P9-CMV-432

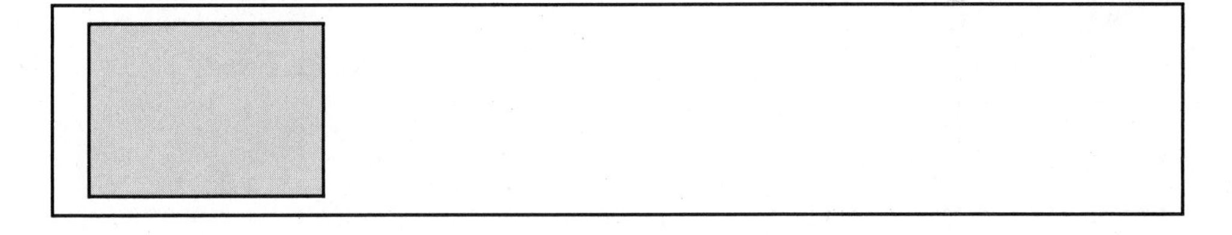

Microelectronic Circuits

The Oxford Series in Electrical Engineering

M.E. Van Valkenburg, *Senior Consulting Editor*
Adel S. Sedra, *Series Editor/Electrical Engineering*
Michael R. Lightner, *Series Editor/Computer Engineering*

ALLEN AND HOLBERG CMOS Analog Circuit Design
BELANGER, ADLER AND RUMIN Introduction to Circuits with Electronics: An Integrated Approach
BOBROW Elementary Linear Circuit Analysis, 2/e
BOBROW Fundamentals of Electrical Engineering
CHEN Linear System Theory and Design
CHEN System and Signal Analysis
COMER Digital Logic and State Machine Design, 2/e
COMER Microprocessor Based System Design
COOPER AND MCGILLEM Probalistic Methods of Signal and System Analysis, 2/e
GHAUSI Electronic Devices and Circuits: Discrete and Integrated
HOSTETTER, SAVANT AND STEPHANI Design of Feedback Control Systems, 2/e
HOUTS Signal Analysis in Linear Systems
JONES Introduction to Optical Fiber Communication Systems
KARNI AND BYATT Mathematical Methods in Continuous and Discrete Systems
KENNEDY Operational Amplifier Circuits: Theory and Application
KUO Digital Control Systems
LASTMAN AND SINHA Microcomputer-Based Numerical Methods for Science and Engineering
LATHI Modern Digital and Analog Communication Systems, 2/e
LEVENTHAL Microcomputer Experimentation with the IBM PC
LEVENTHAL Microcomputer Experimentation with the Intel SDK- 86
LEVENTHAL Microcomputer Experimentation with the Motorola Mc6800 ECB
MCGILLEM AND COOPER Continuous and Discrete Signal and System Analysis, 3/e
NAVON Semiconductor Microdevices and Materials
PAPOULIS Circuits and Systems: A Modern Approach
RAMSHAW AND VAN HEESWIJK Energy Conversion
SADIKU Elements of Electromagnetics
SCHWARZ Electromagnetics for Engineers
SCHWARZ AND OLDHAM Electrical Engineering: An Introduction
SEDRA AND SMITH Microelectronic Circuits, 3/e
SINHA Control Systems
VAN VALKENBURG Analog Filter Design
VRANESIC AND ZAKY Microcomputer Structures
WARNER AND GRUNG Semiconductor Device Electronics
WASSER AND FLYNN Introduction to Arithmetic for Digital Systems Designers
WOLOVICH Robotics: Basic Analysis and Design
YARIV Optical Electronics, 4/e

THIRD EDITION

Microelectronic Circuits

Adel S. Sedra
UNIVERSITY OF TORONTO

Kenneth C. Smith
UNIVERSITY OF TORONTO

New York Oxford
Oxford University Press

Oxford University Press

Oxford New York
Athens Auckland Bangkok Bombay Calcutta Cape Town
Dar es Salaam Delhi Florence Hong Kong Istanbul Karachi
Kuala Lumpur Madras Madrid Melbourne Mexico City
Nairobi Paris Singapore Taipei Tokyo Toronto

and associated companies in

Berlin Ibadan

Published by Oxford University Press, Inc.,
198 Madison Avenue, New York, New York 10016

Oxford is a registered trademark of Oxford University Press

ABOUT THE COVER

In the background is a processed wafer of silicon on which a large number of identical circuit chips are fabricated. When cut from such a wafer, a chip can be housed in a package such as the one shown here with its cover removed. This single chip is a segmented memory array with a very large number of active devices in a variety of intricate circuit structures. The growing need for efficient and reliable low-cost components such as these underlies the importance of the various circuit design and circuit analysis techniques discussed in this text.

Cover Credit: Computer Chip/The Stock Market © 1989 Ted Horowitz

Printed in the United States of America on acid-free paper

0-19-510370-X

10 9 8 7 6 5 4 3

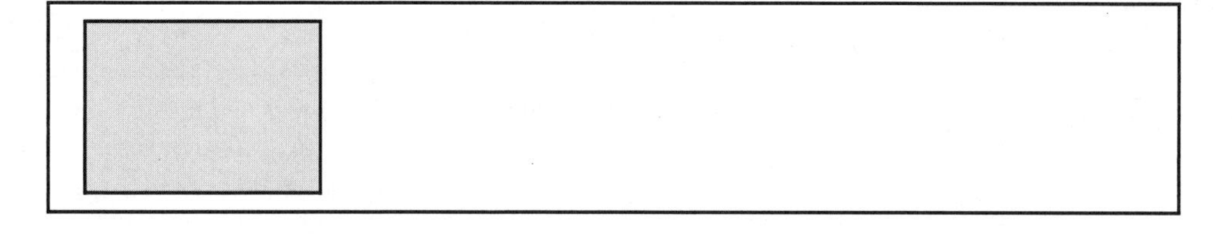

Preface

Microelectronic Circuits, third edition, is intended as a text for the core courses in electronic circuits taught to majors in electrical and computer engineering. It should also prove useful to engineers and other professionals wishing to update their knowledge through self-study.

As was the case with the first two editions, the objective of this book is to develop in the reader the ability to analyze and design electronic circuits, both analog and digital, discrete and integrated. While the application of integrated circuits is covered, emphasis is placed on transistor circuit design. This is done because of our belief that even if the majority of those studying the book were not to pursue a career in IC design, knowledge of what is inside the IC package would enable intelligent and innovative application of such chips. Furthermore, with the advances in VLSI technology and design methodology, IC design itself is becoming accessible to an increasing number of engineers.

Prerequisites

The prerequisite for studying the material in this book is a first course in circuit analysis. As a review, some linear circuits material is included here in appendixes: specifically, two-port network parameters in Appendix B, some useful network theorems in Appendix D, and single-time-constant circuit responses in Appendix E. No prior knowledge of physical electronics is assumed. All required device physics is included, and Appendix A provides a brief description of IC fabrication.

Organization

Although the philosophy and pedagogical approach of the first two editions have been retained, several changes have been made in both organization and coverage. In organization, the size of the ''front end'' of the book has been reduced in order to enable the reader to get to the heart of the subject much more quickly. This has been achieved by deleting some material and moving other material to where it is needed (e.g., Miller's theorem) or to an appendix (e.g., single-time-constant circuits) so it is readily available to those who need it whenever they need it.

Following an introductory chapter that presents some of the basic electronics concepts and establishes notation and conventions, the book is divided into three parts. Part I, Devices and Basic Circuits, is composed of Chapters 2 through 5 and deals with the op amp, the diode, the bipolar junction transistor (BJT), and the field-effect transistor (FET).

It constitutes the bulk of a first course in electronics and most of the material is considered prerequisite to the study of further electronic circuits topics.*

Part II (Chapters 6–12) deals with analog circuits, and Part III (Chapters 13 and 14) deals with digital circuits. Except for requiring some knowledge of the differential pair from Chapter 6, the order of Parts II and III can be reversed. Thus it is possible to study the digital electronics topics of Part III immediately after coverage of the basic devices and circuits of Part I. Such an order of coverage might be preferred for computer engineering students.

Although we recognize that certain economies can be achieved by presenting the BJT and the MOSFET together from the outset, as special cases of a general three-terminal device, we have opted to introduce them separately in Part I and to combine them as much as possible in Part II. It has been our experience that the two devices are sufficiently different that in a first encounter, a separate presentation of each device—its structure and physical operation, its characteristics, and its basic circuit applications—is appropriate. Specifically, in the first course the student needs to ''live'' with each of the two basic devices for a while in order to become comfortable with it. Subsequently, however, a combined treatment is possible and indeed desirable.

Although we have chosen in this edition to place the BJT (Chapter 4) before the MOSFET (Chapter 5), the order of these two topics can be easily reversed. This flexibility is obtained at the expense of a slight redundancy. However, the redundancy can be used to reinforce learning or it can be minimized by a quicker coverage of whichever device is studied second.

Major Changes in Coverage

In addition to the reorganization outlined above, important changes in coverage have been made in the third edition:

1. The subject of diode modeling is more carefully presented, and the treatment of diode application in power-supply design has been expanded (Chapter 3).
2. The analysis of single-stage BJT amplifiers has been made more systematic (Chapter 4).
3. The material on FET devices has been completely rewritten, with the different FET types presented in a more coherent and unified manner (Chapter 5). The JFET whose relative significance has diminished is no longer presented in a separate chapter, rather it is relegated to one section (Section 5.4).
4. Coverage of CMOS op amps has been expanded (Section 10.8).
5. New sections have been added on data converters (Sections 10.9–10.11).
6. Analog filters are covered in a more comprehensive manner in an almost-standalone chapter (Chapter 11).

---

*A possible exception is Chapter 2 on op amps whose study can be postponed in whole or in part to a later stage. More details are given in the section on course organization.

7. Sinusoidal oscillators have been combined with nonlinear waveform generators in a separate chapter (Chapter 12) with new material on IC timers included.

8. New sections on semiconductor memories have been added (Sections 13.9 and 13.10).

Coverage of New Technologies

Since the publication of the previous edition, two IC technologies have gained prominence: gallium arsenide (GaAs) and BiCMOS. The growing interest in these new technologies is reflected in this edition by the inclusion of five sections dealing with their analog and digital applications. Specifically, the GaAs MESFET is introduced in Section 5.11; GaAs analog circuits are studied in Section 6.9; and GaAs digital circuits are presented in Section 13.11. BiCMOS amplifiers are studied in Section 6.8; BiCMOS op amps in Section 10.8; and BiCMOS digital circuits in Section 14.7.

Increased Emphasis on Design

It has been our philosophy that circuit design is best taught by pointing out the various tradeoffs available in selecting a circuit configuration and in selecting component values for a given configuration. The emphasis on design has been increased in this edition by including more design examples, exercise problems, and end-of-chapter problems. Those exercises and end-of-chapter problems that are considered ''design-oriented'' are indicated with a D.

Exercises, End-of-Chapter Problems, and Additional Solved Problems

Over 400 exercises are integrated throughout the text. The answer to each exercise is given below the exercise so students can check their understanding of the material as they read. Solving these exercises should enable the reader to gauge his or her grasp of the preceding material. In addition, more than 1000 end-of-chapter problems, 80% of which are new to this edition, are provided. The problems are keyed to the individual sections (a new feature of this edition) and their degree of difficulty is indicated by a rating system: difficult problems are marked with an asterisk (*); more difficult problems with two asterisks (**); and very difficult (and/or time consuming) problems with three asterisks (***). We must admit, however, that this classification is by no means exact. Our rating no doubt has depended to some degree on our thinking (and mood!) at the time a particular problem was created. However, the number of medium-difficulty problems has been increased in response to users' requests. Answers to about half the problems are given in Appendix G. Complete solutions for all exercises and problems are included in the *Instructor's Manual,* which is available from the publisher for those instructors who adopt the book.

As in the previous two editions, many examples are included. The examples, and indeed most of the problems and exercises, are based on real circuits and anticipate the applications encountered in designing real-life circuits. A new feature of this edition is the inclusion of the numbered solution steps in the figures for many examples, as an attempt to recreate the dynamics of the classroom.

A recurring request from many of the students who used earlier editions of the book has been for solved problems. To satisfy this need, a book of additional problems with solutions is available with this edition (see the list of available ancillaries later in this preface).

Computer-Aided Design (SPICE)

There is no doubt that computer aids play an essential role in the design of electronic circuits. It is our belief, however, that the use of computer aids in the introductory electronics course should be handled with care. Good circuits are still designed by people and not by computers and there is no substitute for the insight that good circuit designers provide in the initial phases of a design. Such insight is acquired, at least in the introductory course, by focusing attention on the fundamental aspects of device models and circuit configurations, and by developing the ability to perform rapid, approximate manual analysis. Computer analysis can help in this process but has to be used judiciously. We have, therefore, chosen *not* to integrate the use of SPICE, the most popular circuit simulator, throughout the text. Rather, SPICE is presented in Appendix C and, more importantly and new to this edition, a new manual is available on the use of SPICE in electronic circuit design. This manual follows the same sequence as the text and is keyed to the text. It presents SPICE solutions to many of the text examples and provides commentary on the additional information obtained in each case as a result of the use of SPICE. The manual also includes analysis and design problems in which the use of SPICE is beneficial. The manual can be used to include and indeed integrate as much SPICE material into the course as the instructor deems necessary without making the text and the course highly dependent on SPICE, which after all is a tool rather than an end in itself.

An Outline for the Reader

The book starts with an introduction to the basic concepts of electronics in Chapter 1. Signals, their frequency spectra, and their analog and digital forms are presented. Amplifiers are introduced as circuit-building blocks and their various types and models are studied. This chapter also establishes some of the terminology and conventions used throughout the text.

The next four chapters are devoted to the study of electronic devices and basic circuits and constitute Part I of the text. Chapter 2 deals with operational amplifiers, their terminal characteristics, simple applications, and limitations. We have chosen to discuss the op amp as a circuit building block at this early stage simply because it is easy to deal with and because the student can experiment with op-amp circuits that perform non-trivial tasks with relative ease and with a sense of accomplishment. We have found this approach to be highly motivating to the student. We should point out, however, that part or all of this chapter can be skipped and studied at a later stage (for instance in conjunction with Chapter 6 or Chapter 8) with no loss of continuity.

Chapter 3 is devoted to the study of the most fundamental electronic device, the *pn* junction diode. The diode terminal characteristics and its hierarchy of models are presented. Also, some of the fundamental applications of diodes, especially those related to power-supply design, are studied. The chapter concludes with an introduction to semiconductors and the physical operation of the *pn* junction.

Chapter 4 introduces the bipolar junction transistor (BJT): its structure, physical operation, terminal characteristics, large- and small-signal models, its operation as an amplifier and as a switch, and the basic configurations of single-stage BJT amplifiers.

The field-effect transistor (FET) family of devices is covered in Chapter 5 where the emphasis, however, is placed on the MOS transistor. Here again the structure, physical operation, terminal characteristics, models, and basic circuit applications of the various FET types are presented. As mentioned earlier, this chapter can, if desired, be studied before the BJT chapter. Our hope is that each of these chapters will make the reader thoroughly familiar and intimately comfortable with the device treated.

By the end of Chapter 5 the reader will have learned about the basic building blocks of electronic circuits and will be ready to consider the more advanced topics of Part II (analog circuits) and Part III (digital circuits). As mentioned earlier, the order of study of Parts II and III can be easily reversed.

Chapter 6 is the first of a sequence of five chapters dealing with more advanced topics in amplifier design. The main topic of Chapter 6 is the differential amplifier, in both its bipolar and FET forms.

In Chapter 7 we study the frequency response of amplifiers. Here emphasis is placed on the choice of configuration to obtain wideband operation.

Chapter 8 deals with the important topic of feedback. Practical circuit applications of negative feedback are presented. We also discuss the stability problem in feedback amplifiers and treat frequency compensation in some detail.

Chapter 9 deals with various types of amplifier output stages. Thermal design is studied and examples of IC power amplifiers are presented.

Chapter 10 presents an introduction to analog integrated circuits. Bipolar, CMOS, and BiCMOS op amps are discussed. Also, basic circuits for the design of data converters are studied. This chapter ties together many of the ideas and methods presented in the previous chapters.

The last two chapters of Part II, Chapters 11 and 12, are application oriented. Chapter 11 is devoted to the study of analog filter design and tuned amplifiers. Chapter 12 presents a study of sinusoidal oscillators, waveform generators, and other nonlinear signal processing circuits.

The last two chapters of the book, Chapters 13 and 14, constitute Part III, digital circuits. They present a concise, modern treatment of digital electronics and should serve as the basis for a more detailed study of digital circuits and systems and/or VLSI design.

Course Organization

The book contains sufficient material for a sequence of two single-semester courses (each of 40 to 50 lecture hours). The organization of the book provides considerable flexibility in course design.

Three possibilities for the first course are

a. Chapters 1 through 5. If time is limited, Sections 2.8 through 2.12 can be postponed to the second course.

b. Chapters 1, 3, 4, 5, and selected topics of Chapters 6 and 7 (e.g., Sections 6.1, 6.2, 6.7, and 7.1 through 7.7).

c. Chapters 1, 3, 4, 5, and selected topics of Chapters 13 and 14 as time permits.

Two possibilities for the second course are

a. Chapters 6 through 12. If time is limited, some sections of Chapters 9, 10, 11, and 12 can be postponed to a third course dealing with analog circuits.

b. Chapters 6, 7, 8, 13, and 14.

Ancillaries

A complete set of ancillary materials is available with this text to support your course:

For the Instructor. The *Instructor's Manual with Transparency Masters,* written by Sedra and Smith, provides complete solutions to all the exercises and problems in the text. It also contains 200 transparency masters that duplicate important figures in the text, the ones most often used in class.

For the Student and the Instructor. The *Laboratory Manual,* written by K. C. Smith, contains approximately 35 experiments. It follows the organization of the text and includes experiments for all major topics. Each experiment includes a list of the equipment required, background material to prepare for the experiment, procedural techniques, and instructions for presenting the results. To help instructors choose and prepare for the experiments, this manual identifies the core experiments all students should perform and includes manufacturers' data sheets for the most common components.

The *Students Problems Book,* also written by the text authors, provides approximately 600 additional problems with complete solutions for students who want more practice.

The *SPICE Manual,* by Gordon Roberts of McGill University and Adel Sedra, written specifically for *Microelectronic Circuits,* third edition, teaches students the techniques, advantages, and limitations of computer-aided design. This valuable supplement is keyed to the text to help instructors.

Acknowledgments

Many of the changes in this edition were made in response to feedback received from some of the instructors who adopted the second edition. We are grateful to all those who took the time to write to us. In particular we wish to thank W. L. Brown of San Diego State University; M. Ghorab of Ryerson Polytechnic Institute; K. S. Imrie of Macquarie University (Australia); G. W. Roberts of McGill University (who is also co-author of the SPICE Manual); and A. T. Tiedemann of the University of Wisconsin-Madison. We continue to be indebted to our former colleague Dr. Frank Holmes, who supplied us with Appendix A on IC fabrication.

Various portions of this edition were reviewed by several people. We thank the following reviewers for their helpful suggestions and hope that they will be pleased with the final results: Ali Akansu of New Jersey Institute of Technology; J. Alvin Connelly of Georgia Institute of Technology; Arnold W. Dipert of University of Illinois–Urbana-Champaign; Glen C. Gerhard of University of New Hampshire; Alfred Johnson of Widener University; W. Marshall Leach of Georgia Institute of Technology; Bryen E. Lorenz of Widener University; N. R. Malik of University of Iowa; John Oristian of U. S.

Military Academy–West Point; Edwyn Smith of University of Toledo; and Sidney Wielin of University of Southern California.

We also remain grateful to the reviewers of the two previous editions: Frank Barnes of University of Colorado; Douglas Brumm of Michigan Technical University; Eugene Chenette of University of Florida; Artice M. Davis of San Jose State University; Randall L. Geiger of Iowa State University; Doug Hamilton of University of Arizona; Richard Jaeger of Auburn University; Alan B. MacNee of University of Michigan; Paul McGrath of Clarkson College of Technology; Jamie Ramirez-Angulo of New Mexico State University; William Sayle, III of Georgia Institute of Technology; Rolf Schaumann of Portland State University; Bernhard M. Schmidt of University of Dayton; Yuh Sun of California State University; and Darrel Vines of Texas Tech University.

A number of individuals helped us in the preparation of this edition. We are grateful to Margaret Tompsett for her skillful typing of the manuscript. Laura Fujino assisted in the preparation of the index. Mei Sum Kwan kept the paper flowing in the right direction and assisted in various other ways. Paul and Mark Sedra helped with a variety of tasks.

We are extremely grateful to Alexa Barnes, our developmental editor, and to Barbara Gingery, our editor, who were the source of numerous good ideas. It has been a pleasure working with them. Last but not least, we wish to thank our families for their support.

Adel S. Sedra
Kenneth C. Smith

Condensed Table of Contents

APPENDIXES

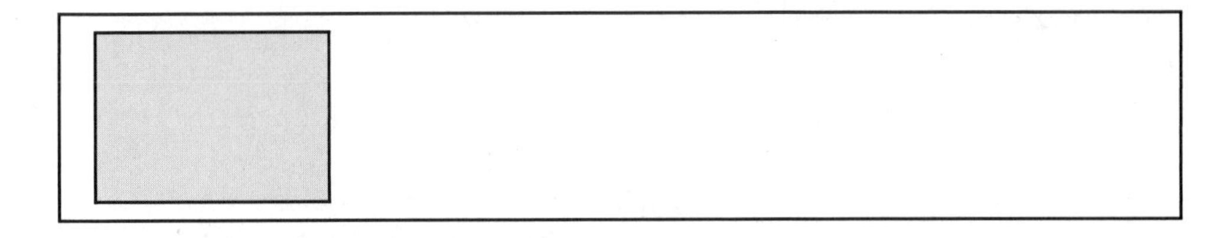

Detailed Table of Contents

CHAPTER 3 DIODES *116*

CHAPTER 4 BIPOLAR JUNCTION TRANSISTORS (BJTs) *191*

CHAPTER 5 FIELD-EFFECT TRANSISTORS (FETs) *298*

CHAPTER 11 FILTERS AND TUNED AMPLIFIERS *762*

CHAPTER 12 SIGNAL GENERATORS AND WAVEFORM-SHAPING CIRCUITS *841*

PART III DIGITAL CIRCUITS *904*

CHAPTER 13 MOS DIGITAL CIRCUITS *906*

CHAPTER 1

Introduction to Electronics

INTRODUCTION

The subject of this book is modern electronics, a field that has come to be known as **microelectronics. Microelectronics** refers to the integrated-circuit (IC) technology that at the time of this writing is capable of producing circuits that contain more than 1 million components in a small piece of silicon (known as a **silicon chip**) whose area is in the order of 60 mm^2. One such microelectronic circuit, for example, is a complete digital computer, which accordingly is known as a **microcomputer** or, more generally, a **microprocessor.**

In this book we shall study electronic devices that can be used singly (in the design of **discrete circuits**) or as components of an integrated-circuit chip. We shall study the design and analysis of interconnections of these devices, which form discrete and integrated circuits of varying complexity and perform a wide variety of functions. We shall also learn about available IC chips and their application in the design of electronic systems.

The purpose of this first chapter is to introduce some basic concepts and terminology. In particular, we shall learn about signals and about one of the most important signal-processing functions electronic circuits are designed to perform, namely, signal amplification. We shall then look at models for linear amplifiers. These models will be employed in subsequent chapters in the design and analysis of actual amplifier circuits.

In addition to motivating the study of electronics, this chapter serves as a bridge between the study of linear circuits and that of the subject of this book: the design and analysis of electronic circuits.

1.1 SIGNALS

Signals contain information about a variety of things and activities in our physical world. Examples abound: Information about the weather is contained in signals that represent the air temperature, pressure, wind speed, etc. The voice of a radio announcer reading the news into a microphone provides an acoustic signal that contains information about world affairs. To monitor the status of a nuclear reactor, instruments are used to measure a multitude of relevant parameters, each instrument producing a signal.

To extract required information from a set of signals, the observer (be it a human or a machine) invariably needs to **process** the signals in some predetermined manner. This **signal processing** is usually most conveniently performed by electronic systems. For this to be possible, however, the signal must first be converted into an electric signal, that is, a voltage or a current. This process is accomplished by devices known as **transducers.** A variety of transducers exist, each suitable for one of the various forms of physical signals. For instance, the sound waves generated by a human can be converted to electric signals using a microphone, which is in effect a pressure transducer. It is not our purpose here to study transducers; rather, we shall assume that the signals of interest already exist in the electrical domain and represent them by one of the two equivalent forms shown in Fig. 1.1. In Fig. 1.1(a) the signal is represented by a voltage source $v_s(t)$ having a source resistance R_s. In the alternate representation of Fig. 1.1(b) the signal is represented by a current source $i_s(t)$ having a source resistance R_s. Although the two representations are equivalent, that in Fig. 1.1(a) (known as the Thévenin form) is preferred when R_s is low. The representation of Fig. 1.1(b) (known as the Norton form) is preferred when R_s is high.

From the discussion above it should be apparent that a signal is a time-varying quantity that can be represented by a graph such as that shown in Fig. 1.2. In fact, the information content of the signal is represented by the changes in its magnitude as time progresses; that is, the information is contained in the ''wiggles'' in the signal waveform. In general, such waveforms are difficult to characterize mathematically. In other words, it is not easy to describe succinctly an arbitrary looking waveform such as that of Fig. 1.2. Of course, such a description is of great importance for the purpose of designing appropriate signal-processing circuits that perform desired functions on the given signal.

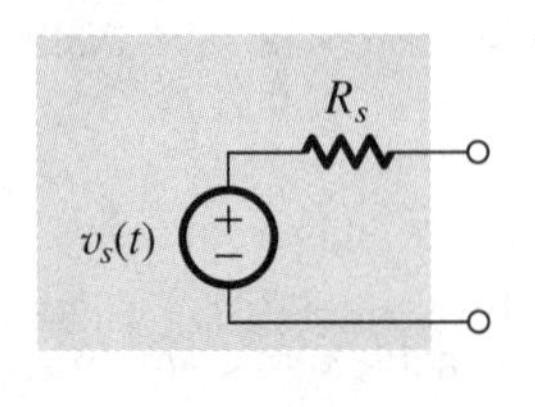

(a)

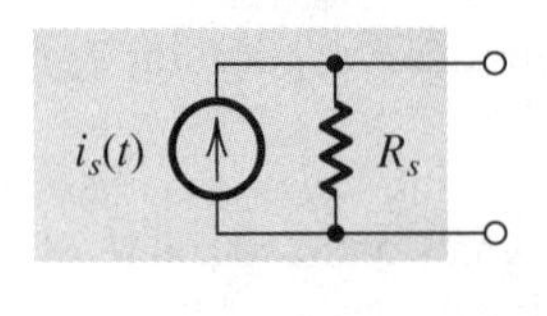

(b)

Fig. 1.1 Two alternative representations of a signal source: **(a)** the Thévenin form, and **(b)** the Norton form.

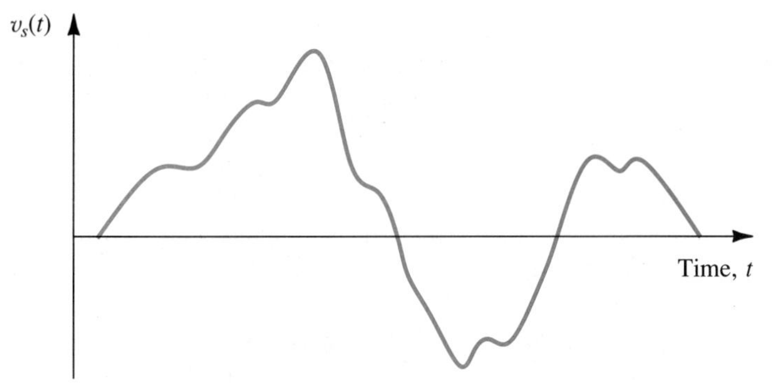

Fig. 1.2 An arbitrary voltage signal $v_s(t)$.

1.2 FREQUENCY SPECTRUM OF SIGNALS

An extremely useful characterization of a signal, and for that matter of any arbitrary function of time, is in terms of its **frequency spectrum.** Such a description of signals is obtained through the mathematical tools of **Fourier series** and **Fourier transform.**[1] We are not interested at this point in the details of these transformations; suffice it to say that they provide the means for representing a voltage signal $v_s(t)$ or a current signal $i_s(t)$ as the sum of sine-wave signals of different frequencies and amplitudes. This makes the sine wave a very important signal in the analysis, design, and testing of electronic circuits. Therefore, we shall briefly review the properties of the sinusoid.

Figure 1.3 shows a sine-wave voltage signal $v_a(t)$,

$$v_a(t) = V_a \sin \omega t \tag{1.1}$$

where V_a denotes the peak value or amplitude in volts and ω denotes the angular frequency in radians per second; that is, $\omega = 2\pi f$ rad/s, where f is the frequency in hertz, $f = 1/T$ Hz, and T is the period in seconds.

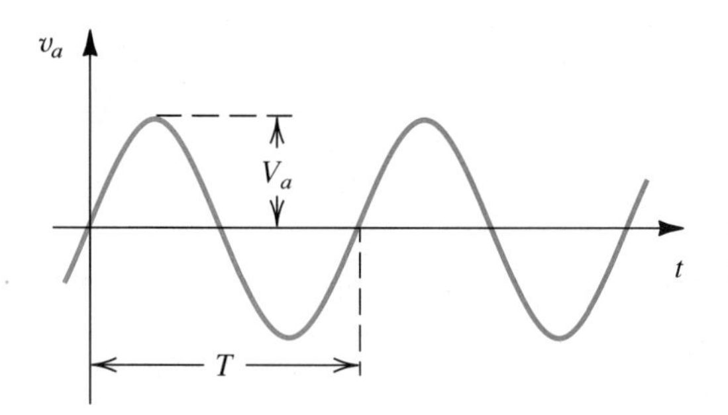

Fig. 1.3 Sine-wave voltage signal of amplitude V_a and frequency $f = 1/T$ Hz. The angular frequency $\omega = 2\pi f$ rad/s.

[1] The reader who has not yet studied these topics should not be alarmed. No detailed application of this material will be made until Chapter 7. Nevertheless, a general understanding of Section 1.2 should be very helpful when studying early parts of the book.

The sine-wave signal is completely characterized by its peak value V_a, its frequency ω, and its phase with respect to an arbitrary reference time. In this case the time origin has been chosen so that the phase angle is 0. It should be mentioned that it is common to express the amplitude of a sine-wave signal in terms of its root-mean-square (rms) value, which is equal to the peak value divided by $\sqrt{2}$. Thus the rms value of the sinusoid $v_a(t)$ of Fig. 1.2 is $V_a/\sqrt{2}$. For instance, when we speak of the wall power supply in our homes as being 120 V, we mean that it has a sine waveform of $120\sqrt{2}$ volts peak value.

Returning now to the representation of signals as the sum of sinusoids, we note that the Fourier series is utilized to accomplish this task for the special case when the signal is a periodic function of time. On the other hand, the Fourier transform is more general and can be used to obtain the frequency spectrum of a signal whose waveform is an arbitrary function of time.

The Fourier series allows us to express a given periodic function of time as the sum of an infinite number of sinusoids whose frequencies are harmonically related. For instance, the symmetrical square-wave signal in Fig. 1.4 can be expressed as

$$v(t) = \frac{4V}{\pi}(\sin \omega_0 t + \tfrac{1}{3} \sin 3\omega_0 t + \tfrac{1}{5} \sin 5\omega_0 t + \cdots) \tag{1.2}$$

where V is the amplitude of the square wave and $\omega_0 = 2\pi/T$ (T is the period of the square wave) is called the **fundamental frequency.** Note that because the amplitudes of the harmonics progressively decrease, the infinite series can be truncated, with the truncated series providing an approximation to the square waveform.

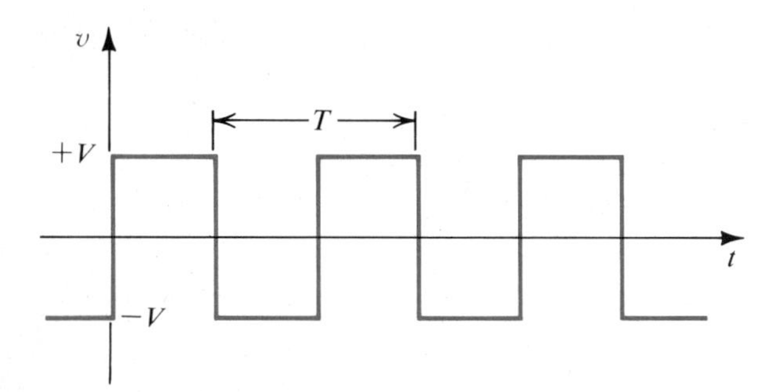

Fig. 1.4 A symmetrical square-wave signal of amplitude V.

The sinusoidal components in the series of Eq. (1.2) constitute the frequency spectrum of the square-wave signal. Such a spectrum can be graphically represented as in Fig. 1.5, where the horizontal axis represents the angular frequency ω in radians per second.

The Fourier transform can be applied to a nonperiodic function of time, such as that depicted in Fig. 1.2, and provides its frequency spectrum as a continuous function of frequency, as indicated in Fig. 1.6. Unlike the case of periodic signals, where the spectrum consists of discrete frequencies (at ω_0 and its harmonics), the spectrum of a nonperiodic signal contains in general all possible frequencies. Nevertheless, the essential parts of

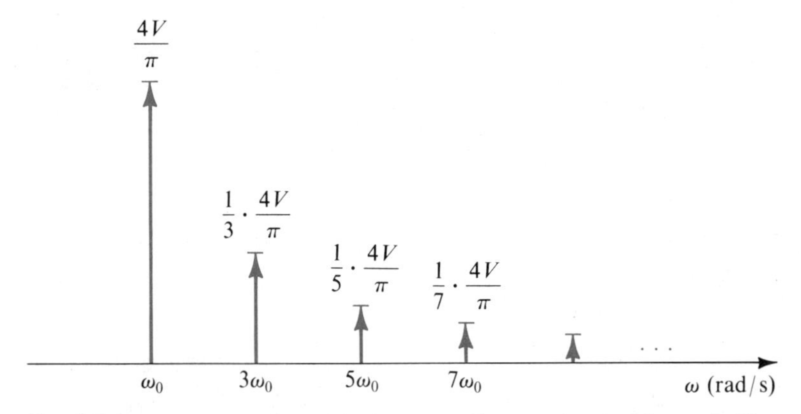

Fig. 1.5 Frequency spectrum (also known as line spectrum) of the periodic square wave of Fig. 1.4.

the spectra of practical signals are usually confined to relatively short segments of the frequency (ω) axis—an observation that is very useful in the processing of such signals. For instance, the spectrum of audible sounds such as speech and music extends from about 20 Hz to about 20 kHz—a frequency range known as the **audio band.** Here we should note that although some musical tones have frequencies above 20 kHz, the human ear is incapable of hearing frequencies that are much above 20 kHz.

We conclude this section by noting that a signal can be represented either by the manner in which its waveform varies with time, as for the voltage signal $v_a(t)$ shown in Fig. 1.2, or in terms of its frequency spectrum, as in Fig. 1.6. The two alternative representations are known as the time-domain representation and the frequency-domain representation, respectively. The frequency-domain representation of $v_a(t)$ will be denoted by the symbol $V_a(\omega)$.

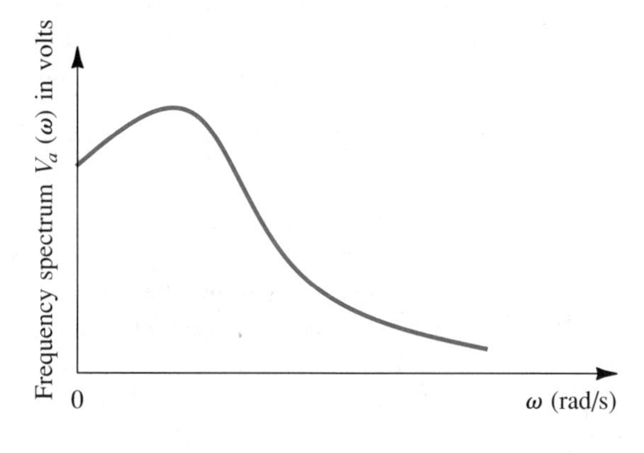

Fig. 1.6 Frequency spectrum of an arbitrary waveform such as that in Fig. 1.2.

Exercises 1.1 Find the frequencies f and ω of a sine-wave signal with a period of 1 ms.

Ans. $f = 1000$ Hz; $\omega = 2\pi \times 10^3$ rad/s

1.2 What is the period T of sine waveforms characterized by frequencies of (a) $f = 60$ Hz? (b) $f = 10^{-3}$ Hz? (c) $f =$ 1 MHz?

Ans. 16.7 ms; 1000 s; 1 μs

1.3 When the square-wave signal of Fig. 1.4, whose Fourier series is given in Eq. (1.2), is applied to a resistor, the total power dissipated may be calculated directly using the relationship $P = 1/T \int_0^T (v^2/R)\, dt$, or indirectly by summing the contribution of each of the harmonic components, that is, $P = P_1 + P_3 + P_5 + \cdots$, which may be found directly from rms values. Verify that the two approaches are equivalent. What fraction of the energy of a square wave is in its fundamental? In its first five harmonics? In its first seven? First nine? In what number of harmonics is 90% of the energy? (Note that in counting harmonics, the fundamental at ω_0 is the first, the one at $2\omega_0$ is the second, etc.)

Ans. 0.81; 0.93; 0.95; 0.96; 3

1.3 ANALOG AND DIGITAL SIGNALS

The voltage signal depicted in Fig. 1.2 is called an **analog signal.** The name derives from the fact that such a signal is analogous to the physical signal that it represents. The magnitude of an analog signal can take on any value; that is, the amplitude of an analog signal exhibits a continuous variation over its range of activity. The vast majority of signals in the world around us are analog. Electronic circuits that process such signals are known as **analog circuits.** A variety of analog circuits will be studied in this book.

An alternative form of signal representation is that of a sequence of numbers, each number representing the signal magnitude at an instant of time. The resulting signal is called a **digital signal.** To see how a signal can be represented in this form—that is, how signals can be converted from analog to digital form—consider Fig. 1.7(a). Here the curve represents a voltage signal, identical to that in Fig. 1.2. At equal intervals along the time axis we have marked the time instants t_0, t_1, t_2, and so on. At each of these time instants the magnitude of the signal is measured, a process known as **sampling.** Figure 1.7(b) shows a representation of the signal of Fig. 1.7(a) in terms of its samples. The signal of Fig. 1.7(b) is defined only at the sampling instants; it no longer is a continuous function of time, but rather, it is a **discrete-time signal.** However, since the magnitude of each sample can take any value in a continuous range, the signal in Fig. 1.7(b) is still an analog signal.

Now if we represent the magnitude of each of the signal samples in Fig. 1.7(b) by a number having a finite number of digits, then the signal amplitude will no longer be continuous; rather, it is said to be **quantized, discretized,** or **digitized.** The resulting digital signal then is simply a sequence of numbers that represent the magnitudes of the successive signal samples.

Electronic circuits that process digital signals are called **digital circuits.** The digital computer is a system constructed of digital circuits. All the internal signals in a digital

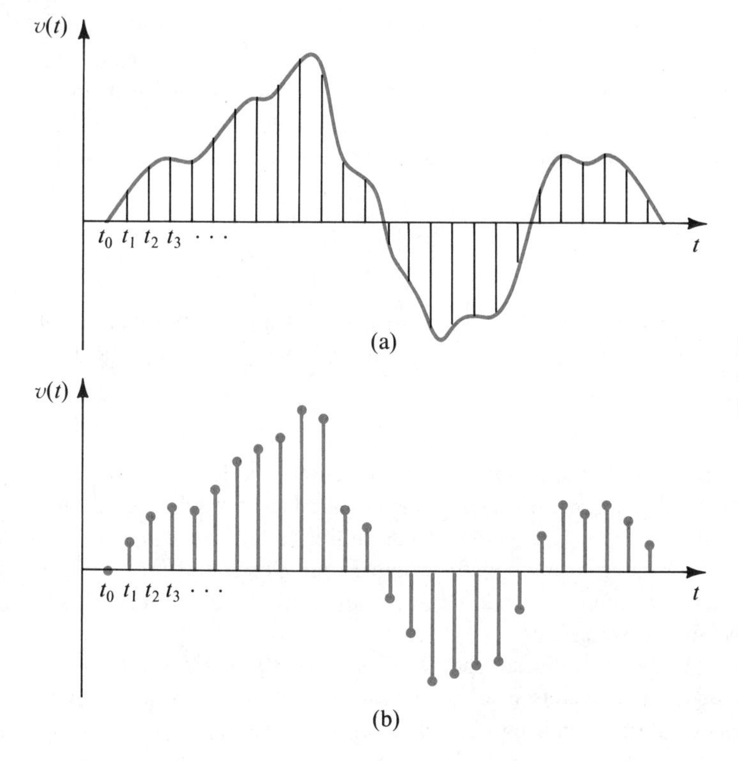

Fig. 1.7 Sampling the continuous-time analog signal in **(a)** results in the discrete-time signal in **(b)**.

computer are digital signals. Digital processing of signals has become quite popular primarily because of the tremendous advances made in the design and fabrication of digital circuits. Another reason for the popularity of digital signal processing is that one generally prefers to deal with numbers. For instance, there is little doubt that the majority of us find the digital display of time (as in a digital watch) much more convenient than the analog display (hands moving relative to a graduated dial). While the latter form of display calls for interpretation on the part of the observer, the former is explicit, eliminating any subjective judgment. This is an important point that perhaps is better appreciated in the context of an instrumentation system, such as that for monitoring the status of a nuclear reactor. In such a system human interpretation of instrument readings and the associated inevitable lack of consistency could be hazardous. Furthermore, in such an instrumentation system the measurement results usually have to be fed to a digital computer for further analysis. It would be convenient therefore if the signals obtained by the measuring instruments were already in digital form.

Digital processing of signals is economical and reliable. Furthermore, it allows a wide variety of processing functions to be performed—functions that are either impossible or impractical to implement by analog means. Nevertheless, as already mentioned, most of the signals in the physical world are analog. Also, there remain many signal-processing tasks that are best performed by analog circuits. It follows that a good electronics engineer must be proficient in both forms of signal processing. Such is the philosophy adopted in this text.

Before leaving this discussion of signals we should point out that not all the signals with which electronic systems deal originate in the physical world. For instance, the electronic calculator and the digital computer perform mathematical and logical operations to solve problems. The internal digital signals represent the variables and parameters of these problems and are obviously not supplied directly from external physical signals.

1.4 AMPLIFIERS

In this section we shall introduce a fundamental signal-processing function that is employed in some form in almost every electronic system, namely, signal amplification.

Signal Amplification

From a conceptual point of view the simplest signal processing task is that of **signal amplification.** The need for amplification arises because transducers provide signals that are said to be ''weak,'' that is, in the microvolt (μV) or millivolt (mV) range and possessing little energy. Such signals are too small for reliable processing, and processing is much easier if the signal magnitude is made larger. The functional block that accomplishes this task is the **signal amplifier.**

It is appropriate at this point to discuss the need for **linearity** in amplifiers. When amplifying a signal, care must be exercised so that the information contained in the signal is not changed and no new information is introduced. Thus when feeding the signal shown in Fig. 1.2 to an amplifier we want the output signal of the amplifier to be an exact replica of that at the input, except of course for having larger magnitude. In other words, the ''wiggles'' in the output waveform must be identical to those in the input waveform. Any change in waveform is considered to be **distortion** and is obviously undesirable.

An amplifier that preserves the details of the signal waveform is characterized by the relationship

$$v_o(t) = Av_i(t) \tag{1.3}$$

where v_i and v_o are the input and output signals, respectively, and A is a constant representing the magnitude of amplification, known as **amplifier gain.** Equation (1.3) is a linear relationship; hence the amplifier it describes is a **linear amplifier.** It should be easy to see that if the relationship between v_o and v_i contains higher powers of v_i, then the waveform of v_o will no longer be identical to that of v_i. The amplifier is then said to exhibit **nonlinear distortion.**

The amplifiers discussed so far are primarily intended to operate on very small input signals. Their purpose is to make the signal magnitude larger and therefore are thought of as **voltage amplifiers.** The **preamplifier** in the home stereo system is an example of a voltage amplifier. However, it usually does more than just amplify the signal; specifically, it performs some shaping of the frequency spectrum of the input signal. This topic, however, is beyond our need at this moment.

At this time we wish to mention another type of amplifier, namely, the power amplifier. Such an amplifier may provide only a modest amount of voltage gain but substantial current gain. Thus while absorbing little power from the input signal source to which it is connected, often a preamplifier, it delivers large amounts of power to its load. An example is found in the power amplifier of the home stereo system, whose purpose is to provide sufficient power to drive the loudspeaker. Here we should note that the loudspeaker is the output transducer of the stereo system; it converts the electric output signal of the system into an acoustic signal. A further appreciation of the need for linearity can be acquired by reflecting on the power amplifier. A linear power amplifier causes both soft and loud music passages to be reproduced without distortion.

Amplifier Circuit Symbol

The signal amplifier is obviously a two-port network. Its function is conveniently represented by the circuit symbol of Fig. 1.8(a). This symbol clearly distinguishes the input and output ports and indicates the direction of signal flow. Thus, in subsequent diagrams it will not be necessary to label the two ports ''input'' and ''output.'' For generality we have shown the amplifier to have two input terminals that are distinct from the two output terminals. A more common situation is illustrated in Fig. 1.8(b), where a common terminal exists between the input and output ports of the amplifier. This common terminal is used as a reference point and is called the **circuit ground.**

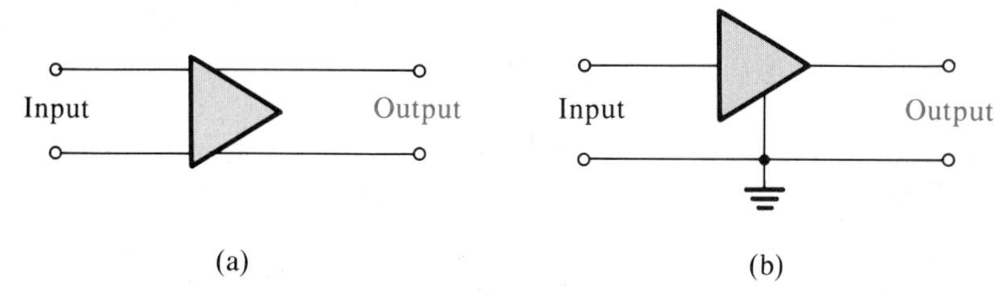

Fig. 1.8 (a) Circuit symbol for amplifier. **(b)** An amplifier with a common terminal (ground) between the input and output points.

Voltage Gain

A linear amplifier accepts an input signal $v_I(t)$ and provides at the output, across a load resistance R_L (see Fig. 1.9(a)), an output signal $v_O(t)$ that is a magnified replica of $v_I(t)$. The **voltage gain** of the amplifier is defined by

$$\text{Voltage gain } (A_v) \equiv \frac{v_O}{v_I} \tag{1.4}$$

Fig. 1.9(b) shows the transfer characteristic of a linear amplifier. If we apply to the input of this amplifier a sinusoidal voltage of amplitude $\hat{V}$, we obtain at the output a sinusoid of amplitude $A_v\hat{V}$.

Power Gain and Current Gain

An amplifier increases the signal power, an important feature that distinguishes an amplifier from a transformer. In the case of a transformer, although the voltage delivered to the

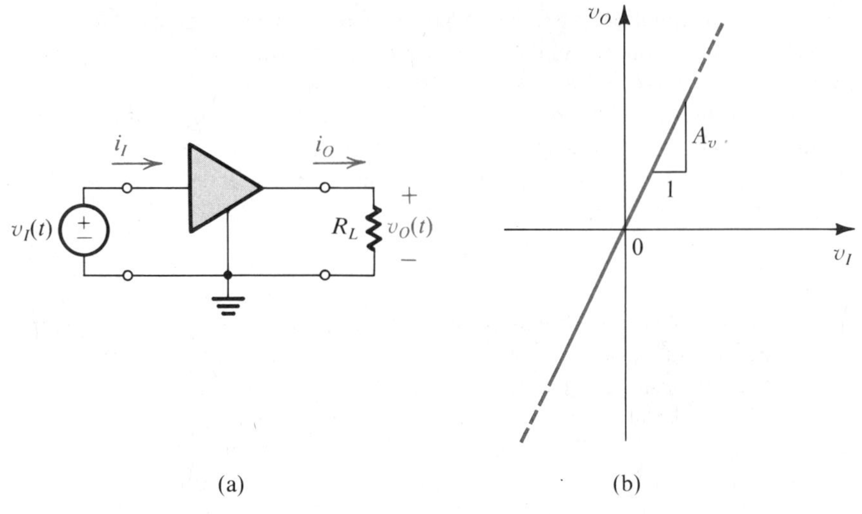

Fig. 1.9 (a) A voltage amplifier fed with a signal $v_I(t)$ and connected to a load resistance R_L. **(b)** Transfer characteristic of a linear voltage amplifier with voltage gain A_v.

load could be greater than the voltage feeding the input side (primary), the power delivered to the load is less than or at most equal to the power supplied by the signal source. On the other hand, an amplifier provides the load with power greater than that obtained from the signal source. That is, amplifiers have power gain. The **power gain** of the amplifier in Fig. 1.9(a) is defined as

$$\text{Power gain } (A_p) \equiv \frac{\text{load power } (P_L)}{\text{input power } (P_I)} \tag{1.5}$$

$$= \frac{v_O i_O}{v_I i_I} \tag{1.6}$$

where i_O is the current that the amplifier delivers to the load (R_L), $i_O = v_O/R_L$, and i_I is the current the amplifier draws from the signal source. The **current gain** of the amplifier is defined as

$$\text{Current gain } (A_i) \equiv \frac{i_O}{i_I} \tag{1.7}$$

From Eqs. (1.4) to (1.7) we note that

$$A_p = A_v A_i \tag{1.8}$$

Expressing Gain in Decibels

The amplifier gains defined above are ratios of similarly dimensioned quantities. Thus they will be expressed either as dimensionless numbers or, for emphasis, as V/V for the voltage gain, A/A for the current gain, and W/W for the power gain. Alternatively, for a number of reasons, some of them historic, electronics engineers express amplifier gain with a logarithmic measure. Specifically the voltage gain A_v can be expressed as

$$\text{Voltage gain in decibels} = 20 \log|A_v| \qquad \text{dB}$$

and the current gain A_i can be expressed as

$$\text{Current gain in decibels} = 20 \log|A_i| \qquad \text{dB}$$

Since power is related to voltage (or current) squared, the power gain A_p can be expressed in decibels as follows:

$$\text{Power gain in decibels} = 10 \log A_p \qquad \text{dB}$$

The absolute values of the voltage and current gains are used because in some cases A_v or A_i may be negative numbers. A negative gain A_v simply means that there is a 180° phase difference between input and output signals; it does not imply that the amplifier is **attenuating** the signal. On the other hand, an amplifier whose voltage gain is, say, -20 dB is in fact attenuating the input signal by a factor of 10 (that is, $A_v = 0.1$).

The Amplifier Power Supplies

Since the power delivered to the load is greater than the power drawn from the signal source, the question arises as to the source of this additional power. The answer is found by observing that amplifiers need dc power supplies for their operation. These dc sources supply the extra power delivered to the load as well as any power that might be dissipated in the internal circuit of the amplifier (such power is converted to heat). In Fig. 1.9(a) we have not explicitly shown these dc sources.

Figure 1.10(a) shows an amplifier that requires two dc sources: one positive of value V_1 and one negative of value V_2. The amplifier has two terminals, labeled V^+ and V^-, for connection to the dc supplies. For the amplifier to operate, the terminal labeled V^+ has to be connected to the positive side of a dc source whose voltage is V_1 and whose negative side is connected to the circuit ground. Also, the terminal labeled V^- has to be connected

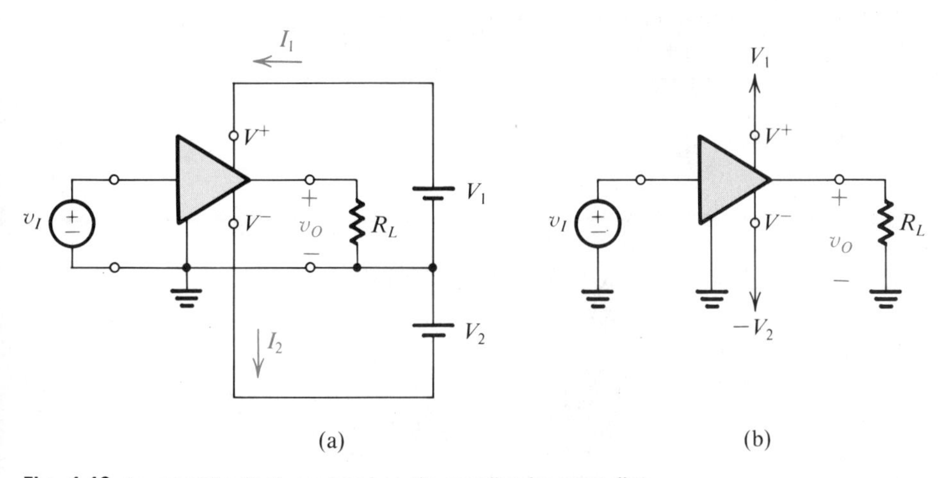

Fig. 1.10 An amplifier that requires two dc supplies for operation.

to the negative side of a dc source whose voltage is V_2 and whose positive side is connected to the circuit ground. Now, if the current drawn from the positive supply is denoted I_1 and that from the negative supply is I_2 (see Fig. 1.10(a)), then the dc power delivered to the amplifier is

$$P_{dc} = V_1 I_1 + V_2 I_2$$

If the power dissipated in the amplifier circuit is denoted $P_{dissipated}$, the power balance equation for the amplifier can be written as

$$P_{dc} + P_I = P_L + P_{dissipated}$$

Since the power drawn from the signal source (P_I) is usually small, the amplifier **efficiency** is defined as

$$\eta \equiv \frac{P_L}{P_{dc}} \times 100 \tag{1.9}$$

The power efficiency is an important performance parameter for amplifiers that handle large amounts of power. Such amplifiers, called power amplifiers, are used, for example, as output amplifiers of stereo systems.

In order to simplify circuit diagrams, we shall adopt the convention illustrated in Fig. 1.10(b). Here the V^+ terminal is shown connected to an arrowhead pointing upward and the V^- terminal to an arrowhead pointing downward. The corresponding voltage is indicated next to each arrowhead. Note that in many cases we will not explicitly show the connections of the amplifier to the dc power sources. Finally, we note that some amplifiers require only one power supply.

EXAMPLE 1.1

Consider an amplifier operating from $\pm$10-V power supplies. It is fed with a sinusoidal voltage having 1 V peak and delivers a sinusoidal voltage output of 9 V peak to a 1-kΩ load. The amplifier draws a current of 9.5 mA from each of its two power supplies. The input current of the amplifier is found to be sinusoidal with 0.1 mA peak. Find the voltage gain, the current gain, the power gain, the power drawn from the dc supplies, the power dissipated in the amplifier, and the amplifier efficiency.

Solution

$$A_v = \frac{9}{1} = 9 \text{ V/V}$$

or

$$A_v = 20 \log 9 \simeq 19.1 \text{ dB}$$

$$\hat{I}_o = \frac{9 \text{ V}}{1 \text{ k}\Omega} = 9 \text{ mA}$$

$$A_i = \frac{\hat{I}_o}{\hat{I}_i} = \frac{9}{0.1} = 90 \text{ A/A}$$

or

$$A_i = 20 \log 90 = 39.1 \text{ dB}$$

$$P_L = V_{o_{\text{rms}}} I_{o_{\text{rms}}}$$

$$= \frac{9}{\sqrt{2}} \frac{9}{\sqrt{2}} = 40.5 \text{ mW}$$

$$P_I = V_{i_{\text{rms}}} I_{i_{\text{rms}}} = \frac{1}{\sqrt{2}} \frac{0.1}{\sqrt{2}} = 0.05 \text{ mW}$$

$$A_p = \frac{P_L}{P_I} = \frac{40.5}{0.05} = 810 \text{ W/W}$$

or

$$A_p = 10 \log 810 = 29.1 \text{ dB}$$

$$P_{\text{dc}} = 10 \times 9.5 + 10 \times 9.5 = 190 \text{ mW}$$

$$P_{\text{dissipated}} = P_{\text{dc}} + P_I - P_L$$

$$= 190 + 0.05 - 40.5 = 149.6 \text{ mW}$$

$$\eta = \frac{P_L}{P_{\text{dc}}} \times 100 = 21.3\%$$

From the above example we observe that the amplifier converts some of the dc power it draws from the power supplies to signal power that it delivers to the load.

Amplifier Saturation

The amplifier transfer characteristic remains linear over only a limited range of input and output voltages. For an amplifier operated from two power supplies the output voltage cannot exceed a specified positive limit and cannot decrease below a specified negative limit. The resulting transfer characteristic is shown in Fig. 1.11, with the positive and negative saturation levels denoted L_+ and L_-, respectively. Each of the two saturation levels is usually within 1 or 2 volts of the voltage of the corresponding power supply.

Obviously, in order to avoid distorting the output signal waveform, the input signal swing must be kept within the linear range of operation.

$$\frac{L_-}{A_v} \leq v_I \leq \frac{L_+}{A_v}$$

Figure 1.11 shows two input waveforms and the corresponding output waveforms. We note that the peaks of the larger waveform have been clipped off because of amplifier saturation.

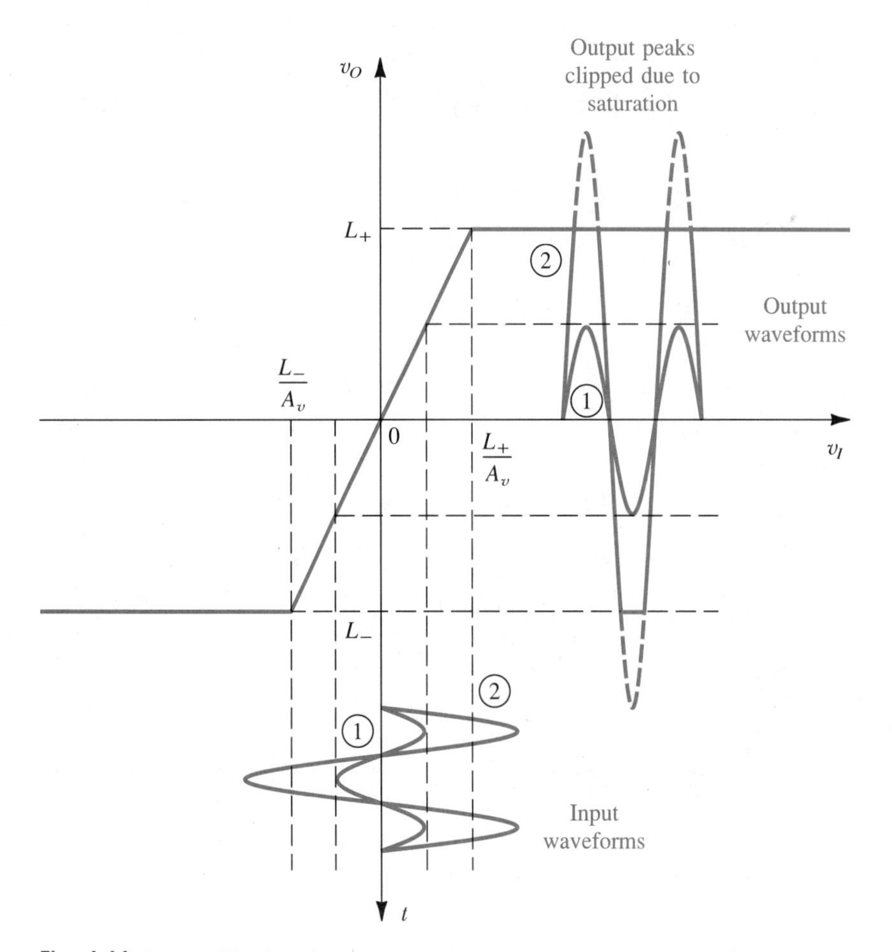

Fig. 1.11 An amplifier transfer characteristic that is linear except for output saturation.

Nonlinear Transfer Characteristics and Biasing

Except for the output saturation effect discussed above, the amplifier transfer characteristics have been assumed to be perfectly linear. In practical amplifiers the transfer characteristic may exhibit nonlinearities of various magnitudes, depending on how elaborate the amplifier circuit is and how much effort has been expended in the design to ensure linear operation. Consider as an example the transfer characteristic depicted in Fig. 1.12. Such a characteristic is typical of simple amplifiers that are operated from a single (positive) power supply. The transfer characteristic is obviously nonlinear and, because of the single-supply operation, is not centered around the origin. Fortunately, a simple technique exists for obtaining linear amplification from an amplifier with such a nonlinear transfer characteristic.

The technique consists of first **biasing** the circuit to operate at a point near the middle of the transfer characteristic. This is achieved by applying a dc voltage V_I, as indicated in Fig. 1.12, where the operating point is labeled Q and the corresponding dc voltage at the

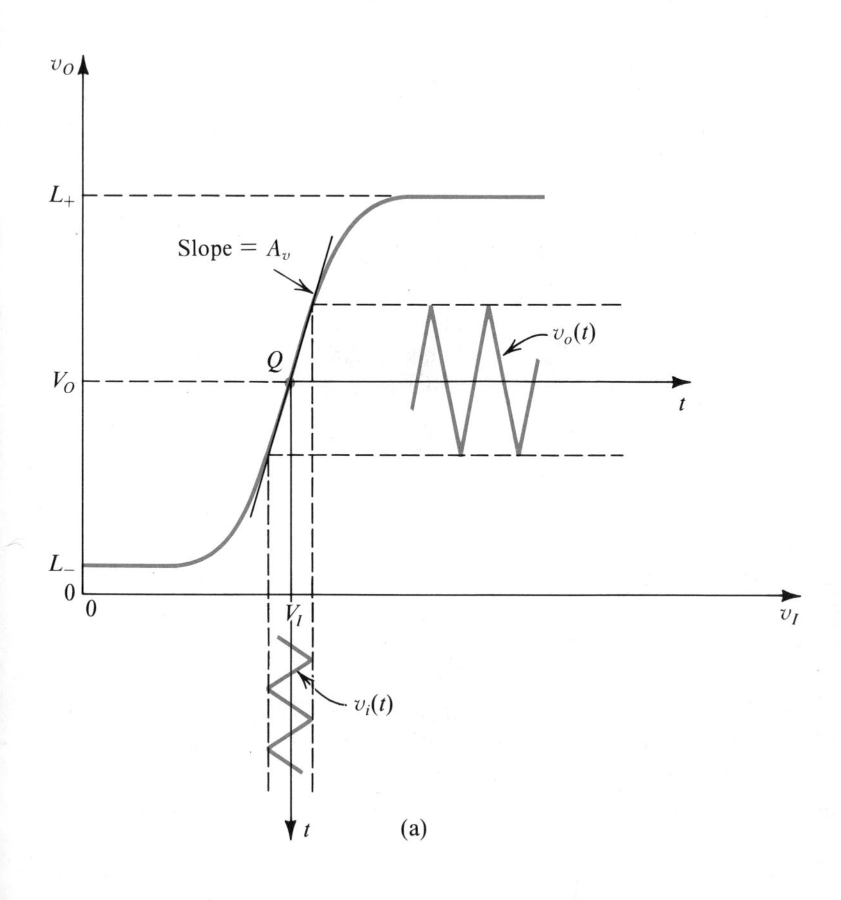

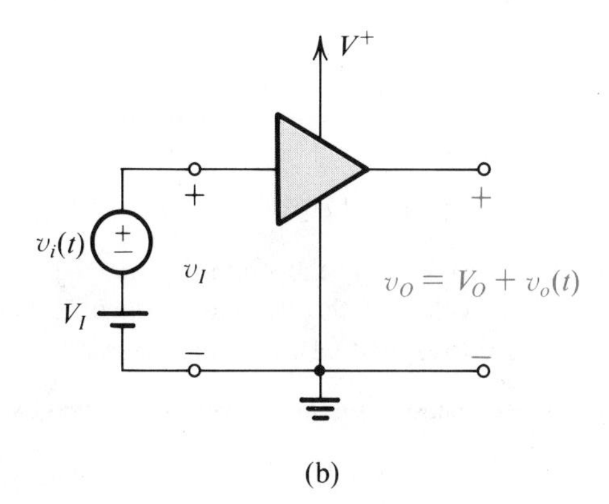

Fig. 1.12 (a) An amplifier transfer characteristic that shows considerable nonlinearity. **(b)** To obtain linear operation the amplifier is biased as shown and the signal amplitude is kept small.

output is V_O. The point Q is known as the quiescent point, the dc bias point, or simply the operating point. The time-varying signal to be amplified, $v_i(t)$, is then superimposed on the dc bias voltage V_I as indicated in Fig. 1.12. Now, as the total instantaneous input $v_I(t)$,

$$v_I(t) = V_I + v_i(t)$$

varies around V_I, the instantaneous operating point moves up and down the transfer curve around the operating point Q. In this way, one can determine the waveform of the total instantaneous output voltage $v_O(t)$. It can be seen that by keeping the amplitude of $v_i(t)$ sufficiently small, the instantaneous operating point can be confined to an almost linear segment of the transfer curve centered about Q. This in turn results in the time-varying portion of the output being proportional to $v_i(t)$; that is,

$$v_O(t) = V_O + v_o(t)$$

with

$$v_o(t) = A_v v_i(t)$$

where A_v is the slope of the almost linear segment of the transfer curve; that is,

$$A_v = \left.\frac{dv_O}{dv_I}\right|_{\text{at}\,Q}$$

In this manner linear amplification is achieved. Of course, there is a limitation: The input signal must be kept sufficiently small. Increasing the amplitude of the input signal can cause the operation to be no longer restricted to an almost linear segment of the transfer curve. This in turn results in a distorted output signal waveform. Such nonlinear distortion is undesirable: The output signal contains additional spurious information that is not part of the input. We shall use this biasing technique and the associated small-signal approximation frequently in the design of transistor amplifiers.

EXAMPLE 1.2

A transistor amplifier has the transfer characteristic

$$v_O = 10 - 10^{-11} e^{40 v_I} \tag{1.10}$$

which applies for $v_I \geq 0$ V and $v_O \geq 0.3$ V. Find the limits L_- and L_+ and the corresponding values of v_I. Also, find the value of the dc bias voltage V_I that results in $V_O = 5$ V and the voltage gain at the corresponding operating point.

Solution

The limit L_- is obviously 0.3 V. The corresponding value of v_I is obtained by substituting $v_O = 0.3$ V in Eq. (1.10); that is,

$$v_I = 0.690 \text{ V}$$

The limit L_+ is determined by $v_I = 0$ and is thus given by

$$L_+ = 10 - 10^{-11} \simeq 10 \text{ V}$$

To bias the device so that $V_O = 5$ V we require a dc input V_I whose value is obtained by substituting $v_O = 5$ V in Eq. (1.10) as follows:

$$V_I = 0.673 \text{ V}$$

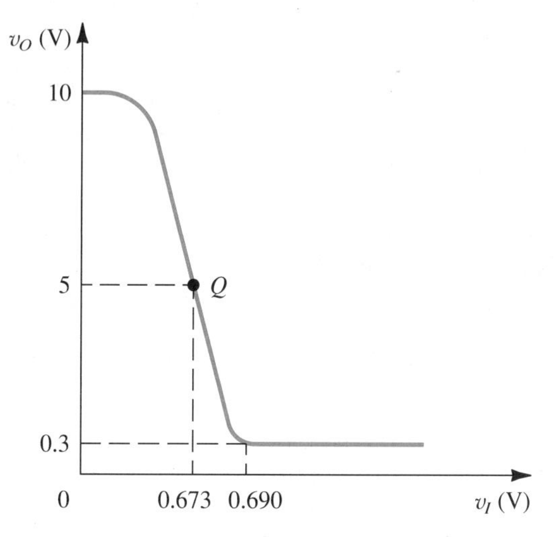

Fig. 1.13 A sketch of the transfer characteristic of the amplifier of Example 1.2.

The gain at the operating point is obtained by evaluating the derivative dv_O/dv_I at v_I = 0.673 V. The result is

$$A_v = -200 \text{ V/V}$$

A sketch of the amplifier transfer characteristic (not to scale) is shown in Fig. 1.13.

At this point we draw the reader's attention to the terminology used above. Total instantaneous quantities are denoted by a lowercase symbol with an uppercase subscript, for example, $i_A(t)$, $v_C(t)$. Direct-current (dc) quantities will be denoted by an uppercase symbol with an uppercase subscript, for example, I_A, V_C. Finally, incremental signal quantities will be denoted by a lowercase symbol with a lowercase subscript, for example $i_a(t)$, $v_c(t)$.

Once an amplifier is properly biased and the input signal is kept sufficiently small, the operation is assumed to be linear. We can then employ the techniques of linear circuit analysis to analyze the signal operation of the amplifier circuit. The remainder of this chapter provides a review and application of these analysis techniques.

Exercises 1.4 An amplifier has a voltage gain of 100 V/V and a current gain of 1000 A/A. Express the voltage and current gains in decibels and find the power gain.

Ans. 40 dB; 60 dB; 50 dB

1.5 An amplifier operating from a single 15-V supply provides a 12-V peak-to-peak sine-wave signal to a 1-kΩ load, and draws negligible input current from the signal source. The dc current drawn from the 15-V supply is 8 mA. What is the power dissipated in the amplifier and what is the amplifier efficiency?

Ans. 102 mW; 15%

1.6 The object of this exercise is to investigate the limitation of the small-signal approximation. Consider the amplifier of Example 1.2 with a positive input signal of 2 mV superimposed on the dc bias voltage V_I. Find the corresponding signal at the output: (a) Assume the amplifier is linear around the operating point; that is, use the value of gain evaluated in Example 1.2. (b) Use the transfer characteristic of the amplifier. Repeat for input signals of 5 mV and 10 mV.

Ans. −0.4 V, −0.409 V; −1 V, −1.087 V; −2 V, −2.416 V

1.5 CIRCUIT MODELS FOR AMPLIFIERS

A good part of this book is concerned with the design of amplifier circuits using transistors of various types. Such circuits will vary in complexity from those using a single transistor to those with 20 or more devices. In order to be able to apply the resulting amplifier circuit as a building block in a system, one must be able to characterize, or model, its terminal behavior. In this section we study simple but effective amplifier models. These models apply irrespective of the complexity of the internal circuit of the amplifier. The values of the model parameters can be found either by analyzing the amplifier circuit or by performing measurements at the amplifier terminals.

Voltage Amplifiers

Figure 1.14(a) shows a circuit model for the voltage amplifier. The model consists of a voltage-controlled voltage source having a gain factor A_{vo}, an input resistance R_i that accounts for the fact that the amplifier draws an input current from the signal source, and an output resistance R_o that accounts for the change in output voltage as the amplifier is called upon to supply output current to a load. To be specific, we show in Fig. 1.14(b) the amplifier model fed with a signal voltage source v_s having a resistance R_s and connected at the output to a load resistance R_L. The nonzero output resistance R_o causes only a fraction of $A_{vo}v_i$ to appear across the output. Using the voltage divider rule we obtain

$$v_o = A_{vo}v_i \frac{R_L}{R_L + R_o}$$

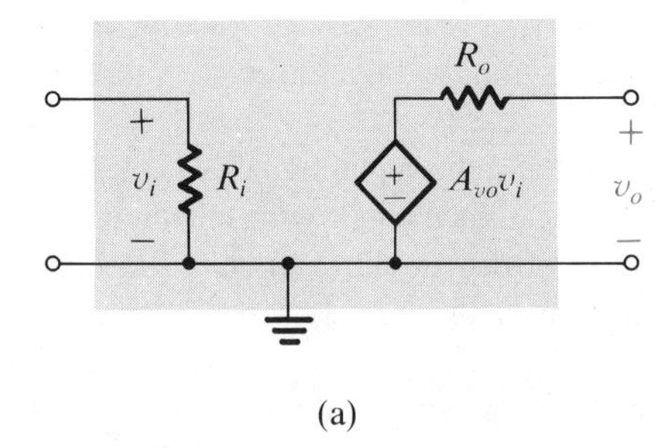

(a)

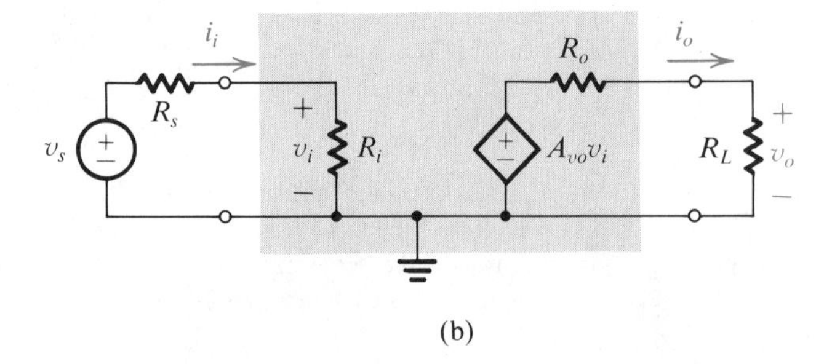

(b)

Fig. 1.14 (a) Circuit model for the voltage amplifier. **(b)** The voltage amplifier with input signal source and load.

Thus the voltage gain is given by

$$A_v \equiv \frac{v_O}{v_i} = A_{vo}\frac{R_L}{R_L + R_o} \tag{1.11}$$

It follows that in order not to lose gain in coupling the amplifier output to a load, the output resistance R_o should be much smaller than the load resistance R_L. In other words, for a given R_L one must design the amplifier so that its R_o is much smaller than R_L. An ideal voltage amplifier is one with $R_o = 0$. Equation (1.11) indicates also that for $R_L = \infty$, $A_v = A_{vo}$. Thus A_{vo} is the voltage gain of the unloaded amplifier, or the **open-circuit voltage gain.** It should also be clear that in specifying the voltage gain of an amplifier, one must also specify the value of load resistance at which this gain is measured or calculated. If a load resistance is not specified, it is normally assumed that the given voltage gain is the open-circuit gain A_{vo}.

The finite input resistance R_i introduces another voltage-divider action at the input, with the result that only a fraction of the source signal v_s actually reaches the input terminals of the amplifier; that is,

$$v_i = v_s\frac{R_i}{R_i + R_s} \tag{1.12}$$

It follows that in order not to lose a significant portion of the input signal in coupling the signal source to the amplifier input, the amplifier must be designed to have an input resistance R_i much greater than the resistance of the signal source, $R_i \gg R_s$. An ideal voltage amplifier is one with $R_i = \infty$. In this ideal case both the current gain and power gain become infinite.

There are situations in which one is interested not in the voltage gain but in a significant power gain. For instance, the source signal can be of a respectable voltage but the source resistance can be much greater than the load resistance. Connecting the source directly to the load would result in significant signal attenuation. In such a case one requires an amplifier with a high input resistance (much greater than the source resistance) and a low output resistance (much smaller than the load resistance) but with a modest voltage gain (or even unity gain). Such an amplifier is referred to as a **buffer amplifier.** We shall encounter buffer amplifiers often throughout this text.

EXAMPLE 1.3

Figure 1.15 depicts an amplifier composed of a cascade of three stages. The amplifier is fed by a signal source with a source resistance of 100 kΩ and delivers its output into a load resistance of 100 Ω. The first stage has a relatively high input resistance and a modest gain factor of 10. The second stage has a higher gain factor but lower input resistance. Finally, the last, or output, stage has unity gain but a low output resistance. We wish to evaluate the overall voltage gain, that is, v_L/v_s, the current gain, and the power gain.

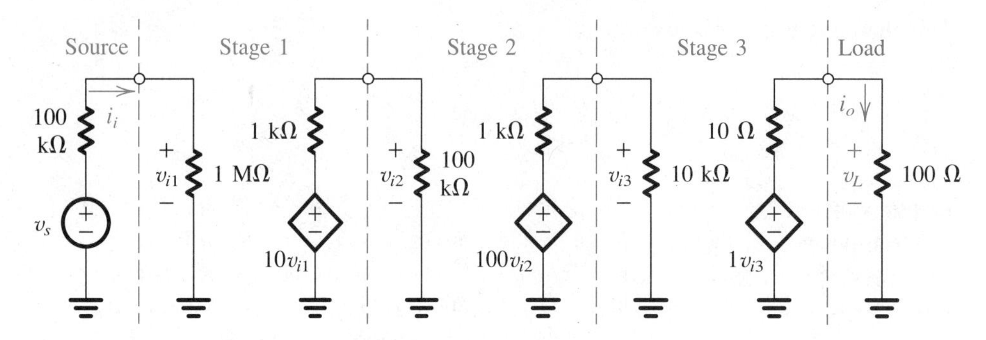

Fig. 1.15 Three-stage amplifier for Example 1.3.

Solution The fraction of source signal appearing at the input terminals of the amplifier is obtained using the voltage-divider rule at the input, as follows:

$$\frac{v_{i1}}{v_s} = \frac{1\ \text{M}\Omega}{1\ \text{M}\Omega + 100\ \text{k}\Omega} = 0.909$$

The voltage gain of the first stage is obtained by considering the input resistance of the second stage to be the load of the first stage; that is,

$$A_{v1} \equiv \frac{v_{i2}}{v_{i1}} = 10\frac{100\ \text{k}\Omega}{100\ \text{k}\Omega + 1\ \text{k}\Omega} = 9.9$$

Similarly, the voltage gain of the second stage is obtained by considering the input resistance of the third stage to be the load of the second stage,

$$A_{v2} \equiv \frac{v_{i3}}{v_{i2}} = 100\frac{10\ \text{k}\Omega}{10\ \text{k}\Omega + 1\ \text{k}\Omega} = 90.9$$

Finally, the voltage gain of the output stage is as follows:

$$A_{v3} \equiv \frac{v_L}{v_{i3}} = 1\frac{100\ \Omega}{100\ \Omega + 10\ \Omega} = 0.909$$

The total gain of the three stages in cascade can be now found from

$$A_v \equiv \frac{v_L}{v_{i1}} = A_{v1}A_{v2}A_{v3} = 818$$

or 58.3 dB.

To find the voltage gain from source to load we multiply A_v by the factor representing the loss of gain at the input; that is,

$$\frac{v_L}{v_s} = \frac{v_L}{v_{i1}}\frac{v_{i1}}{v_s} = A_v\frac{v_{i1}}{v_s}$$

$$= 818 \times 0.909 = 743.6\ \text{V/V}$$

or 57.4 dB.

The current gain is found as follows:

$$A_i \equiv \frac{i_o}{i_i} = \frac{v_L/100\ \Omega}{v_{i1}/1\ \text{M}\Omega}$$
$$= 10^4 \times A_v = 8.18 \times 10^6\ \text{A/A}$$

or 138.3 dB.

The power gain is found from

$$A_p \equiv \frac{P_L}{P_I} = \frac{v_L i_o}{v_{i1} i_i}$$
$$= A_v A_i = 818 \times 8.18 \times 10^6 = 66.9 \times 10^8\ \text{W/W}$$

or 98.3 dB. Note that

$$A_p(\text{dB}) = \tfrac{1}{2}[A_v(\text{dB}) + A_i(\text{dB})]$$

Exercises 1.7 A phonograph cartridge characterized by a voltage of 1 V rms and a resistance of 1 MΩ is available to drive a 10-Ω loudspeaker. If connected directly, what voltage and power levels result at the loudspeaker? If a unity-gain (that is, $A_{vo} = 1$) buffer amplifier with 1-MΩ input resistance and 10-Ω output resistance is interposed between source and load, what do the output voltage and power levels become? For the new arrangement find the voltage gain from source to load, and the power gain (both expressed in decibels).

Ans. 10 μV rms; 10^{-11} W; 0.25 V; 6.25 mW; −12 dB; 44 dB

1.8 The output voltage of a voltage amplifier has been found to decrease by 20% when a load resistance of 1 kΩ is connected. What is the value of the amplifier output resistance?

Ans. 250 Ω

1.9 An amplifier with a voltage gain of +40 dB, an input resistance of 10 kΩ, and an output resistance of 1 kΩ is used to drive a 1-kΩ load. What is the value of A_{vo}? Find the value of power gain in dB.

Ans. 100 V/V; 44 dB

Other Amplifier Types

In the design of an electronic system, the signal of interest—whether at the system input, at an intermediate stage, or at the output—can be either a voltage or a current. For instance, some transducers have very high output resistances and can be more appropriately modeled as current sources. Similarly, there are applications in which the output current rather than the voltage is of interest. Thus, although it is the most popular, the voltage amplifier considered above is just one of four possible amplifier types. The other three are the current amplifier, the transconductance amplifier, and the transresistance amplifier.

Figure 1.16(a) shows a circuit model for the current amplifier. It consists of a current-controlled current source with a current-gain factor A_{is}, an input resistance R_i, and an output resistance R_o. Figure 1.16(b) shows the current amplifier fed with a signal current

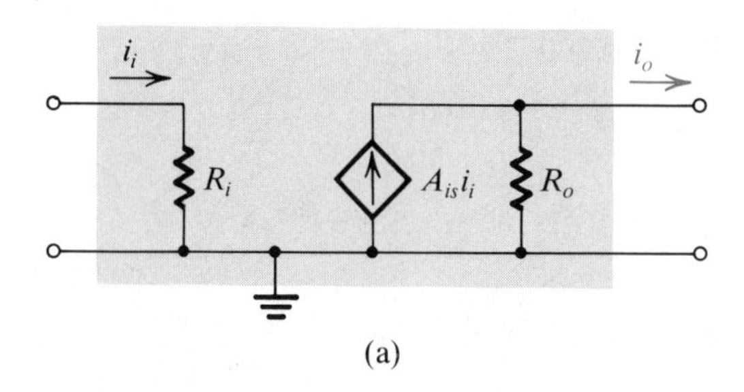

(a)

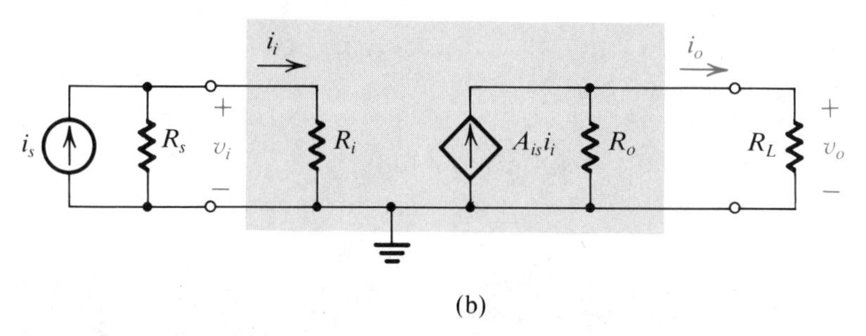

(b)

Fig. 1.16 (a) Circuit model for the current amplifier. **(b)** The current amplifier with input signal source and load.

source i_s having a resistance R_s, and with a load resistance R_L connected at the output. Using the current-divider rule at the output we find i_o as

$$i_o = A_{is}i_i\frac{R_o}{R_o + R_L}$$

Thus, the current gain of the loaded amplifier is given by

$$A_i \equiv \frac{i_o}{i_i} = A_{is}\frac{R_o}{R_o + R_L} \tag{1.13}$$

It follows that to avoid loss of gain in coupling the current amplifier to its load, the amplifier must be designed so that its output resistance R_o is much greater than the load resistance R_L. An ideal current amplifier has an infinite output resistance. Also note that with $R_L = 0$, the current gain is equal to A_{is}. Thus A_{is} is called the **short-circuit current gain.**

At the input side, the input resistance R_i causes current-divider action, with the result that only a fraction of i_s reaches the input of the amplifier; that is,

$$i_i = i_s\frac{R_s}{R_s + R_i} \tag{1.14}$$

To reduce the signal loss at the input side, the current amplifier must be designed so that $R_i \ll R_s$. The ideal current amplifier has $R_i = 0$.

Figure 1.17(a) shows a circuit model for the transconductance amplifier. This type of

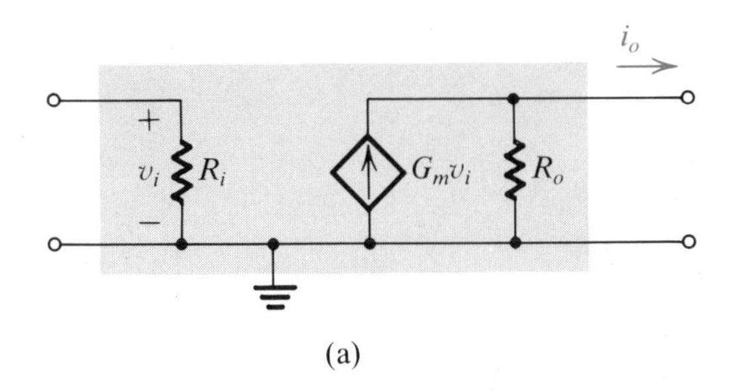

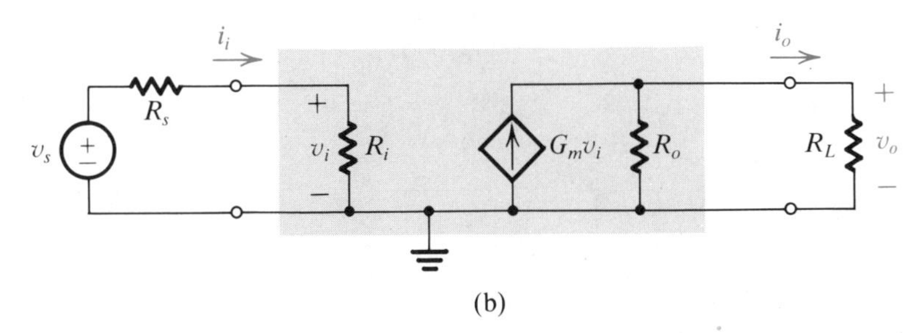

Fig. 1.17 (a) Circuit model for the transconductance amplifier. **(b)** The transconductance amplifier with input signal source and load.

amplifier is intended to work with a voltage input signal and to provide an output current signal, as indicated in Fig. 1.17(b). The gain parameter G_m is the ratio of the short-circuit output current to the input voltage. Thus G_m represents the transfer conductance of the amplifier and is called the **short-circuit transconductance** and has the dimension of ohms or A/V. An ideal transconductance amplifier has an infinite input resistance and an infinite output resistance. We will employ the transconductance amplifier extensively to model the operation of transistor amplifiers.

Finally, we show in Fig. 1.18(a) an equivalent circuit model for the transresistance amplifier. As indicated in Fig. 1.18(b) this type of amplifier is intended to operate with an input current signal and to provide an output voltage signal. The gain parameter R_m is the ratio of the open-circuit output voltage to the input current. It represents the transfer resistance of the amplifier and is called the **open-circuit transresistance** and has the dimension of ohms or V/A. An ideal transresistance amplifier has a zero input resistance and a zero output resistance.

Relationships Between the Four Amplifier Models

Although for a given amplifier a particular one of the four models above is most preferable, *any of the four can be used to model the amplifier*. In fact, simple relationships can be derived to relate the parameters of the various models. For instance, the open-circuit voltage gain A_{vo} can be related to the short-circuit current gain A_{is} as follows: The open-circuit output voltage given by the voltage amplifier model of Fig. 1.14(a) is $A_{vo}v_i$. The

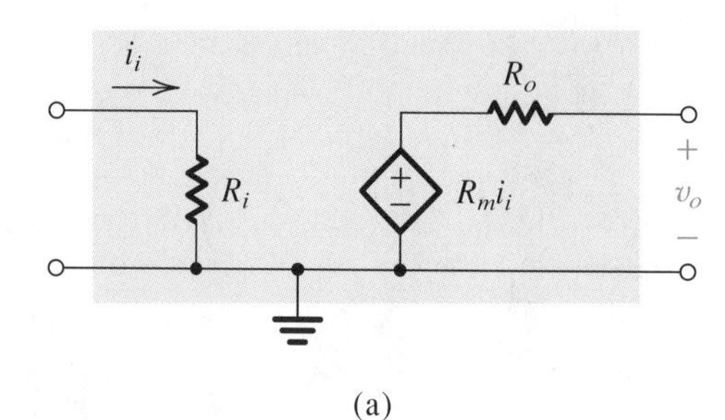

(a)

(b)

Fig. 1.18 (a) Circuit model for the transresistance amplifier. **(b)** The transresistance amplifier with signal source and load.

current amplifier model of Fig. 1.16(a) gives an open-circuit output voltage of $A_{is} i_i R_o$. Equating these two values and noting that $i_i = v_i/R_i$ gives

$$A_{vo} v_i = A_{is}\left(\frac{v_i}{R_i}\right)R_o$$

Thus

$$A_{vo} = A_{is}\left(\frac{R_o}{R_i}\right) \tag{1.15}$$

Similarly, we can show that

$$A_{vo} = G_m R_o \tag{1.16}$$

and

$$A_{vo} = \frac{R_m}{R_i} \tag{1.17}$$

The expressions in Eqs. (1.15) to (1.17) can be used to relate any two of the gain parameters A_{vo}, A_{is}, G_m, and R_m.

From the amplifier circuit models given in Figs. 1.14 to 1.18 we observe that the input resistance R_i of the amplifier can be determined by applying an input voltage v_i and measuring (or calculating) the input current i_i; that is, $R_i = v_i/i_i$. The output resistance is found as the ratio of the open-circuit output voltage to the short-circuit output current.

Alternatively, the output resistance can be found by eliminating the input signal source (then i_i and v_i will both be zero) and applying a voltage signal v_x to the output of the amplifier. If we denote the current drawn from v_x *into* the output terminals by i_x (note that i_x is opposite in direction to i_o), then $R_o = v_x/i_x$. Although these techniques are conceptually correct, in actual practice more refined methods are employed in measuring R_i and R_o.

The amplifier models considered above are **unilateral;** that is, signal flow is unidirectional, from input to output. Most real amplifiers show some reverse transmission, which is usually undesirable but must nonetheless be modeled. We shall not pursue this point further at this time except to mention that more complete models for linear two-port networks are given in Appendix B.

EXAMPLE 1.4

The **bipolar junction transistor (BJT)** is a three-terminal device having the circuit symbol shown in Fig. 1.19(a) with the terminals called **emitter** (E), **base** (B), and **collector** (C). The device is basically nonlinear; thus the relationships between the total instantaneous terminal currents and voltages are generally nonlinear. Nevertheless, the linearization technique discussed in Section 1.4 can be used to provide linear operation for small signals—that is, incremental variations in currents and voltages around a bias or quiescent

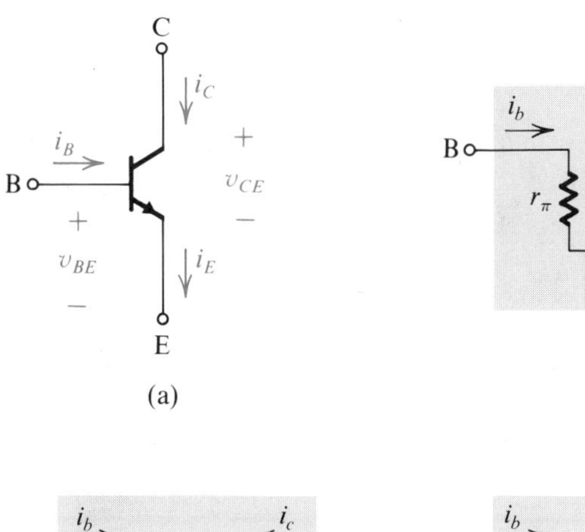

i_b i_c B C + r_π v_{be} − $g_m v_{be}$ i_e E

(c)

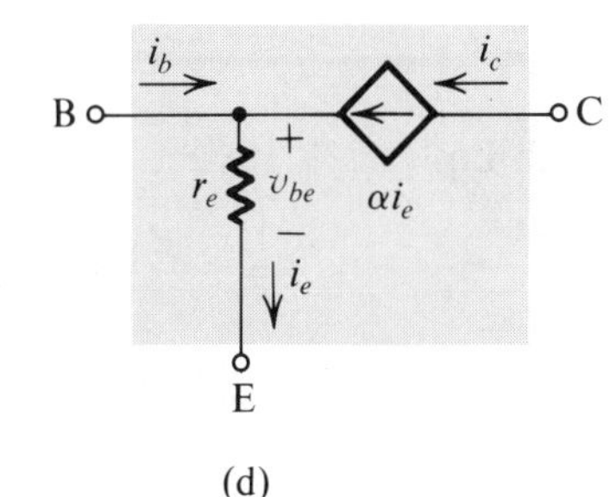

Fig. 1.19 (a) Circuit symbol for the bipolar junction transistor (BJT), illustrating the definition of the total instantaneous terminal quantities. **(b, c, d)** Equivalent circuit models for the small-signal linear operation of the BJT.

point. Thus the total instantaneous quantities can be expressed as the sums of dc or bias quantities and signal quantities as

$$v_{BE} = V_{BE} + v_{be} \qquad i_B = I_B + i_b$$
$$v_{CE} = V_{CE} + v_{ce} \qquad i_C = I_C + i_c$$
$$i_E = I_E + i_e$$

It will be shown in Chapter 4 that under the small-signal approximation the relationships between the signal quantities are linear and that the three-terminal transistor can be represented by one of the equivalent circuit models of Fig. 1.19(b), (c), and (d). If these models are equivalent, we wish to find expressions for the parameters of the models in Figs. 1.19(c) and (d) in terms of the parameters of the model in Fig. 1.19(b).

Solution For the model in Fig. 1.19(b) we have

$$i_b = \frac{v_{be}}{r_\pi} \tag{1.18}$$

$$i_c = \beta i_b \tag{1.19}$$

$$i_e = (\beta + 1)i_b \tag{1.20}$$

For the model in Fig. 1.19(c) we have

$$i_c = g_m v_{be} \tag{1.21}$$

Use of Eqs. (1.18), (1.19), and (1.21) gives

$$g_m = \frac{\beta}{r_\pi}$$

For the model in Fig. 1.19(d) we have

$$i_b = (1 - \alpha)i_e \tag{1.22}$$

$$i_c = \alpha i_e \tag{1.23}$$

$$i_e = \frac{v_{be}}{r_e} \tag{1.24}$$

Use of Eqs. (1.19) and (1.20) gives

$$\frac{i_c}{i_e} = \frac{\beta}{\beta + 1}$$

Comparing this result with Eq. (1.23), we get

$$\alpha = \frac{\beta}{\beta + 1}$$

Equations (1.18) and (1.20) can be combined to yield

$$i_e = \frac{\beta + 1}{r_\pi} v_{be}$$

Comparison of this result with Eq. (1.24) provides

$$r_e = \frac{r_\pi}{\beta + 1}$$

We have thus obtained expressions for the parameters of the models in Fig. 1.19(c) and (d) in terms of those for the model in Fig. 1.19(b).

EXAMPLE 1.5

A **gyrator** is a two-port network that can be realized by connecting in parallel and back-to-back two voltage-controlled current sources of opposite polarities. Such an arrangement is shown in Fig. 1.20(a) (we assume ideal voltage-controlled current sources). It is required to show that if port 2 of the gyrator is terminated in a capacitance, as shown in Fig. 1.20(b), then input port 1 will have the i-v relationship of an inductance.

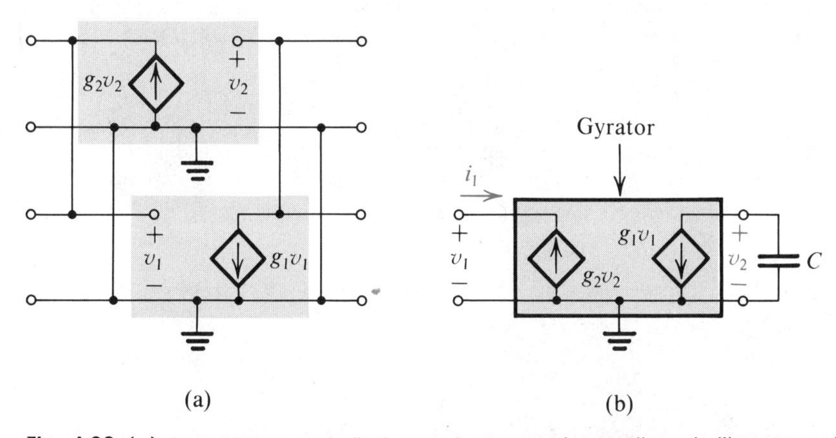

Fig. 1.20 (a) Two voltage-controlled current sources of opposite polarities connected back-to-back in parallel to form a gyrator. **(b)** The gyrator of (a) terminated at port 2 in a capacitance.

Solution We must show that in the circuit of Fig. 1.20(b)

$$v_1 = L\frac{di_1}{dt}$$

where L is a constant representing an inductance. To obtain the relationship between i_1 and v_1, consider first port 2, where we have

$$v_2 = -\frac{1}{C}\int_0^t g_1v_1\,dt$$

where it has been assumed that C was originally (at $t = 0$) uncharged. The current i_1 is given by

$$i_1 = -g_2v_2 = \frac{g_1g_2}{C}\int_0^t v_1\,dt$$

which can be written in differential form as

$$v_1 = \frac{C}{g_1 g_2} \frac{di_1}{dt}$$

Thus the i_1-v_1 relationship is of the form found in an inductor, with the inductance L given by

$$L = \frac{C}{g_1 g_2}$$

Thus the gyrator can be used to realize an inductance using active elements (controlled sources) and a capacitor. This is a significant and useful result because physical inductors are almost impossible to fabricate using integrated circuit (IC) technology.

Exercises 1.10 A current amplifier with an input resistance of 100 Ω, an output resistance of 100 kΩ, and a current gain of 10,000 is connected between a signal current source with a 100-kΩ source resistance and a 100-ohm load. Find the voltage gain and the power gain.

Ans. 80 dB; 80 dB

1.11 A transconductance amplifier with an input resistance of 10 kΩ, an output resistance of 10 kΩ, and a transconductance of 1000 mA/V is connected between a 10-kΩ voltage source and a 10-kΩ load. Find the voltage gain from source to load (that is, v_o/v_s).

Ans. 2500 V/V or 68 dB

1.12 Consider a transistor modeled as in Fig. 1.19(b) to be connected between a voltage source with a 1-kΩ source resistance and a load resistance of 1 kΩ. Let $r_\pi = 2$ kΩ and $\beta = 90$. Calculate the voltage gain (v_c/v_s) and the current gain (i_c/i_b) as ratios. Also give the power gain in decibels.

Ans. −30 V/V; 90 A/A; 36.1 dB

1.13 Find the input resistance between terminals B and G in the circuit shown in Fig. E1.13. The voltage v_x is a test voltage

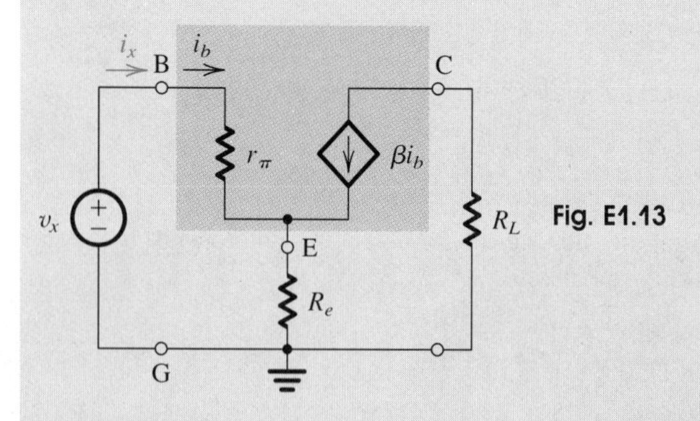

Fig. E1.13

with the input resistance R_{in} defined as $R_{in} \equiv v_x/i_x$.

Ans. $R_{in} = r_\pi + (\beta + 1)R_e$

1.6 FREQUENCY RESPONSE OF AMPLIFIERS

From Section 1.2 we know that the input signal to an amplifier can always be expressed as the sum of sinusoidal signals. It follows that an important characterization of an amplifier is in terms of its response to input sinusoids of different frequencies. Such characterization of amplifier performance is known as the amplifier frequency response.

Measuring the Amplifier Frequency Response

We shall introduce the subject of amplifier frequency response by showing how it can be measured. Figure 1.21 depicts a linear voltage amplifier fed at its input with a sine-wave signal of amplitude V_i and frequency ω. As the figure indicates, the signal measured at the amplifier output also is sinusoidal with exactly the same frequency ω. This is an important point to note: Whenever a sine-wave signal is applied to a linear circuit, the resulting output is sinusoidal with the same frequency as the input. In fact, the sine wave is the only signal that does not change shape as it passes through a linear circuit. Observe, however,

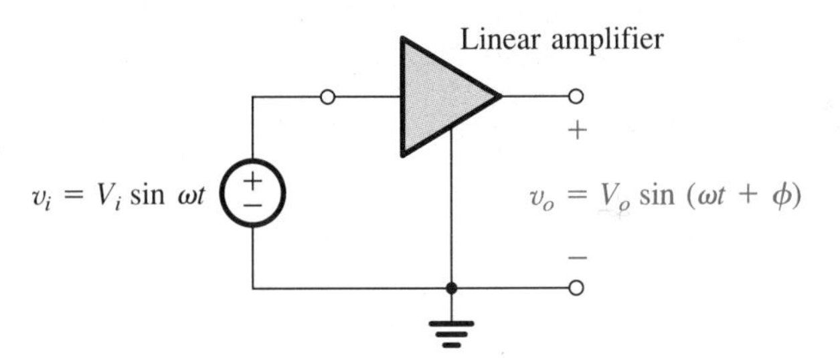

Fig. 1.21 Measuring the frequency response of a linear amplifier. At the test frequency ω, the amplifier gain is characterized by its magnitude (V_o/V_i) and phase ϕ.

that the output sinusoid will in general have a different amplitude and will be shifted in phase relative to the input. The ratio of the amplitude of the output sinusoid (V_o) to the amplitude of the input sinusoid (V_i) is the magnitude of the amplifier gain or transmission at the test frequency ω. Also, the angle ϕ is the phase of the amplifier transmission at the test frequency ω. If we denote the **amplifier transmission,** or **transfer function** as it is more commonly known, by $T(\omega)$ then

$$|T(\omega)| = \frac{V_o}{V_i}$$

$$\angle T(\omega) = \phi$$

The response of the amplifier to a sinusoid of frequency ω is completely described by $|T(\omega)|$ and $\angle T(\omega)$. Now, to obtain the complete frequency response of the amplifier we simply change the frequency of the input sinusoid and measure the new value for $|T|$ and $\angle T$. The end result will be a table and/or graph of gain magnitude [$|T(\omega)|$] versus frequency, and a table and/or graph of phase angle [$\angle T(\omega)$] versus frequency. These two plots together constitute the frequency response of the amplifier; the first is known as the magnitude or amplitude response, and the second is the phase response.

Amplifier Bandwidth

Figure 1.22 shows the magnitude response of an amplifier. It indicates that the gain is almost constant over a wide frequency range, roughly between ω_1 and ω_2. Signals whose frequencies are below ω_1 or above ω_2 will experience lower gain, with the gain decreasing as we move farther away from ω_1 and ω_2. The band of frequencies over which the gain of the amplifier is almost constant, to within a certain number of decibels (usually 3 dB), is called the **amplifier bandwidth.** Normally the amplifier is designed so that its bandwidth coincides with the spectrum of the signals it is required to amplify. If this were not the case, the amplifier would *distort* the frequency spectrum of the input signal, with different components of the input signal being amplified by different amounts.

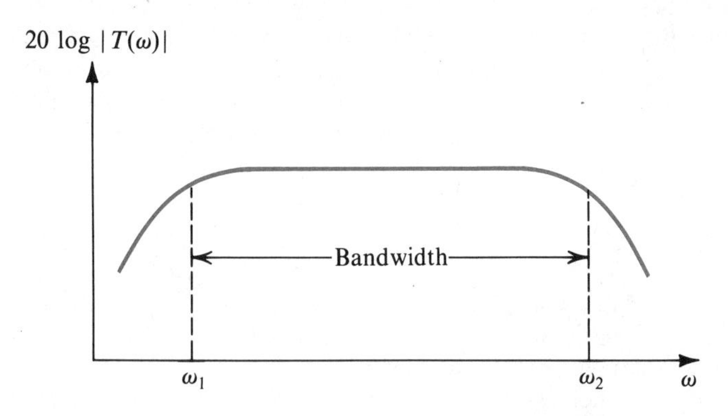

Fig. 1.22 Typical magnitude response of an amplifier. $T(\omega)$ is the amplifier transfer function—that is, the ratio of the output $V_o(\omega)$ to the input $V_i(\omega)$.

Evaluating the Frequency Response of Amplifiers

Above, we described the method used to measure the frequency response of an amplifier. We now briefly discuss the method for analytically obtaining an expression for the frequency response. What we are about to say is just a preview of this important subject, whose detailed study starts in Chapter 7.

To evaluate the frequency response of an amplifier one has to analyze the amplifier equivalent circuit model, taking into account all reactive components.[2] Circuit analysis proceeds in the usual fashion but with inductances and capacitances represented by their reactances. An inductance L has a reactance or impedance $j\omega L$, and a capacitance C has a reactance or impedance $1/j\omega C$ or, equivalently, a susceptance or admittance $j\omega C$. Thus in a *frequency-domain* analysis we deal with impedances and/or admittances. The result of the analysis is the amplifier transfer function $T(\omega)$,

$$T(\omega) = \frac{V_o(\omega)}{V_i(\omega)}$$

where $V_i(\omega)$ and $V_o(\omega)$ denote the input and output signals, respectively. $T(\omega)$ is generally a complex function whose magnitude $|T(\omega)|$ gives the magnitude of transmission or the

[2]Note that in the models considered in previous sections no reactive components were included. These were simplified models and cannot be used alone to predict the amplifier frequency response.

magnitude response of the amplifier. The phase of $T(\omega)$ gives the phase response of the amplifier.

In the analysis of a circuit to determine its frequency response, the algebraic manipulations can be considerably simplified by using the **complex frequency variable** s. In terms of s, the impedance of an inductance L is sL and that of a capacitor C is $1/sC$. Replacing the reactive elements with their impedances and performing standard circuit analysis, we obtain the transfer function $T(s)$ as

$$T(s) \equiv \frac{V_o(s)}{V_i(s)}$$

Subsequently, we replace s by $j\omega$ to determine the network transfer function for **physical frequencies,** $T(j\omega)$. Note that $T(j\omega)$ is the same function we called $T(\omega)$ above;[3] the additional j is included in order to emphasize that $T(j\omega)$ is obtained from $T(s)$ by replacing s with $j\omega$.

Single-Time-Constant Networks

In analyzing amplifier circuits to determine their frequency response, one is greatly aided by knowledge of the frequency response characteristics of single-time-constant (STC) networks. An STC network is one that is composed of, or can be reduced to, one reactive component (inductance or capacitance) and one resistance. Examples are shown in Fig. 1.23. An STC network formed of an inductance L and a resistance R has a time constant $\tau = L/R$. The time constant τ of an STC network composed of a capacitance C and a resistance R is given by $\tau = CR$.

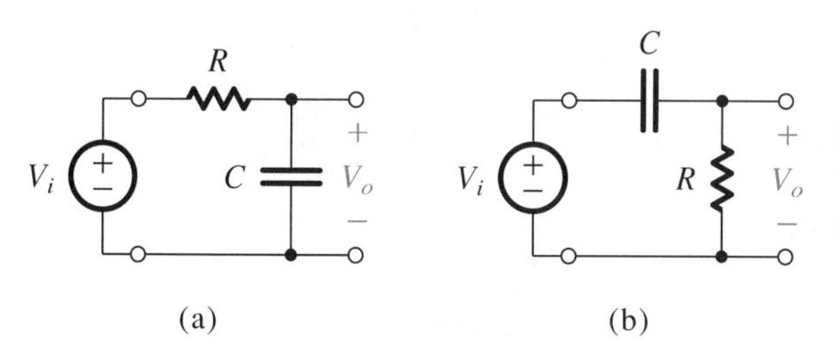

Fig. 1.23 Two examples of STC networks: **(a)** a low-pass network and **(b)** a high-pass network.

Appendix E presents a study of STC networks and their responses to sinusoidal, step, and pulse inputs. Knowledge of this material will be needed at various points throughout this text, and the reader will be encouraged to refer to the Appendix. At this point we need in particular the frequency response results; we will, in fact, briefly discuss this important topic.

Most STC networks can be classified into two categories:[4] **low pass (LP)** and **high pass (HP),** with each of the two categories displaying distinctly different signal re-

[3] At this stage we are using s simply as a shorthand for $j\omega$. We shall not require detailed knowledge of s-plane concepts until Chapter 7.

[4] An important exception is the *all-pass* STC network studied in Problem 1.31 and later in Chapter 11.

sponses. As an example, the STC network shown in Fig. 1.23(a) is of the *low-pass* type, and that in Fig. 1.23(b) is of the *high-pass* type. To see the reasoning behind this classification, observe that the transfer function of each of these two circuits can be expressed as a voltage divider ratio, with the divider composed of a resistor and a capacitor. Now recalling how the impedance of a capacitor varies with frequency ($Z = 1/j\omega C$) it is easy to see that the transmission of the circuit in Fig. 1.23(a) will decrease with frequency and approach zero as ω approaches ∞. Thus the circuit of Fig. 1.23(a) acts as a **low-pass filter;**[5] it passes low-frequency sine-wave inputs with little or no attenuation (at $\omega = 0$, the transmission is unity) and attenuates high-frequency input sinusoids. The circuit of Fig. 1.23(b) does the opposite; its transmission is unity at $\omega = \infty$ and decreases as ω is reduced, reaching 0 for $\omega = 0$. The latter circuit, therefore, performs as a **high-pass filter.**

Table 1.1 provides a summary of the frequency response results for STC networks of both types. Also, sketches of the magnitude and phase responses are given in Figs. 1.24 and 1.25. These frequency response diagrams are known as Bode plots and the 3-dB

Table 1.1 FREQUENCY RESPONSE OF STC NETWORKS

	Low-Pass (LP)	High-Pass (HP)
Transfer Function $T(s)$	$\dfrac{K}{1 + (s/\omega_0)}$	$\dfrac{Ks}{s + \omega_0}$
Transfer Function (for physical frequencies) $T(j\omega)$	$\dfrac{K}{1 + j(\omega/\omega_0)}$	$\dfrac{K}{1 - j(\omega_0/\omega)}$
Magnitude Response $\lvert T(j\omega)\rvert$	$\dfrac{\lvert K\rvert}{\sqrt{1 + (\omega/\omega_0)^2}}$	$\dfrac{\lvert K\rvert}{\sqrt{1 + (\omega_0/\omega)^2}}$
Phase Response $\angle T(j\omega)$	$-\tan^{-1}(\omega/\omega_0)$	$\tan^{-1}(\omega_0/\omega)$
Transmission at $\omega = 0$ (dc)	K	0
Transmission at $\omega = \infty$	0	K
3-dB Frequency	$\omega_0 = 1/\tau$; $\tau \equiv$ time constant $\tau = CR$ or L/R	
Bode Plots	in Fig. 1.24	in Fig. 1.25

[5]A filter is a circuit that passes signals in a specified frequency band (the filter passband) and stops or severely attenuates (filters out) signals in another frequency band (the filter stopband). Filters will be studied in Chapter 11.

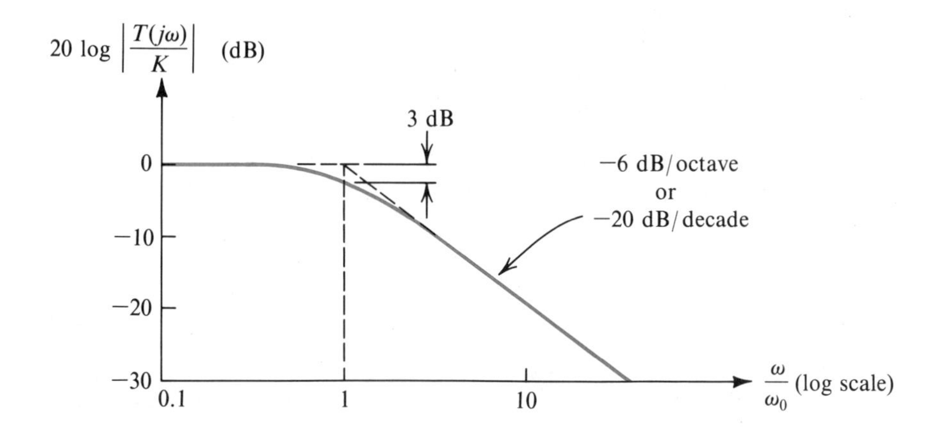

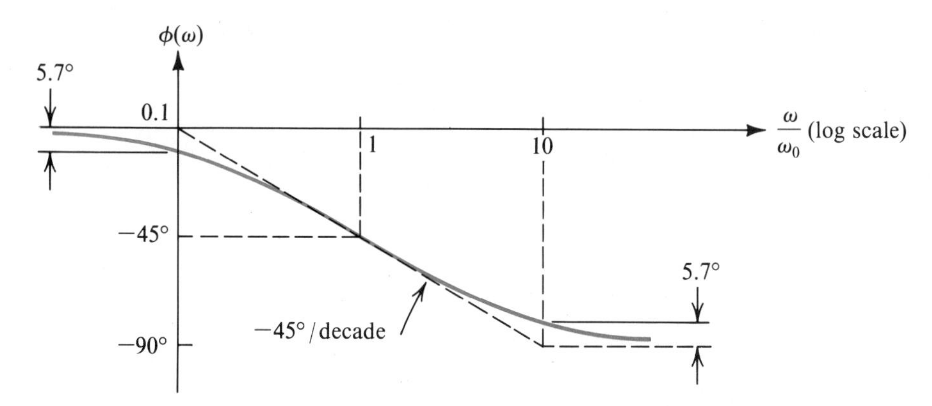

Fig. 1.24 (a) Magnitude and **(b)** phase response of STC networks of the low-pass type.

frequency (ω_0) is also known as the **corner frequency** or **break frequency.** The reader is urged to become familiar with this information and to consult Appendix E if further clarifications are needed.

EXAMPLE 1.6

Figure 1.26 shows a voltage amplifier having an input resistance R_i, an input capacitance C_i, a gain factor μ, and an output resistance R_o. The amplifier is fed with a voltage source V_s having a source resistance R_s, and a load of resistance R_L is connected to the output.

(a) Derive an expression for the amplifier voltage gain V_o/V_s as a function of frequency. From this find expressions for the dc gain and the 3-dB frequency.

(b) Calculate the values of the dc gain, the 3-dB frequency, and the frequency at which the gain becomes 0 dB (i.e., unity) for the case $R_s = 20\ \text{k}\Omega$, $R_i = 100\ \text{k}\Omega$, $C_i = 60\ \text{pF}$, $\mu = 144$, $R_o = 200\ \Omega$, and $R_L = 1\ \text{k}\Omega$.

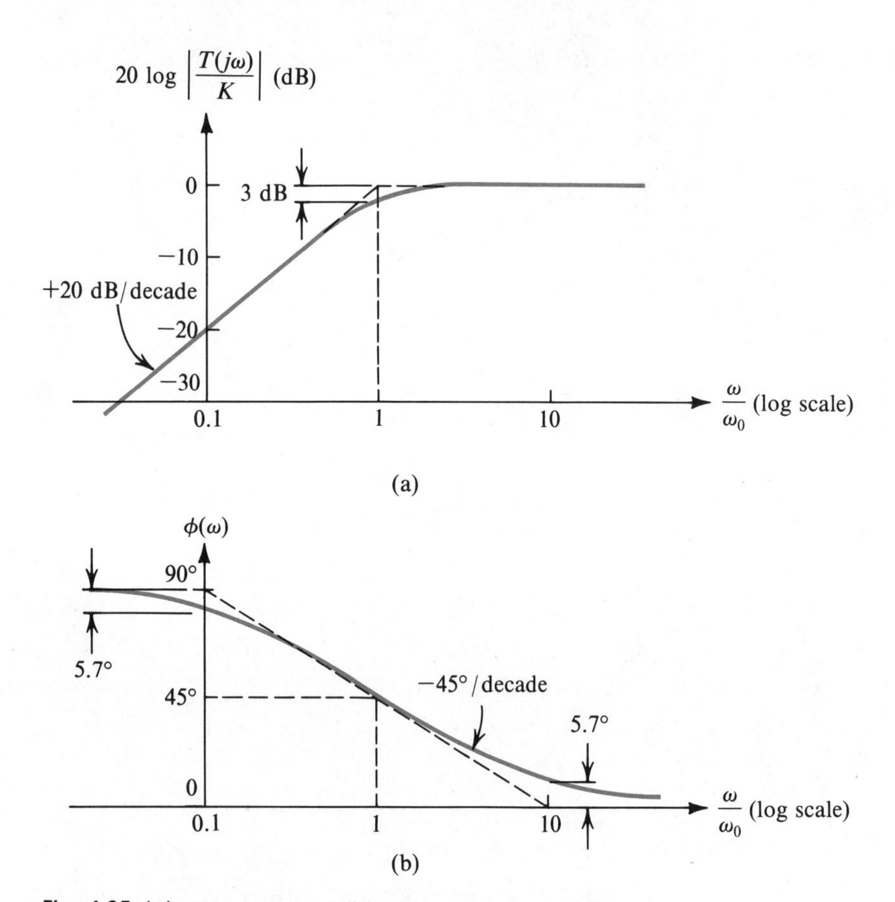

Fig. 1.25 (a) Magnitude and **(b)** phase response of STC networks of the high-pass type.

(c) Find $v_o\ (t)$ for each of the following inputs:

(i) $v_i = 0.1 \sin 10^2\ t$, V

(ii) $v_i = 0.1 \sin 10^5\ t$, V

(iii) $v_i = 0.1 \sin 10^6\ t$, V

(iv) $v_i = 0.1 \sin 10^8\ t$, V

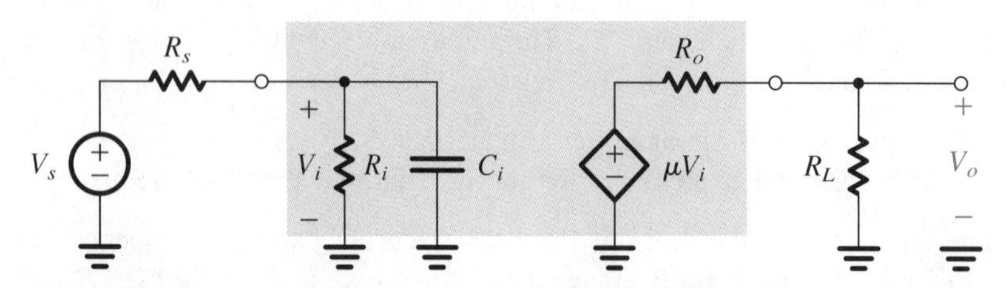

Fig. 1.26 Circuit for Example 1.6.

Solution

(a) Utilizing the voltage divider rule, we can express V_i in terms of V_s as follows

$$V_i = V_s \frac{Z_i}{Z_i + R_s}$$

where Z_i is the amplifier input impedance. Since Z_i is composed of two parallel elements it is obviously easier to work in terms of Y_i,

$$V_i = V_s \frac{1}{1 + R_s Y_i}$$

$$= V_s \frac{1}{1 + R_s[(1/R_i) + sC_i]}$$

Thus,

$$\frac{V_i}{V_s} = \frac{1}{1 + (R_s/R_i) + sC_i R_s}$$

This expression can be put in the standard form for a low-pass STC network (see top of Table 1.1) by extracting $[1 + (R_s/R_i)]$ from the denominator; thus we have

$$\frac{V_i}{V_s} = \frac{1}{1 + (R_s/R_i)} \frac{1}{1 + sC_i[(R_s R_i)/(R_s + R_i)]} \tag{1.25}$$

At the output side of the amplifier we can write

$$V_o = \mu V_i \frac{R_L}{R_L + R_o}$$

This equation can be combined with Eq. (1.25) to obtain the amplifier transfer function as

$$\frac{V_o}{V_s} = \mu \frac{1}{1 + (R_s/R_i)} \frac{1}{1 + (R_o/R_L)} \frac{1}{1 + sC_i[(R_s R_L)/(R_s + R_i)]} \tag{1.26}$$

We note that only the last factor in this expression is new (compared to the expression derived in the last section). This factor is a result of the input capacitance C_i, with the time constant being

$$\tau = C_i \frac{R_s R_i}{R_s + R_i}$$
$$= C_i(R_s // R_i) \tag{1.27}$$

We could have obtained this result by inspection: From Fig. 1.26 we see that the input circuit is an STC network and that its time constant can be found by reducing V_s to zero, with the result that the resistance seen by C_i is R_i in parallel with R_s. From Eq. (1.26), the dc gain is found as

$$K \equiv \frac{V_o}{V_i}(0) = \mu \frac{1}{1 + (R_s/R_i)} \frac{1}{1 + (R_o/R_L)} \tag{1.28}$$

The 3-dB frequency ω_0 can be found from

$$\omega_0 = \frac{1}{\tau} = \frac{1}{C_i(R_s//R_i)} \tag{1.29}$$

Since the frequency response of this amplifier is of the low-pass STC type, the Bode plots for the gain magnitude and phase will take the form shown in Fig. 1.24, where K is given by Eq. (1.28) and ω_0 is given by Eq. (1.29).

(b) Substituting the numerical values given into Eq. (1.28) results in

$$K = 144 \frac{1}{1 + (20/100)} \frac{1}{1 + (200/1000)} = 100 \text{ V/V}$$

Thus the amplifier has a dc gain of 40 dB. Substituting the numerical values into Eq. (1.29) gives the 3 dB frequency

$$\omega_0 = \frac{1}{60 \text{ pF} \times (20 \text{ k}\Omega//100 \text{ k}\Omega)}$$

$$= \frac{1}{60 \times 10^{-12} \times (20 \times 100/(20 + 100)) \times 10^3} = 10^6 \text{ rad/s}$$

Thus,

$$f_0 = \frac{10^6}{2\pi} = 159.2 \text{ kHz}$$

Since the gain falls off at the rate of -20 dB/decade, starting at ω_0 (see Fig. 1.24a) the gain will reach 0 dB in two decades; thus we have

$$\text{Unity-gain frequency} = 10^8 \text{ rad/s or } 15.92 \text{ MHz}$$

(c) To find $v_o(t)$ we need to determine the gain magnitude and phase at 10^2, 10^5, 10^6, and 10^8 rad/s. This can be done either approximately utilizing the Bode plots of Fig. 1.24, or exactly utilizing the expression for the amplifier transfer function,

$$T(j\omega) \equiv \frac{V_o}{V_i}(j\omega) = \frac{100}{1 + j(\omega/10^6)}$$

We shall do both.

(i) For $\omega = 10^2$ rad/s, which is $(\omega_0/10^4)$, the Bode plots of Fig. 1.24 suggest that $|T| \simeq K = 100$ and $\phi = 0°$. The transfer function expression gives $|T| \simeq 100$ and $\phi = -\tan^{-1} 10^{-4} \simeq 0°$. Thus

$$v_o(t) = 10 \sin 10^2 t, \text{ V}$$

(ii) For $\omega = 10^5$ rad/s, which is $(\omega_0/10)$, the Bode plots of Fig. 1.24 suggest that $|T| \simeq K = 100$ and $\phi = -5.7°$. The transfer function expression gives $|T| = 99.5$ and $\phi = -\tan^{-1} 0.1 = -5.7°$. Thus

$$v_o(t) = 9.95 \sin(10^5 t - 5.7°), \text{ V}$$

(iii) For $\omega = 10^6$ rad/s $= \omega_0$, $|T| = 100/\sqrt{2} = 70.7$ V/V or 37 dB and $\phi = -45°$. Thus

$$v_o(t) = 7.07 \sin(10^6 t - 45°), \text{ V}$$

(iv) For $\omega = 10^8$ rad/s, which is ($100\ \omega_0$), the Bode plots suggest that $|T| = 1$ and $\phi = -90°$. The transfer function expression gives

$$|T| \simeq 1 \text{ and } \phi = -\tan^{-1} 100 = -89.4°,$$

Thus

$$v_o(t) = 0.1 \sin(10^8 t - 89.4°), \text{ V}$$

Classification of Amplifiers Based on Frequency Response

Amplifiers can be classified based on the shape of their magnitude-response curve. Figure 1.27 shows typical frequency response curves for various amplifier types. In Figure 1.27(a) the gain remains constant over a wide frequency range but falls off at low and high frequencies. This is a common type of frequency response found in audio amplifiers.

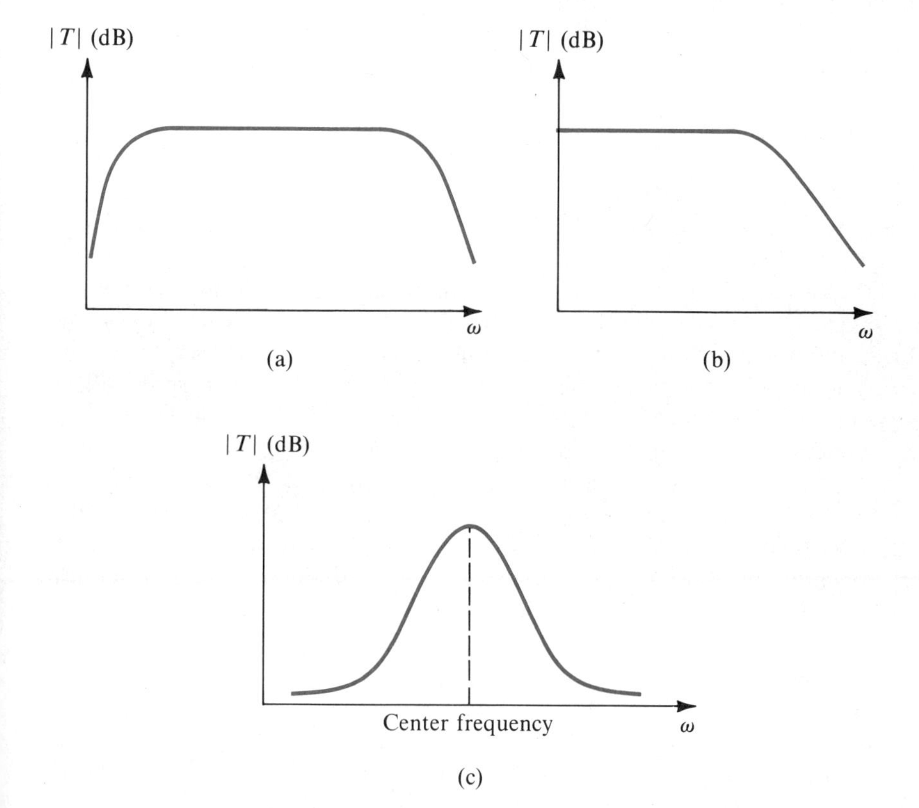

Fig. 1.27 Frequency response for **(a)** a capacitively coupled amplifier, **(b)** a direct-coupled amplifier, and **(c)** a tuned or bandpass amplifier.

As will be shown in Chapter 7, **internal capacitances** in the device (transistor) cause the falloff of gain at high frequencies, just as C_i did in the circuit of Example 1.6. On the other hand, the falloff of gain at low frequencies is usually caused by **coupling capacitors** used to connect one amplifier stage to another, as indicated in Fig. 1.28. This practice is usually adopted to simplify the design process of the different stages. The coupling capacitors are usually chosen quite large (a fraction of a microfarad to a few tens of microfarads) so that their reactance (impedance) is small at the frequencies of interest. Nevertheless, at sufficiently low frequencies the reactance of a coupling capacitor will become large enough to cause part of the signal being coupled to appear as a voltage drop across the coupling capacitor and thus not reach the subsequent stage. Coupling capacitors will thus cause loss of gain at low frequencies and cause the gain to be zero at dc.

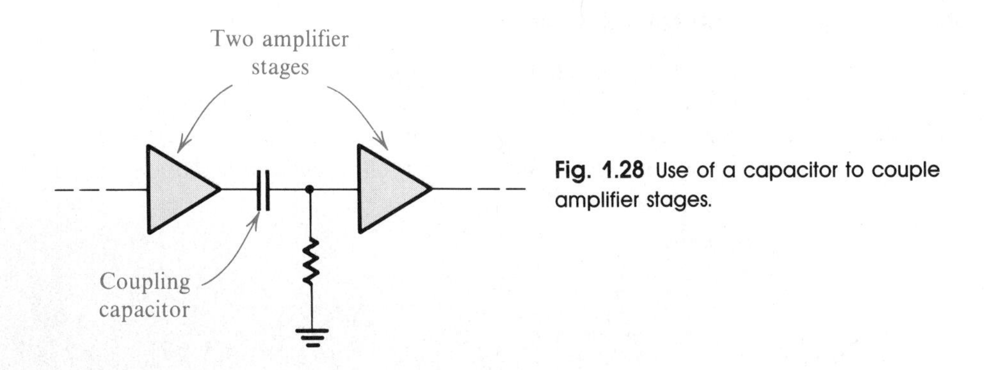

Fig. 1.28 Use of a capacitor to couple amplifier stages.

There are many applications in which it is important that the amplifier maintain its gain at low frequencies down to dc. Furthermore, monolithic integrated circuit (IC) technology does not allow the fabrication of large coupling capacitors. Thus IC amplifiers are usually designed as **directly coupled** or **dc amplifiers** (as opposed to **capacitively coupled** or **ac amplifiers**). Figure 1.27(b) shows the frequency response of a dc amplifier. Such a frequency response characterizes what is referred to as a **low-pass amplifier.** Although it is not very appropriate, the term low-pass amplifier is also often used to refer to the amplifier whose response is shown in Fig. 1.27(a).

In a number of applications, such as in the design of radio and TV receivers, the need arises for an amplifier whose frequency response peaks around a certain frequency (called the **center frequency**) and falls off on both sides of this frequency, as shown in Fig. 1.27(c). Amplifiers with such a response are called **tuned amplifiers, bandpass amplifiers,** or **bandpass filters.** A tuned amplifier forms the heart of the front-end or tuner of a communication receiver; by adjusting its center frequency to coincide with the frequency of a desired communications channel, the signal of this particular channel can be received while those of other channels are attenuated or filtered out.

Exercises 1.14 Consider a voltage amplifier having a frequency response of the low-pass STC type with a dc gain of 60 dB and a 3-dB frequency of 1000 Hz. Find the gain in dB at $f = 10$ Hz, 10 kHz, 100 kHz, and 1 MHz.

Ans. 60 dB; 40 dB; 20 dB; 0 dB

1.15 Consider a transconductance amplifier having the model shown in Fig. 1.17(a) with $R_i = 5$ kΩ, $R_o = 50$ kΩ, and $G_m = 10$ mA/V. If the amplifier load consists of a resistance R_L in parallel with a capacitance C_L, convince yourself that the voltage transfer function realized, V_o/V_i, is of the low-pass STC type. What is the lowest value that R_L can have while a dc gain of at least 40 dB is obtained? With this value of R_L connected, find the highest value that C_L can have while a 3-dB bandwidth of at least 100 kHz is obtained.

Ans. 12.5 kΩ; 159.2 pF

1.16 Consider the situation illustrated in Fig. 1.28. Let the output resistance of the first voltage amplifier be 1 kΩ and the input resistance of the second voltage amplifier (including the resistor shown) be 9 kΩ. The resulting equivalent circuit is shown in Fig. E1.16 where V_s and R_s are the output voltage and output resistance of the first amplifier, C is a coupling

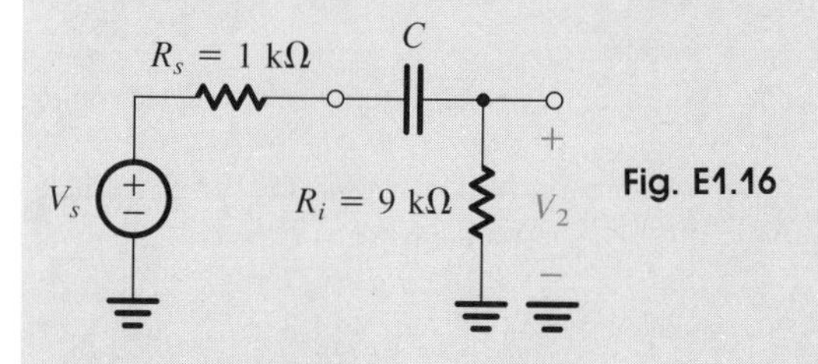

Fig. E1.16

capacitor, and R_i is the input resistance of the second amplifier. Convince yourself that V_2/V_s is a high-pass STC function. What is the smallest value for C that will ensure that the 3-dB frequency is not higher than 100 Hz?

Ans. 0.16 μF

SUMMARY

- An electrical signal source can be represented by either the Thévenin form (a voltage source in series with a source impedance) or the Norton form (a current source in parallel with a source impedance).

- The sine-wave signal is completely characterized by its peak value (or rms value which is the peak/$\sqrt{2}$), its frequency (ω in rad/s or f in Hz; $\omega = 2\pi f$ and $f = 1/T$ where T is the period in seconds), and its phase with respect to an arbitrary reference time.

- A signal can be represented either by its waveform versus time, or as the sum of sinusoids. The latter representation is known as the frequency spectrum of the signal.

- Analog signals have magnitudes that can assume any value. Electronic circuits that process analog signals are called analog circuits. Sampling the magnitude of an analog signal at discrete instants of time and representing each signal sample by a number, results in a digital signal. Digital signals are processed by digital circuits.

- The transfer characteristic, v_O versus v_I, of a linear amplifier is a straight line with a slope equal to the voltage gain.

- Amplifiers increase the signal power and thus require dc power supplies for their operation.

- Linear amplification can be obtained from a device having a nonlinear transfer characteristic by employing dc biasing and keeping the input signal amplitude small.

- Depending on the signal to be amplified (voltage or current) and on the desired form of output signal (voltage or current), there are four basic amplifier types: voltage, current, transconductance, and transresistance amplifiers.

- Sinusoidal signals are used to measure the frequency response of amplifiers.

- The transfer function $T(s) \equiv V_o(s)/V_i(s)$ of a voltage amplifier can be determined from circuit analysis. Substituting $s = j\omega$ gives $T(j\omega)$, whose magnitude $|T(j\omega)|$ is the magnitude response, and whose phase $\phi(\omega)$ is the phase response, of the amplifier.

▸ Amplifiers are classified according to the shape of their frequency response, $|T(j\omega)|$.

▸ Single-time-constant (STC) networks are those networks that are composed of, or can be reduced to, one reactive component (L or C) and one resistance (R). The time constant τ is either L/R or CR.

▸ STC networks can be classified into two categories: low-pass (LP) and high-pass (HP). LP networks pass dc and low frequencies and attenuate high frequencies. The opposite is true for HP networks.

▸ The gain of an LP (HP) STC circuit drops by 3 dB below the zero-frequency (infinite-frequency) value at a frequency $\omega_0 = 1/\tau$. At high frequencies (low frequencies) the gain falls off at the rate of 6 dB/octave or 20 dB/decade.

BIBLIOGRAPHY

E. F. Angelo, Jr., *Electronics: BJTs, FETs, and Microcircuits,* New York: McGraw-Hill, 1969.

L. S. Bobrow, *Elementary Linear Circuit Analysis,* 2nd ed., New York: Holt, Rinehart and Winston, 1987.

W. H. Hayt and J. E. Kemmerly, *Engineering Circuit Analysis,* 3rd ed., New York: McGraw-Hill, 1978.

Scientific American, Special Issue on microelectronics, Sept. 1977.

M. E. Van Valkenburg, *Network Analysis,* 3rd ed., Englewood Cliffs, N.J.: Prentice-Hall, 1974.

PROBLEMS

Section 1.2: Frequency Spectrum of Signals

1.1 A sinusoidal signal source has an open-circuit voltage of 10 mV and a short-circuit current of 10 μA. What is the source resistance?

1.2 Give expressions for the sine-wave voltage signals having:
(a) 10-V peak amplitude and 10-kHz frequency.
(b) 120-V rms and 60-Hz frequency.
(c) 0.2-V peak-to-peak and 1000-rad/s frequency.
(d) 100-mV peak and 1-ms period.

1.3 Illustrate the composition of a square-wave signal by sketching the first four terms of the series given in Eq. (1.2) and then performing a graphical summation.

1.4 For a square-wave audio signal of 10 kHz, what fraction of the available energy is perceived by an average adult listener of age 40 whose hearing extends only to 16 kHz?

***1.5** Numerically evaluate the series expansion for a square wave given in Eq. (1.2) truncated after four terms at $t = T/8$, $T/4$, $5T/8$, and $3T/4$.[1]

1.6 What fraction of the energy contained in a square wave of frequency f and peak-to-peak amplitude 2 V is contained in the harmonic at frequency $9f$?

Section 1.4: Amplifiers

1.7 An amplifier fed by a sine-wave signal of 5-mV peak delivers a sine-wave output of 1-V peak to a load resistance of 2 kΩ. The input current of the amplifier is found to be a sine wave of 5-μA peak. Calculate the voltage gain, current gain, and power gain as ratios and in decibels.

1.8 An amplifier has a current gain of 80 dB and a power gain of 50 dB. Find its voltage gain.

1.9 An amplifier operating from ±15-V power supplies delivers an output sine-wave signal of 20 V peak-to-peak to a 2-kΩ load resistance. The amplifier draws an average current of 1.7 mA from each of its two power supplies and a negligible current from the input signal source. Find the power dissipated in the amplifier circuit.

---

[1] Somewhat difficult problems are marked with an asterisk (*); more difficult problems are marked with two asterisks (**); and very difficult (and/or time consuming) problems are marked with three asterisks (***).

[2] Design-oriented problems are marked with a D.

1.10 An amplifier operating from ± 15-V power supplies has a linear transfer characteristic except for output saturation at ± 13 V. If the amplifier gain is 100 V/V, find the rms value of the largest sine wave signal that can be applied at the input without output clipping.

D*1.11 An amplifier designed using a single metal-oxide-semiconductor (MOS) transistor has the transfer characteristic[2]

$$v_O = 10 - 5(v_I - 2)^2$$

where v_I and v_O are in volts. This transfer characteristic applies for $2 \leq v_I \leq v_O + 2$ and v_O positive. At the limits of this region the amplifier saturates.

(a) Sketch and clearly label the transfer characteristic. What are the saturation levels L_+ and L_- and the corresponding values of v_I?

(b) Bias the amplifier to obtain a dc output voltage of 5 V. What value of input dc voltage V_I is required?

(c) Calculate the value of the small-signal voltage gain at the bias point.

(d) If a sinusoidal input signal is superimposed on the dc bias voltage V_I, that is,

$$v_I = V_I + V_i \cos \omega t$$

find the resulting v_O. Using the trigonometric identity $\cos^2 \Theta = \frac{1}{2} + \frac{1}{2} \cos 2\Theta$, express v_O as the sum of a dc component, a signal component with frequency ω, and a sinusoidal component with frequency 2ω. The latter component is undesirable and is a result of the nonlinear transfer characteristic of the amplifier. If it is required to limit the ratio of the second harmonic component to the fundamental component to 1% (this ratio is known as the second-harmonic distortion) what is the corresponding upper limit on V_i?

D1.12 A transistor amplifier has a transfer characteristic that can be described by the equation

$$v_O = 17 - 10\, v_I$$

for $0.3 \text{ V} \leq v_O \leq 10$ V. At the limits of this range, the amplifier saturates. Sketch the transfer characteristic and find the value of the dc input voltage V_I required to bias the amplifier so that the dc output voltage is 5 V. For signals v_i superimposed on the bias voltage V_I, find the amplifier voltage gain. If v_i is sinusoidal find its largest possible amplitude without output clipping.

Section 1.5: Circuit Models for Amplifiers

1.13 A signal source whose open-circuit voltage is 10 mV rms and whose short-circuit current is 1 μA rms, when connected to the input of a voltage amplifier, supplies a current of 0.5 μA rms. The resulting output of the amplifier when connected to a 1-kΩ load is 1 V rms. Find the amplifier current gain, voltage gain, and power gain, and the overall voltage gain (from source to load). Also find the amplifier input resistance. If, when the 1-kΩ load is disconnected from the output of the amplifier, the output voltage rises to 1.1 V rms, what is the output resistance of the amplifier? Also find its open-circuit voltage gain A_{vo}.

D1.14 A designer has available voltage amplifiers having an input resistance of 10 kΩ, an output resistance of 1 kΩ, and an open-circuit voltage gain of 10. The signal source has a 10-kΩ resistance and provides 10 mV rms signal, and it is required to provide a signal of at least 2 V rms to a 1 kΩ load. How many amplifier stages are required? What is the output voltage actually obtained?

D*1.15 Design an amplifier that provides 0.5 W of signal power to a 100-Ω load resistance. The signal source provides a 30-mV rms signal and has a resistance of 0.5 MΩ. Three types of voltage amplifier stages are available:

(1) a high-input-resistance type with $R_i = 1$ MΩ, $A_{vo} = 10$, and $R_o = 10$ kΩ;

(2) a high-gain type with $R_i = 10$ kΩ, $A_{vo} = 100$, and $R_o = 1$ kΩ;

(3) a low-output-resistance type with $R_i = 10$ kΩ, $A_{vo} = 1$, and $R_o = 20$ Ω.

Design a suitable amplifier using a combination of these stages. Your design should utilize the minimum number of stages. Find the load voltage and power output realized.

1.16 A transconductance amplifier having $R_i = 2$ kΩ, $G_m = 40$ mA/V, and $R_o = 20$ kΩ is fed with a voltage source having a source resistance of 2 kΩ and is loaded with a 1-kΩ resistance. Find the voltage gain realized.

1.17 Figure P1.17 shows a transconductance amplifier whose output is *fed back* to its input. Find the input resistance R_{in} of the resulting one-port network. (*Hint:* Apply a test voltage v_x between the two input terminals and find the current i_x drawn from the source; then $R_{in} \equiv v_x / i_x$.)

1.18 Figure P1.18 shows a current amplifier whose output is fed back to the input. Find the input resistance R_{in} of

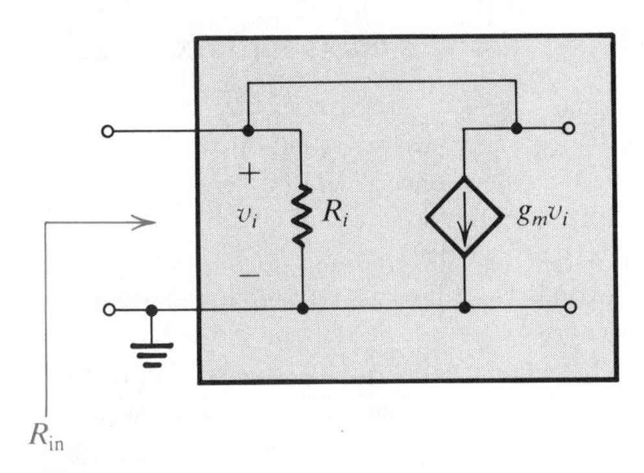

Fig. P1.17

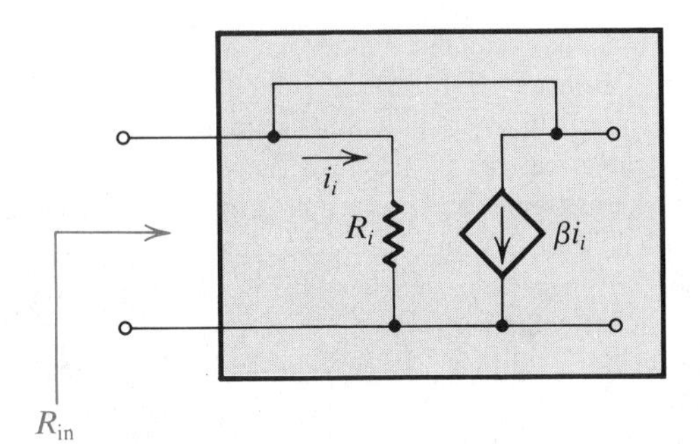

Fig. P1.18

the resulting one-port network. (See hint given in Problem 1.17.)

1.19 Refer to Fig. 1.19(c). A BJT, appropriately biased, is used as a so-called common- (or grounded-) emitter amplifier with E grounded, a source voltage v_s with resistance R_s connected to B, and a load resistance R_L connected between C and ground. Derive an expression for the voltage gain v_o/v_s, where v_o is the voltage measured across R_L. Evaluate the voltage gain for the case of $R_s = 5\ \text{k}\Omega$, $r_\pi = 5\ \text{k}\Omega$, $g_m = 40$ mA/V, and $R_L = 5\ \text{k}\Omega$.

1.20 Consider the capacitively terminated gyrator of Fig. 1.20(b). Apply an input test voltage V_x at port 1 and find the current I_x that is drawn from V_x. Show that the input impedance is given by

$$Z_{in}(s) \equiv \frac{V_x(s)}{I_x(s)} = \frac{sC}{g_1 g_2}$$

Find the value of the input inductance realized for $g_1 = g_2 = 5$ mA/V and $C = 0.01\ \mu$F.

***1.21** In the gyrator circuit of Fig. 1.20(a), let each of the controlled sources have an input resistance R_i and an output resistance R_o. Now for the capacitively terminated gyrator of Fig. 1.20(b) with these input and output resistances included, show that the input admittance is given by

$$Y_{in}(s) = \frac{1}{R} + \frac{g_1 g_2 R}{sCR + 1}$$

where $R = R_i // R_o$. Also show that this admittance can be represented by the circuit shown in Fig. P1.21 and find the values of L, R_1, and R_2 in terms of g_1, g_2, C, and R.

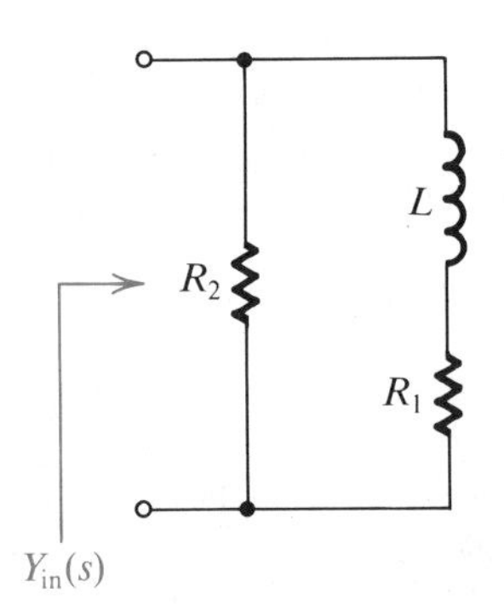

Fig. P1.21

1.22 An amplifier with an input resistance of 10 kΩ, when driven by a current source of 1 μA and a source resistance of 100 kΩ, has a short-circuit output current of 10 mA and an open-circuit output voltage of 10 V. When driving a 4-kΩ load, what are the voltage gain, current gain, and power gain expressed as ratios and in dB?

1.23 For the circuit shown in Fig. P1.23 show that

$$\frac{v_c}{v_b} = \frac{-\beta R_L}{r_\pi + (\beta + 1)R_E}$$

and,

$$\frac{v_e}{v_b} = \frac{R_E}{R_E + [r_\pi/(\beta + 1)]}$$

Evaluate these voltage gains for $\beta = 100$, $r_\pi = 2.5\ \text{k}\Omega$, $R_E = 1\ \text{k}\Omega$, and $R_L = 10\ \text{k}\Omega$.

1.24 Figure P1.24(a) shows two transconductance amplifiers connected in a special configuration. Find v_o in terms of v_1 and v_2. Let $g_m = 100$ mA/V and $R = 5\ \text{k}\Omega$. If $v_1 = v_2 = 1$ V, find the value of v_o. Also find v_o for the case $v_1 = 1.01$ V and $v_2 = 0.99$ V. (*Note:* This circuit is called a **differential amplifier**

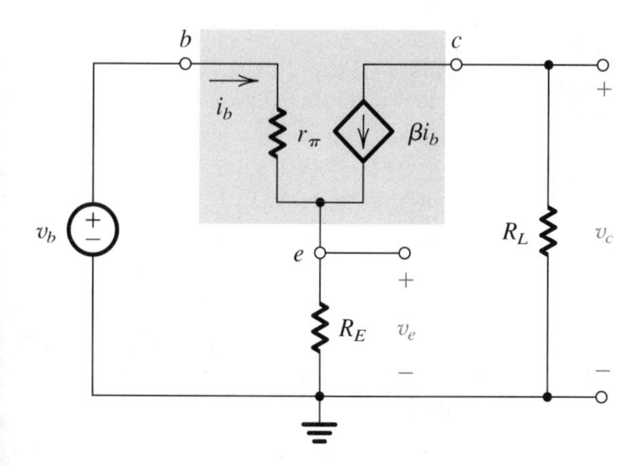

Fig. P1.23

and is given the symbol shown in Fig. 1.24(b).) A particular type of differential amplifier known as an operational amplifier will be studied in Chapter 2.

Section 1.6: Frequency Response of Amplifiers

1.25 Using the voltage-divider rule derive the transfer functions $T(s) \equiv V_o(s)/V_i(s)$ of the circuits shown in Fig. 1.23, and show that the transfer functions are of the form given at the top of Table 1.1.

1.26 Figure P1.26 shows a signal source connected to the input of an amplifier. Here R_s is the source resistance,

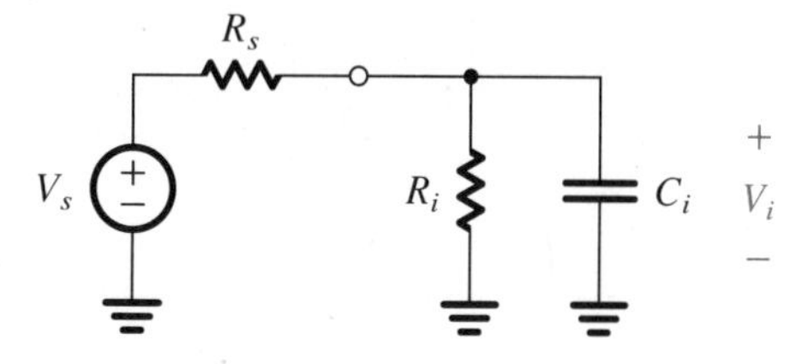

Fig. P1.26

and R_i and C_i are the input resistance and input capacitance, respectively, of the amplifier. Derive an expression for $V_i(s)/V_s(s)$ and show that it is of the low-pass STC type. Find the 3-dB frequency for the case $R_s = 10$ kΩ, $R_i = 100$ kΩ, and $C_i = 10$ pF.

D1.27 It is required to couple a voltage source V_s having a resistance R_s to a load R_L via a capacitor C. Derive an expression for the transfer function from source to load (i.e., V_L/V_s) and show that it is of the high-pass STC type. For $R_s = 10$ kΩ and $R_L = 40$ kΩ find the smallest coupling capacitor that will result in a 3-dB frequency no greater than 10 Hz.

*1.28 The unity-gain voltage amplifiers in the circuit of Fig. P1.28 have infinite input resistances and zero output resistances, and thus they function as perfect buffers. Convince yourself that the overall gain V_o/V_i will drop by 3-dB below the value at dc at the frequency for which the gain of each RC circuit is 0.75 dB down. What is this frequency?

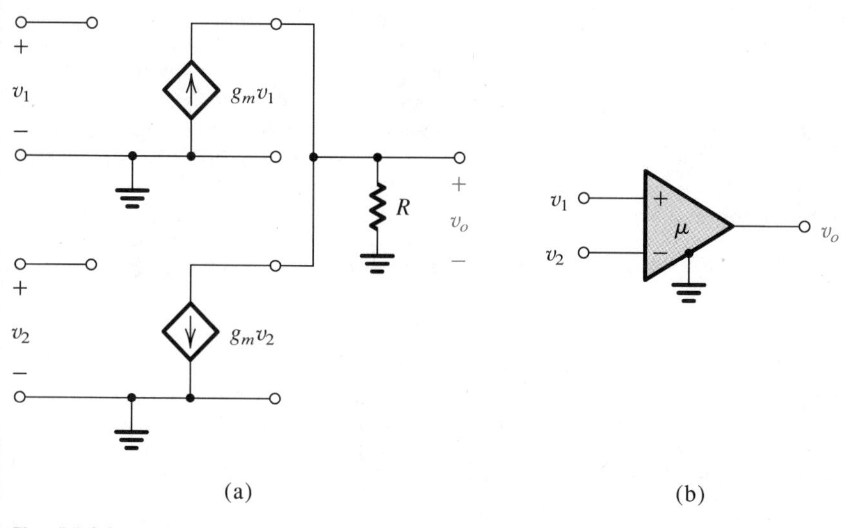

Fig. P1.24

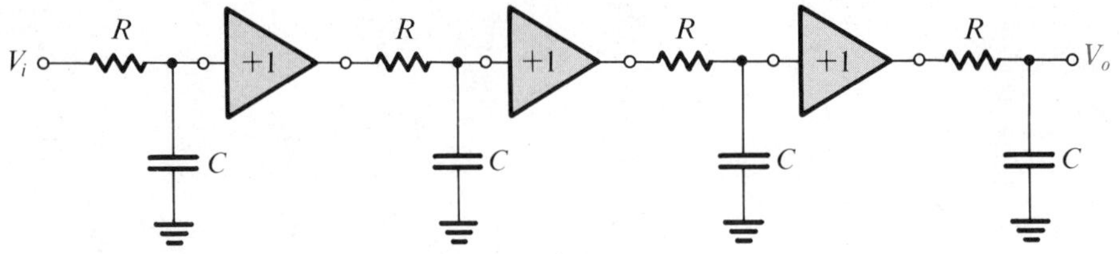

Fig. P1.28

*1.29 A voltage amplifier has the transfer function

$$A_v = \frac{100}{\left(1 + j\frac{f}{10^4}\right)\left(1 + \frac{10^2}{jf}\right)}$$

Using the Bode plots for low-pass and high-pass STC networks (Figs. 1.24 and 1.25) sketch a Bode plot for $|A_v|$. Give approximate values for the gain magnitude at $f = 10$, 10^2, 10^3, 10^4, 10^5, 10^6, and 10^7 Hz. Find the bandwidth of the amplifier (defined as the frequency range over which the gain remains within 3 dB of the maximum value).

D**1.30 A transconductance amplifier having the equivalent circuit shown in Fig. 1.17(b) is fed with a voltage source V_s having a source resistance R_s, and its output is connected to a load consisting of a resistance R_L in parallel with a capacitance C_L. For given values of R_s, R_o, and C_L it is required to specify the values of R_i, G_m, and R_L to meet the following design constraints:

(1) At most $x\%$ of the input signal is lost in coupling the signal source to the amplifier (i.e., $V_i \geq [1 - (x/100)]V_s$).

(2) The 3-dB frequency of the amplifier is equal to or greater than a specified value f_{3dB}.

(3) The dc gain V_o/V_s is equal to or greater than a specified value A_0.

Show that these constraints can be met by selecting

$$R_i \geq \left(\frac{100}{x} - 1\right)R_s$$

$$R_L \leq \frac{1}{2\pi f_{3dB}C_L - (1/R_o)}$$

$$G_m \geq \frac{A_0/[1 - (x/100)]}{(R_L//R_o)}$$

Find R_i, R_L, and G_m for $R_s = 10\ \text{k}\Omega$, $x = 20\%$, $A_0 = 80$, $R_o = 10\ \text{k}\Omega$, $C_L = 10$ pF, and $f_{3dB} = 3$ MHz.

1.31 Show that the transfer function of the circuit in Fig. P1.31 is given by

$$\frac{V_o}{V_i} = \frac{1}{2}\frac{s - 1/RC}{s + 1/RC}$$

For $s = j\omega$, sketch the magnitude and phase versus frequency of this transfer function. *Note:* This circuit is an STC network of the **all-pass** type.

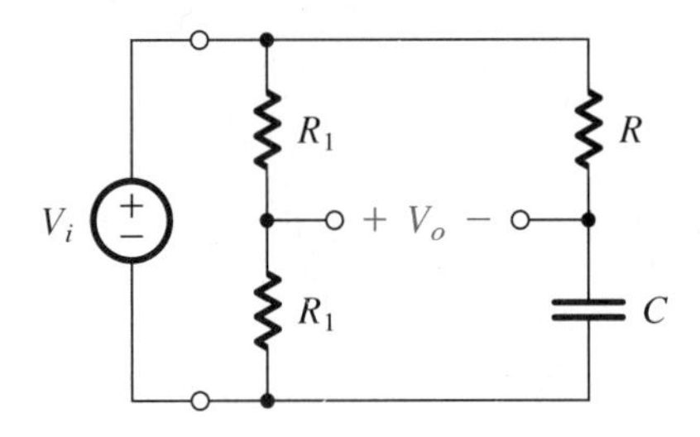

Fig. P1.31

1.32 Use the voltage-divider rule to find the transfer function $V_o(s)/V_i(s)$ of the circuit in Fig. P1.32. Show that the transfer function can be made independent of frequency if the condition $C_1R_1 = C_2R_2$ applies. Under such condition the circuit is called a **compensated attenuator.** Find the transmission of the compensated attenuator in terms of R_1 and R_2.

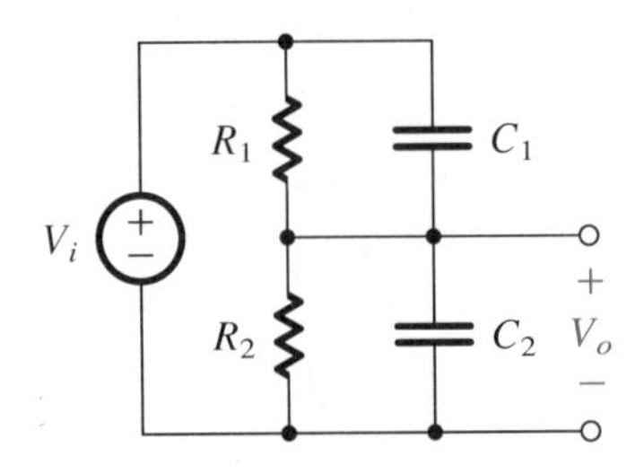

Fig. P1.32

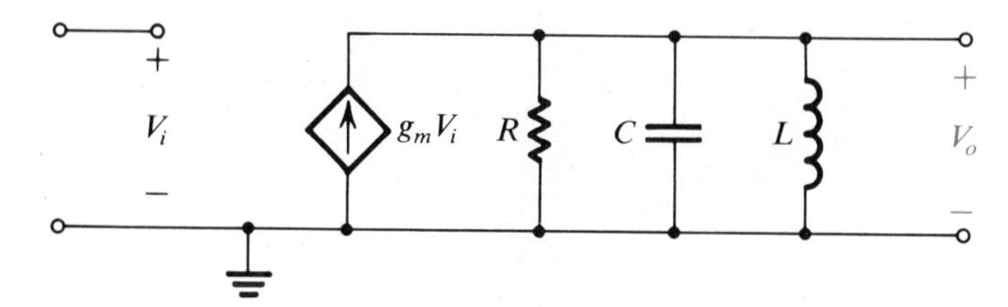

Fig. P1.34

*1.33 An amplifier with a frequency response of the type shown in Fig. 1.22 is specified to have a phase shift no greater than 11.4° over its bandwidth, which extends from 100 Hz to 1 kHz. It has been found that the gain falloff at the low-frequency end is determined by the response of a high-pass STC circuit, and that at the high-frequency end it is determined by a low-pass STC circuit. What do you expect the corner frequencies of these two circuits to be? What is the drop in gain in decibels (relative to the maximum gain) at the two frequencies that define the amplifier bandwidth? What are the frequencies at which the drop in gain is 3 dB?

***1.34 Figure P1.34 shows a tuned, or bandpass, amplifier. Derive its transfer function $T(s) \equiv V_o(s)/V_i(s)$ and find $T(j\omega)$ and $|T(j\omega)|$. What is the value of gain at $\omega = 0$ and at $\omega = \infty$? At what frequency is the gain maximum? What is the value of the maximum gain? Find the two frequencies at which the gain drops by 3 dB below the maximum value. The 3-dB bandwidth of the amplifier is the difference between these two frequencies. Find its value. (Hint: the maximum output occurs at the frequency for which the parallel equivalent impedance of L and C is infinite.)

PART 

DEVICES and BASIC CIRCUITS

INTRODUCTION

As well as passive components such as resistors and capacitors, with which we assume the reader is already familiar, electronic circuits, both analog and digital, are constructed using diodes and transistors. These electronic devices are made of the semiconductor material silicon. The study of silicon diodes and transistors constitutes the bulk of Part I.

An additional circuit building block that is not an electronic device in the most fundamental sense, but is commercially available as an integrated circuit (IC) package and has well defined terminal characteristics, is the operational amplifier (op amp). Although the op amp internal circuit is complex, typically incorporating 20 or more transistors, its almost ideal terminal behavior makes it possible to treat the op amp as a circuit element and to use it in the design of powerful circuits without any knowledge of its internal construction. Our study of electronic circuits begins in Chapter 2 with the IC op amp.[1]

Each of the three succeeding chapters deals with one of the basic semiconductor devices: the diode in Chapter 3, the bipolar junction transistor (BJT) in Chapter 4, and the field-effect transistor (FET) in Chapter 5. Each chapter begins with a study of the terminal characteristics of the respective device and includes a description of the device's physical operation. Appropriate models are then developed to represent the device's terminal characteristics. This is followed with a study of the design and analysis of the fundamental circuit configurations in which the device is commonly utilized.

The objective of Part I is to develop in the reader a high degree of familiarity with the basic electronic devices (op amps, diodes, BJTs, and FETs) and with their use in the design of simple but fundamental circuits. By the end of Part I the reader should be able to proceed with the more advanced aspects of designing analog circuits (in Part II) and digital circuits (in Part III).

[1]Readers who prefer to study op amps at a later stage (perhaps after Chapter 7 or in conjunction with Chapter 10) can skip Chapter 2 with no loss in continuity.

CHAPTER 2

Operational Amplifiers

Introduction

INTRODUCTION

Having learned basic amplifier concepts and terminology, we are now ready to undertake the study of a circuit building block of universal importance: the operational amplifier (op amp). Although op amps have been in use for a long time, their applications were initially in the areas of analog computation and instrumentation. Early op amps were constructed from discrete components (vacuum tubes and then transistors and resistors), and their cost was prohibitively high (tens of dollars). In the mid-1960s the first integrated-circuit (IC) op amp was produced. This unit (the μA 709) was made up of a relatively large number of transistors and resistors all on the same silicon chip. Although its characteristics were poor (by today's standards) and its price was still quite high, its appearance signaled a new era in electronic circuit design. Electronics engineers started using op amps in large quantities, which caused their price to drop dramatically. They also demanded better-quality op amps. Semiconductor manufacturers responded quickly; and within the span of a few years, high-quality op amps became available at extremely low prices (tens of cents) from a large number of suppliers.

One of the reasons for the popularity of the op amp is its versatility. As we will shortly see, one can do almost anything with op amps! Equally important is the fact that the IC op amp has characteristics that closely approach the assumed ideal. This implies that it is quite easy to design circuits using the IC op amp. Also, op amp circuits work at levels that are quite close to their predicted theoretical performance. It is for this reason that we are studying op amps at this early stage. It is expected that by the end of this chapter the reader should be able to design nontrivial circuits successfully using op amps.

As already implied, an IC op amp is made up of a large number of transistors, resistors, and (sometimes) one capacitor connected in a rather complex circuit. Since we have not yet studied transistor circuits, the circuit inside the op amp will not be discussed in this chapter. Rather, we will treat the op amp as a circuit building block and study its terminal characteristics and its applications. This approach is quite satisfactory in many op amp applications. Nevertheless, for the more difficult and demanding applications it is quite useful to know what is inside the op amp package. This topic will be studied in Chapter 10. Finally, it should be mentioned that more advanced applications of op amps will appear in later chapters.

2.1 THE OP AMP TERMINALS

From a signal point of view the op amp has three terminals: two input terminals and one output terminal. Figure 2.1 shows the symbol we shall use to represent the op amp. Terminals 1 and 2 are input terminals, and terminal 3 is the output terminal. As explained in Section 1.4, amplifiers require dc power to operate. Most IC op amps require two dc power supplies, as shown in Fig. 2.2. Two terminals, 4 and 5, are brought out of the op amp package and connected to a positive voltage V^+ and a negative voltage V^-, respectively. In Fig. 2.2(b) we explicitly show the two dc power supplies as batteries with a

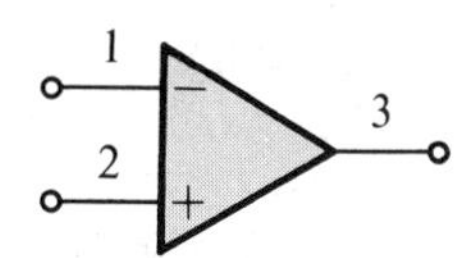

Fig. 2.1 Circuit symbol for the op amp.

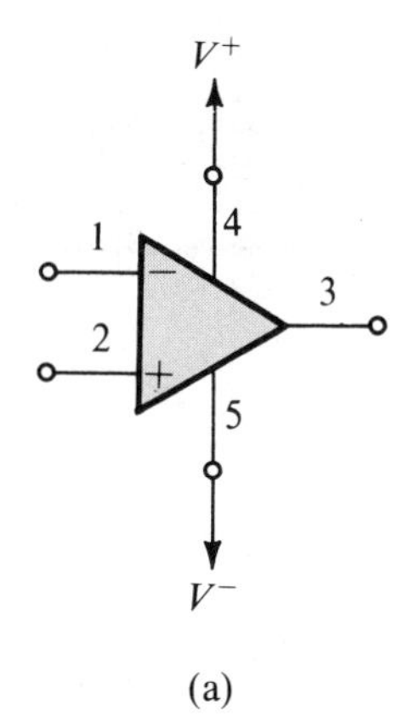

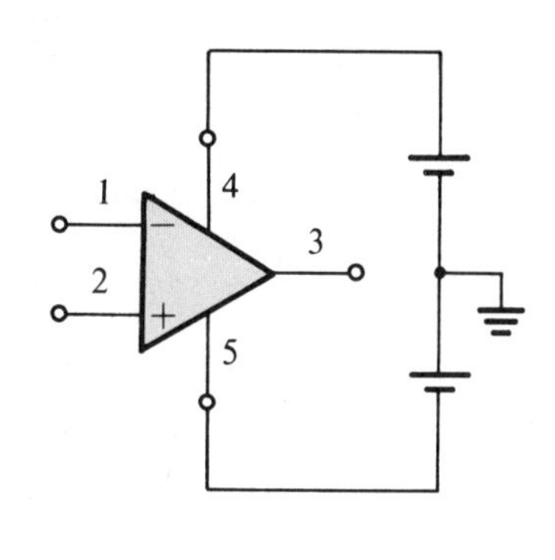

Fig. 2.2 The op amp shown connected to dc power supplies.

common ground. It is interesting to note that the reference grounding point in op amp circuits is just the common terminal of the two power supplies; that is, no terminal of the op amp package is physically connected to ground. In what follows we will not explicitly show the op amp power supplies.

In addition to the three signal terminals and the two power-supply terminals, an op amp may have other terminals for specific purposes. These other terminals can include terminals for frequency compensation and terminals for offset nulling: both functions will be explained in later sections.

2.2 THE IDEAL OP AMP

We now consider the circuit function of the op amp. The op amp is supposed to sense the difference between the voltage signals applied at its two input terminals (that is, the quantity $v_2 - v_1$), multiply this by a number A, and cause the resulting voltage $A(v_2 - v_1)$ to appear at output terminal 3. Here it should be emphasized that when we talk about the voltage at a terminal we mean the voltage between that terminal and ground; thus v_1 means the voltage applied between terminal 1 and ground.

The ideal op amp is not supposed to draw any input current; that is, the signal current into terminal 1 and the signal current into terminal 2 are both zero. In other words, the input impedance of an ideal op amp is supposed to be infinite.

How about the output terminal 3? This terminal is supposed to act as the output terminal of an ideal voltage source. That is, the voltage between terminal 3 and ground will always be equal to $A(v_2 - v_1)$ and will be independent of the current that may be drawn from terminal 3 into a load impedance. In other words, the output impedance of an ideal op amp is supposed to be zero.

Putting together all of the above we arrive at the equivalent circuit model shown in Fig. 2.3. Note that the output is in phase with (has the same sign as) v_2 and out of phase with (has the opposite sign of) v_1. For this reason input terminal 1 is called the **inverting input terminal** and is distinguished by a "−" sign, while input terminal 2 is called the **noninverting input terminal** and is distinguished by a "+" sign.

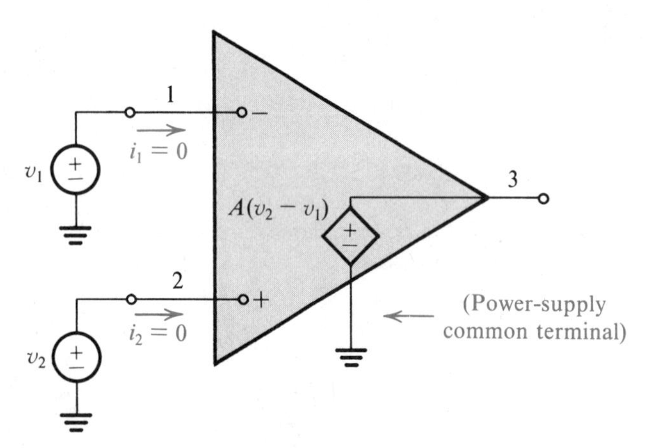

Fig. 2.3 Equivalent circuit of the ideal op amp.

As can be seen from the above description, the op amp responds only to the *difference* signal $v_2 - v_1$ and hence ignores any signal *common* to both inputs. That is, if $v_1 = v_2 =$ 1 V, then the output will—ideally—be zero. We call this property **common-mode rejection,** and we conclude that an ideal op amp has infinite common-mode rejection. We will have more to say about this point later. For the time being note that the op amp is a **differential-input, single-ended-output** amplifier, with the latter term referring to the fact that the output appears between terminal 3 and ground. Furthermore, gain A is called the **differential gain,** for obvious reasons. Perhaps not so obvious is another name that we will attach to A: the **open-loop gain.** The reason for this name will become obvious later on when we "close the loop" around the op amp and define another gain, the closed-loop gain.

An important characteristic of op amps is that they are direct-coupled devices or dc amplifiers, where dc stands for direct-coupled (it could equally well stand for direct current, since a direct-coupled amplifier is one that amplifies signals whose frequency is as low as zero). The fact that op amps are direct-coupled devices will allow us to use them in many important applications. Unfortunately, though, the direct-coupling property can cause some serious problems, as will be discussed in a later section.

How about bandwidth? The ideal op amp has a gain A that remains constant down to zero frequency and up to infinite frequency. That is, ideal op amps will amplify signals of any frequency with equal gain.

We have discussed all of the properties of the ideal op amp except for one, which in fact is the most important. This has to do with the value of A. The ideal op amp should have a gain A whose value is very large and ideally infinite. One may justifiably ask: If the gain A is infinite, how are we going to use the op amp? The answer is very simple: In almost all applications the op amp will *not* be used in an open-loop configuration. Rather, we will apply feedback to close the loop around the op amp, as will be illustrated in detail in Section 2.3.

Exercises 2.1 Consider an op amp that is ideal except that its open-loop gain $A = 10^3$. The op amp is used in a feedback circuit and the voltages appearing at two of its three signal terminals are measured. In each of the following cases, use the measured values to find the expected value of the voltage at the third terminal. (a) $v_2 = 0$ V and $v_3 = 2$ V; (b) $v_2 = +5$ V and $v_3 = -10$ V; (c) $v_1 = 1.002$ V and $v_2 = 0.998$ V; (d) $v_1 = -3.6$ V and $v_3 = -3.6$ V.

Ans. (a) $v_1 = -0.002$ V; (b) $v_1 = +5.01$ V; (c) $v_3 = -4$ V; (d) $v_2 = -3.6036$ V

2.2 The internal circuit of a particular op amp can be modeled by the circuit shown in Fig. E2.2. Express v_3 as a function of v_1 and v_2. For the case $G_m = 10$ mA/V, $R = 10$ kΩ, and $\mu = 100$, find the value of the open-loop gain A.

Ans. $v_3 = \mu G_m R(v_2 - v_1)$; $A = 10{,}000$ V/V

2.3 ANALYSIS OF CIRCUITS CONTAINING IDEAL OP AMPS—THE INVERTING CONFIGURATION

Consider the circuit shown in Fig. 2.4, which consists of one op amp and two resistors R_1 and R_2. Resistor R_2 is connected from the output terminal of the op amp, terminal 3, *back* to the *inverting* or *negative* input terminal, terminal 1. We speak of R_2 as applying **negative feedback;** if R_2 were connected between terminals 3 and 2 we would have called

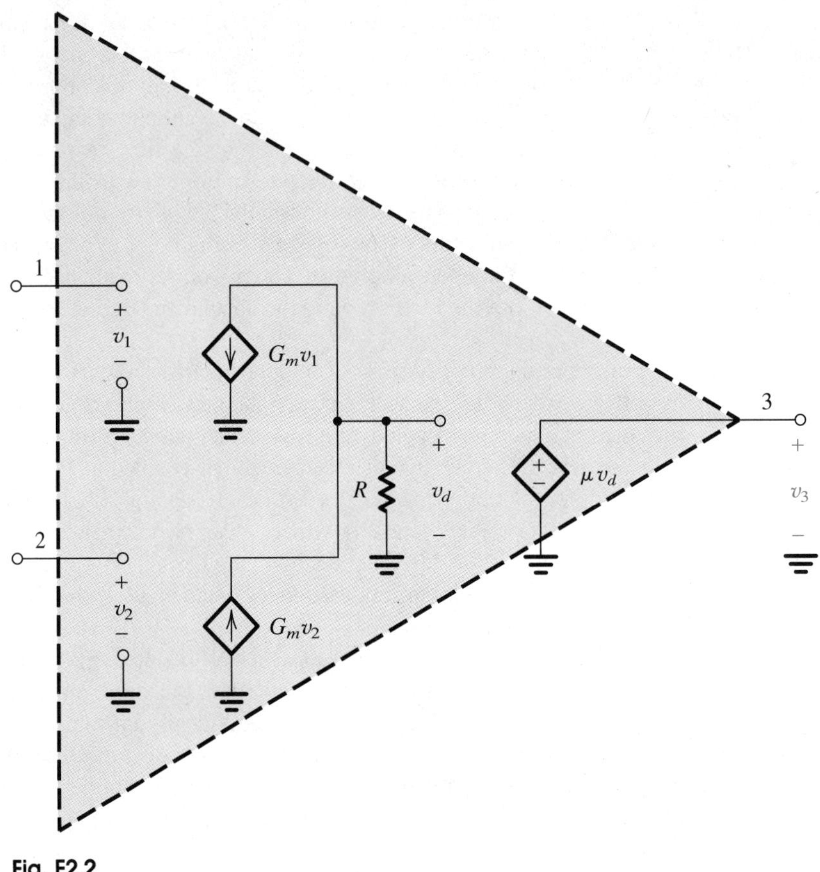

Fig. E2.2

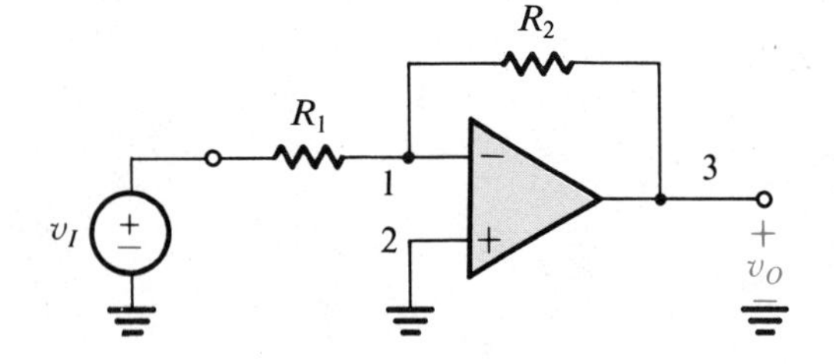

Fig. 2.4 The inverting closed-loop configuration.

this **positive feedback.** Note also that R_2 closes the loop around the op amp. In addition to adding R_2, we have grounded terminal 2 and connected a resistor R_1 between terminal 1 and an input signal source with a voltage v_I. The output of the overall circuit is taken at terminal 3 (that is, between terminal 3 and ground). Terminal 3 is, of course, a convenient point to take the output, since the impedance level there is ideally zero. Thus the voltage v_O will not depend on the value of the current that might be supplied to a load impedance connected between terminal 3 and ground.

The Closed-Loop Gain

We now wish to analyze the circuit in Fig. 2.4 to determine the **closed-loop gain *G*,** defined as

$$G \equiv \frac{v_O}{v_I}$$

We will do so assuming the op amp to be ideal. Figure 2.5(a) shows the equivalent circuit, and the analysis proceeds as follows: The gain *A* is very large (ideally infinite). If we

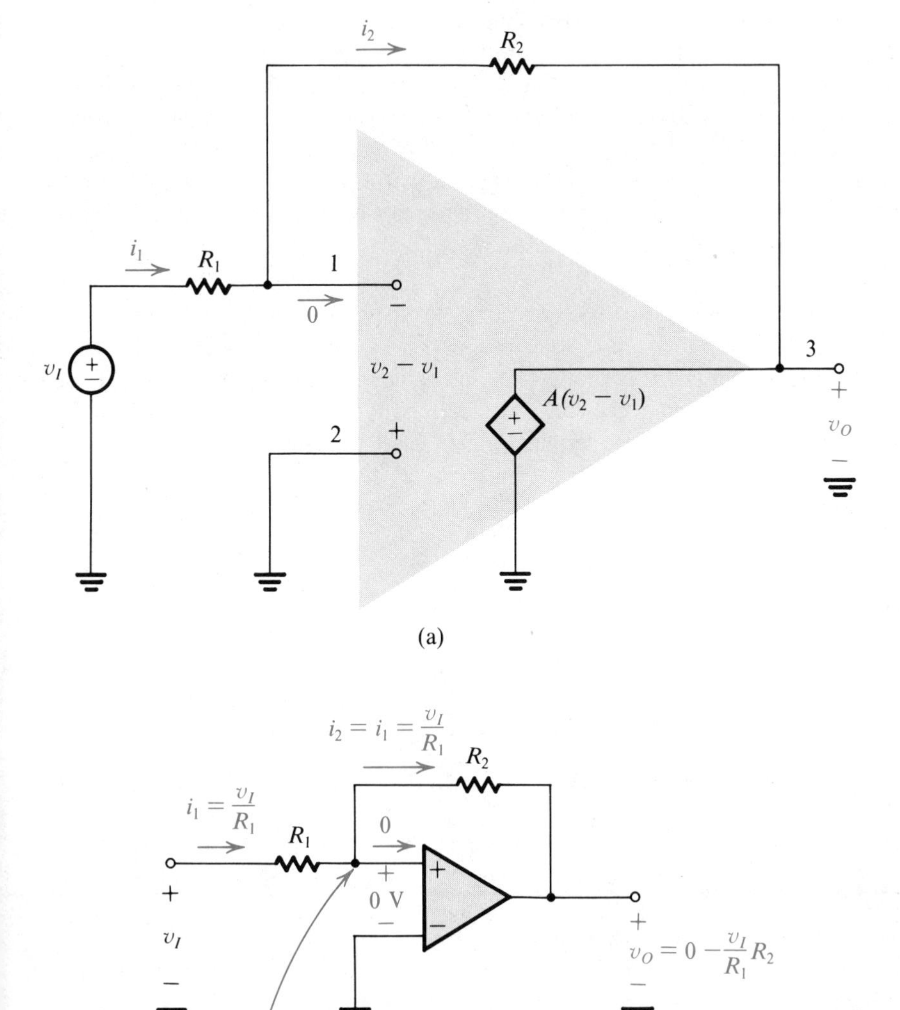

Fig. 2.5 Analysis of the inverting configuration.

assume that the circuit is "working" and producing a finite output voltage at terminal 3, then the voltage between the op amp input terminals should be negligibly small. Specifically, if we call the output voltage v_O, then, by definition,

$$v_2 - v_1 = \frac{v_O}{A} \simeq 0$$

It follows that the voltage at the inverting input terminal (v_1) is given by $v_1 \simeq v_2$. That is, because the gain A approaches infinity, the voltage v_1 approaches v_2. We speak of this as the two input terminals "tracking each other in potential." We also speak of a "virtual short circuit" that exists between the two input terminals. Here the word *virtual* should be emphasized, and one should *not* make the mistake of physically shorting terminals 1 and 2 together while analyzing a circuit. A **virtual short circuit** means that whatever voltage is at 2 will automatically appear at 1 because of the infinite gain A. But terminal 2 happens to be connected to ground; thus $v_2 = 0$ and $v_1 \simeq 0$. We speak of terminal 1 as being a **virtual ground**—that is, having zero voltage but not physically connected to ground.

Now that we have determined v_1 we are in a position to apply Ohm's law and find the current i_1 through R_1 (see Fig. 2.5) as follows:

$$i_1 = \frac{v_I - v_1}{R_1} \simeq \frac{v_I}{R_1}$$

Where will this current go? It cannot go into the op amp, since the ideal op amp has an infinite input impedance and hence draws zero current. It follows that i_1 will have to flow through R_2 to the low-impedance terminal 3. We can then apply Ohm's law to R_2 and determine v_O; that is,

$$\begin{aligned} v_O &= v_1 - i_1 R_2 \\ &= 0 - \frac{v_I}{R_1} R_2 \end{aligned}$$

Thus

$$\frac{v_O}{v_I} = -\frac{R_2}{R_1}$$

which is the required closed-loop gain. Figure 2.5(b) illustrates some of these analysis steps.

We thus see that the closed-loop gain is simply the ratio of the two resistances R_2 and R_1. The minus sign means that the closed-loop amplifier provides signal inversion. Thus if $R_2/R_1 = 10$ and we apply at the input (v_I) a sine-wave signal of 1 V peak-to-peak, then the output v_O will be a sine wave of 10 V peak-to-peak and phase-shifted 180° with respect to the input sine wave. Because of the minus sign associated with the closed-loop gain this configuration is called the **inverting configuration.**

The fact that the closed-loop gain depends entirely on external passive components (resistors R_1 and R_2) is a very interesting one. It means that we can make the closed-loop gain as accurate as we want by selecting passive components of appropriate accuracy. It also means that the closed-loop gain is (ideally) independent of the op amp gain. This is a dramatic illustration of negative feedback: We started out with an amplifier having very large gain A, and through applying negative feedback we have obtained a closed-loop gain R_2/R_1 that is much smaller than A but is stable and predictable. That is, we are trading gain for accuracy.

Effect of Finite Open-Loop Gain

The points just made are more clearly illustrated by deriving an expression for the closed-loop gain under the assumption that the op amp open-loop gain A is finite. Figure 2.6 shows the analysis. If we denote the output voltage v_O, then the voltage between the two input terminals of the op amp will be v_O/A. Since the positive input terminal is grounded,

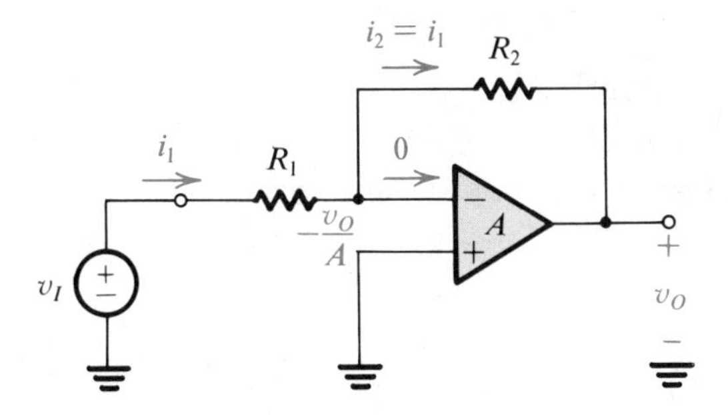

Fig. 2.6 Analysis of the inverting configuration taking into account the finite open-loop gain of the op amp.

the voltage at the negative input terminal must be $-v_O/A$. The current i_1 through R_1 can now be found from

$$i_1 = \frac{v_I - (-v_O/A)}{R_1} = \frac{v_I + v_O/A}{R_1}$$

The infinite input impedance of the op amp forces the current i_1 to flow entirely through R_2. The output voltage v_O can thus be determined from

$$\begin{aligned} v_O &= -\frac{v_O}{A} - i_1 R_2 \\ &= -\frac{v_O}{A} - \left(\frac{v_I + v_O/A}{R_1}\right) R_2 \end{aligned}$$

Collecting terms, the closed-loop gain G is found as

$$G \equiv \frac{v_O}{v_I} = \frac{-R_2/R_1}{1 + (1 + R_2/R_1)/A} \tag{2.1}$$

We note that as A approaches ∞, G approaches the ideal value of $-R_2/R_1$. Also from Fig. 2.6 we see that as A approaches ∞, the voltage at the inverting input terminal approaches zero. This is the virtual ground assumption we used in our earlier analysis when the op amp was assumed to be ideal. Finally note that Eq. (2.1) in fact indicates that to minimize the dependence of the closed-loop gain G on the value of the open-loop gain A, we should make

$$1 + \frac{R_2}{R_1} \ll A$$

EXAMPLE 2.1 Consider the inverting configuration with $R_1 = 1$ kΩ and $R_2 = 100$ kΩ.

(a) Find the closed-loop gain for the cases $A = 10^3$, 10^4, and 10^5. In each case determine the percentage error in the magnitude of G relative to the ideal value of R_2/R_1 (obtained with $A = \infty$). Also determine the voltage v_1 that appears at the inverting input terminal when $v_I = 0.1$ V.

(b) If the open-loop gain A changes from 100,000 to 50,000, what is the corresponding percentage change in the magnitude of the closed-loop gain G?

Solution (a) Substituting the given values in Eq. (2.1), we obtain the values given in the following table where the percentage error ε is defined as

$$\varepsilon \equiv \frac{|G| - (R_2/R_1)}{(R_2/R_1)} \times 100$$

The values of v_1 are obtained from $v_1 = -v_O/A = Gv_I/A$ with $v_I = 0.1$ V.

A	$\|G\|$	ε	v_1
10^3	90.83	−9.17%	−9.08 mV
10^4	99.00	−1.00%	−0.99 mV
10^5	99.90	−0.10%	−0.10 mV

(b) Using Eq. (2.1), we find that for $A = 50{,}000$, $|G| = 99.80$. Thus halving the open-loop gain results in a change of −0.1% in the closed-loop gain!

Input and Output Resistances

Assuming an ideal op amp with infinite open-loop gain, the input resistance of the closed-loop inverting amplifier of Fig. 2.4 is simply equal to R_1. This can be seen from Fig. 2.5(b), where

$$R_{in} \equiv \frac{v_I}{i_1} = \frac{v_I}{v_I/R_1} = R_1$$

Thus to make R_{in} high we should select a high value for R_1. However, if the required gain R_2/R_1 is also high, then R_2 could become impractically large (e.g., greater than a few mega ohms). We may conclude that the inverting configuration suffers from a low input resistance. A solution to this problem is discussed in Example 2.2 below. It should also be mentioned that the finite open-loop gain A has a negligible effect on the value of the input resistance of the inverting amplifier configuration (see Problem 2.12).

Since the output of the inverting configuration is taken at the terminals of the ideal voltage source $A\,(v_2 - v_1)$ (see Fig. 2.5a) it follows that the output resistance of the closed-loop amplifier is zero.

Putting all of the above together we obtain the circuit shown in Fig. 2.7 as the equivalent circuit model of the inverting amplifier configuration of Fig. 2.4 (under the assumption that the op amp is ideal).

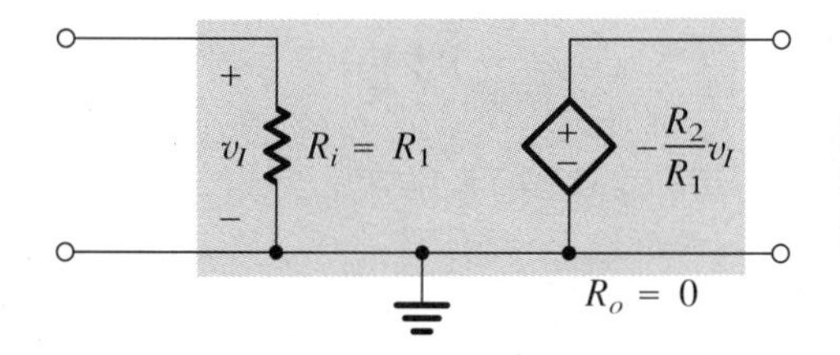

Fig. 2.7 Equivalent circuit model of the inverting amplifier configuration of Fig. 2.4 (assuming the op amp is ideal).

EXAMPLE 2.2

Assuming the op amp to be ideal, derive an expression for the closed-loop gain v_O/v_I of the circuit shown in Fig. 2.8. Use this circuit to design an inverting amplifier with a gain of 100 and input resistance of 1 MΩ. Assume that for practical reasons it is required not to use resistors greater than 1 MΩ. Compare your design with that based on the inverting configuration of Fig. 2.4.

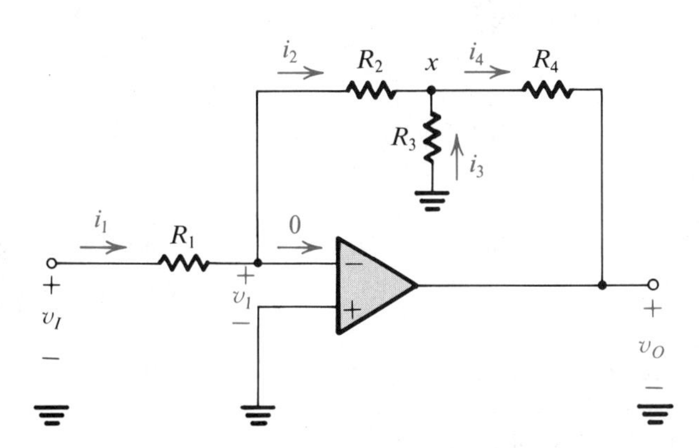

Fig. 2.8 Circuit for Example 2.2.

Solution The analysis begins at the inverting input terminal of the op amp, where the voltage is

$$v_1 = \frac{-v_O}{A} = \frac{-v_O}{\infty} = 0$$

Here we have assumed that the circuit is "working" and producing a finite output voltage v_O. Knowing v_1 we can determine the current i_1, as follows:

$$i_1 = \frac{v_I - v_1}{R_1} = \frac{v_I - 0}{R_1} = \frac{v_I}{R_1}$$

Since zero current flows into the inverting input terminal, all of i_1 will flow through R_2, and thus

$$i_2 = i_1 = \frac{v_I}{R_1}$$

Now we can determine the voltage at node x:

$$v_x = v_1 - i_2 R_2 = 0 - \frac{v_I}{R_1} R_2 = -\frac{R_2}{R_1} v_I$$

This in turn enables us to find the current i_3,

$$i_3 = \frac{0 - v_x}{R_3} = \frac{R_2}{R_1 R_3} v_I$$

Next, a node equation at x yields i_4,

$$i_4 = i_2 + i_3 = \frac{v_I}{R_1} + \frac{R_2}{R_1 R_3} v_I$$

Finally, we can determine v_O from

$$v_O = v_x - i_4 R_4$$

$$= -\frac{R_2}{R_1} v_I - \left(\frac{v_I}{R_1} + \frac{R_2}{R_1 R_3} v_I \right) R_4$$

Thus, the voltage gain is given by

$$\frac{v_O}{v_I} = -\left[\frac{R_2}{R_1} + \frac{R_4}{R_1} \left(1 + \frac{R_2}{R_3} \right) \right]$$

which can be written in the form

$$\frac{v_O}{v_I} = -\frac{R_2}{R_1} \left(1 + \frac{R_4}{R_2} + \frac{R_4}{R_3} \right)$$

Now since an input resistance of 1 MΩ is required, we select $R_1 = 1$ MΩ. Then, with the limitation of using resistors no greater than 1 MΩ, the maximum value possible for the first factor in the gain expression is 1 and is obtained by selecting $R_2 = 1$ MΩ. To obtain a gain of -100, R_3 and R_4 must be selected so that the second factor in the gain expression is 100. If we select the maximum allowed (in this example) value of 1 MΩ for R_4 then the required value of R_3 can be calculated to be 10.2 kΩ. Thus this circuit utilizes three 1-MΩ resistors and a 10.2 kΩ resistor. In comparison, if the inverting configuration were used with $R_1 = 1$ MΩ we would have required a feedback resistor of 100 MΩ, an impractically large value!

Exercises D2.3 Use the circuit of Fig. 2.4 to design an inverting amplifier having a gain of -10 and an input resistance of 100 kΩ. Give the values of R_1 and R_2.

Ans. $R_1 = 100$ kΩ; $R_2 = 1$ MΩ

2.4 The circuit shown in Fig. E2.4(a) can be used to implement a transresistance amplifier (see Section 1.5). Find the value of the input resistance R_i, the transresistance R_m, and the output resistance R_o of the transresistance amplifier. If the signal source shown in Fig. E2.4(b) is connected to the input of the transresistance amplifier, find its output voltage.

Ans. $R_i = 0$; $R_m = -10$ kΩ; $R_o = 0$; $v_o = -5$ V

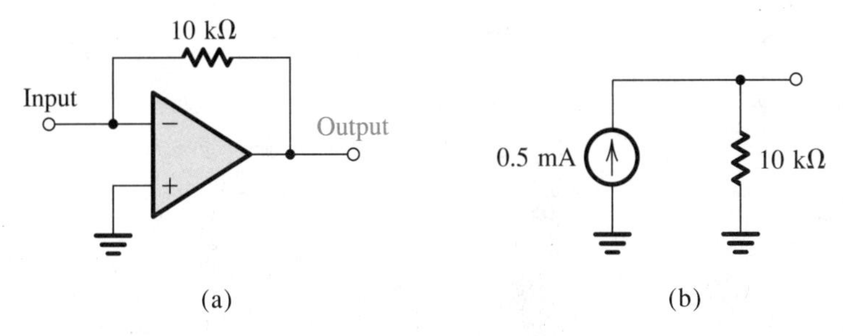

Fig. E2.4

2.4 OTHER APPLICATIONS OF THE INVERTING CONFIGURATION

Rather than using two resistors R_1 and R_2, we may use two impedances Z_1 and Z_2, as shown in Fig. 2.9. The closed-loop gain, or more appropriately the closed-loop transfer

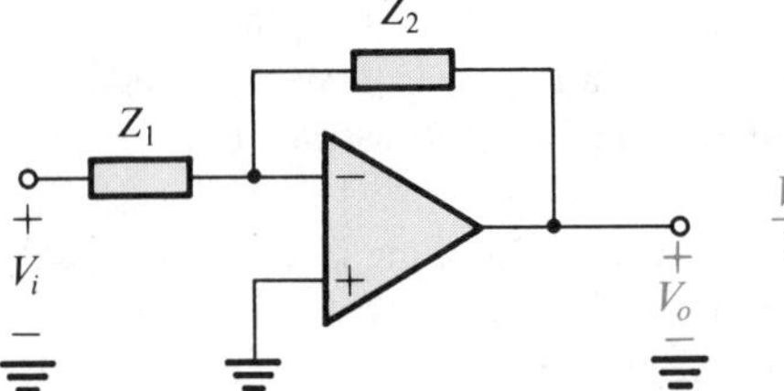

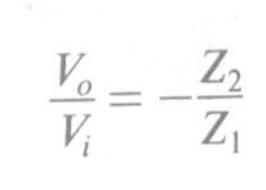

Fig. 2.9 The inverting configuration with general impedances in the feedback and at the input.

function, is given by

$$\frac{V_o}{V_i} = -\frac{Z_2}{Z_1}$$

As a special case consider first the following:

$$Z_1 = R \quad \text{and} \quad Z_2 = \frac{1}{sC}$$

$$\frac{V_o}{V_i} = -\frac{1}{sCR} \tag{2.2a}$$

which for physical frequencies, $s = j\omega$, becomes

$$\frac{V_o}{V_i} = -\frac{1}{j\omega CR} \tag{2.2b}$$

This transfer function can be shown to correspond to integration; that is, $v_O(t)$ will be the integral of $v_I(t)$. To see this in the time domain, consider the corresponding circuit shown in Fig. 2.10. It is easy to see that the current i_1 is given by

$$i_1 = \frac{v_I(t)}{R}$$

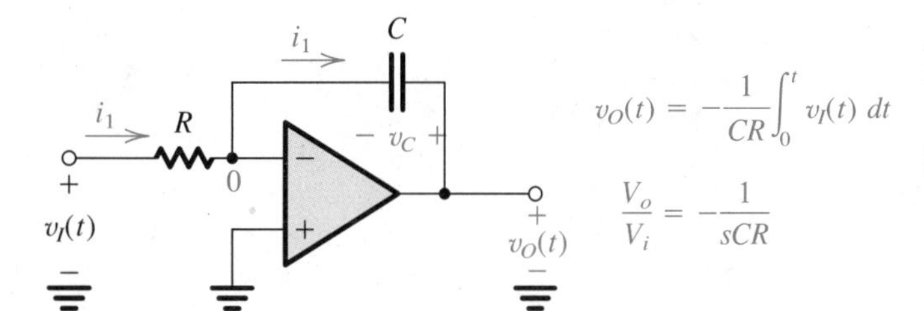

Fig. 2.10 The Miller or inverting integrator.

If at time $t = 0$ the voltage across the capacitor (measured in the direction indicated) is V_C, then

$$v_O(t) = V_C - \frac{1}{C}\int_0^t i_1(t)\,dt$$
$$= V_C - \frac{1}{CR}\int_0^t v_I(t)\,dt$$

Thus $v_O(t)$ is the time integral of $v_I(t)$, and voltage V_C is the initial condition of this integration process. The time constant CR is called the **integration time constant.** This integrator circuit is inverting because of the minus sign associated with its transfer function; it is known as the **Miller integrator.**

From the transfer function in Eq. (2.2) and our discussion of the frequency response of low-pass single-time-constant networks in Chapter 1 (see also Appendix E), it is easy to see that the Miller integrator will have the magnitude response shown in Fig. 2.11, which is identical to that of a low-pass network with zero corner frequency. It is important to note that at zero frequency the closed-loop gain is infinite. That is, at dc the op amp is operating as an open loop, which could be easily seen when we recall that capacitors behave as open circuits for dc. When op amp imperfections are taken into account, we

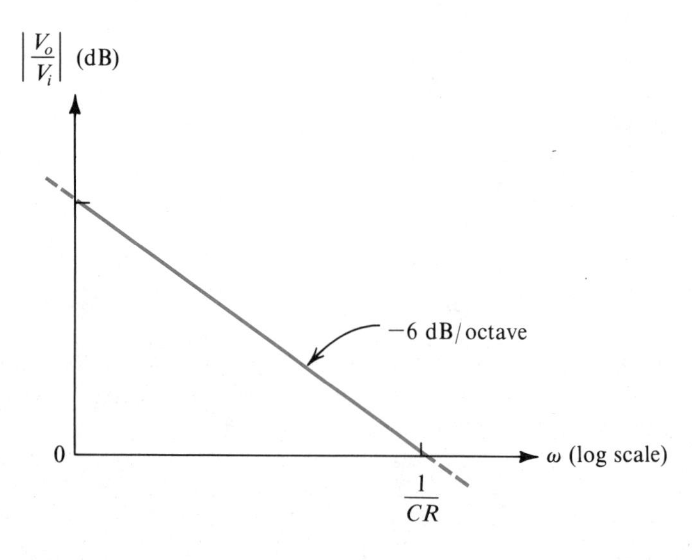

Fig. 2.11 Frequency response of an ideal integrator with a time constant CR.

will find it necessary to modify the integrator circuit to make the closed-loop gain at dc finite. This can be done by connecting a large resistance in parallel with the integrator capacitor, as will be illustrated in a later section. This modification, which is necessary to make the circuit work, unfortunately turns it into a nonideal integrator. Another point to note is that the integrator magnitude response crosses the 0 dB (unity-gain) line at a frequency equal to the inverse of the time constant ($1/CR$). An important application of op amp integrators is their use to convert square waveforms into triangular waves. This application is the subject of Exercise 2.5 below.

As another special case, consider

$$Z_1 = \frac{1}{sC} \quad \text{and} \quad Z_2 = R$$

$$\frac{V_o}{V_i} = -sCR$$

or in terms of physical frequencies,

$$\frac{V_o}{V_i} = -j\omega CR$$

which corresponds to a differentiation operation, that is

$$v_O(t) = -CR\frac{dv_I(t)}{dt}$$

The reader can easily verify that the circuit of Fig. 2.12(a) indeed implements this differentiation operation. Figure 2.12(b) shows a plot for the magnitude of the transfer function of the differentiator, which is identical to that of an STC high-pass network with an infinite corner frequency. Note that the response crosses the 0 dB line at $\omega = 1/CR$.

The very nature of a differentiator circuit causes it to be a "noise magnifier." This is due to the spike introduced at the output every time there is a sharp change in $v_I(t)$; such a change could be a "picked up" interference. For this reason and because they suffer from stability problems (Chapter 8), differentiator circuits are generally avoided in practice. When the circuit of Fig. 2.12(a) is used, it is usually necessary to connect a small resistance in series with the capacitor. This modification, unfortunately, turns the circuit into a nonideal differentiator.

As a final application of the inverting configuration consider the circuit shown in Fig. 2.13. Here we have a resistance R_f in the negative-feedback path (as before), but we have a number of input signals $v_1, v_2, \ldots, v_n$ each applied to a corresponding resistor $R_1, R_2, \ldots, R_n$, which are connected to the inverting terminal of the op amp. From our previous discussion, the ideal op amp will have a virtual ground appearing at its negative input terminal. Ohm's law then tells us that the currents $i_1, i_2, \ldots, i_n$ are given by

$$i_1 = \frac{v_1}{R_1}, \qquad i_2 = \frac{v_2}{R_2}, \qquad \ldots, \qquad i_n = \frac{v_n}{R_n}$$

All these currents sum together to produce the current i; that is,

$$i = i_1 + i_2 + \cdots + i_n$$

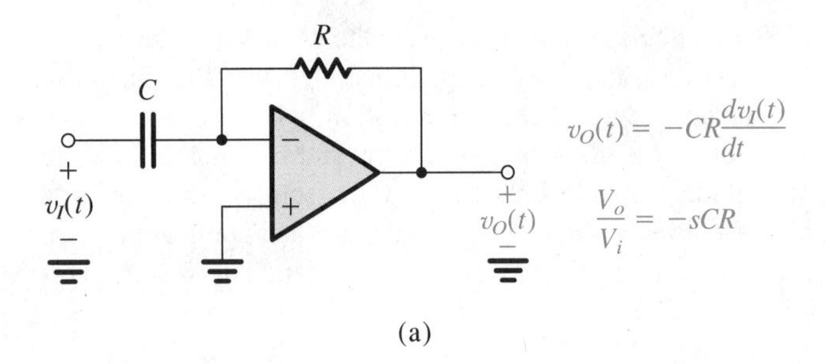

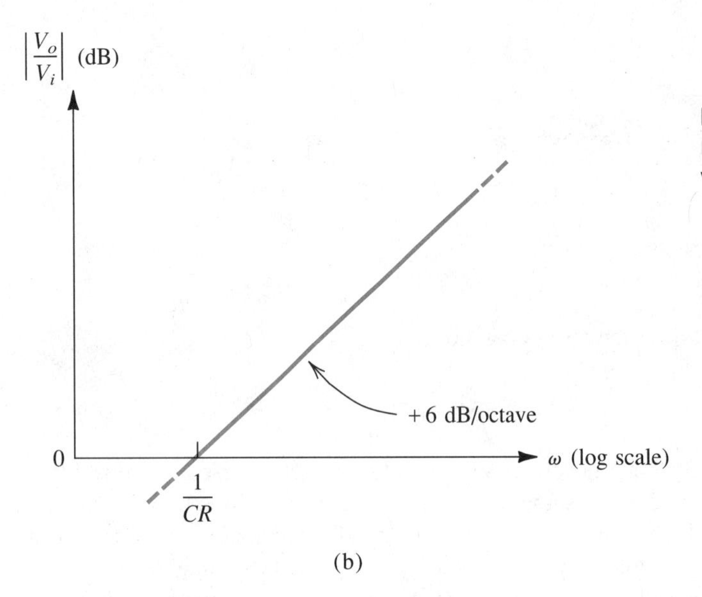

Fig. 2.12 (a) A differentiator. **(b)** Frequency response of a differentiator with a time constant CR.

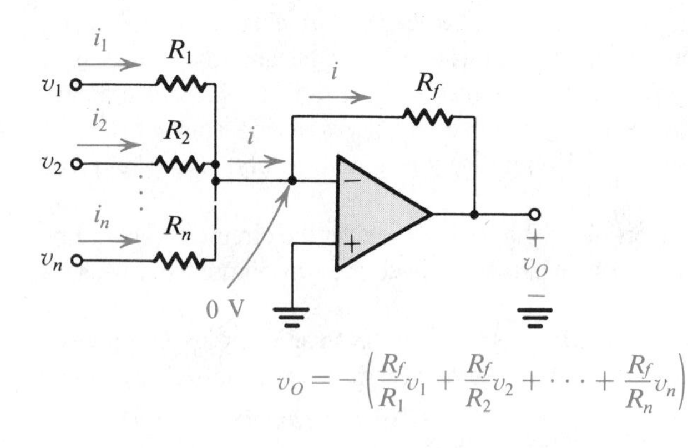

Fig. 2.13 A weighted summer.

will be forced to flow through R_f (since no current flows into the input terminals of an ideal op amp). The output voltage v_O may now be determined by another application of Ohm's law,

$$v_O = 0 - iR_f = -iR_f$$

Thus

$$v_O = -\left(\frac{R_f}{R_1}v_1 + \frac{R_f}{R_2}v_2 + \cdots + \frac{R_f}{R_n}v_n\right)$$

That is, the output voltage is a weighted sum of the input signals $v_1, v_2, \ldots, v_n$. This circuit is therefore called a **weighted summer.** Note that each summing coefficient may be independently adjusted by adjusting the corresponding ''feed-in'' resistor (R_1 to R_n). This nice property, which greatly simplifies circuit adjustment, is a direct consequence of the virtual ground that exists at the inverting op amp terminal. As the reader will soon come to appreciate, virtual grounds are extremely ''handy.''

We have seen above that op amps can be used to multiply a signal by a constant, integrate it, differentiate it, and sum a number of signals with prescribed weights. These are all mathematical operations—hence the name operational amplifier. In fact, the circuits above are functional building blocks needed to perform analog computation. For this reason the op amp has been the basic element of analog computers. Op amps, however, can do much more than just perform the mathematical operations required in analog computation. In this chapter we will get a taste of this versatility, with other applications presented in later chapters.

Exercises 2.5 Consider a symmetrical square wave of 20-V peak-to-peak, 0 average, and 2-ms period applied to a Miller integrator. Find the value of the time constant CR such that the triangular waveform at the output has a 20-V peak-to-peak amplitude.

Ans. 0.5 ms

2.6 Show that the circuit shown in Fig. E2.6 has an STC low-pass transfer function. For the case $R_1 = 1\ \text{k}\Omega$, $R_2 = 100\ \text{k}\Omega$, and $C_2 = 1$ nF, find the dc gain and the 3-dB frequency.

Ans. -100 V/V; 10^4 rad/s

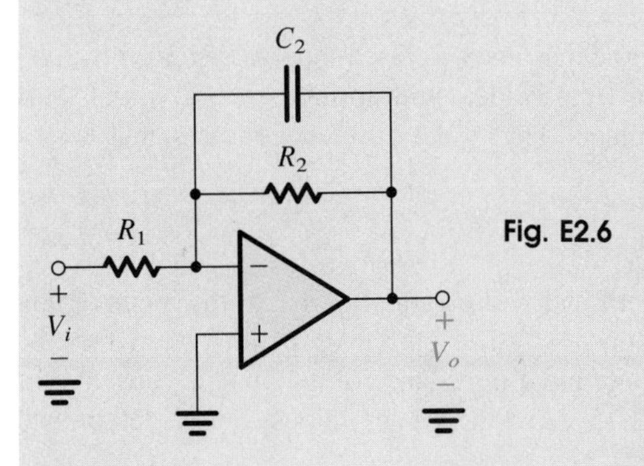

Fig. E2.6

D2.7 Using an ideal op amp, design an inverting integrator with an input resistance of 10 kΩ and an integration time constant of 10^{-3} s. What is the gain magnitude and phase angle of this circuit at 10 rad/s and at 1 rad/s? What is the frequency at which the gain magnitude is unity?

Ans. $R = 10\ \text{k}\Omega$, $C = 0.1\ \mu\text{F}$; at $\omega = 10$ rad/s: $|V_o/V_i| = 100$ V/V and $\phi = +90°$; at $\omega = 1$ rad/s: $|V_o/V_i| = 1{,}000$ V/V and $\phi = +90°$; 1000 rad/s

D2.8 Design a differentiator to have a time constant of 10^{-2} s and an input capacitance of 0.01 μF. What is the gain magnitude and phase of this circuit at 10 rad/s, and at 10^3 rad/s? In order to limit the high-frequency gain of the differentiator circuit to 100, a resistor is added in series with the capacitor. Find the required resistor value.

Ans. $C = 0.01\ \mu$F; $R = 1$ MΩ; at $\omega = 10$ rad/s: $|V_o/V_i| = 0.1$ and $\phi = -90°$; at $\omega = 1000$ rad/s: $|V_o/V_i| = 10$ and $\phi = -90°$; 10 kΩ

D2.9 Design an inverting op amp circuit to form the weighted sum v_O of two inputs v_1 and v_2. It is required that $v_O = -(v_1 + 5v_2)$. Choose values for R_1, R_2, and R_f so that for a maximum output voltage of 10 V the current in the feedback resistor will not exceed 1 mA.

Ans. A possible choice: $R_1 = 10$ kΩ, $R_2 = 2$ kΩ, and $R_f = 10$ kΩ

2.5 THE NONINVERTING CONFIGURATION

The second closed-loop configuration we shall study is shown in Fig. 2.14. Here the input signal v_I is applied directly to the positive input terminal of the op amp while one terminal of R_1 is connected to ground.

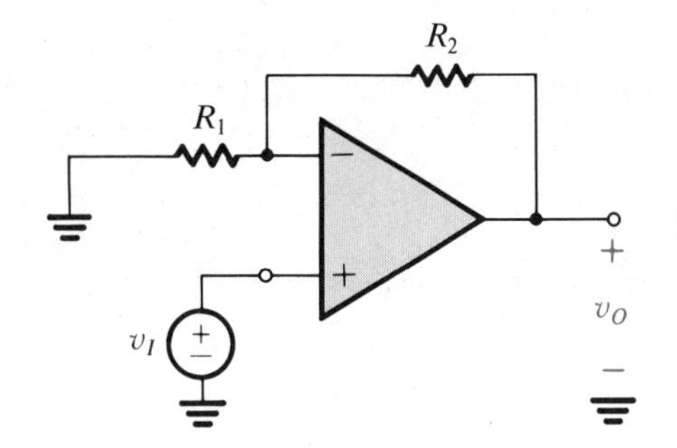

Fig. 2.14 The noninverting configuration.

Analysis of the noninverting circuit to determine its closed-loop gain (v_O/v_I) is illustrated in Fig. 2.15. Assuming that the op amp is ideal with infinite gain, a virtual short circuit exists between its two input terminals. Hence the difference input signal is

$$v_2 - v_1 = \frac{v_O}{A} = 0 \qquad \text{for } A = \infty$$

Thus the voltage at the inverting input terminal will be equal to that at the noninverting input terminal, which is the applied voltage v_I. The current through R_1 can then be determined as v_I/R_1. Because of the infinite input impedance of the op amp, this current will flow through R_2, as shown in Fig. 2.15. Now the output voltage can be determined from

$$v_O = v_I + \left(\frac{v_I}{R_1}\right)R_2$$

which yields

$$\frac{v_O}{v_I} = 1 + \frac{R_2}{R_1} \tag{2.3}$$

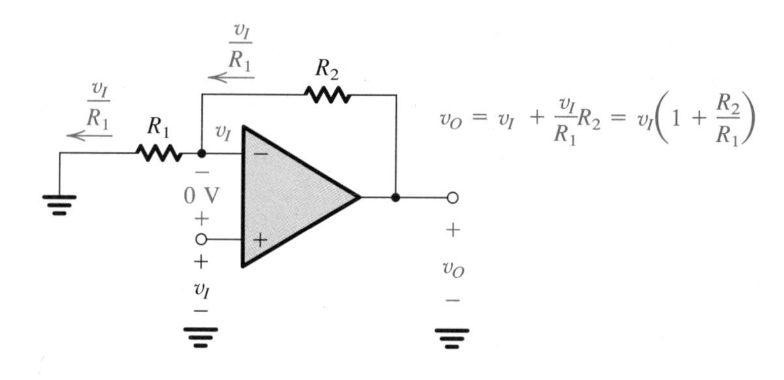

Fig. 2.15 Analysis of the noninverting circuit.

Further insight into the operation of the noninverting configuration can be obtained by considering the following: The voltage divider in the negative-feedback path causes a fraction of the output voltage to appear at the inverting input terminal of the op amp; that is,

$$v_1 = v_O\left(\frac{R_1}{R_1 + R_2}\right)$$

Then the infinite op amp gain and the resulting virtual short circuit between the two input terminals of the op amp forces this voltage to be equal to that applied at the positive input terminal; thus

$$v_O\left(\frac{R_1}{R_1 + R_2}\right) = v_I$$

which yields the gain expression given in Eq. (2.3).

The gain of the noninverting configuration is positive—hence the name *noninverting*. The input impedance of this closed-loop amplifier is ideally infinite, since no current flows into the positive input terminal of the op amp. The output of the noninverting amplifier is taken at the terminals of the ideal voltage source $A\,(v_2 - v_1)$ (see the op amp equivalent circuit in Fig. 2.3), thus the output resistance of the noninverting configuration is zero. Putting these properties together, we arrive at the equivalent circuit model of the noninverting amplifier configuration, shown in Fig. 2.16. This model is obtained under the assumption that the op amp is ideal.

The property of high input impedance is a very desirable feature of the noninverting configuration. It enables using this circuit as a buffer amplifier to connect a source with a

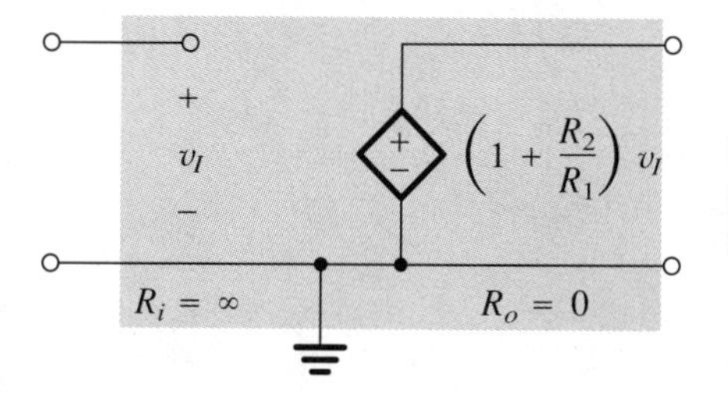

Fig. 2.16 Equivalent circuit model of the noninverting amplifier configuration of Fig. 2.14 (assuming that the op amp is ideal).

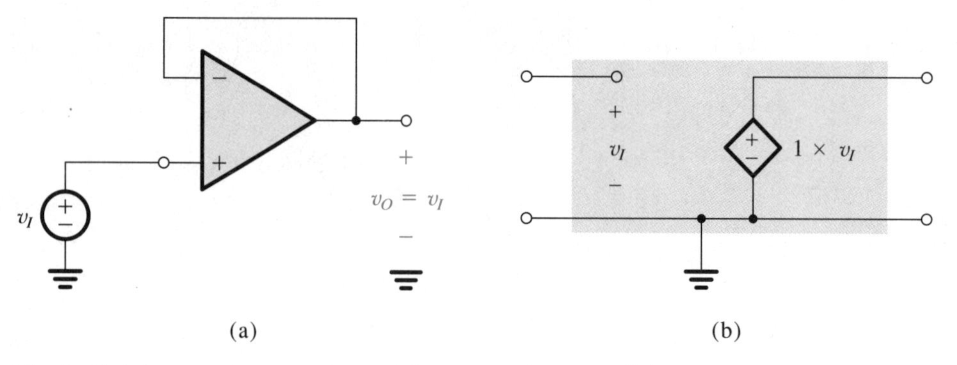

Fig. 2.17 (a) The unity-gain buffer or follower amplifier, and **(b)** its equivalent circuit model.

high impedance to a low-impedance load. We have discussed the need for buffer amplifiers in Chapter 1. In many applications the buffer amplifier is not required to provide any voltage gain; rather, it is used mainly as an impedance transformer or a power amplifier. In such cases we may make $R_2 = 0$ and $R_1 = \infty$ to obtain the unity-gain amplifier shown in Fig. 2.17(a). This circuit is commonly referred to as a **voltage follower,** since the output "follows" the input. In the ideal case, $v_O = v_I$, $R_{in} = \infty$, and $R_{out} = 0$.

Since the noninverting configuration has a gain greater than or equal to unity, depending on the choice of R_2/R_1, some prefer to call it "a follower with gain."

Exercises 2.10 Use the superposition principle to find the output voltage of the circuit shown in Fig. E2.10.

Ans. $v_O = 6v_1 + 4v_2$

2.11 If in the circuit of Fig. E2.10 the 1-kΩ resistor is disconnected from ground and connected to a third signal source v_3, use superposition to determine v_O in terms of v_1, v_2, and v_3.

Ans. $v_O = 6v_1 + 4v_2 - 9v_3$

D2.12 Design a noninverting amplifier with a gain of 2. At the maximum output voltage of 10 V the current in the voltage divider is to be 10 μA.

Ans. $R_1 = R_2 = 0.5\ \text{M}\Omega$

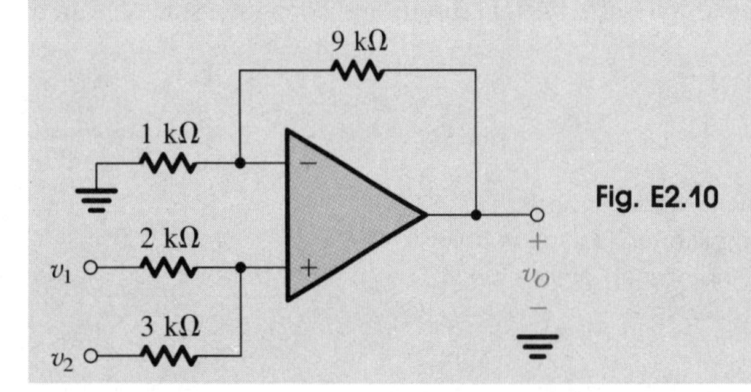

Fig. E2.10

2.13 (a) Show that if the op amp in the circuit of Fig. 2.14 has a finite open-loop gain A, then the closed-loop gain is given by

$$G \equiv \frac{v_O}{v_I} = \frac{1 + R_2/R_1}{1 + (1 + R_2/R_1)/A}$$

(b) For $R_1 = 1\ \text{k}\Omega$ and $R_2 = 9\ \text{k}\Omega$ find the percentage deviation ε of the closed-loop gain from the ideal value of $(1 + R_2/R_1)$ for the cases $A = 10^3$, 10^4, and 10^5. In each case find the voltage between the two input terminals of the op amp assuming that $v_I = 1$ V.

Ans. $\varepsilon = -1\%, -0.1\%, -0.01\%$; $v_2 - v_1 = 9.9$ mV, 1 mV, 0.1 mV

2.6 EXAMPLES OF OP AMP CIRCUITS

Now that we have studied the two most common closed-loop configurations of op amps, we present a number of examples. Our objective is twofold: first, to enable the reader to gain experience in analyzing circuits containing op amps; second, to introduce the reader to some of the many interesting and exciting applications of op amps.

EXAMPLE 2.3

Figure 2.18 shows a circuit for an analog voltmeter of very high input resistance that uses an inexpensive moving-coil meter. As shown, the moving-coil meter is connected in the negative-feedback path of the op amp. The voltmeter measures the voltage v applied between the op amp's positive input terminal and ground. Assume that the moving coil produces full-scale deflection when the current passing through it is 100 μA; we wish to find the value of R such that the full-scale reading for v is +10 V.

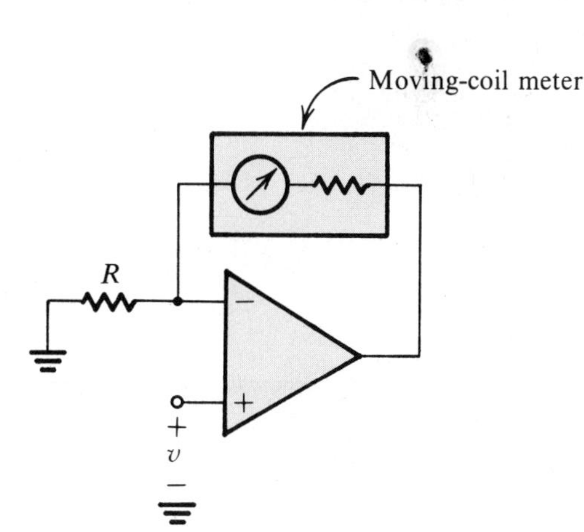

Fig. 2.18 An analog voltmeter with a high input resistance.

Solution The current in the moving-coil meter is v/R because of the virtual short circuit at the op amp input and the infinite input impedance of the op amp. Thus we have to choose R such that $10/R = 100\ \mu\text{A}$. Thus $R = 100\ \text{k}\Omega$.

Note that the resulting voltmeter will produce readings directly proportional to the value of v, irrespective of the value of the internal resistance of the moving-coil meter—a very desirable property.

EXAMPLE 2.4

We need to find an expression for the output voltage v_O in terms of the input voltages v_1 and v_2 for the circuit in Fig. 2.19.

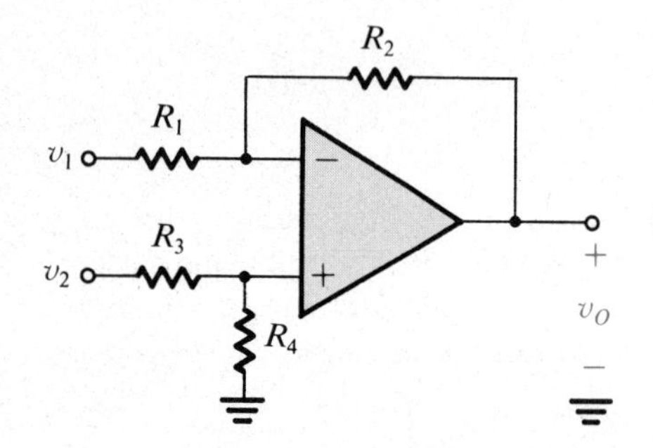

Fig. 2.19 A difference amplifier.

Solution

There are a number of ways to solve this problem; perhaps the easiest is using the principle of superposition. Obviously superposition may be employed here, since the network is linear. To apply superposition we first reduce v_2 to zero—that is, ground the terminal to which v_2 is applied—and then find the corresponding output voltage, which will be due entirely to v_1. We denote this output voltage v_{O1}. Its values may be found from the circuit in Fig. 2.20(a), which we recognize as that of the inverting configuration. The existence of R_3 and R_4 does not affect the gain expression, since no current flows through either of them. Thus

$$v_{O1} = -\frac{R_2}{R_1}v_1$$

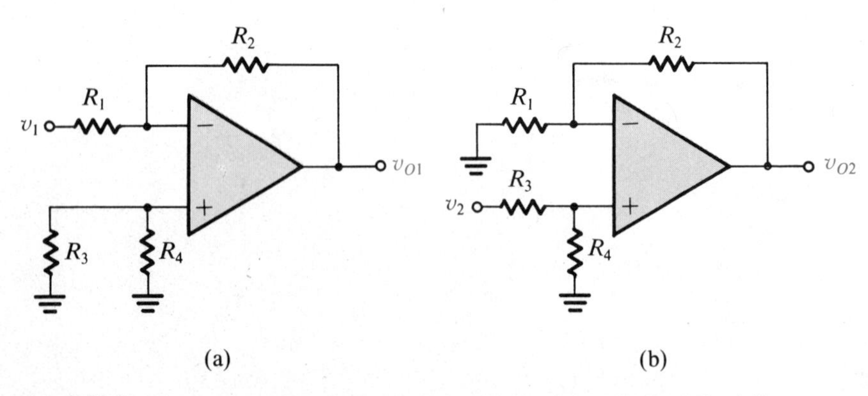

Fig. 2.20 Application of superposition to the analysis of the circuit of Fig. 2.19.

Next, we reduce v_1 to zero and evaluate the corresponding output voltage v_{O2}. The circuit will now take the form shown in Fig. 2.20(b), which we recognize as the noninverting configuration with an additional voltage divider, made up of R_3 and R_4, connected across the input v_2. The output voltage v_{O2} is therefore given by

$$v_{O2} = v_2 \frac{R_4}{R_3 + R_4}\left(1 + \frac{R_2}{R_1}\right)$$

The superposition principle tells us that the output voltage v_O is equal to the sum of v_{O1} and v_{O2}. Thus we have

$$v_O = -\frac{R_2}{R_1}v_1 + \frac{1 + R_2/R_1}{1 + R_3/R_4}v_2 \tag{2.4}$$

This completes the analysis of the circuit in Fig. 2.19. However, because of the practical importance of this circuit we shall pursue it further. We will ask: What is the condition under which this circuit will act as a differential amplifier? In other words, we wish to make the circuit respond (produce an output) in proportion to the difference signal $v_2 - v_1$ and reject common-mode signals (that is, produce zero output when $v_1 = v_2$). The answer can be obtained from the expression we have just derived (Eq. 2.4). Let us set $v_1 = v_2$ and require that $v_O = 0$. It is easy to see that this process leads to the condition $R_2/R_1 = R_4/R_3$. Substituting in Eq. (2.4) results in the output voltage

$$v_O = \frac{R_2}{R_1}(v_2 - v_1)$$

which is clearly that of a differential amplifier with a gain of R_2/R_1.

We next inquire about the input resistance seen between the two input terminals. The circuit is redrawn in Fig. 2.21 with the condition $R_2/R_1 = R_4/R_3$ imposed. In fact, to simplify matters and for other practical considerations, we have made $R_3 = R_1$ and $R_4 = R_2$. We wish to evaluate the input differential resistance R_{in}, defined as

$$R_{in} \equiv \frac{v_2 - v_1}{i}$$

Since the two input terminals of the op amp track each other in potential, we may write a loop equation and obtain

$$v_2 - v_1 = R_1 i + 0 + R_1 i$$

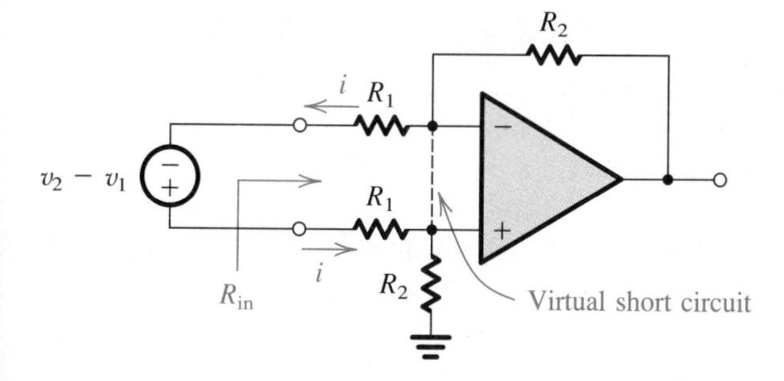

Fig. 2.21 Finding the input resistance of the difference amplifier.

Thus $R_{in} = 2R_1$. Note that if the amplifier is required to have a large differential gain, then R_1, of necessity, will be relatively small and the input resistance will be correspondingly small, a drawback of this circuit.

Differential amplifiers find application in many areas, most notably in the design of instrumentation systems. As an example, consider the case of a transducer that produces between its two output terminals a relatively small signal, say 1 mV. However, between each of the two wires (leading from the transducer to the instrumentation system) and ground there may be large picked-up interference, say 1 V. The required amplifier, known as an **instrumentation amplifier,** must reject this large interference signal, which is common to the two wires (a common-mode signal), and amplify the small difference (or differential) signal. This situation is illustrated in Fig. 2.22 where v_{CM} denotes the common-mode signal and v_d denotes the differential signal.

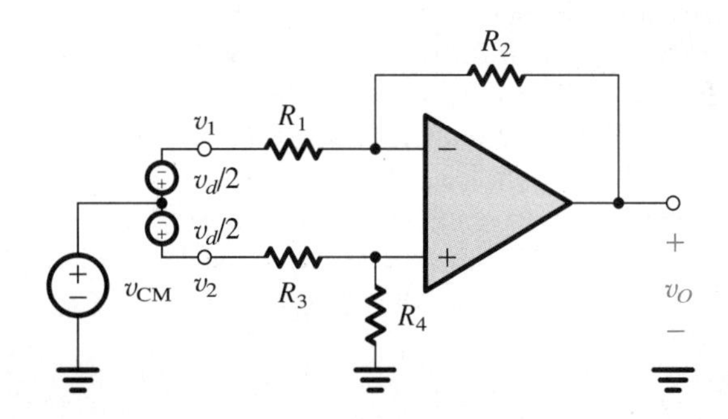

Fig. 2.22 Representation of the common-mode and differential components of the input signal to a difference amplifier. Note that $v_1 = v_{CM} - v_d/2$ and $v_2 = v_{CM} + v_d/2$.

EXAMPLE 2.5

The difference amplifier studied in the previous example is not entirely satisfactory as an instrumentation amplifier. Its major drawbacks are its low input resistance and that its gain cannot be easily varied. A much superior instrumentation amplifier circuit is shown in Fig. 2.23(a). Analyze the circuit to determine v_O as a function of v_1 and v_2, and determine the differential gain. Suggest a way for making the gain variable. Also find the input resistance. Design the circuit to provide a gain that can be varied over the range 2 to 1000 utilizing a 100-kΩ variable resistance (a potentiometer, or "pot" for short).

Solution The circuit consists of two stages: The first stage is formed by op amps A_1 and A_2 and their associated resistors, and the second stage is formed by op amp A_3 together with its four associated resistors. We recognize the second stage as that of the difference amplifier studied in Example 2.4.

Analysis of the circuit, assuming ideal op amps, is straightforward, as is illustrated in Fig. 2.23(b). The key point is that the virtual short circuits at the inputs of op amps A_1 and A_2 cause the input voltages v_1 and v_2 to appear at the two terminals of resistor R_1. Thus the differential input voltage $(v_1 - v_2)$ appears across R_1 and causes a current $i = (v_1 - v_2)/R_1$ to flow through R_1 and the two resistors labeled R_2. This current in turn produces a voltage difference between the output terminals of A_1 and A_2 given by

$$v_{O1} - v_{O2} = \left(1 + \frac{2R_2}{R_1}\right)(v_1 - v_2) \tag{2.5}$$

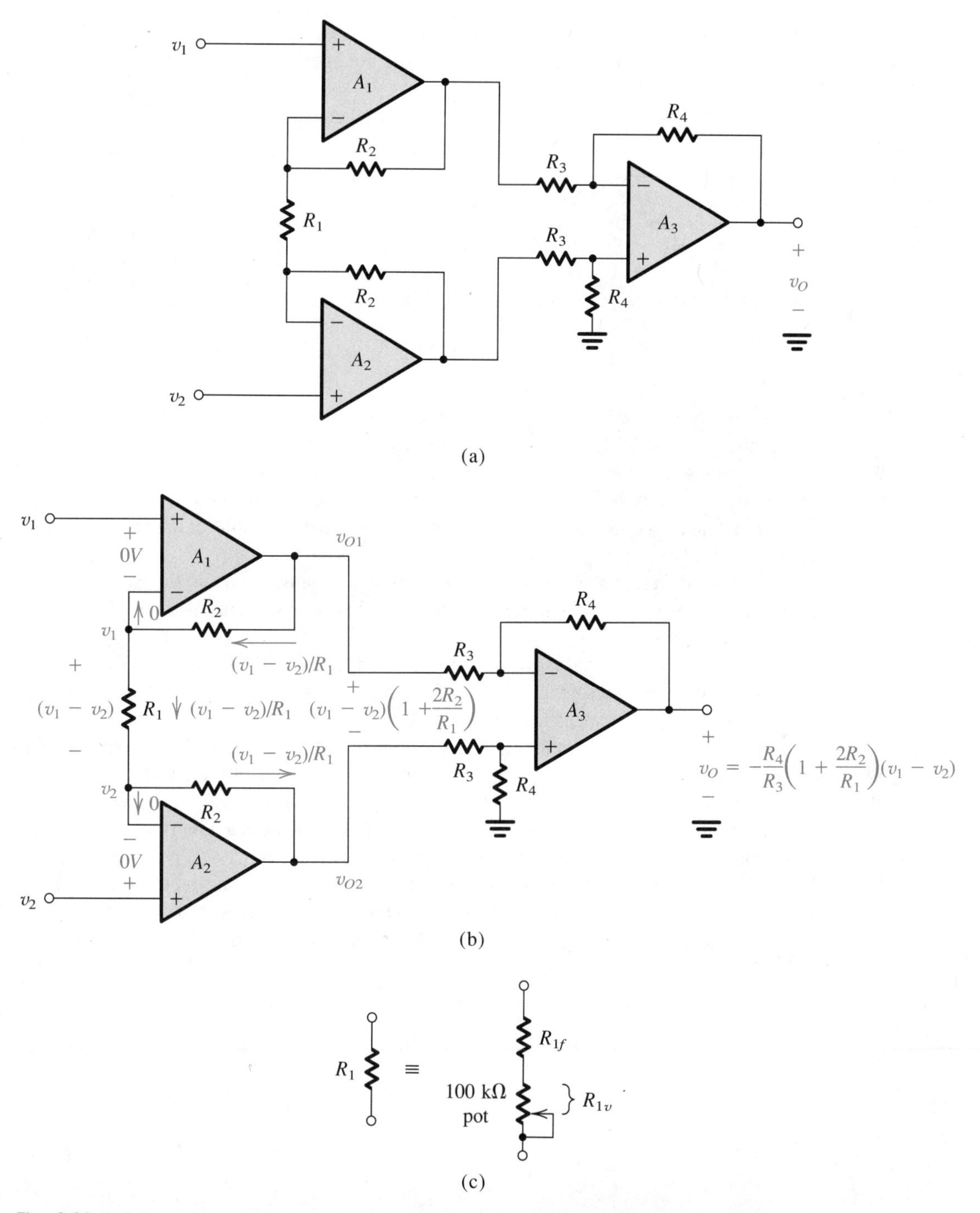

Fig. 2.23 (a) A popular circuit for an instrumentation amplifier. **(b)** Analysis of the circuit in (a) assuming ideal op amps. **(c)** To make the gain variable, R_1 is implemented as the series combination of a fixed resistor R_{1f} and a variable resistor R_{1v}. Resistor R_{1f} ensures that the maximum available gain is limited.

The difference amplifier formed around op amp A_3 senses the voltage difference $(v_{O1} - v_{O2})$ and provides a proportional output voltage v_O,

$$v_O = -\frac{R_4}{R_3}(v_{O1} - v_{O2}) \tag{2.6}$$

Combining Eqs. (2.5) and (2.6) results in

$$v_O = \frac{R_4}{R_3}\left(1 + \frac{2R_2}{R_1}\right)(v_2 - v_1)$$

Thus the instrumentation amplifier has a differential voltage gain

$$A_d \equiv \frac{v_O}{v_2 - v_1} = \left(1 + \frac{2R_2}{R_1}\right)\frac{R_4}{R_3} \tag{2.7}$$

It can be easily shown that an input common-mode signal v_{CM} (applied to both input terminals; see Fig. 2.22) will propagate through the first stage resulting in $v_{O1} = v_{O2} = v_{CM}$ (assuming that for now we have made $v_d = 0$). Thus if the second-stage difference amplifier is properly balanced it will produce a zero output voltage in response to v_{CM}, indicating that the common-mode gain of the instrumentation amplifier has the ideal value of zero.

From the differential-gain expression in Eq. (2.7) we observe that the gain value can be varied by varying the single resistor R_1; any other arrangement involves varying two resistors simultaneously.

Since both of the input-stage op amps are connected in the noninverting configuration, the input impedance seen by each of v_1 and v_2 is (ideally) infinite. This is a major advantage of this instrumentation amplifier configuration.

We now turn to the particular design problem. It is usually preferable to obtain all the required gain in the first stage, leaving the second stage to perform the task of taking the difference between the outputs of the first stage and thus rejecting the common-mode signal. In other words, the second stage is usually designed for a gain of one. Adopting this approach, we select all the second-stage resistors to be equal to a practically convenient value, say 10 kΩ. The problem then reduces to designing the first stage to realize a gain adjustable over the range 2 to 1000. Implementing R_1 as the series combination of a fixed resistor R_{1f} and the variable resistor R_{1v} obtained using the 100-kΩ pot (see Fig. 2.23c) we can write

$$1 + \frac{2R_2}{R_{1f} + R_{1v}} = 2 \text{ to } 1000$$

Thus,

$$1 + \frac{2R_2}{R_{1f}} = 1000$$

and

$$1 + \frac{2R_2}{R_{1f} + 100\text{ k}\Omega} = 2$$

These two equations yield $R_{1f} = 100.2\ \Omega$ and $R_2 = 50.050\ \mathrm{k}\Omega$. Other practical values may be selected; for instance, $R_{1f} = 100\ \Omega$ and $R_2 = 49.9\ \mathrm{k}\Omega$ (both values are available as standard 1%-tolerance metal-film resistors; see Appendix F) results in a gain covering approximately the required range.

EXAMPLE 2.6 We wish to find the input resistance R_{in} of the circuit in Fig. 2.24.

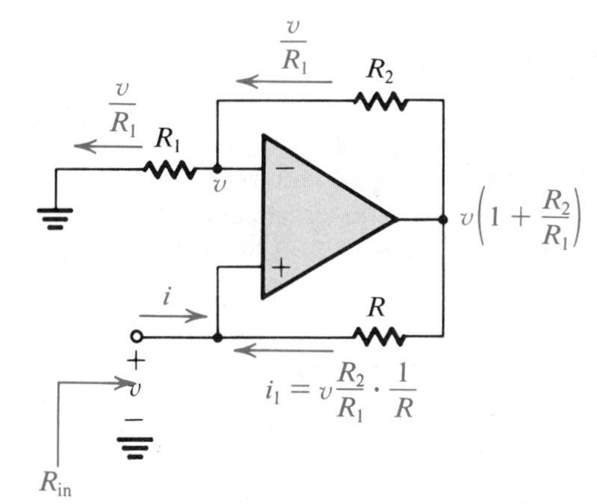

Fig. 2.24 Circuit for Example 2.6.

Solution To find R_{in} we apply an input voltage v and evaluate the input current i. Then R_{in} may be found from its definition, $R_{in} \equiv v/i$.

Owing to the virtual short circuit between the op-amp input terminals, the voltage at the inverting terminal will be equal to v. The current through R_1 will therefore be v/R_1. Owing to the infinite input impedance of the op amp, the current through R_2 will also be v/R_1. Thus the voltage at the op amp output will be

$$v + \frac{v}{R_1}R_2 = \left(1 + \frac{R_2}{R_1}\right)v$$

We may now apply Ohm's law to R and obtain the current through it as

$$i_1 = \frac{v(1 + R_2/R_1) - v}{R} = v\frac{R_2}{R_1}\frac{1}{R}$$

Since no current flows into the positive input terminal of the op amp, we have

$$i = -i_1 = -\frac{v}{R}\frac{R_2}{R_1}$$

Thus

$$R_{in} = -R\frac{R_1}{R_2}$$

that is, the input resistance is negative with a magnitude equal to R, the resistance in the positive-feedback path, multiplied by the ratio R_1/R_2. This circuit is therefore called a **negative impedance converter** (NIC), where R may in general be replaced by an arbitrary impedance Z.

Let us investigate the application of this circuit further. Consider the case $R_1 = R_2 = r$, where r is an arbitrary value; it follows that $R_{in} = -R$. Let the input be fed with a voltage source V_s having a source resistance equal to R, as shown in Fig. 2.25(a). We want to evaluate the current I_l that flows in an impedance Z_L connected as shown. In Fig. 2.25(b) we have utilized the information gained above and replaced the circuit in the dashed box by a resistance, $-R$. Figure 2.25(c) illustrates the conversion of the voltage source to its Norton's equivalent (for a review of Norton's theorem, see Appendix D). Finally, the two parallel resistances R and $-R$ are combined to produce an infinite resistance, resulting in the circuit in Fig. 2.25(d), from which we see that the load current I_l is given by $I_l = V_s/R$, independent of the value of Z_L! This is an interesting result; it tells us that the circuit of Fig. 2.25(a) acts as a **voltage-to-current converter,** providing a current I_l that is directly proportional to V_s ($I_t = V_s/R$) and is independent of the value of the load impedance. That is, terminal 2 acts as a current-source output, with the resistance

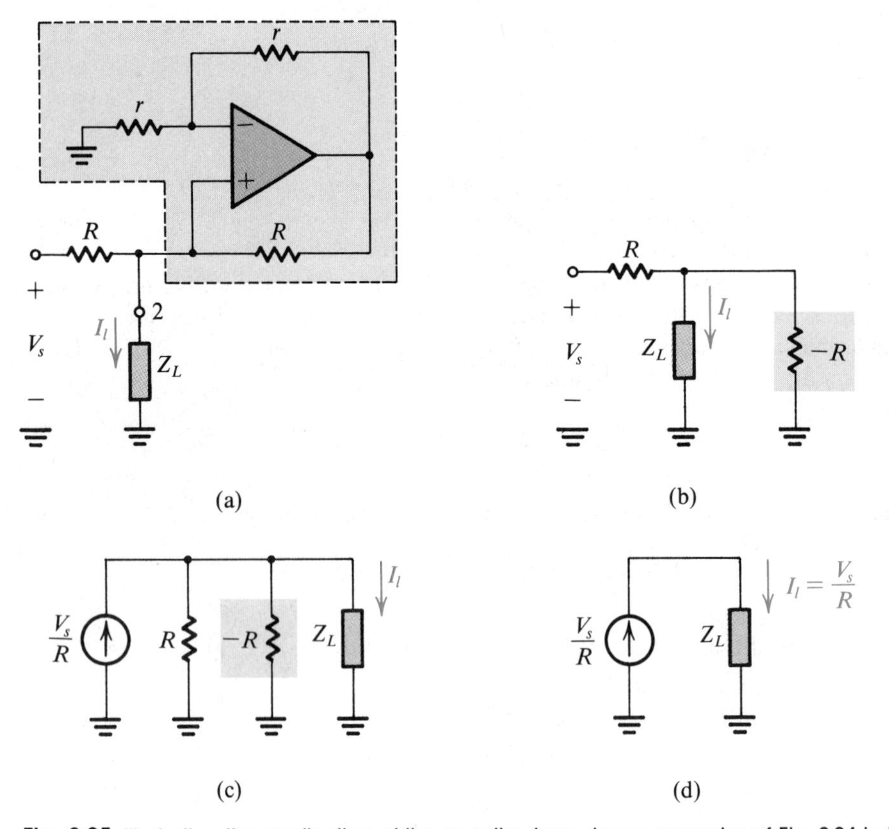

Fig. 2.25 Illustrating the application of the negative impedance converter of Fig. 2.24 in the design of a voltage-to-current converter or a controlled current source.

looking back into 2 equal to infinity. Note that this infinite resistance is obtained via the cancellation of the positive source resistance R with the negative input resistance $-R$.

As it is, the circuit is quite useful and has practical applications where it is necessary to generate a current signal in proportion to a given voltage signal. A specific application is illustrated in Fig. 2.26, where a capacitor C is used as a load. From the analysis above we conclude that capacitor C will be supplied by a current $I = V_i/R$, and thus its voltage V_2 will be given by

$$V_2 = \frac{I}{sC} = \frac{V_i}{sCR}$$

That is,

$$\frac{V_2}{V_i} = \frac{1}{sCR}$$

which is the transfer function of an integrator,

$$v_2 = \frac{1}{CR}\int_0^t v_i \, dt + V$$

where V is the voltage across the capacitor at $t = 0$. This integrator has some interesting properties. The transfer function does not have an associated negative sign, as is the case with the Miller integrator. Noninverting, or positive, integrators are required in many applications. Another useful property is the fact that one terminal of the capacitor is grounded. Among other things, this would simplify the initial charging of the capacitor, as may be necessary to simulate an initial condition in the solution of a differential equation on an analog computer.

The integrator circuit of Fig. 2.26 has a serious problem though. We cannot "take the output" at terminal 2 as indicated, since terminal 2 is a high-impedance point, which means that connecting any load resistance there will change the transfer function V_2/V_i. Fortunately, however, a low-impedance point exists where the signal is proportional to V_2.

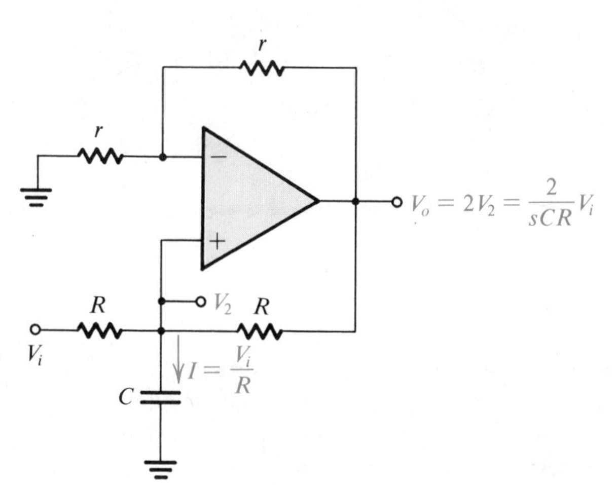

Fig. 2.26 Application of the voltage-to-current converter in the design of a noninverting integrator.

We are speaking of the op amp's output terminal, where, as the reader can easily verify, $V_o = 2V_2$. Thus

$$\frac{V_o}{V_i} = \frac{2}{sCR}$$

Exercises 2.14 Find values for the resistances in the circuit of Fig. 2.19 such that the circuit behaves as a differential amplifier with an input resistance of 20 kΩ and a gain of 100.

Ans. $R_1 = R_3 = 10$ kΩ; $R_2 = R_4 = 1$ MΩ

2.15 For the circuit shown in Fig. E2.15 derive an expression for the transfer function V_o/V_i. Find expressions for the magnitude and phase of the response. *Note:* This circuit functions as a phase shifter. It is also known as a first-order all-pass filter.

Ans. $V_o/V_i = (s - 1/CR)/(s + 1/CR)$; $|V_o/V_i| = 1$; $\phi = 180° - 2\tan^{-1}(\omega CR)$

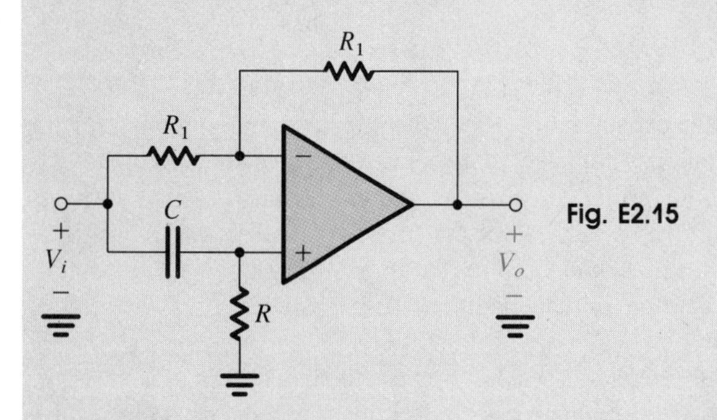

Fig. E2.15

2.16 Consider the difference amplifier circuit in Fig. 2.19. Replace R_2 and R_4 with two equal capacitors C, and let $R_1 = R_3 = R$. Use superposition to show that the circuit becomes a differential integrator with $V_o = (V_2 - V_1)/sCR$.

2.17 This exercise illustrates the use of the negative impedance converter circuit of Fig. 2.24 as an amplifier. Let $R_1 = R_2 = 100$ kΩ, and let the input be connected to a source $v_s = 10$ mV having a resistance $R_s = 0.9$ MΩ. Find the value of R so that a signal of 100 mV appears at the noninverting input terminal of the op amp. What is the value of the signal at the output terminal of the op amp?

Ans. 1 MΩ; 200 mV

2.7 NONIDEAL PERFORMANCE OF OP AMPS

Above we defined the ideal op amp, and we presented a number of circuit applications of op amps. The analysis of these circuits assumed the op amps to be ideal. Although in many applications such an assumption is not a bad one, a circuit designer has to be thoroughly familiar with the characteristics of practical op amps and the effects of such characteristics on the performance of op amp circuits. Only then will the designer be able to use the op amp intelligently, especially if the application at hand is not a straightforward one. The nonideal properties of op amps will, of course, limit the range of operation of the circuits analyzed in the previous examples.

In the remainder of this chapter we consider the nonideal properties of the op amp. We do this by treating one parameter at a time, beginning in this section with the most serious op amp nonideality, its finite gain and bandwidth.

Finite Open-Loop Gain and Bandwidth

The differential open-loop gain of an op amp is not infinite; rather, it is finite and decreases with frequency. Figure 2.27 shows a plot for $|A|$, with the numbers typical of most general-purpose op amps (such as the 741-type op amp, which is available from many semiconductor manufacturers and whose internal circuit is studied in Chapter 10).

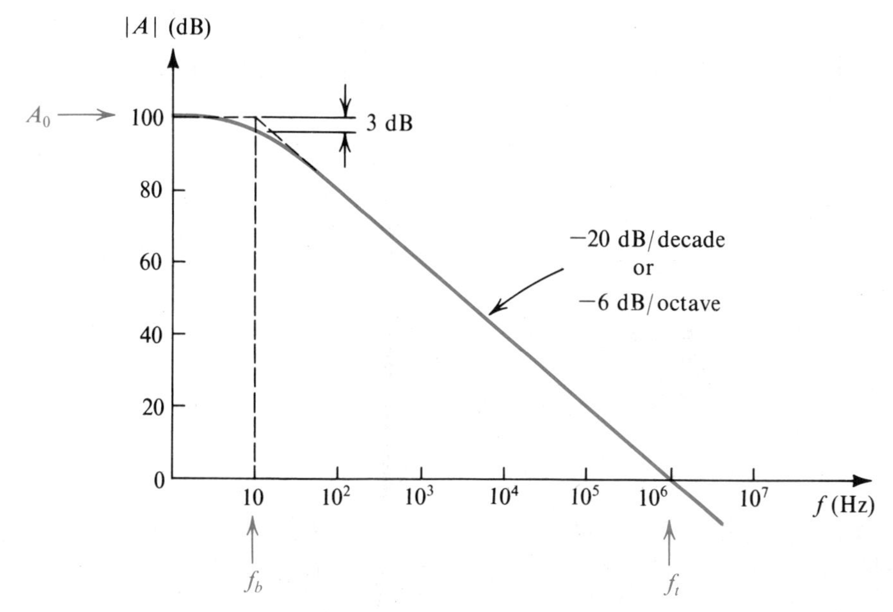

Fig. 2.27 Open-loop gain of a typical general-purpose internally compensated op amp.

Note that although the gain is quite high at dc and low freqencies, it starts to fall off at a rather low frequency (10 Hz in our example). The uniform −20-dB/decade gain rolloff shown is typical of **internally compensated** op amps. These are units that have a network (usually a single capacitor) included on the same IC chip whose function is to cause the op-amp gain to have the single-time-constant low-pass response shown. This process of modifying the open-loop gain is termed **frequency compensation,** and its purpose is to ensure that op amp circuits will be stable (as opposed to oscillatory). The subject of stability of op amp circuits—or, more generally, of feedback amplifiers—will be studied in Chapter 8.

By analogy to the response of low-pass STC circuits (see Section 1.6 and, for more detail, Appendix E), the gain $A(s)$ of an internally compensated op amp may be expressed as

$$A(s) = \frac{A_0}{1 + s/\omega_b} \tag{2.8}$$

which for physical frequencies, $s = j\omega$, becomes

$$A(j\omega) = \frac{A_0}{1 + j\omega/\omega_b} \tag{2.9}$$

where A_0 denotes the dc gain and ω_b is the 3-dB frequency (or "break" frequency). For the example shown in Fig. 2.27, $A_0 = 10^5$ and $\omega_b = 2\pi \times 10$ rad/s. For frequencies $\omega \gg \omega_b$ (about ten times and higher) Eq. (2.9) may be approximated by

$$A(j\omega) \simeq \frac{A_0\omega_b}{j\omega} \tag{2.10}$$

from which it can be seen that the gain $|A|$ reaches unity (0 dB) at a frequency denoted by ω_t and given by

$$\omega_t = A_0\omega_b \tag{2.11}$$

Substituting in Eq. (2.10) gives

$$A(j\omega) \simeq \frac{\omega_t}{j\omega} \tag{2.12}$$

where ω_t is called the **unity-gain bandwidth.**[1] The unity-gain bandwidth $f_t = \omega_t/2\pi$ is usually specified on the data sheets of op amps. Also note that for $\omega \gg \omega_b$ the open-loop gain in Eq. (2.8) becomes

$$A(s) \simeq \frac{\omega_t}{s} \tag{2.13}$$

Thus the op amp behaves as an integrator with time constant $\tau = 1/\omega_t$. This correlates with the -6-dB/octave frequency response indicated in Fig. 2.27.

The gain magnitude can be obtained from Eq. (2.12) as

$$|A(j\omega)| \simeq \frac{\omega_t}{\omega} = \frac{f_t}{f} \tag{2.14}$$

Thus if f_t is known (10^6 Hz in our example), one can easily estimate the magnitude of the op amp gain at a given frequency f. As a matter of practical importance, we note that the production spread in the value of ω_t between op amps of the same type is much smaller than that observed for A_0 and ω_b. For this reason, ω_t (or $f_t = \omega_t/2\pi$) is preferred as a specification parameter. Finally, it should be mentioned that an op amp having this uniform -6-dB/octave gain rolloff is said to have a "single-pole" model. Also, since this single pole dominates the amplifier frequency response, it is called a **dominant pole.** More will be said about poles and zeros in Chapter 7.

[1] Since ω_t is the product of the dc gain A_0 and the 3-dB bandwidth ω_b, it is also known as the **gain–bandwidth product** (GB).

Frequency Response of Closed-Loop Amplifiers

We next consider the effect of limited op amp gain and bandwidth on the closed-loop transfer functions of the two basic configurations: the inverting circuit of Fig. 2.4 and the noninverting circuit of Fig. 2.14. The closed-loop gain of the inverting amplifier, assuming a finite op amp open-loop gain A, was derived in Section 2.3 and given in Eq. (2.1), which we repeat here as

$$\frac{V_o}{V_i} = \frac{-R_2/R_1}{1 + (1 + R_2/R_1)/A} \tag{2.15}$$

Substituting for A from Eq. (2.8) gives

$$\frac{V_o(s)}{V_i(s)} = \frac{-R_2/R_1}{1 + \dfrac{1}{A_0}\left(1 + \dfrac{R_2}{R_1}\right) + \dfrac{s}{\omega_t/(1 + R_2/R_1)}} \tag{2.16}$$

For $A_0 \gg 1 + R_2/R_1$, which is usually the case,

$$\frac{V_o(s)}{V_i(s)} \simeq \frac{-R_2/R_1}{1 + \dfrac{s}{\omega_t/(1 + R_2/R_1)}} \tag{2.17}$$

which is of the same form as that for a low-pass single-time-constant network (see Table 1.1). Thus the inverting amplifier has an STC low-pass response with a dc gain of magnitude equal to R_2/R_1. The closed-loop gain rolls off at a uniform -20-dB decade slope with a corner frequency (3-dB frequency) given by

$$\omega_{3\text{dB}} = \frac{\omega_t}{1 + R_2/R_1} \tag{2.18}$$

Similarly, analysis of the noninverting amplifier of Fig. 2.14, assuming a finite open-loop gain A, yields the closed-loop transfer function

$$\frac{V_o}{V_i} = \frac{1 + R_2/R_1}{1 + (1 + R_2/R_1)/A} \tag{2.19}$$

Substituting for A from Eq. (2.8) and making the approximation $A_0 \gg 1 + R_2/R_1$ results in

$$\frac{V_o(s)}{V_i(s)} \simeq \frac{1 + R_2/R_1}{1 + \dfrac{s}{\omega_t/(1 + R_2/R_1)}} \tag{2.20}$$

Thus the noninverting amplifier has an STC low-pass response with a dc gain of $(1 + R_2/R_1)$ and a 3-dB frequency given also by Eq. (2.18).

EXAMPLE 2.7

Consider an op amp with $f_t = 1$ MHz. Find the 3-dB frequency of closed-loop amplifiers with nominal gains of +1000, +100, +10, +1, −1, −10, −100, and −1000. Sketch the magnitude frequency response for the amplifiers with closed-loop gains of +10 and −10.

Solution Using Eq. (2.18), we obtain the results given in the following table:

Closed-Loop Gain	$\frac{R_2}{R_1}$	$f_{3dB} = f_t/(1 + R_2/R_t)$
+1000	999	1 kHz
+100	99	10 kHz
+10	9	100 kHz
+1	0	1 MHz
−1	1	0.5 MHz
−10	10	90.9 kHz
−100	100	9.9 kHz
−1000	1000	≃1 kHz

Figure 2.28 shows the frequency response for the amplifier whose nominal dc gain is +10, and Fig. 2.29 shows the frequency response for the −10 case. An interesting observation follows from the table above: The unity-gain inverting amplifier has a 3-dB frequency of $f_t/2$ as compared to f_t for the unity-gain noninverting amplifier.

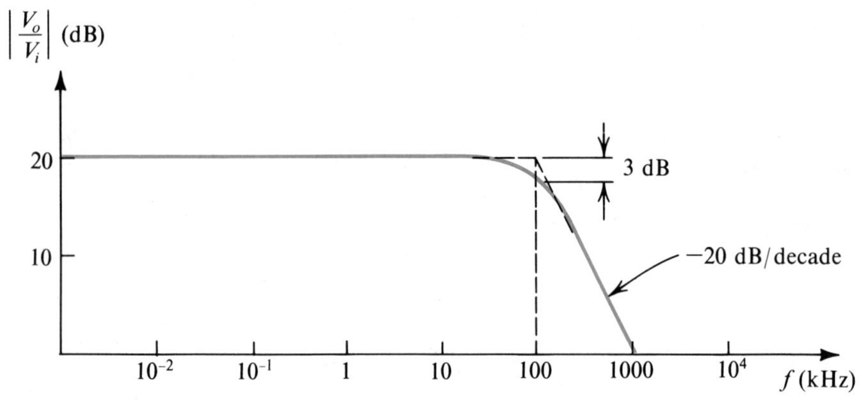

Fig. 2.28 Frequency response of an amplifier with a nominal gain of +10.

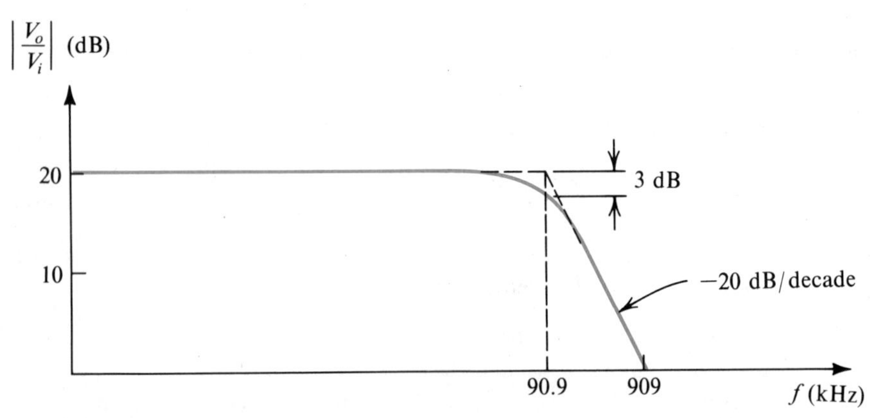

Fig. 2.29 Frequency response of an amplifier with a nominal gain of −10.

An Interpretation in Terms of Feedback

Example 2.7 clearly illustrates the trade-off between gain and bandwidth. For instance, the noninverting configuration exhibits a constant **gain–bandwidth product.** An interpretation of these results in terms of feedback theory will be given in Chapter 8. For the time being, note that both the inverting and the noninverting configurations have identical "feedback loops." This can be seen by eliminating the excitation (that is, short-circuiting the input voltage source), resulting in both cases in the feedback loop shown in Fig. 2.30. Since their feedback loops are identical, the two configurations have the same dependence on finite op amp gain and bandwidth (for example, identical expressions for f_{3dB}).

Before leaving this section we wish to examine the feedback loop of Fig. 2.30 a bit more closely. The forward path of the loop (from node 1 to node 3) consists of an

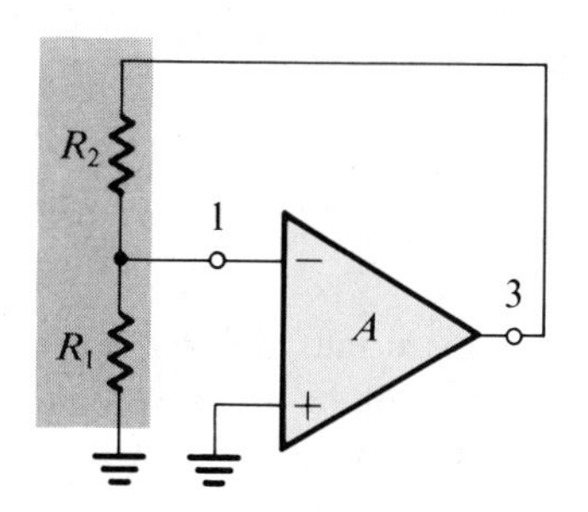

Fig. 2.30 The feedback loop of the inverting and the noninverting configurations.

amplifier of gain $-A$. The feedback path (from node 3 to node 1) consists of a voltage divider of transmission ratio $R_1/(R_1 + R_2)$. This is usually called the feedback ratio (or feedback factor) and is denoted β; that is,

$$\beta = \frac{R_1}{R_1 + R_2} \tag{2.21}$$

The loop gain is given by

$$\text{Loop gain} = -A\beta \tag{2.22}$$

Note that the negative loop gain signifies the fact that the feedback is negative. An important quantity in feedback amplifiers is the amount of feedback, defined by

$$\text{Amount of feedback} \equiv 1 - \text{loop gain} = 1 + A\beta \tag{2.23}$$

Later on in this chapter we shall see that this quantity plays an important role.

Finally, note that for both the inverting and the noninverting amplifier configurations, the 3-dB bandwidth is given by

$$f_{3dB} = \beta f_t \tag{2.24}$$

Exercises 2.18 Consider an op amp having a 106-dB gain at dc and a single-pole frequency response with f_t = 2 MHz. Find the magnitude of gain at f = 1 kHz, 10 kHz, and 100 kHz.

Ans. 2000; 200; 20

2.19 If the op amp in Exercise 2.18 is used to design a noninverting amplifier with nominal dc gain of 100, find the 3-dB frequency of the closed-loop gain.

Ans. 20 kHz

2.20 An op amp with a dc gain $A_0 = 10^4$ is connected in the noninverting configuration with $R_1 = 1\ \text{k}\Omega$ and $R_2 = 99\ \text{k}\Omega$. Find the values of the feedback factor and the loop gain.

Ans. 0.01; −100

2.21 What is the value of the feedback ratio β for a unity-gain follower?

Ans. $\beta = 1$ (100% feedback)

2.22 Consider a unity-gain follower constructed using an op amp with infinite bandwidth but finite dc gain A_0. Find an expression for the closed-loop gain and evaluate it for $A_0 = 10^3$ and 10^5.

Ans. $G = \dfrac{1}{1 + 1/A_0}$; 0.9990; 0.999990

2.8 THE INTERNAL STRUCTURE OF IC OP AMPS

Further insight into the frequency response of op amps, and indeed into their dynamic behavior in general, can be gained by considering the internal circuit of the op amp.[2] This will be done in detail in Chapter 10. For the time being let us consider Fig. 2.31, which shows the internal structure, in block diagram form, of most modern integrated-circuit (IC) op amps. The op amp is seen to consist of three stages: an input stage, which is basically a differential-input transconductance amplifier; an intermediate stage, which is a voltage amplifier with high negative voltage gain ($-\mu$) and with a feedback capacitor C; and an output stage, which is a unity-gain buffer whose purpose is to provide the op amp with a low output resistance. For the purpose of the present discussion, we shall assume that the output buffer is an ideal unity-gain amplifier and consider it no further.

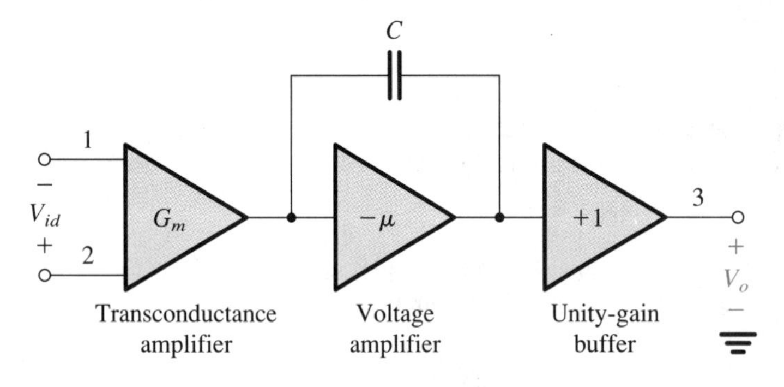

Fig. 2.31 The internal structure of modern IC op amps.

[2] Study of this section will enable us to provide (in the next section) a reasonable explanation of op amp slewing. Nevertheless, if desired, this section may be postponed to a later point in the course (with corresponding modification to the coverage of slew-rate limitation).

Figure 2.32(a) shows a simplified small-signal equivalent circuit model of the op amp with the output stage eliminated. The input stage is shown as having infinite input impedance. This stage senses the differential input voltage V_{id} ($V_{id} = V_2 - V_1$) and provides a proportionate current $G_m V_{id}$. It has an output resistance R_{o1}. The second stage is shown to have an input resistance R_{i2}, a voltage gain $-\mu$, and a zero output resistance. The feedback capacitor C is included for the purpose of ensuring that the op amp will be stable (does not oscillate and provide unwanted output signals) when it is connected in a feedback circuit. The topic of stability will be studied in Chapter 8. For the time being we will show that the capacitor C, known as the frequency-compensation capacitor, provides the op amp with the STC low-pass response studied in the previous section.

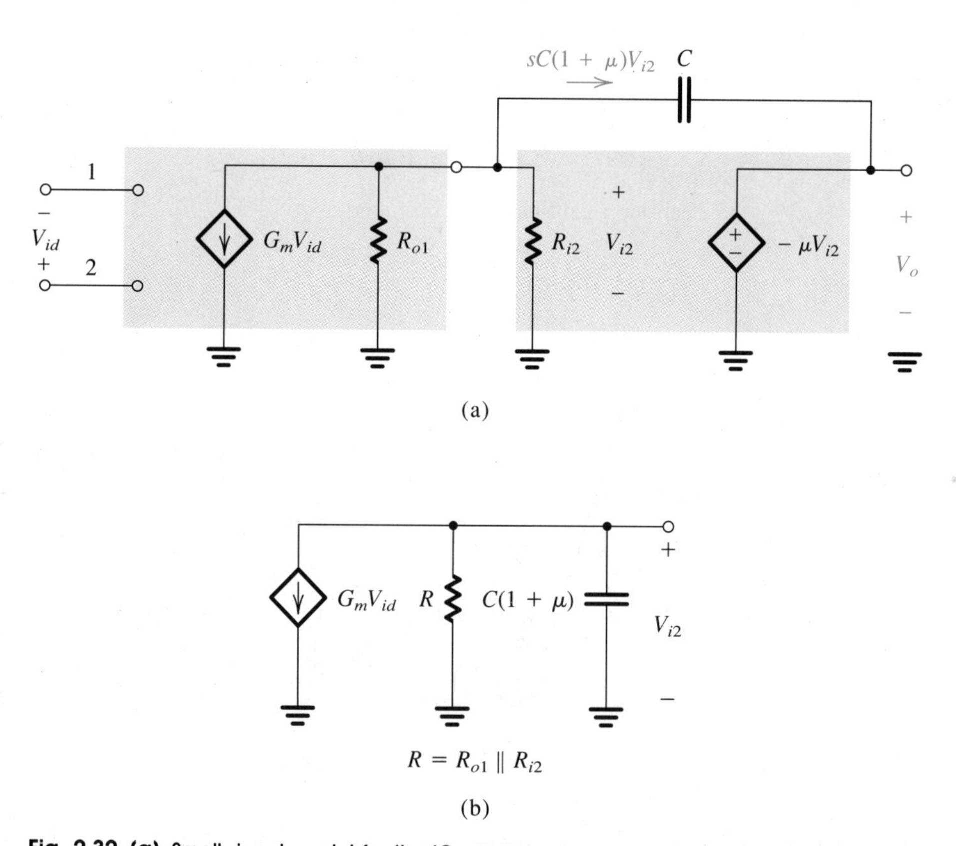

Fig. 2.32 (a) Small-signal model for the IC op amp whose internal structure is shown in Fig. 2.31. Here the output buffer is assumed ideal and, to simplify the diagram, is eliminated. **(b)** Equivalent circuit of the interface between the first and second stages.

We wish to analyze the equivalent circuit of Fig. 2.32(a) to determine the open-loop gain $A \equiv V_o(s)/V_{id}(s)$. Toward that end we observe that the current through the capacitor C is $sC(1 + \mu)V_{i2}$. Since this current is drawn away from a node whose voltage is V_{i2}, the bridging capacitor C can be replaced with a grounded capacitor equal to $C(1 + \mu)$. This replacement is shown in Fig. 2.32(b) and will be seen to simplify the analysis.[3] This figure shows the equivalent circuit at the interface between the first and second stages. Note that we have combined R_{o1} and R_{i2} into a single resistor R,

$$R = R_{o1} \| R_{i2}$$

For the circuit in Fig. 2.32(b), we can write

$$V_{i2} = \frac{-G_m V_{id}}{Y}$$

where

$$Y = \frac{1}{R} + sC(1 + \mu)$$

Thus

$$V_{i2} = -V_{id} \frac{G_m R}{1 + sC(1 + \mu)R}$$

Since $V_o = -\mu V_{i2}$, the open-loop gain can now be found as

$$A(s) \equiv \frac{V_o(s)}{V_{id}(s)}$$

$$= \frac{\mu G_m R}{1 + sC(1 + \mu)R} \tag{2.25}$$

Comparing this expression to that in Eq. (2.8), we find that

$$A_0 = \mu G_m R \tag{2.26}$$

$$\omega_b = \frac{1}{C(1 + \mu)R} \tag{2.27}$$

Now, since the unity-gain bandwidth ω_t is given by $\omega_t = A_0 \omega_b$, it follows that

$$\omega_t = \frac{\mu G_m R}{C(1 + \mu)R}$$

Usually $\mu \gg 1$ and ω_t can be approximated by

$$\omega_t \simeq \frac{G_m}{C} \tag{2.28}$$

We also note that at frequencies much greater than the 3-dB frequency ω_b, the open-loop gain $A(s)$ in Eq. (2.25) can be approximated by

$$A(s) \simeq \frac{\mu G_m R}{sC(1 + \mu)R}$$

---

[3]The replacement we have just performed is an application of Miller's theorem which will be studied in Chapter 7.

Again, for $\mu \gg 1$,

$$A(s) \simeq \frac{G_m}{sC} \tag{2.29}$$

At such frequencies the op-amp internal circuit can be represented by the structure in Fig. 2.33. Here we have assumed that μ approaches ∞, and thus the second stage together with

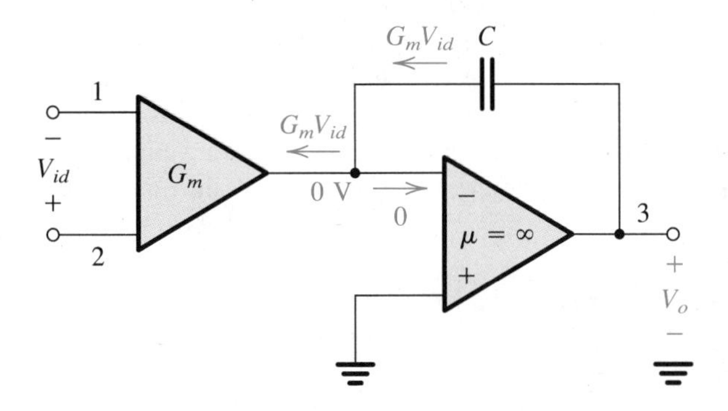

Fig. 2.33 Approximate equivalent circuit of the op amp innards at frequencies $f \gg f_b$ and assuming that the gain of the second stage is large (approaching ∞).

the compensating capacitor acts as an integrator. A virtual ground appears at the input terminal of the second stage, and all the current of the first stage (G_mV_{id}) flows through the feedback capacitor C. Thus

$$V_o = \frac{G_mV_{id}}{sC}$$

resulting in the open-loop gain given by Eq. (2.29). This simplified equivalent circuit will prove useful in our discussion of slew-rate limitation in the next section.[4]

To conclude this section consider the popular 741-type op amp for which $G_m =$ 0.19 mA/V, $R_{o1} = 6.7\ \text{M}\Omega$, $R_{i2} = 4\ \text{M}\Omega$, $\mu = 529$, $C = 30$ pF. Thus

$$A_0 = \mu G_m(R_{o1} // R_{i2}) = 529 \times 0.19 \times 10^{-3} \times (6.7//4) \times 10^6$$
$$= 2.52 \times 10^5$$

$$\omega t = \frac{G_m}{C} = \frac{0.19 \times 10^{-3}}{30 \times 10^{-12}} = 6.33\ \text{Mrad/s}$$

$$f_t = \frac{\omega_t}{2\pi} \simeq 1\ \text{MHz}$$

[4]Note that R_{o1} and R_{i2} are not included in this analysis because the currents through them are equal to zero due to the virtual ground present at the input terminal of the second stage. Also note that the circuit in Fig. 2.33 implies that the dc gain of the op amp is infinite. This is a consequence of the assumption that the gain of the second stage $\mu = \infty$.

2.9 LARGE-SIGNAL OPERATION OF OP AMPS

In this section we study the limitations on the performance of op-amp circuits when large output signals are present.

Output Saturation

Similar to all other amplifiers, op amps operate linearly over a limited range of output voltages. Specifically, the op amp output saturates in the manner shown in Fig. 1.11 with L_+ and L_- within 1 to 3 volts of the positive and negative power supplies, respectively. Thus, an op amp that is operating from $\pm$15-V supplies will saturate when the output voltage reaches about +12 V in the positive direction and −12 V in the negative direction. For this op amp the **rated output voltage** is said to be $\pm$12 V. To avoid clipping off the peaks of the output waveform, and the resulting waveform distortion, the input signal must be kept correspondingly small.

Exercise 2.23 The rated output voltage of a given op amp is $\pm$10 V. If the op amp is used to design a noninverting amplifier with a gain of 200, what is the largest sine-wave input that can be handled without output clipping?

Ans. 0.1 V peak-to-peak

Slew Rate

Another phenomenon that can cause nonlinear distortion when large output signals are present is that of slew-rate limiting. We shall first describe slew-rate limiting and then explain the reason it occurs.

Consider the unity-gain follower shown in Fig. 2.34(a). Let the input voltage v_I be the step of height V shown in Fig. 2.34(b). Now the closed-loop gain of the amplifier is given by Eq. (2.20) with $R_2 = 0$ and $R_1 = \infty$; that is,

$$\frac{V_o}{V_i} = \frac{1}{1 + s/\omega_t} \tag{2.30}$$

which is a low-pass STC response. We would therefore expect the step input to produce the output waveform.

$$v_O(t) = V(1 - e^{-t/\tau}) \tag{2.31}$$

(For a review of the step response of STC networks, see Appendix E) where $\tau = 1/\omega_t$. Figure 2.34(c) shows a sketch of this exponentially rising waveform.

In practice, however, such a response is obtained only if the step size V is "small," where a more precise definition of "small" will follow shortly. For large step inputs (5 V, for example) the output waveform will be the linearly ramping signal shown in Fig. 2.34(d). It is important to note that the slope of the linear ramp is smaller than the initial slope (at $t = 0$) of the exponentially rising waveform of the same magnitude V (shown in Fig. 2.34(c)), which is V/τ. The linear ramping response shown in Fig. 2.34(d) indicates that the op amp output is unable to rise at the rate predicted by Eq. (2.31). When this occurs, the op amp is said to be slew-rate limited (or slewing), and the slope of the linear

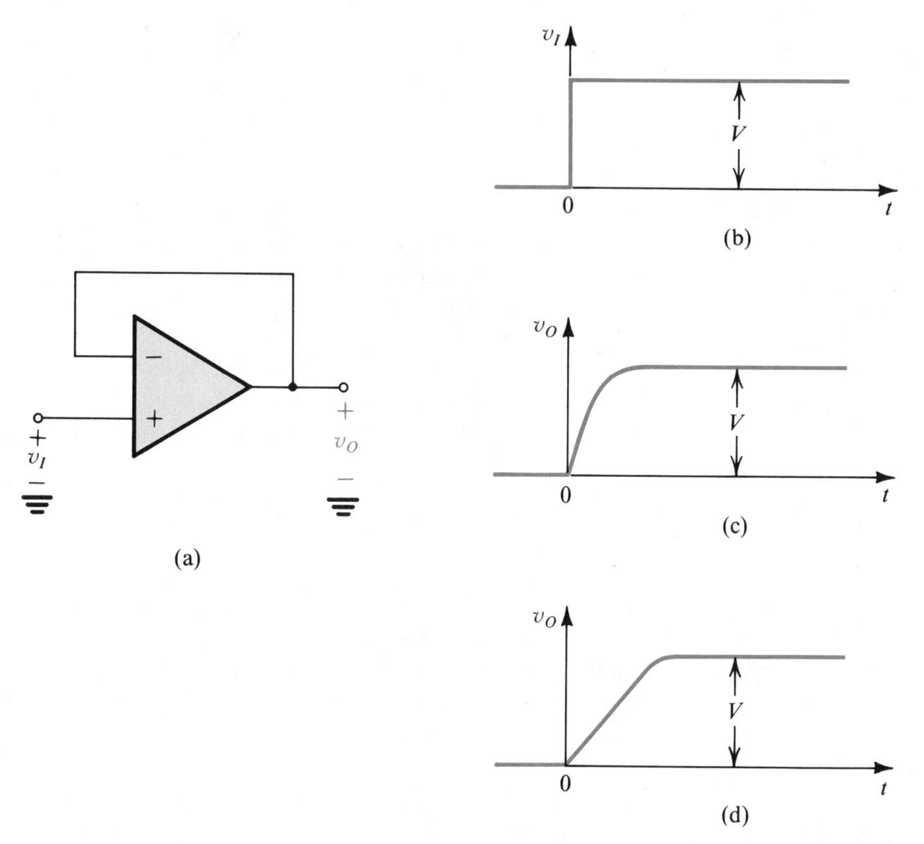

Fig. 2.34 **(a)** Unity-gain follower. **(b)** Input step waveform. **(c)** Exponentially rising output waveform obtained when V is small. **(d)** Linearly rising output waveform obtained when V is large (here the amplifier is slew-rate limited).

ramp at the output is called the slew rate. The slew rate (SR) is the maximum possible rate of change of the op amp output voltage,

$$\mathrm{SR} = \left.\frac{dv_O}{dt}\right|_{\max} \tag{2.32}$$

and is usually specified on the op-amp data sheet in V/μs. It follows that the op amp in Fig. 2.34(a) will begin to slew for a signal V for which the initial slope of the exponentially rising ramp, V/τ, exceeds the op amp slew rate.

We next investigate the origin of the slew-rate limitation. Consider once more the unity-gain follower in Fig. 2.34(a) with an input step of few volts. We see that at time $t = 0$, as the input rises to V volts, the output remains at zero volts. Thus the full size of the step appears between the two input terminals of the op amp. It follows that the input differential voltage V_{id} will be large and the input transconductance amplifier (See Fig. 2.33) will saturate in the manner indicated in Fig. 2.35. Under these conditions the

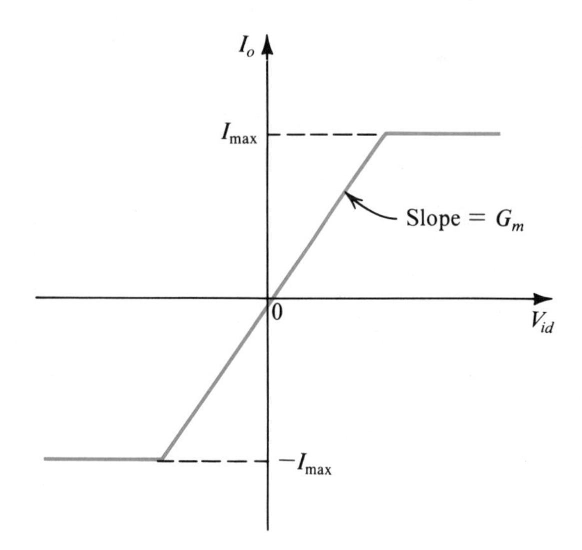

Fig. 2.35 Transfer characteristic of the input transconductance amplifier present in every modern IC op amp.

transconductance amplifier supplies its maximum possible output current I_{max} to the second stage. (Note that I_{max} is smaller than $G_m V_{id}$ because the input stage has saturated.) The constant current I_{max} flows through the frequency-compensation capacitor C of the second stage and causes the output voltage to rise linearly with a slope equal to I_{max}/C. This, the highest possible rate of change of the output voltage, is the op amp slew rate. Thus,

$$\mathrm{SR} = \frac{I_{max}}{C} \tag{2.33}$$

Exercises 2.24 An op amp that has a slew rate of 1 V/μs and a unity-gain bandwidth f_t of 1 MHz is connected in the unity-gain follower configuration. Find the largest possible input voltage step for which the output waveform will still be given by the exponential ramp of Eq. (2.31). For this input voltage what is the 10% to 90% rise time of the output waveform? If an input step 10 times as large is applied, find the 10% to 90% rise time of the output waveform.

Ans. 0.16 V; 0.35 μs; 1.28 μs

2.25 For the 741-type op amp the maximum current that the first stage can supply is 19 μA and the compensation capacitor C is 30 pF. Find the slew rate.

Ans. 0.63 V/μs

Full-Power Bandwidth

Op amp slew-rate limiting can cause nonlinear distortion in sinusoidal waveforms. Consider once more the unity-gain follower with a sine wave input given by

$$v_I = \hat{V}_i \sin \omega t$$

The rate of change of this waveform is given by

$$\frac{dv_I}{dt} = \omega \hat{V}_i \cos \omega t$$

and has a maximum value of $\omega \hat{V}_i$. This maximum occurs at the zero crossings of the input sinusoid. Now if $\omega \hat{V}_i$ exceeds the slew rate of the op amp, the output waveform will be distorted in the manner shown in Fig. 2.36. Observe that the output cannot keep up with the large rate of change of the sinusoid at its zero crossings, and the op amp slews.

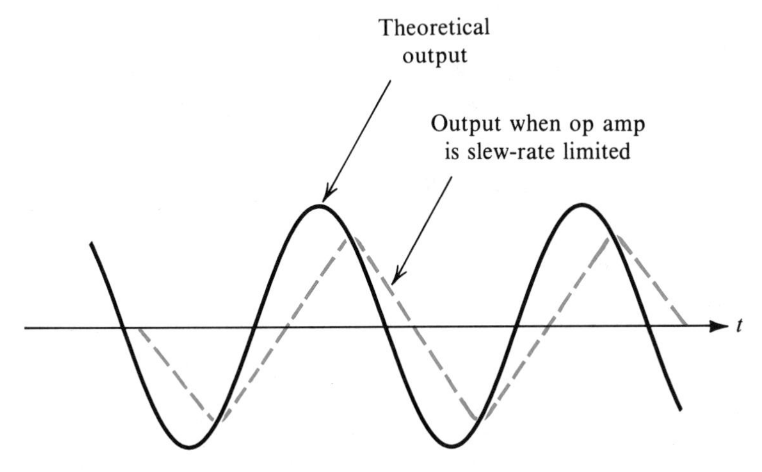

Fig. 2.36 Effect of slew-rate limiting on output sinusoidal waveforms.

The op amp data sheets usually specify a frequency f_M called the **full-power bandwidth.** It is the frequency at which an output sinusoid with amplitude equal to the rated output voltage of the op amp begins to show distortion due to slew-rate limiting. If we denote the rated output voltage V_{omax}, then f_M is related to SR as follows:

$$\omega_m V_{omax} = \text{SR}$$

Thus,

$$f_M = \frac{\text{SR}}{2\pi V_{omax}} \tag{2.34}$$

It should be obvious that output sinusoids of amplitudes smaller than V_{omax} will show slew-rate distortion at frequencies higher than ω_M. In fact, at a frequency ω higher than ω_M, the maximum amplitude of the undistorted output sinusoid is given by:

$$V_o = V_{omax}\left(\frac{\omega_{\text{M}}}{\omega}\right) \tag{2.35}$$

Finally, we note that slew-rate limiting is a phenomenon distinct from the small-signal frequency limitation studied in Section 2.7.

Exercise 2.26 An op amp has a rated output voltage of ± 10 V and a slew rate of 1 V/μs. What is its full-power bandwidth? If an input sinusoid with frequency $f = 5f_M$ is applied to a unity-gain follower constructed using this op amp, what is the maximum possible amplitude that can be accommodated at the output without incurring slew-induced distortion?

Ans. 15.9 kHz; 2 V (peak)

2.10 COMMON-MODE REJECTION

Practical op amps have finite nonzero **common-mode gain;** that is, if the two input terminals are tied together and a signal v_{Icm} is applied, the output will not be zero. The ratio of the output voltage v_O to the input voltage v_{Icm} is called the common-mode gain A_{cm}. Figure 2.37 illustrates this definition.

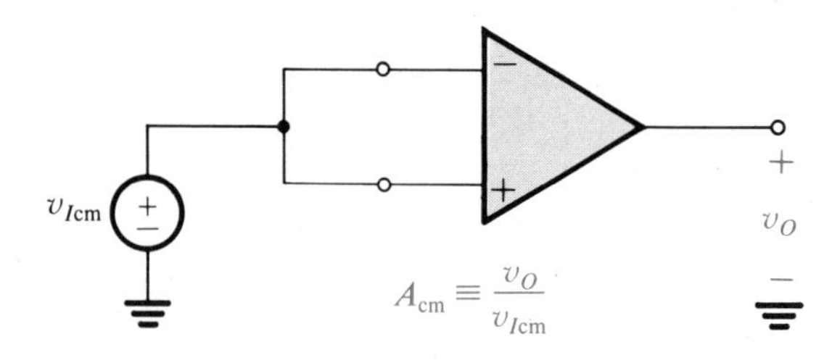

Fig. 2.37 Illustrating the definition of the common-mode gain of the op amp.

To be precise, consider an op amp with signals v_1 and v_2 applied to its inverting and noninverting input terminals, respectively. The difference between the two input signals is the differential-mode, or simply differential, input signal v_{id},

$$v_{id} = v_2 - v_1 \tag{2.36}$$

The average of the two input signals is the common-mode input signal v_{Icm},

$$v_{Icm} = \frac{v_1 + v_2}{2} \tag{2.37}$$

Now the output voltage v_O can be expressed as

$$v_O = Av_{id} + A_{cm}v_{Icm} \tag{2.38}$$

where A is the differential gain and A_{cm} is the common-mode gain.

The ability of an op amp to reject common-mode signals is specified in terms of the **common-mode rejection ratio** (CMRR), defined as

$$\text{CMRR} = \frac{|A|}{|A_{cm}|} \tag{2.39}$$

Usually the CMRR is expressed in decibels:

$$\text{CMRR} = 20 \log \frac{|A|}{|A_{cm}|} \tag{2.40}$$

The CMRR is a function of frequency, decreasing as the frequency is increased. Typical values of CMRR at low frequencies range from 80 to 100 dB.

The finite CMRR of op amps is unimportant in the case of the inverting configuration, since the positive input terminal is grounded and hence the common-mode input signal is approximately zero. On the other hand, in the noninverting configuration the common-mode input signal is nearly equal to the applied input signal, and thus the finite CMRR of the op amp may have to be taken into account in applications that demand high accuracy. The closed-loop configuration that is most adversely affected by the finite

CMRR of the op amp is the differential amplifier of Fig. 2.19. We have found in Example 2.4 that by the appropriate selection of resistor values, the circuit can be made to respond only to differential input signals. This will no longer be true if the finite CMRR of the op amp is taken into account.

A simple method for taking into account the effect of the finite CMRR in calculating closed-loop gain is as follows: A common-mode input signal v_{Icm} gives rise to an output component of value $A_{cm}v_{Icm}$. The same output component can be obtained if a differential input signal

$$v_{\text{error}} = \frac{A_{cm}v_{Icm}}{A} = \frac{v_{Icm}}{\text{CMRR}} \tag{2.41}$$

is applied to an op amp with zero common-mode gain. Thus, in a given circuit, once the input common-mode signal is found, we simply add a signal generator v_{error} in series with one of the op amp input leads and carry out the rest of the analysis assuming the op amp to be ideal.

As an example, Fig. 2.38 shows the analysis of the noninverting configuration taking into account the finite CMRR of the op amp. From Fig. 2.38(a) we observe that

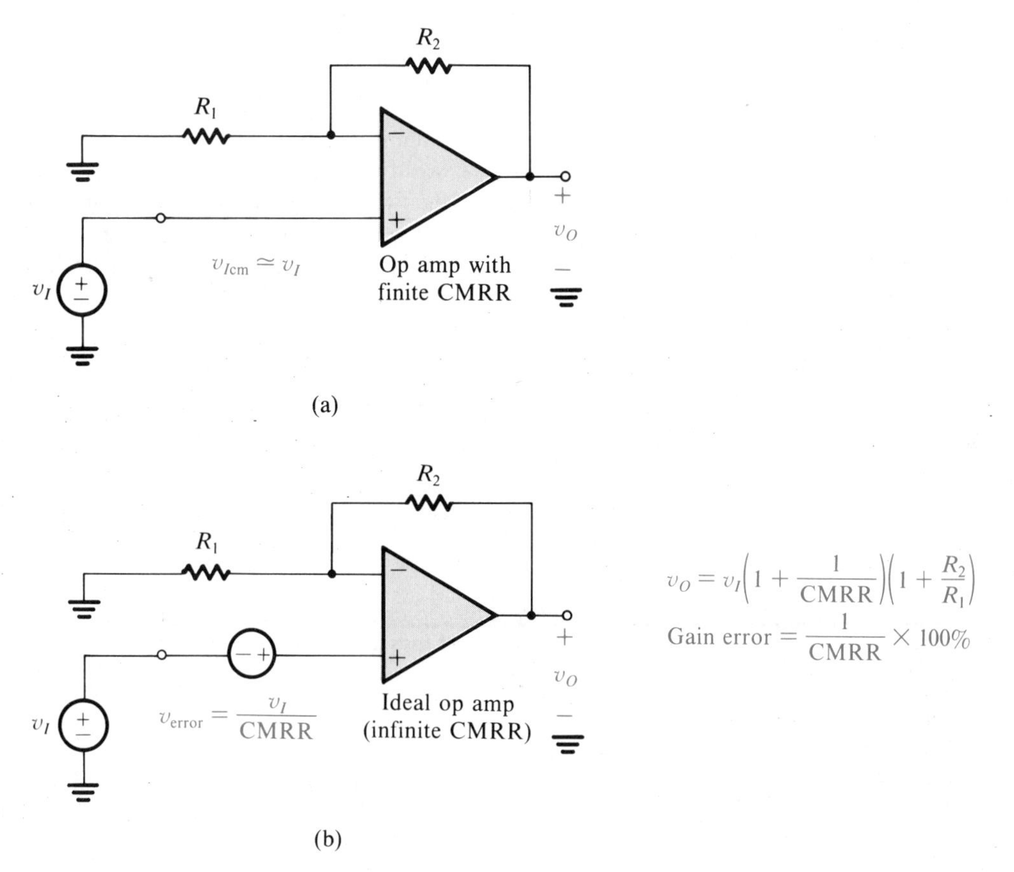

Fig. 2.38 Analysis of the noninverting configuration with the finite CMRR of the op amp taken into account.

$v_{Icm} \simeq v_I$. Thus the op amp is replaced with an ideal one together with the $v_{\text{error}} = v_I/\text{CMRR}$ generator, as shown in Fig. 2.38(b). Analysis of the latter circuit is straightforward, and the result is given in Fig. 2.38. Note that although the error-voltage generator is assigned a polarity, CMRR can be either positive or negative; its sign is generally not known.

Exercise 2.27 We wish to use the method described above to find the common-mode gain of the difference amplifier circuit in Fig. 2.19 under the conditions that $R_2/R_1 = R_4/R_3$ and the op amp has a finite CMRR. (a) With the two input terminals tied together and a common-mode input signal v_{CM} applied, show that

$$v_{\text{error}} = v_{CM}\frac{R_4}{R_3+R_4}\frac{1}{\text{CMRR}}$$

(b) Then, with the input signal source short-circuited and the v_{error} generator inserted in series with the positive input terminal of the op amp (now assumed ideal), find the output voltage, and show that the common-mode gain of the difference amplifier is given by

$$\text{CM gain} = \frac{R_2}{R_1}\frac{1}{\text{CMRR}}$$

2.11 INPUT AND OUTPUT RESISTANCES

Figure 2.39 shows an equivalent circuit of the op amp, incorporating its finite input and output resistances. As shown, the op amp has a differential input resistance R_{id} seen between the two input terminals. In addition, if the two input terminals are tied together and the input resistance (to ground) is measured, the result is the common-mode input resistance R_{icm}. In the equivalent circuit we have split R_{icm} into two equal parts ($2R_{icm}$), each connected between one of the input terminals and ground.

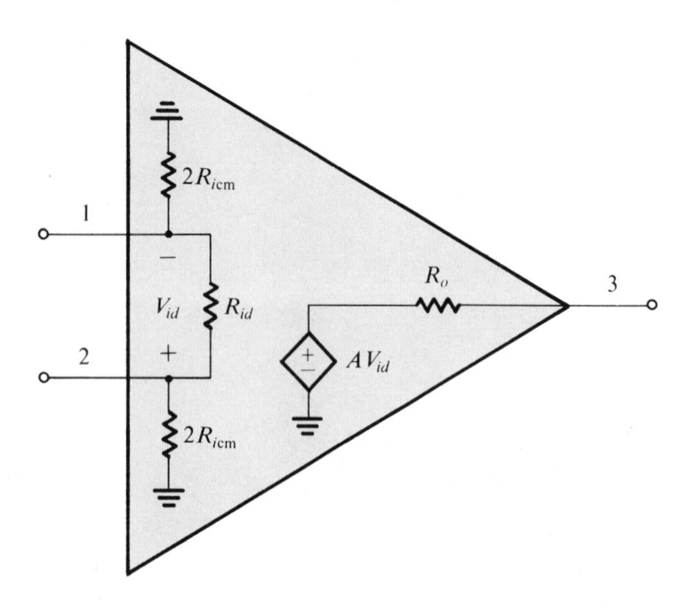

Fig. 2.39 Op-amp model with the input and output resistances shown.

Input Resistance

Typical values for the input resistances of general-purpose op amps using bipolar junction transistors are $R_{id} = 1\ \text{M}\Omega$ and $R_{icm} = 100\ \text{M}\Omega$. Op amps that utilize field-effect transistors in the input stage have much higher input resistances. The value of the input resistance of a particular closed-loop circuit will depend on the values of R_{id} and R_{icm} as well as on the circuit configuration. For the inverting configuration the input resistance is approximately equal to R_1. Detailed analysis shows that taking R_{id} and R_{icm} into account has a negligible effect on the value of the input resistance of the inverting circuit. On the other hand, the input resistance of the noninverting configuration is strongly dependent on the values of R_{id} and R_{icm} as well as on the value of A and R_2/R_1. Straightforward analysis of the noninverting circuit using the op amp model of Fig. 2.39 and assuming that

$$R_o = 0, \qquad R_1 \ll R_{icm}, \qquad \frac{R_2}{R_{id}} \ll A$$

results in the following approximate expression for the input resistance of the noninverting circuit:

$$R_{\text{in}} \simeq \{2R_{icm}\} \| \{(1 + A\beta)R_{id}\} \tag{2.42}$$

where β is the feedback ratio, given by

$$\beta = \frac{R_1}{R_1 + R_2} \tag{2.43}$$

We observe that the input resistance consists of two components in parallel: $(2R_{icm})$, which is very large, and $(1 + A\beta)R_{id}$, which is also large because R_{id} is multiplied by the amount of feedback $(1 + A\beta)$. At low frequencies, $A = A_0$ and the amount of feedback $1 + A_0\beta$ is usually a large number. At higher frequencies, the dependence of A on frequency must be taken into account as illustrated in the following example.

EXAMPLE 2.8

Consider an op amp having $f_t = 1$ MHz, $R_{id} = 1\ \text{M}\Omega$, and $R_{icm} = 100\ \text{M}\Omega$. Find the components of the input impedance of a noninverting amplifier with a nominal gain of 100.

Solution The input impedance can be obtained by substituting in Eq. (2.42) the following:

$$R_{icm} = 10^8\ \Omega$$

$$R_{id} = 10^6\ \Omega$$

Since

$$1 + \frac{R_2}{R_1} = 100$$

we find that $\beta = 0.01$. Also

$$A \simeq \frac{\omega_t}{s} = \frac{2\pi \times 10^6}{s}$$

The result is

$$Z_{in} = (2 \times 10^8)//\left(10^6 + \frac{2\pi \times 10^6 \times 10^{-2} \times 10^6}{s}\right)$$

Note that the second component of Z_{in} consists of a 10^6-Ω resistance in series with a capacitor of value $(1/2\pi \times 10^{-10})$ farads. Figure 2.40 shows an equivalent circuit of the input impedance obtained above.

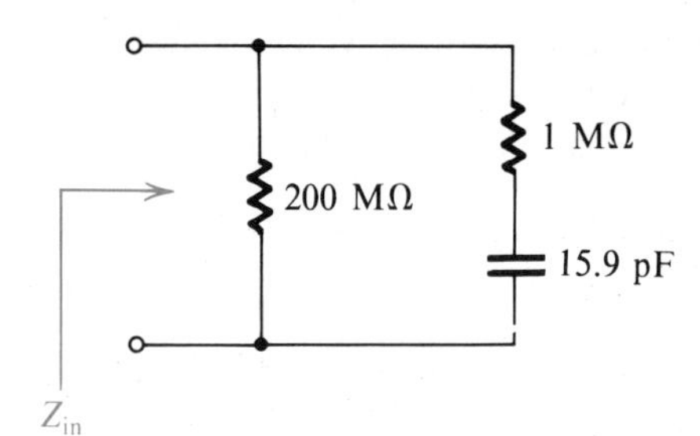

Fig. 2.40 Circuit representation of the input impedance of the amplifier in Example 2.8.

Output Resistance

We now turn to the effect of the finite output resistance R_o shown in the model of Fig. 2.39. Typical values for the open-loop output resistance R_o are 75 to 100 Ω, although there exist amplifiers with much higher output resistances. We wish to find the output resistance of a closed-loop amplifier. To do this, we short the signal source, which makes the inverting and noninverting configurations identical and apply a test voltage V_x to the output as shown in Fig. 2.41. Then the output resistance $R_{out} \equiv V_x/I$ can be obtained by straightforward analysis of the circuit in Fig. 2.41 as follows:

$$V = -V_x \frac{R_1}{R_1 + R_2} = -\beta V_x$$

$$I = \frac{V_x}{R_1 + R_2} + \frac{V_x - AV}{R_o}$$

$$= \frac{V_x}{R_1 + R_2} + \frac{(1 + A\beta)V_x}{R_o}$$

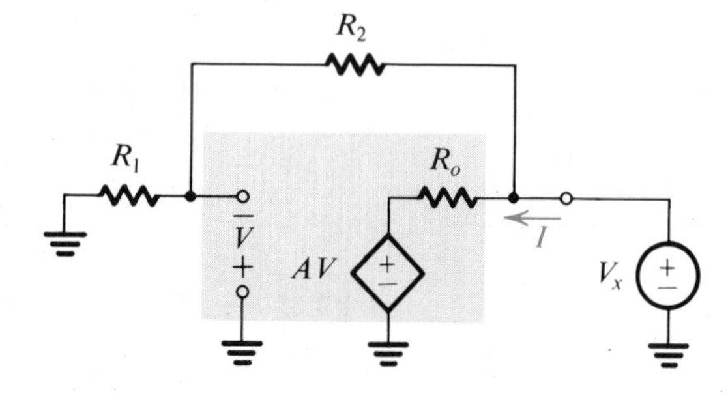

Fig. 2.41 Derivation of the closed-loop output resistance.

Thus

$$\frac{1}{R_{out}} \equiv \frac{I}{V_x} = \frac{1}{R_1 + R_2} + \frac{1 + A\beta}{R_o}$$

where the constant β is defined as

$$\beta \equiv \frac{R_1}{R_1 + R_2}$$

This means that the closed-loop output resistance is composed of two parallel components,

$$R_{out} = [R_1 + R_2]//[R_o/(1 + A\beta)] \tag{2.44}$$

Normally R_o is much smaller than $R_1 + R_2$, resulting in

$$R_{out} \simeq \frac{R_o}{1 + A\beta} \tag{2.45}$$

We note that the closed-loop output resistance is smaller than the op amp open-loop output resistance by a factor equal to the amount of feedback, $1 + A\beta$. A further simplification of the expression for R_{out} is obtained by noting that normally $A\beta \gg 1$, which gives

$$R_{out} \simeq \frac{R_o}{A\beta} \tag{2.46}$$

At very low frequencies A is real and large, resulting in a very small R_{out}. As an example, a voltage follower ($\beta = 1$) designed using an op amp with $R_o = 100\ \Omega$ and $A_0 = 10^5$ will have

$$R_{out} = \frac{100}{10^5 \times 1} = 1\ \text{m}\Omega$$

It is interesting to inquire about the effect of the finite op amp bandwidth on the closed-loop output impedance. Substituting $A = \omega_t/s$ in Eq. (2.45) results in

$$Z_{out} = \frac{R_o}{1 + \beta\omega_t/s} \tag{2.47}$$

Thus,

$$Y_{out} = \frac{1}{Z_{out}} = \frac{1}{R_o} + \frac{\beta\omega_t}{sR_o} \tag{2.48}$$

which indicates that the output impedance consists of a resistance equal to R_o in parallel with an inductance of value $L = R_o/\beta\omega_t$. An equivalent circuit representation of the closed-loop output impedance is shown in Fig. 2.42(a). Note that for this equivalent circuit we have used the more accurate gain expression

$$A = \frac{A_0}{1 + s/\omega_t}$$

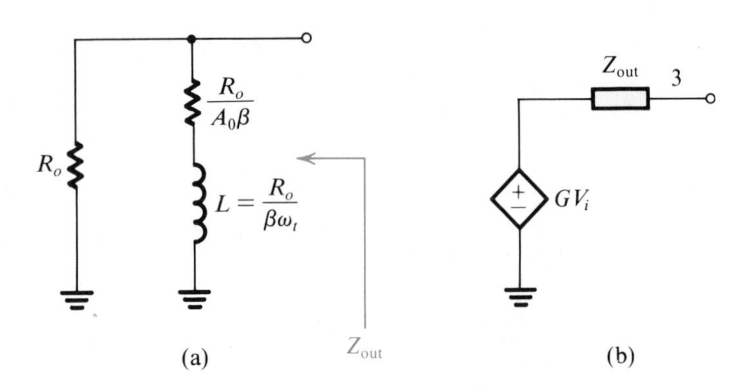

Fig. 2.42 **(a)** Equivalent circuit representation of the closed-loop output impedance of both the inverting and the noninverting configurations. **(b)** Thévenin equivalent of the output of a closed-loop circuit with gain G and output impedance Z_{out}.

Finally we remind the reader of the meaning of the closed-loop output impedance just evaluated by showing in Fig. 2.42(b) the Thévenin equivalent of the output of a closed-loop amplifier having a closed-loop gain G and a closed-loop output impedance Z_{out}.

Exercises 2.28 Find the input resistance, at low frequencies, of a unity-gain buffer constructed using an op amp with $A_0 = 10^4$, $R_{icm} = 100\ \text{M}\Omega$ and $R_{id} = 1\ \text{M}\Omega$.

Ans. $R_{in} \simeq 200\ \text{M}\Omega$

2.29 Repeat Exercise 2.28 for a noninverting amplifier with a nominal closed-loop gain of 100.

Ans. $R_{in} \simeq 67\ \text{M}\Omega$

2.30 Find the values of the various components in the output impedance (Fig. 2.42(a)) of a noninverting amplifier with a closed-loop gain of 100. Let the op amp have $R_o = 100\ \Omega$, $A_0 = 10^5$, and $f_t = 1$ MHz.

Ans. 100 Ω; 0.1 Ω; 1.59 mH

2.12 DC PROBLEMS

Offset Voltage

Because op amps are direct-coupled devices with large gains at dc, they are prone to dc problems. The first such problem is the dc offset voltage. To understand this problem consider the following *conceptual* experiment: If the two input terminals of the op amp are tied together and connected to ground, it will be found that a finite dc voltage exists at the output. In fact, if the op amp has a high dc gain, the output will be at either the positive or negative saturation level. The op amp output can be brought back to its ideal value of 0 V by connecting a dc voltage source of appropriate polarity and magnitude between the two input terminals of the op amp. This external source balances out the input offset voltage of the op amp. It follows that the **input offset voltage** (V_{OS}) must be of equal magnitude and of opposite polarity to the voltage we applied externally.

The input offset voltage arises as a result of the unavoidable mismatches present in the input differential stage inside the op amp. In later chapters we shall study this topic in detail. Here, however, our concern is to investigate the effect of V_{OS} on the operation of closed-loop op amp circuits. Toward that end we note that general-purpose op amps exhibit V_{OS} in the range of 1 to 5 mV. Also, the value of V_{OS} depends on temperature. The op-amp data sheets usually specify typical and maximum values for V_{OS} at room temperature as well as the temperature coefficient of V_{OS} (usually in μV/°C). They do not, however, specify the polarity of V_{OS} because the component mismatches that give rise to V_{OS} are obviously not known *a priori;* different units of the same op amp type may exhibit either a positive or a negative V_{OS}.

To analyze the effect of V_{OS} on the operation of op amp circuits, we need a circuit model for the op amp with input offset voltage. Such a model is shown in Fig. 2.43. It consists of a dc source of value V_{OS} placed in series with the positive input lead of an offset-free op amp. The justification for this model follows from the description above.

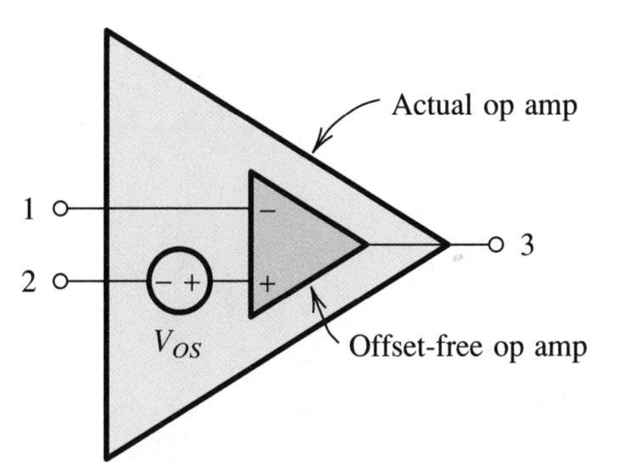

Fig. 2.43 Circuit model for an op amp with input offset voltage V_{OS}.

Exercise 2.31 Use the model of Fig. 2.43 to sketch the transfer characteristic v_O versus v_{Id} ($v_O \equiv v_3$ and $v_{Id} \equiv v_2 - v_1$) of an op amp having $A_0 = 10^4$, output saturation levels of ± 10 V, and V_{OS} of +5 mV.

Ans. See Fig. E2.31.

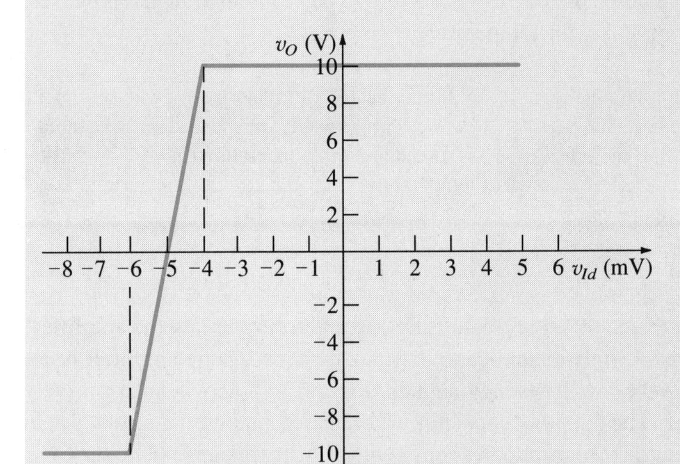

Fig. E2.31 Transfer characteristic of an op amp with V_{OS} = 5 mV.

Analysis of op amp circuits to determine the effect of the op amp V_{OS} on their performance is straightforward: The input voltage signal source is short circuited and the op amp is replaced with the model of Fig. 2.43. (Eliminating the input signal, done to simplify matters, is based on the principle of superposition.) Following this procedure we find that both the inverting and the noninverting amplifier configurations result in the circuit shown in Fig. 2.44, from which the output dc voltage due to V_{OS} is found to be

$$V_O = V_{OS}\left[1 + \frac{R_2}{R_1}\right] \tag{2.49}$$

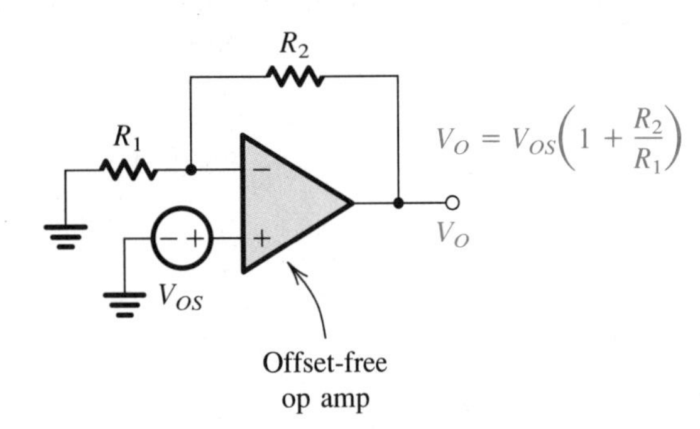

Fig. 2.44 Evaluating the output dc offset due to V_{OS} in a closed-loop amplifier.

This output dc voltage can have a large magnitude. For instance, a noninverting amplifier with a closed-loop gain of 1000, when constructed from an op amp with 5-mV input offset voltage, will have a dc output voltage of +5 V or −5 V (depending on the polarity of V_{OS}), rather than the ideal value of 0 V. Now, when an input signal is applied to the amplifier, the corresponding signal output will be superimposed on the 5-V dc. Obviously then, the allowable signal swing at the output will be reduced. Even worse, if the signal to be amplified is dc, we would not know whether the output is due to V_{OS} or to the signal!

Some op amps are provided with two additional terminals to which a specified circuit can be connected to trim the output dc due to V_{OS} to zero. The problem remains, however, of the variation (or drift) of V_{OS} with temperature.

Exercise 2.32 Consider an inverting amplifier with a nominal gain of 1000 constructed from an op amp with an input offset voltage of 3 mV and with output saturation levels of ±10 V. (a) What is (approximately) the peak sine-wave input signal that can be applied without output clipping? (b) If the effect of V_{OS} is nulled at room temperature (25°C) how large an input can one now apply if: (i) the circuit is to operate at a constant temperature? (ii) the circuit is to operate at a temperature in the range 0°C to 75°C and the temperature coefficient of V_{OS} is 10 μV/°C?

Ans. (a) 7 mV; (b) 10 mV, 9.5 mV

One way to overcome the dc offset problem is by capacitively coupling the amplifier. This, however, will be possible only in applications where the closed-loop amplifier is not required to amplify dc or very-low-frequency signals. Figure 2.45 shows a capacitively coupled inverting amplifier. The coupling capacitor will cause the gain to be zero at dc. In fact, the circuit will have an STC high-pass response with a 3-dB frequency $\omega_0 = 1/CR_1$, and the gain will be $-R_2/R_1$ for frequencies $\omega \gg \omega_0$. The advantage of this arrangement is that V_{OS} will not be amplified. Thus the output dc voltage will be equal to V_{OS} rather

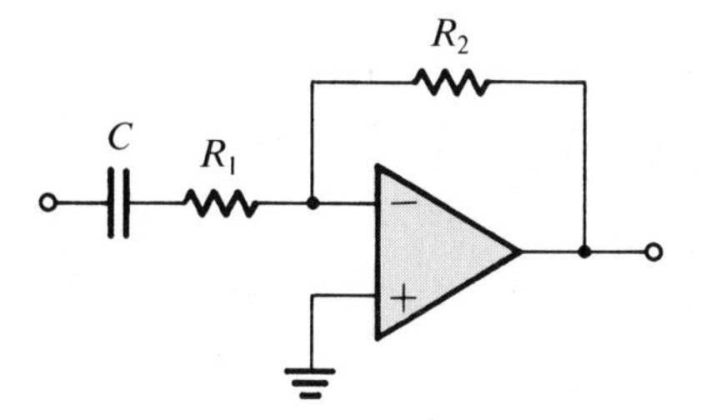

Fig. 2.45 A capacitively coupled inverting amplifier.

than $V_{OS}(1 + R_2/R_1)$, which is the case without the coupling capacitor. Since the capacitor behaves as an open circuit at dc, it is easy to see from Fig. 2.45 that the V_{OS} generator indeed sees a unity-gain follower.

Another op amp circuit that is adversely affected by the op amp input offset voltage is the Miller integrator. Figure 2.46 shows the integrator circuit with the input signal reduced to zero and the op amp replaced with the model of Fig. 2.43. Analysis of the circuit is straightforward and is shown in Fig. 2.46. Assuming that at time $t = 0$ the voltage across the capacitor is zero, the output voltage as a function of time is given by

$$v_O = V_{OS} + \frac{V_{OS}}{CR}t \tag{2.50}$$

Thus v_O increases linearly with time until the op amp saturates—clearly an unacceptable situation! The problem can be alleviated by connecting a resistor R_F across the integrator capacitor C. Such a resistor provides a dc path through which the dc current (V_{OS}/R) can flow, with the result that v_O will now have a dc component of $V_{OS}[1 + (R_F/R)]$ (instead of rising linearly). To keep the dc offset at the output small, one would select a low value for R_F. Unfortunately, however, the lower the value of R_F, the less ideal the integrator circuit becomes. This is another example of the trade-offs that a designer has to consider in creating working circuits from imperfect components.

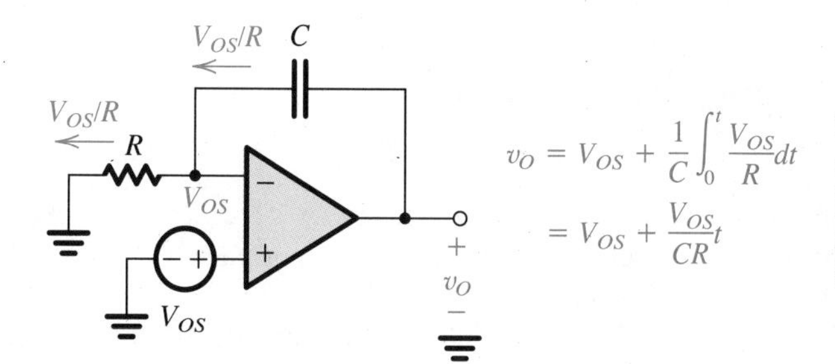

$$v_O = V_{OS} + \frac{1}{C}\int_0^t \frac{V_{OS}}{R}dt$$
$$= V_{OS} + \frac{V_{OS}}{CR}t$$

Fig. 2.46 Determining the effect of the op amp input offset voltage V_{OS} on the Miller integrator circuit. Note that since the output rises with time, the op amp eventually saturates.

Exercise D2.33 Consider a Miller integrator with a time constant of 1 ms and an input resistance of 10 kΩ. Let the op amp have $V_{OS} = 2$ mV and output saturation voltages of ±12 V. (a) Assuming that when the power supply is turned on the capacitor voltage is zero, how long does it take for the amplifier to saturate? (b) Select the largest possible value for a feedback resistor R_F so that at least ±10 V of output signal swing remain available. What is the corner frequency of the resulting STC network?

Ans. (a) 6 s; (b) 10 MΩ, 0.16 Hz

Input Bias Currents

The second dc problem encountered in op amps is illustrated in Fig. 2.47. In order for the op amp to operate, its two input terminals have to be supplied with dc currents, termed the **input bias currents.** In Fig. 2.47 these two currents are represented by two current sources, I_{B1} and I_{B2}, connected to the two input terminals. It should be emphasized that

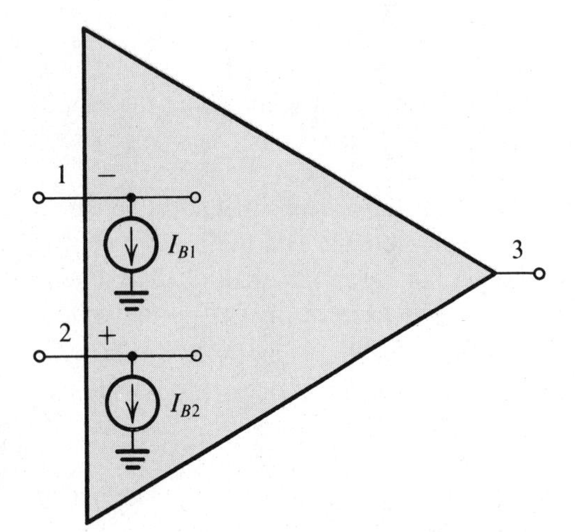

Fig. 2.47 The op-amp input bias currents represented by two current sources I_{B1} and I_{B2}.

the input bias currents are independent of the fact that the op amp has finite input resistance (not shown in Fig. 2.47). The op amp manufacturer usually specifies the average value of I_{B1} and I_{B2} as well as their expected difference. The average value I_B is called the **input bias current,**

$$I_B = \frac{I_{B1} + I_{B2}}{2}$$

and the difference is called the **input offset current** and is given by

$$I_{OS} = |I_{B1} - I_{B2}|$$

Typical values for general-purpose op amps that use bipolar transistors are $I_B = 100$ nA and $I_{OS} = 10$ nA. Op amps that utilize field-effect transistors in the input stage have much smaller input bias current (of the order of picoamperes).

We now wish to find the dc output voltage of the closed-loop amplifier due to the input bias currents. To do this we ground the signal source and obtain the circuit shown in Fig. 2.48 for both the inverting and noninverting configurations. As shown in Fig. 2.48, the output dc voltage is given by

$$V_O = I_{B1}R_2 \simeq I_B R_2 \tag{2.51}$$

This obviously places an upper limit on the value of R_2. Fortunately, however, a technique exists for reducing the value of the output dc voltage due to the input bias currents. The method consists of introducing a resistance R_3 in series with the noninverting input

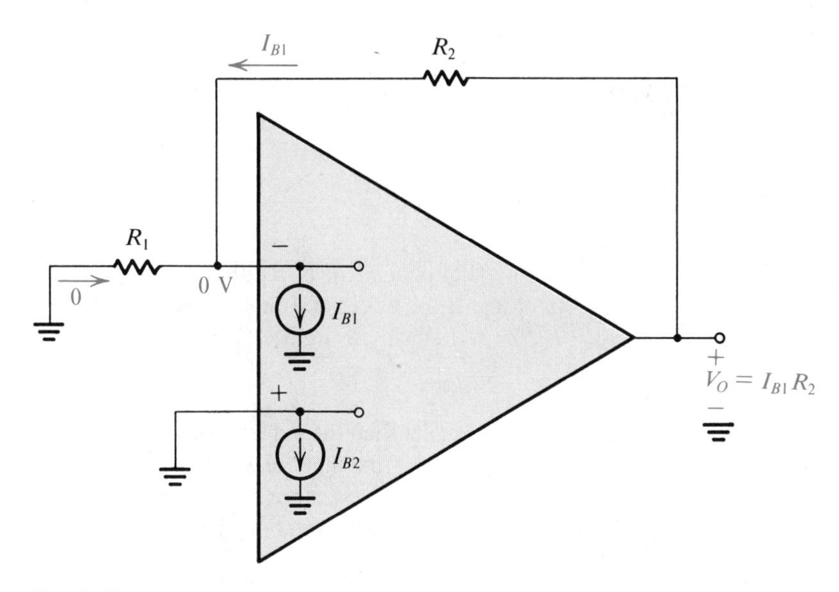

Fig. 2.48 Analysis of the closed-loop amplifier, taking into account the input bias currents.

lead, as shown in Fig. 2.49. From a signal point of view, R_3 has a negligible effect. The appropriate value for R_3 can be determined by analyzing the circuit in Fig. 2.49, where analysis details are shown and the output voltage is given by

$$V_O = -I_{B2}R_3 + R_2(I_{B1} - I_{B2}R_3/R_1) \tag{2.52}$$

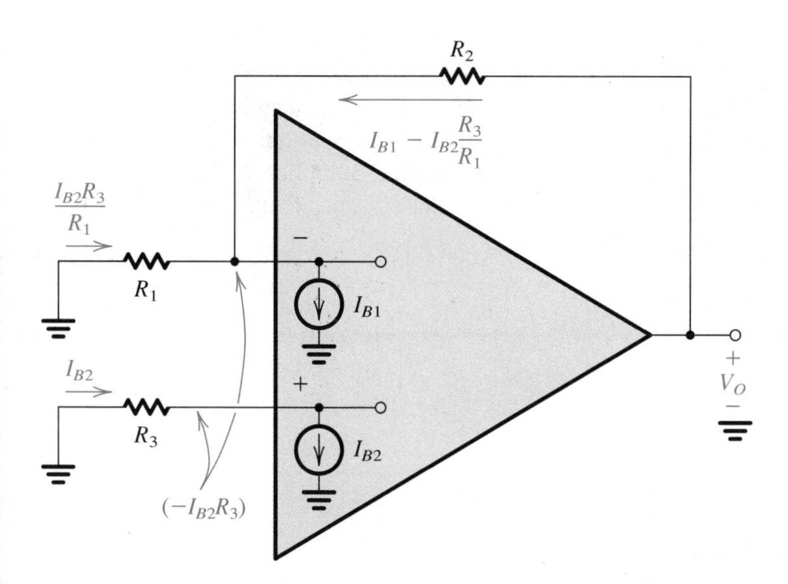

Fig. 2.49 Reducing the effect of the input bias currents by introducing a resistor R_3.

Consider first the case $I_{B1} = I_{B2} = I_B$, which results in

$$V_O = I_B[R_2 - R_3(1 + R_2/R_1)]$$

Thus we may reduce V_O to zero by selecting R_3 such that

$$R_3 = \frac{R_2}{1 + R_2/R_1} = \frac{R_1R_2}{R_1 + R_2} \tag{2.53}$$

That is, R_3 should be made equal to the parallel equivalent of R_1 and R_2.

Having selected R_3 as above, let us evaluate the effect of a finite offset current I_{OS}. Let $I_{B1} = I_B + I_{OS}/2$ and $I_{B2} = I_B - I_{OS}/2$, and substitute in Eq. (2.52). The result is

$$V_O = I_{OS}R_2 \tag{2.54}$$

which is usually about an order of magnitude smaller than the value obtained without R_3 (Eq. 2.51). We conclude that to minimize the effect of the input bias currents one should place in the positive lead a resistance equal to the dc resistance seen by the inverting terminal. We should emphasize the word dc in the last statement; note that if the amplifier is ac-coupled, we should select $R_3 = R_2$, as shown in Fig. 2.50.

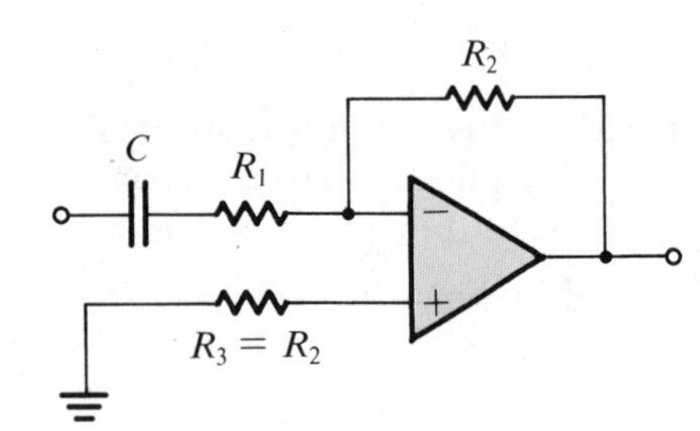

Fig. 2.50 In an ac-coupled amplifier the dc resistance seen by the inverting terminal is R_2; hence R_3 is chosen equal to R_2.

While we are on the subject of ac-coupled amplifiers we should note that one must always provide a continuous dc path between each of the input terminals of the op amp and ground. For this reason the ac-coupled noninverting amplifier of Fig. 2.51 will *not* work without the resistance R_3 to ground. Unfortunately, including R_3 lowers considerably the input resistance of the closed-loop amplifier.

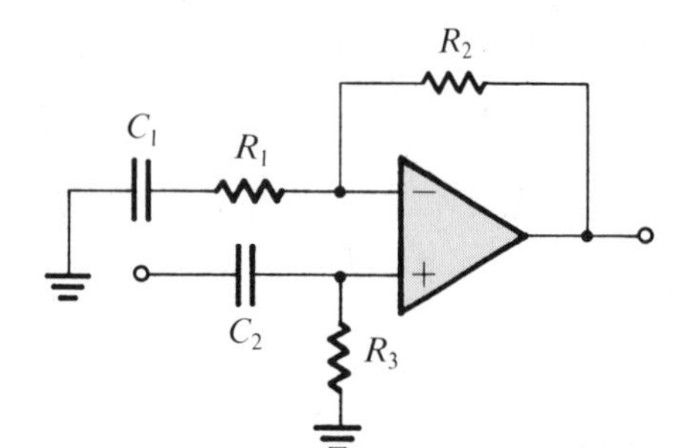

Fig. 2.51 Illustrating the need for a continuous dc path for each of the op amp input terminals.

Exercise 2.34 Consider a Miller integrator circuit with an input resistance R and an integrator capacitor C and with another resistor R connected between the positive input terminal of the op amp and ground. Ground the integrator input terminal and replace the op amp with the equivalent circuit of Fig. 2.47 with $I_{B1} = I_B + I_{OS}/2$ and $I_{B2} = I_B - I_{OS}/2$.
(a) Show that if the capacitor voltage at $t = 0$ is zero then

$$v_O = -\left[I_B - \frac{I_{OS}}{2}\right]R + \frac{I_{OS}}{C}t$$

(b) With a resistor R_F connected in parallel with C and assuming that $R_F \gg R$, show that v_O becomes $v_O \simeq I_{OS}R_F$.

SUMMARY

▸ The IC op amp is a versatile circuit building block. It is easy to apply, and the performance of op-amp circuits closely matches theoretical predictions.

▸ Most op amps are differential amplifiers having a large differential gain (called open-loop gain, A) and a small common-mode gain. The ratio of the two gains is the CMRR.

▸ An ideal op amp has an infinite gain, an infinite input resistance, and a zero output resistance.

▸ In most applications, negative feedback is applied around the op amp, resulting in a closed-loop amplifier with gain determined almost entirely by external components.

▸ The noninverting closed-loop configuration features a very high input resistance. A special case is the unity-gain follower, frequently employed as a buffer amplifier to connect a high-resistance source to a low-resistance load.

▸ In the analysis of op amp circuits, if the op amp is assumed ideal, a virtual short circuit appears between its two input terminals. This, together with the fact that zero currents flow into the input terminals of an ideal op amp, simplifies analysis considerably.

▸ For most internally compensated op amps, the open-loop gain falls off with frequency at a rate of -20 dB/decade, reaching unity at a frequency f_t (the unity-gain bandwidth).

▸ For both the inverting and the noninverting closed-loop configurations, the 3-dB frequency is equal to $f_t/(1 + R_2/R_1)$.

▸ The maximum rate at which the op-amp output voltage can change is called the slew rate. Op amp slewing can result in nonlinear distortion of output signal waveforms.

▸ The full-power bandwidth, f_M, is the maximum frequency at which an output sinusoid with an amplitude equal to the op-amp rated output voltage can be produced without distortion.

▸ The finite op amp CMRR limits the performance of the difference-amplifier closed-loop configuration. The effect of finite CMRR can be taken into account in analysis by including a signal source equal to (v_{Icm}/CMRR) in series with the op amp positive input lead.

▸ The input resistance of the noninverting configuration is approximately equal to $R_{id}(1 + A\beta)$, where β is the feedback factor.

▸ For both the inverting and the noninverting configurations the output resistance is equal to $R_o/(1 + A\beta)$.

▸ The input offset voltage, V_{OS}, is the magnitude of dc voltage that when applied between the op amp input terminals, with appropriate polarity, reduces the dc offset voltage at the output to zero.

▸ The effect of V_{OS} on performance can be evaluated by including in the analysis a dc source V_{OS} in series with the op-amp positive input lead.

▸ Capacitively coupling an op amp reduces the dc offset voltage at the output considerably.

▸ The average of the two dc currents, I_{B1} and I_{B2}, that flow in the input terminals of the op amp, is called the input bias current, I_B. In a closed-loop amplifier, I_B gives rise to a dc offset voltage at the output of magnitude I_BR_2. This voltage can be reduced to $I_{OS}R_2$ by connecting a resistance in series with the positive input terminal equal to the total dc resistance seen by the negative input terminal. I_{OS} is the input offset current; that is, $I_{OS} = |I_{B1} - I_{B2}|$.

▸ Connecting a large resistance in parallel with the capacitor of an op amp integrator prevents op amp saturation (due to the effect of V_{OS} and I_B).

BIBLIOGRAPHY

G. B. Clayton, *Experimenting with Operational Amplifiers,* London: Macmillan, 1975.

G. B. Clayton, *Operational Amplifiers,* 2nd ed., London: Newnes-Butterworths, 1979.

S. Franco, *Design with Operational Amplifiers and Analog Integrated Circuits,* New York: McGraw-Hill, 1988.

J. G. Graeme, G. E. Tobey, and L. P. Huelsman, *Operational Amplifiers: Design and Applications,* New York: McGraw-Hill, 1971.

W. Jung, *IC Op Amp Cookbook,* Indianapolis: Howard Sams, 1974.

E. J. Kennedy, *Operational Amplifier Circuits: Theory and Applications,* New York: Holt, Rinehart and Winston, 1988.

J. K. Roberge, *Operational Amplifiers: Theory and Practice,* New York: Wiley, 1975.

J. I. Smith, *Modern Operational Circuit Design,* New York: Wiley-Interscience, 1971.

J. E. Solomon, "The monolithic op amp: A tutorial study," *IEEE Journal of Solid-State Circuits,* vol. SC-9, no. 6, pp. 314–322, Dec. 1974.

J. V. Wait, L. P. Huelsman, and G. A. Korn, Intrododuction to Operational Amplifier Theory and Applications, New York: McGraw-Hill, 1975.

PROBLEMS

Section 2.1: The Op Amp Terminals

2.1 What is the minimum number of terminals required by a single op amp? What is the minimum number of terminals required on an integrated-circuit package containing four op amps (called a quad op amp)?

Section 2.2: The Ideal Op Amp

2.2 The circuit of Fig. P2.2 uses an op amp that is ideal except for having a finite gain A. Measurements indicate $v_O = 3.5$ V when $v_I = 3.5$ V. What is the op amp gain A?

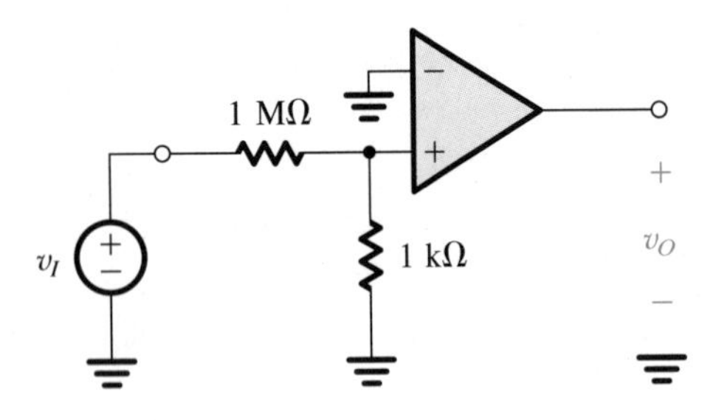

Fig. P2.2

2.3 A set of experiments are run on an op amp that is ideal except for having a finite gain A. The results are tabulated below. Are the results consistent? If not, are they reasonable, in view of the possibility of experimental error? What do they show the gain to be? Using this value, predict values of the measurements that were accidentally omitted (the blank entries).

Experiment #	v_1	v_2	v_O
1	0.00	0.00	0.00
2	1.00	1.00	0.00
3		1.00	1.00
4	1.00	1.10	10.1
5	2.01	2.00	−0.99
6	1.99	2.00	1.00
7	5.10		−5.10

2.4 Refer to Exercise 2.2. This problem explores an alternative internal structure for the op amp. In particular, we wish to model the internal structure of a particular op amp using two transconductance amplifiers and one transresistance amplifier. Suggest an appropriate topology. For equal transconductances G_m and a transresistance R_m, find an expression for the open-loop gain A. For $G_m = 50$ mA/V and $R_m = 10^6\ \Omega$, what value of A results?

Section 2.3: Analysis of Circuits Containing Ideal Op Amps—The Inverting Configuration

2.5 An inverting amplifier with the topology shown in Fig. 2.4 uses an ideal op amp with $R_1 = 33$ kΩ and $R_2 = 330$ kΩ. What is the closed-loop gain you would expect? A second 33-kΩ resistor is connected at the input:
(a) in series with the existing 33 kΩ.
(b) in parallel with the existing 33 kΩ.
What values of gain result?

2.6 Assuming ideal op amps, find the voltage gain v_o/v_i and input resistance R_{in} of each of the circuits in Fig. P2.6.

D2.7 Design an inverting op amp circuit for which the gain is -4 V/V and the total resistance used is 100 kΩ.

D2.8 Using the circuit of Fig. 2.4 and assuming an ideal op amp, design an amplifier with a gain of -50 V/V having the largest possible input resistance under the constraint of having to use resistors no larger than 10 MΩ. What is the input resistance of your design?

2.9 An inverting op amp circuit is fabricated with the resistors R_1 and R_2 having $x\%$ tolerance (that is, the value of each resistance can deviate from the nominal value by as much as $\pm x\%$). What is the tolerance on the realized closed-loop gain? Assume the op amp to be ideal. If the nominal closed-loop gain is -100 V/V and $x = 5$, what is the range of gain values expected from such a circuit?

2.10 An op amp with an open-loop gain of 1000 V/V is used in the inverting configuration. If in this application the output voltage ranges from -10 V to $+10$ V, what is the maximum voltage by which the "virtual ground node" departs from its ideal value?

2.11 The circuit in Fig. P2.11 is frequently used to provide an output voltage v_o proportional to an input signal current i_i. Derive expressions for the transresistance

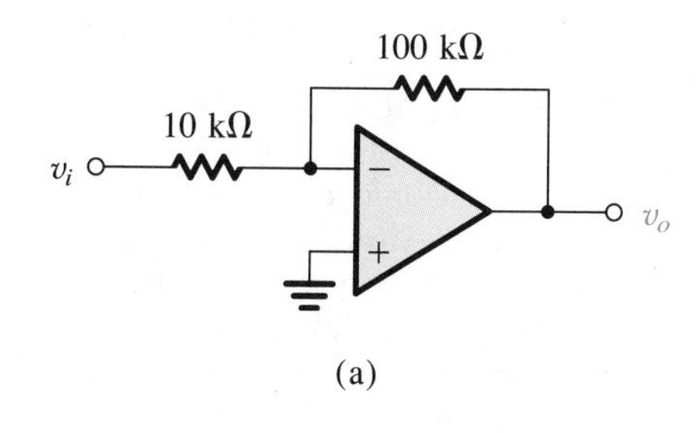

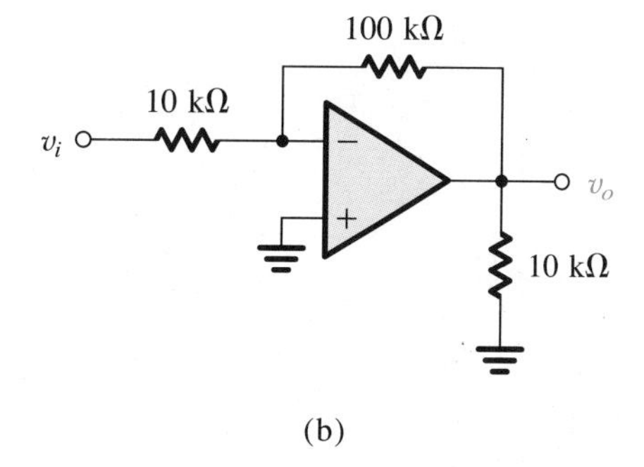

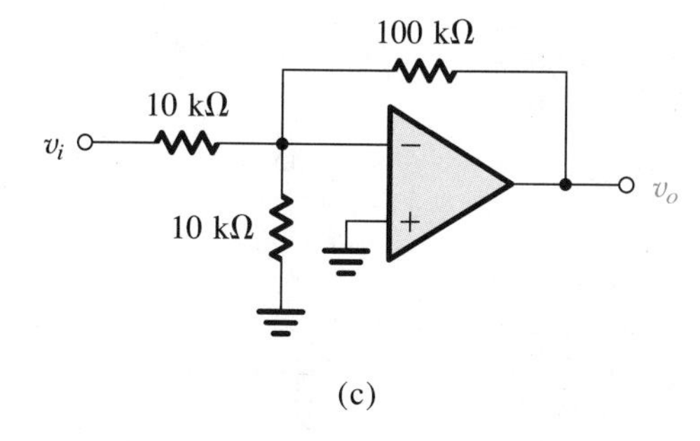

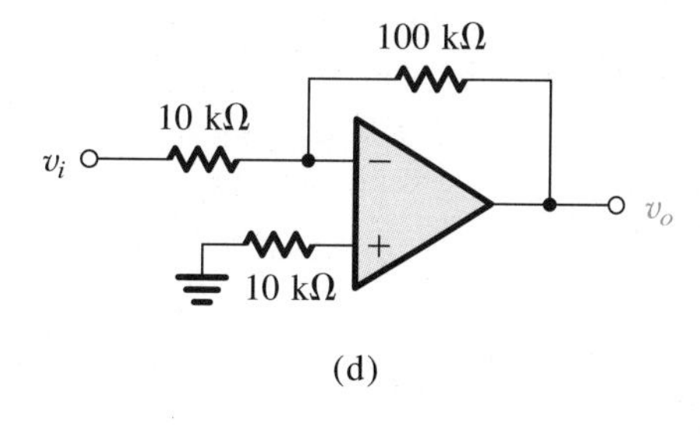

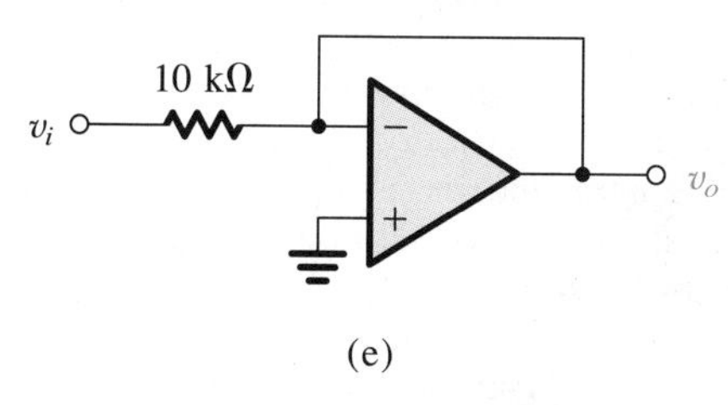

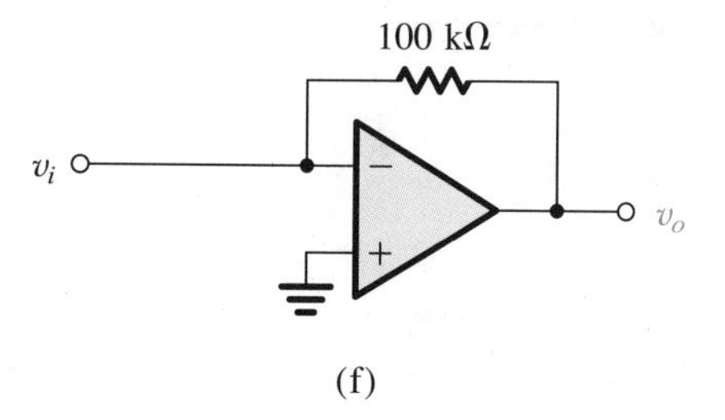

Fig. P2.6

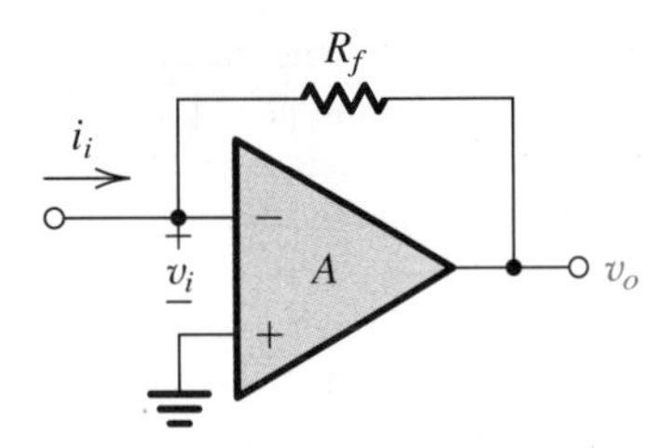

Fig. P2.11

$R_m \equiv v_o/i_i$ and the input resistance $R_i \equiv v_i/i_i$ for the two cases:

(a) A is infinite, and
(b) A is finite.

2.12 Derive an expression for the input resistance of the inverting amplifier of Fig. 2.4 taking into account the finite open-loop gain A of the op amp.

***2.13** Rearrange Eq. (2.1) to give the amplifier open-loop gain A required to realize a specified closed-loop gain ($G_{\text{nominal}} = -R_2/R_1$) within a specified gain error ε,

$$\varepsilon \equiv \left| \frac{G - G_{\text{nominal}}}{G_{\text{nominal}}} \right|$$

For a closed-loop gain of −100 and a gain error of ≤10%, what is the minimum A required?

***2.14** Using Eq. (2.1), determine the value of A for which a reduction of A by x% results in a reduction in $|G|$ by (x/k)%. Find the value of A required for the case in which the nominal closed-loop gain is 100, x is 50, and k is 100.

2.15 Consider the circuit in Fig. 2.8 with $R_1 = R_2 = R_4 = 1\ \text{M}\Omega$, and assume the op amp to be ideal. Find values for R_3 to obtain the following gains:

(a) −10V/V, and
(b) −100 V/V.

2.16 For the circuit in Fig. 2.8, what gain results when all resistors are equal? An extension of this circuit is shown in Fig. P2.16; determine its gain.

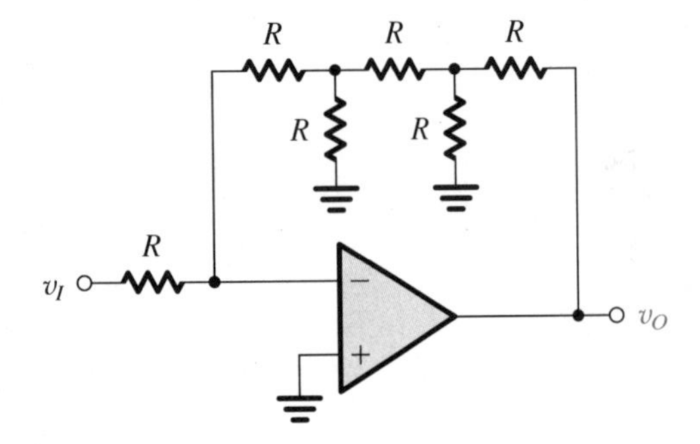

Fig. P2.16

***2.17** Repeat Problem 2.6 for the case the op amps have finite gain $A = 1000$.

D*2.18 (a) For the circuit in Fig. 2.8, taking the finite open-loop gain A of the op amp into account, show that

$$\frac{v_O}{v_I} = \frac{-G_0}{1 + [1 + G_0 + (R_4/R_3)]/A}$$

where G_0 is the nominal magnitude of the closed-loop gain (see Example 2.2),

$$G_0 = \frac{R_2}{R_1}\left[1 + \frac{R_4}{R_2} + \frac{R_4}{R_3}\right]$$

(b) Apply this result to the case $G_0 = 100$, $A = 1000$, and $R_4 = R_2 = R_1$. Find v_O/v_1.
(c) Repeat (b) for the same G_0 and A values but with $R_4 = R_2 = 10\ R_1$.
(d) For comparison, find v_O/v_I for the inverting configuration with the same G_0 and A values.

Note: The expression for v_O/v_I above suggests that the effect of finite A can be made approximately equal to that of the inverting configuration by selecting $R_4 \ll R_3$. This component selection, however, defeats the purpose of using the T network in the feedback. Why? (You need to study the design process in Example 2.2 to be able to answer this question.)

Section 2.4: Other Applications of the Inverting Configuration

2.19 A Miller integrator incorporates an ideal op amp, a resistor R of 100 kΩ, and a capacitor C of 0.1 μF. A sine-wave signal is applied to its input.

(a) At what frequency (in Hz) are the input and output signals equal in amplitude?
(b) At that frequency how does the phase of the output sine wave relate to that of the input?
(c) If the frequency is lowered by a factor of 10 from that found in (a), by what factor does the output voltage change, and in what direction (smaller or larger)?
(d) What is the phase relation between the input and output in situation (c)?

2.20 A Miller integrator whose input and output voltages are initially zero and whose time constant is 1 ms is driven by the signal shown in Fig. P2.20. Sketch and label the output waveform that results. Indicate what happens if the input levels are ±2 V, with the time constant the same (1 ms) and with the time constant raised to 2 ms.

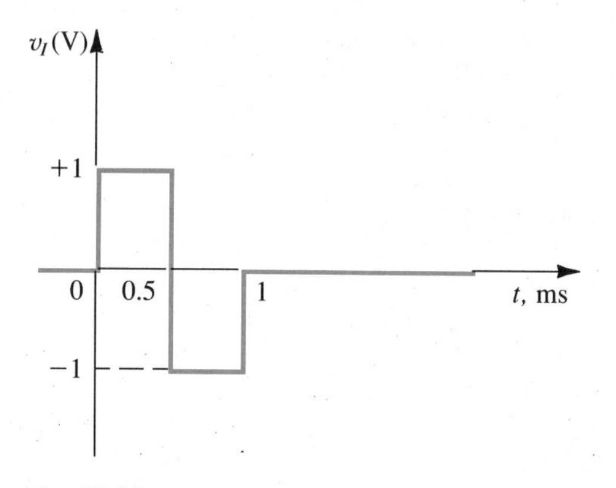

Fig. P2.20

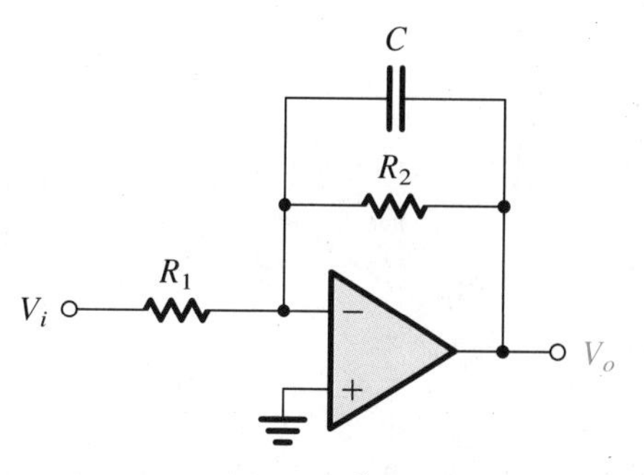

Fig. P2.22

2.21 Consider a Miller integrator having a time constant of 1 ms, and whose output is initially zero, when fed with a string of pulses of 10-μs duration and 1-V amplitude rising from 0 V (see Fig. P2.21). Sketch and label the output waveform resulting. How many pulses are required for an output voltage change of 1 V?

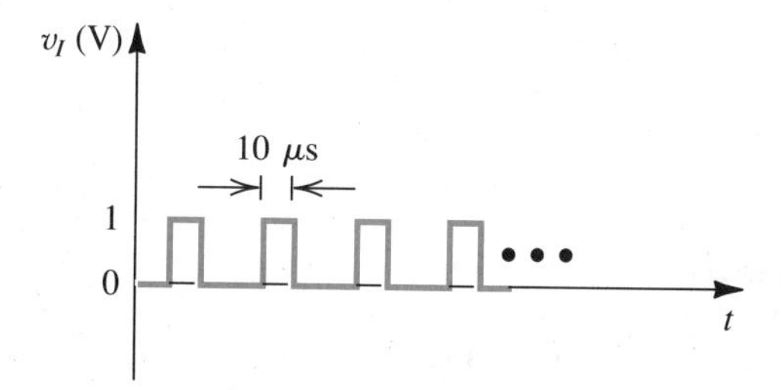

Fig. P2.21

D2.22 Figure P2.22 shows a circuit that performs a low-pass single-time-constant function. Such a circuit is known as a first-order low-pass active filter. Derive the transfer function and show that the dc gain is $(-R_2/R_1)$ and the 3-dB frequency $\omega_0 = 1/CR_2$. Design the circuit to obtain an input resistance of 1 kΩ, a dc gain of 40 dB, and a 3-dB frequency of 4 kHz. At what frequency does the magnitude of the transfer function reduce to unity?

****2.23** In order to limit the low-frequency gain of a Miller integrator, a resistor is often shunted across the integrating capacitor. Consider the case when the input resistor is 100 kΩ, the capacitor is 0.1 μF, and the shunt resistor is 10 MΩ.

(a) Sketch and label a Bode plot for the magnitude response of the resulting circuit and contrast it with that of an ideal integrator (that is, without the shunt resistor). At what frequency does the circuit begin to behave less as an integrator and more as an amplifier?

(b) Sketch and clearly label the output waveform resulting when an input pulse of 0.1-V height and 1-ms duration is applied. Consider the cases without and with the shunt resistor. (*Note:* The pulse response of STC networks is discussed in Appendix E.)

2.24 A differentiator utilizes an ideal op amp, a 10-kΩ resistor, and a 0.01-μF capacitor. What is the frequency f_0 (in Hz) at which its input and output sine-wave signals have equal magnitude? What is the output signal for a 1-V peak-to-peak sine-wave input with frequency equal to 10 f_0?

2.25 An op amp differentiator with 1-ms time constant is driven by the rate-controlled step shown in Fig. P2.25. Assuming v_O to be zero initially, sketch and label its waveform.

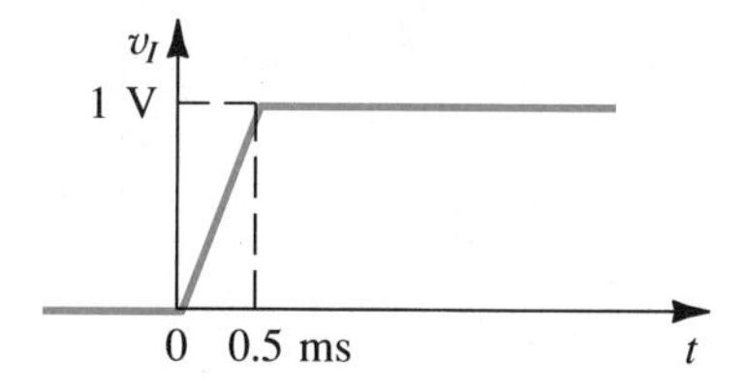

Fig. P2.25

D2.26 Figure P2.26 shows a circuit that performs the high-pass single-time-constant function. Such a circuit is known as a first-order high-pass active filter. Derive the transfer function and show that the high-frequency gain is $(-R_2/R_1)$ and the 3-dB frequency $\omega_0 = 1/CR_1$. Design the circuit to obtain a high-frequency input resistance of 1 kΩ, a high-frequency gain of 40 dB, and a 3-dB frequency of 100 Hz. At what frequency does the magnitude of the transfer function reduce to unity?

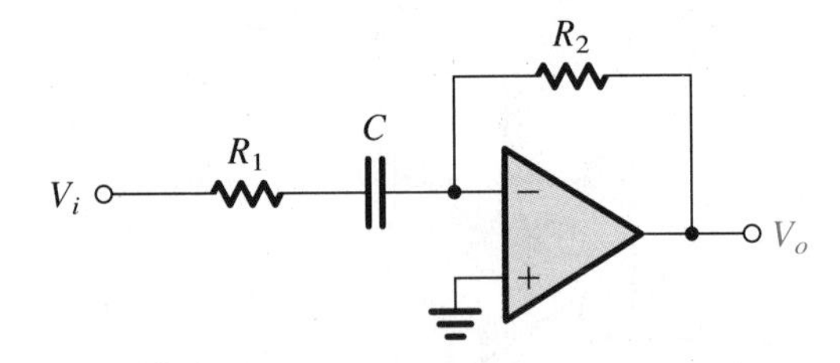

Fig. P2.26

D**2.27 Derive the transfer function of the circuit in Fig. P2.27 (for an ideal op amp) and show that it can be written in the form

$$\frac{V_o}{V_i} = \frac{-R_2/R_1}{[1 + (\omega_1/j\omega)][1 + j(\omega/\omega_2)]}$$

where $\omega_1 = 1/C_1R_1$ and $\omega_2 = 1/C_2R_2$. Assuming that the circuit is designed such that $\omega_2 \gg \omega_1$, find approximate expressions for the transfer function in the following frequency regions:

(a) $\omega \ll \omega_1$

(b) $\omega_1 \ll \omega \ll \omega_2$

(c) $\omega \gg \omega_2$

Use these approximations to sketch a Bode plot for the magnitude response. Observe that the circuit performs as an amplifier whose gain rolls off at the low-frequency end in the manner of a high-pass STC network, and at the high-frequency end in the manner of a low-pass STC network. Design the circuit to provide a gain of 40 dB in the "middle frequency range," a low-frequency 3-dB point at 10 Hz, a high-frequency 3-dB point at 10 kHz, and an input resistance (at $\omega \gg \omega_1$) of 5 kΩ.

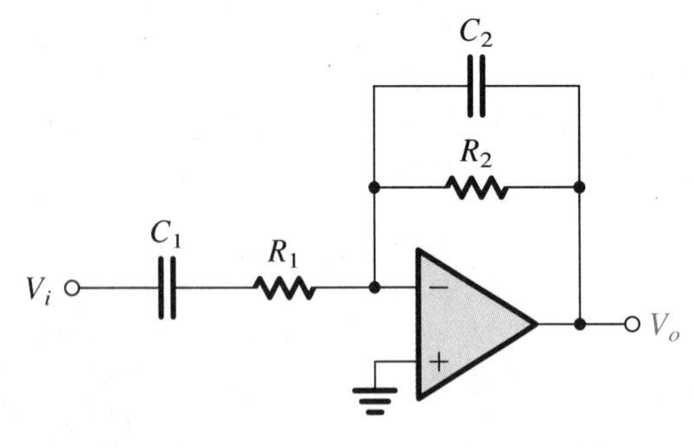

Fig. P2.27

2.28 A weighted summer circuit using an ideal op amp has three inputs using 100-kΩ resistors and a feedback resistor of 50 kΩ. A signal v_1 is connected to two of the inputs, while a signal v_2 is connected to the third. Express v_O in terms of v_1 and v_2. If $v_1 = 3$ V and $v_2 = -3$ V, what is v_O?

*2.29 We wish to investigate the effect of finite op amp open-loop gain A on the performance of the weighted summer. First consider a two-input summer with inputs v_a and v_b connected to R_{1a} and R_{1b}, respectively, and a feedback resistor R_2. Utilize the principle of superposition together with Eq. (2.1) to show that

$$v_O = \frac{1}{1 + \dfrac{(1 + R_2/R_{\text{parallel}})}{A}}\left[v_a\frac{R_2}{R_{1a}} + v_b\frac{R_2}{R_{1b}}\right]$$

where R_{parallel} is the parallel equivalent of the input resistors. Then extend this result to the case of an arbitrary number of inputs.

D2.30 Design an op amp circuit to provide an output $v_O = -[3v_1 + (v_2/2)]$. Choose relatively low values of resistors but ones for which the input current (from each input signal source) does not exceed 0.1 mA for 2-V input signals.

D*2.31 In an instrumentation system, there is a need to take the difference between two signals, one, $v_1 = 3\sin(2\pi \times 60t) + 0.01\sin(2\pi \times 1000t)$, volts, and another $v_2 = 3\sin(2\pi \times 60t) - 0.01\sin(2\pi \times 1000t)$, volts. Draw a circuit that finds the required difference using two op amps and mainly 10-kΩ resistors. Since it is desirable to amplify the 1000-Hz component in the process, arrange to provide overall gain of 10 as well. The op amps available are ideal except that their output voltage swing is limited to ±10 V.

*2.32 Figure P2.32 shows a circuit for a digital-to-analog converter (DAC). The circuit accepts a four-bit input binary word $a_3a_2a_1a_0$, where a_0, a_1, a_2, and a_3 take the values of 0 or 1, and it provides an analog output voltage v_O proportional to the value of the digital input. Each of the bits of the input word controls the correspondingly numbered switch. For instance, if a_2 is 0 then switch S_2 connects the 20-kΩ resistor to ground, while if a_2 is 1 then S_2 connects the 20-kΩ resistor to the +5-V power supply. Show that v_O is given by

$$v_O = -\frac{R_f}{16}[2^0a_0 + 2^1a_1 + 2^2a_2 + 2^3a_3]$$

where R_f is in kΩ. Find the value of R_f so that v_O ranges from 0 to −10 (15/16) volts.

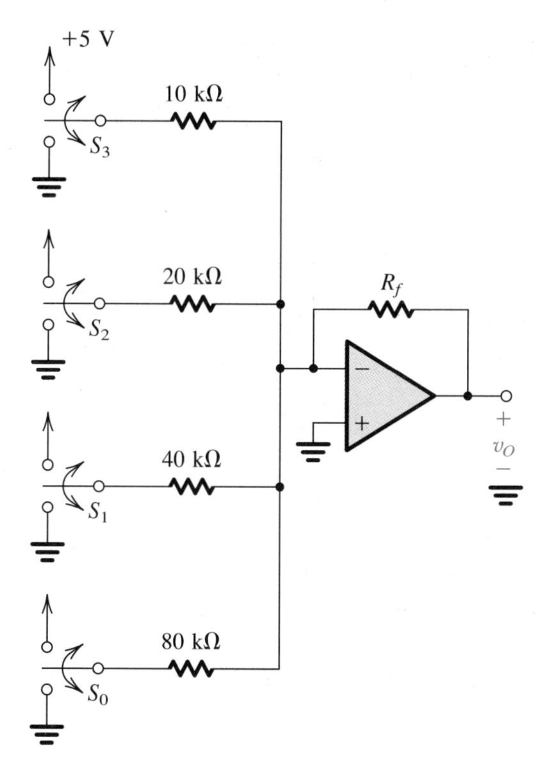

Fig. P2.32

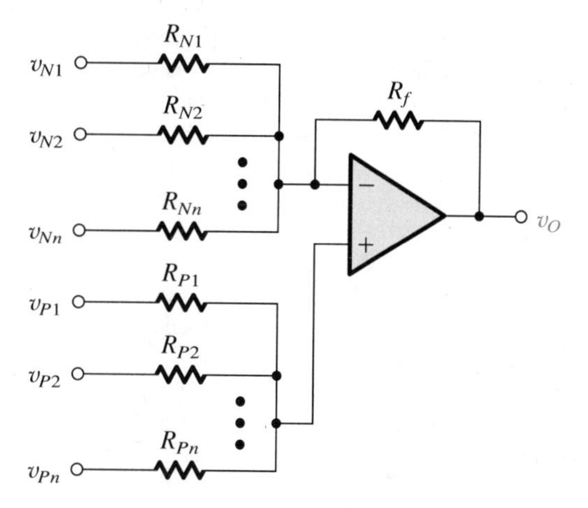

Fig. P2.33

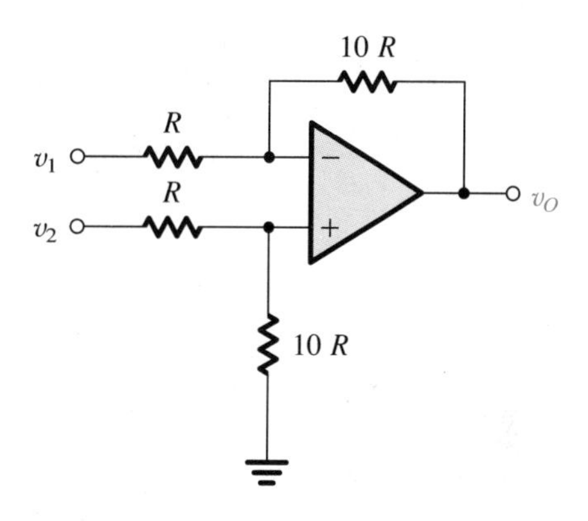

Fig. P2.34

Section 2.5: The Noninverting Configuration

D2.33 (a) Use superposition to show that the output of the circuit in Fig. P2.33 is given by

$$v_O = -\left[\frac{R_f}{R_{N1}}v_{N1} + \frac{R_f}{R_{N2}}v_{N2} + \cdots + \frac{R_f}{R_{Nn}}v_{Nn}\right]$$

$$+\left[1 + \frac{R_f}{R_N}\right]\left[\frac{R_P}{R_{P1}}v_{P1} + \frac{R_P}{R_{P2}}v_{P2} + \cdots + \frac{R_P}{R_{Pn}}v_{Pn}\right]$$

where $R_N = R_{N1} // R_{N2} // \cdots // R_{Nn}$ and $R_P = R_{P1} // R_{P2} // \cdots // R_{Pn}$.

(b) Design a circuit to obtain

$$v_O = -2\, v_{N1} + v_{P1} + 2\, v_{P2}$$

The smallest resistor used should be 10 kΩ.

2.34 For the circuit in Fig. P2.34 use superposition to find v_O in terms of the input voltages v_1 and $v5_2$. Assume an ideal op amp. For

$v_1 = 10\sin(2\pi \times 60t) - 0.1\sin(2\pi \times 1000t)$, volts

$v_2 = 10\sin(2\pi \times 60t) + 0.1\sin(2\pi \times 1000t)$, volts

find v_O.

D2.35 The circuit shown in Fig. P2.35 utilizes a 10-kΩ potentiometer to realize an adjustable-gain amplifier. Derive an expression for the gain as a function of the potentiometer setting x. Assume the op amp to be ideal. What is the range of gains obtained? Show how

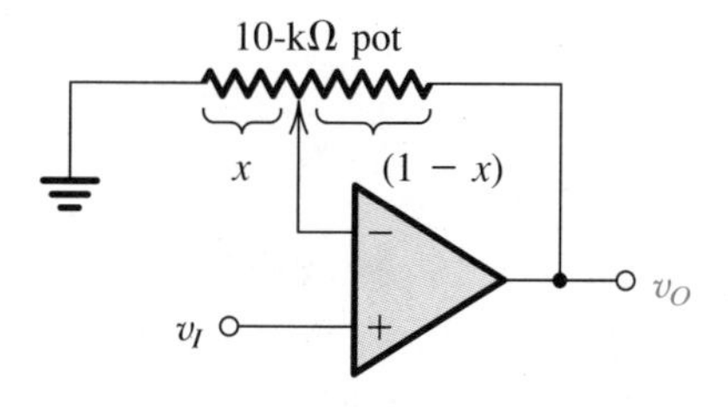

Fig. P2.35

to add a fixed resistor so that the gain range can be 1 to 11 V/V. What should the resistor value be?

2.36 It is required to connect a 10-V source with a source resistance of 100 kΩ to a 1-kΩ load. Find the voltage that will appear across the load if:

(a) the source is connected directly to the load.

(b) an op-amp unity-gain buffer is inserted between the source and the load.

In each case find the load current and the current supplied by the source. Where does the load current come from in case (b)?

2.37 Derive an expression for the gain of the voltage follower of Fig. 2.17 assuming the op amp to be ideal except for having a finite gain A. Calculate the value of the closed-loop gain for $A = 1000$, 100, and 10. In each case find the percentage error in gain magnitude from the nominal value of unity.

2.38 Figure P2.38 shows a circuit that provides an output voltage v_O whose value can be varied by turning the wiper of the 100-kΩ potentiometer. Find the range over which v_O can be varied. If the potentiometer is a "20-turn" device, find the change in v_O corresponding to each turn of the pot.

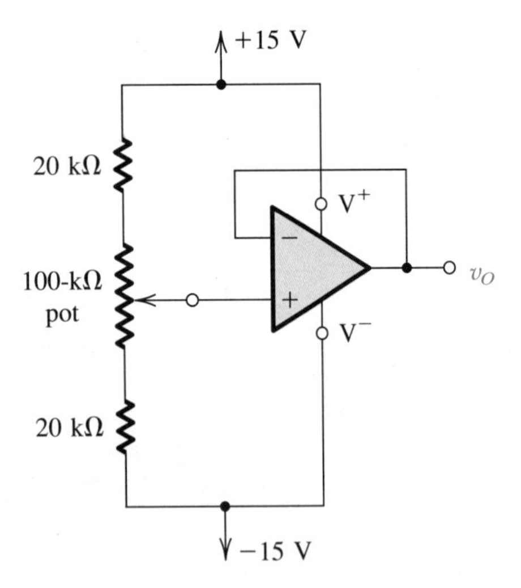

Fig. P2.38

Section 2.6: Examples of Op Amp Circuits

D2.39 For the high-input-resistance metering circuit of Fig. 2.18, using a 1-mA meter movement, find the value of resistor R such that full-scale reading is obtained for $v_I = 2.5$ V. If the meter resistance is 50 Ω, what is the op amp output voltage at half scale?

2.40 For the circuit shown in Fig. P2.40, express v_O as a function of v_1 and v_2. What is the input resistance seen by v_1 alone? By v_2 alone? By a source connected between the two input terminals? By a source connected to both input terminals simultaneously?

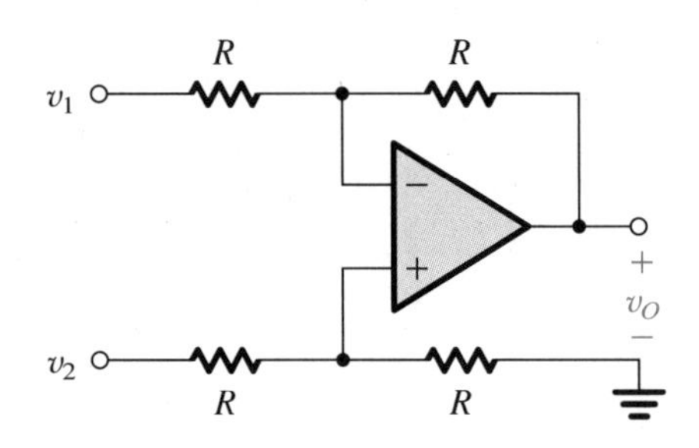

Fig. P2.40

D*2.41 Consider the differential amplifier in Fig. 2.22. It is common to express the output voltage in the form

$$v_O = G_d v_d + G_{cm} v_{CM}$$

where G_d is the differential gain and G_{cm} is the common-mode gain. Using the expression for v_O in Eq. (2.4), find expressions for G_d and G_{cm}, and show that the common-mode rejection ratio (CMRR) of the closed-loop amplifier is given by

$$\text{CMRR} \equiv 20 \log \frac{|G_d|}{|G_{cm}|}$$

$$= 20 \log \frac{1 + \frac{1}{2}\left[\frac{R_1}{R_2} + \frac{R_3}{R_4}\right]}{\left|\frac{R_1}{R_2} - \frac{R_3}{R_4}\right|}$$

Ideally the circuit is designed with $R_1/R_2 = R_3/R_4$, which results in an infinite CMRR. The finite tolerances of the resistor values, however, will make the CMRR finite. Show that if each resistor has a tolerance of $\pm 100\ \varepsilon\%$ then the worst-case CMRR is given by

$$\text{CMRR} = 20 \log \left[\frac{K + 1}{4\varepsilon}\right]$$

where K is the nominal (ideal) value of the ratios (R_2/R_1) and (R_4/R_3). Calculate the value of worst-case CMRR for an amplifier designed to have a differential gain of ideally 100 assuming that the op amp is ideal and that 1% resistors are used.

***2.42** Figure P2.42 shows a modified version of the difference amplifier studied in Example 2.4. The modified

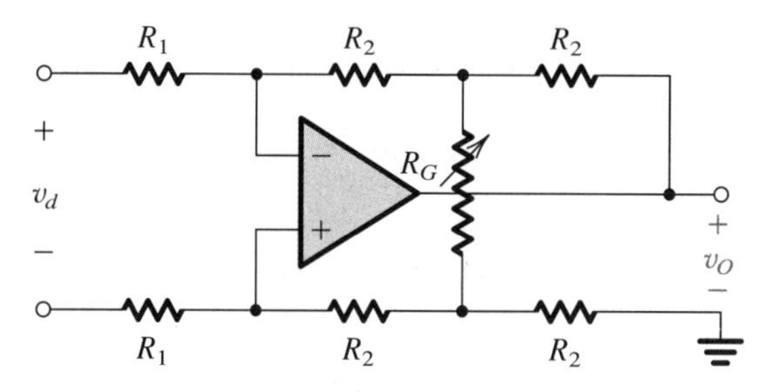

Fig. P2.42

circuit includes a resistor R_G, which can be used to vary the gain. Show that the differential voltage gain is given by

$$\frac{v_O}{v_d} = -2\frac{R_2}{R_1}\left[1 + \frac{R_2}{R_G}\right]$$

Hint: The virtual short circuit at the op amp input causes the current through the R_1 resistors to be $v_d/2R_1$.

2.43 Consider the instrumentation amplifier of Fig. 2.23(a) with a common-mode input voltage of +5 V (dc) and a differential input signal of 10-mV peak sine wave. Let $R_1 = 1\text{ k}\Omega$, $R_2 = 0.5\text{ M}\Omega$, $R_3 = R_4 = 10\text{ k}\Omega$. Find the voltage at every node in the circuit.

D2.44 Design the instrumentation amplifier circuit of Fig. 2.23(a) to realize a differential gain, variable in the range 1 to 100, utilizing a 100-kΩ pot as variable resistor. (*Hint:* Design the second stage for a gain of 0.5.)

2.45 For the negative impedance converter (NIC) circuit of Fig. 2.25(a) with $R = 1\text{ k}\Omega$ and $V_s = 1$ V, find the voltages across the load and at the output of the op amp for load resistances of 0 Ω, 100 Ω, 1 kΩ, and 2 kΩ. For an op amp whose output saturates at ±13 V (in the manner shown in Fig. 1.11), what is the highest value of load resistance that can be tolerated for $V_s \leq$ 1 V?

D*2.46 The circuit in Fig. 2.25(a) (Example 2.6) is a voltage-to-current converter (or alternatively, a voltage-controlled current source). Two alternative circuits for implementing this function are shown in Fig. P2.46. Analyze these circuits to determine i_O as a function of v_I. Comment on the advantages and disadvantages of these two circuits in comparison to that of Fig. 2.25(a).

***2.47** The circuit shown in Fig. P2.47 is intended to supply current to floating loads (those for which both termi-

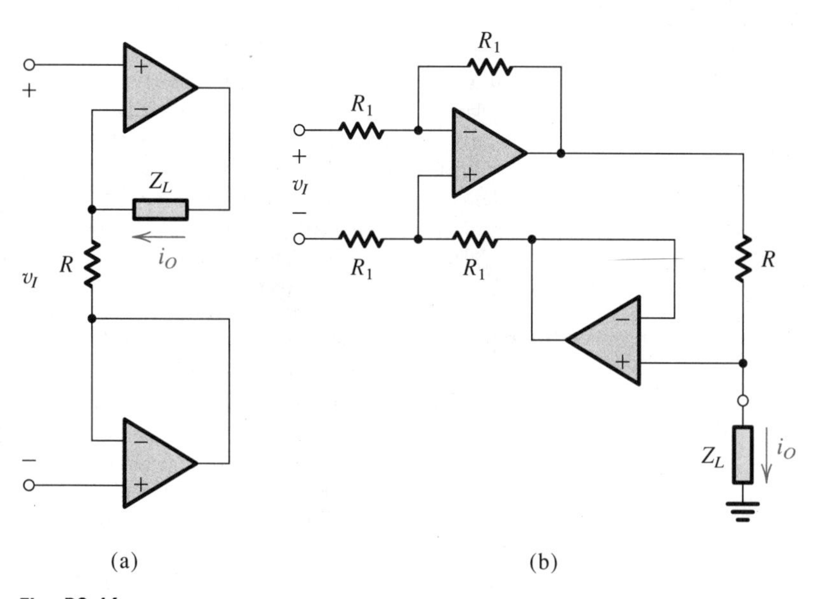

Fig. P2.46

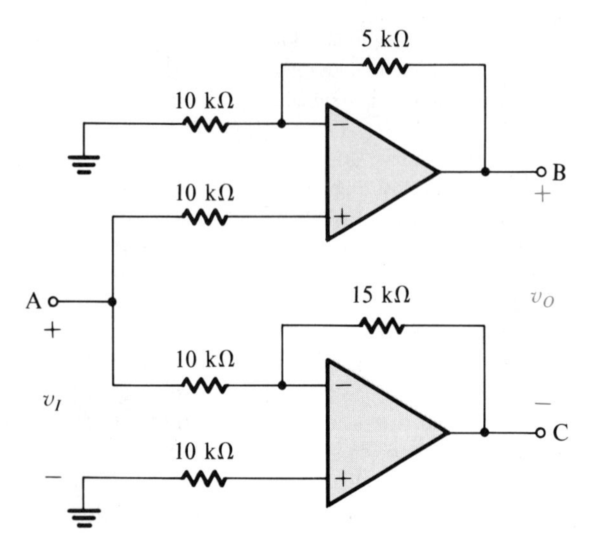

Fig. P2.47

nals are ungrounded) while making greatest possible use of the available power supply.

(a) Assuming ideal op amps, sketch the voltage waveforms at nodes B and C for a 1-V peak-to-peak sine wave applied at A. Also sketch v_O.

(b) What is the voltage gain v_O/v_I?

(c) Assuming that the op amps operate from ±15-V power supplies and that their output saturates at ±14 V (in the manner shown in Fig. 1.11), what is the largest sine wave output that can be accommodated? Specify both its peak-to-peak and rms values.

Section 2.7: Nonideal Performance of Op Amps

2.48 The data in the following table apply to internally compensated op amps. Fill in the blank entries.

A_0	f_b (Hz)	f_t (Hz)
10^6	1	
10^6		10^6
	10^6	10^8
	10^{-1}	10^6
2×10^5	10	

2.49 A measurement of the open-loop gain of an internally compensated op amp at very low frequencies shows it to be 4.2×10^4 V/V; at 100 kHz shows it is 76 V/V. Estimate values for A_0, f_b, and f_t.

2.50 Measurements of the open-loop gain of a compensated op amp intended for high-frequency operation indicate that the gain is 5.1×10^3 at 100 kHz and 8.3×10^3 at 10 kHz. Estimate its 3-dB frequency, its unity-gain frequency, and its dc gain.

2.51 An inverting amplifier with nominal gain of −20V/V employs an op amp having a dc gain of 10^4 and a unity-gain frequency of 10^6 Hz. What is the 3-dB frequency f_{3dB} of the closed-loop amplifier? What is its gain at 0.1 f_{3dB} and at 10 f_{3dB}?

2.52 Consider a unity-gain follower utilizing an internally compensated op amp with $f_t = 1$ MHz. What is the 3-dB frequency of the follower? At what frequency is the gain of the follower 1% below its low-frequency magnitude? If the input to the follower is a 1-V step, find the 10% to 90% rise time of the output voltage. (*Note:* The step response of STC low-pass networks is discussed in Appendix E.)

D*2.53 This problem illustrates the use of cascaded closed-loop amplifiers to obtain an overall bandwidth greater than can be achieved using a single-stage amplifier with the same overall gain.

(a) Show that cascading two identical amplifier stages, each having a low-pass STC frequency response with a 3-dB frequency f_1, results in an overall amplifier with a 3-dB frequency given by

$$f_{3dB} = \sqrt{\sqrt{2} - 1}\, f_1$$

(b) It is required to design a noninverting amplifier with a dc gain of 40 dB utilizing a single internally-compensated op amp with $f_t = 1$ MHz. What is the 3-dB frequency obtained?

(c) Redesign the amplifier of (b) by cascading two identical noninverting amplifiers each with a dc gain of 20 dB. What is the 3-dB frequency of the overall amplifier? Compare to the value obtained in (b) above.

2.54 Consider the use of an op amp with a unity-gain frequency f_t in the realization of

(a) an inverting amplifier with dc gain of magnitude K.

(b) a noninverting amplifier with a dc gain of K.

In each case find the 3-dB frequency and the gain-bandwidth product (GBP ≡ |Gain| × f_{3dB}). Comment on the results.

*2.55 Figure P2.55 shows a generalization of the noninverting configuration, where a resistive network having a

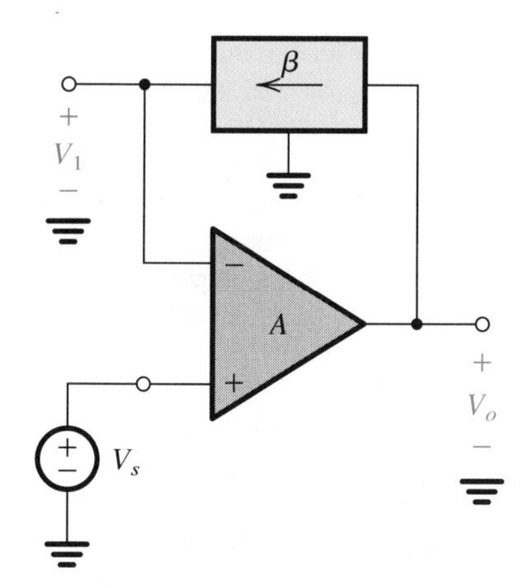

Fig. 2.55

transfer function $\beta \equiv V_1/V_o$ is connected in the negative feedback path of the op amp.

(a) Show that if the open-loop gain A is infinite, then the closed-loop gain $V_o/V_s = 1/\beta$.

(b) Derive an expression for the closed-loop gain assuming A is finite.

(c) If the op amp is internally compensated such that $A \simeq \omega_t/s$, show that the 3-dB frequency of the closed-loop gain is $f_{3\text{dB}} = \beta f_t$.

2.56 Consider an inverting summer with two inputs V_1 and V_2 and with $V_o = -(V_1 + V_2)$. Find the 3-dB frequency of each of the gain functions V_o/V_1 and V_o/V_2 in terms of the op amp f_t. (*Hint:* In each case, the other input to the summer can be set to zero—an application of superposition.)

***2.57** (a) Show that the transfer function of a Miller integrator realized using an internally compensated op amp with a unity-gain frequency ω_t is given approximately by

$$\frac{V_o}{V_i} \simeq -\frac{1}{j\omega CR} \frac{1}{1 + j(\omega/\omega_t)}$$

where it has been assumed that ω_t is much higher than the integrator frequency ω_0 ($\omega_0 = 1/CR$).

(b) What is the "excess phase" that the integrator has due to the op amp ω_t at $\omega = \omega_t/100$? Is the excess phase of the lag or lead type?

2.58 A noninverting amplifier using precision resistors of values 190 kΩ and 10 kΩ has a measured closed-loop gain of 19. In terms of its description as a feedback network, find values for the open-loop gain A, the feedback factor β, the loop gain, and the amount of feedback.

Section 2.8: The Internal Structure of IC Op Amps

2.59 A particular family of op amps whose internal structure can be modeled as in Fig. 2.32 is being considered. Four designs are as follows:

G_m (mA/V)	R_{o1} (MΩ)	R_{i2} (MΩ)	μ	C (pF)
0.2	10	5	500	30
0.02	20	10	10^4	1
20	0.1	0.1	10^3	30
2	1	4	700	20

For each design find the dc open-loop gain and the unity-gain frequency.

Section 2.9: Large-Signal Operation of Op Amps

2.60 A particular op amp using ±15-V supplies operates linearly for outputs in the range −13 V to +13 V. If used in an inverting amplifier configuration of gain −1000, what is the rms value of the largest possible sine wave that can be applied at the input without output clipping?

2.61 For an op-amp differentiator circuit having a time constant of 1 ms, using an op amp whose linear output range is ±11 V, what is the maximum rate of rise of acceptable input signals?

2.62 An op amp having a slew rate of 10 V/μs is to be used in the unity-gain follower configuration, with input pulses that rise from 0 to 5 V. What is the shortest pulse that can be used while ensuring full-amplitude output? For such a pulse, describe the output resulting.

***2.63** For operation with 10-V output pulses with the requirement that the sum of the rise and fall times should represent only 20% of the pulse width (at half amplitude), what is the slew-rate requirement for an op amp to handle pulses 1 μs wide? (*Note:* The rise and fall times of a pulse signal are usually measured between the 10%- and 90%-height points).

2.64 Show that for an op amp having the internal structure shown in Fig. 2.33, the relationship between the slew rate and the unity-gain frequency is given by

$$\text{SR} = \left[\frac{I_{\max}}{G_m}\right] \omega_t$$

where I_{max} is the maximum current available from the input stage and G_m is the transconductance of the input stage.

2.65 For an amplifier having a slew rate of 10 V/μs, what is the highest frequency at which a 20-V peak-to-peak sine wave can be produced at the output?

D*2.66 In designing with op amps one has to check the limitations on the voltage and frequency ranges of operation of the closed-loop amplifier, imposed by the op amp finite bandwidth (f_t), slew rate (SR), and output saturation (V_{omax}). This problem illustrates the point by considering the use of an op amp with $f_t = 2$ MHz, SR = 1 V/μs, and $V_{omax} = 10$ V in the design of a noninverting amplifier with a nominal gain of 10. Assume a sine-wave input with peak amplitude V_i.

(a) If $V_i = 0.5$ V, what is the maximum frequency before the output distorts?

(b) If $f = 20$ kHz, what is the maximum value of V_i before the output distorts?

(c) If $V_i = 50$ mV, what is the useful frequency range of operation?

(d) If $f = 5$ kHz, what is the useful input voltage range?

Section 2.10: Common-Mode Rejection

2.67 A differential amplifier for which the input signals are

$$v_1 = 10.00 \sin(2\pi\, 60t) + 0.01 \sin(2\pi\, 1000t)$$

and

$$v_2 = 10.00 \sin(2\pi\, 60t) - 0.01 \sin(2\pi\, 1000t)$$

has an output

$$v_O = 0.1 \sin(2\pi\, 60t) + 5 \sin(2\pi\, 1000t)$$

For this situation, calculate the common-mode gain, the difference-mode (or differential) gain, and the CMRR both as a ratio and in dB.

***2.68** In somewhat more complex situations than prevail in Problem 2.67, the major (common) interfering signals may not be totally balanced at the two inputs. Such is the case in which

$$v_1 = 10.00 \sin(2\pi\, 60t) + 0.04 \sin(2\pi\, 1000t)$$

$$v_2 = 10.01 \sin(2\pi\, 60t) - 0.04 \sin(2\pi\, 1000t)$$

and

$$v_O = \sin(2\pi\, 60t) + 4 \sin(2\pi\, 1000t)$$

Calculate the difference-mode gain, the common-mode gain, and the CMRR.

2.69 For the difference amplifier analyzed in Exercise 2.27 find the common-mode gain and the closed-loop CMRR for the case $R_2/R_1 = 1000$ and the op amp CMRR is 80 dB.

2.70 An op amp with a large value of A_0 but with a CMRR of only 40 dB is used in a noninverting configuration with a closed-loop gain of 2. What output would you expect for an input sine-wave signal of 10 volts peak-to-peak?

Section 2.11: Input and Output Resistances

2.71 A particular op amp, for which $A_0 = 10^4$, $R_{icm} = 10$ MΩ, and $R_{id} = 10$ kΩ, is connected in the noninverting configuration with a closed-loop gain of 10 (ideally). What is the input resistance presented to the driving signal source at low frequencies?

2.72 An op amp for which $R_{icm} = 50$ MΩ, $R_{id} = 10$ kΩ, $A_0 = 10^4$, and $f_t = 10^6$ Hz is used to design a noninverting amplifier with a nominal closed-loop gain of 10. Find the component values of the input impedance, and calculate $|Z_{in}|$ at $f = 10$ kHz.

***2.73** Show that the input impedance of the op-amp circuit in Fig. P2.73a is as shown in Fig. P2.73b. (Use the op-amp equivalent circuit of Fig. 2.39.) For a particular op amp having $A_0 = 10^5$, $R_{id} = 1$ MΩ, $R_{icm} = 100$ MΩ, $R_o = 100$ Ω, and $\omega_t = 10^6$ rad/s, and for $R_2 = 100$ kΩ, find the values of the components of the input impedance. Evaluate $|Z_{in}|$ at $\omega = 0$, 10, 10^3, 10^4, and 10^6 rad/s.

***2.74** Consider the noninverting op-amp configuration with an external voltage v_i.

(a) Find v_o in terms of v_i, the open-loop gain A, and the feedback ratio $\beta \equiv R_1/(R_1 + R_2)$. Neglect the effects of the op amp input and output resistances.

(b) Draw an equivalent circuit for the noninverting configuration by replacing the op amp with the equivalent circuit in Fig. 2.39 with R_o set to zero. Now, use the v_o obtained in (a) to find v_{id}.

(c) Use v_{id} obtained in (b) to help find the input current drawn from v_i, and thus find the closed-loop input resistance R_{in}. Compare your result to Eq. (2.42).

2.75 An inverting amplifier for which $R_1 = 10$ kΩ and $R_2 = 100$ kΩ is constructed with an op amp whose open-loop output resistance is 1 kΩ, whose dc gain is 10^4, and whose 3-dB frequency is 100 Hz. Find the values of the components of its output impedance, and evaluate $|Z_{out}|$ at $f = 0$, 10^2, 10^4, and 10^6 Hz.

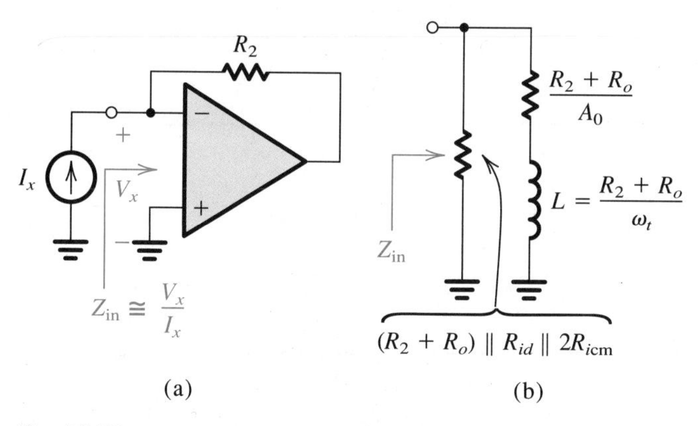

Fig. P2.73

Section 2.12: DC Problems

2.76 An op amp wired in the inverting configuration with the input grounded, having $R_2 = 100$ kΩ and $R_1 = 1$ kΩ, has an output dc voltage of -0.5 V. If the input bias current is known to be very small, find the input offset voltage.

2.77 A noninverting amplifier with a gain of 100 uses an op amp having an input offset voltage of ±2 mV. Find the output when the input is 0.01 sin ωt, volts.

2.78 A noninverting amplifier with a closed-loop gain of 1000 is designed using an op amp having an input offset voltage of 4 mV and output saturation levels of ±12 V. What is the maximum amplitude of the sine wave that can be applied at the input without the output clipping? If the amplifier is capacitively coupled in the manner indicated in Fig. 2.51, what would the maximum possible amplitude be?

2.79 Consider the differential amplifier circuit in Fig. 2.19. Let $R_1 = R_3 = 10$ kΩ and $R_2 = R_4 = 1$ MΩ. If the op amp has $V_{OS} = 3$ mV, $I_B = 0.2$ μA, and $I_{OS} = 50$ nA, find the worst-case (largest) dc offset voltage at the output.

2.80 The circuit shown in Fig. P2.80 uses an op amp having a ±5-mV offset. What is its output offset voltage? What does the output offset become with the input ac coupled through a capacitor C? If, instead, the 1-kΩ resistor is capacitively coupled to ground, what does the output offset become?

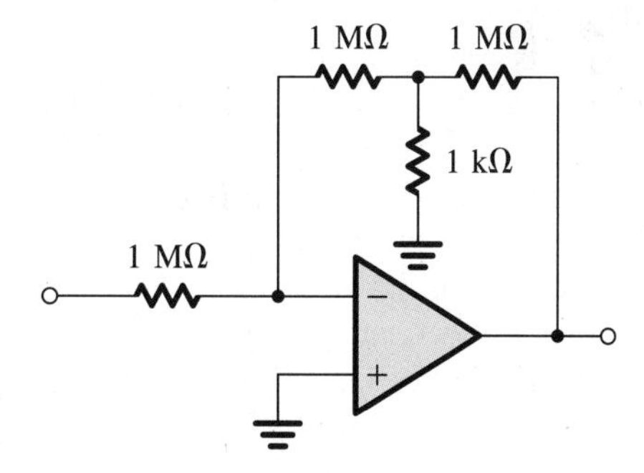

Fig. P.2.80

2.81 Using offset-nulling facilities provided for the op amp, a closed-loop amplifier with gain of +1000 is adjusted at 25°C to produce zero output with the input grounded. If the input offset-voltage drift of the op amp is specified to be 10 μV/°C, what output would you expect at 0°C and at 75°C? While nothing can be said separately about the polarity of the output offset at either 0 or 75°C, what would you expect their relative polarities to be?

2.82 An op amp is connected in a closed loop with gain of +100 utilizing a feedback resistor of 1 MΩ.

(a) If the input bias current is 100 nA, what output voltage results with the input grounded?

(b) If the input offset voltage is ±1 mV, and the input bias current as in (a), what is the largest possible output that can be observed with the input grounded?

(c) If bias-current compensation is used, what is the value of the required resistor? If the offset current is no more than one tenth the bias current, what is the resulting output offset voltage (due to offset current alone)?

(d) With bias-current compensation, as in (c), in place what is the largest dc voltage at the output due to the combined effect of offset voltage and offset current?

CHAPTER 3

Diodes

INTRODUCTION

In the previous chapters we dealt almost entirely with linear circuits; any nonlinearity, such as that introduced by amplifier output saturation, was considered a problem to be solved by the circuit designer. However, there are many other signal-processing functions that can be implemented only by nonlinear circuits. Examples include the generation of dc voltages from the ac power supply and the generation of signals of various waveforms (e.g., sinusoids, square waves, pulses, etc). Also, digital logic and memory circuits constitute a special class of nonlinear circuits.

The simplest and most fundamental nonlinear circuit element is the diode. Just like a resistor, the diode has two terminals; but unlike the resistor, which has a linear (straight-line) relationship between the current flowing through it and the voltage appearing across it, the diode has a nonlinear i–v characteristic.

This chapter is concerned with the study of diodes. In order to understand the essence of diode function, we begin with a fictitious element, the ideal diode. We then introduce the silicon junction diode, explain its terminal characteristics, and provide techniques for the analysis of diode circuits. The latter task involves the important subject of device modeling.

Of the many applications of diodes, their use in the design of rectifiers (which convert ac to dc) is the most common. Therefore, we shall study rectifier circuits in some detail and briefly look at a number of other diode applications. Further nonlinear circuits that utilize diodes and other devices will be found throughout the text and in particular in Chapter 12.

The chapter concludes with an introduction to the physical operation of the *pn junction*. In addition to being a diode, the *pn* junction is the basis of many other solid-state devices, including the bipolar junction transistor studied in the next chapter. Thus an understanding of the *pn* junction is essential to the study of modern electronics. Though simplistic, the qualitative description presented here provides sufficient background for the use of devices (diodes and transistors) in electronic circuit design. A more detailed study of the *pn* junction can be found in texts dealing with device physics.

3.1 THE IDEAL DIODE

The ideal diode may be considered the most fundamental nonlinear element. It is a two-terminal device having the circuit symbol of Fig. 3.1(a) and the i–v characteristic shown in Fig. 3.1(b). The terminal characteristic of the ideal diode can be interpreted as follows: If a negative voltage (relative to the reference direction indicated in Fig. 3.1(a) is applied to the diode, no current flows and the diode behaves as an open circuit (Fig. 3.1(c)). Diodes operated in this mode are said to be **reverse-biased,** or operated in the reverse

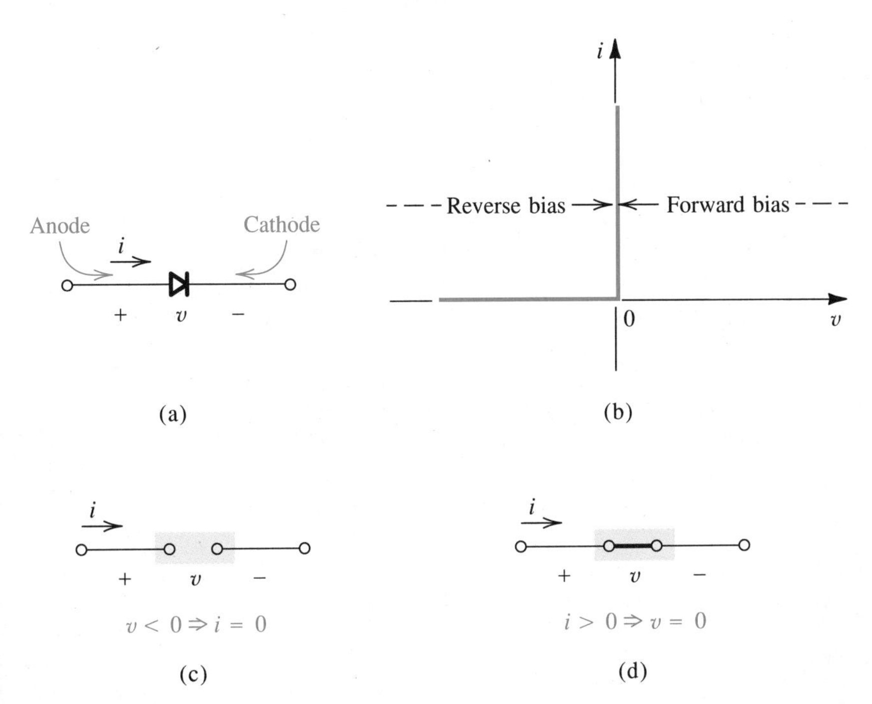

Fig. 3.1 The ideal diode: **(a)** diode circuit symbol; **(b)** i–v characteristic; **(c)** equivalent circuit in the reverse direction; **(d)** equivalent circuit in the forward direction.

direction. An ideal diode has zero current when operated in the reverse direction and is said to be **cut off.**

On the other hand, if a positive current (relative to the reference direction indicated in Fig. 3.1(a) is applied to the ideal diode, zero voltage drop appears across the diode. In other words, the ideal diode behaves as a short circuit in the *forward* direction (Fig. 3.1(d); it passes any current with zero voltage drop. A forward-conducting diode is said to be **turned on,** or simply, **on.**

From the above description it should be noted that the external circuit must be designed so as to limit the forward current through a conducting diode, and the reverse voltage across a cutoff diode, to predetermined values. Figure 3.2 shows two diode circuits that illustrate this point. In the circuit of Fig. 3.2(a) the diode is obviously conducting. Thus its voltage drop will be zero, and the current through it will be determined by the +10-V supply and the 1-kΩ resistor as 10 mA. The diode in the circuit of Fig. 3.2(b) is obviously cut off, and thus its current will be zero, which in turn means that the entire 10-V supply will appear as reverse bias across the diode.

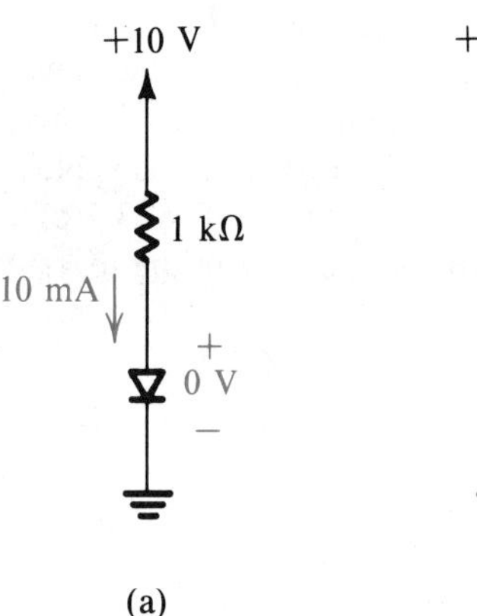

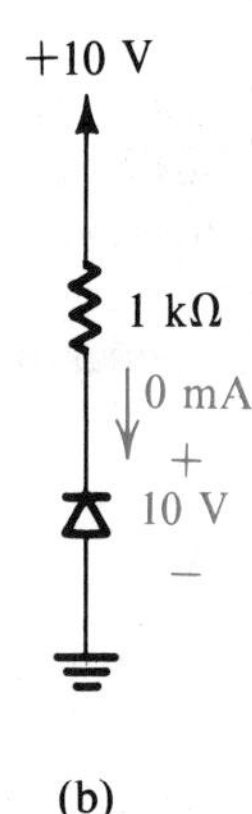

Fig. 3.2 The two modes of operation of ideal diodes and the use of an external circuit to limit the forward current and the reverse voltage.

The positive terminal of the diode is called the **anode** and the negative terminal the **cathode**, a carryover from the days of vacuum-tube diodes. The i–v characteristic of the ideal diode (conducting in one direction and not in the other) should explain the choice of its arrowlike circuit symbol.

As should be evident from the above, the i–v characteristic of the ideal diode is highly nonlinear; it consists of two straight-line segments at 90° to one another. A nonlinear curve that consists of straight-line segments is said to be **piecewise linear.** If a device having a piecewise-linear characteristic is used in a particular application in such a way that the signal across its terminals swings only along one of the linear segments, then the device can be considered a linear circuit element as far as that particular circuit application is concerned. On the other hand, if signals swing past one or more of the break points in the characteristic, linear analysis is no longer possible.

A Simple Application: The Rectifier

A fundamental application of the diode, one that makes use of its severely nonlinear i–v curve, is the rectifier circuit shown in Fig. 3.3(a). The circuit consists of the series

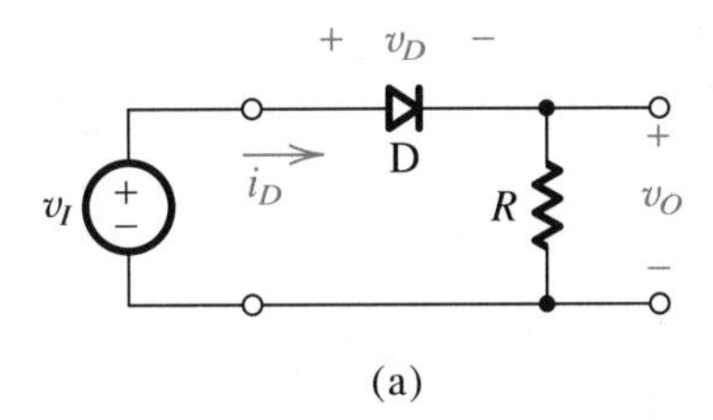

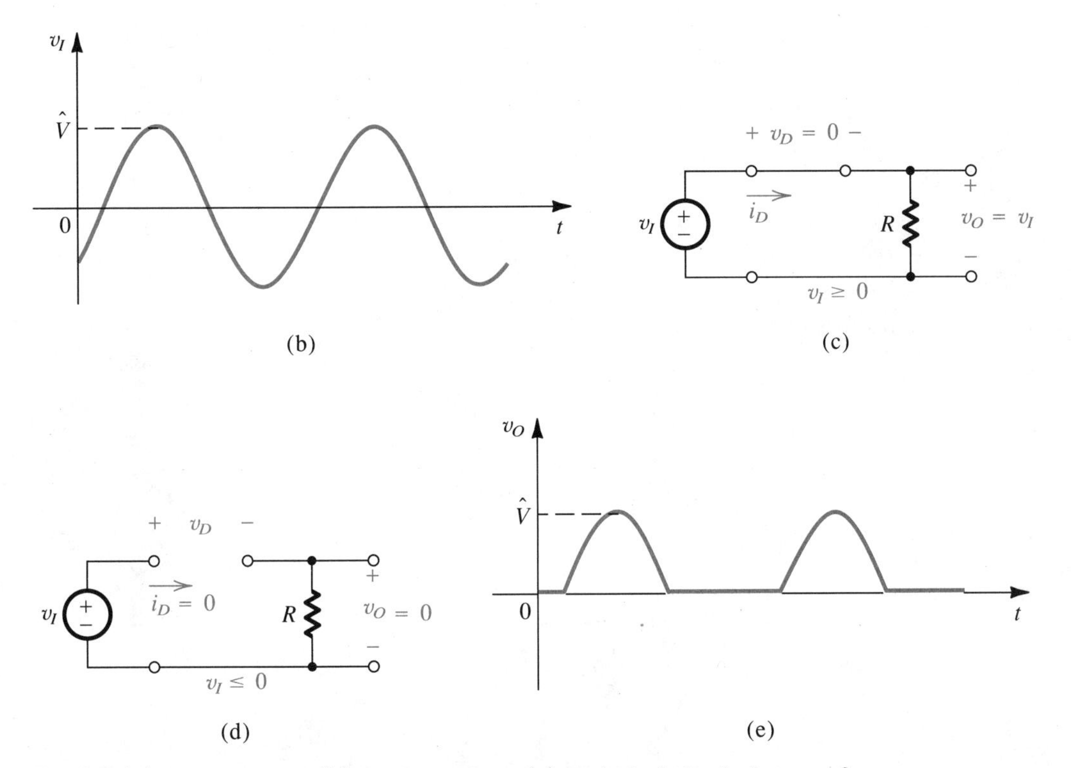

Fig. 3.3 (a) Rectifier circuit. **(b)** Input waveform. **(c)** Equivalent circuit when $v_I \geq 0$. **(d)** Equivalent circuit when $v_I \leq 0$. **(e)** Output waveform.

connection of a diode D and a resistor R. Let the input voltage v_I be the sinusoid shown in Fig. 3.3(b), and assume the diode to be ideal. During the positive half-cycles of the input sinusoid, the positive v_I will cause current to flow through the diode in its forward direction. It follows that the diode voltage v_D will be very small—ideally zero. Thus the circuit will have the equivalent shown in Fig. 3.3(c), and the output voltage v_O will be equal to the input voltage v_I. On the other hand, during the negative half-cycles of v_I, the diode will not conduct. Thus the circuit will have the equivalent shown in Fig. 3.3(d), and v_O will be zero. Thus, the output voltage will have the waveform shown in Fig. 3.3(e). Note that while v_I alternates in polarity and has a zero average value, v_O is unidirectional and has a finite average value or a dc component. Thus the circuit of Fig. 3.3(a) **rectifies** the signal and hence is called a **rectifier.** It can be used to generate dc from ac. We will study rectifier circuits in detail in Section 3.6.

Exercises 3.1 For the circuit in Fig. 3.3(a) sketch the transfer characteristic v_O versus v_I.

Ans. See Fig. E3.1.

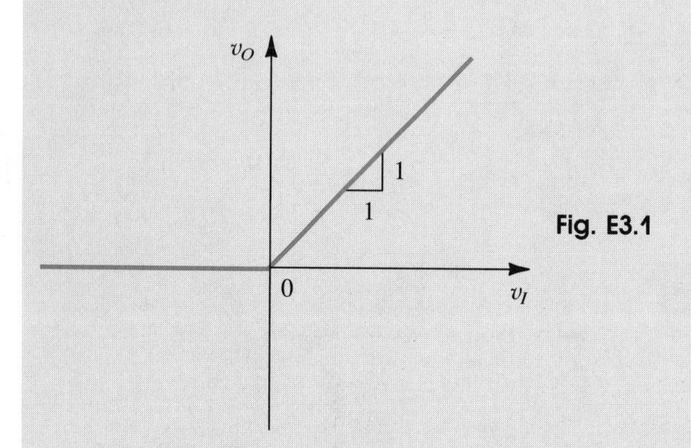

Fig. E3.1

3.2 For the circuit in Fig. 3.3(a) sketch the waveform of v_D.

Ans. See Fig. E3.2.

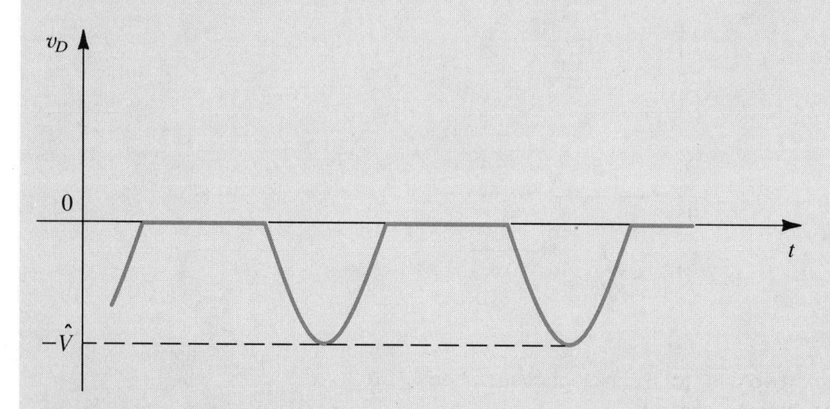

Fig. E3.2

3.3 In the circuit of Fig. 3.3(a) let v_I have a peak value of 10 V and let $R = 1$ kΩ. Find the peak value of i_D and the dc component of v_O.

Ans. 10 mA; 3.18 V

EXAMPLE 3.1

Figure 3.4(a) shows a circuit for charging a 12-V battery. If v_S is a sinusoid with 24-V peak amplitude, find the fraction of each cycle during which the diode conducts. Also find the peak value of the diode current and the maximum reverse-bias voltage that appears across the diode.

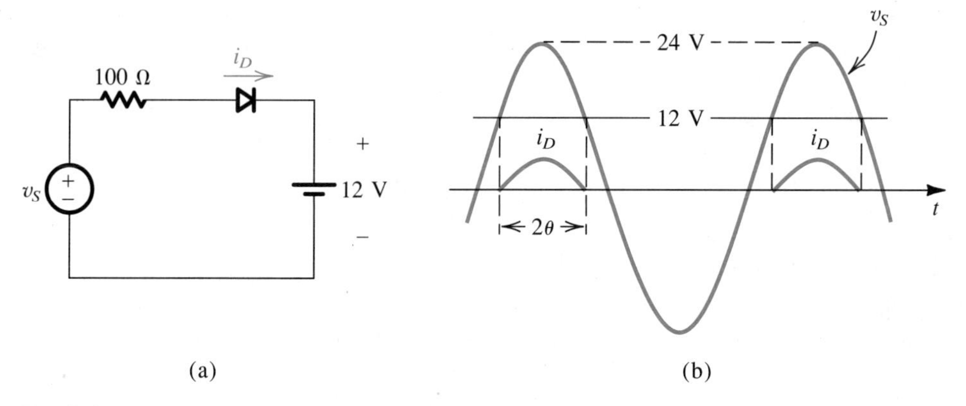

Fig. 3.4 Circuit and waveforms for Example 3.1.

Solution The diode conducts when v_S exceeds 12 V, as shown in Fig. 3.4(b). The conduction angle is 2θ, where θ is given by

$$24 \cos \theta = 12$$

Thus $\theta = 60°$ and the conduction angle is 120° or one-third of a cycle.

The peak value of the diode current is given by

$$I_d = \frac{24 - 12}{100} = 0.12 \text{ A}$$

The maximum reverse voltage across the diode occurs when v_S is at its negative peak and is equal to $24 + 12 = 36$ V.

Another Application: Diode Logic Gates

Diodes together with resistors can be used to implement digital logic functions. Figure 3.5 shows two diode logic gates. To see how these circuits function consider a positive logic

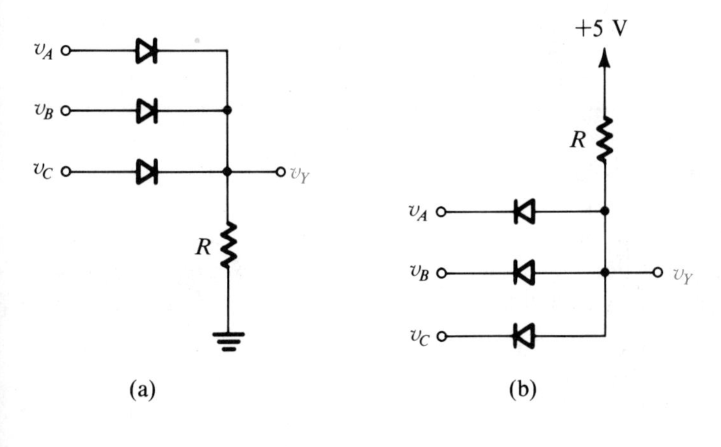

Fig. 3.5 Diode logic gates: **(a)** OR gate; **(b)** AND gate (in a positive logic system).

system in which voltage values close to 0 V correspond to logic 0 (or low) and voltage values close to +5 V correspond to logic 1 (or high). The circuit in Fig. 3.5(a) has three inputs, v_A, v_B, and v_C. It is easy to see that diodes connected to +5-V inputs will conduct, thus clamping the output v_Y to a value equal to +5 V. This positive voltage at the output will keep the diodes whose inputs are low (around 0 V) cut off. Thus the output will be high if one or more of the inputs are high. The circuit therefore implements the logic OR function, which in Boolean notation is expressed as

$$Y = A + B + C$$

Similarly, the reader is encouraged to show that using the same logic system mentioned above, the circuit of Fig. 3.5(b) implements the logic AND function,

$$Y = A \cdot B \cdot C$$

EXAMPLE 3.2 Assuming the diodes to be ideal, find the values of I and V in the circuits of Fig. 3.6.

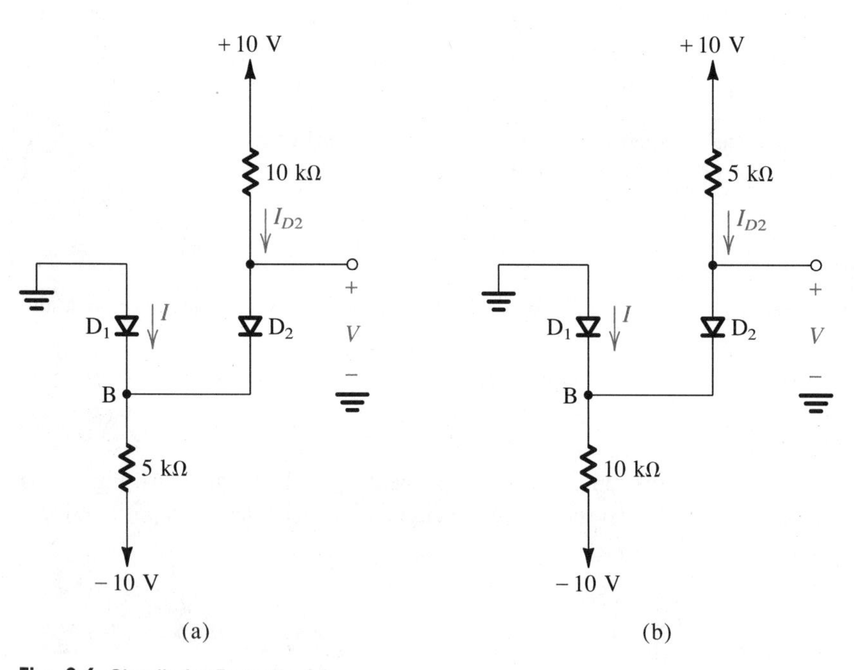

Fig. 3.6 Circuits for Example 3.2.

Solution In these circuits it might not be obvious at first sight whether none, one, or both diodes are conducting. In such a case we make a plausible assumption, proceed with the analysis, and then check whether we end up with a consistent solution. For the circuit in Fig. 3.6(a), we shall assume that both diodes are conducting. It follows that $V_B = 0$ and $V = 0$. The current through D_2 can now be determined from

$$I_{D2} = \frac{10 - 0}{10} = 1 \text{ mA}$$

Writing a node equation at B,

$$I + 1 = \frac{0 - (-10)}{5}$$

results in $I = 1$ mA. Thus D_1 is conducting as originally assumed, and the final result is $I = 1$ mA and $V = 0$ V.

For the circuit in Fig. 3.6(b), if we assume that both diodes are conducting, then $V_B = 0$ and $V = 0$. The current in D_2 is obtained from

$$I_{D2} = \frac{10 - 0}{5} = 2 \text{ mA}$$

The node equation at B is

$$I + 2 = \frac{0 - (-10)}{10}$$

which yields $I = -1$ mA. Since this is not possible, our original assumption is not correct. We start again, assuming that D_1 is off and D_2 is on. The current I_{D2} is given by

$$I_{D2} = \frac{10 - (-10)}{15} = 1.33 \text{ mA}$$

and the voltage at node B is

$$V_B = -10 + 10 \times 1.33 = +3.3 \text{ V}$$

Thus D_1 is reverse-biased as assumed, and the final result is $I = 0$ and $V = 3.3$ V.

Exercises 3.4 Find the values of I and V in the circuits shown in Fig. E3.4.

Ans. (a) 2 mA, 0 V; (b) 0 mA, 5 V; (c) 0 mA, 5 V; (d) 2 mA, 0 V; (e) 3 mA, +3 V; (f) 4 mA, +1 V.

3.5 Figure E3.5 shows a circuit for an ac voltmeter. It utilizes a moving-coil meter that gives a full-scale reading when the *average* current flowing through it is 1 mA. The moving-coil meter has a 50-Ω resistance. Find the value of R that results in the meter indicating a full-scale reading when the input sine-wave voltage v_I is 20 V peak-to-peak. (*Hint:* The average value of half-sine waves is V_p/π.)

Ans. 3.133 kΩ

3.2 TERMINAL CHARACTERISTICS OF JUNCTION DIODES

In this section we study the characteristics of real diodes—specifically, semiconductor junction diodes made of silicon. The physical processes that give rise to the diode terminal characteristics, and to the name "junction diode," will be studied in the latter part of this chapter.

Figure 3.7 shows the i–v characteristic of a silicon junction diode. The same characteristic is shown in Fig. 3.8 with some scales expanded and others compressed, so as to

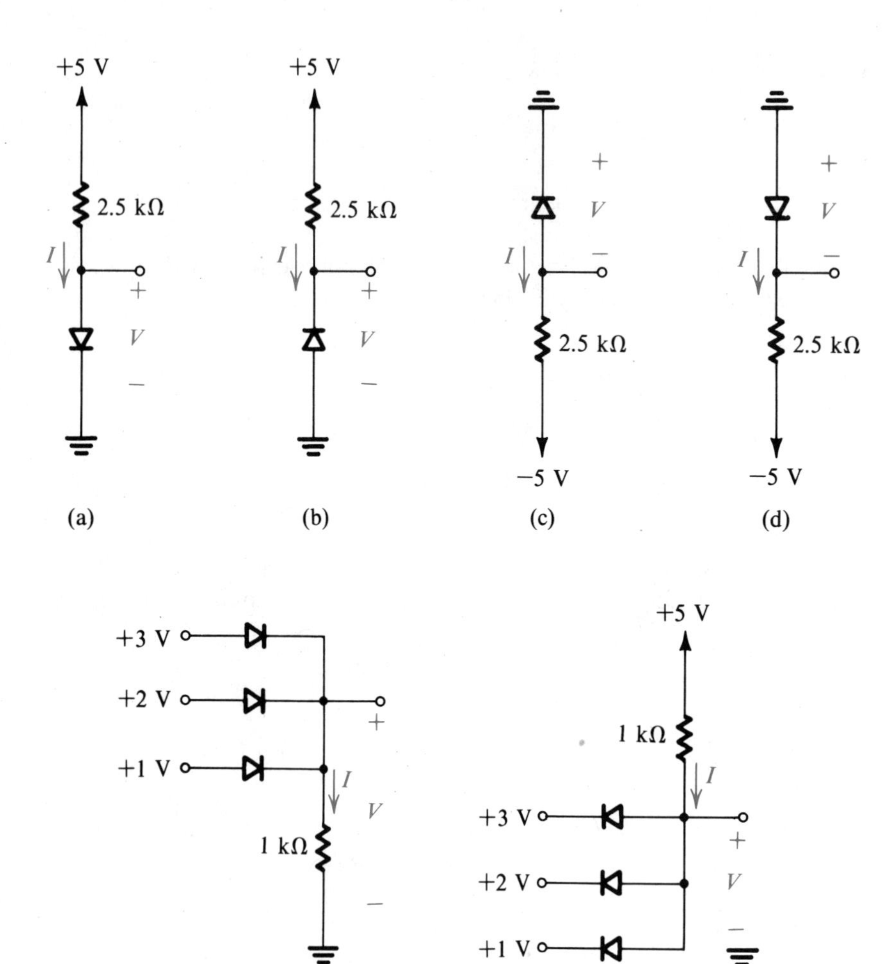

Fig. E3.4

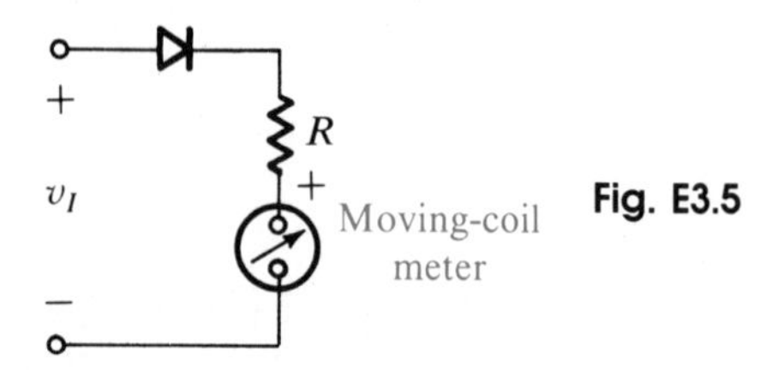

Fig. E3.5

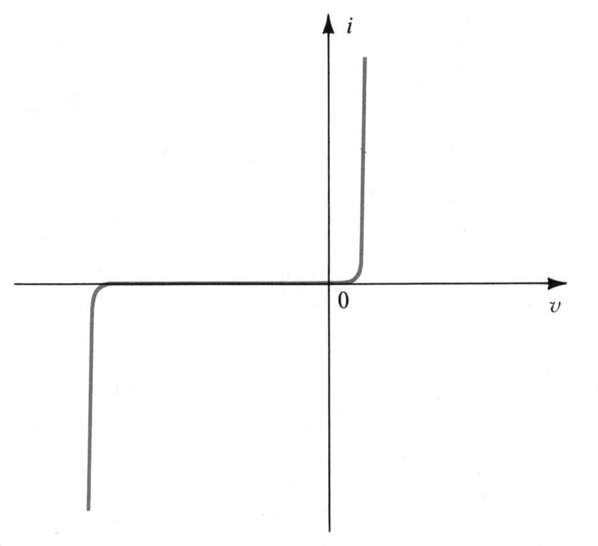

Fig. 3.7 The i–v characteristic of a silicon junction diode.

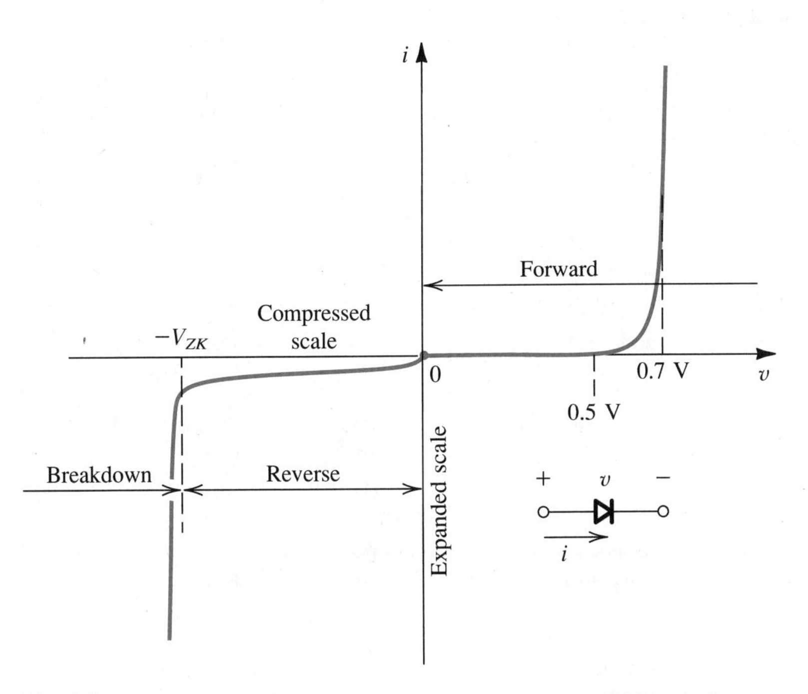

Fig. 3.8 The diode i–v relationship with some scales expanded and others compressed in order to reveal details.

portray details. Note that the scale changes have resulted in the apparent discontinuity at the origin.

As indicated, the characteristic curve consists of three distinct regions:

1. The forward-bias region, determined by $v > 0$
2. The reverse-bias region, determined by $v < 0$
3. The breakdown region, determined by $v < -V_{ZK}$

These three regions of operation are described in the following.

The Forward-Bias Region

The forward-bias—or, simply, forward—region of operation is entered when the terminal voltage v is positive. In the forward region the i–v relationship is closely approximated by

$$i = I_S(e^{v/nV_T} - 1) \tag{3.1}$$

In this equation I_S is a constant for a given diode at a given temperature. The current I_S is usually called the **saturation current** (for reasons that will become apparent shortly). A better name, one that we will use for I_S, is the **scale current.** This name arises from the fact that I_S is directly proportional to the cross-sectional area of the diode. Thus doubling of the junction area results in a diode with double the value of I_S and, as the diode equation indicates, double the value of current i for a given forward voltage v. For "small-signal" diodes, which are small-size diodes intended for low-power applications, I_S is of the order of 10^{-15} A. The value of I_S is, however, a very strong function of temperature. As a rule of thumb, I_S doubles in value for every 5°C rise in temperature.[1]

The voltage V_T in Eq. (3.1) is a constant called the **thermal voltage,** given by

$$V_T = \frac{kT}{q} \tag{3.2}$$

where k = Boltzmann's constant $= 1.38 \times 10^{-23}$ joules/kelvin
T = the absolute temperature in kelvins = 273 + temperature in °C
q = the magnitude of electronic charge = 1.602×10^{-19} coulomb

At room temperature (20°C) the value of V_T is 25.2 mV. In rapid, approximate circuit analysis we shall use $V_T \simeq 25$ mV at room temperature.[2]

In the diode equation the constant n has a value between 1 and 2, depending on the material and the physical structure of the diode. Diodes made using the standard integrated-circuit fabrication process exhibit $n = 1$ when operated under normal conditions.[3] Diodes available as discrete two-terminal components generally exhibit $n = 2$.

[1] An excellent discussion of the temperature dependence of diode characteristics is given in Hodges and Jackson (1988), pp. 146–148. Also given is a derivation for the temperature coefficient of I_S.

[2] A slightly higher ambient temperature (25°C or so) is usually assumed for electronic equipment operating inside a cabinet. At this temperature, $V_T \simeq 25.8$ mV. Nevertheless, for the sake of simplicity and to promote rapid circuit analysis, we shall use $V_T \simeq 25$ mV throughout this book.

[3] On an integrated circuit, diodes are usually obtained by connecting a bipolar junction transistor (BJT) as a two-terminal device, as will be seen in Chapter 6.

For appreciable current i in the forward direction, specifically for $i \gg I_S$, Eq. (3.1) can be approximated by the exponential relationship

$$i \simeq I_S e^{v/nV_T} \tag{3.3}$$

This relationship can be expressed alternatively in the logarithmic form

$$v = nV_T \ln \frac{i}{I_S} \tag{3.4}$$

where ln denotes the natural (base e) logarithm.

The exponential relationship of the current i to the voltage v holds over many decades of current (a span of as many as seven decades—that is, a factor of 10^7—can be found). This is quite a remarkable property of junction diodes, one that is also found in bipolar junction transistors and one that has been exploited in many interesting applications.

Let us consider the forward i–v relationship in Eq. (3.3) and evaluate the current I_1 corresponding to a diode voltage V_1:

$$I_1 = I_S e^{V_1/nV_T}$$

Similarly, if the voltage is V_2, the diode current I_2 will be

$$I_2 = I_S e^{V_2/nV_T}$$

These two equations can be combined to produce

$$\frac{I_2}{I_1} = e^{(V_2 - V_1)/nV_T}$$

which can be rewritten

$$V_2 - V_1 = nV_T \ln \frac{I_2}{I_1}$$

or, in terms of base-10 logarithms,

$$V_2 - V_1 = 2.3nV_T \log \frac{I_2}{I_1} \tag{3.5}$$

This equation simply states that for a decade (factor of 10) change in current the diode voltage drop changes by $2.3nV_T$, which is approximately 60 mV for $n = 1$ and 120 mV for $n = 2$. This also suggests that the diode i–v relationship is most conveniently plotted on a semilog paper. Using the vertical, linear axis for v and the horizontal, log axis for i, one obtains a straight line with a slope of $2.3nV_T$ per decade of current. Finally, it should be mentioned that not knowing the exact value of n (which can be obtained from a simple experiment), circuit designers use the convenient approximate number of 0.1 V/decade for the slope of the diode logarithmic characteristic.

A glance at the i–v characteristic in the forward region (Fig. 3.8) reveals that the current is negligibly small for v smaller than about 0.5 V. This value is usually referred to as the **cut-in voltage.** It should be emphasized, however, that this apparent threshold in the characteristic is simply a consequence of the exponential relationship. Another consequence of this relationship is the rapid increase of i. Thus for a "fully conducting" diode the voltage drop lies in a narrow range, approximately 0.6 to 0.8 V. This gives rise to a simple ''model'' for the diode where it is assumed that a conducting diode has approximately a 0.7-V drop across it. Diodes with different current ratings (that is, different areas and correspondingly different I_S) will exhibit the 0.7-V drop at different currents. For instance, a small-signal diode may be considered to have a 0.7-V drop at $i = 1$ mA, while a higher-power diode may have a 0.7-V drop at $i = 1$ A. We will return to the topics of diode-circuit analysis and diode models in Section 3.3.

EXAMPLE 3.3

A silicon diode said to be a 1-mA device displays a forward voltage of 0.7 V at a current of 1 mA. Evaluate the junction scaling constant I_S in the event that n is either 1 or 2. What scaling constants would apply for a 1-A diode of the same manufacture that conducts 1 A at 0.7 V?

Solution Since

$$i = I_S e^{v/nV_T}$$

then

$$I_S = ie^{-v/nV_T}$$

For the 1-mA diode:

$$\text{If } n = 1: \quad I_S = 10^{-3}e^{-700/25} = 6.9 \times 10^{-16} \text{ A}, \quad \text{or about } 10^{-15} \text{ A}$$

$$\text{If } n = 2: \quad I_S = 10^{-3}e^{-700/50} = 8.3 \times 10^{-10} \text{ A}, \quad \text{or about } 10^{-9} \text{ A}$$

The diode conducting 1 A at 0.7 V corresponds to 1000 1-mA diodes in parallel with a total junction area 1000 times greater. Thus I_S is also 1000 times greater, being 1 pA and 1 μA, respectively for $n = 1$ and $n = 2$.

From this example it should be apparent that the value of n used can be quite important.

Since both I_S and V_T are functions of temperature, the forward i–v characteristic varies with temperature as illustrated in Fig. 3.9. At a given constant diode current the voltage

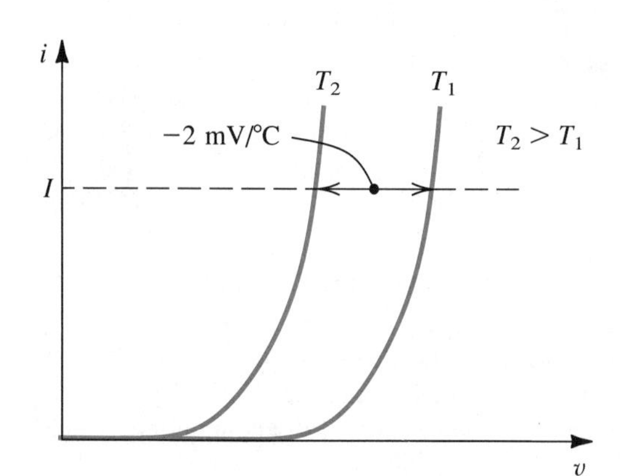

Fig. 3.9 Illustrating the temperature dependence of the diode forward characteristics. At a constant current, the voltage drop decreases by approximately 2 mV for every 1°C increase in temperature.

drop across the diode decreases by approximately 2 mV for every 1°C increase in temperature. The change in diode voltage with temperature has been exploited in the design of electronic thermometers.

Exercises 3.6 Consider a silicon diode with $n = 1.5$. Find the change in voltage if the current changes from 0.1 mA to 10 mA.

Ans. 172.5 mV

3.7 A silicon junction diode with $n = 1$ has $v = 0.7$ V at $i = 1$ mA. Find the voltage drop at $i = 0.1$ mA and $i = 10$ mA.

Ans. 0.64 V; 0.76 V

3.8 Using the fact that a silicon diode has $I_S = 10^{-14}$ A at 25°C and that I_S increases by 15% per °C rise in temperature, find the value of I_S at 125°C.

Ans. 1.17×10^{-8} A

The Reverse-Bias Region

The reverse-bias region of operation is entered when the diode voltage v is made negative. Equation (3.1) predicts that if v is negative and a few times larger than V_T (25 mV) in magnitude, the exponential term becomes negligibly small compared to unity and the diode current becomes

$$i \simeq -I_S$$

that is, the current in the reverse direction is constant and equal to I_S. This is the reason behind the term *saturation current.*

Real diodes exhibit reverse currents that, though quite small, are much larger than I_S. For instance, a small-signal or a 1-mA diode whose I_S is of the order of 10^{-14} to 10^{-15} A could show a reverse current of the order of 1 nA. The reverse current also increases somewhat with the increase in magnitude of the reverse voltage. Note that because of the very small magnitude of the current these details are not clearly evident on the diode i–v characteristic of Fig. 3.8.

A good part of the reverse current is due to leakage effects. These leakage currents are proportional to the junction area, just as I_S is. Finally, it should be mentioned that the reverse current is a strong function of temperature, with the rule of thumb being that it doubles for every 10°C rise in temperature.

Exercise 3.9 The diode in the circuit of Fig. E3.9 is a large, high-current device whose reverse leakage is reasonably independent of voltage. If $V = 1$ V at 20°C, find the value of V at 40°C and at 0°C.

Ans. 4 V; 0.25 V

The Breakdown Region

The third distinct region of diode operation is the breakdown region, which can be easily identified on the diode i–v characteristic in Fig. 3.8. The breakdown region is entered when the magnitude of the reverse voltage exceeds a threshold value specific to the particular diode and called the **breakdown voltage.** This is the voltage at the "knee" of

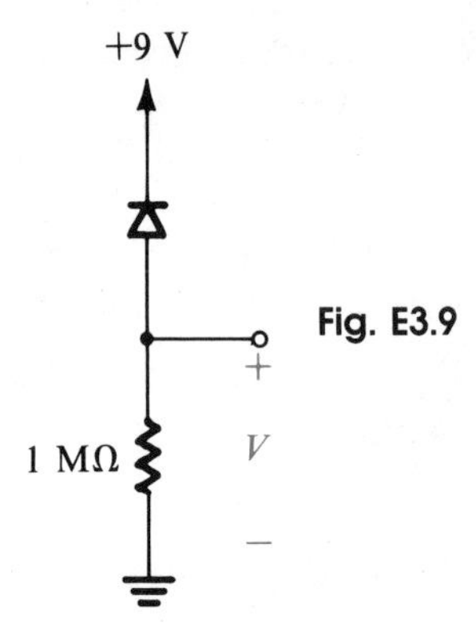

Fig. E3.9

the i–v curve in Fig. 3.8 and is denoted V_{ZK}, where the subscript Z stands for zener (to be explained shortly) and K denotes knee.

As can be seen from Fig. 3.8, in the breakdown region the reverse current increases rapidly, with the associated increase in voltage drop being very small. Diode breakdown is normally not destructive provided that the power dissipated in the diode is limited by external circuitry to a ''safe'' level. This safe value is normally specified on the device data sheets. It therefore is necessary to limit the reverse current in the breakdown region to a value consistent with the permissible power dissipation.

The fact that the diode i–v characteristic in breakdown is almost a vertical line enables it to be used in voltage regulation. This subject will be studied in Section 3.5.

3.3 ANALYSIS OF DIODE CIRCUITS

In this section we shall study methods for the analysis of diode circuits. We shall concentrate on circuits in which the diodes are operating in the forward-bias region. Operation in the other region of interest, the breakdown region, is considered in Section 3.5.

Consider the circuit shown in Fig. 3.10 consisting of a dc source V_{DD}, a resistor R, and a diode. We wish to analyze this circuit to determine the diode current I_D and voltage V_D.

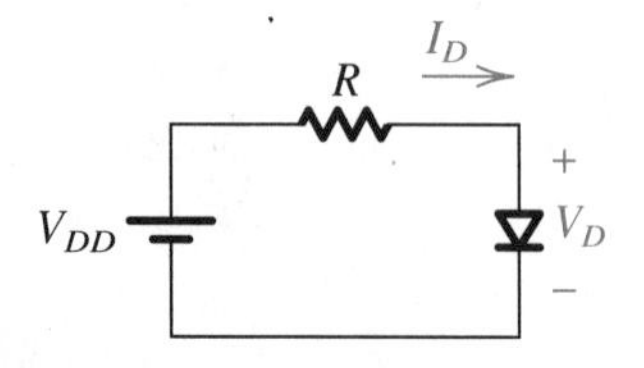

Fig. 3.10 A simple diode circuit.

The diode is obviously biased in the forward direction. Assuming that V_{DD} is greater than 0.5 V or so, the diode current will be much greater than I_S and we can represent the diode i–v characteristic by the exponential relationship, resulting in

$$I_D = I_S e^{V_D/nV_T} \tag{3.6}$$

The other equation that governs circuit operation is obtained by writing a Kirchhoff loop equation, resulting in

$$I_D = \frac{V_{DD} - V_D}{R} \tag{3.7}$$

Assuming that the diode parameters I_S and n are known, Eqs. (3.6) and (3.7) are two equations in the two unknown quantities I_D and V_D. Two alternative ways for obtaining the solution are graphical analysis and iterative analysis.

Graphical Analysis

Graphical analysis is performed by plotting the relationships of Eqs. (3.6) and (3.7) on the i–v plane. The solution can then be obtained as the coordinates of the point of intersection of the two graphs. A sketch of the graphical construction is shown in Fig. 3.11; the curve represents the exponential diode equation (Eq. 3.6) and the straight line represents Eq. (3.7). Such a straight line is known as the **load line,** a name that will become more meaningful in later chapters. The load line intersects the diode curve at point Q, which represents the **operating point** of the circuit. Its coordinates give the values of I_D and V_D.

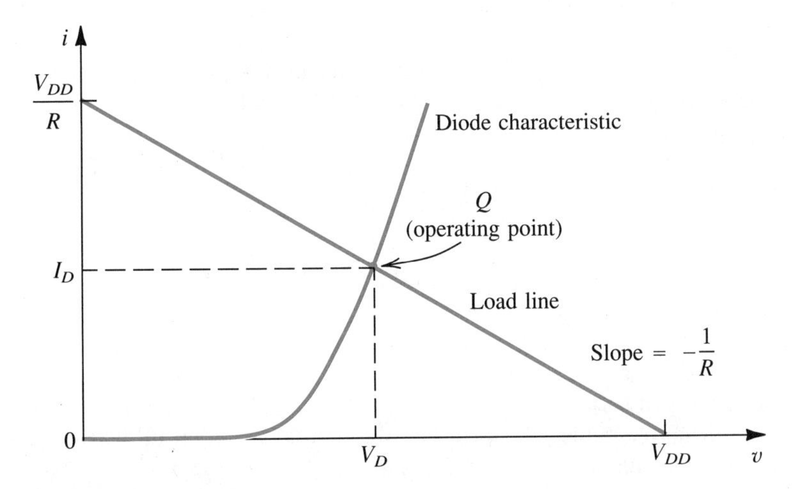

Fig. 3.11 Graphical analysis of the circuit in Fig. 3.10.

Graphical analysis aids in the visualization of circuit operation. However, the effort involved in performing such an analysis, particularly for complex circuits, is too great to be justified in practice.

Iterative Analysis

Equations (3.6) and (3.7) can be solved using a simple iterative procedure, as is illustrated in the following example.

EXAMPLE 3.4 Determine the current I_D and the diode voltage V_D for the circuit in Fig. 3.10 with $V_{DD} = 5$ V and $R = 1$ kΩ. Assume that the diode has a current of 1 mA at a voltage of 0.7 V, and that its voltage drop changes by 0.1 V for every decade change in current.

Solution To begin the iteration we assume that $V_D = 0.7$ V and use Eq. (3.7) to determine the current,

$$I_D = \frac{V_{DD} - V_D}{R}$$

$$= \frac{5 - 0.7}{1} = 4.3 \text{ mA}$$

We then use the diode equation to obtain a better estimate for V_D. This can be done by employing Eq. (3.5), namely,

$$V_2 - V_1 = 2.3\, nV_T \log \frac{I_2}{I_1}$$

For our case, $2.3\, nV_T = 0.1$ V; then

$$V_2 = V_1 + 0.1 \log \frac{I_2}{I_1}$$

Substituting $V_1 = 0.7$ V, $I_1 = 1$ mA, and $I_2 = 4.3$ mA results in $V_2 = 0.763$ V. Thus the results of the first iteration are $I_D = 4.3$ mA and $V_D = 0.763$ V. The second iteration proceeds in a similar manner:

$$I_D = \frac{5 - 0.763}{1} = 4.237 \text{ mA}$$

$$V_2 = 0.763 + 0.1 \log \left[\frac{4.237}{4.3}\right]$$

$$= 0.762 \text{ V}$$

Thus, the second iteration yields $I_D = 4.237$ mA and $V_D = 0.762$ V. Since these values are not much different from the values obtained after the first iteration, no further iterations are necessary, and the solution is $I_D = 4.237$ mA and $V_D = 0.762$ V.

The Need for Rapid Analysis

The iterative analysis procedure utilized in the example above is simple and yields accurate results after two or three iterations. Nevertheless, there are situations in which the effort and time required are still greater than can be justified. Specifically, if one is doing a pencil-and-paper design of a relatively complex circuit, rapid circuit analysis is a necessity. Through quick analysis, the designer is able to evaluate various possibilities before deciding on a suitable circuit. To speed up the analysis process one must be content with less precise results. This, however, is seldom a problem, as the more accurate analysis can be postponed until a final or almost final design is obtained. Accurate analysis of the almost final design can be performed with the aid of a computer circuit-analysis program such as SPICE (see Appendix C and, for more detail the SPICE supplement to this book). The results of such an analysis can then be used to further refine or "fine-tune" the design.

Simplified Diode Models

Although the exponential i–v relationship is an accurate model of the diode characteristic in the forward region, its nonlinear nature complicates the analysis of diode circuits. The analysis can be greatly simplified if we can find linear relationships to describe the diode terminal characteristics. An attempt in this direction is illustrated in Fig. 3.12, where the exponential curve is approximated by two straight lines, line A with zero slope and line B with a slope of $1/r_D$. It can be seen that for this particular diode, over the current range of 0.1 mA to 10 mA the voltages predicted by the straight-lines model differ from those predicted by the exponential model by less than 50 mV. Obviously the choice of these two straight lines is not unique; one can obtain a closer approximation by restricting the current range over which the approximation is required.

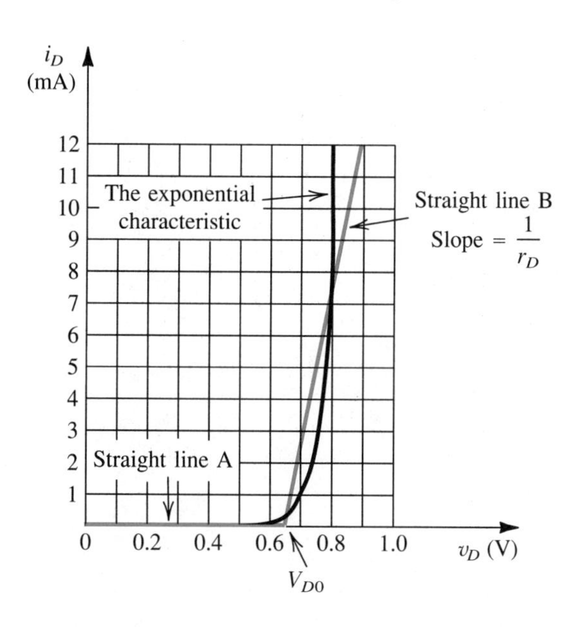

Fig. 3.12 Approximating the diode forward characteristic with two straight lines.

The straight-lines (or piecewise-linear) model of Fig. 3.12 can be described by

$$i_D = 0, \quad v_D \le V_{D0}$$
$$i_D = (v_D - V_{D0})/r_D, \qquad v_D \ge V_{D0} \tag{3.8}$$

where V_{D0} is the intercept of line B on the voltage axis and r_D is the inverse of the slope of line B. For the particular example shown, $V_{D0} = 0.65$ V and $r_D = 20\ \Omega$.

The piecewise-linear model described by Eqs. (3.8) can be represented by the equivalent circuit shown in Fig. 3.13. Note that an ideal diode is included in this model to constrain i_D to flow in the forward direction only. This model is also known as the "battery-plus-resistance" model.

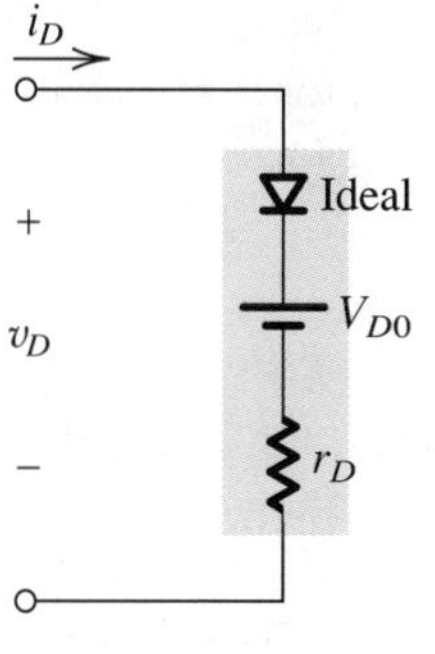

Fig. 3.13 Piecewise-linear model of the diode forward characteristics and its equivalent circuit representation.

EXAMPLE 3.5 Repeat the problem in Example 3.4 utilizing the piecewise-linear model whose parameters are given in Fig. 3.12 ($V_{D0} = 0.65$ V, $r_D = 20\Omega$). Note that the characteristics depicted in this figure are those of the diode described in Example 3.4 (1 mA at 0.7 V and 0.1 V/decade).

Solution Replacing the diode in the circuit of Fig. 3.10 with the equivalent circuit model of Fig. 3.13 results in the circuit in Fig. 3.14, from which we can write for the current I_D,

$$I_D = \frac{V_{DD} - V_{D0}}{R + r_D}$$

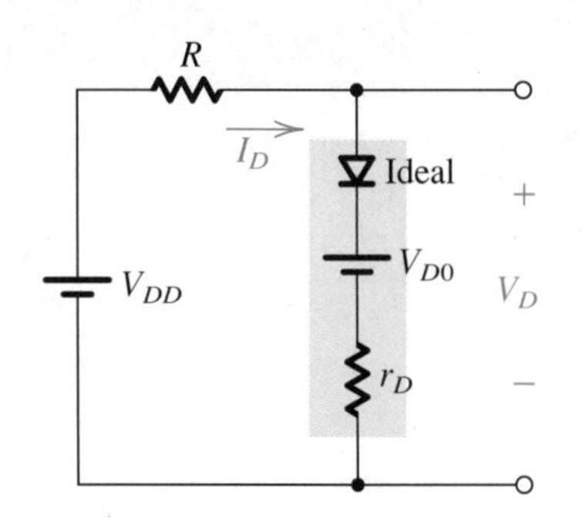

Fig. 3.14 The circuit of Fig. 3.10 with the diode replaced with its piecewise-linear model of Fig. 3.13.

where the model parameters V_{D0} and r_D are seen from Fig. 3.12 to be $V_{D0} = 0.65$ V and $r_D = 20\ \Omega$. Thus

$$I_D = \frac{5 - 0.65}{1 + 0.02} = 4.26 \text{ mA}$$

The diode voltage V_D can now be computed:

$$V_D = V_{D0} + I_D r_D$$
$$= 0.65 + 4.26 \times 0.02 = 0.735 \text{ V}$$

The Constant-Voltage-Drop Model

An even simpler model of the diode forward characteristics can be obtained if we use a vertical straight line to approximate the fast-rising part of the exponential curve, as shown in Fig. 3.15. The resulting model simply says that a forward-conducting diode exhibits a constant voltage drop V_D. The value of V_D is usually taken to be 0.7 V. Note that for the particular diode whose characteristics are depicted in Fig. 3.15, this model predicts the diode voltage to within ± 0.1 V over the current range 0.1 to 10 mA. The constant-voltage-drop model can be represented by the equivalent circuit shown in Fig. 3.16.

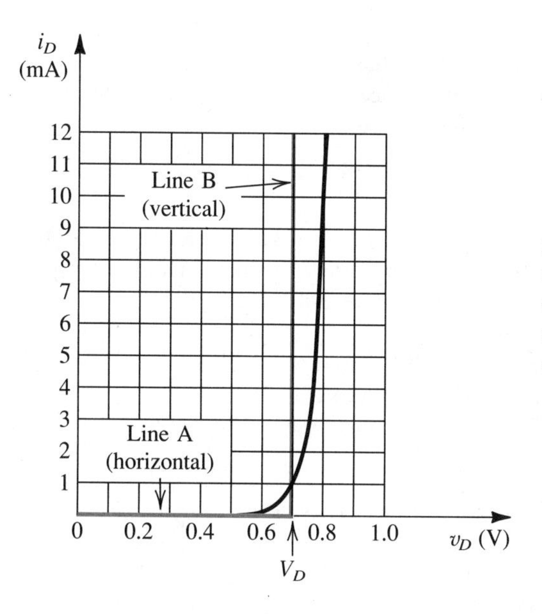

Fig. 3.15 Development of the constant-voltage-drop model of the diode forward characteristics. A vertical straight line (B) is used to approximate the fast-rising exponential.

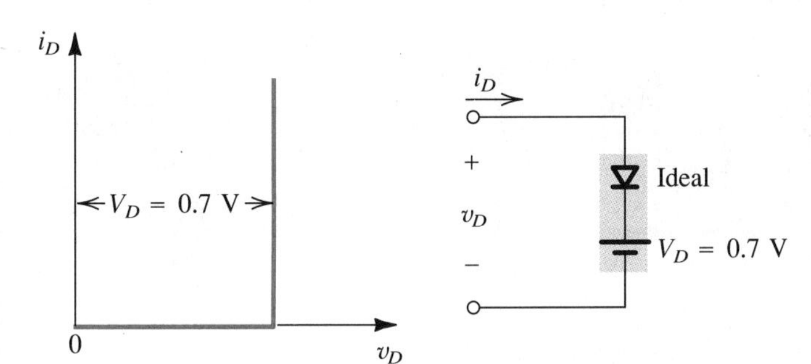

Fig. 3.16 The constant-voltage-drop model of the diode forward characteristics and its equivalent circuit representation.

The constant-voltage-drop model is the one most frequently employed in the initial phases of analysis and design. This is especially true if at these stages one does not have detailed information about the diode characteristics, which is often the case.

Finally, note that if we employ the constant-voltage-drop model to solve the problem in Examples 3.4 and 3.5 we obtain

$$I_D = \frac{V_{DD} - 0.7}{R}$$

$$= \frac{5 - 0.7}{1} = 4.3 \text{ mA}$$

which is not too different from the values obtained before with the more elaborate models.

The Ideal Diode Model

In applications that involve voltages much greater than the diode voltage drop (0.6–0.8 V), we may neglect the diode voltage drop altogether while calculating the diode current. The result is the ideal diode model, which we studied in Section 3.1.

A Concluding Remark

The question of which model to use in a particular application is one that a circuit designer faces repeatedly, not just with diodes but with every circuit element. One's ability to select appropriate device models improves with practice and experience.

Exercises 3.10 For the circuit in Fig. 3.10 find I_D and V_D for the case $V_{DD} = 5$ V and $R = 10$ kΩ. Assume that the diode has a voltage of 0.7 V at 1-mA current and that the voltage changes by 0.1 V/decade of current change. Use (a) iteration (b) the piecewise-linear model with $V_{D0} = 0.65$ V and $r_D = 20$ Ω (c) the constant-voltage-drop model with $V_D = 0.7$ V

Ans. (a) 0.434 mA, 0.663 V; (b) 0.434 mA, 0.659 V; (c) 0.43 mA, 0.7 V

3.11 Consider a diode that is 100 times as large (in junction area) as that whose characteristics are displayed in Fig. 3.12. If we approximate the characteristics in a manner similar to that in Fig. 3.12 (but over a current range 100 times as large) how would the model parameter V_{D0} and r_D change?

Ans. V_{D0} does not change; r_D decreases by a factor of 100 to 0.2 Ω

D3.12 Design the circuit in Fig. E3.12 to provide an output voltage of 2.4 V. Assume that the diodes available have 0.7-V drop at 1 mA and that $\Delta V = 0.1$ V/decade change in current.

Ans. $R = 760$ Ω

3.13 Repeat Exercise 3.4 to obtain better estimates of I and V (than those found in Exercise 3.4) but assuming that nothing much is known about the diodes other than that they are small-signal diodes intended to work in the mA range.

Ans. (a) 1.72 mA, 0.7 V; (b) 0 mA, 5 V; (c) 0 mA, 5 V; (d) 1.72 mA, 0.7 V; (e) 2.3 mA, +2.3 V; 3.3 mA, +1.7 V

3.4 THE SMALL-SIGNAL MODEL AND ITS APPLICATION

There are applications in which a diode is biased to operate at a point on the forward i–v characteristic and a small ac signal is superimposed on the dc quantities. For this situation the diode is best modeled by a resistance equal to the inverse of the slope of the tangent to

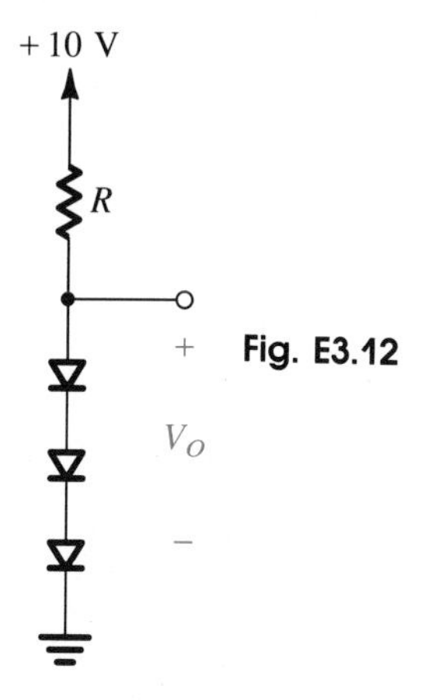

Fig. E3.12

the i–v characteristic at the bias point. The concept of biasing a nonlinear device and restricting signal excursion to a short, almost-linear segment of its characteristic around the bias point was introduced in Section 1.4 for two-port networks. In the following we develop such a small-signal model for the junction diode and illustrate its application.

Consider the conceptual circuit in Fig. 3.17(a) and the corresponding graphical representation in Fig. 3.17(b). A dc voltage V_D, represented by a battery, is applied to the diode; and a time-varying signal $v_d(t)$, assumed (arbitrarily) to have a triangular waveform, is superimposed on the dc voltage V_D. In the absence of the signal $v_d(t)$ the diode voltage is equal to V_D, and correspondingly the diode will conduct a dc current I_D given by

$$I_D = I_S e^{V_D/nV_T} \tag{3.9}$$

When the signal $v_d(t)$ is applied, the total instantaneous diode voltage $v_D(t)$ will be given by

$$v_D(t) = V_D + v_d(t) \tag{3.10}$$

Correspondingly, the total instantaneous diode current $i_D(t)$ will be

$$i_D(t) = I_S e^{v_D/nV_T} \tag{3.11}$$

Substituting for v_D from Eq. (3.10) gives

$$i_D(t) = I_S e^{(V_D+v_d)/nV_T}$$

which can be rewritten

$$i_D(t) = I_S e^{V_D/nV_T} e^{v_d/nV_T}$$

Using Eq. (3.9) we obtain

$$i_D(t) = I_D e^{v_d/nV_T} \tag{3.12}$$

Now if the amplitude of the signal $v_d(t)$ is kept sufficiently small such that

$$\frac{v_d}{nV_T} \ll 1 \tag{3.13}$$

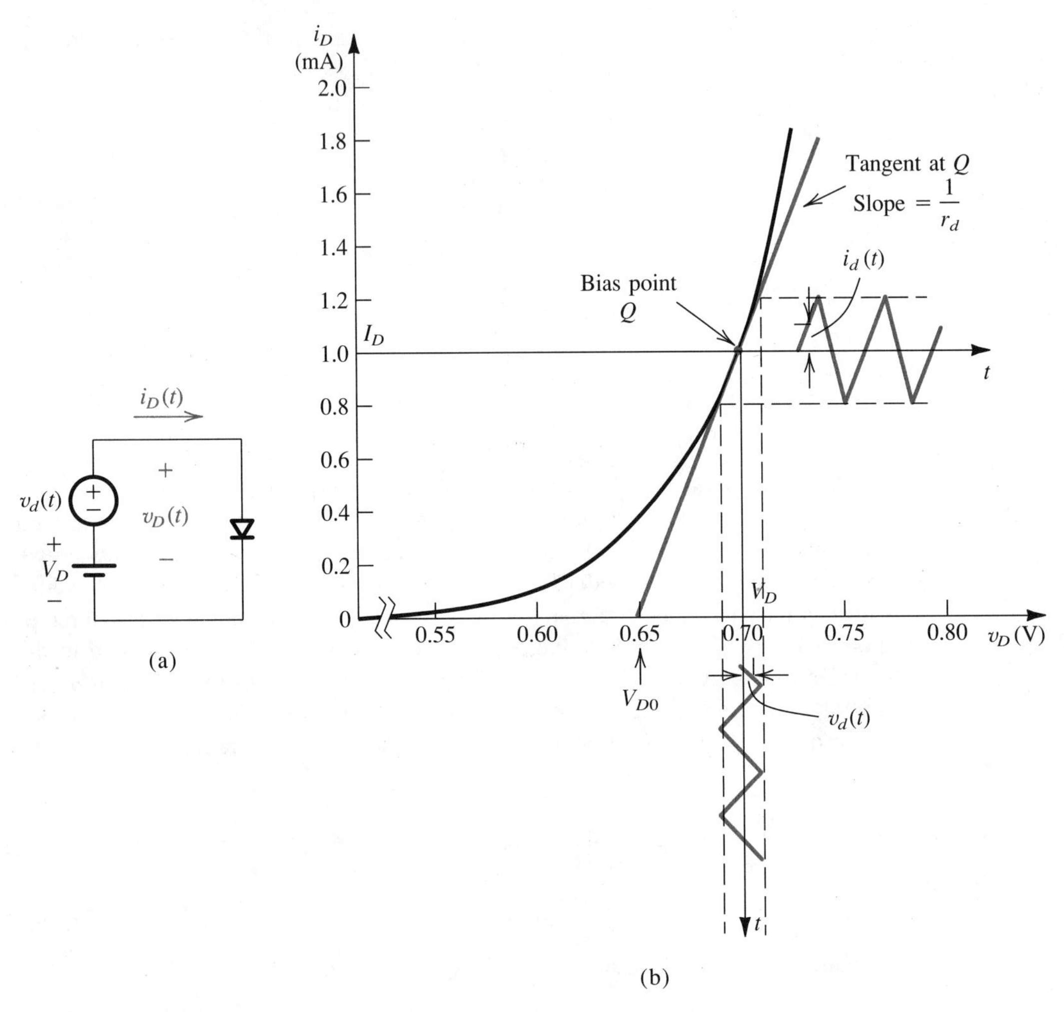

Fig. 3.17 Development of the diode small-signal model. Note that the numerical values shown are for a diode with $n = 2$.

then we may expand the exponential of Eq. (3.12) in a series and truncate the series after the first two terms to obtain the approximate expression

$$i_D(t) \simeq I_D\left(1 + \frac{v_d}{nV_T}\right) \tag{3.14}$$

This is the **small-signal approximation.** It is valid for signals whose amplitudes are smaller than about 10 mV (see Eq. 3.13 and recall that $V_T = 25$ mV).

From Eq. (3.14) we have

$$i_D(t) = I_D + \frac{I_D}{nV_T}v_d \tag{3.15}$$

Thus superimposed on the dc current I_D we have a signal current component directly proportional to the signal voltage v_d. That is,

$$i_D = I_D + i_d \tag{3.16}$$

where

$$i_d = \frac{I_D}{nV_T} v_d \tag{3.17}$$

The quantity relating the signal current i_d to the signal voltage v_d has the dimensions of conductance, mhos (℧), and is called the **diode small-signal conductance.** The inverse of this parameter is the **diode small-signal resistance,** or **incremental resistance,** r_d,

$$r_d = \frac{nV_T}{I_D} \tag{3.18}$$

Note that the value of r_d is inversely proportional to the bias current I_D.

Let us return to the graphical representation in Fig. 3.17(b). It is easy to see that using the small-signal approximation is equivalent to assuming that the signal amplitude is sufficiently small such that the excursion along the i–v curve is limited to a short, almost linear segment. The slope of this segment, which is equal to the slope of the i–v curve at the operating point Q, is equal to the small-signal conductance. The reader is encouraged to prove that the slope of the i–v curve at $i = I_D$ is equal to I_D/nV_T, which is $1/r_d$, that is,

$$r_d = 1 \Big/ \left[\frac{\partial i_D}{\partial v_D}\right]_{i_D = I_D} \tag{3.19}$$

Now, if we denote the point at which the tangent intersects the v_D axis by V_{D0}, we can describe the tangent by the equation

$$i_D = \frac{1}{r_d}(v_D - V_{D0}) \tag{3.20}$$

This equation is a model for the diode operation for small variations around the bias or quiescent point Q. The model can be represented by the equivalent circuit shown in Fig. 3.18, from which we can write

$$\begin{aligned} v_D &= V_{D0} + i_D r_d \\ &= V_{D0} + (I_D + i_d) r_d \\ &= (V_{D0} + I_D r_d) + i_d r_d \\ &= V_D + i_d r_d \end{aligned}$$

Thus, as expected, the incremental or signal voltage across the diode is given by $v_d = i_d r_d$. To illustrate the application of the diode small-signal model, consider the circuit shown in Fig. 3.19(a). Here we have a signal voltage v_s connected in series with the dc source V_{DD}. With $v_s = 0$, the dc current is denoted I_D and the diode dc voltage is denoted V_D. We want to determine the signal current i_d and the signal voltage across the diode v_d.

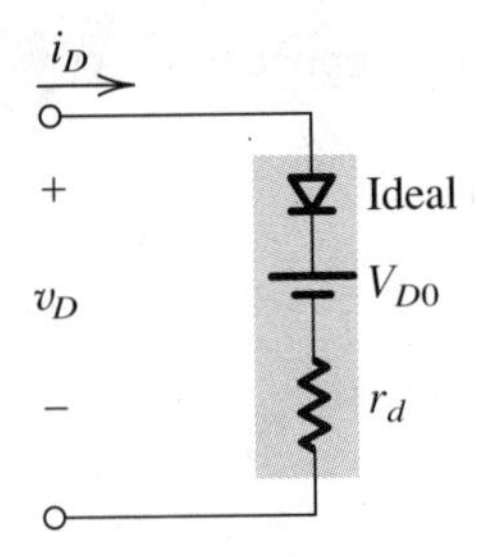

Fig. 3.18 Equivalent circuit model for the diode for small changes around a bias point Q. The incremental resistance r_d is the inverse of the slope of the tangent at Q, and V_{D0} is the intercept of the tangent on the v_D axis (see Fig. 3.17).

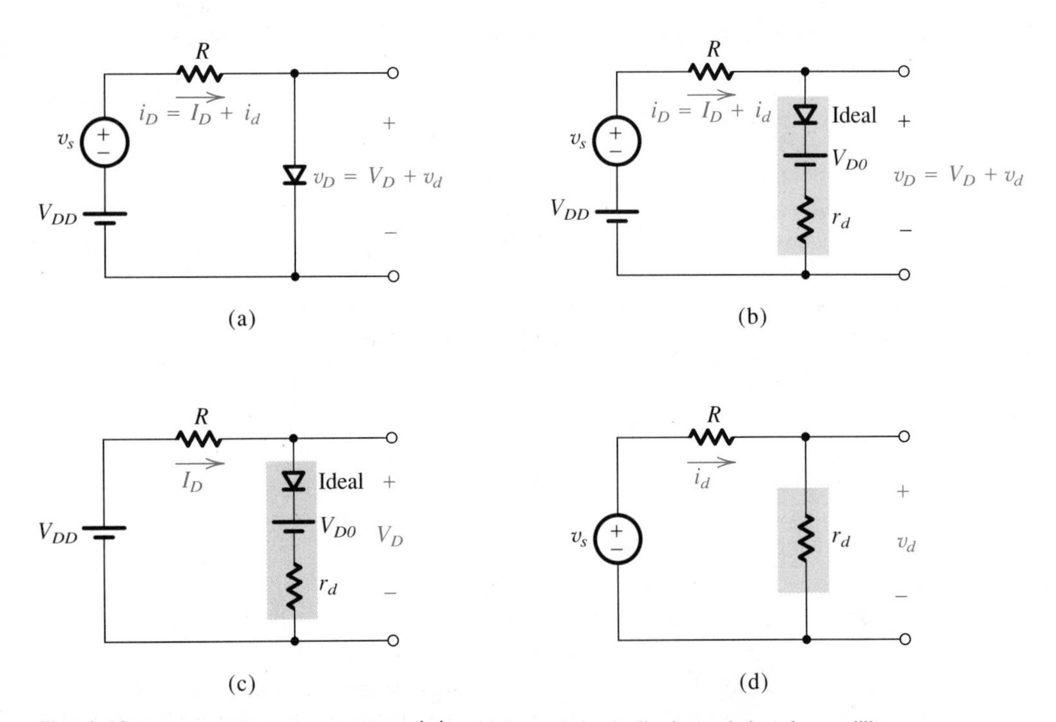

Fig. 3.19 The analysis of the circuit in **(a)**, which contains both dc and signal quantities, can be performed by replacing the diode with the model of Fig. 3.18, as shown in **(b)**. This allows separating the dc analysis [the circuit in **(c)**] from the signal analysis [the circuit in **(d)**].

To do this we replace the diode by the model of Fig. 3.18, thus obtaining the equivalent circuit shown in Fig. 3.19(b). A loop equation for this circuit yields

$$\begin{aligned} V_{DD} + v_s &= i_D R + V_{D0} + i_D r_d \\ &= (I_D + i_d)R + V_{D0} + (I_D + i_d)r_d \\ &= I_D R + (V_{D0} + I_D r_d) + i_d(R + r_d) \\ &= I_D R + V_D + i_d(R + r_d) \end{aligned}$$

Separating the dc and signal quantities on both sides of this equation gives for dc

$$V_{DD} = I_D R + V_D$$

which is represented by the circuit in Fig. 3.19(c), and for the signal,

$$v_s = i_d(R + r_d)$$

which is represented by the circuit in Fig. 3.19(d). We conclude that the small-signal approximation allows one to separate the dc analysis from the signal analysis. The signal analysis is performed by eliminating all dc sources and replacing the diode with its small-signal resistance r_d. From the small-signal equivalent circuit, the diode signal voltage can be simply found using the voltage-divider rule

$$v_d = v_s \frac{r_d}{R + r_d}$$

Separating the dc or bias analysis and the signal analysis is a very useful technique that we will employ often throughout this book.

EXAMPLE 3.6

Consider the circuit shown in Fig. 3.20 for the case $R = 10\ \text{k}\Omega$. The power supply V^+ has a dc value of 10 V on which is superimposed a 60-Hz sinusoid of 1-V peak amplitude. (This "signal" component of the power supply voltage is an imperfection in the power supply design. It is known as the **power supply ripple.** More on this later.) Calculate both the dc voltage of the diode and the sine-wave signal appearing across it. Assume the diode to have a 0.7-V drop at 1-mA current and $n = 2$.

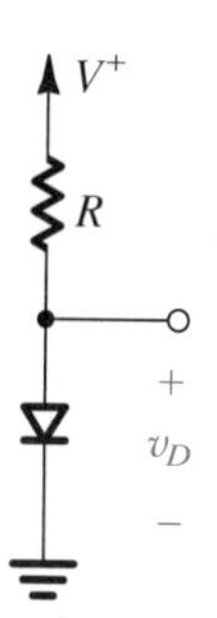

Fig. 3.20 Circuit for Example 3.6.

Solution Considering dc quantities only, we assume $V_D \simeq 0.7$ V and calculate the diode dc current

$$I_D = \frac{10 - 0.7}{10} = 0.93\ \text{mA}$$

Since this value is very close to 1 mA, the diode voltage will be very close to the assumed value of 0.7 V. At this operating point, the diode incremental resistance r_d is

$$r_d = \frac{nV_T}{I_D} = \frac{2 \times 25}{0.93} = 53.8\ \Omega$$

The peak-to-peak signal voltage across the diode can be found by using the voltage divider rule as follows:

$$v_d(\text{peak-to-peak}) = 2\frac{r_d}{R + r_d}$$

$$= 2\frac{0.0538}{10 + 0.0538} = 10.7 \text{ mV}$$

Thus the amplitude of the sinusoidal signal across the diode is 5.35 mV. Since this value is quite small, our use of the small-signal model of the diode is justified.

Use of the Diode Forward Drop in Voltage Regulation

A voltage regulator is a circuit whose purpose is to provide a constant dc voltage between its output terminals. The output voltage is required to remain as constant as possible in spite of (a) changes in the load current drawn from the regulator output terminal and (b) changes in the dc power-supply voltage that feeds the regulator circuit. Since the forward voltage drop of the diode remains almost constant at approximately 0.7 V while the current though it varies by relatively large amounts, a forward-biased diode can make a simple voltage regulator. For instance, we have seen in Example 3.6 that while the 10-V dc supply voltage had a ripple of 2 V peak-to-peak (±10% variation) the corresponding ripple in the diode voltage was only approximately ±5.4 mV (or ±0.8% variation). Regulated voltages greater than 0.7 V can be obtained by connecting a number of diodes in series. For example, the use of three forward-biased diodes in series provides a voltage of about 2 V. One such circuit is investigated in the following example.

EXAMPLE 3.7

Consider the circuit shown in Fig. 3.21. A string of three diodes is used to provide a constant voltage of about 2.1 V. We want to calculate the percentage change in this regulated voltage caused by (a) a ±10% change in the power-supply voltage and (b) connection of a 1-kΩ load resistance. Assume $n = 2$.

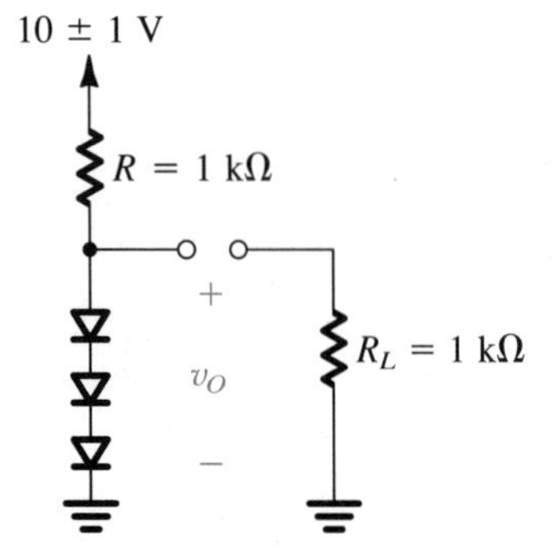

Fig. 3.21 Circuit for Example 3.7.

Solution With no load, the nominal value of the current in the diode string is given by

$$I = \frac{10 - 2.1}{1} = 7.9 \text{ mA}$$

Thus each diode will have an incremental resistance of

$$r_d = \frac{nV_T}{I}$$

Using $n = 2$ gives

$$r_d = \frac{2 \times 25}{7.9} = 6.3 \ \Omega$$

The three diodes in series will have a total incremental resistance of

$$r = 3r_d = 18.9 \ \Omega$$

This resistance along with the resistance R forms a voltage divider whose ratio can be used to calculate the change in output voltage due to a $\pm 10\%$ (that is, ± 1-V) change in supply voltage. Thus the peak-to-peak change in output voltage will be

$$\Delta v_o = 2\frac{r}{r + R} = 2\frac{0.0189}{0.0189 + 1} = 37.1 \text{ mV}$$

that is, corresponding to the ± 1-V ($\pm 10\%$) change in supply voltage the output voltage will change by ± 18.5 mV or $\pm 0.9\%$. Since this implies a change of about ± 6.2 mV per diode, our use of the small-signal model is justified.

When a load resistance of 1 kΩ is connected across the diode string it draws a current of approximately 2.1 mA. Thus the current in the diodes decreases by 2.1 mA, resulting in a decrease in voltage across the diode string given by

$$\Delta v_o = -2.1 \times r = -2.1 \times 18.9 = -39.7 \text{ mV}$$

Since this implies that the voltage across each diode decreases by about 13.2 mV, our use of the small-signal model is not entirely justified. Nevertheless, a detailed calculation of the voltage change using the exponential model results in $\Delta v_o = -35.5$ mV, which is not too different from the approximate value obtained using the incremental model.

Exercises Find the value of the diode small-signal resistance r_d at bias currents of 0.1, 1, and 10 mA. Assume $n = 1$.

Ans. 250 Ω; 25 Ω; 2.5 Ω

3.15 For a diode that conducts 1 mA at a forward voltage drop of 0.7 V and whose $n = 1$, find the equation of the straight-line tangent at $I_D = 1$ mA.

Ans. $i_D = (1/25 \ \Omega)(v_D - 0.675)$

3.16 Consider a diode with $n = 2$ biased at 1 mA. Find the change in current as a result of changing the voltage by (a) −20 mV; (b) −10 mV; (c) −5 mV; (d) +5 mV; (e) +10 mV; (f) +20 mV

In each case, do the calculations **(i)** using the small-signal model and **(ii)** using the exponential model.

Ans. (a) −0.40, −0.33 mA; (b) −0.20, −0.18 mA; (c) −0.10, −0.10 mA; (d) +0.10, +0.11 mA; (e) +0.20, +0.22 mA; (f) +0.40, +0.49 mA

D3.17 Design the circuit of Fig. E3.17 so that $V_O = 3$ V when $I_L = 0$, and V_O changes by 40 mV per 1 mA of load current. Find the value of R and the junction area of each diode (assume all four diodes are identical) relative to a diode with 0.7-V drop at 1-mA current. Assume $n = 1$.

Ans. $R = 4.8$ kΩ; 0.34

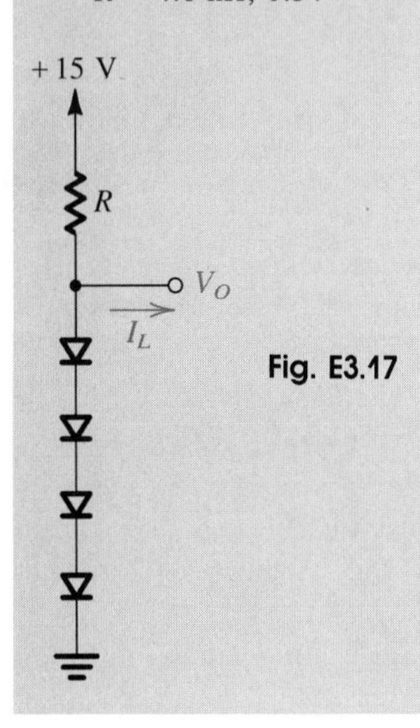

Fig. E3.17

3.5 OPERATION IN THE REVERSE BREAKDOWN REGION—ZENER DIODES

The very steep i–v curve that the diode exhibits in the breakdown region (Fig. 3.8) and the almost constant voltage drop that this indicates suggest that diodes operating in the breakdown region can be used in the design of voltage regulators. This in fact turns out to be a very important application of diodes. Special diodes are manufactured to operate specifically in the breakdown region. Such diodes are called **breakdown diodes** or, more commonly, **zener diodes,** after an early worker in the area.

Figure 3.22 shows the circuit symbol of the zener diode. In normal applications of zener diodes, current flows into the cathode, and the cathode is positive with respect to the anode; thus I_Z and V_Z in Fig. 3.22 have positive values.

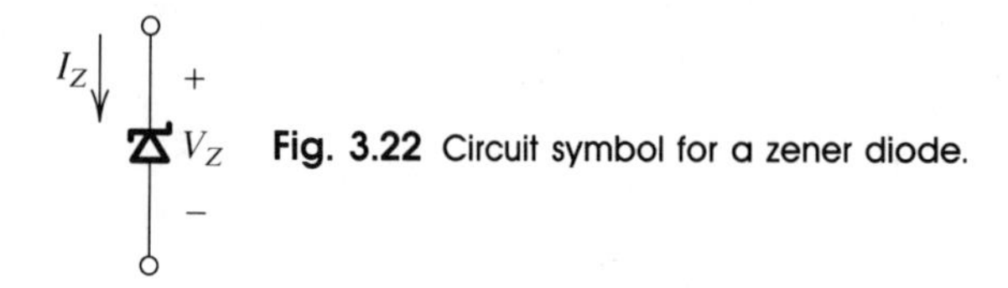

Fig. 3.22 Circuit symbol for a zener diode.

Specifying and Modeling the Zener Diode

Figure 3.23 shows details of the diode i–v characteristics in the breakdown region. We observe that for currents greater than the knee current (I_{ZK}, specified on the data sheet of the zener diode), the i–v characteristic is almost a straight line. The manufacturer usually specifies the voltage across the zener diode V_Z at a specified test current, I_{ZT}. We have

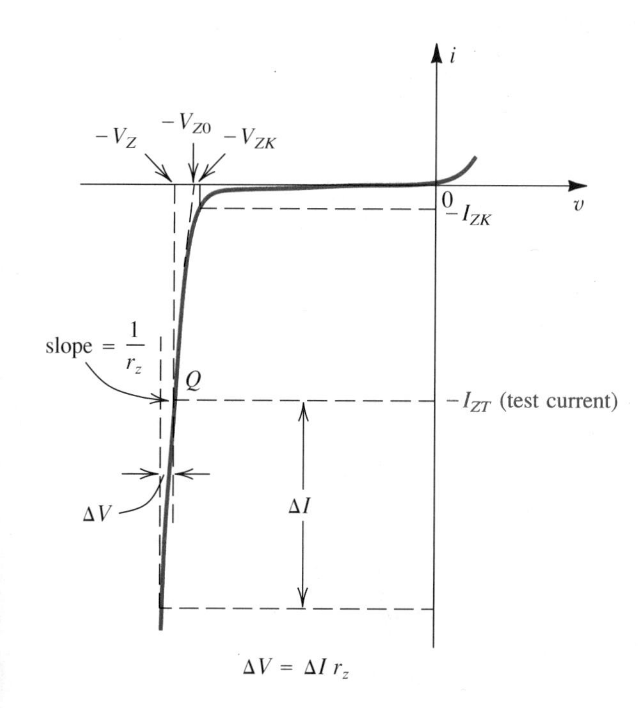

Fig. 3.23 The diode i–v characteristics with the breakdown region shown in some detail.

indicated these parameters in Fig. 3.23 as the coordinates of the point labeled Q. Thus a 6.8-V zener diode will exhibit a 6.8-V drop at a specified test current of, say, 10 mA. As the current through the zener deviates from I_{ZT}, the voltage across it will change, though slightly. Figure 3.23 shows that corresponding to current change ΔI the zener voltage changes by ΔV, which is related to ΔI by

$$\Delta V = r_z\, \Delta I$$

where r_z is the inverse of the slope of the almost-linear i–v curve at point Q. Resistance r_z is the **incremental resistance** of the zener diode at operating point Q. It is also known as the **dynamic resistance** of the zener, and its value is specified on the device data sheet. Typically, r_z is in the range of a few ohms to a few tens of ohms. Obviously, the lower the value of r_z is, the more constant the zener voltage remains as its current varies and thus the more ideal its performance becomes. In this regard, we observe from Fig. 3.23 that while r_z remains low and almost constant over a wide range of current, its value increases considerably in the vicinity of the knee. Therefore, as a general design guideline one should avoid operating the zener in this low-current region.

Zener diodes are fabricated with voltages V_Z in the range of a few volts to a few hundred volts. In addition to specifying V_Z (at a particular current I_{ZT}), r_z, and I_{ZK}, the manufacturer also specifies the maximum power that the device can safely dissipate. Thus a 0.5-W, 6.8-V zener diode can operate safely at currents up to a maximum of about 70 mA.

The almost linear i–v characteristic of the zener diode suggests that the device can be modeled as indicated in Fig. 3.24. Here V_{Z0} denotes the point at which the straight line of slope $1/r_z$ intersects the voltage axis (refer to Fig. 3.23). Although V_{Z0} is shown to be slightly different from the knee voltage V_{ZK}, in practice their values are almost equal. The equivalent circuit model of Fig. 3.24 can be analytically described by

$$V_Z = V_{Z0} + r_z I_Z \tag{3.21}$$

and it applies for $I_Z > I_{ZK}$ and, obviously, $V_Z > V_{Z0}$. The use of the zener model in analysis is illustrated by the following example.

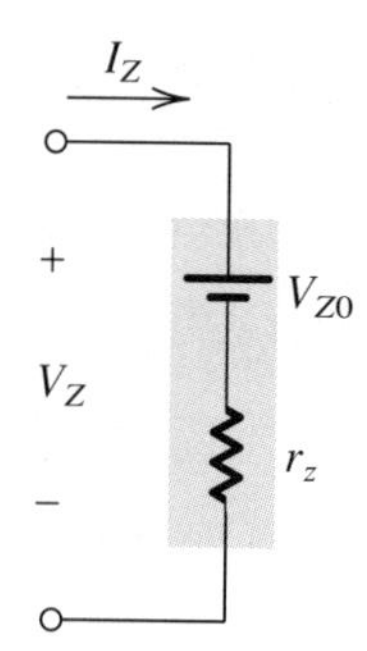

Fig. 3.24 Model for the zener diode.

EXAMPLE 3.8

The 6.8-V zener diode in the circuit of Fig. 3.25(a) is specified to have $V_Z = 6.8$ V at $I_Z = 5$ mA, $r_z = 20\ \Omega$, and $I_{ZK} = 0.2$ mA. The supply voltage V^+ is nominally 10 V but can vary by ± 1 V.

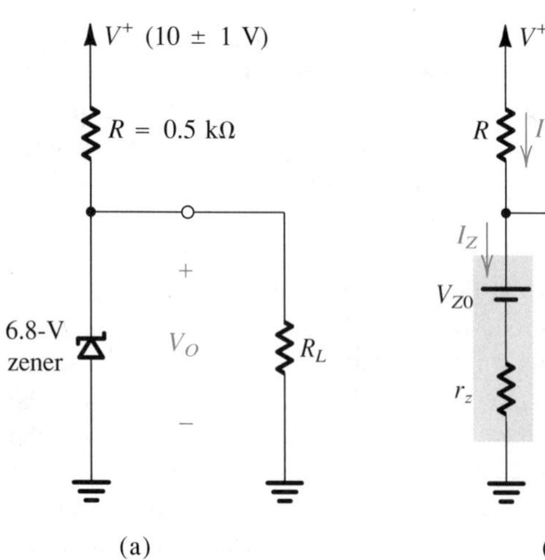

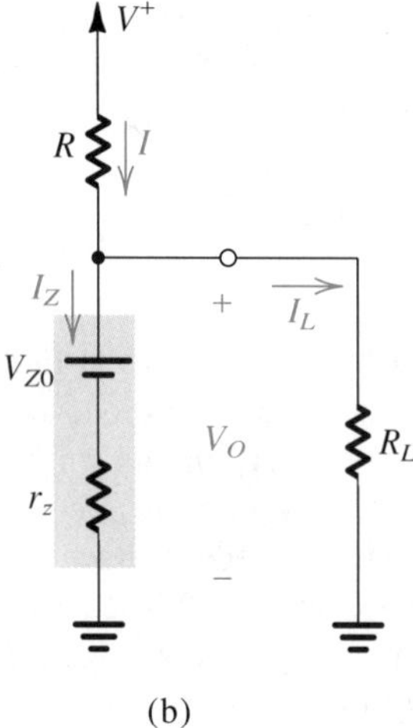

Fig. 3.25 (a) Circuit for Example 3.8. **(b)** The circuit with the zener diode replaced with its equivalent circuit model.

(a) Find V_O with no load and with V^+ at its nominal value.

(b) Find the change in V_O resulting from the ± 1-V change in V^+.

(c) Find the change in V_O resulting from connecting a load resistance $R_L = 2\ \text{k}\Omega$.

(d) Find the value of V_O when $R_L = 0.5\ \text{k}\Omega$.

(e) What is the minimum value of R_L for which the diode still operates in the breakdown region?

Solution First we must determine the value of the parameter V_{Z0} of the zener diode model. Substituting $V_Z = 6.8$ V, $I_Z = 5$ mA, and $r_z = 20\ \Omega$ in Eq. (3.21) yields $V_{Z0} = 6.7$ V. Figure 3.25(b) shows the circuit with the zener diode replaced with its model.

(a) With no load connected, the current through the zener is given by

$$I_Z = I = \frac{V^+ - V_{Z0}}{R + r_z}$$

$$= \frac{10 - 6.7}{0.5 + 0.02} = 6.35 \text{ mA}$$

Thus,

$$V_O = V_{Z0} + I_Z r_z$$

$$= 6.7 + 6.35 \times 0.02 = 6.83 \text{ V}$$

(b) For a ± 1-V change in V^+, the change in output voltage can be found from

$$\Delta V_O = \Delta V^+ \frac{r_z}{R + r_z}$$

$$= \pm 1 \times \frac{20}{500 + 20} = \pm 38.5 \text{ mV}$$

(c) When a load resistance of 2 kΩ is connected, the load current will be approximately 6.8 V/2 kΩ = 3.4 mA. Thus the change in zener current will be $\Delta I_Z = -3.4$ mA, and the corresponding change in zener voltage (output voltage) will thus be

$$\Delta V_O = r_z\, \Delta I_Z$$

$$= 20 \times -3.4 = -68 \text{ mV}$$

A more accurate estimate of ΔV_O can be obtained by analyzing the circuit in Fig. 3.25(b). The result of such an analysis is $\Delta V_O = -70$ mV.

(d) An R_L of 0.5 kΩ would draw a load current of 6.8/0.5 = 13.6 mA. This is not possible because the current I supplied through R is only 6.4 mA (for $V^+ = 10$ V). Therefore, the zener must be cut off. If this is indeed the case, then

V_O is determined by the voltage divider formed by R_L and R,

$$V_O = V^+ \frac{R_L}{R + R_L}$$

$$= 10 \frac{0.5}{0.5 + 0.5} = 5 \text{ V}$$

Since this voltage is lower than the breakdown voltage of the zener, the diode is indeed no longer operating in the breakdown region.

(e) For the zener to be at the edge of the breakdown region, $I_Z = I_{ZK} = 0.2$ mA and $V_Z \simeq V_{ZK} \simeq 6.7$ V. At this point the lowest (worst case) current supplied through R is $(9 - 6.7)/0.5 = 4.6$ mA, and thus the load current is $4.6 - 0.2 = 4.4$ mA. The corresponding value of R_L is

$$R_L = \frac{6.7}{4.4} \simeq 1.5 \text{ k}\Omega$$

Design of the Zener Shunt Regulator

The function of the voltage regulator was described in the previous section. A voltage regulator circuit using a zener diode is shown in Fig. 3.26. This circuit is known as a shunt regulator because the zener diode is connected in parallel (shunt) with the load. The regulator is fed with a supply voltage that, as indicated in the figure, is not very constant; it includes a large ripple component. Such a *raw* supply voltage can be obtained as the output of a rectifier circuit, as will be seen in later sections. The load can be a simple resistor or a complex electronic circuit.

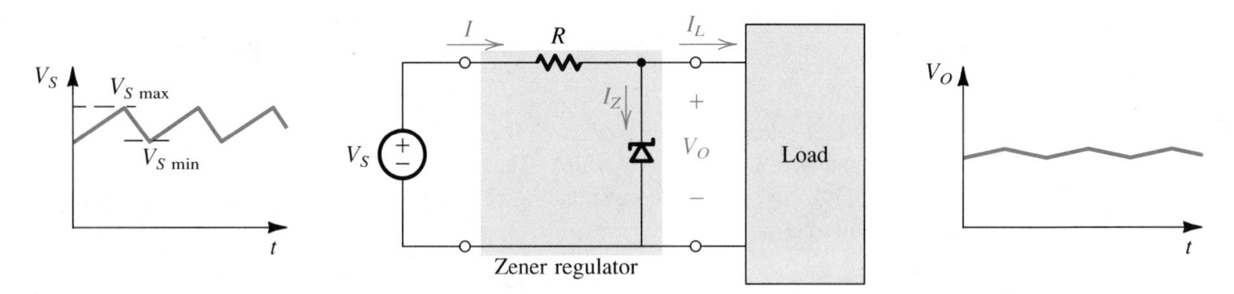

Fig. 3.26 A zener shunt regulator. Observe that while the raw supply V_S has a large ripple component, the regulated voltage V_O has a very small ripple.

The function of the regulator is to provide an output voltage V_O that is as constant as possible in spite of the ripples in V_S and the variations in the load current I_L. Two parameters can be used to measure how well the regulator is performing its function: the **line**

regulation and the **load regulation.** The line regulation is defined as the change in V_O corresponding to a 1-V change in V_S,

$$\text{Line regulation} \equiv \frac{\Delta V_O}{\Delta V_S} \tag{3.22}$$

and is usually expressed in mV/V. The load regulation is defined as the change in V_O corresponding to a 1-mA change in I_L,

$$\text{Load regulation} \equiv \frac{\Delta V_O}{\Delta I_L} \tag{3.23}$$

Expressions for these performance measures can be derived for the shunt regulator of Fig. 3.26 by replacing the zener with its equivalent circuit model, thus obtaining the circuit in Fig. 3.27. Straightforward analysis of this circuit yields

$$V_O = V_{ZO}\frac{R}{R + r_z} + V_S\frac{r_z}{R + r_z} - I_L(r_z//R) \tag{3.24}$$

In the equation, only the first term on the right-hand side is a desirable one. The second and third terms represent the dependence on the supply voltage and the load current, respectively, and thus should be minimized. In fact from Eq. (3.24) and the definitions in Eqs. (3.22) and (3.23) we obtain

$$\text{Line regulation} = \frac{r_z}{R + R_z} \tag{3.25}$$

and

$$\text{Load regulation} = -(r_z//R) \tag{3.26}$$

Note that both of these results could have been derived by inspection of the circuit.

An important consideration in the design of the shunt regulator circuit is to ensure that the current through the zener diode never becomes too low; otherwise, r_z increases and the performance degrades. We note that the minimum zener current occurs when V_S is at its minimum and I_L is at its maximum load current. This can be achieved by the proper selection of the value of R. Analysis of the circuit in Fig. 3.27 with $V_S = V_{S\text{min}}$, $I_Z = I_{Z\text{min}}$ and $I_L = I_{L\text{max}}$ yields

$$R = \frac{V_{S\text{min}} - V_{ZO} - r_z I_{Z\text{min}}}{I_{Z\text{min}} + I_{L\text{max}}} \tag{3.27}$$

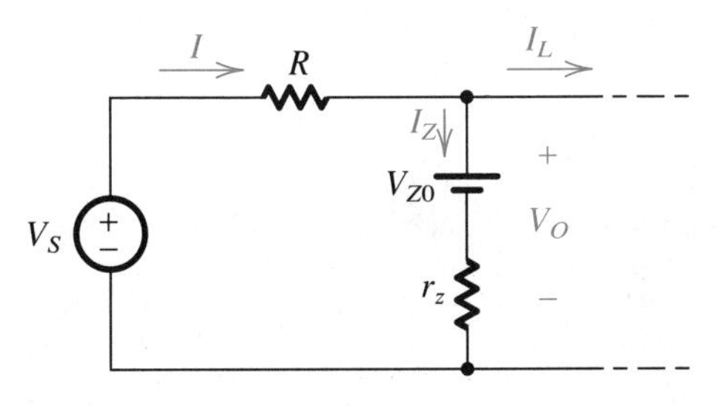

Fig. 3.27 The shunt regulator circuit with the zener diode replaced with its circuit model.

EXAMPLE 3.9 It is required to design a zener shunt regulator to provide an output voltage of approximately 7.5 V. The raw supply varies between 15 and 25 V and the load current varies over the range 0 to 15 mA. The zener diode available has $V_Z = 7.5$ V at a current of 20 mA, and its $r_z = 10\ \Omega$. Find the required value of R and determine the line and load regulation. Also determine the percentage change in V_O corresponding to the full change in V_S and the full change in I_L.

Solution First we determine the value of the parameter V_{Z0} of the zener diode model by substituting $V_Z = 7.5$ V, $I_Z = 20$ mA, and $r_z = 10\ \Omega$ in Eq. (3.21). The result is $V_{Z0} = 7.3$ V. Next we use Eq. (3.27) to determine the value of R by substituting $V_{S\min} = 15$ V and $I_{L\max} = 15$ mA and designing for $I_{Z\min} = (1/3)I_{L\max} = 5$ mA. Thus

$$R = \frac{15 - 7.3 - 0.01 \times 5}{5 + 15} = 383\ \Omega$$

The line regulation can be determined using Eq. (3.25),

$$\text{Line regulation} = \frac{r_z}{r_z + R} = \frac{10}{10 + 383} = 25.4\ \text{mV/V}$$

and the load regulation can be found using Eq. (3.26),

$$\text{Load regulation} = -(r_z//R) = -(10//383) = -9.7\ \text{mV/mA}$$

The full change in V_S (15 to 25 V) results in

$$\Delta V_O = 25.4 \times 10 = 0.254\ \text{V} \quad \text{or} \quad 3.4\%$$

and the full change in I_L (0 to 15 mA) results in

$$\Delta V_O = -9.7 \times 15 \simeq -0.15\ \text{V} \quad \text{or} \quad -2\%$$

Temperature Effects

The dependence of the zener voltage V_Z on temperature is specified in terms of its temperature coefficient TC, or **temco** as it is commonly known, which is usually expressed in mV/°C. The value of TC depends on the zener voltage, and for a given diode the TC varies with the operating current. Zener diodes whose V_Z are lower than about 5 V exhibit a negative TC. On the other hand, zeners with higher voltages exhibit positive TC. The TC of a zener diode with a V_Z of about 5 V can be made zero by operating the diode at a specified current. Another commonly used technique for obtaining a reference voltage with low temperature coefficient is to connect a zener diode with a positive temperature coefficient of about 2 mV/°C in series with a forward conducting diode. Since the forward conducting diode has a voltage drop of $\simeq$0.7 V and a TC of about -2 mV/°C, the series combination will provide a voltage of $(V_Z + 0.7)$ with a TC of about 0.

Exercises 3.18 A zener diode whose nominal voltage is 10 V at 10 mA has an incremental resistance of 50 Ω. What voltage do you expect if the diode current is halved? doubled? What is the value of V_{Z0} of the zener model?

Ans. 9.75 V; 10.5 V; 9.5 V

D3.19 A zener diode exhibits a constant voltage of 5.6 V for currents greater than five times the knee current. I_{ZK} is specified to be 1 mA. It is to be used in the design of a shunt regulator fed from a 15-V supply. The load current varies over the range 0 to 15 mA. Find a suitable value for the resistor R. What is the maximum power dissipation of the zener diode?

Ans. 470 Ω; 112 mW

3.20 A shunt regulator utilizes a zener diode whose voltage is 5.1 V at a current of 50 mA and whose incremental resistance is 7 Ω. The diode is fed from a supply of 15-V nominal voltage through a 200-Ω resistor. What is the output voltage at no load? Find the line regulation and the load regulation.

Ans. 5.1 V; 33.8 mV/V; 6.8 mV/mA

3.6 RECTIFIER CIRCUITS

One of the most important applications of diodes is in the design of rectifier circuits. A diode rectifier forms an essential building block of the dc power supplies required to power electronic equipment. A block diagram of such a power supply is shown in Fig. 3.28. As indicated, the power supply is fed from the 120-V (rms) 60-Hz ac line, and it delivers a dc voltage V_O (usually in the range of 5 to 20 V) to an electronic circuit represented by the *load* block. The dc voltage V_O is required to be as constant as possible in spite of variations in the ac line voltage and in the current drawn by the load.

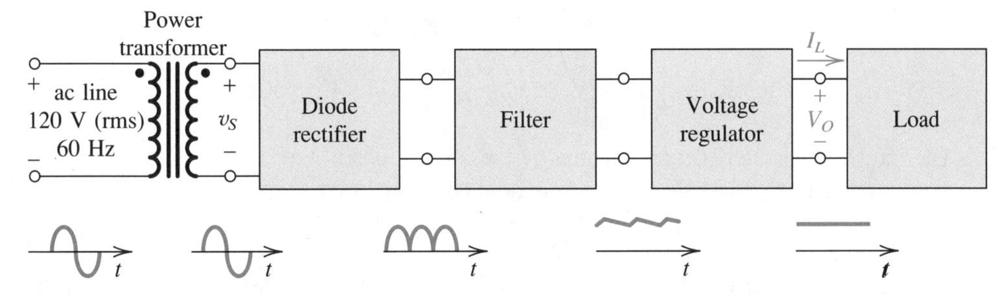

Fig. 3.28 Block diagram of a dc power supply.

The first block in a dc power supply is the **power transformer.** It consists of two separate coils wound around an iron core that magnetically couples the two windings. The **primary winding,** having N_1 turns, is connected to the 120-V ac supply; and the **secondary winding,** having N_2 turns, is connected to the circuit of the dc power supply. Thus an ac voltage v_S of $120(N_2/N_1)$ volts (rms) develops between the two terminals of the secondary winding. By selecting an appropriate turns ratio (N_1/N_2) for the transformer, the designer can step the line voltage down to the value required to yield the particular dc voltage output of the supply. For instance, a secondary voltage of 8-V rms is usually required for a dc output of 5 V. This can be achieved with a 1:15 turns ratio.

In addition to providing the appropriate sinusoidal amplitude for the dc power supply, the power transformer provides electrical isolation between the electronic equipment and the power-line circuit. This isolation minimizes the risk of electric shock to the equipment user.

The diode rectifier converts the input sinusoid v_S to a unipolar output, which can have the pulsating waveform indicated in Fig. 3.28. Although this waveform has a nonzero average or a dc component, its pulsating nature makes it unsuitable as a dc source for electronic circuits, hence the need for a filter. The variations in the magnitude of the rectifier output are considerably reduced by the filter block in Fig. 3.28. In the following we shall study a number of rectifier circuits and a simple implementation of the output filter.

The output of the rectifier filter, though much more constant than without the filter, still contains a time-dependent component, known as **ripple.** To reduce the ripple and to stabilize the magnitude of the dc output voltage of the supply against variations caused by changes in load current, a voltage regulator is employed. Such a regulator can be implemented using the zener shunt regulator configuration studied in Section 3.5. Alternatively, an integrated-circuit (IC) regulator can be used [see, for example, Soclof (1985)].

The Half-Wave Rectifier

The half-wave rectifier utilizes alternate half cycles of the input sinusoid. Figure 3.29(a) shows the circuit of a half-wave rectifier. This circuit was analyzed in Section 3.1 (see Fig. 3.3) assuming an ideal diode. Using the more realistic piecewise-linear diode model, we obtain the equivalent circuit shown in Fig. 3.29(b) from which we can write

$$V_O = 0, \qquad V_S < V_{D0} \tag{3.28a}$$

$$V_O = \frac{R}{R + r_D} V_S - V_{D0} \frac{R}{R + r_D}, \qquad V_S \geq V_{D0} \tag{3.28b}$$

The transfer characteristic represented by these equations is sketched in Fig. 3.29(c). In many applications $r_D \ll R$ and the second equation can be simplified to

$$v_O \simeq v_S - V_{D0} \tag{3.29}$$

where $V_{D0} = 0.7$ or 0.8 V. Figure 3.29(d) shows the output voltage obtained when the input v_S is a sinusoid.

In selecting diodes for rectifier design, two important parameters must be specified: the current-handling capability required of the diode, determined by the largest current the diode is expected to conduct, and the **peak inverse voltage** (PIV) that the diode must be able to withstand without breakdown, determined by the largest reverse voltage that is expected to appear across the diode. In the rectifier circuit of Fig. 3.29(a) we observe that when v_S is negative the diode will be cut off and v_O will be zero. It follows that the PIV is equal to the peak of v_S,

$$\text{PIV} = V_s$$

It is usually prudent, however, to select a diode that has a reverse breakdown voltage at least 50% greater than the expected PIV.

Before leaving the half-wave rectifier the reader should note two points. First, it is possible to use the diode exponential characteristic to determine the exact transfer charac-

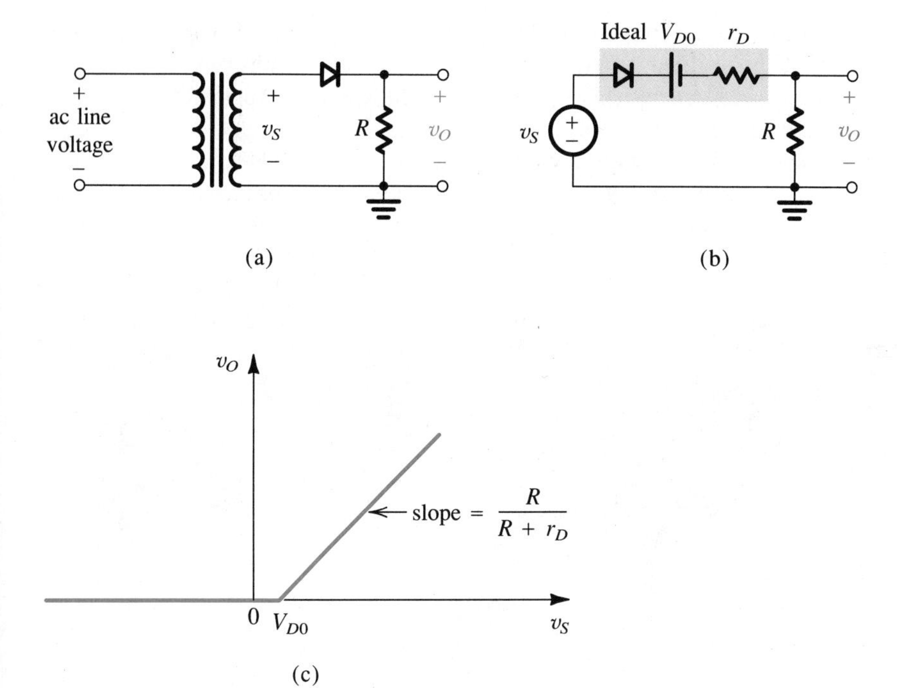

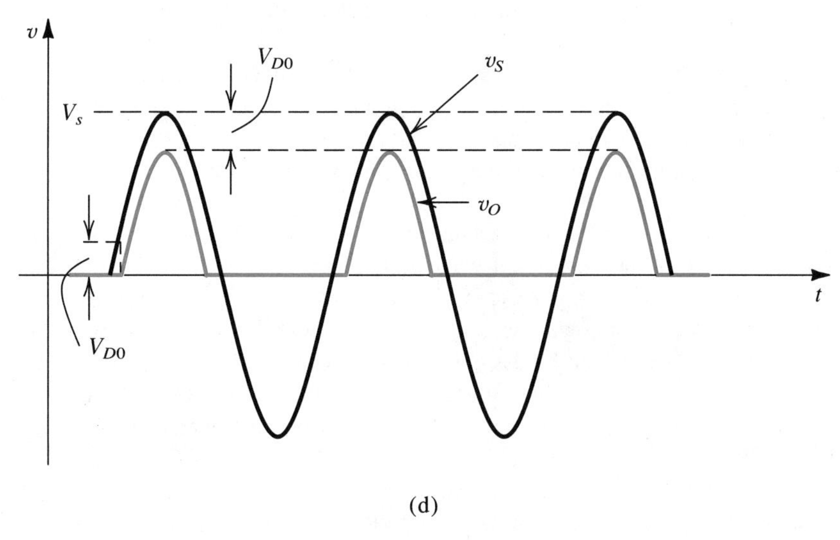

Fig. 3.29 (a) Half-wave rectifier. **(b)** Equivalent circuit of the half-wave rectifier with the diode replaced with its piecewise-linear model. **(c)** Transfer characteristic of the rectifier circuit. **(d)** Input and output waveforms, assuming that $r_D \ll R$.

teristic of the rectifier (see Problem 3.47). However, the amount of work involved is usually too great to be justified in practice. Of course, such an analysis can be easily done using a computer circuit-analysis program such as SPICE (see Appendix C and, for greater detail, the SPICE Supplement to this book).

Second, whether we analyze the circuit accurately or not it should be obvious that this circuit does not function properly when the input signal is small. For instance, this circuit cannot be used to rectify an input sinusoid of 100-mV amplitude. For such an application one resorts to one of the precision circuits presented in Chapter 12.

Exercise 3.21 For the half-wave rectifier circuit in Fig. 3.29(a), neglecting the effect of r_D, show the following. (a) For the half cycles during which the diode conducts, conduction begins at an angle $\Theta = \sin^{-1}(V_{D0}/V_s)$ and terminates at $(\pi - \Theta)$, for a total conduction angle of $(\pi - 2\Theta)$. (b) The average value (dc component) of v_O is $V_O = (1/\pi)V_s - V_{D0}/2$. (c) The peak diode current is $(V_s - V_{D0})/R$.

Find numerical values for these quantities for the case of 12-V (rms) sinusoidal input, $V_{D0} \simeq 0.7$ V, and $R = 100\ \Omega$. Also give the value for PIV.

Ans. (a) $\Theta = 2.4°$, conduction angle = 175°; (b) 5.06 V; (c) 163 mA; 17 V

Full-Wave Rectifier

The full-wave rectifier utilizes both halves of the input sinusoid. To provide a unipolar output, it inverts the negative halves of the sine wave. One possible implementation is shown in Fig. 3.30(a). Here the transformer secondary winding is **center-tapped** to provide two equal voltages v_S across the two halves of the secondary winding with the polarities indicated. Note that when the input line voltage (feeding the primary) is positive, both of the signals labeled v_S will be positive. In this case D_1 will conduct and D_2 will be reverse biased. The current through D_1 will flow through R and back to the center tap of the secondary. The circuit then behaves like a half-wave rectifier, and the output during the positive half cycles will be identical to that produced by the half-wave rectifier.

Now during the negative half cycle of the ac line voltage, both of the voltages labeled v_S will be negative. Thus D_1 will be cut off while D_2 will conduct. The current conducted by D_2 will flow through R and back to the center tap. It follows that during the negative half cycles also the circuit behaves as a half-wave rectifier except that diode D_2 is the one conducting. The important point, however, is that the current through R always flows in the same direction, and thus v_O will be unipolar, as indicated in Fig. 3.30(c). The output waveform shown is obtained by assuming that a conducting diode has a constant voltage drop V_{D0}. In other words we have, for simplicity, neglected the effect of the diode resistance r_D. Thus the transfer characteristic of the full-wave rectifier takes the shape shown in Fig. 3.30(b).

The full-wave rectifier obviously produces a more "energetic" waveform than that provided by the half-wave rectifier. In almost all rectifier applications one opts for the full-wave type.

To find the PIV of the diodes in the full-wave rectifier circuit, consider the situation during the positive half cycles. Diode D_1 is conducting, and D_2 is cut off. The voltage at the cathode of D_2 is v_O, and that at its anode is $-v_S$. Thus the reverse bias of D_2 will be at its maximum when v_O is at its peak value of $(V_s - V_{D0})$ and v_S at its peak value of V_s;

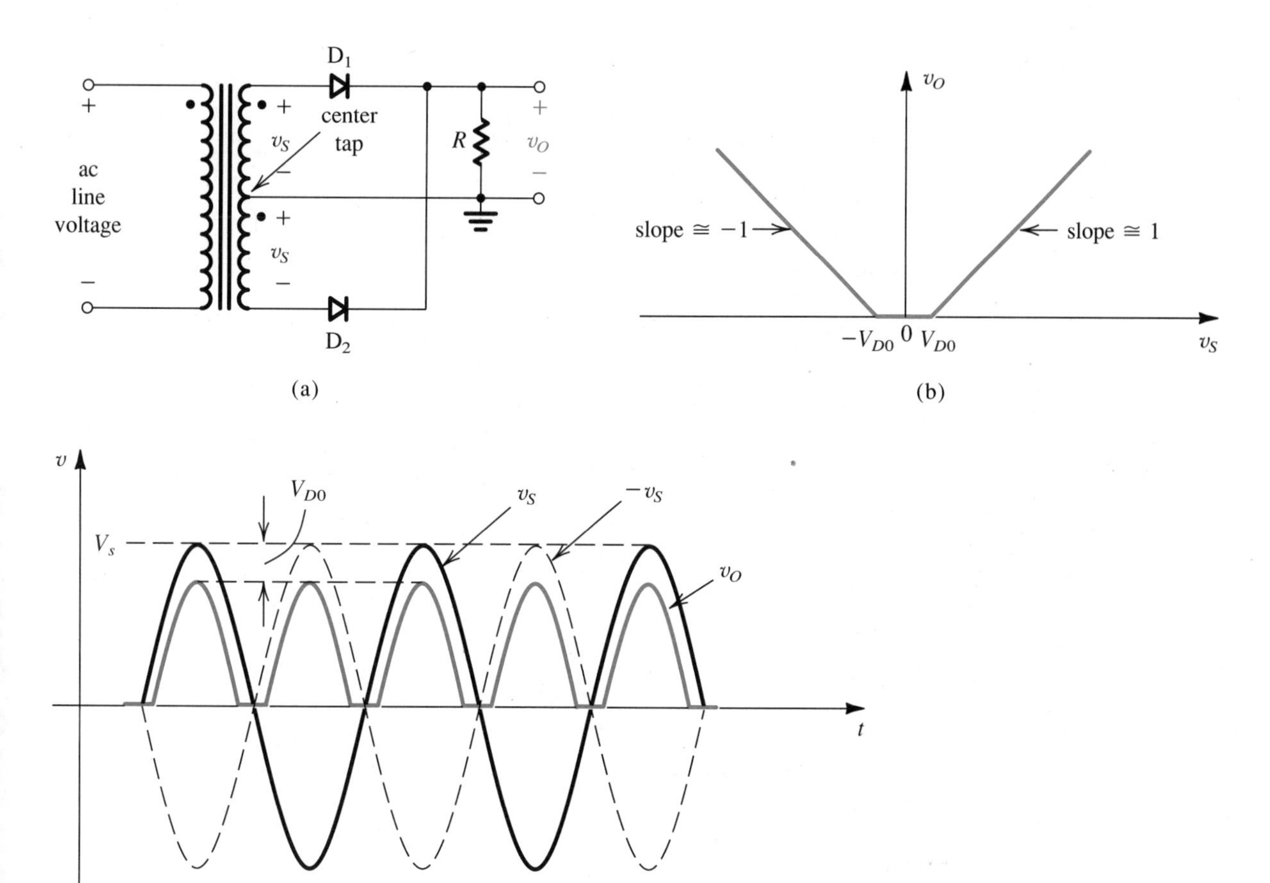

Fig. 3.30 Full-wave rectifier utilizing a transformer with a center-tapped secondary winding. **(a)** Circuit. **(b)** Transfer characteristic assuming a constant-voltage-drop model for the diodes. **(c)** Input and output waveforms.

thus

$$\text{PIV} = 2V_s - V_{D0}$$

which is approximately twice that for the case of the half-wave rectifier.

Exercise 3.22 For the full-wave rectifier circuit in Fig. 3.30(a), neglecting the effect of r_D, show the following: (a) The output is zero for an angle of $2\sin^{-1}(V_{D0}/V_s)$ centered around the zero crossing points of the sine-wave input. (b) The average value (dc component) of v_O is $V_O = (2/\pi)V_s - V_{D0}$. (c) The peak current of each diode is $(V_s - V_{D0})/R$.

Find the fraction (in percent) of each cycle during which $v_O > 0$, the value of V_O, the peak diode current, and the value of PIV, for the case v_S in which is a 12-V (rms) sinusoid, $V_{D0} \simeq 0.7$ V, and $R = 100\ \Omega$.

Ans. 97.4%; 10.1 V; 163 mA; 33.3 V.

The Bridge Rectifier

An alternative implementation of the full-wave rectifier is shown in Fig. 3.31(a). The circuit, known as the bridge rectifier because of the similarity of its configuration to that of the Wheatstone bridge, does not require a center-tapped transformer, a distinct advantage over the full-wave rectifier circuit of Fig. 3.30. The bridge rectifier, however, requires four diodes as compared to two in the previous circuit. This is not much of a disadvantage, as diodes are inexpensive and one can buy a diode bridge in one package.

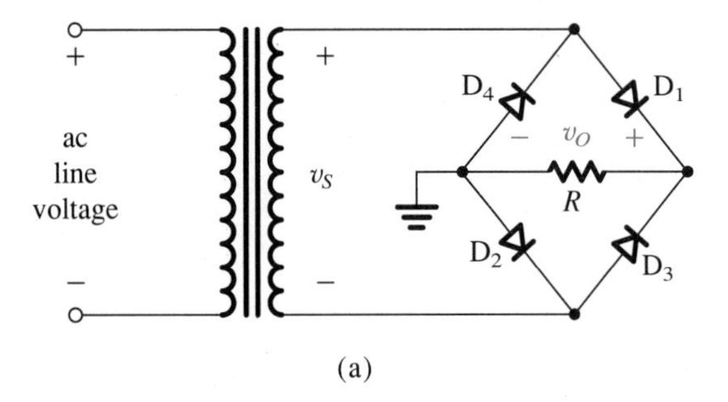

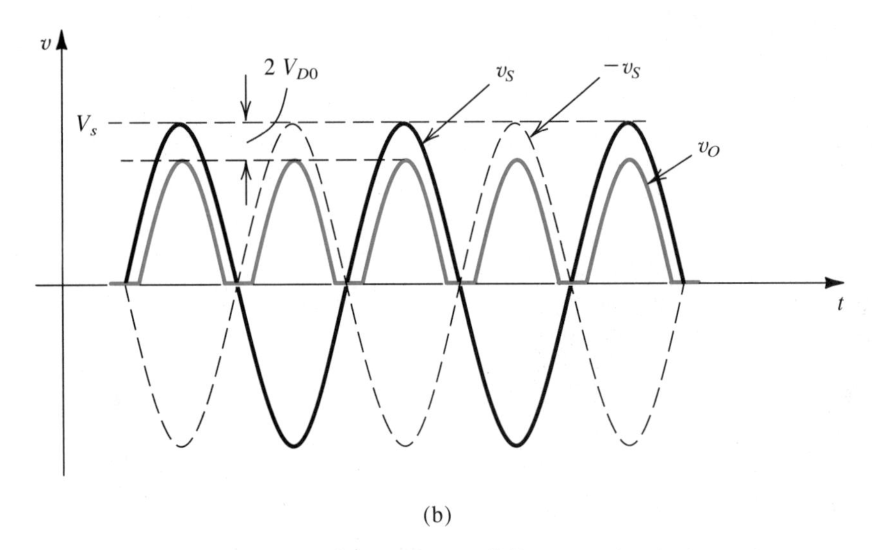

Fig. 3.31 The bridge rectifier; **(a)** circuit and **(b)** input and output waveforms.

The bridge rectifier circuit operates as follows. During the positive half cycles of the input voltage, v_S is positive, and thus current is conducted through diode D_1, resistor R, and diode D_2. Meanwhile diodes D_3 and D_4 will be reverse biased. Observe that there are two diodes in series in the conduction path, and thus v_O will be lower than v_S by two diode drops (compared to one drop in the circuit previously discussed). This is somewhat of a disadvantage of the bridge rectifier.

Next, consider the situation during the negative half cycles of the input voltage. The secondary voltage v_S will be negative, and thus $-v_S$ will be positive, forcing current through D_3, R, and D_4. Meanwhile diodes D_1 and D_2 will be reverse biased. The important point to note, though, is that during both half cycles, current flows through R in the same direction (from right to left) and thus v_O will always be positive, as indicated in Fig. 3.31(b).

To determine the peak inverse voltage (PIV) of each diode, consider the circuit during the positive half cycles. The reverse voltage across D_3 can be determined from the loop formed by D_3, R, and D_2 as

$$v_{D3}\text{ (reverse)} = v_O + v_{D2}\text{ (forward)}$$

Thus the maximum value of v_{D3} occurs at the peak of v_O and is given by

$$\text{PIV} = V_s - 2\,V_{D0} + V_{D0} = V_s - V_{D0}$$

Observe that here the PIV is about half the value for the full-wave rectifier with a center-tapped transformer. This is another advantage of the bridge rectifier.

Yet another advantage of the bridge rectifier circuit over that utilizing a center-tapped transformer is that only about half as many turns are required for the secondary winding of the transformer. Another way of looking at this point can be obtained by observing that each half of the secondary winding of the center-tapped transformer is utilized for only half the time. In conclusion, the bridge rectifier is the most popular rectifier circuit configuration.

Exercise 3.23 For the bridge rectifier circuit of Fig. 3.31(a), use the constant-voltage-drop diode model to show the following. (a) The average (or dc component) of the output voltage is $V_O = (2/\pi)\,V_s - 2\,V_{D0}$. (b) The peak diode current is $(V_s - 2\,V_{D0})/R$.

Find numerical values for the quantities in (a) and (b) and the PIV for the case in which v_S is a 12-V (rms) sinusoid, $V_{D0} \simeq 0.7$ V, and $R = 100\ \Omega$.

Ans. 9.4 V; 156 mA; 16.3 V

The Rectifier with a Filter Capacitor—The Peak Rectifier

The pulsating nature of the output voltage produced by the rectifier circuits discussed above makes it unsuitable as a dc supply for electronic circuits. A simple way to reduce the variation of the output voltage is to place a capacitor across the load resistor. It will be shown that this **filter capacitor** serves to reduce substantially the variations in the rectifier output voltage.

To see how the rectifier circuit with a filter capacitor works, consider first the simple circuit shown in Fig. 3.32. Let the input v_I be a sinusoid with a peak value V_p, and assume the diode to be ideal. As v_I goes positive, the diode conducts and the capacitor is charged so that $v_O = v_I$. This situation continues until v_I reaches its peak value V_p. Beyond the peak, as v_I decreases, the diode becomes reverse-biased and the output voltage remains constant at the value V_p. In fact, theoretically speaking, the capacitor voltage will retain its charge and hence its voltage indefinitely as there is no way for it to discharge. Thus the circuit provides a dc voltage output equal to the peak of the input sine wave. This is a very encouraging situation in view of our desire to produce a dc output.

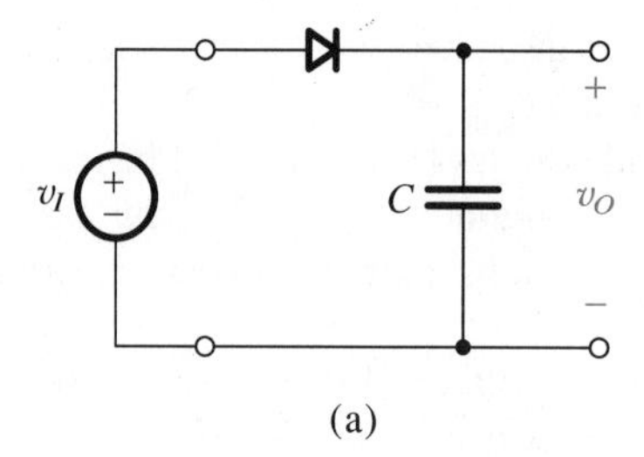

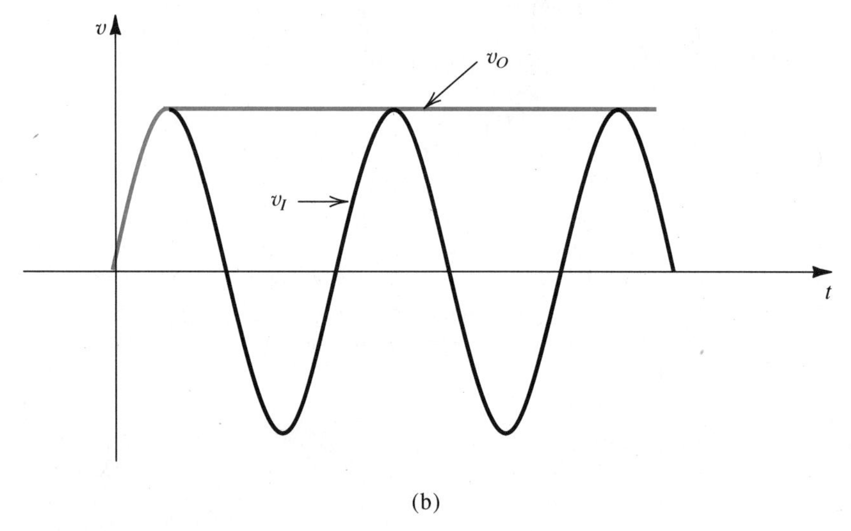

Fig. 3.32 (a) A simple circuit used to illustrate the effect of a filter capacitor. **(b)** Input and output waveforms assuming an ideal diode. Note that the circuit provides a dc voltage equal to the peak of the input sine wave. The circuit is therefore known as a peak rectifier or a peak detector.

We next consider the more practical situation where a load resistance R is connected across the capacitor C, as depicted in Fig. 3.33(a). However, we will continue to assume the diode to be ideal. As before, for a sinusoidal input, the capacitor charges to the peak of the input V_p. Then the diode cuts off and the capacitor discharges through the load resistance R. The capacitor discharge will continue for almost the entire cycle, until the time at which v_I exceeds the capacitor voltage. Then the diode turns on again, charges the capacitor up to the peak of v_I, and the process repeats itself. Observe that to keep the output voltage from decreasing too greatly during capacitor discharge, one selects a value for C so that the time constant CR is much greater than the discharge interval.

We are now ready to analyze the circuit in detail. Figure 3.33(b) shows the steady-state input and output voltage waveforms under the assumption that $CR \gg T$, where T is the period of the input sinusoid. The waveforms of the load current

$$i_L = v_O/R \tag{3.30}$$

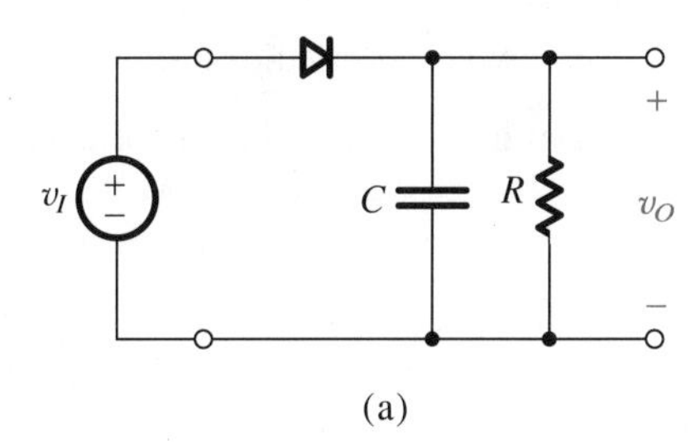

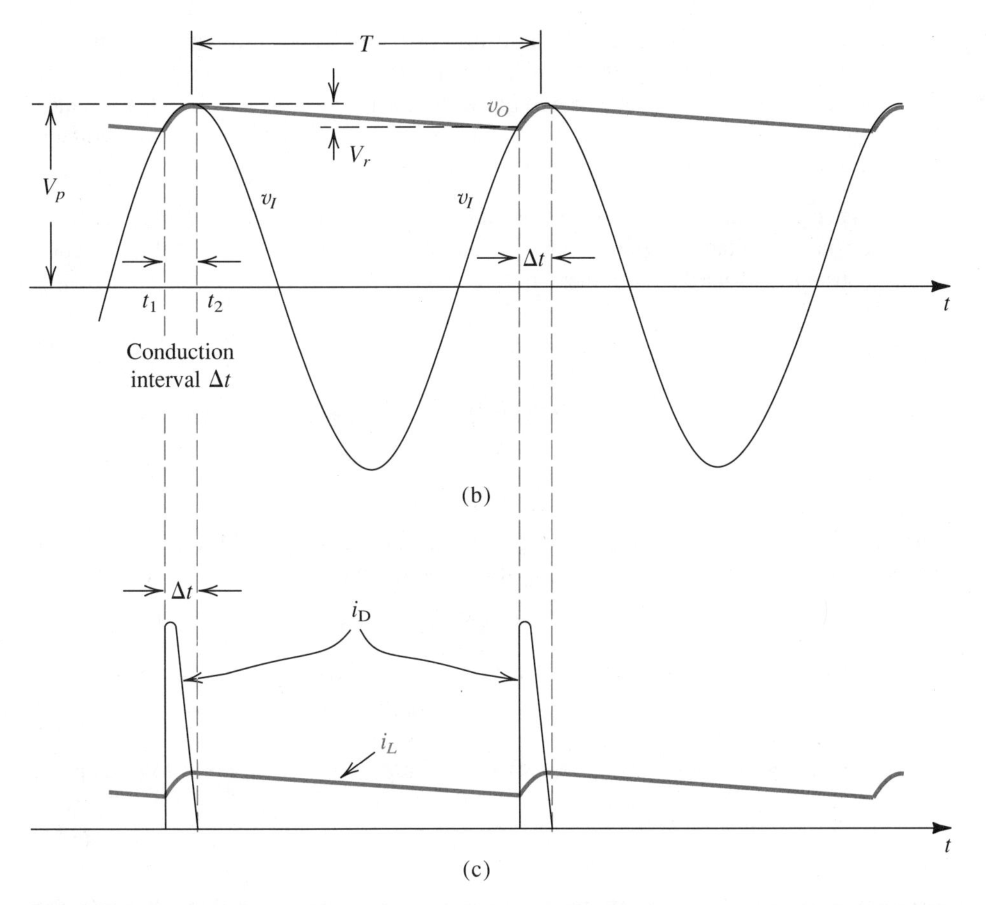

Fig. 3.33 Voltage and current waveforms in the peak rectifier circuit with $CR \gg T$. The diode is assumed ideal.

and of the diode current

$$i_D = i_C + i_L \tag{3.31}$$

$$= C\frac{dv_I}{dt} + i_L \tag{3.32}$$

are shown in Fig. 3.33(c). The following observations are in order:

1. The diode conducts for a brief interval, Δt, near the peak of the input sinusoid and supplies the capacitor with charge equal to that lost during the much longer discharge interval. The latter is approximately equal to the period T.

2. Assuming an ideal diode, the diode conduction begins at time t_1, at which the input v_I equals the exponentially decaying output v_O. Conduction stops at t_2 shortly after the peak of v_I; the exact value of t_2 can be determined by setting $i_D = 0$ in Eq. (3.32).

3. During the diode-off interval, the capacitor C discharges through R and thus v_O decays exponentially with a time constant CR. The discharge interval begins almost at the peak of v_I. At the end of the discharge interval, which lasts for almost the entire period T, $v_O = V_p - V_r$, where V_r is the peak-to-peak ripple voltage. When $CR \gg T$ the value of V_r is small.

4. When V_r is small, v_O is almost constant and equal to the peak value of v_I. Thus the dc output voltage is approximately equal to V_p. Similarly, the current i_L is almost constant and its dc component I_L is given by

$$I_L = \frac{V_p}{R} \tag{3.33}$$

A more accurate expression for the output dc voltage can be obtained by taking the average of the extreme values of v_O,

$$V_O = V_p - \tfrac{1}{2}V_r \tag{3.34}$$

With these observations in hand we now derive expressions for V_r and for the average and peak values of the diode current. During the diode-off interval, v_O can be expressed as

$$v_O = V_p e^{-t/CR}$$

At the end of the discharge interval we have

$$V_p - V_r \simeq V_p e^{-T/CR}$$

Now since $CR \gg T$, we can use the approximation $e^{-T/CR} \simeq 1 - T/CR$ to obtain

$$V_r \simeq V_p \frac{T}{CR} \tag{3.35}$$

We observe that to keep V_r small we must select a capacitance C so that $CR \gg T$. The ripple voltage V_r in Eq. (3.35) can be expressed in terms of the frequency $f = 1/T$ as

$$V_r = \frac{V_p}{f\,CR} \tag{3.36}$$

Note that an alternative interpretation of the approximation made above is that the capacitor discharges by means of a constant current $I_L = V_p/R$. This approximation is valid as long as $V_r \ll V_p$.

Using Fig. 3.33(b) and assuming that diode conduction ceases almost at the peak of v_I, we can determine the conduction interval Δt from

$$V_p \cos(\omega \, \Delta t) = V_p - V_r$$

where $\omega = 2\pi f = 2\pi/T$ is the angular frequency of v_I. Since $(\omega \, \Delta t)$ is a small angle we can employ the approximation $\cos(\omega \, \Delta t) \simeq 1 - \frac{1}{2}(\omega \, \Delta t)^2$ to obtain

$$\omega \, \Delta t \simeq \sqrt{2V_r/V_p} \tag{3.37}$$

We note that when $V_r \ll V_p$, the conduction angle $\omega \, \Delta t$ will be small, as assumed.

To determine the average diode current during conduction, i_{Dav}, we equate the charge that the diode supplies the capacitor,

$$Q_{\text{supplied}} = i_{Cav} \, \Delta t$$

to the charge that the capacitor loses during the discharge interval,

$$Q_{\text{lost}} = CV_r$$

to obtain

$$i_{Dav} = I_L(1 + \pi\sqrt{2V_p/V_r}) \tag{3.38}$$

In deriving this expression we made use of Eq. (3.31) and assumed that i_{Lav} is given by Eq. (3.33). We also used Eqs. (3.36) and (3.37). Observe that when $V_r \ll V_p$, the average diode current during conduction is much greater than the dc load current. This is not surprising since the diode conducts for a very short interval and must replenish the charge lost by the capacitor during the much longer interval in which it is discharged by I_L.

The peak value of the diode current, $i_{D\max}$, can be determined by evaluating the expression in Eq. (3.32) at the onset of diode conduction—that is, at $t = t_1 = -\Delta t$ (where $t = 0$ is at the peak). Assuming that i_L is almost constant at the value given by Eq. (3.33), we obtain

$$i_{D\max} = I_L(1 + 2\pi\sqrt{2V_p/V_r}) \tag{3.39}$$

From Eqs. (3.38) and (3.39) we see that for $V_r \ll V_p$, $i_{D\max} \simeq 2i_{Dav}$, which correlates with the fact that the waveform of i_D is almost a right-angle triangle (see Fig. 3.33c).

EXAMPLE 3.10

Consider a peak rectifier fed by a 60-Hz sinusoid having a peak value $V_p = 100$ V. Let the load resistance $R = 10$ kΩ. Find the value of the capacitance C that will result in a peak-to-peak ripple of 2 V. Also calculate the fraction of the cycle during which the diode is conducting and the average and peak values of the diode current.

Solution

From Eq. (3.36) we obtain the value of C as

$$C = \frac{V_p}{V_r f R} = \frac{100}{2 \times 60 \times 10 \times 10^3} = 83.3 \; \mu\text{F}$$

The conduction angle $\omega\,\Delta t$ is found from Eq. (3.37) as

$$\omega\,\Delta t = \sqrt{2 \times 2/100} = 0.2 \text{ rad}$$

Thus the diode conducts for $(0.2/2\pi) \times 100 = 3.18\%$ of the cycle. The average diode current is obtained from Eq. (3.38), where $I_L = 100/10 = 10$ mA,

$$i_{Dav} = 10(1 + \pi\sqrt{2 \times 100/2}) = 324 \text{ mA}$$

The peak diode current is found using Eq. (3.39),

$$i_{Dmax} = 10(1 + 2\pi\sqrt{2 \times 100/2}) = 638 \text{ mA}$$

The circuit of Fig. 3.33(a) is known as a half-wave **peak rectifier.** The full-wave rectifier circuits of Figs. 3.30(a) and 3.31(a) can be converted to peak rectifiers by including a capacitor across the load resistor. As in the half-wave case the output dc voltage will be almost equal to the peak value of the input sine wave (see Fig. 3.34). The ripple frequency, however, will be twice that of the input. The peak-to-peak ripple voltage, for this case, can be derived using a procedure identical to that above but with the discharge period T replaced by $T/2$, resulting in

$$V_r = \frac{V_p}{2fCR} \tag{3.40}$$

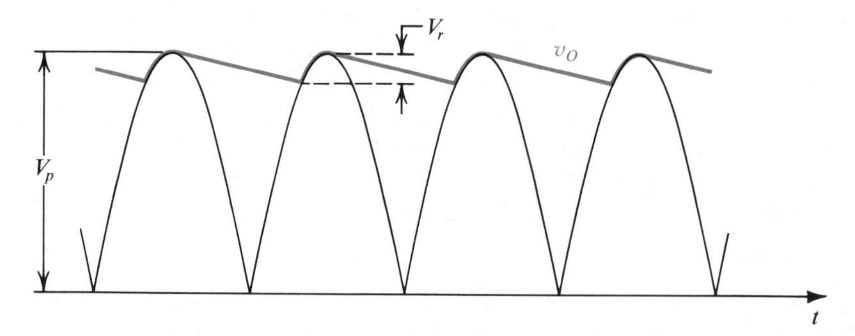

Fig. 3.34 Waveforms in the full-wave peak rectifier.

While the diode conduction interval, Δt, will still be given by Eq. (3.37), the average and peak currents in each of the diodes will be given by

$$i_{Dav} = I_L(1 + \pi\sqrt{V_p/2V_r}) \tag{3.41}$$

$$i_{Dmax} = I_L(1 + 2\pi\sqrt{V_p/2V_r}) \tag{3.42}$$

Comparing these expressions with the corresponding ones for the half-wave case, we note that for the same values of V_p, f, R, and V_r (and thus the same I_L) we need a capacitor half the size of that required in the half-wave rectifier. Also the current in each diode in the full-wave rectifier is approximately half that which flows in the diode of the half-wave circuit.

The analysis above assumed ideal diodes. The accuracy of the results can be improved by taking the diode voltage drop into account. This can be easily done by replacing the peak voltage V_p to which the capacitor charges with $(V_p - V_{D0})$ for the half-wave circuit and the full-wave circuit using a center-tapped transformer, and with $(V_p - 2V_{D0})$ for the bridge rectifier case.

We conclude this section by noting that the peak rectifier circuits find application in signal processing systems where it is required to detect the peak of an input signal. In such a case the circuit is referred to as a **peak detector.** A particularly popular application of the peak detector is in the design of a demodulator for amplitude-modulated (AM) signals. We shall not discuss this application further here.

Exercise D3.24 Consider a bridge rectifier circuit with a filter capacitor C placed across the load resistor R, for the case in which the transformer secondary delivers a sinusoid of 12 V (rms) having a 60-Hz frequency, and assuming $V_{D0} =$ 0.8 V and a load resistance $R = 100\ \Omega$. Find the value of C that results in a ripple voltage no larger than 1 V peak-to-peak. What is the dc voltage at the output? Find the load current. Find the diodes' conduction angle. What is the average diode current? What is the peak reverse voltage across each diode? Specify the diode in terms of its peak current and its PIV.

Ans. 1283 μF; 15.4 V or (a better estimate) 14.9 V; 0.15 A; 0.36 rad (20.6°); 1.44 A; 2.72 A; 16.2 V. Thus select a diode with 3.5 to 4 A peak current and a 20 V PIV rating.

3.7 LIMITING AND CLAMPING CIRCUITS

In this section we shall present additional nonlinear circuit applications of diodes.

Limiter Circuits

Figure 3.35 shows the general transfer characteristic of a limiter circuit. As indicated, for inputs in a certain range, $L_-/K \leq v_I \leq L_+/K$, the limiter acts a linear circuit, providing an output proportional to the input, $v_O = K\ v_I$. Although in general K can be greater than 1,

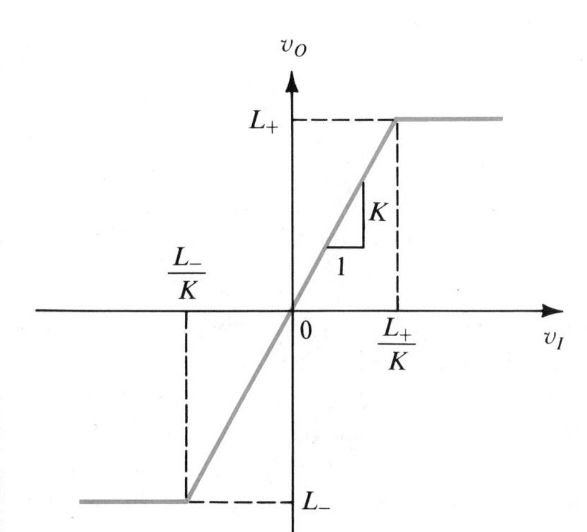

Fig. 3.35 General transfer characteristics for a limiter circuit.

the circuits discussed in this section have $K \leq 1$ and are known as passive limiters. (Examples of active limiters will be presented in Chapter 12.) If v_I exceeds the upper *threshold* (L_+/K), the output voltage is *limited* or clamped to the upper limiting level L_+. On the other hand, if v_I is reduced below the lower limiting threshold (L_-/K), the output voltage v_O is limited to the lower limiting level L_-.

The general transfer characteristic of Fig. 3.35 describes a **double limiter**—that is, a limiter that works on both the positive and negative peaks of an input waveform. **Single limiters,** of course, exist. Finally, note that if an input waveform such as that shown in Fig. 3.36 is fed to a double limiter its two peaks will be *clipped off*. Limiters therefore are sometimes referred to as **clippers.**

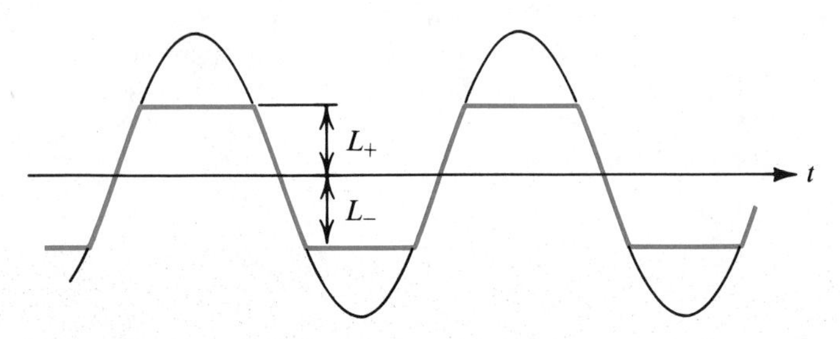

Fig. 3.36 Applying a sine wave to a limiter can result in clipping off its two peaks.

The limiter whose characteristics are depicted in Fig. 3.35 is described as a **hard limiter. Soft limiting** is characterized by smoother transitions between the linear region and the saturation regions and a slope greater than zero in the saturation regions, as illustrated in Fig. 3.37. Depending on the application, either hard or soft limiting may be preferred.

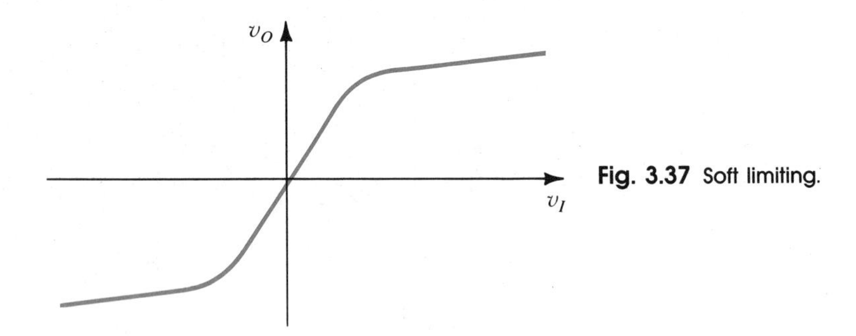

Fig. 3.37 Soft limiting.

Limiters find application in a variety of signal processing systems. One of their simplest applications is in limiting the voltage between the two input terminals of an op amp to a value lower than the breakdown voltage of the transistors that make up the input stage of the op amp circuit. We will have more to say on this and other limiter applications at later points in this book.

Diodes can be combined with resistors to provide simple realizations of the limiter function. A number of examples are depicted in Fig. 3.38. In each part of the figure both the circuit and its transfer characteristic are given. The transfer characteristics are obtained using the constant-voltage-drop ($V_D = 0.7$ V) diode model but assuming a smooth transition between the linear and saturation regions of the transfer characteristics. Better approximations for the transfer characteristics can be obtained using the piecewise-linear diode model. If this is done, the saturation region of the characteristics acquires a slight slope (due to the effect of r_D).

The circuit in Fig. 3.38(a) is that of the half-wave rectifier except that here the output is taken across the diode. For $v_I < 0.5$ V, the diode is cut off, no current flows, and the voltage drop across R is zero; thus $v_O = v_I$. As v_I exceeds 0.5 V the diode turns on, eventually limiting v_O to one diode drop (0.7 V). The circuit of Fig. 3.38(b) is similar to that in (a) except that the diode is reversed.

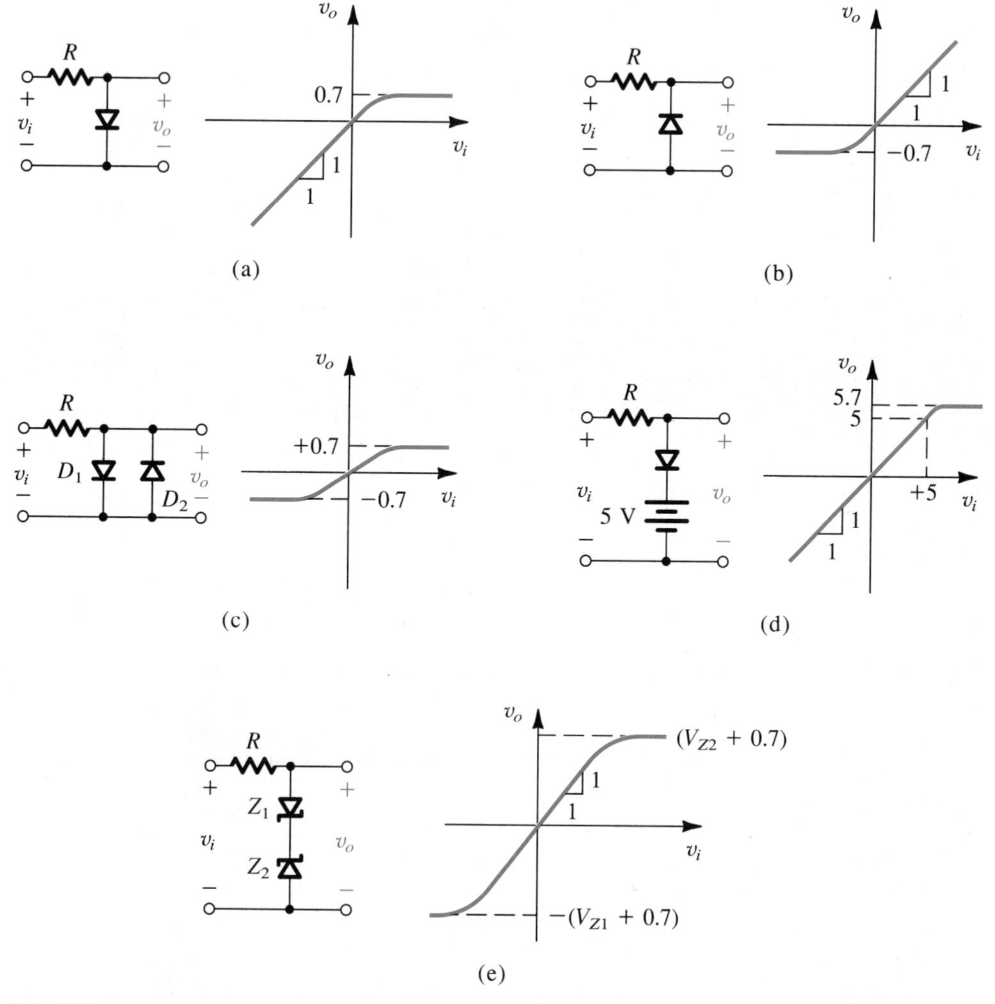

Fig. 3.38 A variety of basic limiting circuits.

Double limiting can be implemented by placing two diodes of opposite polarity in parallel, as shown in Fig. 3.38(c). Here the linear region of the characteristic is obtained for $-0.5\text{ V} \le v_I \le 0.5\text{ V}$. For this range of v_I, both diodes are off and $v_O = v_I$. As v_I exceeds 0.5 V, D_1 turns on and eventually limits v_O to +0.7 V. Similarly, as v_I goes more negative than −0.5 V, D_2 turns on and eventually limits v_O to −0.7 V.

The thresholds and saturation levels of diode limiters can be controlled by using strings of diodes and/or by connecting a dc voltage in series with the diode(s). The latter idea is illustrated in Fig. 3.38(d). Finally, rather than strings of diodes we may use two zener diodes in series, as shown in Fig. 3.38(e). In this circuit limiting occurs in the positive direction at a voltage of $V_{Z2} + 0.7$, where 0.7 V represents the voltage drop across zener diode Z_1 when conducting in the *forward* direction. For negative inputs, Z_1 acts as a zener, while Z_2 conducts in the forward direction. It should be mentioned that pairs of zener diodes connected in series are available commercially for applications of this type under the name **double-anode zener.**

More flexible limiter circuits are possible if op amps are combined with diodes and resistors. Examples of such circuits are discussed in Chapter 12.

Exercise 3.25 Assuming the diodes to be ideal, describe the transfer characteristic of the circuit shown in Fig. E3.25.

Ans. $v_O = v_I$ for $-5 \le v_I \le +5$
$v_O = \frac{1}{2}v_I - 2.5$ for $v_I \le -5$
$v_O = \frac{1}{2}v_I + 2.5$ for $v_I \ge +5$

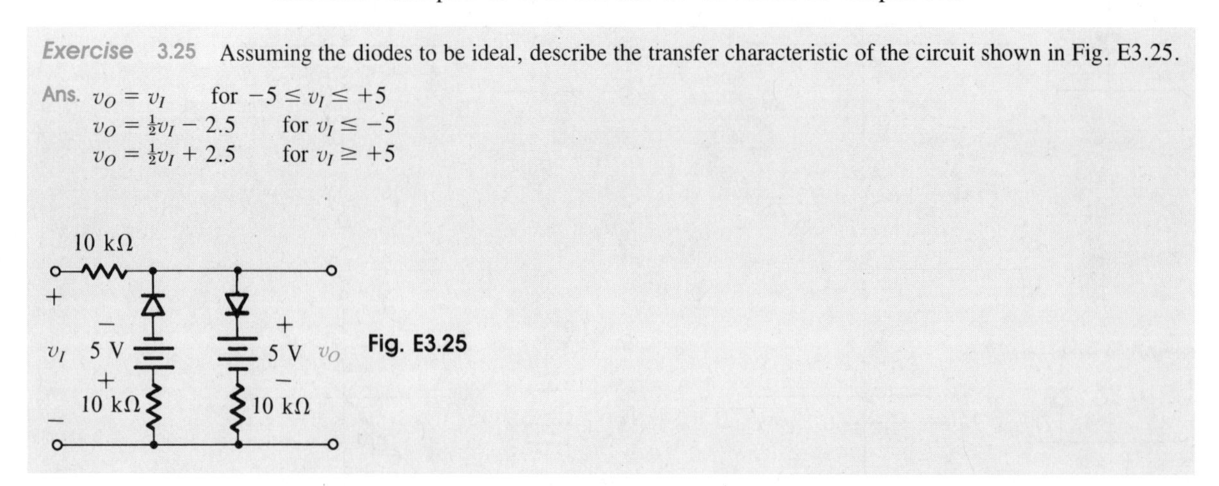

Fig. E3.25

The Clamped Capacitor or dc Restorer

If in the basic peak-rectifier circuit the output is taken across the diode rather than across the capacitor, an interesting circuit with important applications results. The circuit, called a dc restorer, is shown in Fig. 3.39 fed with a square wave. Because of the polarity in which the diode is connected, the capacitor will charge to a voltage v_C (see Fig. 3.39) equal to the magnitude of the most negative peak of the input signal. Subsequently, the diode turns off and the capacitor retains its voltage indefinitely. If, for instance, the input square wave has the arbitrary levels −6 V and +4 V, then v_C will be equal to 6 V. Now, since the output voltage v_O is given by

$$v_O = v_I + v_C$$

it follows that the output waveform will be identical to that of the input, except that it is shifted upwards by v_C volts. In our example the output will be a square wave with levels of 0 V and +10 V.

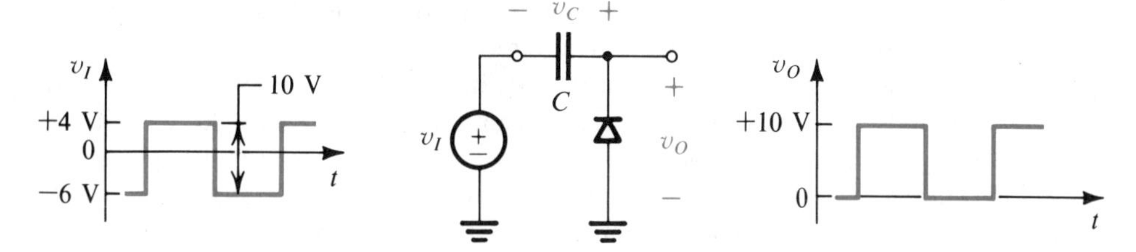

Fig. 3.39 The clamped capacitor or dc restorer with a square-wave input and no load.

Another way of visualizing the operation of the circuit in Fig. 3.39 is to note that because the diode is connected across the output with the polarity shown, it prevents the output voltage from going below 0 V (by conducting and charging up the capacitor, thus causing the output to rise to 0 V), but this connection will not constrain the positive excursion of v_O. The output waveform will therefore have its lowest peak *clamped* to 0 V, which is why the circuit is called a **clamped capacitor.** It should be obvious that reversing the diode polarity will provide an output waveform whose highest peak is clamped to 0 V. In either case the output waveform will have a finite average value or dc component. This dc component is entirely unrelated to the average value of the input waveform. As an application, consider a pulse signal being transmitted through a capacitively coupled or ac-coupled system. The capacitive coupling will cause the pulse train to lose whatever dc component it originally had. Feeding the resulting pulse waveform to a clamping circuit provides it with a well-determined dc component, a process known as **dc restoration.** This is why the circuit is also called a **dc restorer.**

Restoring dc is useful because the dc component of a pulse waveform is an effective measure of its duty cycle. The duty cycle of a pulse waveform can be modulated (in a process called pulsewidth modulation) and made to carry information. In such a system, detection or demodulation could be achieved simply by feeding the received pulse waveform to a dc restorer and then using a simple RC low-pass filter to separate the average of the output waveform from the superimposed pulses.

When a load resistance R is connected across the diode in a clamping circuit, as shown in Fig. 3.40, the situation changes significantly. While the output is above ground,

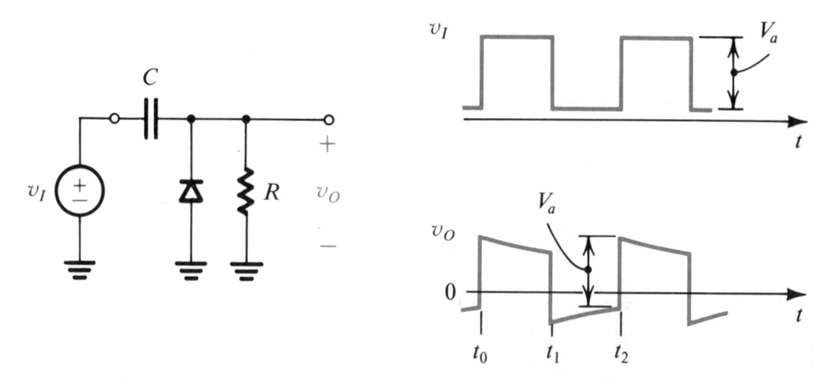

Fig. 3.40 The clamped capacitor with a load resistance *R*.

a net dc current must flow in R. Since at this time the diode is off, this current obviously comes from the capacitor, thus causing the capacitor to discharge and the output voltage to fall. This is shown in Fig. 3.40 for a square-wave input. During the interval t_0 to t_1 the output voltage falls exponentially with time constant CR. At t_1 the input decreases by V_a volts and the output attempts to follow. This causes the diode to conduct heavily and to quickly charge the capacitor. At the end of the interval t_1 to t_2 the output voltage would normally be a few tenths of a volt negative (say, -0.5 V). Then as the input rises by V_a volts (at t_2), the output follows, and the cycle repeats itself. In a steady state the charge lost by the capacitor during the interval t_0 to t_1 is recovered during the interval t_1 to t_2. This charge equilibrium enables us to calculate the average diode current as well as the details of the output waveform.

Voltage Doubler

Figure 3.41(a) shows a circuit composed of two sections in cascade: a clamp formed by C_1 and D_1, and a peak rectifier formed by D_2 and C_2. When excited by a sinusoid of amplitude V_p the clamping section provides the voltage waveform shown in Fig. 3.41(b). Note that while the positive peaks are clamped to zero volts, the negative peak reaches $-2V_p$. In response to this waveform, the peak-detector section provides across capacitor C_2 a negative dc voltage of magnitude $2V_p$. Because the output voltage is double the input peak, the circuit is known as a voltage doubler. The technique can be extended to provide output dc voltages that are higher multiples of V_p.

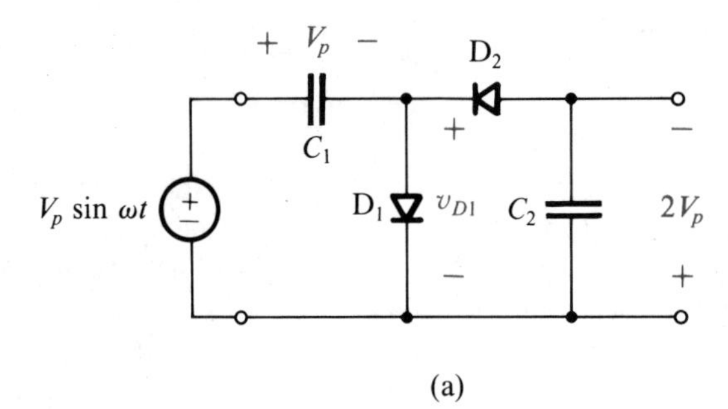

(a)

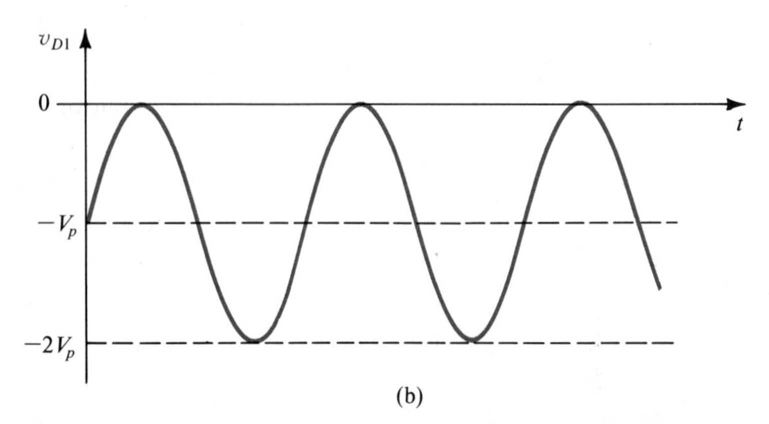

(b)

Fig. 3.41 Voltage doubler: **(a)** circuit; **(b)** waveform of the voltage across D_I.

Exercise 3.26 If the diode in the circuit of Fig. 3.39 is reversed, what will the dc component of v_O become?

Ans. -5 V

3.8 PHYSICAL OPERATION OF DIODES—BASIC SEMICONDUCTOR CONCEPTS

Having studied the terminal characteristics of junction diodes, we will now briefly consider the physical processes that give rise to these characteristics. The following treatment of device physics is qualitative; nevertheless, it should provide sufficient background for the design of diode and other semiconductor circuits.

The *pn* Junction

The semiconductor diode is basically a *pn* junction, as shown schematically in Fig. 3.42. As indicated, the *pn* junction consists of *p*-type semiconductor material (such as silicon) brought into close contact with *n*-type semiconductor material (silicon). In actual practice both the *p* and *n* regions are part of the same silicon crystal; that is, the *pn* junction is formed within a single silicon crystal by creating regions of different ''dopings'' (*p* and *n* regions). Appendix A provides a brief description of the process employed in the fabrication of *pn* junctions. As indicated in Fig. 3.42, external wire connections to the *p* and *n* regions (that is, diode terminals) are made through metal (aluminum) contacts.

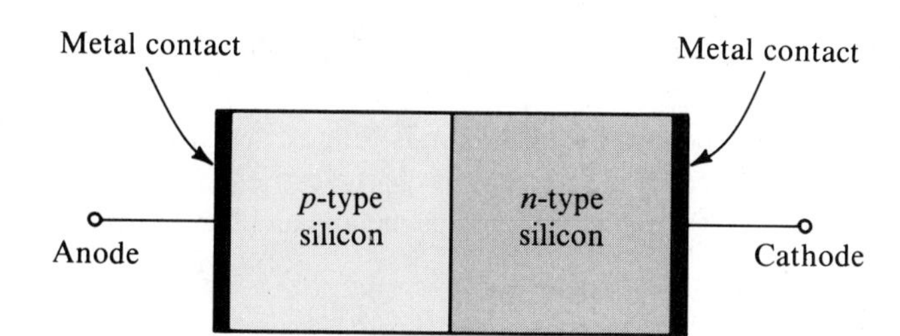

Fig. 3.42 Physical structure of the junction diode. (Actual geometries are given in Appendix A.)

In addition to being essentially a diode, the *pn* junction is the basic component of field-effect transistors (FETs) and bipolar-junction transistors (BJTs). Thus an understanding of the physical operation of *pn* junctions is important to the understanding of the operation and terminal characteristics of diodes and transistors.

Intrinsic Silicon

Although either silicon or germanium can be used to manufacture semiconductor devices—indeed earlier diodes and transistors were made of germanium—today's integrated-circuit technology is based almost entirely on silicon. For this reason we will deal mostly with silicon devices throughout this book.[4]

[4] An exception is the gallium arsenide (GaAs) circuits studied in Chapters 5, 6 and 13.

A crystal of pure or intrinsic silicon has a regular lattice structure where the atoms are held in their positions by bonds, called **covalent bonds,** formed by the four valence electrons associated with each silicon atom. At sufficiently low temperatures all covalent bonds are intact and no (or very few) **free electrons** are available to conduct electric current. However, at room temperature some of the bonds are broken by thermal ionization and some electrons are freed. When a covalent bond is broken an electron leaves its parent atom; thus a positive charge, equal to the magnitude of the electron charge, is left with the parent atom. An electron from a neighboring atom may be attracted to this positive charge, leaving its parent atom. This action fills up the ''hole'' that existed in the ionized atom but creates a new hole in the other atom. This process may repeat itself with the result that we effectively have a positively charged carrier, or **hole,** moving through the silicon crystal structure and being available to conduct electric current. The charge of a hole is equal in magnitude to the charge of an electron.

Thermal ionization results in free electrons and holes in equal numbers and hence equal concentrations. These free electrons and holes move randomly through the silicon crystal structure, and in the process some electrons may fill some of the holes. This process, called **recombination,** results in the disappearance of free electrons and holes. The recombination rate is proportional to the number of free electrons and holes, which is in turn determined by the ionization rate. The ionization rate is a strong function of temperature. In thermal equilibrium the recombination rate is equal to the ionization or thermal-generation rate, and one can calculate the concentration of free electrons n, which is equal to the concentration of holes p,

$$n = p = n_i$$

where n_i denotes the concentration of free electrons or holes in intrinsic silicon at a given temperature.

Finally, it should be mentioned that the reason that silicon is called a **semiconductor** is that its conductivity, which is determined by the number of charge carriers available to conduct electric current, is between that of conductors (such as metals) and that of insulators (such as glass).

Diffusion and Drift

There are two mechanisms by which holes and electrons move through a silicon crystal—**diffusion** and **drift.** Diffusion is associated with random motion due to thermal agitation. In a piece of silicon with uniform concentrations of free electrons and holes this random motion does not result in a net flow of charge (that is, current). On the other hand, if by some mechanism the concentration of, say, free electrons is made higher in one part of the piece of silicon than in another, then electrons will diffuse from the region of high concentration to the region of low concentration. This diffusion process gives rise to a net flow of charge, or **diffusion current.** As an example, consider the bar of silicon shown in Fig. 3.43(a), in which the hole **concentration profile** shown in Fig. 3.43(b) has been created along the x-axis by some unspecified mechanism. The existence of such a concentration profile results in a hole diffusion current in the x direction, with the magnitude of the current at any point being proportional to the slope of the concentration curve, or the concentration gradient, at that point.

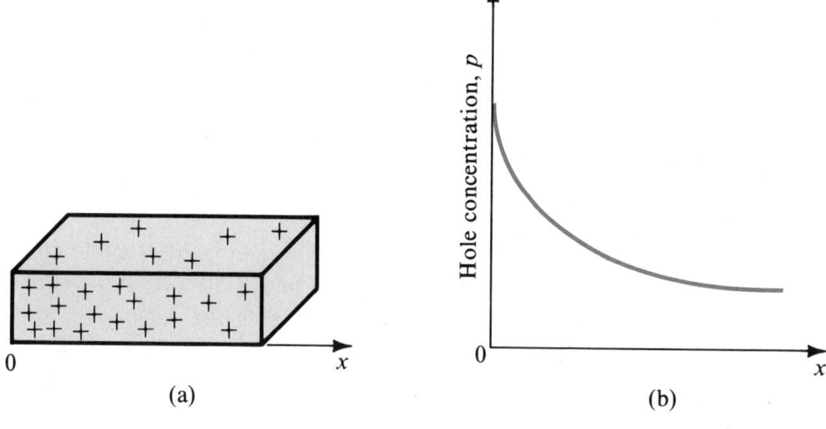

Fig. 3.43 A bar of intrinsic silicon **(a)** in which the hole concentration profile shown in **(b)** has been created along the x-axis by some unspecified mechanism.

The other mechanism for carrier motion in semiconductors is drift. Carrier drift occurs when an electric field is applied across a piece of silicon. Free electrons and holes are accelerated by the electric field and acquire a velocity component (superimposed on the velocity of their thermal motion) called **drift velocity.** The resulting hole and electron current components are called **drift currents.** The relationship between the drift current and the applied electric field represents one form of Ohm's law.

Doped Semiconductors

The intrinsic silicon crystal described above has equal concentrations of free electrons and holes generated by thermal ionization. These concentrations, denoted n_i, are strongly dependent on temperature. Doped semiconductors are materials in which carriers of one kind (electrons or holes) predominate. Doped silicon in which the majority of charge carriers are the *negatively* charged electrons is called ***n* type,** while silicon doped so that the majority of charge carriers are the *positively* charged holes is called ***p* type.**

Doping of a silicon crystal to turn it into n type or p type is achieved by introducing a small number of impurity atoms. For instance, introducing impurity atoms of a pentavalent element such as phosphorus results in n-type silicon because the phosphorus atoms that replace some of the silicon atoms in the crystal structure have five valence electrons, four of which form bonds with the neighboring silicon atoms while the fifth becomes a free electron. Thus each phosphorus atom *donates* a free electron to the silicon crystal, and the phosphorus impurity is called a **donor.** It should be clear, though, that no holes are generated by this process; hence the majority of charge carriers in the phosphorus-doped silicon will be electrons. In fact, if the concentration of donor atoms (phosphorus) is N_D, in thermal equilibrium the concentration of free electrons in the n-type silicon, n_{n0}, will be

$$n_{n0} \simeq N_D$$

where the additional subscript 0 denotes thermal equilibrium. In this n-type silicon the concentration of holes, p_{n0}, that are generated by thermal ionization will be

$$p_{n0} \simeq \frac{n_i^2}{N_D}$$

Since n_i is a function of temperature, it follows that the concentration of the **minority** holes will be a function of temperature, whereas that of the **majority** electrons is independent of temperature.

To produce a p-type semiconductor, silicon has to be doped with a trivalent impurity such as boron. Each of the impurity boron atoms *accepts* one electron from the silicon crystal, so that they may form covalent bonds in the lattice structure. Thus each boron atom gives rise to a hole, and the concentration of the majority holes in p-type silicon, under thermal equilibrium, is approximately equal to the concentration N_A of the **acceptor** (boron) impurity,

$$p_{p0} \simeq N_A$$

In this p-type silicon the concentration of the minority electrons, which are generated by thermal ionization, will be

$$n_{p0} \simeq \frac{n_i^2}{N_A}$$

It should be mentioned that a piece of n-type or p-type silicon is electrically neutral; the majority free carriers (electrons in n-type silicon and holes in p-type silicon) are neutralized by **bound charges** associated with the impurity atoms.

3.9 THE *pn* JUNCTION UNDER OPEN-CIRCUIT CONDITIONS

Figure 3.44 shows a pn junction under open-circuit conditions—that is, the external terminals are left open. The "+" signs in the p-type material denote the majority holes. The charge of these holes is neutralized by an equal amount of bound negative charge associated with the acceptor atoms. For simplicity these bound charges are not shown in the diagram. Also not shown are the minority electrons generated in the p-type material by thermal ionization.

In the n-type material the majority electrons are indicated by "−" signs. Here also the bound positive charge, which neutralizes the charge of the majority electrons, is not shown in order to keep the diagram simple. The n-type material also contains minority holes generated by thermal ionization and not shown on the diagram.

The Diffusion Current I_D

Because the concentration of holes is high in the p region and low in the n region, holes diffuse across the junction from the p side to the n side; similarly, electrons diffuse across the junction from the n side to the p side. These two current components add together to form the diffusion current I_D, whose direction is from the p side to the n side, as indicated in Fig. 3.44.

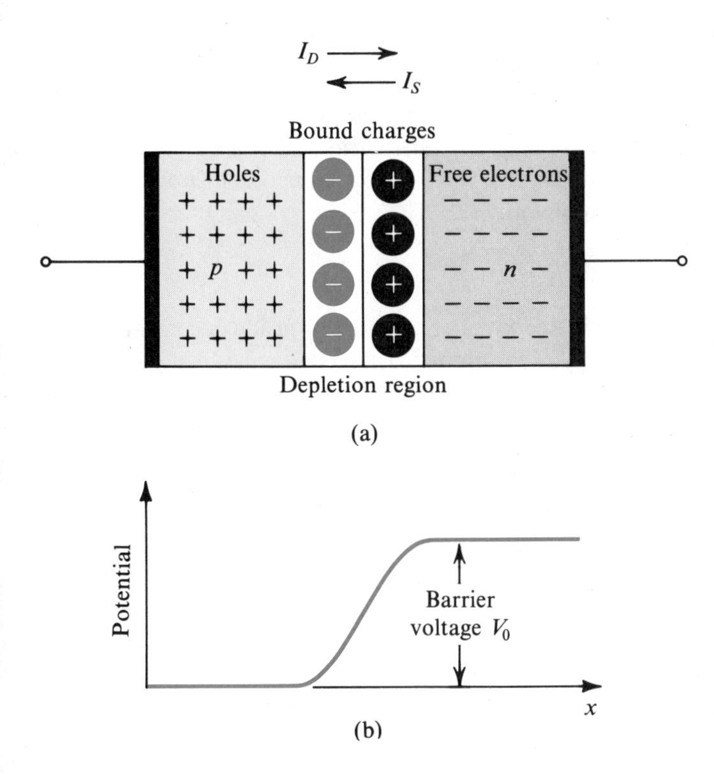

Fig. 3.44 (a) The *pn* junction with no applied voltage (open-circuited terminals). **(b)** The potential distribution along an axis perpendicular to the junction.

The Depletion Region

The holes that diffuse across the junction into the *n* region quickly recombine with some of the majority electrons present there and thus disappear from the scene. This recombination process results in the disappearance of some free electrons from the *n*-type material. Thus some of the bound positive charge will no longer be neutralized by free electrons, and this charge is said to have been **uncovered.** Since recombination takes place close to the junction, there will be a region close to the junction that is depleted of free electrons and contains uncovered bound positive charge, as indicated in Fig. 3.44.

The electrons that diffuse across the junction into the *p* region quickly recombine with some of the majority holes present there and thus disappear from the scene. This results also in the disappearance of some majority holes, causing some of the bound negative charge to be uncovered (that is, no longer neutralized by holes). Thus in the *p* material close to the junction there will be a region depleted of holes and containing uncovered bound negative charge, as indicated in Fig. 3.44.

From the above it follows that a **carrier-depletion region** will exist on both sides of the junction, with the *n* side of this region positively charged and the *p* side negatively charged. This carrier-depletion region—or, simply, **depletion region**—is also called the **space-charge region.** The charge on both sides of the depletion region cause an electric field to be established across the region; hence a potential difference results across the depletion region, with the *n* side at a positive voltage relative to the *p* side, as shown in Fig. 3.44(b). Thus the resulting electric field opposes the diffusion of holes into the *n* region and electrons into the p region. In fact, the voltage drop across the depletion region acts as a **barrier** that has to be overcome for holes to diffuse into the n region and electrons to diffuse into the p region. The larger the barrier voltage, the smaller the number of carriers that will be able to overcome the barrier, and hence the lower the magnitude of diffusion current. Thus the diffusion current I_D depends strongly on the voltage drop V_0 across the depletion region.

The Drift Current I_S and Equilibrium

In addition to the current component I_D due to majority carrier diffusion, a component due to minority carrier drift exists across the junction. Specifically, some of the thermally generated holes in the n material diffuse through the n material to the edge of the depletion region. There they experience the electric field in the depletion region, which sweeps them across that region into the p side. Similarly, some of the minority thermally generated electrons in the p material diffuse to the edge of the depletion region and get swept by the electric field in the depletion region across that region into the n side. These two current components—electrons moved by drift from p to n and holes moved by drift from n to p—add together to form the drift current I_S, whose direction is from the n side to the p side of the junction, as indicated in Fig. 3.44. Since the current I_S is carried by thermally generated minority carriers, its value is strongly dependent on temperature; however, it is independent of the value of the depletion layer voltage V_0.

Under open-circuit conditions (Fig. 3.44) no external current exists; thus the two opposite currents across the junction should be equal in magnitude:

$$I_D = I_S$$

This equilibrium condition is maintained by the barrier voltage V_0. Thus if for some reason I_D exceeds I_S, then more bound charge will be uncovered on both sides of the junction, the depletion layer will widen, and the voltage across it (V_0) will increase. This in turn causes I_D to decrease until equilibrium is achieved with $I_D = I_S$. On the other hand, if I_S exceeds I_D, then the amount of uncovered charge will decrease, the depletion layer will narrow, and the voltage across it will decrease. This causes I_D to increase until equilibrium is achieved with $I_D = I_S$.

The Terminal Voltage

When the pn junction terminals are left open-circuited the voltage measured between them will be zero. That is, the voltage V_0 across the depletion region *does not* appear between the diode terminals. This is because of the contact voltages existing at the metal-semiconductor junctions at the diode terminals, which counter and exactly balance the barrier voltage. If this were not the case, we would have been able to draw energy from the isolated pn junction, which would clearly violate the principle of conservation of energy.

Width of the Depletion Region

From the above it should be apparent that the depletion region exists in both the p and n materials and that equal amounts of charge exist on both sides. However, since usually the doping levels are not equal in the p and n materials, one can reason that the width of the depletion region will not be the same on the two sides. Rather, in order to uncover the same amount of charge the depletion layer will extend deeper into the more lightly doped material. In actual practice it is usual for one side of the junction to be much more lightly doped than the other, with the result that the depletion region exists almost entirely in one of the two semiconductor materials.

3.10 THE *pn* JUNCTION UNDER REVERSE-BIAS CONDITIONS

The behavior of the *pn* junction in the reverse direction is more easily explained on a microscopic scale if we consider exciting the junction with a constant-current source (rather than with a voltage source), as shown in Fig. 3.45. The current source I is obviously in the reverse direction. For the time being let the magnitude of I be less than I_S; if I is greater than I_S, breakdown will occur, as explained in Section 3.11.

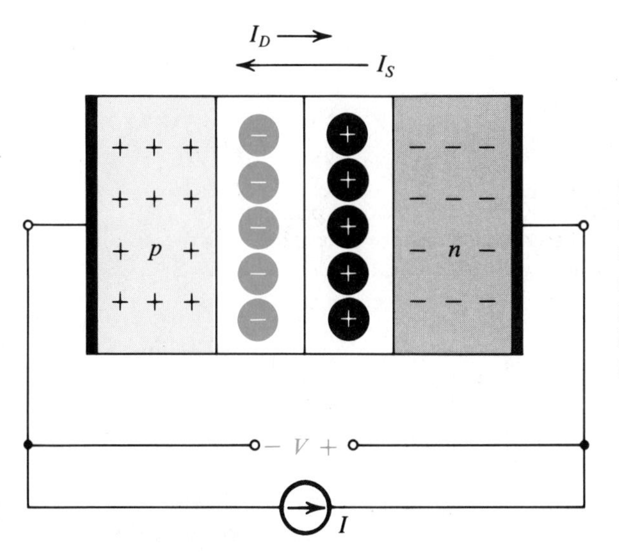

Fig. 3.45 The *pn* junction excited by a constant-current source I in the reverse direction. To avoid breakdown, I is kept smaller than I_S. Note that the depletion layer widens and the barrier voltage increases by V volts, which appears between the terminals as a reverse voltage.

The current I will be carried by electrons flowing in the external circuit from the *n* material to the *p* material (that is, in the direction opposite to that of I). This will cause electrons to leave the *n* material and holes to leave the *p* material. The free electrons leaving the *n* material cause the uncovered positive bound charge to increase. Similarly, the holes leaving the *p* material result in an increase in the uncovered negative bound charge. Thus the reverse current I will result in an increase in the width of, and the charge stored in, the depletion layer. This, in turn, will result in a higher voltage across the depletion region—that is, a greater barrier voltage V_0—which causes the diffusion current I_D to decrease. The drift current I_S, being independent of the barrier voltage, will remain constant. Finally, equilibrium (steady state) will be reached when

$$I_S - I_D = I$$

In equilibrium, the increase in depletion-layer voltage will appear as an external voltage between the diode terminals, with n being positive with respect to p.

We can now consider exciting the *pn* junction by a reverse voltage V, where V is less than the breakdown voltage V_{ZK}. When the voltage V is first applied, a reverse current flows in the external circuit from p to n. This current causes the increase in width and charge of the depletion layer. Eventually the voltage across the depletion layer will increase by the magnitude of the external voltage V, at which time an equilibrium is reached with the external reverse current I equal to $(I_S - I_D)$. Note, however, that initially the external current can be much greater than I_S. The purpose of this initial transient is to *charge* the depletion layer and increase the voltage across it by V volts.

From the above we observe the analogy between the depletion layer of a *pn* junction and a capacitor. As the voltage across the *pn* junction changes, the charge stored in the depletion layer changes accordingly. Figure 3.46 shows a sketch of typical charge-versus-external-voltage characteristic of a *pn* junction. Since this q–v characteristic is nonlinear,

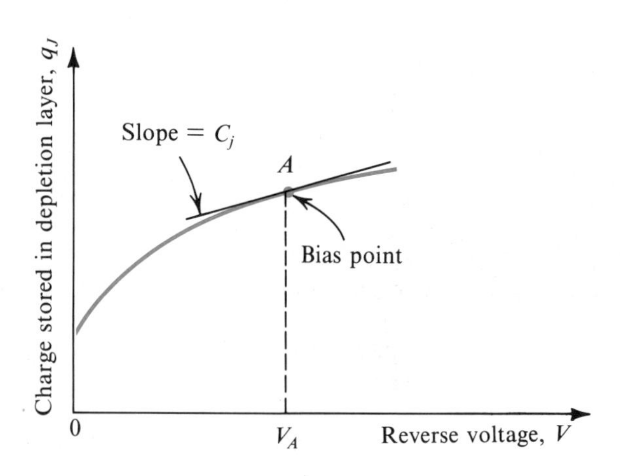

Fig. 3.46 The charge stored on either side of the depletion layer as a function of the reverse voltage V.

one has to be careful in speaking of the "depletion capacitance." As we did in Section 3.4 with the nonlinear i–v characteristic, we may consider operation around a bias point on the q–v curve, such as point A in Fig. 3.46, and define the small-signal or incremental **depletion capacitance** C_j as the slope of the q–v curve at the operating point.

$$C_j = \left.\frac{dq_J}{dv}\right|_{v=V_A}$$

It can be shown that

$$C_j = \frac{K}{(V_0 - V_D)^m}$$

where V_0 = the depletion-layer voltage with zero external voltage,
V_D = the voltage between the diode terminals (V_D is negative in the reverse direction),

K = a constant depending on the area of the junction and impurity concentrations, and

m = a constant depending on the distribution of impurity near the junction. The value of m ranges between $\frac{1}{3}$ and 4 for junctions of various types.

To recap, as a reverse-bias voltage is applied to a *pn* junction, a transient occurs during which the depletion capacitance is charged to the new bias voltage. After the transient dies, the steady-state reverse current is simply equal to $I_S - I_D$. Usually I_D is very small when the diode is reverse-biased and the reverse current is approximately equal to I_S. This, however, is only a theoretical model that does not apply very well. In actual fact, currents as high as few nanoamperes (10^{-9} A) flow in the reverse direction, in devices for which I_S is of the order of 10^{-15} A. This large difference is due to leakage and other effects. Furthermore, the reverse current is dependent to a certain extent on the magnitude of the reverse voltage, contrary to the theoretical model, which states that $I \simeq I_S$ independent of the value of the reverse voltage applied. Nevertheless, because of the very low currents involved, one is usually not interested in the details of the diode i–v characteristic in the reverse direction.

3.11 THE *pn* JUNCTION IN THE BREAKDOWN REGION

In considering diode operation in the reverse-bias region in Section 3.10 it was assumed that the reverse-current source I (Fig. 3.45) is smaller than I_S or, equivalently, that the reverse voltage V is smaller than the breakdown voltage V_{ZK}. We now wish to consider the breakdown mechanisms in *pn* junctions and explain the reasons behind the almost vertical line representing the i–v relationship in the breakdown region. For this purpose, let the *pn* junction be excited by a current source that causes a constant current I greater than I_S to flow in the reverse direction, as shown in Fig. 3.47. This current source will move holes from the p material through the external circuit[5] into the n material and electrons from the n material through the external circuit into the p material. This action results in more and more of the bound charge being uncovered; hence the depletion layer widens and the barrier voltage rises. This latter effect causes the diffusion current to decrease; eventually

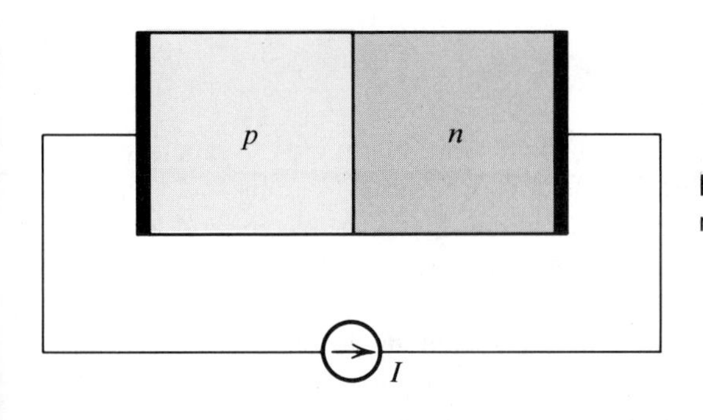

Fig. 3.47 The *pn* junction excited by a reverse-current source I, where $I > I_S$.

[5]The current in the external circuit will, of course, be carried entirely by electrons.

it will be reduced to almost zero. Nevertheless, this is not sufficient to reach a steady state, since I is greater than I_S. Therefore the process leading to the widening of the depletion layer continues until a sufficiently high junction voltage develops, at which point a new mechanism sets in to supply the charge carriers needed to support the current I. As will be now explained, this mechanism for supplying reverse currents in excess of I_S can take one of two forms depending on the *pn* junction material, structure, and so on.

The two possible breakdown mechanisms are the **zener effect** and the **avalanche effect.** If a *pn* junction breaks down with $V_Z < 5$ V, the breakdown mechanism is usually the zener effect. Avalanche breakdown occurs when V_Z is greater than about 7 V. For junctions that break down between 5 and 7 V, the breakdown mechanism can be either the zener or the avalanche effect or a combination of the two.

Zener breakdown occurs when the electric field in the depletion layer increases to the point where it can break covalent bonds and generate electron–hole pairs. The electrons generated this way will be swept by the electric field into the *n* side and the holes into the *p* side. Thus these electrons and holes constitute a reverse current across the junction that helps support the external current I. Once the zener effect starts, a large number of carriers can be generated, with a negligible increase in the junction voltage. Thus the reverse current in the breakdown region will be determined by the external circuit, while the reverse voltage appearing between the diode terminals will remain close to the rated breakdown voltage V_Z.

The other breakdown mechanism is avalanche breakdown, which occurs when the minority carriers that cross the depletion region under the influence of the electric field gain sufficient kinetic energy to be able to break covalent bonds in atoms with which they collide. The carriers liberated by this process may have sufficiently high energy to be able to cause other carriers to be liberated in another ionizing collision. This process occurs in the fashion of an avalanche, with the result that many carriers are created that are able to support any value of reverse current, as determined by the external circuit, with a negligible change in the voltage drop across the junction.

As mentioned before, *pn* junction breakdown is not a destructive process, provided that the maximum specified power dissipation is not exceeded. This maximum power dissipation rating, in turn, implies a maximum value for the reverse current.

3.12 THE *pn* JUNCTION UNDER FORWARD-BIAS CONDITIONS

We next consider operation of the *pn* junction in the forward-bias region. Again it is easier to explain physical operation if we excite the junction by a constant-current source supplying a current I in the forward direction, as shown in Fig. 3.48. The electrons carrying the current I in the external circuit from the *p* material to the *n* material cause holes to be extracted from the *n* region and electrons to be extracted from the *p* region. This results in majority carriers being supplied to both sides of the junction by the external circuit: holes to the *p* material and electrons to the *n* material. These majority carriers will neutralize some of the uncovered bound charge, causing less charge to be stored in the depletion layer. Thus the depletion layer narrows and the depletion barrier voltage reduces. The reduction in barrier voltage enables more holes to cross the barrier from the *p* material into the *n* material and more electrons from the *n* side to cross into the *p* side. Thus the

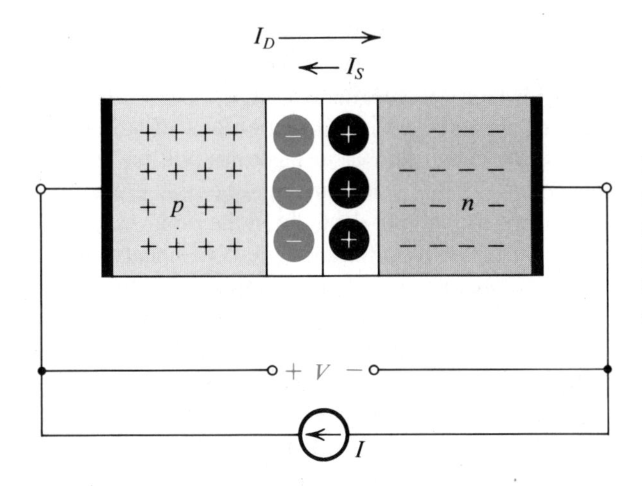

Fig. 3.48 The pn junction excited by a constant-current source supplying a current I in the forward direction. The depletion layer narrows and the barrier voltage decreases by V volts, which appears as an external voltage in the forward direction.

diffusion current I_D increases until equilibrium is achieved with $I_D - I_S = I$, the externally supplied forward current.

Let us now examine closely the current flow across the forward-biased pn junction in the steady state. The barrier voltage is now lower than V_0 by an amount V that appears between the diode terminals as a forward voltage drop (that is, the anode of the diode will be more positive than the cathode by V volts). Owing to the decrease in the barrier voltage or, alternatively, because of the forward voltage drop V, holes are **injected** across the junction into the n region and electrons are injected across the junction into the p region. The holes injected into the n region will cause the minority-carrier concentration there, p_n, to exceed the thermal equilibrium value, p_{n0}. The *excess* concentration $p_n - p_{n0}$ will be highest near the edge of the depletion layer and will decrease (exponentially) as one moves away from the junction, eventually reaching zero. Figure 3.49 shows such a minority-carrier distribution.

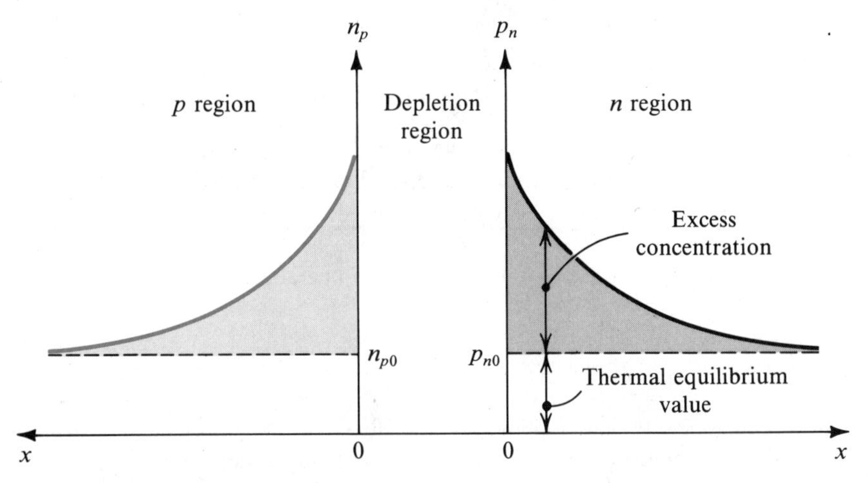

Fig. 3.49 Minority-carrier distribution as a function of the distance from the edges of the depletion layer.

In the steady state the concentration profile of **excess minority carriers** remains constant, and indeed it is such a distribution that gives rise to the increase of diffusion current I_D above the value I_S. This is because the distribution shown causes injected minority holes to diffuse away from the junction into the n region and disappear by recombination. To maintain equilibrium, an equal number of electrons will have to be supplied by the external circuit, thus replenishing the electron supply in the n material.

Similar statements can be made about the minority electrons in the p material. The diffusion current I_D is, of course, the sum of the electron and hole components.

Thus in the steady state, excess minority carrier distributions as shown in Fig. 3.49 exist in both the n and p materials. It can be shown that the hole current crossing the junction is proportional to the total excess hole charge stored in the n material. This charge is proportional to the area under the hole-concentration curve, which is shown shaded in Fig. 3.49 and which, in turn, is proportional to the excess hole concentration at the edge of the junction. Similarly, it can be shown that the current component carried by electrons is proportional to the excess stored electron charge in the p material. Thus the electron-current component is proportional to the area under the minority carrier distribution curve and hence is proportional to the excess concentration of electrons at the edge of the depletion layer.

Finally, we note that if we change the value of the external current I or, alternatively, change the forward voltage drop V, the minority carrier charge stored in the p and n materials will have to be changed for a new steady state to be established (this, of course, is in addition to the change in the charge stored in the depletion region). Thus the pn junction exhibits a capacitive effect—in addition to the depletion capacitance—related to the storage of minority carrier charges. Since these charges q_M are proportional to the current flowing across the junction I, it follows from the diode equation (Eq. 3.1) that q_M is related to the forward voltage drop by a relationship of the form

$$q_M = q_0(e^{v/nV_T} - 1)$$

where q_0 is a constant charge proportional to the current I_S. Thus the q–v curve of this capacitive effect is clearly a nonlinear one. We may model this capacitive behavior, however, by a small-signal capacitance C_d,

$$C_d = \left. \frac{dq_M}{dv} \right|_{v=V_A}$$

where V_A is the dc diode voltage at the operating point A, around which the small-signal model is valid. The capacitance C_d is called the **diffusion capacitance.** From the above relationships one can easily establish that C_d is proportional to the value of $q_M + q_0$. Whereas in the reverse-bias region C_d is zero, its value in the forward-bias region is approximately proportional to the bias current I_A at the operating point.

The Complete Small-Signal Model

From the above we conclude that an appropriate small-signal model for the pn junction consists of the diode incremental resistance r_d in parallel with the depletion layer capacitance C_j, in parallel with the diffusion capacitance C_d. This model is depicted in Fig.

3.50. If the diode is biased in the forward region at a point A on the i–v curve, then

$$r_d = \frac{nV_T}{I_A}$$

$$C_d = k_c I_A$$

$$C_j = \frac{K}{(V_0 - V_A)^m}$$

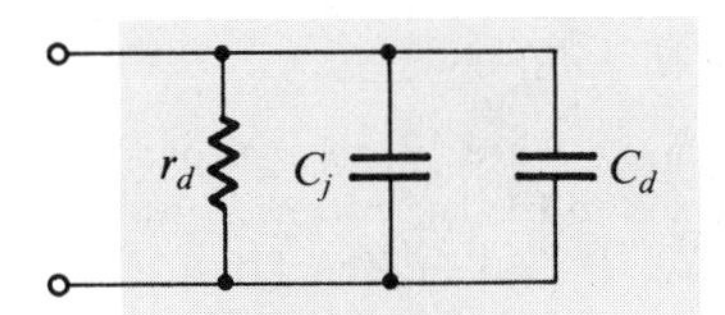

Fig. 3.50 Complete small-signal model of a *pn* junction.

where k_c is a constant and V_A is the forward bias voltage (a positive number). On the other hand, if the diode is reverse-biased with a reverse-bias voltage $v = -V_A$ (V_A is a positive number), then

$$r_d \simeq \infty$$

$$C_d \simeq 0$$

$$C_j = \frac{K}{(V_0 + V_A)^m}$$

The values of C_d and C_j are, of course, dependent on the size of the junction, being directly proportional to the cross-sectional area. For a small-signal diode reverse-biased by few volts, C_j is typically of the order of 1 pF. The same diode, forward-biased with a current of few milliamperes, has a diffusion capacitance of the order of 10 pF.

SUMMARY

- In the forward direction, the ideal diode conducts any current forced by the external circuit while displaying a zero voltage drop. The ideal diode does not conduct in the reverse direction; any applied voltage appears as reverse bias across the diode.

- The unidirectional-current-flow property makes the diode useful in the design of rectifier circuits.

- The forward conduction of practical silicon diodes is accurately characterized by the relationship $i = I_s e^{v/nV_T}$.

- A silicon diode conducts a negligible current until the forward voltage is at least 0.5 V. Then the current increases rapidly, with the voltage drop increasing by 60 to 120 mV (depending on the value of n) for every decade of current change.

- In the reverse direction, a silicon diode conducts a current of the order of 10^{-9} A. This current is much greater than I_S and increases with the magnitude of reverse voltage.

- Beyond a certain value of reverse voltage (that depends on the diode) breakdown occurs, and current increases rapidly with a small corresponding increase in voltage.

- Diodes designed to operate in the breakdown region are called zener diodes. They are employed in the design of voltage regulators whose function is to provide a constant dc voltage that varies little with variations in power supply voltage and/or load current.

▶ A hierarchy of diode models exists, with the selection of an appropriate model dictated by the application.

▶ In many applications, a conducting diode is modeled as having a constant voltage drop, usually about 0.7 V.

▶ A diode biased to operate at a dc current I_D has a small-signal resistance $r_d = nV_T/I_D$.

▶ The silicon junction diode is basically a *pn* junction. Such a junction is formed in a single silicon crystal.

▶ In *p*-type silicon there is an overabundance of holes (positively charged carriers), while in *n*-type silicon electrons are abundant.

▶ A carrier-depletion region develops at the interface in a *pn* junction, with the *n* side positively charged and the *p* side negatively charged. The voltage difference resulting is called the barrier voltage.

▶ A diffusion current I_D flows in the forward direction (carried by holes from the *p* side and electrons from the *n* side), and a current I_S flows in the reverse direction (carried by thermally generated minority carriers). In an open-circuited junction, $I_D = I_S$ and the barrier voltage is denoted V_0.

▶ Applying a reverse-bias voltage $|V|$ to a *pn* junction causes the depletion region to widen, and the barrier voltage increases to $(V_0 + |V|)$. The diffusion current decreases and a net reverse current of $(I_S - I_D)$ flows.

▶ Applying a forward-bias voltage $|V|$ to a *pn* junction causes the depletion region to become narrower, and the barrier voltage decreases to $(V_0 - |V|)$. The diffusion current increases, and a net forward current of $(I_D - I_S)$ flows.

BIBLIOGRAPHY

E. J. Angelo, Jr., *Electronics: BJTs, FETs and Microcircuits,* New York: McGraw-Hill, 1969.

S. B. Burns and P. R. Bond, *Principles of Electronic Circuits,* St. Paul: West, 1987.

S. Franco, *Design with Operational Amplifiers and Analog Integrated Circuits,* New York: McGraw-Hill, 1988.

P. E. Gray and C. L. Searle, *Electronic Principles,* New York: Wiley, 1969.

D. A. Hodges and H. G. Jackson, *Analysis and Design of Digital Integrated Circuits,* Second Edition, New York: McGraw-Hill, 1988.

D. H. Navon, *Semiconductor Microdevices and Materials,* New York: Holt, Rinehart and Winston, 1986.

S. Soclof, *Applications of Analog Integrated Circuits,* Englewood Cliffs, NJ: Prentice-Hall, 1985.

S. M. Sze, *Semiconductor Devices, Physics and Technology,* New York: Wiley, 1985.

PROBLEMS

Section 3.1: The Ideal Diode

3.1 For the circuits shown in Fig. P3.1 using ideal diodes, find the values of the voltages and currents indicated.

3.2 For the circuits shown in Fig. P3.2 using ideal diodes find the values of the labeled voltages and currents.

D3.3 For the logic gate of Fig. 3.5(a) assume ideal diodes and input voltage levels of 0 and +5 V. Find a suitable value for R so that the current required from each of the input signal sources does not exceed 0.2 mA.

D3.4 Repeat Problem 3.3 for the logic gate of Fig. 3.5(b).

3.5 Assuming that the diodes in the circuits of Fig. P3.5 are ideal, find the values of the labeled voltages and currents.

3.6 Assuming that the diodes in the circuits of Fig. P3.6 are ideal, utilize Thévenin's theorem to simplify the circuits and thus find the values of the labeled currents and voltages.

D3.7 For the rectifier circuit of Fig. 3.3(a) let the input sine wave have 120-V rms value and assume the diode to be ideal. Select a suitable value for R so that the peak diode current does not exceed 0.1 A. What is the greatest reverse voltage that will appear across the diode?

3.8 Consider the rectifier circuit of Fig. 3.3 in the event that the input source v_I has a source resistance R_s. For

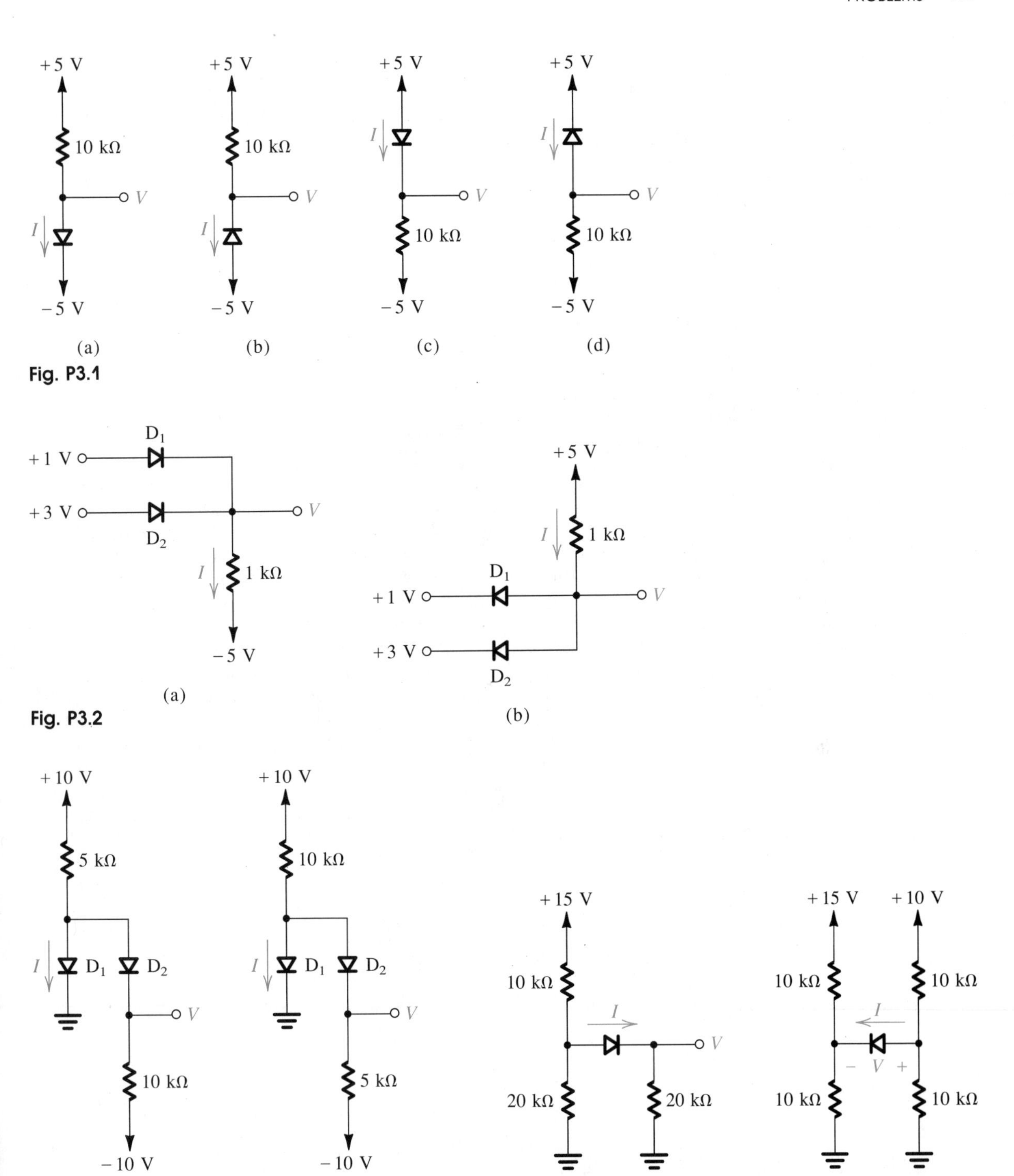

Fig. P3.1

Fig. P3.2

Fig. P3.5

Fig. P3.6

the case $R_s = R$ and assuming the diode to be ideal, sketch and clearly label the transfer characteristic v_O versus v_I.

3.9 Consider the battery charger described in Example 3.1. If the sinusoidal source is replaced with a square wave of the same amplitude, for what fraction of the cycle does the diode conduct? What is the average diode current?

3.10 Repeat Problem 3.9 for a symmetrical triangular wave of 24-V amplitude.

3.11 The circuit of Fig. P3.11 can be used in a signalling system using one wire, plus a common ground return. At any moment, the input has one of three values: +3 V, 0, −3 V. What is the status of the lamps for each input value? (Note that the lamps can be located apart from each other, and that there may be several of each type of connection, all on one wire!).

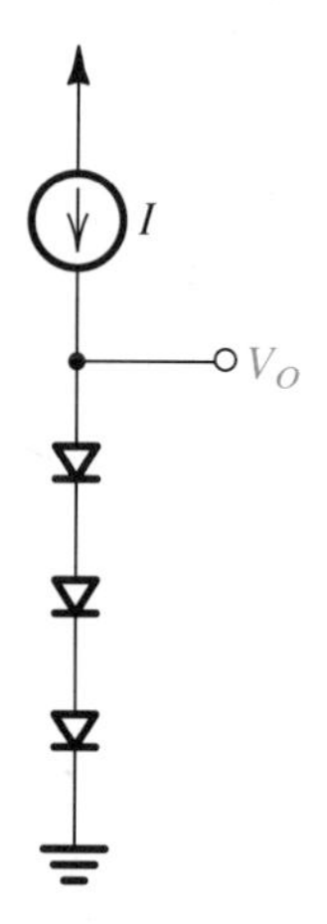

Fig. P3.15

Section 3.2: Terminal Characteristics of Junction Diodes

3.12 At about what current and voltage does the exponential approximation for diode current (Eq. 3.3) provide less than 1% error? Use $n = 2$ and $V_T = 25$ mV.

3.13 At what forward voltage does a diode for which $n = 2$ conduct a current equal to 1000 I_S? In terms of I_S, what current flows in the same diode when its forward voltage is 0.7 V?

3.14 A diode for which the forward voltage drop is 0.7 V at 1.0 mA and for which $n = 1$ is operated at 0.5 V. What is the value of the current?

3.15 The circuit in Fig. P3.15 utilizes three identical diodes having $n = 1$ and $I_S = 10^{-14}$ A. Find the value of the current I required to obtain an output voltage $V_O = 2$ V. If a current of 1 mA is drawn away from the output terminal by a load, what is the change in output voltage?

3.16 A junction diode is operated in a circuit in which it is supplied with a constant current I. What is the effect on the forward voltage of the diode if an identical diode is connected in parallel? Assume $n = 1$.

3.17 A diode measured at two operating currents, 0.2 mA and 10 mA, is found to have corresponding voltages of 0.650 and 0.750 V. Find the values of n and I_S.

3.18 When a 10-A current is applied to a particular diode it is found that the junction voltage immediately becomes 700 mV. However, as the power being dissipated in the diode raises its temperature, it is found

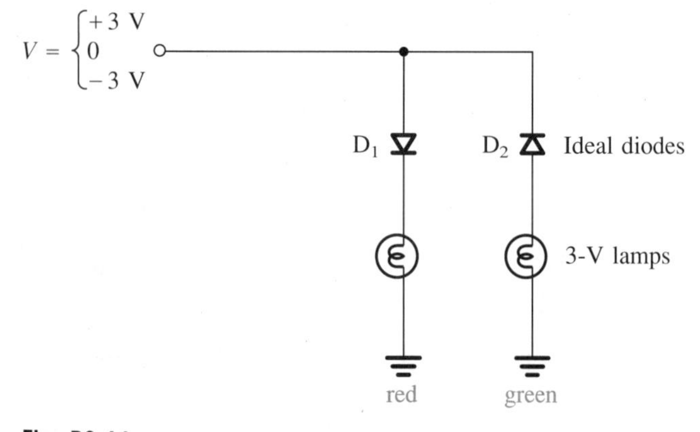

Fig. P3.11

that the voltage decreases and eventually reaches 600 mV. What is the apparent rise in junction temperature? What is the power dissipated in the diode in its final state? What is the temperature rise per watt of power dissipation?

Section 3.3: Analysis of Diode Circuits

3.19 Consider the graphical analysis of the diode circuit of Fig. 3.10 with $V_{DD} = 1$ V, $R = 1$ kΩ, and a diode having $I_S = 10^{-15}$ A and $n = 1$. Calculate a small number of points on the diode characteristic in the vicinity of where you expect the load line to intersect it, and use a graphical process to refine your estimate of diode current. What value of diode current and voltage do you find? Analytically, find the voltage corresponding to your estimate of current. By how much does it differ from the graphically estimated value?

3.20 Use the iterative analysis procedure to determine the diode current and voltage in the circuit of Fig. 3.10 for $V_{DD} = 1$ V, $R = 1$ kΩ, and a diode having $I_S = 10^{-15}$ A and $n = 1$.

3.21 A 1-mA diode (i.e., one that has $v_D = 0.7\ V$ at $i_D = 1$ mA) is connected in series with a 200-Ω resistor to a 1.0-V supply.

(a) Provide a rough estimate of the diode current you would expect.

(b) If the diode is characterized by $n = 2$, estimate the diode current more closely using iterative analysis.

D3.22 Assuming the availability of diodes for which $v_D = 0.7$ V at $i_D = 1$ mA and $n = 1$, design a circuit that utilizes four diodes connected in series, in series with a resistor R connected to a 15-V power supply. The voltage across the string of diodes is to be 3.0 V.

3.23 Find the parameters of a piecewise-linear model of a diode for which $v_D = 0.7$ V at $i_D = 1$ mA and $n = 2$. The model is to fit exactly at 1 mA and 10 mA. Calculate the error in millivolts in predicting v_D using the piecewise-linear model at $i_D = 0.5$, 5, and 14 mA.

3.24 Consider the half-wave rectifier circuit of Fig. 3.3(a) with $R = 1$ kΩ and the diode having the characteristics and the piecewise-linear model shown in Fig. 3.12 ($V_{D0} = 0.65$ V, $r_D = 20$ Ω). Analyze the rectifier circuit using the piecewise-linear model for the diode and thus find the output voltage v_O as a function of v_I. Sketch the transfer characteristic v_O versus v_I for $0 \leq v_I \leq 10$ V. For v_I being a sinusoid with 10-V peak amplitude, sketch and clearly label the waveform of v_O.

3.25 Solve the problems in Example 3.2 using the constant-voltage-drop diode model ($V_D = 0.7$ V).

3.26 For the circuits shown in Fig. P3.1, using the constant-voltage-drop ($V_D = 0.7$ V) diode model, find the voltages and currents indicated.

3.27 For the circuits shown in Fig. P3.2, using the constant-voltage-drop ($V_D = 0.7$ V) diode model, find the voltages and currents indicated.

3.28 For the circuits in Fig. P3.5, using the constant-voltage-drop ($V_D = 0.7$ V) diode model, find the values of the labeled currents and voltages.

3.29 For the circuits in Fig. P3.6, utilize Thévenin's theorem to simplify the circuits and find the values of the labeled currents and voltages. Assume that the diodes can be represented by the constant-voltage-drop model ($V_D = 0.7$ V).

D3.30 Repeat Problem 3.7, representing the diode by its constant-voltage-drop model ($V_D = 0.7$). How different is the resulting design?

3.31 Repeat the problem in Example 3.1 assuming that the diode has 10 times the area of the device whose characteristics and piecewise-linear model are displayed in Fig. 3.12. Represent the diode by its piecewise-linear model ($v_D = 0.65 + 2i_D$).

**3.32 For the circuit shown in Fig. P3.32, utilize the constant-voltage-drop model (0.7 V) for each conducting

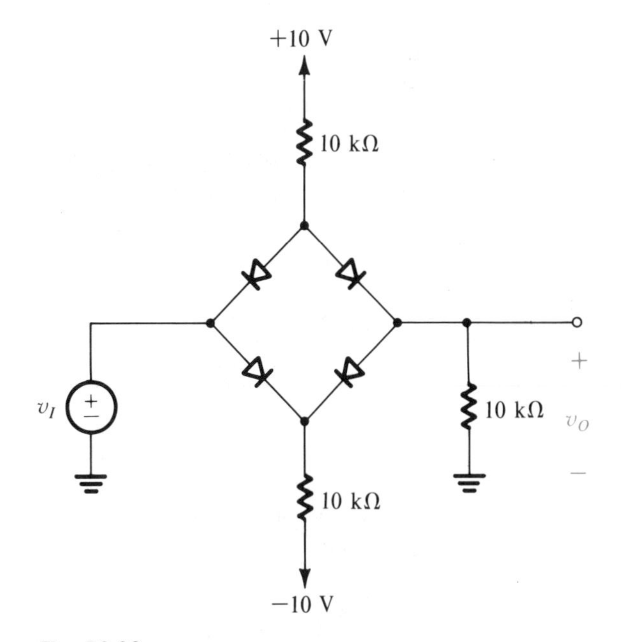

Fig. P3.32

diode and show that the transfer characteristic can be described by

$$\text{for } -4.65 \le v_I \le 4.65 \text{ V}, \qquad v_O = v_I;$$

$$\text{for } v_I \ge +4.65 \text{ V}, \qquad v_O = +4.65 \text{ V};$$

$$\text{for } v_I \le -4.65 \text{ V}, \qquad v_O = -4.65 \text{ V}.$$

Section 3.4: The Small-Signal Model and Its Application

3.33 The small-signal model is said to be valid for voltage variations of about 10 mV. To what percent current change does this correspond (consider both positive and negative signals) for
(a) $n = 1$?
(b) $n = 2$?

3.34 What is the incremental resistance of 10 1-mA diodes connected in parallel and fed with a dc current of 10 mA. Let $n = 2$. (A 1-mA diode is one that has a 0.7-V drop at a 1-mA current.)

*3.35 Consider the voltage regulator circuit shown in Fig. 3.20. The value of R is selected to obtain an output voltage V_O (across the diode) of 0.7 V.
(a) Use the diode small-signal model to show that the change in output voltage corresponding to a change of 1 V in V^+ is

$$\frac{\Delta V_O}{\Delta V^+} = \frac{nV_T}{V^+ + nV_T - 0.7}$$

This quantity is known as the line regulation and is usually expressed in mV/V.
(b) Generalize the expression above for the case of m diodes connected in series and the value of R adjusted so that the voltage across each diode is 0.7 V (and $V_O = 0.7m$ volts).
(c) Calculate the value of line regulation for the case $V^+ = 10$ V (nominally) and (i) m = 1; (ii) m = 3. Use $n = 2$.

D*3.36 Consider the voltage-regulator circuit shown in Fig. 3.20 under the condition that a load current I_L is drawn from the output terminal. Denote the output voltage (across the diode) by V_O.
(a) If the value of I_L is sufficiently small so that the corresponding change in regulator output voltage ΔV_O is small enough to justify using the diode small-signal model, show that

$$\frac{\Delta V_O}{I_L} = -(r_d // R)$$

This quantity is known as the load regulation and is usually expressed in mV/mA.
(b) If the value of R is selected such that at no load the voltage across the diode is 0.7 V and the diode current is I_D, show that the expression derived in (a) becomes

$$\frac{\Delta V_O}{I_L} = -\frac{nV_T}{I_D} \frac{V^+ - 0.7}{V^+ - 0.7 + nV_T}$$

Select the lowest possible value for I_D that results in a load regulation ≤ 5 mV/mA. Assume $n = 2$. If V^+ is nominally 10 V, what value of R is required? Also specify the diode required.
(c) Generalize the expression derived in (b) for the case of m diodes connected in series and R adjusted to obtain $V_O = 0.7m$ volts at no load.

*3.37 In the circuit shown in Fig. P3.37, I is a dc current and v_s is a sinusoidal signal. Capacitor C is very large; its function is to couple the signal to the diode but block the dc current from flowing into the signal source. Use the diode small-signal model to show that the signal component of the output voltage is

$$v_o = v_s \frac{nV_T}{nV_T + IR_s}$$

If $v_s = 10$ mV, find v_o for $I = 1$ mA, 0.1 mA, and 1 μA. Let $R_s = 1$ kΩ and $n = 2$. At what value of I does v_o become one-half of v_s? Note that this circuit functions as a signal attenuator with the attenuation factor controlled by the value of the dc current I.

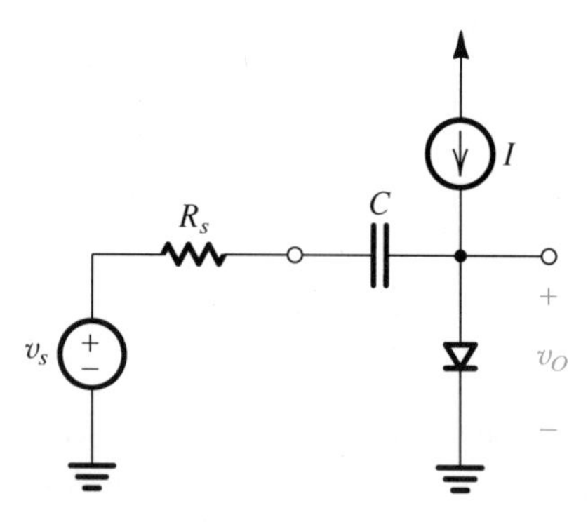

Fig. P3.37

D*3.38 A voltage regulator consisting of two diodes in series fed with a constant-current source is used as a replacement for a single carbon-zinc (battery) cell of nominal voltage 1.5 V. The regulator load current varies from

2 to 7 mA. Constant-current supplies of 5, 10, and 15 mA are available. Which would you choose and why? What change in output voltage would result when the load current varies over its full range? Assume that the diodes have a 0.7 V drop at a 1-mA current and $n = 2$.

Section 3.5: Operation in the Reverse Breakdown Region—Zener Diodes

3.39 A voltage regulator consisting of a 6.8-V zener diode and a 100-Ω resistor and intended for operation with a 9-V supply is accidentally connected to 15-V supply instead. Assuming that r_z is very small, calculate the expected values of zener current and the power dissipation in both the zener diode and the resistor, for both the normal and aberrant situations, and compare the ratios.

D3.40 A zener diode for which $V_Z = 6.8$ V exhibits an almost constant voltage for $I_Z > I_{ZK} = 0.1$ mA. (That is, r_z is negligibly small for $I_Z > I_{ZK}$.) It is employed in a shunt-regulator circuit supplied by a 10-V supply via a resistor R of 1 kΩ value. For what load-current range will the output retain its regulated value? What is the value of load resistance below which the output voltage will no longer be regulated? Redesign the circuit to provide a regulated output voltage with a load resistor of as low as 1 kΩ and with a minimum zener current 10 times I_{ZK}.

D3.41 A designer requires a shunt regulator of about 20 V. Two kinds of zener diodes are available: 6.8-V devices with r_z of 10 Ω and 5.1-V devices with r_z of 30 Ω. For the two major choices possible, find the load regulation. In this calculation neglect the effect of the regulator resistance R.

3.42 A shunt regulator utilizing a zener with an incremental resistance of 4 Ω is fed through an 82-Ω resistor. If the raw supply changes by 1.4 V, what is the corresponding change in the regulated output voltage?

3.43 A 9.1-V zener diode exhibits its nominal voltage at a test current of 28 mA. At this current the incremental resistance is specified as 5 Ω. Find V_{Z0} of the zener model. Find the zener voltage at a current of 10 mA and at 100 mA.

D*3.44 Provide two designs of shunt regulators utilizing the 1N5235 zener diode, which is specified as follows: $V_Z = 6.8$ V and $r_z = 5$ Ω for $I_Z = 20$ mA; at $I_Z = 0.25$ mA (nearer the knee) $r_z = 750$ Ω. For both designs the supply voltage is nominally 9 V and varies by ±1 V. For the first design, assume that the availability of supply current is not a problem, and thus operate the diode at 20 mA. For the second design, assume that the current from the raw supply is limited, and therefore you are forced to operate the diode at 0.25 mA. For the purpose of these initial designs assume no load. For each design find the value of R and the line regulation.

D*3.45 It is required to design a zener shunt regulator to provide a regulated voltage of about 10 V. The available 10-V, 1-W zener of type 1N4740 is specified to have a 10-V drop at a test current of 25 mA. At this current its r_z is 7 Ω. The raw supply available has a nominal value of 20 V but can vary by as much as ±25%. The regulator is required to supply a load current of 0 to 20 mA. Design for a minimum zener current of 5 mA.
(a) Find V_{Z0}.
(b) Calculate the required value of R.
(c) Find the line regulation. What is the change in V_O expressed as a percentage, corresponding to the ±25% change in V_S?
(d) Find the load regulation. By what percentage does V_O change from the no-load to the full-load condition?
(e) What is the maximum current that the zener in your design should be able to conduct? What is the zener power dissipation under this condition?

Section 3.6: Rectifier Circuits

3.46 Consider the half-wave rectifier circuit of Fig. 3.29(a) with the diode reversed. Let v_S be a sinusoid with 10-V peak amplitude, and let $R = 1$ kΩ. Use the constant-voltage-drop diode model with $V_D = 0.7$ V.
(a) Sketch the transfer characteristic.
(b) Sketch the waveform of v_O.
(c) Find the average value of v_O.
(d) Find the peak current in the diode.
(e) Find the PIV of the diode.

3.47 Using the exponential diode characteristic, show that for v_S and v_O both greater than zero, the circuit of Fig. 3.29(a) has the transfer characteristic

$$v_O = v_S - v_D \text{ (at } i_D = 1 \text{ mA)} - n\,V_T \ln (v_O/R)$$

where v_S and v_O are in volts and R is in kilohms.

3.48 Consider a half-wave rectifier circuit with a triangular-wave input of 20-V peak-to-peak amplitude and zero average and with $R = 1$ kΩ. Assume that the diode can be represented with the piecewise-linear model with $V_{D0} = 0.65$ V and $r_D = 20$ Ω. Find the average value of v_O.

3.49 For a half-wave rectifier circuit with $R = 1\ \text{k}\Omega$ and utilizing a diode whose voltage drop is 0.7 V at a current of 1 mA and exhibiting a 0.1-V change per decade of current variation, find the values of the input voltage to the rectifier corresponding to $v_O = 0.1$, 0.5, 1, 2, 5, and 10 V. Plot the rectifier transfer characteristic.

3.50 A half-wave rectifier circuit using a diode with $v_D =$ 0.7 V at $i_D = 1$ mA and a load resistance of 10 kΩ is supplied by a 1-kHz symmetrical square wave having a peak-to-peak amplitude of V volts and a zero average. Calculate the average output voltage for three values of V: 200 V, 20 V, and 2 V. Assume that the diode voltage changes by 0.1 V for every decade change in current.

D3.51 It is required to design a full-wave rectifier circuit using the circuit of Fig. 3.30 to provide an average output voltage of
(a) 10 V,
(b) 100 V.
In each case find the required turns ratio of the transformer. Assume that a conducting diode has a voltage drop of 0.7 V. The ac line voltage is 120 V rms.

D3.52 Repeat Problem 3.51 for the bridge rectifier circuit of Fig. 3.31.

D3.53 Consider the full-wave rectifier in Fig. 3.30 when the transformer turns ratio is such that the voltage across the entire secondary winding is 24 V rms. If the input ac line voltage (120 V rms) fluctuates by as much as ±10%, find the required PIV of the diodes. (Remember to use a factor of safety in your design.)

***3.54** The circuit in Fig. P3.54 implements a complementary-output rectifier. Sketch and clearly label the waveforms of v_O^+ and v_O^-. Assume a 0.7-V drop across each conducting diode. If the magnitude of the average of each output is to be 15 V, find the required amplitude of the sine wave across the entire secondary winding. What is the PIV of each diode?

D*3.55 It is required to use a peak rectifier to design a dc power supply that provides an average dc output voltage of 15 V on which a maximum of ±1-V ripple is allowed. The rectifier feeds a load of 150-Ω resistance. The rectifier is fed from the line voltage (120 V rms, 60 Hz) through a transformer. The diodes available have 0.7-V drop when conducting. If the designer opts for the half-wave circuit:
(a) Specify the rms voltage that must appear across the transformer secondary.
(b) Find the required value of the filter capacitor.
(c) Find the maximum reverse voltage that will appear across the diode, and specify the PIV rating of the diode.
(d) Calculate the average current through the diode during conduction.
(e) Calculate the peak diode current.

D*3.56 Repeat Problem 3.55 for the case the designer opts for a full-wave circuit utilizing a center-tapped transformer.

D*3.57 Repeat Problem 3.55 for the case the designer opts for a full-wave bridge rectifier circuit.

3.58 Consider a half-wave peak rectifier fed with a voltage v_S having a triangular waveform with 20-V peak-to-peak amplitude, zero average and 1-kHz frequency. Assume that the diode has a 0.7-V drop when conducting. Let the load resistance $R = 100\ \Omega$ and the filter capacitor $C = 100\ \mu\text{F}$. Find the average dc output voltage, the time interval during which the diode conducts, the average diode current during conduction, and the maximum diode current.

D*3.59 Consider the circuit in Fig. P3.54 with two equal filter capacitors placed across the load resistors R. Assume that the diodes available exhibit a 0.7-V drop when

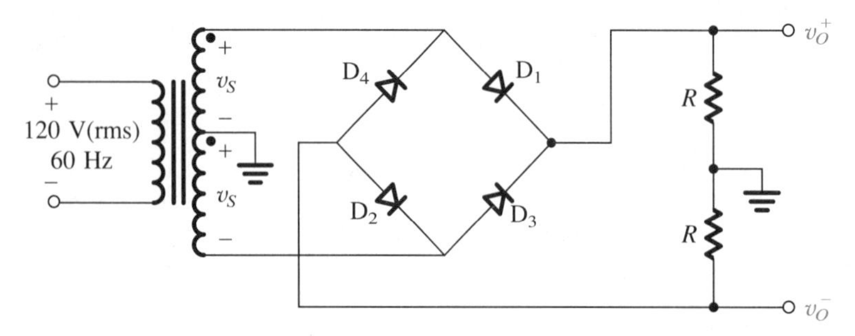

Fig. P3.54

conducting. Design the circuit to provide ±15-V dc output voltage with a peak-to-peak ripple no greater than 1 V. Each supply should be capable of providing 200 mA dc current to its load resistor R. Completely specify the diodes and the transformer.

Section 3.7: Limiting and Clamping Circuits

3.60 Sketch the transfer characteristic v_O versus v_I for the limiter circuits shown in Fig. P3.60. All diodes begin conducting at a forward voltage drop of 0.5 V and display voltage drops of 0.7 V when fully conducting.

3.61 Repeat Problem 3.60 assuming that the diodes are modeled with the piecewise-linear model with $V_{D0} = 0.65$ V and $r_D = 20\ \Omega$.

3.62 The circuits in Fig. P3.60(a) and (d) are connected as follows: The two input terminals are tied together, and the two output terminals are tied together. Sketch the transfer characteristic of the circuit resulting, assuming that the cut-in voltage of the diodes is 0.5 V and their voltage drop when fully conducting is 0.7 V.

3.63 Repeat Problem 3.62 for the two circuits in Fig. P3.60(a) and (b) connected together as follows: The two input terminals are tied together, and the two output terminals are tied together.

*3.64 Sketch and clearly label the transfer characteristic of the circuit in Fig. P3.64 for $-20\text{ V} \le v_I \le +20$ V.

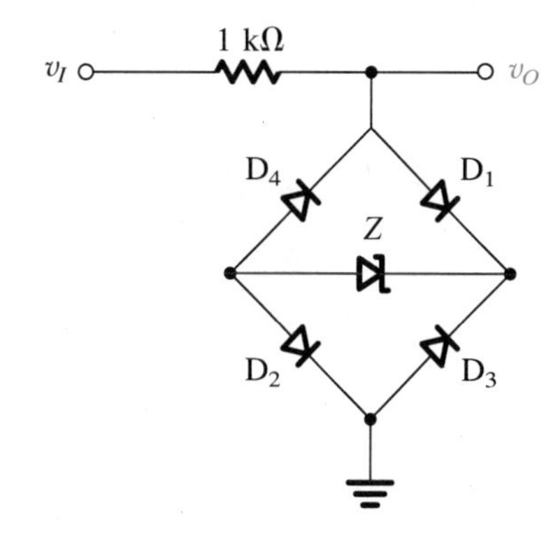

Fig. P3.64

Assume that the diodes can be represented by a piecewise-linear model with $V_{D0} = 0.65$ V and $r_D = 20\ \Omega$. Assuming that the specified zener voltage (8.2 V) is measured at a current of 10 mA and that $r_z = 20\ \Omega$, represent the zener by a piecewise-linear model.

*3.65 Plot the transfer characteristic of the circuit in Fig. P3.65 by evaluating v_I corresponding to $v_O = 0.5$, 0.6, 0.7, 0.8, 0, −0.5, −0.6, −0.7, and −0.8 V. Assume that the diodes are 1-mA units having a 0.1-V/decade logarithmic characteristic. Characterize the

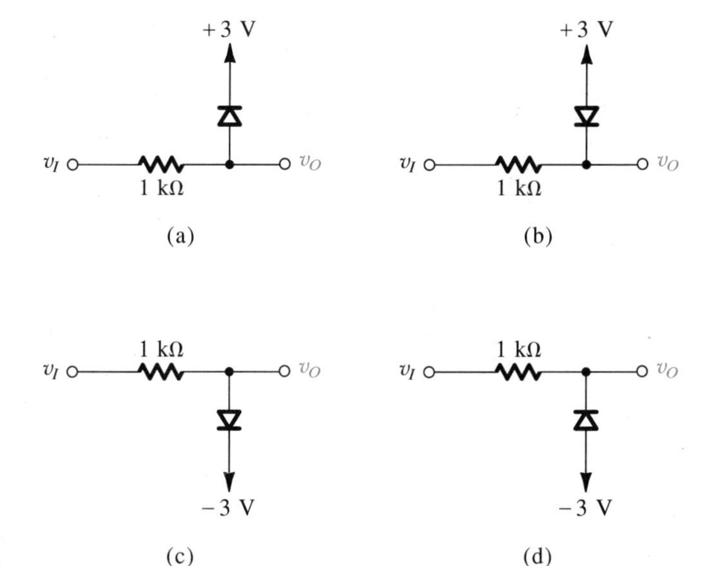

Fig. P3.60

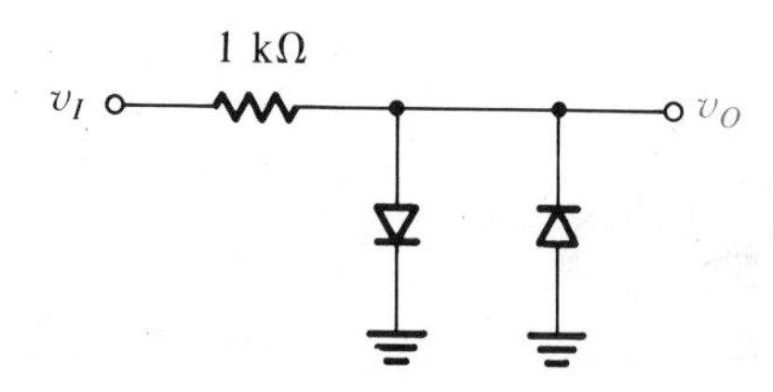

Fig. P3.65

circuit as a hard or soft limiter. What is the value of K? Estimate L_+ and L_-.

D3.66 Design limiter circuits using only diodes and 10-kΩ resistors to provide an output signal limited to the range:
(a) -0.7 V and above,
(b) -2.1 V and above, and
(c) ± 1.4 V.
Assume that each diode has a 0.7-V drop when conducting.

3.67 A clamped capacitor using an ideal diode is supplied with a sine wave of 10-V rms. What is the average (dc) value of the resulting output?

**3.68 For the circuits in Fig. P3.68, each utilizing an ideal diode (or diodes) sketch the output for the input shown. Label the most positive and most negative output levels. Assume $CR \gg T$.

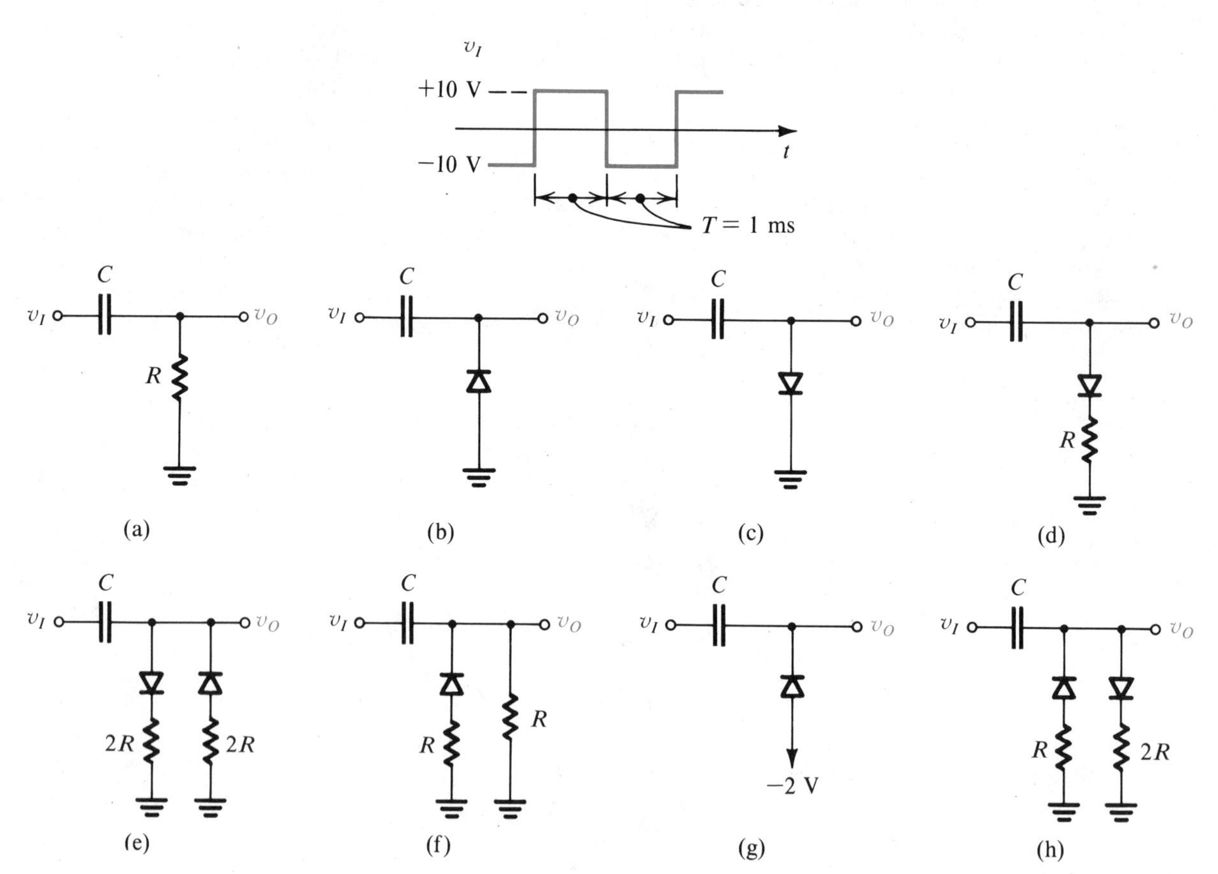

Fig. P3.68

CHAPTER 4

Bipolar Junction Transistors (BJTs)

INTRODUCTION

Having studied the junction diode, which is the most basic two-terminal semiconductor device, we now turn our attention to three-terminal semiconductor devices. Three-terminal devices are far more useful than two-terminal ones because they can be used in a multitude of applications ranging from signal amplification to the design of digital logic and memory circuits. The basic principle involved is the use of the voltage between two terminals to control the current flowing in the third terminal. In this way a three-terminal device can be used to realize a controlled source, which, as we learned in Chapter 1, is the basis for amplifier design. Also, in the extreme, the control signal can be used to cause the current in the third terminal to change from zero to a large value, thus allowing the device to act as a switch. The switch is the basic element of digital circuits.

There are two major types of three-terminal semiconductor devices: the bipolar junction transistor (BJT), which is the subject of this chapter, and the field-effect transistor (FET), which we shall study in Chapter 5. The two transistor types are equally important, and each offers distinct advantages and has unique areas of application.

The bipolar transistor consists of two *pn* junctions constructed in a special way and connected in series, back to back. Current is conducted by both electrons and holes, hence the name bipolar.

The BJT, often referred to simply as ''the transistor,'' is widely used in discrete circuits as well as in IC design, both analog and digital. The device characteristics are so well understood that one is able to design transistor circuits whose performance is remarkably predictable and quite insensitive to variations in device parameters.

We shall start by presenting a simple qualitative description of the operation of the transistor. Though simple, this physical description provides considerable insight into the performance of the transistor as a circuit element. We will quickly move from describing current flow in terms of holes and electrons to a study of transistor terminal characteristics. First-order models for transistor operation in different modes will be developed and utilized in the analysis of transistor circuits. One of the main objectives of this chapter is to develop in the reader a high degree of familiarity with the transistor. Thus by the end of the chapter the reader should be able to perform rapid first-order analysis of transistor circuits, as well as design single-stage transistor amplifiers.

4.1 PHYSICAL STRUCTURE AND MODES OF OPERATION

Figure 4.1 shows a simplified structure for a BJT. A practical transistor structure will be shown later (see also Appendix A, which deals with fabrication technology).

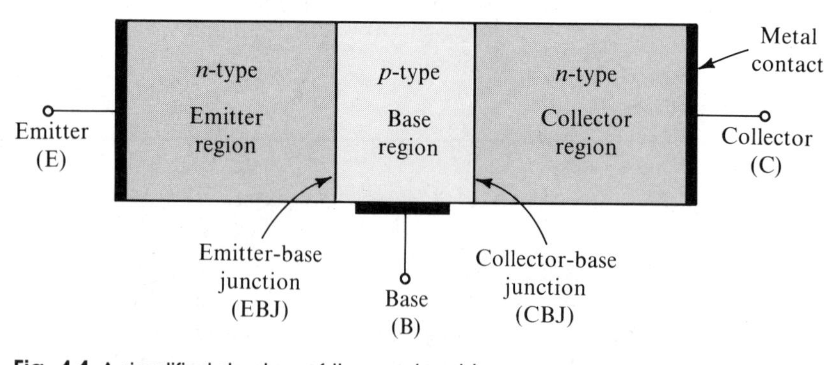

Fig. 4.1 A simplified structure of the *npn* transistor.

As shown in Fig. 4.1, the BJT consists of three semiconductor regions: the emitter region (*n* type), the base region (*p* type), and the collector region (*n* type). Such a transistor is called an *npn* transistor. Another transistor, a dual of the *npn* as shown in Fig. 4.2, has a *p*-type emitter, an *n*-type base, and a *p*-type collector and is appropriately called a *pnp* transistor.

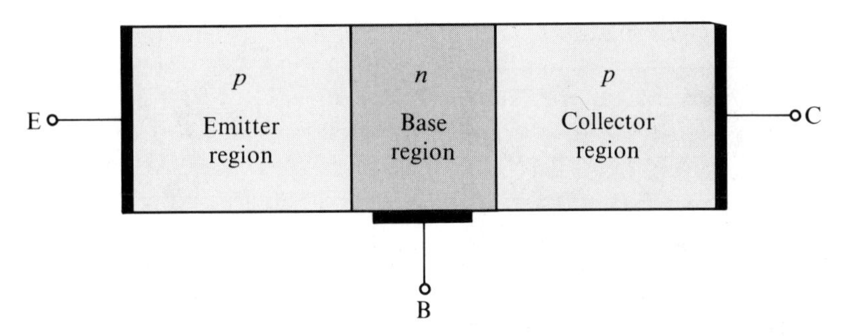

Fig. 4.2 A simplified structure of the *pnp* transistor.

A terminal is connected to each of the three semiconductor regions of a transistor, with the terminals labeled **emitter** (E), **base** (B), and **collector** (C).

The transistor consists of two *pn* junctions, the emitter–base junction (EBJ) and the collector–base junction (CBJ). Depending on the bias condition (forward or reverse) of each of these junctions, different modes of operation of the BJT are obtained, as shown in Table 4.1.

The active mode is the one used if the transistor is to operate as an amplifier. Switching applications (for example, logic circuits) utilize both the cutoff and the saturation modes.

Table 4.1 BJT MODES OF OPERATION

Mode	EBJ	CBJ
Cutoff	Reverse	Reverse
Active	Forward	Reverse
Saturation	Forward	Forward

4.2 OPERATION OF THE *npn* TRANSISTOR IN THE ACTIVE MODE

Let us start by considering the physical operation of the transistor in the active mode. This situation is illustrated in Fig. 4.3 for the *npn* transistor. Two external voltage sources (shown as batteries) are used to establish the required bias conditions for active-mode operation. The voltage V_{BE} causes the *p*-type base to be higher in potential than the *n*-type emitter, thus forward-biasing the emitter–base junction. The collector–base voltage V_{CB} causes the *n*-type collector to be higher in potential than the *p*-type base, thus reverse-biasing the collector–base junction.

Current Flow

In the following description of current flow only diffusion-current components are considered. Drift currents due to thermally generated minority carriers are usually very small and can be neglected. We will have more to say about these reverse-current components at a later stage.

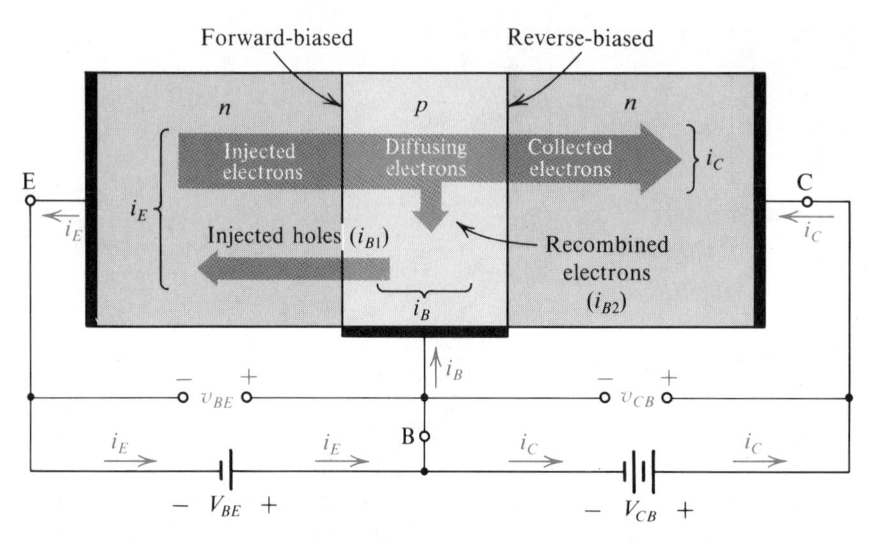

Fig. 4.3 Current flow in an *npn* transistor biased to operate in the active mode. (Reverse current components due to drift of thermally generated minority carriers are not shown.)

The forward bias on the emitter–base junction will cause current to flow across this junction. Current will consist of two components: electrons injected from the emitter into the base, and holes injected from the base into the emitter. As will become apparent shortly, it is highly desirable to have the first component (electrons from emitter to base) at a much higher level than the second component (holes from base to emitter). This can be accomplished by fabricating the device with a heavily doped emitter and a lightly doped base; that is, the device is designed to have a high density of electrons in the emitter and a low density of holes in the base.

The current that flows across the emitter–base junction will constitute the emitter current i_E, as indicated in Fig. 4.3. The direction of i_E is "out of" the emitter lead, which is in the direction of the hole current and opposite to the direction of the electron current, with the emitter current i_E being equal to the sum of these two components. However, since the electron component is much larger than the hole component, the emitter current will be dominated by the electron component.

Let us now consider the electrons injected from the emitter into the base. These electrons will be **minority carriers** in the *p*-type base region. Because the base is usually very thin, in the steady state the excess minority carrier (electron) concentration in the base will have an almost straight-line profile as indicated by the solid straight line in Fig. 4.4. The concentration will be highest [denoted by $n_p(0)$] at the emitter side and lowest (zero) at the collector side.[1] As in the case of any forward-biased *pn* junction (Chapter 3), the concentration $n_p(0)$ will be proportional to e^{v_{BE}/V_T}, where v_{BE} is the forward-bias

[1]This minority carrier distribution in the base results from the boundary conditions imposed by the two junctions. It is not a "natural" diffusion-based distribution, which would result if the base region were infinitely thick.

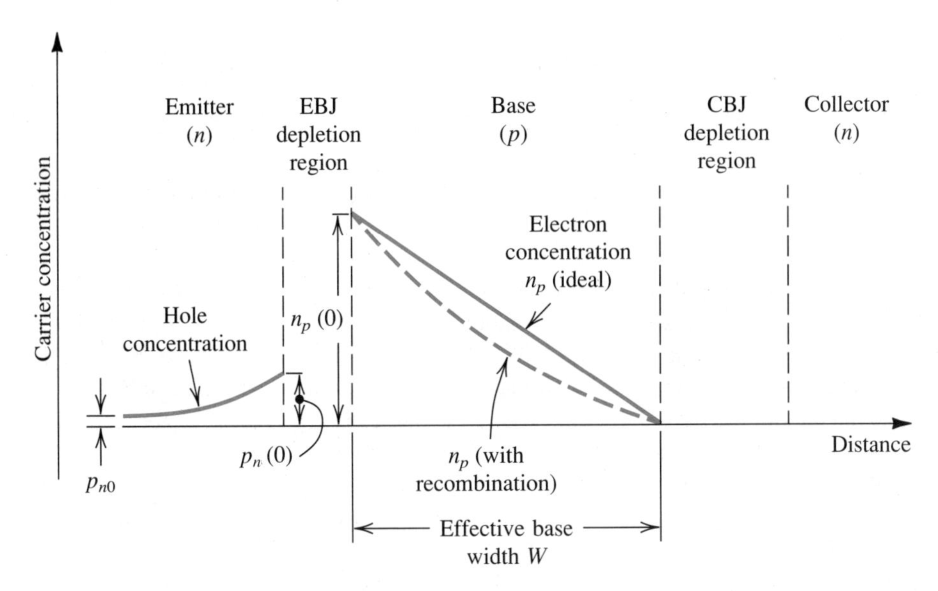

Fig. 4.4 Profiles of minority carrier concentrations in the base and emitter of an *npn* transistor operating in the active mode; $v_{BE} > 0$ and $v_{CB} \geq 0$.

voltage and V_T is the thermal voltage, which is equal to approximately 25 mV at room temperature. The reason for the zero concentration at the collector side of the base is that the positive collector voltage v_{CB} causes the electrons at that end to be swept across the CBJ depletion region.

The tapered minority-carrier concentration profile (Fig. 4.4) causes the electrons injected into the base to diffuse through the base region toward the collector. This diffusion current is directly proportional to the slope of the straight-line concentration profile. Thus the diffusion current will be proportional to the concentration $n_p(0)$ and inversely proportional to the base width W.

Some of the electrons that are diffusing through the base region will combine with holes, which are the majority carriers in the base. However, since the base is usually very thin, the percentage of electrons ''lost'' through this recombination process will be quite small. Nevertheless, the recombination in the base region causes the excess minority carrier concentration profile to deviate from a straight line and take the slightly concave shape indicated by the broken line in Fig. 4.4. The slope of the concentration profile at the EBJ is slightly higher than that at the CBJ, with the difference accounting for the small number of electrons lost in the base region through recombination.

From the above we see that most of the diffusing electrons will reach the boundary of the collector–base depletion region. Because the collector is more positive than the base (by v_{CB} volts), these successful electrons will be swept across the CBJ depletion region into the collector. They will thus get ''collected'' to constitute the collector current i_C. By convention the direction of i_C will be opposite to that of electron flow; thus i_C will flow *into* the collector terminal.

Another important observation to make here is that the magnitude of i_C is independent of v_{CB}. That is, as long as the collector is positive with respect to the base, the electrons that reach the collector side of the base region will be swept into the collector and register as collector current.

The Collector Current

From the above discussion we can see that the collector current i_C may be expressed as

$$i_C = I_S e^{v_{BE}/V_T} \tag{4.1}$$

where I_S is a constant called the saturation current and V_T is the thermal voltage. The reason for the exponential dependence is that the electron diffusion current is proportional to the minority carrier concentration $n_p(0)$, which in turn is proportional to e^{v_{BE}/V_T}. The saturation current I_S is inversely proportional to the base width W and is directly proportional to the area of the EBJ. Typically I_S is in the range of 10^{-12} to 10^{-15} A (depending on the size of the device) and is a function of temperature, approximately doubling for every 5°C rise in temperature.

Because I_S is directly proportional to the junction area (that is, the device size) it will also be referred to as the **current scale factor.** Two transistors that are identical except that one has an EBJ area, say, twice that of the other will have saturation currents with that same ratio (2). Thus for the same value of v_{BE} the larger device will have a collector current twice that in the smaller device. This concept is frequently employed in integrated-circuit design.

The Base Current

The base current i_B is composed of two components. The first and dominant component i_{B1} is due to the holes injected from the base region into the emitter region. This current component is proportional to e^{v_{BE}/V_T} and to the doping density in the base. The second component of base current, i_{B2}, is due to holes that have to be supplied by the external circuit in order to replace the holes lost from the base through the recombination process. The number of electrons (and hence the number of holes) taking part in the recombination process is proportional to the concentration $n_p(0)$ and to the base width W. Thus i_{B2} will be proportional to e^{v_{BE}/V_T} and to W.

From the above discussion we conclude that the total base current $i_B(= i_{B1} + i_{B2})$ will be proportional to $e^{(v_{BE}/V_T)}$. We may therefore express i_B as a fraction of i_C, as follows:

$$i_B = \frac{i_C}{\beta} \tag{4.2}$$

Thus

$$i_B = \frac{I_S}{\beta} e^{v_{BE}/V_T} \tag{4.3}$$

where β is a constant for the particular transistor. For modern *npn* transistors, β is in the range 100 to 200, but it can be as high as 1000 for special devices. For reasons that will become clear later, the constant β is called the **common-emitter current gain.**

As may be seen from the above, the value of β is highly influenced by two factors: the width of the base region and the relative dopings of the emitter region and the base region. To obtain a high β (which is highly desirable since β represents a gain parameter) the base should be thin and lightly doped and the emitter heavily doped. The discussion thus far assumes an idealized situation, where β is a constant for a given transistor.

The Emitter Current

Since the current that enters a transistor should leave it, it can be seen from Fig. 4.3 that the emitter current i_E is equal to the sum of the collector current i_C and the base current i_B,

$$i_E = i_C + i_B \tag{4.4}$$

Use of Eqs. (4.2) and (4.4) gives

$$i_E = \frac{\beta + 1}{\beta} i_C \tag{4.5}$$

that is,

$$i_E = \frac{\beta + 1}{\beta} I_S e^{v_{BE}/V_T} \tag{4.6}$$

Alternatively, we can express Eq. (4.5) in the form

$$i_C = \alpha i_E \tag{4.7}$$

where the constant α is related to β by

$$\alpha = \frac{\beta}{\beta + 1} \tag{4.8}$$

Thus the emitter current in Eq. (4.6) can be written

$$i_E = (I_S/\alpha)e^{v_{BE}/V_T} \tag{4.9}$$

Finally, we can use Eq. (4.8) to express β in terms of α; that is,

$$\beta = \frac{\alpha}{1 - \alpha} \tag{4.10}$$

It can be seen from Eq. (4.8) that α is a constant (for the particular transistor) less than but very close to unity. For instance, if $\beta = 100$, then $\alpha \simeq 0.99$. Equation (4.10) reveals an important fact: Small changes in α correspond to very large changes in β. This mathematical observation manifests itself physically, with the result that transistors of the same type may have widely different values of β. For reasons that will become apparent later, α is called the **common-base current gain.**

Recapitulation

We have presented a first-order model for the operation of the *npn* transistor in the active mode. Basically, the forward-bias voltage v_{BE} causes an exponentially related current i_C to flow in the collector terminal. The collector current i_C is independent of the value of the collector voltage as long as the collector–base junction remains reverse-biased; that is, $v_{CB} \geq 0$. Thus in the active mode the collector terminal behaves as an ideal constant-current source where the value of the current is determined by v_{BE}. The base current i_B is a factor $1/\beta$ of the collector current, and the emitter current is equal to the sum of the collector and base currents. Since i_B is much smaller than i_C (that is, $\beta \gg 1$), $i_E \simeq i_C$. More precisely, the collector current is a fraction α of the emitter current, with α smaller than, but close to, unity.

Equivalent Circuit Models

The first-order model of transistor operation described above can be represented by the equivalent circuit shown in Fig. 4.5(a). Here diode D_E has a current scale factor equal to (I_S/α) and thus provides a current i_E related to v_{BE} according to Eq. (4.9). The current of the controlled source, which is equal to the collector current, is controlled by v_{BE} according to the exponential relationship indicated, a restatement of Eq. (4.1). This model is in essence a nonlinear voltage-controlled current source. It can be converted to the current-controlled current-source model shown in Fig. 4.5(b) by expressing the current of the

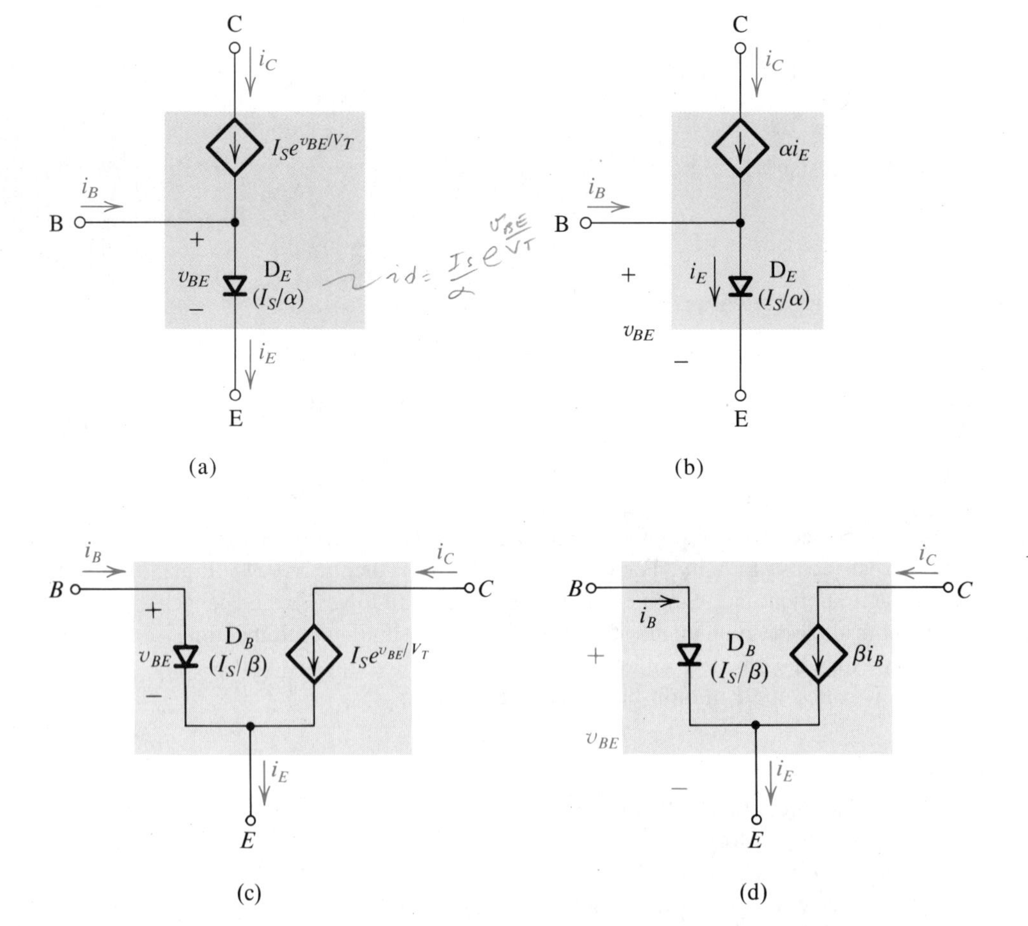

Fig. 4.5 Large-signal equivalent circuit models of the *npn* BJT operating in the active mode.

controlled source as αi_E. Note that this model is also nonlinear because of the exponential relationship of the current i_E through diode D_E and the voltage v_{BE}. From this model we observe that if the transistor is used as a two-port network with the input port between E and B and the output port between C and B (that is, with B as a common terminal), then the current gain observed is equal to α. Thus α is called the common-base current gain.

Two other equivalent circuit models, shown in Fig. 4.5(c) and (d), may be used to represent the large-signal operation of the BJT. The model of Fig. 4.5(c) is essentially a voltage-controlled current source. However, here diode D_B conducts the base current and thus its current scale factor is I_S/β, resulting in the i_B–v_{BE} relationship given in Eq. (4.3). By simply expressing the collector current as βi_B we obtain the current-controlled current-source model shown in Fig. 4.5(d). From this latter model we observe that if the transistor is used as a two-port network with the input port between B and E and the output port between C and E (that is, with E as the common terminal), then the current gain observed is equal to β. Thus β is called the common-emitter current gain.

The Constant *n*

In the diode equation (Chapter 3) we used a constant n in the exponential and mentioned that its value is between 1 and 2. For modern bipolar junction transistors the constant n is close to unity except in special cases: (1) at high currents (that is, high relative to the normal current range of the particular transistor) the i_C–v_{BE} relationship exhibits a value for n that is close to 2, and (2) at low currents, the i_B–v_{BE} relationship shows a value for n of approximately 2. Note that for our purposes we shall assume always that $n = 1$.

The Collector–Base Reverse Current (I_{CBO})

In our discussion of current flow in transistors we ignored the small reverse currents carried by thermally generated minority carriers. Although such currents can be safely neglected in modern transistors, the reverse current across the collector–base junction deserves some mention. This current, denoted I_{CBO}, is the reverse current flowing from collector to base with the emitter open-circuited (hence the subscript O). This current is usually in the nanoampere range, a value that is many times higher than its theoretically predicted value. As with the diode reverse current, I_{CBO} contains a substantial leakage component, and its value is dependent on v_{CB}. I_{CBO} depends strongly on temperature, approximately doubling for every 10°C rise.[2]

The Structure of Actual Transistors

Figure 4.6 shows a simplified but more realistic cross section of an *npn* BJT. Note that the collector virtually surrounds the emitter region, thus making it difficult for the electrons injected into the thin base to escape being collected. In this way, the resulting α is close to unity and β is large. Also, observe that the device is not symmetrical. For more details on the physical structure of actual devices the reader is referred to Appendix A.

[2] The temperature coefficient of I_{CBO} is different from that of I_S because I_{CBO} contains a substantial leakage component.

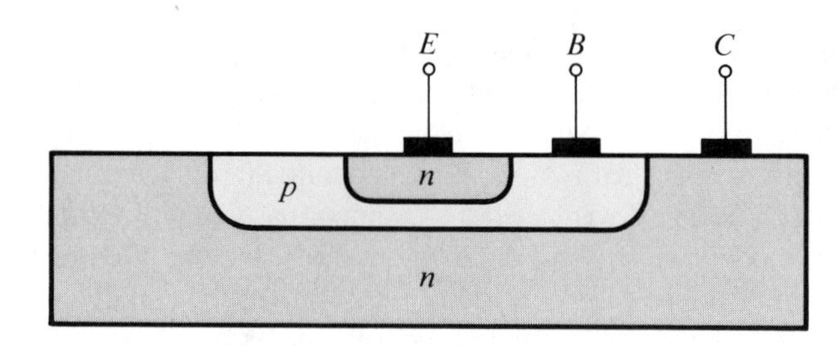

Fig. 4.6 Cross section of an *npn* BJT.

Exercises 4.1 Consider an *npn* transistor with $v_{BE} = 0.7$ V at $i_C = 1$ mA. Find v_{BE} at $i_C = 0.1$ mA and 10 mA.

Ans. 0.64 V; 0.76 V

4.2 Transistors of a certain type are specified to have β values in the range 50 to 150. Find the range of their α values.

Ans. 0.980 to 0.993

4.3 Measurement of an *npn* BJT in a particular circuit shows the base current to be 14.46 μA, the emitter current to be 1.460 mA, and the base–emitter voltage to be 0.7 V. For these conditions calculate α, β, and I_S.

Ans. 0.99; 100; 10^{-15} A

4.4 Calculate β for two transistors for which $\alpha = 0.99$ and 0.98. For collector currents of 10 mA, find the base current of each transistor.

Ans. 99; 49; 0.1 mA; 0.2 mA

4.5 Consider the BJT model shown in Fig. 4.5(d). Find the value of the scale current of D_B given that $I_S = 10^{-14}$ A and $\beta = 100$. If this transistor is operated in the common-emitter configuration, with the base fed with a constant-current source supplying a 10-μA current and with the collector connected to a +10-V dc supply (E is at ground), find V_{BE} and I_C.

Ans. 10^{-16} A; 0.633 V; 1 mA

4.3 THE *pnp* TRANSISTOR

The *pnp* transistor operates in a manner similar to that of the *npn* device described in Section 4.2. Figure 4.7 shows a *pnp* transistor biased to operate in the active mode. Here the voltage V_{EB} causes the *p*-type emitter to be higher in potential than the *n*-type base, thus forward-biasing the base–emitter junction. The collector–base junction is reverse-biased by the voltage V_{BC}, which keeps the *n*-type base higher in potential than the *p*-type collector.

Unlike the *npn* transistor, current in the *pnp* device is mainly conducted by holes injected from the emitter into the base as a result of the forward-bias voltage V_{EB}. Since the component of emitter current contributed by electrons injected from base to emitter is kept small by using a lightly doped base, most of the emitter current will be due to holes. The electrons injected from base to emitter give rise to the dominant component of base current, i_{B1}. Also, a number of the holes injected into the base will recombine with the majority carriers in the base (electrons) and will thus be lost. The disappearing base electrons will have to be replaced from the external circuit, giving rise to the second component of base current, i_{B2}. The holes that succeed in reaching the boundary of the

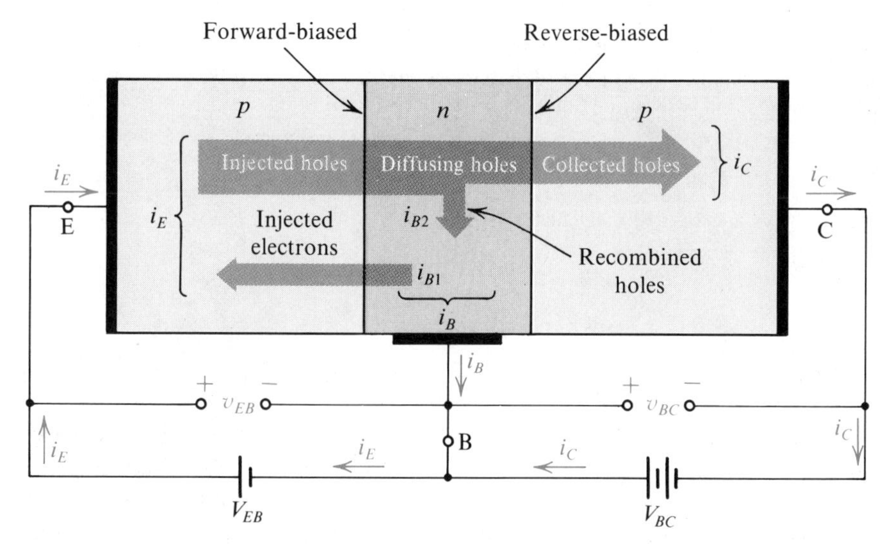

Fig. 4.7 Current flow in a *pnp* transistor biased to operate in the active mode.

depletion region of the collector–base junction will be attracted by the negative voltage on the collector. Thus these holes will be swept across the depletion region into the collector and appear as collector current.

It can easily be seen from the above description that the current–voltage relationships of the *pnp* transistor will be identical to those of the *npn* transistor except that v_{BE} has to be replaced by v_{EB}. Also, the large-signal operation of the *pnp* transistor can be modeled by any one of four possible equivalent circuit models that parallel those given for the *npn* transistor in Fig. 4.5. For illustration, two of the four *pnp* models are depicted in Fig. 4.8.

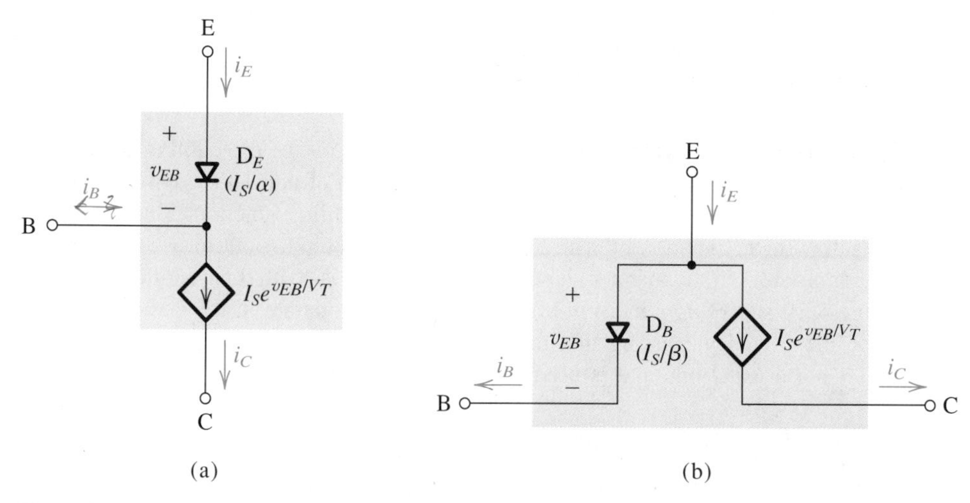

Fig. 4.8 Two large-signal models for the *pnp* transistor operating in the active mode.

Exercises 4.6 Consider the model in Fig. 4.8(a) applied in the case of a *pnp* transistor whose base is grounded, the emitter is fed by a constant-current source that supplies a 2-mA current into the emitter terminal, the collector is connected to a −10-V dc supply. Find the emitter voltage, the base current, and the collector current if for this transistor $\beta = 50$ and $I_S = 10^{-14}$ A.

Ans. 0.650 V; 39.2 μA; 1.96 mA

4.7 For a *pnp* transistor having $I_S = 10^{-11}$ A and $\beta = 100$, calculate v_{EB} for $i_C = 1.5$ A.

Ans. 0.643 V

4.4 CIRCUIT SYMBOLS AND CONVENTIONS

The physical structure used thus far to explain transistor operation is rather cumbersome to employ in drawing the schematic of a multitransistor circuit. Fortunately, a very descriptive and convenient circuit symbol exists for the BJT. Figure 4.9(a) shows the symbol for the *npn* transistor; the *pnp* symbol is given in Fig. 4.9(b). In both symbols the emitter is distinguished by an arrowhead. This distinction is important because practical BJTs are not symmetric devices. That is, interchanging the emitter and collector will result in a different, and much lower, value for α, a value called the **inverse** or **reverse α.**[3]

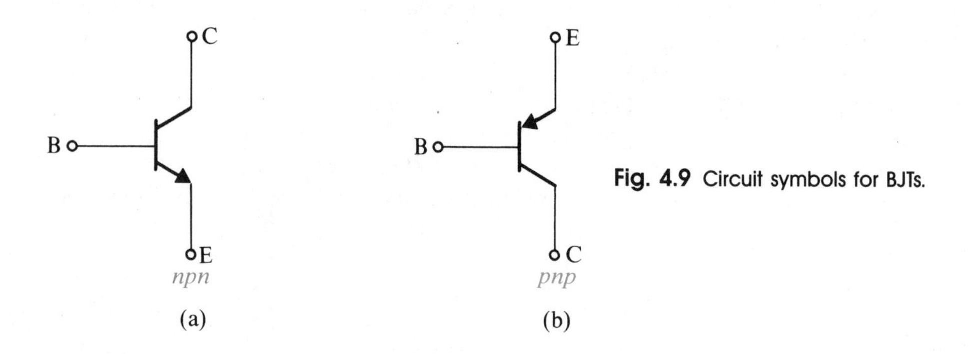

Fig. 4.9 Circuit symbols for BJTs.

The polarity of the device—*npn* or *pnp*—is indicated by the direction of the arrowhead on the emitter. This arrowhead points in the direction of normal current flow in the emitter, which is also the forward direction of the base–emitter junction. Since we have adopted a drawing convention by which currents flow from top to bottom, we will always draw *pnp* transistors in the manner shown in Fig. 4.9 (that is, with their emitters on top).

Figure 4.10 shows *npn* and *pnp* transistors biased to operate in the active mode. It should be mentioned in passing that the biasing arrangement shown, utilizing two dc sources, is not a usual one and is used merely to illustrate operation. Practical biasing

[3]The inverse mode of operation of the BJT will be studied in Chapter 14.

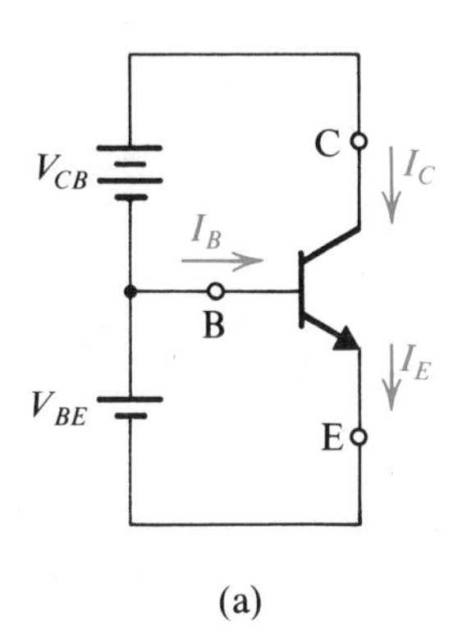

(a)

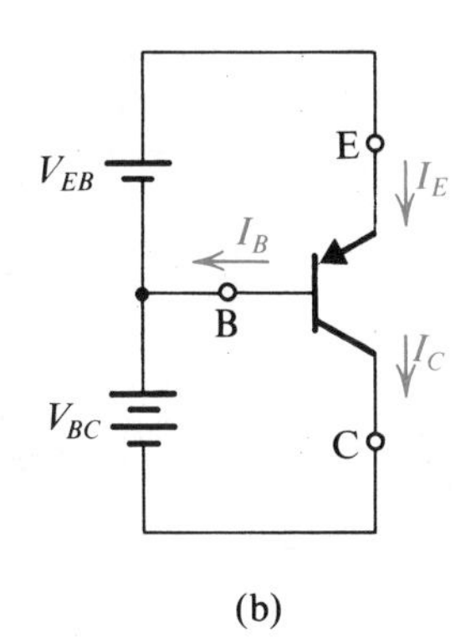

(b)

Fig. 4.10 Voltage polarities and current flow in transistors biased in the active mode.

schemes will be presented in Section 4.8. Figure 4.10 also indicates the reference and actual directions of current flow throughout the transistor. Our convention will be to take the reference direction to coincide with the normal direction of current flow. Hence, normally, we should not encounter a negative value for i_E, i_B, or i_C.

The convenience of the circuit drawing convention that we have adopted should be obvious from Fig. 4.10. Note that currents flow from top to bottom and that voltages are higher at the top and lower at the bottom. The arrowhead on the emitter also implies the polarity of the emitter–base voltage that should be applied in order to forward-bias the emitter–base junction. Just a glance at the circuit symbol of the *pnp* transistor, for example, indicates that we should make the emitter higher in voltage than the base (by v_{EB}) in order to cause current to flow into the emitter (downward). Note that the symbol v_{EB} means the voltage by which the emitter (E) is higher than the base (B). Thus for a *pnp* transistor operating in the active mode v_{EB} is positive, while in an *npn* transistor v_{BE} is positive.

From the discussion of Section 4.3 it follows that an *npn* transistor whose EBJ is forward-biased will operate in the active mode *as long as the collector is higher in potential than the base*. Active-mode operation will be maintained even if the collector voltage falls to equal that of the base, since a silicon *pn* junction is essentially nonconducting when the voltage across it is zero. The collector voltage should not be allowed, however, to fall below that of the base if active-mode operation is required. If it does fall below the base voltage, the collector–base junction could become forward-biased, with the transistor entering a new mode of operation, saturation. We will discuss the saturation mode later.

In a parallel manner, the *pnp* transistor will operate in the active mode *if the potential of the collector is lower than (or equal to) that of the base*. The collector voltage should not be allowed to rise above that of the base if active-mode operation is to be maintained.

EXAMPLE 4.1

The transistor in the circuit of Fig. 4.11(a) has $\beta = 100$ and exhibits a v_{BE} of 0.7 V at $i_C = 1$ mA. Design the circuit so that a current of 2 mA flows through the collector and a voltage of +5 V appears at the collector.

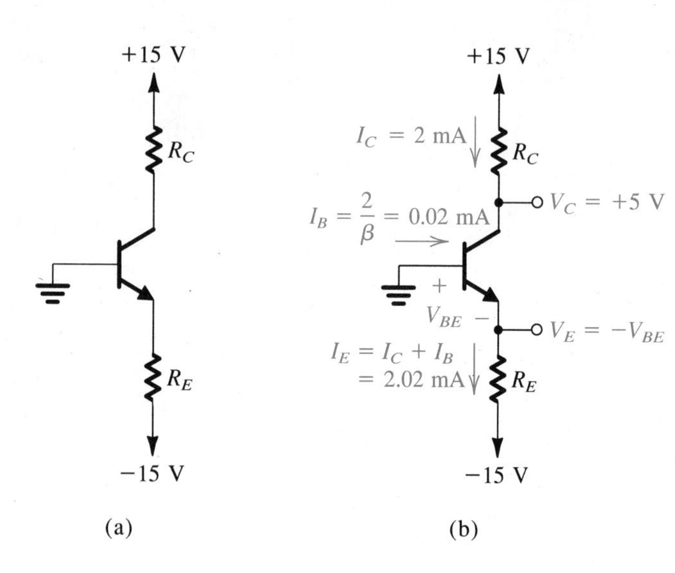

Fig. 4.11 Circuit for Example 4.1.

Solution Refer to Fig. 4.11(b). To obtain a voltage $V_C = +5$ V the voltage drop across R_C must be $15 - 5 = 10$ V. Now since $I_C = 2$ mA, the value of R_C should be selected according to

$$R_C = \frac{10 \text{ V}}{2 \text{ mA}} = 5 \text{ k}\Omega$$

Since $v_{BE} = 0.7$ V at $i_C = 1$ mA, the value of v_{BE} at $i_C = 2$ mA is

$$V_{BE} = 0.7 + V_T \ln\left(\frac{2}{1}\right) = 0.717 \text{ V}$$

Since the base is at 0 V the emitter voltage should be at

$$V_E = -0.717 \text{ V}$$

For $\beta = 100$, $\alpha = 100/101 = 0.99$. Thus the emitter current should be

$$I_E = \frac{I_C}{\alpha} = \frac{2}{0.99} = 2.02 \text{ mA}$$

Now the value required for R_E can be determined from

$$R_E = \frac{V_E - (-15)}{I_E}$$

$$= \frac{-0.717 + 15}{2.02} = 7.07 \, k\Omega$$

This completes the design. We should note, however, that the calculations above were made with a degree of accuracy that is usually neither necessary nor justified in practice in view, for instance, of the expected tolerances of component values. Nevertheless, we chose to do the design precisely in order to illustrate the various steps involved.

Exercises 4.8 In the circuit shown in Fig. E4.8 the voltage at the emitter was measured and found to be -0.7 V. If $\beta = 50$, find I_E, I_B, I_C, and V_C.

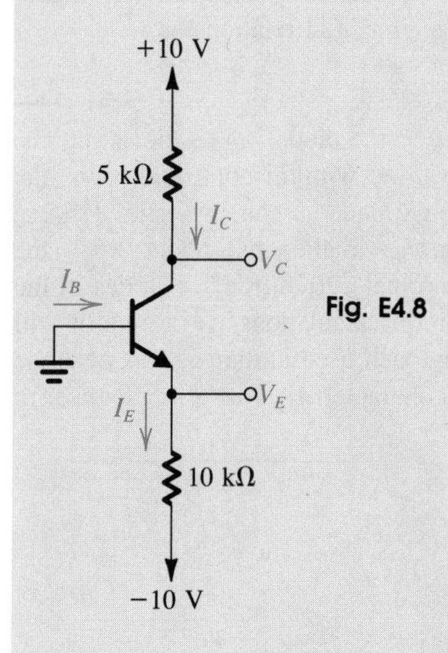

Fig. E4.8

Ans. 0.93 mA; 18.2 μA; 0.91 mA; +5.44 V

4.9 In the circuit shown in Fig. E4.9, measurement indicates V_B to be +1.0 V and V_E to be +1.7 V. What are α and β for this transistor? What voltage V_C do you expect at the collector?

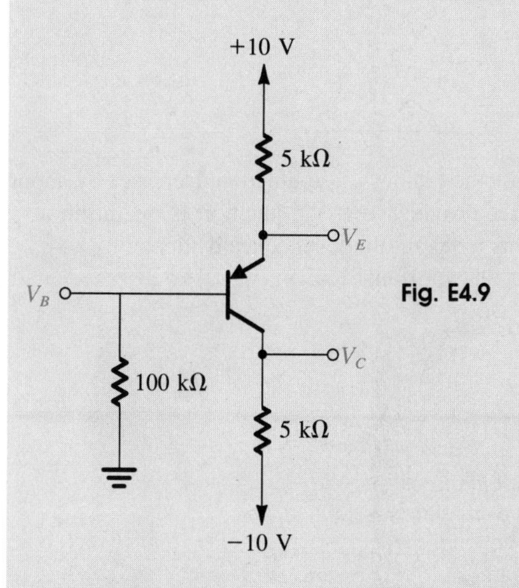

Fig. E4.9

Ans. 0.994; 165; -1.75 V

4.5 GRAPHICAL REPRESENTATION OF TRANSISTOR CHARACTERISTICS

It is sometimes useful to describe the transistor i–v characteristics graphically. Figure 4.12 shows the i_C–v_{BE} characteristic, which is the exponential relationship

$$i_C = I_S e^{v_{BE}/V_T}$$

which is identical (except for the value of constant n) to the diode i–v relationship. The i_E–v_{BE} and i_B–v_{BE} characteristics are also exponential but with different scale currents: I_S/α for i_E and I_S/β for i_B. Since the constant of the exponential characteristic, $1/V_T$, is quite high ($\simeq$40), the curve rises very sharply. For v_{BE} smaller than about 0.5 V the current is negligibly small. Also, over most of the normal current range v_{BE} lies in the range 0.6 to 0.8 V. In performing rapid first-order dc calculations we normally will assume that $V_{BE} \simeq 0.7$ V, which is similar to the approach used in the analysis of diode circuits (Chapter 3). For a *pnp* transistor the i_C–v_{EB} characteristic will look identical to that of Fig. 4.12.

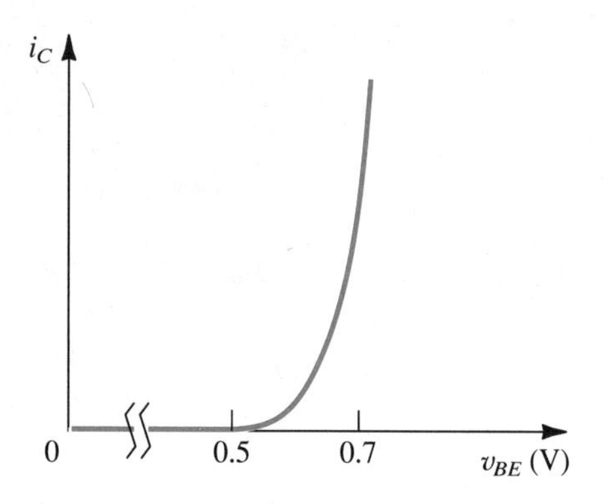

Fig. 4.12 The i_C–v_{BE} characteristic for an *npn* transistor.

As in silicon diodes, the voltage across the emitter–base junction decreases by about 2 mV for each rise of 1°C in temperature, provided that the junction is operating at a constant current. Figure 4.13 illustrates this temperature dependence by depicting i_C–v_{BE} curves at three different temperatures for an *npn* transistor.

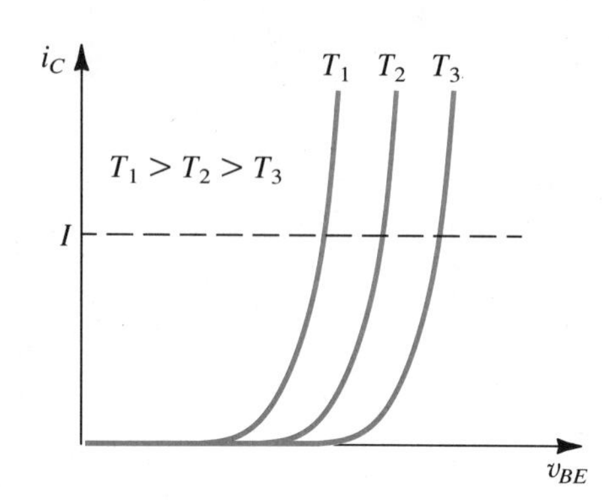

Fig. 4.13 Effect of temperature on the i_C–v_{BE} characteristic. At a constant emitter current (broken line) v_{BE} changes by −2 mV/°C.

Figure 4.14(b) shows the i_C versus v_{CB} characteristics of an *npn* transistor for various values of the emitter current i_E. These characteristics can be measured using the circuit shown in Fig. 4.14(a). Only active-mode operation is shown, since only the portion of the characteristics for $v_{CB} \geq 0$ is drawn. As can be seen, the curves are horizontal straight lines, corroborating the fact that the collector behaves as a constant-current source. In this case the value of the collector current is controlled by that of the emitter current ($i_C = \alpha i_E$), and the transistor may be thought of as a current-controlled current source (see the model of Fig. 4.5b).

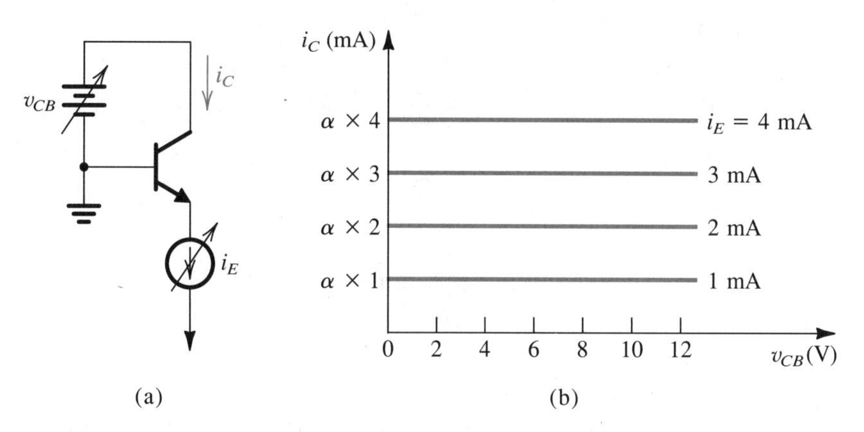

Fig. 4.14 The i_C–v_{CB} characteristics for an *npn* transistor in the active mode.

When operated in the active region, practical BJTs show some dependence of the collector current on the collector voltage, with the result that their i_C–v_{CB} characteristics are not perfectly horizontal straight lines. To see this dependence more clearly, consider the conceptual circuit shown in Fig. 4.15(a). The transistor is connected in the common-emitter configuration, and its V_{BE} can be set to any desired value by adjusting the dc source connected between base and emitter. At each value of V_{BE}, the corresponding i_C–v_{CE} characteristic curve can be measured point-by-point by varying the dc source connected between collector and emitter and measuring the corresponding collector current. The result is the family of i_C–v_{CE} characteristic curves shown in Fig. 4.15(b).

At low values of v_{CE}, as the collector voltage drops below that of the base, the collector–base junction becomes forward-biased and the transistor leaves the active mode and enters the saturation mode. We shall study the saturation mode of operation in a later section. At this time we wish to examine the characteristic curves in the active region in detail. We observe that the characteristic curves, though still straight lines, have finite slope. In fact, when extrapolated, the characteristic lines meet at a point on the negative v_{CE} axis, at $v_{CE} = -V_A$. The voltage V_A, a positive number, is a parameter for the particular BJT, with typical values in the range of 50 to 100 V. It is called the Early voltage, after the scientist who first studied this phenomenon.

At a given value of v_{BE}, increasing v_{CE} increases the reverse-bias voltage on the collector–base junction and thus increases the width of the depletion region of this junc-

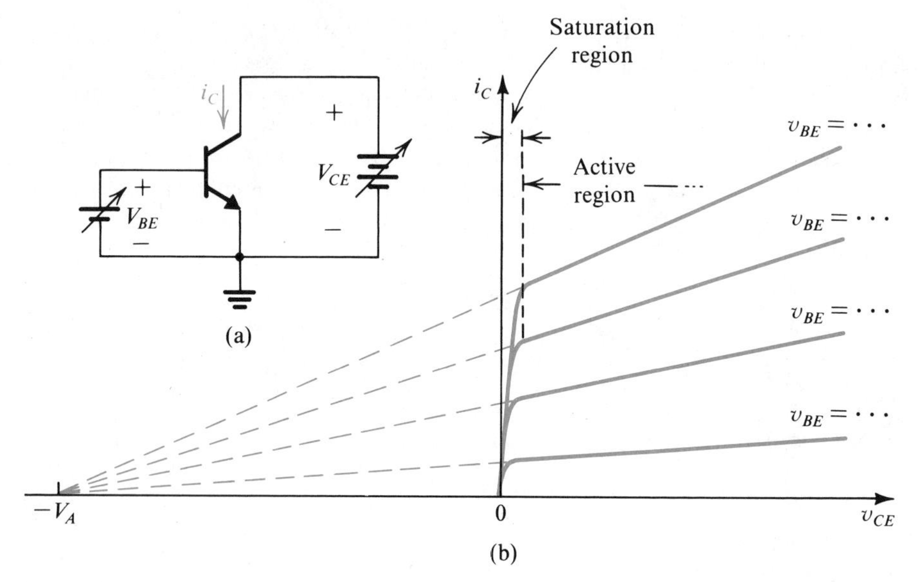

Fig. 4.15 (a) Conceptual circuit for measuring the i_C–v_{CE} characteristics of the BJT. **(b)** The i_C–v_{CE} characteristics of a practical BJT.

tion (refer to Fig. 4.3). This in turn results in a decrease in the **effective base width** W. Recalling that I_S is inversely proportional to W, we see that I_S will increase and i_C increases proportionally. This is the Early effect.

The linear dependence of i_C on v_{CE} can be accounted for by assuming that I_S remains constant and including the factor $(1 + v_{CE}/V_A)$ in the equation for i_C as follows:

$$i_C = I_S e^{v_{BE}/V_T}\left(1 + \frac{v_{CE}}{V_A}\right) \tag{4.11}$$

The nonzero slope of the i_C–v_{CE} straight lines indicates that the **output resistance** looking into the collector is not infinite. Rather, it is finite and defined by

$$r_o \equiv \left[\frac{\partial i_C}{\partial v_{CE}}\bigg|_{v_{BE}=\text{constant}}\right]^{-1} \tag{4.12}$$

Using Eq. (4.11) we can show that

$$r_o \simeq \frac{V_A}{I_C} \tag{4.13}$$

where I_C is the current level corresponding to the constant value of v_{BE}, near the boundary of the active region.

It is rarely necessary to include the dependence of i_C on v_{CE} in dc bias design and analysis. However, the finite output resistance r_o can have a significant effect on the gain of transistor amplifiers, as will be seen in later sections and chapters.

Exercises 4.10 Consider a *pnp* transistor with $v_{EB} = 0.7$ V at $i_E = 1$ mA. Let the base be grounded, the emitter be fed by a 2-mA constant-current source, and the collector be connected to a −5-V supply through a 1-kΩ resistance. If the temperature increases by 30°C, find the changes in emitter and collector voltages. Neglect the effect of I_{CBO}.

Ans. −60 mV; 0 V

4.11 Find the output resistance of a BJT for which $V_A = 100$ V, at $I_C = 0.1$, 1, and 10 mA.

Ans. 1 MΩ; 100 kΩ; 10 kΩ

4.12 Consider the circuit in Fig. 4.15(a). At $V_{CE} = 1$ V, V_{BE} is adjusted to yield a collector current of 1 mA. Then, while V_{BE} is kept constant, V_{CE} is raised to 11 V. Find the new value of I_C. For this transistor $V_A = 100$ V.

Ans. 1.1 mA

4.6 ANALYSIS OF TRANSISTOR CIRCUITS AT DC

We are now ready to consider the analysis of some simple transistor circuits to which only dc voltages are applied. In the following examples we will use the simple constant-V_{BE} model, which is similar to that discussed in Section 3.3 for the junction diode. Specifically we will assume that $V_{BE} = 0.7$ V irrespective of the exact value of current. If it is desired, this approximation can be refined using techniques similar to those employed in the diode case. The emphasis here, however, is on the essence of transistor circuit analysis.

EXAMPLE 4.2

Consider the circuit shown in Fig. 4.16(a), which is redrawn in Fig. 4.16(b) to remind the reader of the convention employed throughout this book for indicating connections to dc sources. We wish to analyze this circuit to determine all node voltages and branch currents. We will assume that β is specified to be 100.

Solution We do not know initially whether the transistor is in the active mode or not. A simple approach would be to assume that the device is in the active mode, proceed with the solution, and finally check whether or not the transistor is in fact in the active mode. If we find that the conditions for active-mode operation are met, then our work is completed. Otherwise, the device is in another mode of operation and we have to solve the problem again. Obviously, at this stage we have learned only about the active mode of operation and thus will not be able to deal with circuits that we find are not in the active mode.

Glancing at the circuit in Fig. 4.16(a), we note that the base is connected to +4 V and the emitter is connected to ground through a resistance R_E. It therefore is safe to conclude that the base–emitter junction will be forward-biased. Assuming that this is the case and assuming that V_{BE} is approximately 0.7 V, it follows that the emitter voltage will be

$$V_E = 4 - V_{BE} \simeq 4 - 0.7 = 3.3 \text{ V}$$

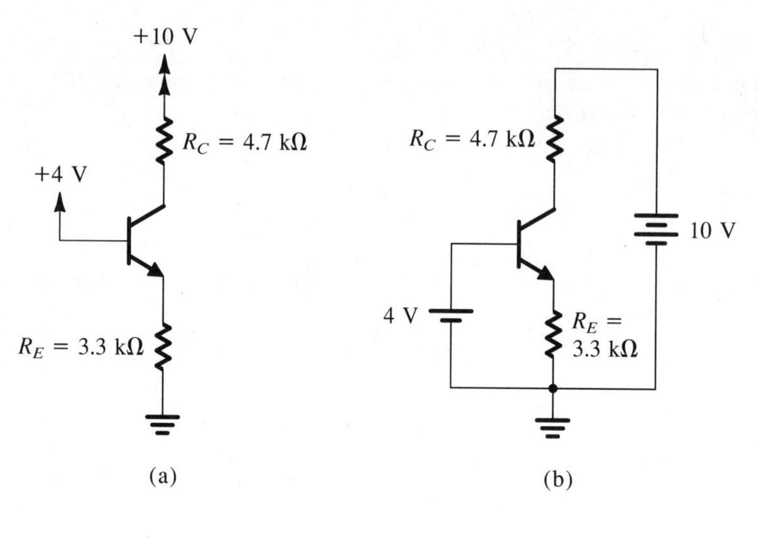

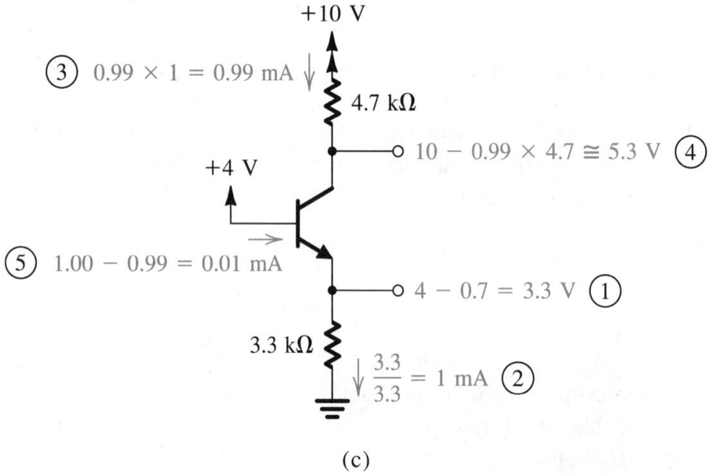

Fig. 4.16 Analysis of the circuit for Example 4.2: **(a)** circuit; **(b)** circuit redrawn to remind the reader of the convention used in this book to show connections to the power supply; **(c)** analysis with the steps numbered.

We are now in an opportune position; we know the voltages at the two ends of R_E and thus can determine the current I_E through it,

$$I_E = \frac{V_E - 0}{R_E} = \frac{3.3}{3.3} = 1 \text{ mA}$$

Since the collector is connected through R_C to the +10-V power supply, it appears possible that the collector voltage will be higher than the base voltage, which is essential for active-mode operation. Assuming that this is the case, we can evaluate the collector current from

$$I_C = \alpha I_E$$

The value of α is obtained from

$$\alpha = \frac{\beta}{\beta + 1} = \frac{100}{101} \simeq 0.99$$

Thus I_C will be given by

$$I_C = 0.99 \times 1 = 0.99 \text{ m}A$$

We are now in a position to use Ohm's law to determine the collector voltage V_C,

$$V_C = 10 - I_C R_C = 10 - 0.99 \times 4.7 \simeq +5.3 \text{ V}$$

Since the base is at +4 V, the collector–base junction is reverse-biased by 1.3 V, and the transistor is indeed in the active mode as assumed.

It remains only to determine the base current I_B, as follows:

$$I_B = \frac{I_E}{\beta + 1} = \frac{1}{101} \simeq 0.01 \text{ mA}$$

Before leaving this example we wish to emphasize strongly the value of carrying out the analysis directly on the circuit diagram. Only in this way will one be able to analyze complex circuits in a reasonable length of time. Figure 4.16(c) illustrates the above analysis on the circuit diagram with the order of the analysis steps indicated by the circled numbers.

EXAMPLE 4.3

We wish to analyze the circuit of Fig. 4.17(a) to determine the voltages at all nodes and the currents through all branches. Note that this circuit is identical to that of Fig. 4.16 except that the voltage at the base is now +6 V.

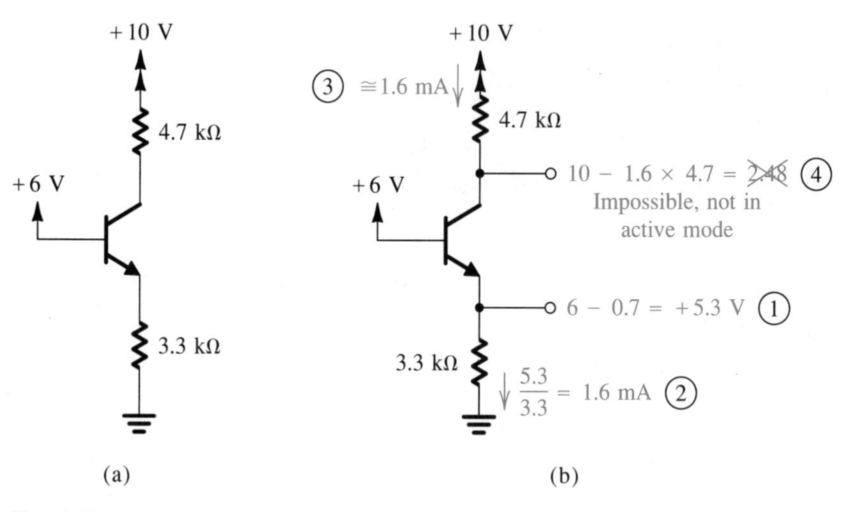

Fig. 4.17 Analysis of the circuit for Example 4.3. Note that the circled numbers indicate the order of the analysis steps.

Solution Assuming active-mode operation, we have

$$V_E = +6 - V_{BE} \simeq 6 - 0.7 = 5.3 \text{ V}$$

$$I_E = \frac{5.3}{3.3} = 1.6 \text{ mA}$$

$$V_C = +10 - 4.7 \times I_C \simeq 10 - 7.52 = 2.48 \text{ V}$$

Since the collector voltage calculated appears to be less than the base voltage by 3.52 V, it follows that our original assumption of active-mode operation is incorrect. In fact, the transistor has to be in the *saturation* mode. Since we have not yet studied the saturation mode of operation, we shall defer the analysis of this circuit to a later section.

The details of the analysis performed above are illustrated in Fig. 4.17(b).

EXAMPLE 4.4

We wish to analyze the circuit in Fig. 4.18(a) to determine the voltages at all nodes and the currents through all branches. Note that this circuit is identical to that considered in Examples 4.2 and 4.3 except that now the base voltage is zero.

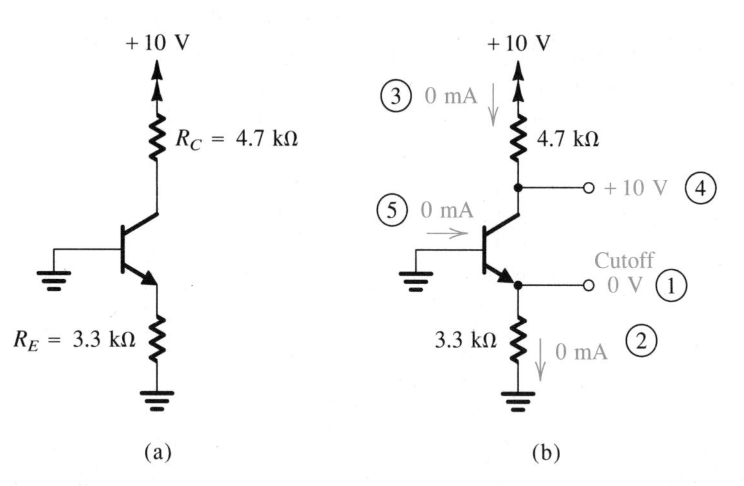

Fig. 4.18 Example 4.3: **(a)** circuit; **(b)** analysis with the order of the analysis steps indicated by circled numbers.

Solution Since the base is at zero volts, the emitter–base junction cannot conduct and the emitter current is zero. Also, the collector–base junction cannot conduct, since the *n*-type collector is connected through R_C to the positive power supply while the *p*-type base is at ground. It follows that the collector current will be zero. The base current will also have to be zero, and the transistor is in the *cutoff* mode of operation.

The emitter voltage will obviously be zero, while the collector voltage will be equal to +10 V, since the voltage drop across R_C is zero. Figure 4.18(b) shows the analysis details.

Exercises D4.13 For the circuit in Fig. 4.16(a), find the highest voltage to which the base can be raised while the transistor remains in the active mode. Assume $\alpha \simeq 1$.

Ans. +4.54 V

D4.14 Redesign the circuit of Fig. 4.16(a) (i.e., find new values for R_E and R_C) to establish a collector current of 0.5 mA and a reverse bias voltage on the collector–base junction of 2 V. Assume $\alpha \simeq 1$.

Ans. $R_E = 6.6$ kΩ; $R_C = 8$ kΩ

EXAMPLE 4.5

We desire to analyze the circuit of Fig. 4.19(a) to determine the voltages at all nodes and the currents through all branches.

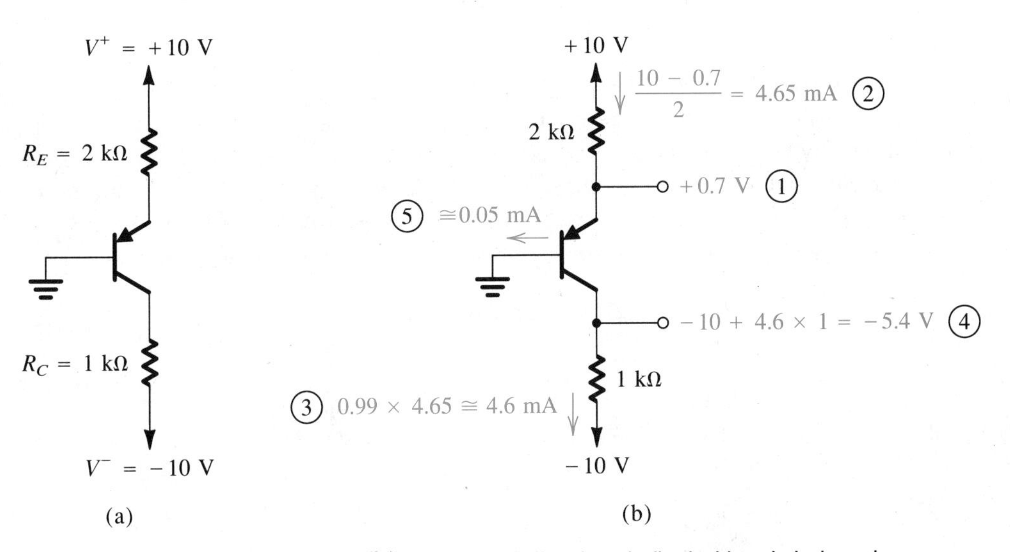

Fig. 4.19 Example 4.4: **(a)** circuit; **(b)** analysis with the steps indicated by circled numbers.

Solution The base of this *pnp* transistor is grounded, while the emitter is connected to a positive supply ($V^+ = +10$ V) through R_E. It follows that the emitter–base junction will be forward-biased with

$$V_E = V_{EB} \simeq 0.7 \text{ V}$$

Thus the emitter current will be given by

$$I_E = \frac{V^+ - V_E}{R_E} = \frac{10 - 0.7}{2} = 4.65 \text{ mA}$$

Since the collector is connected to a negative supply (more negative than the base voltage) through R_C, it is *possible* that this transistor is operating in the active mode. Assuming this to be the case, we obtain

$$I_C = \alpha I_E$$

Since no value for β has been given, we shall assume $\beta = 100$, which results in $\alpha = 0.99$. Since large variations in β result in small differences in α, this assumption will not be critical as far as determining the value of I_C is concerned. Thus

$$I_C = 0.99 \times 4.65 = 4.6 \text{ mA}$$

The collector voltage will be

$$\begin{aligned} V_C &= V^- + I_C R_C \\ &= -10 + 4.6 \times 1 = -5.4 \text{ V} \end{aligned}$$

Thus the collector–base junction is reverse-biased by 5.4 V and the transistor is indeed in the active mode, which supports our original assumption.

It remains only to calculate the base current,

$$I_B = \frac{I_E}{\beta + 1} = \frac{4.65}{101} \simeq 0.05 \text{ mA}$$

Obviously, the value of β critically affects the base current. Note, however, that in this circuit the value of β will have no effect on the mode of operation of the transistor. Since β is generally an ill-specified parameter, this circuit represents a good design. As a rule, one should strive to *design the circuit such that its performance is as insensitive to the value of β as possible*. The analysis details are illustrated in Fig. 4.19(b).

Exercises **D4.15** For the circuit in Fig. 4.19(a), find the largest value to which R_C can be raised while the transistor remains in the active mode.

Ans. 2.17 kΩ

D4.16 Redesign the circuit of Fig. 4.19(a) (i.e., find new values for R_E and R_C) to establish a collector current of 1 mA and a reverse bias on the collector–base junction of 4 V. Assume $\alpha \simeq 1$

Ans. $R_E = 9.3$ kΩ; $R_C = 6$ kΩ

EXAMPLE 4.6

We want to analyze the circuit in Fig. 4.20(a) to determine the voltages at all nodes and the currents in all branches. Assume $\beta = 100$.

Solution

The base–emitter junction is clearly forward-biased. Thus

$$I_B = \frac{+5 - V_{BE}}{R_B} \simeq \frac{5 - 0.7}{100} = 0.043 \text{ mA}$$

Assume that the transistor is operating in the active mode. We now can write

$$I_C = \beta I_B = 100 \times 0.043 = 4.3 \text{ mA}$$

The collector voltage can now be determined as

$$V_C = +10 - I_C R_C = 10 - 4.3 \times 2 = +1.4 \text{ V}$$

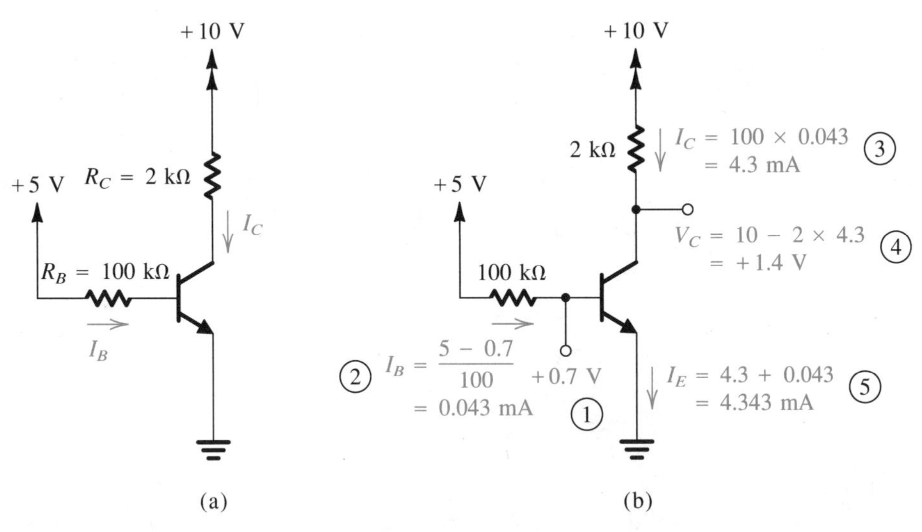

Fig. 4.20 Example 4.5: **(a)** circuit; **(b)** analysis with the steps indicated by the circled numbers.

Since the base voltage V_B is

$$V_B = V_{BE} \simeq +0.7 \text{ V}$$

it follows that the collector–base junction is reverse-biased by 0.7 V and the transistor is indeed in the active mode. The emitter current will be given by

$$I_E = (\beta + 1)I_B = 101 \times 0.043 \simeq 4.3 \text{ mA}$$

We note from this example that the collector and emitter currents depend critically on the value of β. In fact, if β were 10% higher, the transistor would leave the active mode and enter saturation. Therefore this clearly is a bad design. The analysis details are illustrated in Fig. 4.20(b).

Exercise D4.17 The circuit of Fig. 4.20(a) is to be fabricated using a transistor type whose β is specified to be in the range of 50 to 150. (That is, units of this same transistor type can have β values anywhere in this range.) Redesign the circuit by selecting a new value for R_C so that all fabricated circuits are guaranteed to be in the active mode. What is the range of collector voltages that the fabricated circuits may exhibit?

Ans. $R_C = 1.44$ kΩ; $V_C = 0.7$ V to 6.9 V

EXAMPLE 4.7

We want to analyze the circuit of Fig. 4.21(a) to determine the voltages at all nodes and the currents through all branches. Assume $\beta = 100$.

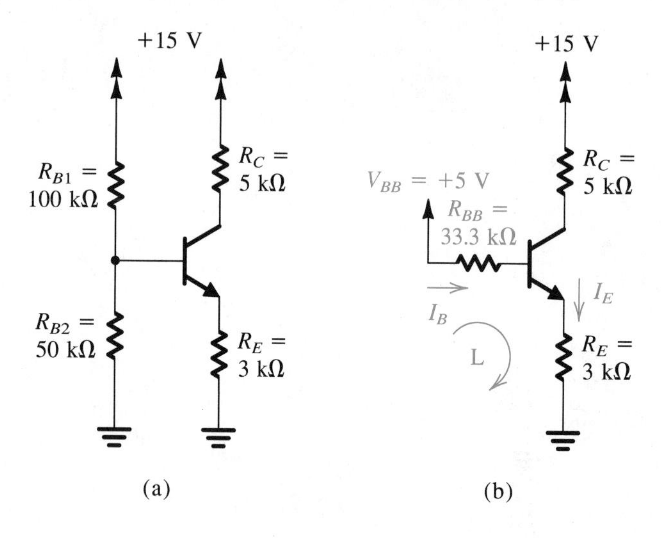

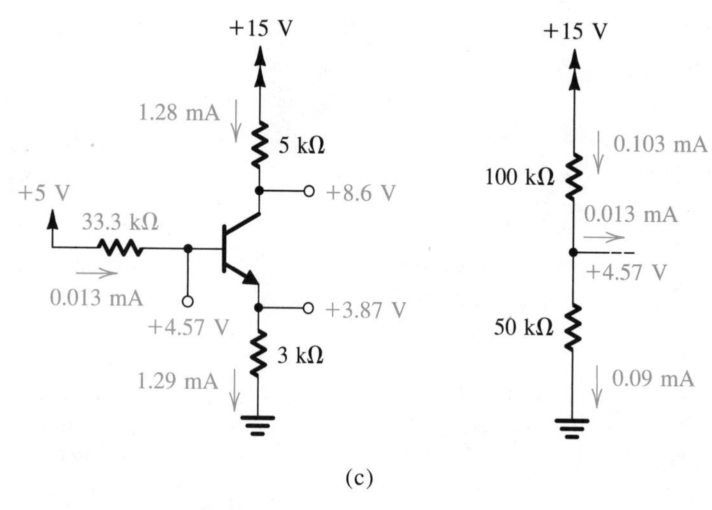

Fig. 4.21 Circuits for Example 4.7.

Solution The first step in the analysis consists of simplifying the base circuit using Thévenin's theorem. The result is shown in Fig. 4.21(b), where

$$V_{BB} = +15 \frac{R_{B2}}{R_{B1} + R_{B2}} = 15 \frac{50}{100 + 50} = +5 \text{ V}$$

$$R_{BB} = (R_{B1} \mathbin{//} R_{B2}) = (100 \mathbin{//} 50) = 3.33 \text{ k}\Omega$$

To evaluate the base or the emitter current we have to write a loop equation around the loop marked L in Fig. 4.21(b). Note, though, that the current through R_{BB} is different from the current through R_E. The loop equation will be

$$V_{BB} = I_B R_{BB} + V_{BE} + I_E R_E$$

Substituting for I_B by

$$I_B = \frac{I_E}{\beta + 1}$$

and rearranging the equation gives

$$I_E = \frac{V_{BB} - V_{BE}}{R_E + [R_{BB}/(\beta + 1)]}$$

For the numerical values given we have

$$I_E = \frac{5 - 0.7}{3 + (33.3/101)} = 1.29 \text{ mA}$$

The base current will be

$$I_B = \frac{1.29}{101} = 0.0128 \text{ mA}$$

The base voltage is given by

$$\begin{aligned} V_B &= V_{BE} + I_E R_E \\ &= 0.7 + 1.29 \times 3 = 4.57 \text{ V} \end{aligned}$$

Assume active-mode operation. We can evaluate the collector current as

$$I_C = \alpha I_E = 0.99 \times 1.29 = 1.28 \text{ mA}$$

The collector voltage can now be evaluated as

$$V_C = +15 - I_C R_C = 15 - 1.28 \times 5 = 8.6 \text{ V}$$

It follows that the collector is higher in potential than the base by 4.03 V, which means that the transistor is in the active mode, as had been assumed. The results of the analysis are given in Fig. 4.21c.

Exercise 4.18 If the transistor in the circuit of Fig. 4.21(a) is replaced with another having half the value of β (i.e., $\beta = 50$), find the new value of I_C and express the change in I_C in percent form.

Ans. $I_C = 1.15$ mA; -10%

EXAMPLE 4.8 We wish to analyze the circuit in Fig. 4.22(a) to determine the voltages at all nodes and the currents through all branches.

Solution We first recognize that part of this circuit is identical to the circuit we analyzed in Example 4.7—namely, the circuit of Fig. 4.21(a). The difference, of course, is that in the new circuit we have an additional transistor Q_2 together with its associated resistors R_{E2} and

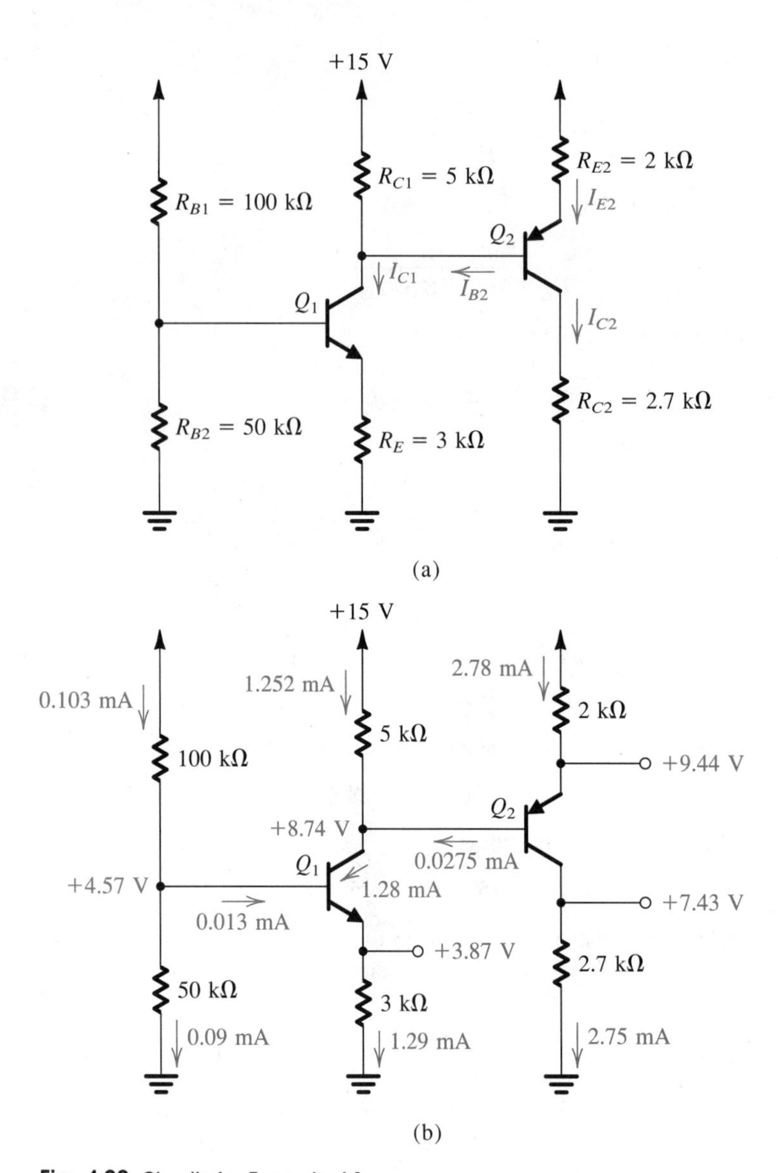

Fig. 4.22 Circuits for Example 4.8.

R_{C2}. Assume that Q_1 is still in the active mode. The following values will be identical to those obtained in the previous example:

$$V_{B1} = +4.57 \text{ V} \qquad I_{E1} = 1.29 \text{ mA}$$

$$I_{B1} = 0.0128 \text{ mA} \qquad I_{C1} = 1.28 \text{ mA}$$

However, the collector voltage will be different than previously calculated, since part of the collector current I_{C1} will flow in the base lead of Q_2 (I_{B2}). As a first approximation we may assume that I_{B2} is much smaller than I_{C1}; that is, we may assume that the current through R_{C1} is almost equal to I_{C1}. This will enable us to calculate V_{C1}:

$$V_{C1} \simeq +15 - I_{C1}R_{C1}$$
$$= 15 - 1.28 \times 5 = +8.6 \text{ V}$$

Thus Q_1 is in the active mode, as had been assumed.

As far as Q_2 is concerned, we note that its emitter is connected to +15 V through R_{E2}. It is therefore safe to assume that the emitter–base junction of Q_2 will be forward-biased. Thus the emitter of Q_2 will be at a voltage V_{E2} given by

$$V_{E2} = V_{C1} + V_{EB}|_{Q_2} \simeq 8.6 + 0.7 = +9.3 \text{ V}$$

The emitter current of Q_2 may now be calculated as

$$I_{E2} = \frac{+15 - V_{E2}}{R_{E2}} = \frac{15 - 9.3}{2} = 2.85 \text{ mA}$$

Since the collector of Q_2 is returned to ground via R_{C2}, it is possible that Q_2 is operating in the active mode. Assume this to be the case. We now find I_{C2} as

$$I_{C2} = \alpha_2 I_{E2}$$
$$= 0.99 \times 2.85 = 2.82 \text{ mA} \qquad (\text{assuming } \beta_2 = 100)$$

The collector voltage of Q_2 will be

$$V_{C2} = I_{C2}R_{C2} = 2.82 \times 2.7 = 7.62 \text{ V}$$

which is lower than V_{B2} by 0.98 V. Thus Q_2 is in the active mode, as assumed.

It is important at this stage to find the magnitude of the error incurred in our calculations by the assumption that I_{B2} is negligible. The value of I_{B2} is given by

$$I_{B2} = \frac{I_{E2}}{\beta_2 + 1} = \frac{2.85}{101} = 0.028 \text{ mA}$$

which is indeed much smaller than I_{C1} (1.28 mA). If desired, we can obtain more accurate results by iterating one more time, assuming I_{B2} to be 0.028 mA. The new values will be

$$\text{Current in } R_{C1} = I_{C1} - I_{B2} = 1.28 - 0.028 = 1.252 \text{ mA}$$
$$V_{C1} = 15 - 5 \times 1.252 = 8.74 \text{ V}$$
$$V_{E2} = 8.74 + 0.7 = 9.44 \text{ V}$$
$$I_{E2} = \frac{15 - 9.44}{2} = 2.78 \text{ mA}$$
$$I_{C2} = 0.99 \times 2.78 = 2.75 \text{ mA}$$
$$V_{C2} = 2.75 \times 2.7 = 7.43 \text{ V}$$
$$I_{B2} = \frac{2.78}{101} = 0.0275 \text{ mA}$$

Note that the new value of I_{B2} is very close to the value used in our iteration, and no further iterations are warranted. The final results are indicated in Fig. 4.22(b).

The reader justifiably might be wondering about the necessity for using an iterative scheme in solving a linear (or linearized) problem. Indeed, we can obtain the exact solution (if we can call anything we are doing with a first-order model exact!) by writing appropriate equations. The reader is encouraged to find this solution and compare the results with those obtained above. It is important to emphasize, however, that in most such problems it is quite sufficient to obtain an approximate solution, provided that we can obtain it quickly and, of course, correctly.

Important Note

In Examples 4.2 through 4.8 we frequently used a precise value of α to calculate the collector current. Since $\alpha \simeq 1$, the error in such calculations will be very small if one assumes $\alpha = 1$ and $i_C = i_E$. Therefore, except in calculations that depend critically on the value of α (such as the calculation of base current), one usually assumes $\alpha \simeq 1$.

Exercises 4.19 For the circuit in Fig. 4.22, find the total current drawn from the power supply. Hence find the power dissipated in the circuit.

Ans. 4.135 mA; 62 mW

4.20 The circuit in Fig. E4.20 is to be connected to the circuit in Fig. 4.22(a) as indicated; specifically, the base of Q_3 is to be connected to the collector of Q_2. If Q_3 has $\beta = 100$, find the new value of V_{C2} and the values of V_{E3} and I_{C3}.

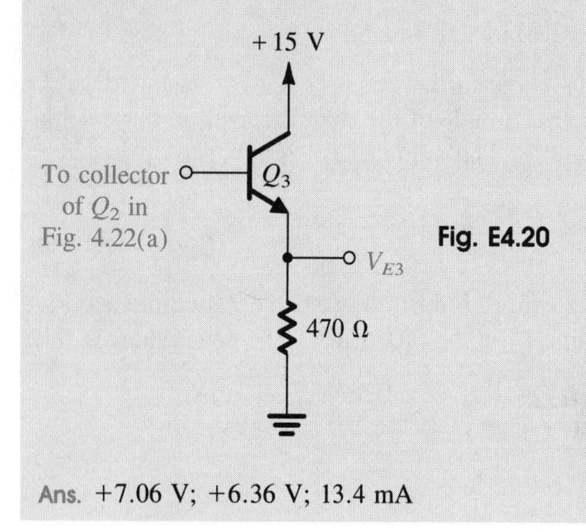

Fig. E4.20

Ans. +7.06 V; +6.36 V; 13.4 mA

4.7 THE TRANSISTOR AS AN AMPLIFIER

To operate as an amplifier a transistor must be biased in the active region. The biasing problem is that of establishing a constant dc current in the emitter (or the collector). This current should be predictable and insensitive to variations in temperature, value of β, and so on. While deferring the study of biasing techniques to Section 4.10, we will demonstrate in the following the need to bias the transistor at a constant collector current. This requirement stems from the fact that the operation of the transistor as an amplifier is highly influenced by the value of the quiescent (or bias) current, as shown below.

To understand how the transistor operates as an amplifier, consider the *conceptual* circuit shown in Fig. 4.23(a). Here the base–emitter junction is forward biased by a dc voltage V_{BE} (battery). The reverse bias of the collector–base junction is established by connecting the collector to another power supply of voltage V_{CC} through a resistor R_C. The input signal to be amplified is represented by the voltage source v_{be} that is superimposed on V_{BE}.

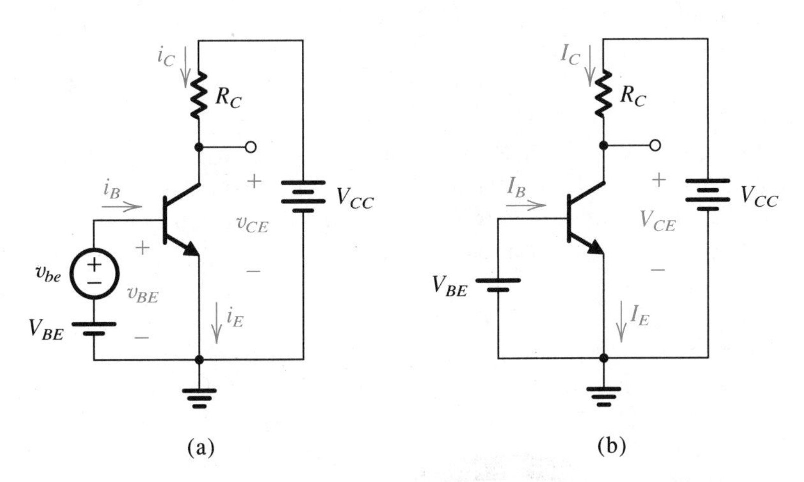

Fig. 4.23 (a) Conceptual circuit to illustrate the operation of the transistor as an amplifier. **(b)** The circuit of (a) with the signal source v_{be} eliminated, to perform dc (bias) analysis.

DC Conditions

We consider first the dc bias conditions by setting the signal v_{be} to zero. The circuit reduces to that in Fig. 4.23(b), and we can write the following relationships for the dc currents and voltages:

$$I_C = I_S e^{V_{BE}/V_T} \tag{4.14}$$

$$I_E = I_C/\alpha \tag{4.15}$$

$$I_B = I_C/\beta \tag{4.16}$$

$$V_C = V_{CE} = V_{CC} - I_C R_C \tag{4.17}$$

Obviously, for active-mode operation, V_C should be greater than V_B by an amount that allows for a reasonable signal swing at the collector yet maintains the transistor in the active region at all times. We shall return to this point later.

The Collector Current and the Transconductance

If a signal v_{be} is applied as shown in Fig. 4.23(a), the total instantaneous base–emitter voltage v_{BE} becomes

$$v_{BE} = V_{BE} + v_{be} \tag{4.18}$$

Correspondingly, the collector current becomes

$$\begin{aligned} i_C &= I_S e^{v_{BE}/V_T} = I_S e^{(V_{BE}+v_{be})/V_T} \\ &= I_S e^{(V_{BE}/V_T)} e^{(v_{be}/V_T)} \end{aligned} \tag{4.19}$$

Use of Eq. (4.14) yields

$$i_C = I_C e^{v_{be}/V_T} \tag{4.20}$$

Now, if $v_{be} \ll V_T$, we may approximate Eq. (4.20) as

$$i_C \simeq I_C \left(1 + \frac{v_{be}}{V_T}\right) \tag{4.21}$$

Here we have expanded the exponential in Eq. (4.20) in a series and retained only the first two terms. This approximation, which is valid only for v_{be} less than about 10 mV, is referred to as the **small-signal approximation.** Under this approximation the total collector current is given by Eq. (4.21) and can be rewritten

$$i_C = I_C + \frac{I_C}{V_T} v_{be} \tag{4.22}$$

Thus the collector current is composed of the dc bias value I_C and a signal component i_c,

$$i_c = \frac{I_C}{V_T} v_{be} \tag{4.23}$$

This equation relates the signal current in the collector to the corresponding base–emitter signal voltage. It can be rewritten as

$$i_c = g_m v_{be} \tag{4.24}$$

where g_m is called the **transconductance,** and from Eq. (4.23) it is given by

$$g_m = \frac{I_C}{V_T} \tag{4.25}$$

We observe that the transconductance of the BJT is directly proportional to the collector bias current I_C. Thus to obtain a constant, predictable value for g_m we need a constant, predictable I_C. Finally, we note that BJTs have relatively high transconductance (as compared to FETs, which are studied in the next chapter); for instance, at $I_C = 1$ mA, $g_m \simeq 40$ mA/V.

A graphical interpretation for g_m is given in Fig. 4.24, where it is shown that g_m is equal to the slope of the i_C–v_{BE} characteristic curve at $i_C = I_C$ (i.e., at the bias point Q). Thus

$$g_m = \left.\frac{\partial i_C}{\partial v_{BE}}\right|_{i_C = I_C} \tag{4.26}$$

The small-signal approximation implies keeping the signal amplitude sufficiently small so that operation is restricted to an almost linear segment of the i_C–v_{BE} exponential curve. Increasing the signal amplitude will result in the collector current having components

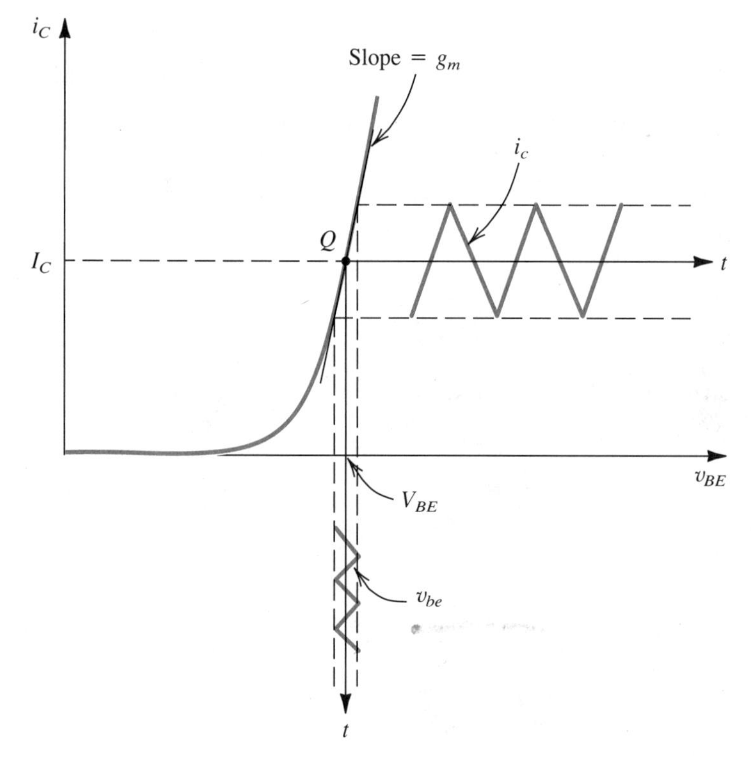

Fig. 4.24 Linear operation of the transistor under the small-signal condition: A small signal v_{be} with a triangular waveform is superimposed on the dc voltage V_{BE}. It gives rise to a collector signal current i_c, also of triangular waveform, superimposed on the dc current I_C. $i_c = g_m v_{be}$, where g_m is the slope of the i_C–v_{BE} curve at the bias point Q.

nonlinearly related to v_{be}. A similar type of approximation was employed for diodes (Section 3.4).

The analysis above suggests that for small signals ($v_{be} \ll V_T$), the transistor behaves as a voltage-controlled current source. The input port of this controlled source is between base and emitter, and the output port is between collector and emitter. The transconductance of the controlled source is g_m, and the output resistance is infinite. The latter ideal property is a result of our first-order model of transistor operation in which the collector voltage has no effect on the collector current in the active mode. As we have seen in Section 4.5, practical BJTs have finite output resistance. The effect of the output resistance on amplifier performance will be considered later.

Exercise 4.21 Use Eq. (4.26) to derive the expression for g_m in Eq. (4.25).

The Base Current and the Input Resistance at the Base

To determine the resistance seen by v_{be}, we first evaluate the total base current i_B using Eq. (4.22), as follows:

$$i_B = \frac{i_C}{\beta} = \frac{I_C}{\beta} + \frac{1}{\beta}\frac{I_C}{V_T} v_{be}$$

Thus

$$i_B = I_B + i_b \tag{4.27}$$

where I_B is equal to I_C/β and the signal component i_b is given by

$$i_b = \frac{1}{\beta}\frac{I_C}{V_T}v_{be} \tag{4.28}$$

Substituting for I_C/V_T by g_m gives

$$i_b = \frac{g_m}{\beta}v_{be} \tag{4.29}$$

The small-signal input resistance between base and emitter, *looking into the base,* is denoted by r_π and is defined as

$$r_\pi \equiv \frac{v_{be}}{i_b} \tag{4.30}$$

Using Eq. (4.29) gives

$$r_\pi = \frac{\beta}{g_m} \tag{4.31}$$

Thus r_π is directly dependent on β and is inversely proportional to the bias current I_C. Substituting for g_m in Eq. (4.31) from Eq. (4.25) and replacing I_C/β by I_B gives an alternative expression for r_π,

$$r_\pi = \frac{V_T}{I_B} \tag{4.32}$$

The Emitter Current and the Input Resistance at the Emitter

The total emitter current i_E can be determined using Eq. (4.22) and

$$i_E = \frac{i_C}{\alpha} = \frac{I_C}{\alpha} + \frac{i_c}{\alpha}$$

Thus

$$i_E = I_E + i_e \tag{4.33}$$

where I_E is equal to I_C/α and the signal current i_e is given by

$$i_e = \frac{i_c}{\alpha} = \frac{I_C}{\alpha V_T}v_{be} = \frac{I_E}{V_T}v_{be} \tag{4.34}$$

If we denote the small-signal resistance between base and emitter, *looking into the emitter,* by r_e, it can be defined as

$$r_e \equiv \frac{v_{be}}{i_e} \tag{4.35}$$

Using Eq. (4.34) we find that r_e, called the **emitter resistance,** is given by

$$r_e = \frac{V_T}{I_E} \tag{4.36}$$

Comparison with Eq. (4.25) reveals that

$$r_e = \frac{\alpha}{g_m} \simeq \frac{1}{g_m} \tag{4.37}$$

The relationship between r_π and r_e can be found by combining their respective definitions in Eqs. (4.30) and (4.35) as

$$v_{be} = i_b r_\pi = i_e r_e$$

Thus

$$r_\pi = (i_e/i_b) r_e$$

which yields

$$r_\pi = (\beta + 1) r_e \tag{4.38}$$

Exercise 4.22 A BJT having $\beta = 100$ is biased at a dc collector current of 1 mA. Find the value of g_m, r_e, and r_π at the bias point.

Ans. 40 mA/V; 25 Ω; 2.5 kΩ

Voltage Gain

Up until now we have established only that the transistor senses the base–emitter signal v_{be} and causes a proportional current $g_m v_{be}$ to flow in the collector lead at a high (ideally infinite) impedance level. In this way the transistor is acting as a voltage-controlled current source. To obtain an output voltage signal we may force this current to flow through a resistor, as is done in Fig. 4.23(a). Then the total collector voltage v_C will be

$$\begin{aligned} v_C &= V_{CC} - i_C R_C \\ &= V_{CC} - (I_C + i_c) R_C \\ &= (V_{CC} - I_C R_C) - i_c R_C \\ &= V_C - i_c R_C \end{aligned} \tag{4.39}$$

Here the quantity V_C is the dc bias voltage at the collector, and the signal voltage is given by

$$\begin{aligned} v_c &= -i_c R_C = -g_m v_{be} R_C \\ &= (-g_m R_C) v_{be} \end{aligned} \tag{4.40}$$

Thus the voltage gain of this amplifier is

$$\text{Voltage gain} \equiv \frac{v_c}{v_{be}} = -g_m R_C \tag{4.41}$$

Here again we note that because g_m is directly proportional to the collector bias current, the gain will be as stable as the collector bias current is made.

Exercise 4.23 In the circuit of Fig. 4.23(a), V_{BE} is adjusted to yield a dc collector current of 1 mA. Let $V_{CC} = 15$ V, $R_C = 10$ kΩ, and $\beta = 100$. Find the voltage gain v_c/v_{be}. If $v_{be} = 0.005 \sin \omega t$ volts, find $v_C(t)$ and $i_B(t)$.

Ans. -400 V/V; $5 - 2 \sin \omega t$ volts; $10 + 2 \sin \omega t$ μA

4.8 SMALL-SIGNAL EQUIVALENT CIRCUIT MODELS

The analysis in the previous section indicates that every current and voltage in the amplifier circuit of Fig. 4.23(a) is composed of two components: a dc component and a signal component. For instance, $v_{BE} = V_{BE} + v_{be}$, $I_C = I_C + i_c$, and so on. The dc components are determined from the dc circuit given in Fig. 4.23(b) and from the relationships imposed by the transistor (Eqs. 4.14 through 4.16). On the other hand, a representation of the signal operation of the BJT can be obtained by eliminating the dc sources, as shown in Fig. 4.25. Observe that since the voltage of an ideal dc supply does not change, the signal voltage across it will be zero. For this reason we have replaced V_{CC} and V_{BE} with short circuits. Had the circuit contained ideal dc current sources, these would have been replaced by open circuits. Note, however, that the circuit of Fig. 4.25 is useful only in so far as it shows the various signal currents and voltages; it is *not* an actual amplifier circuit since the dc bias circuit is not shown.

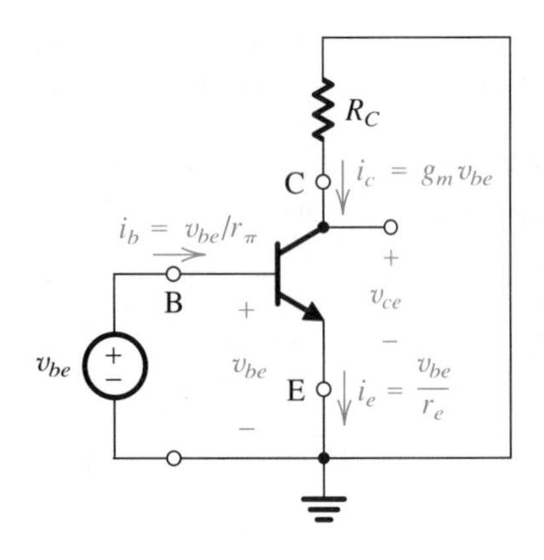

Fig. 4.25 The amplifier circuit of Fig. 4.23(a) with the dc sources V_{BE} and V_{CC} eliminated (short circuited). Thus only the signal components are present. Note that this is a representation of the signal operation of the BJT and not an actual amplifier circuit.

Figure 4.25 also shows the expressions for the current increments i_c, i_b, and i_e obtained when a small signal v_{be} is applied. These relationships can be represented by a circuit. Such a circuit should have three terminals, C, B, and E, and should yield the same terminal currents indicated in Fig. 4.25. The resulting circuit is then equivalent to the transistor as far as small-signal operation is concerned, and thus it can be considered an equivalent small-signal circuit model.

The Hybrid-π Model

An equivalent circuit model for the BJT is shown in Fig. 4.26(a). This model represents the BJT as a voltage-controlled current source and explicitly includes the input resistance looking into the base, r_π. The model obviously yields $i_c = g_m v_{be}$ and $i_b = v_{be}/r_\pi$. Not so obvious, however, is the fact that the model also yields the correct expression for i_e. This can be shown as follows: At the emitter node we have

$$i_e = \frac{v_{be}}{r_\pi} + g_m v_{be} = \frac{v_{be}}{r_\pi}(1 + g_m r_\pi)$$

$$= \frac{v_{be}}{r_\pi}(1 + \beta) = v_{be}\Big/\left(\frac{r_\pi}{1+\beta}\right)$$

$$= v_{be}/r_e$$

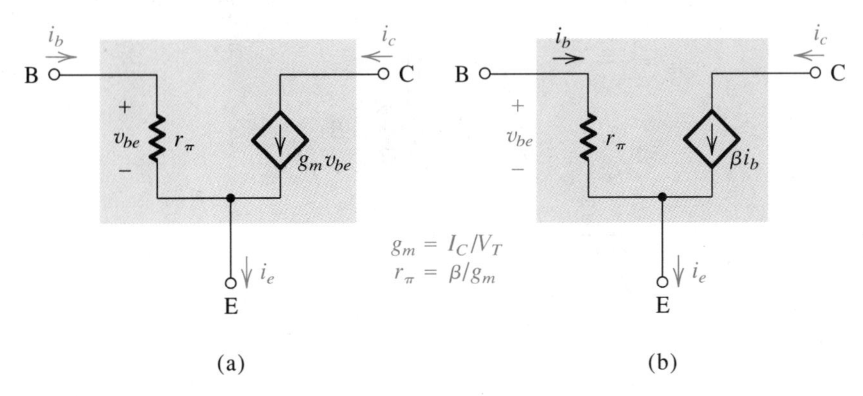

Fig. 4.26 Two slightly different versions of the simplified hybrid-π model for the small-signal operation of the BJT. The equivalent circuit in **(a)** represents the BJT as a voltage-controlled current source (a transconductance amplifier) and that in **(b)** represents the BJT as a current-controlled current source (a current amplifier).

A slightly different equivalent circuit model can be obtained by expressing the current of the controlled source ($g_m v_{be}$) in terms of the base current i_b as follows:

$$g_m v_{be} = g_m(i_b r_\pi)$$

$$= (g_m r_\pi) i_b = \beta i_b$$

This results in the alternative equivalent circuit model shown in Fig. 4.26(b). Here the transistor is represented as a current-controlled current source, with the control current being i_b.

The two models of Fig. 4.26 are simplified versions of what is known as the hybrid-π model. This is the most widely used model for the BJT. The complete hybrid-π model, as will be shown later in this chapter and in Chapter 7, includes additional components that model second-order effects in the BJT.

It is important to note that the small-signal equivalent circuits of Fig. 4.26 model the operation of the BJT at a given bias point. This should be obvious from the fact that the model parameters g_m and r_π depend on the value of the dc bias current I_C, as indicated in Fig. 4.26. Finally, although the models have been developed for an *npn* transistor, they apply equally well to *pnp* transistors *with no change of polarities*.

The T Model

Although we will almost always use the hybrid-π model (in one of its two variants shown in Fig. 4.26) to carry out small-signal analysis of transistor circuits, there are situations in which an alternative model, shown in Fig. 4.27, is somewhat more convenient. This model, called the **T model,** is shown in two versions in Fig. 4.27. The model of Fig. 4.27(a) represents the BJT as a voltage-controlled current source with the control voltage being v_{be}. Here, however, the resistance between base and emitter, looking into the emitter, is explicitly shown. From Fig. 4.27(a) we see clearly that the model yields the correct expressions for i_c and i_e. For i_b we note that at the base node we have

$$i_b = \frac{v_{be}}{r_e} - g_m v_{be} = \frac{v_{be}}{r_e}(1 - g_m r_e)$$

$$= \frac{v_{be}}{r_e}(1 - \alpha) = \frac{v_{be}}{r_e}\left(1 - \frac{\beta}{\beta + 1}\right)$$

$$= \frac{v_{be}}{(\beta + 1)r_e} = \frac{v_{be}}{r_\pi}$$

as should be the case.

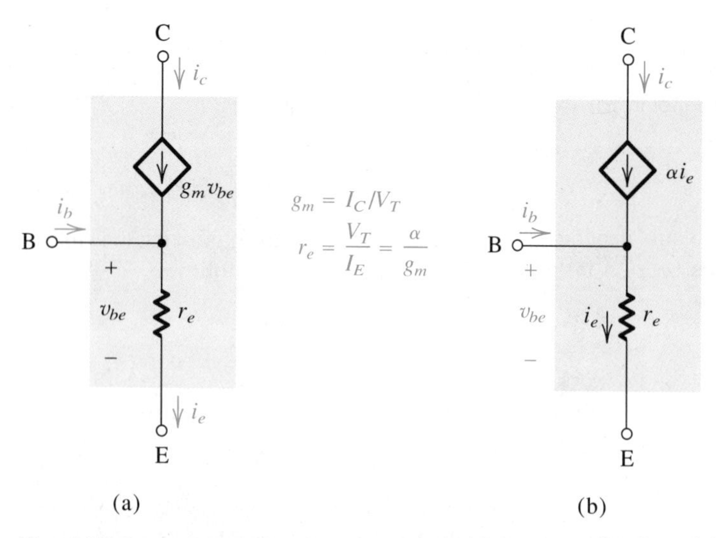

Fig. 4.27 Two slightly different versions of what is known as the *T model* of the BJT. The circuit in **(a)** is a voltage-controlled current source representation and that in **(b)** is a current-controlled current source representation. Those models explicitly show the emitter resistance r_e, unlike the hybrid-π model, which shows r_π.

If in the model of Fig. 4.27(a) the current of the controlled source is expressed in terms of the emitter current as follows:

$$g_m v_{be} = g_m(i_e r_e)$$
$$= (g_m r_e)i_e = \alpha i_e$$

we obtain the alternative T model shown in Fig. 4.27(b). Here the BJT is represented as a current-controlled current source but with the control signal being i_e.

Application of the Small-Signal Equivalent Circuits

The availability of the small-signal BJT circuit models makes the analysis of transistor amplifier circuits a systematic process. First, the dc operating point is determined and the model parameters are calculated. Then the dc sources are eliminated, the BJT is replaced with an equivalent circuit model (usually a hybrid-π model), and the resulting circuit is analyzed to determine the required quantity, e.g., voltage gain, input resistance, etc. The process will be illustrated by the following examples.

EXAMPLE 4.9

We wish to analyze the transistor amplifier shown in Fig. 4.28(a) to determine its voltage gain. Assume $\beta = 100$.

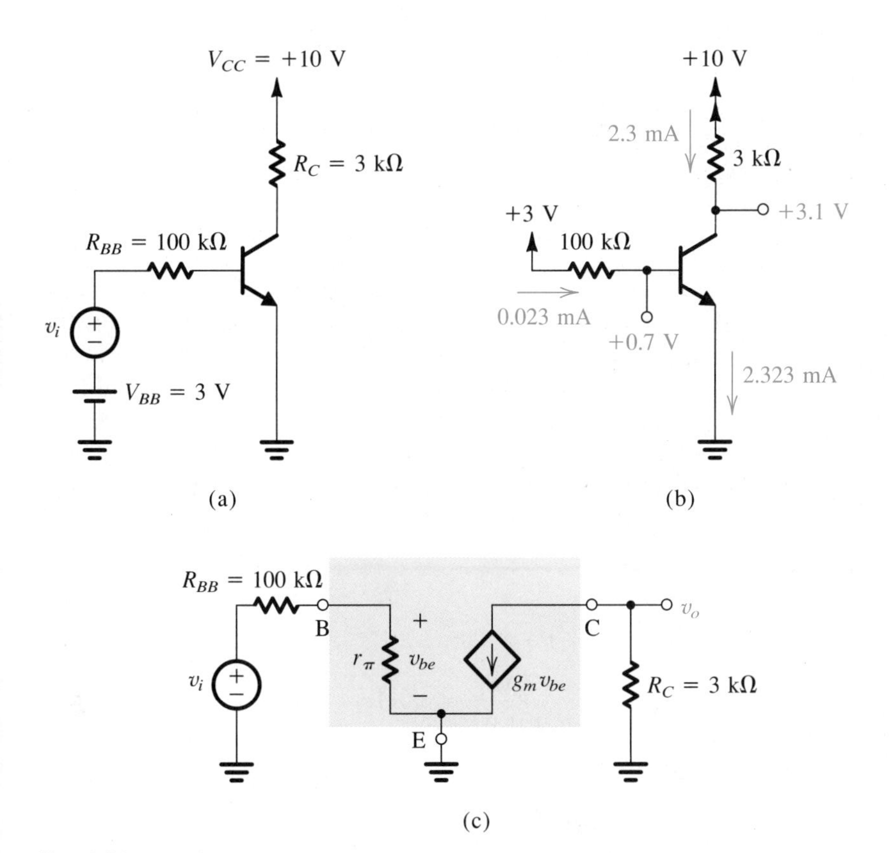

Fig. 4.28 Example 4.9: (a) circuit; (b) dc analysis; (c) small-signal model.

Solution The first step in the analysis consists of determining the quiescent operating point. For this purpose we assume that $v_i = 0$. The dc base current will be

$$I_B = \frac{V_{BB} - V_{BE}}{R_{BB}}$$

$$\simeq \frac{3 - 0.7}{100} = 0.023 \text{ mA}$$

The dc collector current will be

$$I_C = \beta I_B = 100 \times 0.023 = 2.3 \text{ mA}$$

The dc voltage at the collector will be

$$V_C = V_{CC} - I_C R_C$$

$$= +10 - 2.3 \times 3 = +3.1 \text{ V}$$

Since $V_B = +0.7$ V, it follows that in the quiescent condition the transistor will be operating in the active mode. The dc analysis is illustrated in Fig. 4.28(b).

Having determined the operating point, we may now proceed to determine the small-signal model parameters:

$$r_e = \frac{V_T}{I_E} = \frac{25 \text{ mV}}{(2.3/0.99) \text{ mA}} = 10.8 \ \Omega$$

$$g_m = \frac{I_C}{V_T} = \frac{2.3 \text{ mA}}{25 \text{ mV}} = 92 \text{ mA/V}$$

$$r_\pi = \frac{\beta}{g_m} = \frac{100}{92} = 1.09 \text{ k}\Omega$$

To carry out the small-signal analysis it is equally convenient to employ either of the two hybrid-π equivalent circuit models of Fig. 4.26. Using the first results in the amplifier equivalent circuit given in Fig. 4.28(c). Note that no dc quantities are included in this equivalent circuit. It is most important to note that the dc supply voltage V_{CC} has been replaced by a *short circuit* in the signal equivalent circuit because the circuit terminal connected to V_{CC} will always have a constant voltage. That is, the signal voltage at this terminal will be zero. In other words, *a circuit terminal connected to a constant dc source can always be considered as a signal ground.*

Analysis of the equivalent circuit in Fig. 4.28(c) proceeds as follows:

$$v_{be} = v_i \frac{r_\pi}{r_\pi + R_{BB}}$$

$$= v_i \frac{1.09}{101.09} = 0.011 v_i \tag{4.42}$$

The output voltage v_o is given by

$$v_o = -g_m v_{be} R_C$$

$$= -92 \times 0.011 v_i \times 3 = -3.04 v_i$$

Thus the voltage gain will be

$$\frac{v_o}{v_i} = -3.04 \tag{4.43}$$

where the minus sign indicates a phase reversal.

EXAMPLE 4.10

To gain more insight into the operation of transistor amplifiers, we wish to consider the waveforms at various points in the circuit analyzed in the previous example. For this purpose assume that v_i has a triangular waveform. First determine the maximum amplitude that v_i is allowed to have. Then, with the amplitude of v_i set to this value, give the waveforms of $i_B(t)$, $v_{BE}(t)$, $i_C(t)$, and $v_C(t)$.

Solution

One constraint on signal amplitude is the small-signal approximation, which stipulates that v_{be} should not exceed about 10 mV. If we take the triangular waveform v_{be} to be 20 mV peak-to-peak and work backward, Eq. (4.42) can be used to determine the maximum possible peak of v_i,

$$\hat{V}_i = \frac{\hat{V}_{be}}{0.011} = 0.91 \text{ V}$$

To check whether or not the transistor remains in the active mode with v_i having a peak value $\hat{V}_i = 0.91$ V, we have to evaluate the collector voltage. The voltage at the collector will consist of a triangular wave v_c superimposed on the dc value $V_C = 3.1$ V. The peak voltage of the triangular waveform will be

$$\hat{V}_c = \hat{V}_i \times \text{gain} = 0.91 \times 3.04 = 2.77 \text{ V}$$

It follows that when the output swings negative, the collector voltage reaches a minimum of $3.1 - 2.77 = 0.33$ V, which is lower than the base voltage $\simeq 0.7$ V. Thus the transistor will not remain in the active mode with v_i having a peak value of 0.91 V. We can easily determine, though, the maximum value of the peak of the input signal such that the transistor remains active at all times. This can be done by finding the value of $\hat{V}_i$ that corresponds to the minimum value of the collector voltage being equal to the base voltage, which is approximately 0.7 V. Thus

$$\hat{V}_i = \frac{3.1 - 0.7}{3.04} = 0.79 \ V$$

Let us choose $\hat{V}_i$ to be approximately 0.8 V, as shown in Fig. 4.29(a), and complete the analysis of this problem. The signal current in the base will be triangular, with a peak value $\hat{I}_b$ of

$$\hat{I}_b = \frac{\hat{V}_i}{R_{BB} + r_\pi} = \frac{0.8}{100 + 1.09} = 0.008 \text{ mA}$$

This triangular-wave current will be superimposed on the quiescent base current I_B, as shown in Fig. 4.29(b). The base–emitter voltage will consist of a triangular-wave compo-

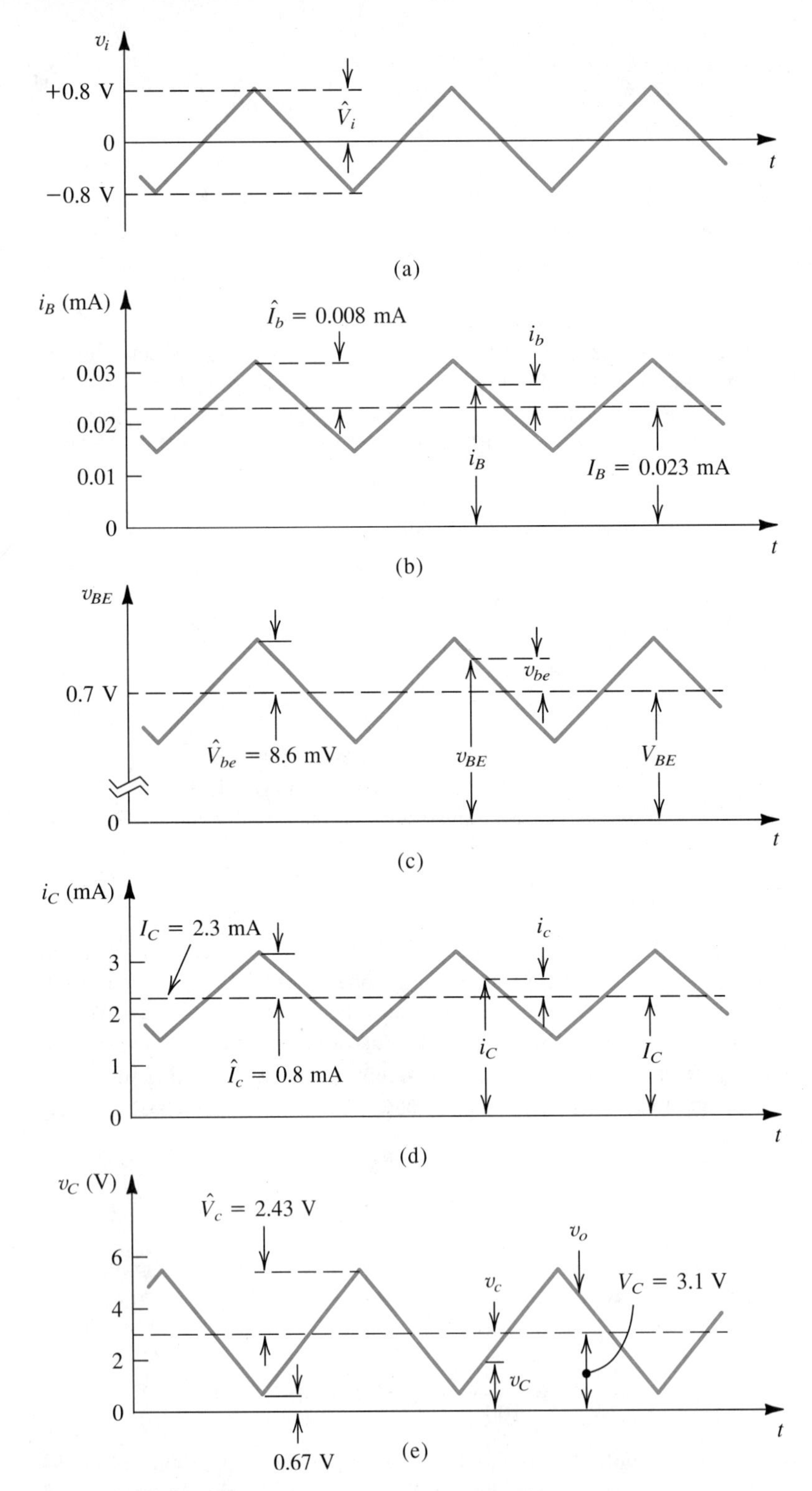

Fig. 4.29 Signal waveforms in the circuit of Fig. 4.28.

nent superimposed on the dc V_{BE} that is approximately 0.7 V. The peak value of the triangular waveform will be

$$\hat{V}_{be} = \hat{V}_i \frac{r_\pi}{r_\pi + R_{BB}} = 0.8 \frac{1.09}{100 + 1.09} = 8.6 \text{ mV}$$

The total v_{BE} is sketched in Fig. 4.29(c).

The signal current in the collector will be triangular in waveform, with a peak value $\hat{I}_c$ given by

$$\hat{I}_c = \beta \hat{I}_b = 100 \times 0.008 = 0.8 \text{ mA}$$

This current will be superimposed on the quiescent collector current I_C (=2.3 mA), as shown in Fig. 4.29(d).

Finally, the signal voltage at the collector can be obtained by multiplying v_i by the voltage gain; that is,

$$\hat{V}_c = 3.04 \times 0.8 = 2.43 \text{ V}$$

Figure 4.29(e) shows a sketch of the total collector voltage v_C versus time. Note the phase reversal between the input signal v_i and the output signal v_c. Also note that although the minimum collector voltage is slightly lower than the base voltage, the transistor will remain in the active mode. In fact, BJTs remain in the active mode even with their collector–base junctions forward-biased by as much as 0.3 or 0.4 V.

EXAMPLE 4.11

We need to analyze the circuit of Fig. 4.30(a) to determine the voltage gain and the signal waveforms at various points. The capacitor C is a coupling capacitor whose purpose is to couple the signal v_i to the emitter while blocking dc. In this way the dc bias established by V^+ *and* V^- together with R_E and R_C will not be disturbed when the signal v_i is connected. For the purpose of this example, C will be assumed to be infinite—that is, acting as a perfect short circuit at signal frequencies of interest. Similarly, another very large capacitor is used to couple the output signal v_o to other parts of the system.

Solution We shall start by determining the dc operating point as follows (see Fig. 4.30b):

$$I_E = \frac{+10 - V_E}{R_E} \simeq \frac{+10 - 0.7}{10} = 0.93 \text{ mA}$$

Assuming $\beta = 100$, then $\alpha = 0.99$ and

$$I_C = 0.99 I_E = 0.92 \text{ mA}$$

$$V_C = -10 + I_C R_C$$

$$= -10 + 0.92 \times 5 = -5.4 \text{ V}$$

Thus the transistor is in the active mode. Furthermore, the collector signal can swing from −5.4 V to zero (which is the base voltage) without the transistor going into saturation. However, a negative 5.4-V swing in the collector voltage will (theoretically) cause the

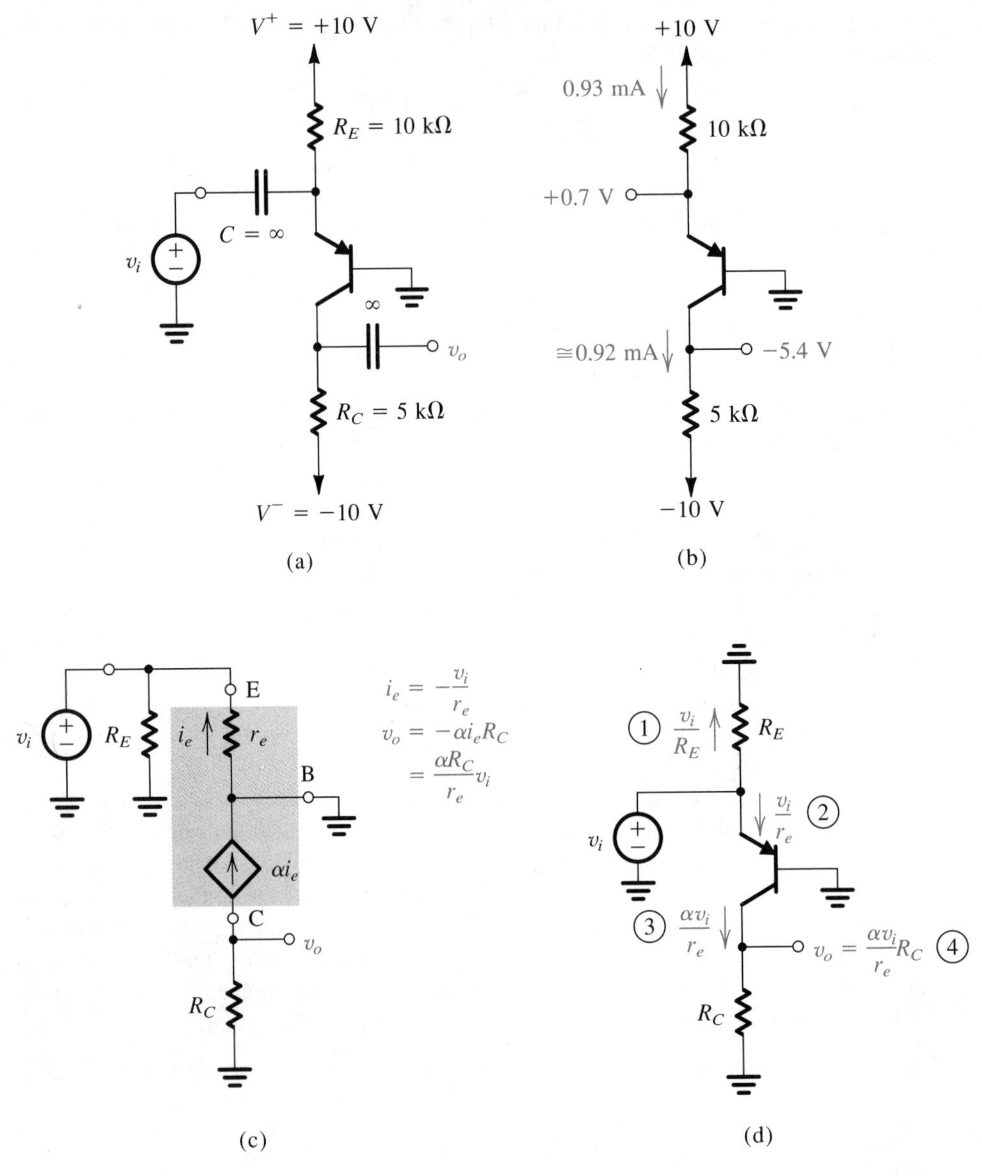

Fig. 4.30 Example 4.11: **(a)** circuit; **(b)** dc analysis; **(c)** small-signal model; **(d)** small-signal analysis performed directly on circuit.

minimum collector voltage to be −10.8 V, which is more negative than the power-supply voltage. It follows that if we attempt to apply an input that results in such an output signal, the transistor will cut off and the negative peaks of the output signal will be clipped off, as illustrated in Fig. 4.31. The waveform in Fig. 4.31, however, is shown to be linear (except for the clipped peak); that is, the effect of the nonlinear i_C–v_{BE} characteristic is not taken into account. This is not correct, since if we are driving the transistor into cutoff

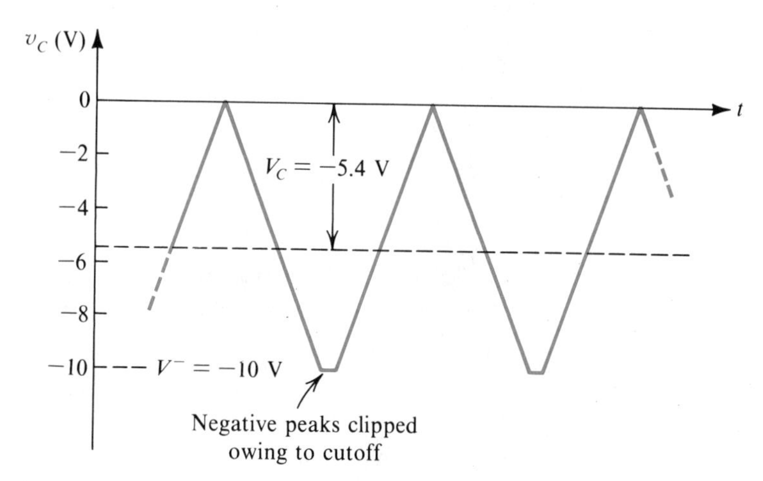

Fig. 4.31 Distortion in output signal due to transistor cutoff. Note that it is assumed that no distortion due to transistor nonlinear characteristics is occurring.

at the negative signal peaks, then we will surely be exceeding the small-signal limit, as will be shown later.

Let us now proceed to determine the small-signal voltage gain. Toward that end, we eliminate the dc sources and replace the BJT with its T equivalent circuit of Fig. 4.27(b). Note that because the base is grounded, the T model is somewhat more convenient than the hybrid-π model. Nevertheless, identical results can be obtained using the latter.

Figure 4.30(c) shows the resulting small-signal equivalent circuit of the amplifier. The model parameters are

$$\alpha = 0.99$$

$$r_e = \frac{V_T}{I_E} = \frac{25 \text{ mV}}{0.93 \text{ mA}} = 27 \ \Omega$$

Analysis of the circuit in Fig. 4.30(c) to determine the output voltage v_o and hence the voltage gain v_o/v_i is straightforward and is given in the figure. The result is

$$\frac{v_o}{v_i} = 183.3 \text{ V/V}$$

Note that the voltage gain is positive, indicating that the output is in phase with the input signal. This property is due to the fact that the input signal is applied to the emitter rather than to the base, as was done in Example 4.9. We should emphasize that the positive gain has nothing to do with the fact that the transistor used in this example is of the *pnp* type.

Returning to the question of allowable signal magnitude, we observe from Fig. 4.30(c) that $v_{eb} = v_i$. Thus, if small-signal operation is desired (for linearity), then the peak of v_i should be limited to about 10 mV. With $\hat{V}_i$ set to this value, as shown for a sine-wave input in Fig. 4.32, the peak amplitude at the collector, $\hat{V}_c$, will be

$$\hat{V}_c = 183.3 \times 0.01 = 1.833 \text{ V}$$

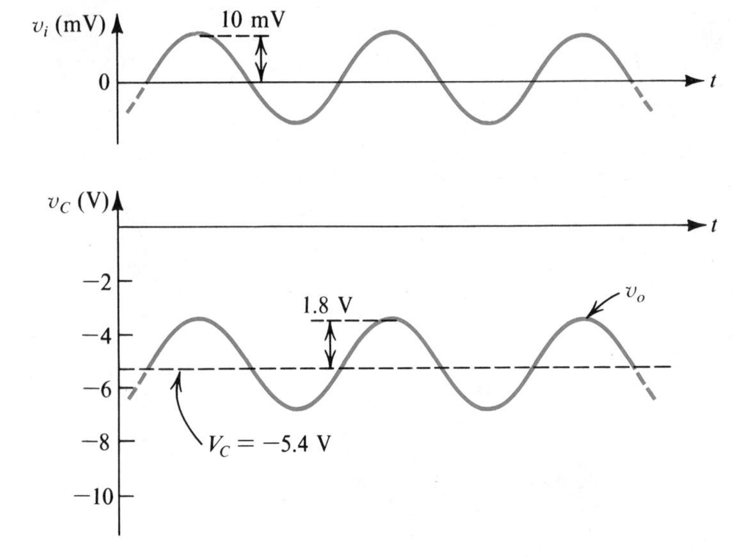

Fig. 4.32 Input and output waveforms for the circuit of Fig. 4.30.

and the total instantaneous collector voltage $v_C(t)$ will be as depicted in Fig. 4.32(b).

Performing Small-Signal Analysis Directly on the Circuit Diagram

In most cases one should explicitly replace each BJT with its small-signal model and analyze the resulting circuit, as we have done in the examples above. This systematic procedure is particularly recommended for beginning students. Experienced circuit designers, however, often perform a first-order analysis directly on the circuit. Figure 4.30(d) illustrates this process for the circuit we have just analyzed. The reader is urged to follow this direct analysis procedure (the steps are numbered). Observe that the equivalent circuit model is *implicitly* utilized; we are only saving the step of redrawing the circuit with the BJT replaced with its model. Direct analysis, however, has an additional very important benefit: It provides insight into the signal transmission through the circuit. Such insight can prove invaluable in design, particularly at the stage of selecting a circuit configuration appropriate for a given application.

Augmenting the Hybrid-π Model to Account for the Early Effect

The Early effect, discussed in Section 4.5, causes the collector current to depend not only on v_{BE} but also on v_{CE}. The dependence on v_{CE} can be modeled by assigning a finite output resistance to the controlled-current source in the hybrid-π model, as shown in Fig. 4.33. The output resistance r_o was defined in Eq. (4.12); its value is given by $r_o \simeq V_A/I_C$, where V_A is the Early voltage and I_C is the collector dc bias current. Note that in the models of Fig. 4.33 we have renamed v_{be} as v_π, in order to conform with the literature.

The question arises as to the effect of r_o on the operation of the transistor as an amplifier. In amplifier circuits in which the emitter is grounded (as in the circuit of Fig. 4.28), r_o simply appears in parallel with R_C. Thus if we include r_o in the equivalent circuit of Fig. 4.28(c), for example, the output voltage v_o becomes

$$v_o = -g_m v_{be}(R_C // r_o)$$

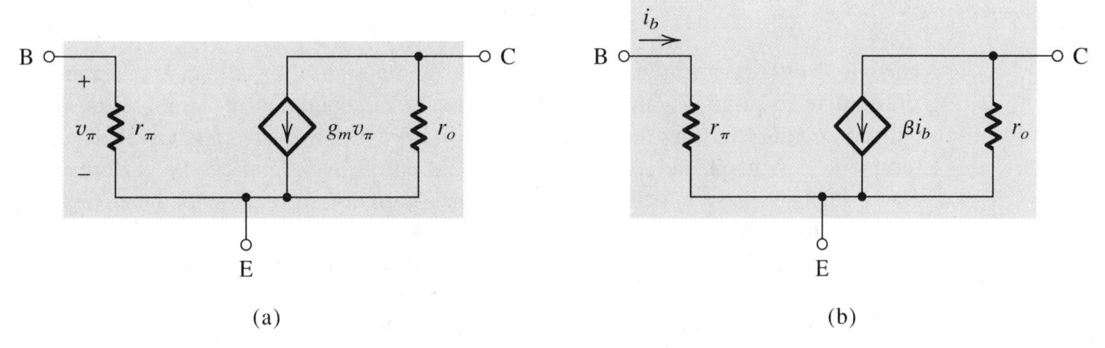

Fig. 4.33 The hybrid-π small-signal model, in its two versions, with the resistance r_o included. Note that $r_o = \left[\frac{(\partial i_C)}{(\partial v_{CE})}\right]^{-1}_{v_{be}=0} = \frac{V_A}{I_C}$, where V_A is the Early voltage and I_C is the collector dc bias current. Note that v_{be} has been renamed. v_π in conformity with the literature.

Thus the gain will be somewhat reduced. Obviously if $r_o \gg R_C$ the reduction in gain will be negligible and one can ignore the effect of r_o. In general, in such a configuration r_o can be neglected if it is greater than 10 R_C.

When the emitter of the transistor is not grounded, including r_o in the model can complicate the analysis. We will make comments regarding r_o and its inclusion or exclusion at frequent occasions throughout the book. Of course, if one is performing an accurate analysis of an almost final design using computer-aided analysis, then r_o can be easily included (see the SPICE supplement to this text).

Exercise 4.24 The transistor in Fig. E4.24 is biased with a constant current source $I = 1$ mA and has $\beta = 100$ and $V_A = 100$ V. (a) Find the dc voltages at the base, emitter, and collector. (b) Find g_m, r_π, and r_o. (c) If terminal Z is connected to ground, X to signal source v_s with a source resistance $R_s = 2$ kΩ, and Y to an 8-kΩ load resistance, use the hybrid-π model of Fig. 4.33(a), to draw the small-signal equivalent circuit of the amplifier. (Note that the current source I should be replaced with an open circuit.) Calculate the voltage gain v_y/v_i. If r_o were neglected what would the error in estimating the gain magnitude be?

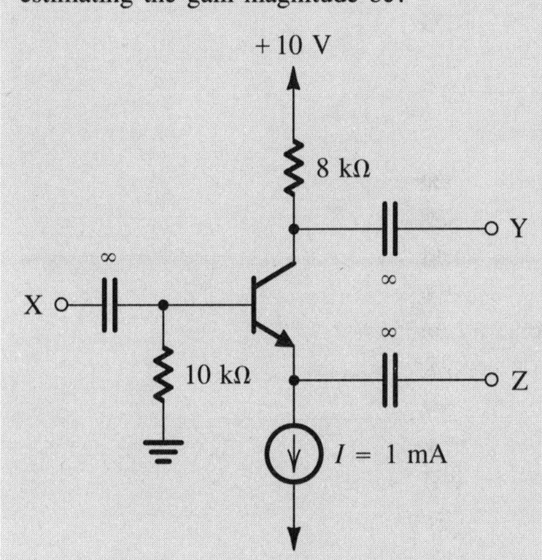

+10 V
8 kΩ
Y
∞
X
∞
∞
10 kΩ
Z
I = 1 mA

Fig. E4.24

Ans. (a) −0.1 V, −0.8 V, +2 V; (b) 40 mA/V, 2.5 kΩ, 100 kΩ; (c) −77 V/V, +3.9%

4.9 GRAPHICAL ANALYSIS

Although of little practical value in the analysis and design of most transistor circuits, it is illustrative to portray with graphical techniques the operation of a simple transistor amplifier circuit. Consider the circuit of Fig. 4.34, which we have already analyzed in Example 4.9. A graphical analysis of the operation of this circuit can be performed as follows: First, we have to determine the base bias current I_B by setting the input signal v_i to zero and using the technique illustrated in Fig. 4.35 (we have employed this technique in the analysis of diode circuits in Chapter 3). We next move to the i_C–v_{CE} characteristics, shown in Fig. 4.36. Observe that each of these characteristic curves is obtained by setting the base current i_B to a constant value, varying v_{CE}, and measuring the corresponding i_C. This family of i_C–v_{CE} characteristic curves should be contrasted to that shown in Fig. 4.15; the latter was obtained by setting v_{BE} constant.

Having determined the base bias current I_B, we know that the operating point will lie on the i_C–v_{CE} curve corresponding to this value of base current (the curve for $i_B = I_B$). Where it lies on the curve will be determined by the collector circuit. Specifically, the collector circuit imposes the constraint

$$v_{CE} = V_{CC} - i_C R_C$$

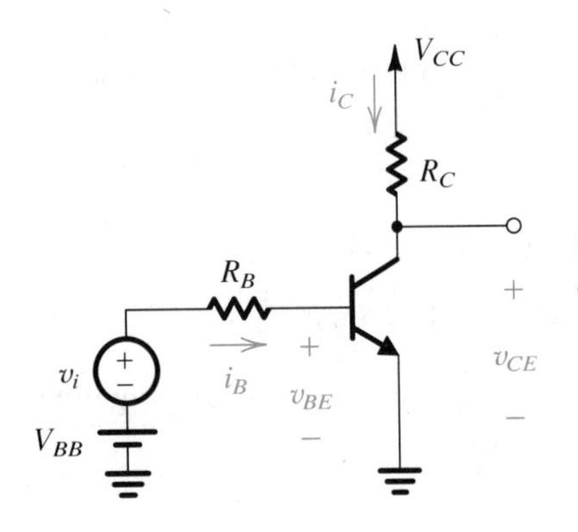

Fig. 4.34 Circuit whose operation is to be analyzed graphically.

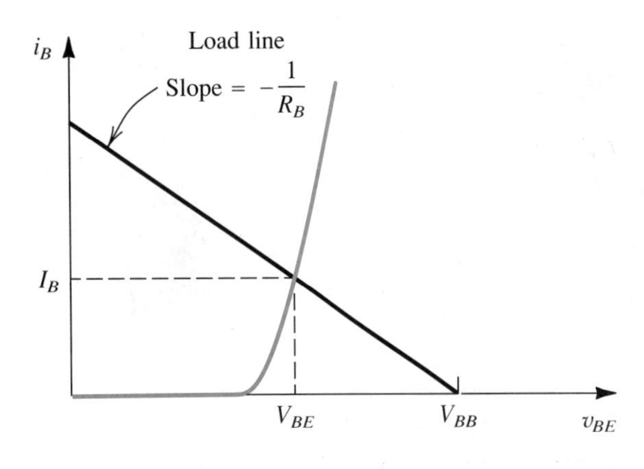

Fig. 4.35 Graphical construction for the determination of the dc base current in the circuit of Fig. 4.34.

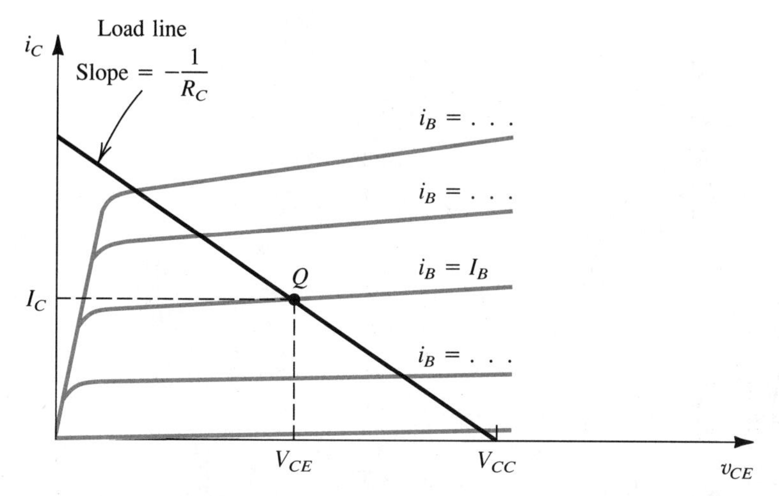

Fig. 4.36 Graphical construction for determining the dc collector current I_C and the collector-to-emitter voltage V_{CE} in the circuit of Fig. 4.34.

which can be rewritten

$$i_C = \frac{V_{CC}}{R_C} - \frac{1}{R_C} v_{CE}$$

and which represents a linear relationship between v_{CE} and i_C. This relationship can be represented by a straight line, as shown in Fig. 4.36. Since R_C can be considered the amplifier load, the straight line of slope $-1/R_C$ is known as the load line.[4] The dc bias point, or quiescent point, Q will be at the intersection of the load line and the i_C–v_{CE} curve corresponding to the base current I_B. The coordinates of point Q give the dc collector current I_C and the dc collector-to-emitter voltage V_{CE}. Observe that for amplifier operation, Q should be in the active region and furthermore should be in the middle of the active region to allow for a reasonable signal swing as the input signal v_i is applied.

The situation when v_i is applied is illustrated in Fig. 4.37. Consider first Fig. 4.37(a), which shows a signal v_i having a triangular waveform being superimposed on the dc voltage V_{BB}. Corresponding to each instantaneous value of $V_{BB} + v_i(t)$, one can draw a straight line with slope $-1/R_B$. Such an "instantaneous load line" intersects the i_B–v_{BE} curve at a point whose coordinates give the total instantaneous values of i_B and v_{BE} corresponding to the particular value of $V_{BB} + v_i(t)$. As an example, Fig. 4.37(a) shows the straight lines corresponding to $v_i = 0$, v_i at its positive peak, and v_i at its negative peak. Now, if the amplitude of v_i is sufficiently small so that the instantaneous operating point is confined to an almost linear segment of the i_B–v_{BE} curve, then the resulting signals i_b and v_{be} will be triangular in waveform, as indicated in the figure. This, of course, is the small-signal approximation. In summary, the graphical construction in Fig.

[4]The term *load line* is also employed for the straight line in Fig. 4.35.

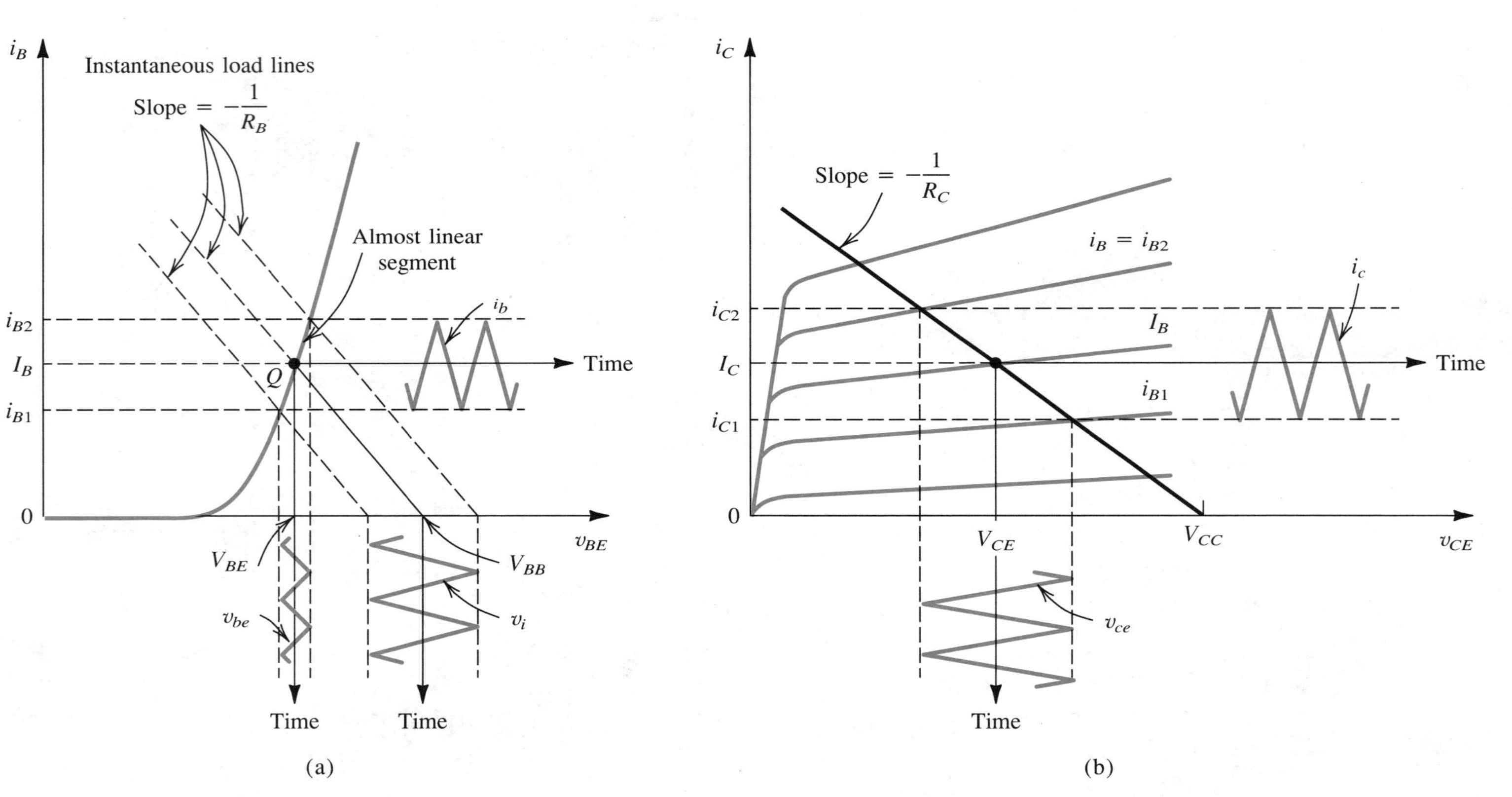

Fig. 4.37 Graphical determination of the signal components v_{be}, i_b, i_c, and v_{ce} when a signal component v_i is superimposed on the dc voltage V_{BB} (see Fig. 4.34).

4.37(a) can be used to determine the total instantaneous value of i_B corresponding to each value of v_i.

Next we move to the i_C–v_{CE} characteristics of Fig. 4.37(b). The operating point will move along the load line of slope $-1/R_C$ as i_B goes through the instantaneous values determined from Fig. 4.37(a). For example, when v_i is at its positive peak, $i_B = i_{B2}$ (from Fig. 4.37a), and the instantaneous operating point in the i_C–v_{CE} plane will be at the intersection of the load line and the curve corresponding to $i_B = i_{B2}$. In this way, one can determine the waveforms of i_C and v_{CE} and hence of the signal components i_c and v_{ce}, as indicated in Fig. 4.37(b).

Exercises 4.25 In terms of the parameters of the hybrid-π equivalent circuit model, what is the slope of the i_B–v_{BE} curve at the bias point? Find an expression for the slope in terms of the dc bias current I_B.

Ans. $1/r_\pi$; I_B/V_T

4.26 Consider the circuit of Fig. 4.34 with $V_{BB} = 1.7$ V, $R_B = 100$ kΩ, $V_{CC} = 10$ V, and $R_C = 5$ kΩ. Let the transistor $\beta = 100$. The input signal v_i is a triangular wave of 0.4 V peak-to-peak. Refer to Fig. 4.37 and use the geometry of the graphical construction shown to answer the following questions. (a) If $V_{BE} = 0.7$ V, find I_B. (b) Assuming operation on a straight line segment of the i_B–v_{BE} curve, find the inverse of its slope (use the result of Exercise 4.25). (c) Find approximate values for the peak-to-peak amplitude of i_b and of v_{be}. (d) Assuming the i_C–v_{CE} curves to be horizontal (i.e., ignoring the Early effect) find I_C and V_{CE}. (e) Find the peak-to-peak amplitude of i_c and of v_{ce}. (f) What is the voltage gain of this amplifier?

Ans. (a) 10 μA; (b) 2.5 kΩ; (c) 4 μA, 10 mV; (d) 1 mA, 5 V; (e) 0.4 mA, 2 V; (f) −5 V/V

4.10 BIASING THE BJT FOR DISCRETE-CIRCUIT DESIGN

The biasing problem is that of establishing a constant dc current in the emitter of the BJT. This current has to be calculable, predictable, and insensitive to variations in temperature and to the large variations in the value of β encountered among transistors of the same type. In this section we shall deal with the classical approaches to solving the bias problem in transistor circuits designed from discrete devices. Bias methods for integrated-circuit design are presented in Chapter 6.

Bias Arrangement Using a Single Power Supply

Figure 4.38(a) shows the arrangement most commonly used for biasing a transistor amplifier if only a single power supply is available. The technique consists of supplying the base of the transistor with a fraction of the supply voltage V_{CC} through the voltage divider R_1, R_2. In addition, a resistor R_E is connected to the emitter.

Figure 4.38(b) shows the same circuit with the voltage-divider network replaced by its Thévenin equivalent.

$$V_{BB} = \frac{R_2}{R_1 + R_2} V_{CC} \tag{4.44}$$

$$R_B = \frac{R_1 R_2}{R_1 + R_2} \tag{4.45}$$

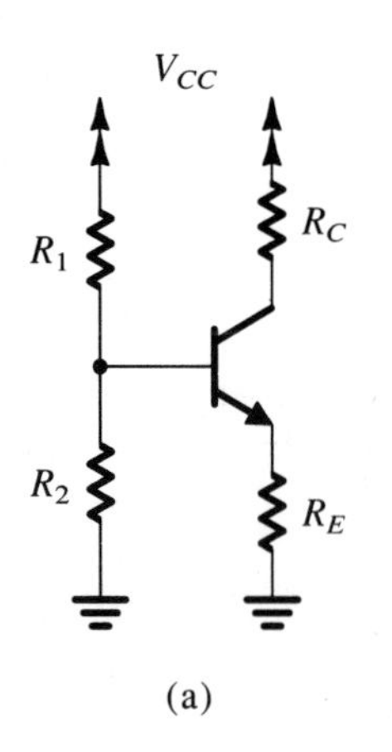

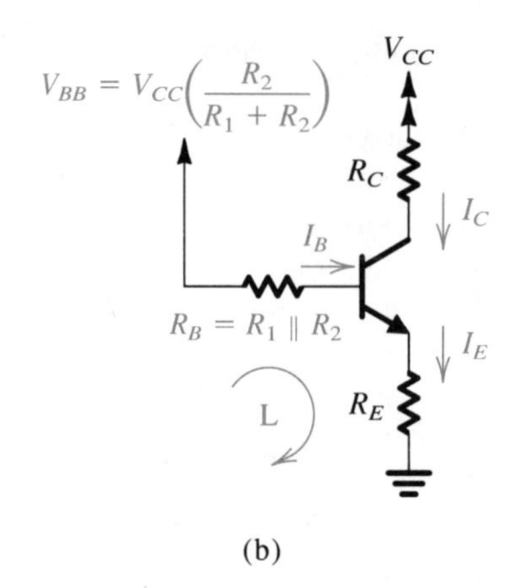

Fig. 4.38 Classical biasing for BJTs using a single power supply: **(a)** circuit; **(b)** circuit with the voltage divider supplying the base replaced with its Thévenin equivalent.

The current I_E can be determined by writing a Kirchhoff loop equation for the base–emitter–ground loop and substituting $I_B = I_E/(\beta + 1)$:

$$I_E = \frac{V_{BB} - V_{BE}}{R_E + R_B/(\beta + 1)} \tag{4.46}$$

To make I_E insensitive to temperature and β variation, we design the circuit to satisfy the following two constraints:

$$V_{BB} \gg V_{BE} \tag{4.47}$$

$$R_E \gg \frac{R_B}{\beta + 1} \tag{4.48}$$

Condition (4.47) ensures that small variations in V_{BE} (around 0.7 V) will be swamped by the much larger V_{BB}. There is a limit, however, on how large V_{BB} can be: For a given value of the supply voltage V_{CC}, the higher the value we use for V_{BB}, the lower will be the sum of voltages across R_C and the collector–base junction (V_{CB}). On the other hand, we want the voltage across R_C to be large in order to obtain high voltage gain and large signal swing (before transistor cutoff). We also want V_{CB} (or V_{CE}) to be large to provide a large signal swing (before transistor saturation). Thus, as is the case in any design problem, we have a set of conflicting requirements, and the solution must be a compromise. As a rule of thumb one designs for V_{BB} about $\frac{1}{3}V_{CC}$, V_{CB} (or V_{CE}) about $\frac{1}{3}V_{CC}$, and $I_C R_C$ about $\frac{1}{3}V_{CC}$.

Condition (4.48) makes I_E insensitive to variations in β and could be satisfied by selecting R_B small. This in turn is achieved by using low values for R_1 and R_2. Lower values for R_1 and R_2, however, will mean a higher current drain from the power supply and normally will result in a lowering of the input resistance of the amplifier (if the input signal is coupled to the base), which is the trade-off involved in this part of the design problem. It should be noted that Condition (4.48) means that we want to make the base voltage independent of the value of β and determined solely by the voltage divider. This will obviously be satisfied if the current in the divider is made much larger than the base current. Typically one selects R_1 and R_2 such that their current is in the range of I_E to $0.1I_E$.

Further insight into the mechanism by which the bias arrangement of Fig. 4.38(a) stabilizes the dc emitter (and hence collector) current is obtained by considering the feedback action provided by R_E. Consider that for some reason the emitter current increases. The voltage drop aross R_E and hence V_E will increase correspondingly. Now, if the base voltage is determined primarily by the voltage divider R_1, R_2, which is the case if R_B is small, the increase in V_E will result in a corresponding decrease in V_{BE}. This in turn reduces the collector (and emitter) current, a change opposite to that originally assumed. Thus R_E provides a *negative feedback* action that stabilizes the bias current. We shall study negative feedback formally in Chapter 8.

EXAMPLE 4.12

We wish to design the bias network of the amplifier in Fig. 4.38 to establish a current $I_E = 1$ mA using a power supply $V_{CC} = +12$ V.

Solution

We shall follow the rule of thumb mentioned above and allocate one-third of the supply voltage to the voltage drop across R_2 and another third to the voltage drop across R_C, leaving one-third for possible signal swing at the collector. Thus

$$V_B = +4 \text{ V}$$

$$V_E = 4 - V_{BE} \simeq 3.3 \text{ V}$$

and R_E is determined from

$$R_E = \frac{V_E}{I_E} = 3.3 \text{ k}\Omega$$

From the discussion above we select a voltage-divider current of $0.1I_E$. Neglecting the base current, we find

$$R_1 + R_2 = \frac{12}{0.1I_E} = 120 \text{ k}\Omega$$

$$\frac{R_2}{R_1 + R_2} V_{CC} = 4 \text{ V}$$

Thus $R_2 = 40$ kΩ and $R_1 = 80$ kΩ.

At this point it is desirable to find a more accurate estimate for I_E, taking into account the nonzero base current. Using Eq. (4.46), and assuming that β is specified to be 100, we obtain

$$I_E = \frac{3.3}{3.3 + 0.267} = 0.93 \text{ mA}$$

We could, of course, have obtained a value much closer to the desired 1 mA by designing with exact equations. However, since our work is based on first-order models, it does not make sense to strive for accuracy better than 5% or 10%.

It should be noted that if we are willing to draw a higher current from the power supply and if we are prepared to accept a lower input resistance for the amplifier, then we may use a voltage-divider current equal, say, to I_E, resulting in $R_1 = 8\ \text{k}\Omega$ and $R_2 = 4\ \text{k}\Omega$. The effect of this on the amplifier input resistance will be analyzed in Section 4.11. We shall refer to the circuit using these latter values as design 2, for which the actual value of I_E will be

$$I_E = \frac{3.3}{3.3 + 0.026} \simeq 1\ \text{mA}$$

The value of R_C can be determined from

$$R_C = \frac{12 - V_C}{I_C}$$

Thus for design 1 we have

$$R_C = \frac{12 - 8}{0.99 \times 0.93} = 4.34\ \text{k}\Omega$$

whereas for design 2 we have

$$R_C = \frac{12 - 8}{0.99 \times 1} = 4.04\ \text{k}\Omega$$

For simplicity we shall select $R_C = 4\ \text{k}\Omega$ for both designs.

Exercise 4.27 For design 1 in Example 4.12 calculate the expected range of I_E if the transistor used has β in the range 50 to 150. Express the range of I_E as a percentage of the nominal value ($I_E = 1$ mA) obtained for $\beta = \infty$. Repeat for design 2.

Ans. For design 1, 0.86 to 0.95 mA, a 9% range; for design 2, 0.984 to 0.995 mA, a 1.1% range

Biasing Using Two Power Supplies

A somewhat simpler bias arrangement is possible if two power supplies are available, as shown in Fig. 4.39. Writing a loop equation for the loop labeled L gives

$$I_E = \frac{V_{EE} - V_{BE}}{R_E + R_B/(\beta + 1)} \tag{4.49}$$

This equation is identical to Eq. (4.46) except for V_{EE} replacing V_{BB}. Thus the two constraints of Eqs. (4.47) and (4.48) apply here as well. Note that if the transistor is to be used with the base grounded (that is, in the common-base configuration discussed in the next section), then R_B can be eliminated altogether. On the other hand, if the input signal is to be coupled to the base, then R_B is needed.

Exercise D4.28 The bias arrangement of Fig. 4.39 is to be used for a common-base amplifier. Design the circuit to establish a dc emitter current of 1 mA and allowing for a maximum signal swing at the collector of ±2 V. Use +10-V and −5-V power supplies.

Ans. $R_B = 0$; $R_E = 4.3\ \text{k}\Omega$; $R_C = 8\ \text{k}\Omega$

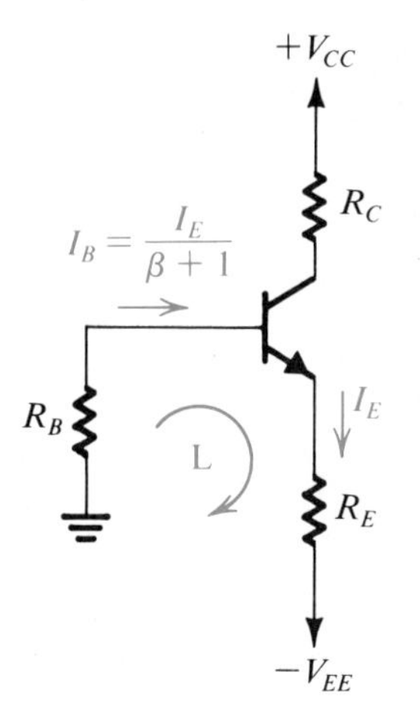

Fig. 4.39 Biasing the BJT using two power supplies. Resistor R_B is needed only if the signal is to be coupled to the base. Otherwise, the base can be connected directly to ground, resulting in almost total independence of the bias current with regard to the value of β.

An Alternative Biasing Arrangement

Figure 4.40(a) shows a simple but effective alternative biasing arrangement suitable for common-emitter amplifiers. Analysis of the circuit is shown in Fig. 4.40(b), from which we can write

$$V_{CC} = I_E R_C + I_B R_B + V_{BE}$$

$$= I_E R_C + \frac{I_E}{\beta + 1} R_B + V_{BE}$$

Thus the emitter bias current is given by

$$I_E = \frac{V_{CC} - V_{BE}}{R_C + R_B/(\beta + 1)} \tag{4.50}$$

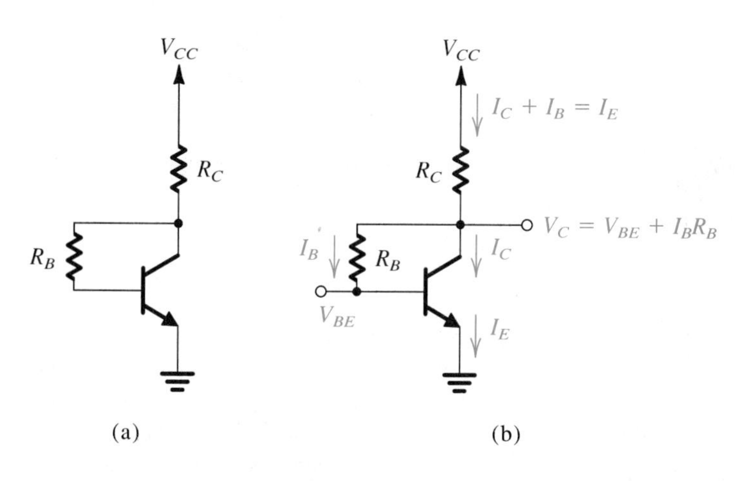

Fig. 4.40 (a) A simple alternative biasing arrangement suitable for common-emitter amplifiers. **(b)** Analysis of the circuit in (a).

To obtain a value of I_E that is insensitive to variation of β, we select $R_B/(\beta + 1) \ll R_C$. Note, however, that the value of R_B determines the allowable signal swing at the collector since

$$V_{CB} = I_B R_B = I_E \frac{R_B}{\beta + 1} \tag{4.51}$$

Bias stability in this circuit is achieved by the negative feedback action of resistor R_B. We will encounter circuits of this type in our study of feedback in Chapter 8.

Exercise D4.29 Design the circuit of Fig. 4.40 to obtain a dc emitter current of 1 mA and to ensure a ±2-V signal swing at the collector. Let $V_{CC} = 10$ V and $\beta = 100$.

Ans. $R_B = 202$ kΩ; $R_C = 7.3$ kΩ. Note that if standard 5% resistor values are used (Appendix F) we select $R_B = 200$ kΩ and $R_C = 7.5$ kΩ. This results in $I_E = 0.98$ mA and $V_C = 2.64$ V.

Biasing Using a Current Source

The BJT can be biased using a constant current source I as indicated in the circuit of Fig. 4.41. This circuit has the advantage that the emitter current is independent of the values of β and R_B. Thus R_B can be made large, enabling an increase in the input resistance at the base without adversely affecting bias stability. The constant current source I can be easily implemented using another BJT, as will be shown in Chapter 6.

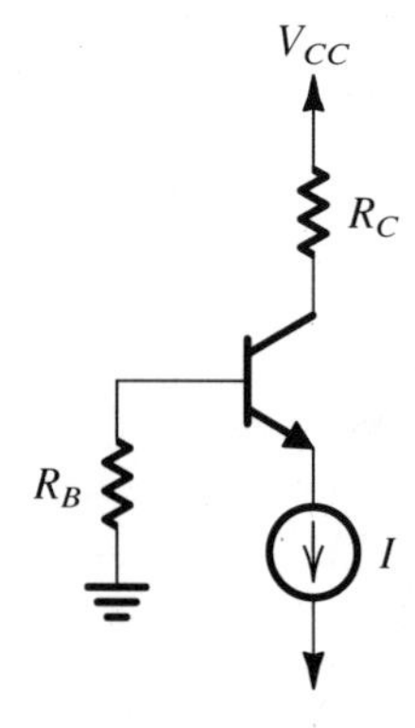

Fig. 4.41 A BJT biased using a constant-current source I.

Exercise 4.30 For the circuit in Fig. 4.41 with $V_{CC} = 10$ V, $I = 1$ mA, $\beta = 100$, $R_B = 100$ kΩ, and $R_C = 7.5$ kΩ, find the dc voltage at the base, the emitter, and the collector.

Ans. −1 V; −1.7 V; +2.5 V

4.11 BASIC SINGLE-STAGE BJT AMPLIFIER CONFIGURATIONS

In this section we study the three basic configurations of BJT amplifiers: the common-emitter (CE), the common-base (CB), and the common-collector (CC) circuits. To enable comparisons to be made between the characteristics of the three configurations, all three

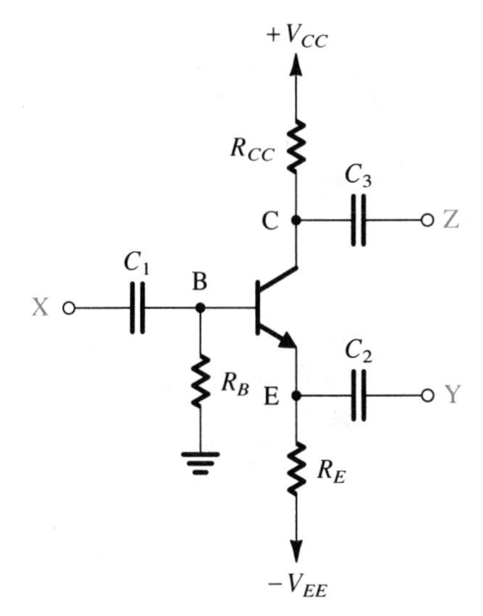

Fig. 4.42 A universal BJT amplifier configuration. Large-valued capacitors C_1, C_2, and C_3 will be used to connect the corresponding BJT terminals to a signal source, a load, or ground. In this way the three basic BJT amplifier configurations, CE, CB, and CC, will be generated.

circuits will be derived from the circuit shown in Fig. 4.42, which may thus be considered a **universal amplifier** configuration.

As indicated, two power supplies are employed for biasing, and thus only one resistor, R_B, is required to establish the dc voltage at the base. Bias design is straightforward and follows the method studied in the previous section. Three large-valued capacitors, C_1, C_2, and C_3, are connected to the base, emitter, and collector terminals, respectively. As will be seen shortly, these capacitors are used to connect the BJT terminals to a signal source, a load resistance, or ground. Since capacitors block dc, such connections do not affect the bias of the BJT, which is the major advantage of ac coupling. The major disadvantage is the requirement for large capacitors, making the circuit suitable only for discrete component design. It should be noted, however, that although the basic amplifier configurations are studied in this section in their capacitor-coupled form, their characteristics remain unchanged when the direct-coupled forms are used. The latter are utilized in IC design, as will be shown in Chapters 6 and 10.

Exercise 4.31 For the circuit in Fig. 4.42 let $R_B = 100\ \text{k}\Omega$, $R_E = 10\ \text{k}\Omega$, $R_C = 10\ \text{k}\Omega$, $V_{CC} = V_{EE} = 10$ V, and let the BJT have $\beta = 100$ and $V_A = 100$ V. Find the values of V_B, V_E, I_C, and V_C. Also find the values of the small-signal model parameters g_m, r_e, r_π, and r_o at the bias point.

Ans. $V_B = -0.84$ V; $V_E = -1.54$ V; $I_C = 0.84$ mA; $V_C = +1.6$ V; $g_m = 33.6$ mA/V; $r_e \simeq 30\ \Omega$; $r_\pi \simeq 3\ \text{k}\Omega$; $r_o = 119\ \text{k}\Omega$

The Mid-Frequency Band

In the analysis to follow we shall assume that capacitors C_1, C_2, and C_3 are very large (ideally infinite), thus postponing the study of the effect of these capacitors on the gain of the amplifier until Chapter 7. In Chapter 7, it will be shown that the finite values of C_1, C_2, and C_3 cause the amplifier gain to fall off at low frequencies. It will also be shown that the amplifier gain falls off at high frequencies as well. The latter effect, however, is due to the transistor internal capacitances (or more properly, the physical phenomena that are modeled by capacitances). In the present chapter, we shall neglect the effect of these internal capacitances. In other words, the analysis of the present chapter assumes the signal frequency is sufficiently high so as to assume that C_1, C_2, and C_3 act as perfect short circuits, but is sufficiently low so that the transistor internal capacitances can be neglected. Such a frequency is said to be in the **mid-frequency range** or **midband** of the amplifier.

The Common-Emitter Amplifier

The common-emitter amplifier configuration can be generated from the circuit of Fig. 4.42 by connecting terminal Y to ground, terminal X to the input signal source, and terminal Z to the load resistance. The resulting circuit is shown in Fig. 4.43(a). The large capacitor C_2 establishes a signal ground on the emitter. Thus, the circuit can be viewed as a two-port network with the input port between base and emitter and the output port between collector and emitter. The emitter at *ground* potential is a *common* terminal between input and output; hence the name common-emitter or grounded-emitter amplifier.

Observe that while capacitors C_1 and C_3 serve to couple the signal source and the load to the BJT, capacitor C_2 serves to short circuit the emitter to ground at signal frequencies. Thus the signal current in the emitter flows through C_2 to ground, thus *bypassing* resistor R_E. Capacitor C_2 is thus called a **bypass capacitor.**

Before we begin our analysis of the CE amplifier, a comment is in order on the signal source and the load. In a multistage amplifier, the signal source can represent the Théve-nin equivalent of the output side of the preceding amplifier stage; that is, v_s is the open-circuit output voltage, and R_s is the output resistance of the preceding stage. Similarly, R_L can represent the input resistance of the succeeding amplifier stage.

Replacing the BJT with its hybrid-π equivalent circuit model results in the CE amplifier equivalent circuit shown in Fig. 4.43(b). We can model the amplifier with any of the four models studied in Section 1.5. Here the most convenient model is the transconductance amplifier (Fig. 1.17a), which is characterized by the input resistance R_i, the output resistance R_o, and the short-circuit transconductance G_m. Figure 4.43(c) shows the CE circuit with the amplifier replaced with its transconductance model. The input resistance R_i is determined from inspection of the circuit in Fig. 4.43(b) to be

$$R_i = R_B // r_\pi \tag{4.52}$$

from which we clearly see the need to select R_B as large as possible, a requirement that conflicts with that for bias stability. For $R_B \gg r_\pi$, the input resistance of the common-emitter amplifier becomes

$$R_i \simeq r_\pi \tag{4.53}$$

The short-circuit transconductance G_m can be determined from the circuit of Fig. 4.43(b) as follows:

$$G_m \equiv \left.\frac{i_o}{v_i}\right|_{R_L=0} = \frac{-g_m v_\pi}{v_\pi}$$

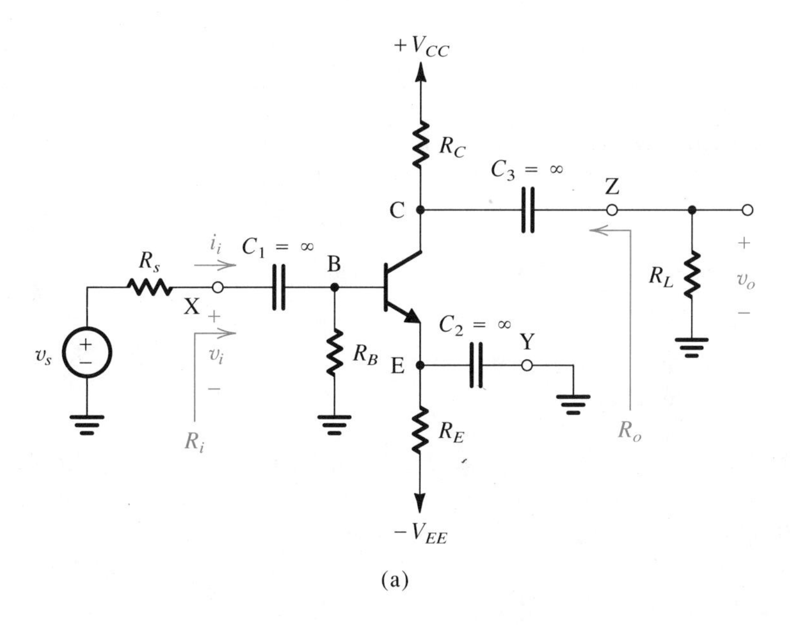

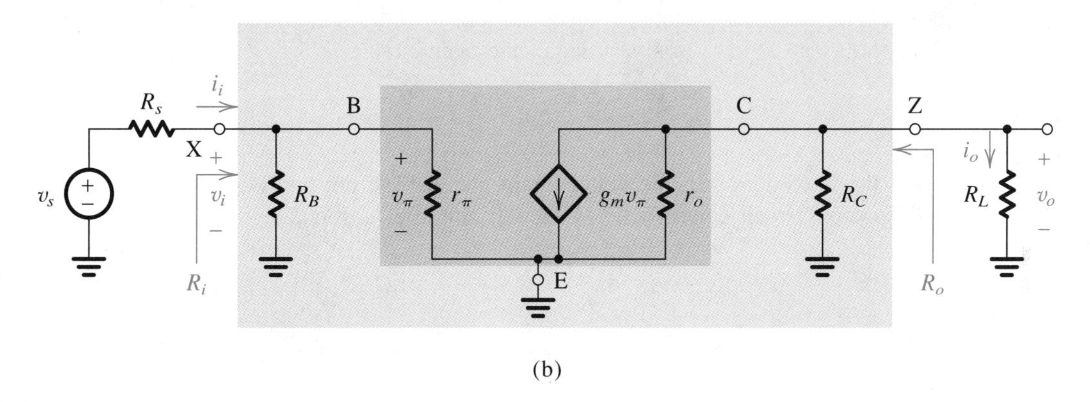

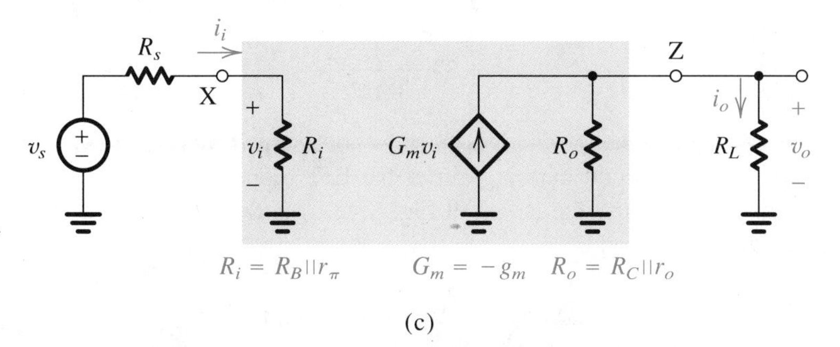

Fig. 4.43 The common-emitter amplifier: **(a)** circuit derived from that in Fig. 4.42; **(b)** equivalent circuit obtained by replacing the BJT with its hybrid-π model; **(c)** equivalent circuit model as a transconductance amplifier (the model parameters are indicated).

Thus,

$$G_m = -g_m \tag{4.54}$$

To determine the output resistance of the CE amplifier we set $v_s = 0$. This results in $v_\pi = 0$ and the controlled source $g_m v_\pi = 0$. Thus

$$R_o = R_C // r_o \tag{4.55}$$

For discrete-component circuits, $R_C \ll r_o$ and

$$R_o \simeq R_C \tag{4.56}$$

It should be noted, however, that in IC amplifiers one can arrange for R_C to be very large, in which case R_o approaches the output resistance of the BJT, r_o. This will be demonstrated in Chapter 6.

The model parameters of the CE amplifier can be used to determine its various gain parameters utilizing the equivalent circuit of Fig. 4.43(c). Thus the open-circuit voltage gain A_{vo} can be found as follows:

$$A_{vo} \equiv \left.\frac{v_o}{v_i}\right|_{R_L=\infty} = G_m R_o = -g_m(R_C // r_o) \tag{4.57}$$

For $r_o \gg R_C$ we observe that r_o reduces the gain slightly. We also note that in the case of IC amplifiers in which R_C is very large, the open-circuit voltage gain approaches

$$\left.A_{vo}\right|_{\max} = -g_m r_o = -\frac{I_C}{V_T}\frac{V_A}{I_C} = -\frac{V_A}{V_T} \tag{4.58}$$

This is the maximum possible voltage gain obtainable from a CE amplifier.

The short-circuit current gain of the CE amplifier is found from

$$A_{is} \equiv \left.\frac{i_o}{i_i}\right|_{R_L=0} = \frac{G_m v_i}{v_i/R_i} = G_m R_i$$

Thus

$$A_{is} = -g_m(R_B // r_\pi) = -\frac{g_m r_\pi R_B}{R_B + r_\pi}$$

$$= -\beta \frac{1}{1 + r_\pi/R_B} \tag{4.59}$$

Note that A_{is} approaches β when $R_B \gg r_\pi$—not a surprising result in view of the fact that β is the CE short-circuit current gain of the BJT proper.

Finally, we can use the model of Fig. 4.43(c) to determine the overall voltage gain, $A_v \equiv v_o/v_s$, as follows:

$$A_v = \frac{v_i}{v_s}\frac{v_o}{v_i} = \frac{R_i}{R_i + R_s} G_m(R_o // R_L)$$

$$= -\frac{(R_B // r_\pi)}{(R_B // r_\pi) + R_s} g_m(R_C // r_o // R_L) \tag{4.60}$$

For $R_B \gg r_\pi$ this expression simplifies to

$$A_v \simeq -\frac{r_\pi}{r_\pi + R_s} g_m(R_C//R_L//r_o)$$

Substituting $g_m r_\pi = \beta$ gives

$$A_v = -\frac{\beta(R_C//R_L//r_o)}{r_\pi + R_s} \tag{4.61}$$

This expression clearly indicates the degree of dependence of the voltage gain on β: For large R_s the gain is highly dependent on the value of β. The dependence on β decreases as R_s is reduced; and in the limit, for $R_s = 0$, the gain is independent of β since $r_\pi = \beta/g_m$.

In summary, the common-emitter amplifier provides an input resistance of moderate value ($\simeq r_\pi$), a high transconductance (equal to g_m of the BJT), a high output resistance ($R_C//r_o$), and high voltage and current gains. In designing an amplifier by cascading several stages, the common-emitter configuration is usually utilized to provide the bulk of the required overall voltage gain. The common-emitter amplifier, however, has a serious drawback: Its high-frequency performance is severely limited by the collector-to-base capacitance of the BJT, as will be shown in Chapter 7.

Exercise 4.32 Consider the amplifier whose bias design was analyzed in Exercise 4.31 when connected in the common-emitter configuration with $R_s = R_L = 10$ kΩ. Find the values of

R_i, G_m, R_o, A_{vo}, A_{is}, A_v, and A_i. (*Note*: $A_i \equiv \dfrac{i_o}{i_i} = \dfrac{R_s + R_i}{R_L} A_v$.)

Ans. 2.9 kΩ; −33.6 mA/V; 9.2 kΩ; −309 V/V; −97.1 A/A; −36.2 V/V, −46.7 A/A

The Common-Emitter Amplifier with a Resistance in the Emitter

Including a resistance in the signal path between emitter and ground, as shown in Fig. 4.44(a), can lead to significant changes in the amplifier characteristics. Thus such a resistor can be utilized by the designer as an effective design tool to tailor the amplifier characteristics so as to fit the design requirements. Observe that in Fig. 4.44(a) the resistance between emitter and ground is R_e, the unbypassed part of the bias resistance R_E.

To analyze the amplifier of Fig. 4.44(a) we replace the BJT with its hybrid-π model, as shown in Fig. 4.44(b). Note that the BJT output resistance r_o connects the output side of the amplifier to the input side. This destroys the unilateral nature of the amplifier and complicates the analysis considerably. Fortunately, however, because r_o is large its effect on the amplifier input resistance R_i and transconductance G_m can be neglected. However, it should be obvious that r_o has an important effect on the value of the output resistance at the collector (R_{oc} in Fig. 4.44b). This will be demonstrated shortly.

Neglecting r_o results in the equivalent circuit shown in Fig. 4.44(c). To determine the input resistance R_i we first note that

$$R_i \equiv \frac{v_i}{i_i} = R_B//R_{ib} \tag{4.62}$$

(a)

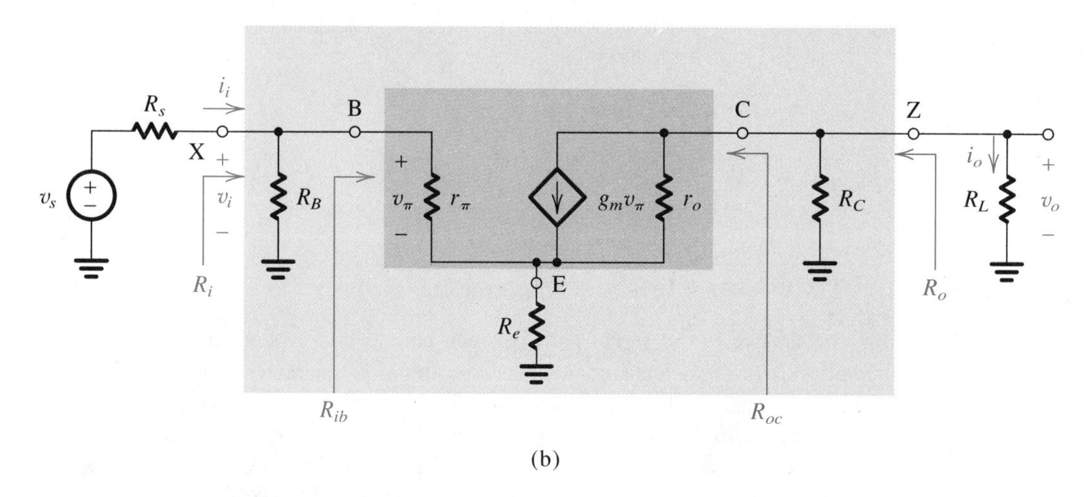
i_i
R_s
B
C
Z
X
v_s
v_i
R_B
v_π
r_π
$g_m v_\pi$
r_o
R_C
i_o
R_L
v_o
R_i
E
R_e
R_o
R_{ib}
R_{oc}

(b)

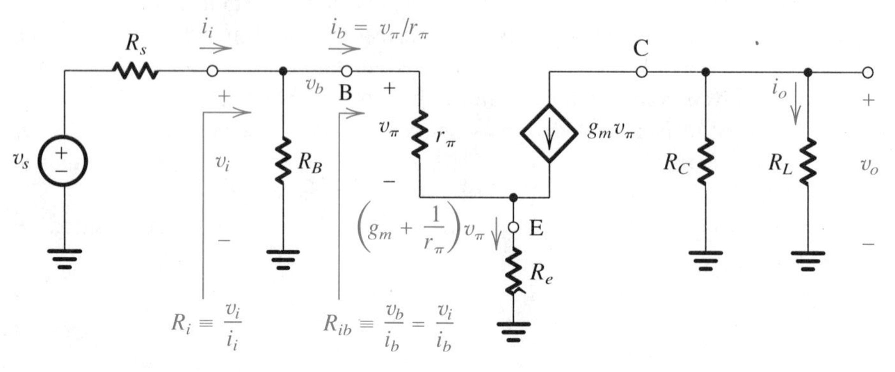
R_s
i_i
$i_b = v_\pi / r_\pi$
C
v_b
B
i_o
v_s
v_i
R_B
v_π
r_π
$g_m v_\pi$
R_C
R_L
v_o
$\left(g_m + \frac{1}{r_\pi}\right) v_\pi$
E
R_e
$R_i \equiv \frac{v_i}{i_i}$
$R_{ib} \equiv \frac{v_b}{i_b} = \frac{v_i}{i_b}$

(c)

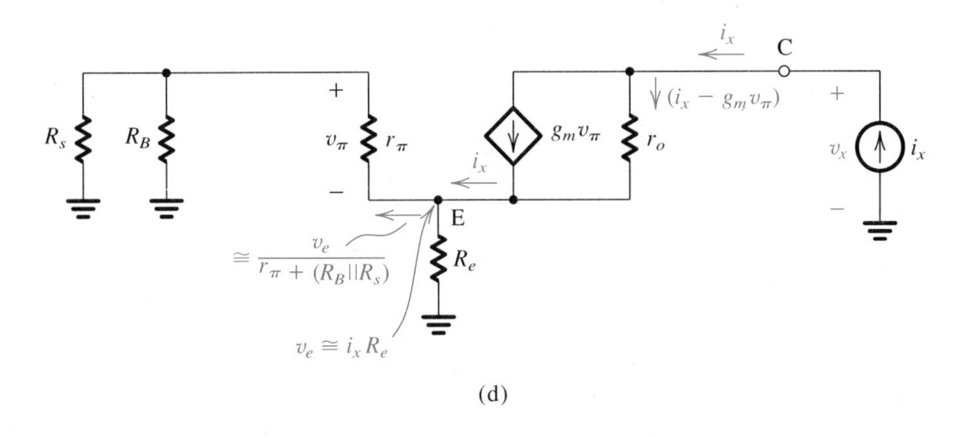

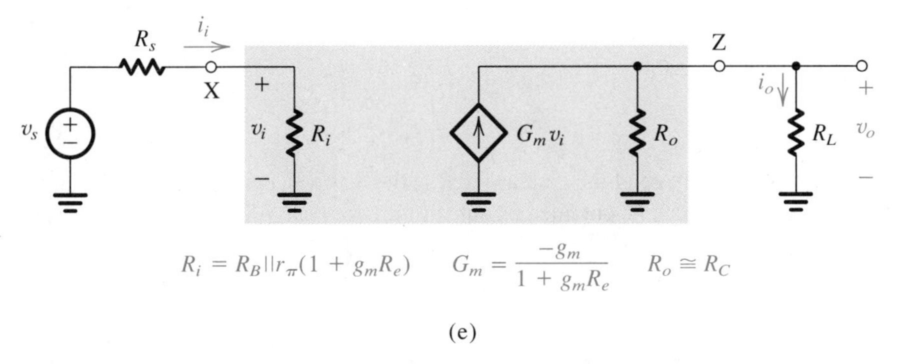

Fig. 4.44 The common-emitter amplifier with a resistance R_e between emitter and ground: **(a)** circuit; **(b)** equivalent circuit obtained by replacing the BJT with its hybrid-π model; **(c)** the equivalent circuit of (b) with r_o neglected; **(d)** circuit for determining the output resistance at the collector, R_{oc}; and **(e)** equivalent circuit using the transconductance amplifier model (approximate expressions are shown for the model parameters).

where R_{ib} is the input resistance looking into the base,

$$R_{ib} \equiv \frac{v_b}{i_b} \tag{4.63}$$

where $v_b = v_i$. Now from the analysis indicated on the circuit diagram in Fig. 4.44(c), we can write

$$i_b = v_\pi / r_\pi \tag{4.64}$$

$$v_b = v_\pi + \left(g_m + \frac{1}{r_\pi}\right) v_\pi R_e$$

Replacing $g_m + (1/r_\pi)$ by $1/r_e$ results in

$$v_b = v_\pi \left(1 + \frac{R_e}{r_e}\right) \tag{4.65}$$

Eqs. (4.64) and (4.65) can be combined to determine R_{ib} according to the definition in Eq. (4.63).

$$R_{ib} = \frac{v_\pi(1 + R_e/r_e)}{v_\pi/r_\pi}$$

Thus

$$R_{ib} = r_\pi \left(1 + \frac{R_e}{r_e}\right) \tag{4.66}$$

which indicates that including the resistance R_e results in increasing the input resistance at the base by the factor

$$[1 + (R_e/r_e)] \simeq 1 + g_m R_e,$$

$$R_{ib} \simeq r_\pi(1 + g_m R_e) \tag{4.67}$$

and the amplifier input resistance R_i becomes

$$R_i = R_B // r_\pi(1 + g_m R_e)$$

Obviously, to take advantage of the increase in R_{ib}, the bias resistance R_B should be large.

A very useful result can be obtained by substituting for r_π in Eq. (4.66) by $(\beta + 1)r_e$,

$$R_{ib} = (\beta + 1)r_e\left(1 + \frac{R_e}{r_e}\right)$$

$$= (\beta + 1)(r_e + R_e) \tag{4.68}$$

which says that the *total resistance looking into the base is* $(\beta + 1)$ *times the total resistance in the emitter circuit*. The latter is equal to the sum of the emitter resistance r_e and the external resistance R_e. The factor $(\beta + 1)$ obviously arises because the base current is $1/(\beta + 1)$ times the emitter current. This **resistance reflection rule** is frequently employed in BJT circuit design.

To determine the short-circuit transconductance G_m, we refer to Fig. 4.44(c),

$$G_m \equiv \left.\frac{i_o}{v_i}\right|_{R_L=0} = \frac{-g_m v_\pi}{v_i}$$

Substituting for v_i from Eq. (4.65) (recall that $v_i = v_b$) gives

$$G_m = -\frac{g_m}{1 + (R_e/r_e)} \tag{4.69}$$

which can be approximated as

$$G_m \simeq -\frac{g_m}{1 + g_m R_e} \tag{4.70}$$

Thus the transconductance is reduced by the factor $(1 + g_m R_e)$, the same factor by which R_{ib} is increased. This factor is the amount of feedback introduced by R_e (Chapter 8).

To determine the output resistance R_o we refer back to Fig. 4.44(b) and observe that

$$R_o = R_C // R_{oc} \tag{4.71}$$

where R_{oc} is the resistance looking into the collector of the BJT. Figure 4.44(d) shows a circuit for determining R_{oc}. Here v_s is set to zero, and a test current source i_x is connected to the collector. If the voltage that develops at the collector is denoted v_x, then

$$R_{oc} \equiv \frac{v_x}{i_x} \tag{4.72}$$

Some of the details of the analysis are shown in Fig. 4.44(d). Specifically, the node equation at the collector shows that the current through r_o is $i_x - g_m v_\pi$ and the current into the emitter node is i_x. Between the emitter node and ground there are two parallel paths; R_e and $r_\pi + (R_s // R_B)$. Since usually $R_e \ll r_\pi + (R_s // R_B)$, most of the current i_x will flow through R_e, with the result that

$$v_e \simeq i_x R_e \tag{4.73}$$

The current through the $r_\pi + (R_s // R_B)$ branch will be approximately $i_x R_e / [r_\pi + (R_s // R_B)]$. Thus v_π will be

$$v_\pi = -\frac{i_x R_e r_\pi}{r_\pi + (R_s // R_B)} \tag{4.74}$$

Finally, the voltage from collector to ground, v_x, can be obtained from

$$v_x = (i_x - g_m v_\pi) r_o + v_e \tag{4.75}$$

Substituting for v_π from Eq. (4.74) and for v_e from Eq. (4.73) and using the definition in Eq. (4.72) results in

$$R_{oc} = r_o \left[1 + \frac{g_m R_e r_\pi}{r_\pi + (R_s // R_B)} \right] + R_e$$

Usually $R_e \ll r_o$, enabling us to neglect the second term on the right-hand side and to express R_{oc} as

$$R_{oc} \simeq r_o \left[1 + \frac{g_m R_e}{1 + (R_s // R_B)/r_\pi} \right] \tag{4.76}$$

Observe that R_{oc} is increased by the factor between the square brackets. For $r_\pi \gg (R_s // R_B)$ (which is *not* always the case) this factor approaches $1 + g_m R_e$. Substituting for R_{oc} from Eq. (4.76) into Eq. (4.71) gives

$$\begin{aligned} R_o &= R_C // r_o \left[1 + \frac{g_m R_e}{1 + (R_s // R_B)/r_\pi} \right] \\ &\simeq R_C \end{aligned} \tag{4.77}$$

We have thus far determined the parameters of the transconductance model of the amplifier, shown in Fig. 4.44(e). This model can be used to determine other gain parameters of interest—for example, A_{vo} and A_{is}—in exactly the same way used for the CE circuit. The overall voltage gain is obtained as follows:

$$\begin{aligned} A_v \equiv \frac{v_o}{v_s} &= \frac{v_i}{v_s}\frac{v_o}{v_i} \\ &= \frac{R_i}{R_i + R_s} \times -G_m(R_o//R_L) \\ &= -\frac{[R_B//r_\pi(1 + g_m R_e)]}{[R_B//r_\pi(1 + g_m R_e)] + R_s}\frac{g_m}{1 + g_m R_e}(R_C//R_L) \end{aligned} \tag{4.78}$$

We observe that although the first factor in the gain expression increases (because of the increase in R_i), the second factor decreases, and the overall effect is a decrease in gain. This is the price paid for the increased input resistance and for the following three additional advantages that accrue when a resistance R_e is included in the emitter:

1. The gain A_v becomes less dependent on the value of β. To see this point, assume that R_B is large so that we may neglect its effect and write Eq. (4.78) as

$$A_v \simeq -\frac{\beta(R_C//R_L)}{r_\pi(1 + g_m R_e) + R_s} \tag{4.79}$$

   Now if $r_\pi(1 + g_m R_e) \gg R_s$, then A_v will be

$$A_v \simeq -\frac{g_m(R_C//R_L)}{1 + g_m R_e} \tag{4.80}$$

   which is totally independent of the value of β. This expression can be written in the alternative form

$$A_v \simeq -\frac{R_C//R_L}{r_e + R_e} \tag{4.81}$$

   which has an intuitive appeal: For small R_s, the voltage gain is approximately the ratio of the total resistance in the collector circuit to the total resistance in the emitter circuit. Even if the condition of small R_s is not satisfied, the value of A_v in Eq. (4.79) is less dependent on β than in the case of the CE amplifier in Eq. (4.61).

2. A larger signal can be applied at the input without risking nonlinear distortion. Specifically, since v_π is a fraction $r_e/(r_e + R_e)$ of v_i (see Eq. 4.65), the input signal can be increased by the factor $1 + R_e/r_e \simeq 1 + g_m R_e$ while keeping v_π unchanged.

3. The high-frequency response is significantly improved, as will be seen in Chapter 7.

Finally, we note that the resistance R_e introduces negative feedback in the amplifier circuit. Indeed, it is this negative feedback that gives rise to all of the characteristics observed above. We shall study feedback formally in Chapter 8.

Exercise 4.33 Consider the amplifier whose bias design was analyzed in Exercise 4.31 when connected in the common-emitter configuration with a resistance of 170 Ω in the signal path between emitter and ground and with $R_s = R_L = 10\ \text{k}\Omega$. Find the values of R_{ib}, R_i, G_m, R_{oc}, R_o, A_v, and A_i. Compare the results with those obtained in Exercise 4.32 for the same amplifier when connected in the common-emitter configuration.

Ans. 20 kΩ; 16.7 kΩ; −5 mA/V; 288 kΩ; 9.7 kΩ; −15.6 V/V; −41.7 A/A

The Common-Base Amplifier

The common-base (CB) amplifier can be derived from the universal circuit of Fig. 4.42 as indicated in Fig. 4.45(a). Here the base is grounded (from a signal standpoint), the input

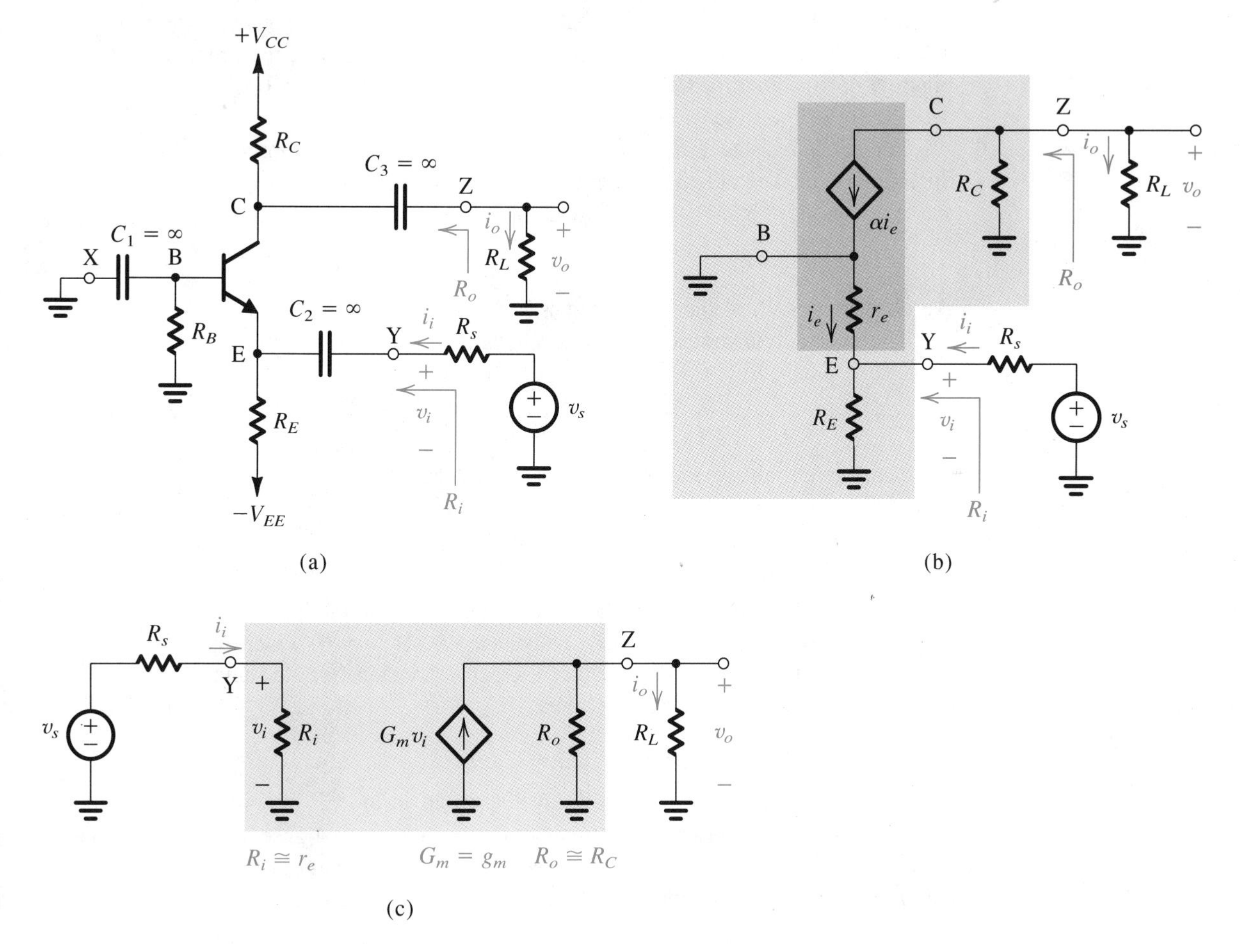

Fig. 4.45 The common-base amplifier: **(a)** circuit as derived from the universal circuit in Fig. 4.42 (note that R_B can be eliminated and the base directly connected to ground); **(b)** equivalent circuit obtained by replacing the BJT with its T model (the transistor output resistance r_o is not included as it has negligible effect on the operation of the CB amplifier); **(c)** equivalent circuit of the CB amplifier utilizing the transconductance amplifier model (expressions for the model parameters are shown).

signal source is coupled to the emitter, and the load is coupled to the collector. Thus, the CB amplifier can be viewed as a two-port network with the input port between the emitter and the base and the output port between the collector and the base. The fact that the base, which is at signal ground, is the *common* terminal between the input and output ports gives the configuration the name common-base or grounded-base amplifier.

It should be obvious from the circuit in Fig. 4.45(a) that R_B does not serve a useful purpose; it can be eliminated and the base connected directly to ground, thus obviating the need for C_1. Nevertheless, we have chosen to show the circuit in the form in Fig. 4.45(a) to show how it can be derived from the universal configuration and, more importantly, to allow a fair comparison of the results with those of the other amplifier configurations.

It can be shown that the BJT output resistance r_o has a negligible effect on the operation of the CB amplifier. We shall therefore neglect r_o in the analysis to follow. This analysis is most conveniently performed using the T equivalent circuit model of the BJT. Figure 4.45(b) shows the CB circuit with the BJT replaced with its T model. We shall now derive expressions for R_i, G_m, and R_o of the amplifier.

The input resistance R_i can be found by inspection of the circuit in Fig. 4.45(b). Specifically, between the input terminal and ground there are two parallel resistances: R_E and r_e. Thus $R_i = R_E // r_e$. Since almost always $R_E \gg r_e$,

$$R_i \simeq r_e \tag{4.82}$$

We observe that R_i of the CB circuit is very low.

The short-circuit transconductance G_m is determined as follows:

$$G_m \equiv \left.\frac{i_o}{v_i}\right|_{R_L=0} = \frac{-\alpha i_e}{v_i}$$

but from the circuit we see that $i_e = -(v_i/r_e)$. Thus

$$G_m = \frac{\alpha}{r_e} = g_m \tag{4.83}$$

which is equal in magnitude to the value obtained in the CE circuit.

Finally, to determine the output resistance we set $v_s = 0$, which makes $i_e = 0$; and thus the controlled source αi_e is reduced to zero. Looking from terminal Z back into the collector we see the resistance R_C; thus

$$R_o = R_C \tag{4.84}$$

which is only slightly greater than the value found in the CE circuit.

Figure 4.45(c) shows the CB circuit modeled utilizing the transconductance-amplifier model whose parameters we have just determined. This circuit can be used as follows to determine other gain parameters of interest. The open-circuit voltage gain A_{vo} is

$$A_{vo} \equiv \left.\frac{v_o}{v_i}\right|_{R_L=\infty} = G_m R_o$$

Thus,

$$A_{vo} = g_m R_C \tag{4.85}$$

which is almost equal in magnitude (but opposite in sign) to the corresponding value of the CE amplifier.

The short-circuit current gain is found as follows:

$$A_{is} \equiv \left.\frac{i_o}{i_i}\right|_{R_L=0} = \frac{G_m v_i}{v_i/R_i} = G_m R_i$$

Thus,

$$A_{is} = g_m r_e = \alpha$$

This is not a surprising result as α is by definition the short-circuit current gain of the BJT in the CB configuration. However, since $\alpha < 1$, we observe that the CB amplifier does not in fact provide current gain. This, together with its very low ($\simeq r_e$) input resistance, severely limits the range of application of the CB amplifier.

The overall voltage gain is

$$A_v = \frac{R_i}{R_i + R_s} G_m(R_C//R_L)$$

$$= \frac{r_e}{r_e + R_s} g_m(R_C//R_L)$$

which will be low if R_s is large.

The common-base circuit finds application as a current buffer, accepting an input current signal at a low input impedance level ($\simeq r_e$) and delivering an almost equal (but slightly smaller) current at the collector at a very high impedance level (infinite if we exclude R_C). It will be shown in Chapter 7 that the CB amplifier exhibits a much wider bandwidth than is obtained in the CE amplifier.

Exercise 4.34 Consider the amplifier whose bias design was analyzed in Exercise 4.31, when connected in the CB configuration. Find the values of R_i, G_m, R_o, A_{vo}, and A_{is}. Also find the overall voltage gain A_v and the overall current gain A_i for the two cases: (a) $R_s = R_L = 10$ kΩ (same case used for the CE amplifier in Exercise 4.32, hence you can compare results), and (b) $R_s = 50$ Ω, $R_L = 10$ kΩ.

Ans. 30 Ω; +33.6 mA/V; 10 kΩ; +336 V/V; 0.99 A/A; (a) $A_v = 0.5$ V/V, $A_i = 0.5$ A/A (b) $A_v = 63$ V/V, $A_i = 0.5$ A/A

The Common-Collector Amplifier or Emitter Follower

The common-collector (CC) configuration can be derived from the universal amplifier of Fig. 4.42 as shown in Fig. 4.46(a). Here capacitor C_3 establishes a signal ground at the collector. The input signal source v_s is coupled to the base via capacitor C_1, and the load resistor R_L is coupled to the emitter via capacitor C_2. The amplifier can be viewed as a two-port network with the input port between base and collector (ground) and the output port between emitter and collector (ground). Thus the collector, at signal ground, is the common terminal between the input and output ports, hence the name common-collector or grounded-collector amplifier.

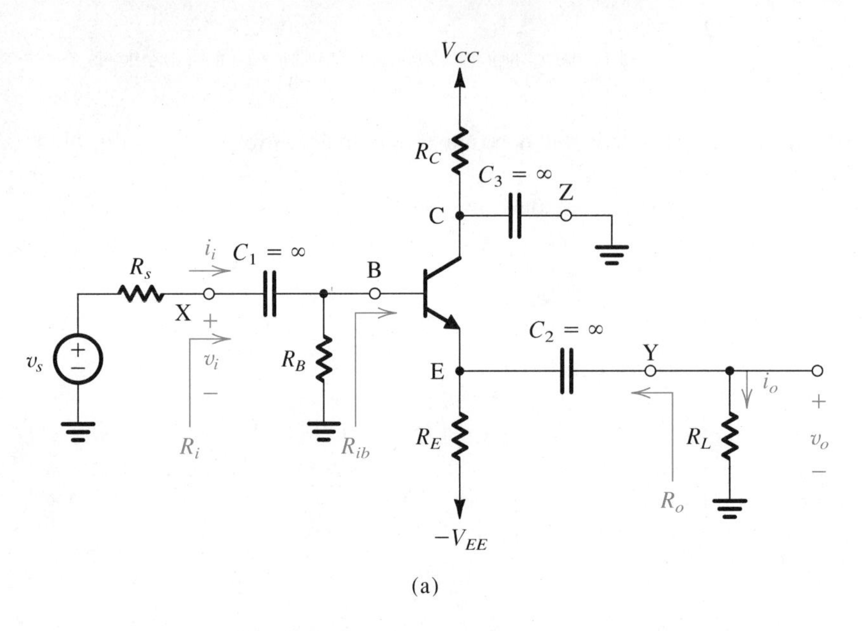

(a)

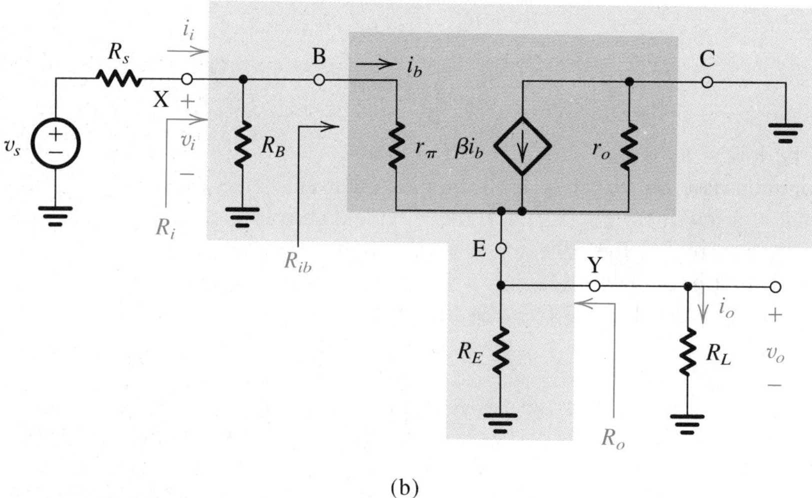

(b)

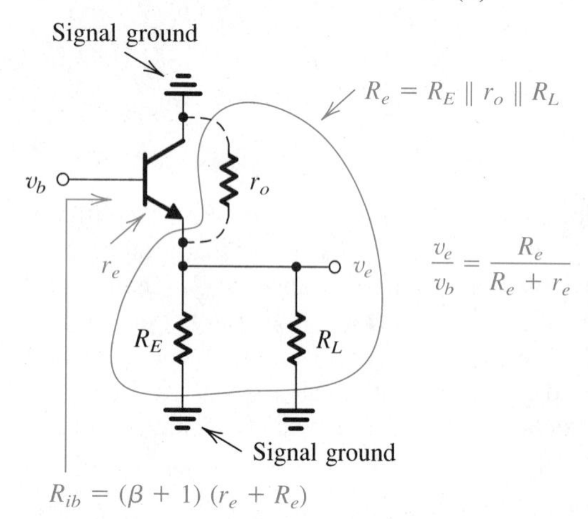

(c)

$$\frac{v_e}{v_b} = \frac{R_e}{R_e + r_e}$$

Fig. 4.46 The common-collector or emitter-follower amplifier configuration: **(a)** circuit obtained from the general configuration of Fig. 4.42; **(b)** equivalent circuit obtained by replacing the BJT with its hybrid-π model; **(c)** analysis can be performed in this manner directly on the circuit with the equivalent circuit model implicitly utilized (rather than explicitly drawn).

Although we have chosen, for unity of presentation, to derive the CC circuit from the universal amplifier configuration, the collector resistor R_C does not serve a useful purpose and can be dispensed with and the collector connected directly to V_{CC}. This will furthermore obviate the need for capacitor C_3. Whether R_C is used or not has no effect on the analysis to follow.

Replacing the BJT with its hybrid-π model results in the equivalent circuit of Fig. 4.46(b). Observe that we have used the current-controlled form of the hybrid-π model for variety and because it is somewhat more convenient in this case. We shall now analyze the circuit to determine the input resistance R_i, voltage gain (v_o/v_s), and output resistance R_o. As will be seen, the input resistance is strongly dependent on the value of R_L, and the output resistance is strongly dependent on the value of R_s. Thus the CC amplifier is *not* unilateral, and the amplifier models studied in Section 1.5 do *not* apply. Therefore, rather than deriving a model for the CC amplifier, we shall determine its characteristics by direct analysis of the circuit in Fig. 4.46(b).

The input resistance is seen to be the parallel equivalent of the bias resistance, R_B, and the input resistance looking into the base, R_{ib},

$$R_i = R_B // R_{ib} \tag{4.86}$$

To determine R_{ib} we use the resistance reflection rule and note that r_o appears in effect in parallel with R_E and R_o. Thus the effective resistance in the emitter lead is

$$R_e = R_E // r_o // R_L$$

resulting in a total emitter resistance of $r_e + R_e$ and an input resistance at the base of

$$\begin{aligned} R_{ib} &= (\beta + 1)(r_e + R_e) \\ &= (\beta + 1)[r_e + (R_E // r_o // R_L)] \end{aligned} \tag{4.87}$$

Substituting in Eq. (4.86) gives

$$R_i = R_B // (\beta + 1)[r_e + (R_E // r_o // R_L)] \tag{4.88}$$

We observe that R_i will normally be very high because of the $(\beta + 1)$ multiplication factor. This is a very important characteristic of the CC amplifier. Indeed, the most important application of this amplifier configuration is to connect a source with a high source resistance to a low-resistance load without significant loss of signal strength. Observe, however, that to realize a high input resistance one must select a high value for R_B. Although a high R_B increases the dependence of the bias current on the value of β, this is not a major concern in the design of the CC circuit as its voltage gain is not strongly affected by the value of I_E, as will be seen shortly. If R_B is sufficiently high to be neglected (or if R_B is not present in the circuit, as is the case for the directly coupled circuit) the input resistance expression simplifies to

$$R_i \simeq R_{ib} = (\beta + 1)[r_e + (R_E // r_o // R_L)] \tag{4.89}$$

A practically useful case occurs when $R_L \ll (R_E // r_o)$, for which

$$\begin{aligned} R_i &\simeq (\beta + 1)(r_e + R_L) \\ &= r_\pi + (\beta + 1)R_L \end{aligned} \tag{4.90}$$

Having determined the input resistance R_i, we now can find the voltage transmission from the signal source to the transistor base by employing the voltage divider rule,

$$\frac{v_i}{v_s} = \frac{R_i}{R_i + R_s} \tag{4.91}$$

Again we observe that for large R_i, the value of v_i/v_s will be close to unity.

From the analysis of the CE amplifier with resistance R_e in the emitter, we found that

$$\frac{v_e}{v_b} = \frac{R_e}{R_e + r_e}$$

We can apply this result to the CC circuit by noting that $v_b = v_i$, $v_e = v_o$, and $R_e = R_E//r_o//R_L$; thus

$$\frac{v_o}{v_i} = \frac{(R_E//r_o//R_L)}{(R_E//r_o//R_L) + r_e} \tag{4.92}$$

Since r_e is usually small, $v_o/v_i \simeq 1$. Thus the output voltage at the emitter *follows* the input voltage at the base, a fact that gives the circuit the name **emitter follower.**

This is a suitable juncture to note that with a bit of experience one can perform the analysis directly on the circuit. For example, one can write the gain expression in Eq. (4.92) directly from inspection of the circuit in Fig. 4.46(a). This view is portrayed in Fig. 4.46(c).

The overall voltage gain of the emitter follower can be obtained by combining Eqs. (4.91) and (4.92):

$$A_v \equiv \frac{v_o}{v_s} = \frac{R_i}{R_i + R_s} \frac{(R_E//r_o//R_L)}{(R_E//r_o//R_L) + r_e} \tag{4.93}$$

We observe that the voltage gain is lower than unity. The gain value, however, is normally close to unity due to the large R_i and the normally small value of r_e. An approximate but insightful expression for the voltage gain can be obtained for the case $R_L \ll (R_E//r_o)$ and R_B large,

$$\frac{v_o}{v_s} \simeq \frac{(\beta + 1)R_L}{(\beta + 1)R_L + r_\pi + R_s} \tag{4.94}$$

The fact that the voltage gain of the emitter follower is less than unity does not detract from the usefulness of the circuit. Its prime function is to connect a high-resistance source to a low-resistance load without significant signal attenuation; that is, it is a **buffer amplifier.** The emitter follower provides a high current gain whose value can be found as follows:

$$A_i \equiv \frac{i_o}{i_i} = \frac{v_o/R_L}{v_s/(R_s + R_i)}$$

$$= \frac{R_i}{R_L} \frac{R_E//r_o//R_L}{(R_E//r_o//R_L) + r_e} \simeq \frac{R_i}{R_L} \tag{4.95}$$

For the case $R_L \ll (R_E//r_o)$, $R_B \gg R_{ib}$, R_i is given approximately by Eq. (4.90) and the current gain becomes approximately equal to $(\beta + 1)$.

To determine the output resistance of the emitter follower we set $v_s = 0$ and apply a test voltage source v_x to the output (emitter) terminal, as shown in Fig. 4.47. We see that

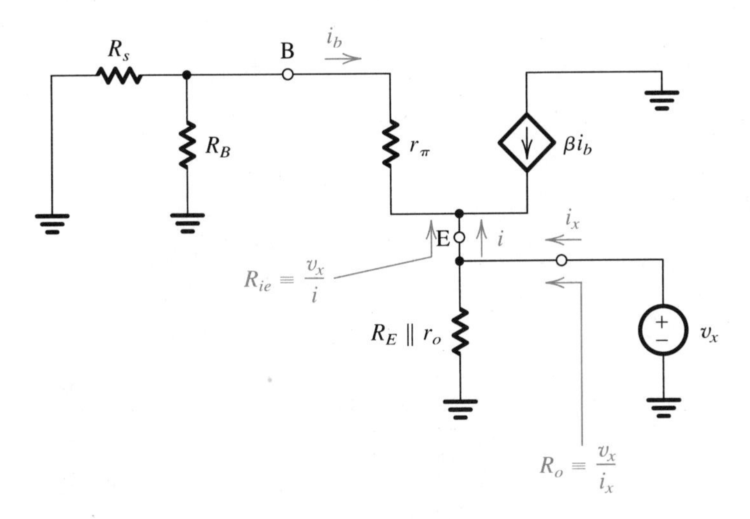

Fig. 4.47 Determining the output resistance R_o of the emitter follower. Note that R_o is the parallel equivalent of $(R_E//r_o)$ and the resistance R_{ie}, looking into the emitter of the BJT. The latter is equal to the total resistance in the base $[r_\pi + (R_s//R_B)]$ divided by $(\beta + 1)$.

$R_o \equiv v_x/i_x$ is the parallel equivalent of $(R_E//r_o)$ and the resistance looking into the emitter, R_{ie},

$$R_{ie} \equiv \frac{v_x}{i}$$

Since

$$i = -i_b - \beta i_b = -(\beta + 1)i_b$$

and

$$i_b = -\frac{v_x}{r_\pi + (R_s//R_B)}$$

we obtain

$$R_{ie} = \frac{r_\pi + (R_s//R_B)}{\beta + 1} \tag{4.96}$$

which says that *looking into the emitter we see the total resistance in the base circuit divided by (β + 1)*. This is an important result and is the inverse of the resistance reflection rule; that is, resistances are reflected from the emitter to the base circuit by multiplying them by $(\beta + 1)$, and from the base to the emitter circuit by dividing them by $(\beta + 1)$.

We conclude that the output resistance is given by

$$R_o = R_E//r_o//\frac{r_\pi + (R_s//R_B)}{\beta + 1} \tag{4.97}$$

Normally, the reflected component of R_o is much smaller than R_E and r_o, thus permitting the approximation

$$R_o \simeq \frac{r_\pi + (R_s//R_B)}{\beta + 1} \tag{4.98}$$

which can also be written in the form

$$R_o = r_e + \frac{(R_s//R_B)}{\beta + 1} \tag{4.99}$$

If R_B is large (or absent) we have

$$R_o = r_e + \frac{R_s}{\beta + 1} \tag{4.100}$$

Thus the emitter follower exhibits a low output resistance, a desirable characteristic in a buffer amplifier.

From Thévenin's theorem we know that the output resistance R_o can be used together with the open-circuit, overall voltage gain,

$$A_v\Big|_{R_L=\infty} \equiv \frac{v_o}{v_s}\Big|_{R_L=\infty} \tag{4.101}$$

to determine the overall voltage gain for any load resistance R_L,

$$A_v = A_v\Big|_{R_L=\infty} \frac{R_L}{R_L + R_o} \tag{4.102}$$

This should yield results identical to those found by direct analysis.

Finally, we consider the question of maximum allowed input signal swing in the emitter-follower circuit. Since only a small fraction of the input signal appears across the base–emitter junction, the emitter follower exhibits linear performance for a large range of input-signal amplitude. In fact the upper limit on the value of the input-signal amplitude is usually imposed by transistor cutoff. To see how this comes about, consider the circuit of Fig. 4.46(a) when the input signal is a sine wave. As the input goes negative, the output v_o will also go negative, and the current in R_L will be flowing from ground into the emitter terminal. The transistor will cut off when this current becomes equal to the current flowing through R_E. Denoting by $\hat{V}_e$ the peak of emitter signal voltage at which cutoff happens, we can write

$$\frac{V_E - \hat{V}_e + V_{EE}}{R_E} = \frac{\hat{V}_e}{R_L} \tag{4.103}$$

where V_E is the dc voltage at the emitter. The value of $\hat{V}_e$ can be obtained from Eq. (4.103) as

$$\hat{V}_e = \frac{V_E + V_{EE}}{1 + R_E/R_L} \tag{4.104}$$

The corresponding value of the amplitude of v_s can be obtained from

$$\hat{V}_s = \frac{\hat{V}_e}{A_v} \tag{4.105}$$

Increasing the amplitude of v_s above this value results in the transistor becoming cut off, and the negative peaks of the output-signal waveform will be clipped off.

In conclusion, we note that the emitter follower is usually employed as the last stage in a multistage amplifier. Its purpose in this application is to provide the amplifier with a low output resistance, which in turn permits the amplifier to be connected to low-valued load resistances without a severe reduction in gain. We shall study the design of amplifier output stages in detail in Chapter 9.

Exercise 4.35 Consider the amplifier whose bias design was analyzed in Exercise 4.31, when connected in the CC configuration. Find the values of R_{ib}, R_i, A_v, A_i, and R_o for the following two cases. (a) $R_s = R_L = 10$ kΩ (the same case used for the CE configuration in Exercise 4.32 and for the CB configuration in Exercise 4.34; hence you can compare results) (b) $R_s = 100$ kΩ and $R_L = 1$ kΩ (a case that illustrates the need for a voltage buffer)

Ans. (a) 488 kΩ, 83 kΩ, 0.89 V/V, 8.3 A/A, 118 Ω; (b) 94 kΩ, 48.5 kΩ, 0.32 V/V, 47 A/A, 497 Ω

Comparing the Various Configurations

Throughout this section, comments have been made on the distinctive features of each of the amplifier configurations and on its area of applicability. To highlight the differences among the various BJT basic amplifier configurations, we present a summary of the results in Table 4.2. Also, numerical values for a typical design are given in Table 4.3. Note that these data are for the design investigated in Exercises 4.31 to 4.35.

4.12 THE TRANSISTOR AS A SWITCH—CUTOFF AND SATURATION

Having studied the active mode of operation in detail, we are now ready to complete the picture by considering what happens when the transistor leaves the active region. At one extreme the transistor will enter the cutoff region, while at the other extreme the transistor enters the saturation region. These two extreme modes of operation are very useful if the transistor is to be used as a switch, such as in digital logic circuits. In the present section we shall study the cutoff and saturation modes of operation. The application of the BJT in switching circuits is presented in Chapter 14, which also includes a detailed study of the saturation mode of operation.

Cutoff Region

To help introduce cutoff and saturation, we consider the simple circuit shown in Fig. 4.48, which is fed with a voltage source v_I. We wish to analyze this circuit for different values of v_I.

Table 4.2 SUMMARY OF THE CHARACTERISTICS OF THE BASIC BJT AMPLIFIER CONFIGURATIONS

	Common Emitter	Common Emitter with a Resistance R_e	Common Base	Common Collector (Emitter Follower)
R_i	$R_B//r_\pi \simeq r_\pi$	$R_B//(\beta+1)(r_e+R_e)$ $\simeq R_B//r_\pi(1+g_mR_e)$	$\simeq r_e$	$R_B//(\beta+1)[r_e+(R_E//r_o//R_L)]$
R_o	$R_C//r_o \simeq R_C$	$R_C//r_o\left[1+\dfrac{g_mR_e}{1+(R_s//R_B)/r_\pi}\right] \simeq R_C$	R_C	$R_E//\left[r_e+\dfrac{R_s//R_B}{\beta+1}\right]$
$G_m = \left.\dfrac{i_o}{v_i}\right\|_{R_L=0}$	$-g_m$	$-g_m/(1+g_mR_e)$	g_m	Not useful
$A_v \equiv \dfrac{v_o}{v_s}$	$-\dfrac{R_B//r_\pi}{(R_B//r_\pi)+R_s}\, g_m(R_C//r_o//R_L)$ $\simeq -\dfrac{\beta(R_C//R_L//r_o)}{r_\pi+R_s}$	$-\dfrac{R_B//r_\pi(1+g_mR_e)}{[R_B//r_\pi(1+g_mR_e)]+R_s}\,\dfrac{g_m}{1+g_mR_e}\,(R_C//R_L)$ $\simeq -\dfrac{\beta(R_C//R_L)}{r_\pi(1+g_mR_e)+R_s}$ $\simeq -(R_C//R_L)/(r_e+R_e)$	$\simeq \dfrac{R_C//R_L}{r_e+R_s}$	$\dfrac{R_i}{R_i+R_s}\,\dfrac{R_E//r_o//R_L}{(R_E//r_o//R_L)+r_e}$ $\simeq \dfrac{(\beta+1)R_L}{(\beta+1)R_L+r_\pi+R_s}$
$A_i \equiv \dfrac{i_o}{i_i}$	$-\dfrac{R_B//r_\pi}{R_L}\, g_m(R_C//r_o//R_L)$ $\simeq -\beta\,\dfrac{R_C//r_o//R_L}{R_L}$	$-\dfrac{R_B//r_\pi(1+g_mR_e)}{R_L}\,\dfrac{g_m}{1+g_mR_e}\,(R_C//R_L)$ $\simeq -\beta\,\dfrac{R_C//R_L}{R_L}$	$\alpha\,\dfrac{R_C//R_L}{R_L}$	$\simeq R_i/R_L$ $\simeq (\beta+1)\,\dfrac{R_E//R_L}{R_L}$

Table 4.3 TYPICAL NUMERICAL VALUES[1]

	Common Emitter	Common Emitter with $R_e = 170\ \Omega$	Common Base[2]	Common Collector[3] (Emitter Follower)
R_i (kΩ)	2.9	16.7	0.03	83
R_o (kΩ)	9.2	9.7	10	0.118
G_m (mA/V)	−33.6	−5	33.6	Not useful
A_v (V/V)	−36.2	−15.6	0.5	0.89
A_i (A/A)	−46.7	−41.7	0.5	8.3

[1] For all cases the BJT is biased as in Exercise 4.31 ($I_C = 0.84$ mA) and $R_s = R_L = 10$ kΩ.
[2] Both the voltage and current gain obtained are less than unity. Normally, the CB configuration is used either as a current buffer, or as a voltage amplifier in case R_s is small.
[3] The characteristics of this emitter follower are impaired by the existence of the bias resistor R_B.

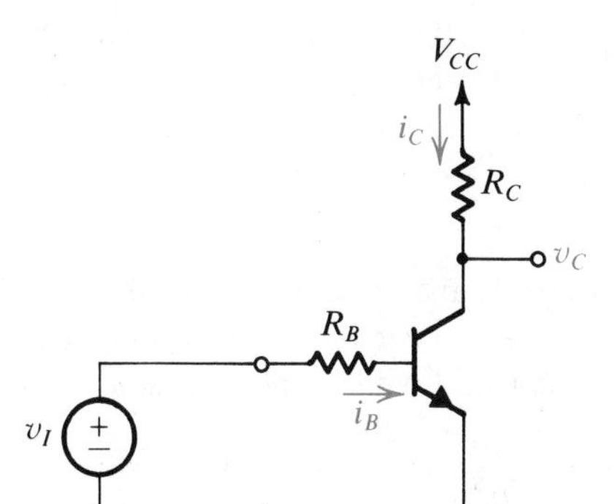

Fig. 4.48 A simple circuit used to illustrate the different modes of operation of the BJT.

If v_I is smaller than about 0.5 V, the EBJ will conduct negligible current. In fact, the EBJ could be considered "reverse"-biased, and the device will be in the cutoff mode. It follows that

$$i_B = 0 \qquad i_E = 0 \qquad i_C = 0 \qquad v_C = V_{CC}$$

Note that the CBJ also is reverse-biased.

Active Region

To turn the transistor on, we have to increase v_I above 0.5 V. In fact, for appreciable currents to flow v_{BE} should be about 0.7 V and v_I should be higher. For $v_I > 0.7$ V we have

$$i_B = \frac{v_I - V_{BE}}{R_B} \tag{4.106}$$

which may be approximated by

$$i_B \simeq \frac{v_I - 0.7}{R_B} \tag{4.107}$$

provided that $v_I \gg 0.7$ V (for example, ≥ 2 V) and that the resulting collector current is in the normal range for this particular transistor. The collector current is given by

$$i_C = \beta i_B \tag{4.108}$$

which applies only if the device is in the active mode. How do we know that the device is in the active mode? We do not know; therefore we assume that it is in the active mode, calculate i_C using Eq. (4.108) and v_C from

$$v_C = V_{CC} - R_C i_C \tag{4.109}$$

and then check whether $v_{CB} \geq 0$ or not. In our case we merely check whether $v_C \geq 0.7$ V or not. If $v_C \geq 0.7$ V, then our original assumption is correct, and we have completed the analysis for the particular value of v_I. On the other hand, if v_C is found to be less than 0.7 V, then the device has left the active region and entered the saturation region.

Obviously as v_I is increased, i_B will increase (Eq. 4.107), i_C will correspondingly increase (Eq. 4.108), and v_C will decrease (Eq. 4.109). Eventually, v_C will become less than v_B (0.7 V), and the device will enter the saturation region.

Saturation Region

Saturation occurs when we attempt to force a current in the collector higher than the collector circuit can support while maintaining active-mode operation. For the circuit in Fig. 4.48 the maximum current that the collector "can take" without the transistor leaving the active mode can be evaluated by setting $v_{CB} = 0$, which results in

$$\hat{I}_C = \frac{V_{CC} - V_B}{R_C} \simeq \frac{V_{CC} - 0.7}{R_C} \tag{4.110}$$

This collector current is obtained by forcing a base current $\hat{I}_B$, given by

$$\hat{I}_B = \frac{\hat{I}_C}{\beta} \tag{4.111}$$

and the corresponding required value of v_I can be obtained from Eq. (4.107). Now, if we increase i_B above $\hat{I}_B$, the collector current will increase and the collector voltage will fall below that of the base. This will continue until the CBJ becomes forward-biased with a forward-bias voltage of about 0.4 to 0.5 V. Note that the forward voltage drop of the collector–base junction is small because the CBJ has a relatively large area (see Fig. 4.6). This situation is referred to as saturation, since any further increase in the base current will result in a very small increase in the collector current and a corresponding small decrease in the collector voltage. This means that in saturation the *incremental* β ($\Delta i_C/\Delta i_B$) is negligibly small. Any "extra" current that we force into the base terminal mostly will flow through the emitter terminal. Thus the ratio of the collector current to the base current of a saturated transistor is *not* equal to β and can be set to any desired value—smaller than β—simply by pushing more current into the base.

Let us now return to the circuit of Fig. 4.48, which we have redrawn in Fig. 4.49 with the assumption that the transistor is in saturation. The value of V_{BE} of a saturated transistor is usually slightly higher than that of the device operating in the active mode.[5] Nevertheless, for simplicity, we shall assume V_{BE} to remain around 0.7 V even if the device is in saturation.

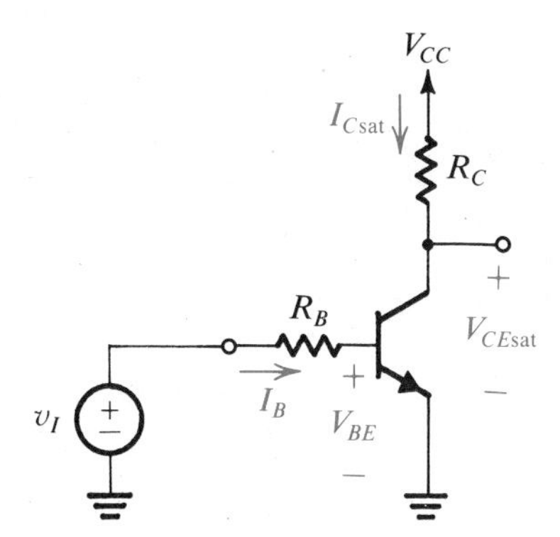

Fig. 4.49 A saturated transistor.

Since in saturation the base voltage is higher than the collector voltage by about 0.4 or 0.5 V, it follows that the collector voltage will be higher than the emitter voltage by 0.3 or 0.2 V. This quantity is referred to as V_{CEsat}, and we will normally assume that $V_{CEsat} \simeq 0.3$ V. Note, however, that if we push more current into the base, we drive the transistor ''deeper'' into saturation and the CBJ forward-bias increases, which means that V_{CEsat} will decrease.

The value of the collector current in saturation will be almost constant. We denote this value by I_{Csat}. It follows that for the circuit in Fig. 4.49 we have

$$I_{Csat} = \frac{V_{CC} - V_{CEsat}}{R_C} \tag{4.112}$$

In order to ensure that the transistor is driven into saturation, we have to force a base current of at least

$$I_{Bsat} = \frac{I_{Csat}}{\beta} \tag{4.113}$$

Normally one designs the circuit such that I_B is higher than I_{Bsat} by a factor of 2 to 10 (called the **overdrive factor**). The ratio of I_{Csat} to I_B is called the **forced β** (β_{forced}), since its value can be set at will,

$$\beta_{forced} = \frac{I_{Csat}}{I_B} \tag{4.114}$$

[5] The increase in V_{BE} is due to the increased base current producing a sizable ohmic (IR) voltage drop across the bulk resistance of the base region. In other words, part of V_{BE} will appear across the base semiconductor material as an IR drop, and the remainder will appear across the EBJ.

Transistor Inverter

Figure 4.50 shows the transfer characteristic of the circuit of Fig. 4.48. The shape of this curve should be obvious to the reader who has followed the above discussion. The three regions of operation—cutoff, active, and saturation—are indicated in Fig. 4.50. For the transistor to be operated as an amplifier, it should be biased somewhere in the active region, such as at the point marked X. The voltage gain of the amplifier is equal to the slope of the transfer characteristic at this point.

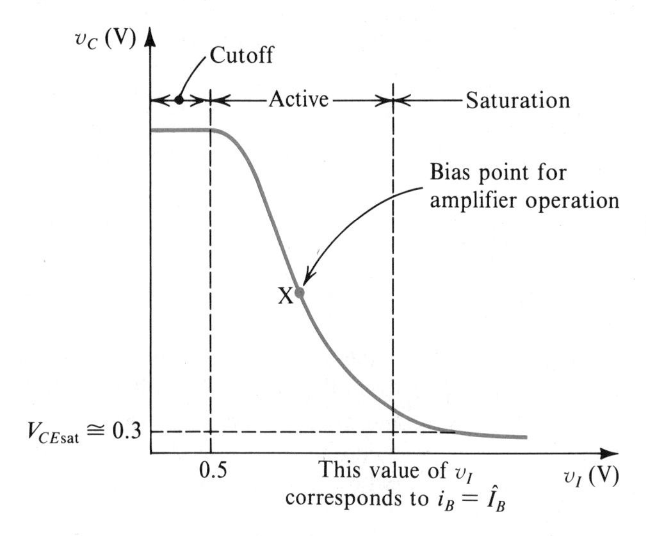

Fig. 4.50 Transfer characteristic for the circuit in Fig. 4.48.

For switching applications the transistor is usually operated in cutoff and saturation.[6] That is, one state of the switch corresponds to the transistor being cut off, and the other state corresponds to the transistor in saturation. There are a number of reasons for choosing these two extreme modes of operation. One reason is that in both cutoff and saturation the currents and voltages in the transistor are well defined and do not depend on such ill-specified parameters as β. Another reason is that in both cutoff and saturation the power dissipated in the transistor is minimal. Although this is obvious for the cutoff mode, it is also the case in the saturation mode, since the voltage V_{CEsat} is small. Finally we note that the circuit we have been using as an example is in fact the basic transistor logic inverter. More will be said about BJT logic circuits in Chapter 14.

Model for the Saturated BJT

From the discussion above we obtain a simple model for transistor operation in the saturation mode, as shown in Fig. 4.51. Normally, we use such a model implicitly in the analysis of a given circuit.

For quick approximate calculations one may consider V_{BE} and V_{CEsat} to be zero and use the three-terminal short circuit shown in Fig. 4.52 to model a saturated transistor.

[6] An exception to this is found in emitter-coupled logic (ECL), which is studied in detail in Chapter 14.

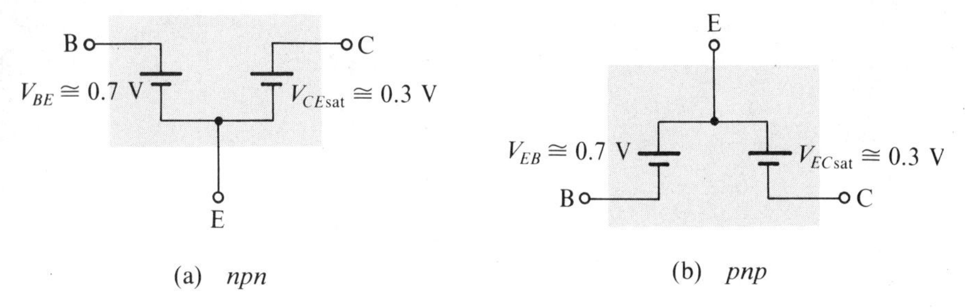

Fig. 4.51 Model for the saturated BJT.

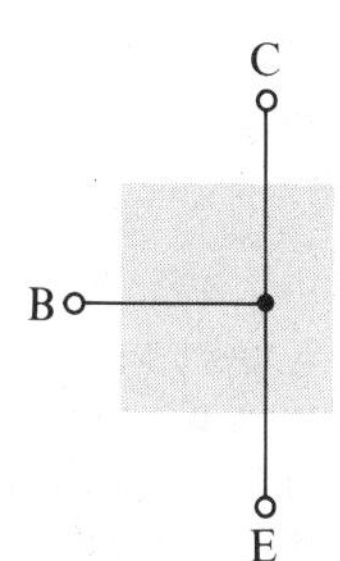

Fig. 4.52 An approximate model for the saturated BJT.

EXAMPLE 4.13

We wish to analyze the circuit in Fig. 4.53(a) to determine the voltages at all nodes and the currents in all branches. Assume the transistor β is specified to be *at least* 50.

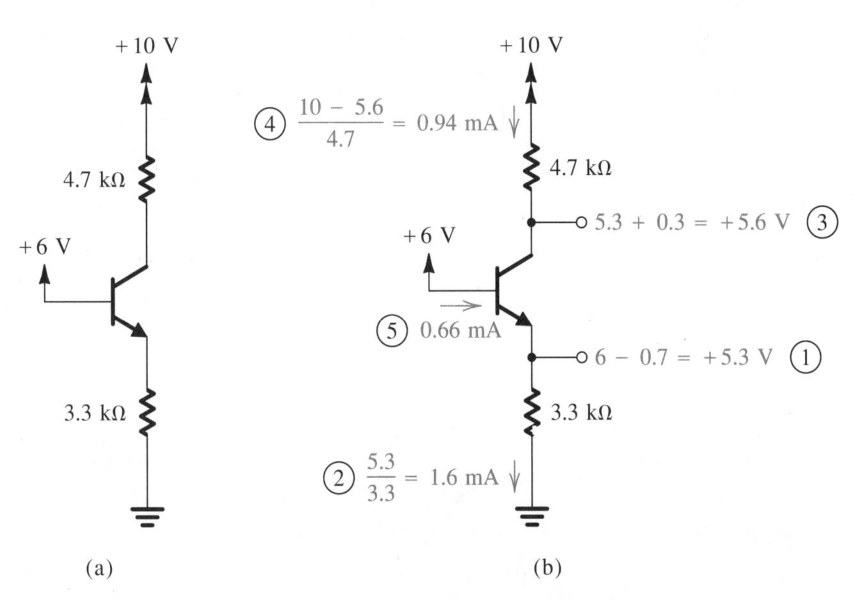

Fig. 4.53 Example 4.13: **(a)** circuit; **(b)** analysis with the order of the steps numbered.

Solution We have already considered this circuit in Example 4.3 and discovered that the transistor has to be in saturation. Assuming this to be the case, we have

$$V_E = +6 - 0.7 = +5.3 \text{ V}$$

$$I_E = \frac{V_E}{3.3} = \frac{5.3}{3.3} = 1.6 \text{ mA}$$

$$V_C = V_E + V_{CE\text{sat}} \simeq +5.3 + 0.3 = +5.6 \text{ V}$$

$$I_C = \frac{+10 - 5.6}{4.7} = 0.94 \text{ mA}$$

$$I_B = I_E - I_C = 1.6 - 0.94 = 0.66 \text{ mA}$$

Thus the transistor is operating at a forced β of

$$\beta_{\text{forced}} = \frac{I_C}{I_B} = \frac{0.94}{0.66} = 1.4$$

Since β_{forced} is less than the *minimum* specified value of β, the transistor is indeed saturated. We should emphasize here that in testing for saturation the minimum value of β should be used. By the same token, if we are designing a circuit in which a transistor is to be saturated, the design should be based on the minimum specified β. Obviously, if a transistor with this minimum β is saturated, then transistors with higher values of β will also be saturated. The details of the analysis are shown in Fig. 4.53(b), where the order of the steps used is indicated by the circled numbers.

EXAMPLE 4.14

The transistor in Fig. 4.54 is specified to have β in the range 50 to 150. Find the value of R_B that results in saturation with an overdrive factor of at least 10.

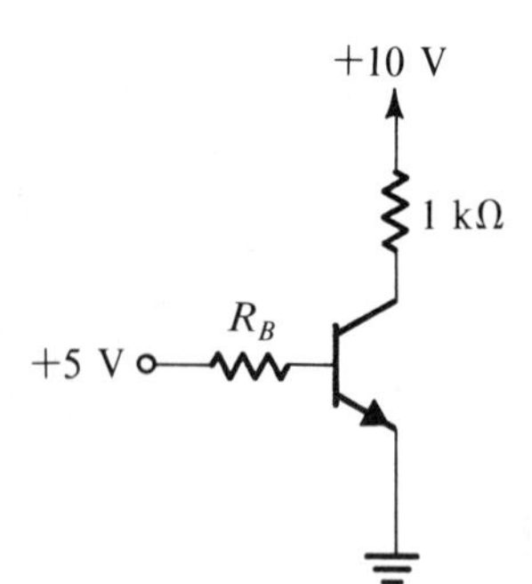

Fig. 4.54 Circuit for Example 4.14.

Solution When the transistor is saturated the collector voltage will be

$$V_C = V_{CE\text{sat}} \simeq 0.3 \text{ V}$$

Thus the collector current is given by

$$I_{C\text{sat}} = \frac{+10 - 0.3}{1} = 9.7 \text{ mA}$$

To saturate the transistor with the lowest β we need to provide a base current of at least

$$I_{B\text{sat}} = \frac{I_{C\text{sat}}}{\beta_{\min}} = \frac{9.7}{50} = 0.194 \text{ mA}$$

For an overdrive factor of 10, the base current should be

$$I_B = 10 \times 0.194 = 1.94 \text{ mA}$$

Thus we require a value of R_B such that

$$\frac{+5 - 0.7}{R_B} = 1.94$$

$$R_B = \frac{4.3}{1.94} = 2.2 \text{ k}\Omega$$

EXAMPLE 4.15

We want to analyze the circuit of Fig. 4.55 to determine the voltages at all nodes and the currents through all branches. The minimum value of β is specified to be 30.

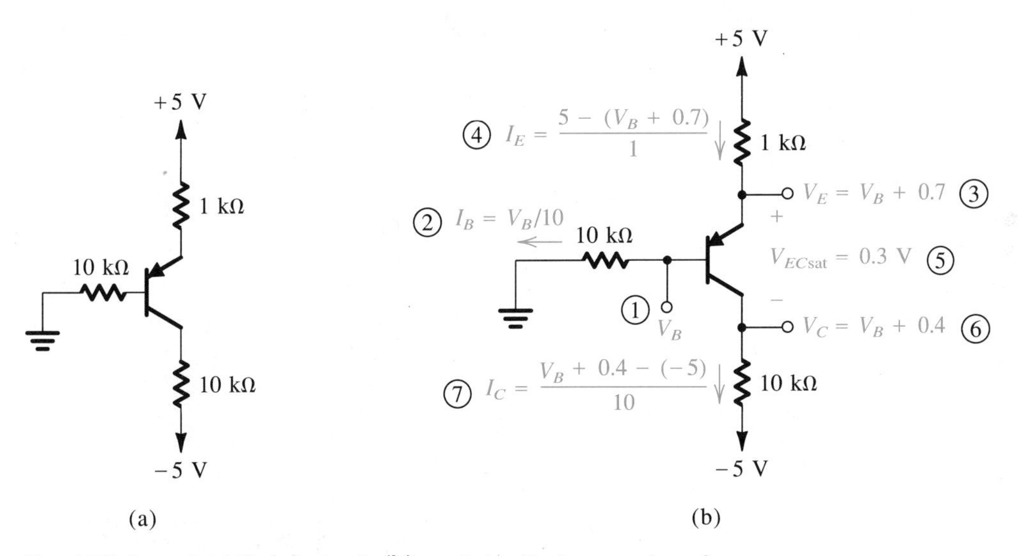

Fig. 4.55 Example 4.15: **(a)** circuit; **(b)** analysis with steps numbered.

Solution A quick glance at this circuit reveals that the transistor will be either active or saturated. Assuming active-mode operation and neglecting the base current, we see that the base voltage will be approximately zero volts, the emitter voltage will be approximately +0.7 V, and the emitter current will be approximately 4.3 mA. Since the maximum current that the collector can support while the transistor remains in the active mode is approximately 0.5 mA, it follows that the transistor is definitely saturated.

Assuming that the transistor is saturated and denoting the voltage at the base by V_B (refer to Fig. 4.55b), it follows that

$$V_E = V_B + V_{EB} \simeq V_B + 0.7$$

$$V_C = V_E - V_{EC\text{sat}} \simeq V_B + 0.7 - 0.3 = V_B + 0.4$$

$$I_E = \frac{+5 - V_E}{1} = \frac{5 - V_B - 0.7}{1} = 4.3 - V_B \quad \text{mA}$$

$$I_B = \frac{V_B}{10} = 0.1\ V_B \quad \text{mA}$$

$$I_C = \frac{V_C - (-5)}{10} = \frac{V_B + 0.4 + 5}{10} = 0.1 V_B + 0.54 \quad \text{mA}$$

Using the relationship $I_E = I_B + I_C$ we obtain

$$4.3 - V_B = 0.1 V_B + 0.1 V_B + 0.54$$

which results in

$$V_B = \frac{3.76}{1.2} \simeq 3.13\ \text{V}$$

Substituting in the equations above, we obtain

$$V_E = 3.83\ \text{V}$$

$$V_C = 3.53\ \text{V}$$

$$I_E = 1.17\ \text{mA}$$

$$I_C = 0.853\ \text{mA}$$

$$I_B = 0.313\ \text{mA}$$

(note that I_E does not exactly equal $I_B + I_C$ because the value of V_B is approximate). It is clear that the transistor is saturated, since the value of forced β is

$$\beta_{\text{forced}} = \frac{0.853}{0.313} \simeq 2.7$$

which is much smaller than the specified minimum β.

EXAMPLE 4.16

We desire to evaluate the voltages at all nodes and the currents through all branches in the circuit of Fig. 4.56(a). Assume $\beta = 100$.

Solution

By examining the circuit we conclude that the two transistors Q_1 and Q_2 cannot be simultaneously conducting. Thus if Q_1 is on, Q_2 will be off, and vice versa. Assume that

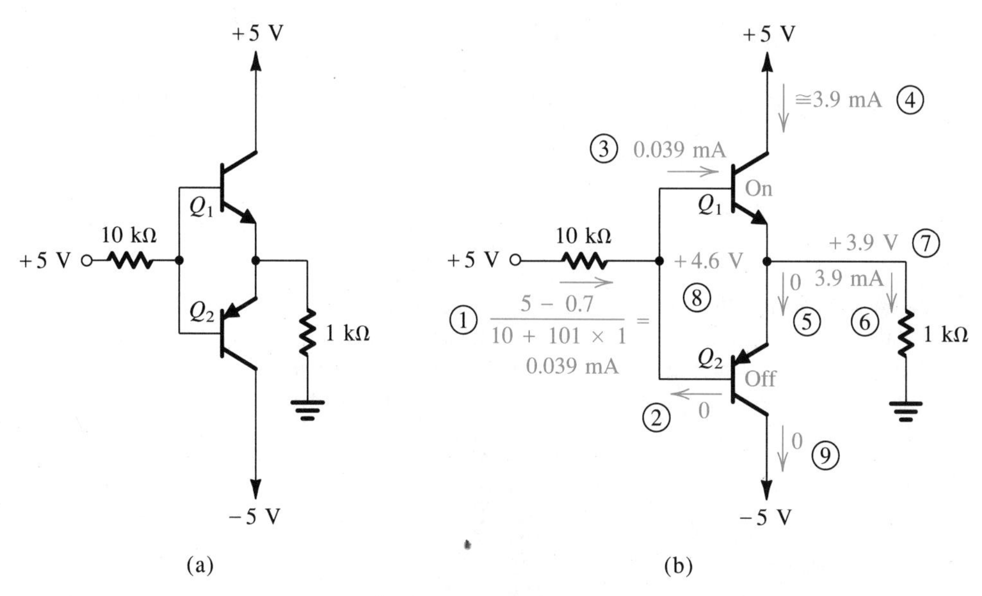

Fig. 4.56 Example 4.16: **(a)** circuit; **(b)** analysis with the steps numbered.

Q_2 is on. It follows that current will flow from ground through the 1-kΩ load resistor into the emitter of Q_2. Thus the base of Q_2 will be at a negative voltage, and base current will be flowing out of the base through the 10-kΩ resistor and into the +5-V supply. This is impossible, since if the base is negative, current in the 10-kΩ resistor will have to flow into the base. Thus we conclude that our original assumption—that Q_2 is on—is incorrect. It follows that Q_2 will be off and Q_1 will be on.

The question now is whether Q_1 is active or saturated. The answer in this case is obvious. Since the base is fed with a +5-V supply and since base current flows into the base of Q_1, it follows that the base of Q_1 will be at a voltage lower than +5 V. Thus the collector–base junction of Q_1 is reverse-biased and Q_1 is in the active mode. It remains only to determine the currents and voltages using techniques already described in detail. The results are given in Fig. 4.56(b).

Exercises 4.36 Consider the circuit of Fig. 4.49 with the input connected to +5 V. Let $V_{CC} = 5$ V, $R_C = 1$ kΩ, $R_B = 10$ kΩ, and $\beta = 50$. What is the value of forced β? Find the value of v_I required to establish $\beta_{\text{forced}} = \beta/2$.

Ans. 10.9; 2.6 V

4.37 Repeat Example 4.14 with an overdrive factor of 5.

Ans. $R_B = 4.4$ kΩ

4.38 Solve the problem in Example 4.16 with the voltage feeding the bases changed to +10 V. Assume that $\beta_{\min} = 30$, and find V_E, V_B, I_{C1}, and I_{C2}.

Ans. +4.7 V; +5.4 V; 4.24 mA; 0

4.13 COMPLETE STATIC CHARACTERISTICS AND SECOND-ORDER EFFECTS

We conclude this chapter with a discussion of the complete static characteristics of the BJT as they appear on its data sheets. We also study a number of secondary characteristics that limit the operation of the BJT in the more advanced circuits presented in subsequent chapters.

Common-Base Characteristics

Figure 4.57 shows the complete set of i_C–v_{CB} characteristic curves for an *npn* transistor. As mentioned in Section 4.5, the i_C–v_{CB} characteristics are measured at constant values of emitter current i_E (see Fig. 4.14a). Since in such an arrangement the base is connected to a constant voltage, the i_C–v_{CB} curves are called the **common-base characteristics.**

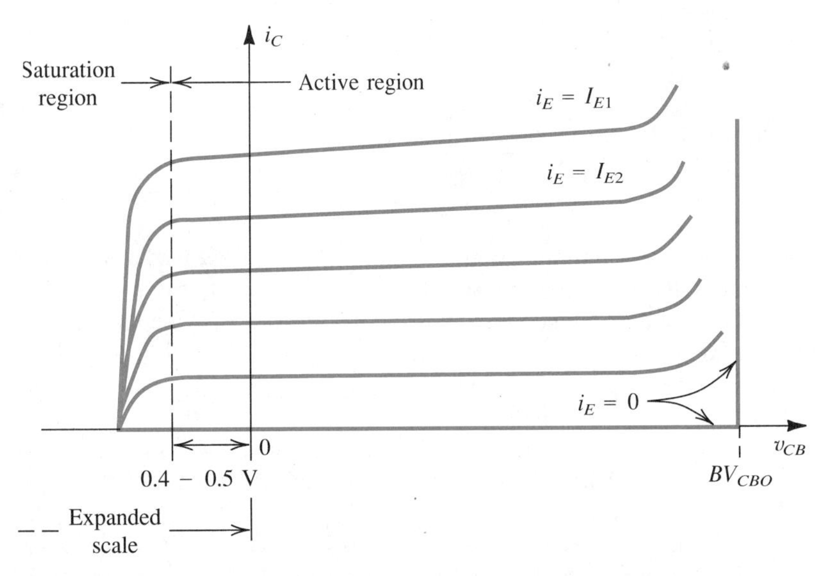

Fig. 4.57 The i_C–v_{CB} or common-base characteristics of an *npn* transistor. Note that in the active region there is a slight dependence of i_C on the value of v_{CB}. The result is a finite output resistance that decreases as the current level in the device is increased.

The curves in Fig. 4.57 differ from those presented in Fig. 4.14(b) in three aspects. First, the avalanche breakdown of the CBJ at large voltages is indicated and will be briefly explained at a later stage. Second, the saturation-region characteristics are included. As indicated, as v_{CB} goes negative the CBJ becomes forward-biased and the collector current decreases. Since for each curve, i_E is held constant, the decrease in i_C results in an equal increase in i_B. The large effect that v_{CB} has on the collector current in saturation is evident from Fig. 4.57 and is consistent with our earlier description of the saturation mode of operation.

The third difference between the characteristic curves of Fig. 4.57 and the curves presented earlier is that in the active region the characteristic curves are shown to have a very small slope. This slope indicates that in the common-base configuration the collector current depends to a small extent on the collector–base voltage, which is a manifestation of the Early effect discussed previously. It should be noted, however, that the slope of the i_C–v_{CB} curves measured with a constant i_E is much smaller than the slope of the i_C–v_{CE} curves measured with a constant v_{BE}. In other words, the output resistance of the common-base configuration is much greater than that of the common-emitter circuit with a constant v_{BE} (that is, r_o). Another important point to note here is that since each i_C–v_{CB} curve is measured at a constant i_E, the increase in i_C with v_{CB} implies a corresponding decrease in i_B. The dependence of i_B on v_{CB} can be modeled by the addition of a resistor r_μ between collector and base in the hybrid-π model, resulting in the augmented model shown in Fig. 4.58. The resistance r_μ is very large, greater than βr_o.

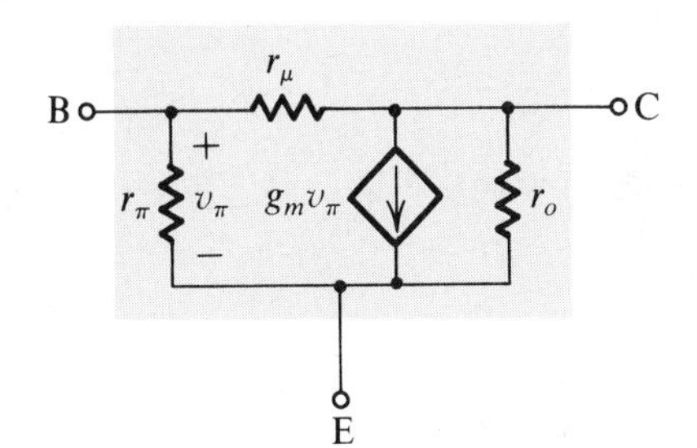

Fig. 4.58 The hybrid-π model, including the resistance r_μ, which models the effect of v_C on i_b.

The augmented hybrid-π model of Fig. 4.58 can be used to find the output resistance of the common-base configuration, which is the inverse of the slope of the i_C–v_{CB} characteristic lines in Fig. 4.57. To do that, simply ground the base, leave the emitter open-circuited (because i_E is constant), apply a test voltage between collector and ground, and find the current drawn from the test voltage. The result (Problem 4.83) is that the output resistance is approximately equal to the parallel equivalent of r_μ and βr_o, and thus is very large.

Common-Emitter Characteristics

An alternative way of graphically displaying the transistor characteristics is shown in Fig. 4.59, where i_C is plotted versus v_{CE} for various values of the base current i_B. These characteristics are measured in a different way from those of Fig. 4.15. While in the latter case v_{BE} is held constant for every curve, here i_B is held constant. As a result the slope in the active region is different from $1/r_o$; in fact, the slope here is greater. It can be shown using the hybrid-π model of Fig. 4.58 that the output resistance of the common-emitter configuration with i_B held constant is approximately equal to $[r_o // (r_\mu/\beta)]$; see Problem 4.84.

The saturation region is evident also on the common-emitter characteristics of Fig. 4.59. We note that whereas the transistor in the active region acts as a current source with a high (but finite) output resistance, in the saturation region it behaves as a ''closed

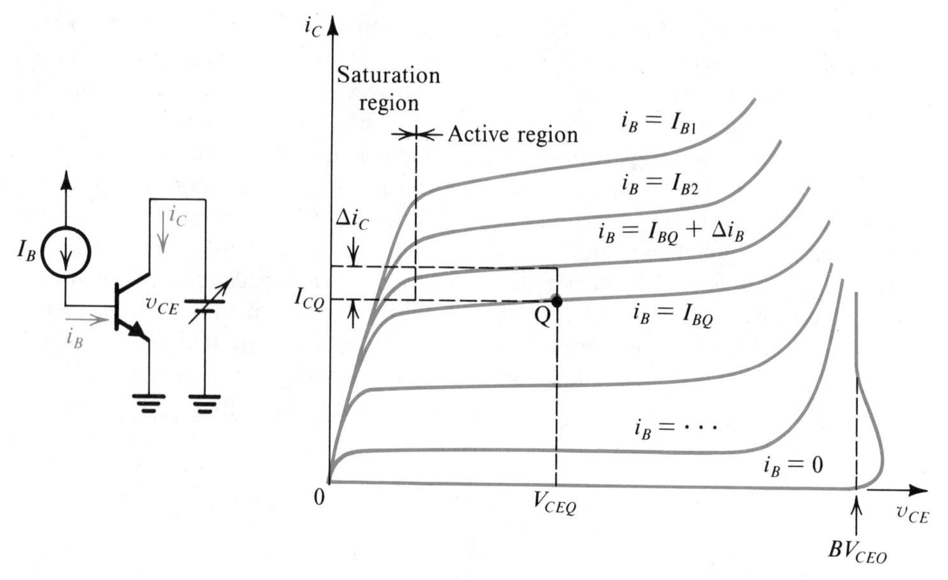

Fig. 4.59 Common-emitter characteristics. Note that the horizontal scale is expanded around the origin to show the saturation region in some detail.

switch'' with a small ''closure resistance'' $R_{CE\text{sat}}$. Since the characteristic curves are all ''bunched together'' in saturation, we show an expanded view of the saturation portion of the characteristics in Fig. 4.60. Note that the curves do not extend directly to the origin. In fact, for a given value of i_B the i_C–v_{CE} characteristic in saturation can be approximated by a straight line intersecting the v_{CE} axis at a point $V_{CE\text{off}}$, as illustrated in Fig. 4.61. The voltage $V_{CE\text{off}}$ is called the **offset voltage** of the transistor switch. Field-effect transistors

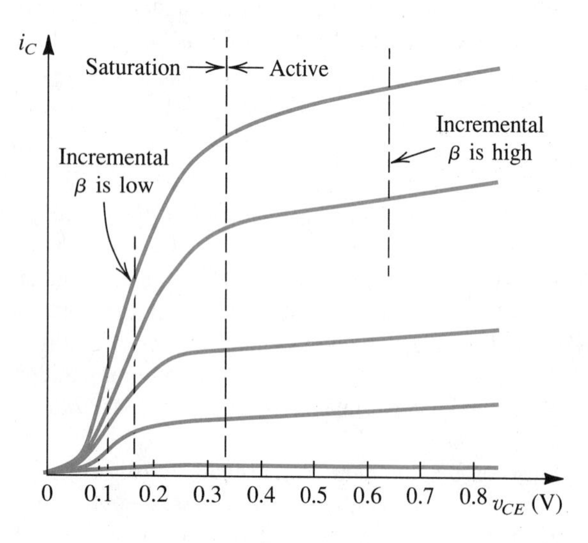

Fig. 4.60 An expanded view of the common-emitter characteristics in the saturation region.

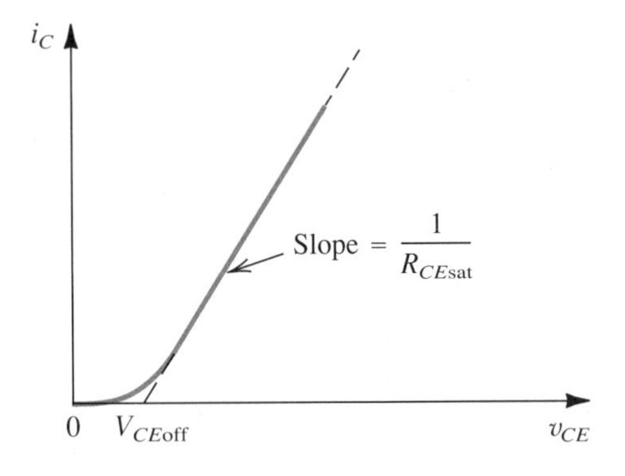

Fig. 4.61 One of the i_C–v_{CE} characteristic curves in the saturation region. Note that the characteristic can be modeled by an offset voltage V_{CEoff} and a small resistance R_{CEsat}.

(Chapter 5) do not exhibit such offset voltages and hence make superior switches. FETs, however, display higher values of closure resistance.

The Transistor β

Earlier we defined β as the ratio of the total current in the collector to the total current in the base when the transistor is operating in the active mode. Let us be more specific. Assume that the transistor is operating at a base current I_{BQ}, a collector current I_{CQ}, and a collector-to-emitter voltage V_{CEQ}. These quantities define the operating or bias point Q in Fig. 4.59. The ratio of I_{CQ} to I_{BQ} is called the dc β or h_{FE} (the reason for the latter name will be explained in Chapter 7),

$$h_{FE} \equiv \beta_{\text{dc}} \equiv \frac{I_{CQ}}{I_{BQ}} \tag{4.115}$$

When the transistor is used as an amplifier it is first biased at a point such as Q. Applied signals then cause incremental changes in i_B, i_C, and v_{CE} around the bias point. We may therefore define an *incremental* or ac β as follows: Let the collector-to-emitter voltage be maintained constant at V_{CEQ} (in order to eliminate the Early effect), and change the base current by an increment Δi_B. If the collector current changes by an increment Δi_C (see Fig. 4.59), then β_{ac} (or h_{fe}, as it is usually called) at the operating point Q is defined as

$$h_{fe} \equiv \beta_{\text{ac}} \equiv \left.\frac{\Delta i_C}{\Delta i_B}\right|_{v_{CE}=\text{constant}} \tag{4.116}$$

The fact that V_{CE} is held constant implies that the incremental voltage v_{ce} is zero; therefore h_{fe} is called the **short-circuit current gain.**

When we perform small-signal analysis, the β should be the ac β (h_{fe}). On the other hand, if we are analyzing or designing a switching circuit, β_{dc} (h_{FE}) is the appropriate β. The difference in value between β_{dc} and β_{ac} is usually small, and we will not normally distinguish between the two. A point worth mentioning, however, is that the value of β depends on the current level in the device, and the relationship takes the form shown in Fig. 4.62. This figure also shows the temperature dependence of β.

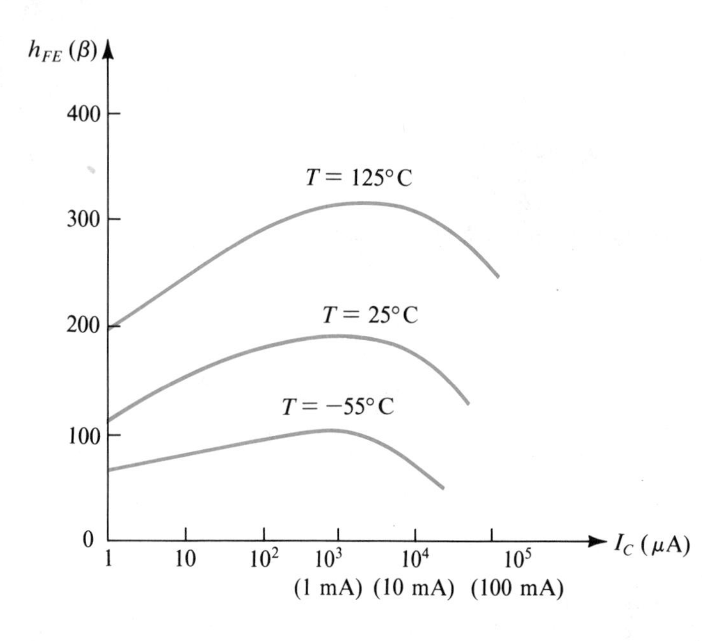

Fig. 4.62 Typical dependence of β on I_C and on temperature in a modern integrated-circuit *npn* silicon transistor intended for operation around 1 mA.

In Section 4.9 we discussed the graphical analysis of transistor circuits. It was shown that for amplifier applications the BJT is biased at a quiescent point in the middle of its active region. Here we wish to show graphically a BJT operating in the saturation region. Figure 4.63 shows a load line that intersects the i_C–v_{CE} characteristic at a point in the saturation region. Note that in this case changes in the base current result in very small changes in i_C and v_{CE} and that in saturation the incremental β (β_{ac}) is very small.

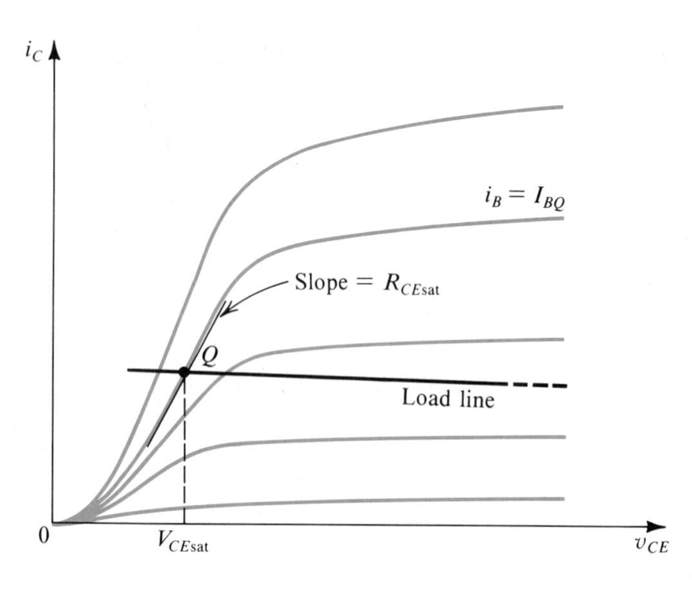

Fig. 4.63 An expanded view of the saturation portion of the characteristics together with a load line that results in operation at a point Q in the saturation region.

Transistor Breakdown

The maximum voltages that can be applied to a BJT are limited by the EBJ and CBJ breakdown effects that follow the avalanche multiplication mechanism described in Section 3.11. Consider first the common-base configuration. The i_C–v_{CB} characteristics in Fig. 4.57 indicate that for $i_E = 0$ (that is, with the emitter open-circuited) the collector–base junction breaks down at a voltage denoted by BV_{CBO}. For $i_E > 0$, breakdown occurs at voltages smaller than BV_{CBO}. Typically BV_{CBO} is greater than 50 V.

Next consider the common-emitter characteristics of Fig. 4.59, which show breakdown occurring at a voltage BV_{CEO}. Here, although breakdown is still of the avalanche type, the effects on the characteristics are more complex than in the common-base configuration. We will not explain these details; it is sufficient to point out that typically BV_{CEO} is about half BV_{CBO}. On the transistor data sheets BV_{CEO} is sometimes referred to as the **sustaining voltage** LV_{CEO}.

Breakdown of the CBJ in either the common-base or common-emitter configuration is not destructive as long as the power dissipation in the device is kept within safe limits. This, however, is not the case with the breakdown of the emitter–base junction. The EBJ breaks down in an avalanche manner at a voltage BV_{EBO} much smaller than BV_{CBO}. Typically BV_{EBO} is in the range 6 to 8 V, and the breakdown is destructive in the sense that the β of the transistor is permanently reduced. This does not prevent use of the EBJ as a zener diode to generate reference voltages in IC design. In such applications, however, one is not concerned with the β-degradation effect. A circuit arrangement to prevent EBJ breakdown in IC amplifiers will be discussed in Chapter 10. Transistor breakdown and maximum allowable power dissipation are important parameters in the design of power amplifiers (Chapter 9).

Exercises 4.39 What is the output voltage of the circuit in Fig. E4.39 if the transistor $BV_{BCO} = 70$ V?

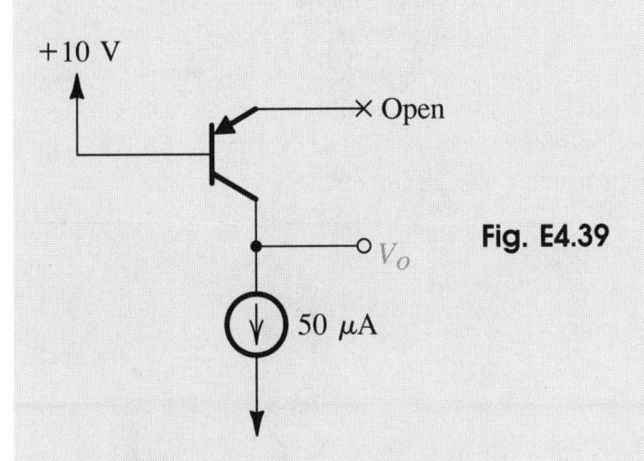

Fig. E4.39

Ans. −60 V

4.40 Measurements made on a BJT switch having a constant base-current drive at low values of v_{CE} provide the following data: at $i_C = 5$ mA, $v_{CE} = 170$ mV; at $i_C = 2$ mA, $v_{CE} = 110$ mV. What are the values of the offset voltage and saturation resistance of this switch?

Ans. 70 mV; 20 Ω

SUMMARY

- Depending on the bias conditions on its two junctions, the BJT can operate in one of three possible modes: cutoff (both junctions reverse-biased), active (the EBJ forward-biased and the CBJ reverse-biased), and saturation (both junctions forward-biased).

- For amplifier applications the BJT is operated in the active mode. Switching applications make use of the cutoff and saturation modes.

- A BJT operating in the active mode provides a collector current $i_C = I_S e^{|v_{BE}|/V_T}$. The base current $i_B = i_C/\beta$, and the emitter current $i_E = i_C + i_B$. Also, $i_C = \alpha i_E$ and thus $\beta = \alpha/(1 - \alpha)$ and $\alpha = \beta/(\beta + 1)$.

- To ensure operation in the active mode, the collector voltage of an *npn* transistor must be kept greater than the base voltage. For a *pnp* transistor the collector voltage must be lower than the base voltage.

- At a constant collector current the magnitude of the base-emitter voltage decreases by about 2 mV for every °C rise in temperature.

- The dc analysis of transistor circuits is greatly simplified by assuming that $|V_{BE}| \simeq 0.7$ V.

- To operate as a linear amplifier the BJT is biased in the active region and the signal v_{be} is kept small ($v_{be} \ll V_T$).

- For small signals, the BJT functions as a linear voltage-controlled current source with a transconductance $g_m = I_C/V_T$. The input resistance between base and emitter, looking into the base, is $r_\pi = \beta/g_m$.

- Bias design seeks to establish a dc collector current that is as independent of the value of β as possible.

- In the common-emitter configuration, the emitter is at signal ground, the input signal is applied to the base, and the output is taken at the collector. A high voltage gain and a reasonably high input resistance are obtained, but the high-frequency response is limited.

- The input resistance of the common-emitter amplifier can be increased by including an unbypassed resistance in the emitter lead.

- In the common-base configuration, the base is at signal ground, the input signal is applied to the emitter, and the output is taken at the collector. A high voltage gain (from emitter to collector) and an excellent high-frequency response are obtained, but the input resistance is very low. The CB amplifier is useful as a current buffer.

- In the emitter follower the collector is at signal ground, the input signal is applied to the base, and the output taken at the emitter. Although the voltage gain is less than unity, the input resistance is very high and the output resistance is very low. The circuit is useful as a voltage buffer.

- In a saturated transistor, $|V_{CE\text{sat}}| \simeq 0.3$ V and $I_{C\text{sat}} = (V_{CC} - V_{CE\text{sat}})/R_C$. The ratio of $I_{C\text{sat}}$ to the base current is the forced β, which is lower than β. The collector-to-emitter resistance, $R_{CE\text{sat}}$, is small (few tens of ohms).

- With the emitter open-circuited ($i_E = 0$), the CBJ breaks down at a reverse voltage BV_{CBO} that is typically > 50 V. For $i_E > 0$, the breakdown voltage is less than BV_{CBO}. In the common-emitter configuration the breakdown voltage specified is BV_{CEO}, which is about half BV_{CBO}. The emitter–base junction breaks down at a reverse bias of 6–8 V. This breakdown usually has an adverse effect on β.

BIBLIOGRAPHY

E. J. Angelo, Jr., *Electronics: BJTs, FETs, and Microcircuits,* New York: McGraw-Hill, 1969.

I. Getreu, *Modeling the Bipolar Transistor,* Beaverton, Ore.: Tektronix, Inc., 1976.

P. R. Gray and R. G. Meyer, *Analysis and Design of Analog Integrated Circuits,* 2nd ed., New York: Wiley, 1984.

P. E. Gray and C. L. Searle, *Electronic Principles,* New York: Wiley, 1971.

J. Millman and A. Grabel, *Microelectronics,* 2nd ed., New York: McGraw-Hill, 1987.

C. L. Searle, A. R. Boothroyd, E. J. Angelo, Jr., P. E. Gray, and D. O. Pederson, *Elementary Circuit Properties of Transistors,* Vol. 3 of the SEEC Series, New York: Wiley, 1964.

PROBLEMS

Section 4.2: Operation of the *npn* Transistor in the Active Mode

4.1 Two transistors, fabricated with the same technology but having different junction areas, when operated at a base-emitter voltage of 0.69 V, have collector currents of 0.13 and 10.9 mA. Find I_S for each device. What are the relative junction areas?

4.2 In a particular BJT, the base current is 7.5 μA, and the collector current is 940 μA. Find β and α for this device.

4.3 Measurements taken on a variety of transistors are found to be incomplete (or possibly in error). Available data are as shown. Provide the missing data and calculations, and detect inconsistencies if any.

Device	i_C (mA)	i_B (mA)	i_E (mA)	α	β
a	10.0		10.1		100
b		0.02	1.12		
c	0.63			0.984	63
d	98.0		99.0	0.990	98
e		0.001	0.011		10
f	10.0	0.2	10.1		100
g	10.1	0.1	10.0	0.99	
h	0.990	0.010			99
i		0.015		0.995	193

4.4 Show that for a transistor with α close to unity, if α changes by a small per-unit amount $(\Delta\alpha/\alpha)$ the corresponding per-unit change in β is given approximately by

$$\frac{\Delta\beta}{\beta} \simeq \beta\left(\frac{\Delta\alpha}{\alpha}\right)$$

Find $\Delta\beta/\beta$ for $\beta = 100$ and α changes by 0.1%.

4.5 Consider the large-signal BJT models shown in Figs. 4.5(b) and (d). What are the relative sizes of the diodes D_E and D_B for transistors for which $\beta = 10$? $\beta = 1000$?

4.6 A particular BJT when conducting a collector current of 10 mA is known to have $v_{BE} = 0.70$ V and $i_B = 100\ \mu$A. Use these data to create specific transistor models of the form shown in Figs. 4.5(a) and (d).

4.7 Using the *npn* transistor model of Fig. 4.5(b), consider the case of a transistor whose base is connected to ground, the collector is connected to a 10-V dc source through a 1-kΩ resistor, and a 5-mA current source is connected to the emitter with the polarity so that current is drawn out of the emitter terminal. If $\beta = 100$ and $I_S = 10^{-14}$ A, find the voltages at the emitter and the collector and calculate the base current.

4.8 The current I_{CBO} of a small transistor is measured to be 15 nA at 25°C. If the temperature of the device is raised to 75°C, what do you expect I_{CBO} current to become?

***4.9** Augment the model of the *npn* BJT shown in Fig. 4.5(c) by a current source representing I_{CBO}. In terms of this addition, what do the terminal currents i_B, i_C, and i_E become? If the base lead is open circuited while the emitter is connected to ground, and the collector is connected to a positive supply, find the emitter and collector currents.

4.10 From Fig. 4.6 we note that the transistor is not a symmetrical device. Thus interchanging the collector and emitter terminals will result in a device with different values of α and β, called the inverse or reverse values and denoted α_R and β_R. An *npn* transistor is accidentally connected with collector and emitter leads interchanged. The resulting currents in the normal emitter and base leads are 5 mA and 1 mA, respectively. What are the values of α_R and β_R?

Section 4.3: The *pnp* Transistor

4.11 Fig. 4.8 shows two large-signal models for the *pnp* transistor operating in the active mode. Sketch two additional models that parallel those given for the *npn* transistor in Fig. 4.5(b) and (d).

4.12 Consider the *pnp* large-signal model of Fig. 4.8(b) applied for a transistor having $I_S = 10^{-13}$ A and $\beta = 40$. If the emitter is connected to ground, the base is connected to a current source that pulls out of the base terminal a current of 10 μA, and the collector is connected to a negative supply of -10 V via a 10-kΩ resistor, find the base voltage, the collector voltage, and the emitter current.

4.13 A *pnp* transistor has $v_{EB} = 0.8$ V at a collector current of 1 A. What do you expect v_{EB} to become at $i_C = 10$ mA? at $i_C = 5$ A?

4.14 A *pnp* transistor has a common-emitter current gain of 50. What is its common-base current gain?

Section 4.4: Circuit Symbols and Conventions

4.15 For the circuits in Fig. P4.15 assume that the transistors have very large β. Some measurements have been made on these circuits, the results are indicated in the figure. Find the values of the other labeled voltages and currents.

4.16 Measurements on the circuits of Fig. P4.16 produce labeled voltages as indicated. Find the value of β for each transistor.

D4.17 Examination of the table of standard values for resistors with 5% tolerance reveals that the closest values to those found in the design of Example 4.1 are 5.1 kΩ and 6.8 kΩ. For these values use approximate calculations (e.g., $V_{BE} \simeq 0.7V$ and $\alpha \simeq 1$) to determine the values of collector current and collector voltage that are likely to result.

D4.18 Redesign the circuit in Example 4.1 to provide $V_C = +7.5$ V and $I_C = 1$ mA.

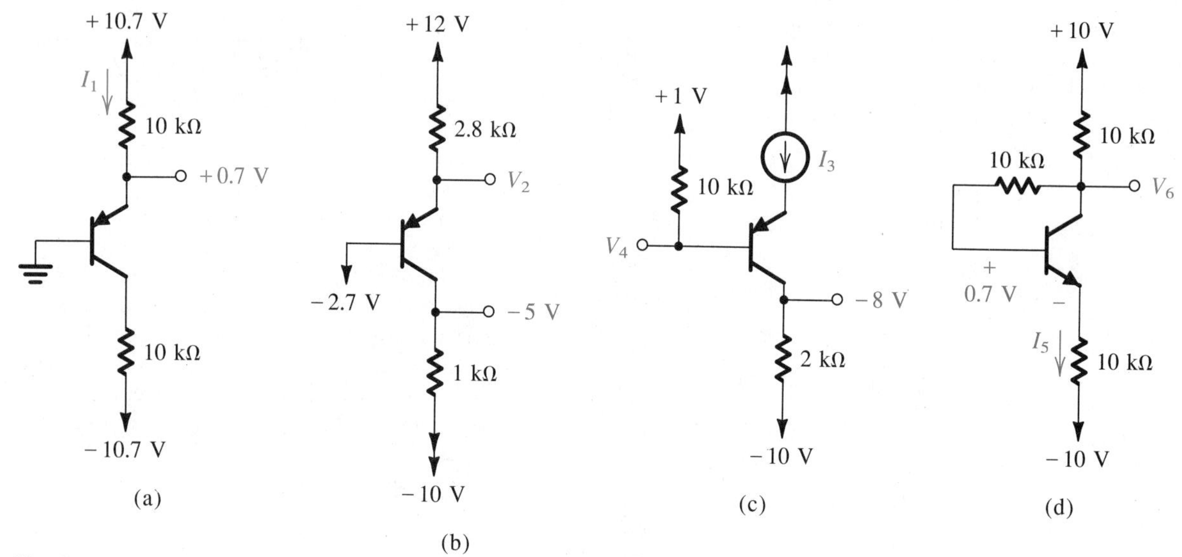

Fig. P4.15

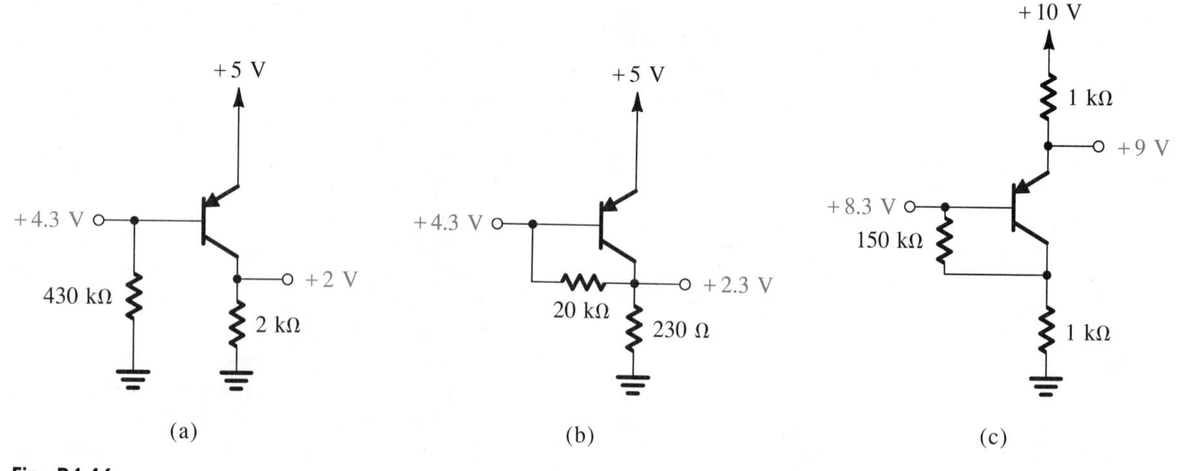

Fig. P4.16

Section 4.5: Graphical Representation of Transistor Characteristics

4.19 Use Eq. (4.11) to plot i_C versus v_{CE} for an *npn* transistor having $I_S = 10^{-15}$ A and $V_A = 100$ V. Provide curves for $v_{BE} = 0.65$, 0.70, 0.72, 0.73, and 0.74 volts. Show the characteristics for v_{CE} up to 15 V.

4.20 For a BJT whose collector current is 12 mA at $v_{BE} = 0.7$ V, what collector current results when v_{BE} is reduced to 0.5 V?

4.21 A BJT whose emitter current is fixed at 1 mA has a base–emitter voltage of 0.67 V at 25°C. What base–emitter voltage would you expect at 0°C? at 100°C?

4.22 For a BJT having an Early voltage of 200 V, what is its output resistance at 1 mA? at 100 μA?

4.23 For a BJT having an output resistance of 10 MΩ at 10 μA, what must its Early voltage be? If the current is raised to 10 mA, what does the output resistance become?

4.24 Consider the circuit of Fig. 4.15(a). Let V_{BE} be adjusted to yield a current $I_C = 1$ mA at $V_{CE} = 1$ V. Then, while keeping V_{BE} constant, V_{CE} is raised to +11 V and I_C is measured to be 1.2 mA. Find V_A for this transistor and the value of r_o at $I_C = 1$ mA.

Section 4.6: Analysis of Transistor Circuits at DC

4.25 The transistor in the circuit of Fig. P4.25 has a very high β. Find V_E and V_C for V_B (a) +3 V, (b) +1 V, and (c) 0 V. Assume $V_{BE} \simeq 0.7$ V.

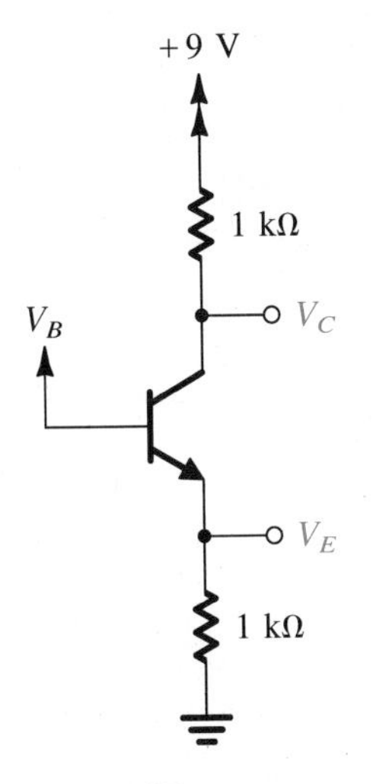

Fig. P4.25

4.26 The transistor in the circuit of Fig. P4.25 has a very high β. Find the highest value of V_B for which the transistor still operates in the active mode.

D4.27 Consider the circuit in Fig. P4.25 with the base voltage V_B obtained using a voltage divider across the 9-V supply. Assuming the transistor β to be very large (that is, ignoring the base current), design the voltage divider to obtain $V_B = 3$ V. Design for a 0.2-mA current in the voltage divider. Now if the BJT $\beta = 100$, analyze the circuit to determine the collector current and the collector voltage.

4.28 A single measurement indicates the emitter voltage of the transistor in the circuit of Fig. P4.28 to be 1.0 V. Under the assumption that $V_{BE} = 0.7$ V, what are V_B, I_B, I_E, I_C, β, and α? (Isn't it surprising what a little measurement can lead to?)

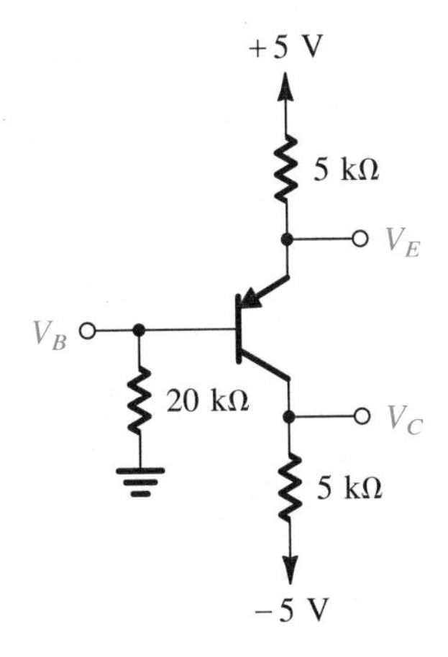

Fig. P4.28

D4.29 Redesign the circuit of Fig. 4.19(a) (i.e., find new values for R_E and R_C) to establish a collector current of 2 mA and a collector voltage of −5 V. Assume $\alpha \simeq 1$.

D**4.30 It is required to design the circuit in Fig. P4.30 so that a current of 1 mA is established in the emitter and a voltage of +5 V appears at the collector. The transistor type used has a nominal β of 100. However, the β value can be as low as 50 and as high as 150. Your design should ensure that the specified emitter current is obtained when $\beta = 100$ and that at the extreme values of β the emitter current does not change by more than 10% of its nominal value. Also, design for as large a value for R_B as possible. Give the values of R_B, R_E, and R_C to the nearest kilohm. What is the expected

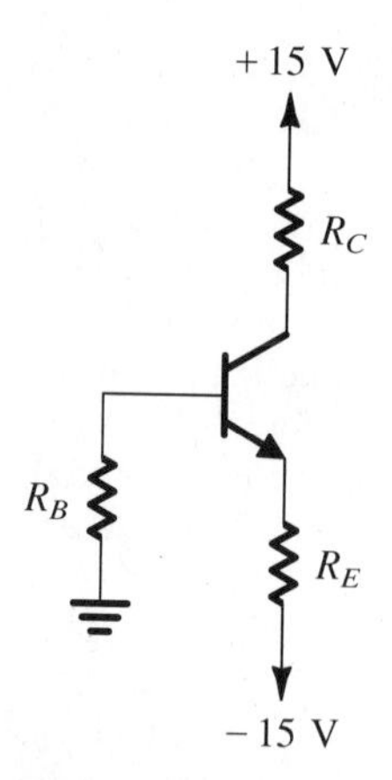

Fig. P4.30

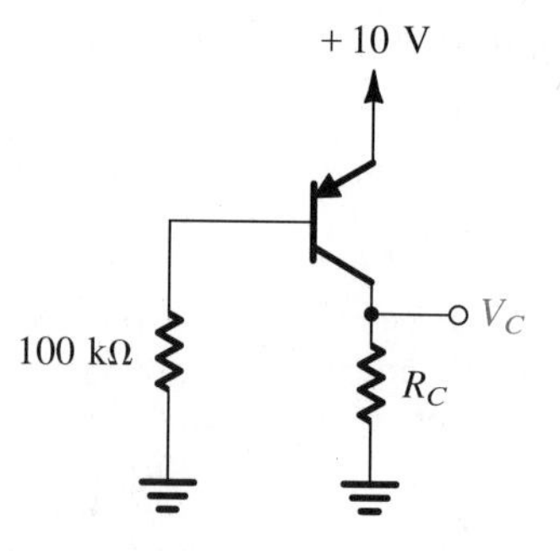

Fig. P4.32

range of collector current and collector voltage corresponding to the full range of β values?

*4.31 For the circuit in Fig. P4.31, find all node voltages for (a) β very high, and (b) $\beta = 99$.

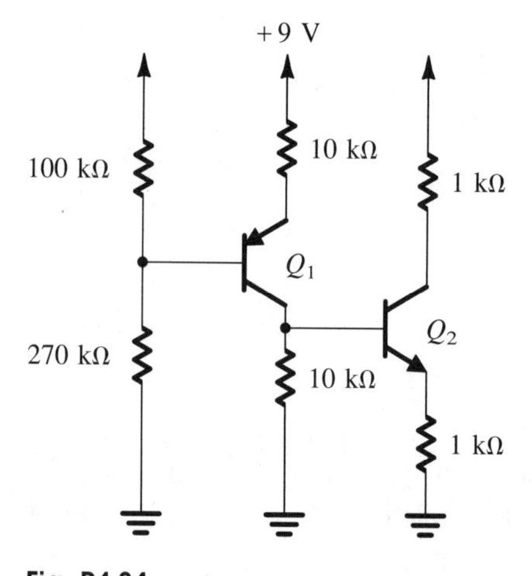

Fig. P4.31

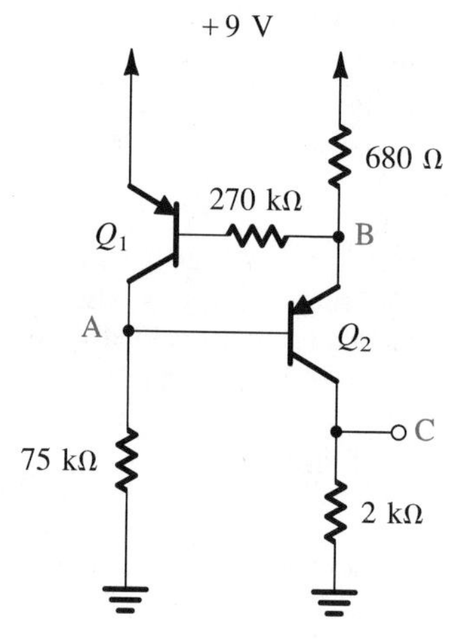

Fig. P4.33

D4.32 The *pnp* transistor in the circuit of Fig. P4.32 has $\beta = 50$. Find the value for R_C to obtain $V_C = +5$ V. What happens if the transistor is replaced with another having $\beta = 100$?

**4.33 For the circuit in Fig. P4.33 in which $|V_{BE}| = 0.7$ V, find the voltages at nodes A, B, and C for (a) $\beta = \infty$, and (b) $\beta = 100$.

Section 4.7: The Transistor as an Amplifier

4.34 Consider a transistor biased to operate in the active mode at a dc collector current I_C. Calculate the collector signal current as a fraction of I_C (that is i_c/I_C) for input signals v_{be} of +1, −1, +2, −2, +5, −5, +8, −8, +10, −10, +12, −12 mV. In each case do the calculation two ways:

(a) using the exponential characteristic, and

(b) using the small-signal approximation.

Present your results in the form of a table. Comment on the range of validity of the small-signal approximation.

4.35 A transistor with $\beta = 150$ is biased to operate at a dc collector current of 2 mA. Find the values of g_m, r_π, and r_e. Repeat for a bias current of 0.2 mA.

D4.36 A designer, in considering a single-stage BJT amplifier, wishes to obtain a g_m of 1 A/V while limiting the base–emitter dc voltage to 0.8 V. Specify the size of the transistor required in terms of its scale current I_S.

D4.37 For reasonably linear small-signal operation of a BJT, v_{be} must be limited to no larger than 10 mV. To what percentage change of bias current does this correspond? For a design in which the required output signal is 10-mA peak, what bias current is required? What is the corresponding value of g_m?

4.38 A transistor operating with nominal g_m of 80 mA/V has a β that ranges from 50 to 200. Also, the bias circuit, being less than ideal, allows a $\pm 25\%$ variation in I_C. What are the extreme values found of the resistance, looking into the base?

4.39 In the circuit of Fig. 4.23, V_{BE} is adjusted so that $V_C = 2$ V. If $V_{CC} = 10$ V, $R_C = 2$ kΩ, and a signal $v_{be} = 0.004 \sin \omega t$ volts is applied, find expressions for the total instantaneous quantities $i_C(t)$, $v_C(t)$, and $i_B(t)$. The transistor has $\beta = 100$. What is the voltage gain?

D*4.40 We wish to design the amplifier circuit of Fig. 4.23 under the constraint that V_{CC} is fixed. Let the input signal $v_{be} = \hat{V}_{be} \sin \omega t$ where $\hat{V}_{be}$ is the maximum value for acceptable linearity. Show that for the design that results in the largest signal at the collector without the BJT leaving the active region,

$$R_C I_C = (V_{CC} - V_{BE} - \hat{V}_{be}) \Big/ \left(1 + \frac{\hat{V}_{be}}{V_T}\right)$$

and find an expression for the voltage gain obtained. For $V_{CC} = 10$ V, $V_{BE} = 0.7$ V, and $\hat{V}_{be} = 5$ mV, find the dc voltage at the collector, the amplitude of the output voltage signal, and the voltage gain.

Section 4.8: Small-Signal Equivalent Circuit Models

4.41 A BJT is biased to operate in the active mode at a dc collector current of 2 mA. It has a β of 120. Give the four small-signal models (Figs. 4.26 and 4.27) of the BJT complete with the values of their parameters.

4.42 The transistor amplifier in Fig. P4.42 is biased with a current source I and has a very high β. Find the dc voltage at the collector, V_C. Also find the value of g_m. Replace the transistor with its simplified hybrid-π model of Fig. 4.26a (note that the dc current source I should be replaced with an open circuit). Hence find the voltage gain v_c/v_i.

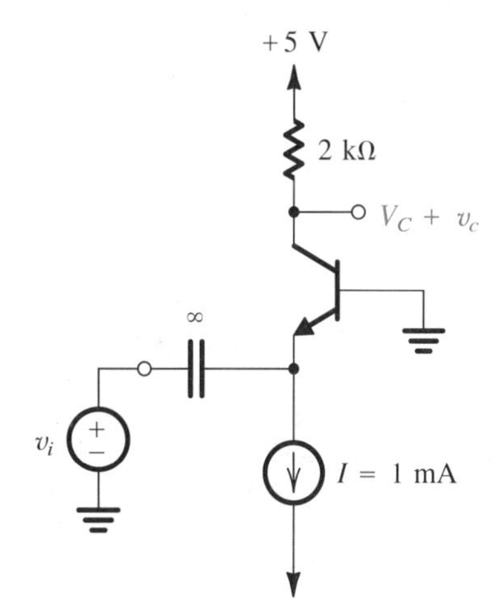

Fig. P4.42

4.43 Using the BJT equivalent circuit model of Fig. 4.27a sketch the equivalent circuit of a transistor amplifier for which a resistance R_e is connected between the emitter and ground, the collector is grounded, and an input signal source v_b is connected between the base and ground. (It is assumed that the transistor is properly biased to operate in the active region.) Show that:

(a) the voltage gain between base and emitter, that is v_e/v_b, is given by

$$\frac{v_e}{v_b} = \frac{R_e}{R_e + r_e}$$

(b) the input resistance,

$$R_{in} \equiv \frac{v_b}{i_b} = (\beta + 1)(R_e + r_e)$$

Find the numerical values for (v_e/v_b) and R_{in} for the case $R_e = 1$ kΩ, $\beta = 100$ and the emitter bias current $I_E = 1$ mA.

D4.44 For the rudimentary amplifier design shown in Fig. 4.28a, show using the equivalent circuit in Fig. 4.28c that the overall voltage gain is given by

$$\frac{v_o}{v_i} = -\frac{\beta R_C}{R_{BB} + r_\pi}$$

Hence find the collector bias current required to obtain a voltage gain of -10 for the case $\beta = 150$, $R_C = 5$ kΩ and $R_{BB} = 20$ kΩ. If the amplitude of the base-emitter signal is limited to 10 mV, find the minimum required value of V_{CC} so that the BJT operates in the active mode at all times.

4.45 Refer to the circuit in Fig. 4.30(a). What is the input resistance of this amplifier? If the signal source v_i has a resistance of 50 Ω what will the overall voltage gain v_c/v_i become?

4.46 When the collector of a transistor is connected to its base, the transistor still operates (internally) in the active region because the collector–base junction is still in effect reverse-biased. Use the simplified hybrid-π model to find the incremental (small-signal) resistance of the resulting two-terminal device.

D*4.47 Design an amplifier using the configuration of Fig. 4.30(a). The power supplies available are ±10 V. The input signal source has a resistance of 100 Ω, and it is required that the amplifier input resistance matches this value. (Note that $R_{in} = r_e // R_E \simeq r_e$.) The amplifier is to have the greatest possible voltage gain and the largest possible output signal, but small-signal linear operation (i.e., the signal component across the base–emitter junction should be limited to no more than 10 mV). Find values for R_E and R_C. What is the value of voltage gain realized?

***4.48** The transistor in the circuit shown in Fig. P4.48 is biased to operate in the active mode. Assuming that β is very large, find the collector bias current I_C. Replace the transistor with the small-signal equivalent circuit model of Fig. 4.27(b) (remember to replace the dc power supply with a short circuit). Analyze the resulting amplifier equivalent circuit to show that

$$\frac{v_{o1}}{v_i} = \frac{R_E}{R_E + r_e}$$

$$\frac{v_{o2}}{v_i} = \frac{-\alpha R_C}{R_E + r_e}$$

Find the values of these voltage gains ($\alpha \simeq 1$).

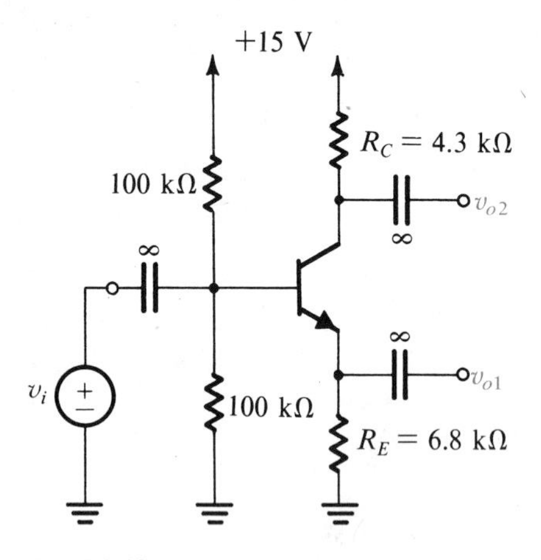

Fig. P4.48

****4.49** The transistor shown in the circuit of Fig. P4.49 has $\beta = 100$ and $V_A = 80$ V

(a) Find the dc voltages at the base, emitter, and collector.

(b) Find g_m and r_π.

(c) If terminal Z is connected to ground, X to signal source v_s having a 10-kΩ source resistance, and Y to a 10-kΩ load resistance, use the small-signal equivalent circuit model of Fig. 4.26(a) to find the voltage gain v_y/v_s.

(d) If terminal X is connected to ground, Z to an input signal source v_s having a 200-Ω source resistance, and Y to a load resistance of 10 kΩ, find the voltage gain v_y/v_s.

(e) If Y is connected to ground, X to an input signal source v_s having a 100-kΩ source resistance, and Z to a load resistance of 1 kΩ, find the voltage gain v_z/v_s.

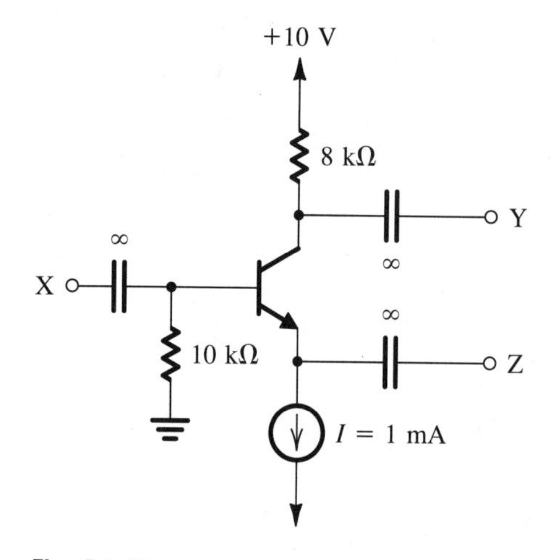

Fig. P4.49

4.50 In the design of BJT integrated-circuit amplifiers the arrangement shown in Fig. P4.50 is often used. Here the transistor is biased with a constant-current source feeding the collector. The external circuit (not shown) is arranged so that a stable dc voltage develops at the collector. Using the hybrid-π equivalent circuit model, including r_o, show that the small-signal volt-

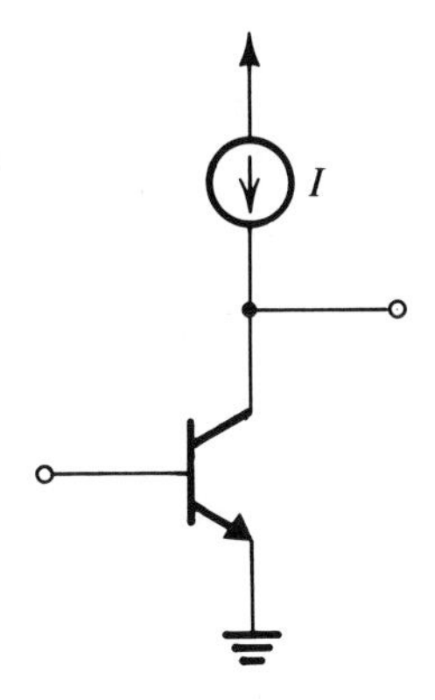

Fig. P4.50

age gain obtained from base to collector is equal to $-(V_A/V_T)$. Find the value of the gain for $V_A = 100$ V.

Section 4.9: Graphical Analysis

4.51 Consider the characteristic curves shown in Fig. 4.36 with the following additional calibration data: From the lowest colored line, $i_B = 1$, 10, 20, 30, and 40 μA. Assume the lines to be horizontal, and let $\beta = 100$. For $V_{CC} = 5$ V, $R_C = 1$ kΩ what peak-to-peak collector voltage swing will result for i_B varying over the range 10 to 40 μA? If at the bias point (not the one shown in the figure) $V_{CE} = \frac{1}{2}V_{CC}$, find the value of I_C and I_B. If at this current $V_{BE} = 0.7$ V, and if $R_B = 100$ kΩ, find the required value of V_{BB}.

*4.52 Sketch the i_C–v_{CE} characteristics for an *npn* transistor having $\beta = 100$ and $V_A = 100$ V. Sketch characteristic curves for $i_B = 20$, 50, 80, and 100 μA. For the purpose of this sketch assume that $i_C = \beta i_B$ at $v_{CE} = 0$. Also sketch the load line obtained for $V_{CC} = 10$ V and $R_C = 1$ kΩ. If the dc bias current into the base is 50 μA, write the equation for the corresponding i_C–v_{CE} curve. Also write the equation for the load line, and solve the two equations to obtain V_{CE} and I_C. If the input signal causes a sinusoidal signal of 30-μA peak amplitude to be superimposed on I_B, find the corresponding signal components of i_C and v_{CE}.

Section 4.10: Biasing the BJT for Discrete-Circuit Design

D4.53 Design the circuit of Fig. 4.38 for the case $V_{CC} = 9$ V to provide $\frac{1}{3}V_{CC}$ across each of R_E and R_C, $I_E = 0.5$ mA, and the current in the voltage divider $0.2\, I_E$. Design assuming a very large β, then find the actual value obtained for I_E with a BJT having $\beta = 100$.

D**4.54 It is required to design the bias circuit of Fig. 4.38 for a BJT whose nominal $\beta = 100$.

(a) Find the largest ratio (R_B/R_E) that will guarantee I_E to remain within $\pm 5\%$ of its nominal value for β as low as 50 and as high as 150.

(b) If the resistance ratio found in (a) is used, find an expression for the voltage $V_{BB} \equiv V_{CC}\, R_2/(R_1 + R_2)$ that will result in a voltage drop of $V_{CC}/3$ across R_E.

(c) For $V_{CC} = 10$ V, find the required values of R_1, R_2, and R_E to obtain $I_E = 2$ mA and to satisfy the requirement for stability of I_E in (a).

(d) Find R_C so that $V_{CE} = 3$ V for β equal to its nominal value.

D4.55 Design a version of the circuit in Fig. 4.39 using ± 9-V supplies for operation at 10 mA (for high β), such that the total variation in I_E is less than 5% for β as low as 50. To obtain the highest possible voltage gain, select the largest possible value for R_C; however, you must ensure that v_{CE} is never less than 2 V.

D*4.56 Utilizing ± 5-V power supplies, it is required to design a version of the circuit in Fig. 4.39 in which the signal will be coupled to the emitter and thus R_B can be set to zero. Find values for R_E and R_C so that a dc emitter current of 1 mA is obtained and so that the gain to the collector is maximized while allowing ± 1 V of signal swing at the collector. If temperature increases from the nominal value of 25°C to 125°C, estimate the percentage change in collector bias current and reduction in signal swing. In addition to the -2mV/°C change in V_{BE}, assume that the transistor β changes over this temperature range from 50 to 150.

D4.57 Using a 5-V power supply design a version of the circuit of Fig. 4.40 to provide a dc emitter current of 0.5 mA and to allow a ± 1-V signal swing at the collector. The BJT has a nominal $\beta = 100$. If the actual BJT used has $\beta = 50$, what emitter current is obtained? Also, what is the allowable signal swing at the collector? Repeat for $\beta = 150$.

Section 4.11: Basic Single-Stage BJT Amplifier Configurations

4.58 For the common-emitter amplifier shown in Fig. P4.58, let $V_{CC} = 9$ V, $R_1 = 27$ kΩ, $R_2 = 15$ kΩ, $R_E = 1.2$ kΩ, and $R_C = 2.2$ kΩ. The transistor has $\beta = 100$ and $V_A = 100$ V. Calculate the dc bias current I_E. If the amplifier operates between a source for

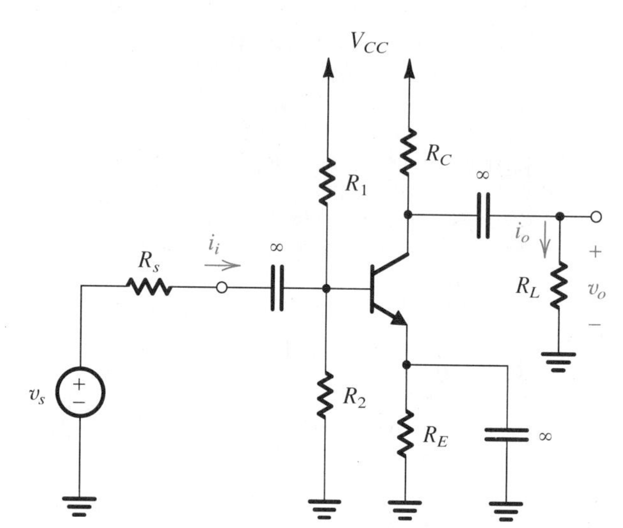

Fig. P4.58

which $R_s = 10$ kΩ and a load of 2 kΩ, replace the transistor with its hybrid-π model, and draw the amplifier equivalent circuit. Find the values of R_i, G_m, R_o, A_v, and A_i.

D4.59 Using the topology of Fig. P4.58, design an amplifier to operate between a 10-kΩ source and a 2-kΩ load with a gain v_o/v_s of -8 V/V. The power supply available is 9 V. Use an emitter current of about 2 mA and a current of about one-tenth that in the voltage divider that feeds the base, with V_{BB} of about one-third of the supply. The transistor available has $\beta = 100$ and $V_A = 100$ V.

4.60 A designer, having examined the situation described in Problem 4.58, and estimating the available gain to be about -8 V/V, wishes to explore the possibility of improvement by reducing the loading of the source by the amplifier input. As an experiment, he varies the resistance levels by a factor of about 3: R_1 to 82 kΩ, R_2 to 47 kΩ, R_E to 3.6 kΩ, and R_C to 6.8 kΩ (standard values of 5%-tolerance resistors). With $V_{CC} = 9$ V, $R_s = 10$ kΩ, $R_L = 2$ kΩ, $\beta = 100$, and $V_A = 100$ V, what does the gain become?

D4.61 Use the approximate voltage gain expression given in Table 4.2 for the CE amplifier of Fig. 4.43 to find the dc collector current that will result in an overall voltage gain of 10. The source has a resistance of 10 kΩ and the load resistance also is 10 kΩ. The BJT has $\beta = 100$ and $V_A = 100$ V. Design for a dc voltage drop across R_C of 4 V. Note that there are two possible values for I_C. It is known that the BJT maintains its specified β over the collector current range of 0.1 to 10 mA only, select an appropriate value for I_C. Select an appropriate value for R_B that will not change the input resistance by more than 10%. Using ±5 V power supplies find values for R_E and R_C.

D4.62 In the circuit of Fig. P4.62, v_s is a small sine-wave signal with zero average. The transistor β is 100.

(a) Find the value of R_E to establish a dc emitter current of about 1 mA.

(b) Find R_C to establish a dc collector voltage of about +5 V.

(c) For $R_L = 5$ kΩ and the transistor $r_o = 100$ kΩ, draw the small-signal equivalent circuit of the amplifier and determine its overall voltage gain.

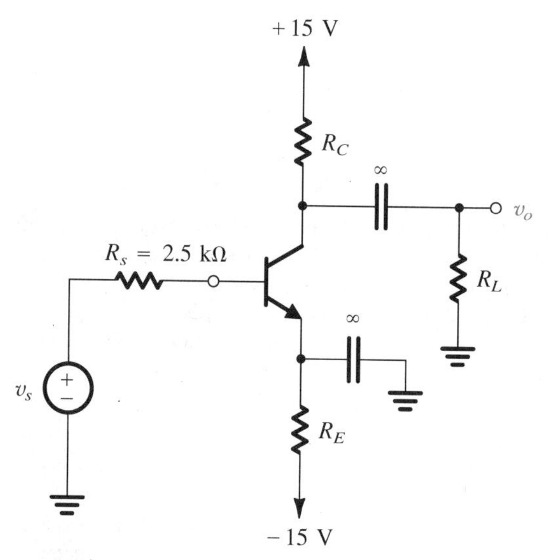

Fig. P4.62

***4.63** The amplifier of Fig. P4.63 consists of two identical common-emitter amplifiers connected in cascade. Observe that the input resistance of the second stage, R_{in2}, constitutes the load resistance of the first stage.

(a) For $V_{CC} = 15$ V, $R_1 = 100$ kΩ, $R_2 = 47$ kΩ, $R_E = 3.9$ kΩ, $R_C = 6.8$ kΩ, and $\beta = 100$, determine the dc collector current and collector voltage of each transistor.

(b) Draw the small-signal equivalent circuit of the entire amplifier and give the values of all its components.

(c) Find R_{in1} and v_{b1}/v_s for $R_s = 5$ kΩ.

(d) Find R_{in2} and v_{b2}/v_{b1}.

(e) For $R_L = 2$ kΩ, find v_o/v_{b2}.

(f) Find the overall voltage gain v_o/v_s.

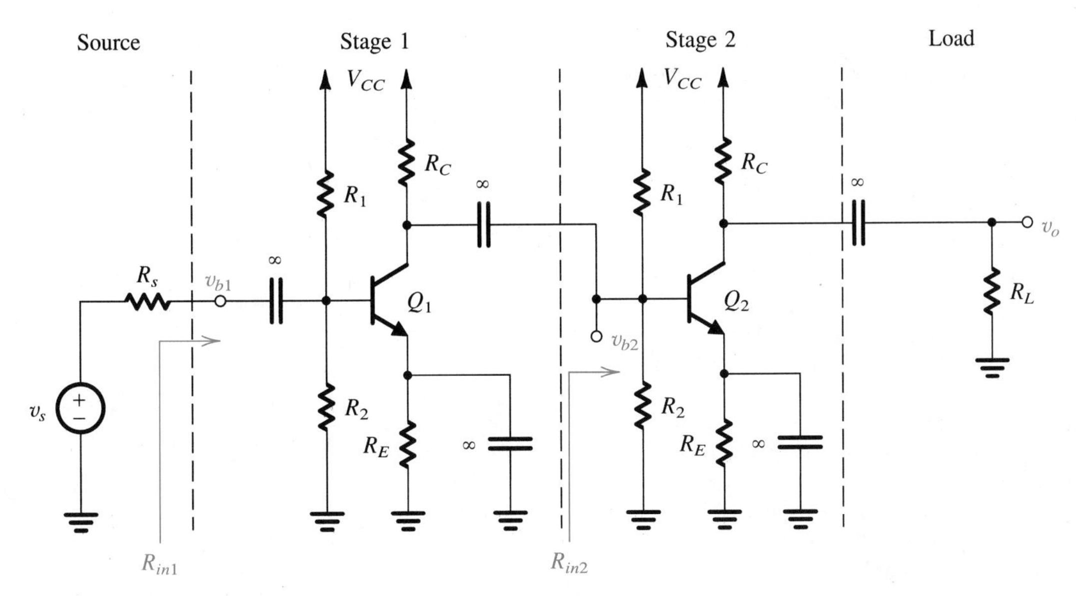

Fig. P4.63

4.64 In the circuit of Fig. P4.64, v_s is a small sine-wave signal with zero average value. Find R_{in} and the gain v_o/v_s. Assume $\beta = 50$. If the amplitude of the signal v_{be} is to be limited to 5 mV, what is the largest signal at the input? What is the corresponding signal at the output?

*4.65 The BJT in the circuit of Fig. P4.65 has $\beta = 100$.

(a) Find the dc collector current and the dc voltage at the collector.

(b) Replacing the transistor by its T model, draw the small-signal equivalent circuit of the amplifier. Analyze the resulting circuit to determine the voltage gain v_o/v_i.

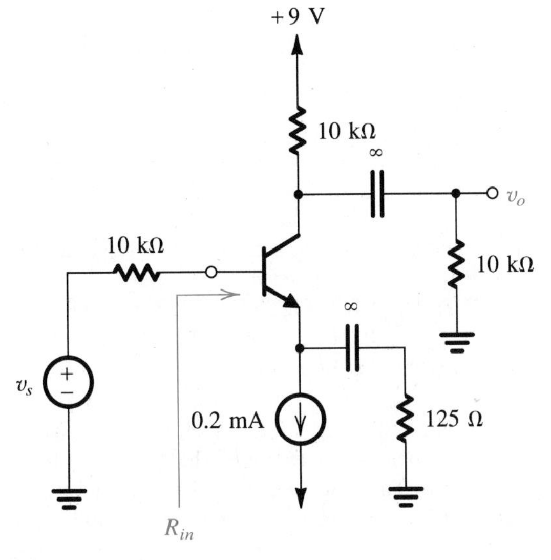

Fig. P4.64

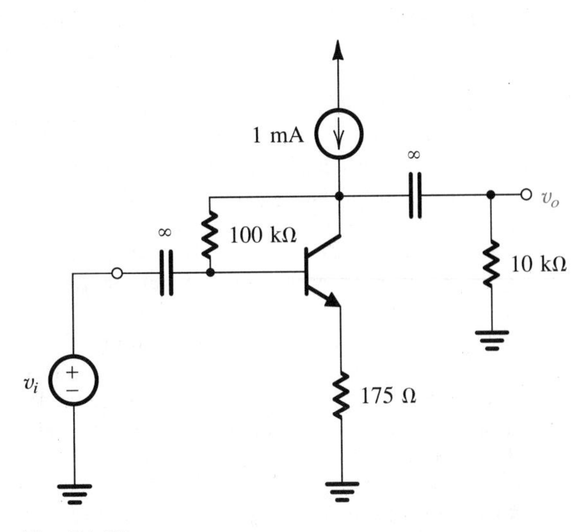

Fig. P4.65

4.66 Replace the BJT in the amplifier of Fig. 4.44(a) with its T model and neglect r_o. Analyze the circuit to show that

$$R_{ib} = (\beta + 1)(r_e + R_e)$$

$$\frac{v_o}{v_i} = -\frac{\alpha(R_C // R_L)}{r_e + R_e}$$

If the transistor is biased at a dc collector current of 0.5 mA and has $\beta = 100$, find the value of R_e that will result in an input resistance R_{ib} of 10 kΩ. If $R_L = 2$ kΩ, what should R_C be to obtain a gain v_o/v_i of -10?

****4.67** Refer to the approximate voltage-gain expressions (in terms of the transistor β) given in Table 4.2 for the CE amplifier with and without a resistance R_e in the emitter. Let the BJT be biased at an emitter current of 1 mA. The source resistance R_s is 10 kΩ. The BJT β is specified to lie in the range 50 to 150 with a nominal value of 100.

(a) What is the ratio of maximum to minimum voltage gain obtained without R_e?

(b) What value of R_e should be used to limit the ratio of maximum to minimum gain to 1.2?

(c) If the R_e found in (b) is used, by what factor is the gain reduced (compared to the case without R_e) for a BJT with a nominal β?

D4.68** Consider the CB amplifier of Fig. 4.45 with R_B set to zero. Let the power supplies be ±5 V. The source has a resistance of 50 Ω, and the load resistance is 1 kΩ. Design the circuit so that the amplifier input resistance is matched to that of the source, and the output signal swing is as large as possible with relatively low distortion (v_{be} limited to 10 mV). Find R_E and R_C and calculate the overall voltage gain obtained and the output signal swing.

4.69 For the circuit in Fig. P4.69 find the input resistance R_i and the voltage gain v_o/v_s. Assume that the source provides a small signal v_s and that β is high.

4.70 For the common-base amplifier with $R_s \gg r_e$ and $R_L \ll R_C$, find an approximate expression for the voltage gain v_o/v_s.

4.71 For the emitter-follower circuit shown in Fig. P4.71 the BJT used is specified to have β values in the range of 20 to 200 (a distressing situation for the circuit designer). For the two extreme values of β ($\beta = 20$ and $\beta = 200$) find:

(a) I_E, V_E, and V_B.

(b) the input resistance R_i.

(c) the voltage gain v_o/v_s.

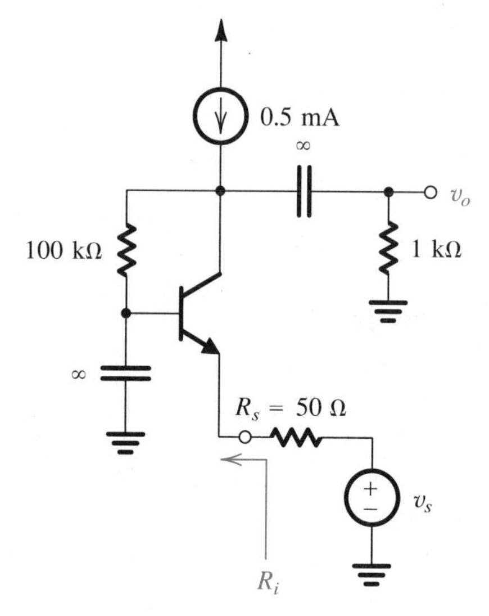

Fig. P4.69

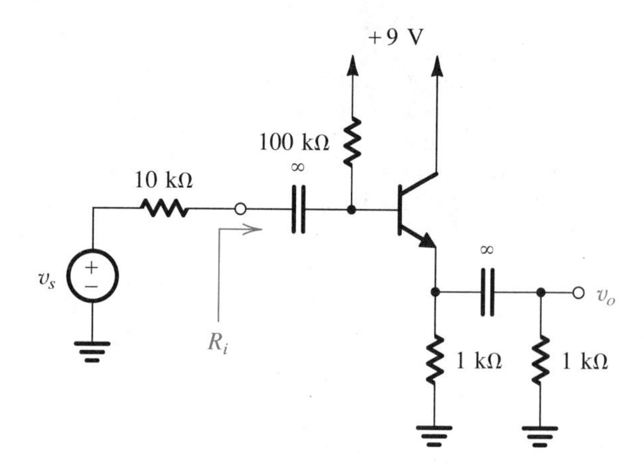

Fig. P4.71

4.72 In the emitter follower in Fig. P4.72, the signal source is directly coupled to the transistor base. If the dc component of v_s is zero, find the dc emitter current. Assume $\beta = 120$. Neglecting r_o, find R_i, the voltage gain v_o/v_s, the current gain i_o/i_i, and the output resistance R_o.

***4.73** In the emitter follower of Fig. P4.73, the signal source is directly connected to the base. The dc component of v_s is zero. The transistor has $\beta = 100$ and $V_A = 125$ V. What is the output resistance of the follower? Find the gain v_o/v_s with no load and with a load of 1 kΩ. With the 1-kΩ load connected, find the largest

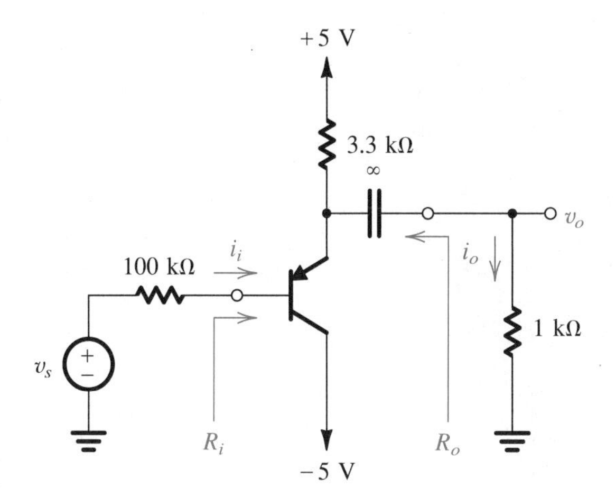

Fig. P4.72

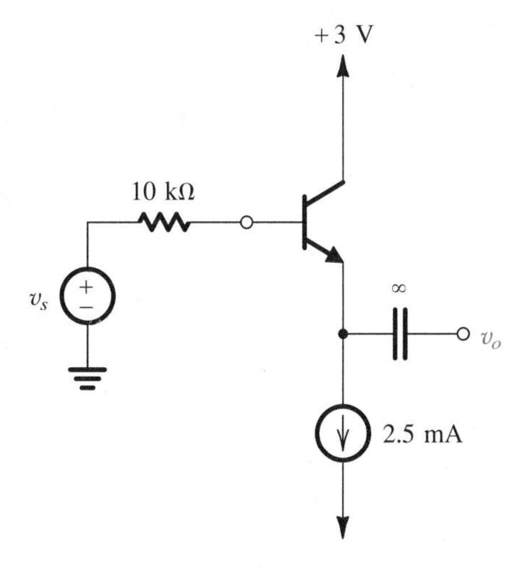

Fig. P4.73

possible negative output signal. What is the largest possible positive output signal if operation is satisfactory up to the point that the base–collector junction is forward biased by 0.2 V?

*4.74 Consider the emitter follower of Fig. 4.46(a) with R_C set to zero, $V_{CC} = V_{EE} = 5$ V, $R_B = 10$ kΩ, and $R_E = 330$ Ω. Let the BJT have $\beta = 100$ and $V_A = 100$ V. The source resistance is 6.8 kΩ and the load is 600 Ω.

(a) Find the dc emitter current and the dc voltage at the emitter.

(b) Find R_i, v_o/v_s, i_o/i_i, and R_o.

(c) What is the maximum possible negative output voltage?

(d) What is the maximum possible positive output voltage? Assume that the BJT enters saturation when v_{CE} is reduced below about 0.5 V.

(e) For acceptable linearity v_{be} is to be limited to 10 mV. What is the corresponding amplitude of the output sinusoid? of the input sinusoid?

*4.75 Consider an emitter follower for which the effect of the bias resistor R_B on signal performance can be neglected. When driven from a 10-kΩ source the open-circuit voltage gain was found to be 0.98 and the output resistance 100 Ω. The output resistance increased to 175 Ω when the source resistance was increased to 20 kΩ. Find the voltage gain when the follower is driven by a 30-kΩ source and loaded by a 1-kΩ resistor.

4.76 For the circuit in Fig. P4.76, called a **bootstrapped follower:

(a) Find the dc emitter current and g_m, r_e, and r_π. Use $\beta = 100$.

(b) Replace the BJT with its hybrid-π model (neglecting r_o) and analyze the circuit to determine the input resistance R_i and the voltage gain v_o/v_s.

(c) Repeat (b) for the case when capacitor C_B is open circuited. Compare the results with those obtained in (b) to find the advantages of bootstrapping.

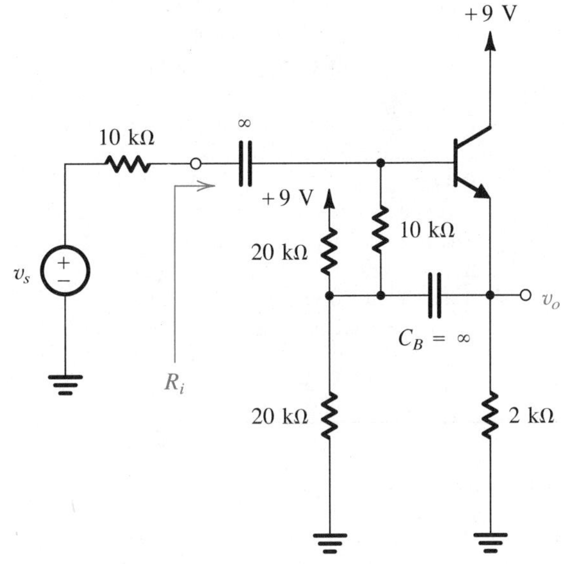

Fig. P4.76

**4.77 For the follower circuit in Fig. P4.77 let transistor Q_1 have $\beta = 20$ and transistor Q_2 have $\beta = 200$, and neglect the effect of r_o. Use $V_{BE} = 0.7$ V.

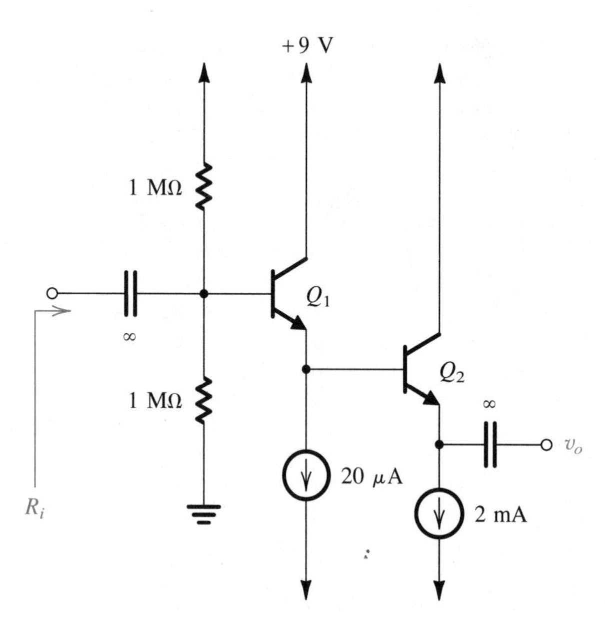

Fig. P4.77

(a) Find the dc emitter current of Q_1 and Q_2. Also find the dc voltages V_{B1} and V_{B2}.

(b) If a load resistance $R_L = 1$ kΩ is connected to the output terminal, find the voltage gain from the base to the emitter of Q_2, v_o/v_{b2}, and find the input resistance looking into the base of Q_2, R_{ib2}. (*Hint:* Consider Q_2 as an emitter follower fed by a voltage v_{b2} at its base.)

(c) Replacing Q_2 with its input resistance R_{ib2} found in (b), analyze the circuit of emitter follower Q_1 to determine its input resistance R_i, and the gain from its base to its emitter, v_{e1}/v_{b1}.

(d) If the circuit is fed with a source having a 100-kΩ resistance, find the transmission to the base of Q_1, v_{b1}/v_s.

(e) Find the overall voltage gain v_o/v_s.

Section 4.12: The Transistor as a Switch—Cutoff and Saturation

D4.78 For the circuit in Fig. P4.78 select a value for R_B so that the transistor saturates with an overdrive factor of 10. The BJT is specified to have a minimum β of 30 and $V_{CEsat} = 0.3$ V. What is the value of forced β obtained?

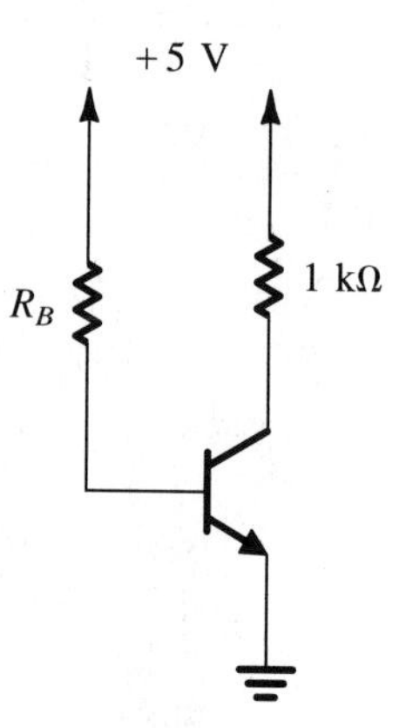

Fig. P4.78

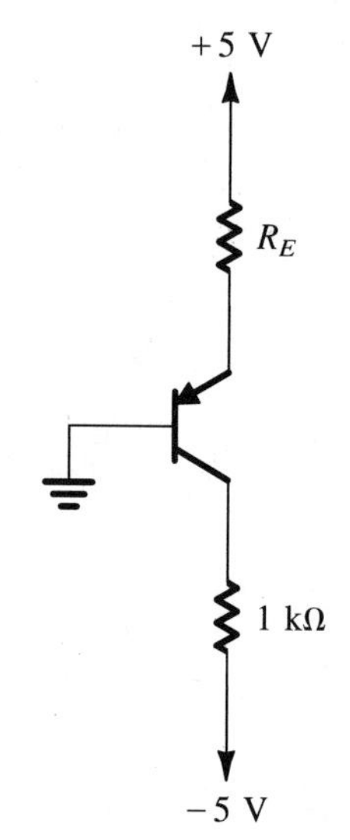

Fig. P4.79

D4.79 For the circuit in Fig. P4.79 select a value for R_E so that the transistor saturates with a forced β of 5.

4.80 For the circuit in Fig. P4.80 find V_B, V_E, and V_C for $R_B = 100$ kΩ, 10 kΩ, and 1 kΩ. Let $\beta = 100$.

4.81 For the circuit in Fig. P4.81 find V_B and V_E for $v_I = 0$, +3 V, −5 V, and −10 V. The BJTs have $\beta = 100$.

*4.82 Using the three-terminal-short model for a saturated transistor, find the approximate collector voltages in the circuits of Fig. P4.82. Also calculate the forced β for each of the transistors.

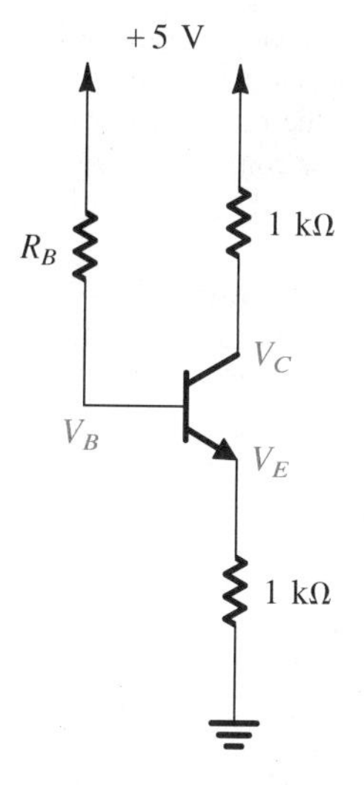

Fig. P4.80

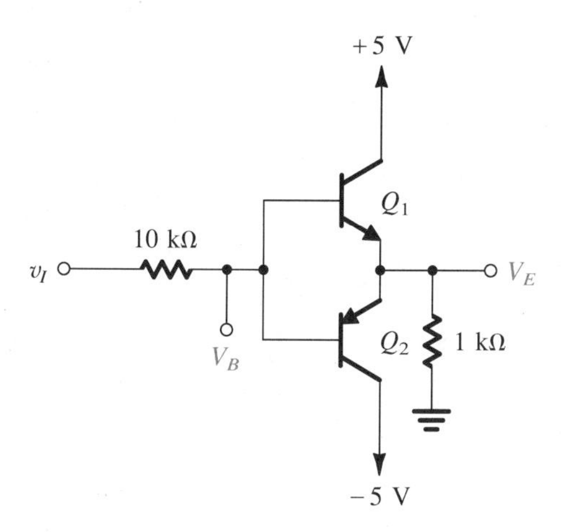

Fig. P4.81

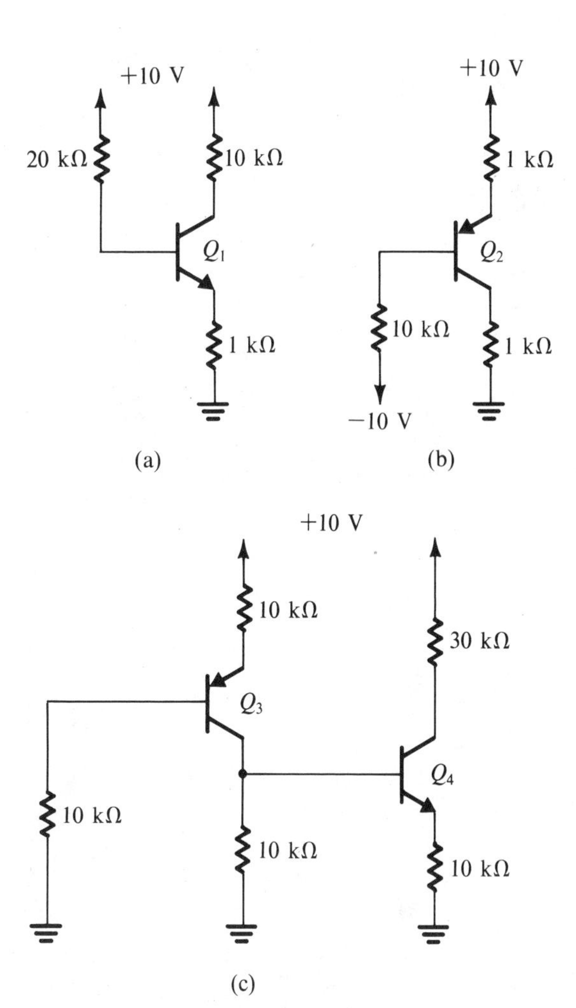

Fig. P4.82

Section 4.13: Complete Static Characteristics and Second-Order Effects

*4.83 Use the hybrid-π model of Fig. 4.58 to obtain an expression for the output resistance of the BJT in the common-base configuration. To do that, ground the base, leave the emitter open-circuited, and apply a test voltage v_x between the collector and ground. Find the current i_x drawn from v_x, and show that the output resistance v_x/i_x is $r_\mu // [r_\pi + (\beta + 1)r_o]$, which is approximately $r_\mu // \beta r_o$.

4.84 Use the hybrid-π model of Fig. 4.58 to obtain an expression for the output resistance of the BJT in the common-emitter configuration when the base is fed with a constant current source (see Fig. 4.59). To do that, ground the emitter, leave the base open-circuited, and apply a test voltage v_x between the collector and ground. Find the current i_x drawn from v_x, and show that the output resistance v_x/i_x is given by $r_o // [(r_\mu + r_\pi)/\beta]$, which is approximately $r_o // (r_\mu/\beta)$.

4.85 Consider a transistor modeled as in Fig. 4.58 with $r_\mu = 10\beta r_o$, and the base driven from a constant current source that produces a collector current of 2.5 mA. The collector voltage is raised by 10 V (without breakdown). What increase in collector current would you expect if $V_A = 200$ V? (*Hint:* The output resistance of a common-emitter transistor whose base is fed with a constant-current source is approximately $r_o // (r_\mu / \beta)$.

*4.86 In the circuit of Fig. P4.86, transistor Q_1 is intended to act as a switch to connect the input and output terminals. The status of the switch (open or closed) is controlled by the voltage v_C through diode D_1, which is identical to the base–emitter junction of Q_1. For $v_I = -2$ V, what value of v_C ensures that the switch closes? opens? For $\beta = 100$, what value of I is minimally necessary? What value of I establishes a forced β of 5? Assuming $V_{CEsat} \leq 0.1$ V what value of v_O might you expect? What current flows in the input?

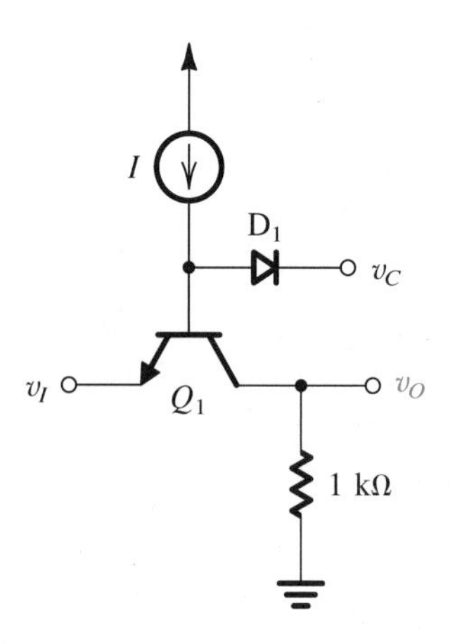

Fig. P4.86

4.87 A BJT switch characterized by an offset voltage of 50 mV and a saturation resistance of 70 Ω is used to connect a circuit node to ground. The Thévenin equivalent at this node is 4.4 V and 2.3 kΩ. What node voltage would you expect with the switch closed? Between this circuit node and ground there is a capacitance of 0.023 μF; thus when the switch is opened, the node voltage rises exponentially from the closure voltage you found to 4.4 V. How long does it take for the voltage to settle to within 0.1 V of its final value?

4.88 A BJT operating at $i_B = 8$ μA and $i_C = 1.2$ mA undergoes a reduction in base current of 0.8 μA. It is found that when v_{CE} is held constant the corresponding reduction in collector current is 0.1 mA. What are the values of h_{FE} and h_{fe} that apply? If the base current is increased from 8 μA to 10 μA and v_{CE} is increased from 8 to 10 V, what collector current results? (Assume $V_A = 100$ V and neglect the effect of r_μ.)

4.89 A BJT for which BV_{CBO} is 30 V is connected as shown in Fig. P4.89. What voltages would you measure on the collector, base, and emitter?

*4.90 In the circuit of Fig. P4.90, for modest values of the voltage V_C, V_E is 5 V. Then as V_C is raised, V_E begins to follow. For $V_C = 40$ V, $V_E = 15$ V. What is happening? Characterize the transistor as completely as you can.

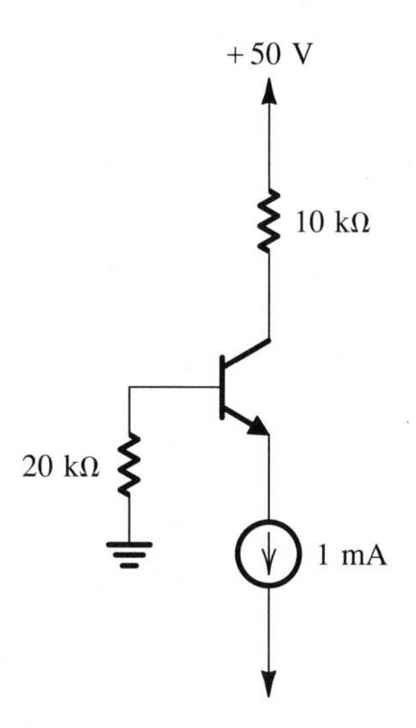

Fig. P4.89

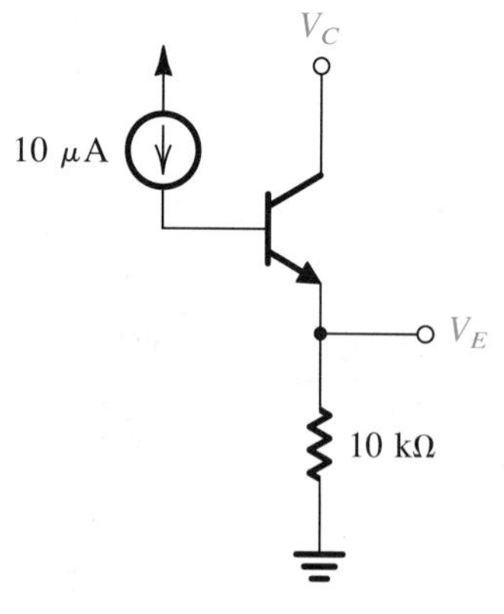

Fig. P4.90

General Problems

*4.91 For the two-terminal circuit of Fig. P4.91, known as the V_{BE}-multiplier, assume I to be sufficiently large and β to be very high and find expressions for the dc voltage drop V and the incremental resistance between X and ground. Find the values of V and the incremental resistance for $R_1 = R_2 = 1\ \text{k}\Omega$, $I = 10$ mA, and $V_{BE} = 0.7$ V. Repeat for $\beta = 100$. In general, what value of I is "sufficiently large" to meet the conditions implied?

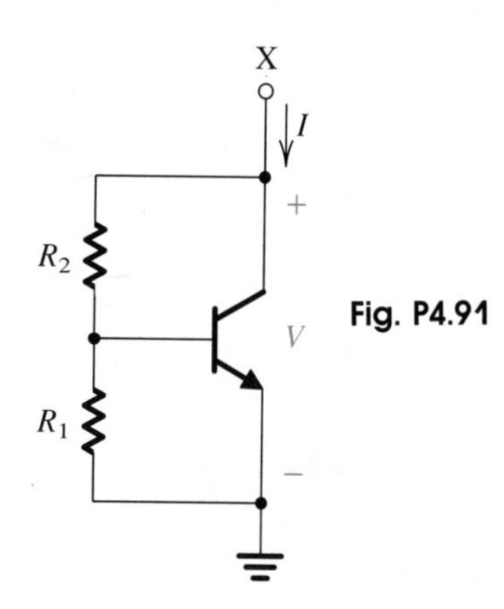

Fig. P4.91

CHAPTER 5

Field-Effect Transistors (FETs)

INTRODUCTION

In this chapter we study the other major type of transistor, the field-effect transistor (FET). As in the case of the BJT (Chapter 4), the voltage between two terminals of the FET controls the current flow in the third terminal. Correspondingly, the FET can be used both as an amplifier and as a switch.

The field-effect transistor derives its name from the essence of its physical operation. Specifically, it will be shown that the current-control mechanism is based on an electric field established by the voltage applied to the control terminal. It will also be shown that current is conducted by only one type of carrier (electrons or holes) depending on the type of FET (*n* channel or *p* channel), which gives the FET another name, the **unipolar transistor.**

Although the basic concept of the FET has been known since the 1930s, the device became a practical reality only in the 1960s. Since the late 1970s, a particular kind of FET, the **metal-oxide semiconductor field-effect transistor** (MOSFET), has been extremely popular. Compared to BJTs, MOS transistors can be made quite small (that is, occupying a small silicon area on the IC chip), and their manufacturing process is relatively simple (see Appendix A). Furthermore, digital logic and memory functions can be implemented with circuits that exclusively use MOSFETs (that is, no resistors or diodes are needed). For these reasons, most very-large-scale integrated (VLSI) circuits are made at the present time using MOS technology. Examples include microprocessor and memory chips. MOS technology has also been applied extensively in the design of analog integrated circuits.

Although the FET family of devices has many different types, and we shall discuss a few, most of the chapter is devoted to the enhancement-type MOSFET, which is by far the most significant field-effect transistor. Its significance is on par with that of the bipolar junction transistor, with each having its own areas of application.

Field-effect transistors are available in discrete form, and we shall study their application in discrete circuit design. Their most important use, however, is in the design of integrated circuits. The study of MOS integrated circuits begins in this chapter and is continued in subsequent chapters.

5.1 STRUCTURE AND PHYSICAL OPERATION OF THE ENHANCEMENT-TYPE MOSFET

The enhancement-type MOSFET is the most widely used field-effect transistor. In this section we shall study its structure and physical operation. This will lead to the current-voltage characteristics of the device, studied in the next section.

Device Structure

Figure 5.1 shows the physical structure of the *n*-channel enhancement-type MOSFET. The meaning of the names "enhancement" and "*n*-channel" will become apparent shortly. The transistor is fabricated on a *p*-type substrate, which is a single-crystal silicon wafer that provides physical support for the device (and for the entire circuit in the case of an integrated circuit). Two heavily doped *n*-type regions, indicated in the figure as the n^+ **source**[1] and the n^+ **drain** regions, are created in the substrate. A thin (about 0.1 μm) layer[2] of silicon dioxide (SiO_2), which is an excellent electrical insulator, is grown on the surface of the substrate, covering the area between the source and drain regions. Metal is deposited on top of the oxide layer to form the **gate electrode** of the device. Metal contacts are also made to the source region, the drain region, and the substrate, also known as the **body.** Thus, four terminals are brought out: the gate terminal (G), the source terminal (S), the drain terminal (D), and the substrate or body terminal (B).

At this point it should be clear that the name of the device (metal-oxide-semiconductor FET) is derived from its physical structure. The name, however, has become a general one and is used also for FETs that do not use metal for the gate electrode. In fact, most

[1]The notation n^+ indicates heavily doped *n*-type silicon. Conversely, n^- is used to denote lightly doped *n*-type silicon. Similar notation applies for *p*-type silicon.

[2]A micrometer (μm), also called micron, is 10^{-6} m.

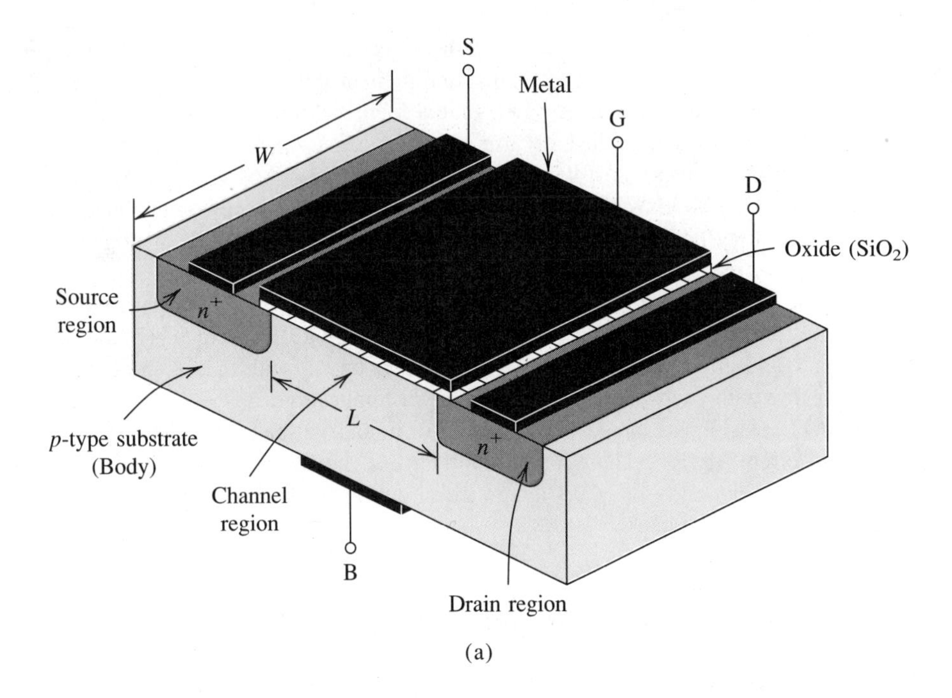

(a)

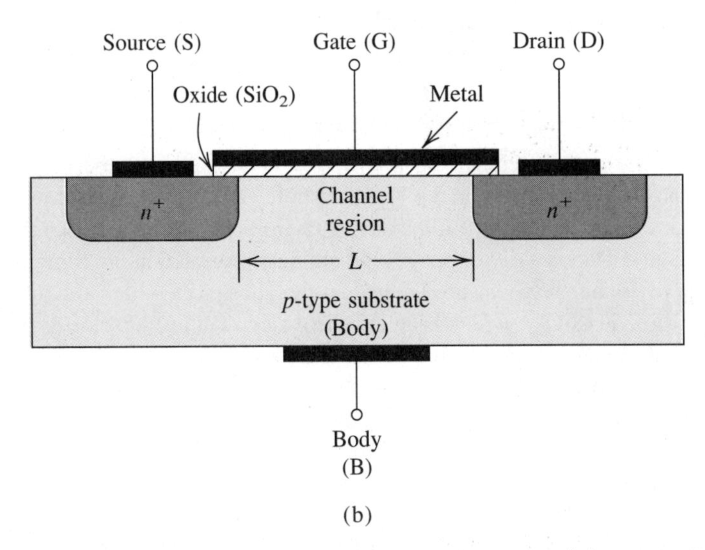

(b)

Fig. 5.1 Physical structure of enhancement-type NMOS transistor: **(a)** perspective view; **(b)** cross section. Typically $L = 1$ to $10\ \mu$m, $W = 2$ to $500\ \mu$m, and the thickness of the oxide layer is of the order of $0.1\ \mu$m.

modern MOSFETs are fabricated using a process known as silicon-gate technology, in which a certain type of silicon, called polysilicon, is used to form the gate electrode (see Appendix A). Our description of MOSFET operation and characteristics apply irrespective of the type of gate electrode.

Another name for the MOSFET is the **insulated-gate FET** or **IGFET.** This name also arises from the physical structure of the device, emphasizing the fact that the gate electrode is electrically insulated from the device body (by the oxide layer). It is this insulation that causes the current in the gate terminal to be extremely small (of the order of 10^{-15} A).

Observe that the substrate forms *pn* junctions with the source and drain regions. In normal operation these *pn* junctions are kept reverse-biased at all times. Since the drain will be at a positive voltage relative to the source, the two *pn* junctions can be effectively cut off by simply connecting the substrate terminal to the source terminal. We shall assume this to be the case in the following description of MOSFET operation. Thus, here, the substrate will be considered as having no effect on device operation, and the MOSFET will be treated as a three-terminal device, with the terminals being the gate (G), the source (S), and the drain (D). It will be shown that a voltage applied to the gate controls current flow between source and drain. This current will flow in the longitudinal direction from drain to source in the region labeled ''channel region.'' Note that this region has a length L and a width W, two important parameters of the MOSFET. Typically, L is in the range 1 to 10 μm, and W is in the range 2 to 500 μm. Finally note that, unlike the BJT, the MOSFET is normally constructed as a symmetrical device. Thus its source and drain can be interchanged with no change in device characteristics.

Operation with No Gate Voltage

With no bias voltage applied to the gate, two back-to-back diodes exist in series between drain and source. One diode is formed by the *pn* junction between the n^+ drain region and the *p*-type substrate, and the other diode is formed by the *pn* junction between the *p*-type substrate and the n^+ source region. These back-to-back diodes prevent current conduction from drain to source when a voltage v_{DS} is applied. In fact, the path between drain and source has a very high resistance (of the order of 10^{12} Ω).

Creating a Channel for Current Flow

Consider next the situation depicted in Fig. 5.2. Here we have grounded the source and the drain and applied a positive voltage to the gate. Since the source is grounded, the gate voltage appears in effect between gate and source and thus is denoted v_{GS}. The positive voltage on the gate causes in the first instance the free holes (which are positively charged) to be repelled from the region of the substrate under the gate (the channel region). These holes are pushed downward into the substrate, leaving behind a carrier-depletion region. The depletion region is populated by the bound negative charge associated with the acceptor atoms. These charges are ''uncovered'' because the neutralizing holes have been pushed downward into the substrate.

As well, the positive gate voltage attracts electrons from the n^+ source and drain regions (where they are in abundance) into the channel region. When a sufficient number of electrons accumulate near the surface of the substrate under the gate, an *n* region is in

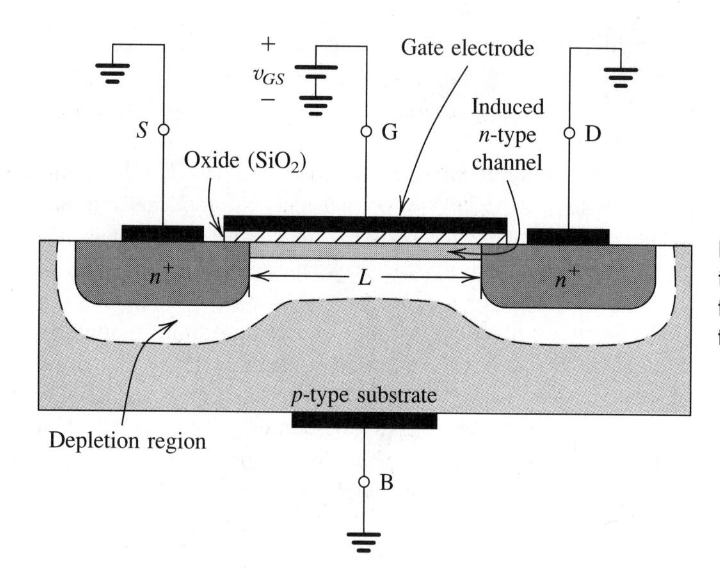

Fig. 5.2 The enhancement-type NMOS transistor with a positive voltage applied to the gate. An n channel is induced at the top of the substrate beneath the gate.

effect created, connecting the source and drain regions, as indicated in Fig. 5.2. Now if a voltage is applied between drain and source, current flows through this induced n region, carried by the mobile electrons. The induced n region thus forms a **channel** for current flow from drain to source and is aptly called so. Correspondingly, the MOSFET of Fig. 5.2 is called an ***n*-channel MOSFET** or, alternatively, an **NMOS transistor.** Note that an n-channel MOSFET is formed in a p-type substrate: The channel is created by inverting the substrate surface from p type to n type. Hence the induced channel is called an **inversion layer.**

The value of v_{GS} at which a sufficient number of mobile electrons accumulate in the channel region to form a conducting channel is called the **threshold voltage** and is denoted V_t.[3] Obviously, V_t for an n-channel FET is positive. The value of V_t is controlled during device fabrication and typically lies in the range 1 to 3 V.

The gate and body of the MOSFET form a parallel-plate capacitor with the oxide layer acting as the capacitor dielectric. The positive gate voltage causes positive charge to accumulate on the top plate of the capacitor (the gate electrode). The corresponding negative charge of the bottom plate is formed by the electrons in the induced channel. An electric field thus develops in the vertical direction. It is this field that controls the amount of charge in the channel, and thus it determines the channel conductivity and, in turn, the current that will flow through the channel when a voltage v_{DS} is applied.

Applying a Small v_{DS}

Having induced a channel, we now apply a positive voltage v_{DS} between drain and source, as shown in Fig. 5.3. We first consider the case where v_{DS} is small (say, 0.1 or 0.2 V).

[3]Many texts use V_T to denote the threshold voltage. We use V_t to avoid confusion with the thermal voltage V_T.

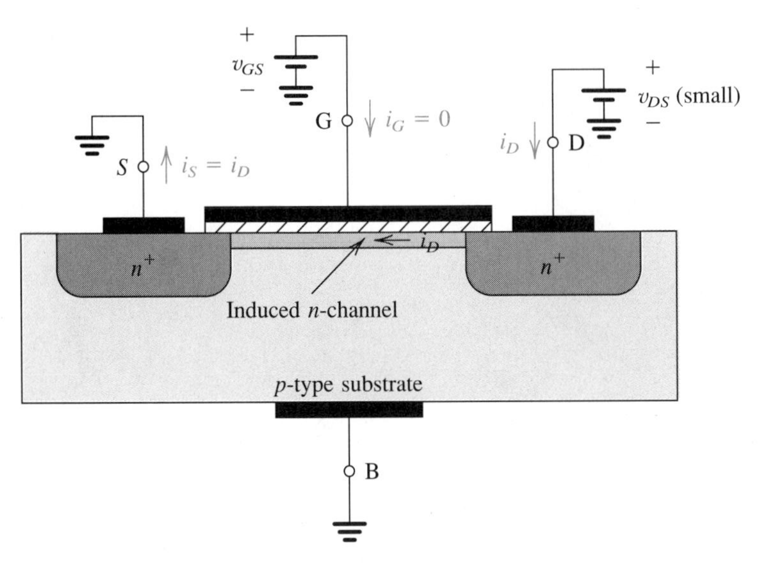

Fig. 5.3 An NMOS transistor with $v_{GS} > V_t$ and with a small v_{DS} applied. The device acts as a conductance whose value is determined by v_{GS}. Specifically, the channel conductance is proportional to $v_{GS} - V_t$, and thus i_D is proportional to $(v_{GS} - V_t)v_{DS}$. Note that the depletion region is not shown (for simplicity).

The voltage v_{DS} causes a current i_D to flow through the induced n channel. Current is carried by free electrons traveling from source to drain (hence the names source and drain). The magnitude of i_D depends on the density of electrons in the channel, which in turn depends on the magnitude of v_{GS}. Specifically, for $v_{GS} = V_t$ the channel is just induced and the current conducted is still negligibly small. As v_{GS} exceeds v_t, more electrons are attracted into the channel. We may visualize the increase in charge carriers in the channel as an increase in the channel depth. The result is a channel of increased conductance or equivalently reduced resistance. In fact, the conductance of the channel is proportional to the **excess gate voltage** $(v_{GS} - V_t)$. It follows that the current i_D will be proportional to $v_{GS} - V_t$ and, of course, to the voltage v_{DS} that causes i_D to flow.

Figure 5.4 shows a sketch of i_D versus v_{DS} for various values of v_{GS}. We observe that the MOSFET is operating as a linear resistance whose value is controlled by v_{GS}. The resistance is infinite for $v_{GS} \leq V_t$, and its value decreases as v_{GS} exceeds V_t.

The description above indicates that for the MOSFET to conduct, a channel has to be induced. Then, increasing v_{GS} above the threshold voltage V_t enhances the channel, hence the names **enhancement-mode operation** and **enhancement-type MOSFET.** Finally, we note that the current that leaves the source terminal (i_S) is equal to the current that enters the drain terminal (i_D), and the gate current $i_G = 0$.

Operation as v_{DS} is Increased

We next consider the situation as v_{DS} is increased. For this purpose let v_{GS} be held constant at a value greater than V_t. Refer to Fig. 5.5, and note that v_{DS} appears as a voltage drop across the length of the channel. That is, as we travel along the channel from source

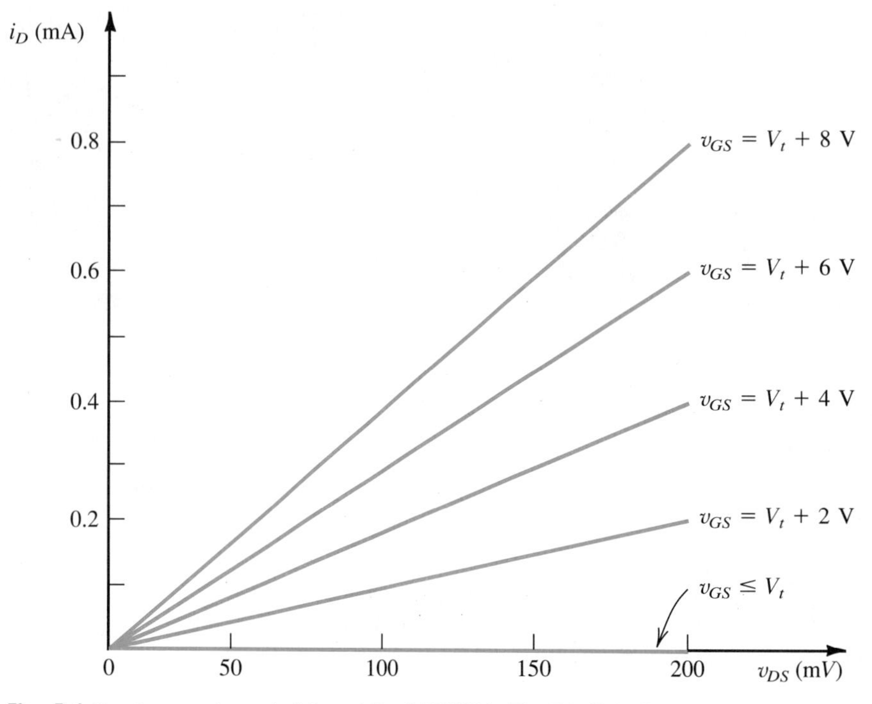

Fig. 5.4 The i_D–v_{DS} characteristics of the MOSFET in Fig. 5.3. Note that v_{DS} is small. The device operates as a linear resistor whose value is controlled by v_{GS}.

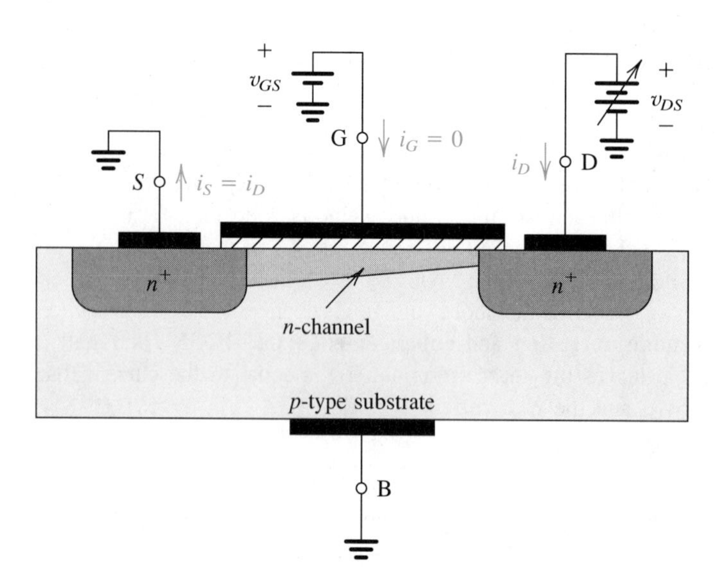

Fig. 5.5 Operation of the enhancement NMOS' transistor as v_{DS} is increased. The induced channel acquires a tapered shape and its resistance increases as v_{DS} is increased. Here v_{GS} is kept constant at a value $> V_t$.

to drain, the voltage (measured relative to the source) increases from 0 to v_{DS}. Thus the voltage between the gate and points along the channel decreases from v_{GS} at the source end to $v_{GS} - v_{DS}$ at the drain end. Since the channel depth depends on this voltage, we find that the channel is no longer of uniform depth; rather, the channel will take the tapered form shown in Fig. 5.5, being deepest at the source end and shallowest at the drain end. As v_{DS} is increased, the channel becomes more tapered and its resistance increases correspondingly. Thus the i_D–v_{DS} curve does not continue as a straight line but bends as shown in Fig. 5.6. Eventually, when v_{DS} is increased to the value that reduces

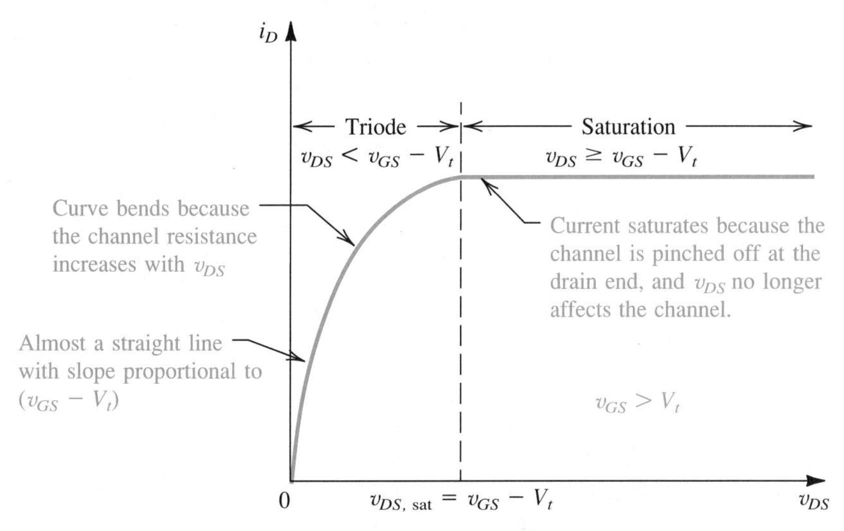

Fig. 5.6 The drain current i_D versus the drain-to-source voltage v_{DS} for an enhancement-type NMOS transistor operated with $v_{GS} > V_t$.

the voltage between gate and channel at the drain end to V_t—that is, $v_{GS} - v_{DS} = V_t$ or $v_{DS} = v_{GS} - V_t$—the channel depth at the drain end decreases to almost zero, and the channel is said to be **pinched off.** Increasing v_{DS} beyond this value has little effect (theoretically, no effect) on the channel shape, and the current through the channel remains constant at the value reached for $v_{DS} = v_{GS} - V_t$. The drain current thus **saturates** at this value, and the MOSFET is said to have entered the **saturation region**[4] of operation. The voltage v_{DS} at which saturation occurs is denoted $v_{DS,\text{sat}}$,

$$v_{DS,\text{sat}} = v_{GS} - V_t \tag{5.1}$$

Obviously, for every value of $v_{GS} \geq V_t$, there is a corresponding value of $v_{DS,\text{sat}}$. The device operates in the saturation region if $v_{DS} \geq v_{DS,\text{sat}}$. The region of the i_D–v_{DS} charac-

[4] Saturation in a MOSFET means a very different thing than in a BJT. This is a rather unfortunate situation but cannot be changed at this time: Virtually the entire electronics literature uses this terminology!

teristic obtained for $v_{DS} < v_{DS,\text{sat}}$ is called the **triode region,** a carryover from the days of vacuum-tube devices whose operation an FET resembles.

To help further in visualizing the effect of v_{DS}, we show in Fig. 5.7 sketches of the channel as v_{DS} is increased while v_{GS} is kept constant. Theoretically, any increase in v_{DS} above $v_{DS,\text{sat}}$ (which is equal to $v_{GS} - V_t$) has no effect on the channel shape and simply appears across the depletion region surrounding the channel and the n^+ drain region.

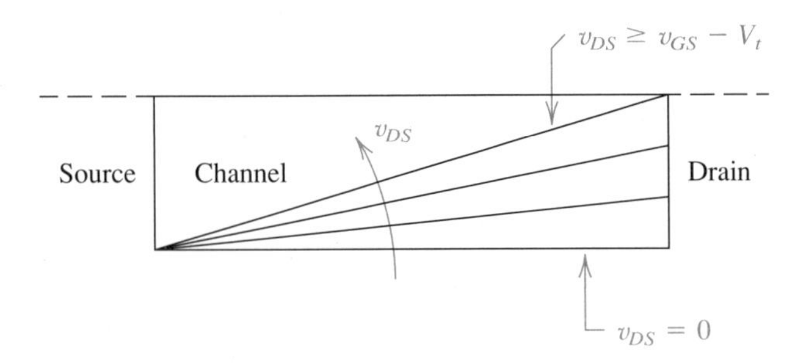

Fig. 5.7 Increasing v_{DS} causes the channel to acquire a tapered shape. Eventually, as v_{DS} reaches $v_{GS} - V_t$, the channel is pinched off at the drain end. Increasing v_{DS} above $v_{GS} - V_t$ has little effect (theoretically, no effect) on the channel's shape.

The *p*-channel MOSFET

A *p*-channel enhancement-type MOSFET (PMOS transistor) is fabricated on an *n*-type substrate with p^+ regions for the drain and source, and holes as charge carriers. The device operates in the same manner as the *n*-channel device except that v_{GS} and v_{DS} are negative and the threshold voltage V_t is negative. Also the current i_D enters the source terminal and leaves through the drain terminal.

PMOS technology was originally the dominant one. However, because NMOS devices can be made smaller and thus operate faster, and because NMOS requires lower supply voltages than PMOS, NMOS technology has virtually replaced PMOS. Nevertheless, it is important to be familiar with the PMOS transistor for two reasons: PMOS devices are still available for discrete-circuit design, and more importantly, both PMOS and NMOS transistors are utilized in **complementary MOS** or **CMOS** circuits.

Complementary MOS or CMOS

As the name implies, complementary MOS technology employs MOS transistors of both polarities. Although CMOS circuits are somewhat more difficult to fabricate than NMOS, the availability of complementary devices makes possible many powerful circuit-design possibilities. Indeed, at the present time CMOS is the most useful of all the integrated-circuit MOS technologies. This statement applies to both analog and digital circuits. Throughout this book we will study many CMOS circuit techniques.

Figure 5.8 shows a cross section of a CMOS chip illustrating how the PMOS and NMOS transistors are fabricated. Observe that while the PMOS transistor is implemented directly in the *n*-type substrate, the NMOS transistor is fabricated in a specially created *p* region, known as a *p* well. The two devices are isolated from each other by a thick region of oxide that functions as an insulator.

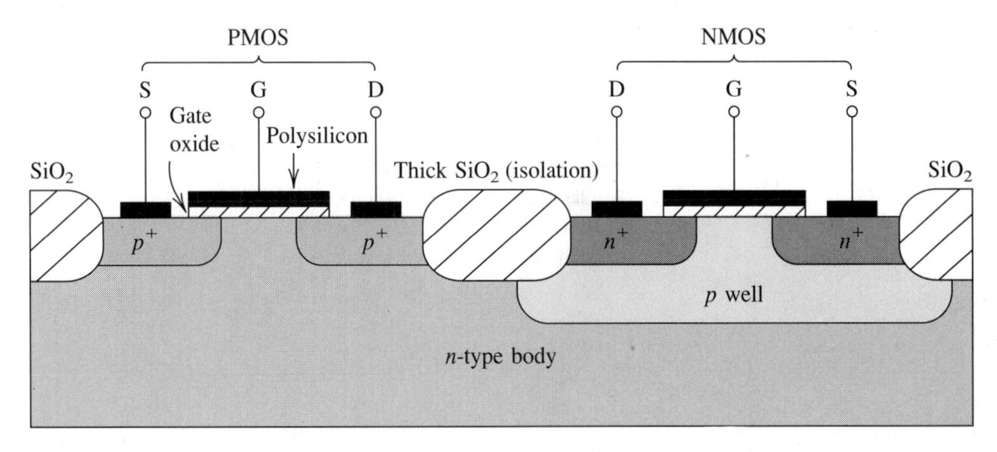

Fig. 5.8 Cross section of a CMOS integrated circuit. Note that the NMOS transistor is formed in a separate *p*-type region, known as a *p* well. Another arrangement is also possible in which a *p*-type body is used and the *p* device is formed in an *n* well.

Operating the MOS Transistor in the Subthreshold Region

The above description of MOSFET operation implies that for $v_{GS} < V_t$, no current flows and the device is cut off. This is not entirely true for it has been found that for values of v_{GS} smaller but close to V_t, a small drain current flows. In this **subthreshold region** of operation the drain current is exponentially related to v_{GS}, much like the i_C–v_{BE} relationship of a BJT.

Although in most applications the MOS transistor is operated with $v_{GS} > V_t$, there are special, but a growing number of, applications that make use of subthreshold operation. In this book we will not consider subthreshold operation any further and refer the reader to the references listed at the end of the chapter [see in particular Mead (1988) and Tsividis (1987)].

5.2 CURRENT–VOLTAGE CHARACTERISTICS OF THE ENHANCEMENT MOSFET

Building on the physical foundation established in the previous section for the operation of the enhancement MOS transistor, we present in this section its complete current–voltage characteristics. These characteristics can be measured at dc or at low frequencies and thus are called static characteristics. The dynamic effects that limit the operation of the MOSFET at high frequencies and high switching speeds will be discussed in Chapter 7 for amplifier circuits, and in Chapter 13 for digital circuits.

Circuit Symbol

Figure 5.9(a) shows the circuit symbol for the *n*-channel enhancement-type MOSFET. The symbol is very descriptive: The vertical solid line denotes the gate electrode; the vertical broken line denotes the channel—the line is broken to indicate that the device is of the enhancement type whose channel does not exist without the application of an

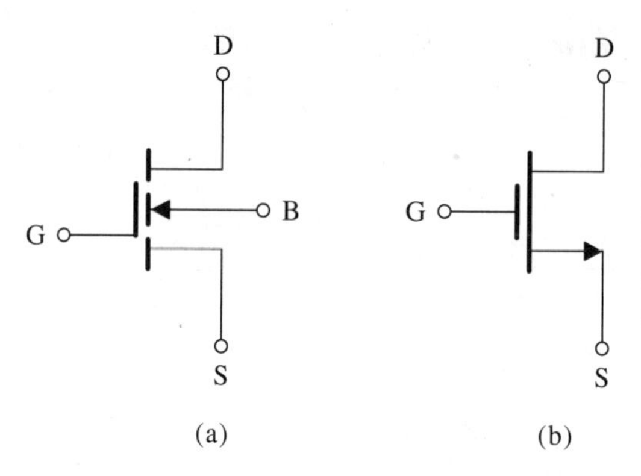

Fig. 5.9 (a) Circuit symbol for the *n*-channel enhancement-type MOSFET. **(b)** Simplified circuit symbol for the enhancement *n*-channel MOSFET with the substrate (B) connected to the source (S).

appropriate gate voltage; and the spacing between the gate line and the channel line represents the fact that the gate electrode is insulated from the body of the device. The polarity of the *pn* junction between the *p*-type substrate and the *n* channel is indicated by the arrowhead on the line representing the substrate (body). This arrowhead also indicates the polarity of the transistor; namely, that it is an *n*-channel device.

Although the MOSFET is a symmetrical device, it is useful in circuit applications to designate one terminal as the source and the other as the drain (without having to write S and D beside the terminals). This goal is achieved in the circuit symbol by drawing the gate terminal line closer to the source than to the drain.[5]

Although the symbol of Fig. 5.9(a) is descriptive it is rather complex and makes drawing a large circuit an awkward task. A simplified circuit symbol that applies for the usual case where the substrate is connected to the source is shown in Fig. 5.9(b). In this symbol, the arrowhead on the source terminal points in the normal direction of current flow and accomplishes two purposes: It distinguishes the source from the drain, and it indicates the polarity of the device (i.e., *n* channel).

The i_D–v_{DS} Characteristics

Figure 5.10(a) shows an *n*-channel enhancement-type MOSFET with voltages v_{GS} and v_{DS} applied and with the normal directions of current flow indicated. This conceptual circuit can be used to measure the i_D–v_{DS} characteristics, which are a family of curves, each measured at a constant v_{GS}. From the study of physical operation in the previous section, we expect each of the i_D–v_{DS} curves to have the shape shown in Fig. 5.6. This indeed is the case, as is evident from Fig. 5.10(b) which shows a typical set of i_D–v_{DS} characteristics. A thorough understanding of the MOSFET terminal characteristics is essential for the reader who intends to design MOS circuits.

The characteristic curves in Fig. 5.10(b) indicate that there are three distinct regions of operation: the cutoff region, the triode region, and the saturation region. The saturation region is used if the FET is to operate as an amplifier. For operation as a switch, the cutoff

[5] In practice, it is the polarity of the voltage impressed across the device that determines source and drain; the drain is always positive relative to the source in an *n*-channel FET.

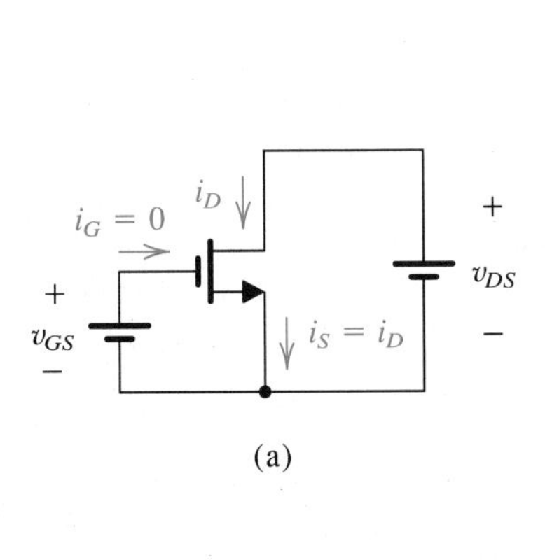

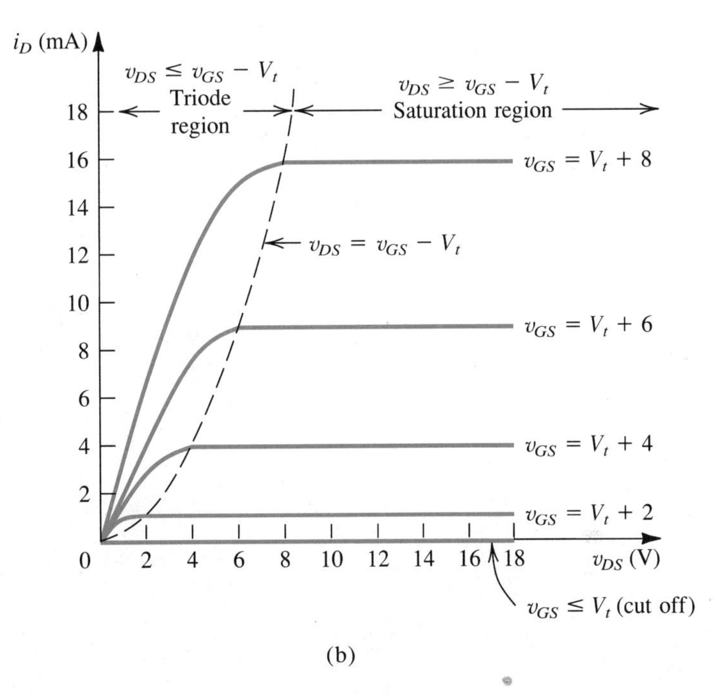

Fig. 5.10 (a) An n-channel enhancement-type MOSFET with v_{GS} and v_{DS} applied and with the normal directions of current flow indicated. **(b)** The i_D–v_{DS} characteristics for a device with $V_t = 2$ V and $K = 0.25$ mA/V^2.

and triode regions are utilized. The device is cut off when $v_{GS} < V_t$. To operate the MOSFET in the triode region we must first induce a channel,

$$v_{GS} \geq V_t \quad \text{(Induced channel)} \tag{5.2}$$

and then keep v_{DS} small enough so that the channel remains continuous. This is achieved by ensuring that the gate-to-drain voltage is

$$v_{GD} > V_t \quad \text{(Continuous channel)} \tag{5.3}$$

This condition can be stated explicitly in terms of v_{DS} by writing $v_{GD} = v_{GS} + v_{SD} = v_{GS} - v_{DS}$; thus

$$v_{GS} - v_{DS} > V_t$$

which can be rearranged to yield

$$v_{DS} < v_{GS} - V_t \quad \text{(Continuous channel)} \tag{5.4}$$

Either Eq. (5.3) or Eq. (5.4) can be used to ascertain triode-region operation. In words, the n-channel enhancement-type MOSFET operates in the triode region when v_{GS} is greater than V_t and the drain voltage is lower than the gate voltage by at least V_t volts.

In the triode region, the i_D–v_{DS} characteristics can be approximately described by the relationship

$$i_D = K[2(v_{GS} - V_t)v_{DS} - v_{DS}^2] \tag{5.5}$$

in which K is a device parameter given by

$$K = \frac{1}{2}\,\mu_n C_{OX}\left(\frac{W}{L}\right) \tag{5.6}$$

where μ_n is a physical constant known as the **electron mobility** (its value in this case applies for the electrons in the induced n channel), C_{OX} (called the **oxide capacitance**) is the capacitance per unit area of the gate-to-body capacitor for which the oxide layer serves as dielectric, L is the length of the channel, and W is its width (refer to Fig. 5.1). Since for a given fabrication process the quantity ($\frac{1}{2}\,\mu_n C_{OX}$) is a constant (approximately 10 μA/V^2 for the standard NMOS process with a 0.1-μm oxide thickness), the **aspect ratio** of the device (W/L) determines its conductivity parameter K. Note from Eq. (5.5) that K has the units A/V^2.

If v_{DS} is sufficiently small so that we can neglect the v_{DS}^2 term in Eq. (5.5), we obtain for the i_D–v_{DS} characteristics near the origin the relationship

$$i_D \simeq 2K(v_{GS} - V_t)v_{DS} \tag{5.7}$$

This linear relationship represents the operation of the MOS transistor as a linear resistance r_{DS},

$$r_{DS} \equiv \frac{v_{DS}}{i_D} = [2K(v_{GS} - V_t)]^{-1} \tag{5.8}$$

whose value is controlled by v_{GS}. We discussed this region of operation in the previous section (refer to Fig. 5.4).

To operate the MOSFET in the saturation region a channel has to be induced,

$$v_{GS} \geq V_t \quad \text{(Induced channel)} \tag{5.9}$$

and pinched off at the drain end by raising v_{DS} to a value that results in the gate-to-drain voltage falling below V_t,

$$v_{GD} \leq V_t \quad \text{(Pinched-off channel)} \tag{5.10}$$

This condition can be expressed explicitly in terms of v_{DS} as

$$v_{DS} \geq v_{GS} - V_t \quad \text{(Pinched-off channel)} \tag{5.11}$$

In words, the n-channel enhancement-type MOSFET operates in the saturation region when v_{GS} is greater than V_t and the drain voltage does not fall below the gate voltage by more than V_t volts.

The boundary between the triode region and the saturation region is characterized by

$$v_{DS} = v_{GS} - V_t \quad \text{(Boundary)} \tag{5.12}$$

Substituting this value of v_{DS} into Eq. (5.5) gives the saturation value of the current i_D as

$$i_D = K(v_{GS} - V_t)^2 \tag{5.13}$$

Thus in saturation the MOSFET provides a drain current whose value is independent of the drain voltage v_{DS} and is determined by the gate voltage v_{GS} according to the square-law relationship in Eq. (5.13), a sketch of which is shown in Fig. 5.11. Thus the saturated

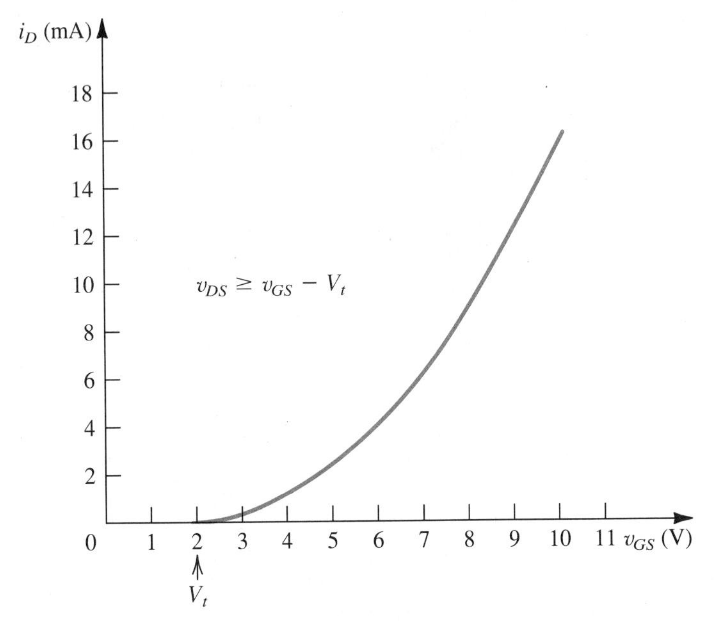

Fig. 5.11 The i_D–v_{GS} characteristic for an enhancement-type NMOS transistor in saturation ($V_t = 2$ V, $K = 0.25$ mA/V²).

MOSFET behaves as an ideal current source whose value is controlled by v_{GS} according to the nonlinear relationship in Eq. (5.13). Figure 5.12 shows a circuit representation of this view of the MOSFET operation in the saturation region. Note that this is a large-signal equivalent circuit model.

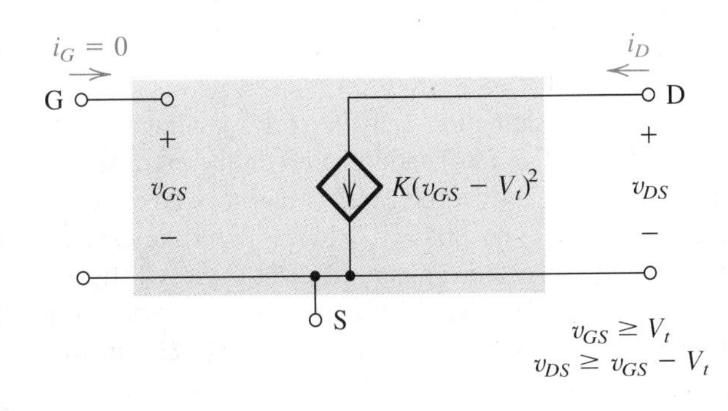

Fig. 5.12 Large-signal equivalent circuit model of an *n*-channel MOSFET operating in the saturation region.

Referring back to the i_D–v_{DS} characteristics in Fig. 5.10(b), we note that the boundary between the triode and the saturation regions is shown as a broken-line curve. Since this curve is characterized by $v_{DS} = v_{GS} - V_t$, its equation can be found by substituting for $v_{GS} - V_t$ by v_{DS} in either the triode-region equation (Eq. 5.5) or the saturation-region equation (Eq. 5.13). The result is

$$i_D = K v_{DS}^2 \tag{5.14}$$

It should be noted that the characteristics depicted in Figs. 5.10, 5.11, and 5.4 are for a MOSFET with $K = 0.25$ mA/V^2 and $V_t = 2$ V.

Finally, the chart in Fig. 5.13(a) shows the relative levels that the terminal voltages of the enhancement-type NMOS transistor must have for operation in the triode region and in the saturation region. The same information is presented in a different way in Fig. 5.13(b).

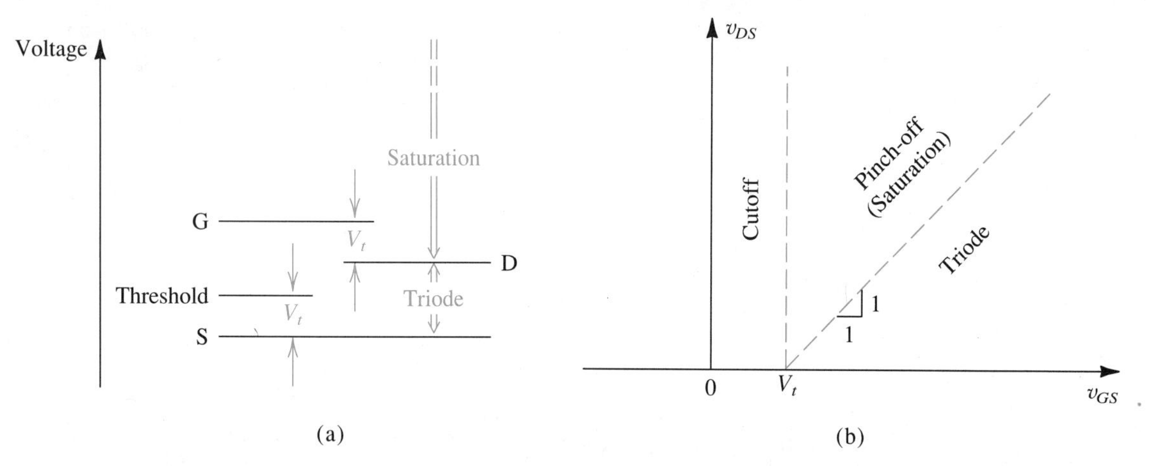

Fig. 5.13 (a) The relative levels of the terminal voltages of the enhancement NMOS transistor for operation in the triode region and in the saturation region. **(b)** Illustrating the different regions of operation on the v_{DS}–v_{GS} plane.

Finite Output Resistance in Saturation

The complete independence of i_D on v_{DS} in saturation and the corresponding infinite output resistance at the drain (see Fig. 5.12) is an idealization based on the premise that once the channel is pinched off at the drain end, further increases in v_{DS} have no effect on the channel's shape. In practice, increasing v_{DS} beyond $v_{DS,\text{sat}}$ does affect the channel somewhat. Specifically, as v_{DS} is increased, the channel pinch-off point is moved slightly away from the drain toward the source. Thus the effective channel length is reduced, a phenomenon known as **channel-length modulation.** Now since K is inversely proportional to the channel length (Eq. 5.6), K and, correspondingly, i_D, increases with v_{DS}.

A typical set of i_D–v_{DS} characteristics showing the effect of channel length modulation is displayed in Fig. 5.14. The slight linear dependence of i_D on v_{DS} in the saturation region can be analytically accounted for by incorporating the factor $1 + \lambda\, v_{DS}$ in the i_D equation as follows:

$$i_D = K(v_{GS} - V_t)^2(1 + \lambda\, v_{DS}) \tag{5.15}$$

where the positive constant λ is a MOSFET parameter. From Fig. 5.14 we observe that the straight-line i_D–v_{DS} characteristics in saturation, when extrapolated, intercept the v_{DS}-axis at the point $v_{DS} = -1/\lambda \equiv -V_A$, where V_A is a positive voltage similar to the

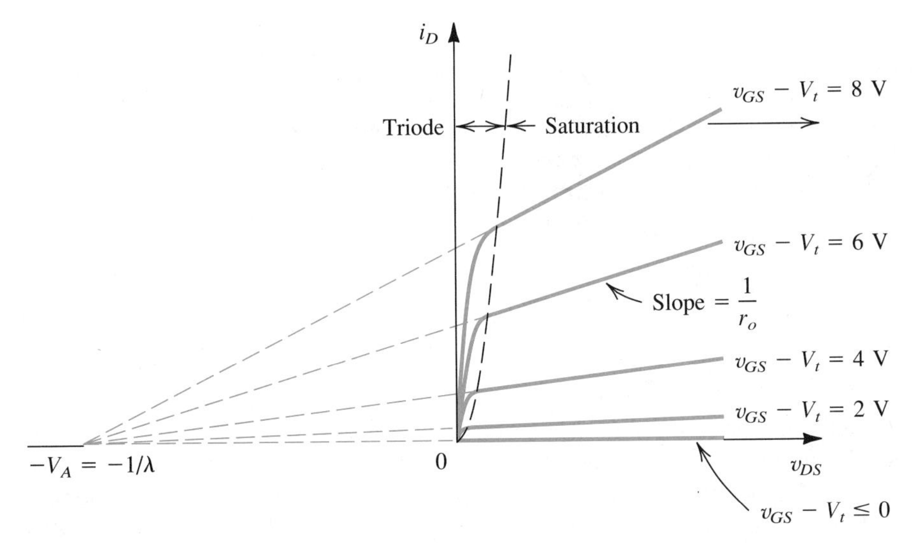

Fig. 5.14 Effect of v_{DS} on i_D in the saturation region. The MOSFET parameter V_A is typically in the range 30 to 200 V.

Early voltage in a BJT. Typically $\lambda = 0.005$ to $0.03\ \text{V}^{-1}$, and, correspondingly, V_A is in the range 200 to 30 volts.

It should be obvious that channel-length modulation makes the output resistance in saturation finite. Defining the output resistance r_o as[6]

$$r_o \equiv \left[\frac{\partial i_D}{\partial v_{DS}} \right]^{-1}_{v_{GS}=\text{constant}} \tag{5.16}$$

results in

$$r_o = [\lambda\ K(V_{GS} - V_t)^2]^{-1}$$

which can be approximated by

$$r_o \simeq [\lambda\ I_D]^{-1} \tag{5.17}$$

where I_D is the current corresponding to the particular value of v_{GS} for which r_o is being evaluated. The approximation in Eq. (5.17) is based on neglecting the effect of the factor $1 + \lambda\ v_{DS}$ in Eq. (5.15) on the value of r_o—a second-order approximation. Equation (5.17) can be written in the alternative form

$$r_o \simeq V_A/I_D \tag{5.18}$$

Thus the output resistance is inversely proportional to the dc bias current I_D. Finally, we show in Fig. 5.15 the large-signal equivalent circuit model incorporating r_o.

[6] In this book we use r_o to represent the incremental output resistance in saturation and r_{DS} to represent the drain-to-source resistance in the triode region for small v_{DS}.

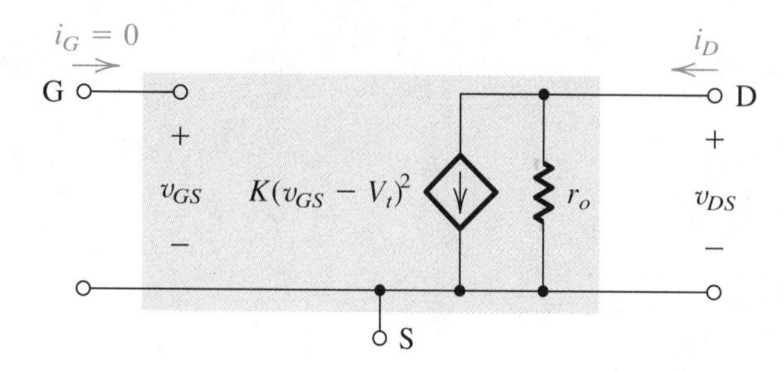

Fig. 5.15 Large-signal equivalent circuit model of the n-channel MOSFET in saturation, incorporating the output resistance r_o. The output resistance models the linear dependence of i_D on v_{DS} and is given by $r_o = V_A/[K(V_{GS} - V_t)^2] \simeq V_A/I_D$.

Characteristics of the p-Channel MOSFET

The circuit symbol for the p-channel enhancement-type MOSFET is shown in Fig. 5.16(a). For the case where the substrate is connected to the gate, the simplified symbol of Fig. 5.16(b) is usually used. The voltage and current polarities for normal operation are indicated in Fig. 5.16(c). Recall that for the p-channel device the threshold voltage V_t is negative. To induce a channel we apply a gate voltage that is more negative than V_t,

$$v_{GS} \leq V_t \quad \text{(Induced channel)} \tag{5.19}$$

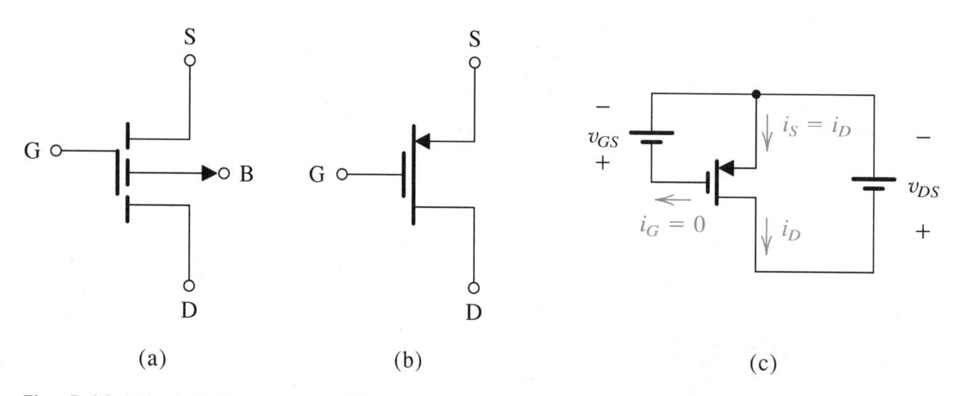

Fig. 5.16 (a) Circuit symbol for the p-channel enhancement-type MOSFET. **(b)** Simplified circuit symbol for the case where the substrate (body, B) is connected to the source. **(c)** The MOSFET with voltages applied and the directions of current flow indicated. Note that v_{GS} and v_{DS} are negative and i_D flows out of the drain terminal.

and apply a drain voltage that is more negative than the source voltage (i.e., v_{DS} is negative or, equivalently, v_{SD} is positive). The current i_D flows out of the drain terminal, as indicated in the figure. To operate in the triode region v_{DS} must satisfy

$$v_{DS} \geq v_{GS} - V_t \quad \text{(Continuous channel)} \tag{5.20}$$

that is, the drain voltage must be higher than the gate voltage by at least $|V_t|$. The current i_D is given by the same equation as for NMOS, Eq. (5.5),

$$i_D = K[2(v_{GS} - V_t)v_{DS} - v_{DS}^2] \tag{5.21}$$

where v_{GS}, V_t, and v_{DS} are negative and K is given by

$$K = \frac{1}{2}\mu_p C_{OX}\left(\frac{W}{L}\right) \tag{5.22}$$

where μ_p is the mobility of holes in the induced p channel. Typically, $\mu_p \simeq \frac{1}{2}\mu_n$, with the result that for the same W/L ratio a PMOS transistor has half the value of K as the NMOS device.

To operate in saturation, v_{DS} must satisfy

$$v_{DS} \leq v_{GS} - V_t \quad \text{(Pinched-off channel)} \tag{5.23}$$

that is, the drain voltage must be lower than (gate voltage $+|V_t|$). The current i_D is given by the same equation used for NMOS, Eq. (5.15),

$$i_D = K(v_{GS} - V_t)^2(1 + \lambda\, v_{DS}) \tag{5.24}$$

where v_{GS}, V_t, λ, and v_{DS} are all negative.

Finally, the chart in Fig. 5.17 shows the relative levels that the terminal voltages of the enhancement-type PMOS transistor must have for operation in the triode region and in the saturation region.

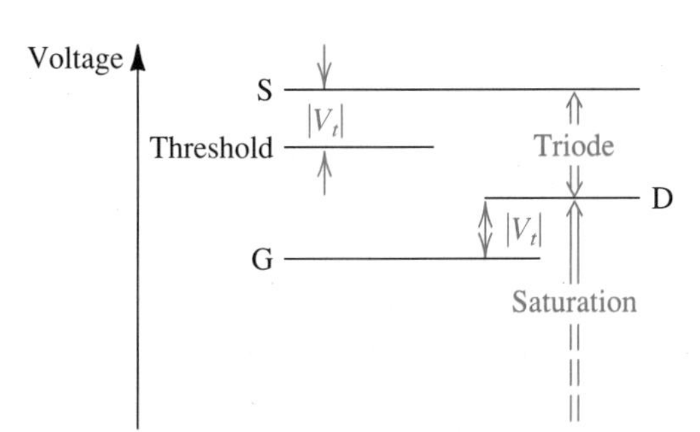

Fig. 5.17 The relative levels of the terminal voltages of the enhancement-type PMOS transistor for operation in the triode region and in the saturation region.

The Role of the Substrate—The Body Effect

In many applications the substrate (or body) terminal B is connected to the source terminal, which results in the *pn* junction between the substrate and the induced channel (see Fig. 5.5) having a constant reverse bias. In such a case the substrate does not play any role in circuit operation and its existence can be ignored altogether.

In integrated circuits, however, the substrate is usually common to many MOS transistors. In order to maintain the reverse-bias condition on the substrate-to-channel junction, the substrate is usually connected to the most negative power supply in an NMOS circuit (the most positive in a PMOS circuit). The resulting reverse-bias voltage between source and body (V_{SB} in an n-channel device) will have an effect on device operation. To appreciate this fact consider an NMOS transistor and let its substrate be made negative relative to the source. The reverse bias voltage will widen the depletion region (refer to Fig. 5.2). This in turn reduces the channel depth. To return the channel to its former state, v_{GS} has to be increased.

The effect of V_{SB} on the channel can be most conveniently represented as a change in the threshold voltage V_t. Specifically, it can be shown [see Tsividis (1987)] that increasing the reverse substrate bias voltage V_{SB} results in an increase in V_t according to the relationship

$$V_t = V_{t0} + \gamma\left[\sqrt{2\phi_f + V_{SB}} - \sqrt{2\phi_f}\right] \tag{5.25}$$

where V_{t0} is the threshold voltage for $V_{SB} = 0$; γ is a process parameter, typically 0.5 $V^{1/2}$; ϕ_f is a physical parameter with $(2\phi_f)$ typically 0.6 V.

Equation (5.25) indicates that an incremental change in V_{SB} gives rise to an incremental change in V_t, which in turn results in an incremental change in i_D even though v_{GS} might have been kept constant. It follows that the body voltage controls i_D; thus the body acts as another gate for the MOSFET, a phenomenon known as the **body effect.** The body effect can cause considerable degradation in circuit performance, as will be shown in Section 5.9.

Temperature Effects

Both V_t and K are temperature sensitive. The magnitude of V_t decreases by about 2 mV for every 1°C rise in temperature. This decrease in $|V_t|$ gives rise to a corresponding increase in drain current as temperature is increased. However, because K decreases with temperature and its effect is a dominant one, the overall observed effect of a temperature increase is a *decrease* in drain current. This very interesting result is put to use in applying the MOSFET in power circuits (Chapter 9).

Breakdown and Input Protection

As the voltage on the drain is increased, a value is reached at which the *pn* junction between the drain region and substrate suffers avalanche breakdown (see Section 3.11). This breakdown usually occurs at voltages of 50 to 100 V and results in a rapid increase in current.

Another breakdown effect that occurs at lower voltages (about 20 V) in modern devices is called **punch-through.** It occurs in devices with relatively short channels when the drain voltage is increased to the point that the depletion region surrounding the drain region extends through the channel to the source. The drain current then increases rapidly. Punch-through does not normally result in permanent damage to the device.

Yet another kind of breakdown occurs when the gate-to-source voltage exceeds about 50 V. This is the breakdown of the gate oxide and results in permanent damage to the device. Although 50 V is high, it must be remembered that the MOSFET has a very high input impedance, and thus small amounts of static charge accumulating on the gate capacitor can cause this breakdown voltage to be exceeded.

To prevent the accumulation of static charge on the gate capacitor of a MOSFET, gate protection devices are usually included at the input of MOS integrated circuits. The protection mechanism invariably makes use of clamping diodes (see Fig. 13.32).

For future reference and to permit comparison of the various types of FETs studied in this chapter, Table 5.1 provides a summary of the important formulas and voltage polarities. (Depletion MOSFETs and JFETs will be studied shortly.)

Table 5.1 SUMMARY OF THE CHARACTERISTICS OF FIELD-EFFECT TRANSISTORS

	n-Channel			*p*-Channel		
	Enhancement MOSFET	Depletion MOSFET	JFET	Enhancement MOSFET	Depletion MOSFET	JFET
Circuit Symbol	i_D	i_D	i_D	i_D	i_D	i_D
V_t	+	−	−	−	+	+
K	$\frac{1}{2}\mu_n C_{OX}(W/L)$		I_{DSS}/V_P^2	$\frac{1}{2}\mu_p C_{OX}(W/L)$		I_{DSS}/V_P^2
To turn transistor on	$v_{GS} > V_t$			$v_{GS} < V_t$		
v_{DS}	+			−		
To operate in the triode region	$v_{DS} \leq v_{GS} - V_t$			$v_{DS} \geq v_{GS} - V_t$		
To operate in the saturation region	$v_{DS} \geq v_{GS} - V_t$			$v_{DS} \leq v_{GS} - V_t$		
$\lambda = 1/V_A$	+			−		
In triode region	$i_D = K[2(v_{GS} - V_t)v_{DS} - v_{DS}^2]$					
In saturation region	$i_D = K(v_{GS} - V_t)^2(1 + \lambda\, v_{DS})$					
r_o	$\lvert V_A \rvert / I_D$					

Exercises 5.1 An enhancement-type NMOS transistor with $V_t = 2$ V has its source grounded and a 3-V dc source connected to the gate. In what region of operation does the device operate for (a) $V_D = +0.5$ V? (b) $V_D = 1$ V? (c) $V_D = 5$ V?

Ans. (a) Triode; (b) Saturation; (c) Saturation

5.2 If the NMOS device in Exercise 5.1 has $\mu_n C_{OX} = 20\ \mu\text{A/V}^2$, $W = 100\ \mu$m, and $L = 10\ \mu$m, find the value of K. Also find the value of drain current that results in each of the three cases (a), (b), and (c) specified in Exercise 5.1. Neglect the dependence of i_D on v_{DS} in saturation.

Ans. 0.1 mA/V^2; (a) 75 μA; (b) 100 μA; (c) 100 μA

5.3 An enhancement-type NMOS transistor with $V_t = 2$ V conducts a current $i_D = 1$ mA when $v_{GS} = v_{DS} = 3$ V. Neglecting the dependence of i_D on v_{DS} in saturation, find the value of i_D for $v_{GS} = 4$ V and $v_{DS} = 5$ V. Also, calculate the value of the drain-to-source resistance r_{DS} for small v_{DS} and $v_{GS} = 4$ V.

Ans. 4 mA; 250 Ω

5.4 An enhancement-type MOSFET with $K = 0.1\ \text{mA/V}^2$, $V_t = 1.5$ V, and $\lambda = 0.02\ \text{V}^{-1}$ is operated at $v_{GS} = 3.5$ V. Find the drain current obtained at $v_{DS} = 2$ V and at $v_{DS} = 10$ V. Determine the output resistance r_o at this value of v_{GS}.

Ans. 0.416 mA; 0.480 mA; 125 kΩ

5.5 An IC NMOS transistor has $W = 100\ \mu\text{m}$, $L = 10\ \mu\text{m}$, $\mu_n C_{OX} = 20\ \mu\text{A/V}^2$, $V_A = 100$ V, $\gamma = \frac{1}{2}\ \text{V}^{1/2}$, $V_{t0} = 1$ V, and $2\phi_f = 0.6$ V. Calculate the values of K and of V_t at $V_{SB} = 4$ V. Also, for $V_{GS} = 3$ V and $V_{DS} = 5$ V, calculate I_D for $V_{SB} = 0$ V and for $V_{SB} = 4$ V. What is the output resistance r_o for each of the two cases?

Ans. 0.1 mA/V^2; 1.7 V; 0.420 mA; 0.177 mA; 250 kΩ; 592 kΩ

5.6 An enhancement-type PMOS transistor has $K = 50\ \mu\text{A/V}^2$ and $V_t = -2$ V. If the gate is grounded and the source is connected to +5 V, what is the highest voltage that can be applied to the drain while the device operates in saturation? Neglecting the finite r_o, find the drain current for $v_D = -5$ V.

Ans. +2 V; 0.45 mA

5.3 THE DEPLETION-TYPE MOSFET

In this section we briefly discuss another type of MOSFET, the depletion-type MOSFET. Its structure is similar to that of the enhancement-type MOSFET with one important difference: The depletion MOSFET has a physically implanted channel. Thus an n-channel depletion type MOSFET has an n-type silicon region connecting the n^+ source and the n^+ drain regions at the top of the p-type substrate. Thus if a voltage v_{DS} is applied between drain and source, a current i_D flows for $v_{GS} = 0$. In other words, there is no need to induce a channel, unlike the case of the enhancement MOSFET.

The channel depth and hence its conductivity can be controlled by v_{GS} in exactly the same manner as in the enhancement-type device. Applying a positive v_{GS} enhances the channel by attracting more electrons into it. Here, however, we also can apply a negative v_{GS}, which causes electrons to be repelled from the channel; and thus the channel becomes shallower and its conductivity decreases. The negative v_{GS} is said to **deplete** the channel of its charge carriers, and this mode of operation (negative v_{GS}) is called **depletion mode.** As the magnitude of v_{GS} is increased in the negative direction, a value is reached at which the channel is completely depleted of charge carriers and i_D is reduced to zero even though v_{DS} may be still applied. This negative value of v_{GS} is the threshold voltage of the n-channel depletion-type MOSFET.

The description above suggests (correctly) that a depletion-type MOSFET can be operated in the enhancement mode by applying a positive v_{GS} and in the depletion mode by applying a negative v_{GS}. The i_D–v_{DS} characteristics are similar to those for the enhancement device except that V_t of the n-channel depletion device is negative.

Figure 5.18(a) shows the circuit symbol for the n-channel depletion-type MOSFET. This symbol differs from that of the enhancement-type device in only one respect: The vertical line representing the channel is solid, signifying that a physical channel exists. When the body (B) is connected to the source (S) the simplified symbol shown in Fig. 5.18(b) can be used. This symbol differs from the corresponding one for the enhancement device in that a shaded area is included to denote the implanted channel.

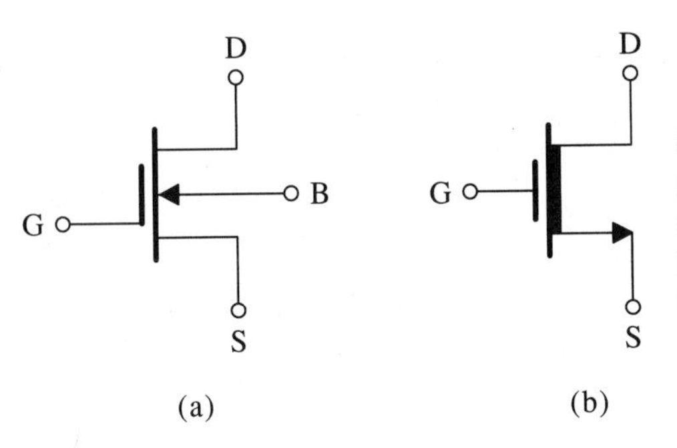

Fig. 5.18 (a) Circuit symbol for the *n*-channel depletion-type MOSFET. **(b)** Simplified circuit symbol applicable for the case the substrate (B) is connected to the source (S).

The i_D–v_{DS} characteristics of a depletion-type *n*-channel MOSFET for which $V_t = -4$ V and $K = 1$ mA/V^2 are sketched in Fig. 5.19(b). Although these characteristics do not show the dependence of i_D on v_{DS} in saturation, such dependence exists and is identical to the case for the enhancement-type device. Observe that because the threshold voltage V_t is negative, the depletion NMOS will operate in the triode region as long as the drain voltage does not exceed the gate voltage by more than $|V_t|$. For it to operate in saturation, the drain voltage must be greater than the gate voltage by at least $|V_t|$ volts. The chart in Fig. 5.20(a) shows the relative levels of the terminal voltages of the depletion NMOS transistor for the two regions of operation. Another way of representing this information is given in Fig. 5.20(b).

Figure 5.19(c) shows the i_D–v_{GS} characteristics in saturation, indicating both the depletion and enhancement modes of operation.

The current–voltage characteristics of the depletion-type MOSFET are described by the equations given in the previous section for the enhancement device except that, for an *n*-channel depletion device, V_t is negative.

Another parameter for the depletion MOSFET is the value of drain current obtained in saturation with $v_{GS} = 0$. It is denoted I_{DSS} and is indicated in Fig. 5.19(b) and (c). It can be shown that

$$I_{DSS} = K\, V_t^2 \tag{5.26}$$

Depletion-type MOSFETs can be fabricated on the same integrated-circuit chip as enhancement-type devices, resulting in circuits with improved characteristics, as will be shown in a later section.

In the above we discussed only *n*-channel depletion devices. Depletion PMOS transistors are available in discrete form and operate in a manner similar to their *n*-channel counterparts except that the polarities of all voltages (including V_t) are reversed. Also, in a *p*-channel device, i_D flows from source to drain, entering the source terminal and leaving by way of the drain terminal. As a summary, we show in Fig. 5.21 sketches of the $i_D - v_{GS}$ characteristics of enhancement and depletion MOSFETs of both polarities (operating in saturation). Finally, we draw the reader's attention to the summary of the characteristics of depletion MOSFETs, presented in Table 5.1.

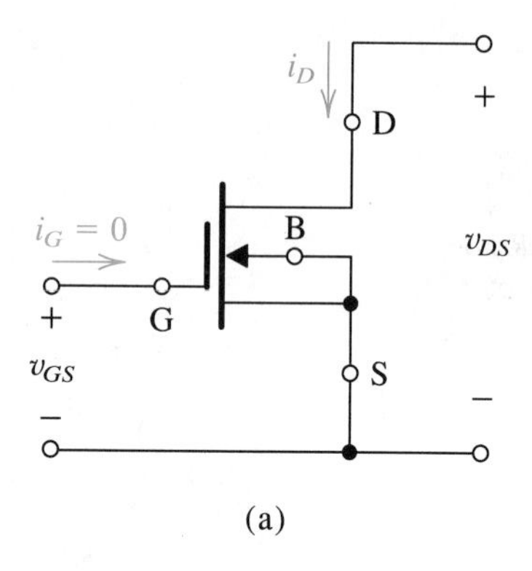

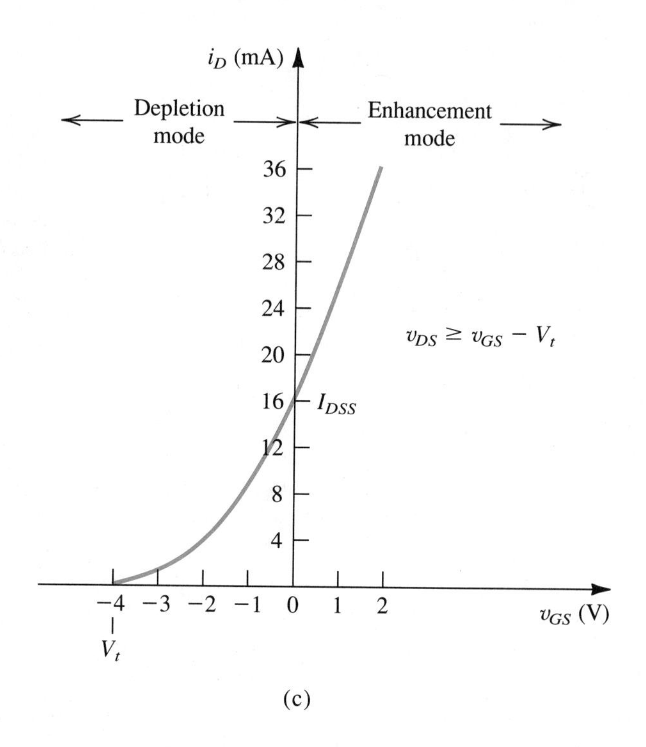

Fig. 5.19 The current–voltage characteristics of a depletion-type *n*-channel MOSFET for which $V_t = -4$ V and $K = 1$ mA/V²: **(a)** transistor with current and voltage polarities indicated; **(b)** the i_D–v_{DS} characteristics; **(c)** the i_D–v_{GS} characteristics in saturation.

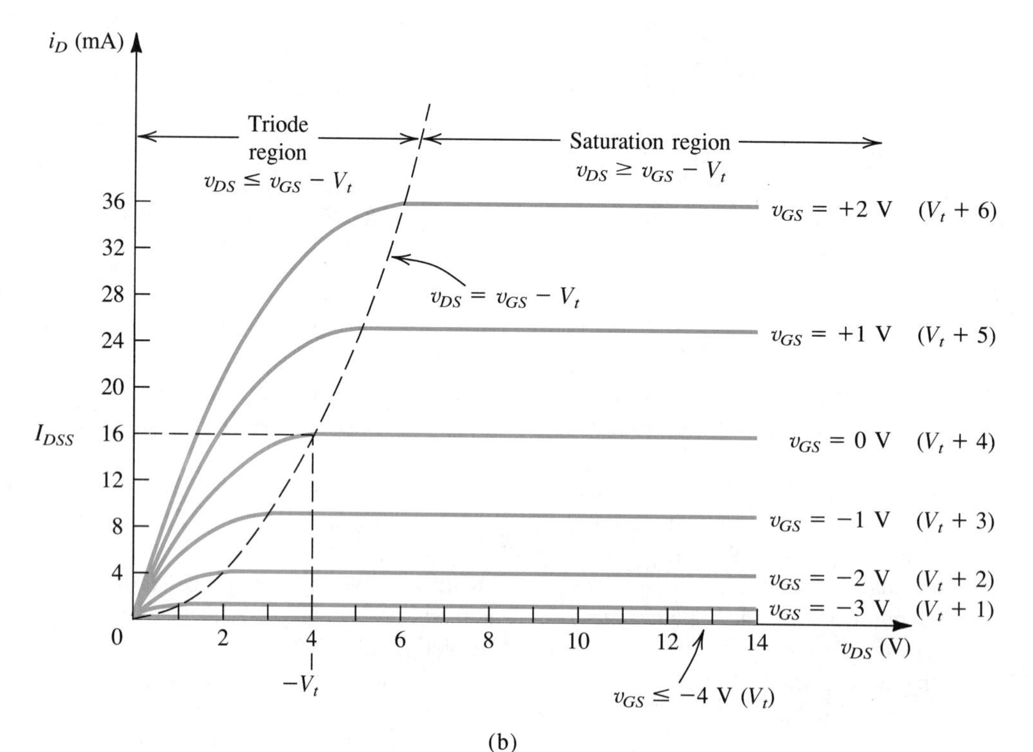

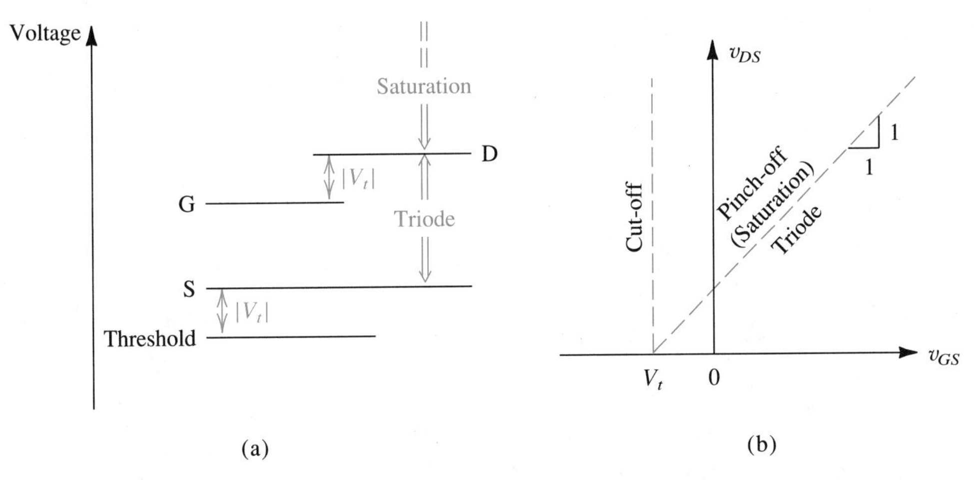

Fig. 5.20 (a) The relative levels of terminal voltages of a depletion-type NMOS transistor for operation in the triode and the saturation regions. The case shown is for operation in the enhancement mode (v_{GS} is positive). **(b)** Representation of the regions of operation on the v_{DS}–v_{GS} plane.

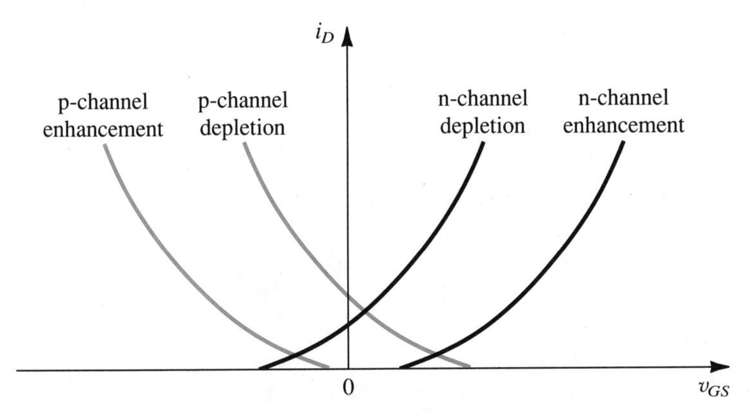

Fig. 5.21 Sketches of the i_D–v_{GS} characteristics for MOSFETs of enhancement and depletion types, of both polarities (operating in saturation). Note that the characteristic curves intersect the v_{GS} axis at V_t. Also note that somewhat different values of $|V_t|$ are shown for *n*-channel and *p*-channel devices.

Exercises 5.7 For a depletion-type NMOS transistor with $V_t = -2$ V and $K = 2$ mA/V^2, find the minimum v_{DS} required to operate in the saturation region when $v_{GS} = +1$ V. What is the corresponding value of i_D?

Ans. 3 V; 18 mA

5.8 The depletion-type MOSFET in Fig. E5.8 has $K = 2$ mA/V^2 and $V_t = -2$ V. Neglecting the effect of v_{DS} on i_D in the saturation region, find the voltage that will appear at the source terminal.

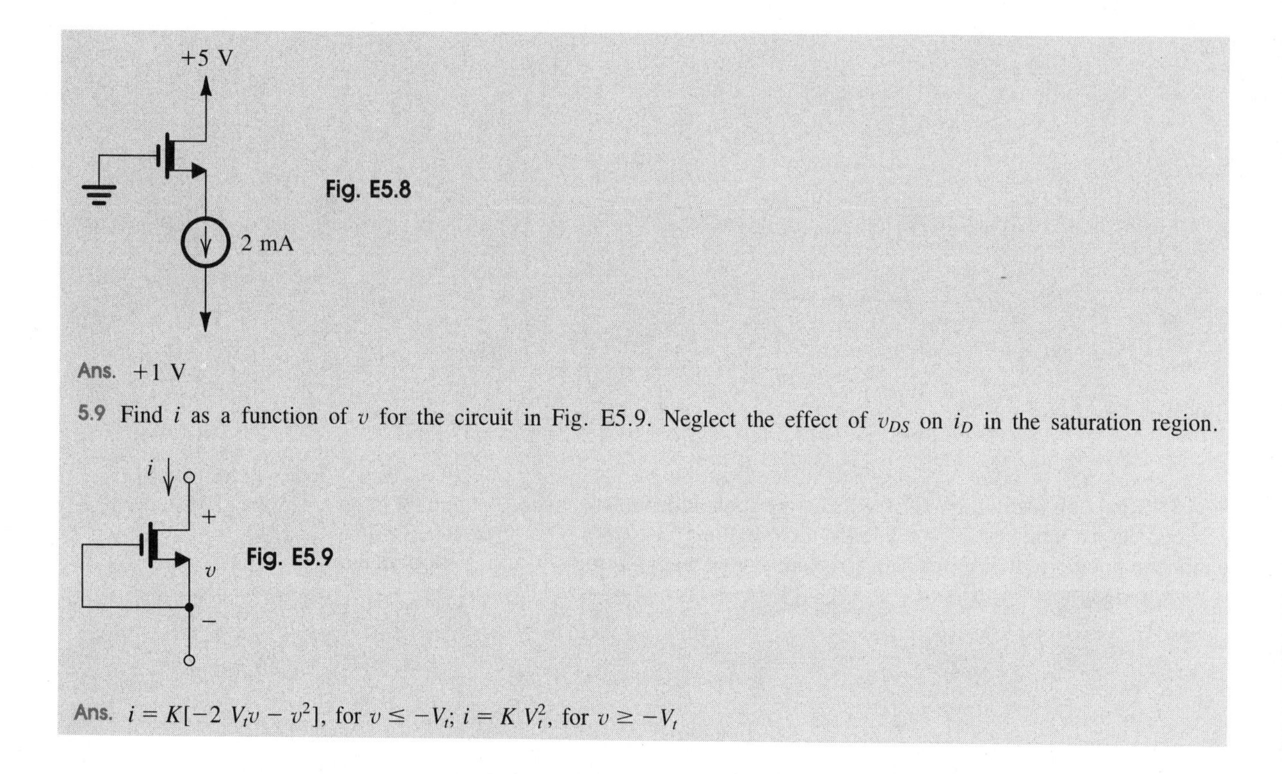

Fig. E5.8

Ans. $+1$ V

5.9 Find i as a function of v for the circuit in Fig. E5.9. Neglect the effect of v_{DS} on i_D in the saturation region.

Fig. E5.9

Ans. $i = K[-2\ V_t v - v^2]$, for $v \leq -V_t$; $i = K\ V_t^2$, for $v \geq -V_t$

5.4 THE JUNCTION FIELD-EFFECT TRANSISTOR (JFET)

The **junction field-effect transistor,** or JFET, is perhaps the simplest transistor available. It has some important characteristics, notably its very high input resistance. Unfortunately, however (for the JFET), the MOSFET has an even higher input resistance. This, together with the many other advantages of MOS transistors, has made MOS the technology of choice for the implementation of very-large-scale integrated (VLSI) circuits. Thus, apart from excelling in a few special applications, the JFET is at the present time not a very significant device. One of these applications involves using JFETs to implement the input stage of an integrated-circuit op amp where the remainder of the circuit is implemented with BJTs. The JFET input stage takes advantage of the high-input-resistance property of JFETs and provides the op amp with an input resistance that is much greater than is possible with BJTs. The JFET is also utilized in discrete-circuit design both as an amplifier and as a switch. In this section we provide a brief introduction to the JFET structure, its operation and terminal characteristics. A number of JFET circuits will be encountered throughout the text.

Device Structure

As with other FET types, the JFET is available in two polarities: n-channel and p-channel. Fig. 5.22(a) shows a simplified[7] structure of the n-channel JFET. It consists of a slab of

[7] For actual geometries and fabrication methods, the reader is referred to Appendix A.

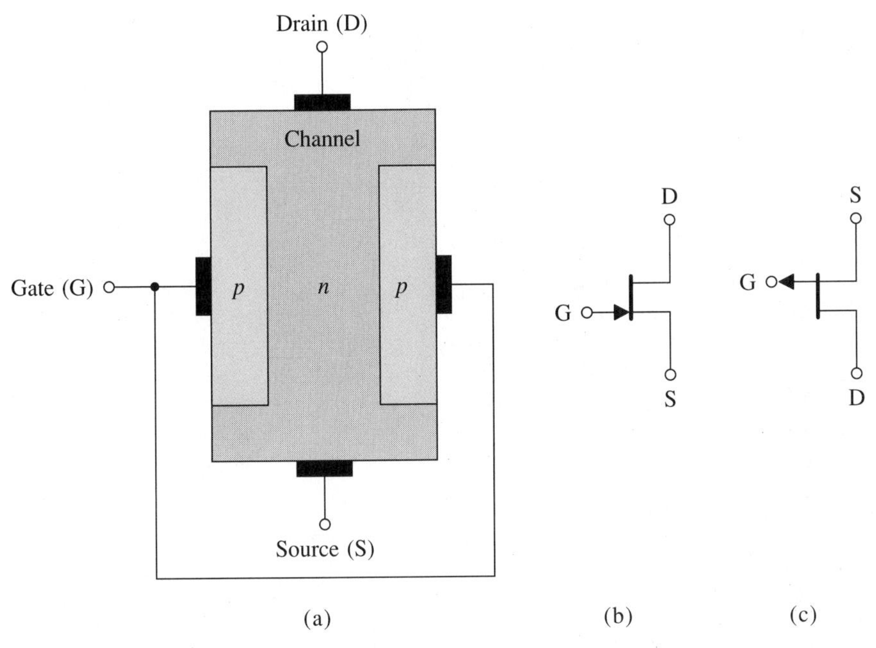

Fig. 5.22 (a) Basic structure of *n*-channel JFET. This is a simplified structure utilized to explain device operation. **(b)** Circuit symbol for the *n*-channel JFET. **(c)** Circuit symbol for the *p*-channel JFET.

n-type silicon with *p*-type regions diffused on its two sides. The *n* region is the channel, and the *p*-type regions are electrically connected together and form the gate. The device operation is based on reverse-biasing the *pn* junction between gate and channel. Indeed, it is the reverse bias on this junction that is used to control the channel width and hence the current flow from drain to source. The major role that this *pn* junction plays in the operation of this FET has given rise to its name: Junction Field-Effect Transistor (JFET).

It should be obvious that a *p*-channel device can be fabricated by simply reversing all the semiconductor types, thus using *p*-type silicon for the channel and *n*-type silicon for the gate regions.

Figures 5.22(b) and (c) show the circuit symbols for JFETs of both polarities. Observe that the device polarity (*n*-channel or *p*-channel) is indicated by the direction of the arrowhead on the gate line. This arrowhead points in the forward direction of the gate–channel *pn* junction. Although the JFET is a symmetrical device whose source and drain can be interchanged, it is useful in circuit design to designate one of these two terminals as source and the other as drain. The circuit symbol achieves this designation by placing the gate closer to the source than to the drain.

Physical Operation

Consider an *n*-channel JFET and refer to Fig. 5.23(a). (Note that to simplify matters we will not show the electrical connection between the gate terminals; it is assumed, however, that the two terminals labeled G are joined together.) With $v_{GS} = 0$, the application of a voltage v_{DS} causes current to flow from the drain to the source. When a negative v_{GS}

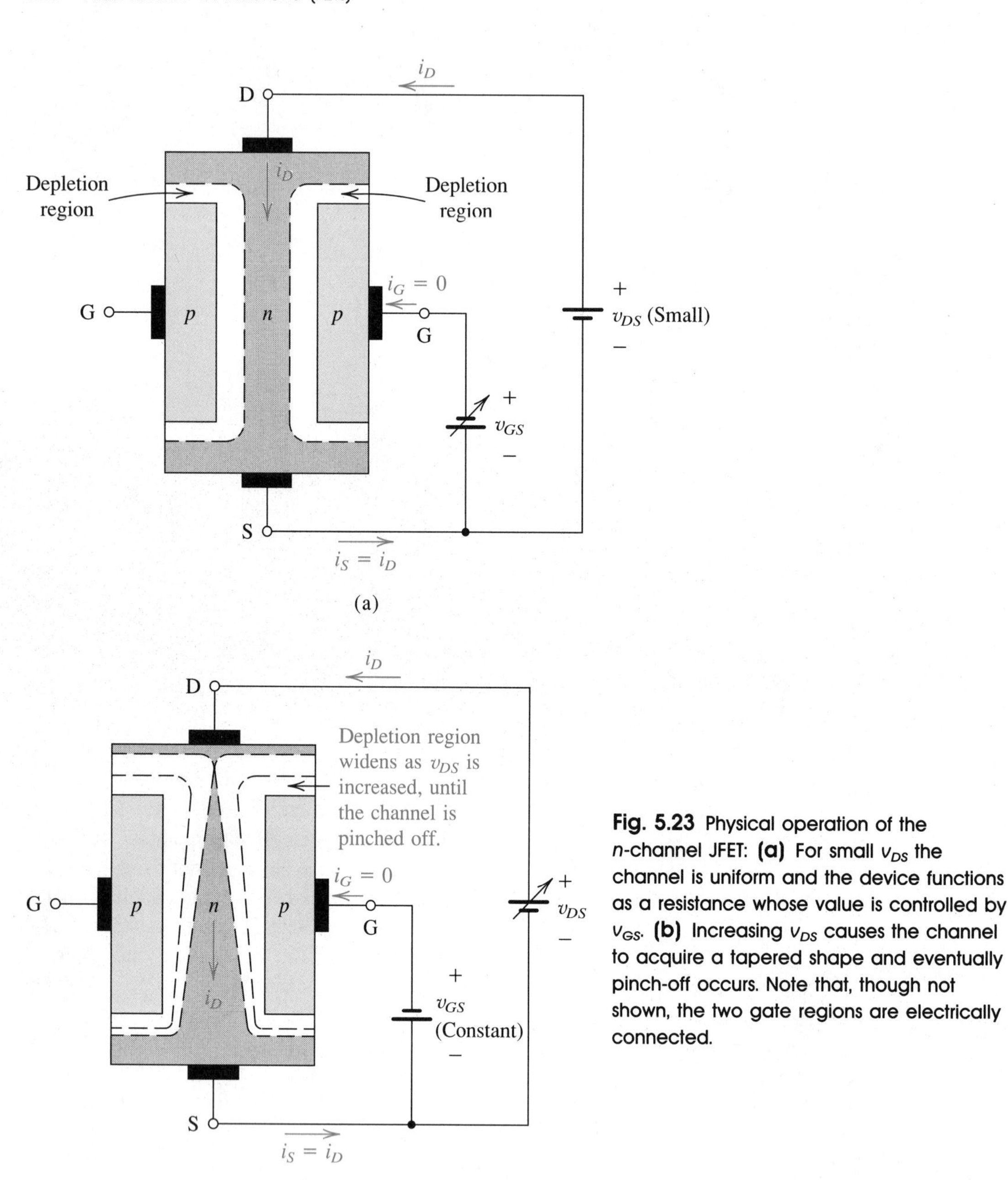

Fig. 5.23 Physical operation of the n-channel JFET: **(a)** For small v_{DS} the channel is uniform and the device functions as a resistance whose value is controlled by v_{GS}. **(b)** Increasing v_{DS} causes the channel to acquire a tapered shape and eventually pinch-off occurs. Note that, though not shown, the two gate regions are electrically connected.

is applied, the depletion region of the gate–channel junction widens and the channel becomes correspondingly narrower; thus the channel resistance increases and the current i_D (for a given v_{DS}) decreases. Because v_{DS} is small, the channel is almost of uniform width. The JFET is simply operating as a resistance whose value is controlled by v_{GS}. If we keep increasing v_{GS} in the negative direction, a value is reached at which the depletion region occupies the entire channel. At this value of v_{GS} the channel is completely depleted of charge carriers (electrons); the channel has in effect disappeared. This value of v_{GS} is therefore the threshold voltage of the device, V_t, which is obviously negative for an n-channel JFET.

Consider next the situation depicted in Fig. 5.23(b). Here v_{GS} is held constant at a value greater (that is, less negative) than V_t, and v_{DS} is increased. Since v_{DS} appears as a voltage drop across the length of the channel, the voltage increases as we move along the channel from source to drain. It follows that the reverse-bias voltage between gate and channel varies at different points along the channel and is highest at the drain end. Thus the channel acquires a tapered shape and the i_D–v_{DS} characteristic becomes nonlinear. When the reverse bias at the drain end, v_{GD}, falls below the threshold voltage V_t, the channel is pinched off at the drain end and the drain current saturates. The remainder of the description of JFET operation follows closely that given for the MOSFET.

The description above clearly indicates that the JFET is a depletion-type device. Its characteristics should therefore be similar to those of the depletion-type MOSFET. This is true with a very important exception: While it is possible to operate the depletion-type MOSFET in the enhancement mode (by simply applying a positive v_{GS} if the device is n channel) this is impossible in the JFET case. If we attempt to apply a positive v_{GS}, the gate–channel pn junction becomes forward biased and the gate ceases to control the channel. Thus the maximum v_{GS} is limited to 0 V, though it is possible to go as high as 0.3 V or so since a pn junction remains essentially cut off at such a small forward voltage.

Current–Voltage Characteristics

The JFET characteristics are displayed in Fig. 5.24. Note that for a JFET the saturation region is usually called the **pinch-off region.** Although Fig. 5.24(b) shows i_D to be independent of v_{DS} in the pinch-off region, this is an ideal situation. Actual JFETs suffer from channel-length modulation in a manner very similar to that for the MOSFETs.

The JFET characteristics shown in Fig. 5.24 are for a device with $V_t = -4$ V and $K = 1$ mA/V^2. For JFETs the threshold voltage is usually called the **pinch-off voltage** and is denoted by V_P, thus $V_P = V_t$. Also, rather than specifying the conductance parameter K, the JFET manufacturer usually specifies the value of drain current at the onset of saturation for $v_{GS} = 0$, denoted I_{DSS}. It can be easily shown that

$$I_{DSS} = K\,V_t^2 = K\,V_P^2 \tag{5.27}$$

The JFET characteristics can be described by the same equations used for MOSFETs. Specifically, relabeling V_t as V_P, we can write: The n-channel JFET will be cut off for

$$v_{GS} \leq V_P$$

where V_P is negative. To turn the device on, we apply a gate-to-source voltage v_{GS},

$$V_P < v_{GS} \leq 0 \tag{5.28}$$

and a positive drain-to-source voltage v_{DS}. The JFET operates in the triode region for

$$v_{DS} \leq v_{GS} - V_P \tag{5.29}$$

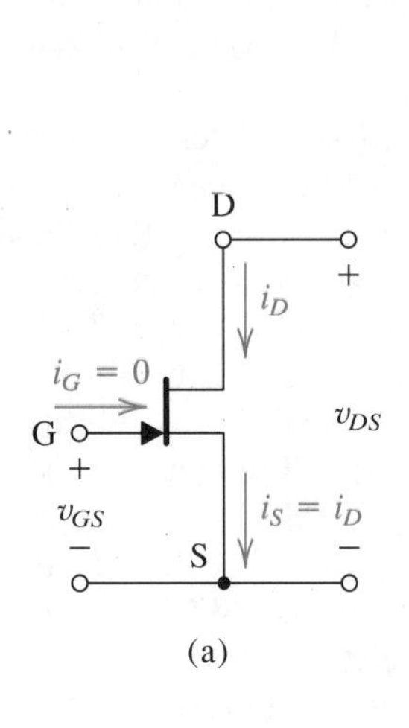

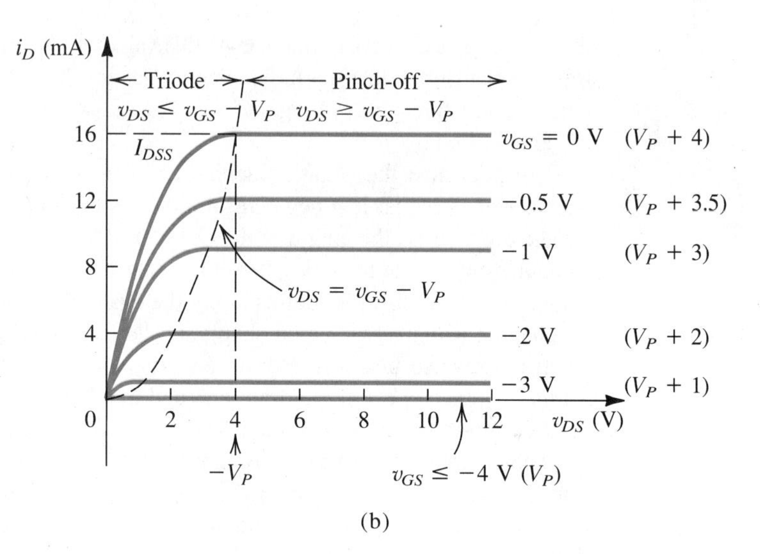

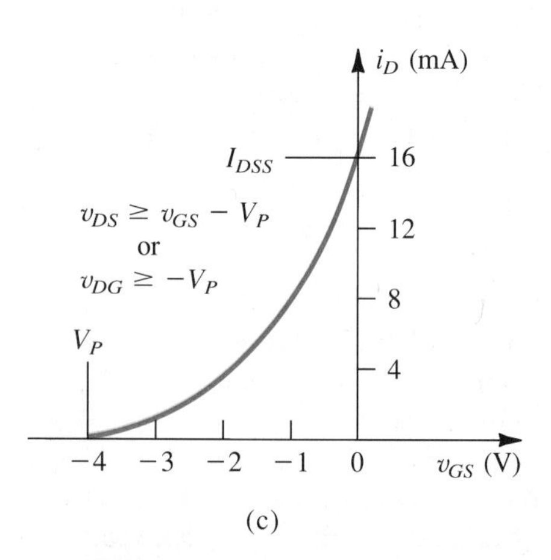

Fig. 5.24 (a) The *n*-channel JFET with the directions of current flow indicated; **(b)** the i_D–v_{DS} characteristics; and **(c)** the i_D–v_{GS} characteristic in pinch-off (saturation).

in which case the drain current is given by

$$i_D = K[2(v_{GS} - V_P)v_{DS} - v_{DS}^2]$$

If we extract a factor V_P^2 from the quantity between the square brackets and replace $K\,V_P^2$ by I_{DSS}, we can express i_D in the form

$$i_D = I_{DSS}\left[2\left(1 - \frac{v_{GS}}{V_P}\right)\left(\frac{v_{DS}}{-V_P}\right) - \left(\frac{v_{DS}}{V_P}\right)^2\right] \tag{5.30}$$

The JFET operates in saturation (pinch-off) for

$$v_{DS} \geq v_{GS} - V_P \tag{5.31}$$

In words, for the JFET to operate in pinch-off the drain voltage must be greater than the gate voltage by at least $|V_P|$. In pinch-off, the drain current is given by

$$i_D = I_{DSS}\left(1 - \frac{v_{GS}}{V_P}\right)^2 (1 + \lambda\, v_{DS}) \tag{5.32}$$

where $\lambda \equiv 1/V_A$ is a positive constant included to account for the dependence of i_D on v_{DS} in pinch-off.

Since the gate–channel junction is always reverse-biased, only a leakage current flows through the gate terminal. From Chapter 3, we know that such a current is of the order of 10^{-9} A. Although i_G is very small, and is assumed zero in almost all applications, it should be noted that the gate current in a JFET is many orders of magnitude greater than the gate current in a MOSFET. Of course the latter is so tiny because of the insulated gate structure. Another complication arises in the JFET because of the strong dependence of gate leakage current on temperature—approximately doubling for every 10°C rise in temperature, just as in the case of a reverse-biased diode (see Chapter 3).

The *p*-Channel JFET

The current–voltage characteristics of the *p*-channel JFET are described by the same equations as the *n*-channel JFET. Note, however, that for the *p*-channel JFET V_P is positive, $0 \le v_{GS} \le V_P$, v_{DS} is negative, λ and V_A are negative, and the current i_D flows out of the drain terminal. To operate the *p*-channel JFET in pinch-off, $v_{DS} \le v_{GS} - V_P$, which in words means that the drain voltage must be lower than the gate voltage by at least $|V_P|$. Otherwise, with $v_{DS} \ge v_{GS} - V_P$, the *p*-channel JFET operates in the triode region. Finally, we draw the reader's attention to the summary of JFET characteristics presented in Table 5.1 (page 317).

Exercises In Exercises 5.10 to 5.14, let the *n*-channel JFET have $V_P = -4$ V and $I_{DSS} = 10$ mA, and unless otherwise specified assume that in pinch-off (saturation) the output resistance is infinite.

5.10 Find the values of V_t and K.

Ans. -4V; $\frac{5}{8}$ mA/V^2

5.11 For $v_{GS} = -2V$, find the minimum v_{DS} for the device to operate in pinch-off. Calculate i_D for $v_{GS} = -2$ V and $v_{DS} = 3$ V.

Ans. 2 V; 2.5 mA

5.12 For $v_{DS} = 3$ V, find the change in i_D corresponding to a change in v_{GS} from -2 to -1.6 V.

Ans. 1.1 mA

5.13 For small v_{DS} calculate the value of r_{DS} at $v_{GS} = 0$ V and at $v_{GS} = -3$ V.

Ans. 200 Ω; 800 Ω

5.14 If $V_A = 400$ V, find the JFET output resistance r_o when operating in pinch-off at a current of 1 mA, 2.5 mA, and 10 mA.

Ans. 400 kΩ; 160 kΩ; 40 kΩ

5.15 Consider a p-channel JFET with $V_P = 5$ V and $I_{DSS} = 10$ mA. If $v_{GS} = 3$ V, find i_D for $v_{DS} = -1$ V and for $v_{DS} = -2$ V. Assume $\lambda = 0$.

Ans. 1.2 mA; 1.6 mA

5.16 A p-channel JFET fabricated with the standard IC process has $V_P = 1$ V, $I_{DSS} = 1$ mA, and $V_A = -100$ V. For operation in the triode region near the origin, find the value of r_{DS} at $v_{GS} = 0$ V. Also find the output resistance when the device is operated in pinch-off at $v_{GS} = 0$ V.

Ans. 500 Ω; 100 kΩ

5.17 Consider the p-channel JFET specified in Exercise 5.16. If the source and gate are grounded, what is the highest voltage that can be applied to the drain while the device operates in pinch-off? What current flows in the drain at this voltage? If the drain voltage is decreased by 4 V, find the change in drain current.

Ans. −1 V; 1 mA; +0.04 mA

5.5 FET CIRCUITS AT DC

Having studied the current–voltage characteristics of various types of FETs, we now consider FET circuits in which only dc quantities are present. Specifically, we shall present a series of design and analysis examples of FET circuits at dc.

EXAMPLE 5.1 Design the circuit of Fig. 5.25 so that the transistor operates at $I_D = 0.4$ mA and $V_D = +1$ V. The NMOS transistor has $V_t = 2$ V, $\mu_n C_{OX} = 20\ \mu A/V^2$, $L = 10\ \mu m$, and $W = 400\ \mu m$. Neglect the channel-length modulation effect (i.e., assume that $\lambda = 0$).

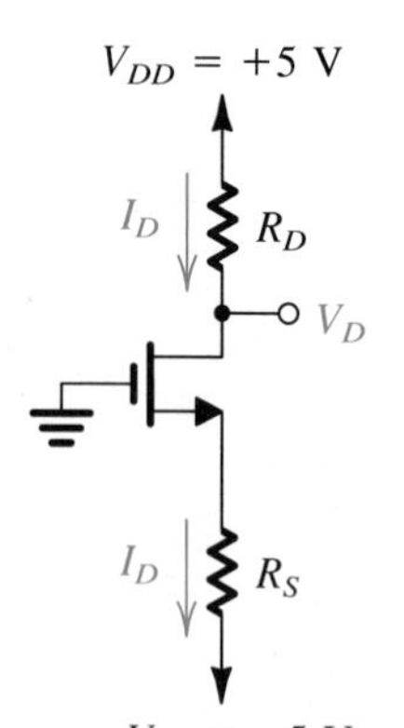

Fig. 5.25 Circuit for Example 5.1.

Solution The conductance parameter K is

$$K = \frac{1}{2}(20)(400/10) = 400\ \mu A/V^2 = 0.4\ mA/V^2$$

Since $V_D = 1$ V means operation in the saturation region, we use the saturation-region expression of i_D to determine the required value of v_{GS},

$$I_D = K(V_{GS} - V_t)^2$$

$$0.4 = 0.4(V_{GS} - 2)^2$$

This equation yields two values for V_{GS}, 1 V and 3 V. The first value does not make physical sense since it is lower than V_t. Thus $V_{GS} = 3$ V. Referring to Fig. 5.25, we note that the gate is at ground potential; thus the source must be at -3 V, and the required value of R_S can be determined from

$$R_S = \frac{V_S - (-V_{SS})}{I_D}$$

$$= \frac{-3 - (-5)}{0.4} = 5 \text{ k}\Omega$$

To establish a dc voltage of $+1$ V at the drain we must select R_D as follows:

$$R_D = \frac{V_{DD} - V_D}{I_D}$$

$$= \frac{5 - 1}{0.4} = 10 \text{ k}\Omega$$

EXAMPLE 5.2

Design the circuit in Fig. 5.26 to obtain a current I_D of 0.4 mA. Give the value required for R and find the dc voltage V_D. Let the NMOS transistor have $V_t = 2$ V, $\mu n\, C_{OX} = 20\ \mu A/V^2$, $L = 10\ \mu m$, and $W = 100\ \mu m$. Neglect the channel-length modulation effect (that is, assume $\lambda = 0$).

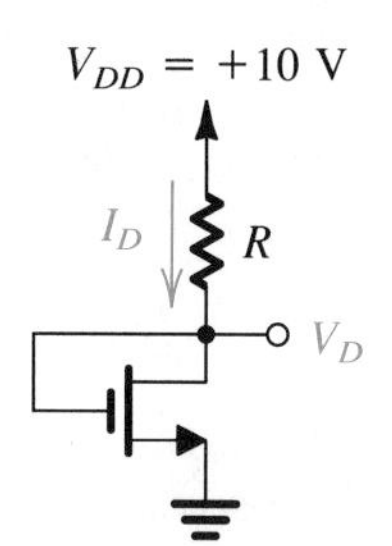

Fig. 5.26 Circuit for Example 5.2.

Solution Because $V_{DG} = 0$, the FET is operating in the saturation region. Thus

$$I_D = K(V_{GS} - V_t)^2$$

$$0.4 = \frac{1}{2}(20)(10^{-3})(100/10)(V_{GS} - 2)^2$$

which yields two values for V_{GS}, namely 4 and 0. The second value obviously does not make physical sense since it is lower than V_t. Thus $V_{GS} = 4$ V, and the drain voltage will be

$$V_D = +4 \text{ V}$$

The required value for R can be found as follows:

$$R = \frac{V_{DD} - V_D}{I_D}$$

$$= \frac{10 - 4}{0.4} = 15 \text{ k}\Omega$$

EXAMPLE 5.3

Design the circuit in Fig. 5.27 to establish a drain voltage of 0.1 V. What is the effective resistance between drain and source at this operating point? Let $V_t = 1$ V and $K = 0.5$ mA/V^2.

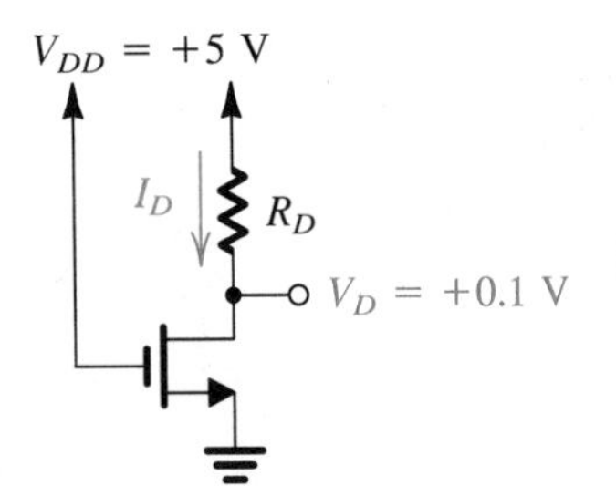

Fig. 5.27 Circuit for Example 5.3.

Solution

Since the drain voltage is lower than the gate voltage by 4.9 V and $V_t = 1$ V, the MOSFET is operating in the triode region. Thus the current I_D is given by

$$I_D = 0.5[2(5 - 1) \times 0.1 - 0.01]$$

$$= 0.395 \text{ mA}$$

The required value for R_D can be found as follows:

$$R_D = \frac{V_{DD} - V_D}{I_D}$$

$$= \frac{5 - 0.1}{0.395} = 12.4 \text{ k}\Omega$$

(Obviously, in a practical design problem one selects the closest standard value available for, say, 5% resistors—in this case, 12 kΩ; see Appendix F). The effective drain-to-source resistance can be determined as follows:

$$r_{DS} = \frac{V_{DS}}{I_D}$$

$$= \frac{0.1}{0.395} = 253 \ \Omega$$

EXAMPLE 5.4 Analyze the circuit shown in Fig. 5.28(a) to determine the voltages at all nodes and the currents through all branches. Let $V_t = 1$ V and $K = 0.5$ mA/V^2. Neglect the channel-length modulation effect (i.e., assume $\lambda = 0$).

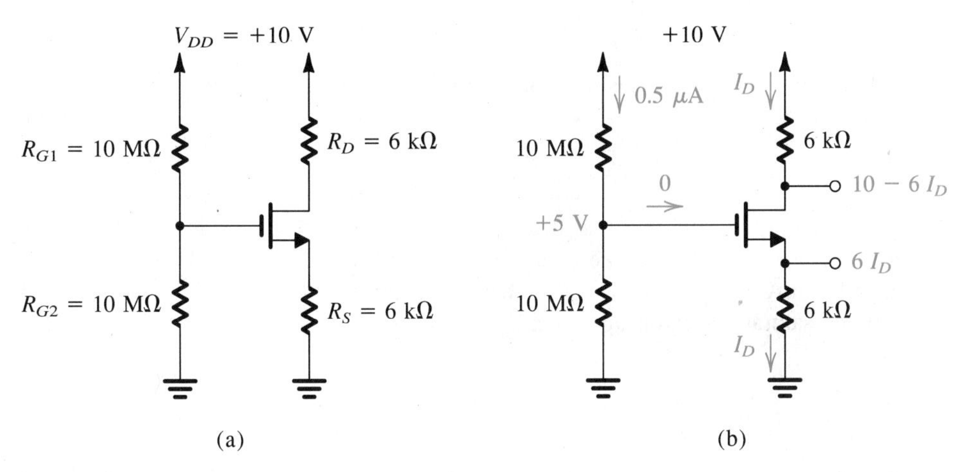

Fig. 5.28 **(a)** Circuit for Example 5.4; **(b)** the circuit with some of the analysis details shown.

Solution Since the gate current is zero, the voltage at the gate is simply determined by the voltage divider formed by the two 10-MΩ resistors,

$$V_G = 10 \times \frac{10}{10 + 10} = +5 \text{ V}$$

With this positive voltage at the gate the NMOS transistor will be turned on. We do not know, however, whether the transistor will be operating in the saturation region or in the triode region. We shall assume saturation-region operation, solve the problem, and then check the validity of our assumption. Obviously, if our assumption turns out not to be valid we will have to solve the problem again for triode-region operation.

Refer to Fig. 5.28(b). Since the voltage at the gate is 5 V and the voltage at the source is $I_D \times 6 = 6\,I_D$ we have

$$V_{GS} = 5 - 6\,I_D$$

Thus I_D is given by

$$\begin{aligned} I_D &= K(V_{GS} - V_t)^2 \\ &= 0.5(5 - 6\,I_D - 1)^2 \end{aligned}$$

which yields the following quadratic equation in I_D:

$$18\,I_D^2 - 25\,I_D + 8 = 0$$

This equation yields two values for I_D: 0.89 mA and 0.5 mA. The first answer results in a source voltage of $6 \times 0.89 = 5.34$, which is greater than the gate voltage and does not make physical sense. Thus

$$I_D = 0.5 \text{ mA}$$

$$V_S = 0.5 \times 6 = +3 \text{ V}$$

$$V_{GS} = 5 - 3 = 2 \text{ V}$$

$$V_D = 10 - 6 \times 0.5 = +7 \text{ V}$$

Since $V_D > V_G - V_t$, the transistor is operating in saturation as initially assumed.

EXAMPLE 5.5 Design the circuit of Fig. 5.29 so that the transistor operates in saturation with $I_D = 0.5$ mA and $V_D = +3$ V. Let the enhancement-type PMOS transistor have $V_t = -1$ V and $K = 0.5\text{ mA/V}^2$. Assume $\lambda = 0$. What is the largest value that R_D can have while maintaining saturation-region operation?

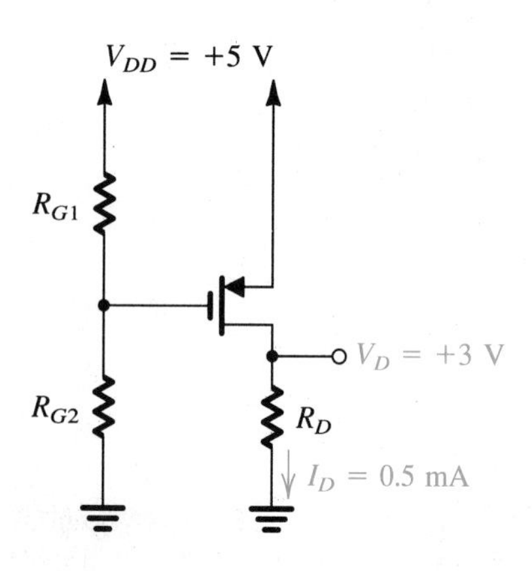

Fig. 5.29 Circuit for Example 5.5.

Solution Since the MOSFET is in saturation and $I_D = 0.5$ mA, we can write

$$0.5 = 0.5[V_{GS} - (-1)]^2$$

Recalling that V_{GS} has to be negative ($V_{GS} < V_t$), we find that the only solution to this equation that makes physical sense is $V_{GS} = -2$ V. Now since the source is at $+5$ V, the gate voltage must be set to $+3$ V. This can be achieved by the appropriate selection of the values of R_{G1} and R_{G2}. A possible selection is $R_{G1} = 2\text{ M}\Omega$ and $R_{G2} = 3\text{ M}\Omega$.

The value of R_D can be found from

$$R_D = \frac{V_D}{I_D} = \frac{3}{0.5} = 6 \text{ k}\Omega$$

Saturation-mode operation will be maintained up to the point that V_D exceeds V_G by $|V_t|$, that is

$$V_{D_{\max}} = 3 + 1 = 4 \text{ V}$$

This value of drain voltage is obtained with R_D given by

$$R_D = \frac{4}{0.5} = 8 \text{ k}\Omega$$

EXAMPLE 5.6 Analyze the circuit in Fig. 5.30 to determine I_D and V_D. Let the depletion-mode PMOS transistor have $V_t = 1$ V, $K = 0.5$ mA/V^2, and $\lambda = 0$.

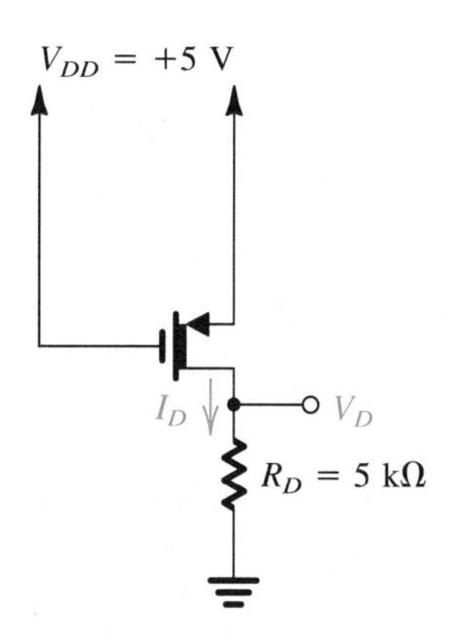

Fig. 5.30 Circuit for Example 5.6.

Solution Since this is a depletion device, it conducts for $V_{GS} = 0$. Assuming operation in the saturation region, we find I_D from

$$I_D = 0.5(0 - 1)^2 = 0.5 \text{ mA}$$

and the voltage V_D becomes

$$V_D = I_D \times 5 = 0.5 \times 5 = +2.5 \text{ V}$$

Since V_D is lower than V_G by 2.5 V (which is greater than V_t), the transistor is operating in the saturation region, as assumed.

EXAMPLE 5.7 For the circuit analyzed in the previous example, find the largest value that R_D can have while the transistor remains in the saturation region.

Solution Since this is a depletion device, saturation-region operation is maintained up to the point that V_D reaches $V_G - V_t = 5 - 1 = 4$ V. Since $I_D = 0.5$ mA (from Example 5.6) the largest R_D possible is

$$R_D = \frac{4}{0.5} = 8 \text{ k}\Omega$$

EXAMPLE 5.8 Design the circuit in Fig. 5.31 to establish a dc voltage of +9.9 V at the source. At this operating point, what is the effective resistance between source and drain of the transistor? Let $V_t = -1$ V and $K = 0.5$ mA/V^2.

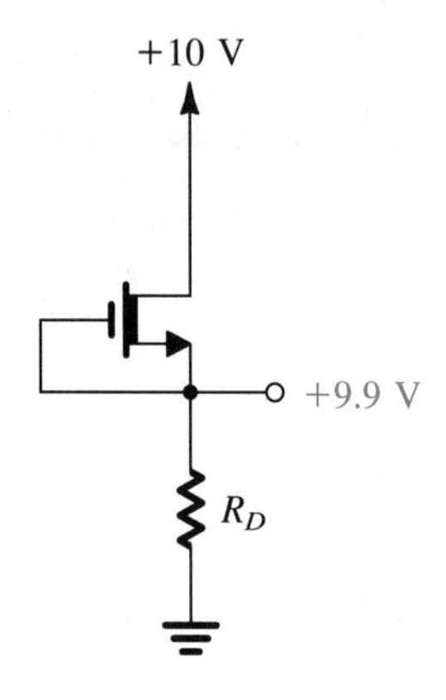

Fig. 5.31 Circuit for Example 5.8.

Solution Here V_G is lower than V_D by only 0.1 V, which is lower than the $|V_t|$ required for saturation-region operation. Thus the depletion-type NMOS transistor is operating in the triode region with $V_{DS} = 0.1$ V and $V_{GS} = 0$ V. Thus the drain current I_D is given by

$$I_D = 0.5[2(0 - (-1)) \times 0.1 - 0.01]$$
$$\simeq 0.1 \text{ mA}$$

We select R_D according to

$$R_D = \frac{9.9 \text{ V}}{0.1 \text{ mA}} = 99 \text{ k}\Omega \simeq 100 \text{ k}\Omega$$

The effective source-to-drain resistance is

$$r_{DS} = \frac{V_{DS}}{I_D} = \frac{0.1 \text{ V}}{0.1 \text{ mA}} = 1 \text{ k}\Omega$$

EXAMPLE 5.9 Design the circuit of Fig. 5.32 to obtain $I_D = 4$ mA and $V_D = 6$ V. Let the n-channel JFET have $V_P = -4$ V and $I_{DSS} = 16$ mA. Assume $\lambda = 0$.

Solution The JFET has $V_t = V_P = -4$ V and a conductance parameter $K = I_{DSS}/V_t^2 = 16/16 = 1$ mA/V^2. Since the required $V_D = 6$ V and $V_G = 0$ V, the FET will be operating in the pinch-off (saturation) region. Thus

$$I_D = K(V_{GS} - V_t)^2$$
$$4 = 1[V_{GS} - (-4)]^2$$

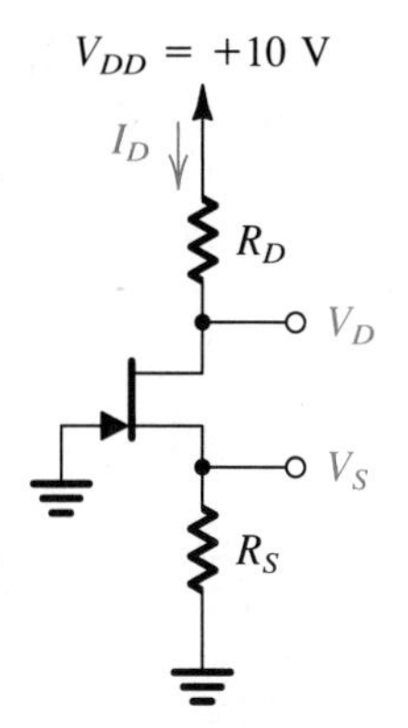

Fig. 5.32 Circuit for Example 5.9.

This equation leads to two answers: $V_{GS} = -6$ V and -2 V. Since the first value is $<V_P$, it implies cutoff and is thus physically meaningless. It follows that

$$V_{GS} = -2 \text{ V}$$

$$V_S = +2 \text{ V}$$

$$R_S = \frac{V_S}{I_D} = \frac{2}{4} = 0.5 \text{ k}\Omega$$

$$R_D = \frac{V_{DD} - V_D}{I_D} = \frac{10 - 6}{4} = 1 \text{ k}\Omega$$

EXAMPLE 5.10 Analyze the circuit of Fig. 5.33 to determine V_S and V_D. The p-channel JFET has $V_P = +2$ V, $I_{DSS} = 4$ mA, and $\lambda = 0$.

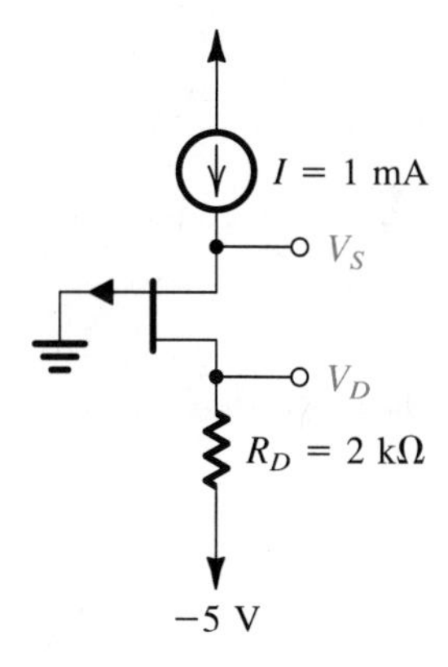

Fig. 5.33 Circuit for Example 5.10.

Solution The JFET has $V_t = V_P = 2$ V and $K = I_{DSS}/V_t^2 = 1$ mA/V^2. The drain voltage can be found from

$$V_D = -5 + I_D R_D = -5 + 1 \times 2 = -3 \ V$$

Thus the drain voltage is lower than the gate voltage by an amount greater than V_t, which guarantees operation in the pinch-off (saturation) region. Thus we can use the drain-current equation to write

$$1 = 1(V_{GS} - 2)^2$$

which yields two values for V_{GS}: +3 V and +1 V. The first value is greater than V_P and thus implies cut-off operation, which is impossible in this case. Thus $V_{GS} = +1$ V and $V_S = -1$ V.

Exercises 5.18 For the circuit designed in Example 5.1, find the largest value that R_D can have while the MOSFET remains in saturation.

Ans. 17.5 kΩ

5.19 Consider the circuit of Fig. 5.26, which is designed in Example 5.2 (to which you should refer before solving this problem). Let the voltage V_D be applied to the gate of another transistor Q_2, as shown in Fig. E5.19. Assume that Q_2 is identical to Q_1. Find the drain current and voltage of Q_2. (Assume $\lambda = 0$.)

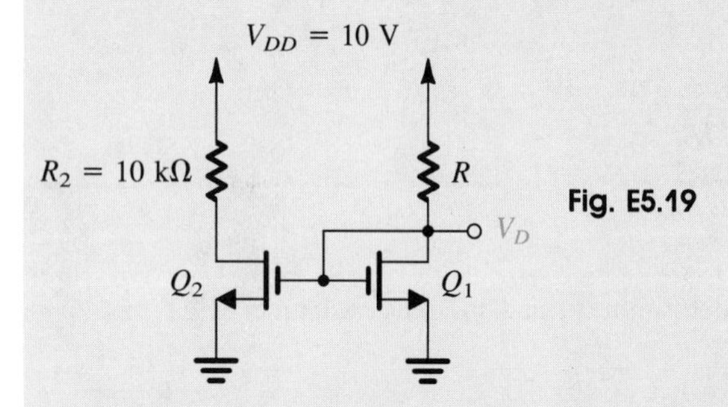

Fig. E5.19

Ans. 0.4 mA; +6 V

5.20 Consider the circuit of Fig. 5.27, which was designed in Example 5.3. If the value of R_D is made double that found in Example 5.3, find the new values of V_D, I_D, and r_{DS}.

Ans. 0.05 V; 0.2 mA; 250 Ω

D5.21 Consider the circuit in Fig. 5.28(a) with new values for the resistors. Design the circuit to obtain approximately 4 V at the gate, a drain current of about 1 mA, and a drain voltage of about 4 V. The transistor has $V_t = 2$ V, $K = 1$ mA/V^2, and $\lambda = 0$.

Ans. Possible values for R_{G1} and R_{G2} are 6.2 MΩ and 3.9 MΩ, respectively; $R_S = 820$ Ω; $R_D = 5.6$ kΩ. (Note that these are standard values for resistors of 5% tolerance; see Appendix F. The result is a gate voltage of 3.86 V, a drain current of 1.03 mA, and a drain voltage of 4.23 V.)

5.22 Analyze the circuit in Fig. E5.22 to determine I_D and V_D. The depletion MOSFET has $V_t = -1$ V, $K = 0.5$ mA/V^2, and $\lambda = 0$.

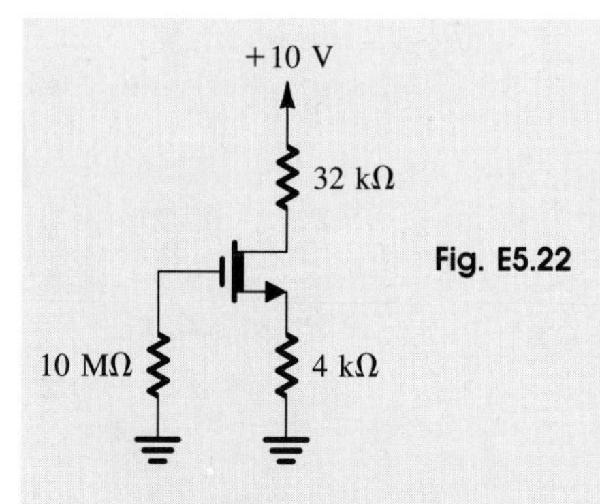

Fig. E5.22

Ans. 0.125 mA; 6 V

D5.23 The JFET in the circuit of Fig. E5.23 has $V_P = -3$ V, $I_{DSS} = 9$ mA, and $\lambda = 0$. Find the values of all resistors so that $V_G = 5$ V, $I_D = 4$ mA, and $V_D = 11$ V. Design for 0.05 mA in the voltage divider.

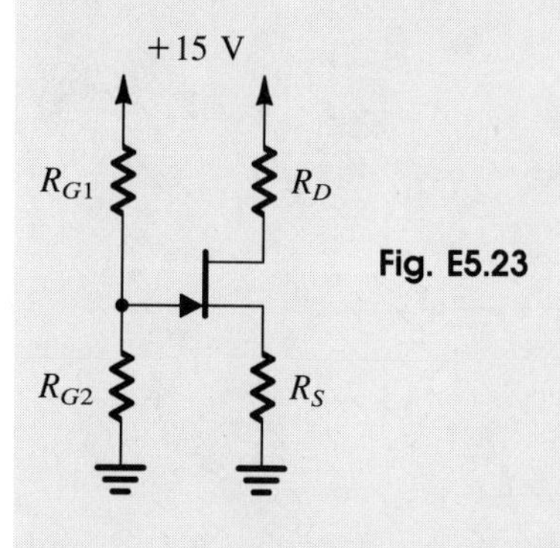

Fig. E5.23

Ans. $R_{G1} = 200$ kΩ; $R_{G2} = 100$ kΩ; $R_S = 1.5$ kΩ; $R_D = 1$ kΩ

5.6 THE FET AS AN AMPLIFIER

In this section we study the operation of the field-effect transistor as an amplifier. Although we shall use an enhancement-type n-channel MOSFET, the results apply equally well to the other FET types.

Graphical Analysis

Since FETs are well characterized by equations, it is rarely necessary to apply graphical techniques in the analysis of FET circuits. Nevertheless, it is illustrative and instructive to begin the study of FET amplifiers with the graphical analysis of the conceptual MOS amplifier circuit shown in Fig. 5.34(a). The enhancement-type n-channel MOSFET is shown biased with a battery V_{GS}, clearly an impractical arrangement but one that should

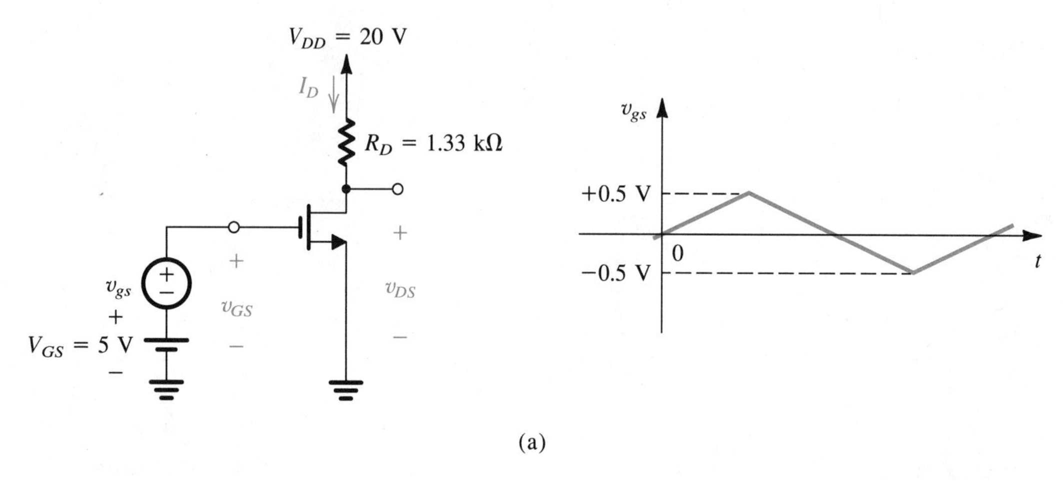

(a)

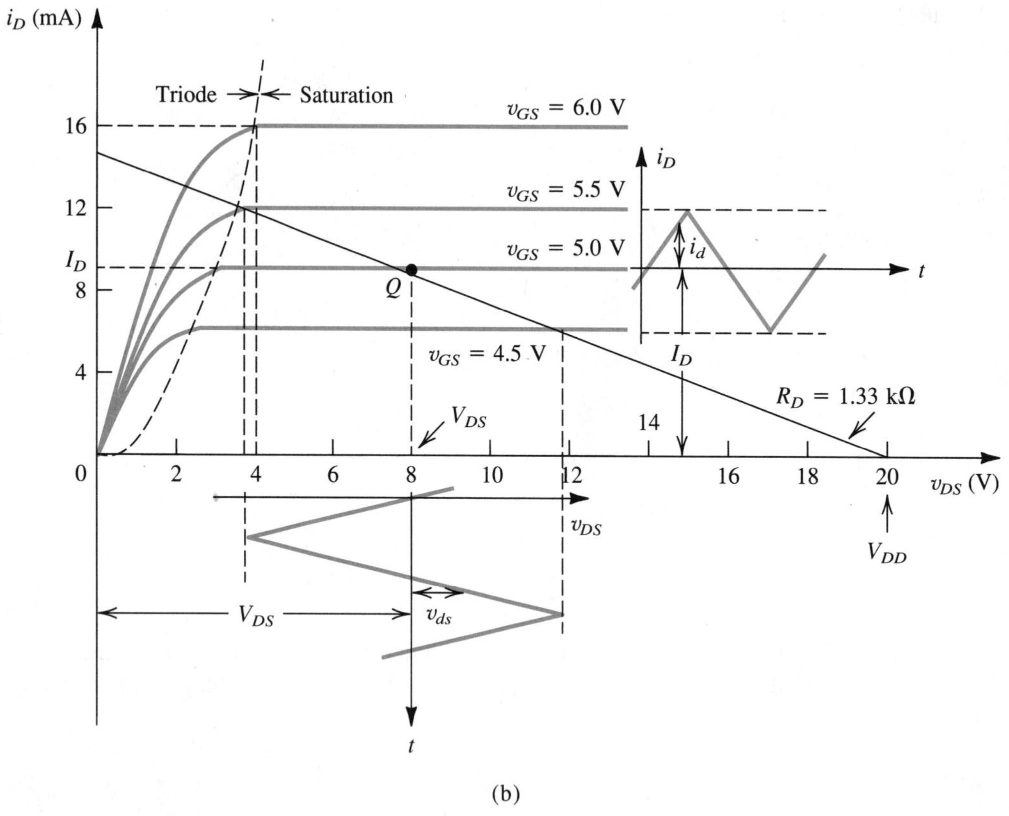

(b)

Fig. 5.34 Graphical analysis of the conceptual MOS amplifier circuit. **(a)** The MOS transistor has $V_t = 2$ V and $K = 1$ mA/V² and is biased using a battery $V_{GS} = 5$ V. A triangular-wave input signal v_{gs} with 1-V peak-to-peak amplitude is applied as shown. **(b)** The result when the instantaneous operating point is confined to the saturation region: Reasonably linear operation with a voltage gain of 8 is obtained. **(c)** A larger load resistance is used, and the operating point is allowed to enter the triode region. The result is severe nonlinear distortion.

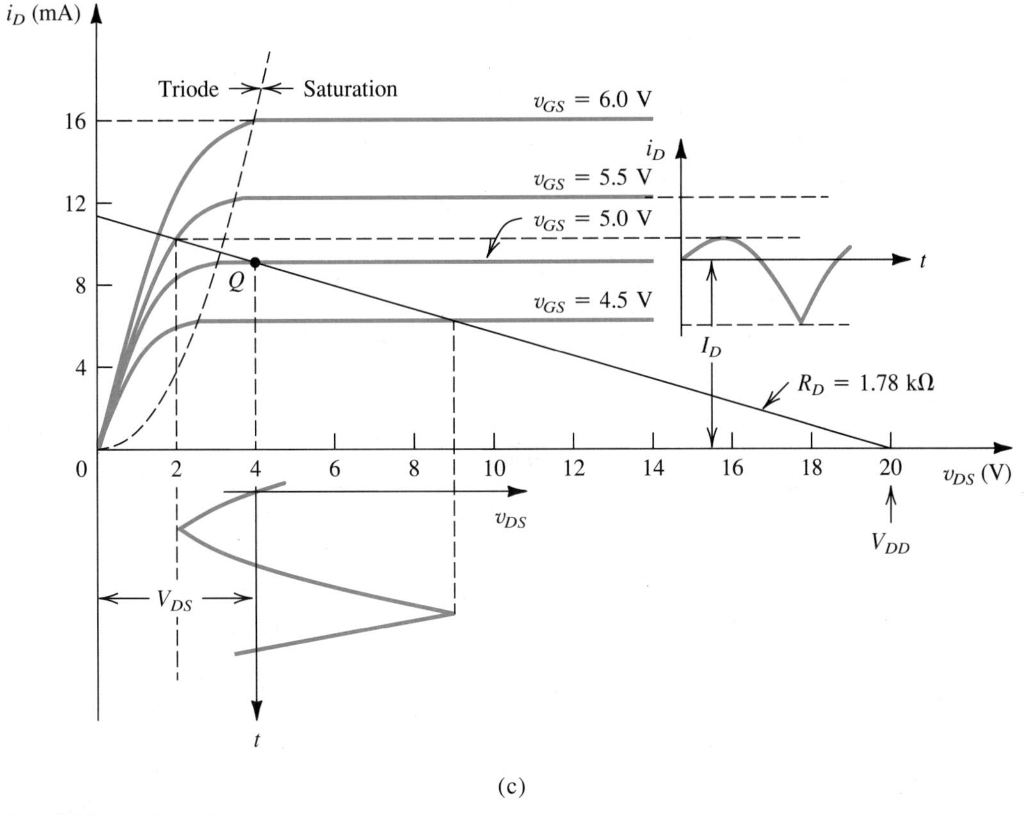

(c)

Fig. 5.34 *(continued)*

serve our present purpose. Superimposed on the gate-to-source dc bias V_{GS} is a time-varying signal v_{gs} that we wish to amplify; thus the total instantaneous gate-to-source voltage is

$$v_{GS} = V_{GS} + v_{gs} \tag{5.33}$$

At any instant of time the operating point will be located on the i_D–v_{DS} curve corresponding to the specific value of v_{GS}. Where it will be on that curve will be determined by V_{DD} and R_D from

$$v_{DS} = V_{DD} - R_D i_D \tag{5.34}$$

which may be rewritten

$$i_D = \frac{V_{DD}}{R_D} - \frac{1}{R_D} v_{DS} \tag{5.35}$$

This is a linear equation in the variables i_D and v_{DS} and can be represented by a straight line on the i_D–v_{DS} plane. Such a line intercepts the v_{DS} axis at V_{DD} and has a slope equal to $-1/R_D$. Since R_D represents the **load resistance** of the amplifier, the straight line representing Eq. (5.35) is called the **load line.** The instantaneous operating point of the MOSFET will lie at the intersection of the load line and the i_D–v_{DS} curve corresponding to the instantaneous value of v_{GS}. The coordinates of the operating point are the instantaneous values of i_D and v_{DS}.

To give concreteness to our analysis we use numerical values. Specifically, the MOSFET is assumed to have a threshold voltage $V_t = 2$ V and a conductance parameter $K = 1\ \text{mA/V}^2$. As indicated in Fig. 5.34(a), the transistor is biased at $V_{GS} = 5$ V and its drain is connected to a positive supply $V_{DD} = 20$ V through a resistance $R_D = 1.33\ \text{k}\Omega$. The signal to be amplified has a triangular waveform with 1-V peak-to-peak amplitude. Figure 5.34(b) shows the i_D–v_{DS} characteristics of the MOSFET together with the load line corresponding to $R_D = 1.33\ \text{k}\Omega$. In the absence of the input signal v_{gs}, the MOSFET will operate at the point labeled Q, which is at the intersection of the curve for $v_{GS} = 5$ V and the load line. This is the **dc bias point, dc operating point,** or **quiescent point,** as it is variously called. The coordinates of Q determine the dc current in the drain, $I_D = 9$ mA, and the dc voltage at the drain, $V_{DS} = 8$ V.

When the triangular-wave signal v_{gs} is applied, the instantaneous operating point will move along the load line in correspondence with the total instantaneous voltage v_{GS}. This is shown in Fig. 5.34(b), from which we observe that, for instance, at the positive peak of the input signal, $v_{gs} = 0.5$ V, $v_{GS} = 5 + 0.5 = 5.5$ V, the corresponding drain current is 12.25 mA, and the corresponding drain voltage is 3.7 V. It follows that the total instantaneous drain current i_D and the total instantaneous drain voltage v_{DS} can be determined in this manner, point by point. Figure 5.34(b) shows the resulting waveforms. We note that superimposed on the dc current I_D we obtain a time-varying component i_d that is almost perfectly triangular in shape. Also, superimposed on the dc voltage v_{DS} we obtain a signal component that also is almost perfectly triangular in shape. This is the output voltage signal and, except for phase inversion, is an amplified replica of the input signal: It has a peak-to-peak amplitude of about 8 V, and thus the gain of the amplifier is -8 V/V.

In the example above we were able to obtain almost linear amplification from the nonlinear MOSFET by properly choosing the dc bias point Q and by keeping the input signal amplitude small. In the following sections we will study the analysis and design of FET amplifiers in detail. At this stage, however, it is important to note that the instantaneous operating point was confined to the saturation region. This is done so that the MOSFET operates as a current source whose magnitude is controlled by v_{gs}. If the instantaneous operating point is allowed to leave the saturation region, the FET no longer operates as a linear-controlled current source and severe nonlinear distortion may result. This is illustrated in Fig. 5.34(c), where the MOSFET is still biased at $V_{GS} = 5$ V but a larger resistance $R_D = 1.78\ \text{k}\Omega$ is used. As a result of using the larger load resistance the instantaneous operating point enters the triode region during most of the positive halves of v_{gs}. As can be seen, the signal component of i_D and the output signal voltage v_{ds} are severely distorted.

In summary, for the FET to operate as a linear amplifier it must be biased at a point in the middle of the saturation region, the instantaneous operating point must at all times be confined to the saturation region, and the input signal must be kept sufficiently small. The last point will be elaborated on in the following.

Algebraic Analysis

We next present an algebraic analysis of the conceptual amplifier circuit of Fig. 5.34(a). With the input signal v_{gs} set to zero we obtain for the dc drain current I_D and the dc drain voltage V_{DS} (or simply V_D since the source is grounded) the relationship

$$I_D = K(V_{GS} - V_t)^2 \tag{5.36}$$

$$V_D = V_{DD} - R_D I_D \tag{5.37}$$

where we have neglected the effect of channel-length modulation. With the signal v_{gs} superimposed on V_{GS}, the total instantaneous gate-to-source voltage v_{GS} is given by

$$v_{GS} = V_{GS} + v_{gs} \tag{5.38}$$

Correspondingly, the total instantaneous current i_D will be

$$\begin{aligned} i_D &= K(v_{GS} - V_t)^2 \\ &= K(V_{GS} + v_{gs} - V_t)^2 \\ &= K(V_{GS} - V_t)^2 + 2K(V_{GS} - V_t)v_{gs} + Kv_{gs}^2 \end{aligned} \tag{5.39}$$

The first term on the right-hand side of Eq. (5.39) can be recognized as the dc or quiescent current I_D [Eq. (5.36)]. The second term represents a current component that is directly proportional to the input signal v_{gs}. The last term is a current component that is proportional to the square of the input signal. This last component is undesirable because it represents nonlinear distortion. To reduce the nonlinear distortion introduced by the MOSFET, the input signal should be kept small,

$$v_{gs} \ll 2(V_{GS} - V_t) \tag{5.40}$$

If this *small-signal condition* is satisfied, we may neglect the last term in Eq. (5.39) and express i_D as

$$i_D \simeq I_D + i_d \tag{5.41}$$

where the signal current i_d is given by

$$i_d = 2K(V_{GS} - V_t)v_{gs}$$

The constant relating i_d and v_{gs} is the transconductance g_m,

$$g_m \equiv \frac{i_d}{v_{gs}} = 2K(V_{GS} - V_t) \tag{5.42}$$

Figure 5.35 presents a graphical interpretation of the small-signal operation of the enhancement MOSFET amplifier. Note that g_m is equal to the slope of the i_D–v_{GS} characteristic at the operating point,

$$g_m = \left.\frac{\partial i_D}{\partial v_{GS}}\right|_{v_{GS}=V_{GS}} \tag{5.43}$$

The Transconductance g_m

We shall now take a closer look at MOSFET transconductance. Substituting for K from Eq. (5.6) into Eq. (5.42) gives

$$g_m = (\mu_n C_{OX})(W/L)(V_{GS} - V_t) \tag{5.44}$$

This relationship indicates that g_m depends on the W/L ratio of the MOS transistor; hence to obtain relatively large transconductance the device must be short and wide. We also

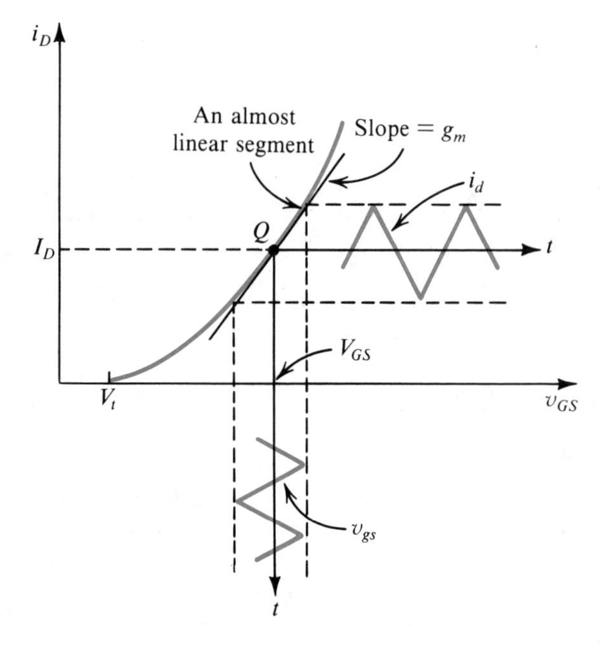

Fig. 5.35 Small-signal operation of the enhancement MOSFET amplifier.

observe that for a given device the transconductance is proportional to the excess voltage $\Delta V = V_{GS} - V_t$, the amount by which the bias voltage V_{GS} exceeds the threshold voltage V_t. Note, however, that increasing g_m by biasing the device at a larger V_{GS} has the disadvantage of reducing the allowable voltage signal swing at the drain.

Another useful expression for g_m can be obtained by substituting for $(V_{GS} - V_t)$ in Eq. (5.44) by $\sqrt{I_D/K}$ (from Eq. 5.36) and again substituting for K by $\frac{1}{2}\mu_n C_{OX}(W/L)$. The result is

$$g_m = \sqrt{2\mu_n C_{OX}}\sqrt{W/L}\sqrt{I_D} \tag{5.45}$$

This expression shows that

1. For a given MOSFET, g_m is proportional to the square root of the dc bias current.
2. At a given bias current, g_m is proportional to $\sqrt{W/L}$.

In contrast, the transconductance of the bipolar junction transistor (BJT) studied in Chapter 4 is proportional to the bias current and is independent of the physical size and geometry of the device.

To gain some insight into the values of g_m obtained in MOSFETs consider an integrated-circuit device operating at $I_D = 1$ mA and having $\mu_n C_{OX} = 20\ \mu A/V^2$. Equation (5.45) shows that for $W/L = 1$, $g_m = 0.2$ mA/V, whereas a device for which $W/L = 100$ has $g_m = 2$ mA/V. In contrast, a BJT operating at a collector current of 1 mA has $g_m = 40$ mA/V. However, in spite of their low g_m, MOSFETs have many other advantages, including high input impedance, small size, low power dissipation, and ease of fabrication.

Voltage Gain

Returning to the circuit of Fig. 5.34(a), we can express the total instantaneous drain voltage v_D as follows:

$$v_D = V_{DD} - R_D i_D$$

Under the small-signal condition we have

$$v_D = V_{DD} - R_D(I_D + i_d)$$

which can be rewritten

$$v_D = V_D - R_D i_d$$

Thus the signal component of the drain voltage is

$$v_d = -i_d R_D = -g_m R_D v_{gs}$$

which indicates that the voltage gain is given by

$$\frac{v_d}{v_{gs}} = -g_m R_D \tag{5.46}$$

The minus sign in Eq. (5.46) indicates that the output signal v_d is 180° out of phase with respect to the input signal v_{gs}. This is illustrated in Fig. 5.36, which shows v_{GS} and v_D. The input signal is assumed to have a triangular waveform with an amplitude much smaller than $2(V_{GS} - V_t)$, the small-signal condition in Eq. (5.40), to ensure linear operation. For operation in the saturation region at all times, the minimum value of v_D should not fall below the corresponding value of v_G by more than V_t. Also, the maximum value of v_D should be smaller than V_{DD}; otherwise the FET will enter the cutoff region and the peaks of the output signal waveform will be clipped off.

Separating the DC Analysis and the Signal Analysis

From the above analysis we see that under the small-signal approximation, signal quantities are superimposed on dc quantities. For instance, the total drain current i_D equals the dc current I_D plus the signal current i_d, the total drain voltage $v_D = V_D + v_d$, and so on. It follows that the analysis and design can be greatly simplified by separating dc or bias calculations from small-signal calculations. That is, once a stable dc operating point has been established and all dc quantities calculated, we may then perform signal analysis ignoring dc quantities.

Small-Signal Equivalent Circuit Models

From a signal point of view the FET behaves as a voltage-controlled current source. It accepts a signal v_{gs} between gate and source and provides a current $g_m v_{gs}$ at the drain terminal. The input resistance of this controlled source is very high—ideally, infinite. The output resistance—that is, the resistance looking into the drain—also is high, and we have assumed it to be infinite thus far. Putting all of this together we arrive at the circuit in Fig. 5.37(a), which represents the small-signal operation of the FET and is thus a small-signal model or a small-signal equivalent circuit.

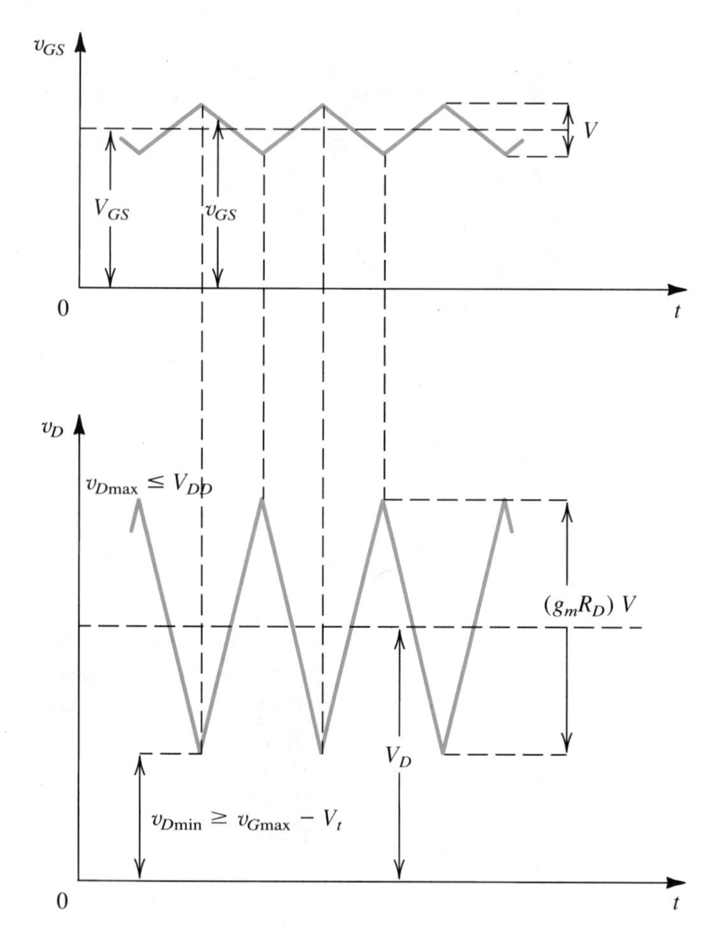

Fig. 5.36 Total instantaneous voltages v_{GS} and v_D for the circuit in Fig. 5.34(a).

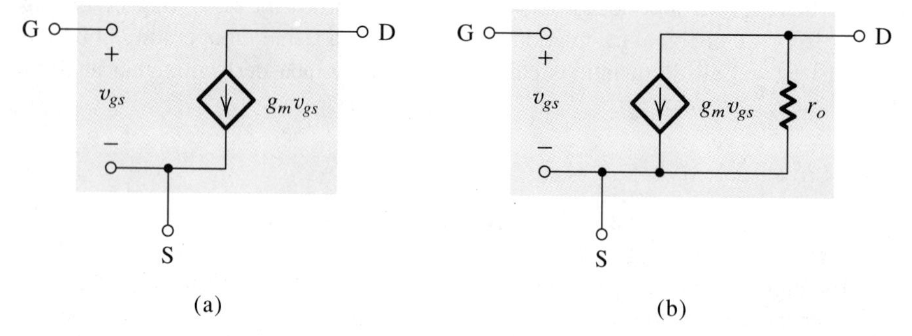

Fig. 5.37 Small-signal models for the FET: **(a)** neglecting the dependence of i_D on v_{DS} in saturation (channel-length modulation effect); and **(b)** including the effect of channel-length modulation, modeled by output resistance $r_o = |V_A|/I_D$.

In the analysis of a FET amplifier circuit, the FET can be replaced by the equivalent circuit model shown in Fig. 5.37(a). The rest of the circuit remains unchanged except that *dc voltage sources are replaced by short circuits*. This is a result of the fact that the voltage across an ideal dc voltage source does not change and thus there will always be a zero voltage signal across a dc voltage source. The circuit resulting can then be used to perform any required signal analysis, such as calculating voltage gain.

The most serious shortcoming of the small-signal model of Fig. 5.37(a) is that it assumes that the drain current in saturation is independent of the drain voltage. From our study of the FET characteristics in saturation we know that the drain current does in fact depend on v_{DS} in a linear manner. Such dependence was modeled by a finite resistance r_o between drain and source, whose value is given approximately by

$$r_o \simeq \frac{|V_A|}{I_D} \tag{5.47}$$

where $V_A = 1/\lambda$ is a FET parameter that either is specified or can be measured. Typically, r_o is in the range 10 to 1000 kΩ. It follows that the accuracy of the small-signal model can be improved by including r_o in parallel with the controlled source, as shown in Fig. 5.37(b).

It is important to note that the small-signal model parameters g_m and r_o depend on the dc bias point of the FET.

Returning to the amplifier of Fig. 5.34(a), we find that replacing the FET with the small-signal model of Fig. 5.37b results in the voltage-gain expression

$$\frac{v_d}{v_{gs}} = -g_m(R_D//r_o) \tag{5.48}$$

Thus the finite output resistance r_o results in a reduction in the magnitude of the voltage gain.

The small-signal equivalent circuit model of Fig. 5.37(b) applies to all types of FETs including *n*-channel and *p*-channel. In the case of the JFET the following alternative expression for g_m is often employed

$$g_m = \frac{2I_{DSS}}{|V_P|}\sqrt{\frac{I_D}{I_{DSS}}}$$

Finally, it should be noted that the small-signal model derived above applies only at low and medium frequencies. High-frequency FET models will be studied in Chapter 7.

EXAMPLE 5.11

Figure 5.38(a) shows an enhancement MOSFET amplifier in which the input signal v_i is coupled to the gate via a large capacitor, and the output signal at the drain is coupled to the load resistance R_L via another large capacitor. We wish to analyze this amplifier circuit to determine its small-signal voltage gain and its input resistance. The transistor has $V_t = 1.5$ V, $K = 0.125$ mA/V^2, and $V_A = 50$ V. Assume the coupling capacitors to be sufficiently large as to act as short circuits at the signal frequencies of interest.

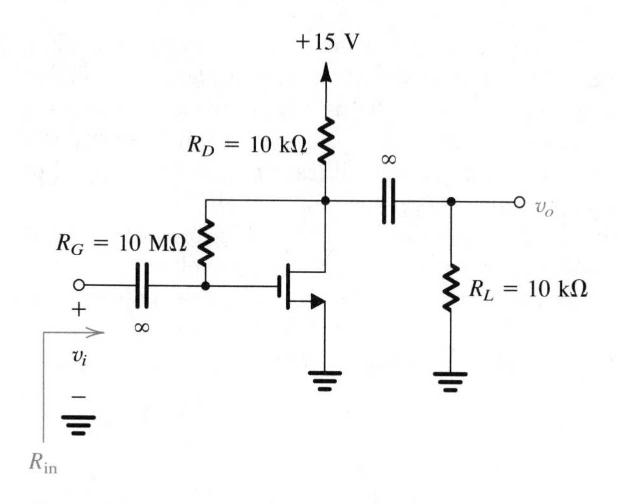

(a)

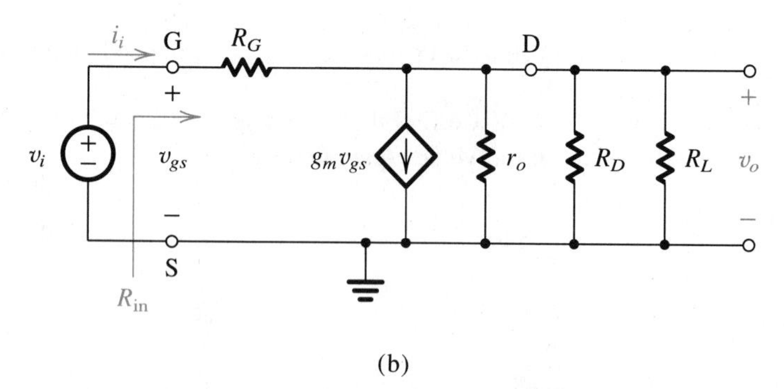

(b)

Fig. 5.38 Example 5.11: **(a)** amplifier circuit; **(b)** equivalent circuit model.

Solution We first evaluate the dc operating point as follows:

$$I_D = 0.125(V_{GS} - 1.5)^2 \tag{5.49}$$

where, for simplicity, we have neglected the channel-length modulation effect. Since the dc gate current is zero, there will be no dc voltage drop across R_G; thus $V_{GS} = V_D$, which, when substituted in Eq. (5.49), yields

$$I_D = 0.125(V_D - 1.5)^2 \tag{5.50}$$

Also,

$$V_D = 15 - R_D I_D = 15 - 10I_D \tag{5.51}$$

Solving Eqs. (5.50) and (5.51) together gives

$$I_D = 1.06 \text{ mA} \quad \text{and} \quad V_D = 4.4 \text{ V}$$

(Note that the other solution to the quadratic equation is not physically meaningful.)

The value of g_m is given by

$$g_m = 2K(V_{GS} - V_t)$$
$$= 2 \times 0.125(4.4 - 1.5) = 0.725 \text{ mA/V}$$

The output resistance r_o is given by

$$r_o = \frac{V_A}{I_D} = \frac{50}{1.06} = 47 \text{ k}\Omega$$

Figure 5.38(b) shows the small-signal equivalent circuit of the amplifier. Since R_G is very large (10 MΩ) the current through it can be neglected compared to that of the controlled source $g_m v_{gs}$, enabling us to write for the output voltage

$$v_o \simeq -g_m v_{gs}(R_D // R_L // r_o)$$

Since $v_{gs} = v_i$, the voltage gain is

$$\frac{v_o}{v_i} = -g_m(R_D // R_L // r_o)$$
$$= -0.725(10//10//47) = -3.3 \text{ V/V}$$

To evaluate the input resistance R_{in}, we note that the input current i_i is given by

$$i_i = (v_i - v_o)/R_G$$
$$= \frac{v_i}{R_G}\left(1 - \frac{v_o}{v_i}\right)$$
$$= \frac{v_i}{R_G}[1 - (-3.3)] = \frac{4.3\ v_i}{R_G}$$

Thus,

$$R_{in} \equiv \frac{v_i}{i_i} = \frac{R_G}{4.3} = \frac{10}{4.3} = 2.33 \text{ M}\Omega$$

The T Equivalent Circuit Model

Through a simple circuit transformation it is possible to develop an alternative equivalent circuit model for the FET. The development of such a model, known as the T model, is illustrated in Fig. 5.39. Figure 5.39(a) shows the equivalent circuit studied above without r_o. In Fig. 5.39(b) we have added a second $g_m v_{gs}$ current source in series with the original controlled source. This addition obviously does not change the terminal currents and is thus allowed. The newly created circuit node, labeled X, is joined to the gate terminal G in Fig. 5.39(c). Observe that the gate current does not change; that is, it remains equal to zero, and thus this connection does not alter the terminal characteristics. We now note that we have a controlled current source $g_m v_{gs}$ connected across its control voltage v_{gs}. We can replace this controlled source by a resistance as long as this resistance draws an equal

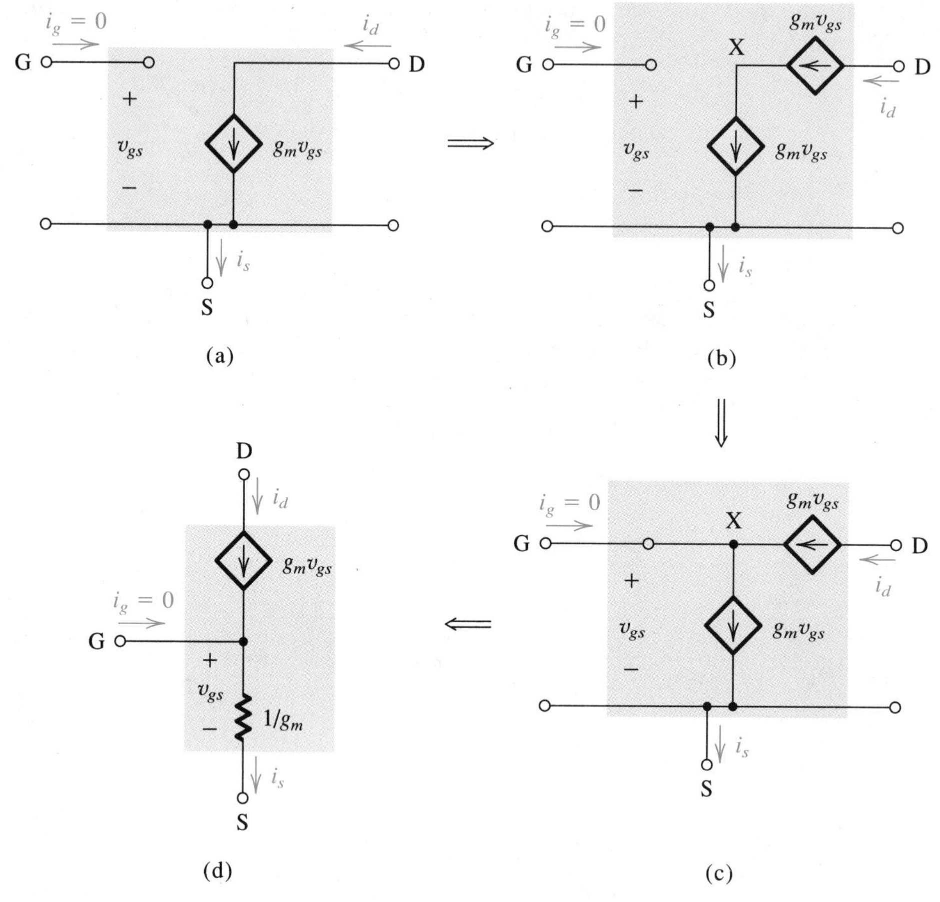

Fig. 5.39 Development of the T equivalent circuit model for the FET. For simplicity, r_o has been omitted but can be added between D and S in the T model of (d).

current as the source. (See the source-absorption theorem in Appendix D.) Thus the value of the resistance is $v_{gs}/g_m v_{gs} = 1/g_m$. This replacement is shown in Fig. 5.39(d), which depicts the alternative model. Observe that i_g is still zero, $i_d = g_m v_{gs}$, and $i_s = v_{gs}/(1/g_m) = g_m v_{gs}$, all the same as in the original model in Fig. 5.39(a).

The model of Fig. 5.39(d) shows that the resistance between gate and source looking into the source is $1/g_m$. This observation and the T model prove useful in some applications. Note that the resistance between gate and source, looking into the gate, is infinite.

In developing the T model we did not include r_o. If desired, this can be done by incorporating in the circuit of Fig. 5.39(d) a resistance r_o between drain and source.

Exercises 5.24 Consider an algebraic analysis of the amplifier circuit of Fig. 5.34(a), which we have analyzed graphically. The MOSFET has $K = 1$ mA/V^2 and $V_t = 2$ V. Also, $V_{GS} = 5$ V, $R_D = 1.33$ kΩ, and $V_{DD} = 20$ V. (a) Find the dc current I_D and the dc voltage V_D. (b) Find g_m. (c) Find the voltage gain. (d) If $v_{gs} = 0.5 \sin \omega t$, find v_d assuming that the small signal approximation holds. What are the minimum and maximum values of v_D? (e) Use Eq. (5.39) to determine the various components of i_D. Using the identity $\sin^2 \omega t = \frac{1}{2} - \frac{1}{2}\cos 2\omega t$ show that there is a slight shift in I_D (how much?) and that there is a second-harmonic component (i.e., a component with frequency 2ω). Express the amplitude of the second-harmonic component as a percentage of the amplitude of the fundamental. (This is known as the second-harmonic distortion.)

Ans. (a) 9 mA, 8 V; (b) 6 mA/V; (c) −8 V/V; (d) $v_d = -4 \sin \omega t$, 4 V, 12 V; (e) $i_D = 9.125 + 3 \sin \omega t - 0.125 \cos 2\omega t$, shift in dc current is 0.125 mA, 4.16%

5.25 Using Eq. (5.42) show that g_m for an n-channel JFET is given by

$$g_m = \frac{2I_{DSS}}{-V_P}\left(1 - \frac{V_{GS}}{V_P}\right)$$

Find the value of g_m for a JFET having $V_P = -4$ V and $I_{DSS} = 16$ mA when operating at $V_{GS} = 0$.

Ans. 8 mA/V

5.26 An enhancement NMOS transistor has $\mu_n C_{OX} = 20\ \mu A/V^2$, $W/L = 64$, $V_t = 1$ V, and $\lambda = 0.01$. Find g_m and r_o when (a) the bias voltage $V_{GS} = 2$ V (b) the bias current $I_D = 1$ mA

Ans. (a) 1.28 mA/V, 156 kΩ; (b) 1.6 mA/V; 100 kΩ

5.7 BIASING THE FET IN DISCRETE CIRCUITS

The first step in the design of a FET amplifier involves establishing a dc operating point that is predictable and stable. Here bias stability refers to the requirement that the dc current I_D remain as constant as possible in the face of changes in operating conditions—for example, temperature—and the variations normally found in the values of device parameters (K and V_t) among transistors of the same type.

To minimize the possibility of nonlinear distortion, the dc operating point should be located in the middle of the saturation region, thus allowing for the required signal swing without the device ever entering the triode region.

In this section we study two biasing schemes commonly used in the design of discrete FET amplifiers. Integrated-circuit biasing techniques will be studied in a later section.

Biasing Using Source-Resistance Feedback

Figure 5.40(a) shows a biasing arrangement that is most commonly employed when the circuit is to be operated from a single power supply. The two-supply version of the circuit is shown in Fig. 5.40(b). Although the circuits are shown for the case of an enhancement MOSFET, this biasing arrangement works equally well for depletion devices, for JFETs, and, as we have seen in Chapter 4, for BJTs also.

Consider first the circuit of Fig. 5.40(a). The voltage divider R_{G1}-R_{G2} supplies the gate with a constant dc voltage V_{GG},

$$V_{GG} = V_{DD}\frac{R_{G2}}{R_{G1} + R_{G2}}$$

In the absence of R_S this voltage appears directly between gate and source, and the corresponding current I_D will be highly dependent on the exact value of V_{GG} and on the

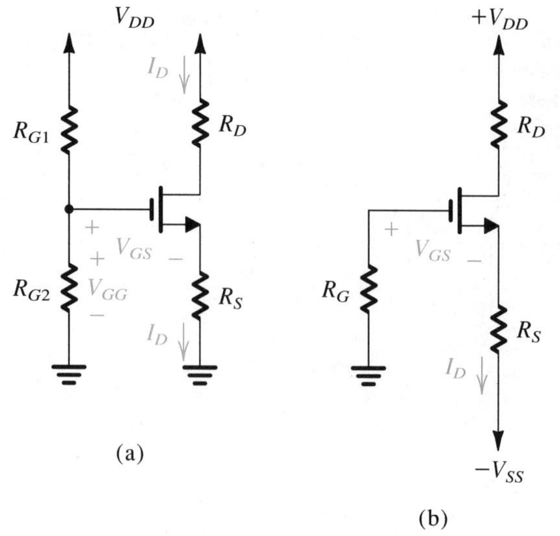

Fig. 5.40 A popular biasing arrangement that can be used for FETs of both the enhancement and the depletion type, for JFETs and for BJTs: **(a)** single-supply version; **(b)** two-supply version. Bias stability is provided by the negative-feedback action of R_S.

device parameters K and V_t. This point is illustrated in Fig. 5.41, where we show the i_D–v_{GS} characteristic curve for two extreme devices of the same type. The large difference in I_D between the two devices should be evident.

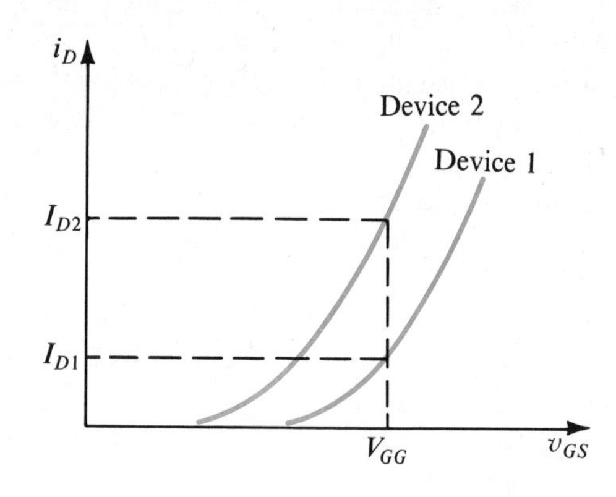

Fig. 5.41 High dependence of the bias current on device parameters if the resistance R_S is not used. Devices 1 and 2 represent two extremes among units of the same type.

Including the resistance R_S results in the following describing equation:

$$V_{GG} = V_{GS} + I_D R_S$$

which can be rewritten

$$I_D = \frac{V_{GG}}{R_S} - \frac{1}{R_S} V_{GS} \tag{5.52}$$

This is the equation of the straight line shown in Fig. 5.42, where again we have shown the characteristics of two extreme devices. Note that the difference in the value of I_D between the two devices is much less than that observed without R_S.

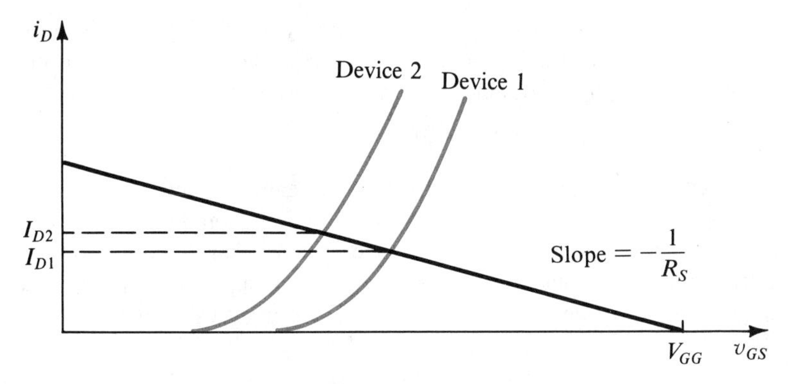

Fig. 5.42 Bias stability obtained by including resistance R_S. Devices 1 and 2 represent two extremes among units of the same type.

The dual-supply circuit shown in Fig. 5.40(b) operates in exactly the same manner as the single-supply version. The reader can easily show that Eq. (5.52) applies to the circuit of Fig. 5.40(b) if V_{GG} is replaced with V_{SS}.

Bias stability in the circuits of Fig. 5.40 is achieved via the negative-feedback action of resistor R_S. To see how this comes about consider the circuit of Fig. 5.40(a) and assume that for some reason (such as a change in temperature) the drain current increases by an amount Δi_D. This incremental change in the bias current results in an incremental increase in the voltage at the source, Δv_S,

$$\Delta v_S = R_S\, \Delta i_D$$

Since the gate is held at a constant voltage V_{GG}, an increase in the source voltage results in an equal decrease in V_{GS},

$$\Delta v_{GS} = -\Delta v_S = -R_S\, \Delta i_D$$

Since a decrease in V_{GS} results in a decrease in I_D, the net increase in I_D will be less than the original value Δi_D, indicating the presence of a negative-feedback mechanism.

In the above it has been assumed implicitly that the value of R_D is chosen such that the device operates in the saturation region. This is accomplished by maintaining the drain voltage greater than $v_G - V_t$ at all times.

EXAMPLE 5.12 Consider an enhancement MOSFET with $K = 0.25$ mA/V^2 and $V_t = 2$ V. We wish to bias the device at $I_D = 1$ mA using the biasing arrangement of Fig. 5.40(a) with $V_{DD} =$ 20 V.

Solution To determine the required value of V_{GS} we use the relationship

$$I_D = K(V_{GS} - V_t)^2$$

For $K = 0.25$ mA/V^2, $V_t = 2$ V, and $I_D = 1$ mA, the value of V_{GS} should be $V_{GS} = 4$ V. If we choose a 4-V drop across R_S, then the gate voltage should be $V_{GG} = 8$ V. This voltage can be established by choosing $R_{G1} = 1.2\ \text{M}\Omega$ and $R_{G2} = 0.8\ \text{M}\Omega$. Of course we should choose values for R_{G1} and for R_{G2} as large as practicable in order to keep the amplifier input resistance as high as possible. The value of R_S is given by

$$R_S = \tfrac{4}{1} = 4\ \text{k}\Omega$$

The choice of a value for R_D is governed by the required gain and signal swing. The higher the value of R_D the higher the gain will be. However, we should ensure that the drain voltage will at no time fall below the gate voltage by more than V_t volts. For this example let us assume that a maximum signal swing of ± 4 V is required at the drain. It follows that we may choose R_D such that $V_D = +10$ V. The required value of R_D will be

$$R_D = \frac{20 - 10}{1} = 10\ \text{k}\Omega$$

Biasing Using Drain-to-Gate Feedback

The second biasing arrangement we shall study is depicted in Fig. 5.43. As shown, a resistance R_G, usually quite large, is connected between drain and gate of an enhancement type MOSFET. Since the gate current is almost zero, the dc gate voltage will be equal to the dc drain voltage. This condition means that the device is still operating in the saturation region. It should be obvious, however, that this biasing arrangement would not work directly with depletion-type devices, including JFETs.

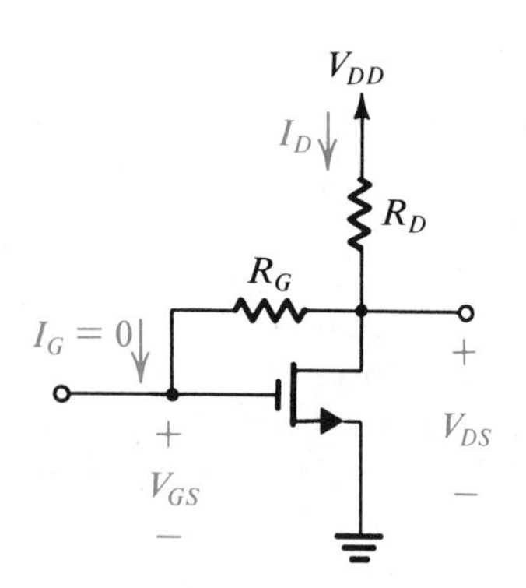

Fig. 5.43 Another popular and simple biasing arrangement for enhancement MOS discrete-circuit amplifiers. Bias stability is provided by the drain-to-gate feedback resistor R_G. Since $V_D = V_G$, this biasing scheme is not directly suitable for depletion-type devices including JFETs.

To evaluate the dc operating point of the circuit of Fig. 5.43 consider Fig. 5.44, which shows the i_D–v_{DS} characteristics of the MOSFET. The parabolic boundary between the triode region and the saturation region is shown as a broken line. This boundary curve is the locus of the points at which $v_{DS} = v_{GS} - V_t$. If one assumes ideal characteristics, then shifting this curve laterally by V_t volts gives the locus of the points for which $v_{DS} = v_{GS}$. Clearly the operating point Q lies on this latter curve, which is represented by the solid line in Fig. 5.44. From the circuit in Fig. 5.43 we can write

$$v_{DS} = V_{DD} - R_D i_D$$

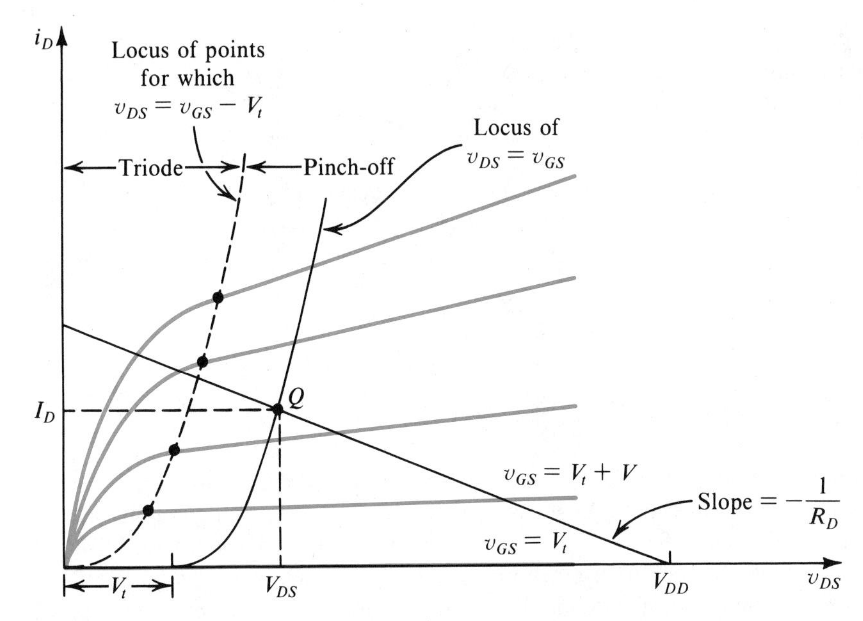

Fig. 5.44 Evaluation of the dc operating point Q of the circuit in Fig. 5.43.

or, equivalently,

$$i_D = \frac{V_{DD}}{R_D} - \frac{1}{R_D} v_{DS} \tag{5.53}$$

which is the equation of the load line shown in Fig. 5.44. The operating point Q will lie at the intersection of the load line and the solid-line parabola. Though quite illustrative, the above graphical procedure is seldom used in practice. It is far more expedient to determine i_D and v_{DS} at the operating point Q by solving Eq. (5.53) together with the equation describing the locus of the points $v_{DS} = v_{GS}$,

$$i_D = K(v_{DS} - V_t)^2$$

where for simplicity we have neglected the channel-length modulation effect.

It should be noted that bias stability in the circuit of Fig. 5.43 is achieved by the negative-feedback action provided by the connection of R_G. To see how this works, assume that for some reason the drain current increases by an increment Δi_D. From the circuit we see that the drain voltage will decrease by $R_D \Delta i_D$. Since the current in R_G is nearly zero, the gate voltage, and hence V_{GS}, will decrease by an equal amount, namely, $R_D \Delta i_D$. As a result of the decrease in V_{GS} the drain current will decrease. Thus the overall increase in drain current will be much smaller than the value originally assumed, Δi_D.

Exercises 5.27 In the circuit of Example 5.12 find the percentage change in I_D if the FET is replaced by another having the same K value but $V_t = 3$ V.

Ans. -20%

D5.28 Consider the bias arrangement of Fig. 5.43. Find the value of R_D required to establish a drain current of 1 mA provided that $V_{DD} = 20$ V, $V_t = 2$ V, and $K = 0.25$ mA/V^2. What is the percentage change in I_D if the device is replaced by another having the same K value but $V_t = 3$ V?

Ans. 16 kΩ; −6%

D5.29 Design a bias circuit of the type shown in Fig. 5.40(b) for a JFET having $V_P = -4$ V and $I_{DSS} = 16$ mA. Let $V_{DD} = V_{SS} = 10$ V. Design for a dc drain current of 4 mA and use the largest possible R_D that allows for a drain signal swing of ±2 V. Give the values of R_S and R_D.

Ans. $R_S = 3$ kΩ; $R_D = 1$ kΩ

5.8 BASIC CONFIGURATIONS OF SINGLE-STAGE FET AMPLIFIERS

In this section we shall study the basic configurations in which a single FET can be applied to provide amplification. As will be seen, each of the three configurations considered offers some unique features. These features are most clearly illustrated using a capacitively coupled amplifier in order to separate the signals from the dc bias. The results, however, apply equally well to direct-coupled amplifier stages, as will be shown in the next section, which deals with IC MOS amplifiers. Also, although all the circuits in this section are presented utilizing an n-channel enhancement-type MOSFET, the results apply equally well to other FET types.

In order to distinguish clearly between the three basic amplifier configurations we shall employ the same biasing arrangement in all cases. Figure 5.45 shows the basic circuit that will be used to implement each of the three basic configurations.

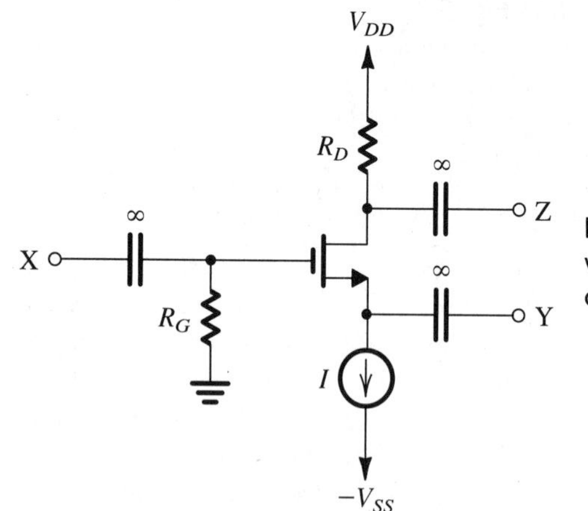

Fig. 5.45 Basic circuit for studying the various FET single-stage amplifier configurations.

The MOSFET is biased by a dc current source connected to the negative supply. Although the device could be biased by connecting a resistor R_S to the negative supply, constant-current biasing is employed in order to simplify the analysis and thus focus attention on the salient features of the various amplifier configurations. Moreover, using current sources for biasing is a common practice in the design of integrated circuits. A resistor R_G

connects the gate to ground, thus establishing dc continuity and fixing the dc voltage at the gate at zero volts. Because the gate current is extremely small, a large resistor R_G (in the megohm range) can be easily employed. A resistor R_D connects the drain to the positive supply voltage, V_{DD}, and thus establishes the dc drain voltage at a value that ensures saturation-region operation at all times while allowing for the required signal swing at the drain. Finally, three large-valued capacitors are used to couple the gate, source, and drain to signal source, load resistance, or ground, as required to configure the circuit in one of the three amplifier configurations. In the analysis to follow we assume that these capacitors act as perfect short circuits.

The Common-Source Amplifier

The common-source amplifier configuration is obtained by connecting terminal Y to ground, thus establishing a signal ground at the source. The input signal is connected to the gate and the load resistance to the drain, resulting in the configuration of Fig. 5.46(a). The circuit can be viewed as a two-port network with the input port between the gate and

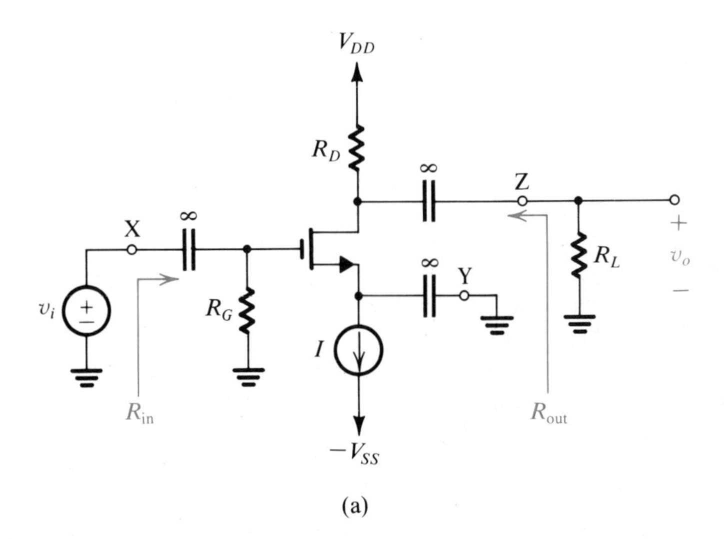

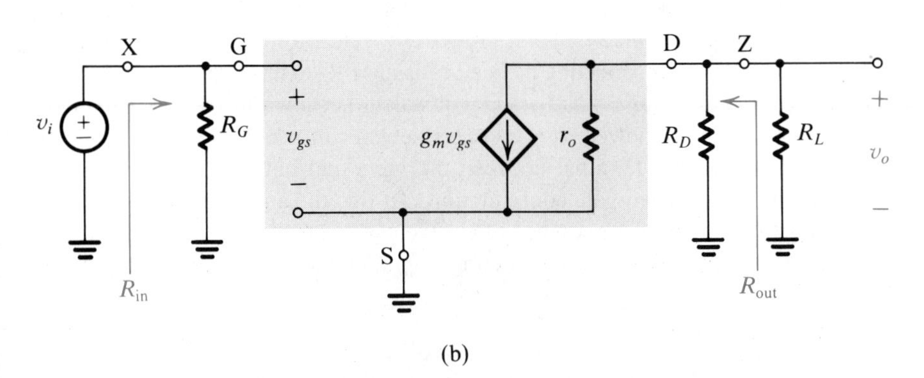

Fig. 5.46 (a) A MOSFET common-source amplifier. **(b)** Its small-signal equivalent circuit.

the source (ground) and the output port between the drain and the source (ground)—hence the name common-source or grounded-source configuration. Replacing the MOSFET with its small-signal equivalent circuit model leads to the circuit shown in Fig. 5.46(b). (Note that the bias current source is replaced by an open circuit.)

The amplifier input resistance R_{in}, output resistance R_{out}, and voltage gain A_v can be found by inspection of the circuit in Fig. 5.46(b), as follows:

$$R_{\text{in}} = R_G \tag{5.55}$$

$$R_{\text{out}} = R_D // r_o \tag{5.56}$$

$$A_v \equiv \frac{v_o}{v_i} = -g_m(R_L // R_D // r_o) \tag{5.57}$$

Note that the voltage gain in Eq. (5.57) includes the effect of the load resistance R_L. Alternatively, the amplifier can be characterized by its output resistance R_{out} in Eq. (5.56) and its open-circuit voltage gain A_{vo} obtained by setting R_L in Eq. (5.57) to ∞; that is,

$$A_{vo} \equiv \left.\frac{v_o}{v_i}\right|_{R_L=\infty} = -g_m[R_D // r_o] \tag{5.58}$$

The gain A_v for a particular R_L can then be evaluated using the voltage-divider rule:

$$A_v = A_{vo}\frac{R_L}{R_L + R_{\text{out}}} \tag{5.59}$$

From the above we note that the common-source amplifier provides a high input resistance, limited only by the value of the biasing resistor R_G, a large negative voltage gain, and a large output resistance. The last property is of course not a desirable one for voltage amplifiers (see Chapter 1).

A major drawback of the common-source configuration is its limited high-frequency response, as will be shown in Chapter 7.

The Common-Gate Amplifier

The common-gate amplifier configuration is obtained by connecting terminal X of the circuit in Fig. 5.45 to ground. This establishes a signal ground at the gate.[8] The input signal is then applied to the source by connecting the v_i generator to terminal Y, and the output is taken at the drain by connecting the load resistance R_L to terminal Z. The result is the circuit in Fig. 5.47(a), whose small-signal equivalent is shown in Fig. 5.47(b). Observe that the circuit can be viewed as a two-port network in which the gate, at signal ground, serves as a common terminal between the input and output ports—hence the name common-gate or grounded-gate configuration. To find the amplifier characteristics let us neglect for the moment the existence of r_o. We can see that $v_{gs} = -v_i$; thus at the input terminal the signal generator v_i sees a current $g_m v_i$ drawn away from v_i. Therefore

---

[8]Observe that the resistor R_G is not serving a useful purpose in this circuit. Thus R_G can be disposed with and the gate connected directly to ground. This also obviates the need for the gate coupling capacitor.

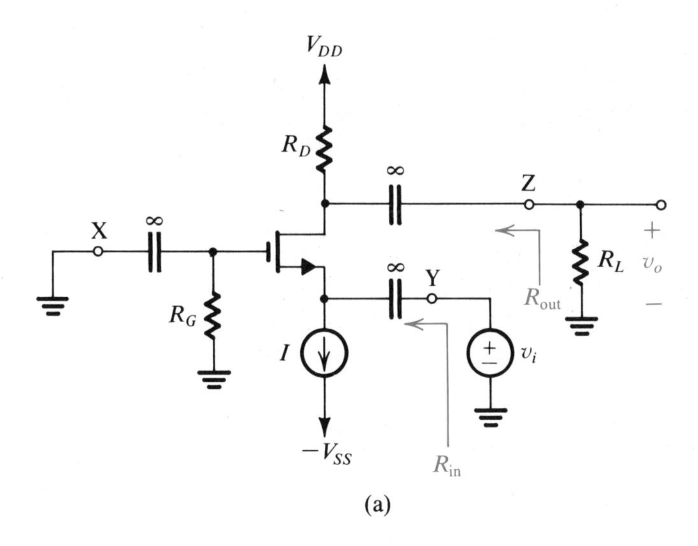

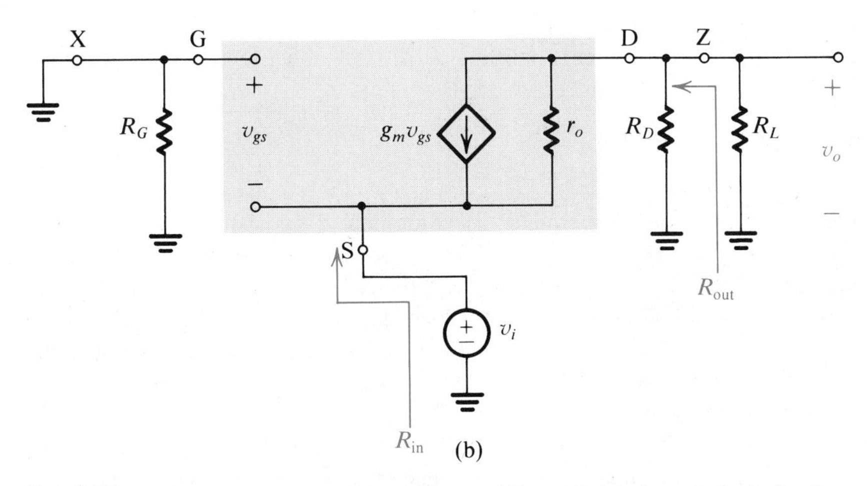

Fig. 5.47 A MOSFET common-gate amplifier and its small-signal equivalent circuit.

the input resistance is simply $1/g_m$ (see the source-absorption theorem in Appendix D). This is consistent with the fact that the resistance between gate and source, looking into the source, is equal to $1/g_m$. Thus

$$R_{in} \simeq 1/g_m \tag{5.60}$$

The voltage gain is found by noting that the drain current is (approximately, because r_o is neglected) $g_m v_{gs} = -g_m v_i$. Thus, $v_o = g_m v_i (R_L // R_D)$ and

$$A_v \equiv \frac{v_o}{v_i} \simeq g_m (R_L // R_D) \tag{5.61}$$

Finally, we can see by inspection of the circuit in Fig. 5.47(b) that

$$R_{out} = R_D // r_o \tag{5.62}$$

The effect of r_o on R_{in} and A_v can be found by analyzing the circuit in Fig. 5.47(b) with r_o in place. If this is done it will be found that R_{in} remains to a good approximation unchanged, although a better approximation for the voltage gain is

$$A_v \simeq g_m(R_L // R_D // r_o) \tag{5.63}$$

From the above we observe that the common-gate configuration provides a voltage gain that is almost equal to that obtained in the common-source amplifier except that here there is no signal inversion. An important difference between the two configurations is that the input resistance of the common-gate circuit is much smaller than that of the common-source circuit. Although this is a drawback in the case of a voltage amplifier, the common-gate circuit is almost always fed with a current signal. In this case the low input resistance becomes an advantage and the common-gate circuit acts simply as a unity-gain current amplifier or a **current follower.** It provides a drain signal current equal to the signal current fed to the source but at a much higher impedance level. The drain signal current is then fed to the parallel equivalent of R_L and R_D to produce the amplifier output voltage. This application of the common-gate circuit will be illustrated at a later stage.

The major advantage of the common-gate amplifier is that it has a much wider bandwidth than the common-source amplifier, as will be shown in Chapter 7.

The Common-Drain Amplifier or Source Follower

The third and only remaining possibility for creating a two-port amplifier from the configuration of Fig. 5.45 is shown in Fig. 5.48(a). Here terminal Z is connected to ground, thus establishing a signal ground at the drain. Obviously we can dispense with R_D altogether and connect the drain directly to V_{DD}, which is a signal ground. The input signal source v_i is connected to X and thus to the gate, and the load resistance R_L is connected to Y and thus to the source. Thus the drain, at signal ground, is the common terminal between the input and output ports of this two-port amplifier—hence the name common-drain or grounded-drain configuration.

Replacing the MOSFET with its small-signal model results in the equivalent circuit shown in Fig. 5.48(b). The amplifier input resistance is found by inspection to be $R_{in} = R_G$, which is very high since R_G can be selected very high. The input resistance can be made even higher by eliminating R_G altogether and connecting v_i directly (that is, without the coupling capacitor) to the gate. This is possible, however, only if the input signal source provides dc continuity.

The output resistance of the common-drain amplifier can be found by short-circuiting the signal source v_i and applying a test voltage v_y to the output terminal (the source) as shown in Fig. 5.48(c). We note that now $v_{gs} = -v_y$ and the current i_y is given by

$$\begin{aligned} i_y &= -g_m v_{gs} + (v_y/r_o) \\ &= g_m v_y + (v_y/r_o) \end{aligned}$$

Thus the output resistance R_{out} is

$$R_{out} \equiv \frac{v_y}{i_y} = 1\Big/\left(g_m + \frac{1}{r_o}\right)$$

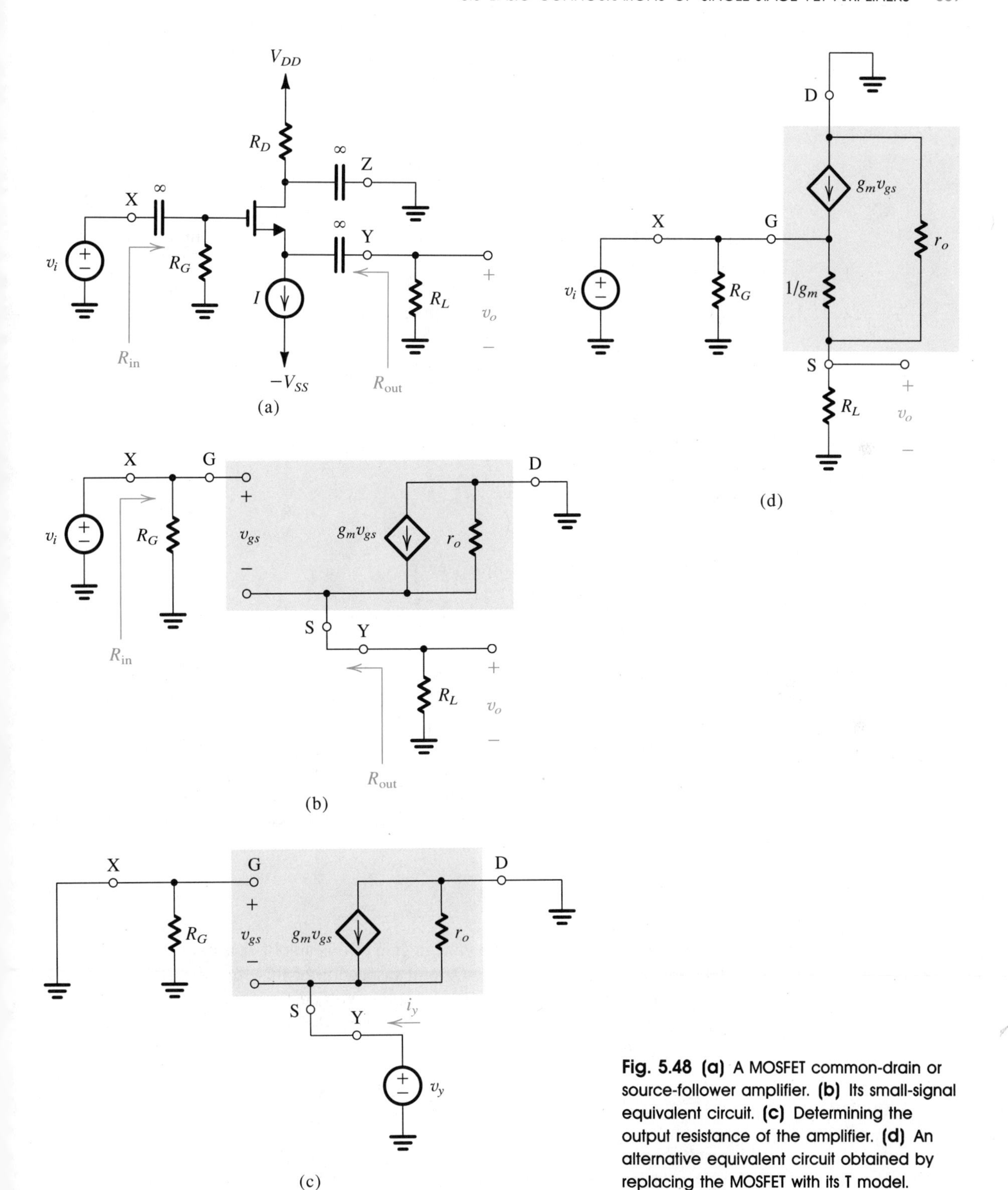

Fig. 5.48 (a) A MOSFET common-drain or source-follower amplifier. **(b)** Its small-signal equivalent circuit. **(c)** Determining the output resistance of the amplifier. **(d)** An alternative equivalent circuit obtained by replacing the MOSFET with its T model.

or alternatively

$$R_{\text{out}} = \frac{1}{g_m} // r_o \tag{5.64}$$

Usually $r_o \gg 1/g_m$, and R_{out} is approximated as

$$R_{\text{out}} \simeq 1/g_m \tag{5.65}$$

which is normally low. In fact the low output-resistance is the most important feature of this amplifier configuration.

To complete the characterization of the common-drain amplifier we return to Fig. 5.48(b) to determine the open-circuit voltage gain,

$$A_{vo} \equiv \left.\frac{v_o}{v_i}\right|_{R_L=\infty}$$

With R_L disconnected we see that

$$v_o = v_s = g_m v_{gs} r_o \tag{5.66}$$

From the input side we can write

$$v_i = v_{gs} + v_o$$

Substituting for v_{gs} from Eq. (5.66) yields

$$v_i = \frac{v_o}{g_m r_o} + v_o$$

Thus

$$A_{vo} = \frac{1}{1 + (1/g_m r_o)} \tag{5.67}$$

from which we see that the open-circuit voltage gain is less than unity. Usually, however, $g_m r_o$ is large and A_{vo} is very close to unity. The open-circuit voltage gain (Eq. 5.67) can be used together with the output resistance (Eq. 5.64) to determine the voltage gain for any load resistance R_L as

$$A_v = A_{vo} \frac{R_L}{R_L + R_{\text{out}}} \tag{5.68}$$

A much quicker way for finding the voltage gain of this amplifier is to use the T model of the FET, as shown in Fig. 5.48(d). Careful inspection of this equivalent circuit reveals that R_L appears in parallel with r_o, and the output voltage v_o is taken across this parallel combination. Also, $R_L // r_o$ is in series with $1/g_m$. Thus the voltage divider rule can be used to find A_v directly,

$$A_v \equiv \frac{v_o}{v_i} = \frac{R_L // r_o}{(R_L // r_o) + (1/g_m)} \tag{5.69}$$

This result should be identical to that found using Eq. (5.68). Now we can go even further and suggest that the result in Eq. (5.69) could have been written by inspection of the original circuit of Fig. 5.48(a): Since the drain is at signal ground, r_o appears in parallel with R_L, and the parallel combination is in series with the resistance looking into the FET source ($1/g_m$). The signal v_i appears across this series combination, and thus we can use the voltage divider rule to write Eq. (5.69).

From Eq. (5.69) we see that as expected the overall voltage gain is less than unity. The gain approaches unity for $R_L \gg 1/g_m$. Since the signal at the source (the output) follows that at the input, this amplifier configuration is most commonly known as the source follower. The source follower finds application as a buffer amplifier or as output stage of a multistage amplifier, where its function is to provide the overall amplifier with a low output resistance ($\simeq 1/g_m$).

Exercise 5.31 Consider the circuit of Fig. 5.45 with $\mu_n C_{OX} = 20 \mu A/V^2$, $W/L = 100$, $V_t = 1$ V, $V_A = 50$ V, $I = 1$ mA, $V_{DD} = V_{SS} = 5$ V, $R_G = 1$ MΩ, and $R_D = 3$ kΩ. Find the following: (a) The dc voltages at the gate, source, and drain. (b) g_m. (c) R_{in}, R_{out}, and A_v for the common-source amplifier with $R_L = 3$ kΩ. (d) R_{in}, R_{out}, and A_v for the common-gate amplifier with $R_L = 3$ kΩ. (e) R_{in}, R_{out}, and A_v for the source follower with $R_L = 3$ kΩ. (f) The values for A_v for the three amplifier configurations when the signal source v_i has a 100-kΩ resistance.

Ans. (a) 0 V, −2 V, +2 V; (b) 2 mA/V; (c) 1 MΩ, 2.83 kΩ, −2.91 V/V; (d) 0.5 kΩ, 2.83 kΩ, +2.91 V/V; (e) 1 MΩ, 495 Ω, 0.85 V/V; (f) −2.65 V/V, 0.01 V/V(!), 0.77 V/V

5.9 INTEGRATED-CIRCUIT MOS AMPLIFIERS

In this section we begin the study of integrated-circuit (IC) MOS amplifiers.[9] The most important feature of IC MOS amplifiers is their use of MOS transistors as load elements, in place of resistors. Since a MOSFET requires a much smaller silicon area than a resistor, the resulting circuits are very efficient in their use of silicon "real estate."

At the present time there are two different MOS integrated-circuit technologies: NMOS and CMOS. NMOS refers to MOS integrated circuits that are based entirely on *n*-channel MOS transistors. The majority of these transistors are of the enhancement type; depletion-type transistors are used only as load devices, as will be explained below. By contrast, CMOS technology is based on using both *n*-channel and *p*-channel devices, all of which are of the enhancement type. The availability of both device polarities makes it easier to design high-quality circuits in CMOS. In fact, at the present time CMOS is by far the most popular technology for digital integrated circuits, and is rivaling bipolar technology (Chapter 4) for analog applications. The NMOS technology, though not as convenient for the circuit designer, currently offers the highest possible functional density (highest number of devices, and hence circuit functions, per chip), and it requires fewer processing steps than CMOS. Thus NMOS allows very high levels of integration. Both CMOS and NMOS are used extensively in the design of very-large-scale integrated (VLSI) circuits.

In this section we shall study some of the circuit techniques employed in the design of NMOS and CMOS amplifiers. More advanced MOS analog circuit design techniques will

[9] Study of this section may be deferred and undertaken later in conjunction with the material on IC MOS amplifiers in Chapters 6 and 10.

be studied in Chapters 6 and 10. Digital NMOS and CMOS circuits are studied in Chapter 13.

NMOS Load Devices

In NMOS technology two types of load elements are used: the enhancement MOSFET with the drain connected to the gate, and the depletion MOSFET with the gate connected to the source. Figure 5.49 shows the "diode-connected" enhancement transistor, together

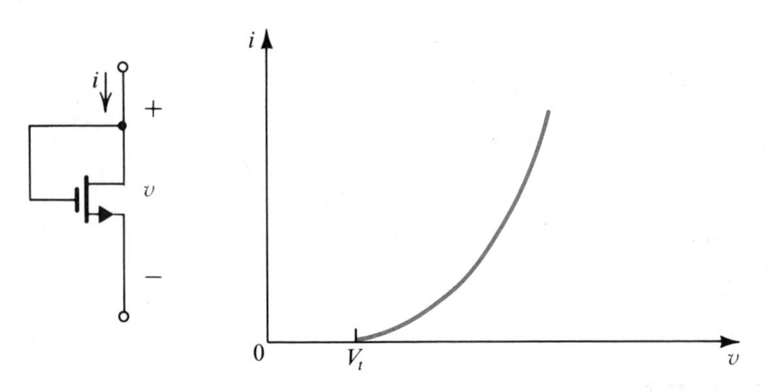

Fig. 5.49 A diode-connected enhancement MOS transistor and its i–v characteristic.

with its i–v characteristic, which is described by

$$i = K(v - V_t)^2 \tag{5.70}$$

Observe that the transistor will always be operating in the saturation region. If the diode-connected transistor is biased at a voltage V, then its incremental or small-signal resistance will be equal to $1/g_m$, with the value of g_m evaluated at the bias point.

The diode-connected depletion MOSFET is shown in Fig. 5.50, together with its i–v characteristic. To operate in the saturation region, the voltage across the two-terminal

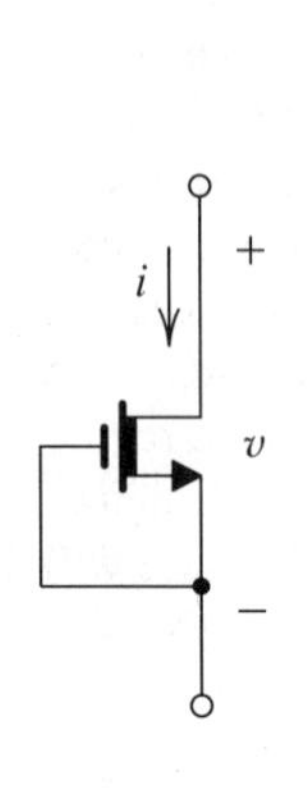

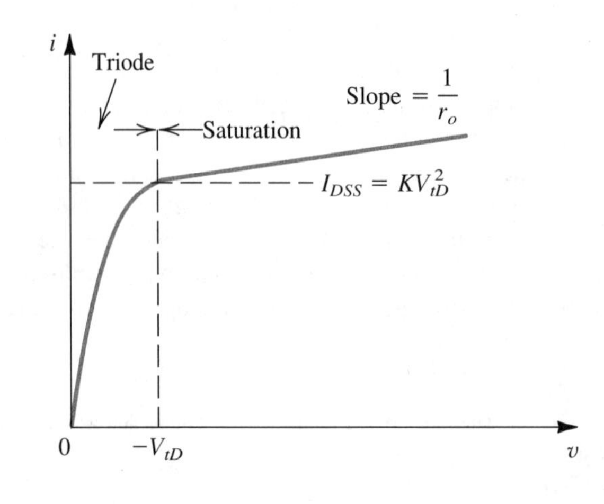

Fig. 5.50 A diode-connected depletion MOS transistor and its i–v characteristic.

device must exceed $-V_{tD}$, where V_{tD} is the threshold voltage of the depletion device and is negative, typically -1 to -4 V. In the triode region the i–v characteristic is described by

$$i = K(-2V_{tD}v - v^2) \tag{5.71}$$

At the onset of saturation, $v = -V_{tD}$ and

$$i = KV_{tD}^2 = I_{DSS} \tag{5.72}$$

where we use the symbol I_{DSS} the same way we did for JFETs. Ideally, in saturation, the i–v characteristic is a horizontal line. However, because of channel-length modulation, the i–v characteristic is a straight line with a finite slope and is described by

$$i \simeq KV_{tD}^2\left(1 + \frac{v}{V_A}\right) \tag{5.73}$$

Obviously, to provide the large resistance required of the load in amplifier applications (in order to obtain high gain), the diode-connected depletion transistor must be operated in the saturation region.

Exercises D5.32 Figure E5.32 shows a voltage divider composed of three diode-connected enhancement MOSFETs. Utilizing a current $I = 90$ μA, find the W/L ratios of the three transistors so that the divider provides $V_1 = +1$ V and $V_2 = -1$ V. Let $V_t = 1$ V and $\mu_n C_{OX} = 20$ μA/V^2. Neglect the small effect of r_o of each of the three devices.

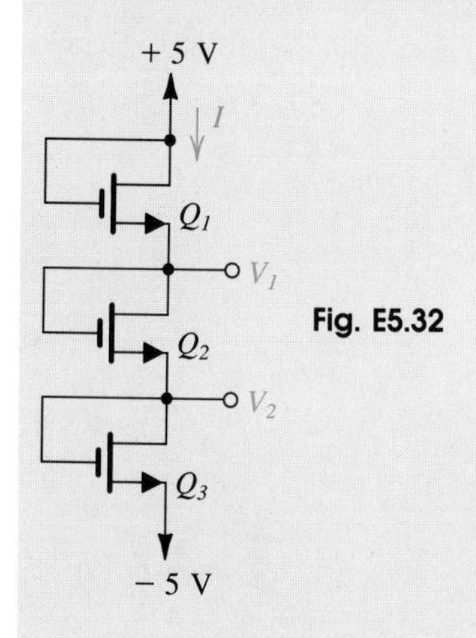

Fig. E5.32

Ans. $W_1/L_1 = W_3/L_3 = 1$; $W_2/L_2 = 9$

5.33 For the diode-connected enhancement MOSFET of Fig. 5.49 find an expression for its incremental (small-signal) resistance at $v = 2V_t$. Neglect the small effect of r_o. Evaluate r for the case $V_t = 1$ V, $\mu_n C_{ox} = 20$ μA/V^2, $W = 6$ μm, and $L = 30$ μm.

Ans. $r = 1/[(\mu_n C_{OX})(W/L)V_t]$; 250 kΩ

NMOS Amplifier With Enhancement Load

Figure 5.51(a) shows an enhancement MOSFET amplifier with an enhancement load. This circuit represents the simplest way of implementing an amplifier and a logic inverter

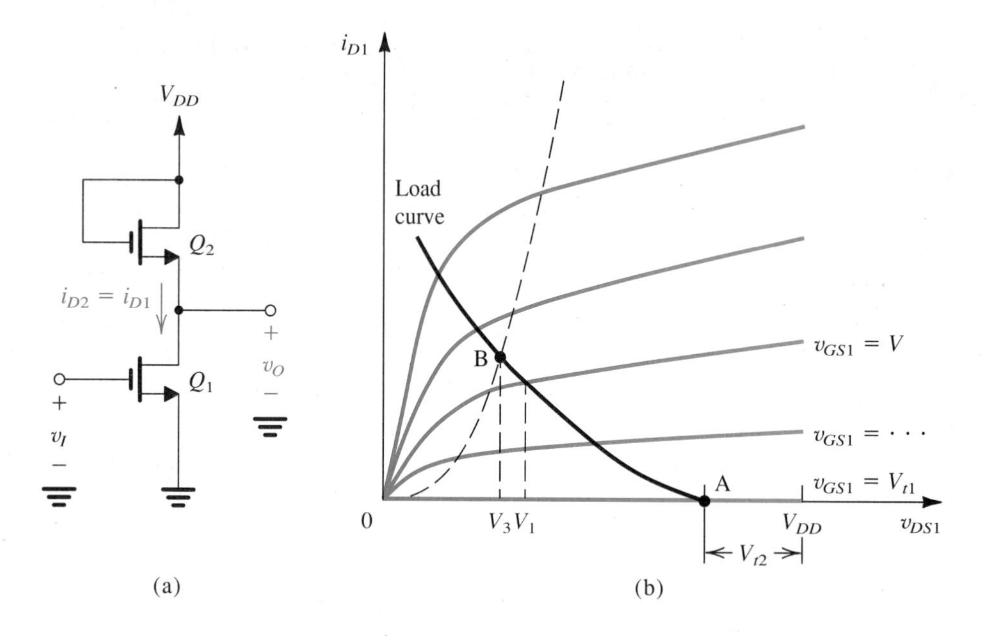

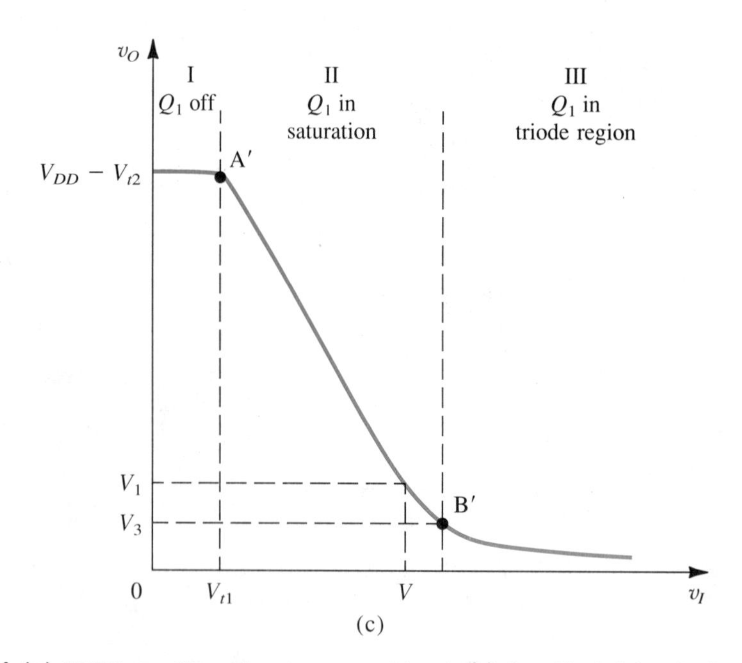

Fig. 5.51 (a) NMOS amplifier with enhancement load. **(b)** Graphical determination of the transfer characteristic. **(c)** Transfer characteristic.

in NMOS technology. We wish to derive the transfer characteristic v_O versus v_I. This can be done graphically, as illustrated in Fig. 5.51(b), which shows a plot of the i_D–v_{DS} characteristics of the driving (amplifying) transistor Q_1. Note that the current i_{D1} is the same current that flows in the load device Q_2. Also note that $v_{DS1} = v_O$, that each of the characteristic curves corresponds to a constant value of v_{GS1}, and that $v_{GS1} = v_I$.

Superimposed on the static characteristics of Q_1 is the **load curve,** which is drawn in the same manner used to draw a load line; namely, we locate the V_{DD} point on the v_{DS1}-axis and draw a mirror image of the i–v characteristic of the load device (Fig. 5.49). Now the transfer characteristic is determined by the intersection points of the load curve and the i_{D1}–v_{DS1} characteristic curves. For instance, for $v_I = V$ we find the intersection of the curve corresponding to $v_{GS1} = V$ and the load curve. As shown, at this point $v_{DS1} = V_1$; thus $v_O = V_1$. This process should be repeated for all possible values of v_I. The result is the transfer characteristic shown in Fig. 5.51(c).

The transfer characteristic displays three well-defined regions. In region I, the driving transistor Q_1 is off, since $v_I < V_{t1}$. Nevertheless, Q_2 is in the saturation region and is conducting a negligible current; thus the voltage across Q_2 is equal to V_{t2}, and hence the output voltage is $V_{DD} - V_{t2}$. (Note that, in fact, Q_2 is always in saturation.) In region II, Q_1 is conducting and is operating in saturation, and, as will be shown analytically, the transfer curve in region II is linear. Therefore, this region is very useful for amplifier operation. Finally, in region III, Q_1 leaves the saturation region and enters the triode region. The onset of this region, point B′, corresponds to the intersection of the load curve and the boundary curve between the saturation and triode regions (point B in Fig. 5.51b).

We shall now derive the equation describing the transfer characteristic under the assumption that both devices have infinite resistance (that is, horizontal characteristic lines) in saturation. Furthermore, the two devices will be assumed to have equal threshold voltages V_t but different values of K (K_1 and K_2), a situation that corresponds to actual practice.

When Q_1 is in saturation we have

$$i_{D1} = K_1(v_{GS1} - V_t)^2 \tag{5.74}$$

Since $i_{D1} = i_{D2} = i_D$ and $v_{GS1} = v_I$, this equation can be rewritten

$$i_D = K_1(v_I - V_t)^2 \tag{5.75}$$

The operation of Q_2 is described by

$$i_D = K_2(v_{GS2} - V_t)^2$$

Since $v_{GS2} = V_{DD} - v_O$, this equation can be rewritten

$$i_D = K_2(V_{DD} - v_O - V_t)^2 \tag{5.76}$$

Combining Eqs. (5.75) and (5.76) and with some simple manipulations we obtain

$$v_O = \left(V_{DD} - V_t + \sqrt{\frac{K_1}{K_2}}V_t\right) - \sqrt{\frac{K_1}{K_2}}v_I \tag{5.77}$$

which is a linear equation between v_O and v_I. This is obviously the equation of the straight-line portion of the transfer characteristic (region II) of Fig. 5.51(c).

From Eq. (5.77) we see that the circuit behaves as a linear amplifier for large signals. The gain of the amplifier is

$$A_v = -\sqrt{\frac{K_1}{K_2}} \tag{5.78}$$

Substituting for K_1 and K_2 using the expression in Eq. (5.6) gives

$$A_v = -\sqrt{\frac{(W/L)_1}{(W/L)_2}} \tag{5.79}$$

Thus the gain is determined by the geometries of the two devices and is fixed for given devices. To obtain relatively large gains, $(W/L)_2$ must be made smaller than $(W/L)_1$. Thus the usual practice is to make the amplifier transistor Q_1 short and wide and the load transistor Q_2 long and narrow. Nevertheless, it is difficult to realize gains greater than about 10.

We shall next consider a small-signal analysis of the amplifier circuit of Fig. 5.51(a), assuming that the circuit is biased to operate somewhere in region II of the transfer characteristic. Figure 5.52 shows the amplifier equivalent circuit obtained by replacing

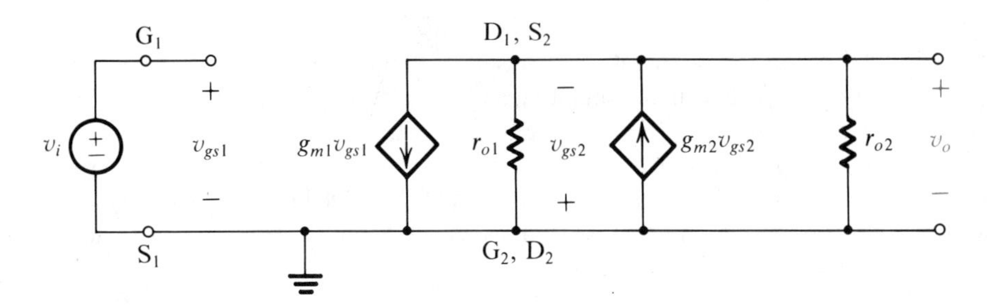

Fig. 5.52 Small-signal equivalent circuit of the enhancement-load amplifier of Fig. 5.51(a).

each of the two transistors by its equivalent circuit model. Since the voltage across the controlled current source $g_{m2}v_{gs2}$ is v_{gs2}, the source can be replaced with a resistance $1/g_{m2}$ (see the source-absorption theorem in Appendix D). Then for v_o we can write

$$v_o = -g_{m1}v_{gs1}[(1/g_{m2})//r_{o1}//r_{o2}]$$

Substituting $v_{gs1} = v_i$, we obtain the voltage gain as follows:

$$A_v = \frac{v_o}{v_i} = \frac{-g_{m1}}{g_{m2} + 1/r_{o1} + 1/r_{o2}} \tag{5.80}$$

Now if r_{o1} and r_{o2} are much larger than $(1/g_{m2})$, the gain expression in Eq. (5.80) reduces to

$$A_v \simeq -\frac{g_{m1}}{g_{m2}} \tag{5.81}$$

which can easily be shown to lead to the expression in Eq. (5.78).

In the above analysis it was implicitly assumed that the substrate of each transistor is connected to its source. However, since this amplifier configuration is intended for integrated-circuit implementation, Q_1 and Q_2 share the same substrate, which is normally connected to the most negative supply voltage in the circuit—ground in this case. It follows that for Q_2, the substrate is at ground potential while the source is not. Thus Q_2 suffers from the body effect mentioned earlier in Section 5.2. We now digress briefly to consider how to model the body effect in general. Then we will return to the NMOS amplifier and reanalyze it, taking the body effect of Q_2 into account.

Modeling the Body Effect

As mentioned in Section 5.2 the body effect occurs in a MOSFET when the substrate is not tied to the source but is connected to the most negative power supply in the integrated circuit. Thus the substrate (body) will be at signal ground, but since the source is not, a signal voltage v_{bs} develops between the body (B) and the source (S). In Section 5.2 it was mentioned that the substrate acts as a ''second gate'' for the MOSFET, much like the gate of a JFET. Thus the signal v_{bs} gives rise to a drain-current component which we shall write as $g_{mb}v_{bs}$, where g_{mb} is the **body transconductance,** defined as

$$g_{mb} \equiv \left. \frac{\partial i_D}{\partial v_{BS}} \right|_{\substack{v_{GS} = \text{constant} \\ v_{DS} = \text{constant}}} \tag{5.82}$$

Recalling that i_D depends on v_{BS} through the dependence of V_t on V_{BS}, Eqs. (5.13), (5.25) and (5.42) can be used to obtain

$$g_{mb} = \chi\, g_m \tag{5.83}$$

where

$$\chi \equiv \frac{\partial V_t}{\partial V_{SB}} = \frac{\gamma}{2\sqrt{2\phi_f + V_{SB}}} \tag{5.84}$$

Typically the value of χ lies in the range 0.1 to 0.3.

Figure 5.53 shows the MOSFET model augmented to include the controlled source $g_{mb}v_{bs}$ that models the body effect. This is the model to be used whenever the substrate is not connected to the source.

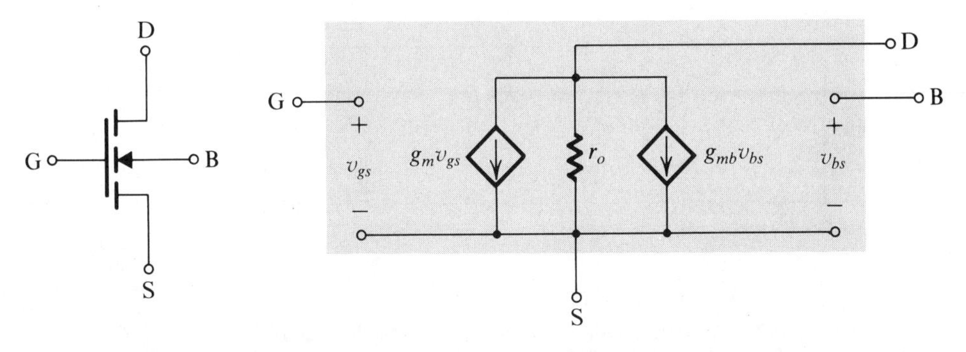

Fig. 5.53 Small-signal equivalent circuit model of a MOSFET in which the body is not connected to the source.

Exercise 5.33 For an NMOS transistor with $2\phi_f = 0.6$ V, $\gamma = 0.5$ V$^{1/2}$, and $V_{SB} = 4$ V, find the value of χ.

Ans. 0.12

Analysis of the Enhancement-Load Amplifier Including the Body Effect

Figure 5.54(a) shows the enhancement-load amplifier with the substrate connections explicitly indicated. Replacing each of the two transistors with the small-signal equivalent circuit of Fig. 5.53 gives rise to the equivalent circuit of Fig. 5.54(b). The only difference between this equivalent circuit and that in Fig. 5.52 is the inclusion of the body effect of Q_2, modeled by the controlled source $g_{mb2}v_{bs2}$. Observing that the voltage across this current source is v_{bs2}, we can replace it by a resistance equal to $1/g_{mb2}$. We thus obtain

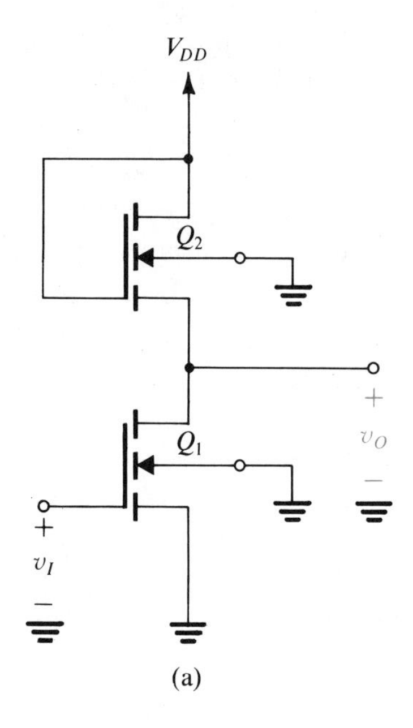

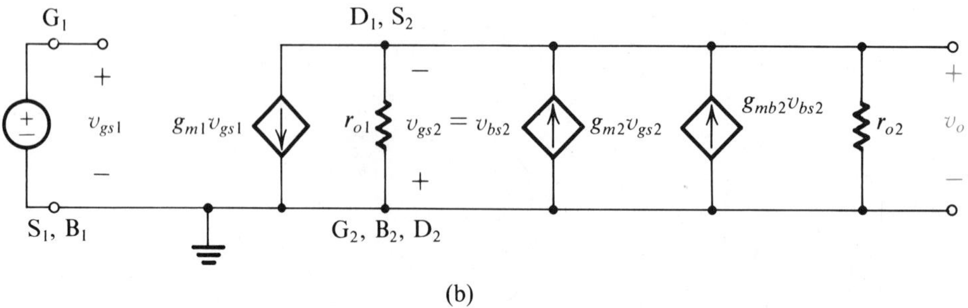

Fig. 5.54 **(a)** The enhancement-load amplifier with the substrate connections explicitly shown. **(b)** Small-signal equivalent circuit of the amplifier in (a), including the body effect of Q_2.

for v_o

$$v_o = -g_{m1}v_{gs1}\left[\left(\frac{1}{g_{m2}}\right)//\left(\frac{1}{g_{mb2}}\right)//r_{o1}//r_{o2}\right]$$

Substituting $v_{gs1} = v_i$ results in the voltage-gain expression

$$A_v = -\frac{g_{m1}}{g_{m2} + g_{mb2} + 1/r_{o1} + 1/r_{o2}} \tag{5.85}$$

Assuming that r_{o1} and r_{o2} are large in comparison to $1/g_{m2}$, we can approximate Eq. (5.85) by

$$A_v \simeq \frac{-g_{m1}}{g_{m2} + g_{mb2}} \tag{5.86}$$

Substituting for g_{mb2} from Eq. (5.83) gives

$$A_v = -\frac{g_{m1}}{g_{m2}}\frac{1}{1+\chi} \tag{5.87}$$

Comparison of this expression with that in Eq. (5.81) reveals that the body effect in the load device results in a reduction in gain by a factor $1/(1 + \chi)$.

A drawback of the enhancement-load amplifier is its rather limited signal swing. Specifically, it can be seen from Fig. 5.51(c) that the output voltage cannot exceed $V_{DD} - V_t$.

EXAMPLE 5.13

The enhancement-load MOSFET amplifier can be also used in discrete circuits to design a linear amplifier for large input signals. As an example, consider the capacitively coupled amplifier shown in Fig. 5.55(a). Resistance R_G establishes a dc operating point on the

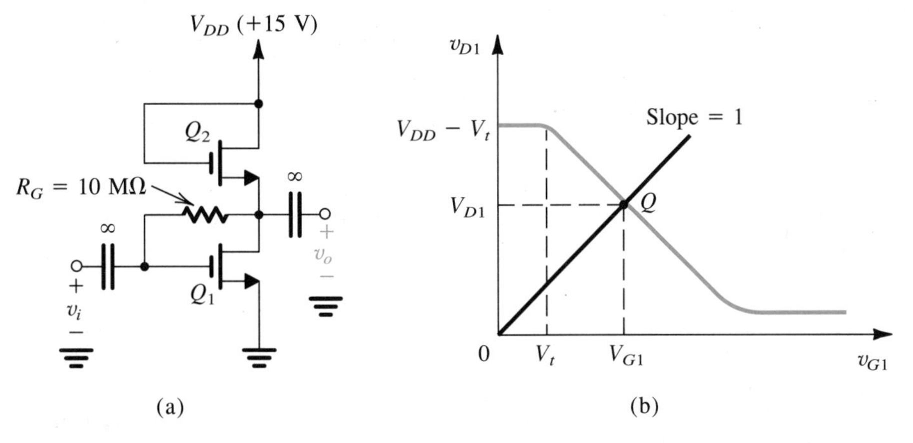

Fig. 5.55 (a) A capacitively coupled MOS amplifier with enhancement load. **(b)** Illustrating the determination of the dc operating point Q.

linear segment of the transfer curve. Figure 5.55(b) illustrates the process of determining the bias point Q, where the 45° straight line represents the constraint that R_G imposes: $V_{D1} = V_{G1}$. Thus the operating point lies at the intersection of this straight line with the transfer curve. Analytically, the straight-line portion of the transfer curve is described by Eq. (5.77). Thus

$$V_{D1} = \left(V_{DD} - V_t + \sqrt{\frac{K_1}{K_2}}V_t\right) - \sqrt{\frac{K_1}{K_2}}V_{G1}$$

Substitution of $V_{G1} = V_{D1}$ gives

$$V_{D1} = \frac{V_{DD} - V_t + V_t\sqrt{K_1/K_2}}{1 + \sqrt{K_1/K_2}}$$

Consider the case where $V_t = 2$ V, $V_{DD} = 15$ V, $K_1 = 270\ \mu\text{A/V}^2$, and $K_2 = 30\ \mu\text{A/V}^2$. The value of V_{D1} will be 4.75 V, and the dc drain current I_D will be given by

$$I_D = 0.27(4.75 - 2)^2 \simeq 2 \text{ mA}$$

At this operating point the voltage gain will be

$$\frac{v_o}{v_i} = -\sqrt{\frac{K_1}{K_2}} = -3$$

The actual voltage gain will be slightly lower than this value because of the finite output resistance r_o of each of the two MOSFETs. Also, if a load resistance is connected to the output, the voltage gain will be further reduced.

Exercises 5.34 For the enhancement-load amplifier of Fig. 5.51(a) let $W_1 = 100\ \mu\text{m}$, $L_1 = 6\ \mu\text{m}$, $W_2 = 6\ \mu\text{m}$, and $L_2 = 30\ \mu\text{m}$. If the body-effect parameter $\chi = 0.1$, find the voltage gain without and with the body effect taken into account. Neglect the effect of r_o.

Ans. −9.12 V/V; −8.3 V/V

5.35 For the enhancement-load amplifier of Fig. 5.51(a) let $W_1 = 100\ \mu\text{m}$, $L_1 = 6\ \mu\text{m}$, $W_2 = 6\ \mu\text{m}$, $L_2 = 30\ \mu\text{m}$, $V_{DD} = 10$ V, and $V_t = 1$ V. Find the coordinates of points A′ and B′ that define the linear region of the transfer characteristic in Fig. 5.51(c).

Ans. $V_{IA'} = 1$ V, $V_{OA'} = 9$ V; $V_{IB'} = 1.9$ V, $V_{OB'} = 0.9$ V

NMOS Amplifier with Depletion Load

Modern NMOS technology allows the fabrication of both enhancement and depletion devices on the same chip. As will be now shown, using the depletion MOSFET as a load device results in an amplifier with performance superior to that of the enhancement-load circuit. The same holds true if the circuit is to be used as a logic inverter (Chapter 13).

The depletion-load amplifier is shown in Fig. 5.56(a). Figure 5.56(b) shows the i–v characteristic of the depletion load. The transfer characteristic of the amplifier can be determined using the graphical technique illustrated in Fig. 5.56(c). Here the i–v load curve is superimposed on the i_D–v_{DS} characteristics of the enhancement transistor Q_1. The

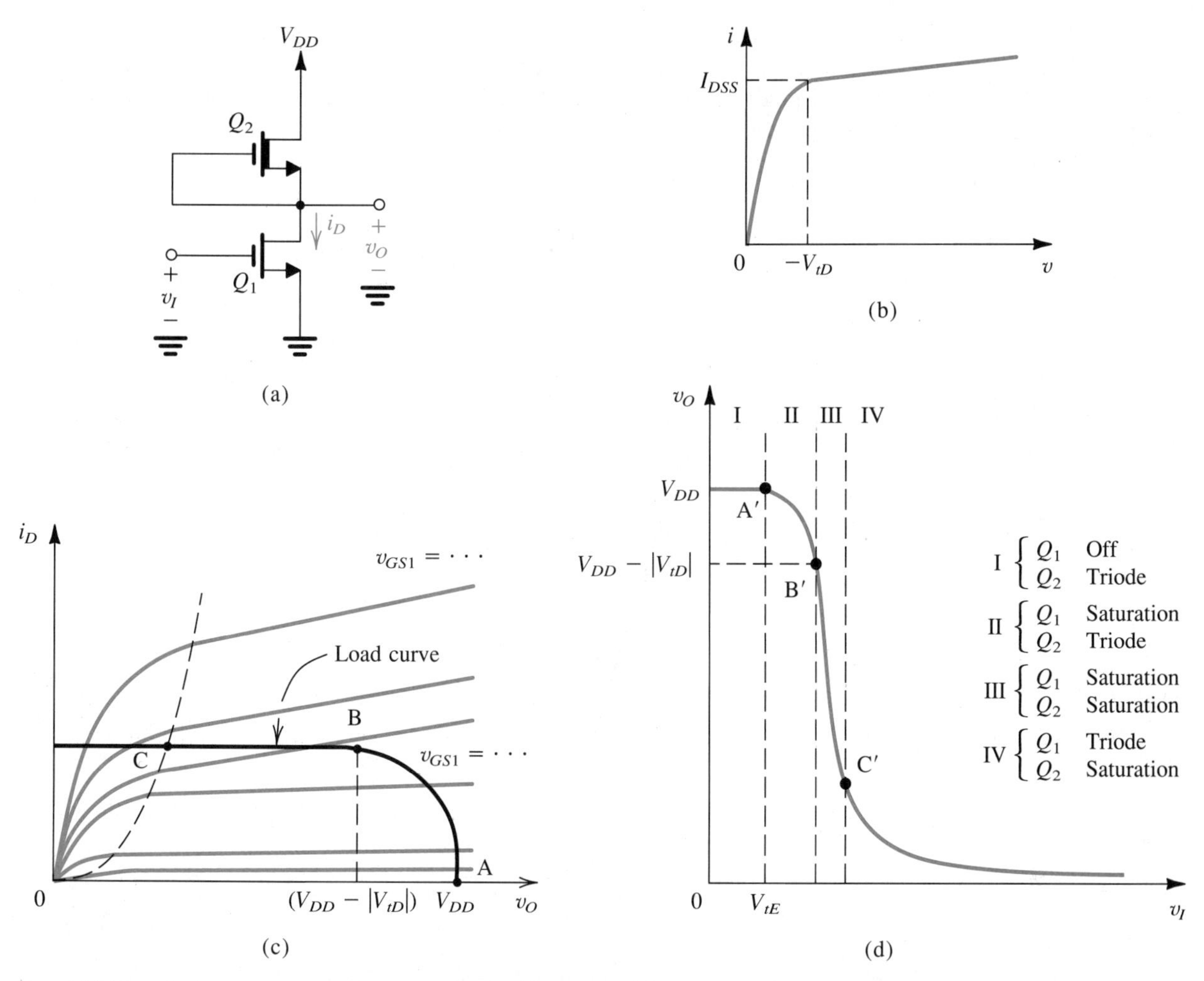

Fig. 5.56 The NMOS amplifier with depletion load: **(a)** circuit; **(b)** *i–v* characteristic of the depletion load; **(c)** graphical construction to determine the transfer characteristic; and **(d)** transfer characteristic.

transfer characteristic can be determined point-by-point in the same manner used for the enhancement-load amplifier. The resulting characteristic is shown in Fig. 5.56(d). We observe four distinct regions. For $v_I < V_{tE}$, the threshold voltage of the enhancement transistor Q_1, transistor Q_1 is off and $v_O = V_{DD}$. Here we note an important difference from the enhancement-load case, where the maximum output is one threshold voltage lower than V_{DD}. Region II is obtained when v_I exceeds V_{tE} and Q_1 turns on; but because the output voltage is high, Q_2 is in the triode region. In fact, Q_2 remains in the triode region until v_O becomes lower than V_{DD} by $|V_{tD}|$. At this point the amplifier enters region III of its transfer characteristic. Here both Q_1 and Q_2 are operating in the saturation region and thus have large output resistances, which gives rise to the large gain indicated by the sharp transfer curve in region III. Region III is the one of interest for amplifier operation;

that is, the amplifier will be biased to operate in region III. Finally, region IV is entered when v_O becomes V_{tE} volts lower than v_I, at which point Q_1 enters the triode region.

If the depletion-load amplifier is biased to operate in region III, then the small-signal gain will be given by

$$A_v \equiv \frac{v_o}{v_i} = -g_{m1}[r_{o1}//r_{o2}] \tag{5.88}$$

In practice, however, a gain much lower than this is realized due to the body effect on transistor Q_2. Specifically, note that because the substrate of Q_2 will be connected to ground, a voltage signal equal to $-v_o$ appears between body and source. The resulting small-signal equivalent circuit is shown in Fig. 5.57, from which we see that the con-

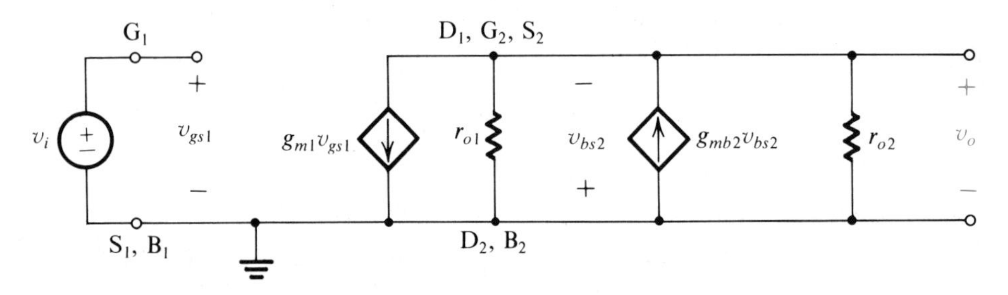

Fig. 5.57 Small-signal equivalent circuit of the depletion-load amplifier of Fig. 5.56(a), incorporating the body effect on Q_2.

trolled current source $g_{mb2}v_{bs2}$ can be replaced by a resistance $1/g_{mb2}$. Then the output voltage can be obtained as

$$v_o = -g_{m1}v_{gs1}\left[\left(\frac{1}{g_{mb2}}\right)//\, r_{o1}//\, r_{o2}\right]$$

Since $v_{gs1} = v_i$, the voltage gain is given by

$$A_v \equiv \frac{v_o}{v_i} = -g_{m1}\left[\left(\frac{1}{g_{mb2}}\right)//\, r_{o1}//\, r_{o2}\right] \tag{5.89}$$

It is usually the case that $1/g_{mb2}$ is much smaller than r_{o1} and r_{o2}, resulting in the approximate gain expression

$$A_v \simeq -\frac{g_{m1}}{g_{mb2}} \tag{5.90}$$

Expressing g_{mb2} as $\chi\, g_{m2}$ gives

$$A_v = -\frac{g_{m1}}{g_{m2}}\left(\frac{1}{\chi}\right) \tag{5.91}$$

or, alternatively,

$$A_v = -\sqrt{\frac{(W/L)_1}{(W/L)_2}}\left(\frac{1}{\chi}\right) \tag{5.92}$$

Comparison of Eq. (5.91) with the gain expression for the enhancement-load amplifier (Eq. 5.87) reveals that the gain of the depletion-load amplifier is a factor $(1 + \chi)/\chi$ greater. Since χ is typically 0.1 to 0.3, the gain increase can be as much as a factor 10.

Finally we observe that the dc bias current of the depletion-load amplifier is approximately equal to I_{DSS} of the depletion load, which is given by Eq. (5.72) as

$$I_D \simeq I_{DSS} = K_D V_{tD}^2 \tag{5.93}$$

Thus the bias current is determined by the technology and by the device geometry and cannot be changed by the circuit designer.

Exercise 5.36 For the depletion-load amplifier of Fig. 5.56(a) let $W_1 = 100\ \mu\text{m}$, $L_1 = 6\ \mu\text{m}$, $W_2 = 6\ \mu\text{m}$, $L_2 = 30\ \mu\text{m}$, $V_{tE} = 1.5$ V, $V_{tD} = -3$ V, $\mu_n C_{OX} = 100\ \mu\text{A/V}^2$, $V_{DD} = 10$ V, $|V_A| = 90$ V, and $\chi = 0.1$. (a) Calculate I_{DSS} for the depletion device. (b) Neglecting the finite output resistance in saturation and the body effect, find the coordinates of points A′, B′, and C′ that define the transfer characteristic of Fig. 5.56(d). (Note that in this case the segment B′C′ is a vertical straight line.) (c) For operation in region III, calculate g_{m1}, g_{m2}, g_{mb2}, r_{o1}, and r_{o2}. (d) Taking into account the bodv effect and the finite output resistance, find the small-signal voltage gain in region III.

Ans. (a) 90 μA; (b) (1.5 V, 10 V), (1.83 V, 7 V), (1.83 V, 0.33 V); (c) 0.55 mA/V, 60 μA/V, 6 μA/V, 1 MΩ, 1 MΩ; (d) −68.8 V/V

The Current Mirror

In both NMOS and CMOS analog integrated circuits a stable and predictable dc reference current is generated and is then used to generate proportional dc currents for biasing the various transistors in the circuit. In the following we discuss the circuit building block that is universally employed to generate dc currents which are constant multiples of the reference current source. The circuit, appropriately called a **current mirror,** is shown in its simplest form in Fig. 5.58(a).

The current mirror consists of two enhancement MOSFETs, Q_1 and Q_2, having equal

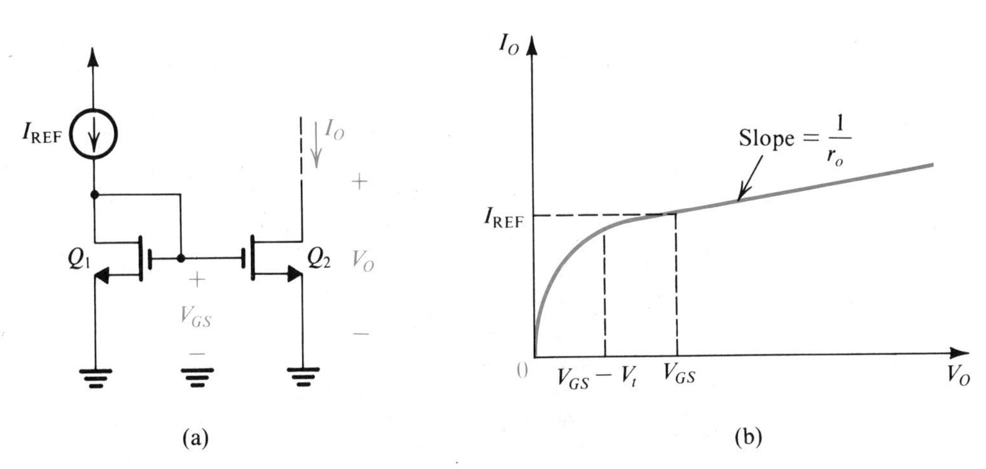

Fig. 5.58 (a) The basic current-mirror circuit. **(b)** The output characteristic of the current mirror for the case of matched transistors Q_1 and Q_2.

threshold voltages V_t, but possibly different (W/L) ratios. Transistor Q_1 is fed with the reference current I_{REF}. The output current I_O is taken at the drain of Q_2, which must be operated in the saturation region. For Q_1 we can write

$$I_{REF} = K_1(V_{GS} - V_t)^2 \tag{5.94}$$

where V_{GS} is the gate-to-source voltage corresponding to a drain current of I_{REF}. Since Q_2 is connected in parallel with Q_1, it will have the same V_{GS}; thus

$$I_O = K_2(V_{GS} - V_t)^2 \tag{5.95}$$

where we have neglected the finite output resistance of Q_2. Equations (5.94) and (5.95) can be combined to obtain

$$I_O = I_{REF}\left(\frac{K_2}{K_1}\right) \tag{5.96}$$

Expressing K_1 and K_2 in terms of the devices' (W/L) ratios gives

$$I_O = I_{REF}\frac{(W/L)_2}{(W/L)_1} \tag{5.97}$$

Thus, ideally, I_O will be a multiple of I_{REF} whose value is determined by device geometry. In practice this value of I_O will be obtained only when the voltage at the drain of Q_2 is equal to V_{GS}. Variation of the drain voltage will result in corresponding changes in I_O due to the finite output resistance r_o of Q_2. Figure 5.58(b) shows I_O as a function of V_O. This is simply the i_D–v_{DS} characteristic curve for Q_2 corresponding to the value of V_{GS} established by passing I_{REF} through Q_1. More elaborate current-mirror circuits will be presented in Chapter 6.

Exercise 5.37 For the current-mirror circuit of Fig. 5.58(a), let $L_1 = L_2 = 6\ \mu\text{m}$, $W_1 = 6\ \mu\text{m}$, $V_t = 1$ V, $\mu_n C_{OX} = 20\ \mu\text{A/V}^2$, $V_A = 50$ V, and $I_{REF} = 10\ \mu\text{A}$. (a) Calculate the value of V_{GS}. (b) Find the value of W_2 that will result in an output current of 100 μA when the output voltage is equal to the voltage at the gate. (c) If the output voltage increases by 5 V, find the resulting value of I_O.

Ans. (a) 2 V; (b) 60 μm; (c) 110 μA

The CMOS Amplifier

In CMOS technology both n-channel and p-channel devices are available, thus making possible a greater variety of circuit design techniques. Furthermore, the devices are usually fabricated in a way that eliminates the body effect, which we have found to cause considerable degradation in the performance of NMOS circuits. The basic CMOS amplifier is shown in Fig. 5.59(a). Here Q_2 and Q_3 are a matched pair of p-channel devices connected as a current mirror that is fed with the reference dc current I_{REF}. Thus Q_2 behaves as a current source and has the i–v characteristic shown in Fig. 5.59(b). Note that Q_2 will be operating in the saturation region when the voltage at its drain is lower than that at its source (V_{DD}) by at least $V_{SG} - |V_{tp}|$, where V_{SG} is the dc bias voltage corresponding to a drain current of I_{REF}. When in saturation, Q_2 has a high output resistance r_{o2},

$$r_{o2} = \frac{|V_A|}{I_{REF}} \tag{5.98}$$

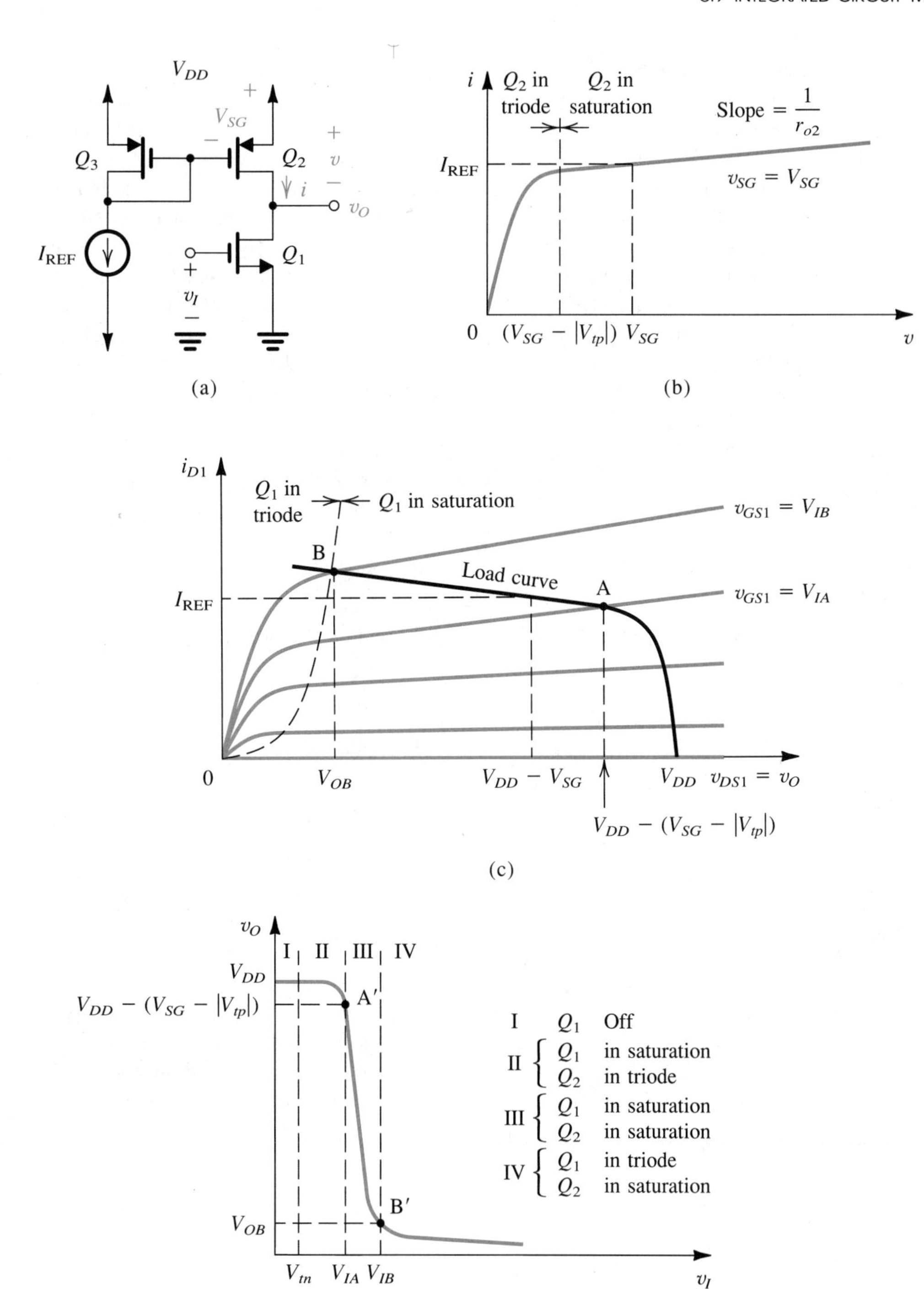

Fig. 5.59 The basic CMOS amplifier: **(a)** circuit; **(b)** *i–v* characteristic of the active load Q_2; **(c)** graphical construction to determine the transfer characteristic; and **(d)** transfer characteristic.

Transistor Q_2 is used as the load resistance for the amplifying transistor Q_1 and is called an **active load.** It follows that when Q_1 is operating in saturation, the small-signal voltage gain will be equal to g_{m1} multiplied by the total resistance between the output and ground, which is $r_{o1}//r_{o2}$. Thus a large voltage gain is obtained in the CMOS amplifier.

Before we consider the voltage gain in more detail we wish to examine the transfer characteristic of the CMOS amplifier. Figure 5.59(c) shows the i_D–v_{DS} characteristics of Q_1, with the load curve corresponding to the active load device Q_2 superimposed. Since $v_{GS1} = v_I$, the transfer characteristic can be determined point-by-point by finding the intersections of the Q_1 characteristic curves, corresponding to different values of v_I, and the load curve. The horizontal coordinate of each intersection point gives the value of v_{DS1} that is equal to v_O. The resulting transfer characteristic is sketched in Fig. 5.59(d), and its four distinct regions are indicated. For amplifier operation, region III is the one of interest. It can be shown (see Example 5.14) that the transfer characteristic in region III is almost linear and, because of the high gain, very steep.

As already mentioned, in region III of the transfer characteristic the small-signal voltage gain is given by

$$A_v \equiv \frac{v_o}{v_i} = -g_{m1}[r_{o1}//r_{o2}] \tag{5.99}$$

Since Q_1 is operating at a dc bias current equal to I_{REF}, g_{m1} can be expressed using Eq. (5.45) as

$$g_{m1} = \sqrt{2(\mu_n C_{OX})(W/L)_1 I_{REF}} \tag{5.100}$$

Substituting in Eq. (5.99) for g_{m1} from Eq. (5.100) and using $r_{o1} = r_{o2} = |V_A|/I_{REF}$ we obtain

$$A_v = -\frac{\sqrt{K_n}|V_A|}{\sqrt{I_{REF}}} \tag{5.101}$$

Thus the voltage gain is inversely proportional to the square root of the bias current.

EXAMPLE 5.14

Consider a CMOS amplifier for which $V_{DD} = 10$ V, $V_{tn} = |V_{tp}| = 1$ V, $\mu_n C_{OX} = 2\mu_p C_{OX} = 20\ \mu\text{A/V}^2$, $W = 100\ \mu$m, and $L = 10\ \mu$m for both n and p devices, $|V_A| = 100$ V and $I_{REF} = 100\ \mu$A. Find the small-signal voltage gain. Also find the coordinates of the extremities of the amplifier region of the transfer characteristic, that is, points A′ and B′.

Solution

$$K_n = \tfrac{1}{2}\mu_n C_{OX}(W/L)$$
$$= \tfrac{1}{2} \times 20 \times (100/10) = 100\ \mu\text{A/V}^2$$

From Eq. (5.101) we obtain

$$A_v = -\frac{\sqrt{100 \times 10^{-6}} \times 100}{\sqrt{10^{-4}}} = -100 \text{ V/V}$$

We observe that the gain is much greater than the values obtained in NMOS amplifiers.

The extremities of the amplifier region of the transfer characteristic are found as follows (refer to Fig. 5.59): First we determine V_{SG} of Q_2 and Q_3 corresponding to $I_D = I_{\text{REF}} = 100\ \mu\text{A}$ using

$$I_D = K_p(V_{SG} - |V_{tp}|)^2\left(1 + \frac{V_{SD}}{|V_A|}\right)$$

Substituting $K_p = \frac{1}{2}\mu_p C_{\text{OX}}(W/L)$ and $V_{SD} = V_{SG}$ and neglecting for simplicity the factor $1 + V_{SG}/|V_A|$, we obtain $V_{SG} \simeq 2.414$ V. Thus for point A′ we have

$$V_{OA} = V_{DD} - (V_{SG} - |V_{tp}|) = 8.586\ V$$

To find the corresponding value of v_I, V_{IA}, we equate the drain currents of Q_1 and Q_2,

$$i_{D1} = K_n(v_I - V_{tn})^2\left(1 + \frac{v_O}{|V_A|}\right)$$

$$i_{D2} = K_p(V_{SG} - |V_{tp}|)^2\left(1 + \frac{V_{DD} - v_O}{|V_A|}\right)$$

and substitute for $K_p(V_{SG} - |V_{tp})^2 \simeq I_{\text{REF}}$ to obtain

$$\begin{aligned} K_n(v_I - V_{tn})^2 &= I_{\text{REF}}\frac{1 + (V_{DD} - v_O)/|V_A|}{1 + (v_O/|V_A|)} \\ &\simeq I_{\text{REF}}\left(1 + \frac{V_{DD}}{|V_A|} - \frac{2v_O}{|V_A|}\right) \end{aligned}$$

which yields

$$v_O = \frac{|V_A|}{2I_{\text{REF}}}\left[I_{\text{REF}}\left(1 + \frac{V_{DD}}{|V_A|}\right) - K_n(v_I - V_{tn})^2\right] \tag{5.102}$$

Substituting $v_O = V_{OA} = 8.586$ V gives the corresponding value of v_I; that is, $V_{IA} = 1.963$ V. Since the width of region III is narrow, we may assume that $V_{IB} \simeq V_{IA} \simeq 2$ V, and thus $V_{OB} = V_{IB} - V_{tn} \simeq 2 - 1 = 1$ V. Substituting this value in Eq. (5.102), we obtain $V_{IB} = 2.039$ V. Thus a more exact value for V_{OB} is 1.039 V. The width of the amplifier region is therefore

$$\Delta V_I = V_{IB} - V_{IA} = 0.076\ V$$

The corresponding output signal swing is

$$\Delta V_O = V_{OA} - V_{OB} = 7.547\ V$$

The large-signal voltage gain is

$$\frac{\Delta V_O}{\Delta V_I} = \frac{7.547}{0.076} = 99.3$$

which is very close to the small-signal value of 100, indicating that the transfer characteristic is quite linear.

Exercises 5.38 For the CMOS amplifier of Example 5.14, find the small-signal voltage gain for $I_{REF} = 25\ \mu A$ and 400 μA.

Ans. -200 V/V; -50 V/V

5.39 For small-signal operation, the CMOS amplifier of Fig. 5.59(a) can be represented as a transconductance amplifier (Section 1.5). Sketch the amplifier equivalent circuit.

Ans. See Fig. E5.39

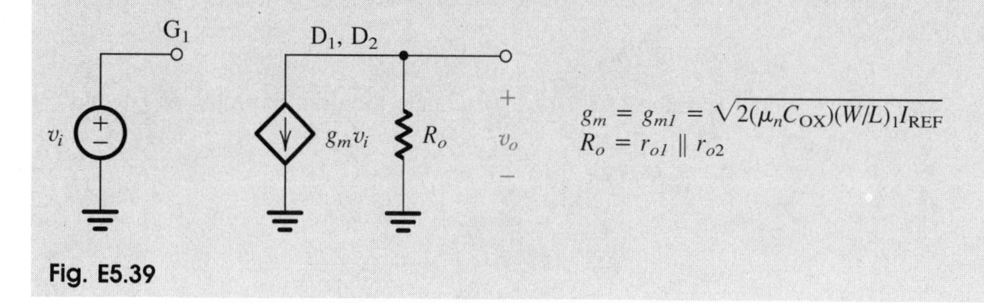

Fig. E5.39

The Source Follower

The source-follower configuration was studied in Section 5.8. In the design of MOS IC amplifiers, the source follower is used as a buffer to obtain a low output resistance. Figure 5.60(a) shows a source follower as it is commonly connected in an IC amplifier. To calculate the small-signal voltage gain and output resistance, we show in Fig. 5.60(b) the circuit with the dc voltage sources replaced with grounds and the dc current source replaced with an open circuit. Also shown are the resistance $1/g_m$, which is the equivalent resistance seen between source and gate, looking into the source; the resistance $1/g_{mb}$,

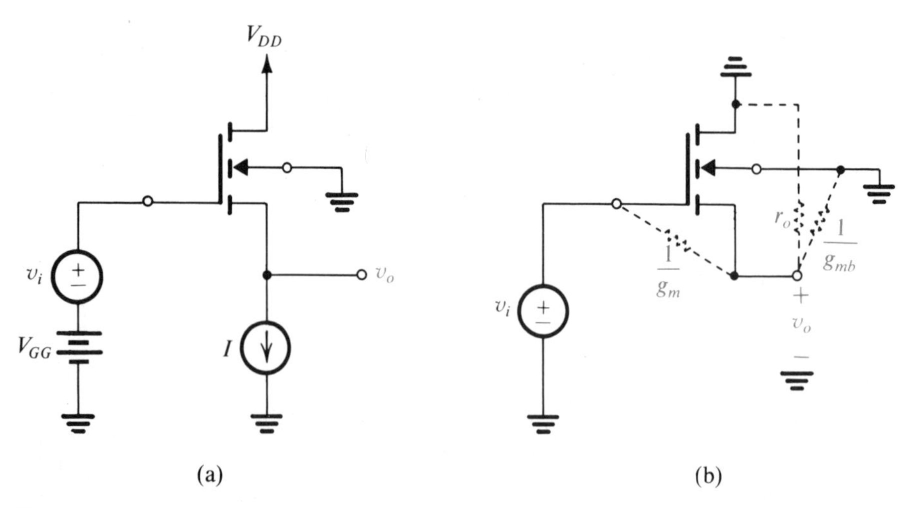

Fig. 5.60 (a) Source-follower circuit. **(b)** Source follower with the source-to-gate equivalent resistance (looking into the source), the source-to-body equivalent resistance (looking into the source), and the drain-to-source resistance shown.

which is the equivalent resistance between source and body, looking into the source; and the source-to-drain resistance r_o. Extreme caution should be exercised in using this equivalent circuit: $1/g_m$ in the resistance looking into the source; the resistance looking into the gate is infinite, since the gate current is zero. Thus the input resistance of the source follower is infinite.

The output voltage v_o appears across the total resistance between source and ground. This is the parallel equivalent of r_o and $1/g_{mb}$. To find the voltage gain we use the voltage-divider rule

$$\frac{v_o}{v_i} = \frac{[(1/g_{mb})//r_o]}{(1/g_m) + [(1/g_{mb})//r_o]} \tag{5.103}$$

If $r_o \gg 1/g_{mb}$, we obtain

$$\frac{v_o}{v_i} \simeq \frac{g_m}{g_m + g_{mb}} \tag{5.104}$$

Substituting $g_{mb} = \chi g_m$ gives

$$\frac{v_o}{v_i} = \frac{1}{1+\chi} \tag{5.105}$$

Thus the body effect reduces the gain from approximately unity to the value given by Eq. (5.105). Note that this gain value is obtained with no load; it is the open-circuit voltage gain. It can be used together with the output resistance R_o of the source follower, to obtain the gain when a load is connected. The output resistance is the resistance between the source and ground with v_i reduced to zero. Short-circuiting the signal source v_i in Fig. 5.60(b), we see that

$$R_o = (1/g_m)//(1/g_{mb})//r_o \tag{5.106}$$

Exercise 5.40 For the source-follower circuit of Fig. 5.60(a), let $W = 100\ \mu\text{m}$, $L = 8\ \mu\text{m}$, $\mu_n C_{OX} = 100\ \mu\text{A/V}^2$, $V_t = 1$ V, $V_A = 100$ V, $I = 0.5$ mA, and $\chi = 0.1$. Calculate the open-circuit voltage gain and the output resistance.

Ans. 0.91 V/V; 810 Ω

5.10 FET SWITCHES

Field-effect transistors can be used as switches for both analog and digital signals. To understand the basis for the operation of the FET as a switch, consider the circuit in Fig. 5.61(a). Here an NMOS transistor is to be operated as a switch between node X and ground. That is, when the switch is open, node X is disconnected from ground, and when the switch is closed node X is connected to ground. The NMOS switch is controlled by its gate voltage v_G, which takes one of two levels: $V_1 < V_t$ and $V_2 > V_t$. Figure 5.61(b) shows a graphical construction to determine the operating point corresponding to each of the two levels of the control signal.

When $v_G = V_1$, the transistor will be cut off and the operating point will be that

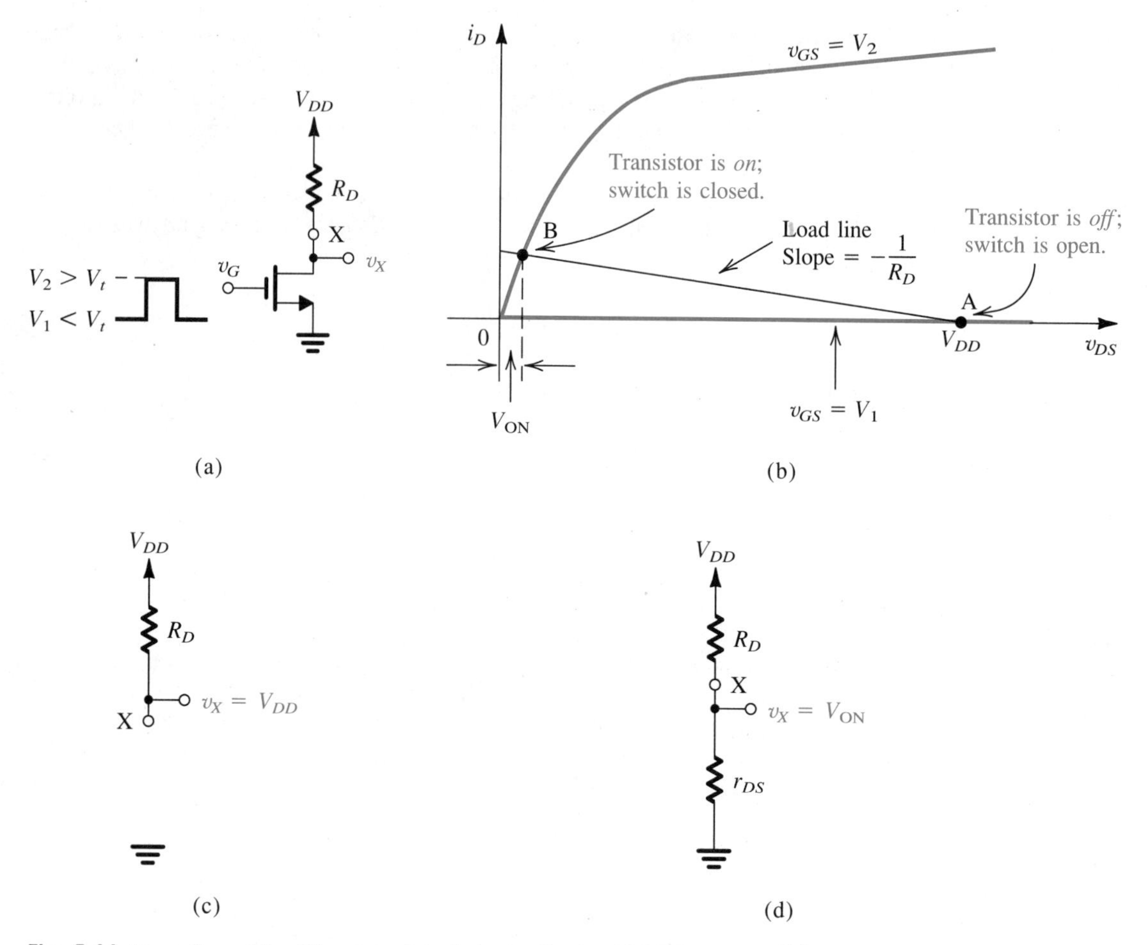

Fig. 5.61 Operation of the FET as a switch: **(a)** circuit using NMOS transistor; **(b)** graphical construction to determine the operating points corresponding to the two levels of the control signal v_G; **(c)** equivalent circuit when the switch is open; and **(d)** equivalent circuit when the switch is closed.

labeled A in Fig. 5.61(b). In this case the switch will be open, and the equivalent circuit corresponding is shown in Fig. 5.61(c).

With v_G at the high level V_2, the transistor will be conducting (on) and operating at point B in Fig. 5.61(b). Observe that the value of R_D is selected so that the device operates in the triode region. Thus the voltage between X and ground, called the **closure voltage,** or the **on voltage,** (V_{ON}) will be small. Also the resistance through the switch, between X and ground, which is r_{DS} of the FET, will be small. Obviously in this state the switch is closed, and the equivalent circuit corresponding is shown in Fig. 5.61(d). Although node X is not short-circuited to ground, as an ideal switch would do, it is connected to ground via a relatively low **on resistance,** r_{DS}.

The JFET switch operates in a similar manner except that one must choose the high level V_2 of the control signal such that the gate-to-channel junction does not conduct; thus $V_2 \leq 0$.

There are two kinds of switches: digital and analog. The digital switch is mainly used to realize a logic inverter: Observe that when v_G is low, v_X is high and vice versa. The proper operation of a digital switch does not depend critically on the exact values of V_{ON} or r_{DS}. We shall study MOSFET digital switches in the context of our study of digital MOS circuits in Chapter 13.

The analog switch, on the other hand, is required to operate with analog signals (i.e., V_{DD} in Fig. 5.61(a) is replaced with an analog signal) that can be small in value and furthermore can be either positive or negative. Performance of the analog switch is sensitive to the value of the on resistance r_{DS}. In the remainder of this section we study the use of the MOSFET in the design of analog switches. MOS analog switches find application in many systems including data converters (Chapter 10) and switched-capacitor filters (Chapter 11).

A final remark: The FET is much better suited to the design of analog switches than the BJT. This is mainly because the FET i_D–v_{DS} characteristics pass right through the origin and extend symmetrically into the third quadrant. In constrast, the i_C–v_{CE} characteristics of the BJT intercept the v_{CE} axis at finite, though small ($\simeq 0.1$ V), voltage. Thus the BJT switch exhibits an offset voltage (see Section 4.12).

An NMOS Analog Switch

When the MOSFET is used to switch (or gate) analog signals, a specific configuration using CMOS circuitry is the most convenient. To appreciate this fact, consider first the analog switch formed by a single enhancement NMOS transistor, as shown in Fig. 5.62. Here v_A is the analog input signal and is assumed to be in the range -5 V to $+5$ V. The load to which this analog signal is to be connected is assumed to be a resistance R_L in parallel with a capacitance C_L. In order to keep the substrate-to-source and substrate-to-drain *pn* junctions reverse-biased at all times, the substrate terminal is connected to -5 V.

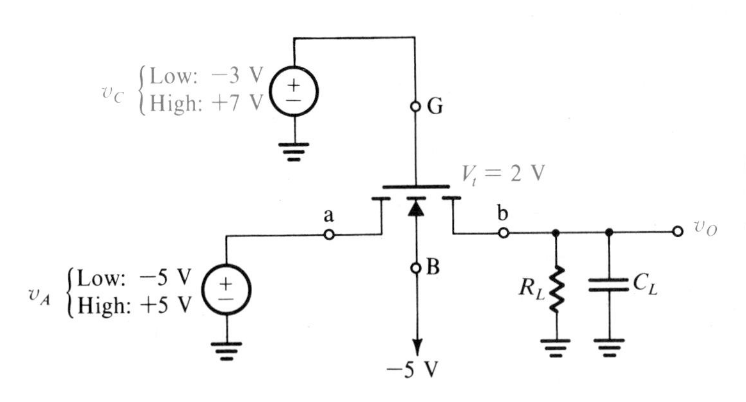

Fig. 5.62 Application of the enhancement MOSFET as an analog switch. The parallel combination of R_L and C_L represents the load.

The purpose of the control signal v_C is to turn the switch on and off. Assume that the device has a threshold voltage $V_t = 2$ V. It follows that in order to turn the transistor on for all possible input signal levels the high value of v_C should be at least $+7$ V. Similarly, to turn the transistor off for all possible input signal levels the low value of v_C should be a maximum of -3 V. Note, however, that these levels are not sufficient in practice, since the transistor will be barely on and barely off at the limits. In any case we see that the range of the control voltage has to be at least equal to the range of the analog input signal being switched. Unfortunately, though, the "on" resistance of the switch will depend to a large extent on the value of the analog input signal. Thus the transient response of the switch will depend on the value of input signal. Another obvious disadvantage is the rather inconvenient levels required for v_C.

The reader will observe that we have not indicated in Fig. 5.62 which terminal is the source and which is the drain. We have avoided labeling them first of all because the MOSFET is a symmetric device, with the source and the drain interchangeable. More important, the operation of the device as a switch is based on this interchangeability of roles. Specifically, if the analog input signal is positive, say, $+4$ V, then it is most convenient to think of terminal *a* as the drain and of terminal *b* as the source. For this case the circuit (when v_C is high) takes the familiar form shown in Fig. 5.63(a). It is easy to see that the device will be operating in the triode region and that v_O will be very close to the input analog signal level of $+4$ V. On the other hand, if the input signal is negative, say, -4 V, then it is most convenient to think of terminal *a* as the source and of terminal *b* as the drain. In this case the circuit (in the case of v_C high) can be redrawn in the familiar form of Fig. 5.63(b). Here again it should be clear that the device operates in the triode region, and v_O will be only slightly higher than the input analog signal level of -4 V.

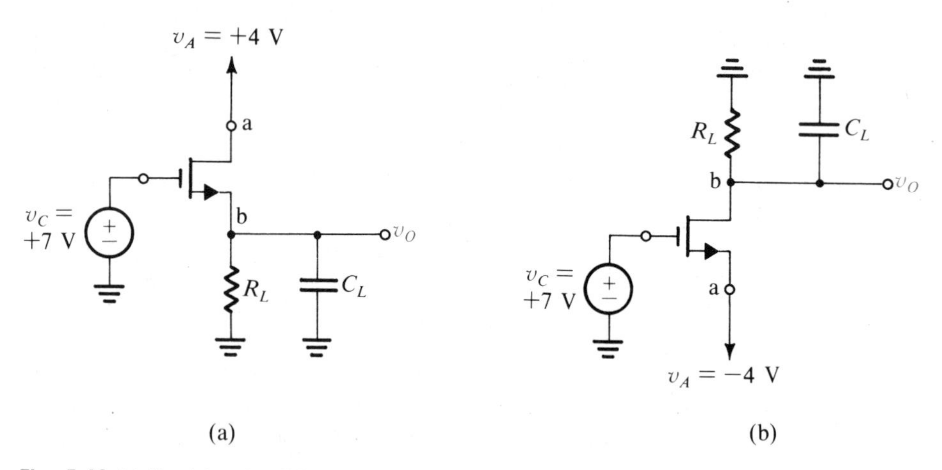

Fig. 5.63 Method for visualizing the operation of the circuit in Fig. 5.62 when v_C is high: **(a)** v_A is positive; **(b)** v_A is negative.

The CMOS Transmission Gate

Let us consider next the more elaborate CMOS analog switch shown in Fig. 5.64. This circuit is commonly known as a **transmission gate.** As in the previous case, we are assuming that the analog signal being switched lies in the range -5 V to $+5$ V. In order

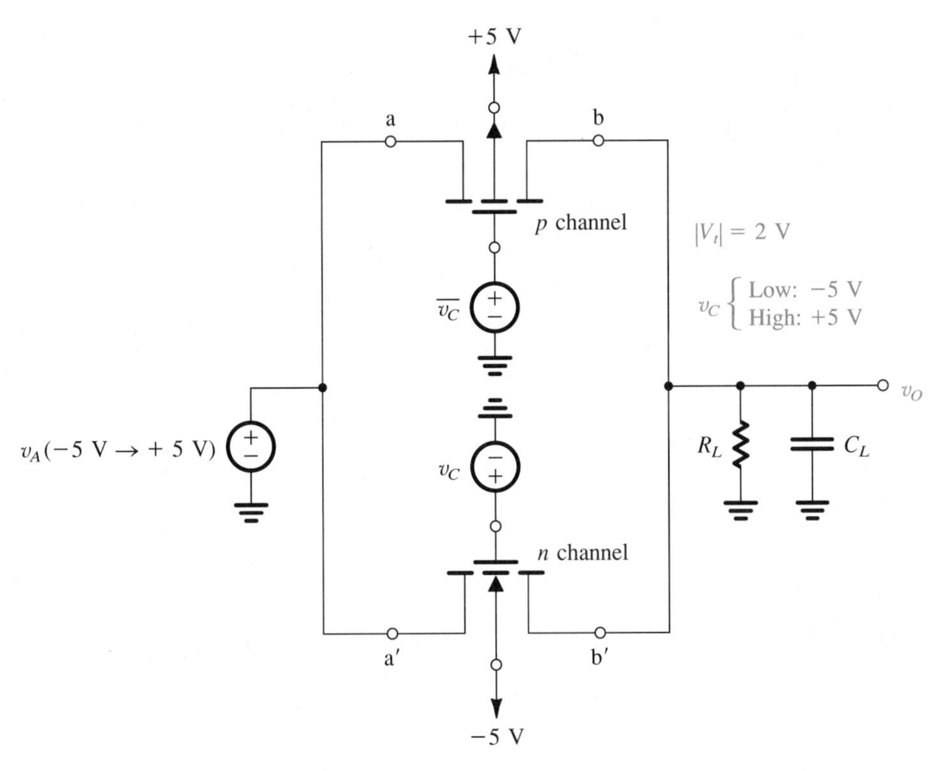

Fig. 5.64 The CMOS transmission gate.

to prevent the substrate junctions from becoming foward-biased at any time, the substrate of the p-channel device is connected to the most positive voltage level (+5 V) and that of the n-channel device is connected to the most negative voltage level (−5 V). The transistor gates are controlled by two complementary signals denoted v_C and $\overline{v_C}$. Unlike the single NMOS switch, here the levels of v_C can be the same as the extremes of the analog signal, +5 V and −5 V. When v_C is at the low level, the gate of the n-channel device will be at −5 V, thus preventing the n-channel device from conducting for any value of v_A (in the range −5 V to +5 V). Simultaneously, the gate of the p-channel device will be at +5 V, which prevents that device from conducting for any value of v_A (in the range −5 V to +5 V). Thus with v_C low the switch is open.

In order for us to close the switch, we have to raise the control signal v_C to the high level of +5 V. Correspondingly, the n-channel device will have its gate at +5 V and will thus conduct for any value of v_A in the range of −5 V to +3 V. Simultaneously, the p-channel device will have its gate at −5 V and will thus conduct for any value of v_A in the range −3 V to +5 V. We thus see that for v_A less than −3 V only the n-channel device will be conducting, while for v_A greater than +3 V only the p-channel device will be conducting. For the range $v_A = -3$ V to +3 V both devices will be conducting. Furthermore, we can see that as one device conducts more heavily, conduction in the other device is reduced. Thus as the resistance r_{DS} of one device decreases, the resistance of the other device increases, with the parallel equivalent, which is the "on" resistance of the switch, remaining approximately constant. This is clearly an advantage of the CMOS transmission gate over the single NMOS switch previously considered. This advantage applies in addition to the obviously more convenient levels required for the control signal in the CMOS switch.

The operation of the transmission gate (in the closed position) can be better understood from the two equivalent circuits shown in Fig. 5.65. Here the circuit in Fig. 5.65(a) applies when v_A is positive, while that in Fig. 5.65(b) applies when v_A is negative. Note the interchangeability of the roles played by the source and drain of each of the two devices. Also note that in thinking about the operation of the CMOS gate one should consider that the drain of one device is connected to the source of the other, and vice versa.

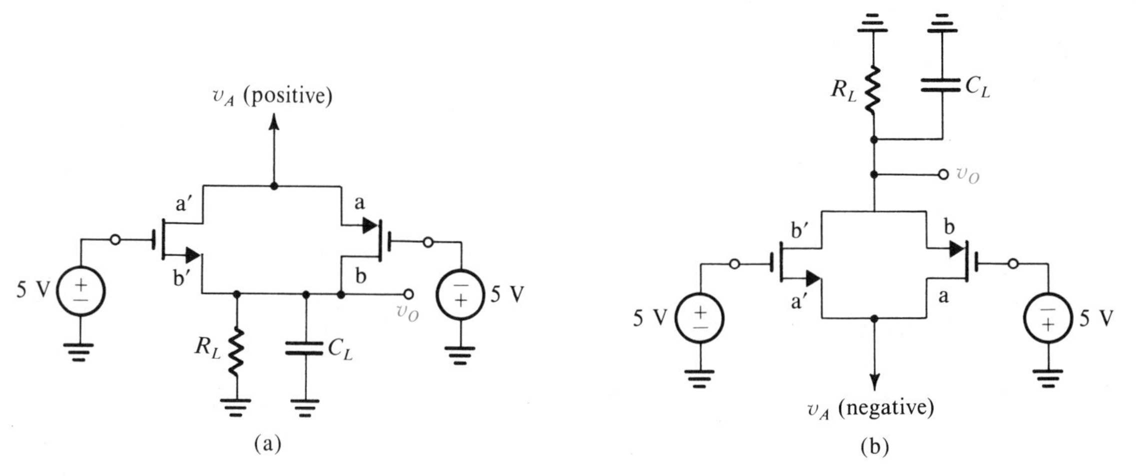

Fig. 5.65 Equivalent circuits for visualizing the operation of the transmission gate in the closed (on) position: **(a)** v_A is positive; **(b)** v_A is negative.

Exercise 5.41 Consider the CMOS transmission gate and its equivalent circuit shown in Fig. 5.65(b). Let the two devices have $|V_t| = 2$ V and $K = 50\ \mu A/V^2$, and let $R_L = 50$ kΩ. For (a) $v_A = -5$ V, (b) $v_A = -2$ V, and (c) $v_A = 0$ V, calculate v_O and the total resistance of the switch. [*Hint:* Since both devices will be operating in the triode region and $|v_{DS}|$ will be small, use $i_D \simeq 2K(v_{GS} - V_t)v_{DS}$ for the n-channel device and $i_D \simeq 2K(v_{SG} - |V_t|)v_{SD}$ *for* the p-channel device.]

Ans. (a) −4.878 V, 1.25 kΩ; (b) −1.986 V, 1.649 kΩ; (c) 0 V, 1.667 kΩ

5.11 GALLIUM ARSENIDE (GaAs) DEVICES—THE MESFET

The devices discussed thus far, and indeed the devices used in most of the circuits studied in this book, are made of silicon.[10] This reflects the situation that has existed in the

[10]The material in this section is required only for the study of the GaAs circuits in Sections 6.9 and 13.11. Otherwise, this section can be skipped without loss of continuity.

microelectronics industry for at least three decades. Furthermore, owing to the advances that are continually being made in silicon device and circuit technology, the dominance of silicon as the most useful semiconductor material is expected to continue for many years to come. Nevertheless, another semiconductor material has been making inroads into digital applications that require extremely high speeds of operation and analog applications that require very high operating frequencies. We refer to gallium arsenide (GaAs), a compound semiconductor formed of gallium, which is in the third column of the periodic table of elements, and arsenic, which is in the fifth column; thus GaAs is known as a III-V semiconductor.

The major advantage that GaAs offers over silicon is that electrons travel much faster in n-type GaAs than in silicon. This is a result of the fact that the electron drift mobility μ_n (which is the constant that relates the electron drift velocity to the electric field; velocity = $\mu_n E$) is five to ten times higher in GaAs than in silicon. Thus for the same input voltages GaAs devices have higher output currents, and thus higher g_m, than the corresponding silicon devices. The larger output currents enable faster charging and discharging of load and parasitic capacitances and thus result in increased speeds of operation.

Gallium arsenide devices have been used for some years in the design of discrete-component amplifiers for microwave applications (in the 10^9 Hz or GHz frequency range). More recently, GaAs has begun to be employed in the design of very-high-speed digital integrated circuits and in analog ICs, such as op amps, that operate in the hundreds of MHz frequency range. Although the technology is still relatively immature, suffering from yield and reliability problems and generally limited to low levels of integration, it offers great potential. Therefore, this book includes a brief study of GaAs devices and circuits. Specifically, the basic GaAs devices are studied in this section; their basic amplifier circuit configurations are discussed in Section 6.9; and GaAs digital circuits are studied in Section 13.11.

The Basic GaAs Devices

Although there are a number of GaAs technologies currently in various stages of development, we shall study the most mature of these technologies. The active device available in this technology is an n-channel field effect transistor known as the **metal semiconductor FET** or **MESFET.** The technology also provides a type of diode known as the **Schottky-barrier diode (SBD).** The structure of these two basic devices is illustrated by their cross sections, depicted in Fig. 5.66. The GaAs circuit is formed on an undoped GaAs substrate. Since the conductivity of undoped GaAs is very low, the substrate is said to be semi-insulating. This turns out to be an advantage of GaAs technology as it simplifies the process of isolating the devices on the chip from one another, as well as resulting in smaller parasitic capacitances between the devices and the circuit ground.

As indicated in Fig. 5.66, a Schottky-barrier diode consists of a metal–semiconductor junction. The metal, referred to as the Schottky-barrier metal to distinguish it from the different kind of metal used to make a contact (see Long and Butner (1990) for a detailed explanation of the difference), forms the anode of the diode. The n-type GaAs forms the cathode. Note that heavily doped n-type GaAs (indicated by n^+) is used between the n region and the cathode metal contact in order to keep the parasitic series resistance low.

The gate of the MESFET is formed by Schottky-barrier metal in direct contact with

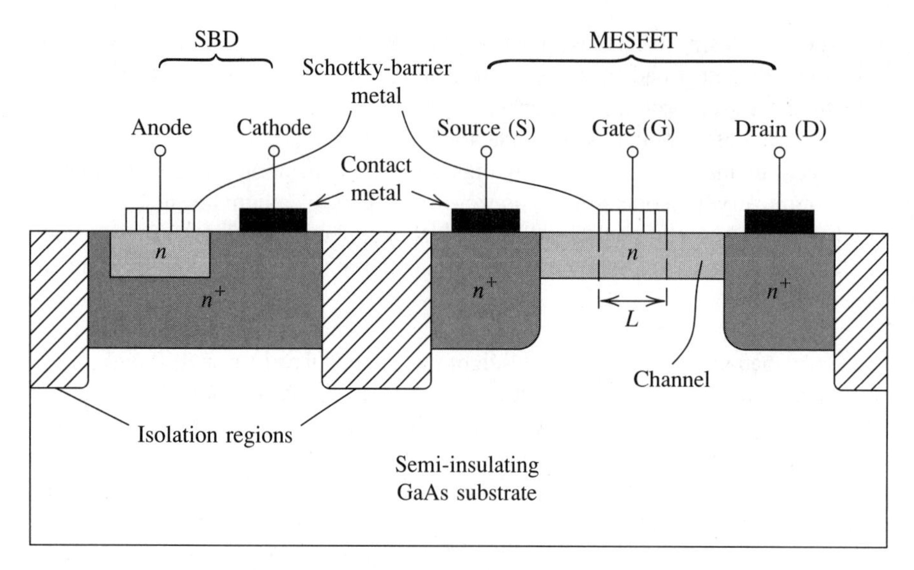

Fig. 5.66 Cross section of a GaAs Schottky-barrier diode (SBD) and a MESFET.

the n-type GaAs that forms the channel region. The channel length L is defined by the length of the gate electrode, and similarly for the width W (in the direction perpendicular to the page). To reduce the parasitic resistances between the drain and source contacts and the channel, the two contacts are surrounded with heavily doped (n^+) GaAs.

Since the main reason for using GaAs circuits is to achieve high speed/frequency of operation, the channel length is made as small as possible. Typically $L = 0.2\text{–}2\ \mu\text{m}$. Also, usually all the transistors on the IC chip are made to have the same length, leaving only the width W of each device to be specified by the circuit designer.

Only n-channel MESFETs are available in GaAs technology. This is because holes have a relatively low drift mobility in GaAs, making p-channel MESFETs unattractive. The lack of complementary transistors is a definite disadvantage of GaAs technology. However, it makes the task of the circuit designer even more challenging than usual.

Device Operation

The MESFET operates in a very similar manner to the JFET, with the Schottky metal playing the role of the p-type gate of the JFET (refer to Fig. 5.22). Basically, a depletion region forms in the channel below the gate surface, and the thickness of the depletion region is controlled by the gate voltage v_{GS}. This in turn effects control over the channel dimensions and thus on the current that flows from drain to source in response to an applied v_{DS}. The latter voltage causes the channel to have a tapered shape, with pinch-off eventually occurring at the drain end of the channel.

The most common GaAs MESFETs available are of the depletion type with a threshold voltage V_t (or, equivalently, pinch-off voltage V_P) in the range of -0.5 to -2.5 V. These devices can be operated with v_{GS} values ranging from the negative V_t to positive values as high as a few tenths of a volt. However, as v_{GS} reaches 0.7 V or so, the

Schottky-barrier diode between gate and channel conducts heavily and the gate voltage no longer effectively controls the drain-to-source current. Gate conduction, which is not possible in MOSFETs, is another definite disadvantage of the MESFET.

Although less common, enhancement-mode MESFETs are available in certain technologies. These normally-off devices are obtained by arranging that the depletion region existing at $v_{GS} = 0$ extends through the entire channel depth, thus blocking the channel and causing $i_D = 0$. To cause current to flow from drain to source the channel must be opened by applying to the gate a positive voltage of sufficient magnitude to reduce the thickness of the depletion region below that of the channel region. Typically, the threshold voltage V_t is between 0.1 and 0.3 V.

The above description of MESFET operation suggests that the i_D–v_{DS} characteristics should saturate at $v_{DS} = v_{GS} - V_t$, as is the case in a silicon JFET. It has been observed, however, that the i_D–v_{DS} characteristics of GaAs MESFETs saturate at lower values of v_{DS} and, furthermore, that the saturation voltages $v_{DS,sat}$ do not depend strongly on the value of v_{GS}. This ''early saturation'' phenomenon comes about because the velocity of the electrons in the channel does not remain proportional to the electric field (which in turn is determined by v_{DS} and L; $E = v_{DS}/L$) as is the case in silicon; rather, the electron velocity reaches a high peak value and then saturates (that is, becomes constant independent of v_{DS}). The velocity saturation effect is even more pronounced in short-channel devices ($L \leq 1$ μm), occurring at values of v_{DS} lower than $(v_{GS} - V_t)$.

Finally, a few words about the operation of the Schottky-barrier diode. Forward current is conducted by the majority carriers (electrons) flowing into the Schottky-barrier metal (the anode). Unlike the *pn*-junction diode, minority carriers play no role in the operation of the SBD. As a result, the SBD does not exhibit minority-carrier storage effects, which gave rise to the diffusion capacitance of the *pn*-junction diode (refer to Section 3.12 and to Fig. 3.49 in particular). Thus, the SBD has only one capacitive effect, that associated with the depletion-layer capacitance C_j.

Device Characteristics and Models

A first-order model for the MESFET, suitable for hand calculations, is obtained by neglecting the velocity saturation effect, and thus the resulting model is almost identical to that of the JFET:

$$i_D = 0 \qquad \text{for } v_{GS} < V_t$$

$$i_D = \beta[2(v_{GS} - V_t)v_{DS} - v_{DS}^2](1 + \lambda\, v_{DS}) \qquad \text{for } v_{GS} \geq V_t,\ v_{DS} < v_{GS} - V_t$$

$$i_D = \beta(v_{GS} - V_t)^2(1 + \lambda\, v_{DS}) \qquad \text{for } v_{GS} \geq V_t,\ v_{DS} \geq v_{GS} - V_t \tag{5.107}$$

The only differences between these equations and those for the JFETs are (1) the channel-length modulation factor, $1 + \lambda\, v_{DS}$, is included also in the equation describing the triode region (also called the ohmic region) simply because λ of the MESFET is rather large and including this factor results in a better fit to measured characteristics; and (2) the conductance parameter K is renamed β to correspond with the MESFET literature (note, however, that this β has absolutely nothing to do with β of the BJT!).

A modification of this model to account for the early saturation effects is given in Hodges and Jackson (1988).

Figure 5.67(a) shows the circuit symbol for the depletion-type n-channel GaAS MESFET. Since only one type of transistor (n channel) is available, all devices will be drawn the same way, and there should be no confusion as to which terminal is the drain and which is the source.

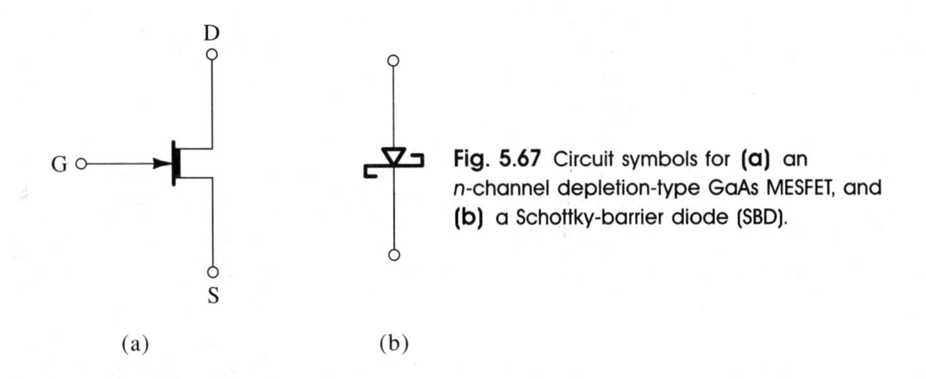

Fig. 5.67 Circuit symbols for **(a)** an n-channel depletion-type GaAs MESFET, and **(b)** a Schottky-barrier diode (SBD).

The circuit symbol of the Schottky-barrier diode is depicted in Fig. 5.67(b). In spite of the fact that the physical operation of the SBD differs from that of the pn-junction diode, their i–v characteristics are identical. Thus the i–v characteristic of the SBD is given by the same exponential relationship studied in Chapter 3. For the GaAs SBD, the constant n is typically in the range of 1 to 1.2.

The small-signal model of the MESFET is identical to that of other FET types. The parameter values are given by

$$g_m = 2\beta(V_{GS} - V_t)(1 + \lambda V_{DS}) \tag{5.108}$$

$$r_o \equiv \left[\frac{\partial i_D}{\partial v_{DS}}\right]^{-1}$$

$$= 1/\lambda\beta(V_{GS} - V_t)^2 \tag{5.109}$$

The MESFET, however, has a rather high value for λ (0.1 to 0.3 V^{-1}) which results in a small output resistance r_o. This turns out to be a serious drawback of GaAs MESFET technology, resulting in low voltage gain obtainable from each stage. Furthermore, it has been found that r_o decreases at high frequencies. Circuit design techniques for coping with the low r_o will be presented in Section 6.9.

For easy reference, Table 5.2 gives typical values for device parameters in a GaAs MESFET technology. The devices in this technology have a channel length $L = 1\ \mu$m. The values given are for a device with a width $W = 1\ \mu$m. The parameter values for actual devices can be obtained by appropriately scaling by the width W. This process is illustrated in the following example. Unless otherwise specified, the values of Table 5.2 are to be used for the exercises and the end-of-chapter problems.

Table 5.2 TYPICAL PARAMETER VALUES FOR GaAs MESFETS AND SCHOTTKY DIODES IN $L = 1\ \mu$m TECHNOLOGY, NORMALIZED FOR $W = 1\ \mu$M

V_t	=	-1.0 V
β	=	10^{-4} A/V^2
λ	=	0.1 V^{-1}
I_S	=	10^{-15} A
n	=	1.1

EXAMPLE 5.15 Figure 5.68 shows a simple GaAs MESFET amplifier, with the W values of the transistors indicated. Assume that the dc component of v_I, that is V_{GS1}, biases Q_1 at the current provided by the current source Q_2 so that both devices operate in saturation and that the dc output is at half of the supply voltage. Find:

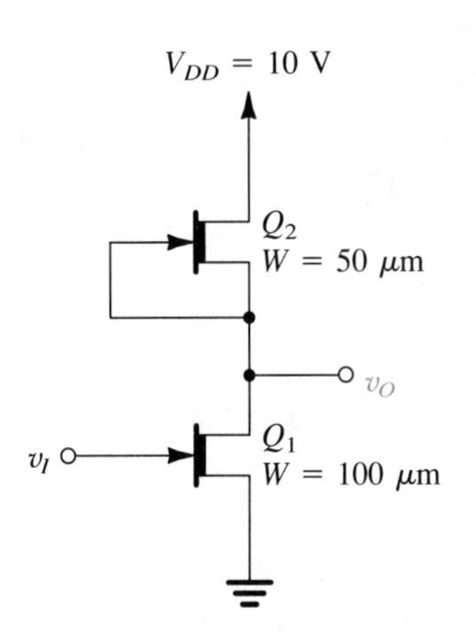

Fig. 5.68 Circuit for Example 5.15: a simple MESFET amplifier.

(a) the β values for Q_1 and Q_2;

(b) V_{GS1};

(c) g_{m1}, r_{o1}, and r_{o2}; and

(d) the small-signal voltage gain.

Solution

(a) The values of β can be obtained by scaling the value given in Table 5.2 using the specified values of W,

$$\beta_1 = 100 \times 10^{-4} = 10^{-2}\ \text{A/V}^2 = 10\ \text{mA/V}^2$$

$$\beta_2 = 50 \times 10^{-4} = 5 \times 10^{-3}\ \text{A/V}^2 = 5\ \text{mA/V}^2$$

(b)

$$I_{D2} = \beta_2(V_{GS2} - V_t)^2(1 + \lambda V_{DS2})$$
$$= 5(0 + 1)^2(1 + 0.1 \times 5)$$
$$= 7.5 \text{ mA}$$
$$I_{D1} = I_{D2} = 7.5 \text{ mA}$$
$$7.5 = \beta_1(V_{GS1} - V_t)^2(1 + \lambda V_{DS1})$$
$$= 10(V_{GS1} + 1)^2(1 + 0.1 \times 5)$$

Thus,

$$V_{GS1} = -0.3 \text{ V}$$

(c)

$$g_{m1} = 2 \times 10(-0.3 + 1)(1 + 0.1 \times 5)$$
$$= 21 \text{ mA/V}$$
$$r_{o1} = \frac{1}{0.1 \times 10(-0.3 + 1)^2} = 2 \text{ k}\Omega$$
$$r_{o2} = \frac{1}{0.1 \times 5(0 + 1)^2} = 2 \text{ k}\Omega$$

(d)

$$A_v = -g_{m1}(r_{o1}//r_{o2})$$
$$= -21 \times (2//2) = -21 \text{ V/V}$$

Exercise 5.42 For a MESFET with the gate shorted to the source and having $W = 10\ \mu$m, find the minimum voltage between drain and source to operate in saturation. For $V_{DS} = 5$ V, find the current I_D. What is the output resistance of this current source?

Ans. 1 V; 1.5 mA; 10 kΩ

SUMMARY

▸ A summary of the characteristics of the various types of field-effect transistors is given in Table 5.1 (page 317).

▸ The relative levels that the FET terminal voltages must have for the different modes of operation are shown in Fig. 5.13 for enhancement NMOS; Fig. 5.17 for enhancement PMOS; and Fig. 5.20 for depletion NMOS.

▸ The $i_D - v_{GS}$ characteristics of the various types of MOSFET are contrasted in Fig. 5.21.

▸ For small signals, a FET biased in the saturation region, functions as a voltage-controlled current source with a transconductance $g_m = 2K(V_{GS} - V_t)$ and an output resistance $r_o = |V_A|/I_D$.

▶ The common-source configuration provides a high voltage gain and a very high input resistance but a limited high-frequency response. A much wider bandwidth is achieved in the common-gate configuration but its input resistance is low.

▶ The source follower provides a voltage gain less than unity but features a low output resistance.

▶ Integrated-circuit MOS amplifiers utilize MOS transistors both as amplifying and as load devices.

▶ In NMOS technology, the load devices can be either enhancement- or depletion-type MOSFETs connected as two-terminal devices. The voltage gain of a depletion-load amplifier is an order of magnitude greater than that of an enhancement-load amplifier.

▶ In CMOS technology, both n- and p-channel enhancement MOSFETs are used, thus providing the circuit designer with considerable flexibility.

▶ In the basic CMOS amplifier, a complementary MOSFET, operated as a constant-current source, is used as a load in an arrangement called an active load.

▶ The voltage gain of the basic CMOS amplifier is approximately $(g_m r_o/2)$.

▶ The CMOS transmission gate is widely used in the switching of analog signals.

▶ GaAs technology offers the advantage of higher speeds and wider bandwidths than is available from silicon devices.

▶ The basic GaAs devices are the n-channel MESFET and the Schottky-barrier diode.

BIBLIOGRAPHY

R. S. C. Cobbold, *Theory and Applications of Field-Effect Transistors,* New York: Wiley, 1969.

P. R. Gray, D. A. Hodges, and R. W. Brodersen, *Analog MOS Integrated Circuits,* New York: IEEE Press, 1980.

P. R. Gray and R. G. Meyer, *Analysis and Design of Analog Integrated Circuits,* 2nd ed; New York: Wiley, 1984.

A. B. Grebene, *Bipolar and MOS Analog Integrated Circuit Design,* New York: Wiley, 1984.

D. J. Hamilton and W. G. Howard, *Basic Integrated Circuit Engineering,* New York: McGraw-Hill, 1975.

D. A. Hodges and H. G. Jackson, *Analysis and Design of Digital Integrated Circuits,* 2nd ed; New York: McGraw-Hill, 1988.

S. I. Long and S. E. Butner, *Gallium Arsenide Digital Integrated Circuit Design,* New York: McGraw-Hill, 1990.

C. Mead, *Analog VLSI and Neural Systems,* Reading, MA: Addison Wesley, 1989.

D. L. Pulfrey and N. G. Tarr, *Introduction to Microelectronic Devices,* Englewood Cliffs, NJ: Prentice-Hall, 1989.

Y. Tsividis, ''Design considerations in single-channel MOS analog integrated circuits—A tutorial,'' *IEEE Journal of Solid-State Circuits,* vol. SC-13, pp. 383–391, June 1978.

Y. Tsividis, *Operation and Modeling of the MOS Transistor,* New York: McGraw-Hill, 1987.

PROBLEMS

Section 5.2: Current–Voltage Characteristics of the Enhancement MOSFET

5.1 Let the n-channel enhancement MOSFET whose characteristics are shown in Fig. 5.10 and for which $V_t = 2$ V and $K = 0.25$ mA/V^2 be operated with $v_{GS} = 6$ V. Find i_D for $v_{DS} = 2$ V and for $v_{DS} = 6$ V.

5.2 A particular enhancement MOSFET for which $V_t = 1$ V and $K = 0.25$ mA/V^2 is to be operated in the saturation region. If i_D is to be 4 mA find the required v_{GS} and the minimum required v_{DS}. Repeat for $i_D = 16$ mA.

5.3 A particular n-channel enhancement MOSFET is measured to have a drain current of 4 mA at $V_{GS} = V_{DS} = 9$ V and of 1 mA at $V_{GS} = V_{DS} = 5$ V. What are the values of K and V_t for this device?

D5.4 For a particular IC fabrication process the quantity $\frac{1}{2}\mu_n C_{OX} = 10$ μA/V^2 and $V_t = 1$ V. In an application in which $v_{GS} = v_{DS} = V_{supply} = 5$ V, a drain current of 0.8 mA is required of a device of minimum length of 2 μm. What value of channel width must the design use?

5.5 In a particular fabrication process for which $\frac{1}{2}\mu_n C_{OX}$ is 30 μA/V^2, layout restricts the critical channel dimensions to the range from 3 to 9 μm. For the same applied voltages, find the range of ratios of the drain currents that flow at the onset of saturation in the four devices having extreme combinations of width and length.

5.6 Consider an enhancement-type NMOS transistor operated at a constant gate-to-source voltage V_{GS}. Show that the drain current decreases to a fraction α of the value at the onset of saturation at

$$v_{DS} = (V_{GS} - V_t)(1 - \sqrt{1 - \alpha})$$

For $V_t = 1$ V and $V_{GS} = 2$ V, find v_{DS} for $\alpha = 0.5$ and $\alpha = 0.1$.

5.7 At the boundary of the saturation region, a particular enhancement MOSFET conducts 8 mA with v_{DS} of 4 V. What is the value of its parameter K? At what value of v_{DS} does the drain current reduce to 2 mA?

5.8 For a particular MOSFET operating in the saturation region at a constant v_{GS}, i_D is found to be 2 mA for $v_{DS} = 4$ V and 2.2 mA for $v_{DS} = 8$ V. What values of r_o, V_A, and λ correspond?

5.9 A particular MOSFET has $V_A = 50$ V. For operation at 1 mA and 10 mA, what are the expected output resistances? In each case, for a change in v_{DS} of 10%, what change in drain current would you expect?

5.10 An enhancement NMOS transistor with V_t of 2 V and K of 0.1 mA/V^2 is to be used as a voltage-controlled linear resistor. What is the required range of v_{GS} to obtain a resistance in the range 0.5 to 5 kΩ?

5.11 A number of enhancement NMOS transistors, whose threshold voltage V_t is 1 V when the source and substrate are joined, are operated with a 10-V supply and a common substrate connected to the negative end of the supply. In operation the dc voltage levels on the source of the devices range from 0 to 9 V. For $\gamma = 0.5$ V$^{1/2}$ what is the expected range of thresholds that will be observed?

*5.12 An enhancement PMOS transistor has $K = 40$ μA/V^2, $V_t = -1.5$ V, and $\lambda = -0.02$ V^{-1}. The gate is connected to ground and the source to +5 V. Find the drain current for (a) $v_D = +4$ V, (b) $v_D = +1.5$ V, (c) $v_D = 0$ V, and (d) $v_D = -5$ V.

*5.13

(a) Using the expression for i_D in saturation and neglecting the channel-length modulation effect (i.e., let $\lambda = 0$) derive an expression for the per unit change in i_D per °C in terms of the per unit change in K per °C, the temperature coefficient of V_t in V/°C, and V_{GS} and V_t.

(b) If V_t decreases by 2 mV for every °C rise in temperature, find the temperature coefficient of K that results in i_D decreasing by 0.2%/°C when the NMOS transistor is operated at $V_{GS} = 5$ V and provided that $V_t = 1$ V.

Section 5.3: The Depletion-Type MOSFET

5.14 A depletion-type n-channel MOSFET with $K = 1$ mA/V^2 and $V_t = -3$ V has its source and gate grounded. Find the region of operation and the drain current for (a) $v_D = 0.1$ V, (b) $v_D = 1$ V, (c) $v_D = 3$ V, and (d) $v_D = 5$ V. Neglect the channel-length modulation effect.

5.15 A depletion-type n-channel MOSFET with $K = 1$ mA/V^2 and $V_t = -3$ V has its gate grounded and a voltage of +1 V applied to the source. Find the minimum drain voltage required in order for the device to operate in saturation. What drain current results at this drain voltage? Neglect the channel-length modulation effect.

*5.16 Neglecting the channel-length modulation effect show that for the depletion-type NMOS transistor of Fig. P5.16 the i–v relationship is given by

$$i = K(v^2 - 2V_t v), \qquad \text{for } v \geq V_t$$

$$i = -KV_t^2, \qquad \text{for } v \leq V_t$$

(Recall that V_t is negative.) Sketch the i–v relationship for the case $V_t = -2$ V and $K = 1$ mA/V^2.

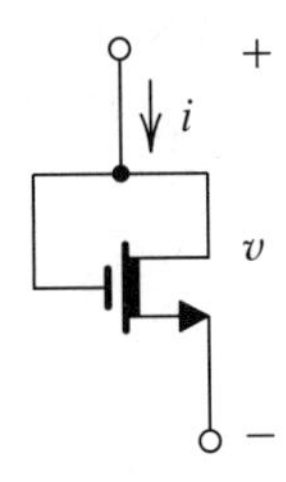

Fig. P5.16

5.17 For the circuit analyzed in Exercise 5.8 (refer to Fig. E5.8) what does the voltage at the source become when the drain voltage is lowered to +1 V?

5.18 A depletion-type NMOS transistor, operating in the triode region with $v_{DS} = 0.1$ V, conducts a drain current of 1 mA at $v_{GS} = -1$ V and 3 mA at $v_{GS} = +1$ V. Find I_{DSS} and V_t.

5.19 A depletion-type NMOS transistor operating in the saturation region with $v_{DS} = 5$ V conducts a drain current of 5 mA at $v_{GS} = -1$ V and 45 mA at $v_{GS} = +1$ V. Find I_{DSS} and V_t. Assume $\lambda = 0$.

Section 5.4: The Junction Field-Effect Transistor (JFET)

5.20 A JFET of unknown type (i.e., n- or p-channel), but for which two of the terminals are known to be either source or drain, has the gate left floating and an ohmmeter connected between source and drain. When the gate is joined to the negative (red) lead of the ohmmeter, the apparent resistance does not change. When the gate is joined to the positive (black) lead of the ohmmeter, the apparent resistance decreases greatly. Is the JFET of the n- or p-channel type?

5.21 Consider an n-channel JFET whose gate and source are joined, thus resulting in a two-terminal device. If the voltage across the device is denoted v and the current through it is denoted i show that

$$i = K(-2\,V_t v - v^2)$$
$$= I_{DSS}\left[2\left(\frac{v}{-V_t}\right) - \left(\frac{v}{V_t}\right)^2\right], \qquad \text{for } 0 \le v \le -V_t$$

$$i = KV_t^2 = I_{DSS}, \qquad \text{for } v \ge -V_t$$

when the channel-length modulation effect is neglected. What is the incremental resistance of this two-terminal device when the JFET is pinched off at the drain? What does the incremental resistance become if the channel-length modulation effect is taken into account?

5.22 An n-channel JFET is to be used as a voltage-controlled resistance that is made almost linear by operating the FET at a very small v_{DS}. Show that the resistance r_{DS} can be expressed in the form

$$r_{DS} = [-V_P/(2I_{DSS})]/(1 - v_{GS}/V_P)$$

For $V_P = -4$ V and $I_{DSS} = 16$ mA find the range of r_{DS} resulting by varying v_{GS} from 0 to 0.9 V_P.

5.23 Consider an n-channel JFET with $I_{DSS} = 4$ mA and $V_P = -2$ V. Find the values of V_t and K. If the source is grounded and a -1-V dc voltage is applied to the gate find the minimum drain voltage that results in pinch-off operation. Find the corresponding value of drain current. If the drain voltage is increased by 10 V and λ for this device is specified to be 0.01 V^{-1} find the new value of drain current.

5.24 A JFET with $V_P = -1$ V and $I_{DSS} = 1$ mA shows an output resistance of 100 kΩ when operated in pinch-off with $v_{GS} = 0$. What is the value of output resistance when the device is operated in pinch-off with $v_{GS} = -0.5$ V?

5.25 For a particular JFET having $V_P = -2$ V and $I_{DSS} = 8$ mA and operating in saturation at $v_{GS} = -1$ V, find the change in v_{GS} required to increase i_D by 0.4 mA. What change in v_{GS} is required to decrease i_D by 0.4 mA from the same original condition? Think about why the changes differ.

5.26 For a JFET having $V_P = -2$ V and $I_{DSS} = 8$ mA operating at $v_{GS} = -1$ V and a very small v_{DS}, what is the value of r_{DS}? Find the value of v_{GS} at which r_{DS} becomes half this value.

5.27 The p-channel JFET in the circuit of Fig. P5.27 has $V_P = 3$ V. Find the range that V_{DD} can have for the device to operate in pinch-off. If V_S is measured with the device in pinch-off and found to be -1 V what do you expect I_{DSS} to be? Assume $\lambda = 0$.

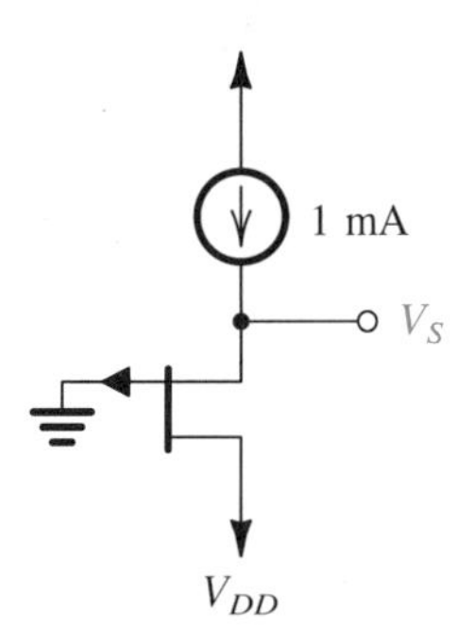

Fig. P5.27

Section 5.5: FET Circuits at DC

D5.28 Design the circuit of Fig. 5.25 to establish a drain current of 1 mA and a drain voltage of 0 V. The MOSFET has $V_t = 2$ V, $\mu_n C_{OX} = 20\ \mu\text{A/V}^2$, $L = 10\ \mu$m, and $W = 400\ \mu$m.

D5.29 Consider the circuit of Fig. E5.19 (page 336). Let Q_1 and Q_2 have $V_t = 2$ V, $\mu_n C_{OX} = 20\ \mu\text{A/V}^2$, $L_1 = L_2 = 10\ \mu$m, $W_1 = 50\ \mu$m, and $\lambda = 0$.

(a) Find the value of R required to establish a current of 0.4 mA in Q_1.

(b) Find W_2 so that Q_2 operates in the saturation region with a current of 0.6 mA.

D5.30 The PMOS transistor in the circuit of Fig. P5.30 has $V_t = -2$ V, $\mu_p C_{OX} = 8\ \mu\text{A/V}^2$, $L = 10\ \mu$m, and

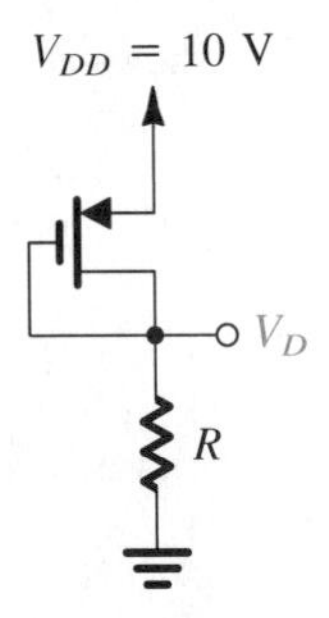

Fig. P5.30

$\lambda = 0$. Find the values required for W and R in order to establish a drain current of 0.1 mA and a voltage V_D of 7 V.

D5.31 The NMOS transistors in the circuit of Fig. P5.31 have $V_t = 2$ V, $\mu_n C_{OX} = 20\ \mu\text{A/V}^2$, $\lambda = 0$, and $L_1 = L_2 = 10\ \mu\text{m}$. Find the required values of gate width for each of Q_1 and Q_2, and the value of R, to obtain the voltages and current values indicated.

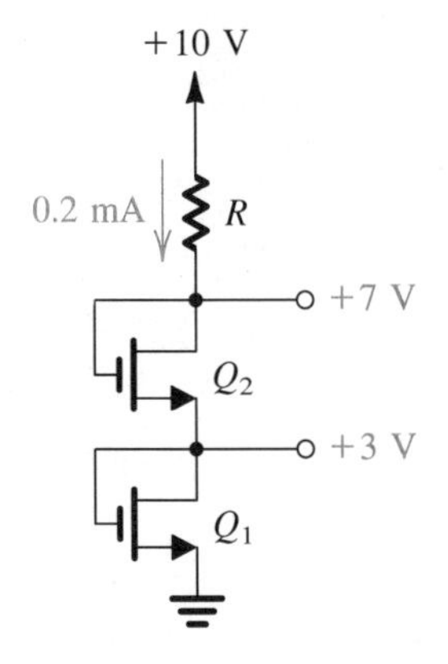

Fig. P5.31

D5.32 The NMOS transistors in the circuit of Fig. P5.32 have $V_t = 2$ V, $\mu_n C_{OX} = 20\mu\text{A/V}^2$, $\lambda = 0$, and $L_1 = L_2 = L_3 = 10\ \mu\text{m}$. Find the required values of gate width for each of Q_1, Q_2, and Q_3 to obtain the voltages and current values indicated.

5.33 The NMOS transistor in the circuit of Fig. P5.33 has $V_t = 1$ V, $K = 0.5\ \text{mA/V}^2$, and $\lambda = 0$. If v_G is a pulse with 0 and 5-V levels, as indicated, find the levels of the pulse signal that will appear at the drain.

5.34 Consider the circuit of Fig. 5.28(a). In Example 5.4 it was found that when $V_t = 1$ V and $K = 0.5\ \text{mA/V}^2$ the drain current is 0.5 mA and the drain voltage is +7 V. If the transistor is replaced with another having $V_t = 2$ V and $K = 1\ \text{mA/V}^2$, find the new values of I_D and V_D. Comment on how tolerant (or intolerant) the circuit is to changes in device parameters.

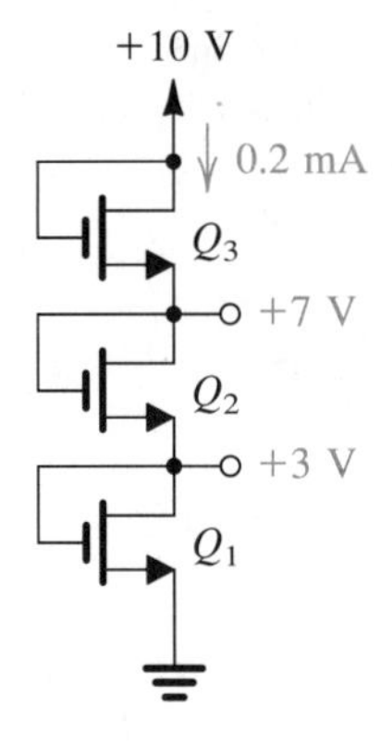

Fig. P5.32

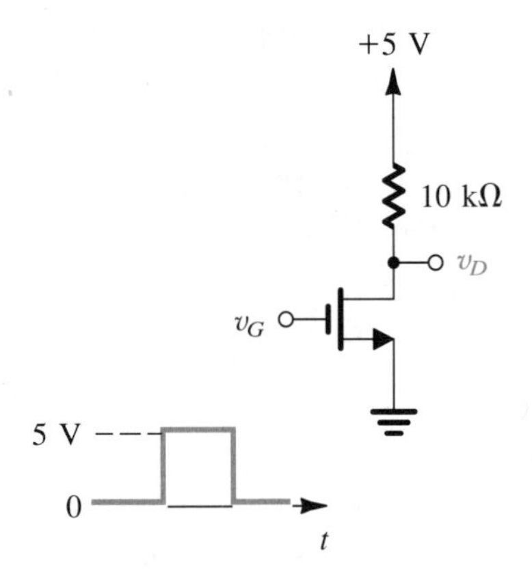

Fig. P5.33

D5.35 Using an enhancement-type PMOS transistor with $V_t = -1.5$ V, $K = 0.5\ \text{mA/V}^2$, and $\lambda = 0$, design a circuit that resembles that in Fig. 5.28(a). Using a 10-V supply design for a gate voltage of +6 V, a drain current of 0.5 mA, and a drain voltage of +5 V. Find the values of R_S and R_D.

5.36 Analyze the circuit in Fig. P5.36 to determine I_D and V_D. Let the NMOS transistor have $V_t = 1$ V, $K = 0.5\ \text{mA/V}^2$, and $\lambda = 0$.

5.37 Analyze the circuit in Fig. P5.37 to determine the drain current and the drain voltage. Assume that the depletion MOSFET has $V_t = -1$ V, $K = 0.5\ \text{mA/V}^2$, and $\lambda = 0$.

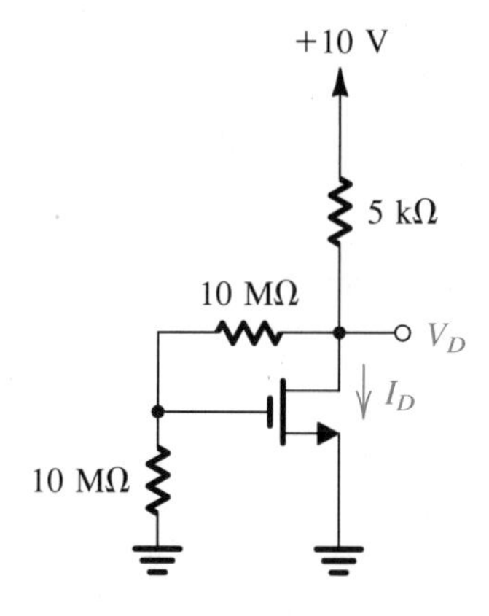

Fig. P5.36

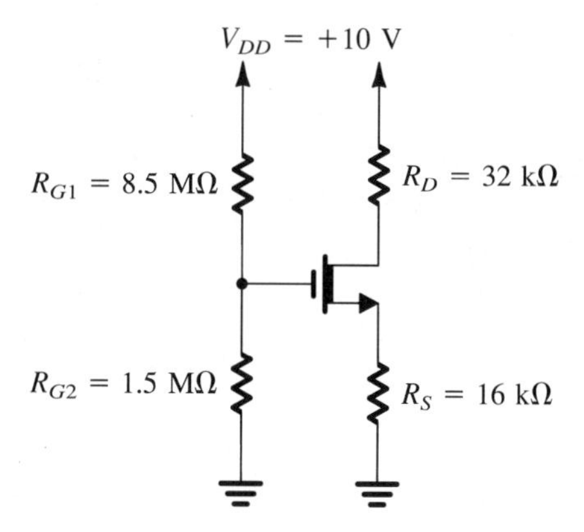

Fig. P5.37

D5.38 While leaving the voltage-divider network in the circuit of Fig. P5.37 unchanged, find new values for R_S and R_D to obtain $I_D = 0.5$ mA and $V_D = 4$ V. The depletion MOSFET has $V_t = -1$ V, $K = 0.5$ mA/V^2, and $\lambda = 0$.

*5.39 For the circuits in Fig. P5.39 calculate the labeled currents and voltages. For all devices $|V_t| = 1$ V, $K = 0.5$ mA/V^2, and $\lambda = 0$.

*5.40 Analyze the circuit in Fig. E5.23 (page 337) to determine I_D and V_D. $R_{G1} = 200$ kΩ, $R_{G2} = 100$ kΩ, $R_S = 1.5$ kΩ, and $R_D = 5$ kΩ. The JFET has $V_P = -4$ V and $I_{DSS} = 12$ mA.

5.41 Consider the circuit of Fig. 5.33. If the JFET has $V_P = 3$ V and $I_{DSS} = 4$ mA, what is the maximum value that the current source I can be set to? At this value of I what is the voltage at the source? At this value of I, what is the maximum value of R_D that still results in pinch-off operation?

Section 5.6: The FET as an Amplifier

*5.42 This problem investigates the nonlinear distortion introduced by a FET amplifier. Let the signal v_{gs} be a sine wave with amplitude V_{gs}, and substitute $v_{gs} = V_{gs} \sin \omega t$ in Eq. (5.39). Using the trigonometric identity $\sin^2 \theta = \frac{1}{2} - \frac{1}{2}\cos 2\theta$, show that the ratio of the signal at frequency 2ω to that at frequency ω, expressed as a percentage, known as the second-harmonic distortion, is

$$\text{Second-harmonic distortion} = \frac{1}{4}\frac{V_{gs}}{V_{GS} - V_t} \times 100$$

If in a particular application $V_t = 2$ V and $V_{GS} = 5$ V, find the largest amplitude of input sine wave for which the second-harmonic distortion does not exceed 1%.

5.43 Consider an NMOS transistor having $V_t = 2$ V and $K = 0.5$ mA/V^2. Let the transistor be biased at $V_{GS} = 4$ V. For operation in saturation what dc bias current I_D results? If a +0.1-V signal is superimposed on V_{GS}, find the corresponding increment in collector current by evaluating the total collector current i_D and subtracting the dc bias current I_D. Repeat for a −0.1-V signal. Use these results to estimate g_m of the FET at this bias point. Compare with the value of g_m obtained using Eq. (5.42).

5.44 Consider the FET amplifier of Fig. 5.34(a) for the case $V_t = 2$ V, $K = 0.5$ mA/V^2, $V_{GS} = 4$ V, $V_{DD} = 10$ V, and $R_D = 3.6$ kΩ.
(a) Find the dc quantities I_D and V_D.
(b) Calculate the value of g_m at the bias point.
(c) Calculate the value of the voltage gain.
(d) If the MOSFET has $\lambda = 0.01$ V^{-1}, find r_o at the bias point and calculate the voltage gain.

5.45 Show that, for a MOSFET biased at a gate-to-source voltage V_{GS} and having a dc drain current I_D, the transconductance g_m can be expressed as

$$g_m = \frac{2I_D}{V_{GS} - V_t}$$

5.46 Consider an NMOS transistor with the drain and gate connected together to create a two-terminal device. As the voltage across the device is varied it ceases conduction at 1 V. When biased at a current of 1 mA, and the two-terminal device exhibits a dc voltage drop of 3 V. Find the incremental or small-signal resistance at this operating condition.

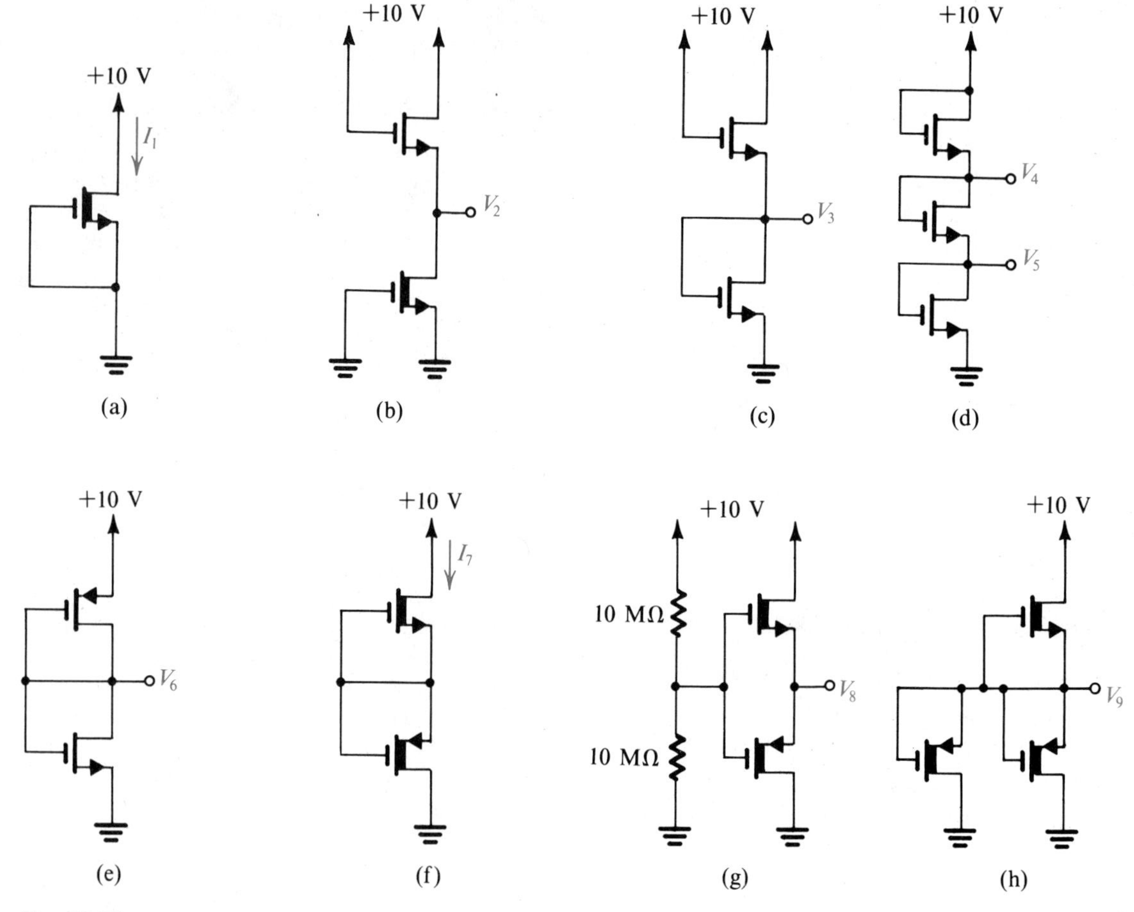

Fig. P5.39

D5.47 An NMOS technology has $\mu_n C_{OX} = 20\ \mu A/V^2$ and $V_t = 1$ V. For a transistor with $L = 10\ \mu m$, find W that results in $g_m = 1$ mA/V at $I_D = 0.5$ mA. Also find the required V_{GS}.

5.48 For the NMOS amplifier in Fig. P5.48, replace the transistor with its T equivalent circuit of Fig. 5.39(d). Derive expressions for the voltage gains v_s/v_i and v_d/v_i.

5.49 The transistor in the circuit of Fig. P5.49 has $V_t = 2$ V, $K = 0.25$ mA/V^2, and $\lambda = 0$. Find the dc voltage V_O. Also, replace the transistor with its T model and find the small-signal resistance R_{out}.

5.50 The transistor in the circuit of Fig. P5.50 has $V_t = 2$ V, $K = 0.25$ mA/V^2, and $\lambda = 0$. Find the dc voltage V_O. Also, replace the transistor with its T model and thus find the small-signal resistance R_{out}.

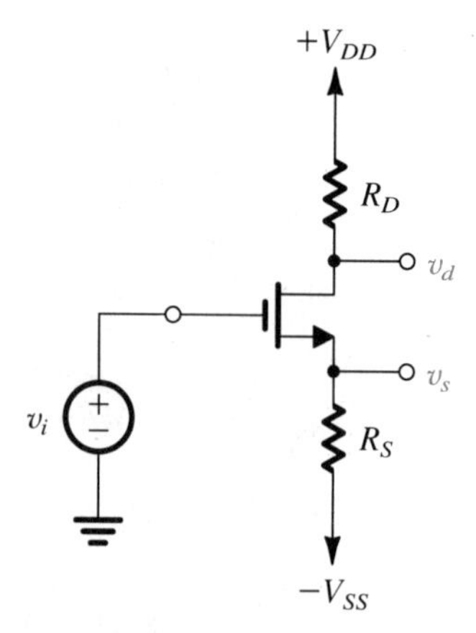

Fig. P5.48

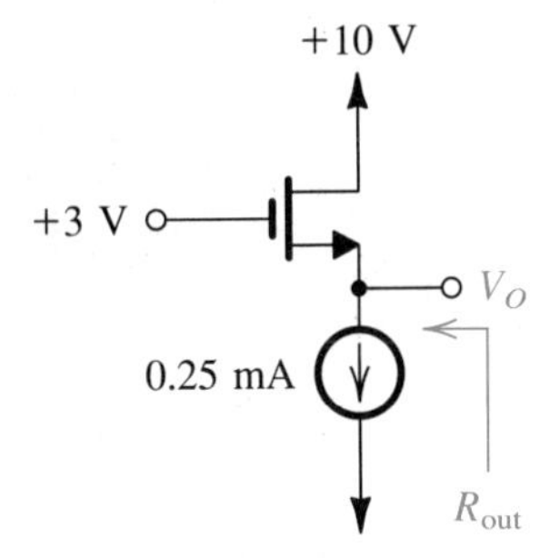

Fig. P5.49

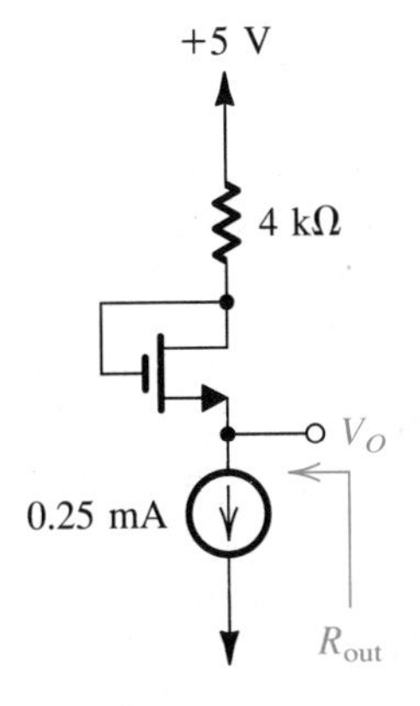

Fig. P5.50

*D5.51 For the NMOS amplifier of Fig. P5.51, replace the transistor with its small-signal equivalent circuit and show that for $R_G \gg r_o$,

$$\frac{v_o}{v_i} \simeq -\frac{2V_A}{V_{GS} - V_t}$$

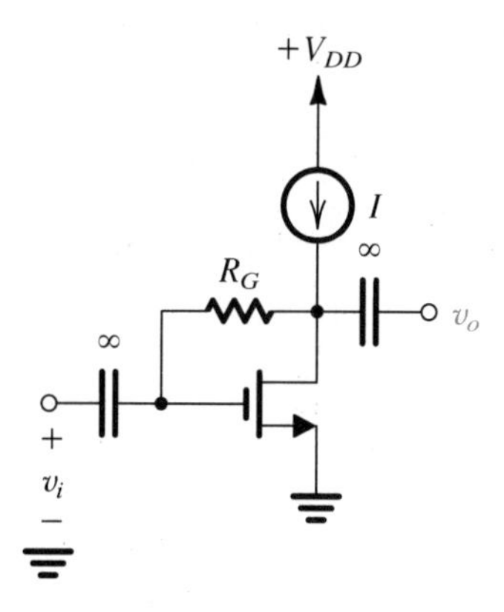

Fig. P5.51

For $V_t = 2$ V, $V_A = 100$ V, find the dc bias voltage at the drain to obtain a voltage gain of −100. If $K = 0.25$ mA/V^2 find the required bias current I. If $R_G = 10$ MΩ, find the input resistance of the amplifier.

Section 5.7: Biasing the FET in Discrete Circuits

5.52 In an electronic instrument using the biasing scheme shown in Fig. 5.40(a), a manufacturing error reduces R_S to zero. Let $V_{DD} = 12$ V, $R_{G1} = 5.6$ MΩ, and $R_{G2} = 2.2$ MΩ. What is the value of V_{GG} created? If supplier specifications allow K to vary from 110 to 190 μA/V^2 and V_t to vary from 1.3 to 2.4 V, what are the extreme values of I_D that may result?

*D5.53 For the situation described in Problem 5.52, what value of R_S should have been installed to limit the maximum value of I_D to 0.15 mA? Choose an appropriate standard 5% resistor value (refer to Appendix F). What extreme values of current now result?

5.54 An enhancement NMOS transistor is connected in the bias circuit of Fig. 5.40(a), with $V_{GG} = 4$ V and $R_S = 1$ kΩ. The transistor has $V_t = 2$ V and $K = 1$ mA/V^2. What bias current results? If a transistor for which K is 50% higher is used, what is the resulting percentage increase in I_D?

5.55 The bias circuit of Fig. 5.40(a) is used in a design with $V_{GG} = 4$ V and $R_S = 1$ kΩ. For an enhancement MOSFET with $K = 1$ mA/V^2, the source voltage was measured and found to be 1 V. What must V_t be for this device? If a device for which V_t is 0.5 V less is used, what does V_S become? What bias current results?

D5.56 Design the circuit of Fig. 5.40(b) for an enhancement MOSFET having $V_t = 2$ V and $K = 1$ mA/V^2. Let $V_{DD} = V_{SS} = 10$ V and design for a dc bias current of 1 mA and for the largest possible voltage gain (and thus the largest possible R_D) consistent with allowing a 2-V peak-to-peak voltage swing at the drain. Assume that the signal voltage on the source terminal of the FET is zero.

*D5.57 It is required to design a bias circuit of the type shown in Fig. 5.40(a) for a JFET. The JFET characteristics are specified to lie between those for the following extreme devices: The "low" device has $V_P = -2$ V and $I_{DSS} = 4$ mA, and the "high" device has $V_P = -8$ V and $I_{DSS} = 16$ mA. Use $V_{DD} = 20$ V. Your design should use the largest possible value for V_{GG} that allows for a voltage drop across R_D of 4 V before the device leaves the saturation region. The design should bias the "low" device at $V_{GS} = 0$. What value should be used for R_S? What is the bias current obtained in the "low" device and in the "high" device?

D5.58 Consider the bias circuit of Fig. 5.40(a) with the MOSFET replaced with a JFET. Find the values of R_D, R_S, R_{G1}, and R_{G2} that will result in $I_D = I_{DSS}/2$ and for which a third of the power-supply voltage will appear across each of R_D, R_S, and the FET (that is, V_{DS}). Let $V_{DD} = 15$ V, $I_{DSS} = 8$ mA, and $V_P = -2$ V. Use 1 μA in the voltage-divider network that feeds the gate.

5.59 An n-channel enhancement MOSFET for which $V_t = 2$ V and $K = 100\ \mu\text{A/V}^2$ is operated from a 9-V supply using a drain resistor R_D of 60 kΩ and the feedback bias scheme depicted in Fig. 5.43. What drain current results? What is V_D? What is the largest sine-wave output signal for which operation in the saturation region is maintained?

D5.60 Using the feedback bias arrangement shown in Fig. 5.43 with a 9-V supply and an NMOS device for which $V_t = 1$ V and $K = 0.2\ \text{mA/V}^2$, find R_D to establish a drain current of 0.2 mA. If resistor values are limited to those on the 5% resistor scale (see Appendix F), what value would you choose? What values of current and V_D result?

5.61 In the circuit of Fig. P5.61, the enhancement PMOS device has $V_t = -2$ V and $K = 0.1\ \text{mA/V}^2$. What value of V results for $I =$ (a) 0.1 mA? (b) 0.2 mA? (c) 0.4 mA?

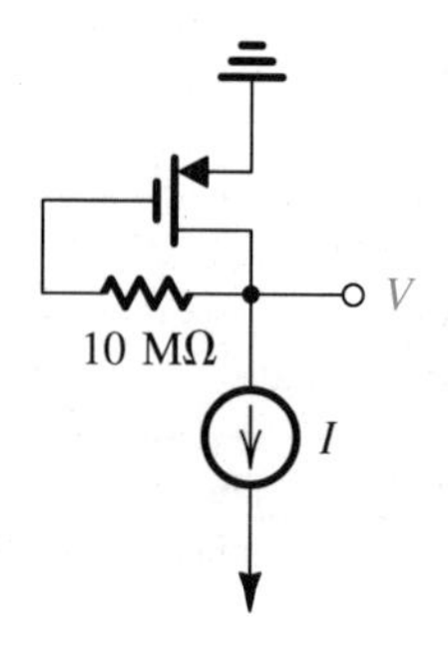

Fig. P5.61

5.62 A MOSFET having $K = 1\ \text{mA/V}^2$ and $V_t = 1$ V operates in a feedback bias arrangement such as that shown in Fig. 5.43, from a 10-V supply with $R_D = 8$ kΩ. What value of I_D results? If the FET is replaced by another with (a) $K = 0.5\ \text{mA/V}^2$ and $V_t = 1$ V, and (b) $K = 1\ \text{mA/V}^2$ and $V_t = 2$ V, what percentage change in I_D results?

Section 5.8: Basic Configurations of Single-Stage FET Amplifiers

5.63 A common-source amplifier having an open-circuit voltage gain of −9 and an output resistance of 10 kΩ is connected to a 20-kΩ load. What is the resulting voltage gain?

5.64 Calculate the voltage gain of a common-source amplifier for which $g_m = 1$ mA/V, $r_o = 20$ kΩ, $R_D = 5$ kΩ, and $R_L = 10$ kΩ.

5.65 A common-gate amplifier using an n-channel enhancement MOS transistor for which $g_m = 5$ mA/V has a 5-kΩ drain resistance (R_D) and a 2-kΩ load resistance (R_L). The amplifier is driven by a voltage source having a 200-Ω resistance. What is the input resistance of the amplifier? What is the overall voltage gain v_o/v_i? If the circuit allows a bias-current increase by a factor of 4 while maintaining linear operation, what do the input resistance and voltage gain become?

5.66 An enhancement MOSFET source follower has $g_m = 5$ mA/V and $r_o = 20$ kΩ. Find the open-circuit voltage gain and the output resistance. What will the gain become when a 1-kΩ load resistance is connected?

***D5.67** The MOSFET in the circuit of Fig. P5.67 has $V_t = 1$ V, $K = 0.4\ \text{mA/V}^2$, and $V_A = 40$ V.

(a) Find the values of R_S, R_D, and R_G so that $I_D = 0.1$ mA, the largest possible value for R_D is used while a maximum signal swing at the drain of ±1 V is possible, and the input resistance at the gate is 10 MΩ.

(b) Find the values of g_m and r_o at the bias point.

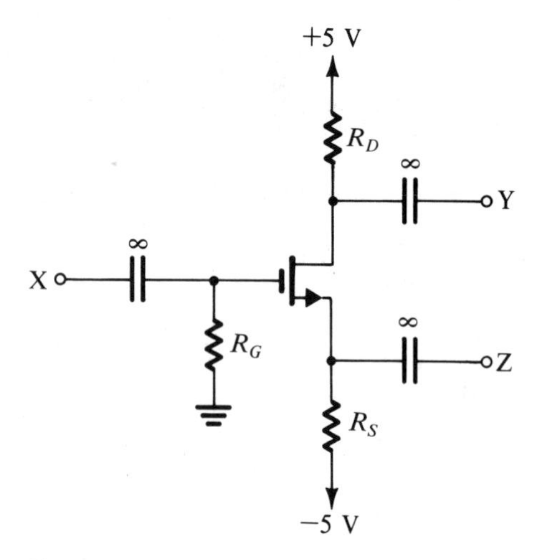

Fig. P5.67

(c) If terminal Z is grounded, terminal X is connected to a signal source having a resistance of 1 MΩ, and terminal Y is connected to a load resistance of 40 kΩ, find the voltage gain from signal source to load.

(d) If terminal Y is grounded, find the voltage gain from X to Z with Z open-circuited. What is the output resistance of the source follower?

(e) If terminal X is grounded and terminal Z is connected to a current source delivering a signal current of 10 μA and having a resistance of 100 kΩ, find the voltage signal that can be measured at Y. For simplicity, neglect the effect of r_o.

*5.68

(a) The NMOS transistor in the source follower circuit of Fig. P5.68(a) has $g_m = 5$ mA/V and a large r_o. Find the open-circuit voltage gain and the output resistance.

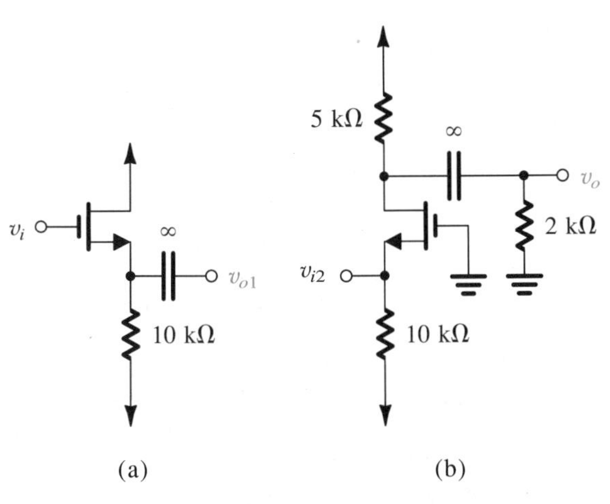

Fig. P5.68

(b) The NMOS transistor in the common-gate amplifier of Fig. P5.68(b) has $g_m = 5$ mA/V and a large r_o. Find the input resistance and the voltage gain.

(c) If the output of the source follower in (a) is connected to the input of the common-gate amplifier in (b), use the results of (a) and (b) to obtain the overall voltage gain v_o/v_i.

*5.69 The JFET in the amplifier circuit in Fig. P5.69 has $V_P = -4$ V and $I_{DSS} = 12$ mA, and at $I_D = 12$ mA the output resistance $r_o = 25$ kΩ.

(a) Determine the dc bias quantities V_G, I_D, V_{GS}, and V_D.

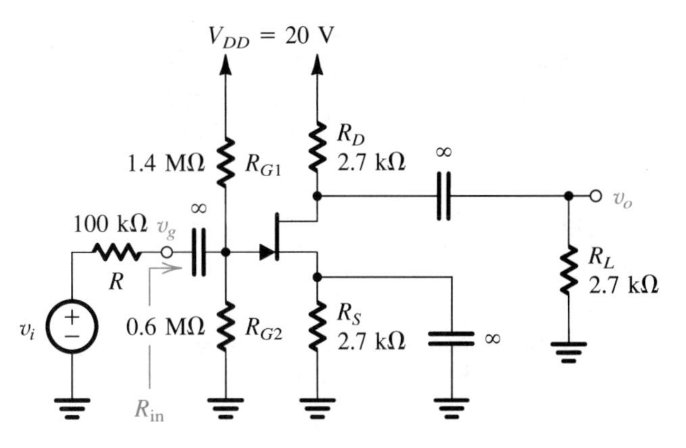

Fig. P5.69

(b) Determine the value of g_m (you can use the same formula as for the enhancement MOSFET). Also determine r_o.

(c) Replace the JFET with its small-signal model, thus obtaining the small-signal equivalent circuit for the amplifier.

(d) Use the equivalent circuit in (c) to determine R_{in} and v_g/v_i.

(e) Use the equivalent circuit to determine v_o/v_g.

(f) Find the overall voltage gain v_o/v_i.

5.70 The JFET source follower of Fig. P5.70 utilizes $R_S = 10$ kΩ and has $r_o = 100$ kΩ. The unloaded (open-circuit) voltage gain from gate to source is found to be 0.9 V/V. Find g_m and R_{out}. Also, find the voltage gain v_o/v_g when $R_L = 910$ Ω.

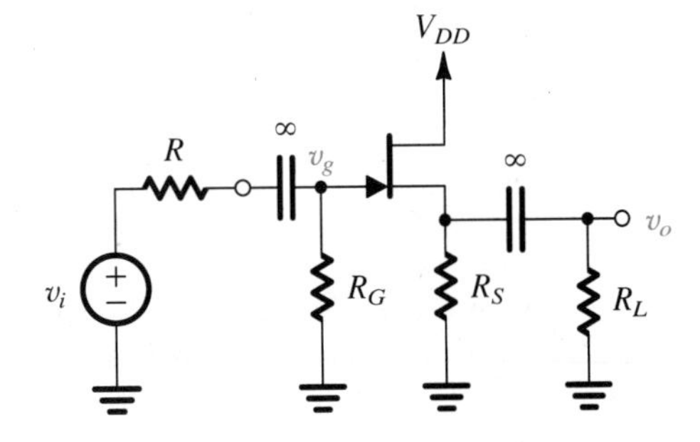

Fig. P5.70

D5.71 Consider the source follower of Fig. P5.70 with R_S returned to a negative supply voltage of −10 V (instead of ground). Let $V_{DD} = +10$ V and assume that the JFET is specified to have $V_P = -4$ V, $I_{DSS} = 12$ mA, and $r_o = \infty$. Design the circuit so that the output resistance is equal to or lower than 200 Ω. What value should be used for R_G to obtain $R_{in} \geq 1$ MΩ? Calculate the open-circuit voltage gain from gate to source of your source follower.

Section 5.9: Integrated-Circuit MOS Amplifiers

D5.72

(a) For a diode-connected enhancement-MOS load, find an expression for the incremental resistance in terms of operating current and device dimensions.

(b) In a particular application utilizing a MOSFET load having the minimum available channel width it is required to raise the load resistance by a factor of 3 while keeping the operating current unchanged. What must be done?

5.73 The enhancement-MOS load of an enhancement-MOS amplifier has a channel width 1/10 and a channel length 10 times that of the amplifier transistor. Ignoring the effects of r_o and the body effect, find the expected voltage gain.

D5.74 The driver transistor in an enhancement-load amplifier has a W/L ratio of 9. Find the W/L ratio of the load to obtain a small-signal voltage gain of -10. Ignore the effect of r_o and assume that $\chi = 0.2$.

5.75 A depletion-load MOSFET amplifier has $K_1/K_2 = 4$. Find the small-signal voltage gain for $\chi = 0.2$. Neglect the effect of r_o.

***5.76** Consider the circuit of an enhancement-load amplifier with a 10-MΩ resistor connected from output to input. Let $V_{DD} = 5$ V and $V_t = 2$ V. For each of the following cases find the dc voltage at the output and the small-signal voltage gain:

(a) $K_2 = K_1$
(b) $K_2 = 0.1K_1$ and
(c) $K_2 = 0.01K_1$.

Neglect the effect of r_o and the body effect.

****5.77** For the circuit in the Fig. P5.77 let $|V_t| = 2$ V. For each of the cases

(a) $K_2 = K_1$
(b) $K_2 = 0.1K_1$, and
(c) $K_2 = 0.01K_1$, find v_O corresponding to $v_I = 0$ V, 3 V, and 6 V.

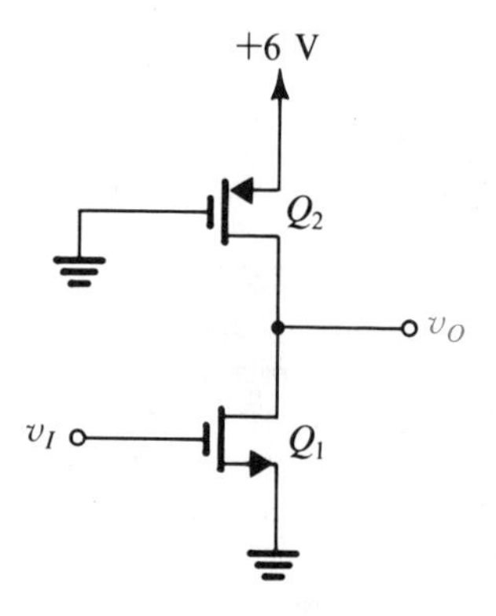

Fig. P5.77

5.78

(a) Figure P5.78(a) shows a MOSFET common-source amplifier biased by a constant drain current source I. Show that the small-signal voltage gain is given by

$$\frac{v_o}{v_i} = -g_m r_o$$

(b) Figure P5.78(b) shows a cascade of two common-source amplifiers, each biased by a constant drain-current source I. Show that the small-signal voltage gain is given by

$$\frac{v_o}{v_i} = (g_{m1} r_{o1})(g_{m2} r_{o2})$$

***5.79** Figure P5.79 shows an IC MOS amplifier known as the cascode configuration. We shall study this circuit in Chapter 6. For the time being, note that it consists of a common-source amplifier Q_1 that feeds its output current to a common-gate amplifier Q_2. V_{BIAS} is a dc bias voltage. Both devices are identical and operate at a dc current I. This amplifier can be modeled as a transconductance amplifier.

(a) Show that the short-circuit transconductance, which is the ratio of the short-circuit output signal current to the input voltage, is equal to g_{m1}.

(b) Short circuit v_i and use equivalent circuit models for Q_1 and Q_2 to show that the output resistance is given by

$$R_o = r_{o1} + r_{o2} + g_{m2} r_{o2} r_{o1} \simeq g_{m2} r_{o2} r_{o1}$$

(c) Use the results of (a) and (b) to obtain the open-circuit voltage gain v_o/v_i.

D5.80 Consider the current mirror of Fig. 5.58(a) when augmented by the addition of a transistor Q_3 whose gate is connected to that of Q_2 and whose source is connected

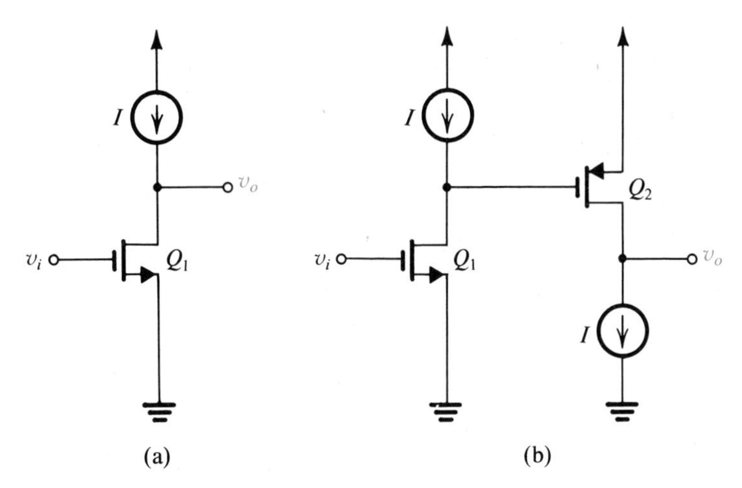

Fig. P5.78

Fig. P5.79

to that of Q_2. The drain of Q_3 provides a second output for the mirror. Label the drain current of Q_2, I_2, and that of Q_3, I_3. Assume that all transistors have equal V_t and equal channel lengths. Transistor Q_1 has $W = 10\ \mu\text{m}$. Neglecting the effects of r_o find the channel width of Q_2 and of Q_3 so that $I_2 = 20\ \mu\text{A}$ and $I_3 = 100\ \mu\text{A}$ when $I_{REF} = 10\mu\text{A}$.

*5.81 The MOSFETs in the circuit of Fig. P5.81 are matched, with $K = 25\ \mu\text{A/V}^2$ and $|V_t| = 2$ V. The resistance $R_2 = 10\ \text{M}\Omega$. For G and D open, what are the drain currents I_{D1} and I_{D2}? For $r_o = \infty$, what is the voltage gain of the amplifier from G to D? For finite r_o ($r_o = |V_A|/I_D$, $|V_A| = 180$ V), what is the voltage gain

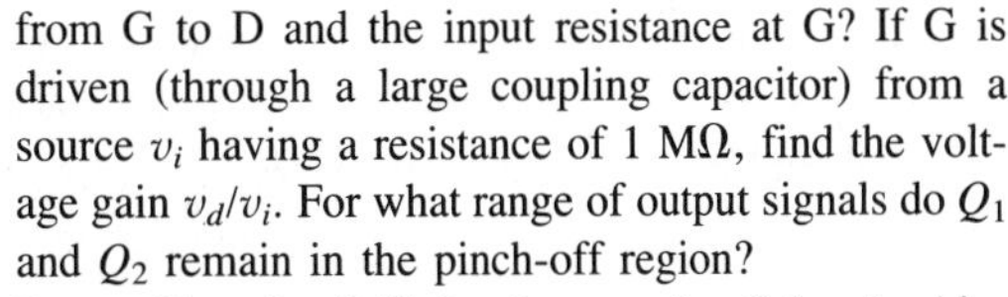
from G to D and the input resistance at G? If G is driven (through a large coupling capacitor) from a source v_i having a resistance of 1 MΩ, find the voltage gain v_d/v_i. For what range of output signals do Q_1 and Q_2 remain in the pinch-off region?

5.82 Repeat Exercise 5.40 for the case $I = 0.1$ mA. Also find the voltage gain when the source follower is loaded by a 10-kΩ resistor.

Section 5.10: FET Switches

5.83 An NMOS switch has a signal input that varies over the range ±5 V and a control input of ±10 V. Calculate the channel resistance at the two signal extremes when the control input is high, turning the switch on. If the switch is to be used in a circuit whose loop resistance is R, how large must R be to ensure at most a 1% signal loss in the switch? For the transistor, $K = 25\ \mu\text{A/V}^2$ and $V_t = 2$ V.

5.84 A CMOS transmission gate for which $K = 25\ \mu\text{A/V}^2$ and $|V_t| = 2$ V utilizes control signals of ±5 V for signals over the range ±5 V. Calculate the switch resistance at the signal extremes and for signals at 0 V. If the switch is to be used in a circuit whose loop resistance is R, how large must R be to ensure at most a 1% signal loss in the switch?

*5.85 A CMOS switch is used to connect a sinusoidal source $0.1 \sin \omega t$ to a load capacitance C. For ±5-V control signals and $K_p = K_n = 25\ \mu\text{A/V}^2$, $V_{tn} = |V_{tp}| = 2$ V, what is the cut-off frequency introduced by the switch if $C = 1000$ pF?

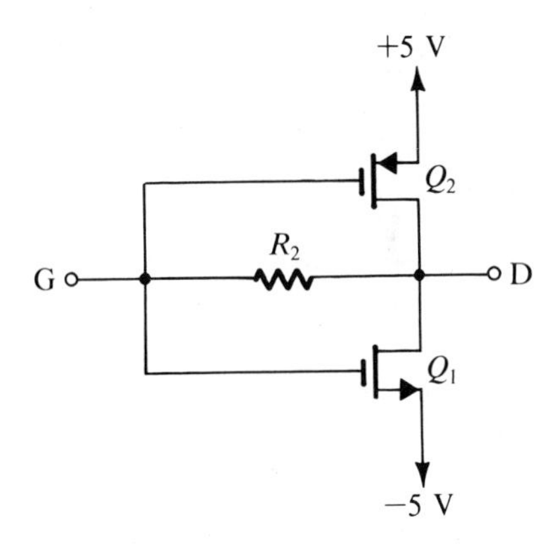

Fig. P5.81

Section 5.11: Gallium Arsenide (GaAs) Devices—The MESFET

5.86 Use the parameter values given in Table 5.2 to find the β value of a MESFET with $W = 100\ \mu\text{m}$. For $V_{GS} = -0.5$ V and $V_{DS} = 1$ V, find I_D. Also find g_m and r_o at this bias point.

D5.87 Use the parameter values given in Table 5.2 to design a MESFET current source (by connecting gate to source) to deliver a current of 6 mA when the voltage across it is 2 V. What transistor width is required? What is the output resistance of the current source? (Recall that for the process specified in Table 5.2 $L = 1\ \mu\text{m}$.)

D5.88 It is required to use three Schottky-barrier diodes, each having $W = 5\ \mu\text{m}$, in series to provide a dc voltage drop of 2.2 V. What bias current is required? Use the parameter values given in Table 5.2. (Recall that for the process specified in Table 5.2 $L = 1\ \mu\text{m}$.)

5.89 If in Example 5.15 the width of each of the two transistors is doubled, find the new values of the bias current and the voltage gain realized.

General Problems:

D**5.90 Show that for the circuit in Fig. P5.90 to operate in pinch-off the following two conditions must be satisfied.

$$I_D R_S \geq 0.5|V_P|$$

$$V_{DD} - I_D R_D \geq 1.5|V_P|$$

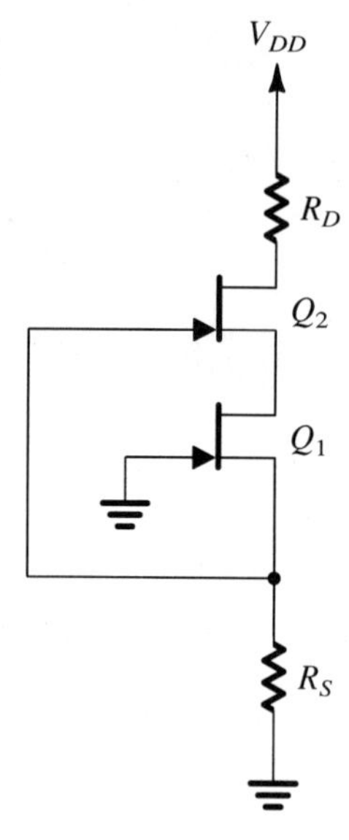

Fig. P5.90

Assume that the two JFETs are matched. For $V_{DD} = 10$ V, $|V_P| = 2$ V, and $I_{DSS} = 4$ mA, design R_S and R_D so that:

(a) $V_{DG1} = |V_P|$ and $V_{DG2} = 2|V_P|$.

(b) $V_{DG1} = 1.5|V_P|$ and $V_{DG2} = 1.5|V_P|$.

***5.91 In the circuit of Fig. P5.91 let $V_{t1} = V_{t2} = 2$ V, $K_1 = 0.5\ \text{mA/V}^2 = 4K_2$, and C_1, C_2 large. Calculate the dc voltage at the output, the voltage gain, and the input resistance. For $C_2 = 0$, what does the input resistance become?

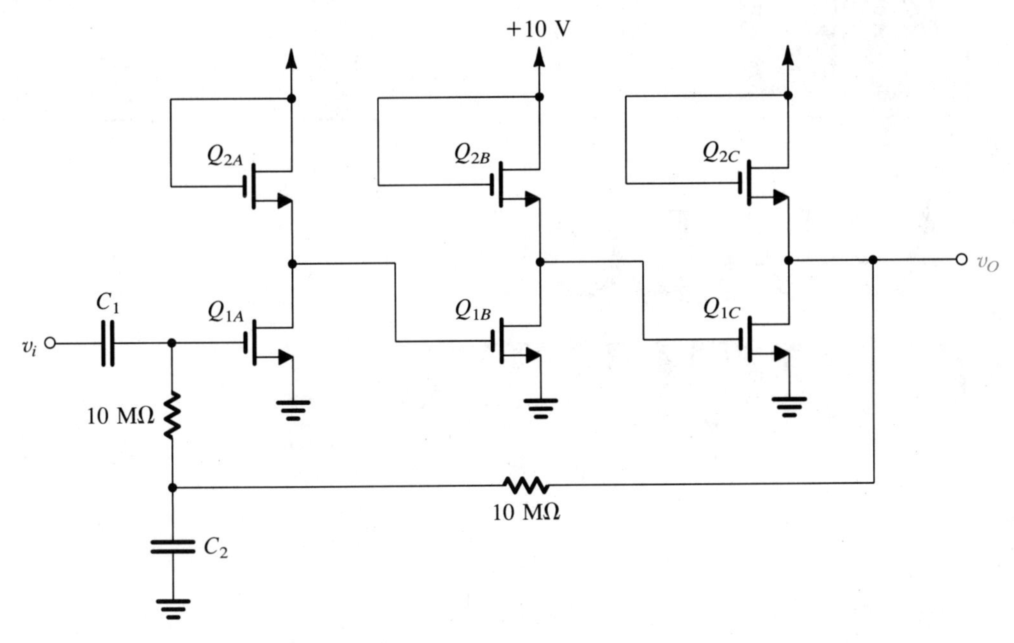

Fig. P5.91

PART 

ANALOG CIRCUITS

INTRODUCTION

Having studied the major electronic devices and their basic circuit configurations, we are now ready to consider the design of more complex analog circuits and systems. Amplifiers constitute the most significant class of analog circuits, and accordingly a large portion of Part II is devoted to the study of various aspects of amplifier design.

Part II begins in Chapter 6 with a widely used amplifier configuration, the differential pair. The analysis and design of BJT, MOSFET, JFET, and MESFET differential amplifiers are presented. Also, BiCMOS, a recently developed technology that combines the advantages of bipolar and CMOS devices, is introduced in the context of amplifier design. To obtain high gain, amplifier stages are placed in cascade, and Chapter 6 concludes with an introduction to multistage amplifiers.

Comments were made in earlier chapters about the limitations that the internal dynamics of the devices place on the frequency response of amplifiers. Chapter 7 presents a detailed study of this topic. High-frequency device models are presented and employed in the frequency-response analysis of the basic amplifier configurations. The mechanisms that limit the frequency response of the basic configurations are identified and more elaborate circuits that circumvent these problems are presented.

Most electronic circuits incorporate some form of feedback; either intentional (that is, designed in) or inevitable (as a result of device and circuit nonidealities). A formal study of the pivotal topic of feedback is presented in Chapter 8. Such a study is essential for the proper application of feedback in the design of amplifiers, to effect desirable properties such as more precise gain value, and to avoid problems such as instability.

The design of amplifiers that are required to deliver large amounts of load power, for example the amplifier that drives the loudspeaker in a stereo system, is based on different considerations than those for small-signal amplifiers. This is the subject of Chapter 9.

The IC op amp was studied in Chapter 2 as a circuit building block. By the time we reach Chapter 10, we will have sufficient knowledge to look inside the op-amp package for a detailed study of its circuit. Chapter 10 presents such a study for bipolar, CMOS, and BiCMOS op amps, with a number of objectives in mind: to tie together many of the ideas and concepts introduced in earlier chapters, to show how a complex circuit can be analyzed and designed by breaking it into smaller parts, to introduce some of the ingenious techniques employed in analog IC design; and to demonstrate that knowledge of what is on an IC chip enables more intelligent application of the chip. Analog-to-digital and digital-to-analog converters form the other major class of analog ICs studied in Chapter 10. These are systems-on-a-chip whose design incorporates techniques that achieve the high levels of precision required.

The last two chapters of Part II have an applications or systems orientation. Chapter 11 deals with the design of filters, which are important building blocks of communications and instrumentation systems. Filter design is one of the rare areas of engineering for which a complete design theory exists, starting from specification and culminating in an actual working circuit.

In the design of electronic systems the need usually arises for signals of various waveforms: sinusoidal, pulse, square wave, etc. The generation of such signals is the subject of Chapter 12.

The designer of analog circuits is usually faced with a vast number of possibilities. This makes the subject both exciting and challenging. It is in the goal of Part II to impart to the readers some of this excitement and to prepare them for the accompanying challenge.

CHAPTER 6

Differential and Multistage Amplifiers

INTRODUCTION

The differential amplifier is the most widely used circuit building block in analog integrated circuits. For instance, the input stage of every op amp is a differential amplifier. Also, the BJT differential amplifier is the basis of a very-high-speed logic circuit family, called emitter-coupled logic (ECL), which we shall study in Chapter 14.

In this chapter we shall study differential amplifiers implemented with BJTs, JFETs, MOSFETs, and MESFETs. Also presented are the biasing techniques employed in the design of bipolar, MOS, and GaAs integrated circuits. The chapter concludes with a discussion of the structure of multistage amplifiers. The analysis of such amplifiers is illustrated by a detailed example.

6.1 THE BJT DIFFERENTIAL PAIR

Qualitative Description of Operation

Figure 6.1 shows the basic BJT differential-pair configuration. It consists of two matched transistors, Q_1 and Q_2, whose emitters are joined together and biased by a constant-current source I. The latter is usually implemented by a transistor circuit of the type described in Section 6.4. Although each collector is connected to the positive supply voltage V_{CC} through a resistance R_C, this connection is not essential to the operation of the differential pair—that is, in some applications the two collectors may be connected to other transistors rather than to resistive loads. It is essential, though, that the collector circuits be such that Q_1 and Q_2 never enter saturation.

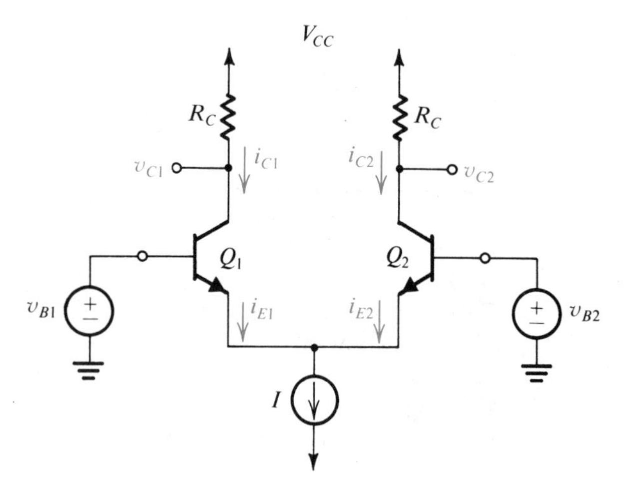

Fig. 6.1 The basic BJT differential-pair configuration.

To see how the differential pair works, consider first the case where the two bases are joined together and connected to a voltage v_{CM}, called the **common-mode voltage.** That is, as shown in Fig. 6.2(a), $v_{B1} = v_{B2} = v_{CM}$. Since Q_1 and Q_2 are matched, it follows from symmetry that the current I will divide equally between the two devices. Thus $i_{E1} = i_{E2} = I/2$, and the voltage at the emitters will be $v_{CM} - V_{BE}$, where V_{BE} is the base–emitter voltage (assumed in Fig. 6.2a to be approximately 0.7 V) corresponding to an emitter current of $I/2$. The voltage at each collector will be $V_{CC} - \frac{1}{2}\alpha I R_C$, and the difference in voltage between the two collectors will be zero.

Now let us vary the value of the common-mode input signal v_{CM}. Obviously, as long as Q_1 and Q_2 remain in the active region the current I will still divide equally between Q_1 and Q_2, and the voltages at the collectors will not change. Thus the differential pair does not respond to (*rejects*) common-mode input signals.

As another experiment, let the voltage v_{B2} be set to a constant value, say, zero (by grounding B_2), and let $v_{B1} = +1$ V (see Fig. 6.2b). With a bit of reasoning it can be seen that Q_1 will be on and conducting all of the current I and that Q_2 will be off. For Q_1 to be on, the emitter has to be at approximately +0.3 V, which keeps the EBJ of Q_2 reverse-biased. The collector voltages will be $v_{C1} = V_{CC} - \alpha I R_C$ and $v_{C2} = V_{CC}$.

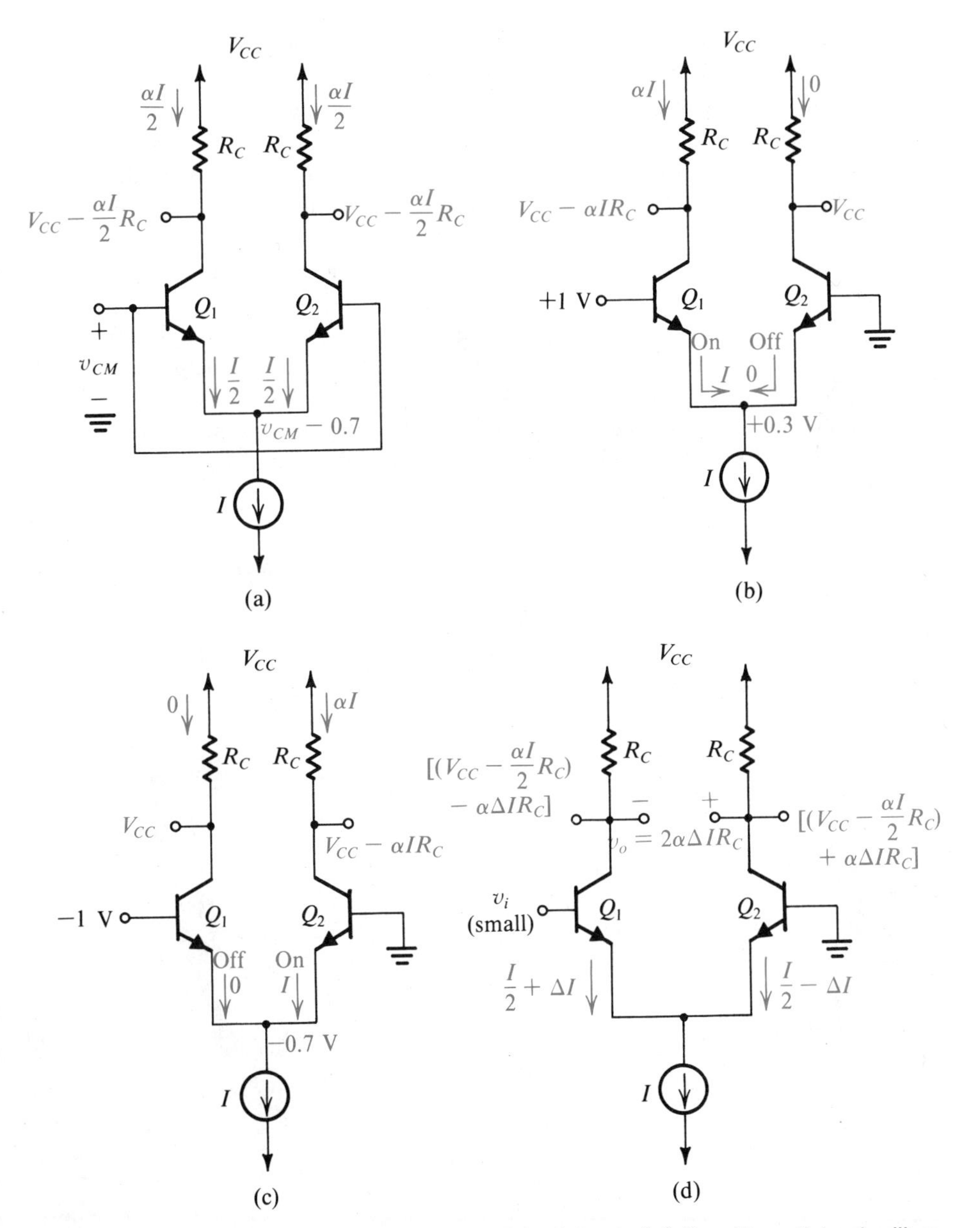

Fig. 6.2 Different modes of operation of the differential pair. **(a)** The differential pair with a common-mode input signal v_{CM}. **(b)** The differential pair with a "large" differential input signal. **(c)** The differential pair with a large differential input signal of polarity opposite to that in (b). **(d)** The differential pair with a small differential input signal v_i.

Let us now change v_{B1} to -1 V (Fig. 6.2c). Again with some reasoning it can be seen that Q_1 will turn off, and Q_2 will carry all the current I. The common emitter will be at -0.7 V, which means that the EBJ of Q_1 will be reverse-biased by 0.3 V. The collector voltages will be $v_{C1} = V_{CC}$ and $v_{C2} = V_{CC} - \alpha I R_C$.

From the above we see that the differential pair certainly responds to **difference-mode or differential signals.** In fact, with relatively small difference voltages we are able to steer the entire bias current from one side of the pair to the other. This current-steering property of the differential pair allows it to be used in logic circuits, as will be demonstrated in Chapter 14.

To use the differential pair as a linear amplifier we apply a very small differential signal (a few millivolts), which will result in one of the transistors conducting a current of $I/2 + \Delta I$; the current in the other transistor will be $I/2 - \Delta I$, with ΔI being proportional to the difference input voltage (see Fig. 6.2d). The output voltage taken between the two collectors will be $2\alpha\, \Delta I R_C$, which is proportional to the differential input signal v_i. The small-signal operation of the differential pair will be studied in Section 6.2.

Exercise 6.1 Find v_E, v_{C1}, and v_{C2} in the circuit of Fig. E6.1. Assume that $|v_{BE}|$ of a conducting transistor is approximately 0.7 V and that $\alpha \simeq 1$.

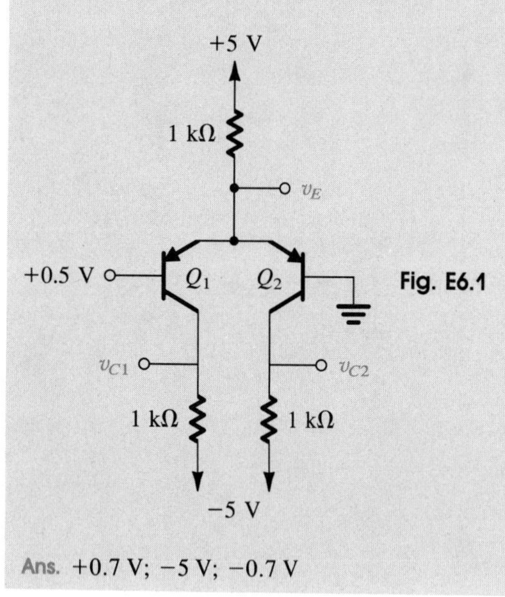

Fig. E6.1

Ans. +0.7 V; −5 V; −0.7 V

Large-Signal Operation of the BJT Differential Pair

We now present a general analysis of the BJT differential pair of Fig. 6.1. If we denote the voltage at the common emitter by v_E, the exponential relationship applied to each of the two transistors may be written

$$i_{E1} = \frac{I_S}{\alpha} e^{(v_{B1} - v_E)/V_T} \tag{6.1}$$

$$i_{E2} = \frac{I_S}{\alpha} e^{(v_{B2} - v_E)/V_T} \tag{6.2}$$

These two equations can be combined to obtain

$$\frac{i_{E1}}{i_{E2}} = e^{(v_{B1} - v_{B2})/V_T} \tag{6.3}$$

which can be manipulated to yield

$$\frac{i_{E1}}{i_{E1} + i_{E2}} = \frac{1}{1 + e^{(v_{B2} - v_{B1})/V_T}} \tag{6.4}$$

$$\frac{i_{E2}}{i_{E1} + i_{E2}} = \frac{1}{1 + e^{(v_{B1} - v_{B2})/V_T}} \tag{6.5}$$

The circuit imposes the additional constraint

$$i_{E1} + i_{E2} = I \tag{6.6}$$

Using Eq. (6.6) together with Eqs. (6.4) and (6.5) gives

$$i_{E1} = \frac{I}{1 + e^{(v_{B2} - v_{B1})/V_T}} \tag{6.7}$$

$$i_{E2} = \frac{I}{1 + e^{(v_{B1} - v_{B2})/V_T}} \tag{6.8}$$

The collector currents i_{C1} and i_{C2} can be obtained simply by multiplying the emitter currents in Eqs. (6.7) and (6.8) by α, which is normally very close to unity.

The fundamental operation of the differential amplifier is illustrated by Eqs. (6.7) and (6.8). First, note that the amplifier responds only to the difference voltage $v_{B1} - v_{B2}$. That is, if $v_{B1} = v_{B2} = v_{CM}$, the current I divides equally between the two transistors irrespective of the value of the common-mode voltage v_{CM}. This is the essence of differential-amplifier operation, which also gives rise to its name.

Another important observation is that a relatively small difference voltage $v_{B1} - v_{B2}$ will cause the current I to flow almost entirely in one of the two transistors. Figure 6.3 shows a plot of the two collector currents (assuming $\alpha \simeq 1$) as a function of the difference signal. This is a normalized plot that can be used universally. Note that a difference voltage of about $4V_T$ ($\simeq$ 100 mV) is sufficient to switch the current almost entirely to one side of the pair.

The nonlinear transfer characteristics of the differential pair, shown in Fig. 6.3, will not be utilized any further in this chapter. In the following we shall be interested specifically in the application of the differential pair as a small-signal amplifier. For this purpose the difference input signal is limited to less than about $V_T/2$ in order that we may operate on a linear segment of the characteristics around the midpoint x.

Exercise 6.2 Find the value of input differential signal sufficient to cause $i_{E1} = 0.99I$.

Ans. 115 mV

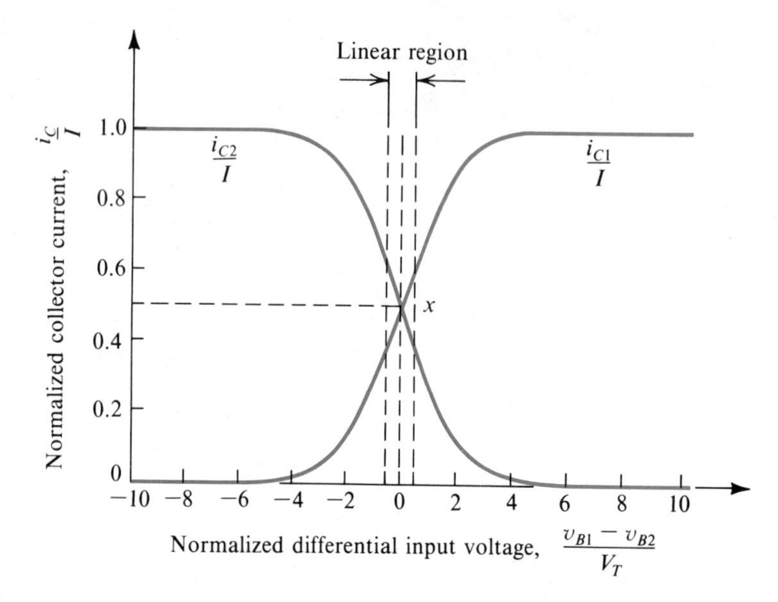

Fig. 6.3 Transfer characteristics of the BJT differential pair of Fig. 6.1 assuming $\alpha \simeq 1$.

6.2 SMALL-SIGNAL OPERATION OF THE BJT DIFFERENTIAL AMPLIFIER

In this section we shall study the application of the BJT differential pair in small-signal amplification. Figure 6.4 shows the differential pair with a difference voltage signal v_d applied between the two bases. Implied is that the dc level at the input—that is, the common-mode input signal—has been somehow established. For instance, one of the two

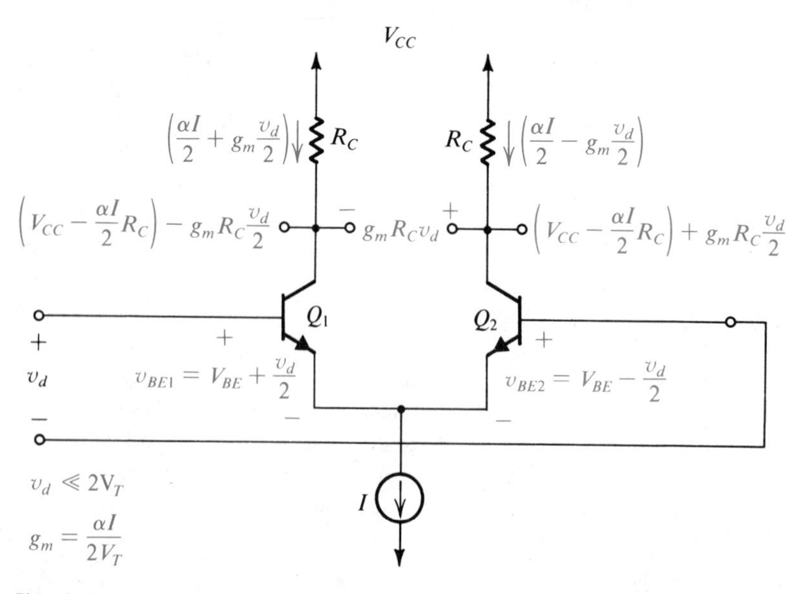

Fig. 6.4 The currents and voltages in the differential amplifier when a small difference signal v_d is applied.

input terminals can be grounded and v_d applied to the other input terminal. Alternatively, the differential amplifier may be fed from the output of another differential amplifier. In this case the voltage at one of the input terminals will be $v_{CM} + v_d/2$ while that at the other input terminal will be $v_{CM} - v_d/2$. We will consider common-mode operation at a later stage.

The Collector Currents When v_d Is Applied

Return now to the circuit of Fig. 6.4. We may use Eqs. (6.7) and (6.8) to find the total currents i_{C1} and i_{C2} as functions of the differential signal v_d by substituting $v_{B1} - v_{B2} = v_d$.

$$i_{C1} = \frac{\alpha I}{1 + e^{-v_d/V_T}} \tag{6.9}$$

$$i_{C2} = \frac{\alpha I}{1 + e^{v_d/V_T}} \tag{6.10}$$

Multiplying the numerator and the denominator of the right-hand side of Eq. (6.9) by $e^{(v_d/2V_T)}$ gives

$$i_{C1} = \frac{\alpha I e^{(v_d/2V_T)}}{e^{(v_d/2V_T)} + e^{(-v_d/2V_T)}} \tag{6.11}$$

Assume that $v_d \ll 2V_T$. We may thus expand the exponential $e^{(\pm v_d/2V_T)}$ in a series and retain only the first two terms:

$$i_{C1} \simeq \frac{\alpha I(1 + v_d/2V_T)}{1 + v_d/2V_T + 1 - v_d/2V_T}$$

Thus

$$i_{C1} = \frac{\alpha I}{2} + \frac{\alpha I}{2V_T}\frac{v_d}{2} \tag{6.12}$$

Similar manipulations can be applied to Eq. (6.10) to obtain

$$i_{C2} = \frac{\alpha I}{2} - \frac{\alpha I}{2V_T}\frac{v_d}{2} \tag{6.13}$$

Equations (6.12) and (6.13) tells us that when $v_d = 0$, the bias current I divides equally between the two transistors of the pair. Thus each transistor is biased at an emitter current of $I/2$. When a "small-signal" v_d is applied differentially (that is, between the two bases), the collector current of Q_1 increases by an increment i_c and that of Q_2 decreases by an equal amount. This ensures that the sum of the total currents in Q_1 and Q_2 remains constant, as constrained by the current-source bias. The incremental or signal current component i_c is given by

$$i_c = \frac{\alpha I}{2V_T}\frac{v_d}{2} \tag{6.14}$$

Equation (6.14) has an easy interpretation. First, note from the symmetry of the circuit (Fig. 6.4) that the differential signal v_d should divide equally between the base–emitter junctions of the two transistors. Thus the total base–emitter voltages will be

$$v_{BE}\big|_{Q_1} = V_{BE} + \frac{v_d}{2} \tag{6.15}$$

$$v_{BE}\big|_{Q_2} = V_{BE} - \frac{v_d}{2} \tag{6.16}$$

where V_{BE} is the dc BE voltage corresponding to an emitter current of $I/2$. Therefore, the collector current of Q_1 will increase by $g_m v_d/2$ and the collector current of Q_2 will decrease by $g_m v_d/2$. Here g_m denotes the transconductance of Q_1 and of Q_2, which are equal and given by

$$g_m = \frac{I_C}{V_T} = \frac{\alpha I/2}{V_T} \tag{6.17}$$

Thus Eq. (6.14) simply states that $i_c = g_m v_d/2$.

An Alternate Viewpoint

There is an extremely useful alternative interpretation of the above results. Assume the current source I to be ideal. Its incremental resistance then will be infinite. Thus the voltage v_d appears across a total resistance of $2r_e$, where

$$r_e = \frac{V_T}{I_E} = \frac{V_T}{I/2} \tag{6.18}$$

Correspondingly there will be a signal current i_e, as illustrated in Fig. 6.5, given by

$$i_e = \frac{v_d}{2r_e} \tag{6.19}$$

Thus the collector of Q_1 will exhibit a current increment i_c and the collector of Q_2 will exhibit a current decrement i_c:

$$i_c = \alpha i_e = \frac{\alpha v_d}{2r_e} = g_m \frac{v_d}{2} \tag{6.20}$$

Note that in Fig. 6.5 we have shown signal quantities only. It is implied, of course, that each transistor is biased at an emitter current of $I/2$.

This method of analysis is particularly useful when resistances are included in the emitters, as shown in Fig. 6.6. For this circuit we have

$$i_e = \frac{v_d}{2r_e + 2R_E} \tag{6.21}$$

Input Differential Resistance

The input differential resistance is the resistance seen between the two bases; that is, it is the resistance seen by the differential input signal v_d. For the differential amplifier in

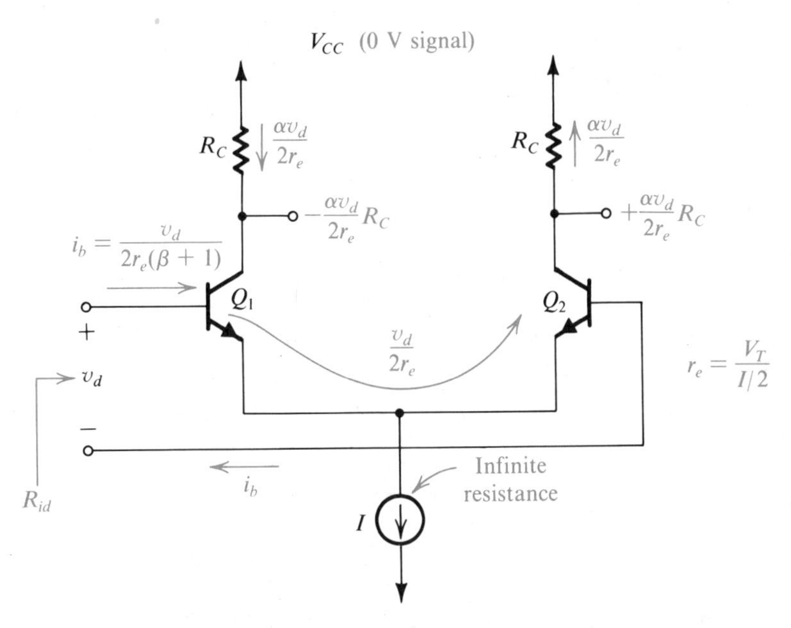

Fig. 6.5 Simple technique for determining the signal currents in a differential amplifier excited by a differential voltage signal v_d; dc quantities are not shown.

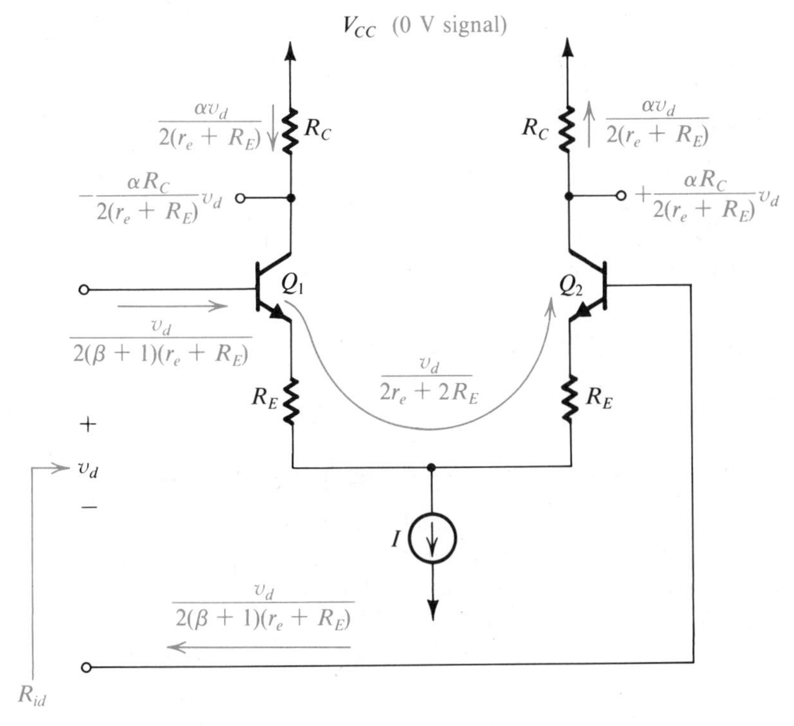

Fig. 6.6 Differential amplifier with emitter resistances.

Figs. 6.4 and 6.5 it can be seen that the base current of Q_1 shows an increment i_b and the base current of Q_2 shows an equal decrement,

$$i_b = \frac{i_e}{\beta + 1} = \frac{v_d/2r_e}{\beta + 1} \tag{6.22}$$

Thus the differential input resistance R_{id} is given by

$$R_{id} \equiv \frac{v_d}{i_b} = (\beta + 1)2r_e = 2r_\pi \tag{6.23}$$

This result is just a restatement of the familiar resistance-reflection rule; namely, *the resistance seen between the two bases is equal to the total resistance in the emitter circuit multiplied by* $\beta + 1$. We can employ this rule to find the input differential resistance for the circuit in Fig. 6.6 as

$$R_{id} = (\beta + 1)(2r_e + 2R_e) \tag{6.24}$$

Differential Voltage Gain

We have established that for small difference input voltages ($v_d \ll 2\,V_T$; that is, v_d smaller than about 20 mV) the collector currents are given by

$$i_{C1} = I_C + g_m\frac{v_d}{2} \tag{6.25}$$

$$i_{C2} = I_C - g_m\frac{v_d}{2} \tag{6.26}$$

where

$$I_C = \frac{\alpha I}{2} \tag{6.27}$$

Thus the total voltages at the collectors will be

$$v_{C1} = (V_{CC} - I_C R_C) - g_m R_C \frac{v_d}{2} \tag{6.28}$$

$$v_{C2} = (V_{CC} - I_C R_C) + g_m R_C \frac{v_d}{2} \tag{6.29}$$

The quantities in parentheses are simply the dc voltages at each of the two collectors.

The output voltage signal of a differential amplifier can be taken either *differentially* (that is, between the two collectors) or *single-ended* (that is, between one collector and ground). If the output is taken differentially, then the differential gain (as opposed to the common-mode gain) of the differential amplifier will be

$$A_d = \frac{v_{c1} - v_{c2}}{v_d} = -g_m R_C \tag{6.30}$$

On the other hand, if we take the output single-ended (say, between the collector of Q_1 and ground), then the differential gain will be given by

$$A_d = \frac{v_{c1}}{v_d} = -\tfrac{1}{2} g_m R_C \tag{6.31}$$

For the differential amplifier with resistances in the emitter leads (Fig. 6.6) the differential gain when the output is taken differentially is given by

$$A_d = -\frac{\alpha(2R_C)}{2r_e + 2R_E} \simeq -\frac{R_C}{r_e + R_E} \tag{6.32}$$

This equation is a familiar one: It states that *the voltage gain is equal to the ratio of the total resistance in the collector circuit ($2R_C$) to the total resistance in the emitter circuit ($2r_e + 2R_E$).*

Equivalence of the Differential Amplifier to a Common-Emitter Amplifier

The above analysis and results are quite similar to those obtained in the case of a common-emitter amplifier stage. That the differential amplifier is in fact equivalent to a common-emitter amplifier is illustrated in Fig. 6.7. Figure 6.7(a) shows a differential amplifier fed by a differential signal v_d with the differential signal applied in a **complementary (push-pull** or **balanced)** manner. That is, while the base of Q_1 is raised by $v_d/2$, the base of Q_2

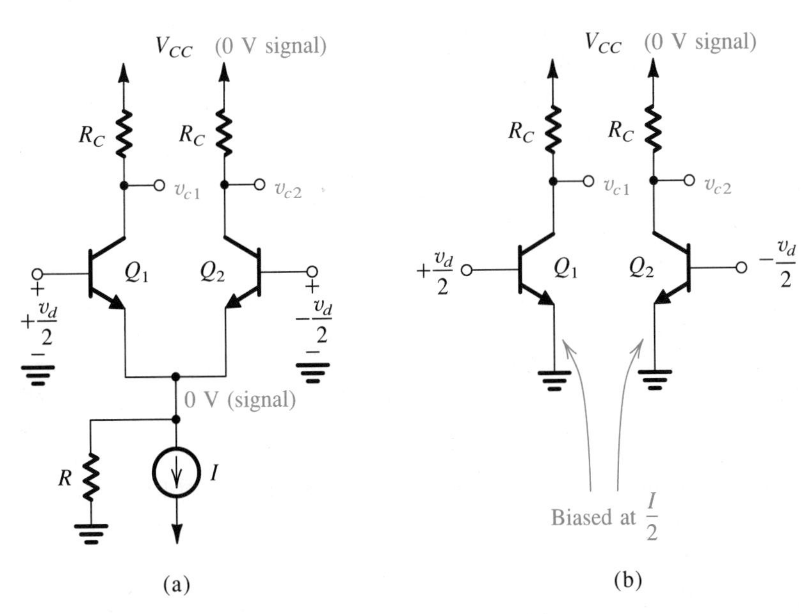

Fig. 6.7 Equivalence of the differential amplifier in **(a)** to the two common-emitter amplifiers in **(b).** This equivalence applies only for differential input signals. Either of the two common-emitter amplifiers in (b) can be used to evaluate the differential gain, input differential resistance, frequency response, and so on, of the differential amplifier.

is lowered by $v_d/2$. We have also included the output resistance R of the bias current source. From symmetry it follows that the signal voltage at the common emitter will be zero. Thus the circuit is equivalent to the two common-emitter amplifiers shown in Fig. 6.7(b), where each of the two transistors is biased at an emitter current of $I/2$. Note that the finite output resistance R of the current source will have no effect on the operation. The equivalent circuit in Fig. 6.7(b) is valid for differential operation only.

In many applications the differential amplifier is not fed in a complementary fashion; rather, the input signal may be applied to one of the input terminals while the other terminal is grounded, as shown in Fig. 6.8. In this case the signal voltage at the emitters will not be zero, and thus the resistance R will have an effect on the operation. Nevertheless, if R is large ($R \gg r_e$), as is usually the case, then v_d will still divide equally (approximately) between the two junctions, as shown in Fig. 6.8. Thus the operation of the differential amplifier in this case will be almost identical to that in the case of symmetric feed, and the common-emitter equivalence can still be employed.

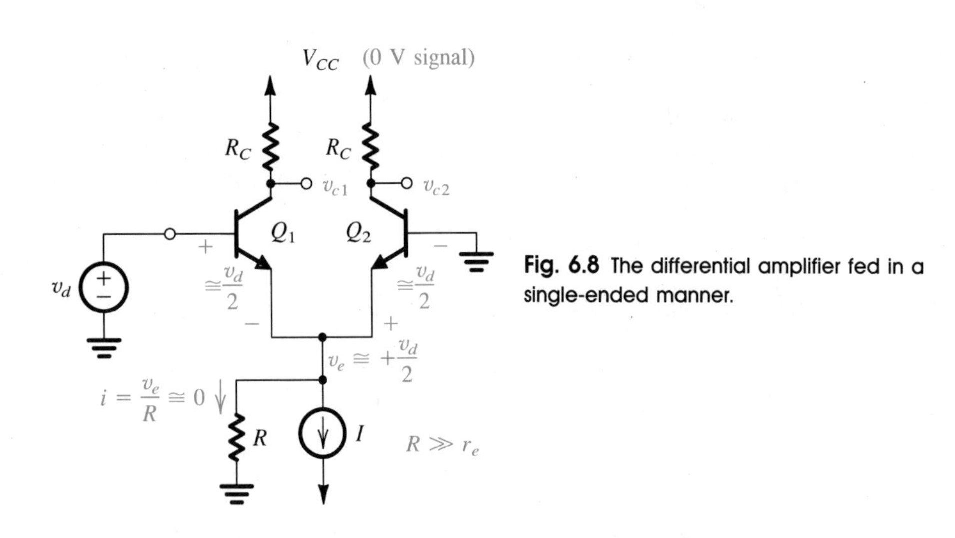

Fig. 6.8 The differential amplifier fed in a single-ended manner.

Since in Fig. 6.7 $v_{c2} = -v_{c1}$, the two common-emitter transistors in Fig. 6.7(b) yield similar results about the performance of the differential amplifier. Thus only one is needed to analyze the differential small-signal operation of the differential amplifier, and is known as the **differential half-circuit.** If we take the common-emitter transistor fed with $+v_d/2$ as the differential half-circuit and replace the transistor with its low-frequency equivalent circuit model, the circuit in Fig. 6.9 results. In evaluating the values of the

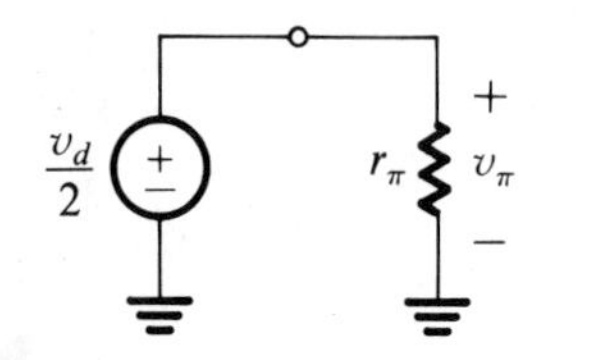

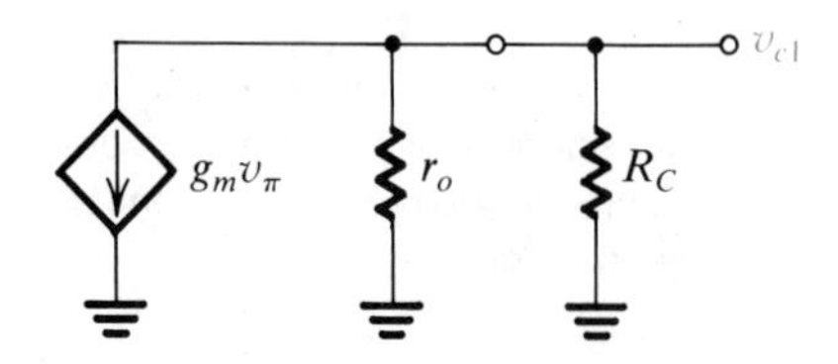

Fig. 6.9 Equivalent circuit model of the differential half-circuit.

model parameters r_π, g_m, and r_o we must recall that the half-circuit is biased at $I/2$. The voltage gain of the differential amplifier (with the output taken differentially) is equal to the voltage gain of the half-circuit—that is, $v_{c1}/(v_d/2)$. Here we note that including r_o will modify the gain expression in Eq. (6.30) to

$$A_d = -g_m(R_C//r_o) \tag{6.33}$$

The input differential resistance of the differential amplifier is twice that of the half-circuit—that is, $2r_\pi$. Finally, we note that the differential half-circuit of the amplifier of Fig. 6.6 is a common-emitter transistor with a resistance R_E in the emitter lead.

Common-Mode Gain

Figure 6.10(a) shows a differential amplifier fed by a common-mode voltage signal v_{CM}. The resistance R is the incremental output resistance of the bias current source. From symmetry it can be seen that the circuit is equivalent to that shown in Fig. 6.10(b), where

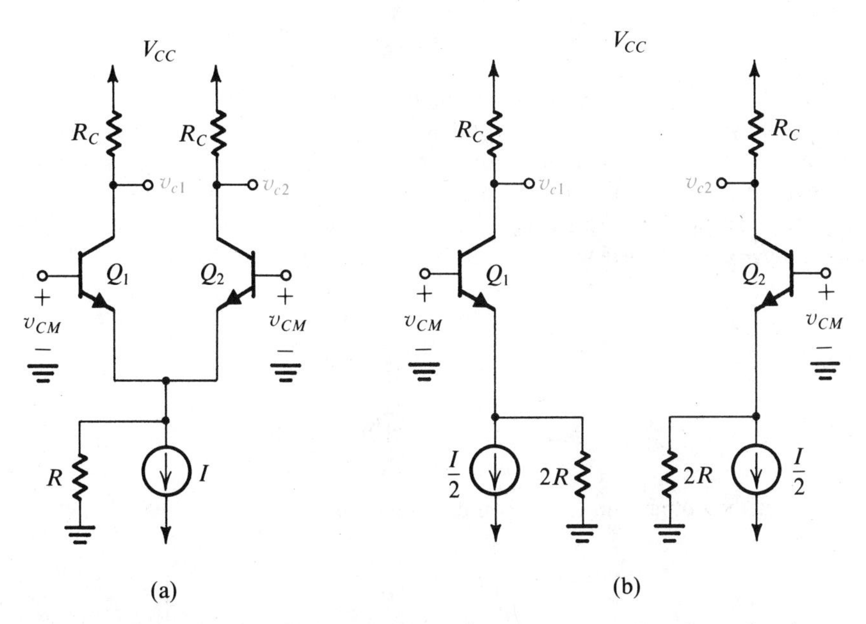

Fig. 6.10 (a) The differential amplifier fed by a common-mode voltage signal. **(b)** Equivalent "half-circuits" for common-mode calculations.

each of the two transistors Q_1 and Q_2 is biased at an emitter current $I/2$ and has a resistance $2R$ in its emitter lead. Thus the common-mode output voltage v_{c1} will be

$$v_{c1} = -v_{\text{CM}}\frac{\alpha R_C}{2R + r_e} \simeq -v_{\text{CM}}\frac{\alpha R_C}{2R} \tag{6.34}$$

At the other collector we have an equal common-mode signal v_{c2},

$$v_{c2} \simeq -v_{\text{CM}}\frac{\alpha R_C}{2R} \tag{6.35}$$

Now, if the output is taken differentially, then the output common-mode voltage $v_{c1} - v_{c2}$ will be zero and the common-mode gain also will be zero. On the other hand, if the output is taken single ended, the common mode gain A_{cm} will be finite and given by[1]

$$A_{cm} = -\frac{\alpha R_C}{2R} \tag{6.36}$$

Since in this case the differential gain is

$$A_d = \tfrac{1}{2} g_m R_C \tag{6.37}$$

the common-mode rejection ratio (CMRR) will be

$$\text{CMRR} = \left|\frac{A_d}{A_{cm}}\right| \simeq g_m R \qquad \alpha \simeq 1 \tag{6.38}$$

Normally the CMRR is expressed in dB:

$$\text{CMRR} = 20 \log \left|\frac{A_d}{A_{cm}}\right| \tag{6.39}$$

Each of the circuits in Fig. 6.10(b) is called the **common-mode half-circuit.**

The above analysis assumes that the circuit is perfectly symmetric. However, practical circuits are not perfectly symmetric, with the result that the common-mode gain will not be zero even if the output is taken differentially. To illustrate, consider the case of perfect symmetry except for a mismatch ΔR_C in the collector resistances. That is, let the collector of Q_1 have a load resistance R_C and that of Q_2 have a load resistance $R_C + \Delta R_C$. It follows that

$$v_{c1} = -v_{CM}\frac{\alpha R_C}{2R + r_e}$$

$$v_{c2} = -v_{CM}\frac{\alpha(R_C + \Delta R_C)}{2R + r_e}$$

Thus the signal at the output due to the common-mode input signal will be

$$v_o = v_{c1} - v_{c2} = v_{CM}\frac{\alpha\, \Delta R_C}{2R + r_e}$$

and the common-mode gain will be

$$A_{cm} = \frac{\alpha\, \Delta R_C}{2R + r_e} \simeq \frac{\Delta R_C}{2R}$$

---

[1] The expressions in Eqs. (6.34) and (6.35) are obtained by neglecting r_o and r_μ. A detailed derivation using the complete hybrid-π equivalent circuit model shows that v_{c1}/v_{CM} and v_{c2}/v_{CM} are approximately

$$\frac{-\alpha R_C}{2R}\left[1 - 2R\left(\frac{1}{\beta r_o} + \frac{1}{\alpha r_\mu}\right)\right]$$

This expression reduces to those in Eqs. (6.34) and (6.35) when $2R \ll \beta r_o$ and $2R \ll \alpha r_\mu$.

This expression can be rewritten as

$$A_{cm} = \frac{R_C}{2R}\frac{\Delta R_C}{R_C} \tag{6.40}$$

Compare the common-mode gain in Eq. (6.40) with that for the case of single-ended output in Eq. (6.36). We see that the common-mode gain is much smaller in the case of differential output. Therefore the input differential stage of an op amp, for example, is usually a balanced one, with the output taken differentially. This ensures that the op amp will have a low common-mode gain or, equivalently, a high CMRR.

The input signals v_1 and v_2 to a differential amplifier usually contain a common-mode component, v_{CM},

$$v_{CM} \equiv \frac{v_1 + v_2}{2} \tag{6.41}$$

and a differential component v_d,

$$v_d \equiv v_1 - v_2 \tag{6.42}$$

Thus the output signal will be given by

$$v_o = A_d(v_1 - v_2) + A_{cm}\left(\frac{v_1 + v_2}{2}\right) \tag{6.43}$$

Input Common-Mode Resistance

The definition of the **common-mode input resistance** R_{icm} is illustrated in Fig. 6.11(a). Figure 6.11(b) shows the equivalent common-mode half-circuit; its input resistance is $2R_{icm}$.

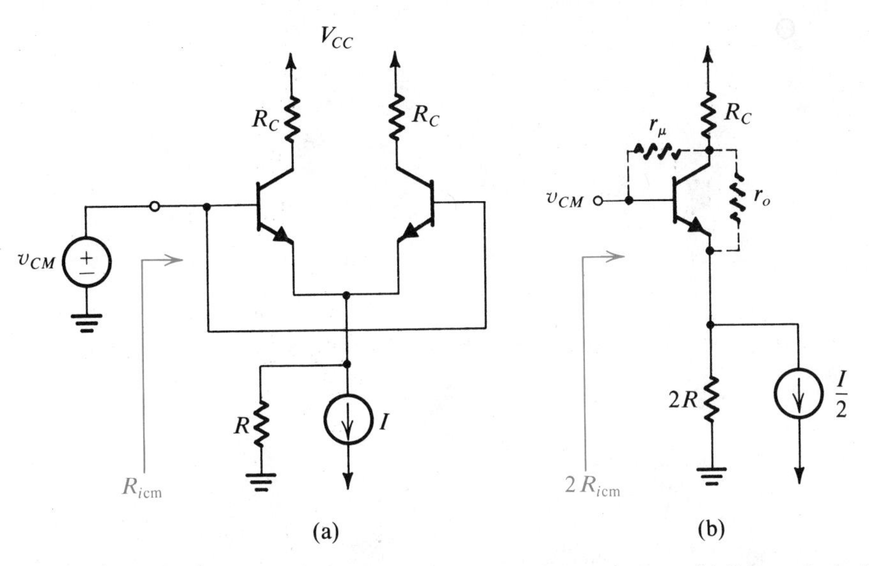

Fig. 6.11 **(a)** Definition of the input common-mode resistance R_{icm}. **(b)** The equivalent common-mode half-circuit.

Since the input common-mode resistance is usually very large, its value will be affected by the transistor resistances r_o and r_μ. These resistances are indicated on the equivalent common-mode half-circuit in Fig. 6.11(b). Now, since the common-mode gain is usually small, the signal at the collector will be very small. We can simplify matters considerably by assuming that the signal at the collector is 0 V; that is, the collector is at signal ground. Under this assumption the input resistance can be found by inspection, as follows:

$$2R_{icm} = r_\mu // [(\beta + 1)(2R)] // [(\beta + 1)r_o]$$

Thus

$$R_{icm} = \left(\frac{r_\mu}{2}\right) // [(\beta + 1)R] // \left[(\beta + 1)\frac{r_o}{2}\right] \tag{6.44}$$

Exercise 6.3 The differential amplifier in Fig. E6.3 uses transistors with $\beta = 100$. Evaluate the following: (a) The input differential resistance R_{id}. (b) The overall voltage gain v_o/v_s (neglect the effect of r_o). (c) The worst-case common-mode gain if the two collector resistances are accurate to within $\pm 1\%$. (d) The CMRR, in dB. (e) The input common-mode resistance (assuming that the Early voltage $V_A = 100$ V and that $r_\mu = 10\ \beta r_o$).

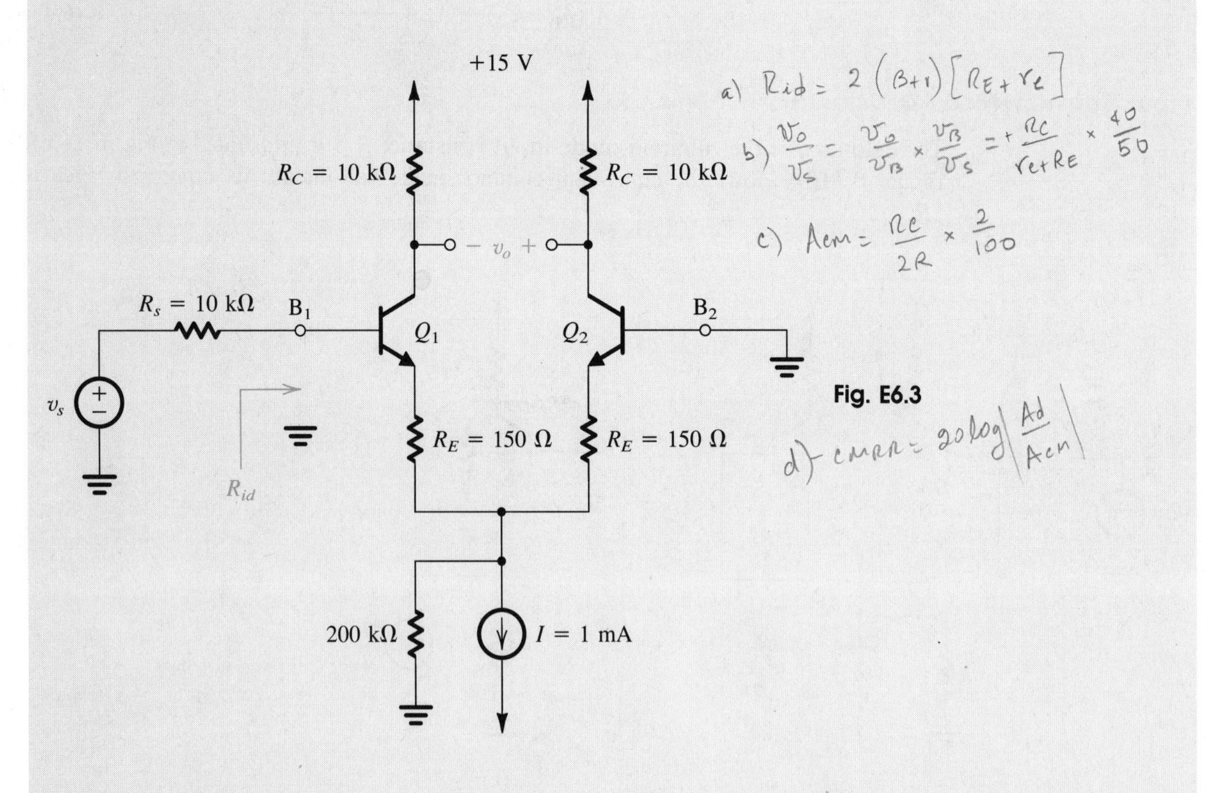

Fig. E6.3

Ans. (a) 40 kΩ; (b) 40 V/V; (c) 5×10^{-4} V/V; (d) 98 dB; (e) 6.3 MΩ

6.3 OTHER NONIDEAL CHARACTERISTICS OF THE DIFFERENTIAL AMPLIFIER

Input Offset Voltage

Consider the basic BJT differential amplifier with both inputs grounded, as shown in Fig. 6.12(a). If the two sides of the differential pair were perfectly matched (that is, Q_1 and Q_2 identical and $R_{C1} = R_{C2} = R_C$), then current I would split equally between Q_1 and

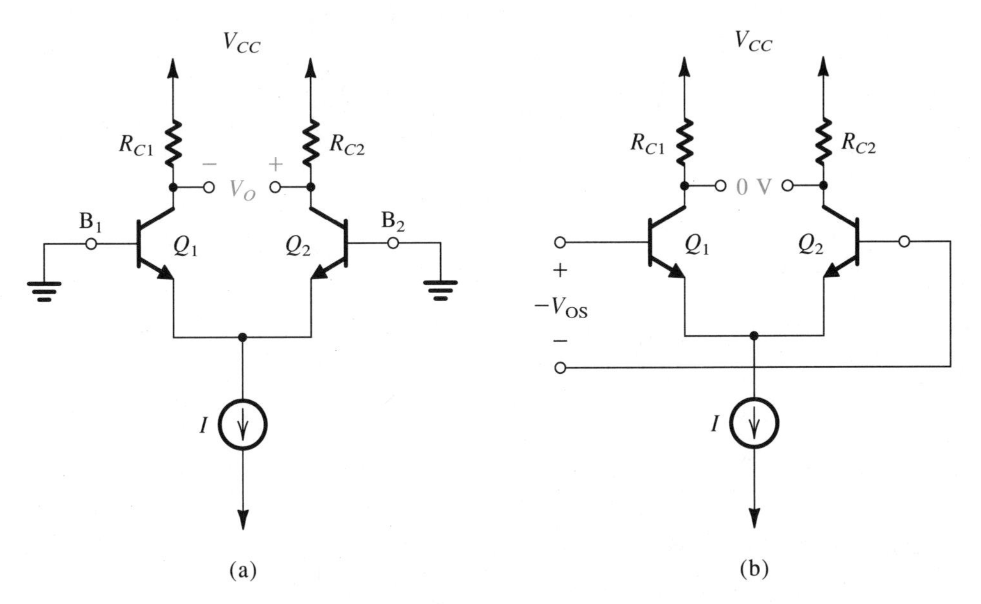

Fig. 6.12 (a) The BJT differential pair with both inputs grounded. Due to device mismatches, a finite dc output V_O results. **(b)** Application of the input offset voltage $V_{OS} \equiv V_O/A_d$ to the input terminals with opposite polarity reduces V_O to zero.

Q_2, and V_O would be zero. Practical circuits exhibit mismatches that result in a dc output voltage V_O even with both inputs grounded. We call V_O the output dc offset voltage. More commonly, we divide V_O by the differential gain of the amplifier, A_d, to obtain a quantity known as the input offset voltage, V_{OS},

$$V_{OS} = V_O/A_d \tag{6.45}$$

Obviously, if we apply a voltage $-V_{OS}$ between the input terminals of the differential amplifier, then the output voltage will be reduced to zero (see Fig. 6.12b). This observation gives rise to the usual definition of the input offset voltage. It should be noted, however, that since the offset voltage is a result of device mismatches, its polarity is not known a priori.

The offset voltage results from mismatches in the load resistances R_{C1} and R_{C2} and from mismatches in Q_1 and Q_2. Consider first the effect of the load mismatch. Let

$$R_{C1} = R_C + \frac{\Delta R_C}{2} \tag{6.46}$$

$$R_{C2} = R_C - \frac{\Delta R_C}{2} \tag{6.47}$$

and assume that Q_1 and Q_2 are perfectly matched. It follows that current I will divide equally between Q_1 and Q_2, and thus

$$V_{C1} = V_{CC} - \left(\frac{\alpha I}{2}\right)\left(R_C + \frac{\Delta R_C}{2}\right)$$

$$V_{C2} = V_{CC} - \left(\frac{\alpha I}{2}\right)\left(R_C - \frac{\Delta R_C}{2}\right)$$

Thus the output voltage will be

$$V_O = V_{C2} - V_{C1} = \alpha\left(\frac{I}{2}\right)(\Delta R_C)$$

and the input offset voltage will be

$$V_{OS} = \frac{\alpha(I/2)(\Delta R_C)}{A_d} \tag{6.48}$$

Substituting $A_d = g_m R_C$ and

$$g_m = \frac{\alpha I/2}{V_T}$$

gives

$$|V_{OS}| = V_T\left(\frac{\Delta R_C}{R_C}\right) \tag{6.49}$$

As an example, consider the situation where the collector resistors are accurate to within $\pm 1\%$. Then the worst case mismatch will be

$$\frac{\Delta R_C}{R_C} = 0.02$$

and the resulting input offset voltage will be

$$|V_{OS}| = 25 \times 0.02 = 0.5 \text{ mV}$$

Next consider the effect of mismatches in transistors Q_1 and Q_2. In particular, let the transistors have a mismatch in their emitter–base junction areas. Such an area mismatch gives rise to a proportional mismatch in the scale currents I_S,

$$I_{S1} = I_S + \left(\frac{\Delta I_S}{2}\right) \tag{6.50}$$

$$I_{S2} = I_S - \left(\frac{\Delta I_S}{2}\right) \tag{6.51}$$

Refer to Fig. 6.12(a) and note that $V_{BE1} = V_{BE2}$. Thus, the current I will split between Q_1 and Q_2 in proportion to their I_S values, resulting in

$$I_{E1} = \frac{I}{2}\left(1 + \frac{\Delta I_S}{2I_S}\right) \tag{6.52}$$

$$I_{E2} = \frac{I}{2}\left(1 - \frac{\Delta I_S}{2I_S}\right) \tag{6.53}$$

It follows that the output offset voltage will be

$$V_O = \alpha\left(\frac{I}{2}\right)\left(\frac{\Delta I_S}{I_S}\right)R_C$$

and the corresponding input offset voltage will be

$$|V_{OS}| = V_T\left(\frac{\Delta I_S}{I_S}\right) \tag{6.54}$$

As an example, an area mismatch of 4% gives rise to $\Delta I_S/I_S = 0.04$ and an input offset voltage of 1 mV.

Finally, we note that since the two contributions to the input offset voltage are not correlated, an estimate of the total input offset voltage can be found as

$$V_{OS} = \sqrt{\left(V_T\frac{\Delta R_C}{R_C}\right)^2 + \left(V_T\frac{\Delta I_S}{I_S}\right)^2}$$

$$= V_T\sqrt{\left(\frac{\Delta R_C}{R_C}\right)^2 + \left(\frac{\Delta I_S}{I_S}\right)^2} \tag{6.55}$$

Input Bias and Offset Currents

In a perfectly symmetric differential pair the two input terminals carry equal dc currents; that is,

$$I_{B1} = I_{B2} = \frac{I/2}{\beta + 1} \tag{6.56}$$

This is the input bias current of the differential amplifier.

Mismatches in the amplifier circuit and most importantly a mismatch in β make the two input dc currents unequal. The resulting difference is the input offset current, I_{OS}, given as

$$I_{OS} = |I_{B1} - I_{B2}| \tag{6.57}$$

Let

$$\beta_1 = \beta + \left(\frac{\Delta\beta}{2}\right)$$

$$\beta_2 = \beta - \left(\frac{\Delta\beta}{2}\right)$$

then

$$I_{B1} = \left(\frac{I}{2}\right)\frac{1}{\beta + 1 + \Delta\beta/2} \simeq \frac{I}{2}\,\frac{1}{\beta + 1}\left(1 - \frac{\Delta\beta}{2\beta}\right) \tag{6.58}$$

$$I_{B2} = \left(\frac{I}{2}\right)\frac{1}{\beta + 1 - \Delta\beta/2} \simeq \frac{I}{2}\,\frac{1}{\beta + 1}\left(1 + \frac{\Delta\beta}{2\beta}\right) \tag{6.59}$$

$$I_{OS} = \frac{I}{2(\beta + 1)}\left(\frac{\Delta\beta}{\beta}\right) \tag{6.60}$$

Formally, the input bias current I_B is defined as follows:

$$I_B \equiv \frac{I_{B1} + I_{B2}}{2} = \frac{I}{2(\beta + 1)} \tag{6.61}$$

Thus

$$I_{OS} = I_B\left(\frac{\Delta\beta}{\beta}\right) \tag{6.62}$$

As an example, a 10% β mismatch results in an offset current one-tenth the value of the input bias current.

Input Common-Mode Range

The input common-mode range of a differential amplifier is the range of the input voltage v_{CM} over which the differential pair behaves as a linear amplifier for differential input signals. The upper limit of the common-mode range is determined by Q_1 and Q_2 leaving the active mode and entering the saturation mode of operation. Thus, the upper limit is approximately equal to the dc collector voltage of Q_1 and Q_2. The lower limit is determined by the transistor that supplies the biasing current I leaving its active region of operation and thus no longer functioning as a constant-current source. Current-source circuits are studied in the next section.

We conclude this section by noting that the definitions presented above are identical to those presented in Chapter 2 for op amps. In fact, as will be seen in Chapter 10, it is the input differential stage in an op amp circuit that primarily determines the op amp dc offset voltage, input bias and offset currents, and input common-mode range.

Exercise 6.4 For a BJT differential amplifier utilizing transistors having $\beta = 100$, matched to 10% or better, and areas that are matched to 10% or better, and collector resistors that are matched to 2% or better, find V_{OS}, I_B, and I_{OS}. The dc bias current is 100 μA.

Ans. 2.5 mV; 0.5 μA; 50 nA

6.4 BIASING IN BJT INTEGRATED CIRCUITS

The BJT biasing techniques discussed in Chapter 4 are not suitable for the design of IC amplifiers. This shortcoming stems from the need for a large number of resistors (one to three per amplifier stage) as well as large coupling and bypass capacitors. With present IC technology it is almost impossible to fabricate large capacitors, and it is uneconomical to manufacture large resistances. On the other hand, IC technology provides the designer with the possibility of using many transistors, which can be produced cheaply. Furthermore, it is easy to make transistors with matched characteristics that track with changes in environmental conditions. The limitations of, and opportunities available in, IC technology dictate a biasing philosophy that is quite different from that employed in discrete BJT amplifiers.

Basically, biasing in integrated-circuit design is based on the use of constant-current sources. We have already seen that the differential pair utilizes constant-current-source bias. On an IC chip with a number of amplifier stages a constant dc current is generated at one location and is then reproduced at various other locations for biasing the various amplifier stages. This approach has the advantage that the bias currents of the various stages track each other in case of changes in power-supply voltage or in temperature.

In this section we shall study a variety of current-source and current-steering circuits. Although these circuits can be used in discrete-unit design, they are primarily intended for application in IC design.

The Diode-Connected Transistor

Shorting the base and collector of a BJT together results in a two-terminal device having an i–v characteristic identical to the i_E–v_{BE} characteristic of the BJT. Figure 6.13 shows two *diode-connected transistors,* one *npn* and the other *pnp*. Observe that since the BJT is still operating in the active mode ($v_{CB} = 0$ results in active-mode operation) the current i divides between base and collector according to the value of the BJT β, as indicated in Fig. 6.13. Thus, internally the BJT still operates as a transistor in the active mode. This is the reason the i–v characteristic of the resulting diode is identical to the i_E–v_{BE} relationship of the BJT.

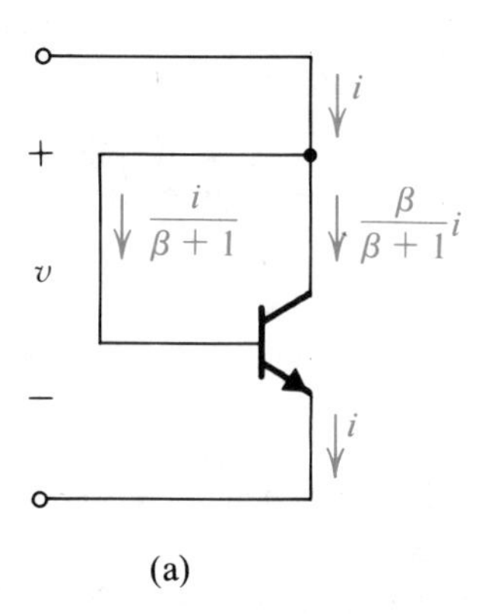

(a)

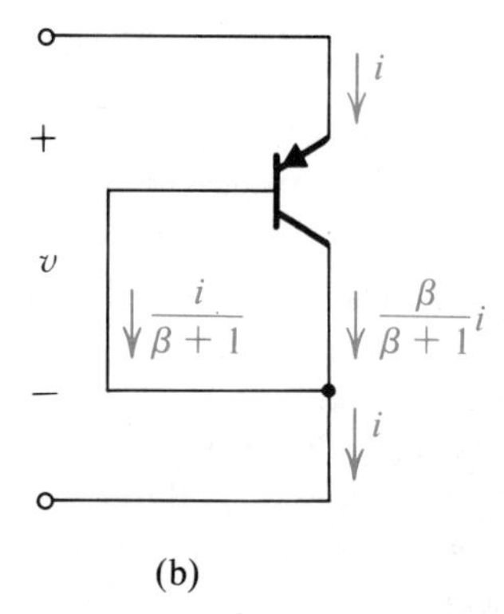

(b)

Fig. 6.13 Diode-connected BJTs.

It can be shown (Exercise 6.5) that the incremental resistance of the diode-connected transistor is approximately equal to r_e. In the following we shall make extensive use of the diode-connected BJT.

Exercise 6.5 Replace the BJT in the diode-connected transistor of Fig. 6.13(a) with its complete low-frequency hybrid-π model. Thus show that the incremental resistance of the two-terminal device is $[r_\pi // (1/g_m) // r_o] \simeq r_e$. Evaluate the incremental resistance for $i = 0.5$ mA.

Ans. 50 Ω

The Current Mirror

The **current mirror,** shown in its simplest form in Fig. 6.14, is the most basic building block in the design of IC current sources and current-steering circuits. (The basic MOS current mirror was studied in Chapter 5.) The current mirror consists of two matched transistors Q_1 and Q_2 with their bases and emitters connected together, and which thus have the same v_{BE}. In addition, Q_1 is connected as a diode by shorting its collector to its base.

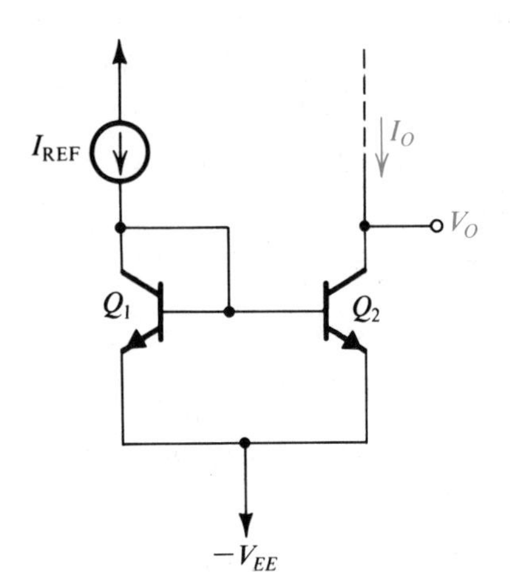

Fig. 6.14 The basic BJT current mirror.

The current mirror is shown fed with a constant-current source I_{REF}, and the output current is taken from the collector of Q_2. The circuit fed by the collector of Q_2 should ensure active-mode operation for Q_2 (by keeping its collector voltage higher than that of the base) at all times. Assume that the BJTs have high β, and thus their base currents are negligibly small. The input current I_{REF} flows through the diode-connected transistor Q_1 and thus establishes a voltage across Q_1 that corresponds to the value of I_{REF}. This voltage in turn appears between the base and emitter of Q_2. Since Q_2 is identical to Q_1, the emitter current of Q_2 will be equal to I_{REF}. It follows that as long as Q_2 is maintained in the active region, its collector current I_O will be approximately equal to I_{REF}. Note that the mirror operation is independent of the value of the voltage $-V_{EE}$ as long as Q_2 remains active.

Next we consider the effect of finite transistor β on the operation of the current mirror. The analysis proceeds as follows: Since Q_1 and Q_2 are matched and since they have equal v_{BE}, their emitter currents will be equal. This is the key point. The rest of the analysis is straightforward and is illustrated in Fig. 6.15. It follows that

$$I_O = \frac{\beta}{\beta + 1} I_E$$

$$I_{REF} = \frac{\beta + 2}{\beta + 1} I_E$$

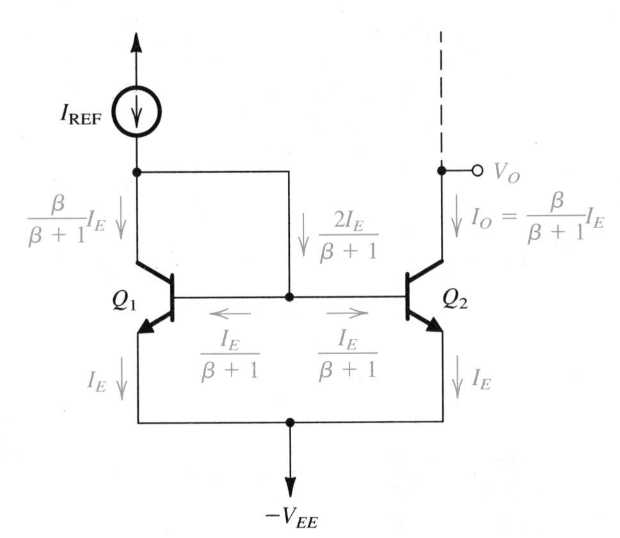

Fig. 6.15 Analysis of the current mirror taking into account the finite β of the BJTs.

Thus the current gain of the mirror is given by

$$\frac{I_O}{I_{REF}} = \frac{\beta}{\beta + 2} = \frac{1}{1 + 2/\beta} \tag{6.63}$$

which approaches unity for $\beta \gg 1$. Note, however, that the deviation of current gain from unity can be relatively high: $\beta = 100$ results in a 2% error.

Another factor that makes I_O unequal to I_{REF} is the linear dependence of the collector current of Q_2, which is I_O, on the collector voltage of Q_2. In fact, even if we ignore the effect of finite β and assume that Q_1 and Q_2 are perfectly matched, the current I_O will be equal to I_{REF} only when the voltage at the collector of Q_2 is equal to the base voltage. As the collector voltage is increased, I_O increases. Since Q_2 is operated at a constant v_{BE} (as determined by I_{REF}) the dependence of I_O on V_O is determined by r_o of Q_2. In other words, the output resistance of the current mirror of Fig. 6.14 is equal to r_o of Q_2.

Taking the effect of finite β and the Early effect together, we find that the output current of the mirror is given by

$$I_O \simeq \frac{I_{REF}}{1 + 2/\beta}\left(1 + \frac{V_O + V_{EE} - V_{BE}}{V_A}\right) \tag{6.64}$$

Exercise 6.6 The current-mirror circuit of Fig. 6.14 utilizes BJTs having $\beta = 100$ and an Early voltage $V_A = 100$ V. Find the output resistance and the output current for $I_{REF} = 1$ mA, $V_{EE} = 5$ V, and $V_O = +5$ V. Assume that $V_{BE} \simeq 0.7$ V.

Ans. 100 kΩ; 1.07 mA

A Simple Current Source

Figure 6.16 shows a simple BJT constant-current-source circuit. It utilizes a pair of matched transistors in a current-mirror configuration, with the input reference current to

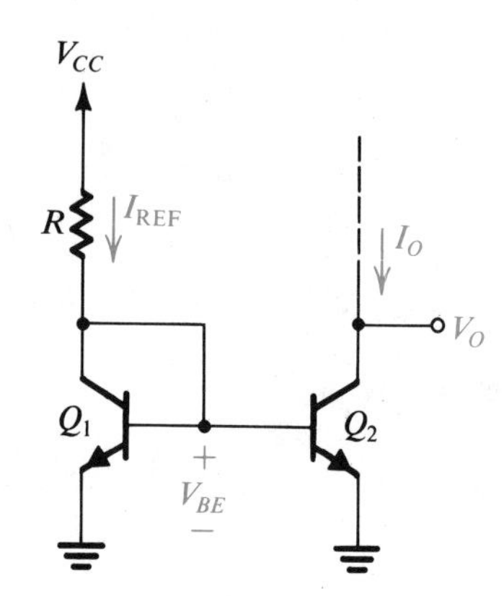

Fig. 6.16 A simple BJT current source.

the mirror, I_{REF}, determined by a resistor R connected to the positive power supply V_{CC}. The current I_{REF} is given by

$$I_{REF} = \frac{V_{CC} - V_{BE}}{R} \tag{6.65}$$

where V_{BE} is the base–emitter voltage corresponding to an emitter current I_{REF}. Neglecting the effect of finite β and the dependence of I_O on V_O, the output current I_O will be equal to I_{REF}. The circuit will operate as a constant-current source as long as Q_2 remains in the active region—that is, for $V_O \geq V_{BE}$. The output resistance of this current source is r_o of Q_2.

Exercise D6.7 For the circuit in Fig. 6.16 find the value of R that results in $I_O = 1$ mA with $V_{CC} = 5$ V. Assume that $V_{BE} \simeq 0.7$ V and neglect the effects of r_o and the finite β.

Ans. 4.3 kΩ

Current-Steering Circuits

As mentioned earlier, in an IC a dc reference current is generated in one location and is then reproduced at other locations for the purpose of biasing the various amplifier stages on the IC. As an example consider the circuit shown in Fig. 6.17. The circuit utilizes two power supplies, V_{CC} and $-V_{EE}$. The dc reference current I_{REF} is generated in the branch

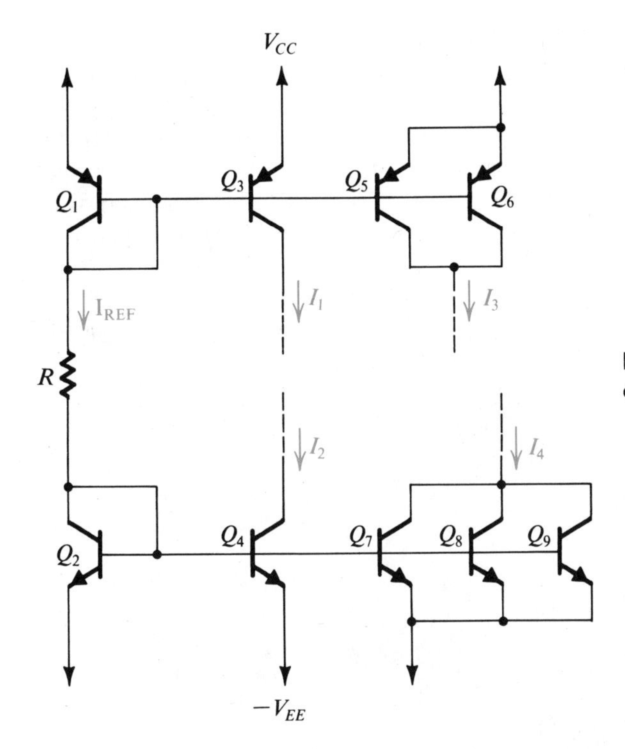

Fig. 6.17 Generation of a number of constant currents.

that consists of the diode-connected transistor Q_1, resistor R, and diode-connected transistor Q_2:

$$I_{\text{REF}} = \frac{V_{CC} + V_{EE} - V_{EB1} - V_{BE2}}{R} \tag{6.66}$$

Now for simplicity assume that all transistors have high β and thus the base currents are negligibly small. Diode-connected transistor Q_1 forms a current mirror with Q_3. Thus Q_3 will supply a constant current I_1 equal to I_{REF}. Transistor Q_3 can supply this current to any load as long as the voltage that develops at the collector does not exceed that at the base $(V_{CC} - V_{EB3})$.

To generate a dc current twice the value of I_{REF}, two transistors Q_5 and Q_6 are connected in parallel, and the combination forms a mirror with Q_1. Thus $I_3 = 2I_{\text{REF}}$. Note that the parallel combination of Q_5 and Q_6 is equivalent to a transistor whose EBJ area is double that of Q_1, which is precisely what would be done if this circuit were to be fabricated in IC form. Current mirrors are indeed used to provide multiples of the reference current by simply designing the transistors to have an area ratio equal to the desired multiple.

Transistor Q_4 forms a mirror with Q_2, and thus Q_4 provides a constant current I_2 equal to I_{REF}. Note an important difference between Q_3 and Q_4: Although both supply equal currents, Q_3 *sources* its current to parts of the circuit whose voltage should not exceed $V_{CC} - V_{EB3}$. On the other hand, Q_4 *sinks* its current from parts of the circuit whose voltage should not decrease below $-V_{EE} + V_{BE4}$. Finally, to generate a current three times the reference, three transistors Q_7, Q_8, and Q_9 are paralleled and the combination placed in a mirror configuration with Q_2. Again, in an IC implementation, Q_7, Q_8, and Q_9 would be replaced with a transistor having a junction area three times that of Q_2.

The above description ignored the effects of the finite transistor β. We have analyzed this effect in the case of a mirror having a single output. The effect of finite β becomes more severe as the number of outputs of the mirror is increased. This is not surprising since the addition of more transistors means that their base currents have to be supplied by the reference current source.

Exercise 6.8 Figure E6.8 shows an N-output current mirror. Assuming all transistors to be matched and have finite β and ignoring the effect of finite output resistances, show that

$$I_1 = I_2 = \cdots = I_N = \frac{1}{1 + (N + 1)/\beta}$$

For $\beta = 100$ find the maximum number of outputs for an error not exceeding 10%.

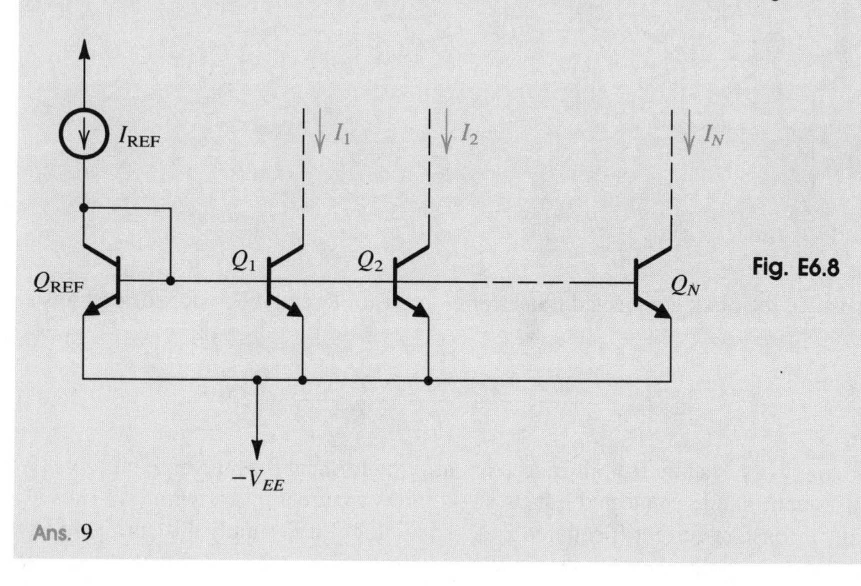

Fig. E6.8

Ans. 9

Improved Current-Source Circuits

Two performance parameters of the BJT current source need improvement. The first is the dependence of I_O on β, which is a result of the error in the mirror current gain introduced by the finite BJT β. The second is the output resistance of the current source, which was found to be equal to the BJT r_o and thus limited to the order of 100 kΩ. The need to increase the current-source output resistance can be seen if we recall that the common-mode gain of the differential amplifier is directly determined by the output resistance of its biasing current source. Also, it will be seen at a later stage that the BJT current sources are usually used in place of the load resistances R_C of the differential amplifier. Thus, to obtain high voltage gain, a large output resistance is required.

We shall now discuss several circuit techniques that result in reduced dependence on β and/or increased output resistance. The first circuit, shown in Fig. 6.18, includes a

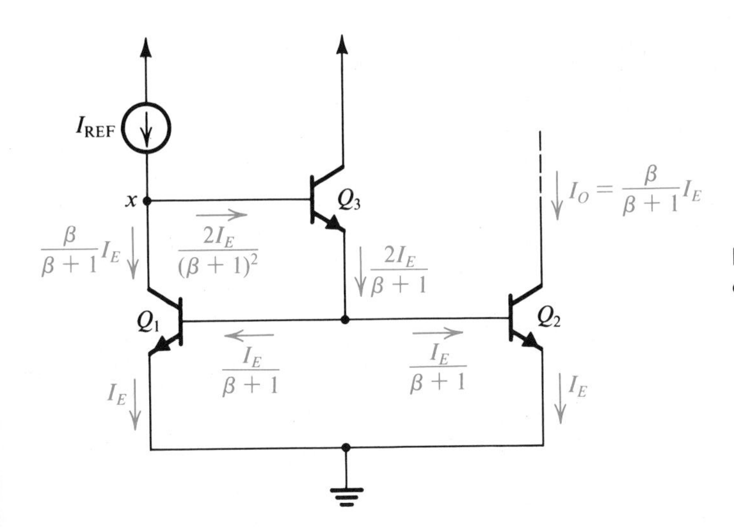

Fig. 6.18 Current mirror with base-current compensation.

transistor Q_3 whose emitter supplies the base currents of Q_1 and Q_2. The sum of the base currents is then divided by $(\beta + 1)$ of Q_3, resulting in a much smaller current that has to be supplied by I_{REF}. Detailed analysis, shown on the circuit diagram, is based on the assumption that Q_1 and Q_2 are matched and thus have equal emitter currents, I_E. A node equation at the node labeled x gives

$$I_{REF} = \left[\frac{\beta}{\beta + 1} + \frac{2}{(\beta + 1)^2}\right] I_E$$

Since

$$I_O = \frac{\beta}{\beta + 1} I_E$$

it follows that the current gain of this mirror is given by

$$\frac{I_O}{I_{REF}} = \frac{1}{1 + 2/(\beta^2 + \beta)} \tag{6.67}$$

$$\simeq \frac{1}{1 + 2/\beta^2} \tag{6.68}$$

which means that the error due to finite β has been reduced from $2/\beta$ to $2/\beta^2$, a tremendous improvement. Finally, note that for simplicity the circuit is shown fed with a current I_{REF}. To use the circuit as a current source we connect a resistance R between node x and the positive supply V_{CC}, then

$$I_{REF} = \frac{V_{CC} - V_{BE1} - V_{BE3}}{R} \tag{6.69}$$

An alternative mirror circuit that achieves both base-current compensation and increased output resistance is the Wilson mirror shown in Fig. 6.19. Analysis of this circuit

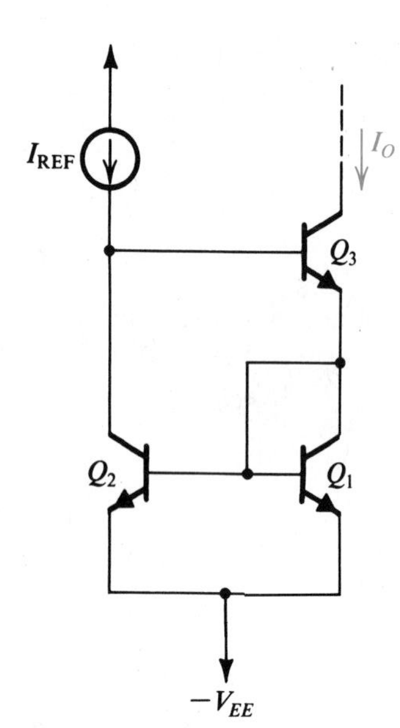

Fig. 6.19 The Wilson current mirror.

taking into account the finite β results in a current-gain expression identical to that in Eq. (6.68). The Wilson current source features an output resistance approximately equal to $\beta r_o/2$, a factor of $\beta/2$ greater than that of the simple current source of Fig. 6.16 (see Problem 6.49).

Exercise 6.9 For the Wilson current mirror in Fig. 6.19 assume all BJTs are matched and have finite β. Denoting the currents in the emitters of Q_1 and Q_2 by I_E, find I_{REF} and I_O in terms of I_E, and show that I_O/I_{REF} is given by Eq. (6.68).

Our final current-source circuit, known as the Widlar current source, is shown in Fig. 6.20. It differs from the basic current mirror circuit in an important way: A resistor R_E is included in the emitter lead of Q_2. Neglecting base currents we can write:

$$V_{BE1} = V_T \ln\left(\frac{I_{\text{REF}}}{I_S}\right) \tag{6.70}$$

and

$$V_{BE2} = V_T \ln\left(\frac{I_O}{I_S}\right) \tag{6.71}$$

where we have assumed that Q_1 and Q_2 are matched devices. Combining Eqs. (6.70) and (6.71) gives

$$V_{BE1} - V_{BE2} = V_T \ln\left(\frac{I_{\text{REF}}}{I_O}\right) \tag{6.72}$$

But from the circuit we see that

$$V_{BE1} = V_{BE2} + I_O R_E \tag{6.73}$$

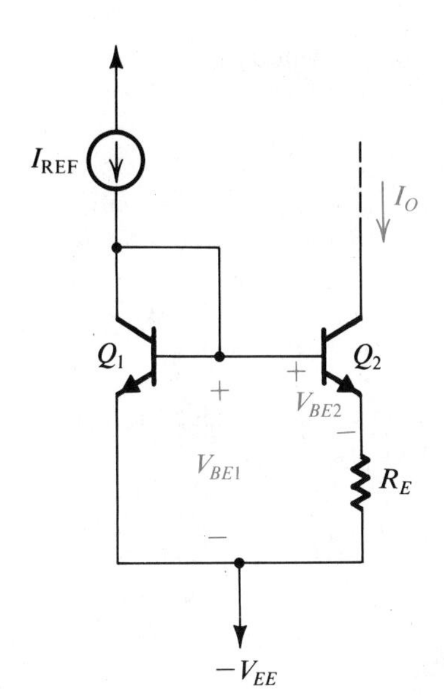

Fig. 6.20 The Widlar current source.

Thus

$$I_O R_E = V_T \ln\left(\frac{I_{REF}}{I_O}\right) \tag{6.74}$$

EXAMPLE 6.1

Figure 6.21 shows two circuits for generating a constant current $I_O = 10\ \mu$A. Determine the values of the required resistors assuming that V_{BE} is 0.7 V at a current of 1 mA and neglecting the effect of finite β.

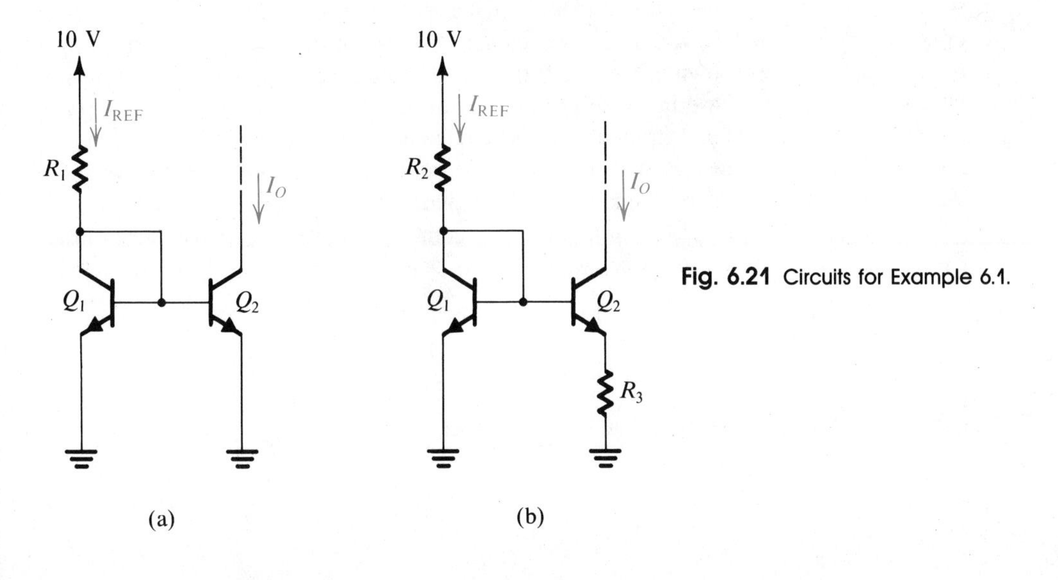

Fig. 6.21 Circuits for Example 6.1.

Solution For the basic current source circuit in Fig. 6.21(a) we choose a value for R_1 to result in $I_{REF} = 10\ \mu A$. At this current, the voltage drop across Q_1 will be

$$V_{BE1} = 0.7 + V_T \ln\left(\frac{10\ \mu A}{1\ mA}\right) = 0.58\ V$$

Thus

$$R_1 = \frac{10 - 0.58}{0.01} = 942\ k\Omega$$

For the Widlar circuit in Fig. 6.21(b) we must first decide on a suitable value for I_{REF}. If we select $I_{REF} = 1$ mA, then $V_{BE1} = 0.7$ V and R_2 is given by

$$R_2 = \frac{10 - 0.7}{1} = 9.3\ k\Omega$$

The value of R_3 can be determined using Eq. (6.74), as follows:

$$10 \times 10^{-6} R_3 = 0.025 \ln\left(\frac{1\ mA}{10\ \mu A}\right)$$

$$R_3 = 11.5\ k\Omega$$

From the above example we observe that using the Widlar circuit allows the generation of a small constant current using relatively small resistors. This is an important advantage that results in considerable savings in chip area. In fact the circuit of Fig. 6.21(a), requiring a 942-kΩ resistance, is totally impractical for implementation in IC form.

Another important characteristic of the Widlar current source is that its output resistance is high. The increase in the output resistance, above that achieved in the basic current source of Fig. 6.16, is due to the emitter resistance R_E. To determine the output resistance of Q_2, we replace the BJT with its low-frequency hybrid-π model and apply a test voltage v_x to the collector, as shown in Fig. 6.22(a). Note that the base of Q_2 is shown grounded, which is not quite the case in the original circuit in Fig. 6.20. Indeed, the base of Q_2 is connected to $-V_{EE}$ (signal ground) via the diode-connected transistor Q_1. The latter, however, has a small incremental resistance (r_e), and thus to simplify matters we shall assume that this resistance is small enough to place the base of Q_2 at signal ground.

The circuit of Fig. 6.22(a) is simplified by combining R_E and r_π in parallel to form R'_E, as illustrated in Fig. 6.22(b), which shows some of the analysis details. A loop equation yields

$$v_x = -v_\pi - \left(g_m + \frac{1}{R'_E}\right) v_\pi r_o \tag{6.75}$$

and a node equation at C provides

$$i_x = g_m v_\pi - \left(g_m + \frac{1}{R'_E}\right) v_\pi \tag{6.76}$$

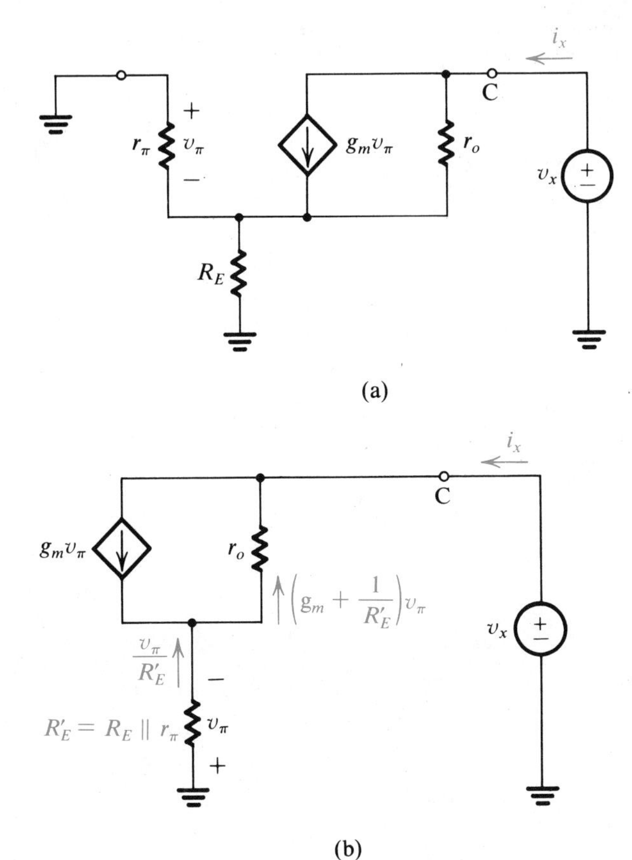

Fig. 6.22 Determination of the output resistance of the Widlar current source of Fig. 6.20.

Dividing Eq. (6.75) by Eq. (6.76) gives the output resistance

$$R_o \equiv \frac{v_x}{i_x} = \frac{1 + \left(g_m + \dfrac{1}{R'_E}\right) r_o}{1/R'_E}$$

which can be rearranged into the form

$$R_o = R'_E + (1 + g_m R'_E) r_o \tag{6.77}$$

$$\simeq (1 + g_m R'_E) r_o \tag{6.78}$$

Thus the output resistance is increased by the factor $1 + g_m R'_E = 1 + g_m(R_E // r_\pi)$. Finally note that in the analysis above we have neglected r_μ. However, it can be easily taken into account since it appears in parallel with the resistance given by Eq. (6.78).

Exercise 6.10 Find the output resistance of each of the two current sources designed in Example 6.1. Let $V_A = 100$ V and $\beta = 100$.

Ans. (a) 10 MΩ; (b) 54 MΩ

EXAMPLE 6.2

Figure 6.23 shows the circuit of a simple operational amplifier. Terminals 1 and 2, shown connected to ground, are the op amp's input terminals, and terminal 3 is the output terminal.

(a) Perform an approximate dc analysis (assuming $\beta \gg 1$ and $|V_{BE}| \simeq 0.7$ V) to calculate the dc currents and voltages everywhere in the circuit. Note that Q_6 has four times the area of each of Q_9 and Q_3.

(b) Calculate the quiescent power dissipation in this circuit.

(c) If transistors Q_1 and Q_2 have $\beta = 100$, calculate the input bias current of the op amp.

(d) What is the common-mode range of this op amp?

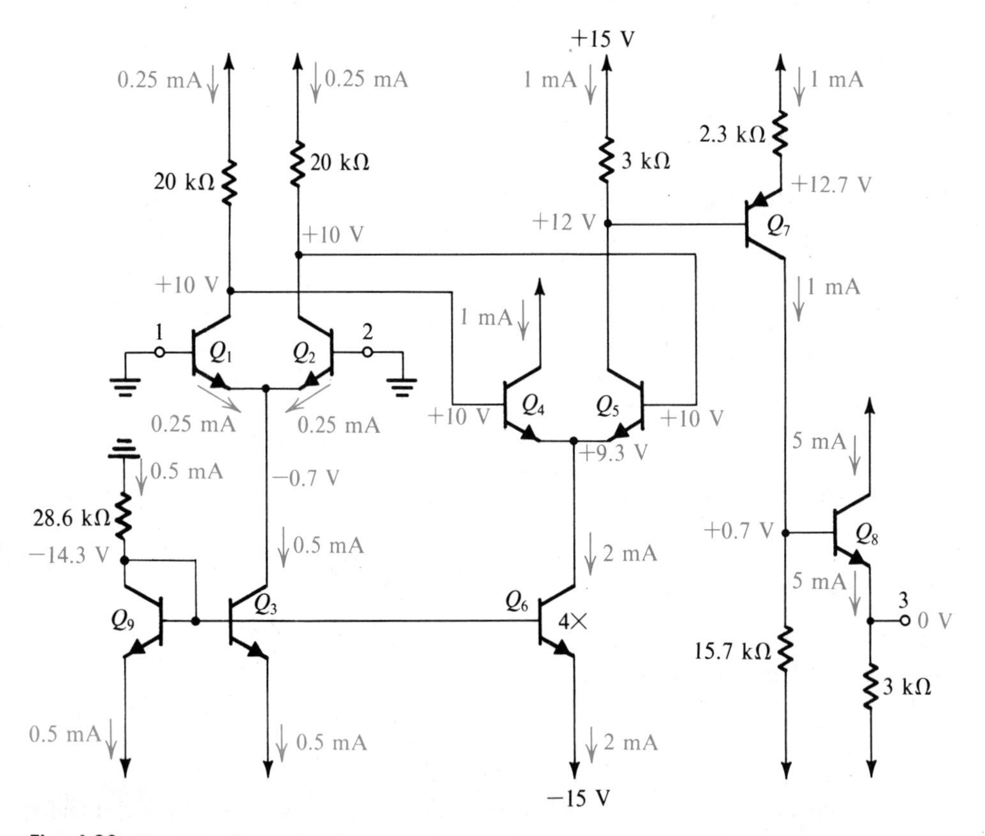

Fig. 6.23 Circuit for Example 6.2.

Solution

(a) The values of all dc currents and voltages are indicated on the circuit diagram. These values were calculated by ignoring the base current of every transistor—that is, by assuming β to be very high. The analysis starts by determining the current through the diode-connected transistor Q_9 to be 0.5 mA. Then we see that transistor Q_3 conducts 0.5 mA and transistor Q_6 conducts 2 mA. The current-source transistor Q_3 feeds the differential pair (Q_1, Q_2) with 0.5 mA. Thus each of Q_1 and Q_2 will be biased at 0.25 mA. The collectors of Q_1 and Q_2 will be at $[+15 - 0.25 \times 20] = +10$ V.

Proceeding to the second differential stage formed by Q_4 and Q_5, we find the voltage at their emitters to be $[+10 - 0.7] = 9.3$ V. This differential pair is biased by the current-source transistor Q_6, which supplies a current of 2 mA; thus Q_4 and Q_5 will each be biased at 1 mA. We can now calculate the voltage at the collector of Q_5 as $[+15 - 1 \times 3] = +12$ V. This will cause the voltage at the emitter of the *pnp* transistor Q_7 to be $+12.7$ V, and the emitter current of Q_7 will be $(+15 - 12.7)/2.3 = 1$ mA.

The collector current of Q_7, 1 mA, causes the voltage at the collector to be $[-15 + 1 \times 15.7] = +0.7$ V. The emitter of Q_8 will be 0.7 V below the base; thus output terminal 3 will be at 0 V. Finally, the emitter current of Q_8 can be calculated to be $[0 - (-15)]/3 = 5$ mA.

(b) To calculate the power dissipated in the circuit in the quiescent state (that is, with zero input signal) we simply evaluate the dc current that the circuit draws from each of the two power supplies. From the +15-V supply the dc current is $I^+ = 0.25 + 0.25 + 1 + 1 + 1 + 5 = 8.5$ mA. Thus the power supplied by the positive power supply is $P^+ = 15 \times 8.5 = 127.5$ mW. The -15-V supply provides a current I^- given by $I^- = 0.5 + 0.5 + 2 + 1 + 5 = 9$ mA. Thus the power provided by the negative supply is $P^- = 15 \times 9 = 135$ mW. Adding P^+ and P^- provides the total power dissipated in the circuit P_D: $P_D = P^+ + P^- = 262.5$ mW.

(c) The input bias current of the op amp is the average of the dc currents that flow in the two input terminals (that is, in the bases of Q_1 and Q_2). These two currents are equal (because we have assumed matched devices); thus the bias current is given by

$$I_B = \frac{I_{E1}}{\beta + 1} \simeq 2.5\ \mu\text{A}$$

(d) The upper limit on the input common-mode voltage is determined by the voltage at which Q_1 and Q_2 leave the active mode and enter saturation. This will happen if the input voltage equals or exceeds the collector voltage, which is +10 V. Thus the upper limit of the common-mode range is +10 V.

The lower limit of the input common-mode range is determined by the voltage at which Q_3 leaves the active mode and thus ceases to act as a constant-current source. This will happen if the collector voltage of Q_3 goes below the voltage at its base, which is -14.3 V. It follows that the input common-mode voltage should not go lower than $-14.3 + 0.7 = -13.6$ V. Thus the common-mode range is -13.6 to $+10$ V.

6.5 THE BJT DIFFERENTIAL AMPLIFIER WITH ACTIVE LOAD

Active devices (transistors) occupy much less silicon area than medium- and large-sized resistors. For this reason we studied in Chapter 5 a variety of MOSFET amplifier circuits that utilize MOSFETs as load devices. Similarly, many practical BJT integrated-circuit amplifiers use BJT loads in place of the resistive loads, R_C. In such circuits the BJT load transistor is usually connected as a constant-current source and thus presents the amplifier transistor with a very-high-resistance load (the output resistance of the current source). Thus amplifiers that utilize *active loads* can achieve higher voltage gains than those with passive (resistive) loads. In this section we study a circuit configuration that has become very popular in the design of BJT ICs.

The active-load differential amplifier circuit is shown in Fig. 6.24. Transistors Q_1 and Q_2 form a differential pair biased with constant current I. The load circuit consists of transistors Q_3 and Q_4 connected in a current mirror configuration. The output is taken single-endedly from the collector of Q_2.

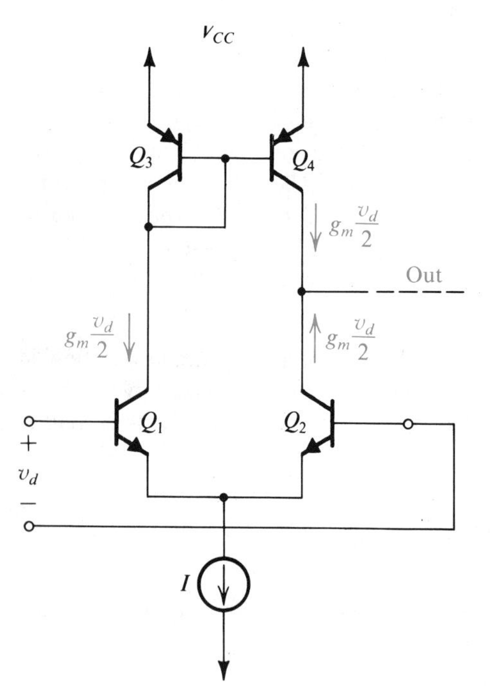

Fig. 6.24 The differential amplifier with an active load.

Consider first the case when no input signal is applied (that is, the two input terminals are grounded). The current I splits equally between Q_1 and Q_2. Thus Q_1 draws a current approximately $I/2$ from the diode-connected transistor Q_3. Assuming $\beta \gg 1$, the mirror supplies an equal current $I/2$ through the collector of Q_4. Since this current is equal to that through the collector of Q_2, no output current flows through the output terminal. It should be noted, however, that in practical circuits the dc quiescent voltage at the output terminal is determined by the subsequent amplifier stage. An example of that will be seen in Chapter 10, where the internal circuit of the 741-type op amp is studied in detail.

Next consider the situation when a differential signal v_d is applied at the input. Current signals $g_m(v_d/2)$ will result in the collectors of Q_1 and Q_2 with the polarities indicated in Fig. 6.24. The current mirror reproduces the current signal $g_m(v_d/2)$ through the collector of Q_4. Thus, at the output node we have two current signals that add together to produce a total current signal of $(g_m v_d)$. Now if the resistance presented by the subsequent amplifier stage is very large, the voltage signal at the output terminal will be determined by the total signal current $(g_m v_d)$ and the total resistance between the output terminal and ground, R_o; that is,

$$v_o = g_m v_d R_o \tag{6.79}$$

The output resistance R_o is the parallel equivalent of the output resistance of Q_2 and the output resistance of Q_4. Since Q_2 is in effect operating in the common-emitter configuration, its output resistance will be equal to r_{o2}. Also, from our study of the basic current mirror circuit in the previous section we know that its output resistance is equal to r_o of Q_4—that is, r_{o4}. Thus,

$$R_o = r_{o2} // r_{o4} \tag{6.80}$$

For the case $r_{o2} = r_{o4} = r_o$,

$$R_o = r_o/2 \tag{6.81}$$

and the output signal voltage will be

$$v_o = g_m v_d (r_o/2) \tag{6.82}$$

leading to a voltage gain

$$\frac{v_o}{v_d} = \frac{g_m r_o}{2} \tag{6.83}$$

Substituting $g_m = I_C/V_T$ *and* $r_o = V_A/I_C$, where $I_C = I/2$, we obtain

$$g_m r_o = \frac{V_A}{V_T} \tag{6.84}$$

which is a constant for a given transistor. Typically, $V_A = 100$ V, leading to $g_m r_o = 4000$ and a stage voltage gain of about 2000.

In some cases the input resistance of the subsequent amplifier stage may be of the same order as R_o and thus must be taken into account in determining voltage gain. In such situations it is convenient to represent the amplifier of Fig. 6.24 by the transconductance amplifier model shown in Fig. 6.25. Here R_i is the differential input resistance, for our

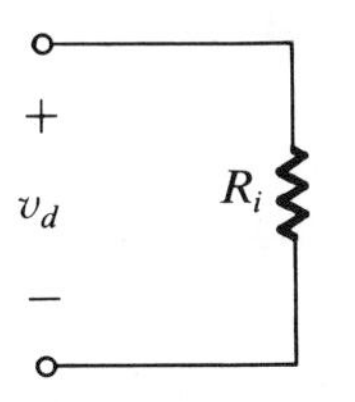

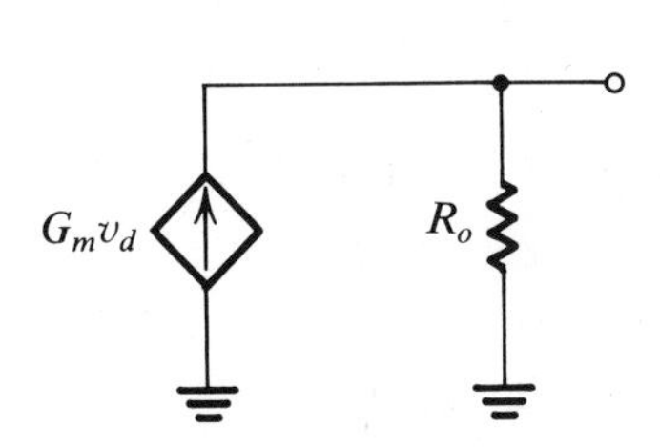

Fig. 6.25 Small-signal model of the differential amplifier of Fig. 6.24.

case $R_i = 2r_\pi$. The amplifier transconductance G_m is the short-circuit transconductance, and for our case

$$G_m = g_m = \frac{I/2}{V_T}$$

Finally, R_o is the output resistance given by Eq. (6.81).

In order to obtain yet higher voltage gains, more elaborate current mirror circuits can be utilized. Also, the basic differential amplifier configuration can be modified to increase the output resistance of Q_2. One such modification involves the use of a circuit arrangement known as the cascode configuration. We shall introduce the cascode configuration shortly.

It is important to observe the role that the current mirror circuit in Fig. 6.24 plays: It inverts the current signal $g_m(v_d/2)$ supplied by the collector of Q_1 and provides an equal current at the collector of Q_4 with such a polarity that it adds to the current signal in the collector of Q_2. Without the current mirror (that is, using only a simple current source) the voltage gain would be half the value found above.

The Cascode Configuration

We shall digress briefly to introduce an important amplifier configuration, the **cascode.** It consists of a common-emitter (CE) stage followed by a common-base (CB) stage, and is shown in differential form in Fig. 6.26(a). Here the pair (Q_1, Q_2) forms the basic differential amplifier which for differential input signals behaves as a common-emitter amplifier. The pair (Q_3, Q_4) forms a differential common-base stage. Although a passive (resistive) load is shown, this can be easily replaced with an active load.

The differential half-circuit is shown in Fig. 6.26(b), from which the CE-CB cascode is readily observable. Also, from this figure we observe that the load resistance seen by the CE transistor Q_1 is no longer R_C but is the much lower input resistance of the CB transistor Q_3, namely its r_e. In Chapter 7 we will find that the reduction in the effective load resistance of Q_1 leads to a tremendous improvement in the amplifier frequency response, which is the most important feature of the cascode amplifier.

Continuing with our examination of the differential half-circuit, we note that the function of the common-base stage is to act as a current buffer; it accepts the signal current $g_m v_d/2$ from the collector of Q_1 at a low input resistance (r_e) and delivers an almost equal current ($\alpha g_m v_d/2$) to the load at a very high output resistance R_o. The high output resistance constitutes the second important feature of the cascode configuration and is particularly useful if the amplifier is to be used with an active load. To find an expression for R_o we observe that it is the output resistance of Q_3 which has an emitter resistance equal to r_o of Q_1. We can utilize Eq. (6.78), substituting $R_E' = r_{\pi 3} // r_{o1} \simeq r_{\pi 3}$ and thus obtaining

$$R_o = r_{o3}(1 + g_{m3} r_{\pi 3})$$

$$= r_{o3}(1 + \beta_3) \simeq \beta_3 r_{o3}$$

Since both devices are biased at the same current ($I/2$) their corresponding small-signal parameters are equal, and thus we can drop the subscripts and express R_o as

$$R_o \simeq \beta r_o \tag{6.85}$$

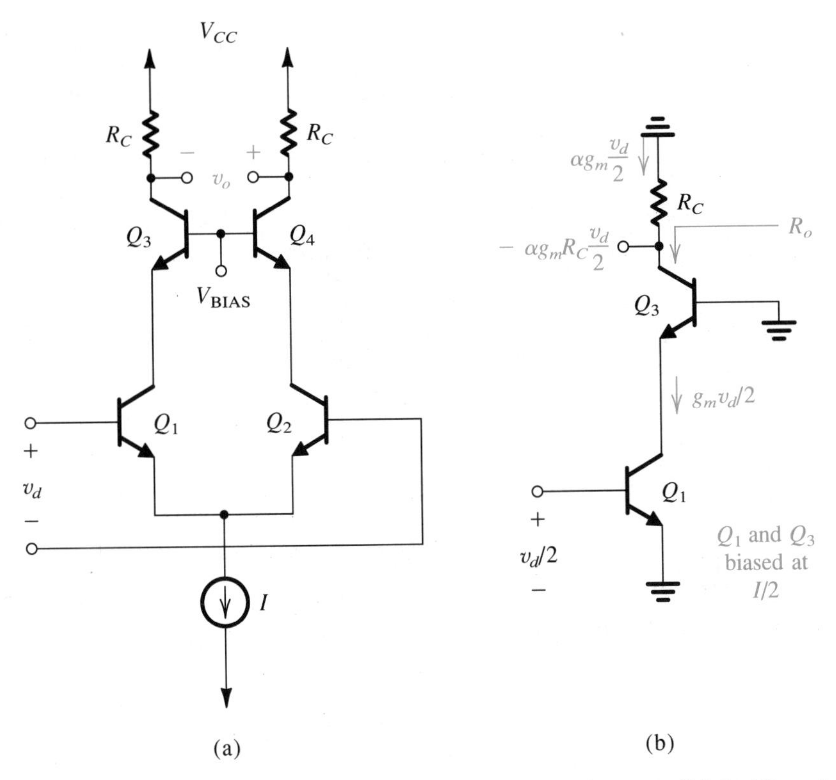

Fig. 6.26 (a) The differential form of the cascode amplifier, and **(b)** its differential half circuit.

Thus the output resistance of the cascode configuration is β times greater than that of the common-emitter amplifier.[2]

Finally, it should be obvious that the cascode configuration can be equally well implemented with MOSFETs, as will be seen at a later point.

Exercises 6.11 For the active-loaded differential amplifier of Fig. 6.24 when biased with a current $I = 0.2$ mA, and if the BJTs have $\beta = 200$ and $V_A = 100$ V, find the values of R_i, G_m, R_o, and the open-circuit voltage gain.

Ans. 100 kΩ; 4 mA/V; 0.5 MΩ; 2000 V/V

6.12 Repeat Exercise 6.11 with the differential pair replaced with a differential cascode amplifier. What is the output resistance of the cascode amplifier?

Ans. 100 kΩ; 4 mA/V; 1 MΩ; 4000 V/V; 200 MΩ

6.13 Repeat Exercise 6.11 with the differential pair replaced with a cascode amplifier and the basic current mirror load replaced with a Wilson current mirror. (Recall that the output resistance of the Wilson mirror is $\beta r_o/2$.)

Ans. 100 kΩ; 4 mA/V; 66.7 MΩ; 2.67×10^5 V/V

---

[2] In this development we have neglected r_μ of Q_3, which appears in parallel with the value of R_o found. Thus a better estimate of R_o is $R_o = \beta r_o // r_\mu$.

6.6 THE JFET DIFFERENTIAL PAIR

Because of their high input resistance, FETs are popular in the design of the differential input stage of op amps. FET-input amplifiers are available commercially and exhibit very small input bias currents (in the picoampere range). In this section we shall study the operation of the JFET differential amplifier.

Figure 6.27 shows the basic JFET differential amplifier fed by two voltage sources v_{G1} and v_{G2} and biased by a constant-current source I. We wish to derive expressions for

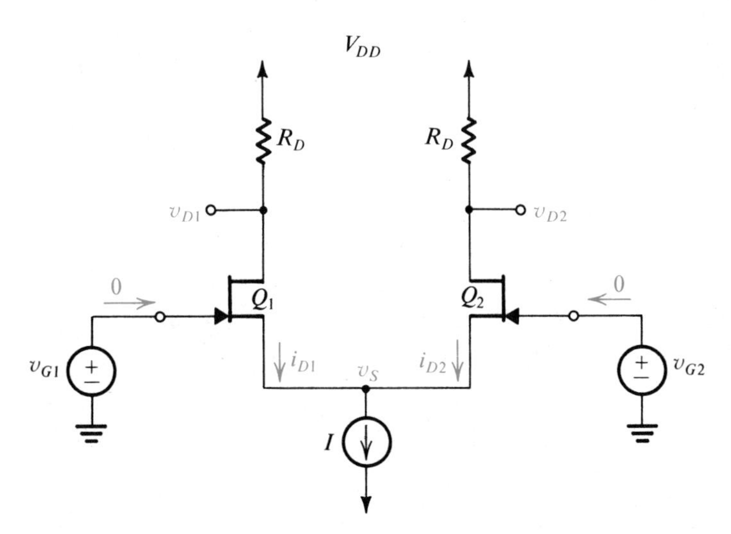

Fig. 6.27 The JFET differential pair.

the currents i_{D1} and i_{D2} in terms of v_{G1} and v_{G2}. To do this we will assume that the JFETs are operating in the pinch-off region. Therefore we may use the square-law i_D–v_{GS} relationship

$$i_D = I_{DSS}\left(1 - \frac{v_{GS}}{V_P}\right)^2$$

where for n-channel devices V_P is negative and we have neglected the channel-length modulation effect (for simplicity). Furthermore, we will assume that the two FETs are perfectly matched. For transistor Q_1 we have

$$i_{D1} = I_{DSS}\left(1 - \frac{v_{G1} - v_S}{V_P}\right)^2$$

which can be rewritten

$$\sqrt{i_{D1}} = \sqrt{I_{DSS}}\left(1 - \frac{v_{G1}}{V_P} + \frac{v_S}{V_P}\right) \tag{6.86}$$

Similarly for Q_2 we have

$$i_{D2} = I_{DSS}\left(1 - \frac{v_{G2} - v_S}{V_P}\right)^2$$

which can be rewritten

$$\sqrt{i_{D2}} = \sqrt{I_{DSS}}\left(1 - \frac{v_{G2}}{V_P} + \frac{v_S}{V_P}\right) \tag{6.87}$$

Subtracting Eq. (6.87) from Eq. (6.86) gives

$$\sqrt{i_{D1}} - \sqrt{i_{D2}} = \sqrt{I_{DSS}}\,\frac{v_{G1} - v_{G2}}{-V_P}$$

Substituting $v_{G1} - v_{G2} = v_{id}$, where v_{id} is the differential input voltage, gives

$$\sqrt{i_{D2}} = \sqrt{i_{D1}} - \sqrt{I_{DSS}}\,\frac{v_{id}}{-V_P} \tag{6.88}$$

The current-source bias imposes the following constraint:

$$i_{D1} + i_{D2} = I \tag{6.89}$$

Substituting for i_{D2} from Eq. (6.88) into Eq. (6.89) results in a quadratic equation that can be solved to yield ultimately

$$i_{D1} = \frac{I}{2} + v_{id}\,\frac{I}{-2V_P}\sqrt{2\frac{I_{DSS}}{I} - \left(\frac{v_{id}}{V_P}\right)^2\left(\frac{I_{DSS}}{I}\right)^2} \tag{6.90}$$

$$i_{D2} = \frac{I}{2} - v_{id}\,\frac{I}{-2V_P}\sqrt{2\frac{I_{DSS}}{I} - \left(\frac{v_{id}}{V_P}\right)^2\left(\frac{I_{DSS}}{I}\right)^2} \tag{6.91}$$

Equations (6.90) and (6.91) indicate that the differential pair responds to a difference signal v_{id} by changing the proportion in which the bias current I divides between the two transistors. For $v_{id} = 0$, $i_{D1} = i_{D2} = I/2$, as should be expected. If v_{id} is positive, the current i_{D1} increases by the amount given by the second term on the right-hand side of Eq. (6.90), and i_{D2} decreases by an equal amount. The maximum value of the increment in i_{D1} (the decrement in i_{D2}) is limited to $I/2$ and is obtained when v_{id} is of such a value that

$$v_{id}\,\frac{I}{-2V_P}\sqrt{2\frac{I_{DSS}}{I} - \left(\frac{v_{id}}{V_P}\right)^2\left(\frac{I_{DSS}}{I}\right)^2} = \frac{I}{2}$$

which results in

$$\left|\frac{v_{id}}{V_P}\right| = \sqrt{\frac{I}{I_{DSS}}} \tag{6.92}$$

That is, the value of v_{id} given by Eq. (6.92) results in the current I being entirely carried by one of the two transistors (Q_1 for positive v_{id} and Q_2 for negative v_{id}). At this point we should note that the bias current I should be smaller than I_{DSS}; otherwise one of the two FETs would carry a current greater than I_{DSS}, which would result in its gate–channel junction becoming forward-biased.

As in the case of the BJT differential pair, the input common-mode range is determined at the low end by the current-source device (which can be a FET or a BJT) leaving the active (pinch-off) region, and at the high end by Q_1 and Q_2 leaving the active (pinch-off) region. This upper limit on v_{CM} will therefore be $|V_P|$ volts below the quiescent voltage at the drain, $V_{DD} - (I/2)R_D$.

Small-Signal Operation

Small-signal analysis of the JFET differential amplifier can be carried out in a manner identical to that given in detail for the BJT circuit. For instance, to evaluate the small-signal differential gain we use Eqs. (6.90) and (6.91) and assume that the terms involving v_{id}^2 are negligibly small; that is,

$$\left|\frac{v_{id}}{V_P}\right| \ll \sqrt{\frac{2I}{I_{DSS}}} \tag{6.93}$$

Note, however, that unlike the BJT case, where v_{id} is limited to about 20 mV, here the small-signal condition limits v_{id} to a volt or so. Under this condition Eqs. (6.90) and (6.91) can be approximated to

$$i_{D1} \simeq \frac{I}{2} + i_d \tag{6.94}$$

$$i_{D2} \simeq \frac{I}{2} - i_d \tag{6.95}$$

where the current signal i_d is given by

$$i_d = \frac{v_{id}}{2}\left(\frac{2I_{DSS}}{-V_P}\sqrt{\frac{I/2}{I_{DSS}}}\right) \tag{6.96}$$

It can be shown that the quantity in parentheses is g_m of Q_1 and of Q_2. Thus Eq. (6.96) simply reaffirms our expectations that the input difference voltage v_{id} divides equally between the two transistors,

$$v_{gs1} = v_{sg2} = \frac{v_{id}}{2} \tag{6.97}$$

and thus the signal current in Q_1 is $g_m v_{id}/2$ and that in Q_2 is $-g_m v_{id}/2$. Alternatively, we can think of the voltage v_{id} as appearing across a total source resistance of $2/g_m$; thus a current

$$i_d = \frac{v_{id}}{2/g_m}$$

flows, as illustrated in Fig. 6.28.

At the drain of Q_1 the voltage signal v_{d1} will be

$$v_{d1} = -i_d R_D = -g_m \frac{v_{id}}{2} R_D \tag{6.98}$$

and at the drain of Q_2 we have

$$v_{d2} = +i_d R_D = +g_m \frac{v_{id}}{2} R_D \tag{6.99}$$

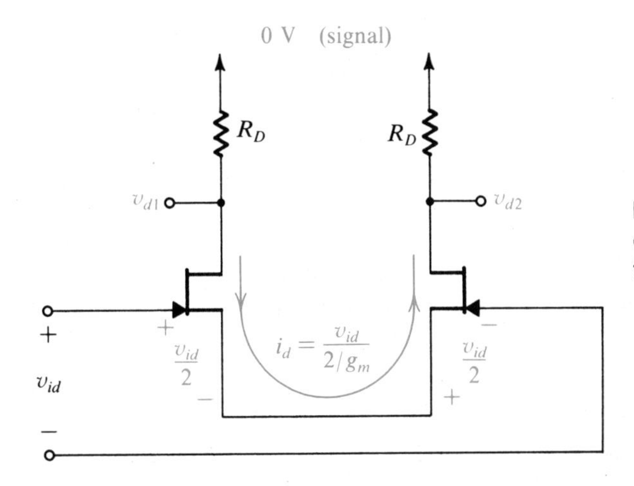

Fig. 6.28 Small-signal analysis of the JFET differential pair. It is assumed that each transistor is biased at a drain current $I/2$.

If the output is taken differentially (that is, between the two drains), then

$$v_o = v_{d1} - v_{d2} = -g_m R_D v_{id} \tag{6.100}$$

and the differential gain will be

$$\frac{v_o}{v_{id}} = -g_m R_D \tag{6.101}$$

If the JFET output resistance r_o is taken into account, the expression for the gain in Eq. (6.101) is modified to

$$\frac{v_o}{v_{id}} = -g_m(\mathrm{R_D}//r_o) \tag{6.102}$$

Finally, it should be pointed out that the current source supplying the bias current I will inevitably have a finite output resistance, which causes the differential amplifier to have a nonzero common-mode gain. The common-mode gain can be evaluated in a manner identical to that used for the BJT case.

Exercise 6.14 Consider the circuit in Fig. 6.27 with $V_{DD} = +15$ V, $I = 1$ mA, $R_D = 10$ kΩ, $I_{DSS} = 2$ mA, and $V_P = -2$ V. Find the value of v_{id} required to switch the current entirely to Q_1. Also calculate the small-signal voltage gain if the output is taken differentially.

Ans. 1.4 V; 10 V/V

6.7 MOS DIFFERENTIAL AMPLIFIERS

During the past decade, the MOS transistor has become very prominent in the design of analog integrated circuits. We have already studied some of the basic MOS IC amplifier circuits in Chapter 5. Building on this material, this section presents the MOS differential pair, which is the most important building block in MOS ICs. We shall also discuss MOS current mirrors, which are used for biasing and as loads for the differential pair. The section concludes with an active-load MOS differential amplifier.

The MOS Differential Pair

Figure 6.29 shows the basic MOS differential pair. It consists of two matched enhancement MOSFETs, Q_1 and Q_2, biased with a constant-current source I. The latter is usually implemented using a current-mirror configuration, much like the case in BJT circuits. Note that the differential amplifier loads are not shown. At this point our purpose is to relate the drain currents to the input voltage. It is, of course, assumed that the load circuit is such that the two MOSFETs in the pair operate in the saturation region.

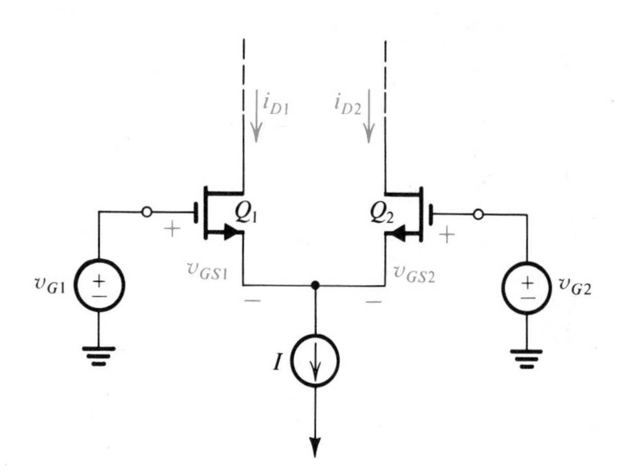

Fig. 6.29 The MOSFET differential pair.

Assuming that the two devices are identical and neglecting the output resistance and body effect, we can express the drain currents as

$$i_{D1} = K(v_{GS1} - V_t)^2 \tag{6.103}$$

$$i_{D2} = K(v_{GS2} - V_t)^2 \tag{6.104}$$

where

$$K = \tfrac{1}{2}\mu_n C_{OX}(W/L) \tag{6.105}$$

Equations (6.103) and (6.104) can be rewritten

$$\sqrt{i_{D1}} = \sqrt{K}(v_{GS1} - V_t) \tag{6.106}$$

$$\sqrt{i_{D2}} = \sqrt{K}(v_{GS2} - V_t) \tag{6.107}$$

Subtracting Eq. (6.107) from Eq. (6.106) and substituting

$$v_{GS1} - v_{GS2} = v_{id}$$

where v_{id} is the differential input voltage, gives

$$\sqrt{i_{D1}} - \sqrt{i_{D2}} = \sqrt{K}v_{id} \tag{6.108}$$

The current-source bias imposes the constraint

$$i_{D1} + i_{D2} = I \tag{6.109}$$

Equations (6.108) and (6.109) are two equations in the two unknowns i_{D1} and i_{D2}. They can be solved together to yield

$$i_{D1} = \frac{I}{2} + \sqrt{2KI}\left(\frac{v_{id}}{2}\right)\sqrt{1 - \frac{(v_{id}/2)^2}{(I/2K)}} \tag{6.110}$$

$$i_{D2} = \frac{I}{2} - \sqrt{2KI}\left(\frac{v_{id}}{2}\right)\sqrt{1 - \frac{(v_{id}/2)^2}{(I/2K)}} \tag{6.111}$$

At the bias (quiescent) point, $v_{id} = 0$, leading to

$$i_{D1} = i_{D2} = \frac{I}{2} \tag{6.112}$$

Correspondingly,

$$v_{GS1} = v_{GS2} = V_{GS}$$

where

$$\frac{I}{2} = K(V_{GS} - V_t)^2 \tag{6.113}$$

This relationship can be used to rewrite Eqs. (6.110) and (6.111) in the form

$$i_{D1} = \frac{I}{2} + \left(\frac{I}{V_{GS} - V_t}\right)\left(\frac{v_{id}}{2}\right)\sqrt{1 - \left(\frac{v_{id}/2}{V_{GS} - V_t}\right)^2} \tag{6.114}$$

$$i_{D2} = \frac{I}{2} - \left(\frac{I}{V_{GS} - V_t}\right)\left(\frac{v_{id}}{2}\right)\sqrt{1 - \left(\frac{v_{id}/2}{V_{GS} - V_t}\right)^2} \tag{6.115}$$

For $v_{id}/2 \ll V_{GS} - V_t$ (small-signal approximation),

$$i_{D1} \simeq \frac{I}{2} + \left(\frac{I}{V_{GS} - V_t}\right)\left(\frac{v_{id}}{2}\right) \tag{6.116}$$

$$i_{D2} \simeq \frac{I}{2} - \left(\frac{I}{V_{GS} - V_t}\right)\left(\frac{v_{id}}{2}\right) \tag{6.117}$$

From Chapter 5 we recall that a MOSFET biased at a drain current I_D has $g_m = 2I_D/(V_{GS} - V_t)$. Thus we see that, for each transistor in the differential pair,

$$g_m = \frac{2(I/2)}{V_{GS} - V_t} = \frac{I}{V_{GS} - V_t} \tag{6.118}$$

and Eqs. (6.116) and (6.117) simply state that for small differential input signals, $v_{id} \ll 2(V_{GS} - V_t)$, the current in Q_1 increases by i_d and that in Q_2 decreases by i_d, where

$$i_d = g_m(v_{id}/2) \tag{6.119}$$

Returning to Eqs. (6.114) and (6.115), we can find the value of v_{id} at which full switching occurs (that is, $i_{D1} = I$ and $i_{D2} = 0$, or vice versa for negative v_{id}) by equating the second term in Eq. (6.114) to $I/2$. The result is

$$|v_{id}|_{\max} = \sqrt{2}(V_{GS} - V_t) \tag{6.120}$$

Figure 6.30 shows plots of the normalized currents i_{D1}/I and i_{D2}/I versus the normalized differential input voltage $v_{id}/(V_{GS} - V_t)$.

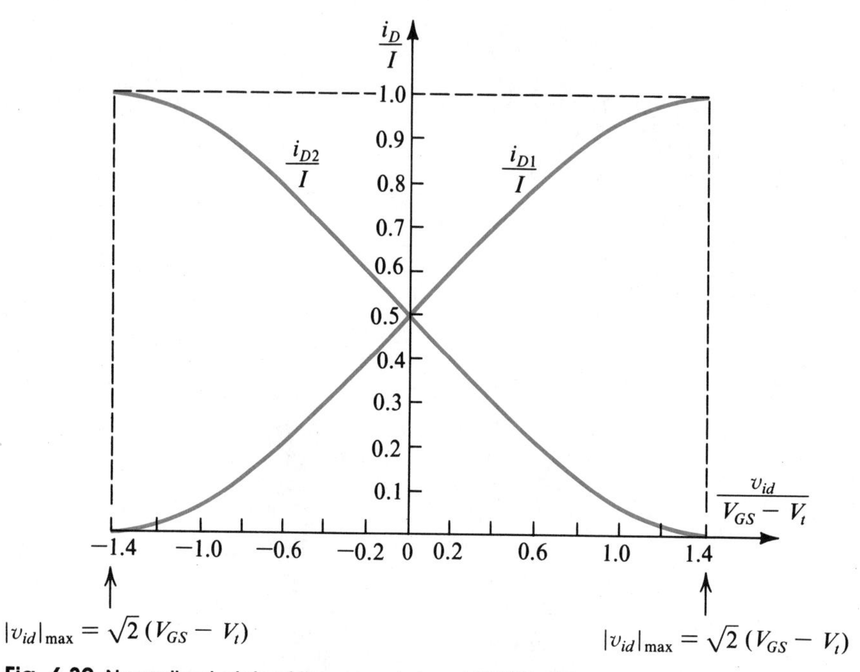

Fig. 6.30 Normalized plots of the currents in a MOSFET differential pair. Note that V_{GS} is the gate-to-source voltage when the drain current is equal to the dc bias current ($I/2$).

Finally, we note that for differential input signals, each MOSFET in the pair operates as a common-source amplifier and thus has an output resistance equal to r_o.

Exercise 6.15 An MOS differential amplifier utilizes a bias current $I = 25\ \mu A$. The devices have $V_t = 1$ V, $W = 120\ \mu m$, $L = 6\ \mu m$, and $(\mu_n C_{OX})$ for this technology is $20\ \mu A/V^2$. Find V_{GS}, g_m, and the value of v_{id} for full current switching.

Ans. 1.25 V; 0.1 mA/V; 0.35 V

Offset Voltage

Three factors contribute to the dc offset voltage of the MOS differential pair: mismatch in load resistances, mismatch in K, and mismatch in V_t. We shall consider the three contributing factors one at a time.

Consider the differential pair shown in Fig. 6.31, in which resistive loads are used in order to simplify the analysis. Since both inputs are grounded, the output voltage V_O is the

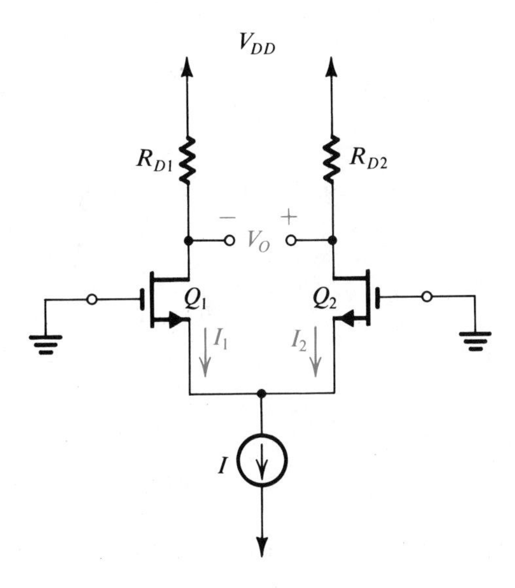

Fig. 6.31 The MOS differential pair with both inputs grounded. Output V_O is the dc output offset voltage.

dc output offset voltage. Consider first the case where Q_1 and Q_2 are perfectly matched but R_{D1} and R_{D2} show a mismatch ΔR_D; that is,

$$R_{D1} = R_D + \left(\frac{\Delta R_D}{2}\right) \tag{6.121}$$

$$R_{D2} = R_D - \left(\frac{\Delta R_D}{2}\right) \tag{6.122}$$

The current I will split equally between Q_1 and Q_2. Nevertheless, because of the mismatch in load resistances, an output voltage V_O develops,

$$V_O = \left(\frac{I}{2}\right)\Delta R_D \tag{6.123}$$

The corresponding input offset voltage is obtained by dividing V_O by the gain $g_m R_D$ and substituting for g_m from Eq. (6.118). The result is

$$V_{OS} = \left(\frac{V_{GS} - V_t}{2}\right)\left(\frac{\Delta R_D}{R_D}\right) \tag{6.124}$$

Next consider the effect of a mismatch in the W/L ratios of Q_1 and Q_2,

$$\left(\frac{W}{L}\right)_1 = \frac{W}{L} + \frac{1}{2}\Delta\left(\frac{W}{L}\right) \tag{6.125}$$

$$\left(\frac{W}{L}\right)_2 = \frac{W}{L} - \frac{1}{2}\Delta\left(\frac{W}{L}\right) \tag{6.126}$$

Such a mismatch gives rise to a proportional mismatch in the conductivity parameter $K = \frac{1}{2}(\mu_n C_{OX})(W/L)$; that is,

$$K_1 = K + \frac{\Delta K}{2} \tag{6.127}$$

$$K_2 = K - \frac{\Delta K}{2} \tag{6.128}$$

The currents I_1 and I_2 will no longer be equal; rather, it can be easily shown that

$$I_1 = \frac{I}{2} + \frac{I}{2}\left(\frac{\Delta K}{2K}\right) \tag{6.129}$$

$$I_2 = \frac{I}{2} - \frac{I}{2}\left(\frac{\Delta K}{2K}\right) \tag{6.130}$$

Dividing the current increment

$$\frac{I}{2}\left(\frac{\Delta K}{2K}\right)$$

by g_m gives half the input offset voltage (due to the mismatch in K values). Thus

$$V_{OS} = \left(\frac{V_{GS} - V_t}{2}\right)\left(\frac{\Delta K}{K}\right) \tag{6.131}$$

Finally, we consider the effect of a mismatch ΔV_t between the two threshold voltages,

$$V_{t1} = V_t + \frac{\Delta V_t}{2} \tag{6.132}$$

$$V_{t2} = V_t - \frac{\Delta V_t}{2} \tag{6.133}$$

The current I_1 will be given by

$$I_1 = K\left(V_{GS} - V_t - \frac{\Delta V_t}{2}\right)^2$$

$$= K(V_{GS} - V_t)^2\left[1 - \frac{\Delta V_t}{2(V_{GS} - V_t)}\right]^2$$

which, for $\Delta V_t \ll 2(V_{GS} - V_t)$, can be approximated as

$$I_1 \simeq K(V_{GS} - V_t)^2\left(1 - \frac{\Delta V_t}{V_{GS} - V_t}\right)$$

Similarly,

$$I_2 \simeq K(V_{GS} - V_t)^2\left(1 + \frac{\Delta V_t}{V_{GS} - V_t}\right)$$

It follows that

$$K(V_{GS} - V_t)^2 = \frac{I}{2}$$

and the current increment (decrement) in Q_2 (Q_1) is

$$\Delta I = \frac{I}{2} \frac{\Delta V_t}{V_{GS} - V_t}$$

Dividing ΔI by g_m gives half the input offset voltage (due to ΔV_t). Thus,

$$V_{OS} = \Delta V_t \tag{6.134}$$

For modern silicon-gate MOS technology ΔV_t can be easily as high as 2 mV. We note that ΔV_t has no counterpart in BJT differential amplifiers. Also, comparison of V_{OS} for the MOS differential pair in Eqs. (6.124) and (6.131) to V_{OS} of the BJT differential pair in Eqs. (6.49) and (6.54) shows that the offset voltage is larger in the MOS pair because $(V_{GS} - V_t)/2$ is usually much greater than V_T. Finally, we observe from Eqs. (6.124) and (6.131) that to keep V_{OS} small, one attempts to operate Q_1 and Q_2 at low values of $V_{GS} - V_t$.

Exercise 6.16 For the MOS differential pair specified in Exercise 6.15, find the three components of input offset voltage. Let $\Delta R_D/R_D = 2\%$, $\Delta K/K = 2\%$, and $\Delta V_t = 2$ mV.

Ans. 2.5 mV; 2.5 mV; 2 mV

Current Mirrors

As in BJT integrated circuits, current mirrors are used in the design of current sources for biasing as well as to operate as active loads. The basic MOS current mirror circuit, shown in Fig. 6.32(a), was studied in Section 5.9. The inaccuracy in current transfer ratio due to the finite β of the BJT has no counterpart in MOS mirrors. Thus the only performance parameter of interest here is the output resistance. For the simple mirror in Fig. 6.32(a), the output resistance is approximately equal to r_{o2}.

The output resistance can be increased by using either the cascode mirror of Fig. 6.32(b) or the Wilson mirror of Fig. 6.32(c). To determine the output resistance of the cascode mirror we use the equivalent circuit shown in Fig. 6.33(a). Note that since the incremental resistance of each of the diode-connected transistors Q_1 and Q_4 is equal to $1/g_m$ and thus is relatively small, we have assumed that the signal voltages at the gates of Q_2 and Q_3 are approximately zero. Replacing Q_2 by its output resistance r_{o2} and replacing Q_3 by its equivalent circuit model leads to the circuit in Fig. 6.33(b). Analysis of the latter circuit is straightforward and leads to

$$R_o \equiv \frac{v_x}{i_x} = r_{o3} + r_{o2} + g_{m3}r_{o3}r_{o2}$$

$$\simeq (g_{m3}r_{o3})r_{o2} \tag{6.135}$$

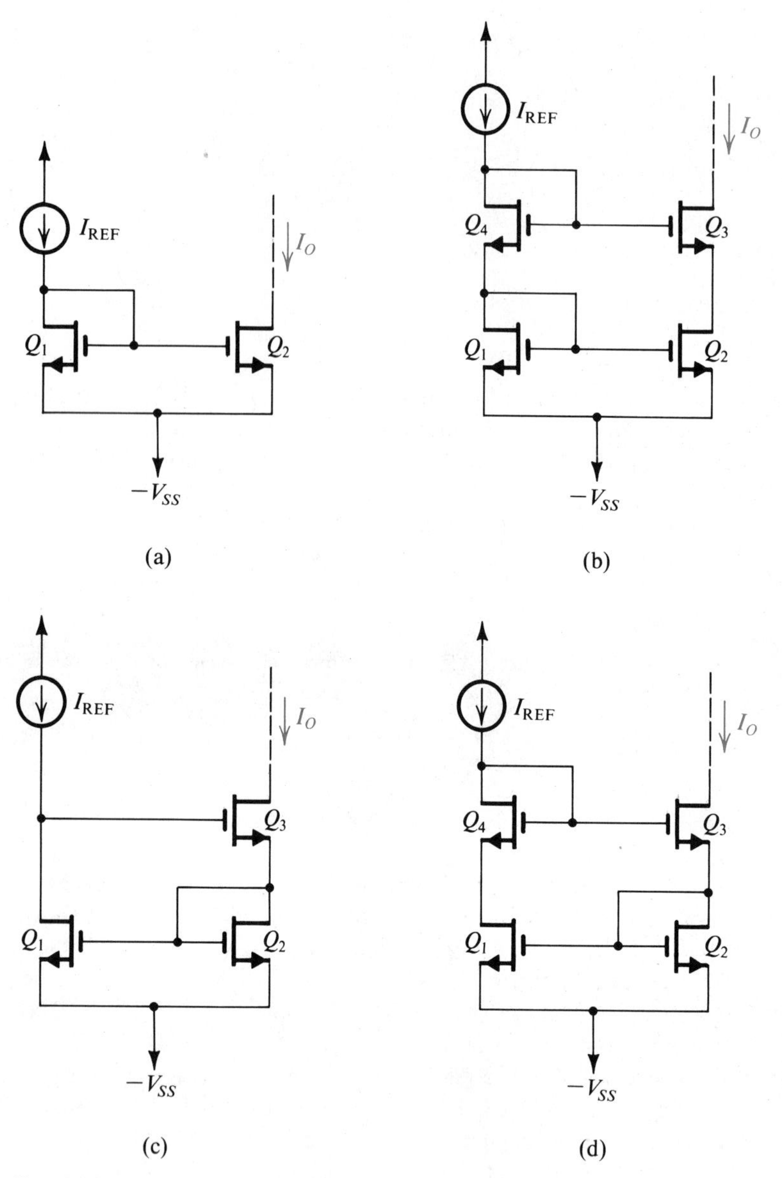

Fig. 6.32 MOS current mirrors: **(a)** basic, **(b)** cascode, **(c)** Wilson, **(d)** modified Wilson.

Thus cascoding transistor Q_2 with transistor Q_3 increases the output resistance from r_{o2} to $(g_{m3}r_{o3})r_{o2}$, an increase by the factor $(g_{m3}r_{o3})$. Similar results are obtained with the Wilson circuit in Fig. 6.32(c). The Wilson circuit, however, suffers from the fact that the drain voltages of Q_1 and Q_2 are not equal, and thus their currents will be unequal. This problem can be solved by including the diode-connected transistor Q_4, as shown in Fig. 6.32(d).

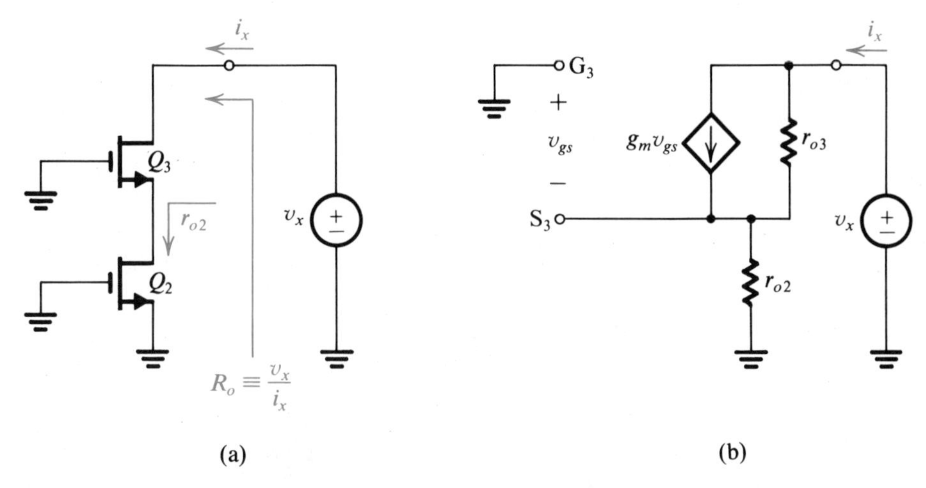

Fig. 6.33 Determining the output resistance of the cascode mirror of Fig. 6.32(b). Note that to simplify matters the diode-connected transistors Q_1 and Q_4 have been assumed to have low resistances and thus are replaced by short circuits.

Exercises 6.17 Consider the current mirror circuit of Fig. 6.32(a) with $V_{SS} = -5$ V and $I_{REF} = 10\ \mu$A. Let Q_1 and Q_2 be identical, with $V_t = 1$ V, $\mu_n C_{OX} = 20\ \mu\text{A/V}^2$, $L = 10\ \mu$m, $W = 40\ \mu$m, and $V_A = 20$ V. Find the output resistance, V_{GS}, and the lowest allowable output voltage.

Ans. 2 MΩ; 1.5 V; −4.5 V

6.18 Repeat Exercise 6.17 for the cascode mirror of Fig. 6.32(b), assuming all devices to be identical.

Ans. 164 MΩ; 1.5 V; −3 V

An Active-Loaded CMOS Amplifier

We conclude this section with a discussion of a popular configuration for a differential amplifier in CMOS technology. The circuit, shown in Fig. 6.34, consists of the differential pair Q_1 and Q_2 loaded by the current mirror formed by Q_3 and Q_4. The dc bias voltage at the output is normally set by the subsequent amplifier stage, as will be shown in Chapter 10.

The circuit is analogous to the BJT version in Fig. 6.24. The signal current i is given by

$$i = g_m(v_{id}/2)$$

where

$$g_m = \frac{I}{V_{GS} - V_t}$$

The output signal voltage is given by

$$v_o = 2i(r_{o2}//r_{o4})$$

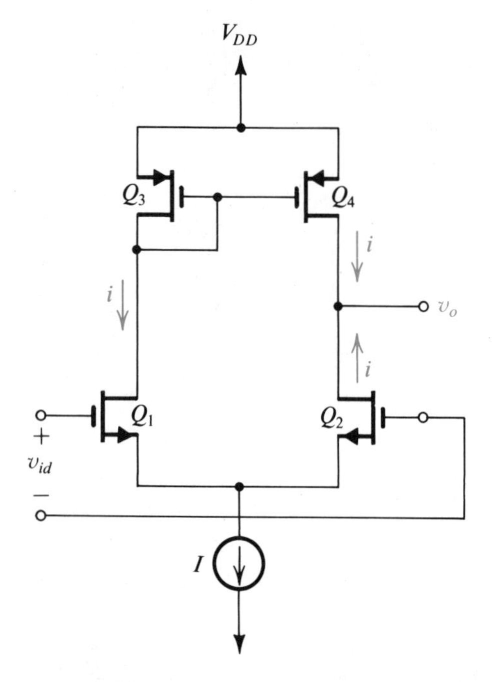

Fig. 6.34 An active-loaded differential amplifier in CMOS technology.

For

$$r_{o2} = r_{o4} = r_o = \frac{V_A}{I/2}$$

the voltage gain becomes

$$A_v \equiv \frac{v_o}{v_{id}} = g_m \frac{r_o}{2}$$

$$= \frac{V_A}{V_{GS} - V_t} \tag{6.136}$$

To obtain higher voltage gains, a cascode current mirror and a cascode differential stage can be used. This, however, reduces the allowable output signal swing. We will have more to say about CMOS differential amplifiers in Chapter 10.

Exercise 6.19 Find the voltage gain of the differential amplifier circuit of Fig. 6.34 under the conditions that $I = 25\ \mu\text{A}$, $V_t = 1$ V, $W_1 = W_2 = 120\ \mu\text{m}$, $L_1 = L_2 = 6\ \mu\text{m}$, $\mu_n C_{OX} = 20\ \mu\text{A/V}^2$, $V_A = 20$ V.

Ans. 80 V/V

6.8 BiCMOS AMPLIFIERS

There are two basic silicon technologies for the design of integrated amplifiers and other analog circuits: bipolar technology, based on the BJT, and CMOS technology, based on

NMOS and PMOS transistors. The BJT has the advantage over the MOSFET of a much higher transconductance (g_m) at the same valuc of dc bias current. Thus much higher voltage gains are realized in a bipolar transistor amplifier stage than in the corresponding MOSFET circuit. Also, as will be seen in the next chapter, bipolar transistor amplifiers have superior high-frequency performance than their MOS counterparts.

On the other hand, the practically infinite input resistance at the gate of a MOSFET makes it possible to design MOS amplifiers with extremely high input resistance and an almost zero input bias current. Also, the MOSFET provides an excellent implementation of a switch; while a saturated BJT exhibits an offset voltage of few tenths of a volt, the i–v characteristics of the MOSFET pass right through the origin, resulting in zero offset. The availability of good switches in CMOS technology makes possible a host of analog circuit techniques that are employed in the design of, among other things, data converters (Chapter 10) and filters (Chapter 11). CMOS is also currently the most widely used digital circuit technology.

It can thus be seen that each of the two circuit technologies, bipolar and CMOS, has its distinct and unique advantages. An IC technology that combines these two device types, allowing each to be utilized in circuit functions for which it is best suited, is now (1990) emerging and is aptly named BiCMOS. This technology is useful in the design of both analog and digital chips as well as chips that combine both analog and digital circuits. Basic BiCMOS amplifier circuits are studied in this section.

Basic Amplifier Stages

Before presenting two BiCMOS amplifier stages, we examine and compare the basic BJT and MOS amplifier stages. Figure 6.35(a) shows an active-loaded BJT common-emitter amplifier. This circuit can be used by itself or can be considered as the differential half-circuit of a differential amplifier. Assuming that the current-source load has an infinite incremental resistance, the total resistance at the collector is the output resistance of the BJT, r_o, and thus the voltage gain realized is

$$\frac{v_o}{v_i} = -g_m r_o \tag{6.138}$$

$$= -\frac{I_C}{V_T}\frac{V_A}{I_C} = -\frac{V_A}{V_T} \tag{6.139}$$

This is the largest gain obtainable from a CE stage. Typically $V_A = 50$ V, and since $V_T = 0.025$ V at room temperature the *intrinsic gain* of the CE stage is about 2000 V/V. A disadvantage of this circuit, however, is its low input resistance, approximately equal to r_π of the BJT.

The corresponding MOSFET amplifier stage is shown in Fig. 6.35(b), for which the voltage gain is

$$\frac{v_o}{v_i} = -g_m r_o \tag{6.140}$$

$$= -\sqrt{2\mu_n C_{OX}(W/L)I}/\lambda I$$

$$= -\frac{\sqrt{2\mu_n C_{OX}(W/L)}}{\lambda\sqrt{I}} \tag{6.141}$$

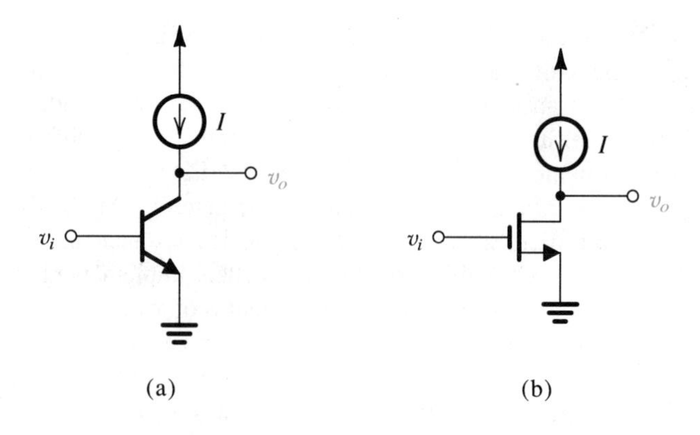

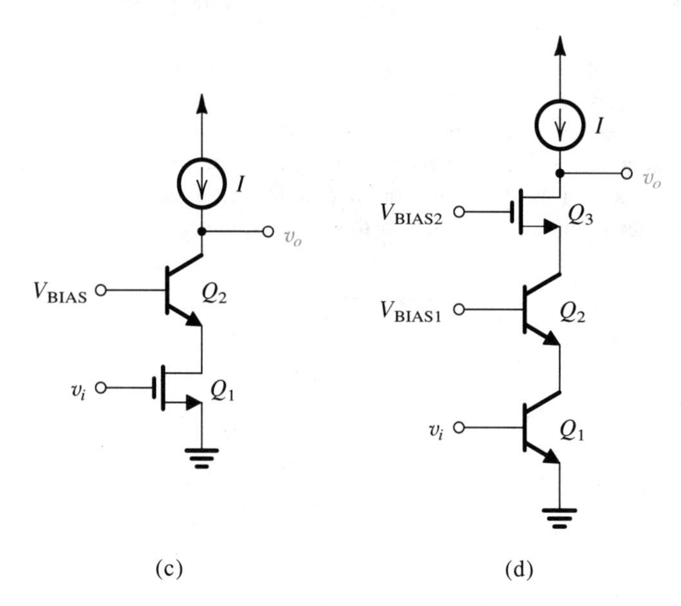

Fig. 6.35 Basic active-loaded amplifier stages: **(a)** bipolar; **(b)** MOS; **(c)** BiCMOS obtained by cascoding Q_1 with a BJT, Q_2; **(d)** BiCMOS double cascode.

where λ is the channel-length modulation factor, $\lambda = 1/V_A$. Observe that unlike the case of the BJT where the gain is independent of the value of the bias current, the gain of the MOSFET amplifier is inversely proportionally to $\sqrt{I}$. Figure 6.36 shows a typical plot of voltage gain versus bias current, from which we note that the gain increases as the current is lowered. At very low currents the MOSFET enters the subthreshold region of operation and the gain becomes constant. It should be noted that although higher gain is obtained by decreasing the level of dc bias current, the price paid is a reduction in amplifier bandwidth (Chapter 7).

Since for the same bias current I, g_m of the MOS transistor is much lower than that of the BJT and since, furthermore, the value of V_A and correspondingly r_o of the BJT is greater than that of the MOSFET (for which V_A is typically 20 V), the intrinsic gain of the MOSFET stage is typically an order of magnitude lower than that for the BJT amplifier. The MOSFET amplifier, however, has the advantage of a practically infinite input resistance.

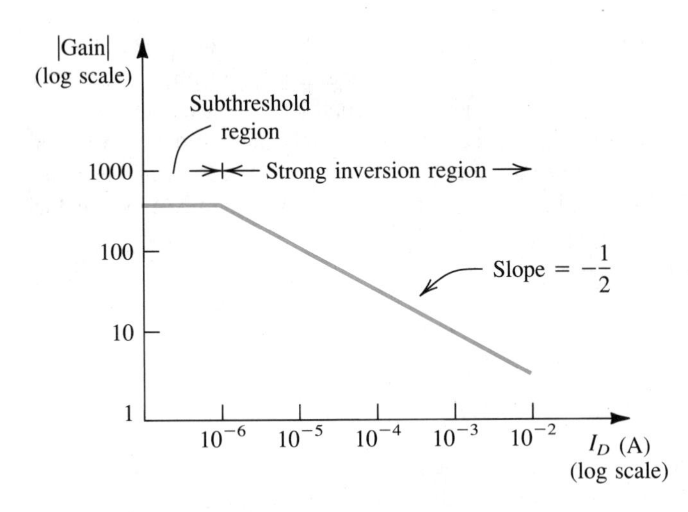

Fig. 6.36 Voltage gain of the active-loaded common-source amplifier versus the bias current I_D. Outside the subthreshold region, this is a plot of Eq. (6.141) for the case $\mu_n C_{OX} = 20\ \mu A/V^2$, $\lambda = 0.05\ V^{-1}$, $L = 2\ \mu m$ and $W = 20\ \mu m$.

Exercise 6.20 For $I = 100\ \mu A$, find g_m, R_i, r_o, and the voltage gain of the CE and CS amplifier stages of Figs. 6.35(a) and (b). For the BJT: $V_A = 50$ V and $\beta = 100$. For the MOSFET: $\lambda = 0.05\ V^{-1}$, $\mu_n C_{OX} = 20\ \mu A/V^2$, $L = 2\ \mu m$, and $W = 20\ \mu m$.

Ans. BJT: $g_m = 4$ mA/V, $R_i = 25$ kΩ, $r_o = 500$ kΩ, gain = −2000 V/V; MOSFET: $g_m = 0.2$ mA/V, $R_i = \infty$, $r_o =$ 200 kΩ, gain = −40 V/V.

The gain of the MOSFET amplifier can be increased by cascoding it with a bipolar transistor Q_2, as shown in the BiCMOS circuit of Fig. 6.35(c). There are two reasons for preferring a bipolar transistor to a MOSFET for the cascode device Q_2: higher output resistance and hence higher voltage gain, and, more importantly, greater bandwidth are achieved with the bipolar transistor. The second point will be appreciated when we study the frequency response of the cascode amplifier in Chapter 7. The gain of the BiCMOS amplifier can be found by first noting that the signal current in the drain of Q_1 is $g_{m1}v_i$. An almost equal current (assuming $\alpha_2 \simeq 1$) flows in the collector of Q_2, resulting in an output voltage of $-(g_{m1}v_i)R_o$ where R_o is the output resistance at the collector of Q_2. Assuming that the bias current-source I has an infinite incremental resistance, the resistance R_o is simply that looking back into the collector of Q_2. Now Q_2 has in its emitter a resistance R_E equal to r_{o1}, and we can use Eq. (6.78) to find the output resistance by substituting $R'_E = r_{o1}//r_{\pi 2} \simeq r_{\pi 2}$. The result is

$$R_o \simeq \beta_2 r_{o2} \tag{6.142}$$

Thus the voltage gain is

$$\frac{v_o}{v_i} = -g_{m1}\beta_2 r_{o2} \tag{6.143}$$

which can be quite high.

Exercise 6.21 Using the device data given in Exercise 6.20, find the voltage gain of the BiCMOS amplifier of Fig. 6.35(c).

Ans. −10,000 V/V

Another BiCMOS amplifier stage is shown in Fig. 6.35(d). Here the emphasis is on obtaining a very high output resistance, R_o, and a correspondingly high gain, $g_{m1}R_o$. This is achieved by employing **double cascoding,** with the second level of cascoding realized using a MOSFET (Q_3). The reason a MOSFET is used is simply that a bipolar transistor would not increase the output resistance any further than the value already available at the collector of Q_2. (Why?) The MOSFET, on the other hand, increases the resistance by the factor $g_{m3}r_{o3}$ (from Eq. 6.135); thus

$$R_o = (g_{m3}r_{o3})(\beta_2 r_{o2}) \tag{6.144}$$

and the voltage gain is

$$\frac{v_o}{v_i} = -g_{m1}g_{m3}r_{o3}\beta_2 r_{o2} \tag{6.145}$$

To realize the full benefit of the extremely high output resistance of this BiCMOS amplifier, the current source I should be realized with a very-high-output-resistance circuit (e.g., a cascode or a double cascode mirror) and the output node should be buffered by a source follower.

Exercise 6.22 Using the device data given in Exercise 6.20 find R_o and the voltage gain of the double cascode BiCMOS amplifier of Fig. 6.35(d).

Ans. 2000 MΩ; -8×10^6 V/V

Current Mirrors

A double cascode BiCMOS current mirror featuring an extremely high output resistance is shown in Fig. 6.37. If we assume that the incremental resistance of each of the three diode-connected transistors (Q_4, Q_5, and Q_6) is small, the output resistance of the mirror will be the same as that found for the amplifier circuit of Fig. 6.35(d), namely that given by Eq. (6.144). Thus the most significant feature of this mirror is its extremely high output resistance.

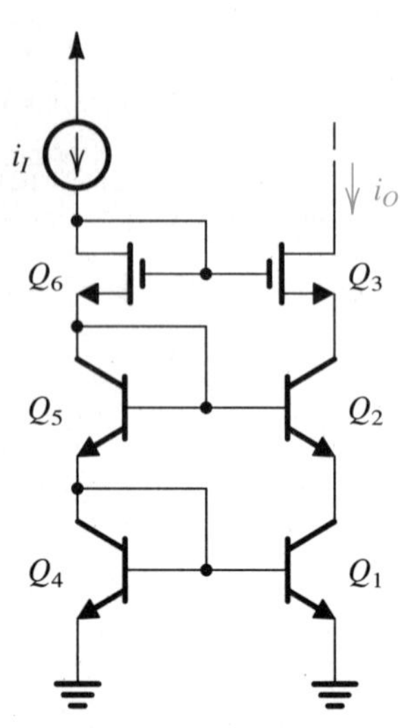

Fig. 6.37 A BiCMOS double cascode current mirror.

Differential Amplifiers

From the analysis of thc offsct voltage of the BJT differential pair in Section 6.3 and of the MOS differential pair in Section 6.7, we know that V_{OS} for the MOS amplifier (typically, 10 mV) is much greater than that for the BJT amplifier (typically, 1 mV). This is due to two factors:

1. The MOS pair suffers from the V_t mismatch, which can be as high as 5 mV. There is no corresponding effect in the BJT pair.
2. The components of V_{OS} of the MOS amplifier (other than that due to ΔV_t) are proportional to $\frac{1}{2}(V_{GS} - V_t)$, which is typically 0.2–0.5 V. The corresponding factor in the BJT amplifier is the thermal voltage V_T, which is 0.025 V at room temperature.

It follows that if a low input offset voltage is a critical requirement a BJT pair is recommended. Otherwise, a MOS differential pair can be used to advantage from the standpoint of input resistance (practically infinite) and input bias current (zero). Figure 6.38 shows the circuit of an active-loaded differential amplifier utilizing an input PMOS pair and a

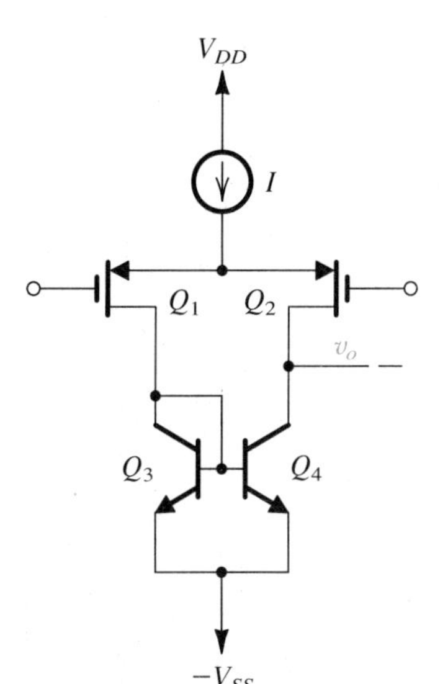

Fig. 6.38 A BiCMOS differential amplifier stage employing a PMOS differential pair. The load circuit and subsequent stages of the amplifier (not shown) utilize BJTs to obtain a wide bandwidth.

bipolar mirror for the load circuit. Such a circuit can serve as the input stage of an op amp. We will have more to say about BiCMOS op amp circuits in Chapter 10. Also, BiCMOS digital circuits will be studied in Chapter 14.

6.9 GaAs AMPLIFIERS

Gallium arsenide (GaAs) technology makes possible the design of amplifiers having very wide bandwidths, in the hundreds of megahertz range. In this section we shall study some of the circuit design techniques that have been developed over the last few years for the design of GaAs amplifiers. As will be seen, these techniques aim to circumvent the major problem of the MESFET, namely, its low output resistance in saturation. Before proceeding with this section the reader is advised to review the material on GaAs devices presented in Section 5.11.

Current Sources

Current sources play a fundamental role in the design of integrated-circuit amplifiers, being employed both for biasing and as active loads. In GaAs technology, the simplest way to implement a current source is to connect the gate of a depletion-type MESFET to its source, as shown in Fig. 6.39(a). Provided that v_{DS} is maintained greater than $|V_t|$, the MESFET will operate in saturation and the current i_D will be

$$i_D = \beta V_t^2(1 + \lambda v_{DS}) \tag{6.146}$$

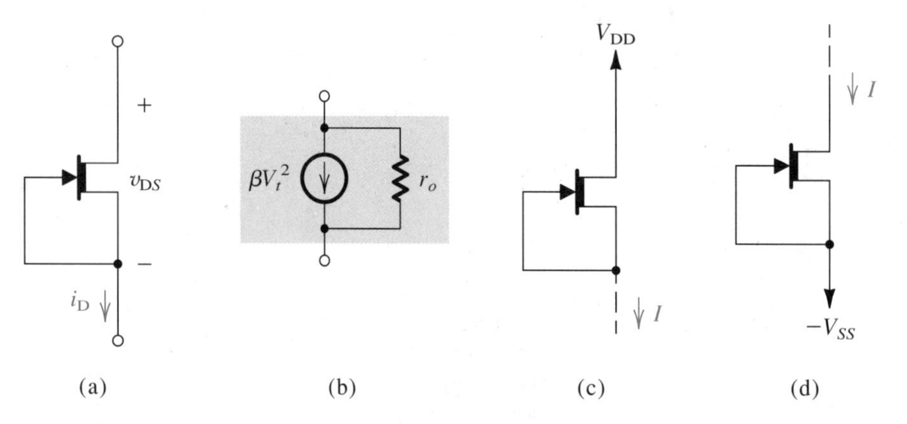

Fig. 6.39 (a) The basic MESFET current source; **(b)** equivalent circuit of the current source; **(c)** the current source connected to a positive power supply to source currents to loads at voltages $\leq V_{DD} - |V_t|$; **(d)** the current source connected to a negative power supply to sink currents from loads at voltages $\geq -V_{SS} + |V_t|$.

Thus the current source will have the equivalent circuit shown in Fig. 6.39(b), where the output resistance is the MESFET r_o,

$$r_o = 1/\lambda\beta V_t^2 \tag{6.147}$$

In JFET terminology, $\beta V_t^2 = I_{DSS}$ and $\lambda = 1/|V_A|$; thus

$$r_o = |V_A|/I_{DSS} \tag{6.148}$$

Since for the MESFET, λ is relatively high (0.1 to 0.3 V^{-1}) the output resistance of the current source of Fig. 6.39(a) is usually low, rendering this current-source realization inadequate for most applications. Before considering means for increasing the effective output resistance of the current source, we show in Fig. 6.39(c) how the basic current source can be connected to *source* currents to a load whose voltage can be as high as

$V_{DD} - |V_t|$. Alternatively, the same device can be connected as shown in Fig. 6.39(d) to *sink* currents from a load whose voltage can be as low as $-V_{SS} + |V_t|$.

Exercise 6.23 Using the device data given in Table 5.2 (page 389), find the current provided by a 10-μm-wide MESFET connected in the current-source configuration. Let the source be connected to a -5-V supply and find the current when the drain voltage is -4V. What is the output resistance of the current source? What change in current occurs if the drain voltage is raised by $+4$V?

Ans. 1.1 mA; 10 kΩ; 0.4 mA

A Cascode Current Source

The output resistance of the current source can be increased by utilizing the cascode configuration as shown in Fig. 6.40. The output resistance R_o of the cascode current source can be found by using Eq. (6.135),

$$R_o \simeq g_{m2} r_{o2} r_{o1} \tag{6.149}$$

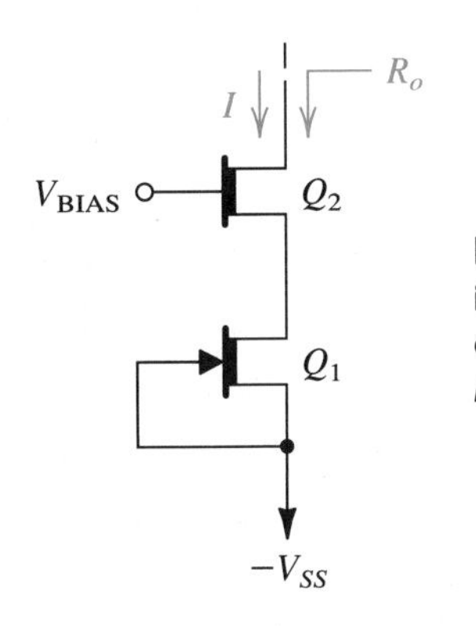

Fig. 6.40 Adding the cascode transistor Q_2 increases the output resistance of the current source by the factor $g_{m2}r_{o2}$; that is, $R_o = g_{m2}r_{o2}r_{o1}$.

Thus, adding the cascode transistor Q_2 raises the output resistance of the current source by the factor $g_{m2}r_{o2}$, which is the intrinsic voltage gain of Q_2. For GaAs MESFETs, $g_{m2}r_{o2}$ is typically 10 to 40. To allow a wide range of voltages at the output of the cascode current source, V_{BIAS} should be the lowest value that results in Q_1 operating in saturation.

Exercise D6.24 For the cascode current source of Fig. 6.40 let $V_{SS} = 5$ V, $W_1 = 10$ μm, and $W_2 = 20$ μm, and assume that the devices have the typical parameter values given in Table 5.2. (a) Find the value of V_{BIAS} that will result in Q_1 operating at the edge of the saturation region (i.e., $V_{DS1} = |V_t|$) when the voltage at the output is -3 V. (b) What is the lowest allowable voltage at the current-source output? (c) What value of output current is obtained for $V_O = -3$ V? (d) What is the output resistance of the current source? (e) What change in output current results when the output voltage is raised from -3 V to $+1$ V?

Ans. (a) -4.3 V; (b) -3.3 V; (c) 1.1 mA; (d) 310 kΩ; (e) 0.013 mA

Increasing the Output Resistance by Bootstrapping

Another technique frequently employed to increase the effective output resistance of a MESFET, including the current-source-connected MESFET, is known as bootstrapping. The bootstrapping idea is illustrated in Fig. 6.41(a). Here the circuit inside the box senses

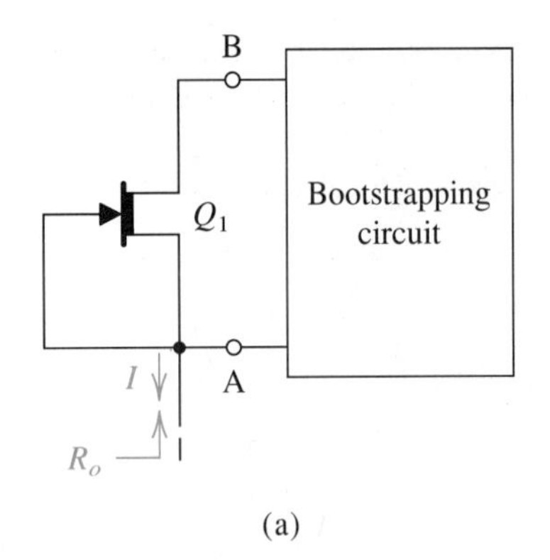

Fig. 6.41 Bootstrapping of a MESFET current source Q_1: **(a)** basic arrangement; **(b)** an implementation; **(c)** small-signal equivalent circuit model of the circuit in (b), for the purpose of determining the output resistance R_o.

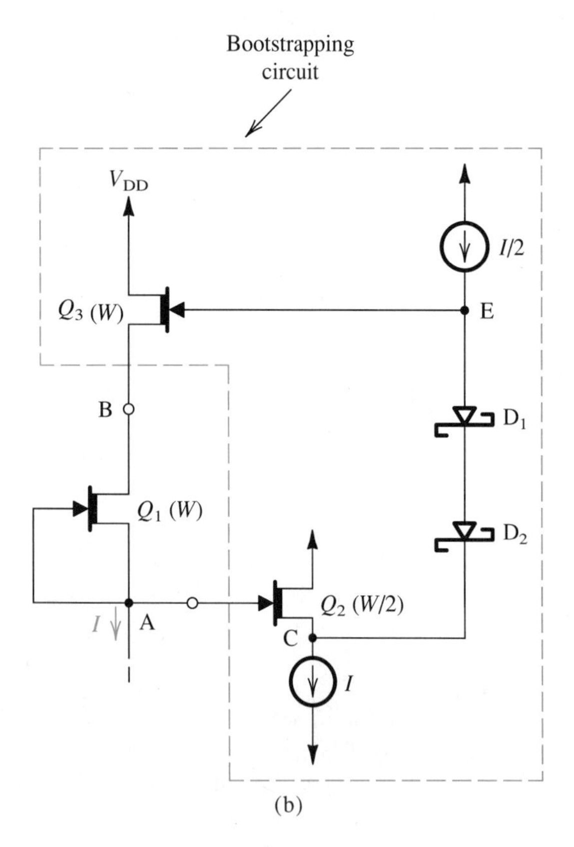

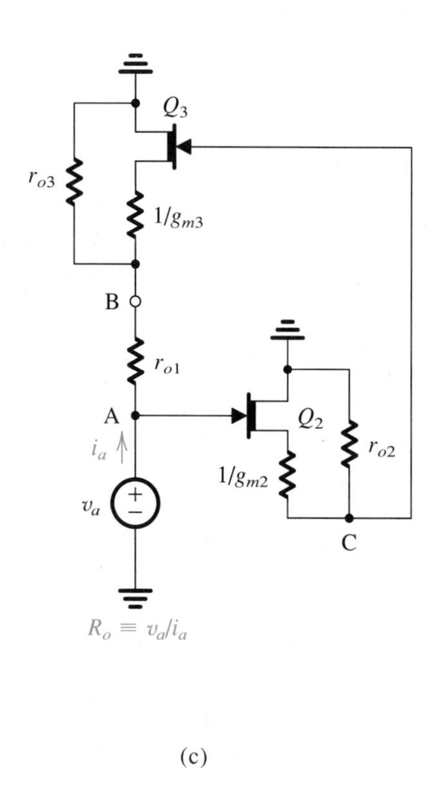

the voltage at the bottom node of the current source, v_A, and causes a voltage v_B to appear at the top node of a value

$$v_B = V_S + \alpha v_A \tag{6.150}$$

where V_S is the dc voltage required to operate the current-source transistor in saturation ($V_S \geq |V_t|$) and α is a constant ≤ 1. The incremental output resistance of the bootstrapped current source can be found by causing the voltage v_A to increase by an increment v_a. From Eq. (6.150) we find that the resulting increment in v_B is $v_b = \alpha v_a$. The incremental current through the current source is therefore $(v_a - v_b)/r_o$ or $(1 - \alpha)v_a/r_o$. Thus the output resistance R_o is

$$R_o = \frac{v_a}{(1 - \alpha)v_a/r_o} = \frac{r_o}{1 - \alpha} \tag{6.151}$$

Thus bootstrapping increases the output resistance by the factor $1/(1 - \alpha)$, which increases as α approaches unity. Perfect bootstrapping is achieved with $\alpha = 1$, resulting in $R_o = \infty$.

From the above we observe that the bootstrapping circuit senses whatever change occurs in the voltage at one terminal of the current source and causes an almost equal change to occur at the other terminal, thus maintaining an almost constant voltage across the current source and minimizing the change in current through the current-source transistor. The action of the bootstrapping circuit can be likened to that of a person who attempts to lift himself off the ground by pulling on the straps of his boots (!), the origin of the name of this circuit technique, which, incidentally, predates GaAs technology. Bootstrapping is a form of positive feedback; the signal v_b that is fed back by the bootstrapping circuit is in phase with (has the same polarity as) the signal that is being sensed, v_a. Feedback will be studied formally in Chapter 8.

An implementation of the bootstrapped current source is shown in Fig. 6.41(b). Here transistor Q_2 is a source follower used to buffer node A, whose voltage is being sensed. The width of Q_2 is half that of Q_1 and is operating at half the bias current. (Transistors Q_1 and Q_2 are said to operate at the same **current density.**) Thus V_{GS} of Q_2 will be equal to that of Q_1—namely, zero—and hence $V_C = V_A$. The two Schottky diodes behave as a battery of approximately 1.4 V, resulting in the dc voltage at node E being 1.4 V higher than V_C . Note that the signal voltage at node C appears intact at node E; only the dc level is shifted. The diodes are said to perform **level shifting,** a common application of Schottky diodes in GaAs MESFET technology.

Transistor Q_3 is a source follower that is operating at the same current density as Q_1, and thus its V_{GS} must be zero, resulting in $V_B = V_E$. The end result is that the bootstrapping circuit causes a dc voltage of 1.4 V to appear across the current-source transistor Q_1. Provided that $|V_t|$ of Q_1 is less than 1.4 V, Q_1 will be operating in saturation as required.

To determine the output resistance of the bootstrapped current source, apply an incremental voltage v_a to node A, as shown in Fig. 6.41(c). Note that this small-signal equivalent circuit is obtained by implicitly using the T model (including r_o) for each FET and assuming that the Schottky diodes act as a perfect level shifter (that is, as an ideal dc voltage of 1.4 V with zero internal resistance). Analysis of this circuit is straightforward and yields

$$\alpha \equiv \frac{v_b}{v_a} = \frac{g_{m3}r_{o3}\dfrac{g_{m2}r_{o2}}{g_{m2}r_{o2} + 1} + \dfrac{r_{o3}}{r_{o1}}}{g_{m3}r_{o3} + \dfrac{r_{o3}}{r_{o1}} + 1} \tag{6.152}$$

which is smaller than but close to unity, as required. The output resistance R_o is then obtained as

$$R_o \equiv \frac{v_a}{i_a} = \frac{r_{o1}}{1-\alpha}$$
$$= r_{o1}\frac{g_{m3}r_{o3} + (r_{o3}/r_{o1}) + 1}{g_{m3}r_{o3}/(g_{m2}r_{o2}+1) + 1} \tag{6.153}$$

For $r_{o3} = r_{o1}$, assuming that $g_{m3}r_{o3}$ and $g_{m2}r_{o2}$ are $\gg 1$, and using the relationships for g_m and r_o for Q_2 and Q_3, one can show that

$$R_o \simeq r_{o1}(g_{m3}r_{o3}/2) \tag{6.154}$$

which represents an increase of about an order of magnitude in output resistance. Unfortunately, however, the circuit is rather complex.

A Simple Cascode Configuration—The Composite Transistor

The rather low output resistance of the MESFET places a severe limitation on the performance of MESFET current sources and various MESFET amplifiers. This problem can be alleviated by using the composite MESFET configuration shown in Fig. 6.42(a) in place of a single MESFET. This circuit is unique to GaAs MESFETs and works only because of the early-saturation phenomenon observed in these devices. Recall from the discussion in Section 5.11 that early saturation refers to the fact that in a GaAs MESFET the drain current saturates at a voltage $v_{DS,\text{sat}}$ that is lower than $v_{GS} - V_t$.

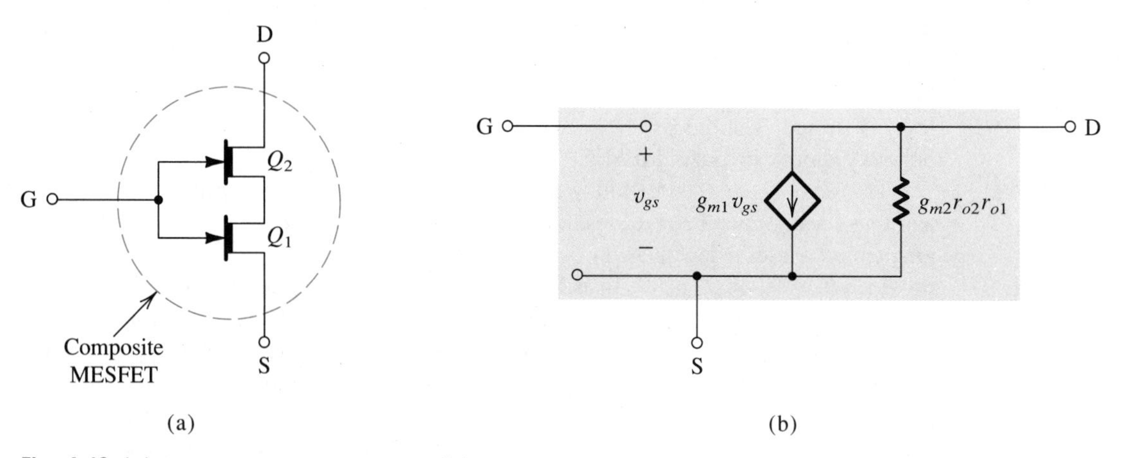

Fig. 6.42 **(a)** The composite MESFET and **(b)** its small-signal model.

In the composite MESFET of Fig. 6.42(a), Q_2 is made much wider than Q_1. It follows that since the two devices are conducting the same current, Q_2 will have a gate-to-source voltage v_{GS2} whose magnitude is much closer to $|V_t|$ than $|v_{GS1}|$ is (thus, $|v_{GS2}| \gg |v_{GS1}|$). For instance, if we use the devices whose typical parameters are given in Table 5.2 and ignore for the moment channel-length modulation ($\lambda = 0$), we find that for

$W_1 = 10\ \mu$m and $W_2 = 90\ \mu$m, at a current of 1 mA, $v_{GS1} = 0$ and $v_{GS2} = -\frac{2}{3}$ V. Now, since the drain-to-source voltage of Q_1 is $v_{DS1} = -v_{GS2} + v_{GS1}$, we see that v_{DS1} will be positive and close to but lower than $v_{GS1} - V_t$ ($\frac{2}{3}$ V in our example compared to 1 V). Thus in the absence of early saturation, Q_1 would be operating in the triode region. With early saturation, however, it has been found that saturation-mode operation is achieved for Q_1 by making Q_2 5 to 10 times wider.

The composite MESFET of Fig. 6.42(a) can be thought of as a cascode configuration, in which Q_2 is the cascode transistor, but without a separate bias line to feed the gate of the cascode transistor (as in Fig. 6.40). By replacing each of Q_1 and Q_2 with their small-signal models one can show that the composite device can be represented with the equivalent circuit model of Fig. 6.42(b). Thus while g_m of the composite device is equal to that of Q_1, the output resistance is increased by the intrinsic gain of Q_2, $g_{m2}r_{o2}$, which is typically in the range 10 to 40. This is a substantial increase and is the reason for the attractiveness of the composite MESFET.

The composite MESFET can be employed in any of the applications that can benefit from its increased output resistance. Some examples are shown in Fig. 6.43. The circuit in Fig. 6.43(a) is that of a current source with increased output resistance. Another view of

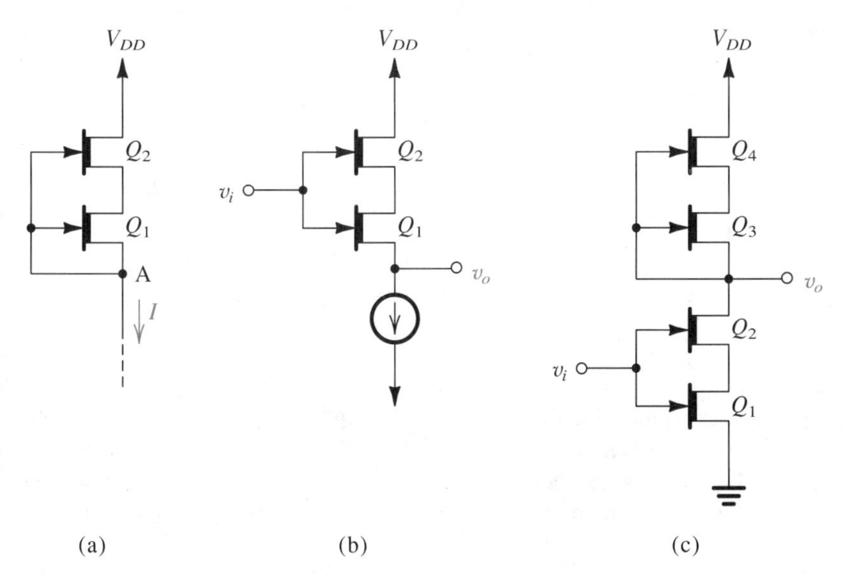

Fig. 6.43 Applications of the composite MESFET: **(a)** as a current source; **(b)** as a source follower; and **(c)** as a gain stage.

the operation of this circuit can be obtained by considering Q_2 as a source follower that causes the drain of Q_1 to follow the voltage changes at the current-source terminal (node A), thereby bootstrapping Q_1 and increasing the effective output resistance of the current source. This alternative interpretation of circuit operation has resulted in its alternative name: the **self-bootstrapped** current source.

The application of the composite MESFET as a source follower is depicted in

Fig. 6.43(b). Assuming the bias-current source I to be ideal, we can write for the gain of this follower

$$\frac{v_o}{v_i} = \frac{r_{o,eff}}{r_{o,eff} + (1/g_{m1})}$$

$$= \frac{g_{m2}r_{o2}r_{o1}}{g_{m2}r_{o2}r_{o1} + (1/g_{m1})} \tag{6.155}$$

which is much closer to the ideal value of unity than the gain of a single MESFET source follower is.

Exercise 6.25 Using the device data given in Table 5.2, contrast the voltage gain of a source follower formed using a single MESFET having $W = 10\ \mu\text{m}$ with a composite MESFET follower with $W_1 = 10\ \mu\text{m}$ and $W_2 = 90\ \mu\text{m}$. In both cases assume biasing at 1 mA and neglect λ while calculating g_m (for simplicity).

Ans. Single: 0.952 V/V; composite: 0.999 V/V

A final example of the application of the composite MESFET is shown in Fig. 6.43(c). The circuit is a gain stage utilizing a composite MESFET (Q_1, Q_2) as a driver and another composite MESFET (Q_3, Q_4) as a current-source load. The small-signal gain is given by

$$\frac{v_o}{v_i} = -g_{m1}R_o \tag{6.156}$$

where R_o is the output resistance,

$$R_o = r_{o,eff}(Q_1, Q_2)//r_{o,eff}(Q_3, Q_4)$$

$$= g_{m2}r_{o2}r_{o1}//g_{m4}r_{o4}r_{o3} \tag{6.157}$$

Differential Amplifiers

The simplest possible implementation of a differential amplifier in GaAs MESFET technology is shown in Fig. 6.44. Here Q_1 and Q_2 form the differential pair, Q_3 forms the bias current source, and Q_4 forms the active (current-source) load. The performance of the circuit is impaired by the low output resistances of Q_3 and Q_4. The voltage gain is given by

$$\frac{v_o}{v_i} = -g_{m2}(r_{o2}//r_{o4}) \tag{6.158}$$

The gain can be increased by using one of the improved current-source implementations discussed above. Also, a rather ingenious technique has been developed for enhancing the gain of the MESFET differential pair. The circuit is shown in Fig. 6.45(a). While the drain of Q_2 is loaded with a current-source load (as before), the output signal developed is fed back to the drain of Q_1 via the source follower Q_3. The small-signal analysis of the circuit is illustrated in Fig. 6.45(b) where the current sources I and $I/2$ have been assumed ideal and thus replaced with open circuits. To determine the voltage gain, we

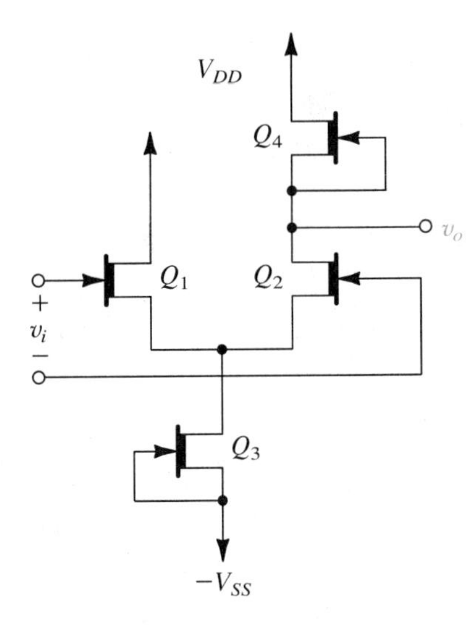

Fig. 6.44 A simple MESFET differential amplifier.

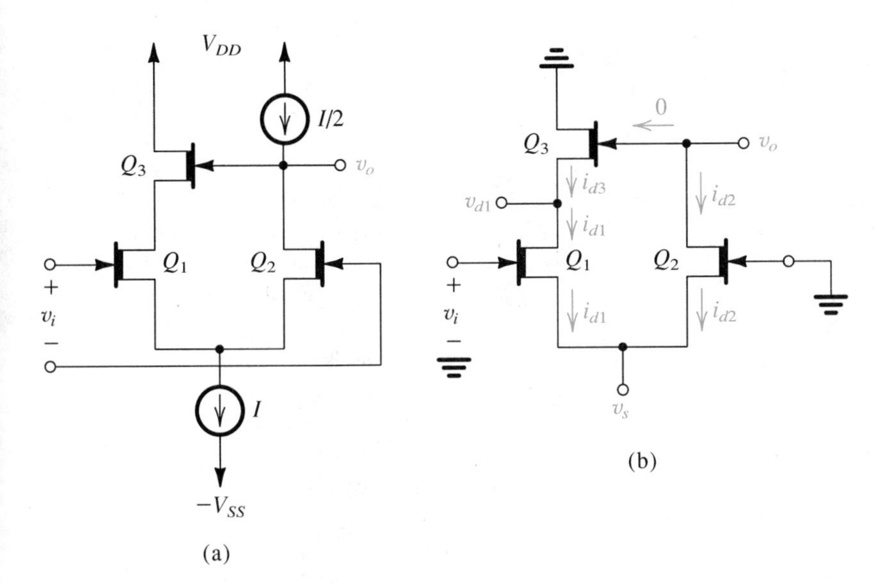

Fig. 6.45 (a) A MESFET differential amplifier whose gain is enhanced by the application of positive feedback through the source follower Q_3; **(b)** small-signal analysis of the circuit in (a).

have grounded the gate terminal of Q_2 and applied the differential input signal v_i to the gate of Q_1. The analysis proceeds along the following steps:

1. From the output node we see that $i_{d2} = 0$.
2. From the sources node, since $i_{d2} = 0$, we find that $i_{d1} = 0$.
3. From the node at the drain of Q_1, since $i_{d1} = 0$, we find that $i_{d3} = 0$.

4. Writing for each transistor

$$i_d = g_m v_{gs} + v_{ds}/r_o = 0$$

we obtain three equations in the three unknowns v_{d1}, v_s, and v_o. The solution yields

$$\frac{v_o}{v_i} = g_{m1}r_{o1} \Bigg/ \left[\frac{g_{m1}r_{o1} + 1}{g_{m2}r_{o2} + 1} - \frac{g_{m3}r_{o3}}{g_{m3}r_{o3} + 1} \right] \tag{6.159}$$

If all three transistors have the same geometry and are operating at equal dc currents, their g_m and r_o values will be equal and the expression in Eq. (6.159) reduces to

$$\frac{v_o}{v_i} \simeq (g_m r_o)^2 \tag{6.160}$$

Thus application of positive feedback through follower Q_3 enables one to obtain a gain equal to the square of that naturally available from a single stage!

Exercise 6.26 Using the device data given in Table 5.2, find the gain of the differential amplifier circuit of Fig. 6.45(a) for $I = 10$ mA and $W_1 = W_2 = W_3 = 100$ μm.

Ans. 400 V/V

6.10 MULTISTAGE AMPLIFIERS

Practical transistor amplifiers usually consist of a number of stages connected in cascade. In addition to providing gain, the first (or input) stage is usually required to provide a high input resistance in order to avoid loss of signal level when the amplifier is fed from a high-resistance source. In a differential amplifier the input stage must also provide large common-mode rejection. The function of the middle stages of an amplifier cascade is to provide the bulk of the voltage gain. In addition, the middle stages provide such other functions as the conversion of the signal from differential mode to single-ended mode and the shifting of the dc level of the signal. These two functions and others will be illustrated later in this section and in greater detail in Chapter 10.

Finally, the main function of the last (or output) stage of an amplifier is to provide a low output resistance in order to avoid loss of gain when a low-valued load resistance is connected to the amplifier. Also, the output stage should be able to supply the current required by the load in an efficient manner—that is, without dissipating an unduly large amount of power in the output transistors. We have already studied one type of amplifier configuration suitable for implementing output stages, namely, the source follower and the emitter follower. It will be shown in Chapter 9 that the source and emitter followers are not optimum from the point of view of power efficiency and that other, more appropriate circuit configurations exist for output stages required to supply large amounts of output power.

To illustrate the structure and method of analysis of multistage amplifiers, we will conclude this chapter with a detailed example. The amplifier circuit to be analyzed is shown in Fig. 6.46. The dc analysis of this simple op amp circuit was presented in Example 6.2, which we urge the reader to review before studying the following material.

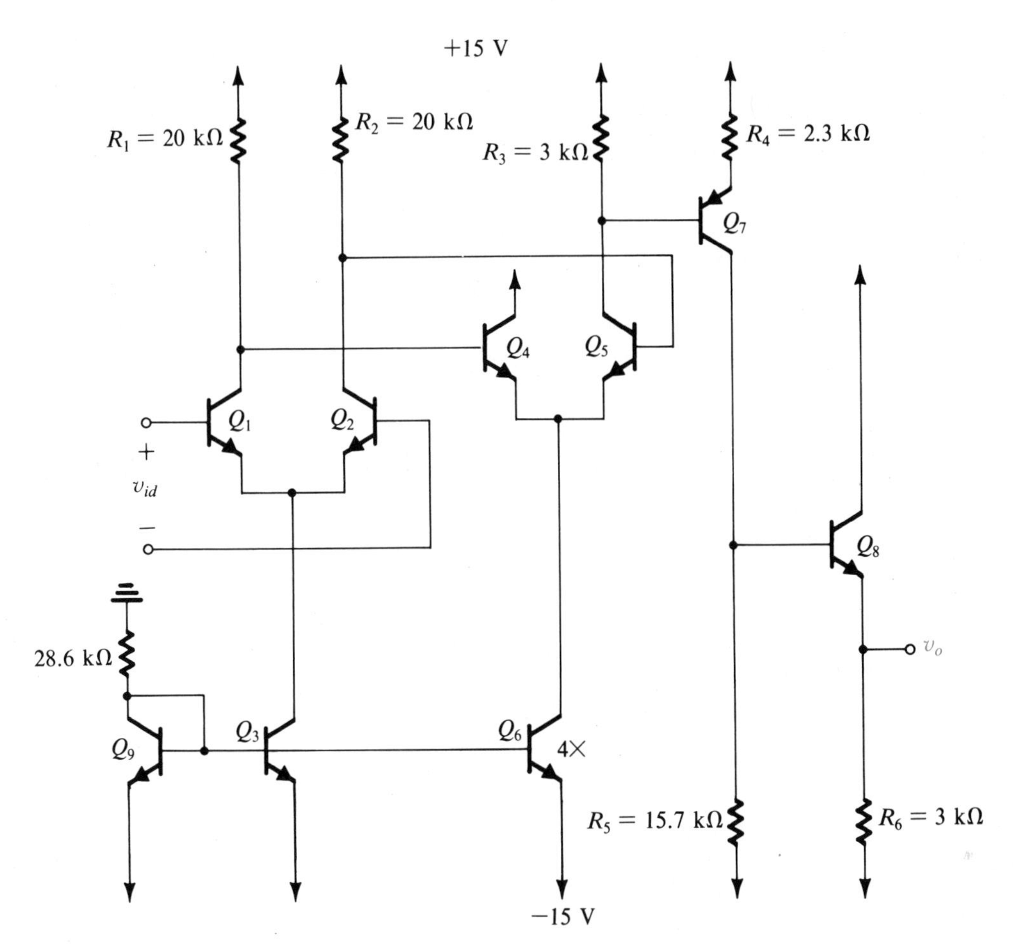

Fig. 6.46 A multistage amplifier circuit (Example 6.3).

The op amp circuit in Fig. 6.46 consists of four stages. The input stage is **differential-in, differential-out** and consists of transistors Q_1 and Q_2, which are biased by current source Q_3. The second stage is also a differential-input amplifier, but its output is taken single-ended at the collector of Q_5. This stage is formed by Q_4 and Q_5, which are biased by the current source Q_6. Note that the conversion from differential to single-ended as performed by the second stage results in a loss of gain of a factor of 2. A more elaborate method for accomplishing this conversion was studied in Sections 6.5 and 6.7; it involves using the current mirror as an active load.

In addition to providing some voltage gain, the third stage, consisting of the *pnp* transistor Q_7, provides the essential function of *shifting the dc level* of the signal. Thus while the signal at the collector of Q_5 is not allowed to swing below the voltage at the base of Q_5 (+10 V), the signal at the collector of Q_7 can swing negative (and positive, of course). From our study of op amps in Chapter 2 we know that the output terminal of the op amp should be capable of positive and negative voltage swings. Therefore every op amp circuit includes a **level-shifting** arrangement. Although the use of the complementary *pnp* transistor provides a simple solution to the level-shifting problem, other forms of level shifters exist, one of which will be discussed in Chapter 10.

Finally, we note that the output stage consists of emitter follower Q_8 and that ideally the dc level at the output is zero volts (as was calculated in Example 6.2).

EXAMPLE 6.3

Use the dc bias quantities evaluated in Example 6.2 and analyze the circuit in Fig. 6.46 to determine the input resistance, the voltage gain, and the output resistance.

Solution

The input differential resistance R_{id} is given by

$$R_{id} = r_{\pi 1} + r_{\pi 2}$$

Since Q_1 and Q_2 are each operating at an emitter current of 0.25 mA, it follows that

$$r_{e1} = r_{e2} = \frac{25}{0.25} = 100\ \Omega$$

Assume $\beta = 100$; then

$$r_{\pi 1} = r_{\pi 2} = 101 \times 100 = 10.1\ \text{k}\Omega$$

Thus $R_{id} = 20.2\ \text{k}\Omega$.

To evaluate the gain of the first stage we first find the input resistance of the second stage, R_{i2},

$$R_{i2} = r_{\pi 4} + r_{\pi 5}$$

Q_4 and Q_5 are each operating at an emitter current of 1 mA; thus

$$r_{e4} = r_{e5} = 25\ \Omega$$

$$r_{\pi 4} = r_{\pi 5} = 101 \times 25 = 2.525\ \text{k}\Omega$$

Thus $R_{i2} = 5.05\ \text{k}\Omega$. This resistance appears between the collectors of Q_1 and Q_2, as shown in Fig. 6.47. Thus the gain of the first stage will be

$$A_1 \equiv \frac{v_{o1}}{v_{id}} \simeq \frac{\text{Total resistance in collector circuit}}{\text{Total resistance in emitter circuit}}$$

$$= \frac{[R_{i2}//(R_1 + R_2)]}{r_{e1} + r_{e2}}$$

$$= \frac{(5.05\ \text{k}\Omega//40\ \text{k}\Omega)}{200\ \Omega} = 22.4\ \text{V/V}$$

Figure 6.48 shows an equivalent circuit for calculating the gain of the second stage. As indicated, the input voltage to the second stage is the output voltage of the first stage, v_{o1}. Also shown is the resistance R_{i3}, which is the input resistance of the third stage formed by Q_7. The value of R_{i3} can be found by multiplying the total resistance in the emitter of Q_7 by $\beta + 1$:

$$R_{i3} = (\beta + 1)(R_4 + r_{e7})$$

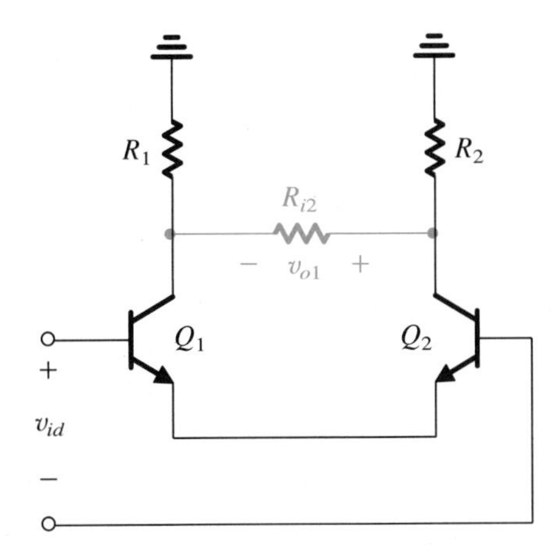

Fig. 6.47 Equivalent circuit for calculating the gain of the input stage of the amplifier in Fig. 6.46.

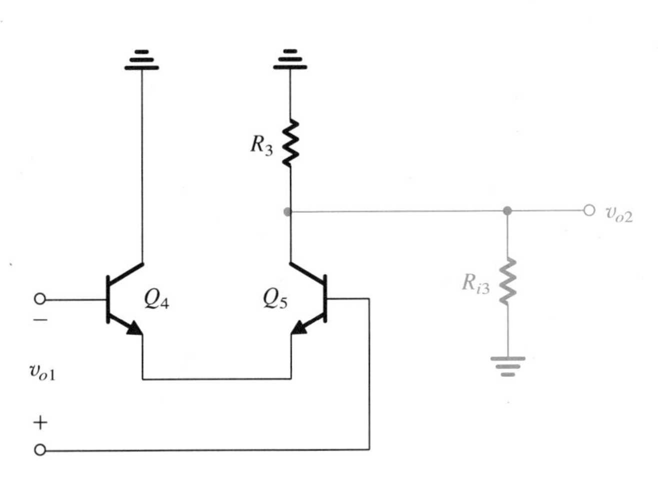

Fig. 6.48 Equivalent circuit for calculating the gain of the second stage of the amplifier in Fig. 6.46.

Since Q_7 is operating at an emitter current of 1 mA,

$$r_{e7} = \frac{25}{1} = 25\ \Omega$$

$$R_{i3} = 101 \times 2.325 = 234.8\ \text{k}\Omega$$

We can now find the gain A_2 of the second stage as the ratio of the total resistance in the collector circuit to the total resistance in the emitter circuit:

$$A_2 \equiv \frac{v_{o2}}{v_{o1}} \simeq -\frac{(R_3//R_{i3})}{r_{e4} + r_{e5}}$$

$$= -\frac{(3\ \text{k}\Omega//234.8\ \text{k}\Omega)}{50\ \Omega} = -59.2\ \text{V/V}$$

To obtain the gain of the third stage we refer to the equivalent circuit shown in Fig. 6.49, where R_{i4} is the input resistance of the output stage formed by Q_8. Using the

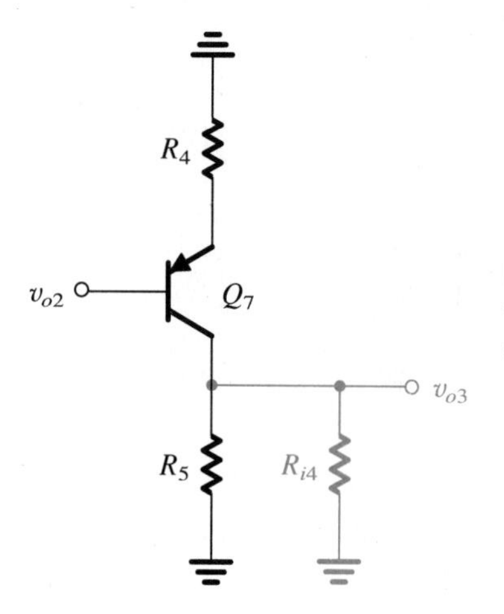

Fig. 6.49 Equivalent circuit for evaluating the gain of the third stage in the amplifier circuit of Fig. 6.46.

resistance-reflection rule we calculate the value of R_{i4} as

$$R_{i4} = (\beta + 1)(r_{e8} + R_6)$$

where

$$r_{e8} = \frac{25}{5} = 5\ \Omega$$

$$R_{i4} = 101(5 + 3000) = 303.5\ \text{k}\Omega$$

The gain of the third stage is given by

$$A_3 \equiv \frac{v_{o3}}{v_{o2}} \simeq -\frac{(R_5 // R_{i4})}{r_{e7} + R_4}$$

$$= -\frac{(15.7\ \text{k}\Omega // 303.5\ \text{k}\Omega)}{2.325\ \text{k}\Omega} = -6.42\ \text{V/V}$$

Finally, to obtain the gain A_4 of the output stage we refer to the equivalent circuit in Fig. 6.50 and write

$$A_4 \equiv \frac{v_o}{v_{o3}} = \frac{R_6}{R_6 + r_{e8}}$$

$$= \frac{3000}{3000 + 5} = 0.998 \simeq 1$$

The overall voltage gain of the amplifier can then be obtained as follows:

$$\frac{v_o}{v_{id}} = A_1 A_2 A_3 A_4 = 8513\ \text{V/V}$$

or 78.6 dB.

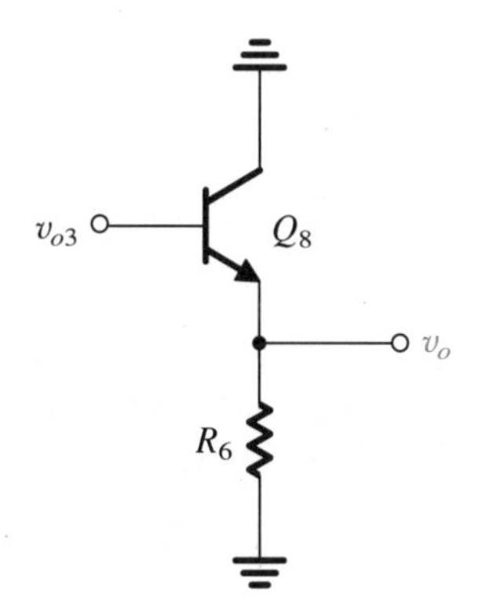

Fig. 6.50 The output stage of the amplifier circuit of Fig. 6.46.

To obtain the output resistance R_o we ''grab hold'' of the output terminal in Fig. 6.46 and look back into the circuit. By inspection we find

$$R_o = \{R_6 // [r_{e8} + R_5/(\beta + 1)]\}$$

which gives $R_o = 152\ \Omega$.

Exercise 6.27 Use the results of Example 6.3 to calculate the overall voltage gain of the amplifier in Fig. 6.46 when it is connected to a source having a resistance of 10 kΩ and a load of 1 kΩ.

Ans. 4943 V/V

SUMMARY

- The differential pair is the most important building block in analog IC design. The input stage of every op amp is a differential amplifier.
- Differential amplifiers are implemented using BJTs, JFETs, MOSFETs, or GaAs MESFETs. FETs offer the advantage of a very small input bias current and very high input resistance. BJTs provide smaller input offset voltages and wider bandwidth. GaAs provides the widest bandwidth.
- The differential amplifier has high differential gain and low common-mode gain; the ratio of the two is the CMRR, usually expressed in decibels.
- In the BJT differential pair, a differential input voltage of about 100 mV is sufficient to steer the bias current to one side of the pair.
- For differential input signals, operation of the differential amplifier can be analyzed using the differential half-circuit. Common-mode gain and input resistance can be obtained from the common-mode half-circuit.
- To obtain low common-mode gain, and thus high CMRR, the bias current source must be designed to have a high output resistance.
- High CMRR is achieved by taking the output differentially and ensuring a high degree of matching between the two sides of the differential amplifier.
- Mismatches between the two sides of the differential amplifier give rise to an input offset voltage and, in BJT amplifiers, an input offset current.
- Biasing in analog ICs is based on using constant-current sources. The basic BJT current-mirror circuit suffers from the dependence on transistor β and has a relatively low output resistance. The β-dependence can be minimized and the output resistance increased by using more elaborate mirror circuits. No β-dependence problems exist in MOS mirrors.
- The differential amplifier utilizing a current-mirror active load is a popular circuit for both bipolar and MOS analog IC design.

▶ The cascode configuration consists of a common-emitter (or a common-source) transistor followed by a common-base (or a common-gate) transistor. It features increased output resistance and wider bandwidth (Chapter 7).

▶ BiCMOS technology enables the designer to combine the high g_m, the high-frequency capability, and the ease of device-matching of bipolar transistors with the infinite input resistance and excellent switching characteristics of MOS transistors.

▶ GaAs MESFET technology offers very high frequencies of operation. The major challenge in designing GaAs amplifiers is to minimize the effects of their low output resistance. Bootstrapping and cascoding are two techniques frequently employed for this purpose.

▶ A multistage amplifier usually consists of an input stage having high input resistance and, if differential, high CMRR, one or more intermediate stages to realize the bulk of the gain, and an output stage having low output resistance. In analyzing a multistage amplifier, the loading effect of each stage on the one that precedes it must be taken into account.

BIBLIOGRAPHY

A. A. Abidi, ''An analysis of bootstrapped gain enhancement techniques,'' *IEEE Journal of Solid-State Circuits,* vol. SC-22, No. 6, pp. 1200–1204, December 1987.

J. N. Giles, *Linear Integrated Circuits Applications Handbook,* Mountain View, Calif.: Fairchild Semiconductors, 1967.

P. R. Gray and R. G. Meyer, *Analysis and Design of Analog Integrated Circuits,* 2nd ed., New York: Wiley, 1984.

A. B. Grebene, *Bipolar and MOS Analog Integrated Circuit Design,* New York: Wiley, 1984.

D. J. Hamilton and W. G. Howard, *Basic Integrated Circuit Engineering,* New York: McGraw-Hill, 1975.

L. E. Larson, K. W. Martin, and G. C. Temes, ''GaAs switched-capacitor circuits for high-speed signal processing,'' *IEEE Journal of Solid State Circuits,* vol. SC-22, No. 6, pp. 971–981, December 1987.

H. S. Lee, ''Analog design,'' Chapter 8 in *BiCMOS Technology and Applications,* A. R. Alvarez, editor, Boston, Mass.: Kluwer Academic Publishers, 1989.

K. W. Martin, ''Ultra-high-speed GaAs analog circuits, systems, and design methodologies,'' Internal Report, Univ. of California, Los Angeles, February 1990.

J. K. Roberge, *Operational Amplifiers: Theory and Practice,* New York: Wiley, 1975.

C. Tamazou and D. Haigh, ''Gallium arsenide analogue integrated circuit design techniques,'' Chapter 8 in *Analog IC Design: the Current-Mode Approach,* C. Toumazou, F. J. Lidgey, and D. G. Haigh, editors, London: Peter Peregrinus Ltd., 1990.

PROBLEMS

Section 6.1: The BJT Differential Pair

6.1 For the differential amplifier of Fig. 6.2(a) let $I = 2$ mA, $V_{CC} = 5$ V, $v_{CM} = -2$ V, $R_C = 2$ kΩ, and $\beta = 100$. Assume that the BJTs have $v_{BE} = 0.7$ V at $i_C = 1$ mA. Find the voltage at the emitters and at the outputs.

6.2 For the circuit of Fig. 6.2(b) with an input of +1 V as indicated, and with $I = 2$ mA, $V_{CC} = 5$ V, $R_C = 2$ kΩ, and $\beta = 100$, find the voltage at the emitters and the collector voltages. Assume that the BJTs have $v_{BE} = 0.7$ V at $i_C = 1$ mA.

6.3 Repeat Exercise 6.1 (page 410) for an input of −0.2 V.

6.4 For the BJT differential amplifier of Fig. 6.1 find the value of the input differential signal, $v_{B1} - v_{B2}$, that causes $i_{E1} = 0.90I$.

D6.5 Consider the differential amplifier of Fig. 6.1 and let the BJT β be very large:

(a) What is the largest input common-mode signal that can be applied while the BJTs remain in the active region?

(b) If an input difference signal is applied that is large enough to steer the current entirely to one side of the pair, what is the change in voltage at each collector (from the condition for which $v_d = 0$)?

(c) If the available power supply V_{CC} is 5 V, what

value of IR_C should you choose in order to allow a common-mode input signal of ±3 V?

(d) For the value of IR_C found in (c), select values for I and R_C. Use the largest possible value for I subject to the constraint that the base current of each transistor (when I divides equally) should not exceed 2 μA. Let $\beta = 100$.

6.6 To provide insight into the possibility of nonlinear distortion resulting from large differential input signals applied to the differential amplifier of Fig. 6.1, evaluate the ratio i_{E1}/I for differential input signals of 5, 10, 20, 30, and 40 mV.

*6.7 For the circuit in Fig. P6.7 in which the transistors have high β, what value would you expect for v_2? If the resistor R_1 is reduced to 2.5 kΩ, what does v_2 become?

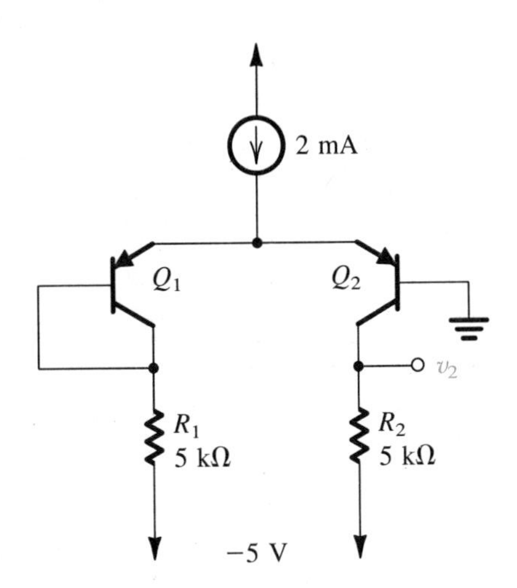

Fig. P6.7

D6.8 Design the circuit of Fig. 6.1 to provide a differential output voltage (that is between the two collectors) of 1 V when the differential input signal is 10 mV. A current source of 1 mA and a positive supply of +10 V are available. What is the largest possible input common-mode voltage for which operation is as required? Assume $\alpha \simeq 1$.

6.9 In a differential amplifier using a 6-mA emitter bias current source the two BJTs are not matched. Rather, one has twice the emitter junction area of the other. For a differential input signal of zero volts, what do the collector currents become? What difference input is needed to equalize the collector currents? Assume $\alpha = 1$.

Section 6.2: Small-Signal Operation of the BJT Differential Amplifier

6.10 For the circuit in Fig. 6.4 let $I = 1$ mA, $V_{CC} = 15$ V, $R_C = 10$ kΩ, and $\alpha = 1$, and let the input voltages be: $v_{B1} = 5 + 0.005 \sin 2\pi \times 1000\,t$, volts, and $v_{B2} = 5 - 0.005 \sin 2\pi \times 1000\,t$, volts.

(a) If the BJTs are specified to have v_{BE} of 0.7 V at a collector current of 1 mA, find the voltage at the emitters. (*Hint:* Observe the symmetry of the circuit.)

(b) Find g_m for each of the two transistors.

(c) Find i_C for each of the two transistors.

(d) Find v_C for each of the two transistors.

(e) Find the voltage between the two collectors.

(f) Find the gain experienced by the 1000-Hz signal.

6.11 A BJT differential amplifier uses a 100-μA bias current. What is the value of g_m of each device? If β is 200, what is the differential input resistance?

D6.12 Design the basic BJT differential amplifier circuit of Fig. 6.4 to provide a differential input resistance of at least 10 kΩ and a differential voltage gain (with the output taken between the two collectors) of 200 V/V. The transistor β is specified to be at least 100. The available power supply is 10 V.

6.13 A BJT differential amplifier is biased from a 2-mA constant-current source and includes a 100-Ω resistor in each emitter. The collectors are connected to V_{CC} via 5-kΩ resistors. A differential input signal of 0.1 V is applied between the two bases.

(a) Find the signal current in the emitters (i_e) and the signal voltage v_{be} for each BJT.

(b) What is the total emitter current in each BJT?

(c) What is the signal voltage at each collector? Assume $\alpha = 1$.

(d) What is the voltage gain realized when the output is taken between the two collectors.

D6.14 Design a BJT differential amplifier to amplify a differential input signal of 0.2 V and provide a differential output signal of 4 V. To ensure adequate linearity, it is required to limit the signal amplitude across each base–emitter junction to a maximum of 5 mV. Another design requirement is that the differential input resistance be at least 80 kΩ. The BJTs available are specified to have $\beta \geq 200$. Give the circuit configuration and specify the values of all its components.

6.15 Find the voltage gain and the input resistance of the amplifier shown in Fig. P6.15 assuming $\beta = 100$.

6.16 Find the voltage gain and input resistance of the amplifier in Fig. P6.16 assuming that $\beta = 100$.

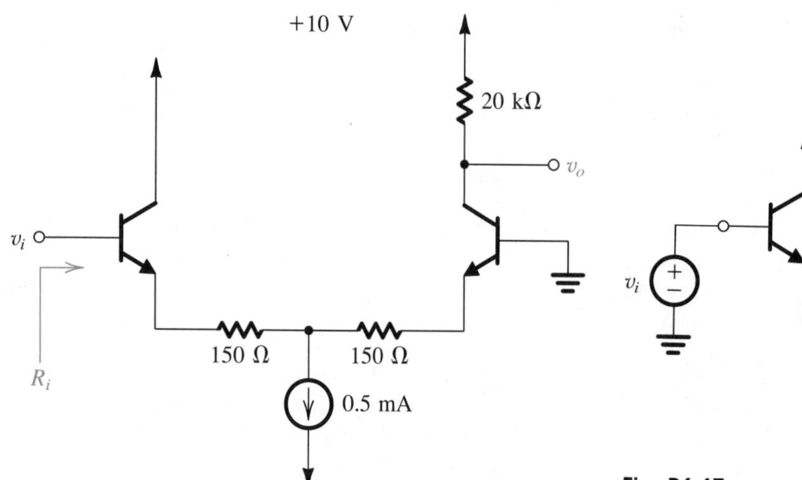

Fig. P6.15

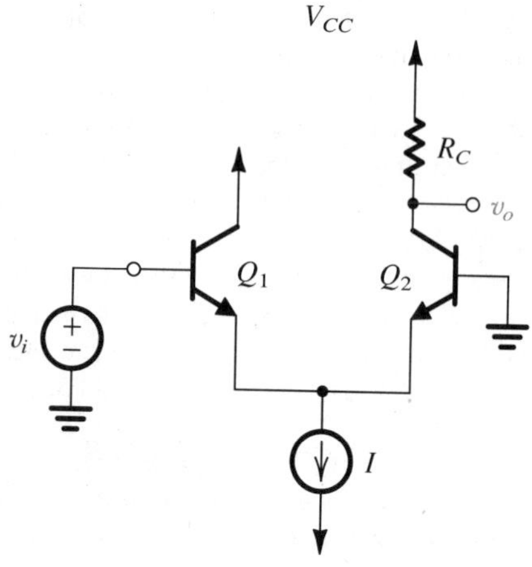

Fig. P6.17

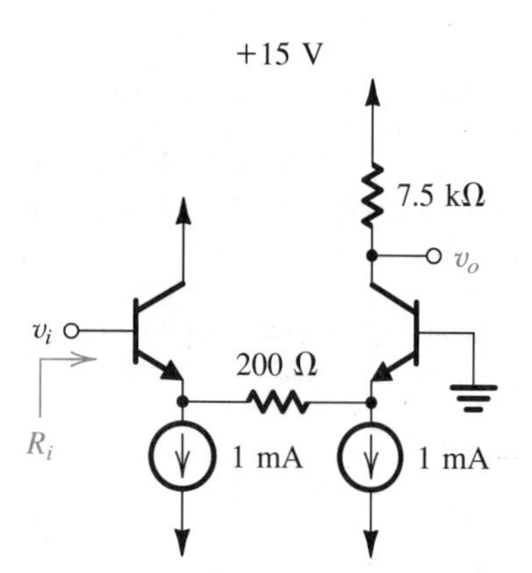

Fig. P6.16

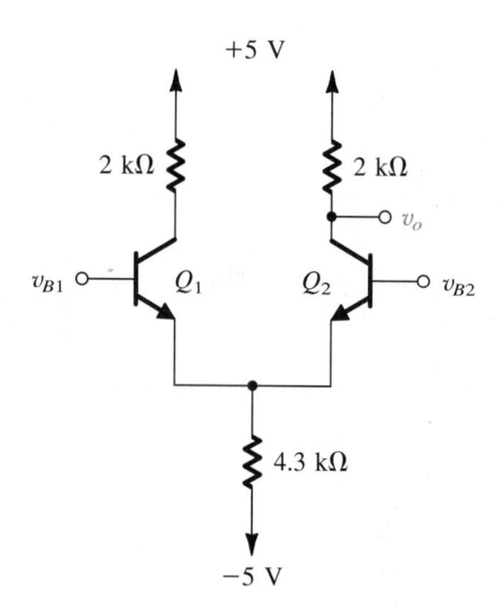

Fig. P6.18

6.17 Derive an expression for the small-signal voltage gain v_o/v_i of the circuit shown in Fig. P6.17 in two different ways:

(a) as a differential amplifier,

(b) as a cascade of a common-collector stage Q_1 and a common-base stage Q_2.

Assume that the BJTs are matched and have a current gain α. Verify that both approaches lead to the same result.

6.18 The differential amplifier circuit of Fig. P6.18 utilizes a resistor connected to the negative power supply to establish the bias current I.

(a) For $v_{B1} = v_d/2$ and $v_{B2} = -v_d/2$, where v_d is a small signal with zero average, find the magnitude of the differential gain, $|v_o/v_d|$.

(b) For $v_{B1} = v_{B2} = v_{CM}$, find the magnitude of the common-mode gain, $|v_o/v_{CM}|$.

(c) Calculate the CMRR.

(d) If $v_{B1} = 0.1 \sin 2\pi \times 60t + 0.005 \sin 2\pi \times 1000t$, volts, $v_{B2} = 0.1 \sin 2\pi \times 60t - 0.005 \sin 2\pi \times 1000t$, volts, find v_o.

***D6.19** Design a BJT differential amplifier that provides two single-ended outputs (at the collectors). The amplifier is to have a differential gain (to each of the two outputs) of at least 100 V/V, a differential input resistance $\geq$10 kΩ, and a common-mode gain (to each of the two outputs) no greater than 0.1 V/V. Use a 2-mA current source for biasing. Give the complete circuit with component values and suitable power supplies that allow for ± 2 V swing at each collector. Specify the minimum value that the output resistance of the bias current source must have. The BJTs available have $\beta \geq 100$. What is the value of the input common-mode resistance when the bias source has the lowest acceptable resistance?

6.20 A BJT differential amplifier with 12-kΩ collector resistances is found to have a common-mode gain for single-ended outputs of -50 dB. Find the value of the output resistance of the bias current source. Assume $\alpha \simeq 1$.

6.21 When the output of a BJT differential amplifier is taken differentially, its CMRR is found to be 40 dB higher than when the output is taken single-ended. If the only source of common-mode gain when the output is taken differentially is the mismatch in collector resistances, what must this mismatch be (in percent)?

*6.22 In a particular BJT differential amplifier a production error results in one of the transistors having an emitter–base junction area that is twice that of the other. With the inputs grounded, how will the emitter bias current split between the two transistors? If the output resistance of the current source is 1 MΩ and the resistance in each collector (R_C) is 12 kΩ, find the common-mode gain obtained when the output is taken differentially. Assume $\alpha \simeq 1$.

6.23 A BJT differential amplifier utilizes a 0.2-mA bias current from a source for which the output resistance is 2 MΩ. The amplifier transistors have $\beta = 200$, $V_A = 100$ V, and $r_\mu = 10\,\beta r_o$. What is the common-mode input resistance?

Section 6.3: Other Nonideal Characteristics of the Differential Amplifier

6.24 A differential amplifier using a 300-μA emitter bias source uses two well matched transistors but collector load resistors that are mismatched by 8%. What input offset voltage is required to reduce the differential output voltage to zero?

6.25 A differential amplifier using a 300-μA emitter bias source uses two transistors whose scale currents I_S differ by 10%. If the two collector resistors are well matched, find the resulting input offset voltage.

6.26 Modify Eq. (6.49) for the case of a differential amplifier having a resistance R_E connected in the emitter of each transistor. Let the bias current source be I.

6.27 A differential amplifier uses two transistors whose β values are 100 and 200. If everything else is matched find the input offset voltage.

*6.28 A differential amplifier uses two transistors having V_A values of 100 V and 300 V. If everything else is matched, find the resulting input offset voltage. Assume that the two transistors are intended to be biased at a V_{CE} of about 10 V.

6.29 A differential amplifier for which the total emitter bias current is 200 μA uses transistors for which β is specified to lie between 75 and 300. What is the largest possible input bias current? the smallest possible input bias current? the largest possible input offset current?

*6.30 A BJT differential amplifier, operating at a bias current of 500 μA, employs collector resistors of 27 kΩ (each) connected to a $+15$-V supply. The emitter current source employs a BJT whose emitter voltage is -5 V. What are the positive and negative limits of the input common-mode range of the amplifier for signals of $\leq$20-mV peak amplitude, applied differentially?

**6.31 In a particular BJT differential amplifier a production error results in one of the transistors having an emitter–base junction area twice that of the other. With both inputs grounded, find the current in each of the two transistors and hence the dc offset voltage at the output, assuming that the collector resistances are equal. Use small-signal analysis to find the input voltage that would restore current balance to the differential pair. Repeat using large-signal analysis and compare results. Also find the input bias and offset currents assuming $I = 0.1$ mA and $\beta_1 = \beta_2 = 100$.

Section 6.4: Biasing in BJT Integrated Circuits

6.32 The circuit in Fig. P6.32 provides a constant current I_O as long as the circuit to which the collector is connected maintains the BJT in the active mode. Show that

$$I_O \equiv \alpha \frac{V_{CC}[R_2/(R_1 + R_2)] - V_{BE}}{R_E + (R_1 // R_2)/(\beta + 1)}$$

*D6.33 For the circuit in Fig. P6.33, assuming all transistors to be identical with β infinite, derive an expression for the output current I_O, and show that by selecting

$$R_1 = R_2$$

and keeping the current in each junction the same, the current I_O will be

$$I_O = \frac{\alpha V_{CC}}{2R_E}$$

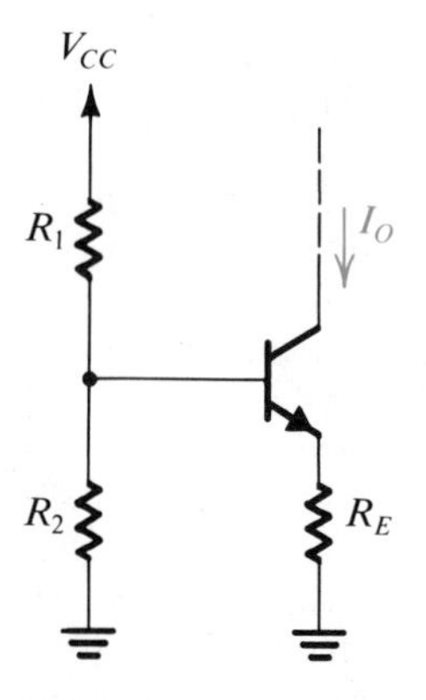

Fig. P6.32

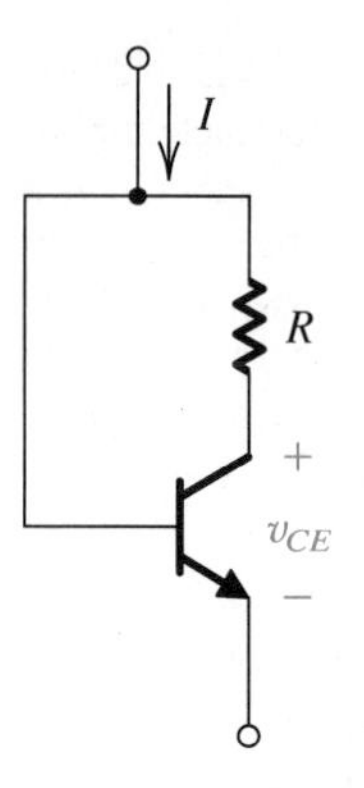

Fig. P6.34

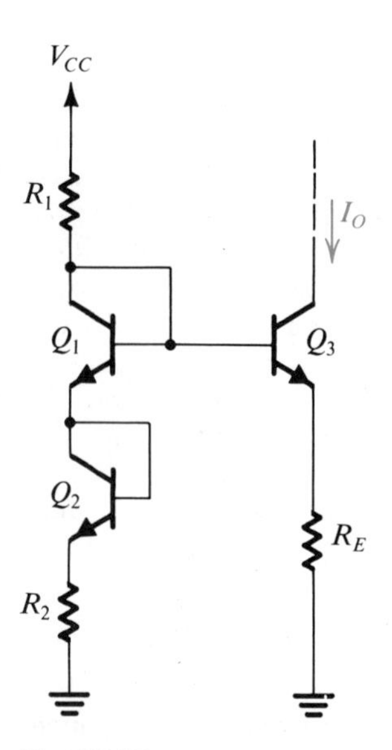

Fig. P6.33

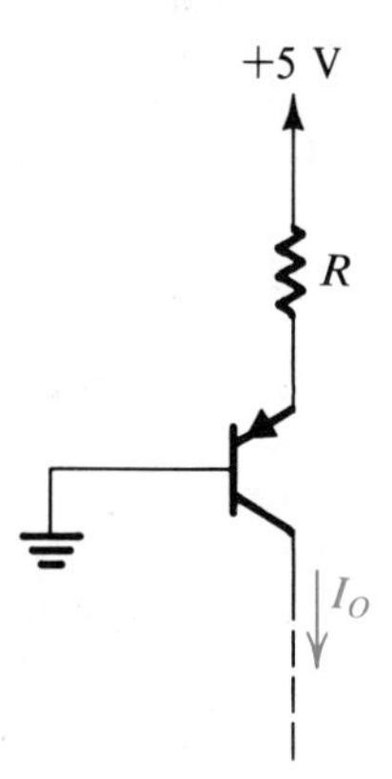

Fig. P6.35

which is independent of V_{BE}. What must the relationship of R_E to R_1 and R_2 be? For $V_{CC} = 15$ V, and assuming $\alpha \simeq 1$ and $V_{BE} = 0.7$ V, design the circuit to obtain an output current of 1 mA. What is the lowest voltage that can be applied to the collector of Q_3?

*6.34 For the circuit in Fig. P6.34 show that by selecting $R = 1/g_m$, v_{CE} is kept constant for small changes in I.

D6.35 For the circuit in Fig. P6.35 find the value of R that will result in $I_O \simeq 1$ mA. What is the largest voltage that can be applied to the collector? Assume $|V_{BE}| =$ 0.7 V.

6.36 The transistors Q_1, Q_2, and Q_3 in the circuit of Fig. P6.36 have emitter–base junction areas in the ratio of 1:2:3, respectively.

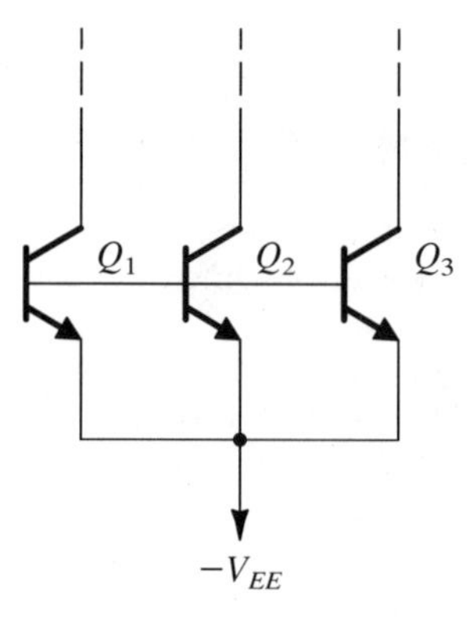

Fig. P6.36

(a) If Q_1 is diode-connected and fed with a 1-mA current source, what currents result in the collectors of Q_2 and Q_3, assuming that the collector voltages are such that active-mode operation is maintained?

(b) Repeat with Q_2 diode-connected and fed with a 1-mA current source.

(c) Repeat with Q_3 diode-connected and fed with a 1-mA current source.

D6.37 Give the circuit for the *pnp* version of the basic current-mirror circuit of Fig. 6.14. If β of the *pnp* transistor is 20, what is the current gain (or transfer ratio) I_O/I_{REF}, neglecting the Early effect?

*6.38 Consider the basic current-mirror circuit of Fig. 6.14 in the case $V_{EE} = 0$ V, and let the BJTs have $\beta = 100$ and $V_A = 100$ V.

(a) Show that the circuit of Fig. P6.38 models the output of the mirror for $V_O \geq 0.7$ V (the active region for Q_2).

(b) Find I_O for $I_{REF} = 1$ mA and $V_O =$ (i) 1 V, (ii) 5 V, (iii) 15 V.

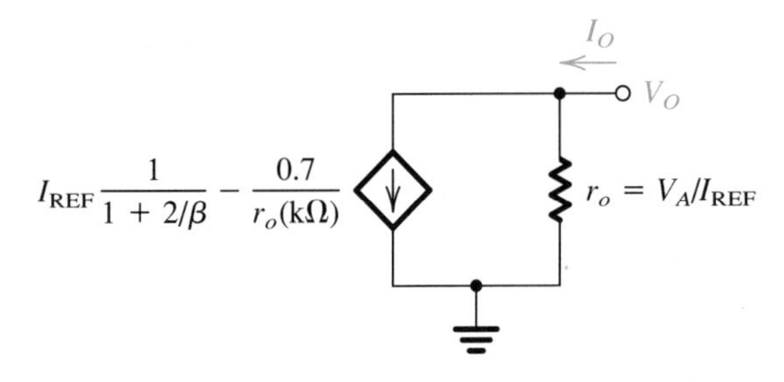

Fig. P6.38

D6.39 Consider the basic current mirror of Fig. 6.14 for the case in which the EBJ of Q_2 has n times the area of that of Q_1. Derive an expression for the current transfer ratio I_O/I_{REF}, assuming finite β but neglecting the Early effect. What is the minimum β that will ensure that the output current be within 5% of its nominal value of nI_{REF}?

D6.40 Assuming a large β and a transistor v_{BE} of 0.7 V at a current of 1 mA, design the circuit of Fig. 6.16 to provide an output current of 2 mA. The power supply available is 15 V. If V_A of the BJT is 100 V, what is the percentage change in I_O obtained as V_O is varied through the range 1 to 15 V?

D6.41 Utilizing a +5-V power supply, design a *pnp* version of the simple current source of Fig. 6.16. The current source is to supply a current of 0.5 mA. Assume that the *pnp* transistors available have high β and a v_{BE} of 0.7 V at $i_E = 1$ mA. What is the maximum possible value of load resistance to which this current source can be connected? (The other terminal of the load is to be grounded.)

D6.42 Using a simple extension of the topology of Fig. 6.16 with a 5-V supply and transistors for which $v_{BE} =$ 0.7 V at 1 mA, design a 3-mA current source using a 1-mA reference current. Assume β to be large.

*6.43 Find the voltages at all nodes and the currents through all branches in the circuit of Fig. P6.43. Assume $|V_{BE}| = 0.7$ V and $\beta = \infty$.

6.44 For the circuit in Fig. P6.44 let $|V_{BE}| = 0.7$ V and $\beta = \infty$. Find I, V_1, V_3, V_4, and V_5 for (a) $R = 10$ kΩ and (b) $R = 100$ kΩ.

D6.45 Using the ideas embodied in Fig. 6.17, design a multiple-mirror circuit using power supplies of ±10 V, to create source currents of 1, 2, and 4 mA and sink currents of 0.5, 1, and 2 mA. Assume that the BJTs have $V_{BE} \simeq 0.7$ V and large β.

*6.46 The circuit shown in Fig. P6.46 is known as a **current conveyor.**

(a) Assuming that Y is connected to a voltage V and that a current I is forced into X, show that a current equal to I flows through terminal Y, that a voltage equal to V appears at terminal X, and that a current equal to I flows through terminal Z. Assume β to be large.

(b) With Y connected to ground, show that a virtual ground appears at X. If X is connected to a +5-V supply through a 10-kΩ resistance, what current flows through Z?

D6.47 Extend the current-mirror circuit of Fig. 6.18 to n outputs. What is the resulting current transfer ratio from the input to each output, I_O/I_{REF}. If the deviation from unity is to be kept ≤0.1%, what is the maximum possible number of outputs for BJTs with $\beta = 100$?

D6.48 Use the *pnp* version of the Wilson current mirror to design a 0.1-mA current source. The current source is required to operate with the voltage at its output terminal as low as −5 V. If the power supplies available are ±5 V, what is the highest voltage possible at the output terminal?

**6.49 For the Wilson current mirror of Fig. 6.19, replace the diode-connected transistor Q_1 with its incremental resistance r_e, and replace Q_2 and Q_3 with their low-frequency hybrid-π models including r_o but excluding r_μ (for simplicity). Note that all three transistors operate at equal dc currents and thus have identical model parameters. Apply a test voltage v_x between the collector of Q_3 and ground and determine the current i_x

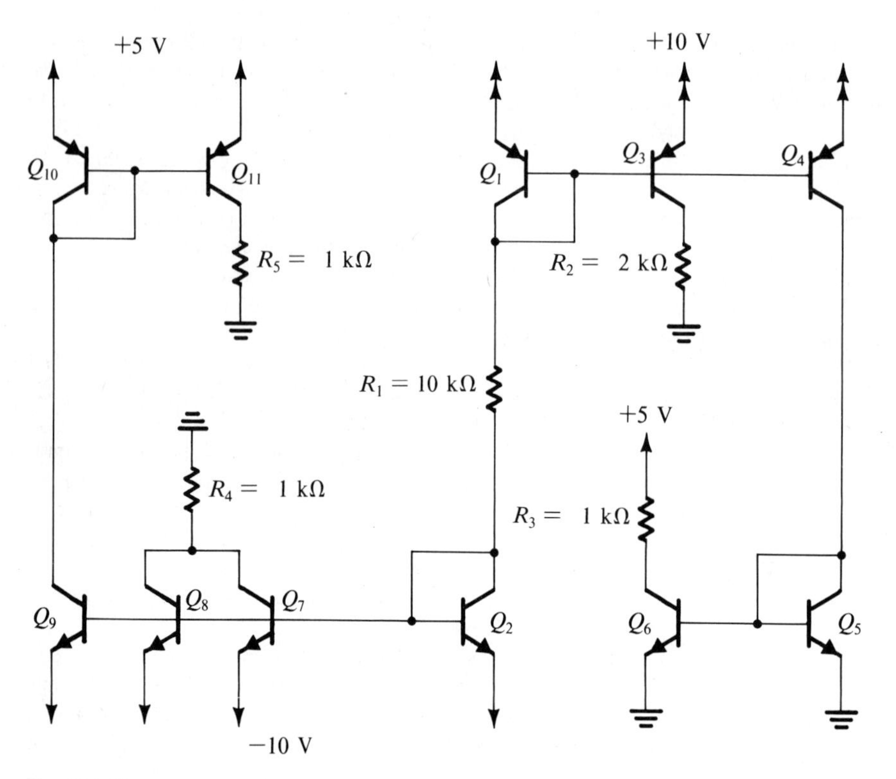

Fig. P6.43

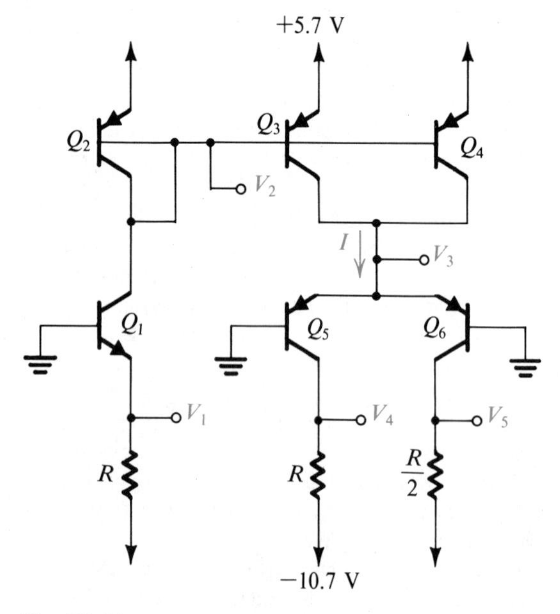

Fig. P6.44

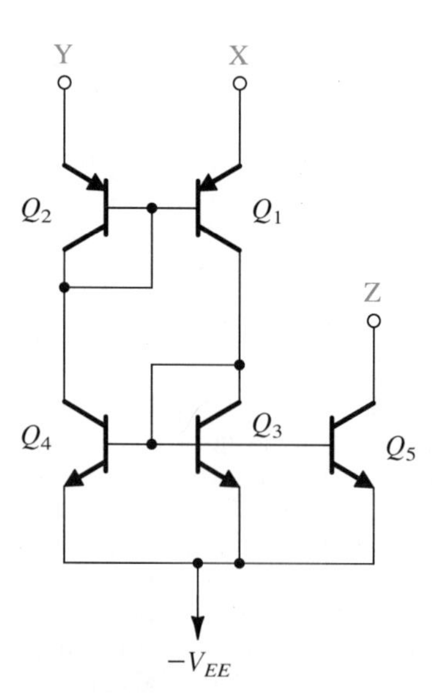

Fig. P6.46

drawn from the source v_x. From this, show that the output resistance $R_o \equiv v_x/i_x \simeq \beta r_o/2$.

6.50 Consider the Wilson current-mirror circuit of Fig. 6.19 when supplied with a reference current I_{REF} of 1 mA. What is the change in I_O corresponding to a change of +10 V in the voltage at the collector of Q_3? Give both the absolute value and percentage change. Let $\beta = 100$, $V_A = 100$ V, and recall that the output resistance of the Wilson circuit is $\beta r_o/2$.

****D6.51** (a) For the circuit in Fig. P6.51, find I_{O1} and I_{O2} in terms of I_{REF}. Assume all transistors to be matched with current gain β.
(b) Use this idea to design a circuit that generates currents of 1, 2, and 4 mA using a reference current source of 7 mA. What are the actual values of the currents generated for $\beta = 50$?

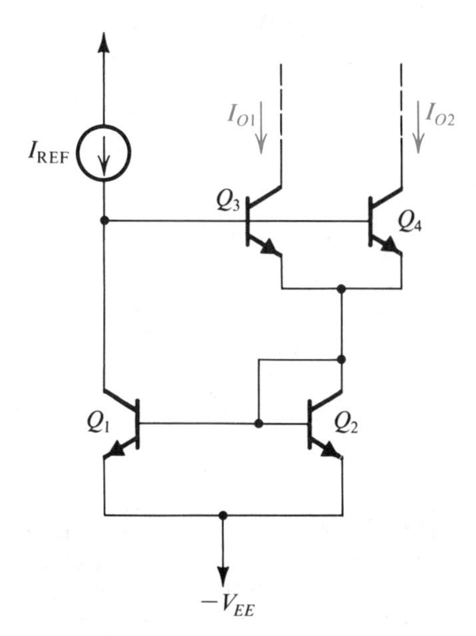

Fig. P6.51

***D6.52** (a) Utilizing a reference current of 100 μA, design a Widlar current source to provide an output current of 10 μA. Let the BJTs have $v_{BE} = 0.7$ V at 1-mA current, and assume β to be high.
(b) If $\beta = 200$ and $V_A = 100$ V, find the value of the output resistance, and find the change in output current corresponding to a 5-V change in output voltage.

6.53 The BJT in the circuit of Fig. P6.53 has $V_{BE} = 0.7$ V, $\beta = 100$, $V_A = 100$ V, and $r_\mu = 10\ \beta r_o$. Find R_o.

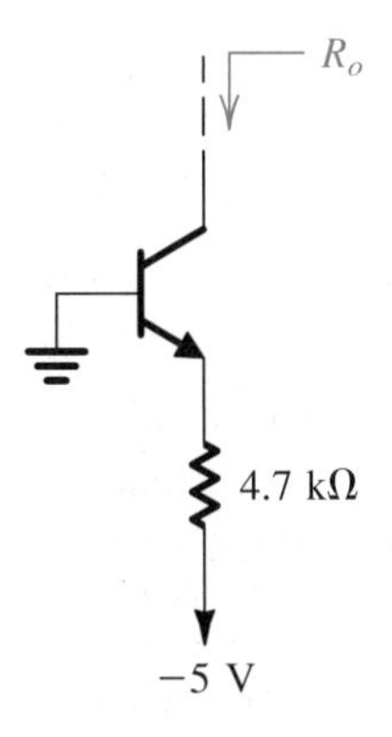

Fig. P6.53

D6.54 (a) For the circuit in Fig. P6.54, assume high β and BJTs having $v_{BE} = 0.7$ V at 1 mA. Find the value of R that will result in $I_O = 10\ \mu$A.
(b) For the design in (a), find R_o assuming $\beta = 100$ and $V_A = 100$ V.

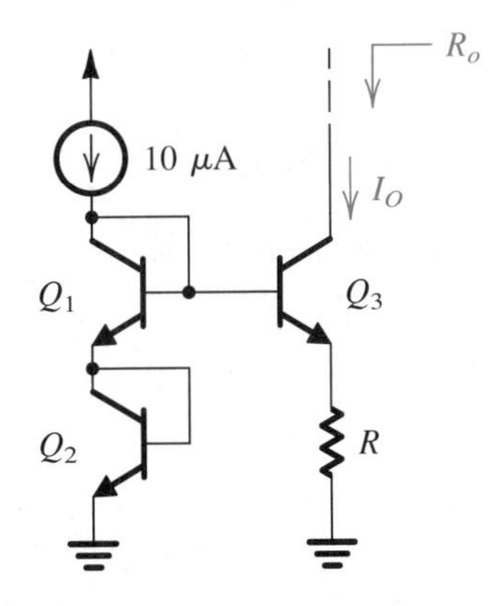

Fig. P6.54

6.55 Consider the effect of power-supply variation on the dc bias of the op amp circuit of Fig. 6.23: If +15 V is lowered to +14 V, what does the output voltage become? If, separately, −15 V is raised to −14 V, what does the output voltage become?

Section 6.5: The BJT Differential Amplifier with Active Load

6.56 The differential amplifier in Fig. 6.24 is operated with $I = 100\ \mu$A, with devices for which $V_A = 200$ V and $\beta = 100$. What differential input resistance, output resistance, equivalent transconductance, and open-circuit voltage gain would you expect? What will the voltage gain be if the input resistance of the subsequent stage is 1 MΩ?

*D6.57 Design the circuit of Fig. 6.24 using a basic current mirror to implement the current source I. It is required that the equivalent transconductance be 4 mA/V. Use ±15-V power supplies and BJTs that have $\beta = 200$ and $V_A = 100$ V. Give the complete circuit with component values and specify the differential input resistance R_i, the output resistance R_o, the open-circuit voltage gain, the input bias current, the input common-mode range, and the common-mode input resistance.

6.58 The finite β (which is usually low for the standard IC process) of the *pnp* transistors that form the current mirror in the circuit of Fig. 6.24 results in a dc offset voltage.

(a) Show that, with both inputs grounded, there will be an output current directed into the amplifier of approximately I/β_p, where β_p is the β of the *pnp* transistors. (Assume that β of the *npn* transistors is high.)

(b) Find the differential input voltage that would reduce the output current found in (a) to zero. This is the input dc offset voltage due to the finite β_p.

(c) For $\beta_p = 25$, find V_{OS}.

6.59 For the cascode amplifier of Fig. 6.26(b), what is the voltage gain from the base to the collector of Q_1?

D6.60 For the cascode amplifier of Fig. 6.26(a), what is the minimum value of V_{BIAS} that results in the upper limit of the input common-mode range being at least +10 V?

*6.61 Consider the differential amplifier circuit of Fig. 6.24 with the two input terminals tied together and an input common-mode signal v_{CM} applied. Let the output resistance of the bias current source be denoted by R, and let the β of the *pnp* transistors be denoted β_p. Assuming that β of the *npn* transistors is high, show that there will be an output current of $v_{CM}/\beta_p R$. Thus show that the common-mode transconductance is $1/\beta_p R$. Use this result together with the differential transconductance G_m (derived in the text) to find the common-mode rejection ratio. Calculate the value of CMRR for the case $I = 0.2$ mA, $R = 1$ MΩ, and $\beta_p = 25$.

Section 6.6: The JFET Differential Pair

6.62 Consider the JFET differential pair of Fig. 6.27 with the gate of Q_2 grounded. Let $I_{DSS} = 2$ mA, $V_P = -2$ V, and the bias current $I = 1$ mA. Find v_{GS1} and v_{GS2} and hence v_{G1} and v_S, that will result in the following current distributions:

(a) $i_{D1} = i_{D2} = 0.5$ mA,

(b) $i_{D1} = 0.75$ mA and $i_{D2} = 0.25$ mA,

(c) $i_{D1} = 1$ mA and $i_{D2} = 0$.

*6.63 A differential amplifier, utilizing JFETs for which $I_{DSS} = 2$ mA, $|V_P| = 2$ V, and $V_A = 100$ V, is biased at a constant current of 2 mA. For drain resistors of 10 kΩ, what is the gain of the amplifier for differential output? If the drain resistors have ±1% tolerance and the current source has an output resistance of 100 kΩ, what is the worst-case common-mode gain and CMRR? Also find the worst-case input offset voltage due to the mismatch in R_D.

6.64 The JFET circuit shown in Fig. P6.64 can be used to implement the current-source bias in a differential amplifier. Find the output resistance if $I_{DSS} = 2$ mA, $V_P = -2$ V, and $V_A = 100$ V.

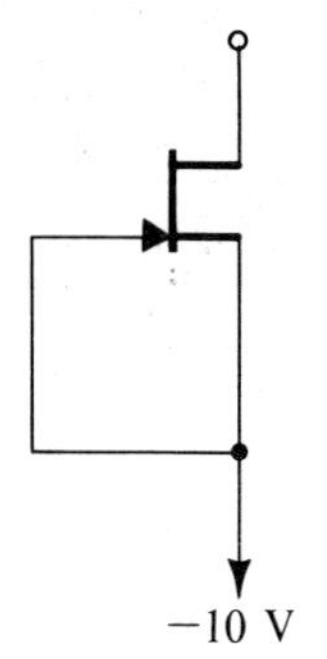

Fig. P6.64

D6.65 It is required to design a JFET differential amplifier utilizing the circuit shown in Fig. P6.65 to implement the bias current source. The JFETs available have $I_{DSS} = 2$ mA and $V_P = -2$ V, and a ±15-V supply is available. Design for a bias current of 1 mA and for

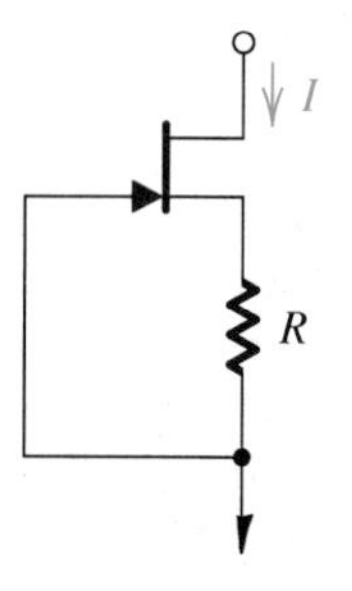

Fig. P6.65

the maximum possible differential gain consistent with allowing the upper limit of the input common-mode range to be at least +5 V. What is the differential gain obtained? What is the lower limit of the input common-mode range?

6.66 A JFET differential amplifier is loaded with the basic BJT current mirror. The JFETs have $V_P = -2$ V, $I_{DSS} = 4$ mA, and $V_A = 100$ V. The BJTs have $|V_A| = 100$ V and β is large. The bias current $I = 2$ mA. Find R_i, G_m, R_o, and the open-circuit voltage gain.

Section 6.7: MOS Differential Amplifiers

6.67 Consider the MOS differential pair of Fig. 6.29 with the gate of Q_2 grounded. Let $\mu_n C_{OX} = 20\ \mu\text{A/V}^2$, $V_t = 1$ V, $W/L = 20$, and $I = 16\ \mu$A. Find v_{GS1}, v_{GS2}, v_S, and v_{G1} that correspond to the following distributions of the current I between Q_1 and Q_2:
(a) $i_{D1} = i_{D2} = 8\ \mu$A.
(b) $i_{D1} = 12\ \mu$A and $i_{D2} = 4\ \mu$A.
(c) $i_{D1} = 16\ \mu$A and $i_{D2} = 0$ (Q_2 *just* cuts off).
Confirm that the value of v_{G1} obtained in (c) is the same as that found using Eq. (6.120).

6.68 An NMOS differential amplifier utilizes a bias current of 8μA. The devices have $V_t = 1$ V, $W = 60\ \mu$m, and $L = 6\ \mu$m, in a technology for which $\mu_n C_{OX} = 20\ \mu\text{A/V}^2$. Find V_{GS}, g_m, and the value of v_{id} for full current switching. To what value should the bias current be changed in order to double the value of v_{id} for full current switching.

D6.69 Design the MOS differential amplifier of Fig. 6.29 to operate at $V_{GS} - V_t = 0.2$ V and to provide a transconductance g_m of 0.1 mA/V. Specify the W/L ratios and the bias current. The technology available provides $V_t = 1$ V and $\mu_n C_{OX} = 20\ \mu\text{A/V}^2$.

D6.70 An NMOS differential pair is to be used in an amplifier whose drain resistors are 100 kΩ ± 1%. For the pair, $K = 100\ \mu\text{A/V}^2$ and $V_t = 1$ V. A decision is to be made concerning the bias current to be used, whether 100 μA or 200 μA. For differential output, contrast the differential gain and input offset voltage for the two possibilities.

D6.71 An NMOS amplifier, whose designed operating point is with V_{GS} one-half volt above the nominal 1-V threshold, is suspected to have a variability of V_t, K, and R_D (independently) of ±5%. What is the worst-case input offset voltage you would expect to find? What is the major contribution to this total offset? If you used a variation of one of the drain resistors to reduce the output offset to zero and thereby compensate for the uncertainties (including that of the other R_D), what percentage change from nominal would you require? If by selection you reduced the contribution of the worst cause of offset by a factor of 10, what change in R_D would be needed?

6.72 For the simple MOS mirror shown in Fig. 6.32(a) the devices nominally have $V_t = 1$ V and $K = 100\ \mu\text{A/V}^2$. Measurement with $V_{DS2} = V_{DS1}$ and $I_{REF} = 1$ mA shows the output current to be 5% low. It is assumed that Q_2 differs from Q_1 in one or more ways. If the difference in current is all due to K_2 being different from K_1, find what this difference must be. If, on the other hand, the difference in current is all due to V_{t2} being different from V_{t1}, find what this difference must be.

6.73 For the simple MOS mirror shown in Fig. 6.32(a), the devices have $V_t = 1$ V, $K = 100\ \mu\text{A/V}^2$, and $V_A = 20$ V. $I_{REF} = 100\ \mu$A, $V_{SS} = 5$ V, and $V_O = +5$ V. What value of I_O results?

6.74 For the cascode current mirror of Fig. 6.32(b) with $V_t = 1$ V, $K = 100\ \mu\text{A/V}^2$, $V_A = 20$ V, $I_{REF} = 100\ \mu$A, $V_{SS} = 5$ V, and $V_O = +5$ V, what value of I_O results? (Note the tremendous improvement over the situation obtained with the simple mirror of Problem 6.73.)

*6.75 It is required to derive an expression for the output resistance of the Wilson current mirror shown in Fig. 6.32(c). Replace each MOSFET with its small-signal model including r_o. For the diode-connected transistor Q_2, the model will reduce to a resistance $(1/g_{m2})$ in parallel with r_{o2}. To simplify matters, neglect r_{o2}. Apply a test voltage v_x between the output terminal and ground and denote the current drawn from v_x by i_x. Analyze the circuit to show that

$$R_o \equiv v_x/i_x \simeq (g_m r_o)r_o$$

*6.76 A Wilson current mirror uses devices for which $V_t = 1$ V, $K = 100\ \mu\text{A/V}^2$, and $V_A = 10$ V. $I_{REF} = 100\ \mu$A and $V_{SS} = 0$ V. What value of I_O results? Now if the circuit is modified to that in Fig. 6.32(d), what value of I_O results?

6.77 Find the output resistance of the current mirror in Fig. P6.77. To simplify matters, assume that the incremental voltage at the gates of Q_1, Q_2, and Q_3 is zero. (*Hint:* Use the relationship in Eq. 6.135).

D6.78 In an active-loaded differential amplifier of the form shown in Fig. 6.34, all transistors are characterized by $|V_t| = 1$ V, $K = 400\ \mu\text{A/V}^2$, and $|V_A| = 20$ V. Find the bias current I for which the gain $v_o/v_{id} = 80$ V/V.

*D6.79 Consider an active-loaded differential amplifier such as that shown in Fig. 6.34 with the bias current source implemented with the modified Wilson mirror of

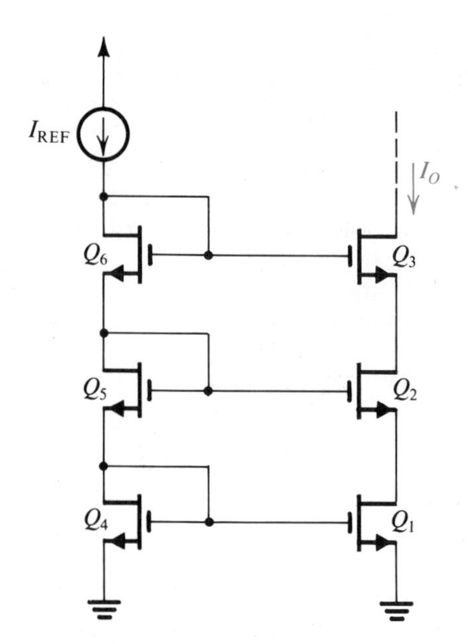

Fig. P6.77

Fig. 6.32(d) with $I_{REF} = 25\ \mu$A. The transistors have $|V_t| = 1$ V and $K = 400\ \mu\text{A/V}^2$. What is the lowest value of the total power supply ($V_{DD} + V_{SS}$) which allows each transistor to operate with $|V_{DS}| \geq |V_{GS}|$?

***D6.80** If the *pnp* transistor in the circuit of Fig. P6.80 is characterized by its exponential relationship with a scale current I_S, show that the dc current I is given by $IR = V_T \ln(I/I_S)$. Assume Q_1 and Q_2 to be matched and Q_3, Q_4, and Q_5 to be matched. Find the value of R that yields a current $I = 10\ \mu$A. For the BJT, $V_{EB} = 0.7$ V at $I_E = 1$ mA.

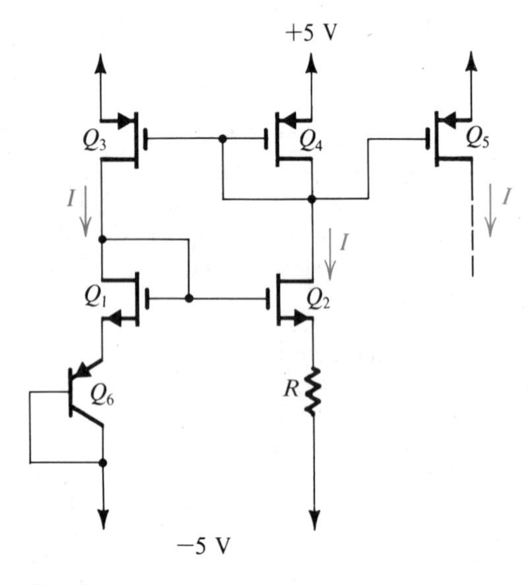

Fig. P6.80

Section 6.8: BiCMOS Amplifiers

For the problems in this section assume that the BiCMOS technology available provides bipolar devices with $V_A = 50$ V and $\beta = 100$; and MOS devices with $L = 2\ \mu$m, $\lambda = 0.05\ \text{V}^{-1}$, $|V_t| = 1$ V, and $\mu_n C_{OX} = 20\ \mu\text{A/V}^2$.

6.81 What W/L ratio must a MOSFET have in order to have a g_m equal to that of a BJT when both are biased at a current I of 100 μA? Comment on the result.

6.82 If the parameter λ of a MOSFET is approximately inversely proportional to the channel length L, how do r_o and the intrinsic voltage gain change with L?

6.83 How does the thickness of the gate oxide affect the intrinsic gain of the common-source amplifier? Discuss the trade-offs that have to be considered in deciding on oxide thickness.

6.84 Repeat Exercise 6.20 for a bias current of 10 μA.

***D6.85** **(a)** Show that for the BiCMOS cascode amplifier of Fig. 6.35(c) the gain can be expressed in the form

$$\frac{v_o}{v_i} = -\frac{2\beta_2 V_A}{V_{GS} - V_t}$$

(b) Design the circuit to realize a gain of −40,000 V/V. In order to obtain acceptable high-frequency performance the bias current should be at least 10 μA. Specify the width of the MOSFET and the value of V_{GS} at which Q_1 should be biased. What is the minimum value for V_{BIAS}? Assume that the BJT has a v_{BE} of 0.7 V at a current of 10 μA.

6.86 If the mirror in Fig. 6.37 has MOSFETs with $W = 20\ \mu$m and is operating at a 100-μA current, what is the minimum voltage required at the output in order for the circuit to exhibit its high output resistance? Assume that $v_{BE} = 0.7$ V at a current of 100 μA. What is the change in I_O resulting from the output voltage rising by 8 V?

Section 6.9: GaAs Amplifiers

Unless otherwise specified, use the device data given in Table 5.2 (page 389).

D6.87 For the current source of Fig. 6.39(c) with $V_{DD} = +5$ V, find the device width that will result in a

source is
e current

.41(b) let
hest volt-
hat is the
he output

. 6.43(a)
Neglect-
ie of cur-
$V_{DS,sat}$ for
ly. If the
, find the
put resist-

let W_1 =
nd W_4 =
r simplic-
he values

elf that if
of Q_2 is
zed is ap-
Q_1 is bi-
for the dc
o find an

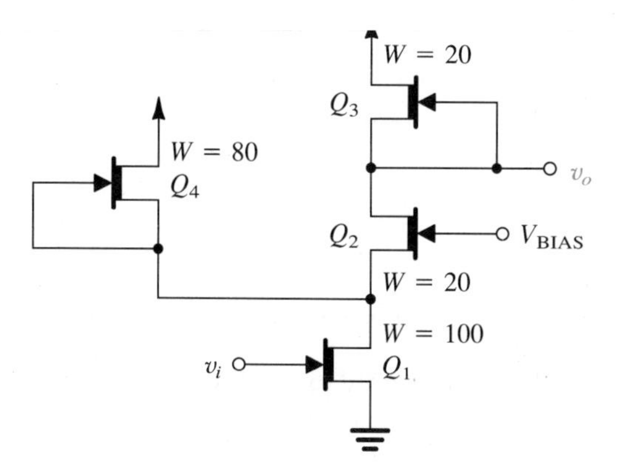

Fig. P6.91

approximate value for the voltage gain. (To simplify matters, neglect λ when calculating the value of g_{m1}.)

**6.92 For the circuit in Fig. 6.45(a), whose small-signal analysis is shown in Fig. 6.45(b), derive expressions for v_s/v_i and v_{d1}/v_i. (*Hint:* Follow the procedure outlined in the text.) For I = 10 mA, $W_1 = W_2 = W_3 =$ 100 μm, calculate the values of v_s/v_i, v_{d1}/v_i, and v_o/v_i.

Section 6.10: Multistage Amplifiers

6.93 A BJT differential amplifier, biased to have r_e = 50 Ω and utilizing two 100-Ω emitter resistors and 5-kΩ loads, drives a second differential stage biased to have r_e = 20 Ω. All BJTs have β = 150. What is the voltage gain of the first stage?

6.94 In the multistage amplifier of Fig. 6.46, emitter resistors are to be introduced—100 Ω in the emitter lead of each of the first-stage transistors and 25 Ω for each of the second-stage transistors—what is the effect on input resistance, the voltage gain of the first stage, and the overall voltage gain? Use the bias values found in Example 6.2.

D6.95 Consider the circuit of Fig. 6.46 and its output resistance. Which resistor has the most effect on the output resistance? What should this resistance be changed to if the output resistance is to be reduced by a factor of 2? What will the amplifier gain become after this change? What other change can you make to restore the amplifier gain to approximately its prior value?

D6.96 If, in the multistage amplifier of Fig 6.46, the resistor R_5 is replaced by a constant-current source ≃ 1 mA, such that the bias situation is essentially unaffected, what does the overall voltage gain of the amplifier become? Assume that the output resistance of the current source is very high. Use the results of Example 6.3.

*D6.97 With the modification suggested in the previous problem to the multistage amplifier, what is the effect of the change on output resistance? What is the overall gain of the amplifier when loaded by 100 Ω to ground? The original amplifier (before modification) has an output resistance of 152 Ω and a voltage gain of 8513 V/V. What is its gain when loaded by 100 Ω? Comment. Use β = 100.

CHAPTER 7

Frequency Response

INTRODUCTION

In Chapter 1 we introduced the topic of amplifier frequency response and briefly mentioned the various frequency-response shapes encountered. In addition, the frequency response of single-time-constant (STC) networks was presented (more details are given in Appendix E). This material was found to be directly applicable to the evaluation of the frequency response of op amp circuits (Chapter 2). Apart from this and the occasional mention of limitations imposed on amplifier frequency response, the detailed study of this important topic has been deferred to the present chapter. This was done for a number of reasons, the most important of which is the need for circuit theoretic concepts and methods that the reader might not have been exposed to at the beginning of the book. We refer specifically to the complex frequency variable s and associated concepts such as poles and zeros. These topics are normally found in introductory texts on circuit analysis [see, for example, Van Valkenburg (1974) and Bobrow (1987)]. In the following it will be assumed that the reader is familiar with this material.

After a brief review of s-domain analysis (Section 7.1) and a study of amplifier transfer functions (Section 7.2), the low-frequency response of the capacitively coupled common-source amplifier is presented (Section 7.3). The FET high-frequency model is then presented and applied in the analysis of the high-frequency response of the common-source amplifier (Section 7.4). The BJT hybrid-π model is studied in Section 7.5 and used to find the frequency response of the common-emitter amplifier (Section 7.6), the common-base amplifier (Section 7.7), and the common-collector amplifier (Section 7.8). A number of two-stage amplifier configurations having the advantage of extended bandwidth are also studied in this chapter. Also, the differential amplifier frequency response is considered in detail.

The single-stage and two-stage amplifiers analyzed in this chapter can be used either singly or as building blocks for more complex multistage amplifiers. Frequency-response analysis of multistage amplifiers uses the methods studied in this chapter as will be illustrated further in Chapter 10 in the context of analyzing op amp circuits.

Although the emphasis in this chapter is on analysis, the material is readily applicable to design. This is achieved by keeping the analysis relatively simple and thus focusing attention on the mechanisms that limit frequency response and on methods for extending amplifier bandwidth. Of course, once an initial design is obtained, its exact frequency response can be found using computer-aided analysis (see Appendix C and the SPICE supplement). The results obtained this way can be used to improve the design further.

7.1 s-DOMAIN ANALYSIS

Most of our work in this chapter will be concerned with finding amplifier voltage gain as a transfer function of the complex frequency s. In this s-domain analysis a capacitance C is replaced by an admittance sC, or equivalently an impedance $1/sC$, and an inductance L is replaced by an impedance sL. Then, using usual circuit-analysis techniques, one derives the voltage transfer function $T(s) \equiv V_o(s)/V_i(s)$.

Exercise 7.1 Find the voltage transfer function $T(s) \equiv V_o(s)/V_i(s)$ for the STC network shown in Fig. E7.1.

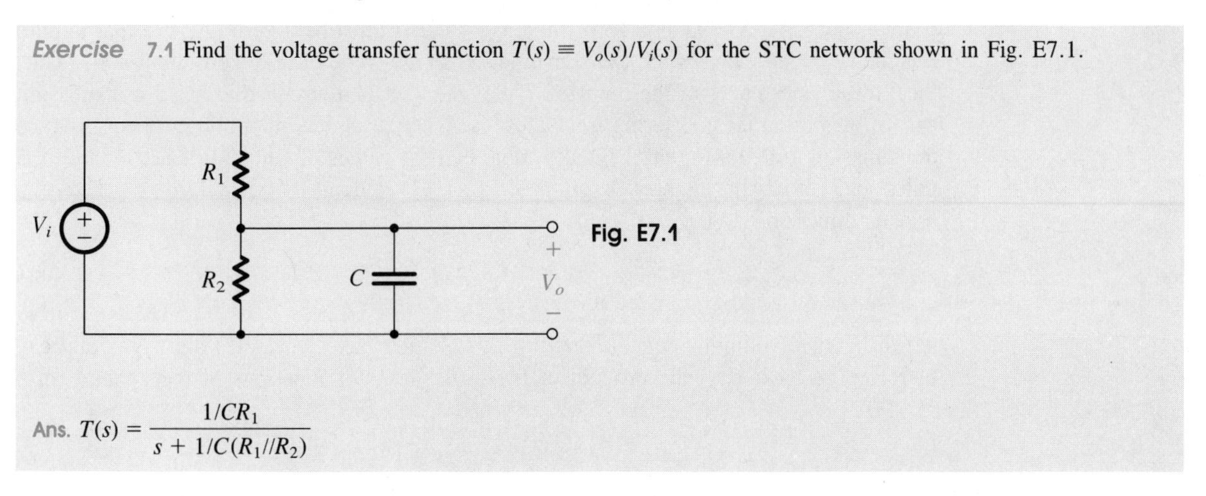

Fig. E7.1

Ans. $T(s) = \dfrac{1/CR_1}{s + 1/C(R_1//R_2)}$

Once the transfer function $T(s)$ is obtained, it can be evaluated for **physical frequencies** by replacing s by $j\omega$. The resulting transfer function $T(j\omega)$ is in general a complex quantity whose magnitude gives the magnitude response (or transmission) and whose angle gives the phase response of the amplifier.

In many cases it will not be necessary to substitute $s = j\omega$ and evaluate $T(j\omega)$; rather, the form of $T(s)$ will reveal many useful facts about the circuit performance. In general, for all the circuits dealt with in this chapter, $T(s)$ can be expressed in the form

$$T(s) = \frac{a_m s^m + a_{m-1}s^{m-1} + \cdots + a_0}{s^n + b_{n-1}s^{n-1} + \cdots + b_0} \tag{7.1}$$

where the coefficients a and b are real numbers and the order m of the numerator is smaller than or equal to the order n of the denominator; the latter is called the **order of the network.** Furthermore, for a **stable circuit**—that is, one that does not generate signals on its own—the denominator coefficients should be such that *the roots of the denominator polynomial all have negative real parts*. We shall study the problem of amplifier stability in Chapter 8.

Poles and Zeros

An alternate form for expressing $T(s)$ is

$$T(s) = a_m \frac{(s - Z_1)(s - Z_2)\cdots(s - Z_m)}{(s - P_1)(s - P_2)\cdots(s - P_n)} \tag{7.2}$$

where a_m is a multiplicative constant (the coefficient of s^m in the numerator), Z_1, $Z_2, \ldots, Z_m$ are the roots of the numerator polynomial, and $P_1, P_2, \ldots, P_n$ are the roots of the denominator polynomial. $Z_1, Z_2, \ldots, Z_m$ are called the **transfer-function zeros** or **transmission zeros,** and $P_1, P_2, \ldots, P_n$ are the **transfer-function poles** or the **natural modes** of the network. A transfer function is completely specified in terms of its poles and zeros together with the value of the multiplicative constant.

The poles and zeros can be either real or complex numbers. However, since the a and b coefficients are real numbers, the complex poles (or zeros) must occur in **conjugate pairs.** That is, if $5 + j3$ is a zero, then $5 - j3$ also must be a zero. A zero that is pure imaginary ($\pm j\omega_Z$) causes the transfer function $T(j\omega)$ to be exactly zero at $\omega = \omega_Z$. Thus the "trap" one places at the input of a television set is a circuit that has a transmission zero at the particular interfering frequency. Real zeros, on the other hand, do not produce transmission nulls (why not?). Finally, note that for values of s much greater than all the poles and zeros, the transfer function in Eq. (7.1) becomes $T(s) \simeq a_m/s^{n-m}$. Thus the transfer function has $(n - m)$ zeros at $s = \infty$.

First-Order Functions

All the transfer functions encountered in this chapter have real poles and zeros and can therefore be written as the product of first-order transfer functions of the general form

$$T(s) = \frac{a_1 s + a_0}{s + \omega_0} \tag{7.3}$$

where $-\omega_0$ is the location of the real pole. The quantity ω_0, called the **pole frequency,** is equal to the inverse of the time constant of this single-time-constant (STC) network (see Section 1.6 and Appendix E). The constants a_0 and a_1 determine the type of STC network. Specifically, we studied in Chapter 1 two types of STC networks, low pass and high pass. For the low-pass first-order network we have

$$T(s) = \frac{a_0}{s + \omega_0} \tag{7.4}$$

In this case the dc gain is a_0/ω_0, and ω_0 is the corner or 3-dB frequency. Note that this transfer function has one zero at $s = \infty$. On the other hand, the first-order high-pass transfer function has a zero at dc and can be written

$$T(s) = \frac{a_1 s}{s + \omega_0} \tag{7.5}$$

At this point the reader is strongly urged to review the material on STC networks and their frequency and pulse responses in Appendix E. Of specific interest are the plots of the magnitude and phase responses of the two special kinds of STC networks. Such plots can be employed to generate the magnitude and phase plots of a high-order transfer function, as explained below.

Bode Plots

A simple technique exists for obtaining an approximate plot of the magnitude and phase of a transfer function given its poles and zeros. The technique is particularly useful in the case of real poles and zeros. The method was developed by H. Bode, and the resulting diagrams are called **Bode plots.**

A transfer function of the form depicted in Eq. (7.2) consists of a product of factors of the form $s + a$, where such a factor appears on top if it corresponds to a zero and on the bottom if it corresponds to a pole. It follows that the magnitude response in decibels of the network can be obtained by summing together terms of the form $20 \log_{10} \sqrt{a^2 + \omega^2}$, and the phase response can be obtained by summing terms of the form $\tan^{-1}(\omega/a)$. In both cases the terms corresponding to poles are summed with negative signs. For convenience we can extract the constant a and write the typical magnitude term in the form $20 \log \sqrt{1 + (\omega/a)^2}$. On a plot of decibels versus log frequency this term gives rise to the curve and straight-line asymptotes shown in Fig. 7.1. Here the low-frequency asymptote is a horizontal straight line at 0-dB level and the high-frequency asymptote is a straight line with a slope of 6 dB/octave or, equivalently, 20 dB/decade. The two asymptotes meet at the frequency $\omega = |a|$, which is called the **corner frequency.** As indicated, the actual magnitude plot differs slightly from the value given by the asymptotes; the maximum difference is 3 dB and occurs at the corner frequency.

For $a = 0$—that is, a pole or a zero at $s = 0$—the plot is simply a straight line of 6 dB/octave slope intersecting the 0-dB line at $\omega = 1$.

In summary, to obtain the Bode plot for the magnitude of a transfer function, the asymptotic plot for each pole and zero is first drawn. The slope of the high-frequency asymptote of the curve corresponding to a zero is +20 dB/decade, while that for a pole is −20 dB/decade. The various plots are then added together, and the overall curve is

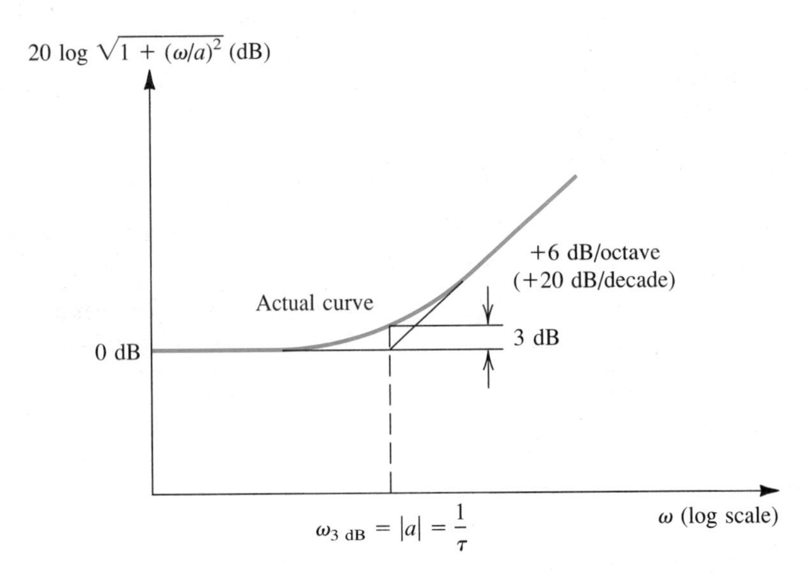

Fig. 7.1 Bode plot for the typical magnitude term. The curve shown applies for the case of a zero. For a pole, the high-frequency asymptote should be drawn with a −6-dB/octave slope.

shifted vertically by an amount determined by the multiplicative constant of the transfer function.

EXAMPLE 7.1

An amplifier has the voltage transfer function

$$T(s) = \frac{10s}{(1 + s/10^2)(1 + s/10^5)}$$

Find the poles and zeros and sketch the magnitude of the gain versus frequency. Find approximate values for the gain at $\omega = 10$, 10^3, and 10^6 rad/s.

Solution The zeros are as follows: one at $s = 0$ and one at $s = \infty$. The poles are as follows: one at $s = -10^2$ rad/s and one at $s = -10^5$ rad/s.

Figure 7.2 shows the asymptotic Bode plots of the different factors of the transfer function. Curve 1, which is a straight line with +20 dB/decade slope, corresponds to the s term (that is, the zero at $s = 0$) in the numerator. The pole at $s = -10^2$ results in curve 2, which consists of two asymptotes intersecting at $\omega = 10^2$. Similarly, the pole at $s = -10^5$ is represented by curve 3, where the intersection of the asymptotes is at $\omega = 10^5$. Finally, curve 4 represents the multiplicative constant of value 10.

Adding the four curves results in the asymptotic Bode diagram of the amplifier gain (curve 5). Note that since the two poles are widely separated, the gain will be very close to 10^3 (60 dB) over the frequency range 10^2 to 10^5 rad/s. At the two corner frequencies (10^2 and 10^5 rad/s) the gain will be approximately 3 dB below the maximum of 60 dB. At the three specific frequencies the values of the gain as obtained from the Bode plot and from

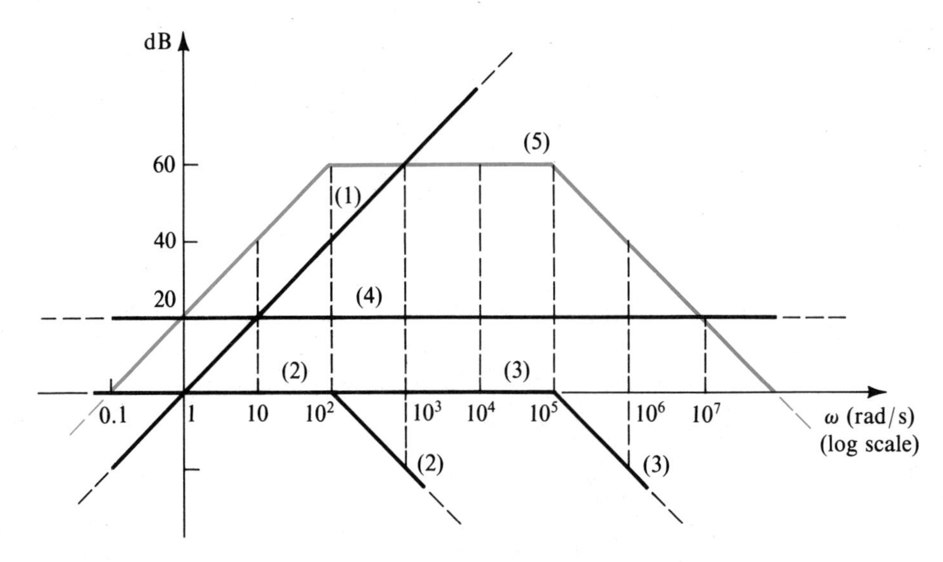

Fig. 7.2 Bode plots for Example 7.1.

exact evaluation of the transfer function are as follows:

ω	Approximate Gain	Exact Gain
10	40 dB	39.96 dB
10^3	60 dB	59.96 dB
10^6	40 dB	39.96 dB

We next consider the Bode phase plot. Figure 7.3 shows a plot of the typical phase term $\tan^{-1}(\omega/a)$, assuming that a is negative. Also shown is an asymptotic straight-line approximation of the arctan function. The asymptotic plot consists of three straight lines. The first is horizontal at $\phi = 0$ and extends up to $\omega = 0.1|a|$. The second line has a slope of $-45°$/decade and extends from $\omega = 0.1|a|$ to $\omega = 10|a|$. The third line has a zero slope and a level of $\phi = -90°$. The complete phase response can be obtained by summing the asymptotic Bode plots of the phase of all poles and zeros.

EXAMPLE 7.2

Find the Bode plot for the phase of the transfer function of the amplifier considered in Example 7.1.

Solution

The zero at $s = 0$ gives rise to a constant $+90°$ phase function represented by curve 1 in Fig. 7.4. The pole at $s = -10^2$ gives rise to the phase function

$$\phi_1 = -\tan^{-1}\frac{\omega}{10^2}$$

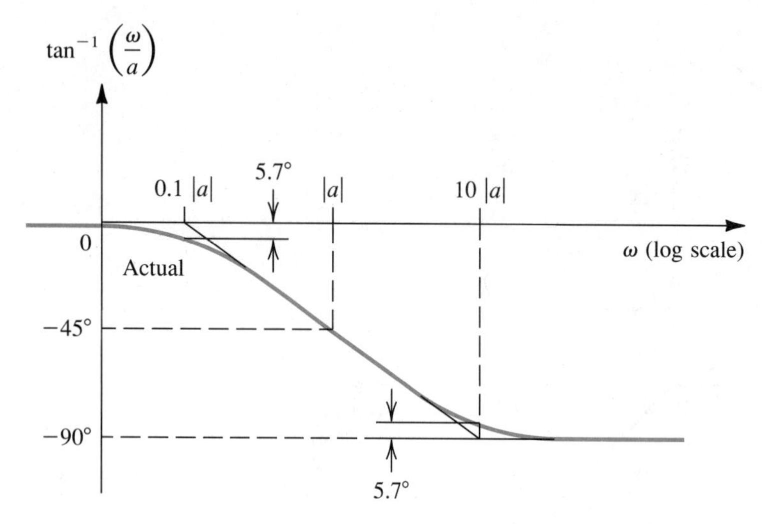

Fig. 7.3 Bode plot of the typical phase term $\tan^{-1}(\omega/a)$ when a is negative.

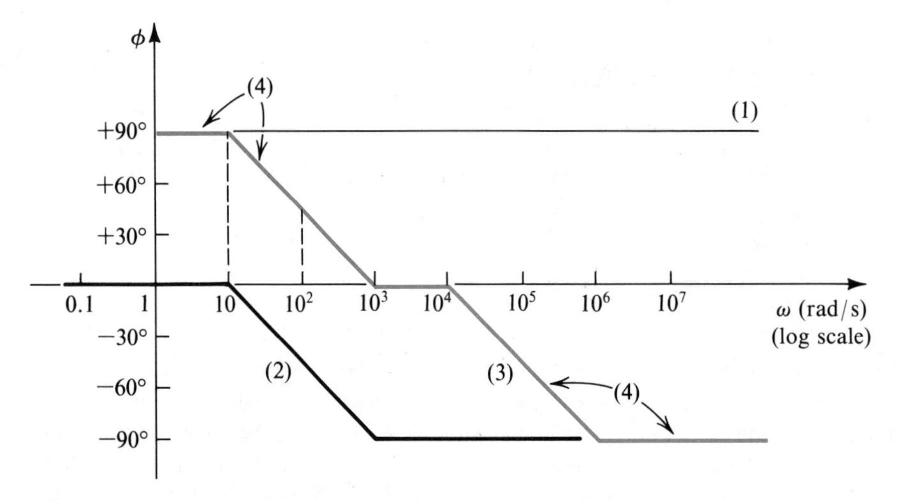

Fig. 7.4 Phase plots for Example 7.2.

(the leading minus sign is due to the fact that this singularity is a pole). The asymptotic plot for this function is given by curve 2 in Fig. 7.4. Similarly, the pole at $s = -10^5$ gives rise to the phase function

$$\phi_2 = -\tan^{-1}\frac{\omega}{10^5}$$

whose asymptotic plot is given by curve 3. The overall phase response (curve 4) is obtained by direct summation of the three plots. We see that at 100 rad/s, the amplifier phase leads by 45° and at 10^5 rad/s the phase lags by 45°.

7.2 THE AMPLIFIER TRANSFER FUNCTION

The amplifiers considered in this chapter have voltage-gain functions of either of the two forms shown in Fig. 7.5. Figure 7.5(a) applies for direct-coupled or dc amplifiers and Fig. 7.5(b) for capacitively coupled or ac amplifiers. The only difference between the two types is that the gain of the ac amplifier falls off at low frequencies. In the following we shall study the more general response shown in Fig. 7.5(b). The response of the dc amplifier follows as a special case.

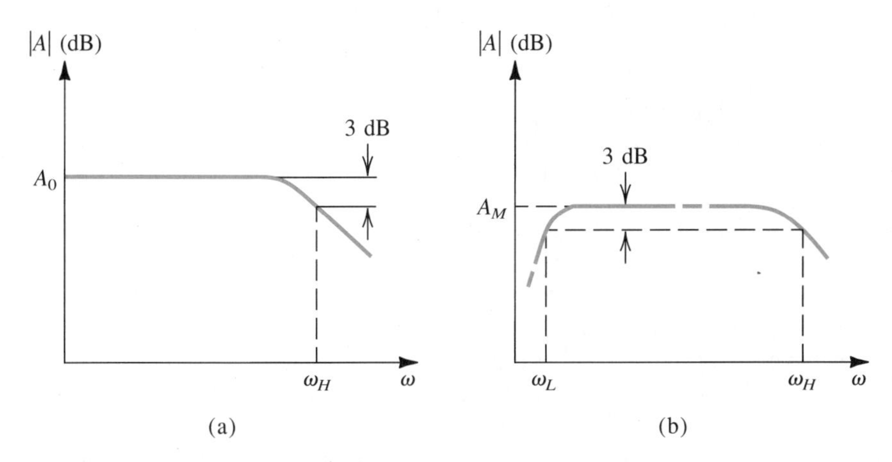

Fig. 7.5 Frequency response for **(a)** a dc amplifier and **(b)** a capacitively coupled amplifier.

The Three Frequency Bands

As can be seen from Fig. 7.5(b) the amplifier gain is almost constant over a wide frequency range called the midband. In this frequency range all capacitances (coupling, bypass, and transistor internal capacitances) have negligible effects and can be ignored in gain calculations. At the high-frequency end of the spectrum the gain drops owing to the effect of the internal capacitances of the device. On the other hand, at the low-frequency end of the spectrum the coupling and bypass capacitances no longer act as perfect short circuits and thus cause the gain to drop. The extent of the midband is usually defined by the two frequencies ω_L and ω_H. These are the frequencies at which the gain drops by 3 dB below the value at midband. The amplifier bandwidth is usually defined as

$$\text{BW} = \omega_H - \omega_L \tag{7.6}$$

and, since, $\omega_L \ll \omega_H$,

$$\text{BW} \simeq \omega_H \tag{7.7}$$

A figure of merit for the amplifier is its **gain-bandwidth product,** defined as

$$GB \equiv A_M \omega_H \tag{7.8}$$

where A_M is the magnitude of midband gain in volts per volt. As will be shown in later sections, it is generally possible to trade off gain for bandwidth.

The Gain Function $A(s)$

The amplifier gain as a function of the complex frequency s can be expressed in the general form

$$A(s) = A_M F_L(s) F_H(s) \tag{7.9}$$

where $F_L(s)$ and $F_H(s)$ are functions that account for the dependence of gain on frequency in the low-frequency band and in the high-frequency band, respectively. For frequencies ω much greater than ω_L the function $F_L(s)$ approaches unity. Similarly, for frequencies ω much smaller than ω_H the function $F_H(s)$ approaches unity. Thus for $\omega_L \ll \omega \ll \omega_H$,

$$A(s) \simeq A_M$$

as should have been expected. It also follows that the gain of the amplifier in the low-frequency band, $A_L(s)$, can be expressed as

$$A_L(s) \simeq A_M F_L(s) \tag{7.10}$$

and the gain in the high-frequency band can be expressed as

$$A_H(s) \simeq A_M F_H(s) \tag{7.11}$$

The midband gain is determined by analyzing the amplifier equivalent circuit with the assumption that the coupling and bypass capacitors are acting as perfect short circuits and the internal capacitors of the transistor model are acting as perfect open circuits. The low-frequency transfer function, $A_L(s)$, is determined from analysis of the amplifier equivalent circuit including the coupling and bypass capacitors but assuming that the transistor-model capacitances behave as perfect open circuits. On the other hand, the high-frequency transfer function, $A_H(s)$, is determined from analysis of the amplifier equivalent circuit including the transistor-model capacitors but assuming that the coupling and bypass capacitors behave as perfect short circuits.

The Low-Frequency Response

The function $F_L(s)$, which characterizes the low-frequency response of the amplifier, takes the general form

$$F_L(s) = \frac{(s + \omega_{Z1})(s + \omega_{Z2}) \cdots (s + \omega_{Zn_L})}{(s + \omega_{P1})(s + \omega_{P2}) \cdots (s + \omega_{Pn_L})} \tag{7.12}$$

where $\omega_{P1}, \omega_{P2}, \ldots, \omega_{Pn_L}$ are positive numbers representing the frequencies of the n_L low-frequency poles and $\omega_{Z1}, \omega_{Z2}, \ldots, \omega_{Zn_L}$ are positive, negative, or zero numbers representing the n_L zeros. It should be noted from Eq. (7.12) that as s approaches infinity (in fact, as $s = j\omega$ approaches midband frequencies), $F_L(s)$ approaches unity.

In many cases the zeros are at such low frequencies (much smaller than ω_L) as to be of little importance in determining the lower 3-dB frequency ω_L. Also, usually one of the poles—say, ω_{P1}—has a much higher frequency than all other poles. It follows that for frequencies ω close to the midband, $F_L(s)$ can be approximated by

$$F_L(s) \simeq \frac{s}{s + \omega_{P1}} \tag{7.13}$$

which is the transfer function of a first-order high-pass network. In this case the low-frequency response of the amplifier is *dominated* by the pole at $s = -\omega_{P1}$ and the lower 3-dB frequency is approximately equal to ω_{P1},

$$\omega_L \simeq \omega_{P1} \tag{7.14}$$

If this **dominant-pole approximation** holds, it becomes a simple matter to determine ω_L. Otherwise one has to find the complete Bode plot for $|F_L(j\omega)|$ and thus determine ω_L. As a rule of thumb the dominant-pole approximation can be made if the highest-frequency pole is separated from the nearest pole or zero by at least two octaves (that is, a factor of four).

If a dominant low-frequency pole does not exist, an approximate formula can be derived for ω_L in terms of the poles and zeros. For simplicity consider the case of a circuit having two poles and two zeros at the low-frequency end; that is,

$$F_L(s) = \frac{(s + \omega_{Z1})(s + \omega_{Z2})}{(s + \omega_{P1})(s + \omega_{P2})} \tag{7.15}$$

Substituting $s = j\omega$ and taking the squared magnitude gives

$$|F_L(j\omega)|^2 = \frac{(\omega^2 + \omega_{Z1}^2)(\omega^2 + \omega_{Z2}^2)}{(\omega^2 + \omega_{P1}^2)(\omega^2 + \omega_{P2}^2)} \tag{7.16}$$

By definition, at $\omega = \omega_L$, $|F_L|^2 = \frac{1}{2}$, and thus

$$\begin{aligned} \frac{1}{2} &= \frac{(\omega_L^2 + \omega_{Z1}^2)(\omega_L^2 + \omega_{Z2}^2)}{(\omega_L^2 + \omega_{P1}^2)(\omega_L^2 + \omega_{P2}^2)} \\ &= \frac{1 + (1/\omega_L^2)(\omega_{Z1}^2 + \omega_{Z2}^2) + (1/\omega_L^4)(\omega_{Z1}^2\omega_{Z2}^2)}{1 + (1/\omega_L^2)(\omega_{P1}^2 + \omega_{P2}^2) + (1/\omega_L^4)(\omega_{P1}^2\omega_{P2}^2)} \end{aligned} \tag{7.17}$$

Since ω_L is usually greater than the frequencies of all the poles and zeros, we may neglect the terms containing $(1/\omega_L^4)$ and solve for ω_L to obtain

$$\omega_L \simeq \sqrt{\omega_{P1}^2 + \omega_{P2}^2 - 2\omega_{Z1}^2 - 2\omega_{Z2}^2} \tag{7.18}$$

This relationship can be extended to any number of poles and zeros. Note that if one of the poles—say, P_1—is dominant, then $\omega_{P1} \gg \omega_{P2}, \omega_{Z1}, \omega_{Z2}$, and Eq. (7.18) reduces to Eq. (7.14).

EXAMPLE 7.3

The low-frequency response of an amplifier is characterized by the transfer function

$$F_L(s) = \frac{s(s + 10)}{(s + 100)(s + 25)}$$

Determine its 3-dB frequency, approximately and exactly.

Solution

Noting that the highest-frequency pole at 100 rad/s is two octaves higher than the second pole and a decade higher than the zero, we find that a dominant-pole situation almost exists and $\omega_L \simeq 100$ rad/s. A better estimate of ω_L can be obtained using Eq. (7.18), as follows:

$$\omega_L = \sqrt{100^2 + 25^2 - 2 \times 10^2} = 102 \text{ rad/s}$$

The exact value of ω_L can be determined from the given transfer function as 105 rad/s. Finally, we show in Fig. 7.6 a Bode plot and an exact plot for the magnitude of the given

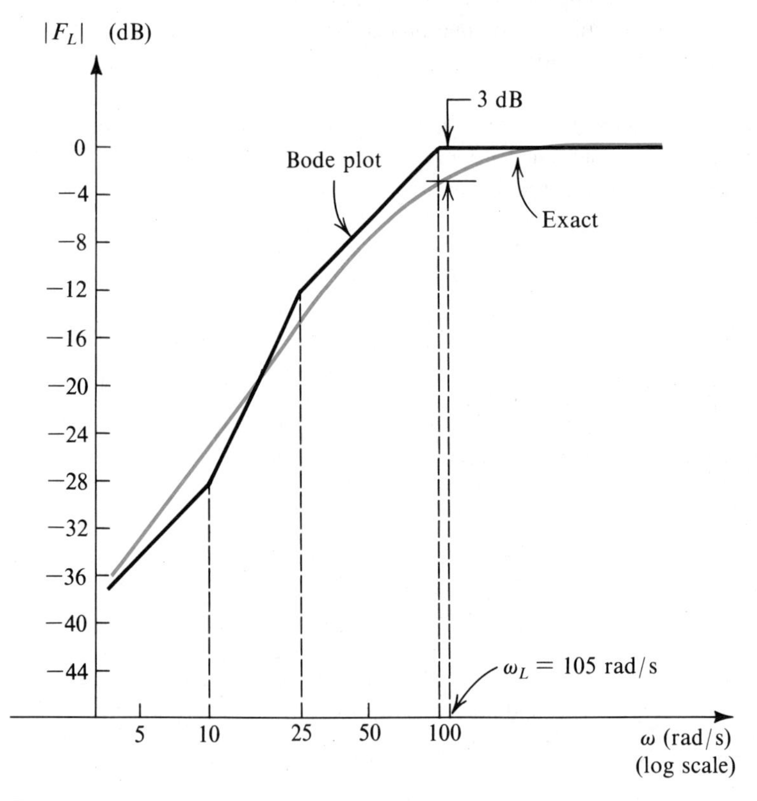

Fig. 7.6 Normalized low-frequency response of the amplifier in Example 7.3.

transfer function. Note that this is a plot of the low-frequency response of the amplifier normalized relative to the midband gain. That is, if the midband gain is 100 dB, then the entire plot should be shifted upward by 100 dB.

The High-Frequency Response

Consider next the high-frequency end. The function $F_H(s)$ can be expressed in the general form

$$F_H(s) = \frac{(1 + s/\omega_{Z1})(1 + s/\omega_{Z2}) \cdots (1 + s/\omega_{Zn_H})}{(1 + s/\omega_{P1})(1 + s/\omega_{P2}) \cdots (1 + s/\omega_{Pn_H})} \tag{7.19}$$

where $\omega_{P1}, \omega_{P2}, \ldots, \omega_{Pn_H}$ are positive numbers representing the frequencies of the n_H high-frequency real poles and $\omega_{Z1}, \omega_{Z2}, \ldots, \omega_{Zn_H}$ are positive, negative, or infinite numbers representing the frequencies of the n_H high-frequency zeros. Note from Eq. (7.19) that as s approaches 0 (in fact as $s = j\omega$ approaches midband frequencies), $F_H(s)$ approaches unity.

In many cases the zeros are either at infinity or at such high frequencies as to be of little significance in determining the upper 3-dB frequency ω_H. If in addition one of the high-frequency poles—say, ω_{P1}—is of much lower frequency than any of the other poles, then the high-frequency response of the amplifier will be *dominated* by this pole, and the function $F_H(s)$ can be approximated by

$$F_H(s) \simeq \frac{1}{1 + s/\omega_{P1}} \tag{7.20}$$

which is the transfer function of a first-order low-pass network. It follows that if a dominant high-frequency pole exists, then the determination of ω_H is greatly simplified:

$$\omega_H \simeq \omega_{P1} \tag{7.21}$$

If a dominant high-frequency pole does not exist, the upper 3-dB frequency ω_H can be determined from a plot of $|F_H(j\omega)|$. Alternatively, an approximate formula for ω_H in terms of the high-frequency poles and zeros can be derived in a manner similar to that used above in deriving Eq. (7.18). The formula for ω_H is

$$\omega_H \simeq 1 \Bigg/ \sqrt{\frac{1}{\omega_{P1}^2} + \frac{1}{\omega_{P2}^2} + \cdots - \frac{2}{\omega_{Z1}^2} - \frac{2}{\omega_{Z2}^2} \cdots} \tag{7.22}$$

Note that if one of the poles, say P_1, is dominant then $\omega_{P1} \ll \omega_{P2}, \omega_{P3}, \ldots, \omega_{Z1}, \omega_{Z2}, \ldots$ and Eq. (7.22) reduces to Eq. (7.21).

EXAMPLE 7.4

The high-frequency response of an amplifier is characterized by the transfer function

$$F_H(s) = \frac{1 - s/10^5}{(1 + s/10^4)(1 + s/4 \times 10^4)}$$

Determine the 3-dB frequency approximately and exactly.

Solution

Noting that the lowest-frequency pole at 10^4 rad/s is two octaves lower than the second pole and a decade lower than the zero, we find that a dominant-pole situation almost exists and $\omega_H \simeq 10^4$ rad/s. A better estimate of ω_H can be obtained using Eq. (7.22), as follows:

$$\omega_H = 1 \Bigg/ \sqrt{\frac{1}{10^8} + \frac{1}{16 \times 10^8} - \frac{2}{10^{10}}}$$

$$= 9800 \text{ rad/s}$$

The exact value of ω_H can be determined from the given transfer function as 9537 rad/s. Finally, we show in Fig. 7.7 a Bode plot and an exact plot for the given transfer function. Note that this is a plot of the high-frequency response of the amplifier normalized relative

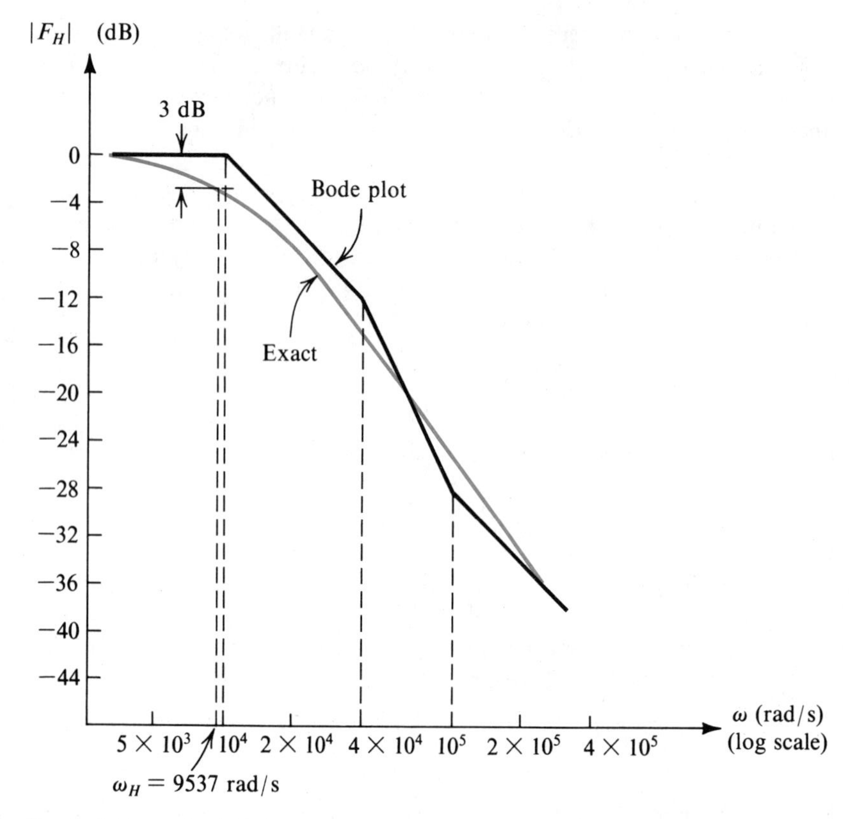

Fig. 7.7 Normalized high-frequency response of the amplifier in Example 7.4.

to its midband gain. That is, if the midband gain is 100 dB, then the entire plot should be shifted upward by 100 dB.

Using Short-Circuit and Open-Circuit Time Constants for the Approximate Determination of ω_L and ω_H

If the poles and zeros of the amplifier transfer function can be determined easily, then we can determine ω_L and ω_H using the techniques described above. In many cases, however, it is not a simple matter to determine the poles and zeros. In such cases, approximate values of ω_L and ω_H can be obtained using the following method.

Consider first the high-frequency response. The function $F_H(s)$ of Eq. (7.19) can be expressed in the alternative form

$$F_H(s) = \frac{1 + a_1 s + a_2 s^2 + \cdots + a_{n_H} s^{n_H}}{1 + b_1 s + b_2 s^2 + \cdots + b_{n_H} s^{n_H}} \tag{7.23}$$

where the coefficients a and b are related to the zero and pole frequencies, respectively. Specifically, the coefficient b_1 is given by

$$b_1 = \frac{1}{\omega_{P1}} + \frac{1}{\omega_{P2}} + \cdots + \frac{1}{\omega_{Pn_H}} \tag{7.24}$$

It can be shown [see Gray and Searle (1969)] that the value of b_1 can be obtained by considering the various capacitances in the high-frequency equivalent circuit one at a time while reducing all other capacitors to zero (or, equivalently, replacing them with open circuits). That is, to obtain the contribution of capacitance C_i we reduce all other capacitances to zero, reduce the input signal source to zero, and determine the resistance R_{io} seen by C_i. This process is then repeated for all other capacitors in the circuit. The value of b_1 is computed by summing the individual time constants, called **open-circuit time constants,**

$$b_1 = \sum_{i=1}^{n_H} C_i R_{io} \tag{7.25}$$

where we have assumed that there are n_H capacitors in the high-frequency equivalent circuit.

This method for determining b_1 is *exact*; the approximation comes about in using the value of b_1 to determine ω_H. Specifically, if the zeros are not dominant and if one of the poles—say, P_1—is dominant, then from Eq. (7.24)

$$b_1 \simeq \frac{1}{\omega_{P1}} \tag{7.26}$$

and the upper 3-dB frequency will be approximately equal to ω_{P1}, leading to

$$\omega_H \simeq \frac{1}{\left[\sum_i C_i R_{io}\right]} \tag{7.27}$$

Here it should be pointed out that in complex circuits we usually do not know whether or not a dominant pole exists. Nevertheless, using Eq. (7.27) to determine ω_H normally yields remarkably good results[1] even if a dominant pole does not exist. The method will be illustrated by an example.

EXAMPLE 7.5

Figure 7.8(a) shows the high-frequency equivalent circuit of a common-source FET amplifier. The amplifier is fed with a signal generator having a resistance R. Resistance R_{in} is due to the biasing network. Resistance R_L' is the parallel equivalent of the load resistance R_L, the drain bias resistance R_D, and the FET output resistance r_o. Capacitors C_{gs} and C_{gd} are the FET internal capacitances, as will be explained in detail in the next section. For

[1] The method of open-circuit time constants yields good results only when all the poles are real, as is the case in this chapter.

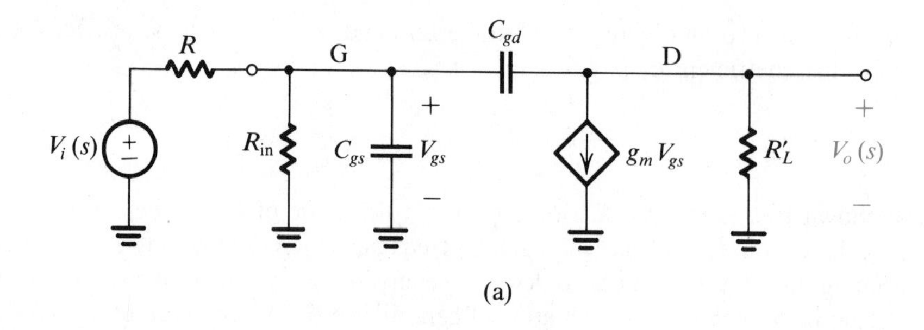

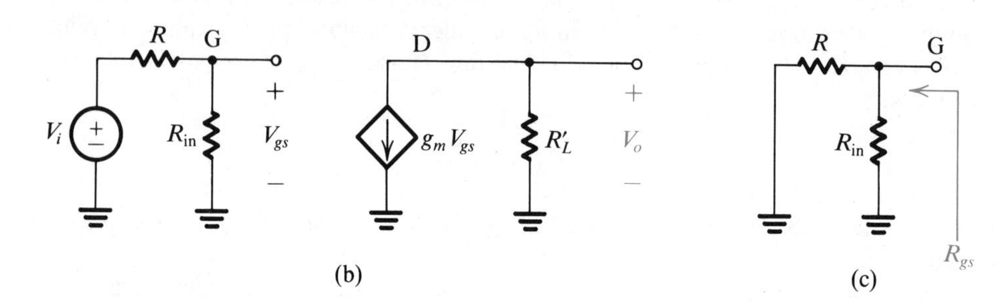

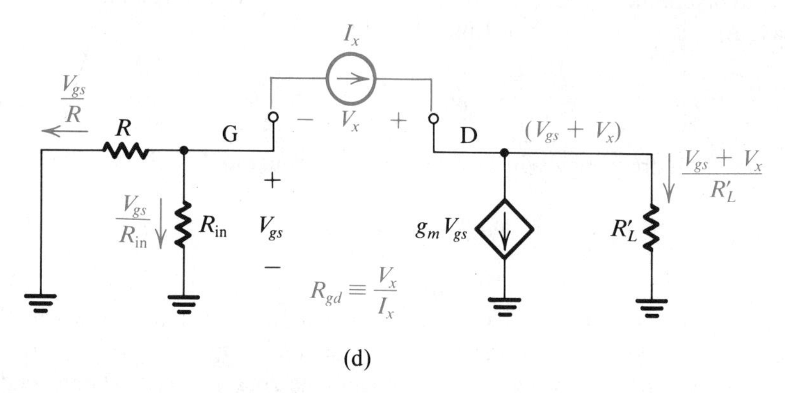

Fig. 7.8 Circuits for Example 7.5.

$R = 100$ kΩ, $R_{\text{in}} = 420$ kΩ, $C_{gs} = C_{gd} = 1$ pF, $g_m = 4$ mA/V, and $R'_L = 3.33$ kΩ, find the midband voltage gain, $A_M = V_o/V_i$, and the upper 3-dB frequency, f_H.

Solution The midband voltage gain is determined by assuming that the capacitors in the FET model are perfect open circuits. This results in the midband equivalent circuit shown in Fig. 7.8(b), from which we find

$$A_M \equiv \frac{V_o}{V_i} = -\frac{R_{\text{in}}}{R_{\text{in}} + R}(g_m R'_L)$$

$$= -\frac{420}{420 + 100} \times 4 \times 3.33 = -10.8 \text{ V/V}$$

We shall determine ω_H using the method of open-circuit time constants. The resistance R_{gs} seen by C_{gs} is found by setting $C_{gd} = 0$ and short-circuiting the signal generator V_i. This results in the circuit of Fig. 7.8(c), from which we find that

$$R_{gs} = R_{\text{in}}//R = 420\ \text{k}\Omega//100\ \text{k}\Omega = 80.8\ \text{k}\Omega$$

Thus the open-circuit time constant of C_{gs} is

$$\tau_{gs} \equiv C_{gs}R_{gs} = 1 \times 10^{-12} \times 80.8 \times 10^3 = 80.8\ \text{ns}$$

The resistance R_{gd} seen by C_{gd} is found by setting $C_{gs} = 0$ and short-circuiting V_i. The result is the circuit in Fig. 7.8(d), to which we apply a test current I_x. Writing a node equation at G gives

$$I_x = -\frac{V_{gs}}{R_{\text{in}}} - \frac{V_{gs}}{R}$$

Thus,

$$V_{gs} = -I_x R' \tag{7.28}$$

where $R' = R_{\text{in}}//R$. A node equation at D provides

$$I_x = g_m V_{gs} + \frac{V_{gs} + V_x}{R'_L}$$

Substituting for V_{gs} from Eq. (7.28) and rearranging terms, yields

$$R_{gd} \equiv \frac{V_x}{I_x} = R' + R'_L + g_m R'_L R' = 1.16\ \text{M}\Omega$$

Thus, the open-circuit time constant of C_{gd} is

$$\begin{aligned}\tau_{gd} &\equiv C_{gd}R_{gd} \\ &= 1 \times 10^{-12} \times 1.16 \times 10^6 = 1160\ \text{ns}\end{aligned}$$

The upper 3-dB frequency ω_H can now be determined from

$$\begin{aligned}\omega_H &\simeq \frac{1}{\tau_{gs} + \tau_{gd}} \\ &= \frac{1}{(80.8 + 1160) \times 10^{-9}} = 806\ \text{krad/s}\end{aligned}$$

Thus

$$f_H = \frac{\omega_H}{2\pi} = 128.3\ \text{kHz}$$

The method of open-circuit time constants has an important advantage in that it tells the circuit designer which of the various capacitances is significant in determining the amplifier frequency response. Specifically, the relative contribution of the various capacitances to the effective time constant b_1 is immediately obvious. For instance, in the above example we see that C_{gd} is the dominant capacitance in determining f_H. We also note that effectively to increase f_H either we use an FET with smaller C_{gd} or, for a given FET, we use a smaller R' or R'_L, thus reducing R_{gd}. If R' is fixed, then for a given FET the only way to increase bandwidth is by reducing the load resistance. Unfortunately, this also decreases the midband gain.

Next we outline the use of short-circuit time constants to determine the lower 3-dB frequency, ω_L. The function $F_L(s)$ of Eq. (7.12) can be expressed in the alternative form

$$F_L(s) = \frac{s^{n_L} + d_1 s^{n_L - 1} + \cdots}{s^{n_L} + e_1 s^{n_L - 1} + \cdots} \tag{7.29}$$

where the coefficients d and e are related to the zero and pole frequencies, respectively. Specifically, the coefficient e_1 is given by

$$e_1 = \omega_{P1} + \omega_{P2} + \cdots + \omega_{Pn_L} \tag{7.30}$$

As shown in Gray and Searle (1969), the exact value of e_1 can be obtained by analyzing the amplifier low-frequency equivalent circuit, considering the various capacitors one at a time while setting all other capacitors to ∞ (or, equivalently, replacing them with short circuits). Thus if capacitor C_i is under consideration, we replace all other capacitors with short circuits, and also reduce the input signal to zero, and determine the resistance R_{is} seen by C_i. The process is then repeated for all other capacitors, and the value of e_1 is computed from

$$e_1 = \sum_{i=1}^{n_L} \frac{1}{C_i R_{is}} \tag{7.31}$$

where it is assumed that there are n_L capacitors in the low-frequency equivalent circuit.

The value of e_1 can be used to obtain an approximate value of the 3-dB frequency ω_L provided that none of the zeros is dominant and that a dominant pole exists. This condition is satisfied if one of the poles—say, P_1—has a frequency ω_{P1} much higher than (at least four times) that of all the other poles and zeros. If this is the case then $\omega_L \simeq \omega_{P1}$ and from Eq. (7.30) we see that $e_1 \simeq \omega_{P1}$, leading to

$$\omega_L \simeq \sum_i \frac{1}{C_i R_{is}} \tag{7.32}$$

Of course, in a complex circuit it is usually difficult to ascertain whether or not a dominant low-frequency pole exists. Nevertheless, the method of short-circuit time constants usually provides a reasonable estimate of ω_L. Such an estimate is quite sufficient for an initial paper-and-pencil design. The method also allows the designer considerable insight into which of the various capacitors most severely limits the low-frequency response. These points will be further illustrated in subsequent sections.

Exercises 7.2 A first-order circuit, having a gain of 10 at dc and a gain of 1 at infinite frequency, has its pole at 10 kHz. Find its transfer function.

Ans. $\dfrac{s + 2\pi \times 10^5}{s + 2\pi \times 10^4}$

7.3 A direct-coupled amplifier has a gain of 1000 and an upper 3-dB frequency of 100 kHz. What is its gain–bandwidth product in hertz?

Ans. 10^8 Hz

7.4 The high-frequency response of an amplifier is characterized by two zeros at $s = \infty$ and two poles at ω_{P1} and ω_{P2}. Expressing $\omega_{P2} = k\omega_{P1}$, find the value of k that results in the exact value of ω_H being $0.9\omega_{P1}$. Repeat for $\omega_H = 0.99\omega_{P1}$.

Ans. 2.78; 9.88

7.5 For the amplifier described in Exercise 7.4, find the exact and approximate values (using Eq. 7.22) of ω_H (as a function of ω_{P1}) for the cases $k = 1, 2, 4$.

Ans. 0.64, 0.71; 0.84, 0.89; 0.95, 0.97

7.6 For the amplifier in Example 7.5, find the gain–bandwidth product in megahertz. Find the value of R_L' that will result in $f_H = 180$ kHz. Find the new values of the midband gain and of the gain–bandwidth product.

Ans. 1.39 MHz; 2.23 kΩ; −7.2 V/V; 1.30 MHz

7.3 LOW-FREQUENCY RESPONSE OF THE COMMON-SOURCE AMPLIFIER

In this section the low-frequency response of the classical capacitor-coupled common-source amplifier stage, shown in Fig. 7.9, will be analyzed. The analysis given applies equally well to MOSFETs.

Figure 7.10 shows the amplifier circuit of Fig. 7.9 prepared to find the voltage gain at low frequencies. To find the gain, we shall start at the signal source and proceed toward the load in a step-by-step manner. Using the voltage-divider rule at the input side, we can find the voltage V_g (between gate and ground) as

$$V_g(s) = V_i(s)\frac{R_{\text{in}}}{R_{\text{in}} + R + 1/sC_{C1}}$$

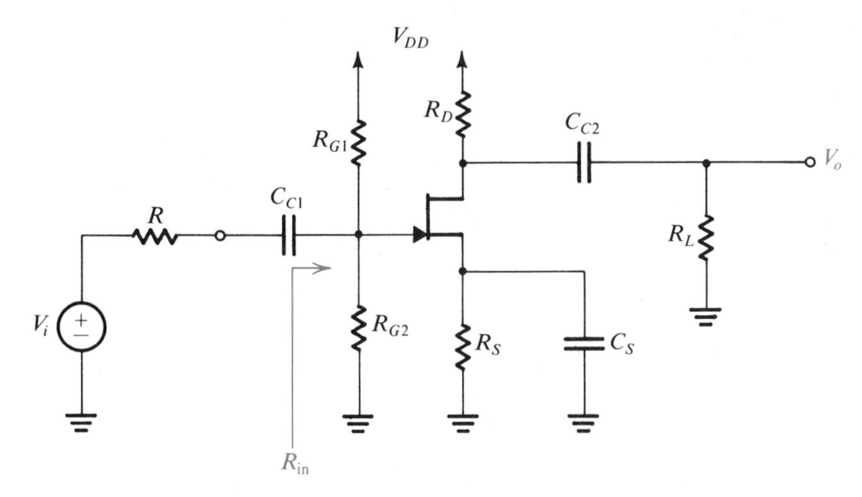

Fig. 7.9 The classical capacitively coupled common-source amplifier.

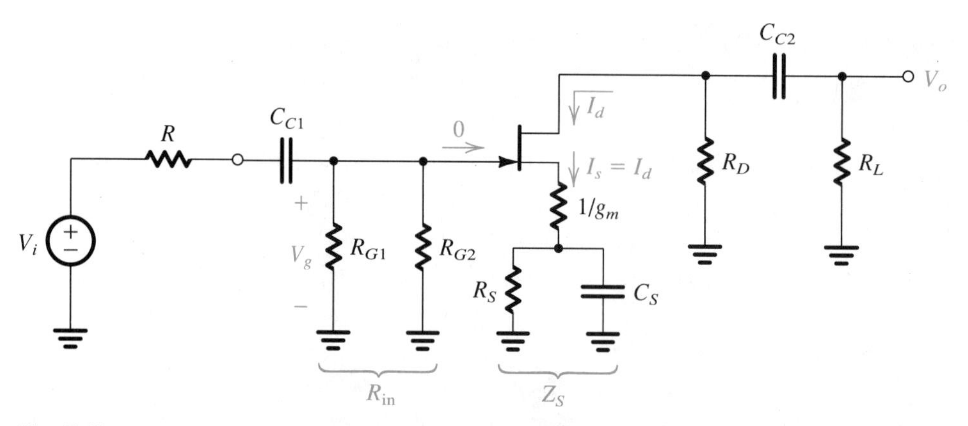

Fig. 7.10 The amplifier circuit of Fig. 7.9 prepared for finding the gain at low frequencies. The resistance $1/g_m$ shown is the FET internal resistance between gate and source looking into the source (i.e., that of the T model).

where $R_{in} = (R_{G1}//R_{G2})$. Thus the transfer function from the input to the gate is given by

$$\frac{V_g(s)}{V_i(s)} = \frac{R_{in}}{R_{in} + R} \frac{s}{s + 1/C_{C1}(R_{in} + R)} \tag{7.33}$$

which is a high-pass function indicating that C_{C1} introduces a zero at zero frequency (dc) and a real pole with a frequency ω_{P1},

$$\omega_{P1} = \frac{1}{C_{C1}(R_{in} + R)} \tag{7.34}$$

Note that we could have arrived at this result by inspection of the input circuit of the amplifier using the techniques of STC network analysis (see Appendix E). Specifically, the input circuit is a high-pass STC network with a time constant equal to C_{C1} multiplied by the total resistance seen by C_{C1}; ω_{P1} is simply the inverse of this time constant.

The next step in the analysis is to find the drain current $I_d(s)$:

$$I_d(s) = I_s(s) = \frac{V_g(s)}{1/g_m + Z_S} \tag{7.35}$$

where we have made use of the fact that the equivalent resistance between gate and source is equal to $1/g_m$ (that is, we have used the T model of the FET implicitly); thus the total impedance between gate and ground, on the source side, is $1/g_m$ in series with Z_S, which denotes the parallel equivalent of R_S and C_S. Equation (7.35) can be rewritten

$$I_d(s) = g_m V_g(s) \frac{Y_S}{g_m + Y_S}$$

where

$$Y_S = \frac{1}{Z_S} = \frac{1}{R_S} + sC_S$$

Thus

$$I_d(s) = g_m V_g(s) \frac{1/R_S + sC_S}{g_m + 1/R_S + sC_S}$$

that is,

$$I_d(s) = g_m V_g(s) \frac{s + 1/C_S R_S}{s + (g_m + 1/R_S)/C_S} \tag{7.36}$$

which indicates that the bypass capacitor C_S introduces a real zero and a real pole. The real zero has a frequency ω_Z,

$$\omega_Z = \frac{1}{C_S R_S} \tag{7.37}$$

while the frequency of the real pole is given by

$$\omega_{P2} = \frac{g_m + 1/R_S}{C_S} = \frac{1}{C_S(R_S//1/g_m)} \tag{7.38}$$

It thus can be seen that ω_Z will always be lower in value than ω_{P2}.

It is instructive to interpret physically the above results regarding the effects of C_S. C_S introduces a zero at the value of s that makes Z_S infinite, which makes physical sense because an infinite Z_S will cause I_d, and hence V_o, to be zero. The pole frequency is the inverse of the time constant formed by multiplying C_S by the resistance seen by the capacitor. To evaluate the latter resistance we ground the signal source (note that the network poles, or natural modes, are independent of the excitation) and grab hold of the terminals of C_S. The resistance seen by C_S will be R_S in parallel with the resistance between source and gate, which is $1/g_m$.

Having determined $I_d(s)$, we can now obtain the output voltage using the output equivalent circuit in Fig. 7.11(a). There is a slight approximation in this equivalent circuit: The resistance r_o is shown connected between drain and ground, rather than between drain and source, in spite of the fact that the source terminal is no longer at ground potential because C_S is not acting as a perfect bypass. However, since the effect of r_o is small anyway (assuming that $r_o \gg R_D$ and R_L), this approximation is valid.

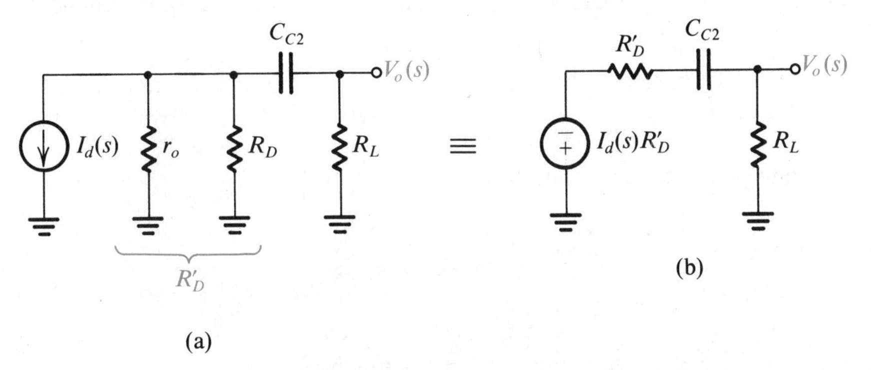

Fig. 7.11 The output equivalent circuit (at low frequency) for the amplifier in Figs. 7.9 and 7.10.

Figure 7.11(b) shows the equivalent output circuit after application of Thévenin's theorem. After some manipulation, we obtain from this figure

$$V_o(s) = -I_d(s)(R_D//r_o//R_L)\frac{s}{s + 1/C_{C2}[R_L + (R_D//r_o)]} \tag{7.39}$$

Thus C_{C2} introduces a zero at zero frequency (dc) and a real pole with a frequency ω_{P3},

$$\omega_{P3} = \frac{1}{C_{C2}[R_L + (R_D//r_o)]} \tag{7.40}$$

Again the frequency of this pole could have been found by inspection of the circuit in Fig. 7.11(a): It equals the inverse of the time constant found by multiplying C_{C2} by the total resistance seen by that capacitor.

The low-frequency amplifier gain $A_L(s)$ can be found by combining Eqs. (7.33), (7.36), and (7.39):

$$A_L(s) = \frac{V_o(s)}{V_i(s)} = A_M\frac{s}{s + \omega_{P1}}\,\frac{s + \omega_Z}{s + \omega_{P2}}\,\frac{s}{s + \omega_{P3}} \tag{7.41}$$

where the midband gain A_M is given by

$$A_M = -\frac{R_{\text{in}}}{R_{\text{in}} + R}g_m(R_D//r_o//R_L) \tag{7.42}$$

and where ω_{P1}, ω_Z, ω_{P2}, and ω_{P3} are given by Eqs. (7.34), (7.37), (7.38), and (7.40), respectively. Note from Eq. (7.41) that as the frequency $s = j\omega$ becomes much larger in magnitude than ω_{P1}, ω_{P2}, ω_{P3}, and ω_Z, the gain approaches the midband value A_M.

Having determined the low-frequency poles and zeros, we can employ the techniques of Section 7.2 to find the lower 3-dB frequency ω_L.

EXAMPLE 7.6

We wish to select appropriate values for the coupling capacitors C_{C1} and C_{C2} and the bypass capacitor C_S of the amplifier in Fig. 7.9, so that the low-frequency response will be dominated by a pole at 100 Hz and that the nearest pole or zero will be at least a decade away. Let $V_{DD} = 20$ V, $R = 100$ kΩ, $R_{G1} = 1.4$ MΩ, $R_{G2} = 0.6$ MΩ, $R_S = 3.5$ kΩ, $R_D = 5$ kΩ, $r_o = \infty$, $R_L = 10$ kΩ, $V_P = -2$ V, and $I_{DSS} = 8$ mA. Also, determine the midband gain.

Solution With the methods of Chapter 5, the following dc operating point is determined:

$$I_D = 2\text{ mA} \qquad V_{GS} = -1\text{ V} \qquad V_D = +10\text{ V}$$

At this operating point the transconductance is

$$g_m = \frac{2I_{DSS}}{-V_P}\sqrt{\frac{I_D}{I_{DSS}}}$$

Thus

$$g_m = \frac{2 \times 8}{2}\sqrt{\frac{2}{8}} = 4\text{ mA/V}$$

The midband voltage gain can be determined as follows: The input resistance R_{in} is given by

$$R_{\text{in}} = \frac{R_{G1}R_{G2}}{R_{G1} + R_{G2}} = \frac{1.4 \times 0.6}{2} = 420 \text{ k}\Omega$$

and the midband voltage gain A_M can be written

$$A_M = \frac{R_{\text{in}}}{R_{\text{in}} + R} \times -g_m(R_D // R_L)$$

$$= -\frac{420}{520} \times 4 \times \frac{5 \times 10}{5 + 10} = -10.8 \text{ V/V}$$

Thus the amplifier has a midband gain of 20.7 dB.

To find which of the three capacitors, C_{C1}, C_S, and C_{C2}, should be made to cause the dominant low-frequency pole at 100 Hz, we first determine the resistance associated with each, as follows:

$$R_{C_{C1}} = R + R_{\text{in}} = 520 \text{ k}\Omega$$

$$R_{C_S} = R_S // (1/g_m) = 0.233 \text{ k}\Omega$$

$$R_{C_{C2}} = R_D + R_L = 15 \text{ k}\Omega$$

We select the smallest resistance, R_{C_S}, as the one to form the highest-frequency (and thus dominant) pole. Thus,

$$C_S = \frac{1}{2\pi f_L R_{C_S}}$$

$$= \frac{1}{2\pi \times 100 \times 0.233 \times 10^3} = 6.83 \ \mu\text{F}$$

The zero due to C_S can be found, using Eq. (7.37), as

$$f_Z = \frac{1}{2\pi C_S R_S} = \frac{1}{2\pi \times 6.83 \times 10^{-6} \times 3.5 \times 10^3} = 6.7 \text{ Hz}$$

To place the two other poles, due to C_{C1} and C_{C2}, at least a decade away from f_L (that is, equal to or lower than 10 Hz) we select the capacitors as follows:

$$C_{C1} \geq \frac{1}{2\pi \times 10 \times 520 \times 10^3} = 0.03 \ \mu\text{F}$$

and

$$C_{C2} \geq \frac{1}{2\pi \times 10 \times 15 \times 10^3} = 1.06 \ \mu\text{F}$$

Note that selecting the smallest resistance to cause the highest-frequency pole results in reasonably small values for the two other capacitors associated with the nondominant poles.

Exercises 7.7 A common-source FET amplifier utilizes a resistance R_S of 1 kΩ bypassed by a capacitor C_S. It is found that the pole and zero due to C_S are at 100 rad/s and 10 rad/s, respectively. Find the values of C_S and g_m.

Ans. 100 μF; 9 mA/V

7.8 If, in the common-source FET amplifier, R_S is replaced with an ideal constant-current source, find the frequencies ω_Z and ω_P caused by C_S.

Ans. $\omega_Z = 0$; $\omega_P = g_m/C_S$

7.4 HIGH-FREQUENCY RESPONSE OF THE COMMON-SOURCE AMPLIFIER

In this section we analyze the high-frequency response of the FET common-source amplifier. The analysis applies to JEETs, MOSFETs, and GaAs MESFETs. It also applies whether the amplifier is capacitively coupled or direct coupled. However, to maintain continuity, the high-frequency response analysis will be performed on the same amplifier circuit considered in the previous section, namely, that in Fig. 7.9.

The FET High-Frequency Model

To investigate the response of FET amplifiers at high frequencies we must have a FET model that takes into account the capacitive effects present in the device. Such a model can be obtained by augmenting the low-frequency model studied in Chapter 5 with appropriate capacitors. For the MOSFET the resulting model is shown in Fig. 7.12(a). To understand the origin of the various capacitances in this model, recall that the dominant capacitive effect in the MOSFET is the gate-to-channel capacitance. When the device operates in the triode region with small v_{DS}, the channel is uniform and the gate-to-channel capacitance is WLC_{OX}. It is a common practice to model this capacitance by two equal capacitors, C_{gs} and C_{gd}, each equal to $\frac{1}{2}WLC_{OX}$. However, in the saturation region, which is the operating region of interest to us here, the channel takes a tapered shape and is pinched off at the drain end. It can be shown [see Gray and Meyer (1984)] that the gate-to-channel capacitance in this case becomes $\frac{2}{3}WLC_{OX}$ and can be modeled by a single capacitor between gate and source. This is the major component of C_{gs} in the model of Fig. 7.12(a). Typically, for a technology with a 0.05-μm oxide thickness, $C_{OX} = 0.7$ fF/μm^2 [a femtofarad (fF) is 10^{-15} F]; thus a device with $L = 10$ μm and $W = 30$ μm has a gate-to-channel capacitance of 0.14 pF.

The other component of C_{gs} arises because the source region extends slightly under the gate electrode (refer to Fig. 5.1). This component of C_{gs} is aptly called **overlap capacitance.** If the length of the overlap is L_d, then the gate-to-source overlap capacitance is WL_dC_{OX}, typically 1 to 10 fF. A similar overlap occurs between the drain region and the gate electrode, giving rise to the model capacitance C_{gd}, which is entirely an overlap capacitance with a typical value of 1 to 10 fF.

The charge storage effects that take place in the depletion region that forms beneath the active area of the device in the substrate (refer to Fig. 5.2) are modeled by the two

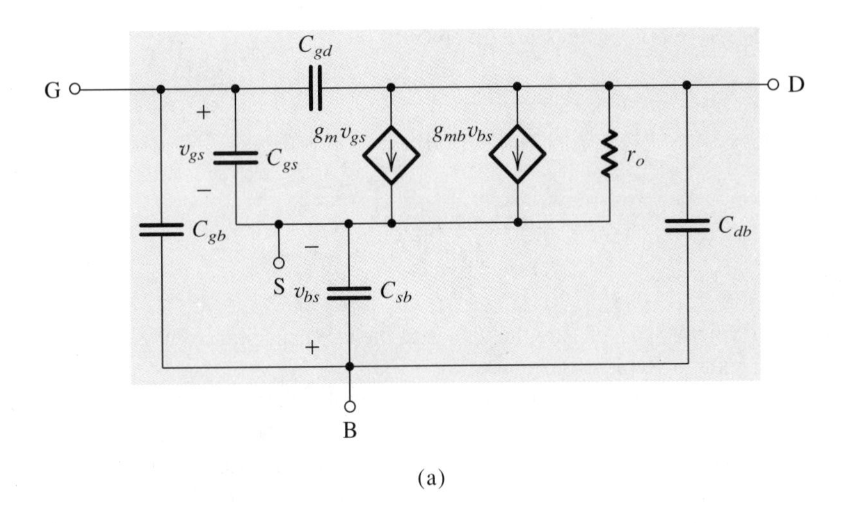

(a)

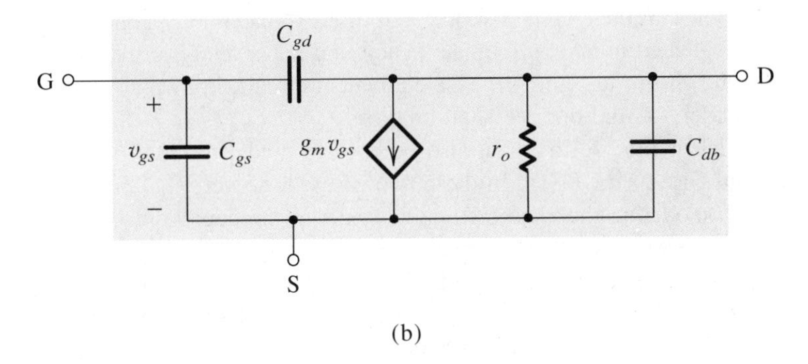

(b)

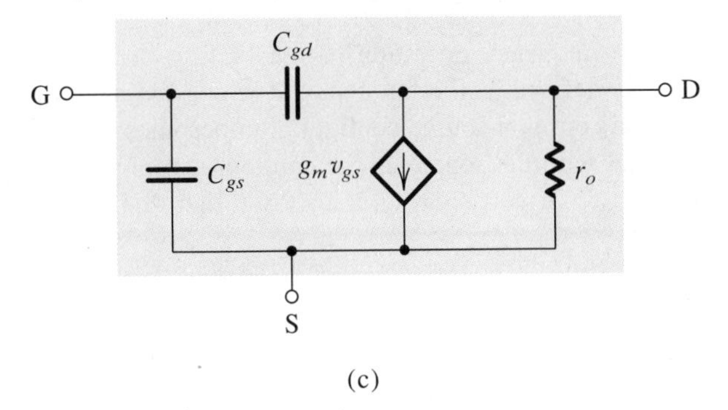

(c)

Fig. 7.12 (a) High-frequency equivalent circuit model for the MOSFET; **(b)** the equivalent circuit for the case the source is connected to the substrate (body). Here C_{gs} includes C_{gb}; **(c)** the equivalent circuit model of (b) with C_{db} neglected (to simplify analysis).

capacitances C_{sb} and C_{db}, whose values are given by[2]

$$C_{sb} = \frac{C_{sb0}}{\left(1 + \frac{|V_{SB}|}{V_0}\right)^{0.5}} \tag{7.43}$$

$$C_{db} = \frac{C_{db0}}{\left(1 + \frac{|V_{DB}|}{V_0}\right)^{0.5}} \tag{7.44}$$

where C_{sb0} and C_{db0} (typically 0.1 pF) are the values of the two capacitances at zero bias, and V_0 is the barrier voltage (typically 0.6 V).

Finally, capacitor C_{gb} models the parasitic capacitance between the gate electrode and the substrate, typically 0.1 pF.

In applications for which the source (S) is connected to the substrate (B) the model simplifies to that shown in Fig. 7.12(b). Here the parasitic capacitance C_{gb} is included in C_{gs}. Typically, C_{gs} is in the range 0.1 to 0.5 pF, C_{gd} is 0.01 to 0.03 pF, and C_{db} is 0.01 to 0.05 pF. It will be shown shortly that C_{gd}, though small, plays a major role in determining the high-frequency response of the common-source amplifier. Capacitor C_{db}, on the other hand, can usually be neglected in an approximate pencil-and-paper analysis, resulting in the simplified model shown in Fig. 7.12(c). This equivalent circuit, known as the high-frequency hybrid-π model,[3] is the one we shall employ.

The hybrid-π model of Fig. 7.12(c) can also be used to model the high-frequency operation of JFETs and GaAs MESFETs. In these two cases, however, C_{gs} and C_{gd} are depletion capacitances, modeling the charge storage effects in the depletion region that forms in the channel under the gate. Thus, the values of C_{gs} and C_{gd} can be found using expressions similar to those in Eqs. (7.43) and (7.44). Typical values for JFETs are $C_{gs} = 1$ to 3 pF and $C_{gd} = 0.1$ to 0.5 pF. For a particular GaAs technology with $L = 1\ \mu\text{m}$, C_{gs} (at $V_{GS} = 0$ V) is 1.6 fF/μm-width and C_{gd} (at $V_{DS} = 2$ V) is 0.16 fF/μm-width. Thus for a MESFET with $W = 100\ \mu\text{m}$, $C_{gs} = 0.16$ pF and $C_{gd} = 0.016$ pF.

The FET Unity-Gain Frequency (f_T)

A figure of merit for the high-frequency operation of the FET as an amplifier is the **unity-gain frequency,** f_T. It is defined as the frequency at which the magnitude of the short-circuit current gain of the common-source configuration becomes unity. Fig. 7.13 shows the FET hybrid-π model with the source as the common terminal between input and output ports and with the output port short-circuited. We find that

$$i_o = g_m v_{gs} \quad \text{and} \quad v_{gs} = i_i / s(C_{gs} + C_{gd})$$

Thus the current gain is given by

$$\frac{i_o}{i_i} = \frac{g_m}{s(C_{gs} + C_{gd})} \tag{7.45}$$

---

[2] See Section 3.10 for a discussion of depletion regions and depletion capacitance.

[3] The name hybrid-π derives from the fact that the model is most conveniently described by the two-port hybrid- or h-parameters (see Appendix B) and from the shape (π) in which it is commonly drawn.

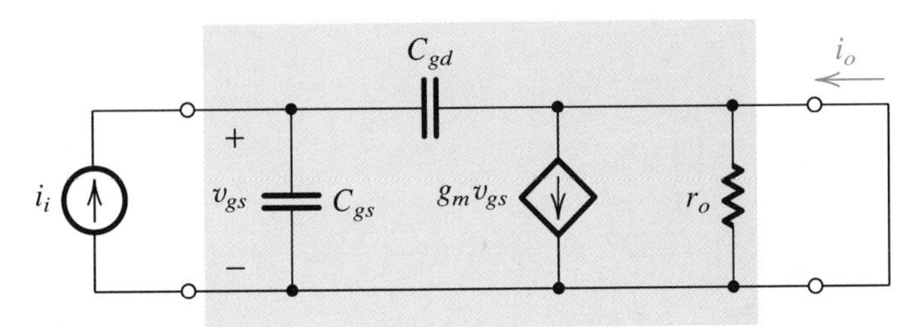

Fig. 7.13 Measurement of the short-circuit current gain i_o/i_i.

For physical frequencies $s = j\omega$, it can be seen that the magnitude of the current gain becomes unity at the frequency

$$\omega_T = g_m/(C_{gs} + C_{gd})$$

Thus, the unity-gain frequency $f_T = \omega_T/2\pi$ is

$$f_T = \frac{g_m}{2\pi(C_{gs} + C_{gd})} \tag{7.46}$$

Note that f_T is a parameter of the device and is independent of the amplifier circuit in which the device is connected. Since f_T is proportional to g_m and inversely proportional to the FET internal capacitances, the higher the value of f_T the more effective the FET as a high-frequency amplifier. Typical values of f_T for various FET types are as follows: For MOSFETs, f_T ranges from about 100 MHz for the older, well established technologies (e.g., a 5-μm CMOS process[4]) to 2 GHz for newer, high-speed technologies (e.g., a 1-μm CMOS process). For JFETs, f_T is typically in the range 20 to 100 MHz. Gallium arsenide technology provides MESFETs with f_T in the range 5 to 15 GHz.

Miller's Theorem

When an FET is connected in the common-source amplifier configuration, capacitor C_{gd} of its hybrid-π model appears in the feedback path from the amplifier output (the FET drain) to its input (the FET gate). The "bridging" nature of C_{gd} complicates the circuit analysis. Fortunately, there is a circuit theorem that allows one to replace a bridging element such as C_{gd} with two grounded elements, one between gate and ground and the other between drain and ground. This replacement not only simplifies circuit analysis, but, more importantly, it also makes clear the significant effect that C_{gd} has on the high-frequency response of the common-source amplifier. This circuit theorem is known as Miller's theorem, and we shall digress briefly to discuss it.

Consider the situation shown in Fig. 7.14(a). We have identified two nodes, 1 and 2, together with the ground terminal of a particular network. As shown, an admittance Y is connected between nodes 1 and 2. In addition nodes 1 and 2 may be connected by other components to other nodes in the network; this is signified by the broken lines emanating from nodes 1 and 2. Miller's theorem provides the means for replacing the "bridging"

[4] A 5-μm process is one in which the minimum allowed channel length is 5 μm. Usually, however, for analog circuits one uses somewhat longer channel lengths (e.g., 10 μm) than the minimum permitted in the particular technology.

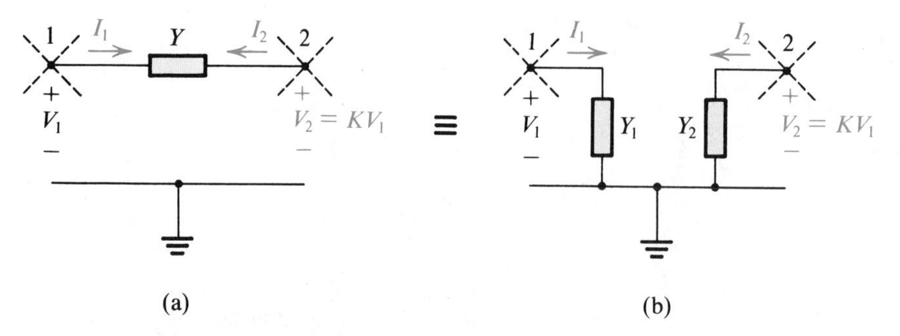

Fig. 7.14 Miller's theorem.

admittance Y with two admittances: Y_1 between node 1 and ground and Y_2 between node 2 and ground, as shown in Fig. 7.14(b).

The Miller replacement illustrated in Fig. 7.14 is based on the premise that it is possible, by independent means, to determine the voltage gain from node 1 to node 2, denoted K, where $K \equiv V_2/V_1$. With the gain K known, the values of Y_1 and Y_2 can be determined as follows: It may be seen from Fig. 7.14(a) that the only way that node 1 "knows of the existence" of the admittance Y is through the current drawn by Y away from node 1. This current, I_1, is given by

$$I_1 = Y(V_1 - V_2) = YV_1(1 - V_2/V_1)$$
$$= YV_1(1 - K)$$

For the circuit of Fig. 7.14(b) to be equivalent to that of Fig. 7.14(a) it is essential that the admittance Y_1 be of such value that the current it draws from node 1 is equal to I_1:

$$Y_1V_1 = I_1$$

which leads to

$$Y_1 = Y(1 - K) \tag{7.47}$$

Similarly, node 2 "feels" the existence of admittance Y only through the current I_2 drawn by Y away from node 2 (note that $I_2 = -I_1$):

$$I_2 = Y(V_2 - V_1)$$
$$= YV_2(1 - V_1/V_2)$$

Thus

$$I_2 = Y(1 - 1/K)V_2$$

For the circuit of Fig. 7.14(b) to be equivalent to that of Fig. 7.14(a) it is essential that the value of Y_2 be such that the current it draws from node 2 is equal to I_2:

$$Y_2V_2 = I_2$$

which leads to

$$Y_2 = Y(1 - 1/K) \tag{7.48}$$

Equations (7.47) and (7.48) are the two necessary and sufficient conditions for the network in Fig. 7.14(b) to be equivalent to that of Fig. 7.14(a). Note that in both networks the voltage gain from node 1 to node 2 is equal to K. We shall apply Miller's theorem quite frequently throughout this book. An important caution, however, is in order. *The Miller equivalent circuit of Fig. 7.14(b) is valid only as long as the conditions that existed in the network when K was determined are not changed.* It follows that although Miller's theorem is very useful in determining the input impedance and the gain of an amplifier, it cannot be used to determine its output resistance. This is because in determining the output resistance in the conventional way, the input signal source is eliminated and a test voltage is applied to the output terminals. This obviously changes the value of K and makes the Miller equivalent invalid.

To appreciate the usefulness of Miller's theorem, consider the case of a common-source amplifier that has been determined, by some means, to have a voltage gain from gate to drain of, say, -100 V/V. The small capacitance C_{gd} (say, 1 pF) gives rise to an input capacitance between gate and ground that can be determined using Eq. (7.47) as

$$\begin{aligned} C_1 &= C_{gd}(1 - K) \\ &= 1[1 - (-100)] = 101 \text{ pF} \end{aligned}$$

a rather large input capacitance that surely will limit the high-frequency response of the amplifier! The multiplication effect experienced by C_{gd} is known as the **Miller effect.** The application of Miller's theorem in the analysis of the high-frequency response of the common-source amplifier is presented next.

Analysis of the High-Frequency Response

Armed with a high-frequency FET model and with a technique for simplifying its application, we now consider the analysis of the high-frequency response of the common-source amplifier of Fig. 7.9. Replacing the JFET with its hybrid-π equivalent circuit model and replacing $(R_{G1}//R_{G2})$ with the equivalent resistance R_{in} results in the amplifier high-frequency equivalent circuit shown in Fig. 7.15(a). A simplified version of the equivalent circuit, as shown in Fig. 7.15(b), is obtained by applying Thévenin's theorem at the input side and by combining the three resistances r_o, R_D, and R_L.

There are a number of ways to analyze the high-frequency equivalent circuit of Fig. 7.15(b) to determine the upper 3-dB frequency ω_H. One approach is to use the method of open-circuit time constants. We have in fact already done this in Example 7.5. Another approximate method that yields considerable insight into high-frequency limitations involves the application of Miller's theorem to replace C_{gd} by an equivalent input capacitance between the gate and ground. This method is based on the observation that since C_{gd} is small, the current through it will be much smaller than that of the controlled source $g_m V_{gs}$. Thus, neglecting the current through C_{gd} in determining the output voltage V_o, we can write

$$V_o \simeq -g_m V_{gs} R_L' \tag{7.49}$$

Using the ratio of the voltages at the two sides of C_{gd} enables us to replace C_{gd} at the input (gate) side with the equivalent Miller capacitance

$$C_{\text{eq}} = C_{gd}(1 + g_m R_L') \tag{7.50}$$

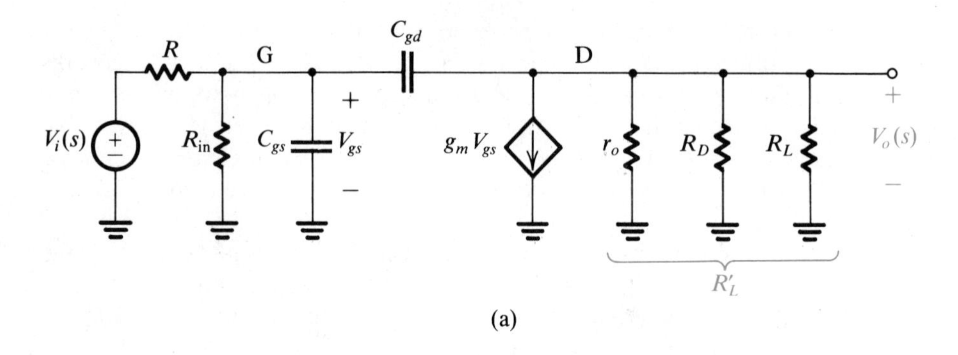

(a)

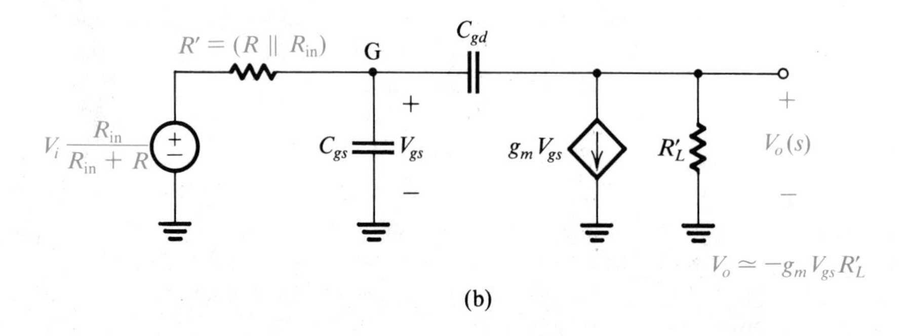

(b)

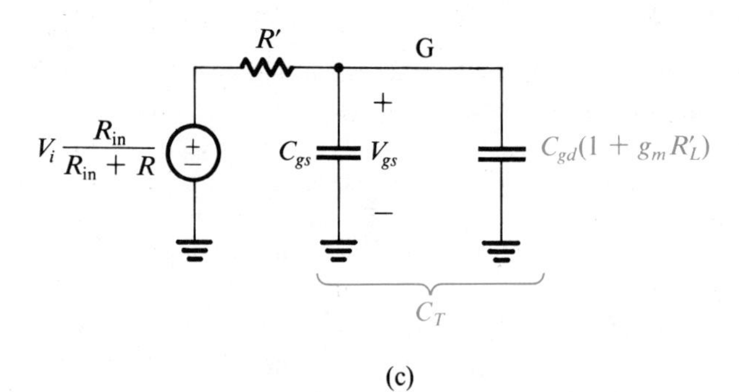

(c)

Fig. 7.15 Equivalent circuits for evaluating the high-frequency response of the amplifier of Fig. 7.9.

as shown in Fig. 7.15(c). We recognize the circuit at the input side as that of a first-order low-pass filter whose time constant is determined by the total input capacitance

$$C_T = C_{gs} + C_{gd}(1 + g_m R'_L) \tag{7.51}$$

and the effective generator resistance

$$R' = R // R_{in} \tag{7.52}$$

This first-order circuit determines the high-frequency response of the common-source amplifier, introducing a dominant high-frequency pole. Thus the upper 3-dB frequency will be

$$\omega_H = \frac{1}{C_T R'} \tag{7.53}$$

We may thus express the high-frequency gain as

$$A_H(s) = A_M \frac{1}{1 + s/\omega_H} \tag{7.54}$$

where, as before, A_M denotes the midband gain given in Eq. (7.42).

From the above, we note the imporant role played by the small feedback capacitance C_{gd} in determining the high-frequency response of the common-source amplifier. Because the voltages at the two sides of C_{gd} are in the ratio of $-g_m R_L'$, which is a large number approximately equal to the midband gain, C_{gd} gives rise to a large capacitance, $C_{gd}(1 + g_m R_L')$, across the input terminals of the amplifier. This is the Miller effect. It follows that to increase the upper 3-dB or cutoff frequency of the amplifier, one has either to reduce $g_m R_L'$, which reduces the midband gain, or to reduce the source resistance, which might not always be possible. Alternatively, one can use circuit configurations that do not suffer from the Miller effect, such as the cascode configuration introduced in Section 6.5. This and other special configurations for *wideband amplifiers* will be studied in later sections.

EXAMPLE 7.7

Use the approximate method based on the Miller effect to find the upper 3-dB frequency of the common-source amplifier whose component values are specified in Example 7.6. Let $C_{gs} = C_{gd} = 1$ pF. Compare the result with that obtained, for the same amplifier, in Example 7.5 using the method of open-circuit time constants.

Solution

The total input capacitance is obtained from Eq. (7.51) as follows:

$$C_T = 1 + 1 \times (1 + 4 \times 3.33) = 15.3 \text{ pF}$$

The effective generator resistance is obtained from Eq. (7.52):

$$R' = 100 \text{ k}\Omega // 420 \text{ k}\Omega = 80.8 \text{ k}\Omega$$

Thus, using Eq. (7.53), we obtain f_H as follows:

$$f_H = \frac{\omega_H}{2\pi} = \frac{1}{2\pi \times 15.3 \times 10^{-12} \times 80.8 \times 10^3} = 128.7 \text{ kHz}$$

which is very close to the value (128.3 kHz) obtained in Example 7.5.

The approximation involved in the above method for determining ω_H is equivalent to assuming that a dominant high-frequency pole exists. To verify that this indeed is the case we shall derive the exact high-frequency transfer function of the common-source amplifier. Converting the input-signal generator to the Norton's form results in the circuit

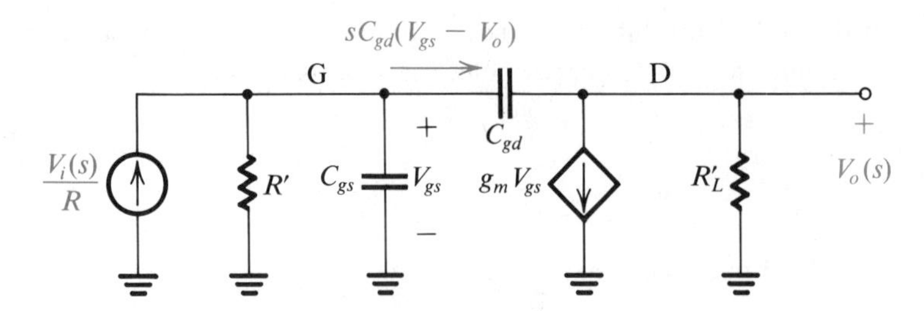

Fig. 7.16 Determination of the exact high-frequency transfer function of the common-source amplifier.

shown in Fig. 7.16. Writing a node equation at G yields

$$\frac{V_i(s)}{R} = \frac{V_{gs}}{R'} + sC_{gs}V_{gs} + sC_{gd}(V_{gs} - V_o) \tag{7.55}$$

Writing a node equation at D gives

$$sC_{gd}(V_{gs} - V_o) = g_m V_{gs} + \frac{V_o(s)}{R'_L} \tag{7.56}$$

Eliminating V_{gs} from Eqs. (7.55) and (7.56) gives the transfer function

$$\frac{V_o(s)}{V_i(s)} = -A_M \frac{1 - \dfrac{s}{(g_m/C_{gd})}}{1 + s[C_{gs} + C_{gd}(1 + g_m R'_L) + C_{gd}(R'_L/R')]R' + s^2 C_{gs} C_{gd} R' R'_L} \tag{7.57}$$

Thus the amplifier has a zero with frequency $\omega_Z = g_m/C_{gd}$ and two poles whose frequencies can be determined from the denominator polynomial. Note that the coefficient of the s term in the denominator is, as expected, equal to the value derived in Example 7.5 using open-circuit time constants. Apart from this observation, the denominator polynomial is unfortunately too complex to draw useful information from directly. We can, however, substitute numerical values and obtain the frequencies of the poles, as is requested in the following exercise.

Exercise 7.9 For the common-source amplifier specified in Examples 7.6 and 7.7 use Eq. (7.57) to determine the frequencies of its finite zero and two poles.

Ans. $f_Z = 637$ MHz; $f_{P1} = 128.4$ kHz; $f_{P2} = 734$ MHz

The answers to Exercise 7.9 show that the zero and second pole are indeed at much higher frequencies than the dominant pole. The fact that the two poles are so widely separated enables us to factor the denominator of Eq. (7.57) as shown below. The denominator polynomial $D(s)$ can be written as

$$\begin{aligned} D(s) &= \left(1 + \frac{s}{\omega_{P1}}\right)\left(1 + \frac{s}{\omega_{P2}}\right) \\ &= 1 + s\left(\frac{1}{\omega_{P1}} + \frac{1}{\omega_{P2}}\right) + \frac{s^2}{\omega_{P1}\omega_{P2}} \\ &\simeq 1 + \frac{s}{\omega_{P1}} + \frac{s^2}{\omega_{P1}\omega_{P2}} \end{aligned} \tag{7.58}$$

Equating the coefficients of the s terms in Eqs. (7.57) and (7.58) gives

$$\omega_{P1} = \frac{1}{[C_{gs} + C_{gd}(1 + g_m R'_L) + C_{gd}(R'_L/R')]R'} \tag{7.59}$$

which is slightly different from the value obtained using the Miller effect but identical to the value of ω_H obtained using open-circuit time constants. Equating the coefficients of s^2 in Eqs. (7.57) and (7.58) and using (7.59) gives the frequency of the second pole

$$\omega_{P2} = \frac{C_{gs} + C_{gd}(1 + g_m R'_L) + C_{gd}(R'_L/R')}{C_{gs} C_{gd} R'_L} \tag{7.60}$$

For $g_m R'_L \gg 1$ and $R'_L < R'$, this expression can be approximated as

$$\omega_{P2} \simeq \frac{g_m}{C_{gs}} \tag{7.61}$$

which shows that ω_{P2} will usually be very high.

EXAMPLE 7.8 If the amplifier analyzed in Examples 7.5 to 7.7 is excited by a pulse of 0.2-V height and 0.2-ms width, find the height, the rise and fall times, and the sag of the output pulse.

Solution The output pulse will have a height V,

$$V = A_M \times 0.2 = 10.8 \times 0.2 = 2.16 \text{ V}$$

The rise and fall times of the output pulse will be determined by the dominant high-frequency pole (see Appendix E),

$$t_r = t_f \simeq 2.2\tau_H$$

where

$$\tau_H = \frac{1}{\omega_H} = \frac{1}{2\pi f_H} = \frac{1}{2\pi \times 128.7 \times 10^3} \simeq 1.2\ \mu\text{s}$$

Thus

$$t_r = t_f = 2.64\ \mu\text{s}$$

On the other hand, the sag or decay in the amplitude of the output pulse will be determined by the dominant low-frequency pole (see Appendix E). Specifically, if the loss in pulse height is denoted by ΔV, then

$$\text{sag} = \frac{\Delta V}{V} = \frac{t_p}{\tau_L} = \omega_L t_p = 2\pi f_L t_p$$

Thus

$$\text{sag} = 2\pi \times 100 \times 0.2 \times 10^{-3} = 0.126 = 12.6\%$$

7.5 THE HYBRID-π EQUIVALENT CIRCUIT MODEL OF THE BJT

Before considering the frequency response analysis of BJT amplifiers, we shall take a closer look at its hybrid-π model at both low and high frequencies and give methods for the determination of the model parameters from terminal measurements or from data-sheet specifications.

The Low-Frequency Model

Figure 7.17 shows the complete low-frequency hybrid-π equivalent circuit model. In addition to the intrinsic model parameters r_π and g_m, this model includes three resistances: r_o, r_μ, and r_x.

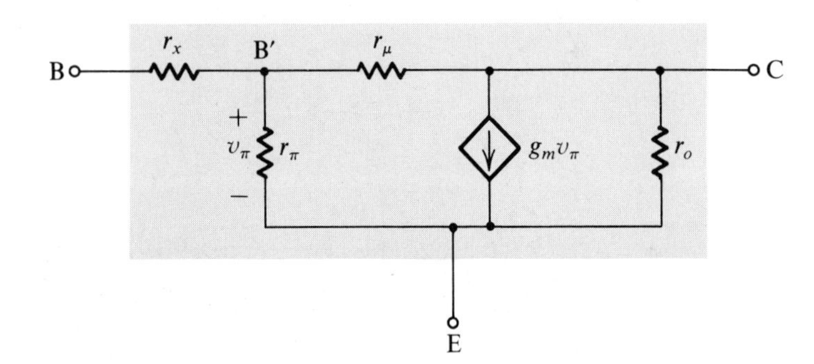

Fig. 7.17 Complete low-frequency hybrid-π model.

Resistance r_o models the slight effect of the collector voltage on the collector current in the active region of operation. Typically r_o is in the range of tens to hundreds of kΩ and its value is inversely proportional to the dc bias current ($r_o = V_A/I_C$). Since g_m is directly proportional to the bias current, the product $g_m r_o$, denoted μ, where $\mu = V_A/V_T$, is a constant for a given transistor, with a value of a few thousand.

The resistance r_μ models the effect of the collector voltage on the base current, and its value is usually much larger than r_o. It can be shown from physical considerations of device operation that r_μ is at least equal to $\beta_0 r_o$ (β_0 denotes the value of β at low frequencies). For modern IC transistors r_μ is closer to $10\beta_0 r_o$. Because of its extremely large value and because including r_μ in the equivalent circuit model destroys its *unilateral* character and thus complicates the analysis, one usually ignores r_μ. There are very special situations, however, where one has to include r_μ. We have already encountered one such case (in evaluating the common-mode input resistance of a differential amplifier).

The resistance r_x models the resistance of the silicon material of the base region between the base terminal B and a fictitious internal, or intrinsic, base terminal B′. The latter node represents the base side of the emitter–base junction. Typically, r_x is a few tens of ohms, and its value depends on the current level in a rather complicated manner. Since r_x is much smaller than r_π, the effect of r_x is negligible at low frequencies. Its presence is felt, however, at high frequencies, where the input impedance of the transistor becomes highly capacitive. This point will become apparent later in this section. In conclusion, r_x can usually be neglected in low-frequency applications.

Exercise 7.10 Each of the common-emitter i_C–v_{CE} curves is measured with the value of base current held constant. This is equivalent to assuming that the signal current $i_b = 0$; that is, from a signal point of view, the base is open-circuited. Use the equivalent circuit in Fig. E7.10 to determine the slope of the i_C–v_{CE} curves in the active region.

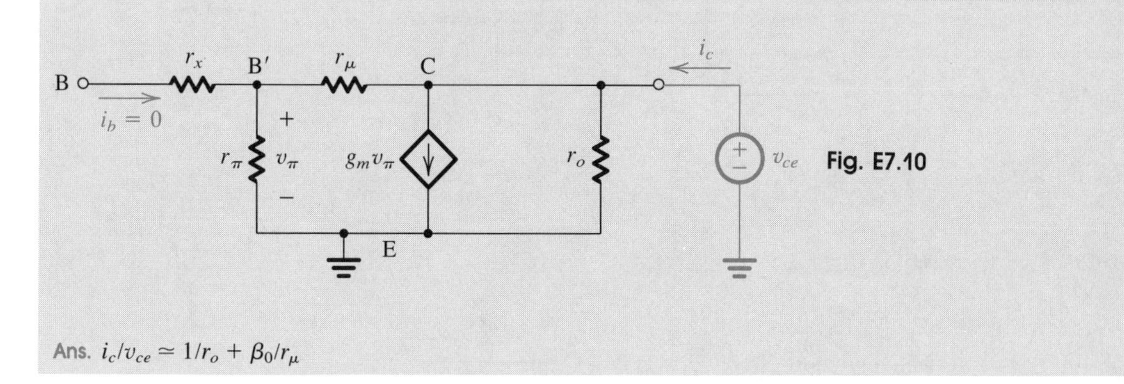

Fig. E7.10

Ans. $i_c/v_{ce} \simeq 1/r_o + \beta_0/r_\mu$

Determination of the Low-Frequency Model Parameters

The transistor is a three-terminal device that can be converted into a two-port network by grounding one of its terminals. It therefore can be characterized by one of the various two-port parameter sets (see Appendix B). For the BJT at low frequencies, the h parameters have been found to be the most convenient. In the following we briefly discuss methods for measuring the h parameters for a transistor biased to operate in the active region. We also derive formulas relating the hybrid-π model parameters to the measured h parameters. These formulas allow us to determine the hybrid-π parameters. It should be emphasized that because the hybrid-π model is closely related to the physical operation of the transistor, its use provides the circuit designer with considerable insight into circuit operation. Thus our interest in h parameters is solely for the purpose of determining the hybrid-π component values.

If the emitter of a transistor biased in the active mode is grounded, port 1 is defined to be between base and emitter, and port 2 is defined to be between collector and emitter, then for small signals around the given bias point we can write

$$v_b = h_{ie} i_b + h_{re} v_c \tag{7.62}$$

$$i_c = h_{fe} i_b + h_{oe} v_c \tag{7.63}$$

These are the defining equations of the common-emitter h parameters where, rather than using the notation h_{11}, h_{12}, and so on, we have assigned more desciptive subscripts to the h parameters: i means input, r means reverse, f means forward, o means output, and the added e denotes a common emitter.

For measurement of h_{ie} and h_{fe}, the circuit shown in Fig. 7.18 can be used. Here R_B is a large resistance that together with V_{BB} determines I_B. The resistance R_C is used to establish the desired dc voltage at the collector, and a very small resistance R_L is used to enable measuring the signal current in the collector. Since R_L is small, the collector is

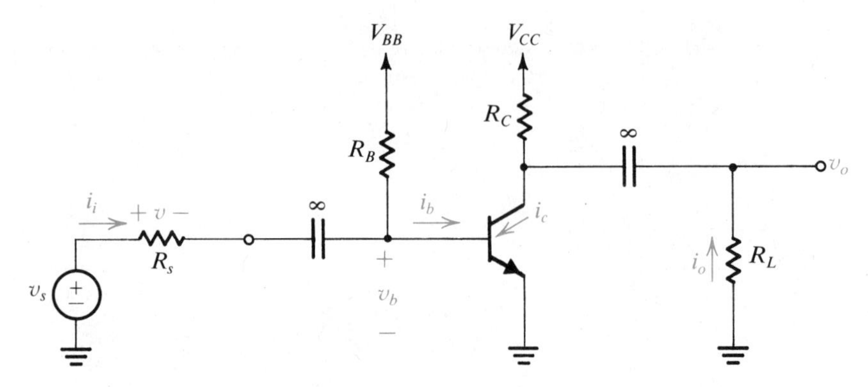

Fig. 7.18 Circuit for measuring h_{ie} and h_{fe}.

effectively short-circuited to ground and

$$i_c \simeq i_o = -\frac{v_o}{R_L}$$

The input signal current i_i is determined by measuring the voltage v across a known resistance R_s. If R_B is large, then

$$i_b \simeq i_i = \frac{v}{R_s}$$

The input signal voltage v_b can be measured directly at the base. Using these measured values one can compute h_{ie} and h_{fe}:

$$h_{ie} = \frac{v_b}{i_b} \quad \text{and} \quad h_{fe} = \frac{i_c}{i_b}$$

To measure h_{re} we use the circuit shown in Fig. 7.19. Here again R_B should be large (much larger than r_π), and the voltmeter used to measure v_b should have a high input

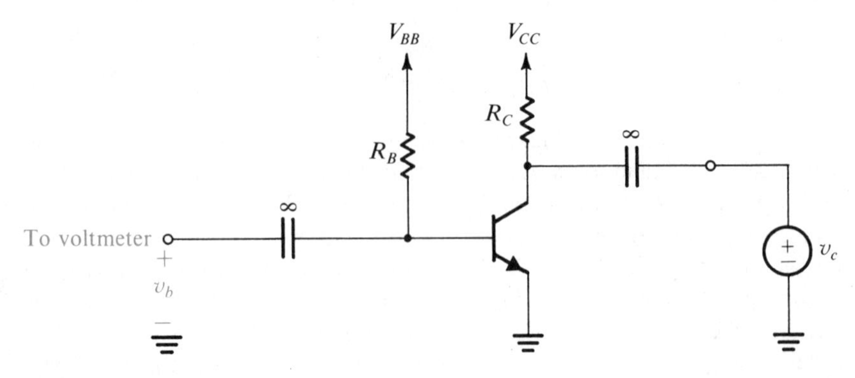

Fig. 7.19 Circuit for measuring h_{re}.

resistance to ensure that the base is effectively open-circuited. The value of h_{re} can then be determined, as follows:

$$h_{re} = \frac{v_b}{v_c}$$

Finally, from Eq. (7.63) we note that h_{oe} is the output conductance with the base open-circuited, which by definition is the slope of the i_C–v_{CE} characteristic curves. Thus h_{oe} can be most conveniently determined from the static common-emitter characteristics.

Expressions for h_{ie} and h_{fe} in terms of the hybrid-π model parameters can be derived by analyzing the equivalent circuit in Fig. 7.20. This analysis yields

$$h_{ie} = r_x + (r_\pi // r_\mu) \tag{7.64}$$

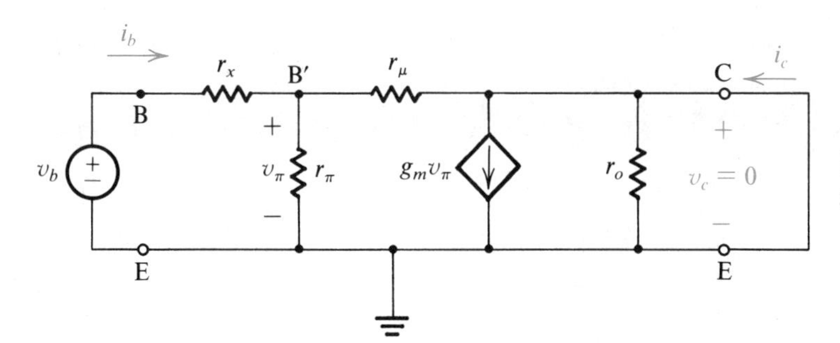

Fig. 7.20 Circuit for deriving expressions for h_{ie} and h_{fe}.

which can be approximated as

$$h_{ie} \simeq r_x + r_\pi \tag{7.65}$$

and

$$h_{fe} = g_m r_\pi \tag{7.66}$$

Here we should note that, by definition, h_{fe} is identical to the ac β, or β_{ac}, and that the value given by Eq. (7.66) indeed corresponds to the formula we used earlier for the low-frequency β, or β_0.

To derive expressions for h_{re} and h_{oe} we use the equivalent circuit model in Fig. 7.21 and obtain

$$h_{re} = \frac{r_\pi}{r_\pi + r_\mu}$$

which can be approximated as

$$h_{re} \simeq \frac{r_\pi}{r_\mu} \tag{7.67}$$

$$h_{oe} \simeq \frac{1}{r_o} + \frac{\beta_0}{r_\mu} \tag{7.68}$$

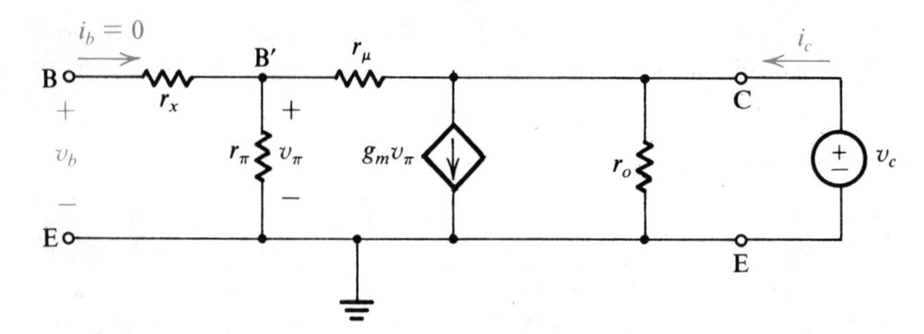

Fig. 7.21 Circuit for deriving expressions for h_{re} and h_{oe}.

This last expression is identical to that given for the slope of the i_C–v_{CE} characteristics with the base open-circuited (Exercise 7.10).

The above expressions can be used to determine the values of the hybrid-π model parameters from the measured h parameters as follows:

$$g_m = I_C/V_T \tag{7.69}$$

$$r_\pi = h_{fe}/g_m \tag{7.70}$$

$$r_x = h_{ie} - r_\pi \tag{7.71}$$

$$r_\mu = r_\pi/h_{re} \tag{7.72}$$

$$r_o = \left(h_{oe} - \frac{h_{fe}}{r_\mu}\right)^{-1} \tag{7.73}$$

$$= V_A/I_C \tag{7.74}$$

It should be noted, however, that since normally $r_x \ll r_\pi$, Eq. (7.71) does not provide an accurate determination of r_x. In fact, there is no accurate way to determine r_x at low frequencies, which should come as no surprise since r_x plays a minor role at low frequencies.

Exercise 7.11 The following parameters were measured on a transistor biased at $I_C = 1$ mA: $h_{ie} = 2.6$ kΩ, $h_{fe} = 100$, $h_{re} = 0.5 \times 10^{-4}$, $h_{oe} = 1.2 \times 10^{-5}$ A/V. Determine the values of g_m, r_π, r_x, r_μ, and r_o, and V_A.

Ans. 40 mA/V; 2.5 kΩ; 100 Ω; 50 MΩ; 100 kΩ; 100 V

The High-Frequency Model

With the exception of the extremely large resistance r_μ, the high-frequency hybrid-π model shown in Fig. 7.22 includes all the resistances of the low-frequency model as well as two capacitances: the emitter–base capacitance C_π and the collector–base capacitance C_μ. The resistance r_μ is omitted because, even at moderate frequencies, the reactance of C_μ is much smaller than r_μ. The emitter–base capacitance C_π is composed of two parts: a diffusion capacitance C_{de}, which is proportional to the dc bias current, and a depletion-layer capacitance C_{je}, which depends on the value of V_{BE}. The collector–base capacitance C_μ is entirely a depletion capacitance (C_{jc}), and its value depends on V_{CB}. Typically C_π is

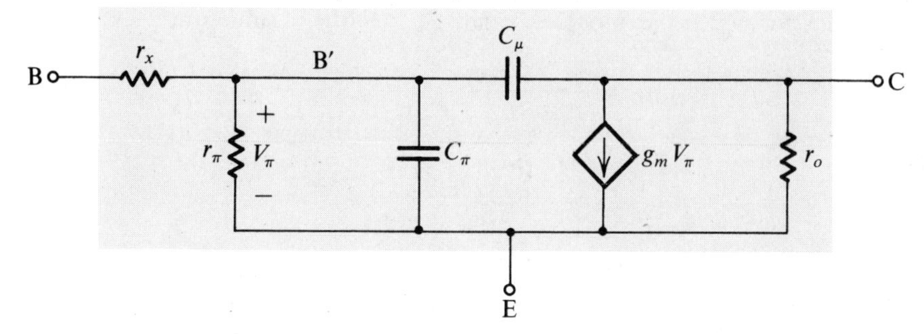

Fig. 7.22 The high-frequency hybrid-π model.

in the range of a few picofarads to a few tens of picofarads, and C_μ is in the range of a fraction of a picofarad to a few picofarads.

The Cutoff Frequency

The transistor data sheets do not usually specify the value of C_π. Rather, the behavior of h_{fe} versus frequency is normally given. In order to determine C_π and C_μ we shall derive an expression for h_{fe} as a function of frequency in terms of the hybrid-π components. For this purpose consider the circuit shown in Fig. 7.23, in which the collector is shorted to the emitter. The short-circuit collector current I_c is

$$I_c = (g_m - sC_\mu)V_\pi \tag{7.75}$$

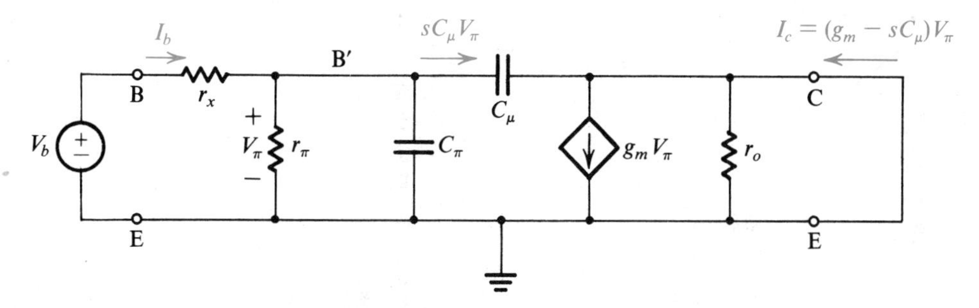

Fig. 7.23 Circuit for deriving an expression for $h_{fe}(s) \equiv I_c/I_b$.

A relationship between V_π and I_b can be established by multiplying I_b by the impedance seen between B′ and E:

$$V_\pi = I_b(r_\pi // C_\pi // C_\mu) \tag{7.76}$$

Thus h_{fe} can be obtained by combining Eqs. (7.75) and (7.76):

$$h_{fe} \equiv \frac{I_c}{I_b} = \frac{g_m - sC_\mu}{1/r_\pi + s(C_\pi + C_\mu)}$$

At the frequencies for which this model is valid, $g_m \gg \omega C_\mu$, resulting in

$$h_{fe} \simeq \frac{g_m r_\pi}{1 + s(C_\pi + C_\mu)r_\pi}$$

Thus

$$h_{fe} = \frac{\beta_0}{1 + s(C_\pi + C_\mu)r_\pi} \tag{7.77}$$

Thus h_{fe} has a single-pole response with a 3-dB frequency at $\omega = \omega_\beta$,

$$\omega_\beta = \frac{1}{(C_\pi + C_\mu)r_\pi} \tag{7.78}$$

Figure 7.24 shows a Bode plot for $|h_{fe}|$. From the -6-dB/octave slope it follows that the frequency at which $|h_{fe}|$ drops to unity, which is called the **unity-gain bandwidth** ω_T, is given by

$$\omega_T = \beta_0 \omega_\beta \tag{7.79}$$

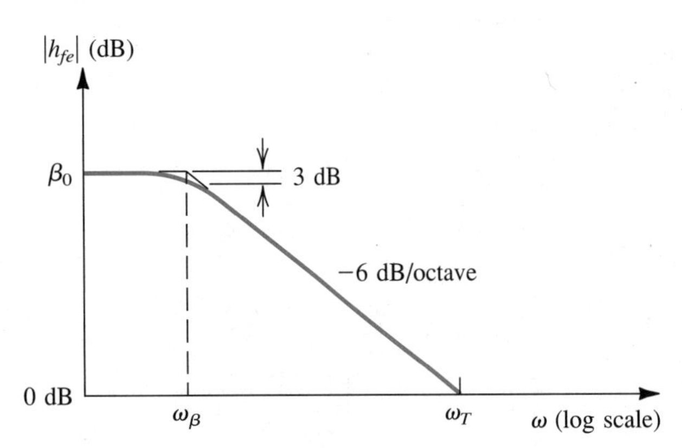

Fig. 7.24 Bode plot for $|h_{fe}|$.

Thus

$$\omega_T = \frac{g_m}{C_\pi + C_\mu} \tag{7.80}$$

and

$$f_T = \frac{g_m}{2\pi(C_\pi + C_\mu)} \tag{7.81}$$

The unity-gain bandwidth f_T is usually specified on the data sheets of the transistor. In some cases f_T is given as a function of I_C and V_{CE}. To see how f_T changes with I_C, recall that g_m is directly proportional to I_C, but only part of C_π (the diffusion capacitance C_d) is directly proportional to I_C. It follows that f_T decreases at low currents, as shown in Fig. 7.25. However, the decrease in f_T at high currents, also shown in Fig. 7.25, can-

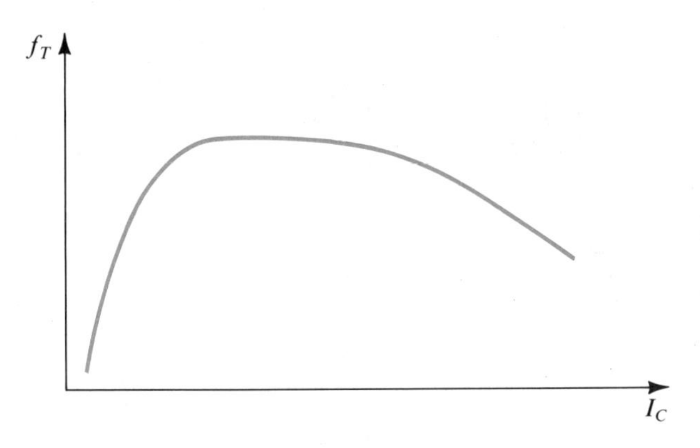

Fig. 7.25 Variation of f_T with I_C.

not be explained by this argument; rather it is due to the same phenomenon that causes β_0 to decrease at high currents. In the region where f_T is almost constant, C_π is dominated by the diffusion part.

Typically, f_T is in the range of 100 MHz to a few GHz. The value of f_T can be used in Eq. (7.81) to determine $C_\pi + C_\mu$. The capacitance C_μ is usually determined separately by measuring the capacitance between base and collector at the desired reverse-bias voltage V_{CB}.

Before leaving this section we should mention that the hybrid-π model of Fig. 7.22 characterizes transistor operation fairly accurately up to a frequency of about $0.2\omega_T$. At higher frequencies one has to add other parasitic elements to the model as well as refine the model to account for the fact that the transistor is in fact a distributed-parameter network that we are trying to model with a lumped-component circuit. One such refinement consists of splitting r_x into a number of parts and replacing C_μ by a number of capacitors each connected between the collector and one of the taps of r_x. This topic is beyond the scope of this book.

An important observation to make from the high-frequency model of Fig. 7.22 is that at frequencies above $5\omega_\beta$ or $10\omega_\beta$ one may ignore the resistance r_π. It can be seen then that r_x becomes the only resistive part of the input impedance at high frequencies. Thus r_x plays an important role in determining the frequency response of transistors at high frequencies. It follows that an accurate determination of r_x should be made from a high-frequency measurement.

Exercises 7.12 For the same transistor considered in Exercise 7.11 and at the same bias point, determine f_T and C_π if $C_\mu = 2$ pF and $|h_{fe}| = 10$ at 50 MHz.

Ans. 500 MHz; 10.7 pF

7.13 If C_π of the BJT in Exercise 7.12 includes a relatively constant depletion-layer capacitance of 2 pF, find f_T of the BJT when operated at $I_C = 0.1$ mA.

Ans. 130.7 MHz

7.6 FREQUENCY RESPONSE OF THE COMMON-EMITTER AMPLIFIER

Analysis of the frequency response of the classical common-emitter amplifier stage in Fig. 7.26 follows a procedure identical to that used for the FET common-source amplifier in Sections 7.3 and 7.4. Thus the overall gain can be written in the form

$$A(s) = A_M \frac{1}{1 + s/\omega_H} \frac{s^2(s + \omega_Z)}{(s + \omega_{P1})(s + \omega_{P2})(s + \omega_{P3})} \tag{7.82}$$

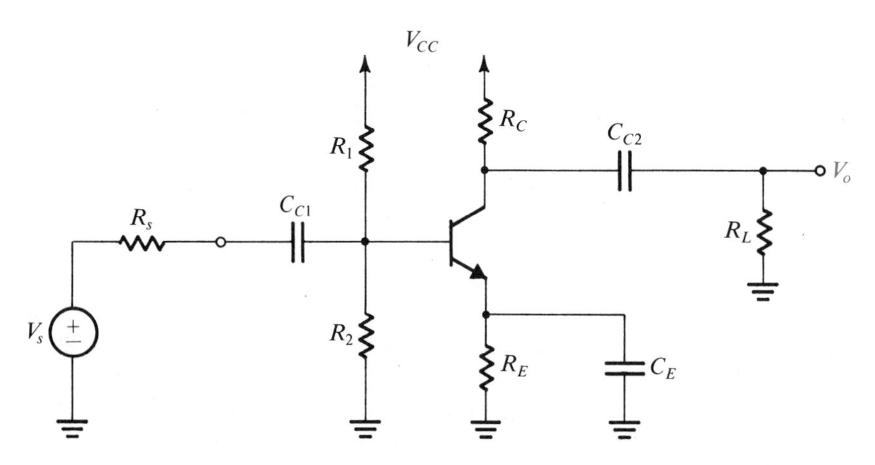

Fig. 7.26 The classical common-emitter amplifier stage.

where the midband gain A_M is evaluated by ignoring all capacitive effects, and where ω_H is the frequency of the dominant high-frequency pole obtained in a manner identical to that used in the FET case, and ω_Z, ω_{P1}, ω_{P2}, and ω_{P3} are the zero and three poles introduced in the low-frequency band by the coupling and the bypass capacitors. Because of the finite input resistance of the BJT, determination of the low-frequency singularities is more complicated than it is for the FET case. This can be seen from the low-frequency equivalent circuit shown in Fig. 7.27. Although we can certainly analyze this circuit and determine its transfer function and hence the poles and zeros, the expressions derived will be too complicated to yield useful insights. Rather, we will make use of the method of short-circuit time constants, described in Section 7.2, to obtain an estimate of the lower 3-dB frequency ω_L.

The determination of ω_L proceeds as follows: First, we set V_s to zero. Then we set C_E and C_{C2} to infinity and find the resistance R_{C1} seen by C_{C1}. From the equivalent circuit in Fig. 7.27, with C_E set to ∞, we find

$$R_{C1} = R_s + [R_B//(r_x + r_\pi)] \tag{7.83}$$

Next, we set C_{C1} and C_{C2} to ∞ and determine the resistance R'_E seen by C_E. Again from the equivalent circuit in Fig. 7.27, or simply using the rule for reflecting resistances from

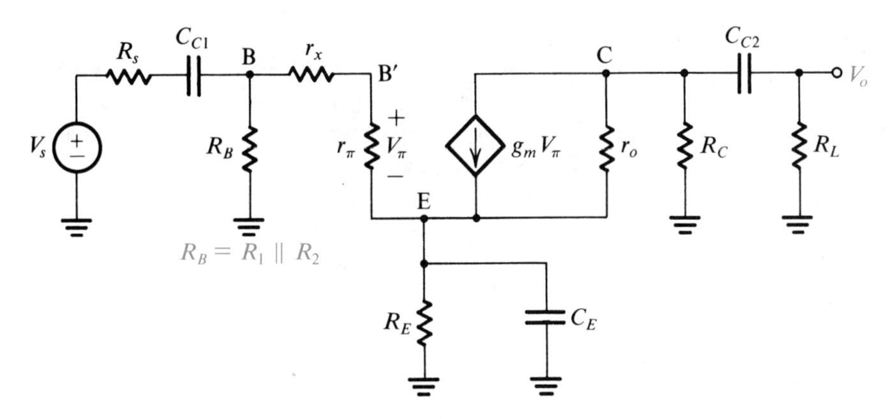

Fig. 7.27 Equivalent circuit for the amplifier of Fig. 7.26 in the low-frequency band.

the base to the emitter circuit, we obtain[5]

$$R'_E = R_E \Big/\!\!\Big/ \frac{r_\pi + r_x + (R_B // R_s)}{\beta_0 + 1} \tag{7.84}$$

Finally, we set both C_{C1} and C_E to infinity and obtain the resistance seen by C_{C2}:

$$R_{C2} = R_L + (R_C // r_o) \tag{7.85}$$

An approximate value for the lower 3-dB frequency can now be determined from

$$\omega_L \simeq \frac{1}{C_{C1}R_{C1}} + \frac{1}{C_E R'_E} + \frac{1}{C_{C2}R_{C2}} \tag{7.86}$$

At this point we should note that the zero introduced by C_E is at the value of s that makes $Z_E = 1/(1/R_E + sC_E)$ infinite,

$$s_Z = -\frac{1}{C_E R_E} \tag{7.87}$$

The frequency of the zero is usually much lower than ω_L, justifying the approximation involved in using the method of short-circuit time constants.

Given a desired value for ω_L, Eq. (7.86) can be used in design as follows: Since R'_E is usually the smallest of the three resistances R_{C1}, R'_E, and R_{C2}, we select a value for C_E so that $(1/C_E R'_E)$ is the dominant term on the right-hand side of Eq. (7.86), say $1/C_E R'_E = 0.8\ \omega_L$. This is equivalent to making C_E form the dominant low-frequency pole; in other words, at $\omega = \omega_L$ the two other capacitors will have small reactances and will thus be playing a minor role. The remaining 20% of ω_L is then split equally between the two other terms in Eq. (7.86). Finally, practical values for the three capacitors are used so that the realized ω_L is equal to or smaller than the specified value.

[5] We also neglect r_o (i.e., let $r_o = \infty$) to simplify matters. The effect of r_o on the value of R'_E is negligibly small.

All the equations derived in the previous section for the high-frequency response of the FET amplifier apply equally well to the BJT amplifier by simply changing symbols (that is, C_{gs} is replaced with C_π, C_{gd} with C_μ, and so on). Thus the comments made, and the conclusions reached, in the previous section apply here as well. Finally, if we assume that the common-emitter amplifier is indeed properly characterized by a dominant low-frequency pole, Eq. (7.82) can be approximated by

$$A(s) \simeq A_M \frac{1}{1 + s/\omega_H} \frac{s}{s + \omega_L} \tag{7.88}$$

Exercises The following exercises pertain to the common-emitter amplifier in Fig. 7.26 with $R_s = 4\ \text{k}\Omega$, $R_1 = 8\ \text{k}\Omega$, $R_2 = 4\ \text{k}\Omega$, $R_E = 3.3\ \text{k}\Omega$, $R_C = 6\ \text{k}\Omega$, $R_L = 4\ \text{k}\Omega$, and $V_{CC} = 12$ V. The dc emitter current can be shown to be $I_E \simeq 1$ mA. At this current the transistor has $\beta_0 = 100$, $C_\pi = 13.9$ pF, $C_\mu = 2$ pF, $r_o = 100\ \text{k}\Omega$, and $r_x = 50\ \Omega$.

7.14 Find the midband gain.

Ans. $A_M = -22.5$ V/V

7.15 Find R_{C1}, R_E', and R_{C2}, and hence f_L, for the case $C_{C1} = C_{C2} = 1\ \mu\text{F}$ and $C_E = 10\ \mu\text{F}$. Also find the frequency of the zero.

Ans. 5.3 kΩ; 40.5 Ω; 9.66 kΩ; 439.5 Hz; 4.8 Hz

7.16 It is required to change the values of C_{C1}, C_{C2}, and C_E so that f_L becomes 100 Hz. Use the design procedure described in the text to obtain the new capacitor values.

Ans. $C_{C1} = 3\ \mu\text{F}$; $C_E = 49.1\ \mu\text{F}$; $C_{C2} = 1.65\ \mu\text{F}$

7.17 Use the Miller-effect method to determine the total input capacitance and hence the dominant high-frequency pole.

Ans. 203.4 pF; 787 kHz.

7.18 Use Eq. (7.59), with the symbols replaced with those for the BJT, to determine a better estimate of the dominant pole. Also, use Eq. (7.60) to determine the second pole.

Ans. 773.5 kHz; 508.5 MHz

7.19 Consider the BJT amplifier with midband gain of −22.5 V/V, a lower 3-dB frequency of 439.5 Hz, and an upper 3-dB frequency of 770 kHz. If a negative input pulse of 10-mV height and 10-μs width is applied at the input, find the rise time, height, and sag of the output pulse waveform.

Ans. 0.45 μs; 225 mV; 6.3 mV

7.7 THE COMMON-BASE, COMMON-GATE, AND CASCODE CONFIGURATIONS

In the previous sections it was shown that the high-frequency response of the common-source amplifier and the common-emitter amplifier is limited by the Miller effect introduced by the feedback capacitance (C_{gd} in the FET and C_μ in the BJT). It follows that to extend the upper frequency limit of a transistor amplifier stage one has to reduce or eliminate the Miller capacitance multiplication. In the following we shall show that this can be achieved in the common-base amplifier configuration. An almost identical analysis can be applied to the common-gate configuration.

We shall also analyze the frequency response of the cascode configuration, and show that it combines the advantages of the common-emitter and the common-base circuits (the common-source and the common-gate circuits in the FET case). To be general and to show the parallels with the common-emitter amplifier, we shall present the circuits in their capacitively coupled form. The high-frequency analysis, however, applies directly to direct-coupled circuits.

Analysis of the Common-Base Amplifier

Figure 7.28 shows a common-base amplifier stage of the capacitively coupled type. Recall that such a configuration was analyzed at midband frequencies (that is, with all capacitive effects neglected) in Chapter 4. In the following we shall be interested specifically in the high-frequency analysis; the low-frequency response can be determined using techniques similar to those of the previous sections.

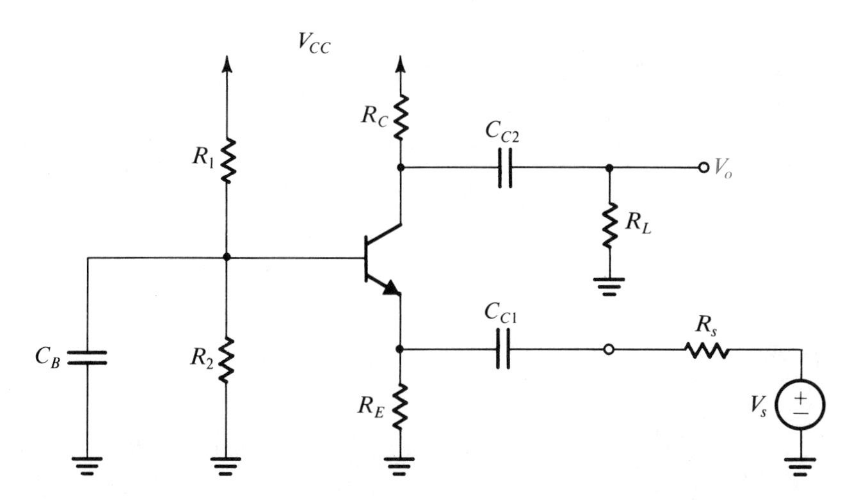

Fig. 7.28 Common-base amplifier stage.

The high-frequency equivalent circuit of the common-base amplifier is shown in Fig. 7.29(a). To simplify matters and to focus attention on the special features of the common-base circuit, r_o and r_x have been omitted.

In the circuit of Fig. 7.29(a) we observe that the voltage at the emitter terminal V_e is equal to $-V_\pi$. We can write a node equation at the emitter terminal that enables us to express the emitter current I_e as

$$I_e = -V_\pi\left(\frac{1}{r_\pi} + sC_\pi\right) - g_m V_\pi = V_e\left(\frac{1}{r_\pi} + g_m + sC_\pi\right)$$

Thus the input admittance looking into the emitter is

$$\frac{I_e}{V_e} = \frac{1}{r_\pi} + g_m + sC_\pi = \frac{1}{r_e} + sC_\pi \tag{7.89}$$

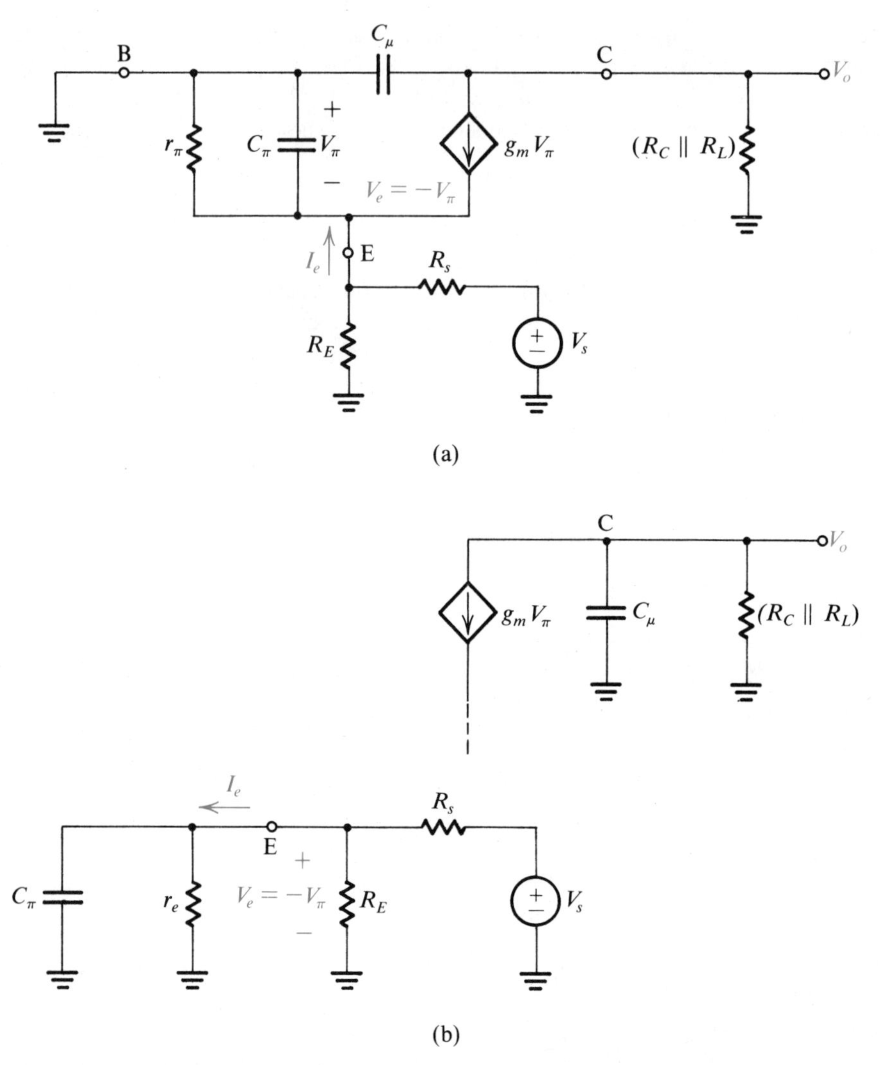

Fig. 7.29 (a) Equivalent circuit of the common-base amplifier in Fig. 7.28; **(b)** simplified version of the circuit in (a).

Therefore at the input of the circuit we may replace the transistor by this input admittance, as shown in Fig. 7.29(b).

At the output side (Fig. 7.29a) we see that V_o is determined by the current source $g_m V_\pi$ feeding $(R_C // R_L // C_\mu)$. This observation is used in drawing the output part in the simplified equivalent circuit shown in Fig. 7.29(b).

The simplified equivalent circuit of Fig. 7.29(b) clearly shows the most important feature of the common-base configuration: the absence of an internal feedback capacitance. Unlike the common-emitter circuit, here C_μ has one terminal grounded, and no Miller effect is present. We therefore expect that the upper cutoff frequency will be much higher than that of the common-emitter configuration.

The high-frequency poles can be directly determined from the equivalent circuit of Fig. 7.29(b). At the input side we have a pole whose frequency ω_{P1} can be written by inspection as

$$\omega_{P1} = \frac{1}{C_\pi(r_e//R_E//R_s)} \tag{7.90}$$

Since r_e is usually very small, the frequency ω_{P1} will be quite high. At the output side there is a pole with frequency ω_{P2} given by

$$\omega_{P2} = \frac{1}{C_\mu(R_C//R_L)} \tag{7.91}$$

Since C_μ is quite small, ω_{P2} also will be quite high. The pole frequencies of a common-gate amplifier can be determined using Eqs. (7.90) and (7.91) with C_π replaced with C_{gs}, r_e replaced with $1/g_m$, and C_μ replaced with C_{gd}.

The question now arises as to the accuracy of the above analysis. Since we are dealing with poles at very high frequencies, we should take into account effects normally thought to be negligible. For instance, the parasitic capacitance usually present between collector and substrate (ground) in an IC transistor will obviously have a considerable effect on the value of ω_{P2}. Also, it is not clear that r_x could be neglected. It follows that for the accurate determination of the high-frequency response of a common-base amplifier a more elaborate transistor model should be used, and one normally employs a computer circuit-analysis program (see Appendix C and the SPICE supplement). Nevertheless, the point that we wish to emphasize here is that the common-base amplifier has a much higher upper cutoff frequency than that of the common-emitter circuit. An analogous statement applies to the common-gate amplifier as compared to the common-source circuit.

The Cascode Configuration

As discussed in Section 6.5, the cascode configuration combines the advantages of the common-emitter and the common-base circuits. Figure 7.30 shows a capacitively coupled cascode amplifier designed using bipolar transistors. Here a large-valued capacitor C_B is used to establish signal ground at the base of Q_2. The following analysis applies equally well to FET cascode circuits and to the BiCMOS cascode circuit of Fig. 6.35(c), in which Q_1 is a MOSFET and Q_2 is a BJT.

In the cascode circuit, Q_1 is connected in the common-emitter configuration and therefore presents a relatively high input resistance to the signal source. The collector signal current of Q_1 is fed to the emitter of Q_2, which is connected in the common-base configuration. Thus the load resistance seen by Q_1 is simply the input resistance r_e of Q_2. This low load resistance of Q_1 considerably reduces the Miller multiplier effect of $C_{\mu 1}$ and thus extends the upper cutoff frequency. This is achieved without reducing the midband gain, since the collector of Q_2 carries a current almost equal to the collector current of Q_1. Furthermore, since it is in the common-base configuration, Q_2 does not suffer from the Miller effect and hence does not limit the high-frequency response. Transistor Q_2 acts essentially as a current buffer or an impedance transformer, faithfully passing on the signal current to the load while presenting a low load resistance to the amplifying device Q_1.

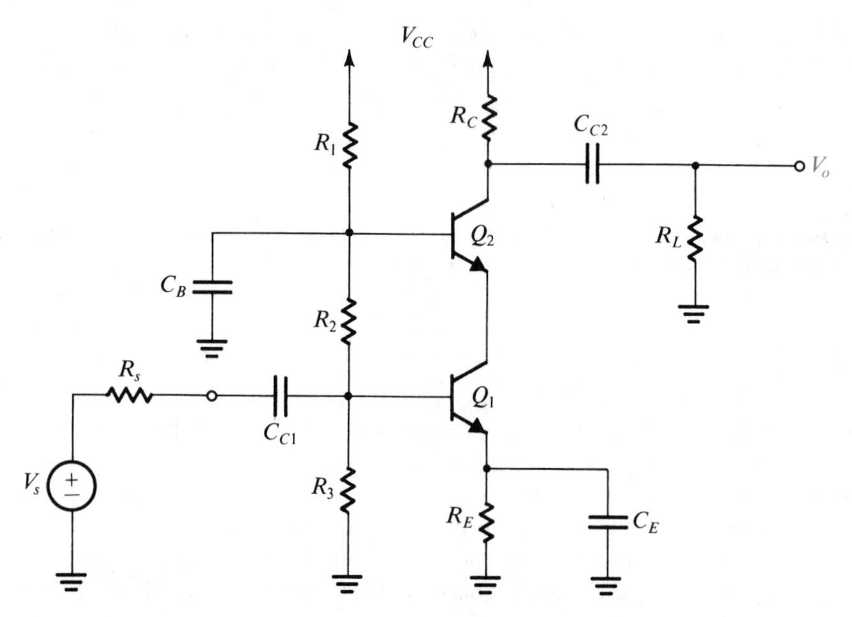

Fig. 7.30 A capacitively-coupled cascode amplifier using bipolar transistors. Capacitor C_B is large, establishing a signal ground at the base of Q_2.

A detailed analysis of the cascode amplifier of Fig. 7.30 will now be presented: Figure 7.31(a) shows the high-frequency equivalent circuit. To simplify matters, r_{x2} and r_{o2} have been omitted. Although the two transistors are operating at equal bias currents and therefore their corresponding parameters are equal, we have for clarity kept the identity of the two sets of parameters separate.

Application of Thévenin's theorem enables us to reduce the circuit to the left of line xx' (Fig. 7.31a) to a source V_s' and a resistance R_s', as shown in Fig. 7.31(b), where

$$V_s' = V_s \frac{(R_2//R_3)}{R_s + (R_2//R_3)} \frac{r_{\pi 1}}{r_{\pi 1} + r_{x1} + (R_2//R_3//R_s)} \tag{7.92}$$

$$R_s' = \{r_{\pi 1}//[r_{x1} + (R_3//R_2//R_s)]\} \tag{7.93}$$

Another important simplification included in the circuit of Fig. 7.31(b) is the replacement of the current source $g_{m2}V_{\pi 2}$ by a resistance $1/g_{m2}$ (see the source absorption theorem in Appendix D). This resistance is then combined with the parallel resistance $r_{\pi 2}$ to obtain r_{e2}. Since $r_{e2} \ll r_{o1}$, we see that between the collector of Q_1 and ground the total resistance is approximately r_{e2}. Capacitance $C_{\pi 2}$ together with resistance r_{e2} produces a transfer-function pole with a frequency

$$\omega_2 = \frac{1}{C_{\pi 2} r_{e2}} \simeq \omega_{T2} \tag{7.94}$$

which is much higher than the frequency of the pole that arises due to the interaction of R_s' and the input capacitance of Q_1. It follows that in the frequency range of interest $C_{\pi 2}$

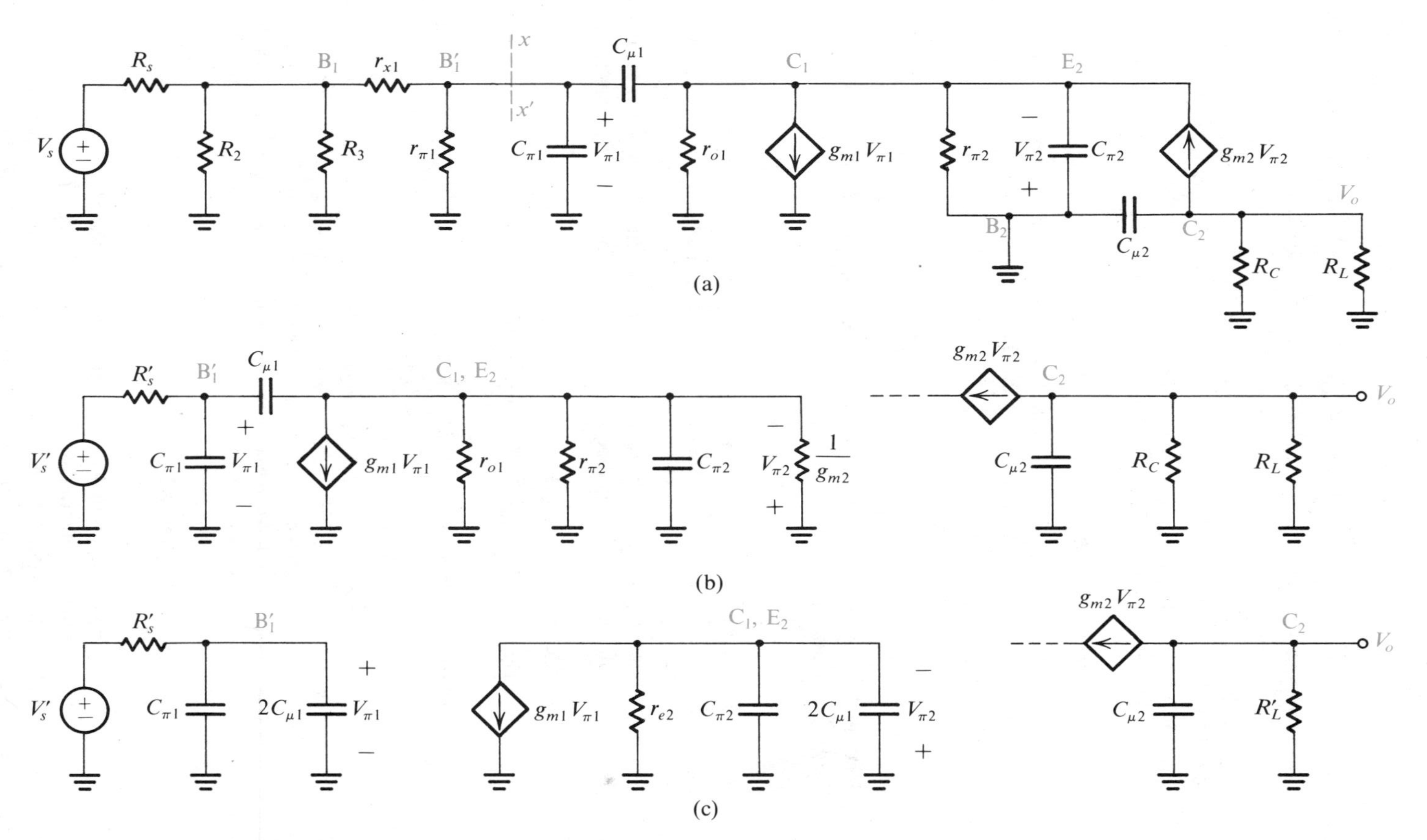

Fig. 7.31 High-frequency analysis of the cascode amplifier in Fig. 7.30. Note that to simplify the analysis, r_{x2} and r_{o2} are not included.

can be ignored in calculating the voltage at the collector of Q_1; that is,

$$V_{c1} \simeq -g_{m1}V_{\pi 1}r_{e2} \simeq -V_{\pi 1}$$

Thus the gain between B_1' and C_1 (Fig. 7.31b) is approximately -1, and we can employ Miller's theorem to replace the bridging capacitance $C_{\mu 1}$ by a capacitance $2C_{\mu 1}$ between B_1' and ground and a capacitance $2C_{\mu 1}$ between C_1 and ground. The resulting equivalent circuit is shown in Fig. 7.31(c), from which we can now evaluate the frequency of the pole due to the RC low-pass circuit at the input as

$$\omega_1 = \frac{1}{R_s'(C_{\pi 1} + 2C_{\mu 1})} \tag{7.95}$$

For the case in which the source resistance is large this frequency will be much lower than both ω_2 (Eq. 7.94) and the frequency of the pole produced by the output part of the circuit,

$$\omega_3 = \frac{1}{C_{\mu 2}R_L'} \tag{7.96}$$

Thus in this case the input circuit produces a dominant high-frequency pole, and the upper 3-dB frequency ω_H is given by

$$\omega_H \simeq \omega_1 \tag{7.97}$$

A slightly better estimate of ω_H can be obtained by combining the frequencies of the three poles using Eq. (7.22). Also, it should be pointed out that the method of open-circuit time constants could have been applied directly to the circuit in Fig. 7.31(b).

The midband gain can be easily evaluated by ignoring the capacitances in the equivalent circuit of Fig. 7.31(c) together with substituting for V_s' and R_s' from Eqs. (7.92) and (7.93):

$$A_M = \frac{V_o}{V_s} = -g_m(R_L//R_C)\frac{(R_2//R_3)}{(R_2//R_3) + R_s}\frac{r_\pi}{r_\pi + r_x + (R_2//R_3//R_s)} \tag{7.98}$$

This expression is identical in form to the expression for the gain of a common-emitter circuit.

Exercise 7.20 Consider the cascode circuit in Fig. 7.30 with the following component values: $R_s = 4\ \text{k}\Omega$, $R_1 = 18\ \text{k}\Omega$, $R_2 = 4\ \text{k}\Omega$, $R_3 = 8\ \text{k}\Omega$, $R_E = 3.3\ \text{k}\Omega$, $R_C = 6\ \text{k}\Omega$, $R_L = 4\ \text{k}\Omega$, $C_{C1} = 1\ \mu\text{F}$, $C_{C2} = 1\ \mu\text{F}$, $C_B = 10\ \mu\text{F}$, $C_E = 10\ \mu\text{F}$, and $V_{CC} = +15$ V. Show that each transistor is operating at $I_E \simeq 1$ mA. Note that this design is identical to that of the common-emitter amplifier in Exercises 7.13–7.18. If we thus assume that the transistors are of the same type, we are able to compare results and draw conclusions. Calculate $A_M, f_1, f_2,$ and f_3. Then, use the sum-of-squares formula in Eq. (7.22) to find f_H.

Ans. -23.1 V/V; 8.95 MHz; 456 MHz; 33 MHz; 8.64 MHz

There is another important case to consider, a case that occurs frequently in the design of active-loaded integrated-circuit cascode amplifiers. In such a case, R_s is usually small, rendering the input pole of frequency ω_1 nondominant. Because of the active load, R_L' is usually very large, causing the pole at the output to dominate. A better estimate of the frequency of this pole can be obtained by taking into account the parasitic load capacitance C_L normally present at the output node, thus modifying ω_3 of Eq. (7.96) to

$$\omega_3 = \frac{1}{(C_{\mu 2} + C_L)R'_L} \tag{7.99}$$

Note to ensure that this pole is indeed dominant one has to arrange that $\omega_2 \simeq \omega_{T2}$ is much higher than ω_3, which is usually the case when Q_2 is a BJT. In a MOS cascode, however, ω_{T2} can be low, thus reducing the amplifier bandwidth. This situation can be considerably improved by using the BiCMOS cascode of Fig. 6.35(c).

7.8 FREQUENCY RESPONSE OF THE EMITTER AND SOURCE FOLLOWERS

The emitter-follower or common-collector configuration was studied in Section 4.11. In the following we consider the high-frequency response of this important circuit configuration. The results apply, with simple modifications, to the FET source follower.

Consider the direct-coupled emitter-follower circuit shown in Fig. 7.32(a), where R_s represents the source resistance and R_E represents the combination of emitter-biasing resistance and load resistance. The high-frequency equivalent circuit is shown in Fig. 7.32(b)[6] and is redrawn in a slightly different form in Fig. 7.32(c). Analysis of this circuit results in the emitter-follower transfer function $V_o(s)/V_s(s)$, which can be shown to have two poles and one real zero:

$$\frac{V_o(s)}{V_s(s)} = A_M \frac{1 + s/\omega_Z}{(1 + s/\omega_{P1})(1 + s/\omega_{P2})} \tag{7.100}$$

where A_M denotes the value of the gain at low and medium frequencies. Unfortunately, though, symbolic analysis of the circuit will not reveal whether or not one of the poles is dominant. To gain more insight we shall take an alternative route.

Writing a node equation at the emitter (Fig. 7.32b) results in

$$V_o = (g_m + y_\pi)V_\pi R_E \tag{7.101}$$

where

$$y_\pi = \frac{1}{r_\pi} + sC_\pi$$

Thus V_o will be zero at the value of s that makes $V_\pi = 0$ and at the value of s that makes $g_m + y_\pi = 0$. In turn, V_π will be zero at the value of s that makes $z_\pi = 0$ or equivalently $y_\pi = \infty$, namely, $s = \infty$. The fact that a transmission zero exists at $s = \infty$ correlates with Eq. (7.100). The other transmission zero is obtained from

$$g_m + y_\pi = 0$$

[6] Although r_o is not shown, it can be easily included and appears in parallel with R_E. Thus the analysis to follow need only be modified by replacing R_E with $(R_E//r_o)$.

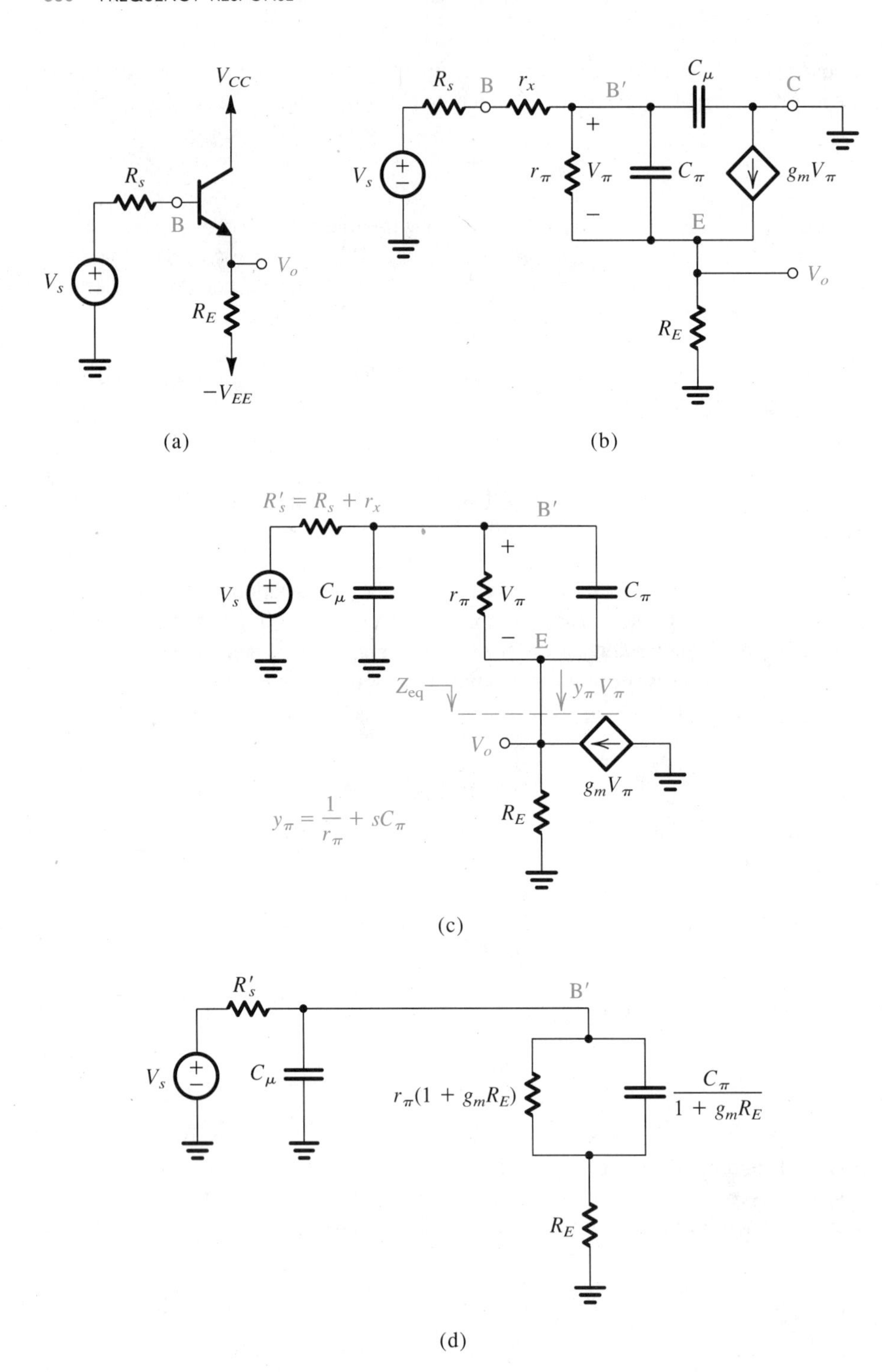

Fig. 7.32 High-frequency analysis of the emitter follower.

that is,

$$g_m + \frac{1}{r_\pi} + s_Z C_\pi = 0$$

which yields

$$s_Z = -\frac{g_m + 1/r_\pi}{C_\pi} = -\frac{1}{C_\pi r_e} \simeq -\omega_T \tag{7.102}$$

Since the frequency of this zero is quite high, it will normally play a minor role in determining the high-frequency response of the emitter follower.

Next we consider the poles. Whether or not one of the two poles is dominant will depend on the particular application—specifically, on the values of R_s and R_E. In most applications R_s is large, and it together with the input capacitance provides a dominant pole. To see this more clearly, consider the equivalent circuit of Fig. 7.32(c). By invoking the source-absorption theorem (Appendix D), we can replace the circuit below the broken line by its equivalent impedance Z_{eq},

$$Z_{eq} \equiv \frac{V_o}{y_\pi V_\pi}$$

Thus

$$Z_{eq} = \frac{(g_m + y_\pi)R_E}{y_\pi} \tag{7.103}$$

Note that Z_{eq} is simply R_E reflected to the base side through use of a generalized form of the reflection rule: R_E is multiplied by $(h_{fe} + 1)$. The total impedance between B′ and ground is

$$Z_{b'} = \frac{1}{y_\pi} + Z_{eq} = \frac{1 + g_m R_E}{y_\pi} + R_E$$

As shown in Fig. 7.32(d), this impedance can be represented by a resistance R_E in series with an RC network consisting of a resistance $(1 + g_m R_E)r_\pi$ in parallel with a capacitance $C_\pi/(1 + g_m R_E)$. Since the impedance of the parallel RC circuit is usually much larger than R_E, we may neglect the latter impedance and obtain a simple STC low-pass network. From this STC circuit it follows that a pole exists at

$$\omega_P = \left[\left(C_\mu + \frac{C_\pi}{1 + g_m R_E}\right)[R_s' // (1 + g_m R_E) r_\pi]\right]^{-1} \tag{7.104}$$

Even though this pole is usually dominant, its frequency is normally quite high, giving the emitter follower a wide bandwidth.[7]

[7]Although we have not considered the emitter follower with a capacitive load, this case deserves a comment: A capacitive load for the emitter follower leads to a *negative* input conductance; and if the source impedance is inductive, the circuit may oscillate. Stability and oscillations are studied in Chapters 8 and 12.

An alternative approach for finding an approximate value of the 3-dB frequency ω_H is to use the open-circuit time-constants method on the equivalent circuit in Fig. 7.32(d).

Finally we note that all the results above can be applied to the case of an FET by replacing R_s' with R_s, r_π with ∞, C_μ with C_{gd}, and C_π with C_{gs}.

Exercises 7.21 For an emitter follower biased at $I_C = 1$ mA and having $R_s = R_E = 1$ kΩ, and using a transistor specified to have $f_T = 400$ MHz, $C_\mu = 2$ pF, $r_x = 100$ Ω, and $\beta_0 = 100$, evaluate the midband-gain A_M and the frequency of the dominant high-frequency pole.

Ans. 0.97 V/V; 62.5 MHz

7.22 For the emitter follower specified in Exercise 7.21, use the method of open-circuit time constants to estimate the upper 3-dB frequency f_H.

Ans. 55.3 MHz

7.9 THE COMMON-COLLECTOR COMMON-EMITTER CASCADE

The excellent high-frequency response of the emitter follower is due to the absence of the Miller capacitance-multiplication effect. The problem with it, though, is that it does not provide voltage gain. It appears possible that we can obtain both gain and wide bandwidth by using a cascade of a common-collector and a common-emitter stage, as shown in Fig. 7.33. Here the emitter-follower transistor Q_1 is shown biased by a current source I, as is usually the case in integrated-circuit design. Because the collector of Q_1 is at a signal ground, $C_{\mu 1}$ does not get multiplied by the stage gain, as is the case in common-emitter amplifier. Thus the pole caused by the interaction of the source resistance and the input capacitance will be at a high frequency.

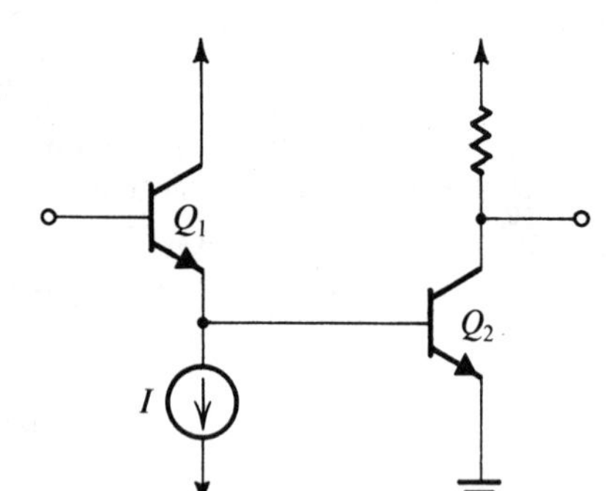

Fig. 7.33 The common-collector common-emitter cascade amplifier.

The voltage gain is provided by the common-emitter transistor Q_2. This transistor suffers from the Miller effect; that is, the total effective capacitance between its base and ground will be large. Nevertheless, this will not be detrimental; the resistance seen by that capacitance will be small because of the low output resistance of the emitter follower Q_1.

Before considering a numerical example, we wish to draw the reader's attention to the similarities between the circuit of Fig. 7.33 and the cascode amplifier studied in

Section 7.7. Both circuits employ a common-emitter amplifier to obtain voltage gain. Both circuits achieve wider bandwidth (than that obtained in a common-emitter amplifier) through minimizing the effect of the Miller multiplier. In the cascode circuit this is achieved by isolating the load resistance from the collector of the common-emitter stage by a low-input-resistance common-base stage. In the present circuit, although Miller multiplication occurs, the resulting large capacitance is isolated from the source resistance by an emitter follower.

EXAMPLE 7.9

Figure 7.34 shows a capacitively coupled amplifier designed as the cascade of a common-collector stage and a common-emitter stage. Assume that the transistors used have $\beta = 100$, $f_T = 400$ MHz, and $C_\mu = 2$ pF, and neglect r_x and r_o. We wish to evaluate the midband gain and the high-frequency response of this circuit. Note that the load and source resistances and the transistor parameters are identical to those used in the common-emitter case (Exercises 7.14 to 7.19) and in the cascode case (Exercise 7.20); hence comparisons can be made.

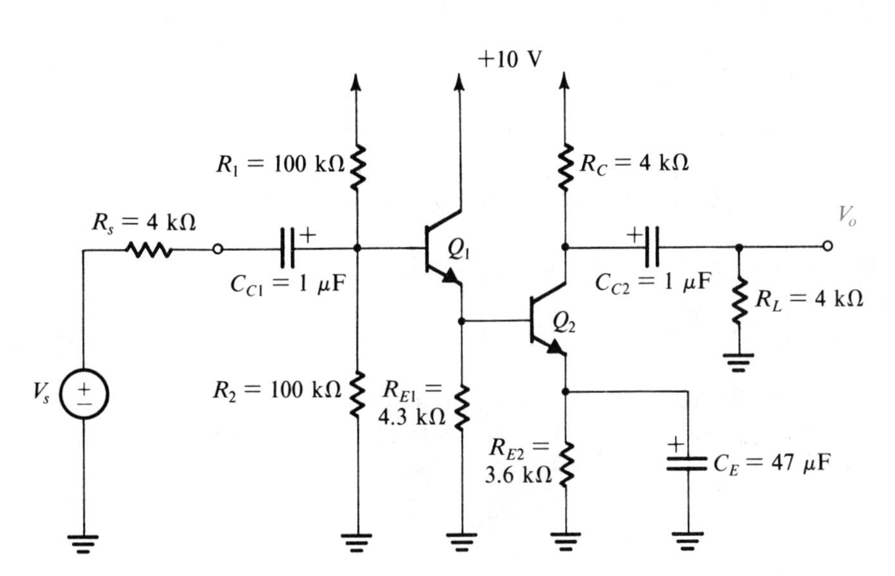

Fig. 7.34 A capacitively coupled amplifier using the common-collector common-emitter cascade configuration.

Solution We first determine the dc bias currents as follows:

$$V_{B1} \simeq 5 \text{ V} \qquad V_{E1} \simeq 4.3 \text{ V} \qquad I_{E1} = \frac{4.3 \text{ V}}{4.3 \text{ k}\Omega} = 1 \text{ mA}$$

$$V_{E2} \simeq 3.6 \text{ V} \qquad I_{E2} = \frac{3.6 \text{ V}}{3.6 \text{ k}\Omega} = 1 \text{ mA}$$

Thus both transistors are operating at emitter currents of approximately 1 mA, and both are in the active mode. At this operating point the equivalent circuit components are

$$g_m \simeq 40 \text{ mA/V} \qquad r_e \simeq 25\ \Omega \qquad r_\pi \simeq 2.5 \text{ k}\Omega$$

$$C_\pi + C_\mu = \frac{g_m}{\omega_T} = 15.9 \text{ pF} \qquad C_\mu = 2 \text{ pF} \qquad C_\pi = 15.9 - 2 = 13.9 \text{ pF}$$

To evaluate the midband gain we shall first determine the value of the input resistance R_{in}. Toward that end note that the input resistance between the base of Q_2 and ground is equal to $r_{\pi 2}$. Thus in the emitter circuit of Q_1 we have R_{E1} in parallel with $r_{\pi 2}$. The input resistance R_{in} will therefore be given by

$$R_{\text{in}} = R_1 // R_2 // \{(\beta_1 + 1)[r_{e1} + (R_{E1} // r_{\pi 2})]\}$$

which leads to $R_{\text{in}} \simeq 38$ kΩ. The transmission from the input to the base of Q_1 is

$$\frac{V_{b1}}{V_s} = \frac{R_{\text{in}}}{R_{\text{in}} + R_s} = 0.9 \tag{7.105}$$

Next, the gain of the emitter follower Q_1 can be obtained as

$$\frac{V_{e1}}{V_{b1}} = \frac{(R_{E1} // r_{\pi 2})}{(R_{E1} // r_{\pi 2}) + r_{e1}} = 0.98 \tag{7.106}$$

Finally, the gain of the common-emitter amplifier Q_2 can be evaluated as

$$\frac{V_o}{V_{e1}} = -g_{m2}(R_C // R_L) = -80 \tag{7.107}$$

The overall voltage gain can be obtained by combining Eqs. (7.105) through (7.107):

$$\frac{V_o}{V_s} = -0.9 \times 0.98 \times 80 = -70.6 \text{ V/V} \tag{7.108}$$

The high-frequency response can be determined from the equivalent circuit shown in Fig. 7.35(a). We apply Miller's theorem to the second stage[8] and perform a number of other simplifications, and we obtain the circuit in Fig. 7.35(b), where

$$R_s' = (R_s // R_1 // R_2)$$

$$V_s' = V_s \frac{(R_1 // R_2)}{(R_1 // R_2) + R_s}$$

$$C_T = C_{\pi 2} + C_{\mu 2}(1 + g_{m2} R_L')$$

$$R_L' = (R_L // R_C)$$

[8] Assuming that $g_{m2} R_L' \gg 1$, application of Miller's theorem results in a capacitance approximately equal to $C_{\mu 2}$ in parallel with R_L'. Note that this is *not* the output capacitance of the amplifier (refer to the discussion of Miller's theorem in Section 7.4). Nevertheless, it can be used in the computation of f_H using the method of open-circuit time constants.

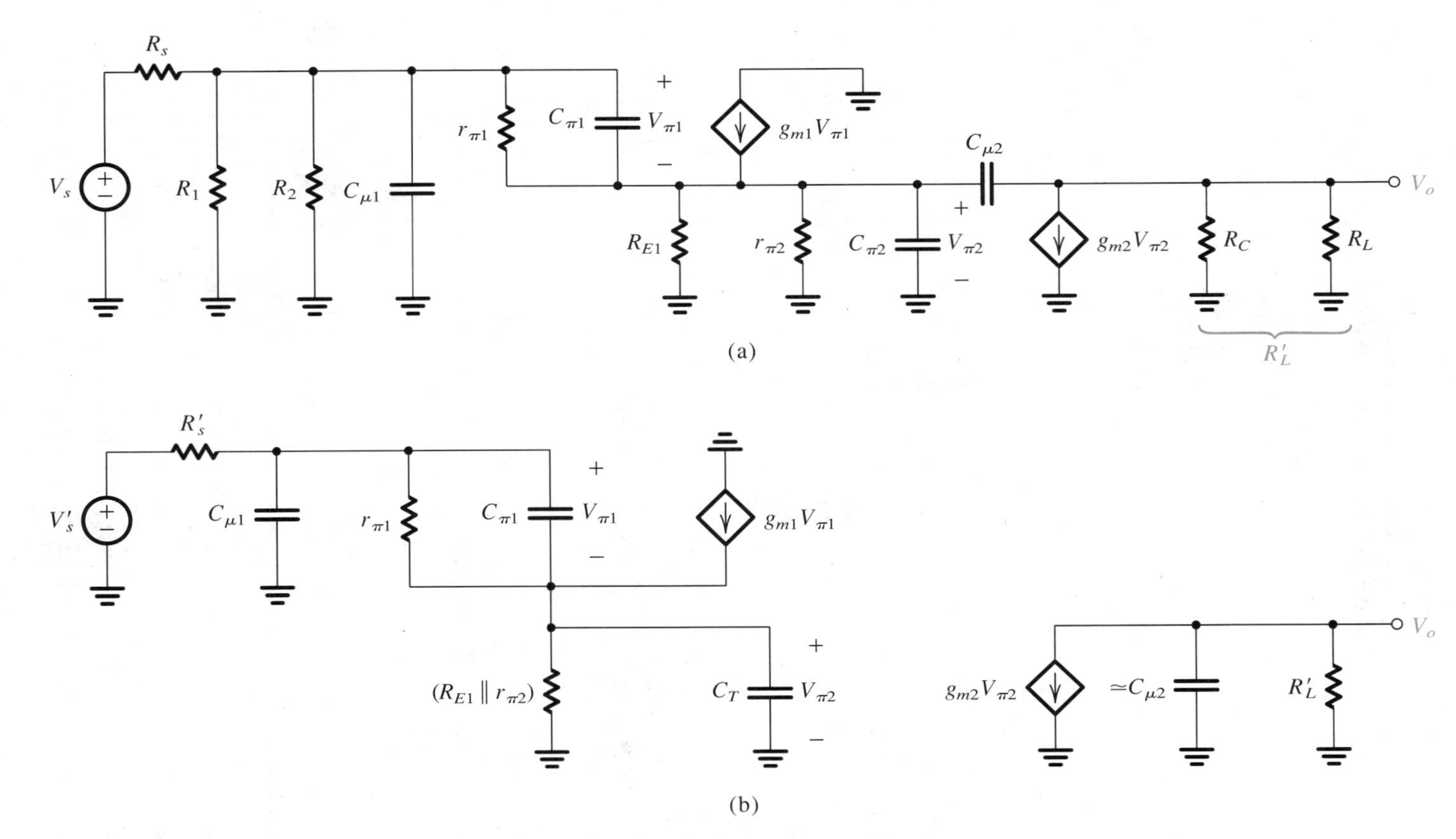

Fig. 7.35 Equivalent circuits for the determination of the high-frequency response of the amplifier in Fig. 7.34.

This circuit is still quite complex, and an exact pencil-and-paper analysis would be quite tedious. Alternatively, we employ the technique of open-circuit time constants discussed in Section 7.2 to determine f_H as follows: Capacitor $C_{\mu 1}$ sees a resistance $R_{\mu 1}$ given by

$$R_{\mu 1} = (R_s // R_{in}) = (4//38) = 3.62 \text{ k}\Omega$$

It can be shown that $C_{\pi 1}$ sees a resistance $R_{\pi 1}$ given by

$$R_{\pi 1} = \left(r_{\pi 1} // \frac{R'_s + R'_{E1}}{1 + g_{m1} R'_{E1}} \right)$$

where $R'_{E1} = (R_{E1} // r_{\pi 2})$. Thus $R_{\pi 1} = 80\ \Omega$.

Capacitance C_T, which is equal to 175.9 pF, sees a resistance R_T given by

$$R_T = \left(R'_{E1} // \frac{r_{\pi 1} + R'_s}{\beta_1 + 1} \right) = 59\ \Omega$$

Capacitance $C_{\mu 2}$ sees a resistance R'_L given by

$$R'_L = (R_L // R_C) = 2 \text{ k}\Omega$$

Thus the effective time constant is given by

$$\tau = C_{\mu 1} R_{\mu 1} + C_{\pi 1} R_{\pi 1} + C_T R_T + C_{\mu 2} R'_L = 22.7 \text{ ns}$$

which corresponds to an upper 3-dB frequency of

$$f_H \simeq \frac{1}{2\pi\tau} = 7 \text{ MHz}$$

Although the upper 3-dB frequency is not as high as the value obtained for the cascode amplifier (8.95 MHz), the midband gain here, 70.6, is higher than that found for the cascode (23.1). A figure of merit for an amplifier is its gain–bandwidth product.

Finally we note that analysis similar to that presented in this section can be applied to the common-drain common-source cascade and to the BiCMOS amplifer stage formed as a common-drain common-emitter cascade.

Exercises 7.23 Use the method of short-circuit time constants to determine an approximate value of the lower 3-dB frequency f_L of the amplifier circuit in Fig. 7.34. Also find the frequency of the zero introduced by C_E.

Ans. 156.8 Hz; 0.94 Hz

7.24 The common-collector common-emitter configuration studied in this section is a modified version of the composite device obtained by connecting two transistors in the form shown in Fig. E7.24(a). This configuration, known as the **Darlington configuration,** is equivalent to a single transistor with $\beta \simeq \beta_1 \beta_2$. It can therefore be used as a high-performance follower, as illustrated in Fig. E7.24(b). For the latter circuit assume that Q_2 is biased at $I_E = 5$ mA and let $R_s = 100$ kΩ, $R_E = 1$ kΩ, and $\beta_1 = \beta_2 = 100$. Find R_{in}, V_o/V_s, and R_{out}.

Ans. 10.3 MΩ; 0.98 V/V; 20 Ω

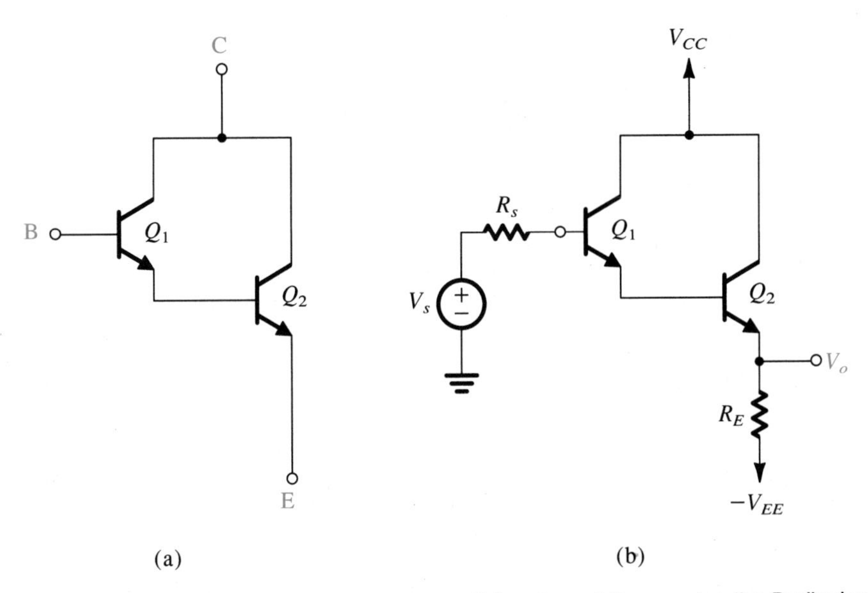

Fig. E7.24 (a) The Darlington configuration; **(b)** voltage follower using the Darlington configuration.

7.10 FREQUENCY RESPONSE OF THE DIFFERENTIAL AMPLIFIER

The differential pair studied in Chapter 6 is the most important building block in analog integrated-circuit design. In this section we analyze its frequency response. Although only the BJT differential pair is considered, the method applies equally well to the FET pair.

Frequency Response in the Case of Symmetric Excitation

Consider the differential amplifier shown in Fig. 7.36(a). The input signal V_s is applied in a complementary (push-pull) fashion, and the source resistance R_s is equally distributed between the two sides of the pair. This situation arises, for instance, if the differential amplifier is fed from the output of another differential stage.

Since the circuit is symmetric and is fed in a complementary fashion, its frequency response will be identical to that of the equivalent common-emitter circuit shown in Fig. 7.36(b). We have analyzed the common-emitter circuit in detail in Section 7.6. Since the differential pair is a direct-coupled amplifier, its gain will extend down to zero frequency with a low-frequency value of

$$\frac{V_o}{V_s} = -\frac{r_\pi}{r_\pi + R_s/2} g_m R_C \tag{7.109}$$

The high-frequency response will be dominated by a real pole with a frequency ω_P,

$$\omega_P = \frac{1}{[(R_s/2)//r_\pi][C_\pi + C_\mu(1 + g_m R_C)]} \tag{7.110}$$

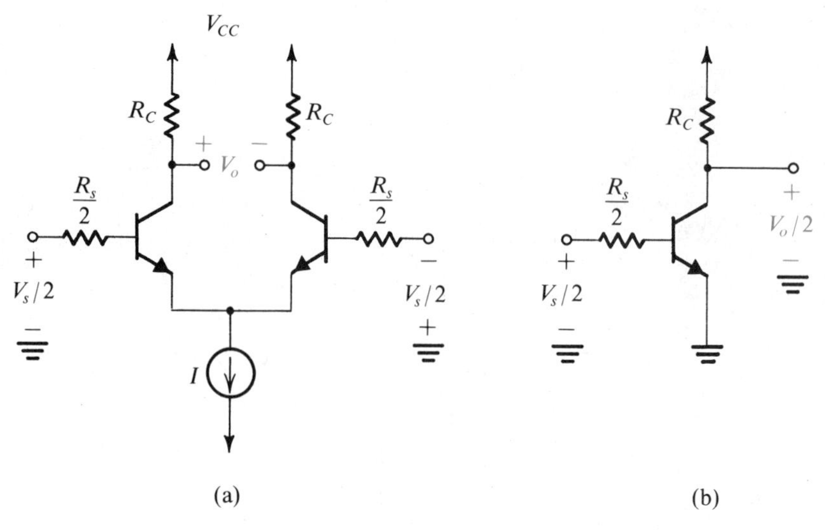

Fig. 7.36 (a) A symmetrically excited differential pair; **(b)** its equivalent half-circuit.

Denote the low-frequency gain given in Eq. (7.109) by A_0. The transfer function of the differential pair will be given by

$$\frac{V_o}{V_s} = \frac{A_0}{1 + s/\omega_P} \tag{7.111}$$

Thus the plot of gain versus frequency will have the standard single-pole shape shown in Fig. 7.37. The 3-dB frequency ω_H is equal to the pole frequency, $\omega_H = \omega_P$. In the above analysis r_x was neglected. It can be easily included simply by replacing $R_s/2$ by $R_s/2 + r_x$ in Eqs. (7.109) and (7.110).

Frequency Response in the Case of Single-Ended Excitation

Figure 7.38(a) shows a differential amplifier driven in a single-ended fashion. Although it is not obvious, the frequency response in this case is almost identical to that of the

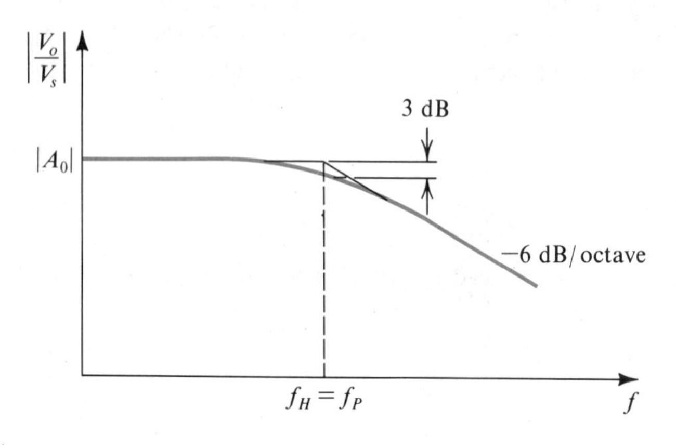

Fig. 7.37 Frequency response of the differential amplifier.

symmetrically driven amplifier considered above. A proof of this assertion is illustrated by a series of equivalent circuits given in Fig. 7.38. Figure 7.38(b) shows the complete equivalent circuit. We can write a node equation at node X and obtain

$$V_{\pi 1}\left(\frac{1}{r_\pi} + sC_\pi + g_m\right) + V_{\pi 2}\left(\frac{1}{r_\pi} + sC_\pi + g_m\right) = 0$$

Thus $V_{\pi 1} = -V_{\pi 2}$. This leads to the equivalent circuit in Fig. 7.38(c), which is further simplified in Fig. 7.38(d).

Consider Fig. 7.38(d). If we neglect $I_{\mu 1}$ and $I_{\mu 2}$ in comparison with $g_m V_\pi$, it follows that

$$V_{c1} \simeq -g_m R_C V_\pi \qquad V_{c2} \simeq +g_m R_C V_\pi$$

Thus $V_o \simeq -g_m R_C(2V_\pi)$. If we further assume that r_x is small so that the voltage at B_2' is approximately zero, and that $g_m R_C \gg 2$, then Miller's theorem now can be applied to obtain the simplified input equivalent circuit shown in Fig. 7.38(e), from which we see that the high-frequency response is dominated by a pole at $s = -\omega_P$, where ω_P is given by

$$\omega_P = \frac{1}{[2r_\pi /\!/ (R_s + 2r_x)][C_\pi/2 + (C_\mu/2)(g_m R_C)]} \tag{7.112}$$

The low-frequency gain A_0 is given by

$$A_0 = \frac{V_o}{V_s} = -g_m R_C \frac{2r_\pi}{2r_\pi + R_s + 2r_x} \tag{7.113}$$

These results are almost identical to those obtained in the case of symmetric excitation.

Effect of Emitter Resistance on the Frequency Response

The bandwidth of the differential amplifier can be widened (that is, ω_H can be increased) by including two equal resistances R_E in the emitters. This is achieved at the expense of a reduction in the low-frequency gain. To evaluate the effect of the emitter resistances on frequency response, consider the equivalent half-circuit shown in Fig. 7.39(a). The low-frequency gain is given by

$$A_0 \equiv \frac{V_o}{V_s} = \frac{-(\beta + 1)(r_e + R_E)}{R_s/2 + r_x + (\beta + 1)(r_e + R_E)} \frac{\alpha R_C}{R_E + r_e} \tag{7.114}$$

The high-frequency equivalent circuit is shown in Fig. 7.39(b). Since it is no longer convenient to apply Miller's theorem, we shall use the technique of open-circuit time constants explained in Section 7.2. The method proceeds as follows: We first eliminate C_μ and determine the resistance seen by C_π, which we shall call R_π. Figure 7.39(c) shows the circuit for finding R_π,

$$R_\pi = \left(r_\pi /\!/ \frac{R_s' + R_E}{1 + g_m R_E}\right) \tag{7.115}$$

Next, the resistance R_μ seen by C_μ can be determined from the circuit in Fig. 7.39(d),

$$R_\mu = R_C + \frac{1 + R_E/r_e + g_m R_C}{1/r_\pi + (1/R_s')(1 + R_E/r_e)} \tag{7.116}$$

The overall effective time constant will be given by

$$\tau = C_\pi R_\pi + C_\mu R_\mu \tag{7.117}$$

and the 3-dB frequency ω_H will be

$$\omega_H \simeq \frac{1}{\tau} \tag{7.118}$$

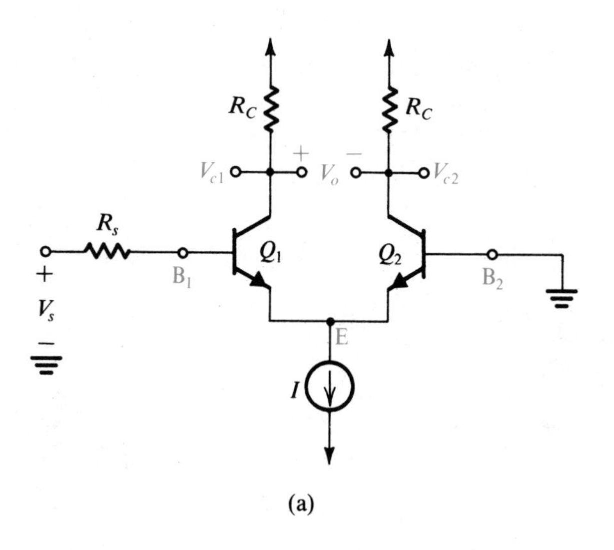

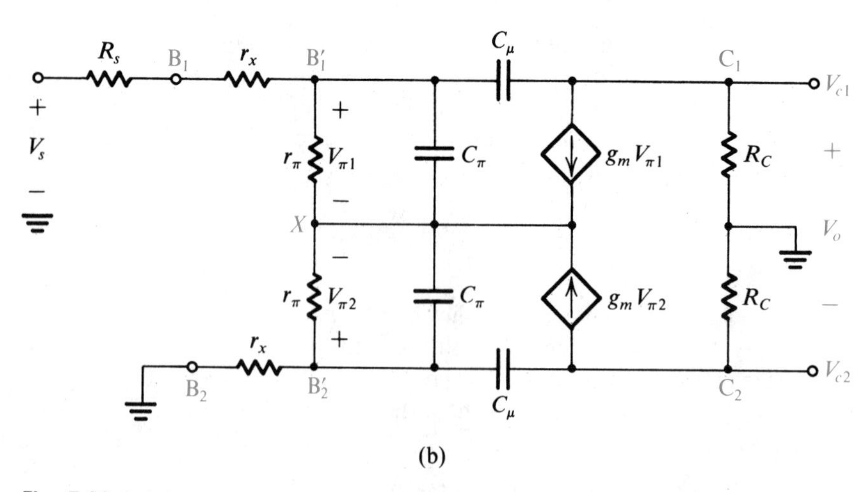

Fig. 7.38 (a) The differential amplifier excited in a single-ended fashion. **(b)** Equivalent circuit of the amplifier in (a).

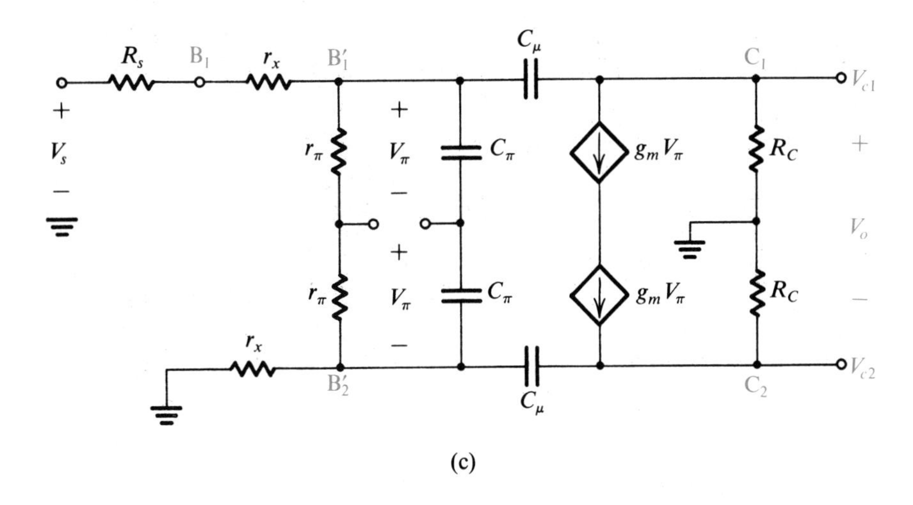

(c)

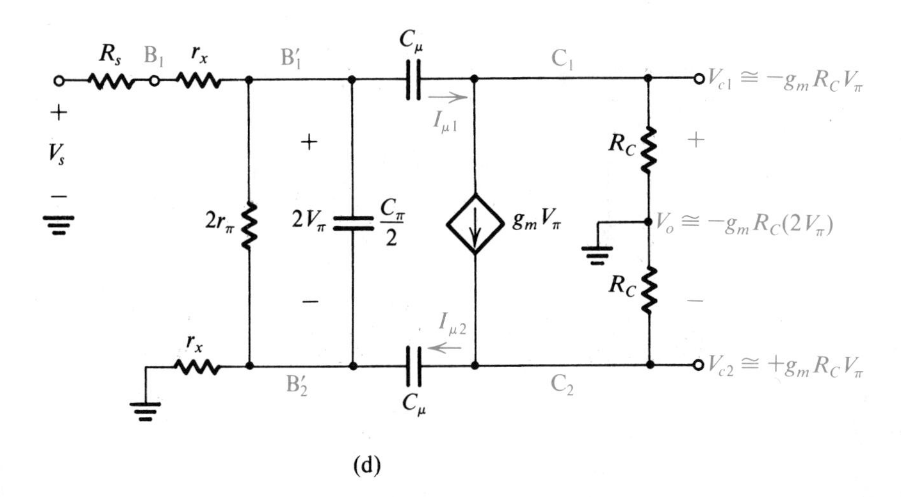

(d)

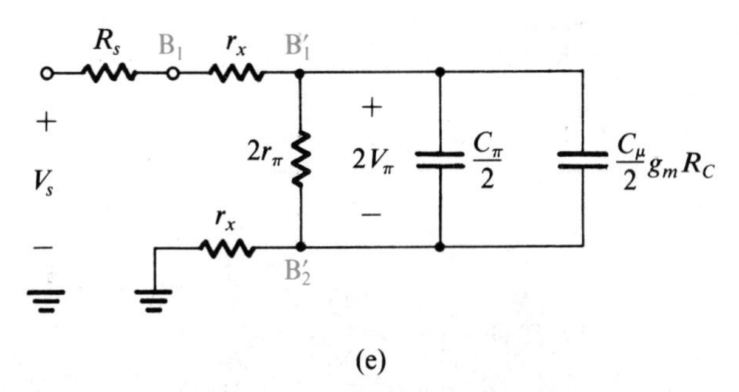

(e)

Fig. 7.38 (c) Simplified equivalent circuit using the fact that $V_{\pi 1} = -V_{\pi 2}$. **(d)** A further simplification of the equivalent circuit. **(e)** The input equivalent circuit after use of Miller's theorem.

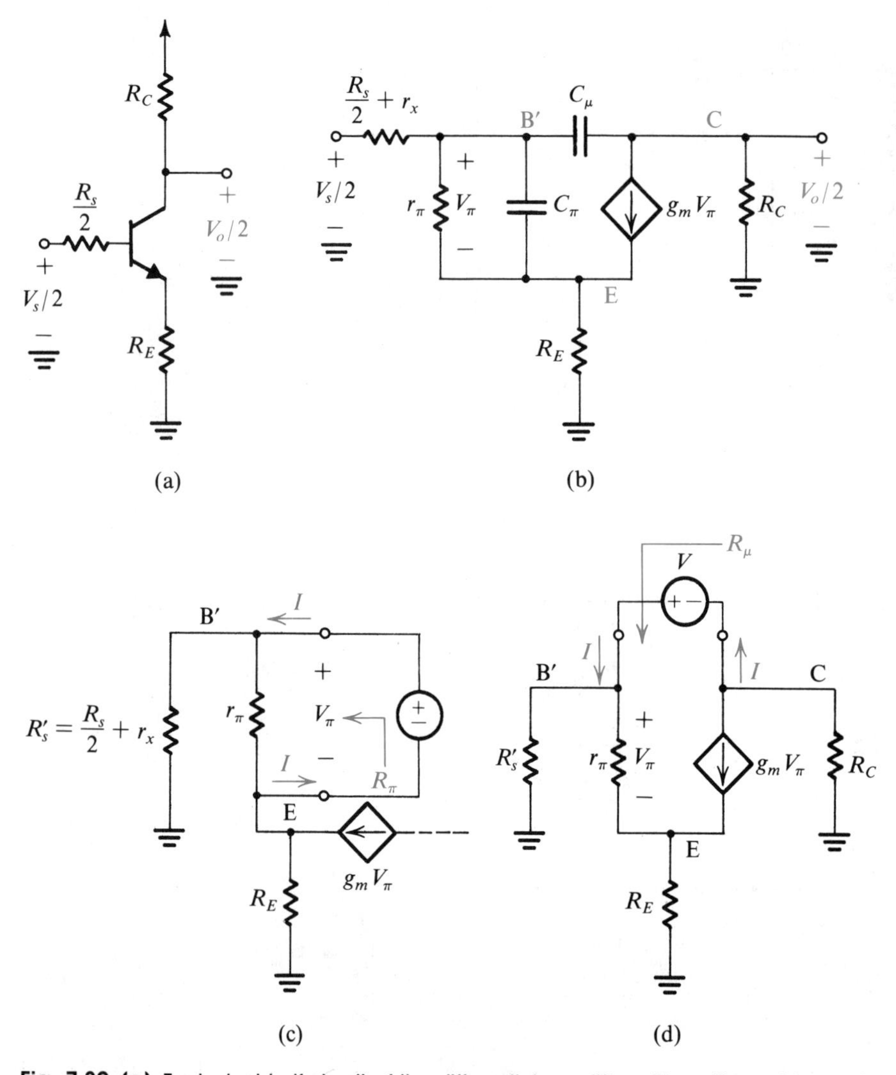

Fig. 7.39 **(a)** Equivalent half-circuit of the differential amplifier with emitter resistance R_E. **(b)** Equivalent circuit of the half-circuit in (a). **(c)** Circuit for determining the resistance R_π seen by C_π. **(d)** Circuit for determining the resistance R_μ seen by C_μ.

Exercises 7.25 Consider a differential amplifier biased with a current source $I = 1$ mA and having $R_C = 10$ kΩ. Let the amplifier be fed with a source having $R_s = 10$ kΩ. Also let the transistors be specified to have $\beta_0 = 100$, $C_\pi = 6$ pF, $C_\mu = 2$ pF, and $r_x = 50$ Ω. Find the dc differential gain A_0, the 3-dB frequency f_H, and the gain–bandwidth product.

Ans. 100 V/V (40 dB); 156 kHz; 15.6 MHz

7.26 Consider the differential amplifier of Exercise 7.25 but with a 150-Ω resistance included in each emitter lead. Find A_0, R_π, R_μ, f_H, and the gain–bandwidth product.

Ans. 40 V/V (32 dB); 1.03 kΩ; 215.6 kΩ; 364 kHz, 14.6 MHz

(*Note:* The difference between the gain–bandwidth products calculated in Exercises 7.25 and 7.26 is due mainly to the different approximations made.)

Variation of the CMRR with Frequency

The common-mode rejection ratio (CMRR) of a differential amplifier falls off at high frequencies because of a number of factors, the most important of which is the increase of the common-mode gain with frequency. To see how this comes about, consider the common-mode equivalent half-circuit shown in Fig. 7.40. Here the resistance R is the output resistance and the capacitance C is the output capacitance of the bias current source. From our study of the frequency response of the common-emitter amplifier in Section 7.6 we know that the components $2R$ and $C/2$ will introduce a zero in the common-mode gain function. This zero will be at a frequency f_Z,

$$f_Z = \frac{1}{2\pi(2R)(C/2)} = \frac{1}{2\pi RC} \tag{7.119}$$

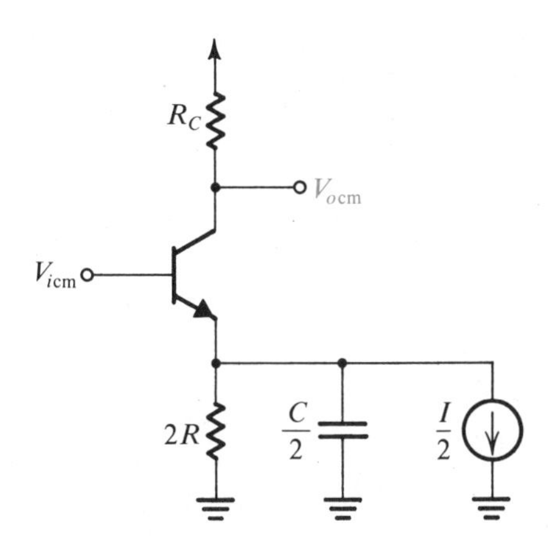

Fig. 7.40 The equivalent common-mode half-circuit.

Since R is usually very large, even a very small output capacitance C will result in f_Z having a relatively low value. The result is that the common-mode gain will start increasing, with a slope of +6 dB/octave at a relatively low frequency as shown in Fig. 7.41(a). The common-mode gain falls off at higher frequencies because of the internal capacitances C_π and C_μ and because of the pole created by $C/2$.

The behavior of the common-mode gain shown in Fig. 7.41(a), together with the high-frequency rolloff of the differential gain (Fig. 7.41(b)), results in the CMRR having the frequency response shown in Fig. 7.41(c).

Exercise 7.27 A differential amplifier is biased by a constant-current source having an output resistance of 30 MΩ and an output capacitance of 2 pF. Find the frequency at which the CMRR decreases by 3 dB.

Ans. 2.65 kHz

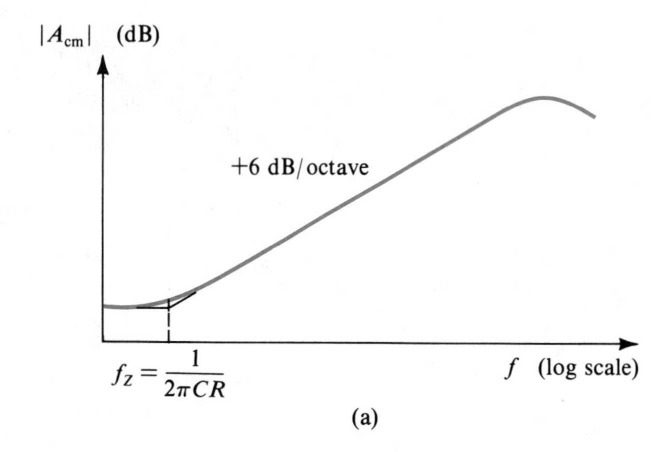

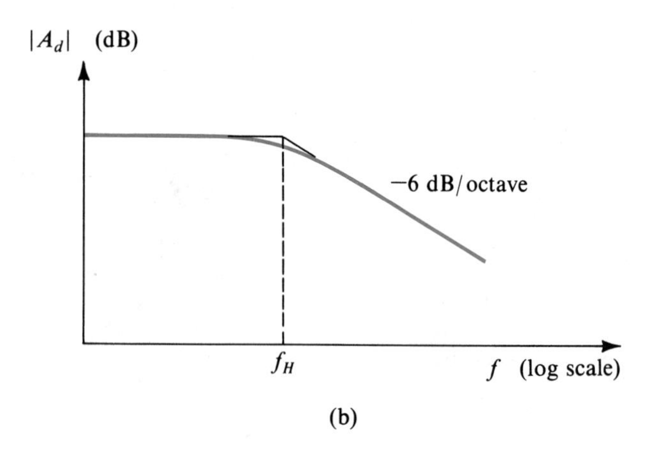

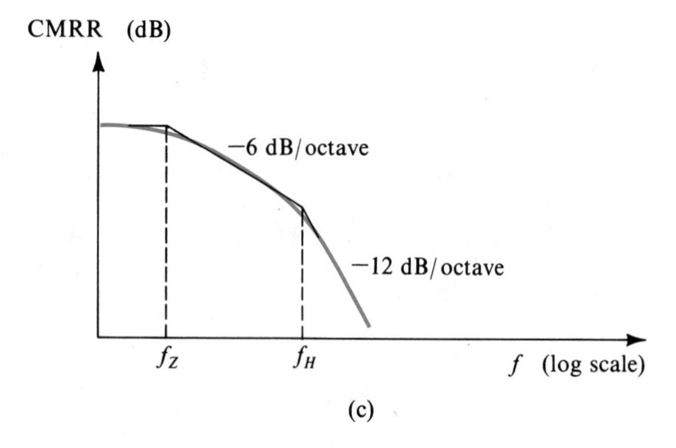

Fig. 7.41 Variation of **(a)** common-mode gain, **(b)** differential gain, and **(c)** common-mode rejection ratio with frequency.

7.11 THE DIFFERENTIAL PAIR AS A WIDEBAND AMPLIFIER— THE COMMON-COLLECTOR COMMON-BASE CONFIGURATION

A slight modification of the differential-amplifier circuit results in a configuration with a much higher bandwidth. The resulting circuit, shown in Fig. 7.42, is obtained by simply eliminating the collector resistance of Q_1. It can be easily seen that this eliminates the Miller capacitance multiplication of $C_{\mu 1}$. Furthermore, since $C_{\mu 2}$ has one of its terminals grounded, there will be no Miller effect in Q_2 either. We should therefore expect the circuit of Fig. 7.42 to have an extended frequency response, and we may add it to our repertoire of wideband amplifiers.

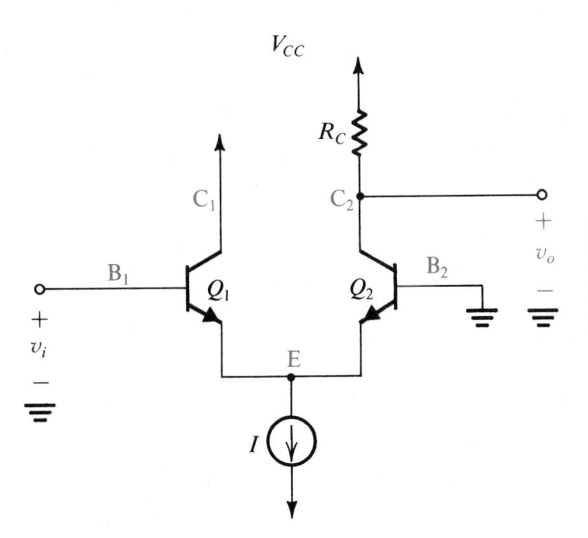

Fig. 7.42 The differential amplifier modified for wideband amplification. Eliminating the collector resistance of Q_1 eliminates the Miller capacitance multiplication.

Low-Frequency Gain

An alternative way of looking at the circuit of Fig. 7.42 is to consider it as a common-collector stage (Q_1) followed by a common-base stage (Q_2). The common-collector transistor Q_1 has in its emitter circuit the emitter resistance r_e of Q_2. Thus the signal current in the emitter of Q_1 is given by

$$i_e = \frac{v_i}{r_{e1} + r_{e2}} = \frac{v_i}{2r_e} \tag{7.120}$$

because

$$r_{e1} = r_{e2} = r_e = \frac{V_T}{I/2} \tag{7.121}$$

This signal current flows in the emitter of Q_2 and appears in its collector multiplied by α,

$$i_{c2} = \alpha i_e = \frac{\alpha v_i}{2r_e} \tag{7.122}$$

Thus the output signal voltage at the collector of Q_2 will be

$$v_o = i_{c2}R_C = \frac{\alpha v_i}{2r_e}R_C \tag{7.123}$$

and the voltage gain will be

$$\frac{v_o}{v_i} = \frac{\alpha R_C}{2r_e} \tag{7.124}$$

Frequency Response

To simplify matters we shall neglect the effect of r_x and thus obtain the equivalent circuit shown in Fig. 7.43(a). It is assumed that the circuit is fed with a signal source having a voltage V_s and a resistance R_s. A node equation at node E reveals that

$$V_{\pi 1} = -V_{\pi 2}$$

This allows us to simplify the circuit to that shown in Fig. 7.43(b), where $V_\pi = V_{\pi 1} = -V_{\pi 2}$. It can be seen that there are two real poles, one at the input with a frequency f_{P1},

$$f_{P1} = \frac{1}{2\pi(R_s//2r_\pi)(C_\pi/2 + C_\mu)} \tag{7.125}$$

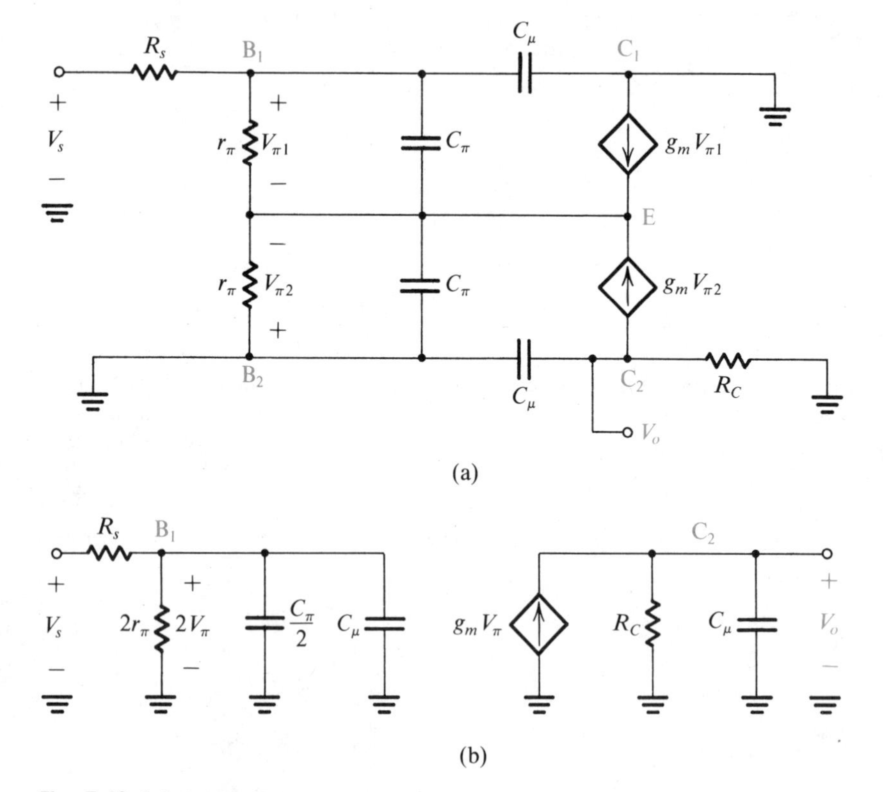

Fig. 7.43 (a) Equivalent circuit for the modified differential amplifier in Fig. 7.42. **(b)** Simplified equivalent circuit.

and one at the output with a frequency f_{P2},

$$f_{P2} = \frac{1}{2\pi R_C C_\mu} \tag{7.126}$$

Whether one of these poles is dominant or not will depend on the particular application. If no dominant pole exists, then one has to use the overall transfer function to evaluate the 3-dB frequency f_H. This is demonstrated in Example 7.10.

EXAMPLE 7.10 Consider the modified differential amplifier discussed above and let $I = 1$ mA, $R_C = 10$ kΩ, $R_s = 10$ kΩ, $f_T = 400$ MHz, and $C_\mu = 2$ pF. Evaluate the low-frequency gain and the 3-dB frequency f_H ($\beta = 100$).

Solution Each transistor is biased at an emitter current of 0.5 mA. Thus

$$r_e \simeq 50\ \Omega \qquad g_m \simeq 20\ \text{mA/V} \qquad r_\pi \simeq 5\ \text{k}\Omega$$

$$C_\pi + C_\mu = \frac{g_m}{\omega_T} = \frac{20 \times 10^{-3}}{2\pi \times 400 \times 10^6} \simeq 8\ \text{pF}$$

$$C_\mu = 2\ \text{pF} \qquad C_\pi = 6\ \text{pF}$$

The low-frequency gain is given by

$$A_0 = \frac{2r_\pi}{R_s + 2r_\pi}\,\frac{\alpha R_C}{2r_e} \simeq \frac{10}{10 + 10}\,\frac{10}{0.1}$$

Thus $A_0 = 50$, or 34 dB. The pole at the input has a frequency ω_{P1},

$$\omega_{P1} = \frac{1}{(R_s // 2r_\pi)(C_\pi/2 + C_\mu)}$$

$$= \frac{1}{5 \times 10^3(3 + 2) \times 10^{-12}} = 40\ \text{Mrad/s}$$

The pole at the output has a frequency ω_{P2}, given by

$$\omega_{P2} = \frac{1}{C_\mu R_C} = \frac{1}{2 \times 10^{-12} \times 10 \times 10^3} = 50\ \text{Mrad/s}$$

Since the two poles are very close to each other, we have to use the overall transfer function to evaluate the 3-dB frequency f_H. The amplifier transfer function is given by

$$A(s) \equiv \frac{V_o}{V_s} = \frac{A_0}{(1 + s/\omega_{P1})(1 + s/\omega_{P2})}$$

Thus

$$|A(j\omega)| = \frac{A_0}{\sqrt{(1 + \omega^2/\omega_{P1}^2)(1 + \omega^2/\omega_{P2}^2)}}$$

At $\omega = \omega_H$, $|A(j\omega_H)| = A_0/\sqrt{2}$. Thus

$$2 = \left(1 + \frac{\omega_H^2}{\omega_{P1}^2}\right)\left(1 + \frac{\omega_H^2}{\omega_{P2}^2}\right)$$

which leads to $\omega_H = 28.53$ Mrad/s. Thus the upper 3-dB frequency f_H is $f_H = \omega_H/2\pi = 4.54$ MHz.

Before leaving this section we should point out a disadvantage of the modified differential-amplifier circuit. Since the output is taken single-ended, the CMRR will be much lower than that of the balanced differential amplifier. For this reason the first stage of an operational amplifier is usually a balanced one. The modified circuit can be used in subsequent stages or in wideband amplifiers, where the requirement of a high CMRR is not so important.

SUMMARY

▶ Bode plots provide a convenient means for sketching the magnitude and phase of amplifier gain versus frequency. They are particularly suited for amplifiers with real poles and zeros.

▶ The amplifier gain remains almost constant over the midfrequency band. It falls off at high frequencies where transistor-model capacitors no longer have very high reactances. For ac amplifiers the gain falls off at low frequencies as well because the coupling and bypass capacitors no longer have very low reactances.

▶ The amplifier bandwidth is the frequency range over which the gain remains within 3 dB of the value at midband. The limits of the bandwidth are the frequencies f_L and f_H (for a dc amplifier only f_H is meaningful).

▶ If the amplifier poles and zeros are known (or easy to find), then f_L and f_H can be determined exactly either graphically or analytically. Alternatively, simple, approximate formulas exist for obtaining reasonably good estimates of f_L and f_H.

▶ If the poles and zeros are not easy to find, an approximate value for f_L (or f_H) can be determined by evaluating the short-circuit (or open-circuit) time constants of the circuit. The accuracy of this method improves if the zeros are not important (that is, if the zeros in the low-frequency band are at very low frequencies and if those in the high-frequency band are at very high frequencies) and if one pole is dominant. The method pinpoints the capacitor(s) that is (are) most significant in determining the frequency response, and is thus useful in design.

▶ The amplifier is said to have a dominant pole in the low-frequency band if the nearest pole or zero is at least two octaves lower in frequency. A dominant pole in the high-frequency band exists if the nearest pole or zero is at least two octaves higher in frequency.

▶ If a dominant low-frequency pole exists, then the low-frequency response is essentially that of an STC high-pass network, and f_L is equal to the pole frequency. A dominant high-frequency pole makes the high-frequency response like that of an STC low-pass network with f_H equal to the pole frequency.

▶ The high-frequency response of the common-source (common-emitter) amplifier is limited by the Miller effect, which multiplies the feedback capacitance C_{gd} (C_μ) and forms a dominant pole at the input. The bandwidth can be extended by reducing the generator resistance and/or the load resistance. The latter action reduces the gain.

▶ The components of the low-frequency hybrid-π model of the BJT can be determined from measurement of the common-emitter h-parameters.

▶ The common-emitter short-circuit current gain h_{fe} is equal to β at low frequencies. Its magnitude begins to fall off around f_β with -6-dB/octave slope, reaching unity at f_T.

▶ The common-base (common-gate) configuration does not suffer from the Miller effect and thus exhibits wide bandwidth. However, it has a low input resistance.

▶ The cascode configuration combines the wide bandwidth of the common-base (common-gate) circuit with the high input resistance of the common-emitter (common-source) circuit.

- The emitter follower (source follower) features a wide bandwidth that approaches the transistor f_T. [For the FET we define f_T as $g_m/2\pi(C_{gs} + C_{gd})$.]

- The common-collector common-emitter cascade provides a wideband amplifier comparable to the cascode.

- The frequency response of the differential pair is obtained by analyzing its equivalent differential half-circuit, which is a common-emitter (common-source) configuration.

- The bandwidth of the differential amplifier can be extended (at the expense of gain reduction) by adding resistors in the emitter (source) leads.

- The CMRR of the differential pair decreases with frequency as a result of the zero introduced in the common-mode gain by the output capacitance of the biasing current source.

- Wideband amplification can be obtained by eliminating the collector resistor of the input transistor in the differential pair, thus converting the circuit into a common-collector common-base cascade.

BIBLIOGRAPHY

L. S. Bobrow, *Elementary Linear Circuit Analysis,* 2nd ed., New York: Holt, Rinehart and Winston, 1987.

P. E. Gray and C. L. Searle, *Electronic Principles,* New York: Wiley, 1969.

P. R. Gray and R. G. Meyer, *Analysis and Design of Analog Integrated Circuits,* 2nd ed., New York: Wiley, 1984.

J. M. Steininger, "Understanding wideband MOS transistors," *IEEE Circuits and Devices,* vol. 6, no. 3, pp. 26–31, May 1990.

M. E. Van Valkenburg, *Network Analysis,* 3rd ed., Englewood Cliffs, N.J.: Prentice-Hall, 1974.

PROBLEMS

Section 7.1: s-Domain Analysis

7.1 Find the transfer function $T(s) = V_o(s)/V_i(s)$ of the circuit in Fig. P7.1. Is this an STC network? If so, of what type? For $C_1 = C_2 = 0.5\ \mu F$ and $R = 10\ k\Omega$, find the location of the pole(s) and zero(s), and sketch Bode plots for the magnitude response and the phase response.

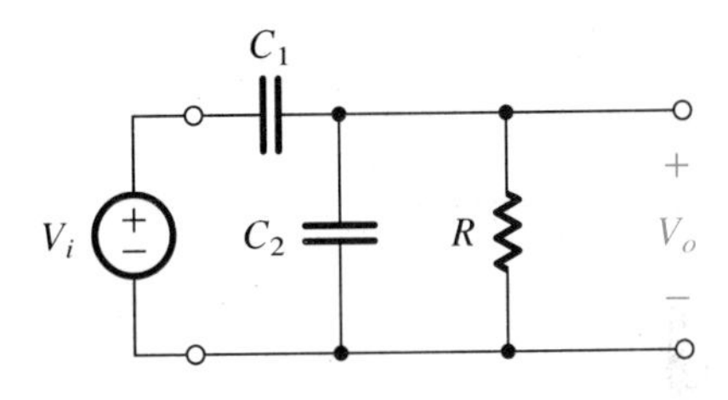

Fig. P7.1

7.2 A transfer function has the following zeros and poles: one zero at $s = 0$ and one zero at $s = \infty$; one pole at $s = -10$ and one pole at $s = -10^5$. The magnitude of the transfer function at $\omega = 10^3$ rad/s is 1000. Find the transfer function $T(s)$ and sketch a Bode plot for its magnitude.

7.3 Sketch Bode plots for the magnitude and phase of the transfer function

$$T(s) = \frac{10^4(1 + s/10^5)}{(1 + s/10^3)(1 + s/10^4)}$$

From your sketches determine approximate values for the magnitude and phase at $\omega = 10^6$ rad/s. What are the exact values determined from the transfer function?

7.4 An amplifier has a low-pass STC frequency response. The magnitude of the gain is 20 dB at dc and 0 dB at 10 kHz. What is the corner frequency? At what frequency is the gain 19 dB? At what frequency is the phase $-6°$?

7.5 A direct-coupled differential amplifier has a differential gain of 100 with poles at 10^6 and 10^8 rad/s, and a common-mode gain of 10^{-3} with a zero at 10^4 rad/s and a pole at 10^8 rad/s. Sketch the Bode magnitude plots for the differential gain, the common-mode gain, and the CMRR. What is the CMRR at 10^7 rad/s? (*Hint:* Division of magnitudes corresponds to subtraction of logarithms.)

Section 7.2: The Amplifier Transfer Function

7.6 An amplifier has the gain transfer function

$$A(s) = 10 \frac{s}{s + 2\pi \times 10} \frac{1}{1 + s/2\pi \times 10^6}$$

Sketch a Bode plot for its magnitude and find the midband gain, the lower 3-dB frequency f_L, and the upper 3-dB frequency f_H. Also, find approximate values for the frequencies at which the gain decreases to unity.

*D7.7 The equivalent circuit of an amplifier is shown in Fig. P7.7. The input signal source is coupled to the amplifier input via coupling capacitor C_C. Capacitor C_L represents a parasitic capacitance appearing across the load resistance R_L.

(a) Derive an expression for the amplifier voltage gain $A(s) \equiv V_o(s)/V_i(s)$.

(b) Noting that C_C is responsible for the frequency dependence of the gain at low frequencies and that C_L causes the gain to fall off at high frequencies, find A_M, $F_L(s)$, and $F_H(s)$.

(c) For $R_s = 10\ \text{k}\Omega$, $R_i = 100\ \text{k}\Omega$, and $R_L = 10\ \text{k}\Omega$, find the required value of G_m to obtain a midband gain of 20 dB.

(d) Find the minimum value of C_C so that f_L is at most 10 Hz.

(e) Find the maximum value that C_L can have while f_H is at least 1 MHz.

7.8 If in the circuit of Fig. P7.8 A is an ideal voltage amplifier of gain 100, find A_M, $F_L(s)$, and $F_H(s)$. Also find ω_L and ω_H.

7.9 Consider an amplifier whose $F_L(s)$ is given by

$$F_L(s) = \frac{s^2}{(s + \omega_{P1})(s + \omega_{P2})}$$

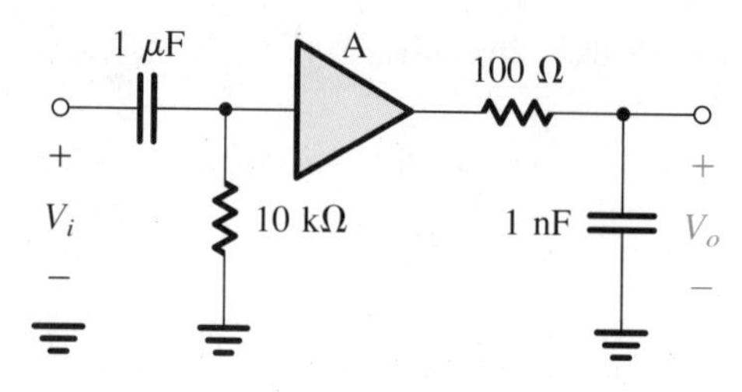

Fig. P7.8

with $\omega_{P1} > \omega_{P2}$. Find the ratio ω_{P1}/ω_{P2} for which the value of the 3-dB frequency ω_L calculated using the dominant pole approximation differs from that calculated using the root-sum-of-squares formula (Eq. 7.18) by (a) 10%, (b) 1%.

7.10 An amplifier is formed by cascading n identical stages, each characterized by a high-pass STC response with a corner frequency ω_0. Find the lower 3-dB frequency of the overall amplifier in terms of ω_0 and n. Find ω_L for $n = 2$, 3, and 4.

7.11 The low-frequency response of an amplifier is characterized by three poles of frequencies 10 Hz, 3 Hz, and 1 Hz, and three zeros at $\omega = 0$. Estimate the lower 3-dB frequency f_L using

(a) the dominant pole approximation,

(b) the root-sum-of-squares (Eq. 7.18) approximation.

7.12 The high-frequency response of a direct-coupled amplifier having a dc gain of -100 incorporates zeros at ∞ and 10^6 rad/s (one at each frequency) and poles at 10^5 and 10^7 rad/s (one at each frequency). Write an expression for the amplifier transfer function. Find ω_H using:

(a) the dominant-pole approximation, and

(b) the root-sum-of-squares approximation (Eq. 7.22).

If a means is found to lower the frequency of the finite zero to 10^5 rad/s, what does the transfer function be-

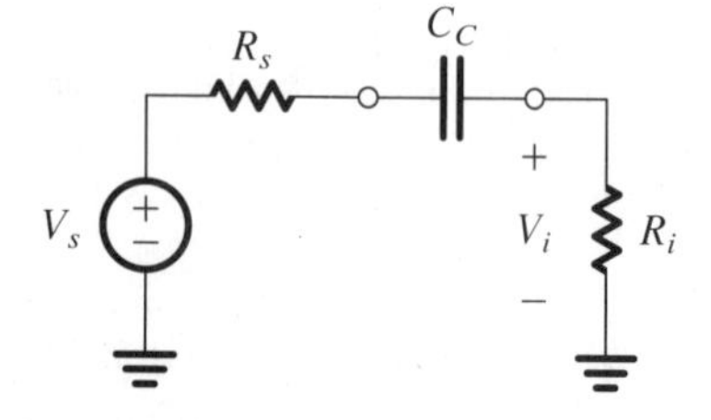

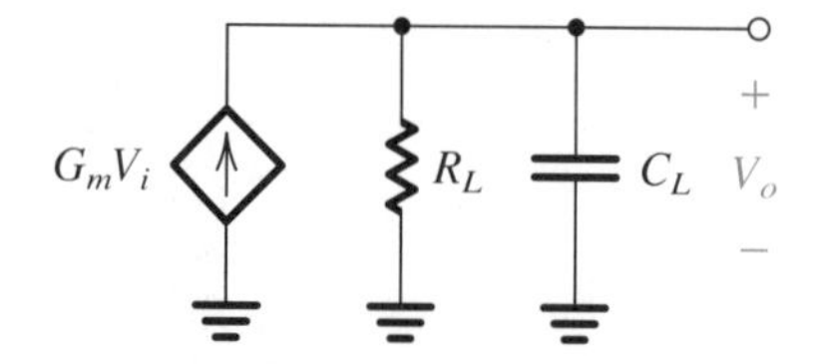

Fig. P7.7

come? What is the upper 3-dB frequency of the resulting amplifier?

7.13 A direct-coupled amplifier has a dominant pole at 100 rad/s and three coincident poles at a much higher frequency. These nondominant poles cause the phase lag of the amplifier at high frequencies to exceed the 90° angle due to the dominant pole. It is required to limit the excess phase at $\omega = 10^6$ rad/s to 30° (i.e., to limit the total phase angle to $-120°$). Find the corresponding frequency of the nondominant poles.

D7.14 Refer to Example 7.5. Give an expression for ω_H in terms of C_{gs}, R' (note that $R' = R_{\text{in}}//R$), C_{gd}, R'_L, and g_m. If all component values except for the generator resistance R are left unchanged, what must R be reduced to in order to raise f_H to 150 kHz?

****D7.15** Figure P7.15 shows the high-frequency equivalent circuit of a FET amplifier with a resistance R_S connected in the source lead. The purpose of this problem is to show that the value of R_S can be used to control the gain and bandwidth of the amplifier, specifically to allow the designer to trade off gain for increased bandwidth.

(a) Derive an expression for the low-frequency voltage gain (set C_{gs} and C_{gd} to zero).

(b) In order to be able to determine ω_H using the open-circuit time constants method, derive expressions for R_{gs} and R_{gd}.

(c) Let $R = 100$ kΩ, $g_m = 4$ mA/V, $R_L = 5$ kΩ, and $C_{gs} = C_{gd} = 1$ pF. Use the expressions in (a) and (b) above to find the low-frequency gain and the upper 3-dB frequency ω_H for the three cases $R_S =$ 0, 100, and 250 Ω. In each case evaluate also the gain–bandwidth product.

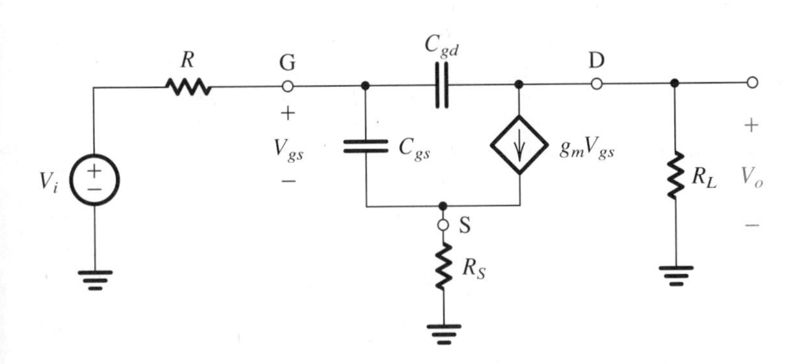

Fig. P7.15

Section 7.3: Low-Frequency Response of the Common-Source Amplifier

D7.16 Consider the common-source amplifier of Fig. 7.9. For a situation in which $R = 1$ MΩ and $R_{\text{in}} \equiv R_{G1}//R_{G2} = 1$ MΩ, what value of C_{C1} must be chosen to place the corresponding pole at 10 Hz? What value would you choose if available capacitors are specified to only one significant digit and the pole frequency is not to exceed 10 Hz? What is the pole frequency, f_{P1}, obtained with your choice? If a designer wishes to lower this by raising R_{in}, what is the most she can expect if available resistors are limited to 10 times those now used?

D7.17 The amplifier in Fig. P7.17 is biased to operate at $I_D = 1$ mA and $g_m = 1$ mA/V. Neglecting r_o, find the

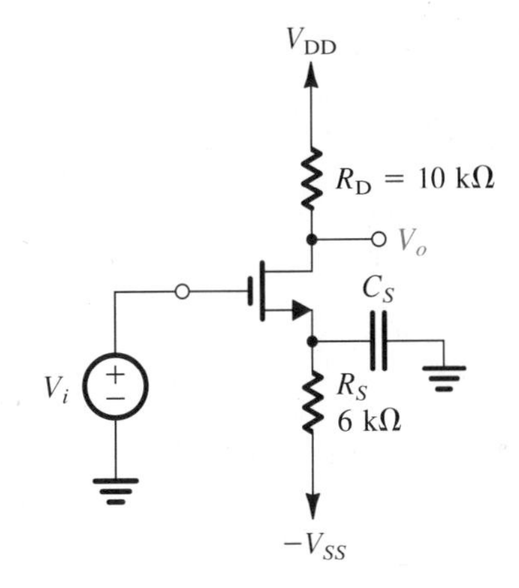

Fig. P7.17

midband gain. Find the value of C_S that places the corresponding pole at 10 Hz. What is the frequency of the transfer-function zero introduced by C_S? Give an expression for the gain function $V_o(s)/V_i(s)$. What is the gain of the amplifier at dc?

7.18 A common-source amplifier employs a resistor $R_S = 4.7\ \text{k}\Omega$ to establish $g_m = 1.2$ mA/V. The source-bypass capacitor chosen is 10 μF. What are the associated pole and zero frequencies? If R_S is replaced with a current source for which g_m remains the same, what do the pole and zero frequencies become?

D7.19 Consider the amplifier of Fig. 7.9, whose equivalent output circuit is depicted in Fig. 7.11. Let $R_D = 15\ \text{k}\Omega$, $r_o = 150\ \text{k}\Omega$, and $R_L = 10\ \text{k}\Omega$. Find the value of C_{C2}, specified to one significant digit, to ensure that the associated pole is at or below 10 Hz. If a higher power design results in doubling I_D, with both R_D and r_o reduced by a factor of 2, what does the corner frequency (due to C_{C2}) become? For increasingly higher power designs, what is the highest corner frequency that can be associated with C_{C2}?

D7.20 Figure P7.20 shows a MOS amplifier whose bias design and midband analysis were performed in Example 5.11. Specifically, the MOSFET is biased at $I_D = 1.06$ mA, and has $g_m = 0.725$ mA/V and $r_{o2} = 47\ \text{k}\Omega$. The midband analysis showed that $V_o/V_i = -3.3$ V/V and $R_{\text{in}} = 2.33\ \text{M}\Omega$. Select appropriate values for the two capacitors so that the low-frequency response is dominated by a pole at 10 Hz with the other pole at least a decade lower. (*Hint:* In determining the pole due to C_{C2}, resistance R_G can be neglected.)

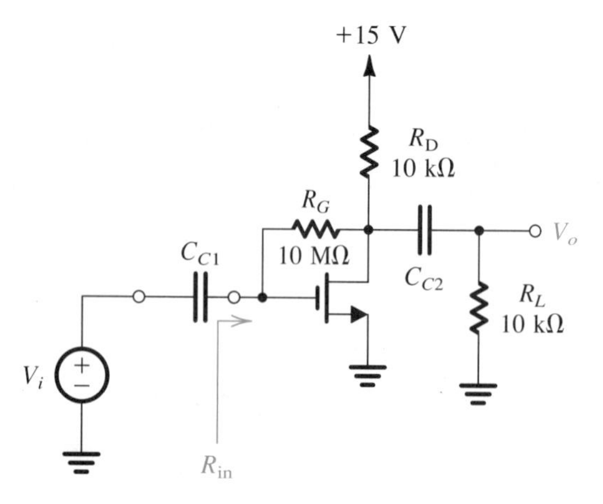

Fig. P7.20

Section 7.4: High-Frequency Response of the Common-Source Amplifier

7.21 Refer to the MOSFET high-frequency model in Fig. 7.12(a). Evaluate the model parameters for an NMOS transistor operating at $I_D = 100\ \mu$A, $V_{SB} = 2$ V, and $V_{DS} = 5$ V. The MOSFET has $W = 30\ \mu$m, $L = 10\ \mu$m, $t_{OX} = 0.1\ \mu$m, $\mu_n = 0.06\ \text{m}^2/\text{V}\cdot\text{s}$, $\gamma = 0.5\ \text{V}^{1/2}$, $2\phi_f = 0.6$ V, $\lambda = 0.02\ \text{V}^{-1}$, $C_{OX} = 0.33\ \text{fF}/\mu\text{m}^2$, $V_0 = 0.6$ V, and $C_{sb0} = C_{db0} = 0.1$ pF. Assume that the overlap capacitance from gate to source and gate to drain is 0.01 pF and that the parasitic capacitance between gate and substrate is 0.05 pF. (Recall that $g_{mb} = \chi g_m$, where $\chi = (\gamma/(2\sqrt{2\phi_f + V_{SB}}).)$

7.22 Find f_T for a MOSFET operating at $I_D = 100\ \mu$A and having $\mu_n C_{OX} = 20\ \mu\text{A/V}^2$, $W = 30\ \mu$m, $L = 10\ \mu$m, $C_{gs} = 0.08$ pF, and $C_{gd} = 0.01$ pF.

7.23 Starting from the definition of ω_T for a MOSFET,

$$\omega_T = \frac{g_m}{C_{gs} + C_{gd}}$$

and making the approximation that $C_{gs} \gg C_{gd}$ and that the overlap component of C_{gs} is negligibly small, show that

$$\omega_T \simeq \frac{3}{L}\sqrt{\frac{\mu_n I_D}{2C_{OX}WL}}$$

Thus note that to obtain a high ω_T from a given device it must be operated at a high current. Also note that faster operation is obtained from smaller devices.

7.24 A GaAs MESFET has a conductance parameter $\beta = 10^{-4}\ \text{A/V}^2$ per μm width, $W = 100\ \mu$m, $L = 1\ \mu$m, $V_t = -1$ V, and C_{gs} (at $V_{GS} = 0$ V) = 1.6 fF per μm width. If the device is operated at $V_{GS} = 0$ V and its f_T is found to be 15 GHz, find C_{gd}.

7.25 Use Miller's theorem to determine the input resistance R_{in} of the amplifier in Fig. P7.20, given that the voltage gain V_o/V_i at midband frequencies has been found to be -3.3 V/V.

7.26 Consider an ideal voltage amplifier with a gain of 0.95 V/V and a resistance $R = 100\ \text{k}\Omega$ connected in the feedback path—that is, between the output and input terminals. Use Miller's theorem to find the input resistance of this circuit.

7.27 Figure P7.27 shows a transconductance amplifier with an infinite input resistance, a 10-kΩ output resistance, and a transconductance $G_m = 0.1$ A/V. A 1-MΩ resistor R_f is connected from the output of the amplifier

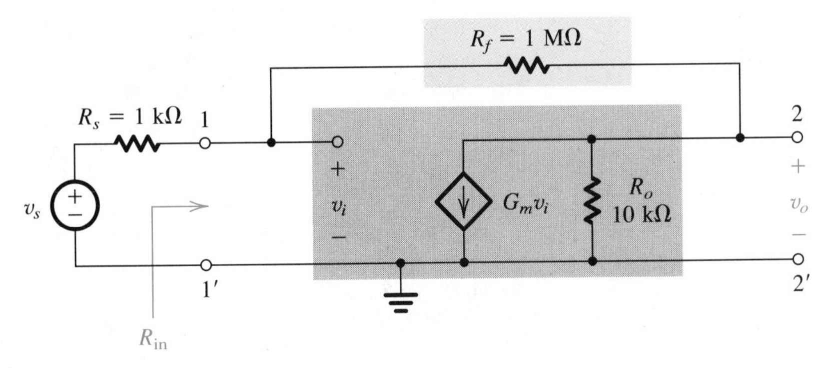

Fig. P7.27

back to its input. The amplifier is fed with a source v_s having a source resistance R_s. Find R_{in}, v_o/v_s, and R_{out}. (*Hint:* Miller's theorem is useful in finding R_{in} but *not* R_{out}.)

7.28 A FET common-source amplifier has $R_{in} = 2\ \text{M}\Omega$, $g_m = 4\ \text{mA/V}$, $r_o = 100\ \text{k}\Omega$, $R_D = 10\ \text{k}\Omega$, $C_{gs} = 2\ \text{pF}$, and $C_{gd} = 0.5\ \text{pF}$. The amplifier is fed from a voltage source with an internal resistance of 500 kΩ and is connected to a 10-kΩ load. Find:
(a) the overall midband gain A_M,
(b) the dominant high-frequency pole, using the Miller approximation,
(c) the location of the two poles and zeros using the transfer function of Eq. (7.57).

7.29 The analysis of the high-frequency response of the common-source amplifier, presented in the text, is based on the assumption that the resistance of the signal source, R, is large enough to form a dominant pole with the input capacitance. In this problem we wish to consider the situation when this assumption is not valid. For this purpose consider the common-source amplifier (Fig. 7.15a) when $R = 0$. Since, as you will find out soon, the load capacitance now plays an important role in determining the high-frequency response, introduce in the model a capacitance C_L between the drain and ground. Such a capacitance would represent the sum of the input capacitance of a possible subsequent amplifier stage, the drain-to-substrate capacitance C_{db} (present in a MOSFET), the inevitable parasitic capacitance present at the output of the amplifier, etc. Analyze the circuit to determine the transfer function $V_o(s)/V_i(s)$ and thus the poles and zeros. Find the frequencies of the pole and zero for the case $C_{gd} = 0.5\ \text{pF}$, $C_L = 2\ \text{pF}$, $g_m = 4\ \text{mA/V}$, and $R'_L = 5\ \text{k}\Omega$.

*7.30 Figure P7.30 shows an active-loaded common-source MOSFET amplifier fed with an input signal source V_i, having a negligible resistance. Capacitance C_L represents the sum of C_{db}, the input capacitance of the subsequent stage, the parasitic capacitance between the output node and ground, etc. Replace the MOSFET with its high-frequency equivalent circuit and analyze the resulting circuit to show that the transfer function is

$$\frac{V_o}{V_i} = -g_m r_o \frac{1 - s/(g_m/C_{gd})}{1 + s(C_L + C_{gd})r_o}$$

Convince yourself that usually the zero will have a much higher frequency than the pole and thus the pole dominates. Sketch a Bode plot for the gain magnitude. Give an approximate expression for the frequency at which the magnitude of gain becomes unity. (This is the gain–bandwidth product of the amplifier.) Evaluate the dc gain, the pole frequency, the zero fre-

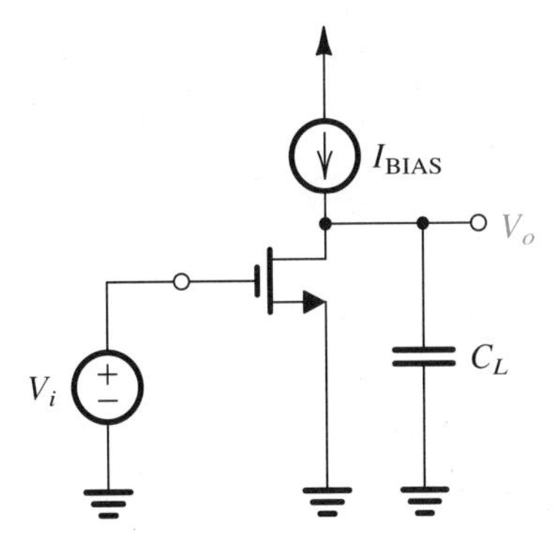

Fig. P7.30

quency, and the gain–bandwidth product for the case of a MOSFET biased at $I_D = 100\ \mu$A that has $g_m = 100\ \mu$A/V, $V_A = 50$ V, $C_{gd} = 0.1$ pF, and $C_L = 0.9$ pF.

***7.31** It is required to analyze the high-frequency response of the CMOS amplifier shown in Fig. P7.31. The dc bias current is 10 μA. For Q_1, $\mu_n C_{OX} = 20\ \mu$A/V^2, $V_A = 50$ V, $W/L = 64$, and $C_{gs} = C_{gd} = 1$ pF. For Q_2, $C_{gd} = 1$ pF and $V_A = 50$ V. Also, there is a 1-pF stray capacitance between the common drain connection and ground. Assume that the resistance of the input signal generator is negligibly small. Also, for simplicity assume that the signal voltage at the gate of Q_2 is zero. Find the frequency of the pole and zero.

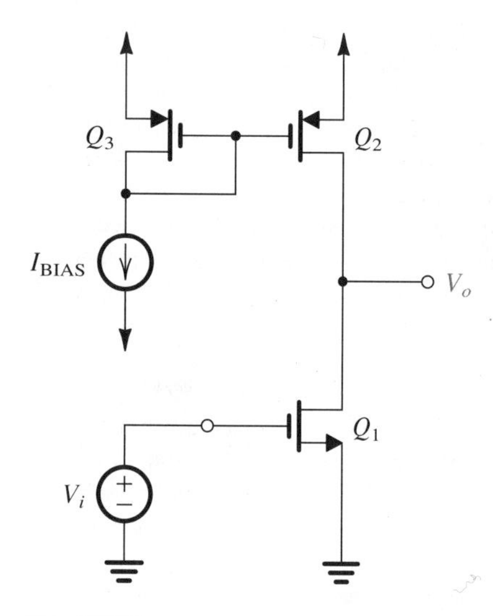

Fig. P7.31

D7.32** This problem investigates the use of MOSFETs in the design of wideband amplifiers (Steininger 1990). Such amplifiers can be realized by cascading low-gain stages.

(a) Show that for the case $C_{gd} \ll C_{gs}$ and the gain of the common-source amplifier is low so that the Miller effect is negligible, the MOSFET can be modeled by the approximate equivalent circuit shown in Fig. P7.32(a).

(b) Figure P7.32(b) shows an amplifier stage suitable for the realization of low gain and wide bandwidth. Transistors Q_1 and Q_2 have the same channel length L but different widths W_1 and W_2. They are biased at the same V_{GS} and have the same f_T. Use the MOSFET equivalent circuit of Fig. P7.32(a) to model this amplifier stage assuming that its output is connected to the input of an identical stage. Show that the voltage gain V_o/V_i is given by

$$\frac{V_o}{V_i} = -\frac{A_0}{1 + \dfrac{s}{\omega_T/(A_0 + 1)}}$$

where

$$A_0 = \frac{g_{m1}}{g_{m2}} = \frac{W_1}{W_2}$$

(c) For $L = 2\ \mu$m, $W_2 = 100\ \mu$m, $f_T = 2$ GHz, $\mu_n C_{OX} = 20\ \mu$A/V^2, and $V_t = 1$ V, design the circuit to obtain a gain of 3 per stage. Bias the MOSFETs at $V_{GS} = 3$ V. Specify the required values of W_1 and I. What is the 3-dB frequency achieved?

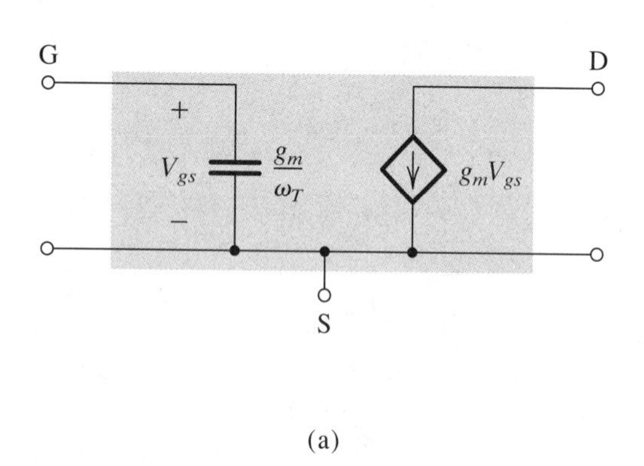

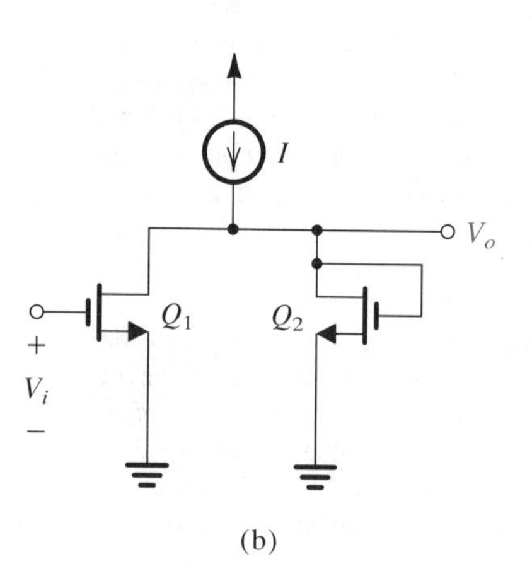

Fig. P7.32

D7.33 A pulse amplifier is required. The output droop should not exceed 2%/ms, and the rise and fall times should not exceed 10 ns. Specify the upper and lower 3-dB frequencies in Hz.

Section 7.5: The Hybrid-π Equivalent Circuit Model of the BJT

7.34 A BJT biased by a 10-mA emitter-current source, and having its emitter bypassed to ground by a large capacitor, is fed a small-signal current through an external 1-kΩ series base resistor. Its collector is connected to a suitable supply via a 100-Ω resistor. Careful ac measurements taken with respect to ground provide the following voltages: on the emitter, 0 mV (approximately); on the base; 3.3 mV; on the collector, 120 mV; on the input resistor remote from the base; 13.3 mV. Find r_e, r_π, h_{fe}, g_m, and r_x.

****7.35** In the circuit of Fig. 7.18, $R_s = 10$ kΩ, $R_B = 100$ kΩ, $R_C = 10$ kΩ, and $R_L = 100$ Ω. AC measurements are taken as follows: $v_s = 15$ mV, $v_b = 5$ mV, and $v_o = 20$ mV.

(a) Calculate h_{ie} and h_{fe} both approximately and exactly (that is, taking the currents through R_B and R_C into account).

(b) If r_x is determined by a separate high-frequency measurement to be 50 Ω, calculate r_π, r_e, g_m, and I_E.

(c) The circuit is then converted to that in Fig. 7.19 and a signal $v_c = 5$ V is applied. Measurement indicates v_b to be 1 mV. What is the value of h_{re}? Note that if the purpose of this measurement is to determine r_μ then it must be done at very low frequencies. Why?

(d) Using the results in (b) and (c), calculate r_μ. At what frequency does the magnitude of the impedance of a 1-pF capacitance equal this value of r_μ?

(e) Consider the circuit of Fig. 7.19 augmented with a resistor of 100 Ω in series with the collector. With $v_c = 5$ V, 8 mV is measured across the 100-Ω resistor. For this situation estimate h_{oe} and hence r_o and V_A.

7.36 The following parameters were measured on a transistor biased at $I_C = 2$ mA: $h_{ie} = 1.350$ kΩ, $h_{fe} = 100$, $h_{re} = 5 \times 10^{-5}$, $h_{oe} = 2.4 \times 10^{-5}$ A/V. Determine the values of g_m, r_π, r_e, r_x, r_o, r_μ, and V_A.

7.37 A particular BJT operating at $I_C = 1$ mA has $C_\mu = 1$ pF, $C_\pi = 10$ pF, and $\beta = 150$. What are ω_T and ω_β for this situation?

7.38 For the transistor described in Problem 7.37, C_π includes a relatively constant depletion-layer capacitance of 2 pF. If the device is operated at $I_C = 0.1$ mA, what does its ω_T become?

7.39 A particular small-geometry BJT has f_T of 5 GHz and $C_\mu = 0.1$ pF when operated at $I_C = 0.5$ mA. What is C_π in this situation? Also find g_m, and for $\beta = 150$ find r_π and f_β.

7.40 For a BJT whose unity-gain bandwidth is 1 GHz and $\beta_0 = 200$, at what frequency does the magnitude of h_{fe} become 10? What is ω_β?

***7.41** For a sufficiently high frequency, measurement of the complex input impedance of a BJT having (ac) grounded emitter and collector yields a real part approximating r_x. For what frequency, defined in terms of ω_β, is such an estimate of r_x good to within 10% under the condition that $r_x \leq r_\pi/10$?

***7.42** Complete the table entries below for transistors (a) through (g) under the conditions indicated. Neglect r_x.

Section 7.6: Frequency Response of the Common-Emitter Amplifier

7.43 Consider the common-emitter amplifier of Fig. 7.26 under the following conditions: $R_s = 5$ kΩ, $R_1 =$

Transistor	I_E (mA)	r_e (Ω)	g_m (mA/V)	r_π (kΩ)	β_0	f_T (MHz)	C_μ (pF)	C_π (pF)	f_β (MHz)
(a)	1				100	400	2		
(b)		25					2	10.7	4
(c)				2.525		400		13.8	4
(d)	10				100	400	2		
(e)	0.1				100	100	2		
(f)	1				10	400	2		
(g)						800	1	9	80

33 kΩ, $R_2 = 22$ kΩ, $R_E = 3.9$ kΩ, $R_C = 4.7$ kΩ, $R_L = 5.6$ kΩ, $V_{CC} = 5$ V. The dc emitter current can be shown to be $I_E \simeq 0.33$ mA, at which $\beta_0 = 120$, $r_o = 300$ kΩ, $r_x = 50$ Ω, $f_T = 700$ MHz, and $C_\mu =$ 1 pF. Find the input resistance, R_{in}, the midband gain, A_M, and an estimate of the frequency of the input pole, f_H.

7.44 For the amplifier described in Problem 7.43, with $C_{C1} = C_{C2} = 1$ μF and $C_E = 10$ μF, estimate the low-frequency 3-dB frequency. Also find the frequency of the zero introduced by C_E.

D7.45 For the amplifier described in Problem 7.43, design the coupling and bypass capacitors for a lower 3-dB frequency of 100 Hz. Design so that the zero cancels the pole introduced by C_{C2}, and that the contribution of C_{C1} to determining f_L is only 1%.

***7.46** For the amplifier described in Problem 7.43, use the BJT equivalent of Eq. (7.57) to determine the high-frequency poles and zero. Find an estimate for f_H.

7.47 Consider the circuit of Fig. 7.26. For $R_s = 10$ kΩ, $R_B \equiv R_1 // R_2 = 10$ kΩ, $r_x = 100$ Ω, $r_\pi = 1$ kΩ, $\beta_0 =$ 100, and $R_E = 1$ kΩ, what is the ratio C_E/C_{C1} that makes their contributions to the determination of ω_L equal?

***D7.48** Consider the circuit of Fig. 7.26 augmented by a resistance r_E in series with C_E. Using the equivalent circuit of Fig. 7.27 determine the resistance R'_E seen by C_E. Comment on the effect of this circuit change on the associated pole and zero frequencies (and the lower 3-dB frequency). What is the effect of including r_E on the midband gain?

***7.49** Consider the active-loaded common-emitter amplifier of Fig. P7.49 (bias details not shown). Let the amplifier be fed with an ideal voltage source V_i and neglect the effect of r_x. Assume that there is a capacitance C_L present between the output node and ground. This capacitance represents the sum of the input capacitance of the subsequent stage and the inevitable parasitic capacitance between collector and ground. Show that the voltage gain is given by

$$\frac{V_o}{V_i} = -g_m r_o \frac{1 - s(C_\mu/g_m)}{1 + s(C_L + C_\mu)r_o}$$

$$\simeq -\frac{g_m r_o}{1 + s(C_L + C_\mu)r_o}, \quad \text{for small } C_\mu$$

If the transistor is biased at $I_C = 200$ μA, and for $V_A = 100$ V, $C_\mu = 0.2$ pF, and $C_L = 1$ pF, find the dc gain, the 3-dB frequency, and the frequency at which the gain reduces to unity. Sketch a Bode plot for the gain magnitude.

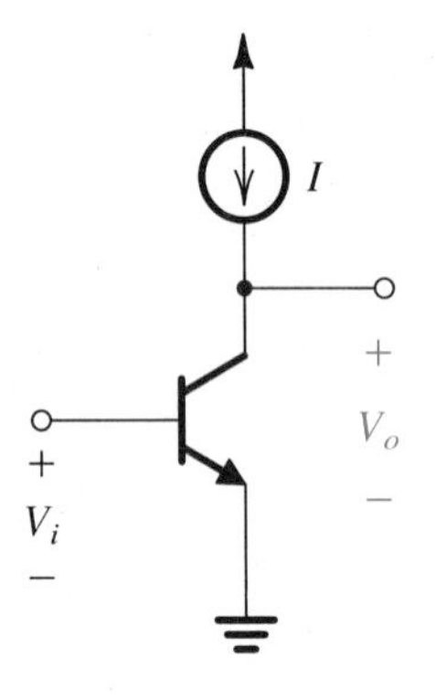

Fig. P7.49

***7.50** Refer to Fig. P7.50. Utilizing the BJT high-frequency hybrid-π model with $r_x = 0$ and $r_o = \infty$, derive an expression for $Z_i(s)$ as a function of r_e and C_π. Find the frequency at which the impedance has a phase angle of 45° for the case in which the BJT has $f_T = 400$ MHz and the bias current is relatively high. What is the frequency when the bias current is reduced so that $C_\pi \simeq C_\mu$? (Assume $\alpha = 1$.)

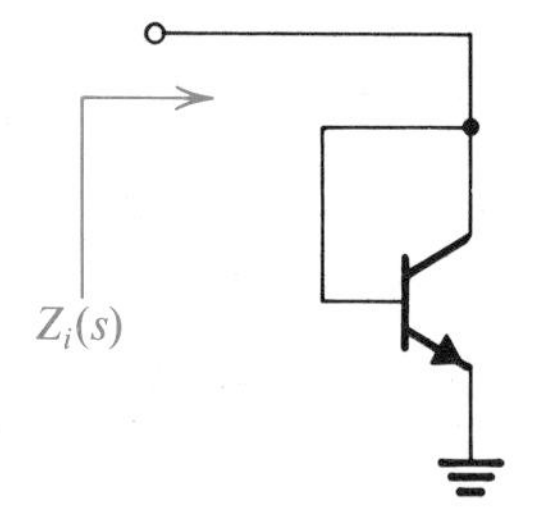

Fig. P7.50

***7.51** For the current mirror in Fig. P7.51 derive an expression for the current transfer function $I_o(s)/I_i(s)$ taking into account the BJT internal capacitances and neglecting r_x and r_o. Assume the BJTs to be identical. Observe that a signal ground appears on the collector of Q_2. If the mirror is biased at 1 mA and the BJTs at this operating point are characterized by $f_T =$ 400 MHz, $C_\mu = 2$ pF, and $\beta_0 = \infty$, find the frequencies of the pole and zero of the transfer function.

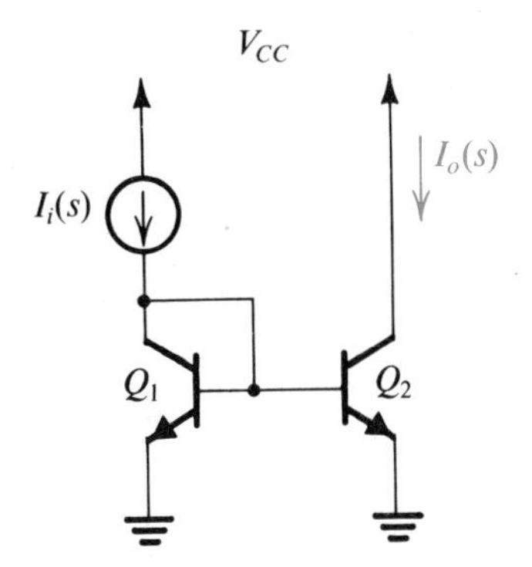

Fig. P7.51

Section 7.7: The Common-Base, Common-Gate, and Cascode Configurations

7.52 Consider the common-base amplifier of Fig. 7.28 under the conditions that $R_1 = 33\ \text{k}\Omega$, $R_2 = 22\ \text{k}\Omega$, $R_E = 3.9\ \text{k}\Omega$, $R_C = 4.7\ \text{k}\Omega$, $R_L = 5.6\ \text{k}\Omega$, $V_{CC} = 5\ \text{V}$, and $R_s = 75\ \Omega$. The dc emitter current can be shown to be 0.33 mA, at which $\beta_0 = 120$, $r_o = 300\ \text{k}\Omega$, $f_T = 700$ MHz, and $C_\mu = 0.5$ pF. Find the input resistance seen by the source and the midband gain. Estimate the location of the two poles and the upper cutoff frequency.

7.53 Consider the cascode configuration of Fig. 7.30 in which $R_s = 5\ \text{k}\Omega$, $R_1 = 22\ \text{k}\Omega$, $R_2 = 11\ \text{k}\Omega$, $R_3 = 22\ \text{k}\Omega$, $R_E = 3.9\ \text{k}\Omega$, $R_C = 4.7\ \text{k}\Omega$, $R_L = 5.6\ \text{k}\Omega$, and $V_{CC} = 5$ V. Analysis indicates that the dc emitter current is 0.33 mA, for which $\beta_0 = 120$, $r_o = 300\ \text{k}\Omega$, $r_x = 50\ \Omega$, $f_T = 700$ MHz, and $C_\mu = 0.5$ pF. Find the midband gain, the high-frequency poles, and the upper 3-dB frequency.

7.54 For the circuit in Fig. P7.54 let $R_C = 5\ \text{k}\Omega$, $I = 1$ mA, $\beta = 100$, $C_\pi = 5$ pF, and $C_\mu = 1$ pF. Find the frequencies of the three high-frequency poles and estimate f_H for two cases: (a) $R_s = 1\ \text{k}\Omega$ and (b) $R_s = 10\ \text{k}\Omega$.

*7.55 Figure P7.55 shows a cascode CMOS amplifier using three devices whose substrates are connected appropriately to the power supplies.

(a) Draw the amplifier small-signal equivalent circuit. It should include three capacitors: C_{gd1}, C_1, and C_2, where C_1 is the total capacitance between the drain of Q_1 and ground, $C_1 = C_{db1} + C_{sb2} + C_{gs2}$; and C_2 is the total capacitance between the drain of Q_2 and ground, $C_2 = C_{db2} + C_{db3} + C_{gd3} + C_L$. Since the resistance between the output node and ground is dominated by r_{o3} you may neglect the effect of r_{o2} and omit it from the equivalent circuit. Also, to simplify matters neglect the body effect in Q_2.

(b) Show that the dc gain is approximately $-g_{m1}r_{o3}$, that the poles have frequencies of $1/2\pi r_{o3}C_2$ and $g_{m2}/2\pi(C_1 + C_{gd1})$, and that the zero has a frequency of $g_{m1}/2\pi C_{gd1}$.

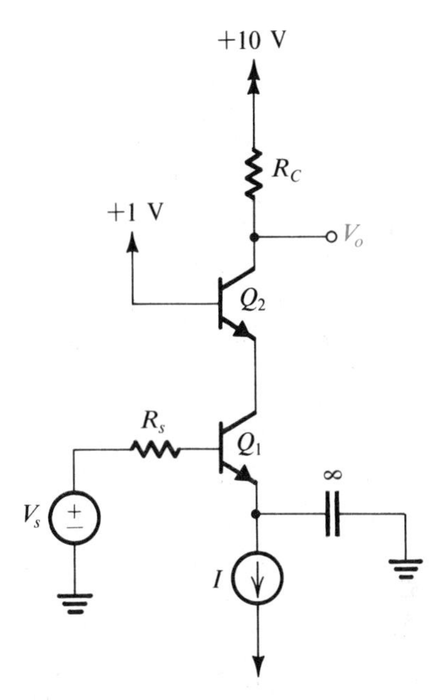

Fig. P7.54

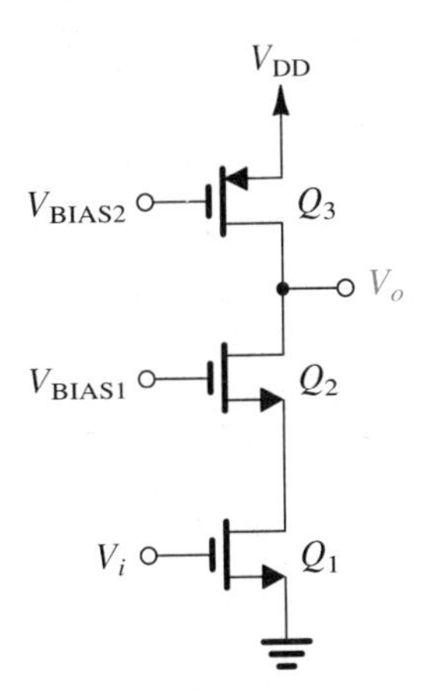

Fig. P7.55

(c) Calculate the values of dc gain and the frequencies of poles and zero for $I_D = 50\ \mu A$, $W/L = 1$ for all devices with $L = 10\ \mu m$, $\mu_n C_{OX} = 20\ \mu A/V^2$, $\lambda_p = 0.02\ V^{-1}$, $C_{gs} = 30$ fF, $C_{gd} = 3.5$ fF, $C_{sbn} = C_{dbn} = 24$ fF, $C_{sbp} = C_{dbp} = 12$ fF and $C_L = 0.1$ pF.

Section 7.8: Frequency Response of the Emitter and Source Followers

7.56 Consider a JFET source follower fed by a signal generator having an internal resistance R. The follower has a resistance in the source lead R_S, a transconductance g_m, and internal capacitances C_{gs} and C_{gd}. Adapt the results of Section 7.8 to this case, and evaluate the midband gain and the frequency of the dominant high-frequency pole for $R = 100\ k\Omega$, $R_S = 10\ k\Omega$, $g_m = 2$ mA/V, $C_{gs} = 2$ pF, and $C_{gd} = 1$ pF.

*7.57 Refer to Fig. 7.32(a). Extract r_x and add it in series with R_s thus changing R_s to R_s'. What is the dc gain from B' to E? Noting that y_π is connected between B' and E and that you know the gain from B' to E, use Miller's theorem to derive the equivalent circuit (or an approximation to it) shown in Fig. 7.32(d).

*D7.58 For the emitter-follower circuit of Fig. P7.58, design the coupling capacitors so that a dominant low-frequency pole at 10 Hz is obtained. Design so that the contributions of the capacitors are in the ratio of 10:1 and that the total capacitance is minimized. Use $\beta_0 = 120$. What is the midband gain? If $C_\pi = 2.5$ pF and $C_\mu = 0.5$ pF, what upper cutoff results?

7.59 For an emitter-follower biased at $I_C = 1$ mA and having $R_s = R_E = 1\ k\Omega$, and using a transistor specified to have $f_T = 2$ GHz, $C_\mu = 0.1$ pF, $r_x = 100\ \Omega$, $\beta_0 = 100$, and $V_A = 20$ V, evaluate the midband gain and the frequency of the dominant high-frequency pole.

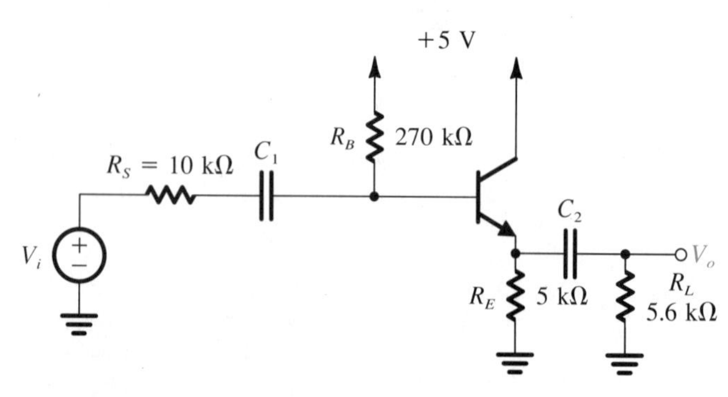

Fig. P7.58

**7.60 For the emitter-follower shown in Fig. P7.60 find the midband gain, the frequency of the high-frequency dominant pole, an estimate of f_H using the method of open-circuit time constants, and the frequency of the transfer-function zero for the cases
(a) $R_s = 1\ k\Omega$,
(b) $R_s = 10\ k\Omega$, and
(c) $R_s = 100\ k\Omega$.
Let $R_L = 1\ k\Omega$, $\beta_0 = 100$, $f_T = 400$ MHz, and $C_\mu = 2$ pF.

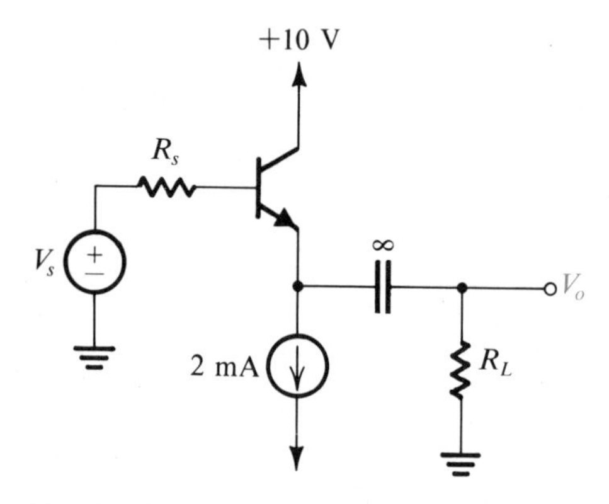

Fig. P7.60

Section 7.9: The Common-Collector Common-Emitter Cascade

*D7.61 The transistors in the circuit of Fig. P7.61 have $\beta_0 = 100$, $V_A = 100$ V, $C_\mu = 0.2$ pF, and $C_{je} = 0.8$ pF. At a bias current of 100 μA, $f_T = 400$ MHz. (Note that the bias details are not shown.)
(a) Find R_{in} and the midband gain.

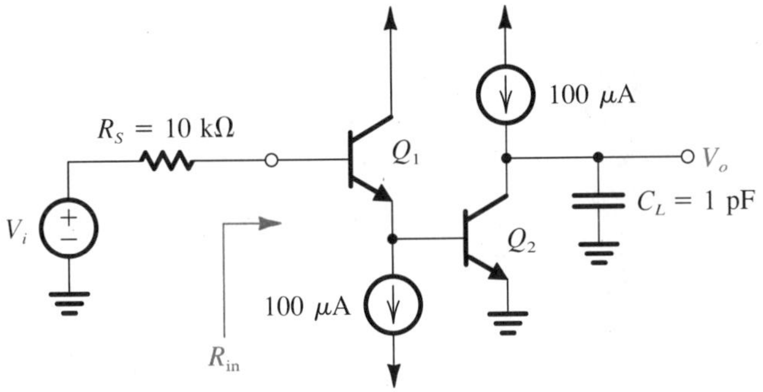

Fig. P7.61

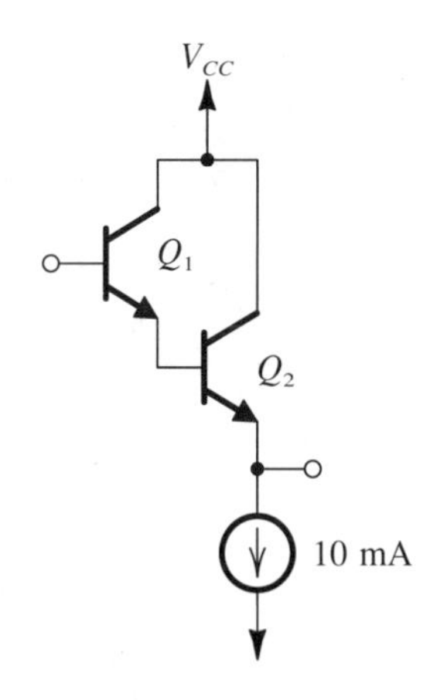

Fig. P7.62

(b) Find an estimate of the upper 3-dB frequency f_H. Which capacitor dominates? Which one is the second most significant?

(c) What are the effects of increasing the bias currents by a factor of 10?

7.62 The BJTs in the Darlington follower of Fig. P7.62 have $\beta_0 = 100$. If the follower is fed with a source having a 100-kΩ resistance and is loaded with 1 kΩ, find the input resistance and the output resistance (excluding the load) and the overall voltage gain, both open circuited and with load.

Section 7.10: Frequency Response of the Differential Amplifier

7.63 A BJT differential amplifier operating with a 2-mA emitter current source uses transistors for which $\beta = 120$, $f_T = 700$ MHz, $C_\mu = 0.5$ pF, and $r_x = 50\ \Omega$. Each of the collector resistances is 10 kΩ, and the amplifier is symmetrically excited with each of the source resistances being 10 kΩ. Find the differential-output dc gain and the 3-dB frequency f_H. Also find the gain–bandwidth product.

7.64 Reconsider the differential amplifier of Problem 7.63 when one of the input terminals is grounded and a signal source with a 20-kΩ resistance is connected to the other input terminal. Find the differential-output dc gain and the 3-dB frequency f_H. Also find the gain–bandwidth product.

7.65 The differential amplifier circuit specified in Problem 7.63 is modified by including a 25-Ω resistor in each of the emitters. Find the differential-output dc gain and the 3-dB frequency f_H. Also find the gain–bandwidth product.

***D7.66** Consider the differential amplifier specified in Problem 7.63. It is required to increase its 3-dB bandwidth to approximately 1 MHz by introducing emitter resistors. Find the required values of the resistors, and evaluate the new value of dc gain and gain–bandwidth product.

***7.67** Figure P7.67 shows a simplified, approximate small-signal model for the CMOS active-loaded differential amplifier studied in Section 6.7 (Fig. 6.34). In this equivalent circuit it is assumed that the amplifier is fed symmetrically, and thus a signal ground appears at the sources of Q_1 and Q_2. Also, it is assumed that the input signal source is ideal, and thus C_{gs1} and C_{gs2} have no effect and therefore have been omitted from the equivalent circuit. The effects of C_{gd1} and C_{gd2} on the circuit poles have been retained by including these capacitances in C_1 and C_2, respectively. The effects of C_{gd1} and C_{gd2} on introducing zeros have been neglected since these zeros are usually at very high frequencies. Since $(r_{o1} // r_{o2}) \ll 1/g_{m3}$ these two resistances have not been included. Also, omitted from the equivalent circuit is capacitance C_{gd4}. This is done to simplify the analysis considerably and because it has been found by computer simulation that its effect is negligibly small. Analyze the circuit to determine its transfer function V_o/V_{id}. Give expressions for the dc gain and the poles. Which pole is dominant?

7.68 A differential amplifier is biased by a current source having an output resistance of 1 MΩ and an output capacitance of 10 pF. The differential gain exhibits a dominant pole of 500 kHz. What are the poles of the CMRR?

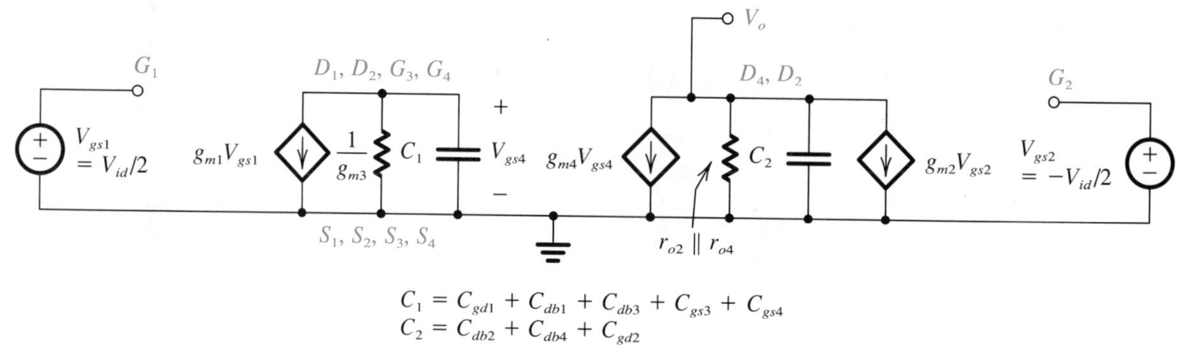

Fig. P7.67

Section 7.11: The Differential Pair as a Wideband Amplifier—The Common-Collector Common-Base Configuration

7.69 For the differential amplifier specified in Problem 7.63, find the dc gain and f_H when the circuit is modified by eliminating the collector resistor of the left-hand-side transistor and the input signal is fed to the base of the left-hand-side transistor while the base of the other transistor in the pair is grounded. Let the source resistance be 20 kΩ and neglect r_x.

***D7.70 Show that including a resistance R_E in each of the emitters of the differential pair of Fig. 7.42 results in the equivalent circuit of Fig. 7.43(b) being modified to resemble that of the emitter-follower in Fig. 7.32(c) (except for a factor of 2 multiplying the impedance between B′ and ground).

(a) Neglecting the resistance $2R_E$ in this equivalent circuit, find the pole frequency f_{P1}. Compare this expression to that in Eq. (7.125).

(b) For a bias current $I = 1$ mA and transistors having $\beta = 100$, $C_\pi = 7.5$ pF and $C_\mu = 0.5$ pF, use the result of (a) to find the value of R_E that results in $\omega_{P1} = 100$ Mrad/s. Assume that $R_s = 10$ kΩ.

(c) From examination of the expression of the other pole f_{P2} in Eq. (7.126), how can one control the frequency of this pole? Find the value of R_C that places the second pole at 400 Mrad/s.

(d) For the amplifier designed in (b) and (c), find the dc gain and f_H.

General Problems

***7.71 In each of the six circuits in Fig. P7.71, let $\beta = 100$, $C_\mu = 2$ pF, and $f_T = 400$ MHz. Calculate the midband gain and the upper 3-dB frequency.

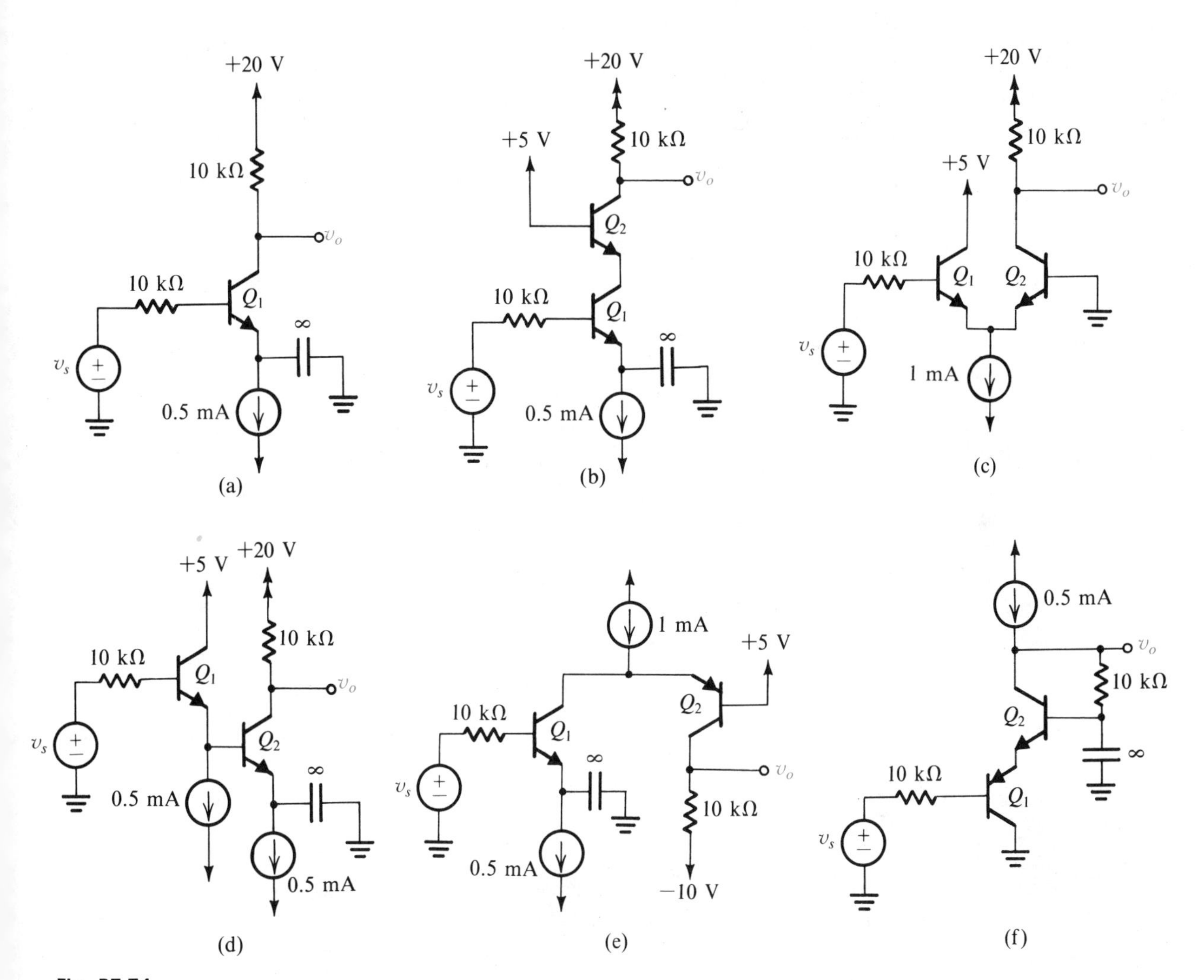
+20 V
10 kΩ
v_o
10 kΩ
Q_1
∞
v_s
0.5 mA
(a)
+20 V
+5 V
10 kΩ
v_o
Q_2
10 kΩ
Q_1
∞
v_s
0.5 mA
(b)
+20 V
10 kΩ
+5 V
v_o
10 kΩ
Q_1
Q_2
v_s
1 mA
(c)
+5 V
+20 V
10 kΩ
10 kΩ
Q_1
v_o
v_s
Q_2
∞
0.5 mA
0.5 mA
(d)
1 mA
+5 V
Q_2
10 kΩ
Q_1
∞
v_o
v_s
10 kΩ
0.5 mA
−10 V
(e)
0.5 mA
v_o
10 kΩ
Q_2
∞
10 kΩ
Q_1
v_s
(f)

Fig. P7.71

CHAPTER 8

Feedback

INTRODUCTION

Most physical systems incorporate some form of feedback. It is interesting to note, though, that the theory of negative feedback has been developed by electronics engineers. In his search for methods for the design of amplifiers with stable gain for use in telephone repeaters, Harold Black, an electronics engineer with the Western Electric Company, invented the feedback amplifier in 1928. Since then the technique has been so widely used that it is almost impossible to think of electronic circuits without some form of feedback, either implicit or explicit. Furthermore, the concept of feedback and its associated theory is currently used in areas other than engineering, such as in the modeling of biological systems.

Feedback can be either **negative (degenerative)** or **positive (regenerative).** In amplifier design, negative feedback is applied to effect one or more of the following properties:

1. *Desensitize the gain;* that is, make the value of the gain less sensitive to variations in the value of circuit components, such as variations that might be caused by changes in temperature.

2. *Reduce nonlinear distortion;* that is, make the output proportional to the input (in other words, make the gain constant independent of signal level).

3. *Reduce the effect of noise;* that is, minimize the contribution to the output of unwanted electric signals generated by the circuit components and extraneous interference.

4. *Control the input and output impedances;* that is, raise or lower input and output impedances by the selection of appropriate feedback topology.

5. *Extend the bandwidth* of the amplifier.

All of the above desirable properties are obtained at the expense of a reduction in gain. It will be shown that the gain-reduction factor, called the **amount of feedback,** is the factor by which the circuit is desensitized, by which the input impedance of a voltage amplifier is increased, by which the bandwidth is extended, and so on. In short, the basic idea of negative feedback is to trade off gain for other desirable properties. This chapter is devoted to the study of negative-feedback amplifiers: their analysis, design, and characteristics.

Under certain conditions the negative feedback in an amplifier can become positive and of such a magnitude as to cause oscillation. In fact, in Chapter 12 we will study the use of positive feedback in the design of oscillators and bistable circuits. In this chapter, however, we are interested in the design of stable amplifiers. We shall therefore study the stability problems of negative-feedback amplifiers.

It should not be implied, however, that positive feedback always leads to instability. In fact, positive feedback is quite useful in a number of applications, such as the design of active filters, which are studied in Chapter 11.

Before we begin our study of negative feedback we wish to remind the reader that we have already encountered negative feedback in a number of applications. Almost all op amp circuits employ negative feedback. Another popular application of negative feedback is the use of the emitter resistance R_E to stabilize the bias point of bipolar transistors and to increase the input resistance and bandwidth of a BJT differential amplifier. In addition, the emitter follower and the source follower employ a large amount of negative feedback. The question then arises as to the need for a formal study of negative feedback. As will be appreciated by the end of this chapter, the formal study of feedback provides an invaluable tool for the analysis and design of electronic circuits. Also, the insight gained by thinking in terms of feedback is extremely profitable.

8.1 THE GENERAL FEEDBACK STRUCTURE

Figure 8.1 shows the basic structure of a feedback amplifier. Rather than showing voltages and currents, Fig. 8.1 is a **signal-flow diagram,** where each x can represent either a voltage or a current signal. The *open-loop* amplifier has a gain A; thus its output x_o is related to the input x_i by

$$x_o = Ax_i \tag{8.1}$$

The output x_o is fed to the load as well as to a feedback network, which produces a sample of the output. This sample x_f is related to x_o by the **feedback factor** β,

$$x_f = \beta x_o \tag{8.2}$$

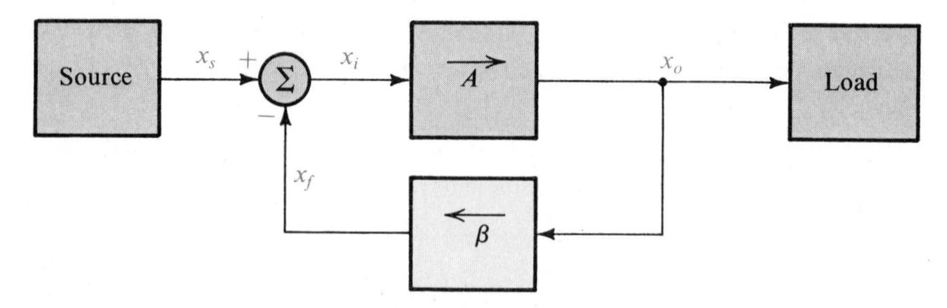

Fig. 8.1 General structure of the feedback amplifier. This is a signal-flow diagram, and the quantities x represent either voltage or current signals.

The feedback signal x_f is *subtracted* from the source signal x_s, which is the input to the complete feedback amplifier, to produce the signal x_i, which is the input to the basic amplifier,

$$x_i = x_s - x_f \tag{8.3}$$

Here we note that it is this subtraction that makes the feedback negative. In essence, negative feedback reduces the signal that appears at the input of the basic amplifier.

Implicit in the above description is that the source, the load, and the feedback network *do not* load the basic amplifier. That is, the gain A does not depend on any of these three networks. In practice this will not be the case, and we shall have to find a method for casting a real circuit into the ideal structure depicted in Fig. 8.1. Also implicit in Fig. 8.1 is that the forward transmission occurs entirely through the basic amplifier and the reverse transmission occurs entirely through the feedback network.

The gain of the feedback amplifier can be obtained by combining Eqs. (8.1) through (8.3):

$$A_f \equiv \frac{x_o}{x_s} = \frac{A}{1 + A\beta} \tag{8.4}$$

The quantity $A\beta$ is called the **loop gain,** a name that follows from Fig. 8.1. For the feedback to be negative, the loop gain $A\beta$ should be positive; that is, the feedback signal x_f should have the same sign as x_s, thus resulting in a smaller difference signal x_i. Equation (8.4) indicates that for positive $A\beta$ the gain with feedback will be smaller than the open-loop gain A by the quantity $1 + A\beta$, which is called the **amount of feedback.**

If, as is the case in many circuits, the loop gain $A\beta$ is large, $A\beta \gg 1$, then from Eq. (8.4) it follows that $A_f \simeq 1/\beta$, which is a very interesting result: *The gain of the feedback amplifier is almost entirely determined by the feedback network.* Since the feedback network usually consists of passive components, which can be chosen to be as accurate as one wishes, the advantage of negative feedback in obtaining accurate, predictable, and stable gain should be apparent. In other words the overall gain will have very little dependence on the gain of the basic amplifier, A, a desirable property because the gain A is usually a function of many parameters, some of which might have wide tolerances. We have seen a dramatic illustration of all of these results in op amp circuits, where the **closed-loop gain** (which is another name for the gain with feedback) is almost entirely determined by the feedback elements.

Equations (8.1) through (8.3) can be combined to obtain the following expression for the feedback signal x_f:

$$x_f = \frac{A\beta}{1 + A\beta} x_s$$

Thus for $A\beta \gg 1$ we see that $x_f \simeq x_s$, which implies that the signal x_i at the input of the basic amplifier is reduced to almost zero. Thus if a large amount of negative feedback is employed, the feedback signal x_f becomes an almost identical replica of the input signal x_s. An outcome of this property is the tracking of the two input terminals of an op amp. The difference between x_s and x_f, which is x_i, is sometimes referred to as the "error signal." Accordingly the input differencing circuit is often also called a **comparison circuit.** (It is also known as a **mixer.**)

Exercise 8.1 The noninverting op amp configuration shown in Fig. E8.1 provides a direct implementation of the feedback loop of Fig. 8.1.

(a) Assume that the op amp has infinite input resistance and zero output resistance. Find an expression for the feedback factor β. (b) If the open-loop voltage gain $A = 10^4$, find R_2/R_1 to obtain a closed-loop voltage gain A_f of 10. (c) What is the amount of feedback in decibels? (d) If $V_s = 1$ V, find V_o, V_f, and V_i. (e) If A decreases by 20%, what is the corresponding decrease in A_f?

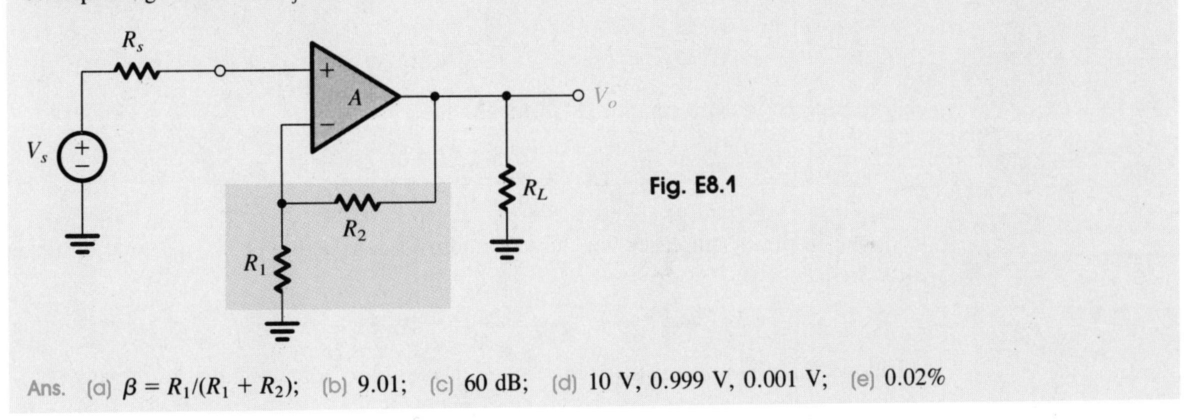

Fig. E8.1

Ans. (a) $\beta = R_1/(R_1 + R_2)$; (b) 9.01; (c) 60 dB; (d) 10 V, 0.999 V, 0.001 V; (e) 0.02%

8.2 SOME PROPERTIES OF NEGATIVE FEEDBACK

The properties of negative feedback were mentioned in the Introduction. In the following we shall consider some of these properties in more detail.

Gain Desensitivity

The effect of negative feedback on desensitizing the closed-loop gain was demonstrated in Exercise 8.1, where we saw that a 20% reduction in the gain of the basic amplifier gave rise to only a 0.02% reduction in the gain of the closed-loop amplifier. This sensitivity-reduction property can be analytically established as follows:

Assume that β is constant. Taking differentials of both sides of Eq. (8.4) results in

$$dA_f = \frac{dA}{(1 + A\beta)^2} \tag{8.5}$$

Dividing Eq. (8.5) by Eq. (8.4) yields

$$\frac{dA_f}{A_f} = \frac{1}{(1 + A\beta)} \frac{dA}{A} \tag{8.6}$$

which says that the percentage change in A_f (due to variations in some circuit parameter) is smaller than the percentage change in A by the amount of feedback. For this reason the amount of feedback, $1 + A\beta$, is also known as the **desensitivity factor.**

Bandwidth Extension

Consider an amplifier whose high-frequency response is characterized by a single pole. Its gain at mid and high frequencies can be expressed as

$$A(s) = \frac{A_M}{1 + s/\omega_H} \tag{8.7}$$

where A_M denotes the midband gain and ω_H is the upper 3-dB frequency. Application of negative feedback, with a frequency-independent factor β, around this amplifier results in a closed-loop gain $A_f(s)$ given by

$$A_f(s) = \frac{A(s)}{1 + \beta A(s)}$$

Substituting for $A(s)$ from Eq. (8.7) results in

$$A_f(s) = \frac{A_M/(1 + A_M\beta)}{1 + s/\omega_H(1 + A_M\beta)}$$

Thus the feedback amplifier will have a midband gain of $A_M/(1 + A_M\beta)$ and an upper 3-dB frequency ω_{Hf} given by

$$\omega_{Hf} = \omega_H(1 + A_M\beta) \tag{8.8}$$

It follows that the upper 3-dB frequency is increased by a factor equal to the amount of feedback.

Similarly, it can be shown that if the open-loop gain is characterized by a dominant low-frequency pole giving rise to a lower 3-dB frequency ω_L, then the feedback amplifier will have a lower 3-dB frequency ω_{Lf},

$$\omega_{Lf} = \frac{\omega_L}{1 + A_M\beta} \tag{8.9}$$

Note that the amplifier bandwidth is increased by the same factor by which its midband gain is decreased, maintaining the gain–bandwidth product constant.

Exercise 8.2 Consider the noninverting op amp circuit of Exercise 8.1. Let the open-loop gain A have a low-frequency value of 10^4 and a uniform -6 dB/octave rolloff at high frequencies with a 3-dB frequency of 100 Hz. Find the low-frequency gain and the upper 3-dB frequency of a closed-loop amplifier with $R_1 = 1$ kΩ and $R_2 = 9$ kΩ.

Ans. 9.99 V/V; 100.1 kHz

Noise Reduction

Negative feedback can be employed to reduce the noise or interference in an amplifier or, more precisely, to increase the ratio of signal to noise. However, as we shall now explain, this noise-reduction process is possible only under certain conditions. Consider the situation illustrated in Fig. 8.2. Figure 8.2(a) shows an amplifier with gain A_1, an input signal V_s, and noise, or interference, V_n. It is assumed that for some reason this amplifier suffers from noise and that the noise can be assumed to be introduced at the input of the amplifier. The **signal-to-noise ratio** for this amplifier is

$$S/N = V_s/V_n$$

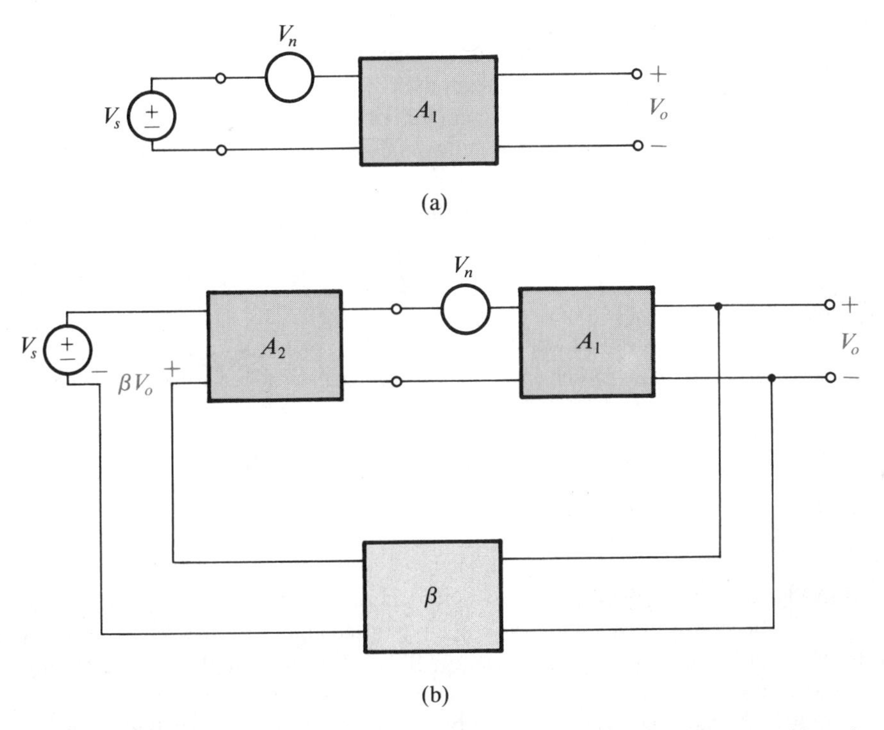

Fig. 8.2 Illustrating the application of negative feedback to improve the signal-to-noise ratio in amplifiers.

Consider next the circuit in Fig. 8.2(b). Here we assume that it is possible to build another amplifier stage with gain A_2 that does not suffer from the noise problem. If this is the case, then we may precede our original amplifier A_1 by the *clean* amplifier A_2 and apply negative feedback around the overall cascade of such an amount as to keep the overall gain constant. The output voltage of the circuit in Fig. 8.2(b) can be found by superposition:

$$V_o = V_s \frac{A_1 A_2}{1 + A_1 A_2 \beta} + V_n \frac{A_1}{1 + A_1 A_2 \beta}$$

Thus the signal-to-noise ratio at the output becomes

$$\frac{S}{N} = \frac{V_s}{V_n} A_2$$

which is A_2 times higher than in the original case.

We should emphasize once more that the improvement in signal-to-noise ratio by the application of feedback is possible only if one can precede the noisy stage by a (relatively) noise-free stage. This situation, however, is not uncommon in practice. The best example is found in the output power-amplifier stage of an audio amplifier. Such a stage usually suffers from a problem known as **power-supply hum.** The problem arises because of the large currents that this stage draws from the power supply and the difficulty in providing adequate power-supply filtering inexpensively.

The power-output stage is required to provide large power gain but little or no voltage gain. We may therefore precede the power-output stage by a small-signal amplifier that provides large voltage gain and apply a large amount of negative feedback, thus restoring the voltage gain to its original value. Since the small-signal amplifier can be fed from another, less hefty (and hence better regulated) power supply, it will not suffer from the hum problem. The hum at the output will then be reduced by the amount of the voltage gain of this added **preamplifier.**

Exercise 8.3 Consider a power-output stage with voltage gain $A_1 = 1$, an input signal $V_s = 1$ V, and a hum V_n of 1 V. Assume that this power stage is preceded by a small-signal stage with gain $A_2 = 100$ V/V and that overall feedback with $\beta = 1$ is applied. If V_s and V_n remain unchanged, find the signal and noise voltages at the output and hence the improvement in S/N.

Ans. $\simeq 1$ V; $\simeq 0.01$ V; 100 (40 dB)

Reduction in Nonlinear Distortion

Curve (a) in Fig. 8.3 shows the transfer characteristic of an amplifier. As indicated, the characteristic is piecewise linear, with the voltage gain changing from 1000 to 100 and then to 0. This nonlinear transfer characteristic will result in this amplifier generating a large amount of nonlinear distortion.

The amplifier transfer characteristic can be considerably **linearized** (that is, made less nonlinear) through the application of negative feedback. That this is possible should not be too surprising, since we have already seen that negative feedback reduces the dependence of the overall closed-loop amplifier gain on the open-loop gain of the basic amplifier. Thus large changes in open-loop gain (1000 to 100 in this case) give rise to much smaller corresponding changes in the closed-loop gain.

To illustrate, let us apply negative feedback with $\beta = 0.01$ to the amplifier whose open-loop voltage transfer characteristic is depicted in Fig. 8.3. The resulting transfer characteristic of the closed-loop amplifier is shown in Fig. 8.3 as curve (b). Here the slope of the steepest segment is given by

$$A_{f1} = \frac{1000}{1 + 1000 \times 0.01} = 90.9$$

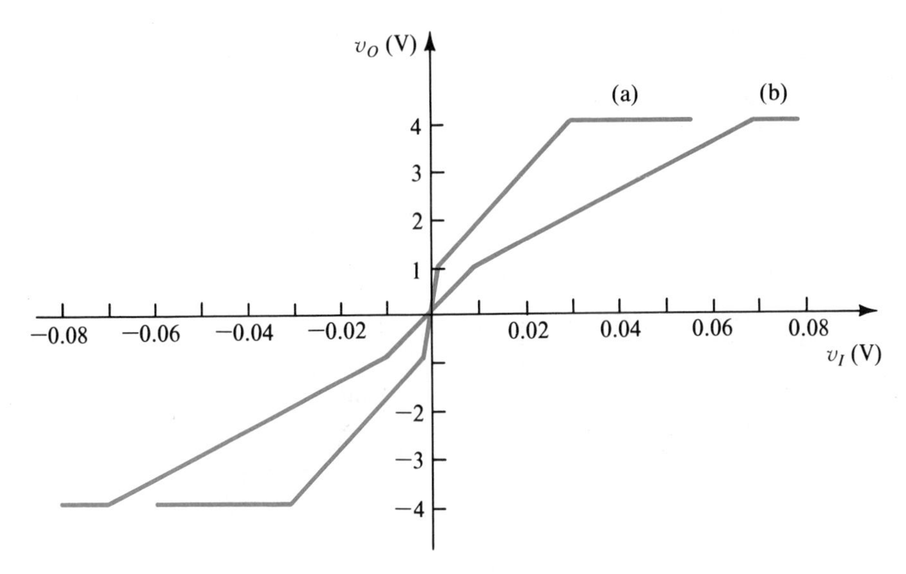

Fig. 8.3 Illustrating the application of negative feedback to reduce the nonlinear distortion in amplifiers. Curve (a) shows the amplifier transfer characteristic without feedback. Curve (b) shows the characteristic with negative feedback ($\beta = 0.01$) applied.

and the slope of the next segment is given by

$$A_{f2} = \frac{100}{1 + 100 \times 0.01} = 50$$

Thus the order-of-magnitude change in slope has been considerably reduced. The price paid, of course, is a reduction in voltage gain. Thus if the overall gain has to be restored, then a preamplifier should be added. This preamplifier should not present a severe nonlinear distortion problem, since it will be dealing with smaller signals.

Finally, it should be noted that negative feedback does nothing about amplifier saturation, since in saturation the gain is very small (almost zero) and hence the amount of feedback is also very small (almost zero).

8.3 THE FOUR BASIC FEEDBACK TOPOLOGIES

Based on the quantity to be amplified (voltage or current) and on the desired form of output (voltage or current), amplifiers can be classified into four categories. These categories were discussed in Chapter 1. In the following we shall review this amplifier classification and point out the feedback topology appropriate in each case.

Voltage Amplifiers

Voltage amplifiers are intended to amplify an input voltage signal and provide an output voltage signal. The voltage amplifier is essentially a voltage-controlled voltage source. The input impedance is required to be high, and the output impedance is required to be

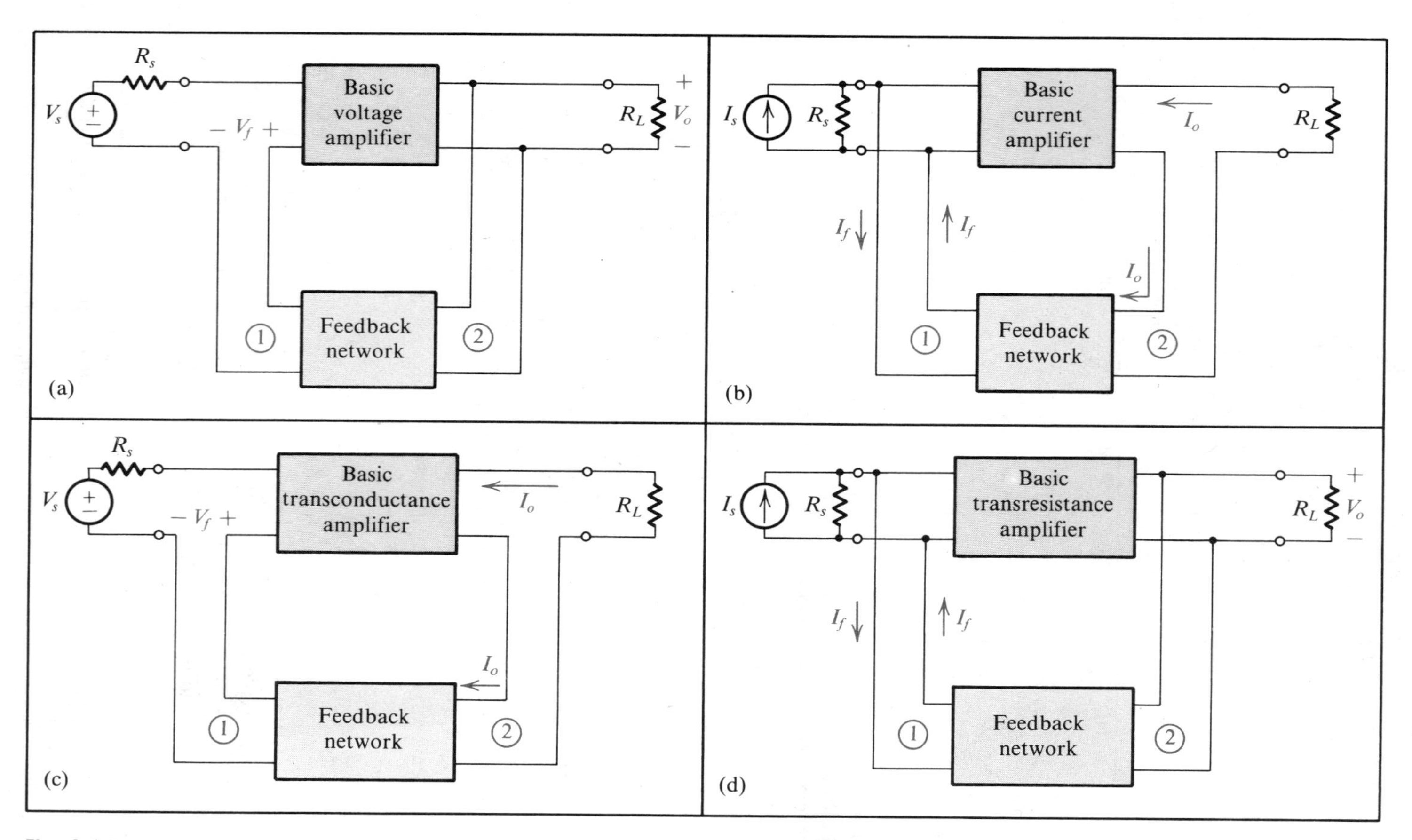

Fig. 8.4 The four basic feedback topologies: **(a)** voltage-sampling series-mixing (series-shunt) topology; **(b)** current-sampling shunt-mixing (shunt-series) topology; **(c)** current-sampling series-mixing (series-series) topology; **(d)** voltage-sampling shunt-mixing (shunt-shunt) topology.

low. Since the signal source is essentially a voltage source, it is convenient to represent it in terms of a Thévenin equivalent circuit. In a voltage amplifier the output quantity of interest is the output voltage. It follows that the feedback network should *sample* the output *voltage*. Also, because of the Thévenin representation of the source, the feedback signal x_f should be a voltage that can be *mixed* with the source voltage in *series*.

A suitable feedback topology for the voltage amplifier is the **voltage-sampling series-mixing** one shown in Fig. 8.4(a). As will be shown, this topology not only stabilizes the voltage gain but also results in a higher input resistance (intuitively, a result of the series connection at the input) and a lower output resistance (intuitively, a result of the parallel connection at the output), which are desirable properties for a voltage amplifier. The noninverting op amp configuration of Fig. E8.1 is an example of this feedback topology. Finally, it should be mentioned that this feedback topology is also known as **series-shunt feedback,** where "series" refers to the connection at the input and "shunt" refers to the connection at the output.

Current Amplifiers

Here the input signal is essentially a current, and thus the signal source is most conveniently represented by its Norton equivalent. The output quantity of interest is current; hence the feedback network should *sample* the output *current*. The feedback signal should be in current form so that it may be *mixed* in *shunt* with the source current. Thus the feedback topology suitable for a current amplifier is the **current-sampling shunt-mixing** topology, illustrated in Fig. 8.4(b). As will be shown, this topology not only stabilizes the current gain but also results in a lower input resistance and a higher output resistance, desirable properties for a current amplifier.

An example of the current-sampling shunt-mixing feedback topology is given in Fig. 8.5. Note that the bias details are not shown. Also note that the current being sampled

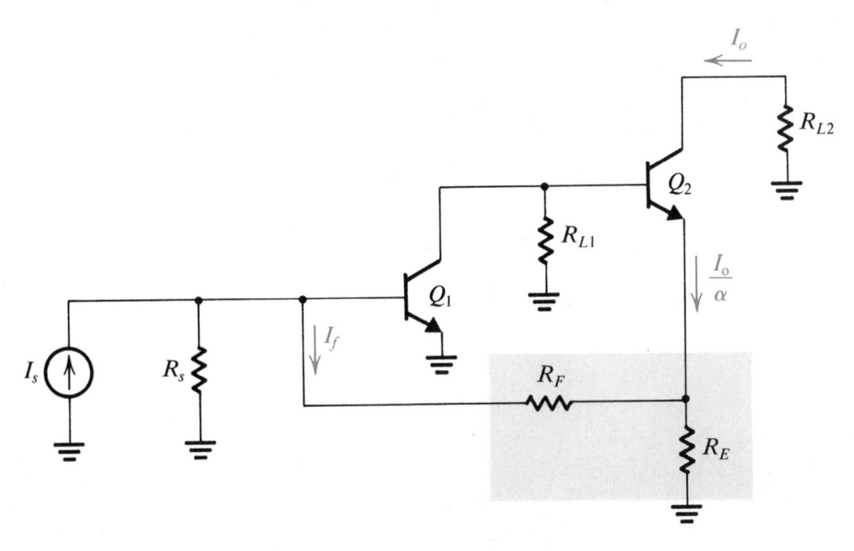

Fig. 8.5 A transistor amplifier with shunt-series feedback.

is not the output current but the almost equal emitter current of Q_2. This is done for circuit design convenience and is quite usual in circuits involving current sampling.

The reference direction indicated in Fig. 8.5 for the feedback current I_f is such that it subtracts from I_s. This reference notation will be followed in all circuits in this chapter, since it is consistent with the notation used in the general feedback structure of Fig. 8.1. Therefore, in all circuits, for the feedback to be negative the loop gain $A\beta$ should be positive. The reader is urged to verify that in the circuit of Fig. 8.5 A is negative and β is negative.

It is of utmost importance to be able to ascertain qualitatively the feedback polarity (positive or negative). This can be done by "following the signal around the loop." For instance, let the current I_s in Fig. 8.5 increase. We see that the base current of Q_1 will increase, and thus its collector current will also increase. This will cause the collector voltage of Q_1 to decrease, and thus the collector current of Q_2, I_o, will decrease. Thus the emitter current of Q_2, I_o/α (where α is the common-base current gain of the BJT), decreases. From the feedback network we see that if I_o/α decreases, then I_f (in the direction shown) will increase. The increase in I_f will subtract from I_s, causing a smaller increment to be seen by the amplifier. Hence the feedback is negative.

Finally, we should mention that this feedback topology is also known as **shunt-series feedback.** Again, the first word in the name (shunt) refers to the connection at the input, and the second word (series) refers to the connection at the output.

Transconductance Amplifiers

Here the input signal is a voltage and the output signal is a current. It follows that the appropriate feedback topology is the **current-sampling series-mixing** topology, illustrated in Fig. 8.4(c).

An example of this feedback topology is given in Fig. 8.6. Here note that as in the

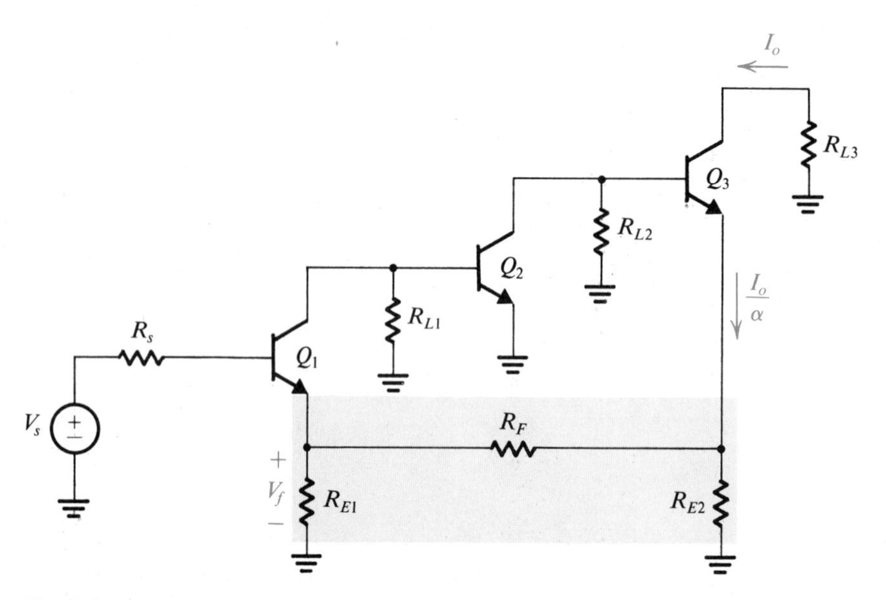

Fig. 8.6 An example of the series-series feedback topology.

circuit of Fig. 8.5 the current sampled is not the output current but the almost equal emitter current of Q_3. In addition, the mixing loop is not a conventional one; it is not a simple series connection, since the feedback signal developed across R_{E1} is in the emitter circuit of Q_1, while the source is in the base circuit of Q_1. These two approximations are done for convenience of circuit design.

Finally, it should be mentioned that the current-sampling series-mixing feedback topology is also known as the **series-series** feedback configuration.

Transresistance Amplifiers

Here the input signal is current and the output signal is voltage. It follows that the appropriate feedback topology is of the **voltage-sampling shunt-mixing** type, shown in Fig. 8.4(d).

An example of this feedback topology is found in the inverting op amp configuration of Fig. 8.7(a). The circuit is redrawn in Fig. 8.7(b) with the source converted to Norton's form.

This feedback topology is also known as **shunt-shunt** feedback.

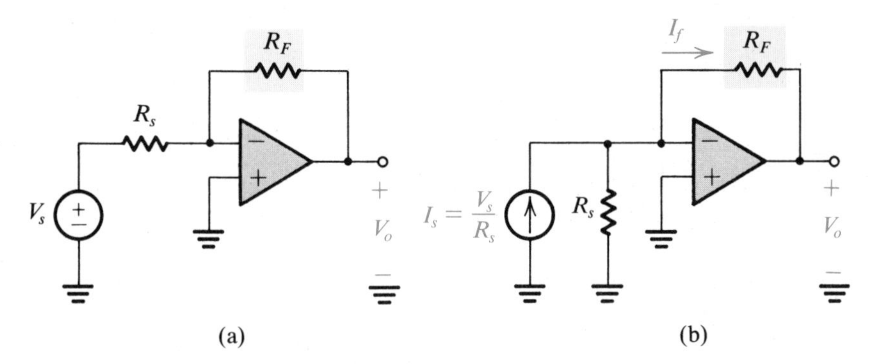

Fig. 8.7 The inverting op amp configuration as an example of shunt-shunt feedback.

8.4 THE SERIES-SHUNT FEEDBACK AMPLIFIER

The Ideal Situation

The ideal structure of the series-shunt feedback amplifier is shown in Fig. 8.8(a). It consists of a *unilateral* open-loop amplifier (the A circuit) and an ideal voltage-sampling series-mixing feedback network (the β circuit). The A circuit has an input resistance R_i, a voltage gain A, and an output resistance R_o. It is assumed that the source and load resistances have been included inside the A circuit (more on this point later). Furthermore, note that the β circuit does *not* load the A circuit; that is, connecting the β circuit does not change the value of A (defined $A \equiv V_o/V_i$).

The circuit of Fig. 8.8(a) exactly follows the ideal feedback model of Fig. 8.1. Therefore the closed-loop voltage gain A_f is given by

$$A_f \equiv \frac{V_o}{V_s} = \frac{A}{1 + A\beta}$$

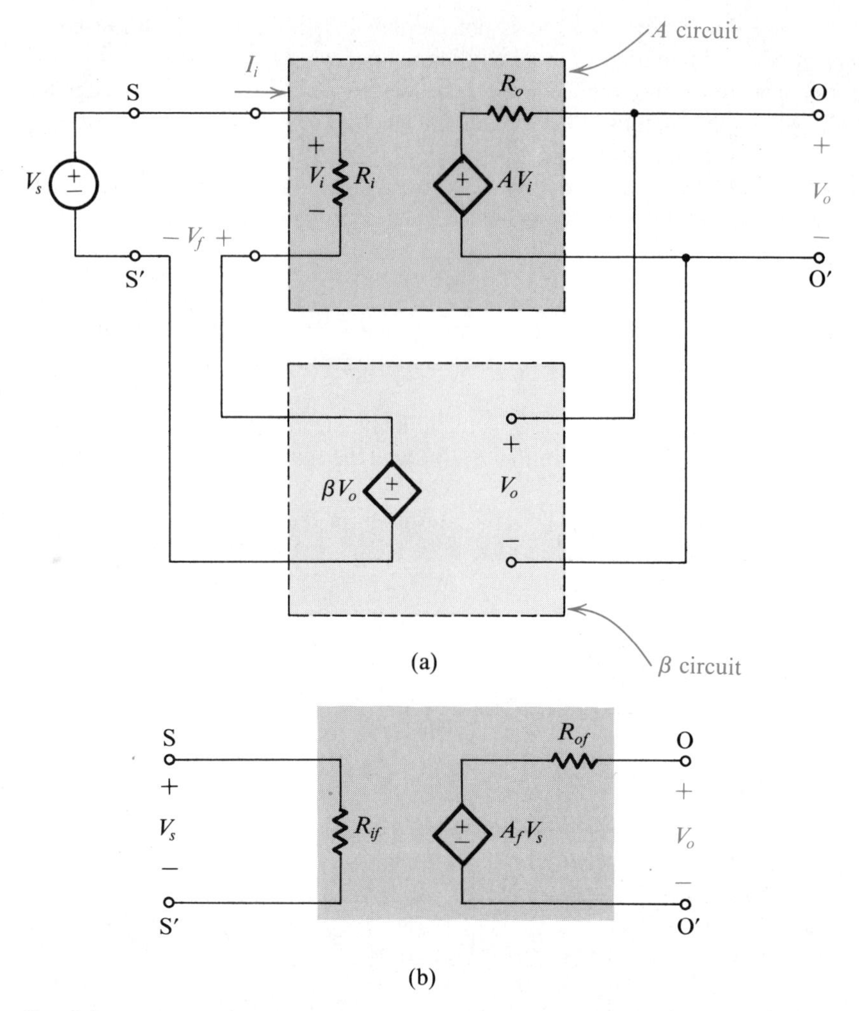

Fig. 8.8 The series-shunt feedback amplifier: **(a)** ideal structure; **(b)** equivalent circuit.

Note that A and β have reciprocal units. This in fact is always the case, resulting in a dimensionless loop gain $A\beta$.

The equivalent circuit model of the series-shunt feedback amplifier is shown in Fig. 8.8(b). Here R_{if} and R_{of} denote the input and output resistances with feedback. The relationship between R_{if} and R_i can be established by considering the circuit in Fig. 8.8(a):

$$R_{if} \equiv \frac{V_s}{I_i} = \frac{V_s}{V_i/R_i}$$

$$= R_i \frac{V_s}{V_i} = R_i \frac{V_i + \beta A V_i}{V_i}$$

Thus

$$R_{if} = R_i(1 + A\beta) \tag{8.10}$$

That is, in this case the negative feedback increases the input resistance by a factor equal to the amount of feedback. Since the above derivation does not depend on the method of sampling (shunt or series), it follows that the relationship between R_{if} and R_i is a function only of the method of mixing. We shall discuss this point further in later sections.

Note, however, that this result is not surprising and is physically intuitive: Since the feedback voltage V_f subtracts from V_s, the voltage that appears across R_i—that is, V_i—becomes quite small. Thus the input current I_i becomes correspondingly small and the resistance seen by V_s becomes large. Finally, it should be pointed out that Eq. (8.10) can be generalized to the form

$$Z_{if}(s) = Z_i(s)[1 + A(s)\beta(s)] \tag{8.11}$$

To find the output resistance, R_{of}, of the feedback amplifier in Fig. 8.8(a) we reduce V_s to zero and apply a test voltage V_t at the output, as shown in Fig. 8.9,

$$R_{of} \equiv \frac{V_t}{I}$$

From Fig. 8.9 we can write

$$I = \frac{V_t - AV_i}{R_o}$$

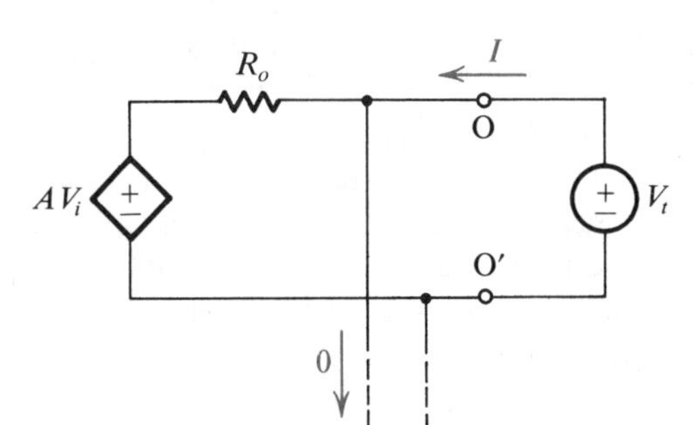

Fig. 8.9 Measuring the output resistance of the feedback amplifier of Fig. 8.8(a): $R_{of} \equiv V_t/I$.

and since $V_s = 0$ it follows from Fig. 8.8(a) that

$$V_i = -V_f = -\beta V_o = -\beta V_t$$

Thus

$$I = \frac{V_t + A\beta V_t}{R_o}$$

leading to

$$R_{of} = \frac{R_o}{1 + A\beta} \tag{8.12}$$

That is, the negative feedback in this case reduces the output resistance by a factor equal to the amount of feedback. With a little thought one can see that the derivation of Eq. (8.12) does not depend on the method of mixing. Thus the relationship between R_{of} and R_o depends only on the method of sampling. Again this result is not surprising and is physically intuitive: With reference to Fig. 8.9 we see that since the feedback samples V_t and is negative, the controlled voltage source will be $-A\beta V_t$. This large negative voltage causes the current I to be large, indicating that the effective output resistance is small. Finally, we note that Eq. (8.12) can be generalized to

$$Z_{of}(s) = \frac{Z_o(s)}{1 + A(s)\beta(s)} \tag{8.13}$$

The Practical Situation

In a practical series-shunt feedback amplifier the feedback network will not be an ideal voltage-controlled voltage source. Rather, the feedback network is usually passive and hence will load the basic amplifier and thus affect the values of A, R_i, and R_o. In addition, the source and load resistances will affect these three parameters. Thus the problem we have is as follows: Given a series-shunt feedback amplifier represented by the block diagram of Fig. 8.10(a), find the A circuit and the β circuit.

Our problem essentially involves representing the amplifier of Fig. 8.10(a) by the ideal structure of Fig. 8.8(a). As a first step toward that end we observe that the source and load resistances should be lumped with the basic amplifier. This, together with representing the two-port feedback network in terms of its h parameters (see Appendix B), is illustrated in Fig. 8.10(b). The choice of h parameters is based on the fact that this is the only parameter set that represents the feedback network by a series network at port 1 and a parallel network at port 2. Such a representation is obviously convenient in view of the series connection at the input and the parallel connection at the output.

Examination of the circuit in Fig. 8.10(b) reveals that the current source $h_{21}I_1$ represents the forward transmission of the feedback network. Since the feedback network is usually passive, its forward transmission can be neglected in comparison to the much larger forward transmission of the basic amplifier. We will therefore assume that $|h_{21}|_{\substack{\text{feedback}\\ \text{network}}} \ll |h_{21}|_{\substack{\text{basic}\\ \text{amplifier}}}$ and thus omit the controlled source $h_{21}I_1$ altogether.

Compare the circuit of Fig. 8.10(b) (after eliminating the current source $h_{21}I_1$) to the ideal circuit of Fig. 8.8(a). We see that by including h_{11} and h_{22} with the basic amplifier we obtain the circuit shown in Fig. 8.10(c), which is very similar to the ideal circuit. Now, if the basic amplifier is unilateral (or almost unilateral), a situation that prevails when

$$|h_{12}|_{\substack{\text{basic}\\ \text{amplifier}}} \ll |h_{12}|_{\substack{\text{feedback}\\ \text{network}}},$$

then the circuit of Fig. 8.10(c) is equivalent (or approximately equivalent) to the ideal circuit. It follows then that the A circuit is obtained by augmenting the basic amplifier at the input with the source impedance R_s and the impedance h_{11} of the feedback network, and augmenting it at the output with the load impedance R_L and the admittance h_{22} of the feedback network.

We conclude that the loading effect of the feedback network on the basic amplifier is

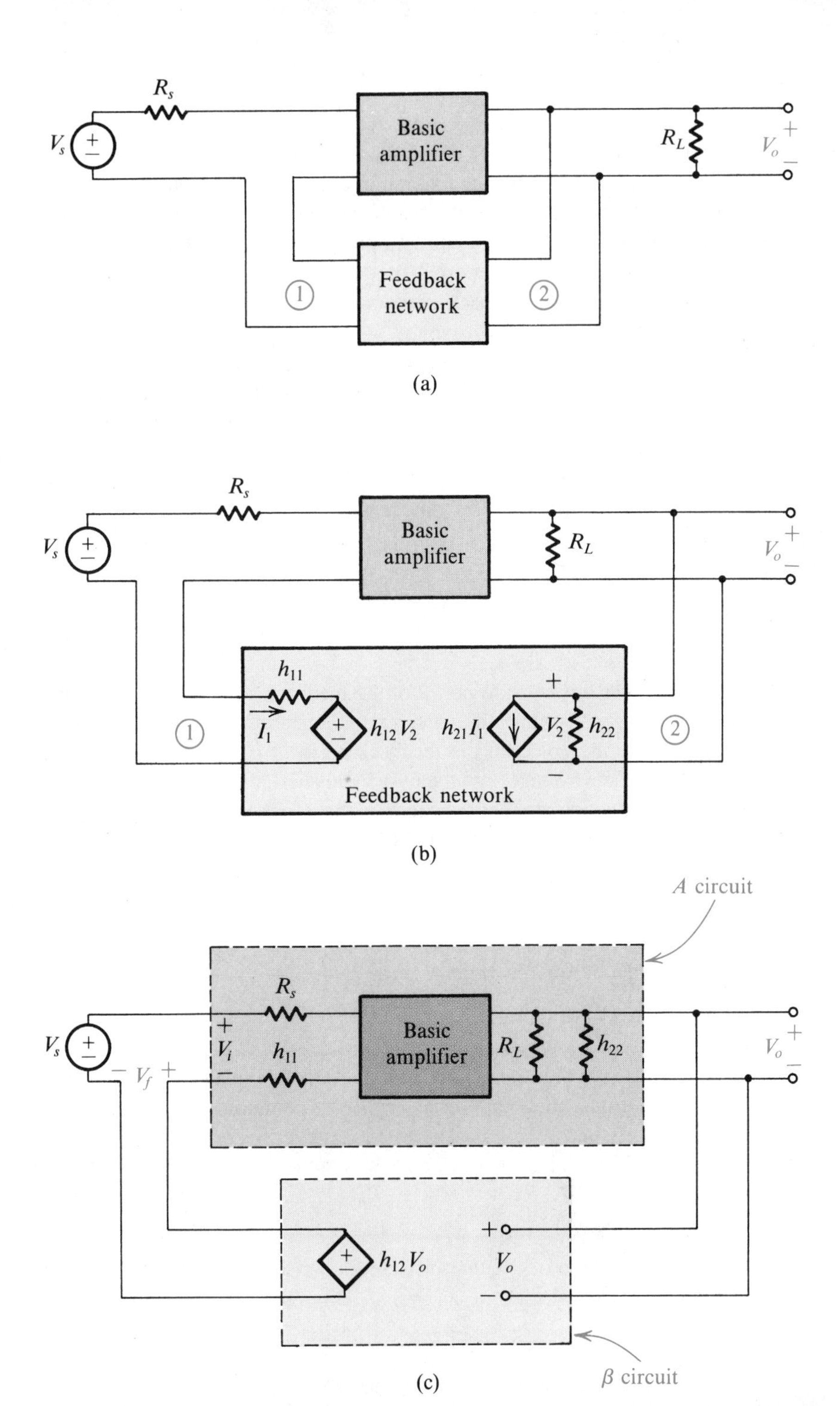

Fig. 8.10 Derivation of the A circuit and β circuit for the series-shunt feedback amplifier. **(a)** Block diagram of a practical series-shunt feedback amplifier. **(b)** The circuit in (a) with the feedback network represented by its h parameters. **(c)** The circuit in (b) after neglecting h_{21}.

represented by the components h_{11} and h_{22}. From the definitions of the h parameters in Appendix B we see that h_{11} is the impedance looking into port 1 of the feedback network with port 2 short-circuited. Since port 2 of the feedback network is connected in *shunt* with the output port of the amplifier, short-circuiting port 2 destroys the feedback. Similarly, h_{22} is the admittance looking into port 2 of the feedback network with port 1 open-circuited. Since port 1 of the feedback network is connected in *series* with the amplifier input, open-circuiting port 1 destroys the feedback.

These observations suggest a simple rule for finding the loading effects of the feedback network on the basic amplifier: The loading effect is found by looking into the appropriate port of the feedback network while the other port is open-circuited or short-circuited so as to destroy the feedback. If the connection is a shunt one, we short-circuit the port; if it is a series one, we open-circuit it. In Sections 8.5 and 8.6 it will be seen that this simple rule applies also to the other three feedback topologies.[1]

We next consider the determination of β. From Fig. 8.10(c), we see that β is equal to h_{12} of the feedback network,

$$\beta = h_{12} \equiv \left.\frac{V_1}{V_2}\right|_{I_1=0}$$

Thus to measure β one applies a voltage to port 2 of the feedback network and measures the voltage that appears at port 1 while the latter port is open-circuited. This result is intuitively appealing because the object of the feedback network is to sample the output voltage ($V_2 = V_o$) and provide a voltage signal ($V_1 = V_f$) that is mixed in series with the input source. The series connection at the input suggests that (as in the case of finding the loading effects of the feedback network) β should be found with port 1 open-circuited.

Summary

A summary of the rules for finding the A circuit and β for a given series-shunt feedback amplifier of the form in Fig. 8.10(a) is given in Fig. 8.11.

EXAMPLE 8.1

Figure 8.12(a) shows an op amp connected in the noninverting configuration. The op amp has an open-loop gain μ, a differential input resistance R_{id}, a common-mode input resistance R_{icm}, and an output resistance r_o. Find expressions for A, β, the closed-loop gain V_o/V_s, the input resistance R'_{if} (see Fig. 8.12a), and the output resistance R'_{of}. Also find numerical values, given $\mu = 10^4$, $R_{id} = 100\text{ k}\Omega$, $R_{icm} = 10\text{ M}\Omega$, $r_o = 1\text{ k}\Omega$, $R_L = 2\text{ k}\Omega$, $R_1 = 1\text{ k}\Omega$, $R_2 = 1\text{ M}\Omega$, and $R_s = 10\text{ k}\Omega$.

Solution

We first observe that the existence of the two resistances labeled $2R_{icm}$ between the op amp inputs and ground will complicate the analysis.[2] This problem, however, can be

---

[1] A simple rule to remember is: If the connection is *sh*unt, *sh*ort it; if *se*ries, *se*ver it.

[2] Refer to Fig. 8.10(c) and let the lower terminal of V_s be grounded. We can see that none of the input terminals of the A circuit can be grounded. Thus the resistance R_i is simply the resistance between the two input terminals of the A circuit. If, however, there exist inside the A circuit resistances with grounded terminals, the determination of R_i becomes quite lengthy.

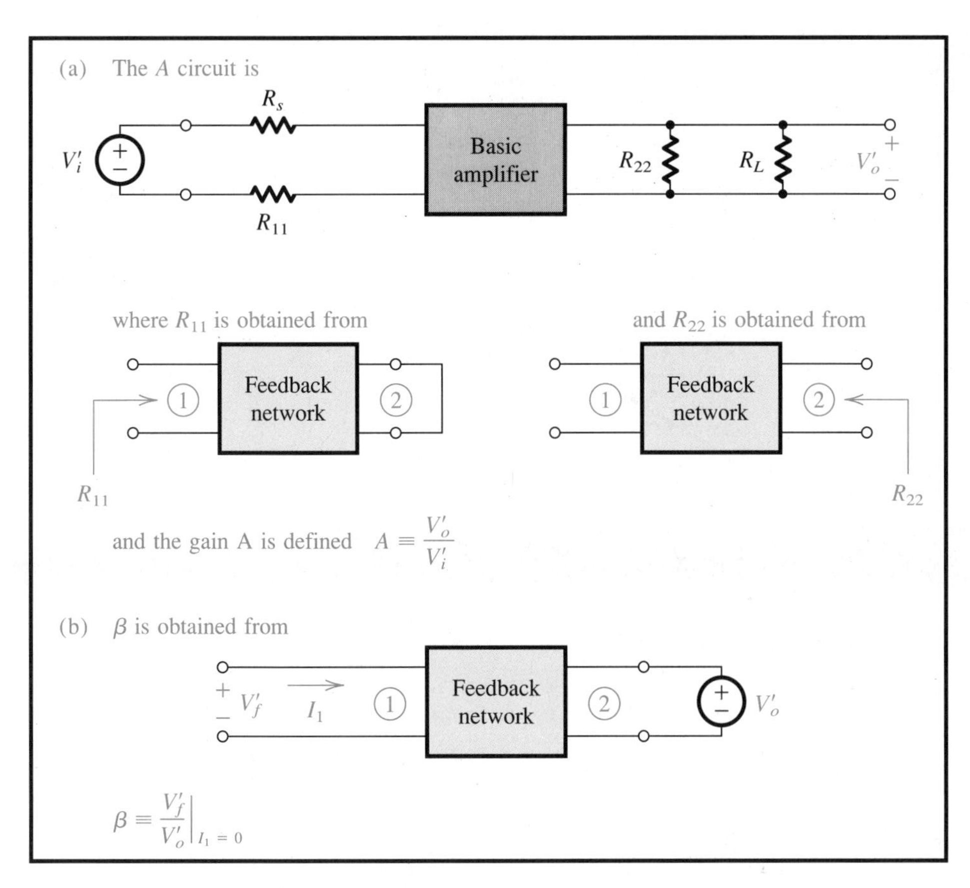

Fig. 8.11 Summary of the rules for finding the *A* circuit and β for the voltage-sampling series-mixing case of Fig. 8.10(a).

easily overcome using Thévenin's theorem, as shown in Fig. 8.12(b). We now observe that the feedback network consists of R_2 and R_1'. This network samples the output voltage V_o and provides a voltage signal (across R_1') that is mixed in series with the input source V_s'.

The *A* circuit is easily obtained with the rules of Fig. 8.11 and is shown in Fig. 8.12(c). For this circuit we can write by inspection

$$A \equiv \frac{V_o'}{V_i'} = \mu \frac{[R_L//(R_1' + R_2)]}{[R_L//(R_1' + R_2)] + r_o} \frac{R_{id}}{R_{id} + R_s' + (R_1'//R_2)}$$

For the values given we find that $R_1' \simeq R_1 = 1\ \text{k}\Omega$ and $R_s' \simeq R_s = 10\ \text{k}\Omega$, leading to $A \simeq 6000$ V/V.

The circuit for obtaining β is shown in Fig. 8.12(d), from which we obtain

$$\beta \equiv \frac{V_f'}{V_o'} = \frac{R_1'}{R_1' + R_2} \simeq 10^{-3}\ \text{V/V}$$

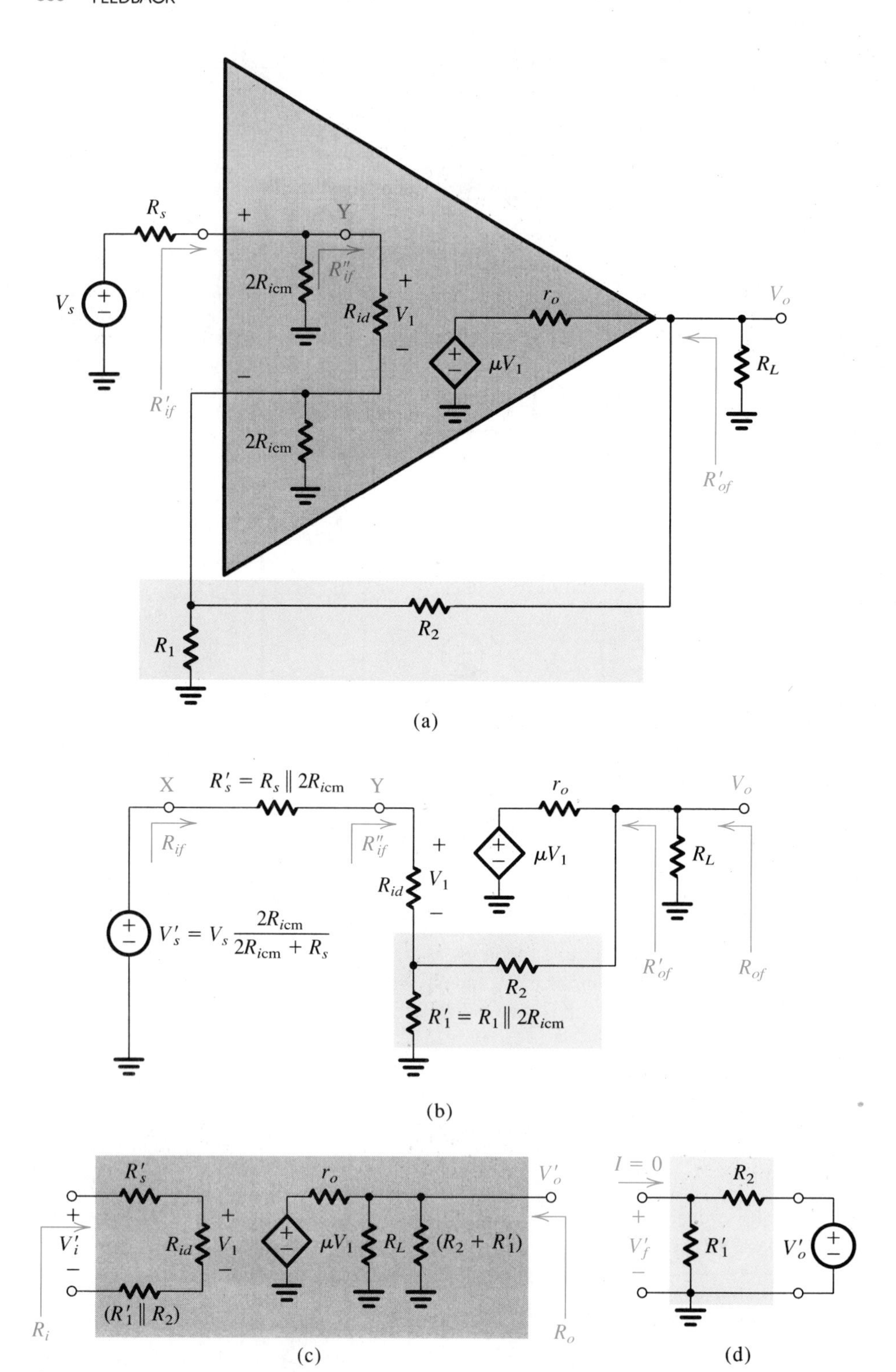

Fig. 8.12 Circuits for Example 8.1.

The voltage gain with feedback is now obtained as

$$A_f \equiv \frac{V_o}{V_s'} = \frac{A}{1 + A\beta} = \frac{6000}{7} = 857 \text{ V/V}$$

This can be used to obtain V_o/V_s:

$$\frac{V_o}{V_s} = \frac{V_o}{V_s'} \frac{2R_{icm}}{2R_{icm} + R_s} \simeq \frac{V_o}{V_s'} = 857 \text{ V/V}$$

The input resistance R_{if} given by the feedback equations is the resistance seen by the external source (in this case V_s'); that is, it is the resistance between node X and ground in Fig. 8.12(b). This resistance is given by

$$R_{if} = R_i(1 + A\beta)$$

where R_i is the input resistance of the A circuit in Fig. 8.12(c):

$$R_i = R_s' + R_{id} + (R_1' // R_2)$$

For the values given, $R_i \simeq 111$ kΩ, resulting in

$$R_{if} = 111 \times 7 = 777 \text{ k}\Omega$$

This, however, is not the resistance asked for. What is required is R_{if}', indicated in Fig. 8.12(a). To obtain R_{if}' we first subtract R_s' from R_{if} and find R_{if}'', indicated in Fig. 8.12(b) and in Fig. 8.12(a). From the latter figure we see that R_{if}' can be obtained by including $2R_{icm}$ in parallel with R_{if}'':

$$R_{if}' = [2R_{icm} // (R_{if} - R_s')]$$

For the values given, $R_{if}' = 739$ kΩ. The resistance R_{of} given by the feedback equations is the output resistance of the feedback amplifier, including the load resistance R_L, as indicated in Fig. 8.12(b). R_{of} is given by

$$R_{of} = \frac{R_o}{1 + A\beta}$$

where R_o is the output resistance of the A circuit. R_o can be obtained by inspection of Fig. 8.12(c) as

$$R_o = [r_o // R_L // (R_2 + R_1')]$$

For the values given, $R_o \simeq 667$ Ω and

$$R_{of} = \frac{667}{7} = 95.3 \ \Omega$$

The resistance asked for, R_{of}', is the output resistance of the feedback amplifier excluding R_L. From Fig. 8.12(b) we see that

$$R_{of} = (R_{of}' // R_L)$$

Thus

$$R_{of}' \simeq 100 \ \Omega$$

Exercises 8.4 If the op amp of Example 8.1 has a uniform −6-dB/octave high-frequency rolloff with $f_{3dB} = 1$ kHz, find the 3-dB frequency of the closed-loop gain V_o/V_s.

Ans. 7 kHz

8.5 The circuit shown in Fig. E8.5 consists of a differential stage followed by an emitter follower, with series-shunt feedback supplied by the resistors R_1 and R_2. Assuming that the dc component of V_s is zero, find the dc operating current of each of the three transistors and show that the dc voltage at the output is approximately zero. Then find the values of A, β, $A_f \equiv V_o/V_s$, R'_{if}, and R'_{of}. Assume that the transistors have $\beta = 100$.

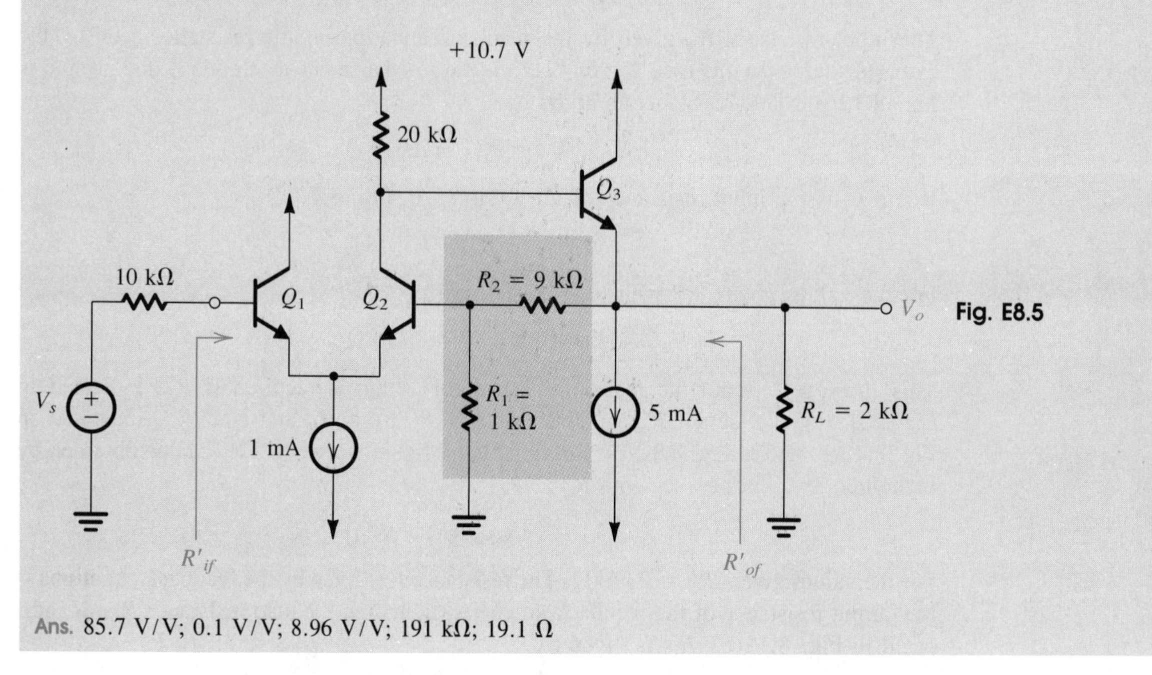

Fig. E8.5

Ans. 85.7 V/V; 0.1 V/V; 8.96 V/V; 191 kΩ; 19.1 Ω

8.5 THE SERIES-SERIES FEEDBACK AMPLIFIER

The Ideal Case

As mentioned in Section 8.3, the series-series feedback topology stabilizes I_o/V_s and is therefore best suited for transconductance amplifiers. Figure 8.13(a) shows the ideal structure for the series-series feedback amplifier. It consists of a unilateral open-loop amplifier (the A circuit) and an ideal feedback network. Note that in this case A is a transconductance,

$$A \equiv \frac{I_o}{V_i} \tag{8.14}$$

while β is a transresistance. Thus the loop gain $A\beta$ remains a dimensionless quantity, as it should always be.

In the ideal structure of Fig. 8.13(a), the load and source resistances have been absorbed inside the A circuit, and the β circuit does not load the A circuit. Thus the circuit

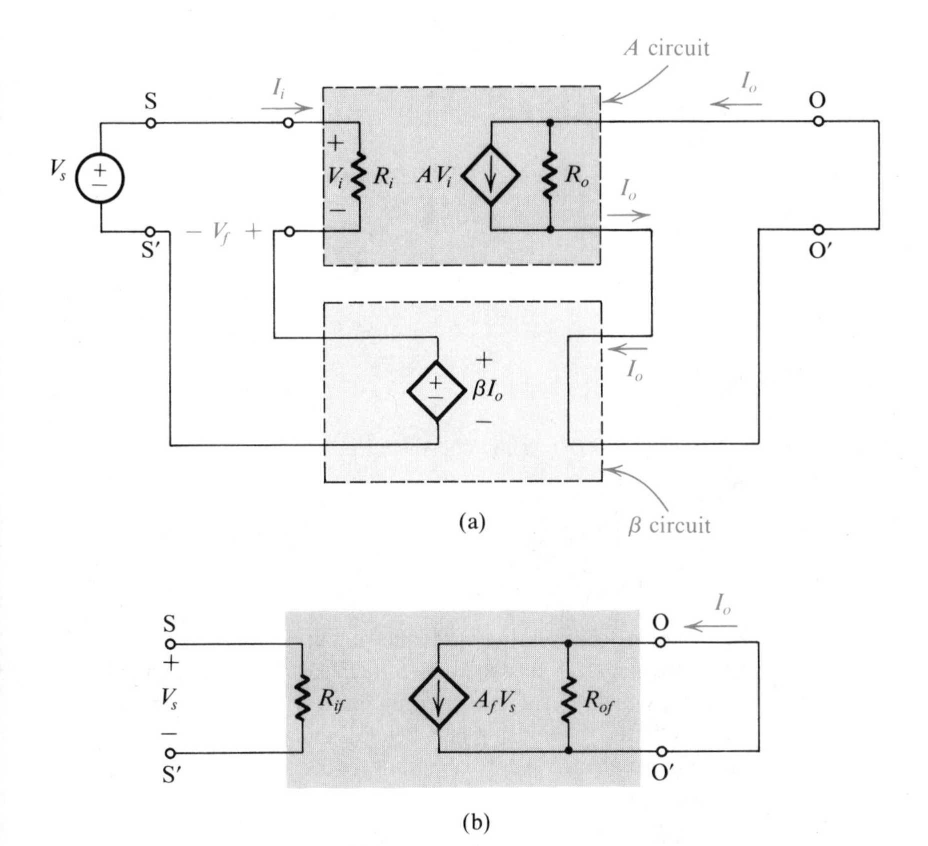

Fig. 8.13 The series-series feedback amplifier: **(a)** Ideal structure; **(b)** equivalent circuit.

follows the ideal feedback model of Fig. 8.1, and we can write

$$A_f \equiv \frac{I_o}{V_s} = \frac{A}{1 + A\beta} \tag{8.15}$$

This transconductance-with-feedback is included in the equivalent circuit model of the feedback amplifier, shown in Fig. 8.13(b). In this model R_{if} is the input resistance with feedback. Using an analysis similar to that in Section 8.4 we can show that

$$R_{if} = R_i(1 + A\beta) \tag{8.16}$$

This relationship is identical to that obtained in the case of series-shunt feedback. This confirms our earlier observation that the relationship between R_{if} and R_i is a function only of the method of mixing. Series mixing therefore always increases the input resistance.

To find the output resistance R_{of} of the series-series feedback amplifier of Fig. 8.13(a) we reduce V_s to zero and break the output circuit to apply a test current I_t, as shown in Fig. 8.14:

$$R_{of} \equiv \frac{V}{I_t} \tag{8.17}$$

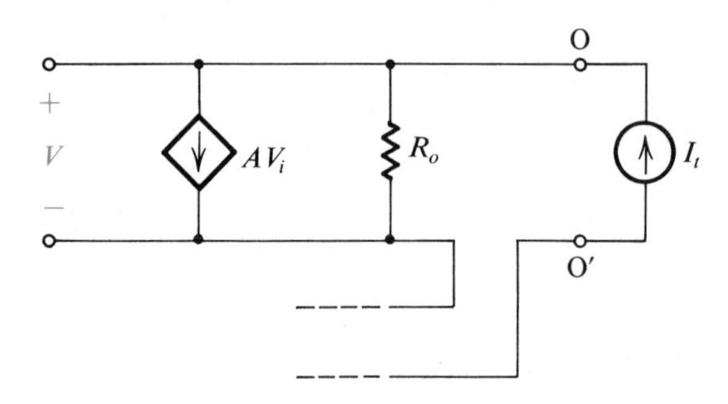

Fig. 8.14 Measuring the output resistance R_{of} of the series-series feedback amplifier.

In this case, $V_i = -V_f = -\beta I_o = -\beta I_t$. Thus for the circuit in Fig. 8.14 we obtain

$$V = (I_t - AV_i)R_o = (I_t + A\beta I_t)R_o$$

Hence

$$R_{of} = (1 + A\beta)R_o \tag{8.18}$$

That is, in this case the negative feedback increases the output resistance. This should have been expected, since the negative feedback tries to make I_o constant in spite of changes in the output voltage, which means increased output resistance. This result also confirms our earlier observation; the relationship between R_{of} and R_o is a function only of the method of sampling. While voltage (shunt) sampling reduces the output resistance, current (series) sampling increases it.

The Practical Case

Figure 8.15(a) shows a block diagram for a practical series-series feedback amplifier. To be able to apply the feedback equations to this amplifier we have to represent it by the ideal structure of Fig. 8.13(a). Our objective therefore is to devise a simple method for finding A and β.

The series-series amplifier of Fig. 8.15(a) is redrawn in Fig. 8.15(b) with R_s and R_L shown closer to the basic amplifier and the two-port feedback network represented by its z parameters (Appendix B). This parameter set has been chosen because it is the only one that provides a representation of the feedback network with a series circuit at the input and a series circuit at the output. This is obviously convenient in view of the series connections at input and output.

As we have done in the case of the series-shunt amplifier, we shall assume that the forward transmission through the feedback network is negligible as compared with that through the basic amplifier; that is, the condition

$$|z_{21}|_{\substack{\text{feedback}\\ \text{network}}} \ll |z_{21}|_{\substack{\text{basic}\\ \text{amplifier}}}$$

is satisfied. We can then dispense with the voltage source $z_{21}I_1$ in Fig. 8.15(b). Doing this, and redrawing the circuit to include z_{11} and z_{22} with the basic amplifier, results in the circuit in Fig. 8.15(c). Now if the basic amplifier is unilateral (or almost unilateral), a

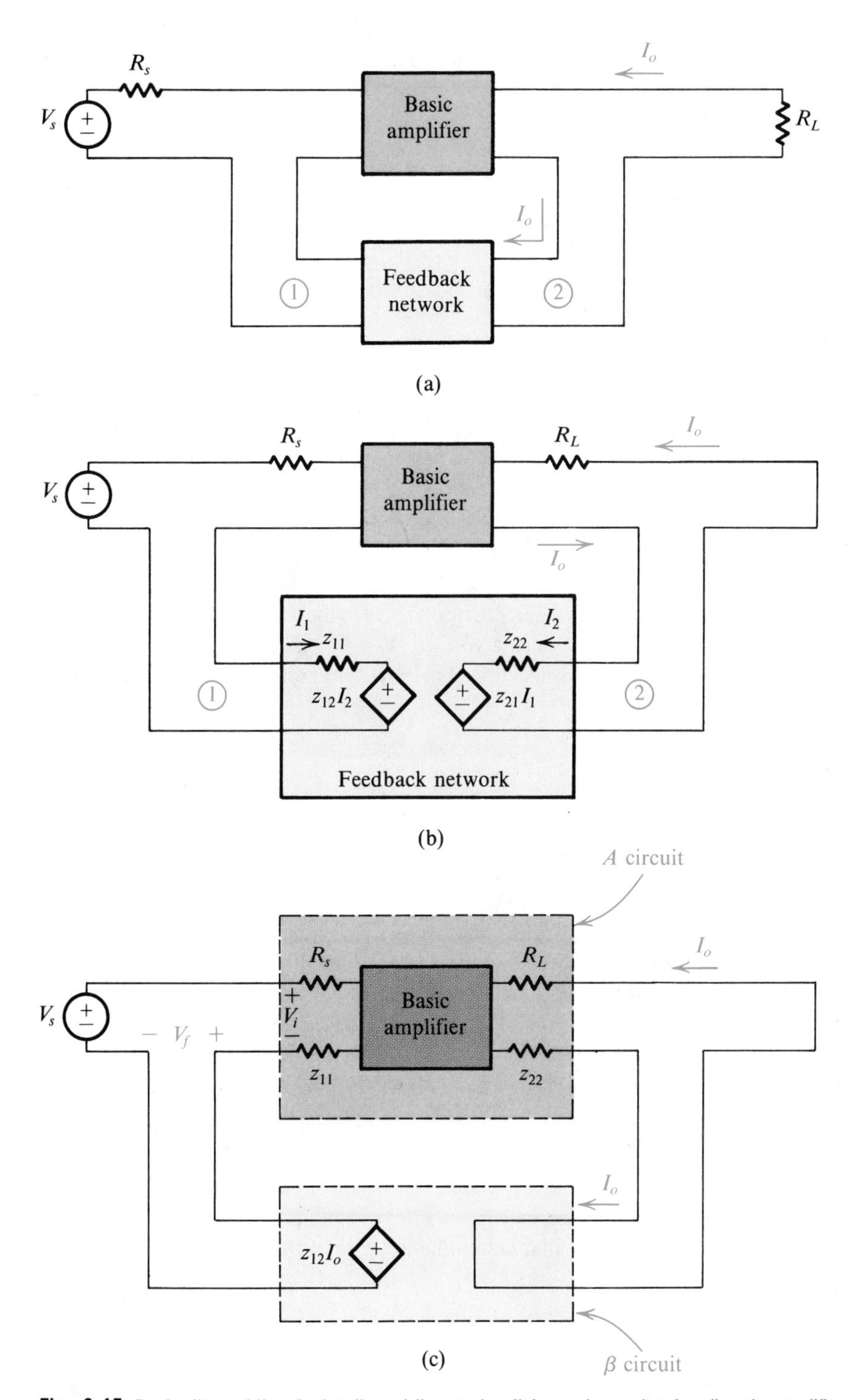

Fig. 8.15 Derivation of the A circuit and the β circuit for series-series feedback amplifiers. **(a)** A series-series feedback amplifier. **(b)** The circuit of (a) with the feedback network represented by its z parameters. **(c)** A redrawing of the circuit in (b) after neglecting z_{21}.

situation that is obtained when

$$|z_{12}|_{\substack{\text{basic}\\ \text{amplifier}}} \ll |z_{12}|_{\substack{\text{feedback}\\ \text{network}}}$$

then the circuit in Fig. 8.15(c) is equivalent (or almost equivalent) to the ideal circuit of Fig. 8.13(a).

It follows that the A circuit is composed of the basic amplifier augmented at the input with R_s and z_{11} and augmented at the output with R_L and z_{22}. Since z_{11} and z_{22} are the impedances looking into ports 1 and 2, respectively, of the feedback network with the other port open-circuited, we see that finding the loading effects of the feedback network on the basic amplifier follows the rule formulated in Section 8.4. That is, we look into one port of the feedback network while the other port is open-circuited or short-circuited so as to destroy the feedback (open if series and short if shunt).

From Fig. 8.15(c) we see that β is equal to z_{12} of the feedback network,

$$\beta = z_{12} \equiv \left.\frac{V_1}{I_2}\right|_{I_1=0} \tag{8.19}$$

This result is intuitively appealing. Recall that in this case the feedback network samples the output current $[I_2 = I_o]$ and provides a voltage $[V_f = V_1]$ that is mixed in series with the input source. Again, the series connection at the input suggests that β is measured with port 1 open.

Summary

For future reference we present in Fig. 8.16 a summary of the rules for finding A and β for a given series-series feedback amplifier of the type shown in Fig. 8.15(a).

EXAMPLE 8.2

Because negative feedback extends the amplifier bandwidth, it is commonly used in the design of broadband amplifiers. One such amplifier is the MC1553. Part of the circuit of the MC1553 is shown in Fig. 8.17(a). The circuit shown (called a *feedback triple*) is composed of three gain stages with series-series feedback provided by the network composed of R_{E1}, R_F, and R_{E2}. Assume that the bias circuit, which is not shown, causes $I_{C1} = 0.6$ mA, $I_{C2} = 1$ mA, and $I_{C3} = 4$ mA. Using these values and assuming $h_{fe} = 100$ and $r_o = \infty$, find the open-loop gain A, the feedback factor β, the closed loop gain $A_f \equiv I_o/V_s$, the voltage gain V_o/V_s, and the input resistance R_{if}.

Solution

Employing the loading rules given in Fig. 8.16, we obtain the A circuit shown in Fig. 8.17(b). To find $A \equiv I_o'/V_i'$ we first determine the gain of the first stage. This can be written by inspection as

$$\frac{V_{c1}}{V_i'} = \frac{-\alpha_1(R_{C1}//r_{\pi 2})}{r_{e1} + [R_{E1}//(R_F + R_{E2})]}$$

Since Q_1 is biased at 0.6 mA, $r_{e1} = 41.7\ \Omega$. Transistor Q_2 is biased as 1 mA; thus $r_{\pi 2} = h_{fe}/g_{m2} = 100/40 = 2.5$ kΩ. Substituting these values together with $\alpha_1 = 0.99$,

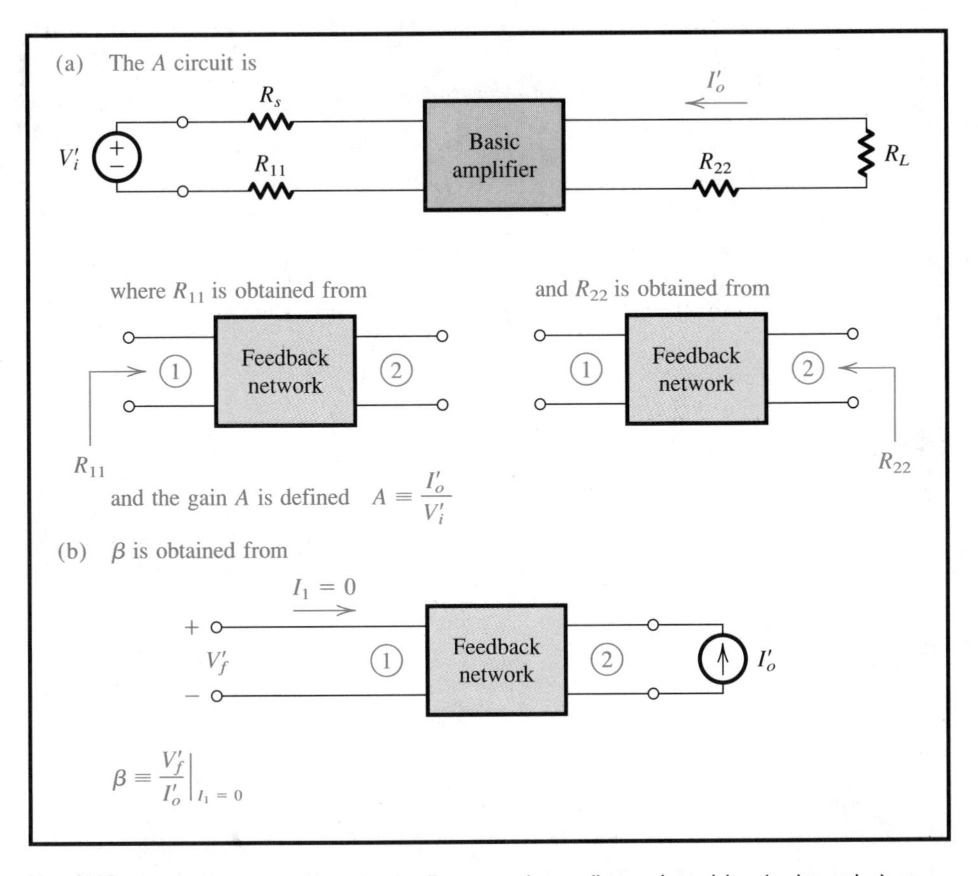

Fig. 8.16 Finding the *A* circuit and *β* for the current-sampling series-mixing (series-series) case.

$R_{C1} = 9\ \text{k}\Omega$, $R_{E1} = 100\ \Omega$, $R_F = 640\ \Omega$, and $R_{E2} = 100\ \Omega$ results in

$$\frac{V_{c1}}{V_i'} = -14.92\ \text{V/V}$$

Next we determine the gain of the second stage, which can be written by inspection as (note that $V_{b2} = V_{c1}$)

$$\frac{V_{c2}}{V_{c1}} = -g_{m2}\{R_{C2}//(h_{fe} + 1)[r_{e3} + (R_{E2}//(R_F + R_{E1}))]\}$$

Substituting $g_{m2} = 40\ \text{mA/V}$, $R_{C2} = 5\ \text{k}\Omega$, $h_{fe} = 100$, $r_{e3} = 25/4 = 6.25\ \Omega$, $R_{E2} = 100\ \Omega$, $R_F = 640\ \Omega$, and $R_{E1} = 100\ \Omega$ results in

$$\frac{V_{c2}}{V_{c1}} = -131.2\ \text{V/V}$$

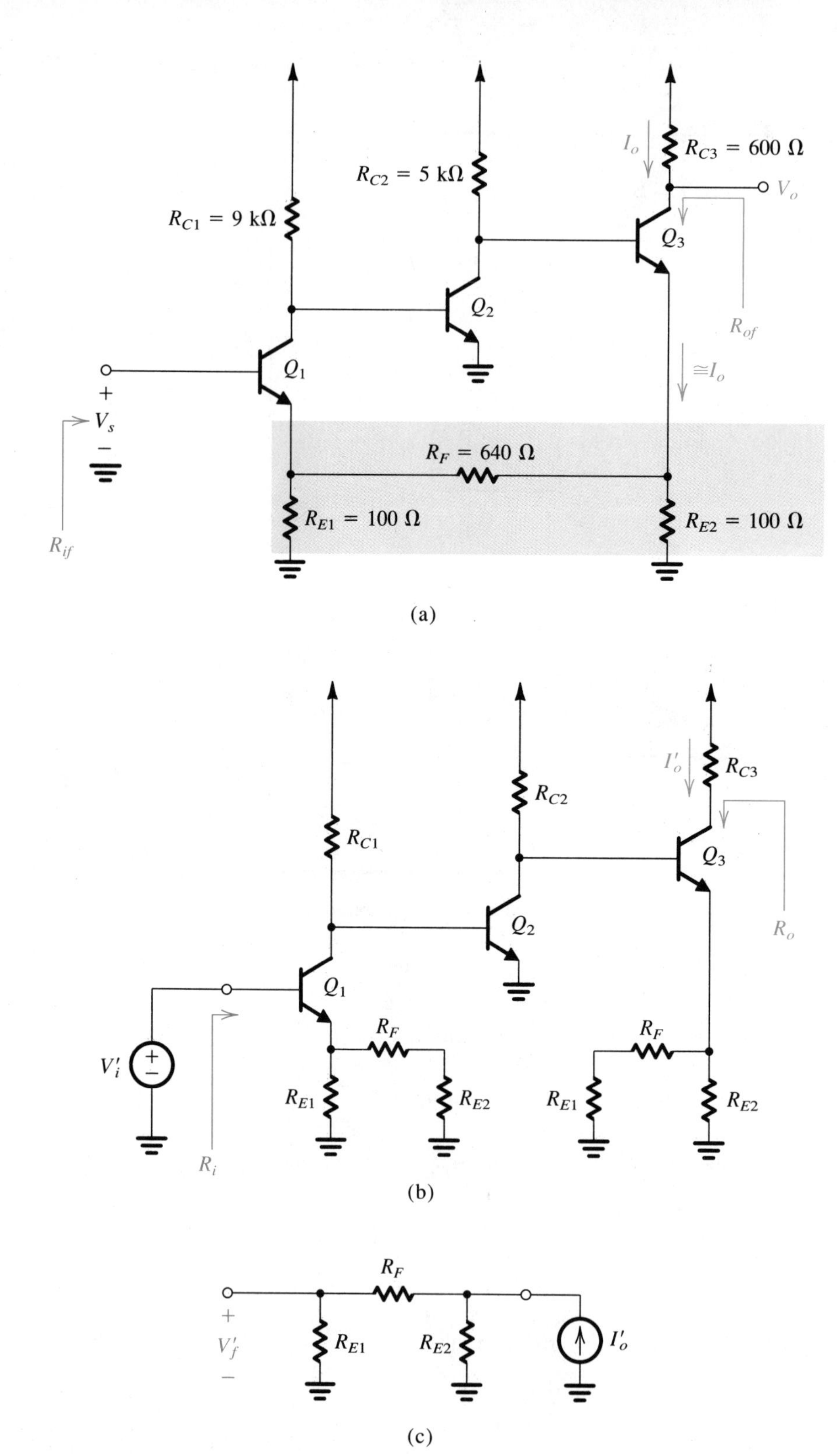

Fig. 8.17 Circuits for Example 8.2.

Finally for the third stage we can write by inspection

$$\frac{I'_o}{V_{c2}} = \frac{\alpha I_{e3}}{V_{b3}} = \frac{\alpha}{r_{e3} + (R_{E2}//(R_F + R_{E1}))}$$

$$= \frac{0.99}{6.25 + (100//740)} = 10.49 \text{ mA/V}$$

Combining the gains of the three stages results in

$$A \equiv \frac{I'_o}{V'_i} = -14.92 \times -131.2 \times 10.49 \times 10^{-3}$$

$$= 20.5 \text{ A/V}$$

The circuit for determining the feedback factor β is shown in Fig. 8.17(c), from which we find

$$\beta \equiv \frac{V'_f}{I'_o} = \frac{R_{E2}}{R_{E2} + R_F + R_{E1}} \times R_{E1}$$

$$= \frac{100}{100 + 640 + 100} \times 100 = 11.9\ \Omega$$

The closed-loop gain A_f can now be found from

$$A_f \equiv \frac{I_o}{V_s} = \frac{A}{1 + A\beta}$$

$$= \frac{20.5}{1 + 20.5 \times 11.9} = 83.7 \text{ mA/V}$$

The voltage gain is found from

$$\frac{V_o}{V_s} = \frac{-I_o R_{C3}}{V_s} = -A_f R_{C3}$$

$$= -83.7 \times 10^{-3} \times 600 = -50.2 \text{ V/V}$$

The input resistance of the feedback amplifier is given by

$$R_{if} = R_i(1 + A\beta)$$

where R_i is the input resistance of the A circuit. The value of R_i can be found from the circuit in Fig. 8.17(b) as follows:

$$R_i = (h_{fe} + 1)[r_{e1} + (R_{E1}//(R_F + R_{E2}))]$$

$$= 13.65 \text{ k}\Omega$$

Thus,

$$R_{if} = 13.65(1 + 20.5 \times 11.9) = 3.34 \text{ M}\Omega$$

Exercise 8.6 The circuit of Fig. 8.17(a) can be considered a voltage-to-current converter, providing an output current $I_o = A_f V_s$ to an arbitrary load R_{C3}. In such an application, the output resistance looking back into the collector of Q_3 becomes of interest. This is the output resistance R_{of} of the feedback amplifier. It is required to determine the value of R_{of}. First, find the value of the output resistance R_o of the A circuit—that is, the resistance looking back into the collector of Q_3—assuming that r_o of Q_3 is 25 kΩ. (*Hint:* Use the formula given in Eq. (4.76) for the output resistance of a common-emitter amplifier with a resistance R_e in the emitter.) Then utilize the feedback formulation to obtain R_{of} (using the results of Example 8.2).

Ans. $R_o = 65.6$ kΩ; $R_{of} = 16.1$ MΩ.

8.6 THE SHUNT-SHUNT AND THE SHUNT-SERIES FEEDBACK AMPLIFIERS

In this section we shall extend—without proof—the method of Sections 8.4 and 8.5 to the two remaining feedback topologies.

The Shunt-Shunt Configuration

Figure 8.18 shows the ideal structure for a shunt-shunt feedback amplifier. Here the A circuit has an input resistance R_i, a transresistance A, and an output resistance R_o. The β circuit is a voltage-controlled current source, and β is a transconductance. The closed-loop gain A_f is defined

$$A_f \equiv \frac{V_o}{I_s} \tag{8.20}$$

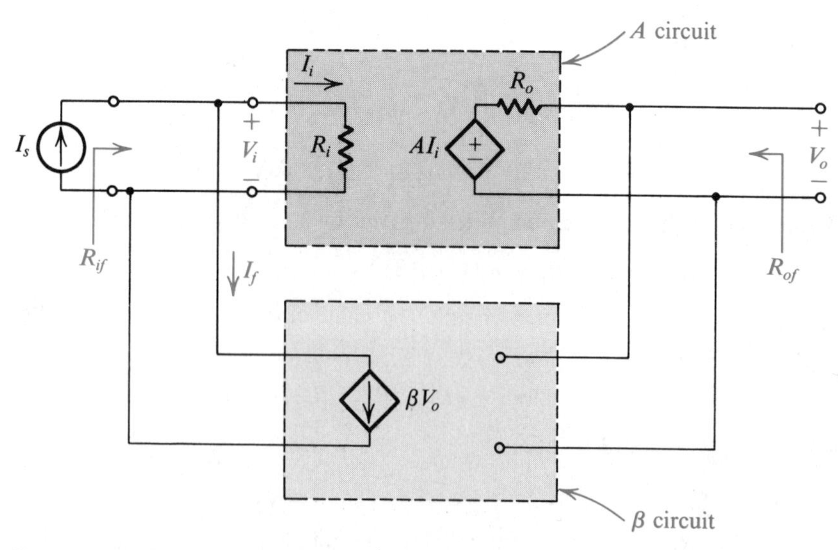

Fig. 8.18 Ideal structure for the shunt-shunt feedback amplifier.

and is given by

$$A_f = \frac{A}{1 + A\beta}$$

The input resistance with feedback is given by

$$R_{if} = \frac{R_i}{1 + A\beta} \tag{8.21}$$

where we note that the shunt connection at the input results in a reduced input resistance. Also note that the resistance R_{if} is the resistance seen by the source I_s, and it includes any source resistance.

The output resistance with feedback is given by

$$R_{of} = \frac{R_o}{1 + A\beta} \tag{8.22}$$

where we note that the shunt connection at the output results in a reduced output resistance. This resistance includes any load resistance.

Given a practical shunt-shunt feedback amplifier having the block diagram of Fig. 8.19, we use the method given in Fig. 8.20 to obtain the A circuit and the circuit for determining β. As in Sections 8.4 and 8.5, the method of Fig. 8.20 assumes that the basic amplifier is almost unilateral and that the forward transmission through the feedback network is negligibly small. The first assumption is justified when the reverse y parameters of the basic amplifier and of the feedback network satisfy the condition

$$|y_{12}|_{\substack{\text{basic}\\\text{amplifier}}} \ll |y_{12}|_{\substack{\text{feedback}\\\text{network}}}$$

The second assumption is justified when the forward y parameters satisfy the condition

$$|y_{21}|_{\substack{\text{feedback}\\\text{network}}} \ll |y_{21}|_{\substack{\text{basic}\\\text{amplifier}}}$$

(For the definition of the y parameters, refer to Appendix B.)

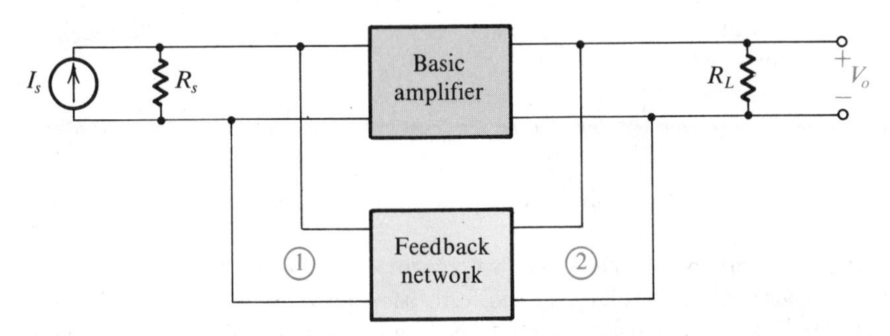

Fig. 8.19 Block diagram for a practical shunt-shunt feedback amplifier.

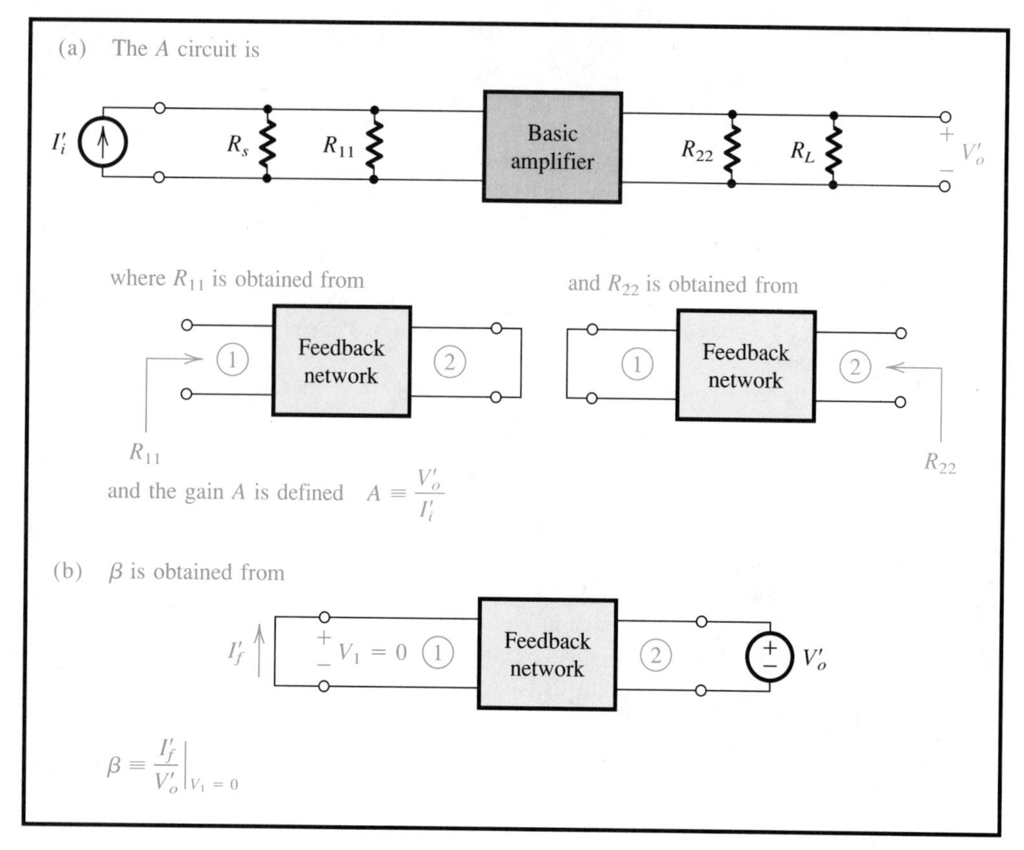

Fig. 8.20 Finding the *A* circuit and β for the voltage-sampling shunt-mixing (shunt-shunt) case.

EXAMPLE 8.3

We want to analyze the circuit of Fig. 8.21(a) to determine the small-signal voltage gain V_o/V_s, the input resistance R'_{if}, and the output resistance R_{of}. The transistor has $\beta = 100$.

Solution

First we determine the transistor dc operating point. The dc analysis is illustrated in Fig. 8.21(b), from which we can write

$$V_C = 0.7 + (I_B + 0.07)47 = 3.99 + 47I_B \qquad \text{and} \qquad \frac{12 - V_C}{4.7} = (\beta + 1)I_B + 0.07$$

These two equations can be solved to obtain $I_B \simeq 0.015$ mA, $I_C \simeq 1.5$ mA, and $V_C = 4.7$ V.

To carry out small-signal analysis we first recognize that the feedback is provided by R_f, which samples the output voltage V_o and feeds back a current that is mixed with the source current. Thus it is convenient to use the Norton source representation, as shown in Fig. 8.21(c). The *A* circuit can be easily obtained using the rules of Fig. 8.20, and it is shown in Fig. 8.21(d). For the *A* circuit we can write by inspection

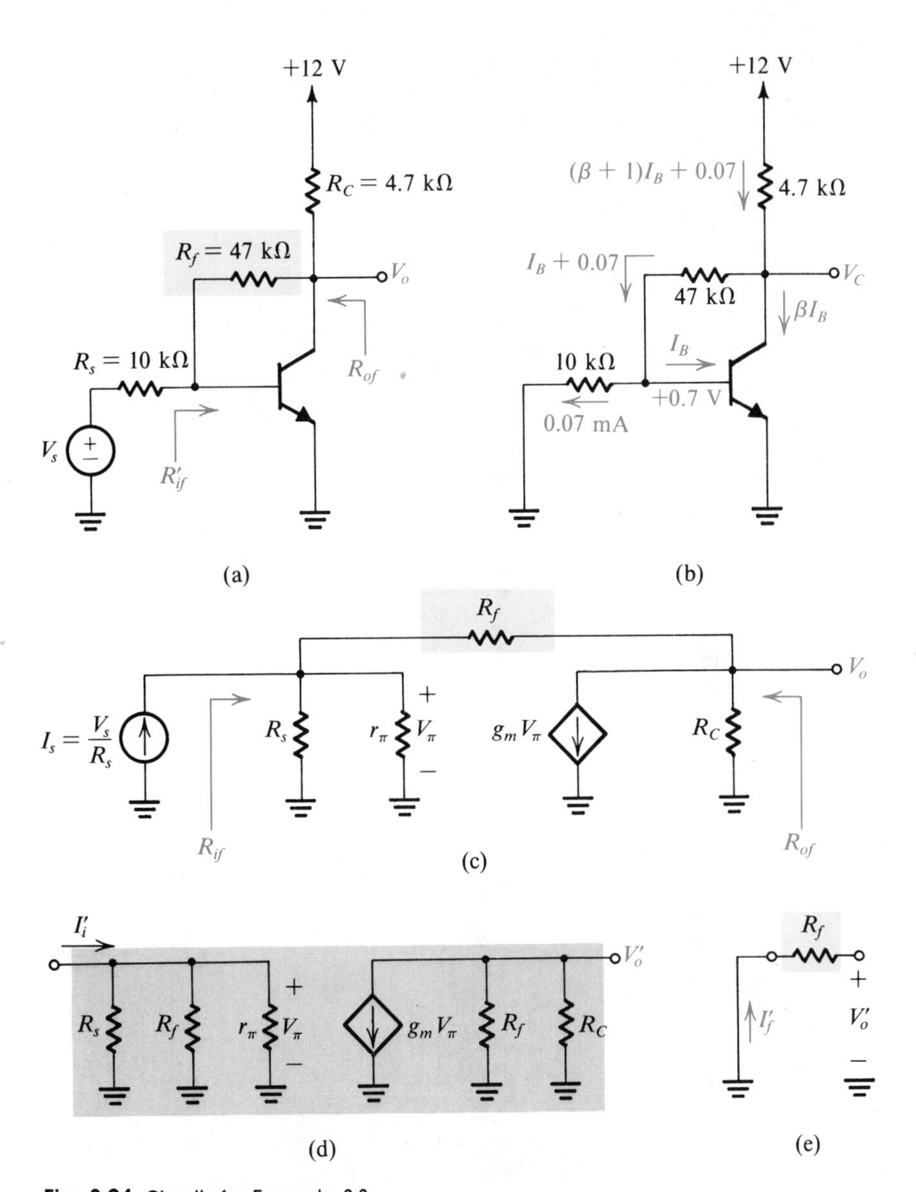

Fig. 8.21 Circuits for Example 8.3.

$$V_\pi = I'_i(R_s // R_f // r_\pi)$$

$$V'_o = -g_m V_\pi (R_f // R_C)$$

Thus

$$A = \frac{V'_o}{I'_i} = -g_m(R_f // R_C)(R_s // R_f // r_\pi)$$

$$= -358.7 \text{ k}\Omega$$

The input and output resistances of the A circuit can be obtained from Fig. 8.21(d) as

$$R_i = (R_s // R_f // r_\pi) = 1.4\ \text{k}\Omega$$

$$R_o = (R_C // R_f) = 4.27\ \text{k}\Omega$$

The circuit for determining β is shown in Fig. 8.21(e), from which we obtain

$$\beta \equiv \frac{I_f'}{V_o'} = -\frac{1}{R_f} = -\frac{1}{47\ \text{k}\Omega}$$

Note that as usual the reference direction for I_f has been selected so that I_f subtracts from I_s. The resulting negative sign of β should cause no concern, since A is also negative, keeping the loop gain $A\beta$ positive, as it should be for the feedback to be negative.

We can now obtain A_f (for the circuit in Fig. 8.21c) as

$$A_f \equiv \frac{V_o}{I_s} = \frac{A}{1 + A\beta}$$

$$\frac{V_o}{I_s} = \frac{-358.7}{1 + 358.7/47} = \frac{-358.7}{8.63} = -41.6\ \text{k}\Omega$$

To find the voltage gain V_o/V_s we note that

$$V_s = I_s R_s$$

Thus

$$\frac{V_o}{V_s} = \frac{V_o}{I_s R_s} = \frac{-41.6}{10} \simeq -4.16\ \text{V/V}$$

The input resistance with feedback is given by

$$R_{if} = \frac{R_i}{1 + A\beta}$$

Thus

$$R_{if} = \frac{1.4}{8.63} = 162.2\ \Omega$$

This is the resistance seen by the current source I_s in Fig. 8.21(c). To obtain the input resistance of the feedback amplifier excluding R_s (that is, the required resistance R_{if}') we subtract $1/R_s$ from $1/R_{if}$ and invert the result; thus $R_{if}' = 165\ \Omega$. Finally, the amplifier output resistance R_{of} is evaluated using

$$R_{of} = \frac{R_o}{1 + A\beta} = \frac{4.27}{8.63} = 495\ \Omega$$

An Important Note

The method we have been employing for the analysis of feedback amplifiers is predicated on two premises: Most of the forward transmission occurs in the basic amplifier, and most of the reverse transmission (feedback) occurs in the feedback network. For each of the three topologies considered thus far, these two assumptions were mathematically expressed as conditions on the relative magnitudes of the forward and reverse two-port parameters of the basic amplifier and the feedback network. Since the circuit considered in Example 8.3 is simple, we have a good opportunity to check the validity of these assumptions.

Reference to Fig. 8.21(d) indicates clearly that the basic amplifier is unilateral; thus *all* of the reverse transmission takes place in the feedback network. The case with forward transmission, however, is not so clear, and we must evaluate the forward y parameters. For the A circuit in Fig. 8.21(d), $y_{21} = g_m$. For the feedback network it can be easily shown that $y_{21} = -1/R_f$. Thus for our analysis method to be valid we must have $g_m \gg 1/R_f$. For the numerical values in Example 8.3, $g_m = 60$ mA/V and $1/R_f = 0.02$ mA/V, indicating that this assumption is more than justified. Nevertheless, in designing feedback amplifiers, care should be taken in choosing component values to ensure that the two basic assumptions are valid.

The Shunt-Series Configuration

Figure 8.22 shows the ideal structure of the shunt-series feedback amplifier. It is a current amplifier whose gain with feedback is defined as

$$A_f \equiv \frac{I_o}{I_s} = \frac{A}{1 + A\beta} \tag{8.23}$$

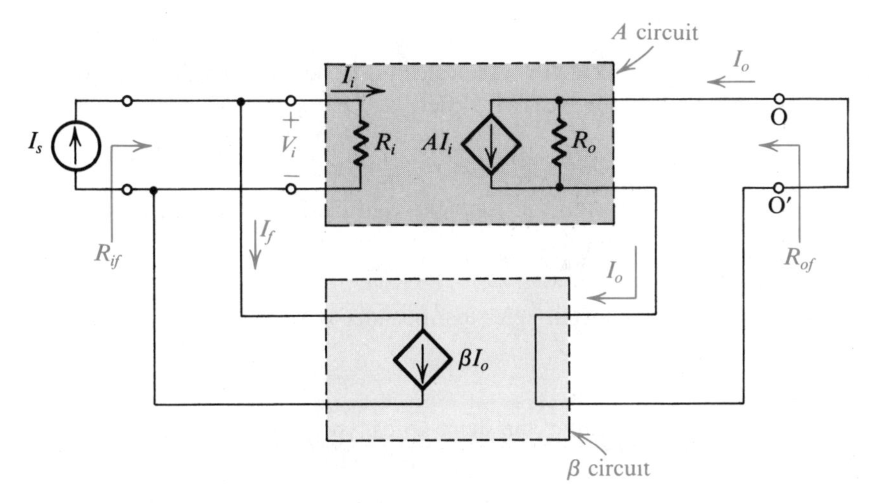

Fig. 8.22 Ideal structure for the shunt-series feedback amplifier.

The input resistance with feedback is the resistance seen by the current source I_s and is given by

$$R_{if} = \frac{R_i}{1 + A\beta} \tag{8.24}$$

Again we note that the shunt connection at the input reduces the input resistance. The output resistance with feedback is the resistance seen by breaking the output circuit, such as between O and O′, and looking between the two terminals thus generated (that is, between O and O′). This resistance, R_{of}, is given by

$$R_{of} = R_o(1 + A\beta) \tag{8.25}$$

where we note that the increase in output resistance is due to the current (series) sampling.

Given a practical shunt-series feedback amplifier, such as that represented by the block diagram of Fig. 8.23, we follow the method given in Fig. 8.24 in order to obtain A

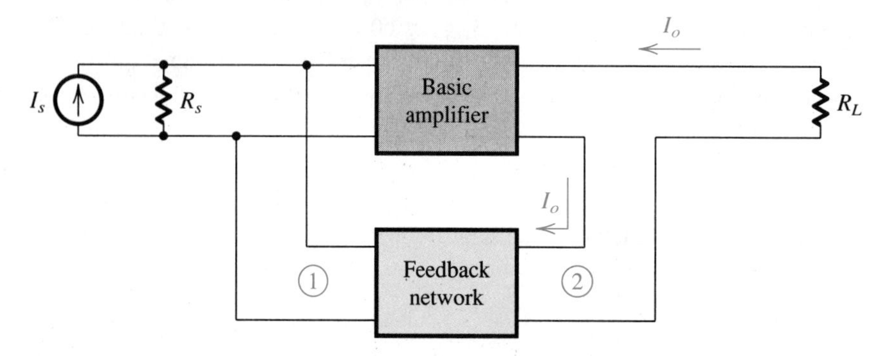

Fig. 8.23 Block diagram for a practical shunt-series feedback amplifier.

and β. Here again the analysis method is predicated on the assumption that most of the forward transmission occurs in the basic amplifier,

$$|g_{21}|_{\substack{\text{feedback}\\\text{network}}} \ll |g_{21}|_{\substack{\text{basic}\\\text{amplifier}}}$$

and that most of the reverse transmission takes place in the feedback network,

$$|g_{12}|_{\substack{\text{basic}\\\text{amplifier}}} \ll |g_{12}|_{\substack{\text{feedback}\\\text{network}}}$$

(For the definition of the g parameters refer to Appendix B.)

EXAMPLE 8.4

Figure 8.25 shows a feedback circuit of the shunt-series type. Find I_{out}/I_{in}, R_{in}, and R_{out}. Assume the transistors to have $\beta = 100$ and $r_o = 100\ \text{k}\Omega$.

Solution We begin by determining the dc operating points. In this regard we note that the feedback signal is capacitively coupled; thus the feedback has no effect on dc bias. The dc analysis proceeds as follows:

$$V_{B1} \simeq 12\frac{15}{100 + 15} = 1.57\ \text{V}$$

$$V_{E1} \simeq 1.57 - 0.7 = 0.87\ \text{V}$$

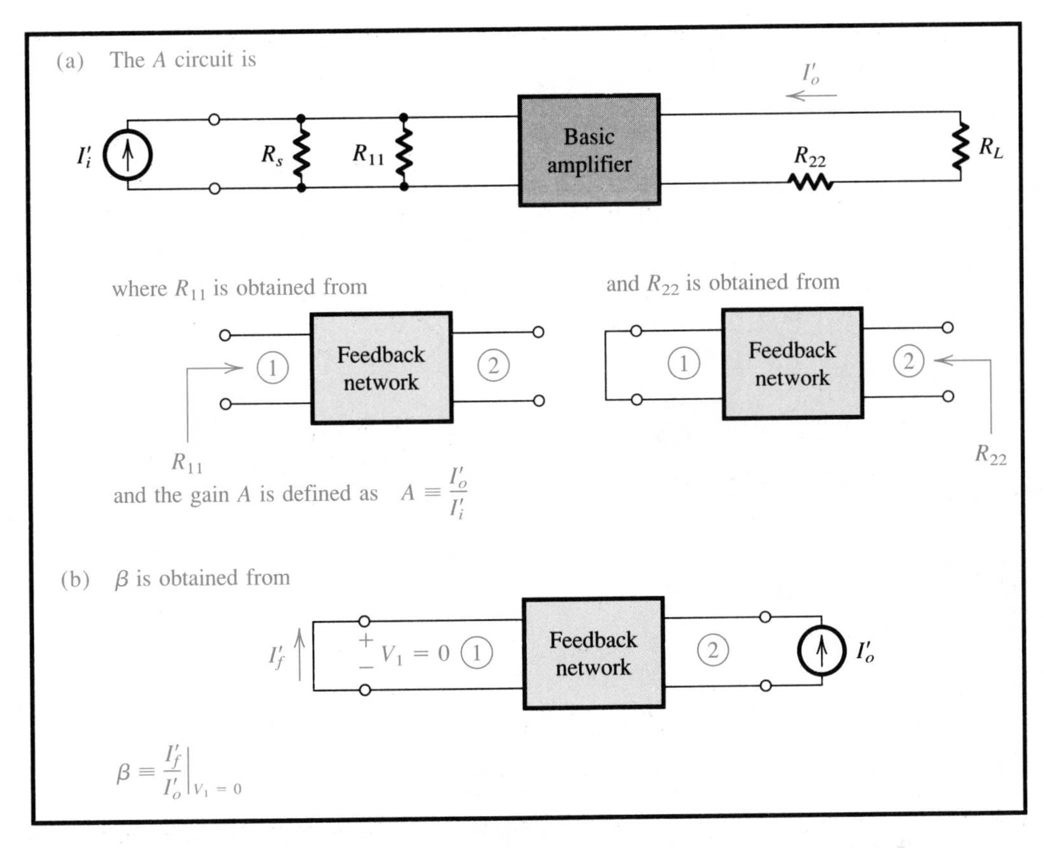

Fig. 8.24 Finding the A circuit and β for the current-sampling shunt-mixing (shunt-series) case.

$$I_{E1} = 0.87/0.87 = 1 \text{ mA}$$

$$V_{C1} \simeq 12 - 10 \times 1 = 2 \text{ V}$$

$$V_{E2} \simeq 2 - 0.7 = 1.3 \text{ V}$$

$$I_{E2} \simeq 1.3/1.3 \simeq 1 \text{ mA}$$

$$V_{C2} \simeq 12 - 1 \times 8 = 4 \text{ V}$$

The amplifier equivalent circuit is shown in Fig. 8.25(b), from which we note that the feedback network is composed of R_{E2} and R_f. The feedback network samples the emitter current of Q_2, which is approximately equal to the collector current I_o. Also note that the required current gain, $I_{\text{out}}/I_{\text{in}}$, will be slightly different than the closed-loop current gain $A_f \equiv I_o/I_s$.

The A circuit is shown in Fig. 8.25(c), where we have obtained the loading effects of the feedback network using the rules of Fig. 8.24. For the A circuit we can write

$$V_{\pi 1} = I_i'[R_s//(R_{E2} + R_f)//R_B//r_{\pi 1}]$$

$$V_{b2} = -g_{m1}V_{\pi 1}\{r_{o1}//R_{C1}//[r_{\pi 2} + (\beta + 1)(R_{E2}//R_f)\}$$

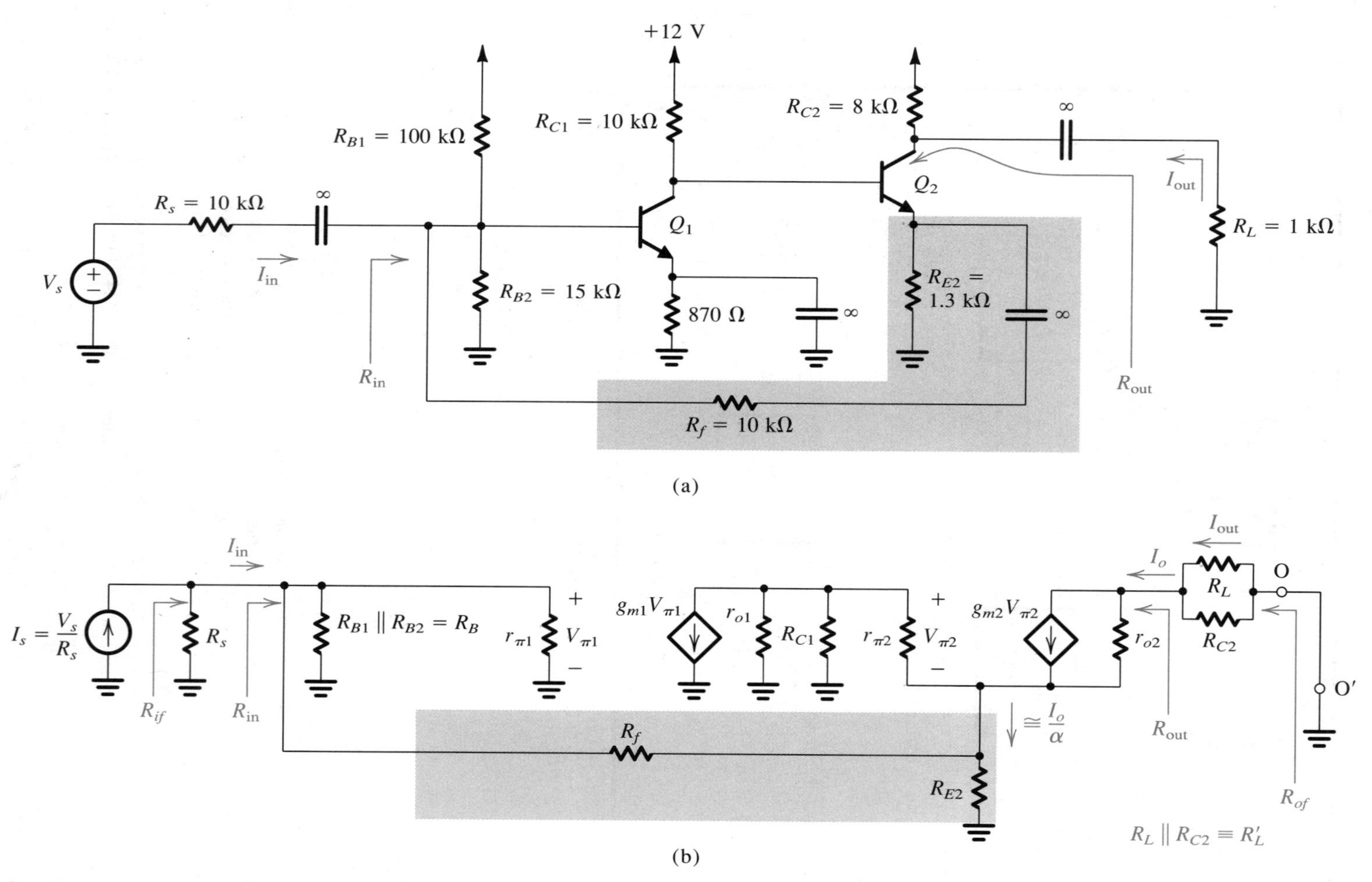

Fig. 8.25 Circuits for Example 8.4.

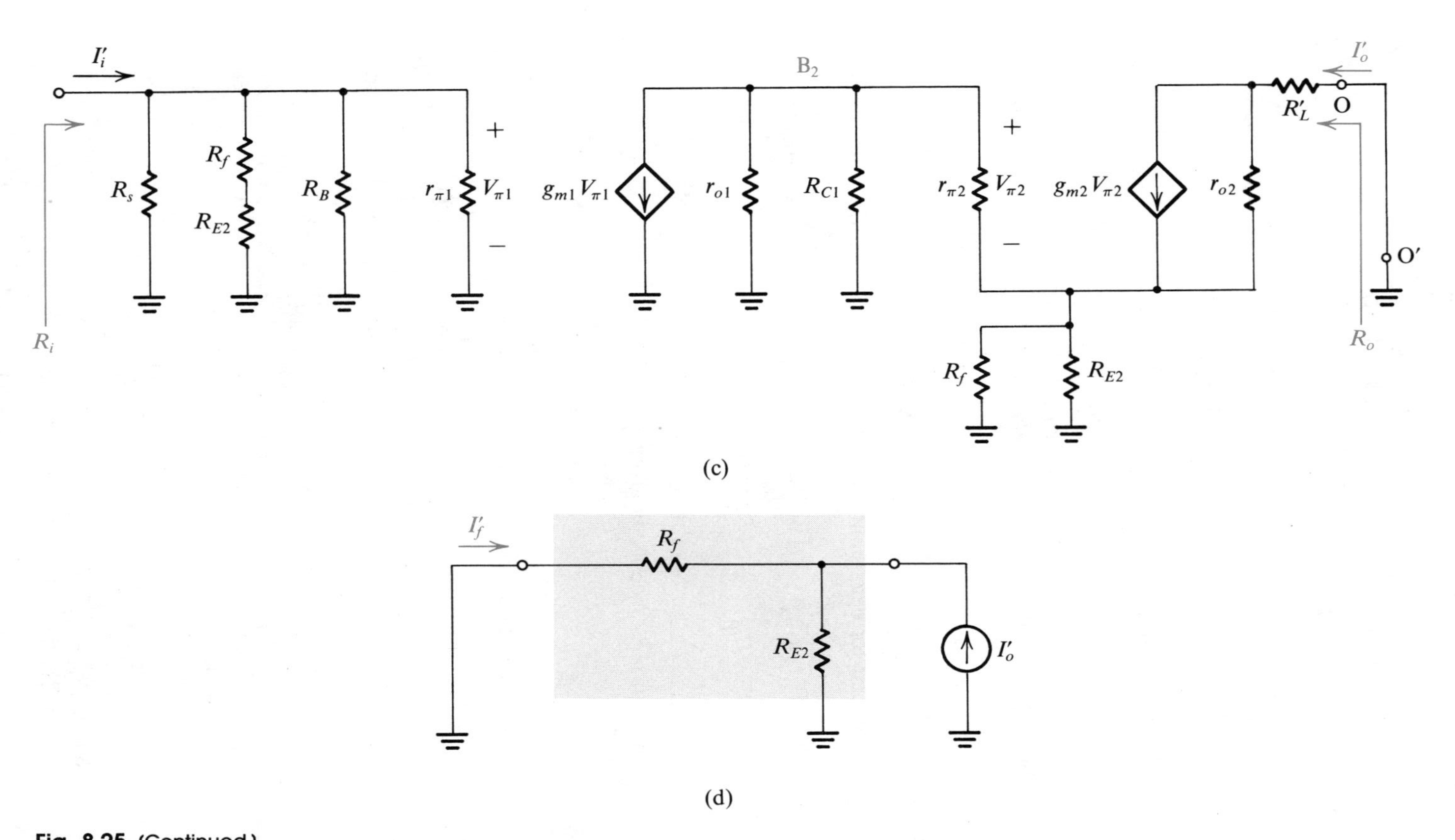

Fig. 8.25 (Continued.)

$$V_{\pi 2} = V_{b2} \frac{r_{\pi 2}}{r_{\pi 2} + (\beta + 1)(R_{E2}//R_f)}$$

$$I_o' \simeq g_{m2} V_{\pi 2}$$

These equations can be combined to obtain the open-loop current gain A,

$$A \equiv \frac{I_o'}{I_i'} \simeq -426.9$$

The input resistance R_i is given by

$$R_i = [R_s//(R_{E2} + R_f)//R_B//r_{\pi 1}] = 1.5 \text{ k}\Omega$$

The output resistance R_o is that found by looking into the output loop of the A circuit (see Fig. 8.25c) with the input excitation I_i' set to zero. It can be shown that

$$R_o = R_L' + r_{o2}\left[1 + \frac{g_{m2}(R_{E2}//R_f)}{1 + (r_{o1}//R_{C1})/r_{\pi 2}}\right]$$
$$= 1.09 \text{ M}\Omega$$

The circuit for determining β is shown in Fig. 8.25(d), from which we find

$$\beta \equiv \frac{I_f'}{I_o'} = -\frac{R_{E2}}{R_{E2} + R_f} = -\frac{1.3}{11.3} = -0.115$$

Thus

$$1 + A\beta = 50.1$$

The input resistance R_{if} is given by

$$R_{if} = \frac{R_i}{1 + A\beta} = 29.9 \ \Omega$$

The required input resistance R_{in} is given by

$$R_{\text{in}} = \frac{1}{1/R_{if} - 1/R_s} = 30 \ \Omega$$

Since $R_{\text{in}} \simeq R_{if}$, it follows from Fig. 8.25(b) that $I_{\text{in}} \simeq I_s$. The current gain A_f is given by

$$A_f \equiv \frac{I_o}{I_s} = \frac{A}{1 + A\beta} = -8.52$$

Note that because $A\beta \gg 1$ the closed-loop gain is approximately equal to $1/\beta$.

Now the required current gain is given by

$$\frac{I_{\text{out}}}{I_{\text{in}}} \simeq \frac{I_{\text{out}}}{I_s} = \frac{R_{C2}}{R_L + R_{C2}} \frac{I_o}{I_s}$$

Thus $I_{\text{out}}/I_{\text{in}} = -7.57$.

Finally, the output resistance R_{of} is given by

$$R_{of} = R_o(1 + A\beta) \simeq 55 \text{ M}\Omega$$

The required output resistance R_{out} can be obtained by subtracting R'_L from R_{of}:

$$R_{out} \simeq R_{of} = 55 \text{ M}\Omega$$

Exercise 8.7 Use the feedback method to find the voltage gain V_o/V_s, the input resistance R'_{if}, and the output resistance R'_{of} of the inverting op amp configuration of Fig. E8.7. Let the op amp have open-loop gain $\mu = 10^4$, $R_{id} = 100 \text{ k}\Omega$, $R_{icm} = 10 \text{ M}\Omega$, and $r_o = 1 \text{ k}\Omega$. (*Hint:* The feedback is of the shunt-shunt type.)

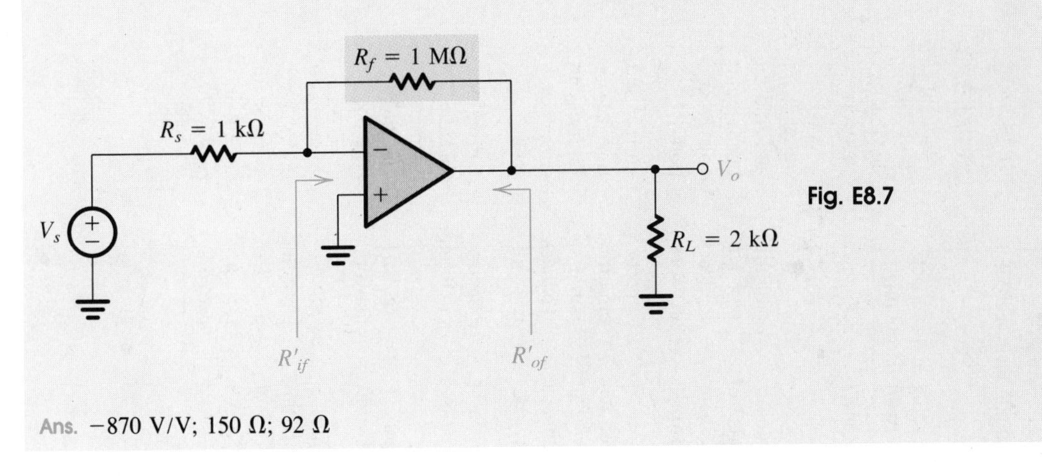

Fig. E8.7

Ans. -870 V/V; 150 Ω; 92 Ω

Summary of Results

Table 8.1 provides a summary of the rules and relationships employed in the analysis of the four types of feedback amplifiers.

8.7 DETERMINING THE LOOP GAIN

We have already seen that the loop gain $A\beta$ is a very important quantity that characterizes a feedback loop. Furthermore, in the following sections it will be shown that $A\beta$ determines whether the feedback amplifier is stable (as opposed to oscillatory). In this section we shall describe an alternative approach to the determination of loop gain.

Consider first the general feedback amplifier shown in Fig. 8.1. Let the external source x_s be set to zero. Open the feedback loop by breaking the connection of x_o to the feedback network and apply a test signal x_t. We see that the signal at the output of the feedback network is $x_f = \beta x_t$; that at the input of the basic amplifier is $x_i = -\beta x_t$; and the signal at the output of the amplifier, where the loop was broken, will be $x_o = -A\beta x_t$. It follows that the loop gain $A\beta$ is given by the negative of the ratio of the *returned* signal to the applied test signal; that is, $A\beta = -x_o/x_t$. It should also be obvious that this applies regardless of where the loop is broken.

In breaking the feedback loop of a practical amplifier circuit, we must ensure that the conditions that existed prior to breaking the loop do not change. This is achieved by terminating the loop where it is opened with an impedance equal to that seen (by the driving circuit) before the loop was broken. To be specific, consider the conceptual

Table 8.1 SUMMARY OF RELATIONSHIPS FOR THE FOUR FEEDBACK-AMPLIFIER TOPOLOGIES

Feedback Amplifier	x_i	x_o	x_f	x_s	A	β	A_f	Source Form	Loading of Feedback network is obtained: At input	Loading of Feedback network is obtained: At output	To find β, apply to port 2 of feedback network	Z_{if}	Z_{of}	Refer to Figs.
Series-shunt (voltage amplifier)	V_i	V_o	V_f	V_s	$\frac{V_o}{V_i}$	$\frac{V_f}{V_o}$	$\frac{V_o}{V_s}$	Thévenin	By short-circuiting port 2 of feedback network	By open-circuiting port 1 of feedback network	A voltage and find the open-circuit voltage at port 1	$Z_i(1 + A\beta)$	$\frac{Z_o}{1 + A\beta}$	8.4(a) 8.8 8.10 8.11
Shunt-series (current amplifier)	I_i	I_o	I_f	I_s	$\frac{I_o}{I_i}$	$\frac{I_f}{I_o}$	$\frac{I_o}{I_s}$	Norton	By open-circuiting port 2 of feedback network	By short-circuiting port 1 of feedback network	A current and find the short-circuit current at port 1	$\frac{Z_i}{1 + A\beta}$	$Z_o(1 + A\beta)$	8.4(b) 8.22 8.23 8.24
Series-series (transconductance amplifier)	V_i	I_o	V_f	V_s	$\frac{I_o}{V_i}$	$\frac{V_f}{I_o}$	$\frac{I_o}{V_s}$	Thévenin	By open-circuiting port 2 of feedback network	By open-circuiting port 1 of feedback network	A current and find the open-circuit voltage at port 1	$Z_i(1 + A\beta)$	$Z_o(1 + A\beta)$	8.4(c) 8.13 8.15 8.16
Shunt-shunt (transresistance amplifier)	I_i	V_o	I_f	I_s	$\frac{V_o}{I_i}$	$\frac{I_f}{V_o}$	$\frac{V_o}{I_s}$	Norton	By short-circuiting port 2 of feedback network	By short-circuiting port 1 of feedback network	A voltage and find the short-circuit current at port 1	$\frac{Z_i}{1 + A\beta}$	$\frac{Z_o}{1 + A\beta}$	8.4(d) 8.18 8.19 8.20

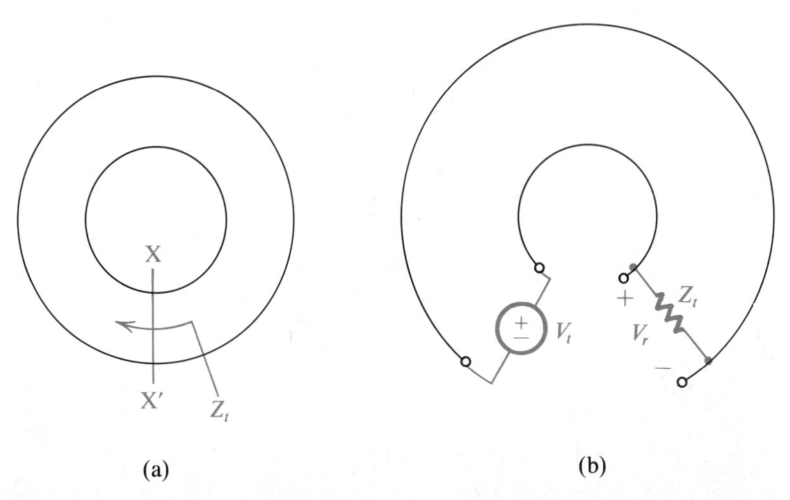

Fig. 8.26 A conceptual feedback loop is broken at XX′ and a test voltage V_t is applied. The impedance Z_t is equal to that previously seen looking to the left of XX′. The loop gain $A\beta = -V_r/V_t$, where V_r is the *returned* voltage.

feedback loop shown in Fig. 8.26(a). If we break the loop at XX′, and apply a test voltage V_t to the terminals thus created to the left of XX′, the terminals at the right of XX′ should be loaded with an impedance Z_t as shown in Fig. 8.26(b). The impedance Z_t is equal to that previously seen looking to the left of XX′. The loop gain $A\beta$ is then determined from

$$A\beta = -\frac{V_r}{V_t}$$

Finally, it should be noted that in some cases it may be convenient to determine $A\beta$ by applying a test current I_t and finding the returned current signal I_r. In this case, $A\beta = -I_r/I_t$.

To illustrate the above procedure we consider the feedback loop shown in Fig. 8.27(a). This feedback loop represents both the inverting and the noninverting op amp configurations. Using a simple equivalent circuit model for the op amp we obtain the circuit of Fig. 8.27(b). Examination of this circuit reveals that a convenient place to break the loop is at the input terminals of the op amp. The loop, broken in this manner, is shown in Fig. 8.27(c) with a test signal V_t applied to the right-hand-side terminals and a resistance R_{id} terminating the left-hand-side terminals. The returned voltage V_r is found by inspection as

$$V_r = -\mu V_1 \frac{\{R_L//[R_2 + R_1//(R_{id} + R)]\}}{\{R_L//[R_2 + R_1//(R_{id} + R)]\} + r_o} \frac{[R_1//(R_{id} + R)]}{[R_1//(R_{id} + R)] + R_2} \frac{R_{id}}{R_{id} + R}$$

This equation can be used directly to find the loop gain $L = A\beta = -V_r/V_t = -V_r/V_1$.

Since the loop gain L is generally a function of frequency, it is usual to call it **loop transmission** and denote it by $L(s)$ or $L(j\omega)$.

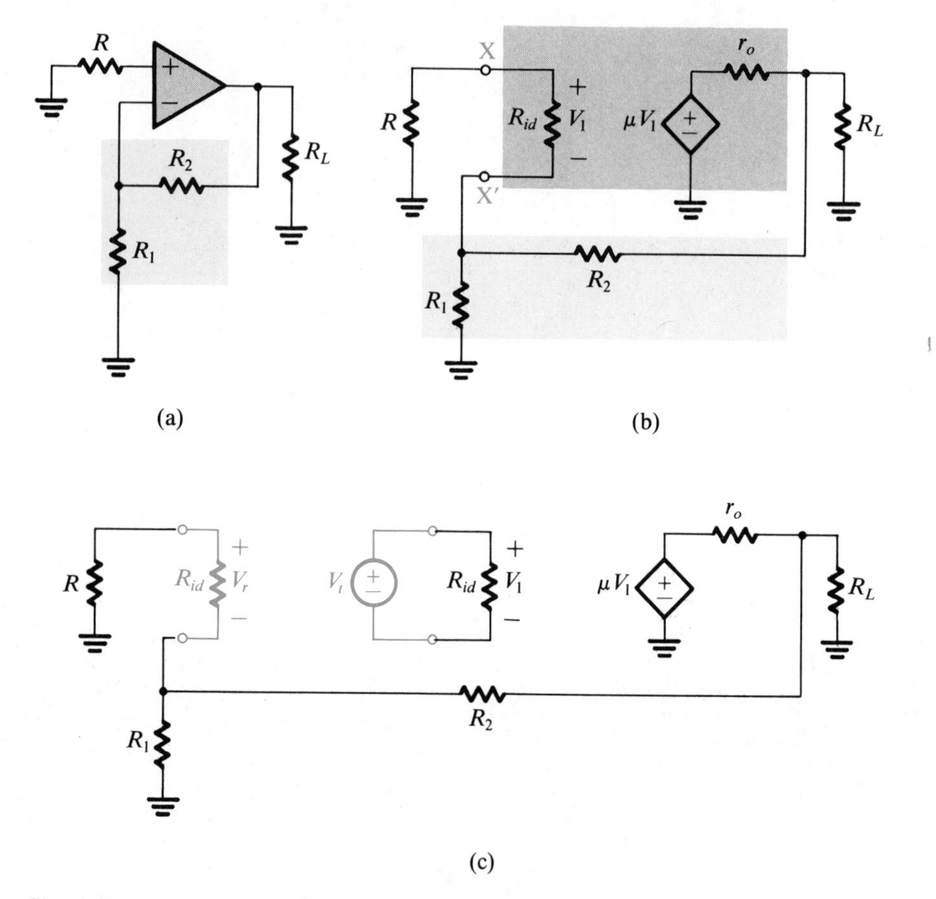

Fig. 8.27 Determination of the loop gain of the feedback loop in (a).

Equivalence of Circuits from a Feedback-Loop Point of View

From the study of circuit theory we know that the poles of a circuit are independent of the external excitation. In fact the poles, or the natural modes (which is a more appropriate name), are determined by setting the external excitation to zero. It follows that the poles of a feedback amplifier depend only on the feedback loop. This will be confirmed in a later section, where we show that the characteristic equation (whose roots are the poles) is completely determined by the loop gain. Thus a given feedback loop may be used to generate a number of circuits having the same poles but different transmission zeros. The closed-loop gain and the transmission zeros depend on how and where the input signal is injected into the loop.

As an example consider the feedback loop of Fig. 8.27(a). This loop can be used to generate the noninverting op amp circuit by feeding the input voltage signal to the terminal of R that is connected to ground; that is, we lift this terminal off ground and connect it to V_S. The same feedback loop can be used to generate the inverting op amp circuit by feeding the input voltage signal to the terminal of R_1 that is connected to ground.

Recognition of the fact that two or more circuits are equivalent from a feedback-loop point of view is very useful because (as will be shown in Section 8.8) stability is a function of the loop. Thus one needs to perform the stability analysis only once for a given loop.

In Chapter 11 we shall employ the concept of loop equivalence in the synthesis of active filters.

Exercises 8.8 Consider the feedback amplifier whose equivalent circuit is shown in Fig. 8.25(b). Break the feedback loop at the input to transistor Q_1; that is, apply a test voltage V_t to the base of Q_1, and find the returned voltage that appears across $r_{\pi 1}$. Thus show that the loop gain is given (neglecting r_{o2} in order to simplify the analysis) by:

$$A\beta = g_{m1}\{r_{o1}//R_{C1}//[r_{\pi 2} + (h_{fe}+1)(R_{E2}//(R_f + (R_B//R_s//r_{\pi 1})))]\}$$
$$\times \frac{R_{E2}//(R_f + (R_B//R_s//r_{\pi 1}))}{R_{E2}//(R_f + (R_B//R_s//r_{\pi 1})) + r_{e2}} \frac{R_s//R_B//r_{\pi 1}}{(R_s//R_B//r_{\pi 1}) + R_f}$$

Using the component values in Example 8.4, find the value of $A\beta$ and compare to the value found in Example 8.4.

Ans. 48.9 compared to 49.1

8.9 Find the numerical value of $A\beta$ for the amplifier in Exercise 8.7.

Ans. 6589 V/V

8.8 THE STABILITY PROBLEM

In a feedback amplifier such as that represented by the general structure of Fig. 8.1, the open-loop gain A is generally a function of frequency, and it should therefore be more accurately called the **open-loop transfer function,** $A(s)$. Also, we have been assuming for the most part that the feedback network is resistive and hence that the feedback factor β is constant, but this need not be always the case. We shall therefore assume that in the general case the **feedback transfer function** is $\beta(s)$. It follows that the **closed-loop transfer function** $A_f(s)$ is given by

$$A_f(s) = \frac{A(s)}{1 + A(s)\beta(s)} \tag{8.26}$$

To focus attention on the points central to our discussion in this section, we shall assume that the amplifier is direct-coupled with constant dc gain A_0 and with poles and zeros occurring in the high-frequency band. Also, for the time being let us assume that at low frequencies $\beta(s)$ reduces to a constant value. Thus at low frequencies the loop gain $A(s)\beta(s)$ becomes a constant, which should be a positive number; otherwise the feedback would not be negative. The question then arises as to what happens at higher frequencies.

For physical frequencies $s = j\omega$, Eq. (8.26) becomes

$$A_f(j\omega) = \frac{A(j\omega)}{1 + A(j\omega)\beta(j\omega)} \tag{8.27}$$

Thus the loop gain $A(j\omega)\beta(j\omega)$ is a complex number that can be represented by its magnitude and phase,

$$L(j\omega) \equiv A(j\omega)\beta(j\omega) = |A(j\omega)\beta(j\omega)|e^{j\phi(\omega)} \tag{8.28}$$

It is the manner in which the loop gain varies with frequency that determines the stability or instability of the feedback amplifier. To appreciate this fact, consider the frequency at which the phase angle $\phi(\omega)$ becomes 180°. At this frequency, ω_{180}, the loop gain $A(j\omega)\beta(j\omega)$ will be a real number with a negative sign. Thus at this frequency the feedback will become positive. If at $\omega = \omega_{180}$ the magnitude of the loop gain is less than unity, then from Eq. (8.27) we see that the closed-loop gain $A_f(j\omega)$ will be greater than the open-loop gain $A(j\omega)$, since the denominator of Eq. (8.27) will be smaller than unity. Nevertheless, the feedback amplifier will be stable.

On the other hand, if at the frequency ω_{180} the magnitude of the loop gain is equal to unity, it follows from Eq. (8.27) that $A_f(j\omega)$ will be infinite. This means that the amplifier will have an output for zero input; this is by definition an **oscillator.** To visualize how this feedback loop may oscillate, consider the general loop of Fig. 8.1 with the external input x_s set to zero. Any disturbance in the circuit, such as the closure of the power-supply switch, will generate a signal $x_i(t)$ at the input to the amplifier. Such a noise signal usually contains a wide range of frequencies, and we shall now concentrate on the component with frequency $\omega = \omega_{180}$, that is, the signal $X_i \sin(\omega_{180} t)$. This input signal will result in a feedback signal given by

$$X_f = A(j\omega_{180})\beta(j\omega_{180})X_i = -X_i$$

Since X_f is further multiplied by -1 in the summer block at the input, we see that the feedback causes the signal X_i at the amplifier input to be *sustained*. That is, from this point on, there will be sinusoidal signals at the amplifier input and output of frequency ω_{180}. Thus the amplifier is said to oscillate at the frequency ω_{180}.

The question now is: What happens if at ω_{180} the magnitude of the loop gain is greater than unity? We shall answer this question, not in general, but for the restricted yet very important class of circuits in which we are interested here. The answer, which is not obvious from Eq. (8.27), is that the circuit will oscillate, and the oscillations will grow in amplitude until some nonlinearity (which is always present in some form) reduces the magnitude of the loop gain to exactly unity, at which point sustained oscillations will be obtained. This mechanism for starting oscillations by using positive feedback with a loop gain greater than unity, and then using a nonlinearity to reduce the loop gain to unity at the desired amplitude, will be exploited in the design of sinusoidal oscillators in Chapter 12. Our objective here is just the opposite: Now that we know how oscillations could occur in a negative-feedback amplifier, we wish to find methods to prevent their occurrence.

The Nyquist Plot

The Nyquist plot is a formalized approach for testing for stability based on the above discussion. It is simply a polar plot of loop gain with frequency used as a parameter.

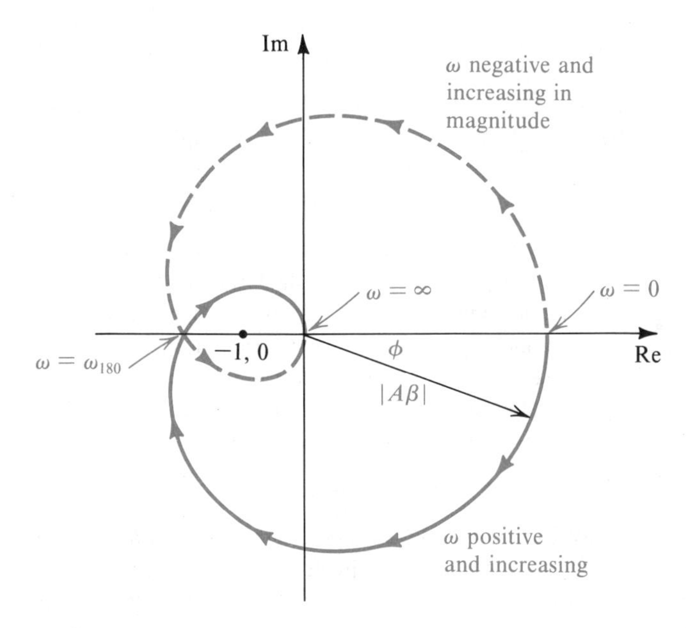

Fig. 8.28 The Nyquist plot.

Figure 8.28 shows such a plot. Note that the radial distance is $|A\beta|$ and the angle is the phase angle ϕ. The solid-line plot is for positive frequencies. Since the loop gain—and for that matter any gain function of a physical network—has a magnitude that is an even function of frequency and a phase that is an odd function of frequency, the $A\beta$ plot for negative frequencies, shown in Fig. 8.28 as a broken line, can be drawn as a mirror image through the Re axis.

The Nyquist plot intersects the negative real axis at the frequency ω_{180}. Thus if this intersection occurs to the left of the point $(-1, 0)$, we know that the magnitude of loop gain at this frequency is greater than unity and the amplifier will be unstable. On the other hand, if the intersection occurs to the right of the point $(-1, 0)$ the amplifier will be stable. It follows that if the Nyquist plot *encircles* the point $(-1, 0)$ then the amplifier will be unstable. It should be mentioned, however, that this statement is a simplified version of the **Nyquist criterion;** nevertheless, it applies to all the circuits in which we are interested. For the full theory behind the Nyquist method and for details on its application, consult Haykin (1970).

Exercise 8.10 Consider a feedback amplifier for which the open-loop transfer function $A(s)$ is given by

$$A(s) = \left(\frac{10}{1 + s/10^4}\right)^3$$

Let the feedback factor β be a constant independent of frequency. Find the frequency ω_{180} at which the phase shift is 180°. Then, show that the feedback amplifier will be stable if the feedback factor β is less than a critical value β_{cr} and unstable if $\beta \geq \beta_{cr}$, and find the value of β_{cr}.

Ans. $\omega_{180} = \sqrt{3} \times 10^4$ rad/s; $\beta_{cr} = 0.008$

8.9 EFFECT OF FEEDBACK ON THE AMPLIFIER POLES

The amplifier frequency response and stability are determined directly by its poles. We shall therefore investigate the effect of feedback on the poles of the amplifier.

Stability and Pole Location

We shall begin by considering the relation between stability and pole location. For an amplifier or any other system to be stable, its poles should lie in the left half of the s plane. A pair of complex conjugate poles on the $j\omega$ axis gives rise to sustained sinusoidal oscillations. Poles in the right half of the s plane give rise to growing oscillations.

To verify the above statement, consider an amplifier with a pole pair at $s = \sigma_0 \pm j\omega_n$. If this amplifier is subjected to a disturbance, such as that caused by closure of the power-supply switch, its transient response will contain terms of the form

$$v(t) = e^{\sigma_0 t}[e^{+j\omega_n t} + e^{-j\omega_n t}] = 2e^{\sigma_0 t}\cos(\omega_n t)$$

This is a sinusoidal signal with an envelope $e^{\sigma_0 t}$. Now if the poles are in the left half of the s plane, then σ_0 will be negative and the oscillations will decay exponentially toward zero, as shown in Fig. 8.29(a), indicating that the system is stable. If, on the other hand, the poles are in the right half-plane, then σ_0 will be positive and the oscillations will grow exponentially (until some nonlinearity limits their growth), as shown in Fig. 8.29(b). Finally, if the poles are on the $j\omega$ axis, then σ_0 will be zero and the oscillations will be sustained, as shown in Fig. 8.29(c).

Although the above discussion is in terms of complex conjugate poles, it can be shown that the existence of any right-half-plane poles results in instability.

Poles of the Feedback Amplifier

From the closed-loop transfer function in Eq. (8.26) we see that the poles of the feedback amplifier are the zeros of $1 + A(s)\beta(s)$. That is, the feedback amplifier poles are obtained by solving the equation

$$1 + A(s)\beta(s) = 0 \tag{8.29}$$

which is called the **characteristic equation** of the feedback loop. It should therefore be apparent that applying feedback to an amplifier changes its poles.

In the following we shall consider how feedback affects the amplifier poles. For this purpose we shall assume that the open-loop amplifier has real poles and no finite zeros (that is, all the zeros are at $s = \infty$). This will simplify the analysis and enable us to focus our attention on the fundamental concepts involved. We shall also assume that the feedback factor β is independent of frequency.

Amplifier With Single-Pole Response

Consider first the case of an amplifier whose open-loop transfer function is characterized by a single pole:

$$A(s) = \frac{A_0}{1 + s/\omega_P} \tag{8.30}$$

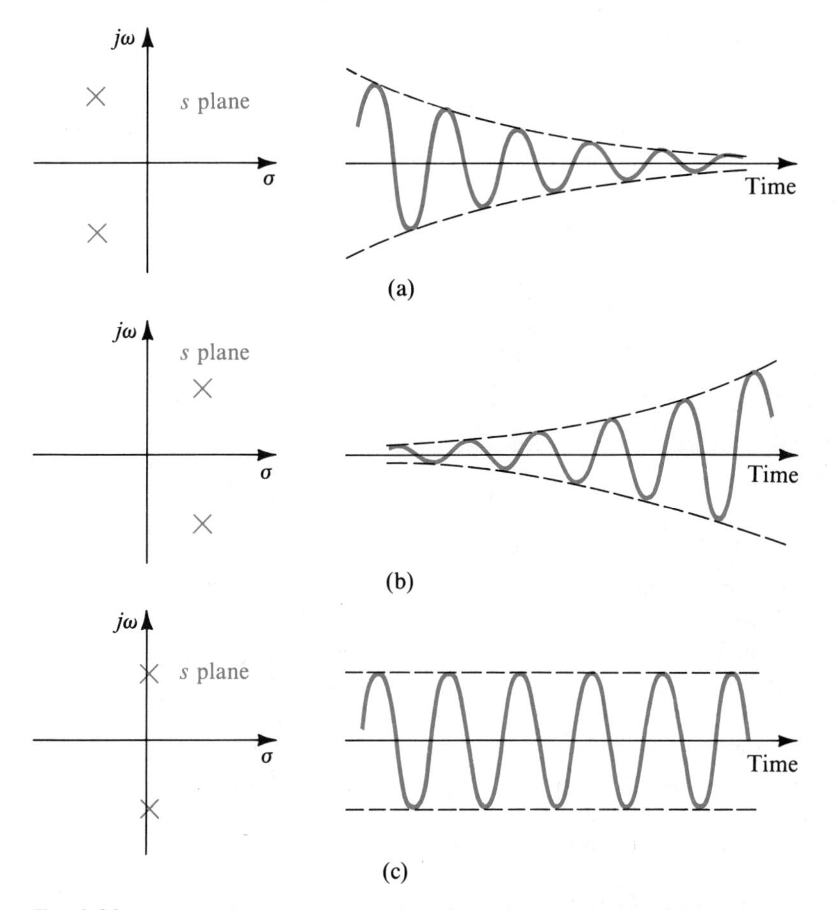

Fig. 8.29 Relationship between pole location and transient response.

The closed-loop transfer function is given by

$$A_f(s) = \frac{A_0/(1 + A_0\beta)}{1 + s/\omega_P(1 + A_0\beta)} \tag{8.31}$$

Thus the feedback moves the pole along the negative real axis to a frequency ω_{Pf},

$$\omega_{Pf} = \omega_P(1 + A_0\beta) \tag{8.32}$$

This process is illustrated in Fig. 8.30(a). Figure 8.30(b) shows Bode plots for $|A|$ and $|A_f|$. Note that while at low frequencies the difference between the two plots is $20 \log(1 + A_0\beta)$, the two curves coincide at high frequencies. One can show that this indeed is the case by approximating Eq. (8.31) for frequencies $\omega \gg \omega_P(1 + A_0\beta)$:

$$A_f(s) \simeq \frac{A_0\omega_P}{s} \simeq A(s) \tag{8.33}$$

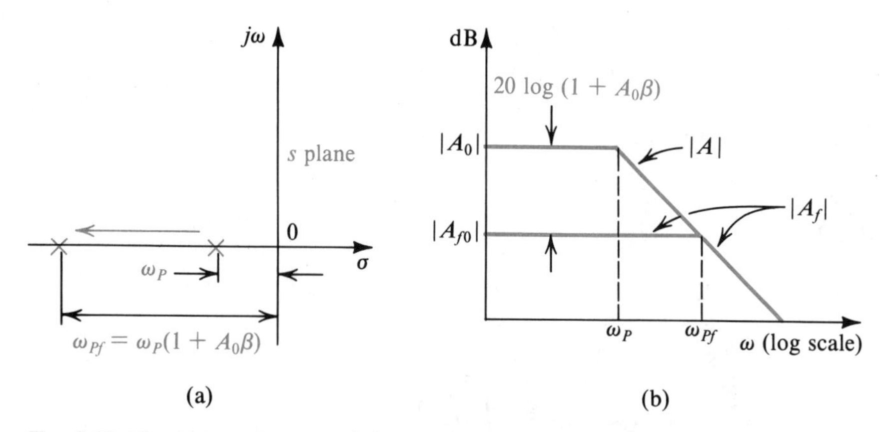

Fig. 8.30 Effect of feedback on **(a)** the pole location, and **(b)** the frequency response of an amplifier having a single-pole open-loop response.

Physically speaking, at such high frequencies the loop gain is much smaller than unity and the feedback is ineffective.

Figure 8.30(b) clearly illustrates the fact that applying negative feedback to an amplifier results in extending its bandwidth at the expense of a reduction in gain. Since the pole of the closed-loop amplifier never enters the right half of the s plane, the single-pole amplifier is stable for any value of β. Thus this amplifier is said to be **unconditionally stable.** This result, however, is hardly surprising, since the phase lag associated with a single-pole response can never be greater than 90°. Thus the loop gain never achieves the 180° phase shift required for the feedback to become positive.

Exercise 8.11 An op amp having a single-pole rolloff at 100 Hz and a low-frequency gain of 10^5 is operated in a feedback loop with $\beta = 0.01$. What is the factor by which feedback shifts the pole? To what frequency? If β is changed to a value that results in a closed-loop gain of +1, to what frequency does the pole shift?

Ans. 1001; 100.1 kHz; 10 MHz

Amplifier with Two-Pole Response

Consider next an amplifier whose open-loop transfer function is characterized by two real-axis poles:

$$A(s) = \frac{A_0}{(1 + s/\omega_{P1})(1 + s/\omega_{P2})} \tag{8.34}$$

In this case the closed-loop poles are obtained from $1 + A(s)\beta = 0$, which leads to

$$s^2 + s(\omega_{P1} + \omega_{P2}) + (1 + A_0\beta)\omega_{P1}\omega_{P2} = 0 \tag{8.35}$$

Thus the closed-loop poles are given by

$$s = -\tfrac{1}{2}(\omega_{P1} + \omega_{P2}) \pm \tfrac{1}{2}\sqrt{(\omega_{P1} + \omega_{P2})^2 - 4(1 + A_0\beta)\omega_{P1}\omega_{P2}} \tag{8.36}$$

From this equation we see that as the loop gain $A_0\beta$ is increased from zero, the poles are brought closer together. Then a value of loop gain is reached at which the poles become coincident. If the loop gain is further increased, the poles become complex conjugate and move along a vertical line. Figure 8.31 shows the locus of the poles for increasing loop gain. This plot is called a **root-locus diagram,** where "root" refers to the fact that the poles are the roots of the characteristic equation.

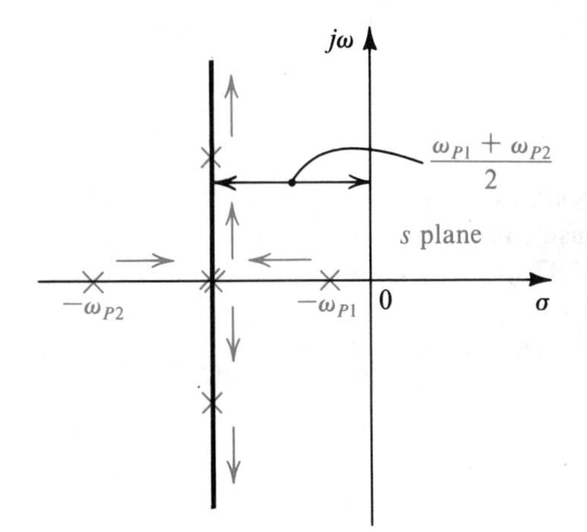

Fig. 8.31 Root-locus diagram for a feedback amplifier whose open-loop transfer function has two real poles.

From the root–locus diagram of Fig. 8.31 we see that this feedback amplifier also is unconditionally stable. Again, this result should come as no surprise; the maximum phase shift of $A(s)$ in this case is 180° (90° per pole), but this value is reached at $\omega = \infty$. Thus there is no finite frequency at which the phase shift reaches 180°.

Another observation to make on the root–locus diagram of Fig. 8.31 is that the open-loop amplifier might have a dominant pole, but this is not necessarily the case for the closed-loop amplifier. The response of the closed-loop amplifier can, of course, always be plotted once the poles are found from Eq. (8.36). As is the case with second-order responses generally, the closed-loop response can show a peak (see Chapter 11). To be more specific, the characteristic equation of a second-order network can be written in the standard form

$$s^2 + s\frac{\omega_0}{Q} + \omega_0^2 = 0 \tag{8.37}$$

where ω_0 is called the **pole frequency** and Q is called **pole Q factor.** The poles are complex if Q is greater than 0.5. A geometric interpretation for ω_0 and Q of a pair of complex conjugate poles is given in Fig. 8.32, from which we note that ω_0 is the radial distance of the poles and that Q indicates the distance of the poles from the $j\omega$ axis. Poles on the $j\omega$ axis have $Q = \infty$.

By comparing Eqs. (8.35) and (8.37) we obtain the Q factor for the poles of the feedback amplifier as

$$Q = \frac{\sqrt{(1 + A_0\beta)\omega_{P1}\omega_{P2}}}{(\omega_{P1} + \omega_{P2})} \tag{8.38}$$

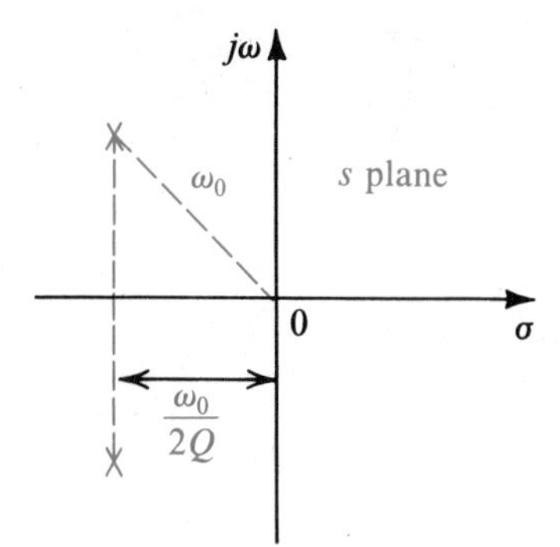

Fig. 8.32 Definition of ω_0 and Q of a pair of complex conjugate poles.

From the study of second-order network responses in Chapter 11 it will be seen that the response of the feedback amplifier under consideration shows no peaking for $Q \leq 0.707$. The boundary case corresponding to $Q = 0.707$ (poles at 45° angles) results in the **maximally flat** response. Figure 8.33 shows a number of possible responses obtained for various values of Q (or, correspondingly, various values of $A_0\beta$).

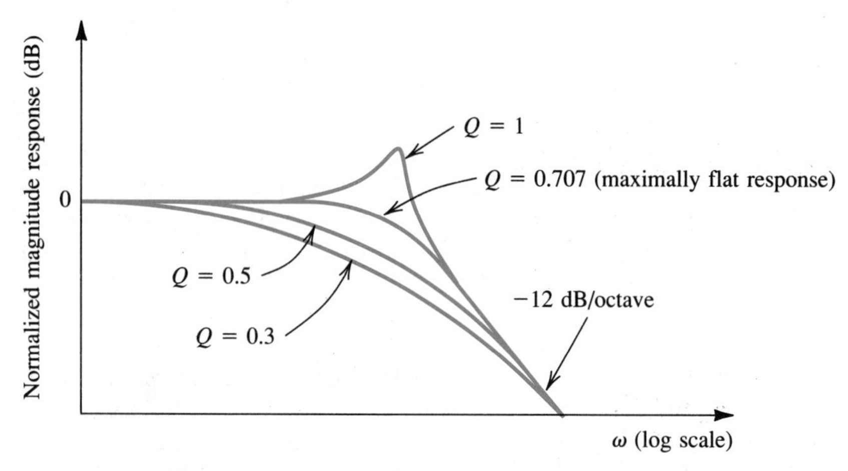

Fig. 8.33 Normalized magnitude response of a two-pole feedback amplifier for various values of Q. Note that Q is determined by the loop gain according to Eq. (8.38).

Exercise 8.12 An amplifier with a low-frequency gain of 100 and poles at 10^4 and 10^6 rad/s is incorporated in a negative-feedback loop with feedback factor β. For what value of β do the poles of the closed-loop amplifier coincide? What is the corresponding Q of the resulting second-order system? For what value of β is a maximally flat response achieved? What is the low-frequency closed-loop gain in the maximally flat case?

Ans. 0.245; 0.5; 0.5; 1.96 V/V

EXAMPLE 8.5

As an illustration of some of the ideas discussed above we consider the positive-feedback circuit shown in Fig. 8.34(a). Find the loop transmission $L(s)$ and the characteristic

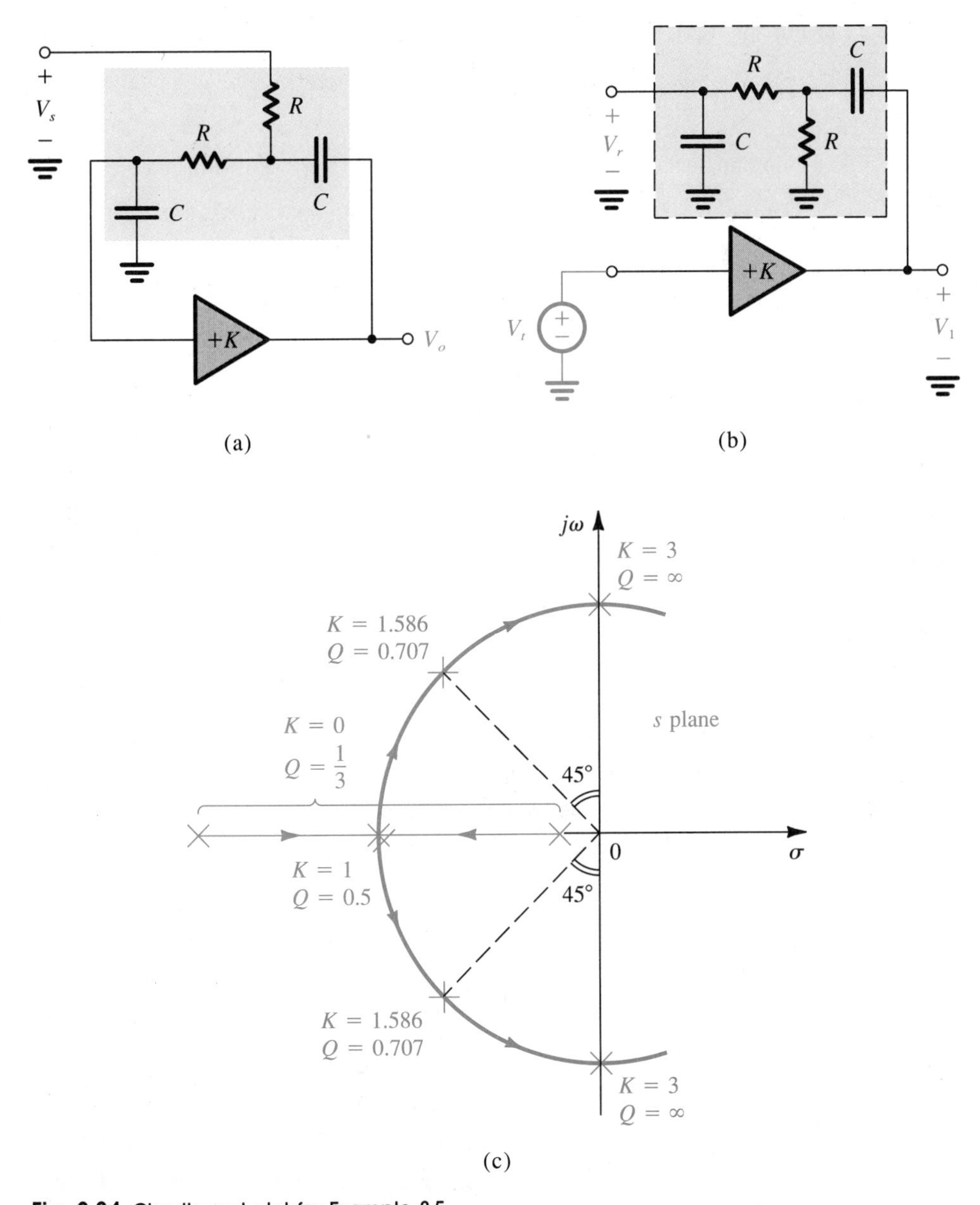

Fig. 8.34 Circuits and plot for Example 8.5.

equation. Sketch a root-locus diagram for varying K, and find the value of K that results in a maximally flat response, and the value of K that makes the circuit oscillate. Assume that the amplifier has infinite input impedance and zero output impedance.

Solution To obtain the loop transmission we short-circuit the signal source and break the loop at the amplifier input. We then apply a test voltage V_t and find the returned voltage V_r, as indicated in Fig. 8.34(b). The loop transmission $L(s) \equiv A(s)\beta(s)$ is given by

$$L(s) = -\frac{V_r}{V_t} = -KT(s) \tag{8.39}$$

where $T(s)$ is the transfer function of the two-port RC network shown inside the broken-line box in Fig. 8.34(b):

$$T(s) \equiv \frac{V_r}{V_1} = \frac{s(1/CR)}{s^2 + s(3/CR) + (1/CR)^2} \tag{8.40}$$

Thus

$$L(s) = \frac{-s(K/CR)}{s^2 + s(3/CR) + (1/CR)^2} \tag{8.41}$$

The characteristic equation is

$$1 + L(s) = 0 \tag{8.42}$$

that is,

$$s^2 + s\frac{3}{CR} + \left(\frac{1}{CR}\right)^2 - s\frac{K}{CR} = 0$$

$$s^2 + s\frac{3-K}{CR} + \left(\frac{1}{CR}\right)^2 = 0 \tag{8.43}$$

By comparing this equation to the standard form of the second-order characteristic equation (Eq. 8.37) we see that the pole frequency ω_0 is given by

$$\omega_0 = \frac{1}{CR} \tag{8.44}$$

and the Q factor is

$$Q = \frac{1}{3-K} \tag{8.45}$$

Thus for $K = 0$ the poles have $Q = \frac{1}{3}$ and are therefore located on the negative real axis. As K is increased the poles are brought closer together and eventually coincide ($Q = 0.5$, $K = 1$). Further increasing K results in the poles becoming complex and conjugate. The root locus is then a circle because the radial distance ω_0 remains constant (Eq. 8.44) independent of the value of K.

The maximally flat response is obtained when $Q = 0.707$, which results when $K = 1.586$. In this case the poles are at 45° angles, as indicated in Fig. 8.34(c). The poles cross the $j\omega$ axis into the right half of the s plane at the value of K that results in $Q = \infty$, that is, $K = 3$. Thus for $K \geq 3$ this circuit becomes unstable. This might appear to contradict our earlier conclusion that the feedback amplifier with a second-order response is unconditionally stable. Note, however, that the circuit in this example is quite different from the negative-feedback amplifier that we have been studying. Here we have an amplifier with a positive gain K and a feedback network whose transfer function $T(s)$ is frequency dependent. This feedback is in fact *positive*, and the circuit will oscillate at the frequency for which the phase of $T(j\omega)$ is zero.

Example 8.5 illustrates the use of feedback (positive feedback in this case) to move the poles of an RC network from their negative real-axis locations to complex conjugate locations. One can accomplish the same task using negative feedback, as the root-locus diagram of Fig. 8.31 demonstrates. The process of pole control is the essence of *active-filter design*, as will be discussed in Chapter 11.

Amplifiers with Three or More Poles

Figure 8.35 shows the root-locus diagram for a feedback amplifier whose open-loop response is characterized by three poles. As indicated, increasing the loop gain from zero moves the highest-frequency pole outward while the two other poles are brought closer together. As $A_0\beta$ is increased further, the two poles become coincident and then become complex and conjugate. A value of $A_0\beta$ exists at which this pair of complex-conjugate poles enters the right half of the s plane, thus causing the amplifier to become unstable.

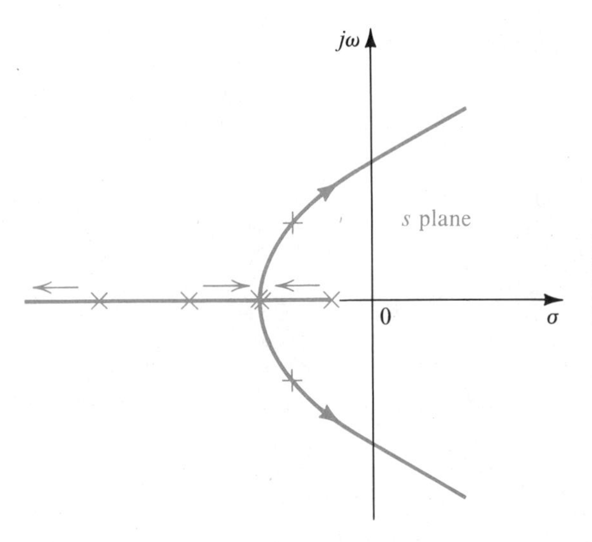

Fig. 8.35 Root-locus diagram for an amplifier with three poles. The arrows indicate the pole movement as $A_0\beta$ is increased.

This result is not entirely unexpected, since an amplifier with three poles has a phase shift that reaches $-270°$ as ω approaches ∞. Thus there exists a finite frequency ω_{180}, at which the loop gain has 180° phase shift.

From the root-locus diagram of Fig. 8.35 we observe that one can always maintain amplifier stability by keeping the loop gain $A_0\beta$ smaller than the value corresponding to the poles entering the right half-plane. In terms of the Nyquist diagram, the critical value of $A_0\beta$ is that for which the diagram passes through the $(-1, 0)$ point. Reducing $A_0\beta$ below this value causes the Nyquist plot to shrink and thus intersect the negative real axis to the right of the $(-1, 0)$ point, indicating stable amplifier performance. On the other hand, increasing $A_0\beta$ above the critical value causes the Nyquist plot to expand, thus encircling the $(-1, 0)$ point and indicating unstable performance.

For a given open-loop gain the above conclusions can be stated in terms of the feedback factor β. That is, there exists a *maximum value* for β above which the feedback amplifier becomes unstable. Alternatively, we can state that there exists a *minimum value* for the closed-loop gain A_{f0} below which the amplifier becomes unstable. To obtain lower values of closed-loop gain one needs therefore to alter the loop transfer function $L(s)$. This is the process known as *frequency compensation*. We shall study the theory and techniques of frequency compensation in Section 8.11.

Before leaving this section we should point out that construction of the root locus diagram for amplifiers having three or more poles as well as finite zeros is an involved process for which a systematic procedure exists. However, such a procedure will not be presented here, and the interested reader should consult Haykin (1970). Although the root-locus diagram provides the amplifier designer with considerable insight, other, simpler techniques based on Bode plots can be effectively employed, as will be explained in Section 8.10.

Exercise 8.13 Consider a feedback amplifier for which the open-loop transfer function $A(s)$ is given by

$$A(s) = \left(\frac{10}{1 + s/10^4}\right)^3$$

Let the feedback factor β be frequency-independent. Find the closed-loop poles as functions of β, and show that the root locus is that of Fig. E8.13. Also find the value of β at which the amplifier becomes unstable. (*Note:* This is the same amplifier that was considered in Exercise 8.10.)

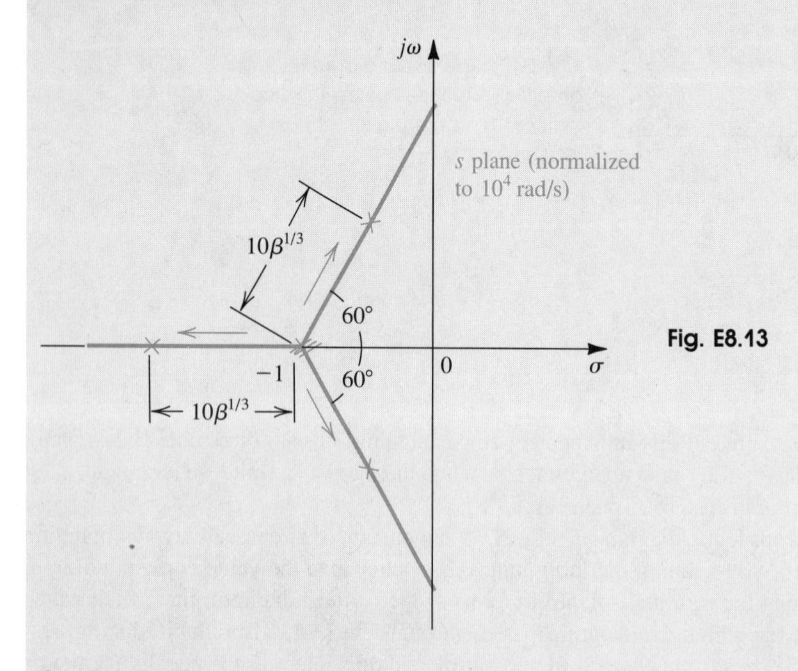

Fig. E8.13

Ans. See Fig. E8.13; $\beta_{\text{critical}} = 0.008$

8.10 STABILITY STUDY USING BODE PLOTS

Gain and Phase Margins

From Sections 8.8 and 8.9 we know that one can determine whether the feedback amplifier is stable or not by examining the loop gain $A\beta$ as a function of frequency. One of the simplest and most effective means for doing this is through the use of a Bode plot for $A\beta$, such as the one shown in Fig. 8.36. (Note that because the phase approaches $-360°$, the network examined is a fourth-order one.) The feedback amplifier whose loop gain is plotted in Fig. 8.36 will be stable, since at the frequency of 180° phase shift, ω_{180}, the magnitude of loop gain is less than unity (negative dB). The difference between the value of $|A\beta|$ at ω_{180} and unity, called the **gain margin,** is usually expressed in dB. The gain margin represents the amount by which the loop gain can be increased while stability is maintained. Feedback amplifiers are usually designed to have sufficient gain margin to allow for the inevitable changes in loop gain with temperature, time, and so on.

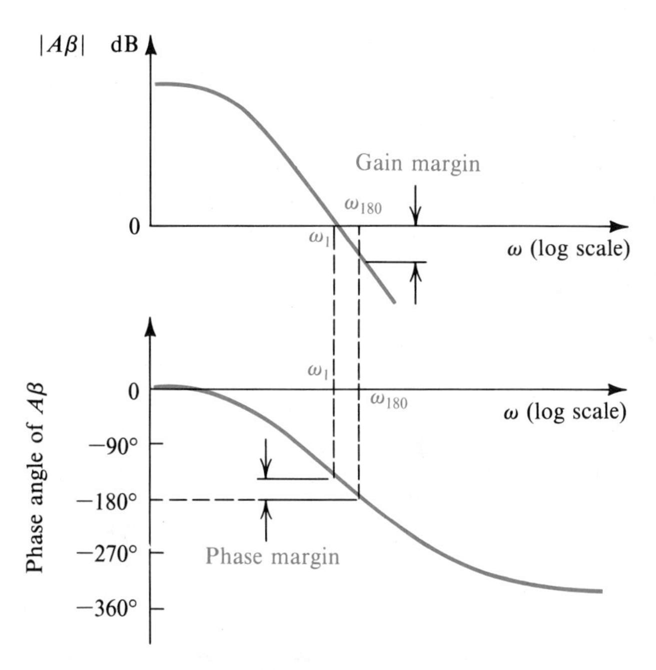

Fig. 8.36 Bode plot for the loop gain $A\beta$ illustrating the definitions of the gain and phase margins.

Another way to investigate the stability and to express its degree is to examine the Bode plot at the frequency for which $|A\beta| = 1$, which is the point at which the magnitude plot crosses the 0-dB line. If at this frequency the phase angle is less (in magnitude) than 180°, then the amplifier is stable. This is the situation illustrated in Fig. 8.36. The difference between the phase angle at this frequency and 180° is termed the **phase margin.** On the other hand, if at the frequency of unity loop-gain magnitude, the phase lag is in excess of 180°, the amplifier will be unstable.

Exercise 8.14 Consider an op amp having a single-pole open-loop response with $A_0 = 10^5$ and $f_P = 10$ Hz. Let the op amp be ideal otherwise (infinite input impedance, zero output impedance, and so on). If this amplifier is connected in the noninverting configuration with a nominal low-frequency closed-loop gain of 100, find the frequency at which $|A\beta| = 1$. Also, find the phase margin.

Ans. 10^4 Hz; 90°

Effect of Phase Margin on Closed-Loop Response

Feedback amplifiers are normally designed with a phase margin of at least 45°. The amount of phase margin has a profound effect on the shape of the closed-loop magnitude response. To see this relationship, consider a feedback amplifier with a large low-frequency loop gain, $A_0\beta \gg 1$. It follows that the closed-loop gain at low frequencies is approximately $1/\beta$. Denoting the frequency at which the magnitude of loop gain is unity by ω_1 we have (refer to Fig. 8.36)

$$A(j\omega_1)\beta = 1 \times e^{-j\theta} \tag{8.46}$$

where

$$\theta = 180° - \text{phase margin} \tag{8.47}$$

At ω_1 the closed-loop gain is

$$A_f(j\omega_1) = \frac{A(j\omega_1)}{1 + A(j\omega_1)\beta}$$

Substituting from Eq. (8.46) gives

$$A_f(j\omega_1) = \frac{(1/\beta)e^{-j\theta}}{1 + e^{-j\theta}}$$

Thus the magnitude of the gain at ω_1 is

$$|A_f(j\omega_1)| = \frac{1/\beta}{|1 + e^{-j\theta}|} \tag{8.48}$$

For a phase margin of 45°, $\theta = 135°$; and we obtain

$$|A_f(j\omega_1)| = 1.3\,\frac{1}{\beta}$$

That is, the gain peaks by a factor of 1.3 above the low-frequency value of $1/\beta$. This peaking increases as the phase margin is reduced, eventually reaching ∞ when the phase margin is zero. Zero phase margin, of course, implies that the amplifier can sustain oscillations [poles on the $j\omega$ axis; Nyquist plot passes through $(-1, 0)$].

Exercise 8.15 Find the closed-loop gain at ω_1 relative to the low-frequency gain when the phase margin is 30°, 60°, and 90°.

Ans. 1.93; 1; 0.707

An Alternative Approach

Investigating stability by constructing Bode plots for the loop gain $A\beta$ can be a tedious and time-consuming process, especially if we have to investigate the stability of a given amplifier for a variety of feedback networks. An alternative approach, which is much simpler, is to construct a Bode plot for the open-loop gain $A(j\omega)$ only. Assuming for the time being that β is independent of frequency, we can plot $20 \log(1/\beta)$ as a horizontal straight line on the same plane used for $20 \log|A|$. The difference between the two curves will be

$$20 \log|A(j\omega)| - 20 \log \frac{1}{\beta} = 20 \log|A\beta| \tag{8.49}$$

which is the loop gain expressed in dB. We may therefore study stability by examining the difference between the two plots. If we wish to evaluate stability for a different feedback factor we simply draw another horizontal straight line at the level $20 \log(1/\beta)$.

To illustrate, consider an amplifier whose open-loop transfer function is characterized by three poles. For simplicity let the three poles be widely separated—say, at 0.1, 1, and 10 MHz, as shown in Fig. 8.37. Note that because the poles are widely separated, the phase is approximately $-45°$ at the first pole frequency, $-135°$ at the second, and $-225°$ at the third. The frequency at which the phase of $A(j\omega)$ is $-180°$ lies on the -40-dB/decade segment, as indicated in Fig. 8.37.

The open-loop gain of this amplifier can be expressed as

$$A = \frac{10^5}{(1 + jf/10^5)(1 + jf/10^6)(1 + jf/10^7)} \tag{8.50}$$

from which $|A|$ can be easily determined for any frequency f (in Hz), and the phase can be obtained as

$$\phi = -[\tan^{-1}(f/10^5) + \tan^{-1}(f/10^6) + \tan^{-1}(f/10^7)] \tag{8.51}$$

The magnitude and phase graphs shown in Fig. 8.37 are obtained using the method for constructing Bode plots (Section 7.1). These graphs provide approximate values for important amplifier parameters, with more exact values obtainable from Eqs. (8.50) and (8.51). For example the frequency f_{180} at which the phase angle is 180° can be found from Fig. 8.37 to be approximately 3.2×10^6 Hz. Using this value as a starting point, a more exact value can be found by trial and error using Eq. (8.51). The result is $f_{180} = 3.34 \times 10^6$ Hz. At this frequency, Eq. (8.50) gives a gain magnitude of 58.2 dB, which is reasonably close to the approximate value of 60 dB given by Fig. 8.37.

Consider next the straight line labeled (a) in Fig. 8.37. This line represents a feedback factor for which $20 \log(1/\beta) = 85$ dB, which corresponds to $\beta = 5.623 \times 10^{-5}$ and a closed-loop gain of 83.6 dB. Since the loop gain is the difference between the $|A|$ curve and the $1/\beta$ line, the point of intersection X_1 corresponds to the frequency at which $|A\beta| = 1$. Using the graphs of Fig. 8.37, this frequency can be found to be approximately 5.6×10^5 Hz. A more exact value of 4.936×10^5 can be obtained using the transfer-function equations. At this frequency the phase angle is approximately $-108°$. Thus the closed-loop amplifier, for which $20 \log(1/\beta) = 85$ dB, will be stable with a phase margin of 72°. The gain margin can be easily obtained from Fig. 8.37; it is 25 dB.

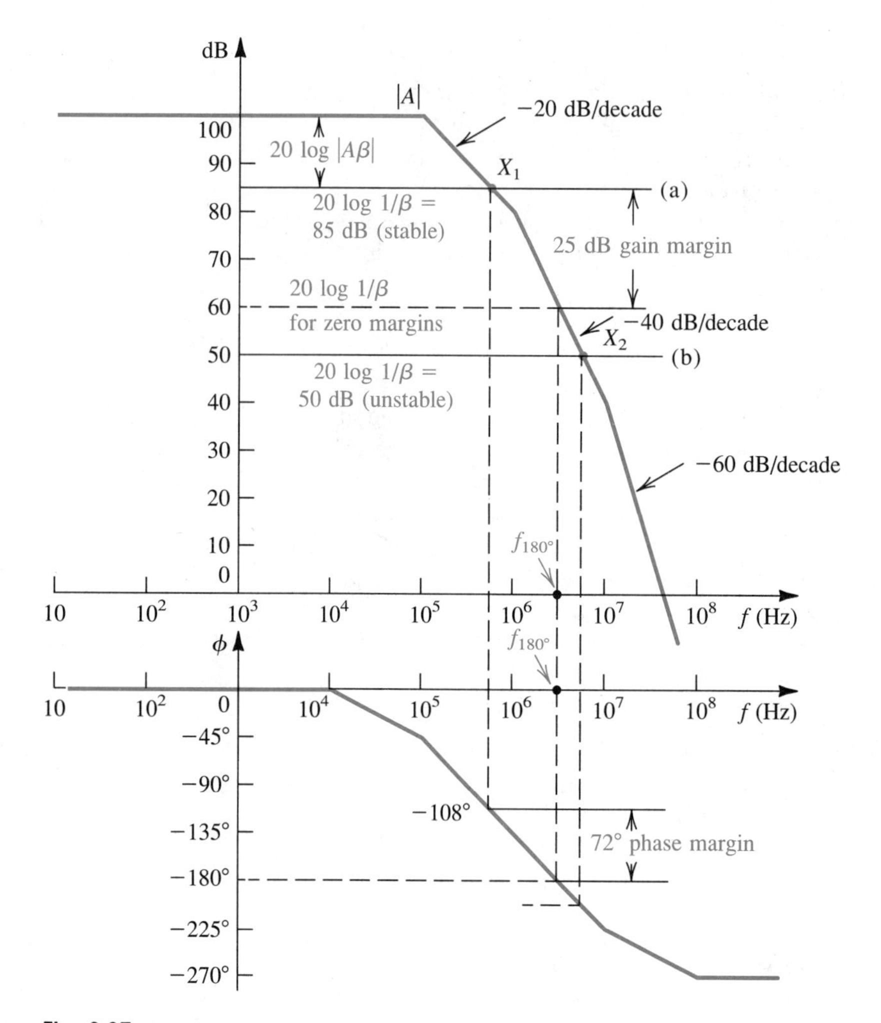

Fig. 8.37 Stability analysis.

Next, suppose that we wish to use this amplifier to obtain a closed-loop gain of 50-dB nominal value. Since $A_0 = 100$ dB, we see that $A_0\beta \gg 1$ and $20 \log(A_0\beta) \simeq 50$ dB, resulting in $20 \log(1/\beta) \simeq 50$ dB. To see whether this closed-loop amplifier is stable or not, we draw line (b) in Fig. 8.37 with a height of 50 dB. This line intersects the open-loop gain curve at point X_2, where the corresponding phase is greater than 180°. Thus the closed-loop amplifier with 50-dB gain will be unstable.

In fact, it can easily be seen from Fig. 8.37 that the *minimum* value of $20 \log(1/\beta)$ that can be used, with the resulting amplifier being stable, is 60 dB. In other words, the minimum value of stable closed-loop gain obtained with this amplifier is approximately

60 dB. At this value of gain, however, the amplifier may still oscillate, since no margin is left to allow for possible changes in gain.

Since the 180°-phase point always occurs on the −40-dB/decade segment of the Bode plot for $|A|$, a rule of thumb to guarantee stability is as follows: The closed-loop amplifier will be stable if the $20 \log(1/\beta)$ line intersects the $20 \log|A|$ curve at a point on the −20-dB/decade segment. Following this rule ensures that a phase margin of at least 45° is obtained. For the example of Fig. 8.37, the rule implies that the maximum value of β is 10^{-4}, which corresponds to a closed-loop gain of approximately 80 dB.

The above rule of thumb can be generalized for the case in which β is a function of frequency. The general rule states that at the intersection of $20 \log[1/|\beta(j\omega)|]$ and $20 \log|A(j\omega)|$ the difference of slopes (called the **rate of closure**) should not exceed 20 dB/decade.

Exercise 8.16 Consider an op amp whose open-loop gain is identical to that of Fig. 8.37. Assume that the op amp is ideal otherwise. Let the op amp be connected as a differentiator. Use the above rule of thumb to show that for stable performance the differentiator time constant should be greater than 159 ms. (*Hint:* Recall that for a differentiator, the Bode plot for $1/|\beta(j\omega)|$ has a slope of +20 dB/decade and intersects the 0-dB line at $1/\tau$, where τ is the differentiator time constant.)

8.11 FREQUENCY COMPENSATION

In this section we shall discuss methods for modifying the open-loop transfer function $A(s)$ of an amplifier having three or more poles so that the closed-loop amplifier is stable for any desired value of closed-loop gain.

Theory

The simplest method of frequency compensation consists of introducing a new pole in the function $A(s)$ at a sufficiently low frequency, f_D, such that the modified open-loop gain, $A'(s)$, intersects the $20 \log(1/|\beta|)$ curve with a slope difference of 20 dB/decade. As an example, let it be required to compensate the amplifier whose $A(s)$ is shown in Fig. 8.38 such that closed-loop amplifiers with β as high as 10^{-2} (that is, closed-loop gains as low as approximately 40 dB) will be stable. First, we draw a horizontal straight line at the 40-dB level to represent $20 \log(1/\beta)$, as shown in Fig. 8.38. We then locate point Y on this line at the frequency of the first pole, f_{P1}. From Y we draw a line with −20-dB/decade slope and determine the point at which this line intersects the dc gain line, point Y'. This latter point gives the frequency f_D of the new pole that has to be introduced in the open-loop transfer function.

The compensated open-loop response $A'(s)$ is indicated in Fig. 8.38. It has four poles: at f_D, f_{P1}, f_{P2}, and f_{P3}. Thus $|A'|$ begins to roll off with a slope of −20 dB/decade at f_D. At f_{P1} the slope changes to −40 dB/decade, at f_{P2} it changes to −60 dB/decade, and so on. Since the $20 \log/(1/\beta)$ line intersects the $20 \log|A'|$ curve at point Y on the −20-dB/decade segment, the closed-loop amplifier with this β value (or lower values) will be stable.

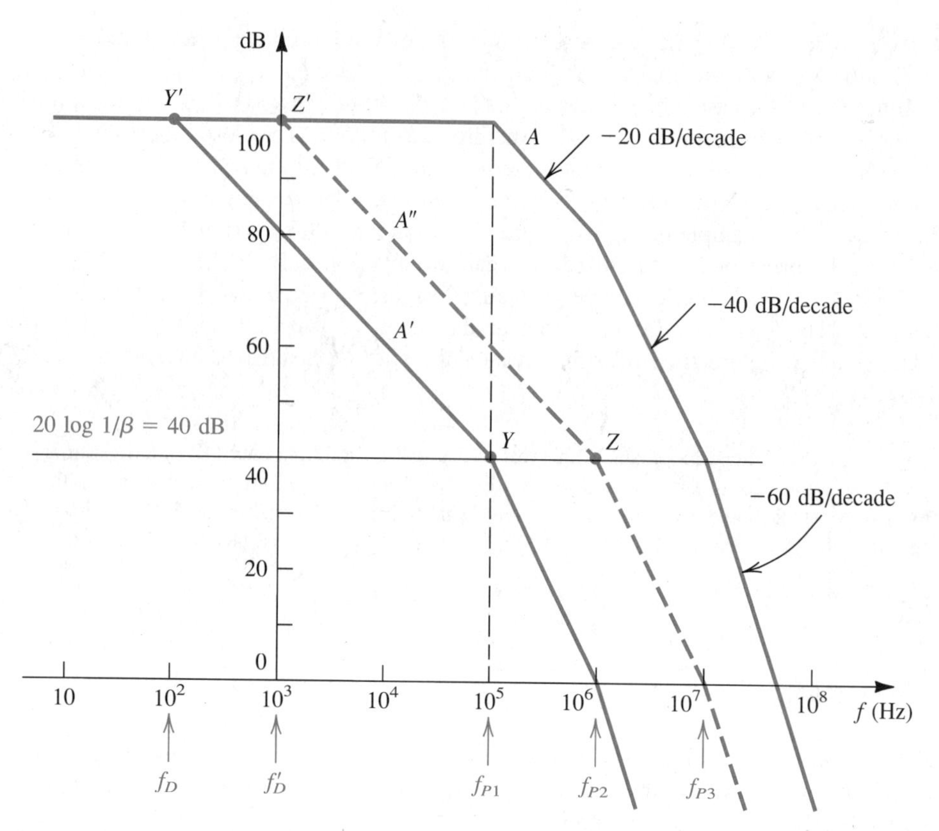

Fig. 8.38 Frequency compensation for $\beta = 10^{-2}$. The response labeled A' is obtained by introducing an additional pole at f_D. The A'' response is obtained by moving the low-frequency pole to f'_D.

A serious disadvantage of this compensation method is that at most frequencies the open-loop gain has been drastically reduced. This means that at most frequencies the amount of feedback available will be small. Since all the advantages of negative feedback are directly proportional to the amount of feedback, the performance of the compensated amplifier has been impaired.

Careful examination of Fig. 8.38 shows that the reason the gain $A'(s)$ is low is the pole at f_{P1}. If we can somehow eliminate this pole, then—rather than locating point Y, drawing YY', and so on—we can now start from point Z (at the frequency of the second pole) and draw the line ZZ'. This would result in the open-loop curve $A''(s)$, which shows considerably higher gain than $A'(s)$.

Although it is not possible to eliminate the pole at f_{P1}, it is usually possible to shift that pole from $f = f_{P1}$ to $f = f'_D$. This makes the pole dominant and eliminates the need for introducing an additional lower-frequency pole, as will be explained next.

Implementation

We shall now address the question of implementing the frequency-compensation scheme discussed above. The amplifier circuit normally consists of a number of cascaded gain stages, with each stage responsible for one or more of the transfer-function poles. Through manual and/or computer analysis of the circuit, one identifies which stage introduces each of the important poles f_{P1}, f_{P2}, and so on. For the purpose of our discussion, assume that the first pole f_{P1} is introduced at the interface between the two cascaded differential stages shown in Fig. 8.39(a). In Fig. 8.39(b) we show a simple small-signal model of the circuit at this interface. Current source I_x represents the output signal current of the $Q_1 - Q_2$ stage. Resistance R_x and capacitance C_x represent the total resistance and capacitance between the two nodes B and B′. It follows that the pole f_{P1} is given by

$$f_{P1} = \frac{1}{2\pi C_x R_x} \tag{8.52}$$

Let us now connect the compensating capacitor C_C between nodes B and B′. This will result in the modified equivalent circuit shown in Fig. 8.39(c) from which we see that the pole introduced will no longer be at f_{P1}; rather, the pole can be at any desired lower frequency f'_D:

$$f'_D = \frac{1}{2\pi(C_x + C_C)R_x} \tag{8.53}$$

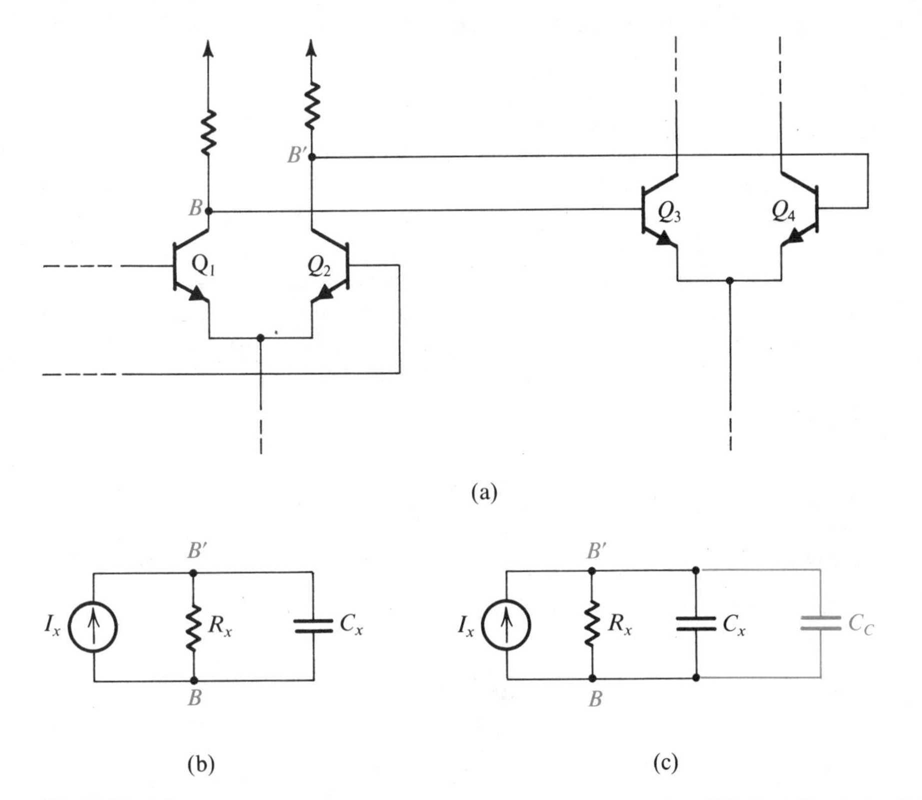

Fig. 8.39 (a) Two cascaded gain stages of a multistage amplifier. **(b)** Equivalent circuit for the interface between the two stages in (a). **(c)** Same circuit as in (b) but with a compensating capacitor C_C added.

We thus conclude that one can select an appropriate value for C_C so as to shift the pole frequency from f_{P1} to the value f'_D determined by point Z' in Fig. 8.38.

At this juncture it should be pointed out that adding the capacitor C_C will usually result in changes in the location of the other poles (those at f_{P2} and f_{P3}). One might therefore need to calculate the new location of f_{P2} and perform a few iterations to arrive at the required value for C_C.

A disadvantage of this implementation method is that the required value of C_C is usually quite large. Thus if the amplifier to be compensated is an IC op amp, it will be difficult, and probably impossible, to include this compensating capacitor on the IC chip. (As pointed out in Chapter 10 and in Appendix A, the maximum practical size of a monolithic capacitor is about 100 pF.) An elegant solution to this problem is to connect the compensating capacitor in the feedback path of an amplifier stage. Because of the Miller effect, the compensating capacitance will be multiplied by the stage gain, resulting in a much larger effective capacitance. Furthermore, as explained below, another unexpected benefit accrues.

Miller Compensation and Pole Splitting

Figure 8.40(a) shows one gain stage in a multistage amplifier. For simplicity, the stage is shown as a common-emitter amplifier, but in practice it can be a more elaborate circuit. In the feedback path of this common-emitter stage we have placed a compensating capacitor C_f.

Figure 8.40(b) shows a simplified equivalent circuit of the gain stage of Fig. 8.40(a). Here R_1 and C_1 represent the total resistance and total capacitance between node B and ground. Similarly, R_2 and C_2 represent the total resistance and total capacitance between node C and ground. Furthermore, it is assumed that C_1 and C_2 include the Miller components due to capacitance C_μ, and C_2 includes the input capacitance of the succeeding amplifier stage. Finally, I_i represents the output signal current of the preceding stage.

In the absence of the compensating capacitor C_f, we can see from Fig. 8.40(b) that there are two poles—one at the input and one at the output. Let us assume that these two

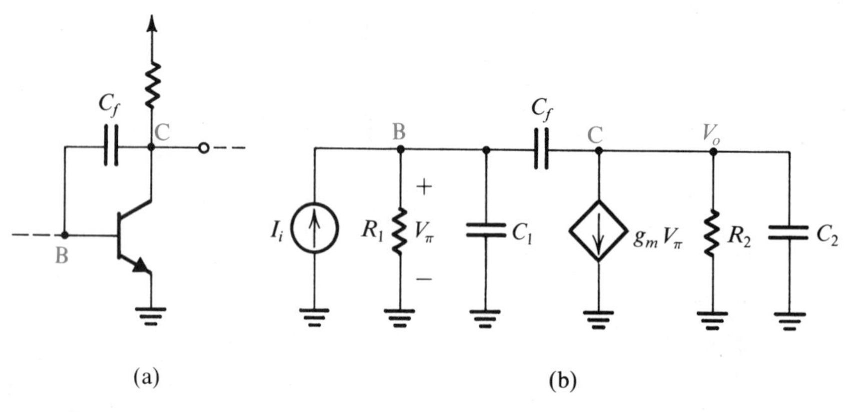

Fig. 8.40 (a) A gain stage in a multistage amplifier with a compensating capacitor connected in the feedback loop, and **(b)** equivalent circuit.

poles are f_{P1} and f_{P2} of Fig. 8.38; thus

$$f_{P1} = \frac{1}{2\pi C_1 R_1} \qquad f_{P2} = \frac{1}{2\pi C_2 R_2} \tag{8.54}$$

With C_f present, analysis of the circuit yields the transfer function

$$\frac{V_o}{I_i} = \frac{(sC_f - g_m)R_1R_2}{1 + s[C_1R_1 + C_2R_2 + C_f(g_mR_1R_2 + R_1 + R_2)] + s^2[C_1C_2 + C_f(C_1 + C_2)]R_1R_2} \tag{8.55}$$

The zero is usually at a much higher frequency than the dominant pole, and we shall neglect its effect. The denominator polynomial $D(s)$ can be written in the form

$$D(s) = \left(1 + \frac{s}{\omega'_{P1}}\right)\left(1 + \frac{s}{\omega'_{P2}}\right) = 1 + s\left(\frac{1}{\omega'_{P1}} + \frac{1}{\omega'_{P2}}\right) + \frac{s^2}{\omega'_{P1}\omega'_{P2}} \tag{8.56}$$

where ω'_{P1} and ω'_{P2} are the new frequencies of the two poles. Normally one of the poles will be dominant; $\omega'_{P1} \ll \omega'_{P2}$. Thus

$$D(s) \simeq 1 + \frac{s}{\omega'_{P1}} + \frac{s^2}{\omega'_{P1}\omega'_{P2}} \tag{8.57}$$

Equating the coefficients of s in the denominator of Eq. (8.55) and in Eq. (8.57) results in

$$\omega'_{P1} = \frac{1}{C_1R_1 + C_2R_2 + C_f(g_mR_1R_2 + R_1 + R_2)}$$

which can be approximated by

$$\omega'_{P1} \simeq \frac{1}{g_mR_2C_fR_1} \tag{8.58}$$

To obtain ω'_{P2} we equate the coefficients of s^2 in the denominator of Eq. (8.55) and in Eq. (8.57) and use Eq. (8.58):

$$\omega'_{P2} \simeq \frac{g_mC_f}{C_1C_2 + C_f(C_1 + C_2)} \tag{8.59}$$

From Eqs. (8.58) and (8.59) we see that as C_f is increased, ω'_{P1} is reduced and ω'_{P2} is increased. This is referred to as **pole splitting.** Note that the increase in ω'_{P2} is highly beneficial; it allows us to move point Z (see Fig. 8.38) further to the right, thus resulting in higher compensated open-loop gain. Finally, note from Eq. (8.58) that C_f is multiplied by the Miller-effect factor g_mR_2, thus resulting in a much larger capacitance, $g_mR_2C_f$. In other words, the required value of C_f will be much smaller than that of C_C in Fig. 8.39.

EXAMPLE 8.6

Consider an op amp whose open-loop transfer function is identical to that shown in Fig. 8.37. We wish to compensate this op amp so that the closed-loop amplifier with resistive feedback is stable for any gain (that is, for β up to unity). Assume that the op amp circuit includes a stage such as that of Fig. 8.40 with C_1 = 100 pF, C_2 = 5 pF, and g_m =

40 mA/V, that the pole at f_{P1} is caused by the input circuit of that stage, and that the pole at f_{P2} is introduced by the output circuit. Find the value of the compensating capacitor if it is connected either between the input node B and ground or in the feedback path of the transistor.

Solution First we determine R_1 and R_2 from

$$f_{P1} = 0.1 \text{ MHz} = \frac{1}{2\pi C_1 R_1}$$

Thus

$$R_1 = \frac{10^5}{2\pi}\ \Omega \qquad f_{P2} = 1 \text{ MHz} = \frac{1}{2\pi C_2 R_2}$$

Thus

$$R_2 = \frac{10^5}{\pi}\ \Omega$$

If a compensating capacitor C_C is connected across the input terminals of the transistor stage, then the frequency of the first pole changes from f_{P1} to f'_D:

$$f'_D = \frac{1}{2\pi(C_1 + C_C)R_1}$$

The second pole remains unchanged. The required value for f'_D is determined by drawing a -20-dB/decade line from the 1-MHz frequency point on the $20 \log(1/\beta) = 20 \log 1 = 0$ dB line. This line will intersect the 100-dB dc gain line at 10 Hz. Thus

$$f'_D = 10 \text{ Hz} = \frac{1}{2\pi(C_1 + C_C)R_1}$$

which results in $C_C \simeq 1\ \mu\text{F}$, which is quite large and which certainly cannot be included on the IC chip.

Next, if a compensating capacitor C_f is connected in the feedback path of the transistor, then both poles change location to the values given by Eqs. (8.58) and (8.59):

$$f'_{P1} \simeq \frac{1}{2\pi g_m R_2 C_f R_1} \qquad f'_{P2} \simeq \frac{g_m C_f}{2\pi[C_1 C_2 + C_f(C_1 + C_2)]} \tag{8.60}$$

To determine where we should locate the first pole we need to know the value of f'_{P2}. As an approximation let us assume that $C_f \gg C_2$, which enables us to obtain

$$f'_{P2} \simeq \frac{g_m}{2\pi(C_1 + C_2)} = 60.6 \text{ MHz}$$

Thus it appears that this pole will move to a frequency higher than f_{P3} (which is 10 MHz). Let us therefore assume that the second pole will be at f_{P3}. This requires that the first pole be located at 100 Hz:

$$f'_{P1} = 100 \text{ Hz} = \frac{1}{2\pi g_m R_2 C_f R_1}$$

which results in $C_f = 78.5$ pF. Although this value is indeed much greater than C_2, we can determine the location of the pole f'_{P2} from Eq. (8.60) which yields $f'_{P2} = 57.2$ MHz, confirming the fact that this pole has indeed been moved past f_{P3}.

We conclude that using Miller compensation not only results in a much smaller compensating capacitor but, owing to pole splitting, also enables us to place the dominant pole a decade higher in frequency. This results in a wider bandwidth for the compensated op amp.

Exercises 8.17 A multipole amplifier having a first pole at 1 MHz and an open-loop gain of 100 dB is to be compensated for closed-loop gains as low as 20 dB by the introduction of a new dominant pole. At what frequency must the new pole be placed?

Ans. 100 Hz

8.18 For the amplifier described in Exercise 8.17, rather than introducing a new dominant pole we can use additional capacitance at the circuit node at which the first pole is formed to reduce the frequency of the first pole. If the frequency of the second pole is 10 MHz and if it remains unchanged while additional capacitance is introduced as mentioned, find the frequency to which the first pole must be lowered so that the resulting amplifier is stable for closed-loop gains as low as 20 dB. By what factor must the capacitance at the controlling node be increased?

Ans. 1000 Hz; 1000

SUMMARY

- Negative feedback is employed to make the amplifier gain less sensitive to component variations, in order to control input and output impedances, to extend bandwidth, to reduce nonlinear distortion, and to enhance signal-to-noise (and signal-to-interference) ratio.

- The above advantages are obtained at the expense of a reduction in gain and at the risk of the amplifier becoming unstable (that is, oscillating). The latter problem is solved by careful design.

- For each of the four basic types of amplifiers, there is an appropriate feedback topology. The four topologies, together with their analysis procedure and their effects on input and output impedances, are summarized in Table 8.1.

- The key feedback parameters are the loop gain ($A\beta$), which for negative feedback must be a positive dimensionless number, and the amount of feedback ($1 + A\beta$). The latter directly determines gain reduction, gain desensitivity, bandwidth extension, and changes in Z_i and Z_o.

- Since A and β are in general frequency-dependent, the poles of the feedback amplifier are obtained by solving the characteristic equation $1 + A(s)\beta(s) = 0$.

- For the feedback amplifier to be stable, its poles must all be in the left half of the s plane.

- Stability is guaranteed if at the frequency for which the phase angle of $A\beta$ is 180°, (that is, ω_{180}), $|A\beta|$ is less than unity; the amount by which it is less than unity, expressed in decibels, is the gain margin. Alternatively, the amplifier is stable if, at the frequency at which $|A\beta| = 1$, the phase angle is less than 180°; the difference is the phase margin.

- The stability of a feedback amplifier can be analyzed by constructing Bode plots for $|A|$ and $1/|\beta|$. Stability is guaranteed if the two plots intersect with a difference in slope no greater than 6 dB/octave.

- To make a given amplifier stable for a given feedback factor β, the open-loop frequency response can be suitably modified by a process known as frequency compensation.

- A popular method for frequency compensation involves connecting a feedback capacitor to an inverting stage in the amplifier. This causes the pole formed at the input of the amplifier stage to shift to a lower frequency and thus become dominant, while the pole formed at the output of the amplifier stage is moved to a very high frequency and thus becomes unimportant. This process is known as pole splitting.

BIBLIOGRAPHY

P. E. Gray and C. L. Searle, *Electronic Principles,* New York: Wiley, 1969.

P. R. Gray and R. G. Meyer, *Analysis and Design of Analog Integrated Circuits,* 2nd ed., New York: Wiley, 1984.

S. S. Haykin, *Active Network Theory,* Reading, Mass.: Addison-Wesley, 1970.

E. S. Kuh and R. A. Rohrer, *Theory of Linear Active Networks,* San Francisco: Holden-Day, Inc., 1967. (This is an advanced-level text.)

Linear Integrated Circuits, Harrison, N.J.: RCA, 1967.

G. S. Moschytz, *Linear Integrated Networks: Design,* New York: Van Nostrand Reinhold, 1974.

E. Renschler, *The MC1539 Operational Amplifier and Its Applications,* Application Note AN-439, Phoenix, Ariz.: Motorola Semiconductor Products.

J. K. Roberge, *Operational Amplifiers: Theory and Practice,* New York: Wiley, 1975.

PROBLEMS

Section 8.1: The General Feedback Structure

8.1 A negative-feedback amplifier has a closed-loop gain $A_f = 100$ and an open-loop gain $A = 10^5$. What is the feedback factor β? If a manufacturing error results in a reduction of A to 10^3, what closed-loop gain results? What is the percentage change in A_f corresponding to this factor of 100 reduction in A?

8.2 Repeat Exercise 8.1, parts (b) through (e), for $A = 100$.

8.3 Repeat Exercise 8.1, parts (b) through (e), for $A_f = 10^3$. For part (d) use $V_s = 0.01$ V.

8.4 The noninverting buffer op amp configuration shown in Fig. P8.4 provides a direct implementation of the feedback loop of Fig. 8.1. Assuming that the op amp has infinite input resistance and zero output resistance, what is β? If $A = 100$, what is the closed-loop voltage gain? What is the amount of feedback in dB? For $V_s = 1$ V, find V_o and V_i. If A decreases by 10%, what is the corresponding decrease in A_f?

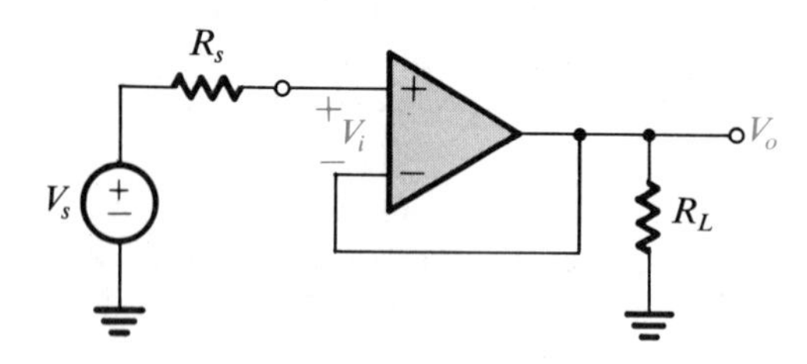

Fig. P8.4

Section 8.2.: Some Properties of Negative Feedback

8.5 For the negative-feedback loop of Fig. 8.1, find the relationship between A and β for which the sensitivity of closed-loop gain to open-loop gain—i.e., $(dA_f/A_f)/(dA/A)$—is -20 dB. For what value of $A\beta$ does the sensitivity become 1/2?

D8.6 It is required to design an amplifier with a gain of 100 that is accurate to within $\pm 1\%$. You have available amplifier stages with a gain of 1000 that is accurate to within $\pm 30\%$. Provide a design that uses a number of these gain stages in cascade, with each stage employing negative feedback of an appropriate amount.

8.7 In a feedback amplifier for which $A = 10^4$ and $A_f = 10^3$, what is the gain-desensitivity factor? Find A_f exactly, and approximately using Eq. (8.6), in the two cases: (a) A drops by 10%; and (b) A drops by 40%.

8.8 Consider an amplifier having a midband gain A_M and a low-frequency response characterized by a pole at $s = -\omega_L$ and a zero at $s = 0$. Let the amplifier be connected in a negative-feedback loop with a feedback factor β. Find an expression for the midband gain and the lower 3-dB frequency of the closed-loop amplifier. By what factor have both changed?

8.9 A capacitively coupled amplifier has a midband gain of 100, a single high-frequency pole at 10 kHz, and a single low-frequency pole at 100 Hz. Negative feedback is employed so that the midband gain is reduced to 10. What are the upper and lower 3-dB frequencies of the closed-loop gain?

D**8.10 It is required to design a dc amplifier with a low-frequency gain of 1000 and a 3-dB frequency of

0.5 MHz. You have available gain stages with a gain of 1000 but with a dominant high-frequency pole at 10 kHz. Provide a design that employs a number of such stages in cascade, each with negative feedback of an appropriate amount. [*Hint:* When negative feedback of an amount $(1 + A\beta)$ is employed around a gain stage, its x-dB frequency is increased by the factor $(1 + A\beta)$.]

D8.11 A power-output stage with voltage gain of 1, fed by a 1-kHz sine wave, has an output consisting of 2V peak-to-peak at 1 kHz contaminated by power-supply hum at 120 Hz of 3V peak-to-peak. Design a feedback circuit utilizing a low-noise preamplifier so that the output noise is reduced to 10 mV peak-to-peak while the signal voltage gain remains approximately unity. Give the value of β and of the gain required from the preamplifier. What is the improvement in signal-to-noise ratio (in dB)?

*8.12 The complementary BJT follower shown in Fig. P8.12(a) has the approximate transfer characteristic shown in Fig. P8.12(b). Observe that for $-0.7\text{ V} \le v_I \le +0.7\text{ V}$, the output is zero. This "dead band" leads to crossover distortion (see Section 9.3). Consider this follower driven by the output of a differential amplifier of gain 100 whose positive input terminal is connected to the input signal source v_S and whose negative input terminal is connected to the emitters of the follower. Sketch the transfer characteristic v_O versus v_S of the resulting feedback amplifier. What are the limits of the dead band and what are the gains outside the dead band?

D8.13 A particular amplifier has a nonlinear transfer characteristic that can be approximated as follows:

(a) for small input signals, $|v_I| \le 10\text{ mV}$, $v_O/v_I = 10^3$;

(b) for intermediate input signals, $10\text{ mV} \le |v_I| \le 50\text{ mV}$, $v_O/v_I = 10^2$;

(c) for large input signals, $|v_I| \ge 50\text{ mV}$, the output saturates.

If the amplifier is connected in a negative-feedback loop, find the feedback factor β that reduces the factor of 10 change in gain (occurring at $|v_I| = 10\text{ mV}$) to only a 10% change. What is the transfer characteristic of the amplifier with feedback?

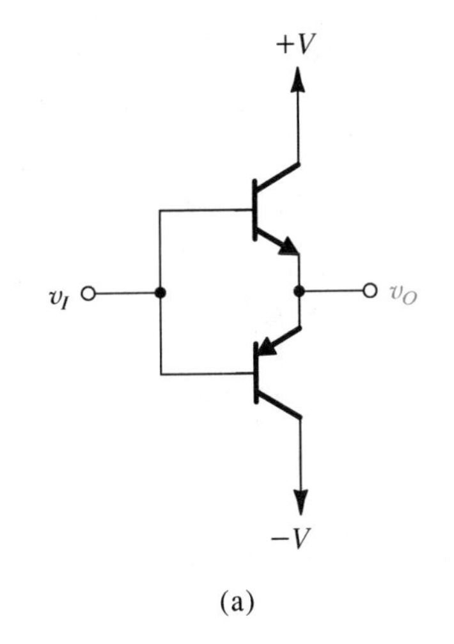

(a)

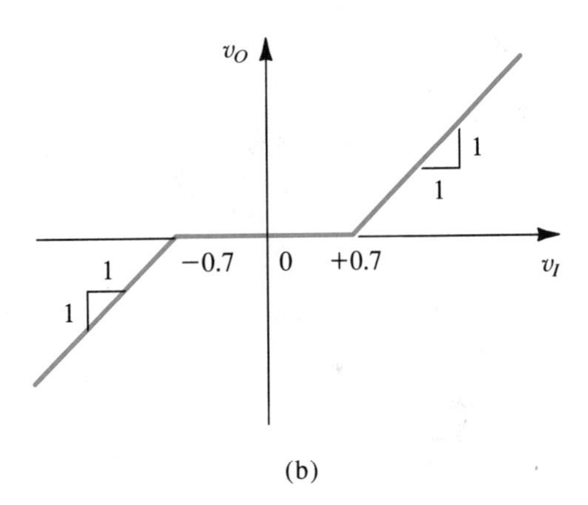

(b)

Fig. P8.12

Section 8.3: The Four Basic Feedback Topologies

8.14 A series-shunt feedback amplifier representable by Fig. 8.4(a) and using an ideal basic voltage amplifier operates with $V_s = 100\text{ mV}$, $V_f = 90\text{ mV}$, and $V_o = 10\text{ V}$. What are the values of A and β which correspond? Include the correct units for each.

8.15 A shunt-series feedback amplifier representable by Fig. 8.4(b) and using an ideal basic current amplifier operates with $I_s = 100\ \mu\text{A}$, $I_f = 90\ \mu\text{A}$, and $I_o = 10\text{ mA}$. What are the values of A and β which correspond? Include the correct units for each.

8.16 Consider the shunt-series feedback amplifier in Fig. 8.5. Assuming the input resistance at the base of Q_1 to be quite low, find an expression for $\beta \equiv I_f/I_o$.

8.17 Consider the series-series feedback circuit of Fig. 8.6. Assuming the resistance looking into the emitter of Q_1 to be relatively high, find an expression for $\beta \equiv V_f/I_o$.

8.18 A series-series feedback circuit representable by Fig. 8.4(c) and using an ideal transconductance amplifier operates with $V_s = 100$ mV, $V_f = 90$ mV, and $I_o = 10$ mA. What are the values of A and β which correspond? Include the correct units for each.

8.19 A shunt-shunt feedback circuit representable by Fig. 8.4(d) and using an ideal transresistance amplifier operates with $I_s = 100\ \mu$A, $I_f = 90\ \mu$A, and $V_o = 10$ V. What are the values of A and β which correspond? Include the correct units for each.

Section 8.4: The Series-Shunt Feedback Amplifier

8.20 A series-shunt feedback amplifier employs a basic amplifier with input and output resistances each of 1 kΩ and gain $A = 1000$ V/V. The feedback factor $\beta = 0.1$ V/V. Find the gain A_f, the input resistance R_{if}, and the output resistance R_{of} of the closed-loop amplifier.

8.21 For a particular amplifier connected in a feedback loop in which the output voltage is sampled, measurement of the output resistance before and after the loop is connected shows a change by a factor of 100. Is the resistance with feedback higher or lower? What is the value of the loop gain $A\beta$? If R_{of} is 100 Ω, what is R_o without feedback?

****8.22** A series-shunt feedback circuit employs a basic voltage amplifier that has a dc gain of 10^4 V/V and an STC frequency response with a unity-gain frequency of 1 MHz. The input resistance of the basic amplifier is 10 kΩ, and its output resistance is 1 kΩ. If the feedback factor $\beta = 0.1$ V/V, find the input impedance Z_{if} and the output impedance Z_{of} of the feedback amplifier. Give equivalent circuit representations of these impedances. Also find the value of each impedance at 10^3 Hz and at 10^5 Hz.

8.23 A series-shunt feedback amplifier utilizes the feedback circuit shown in Fig. P8.23.

(a) Find expressions for the h-parameters of the feedback circuit (see Fig. 8.10b).

(b) If $R_1 = 1$ kΩ and $\beta = 0.01$, what are the values of all four h-parameters? Give the units of each parameter.

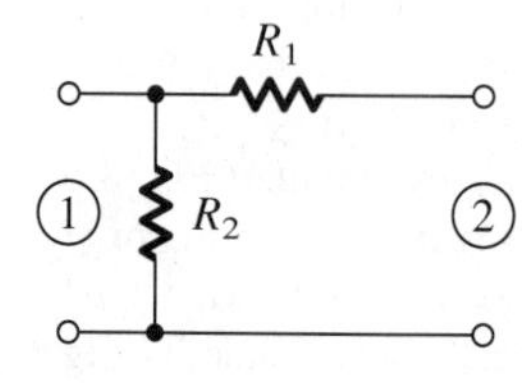

Fig. P8.23

(c) For the case $R_s = 1$ kΩ and $R_L = 1$ kΩ, sketch and label an equivalent circuit following the model in Fig. 8.10(c).

8.24 A feedback amplifier utilizing voltage sampling and employing a basic voltage amplifier with a gain of 100 and an output resistance of 1000 Ω has a closed-loop output resistance of 100 Ω. What is the closed-loop gain? If the basic amplifier is used to implement a unity-gain voltage buffer, what output resistance do you expect?

8.25 The circuit of Fig. E8.5 (page 590) is modified by replacing R_2 with a short circuit. Find the values of A, β, A_f, R'_{if}, and R'_{of}.

8.26 The circuit in Exercise 8.5 is modified by replacing the 20-kΩ resistor with a 0.5-mA current source having an equivalent output resistance of 1 MΩ. Find A, β, A_f, R'_{if}, and R'_{of}.

****8.27** For the circuit in Fig. P8.27, $|V_t| = 1$ V, $K = 0.5$ mA/V^2, $h_{fe} = 100$, and the Early voltage magnitude for all devices (including those that implement the current sources) is 100 V. The signal source V_s has a zero dc component. Find the dc voltage at the output and at the base of Q_3. Find the values of A, β, A_f, R'_{if}, and R'_{of}.

D*8.28 Figure P8.28 shows a series-shunt feedback amplifier without details of the bias circuit.

(a) Sketch the A circuit and the circuit for determining β.

(b) Show that if $A\beta$ is large then the closed-loop voltage gain is given approximately by

$$A_f \equiv \frac{V_o}{V_s} \simeq \frac{R_F + R_E}{R_E}$$

(c) If R_E is selected equal to 50 Ω, find R_F that will result in a closed-loop gain of approximately 25 V/V.

(d) If Q_1 is biased at 1 mA, Q_2 at 2 mA, and Q_3 at 5 mA, and assuming the transistors have $h_{fe} = 100$, find approximate values for R_{C1} and R_{C2} to obtain gains from the stages of the A-circuit as follows: a voltage gain of Q_1 of about -10 and a voltage gain of Q_2 of about -50.

(e) For your design, what is the closed-loop voltage gain realized?

(f) Calculate the input and output resistances of the closed-loop amplifier designed.

D8.29** The transistors of the circuit in Fig. P8.29 have the following parameters: For Q_1, $I_{DSS} = 4$ mA, $V_P = -2$ V; for Q_2, $|V_{BE}| = 0.7$ V, $h_{fe} = 100$.

(a) Find resistor values to operate Q_1 at $I_D = 1$ mA

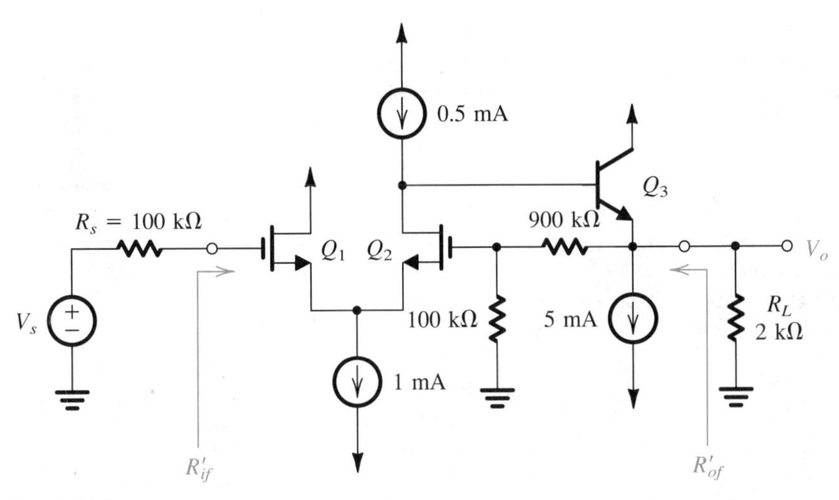

Fig. P8.27

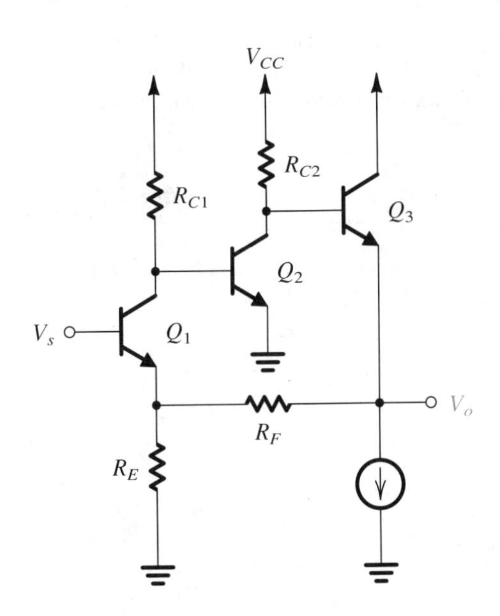

Fig. P8.28

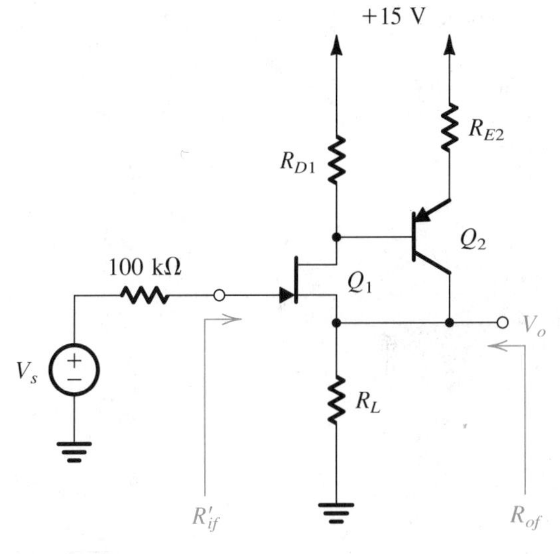

Fig. P8.29

and Q_2 at $I_C = 9$ mA and to establish a dc voltage of +13.5 V at the drain of Q_1. Assume that V_s has a zero dc component.

(b) Calculate the values of A, β, A_f, R_{if}, and R_{of}.

(c) Verify the value found for A_f by direct analysis, that is, without using the feedback method. *Hint:* recall that for the JFET,

$$I_D = I_{DSS}\left(1 - \frac{V_{GS}}{V_P}\right)^2 \text{ and } g_m = \frac{2I_{DSS}}{|V_P|}\sqrt{\frac{I_D}{I_{DSS}}}$$

Section 8.5: The Series-Series Feedback Amplifier

8.30 A series-series feedback amplifier employs a transconductance amplifier having $G_m = 100$ mA/V, input resistance of 10 kΩ, and output resistance of 100 kΩ. The feedback network has $\beta = 0.1$ V/mA, an input resistance (with port 1 open circuited) of 100 Ω, and an input resistance (with port 2 open circuited) of 10 kΩ. The amplifier operates with a signal source having a resistance of 10 kΩ and with a load of 10-kΩ resistance. Find A_f, R'_{if}, and R'_{of}.

D*8.31 Figure P8.31 shows a circuit for a voltage-controlled current source employing series-series feedback through the resistor R_E. (The bias circuit for the transistor is not shown.) Show that if the loop gain $A\beta$ is large,

$$\frac{I_o}{V_s} \simeq \frac{1}{R_E}$$

Then find the value of R_E to obtain a circuit transconductance of 1 mA/V. If the voltage amplifier has a differential input resistance of 100 kΩ, a voltage gain of 100, and an output resistance of 1 kΩ, and if the transistor is biased at a current of 1 mA and has h_{fe} of 100 and r_o of 100 kΩ, find the actual value of transconductance (I_o/V_s) realized. Use R_s = 10 kΩ. Also find the input resistance R_{in} and the output resistance R_{out}. For calculating R_{out} recall that the output resistance of a BJT with an emitter resistance R_E and a resistance in the base circuit of R_B is given by

$$r_o\left[1 + \frac{g_m R_E}{1 + R_B/r_\pi}\right]$$

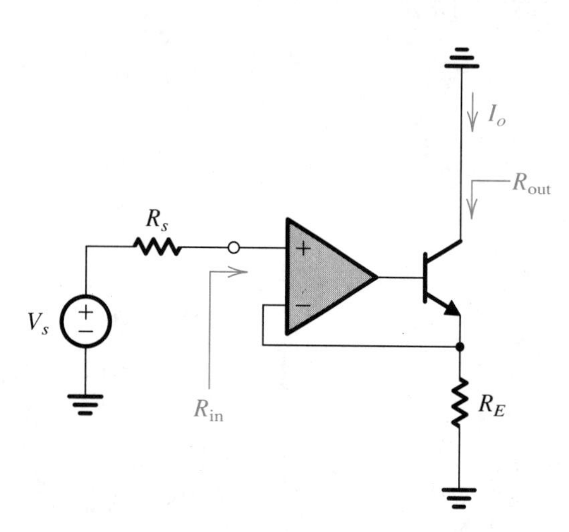

Fig. P8.31

8.32 For the circuit in Fig. 8.17(a) find an approximate value for I_o/V_s assuming that the loop gain is large. Compare to the value found in Example 8.2.

***8.33** Figure P8.33 shows a circuit for a voltage-to-current converter employing series-series feedback via resistor R_F. The MOSFETs have the dimensions shown and $\mu_n C_{OX} = 20\ \mu A/V^2$, $|V_t| = 1$ V, and $|V_A| = 100$ V. What is the value of I_o/V_s obtained for large loop gain? Use feedback analysis to find a more exact value for I_o/V_s.

Fig. P8.33

Section 8.6: The Shunt-Shunt and the Shunt-Series Feedback Amplifiers

D*8.34 For the amplifier topology shown in Fig. 8.21(a) show that for large loop gain,

$$\frac{V_o}{V_s} \simeq -\frac{R_f}{R_s}$$

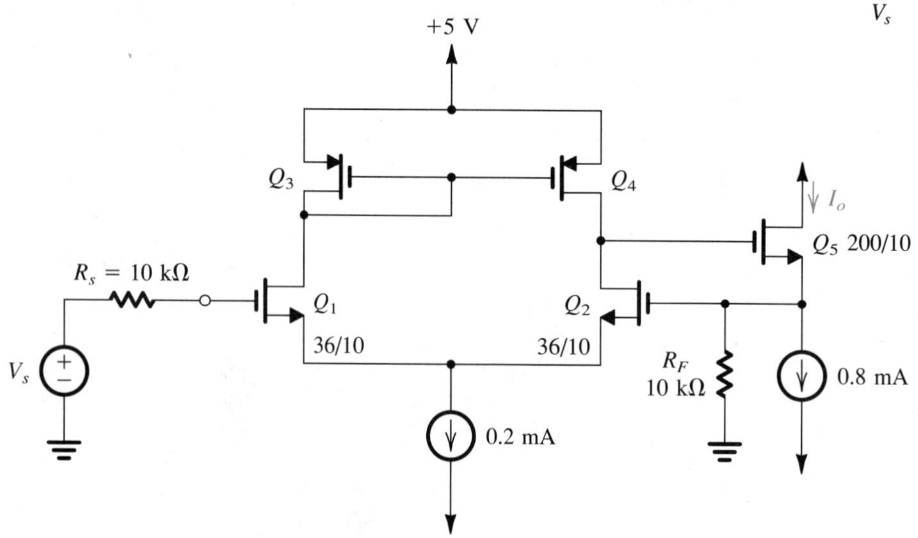

Calculate this value for the component values given on the circuit diagram, and compare the result to that found in Example 8.3. Find a new value for R_f to obtain a voltage gain of approximately -7.5 V/V.

8.35 A transresistance amplifier having an open-circuit "gain" of 100 V/mA, an input resistance of 1 kΩ, and an output resistance of 1 kΩ is connected in a negative-feedback loop employing a shunt-shunt topology. The feedback network has an input resistance (with port 1 short-circuited) of 10 kΩ and an input resistance (with port 2 short-circuited) of 10 kΩ and provides a feedback factor $\beta = 0.1$ mA/V. The amplifier is fed with a current source having $R_s = 10$ kΩ, and a load resistance $R_L = 1$ kΩ is connected at the output. Find the transresistance A_f of the feedback amplifier, its input resistance R'_{if}, and its output resistance R'_{of}.

D8.36**

(a) Show that for the circuit in Fig. P8.36(a) if the loop gain is large, the voltage gain V_o/V_s is given approximately by

$$\frac{V_o}{V_s} \simeq -\frac{R_f}{R_s}$$

(b) Using three cascaded stages of the type shown in Fig. P8.36(b) to implement the amplifier μ, design a feedback amplifier with a voltage gain of approximately -100 V/V. The amplifier is to operate between a source resistance $R_s = 10$ kΩ and a load resistance $R_L = 1$ kΩ. Calculate the actual value of V_o/V_s realized, the input resistance (excluding R_s), and the output resistance (excluding R_L). Assume that the BJTs have h_{fe} of 100. (Note: In practice the three amplifier stages are not made identical, for stability reasons.)

D8.37 Negative feedback is to be used to modify the characteristics of a particular amplifier for various purposes. Identify the feedback topology to be used if:

(a) Input resistance is to be lowered and output resistance raised.

(b) Both input and output resistances are to be raised.

(c) Both input and output resistances are to be lowered.

***8.38** For $V_t = 2$ V and $K = 0.25$ mA/V^2, find the voltage gain (V_o/V_s) and input and output resistances of the circuit in Fig. P8.38 using feedback analysis. Verify by direct analysis.

***8.39** For the circuit of Fig. P8.39, use the feedback method to find the voltage gain V_o/V_s, the input resistance R'_{if}, and the output resistance R'_{of}. The op amp has open-

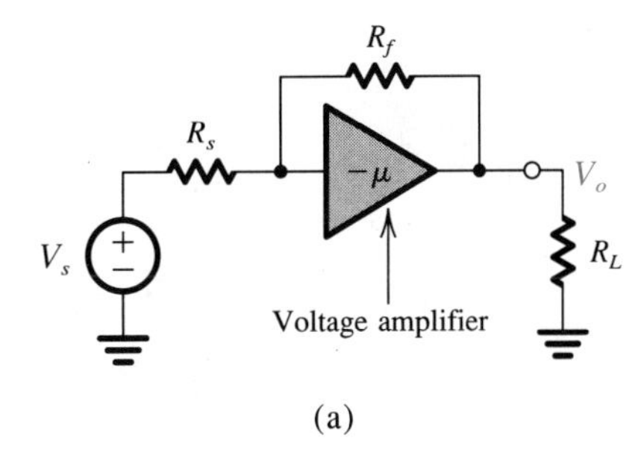

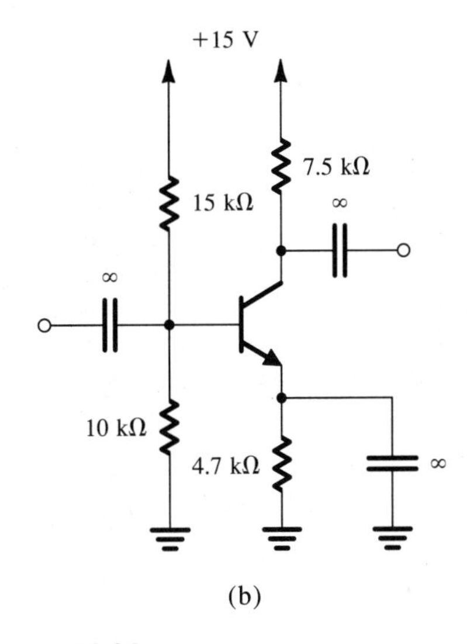

Fig. P8.36

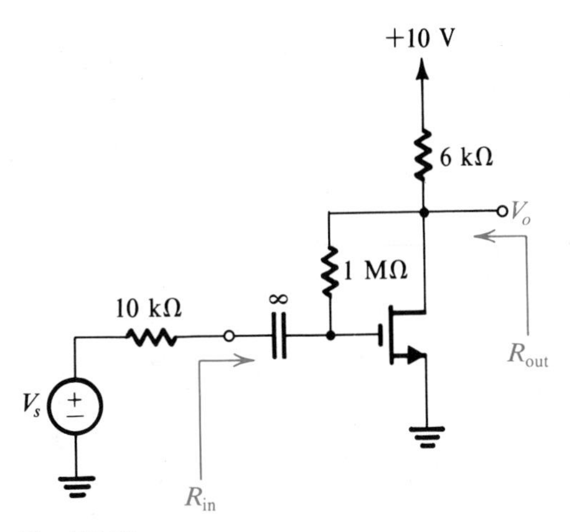

Fig. P8.38

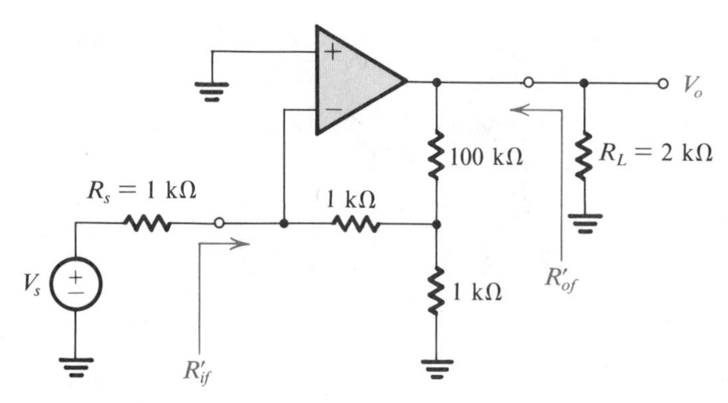

Fig. P8.39

loop gain $\mu = 10^4$ V/V, $R_{id} = 100$ kΩ, $R_{icm} = \infty$, and $r_o = 1$ kΩ.

*8.40 Consider the amplifier of Fig. 8.25(a) to have its output at the emitter of the rightmost transistor Q_2. Use the technique for a shunt-shunt feedback amplifier to calculate (V_{out}/I_{in}) and R_{in}. Using this result, calculate I_{out}/I_{in}. Compare this with the results obtained in Example 8.4.

8.41 A current amplifier with a short-circuit current gain of 100 A/A, an input resistance of 1 kΩ, and an output resistance of 10 kΩ is connected in a negative-feedback loop employing the shunt-series topology. The feedback network provides a feedback factor $\beta = 0.1$ A/A. Lacking complete data about the situation, estimate the current gain, input resistance, and output resistance of the feedback amplifier.

8.42 Figure P8.42 shows how shunt-series feedback can be employed to design a current amplifier utilizing an op amp.

(a) Show that for large loop gain, the current gain is given approximately by

$$\frac{I_o}{I_s} \simeq 1 + \frac{R_f}{r}$$

(b) Using the feedback analysis method find the closed loop gain I_o/I_s, the input resistance (excluding R_s), and the output resistance (excluding R_L) for the case: open-loop voltage gain of op amp = 10^4 V/V, $R_{id} = 100$ kΩ, op-amp output resistance = 1 kΩ, $R_s = R_L = 10$ kΩ, $r = 100$ Ω, and $R_f = 1$ kΩ.

*8.43 For the amplifier circuit in Fig. P8.43, assuming that V_s has a zero dc component find the dc voltages at all nodes and the dc emitter currents of Q_1 and Q_2. Let the BJTs have $\beta = 100$. Use feedback analysis to find V_o/V_s and R_{in}.

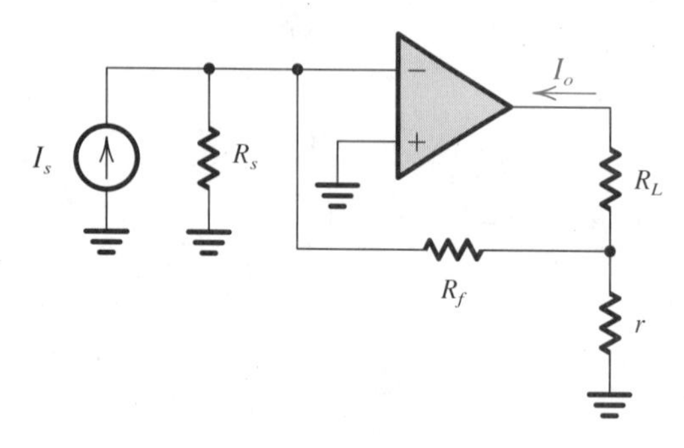

Fig. P8.42

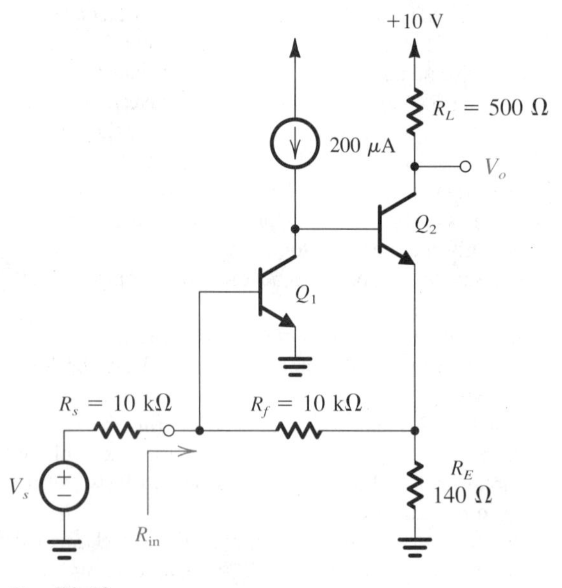

Fig. P8.43

Section 8.7: Determining the Loop Gain

8.44 Determine the loop gain of the amplifier in Fig. P8.27 by breaking the loop at the gate of Q_2 and finding the returned voltage across the 100-kΩ resistor (while setting V_s to zero). The devices have $|V_t| = 1$ V, $K = 0.5$ mA/V^2, and $h_{fe} = 100$. The Early voltage magnitude for all devices (including those that implement the current sources) is 100 V. The signal source V_s has a zero dc component. Determine the output resistance R'_{of}.

8.45 It is required to determine the loop gain of the amplifier circuit shown in Fig. P8.28. The most convenient place to break the loop is at the base of Q_2. Thus, connect a resistance equal to $r_{\pi 2}$ between the collector of Q_1 and ground, apply a test voltage V_t to the base of Q_2, and determine the returned voltage at the collector of Q_1 (with V_s set to zero, of course). Show that

$$A\beta = \frac{g_{m2}R_{C2}(h_{fe3}+1)}{R_{C2}+(h_{fe3}+1)[r_{e3}+R_F+(R_E//r_{e1})]} \times \frac{\alpha_1 R_E}{R_E + r_{e1}}(R_{C1}//r_{\pi 2})$$

8.46 To determine the loop gain of the amplifier circuit shown in Fig. P8.29, the most convenient place to break the loop is between the collector terminal of Q_2 and R_L. Do this, apply a current signal I_t to the source node of Q_1, and determine the returned current signal in the collector terminal of Q_2. As usual, V_s should be set to zero. Show that

$$A\beta = \frac{R_L}{R_L + 1/g_{m1}} \frac{h_{fe2}R_{D1}}{R_{D1}+(h_{fe2}+1)(r_{e2}+R_{E2})}$$

8.47 Show that the loop gain of the amplifier circuit in Fig. P8.33 is

$$A\beta = g_{m1,2}(r_{o2}//r_{o4})\frac{R_F//r_{o5}}{(R_F//r_{o5})+1/g_{m5}}$$

where $g_{m1,2}$ is the g_m of each of Q_1 and Q_2.

8.48 For the circuit in Fig. P8.38 calculate the loop gain. Assume that the FET has $g_m = 1$ mA/V.

8.49 The op amp in the circuit of Fig. P8.42 has a differential input resistance R_{id}, an open-loop gain μ, and an output resistance r_o. Show that the loop gain is

$$A\beta = \frac{\mu}{r_o + R_L + r//[(R_s//R_{id})+R_f]} \times \frac{r(R_s//R_{id})}{r+R_f+(R_s//R_{id})}$$

8.50 For the circuit in Fig. P8.43 calculate the loop gain and then find the input resistance R_{in}. Assume that the BJTs have $h_{fe} = 100$.

Section 8.8: The Stability Problem

8.51 An op amp designed to have a low-frequency gain of 10^5 and a high-frequency response dominated by a single pole at 100 rad/s, acquires, through a manufacturing error, a pair of additional poles at 10,000 rad/s. At what frequency does the total phase shift reach 180°? At this frequency, for what value of β, assumed to be frequency-independent, does the loop gain reach a value of unity? What is the corresponding value of closed-loop gain at low frequencies?

****8.52** For the situation described in Problem 8.51, sketch Nyquist plots for $\beta = 1.0$ and 10^{-3}. (Plot for $\omega = 0$, 100, 10^3, 10^4, and ∞ rad/s.)

8.53 An op amp having a low-frequency gain of 10^3 and a single-pole rolloff at 10^4 rad/s is connected in a negative-feedback loop via a feedback network having a transmission k and a two-pole rolloff at 10^4 rad/s. Find the value of k above which the closed-loop amplifier becomes unstable.

8.54 Consider a feedback amplifier for which the open-loop gain $A(s)$ is given by

$$A(s) = \frac{1000}{(1+s/10^4)(1+s/10^5)^2}$$

If the feedback factor β is independent of frequency, find the frequency at which the phase shift is 180°, and find the critical value of β at which oscillation will commence.

Section 8.9: Effect of Feedback on the Amplifier Poles

8.55 A dc amplifier having a single-pole response with pole frequency 10^4 Hz and unity-gain frequency of 10 MHz is operated in a loop whose frequency-independent feedback factor is 0.1. Find the low-frequency gain, the 3-dB frequency, and the unity-gain frequency of the closed-loop amplifier. By what factor does the pole shift?

***8.56** An amplifier having a low-frequency gain of 10^3 and poles at 10^4 and 10^5 Hz is operated in a closed negative-feedback loop with a frequency-independent β.

(a) For what value of β do the closed-loop poles become coincident? At what frequency?

(b) What is the low-frequency gain corresponding to the situation in (a)? What is the value of the gain at the frequency of the coincident poles?

(c) What is the value of Q corresponding to the situation in (a)?

(d) If β is increased by a factor of 10, what are the new pole locations? What is the corresponding pole-Q?

D8.57 A dc amplifier has an open-loop gain of 1000 and two poles, a dominant one at 1 kHz and a high-frequency one whose location can be controlled. It is required to connect this amplifier in a negative-feedback loop that provides a dc closed-loop gain of 100 and a maximally flat response. Find the required value of β and the frequency at which the second pole should be placed.

8.58 Reconsider Example 8.5 with the circuit in Fig. 8.34 modified to incorporate a so-called tapered network, in which the components immediately adjacent to the amplifier input are raised in impedance to $C/10$ and $10\ R$. Find expressions for the resulting pole frequency ω_0 and Q factor. For what value of K do the poles coincide? For what value of K does the response become maximally flat? For what value of K does the circuit oscillate?

8.59 Three identical logic inverters, each of which can be characterized in its switching region as a linear amplifier having a gain $-K$ and a pole at 10^7 Hz, are connected in a ring. Regarding this as a negative-feedback loop with $\beta = 1$, find the minimum value of K for which the inverter ring must oscillate. What would the frequency of oscillation be for very small signal operation? (Note that in practice such a ring oscillator operates with relatively larger signal (logic levels) at a somewhat lower frequency.)

Section 8.10: Stability Study Using Bode Plots

8.60 Reconsider Exercise 8.14 for the case of the op amp wired as a unity-gain buffer. At what frequency is $|A\beta| = 1$? What is the corresponding phase margin?

8.61 Reconsider Exercise 8.14 for the case of a manufacturing error introducing a second pole at 10^4 Hz. What is now the frequency for which $|A\beta| = 1$? What is the corresponding phase margin? For what values of β is the phase margin 45° or more?

8.62 For what phase margin does the gain peaking have a value of 5%? of 10%? of 0.1 dB? of 1 dB? [Hint: Use the result in Eq. (8.48).]

8.63 An amplifier has a dc gain of 10^5 and poles at 10^5 Hz, 3.16×10^5 Hz, and 10^6 Hz. Find the value of β, and the corresponding closed-loop gain, for which a phase margin of 45° is obtained.

8.64 A two-pole amplifier for which $A_0 = 10^3$ and having poles at 1 MHz and 10 MHz is to be connected as a differentiator. On the basis of the rate-of-closure rule, what is the smallest differentiator time-constant for which operation is stable? What are the corresponding gain and phase margins?

***8.65** For the amplifier described by Fig. 8.37 and with frequency-independent feedback, what is the minimum closed-loop voltage gain that can be obtained for phase margins of 90° and 45°?

Section 8.11: Frequency Compensation

D8.66 A multipole amplifier having a first pole at 2 MHz and a dc open-loop gain of 80 dB is to be compensated for closed-loop gains as low as unity by the introduction of a new dominant pole. At what frequency must the new pole be placed?

D8.67 For the amplifier described in Problem 8.66, rather than introducing a new dominant pole we can use additional capacitance at the circuit node at which the pole is formed to reduce the frequency of the first pole. If the frequency of the second pole is 10 MHz and if it remains unchanged while additional capacitance is introduced as mentioned, find the frequency to which the first pole must be lowered so that the resulting amplifier is stable for closed-loop gains as low as unity. By what factor is the capacitance at the controlling node increased?

8.68 Contemplate the effects of pole splitting by considering Eqs. (8.58) and (8.59) under the conditions that $R_1 \simeq R_2 = R$, $C_2 \simeq C_1/10 = C$, $C_f \gg C$, and $g_m = 100/R$, by calculating ω_{P1}, ω_{P2} and ω'_{P1}, ω'_{P2}.

D8.69 An op amp with open-loop voltage gain of 10^4 and poles at 10^5, 10^6, and 10^7 Hz is to be compensated by the addition of a fourth dominant pole to operate stably with unity feedback ($\beta = 1$). What is the frequency of the required dominant pole? The compensation network is to consist of an RC low-pass network placed in the negative-feedback path of the op amp. The dc bias conditions are such that a 1-MΩ resistor can be tolerated in series with each of the negative and positive input terminals. What capacitor is required between the negative input and ground to implement the required fourth pole?

D*8.70 An op amp with an open-loop voltage gain of 80 dB and poles at 10^5, 10^6, and 2×10^6 Hz is to be compensated to be stable for unity β. Assume that the op amp incorporates an amplifier equivalent to that in Fig. 8.40, with $C_1 = 150$ pF, $C_2 = 5$ pF, and $g_m = 40$ mA/V, and that f_{P1} is caused by the input circuit and f_{P2} by the output circuit of this amplifier. Find the required value of the compensating Miller capacitance and the new frequency of the output pole.

****8.71** The op amp in the circuit of Fig. P8.71 has an open-loop gain of 10^5 and a single-pole rolloff with $\omega_{3dB} = 10$ rad/s.

(a) Sketch a Bode plot for the loop gain.

(b) Find the frequency at which $|A\beta| = 1$, and find the corresponding phase margin.

(c) Find the closed-loop transfer function, including its zero and poles. Sketch a pole-zero plot. Sketch the magnitude of the transfer function versus frequency, and label the important parameters on your sketch.

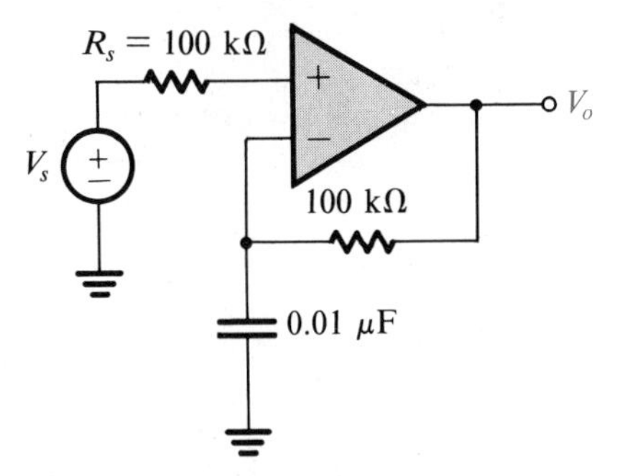

Fig. P8.71

CHAPTER 9

Output Stages and Power Amplifiers

INTRODUCTION

An important function of the output stage is to provide the amplifier with a low output resistance so that it can deliver the output signal to the load without loss of gain. Since the output stage is the final stage of the amplifier, it usually deals with relatively large signals. Thus the small-signal approximations and models either are not applicable or must be used with care. Nevertheless, linearity remains a very important requirement. In fact, a measure of goodness of the design of the output stage is the **total harmonic distortion** (THD) it introduces. This is the rms value of the harmonic components of the output signal, excluding the fundamental, expressed as a percentage of the rms of the fundamental. A high-fidelity audio power amplifier features a THD of the order of a fraction of a percent.

The most challenging requirement in the design of the output stage is that it deliver the required amount of power to the load in an *efficient* manner. This implies that the power *dissipated* in the output-stage transistors must be as low as possible. This requirement stems mainly from the fact that the power dissipated in a transistor raises its internal **junction temperature,** and there is a maximum temperature (in the range of 150°C to 200°C for silicon devices) above which the transistor is destroyed. Other reasons for requiring a high power-conversion efficiency are to prolong the life of batteries employed in battery-powered circuits, to permit a smaller, lower-cost power supply, or to obviate the need for cooling fans.

We begin this chapter with a study of the various output-stage configurations employed in amplifiers that handle both low and high power. In this context, ''high power'' generally means greater than 1 W. We then consider the specific requirements of BJTs employed in the design of high-power output stages, called **power transistors.** Special attention will be paid to the thermal properties of the transistor.

A power amplifier is simply an amplifier with a high-power output stage. Examples of discrete- and integrated-circuit power amplifiers will be presented. The chapter concludes with a brief discussion of MOSFET structures that are currently finding application in power-circuit design.

9.1 CLASSIFICATION OF OUTPUT STAGES

Output stages are classified according to the collector current waveform that results when an input signal is applied. Figure 9.1 illustrates the classification for the case of a sinusoidal input signal. The class A stage, whose associated waveform is shown in Fig. 9.1(a), is biased at a current I_C greater than the amplitude of the signal current, $\hat{I}_c$. Thus the transistor in a class A stage conducts for the entire cycle of the input signal; that is, the conduction angle is 360°. In contrast, the class B stage, whose associated waveform is shown in Fig. 9.1(b), is biased at zero dc current. Thus a transistor in a class B stage conducts for only half of the cycle of the input sine wave, resulting in a conduction angle of 180°. As will be seen later, the negative halves of the sinusoid will be supplied by another transistor that also operates in the class B mode and conducts during the alternate half cycles.

An intermediate class between A and B, appropriately named class AB, involves biasing the transistor at a nonzero dc current much smaller than the peak current of the sine-wave signal. As a result, the transistor conducts for an interval slightly greater than half a cycle, as illustrated in Fig. 9.1(c). The resulting conduction angle is greater than 180° but much less than 360°. The class AB stage has another transistor that conducts for an interval slightly greater than that of the negative half-cycle, and the currents from the two transistors are combined in the load. It follows that, during the intervals near the zero crossings of the input sinusoid, both transistors conduct.

Figure 9.1(d) shows the collector-current waveform for a transistor operated as a class C amplifier. Observe that the transistor conducts for an interval shorter than that of a half-cycle; that is, the conduction angle is less than 180°. The result is the periodically pulsating current waveform shown. To obtain a sinusoidal output voltage, this current is passed through a parallel LC circuit, tuned to the frequency of the input sinusoid. The tuned circuit acts as a bandpass filter and provides an output voltage proportional to the amplitude of the fundamental component in the Fourier series representation of the current waveform.

Class A, AB, and B amplifiers are studied in this chapter. They are employed as output stages of op amps and audio power amplifiers. In the latter application, class AB is the preferred choice, for reasons that will be explained in the following sections. Class C amplifiers are usually employed for radio-frequency (RF) power amplification (required, for example, in radio and TV transmitters). The design of class C amplifiers is a rather specialized topic and is not included in this book.

Although the BJT has been used to illustrate the definition of the various output-stage classes, the same classification applies to output stages implemented with MOSFETs.

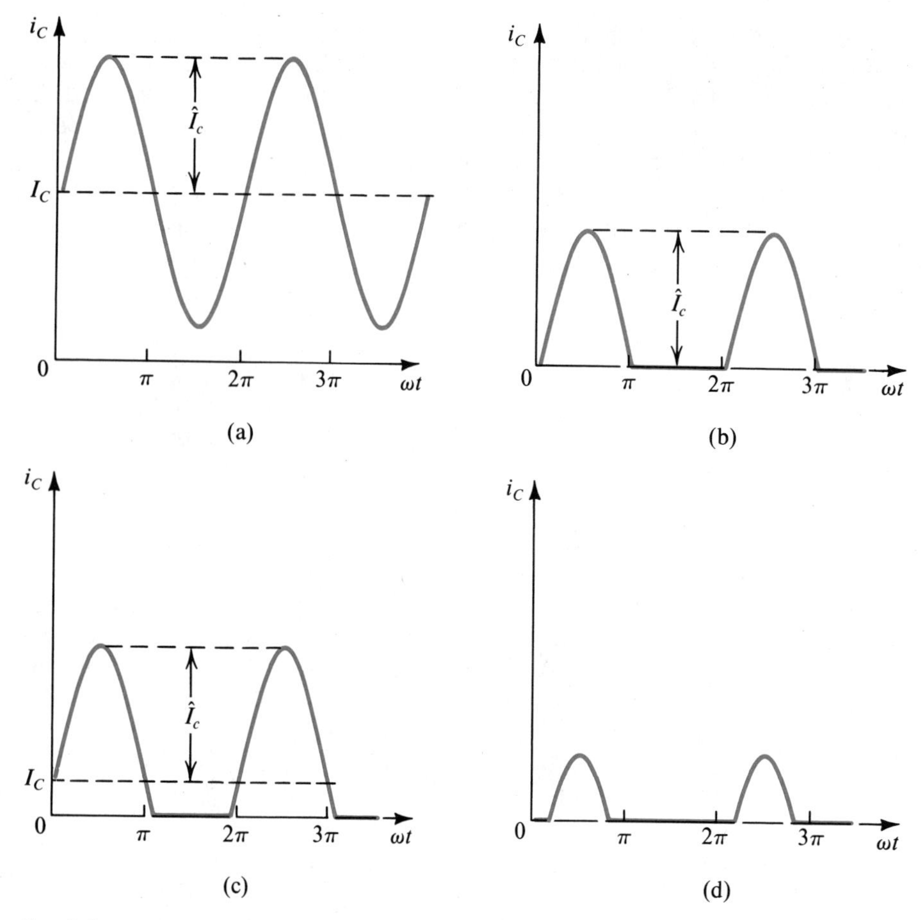

Fig. 9.1 Collector current waveforms for transistors operating in **(a)** class A, **(b)** class B, **(c)** class AB, and **(d)** class C amplifier stages.

Furthermore, the above classification extends to amplifier stages other than those used at the output. In this regard, all the common-emitter, common-base, and common-collector amplifiers (and their FET counterparts) studied in previous chapters fall into the class A category.

9.2 CLASS A OUTPUT STAGE

Because of its low output resistance, the emitter follower is the most popular class A output stage. We have already studied the emitter follower in Chapters 4 and 6; in the following we consider its large-signal operation.

Transfer Characteristic

Figure 9.2 shows an emitter follower Q_1 biased with a constant current I supplied by transistor Q_2. Since the emitter current $i_{E1} = I + i_L$, the bias current I must be greater

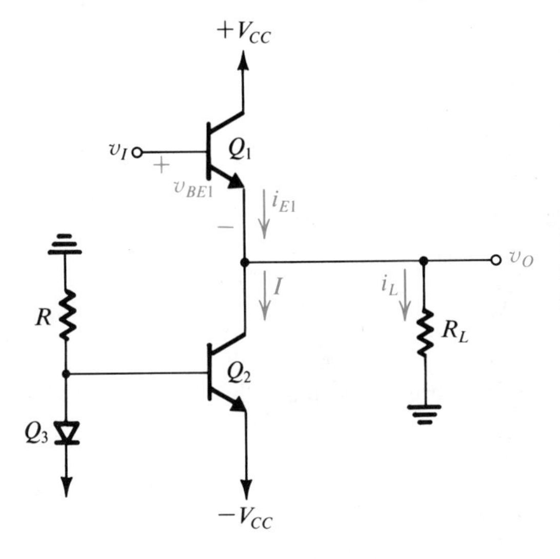

Fig. 9.2 An emitter follower (Q_1) biased with a constant current source I supplied by transistor Q_2.

than the largest negative load current; otherwise, Q_1 cuts off and class A operation will no longer be maintained.

The transfer characteristic of the emitter follower of Fig. 9.2 is described by

$$v_O = v_I - v_{BE1} \tag{9.1}$$

where v_{BE1} depends on the emitter current i_{E1} and thus on the load current i_L. If we neglect the relatively small changes in v_{BE1} (60 mV for every factor of 10 change in emitter current), the linear transfer curve shown in Fig. 9.3 results. As indicated, the

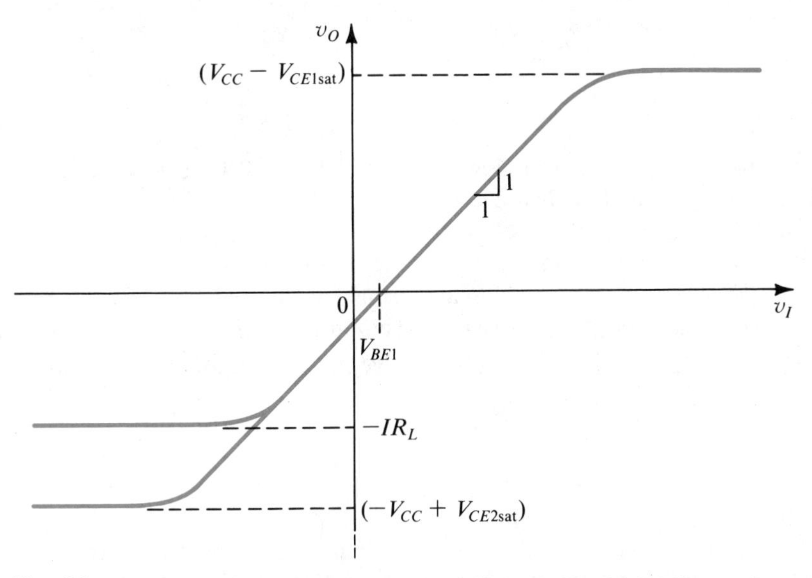

Fig. 9.3 Transfer characteristic of the emitter follower in Fig. 9.2. This linear characteristic is obtained by neglecting the change in v_{BE1} with i_L. The maximum positive output is determined by the saturation of Q_1. In the negative direction, the limit of the linear region is determined either by Q_1 turning off or by Q_2 saturating, depending on the values of I and R_L.

positive limit of the linear region is determined by the saturation of Q_1; thus

$$v_{O\max} = V_{CC} - V_{CE1\mathrm{sat}} \tag{9.2}$$

In the negative direction, the limit of the linear region is determined either by Q_1 turning off,

$$v_{O\min} = -IR_L \tag{9.3}$$

or by Q_2 saturating,

$$v_{O\min} = -V_{CC} + V_{CE2\mathrm{sat}} \tag{9.4}$$

depending on the values of I and R_L. The absolutely lowest output voltage is that given by Eq. (9.4) and is achieved provided that the bias current I is greater than the magnitude of the corresponding load current,

$$I \geq \frac{|-V_{CC} + V_{CE2\mathrm{sat}}|}{R_L} \tag{9.5}$$

Exercises D9.1 For the emitter follower in Fig. 9.2 let $V_{CC} = 15$ V, $V_{CE\mathrm{sat}} = 0.2$ V, $V_{BE} = 0.7$ V and constant, β is very high. Find the value of R that will establish a bias current sufficiently large to allow the largest possible output signal swing for $R_L = 1$ kΩ. Determine the resulting output signal swing and the minimum and maximum emitter currents.

Ans. 0.97 kΩ; -14.8 V to $+14.8$ V; 0 to 29.6 mA

9.2 For the emitter follower of Exercise 9.1, in which $I = 14.8$ mA, consider the case in which v_O is limited to the range -10 V to $+10$ V. Let Q_1 have $v_{BE} = 0.6$ V at $i_C = 1$ mA, and assume $\alpha \simeq 1$. Find v_I corresponding to $v_O = -10$ V, 0, and $+10$ V. At each of these points, use small-signal analysis to determine the voltage gain v_o/v_i. Note that the incremental voltage gain gives the slope of the v_O versus v_I characteristic.

Ans. -9.36 V, 0.67 V, and 10.68 V; 0.995 V/V, 0.998 V/V, and 0.999 V/V

Signal Waveforms

Consider the operation of the emitter-follower circuit of Fig. 9.2 for sine-wave input. Neglecting $V_{CE\mathrm{sat}}$, we see that if the bias current I is properly selected the output voltage can swing from $-V_{CC}$ to $+V_{CC}$ with the quiescent value being zero, as shown in Fig. 9.4(a). Figure 9.4(b) shows the corresponding waveform of $v_{CE1} = V_{CC} - v_O$. Now, assuming that the bias current I is selected to allow a maximum negative load current of V_{CC}/R_L, the collector current of Q_1 will have the waveform shown in Fig. 9.4(c). Finally, Fig. 9.4(d) shows the waveform of the **instantaneous power dissipation** in Q_1,

$$p_{D1} \equiv v_{CE1} i_{C1} \tag{9.6}$$

Power Dissipation

Figure 9.4(d) indicates that the maximum instantaneous power dissipation in Q_1 is $V_{CC}I$. This is equal to the quiescent power dissipation in Q_1. Thus the emitter-follower transistor dissipates the largest amount of power when $v_O = 0$. Since this condition (no input signal) can easily prevail for prolonged periods of time, transistor Q_1 must be able to withstand a continuous power dissipation of $V_{CC}I$.

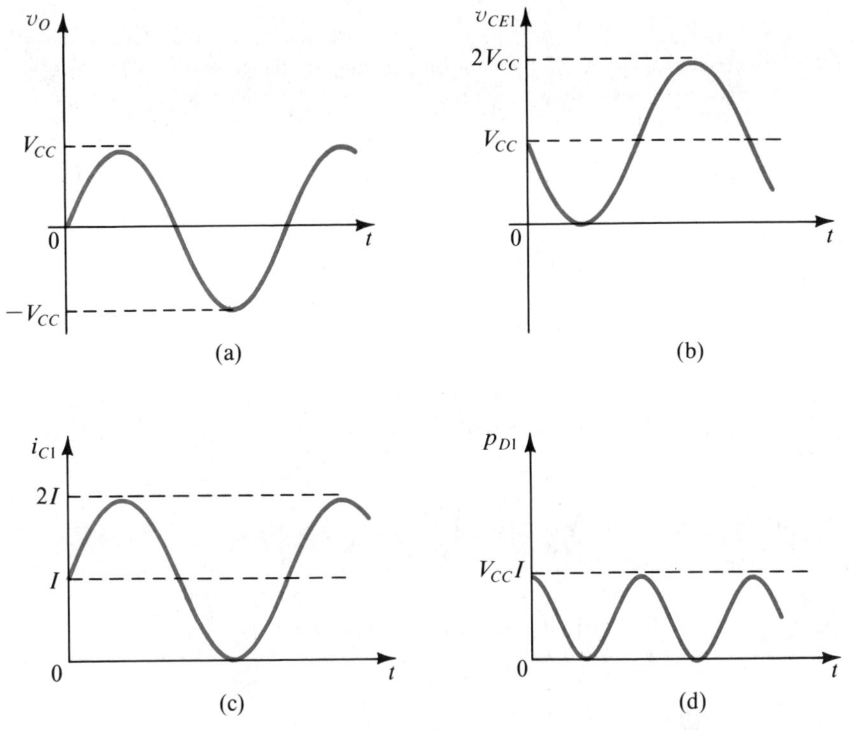

Fig. 9.4 Maximum signal waveforms in the class A output stage of Fig. 9.2 under the condition $I = V_{CC}/R_L$ or, equivalently, $R_L = V_{CC}/I$.

The power dissipation in Q_1 depends on the value of R_L. Consider the extreme case of an output open circuit, that is $R_L = \infty$. In this case $i_{C1} = I$ is constant and the instantaneous power dissipation in Q_1 will depend on the instantaneous value of v_O. The maximum power dissipation will occur when $v_O = -V_{CC}$, for in this case v_{CE1} is a maximum of $2V_{CC}$ and $p_{D1} = 2\ V_{CC}I$. This condition, however, would not normally persist for a prolonged interval, so the design need not be that conservative. Observe that with an open-circuit load the average power dissipation in Q_1 is $V_{CC}I$. A far more dangerous situation occurs at the other extreme of R_L—specifically, $R_L = 0$. In the event of an output short circuit, a positive input voltage would theoretically result in an infinite load current. In practice, a very large current may flow through Q_1, and if the short-circuit condition persists, the resulting large power dissipation in Q_1 can raise its junction temperature beyond the specified maximum, causing Q_1 to burn up. To guard against such a situation, output stages are usually equipped with *short-circuit protection,* as will be explained later.

The power dissipation in Q_2 also must be taken into account in designing an emitter-follower output stage. Since Q_2 conducts a constant current I, and the maximum value of v_{CE2} is $2V_{CC}$, the maximum instantaneous power dissipation in Q_2 is $2V_{CC}I$. This maximum, however, occurs when $v_O = V_{CC}$, a condition that would not normally prevail for a prolonged period of time. A more significant quantity for design purposes is the average power dissipation in Q_2, which is $V_{CC}I$.

Exercise 9.3 Consider the emitter follower in Fig. 9.2 with $V_{CC} = 10$ V, $I = 100$ mA, and $R_L = 100\ \Omega$. Find the power dissipated in Q_1 and Q_2 under quiescent conditions ($v_O = 0$). For a sinusoidal output voltage of maximum possible amplitude (neglecting V_{CEsat}) find the average power dissipation in Q_1 and Q_2. Also find the load power.

Ans. 1 W, 1 W; 0.5 W, 1 W; 0.5 W

Power Conversion Efficiency

The power conversion efficiency of an output stage is defined as

$$\eta \equiv \frac{\text{load power } (P_L)}{\text{supply power } (P_S)} \tag{9.7}$$

For the emitter follower of Fig. 9.2, assuming that the output voltage is a sinusoid with the peak value $\hat{V}_o$, the average load power will be

$$P_L = \frac{1}{2}\frac{\hat{V}_o^2}{R_L} \tag{9.8}$$

Since the current in Q_2 is constant (I), the power drawn from the negative supply[1] is $V_{CC}I$. The *average* current in Q_1 is equal to I, and thus the average power drawn from the positive supply is $V_{CC}I$. Thus the total average supply power is

$$P_S = 2V_{CC}I \tag{9.9}$$

Equations (9.8) and (9.9) can be combined to yield

$$\begin{aligned}\eta &= \frac{1}{4}\frac{\hat{V}_o^2}{IR_LV_{CC}} \\ &= \frac{1}{4}\left(\frac{\hat{V}_o}{IR_L}\right)\left(\frac{\hat{V}_o}{V_{CC}}\right)\end{aligned} \tag{9.10}$$

Since $\hat{V}_o \le V_{CC}$ and $\hat{V}_o \le IR_L$, maximum efficiency is obtained when

$$\hat{V}_o = V_{CC} = IR_L \tag{9.11}$$

The maximum efficiency attainable is 25%. Because this is a rather low figure, the class A output stage is rarely used in large-power applications (of more than 1W). Note also that in practice the output voltage is limited to lower values in order to avoid transistor saturation and associated nonlinear distortion. Thus the efficiency achieved is usually in the 10% to 20% range.

Exercise 9.4 For the emitter follower of Fig. 9.2 let $V_{CC} = 10$ V, $I = 100$ mA, and $R_L = 100\ \Omega$. If the output voltage is an 8-V-peak sinusoid, find the following: (a) The power delivered to the load. (b) The average power drawn from the supplies. (c) The power conversion efficiency.
Ignore the loss in Q_3.

Ans. 0.32 W; 2 W; 16%

[1] This does *not* include the power drawn by the biasing diode-connected transistor Q_3.

9.3 CLASS B OUTPUT STAGE

Figure 9.5 shows a class B output stage. It consists of a complementary pair of transistors (that is, an *npn* and a *pnp*) connected in such a way that both cannot conduct simultaneously.

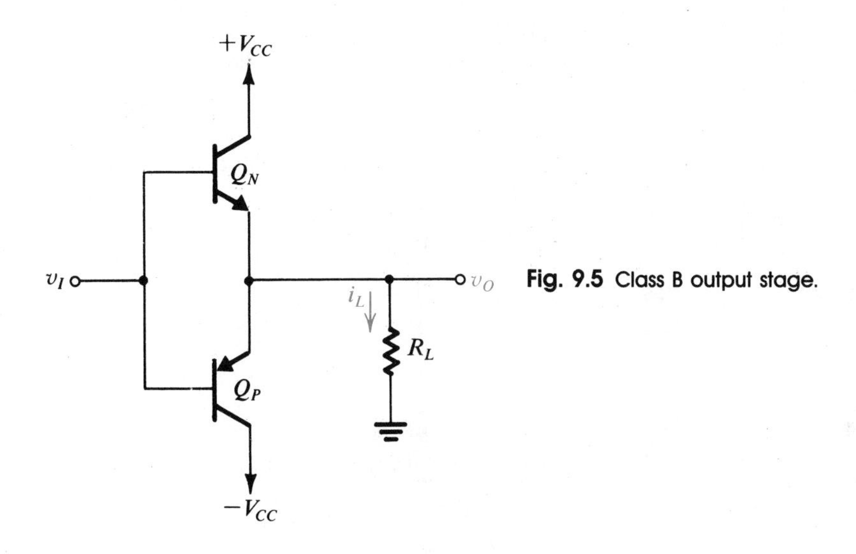

Fig. 9.5 Class B output stage.

Circuit Operation

When the input voltage v_I is zero, both transistors are cut off and the output voltage v_O is zero. As v_I goes positive and exceeds about 0.5 V, Q_N conducts and operates as an emitter follower. In this case v_O follows v_I (that is, $v_O = v_I - v_{BEN}$) and Q_N supplies the load current. Meanwhile, the emitter–base junction of Q_P will be reverse-biased by the V_{BE} of Q_N, which is approximately 0.7 V. Thus Q_P will be cut off.

If the input goes negative by more than about 0.5 V, Q_P turns on and acts as an emitter follower. Again v_O follows v_I (that is, $v_O = v_I + v_{EBP}$), but in this case Q_P supplies the load current and Q_N will be cut off.

We conclude that the transistors in the class B stage of Fig. 9.5 are biased at zero current and conduct only when the input signal is present. The circuit operates in a **push–pull** fashion: Q_N *pushes* (sources) current into the load when v_I is positive, and Q_P *pulls* (sinks) current from the load when v_I is negative.

Transfer Characteristic

A sketch of the transfer characteristic of the class B stage is shown in Fig. 9.6. Note that there exists a range of v_I centered around zero where both transistors are cut off and v_O is zero. This **dead band** results in the **crossover distortion** illustrated in Fig. 9.7 for the case of an input sine wave. The effect of crossover distortion will be most pronounced when the amplitude of the input signal is small. Crossover distortion in audio power amplifiers gives rise to unpleasant sounds.

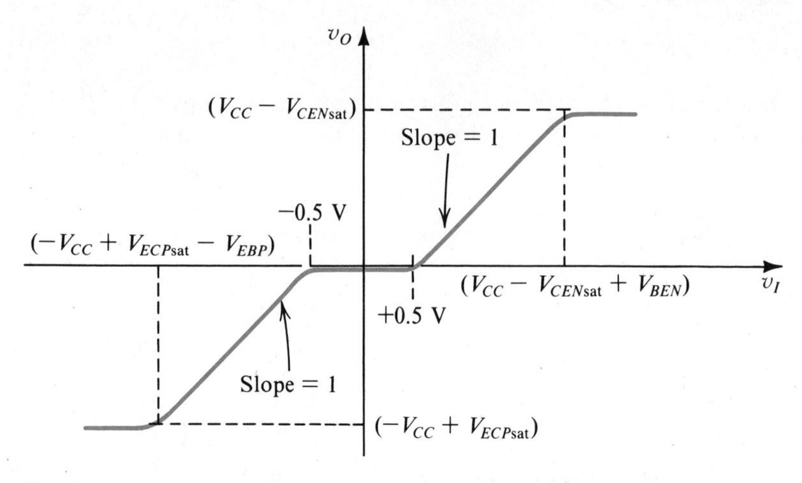

Fig. 9.6 Transfer characteristic for the class B output stage in Fig. 9.5.

Power-Conversion Efficiency

To calculate the power-conversion efficiency, η, of the class B stage, we neglect the crossover distortion and consider the case of an output sinusoid of peak amplitude $\hat{V}_o$. The average load power will be

$$P_L = \frac{1}{2}\frac{\hat{V}_o^2}{R_L} \tag{9.12}$$

The current drawn from each supply will consist of half sine waves of peak amplitude $(\hat{V}_o/R_L)$. Thus the average current drawn from each of the two power supplies will be

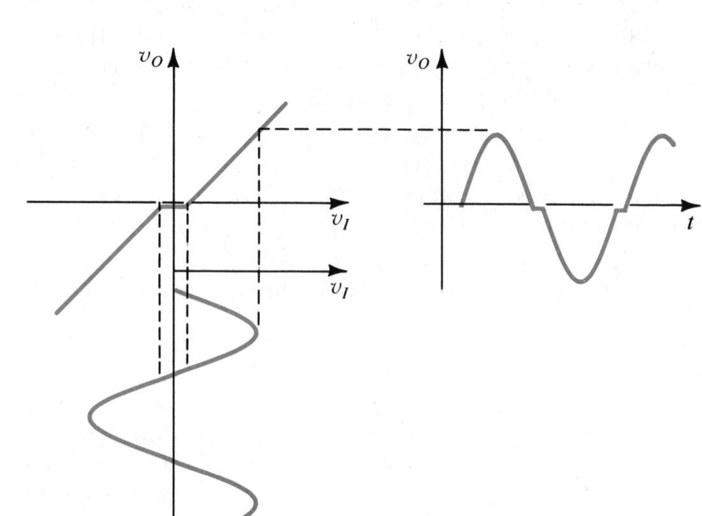

Fig. 9.7 Illustrating how the dead band in the class B transfer characteristic results in crossover distortion.

$\hat{V}_o/\pi R_L$. It follows that the average power drawn from each of the two power supplies will be the same,

$$P_{S+} = P_{S-} = \frac{1}{\pi}\frac{\hat{V}_o}{R_L}V_{CC} \tag{9.13}$$

and the total supply power will be

$$P_S = \frac{2}{\pi}\frac{\hat{V}_o}{R_L}V_{CC} \tag{9.14}$$

Thus the efficiency will be given by

$$\eta = \frac{\pi}{4}\frac{\hat{V}_o}{V_{CC}} \tag{9.15}$$

It follows that the maximum efficiency is obtained when $\hat{V}_o$ is at its maximum. This maximum is limited by the saturation of Q_N and Q_P to $V_{CC} - V_{CEsat} \simeq V_{CC}$. At this value of peak output voltage, the power-conversion efficiency is

$$\eta_{max} = \frac{\pi}{4} = 78.5\% \tag{9.16}$$

This value is much larger than that obtained in the class A stage (25%). Finally, we note that the maximum average power available from a class B output stage is obtained by substituting $\hat{V}_o = V_{CC}$ in Eq. (9.12),

$$P_{Lmax} = \frac{1}{2}\frac{V_{CC}^2}{R_L} \tag{9.17}$$

Power Dissipation

Unlike the class A stage, which dissipates maximum power under quiescent conditions ($v_O = 0$), the quiescent power dissipation of the class B stage is zero. When an input signal is applied, the *average* power dissipated in the class B stage is given by

$$P_D = P_S - P_L \tag{9.18}$$

Substituting for P_S from Eq. (9.14) and for P_L from Eq. (9.12) results in

$$P_D = \frac{2}{\pi}\frac{\hat{V}_o}{R_L}V_{CC} - \frac{1}{2}\frac{\hat{V}_o^2}{R_L} \tag{9.19}$$

From symmetry we see that half of P_D is dissipated in Q_N and the other half in Q_P. Thus Q_N and Q_P must be capable of safely dissipating $\frac{1}{2}P_D$ watts. Since P_D depends on $\hat{V}_o$, we must find the worst-case power dissipation, P_{Dmax}. Differentiating Eq. (9.19) with respect to $\hat{V}_o$ and equating the derivative to zero gives the value of $\hat{V}_o$ that results in maximum average power dissipation as

$$\hat{V}_o|_{P_{Dmax}} = \frac{2}{\pi}V_{CC} \tag{9.20}$$

Substituting this value in Eq. (9.19) gives

$$P_{D\max} = \frac{2V_{CC}^2}{\pi^2 R_L} \tag{9.21}$$

Thus

$$P_{DN\max} = P_{DP\max} = \frac{V_{CC}^2}{\pi^2 R_L} \tag{9.22}$$

At the point of maximum power dissipation the efficiency can be evaluated by substituting for $\hat{V}_o$ from Eq. (9.20) into Eq. (9.15); hence, $\eta = 50\%$.

Figure 9.8 shows a sketch of P_D (Eq. 9.19) versus the peak output voltage $\hat{V}_o$. Curves such as this are usually given on the data sheets of IC power amplifiers. (Usually, however, P_D is plotted versus

$$P_L = \frac{1}{2}\frac{\hat{V}_o^2}{R_L}$$

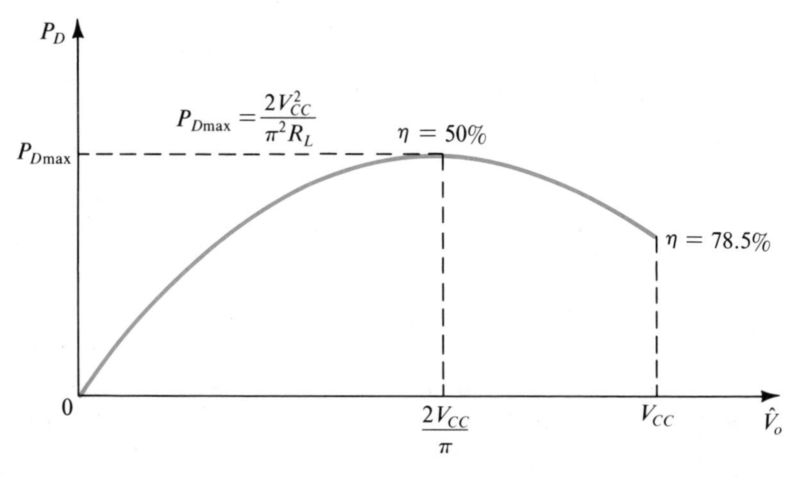

Fig. 9.8 Power dissipation of the class B output stage versus amplitude of the output sinusoid.

rather than $\hat{V}_o$). An interesting observation follows from Fig. 9.8: Increasing $\hat{V}_o$ beyond $2V_{CC}/\pi$ *decreases* the power dissipated in the class B stage while increasing the load power. The price paid is an increase in nonlinear distortion as a result of approaching the saturation region of operation of Q_N and Q_P. Transistor saturation flattens the peaks of the output sine waveform. Unfortunately, this type of distortion cannot be significantly reduced by the application of negative feedback (see Section 8.2), and thus transistor saturation should be avoided in applications requiring low THD.

EXAMPLE 9.1

It is required to design a class B output stage to deliver an average power of 20 W to an 8-Ω load. The power supply is to be selected such that V_{CC} is about 5 V greater than the peak output voltage. This avoids transistor saturation and the associated nonlinear distortion, and allows for including short-circuit protection circuitry. (The latter will be discussed in Section 9.7.) Determine the supply voltage required, the peak current drawn from each supply, the total supply power, and the power-conversion efficiency. Also determine the maximum power that each transistor must be able to dissipate safely.

Solution Since

$$P_L = \frac{1}{2}\frac{\hat{V}_o^2}{R_L}$$

then

$$\hat{V}_o = \sqrt{2P_L R_L}$$

$$= \sqrt{2 \times 20 \times 8} = 17.9 \text{ V}$$

Therefore we select $V_{CC} = 23$ V.

The peak current drawn from each supply is

$$\hat{I}_o = \frac{\hat{V}_o}{R_L} = \frac{17.9}{8} = 2.24 \text{ A}$$

The average power drawn from each supply is

$$P_{S+} = P_{S-} = \frac{1}{\pi} \times 2.24 \times 23 = 16.4 \text{ W}$$

for a total supply power of 32.8 W. The power-conversion efficiency is

$$\eta = \frac{P_L}{P_S} = \frac{20}{32.8} \times 100 = 61\%$$

The maximum power dissipated in each transistor is given by Eq. (9.22); thus

$$P_{DN\max} = P_{DP\max} = \frac{V_{CC}^2}{\pi^2 R_L}$$

$$= \frac{(23)^2}{\pi^2 \times 8} = 6.7 \text{ W}$$

Reducing Crossover Distortion

The crossover distortion of a class B output stage can be reduced substantially by employing a high-gain op amp and overall negative feedback, as shown in Fig. 9.9. The ±0.7-V dead band is reduced to $\pm 0.7/A_0$ volts, where A_0 is the dc gain of the op amp. Nevertheless, the slew-rate limitation of the op amp will cause the alternate turning on and off of the output transistors to be noticeable, especially at high frequencies. A more practical method for reducing and almost eliminating crossover distortion is found in the class AB stage, which will be studied in the next section.

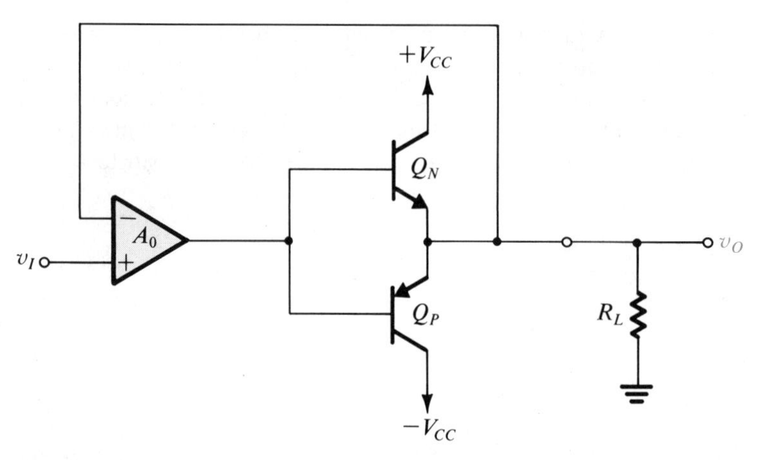

Fig. 9.9 Class B circuit with an op amp connected in a negative-feedback loop to reduce crossover distortion.

Single-Supply Operation

The class B stage can be operated from a single power supply, in which case the load is capacitively coupled, as shown in Fig. 9.10. Note that in order to make the formulas derived above directly applicable, the single power supply is denoted $2V_{CC}$.

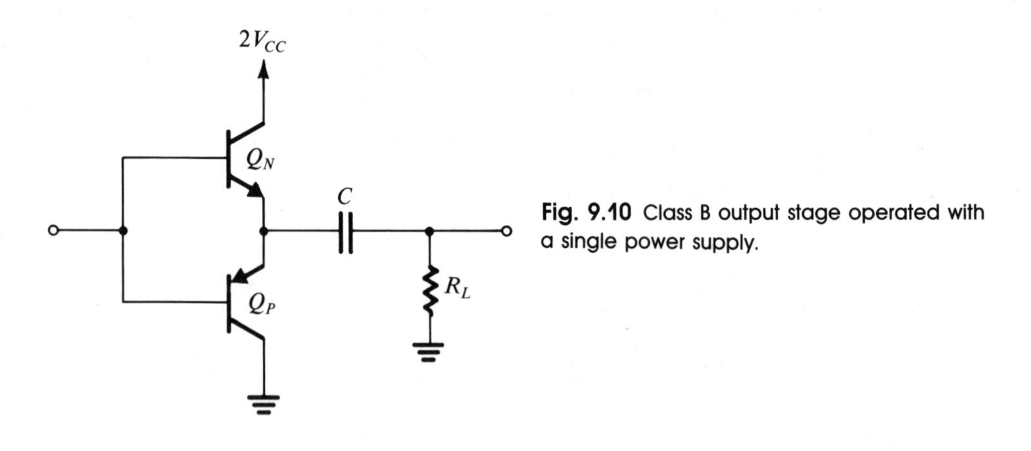

Fig. 9.10 Class B output stage operated with a single power supply.

Exercise 9.5 For the class B output stage of Fig. 9.5 let $V_{CC} = 6$ V and $R_L = 4\ \Omega$. If the output is a sinusoid with 4.5-V peak amplitude, find (a) the output power; (b) the average power drawn from each supply; (c) the power efficiency obtained at this output voltage; (d) the peak currents supplied by v_I, assuming that $\beta_N = \beta_P = 50$; (e) the maximum power that each transistor must be capable of dissipating safely.

Ans. (a) 2.53 W; (b) 2.15 W; (c) 59%; (d) 22.1 mA; (e) 0.91 W

9.4 CLASS AB OUTPUT STAGE

Crossover distortion can be virtually eliminated by biasing the complementary output transistors at a small, nonzero current. The result is the class AB output stage shown in Fig. 9.11. A bias voltage V_{BB} is applied between the bases of Q_N and Q_P. For $v_I = 0$, $v_O = 0$ and a voltage $V_{BB}/2$ appears across the base-emitter junction of each of Q_N and Q_P. Assuming matched devices,

$$i_N = i_P = I_Q = I_S e^{V_{BB}/2V_T} \tag{9.23}$$

The value of V_{BB} is selected so as to yield the required quiescent current I_Q.

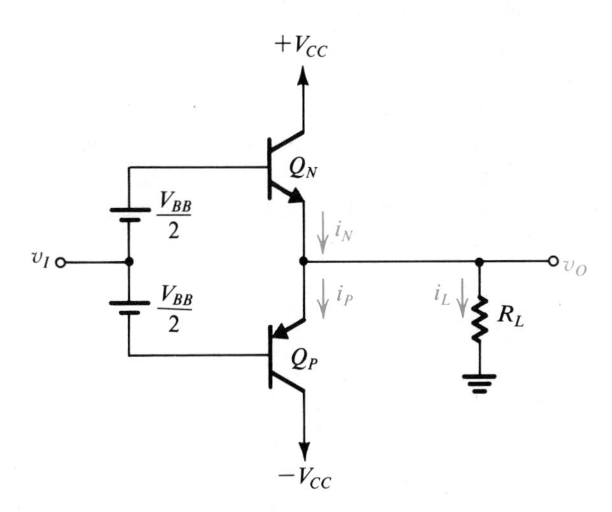

Fig. 9.11 Class AB output stage. A bias voltage V_{BB} is applied between the bases of Q_N and Q_P, giving rise to a bias current I_Q given by Eq. (9.23). Thus for small v_I, both transistors conduct and crossover distortion is almost completely eliminated.

Circuit Operation

When v_I goes positive by a certain amount, the voltage at the base of Q_N increases by the same amount and the output becomes positive at an almost equal value,

$$v_O = v_I + \frac{V_{BB}}{2} - v_{BEN} \tag{9.24}$$

The positive v_O causes a current i_L to flow through R_L, and thus i_N must increase; that is,

$$i_N = i_P + i_L \tag{9.25}$$

The increase in i_N will be accompanied by a corresponding increase in v_{BEN} (above the quiescent value of $V_{BB}/2$). However, since the voltage between the two bases remains constant at V_{BB}, the increase in v_{BEN} will result in an equal decrease in v_{EBP} and hence in i_P. The relationship between i_N and i_P can be derived as follows:

$$v_{BEN} + v_{EBP} = V_{BB}$$

$$V_T \ln\left(\frac{i_N}{I_S}\right) + V_T \ln\left(\frac{i_P}{I_S}\right) = 2V_T \ln\left(\frac{I_Q}{I_S}\right)$$

$$i_N i_P = I_Q^2 \tag{9.26}$$

Thus as i_N increases, i_P decreases by the same ratio while the product remains constant. Equations (9.25) and (9.26) can be combined to yield i_N for a given i_L as the solution to the quadratic equation

$$i_N^2 - i_L i_N - I_Q^2 = 0 \tag{9.27}$$

From the above, we can see that for positive output voltages, the load current is supplied by Q_N, which acts as the output emitter follower. Meanwhile, Q_P will be conducting a current that decreases as v_O increases; for large v_O the current in Q_P can be ignored altogether.

For negative input voltages the opposite occurs; the load current will be supplied by Q_P, which acts as the output emitter follower, while Q_N conducts a current that gets smaller as v_I becomes more negative. Equation (9.26) relating i_N and i_P holds for negative inputs as well.

We conclude that the class AB stage operates in much the same manner as the class B circuit, with one important exception: For small v_I, both transistors conduct, and as v_I is increased or decreased, one of the two transistors takes over the operation. Since the transition is a smooth one, crossover distortion will be almost totally eliminated. Figure 9.12 shows the transfer characteristic of the class AB stage.

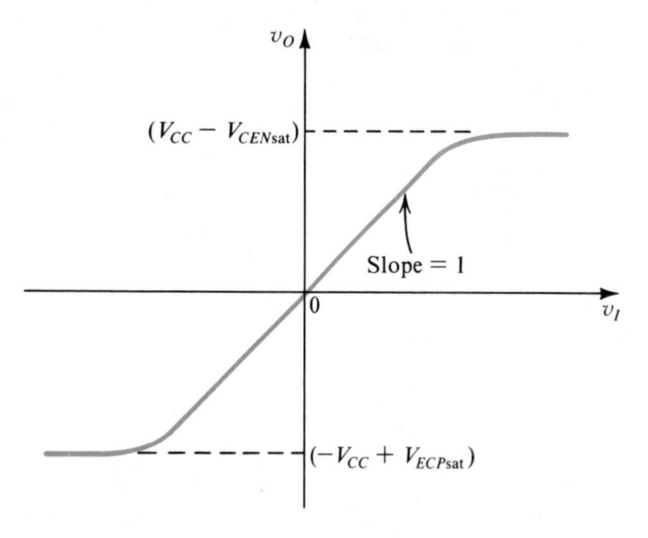

Fig. 9.12 Transfer characteristic of the class AB stage in Fig. 9.11.

The power relationships in the class AB stage are almost identical to those derived for the class B circuit in the previous section. The only difference is that under quiescent conditions the class AB circuit dissipates a power of $V_{CC}I_Q$ per transistor. Since I_Q is usually much smaller than the peak load current, the quiescent power dissipation is usually small. Nevertheless, it can be taken into account easily. Specifically, we can simply add the quiescent dissipation per transistor to its maximum power dissipation with an input signal applied, to obtain the total power dissipation that the transistor must be able to handle safely.

Output Resistance

If we assume that the source supplying v_I is ideal, then the output resistance of the class AB stage can be determined from the circuit in Fig. 9.13 as

$$R_{out} = r_{eN} // r_{eP} \tag{9.28}$$

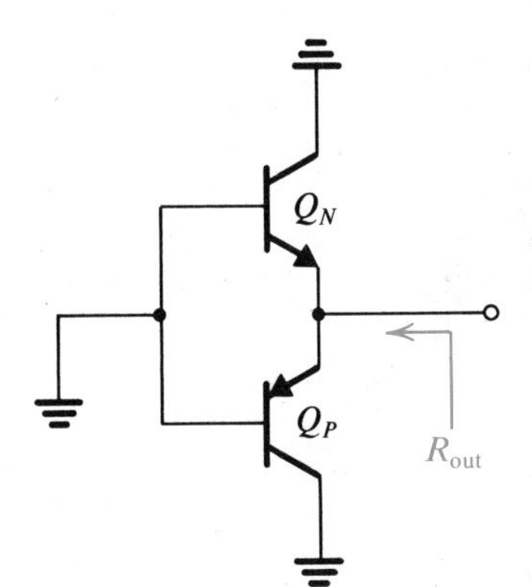

Fig. 9.13 Determining the small-signal output resistance of the class AB circuit of Fig. 9.11.

where r_{eN} and r_{eP} are the small-signal emitter resistances of Q_N and Q_P, respectively. At a given input voltage, the currents i_N and i_P can be determined, and r_{eN} and r_{eP} are given by

$$r_{eN} = \frac{V_T}{i_N} \tag{9.29}$$

$$r_{eP} = \frac{V_T}{i_P} \tag{9.30}$$

Thus

$$R_{out} = \frac{V_T}{i_N} \bigg/ \bigg/ \frac{V_T}{i_P} = \frac{V_T}{i_P + i_N} \tag{9.31}$$

Since as i_N increases, i_P decreases, and vice versa, the output resistance remains approximately constant in the region around $v_I = 0$. This, in effect, is the reason for the virtual absence of crossover distortion. At larger load currents either i_N or i_P will be significant, and R_{out} decreases as the load current increases.

Exercise 9.6 Consider a class AB circuit with $V_{CC} = 15$ V, $I_Q = 2$ mA, and $R_L = 100\ \Omega$. Determine V_{BB}. Construct a table giving i_L, i_N, i_P, v_{BEN}, v_{EBP}, v_I, v_O/v_I, R_{out}, and v_o/v_i versus v_O for $v_O = 0, 0.1, 0.2, 0.5, 1, 5, 10, -0.1, -0.2, -0.5, -1, -5$ and -10 V. Note that v_O/v_I is the large-signal voltage gain and v_o/v_i is the incremental gain obtained as $R_L/(R_L + R_{out})$. The incremental gain is equal to the slope of the transfer curve. Assume Q_N and Q_P to be matched, with $I_S = 10^{-13}$ A.

Ans. $V_{BB} = 1.186$ V

v_O (V)	i_L (mA)	i_N (mA)	i_P (mA)	v_{BEN} (V)	v_{EBP} (V)	v_I (V)	v_O/v_I	R_{out} (Ω)	v_o/v_i
+10.0	100	100.04	0.04	0.691	0.495	10.1	0.99	0.25	1.00
+5.0	50	50.08	0.08	0.673	0.513	5.08	0.98	0.50	1.00
+1.0	10	10.39	0.39	0.634	0.552	1.041	0.96	2.32	0.98
+0.5	5	5.70	0.70	0.619	0.567	0.526	0.95	4.03	0.96
+0.2	2	3.24	1.24	0.605	0.581	0.212	0.94	5.58	0.95
+0.1	1	2.56	1.56	0.599	0.587	0.106	0.94	6.07	0.94
0	0	2	2	0.593	0.593	0	—	6.25	0.94
−0.1	−1	1.56	2.56	0.587	0.599	−0.106	0.94	6.07	0.94
−0.2	−2	1.24	3.24	0.581	0.605	−0.212	0.94	5.58	0.95
−0.5	−5	0.70	5.70	0.567	0.619	−0.526	0.95	4.03	0.96
−1.0	−10	0.39	10.39	0.552	0.634	−1.041	0.96	2.32	0.98
−5.0	−50	0.08	50.08	0.513	0.673	−5.08	0.98	0.50	1.00
−10.0	−100	0.04	100.04	0.495	0.691	−10.1	0.99	0.25	1.00

9.5 BIASING THE CLASS AB CIRCUIT

In this section we discuss two approaches for generating the voltage V_{BB} required for biasing the class AB output stage.

Biasing Using Diodes

Figure 9.14 shows a class AB circuit in which the bias voltage V_{BB} is generated by passing a constant current I_{bias} through a pair of diodes, or diode-connected transistors, D_1 and D_2. In circuits that supply large amounts of power the output transistors are large-geometry devices. The biasing diodes, however, need not be large devices, and thus the quiescent current I_Q established in Q_N and Q_P will be $I_Q = nI_{bias}$, when n is the ratio of the emitter junction area of the output devices to the junction area of the biasing diodes. In other words, the saturation (or scale) current I_S of the output transistors is n times that of

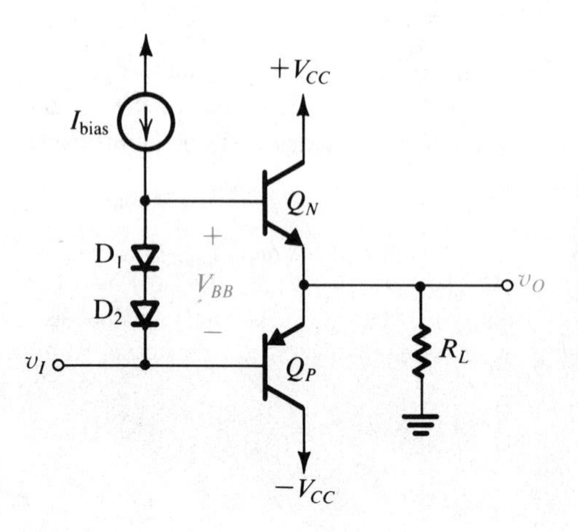

Fig. 9.14 Class AB output stage utilizing diodes for biasing. If the junction area of the output devices, Q_N and Q_P, is n times that of the biasing devices D_1 and D_2, a quiescent current $I_Q = nI_{bias}$ flows in the output devices.

the biasing diodes. Area ratioing is simple to implement in integrated circuits but difficult to realize in discrete-circuit designs.

When the output stage of Fig. 9.14 is sourcing current to the load, the base current of Q_N increases from I_Q/β_N (which is usually small) to approximately i_L/β_N. This base current drive must be supplied by the current source I_{bias}. It follows that I_{bias} must be greater than the maximum anticipated base drive for Q_N. This sets a lower limit on the value of I_{bias}. Now, since $I_Q = nI_{\text{bias}}$ and since I_Q is usually much smaller than the peak load current (less than 10%), we see that we cannot make n a large number. In other words, we cannot make the diodes much smaller than the output devices. This is a disadvantage of the diode biasing scheme.

From the above discussion we see that the current through the biasing diodes will decrease when the output stage is sourcing current to the load. Thus the bias voltage V_{BB} will also decrease, and the analysis of the previous section must be modified to take this effect into account.

The diode biasing arrangement has an important advantage: It can provide thermal stabilization of the quiescent current in the output stage. To appreciate this point recall that the class AB output stage dissipates power under quiescent conditions. Power dissipation raises the internal temperature of the BJTs. From Chapter 4 we know that a rise in transistor temperature results in a decrease in its V_{BE} (approximately -2 mV/°C) if the collector current is held constant. Alternatively, if V_{BE} is held constant and the temperature increases, the collector current increases. The increase in collector current increases the power dissipation, which in turn increases the collector current. Thus a positive feedback mechanism exists that can result in a phenomenon called **thermal runaway.** Unless checked, thermal runaway can lead to the ultimate destruction of the BJT. Diode biasing can be arranged to provide a compensating effect that can protect the output transistors against thermal runaway under quiescent conditions. Specifically, if the diodes are in close thermal contact with the output transistors, their temperature will increase by the same amount as that of Q_N and Q_P. Thus V_{BB} will decrease at the same rate as $V_{BEN} + V_{EBP}$, with the result that I_Q remains constant. Close thermal contact is easily achieved in IC fabrication. It is obtained in discrete circuits by mounting the bias diodes on the metal case of Q_N or Q_P.

EXAMPLE 9.2

Consider the class AB output stage under the conditions that $V_{CC} = 15$ V, $R_L = 100\ \Omega$, and the output is sinusoidal with a maximum amplitude of 10 V. Let Q_N and Q_P be matched with $I_S = 10^{-13}$ A and $\beta = 50$. Assume that the biasing diodes have one-third the junction area of the output devices. Find the value of I_{bias} that guarantees a minimum of 1 mA through the diodes at all times. Determine the quiescent current and the quiescent power dissipation in the output transistors (i.e., at $v_O = 0$). Also find V_{BB} for $v_O = 0$, $+10$ V, and -10 V.

Solution

The maximum current through Q_N is approximately equal to $i_{L\text{max}} = 10\text{ V}/0.1\text{ k}\Omega = 100$ mA. Thus the maximum base current in Q_N is approximately 2 mA. To maintain a minimum of 1 mA through the diodes, we select $I_{\text{bias}} = 3$ mA. The area ratio of 3 yields a quiescent current of 9 mA through Q_N and Q_P. The quiescent power dissipation is

$$P_{DQ} = 2 \times 15 \times 9 = 270\text{ mW}$$

For $v_O = 0$, the base current of Q_N is $9/51 \simeq 0.18$ mA, leaving a current of $3 - 0.18 = 2.82$ mA to flow through the diodes. Since the diodes have $I_S = \frac{1}{3} \times 10^{-13}$ A, the voltage V_{BB} will be

$$V_{BB} = 2V_T \ln \frac{2.82 \text{ mA}}{I_S} = 1.26 \text{ V}$$

At $v_O = +10$ V, the current through the diodes will decrease to 1 mA, resulting in $V_{BB} \simeq 1.21$ V. At the other extreme of $v_O = -10$ V, Q_N will be conducting a very small current; thus its base current will be negligibly small and all of I_{bias} (3 mA) flows through the diodes, resulting in $V_{BB} \simeq 1.26$ V.

Exercises 9.7 For the circuit of Example 9.2 find i_N and i_P for $v_O = +10$ V and $v_O = -10$ V.

Ans. 100.1 mA, 0.1 mA; 0.8 mA, 100.8 mA

9.8 If the collector current of a transistor is held constant, its v_{BE} decreases by 2 mV for every °C rise in temperature. Alternatively, if v_{BE} is held constant, then i_C increases by approximately $g_m \times 2$ mV for every °C rise in temperature. For a device operating at $I_C = 10$ mA find the change in collector current resulting from an increase in temperature of 5°C.

Ans. 4 mA

Biasing Using the V_{BE} Multiplier

An alternative biasing arrangement that provides the designer with considerably more flexibility in both discrete and integrated designs is shown in Fig. 9.15. The bias circuit

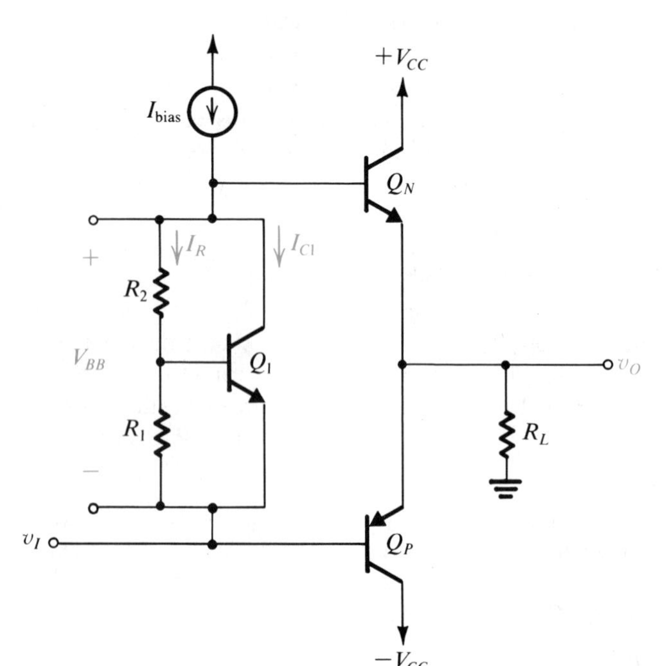

Fig. 9.15 Class AB output stage utilizing a V_{BE} multiplier for biasing.

consists of transistor Q_1 with a resistor R_1 connected between base and emitter and a feedback resistor R_2 connected between collector and base. The resulting two-terminal network is fed with a constant-current source I_{bias}. If we neglect the base current of Q_1, then R_1 and R_2 will carry the same current I_R, given by

$$I_R = \frac{V_{BE1}}{R_1} \tag{9.32}$$

and the voltage V_{BB} across the bias network will be

$$\begin{aligned} V_{BB} &= I_R(R_1 + R_2) \\ &= V_{BE1}\left(1 + \frac{R_2}{R_1}\right) \end{aligned} \tag{9.33}$$

Thus the circuit simply multiplies V_{BE1} by the factor $(1 + R_2/R_1)$, and is known as the "V_{BE} multiplier." The multiplication factor is obviously under the designer's control and can be used to establish the value of V_{BB} required to yield a desired quiescent current I_Q. In IC design it is relatively easy to control accurately the ratio of two resistances. In discrete-circuit design, a potentiometer can be used, as shown in Fig. 9.16, and is manually set to produce the desired value of I_Q.

The value of V_{BE1} in Eq. (9.33) is determined by the portion of I_{bias} that flows through the collector of Q_1; that is,

$$I_{C1} = I_{\text{bias}} - I_R \tag{9.34}$$

$$V_{BE1} = V_T \ln\left(\frac{I_{C1}}{I_{S1}}\right) \tag{9.35}$$

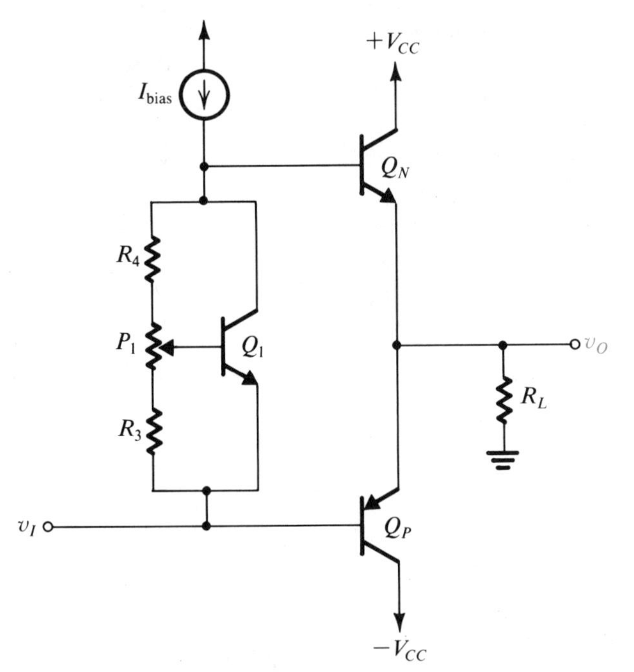

Fig. 9.16 A discrete-circuit class AB output stage with a potentiometer used in the V_{BE} multiplier. The potentiometer is adjusted to yield the desired value of quiescent current in Q_N and Q_P.

where we have neglected the base current of Q_N, which is normally small both under quiescent conditions and when the output voltage is swinging negative. However, for positive v_O, especially at and near its peak value, the base current of Q_N can become sizable and reduces the current available for the V_{BE} multiplier. Nevertheless, since large changes in I_{C1} correspond to only small changes in V_{BE1}, the decrease in current will be mostly absorbed by Q_1, leaving I_R, and hence V_{BB}, almost constant.

Exercise 9.9 Consider a V_{BE} multiplier with $R_1 = R_2 = 1.2\ \text{k}\Omega$, utilizing a transistor that has $V_{BE} = 0.6$ V at $I_C =$ 1 mA, and a very high β. (a) Find the value of the current I that should be supplied to the multiplier to obtain a terminal voltage of 1.2 V. (b) Find the value of I that will result in the terminal voltage changing (from the 1.2-V value) by +50 mV, +100 mV, +200 mV, −50 mV, −100 mV, −200 mV.

Ans. (a) 1.5 mA; (b) 3.24 mA, 7.93 mA, 55.18 mA, 0.85 mA, 0.59 mA, 0.43 mA

Like the diode biasing network, the V_{BE}-multiplier circuit can provide thermal stabilization of I_Q. This is especially true if $R_1 = R_2$, and if Q_1 is in close thermal contact with the output transistors.

EXAMPLE 9.3

It is required to redesign the output stage of Example 9.2 utilizing a V_{BE} multiplier for biasing. Use a small-geometry transistor for Q_1 with $I_S = 10^{-14}$ A and design for a quiescent current $I_Q = 2$ mA.

Solution

Since the peak positive current is 100 mA, the base current of Q_N can be as high as 2 mA. We shall therefore select $I_{\text{bias}} = 3$ mA, thus providing the multiplier with a minimum current of 1 mA.

Under quiescent conditions ($v_O = 0$ and $i_L = 0$) the base current of Q_N can be neglected and all of I_{bias} flows through the multiplier. We now must decide on how this current (3 mA) is to be divided between I_{C1} and I_R. If we select I_R greater than 1 mA, the transistor will be almost cut off at the positive peak of v_O. Therefore, we shall select $I_R = 0.5$ mA, leaving 2.5 mA for I_{C1}.

To obtain a quiescent current of 2 mA in the output transistors, V_{BB} should be

$$V_{BB} = 2V_T \ln \frac{2 \times 10^{-3}}{10^{-13}} = 1.19\ \text{V}$$

We can now determine $R_1 + R_2$ as follows:

$$R_1 + R_2 = \frac{V_{BB}}{I_R} = \frac{1.19}{0.5} = 2.38\ \text{k}\Omega$$

At a collector current of 2.5 mA, Q_1 has

$$V_{BE1} = V_T \ln \frac{2.5 \times 10^{-3}}{10^{-14}} = 0.66\ \text{V}$$

The value of R_1 can now be determined as

$$R_1 = \frac{0.66}{0.5} = 1.32\ \text{k}\Omega$$

and R_2 as

$$R_2 = 2.38 - 1.32 = 1.06 \text{ k}\Omega$$

9.6 POWER BJTs

Transistors that are required to conduct currents in the ampere range and withstand power dissipation in the watts and tens-of-watts range differ in their physical structure, packaging, and specification from the small-signal transistors considered in previous chapters. In this section we consider some of the important properties of power transistors, especially those aspects that pertain to the design of circuits of the type discussed in the previous sections. There are, of course, other important applications of power transistors, such as their use as switching elements in power inverters and motor-control circuits. Such applications are not studied in this book.

Junction Temperature

Power transistors dissipate large amounts of power in their collector-base junctions. The dissipated power is converted into heat, which raises the junction temperature. However, the junction temperature, T_J, must not be allowed to exceed a specified maximum, $T_{J\max}$; otherwise the transistor could suffer permanent damage. For silicon devices $T_{J\max}$ is in the range from 150°C to 200°C.

Thermal Resistance

Consider first the case of a transistor operating in free air—that is, with no special arrangements for cooling. The heat dissipated in the transistor junction will be conducted away from the junction to the transistor case, and from the case to the surrounding environment. In a steady state in which the transistor is dissipating P_D watts, the temperature rise of the junction relative to the surrounding ambience can be expressed as

$$T_J - T_A = \theta_{JA} P_D \tag{9.36}$$

where θ_{JA} is the **thermal resistance** between junction and ambience, having the units of °C per watt. Note that θ_{JA} simply gives the rise in junction temperature over ambient temperature for each watt of dissipated power. Since we wish to be able to dissipate large amounts of power without raising the junction temperature above $T_{J\max}$, it is desirable to have as small a value for the thermal resistance θ_{JA} as possible. For operation in free air, θ_{JA} depends primarily on the type of case in which the transistor is packaged. The value of θ_{JA} is usually specified on the transistor data sheet.

Equation (9.36), which describes the thermal-conduction process, is analogous to Ohm's law, which describes the electrical-conduction process. In this analogy, power dissipation corresponds to current, temperature difference corresponds to voltage difference, and thermal resistance corresponds to electrical resistance. Thus, we may represent the thermal-conduction process by the electric circuit shown in Fig. 9.17.

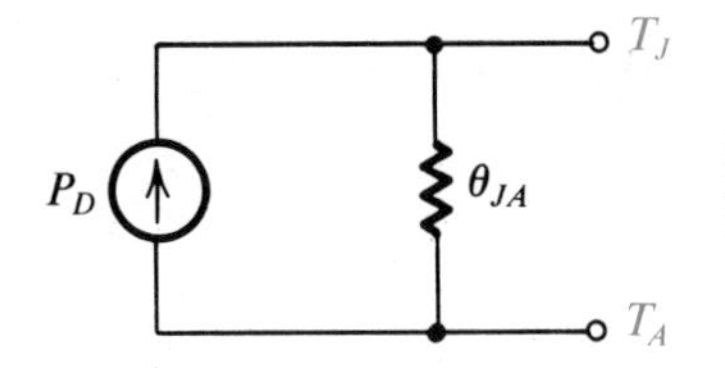

Fig. 9.17 Electrical equivalent circuit of the thermal-conduction process; $T_J - T_A = P_D\theta_{JA}$.

Power Dissipation versus Temperature

The transistor manufacturer usually specifies $T_{J\max}$, the maximum power dissipation at a particular ambient temperature T_{A0} (usually, 25°C), and the thermal resistance θ_{JA}. In addition, a graph such as that shown in Fig. 9.18 is usually provided. The graph simply states that for operation at ambient temperatures below T_{A0}, the device can safely dissipate the rated value of P_{D0} watts. However, if the device is to be operated at higher ambient temperatures, the maximum allowable power dissipation must be **derated** according to the straight line shown in Fig. 9.18. The power-derating curve is a graphical representation of Eq. (9.36). Specifically, note that if the ambient temperature is T_{A0} and the power dissipation is at the maximum allowed (P_{D0}), then the junction temperature will be $T_{J\max}$. Substituting these quantities in Eq. (9.36) results in

$$\theta_{JA} = \frac{T_{J\max} - T_{A0}}{P_{D0}}, \tag{9.37}$$

which is the inverse of the slope of the power-derating straight line. At an ambient temperature T_A, higher than T_{A0}, the maximum allowable power dissipation $P_{D\max}$ can be obtained from Eq. (9.36) by substituting $T_J = T_{J\max}$, thus

$$P_{D\max} = \frac{T_{J\max} - T_A}{\theta_{JA}} \tag{9.38}$$

Observe that as T_A approaches $T_{J\max}$, the allowable power dissipation decreases; the lower thermal gradient limits the amount of heat that can be removed from the junction. In the extreme case of $T_A = T_{J\max}$, no power can be dissipated because no heat can be removed from the junction.

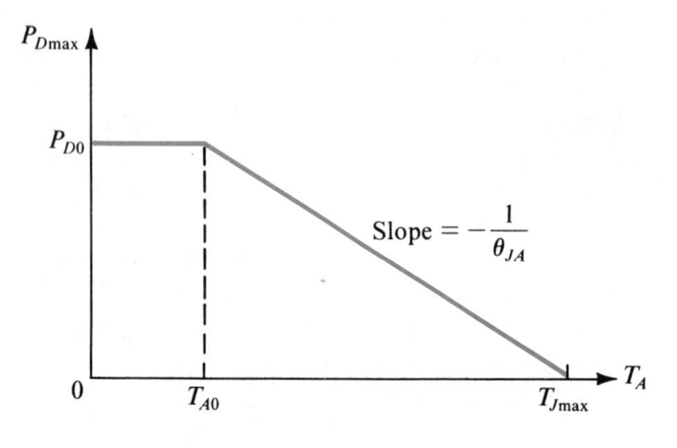

Fig. 9.18 Maximum allowable power dissipation versus ambient temperature for a BJT operated in free air. This is known as a "power-derating" curve.

EXAMPLE 9.4

A BJT is specified to have a maximum power dissipation P_{D0} of 2 watts at an ambient temperature T_{A0} of 25°C, and a maximum junction temperature $T_{J\max}$ of 150°C. Find the following.

(a) The thermal resistance θ_{JA}.

(b) The maximum power that can be safely dissipated at an ambient temperature of 50°C.

(c) The junction temperature if the device is operating at $T_A = 25$°C and is dissipating 1 W.

Solution

(a) $\theta_{JA} = \dfrac{T_{J\max} - T_{A0}}{P_{D0}} = \dfrac{150 - 25}{2} = 62.5\text{°C/W}$

(b) $P_{D\max} = \dfrac{T_{J\max} - T_A}{\theta_{JA}} = \dfrac{150 - 50}{62.5} = 1.6 \text{ W}$

(c) $T_J = T_A + \theta_{JA} P_D = 25 + 62.5 \times 1 = 87.5\text{°C}$

Transistor Case and Heat Sink

The thermal resistance between junction and ambience, θ_{JA}, can be expressed as

$$\theta_{JA} = \theta_{JC} + \theta_{CA} \tag{9.39}$$

where θ_{JC} is the thermal resistance between junction and transistor case (package) and θ_{CA} is the thermal resistance between case and ambience. For a given transistor, θ_{JC} is fixed by the device design and packaging. The device manufacturer can reduce θ_{JC} by encapsulating the device in a relatively large metal case and placing the collector (where most of the heat is dissipated) in direct contact with the case. Most high-power transistors are packaged in this fashion. Fig. 9.19 shows a sketch of a typical package.

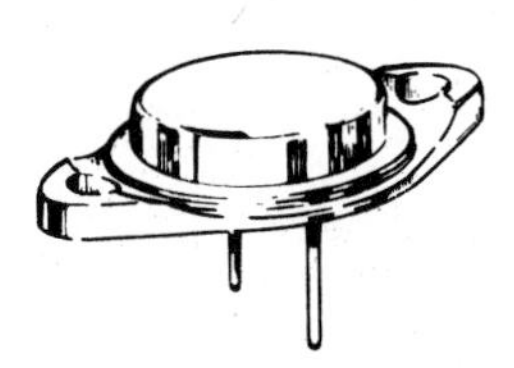

Fig. 9.19 A popular package for power transistors. The case is metal with a diameter of about 2.2 cm; the outside dimension of the "seating plane" is about 4 cm. The seating plane has two holes for screws to bolt it to a heat sink. The collector is electrically connected to the case.

Although the circuit designer has no control over θ_{JC} (once a particular transistor is selected), the designer can considerably reduce θ_{CA} below its free-air value (specified by the manufacturer as part of θ_{JA}). Reduction of θ_{CA} can be effected by providing means to facilitate heat transfer from case to ambience. A popular approach is to bolt the transistor to the chassis or to an extended metal surface. Such a metal surface then functions as a **heat sink.** Heat is easily conducted from the transistor case to the heat sink; that is, the thermal resistance θ_{CS} is usually very small. Also, heat is efficiently transferred (by convection) from the heat sink to the ambience, resulting in a low thermal resistance θ_{SA}. Thus, if a heat sink is utilized, the case-to-ambience thermal resistance will be given by

$$\theta_{CA} = \theta_{CS} + \theta_{SA} \tag{9.40}$$

and can be small because its two components can be made small by the choice of an appropriate heat sink.[2] For example, in very high-power applications the heat sink is usually equipped with fins that further facilitate cooling by radiation.

The electrical analog of the thermal-conduction process when a heat sink is employed is shown in Fig. 9.20, from which we can write

$$T_J - T_A = P_D(\theta_{JC} + \theta_{CS} + \theta_{SA}) \tag{9.41}$$

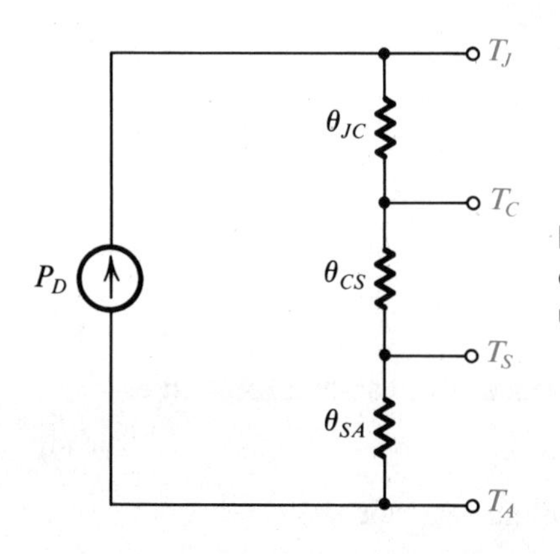

Fig. 9.20 Electrical analog of the thermal conduction process when a heat sink is utilized.

As well as specifying θ_{JC}, the device manufacturer usually supplies a derating curve for $P_{D\max}$ versus the case temperature, T_C. Such a curve is shown in Fig. 9.21. Note that the slope of the power-derating straight line is $-1/\theta_{JC}$. For a given transistor, the maximum

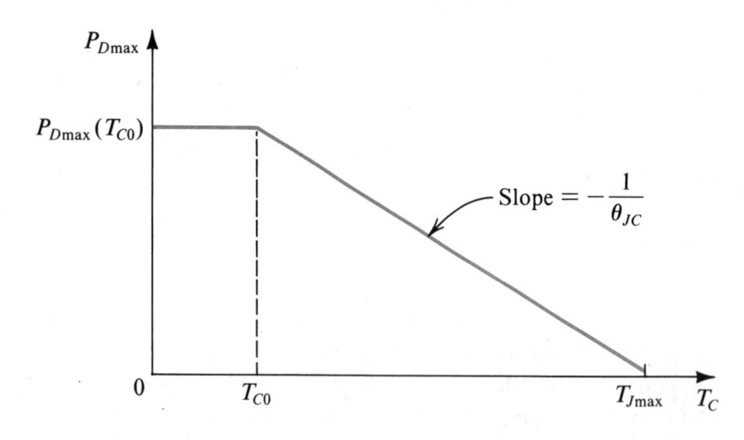

Fig. 9.21 Maximum allowable power dissipation versus the transistor case temperature.

[2] As mentioned earlier, the metal case of a power transistor is electrically connected to the collector. Thus an electrically insulating material such as mica is usually placed between the metal case and the metal heat sink. Also, insulating bushings are generally used in bolting the transistor to the heat sink.

power dissipation at a *case temperature* T_{C0} (usually 25°C) is much greater than that at an *ambient temperature* T_{A0} (usually 25°C). If the device can be maintained at a case temperature T_C, $T_{C0} \leq T_C \leq T_{J\max}$, then the maximum safe power dissipation is obtained when $T_J = T_{J\max}$,

$$P_{D\max} = \frac{T_{J\max} - T_C}{\theta_{JC}} \tag{9.42}$$

EXAMPLE 9.5

A BJT is specified to have $T_{J\max} = 150°\text{C}$ and to be capable of dissipating the following maximum power:

$$40 \text{ W at } T_C = 25°\text{C}$$

$$2 \text{ W at } T_A = 25°\text{C}$$

Above 25°C, the maximum power dissipation is to be derated linearly with $\theta_{JC} = 3.12°\text{C/W}$ and $\theta_{JA} = 62.5°\text{C/W}$. Find the following.

(a) The maximum power that can be dissipated safely by this transistor when operated in free air at $T_A = 50°\text{C}$.

(b) The maximum power that can be dissipated safely by this transistor when operated at an ambient temperature of 50°C, but with a heat sink for which $\theta_{CS} = 0.5°\text{C/W}$ and $\theta_{SA} = 4°\text{C/W}$. In this case find the temperature of the case and of the heat sink.

(c) The maximum power that can be dissipated safely if an *infinite heat sink* is used and $T_A = 50°\text{C}$.

Solution

(a) $P_{D\max} = \dfrac{T_{J\max} - T_A}{\theta_{JA}} = \dfrac{150 - 50}{62.5} = 1.6 \text{ W}$

(b) With a heat sink, θ_{JA} becomes

$$\theta_{JA} = \theta_{JC} + \theta_{CS} + \theta_{SA}$$

$$= 3.12 + 0.5 + 4 = 7.62°\text{C/W}$$

Thus

$$P_{D\max} = \frac{150 - 50}{7.62} = 13.1 \text{ W}$$

Figure 9.22 shows the thermal equivalent circuit with the various temperatures indicated.

(c) An infinite heat sink, if it existed, would cause the case temperature T_C to equal the ambient temperature T_A. The infinite heat sink has $\theta_{CA} = 0$. Obviously, one cannot buy an infinite heat sink; nevertheless, this terminology is used by some manufacturers to describe the power-derating curve of Fig. 9.21. The abcissa is then labeled T_A and the curve is called ''power dissipation versus ambient temper-

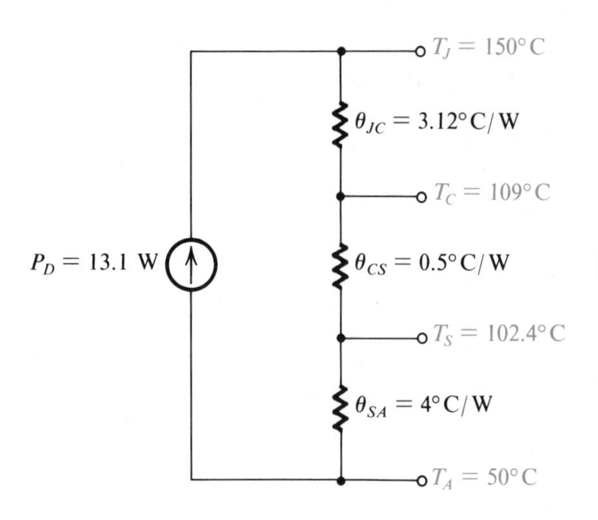

Fig. 9.22 Thermal equivalent circuit for Example 9.5.

ature with infinite heat sink.'' For our example, with infinite heat sink,

$$P_{D\max} = \frac{T_{J\max} - T_A}{\theta_{JC}} = \frac{150 - 50}{3.12} = 32 \text{ W}$$

The advantage of using a heat sink is clearly evident from the above example: With a heat sink, the maximum allowable power dissipation increases from 1.6 W to 13.1 W. Also note that although the transistor considered can be called a ''40-W transistor,'' this level of power dissipation cannot be achieved in practice; it would require an infinite heat sink and an ambient temperature $T_A \leq 25°C$.

Exercise 9.10 The 2N6306 power transistor is specified to have $T_{J\max} = 200°C$ and $P_{D\max} = 125$ W for $T_C \leq 25°C$. For $T_C \geq 25°C$, $\theta_{JC} = 1.4°C/W$. If in a particular application this device is to dissipate 50 W and operate at an ambient temperature of 25°C, find the maximum thermal resistance of the heat sink that must be used (i.e., θ_{SA}). Assume $\theta_{CS} = 0.6°C/W$. What is the case temperature, T_C?

Ans. 1.5°C/W; 130°C

The BJT Safe Operating Area

In addition to specifying the maximum power dissipation at different case temperatures, power transistor manufacturers usually provide a plot of the boundary of the safe operating area (SOA) in the i_C–v_{CE} plane. The SOA specification takes the form illustrated by the sketch in Fig. 9.23. The paragraph numbers below correspond to the boundaries on the sketch.

1. The maximum allowable current $I_{C\max}$. Exceeding this current on a continuous basis can result in melting the wires that bond the device to the package terminals.

2. The maximum power dissipation hyperbola. This is the locus of the points for which $v_{CE} i_C = P_{D\max}$(at T_{C0}). For temperatures $T_C > T_{C0}$, the power-derating

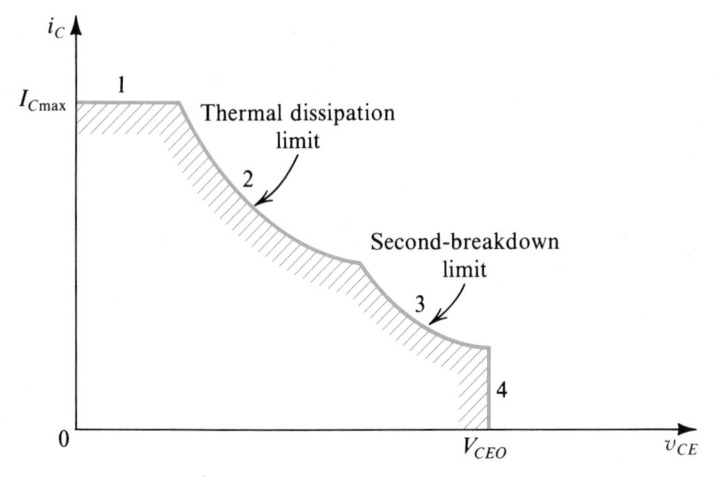

Fig. 9.23 Safe operating area (SOA) of a BJT.

curves described above should be used to obtain the applicable $P_{D\max}$ and thus a correspondingly lower hyperbola. Although the operating point can be allowed to move temporarily above the hyperbola, the *average* power dissipation should not be allowed to exceed $P_{D\max}$.

3. The **second-breakdown** limit. Second breakdown is a phenomenon that results because current flow across the emitter–base junction is not uniform. Rather, the current density is greatest near the periphery of the junction. This ''current crowding'' gives rise to increased localized power dissipation and hence temperature rise (at locations called hot spots). Since a temperature rise causes an increase in current, a form of thermal runaway can occur, leading to junction destruction.

4. The collector-to-emitter breakdown voltage, BV_{CEO}. The instantaneous value of v_{CE} should never be allowed to exceed BV_{CEO}; otherwise, avalanche breakdown of the collector-base junction may occur (see Section 4.13).

Finally, it should be mentioned that logarithmic scales are usually used for i_C and v_{CE}, leading to an SOA boundary that consists of straight lines.

Parameter Values of Power Transistors

Owing to their large geometry and high operating currents, power transistors display typical parameter values that can be quite different from those of small-signal transistors. The important differences are as follows:

1. At high currents, the exponential i_C–v_{BE} relationship exhibits a constant $n = 2$; that is, $i_C = I_S e^{v_{BE}/2V_T}$.

2. β is low, typically 30 to 80, but it can be as low as 5. Here it is important to note that β has a positive temperature coefficient.

3. At high currents, r_π becomes very small and r_x (a few ohms) becomes important (r_x is defined and explained in Section 7.5).

4. f_T is low (a few MHz), C_μ is large (hundreds of pF), and C_π is even larger. (These parameters are defined and explained in Section 7.5.)
5. I_{CBO} is large (a few tens of μA) and, as usual, doubles for every 10°C rise in temperature.
6. BV_{CEO} is typically 50 to 100 V, but it can be as high as 500 V.
7. $I_{C\max}$ is typically in the ampere range, but it can be as high as 100 A.

9.7 VARIATIONS ON THE CLASS AB CONFIGURATION

In this section we discuss a number of circuit improvements and protection techniques for the class AB output stage.

Use of Input Emitter Followers

Figure 9.24 shows a class AB circuit biased using transistors Q_1 and Q_2, which also function as emitter followers, thus providing the circuit with a high input resistance. In effect, the circuit functions as a unity-gain buffer amplifier. Since all four transistors are usually matched, the quiescent current ($v_I = 0$, $R_L = \infty$) in Q_3 and Q_4 is equal to that in Q_1 and Q_2. Resistors R_3 and R_4 are usually very small and are included to compensate for

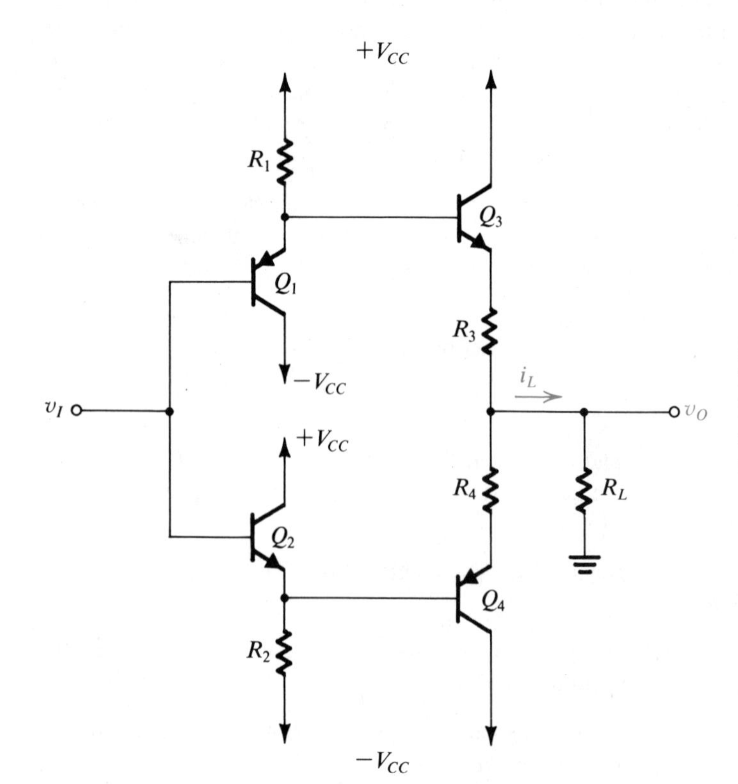

Fig. 9.24 Class AB output stage with an input buffer. In addition to providing a high input resistance, the buffer transistors Q_1 and Q_2 also bias the output transistors Q_3 and Q_4.

possible mismatches between Q_3 and Q_4 and to guard against the possibility of thermal runaway due to temperature differences between the input- and output-stage transistors. The latter point can be appreciated by noting that an increase in the current of, say, Q_3 causes an increase in the voltage drop across R_3 and a corresponding decrease in V_{BE3}. Thus R_3 provides negative feedback that helps stabilize the current through Q_3.

Because the circuit of Fig. 9.24 requires high-quality *pnp* transistors, it is not suitable for implementation in conventional monolithic IC technology. However, excellent results have been obtained with this circuit implemented in hybrid thick-film technology (Wong and Sherwin, 1979). This technology permits component trimming, for instance, to minimize the output offset voltage. The circuit can be used alone or together with an op amp to provide increased output driving capability. The latter application will be discussed in the next section.

Exercise 9.11 (*Note:* Although very instructive, this exercise is rather long.) Consider the circuit of Fig. 9.24 with $R_1 = R_2 = 5\ \text{k}\Omega$, $R_3 = R_4 = 0\ \Omega$, and $V_{CC} = 15$ V. Let the transistors be matched with $I_S = 3.3 \times 10^{-14}$ A, $n = 1$, and $\beta = 200$. (These are the values used in the LH002 manufactured by National Semiconductor, except that $R_3 = R_4 = 2\ \Omega$ there.) (a) For $v_I = 0$ and $R_L = \infty$ find the quiescent current in each of the four transistors and v_O. (b) For $R_L = \infty$, find i_{C1}, i_{C2}, i_{C3}, i_{C4}, and v_O for $v_I = +10$ V and -10 V. (c) Repeat (b) for $R_L = 100\ \Omega$.

Ans. (a) 2.87 mA; 0 V; (b) for $v_I = +10$ V: 0.88 mA, 4.87 mA, 1.95 mA, 1.95 mA, +9.98 V; for $v_I = -10$ V: 4.87 mA, 0.88 mA, 1.95 mA, 1.95 mA, −9.98 V; (c) for $v_I = +10$ V: 0.38 mA, 4.87 mA, 100 mA, 0.02 mA, +9.86 V; for $v_I = -10$ V: 4.87 mA, 0.38 mA, 0.02 mA, 100 mA, −9.86 V

Use of Compound Devices

In order to increase the current gain of the output-stage transistors, and thus reduce the required base current drive, the Darlington configuration shown in Fig. 9.25 is frequently used to replace the *npn* transistor of the class AB stage. The Darlington configuration is equivalent to a single *npn* transistor having $\beta \simeq \beta_1\beta_2$, but almost twice the value of V_{BE}.

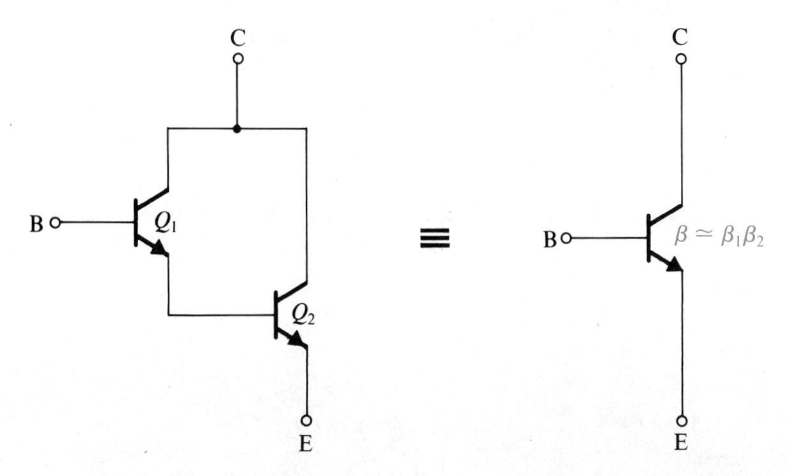

Fig. 9.25 The Darlington configuration.

The Darlington configuration can be also used for *pnp* transistors, and this is indeed done in discrete-circuit design. In IC design, the lack of good-quality *pnp* transistors prompted the use of the alternative compound configuration shown in Fig. 9.26. This compound device is equivalent to a single *pnp* transistor having $\beta \simeq \beta_1\beta_2$. When fabricated with the standard IC technology, Q_1 is usually a lateral *pnp* having a low β ($\beta = 5-10$) and poor high-frequency response ($f_T \simeq 5$ MHz); see Appendix A. The compound device, although it has a relatively high equivalent β, still suffers from a poor high-frequency response. It also suffers from another problem: The feedback loop formed by Q_1 and Q_2 is prone to high-frequency oscillations (with frequency near f_T of the *pnp* device, that is, about 5 MHz). Methods exist for preventing such oscillations. The subject of feedback amplifier stability is studied in Chapter 8.

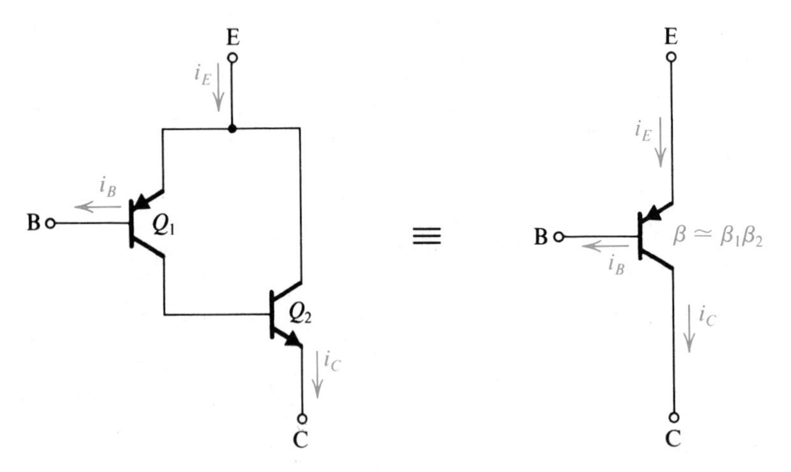

Fig. 9.26 The compound-*pnp* configuration.

To illustrate the application of the Darlington configuration and the compound *pnp*, we show in Fig. 9.27 an output stage utilizing both. The class AB biasing is achieved using a V_{BE} multiplier. Note that the Darlington *npn* adds one more V_{BE} drop, and thus the V_{BE} multiplier is required to provide a bias voltage of about 2 V. The design of this class AB stage is investigated in Problem 9.39.

Exercise 9.12 (a) Refer to Fig. 9.26. Show that, for the composite *pnp* transistor,

$$i_B \simeq \frac{i_C}{\beta_N \beta_P}$$

and

$$i_E \simeq i_C$$

Hence show that

$$i_C \simeq \beta_N I_{SP} e^{v_{EB}/V_T}$$

and thus the transistor has an effective scale current

$$I_S = \beta_N I_{SP}$$

(b) For $\beta_P = 20$, $\beta_N = 50$, $I_{SP} = 10^{-14}$ A, find the effective current gain of the compound device and its v_{EB} when $i_C = 100$ mA. Let $n = 1$.

Ans. (b) 1000; 0.651 V

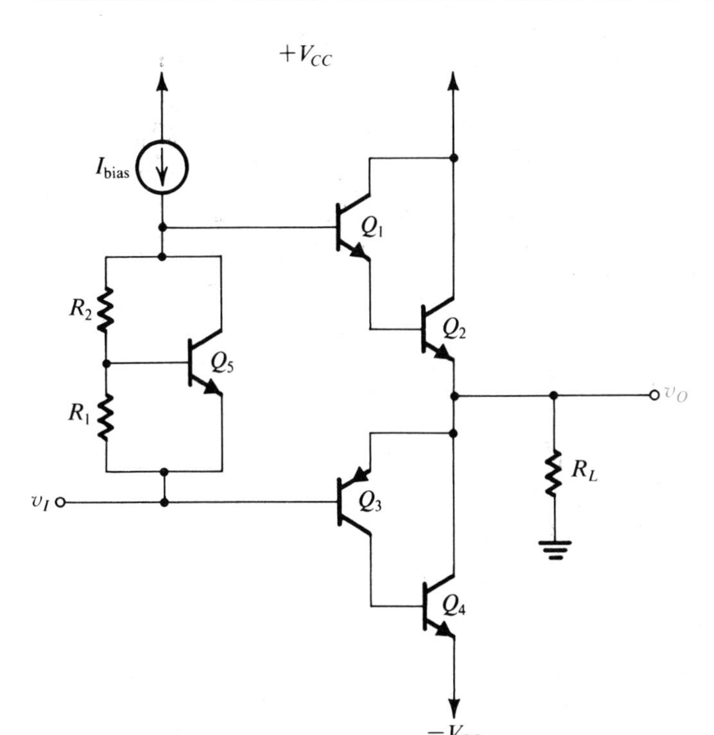

Fig. 9.27 A class AB output stage utilizing a Darlington *npn* and a compound *pnp*. Biasing is obtained using a V_{BE} multiplier.

Short-Circuit Protection

Figure 9.28 shows a class AB output stage equipped with protection against the effect of short-circuiting the output while the stage is sourcing current. The large current that flows through Q_1 in the event of a short circuit will develop a voltage drop across R_{E1} of sufficient value to turn Q_5 on. The collector of Q_5 will then conduct most of the current I_{bias}, robbing Q_1 of its base drive. The current through Q_1 will thus be reduced to a safe operating level.

This method of short-circuit protection is effective in ensuring device safety, but it has the disadvantage that under normal operation about 0.5 V drop might appear across each R_E. This means that the voltage swing at the output will be reduced by that much, in each direction. On the other hand, the inclusion of emitter resistors provides the additional benefit of protecting the output transistors against thermal runaway.

Exercise D9.13 In the circuit of Fig. 9.28 let $I_{bias} = 2$ mA. Find the value of R_{E1} that causes Q_5 to turn on and absorb all 2 mA when the output current being sourced reaches 150 mA. For Q_5, $I_S = 10^{-14}$ A and $n = 1$. If the normal peak output current is 100 mA, find the voltage drop across R_{E1} and the collector current of Q_5.

Ans. 4.3 Ω; 430 mV; 0.3 μA

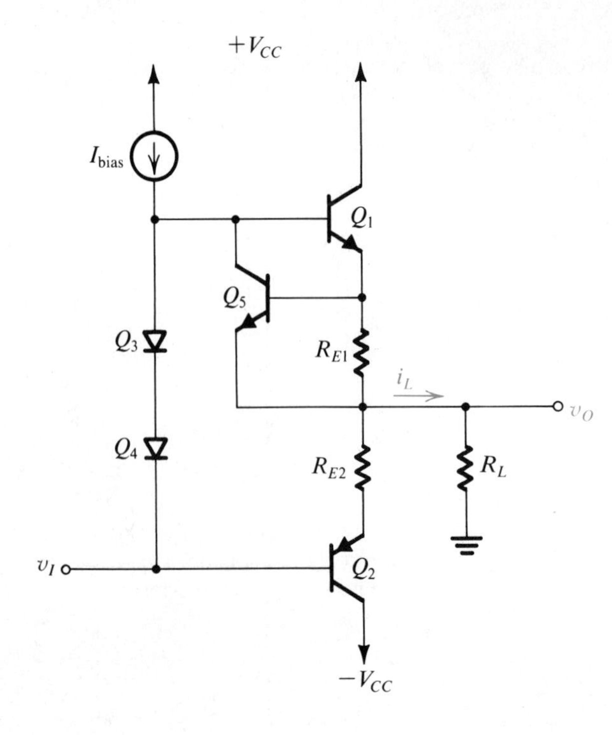

Fig. 9.28 A class AB output stage with short-circuit protection. The protection circuit shown operates in the event of an output short circuit while v_O is positive.

Thermal Shutdown

In addition to short-circuit protection, most IC power amplifiers are usually equipped with a circuit that senses the temperature of the chip and turns on a transistor in the event that the temperature exceeds a safe preset value. The turned-on transistor is connected in such a way that it absorbs the bias current of the amplifier, thus virtually shutting down its operation.

Figure 9.29 shows a thermal shutdown circuit. Here transistor Q_2 is normally off. As the chip temperature rises, the combination of the positive temperature coefficient of zener diode Z_1 and the negative temperature coefficient of V_{BE1} causes the voltage at the emitter of Q_1 to rise. This in turn raises the voltage at the base of Q_2 to the point at which Q_2 turns on.

9.8 IC POWER AMPLIFIERS

A variety of IC power amplifiers are available. Most consist of a high-gain small-signal amplifier followed by a class AB output stage. Some have overall negative feedback already applied, resulting in a fixed closed-loop voltage gain. Others do not have on-chip feedback and are, in effect, op amps with large output-power capability. In fact, the output current driving capability of any general-purpose op amp can be increased by cascading it with a class B or class AB output stage and applying overall negative feedback. The additional output stage can be either a discrete circuit or a hybrid IC such as the buffer

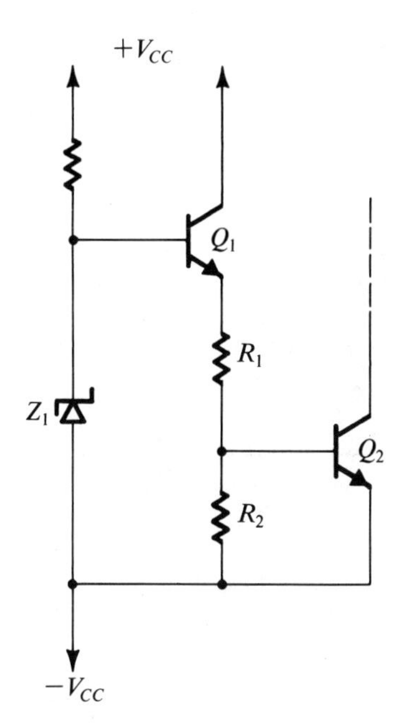

Fig. 9.29 Thermal shutdown circuit.

discussed in the previous section. In the following we discuss some power amplifier examples.

A Fixed-Gain IC Power Amplifier

Our first example is the LM380 (a product of National Semiconductor Corporation), which is a fixed-gain monolithic power amplifier. A simplified version of the internal circuit of the amplifier[3] is shown in Fig. 9.30. The circuit consists of an input differential amplifier utilizing Q_1 and Q_2 as emitter followers for input buffering, and Q_3 and Q_4 as a differential pair with an emitter resistor R_3. The two resistors R_4 and R_5 provide dc paths to ground for the base currents of Q_1 and Q_2, thus enabling the input signal source to be capacitively coupled to either of the two input terminals.

The differential amplifier transistors Q_3 and Q_4 are biased by two separate direct currents: Q_3 is biased by a current from the dc supply V_S through the diode-connected transistor Q_{10}, and resistor R_1; Q_4 is biased by a dc current from the output terminal through R_2. Under quiescent conditions (that is, with no input signal applied) the two bias currents will be equal and the current through, and the voltage across, R_3 will be zero. For the emitter current of Q_3 we can write

$$I_3 \simeq \frac{V_S - V_{EB10} - V_{EB3} - V_{EB1}}{R_1}$$

[3]The main objective of showing this circuit is to point out some interesting design features. The circuit is *not* a detailed schematic diagram of what is actually on the chip.

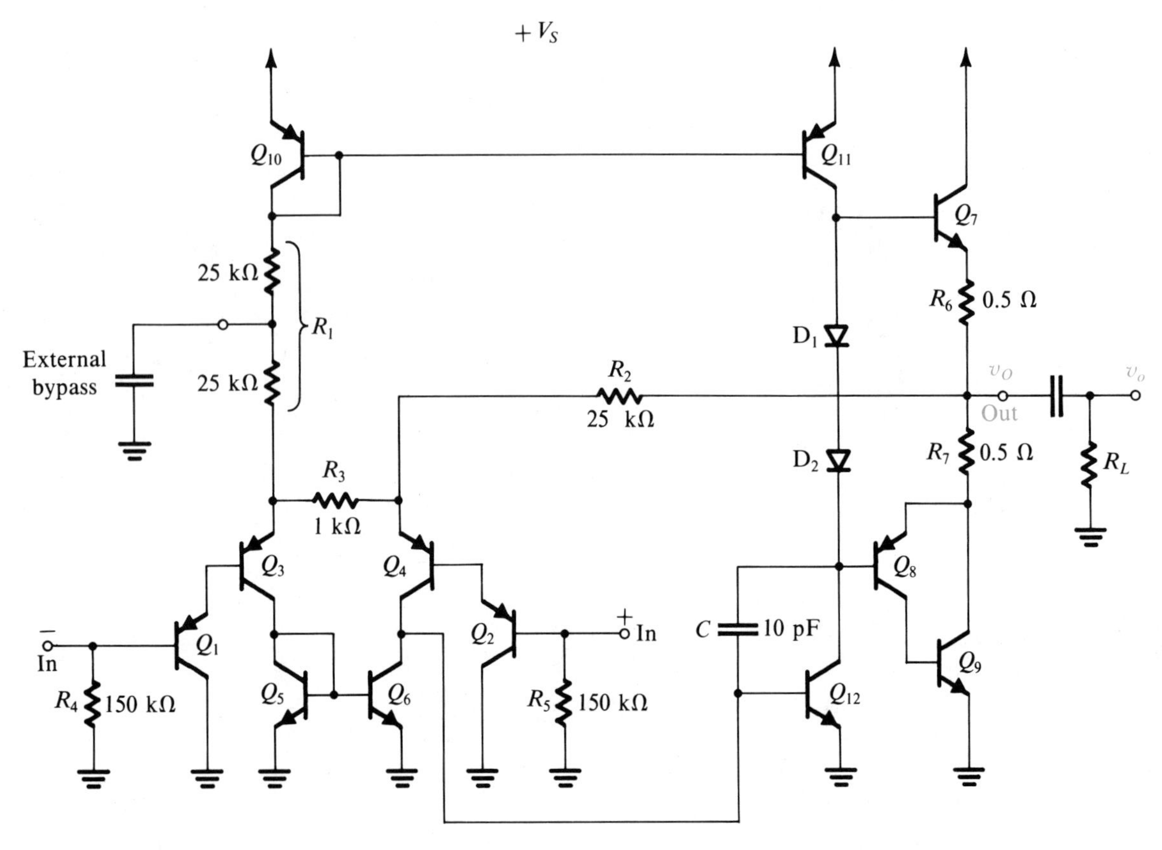

Fig. 9.30 Simplified internal circuit of the LM380 IC power amplifier. (Courtesy National Semiconductor Corporation.)

where we have neglected the small dc voltage drop across R_4. Assuming, for simplicity, all V_{EB} to be equal,

$$I_3 \simeq \frac{V_S - 3V_{EB}}{R_1} \tag{9.43}$$

For the emitter current of Q_4 we have

$$I_4 = \frac{V_O - V_{EB4} - V_{EB2}}{R_2}$$

$$\simeq \frac{V_O - 2V_{EB}}{R_2} \tag{9.44}$$

where V_O is the dc voltage at the output and where we have neglected the small drop across R_5. Equating I_3 and I_4 and using the fact that $R_1 = 2R_2$ results in

$$V_O = \tfrac{1}{2}V_S + \tfrac{1}{2}V_{EB} \tag{9.45}$$

Thus the output is biased at approximately half the power supply voltage, as desired for maximum output voltage swing. An important feature is the dc feedback from the output to the emitter of Q_4, through R_2. This dc feedback acts to stabilize the output dc bias voltage at the value in Eq. (9.45). Qualitatively, the dc feedback functions as follows: If for some reason V_O increases, a corresponding current increment will flow through R_2 and into the emitter of Q_4. Thus the collector current of Q_4 increases, resulting in a positive increment in the voltage at the base of Q_{12}. This, in turn, causes the collector current of Q_{12} to increase, thus bringing down the voltage at the base of Q_7 and hence V_O.

Continuing with the description of the circuit in Fig. 9.30, we observe that the differential amplifier (Q_3, Q_4) has a current-mirror load composed of Q_5 and Q_6 (refer to Section 6.5 for a discussion of active loads). The single-ended output voltage signal of the first stage appears at the collector of Q_6 and thus is applied to the base of the second-stage common-emitter amplifier Q_{12}. Transistor Q_{12} is biased by the constant-current source Q_{11}, which also acts as its active load. In actual operation, however, the load of Q_{12} will be dominated by the reflected resistance due to R_L. Capacitor C provides frequency compensation (see Chapter 8).

The output stage is class AB, utilizing a compound *pnp* transistor (Q_8 and Q_9). Negative feedback is applied from the output to the emitter of Q_4 via resistor R_2. To find the closed-loop gain consider the small-signal equivalent circuit shown in Fig. 9.31. We have replaced the second-stage common-emitter amplifier and the output stage with an inverting amplifier block with gain A. We shall assume that the amplifier A has high gain

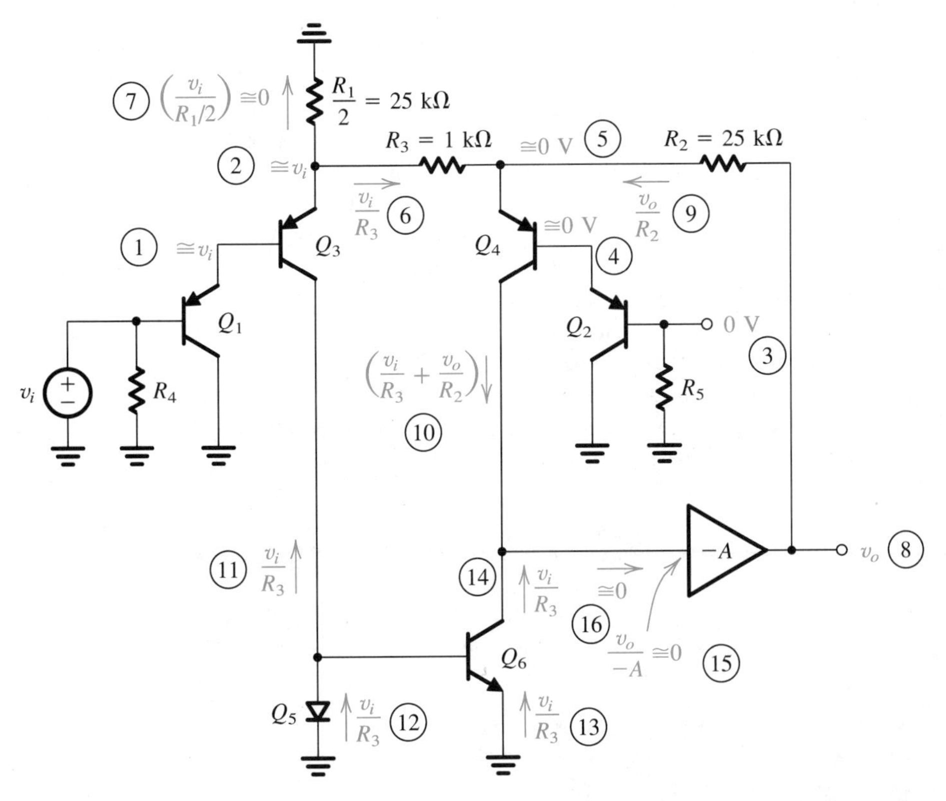

Fig. 9.31 Small-signal analysis of the circuit in Fig. 9.30. The circled numbers indicate the order of the analysis steps.

and high input resistance, and thus the input signal current into A is negligibly small. Under this assumption, Fig. 9.31 shows the analysis details with an input signal v_i applied to the inverting input terminal. The order of the analysis steps is indicated by the circled numbers. Note that since the input differential amplifier has a relatively large resistance, R_3, in the emitter circuit, most of the applied input voltage appears across R_3. In other words, the signal voltages across the emitter–base junctions of Q_1, Q_2, Q_3, and Q_4 are small compared to the voltage across R_3. The voltage gain can be found by writing a node equation at the collector of Q_6:

$$\frac{v_i}{R_3} + \frac{v_o}{R_2} + \frac{v_i}{R_3} = 0$$

which yields

$$\frac{v_o}{v_i} = -\frac{2R_2}{R_3} \simeq -50 \text{ V/V}$$

Exercise 9.14 Denoting the total resistance between the collector of Q_6 and ground R, show, using Fig. 9.31, that

$$\frac{v_o}{v_i} = \frac{-2R_2/R_3}{1 + (R_2/AR)}$$

which reduces to $(-2R_2/R_3)$ under the condition that $AR \gg R_2$.

As was demonstrated in Chapter 8, one of the advantages of negative feedback is the reduction of nonlinear distortion. This is the case in the circuit of the LM380.

The LM380 is designed to operate from a single supply V_S in the range 12–22 V. The selection of supply voltage depends on the value of R_L and the required output power P_L. The manufacturer supplies curves for the device power dissipation versus output power for a given load resistance and different supply voltages. One such set of curves for $R_L = 8\ \Omega$ is shown in Fig. 9.32. Note the similarity to the class B power dissipation curve of Fig. 9.8. In fact, the reader can easily verify that the location and value of the peaks of the curves in Fig. 9.32 are accurately predicted by Eqs. (9.20) and (9.21), respectively (where $V_{CC} = \frac{1}{2}V_S$). The line labeled "3% distortion" in Fig. 9.32 is the locus of the points on the various curves at which the distortion (THD) reaches 3%. A THD of 3% represents the onset of peak clipping due to output transistor saturation.

The manufacturer also supplies curves for maximum power dissipation versus temperature (derating curves) similar to those discussed in Section 9.6 for discrete power transistors.

Exercises 9.15 The manufacturer specifies that for ambient temperatures below 25°C the LM380 can dissipate a maximum of 3.6 W. This is obtained under the condition that the dual-in-line package is soldered onto a printed circuit board in thermal contact with 6 square inches of 2-ounce copper foil. Above $T_A = 25$°C the thermal resistance is $\theta_{JA} = 35$°C/W. $T_{J\max}$ is specified to be 150°C. Find the maximum power dissipation possible if the ambient temperature is to be 50°C.

Ans. 2.9 W

D9.16 It is required to use the LM380 to drive an 8-Ω loudspeaker. Use the curves of Fig. 9.32 to determine the maximum power supply possible while limiting the maximum power dissipation to the 2.9 W determined in Exercise 9.15. If for this application a 3% THD is allowed, find P_L and the peak-to-peak output voltage.

Ans. 20 V; 4.2 W; 16.4 V

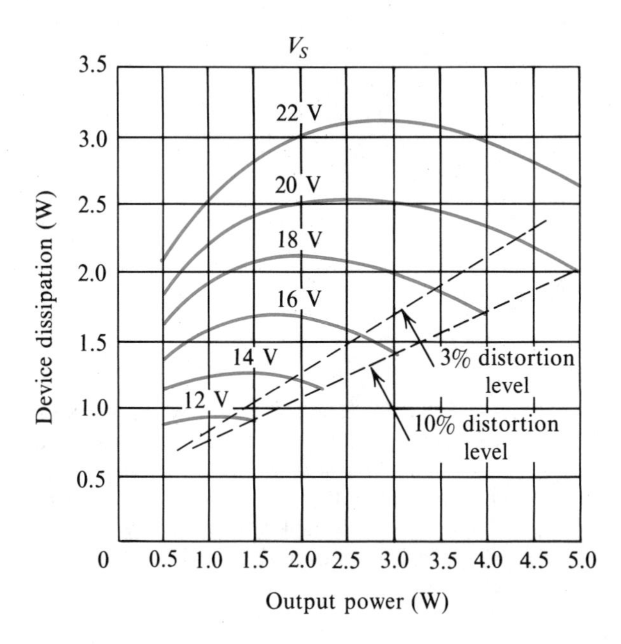

Fig. 9.32 Power dissipation (P_D) versus output power (P_L) for the LM380 with $R_L = 8\ \Omega$. (Courtesy National Semiconductor Corporation.)

Power Op Amps

Figure 9.33 shows the general structure of a power op amp. It consists of an op amp followed by a class AB buffer similar to that discussed in Section 9.7. The buffer consists of transistors Q_1, Q_2, Q_3, and Q_4, with bias resistors R_1 and R_2 and emitter degeneration resistors R_5 and R_6. The buffer supplies the required load current until the current increases to the point that the voltage drop across R_3 (in the current sourcing mode) becomes sufficiently large to turn Q_5 on. Transistor Q_5 then supplies the additional load current required. In the current sinking mode, Q_4 supplies the load current until sufficient voltage develops across R_4 to turn Q_6 on. Then, Q_6 sinks the additional load current. Thus the stage formed by Q_5 and Q_6 acts as a **current booster.** The power op amp is intended to be used with negative feedback in the usual closed-loop configurations. A circuit based on the structure of Fig. 9.33 is commercially available from National Semiconductor as LH0101. This op amp is capable of providing a continuous output current of 2 A, and with appropriate heat sinking can provide 40 W of output power (Wong and Johnson, 1981). The LH0101 is fabricated using hybrid thick-film technology.

The Bridge Amplifier

We conclude this section with a discussion of a circuit configuration that is popular in high-power applications. This is the bridge amplifier configuration shown in Fig. 9.34 utilizing two power op amps, A_1 and A_2. While A_1 is connected in the noninverting configuration with a gain $K = 1 + (R_2/R_1)$, A_2 is connected as an inverting amplifier with a gain of equal magnitude $K = R_4/R_3$. The load R_L is floating and is connected between the output terminals of the two op amps.

If v_I is a sinusoid with amplitude $\hat{V}_i$, the voltage swing at the output of each op amp will be $\pm K\hat{V}_i$, and that across the load will be $\pm 2K\hat{V}_i$. Thus with op amps operated from

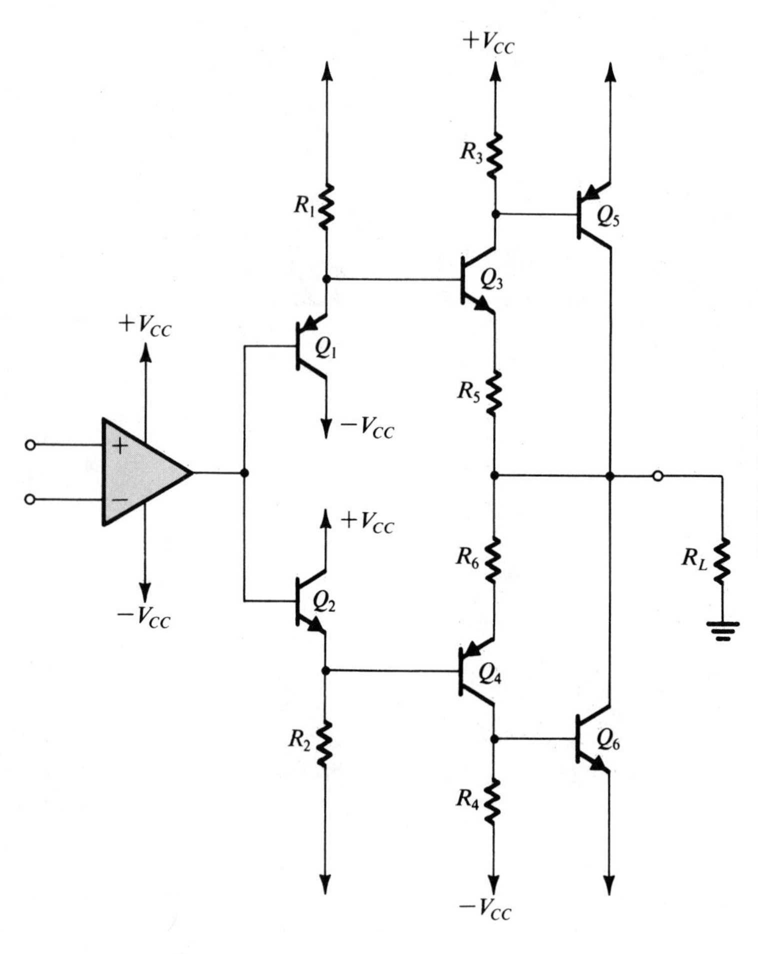

Fig. 9.33 Structure of a power op amp. The circuit consists of an op amp followed by a class AB buffer similar to that discussed in Section 9.7. The output current capability of the buffer, consisting of Q_1, Q_2, Q_3, and Q_4, is further boosted by Q_5 and Q_6.

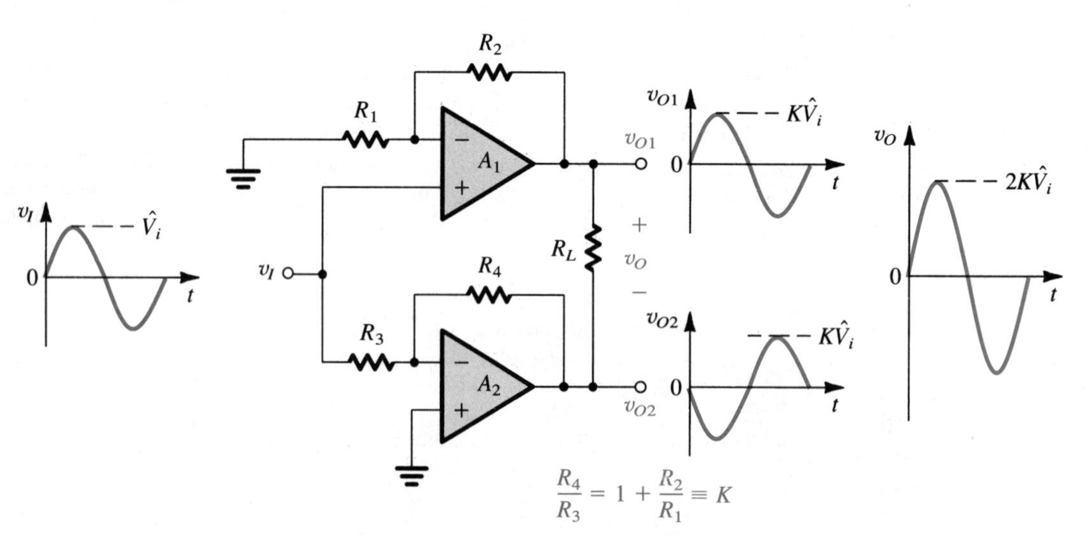

Fig. 9.34 The bridge amplifier configuration.

± 15-V supplies and capable of providing, say a ± 12-V output swing, an output swing of ± 24 V is obtained across the load of the bridge amplifier.

In designing bridge amplifiers, note should be taken of the fact that the peak current drawn from each op amp is $2K\hat{V}_i/R_L$. This effect can be taken into account by considering the load seen by each op amp to be $R_L/2$.

Exercise 9.17 Consider the circuit of Fig. 9.34 with $R_1 = R_3 = 10\ \text{k}\Omega$, $R_2 = 5\ \text{k}\Omega$, $R_4 = 15\ \text{k}\Omega$, and $R_L = 8\ \Omega$. Find the voltage gain and input resistance. The power supply used is ± 18 V. If v_I is a 20-V peak-to-peak sine wave, what is the peak-to-peak output voltage? What is the peak load current? What is the load power?

Ans. 3 V/V; 10 kΩ; 60 V; 3.75 A; 56.25 W

9.9 MOS POWER TRANSISTORS

Thus far in this chapter we have dealt exclusively with BJT circuits. However, recent technological developments have resulted in MOS power transistors with specifications that are quite competitive with those of BJTs. In this section we consider the structure, characteristics, and application of power MOSFETs.

Structure of the Power MOSFET

The enhancement-MOSFET structure studied in Chapter 5 (Fig. 5.1) is not suitable for high-power applications. To appreciate this fact, recall that the drain current of an n-channel MOSFET operating in the saturation region is given by

$$i_D = \frac{1}{2}\mu_n C_{OX}\left(\frac{W}{L}\right)(v_{GS} - V_t)^2 \tag{9.46}$$

It follows that to increase the current capability of the MOSFET its width W should be made large and its channel length L should be made as small as possible. Unfortunately, however, reducing the channel length of the standard MOSFET structure results in a drastic reduction in its breakdown voltage. Specifically, the depletion region of the reverse-biased body-to-drain junction spreads into the short channel, resulting in breakdown at a relatively low voltage. Thus the resulting device would not be capable of handling the high voltages typical of power-transistor applications. For this reason, new structures had to be found for fabricating short-channel (1- to 2-micrometer) MOSFETs with high breakdown voltages.

At the present time the most popular structure for a power MOSFET is the double-diffused or DMOS transistor shown in Fig. 9.35. As indicated, the device is fabricated on a lightly doped n-type substrate with a heavily doped region at the bottom for the drain contact. Two diffusions[4] are employed, one to form the p-type body region and another to form the n-type source region.

The DMOS device operates as follows. Application of a positive gate voltage, v_{GS}, greater than the threshold voltage V_t, induces a lateral n channel in the p-type body region

[4] See Appendix A for a description of the IC fabrication process.

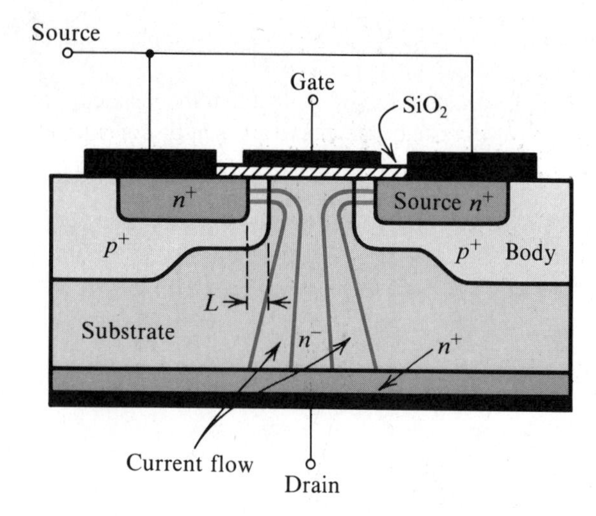

Fig. 9.35 Double-diffused vertical MOS transistor (DMOS).

underneath the gate oxide. The resulting channel is short, its length is denoted L in Fig. 9.35. Current is then conducted by electrons from the source moving through the resulting short channel to the substrate and then vertically down the substrate to the drain. This should be contrasted with the lateral current flow in the standard small-signal MOSFET structure (Chapter 5).

Despite the fact the DMOS transistor has a short channel, its breakdown voltage can be very high (as high as 600 V). This is because the depletion region between the substrate and the body extends mostly in the lightly doped substrate and does not spread into the channel. The result is a MOS transistor that simultaneously has a high current capability (50 A is possible) as well as the high breakdown voltage just mentioned. Finally, we note that the vertical structure of the device provides efficient utilization of the silicon area.

An earlier structure used for power MOS transistors deserves mention. This is the V-groove MOS device (see Severns, 1984). Although still in use, the V-groove MOSFET has lost application ground to the vertical DMOS structure of Fig. 9.35, except possibly for high-frequency applications. Because of space limitations, we shall not describe the V-groove MOSFET.

Characteristics of Power MOSFETs

In spite of their radically different structure, power MOSFETs exhibit characteristics that are quite similar to those of the small-signal MOSFETs studied in Chapter 5. Important differences exist, however, and these are discussed in the following.

Power MOSFETs have threshold voltages in the range of 2 to 4 V. In saturation, the drain current is related to v_{GS} by the square-law characteristic of Eq. (9.46). However, as shown in Fig. 9.36, the i_D–v_{GS} characteristic becomes linear for larger values of v_{GS}. The linear portion of the characteristic occurs as a result of the high electric field along the short channel, causing the velocity of charge carriers to reach an upper limit, a phenomenon known as **velocity saturation.** The drain current is then given by

$$i_D = \tfrac{1}{2} C_{OX} \, W U_{sat}(v_{GS} - V_t) \qquad (9.47)$$

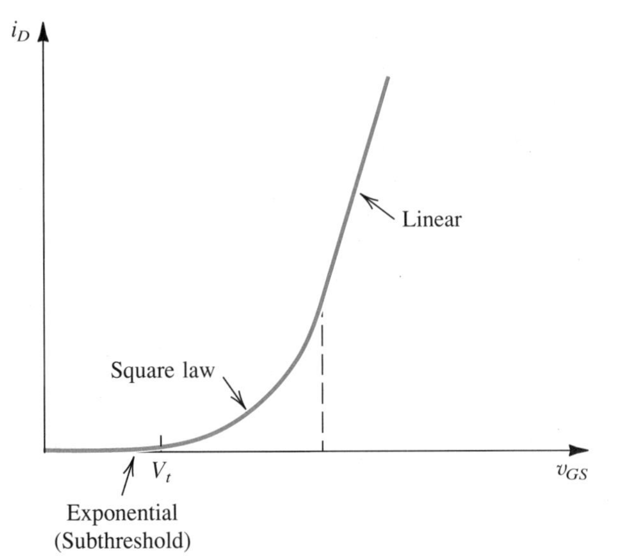

Fig. 9.36 Typical i_D–v_{GS} characteristic for a power MOSFET.

where U_{sat} is the saturated velocity value (5×10^6 cm/s for electrons in silicon). The linear i_D–v_{GS} relationship implies a constant g_m in the velocity-saturation region. It is interesting to note that g_m is proportional to W, which is usually large for power devices; thus power MOSFETs exhibit relatively high transconductance values.

The i_D–v_{GS} characteristic shown in Fig. 9.36 includes a segment labeled ''subthreshold.'' Though of little significance for power devices, the subthreshold region of operation is of interest in very-low-power applications (see Section 5.1).

Temperature Effects

Of most interest in the design of MOS power circuits is the variation of MOSFET characteristics with temperature, illustrated in Fig. 9.37. Observe that there is a value of v_{GS} (in the range of 4–6 V for most power MOSFETs) at which the temperature coefficient of i_D is zero. At higher values of v_{GS}, i_D exhibits a negative temperature coefficient. This is a significant property; it implies that a MOSFET operating beyond the zero-temperature-coefficient point does not suffer from the possibility of thermal runaway. This is *not* the case, however, at low currents (that is, lower than the zero-temperature-coefficient point). In the (relatively) low-current region, the temperature coefficient of i_D is positive, and the power MOSFET can easily suffer thermal runaway (with unhappy consequences). Since class AB output stages are biased at low currents, means must be provided to guard against thermal runaway.

The reason for the positive temperature coefficient of i_D at low currents is that $v_{GS} - V_t$ is relatively low, and the temperature dependence is dominated by the negative temperature coefficient of V_t (in the range of -3 to -6 mV/°C).

Comparison with BJTs

The power MOSFET does not suffer from second breakdown, which limits the safe operating area of BJTs. Also, power MOSFETs do not require the large base-drive cur-

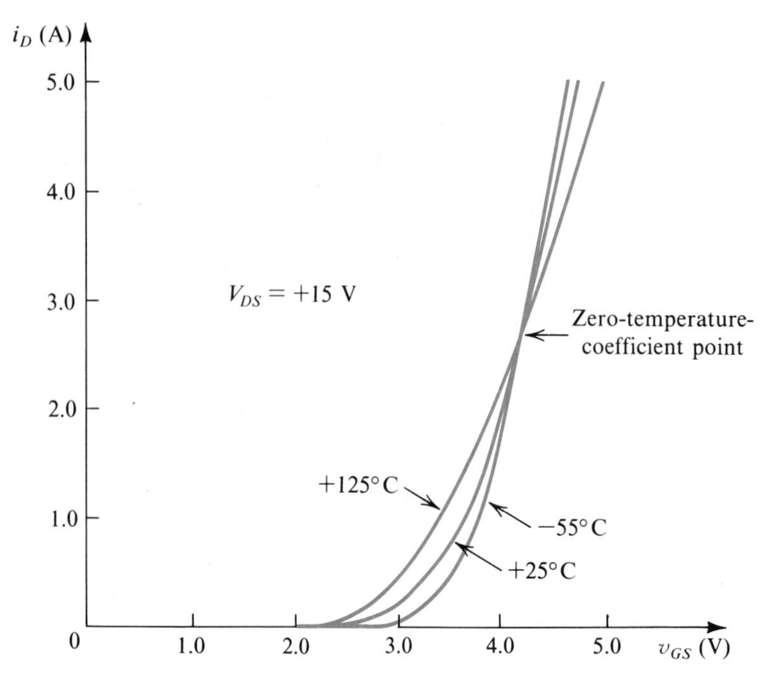

Fig. 9.37 The i_D–v_{GS} characteristic curve of a power MOS transistor (IRF 630, Siliconix) at a case temperature of −55°C, +25°C, and +125°C. (Courtesy Siliconix Inc.)

rents of power BJTs. Note, however, that the driver stage in a MOS power amplifier should be capable of supplying sufficient current to charge and discharge the MOSFET large and nonlinear input capacitance in the time allotted. Finally, the power MOSFET features, in general, a higher speed of operation than the power BJT. This makes MOS power transistors especially suited to switching applications—for instance, in motor-control circuits.

A Class AB Output Stage Utilizing MOSFETs

As an application of power MOSFETs, we show in Fig. 9.38 a class AB output stage utilizing a pair of complementary MOSFETs and employing BJTs for biasing and in the driver stage. The latter consists of complementary Darlington emitter followers formed by Q_1 through Q_4 and has the low output-resistance necessary for driving the output MOSFETs at high speeds.

Of special interest in the circuit of Fig. 9.38 is the bias circuit utilizing two V_{BE} multipliers formed by Q_5 and Q_6 and their associated resistors. Transistor Q_6 is placed in direct thermal contact with the output transistors; this is achieved by simply mounting Q_6 on their common heat sink. Thus, by the appropriate choice of the V_{BE} multiplication factor of Q_6, the bias voltage V_{GG} (between the gates of the output transistors) can be made to decrease with temperature at the same rate as the sum of the threshold voltages ($V_{tN} + |V_{tP}|$) of the output MOSFETs. In this way the quiescent current of the output transistors can be stabilized against temperature variations.

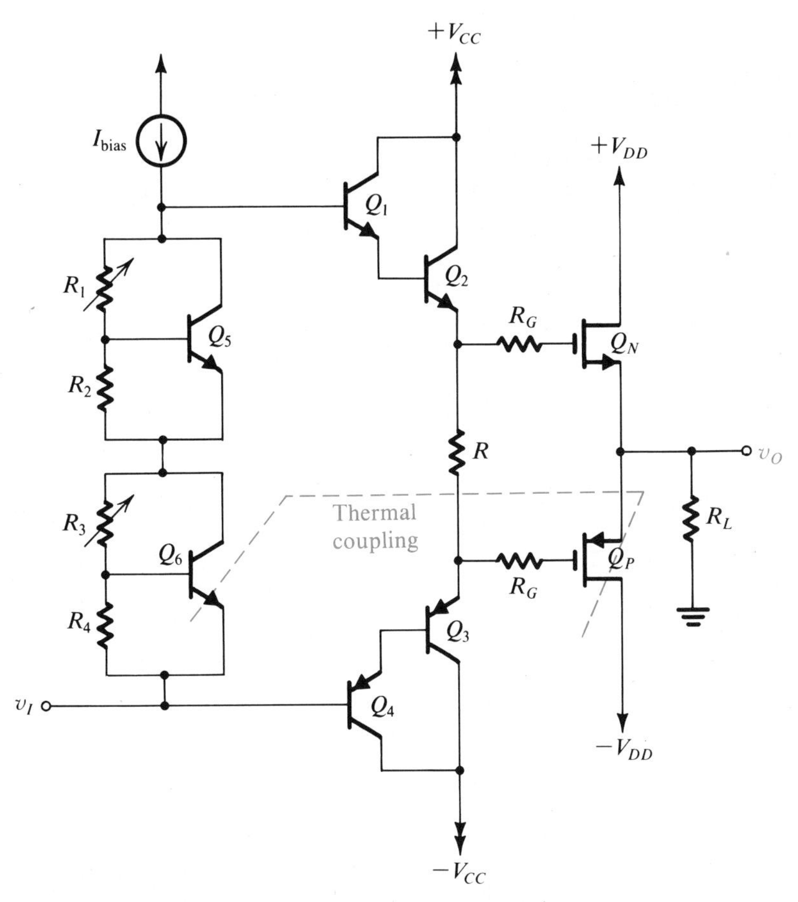

Fig. 9.38 A class AB amplifier with MOS output transistors and BJT drivers. Resistor R_3 is adjusted to provide temperature compensation while R_1 is adjusted to yield the desired value of quiescent current in the output transistors.

Analytically, V_{GG} is given by

$$V_{GG} = \left(1 + \frac{R_3}{R_4}\right) V_{BE6} + \left(1 + \frac{R_1}{R_2}\right) V_{BE5} - 4V_{BE} \tag{9.48}$$

Since V_{BE6} is thermally coupled to the output devices while the other BJTs remain at constant temperature, we have

$$\frac{\partial V_{GG}}{\partial T} = \left(1 + \frac{R_3}{R_4}\right) \frac{\partial V_{BE6}}{\partial T} \tag{9.49}$$

which is the relationship needed to determine R_3/R_4 so that $\partial V_{GG}/\partial T = \partial(V_{tN} + |V_{tP}|/\partial T)$. The other V_{BE} multiplier is then adjusted to yield the required value of V_{GG} and hence the desired quiescent current in Q_N and Q_P.

Exercises 9.18 For the circuit in Fig. 9.38, find the ratio R_3/R_4 that provides temperature stabilization of the quiescent current in Q_N and Q_P. Assume that $|V_t|$ changes at -3 mV/°C and that $\partial V_{BE}/\partial T = -2$ mV/°C.

Ans. 2

9.19 For the circuit in Fig. 9.38 assume that the BJTs have a nominal V_{BE} of 0.7 V and that the MOSFETs have $|V_t| = 3$ V and $K \equiv \frac{1}{2}\mu_n C_{OX}(W/L) = 1$ A/V^2. It is required to establish a quiescent current of 100 mA in the output stage and 20 mA in the driver stage. Find $|V_{GS}|$, V_{GG}, R, and R_1/R_2. Use the value of R_3/R_4 found in Exercise 9.18.

Ans. 3.32 V; 6.64 V; 332 Ω; 9.5

SUMMARY

- Output stages are classified according to the transistor conduction angle: class A (360°), class AB (slightly greater than 180°), class B (180°), and class C (smaller than 180°).

- The most common class A output stage is the emitter follower. It is biased at a current greater than the peak load current.

- The class A output stage dissipates its maximum power under quiescent conditions ($v_O = 0$). It achieves a maximum power conversion efficiency of 25%.

- The class B stage is biased at zero current and thus dissipates no power in quiescence.

- The class B stage can achieve a power conversion efficiency as high as 78.5%. It dissipates its maximum power for $\hat{V}_o = (2/\pi)V_{CC}$.

- The class B stage suffers from crossover distortion.

- The class AB output stage is biased at a small current; thus both transistors conduct for small input signals and crossover distortion is virtually eliminated.

- Except for an additional small quiescent power dissipation, the power relationships of the class AB stage are similar to those in class B.

- To guard against the possibility of thermal runaway, the bias voltage of the class AB circuit is made to vary with temperature in the same manner as does V_{BE} of the output transistors.

- To facilitate the removal of heat from the silicon chip, power devices are usually mounted on heat sinks. The maximum power that can be safely dissipated in the device is given by

$$P_{D\max} = \frac{T_{J\max} - T_A}{\theta_{JC} + \theta_{CS} + \theta_{SA}}$$

where $T_{J\max}$ and θ_{JC} are specified by the manufacturer, while θ_{CS} and θ_{SA} depend on the heat sink design.

- Use of the Darlington configuration in the class AB output stage reduces the base-current drive requirement. In integrated circuits, the compound *pnp* configuration is commonly used.

- Output stages are usually equipped with circuitry that, in the event of a short circuit, can turn on and limit the base-current drive, and hence the emitter current, of the output transistors.

- IC power amplifiers consist of a small-signal voltage amplifier cascaded with a high-power output stage. Overall feedback is applied either on-chip or externally.

- The bridge-amplifier configuration provides, across a floating load, a peak-to-peak output voltage twice that possible from a single amplifier with a grounded load.

- The DMOS transistor is a short-channel power device capable of both high-current and high-voltage operation.

- At low currents, the drain current of a power MOSFET exhibits a positive temperature coefficient, and thus the device can suffer thermal runaway. At high currents the temperature coefficient of i_D is negative.

BIBLIOGRAPHY

C. A. Holt, *Electronic Circuits,* New York: Wiley, 1978.

National Semiconductor Corporation, *Audio/Radio Handbook,* Santa Clara, Calif.: National Semiconductor Corporation, 1980.

D. L. Schilling and C. Belove, *Electronic Circuits,* 2nd ed., New York: McGraw-Hill, 1979.

A. S. Sedra and G. W. Roberts, "Current conveyor theory and practice," Chapter 3 in *Analogue IC Design: The Current-Mode Approach,* C. Toumazou, F. J. Lidgey, and D. G. Haigh, editors, London: Peter Peregrinus, 1990.

R. Severns, (Ed.), *MOSPOWER Applications Handbook,* Santa Clara, Calif.: Siliconix, 1984.

S. Soclof, *Applications of Analog Integrated Circuits,* Englewood Cliffs, N.J.: Prentice-Hall, 1985.

Texas Instruments, Inc., *Power-Transistor and TTL Integrated-Circuit Applications,* New York: McGraw-Hill, 1977.

J. Wong and R. Johnson, "Low-distortion wideband power op amp," *Application Note 261,* Santa Clara, Calif.: National Semiconductor Corporation, July 1981.

J. Wong and J. Sherwin, "Applications of wide-band buffer amplifiers," *Application Note 227,* Santa Clara, Calif.: National Semiconductor Corporation, October 1979.

PROBLEMS

Section 9.2: Class A Output Stage

9.1 A class A emitter follower, biased using the circuit shown in Fig. 9.2, uses $V_{CC} = 5$ V, $R = R_L = 1$ kΩ, with all transistors (including Q_3) identical. Assume $V_{BE} = 0.7$ V, $V_{CEsat} = 0.3$ V, and β very large. For linear operation, what are the upper and lower limits of output voltage, and the corresponding inputs? How do these values change if the emitter–base junction area of Q_3 is made twice as big as that of Q_2? Half as big?

9.2 A source-follower circuit using enhancement NMOS transistors is constructed following the pattern shown in Fig. 9.2. All three transistors used are identical with $V_t = 1$ V and $K = 10$ mA/V^2. $V_{CC} = 5$ V, $R = R_L = 1$ kΩ. For linear operation, what are the upper and lower limits of the output voltage, and the corresponding inputs?

D9.3 Using the follower configuration shown in Fig. 9.2 with ± 9-V supplies, provide a design capable of ± 7-V outputs with a 1-kΩ load, using the smallest possible total supply current. You are provided with four identical, high-β BJTs and a resistor of your choice.

D9.4 An emitter follower using the circuit of Fig. 9.2, for which the output voltage range is ± 5 V, is required using $V_{CC} = 10$ V. The circuit is to be designed such that the current variation in the emitter-follower transistor is no greater than a factor of 10, for load resistances as low as 100 Ω. What is the value of R required? Find the incremental voltage gain of the resulting follower at $v_O = +5$, 0, and -5 V with a 100-Ω load. What is the percentage change in gain over this range of v_O?

***9.5** Consider the operation of the follower circuit of Fig. 9.2 for which $R_L = V_{CC}/I$, when driven by a square wave such that the output ranges from $+V_{CC}$ to $-V_{CC}$ (ignoring V_{CEsat}). For this situation sketch the equivalent of Fig. 9.4 for v_O, i_{C1}, and p_{D1}. Repeat for a square-wave output that has peak levels of $\pm V_{CC}/2$. What is the average power dissipation in Q_1 in each case? Compare these results for sine waves of peak amplitude V_{CC} and $V_{CC}/2$, respectively.

9.6 Consider the situation described in Problem 9.5. For square-wave outputs having peak-to-peak values of $2V_{CC}$ and V_{CC}, and for sine waves of the same peak-to-peak values, find the average power loss in the current-source transistor Q_2.

9.7 Reconsider the situation described in Exercise 9.4 for variation in V_{CC}—specifically for $V_{CC} = 16$, 12, 10, and 8 V. Assume V_{CEsat} is nearly zero. What is the power-conversion efficiency in each case?

9.8 The BiCMOS follower shown in Fig. P9.8 uses devices for which $V_{BE} = 0.7$ V, $V_{CEsat} = 0.3$ V, $K = 10$ mA/V^2, and $V_t = -2$ V. For linear operation, what is the range of output voltages obtained with $R_L = \infty$? With $R_L = 100$ Ω? What is the smallest load resistor allowed for which a 1-V peak sine-wave output is available? What is the corresponding power-conversion efficiency?

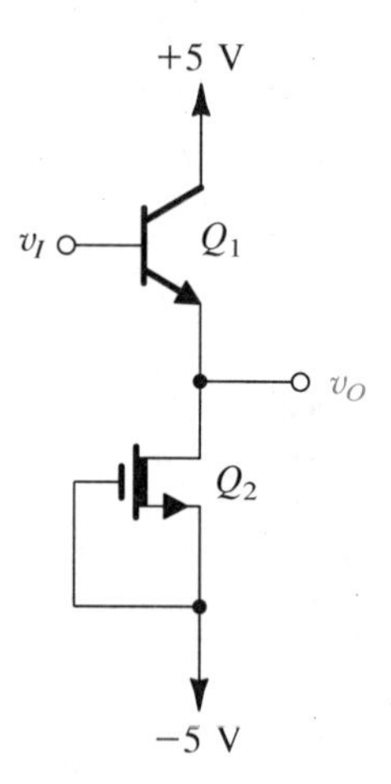

Fig. P9.8

Section 9.3: Class B Output Stage

9.9 Consider the circuit of a complementary BJT class B output stage. For what amplitude of input signal does the crossover distortion represent a 10% loss in peak amplitude?

9.10 Consider the feedback configuration with class B output shown in Fig. 9.9. Let the amplifier gain $A_0 =$ 100 V/V. Derive an expression for v_O versus v_I assuming that $|V_{BE}| = 0.7$ V. Sketch the transfer characteristic v_O versus v_I and compare it to that without feedback.

9.11 Consider the class B output stage using enhancement MOSFETs shown in Fig. P9.11. Let the devices have $|V_t| = 1$ V and $K = 100\ \mu\text{A/V}^2$. With a 10-kHz sine-wave input of 5-V peak and a high value of load resistance, what peak output would you expect? What fraction of the sine-wave period does the crossover interval represent? For what value of load resistor is the peak output voltage reduced to half the input?

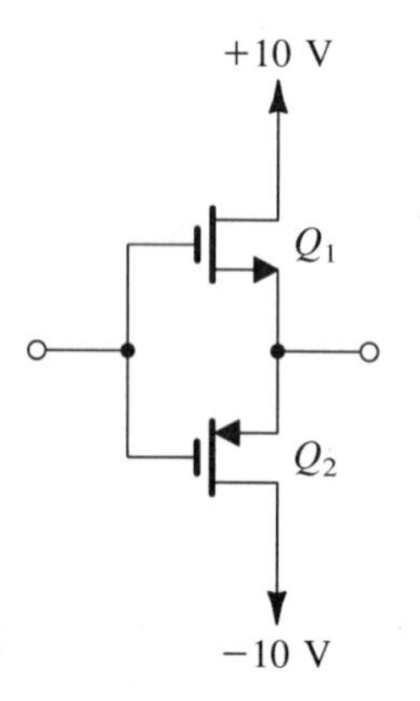

Fig. P9.11

9.12 Consider the complementary BJT class B output stage and neglect the effects of V_{BE} and V_{CEsat}. For ±10-V power supplies and a 100-Ω load resistance, what is the maximum sine-wave output power available? What supply power corresponds? What is the power-conversion efficiency? For output signals of half this amplitude, find the output power, the supply power, and the power-conversion efficiency.

D9.13 A class B output stage operates from ±5-V supplies. Assuming relatively-ideal transistors, what is the output voltage for maximum power-conversion efficiency? What is the output voltage for maximum device dissipation? If each of the output devices is individually rated for 1-W dissipation and a factor of 2 safety margin is to be used, what is the smallest value of load resistance that can be tolerated, if operation is always at full output voltage? If operation is allowed at half the full output voltage, what is the smallest load permitted? What is the greatest possible output power available in each case?

D9.14 A class B output stage is required to deliver an average power of 100 W into a 16-Ω load. The power supply should be 4 V greater than the corresponding peak sine-wave output voltage. Determine the power-supply voltage required (to the nearest volt in the appropriate direction), the peak current from each supply, the total supply power, and the power-conversion efficiency. Also determine the maximum possible power dissipation in each transistor for a sine-wave input.

9.15 Consider the class B BJT output stage with a square-wave output voltage of amplitude $\hat{V}_o$ across a load R_L and employing power supplies $\pm V_{SS}$. Neglecting the effects of V_{BE} and V_{CEsat}, determine the load power, the supply power, the power-conversion efficiency, the maximum attainable power-conversion efficiency and the corresponding value of $\hat{V}_o$, and the maximum available load power. Also find the value of $\hat{V}_o$ at which the power dissipation in the transistors reaches its peak, and the corresponding value of power-conversion efficiency.

Section 9.4: Class AB Output Stage

D9.16 Design the quiescent current of a class AB BJT output stage so that the incremental voltage gain for v_I in the vicinity of the origin is in excess of 0.99 V/V for loads larger than 100 Ω. If the BJTs have V_{BE} of 0.7 V at a current of 100 mA, determine the value of V_{BB} required.

D9.17 The design of a class AB enhancement MOS output stage is being considered. The available devices have $|V_t| = 1$ V and $K = 100$ mA/V^2. What value of gate-to-gate bias voltage, V_{GG}, is required to reduce the incremental output resistance in the quiescent state to 10 Ω?

***9.18** A class AB output stage, resembling that in Fig. 9.11 but utilizing a single supply of +10 V and biased at $V_I = 6$ V, is capacitively coupled to a 100-Ω load. For transistors for which $|V_{BE}| = 0.7$ V at 1 mA and for a bias voltage $V_{BB} = 1.4$ V, what quiescent current results? For a step change in output from 0 to −1 V, what input step is required? Assuming transistor saturation voltages of zero, find the largest possible positive-going and negative-going steps at the output.

Section 9.5: Biasing the Class AB Circuit

D9.19 Consider the diode-biased class AB circuit of Fig. 9.14. For $I_{\text{bias}} = 100$ μA, find the relative size (n) that should be used for the output devices in comparison to the biasing devices to ensure an output resistance of 10 Ω or less.

***D9.20** A class AB output stage using a two-diode bias network as shown in Fig. 9.14 utilizes diodes having the same junction area as the output transistors. For $V_{CC} = 10$ V, $I_{\text{bias}} = 0.5$ mA, $R_L = 100$ Ω, $\beta_N = 50$, and $|V_{CEsat}| = 0$ V, what is the quiescent current? What are the largest possible positive and negative output signal levels? To achieve a positive peak output level equal to the negative peak level, what value of β_N is needed if I_{bias} is not changed? What value of I_{bias} is needed if β_N is held at 50? For this value, what does I_Q become?

****9.21** A class AB output stage using a two-diode bias network as shown in Fig. 9.14, utilizes diodes having the same junction area as the output transistors. At room temperature of about 20°C the quiescent current is 1 mA and $|V_{BE}| = 0.6$ V. Through a manufacturing error, the thermal coupling between the output transistors and the biasing diode-connected transistors is removed. After some output activity, the output devices heat up to 70°C while the biasing devices remain at 20°C. Thus while the V_{BE} of each device remains unchanged, the quiescent current in the output devices increases. To calculate the new current value recall that there are two effects: I_S increases by about 14%/°C and $V_T = kT/g$ changes since T = (273° + temperature in °C) where $V_T = 25$ mV only at 20°C. You may assume that β_N remains almost constant. This assumption is based on the fact that β increases with temperature but decreases with current (see Fig. 4.62). What is the new value of I_Q? If the power supply is ±20 V, what additional power is dissipated? If the temperature of the output transistors increases by 10°C for every watt of additional power dissipation, what additional temperature rise and current increase results? This is the process called thermal runaway.

D9.22 Figure P9.22 shows an enhancement-MOSFET class AB output stage. All transistors have $|V_t| = 1$ V and $K_1 = K_2 = nK_3 = nK_4$. Also, $K_3 = 1$ mA/V^2. For $I_{\text{bias}} = 100$ μA and $R_L = 1$ kΩ find the value of n that results in a small-signal gain of 0.99 for output voltages around zero, and the corresponding value of I_Q.

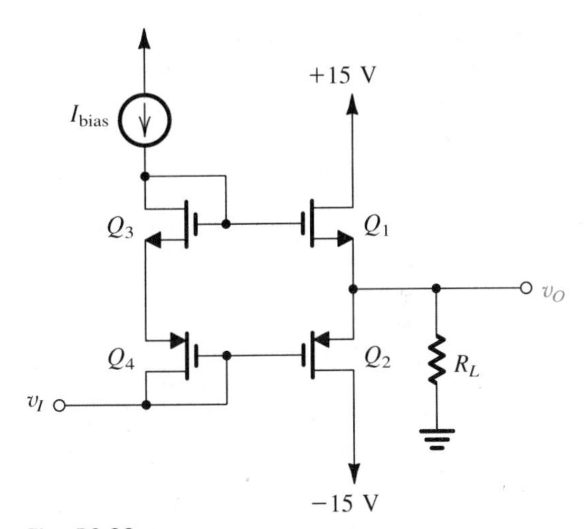

Fig. P9.22

D9.23 Repeat Example 9.3 for the situation in which the peak positive output current is 200 mA. Use the same general approach to safety margins. What are the values of R_1 and R_2 you have chosen?

****9.24** A V_{BE} multiplier is designed with equal resistances for nominal operation at a terminal current of 1 mA, with half the current flowing in the bias network. The initial design is based on $\beta = \infty$ and $V_{BE} = 0.7$ V at 1 mA.

(a) Find the required resistor values and the terminal voltage.

(b) Find the terminal voltage that results when the terminal current increases to 2 mA. Assume $\beta = \infty$.

(c) Repeat (b) for the case the terminal current becomes 10 mA.

(d) Repeat (c) using the more realistic value of $\beta = 100$.

Section 9.6: Power BJTs

D9.25 A particular transistor having a thermal resistance $\theta_{JA} = 2°C/W$ is operating at an ambient temperature of 30°C with a collector–emitter voltage of 20 V. If long life requires a maximum junction temperature of 130°C, what is the corresponding device power rating? What is the greatest average collector current that should be considered?

9.26 A particular transistor has a power rating at 25°C of 200 mW, and a maximum junction temperature of 150°C. What is its thermal resistance? What is its power rating when operated at an ambient temperature of 70°C? What is its junction temperature when dissipating 100 mW at an ambient temperature of 50°C?

9.27 A power transistor operating at an ambient temperature of 50°C, and an average emitter current of 3A, dissipates 30 W. If the thermal resistance of the transistor is known to be less than 3°C/W, what is the greatest junction temperature you would expect? If the transistor V_{BE} measured using a pulsed emitter current of 3 A at a junction temperature of 25°C is 0.80 V, what average V_{BE} would you expect under normal operating conditions? (Use a temperature coefficient of -2 mV/°C.)

9.28 For a particular application of the transistor specified in Example 9.4, extreme reliability is essential. To improve reliability the maximum junction temperature is to be limited to 100°C. What are the consequences of this decision for the conditions specified?

9.29 A power transistor is specified to have a maximum junction temperature of 130°C. When operated at this temperature with a heat sink, the case temperature is found to be 90°C. The case is attached to the heat sink with a bond having a thermal resistance $\theta_{CS} = 0.5°C/W$ and the thermal resistance of the heat sink $\theta_{SA} = 0.1°C/W$. If the ambient temperature is 30°C what is the power being dissipated in the device? What is the thermal resistance of the device, θ_{JC}, from junction to case?

9.30 A power transistor for which $T_{J\max} = 180°C$ can dissipate 50 W at a case temperature of 50°C. If it is connected to a heat sink using an insulating washer for which the thermal resistance is 0.6°C/W, what heat-sink temperature is necessary to ensure safe operation at 30 W? For an ambient temperature of 39°C, what heat-sink thermal resistance is required? If, for a particular extruded-aluminum-finned heat sink, the thermal resistance in still air is 4.5°C/W per cm of length, how long a heat sink is needed?

9.31 An *npn* power transistor operating at $I_C = 10$ A is found to have a base current of 0.5 A and an incremental base input resistance of 0.95 Ω. What value of r_x do you suspect? (At this high current density, $n = 2$.)

9.32 An *npn* power transistor operating at $I_C = 5$ A, with a base–emitter voltage of 1.05 V and a base current of 190 mA, has been measured to have a base spreading resistance (r_x) of 0.8 Ω. Assuming that $n = 2$ for high-current-density operation, what base–emitter voltage would you expect for operation at $I_C = 2$ A?

Section 9.7: Variations on the Class AB Configuration

9.33 Use the results given in the answer to Exercise 9.11 to determine the input current of the circuit in Fig. 9.24 for $v_I = 0$ and ± 10 V with infinite and 100-Ω loads.

*****D9.34** Consider the circuit of Fig. 9.24 in which Q_1 and Q_2 are matched, and Q_3 and Q_4 are matched but have three times the junction area of the others. For $V_{CC} = 10$ V, find values for resistors R_1 through R_4 which allow for a base current of at least 10 mA in Q_3 and Q_4 at $v_I = +5$ V (when a load demands it) with at most a 2 to 1 variation in currents in Q_1 and Q_2, and a no-load quiescent current of 40 mA in Q_3 and Q_4. $\beta_{1,2} \geq 150$; $\beta_{3,4} \geq 50$. For input voltages around 0 V, estimate the output resistance of the overall follower driven by a source having zero resistance. For an input voltage of $+1$ V and a load resistance of 2 Ω, what output voltage results? Q_1 and Q_2 have $|V_{BE}|$ of 0.7 V at a current of 10 mA and exhibit a constant $n = 1$.

9.35 A circuit resembling that in Fig. 9.24 uses four matched transistors for which $|V_{BE}| = 0.7$ V at 10 mA, $n = 1$, and $\beta \geq 50$. Resistors R_1 and R_2 are replaced by 2-mA current sources, and $R_3 = R_4 = 0$. What quiescent current flows in the output transistors? What bias current flows in the bases of the input transistors? Where does it flow? What is the net input current (the offset current) for a β mismatch of 10%? For a load resistance $R_L = 100$ Ω, what is the input resistance? What is the small-signal voltage gain?

9.36 Characterize a Darlington compound transistor formed from two *npn* BJTs for which $\beta \geq 50$, $V_{BE} = 0.7$ V at 1 mA, and $n = 1$. For operation at 10 mA, what values would you expect for β_{eq}, V_{BEeq}, $r_{\pi eq}$, and g_{meq}.

9.37 For the circuit in Fig. P9.37 in which the transistors have $V_{BE} = 0.7$ V and $\beta = 100$, find i_c, g_{meq}, v_o/v_i, and R_{in}.

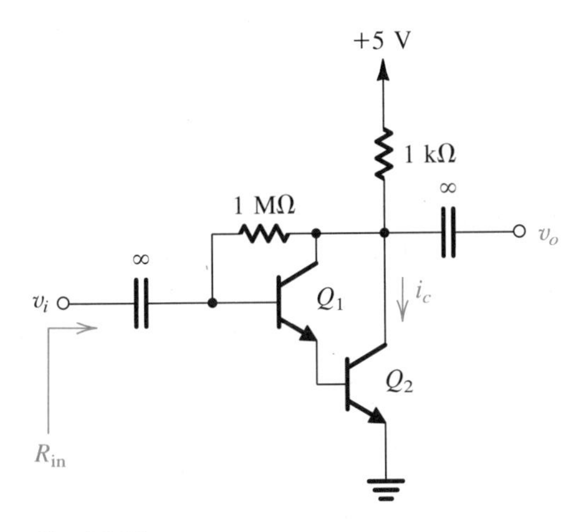

Fig. P9.37

****9.38** The BJTs in the circuit of Fig. P9.38 have $\beta_P = 10$, $\beta_N = 100$, $|V_{BE}| = 0.7$ V, $|V_A| = 100$ V.

(a) Find the dc collector current of each transistor and the value of V_C.

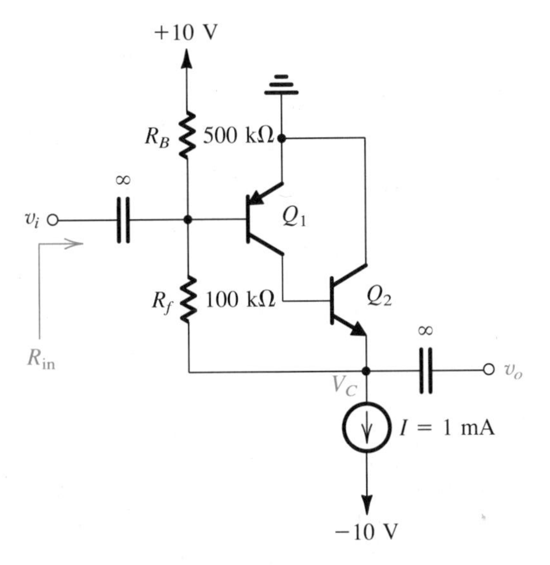

Fig. P9.38

(b) Replacing each BJT with its hybrid-π model, show that

$$\frac{v_o}{v_i} \simeq g_{m1}\,[r_{o1}//\beta_N\,(r_{o2}//R_f)]$$

(c) Find the values of v_o/v_i and R_{in}.

****D9.39** Consider the compound-transistor class AB output stage shown in Fig. 9.27 in which Q_2 and Q_4 are matched transistors with $V_{BE} = 0.7$ V at 10 mA and $\beta = 100$, Q_1 and Q_5 have $V_{BE} = 0.7$ V at 1-mA currents and $\beta = 100$, and Q_3 has $V_{EB} = 0.7$ V at a 1-mA current and $\beta = 10$. All transistors have $n = 1$. Design the circuit for a quiescent current of 2 mA in Q_2 and Q_4, I_{bias} that is 100 times the standby base current in Q_1, and a current in Q_5 that is nine times that in the associated resistors. Find the values of the input voltage required to produce outputs of ± 10 V for a 1-kΩ load. Use V_{CC} of 15 V.

9.40 Repeat Exercise 9.13 for a design variation in which transistor Q_5 is increased in size by a factor of 10, all other conditions remaining the same.

9.41 Repeat Exercise 9.13 for a design in which the limiting output current and normal peak current are 50 mA and 33.3 mA, respectively.

D9.42 The circuit shown in Fig. P9.42 operates in a manner analogous to that in Fig. 9.28 to limit the output current from Q_3 in the event of a short circuit or other mishap. It has the advantage that the current-sensing resistor, R, does not appear directly at the output. Find the value of R that causes Q_5 to turn on and absorb all of $I_{\text{bias}} = 2$ mA, when the current being sourced reaches 150 mA. For Q_5, $I_S = 10^{-14}$ A and $n = 1$. If the normal peak output current is 100 mA, find the voltage drop across R and the collector current in Q_5.

D9.43 Consider the thermal shutdown circuit shown in Fig. 9.29. At 25°C, Z_1 is a 6.8-V zener diode with a TC of 2 mV/°C, and Q_1 and Q_2 are BJTs that display V_{BE} of 0.7 V at a current of 100 μA and have a TC of −2 mV/°C. Design the circuit so that at 125°C, a current of 100 μA flows in each of Q_1 and Q_2. What is the current in Q_2 at 25°C?

Section 9.8: IC Power Amplifiers

D9.44 In the power amplifier circuit of Fig. 9.30 two resistors are important in controlling the overall voltage gain. Which are they? Which controls the gain alone? Which affects both the dc output level and the gain? A new design is being considered in which the output dc level is approximately $\frac{1}{3}V_S$ (rather than approximately

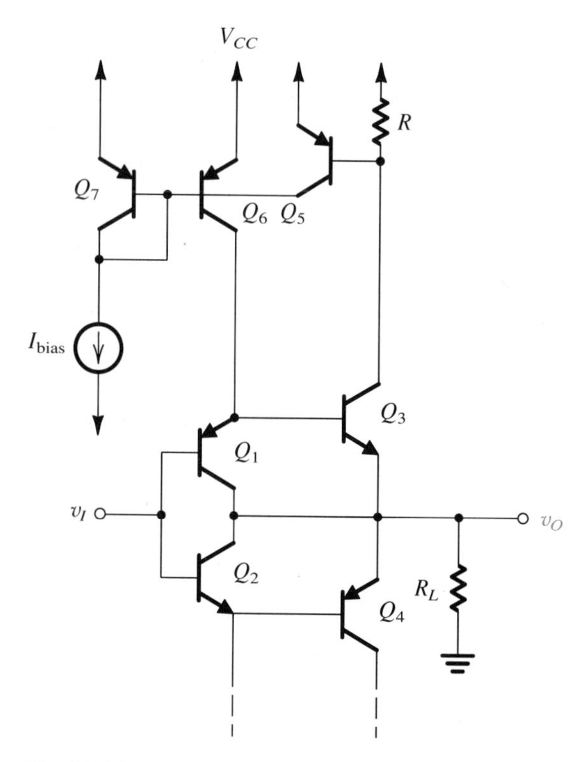

Fig. P9.42

$\frac{1}{2}V_S$) with a gain of 50 (as before). What changes are needed?

9.45 Consider the front end of the circuit in Fig. 9.30. For $V_S = 20$ V, calculate approximate values for the bias currents in Q_1 through Q_6. Assume $\beta_{npn} = 100$, $\beta_{pnp} = 20$, and $|V_{BE}| = 0.7$ V. Also find the dc voltage at the output.

9.46 Assume that the output voltage of the circuit of Fig. 9.30 is at signal ground (and thus the signal feedback is deactivated) and find the differential and common-mode input resistances. For this purpose do not include R_4 and R_5. Let $V_S = 20$ V, $\beta_{npn} = 100$, and $\beta_{pnp} = 20$. Also find the transconductance from the input to the output of the first stage (at the connection of the collectors of Q_4 and Q_6 and the base of Q_{12}).

9.47 It is required to use the LM380 power amplifier to drive an 8-Ω loudspeaker while limiting the maximum possible device dissipation to 1.5 W. Use the graphs of Fig. 9.32 to determine the maximum possible power supply voltage that can be used. (Use only the given graphs; do not interpolate.) If the maximum allowed THD is to be 3%, what is the maximum possible load power? To deliver this power to the load what peak-to-peak output sinusoidal voltage is required?

9.48 Note the resemblance of the structure of the LM380 amplifier to that of the op amp shown in Figs. 2.31 and 2.32. If, when operated with a 20-V supply, the transconductance of the first stage is 1.6 mA/V, find the unity-gain bandwidth (f_t). Since the closed-loop gain is approximately 50 V/V, find its 3-dB bandwidth.

D9.49 Consider the power op amp output stage shown in Fig. 9.33. Using a ±15-V supply, provide a design that provides an output of ±11 V or more, with currents up to ±20 mA provided primarily by Q_3 and Q_4 with a 10% contribution by Q_5 and Q_6, and peak output currents of 1A at full output (+11 V). As the basis of an initial design, use $\beta = 50$ and $|V_{BE}| = 0.7$ V for all devices at all currents. Also use $R_5 = R_6 = 0$.

9.50 For the circuit in Fig. P9.50, assuming all transistors to have large β, show that $i_O = v_I/R$. (This voltage-to-current converter is an application of a versatile circuit building block known as the *current conveyor;* see Sedra and Roberts, 1990.) For $\beta = 100$, by what approximate percentage is i_O actually lower than this?

D9.51 For the bridge amplifier of Fig. 9.34, let $R_1 = R_3 =$ 10 kΩ. Find R_2 and R_4 to obtain an overall gain of 10.

D9.52 An alternative bridge amplifier configuration with high input resistance is shown in Fig. P9.52. (Note the similarity of this circuit to the front end of the instrumentation amplifier circuit shown in Fig. 2.23.) What is the gain v_O/v_I? For op amps using ±15-V supplies that limit at ±13 V, what is the largest sine wave you can provide across R_L? Using 1 kΩ as the smallest resistor, find resistor values that make $v_O/v_I =$ 10 V/V.

Section 9.9: MOS Power Transistors

9.53 A particular power DMOS device for which C_{OX} is 400 μF/m^2, W is 10^5 μm, and $V_t = 2$ V enters velocity saturation at $v_{GS} = 5$ V. Use Eqs. (9.46) and (9.47) to find an expression for L and its value for this transistor. At what value of drain current does velocity saturation begin? For electrons in silicon, $U_{sat} = 5 \times 10^6$ cm/s and $\mu_n = 500$ cm^2/V.s. What is g_m for this device at high currents?

D9.54 Consider the design of the class AB amplifier of Fig. 9.38 under the following conditions: $|V_t| = 2$ V, $K =$ 100 mA/V^2, $|V_{BE}| = 0.7$ V, β is high, $I_{QN} = I_{QP} = I_R = 10$ mA, $I_{bias} = 100$ μA, $I_{Q5} = I_{Q6} = I_{bias}/2$, $R_2 = R_4$, temperature coefficient of $V_{BE} =$

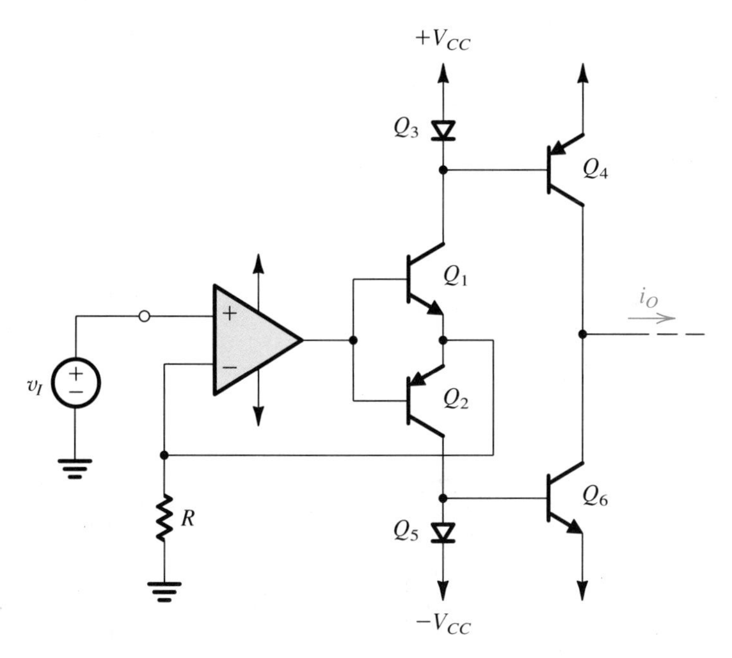

Fig. P9.50

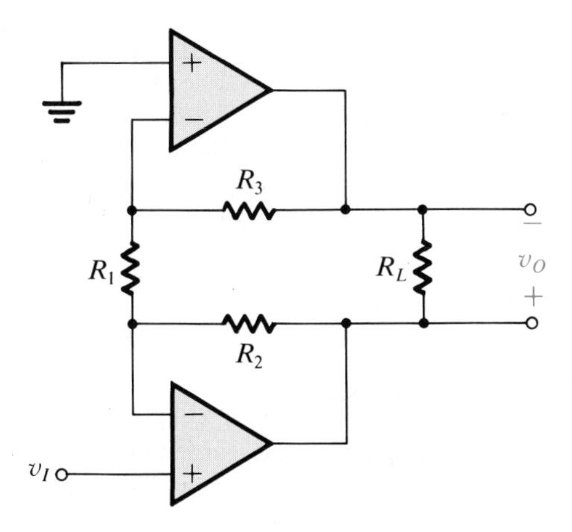

Fig. P9.52

-2 mV/°C, and temperature coefficient of $V_t = -3$ mV/°C in the low-current region. Find the values of R, R_1, R_2, R_3, and R_4. Assume Q_6, Q_P, and Q_N to be thermally coupled. (R_G, used to suppress parasitic oscillation at high frequency, is usually 100 Ω or so.)

CHAPTER 10

Analog Integrated Circuits

INTRODUCTION

Analog ICs include operational amplifiers, analog multipliers, analog-to-digital (A/D) and digital-to-analog (D/A) converters, phase-locked loops, and a variety of other, more specialized functional blocks. All of these analog subsystems are internally constructed using the basic building blocks we have studied in previous chapters, including differential pairs, current mirrors, MOS switches, and others.

In this chapter we shall study the internal circuitry of the most important analog ICs—namely, the operational amplifier and data converters. The terminal characteristics and circuit applications of op amps have already been covered in Chapter 2. Here, our objective is to expose the reader to some of the ingenious techniques that have been evolved over the years for combining elementary analog circuit building blocks so as to realize a complete op amp. Specifically, we shall study in some detail the circuit of the most popular analog IC in production today, the 741 internally compensated op amp. This op amp was introduced in 1966 and is currently produced by almost every manufacturer of analog semiconductors. We shall also study in some detail a popular circuit of a CMOS op amp. Although the performance of CMOS op amps does not match that available in bipolar units, it is more than adequate for their application in VLSI systems. Combining the advantages of bipolar and CMOS devices in the presently evolving BiCMOS technology makes possible some excellent op amp designs. We shall briefly discuss one such circuit.

Analog-to-digital and digital-to-analog converters constitute another important class of analog ICs to which the reader will be introduced in this chapter.

In addition to exposing the reader to some of the ideas that make analog IC design an exciting topic, this chapter should serve to tie together many of the concepts and methods studied in the previous chapters.

10.1 THE 741 OP AMP CIRCUIT

We shall begin with a qualitative study of the 741 op amp circuit, which is shown in Fig. 10.1. Note that in keeping with the IC design philosophy the circuit uses a large number of transistors but relatively few resistors and only one capacitor. This philosophy is dictated by the economics (silicon area, ease of fabrication, quality of realizable components) of the fabrication of active and passive components in IC form (see Appendix A).

As is the case with most modern IC op amps, the 741 requires two power supplies, $+V_{CC}$ and $-V_{EE}$. Normally, $V_{CC} = V_{EE} = 15$ V, but the circuit operates satisfactorily with the power supplies reduced to much lower values (such as ± 5 V). It is important to observe that no circuit node is connected to ground, the common terminal of the two supplies.

With a relatively large circuit such as that shown in Fig. 10.1, the first step in the analysis is the identification of its recognizable parts and their functions. This can be done as follows.

Bias Circuit

The reference bias current of the 741 circuit, I_{REF}, is generated in the branch at the extreme left, consisting of the two diode-connected transistors Q_{11} and Q_{12} and the resistance R_5. Using a Widlar current source formed by Q_{11}, Q_{10}, and R_4, bias current for the first stage is generated in the collector of Q_{10}. Another current mirror formed by Q_8 and Q_9 takes part in biasing the first stage.

The reference bias current I_{REF} is used to provide two proportional currents in the collectors of Q_{13}. This double-collector *lateral*[1] *pnp* transistor can be thought of as two transistors whose base–emitter junctions are connected in parallel. Thus Q_{12} and Q_{13} form a two-output current mirror: One output, the collector of Q_{13B}, provides bias current for Q_{17}, and the other output, the collector of Q_{13A}, provides bias current for the output stage of the op amp.

[1] See Appendix A for a description of lateral *pnp* transistors.

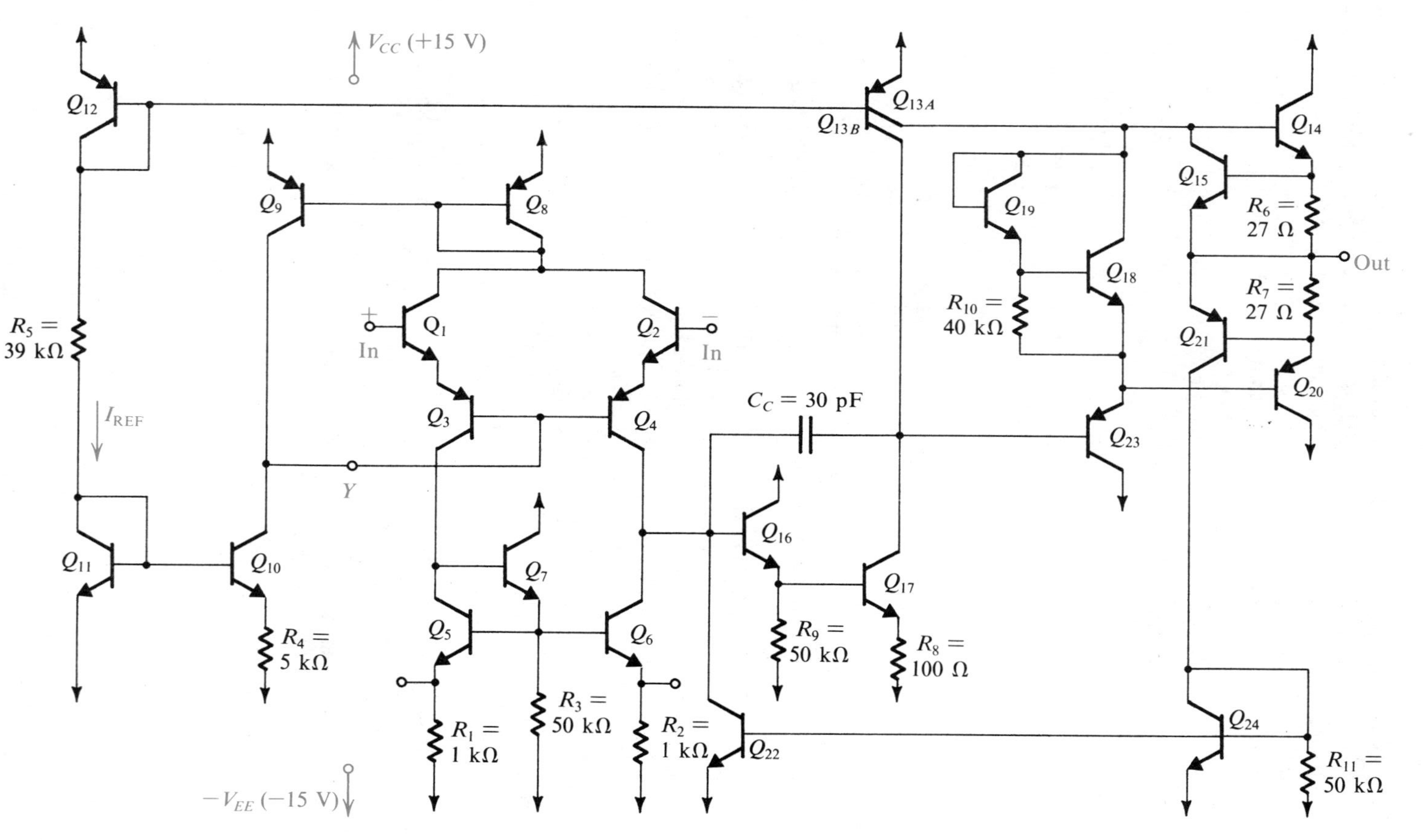

Fig. 10.1 The 741 op amp circuit. Q_{11}, Q_{12}, and R_5 generate a reference bias current, I_{REF}. Q_{10}, Q_9, and Q_8 bias the input stage, which is composed of Q_1 to Q_7. The second gain stage is composed of Q_{16} and Q_{17} with Q_{13B} acting as active load. The class AB output stage is formed by Q_{14} and Q_{20} with biasing devices Q_{18} and Q_{19} and an input buffer Q_{23}. Transisters Q_{15}, Q_{21}, Q_{24}, and Q_{22} serve to protect the amplifier against output short circuit and are normally off.

Two more transistors, Q_{18} and Q_{19}, take part in the dc bias process. The purpose of Q_{18} and Q_{19} is to establish two V_{BE} drops between the bases of the output transistors Q_{14} and Q_{20}.

Short-Circuit Protection Circuitry

The 741 circuit includes a number of transistors that are normally off and that conduct only in the event that one attempts to draw a large current from the op amp output terminal. This would happen, for example, if the output terminal is short-circuited to one of the two supplies. The short-circuit protection network consists of R_6, R_7, Q_{15}, Q_{21}, Q_{24}, and Q_{22}. In the following we shall assume that these transistors are off. Operation of the short-circuit protection network will be explained in Section 10.5.

The Input Stage

The 741 circuit consists of three stages: an input differential stage, an intermediate single-ended high-gain stage, and an output-buffering stage. The input stage consists of transistors Q_1 through Q_7, with biasing performed by Q_8, Q_9, and Q_{10}. Transistors Q_1 and Q_2 act as emitter followers, causing the input resistance to be high and delivering the differential input signal to the differential common-base amplifier formed by Q_3 and Q_4.

Transistors Q_5, Q_6, and Q_7 and resistors R_1, R_2, and R_3 form the load circuit of the input stage. This is an elaborate current-mirror load circuit, which we will analyze in detail in Section 10.3. It will be shown that this load circuit not only provides a high-resistance load but also converts the signal from differential to single-ended form with no loss in gain or common-mode rejection. The output of the input stage is taken single-endedly at the collector of Q_6.

As mentioned in Section 6.10, every op amp circuit includes a *level shifter* whose function is to shift the dc level of the signal so that the signal at the op amp output can swing positive and negative. In the 741, level shifting is done in the first stage using the lateral *pnp* transistors Q_3 and Q_4. Although lateral *pnp* transistors have poor high-frequency performance, their use in the common-base configuration (which is known to have good high-frequency response) does not seriously impair the op amp frequency response.

The use of the lateral *pnp* transistors Q_3 and Q_4 in the first stage results in an added advantage: protection of the input-stage transistors Q_1 and Q_2 against emitter-base junction breakdown. Since the emitter-base junction of an *npn* transistor breaks down at about 7 V of reverse bias (see Section 4.13), regular *npn* differential stages would suffer such a breakdown if, say, the supply voltage is accidentally connected between the input terminals. Lateral *pnp* transistors, however, have high emitter-base breakdown voltages (about 50 V); and because they are connected in series with Q_1 and Q_2, they provide protection of the 741 input transistors, Q_1 and Q_2.

The Second Stage

The second or intermediate stage is composed of Q_{16}, Q_{17}, Q_{13B}, and the two resistors R_8 and R_9. Transistor Q_{16} acts as an emitter follower, thus giving the second stage a high input resistance. This minimizes the loading on the input stage and avoids loss of gain.

Transistor Q_{17} acts as a common-emitter amplifier with a 100-Ω resistor in the emitter. Its load is composed of the high output resistance of the *pnp* current source Q_{13B} in parallel with the input resistance of the output stage (seen looking into the base of Q_{23}). Using a transistor current source as a load resistance is a technique called *active load* (Section 6.5). It enables one to obtain high gain without resorting to the use of high load resistances, which would occupy a large chip area.

The output of the second stage is taken at the collector of Q_{17}. Capacitor C_C is connected in the feedback path of the second stage to provide frequency compensation using the Miller compensation technique studied in Section 8.11. It will be shown in Section 10.6 that the relatively small capacitor C_C gives the 741 a dominant pole at about 4 Hz. Furthermore, pole splitting causes other poles to be shifted to much higher frequencies, giving the op amp a uniform −20 dB/decade gain rolloff with a unity-gain bandwidth of about 1 MHz. It should be pointed out that although C_C is small in value, the chip area that it occupies is about 13 times that of a standard *npn* transistor!

The Output Stage

The purpose of the output stage (Chapter 9) is to provide the amplifier with a low output resistance. In addition, the output stage should be able to supply relatively large load currents without dissipating an unduly large amount of power in the IC. The 741 uses an efficient class AB output stage, which we shall study in detail in Section 10.5.

The output stage consists of the complementary pair Q_{14} and Q_{20}, where Q_{20} is a *substrate pnp* (see Appendix A). Transistors Q_{18} and Q_{19} are fed by current source Q_{13A} and bias the output transistors Q_{14} and Q_{20}. Transistor Q_{23} (which is another substrate *pnp*) acts as an emitter follower, thus minimizing the loading effect of the output stage on the second stage.

Device Parameters

In the following sections we shall carry out a detailed analysis of the 741 circuit. For the standard *npn* and *pnp* transistors, the following parameters will be used:

$$npn\text{:}\quad I_S = 10^{-14}\text{ A},\ \beta = 200,\ V_A = 125\text{ V}$$

$$pnp\text{:}\quad I_S = 10^{-14}\text{ A},\ \beta = 50,\ V_A = 50\text{ V}$$

In the 741 circuit the nonstandard devices are Q_{13}, Q_{14}, and Q_{20}. Transistor Q_{13} will be assumed to be equivalent to two transistors, Q_{13A} and Q_{13B}, with parallel base–emitter junctions and with the following saturation currents:

$$I_{SA} = 0.25 \times 10^{-14}\text{ A} \qquad I_{SB} = 0.75 \times 10^{-14}\text{ A}$$

Transistors Q_{14} and Q_{20} will be assumed to each have an area three times that of a standard device. Output transistors usually have relatively large areas in order to be able to supply large load currents and dissipate relatively large amounts of power with only a moderate increase in the device temperature.

Exercises 10.1 For the standard *npn* transistor whose parameters are given above, find approximate values for the following parameters at $I_C = 1$ mA: V_{BE}, g_m, r_e, r_π, r_o, r_μ. (*Note:* Assume $r_\mu = 10\beta r_o$.)

Ans. 633 mV; 40 mA/V; 25 Ω; 5 kΩ; 125 kΩ; 250 MΩ

10.2 For the circuit in Fig. E10.2, neglect base currents and use the exponential i_C–v_{BE} relationship to show that

$$I_3 = I_1\sqrt{\frac{I_{S3}I_{S4}}{I_{S1}I_{S2}}}$$

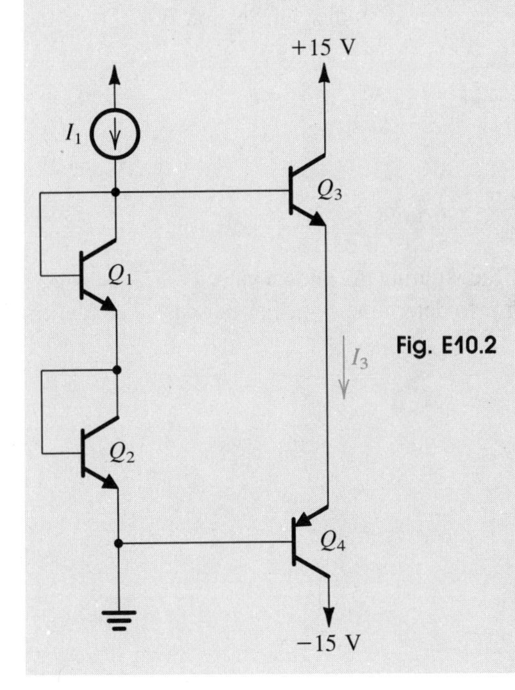

Fig. E10.2

10.2 DC ANALYSIS OF THE 741

In this section we shall carry out a dc analysis of the 741 circuit to determine the bias point of each device. For the dc analysis of an op amp circuit the input terminals are grounded. Theoretically speaking, this should result in zero dc voltage at the output. However, because the op amp has very large gain, any slight approximation in the analysis will show that the output voltage is far from being zero and is close to either $+V_{CC}$ or $-V_{EE}$. In actual practice an op amp left open-loop will have an output voltage saturated close to one of the two supplies. To overcome this problem in the dc analysis, it will be assumed that the op amp is connected in a negative-feedback loop that stabilizes the output dc voltage to zero volts.

Reference Bias Current

The reference bias current I_{REF} is generated in the branch composed of the two diode-connected transistors Q_{11} and Q_{12} and resistor R_5. With reference to Fig. 10.1, we can write

$$I_{REF} = \frac{V_{CC} - V_{EB12} - V_{BE11} - (-V_{EE})}{R_5}$$

For $V_{CC} = V_{EE} = 15$ V and $V_{BE11} = V_{EB12} \simeq 0.7$ V, we have $I_{REF} = 0.73$ mA.

Input Stage Bias

Transistor Q_{11} is biased by I_{REF}, and the voltage developed across it is used to bias Q_{10}, which has a series emitter resistance R_4. This part of the circuit is redrawn in Fig. 10.2 and can be recognized as the Widlar current source studied in Section 6.4. From the circuit we have

$$V_{BE11} - V_{BE10} = I_{C10}R_4$$

Thus

$$V_T \ln \frac{I_{\text{REF}}}{I_{C10}} = I_{C10}R_4 \tag{10.1}$$

where it has been assumed that $I_{S10} = I_{S11}$. Substituting the known values for I_{REF} and R_4, this equation can be solved by trial and error to determine I_{C10}. For our case the result is $I_{C10} = 19\ \mu\text{A}$.

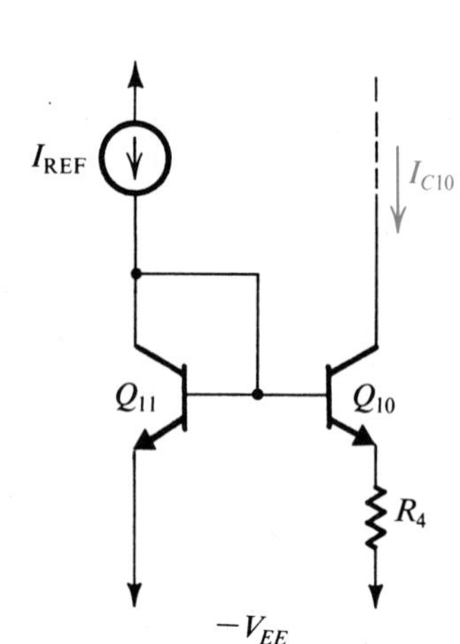

Fig. 10.2 Widlar current source.

Exercise D10.3 Design the Widlar current source of Fig. 10.2 to generate a current $I_{C10} = 10\ \mu\text{A}$ given that $I_{\text{REF}} = 1$ mA. If at a collector current of 1 mA, $V_{BE} = 0.7$ V, find V_{BE11} and V_{BE10}.

Ans. $R_4 = 11.5\ \text{k}\Omega$; $V_{BE11} = 0.7$ V; $V_{BE10} = 0.585$ V

Having determined I_{C10}, we proceed to determine the dc current in each of the input-stage transistors. Part of the input stage is redrawn in Fig. 10.3. From symmetry we see that

$$I_{C1} = I_{C2}$$

Denote this current by I. We see that if the *npn* β is high, then

$$I_{E3} = I_{E4} \simeq I$$

and the base currents of Q_3 and Q_4 are equal, with a value of $I/(\beta_P + 1) \simeq I/\beta_P$, where β_P denotes β of the *pnp* devices.

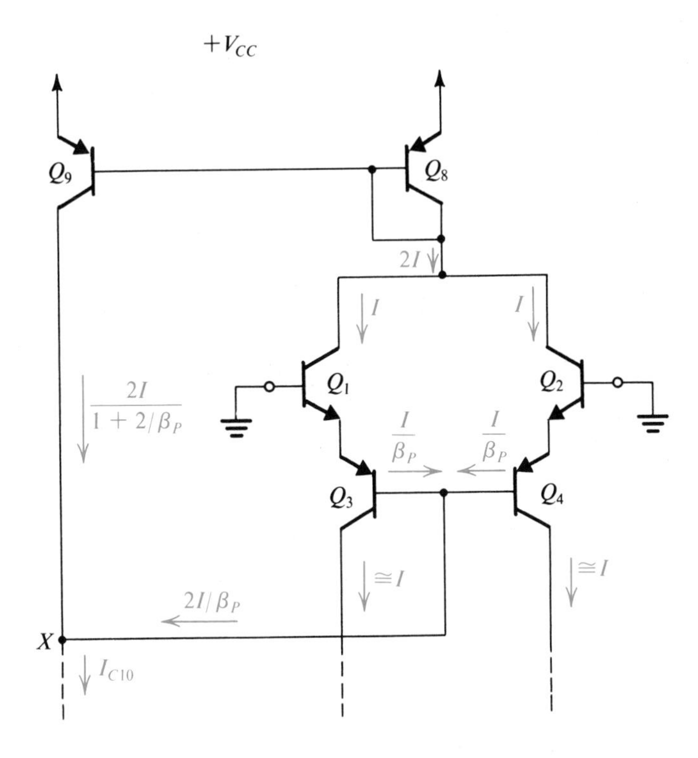

Fig. 10.3 The dc analysis of the 741 input stage.

The current mirror formed by Q_8 and Q_9 is fed by an input current of $2I$. Using the result in Eq. (6.63), we can express the output current of the mirror as

$$I_{C9} = \frac{2I}{1 + 2/\beta_P}$$

We can now write a node equation for node X in Fig. 10.3 and thus determine the value of I. If $\beta_P \gg 1$, then this node equation gives

$$2I \simeq I_{C10}$$

For the 741, $I_{C10} = 19\ \mu\text{A}$; thus $I \simeq 9.5\ \mu\text{A}$. We have thus determined that

$$I_{C1} = I_{C2} \simeq I_{C3} = I_{C4} = 9.5\ \mu\text{A}$$

At this point we should note that transistors Q_1 through Q_4, Q_8, and Q_9 form a **negative-feedback loop,** which works to stabilize the value of I at approximately $I_{C10}/2$. To appreciate this fact, assume that for some reason the current I in Q_1 and Q_2 increases. This will cause the current pulled from Q_8 to increase, and the output current of the Q_8–Q_9 mirror will correspondingly increase. However, since I_{C10} remains constant, node X forces the combined base currents of Q_3 and Q_4 to decrease. This in turn will cause the emitter currents of Q_3 and Q_4, and hence the collector currents of Q_1 and Q_2, to decrease. This is opposite in direction to the change originally assumed. Hence the feedback is negative, and it stabilizes the value of I.

Figure 10.4 shows the remainder of the 741 input stage. If we neglect the base current of Q_{16}, then

$$I_{C6} \simeq I$$

Similarly, neglecting the base current of Q_7 we obtain

$$I_{C5} \simeq I$$

The bias current of Q_7 can be determined from

$$I_{C7} \simeq I_{E7} = \frac{2I}{\beta_N} + \frac{V_{BE6} + IR_2}{R_3} \tag{10.2}$$

where β_N denotes β of the *npn* transistors. To determine V_{BE6} we use the transistor exponential relationship and write

$$V_{BE6} = V_T \ln \frac{I}{I_S}$$

Substituting $I_S = 10^{-14}$ A and $I = 9.5\ \mu$A results in $V_{BE6} = 517$ mV. Then substituting in Eq. (10.2) yields $I_{C7} = 10.5\ \mu$A. Note that the base current of Q_7 is indeed negligible compared to the value of I, as has been assumed.

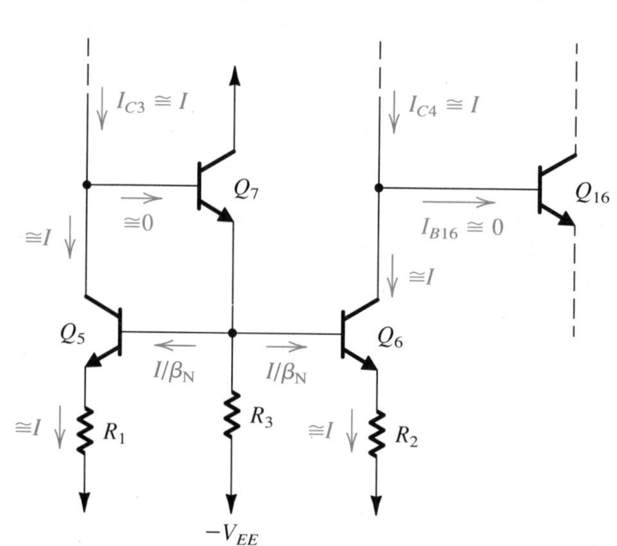

Fig. 10.4 The dc analysis of the 741 input stage, continued.

Input Bias and Offset Currents

The **input bias current** of an op amp is defined (Chapters 2 and 6) as

$$I_B = \frac{I_{B1} + I_{B2}}{2}$$

For the 741 we obtain

$$I_B = \frac{I}{\beta_N}$$

Using $\beta_N = 200$ yields $I_B = 47.5$ nA. Note that this value is reasonably small and is typical of general-purpose op amps that use BJTs in the input stage. Much lower input bias currents (in the picoamp range) can be obtained using an FET input stage. Also, there exist techniques for reducing the input bias current of bipolar-input op amps.

Because of possible mismatches in the β values of Q_1 and Q_2, the input base currents will not be equal. Given the value of the β mismatch, one can use Eq. (6.60) to calculate the **input offset current,** defined as

$$I_{OS} = |I_{B1} - I_{B2}|$$

Input Offset Voltage

From Chapter 6 we know that the input offset voltage is determined primarily by mismatches between the two sides of the input stage. In the 741 op amp the input offset voltage is due to mismatches between Q_1 and Q_2, between Q_3 and Q_4, between Q_5 and Q_6, and between R_1 and R_2. Evaluation of the components of V_{OS} corresponding to the various mismatches follows the method outlined in Section 6.3. Basically, we find the current that results at the output of the first stage due to the particular mismatch being considered. Then we find the differential input voltage that must be applied to reduce the output current to zero.

Input Common-Mode Range

The **input common-mode range** is the range of input common-mode voltages over which the input stage remains in the linear active mode. Refer to Fig. 10.1. We see that in the 741 circuit the input common-mode range is determined at the upper end by saturation of Q_1 and Q_2, and at the lower end by saturation of Q_3 and Q_4.

Exercise 10.4 Neglect the voltage drops across R_1 and R_2 and assume that $V_{CC} = V_{EE} = 15$ V. Show that the input common-mode range of the 741 is approximately -12.6 to $+14.4$ V. (Assume that $V_{BE} \simeq 0.6$ V.)

Second-Stage Bias

If we neglect the base current of Q_{23} then we see from Fig. 10.1 that the collector current of Q_{17} is approximately equal to the current supplied by current source Q_{13B}. Because Q_{13B} has a scale current 0.75 times that of Q_{12}, its collector current will be $I_{C13B} \simeq 0.75 I_{\text{REF}}$, where we have assumed that $\beta_P \gg 1$. Thus $I_{C13B} = 550\ \mu$A and $I_{C17} \simeq 550\ \mu$A. At this current level the base–emitter voltage of Q_{17} is

$$V_{BE17} = V_T \ln \frac{I_{C17}}{I_S} = 618 \text{ mV}$$

The collector current of Q_{16} can be determined from

$$I_{C16} \simeq I_{E16} = I_{B17} + \frac{I_{E17} R_8 + V_{BE17}}{R_9}$$

This calculation yields $I_{C16} = 16.2\ \mu$A. Note that the base current of Q_{16} will indeed be negligible compared to the input-stage bias I, as we have previously assumed.

Output-Stage Bias

Figure 10.5 shows the output stage of the 741 with the short-circuit protection circuitry omitted. Current source Q_{13A} delivers a current of $0.25I_{\text{REF}}$ (because I_S of Q_{13A} is 0.25 times the I_S of Q_{12}) to the network composed of Q_{18}, Q_{19}, and R_{10}. If we neglect the base currents of Q_{14} and Q_{20}, then the emitter current of Q_{23} will also be equal to $0.25I_{\text{REF}}$. Thus

$$I_{C23} \simeq I_{E23} \simeq 0.25I_{\text{REF}} = 180\ \mu\text{A}$$

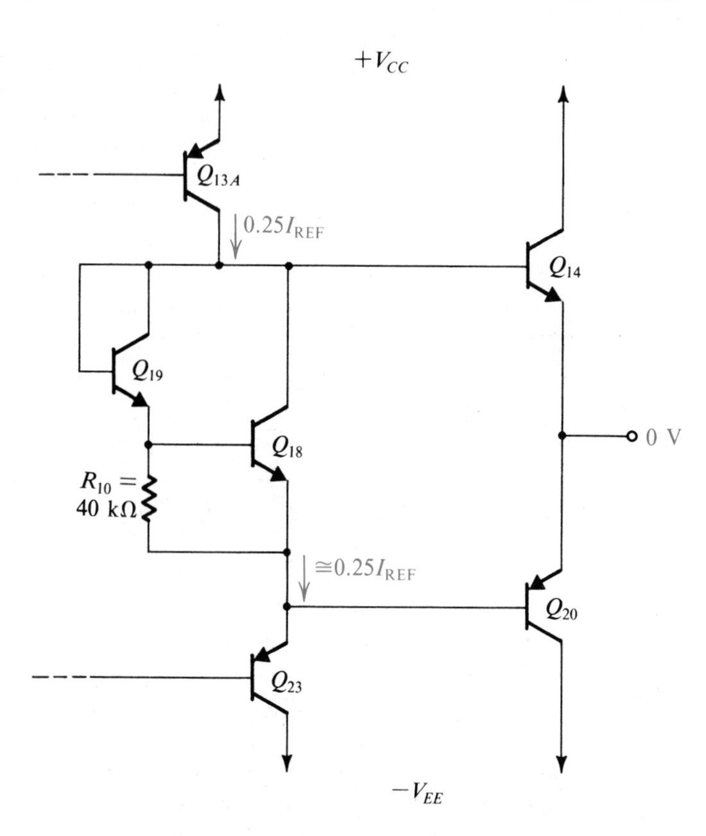

Fig. 10.5 The 741 output stage without the short-circuit protection devices.

Thus we see that the base current of Q_{23} is only 180/50 = 3.6 μA, which is negligible compared to I_{C17}, as we have previously assumed.

If we assume that V_{BE18} is approximately 0.6 V, we can determine the current in R_{10} as 15 μA. The emitter current of Q_{18} is therefore

$$I_{E18} = 180 - 15 = 165\ \mu A$$

Also,

$$I_{C18} \simeq I_{E18} = 165\ \mu\text{A}$$

At this value of current we find that V_{BE18} = 588 mV, which is quite close to the value assumed. The base current of Q_{18} is 165/200 = 0.8 μA, which can be added to the current in R_{10} to determine the Q_{19} current as

$$I_{C19} \simeq I_{E19} = 15.8\ \mu\text{A}$$

The voltage drop across the base–emitter junction of Q_{19} can now be determined as

$$V_{BE19} = V_T \ln \frac{I_{C19}}{I_S} = 530\ \text{mV}$$

As mentioned in Section 10.1, the purpose of the Q_{18}–Q_{19} network is to establish two V_{BE} drops between the bases of the output transistors Q_{14} and Q_{20}. This voltage drop, V_{BB}, can be now calculated as

$$V_{BB} = V_{BE18} + V_{BE19} = 588 + 530 = 1.118\ \text{V}$$

Since V_{BB} appears across the series combination of the base–emitter junctions of Q_{14} and Q_{20}, we can write

$$V_{BB} = V_T \ln \frac{I_{C14}}{I_{S14}} + V_T \ln \frac{I_{C20}}{I_{S20}}$$

Using the calculated value of V_{BB} and substituting $I_{S14} = I_{S20} = 3 \times 10^{-14}$ A, we determine the collector currents as

$$I_{C14} = I_{C20} = 154\ \mu\text{A}$$

Summary

For future reference, Table 10.1 provides a listing of the values of the collector bias currents of the 741 transistors.

Table 10.1 DC COLLECTOR CURRENTS OF THE 741 CIRCUIT (μA)

Q_1	9.5	Q_8	19	Q_{13B}	550	Q_{19}	15.8
Q_2	9.5	Q_9	19	Q_{14}	154	Q_{20}	154
Q_3	9.5	Q_{10}	19	Q_{15}	0	Q_{21}	0
Q_4	9.5	Q_{11}	730	Q_{16}	16.2	Q_{22}	0
Q_5	9.5	Q_{12}	730	Q_{17}	550	Q_{23}	180
Q_6	9.5	Q_{13A}	180	Q_{18}	165	Q_{24}	0
Q_7	10.5						

Exercise 10.5 If in the circuit of Fig. 10.5 the Q_{18}–Q_{19} network is replaced by two diode-connected transistors, find the current in Q_{14} and Q_{20}. (*Hint:* Use the result of Exercise 10.2.)

Ans. 540 μA

10.3 SMALL-SIGNAL ANALYSIS OF THE 741 INPUT STAGE

Figure 10.6 shows part of the 741 input stage for the purpose of performing small-signal analysis. Note that since the collectors of Q_1 and Q_2 are connected to a constant dc

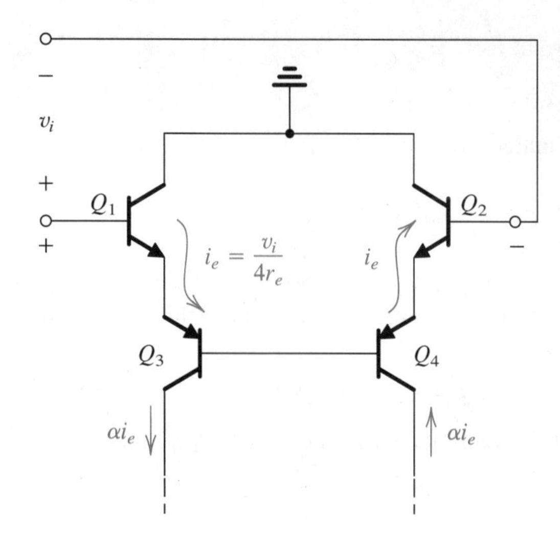

Fig. 10.6 Small-signal analysis of the 741 input stage.

voltage, they are shown grounded. Also, the constant-current biasing of the bases of Q_3 and Q_4 is equivalent to having the common base terminal open-circuited.

The differential signal v_i applied between the input terminals effectively appears across four equal emitter resistances connected in series—those of Q_1, Q_2, Q_3, and Q_4. As a result, emitter signal currents flow as indicated in Fig. 10.6 with

$$i_e = \frac{v_i}{4r_e} \tag{10.3}$$

where r_e denotes the emitter resistance of each of Q_1 through Q_4. Thus

$$r_e = \frac{V_T}{I} = \frac{25\ \text{mV}}{9.5\ \mu\text{A}} = 2.63\ \text{k}\Omega$$

Thus the four transistors Q_1 through Q_4 supply the load circuit with a pair of complementary current signals αi_e, as indicated in Fig. 10.6.

The input differential resistance of the op amp can be obtained from Fig. 10.6 as

$$R_{id} = 4r_\pi = 4(\beta_N + 1)r_e \tag{10.4}$$

For $\beta_N = 200$ we obtain $R_{id} = 2.1\ \text{M}\Omega$.

Proceeding with the input-stage analysis, we show in Fig. 10.7 the load circuit fed with the complementary pair of current signals found above. Neglecting the signal current in the base of Q_7, we see that the collector signal current of Q_5 is approximately equal to the input current αi_e. Now, since Q_5 and Q_6 are identical and their bases are tied together, and since equal resistances are connected in their emitters, it follows that their collector signal currents must be equal. Thus the signal current in the collector of Q_6 is forced to be equal to αi_e. In other words, the load circuit functions as a **current mirror.**

Now consider the output node of the input stage. The output current i_o is given by

$$i_o = 2\alpha i_e \tag{10.5}$$

The factor of two in this equation indicates that conversion from differential to single-ended is performed without losing half the signal. The trick, of course, is the use of the

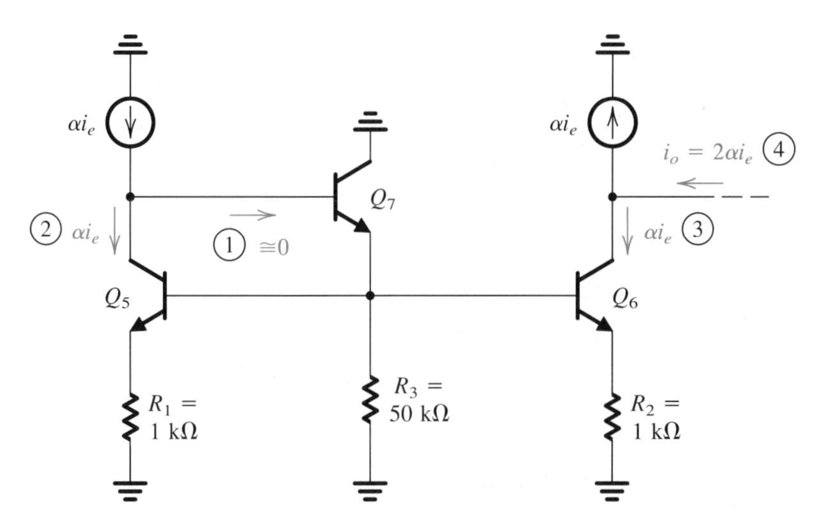

Fig. 10.7 The load circuit of the input stage fed by the two complementary current signals generated by Q_1 through Q_4 in Fig. 10.6. The order of the analysis steps is indicated by the circled numbers.

current mirror to invert one of the current signals and then add the result to the other current signal (see Section 6.5).

Equations (10.3) and (10.5) can be combined to obtain the transconductance of the input stage, G_{m1}:

$$G_{m1} \equiv \frac{i_o}{v_i} = \frac{\alpha}{2r_e} \tag{10.6}$$

Substituting $r_e = 2.63$ kΩ and $\alpha \simeq 1$ yields $G_{m1} = 1/5.26$ mA/V.

Exercise 10.6 For the circuit in Fig. 10.7 find in terms of i_e: (a) The signal voltage at the base of Q_6; (b) The signal current in the emitter of Q_7; (c) The signal current in the base of Q_7; (d) The signal voltage at the base of Q_7; (e) The input resistance seen by the left-hand-side signal current source αi_e.

(*Note:* For simplicity assume that $I_{C7} \simeq I_{C5} = I_{C6}$.)

Ans. (a) 3.63 kΩ × i_e; (b) $0.08 i_e$; (c) $0.0004 i_e$; (d) 3.84 kΩ × i_e; (e) 3.84 kΩ

To complete our modeling of the 741 input stage we must find its output resistance R_{o1}. This is the resistance seen ''looking back'' at the collector terminal of Q_6 in Fig. 10.7. Thus R_{o1} is the parallel equivalent of the output resistance of the current source supplying the signal current αi_e and the output resistance of Q_6. The first component is the resistance looking into the collector of Q_4 in Fig. 10.6. Finding this resistance is considerably simplified if we assume that the common bases of Q_3 and Q_4 are at a *virtual ground*. This of course happens only when the input signal vi is applied in a complementary fashion. Nevertheless, this assumption does not result in a large error.

Assuming that the base of Q_4 is at virtual ground, the resistance we are after is R_{o4}, indicated in Fig. 10.8(a). This is the output resistance of a common-base transistor that

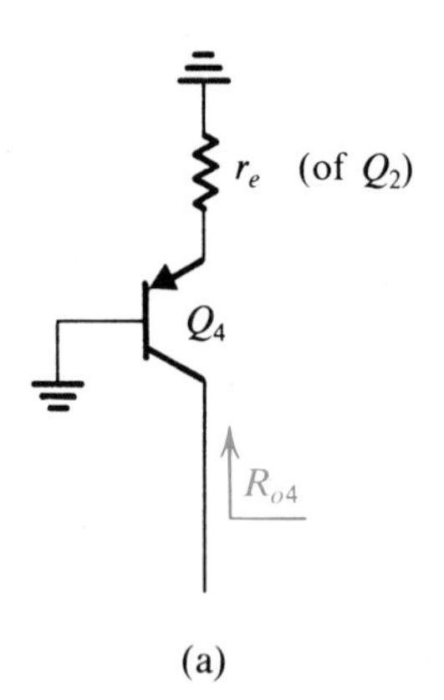

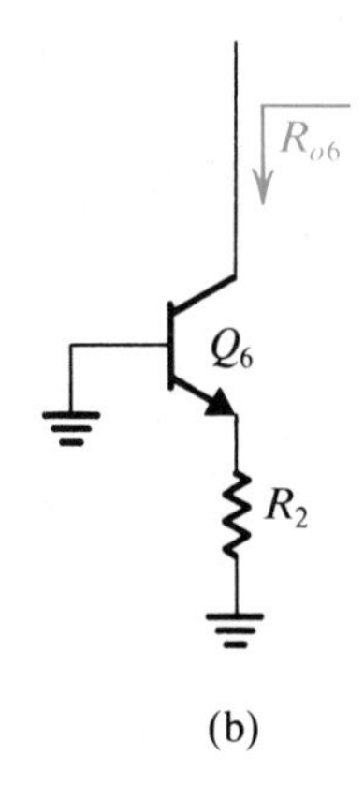

Fig. 10.8 Simplified circuits for finding the two components of the output resistance R_{o1} of the first stage.

has a resistance (r_e of Q_2) in its emitter. To find R_{o4} we may use the expression developed in Chapter 6 (Eq. 6.78):

$$R_o = r_o[1 + g_m(R_E // r_\pi)] \tag{10.7}$$

Substituting $R_E = r_e \equiv 2.63$ kΩ and $r_o = V_A/I$, where $V_A = 50$ V and $I = 9.5$ μA (thus $r_o = 5.26$ MΩ), results in $R_{o4} = 10.5$ MΩ.

The second component of the output resistance is that seen looking into the collector of Q_6 in Fig. 10.7. Although the base of Q_6 is not at signal ground, we shall assume that the signal voltage at the base is sufficiently small to make this approximation valid. The circuit then takes the form in Fig. 10.8(b), and R_{o6} can be determined using Eq. (10.7) with $R_E = R_2$. Thus $R_{o6} \simeq 18.2$ MΩ.

Finally, we combine R_{o4} and R_{o6} in parallel to obtain the output resistance of the input stage, R_{o1}, as $R_{o1} = 6.7$ MΩ.

Figure 10.9 shows the equivalent circuit that we have derived for the input stage. This is a simplified version of the y-parameter model of a two-port network with y_{12} assumed negligible (see Appendix B).

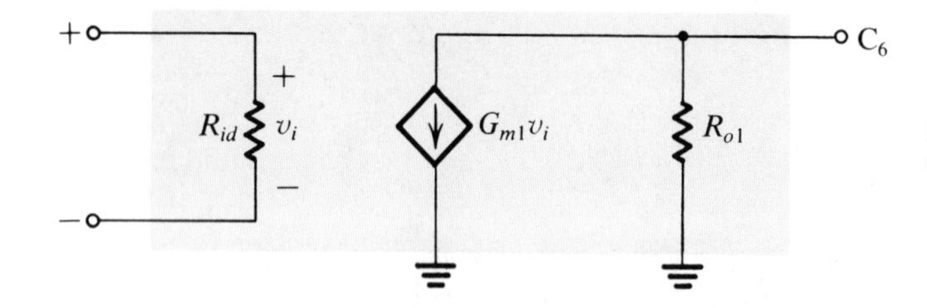

Fig. 10.9 Small-signal equivalent circuit for the input stage of the 741 op amp.

EXAMPLE 10.1 We wish to find the input offset voltage resulting from a 2% mismatch between the resistances R_1 and R_2 in Fig. 10.1.

Solution Consider first the situation when both input terminals are grounded, and assume that $R_1 = R$ and $R_2 = R + \Delta R$, where $\Delta R/R = 0.02$. From Fig. 10.10 we see that while Q_5

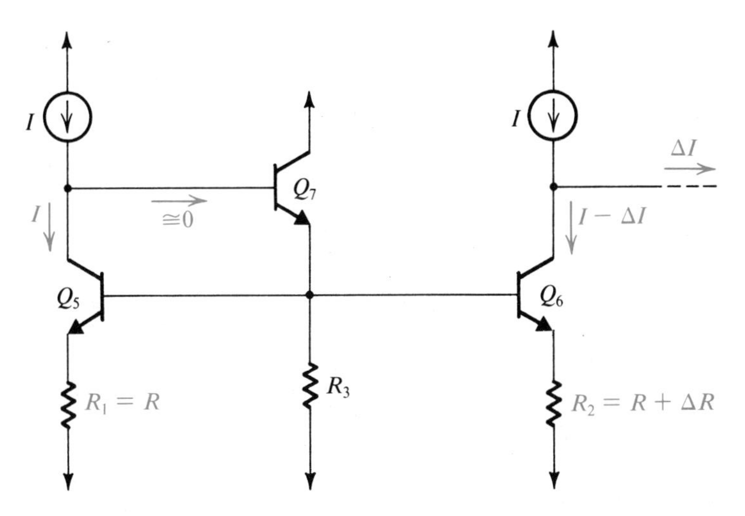

Fig. 10.10 Input stage with both inputs grounded and a mismatch ΔR between R_1 and R_2.

still conducts a current equal to I, the current in Q_6 will be smaller by ΔI. The value of ΔI can be found from

$$V_{BE5} + IR = V_{BE6} + (I - \Delta I)(R + \Delta R)$$

Thus

$$V_{BE5} - V_{BE6} = I\,\Delta R - \Delta I(R + \Delta R) \tag{10.8}$$

The quantity on the left-hand side is in effect the change in V_{BE} due to a change in I_E of ΔI. We may therefore write

$$V_{BE5} - V_{BE6} \simeq \Delta I r_e \tag{10.9}$$

Equations (10.8) and (10.9) can be combined to obtain

$$\frac{\Delta I}{I} = \frac{\Delta R}{R + \Delta R + r_e} \tag{10.10}$$

Substituting $R = 1\ \text{k}\Omega$ and $r_e = 2.63\ \text{k}\Omega$ shows that a 2% mismatch between R_1 and R_2 gives rise to an output current $\Delta I = 5.5 \times 10^{-3} I$. To reduce this output current to zero we have to apply an input voltage V_{OS} given by

$$V_{OS} = \frac{\Delta I}{G_{m1}} = \frac{5.5 \times 10^{-3} I}{G_{m1}} \tag{10.11}$$

Substituting $I = 9.5\ \mu\text{A}$ and $G_{m1} = 1/5.26\ \text{mA/V}$ results in the offset voltage $V_{OS} \simeq 0.3\ \text{mV}$.

It should be pointed out that the offset voltage calculated is only one component of the input offset voltage of the 741. Other components arise because of mismatches in transistor characteristics. The 741 offset voltage is specified to be typically 2 mV.

Exercises The purpose of the following series of exercises is to determine the finite common-mode gain that results from a mismatch in the load circuit of the input stage of the 741 op amp. Figure E10.7 shows the input stage with an input common-mode signal v_{icm} applied and with a mismatch ΔR between the two resistances R_1 and R_2. Note that to simplify matters we have opened the common-mode feedback loop and included a resistance R_o, which is the resistance seen looking to the left of node Y in the circuit of Fig. 10.1. Thus R_o is the parallel equivalent of R_{o9} (the output resistance of Q_9) and R_{o10} (the output resistance of Q_{10}).

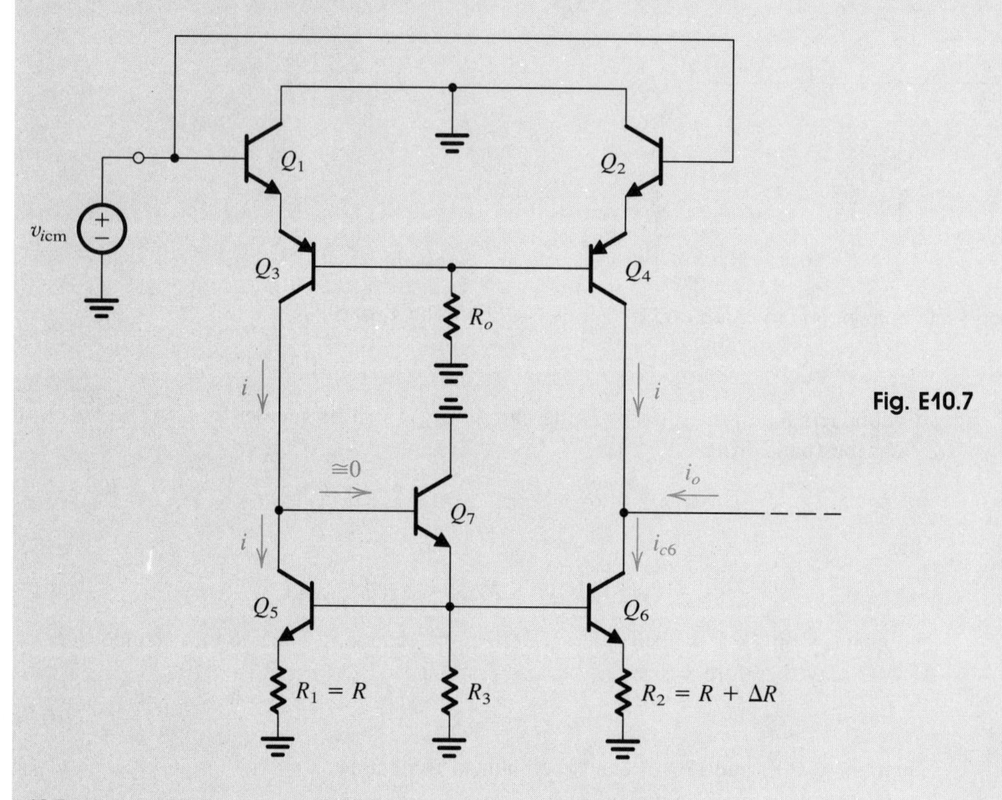

Fig. E10.7

10.7 Show that the current i (Fig. E10.7) is given approximately by

$$i = \frac{v_{icm}}{r_{e1} + r_{e3} + [2R_o/(\beta_P + 1)]}$$

10.8 Show that

$$i_o = -i\frac{\Delta R}{R + r_{e5} + \Delta R}$$

10.9 Using the results of Exercises 10.7 and 10.8, and assuming that $\Delta R \ll (R + r_e)$ and $R_o/(\beta_P + 1) \gg (r_{e1} + r_{e3})$, show that the common-mode transconductance G_{mcm} is given approximately by

$$G_{mcm} \equiv \frac{|i_o|}{v_{icm}} \simeq \frac{\beta_P}{2R_o}\frac{\Delta R}{R + r_{e5}}$$

10.10 Refer to Fig. 10.1 and assume that the bases of Q_9 and Q_{10} are at approximately constant voltages (signal ground). Find R_{o9}, R_{o10} and hence R_o. Use $V_A = 125$ V for *npn* and 50 V for *pnp* transistors and neglect r_μ.

Ans. $R_{o9} = 2.63\ \text{M}\Omega$; $R_{o10} = 31.1\ \text{M}\Omega$; $R_o = 2.43\ \text{M}\Omega$

10.11 For $\beta_P = 50$ and $\Delta R/R = 0.02$ evaluate G_{mcm} obtained in Exercise 10.9.

Ans. 0.057 μA/V

10.12 Use the value of G_{mcm} obtained in Exercise 10.11 and the value of G_{m1} obtained from Eq. (10.6) to find the CMRR, that is the ratio of G_{m1} to G_{mcm}, expressed in decibels.

Ans. 70.5 dB

10.13 Noting that with the common-mode negative-feedback loop in place the common-mode gain will decrease by the amount of feedback, and noting that the loop gain is approximately equal to β_P (see Problem 10.8), find the CMRR with the feedback loop in place.

Ans. 104.6 dB

10.4 SMALL-SIGNAL ANALYSIS OF THE 741 SECOND STAGE

Figure 10.11 shows the 741 second stage prepared for small-signal analysis. In this section we shall analyze the second stage to determine the values of the parameters of the equivalent circuit shown in Fig. 10.12. Again, this is a simplified y-parameter equivalent circuit with y_{12} neglected.

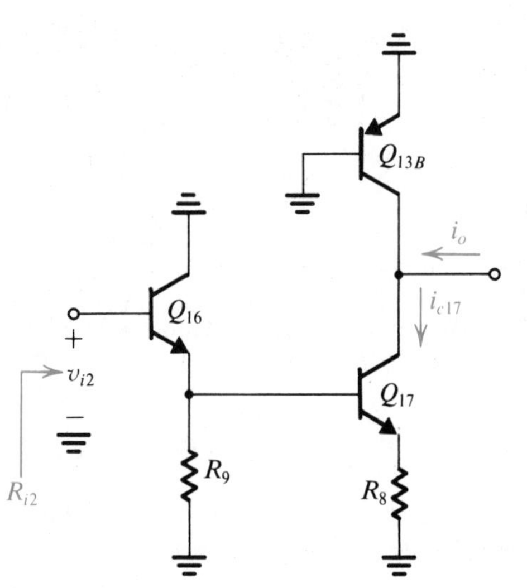

Fig. 10.11 The 741 second stage prepared for small-signal analysis.

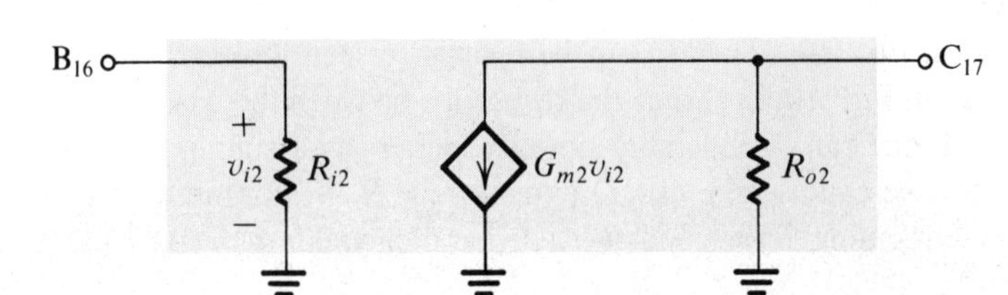

Fig. 10.12 Small-signal equivalent circuit model of the second stage.

Input Resistance

The input resistance R_{i2} can be found by inspection to be

$$R_{i2} = (\beta_{16} + 1)[r_{e16} + R_9//(\beta_{17} + 1)(r_{e17} + R_8)] \tag{10.12}$$

Substituting the appropriate parameter values yields $R_{i2} \simeq 4\ \text{M}\Omega$.

Transconductance

From the equivalent circuit of Fig. 10.12 we see that the transconductance G_{m2} is the ratio of the *short-circuit output current* to the input voltage. Short-circuiting the output terminal of the second stage (Fig. 10.11) to ground makes the signal current through the output resistance of Q_{13B} zero, and the output short-circuit current becomes equal to the collector signal current of Q_{17} (i_{c17}). This latter current can be easily related to v_{i2} as follows:

$$i_{c17} = \frac{\alpha v_{b17}}{r_{e17} + R_8} \tag{10.13}$$

$$v_{b17} = v_{i2}\frac{(R_9//R_{i17})}{(R_9//R_{i17}) + r_{e16}} \tag{10.14}$$

$$R_{i17} = (\beta_{17} + 1)(r_{e17} + R_8) \tag{10.15}$$

These equations can be combined to obtain

$$G_{m2} \equiv \frac{i_{c17}}{v_{i2}} \tag{10.16}$$

which, for the 741 parameter values, is found to be $G_{m2} = 6.5$ mA/V.

Output Resistance

To determine the output resistance R_{o2} of the second stage in Fig. 10.11, we ground the input terminal and find the resistance looking back into the output terminal. It follows that R_{o2} is given by

$$R_{o2} = (R_{o13B}//R_{o17}) \tag{10.17}$$

where R_{o13B} is the resistance looking into the collector of Q_{13B} while its base and emitter are connected to ground. It can be easily shown that

$$R_{o13B} = r_{o13B} \tag{10.18}$$

For the 741 component values we obtain $R_{o13B} = 90.9\ \text{k}\Omega$.

The second component in Eq. (10.17), R_{o17}, is the resistance seen looking into the collector of Q_{17}, as indicated in Fig. 10.13. Since the resistance between the base of Q_{17} and ground is relatively small, one can considerably simplify matters by assuming that the base is grounded. Doing this, we can use Eq. (10.7) to determine R_{o17}. For our case the result is $R_{o17} \cong 787\ \text{k}\Omega$. Combining R_{o13B} and R_{o17} in parallel yields $R_{o2} = 81\ \text{k}\Omega$.

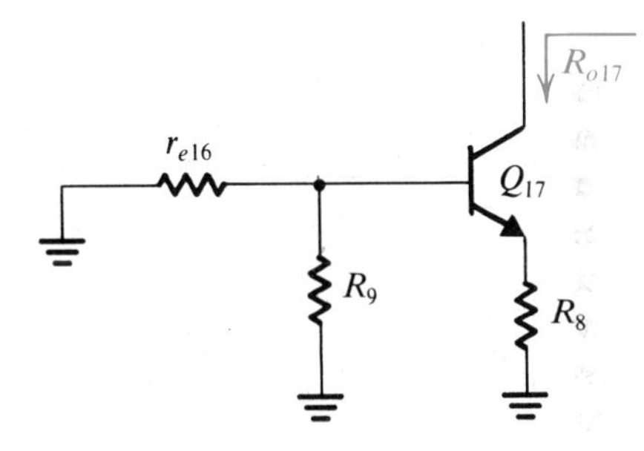

Fig. 10.13 Definition of R_{o17}.

Thévenin Equivalent Circuit

The second-stage equivalent circuit can be converted to the Thévenin form, as shown in Fig. 10.14. Note that the stage open-circuit voltage gain is $-G_{m2}R_{o2}$.

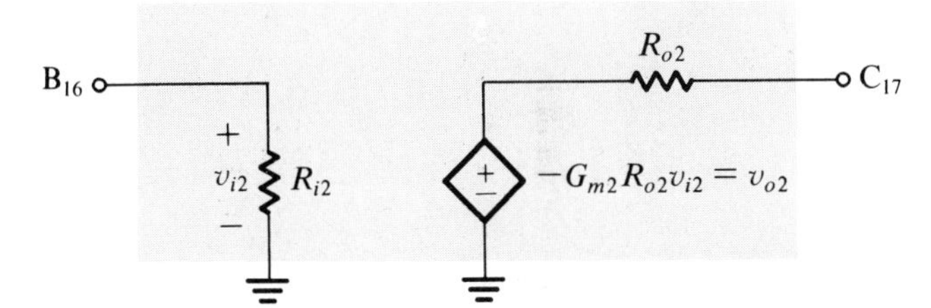

Fig. 10.14 Thévenin form of the small-signal model of the second stage.

Exercises **10.14** Use Eq. (10.12) to show that $R_{i2} \simeq 4\ \text{M}\Omega$.

10.15 Use Eqs. (10.13) to (10.16) to verify that G_{m2} is 6.5 mA/V.

10.16 Verify that $R_{o2} \simeq 81\ \text{k}\Omega$.

10.17 Find the open-circuit voltage gain of the second stage of the 741.

Ans. −526.5 V/V

10.5 ANALYSIS OF THE 741 OUTPUT STAGE

The 741 output stage is shown in Fig. 10.15 without the short-circuit protection circuit. The stage is shown driven by the second-stage transistor Q_{17} and loaded with a 2-kΩ resistance. The circuit is of the AB class (Chapter 9), with the network composed of Q_{18}, Q_{19}, and R_{10} providing the bias of the output transistors Q_{14} and Q_{20}. The use of this network rather than two diode-connected transistors in series enables biasing the output transistors at a low current (0.15 mA) in spite of the fact that the output devices are about three times as large as the standard devices. This is obtained by arranging that the current in Q_{19} is very small and thus its V_{BE} is also small. We analyzed the dc bias in Section 10.2.

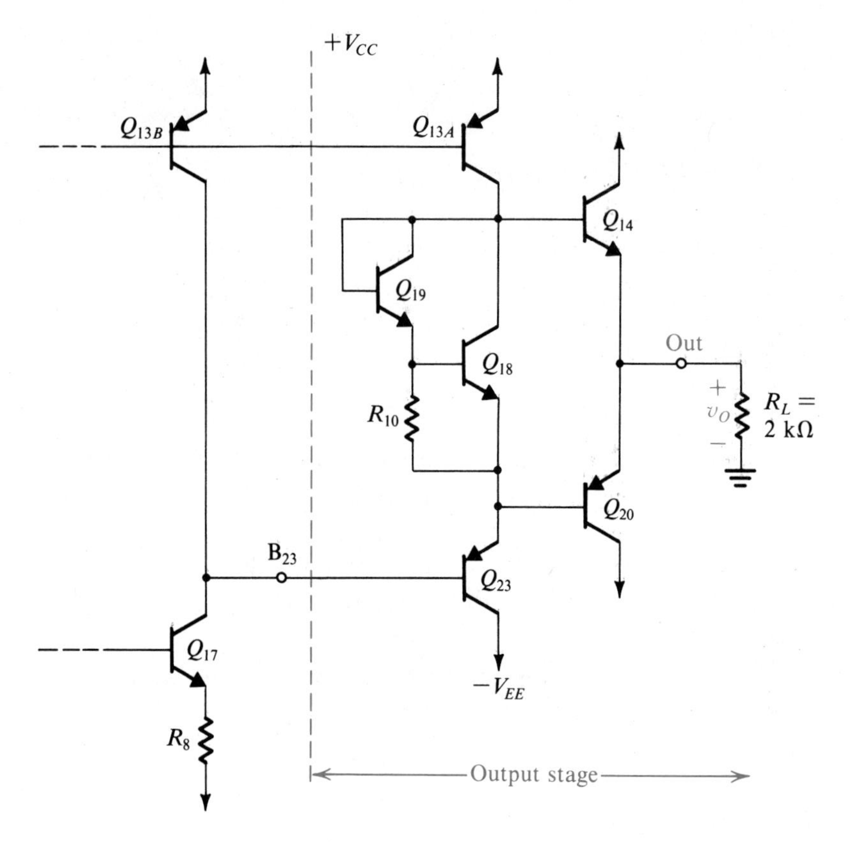

Fig. 10.15 The 741 output stage.

Another feature of the 741 output stage worth noting is that the stage is driven by an emitter follower Q_{23}. As will be shown, this emitter follower provides added buffering, which makes the op amp gain almost independent of the parameters of the output transistors.

Output Voltage Limits

The maximum positive output voltage is limited by the saturation of current-source transistor Q_{13A}. Thus

$$v_{O\max} = V_{CC} - V_{CE\text{sat}} - V_{BE14} \tag{10.19}$$

which is about 1 V below V_{CC}. The minimum output voltage (that is, maximum negative amplitude) is limited by the saturation of Q_{17}. Neglecting the voltage drop across R_8 we obtain

$$v_{O\min} = -V_{EE} + V_{CE\text{sat}} + V_{EB23} + V_{EB20} \tag{10.20}$$

which is about 1.5 V above $-V_{EE}$.

Small-Signal Model

We shall now carry out a small-signal analysis of the output stage for the purpose of determining the values of the parameters of the equivalent circuit model shown in Fig. 10.16. The model is shown fed by v_{o2}, which is the open-circuit output voltage of the second stage. From Fig. 10.14, v_{o2} is given by

$$v_{o2} = -G_{m2}R_{o2}v_{i2} \tag{10.21}$$

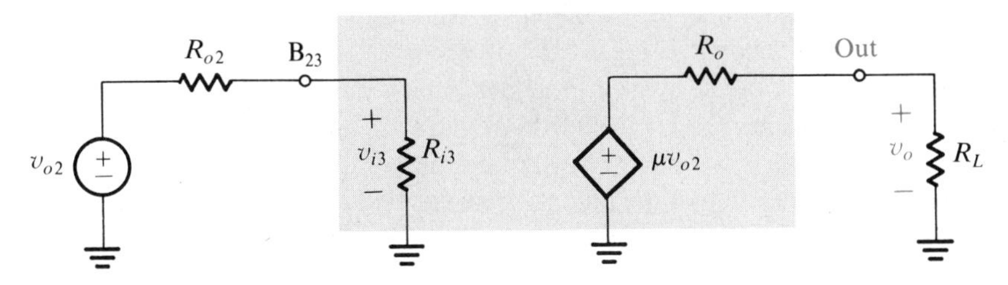

Fig. 10.16 Model for the 741 output stage.

and G_{m2} and R_{o2} were previously determined as $G_{m2} = 6.5$ mA/V and $R_{o2} = 81$ kΩ. Resistance R_{i3} is the input resistance of the output stage determined with the amplifier loaded with R_L. Although the effect of loading an amplifier stage on its input resistance is negligible in the input and second stages, this is not the case in general in an output stage. Defining R_{i3} in this manner enables correct evaluation of the voltage gain of the second stage, A_2, as

$$A_2 \equiv \frac{v_{i3}}{v_{i2}} = -G_{m2}R_{o2}\frac{R_{i3}}{R_{i3} + R_{o2}} \tag{10.22}$$

To determine R_{i3} assume that one of the two output transistors—say, Q_{20}—is conducting a current of, say, 5 mA. It follows that the input resistance looking into the base of Q_{20} is approximately $\beta_{20}R_L$. Assuming $\beta_{20} = 50$, for $R_L = 2$ kΩ the input resistance of Q_{20} is 100 kΩ. This resistance appears in parallel with the series combination of the output resistance of Q_{13A} ($r_{o13A} \simeq 280$ kΩ) and the resistance of the Q_{18}–Q_{19} network. The latter resistance is very small (about 160 Ω; see Exercise 10.18). Thus the total resistance in the emitter of Q_{23} is approximately (100 kΩ//280 kΩ) or 74 kΩ and the input resistance R_{i3} is given by

$$R_{i3} \simeq \beta_{23} \times 74 \text{ k}\Omega$$

which for $\beta_{23} = 50$ is $R_{i3} \simeq 3.7$ MΩ. Since $R_{o2} = 81$ kΩ, we see that $R_{i3} \gg R_{o2}$, and the value of R_{i3} will have little effect on the performance of the op amp. We can use the value obtained for R_{i3} to determine the gain of the second stage in Eq. (10.22) as $A_2 = -515$. The value of A_2 will be needed in Section 10.6 in connection with frequency-response analysis.

Continuing with the determination of the equivalent circuit-model-parameters, we note from Fig. 10.16 that μ is the **open-circuit voltage gain** of the output stage,

$$\mu = \left.\frac{v_o}{v_{o2}}\right|_{R_L=\infty} \tag{10.23}$$

With $R_L = \infty$ the gain of the emitter-follower output transistor (Q_{14} or Q_{20}) will be nearly unity. Also, with $R_L = \infty$ the resistance in the emitter of Q_{23} will be very large. This means that the gain of Q_{23} will be nearly unity and the input resistance of Q_{23} will be very large. We thus conclude that $\mu \simeq 1$.

Next we shall find the value of the output resistance of the op amp, R_o. For this purpose refer to the circuit shown in Fig. 10.17. In accordance with the definition of R_o, the input source feeding the output stage is grounded but its resistance (which is the output resistance of the second stage, R_{o2}) is included. We have assumed that the output voltage v_o is negative, and thus Q_{20} is conducting most of the current; transistor Q_{14} has therefore been eliminated. The exact value of the output resistance will of course depend on which transistor (Q_{14} or Q_{20}) is conducting and on the value of load current. Nevertheless, we wish to find an estimate of R_o.

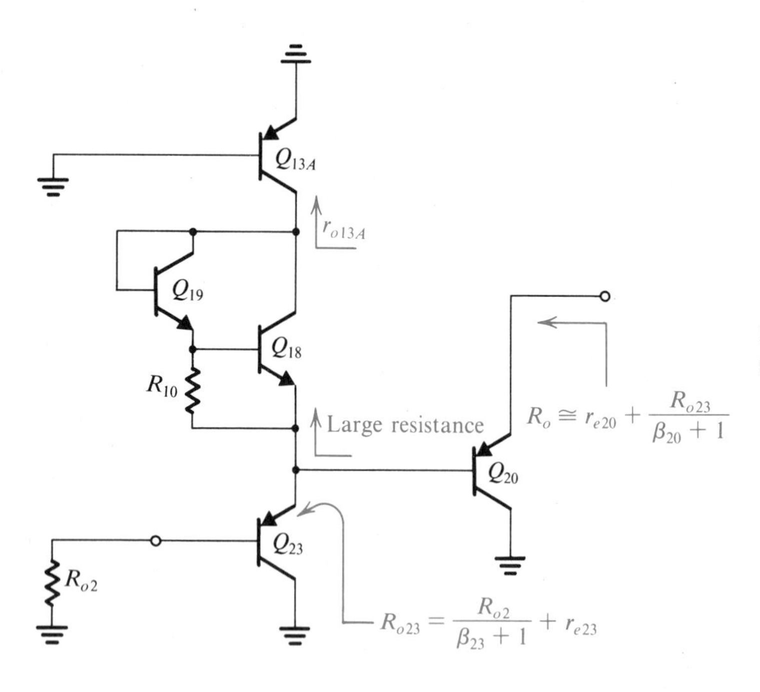

Fig. 10.17 Circuit for finding the output resistance R_o.

As indicated in Fig. 10.17, the resistance seen looking into the emitter of Q_{23} is

$$R_{o23} = \frac{R_{o2}}{\beta_{23}+1} + r_{e23} \tag{10.24}$$

Substituting $R_{o2} = 81\text{ k}\Omega$, $\beta_{23} = 50$, and $r_{e23} = 25/0.18 = 139\ \Omega$ yields $R_{o23} = 1.73\text{ k}\Omega$. This resistance appears in parallel with the series combination of r_{o13A} and the resistance of the Q_{18}–Q_{19} network. Since r_{o13A} alone (0.28 MΩ) is much larger than R_{o23}, the effective resistance between the base of Q_{20} and ground is approximately equal to R_{o23}. Now we can find the output resistance R_o as

$$R_o = \frac{R_{o23}}{\beta_{20} + 1} + r_{e20} \tag{10.25}$$

For $\beta_{20} = 50$ the first component of R_o is 34 Ω. The second component depends critically on the value of output current. For an output current of 5 mA, r_{e20} is 5 Ω and R_o is 39 Ω. To this value we must add the resistance R_7 (27 Ω) (see Fig. 10.1), which is included for short-circuit protection. The output resistance of the 741 is specified to be typically 75 Ω.

Exercises 10.18 Using a simple (r_π, g_m) model for each of the two transistors Q_{18} and Q_{19} in Fig. E10.18, find the small-signal resistance between A and A'. (*Note:* From Table 10.1, $I_{C18} = 165\ \mu A$ and $I_{C19} \simeq 16\ \mu A$.

Ans. 163 Ω

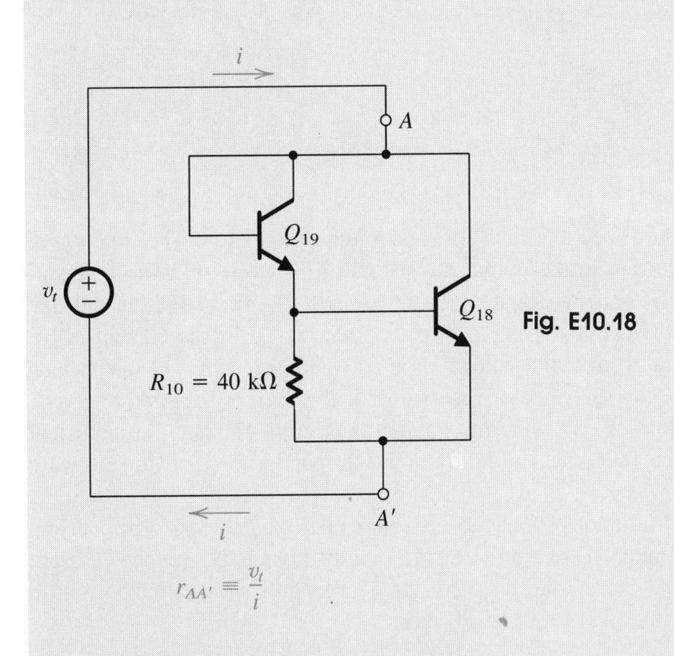

Fig. E10.18

10.19 Figure E10.19 shows the circuit for determining the op amp output resistance when v_O is positive and Q_{14} is conducting most of the current. Using the resistance of the Q_{18}–Q_{19} network calculated in Exercise 10.18 and neglecting the large output resistance of Q_{13A}, find R_o when Q_{14} is sourcing an output current of 5 mA.

Ans. 14.4 Ω

Output Short-Circuit Protection

If the op amp output terminal is short-circuited to one of the power supplies, one of the two output transistors could conduct a large amount of current. Such a large current can result in sufficient heating to cause burnout of the IC. To guard against this possibility, the

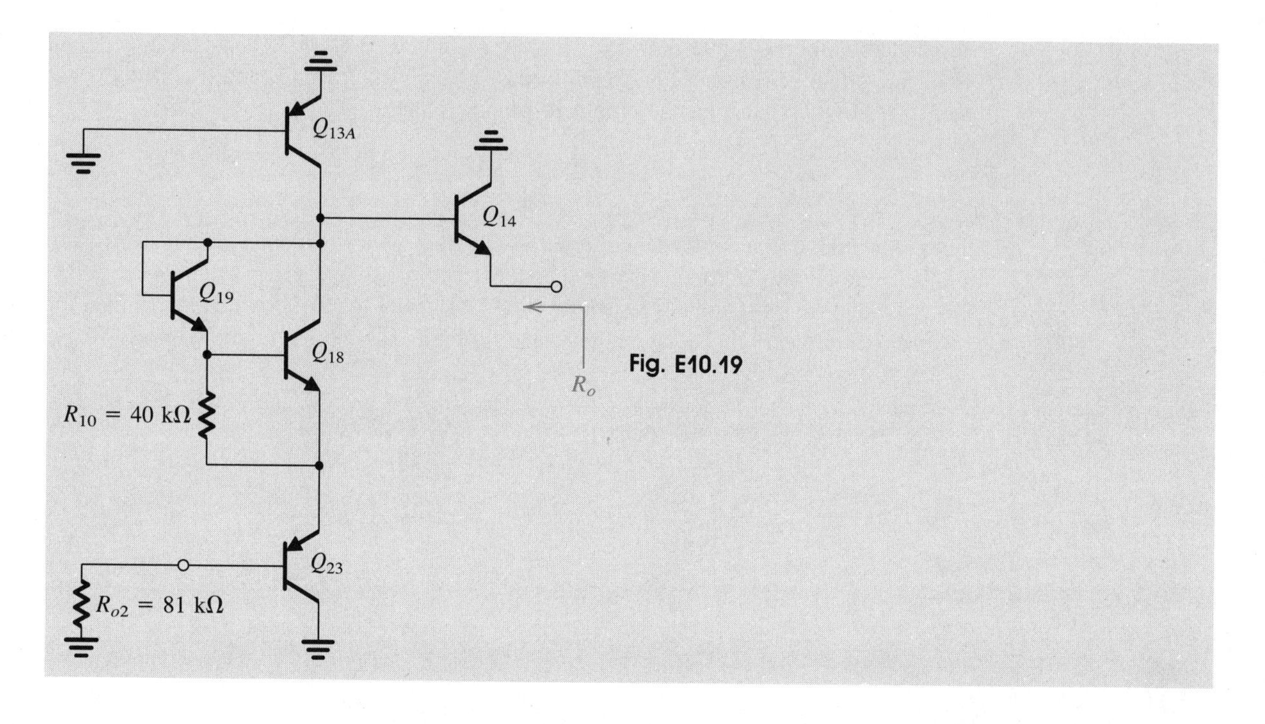

Fig. E10.19

741 op amp is equipped with a special circuit for short-circuit protection. The function of this circuit is to limit the current in the output transistors in the event of a short circuit.

Refer to Fig. 10.1. Resistance R_6 together with transistor Q_{15} limits the current that would flow out of Q_{14} in the event of a short circuit. Specifically, if the current in the emitter of Q_{14} exceeds about 20 mA, the voltage drop across R_6 exceeds 540 mV, which turns Q_{15} on. As Q_{15} turns on, its collector robs some of the current supplied by Q_{13A}, thus reducing the base current of Q_{14}. This mechanism thus limits the maximum current that the op amp can source (that is, supply from the output terminal in the outward direction) to about 20 mA.

Limiting of the maximum current that the op amp can sink, and hence the current through Q_{20}, is done by a mechanism similar to the one discussed above. The relevant circuit is composed of R_7, Q_{21}, Q_{24}, and Q_{22}. For the components shown, the current in the inward direction is limited also to about 20 mA.

10.6 GAIN AND FREQUENCY RESPONSE OF THE 741

In this section we shall evaluate the overall small-signal voltage gain of the 741 op amp. We shall then consider the op amp's frequency response and its slew-rate limitation.

Small-Signal Gain

The overall small-signal gain can be easily found from the cascade of the equivalent circuits derived in the previous sections for the three op amp stages. This cascade is shown in Fig. 10.18, loaded with $R_L = 2\ \text{k}\Omega$, which is the typical value used in measuring and

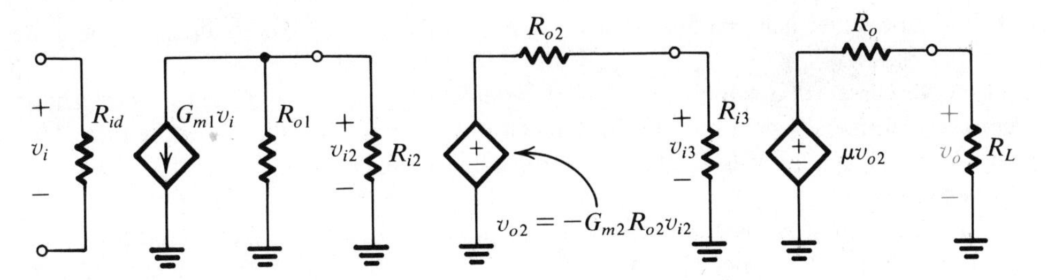

Fig. 10.18 Cascading the small-signal equivalent circuits of the individual stages for the evaluation of the overall voltage gain.

specifying the 741 data. The overall gain can be expressed as

$$\frac{v_o}{v_i} = \frac{v_{i2}}{v_i}\frac{v_{o2}}{v_{i2}}\frac{v_o}{v_{o2}} \tag{10.26}$$

$$= -G_{m1}(R_{o1}//R_{i2})(-G_{m2}R_{o2})\mu\frac{R_L}{R_L + R_o} \tag{10.27}$$

Using the values found in previous sections yields

$$\begin{aligned}\frac{v_o}{v_i} &= -476.1 \times (-526.5) \times 0.97 = 243{,}147 \text{ V/V} \\ &= 107.7 \text{ dB}\end{aligned} \tag{10.28}$$

Frequency Response

The 741 is an internally compensated op amp. It employs the Miller compensation technique, studied in Section 8.11, to introduce a dominant low-frequency pole. Specifically, a 30-pF capacitor (C_C) is connected in the negative-feedback path of the second stage. An approximate estimate of the frequency of the dominant pole can be obtained as follows:

Using Miller's theorem (Section 7.4) the effective capacitance due to C_C between the base of Q_{16} and ground is (see Fig. 10.1)

$$C_i = C_C(1 + |A_2|) \tag{10.29}$$

where A_2 is the second-stage gain. Use of the value calculated for A_2 in Section 10.5, $A_2 = -515$, results in $C_i = 15{,}480$ pF. Since this capacitance is quite large, we shall neglect all other capacitances between the base of Q_{16} and signal ground. The total resistance between this node and ground is

$$\begin{aligned}R_t &= (R_{o1}//R_{i2}) \\ &= (6.7 \text{ M}\Omega//4 \text{ M}\Omega) = 2.5 \text{ M}\Omega\end{aligned} \tag{10.30}$$

Thus the dominant pole has a frequency f_P given by

$$f_P = \frac{1}{2\pi C_i R_t} = 4.1 \text{ Hz} \tag{10.31}$$

It should be noted that this approach is equivalent to using the approximate formula in Eq. (8.58).

As discussed in Section 8.11, Miller compensation produces an additional advantageous effect, namely pole splitting. As a result, the other poles of the circuit are moved to very high frequencies. This has been confirmed by computer-aided analysis [see Gray and Meyer (1984)].

Assuming that all nondominant poles are at high frequencies, the calculated values give rise to the Bode plot shown in Fig. 10.19. The unity-gain bandwidth f_t can be calculated from

$$f_t = A_0 f_{3\text{dB}} \tag{10.32}$$

Thus

$$f_t = 243{,}147 \times 4.1 \simeq 1 \text{ MHz} \tag{10.33}$$

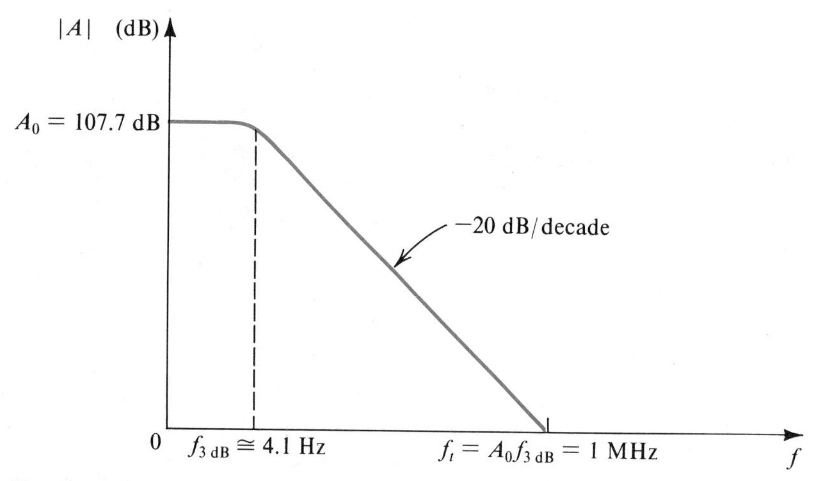

Fig. 10.19 Bode plot for the 741 gain, neglecting nondominant poles.

Although this Bode plot implies that the phase shift at f_t is $-90°$ and thus that the phase margin is 90°, in practice a phase margin of about 80° is obtained. The excess phase shift (about 10°) is due to the nondominant poles. This phase margin is sufficient to provide stable operation of closed-loop amplifiers with any value of feedback factor β. This convenience of use of the internally compensated 741 is achieved at the expense of a great reduction in open-loop gain and hence in the amount of negative feedback. In other words, if one requires a closed-loop amplifier with a gain of 1000, then the 741 is overcompensated for such an application and one would be much better off designing one's own compensation (assuming, of course, the availability of an op amp that is not internally compensated).

A Simplified Model

Figure 10.20 shows a simplified model of the 741 op amp in which the high-gain second stage, with its feedback capacitance C_C, is modeled by an ideal integrator. In this model

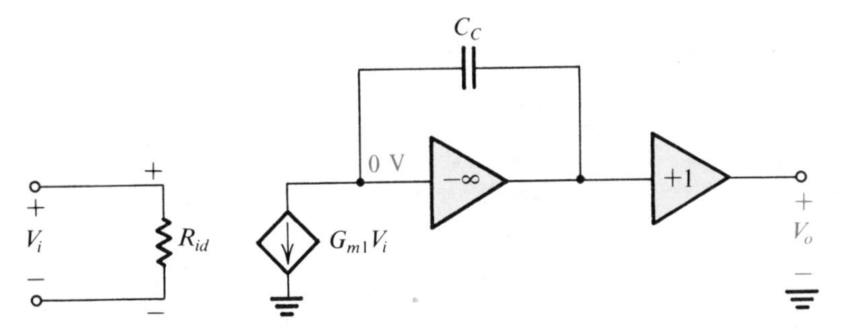

Fig. 10.20 A simple model for the 741 based on modeling the second stage as an integrator.

the gain of the second stage is assumed sufficiently large that a virtual ground appears at its input. For this reason the output resistance of the input stage and the input resistance of the second stage have been omitted. Furthermore, the output stage is assumed to be an ideal unity-gain follower. The reader will recall that this model was used in our study of the op amp terminal characteristics in Chapter 2.

Analysis of the model in Fig. 10.20 gives

$$A(s) \equiv \frac{V_o(s)}{V_i(s)} = \frac{G_{m1}}{sC_C} \tag{10.34}$$

Thus

$$A(j\omega) = \frac{G_{m1}}{j\omega C_C} \tag{10.35}$$

and the magnitude of gain becomes unity at $\omega = \omega_t$, where

$$\omega_t = \frac{G_{m1}}{C_C} \tag{10.36}$$

Substituting $G_{m1} = 1/5.26$ mA/V and $C_C = 30$ pF yields

$$f_t = \frac{\omega_t}{2\pi} \simeq 1 \text{ MHz} \tag{10.37}$$

which is equal to the value calculated before. It should be pointed out, however, that this model is valid only at frequencies $f \gg f_{3dB}$. At such frequencies the gain falls off with a slope of -20 dB/decade, just like an integrator.

Slew Rate

The slew-rate limitation of op amps is discussed in Chapter 2. Here we shall illustrate the origin of the slewing phenomenon in the context of the 741 circuit.

Consider the unity-gain follower of Fig. 10.21 with a step of, say, 10 V applied at the input. Because of amplifier dynamics, its output will not change in zero time. Thus immediately after the input is applied, almost the entire value of the step will appear as a differential signal between the two input terminals. This large input voltage causes the input stage to be **overdriven,** and its small-signal model no longer applies. Rather, half

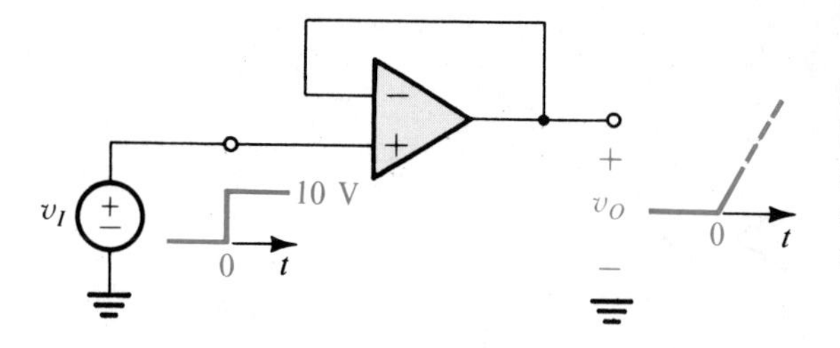

Fig. 10.21 A unity-gain follower with a large step input. Since the output voltage cannot change immediately, a large differential voltage appears between the op amp input terminals.

the stage cuts off and the other half conducts all the current. Specifically, reference to Fig. 10.1 shows that a large differential input voltage causes Q_1 and Q_3 to conduct all the available bias current ($2I$) while Q_2 and Q_4 will be cut off. The current mirror Q_5, Q_6, and Q_7 will still function, and Q_6 will produce a collector current of $2I$.

Using the above observations, and modeling the second stage as an ideal integrator, results in the model of Fig. 10.22. From this circuit we see that the output voltage will be a ramp with a slope of $2I/C_C$:

$$v_O(t) = \frac{2I}{C_C}t \tag{10.38}$$

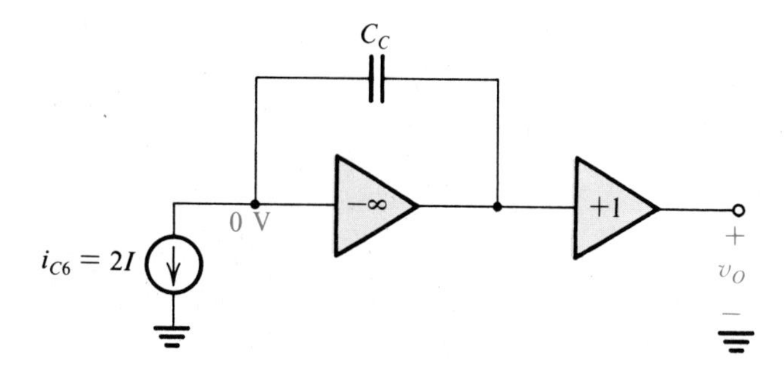

Fig. 10.22 Model for the 741 op amp when a large differential signal is applied.

Thus the slew rate SR is given by

$$\text{SR} = \frac{2I}{C_C} \tag{10.39}$$

For the 741, $I = 9.5\ \mu\text{A}$ and $C_C = 30$ pF, resulting in SR = 0.63 V/μs.

It should be pointed out that this is a rather simplified model of the slewing process. More detail can be found in Gray and Meyer (1984).

Exercise 10.20 Use the value of the slew rate calculated above to find the full-power bandwidth f_M of the 741 op amp. Assume that the maximum output is ±10 V.

Ans. 10 kHz

Relationship Between f_t and SR

A simple relationship exists between the unity-gain bandwidth f_t and the slew rate SR. This relationship is obtained from Eqs. (10.36) and (10.39) together with

$$G_{m1} = 2\frac{1}{4r_e}$$

where r_e is the emitter resistance of each of Q_1 through Q_4. Thus

$$r_e = \frac{V_T}{I}$$

and

$$G_{m1} = \frac{I}{2V_T} \tag{10.40}$$

Substituting in Eq. (10.36) results in

$$\omega_t = \frac{I}{2C_C V_T} \tag{10.41}$$

Substituting for I/C_C from Eq. (10.39) gives

$$\omega_t = \frac{\text{SR}}{4V_T} \tag{10.42}$$

which can be expressed in the alternative form

$$\text{SR} = 4\omega_t V_T \tag{10.43}$$

As a check, for the 741 we have

$$\text{SR} = 4 \times 2\pi \times 10^6 \times 25 \times 10^{-3} = 0.63 \text{ V}/\mu\text{s}$$

which is the result obtained previously.

Exercises 10.21 Consider the integrator model of the op amp in Fig. 10.20. Find the value of the resistor that, when connected across C_C, provides the correct value of the dc gain.

Ans. 1279 MΩ

D10.22 If a resistance R_E is included in each of the emitter leads of Q_3 and Q_4 show that $\text{SR} = 4\omega_t(V_T + IR_E/2)$. Hence find the value of R_E that would double the 741 slew rate while keeping ω_t and I unchanged. What are the new values of C_C, the dc gain, and the 3-dB frequency?

Ans. 5.26 kΩ; 15 pF; 101.7 dB (a 6-dB decrease); 8.2 Hz

(*Note:* This is a viable technique in general for increasing slew rate. It is referred to as the G_m-reduction method.)

10.7 CMOS OP AMPS

Unlike the 741 op amp, which is a general-purpose operational amplifier intended for a variety of applications, most CMOS op amps are designed to be used as part of a VLSI circuit. This constrained application environment implies that the amplifier specifications can be relaxed in return for a simpler circuit that occupies a relatively small silicon area. The most significant specification relaxation is in the load-driving capability of the op amp. Most CMOS op amps are required to drive on-chip capacitive loads of few picofarads. It follows that such an op amp does not need a sophisticated output stage. In fact, most CMOS op amps do not have a low-impedance output stage. On a VLSI chip, however, some of the amplifiers are required to drive off-chip loads, and these few amplifiers are usually equipped with an output stage of a classical type.

Two-Stage Topology

Figure 10.23 shows a popular two-stage CMOS op amp configuration. The circuit utilizes two power supplies, which are usually ±5 V but can be as low as ±2.5 V for advanced, reduced-feature-size technologies. A reference bias current I_{REF} is generated either externally or using on-chip circuits [see Gray and Meyer (1984)]. The current mirror formed by Q_8 and Q_5 supplies the differential pair Q_1–Q_2 with bias current. The W/L ratio of Q_5 is selected to yield the desired input-stage bias. The input differential pair is actively loaded with the current mirror formed by Q_3 and Q_4. Thus the input stage is identical to that studied in Section 6.7.

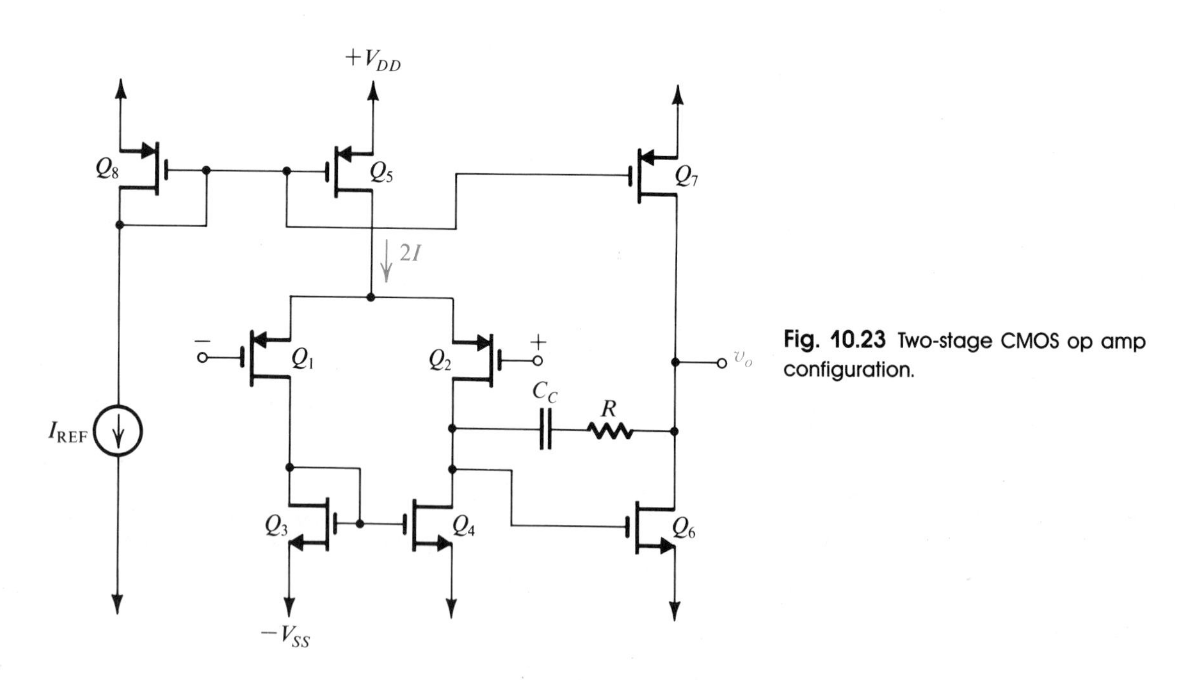

Fig. 10.23 Two-stage CMOS op amp configuration.

The second stage consists of Q_6, which is a common-source amplifier actively loaded with the current-source transistor Q_7. As in the 741, frequency compensation is implemented using a Miller feedback capacitor C_C. Here, however, there is an additional resistor included in series with C_C. The function of this resistor, which is usually implemented using one or two MOS transistors, will be explained shortly.

Voltage Gain

The voltage gain of the first stage was found in Section 6.7 to be given by

$$A_1 = -g_{m1}(r_{o2}//r_{o4}) \tag{10.44}$$

The second stage is an actively loaded common-source amplifier whose voltage gain is given by

$$A_2 = -g_{m6}(r_{o6}//r_{o7}) \tag{10.45}$$

The dc open-loop gain of the op amp is the product of A_1 and A_2.

EXAMPLE 10.2

Consider the circuit in Fig. 10.23 with the following device geometries.

Transistor	Q_1	Q_2	Q_3	Q_4	Q_5	Q_6	Q_7	Q_8
W/L	120/8	120/8	50/10	50/10	150/10	100/10	150/10	150/10

Let $I_{\text{REF}} = 25\ \mu\text{A}$, $|V_t|$ (for all devices) $= 1$ V, $\mu_n C_{\text{OX}} = 20\ \mu\text{A/V}^2$, $\mu_p C_{\text{OX}} = 10\ \mu\text{A/V}^2$, $|V_A|$ (for all devices) $= 25$ V, $V_{DD} = V_{SS} = 5$ V. For all devices evaluate I_D, $|V_{GS}|$, g_m, and r_o. Also find A_1, A_2, the dc open-loop voltage gain, the input common-mode range, and the output voltage range. Neglect the effect of V_A on bias current.

Solution

Refer to Fig. 10.23. Since Q_8 and Q_5 are matched, $2I = I_{REF}$. Thus Q_1, Q_2, Q_3, and Q_4 each conducts a current equal to $I = 12.5\ \mu$A. Since Q_7 is matched to Q_5 and Q_8, the current in Q_7 is equal to $I_{REF} = 25\ \mu$A. Finally, Q_6 conducts an equal current of 25 μA.

With I_D of each device known, we use

$$I_D = \tfrac{1}{2}(\mu C_{OX})(W/L)(|V_{GS}| - |V_t|)^2$$

to determine $|V_{GS}|$. The results are given in Table 10.2.

The transconductance g_m of each device is determined from

$$g_m = 2K(|V_{GS} - |V_t|) = \mu C_{\text{OX}}(W/L)(|V_{GS}| - |V_t|)$$
$$= \sqrt{2(\mu C_{\text{OX}})(W/L)I_D}$$

or alternatively

$$g_m = 2I_D/(|V_{GS}| - |V_t|)$$

Table 10.2

	Q_1	Q_2	Q_3	Q_4	Q_5	Q_6	Q_7	Q_8
$I_D(\mu\text{A})$	12.5	12.5	12.5	12.5	25	25	25	25
$\|V_{GS}\|(\text{V})$	1.4	1.4	1.5	1.5	1.6	1.5	1.6	1.6
$g_m(\mu\text{A/V})$	62.5	62.5	50	50	83.3	100	83.3	83.3
$r_o(\text{M}\Omega)$	2	2	2	2	1	1	1	1

The value of r_o is determined from

$$r_o = |V_A|/I_D$$

The resulting values of g_m and r_o are given in Table 10.2.

The voltage gain of the first stage is determined using Eq. (10.44),

$$A_1 = -g_{m1}(r_{o2}//r_{o4})$$
$$= -62.5(2//2) = -62.5 \text{ V/V}$$

The voltage gain of the second stage is determined using Eq. (10.45),

$$A_2 = -g_{m6}(r_{o6}//r_{o7})$$
$$= -100(1//1) = -50 \text{ V/V}$$

Thus the overall dc open-loop gain is

$$A_0 = A_1A_2 = (-62.5)(-50) = 3125 \text{ V/V}$$

The lower limit of the input common-mode range is the value at which Q_1 and Q_2 leave the saturation region. This occurs when the input voltage falls below the voltage at the drain of Q_1 by $|V_t|$ volts. Since the drain of Q_1 is at $-5 + 1.5 = -3.5$ V, then the lower limit of the input common-mode range is -4.5 V.

The upper limit of the input common-mode range is the value of input voltage at which Q_5 leaves the saturation region. Thus

$$v_{I_{\text{cm/max}}} = V_{DD} - |V_{GS5}| + |V_t| - |V_{GS1}|$$
$$= 5 - 1.6 + 1 - 1.4 = 3 \text{ V}$$

Finally, the output voltage range is determined from Q_7 leaving the saturation region,

$$v_{O\max} = V_{DD} - |V_{GS7}| + |V_t|$$
$$= 5 - 1.6 + 1 = 4.4 \text{ V}$$

and from Q_6 leaving the saturation region,

$$v_{O\min} = -V_{SS} + |V_{GS6}| - |V_t| = -4.5 \text{ V}$$

Input Offset Voltage

The inevitable device mismatches in the input stage give rise to an input offset voltage. The components of this input offset voltage can be calculated using the methods developed in Chapter 6 and applied in previous sections for the 741 op amp. Because device mismatches are random in nature, the resulting offset voltage is referred to as **random offset.** This is to distinguish it from another type of input offset voltage that can be found in CMOS op amps even if all appropriate devices are perfectly matched. This predictable or **systematic offset** can be minimized by careful design. It does not occur in BJT op amps because of the large gain per stage.

To see how systematic offset can occur, consider the circuit of Fig. 10.23 with the two input terminals grounded. If the input stage is perfectly balanced, then the voltage appearing at the drain of Q_4 will be equal to that at the drain of Q_3, which is $(-V_{SS} + V_{GS4})$. Now this is also the voltage that is fed to the gate of Q_6. In other words, a voltage equal to V_{GS4} appears between gate and source of Q_6. If this voltage is different from the value of V_{GS6} that will make $I_6 = I_7$, an output current, and hence an output offset voltage, results. It can be shown that this output offset can be eliminated by selecting device geometries that satisfy the constraint

$$\frac{K_4}{K_6} = \frac{1}{2}\frac{K_5}{K_7} \tag{10.46}$$

where K is the conductivity parameter $[K = \frac{1}{2}\mu C_{OX}(W/L)]$.

Exercise 10.23 Derive Eq. (10.46).

Frequency Response

To appreciate the need for the resistor R placed in series with the Miller compensation capacitor C_C in the circuit of Fig. 10.23, consider first the situation without R. Figure 10.24(a) shows the small-signal equivalent circuit of the op amp with only C_C

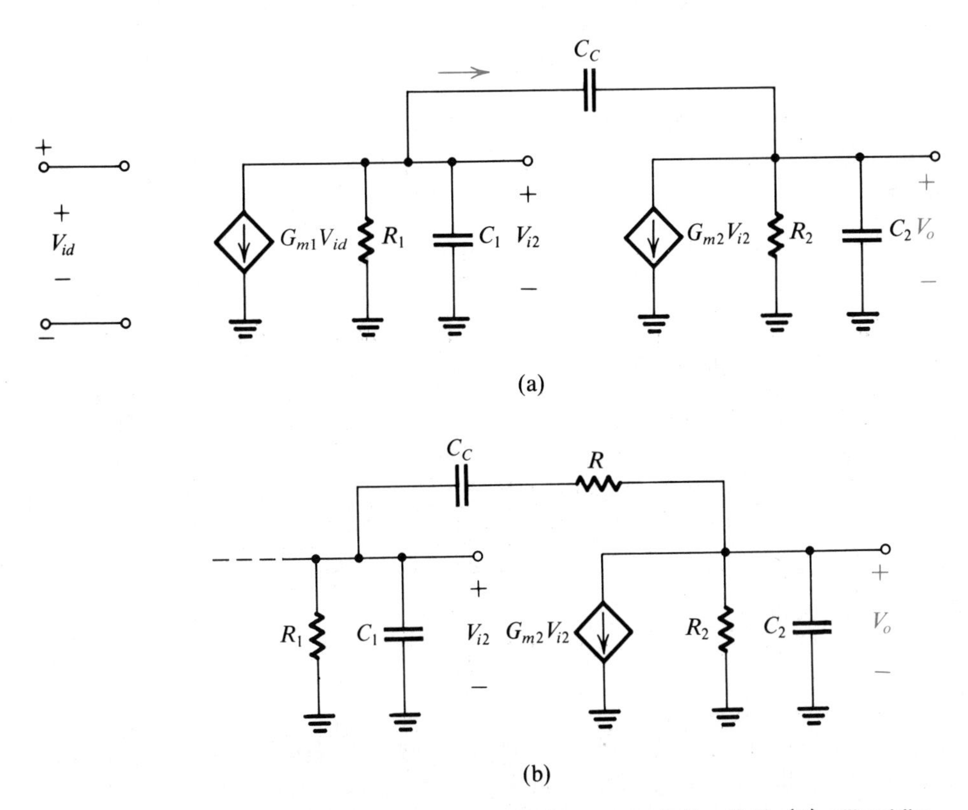

Fig. 10.24 Small-signal equivalent circuit of the CMOS op amp in Fig. 10.23: **(a)** without the resistance R; **(b)** with R included.

included. Note that G_{m1} is the transconductance of the input stage ($G_{m1} = g_{m1} = g_{m2}$), R_1 is the output resistance of the first stage [$R_1 = r_{o2}//r_{o4}$], C_1 is the total capacitance at the interface between the first and second stages, G_{m2} is the transconductance of the second stage ($G_{m2} = g_{m6}$), R_2 is the output resistance of the second stage ($R_2 = r_{o6}//r_{o7}$), and C_2 is the total capacitance at the output node of the op amp. Since C_2 includes the load capacitance, it is usually much larger than C_1.

A circuit similar to that in Fig. 10.24(a) was analyzed in Section 8.11, and the two poles were found to be as follows:

$$\omega_{P1} \simeq \frac{1}{G_{m2}R_2C_CR_1} \tag{10.47}$$

$$\omega_{P2} \simeq \frac{G_{m2}C_C}{C_1C_2 + C_C(C_1 + C_2)} \tag{10.48}$$

We observe that the first pole is due to the Miller capacitance $(1 + G_{m2}R_2)\ C_C \simeq G_{m2}R_2C_C$ (which is much larger than C_1) interacting with R_1. To make ω_{P1} the dominant pole, we select a value for C_C that will result in a value for ω_{P1} that, when multiplied by the dc gain A_0, gives the desired unity-gain frequency ω_t. The value of ω_t is usually selected to be lower than the frequencies of nondominant poles and zeros. Thus, for our case,

$$\begin{aligned} A_0\omega_{P1} &= \omega_t \\ (G_{m1}R_1G_{m2}R_2)\left(\frac{1}{G_{m2}R_2C_CR_1}\right) &= \omega_t \end{aligned} \tag{10.49}$$

which yields

$$\omega_t = \frac{G_{m1}}{C_C} \tag{10.50}$$

The Miller capacitance C_C also introduces a right-half-plane zero in the amplifier transfer function. We paid no attention to this zero in the case of the 741 because it was at a very high frequency. Unfortunately, in the CMOS amplifier this is not the case. The zero location can be most easily determined directly from the circuit in Fig. 10.24(a). We wish to find the value of s at which $V_o = 0$. When we set $V_o = 0$, the current in C_C becomes sC_CV_{i2} in the direction indicated. Now because $V_o = 0$, there will be no current in R_2 and C_2. Thus a node equation at the output provides

$$sC_CV_{i2} = G_{m2}V_{i2}$$

In other words the zero is at

$$s = \frac{G_{m2}}{C_C} \tag{10.51}$$

Since G_{m2} for CMOS amplifiers is of the same order of magnitude as G_{m1}, the zero frequency will be close to ω_t given by Eq. (10.50). Since the zero is in the right half-plane, the phase shift it introduces will decrease the phase margin and thus impair the amplifier stability. Once again we note that this problem is not encountered in BJT op amps because G_{m2} is usually much greater than G_{m1} and thus the zero is at a much higher frequency than ω_t.

The above problem can be solved by including the resistance R in series with the feedback capacitor C_C, as shown in Fig. 10.24(b). To find the new location of the transfer function zero, set $V_o = 0$. Then the current through C_C will be $V_{i2}/(R + 1/sC_C)$, and a node equation at the output yields

$$\frac{V_{i2}}{R + 1/sC_C} = G_{m2}V_{i2}$$

Thus the zero is at

$$s = \frac{1}{C_C(1/G_{m2} - R)} \tag{10.52}$$

We observe that by selecting $R = 1/G_{m2}$, the zero can be placed at infinite frequency. An even better choice would be to select R greater than $1/G_{m2}$, thus placing the zero at a negative real-axis location where the phase it introduces *adds* to the phase margin.

Even with the inclusion of the resistor R, another problem still remains: The frequency of the second pole (Eq. 10.48) is not very far from ω_t. Thus the second pole introduces appreciable phase shift at ω_t, which reduces the phase margin. To see this more clearly, consider the case in which C_2 and C_C are much greater than C_1. Equation (10.48) can be approximated by

$$\omega_{P2} \simeq \frac{G_{m2}}{C_2} \tag{10.53}$$

Now, comparing Eq. (10.50) with Eq. (10.53), we see that for C_2 of the order of C_C (which can happen in the case of a relatively large load capacitance), ω_{P2} will be close to ω_t. This difficulty can be alleviated by increasing C_C and thus decreasing ω_t (see Problem 10.50).

Exercises The following exercises refer to the amplifier analyzed in Example 10.2.

10.24 Find the value of C_C that will result in $f_t = 1$ MHz.

Ans. 10 pF

10.25 Find the value of R that will place the transfer-function zero at infinity.

Ans. 10 kΩ

10.26 Find the frequency of the second pole under the condition that the total capacitance at the output is 10 pF. Hence find the excess phase introduced by the second pole at $\omega = \omega_t$ and the resulting phase margin, assuming that the zero is at $\omega = \infty$.

Ans. 1.59 MHz; 32.2°; 57.8°

Slew Rate

The slew rate of the CMOS op amp of Fig. 10.23 is given by

$$\text{SR} = \frac{2I}{C_C} \tag{10.54}$$

Using $\omega_t = g_{m1}/C_C = 2I/[(|V_{GS}| - |V_t|C_C]$, where $|V_{GS}|$ is the magnitude of the gate-to-source voltage of Q_1 and Q_2, we relate SR and ω_t by

$$SR = (V_{GS} - V_t)\omega_t \tag{10.55}$$

Comparing this relationship with the corresponding one for BJT amplifiers (Eq. 10.43), we see that since $(V_{GS} - V_t)$ is usually much greater than $(4V_T)$, CMOS op amps exhibit higher slew rates than BJT units (for the same ω_t). We also observe that higher SR can be obtained by operating the input-stage devices at a higher V_{GS}. This, however, increases the input offset voltage (see Section 6.7).

Exercise 10.27 Calculate the slew rate of the op amp in Fig. 10.23, whose parameters are specified in Example 10.2, for the compensation designed in Exercise 10.24 (C_C = 10 pF, f_t = 1 MHz).

Ans. 2.5 V/μs

10.8 ALTERNATIVE CONFIGURATIONS FOR CMOS AND BiCMOS OP AMPS

The two-stage CMOS op amp configuration studied in the previous section is by far the most popular topology for the design of op amps for VLSI circuit applications. Perhaps the most important of these applications is the realization of switched-capacitor filters (Section 11.10). The two-stage CMOS op amp works well as long as its load is mostly capacitive and of reasonably low value (<10 pF). A resistive load causes the dc open-loop gain to decrease. A large capacitive load causes the nondominant pole frequency ω_{P2} (Eq. 10.53) to decrease, thus decreasing the phase margin and eventually causing unstable behavior.

If the two-stage CMOS op amp is required to drive larger loads, such as off-chip capacitances, it must be equipped with a low-output-impedance stage. Class AB output stages resembling those studied in Chapter 9 can be implemented in CMOS technology for this purpose. There are many applications, however, in which a low-output resistance is not required; rather, a high open-loop gain and the ability to drive capacitive loads while maintaining a large phase margin are the primary requirements. For such applications the use of the cascode configuration (Sections 6.5 and 6.7) and of BiCMOS technology (Section 6.8) provide attractive design solutions. In this section we shall discuss briefly some of the resulting circuit topologies.

Cascode CMOS Op Amp

Although the gain of the two-stage CMOS op amp can be increased by adding gain stages in cascade, this is not a practical solution: Each additional gain stage increases the phase shift and makes frequency compensation more difficult. An alternative to additional stages is to increase the gain available from existing stages. This can be accomplished by utilizing the cascode configuration as shown in Fig. 10.25 for the case of the input stage. Here the two common-gate transistors Q_{1C} and Q_{2C} are the cascode devices for the differential-pair transistors Q_1 and Q_2. The output resistance looking into the drain of Q_{2C} is (see Eq. 6.135)

$$R_{o2C} \simeq g_{m2C} r_{o2C} r_{o2} \tag{10.56}$$

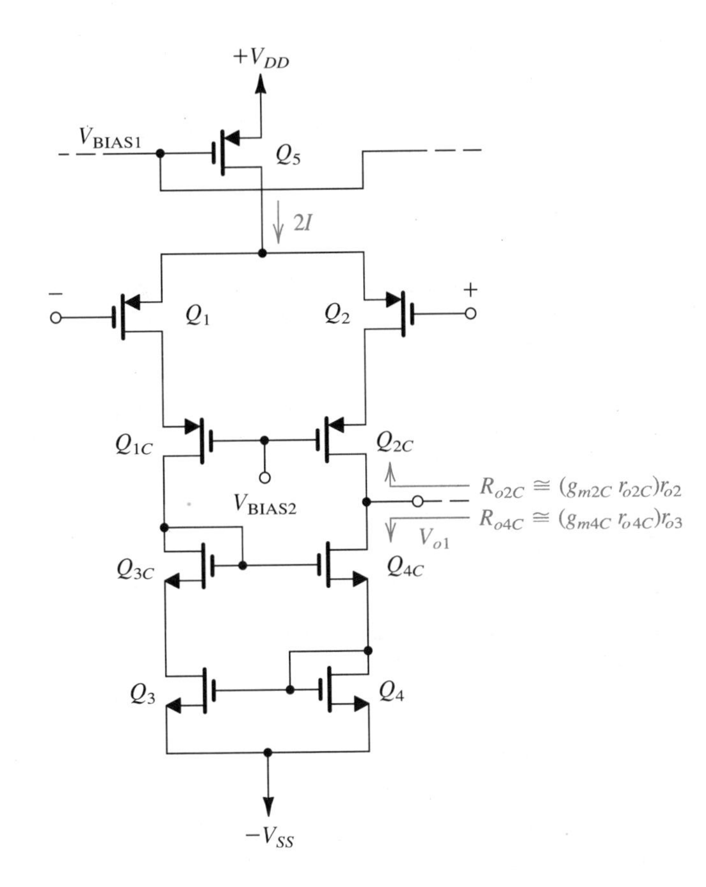

Fig. 10.25 Employing the cascode configuration in the first stage of a CMOS op amp: Transistors Q_{1C} and Q_{2C} are the cascode transistors for the differential amplifier Q_1, Q_2, and raise the output resistance of Q_2 by the factor $g_{m2C}r_{o2C}$. Transistors Q_3, Q_{3C}, Q_4, and Q_{4C} form a Wilson current mirror that provides a high load resistance, R_{o4C}. The total output resistance of the first stage is about two orders of magnitude higher than that of the first stage of the circuit in Fig. 10.23.

which is greater than the value without cascoding by the factor $g_{m2C}r_{o2C}$ (typically about 100). In order to realize the benefit of this increased output resistance, the resistance of the active load also must be raised. This is achieved in the circuit of Fig. 10.25 by employing a Wilson current mirror (Section 6.7). The output resistance of the Wilson mirror, looking into the drain of Q_{4C}, is

$$R_{o4C} \simeq g_{m4C}r_{o4C}r_{o3} \tag{10.57}$$

which is greater than the output resistance of the simple current mirror by the factor $g_{m4C}r_{o4C}$ (also, typically about 100). Thus the output resistance of the first stage becomes

$$\begin{aligned} R_o &= R_{o2C}//R_{o4C} \\ &= (g_{m2C}r_{o2C}r_{o2})//(g_{m4C}r_{o4C}r_{o3}) \end{aligned} \tag{10.58}$$

Since the voltage gain of the first stage is given by

$$A_1 = -g_{m1}R_o \tag{10.59}$$

increasing R_o by about two orders of magnitude increases A_1 by the same factor. Indeed such a gain stage can be designed to provide a voltage gain of 5000 to 10,000.

An important disadvantage of the gain stage of Fig. 10.25 is that the input common-mode range is considerably lower than that obtained in the two-stage amplifier. This is due to the two additional transistors that are placed in the stack between the two power supplies.

The output of the cascoded stage of Fig. 10.25 can be connected to a second gain stage as in the case of the circuit in Fig. 10.23. Here, however, a dc level shifting stage is usually employed [see Gregorian and Temes (1986)]. Obviously the cascode configuration can be employed in the second stage also. This, however, results in decreasing the range of output voltage swing.

Since considerable amount of voltage gain is available from the cascoded gain stage, a single-stage CMOS op amp is a possibility. Such a design, employing a variation on the cascode circuit of Fig. 10.25, will be described next.

Exercise 10.28 For the cascoded gain stage of Fig. 10.25 let $2I = 25\ \mu A$; $\mu_n C_{OX} = 20\ \mu A/V^2$; $\mu_p C_{OX} = 10\ \mu A/V^2$; $|V_t| = 1$ V; $|V_A| = 25$ V; W/L for Q_1, Q_2, Q_{1C}, and $Q_{2C} = 120/8$; W/L for Q_{3C} and $Q_{4C} = 60/8$; and W/L for Q_3 and $Q_4 = 8/8$. Find R_o and A_1.

Ans. 122.5 MΩ; −7520 V/V.

The Folded-Cascode CMOS Op Amp

If in the circuit of Fig. 10.25 each of the six transistors below Q_1 and Q_2 is replaced with its complement, and the group of six devices is disconnected from $-V_{SS}$, ''folded over,'' and connected to $+V_{DD}$, we obtain the circuit shown in Fig. 10.26. Note that in addition to the folding, two current sources, Q_6 and Q_7, have been added. The resulting circuit, aptly named **folded cascode,** operates in much the same manner as the cascode circuit of Fig. 10.25. Here, however, the input common-mode range is larger because only three transistors are now stacked in the input chain between the two power supplies (as compared to five in the original circuit).

Exercise 10.29 For the circuit in Fig. 10.26 let $2I = I_B = 25\ \mu A$, $\mu_n C_{OX} = 2\ \mu_p C_{OX} = 20\ \mu A/V^2$, $|V_t| = 1$ V, $(W/L)_1 = (W/L)_2 = 120/8$, $(W/L)_{1C} = (W/L)_{2C} = 60/8$, $(W/L)_{3C} = (W/L)_{4C} = 120/8$, $(W/L)_3 = (W/L)_4 = 8/8$, $(W/L)_5 = 150/10$, $(W/L)_6 = (W/L)_7 = 8/8$, and $V_{DD} = V_{SS} = 5$ V. Find the values of V_{BIAS1}, V_{BIAS2}, and V_{BIAS3} so that Q_6 and Q_7 operate at the edge of saturation. Find the input common-mode range and the output voltage range.

Ans. +3.4 V, −2 V, −2.4 V; −4.4 V to +3 V; −3 V to +4 V

The folded-cascode circuit is frequently used as a single-stage op amp. Its voltage gain can be determined from

$$A = g_{m1} R_o \tag{10.60}$$

where R_o is the output resistance,

$$\begin{aligned} R_o &= R_{o2C} // R_{o4C} \\ &= [g_{m2C}\, r_{o2C}(r_{o7} // r_{o2})] // (g_{m4C}\, r_{o4C}\, r_{o3}) \end{aligned} \tag{10.61}$$

Exercise 10.30 Show that the voltage gain of the folded-cascode op amp in Fig. 10.26 for the case $I_B = 2I$ and devices matched in pairs is given by

$$A = \frac{|V_A|^2}{I} \frac{2\sqrt{\mu_n\, \mu_p C_{OX}}\sqrt{W_1/L_1}}{3/\sqrt{(W/L)_{2C}} + \sqrt{\mu_n/\mu_p}/\sqrt{(W/L)_{4C}}} \tag{10.62}$$

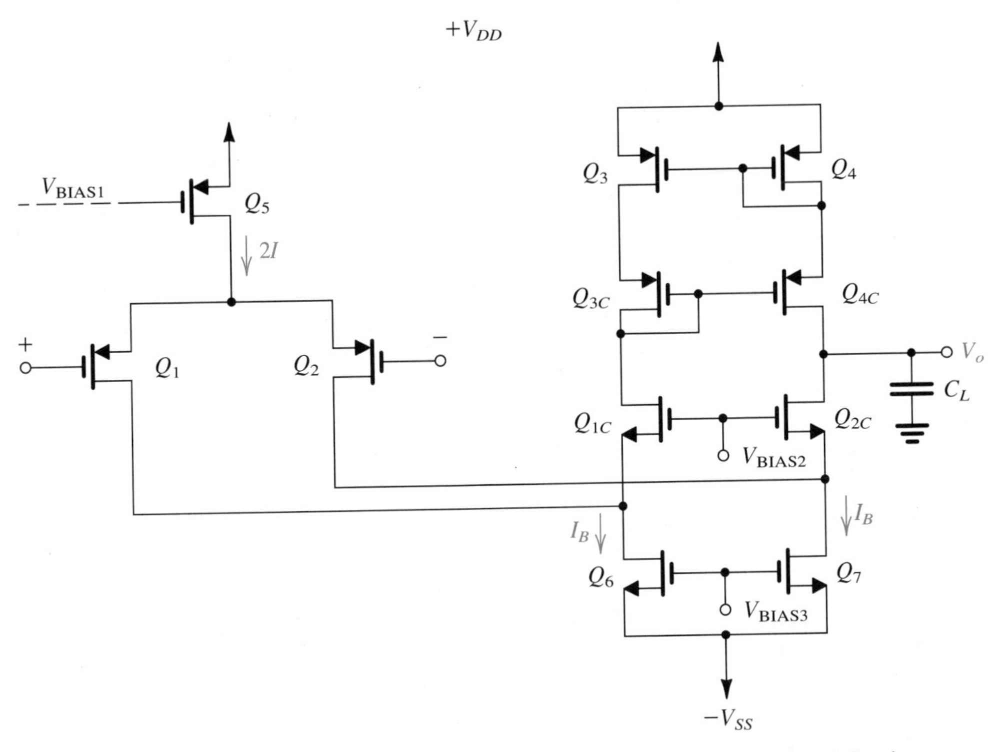

Fig. 10.26 A folded-cascode CMOS op amp. Here the *n*-channel devices Q_{1C} and Q_{2C} form the common-gate stage of the cascoded input stage. Transistors Q_3, Q_4, Q_{3C}, and Q_{4C} form a Wilson current mirror that acts as the load for the gain stage.

A very important feature of the folded cascode circuit is that the dominant pole is established by the total capacitance at the output node, C_L, where C_L includes the load capacitance. Specifically, if we denote the frequency of the dominant pole by ω_D then

$$\omega_D = 1/C_L\, R_o \tag{10.63}$$

Thus the unity-gain frequency ω_t is given by

$$\omega_t = A\ \omega_D = \frac{g_{m1}}{C_L} \tag{10.64}$$

The circuit of course has nondominant poles, but these are usually at frequencies beyond ω_t. If it is found that this is not the case, or if the phase margin at ω_t is not sufficient, one simply increases the load capacitance C_L. Such an increase improves the phase margin at the expense of a reduced ω_t. We observe that here the effect of increasing the load capacitance is opposite to that in the case of the two-stage op amp. This, together with the fact that no special frequency-compensation network is required, makes the folded cascode circuit ideally suited for application in high-frequency switched-capacitor filter design.

The slew rate of the folded-cascode op amp is

$$\text{SR} = 2I/C_L \tag{10.65}$$

Another important advantage of the folded cascode circuit is that it is much less susceptible to the effect of high-frequency noise on the negative power supply ($-V_{SS}$). Such noise inevitably appears on the power supply line in VLSI chips that include switching circuits, e.g., in the clock circuits required for the operation of switched-capacitor filters (see Section 11.10). In some op amp circuits the high-frequency power-supply noise is coupled to the output, thus contaminating the output signal. To see how this can occur refer to the two-stage CMOS op amp circuit in Fig. 10.23. A high-frequency noise signal on the $-V_{SS}$ line appears at the source terminal of Q_6 and is coupled to the gate of Q_6. It is then coupled via capacitor C_C and the small series resistance R to the output. No such coupling path exists in the folded-cascode circuit. Thus the folded cascode circuit is said to have greater **power-supply rejection ratio** (PSRR) than the two-stage op amp circuit. (For a detailed treatment of PSRR refer to Gregorian and Temes, 1986.)

A disadvantage of the folded-cascode circuit is the reduced output voltage swing capability due to the fact that two series transistors exist between the output terminal and each power supply.

A Folded-Cascode BiCMOS Op Amp

To increase the bandwidth of the folded-cascode op amp, the frequencies of the nondominant poles must be increased. Usually the lowest-frequency nondominant pole is that which arises at the input of the common-gate stage Q_{1C}, Q_{2C}. Specifically, the resistance at the node connecting Q_1 and Q_{1C} (and similarly at the node connecting Q_2 and Q_{2C}) is approximately equal to $1/g_{m1C}$ (which is equal to $1/g_{m2C}$). Thus if we denote the total capacitance at each of these two nodes by C_{P1}, the frequency of the resulting pole can be expressed as

$$\omega_P = g_{m1C}/C_{P1} \tag{10.66}$$

Since it is possible to realize much greater values of transconductance using BJTs than is possible with MOSFETs, the frequency ω_P of the nondominant pole can be increased by using a common-base BJT stage in place of the common-gate MOS stage. The result is the BiCMOS folded-cascode circuit shown in Fig. 10.27. Now with ω_P increased, ω_t can be correspondingly increased while C_L remains constant by designing the first stage to have a larger transconductance g_{m1} (see Eq. 10.64).

The BiCMOS circuit combines the increased bandwidth with the advantages of an MOS input stage; namely, a nearly infinite input impedance, a zero input bias current, and a higher slew rate (for the latter point refer to Section 10.7 and to the following exercise.)

Exercise 10.31 We wish to compare the folded-cascode BiCMOS op amp shown in Fig. 10.27 with one in which the first stage is replaced with a BJT differential pair. For both circuits an f_t of 2 MHz is required for $C_L = 10$ pF. (a) For the BiCMOS circuit let $(W/L)_1 = 150/10$, $|V_t| = 1$ V, and $\mu_p C_{OX} = 10\ \mu\text{A/V}^2$. Find the required bias current $2I$ and the slew rate obtained. (b) For the circuit with the BJT input stage, find the required bias current $2I$ and the slew rate obtained.

Ans. (a) 105.3 μA; 10.5 V/μs. (b) 6.3 μA; 0.63 V/μs.

10.9 DATA CONVERTERS—AN INTRODUCTION

In this section we begin the study of another group of analog IC circuits of great importance; namely, data converters.

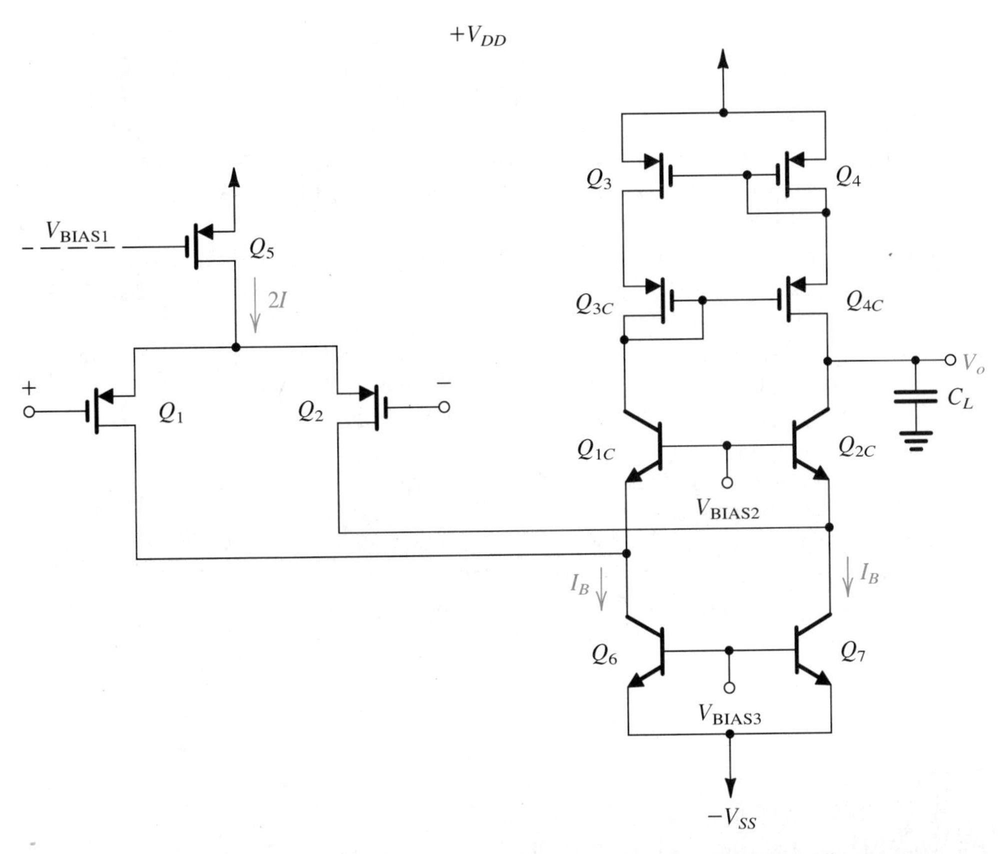

Fig. 10.27 A BiCMOS folded-cascode op amp. BJTs are employed to implement the common-base stage Q_{1C}, Q_{2C}. The high value of $g_{m1C} = g_{m2C}$ raises the frequency of the nondominant pole that results at the input of the common-base stage. Thus for a given value of C_L, $\omega_t = g_{m1}/C_L$ can be increased by operating the first stage at a higher g_m.

Digital Processing of Signals

Most physical signals, such as those obtained at transducer outputs, exist in analog form. Some of the processing required on these signals is most conveniently performed in an analog fashion. For instance, in instrumentation systems it is quite common to use a high-input-impedance, high-gain, high-CMRR differential amplifier right at the output of the transducer. This is usually followed by a filter whose object is to eliminate interference. However, further signal processing is usually required, which can range from simply obtaining a measurement of signal strength to performing some algebraic manipulations on this and related signals to obtain the value of a particular system parameter of interest, as is usually the case in systems intended to provide a complex control function. Another example of signal processing can be found in the common need for transmission of signals to a remote receiver.

All such forms of signal processing can be performed by analog means. In previous chapters we encountered circuits for implementing a number of such tasks. However, an attractive alternative exists: It is to convert, following some initial analog processing, the signal from analog to digital form and then use economical, accurate, and convenient digital ICs to perform **digital signal processing.** Such processing can in its simplest form provide us with a measure of the signal strength as an easy-to-read number (consider, for example, the digital voltmeter). In more involved cases the digital signal processor can perform a variety of arithmetic and logic operations that implement a **filtering algorithm.** The resulting **digital filter** does many of the same tasks that an analog filter performs—namely, eliminate interference and noise. Yet another example of digital signal processing is found in digital communications systems, where signals are transmitted as a sequence of binary pulses, with the obvious advantage that corruption of the amplitudes of these pulses by noise is, to a large extent, of no consequence.

Once digital signal processing has been performed, we might be content to display the result in digital form, such as a printed list of numbers. Alternatively, we might require an analog output. Such is the case in a telecommunications system, where the usual output may be speech. If an analog output is desired, then obviously we need to convert the digital signal back to an analog form.

It is not our purpose here to study the techniques of digital signal processing. Rather, we shall examine the interface circuits between the analog and digital domains. Specifically, we shall study the basic techniques and circuits employed to convert an analog signal to digital form **(analog-to-digital or simply A/D conversion)** and those used to convert a digital signal to analog form **(digital-to-analog or simply D/A conversion).** Digital circuits are studied in Chapters 13 and 14.

Sampling of Analog Signals

The principle underlying digital signal processing is that of **sampling** the analog signal. Figure 10.28 illustrates in a conceptual form the process of obtaining samples of an analog signal. The switch shown closes periodically under the control of a periodic pulse signal (clock). The closure time of the switch, τ, is relatively short, and the samples obtained are stored (held) on the capacitor. The circuit of Fig. 10.28 is known as a **sample-and-hold (S/H) circuit.** As indicated, the S/H circuit consists of an analog switch that can be implemented by a MOSFET transmission gate (Section 5.10), a storage capacitor, and (not shown) a buffer amplifier.

Between the sampling intervals—that is, during the *hold* intervals—the voltage level on the capacitor represents the signal samples we are after. Each of these voltage levels is then fed to the input of an A/D converter, which provides an N-bit binary number proportional to the value of signal sample.

The fact that we can do our processing on a limited number of samples of an analog signal while ignoring the analog-signal details between samples is based on the sampling theorem [see Lathi (1965)].

Signal Quantization

Consider an analog signal whose values range from 0 to +10 V. Let us assume that we wish to convert this signal to digital form and that the required output is a 4-bit[2] signal.

[2] *Bit* stands for *b*inary dig*it*.

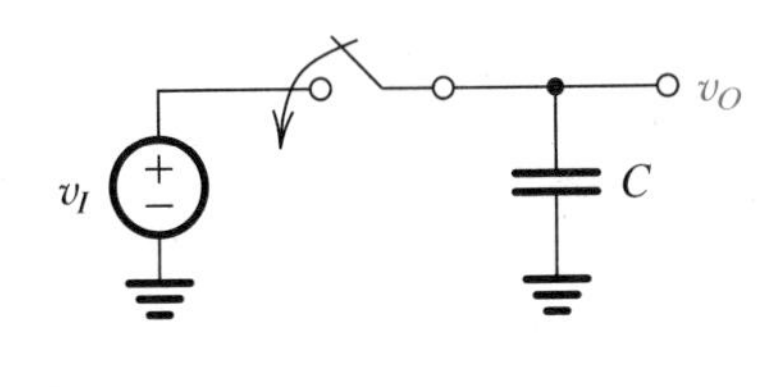

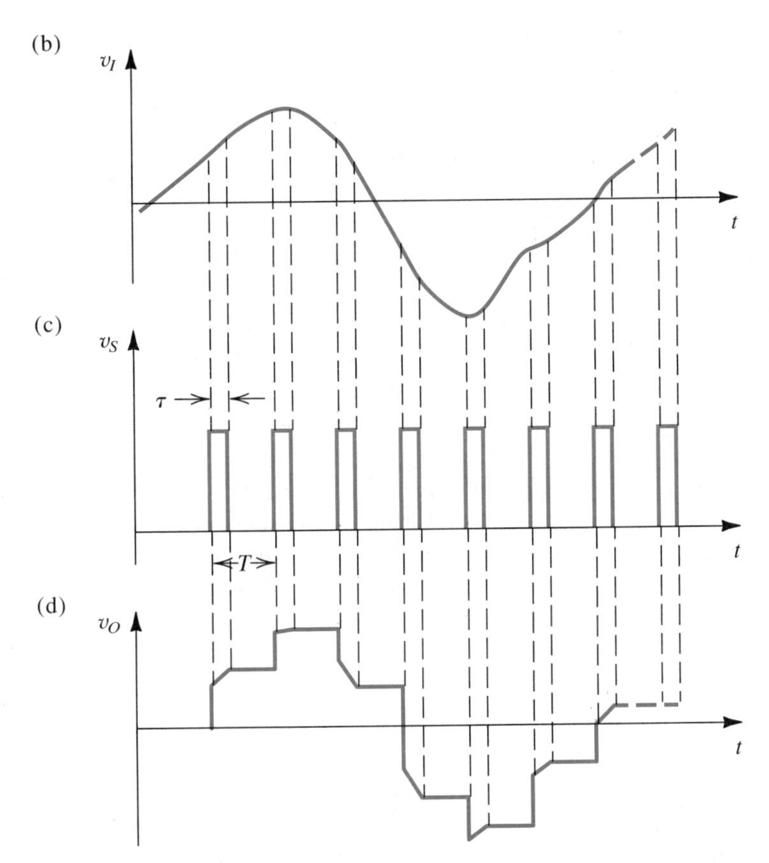

Fig. 10.28 Process of periodically sampling an analog signal. **(a)** Sample-and-hold (S/H) circuit. The switch closes τ seconds every period. **(b)** Input signal waveform. **(c)** Sampling signal (control signal for the switch). **(d)** Output signal (to be fed to A/D converter).

We know that a 4-bit binary number can represent 16 different values, 0 to 15; it follows that the *resolution* of our conversion will be $10\ \text{V}/15 = \frac{2}{3}$ V. Thus an analog signal of 0 V will be represented by 0000, $\frac{2}{3}$ V will be represented by 0001, 6 V will be represented by 1001, and 10 V will be represented by 1111.

All of the above sample numbers were multiples of the basic increment ($\frac{2}{3}$V). A question now arises regarding the conversion of numbers that fall between these successive incremental levels. For instance, consider the case of 6.2-V analog level. This falls between 18/3 and 20/3. However, since it is closer to 18/3 we treat it as if it were 6 V and *code* it as 1001. This process is called **quantization.** Obviously errors are inherent in this process; such errors are called quantization errors. Using more bits to represent (encode or code) an analog signal reduces quantization errors but requires more complex circuitry.

The A/D and D/A Converters as Functional Blocks

Figure 10.29 depicts the functional block representations of A/D and D/A converters. As indicated, the **A/D converter** (also called an ADC) accepts an analog sample v_A and produces an N-bit **digital word.** Conversely, the **D/A converter** (also called a DAC)

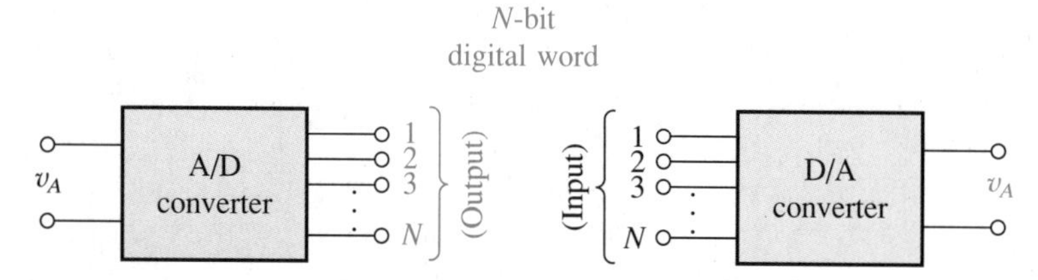

Fig. 10.29 The A/D and D/A converters as circuit blocks.

accepts an n-bit digital word and produces an analog sample. The output samples of the D/A converter are often fed to a sample-and-hold circuit. At the output of the S/H circuit a staircase waveform, such as that in Fig. 10.30, is obtained. The staircase waveform can then be smoothed by a low-pass filter, giving rise to the smooth curve shown in color in Fig. 10.30. In this way an analog output signal is reconstructed. Finally, note that the quantization error of an A/D converter is equivalent to $\pm\frac{1}{2}$ least significant bit (b_N).

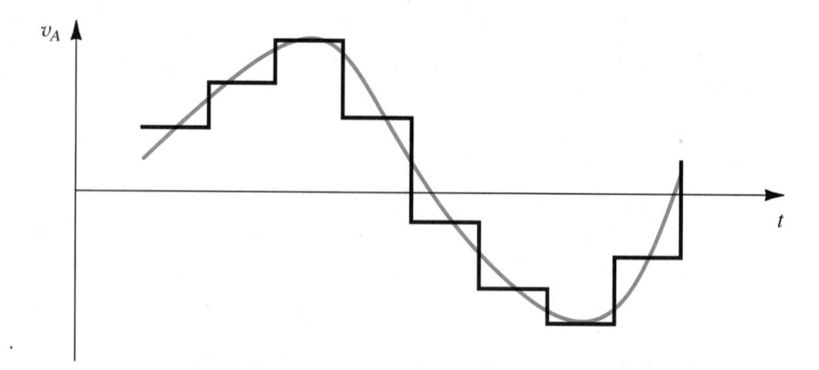

Fig. 10.30 The analog samples at the output of a D/A converter are usually fed to a sample-and-hold circuit to obtain the staircase waveform shown. This waveform can then be filtered to obtain the smooth waveform, shown in color. The time delay usually introduced by the filter is not shown.

Exercise 10.32 An analog signal in the range 0 to +10 V is to be converted to an 8-bit digital signal. What is the resolution of the conversion in volts? What is the digital representation of an input of 6 V? What is the representation of an input of 6.2 V? What is the error made in the quantization of 6.2 V in absolute terms and as a percent of the input? as a percent of full scale? What is the largest possible quantization error as a percent of full scale?

Ans. 0.0392 V; 10011001; 10011110; −0.0064 V; −0.1%; −0.064%; 0.196%

10.10 D/A CONVERTER CIRCUITS

Basic Circuit Using Binary-Weighted Resistors

Figure 10.31 shows a simple circuit for an N-bit D/A converter. The circuit consists of a reference voltage V_{ref}, N binary-weighted resistors R, $2R$, $4R$, $8R$, . . . , $2^{N-1}R$, N single-pole double-throw switches S_1, S_2, . . . , S_N, and an op amp together with its feedback resistance $R_f = R/2$.

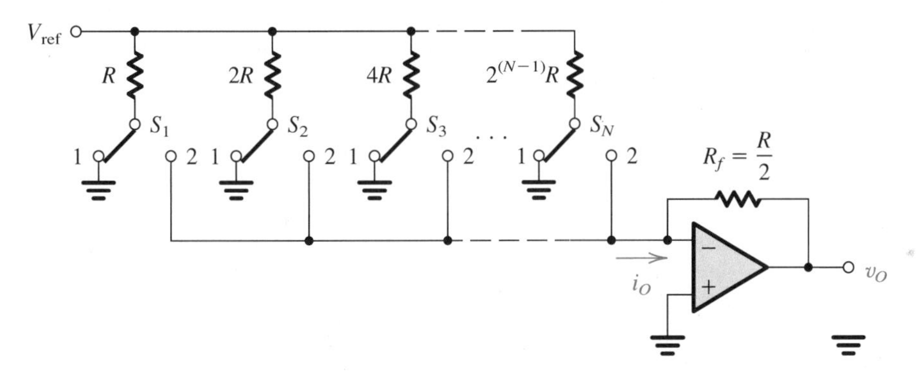

Fig. 10.31 An N-bit D/A converter using a binary-weighted resistive ladder network.

The switches are controlled by an N-bit digital input word D,

$$D = \frac{b_1}{2^1} + \frac{b_2}{2^2} + \cdots + \frac{b_N}{2^N} \tag{10.67}$$

where b_1, b_2, and so on are bit coefficients that are either 1 or 0. Note that the bit b_N is the **least significant bit (LSB)** and b_1 is the **most significant bit (MSB).** In the circuit in Fig. 10.31, b_1 controls switch S_1, b_2 controls S_2, and so on. When b_i is 0, switch S_i is in position 1, and when b_i is 1 switch S_i is in position 2.

Since position 1 of all switches is ground and position 2 is virtual ground, the current through each resistor remains constant. Each switch simply controls where its corresponding current goes: to ground (when the corresponding bit is 0) or to virtual ground (when the corresponding bit is 1). The currents flowing into the virtual ground add up, and the sum flows through the feedback resistance R_f. The total current i_O is therefore given by

$$i_O = \frac{V_{\text{ref}}}{R}b_1 + \frac{V_{\text{ref}}}{2R}b_2 + \cdots + \frac{V_{\text{ref}}}{2^{N-1}R}b_N$$

$$= \frac{2V_{\text{ref}}}{R}\left(\frac{b_1}{2^1} + \frac{b_2}{2^2} + \cdots + \frac{b_N}{2^N}\right)$$

Thus

$$i_O = \frac{2V_{\text{ref}}}{R}D \tag{10.68}$$

and the output voltage v_O is given by

$$v_O = -i_O R_f = -V_{\text{ref}}\,D \tag{10.69}$$

which is directly proportional to the digital word D, as desired.

It should be noted that the accuracy of the DAC depends critically on (1) the accuracy of V_{ref}, (2) the precision of the binary-weighted resistors, and (3) the perfection of the switches. Regarding the third point, we should emphasize that these switches handle analog signals; thus their perfection is of considerable interest. While the offset voltage and the finite on resistance are not of critical significance in a digital switch, these parameters are of immense importance in *analog switches*. The use of FETs to implement analog switches was discussed in Chapter 5. Also, we shall shortly see that in a practical circuit implementation of the DAC the binary-weighted currents are generated by current sources. In this case the analog switch can be realized using the differential-pair circuit, as will be shown.

A disadvantage of the binary-weighted resistor network is that for a large number of bits ($N > 4$) the spread between the smallest and largest resistances becomes quite large. This implies difficulties in maintaining accuracy in resistor values. A more convenient scheme exists utilizing a resistive network called the R-$2R$ ladder.

R-2*R* Ladders

Figure 10.32 shows the basic arrangement of a DAC using an R-$2R$ ladder. Because of the small spread in resistance values, this network is usually preferred to the binary-weighted scheme discussed above, especially for $N > 4$. Operation of the R-$2R$ ladder is straightforward. First, it can be shown, by starting from the right and working toward the left, that the resistance to the right of each ladder node, such as that labeled X, is equal to $2R$. Thus the current flowing to the right, away from each node, is equal to the current flowing downward to ground, and twice that current flows into the node from the left side. It follows that

$$I_1 = 2I_2 = 4I_3 = \cdots = 2^{N-1}I_N$$

Thus, as in the binary-weighted resistive network, the currents controlled by the switches are binary weighted. The output current i_O will therefore be given by

$$i_O = \frac{V_{\text{ref}}}{R}D$$

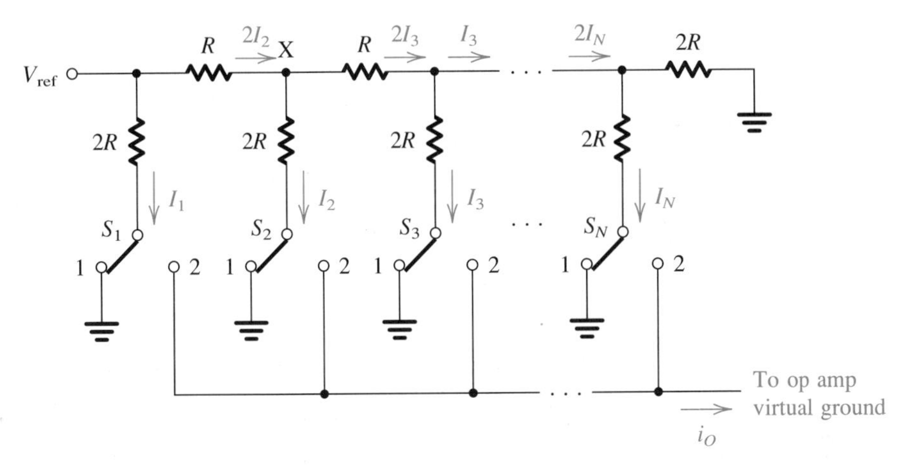

Fig. 10.32 Basic circuit configuration of a DAC utilizing an *R*-2*R* ladder network.

A Practical Circuit Implementation

A practical circuit implementation of the DAC utilizing an *R*-2*R* ladder is shown in Fig. 10.33. The circuit utilizes BJTs to generate binary-weighted constant currents $I_1, I_2, \ldots, I_N$, which are switched between ground and virtual ground of an output summing op amp (not shown). We shall first show that the currents I_1 to I_N are indeed binary weighted, with I_1 corresponding to the MSB and I_N corresponding to the LSB of the DAC.

Starting at the two rightmost transistors, Q_N and Q_t, we see that if they are matched their emitter currents will be equal and are denoted (I_N/α). Transistor Q_t is included to provide proper termination of the *R*-2*R* network. The voltage between the base line of the BJTs and node *N* will be

$$V_N = V_{BE_N} + \left(\frac{I_N}{\alpha}\right)(2R)$$

where V_{BE_N} is the base–emitter voltage of Q_N. Since the current flowing through the resistor *R* connected to node *N* is $(2I_N/\alpha)$, the voltage between node B and node $(N-1)$ will be

$$V_{N-1} = V_N + \left(\frac{2I_N}{\alpha}\right)R = V_{BE_N} + \frac{4I_N}{\alpha}R$$

Assuming, for the moment, that $V_{BE_{N-1}} = V_{BE_N}$, we see that a voltage of $(4I_N/\alpha)R$ appears across the resistance 2*R* in the emitter of Q_{N-1}. Thus Q_{N-1} will have an emitter current of $(2I_N/\alpha)$ and a collector current of $(2I_N)$, twice the current in Q_N. The two transistors will have equal V_{BE} drops if their junction areas are scaled in the same proportion as their currents, which is usually done in practice.

Proceeding in the above manner we can show that

$$I_1 = 2I_2 = 4I_3 = \cdots = 2^{N-1}I_N$$

under the assumption that the EBJ areas of Q_1 to Q_N are scaled in a binary-weighted fashion.

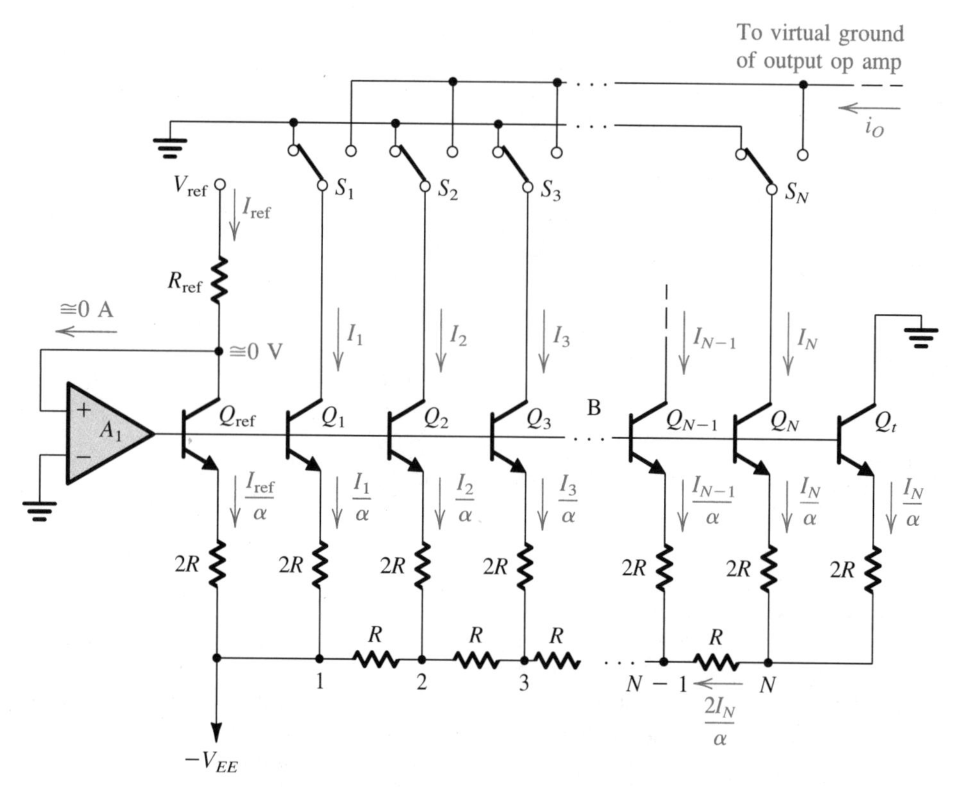

Fig. 10.33 A practical circuit implementation of a DAC utilizing an *R*-2*R* ladder network.

Next consider op amp A_1, which, together with the reference transistor Q_{ref}, forms a negative-feedback loop. (Convince yourself that the feedback is indeed negative.) A virtual ground appears at the collector of Q_{ref} forcing it to conduct a collector current $I_{ref} = V_{ref}/R_{ref}$ independent of whatever imperfections Q_{ref} might have. Now, if Q_{ref} and Q_1 are matched, their collector currents will be equal,

$$I_1 = I_{ref}$$

Thus, the binary-weighted currents are directly related to the reference current, independent of the exact values of V_{BE} and α. Also observe that op amp A_1 supplies the base currents of all the BJTs.

Current Switches

Each of the single-pole double-throw switches in the DAC circuit of Fig. 10.33 can be implemented by a circuit such as that shown in Fig. 10.34 for switch S_m. Here I_m denotes the current flowing in the collector of the mth bit transistor. The circuit is a differential pair with the base of the reference transistor Q_{mr} connected to a suitable dc voltage V_{bias}, and the digital signal representing the mth bit b_m applied to the base of the other transistor Q_{ms}. If the voltage representing b_m is higher than V_{bias} by a few hundred millivolts, Q_{ms}

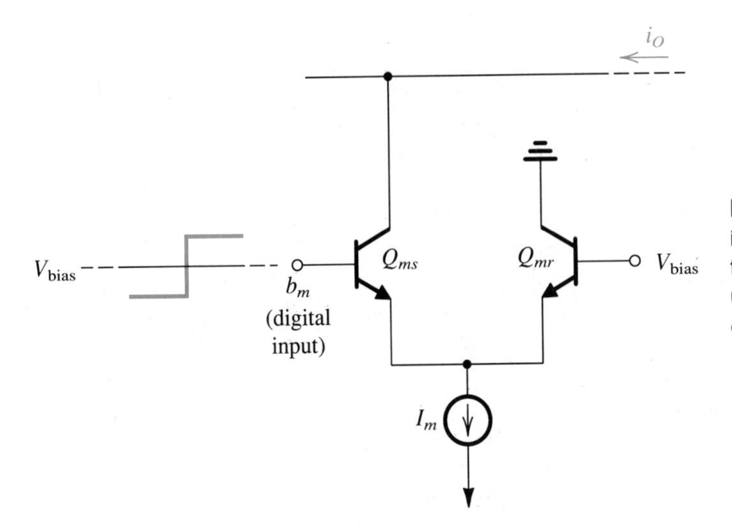

Fig. 10.34 Circuit implementation of switch s_m in the DAC of Fig. 10.33. In a BiCMOS technology, Q_{ms} and Q_{mr} can be implemented using MOSFETs, thus avoiding the inaccuracy caused by the base current of BJTs.

will turn on and Q_{mr} will turn off. The bit current I_m will flow through Q_{ms} and onto the output summing line. On the other hand, when b_m is low, Q_{ms} will be off and I_m flows through Q_{mr} to ground.

The current switch of Fig. 10.34 is simple and features high-speed operation. It suffers, however, from the fact that part of the current I_m flows through the base of Q_{ms} and thus does not appear on the output summing line. More elaborate circuits for current switches can be found in Grebene (1984). Also, in a BiCMOS technology the differential-pair transistors Q_{ms} and Q_{mr} can be replaced with MOSFETs, thus eliminating the base current problem.

Exercises 10.33 What is the maximum resistor ratio required by a 12-bit D/A converter utilizing a binary-weighted ladder?

Ans. 2048

10.34 If the input bias current of an op amp, used as the output summer in a 10-bit DAC, is to be no more than that equivalent to $\frac{1}{4}$ LSB, what is the maximum current required to flow in R_f for an op amp whose bias current is as great as 0.5 μA?

Ans. 2.046 mA

10.11 A/D CONVERTER CIRCUITS

There exist a number of A/D conversion techniques varying in complexity and speed of conversion. In the following we shall discuss two simple but slow schemes, one complex (in terms of the amount of circuitry required) but extremely fast method, and finally a method particularly suited for MOS implementation.

The Feedback-Type Converter

Figure 10.35 shows a simple A/D converter that employs a comparator, an up-down counter, and a D/A converter. The comparator circuit provides an output that assumes one of two distinct values: positive when the difference input signal is positive, and negative when the difference input signal is negative. We shall study comparator circuits in Chapter 12. An up-down counter is simply a counter that can count either up or down depending on the binary level applied at its up-down control terminal. Because the A/D converter of Fig. 10.35 employs a DAC in its feedback loop it is usually called a feedback-type A/D converter. It operates as follows: With a 0 count in the counter, the D/A converter output, v_O, will be zero and the output of the comparator will be high, instructing the counter to count the clock pulses in the up direction. As the count increases, the output of the DAC rises. The process continues until the DAC output reaches the value of the analog input signal, at which point the comparator switches and stops the counter. The counter output will then be the digital equivalent of the input analog voltage.

Operation of the converter of Fig. 10.35 is slow if it starts from zero. This converter however, tracks incremental changes in the input signal quite rapidly.

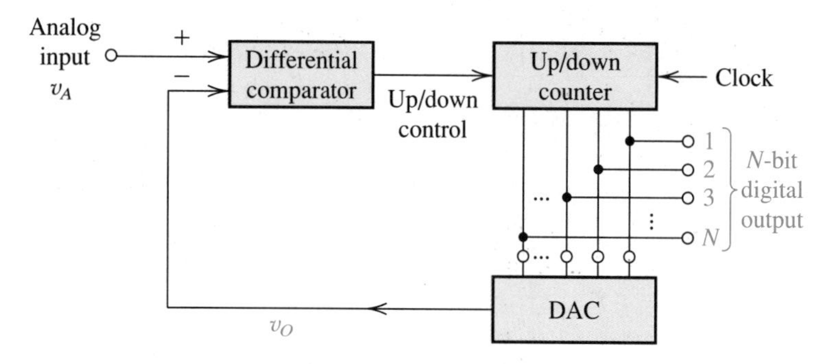

Fig. 10.35 A simple feedback-type A/D converter.

The Dual-Slope A/D Converter

A very popular high-resolution (12- to 14-bit) (but slow) A/D conversion scheme is illustrated in Fig. 10.36. To see how it operates, refer to Fig. 10.36 and assume that the analog input signal v_A is negative. Prior to the start of the conversion cycle, switch S_2 is closed, thus discharging capacitor C and setting $v_1 = 0$. The conversion cycle begins with opening S_2 and connecting the integrator input through switch S_1 to the analog input signal. Since v_A is negative, a current $I = v_A/R$ will flow through R in the direction away from the integrator. Thus v_1 rises linearly with a slope of $I/C = v_A/RC$, as indicated in Fig. 10.36(b). Simultaneously, the counter is enabled and it counts the pulses from a fixed-frequency clock. This phase of the conversion process continues for a fixed duration T_1. It ends when the counter accumulates a fixed count denoted n_{ref}. Usually, for N-bit converter, $n_{\text{ref}} = 2^N$. Denoting the peak voltage at the output of the integrator V_{peak}, we

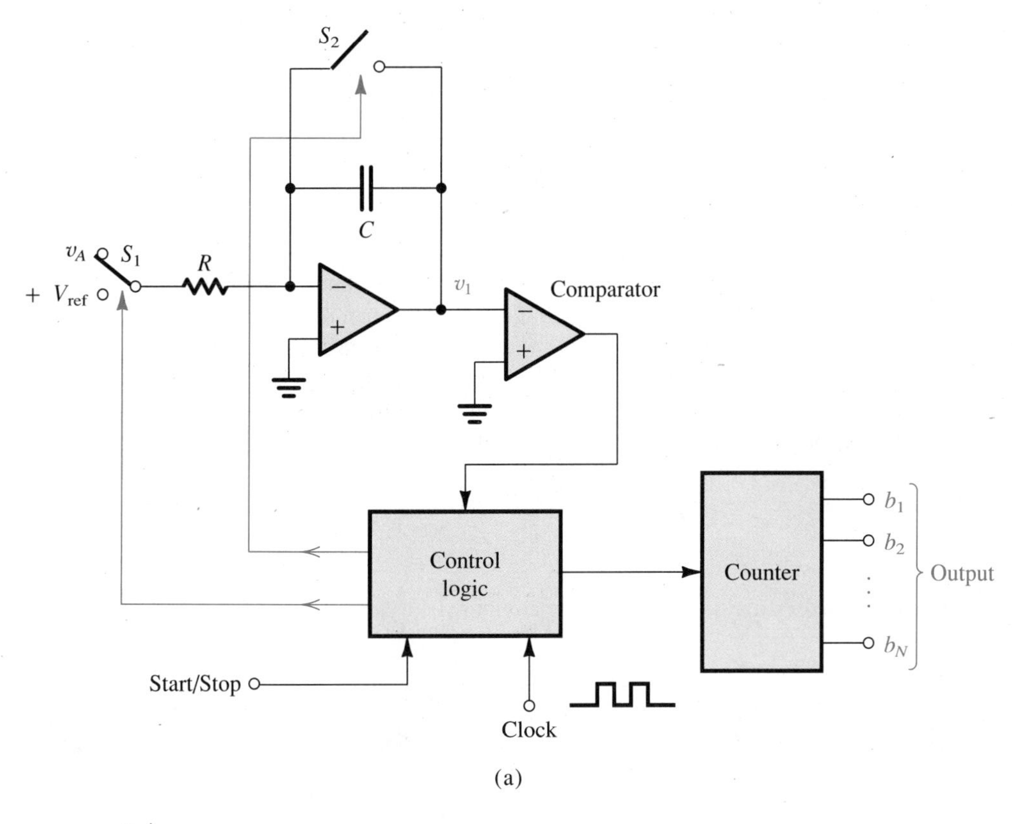

(a)

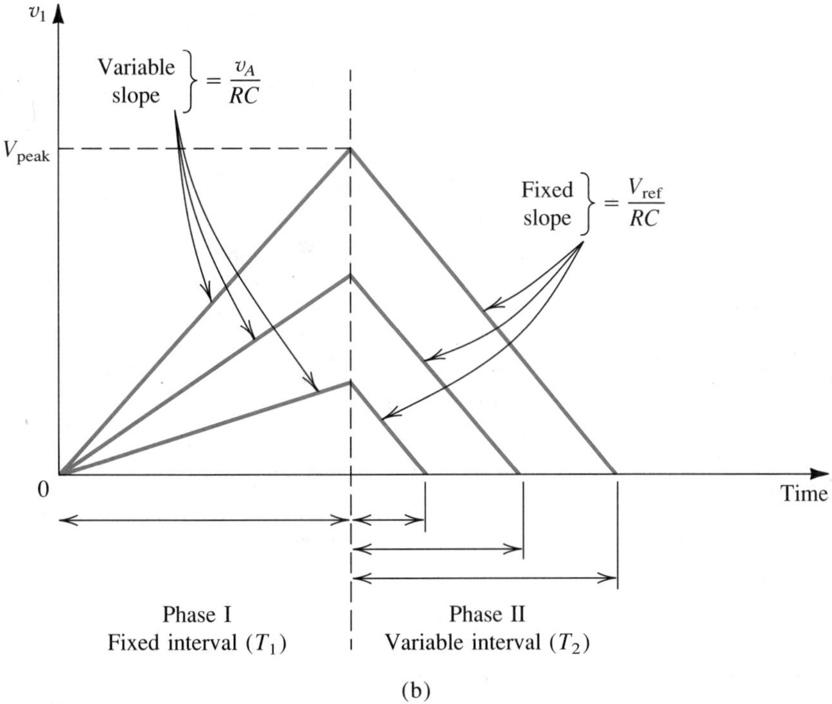

(b)

Fig. 10.36 The dual-slope A/D conversion method. Note that v_A is assumed to be negative.

can write with reference to Fig. 10.36(b)

$$\frac{V_{\text{peak}}}{T_1} = \frac{v_A}{RC} \tag{10.70}$$

At the end of this phase, the counter is reset to zero.

Phase II of the conversion begins at $t = T_1$ by connecting the integrator input through switch S_1 to the positive reference voltage V_{ref}. The current into the integrator reverses direction and is equal to V_{ref}/R. Thus v_1 decreases linearly with a slope of (V_{ref}/RC). Simultaneously the counter is enabled and it counts the pulses from the fixed-frequency clock. When v_1 reaches zero volts, the comparator signals the control logic to stop the counter. Denoting the duration of phase II by T_2, we can write, by reference to Fig. 10.36(b),

$$\frac{V_{\text{peak}}}{T_2} = \frac{V_{\text{ref}}}{RC} \tag{10.71}$$

Equations (10.70) and (10.71) can be combined to yield

$$T_2 = T_1\left(\frac{v_A}{V_{\text{ref}}}\right) \tag{10.72}$$

Since the counter reading, n_{ref}, at the end of T_1 is proportional to T_1 and the reading, n, at the end of T_2 is proportional to T_2, we have

$$n = n_{\text{ref}}\left(\frac{v_A}{V_{\text{ref}}}\right) \tag{10.73}$$

Thus the content of the counter,[3] n, at the end of the conversion process is the digital equivalent of v_A.

The dual-slope converter features high accuracy, since its performance is independent of the exact values of R and C. There exist many commercial implementations of the dual-slope method, some of which utilize CMOS technology.

The Parallel or Flash Converter

The fastest A/D conversion scheme is the simultaneous, parallel, or flash conversion illustrated in Fig. 10.37. Conceptually, flash conversion is very simple. It utilizes $2^N - 1$ comparators to compare the input signal level with each of the $2^N - 1$ possible quantization levels. The outputs of the comparators are processed by an encoding-logic block to provide the N bits of the output digital word. Note that a complete conversion can be obtained within one clock cycle.

Although flash conversion is very fast, the price paid is a rather complex circuit implementation. Variations on the basic technique have been successfully employed in the design of IC converters.

---

[3]Note that n is *not* a continuous function of v_A, as might be inferred from Eq. (10.73). Rather, n takes on discrete values corresponding to the quantized levels of v_A.

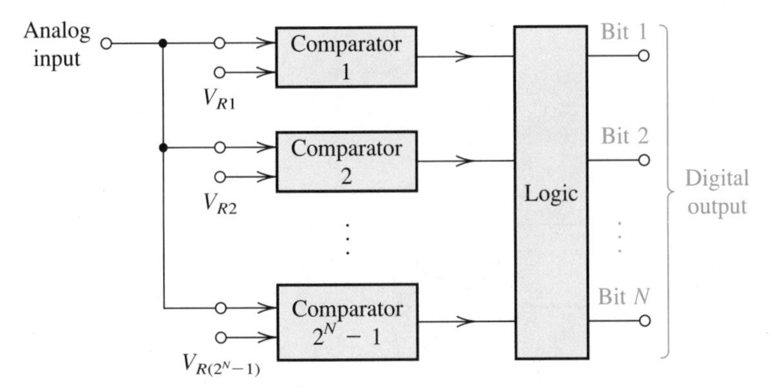

Fig. 10.37 Parallel, simultaneous, or flash A/D conversion.

The Charge-Redistribution Converter

The last A/D conversion technique we shall discuss is particularly suited for CMOS implementation. As shown in Fig. 10.38, the circuit utilizes a binary-weighted capacitor array, a voltage comparator, analog switches, and control logic (not shown). The circuit shown is for a 5-bit converter; capacitor C_T serves the purpose of terminating the capacitor array, making the total capacitance equal to the desired value of $2C$.

Operation of the converter can be divided into three distinct phases, illustrated in Figs. 10.38(a), (b), and (c). In the sample phase (Fig. 10.38a) switch S_B is closed, thus connecting the top plate of all capacitors to ground and setting v_O to zero. Meanwhile, switch S_A is connected to the analog input voltage v_A. Thus the voltage v_A appears across the total capacitance of $2C$, resulting in a stored charge of $2Cv_A$. Thus during this phase a sample of v_A is taken and a proportional amount of charge is stored on the capacitor array.

During the hold phase (Fig. 10.38b), switch S_B is opened and switches S_1 to S_5, and S_T are thrown to the ground side. Thus the top plate of the capacitor array is open-circuited while their bottom plates are connected to ground. Since no discharge path has been provided, the capacitor charges must remain constant with the total equal to $2Cv_A$. It follows that the voltage at the top plate must become $-v_A$. Finally, note that during the hold phase S_A is connected to V_{ref} in preparation for the charge-redistribution phase.

Next we consider the operation during the charge-redistribution phase illustrated in Fig. 10.38(c). First, switch S_1 is connected to V_{ref}. The circuit then consists of V_{ref}, a series capacitor C, and a total capacitance to ground of value C. This capacitive divider causes a voltage increment of $V_{\text{ref}}/2$ to appear on the top plates. Now if v_A is greater than $V_{\text{ref}}/2$, the net voltage at the top plate will remain negative, which means that S_1 will be left in its new position as we move on to switch S_2. If, on the other hand, v_A was smaller than $V_{\text{ref}}/2$, then the net voltage at the top plate would become positive. The comparator will detect this situation and signal the control logic to return S_1 to its ground position and then to move on to S_2.

Next, switch S_2 is connected to V_{ref}, which causes a voltage increment of $V_{\text{ref}}/4$ to appear on the top plate. If the resulting voltage is still negative, S_2 is left in its new position; otherwise, S_2 is returned to its ground position. We then move on to switch S_3, and so on until all the bit switches S_1 to S_5 have been tried.

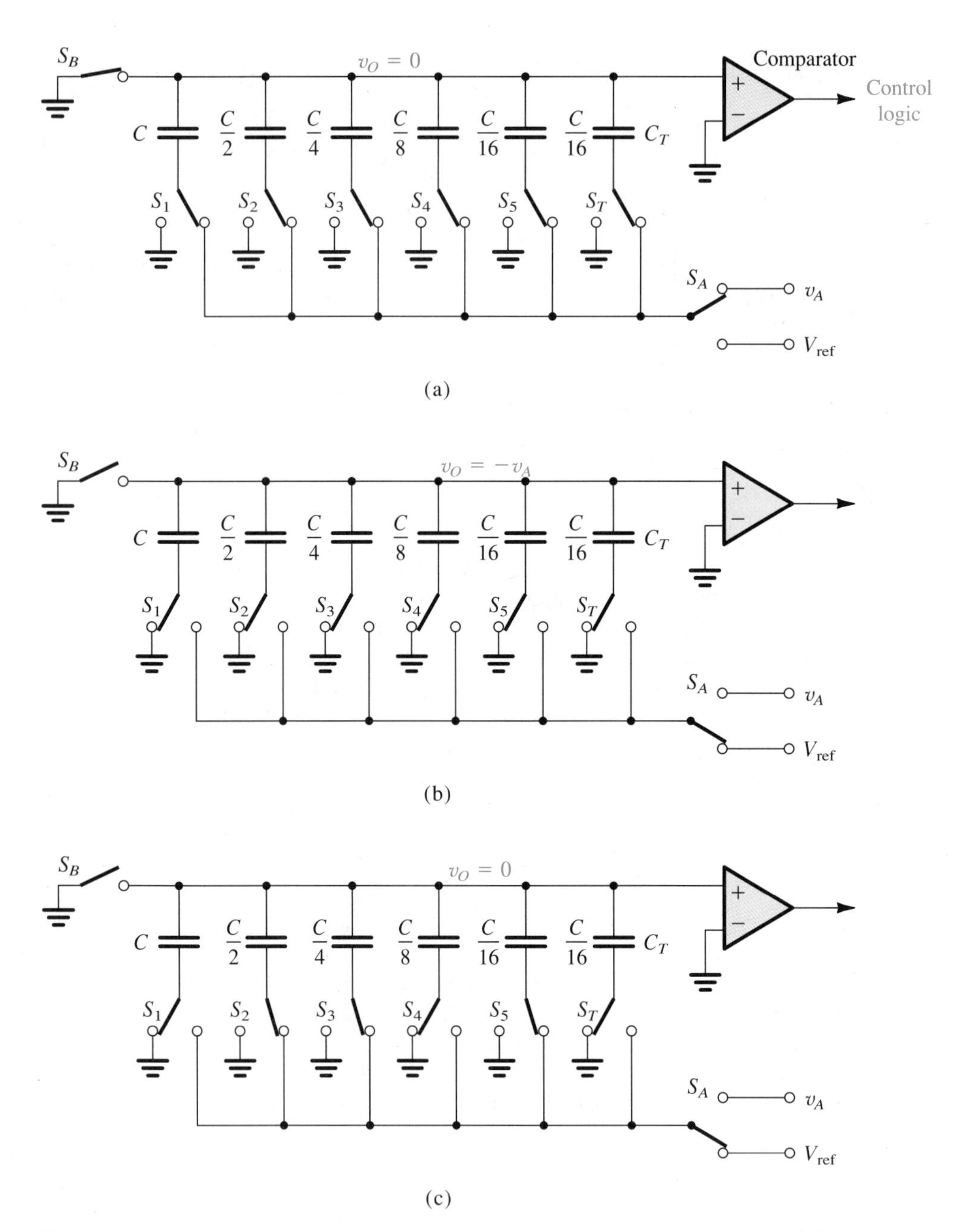

Fig. 10.38 Charge-redistribution A/D converter suitable for CMOS implementation. **(a)** Sample phase; **(b)** hold phase; and **(c)** charge-redistribution phase.

It can be seen that during the charge-redistribution phase the voltage on the top plate will be reduced incrementally to zero. The connection of the bit switches at the conclusion of this phase gives the output digital word; a switch connected to ground indicates a 0 value for the corresponding bit, whereas connection to V_{ref} indicates a 1. The particular switch configuration depicted in Fig. 10.38(c) is for $D = 01101$. Observe that at the end of the conversion process, all the charge is stored in the capacitors corresponding to 1 bits; the capacitors of the 0 bits have been discharged.

The accuracy of this A/D conversion method is independent of the value of stray capacitances from the bottom plate of the capacitors to ground. This is because the bottom plates are connected either to ground or to V_{ref}; thus the charge on the stray capacitances will not flow into the capacitor array. Also, because both the initial and the final voltages on the top plate are zero, the circuit is also insensitive to the stray capacitances between the top plates and ground.[4] The insensitivity to stray capacitances makes the charge-redistribution technique a reasonably accurate method capable of implementing A/D converters with as many as 10 bits.

Exercises 10.35 Consider the 5-bit charge-redistribution converter in Fig. 10.38 with $V_{ref} = 4$ V. What is the voltage increment appearing on the top plate when S_5 is switched? What is the full-scale voltage of this converter? If $v_A = 2.5$ V which switches will be connected to V_{ref} at the end of conversion?

Ans. $\frac{1}{8}$ V; $\frac{31}{8}$ V; S_1 and S_3

10.36 Express the maximum quantization error of an N-bit A/D converter in terms of its least significant bit (LSB) and in terms of its full-scale analog input V_{FS}.

Ans. $\pm\frac{1}{2}$ LSB; $V_{FS}/2(2^N - 1)$

SUMMARY

- The internal circuit of the 741 op amp embodies many of the design techniques employed in bipolar analog integrated circuits.
- The 741 circuit consists of an input differential stage, a high-gain single-ended second stage, and a class AB output stage. This structure is typical of modern op amps and is known as the two-stage design (not counting the output stage). The same structure is used in CMOS op amps.
- To obtain low input offset voltage and current and high CMRR, the 741 input stage is designed to be perfectly balanced. The CMRR is increased by common-mode feedback, which also stabilizes the dc operating point.
- To obtain high input resistance and low input bias current the input stage of the 741 is operated at a very low current level.
- In the 741, output short-circuit protection is accomplished by turning on a transistor that takes away most of the base current drive of the output transistor.
- The use of Miller frequency compensation in the 741 circuit enables locating the dominant pole at a very low frequency, while utilizing a relatively small compensating capacitance.
- Two-stage op amps can be modeled as a transconductance amplifier feeding an ideal integrator with C_C as the integrating capacitor.

[4]The final voltage can deviate from zero by as much as the analog equivalent of the LSB. Thus the insensitivity to top-plate capacitance is not complete.

▶ The slew rate of a two-stage op amp is determined by the first-stage bias current and the frequency-compensation capacitor.

▶ Most CMOS op amps are designed to operate as part of a VLSI circuit and thus are required to drive only small capacitive loads. Therefore, most do not have a low-output-resistance stage.

▶ In the two-stage CMOS op amp, approximately equal gains are realized in the two stages.

▶ The threshold mismatch ΔV_t and the low transconductance of the input stage result in a larger input offset voltage for CMOS op amps as compared to bipolar units.

▶ Miller compensation is employed in CMOS op amps also, but a series resistor is required in order to place the transmission zero at either $s = \infty$ or on the negative real axis.

▶ CMOS op amps have higher slew rates than their bipolar counterparts with comparable f_t values.

▶ Use of the cascode configuration increases the gain of a CMOS amplifier stage by about two orders of magnitude, thus making possible a single-stage op amp.

▶ The folded-cascode circuit of Fig. 10.26 is a popular implementation of CMOS op amps. In addition to the advantages of the cascode configuration, the folded cascode features a larger input common-mode range. The dominant pole of the folded-cascode op amp is determined by the total capacitance at the output node, C_L. Increasing C_L improves the phase margin at the expense of reducing the bandwidth.

▶ The bandwidth of the folded-cascode circuit can be increased by using bipolar transistors for the cascode devices, resulting in the BiCMOS circuit of Fig. 10.27.

▶ A/D and D/A converters constitute an important group of analog ICs.

▶ A DAC consists of: (a) a circuit that generates a reference current, (b) a circuit that assigns binary weights to the value of the reference current, (c) switches that, under the control of the bits of the input digital word, direct the proper combination of binary-weighted currents to an output summing node, and (d) an op amp that converts the current sum to an output voltage. The circuit of (b) can be implemented by either a binary-weighted resistive ladder or an R-$2R$ ladder.

▶ Two simple but slow implementations of the ADC are the feedback-type converter [Fig. 10.35] and the dual-slope converter [Fig. 10.36].

▶ The fastest possible ADC implementation is the parallel or flash converter [Fig. 10.37].

▶ The charge-redistribution method [Fig. 10.38] utilizes switched-capacitor techniques and is particularly suited for the implementation of ADCs in CMOS technology.

BIBLIOGRAPHY

P. E. Allen and D. R. Holberg, *CMOS Analog Circuit Design,* New York: Holt, Rinehart and Winston, 1987.

J. A. Connely (Ed.), *Analog Integrated Circuits,* New York: Wiley-Interscience, 1975.

R. L. Geiger, P. E. Allen, and N. R. Strader, *VLSI Design Techniques for Analog and Digital Circuits,* New York: McGraw Hill, 1990.

P. R. Gray, D. A. Hodges, and R. W. Brodersen, *Analog MOS Integrated Circuits,* New York: IEEE Press, 1980.

P. R. Gray and R. G. Meyer, *Analysis and Design of Analog Integrated Circuits,* 2nd ed. New York: Wiley, 1984.

A. B. Grebene (Ed.), *Analog Integrated Circuits,* New York: IEEE Press, 1978.

A. B. Grebene, *Bipolar and MOS Analog Integrated Circuit Design,* New York: Wiley, 1984.

R. Gregorian and G. C. Temes, *Analog MOS Integrated Circuits for Signal Processing,* New York: Wiley-Interscience, 1986.

IEEE Journal of Solid-State Circuits. The December issue of each year has been devoted to analog ICs.

B. P. Lathi, *Signals, Systems and Communication;* Chapter 11, New York: Wiley, 1965.

H. S. Lee, "Analog Design," Chapter 8, in *BiCMOS Technology and Applications,* A. R. Alvarez, editor, Boston: Kluwer Academic Publishers, 1989.

R. G. Meyer (Ed.), *Integrated-Circuit Operational Amplifiers,* New York: IEEE Press, 1978.

J. E. Solomon, "The monolithic op amp: A tutorial study," *IEEE Journal of Solid-State Circuits,* vol. SC-9, no. 6, pp. 314–332, Dec. 1974.

Staff of Analog Devices, Inc., *Analog-Digital Conversion Notes,* Norwood, Mass: Analog Devices, 1977.

R. J. Widlar, "Some circuit design techniques for linear

integrated circuits,'' *IEEE Transactions on Circuit Theory,* vol. CT-12, pp. 586–590, Dec. 1965.

R. J. Widlar, ''Design techniques for monolithic operational amplifiers,'' *IEEE Journal of Solid-State Circuits,* vol. SC-9, no. 6, pp. 314–322, Dec. 1974.

B. A. Wooley, ''BiCMOS Analog Circuit Techniques,'' Proceedings of the 1990 IEEE International Symposium on Circuits and Systems, pp. 1983–1985, New Orleans, 1990.

PROBLEMS

Section 10.1: The 741 Op Amp Circuit

10.1 In the 741 op amp circuit of Fig. 10.1, Q_1, Q_2, Q_5, and Q_6 are biased at collector currents of 9.5 μA; Q_{16} is biased at a collector current of 16.2 μA; and Q_{17} is biased at a collector current of 550 μA. All these devices are of the ''standard *npn*'' type, having $I_S = 10^{-14}$ A, $\beta = 200$, and $V_A = 125$ V. For each of these transistors find V_{BE}, g_m, r_e, r_π, r_o, and r_μ (assume $r_\mu = 10\ \beta r_o$). Provide your results in a table form. (Note that these parameter values are utilized in the text in the analysis of the 741 circuit.)

D10.2 For the (mirror) bias circuit shown in Fig. E10.2 and the result verified in the associated exercise, find I_1 for the case in which $I_{S3} = 3 \times 10^{-14}$ A, $I_{S4} = 6 \times 10^{-14}$ A, and $I_{S1} = I_{S2} = 10^{-14}$ A and for which a bias current $I_3 = 154$ μA is required.

10.3 Transistor Q_{13} in the circuit of Fig. 10.1 consists, in effect, of two transistors whose emitter–base junctions are connected in parallel and for which $I_{SA} = 0.25 \times 10^{-14}$ A, $I_{SB} = 0.75 \times 10^{-14}$ A, $\beta = 50$, and $V_A = 50$ V. For operation at a total emitter current of 0.73 mA, find values for the parameters V_{EB}, g_m, r_e, r_π, and r_o for the A and B devices.

10.4 In the circuit of Fig. 10.1, Q_1 and Q_2 exhibit emitter–base breakdown at 7 V, while for Q_3 and Q_4 such a breakdown occurs at about 50 V. What differential input voltage would result in the breakdown of the input-stage transistors?

D*10.5 Figure P10.5 shows the CMOS version of the circuit in Fig. E10.2. Find the relationship between I_3 and I_1 in terms of K_1, K_2, K_3, and K_4 of the four transistors, assuming the threshold voltages of all devices to be equal in magnitude. In the event that $K_1 = K_2$ and $K_3 = K_4 = 16\ K_1$, find the required value of I_1 to yield a bias current in Q_3 and Q_4 of 1.6 mA.

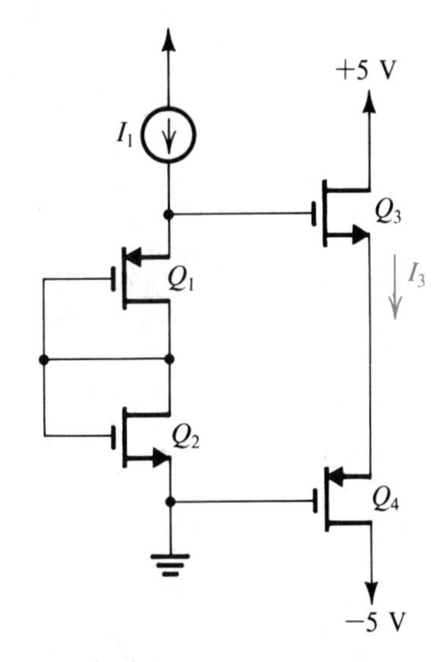

Fig. P10.5

Section 10.2: DC Analysis of the 741

D10.6 For the 741 circuit estimate the input reference current I_{REF} in the event that ±5-V supplies are used. Find a more precise value assuming that for the two BJTs involved $I_S = 10^{-14}$ A. What value of R_5 would be necessary to reestablish the same bias current for ±5-V supplies as exists for ±15 V in the original design?

D10.7 In the reference-bias part of the circuit of Fig. 10.1 consider the replacement of the resistor R_5 by an *n*-channel JFET suitably connected.

(a) If the JFET has an infinite output resistance in pinch-off, find its required I_{DSS} so that $I_{REF} = 0.73$ mA.

(b) If the JFET has an Early voltage $V_A = 50$ V, $I_{DSS} = 0.5$ mA, and $V_P = -2$ V, find I_{REF} for supply voltages of ±5 V and ±15 V.

(c) What is the minimum total supply voltage below which I_{REF} begins to deviate rapidly from the design value?

***10.8** In the 741 circuit consider the common-mode feedback loop comprised of transistors Q_1, Q_2, Q_3, Q_4,

Q_8, Q_9, and Q_{10}. We wish to find the loop gain. This can be conveniently done by breaking the loop between the common collector connection of Q_1 and Q_2, and the diode-connected transistor Q_8. Apply a test current signal I_t to Q_8 and find the returned current signal I_r in the combined collector connection of Q_1 and Q_2. Thus determine the loop gain. Assume that Q_9 and Q_{10} act as ideal current sources. If Q_3 and Q_4 have $\beta = 50$, find the amount of common-mode feedback in dB.

D10.9 Design the Widlar current source of Fig. 10.2 to generate a current $I_{C10} = 20\ \mu A$ given that $I_{REF} = 0.5$ mA. If for the transistors $I_S = 10^{-14}$ A, find V_{BE11} and V_{BE10}. Assume β to be high.

10.10 Consider the analysis of the 741 input stage shown in Fig. 10.3. For what value of β_P do the currents in Q_1 and Q_2 differ from the ideal value of $I_{C10}/2$ by 10%?

D10.11 Consider the analysis of the 741 input stage shown in Fig. 10.3 for the situation in which $I_{S9} = 2I_{S8}$. For $I_{C10} = 19\ \mu A$ and assuming β_P to be high, what does I become? Redesign the Widlar source to reestablish $I_{C1} = I_{C2} = 9.5\ \mu A$.

10.12 For the mirror circuit shown in Fig. 10.4 with the bias and component values given in the text for the 741 circuit, what does the current in Q_6 become if R_2 is shorted?

D10.13 It is required to redesign the circuit of Fig. 10.4 by selecting a new value for R_3 so that when the base currents are *not* neglected, the collector currents of Q_5, Q_6, and Q_7 become all equal assuming that the input current $I_{C3} = 9.4\ \mu A$. Find the new value of R_3 and the three currents. Recall that $\beta_N = 200$.

10.14 Consider the input circuit of the 741 op amp of Fig. 10.1 under the conditions that the emitter current of Q_8 is about 19 μA. If β of Q_1 is 150 and that of Q_2 is 200, find the input bias current I_B and the input offset current I_{OS} of the op amp.

10.15 For a particular application, consideration is being given to selecting 741 ICs for bias and offset currents limited to 40 nA and 4 nA, respectively. Assuming other aspects of the selected units to be normal, what minimum β_N and what β_N variation are implied?

10.16 A manufacturing problem in a 741 op amp causes the current transfer ratio of the mirror circuit that loads the input stage to become 0.9 A/A. For input devices (Q_1 through Q_4) appropriately matched and with high β and normally biased at 9.5 μA, what input offset voltage results?

10.17 The solution to Exercise 10.4 was based on the assumption that a BJT ceases linear operation as soon as its collector–base junction becomes forward-biased. If linear operation continues for as much as 0.3 V of collector–base forward bias, what does the input common-mode range become?

D10.18 Consider the design of the second stage of the 741. What value of R_9 would be needed to reduce I_{C16} to 9.5 μA?

D10.19 Reconsider the 741 output stage as shown in Fig. 10.5, in which R_{10} is adjusted to make $I_{C19} = I_{C18}$. What is the new value of R_{10}? What values of I_{C14} and I_{C20} result?

D*10.20 An alternative approach to providing the voltage drop needed to bias the output transistors is the V_{BE}-multiplier circuit shown in Fig. P10.20. Design the circuit to provide a terminal voltage of 1.118 V (the same as in the 741 circuit). Base your design on half the current flowing through R_1, and assume that $I_S = 10^{-14}$ A and $\beta = 200$. What is the incremental resistance between the two terminals of the V_{BE}-multiplier circuit?

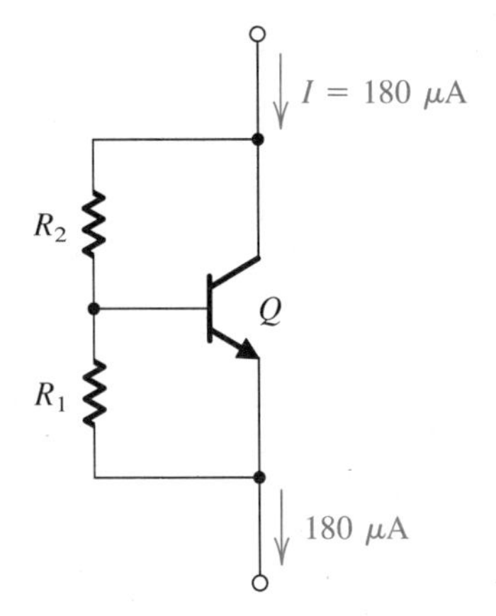

Fig. P10.20

10.21 For the circuit of Fig. 10.1, what is the total current required from the power supplies when the op amp is operated in the linear mode but with no load? Hence estimate the quiescent power dissipation in the circuit. (*Hint:* Use the data given in Table 10.1.)

Section 10.3: Small-Signal Analysis of the 741 Input Stage

10.22 Consider the 741 input stage as modeled in Fig. 10.6, with two additional *npn* diode-connected transistors, Q_{1a} and Q_{2a}, connected between the present *npn* and *pnp* devices, one per side. Convince yourself that the additional devices will each be biased at the same current as Q_1 to Q_4—that is, 9.5 μA. What does R_{id} become? What does G_{m1} become? What is the value of R_{o4} now? What is the output resistance of the first stage, R_{o1}? What is the new open-circuit voltage gain, $G_{m1}\ R_{o1}$? Compare these values to the original ones.

D10.23 What relatively simple change can be made to the mirror load of stage 1 to increase its output resistance, say by a factor of two?

10.24 Repeat Exercise 10.6 with $R_1 = R_2$ replaced by 2-kΩ resistors.

***10.25** In Example 10.1 we investigated the effect of a mismatch between R_1 and R_2 on the input offset voltage of the op amp. Conversely, R_1 and R_2 can be deliberately mismatched (using the circuit shown in Fig. P10.25, for example) to compensate for the op amp input offset voltage.

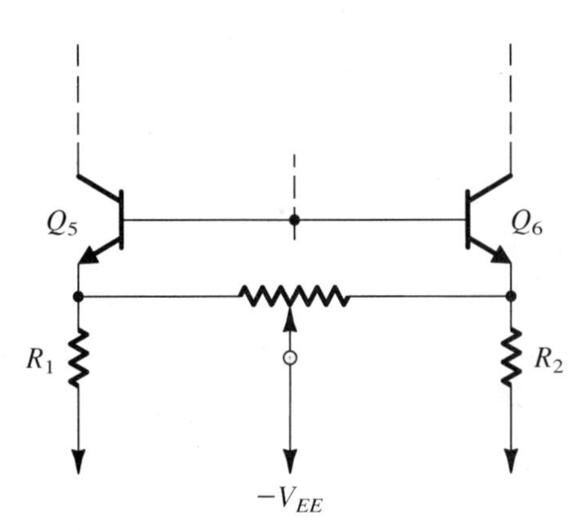

Fig. P10.25

(a) Show that an input offset voltage V_{OS} can be compensated for (i.e., reduced to zero) by creating a relative mismatch $\Delta R/R$ between R_1 and R_2,

$$\frac{\Delta R}{R} = \frac{V_{OS}}{2V_T}\,\frac{1 + r_e/R}{1 - V_{OS}/2V_T}$$

where r_e is the emitter resistance of each of Q_1 to Q_6 and R is the nominal value of R_1 and R_2. (Hint: Use Eq. 10.10)

(b) Find $\Delta R/R$ to trim a 5-mV offset to zero.

(c) What is the maximum offset voltage that can be trimmed this way (corresponding to R_2 completely shorted)?

10.26 Through a processing imperfection, the β of Q_4 in Fig. 10.1 is reduced to 25, while the β of Q_3 remains at its regular value of 50. Find the input offset voltage that this mismatch introduces. (*Hint:* Follow the general procedure outlined in Example 10.1.)

10.27 Consider the circuit of Fig. 10.1 modified to include resistors R in series with the emitters of each of Q_8 and Q_9. What does the resistance looking into the collector of Q_9, R_{o9}, become? For what value of R does it equal R_{o10}? For this case what does R_o of the source at node Y become?

10.28 Refer to Fig. E10.7 and let $R_1 = R_2$. If Q_3 and Q_4 have a β mismatch so that for Q_3 the current gain is β_P and for Q_4 the current gain is $k\beta_P$, find i_o and G_{mcm}. For $R_o = 2.43$ MΩ, $\beta_P = 20$, $0.5 \le k \le 2$, G_{m1} (differential) $= 1/5.26$ kΩ, find the worst case CMRR $\equiv G_{m1}/G_{mcm}$ (in dB) that results. Assume everything else is ideal.

***10.29** What is the effect on the differential gain of the 741 op amp of short-circuiting one, or the other, or both, of R_1 and R_2 in Fig. 10.1? (Refer to Fig. 10.7.) For simplicity, assume $\beta = \infty$.

***10.30** Figure P10.30 shows the equivalent common-mode half-circuit of the input stage of the 741. Here R_o is the resistance seen looking to the left of node Y in Fig. 10.1; its value is approximately 2.4 MΩ. Transistors Q_1 and Q_3 operate at a bias current of 9.5 μA. Find the input resistance of the common-mode half circuit using $\beta_N = 200$, $\beta_P = 50$, $r_\mu = 10\beta r_o$, and $V_A = 125$ V for *npn* and 50 V for *pnp* transistors. To find the common-mode input resistance of the 741 note that it has common-mode feedback that increases the input common-mode resistance. The loop

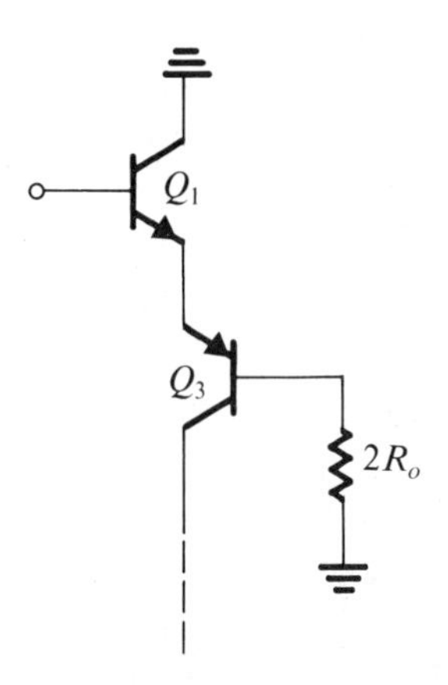

Fig. P10.30

gain is approximately equal to β_P. Find the value of R_{icm}.

Section 10.4: Small-Signal Analysis of the 741 Second Stage

10.31 Consider a variation on the design of the 741 second stage in which $R_8 = 50\ \Omega$. What R_{12} and G_{m2} correspond?

10.32 In the analysis of the 741 second stage, note that R_{o2} is affected most strongly by the low value of R_{o13B}. Consider the effect of placing appropriate resistors in the emitters of Q_{12}, Q_{13A}, and Q_{13B} on this value. What resistor in the emitter of Q_{13B} would be required to make R_{o13B} equal to R_{o17} and thus R_{o2} half as great? What resistors in each of the other emitters would be required?

Section 10.5: Analysis of the 741 Output Stage

10.33 For a 741 employing ± 5 V supplies, $|V_{BE}| = 0.6$ V, and $|V_{CEsat}| = 0.2$ V, find the output voltage limits that apply.

D10.34 Consider an alternative to the present 741 output stage in which Q_{23} is not used, that is in which its base and emitter are joined. Reevaluate the reflection of $R_L = 2\ \text{k}\Omega$ to the collector of Q_{17}. What does A_2 become?

10.35 Consider the positive current-limiting circuit involving Q_{13A}, Q_{15}, and R_6. Find the current in R_6 at which the collector current of Q_{15} equals the current available from Q_{13A} (180 μA) minus the base current of Q_{14}. (You need to perform a couple of iterations.)

*10.36 Consider the 741 sinking-current limit involving R_7, Q_{21}, Q_{24}, R_{11}, and Q_{22}. For what current through R_7 is the current in Q_{22} equal to the maximum current available from the input stage, that is the current in Q_8? What simple change would you make to reduce this current limit to 10 mA?

Section 10.6: Gain and Frequency Response of the 741

10.37 Using the data provided in Eq. (10.28) (alone) for the overall gain of the 741 with a 2-kΩ load, and realizing the significance of the factor 0.97 in relationship to the load, calculate the open-circuit voltage gain, the output resistance, and the gain with a load of 200 Ω. What is the maximum output voltage available for such a load?

10.38 A 741 op amp has a phase margin of 80°. If the excess phase shift is due to a second single pole, what is the frequency of this pole?

10.39 A 741 op amp has a phase margin of 80°. If the op amp has nearly coincident second and third poles, what is their frequency?

D10.40 For a modified 741 whose second pole is at 5 MHz, what dominant-pole frequency is required for 85° phase margin with a closed-loop gain of 100? Assuming C_C continues to control the dominant pole, what value of C_C would be required?

10.41 An internally compensated op amp having an f_t of 5 MHz and dc gain of 10^6 utilizes Miller compensation around an inverting amplifier stage with a gain of 1000. If space exists for at most a 50-pF capacitor, what resistance level must be reached at the input of the Miller amplifier for compensation to be possible?

10.42 Consider the integrator op amp model shown in Fig. 10.20. For $G_{m1} = 10$ mA/V, $C_C = 50$ pF, and a resistance of $10^8\ \Omega$ shunting C_C, sketch and label a Bode plot for the magnitude of the open-loop gain. If G_{m1} is related to the first-stage bias current via Eq. (10.40), find the slew rate of this op amp.

10.43 For an amplifier with a slew rate of 10 V/μs, what is the full-power bandwidth for outputs of ± 10 V? What unity-gain bandwidth, ω_t, would you expect if the topology was similar to that of the 741?

D**10.44 Figure P10.44 shows a circuit suitable for op amp applications. For all transistors $\beta = 100$, $V_{BE} = 0.7$ V, and $r_o = \infty$.

(a) For inputs grounded and output held at 0 V (by negative feedback) find the emitter currents of all transistors.

(b) Calculate the gain of the amplifier with a load of 10 kΩ.

(c) With load as in (b) calculate the value of the capacitor C required for a 3-dB frequency of 1 kHz.

Section 10.7: CMOS Op Amps

D*10.45 In a particular design of the CMOS op amp of Fig. 10.23 the designer wishes to investigate the effects of increasing the W/L ratio of both Q_1 and Q_2 by a factor of 4. Assuming that all other parameters are kept unchanged:

(a) Find the resulting change in ($|V_{GS}| - |V_t|$) and in g_m of Q_1 and Q_2.

(b) What change results in the voltage gain of the input stage? in the overall voltage gain?

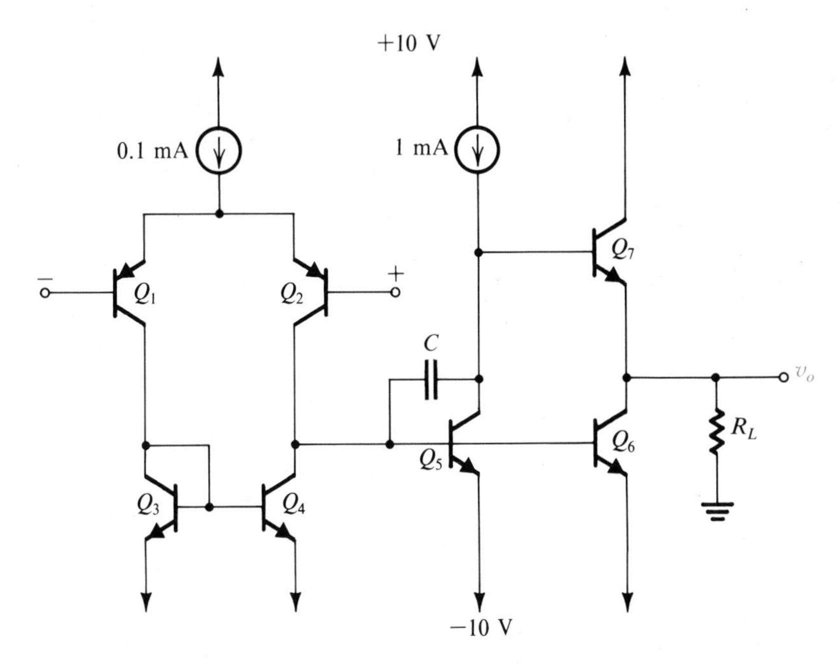

Fig. P10.44

(c) What is the effect on the input offset voltage? (You might wish to refer to Section 6.7.)

(d) If f_t is kept unchanged (so that the phase margin does not decrease), how must C_C be changed? What is the corresponding change in the amplifier slew rate?

10.46 Consider the amplifier of Fig. 10.23, whose parameters are specified in Example 10.2. If a manufacturing error results in the W/L ratio of Q_7 being 180/10, find the current that Q_7 will now conduct. Thus find the systematic offset voltage that will appear at the output. (Use the results of Example 10.2.) Assuming that the open-loop gain will remain approximately unchanged from the value found in Example 10.2, find the corresponding value of input offset voltage.

*10.47 Consider the input stage of the CMOS op amp in Fig. 10.23 with both inputs grounded. Assume that the two sides of the input stage are perfectly matched except that the threshold voltages of Q_3 and Q_4 show a mismatch ΔV_t. Show that a current $g_{m3}\,\Delta V_t$ appears at the output of the first stage. What is the corresponding input offset voltage? Evaluate this offset voltage for the circuit specified in Example 10.2 for $\Delta V_t = 2$ mV. (Use the results of Example 10.2.)

10.48 What is the open-loop output resistance of the amplifier of Fig. 10.23, whose parameters are specified in Example 10.2? For unity-gain-feedback, what does R_{out} become? (Use the results of Example 10.2.)

D10.49 For the amplifier analyzed in Example 10.2:

(a) Find the value of C_C that results in $f_t = 0.5$ MHz.

(b) For $R = 1/G_{m2}$, what is the maximum allowed load capacitance C_2 for which a phase margin of at least 45° is obtained?

D10.50 It is required to design the frequency-compensation network for the amplifier in Fig. 10.23, whose parameters are specified in Example 10.2. The transmission zero is to be placed at infinite frequency and the amplifier is to have 80° of phase margin when the total capacitance at the output is 10 pF. What are the required values of C_C and R? What are the resulting values of f_t and SR?

D*10.51 Redesign the compensation network of Problem 10.50, this time placing the transmission zero on the negative real axis. In this way it introduces a positive phase that increases the phase margin. Design the network so that $f_t = 1$ MHz and the phase margin is 80° with a total capacitance at the output of 10 pF. Find C_C and R and the resulting slew rate assuming that C_2 and C_C are much greater than C_1.

10.52 A two-stage CMOS amplifier resembling that in Fig. 10.23 is found to have a slew rate of 5 V/μs and $f_t = 2$ MHz. If the first-stage bias current $(2I)$ is

50 μA, what value of C_C must be used? If devices with 1-V threshold are used, what gate-to-source bias voltage is used in the input stage? For a process for which $\mu_n C_{OX} = 20\ \mu A/V^2$, what W/L ratio applies for the input-stage devices?

**10.53 Consider a CMOS amplifier that is complementary to that in Fig. 10.23 in which each device is replaced by its complement of the same physical size with the supplies reversed. Use the overall conditions as specified in Example 10.2. For all devices evaluate I_D, g_m, and r_o. Find A_1, A_2, the dc open-loop gain, the input common-mode range, and the output voltage range. Neglect the effect of V_A on bias currents.

Section 10.8: Alternative Configurations for CMOS and BiCMOS Op Amps

D**10.54 Consider the cascoded input stage of Fig. 10.25. Let $2I = 25\ \mu A$, $\mu_p C_{OX} = 10\ \mu A/V^2$, $|V_t| = 1$ V, and W/L for Q_1, Q_2, Q_{1C}, and $Q_{2C} = 120/8$. By how much should V_{BIAS2} be set below the voltage at the common source connection of Q_1 and Q_2, so that Q_1, Q_2, Q_{1C}, and Q_{2C} are operating at the boundary of the saturation region? An arrangement that is usually used to generate V_{BIAS2} via creating a constant voltage difference between the sources of Q_1 and Q_2 and the gates of Q_{1C} and Q_{2C} is shown in Fig. P10.54. If I_{BIAS} is selected to be 5 μA, find the required W/L ratio for Q_B. Also, if W/L for Q_5 is 150/10 what must V_{BIAS1} be? Now draw the complete circuit and calculate V_{GS} for each of Q_3, Q_{3C}, Q_4, and Q_{4C} assuming that $\mu_n C_{OX} = 20\ \mu A/V^2$, W/L for each of Q_{3C} and $Q_4 = 60/8$. Find the input common-mode range.

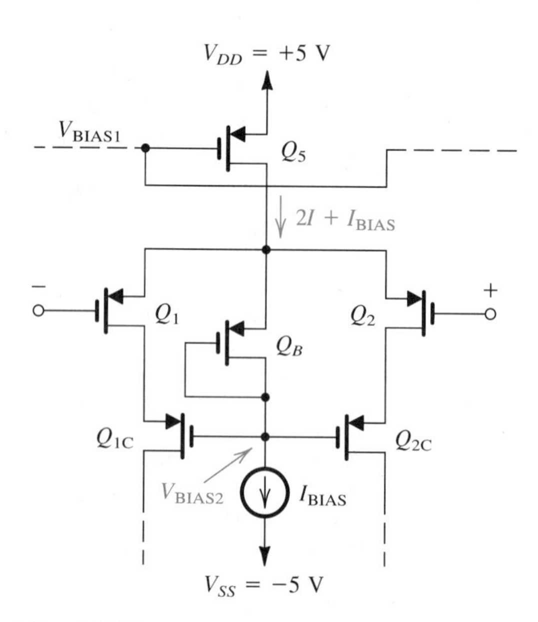

Fig. P10.54

10.55 Sketch the circuit that is complementary to that in Fig. 10.25, that is, one that uses an input n-channel differential pair.

10.56 Find the output resistance and the dc open-loop voltage gain of the folded cascode amplifier of Fig. 10.26 whose parameters are specified in Exercise 10.29. Assume $|V_A| = 25$ V for all devices.

D*10.57 Design the folded cascode circuit of Fig. 10.26 to obtain a dc open-loop voltage gain of 10,000 V/V and a unity-gain bandwidth of 1 MHz when the total capacitance at the output is 10 pF. Design for $I_B = 2I$, $(W/L)_1 = (W/L)_{4C} = 2(W/L)_{2C}$. Specify the required values of I and $(W/L)_1$. Let $\mu_n C_{OX} = 2\mu_p C_{OX} = 20\ \mu A/V^2$ and $|V_A| = 25$ V. (Hint: Use Eq. 10.62.)

D10.58 It is required to design the folded-cascode CMOS op-amp circuit of Fig. 10.26. The load capacitance C_L (including all parasitics) is 10 pF. The total capacitance at the input of each of the common-gate transistors Q_{1C} and Q_{2C} is $C_P = 1$ pF. Design for bias currents $2I = I_B = 100\ \mu A$ and $(W/L)_{1C} = (W/L)_{2C} = 10/10$. To obtain a sufficient phase margin the design should ensure that $f_t \leq f_P/3$, where f_P is the frequency of the nondominant pole due to C_P. Specify the required W/L ratios for the input transistors to obtain the largest possible f_t. What is the value of f_t realized? Assume that $\mu_n C_{OX} = 2\ \mu_p C_{OX} = 20\ \mu A/V^2$.

D10.59 A folded-cascode BiCMOS amplifier having the topology of Fig. 10.27 is designed to operate at high frequencies. The bias currents are $2I = I_B = 400\ \mu A$, and the W/L ratio for the input stage transistors is 300/10. Find f_t for a load capacitance C_L (including all the output node parasitics) of 2 pF. To maintain an acceptable phase margin, the parasitic pole created at the input to the cascode transistors Q_{1C} and Q_{2C} must be at least three times higher in frequency than f_t. What is the largest parasitic capacitance C_P that can be tolerated? Assume $\mu_p C_{OX} = 10\ \mu A/V^2$.

Section 10.9: Data Converters—An Introduction

10.60 An analog signal in the range 0 to +10 V is to be digitized with a quantization error of less than 1% of full scale. What is the number of bits required? What is the resolution of the conversion? If the range is to be extended to ±10 V with the same requirement, what is the number of bits required? For an extension to a range of 0 to +15 V, how many bits are required to provide the same resolution? What is the corresponding resolution and quantization error?

*10.61 Consider Fig. 10.30. On the staircase output of the S/H circuit sketch the output of a simple low-pass RC circuit with a time constant that is (a) one-third of the sampling interval; (b) equal to the sampling interval.

Section 10.10: D/A Converter Circuits

*10.62 Consider the DAC circuit of Fig. 10.31 for the cases $N = 2$, 4, and 8. What is the tolerance, expressed as $\pm x\%$, to which the resistors should be selected so as to limit the resulting output error to the equivalent of $\pm\frac{1}{2}$ LSB?

10.63 The BJTs in the circuit of Fig. P10.63 have their base–emitter junction areas scaled in the ratios indicated. Find I_1 to I_4 in terms of I.

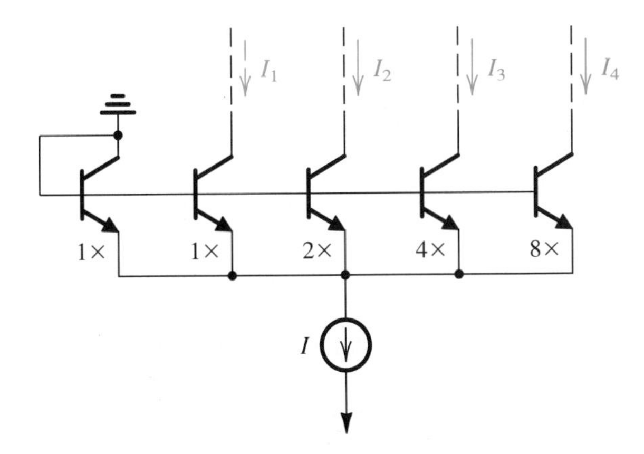

Fig. P10.63

10.64 A problem encountered in the DAC circuit of Fig. 10.33 is the large spread in transistor EBJ areas required when N is large. As an alternative arrangement consider using the circuit in Fig. 10.33 for 4 bits only. Then, feed the current in the collector of the terminating transistor Q_t to the circuit of Fig. P10.63 (in place of the current source I), thus producing currents for 4 more bits. In this way an 8-bit DAC can be implemented with a maximum spread in areas of 8. What is the total area of emitters needed in terms of the smallest device? Contrast this with the usual 8-bit circuit. Give the complete circuit of the converter thus realized.

D*10.65 The circuit in Fig. 10.31 can be used to multiply an analog signal by a digital one by feeding the analog signal to the V_{ref} terminal. In this case the D/A converter is called a **multiplying DAC** or MDAC. Given an input sine-wave signal of 0.1 sin ωt volts, use the circuit of Fig. 10.31 together with an additional op amp to obtain $v_O = 10D \sin \omega t$ where D is the digital word given by Eq. (10.67) and $N = 4$. How many discrete sine-wave amplitudes are available at the output? What is the smallest? What is the largest? To what digital input does a 10-V peak-to-peak output correspond?

10.66 What is the input resistance seen by V_{ref} in the circuit of Fig. 10.32?

Section 10.11: A/D Converter Circuits

10.67 A 12-bit dual-slope ADC of the type illustrated in 10.36 utilizes a 1-MHz clock and has $V_{ref} = 10$ V. Its analog input voltage is in the range 0 to −10 V. The fixed interval T_1 is the time taken for the counter to accumulate a count of 2^N. What is the time required to convert an input voltage equal to the full-scale value? If the peak voltage reached at the output of the integrator is 10 V, what is the integrator time constant? If through aging R increases by 2% and C decreases by 1%, what does V_{peak} become? Does the conversion accuracy change?

10.68 The design of a 4-bit flash ADC as shown in 10.37 is being considered. How many comparators are required? For an input signal in the range of 0 to +10 V, what are the reference voltages needed? Show how they can be generated using a 10-V reference and several 1-kΩ resistors (how many?). If a comparison is possible in 50 ns and the associated logic requires 35 ns, what is the maximum possible conversion rate? Indicate the digital code you expect at the output of the comparators and at the output of the logic for an input of (a) 0 V, (b) +5.1 V, and (c) +10 V.

CHAPTER 11

Filters and Tuned Amplifiers

INTRODUCTION

In this chapter we study the design of an important building block of communication and instrumentation systems, the electronic filter. Filter design is one of the very few areas of engineering for which a complete design theory exists, starting from specification and ending with a circuit realization. A detailed study of filter design requires an entire book, and indeed such textbooks exist. In the limited space available here we shall concentrate on a selection of topics that provide an introduction to the subject as well as a useful arsenal of filter circuits and design methods.

The oldest technology for realizing filters makes use of inductors and capacitors, and the resulting circuits are called **passive LC filters.** Such filters work well at high frequencies; however, in low-frequency applications (dc to 100 kHz) the required inductors are large and physically bulky, and their characteristics are quite nonideal. Furthermore, such inductors are impossible to fabricate in monolithic form and are incompatible with any of the modern techniques for assembling electronic systems. Therefore, there has been considerable interest in finding filter realizations that do not require inductors. Of the various possible types of **inductorless filters,** we shall study **active-RC filters** and **switched-capacitor filters.**

Active-RC filters utilize op amps together with resistors and capacitors and are fabricated using discrete, hybrid thick-film, or hybrid thin-film technology. However, for large-volume production, such technologies do not yield the economies achieved by monolithic fabrication. At the present time the most viable approach for realizing fully integrated monolithic filters is the switched-capacitor technique.

The last topic studied in this chapter is the tuned amplifier commonly employed in the design of radio and TV receivers. Although tuned amplifiers are in effect bandpass filters, they are studied separately because their design is based on somewhat different techniques.

11.1 FILTER TRANSMISSION, TYPES, AND SPECIFICATION

The filters we are about to study are linear circuits that can be represented by the general two-port network shown in Fig. 11.1. The filter **transfer function** $T(s)$ is the ratio of the output voltage $V_o(s)$ to the input voltage $V_i(s)$,

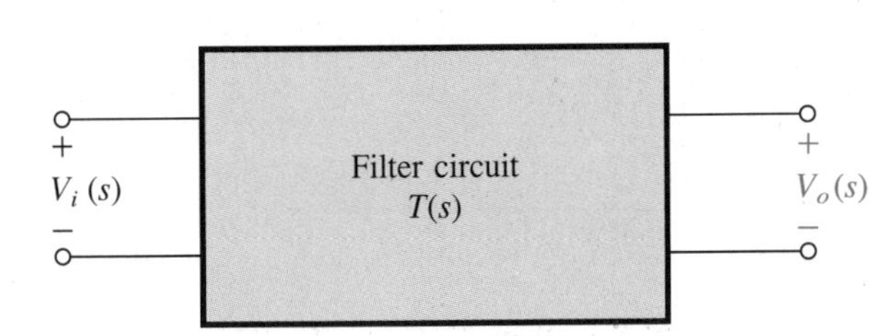

Fig. 11.1 The filters studied in this chapter are linear circuits represented by the general two-port network shown. The filter transfer function $T(s) \equiv V_o(s)/V_i(s)$.

$$T(s) \equiv \frac{V_o(s)}{V_i(s)} \tag{11.1}$$

The filter **transmission** is found by evaluating $T(s)$ for physical frequencies, $s = j\omega$, and can be expressed in terms of its magnitude and phase as

$$T(j\omega) = |T(j\omega)|\, e^{j\phi(\omega)} \tag{11.2}$$

The magnitude of transmission is often expressed in decibels in terms of the **gain function**

$$G(\omega) \equiv 20 \log|T(j\omega)|, \text{ dB} \tag{11.3}$$

or, alternatively, in terms of the **attenuation function**

$$A(\omega) \equiv -20 \log|T(j\omega)|, \text{ dB} \tag{11.4}$$

A filter shapes the frequency spectrum of the input signal, $|V_i(j\omega)|$, according to the magnitude of the transfer function, $|T(j\omega)|$, thus providing an output $V_o(j\omega)$ with a spectrum

$$|V_o(j\omega)| = |T(j\omega)|\,|V_i(j\omega)| \tag{11.5}$$

Also, the phase characteristics of the signal are modified as it passes through the filter according to the filter phase function $\phi(\omega)$.

We are specifically interested here in filters that perform a **frequency selection** function: **passing** signals whose frequency spectrum lies within a specified range, and **stopping** signals whose frequency spectrum falls outside this range. Such a filter has ideally a frequency band (or bands) over which the magnitude of transmission is unity (the filter **passband**) and a frequency band (or bands) over which the transmission is zero (the filter **stopband**). Figure 11.2 depicts the ideal transmission characteristics of the four major filter types: **low-pass** (LP) in Fig. 11.2(a), **high-pass** (HP) in Fig. 11.2(b), **bandpass** (BP) in Fig. 11.2(c) and **bandstop** (BS) or **band-reject** in Fig. 11.2(d). These idealized characteristics, by virtue of their vertical edges, are known as **brick-wall** type responses.

Filter Specification

The filter design process begins with the filter user specifying the transmission characteristics required of the filter. Such a specification cannot be of the form shown in Fig. 11.2 because physical circuits cannot realize these idealized characteristics. Figure 11.3 shows realistic specifications for the transmission characteristics of a low-pass filter. Observe that since a physical circuit cannot provide constant transmission at all passband frequen-

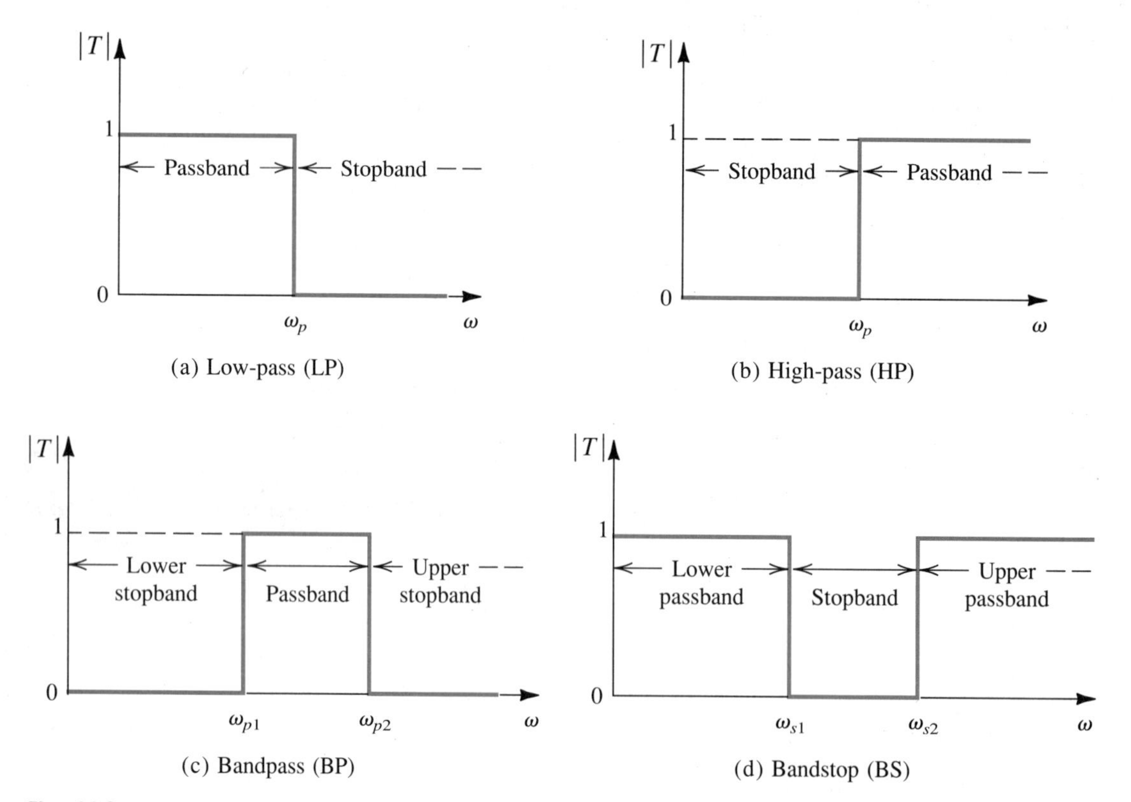

Fig. 11.2 Ideal transmission characteristics of the four major filter types: **(a)** low-pass (LP), **(b)** high-pass (HP), **(c)** bandpass (BP), and **(d)** bandstop (BS).

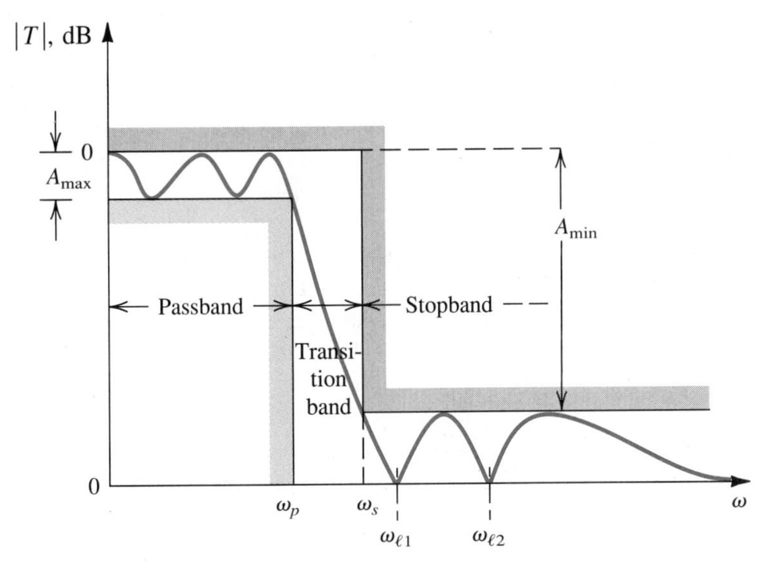

Fig. 11.3 Specification of the transmission characteristics of a low-pass filter. The magnitude response of a filter that just meets specifications is also shown.

cies, the specifications allow for deviation of the passband transmission from the ideal 0 dB, but places an upper bound, A_{max} (dB), on this deviation. Depending on the application, A_{max} typically ranges from 0.05 to 3 dB. Also, since a physical circuit cannot provide zero transmission at all stopband frequencies, the specifications in Fig. 11.3 allow for some transmission over the stopband. However, the specifications require the stopband signals to be attenuated by at least A_{min} (dB) relative to the passband signals. Depending on the filter application, A_{min} can range from 20 to 100 dB.

Since the transmission of a physical circuit cannot change abruptly at the edge of the passband, the specifications of Fig. 11.3 provide for a band of frequencies over which the attenuation increases from near 0 dB to A_{min}. This **transition band** extends from the passband edge ω_p to the stopband edge ω_s. The ratio ω_s/ω_p is usually used as a measure of the sharpness of the low-pass filter response and is called the **selectivity factor.** Finally, observe that for convenience the passband transmission is specified to be 0 dB. The final filter, however, can be given a passband gain, if desired, without changing its selectivity characteristics.

To summarize, the transmission of a low-pass filter is specified by four parameters:

1. the passband edge, ω_p;
2. the maximum allowed variation in passband transmission, A_{max};
3. the stopband edge, ω_s; and
4. the minimum required stopband attenuation, A_{min}.

The more tightly one specifies a filter—that is, lower A_{max}, higher A_{min}, and/or a selectivity ratio ω_s/ω_p closer to unity—the closer the response of the resulting filter to the ideal.

However, the resulting filter circuit must be of higher order and thus more complex and expensive.

In addition to specifying the magnitude of transmission, there are applications in which the phase response of the filter is also of interest. The filter design problem, however, is considerably complicated when both magnitude and phase are specified.

Once the filter specifications have been decided upon, the next step in the design is to find a transfer function whose magnitude meets the specification. To meet specification, the magnitude response curve must lie in the unshaded area in Fig. 11.3. The curve shown in the figure is for a filter that *just* meets specifications. Observe that for this particular filter the magnitude response *ripples* throughout the passband with the ripple peaks being all equal. Since the peak ripple is equal to A_{max} it is usual to refer to A_{max} as the **passband ripple** and to ω_p as the **ripple bandwidth.** The particular filter response shown ripples also in the stopband, again with the ripple peaks all equal and of such a value that the minimum stopband attenuation achieved is equal to the specified value, A_{min}. Thus this particular response is said to be **equiripple** in both the passband and the stopband.

The process of obtaining a transfer function that meets given specifications is known as **filter approximation.** Filter approximation is usually performed using computer programs (Snelgrove, 1982) or filter design tables (Zverev, 1967). In simpler cases, filter approximation can be performed using closed-form expressions, as will be seen in Section 11.3.

Finally, Fig. 11.4 shows transmission specifications for a bandpass filter and the response of a filter that meets these specifications. For this example we have chosen an approximation function that does not ripple in the passband; rather, the transmission

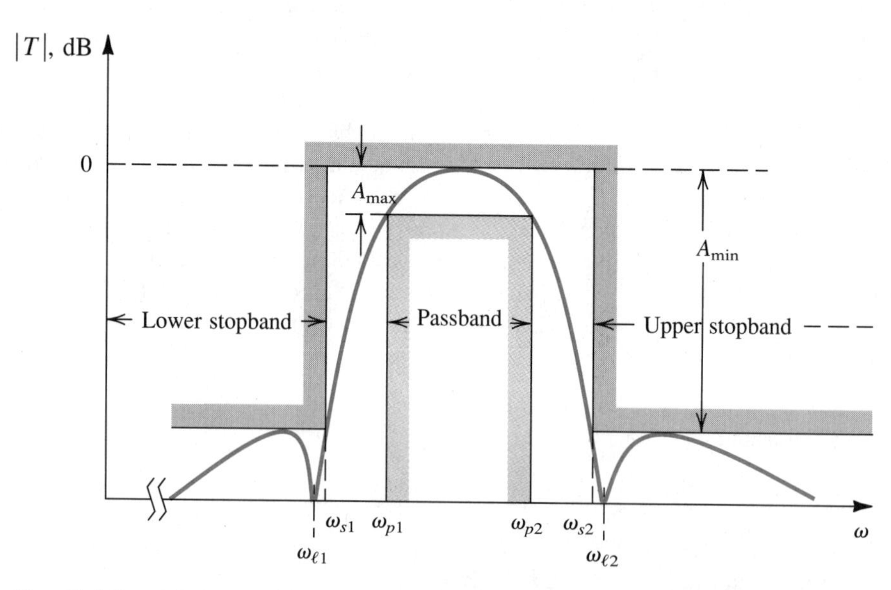

Fig. 11.4 Transmission specifications for a bandpass filter. The magnitude response of a filter that just meets specifications is also shown. Note that this particular filter has a monotonically decreasing transmission in the passband on both sides of the peak frequency.

decreases monotonically on both sides of the center frequency, attaining the maximum allowable deviation at the two edges of the passband.

Exercises 11.1 Find approximate values of attenuation corresponding to filter transmissions of: 1, 0.99, 0.9, 0.8, 0.7, 0.5, 0.1, 0.

Ans. 0, 0.1, 1, 2, 3, 6, 20, ∞ (dB).

11.2 If the magnitude of passband transmission is to remain constant to within $\pm 5\%$, and if the stopband transmission is to be no greater than 1% of the passband transmission, find A_{max} and A_{min}.

Ans. 0.9 dB; 40 dB.

11.2 THE FILTER TRANSFER FUNCTION

The filter transfer function $T(s)$ can be written as the ratio of two polynomials as

$$T(s) = \frac{a_M\, s^M + a_{M-1}\, s^{M-1} + \cdots + a_0}{s^N + b_{N-1}\, s^{N-1} + \cdots + b_0} \tag{11.6}$$

The degree of the denominator, N, is the **filter order.** For the filter circuit to be stable, the degree of the numerator must be less than or equal to that of the denominator; $M \leq N$. The numerator and denominator coefficients, $a_0, a_1, \ldots, a_M$ and $b_0, b_1, \ldots, b_{N-1}$, are real numbers. The polynomials in the numerator and denominator can be factored, and $T(s)$ can be expressed in the form

$$T(s) = \frac{a_M\,(s - z_1)(s - z_2)\cdots(s - z_M)}{(s - p_1)(s - p_2)\cdots(s - p_N)} \tag{11.7}$$

The numerator roots, $z_1, z_2, \ldots, z_M$, are the **transfer-function zeros,** or **transmission zeros;** and the denominator roots, $p_1, p_2, \ldots, p_N$, are the **transfer-function poles,** or the **natural modes**[1]**.** Each transmission zero or pole can be either a real or a complex number. Complex zeros and poles, however, must occur in conjugate pairs. Thus, if $-1 + j2$ happens to be a zero then $-1 - j2$ also must be a zero.

Since in the filter stopband the transmission is required to be zero or small, the filter transmission zeros are usually placed on the $j\omega$ axis at stopband frequencies. This indeed is the case for the filter whose transmission function is sketched in Fig. 11.3. This particular filter can be seen to have infinite attenuation (zero transmission) at two stopband frequencies: ω_{l1} and ω_{l2}. The filter then must have transmission zeros at $s = +j\omega_{l1}$ and $s = +j\omega_{l2}$. However, since complex zeros occur in conjugate pairs, there must also be transmission zeros at $s = -j\omega_{l1}$ and $s = -j\omega_{l2}$. Thus the numerator polynomial of this filter will have the factors $(s + j\omega_{l1})(s - j\omega_{l1})(s + j\omega_{l2})(s - j\omega_{l2})$, which can be written as $(s^2 + \omega_{l1}^2)(s^2 + \omega_{l2}^2)$. For $s = j\omega$ (physical frequencies) the numerator becomes $(-\omega^2 + \omega_{l1}^2)(-\omega^2 + \omega_{l2}^2)$, which indeed is zero at $\omega = \omega_{l1}$ and $\omega = \omega_{l2}$.

Continuing with the example in Fig. 11.3, we observe that the transmission decreases toward $-\infty$ as ω approaches ∞. Thus the filter must have one or more transmis-

[1] Throughout this chapter we use the names *poles* and *natural modes* interchangeably.

sion zeros at $s = \infty$. In general, the number of transmission zeros at $s = \infty$ is the difference between the degree of the numerator polynomial, M, and the degree of the denominator polynomial, N, of the transfer function in Eq. (11.6). This is because as s approaches ∞, $T(s)$ approaches a_M/s^{N-M} and thus is said to have $N - M$ zeros at $s = \infty$.

For a filter circuit to be stable all its poles must lie in the left half of the s-plane, and thus $p_1, p_2, \ldots, p_N$ must all have negative real parts. Figure 11.5 shows typical pole and zero locations for the low-pass filter whose transmission function is depicted in Fig. 11.3. We have assumed that this filter is of fifth order ($N = 5$). It has two pairs of complex conjugate poles and one real-axis pole, for a total of five poles. All the poles lie in the vicinity of the passband, which is what gives the filter its high transmission at passband frequencies. The five transmission zeros are at $s = \pm j\omega_{l1}$, $s = \pm j\omega_{l2}$, and $s = \infty$. Thus the transfer function for this filter is of the form

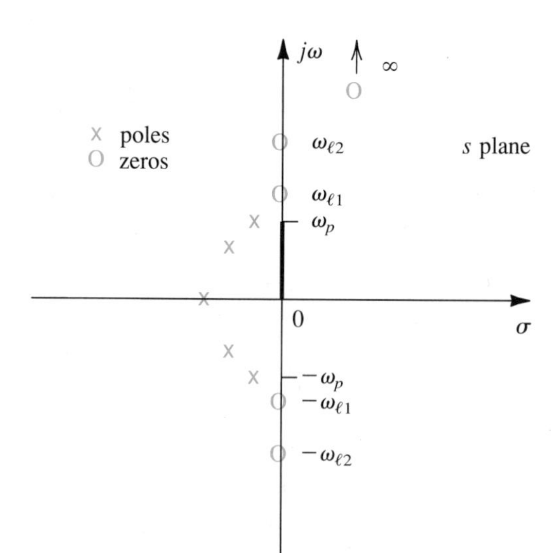

Fig. 11.5 Pole–zero pattern for the low-pass filter whose transmission is sketched in Fig. 11.3. This filter is of fifth order ($N = 5$).

$$T(s) = \frac{a_4\,(s^2 + \omega_{l1}^2)(s^2 + \omega_{l2}^2)}{s^5 + b_4\,s^4 + b_3\,s^3 + b_2\,s^2 + b_1\,s + b_0} \tag{11.8}$$

As another example, consider the bandpass filter whose magnitude response is shown in Fig. 11.4. This filter has transmission zeros at $s = \pm j\omega_{l1}$ and $s = \pm j\omega_{l2}$. It also has one or more zeros at $s = 0$ and one or more zeros at $s = \infty$ (because the attenuation decreases toward $-\infty$ as ω approaches 0 and ∞). Assuming that only one zero exists at each of $s = 0$ and $s = \infty$, the filter must be of sixth order, and its transfer function takes the form

$$T(s) = \frac{a_5\,s\,(s^2 + \omega_{l1}^2)(s^2 + \omega_{l2}^2)}{s^6 + b_5\,s^5 + \cdots + b_0} \tag{11.9}$$

A typical pole–zero plot for such a filter is shown in Fig. 11.6.

As a third and final example, consider the low-pass filter whose transmission function is depicted in Fig. 11.7(a). We observe that in this case there are no finite values of ω at

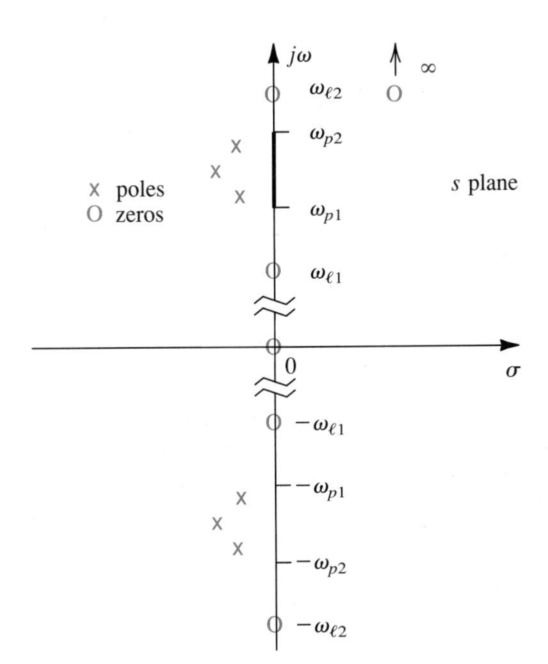

Fig. 11.6 Pole–zero pattern for the bandpass filter whose transmission function is shown in Fig. 11.4. This filter is of sixth order ($N = 6$).

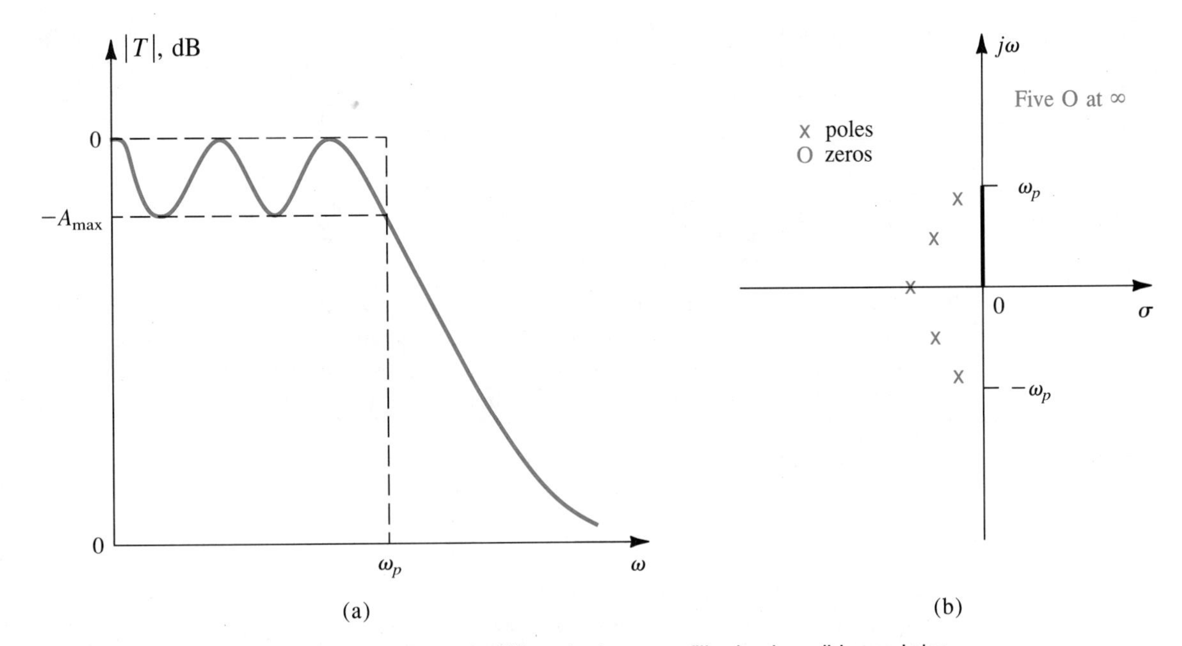

Fig. 11.7 (a) Transmission characteristics of a fifth-order low-pass filter having all transmission zeros at infinity. **(b)** Pole–zero pattern for the filter in (a).

which the attenuation is infinite (zero transmission). Thus it is possible that all the transmission zeros of this filter are at $s = \infty$. If this is the case, the filter transfer function takes the form

$$T(s) = \frac{a_M}{s^N + b_{N-1}\, s^{N-1} + \cdots + b_0} \tag{11.10}$$

Such a filter is known as an **all-pole filter.** Typical pole–zero locations for a fifth-order all-pole low-pass filter are shown in Fig. 11.7(b).

Almost all of the filters studied in this chapter have all of their transmission zeros on the $j\omega$-axis, in the filter stopband(s), including[2] $\omega = 0$ and $\omega = \infty$. Also, to obtain high selectivity all the natural modes will be complex conjugate (except for the case of odd-order filters where one natural mode must be on the real axis). Finally we note that the more selective the required filter response is, the higher its order must be, and the closer its natural modes are to the $j\omega$ axis.

Exercises 11.3 A second-order filter has its poles at $s = -(1/2) \pm j(\sqrt{3}/2)$. The transmission is zero at $\omega = 2$ rad/s and is unity at dc ($\omega = 0$). Find the transfer function.

Ans. $T(s) = \frac{1}{4}\frac{s^2 + 4}{s^2 + s + 1}$

11.4 A fourth-order filter has zero transmission at $\omega = 0$, $\omega = 2$ rad/s, and $\omega = \infty$. The natural modes are $-0.1 \pm j0.8$ and $-0.1 \pm j1.2$. Find $T(s)$.

Ans. $T(s) = \frac{a_3\, s(s^2 + 4)}{(s^2 + 0.2s + 0.65)(s^2 + 0.2s + 1.45)}$

11.5 Find the transfer function $T(s)$ of a third-order all-pole low-pass filter whose poles are at a radial distance of 1 rad/s from the origin and whose complex poles are at 30° angles from the $j\omega$-axis. The dc gain is unity. Show that $|T(j\omega)| = 1/\sqrt{1 + \omega^6}$. Find ω_{3dB} and the attenuation at $\omega = 3$ rad/s.

Ans. $T(s) = 1/(s + 1)(s^2 + s + 1)$; 1 rad/s; 28.6 dB.

11.3 BUTTERWORTH AND CHEBYSHEV FILTERS

In this section we present two functions that are frequently used in approximating the transmission characteristics of low-pass filters. These functions have the advantage that closed-form expressions are available for their parameters. Thus one can use them in filter design without the need for computers or filter design tables. Their utility, however, is limited to relatively simple applications.

Although in this section we discuss the design of low-pass filters only, the approximation functions presented can be applied to the design of other filter types through the use of frequency transformations (see Sedra and Brackett, 1978).

[2]Obviously a low-pass filter should *not* have a transmission zero at $\omega = 0$, and, similarly, a high-pass filter should not have a transmission zero at $\omega = \infty$.

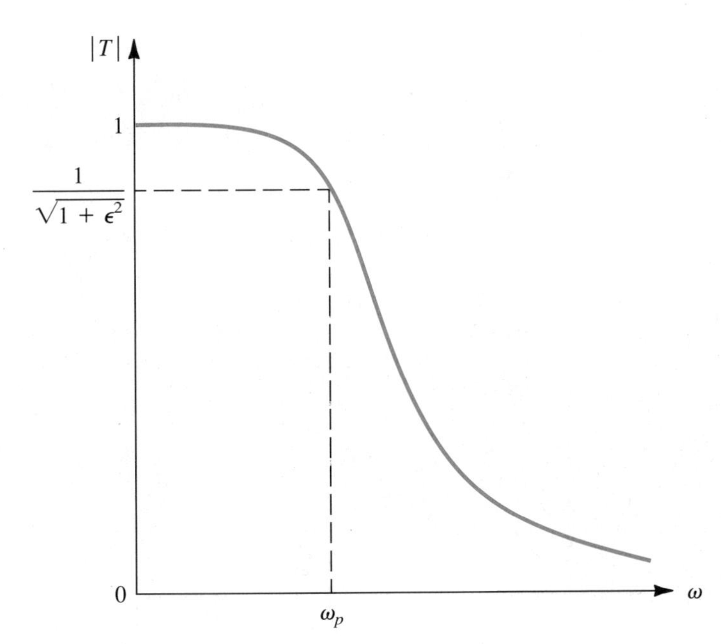

Fig. 11.8 The magnitude response of a Butterworth filter.

The Butterworth Filter

Figure 11.8 shows a sketch of the magnitude response of a Butterworth[3] filter. This filter exhibits a monotonically decreasing transmission with all the transmission zeros at $\omega = \infty$, making it an all-pole filter. The magnitude function for an Nth-order Butterworth filter with a passband edge ω_p is given by

$$|T(j\omega)| = \frac{1}{\sqrt{1 + \epsilon^2 \left(\frac{\omega}{\omega_p}\right)^{2N}}} \tag{11.11}$$

At $\omega = \omega_p$,

$$|T(j\omega_p)| = \frac{1}{\sqrt{1 + \epsilon^2}} \tag{11.12}$$

Thus the parameter ϵ determines the maximum variation in passband transmission, A_{max}, according to

$$A_{max} = 20 \log\sqrt{1 + \epsilon^2} \tag{11.13}$$

Conversely, given A_{max}, the value of ϵ can be determined from

$$\epsilon = \sqrt{10^{A_{max}/10} - 1} \tag{11.14}$$

---

[3] The Butterworth filter approximation is named after S. Butterworth, a British engineer who in 1930 was among the first to employ it.

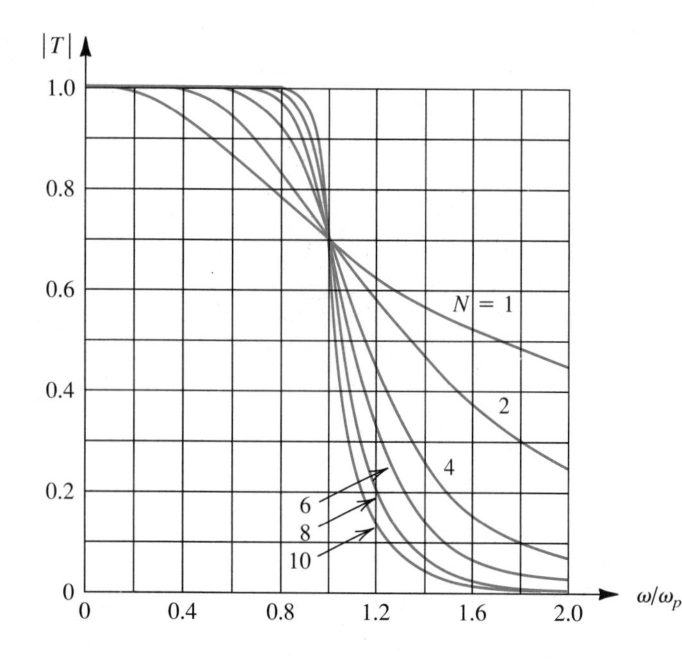

Fig. 11.9 Magnitude response for Butterworth filters of various order with $\epsilon = 1$. Note that as the order increases, the response approaches the ideal brick-wall type transmission.

Observe that in the Butterworth response the maximum deviation in passband transmission (from the ideal value of unity) occurs at the passband edge only. It can be shown that the first $2N - 1$ derivatives of $|T|$ relative to ω are zero at $\omega = 0$ (see Van Valkenburg, 1980). This property makes the Butterworth response very flat near $\omega = 0$ and gives the response the name **maximally flat** response. The degree of passband flatness increases as the order N is increased, as can be seen from Fig. 11.9. This figure indicates also that, as should be expected, as the order N is increased the filter response approaches the ideal brick-wall type response.

At the edge of the stopband, $\omega = \omega_s$, the attenuation of the Butterworth filter is given by

$$\begin{aligned} A(\omega_s) &= -20 \log 1/\sqrt{1 + \epsilon^2 (\omega_s/\omega_p)^{2N}} \\ &= 10 \log[1 + \epsilon^2 (\omega_s/\omega_p)^{2N}] \end{aligned} \tag{11.15}$$

This equation can be used to determine the filter order required, which is the lowest integer value of N that yields $A(\omega_s) \geq A_{\min}$.

The natural modes of an Nth-order Butterworth filter can be determined from the graphical construction shown in Fig. 11.10(a). Observe that the natural modes lie on a circle of radius $\omega_p (1/\epsilon)^{1/N}$ and are spaced by equal angles of π/N, with the first mode at an angle $\pi/2N$ from the $+j\omega$-axis. Since the natural modes all have equal radial distance from the origin they all have the same frequency $\omega_0 = \omega_p (1/\epsilon)^{1/N}$. Figs. 11.10(b), (c), and (d) show the natural modes of Butterworth filters of order $N = 2$, 3, and 4, respectively. Once the N natural modes $p_1, p_2, \ldots, p_N$ are found the transfer function can be

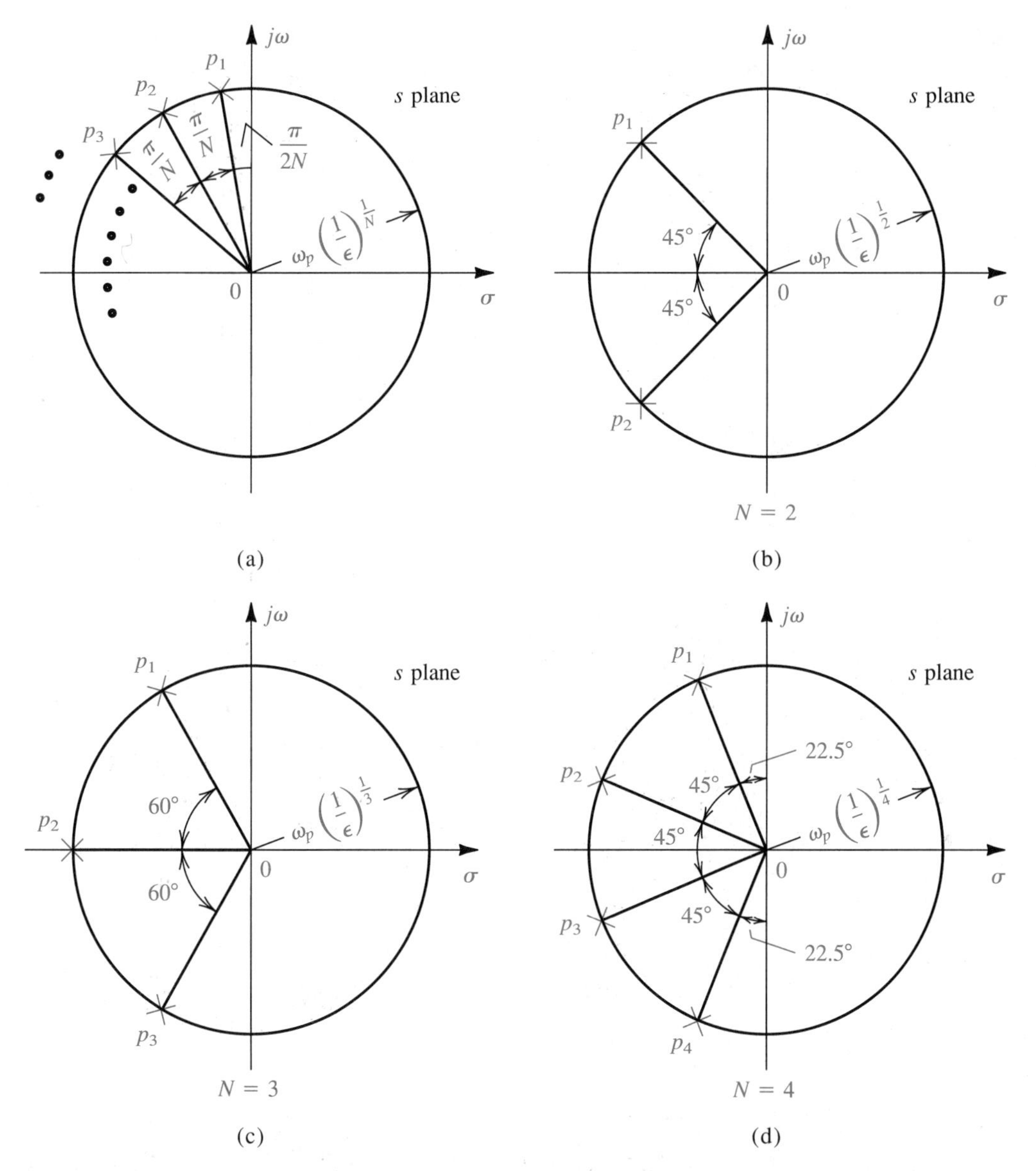

Fig. 11.10 Graphical construction for determining the poles of a Butterworth filter of order N. All the poles lie in the left half of the s-plane on a circle of radius $\omega_0 = \omega_p(1/\epsilon)^{1/N}$, where ϵ is the passband deviation parameter ($\epsilon = \sqrt{10^{A_{max}/10} - 1}$): **(a)** the general case, **(b)** $N = 2$, **(c)** $N = 3$, **(d)** $N = 4$.

written as

$$T(s) = \frac{K\,\omega_0^N}{(s - p_1)(s - p_2)\cdots(s - p_N)} \tag{11.16}$$

where K is a constant equal to the required dc gain of the filter.

To summarize, to find a Butterworth transfer function that meets transmission specifications of the form in Fig. 11.3 we perform the following procedure:

1. Determine ϵ from Eq. (11.14).
2. Use Eq. (11.15) to determine the required filter order as the lowest integer value of N that results in $A(\omega_s) \geq A_{min}$.
3. Use Fig. 11.10(a) to determine the N natural modes.
4. Use Eq. (11.16) to determine $T(s)$.

EXAMPLE 11.1 Find the Butterworth transfer function that meets the following low-pass filter specifications: $f_p = 10$ kHz, $A_{max} = 1$ dB, $f_s = 15$ kHz, $A_{min} = 25$ dB, dc gain = 1.

Solution Substituting $A_{max} = 1$ dB into Eq. (11.14) yields $\epsilon = 0.5088$. Equation (11.15) is then used to determine the filter order by trying various values for N. We find that $N = 8$ yields $A(\omega_s) = 22.3$ dB and $N = 9$ gives 25.8 dB. We thus select $N = 9$.

Figure 11.11 shows the graphical construction for determining the poles. The poles all have the same frequency $\omega_0 = \omega_p\,(1/\epsilon)^{1/N} = 2\pi \times 10 \times 10^3\,(1/0.5088)^{1/9} = 6.773 \times 10^4$ rad/s. The first pole p_1 is given by

$$p_1 = \omega_0\,(-\cos 80^\circ + j\sin 80^\circ) = \omega_0\,(-0.1736 + j0.9848)$$

Combining p_1 with its complex conjugate p_9 yields the factor $(s^2 + s\;0.3472\;\omega_0 + \omega_0^2)$ in the denominator of the transfer function. The same can be done for the other complex poles, and the complete transfer function is obtained using Eq. (11.16),

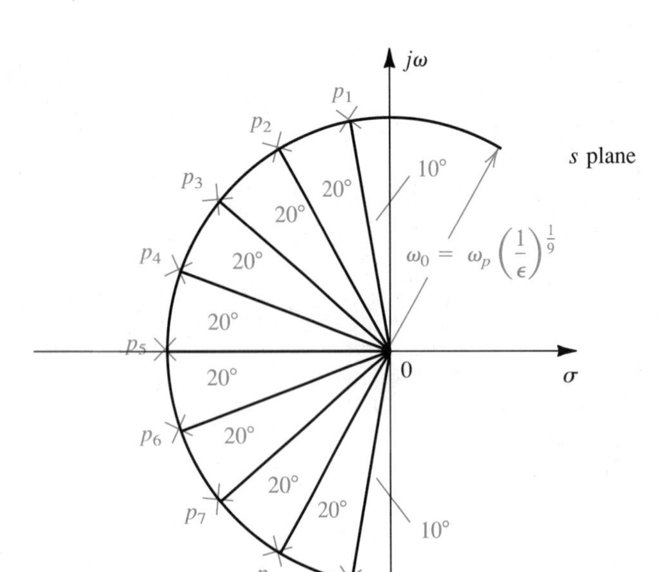

Fig. 11.11 Poles of the ninth-order Butterworth filter of Example 11.1.

$$T(s) = \frac{\omega_0^9}{(s + \omega_0)(s^2 + s\ 1.8794\ \omega_0 + \omega_0^2)(s^2 + s\ 1.5321\ \omega_0 + \omega_0^2)} \times \frac{1}{(s^2 + s\ \omega_0 + \omega_0^2)(s^2 + s\ 0.3472\ \omega_0 + \omega_0^2)} \quad (11.17)$$

The Chebyshev Filter

Figure 11.12 shows representative transmission functions for Chebyshev[4] filters of even and odd order. The Chebyshev filter exhibits an equiripple response in the passband and a monotonically decreasing transmission in the stopband. While the odd-order filter has $|T(0)| = 1$, the even-order filter exhibits its maximum magnitude deviation at $\omega = 0$. In both cases the total number of passband maxima and minima equals the order of the filter, N. All the transmission zeros of the Chebyshev filter are at $\omega = \infty$, making it an all-pole filter.

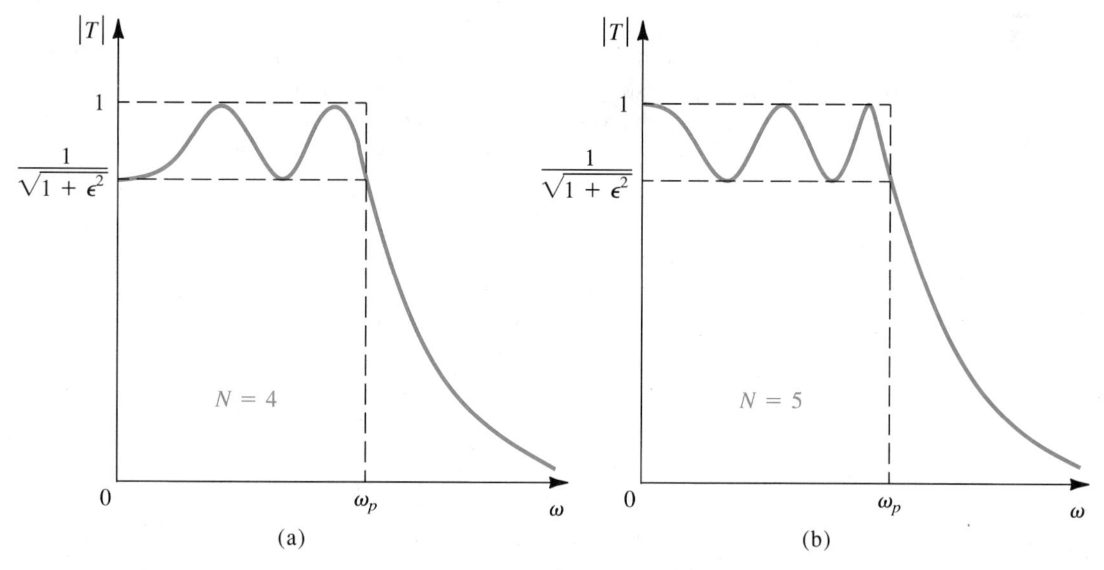

Fig. 11.12 Sketches of the transmission characteristics of representative even- and odd-order Chebyshev filters.

The magnitude of the transfer function of an Nth-order Chebyshev filter with a passband edge (ripple bandwidth) ω_p is given by

$$|T(j\omega)| = \frac{1}{\sqrt{1 + \epsilon^2 \cos^2[N \cos^{-1}(\omega/\omega_p)]}} \qquad \text{for } \omega \leq \omega_p \quad (11.18)$$

and

[4]Named after the Russian mathematician P. L. Chebyshev, who in 1899 used these functions in studying the construction of steam engines.

$$|T(j\omega)| = \frac{1}{\sqrt{1 + \epsilon^2 \cosh^2[N \cosh^{-1}(\omega/\omega_p)]}} \qquad \text{for } \omega \geq \omega_p \tag{11.19}$$

At the passband edge, $\omega = \omega_p$, the magnitude function is given by

$$|T(j\omega_p)| = \frac{1}{\sqrt{1 + \epsilon^2}}$$

Thus, the parameter ϵ determines the passband ripple according to

$$A_{\max} = 10 \log(1 + \epsilon^2) \tag{11.20}$$

Conversely, given $A_{\max}$, the value of ϵ is determined from

$$\epsilon = \sqrt{10^{A_{\max}/10} - 1} \tag{11.21}$$

The attenuation achieved by the Chebyshev filter at the stopband edge ($\omega = \omega_s$) is found using Eq. (11.19) as

$$A(\omega_s) = 10 \log[1 + \epsilon^2 \cosh^2(N \cosh^{-1}\omega_s/\omega_p)] \tag{11.22}$$

With the aid of a calculator this equation can be used to determine the order N required to obtain a specified $A_{\min}$ by finding the lowest integer value of N that yields $A(\omega_s) \geq A_{\min}$. As in the case of the Butterworth filter, increasing the order N of the Chebyshev filter causes its magnitude function to approach the ideal brick-wall low-pass response.

The poles of the Chebyshev filter are given by

$$\begin{aligned} p_k = &-\omega_p \sin\left(\frac{2k-1}{N}\frac{\pi}{2}\right) \sinh\left(\frac{1}{N}\sinh^{-1}\frac{1}{\epsilon}\right) \\ &+ j\omega_p \cos\left(\frac{2k-1}{N}\frac{\pi}{2}\right) \cosh\left(\frac{1}{N}\sinh^{-1}\frac{1}{\epsilon}\right) \qquad k = 1, 2, \ldots, N \end{aligned} \tag{11.23}$$

Finally, the transfer function of the Chebyshev filter can be written as

$$T(s) = \frac{K\,\omega_p^N}{\epsilon\, 2^{N-1}\,(s - p_1)(s - p_2)\cdots(s - p_N)} \tag{11.24}$$

where K is the dc gain that the filter is required to have.

To summarize, given low-pass transmission specifications of the type shown in Fig. 11.3, the transfer function of a Chebyshev filter that meets these specifications can be found as follows:

1. Determine ϵ from Eq. (11.21).
2. Use Eq. (11.22) to determine the order required.
3. Determine the poles using Eq. (11.23).
4. Determine the transfer function using Eq. (11.24).

The Chebyshev filter provides a more efficient approximation than the Butterworth filter. Thus, for the same order and the same $A_{\max}$, the Chebyshev filter provides greater stopband attenuation than the Butterworth filter. Alternatively, to meet identical specifications, one requires a lower order for the Chebyshev than for the Butterworth. This point will be illustrated by the following example.

EXAMPLE 11.2 Find the Chebyshev transfer function that meets the same low-pass filter specifications given in Example 11.1; namely, $f_p = 10$ kHz, $A_{max} = 1$ dB, $f_s = 15$ kHz, $A_{min} = 25$ dB, dc gain = 1.

Solution Substituting $A_{max} = 1$ dB into Eq. (11.21) yields $\epsilon = 0.5088$. By trying various values for N in Eq. (11.22) we find that $N = 4$ yields $A(\omega_s) = 21.6$ dB and $N = 5$ provides 29.9 dB. We thus select $N = 5$ and observe that we required a ninth-order Butterworth filter to meet the same specifications in Example 11.1.

The poles are obtained by substituting in Eq. (11.23) as

$$p_1, p_5 = \omega_p\,(-0.0895 \pm j0.9901)$$

$$p_2, p_4 = \omega_p\,(-0.2342 \pm j0.6119)$$

$$p_5 = \omega_p\,(-0.2895)$$

The transfer function is obtained by substituting these values in Eq. (11.24) as

$$T(s) = \frac{\omega_p^5}{8.1408\,(s + 0.2895\,\omega_p)(s^2 + s\,0.4684\,\omega_p + 0.4293\,\omega_p^2)} \times \frac{1}{s^2 + s\,0.1789\,\omega_p + 0.9883\,\omega_p^2} \qquad (11.25)$$

where $\omega_p = 2\pi \times 10^4$ rad/s.

Exercises D11.6 Determine the order N of a Butterworth filter for which $A_{max} = 1$ dB, $\omega_s/\omega_p = 1.5$, and $A_{min} = 30$ dB. What is the actual value of minimum stopband attenuation realized? If A_{min} is to be exactly 30 dB to what value can A_{max} be reduced?

Ans. $N = 11$; $A_{min} = 32.87$ dB; 0.54 dB

11.7 Find the natural modes and the transfer function of a Butterworth filter with $\omega_p = 1$ rad/s, $A_{max} = 3$ dB ($\epsilon \simeq 1$), and $N = 3$.

Ans. $-0.5 \pm j\sqrt{3}/2$ and -1; $T(s) = 1/(s + 1)(s^2 + s + 1)$

11.8 Observe that Eq. (11.18) can be used to find the passband frequencies at which $|T|$ is at its peaks and at its valleys. (The peaks are reached when the $\cos^2[\;\;]$ term is zero, and the valleys correspond to the $\cos^2[\;\;]$ term equal to unity.) Find these frequencies for a fifth-order filter.

Ans. Peaks at $\omega = 0$, $0.59\,\omega_p$, and $0.95\,\omega_p$; the minima at $\omega = 0.31\,\omega_p$ and $0.81\,\omega_p$

D11.9 Find the attenuation provided at $\omega = 2\omega_p$ by a seventh-order Chebyshev filter with a 0.5-dB passband ripple. If the passband ripple is allowed to increase to 1 dB, by how much does the stopband attenuation increase?

Ans. 64.9 dB; 3.5 dB

D11.10 It is required to design a low-pass filter having $f_p = 1$ kHz, $A_{max} = 1$ dB, $f_s = 1.5$ kHz, $A_{min} = 50$ dB. (a) Find the required order of Chebyshev filter. What is the excess stopband attenuation obtained? (b) Repeat for a Butterworth filter.

Ans. (a) $N = 8$, 5 dB. (b) $N = 16$, 0.5 dB

11.4 FIRST-ORDER AND SECOND-ORDER FILTER FUNCTIONS

In this section we shall study the simplest filter transfer functions, those of first and second order. These functions are useful in their own right in the design of simple filters. First- and second-order filters can also be cascaded to realize a high-order filter. Cascade design is in fact one of the most popular methods for the design of active filters (those utilizing op amps and RC circuits). Because the filter poles occur in complex-conjugate pairs, a high-order transfer function $T(s)$ is factored into the product of second-order functions. If $T(s)$ is odd, there will also be a first-order function in the factorization. Each of the second-order functions (and the first-order function in the case $T(s)$ odd) is then realized using one of the op amp–RC circuits that will be studied in this chapter, and the resulting blocks are placed in cascade. If the output of each block is taken at the output terminal of an op amp where the impedance level is low (ideally zero), cascading does not change the transfer functions of the individual blocks. Thus the overall transfer function of the cascade is simply the product of the transfer functions of the individual blocks, which is the original $T(s)$.

First-Order Filters

The general first-order transfer function is given by

$$T(s) = \frac{a_1 s + a_0}{s + \omega_0} \tag{11.26}$$

This **bilinear transfer function** characterizes a first-order filter with a natural mode at $s = -\omega_0$, a transmission zero at $s = -a_0/a_1$, and a high-frequency gain that approaches a_1. The numerator coefficients, a_0 and a_1, determine the type of filter (e.g., low-pass, high-pass, etc.). Some special cases together with passive (RC) and active (op amp–RC) realizations are shown in Fig. 11.13. Note that the active realizations provide considerably more versatility over their passive counterparts; in many cases the gain can be set to a desired value and some transfer-function parameters can be adjusted without affecting others. The output impedance of the active circuit is also very low, making cascading possible. The op amp, however, limits the high-frequency operation of the active circuits.

Another important special case of the first-order filter function is the all-pass filter shown in Fig. 11.14. Here the transmission zero and the natural mode are symmetrically located relative to the $j\omega$-axis. (They are said to display mirror-image symmetry with respect to the $j\omega$-axis.) Observe that although the transmission of the all-pass filter is (ideally) constant at all frequencies, its phase shows frequency selectivity. All-pass filters are used as phase shifters and in systems that require phase shaping (e.g., in the design of circuits called *delay equalizers,* which cause the overall time delay of a transmission system to be constant with frequency).

Exercises D11.11 Using $R_1 = 10\ \text{k}\Omega$, design the op amp–RC circuit of Fig. 11.13(b) to realize a high-pass filter with a corner frequency of 10^4 rad/s and a high-frequency gain of 10.

Ans. $R_2 = 100\ \text{k}\Omega$ and $C = 0.01\ \mu\text{F}$

D11.12 Design the op amp–RC circuit of Fig. 11.14 to realize an all-pass filter with a 90° phase shift at 10^3 rad/s. Select suitable component values.

Ans. Possible choice: $R = R_1 = R_2 = 10\ \text{k}\Omega$, $C = 0.1\ \mu\text{F}$.

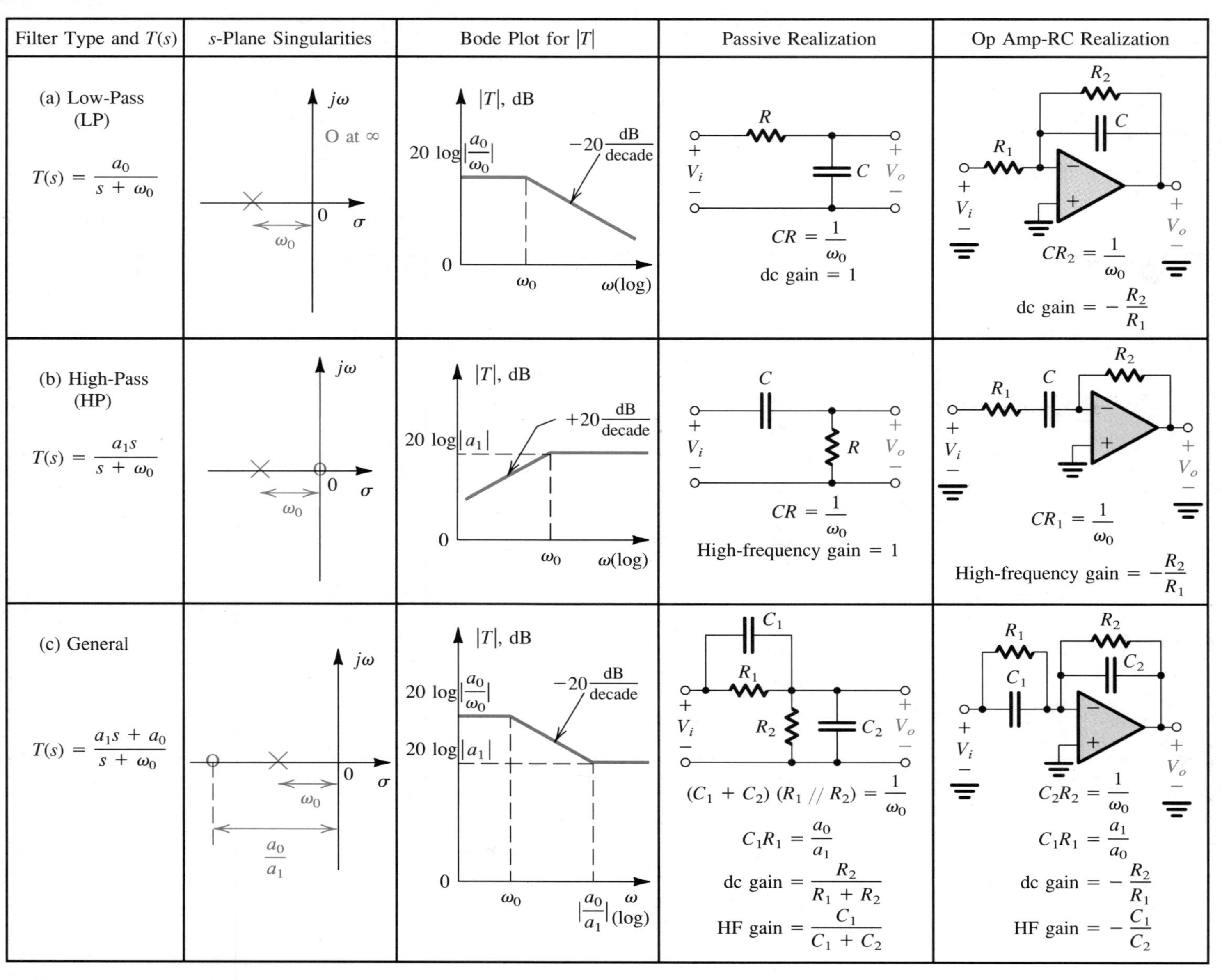

Fig. 11.13 First-order filters.

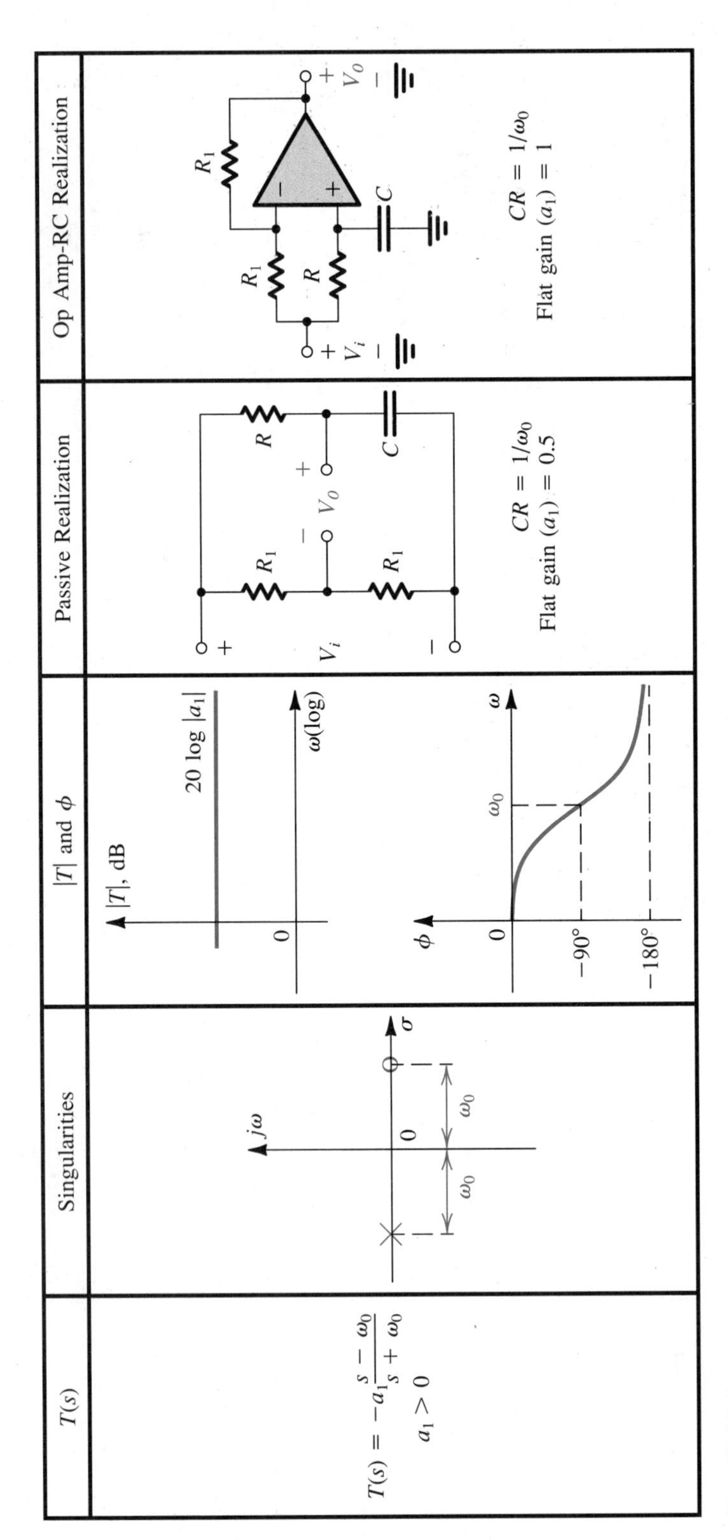

Fig. 11.14 First-order all-pass filter.

Second-Order Filter Functions

The general second-order or **biquadratic** filter transfer function is usually expressed in the standard form

$$T(s) = \frac{a_2 s^2 + a_1 s + a_0}{s^2 + (\omega_0/Q)\, s + \omega_0^2} \tag{11.27}$$

where ω_0 and Q determine the natural modes (poles) according to

$$p_1, p_2 = -\frac{\omega_0}{2Q} \pm j\omega_0 \sqrt{1 - (1/4Q^2)} \tag{11.28}$$

We are usually interested in the case of complex-conjugate natural modes, obtained for $Q > 0.5$. Figure 11.15 shows the location of the pair of complex-conjugate poles in the s-plane. Observe that the radial distance of the natural modes (from the origin) is equal to ω_0, which is known as the **pole frequency.** The parameter Q determines the distance of the poles from the $j\omega$-axis; the higher the value of Q, the closer the poles are to the $j\omega$-axis and the more selective the filter response becomes. An infinite value for Q locates the poles on the $j\omega$-axis and can yield sustained oscillations in the circuit realization. A negative value of Q implies that the poles are in the right half of the s-plane, which certainly produces oscillations. The parameter Q is called the **pole quality factor,** or simply, **pole-Q.**

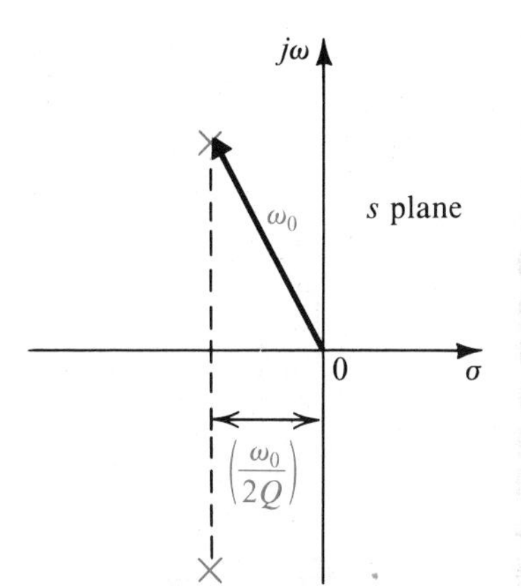

Fig. 11.15 Definition of the parameters ω_0 and Q of a pair of complex conjugate poles.

The transmission zeros of the second-order filter are determined by the numerator coefficients, a_0, a_1, and a_2. It follows that the numerator coefficients determine the type of second-order filter function (i.e., LP, HP, etc.). Seven special cases of interest are illustrated in Fig. 11.16. For each case we give the transfer function, the s-plane locations of the transfer function singularities, and the magnitude response. Circuit realizations for the various second-order filter functions will be given in subsequent sections.

All seven special second-order filters have a pair of complex conjugate natural modes characterized by a frequency ω_0 and a Q factor, Q.

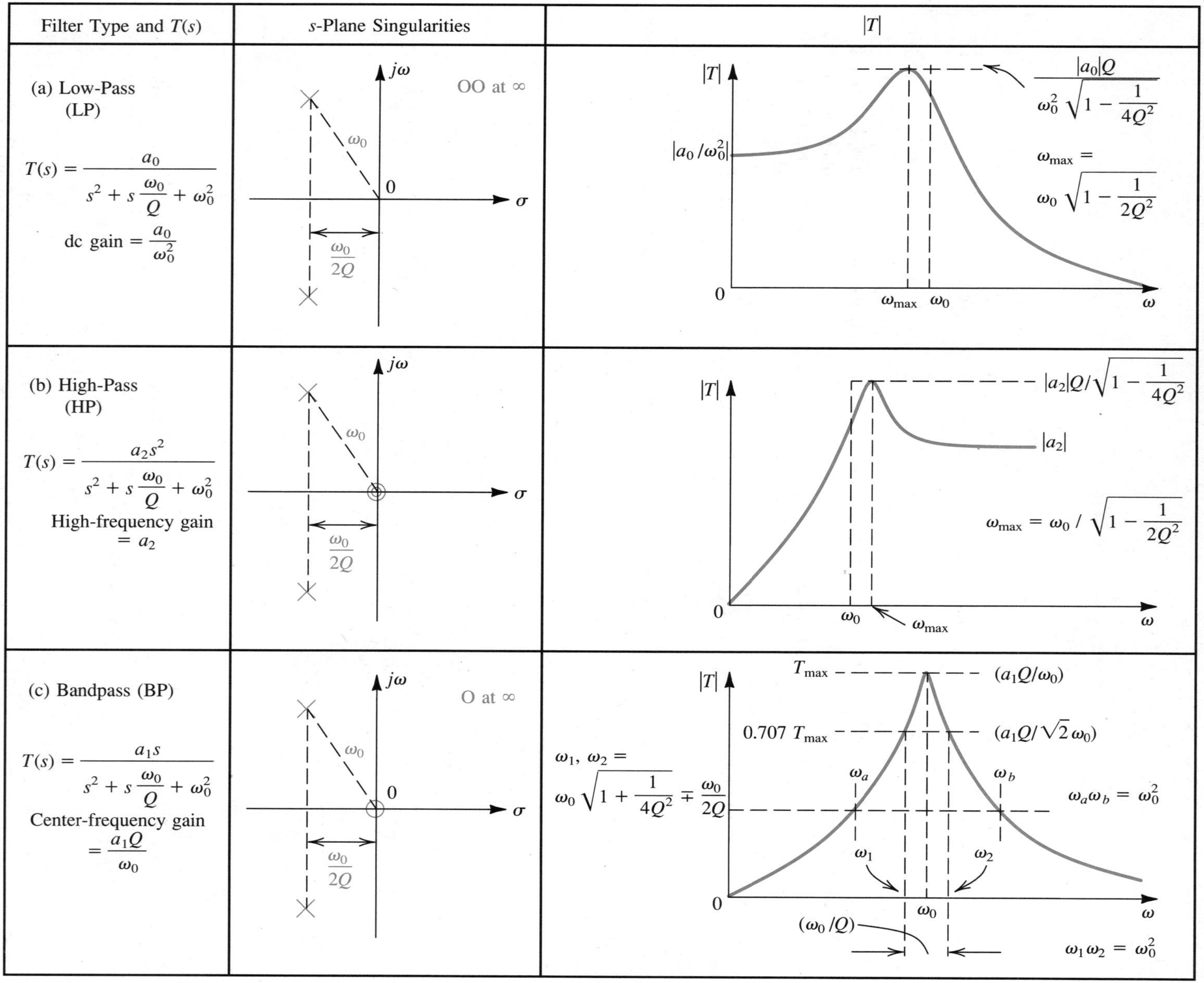

Filter Type and $T(s)$	s-Plane Singularities	$\lvert T\rvert$
(a) Low-Pass (LP) $T(s) = \dfrac{a_0}{s^2 + s\dfrac{\omega_0}{Q} + \omega_0^2}$ dc gain $= \dfrac{a_0}{\omega_0^2}$	$j\omega$, σ, 0, ω_0, $\dfrac{\omega_0}{2Q}$, OO at ∞	$\lvert T\rvert$, $\lvert a_0/\omega_0^2\rvert$, 0, ω_{max}, ω_0, ω; $\dfrac{\lvert a_0\rvert Q}{\omega_0^2\sqrt{1-\dfrac{1}{4Q^2}}}$; $\omega_{max} = \omega_0\sqrt{1-\dfrac{1}{2Q^2}}$
(b) High-Pass (HP) $T(s) = \dfrac{a_2 s^2}{s^2 + s\dfrac{\omega_0}{Q} + \omega_0^2}$ High-frequency gain $= a_2$	$j\omega$, σ, ω_0, $\dfrac{\omega_0}{2Q}$	$\lvert T\rvert$, 0, ω_0, ω_{max}, ω; $\lvert a_2\rvert Q/\sqrt{1-\dfrac{1}{4Q^2}}$; $\lvert a_2\rvert$; $\omega_{max} = \omega_0 / \sqrt{1-\dfrac{1}{2Q^2}}$
(c) Bandpass (BP) $T(s) = \dfrac{a_1 s}{s^2 + s\dfrac{\omega_0}{Q} + \omega_0^2}$ Center-frequency gain $= \dfrac{a_1 Q}{\omega_0}$	$j\omega$, σ, 0, ω_0, $\dfrac{\omega_0}{2Q}$, O at ∞	$\lvert T\rvert$, 0, ω; T_{max} — $(a_1 Q/\omega_0)$; $0.707\ T_{max}$ — $(a_1 Q/\sqrt{2}\,\omega_0)$; $\omega_1, \omega_2 = \omega_0\sqrt{1+\dfrac{1}{4Q^2}} \mp \dfrac{\omega_0}{2Q}$; ω_a, ω_b, $\omega_a\omega_b = \omega_0^2$; ω_1, ω_0, ω_2; (ω_0/Q); $\omega_1\omega_2 = \omega_0^2$

Fig. 11.16 Second-order filtering functions.

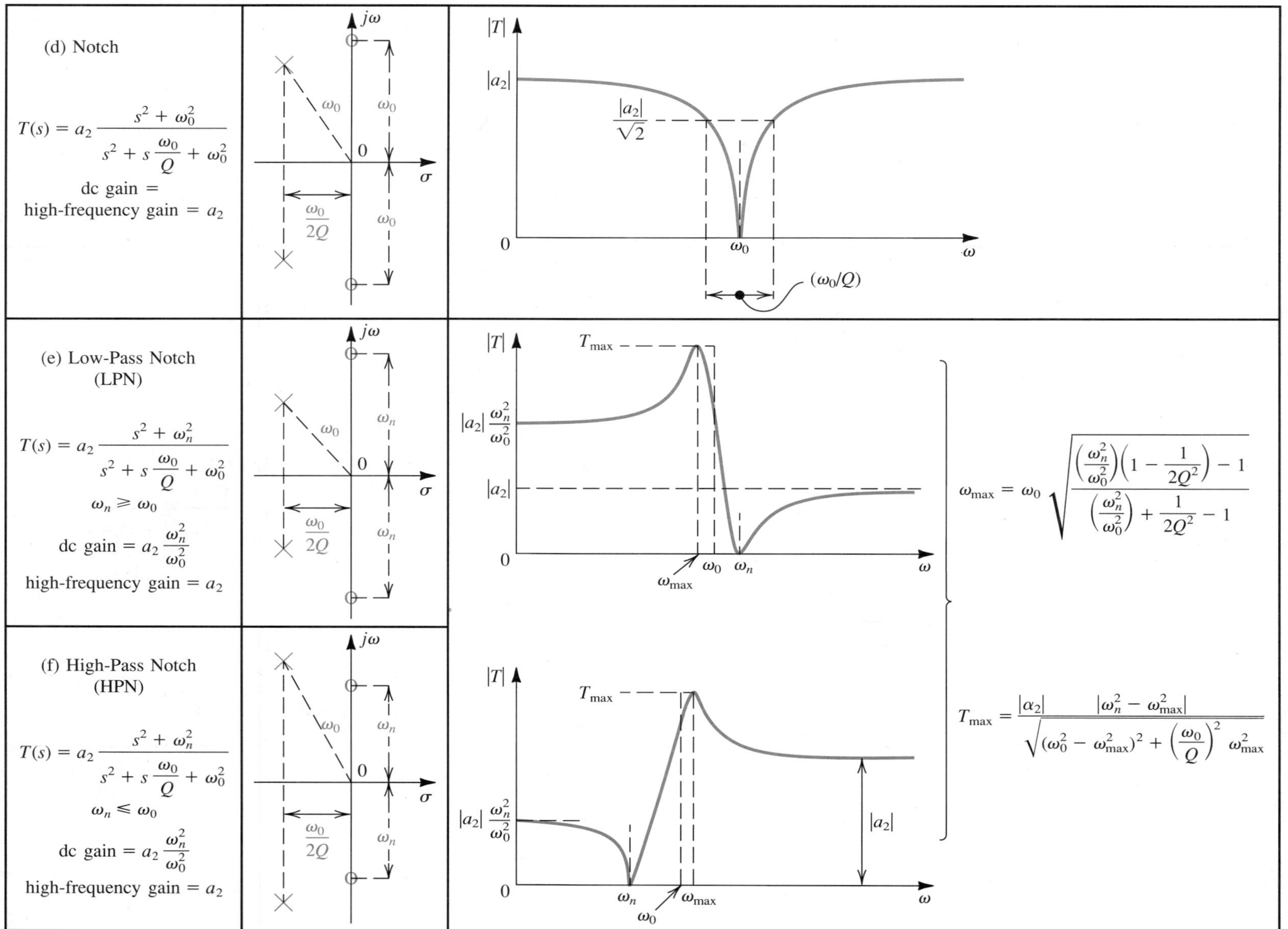

(d) Notch
$T(s) = a_2 \frac{s^2 + \omega_0^2}{s^2 + s\frac{\omega_0}{Q} + \omega_0^2}$
dc gain = high-frequency gain = a_2
(e) Low-Pass Notch (LPN)
$T(s) = a_2 \frac{s^2 + \omega_n^2}{s^2 + s\frac{\omega_0}{Q} + \omega_0^2}$
$\omega_n \geq \omega_0$
dc gain = $a_2 \frac{\omega_n^2}{\omega_0^2}$
high-frequency gain = a_2
(f) High-Pass Notch (HPN)
$T(s) = a_2 \frac{s^2 + \omega_n^2}{s^2 + s\frac{\omega_0}{Q} + \omega_0^2}$
$\omega_n \leq \omega_0$
dc gain = $a_2 \frac{\omega_n^2}{\omega_0^2}$
high-frequency gain = a_2
$j\omega$
σ
0
ω_0
ω_n
$\frac{\omega_0}{2Q}$
$|T|$
$|a_2|$
$\frac{|a_2|}{\sqrt{2}}$
(ω_0/Q)
ω
T_{max}
$|a_2| \frac{\omega_n^2}{\omega_0^2}$
ω_{max}
$\omega_{max} = \omega_0 \sqrt{\frac{\left(\frac{\omega_n^2}{\omega_0^2}\right)\left(1 - \frac{1}{2Q^2}\right) - 1}{\left(\frac{\omega_n^2}{\omega_0^2}\right) + \frac{1}{2Q^2} - 1}}$
$T_{max} = \frac{|\alpha_2| \quad |\omega_n^2 - \omega_{max}^2|}{\sqrt{(\omega_0^2 - \omega_{max}^2)^2 + \left(\frac{\omega_0}{Q}\right)^2 \omega_{max}^2}}$

Fig. 11.16 (continued)

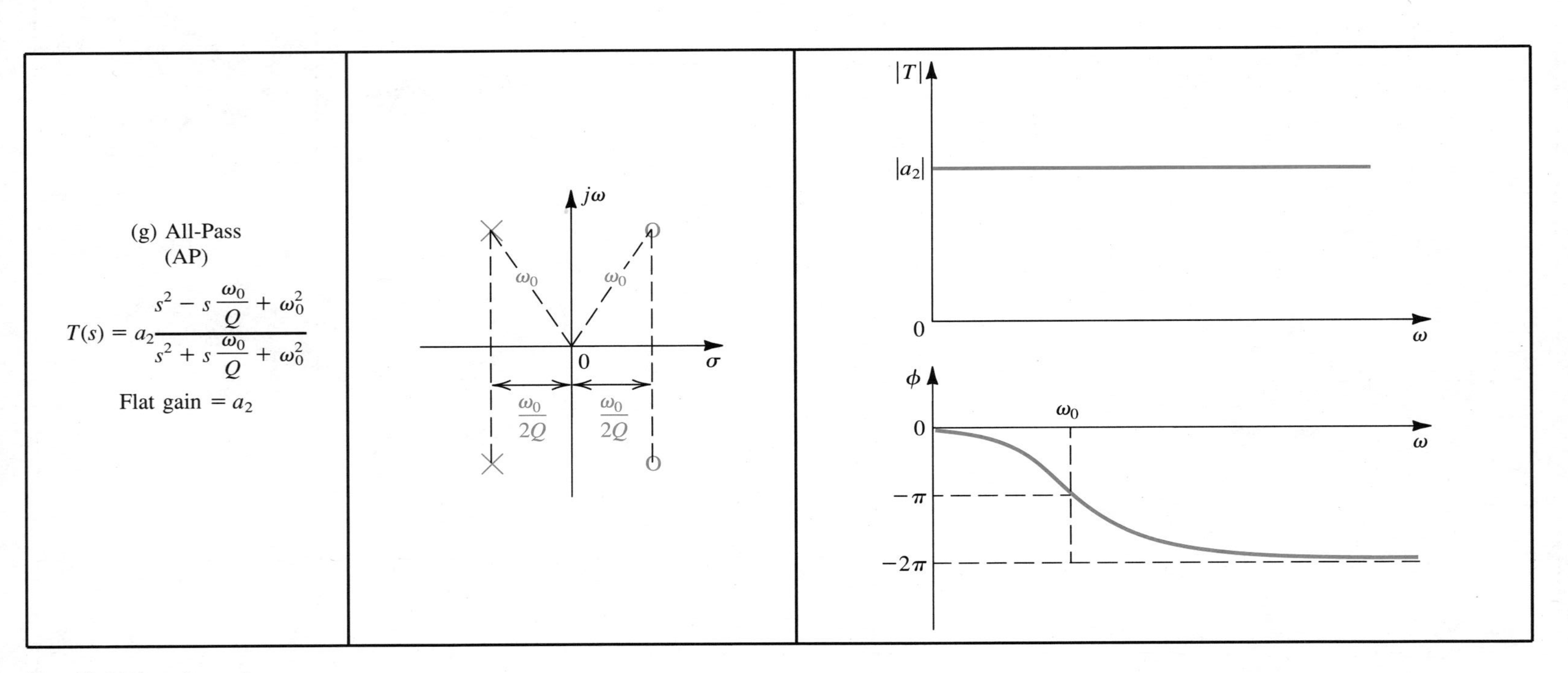

Fig. 11.16 (continued)

In the low-pass (LP) case, shown in Fig. 11.16(a), the two transmission zeros are at $s = \infty$. The magnitude response can exhibit a peak with the details indicated. It can be shown that the peak occurs only for $Q > 1/\sqrt{2}$. The response obtained for $Q = 1/\sqrt{2}$ is the Butterworth, or maximally flat, response.

The high-pass (HP) function shown in Fig. 11.16(b) has both of its transmission zeros at $s = 0$ (dc). The magnitude response shows a peak for $Q > 1/\sqrt{2}$, with the details of the response as indicated. Observe the duality between the LP and HP responses.

Next consider the bandpass filter function shown in Fig. 11.16(c). Here one transmission zero is at $s = 0$ (dc), and the other is at $s = \infty$. The magnitude response peaks at $\omega = \omega_0$. Thus the **center frequency** of the bandpass filter is equal to the pole frequency ω_0. The selectivity of the second-order bandpass filter is usually measured by its 3-dB *bandwidth*. This is the difference between the two frequencies ω_1 and ω_2 at which the magnitude response is 3 dB below its maximum value (at ω_0). It can be shown that

$$\omega_1, \omega_2 = \omega_0\sqrt{1 + (1/4Q^2)} \pm \frac{\omega_0}{2Q} \tag{11.29}$$

Thus

$$BW \equiv \omega_2 - \omega_1 = \omega_0/Q \tag{11.30}$$

Observe that as Q increases, the bandwidth decreases and the bandpass filter becomes more selective.

If the transmission zeros are located on the $j\omega$-axis, at the complex conjugate locations $\pm j\omega_n$, then the magnitude response exhibits zero transmission at $\omega = \omega_n$. Thus a **notch** in the magnitude response occurs at $\omega = \omega_n$, and ω_n is known as the **notch frequency.** Three cases of the second-order notch filter are possible: the regular notch, obtained when $\omega_n = \omega_0$ (Fig. 11.16d); the low-pass notch, obtained when $\omega_n > \omega_0$ (Fig. 11.16e); and the high-pass notch, obtained when $\omega_n < \omega_0$ (Fig. 11.16f). The reader is urged to verify the response details given in these figures (a rather tedious task, though!). Observe that in all notch cases, the transmission at dc and at $s = \infty$ is finite. This is so because there are no transmission zeros at either $s = 0$ or $s = \infty$.

The last special case of interest is the all-pass filter whose characteristics are illustrated in Fig. 11.16(g). Here the two transmission zeros are in the right half of the s-plane, at the mirror-image locations of the poles. (This is the case for all-pass functions of any order.) The magnitude response of the all-pass function is constant over all frequencies; the **flat gain,** as it is called, is in our case equal to $|a_2|$. The frequency selectivity of the all-pass function is in its phase response.

Exercises 11.13 For a maximally flat second-order low-pass filter ($Q = 1/\sqrt{2}$) show that at $\omega = \omega_0$ the magnitude response is 3 dB below the value at dc.

11.14 Give the transfer function of a second-order bandpass filter with a center frequency of 10^5 rad/s, a center-frequency gain of 10, and a 3-dB bandwidth of 10^3 rad/s.

Ans. $T(s) = \dfrac{10^4 s}{s^2 + 10^3 s + 10^{10}}$

11.15 (a) For the second-order notch function with $\omega_n = \omega_0$ show that in order for the attenuation to be greater than A dB over a frequency band BW_a, the value of Q is given by

$$Q = \frac{\omega_0}{BW_a \sqrt{10^{A/10} - 1}}$$

(b) Use the result of (a) to show that the 3-dB bandwidth is ω_0/Q, as indicated in Fig. 11.16(d).

11.16 Consider a low-pass notch with $\omega_0 = 1$ rad/s, $Q = 10$, $\omega_n = 1.2$ rad/s, and a dc gain of unity. Find the frequency and magnitude of the transmission peak. Also find the high-frequency transmission.

Ans. 0.973 rad/s; 3; 0.7

11.5 THE SECOND-ORDER LCR RESONATOR

In this section we shall study the second-order LCR resonator shown in Fig. 11.17(a). The use of this resonator to derive circuit realizations for the various second-order filter functions will be demonstrated. Also, it will be shown in the next section that replacing the inductor L by a simulated inductance obtained using an op amp–RC circuit results in an op amp–RC resonator. The latter forms the basis of an important class of active-RC filters to be studied in the next section.

The Resonator Natural Modes

The natural modes of the parallel resonance circuit of Fig. 11.17(a) can be determined by applying an excitation that does not change the natural structure of the circuit. Two

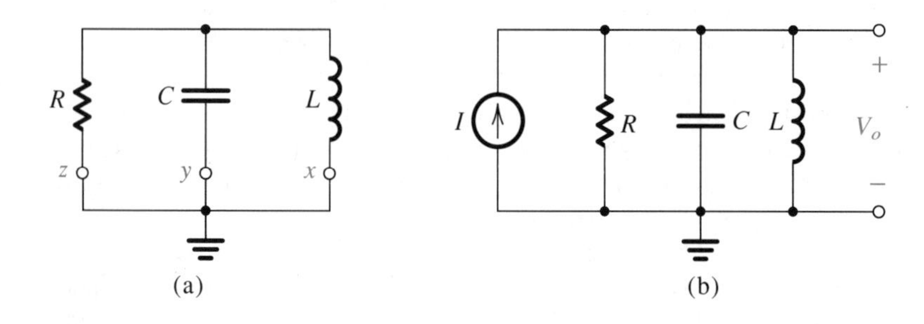

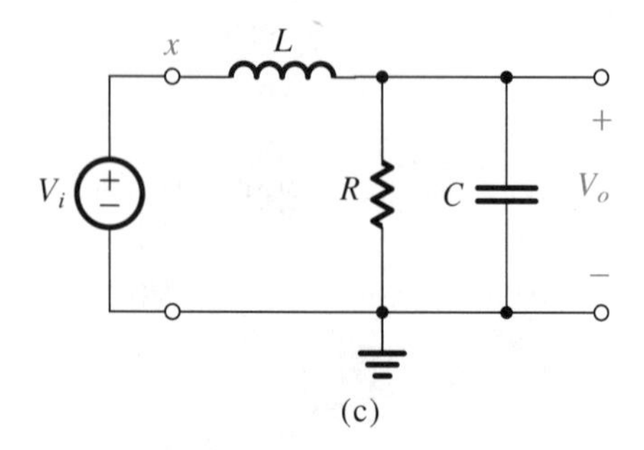

Fig. 11.17 **(a)** The second-order parallel LCR resonator. **(b)** and **(c)** Two ways for exciting the resonator of (a) without changing its *natural structure*. The resonator poles are the poles of V_o/I and V_o/V_i.

possible ways of exciting the circuit are shown in Figs. 11.17(b) and (c). In Fig. 11.17(b) the resonator is excited with a current source I connected in parallel. Since as far as the natural response of a circuit is concerned, an independent ideal current source is equivalent to an open circuit, the excitation of Fig. 11.17(b) does not alter the natural structure of the resonator. Thus the circuit in Fig. 11.17(b) can be used to determine the natural modes of the resonator by simply finding the poles of any response function. We can for instance take the voltage V_o across the resonator as the response and thus obtain the response function $V_o/I = Z$, where Z is the impedance of the parallel resonance circuit. It is obviously more convenient, however, to work in terms of the admittance Y; thus

$$\begin{aligned}\frac{V_o}{I} = \frac{1}{Y} &= \frac{1}{(1/sL) + sC + (1/R)} \\ &= \frac{s/C}{s^2 + s\,(1/CR) + (1/LC)}\end{aligned} \tag{11.31}$$

Equating the denominator to the standard form $s^2 + s\,(\omega_0/Q) + \omega_0^2$ leads to

$$\omega_0^2 = 1/LC \tag{11.32}$$

and

$$\omega_0/Q = 1/CR \tag{11.33}$$

Thus,

$$\omega_0 = 1/\sqrt{LC} \tag{11.34}$$

$$Q = \omega_0\, CR \tag{11.35}$$

These expressions should be familiar to the reader from earlier studies of parallel resonance circuits in introductory courses on circuit theory.

An alternative way of exciting the parallel LCR resonator for the purpose of determining its natural modes is shown in Fig. 11.17(c). Here node x of inductor L has been disconnected from ground and connected to an ideal voltage source V_i. Now, since as far as the natural response of a circuit is concerned, an ideal independent voltage source is equivalent to a short circuit, the excitation of Fig. 11.17(c) does not alter the natural structure of the resonator. Thus we can use the circuit in Fig. 11.17(c) to determine the natural modes of the resonator. These are the poles of any response function. For instance, we can select V_o as the response variable and find the transfer function V_o/V_i. The reader can easily verify that this will lead to the natural modes determined above.

In a design problem, we will be given ω_0 and Q and will be asked to determine L, C, and R. Equations (11.34) and (11.35) are two equations in the three unknowns. The one available degree of freedom can be utilized to set the impedance level of the circuit to a value that results in practical component values.

Realization of Transmission Zeros

Having selected the component values of the LCR resonator so as to realize a given pair of complex-conjugate natural modes, we now consider the use of the resonator to realize a desired filter type (e.g., LP, HP, etc.). Specifically, we wish to find out where to inject the input voltage signal V_i so that the transfer function V_o/V_i is the desired one. Toward that end, note that in the resonator circuit in Fig. 11.17(a) any of the nodes labeled x, y, or z can be disconnected from ground and connected to V_i without altering the circuit's natural modes. When this is done the circuit takes the form of a voltage divider, as shown in Fig. 11.18(a). Thus the transfer function realized is

$$T(s) = \frac{V_o(s)}{V_i(s)} = \frac{Z_2(s)}{Z_1(s) + Z_2(s)} \tag{11.36}$$

We observe that the transmission zeros are the values of s at which $Z_2(s)$ is zero, provided that $Z_1(s)$ is not simultaneously zero, and the values of s at which $Z_1(s)$ is infinite, provided that $Z_2(s)$ is not simultaneously infinite. This statement makes physical sense: The output will be zero either when $Z_2(s)$ behaves as a short circuit or when $Z_1(s)$ behaves as an open circuit. If there is a value of s at which both Z_1 and Z_2 are zero, then V_o/V_i will be finite and no transmission zero is obtained. Similarly, if there is a value of s at which both Z_1 and Z_2 are infinite, then V_o/V_i will be finite and no transmission zero is realized.

Realization of the Low-Pass Function

Using the scheme outlined above we see that to realize a low-pass function, node x is disconnected from ground and connected to V_i, as shown in Fig. 11.18(b). The transmission zeros of this circuit will be at the value of s for which the series impedance becomes infinite (sL becomes infinite at $s = \infty$) and the value of s at which the shunt impedance becomes zero ($1/[sC + (1/R)]$ becomes zero at $s = \infty$). Thus this circuit has two transmission zeros at $s = \infty$, as an LP is supposed to. The transfer function can be written either by inspection or by using the voltage-divider rule. Following the latter approach, we obtain

$$T(s) \equiv \frac{V_o}{V_i} = \frac{Z_2}{Z_1 + Z_2} = \frac{Y_1}{Y_1 + Y_2} = \frac{1/sL}{(1/sL) + sC + (1/R)}$$

$$= \frac{1/LC}{s^2 + s\,(1/CR) + (1/LC)} \tag{11.37}$$

Realization of the High-Pass Function

To realize the second-order high-pass function, node y is disconnected from ground and connected to V_i, as shown in Fig. 11.18(c). Here the series capacitor introduces a transmission zero at $s = 0$ (dc), and the shunt inductor introduces another transmission zero at $s = 0$ (dc). Thus, by inspection, the transfer function may be written as

$$T(s) \equiv \frac{V_o}{V_i} = \frac{a_2\, s^2}{s^2 + s\,(\omega_0/Q) + \omega_0^2} \tag{11.38}$$

where ω_0 and Q are the natural mode parameters given by Eqs. (11.34) and (11.35) and a_2 is the high-frequency transmission. The value of a_2 can be determined from the circuit by observing that as s approaches ∞, the capacitor approaches a short circuit and V_o approaches V_i, resulting in $a_2 = 1$.

Realization of the Bandpass Function

The bandpass function is realized by disconnecting node z from ground and connecting it to V_i, as shown in Fig. 11.18(d). Here the series impedance is resistive and thus does not

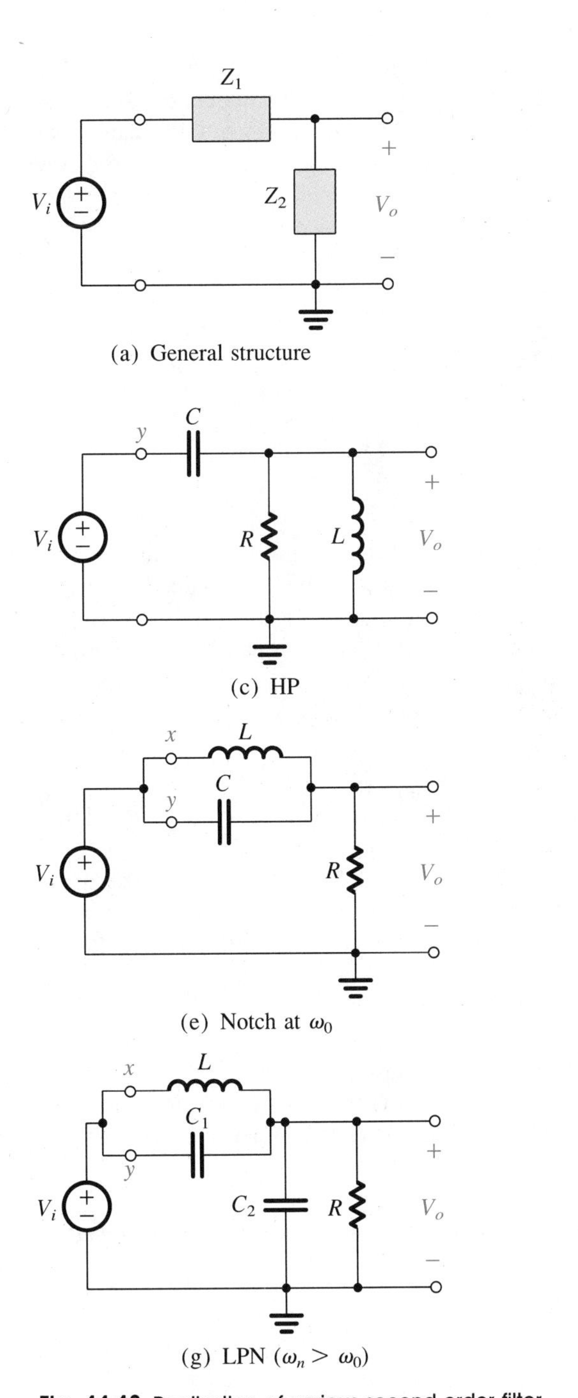

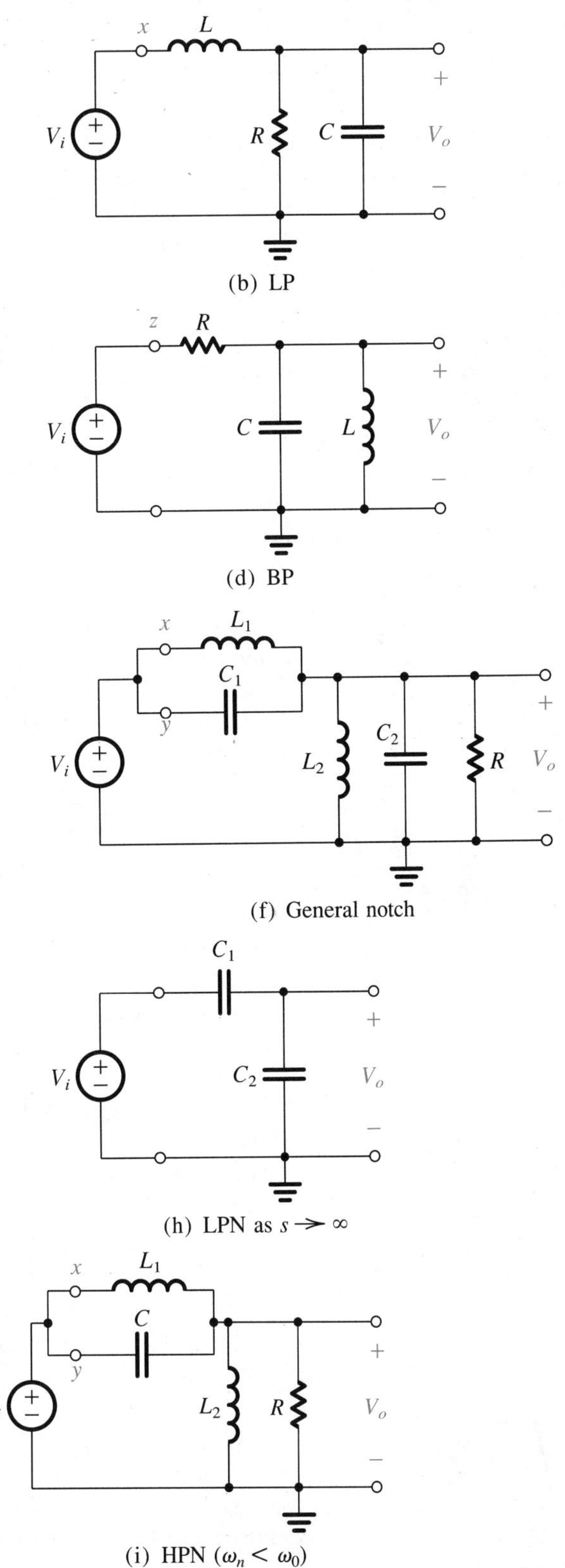

Fig. 11.18 Realization of various second-order filter functions using the LCR resonator of Fig. 11.17(b): **(a)** general structure, **(b)** LP, **(c)** HP, **(d)** BP, **(e)** notch at ω_0, **(f)** general notch, **(g)** LPN ($\omega_n \geq \omega_0$), **(h)** LPN as $s \rightarrow \infty$, **(i)** HPN ($\omega_n < \omega_0$).

introduce any transmission zeros. These are obtained as follows: One zero at $s = 0$ is realized by the shunt inductor, and one zero at $s = \infty$ is realized by the shunt capacitor. At the center-frequency ω_0, the parallel LC tuned circuit exhibits an infinite impedance, and thus no current flows in the circuit. It follows that at $\omega = \omega_0$, $V_o = V_i$. In other words, the center-frequency gain of the bandpass filter is unity. Its transfer function can be obtained as follows:

$$T(s) = \frac{Y_R}{Y_R + Y_L + Y_C} = \frac{1/R}{(1/R) + (1/sL) + sC}$$

$$= \frac{s\,(1/CR)}{s^2 + s\,(1/CR) + (1/LC)} \tag{11.39}$$

Realization of the Notch Functions

To obtain a pair of transmission zeros on the $j\omega$-axis we use a parallel resonance circuit in the series arm, as shown in Fig. 11.18(e). Observe that this circuit is obtained by disconnecting both nodes x and y from ground and connecting them together to V_i. The impedance of the LC circuit becomes infinite at $\omega = \omega_0 = 1/\sqrt{LC}$, thus causing zero transmission at this frequency. The shunt impedance is resistive and thus does not introduce transmission zeros. It follows that the circuit in Fig. 11.18(e) will realize the notch transfer function

$$T(s) = a_2 \frac{s^2 + \omega_0^2}{s^2 + s\,(\omega_0/Q) + \omega_0^2} \tag{11.40}$$

The value of the high-frequency gain a_2 can be found from the circuit to be unity.

To obtain a notch filter realization in which the notch frequency ω_n is arbitrarily placed relative to ω_0, we adopt a variation on the above scheme. We still use a parallel LC circuit in the series branch, as shown in Fig. 11.18(f) where L_1 and C_1 are selected so that

$$L_1C_1 = 1/\omega_n^2 \tag{11.41}$$

Thus the L_1C_1 tank circuit will introduce a pair of transmission zeros at $\pm j\omega_n$, provided that the L_2C_2 tank is not resonant at ω_n. Apart from this restriction, the values of L_2 and C_2 must be selected so as to ensure that the natural modes have not been altered; thus

$$C_1 + C_2 = C \tag{11.42}$$

$$L_1//L_2 = L \tag{11.43}$$

In other words, when V_i is replaced by a short circuit, the circuit should reduce to the original LCR resonator. Another way of thinking about the circuit of Fig. 11.18(f) is that it is obtained from the original LCR resonator by lifting part of L and part of C off ground and connecting them to V_i.

It should be noted that in the circuit of Fig. 11.18(f), L_2 does *not* introduce a zero at $s = 0$ because at $s = 0$, the L_1C_1 circuit also has a zero. In fact at $s = 0$ the circuit reduces to an inductive voltage divider with the dc transmission being $L_2/(L_1 + L_2)$. Similar comments can be made about C_2 and the fact that it does *not* introduce a zero at $s = \infty$.

The LPN and HPN filter realizations are special cases of the general notch circuit of Fig. 11.18(f). Specifically, for the LPN,

$$\omega_n > \omega_0,$$

thus

$$L_1C_1 < (L_1//L_2)(C_1 + C_2)$$

This condition can be satisfied with L_2 eliminated (i.e., $L_2 = \infty$ and $L_1 = L$), resulting in the LPN circuit in Fig. 11.18(g). The transfer function can be written by inspection as

$$T(s) \equiv \frac{V_o}{V_i} = a_2 \frac{s^2 + \omega_n^2}{s^2 + s\,(\omega_0/Q) + \omega_0^2} \tag{11.44}$$

where $\omega_n^2 = 1/LC_1$, $\omega_0^2 = 1/L(C_1 + C_2)$, $\omega_0/Q = 1/CR$, and a_2 is the high-frequency gain. From the circuit we see that as $s \to \infty$, the circuit reduces to that in Fig. 11.18(h), for which

$$\frac{V_o}{V_i} = \frac{C_1}{C_1 + C_2}$$

Thus

$$a_2 = \frac{C_1}{C_1 + C_2} \tag{11.45}$$

To obtain an HPN realization we start with the circuit of Fig. 11.18(f) and use the fact that $\omega_n < \omega_0$ to obtain

$$L_1C_1 > (L_1//L_2)(C_1 + C_2)$$

which can be satisfied while selecting $C_2 = 0$ (i.e., $C_1 = C$). Thus we obtain the reduced circuit shown in Fig. 11.18(i). Observe that as $s \to \infty$, V_o approaches V_i and thus the high-frequency gain is unity. Thus, the transfer function can be expressed as

$$T(s) \equiv \frac{V_o}{V_i} = \frac{s^2 + (1/L_1C)}{s^2 + s\,(1/CR) + [1/(L_1//L_2)\,C]} \tag{11.46}$$

Realization of the All-Pass Function

The all-pass transfer function

$$T(s) = \frac{s^2 - s\,(\omega_0/Q) + \omega_0^2}{s^2 + s\,(\omega_0/Q) + \omega_0^2} \tag{11.47}$$

can be written as

$$T(s) = 1 - \frac{s\,2\,(\omega_0/Q)}{s^2 + s\,(\omega_0/Q) + \omega_0^2} \tag{11.48}$$

The second term on the right-hand side is a bandpass function with a center-frequency gain of 2. We already have a bandpass circuit (Fig. 11.18d) but with a center-frequency gain of unity. We shall therefore attempt an all-pass realization with a flat gain of 0.5, that is,

$$T(s) = 0.5 - \frac{s\ (\omega_0/Q)}{s^2 + s\ (\omega_0/Q) + \omega_0^2}$$

This function can be realized using an attenuator with a transmission ratio of 0.5 together with the bandpass circuit of Fig. 11.18(d). To effect the subtraction, the output of the all-pass circuit is taken between the output terminal of the attenuator and that of the bandpass filter, as shown in Fig. 11.19. Unfortunately this circuit has the disadvantage of lacking a common ground terminal between the input and the output. An op amp–RC realization of the all-pass function will be presented in the next section.

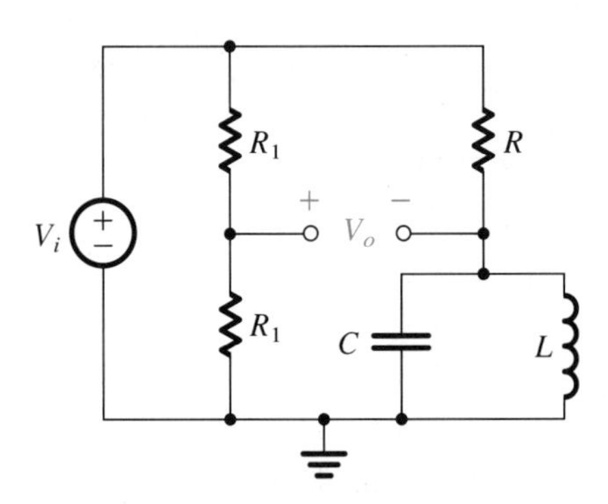

Fig. 11.19 Realization of the second-order all-pass transfer function using a voltage divider and an LCR resonator.

Exercises 11.17 Use the circuit of Fig. 11.18(b) to realize a second-order low-pass function of the maximally flat type with a 3-dB frequency of 100 kHz.

Ans. Selecting $R = 1\ \text{k}\Omega$, we obtain $C = 1125$ pF and $L = 2.25$ mH.

11.18 Use the circuit of Fig. 11.18(e) to design a notch filter to eliminate a bothersome power-supply hum at a 60-Hz frequency. The filter is to have a 3-dB bandwidth of 10 Hz (i.e., the attenuation is greater than 3 dB over a 10-Hz band around the 60-Hz center frequency; see Exercise 11.15 and Fig. 11.16d). Use $R = 1\ \text{M}\Omega$.

Ans. $C = 0.016\ \mu\text{F}$ and $L = 4.42$ H (Note the large inductor required. This is the reason passive filters are not practical in low-frequency applications.)

11.6 SECOND-ORDER ACTIVE FILTERS BASED ON INDUCTOR REPLACEMENT

In this section we study a family of op amp–RC circuits that realize the various second-order filter functions. The circuits are based on an op amp–RC resonator obtained by replacing the inductor L in the LCR resonator with an op amp–RC circuit that has an inductive input impedance.

The Antoniou Inductance-Simulation Circuit

Over the years, many op amp–RC circuits have been proposed for simulating the operation of an inductor. Of these, one circuit invented by A. Antoniou (see Antoniou, 1969) has proved to be the ''best.'' By ''best'' we mean that the operation of the circuit is very tolerant to the nonideal properties of the op amps, in particular their finite gain and bandwidth. Figure 11.20(a) shows the Antoniou inductance simulation circuit. If the circuit is fed at its input (node 1) with a voltage source V_1 and the input current is denoted I_1, then for ideal op amps the input impedance can be shown to be

$$Z_{in} \equiv V_1/I_1 = sC_4R_1R_3R_5/R_2 \tag{11.49}$$

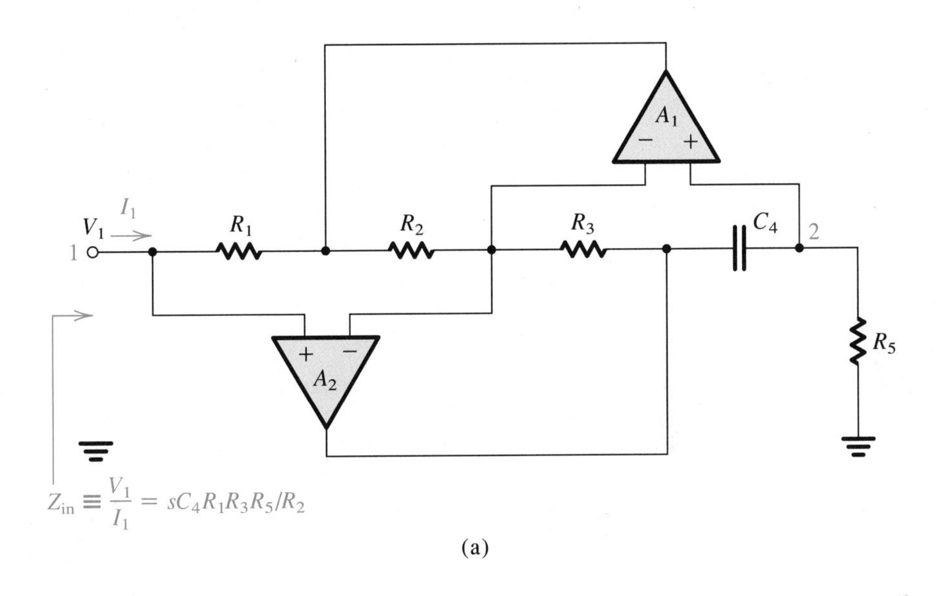

(a)

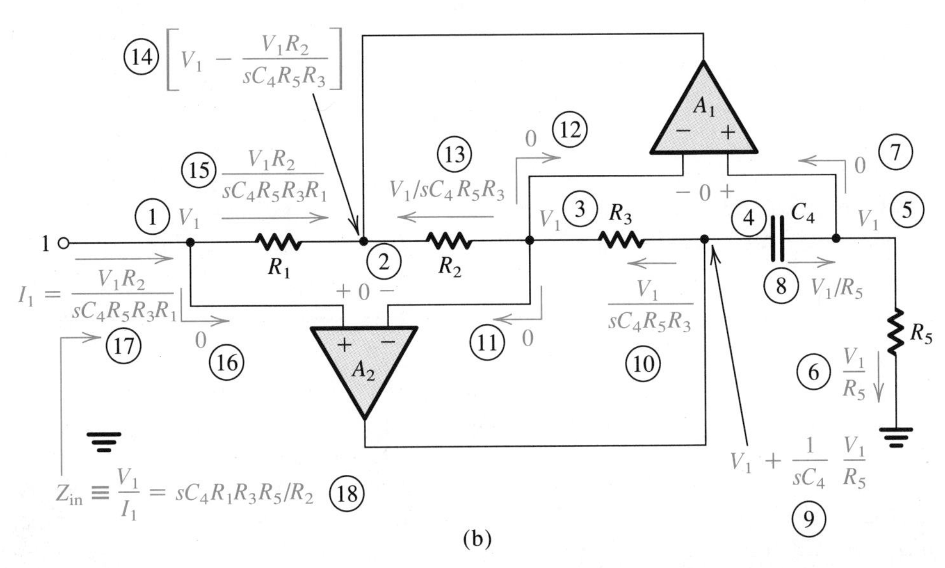

(b)

Fig. 11.20 (a) The Antoniou inductance-simulation circuit. **(b)** Analysis of the circuit assuming ideal op amps. The order of the analysis steps is indicated by the circled numbers.

which is that of an inductance L given by

$$L = C_4 R_1 R_3 R_5 / R_2 \tag{11.50}$$

Figure 11.20(b) shows the analysis of the circuit assuming that the op amps are ideal and thus that a virtual short circuit appears between the two input terminals of each op amp, and assuming also that the input currents of the op amps are zero. The analysis begins at node 1, which is assumed to be fed by a voltage source V_1, and proceeds in a step-by-step manner, with the order of the steps indicated by the circled numbers. The result of the analysis is the expression shown for the input current I_1 from which Z_{in} is found.

The design of this circuit is usually based on selecting $R_1 = R_2 = R_3 = R_5 = R$ and $C_4 = C$, which leads to $L = CR^2$. Convenient values are then selected for C and R to yield the desired inductance value L. More details on this circuit and the effect of the nonidealities of the op amps on its performance can be found in Sedra and Brackett (1978).

The Op Amp–RC Resonator

Figure 11.21(a) shows the LCR resonator we studied in detail in the previous section. Replacing the inductor L with a simulated inductance realized by the Antoniou circuit of Fig. 11.20(a) results in the op amp–RC resonator of Fig. 11.21(b). (Ignore for the moment the additional amplifier shown with broken line.) The circuit of Fig. 11.21(b) is a second-order resonator having a pole frequency

$$\omega_0 = 1/\sqrt{LC_6} = 1/\sqrt{C_4 C_6 R_1 R_3 R_5 / R_2} \tag{11.51}$$

where we have used the expression for L given in Eq. (11.50), and a pole Q factor,

$$Q = \omega_0 \, C_6 R_6 = R_6 \sqrt{\frac{C_6}{C_4} \frac{R_2}{R_1 R_3 R_5}} \tag{11.52}$$

Usually one selects $C_4 = C_6 = C$ and $R_1 = R_2 = R_3 = R_5 = R$, which results in

$$\omega_0 = 1/CR \tag{11.53}$$

$$Q = R_6/R \tag{11.54}$$

Select a practically convenient value for C; use Eq. (11.53) to determine the value of R to realize a given ω_0, and then use Eq. (11.54) to determine the value of R_6 to realize a given Q.

Realization of the Various Filter Types

The op amp–RC resonator of Fig. 11.21(b) can be used to generate circuit realizations for the various second-order filter functions by following the approach described in detail in the previous section in connection with the LCR resonator. Thus to obtain a bandpass function we disconnect node z from ground and connect it to the signal source V_i. A high-pass function is obtained by injecting V_i to node y. To realize a low-pass function using the LCR resonator, the inductor terminal x is disconnected from ground and connected to V_i. The corresponding node in the active resonator is the node at which R_5 is

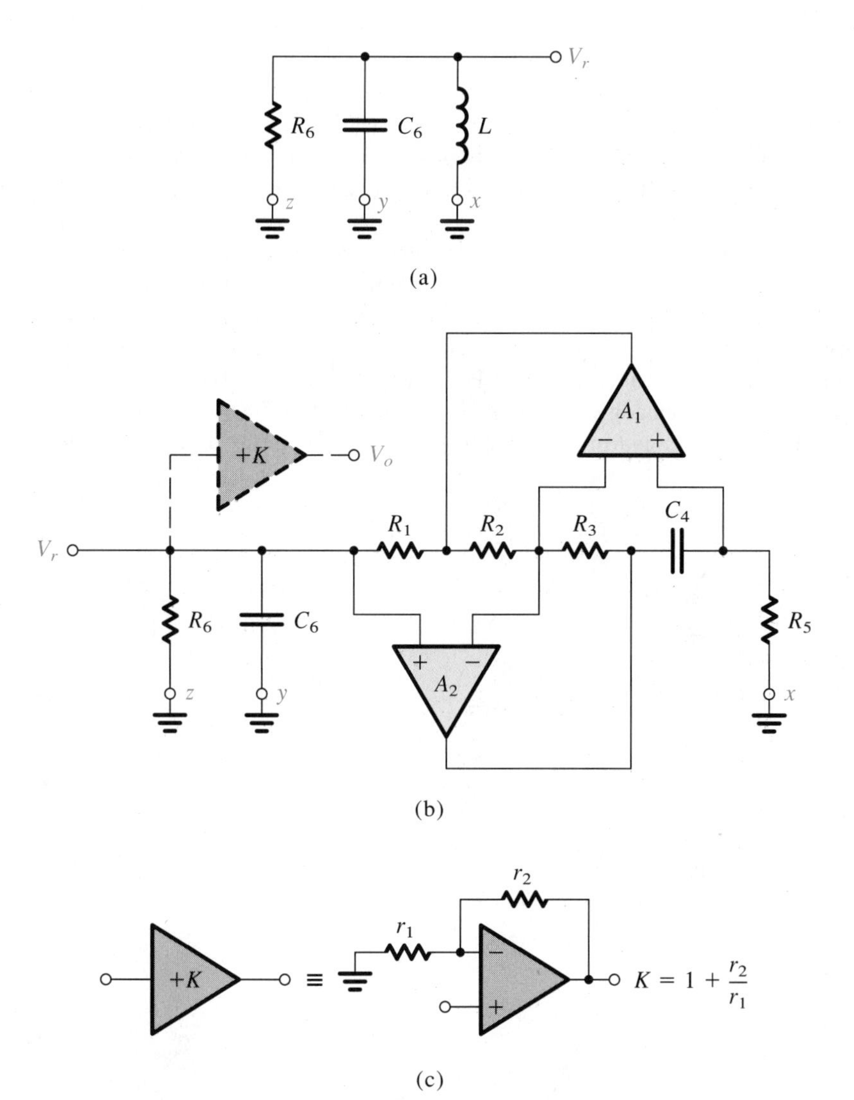

Fig. 11.21 **(a)** An LCR resonator. **(b)** An op amp–RC resonator obtained by replacing the inductor *L* in the LCR resonator of (a) with a simulated inductance realized by the Antoniou circuit of Fig. 11.20(a). **(c)** Implementation of the buffer amplifier *K*.

connected to ground,[5] labeled as node x in Fig. 11.21(b). A regular notch function ($\omega_n = \omega_0$) is obtained by feeding V_i to nodes x and y. In all cases the output can be taken as the voltage across the resonance circuit, V_r. However, this is not a convenient node to use as the filter output terminal since connecting a load would change the filter characteristics.

[5]This point might not be obvious! The reader, however, can show that when V_i is fed to this node the function V_r/V_i is indeed low-pass.

The problem can be solved easily by utilizing a buffer amplifier. This is the amplifier of gain K, drawn with broken lines in Fig. 11.21(b). Figure 11.21(c) shows how this amplifier can be simply implemented using an op amp connected in the noninverting configuration. Note that not only does the amplifier K buffer the output of the filter, but it also allows the designer to set the filter gain to any desired value by appropriately selecting the value of K.

Figure 11.22 shows the various second-order filter circuits obtained from the resonator of Fig. 11.21(b). The transfer functions and design equations for these circuits are given in Table 11.1. Note that the transfer functions can be written by analogy to those of the LCR resonator. We have already commented on the LP, HP, BP, and regular-notch circuits given in Figs. 11.22(a) to (d). The LPN and HPN circuits in Figs. 11.22(e) and (f) are obtained by direct analogy to their LCR counterparts in Figs. 11.18(g) and (i), respectively. The all-pass circuit in Fig. 11.22(g), however, deserves some explanation.

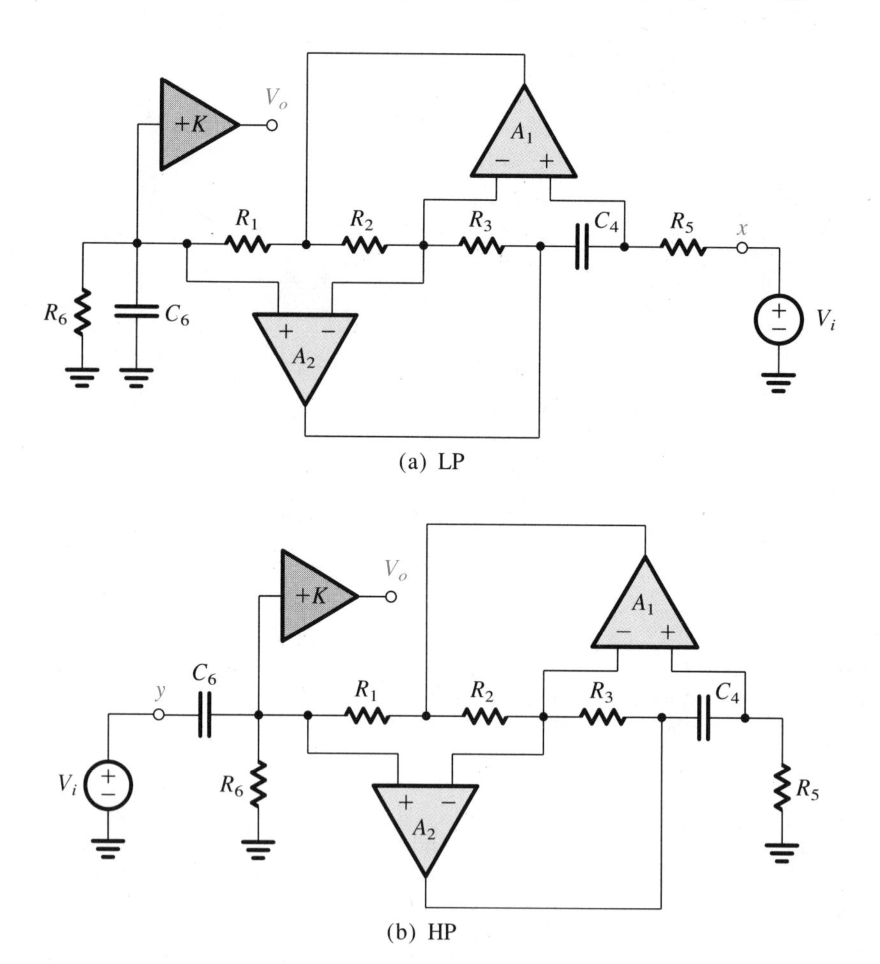

Fig. 11.22 Realizations for the various second-order filter functions using the op amp–RC resonator of Fig. 11.21(b). **(a)** LP; **(b)** HP; **(c)** BP, **(d)** notch at ω_0; **(e)** LPN, $\omega_n \geq \omega_0$; **(f)** HPN, $\omega_n \leq \omega_0$; **(g)** all-pass. The circuits are based on the LCR circuits in Fig. 11.18. Design equations are given in Table 11.1.

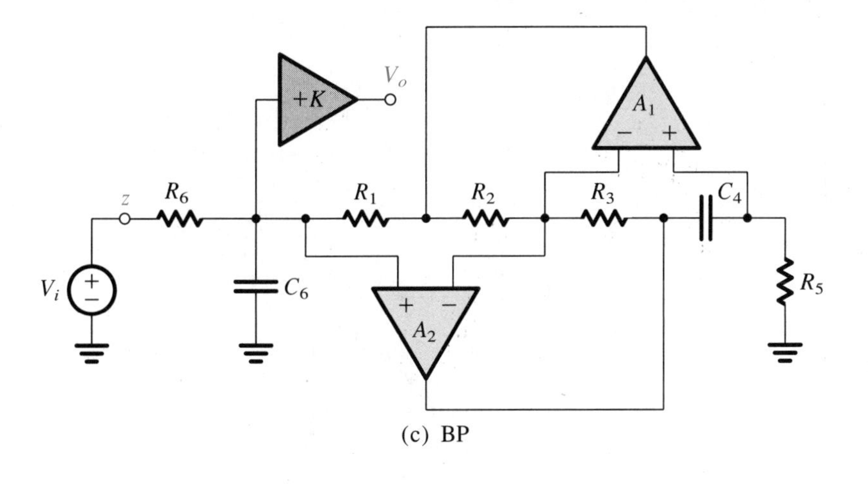

(c) BP

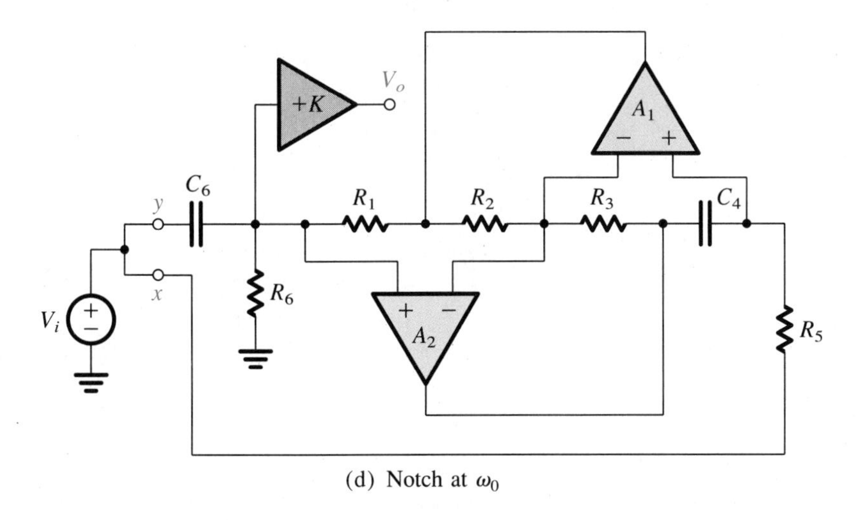

(d) Notch at ω_0

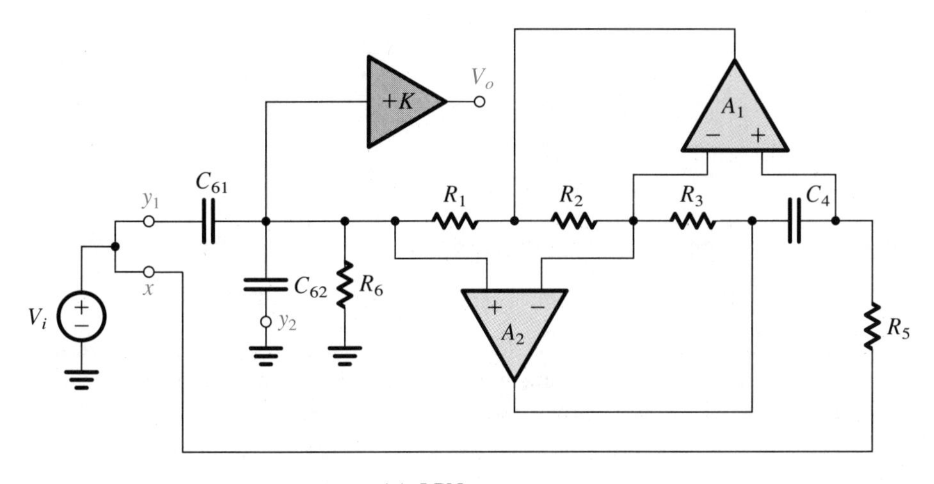

(e) LPN, $\omega_n \geq \omega_0$

Fig. 11.22 (continued)

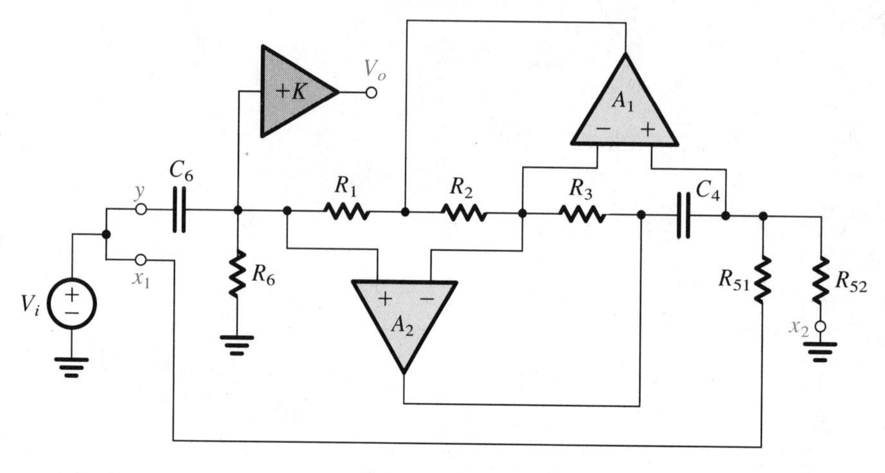

(f) HPN, $\omega_n \leq \omega_0$

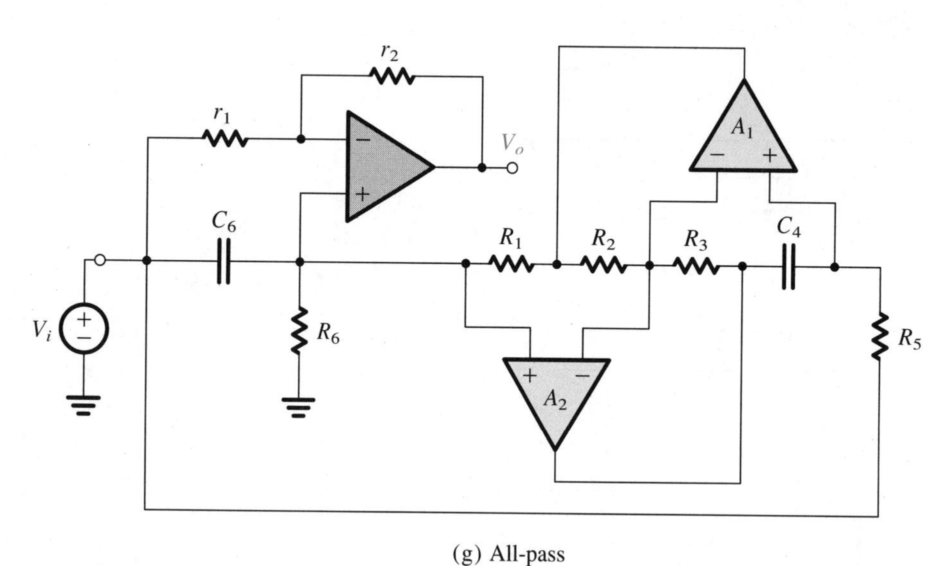

(g) All-pass

Fig. 11.22 (continued)

The All-Pass Circuit

An all-pass function with a flat gain of unity can be written as

$$\text{AP} = 1 - (\text{BP with a center frequency gain of 2}) \tag{11.55}$$

(see Eq. 11.48). Two circuits whose transfer functions are related in this fashion are said to be **complementary.**[6] Thus the all-pass circuit with unity flat gain is the complement of the bandpass circuit with a center-frequency gain of 2. A simple procedure exists for

[6]More about complementary circuits will be presented in conjunction with Fig. 11.31 following.

Table 11.1 DESIGN DATA FOR THE CIRCUITS OF FIG. 11.22

CIRCUIT	TRANSFER FUNCTION AND OTHER PARAMETERS	DESIGN EQUATIONS
Resonator Fig. 11.21(b)	$\omega_0 = 1/\sqrt{C_4C_6R_1R_3R_5/R_2}$ $Q = R_6\sqrt{\frac{C_6}{C_4}\frac{R_2}{R_1R_3R_5}}$	$C_4 = C_6 = C$ (practical value) $R_1 = R_2 = R_3 = R_5 = 1/\omega_0 C$ $R_6 = Q/\omega_0 C$
Low-pass (LP) Fig. 11.22(a)	$T(s) = \dfrac{KR_2/C_4C_6R_1R_3R_5}{s^2 + s\dfrac{1}{C_6R_6} + \dfrac{R_2}{C_4C_6R_1R_3R_5}}$	K = dc gain
High-pass (HP) Fig. 11.22(b)	$T(s) = \dfrac{Ks^2}{s^2 + s\dfrac{1}{C_6R_6} + \dfrac{R_2}{C_4C_6R_1R_3R_5}}$	K = High-frequency gain
Bandpass (BP) Fig. 11.22(c)	$T(s) = \dfrac{Ks/C_6R_6}{s^2 + s\dfrac{1}{C_6R_6} + \dfrac{R_2}{C_4C_6R_1R_3R_5}}$	K = Center-frequency gain
Regular notch (N) Fig. 11.22(d)	$T(s) = \dfrac{K[s^2 + (R_2/C_4C_6R_1R_3R_5)]}{s^2 + s\dfrac{1}{C_6R_6} + \dfrac{R_2}{C_4C_6R_1R_3R_5}}$	K = Low- and high-frequency gain
Low-pass notch (LPN) Fig. 11.22(e)	$T(s) = K\dfrac{C_{61}}{C_{61} + C_{62}}$ $\times \dfrac{s^2 + (R_2/C_4C_{61}R_1R_3R_5)}{s^2 + s\dfrac{1}{C_6R_6} + \dfrac{R_2}{C_4(C_{61} + C_{62})R_1R_3R_5}}$ $\omega_n = 1/\sqrt{C_4C_{61}R_1R_3R_5/R_2}$ $\omega_0 = 1/\sqrt{C_4(C_{61} + C_{62})R_1R_3R_5/R_2}$ $Q = R_6\sqrt{\frac{C_{61} + C_{62}}{C_4}\frac{R_2}{R_1R_3R_5}}$	K = dc gain $C_{61} + C_{62} = C_6 = C$ $C_{61} = C(\omega_0/\omega_n)^2$ $C_{62} = C - C_{61}$
High-pass notch (HPN) Fig. 11.22(f)	$T(s) = K\dfrac{s^2 + (R_2/C_4C_6R_1R_3R_{51})}{s^2 + s\dfrac{1}{C_6R_6} + \dfrac{R_2}{C_4C_6R_1R_3}\left(\dfrac{1}{R_{51}} + \dfrac{1}{R_{52}}\right)}$ $\omega_n = 1/\sqrt{C_4C_6R_1R_3R_{51}/R_2}$ $\omega_0 = \sqrt{\frac{R_2}{C_4C_6R_1R_3}\left(\frac{1}{R_{51}} + \frac{1}{R_{52}}\right)}$ $Q = R_6\sqrt{\frac{C_6}{C_4}\frac{R_2}{R_1R_3}\left(\frac{1}{R_{51}} + \frac{1}{R_{52}}\right)}$	K = High frequency gain $\dfrac{1}{R_{51}} + \dfrac{1}{R_{52}} = \dfrac{1}{R_5} = \omega_0 C$ $R_{51} = R_5(\omega_0/\omega_n)^2$ $R_{52} = R_5/[1 - (\omega_n/\omega_o)^2]$
All-pass (AP) Fig. 11.22(g)	$T(s) = \dfrac{s^2 - s\dfrac{1}{C_6R_6}\dfrac{r_2}{r_1} + \dfrac{R_2}{C_4C_6R_1R_3R_5}}{s^2 + s\dfrac{1}{C_6R_6} + \dfrac{R_2}{C_4C_6R_1R_3R_5}}$ $\omega_z = \omega_0 \quad Q_z = Q(r_1/r_2)$ Flat gain = 1	$r_1 = r_2 = r$ (arbitrary) Adjust r_2 to make $Q_z = Q$.

obtaining the complement of a given linear circuit: Disconnect all the circuit nodes that are connected to ground and connect them to V_i, and disconnect all the nodes that are connected to V_i and connect them to ground. That is, interchanging input and ground in a linear circuit generates a circuit whose transfer function is the complement of that of the original circuit.

Returning to the problem at hand, we first use the circuit of Fig. 11.22(c) to realize a BP with a gain of 2 by simply selecting $K = 2$ and implementing the buffer amplifier with the circuit of Fig. 11.21(c) with $r_1 = r_2$. We then interchange input and ground and thus obtain the all-pass circuit of Fig. 11.22(g).

Finally, in addition to being simple to design, the circuits in Fig. 11.22 exhibit excellent performance. They can be used on their own to realize second-order filter functions, or they can be cascaded to implement high-order filters.

Exercises D11.19 Use the circuit of Fig. 11.22(c) to design a second-order bandpass filter with a center frequency of 10 kHz, a 3-dB bandwidth of 500 Hz, and a center-frequency gain of 10. Use $C = 1.2$ nF.

Ans. $R_1 = R_2 = R_3 = R_5 = 13.26$ kΩ; $R_6 = 265$ kΩ; $C_4 = C_6 = 1.2$ nF; $K = 10$, $r_1 = 10$ kΩ, $r_2 = 90$ kΩ

D11.20 Realize the Chebyshev filter of Example 11.2, whose transfer function is given in Eq. (11.25), as the cascade connection of two circuits of the type shown in Fig. 11.22(a) and one first-order op amp-RC circuit of the type shown in Fig. 11.13(a). Note that you can make the dc gain of all sections unity. Do so. Use as many 10-kΩ resistors as possible.

Ans. First-order section: $R_1 = R_2 = 10$ kΩ, $C = 5.5$ nF. Second-order section with $\omega_0 = 4.117 \times 10^4$ rad/s and $Q = 1.4$: $R_1 = R_2 = R_3 = R_5 = 10$ kΩ, $R_6 = 14$ kΩ, $C_4 = C_6 = 2.43$ nF, $r_1 = \infty$, $r_2 = 0$. Second-order section with $\omega_0 = 6.246 \times 10^4$ rad/s and $Q = 5.56$: $R_1 = R_2 = R_3 = R_5 = 10$ kΩ, $R_6 = 55.6$ kΩ, $C_4 = C_6 = 1.6$ nF, $r_1 = \infty$, $r_2 = 0$.

11.7 SECOND-ORDER ACTIVE FILTERS BASED ON THE TWO-INTEGRATOR-LOOP TOPOLOGY

In this section we study another family of op amp–RC circuits that realize second-order filter functions. The circuits are based on the use of two integrators connected in cascade in an overall feedback loop and are thus known as two-integrator-loop circuits.

Derivation of the Two-Integrator-Loop Biquad

To derive the two-integrator-loop biquadratic circuit, or **biquad** as it is commonly known, consider the second-order high-pass transfer function

$$\frac{V_{hp}}{V_i} = \frac{Ks^2}{s^2 + s(\omega_0/Q) + \omega_0^2} \tag{11.56}$$

where K is the high-frequency gain. Cross multiplying Eq. (11.56) and dividing both sides of the resulting equation by s^2 (in order to get all the terms involving s in the form $1/s$, which is the transfer function of an integrator) gives

$$V_{hp} + \frac{1}{Q}\left(\frac{\omega_0}{s}V_{hp}\right) + \left(\frac{\omega_0^2}{s^2}V_{hp}\right) = K\ V_i \tag{11.57}$$

In this equation we observe that the signal $(\omega_0/s)V_{\text{hp}}$ can be obtained by passing V_{hp} through an integrator with a time constant equal to $1/\omega_0$. Furthermore, passing the resulting signal through another identical integrator results in the third signal involving V_{hp} in Eq. (11.57)—namely, $(\omega_0^2/s^2)V_{\text{hp}}$. Figure 11.23(a) shows a block diagram for such a two-integrator arrangement. Note that in anticipation of the use of the inverting op amp Miller integrator circuit to implement each integrator, the integrator blocks in Fig. 11.23(a) have been assigned negative signs.

The problem still remains, however, of how to form V_{hp}, the input signal feeding the two cascaded integrators. Toward that end we rearrange Eq. (11.57), expressing V_{hp} in terms of its single- and double-integrated versions and of V_i as

$$V_{\text{hp}} = K\,V_i - \frac{1}{Q}\frac{\omega_0}{s}V_{\text{hp}} - \frac{\omega_0^2}{s^2}V_{\text{hp}} \qquad (11.58)$$

which suggests that V_{hp} can be obtained by using the weighted summer of Fig. 11.23(b). Now it should be easy to see that a complete block-diagram realization can be obtained by combining the integrator blocks of Fig. 11.23(a) with the summer block of Fig. 11.23(b), as shown in Fig. 11.23(c).

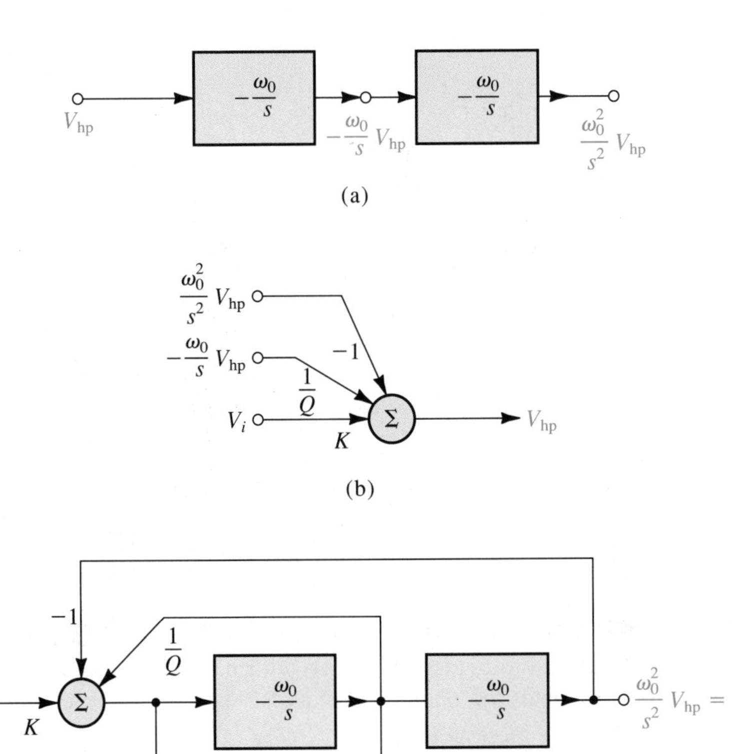

Fig. 11.23 Derivation of a block-diagram realization of the two-integrator-loop biquad.

In the realization of Fig. 11.23(c), V_{hp}, obtained at the output of the summer, realizes the high-pass transfer function $T_{hp} \equiv V_{hp}/V_i$ of Eq. (11.56). The signal at the output of the first integrator is $-(\omega_0/s)V_{hp}$, which is a bandpass function,

$$\frac{(-\omega_0/s)\,V_{hp}}{V_i} = -\frac{K\,\omega_0 s}{s^2 + s(\omega_0/Q) + \omega_0^2} = T_{bp}(s) \tag{11.59}$$

Therefore the signal at the output of the first integrator is labeled V_{bp}. Note that the center-frequency gain of the bandpass filter realized is equal to $-KQ$.

In a similar fashion, we can show that the transfer function realized at the output of the second integrator is the low-pass function,

$$\frac{(\omega_0^2/s^2)V_{hp}}{V_i} = \frac{K\omega_0^2}{s^2 + s(\omega_0/Q) + \omega_0^2} = T_{lp}(s) \tag{11.60}$$

Thus the output of the second integrator is labeled V_{lp}. Note that the dc gain of the low-pass filter realized is equal to K.

We conclude that the two-integrator-loop biquad shown in block-diagram form in Fig. 11.23(c) realizes the three basic second-order filtering functions, LP, BP, and HP, *simultaneously*. This versatility has made the circuit very popular and has given it the name *universal active filter*.

Circuit Implementation

To obtain an op amp circuit implementation of the two-integrator-loop biquad of Fig. 11.23(c), we replace each integrator with a Miller integrator circuit having $CR = 1/\omega_0$, and we replace the summer block with an op amp summing circuit that is capable of assigning both positive and negative weights to its inputs. The resulting circuit, known as the Kerwin–Huelsman–Newcomb or **KHN biquad** after its inventors, is shown in Fig. 11.24(a). Given values for ω_0, Q, and K, the design of the circuit is straightforward: We select suitably practical values for the components of the integrators, C and R, so that $CR = 1/\omega_0$. To determine the values of the resistors associated with the summer, we first use superposition to express the output of the summer in terms of its inputs as

$$V_{hp} = \frac{R_3}{R_2 + R_3}\left(1 + \frac{R_f}{R_1}\right)V_i + \frac{R_2}{R_2 + R_3}\left(1 + \frac{R_f}{R_1}\right)\left(-\frac{\omega_0}{s}V_{hp}\right) - \frac{R_f}{R_1}\left(\frac{\omega_0^2}{s^2}V_{hp}\right) \tag{11.61}$$

Equating the last terms on the right-hand side of Eqs. (11.61) and (11.58) gives

$$R_f/R_1 = 1 \tag{11.62}$$

which implies that we can select arbitrary but practically convenient equal values for R_1 and R_f. Then equating the second-to-last terms on the right-hand side of Eqs. (11.61) and (11.58) and setting $R_1 = R_f$ yields the ratio R_3/R_2 required to realize a given Q as

$$R_3/R_2 = 2Q - 1 \tag{11.63}$$

Thus an arbitrary but convenient value can be selected for either R_2 or R_3, and the value of the other resistance can be determined using Eq. (11.63). Finally, equating the coefficients of V_i in Eqs. (11.61) and (11.58) and substituting $R_f = R_1$ and for R_3/R_2 from Eq. (11.63) results in

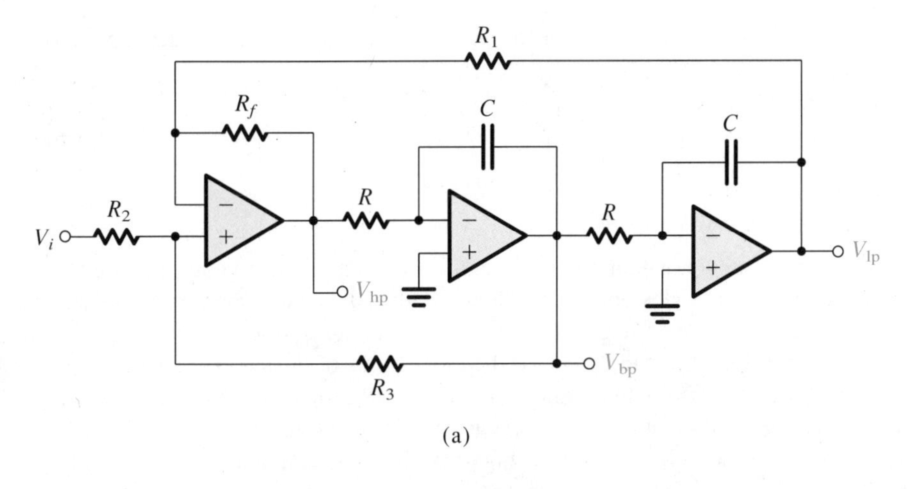

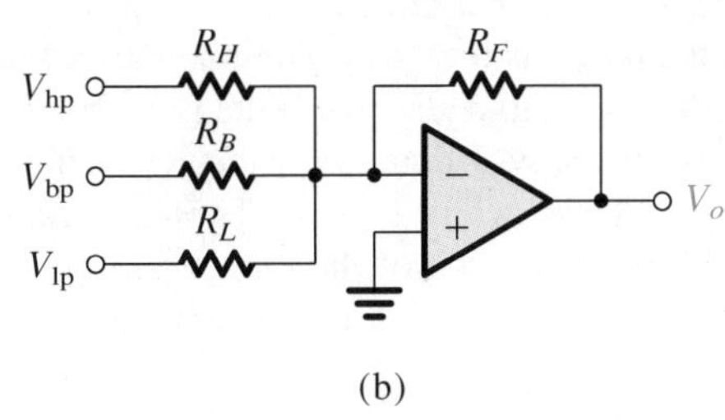

Fig. 11.24 (a) The KHN biquad circuit, obtained as a direct implementation of the block diagram of Fig. 11.23(c). The three basic filtering functions, HP, BP, and LP, are simultaneously realized. **(b)** To obtain notch and all-pass functions, the three outputs are summed with appropriate weights using this op amp summer.

$$K = 2 - (1/Q) \tag{11.64}$$

Thus the gain parameter K is fixed to this value.

The KHN biquad can be used to realize notch and all-pass functions by summing weighted versions of the three outputs, LP, BP, and HP. Such an op amp summer is shown in Fig. 11.24(b); for this summer we can write

$$\begin{aligned} V_o &= -\left(\frac{R_F}{R_H}V_{\text{hp}} + \frac{R_F}{R_B}V_{\text{bp}} + \frac{R_F}{R_L}V_{\text{lp}}\right) \\ &= -V_i\left(\frac{R_F}{R_H}T_{\text{hp}} + \frac{R_F}{R_B}T_{\text{bp}} + \frac{R_F}{R_L}T_{\text{lp}}\right) \end{aligned} \tag{11.65}$$

Substituting for T_{hp}, T_{bp}, and T_{lp} from Eqs. (11.56), (11.59), and (11.60), respectively, gives the overall transfer function

$$\frac{V_o}{V_i} = -K\,\frac{(R_F/R_H)s^2 - s(R_F/R_B)\omega_0 + (R_F/R_L)\omega_0^2}{s^2 + s(\omega_0/Q) + \omega_0^2} \tag{11.66}$$

from which we can see that different transmission zeros can be obtained by the appropriate selection of the values of the summing resistors. For instance, a notch is obtained by selecting $R_B = \infty$ and

$$\frac{R_H}{R_L} = \left(\frac{\omega_n}{\omega_0}\right)^2 \tag{11.67}$$

An Alternative Two-Integrator-Loop Biquad Circuit

An alternative two-integrator-loop biquad circuit in which all three op amps are used in a single-ended mode can be developed as follows: Rather than using the input summer to add signals with positive and negative coefficients, we can introduce an additional inverter, as shown in Fig. 11.25(a). Now all the coefficients of the summer have the same sign, and we can dispense with the summing amplifier altogether and perform the summation at the virtual-ground input of the first integrator. The resulting circuit is shown in Fig. 11.25(b), from which we observe that the high-pass function is no longer available! This is the price paid for obtaining a circuit that utilizes all op amps in a single-ended mode. The circuit of Fig. 11.25(b) is known as the **Tow–Thomas biquad,** after its originators.

Rather than using a fourth op amp to realize the finite transmission zeros required for the notch and all-pass functions, as was done with the KHN biquad, an economical *feedforward* scheme can be employed with the Tow–Thomas circuit. Specifically, the virtual ground available at the input of each of the three op amps in the Tow–Thomas circuit permits the input signal to be fed to all three op amps, as shown in Fig. 11.26. If

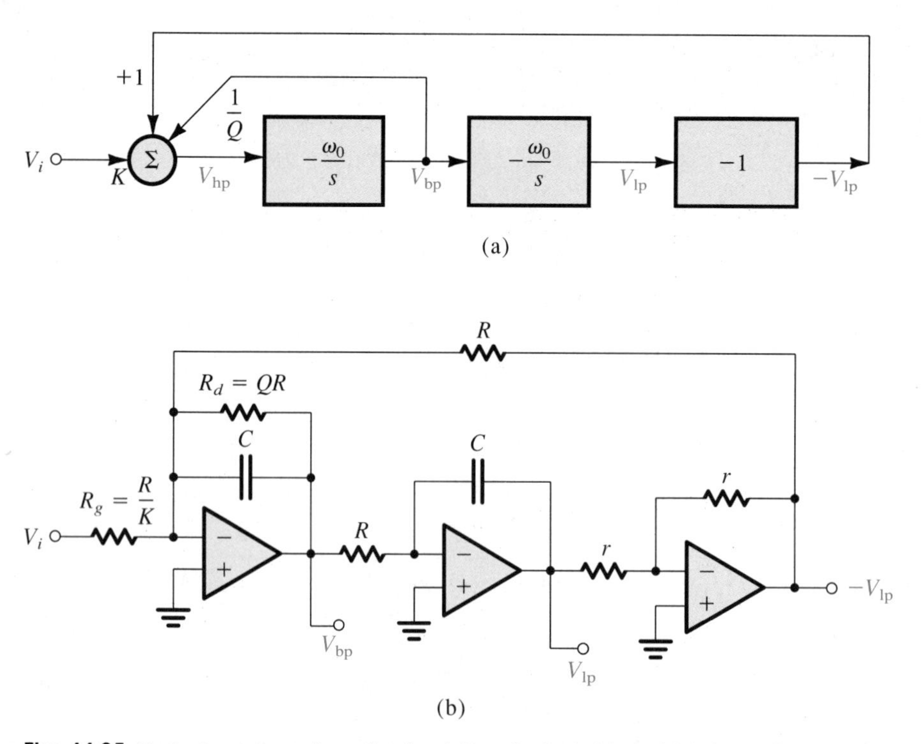

Fig. 11.25 Derivation of an alternative two-integrator-loop biquad in which all op amps are used in a single-ended fashion. The resulting circuit in (b) is known as the Tow–Thomas biquad.

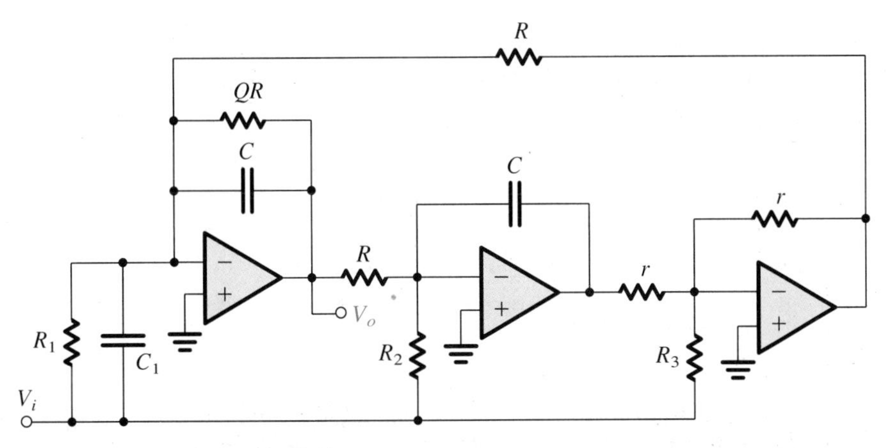

Fig. 11.26 The Tow–Thomas biquad with feedforward. The transfer function of Eq. (11.68) is realized by feeding the input signal through appropriate components to the inputs of the three op amps. This circuit can realize all special second-order functions. The design equations are given in Table 11.2.

V_o is taken at the output of the damped integrator, straightforward analysis yields the filter transfer function

$$\frac{V_o}{V_i} = -\frac{s^2\left(\frac{C_1}{C}\right) + s\frac{1}{C}\left(\frac{1}{R_1} - \frac{r}{RR_3}\right) + \frac{1}{C^2RR_2}}{s^2 + s\,\frac{1}{QCR} + \frac{1}{C^2R^2}} \tag{11.68}$$

which can be used to obtain the design data given in Table 11.2.

Table 11.2 DESIGN DATA FOR THE CIRCUIT IN FIG. 11.26

All cases	C = arbitrary, $R = 1/\omega_0 C$, r = arbitrary
LP	$C_1 = 0$, $R_1 = \infty$, $R_2 = R/\text{dc gain}$, $R_3 = \infty$
Positive BP	$C_1 = 0$, $R_1 = \infty$, $R_2 = \infty$, $R_3 = Qr/\text{center-frequency gain}$
Negative BP	$C_1 = 0$, $R_1 = QR/\text{center-frequency gain}$, $R_2 = \infty$, $R_3 = \infty$
HP	$C_1 = C \times \text{high-frequency gain}$, $R_1 = \infty$, $R_2 = \infty$, $R_3 = \infty$
Notch (all types)	$C_1 = C \times \text{high-frequency gain}$, $R_1 = \infty$, $R_2 = R(\omega_0/\omega_n)^2/\text{high-frequency gain}$, $R_3 = \infty$
AP	$C_1 = C \times \text{flat gain}$, $R_1 = \infty$, $R_2 = R/\text{gain}$, $R_3 = Qr/\text{gain}$

Final Remarks

The two-integrator-loop biquads are extremely versatile and easy to design. However, their performance is adversely affected by the finite bandwidth of the op amps. Special techniques exist for compensating the circuit for such effects (see Sedra and Brackett, 1978).

Exercises D11.21 Design the KHN circuit to realize a high-pass function with $f_0 = 10$ kHz and $Q = 2$. Choose $C = 1$ nF. What is the value of high-frequency gain obtained? What is the center-frequency gain of the bandpass function that is simultaneously available at the output of the first integrator?

Ans. $R = 15.9$ kΩ; $R_1 = R_f = R_2 = 10$ kΩ (arbitrary), $R_3 = 30$ kΩ; 1.5; 3

D11.22 Use the KHN circuit together with an output summing amplifier to design a low-pass notch filter with $f_0 = 5$ kHz, $f_n = 8$ kHz, $Q = 5$, and a dc gain of 3. Select $C = 1$ nF and $R_L = 10$ kΩ.

Ans. $R = 31.83$ kΩ; $R_1 = R_f = R_2 = 10$ kΩ (arbitrary); $R_3 = 90$ kΩ; $R_H = 25.6$ kΩ; $R_F = 42.7$ kΩ; $R_B = \infty$

D11.23 Use the Tow–Thomas biquad (Fig. 11.25b) to design a second-order bandpass filter with $f_0 = 10$ kHz, $Q = 20$, and unity center-frequency gain. If $R = 10$ kΩ, give the values of C, R_d, and R_g.

Ans. 1.59 nF; 200 kΩ; 200 kΩ

D11.24 Use the data of Table 11.2 to design the biquad circuit of Fig. 11.26 to realize an all-pass filter with $\omega_0 = 10^4$ rad/s, $Q = 5$, and flat gain = 1. Use $C = 10$ nF and $r = 10$ kΩ.

Ans. $R = 10$ kΩ; Q-determining resistor = 50 kΩ; $C_1 = 10$ nF; $R_1 = \infty$; $R_2 = 10$ kΩ; $R_3 = 50$ kΩ

11.8 SINGLE-AMPLIFIER BIQUADRATIC ACTIVE FILTERS

The op amp-RC biquadratic circuits studied in the two previous sections provide good performance, are versatile, and are easy to design and to adjust (tune) after final assembly. Unfortunately, however, they are not economic in their use of op amps, requiring three or four amplifiers per second-order section. This can be a problem, especially in applications where power-supply current is to be conserved; for instance, in a battery-operated instrument. In this section we shall study a class of second-order filter circuits that requires only one op amp per biquad. These minimal realizations, however, suffer a greater dependence on the limited gain and bandwidth of the op amp and can also be more sensitive to the unavoidable tolerances in the values of resistors and capacitors than the multiple-op amp biquads of the previous sections. The **single-amplifier biquads** (SABs) are therefore limited to the less stringent filter specifications—for example, pole Q factors less than about 10.

The synthesis of SAB circuits is based on the use of feedback to move the poles of an RC circuit from the negative real axis, where they naturally lie, to the complex conjugate locations required to provide selective filter response. The synthesis of SABs follows a two-step process:

1. Synthesis of a feedback loop that realizes a pair of complex conjugate poles characterized by a frequency ω_0 and a Q factor Q.
2. Injecting the input signal in a way that realizes the desired transmission zeros.

Synthesis of the Feedback Loop

Consider the circuit shown in Fig. 11.27(a), which consists of a two-port RC network n placed in the negative-feedback path of an op amp. We shall assume that, except for having a finite gain A, the op amp is ideal. We shall denote by $t(s)$ the open-circuit voltage

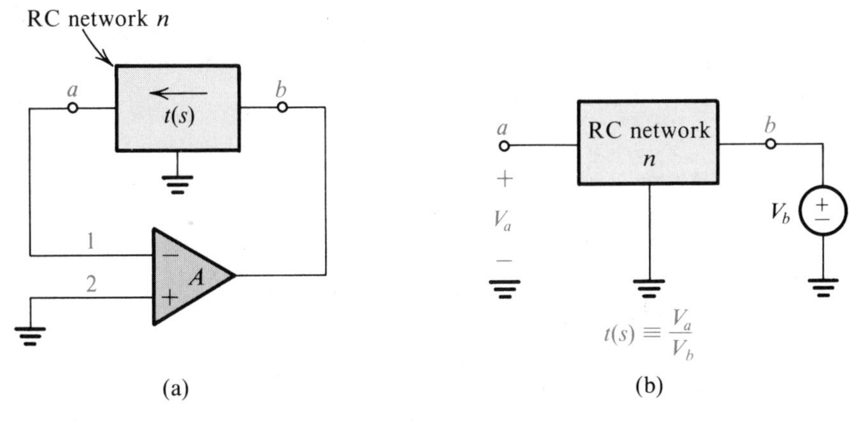

Fig. 11.27 (a) Feedback loop obtained by placing a two-port RC network n in the feedback path of an op amp. **(b)** Definition of the open-circuit transfer function $t(s)$ of the RC network.

transfer function of the RC network n, where the definition of $t(s)$ is illustrated in Fig. 11.27(b). The transfer function $t(s)$ can in general be written as the ratio of two polynomials $N(s)$ and $D(s)$:

$$t(s) = \frac{N(s)}{D(s)}$$

The roots of $N(s)$ are the transmission zeros of the RC network, and the roots of $D(s)$ are its poles. Study of network theory shows that while the poles of an RC network are restricted to lie on the negative real axis, the zeros can in general lie anywhere in the s-plane.

The loop gain $L(s)$ of the feedback circuit in Fig. 11.27(a) can be determined using the method of Section 8.7. It is simply the product of the op amp gain A and the transfer function $t(s)$,

$$L(s) = At(s) = \frac{AN(s)}{D(s)} \tag{11.69}$$

Substituting for $L(s)$ into the characteristic equation

$$1 + L(s) = 0 \tag{11.70}$$

results in the poles s_P of the closed-loop circuit obtained as solutions to the equation

$$t(s_P) = -\frac{1}{A} \tag{11.71}$$

In the ideal case, $A = \infty$ and the poles are obtained from

$$N(s_P) = 0 \tag{11.72}$$

That is, *the poles are identical to the zeros of the RC network.*

Since our objective is to realize a pair of complex conjugate poles, we should select an RC network that has complex conjugate transmission zeros. The simplest such net-

$$t(s) = \frac{s^2 + s\left(\frac{1}{C_1} + \frac{1}{C_2}\right)\frac{1}{R_3} + \frac{1}{C_1 C_2 R_3 R_4}}{s^2 + s\left(\frac{1}{C_1 R_3} + \frac{1}{C_2 R_3} + \frac{1}{C_1 R_4}\right) + \frac{1}{C_1 C_2 R_3 R_4}}$$

(a)

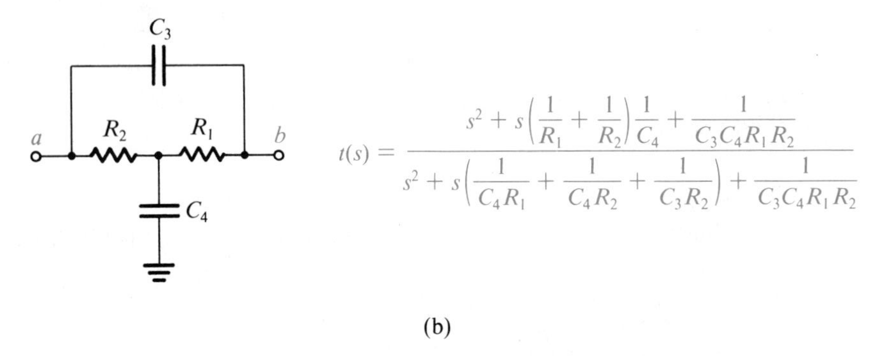

$$t(s) = \frac{s^2 + s\left(\frac{1}{R_1} + \frac{1}{R_2}\right)\frac{1}{C_4} + \frac{1}{C_3 C_4 R_1 R_2}}{s^2 + s\left(\frac{1}{C_4 R_1} + \frac{1}{C_4 R_2} + \frac{1}{C_3 R_2}\right) + \frac{1}{C_3 C_4 R_1 R_2}}$$

(b)

Fig. 11.28 Two RC networks (called bridged-T networks) that have complex transmission zeros. The transfer functions given are from b to a with a open-circuited.

works are the bridged-T networks shown in Fig. 11.28 together with their transfer function $t(s)$ from b to a, with a open-circuited. As an example, consider the circuit generated by placing the bridged-T network of Fig. 11.28(a) in the negative-feedback path of an op amp, as shown in Fig. 11.29. The pole polynomial of the active-filter circuit will be equal to the numerator polynomial of the bridged-T network; thus

$$s^2 + s\frac{\omega_0}{Q} + \omega_0^2 = s^2 + s\left(\frac{1}{C_1} + \frac{1}{C_2}\right)\frac{1}{R_3} + \frac{1}{C_1 C_2 R_3 R_4}$$

which enables us to obtain ω_0 and Q as

$$\omega_0 = \frac{1}{\sqrt{C_1 C_2 R_3 R_4}} \tag{11.73}$$

$$Q = \left[\frac{\sqrt{C_1 C_2 R_3 R_4}}{R_3}\left(\frac{1}{C_1} + \frac{1}{C_2}\right)\right]^{-1} \tag{11.74}$$

If we are designing this circuit, ω_0 and Q are given and Eqs. (11.73) and (11.74) can be used to determine C_1, C_2, R_3, and R_4. It follows that there are two degrees of freedom. Let us exhaust one of these by selecting $C_1 = C_2 = C$. Let us also denote $R_3 = R$ and $R_4 = R/m$. By substituting in Eqs. (11.73) and (11.74) and with some manipulation, we obtain

$$m = 4Q^2 \tag{11.75}$$

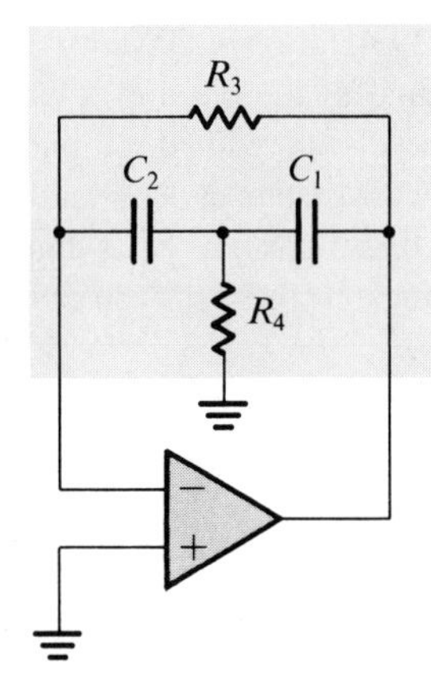

Fig. 11.29 An active-filter feedback loop generated using the bridged-T network of Fig. 11.28(a).

$$CR = \frac{2Q}{\omega_0} \tag{11.76}$$

Thus if we are given the value of Q, Eq. (11.75) can be used to determine the ratio of the two resistances R_3 and R_4. Then the given values of ω_0 and Q can be substituted in Eq. (11.76) to determine the time constant CR. There remains one degree of freedom—the value of C or R can be arbitrarily chosen. In an actual design, this value, which sets the *impedance level* of the circuit, should be chosen so that the resulting component values are practical.

Exercises D11.25 Design the circuit of Fig. 11.29 to realize a pair of poles with $\omega_0 = 10^4$ rad/s and $Q = 1$. Select $C_1 = C_2 = 1$ nF.

Ans. $R_3 = 200$ kΩ; $R_4 = 50$ kΩ

11.26 For the circuit designed in Exercise 11.25, find the location of the poles of the RC network in the feedback loop.

Ans. -0.382×10^4 and -2.618×10^4 rad/s

Injecting the Input Signal

Having synthesized a feedback loop that realizes a given pair of poles, we now consider connecting the input signal source to the circuit. We wish to do this, of course, without altering the poles.

Since, for the purpose of finding the poles of a circuit, an ideal voltage source is equivalent to a short-circuit, it follows that any circuit node that is connected to ground can instead be connected to the input voltage source without causing the poles to change. Thus the method of injecting the input signal into the feedback loop is simply to disconnect a component (or several components) that is (are) connected to ground and connect it (them) to the input source. Depending on the component(s) through which the input signal is injected, different transmission zeros are obtained. This is, of course, the same method we used in Section 11.5 with the LCR resonator and in Section 11.6 with the biquads based on the LCR resonator.

As an example, consider the feedback loop of Fig. 11.29. Here we have two grounded nodes (one terminal of R_4 and the positive input terminal of the op amp) that can serve for injecting the input signal. Figure 11.30 shows the circuit with the input signal injected through part of the resistance R_4. Note that the two resistances R_4/α and $R_4/(1-\alpha)$ have a parallel equivalent of R_4. It can be shown (by direct analysis) that this circuit realizes the second-order bandpass function and that the value of α $(0 < \alpha \leq 1)$ can be used to obtain the desired center-frequency gain.

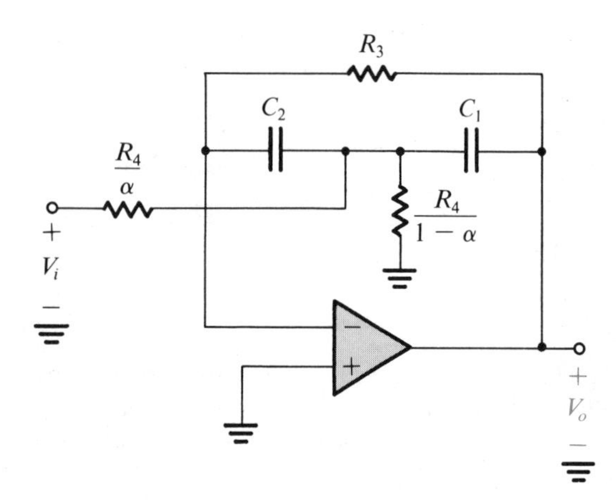

Fig. 11.30 The loop of Fig. 11.29 with the input signal injected through part of resistance R_4. This circuit realizes the bandpass function.

Exercise 11.27 Assume the op amp to be ideal and analyze the circuit in Fig. 11.30 to obtain its transfer function. Thus show that the circuit realizes the bandpass function and that its poles are identical to the zeros of $t(s)$ in Fig. 11.28(a). Using the component values of Exercise 11.25 find the values of R_4/α and $R_4/(1-2)(1-\alpha)$ to obtain unity center-frequency gain.

Ans. 100 kΩ; 100 kΩ

Generation of Equivalent Feedback Loops

The **complementary transformation** of feedback loops is based on the property of linear networks illustrated in Fig. 11.31 for the two-port (three-terminal) network n. In part (a) of the figure, terminal c is grounded and a signal V_b is applied to terminal b. The transfer function from b to a with c grounded is denoted t. Then, in part (b) of the figure, terminal b is grounded and the input signal is applied to terminal c. The transfer function from c to a with b grounded can be shown to be the complement of t—that is, $1-t$. (Recall that we used this property in generating a circuit realization for the all-pass function in Section 11.6.)

Application of the complementary transformation to a feedback loop to generate an equivalent feedback loop is a two-step process:

1. Nodes of the feedback network and any of the op amp inputs that are connected to ground should be disconnected from ground and connected to the op amp output. Conversely, those nodes that were connected to the op amp output should be now connected to ground. That is, we simply interchange the op amp output terminal with ground.

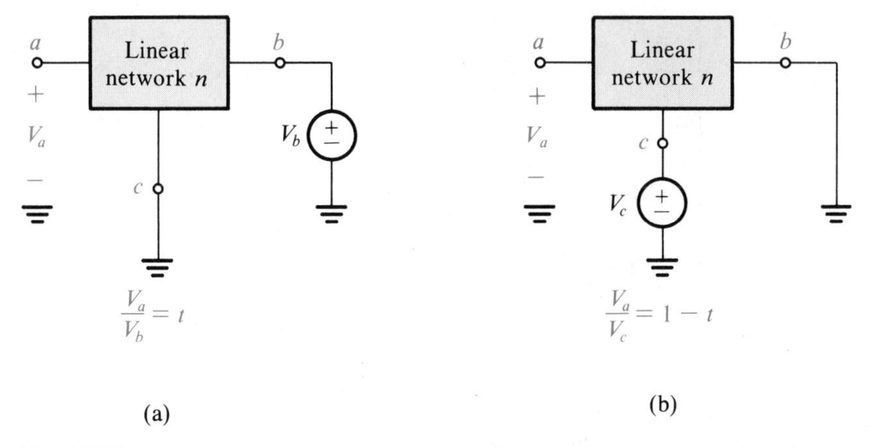

Fig. 11.31 Interchanging input and ground results in the complement of the transfer function.

2. The two input terminals of the op amp should be interchanged.

The feedback loop generated by this transformation has the same characteristic equation, and hence the same poles, as the original loop.

To illustrate, we show in Fig. 11.32(a) the feedback loop formed by connecting a two-port RC network in the negative-feedback path of an op amp. Application of the complementary transformation to this loop results in the feedback loop of Fig. 11.32(b). Note that in the latter loop the op amp is used in the unity-gain follower configuration. We shall now show that the two loops of Fig. 11.32 are equivalent.

If the op amp has an open-loop gain A, the follower in the circuit of Fig. 11.32(b) will have a gain of $A/(A + 1)$. This, together with the fact that the transfer function of network n from c to a is $1 - t$ (see Fig. 11.31), enables us to write for the circuit in Fig. 11.32(b) the characteristic equation

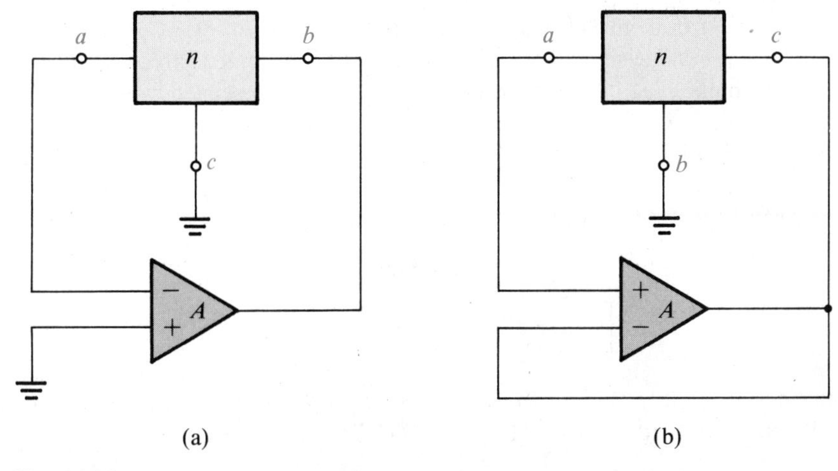

Fig. 11.32 Application of the complementary transformation to the feedback loop in **(a)** results in the equivalent (same poles) loop in **(b)**.

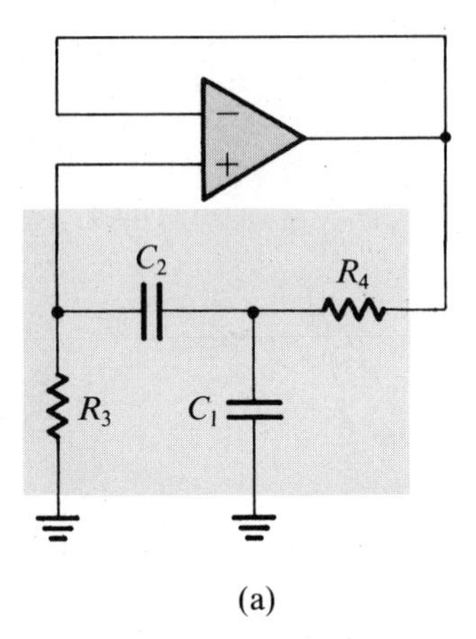

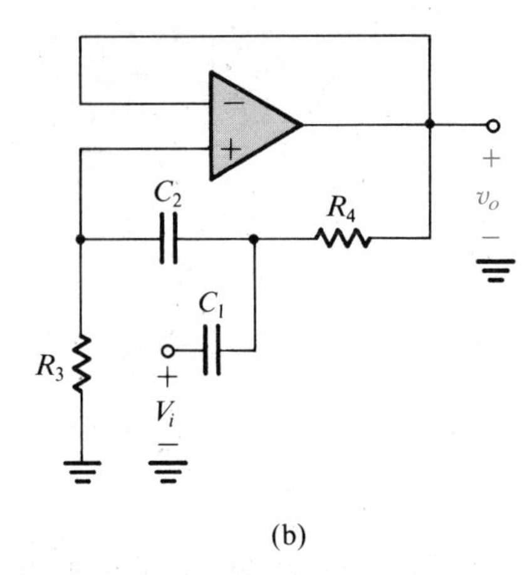

Fig. 11.33 (a) Feedback loop obtained by applying the complementary transformation to the loop in Fig. 11.29. **(b)** Injecting the input signal through C_1 realizes the high-pass function. This is one of the Sallen-and-Key family of circuits.

$$1 - \frac{A}{A+1}(1-t) = 0$$

This equation can be manipulated to the form

$$1 + At = 0$$

which is the characteristic equation of the loop in Fig. 11.32(a). As an example, consider the application of the complementary transformation to the feedback loop of Fig. 11.29: The feedback loop of Fig. 11.33(a) results. Injecting the input signal through C_1 results in the circuit in Fig. 11.33(b), which can be shown (by direct analysis) to realize a second-order high-pass function. This circuit is one of a family of SABs known as the Sallen–and–Key circuits, after their originators. The design of the circuit in Fig. 11.33(b) is based on Eqs. (11.73) through (11.76); namely, $R_3 = R$, $R_4 = R/4Q^2$, $C_1 = C_2 = C$, $CR = 2Q/\omega_0$, and the value of C is arbitrarily chosen to be practically convenient.

As another example, Fig. 11.34(a) shows the feedback loop generated by placing the two-port RC network of Fig. 11.28(b) in the negative-feedback path of an op amp. For an ideal op amp, this feedback loop realizes a pair of complex conjugate natural modes having the same location as the zeros of $t(s)$ of the RC network. Thus, using the expression for $t(s)$ given in Fig. 11.28(b), we can write for the active-filter poles

$$\omega_0 = 1/\sqrt{C_3 C_4 R_1 R_2} \qquad (11.77)$$

$$Q = \left[\frac{\sqrt{C_3 C_4 R_1 R_2}}{C_4}\left(\frac{1}{R_1} + \frac{1}{R_2}\right)\right]^{-1} \qquad (11.78)$$

Normally the design of this circuit is based on selecting $R_1 = R_2 = R$, $C_4 = C$, and $C_3 = C/m$. When substituted in Eqs. (11.77) and (11.78), these yield

$$m = 4Q^2 \qquad (11.79)$$

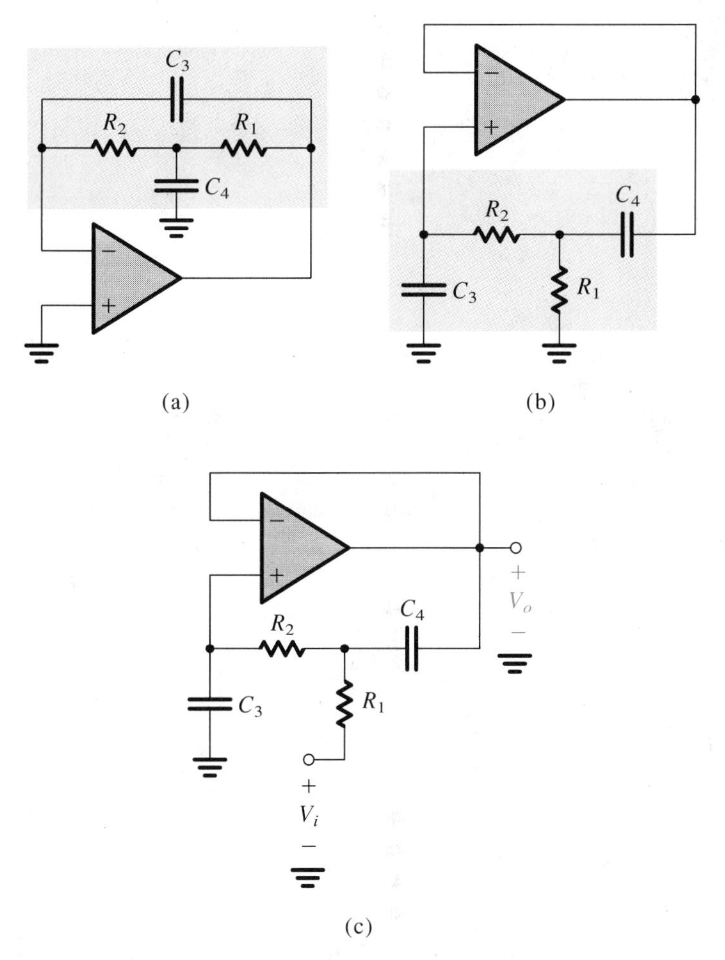

Fig. 11.34 (a) Feedback loop obtained by placing the bridged-T network of Fig. 11.28(b) in the negative-feedback path of an op amp. **(b)** Equivalent feedback loop generated by applying the complementary transformation to the loop in (a). **(c)** A low-pass filter obtained by injecting V_i through R_1 into the loop in (b).

$$CR = 2Q/\omega_0 \tag{11.80}$$

with the remaining degree of freedom (the value of C or R) left to the designer to choose.

Injecting the input signal to the C_4 terminal that is connected to ground can be shown to result in a bandpass realization. If, however, we apply the complementary transformation to the feedback loop in Fig. 11.34(a) we obtain the equivalent loop in Fig. 11.34(b). The loop equivalence means that the circuit of Fig. 11.34(b) has the same poles and thus the same ω_0 and Q and the same design equations (Eqs. 11.77 through 11.80). The new loop in Fig. 11.34(b) can be used to realize a low-pass function by injecting the input signal as shown in Fig. 11.34(c).

Exercises 11.28 Analyze the circuit in Fig. 11.34(c) to determine its transfer function $V_o(s)/V_i(s)$ and thus show that ω_0 and Q are indeed those in Eqs. (11.77) and (11.78). Also show that the dc gain is unity.

D11.29 Design the circuit in Fig. 11.34(c) to realize a low-pass filter with $f_0 = 4$ kHz and $Q = 1/\sqrt{2}$. Use 10-kΩ resistors.

Ans. $R_1 = R_2 = 10$ kΩ; $C_3 = 2.81$ nF; $C_4 = 5.63$ nF

11.9 SENSITIVITY

Because of the tolerances in component values and because of the finite op amp gain, the response of the actual assembled filter will deviate from the ideal response. As a means for predicting such deviations, the filter designer employs the concept of **sensitivity.** Specifically, for second-order filters one is normally interested in finding how *sensitive* their poles are relative to variations (both initial tolerances and future changes) in RC component values and amplifier gain. These sensitivities can be quantified using the **classical sensitivity function** S_x^y, defined

$$S_x^y \equiv \lim_{\Delta x \to 0} \frac{\Delta y/y}{\Delta x/x} \tag{11.81}$$

Thus

$$S_x^y = \frac{\partial y}{\partial x}\frac{x}{y} \tag{11.82}$$

Here x denotes the value of a component (a resistor, a capacitor, or an amplifier gain) and y denotes a circuit parameter of interest (say, ω_0 or Q). For small changes

$$S_x^y \simeq \frac{\Delta y/y}{\Delta x/x} \tag{11.83}$$

Thus we can use the value of S_x^y to determine the per-unit change in y due to a given per-unit change in x. For instance, if the sensitivity of Q relative to a particular resistance R_1 is 5, then a 1% increase in R_1 results in a 5% increase in the value of Q.

EXAMPLE 11.3 For the feedback loop of Fig. 11.29 find the sensitivities of ω_0 and Q relative to all the passive components and the op amp gain. Evaluate these sensitivities for the design considered in the previous section for which $C_1 = C_2$.

Solution To find the sensitivities with respect to the passive components, called **passive sensitivities,** we assume that the op amp gain is infinite. In this case ω_0 and Q are given by Eqs. (11.73) and (11.74). Thus for ω_0 we have

$$\omega_0 = \frac{1}{\sqrt{C_1 C_2 R_3 R_4}}$$

which can be used together with the sensitivity definition of Eq. (11.82) to obtain

$$S_{C_1}^{\omega_0} = S_{C_2}^{\omega_0} = S_{R_3}^{\omega_0} = S_{R_4}^{\omega_0} = -\tfrac{1}{2}$$

For Q we have

$$Q = \left[\sqrt{C_1C_2R_3R_4}\left(\frac{1}{C_1} + \frac{1}{C_2}\right)\frac{1}{R_3}\right]^{-1}$$

on which we apply the sensitivity definition to obtain

$$S_{C_1}^{Q} = \frac{1}{2}\left(\sqrt{\frac{C_2}{C_1}} - \sqrt{\frac{C_1}{C_2}}\right)\left(\sqrt{\frac{C_2}{C_1}} + \sqrt{\frac{C_1}{C_2}}\right)^{-1}$$

For the design with $C_1 = C_2$ we see that $S_{C_1}^{Q} = 0$. Similarly, we can show that

$$S_{C_2}^{Q} = 0, \qquad S_{R_3}^{Q} = \tfrac{1}{2}, \qquad S_{R_4}^{Q} = -\tfrac{1}{2}$$

It is important to remember that the sensitivity expression should be derived *before* substituting values corresponding to a particular design.

Next we consider the sensitivities relative to the amplifier gain. If we assume the op amp to have a finite gain A, the characteristic equation for the loop becomes

$$1 + At(s) = 0 \tag{11.84}$$

where $t(s)$ is given in Fig. 11.28(a). To simplify matters we can substitute for the passive components by their design values. This causes no errors in evaluating sensitivities, since we are now finding the sensitivity with respect to the amplifier gain. Using the design values previously obtained—namely, $C_1 = C_2 = C$, $R_3 = R$, $R_4 = R/4Q^2$, and $CR = 2Q/\omega_0$—we get

$$t(s) = \frac{s^2 + s(\omega_0/Q) + \omega_0^2}{s^2 + s(\omega_0/Q)(2Q^2 + 1) + \omega_0^2} \tag{11.85}$$

where ω_0 and Q denote the nominal or design values of the pole frequency and Q factor. The actual values are obtained by substituting for $t(s)$ in Eq. (11.84):

$$s^2 + s\frac{\omega_0}{Q}(2Q^2 + 1) + \omega_0^2 + A\left(s^2 + s\frac{\omega_0}{Q} + \omega_0^2\right) = 0$$

Assuming the gain A to be real and dividing both sides by $A + 1$, we get

$$s^2 + s\frac{\omega_0}{Q}\left(1 + \frac{2Q^2}{A + 1}\right) + \omega_0^2 = 0 \tag{11.86}$$

From this equation we see that the actual pole frequency, ω_{0a}, and pole-Q, Q_a, are

$$\omega_{0a} = \omega_0 \tag{11.87}$$

$$Q_a = \frac{Q}{1 + 2Q^2/(A + 1)} \tag{11.88}$$

Thus

$$S_A^{\omega_{0a}} = 0$$

$$S_A^{Q_a} = \frac{A}{A + 1}\,\frac{2Q^2/(A + 1)}{1 + 2Q^2/(A + 1)}$$

For $A \gg 2Q^2$ and $A \gg 1$ we obtain

$$S_A^{Q_a} \simeq \frac{2Q^2}{A}$$

It is usual to drop the subscript a in this expression and write

$$S_A^Q \simeq \frac{2Q^2}{A} \tag{11.89}$$

Note that if Q is high ($Q \geq 5$) its sensitivity relative to the amplifier gain can be quite high.[7]

A Concluding Remark

The results of Example 11.3 indicate a serious disadvantage of single-amplifier biquads—the sensitivity of Q relative to the amplifier gain is quite high. Although a technique exists for reducing S_A^Q in SABs (see Sedra, et al., 1980), this is done at the expense of increased passive sensitivities. Nevertheless, the resulting SABs are used extensively in many applications. However, for filters with Q factors greater than about 10, one usually opts for one of the multiamplifier biquads studied in Sections 11.6 and 11.7. For these circuits S_A^Q is proportional to Q, rather than to Q^2 as in the SAB case (Eq. 11.89).

Exercise 11.30 In a particular filter utilizing the feedback loop of Fig. 11.29, with $C_1 = C_2$, find the expected percentage change in ω_0 and Q under the conditions that: (a) R_3 is 2% high. (b) R_4 is 2% high. (c) Both R_3 and R_4 are 2% high. (d) Both capacitors are 2% high and both resistors are 2% low.

Ans. (a) −1%, +1%; (b) −1%, −1%; (c) −2%, 0%; (d) 0%, 0%

11.10 SWITCHED-CAPACITOR FILTERS

The active-RC filter circuits presented above have two properties that make their production in monolithic IC form difficult, if not practically impossible; these are the need for large-valued capacitors and the requirement of accurate RC time constants. The search therefore continued for a method of filter design that would lend itself more naturally to IC implementation. In this section we shall introduce one such method. At the time of this writing it appears to be the best contender for the task.

The Basic Principle

The switched-capacitor filter technique is based on the realization that a capacitor switched between two circuit nodes at a sufficiently high rate is equivalent to a resistor connecting these two nodes. To be specific, consider the active-RC integrator of Fig.

[7]Because the open-loop gain A of op amps usually has wide tolerance, it is important to keep $S_A^{\omega_0}$ and S_A^Q very small.

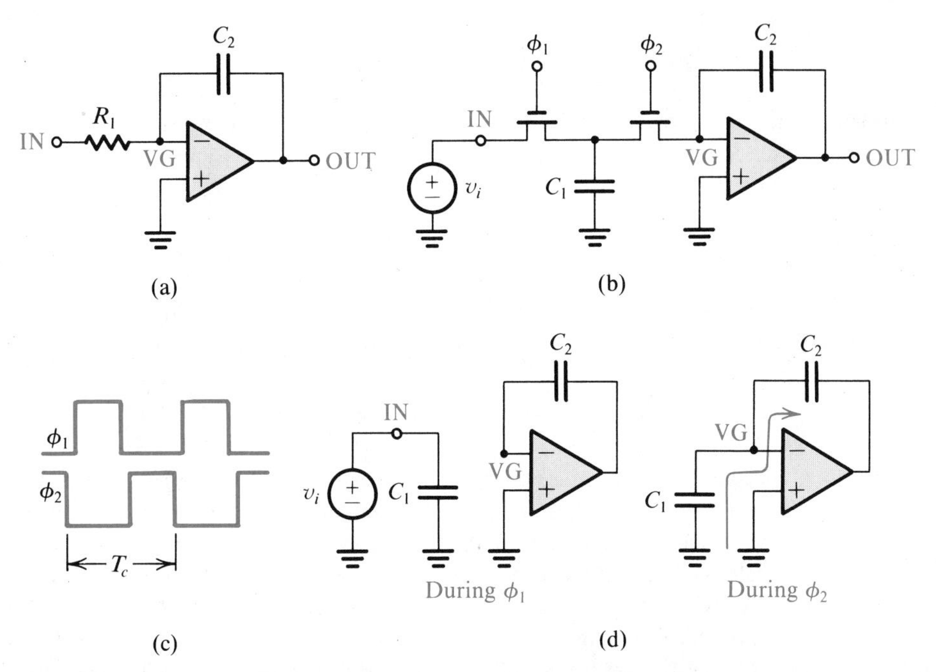

Fig. 11.35 Basic principle of the switched-capacitor filter technique. **(a)** Active-RC integrator. **(b)** Switched-capacitor integrator. **(c)** Two-phase clock (nonoverlapping). **(d)** During ϕ_1, C_1 charges up to the current value of v_i and then, during ϕ_2, discharges into C_2.

11.35(a). This is the familiar Miller integrator, which we used in the two-integrator-loop biquad in Section 11.7. In Fig. 11.35(b) we have replaced the input resistor R_1 by a grounded capacitor C_1 together with two MOS transistors acting as switches. In some circuits more elaborate switch configurations are used, but such details are beyond our present need.

The two MOS switches in Fig. 11.35(b) are driven by a *nonoverlapping* two-phase clock. Figure 11.35(c) shows the clock waveforms. We shall assume in this introductory exposition that the clock frequency f_c ($f_c = 1/T_c$) is much higher than the frequency of the signal being filtered. Thus during clock phase ϕ_1, when C_1 is connected across the input signal source v_i, the variations in the input signal are negligibly small. It follows that during ϕ_1 capacitor C_1 charges up to the voltage v_i,

$$q_{C1} = C_1 v_i$$

Then, during clock phase ϕ_2, capacitor C_1 is connected to the virtual-ground input of the op amp, as indicated in Fig. 11.35(d). Capacitor C_1 is thus forced to discharge, and its previous charge q_{C1} is transferred to C_2, in the direction indicated in Fig. 11.35(d).

From the above description we see that during each clock period T_c an amount of charge $q_{C1} = C_1 v_i$ is extracted from the input source and supplied to the integrator capacitor C_2. Thus the average current flowing between the input node (IN) and the virtual ground node (VG) is

$$i_{av} = \frac{C_1 v_i}{T_c}$$

If T_c is sufficiently short, one can think of this process as almost continuous and define an equivalent resistance R_{eq} that is in effect present between nodes IN and VG:

$$R_{eq} \equiv v_i / i_{av}$$

Thus

$$R_{eq} = T_c / C_1 \tag{11.90}$$

Using R_{eq} we obtain an equivalent time-constant for the integrator:

$$\text{Time constant} = C_2 R_{eq} = T_c \frac{C_2}{C_1} \tag{11.91}$$

Thus the time constant that determines the frequency response of the filter is determined by the clock period T_c and the capacitor ratio C_2/C_1. Both of these parameters can be well controlled in an IC process. Specifically, note the dependence on capacitor ratios rather than on absolute values of capacitors. The accuracy of capacitor ratios in MOS technology can be controlled to within 0.1%.

Another point worth observing is that with a reasonable clocking frequency (such as 100 kHz) and not-too-large capacitor ratios (say, 10) one can obtain reasonably large time constants (such as 10^{-4} s) suitable for audio applications. Since capacitors typically occupy relatively large areas on the IC chip, it is important to note that the ratio accuracies quoted above are obtainable with the smaller capacitor value as low as 0.2 pF.

Practical Circuits

The switched-capacitor (SC) circuit in Fig. 11.35(b) realizes an inverting integrator (note the direction of charge flow through C_2 in Fig. 11.35d). As we saw in Section 11.7, a two-integrator-loop active filter is composed of one inverting and one noninverting integrator.[8] To realize a switched-capacitor biquad filter we therefore need a pair of complementary switched-capacitor integrators. Figure 11.36(a) shows a noninverting, or positive, integrator circuit. The reader is urged to follow the operation of this circuit during the two clock phases and thus show that it operates much the same way as the basic circuit of Fig. 11.35(b), except for a sign reversal.

In addition to realizing a noninverting integrator function, the circuit in Fig. 11.36(a) is insensitive to stray capacitances; however, we shall not explore this point any further. The interested reader is referred to Schaumann, Ghausi, and Laker (1990). By reversal of the clock phases on two of the switches, the circuit in Fig. 11.36(b) is obtained. This circuit realizes the inverting integrator function, like the circuit of Fig. 11.35(b), but is insensitive to stray capacitances (which the original circuit of Fig. 11.35b is not). The pair of complementary integrators of Fig. 11.36 has become the standard building block in the design of switched-capacitor filters.

[8] In the two-integrator loop of Fig. 11.25(b) the noninverting integrator is realized by the cascade of a Miller integrator and an inverter.

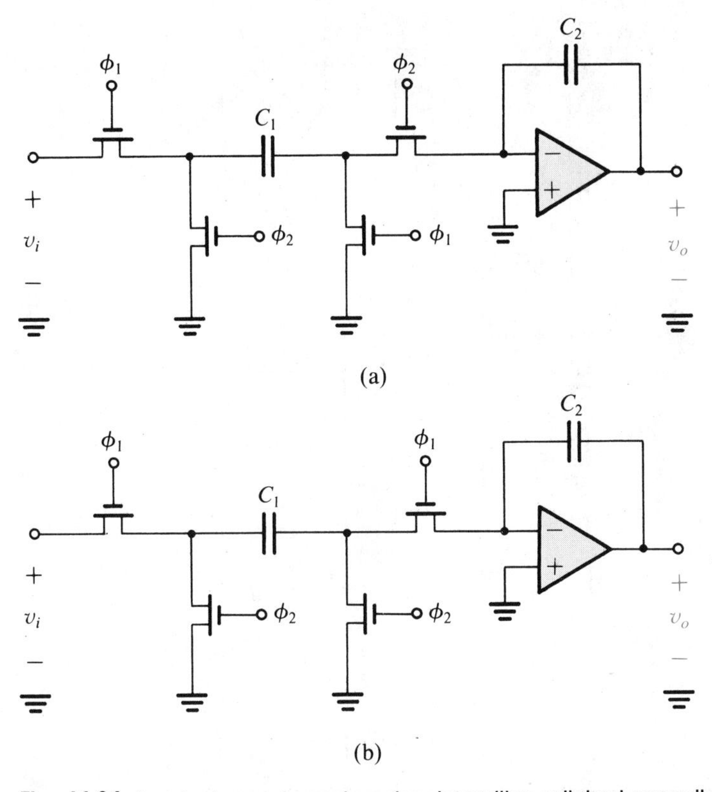

Fig. 11.36 A pair of complementary stray-insensitive switched-capacitor integrators. **(a)** Noninverting switched-capacitor integrator. **(b)** Inverting switched-capacitor integrator.

Let us now consider the realization of a complete biquad circuit. Figure 11.37(a) shows the active-RC two-integrator-loop circuit previously studied. By considering the cascade of integrator 2 and the inverter as a positive integrator, and then simply replacing each resistor by its switched-capacitor equivalent, we obtain the circuit in Fig. 11.37(b). Ignore the damping around the first integrator (that is, the switched capacitor C_5) for the time being and note that the feedback loop indeed consists of one inverting and one noninverting integrator. Then note the phasing of the switched capacitor used for damping. Reversing the phases here would convert the feedback to positive and move the poles to the right half of the s-plane. On the other hand, the phasing of the feed-in switched capacitor (C_6) is not that important; a reversal of phases would result only in an inversion in the sign of the function realized.

Having identified the correspondences between the active-RC biquad and the switched-capacitor biquad, we can now derive design equations. Analysis of the circuit in Fig. 11.37(a) yields

$$\omega_0 = \frac{1}{\sqrt{C_1 C_2 R_3 R_4}} \tag{11.92}$$

Substituting for R_3 and R_4 by their SC equivalent values, that is,

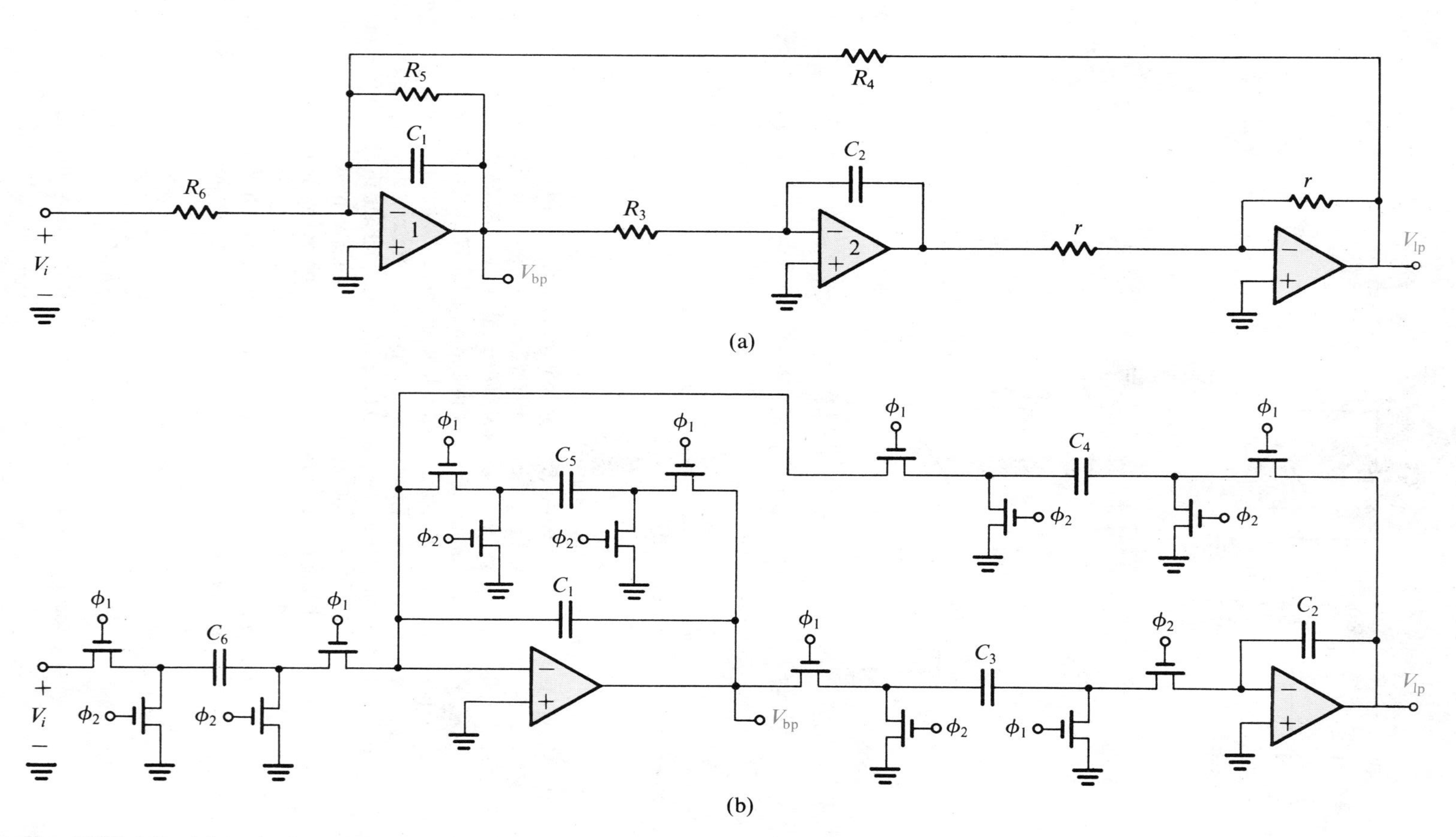

Fig. 11.37 A two-integrator-loop active-RC biquad and its switched-capacitor counterpart.

$$R_3 = T_c/C_3 \quad \text{and} \quad R_4 = T_c/C_4$$

gives ω_0 of the SC biquad as

$$\omega_0 = \frac{1}{T_c}\sqrt{\frac{C_3\,C_4}{C_2\,C_1}} \tag{11.93}$$

It is usual to select the time constants of the two integrators to be equal; that is,

$$\frac{T_c}{C_3}C_2 = \frac{T_c}{C_4}C_1 \tag{11.94}$$

If we further select the two integrating capacitors C_1 and C_2 to be equal,

$$C_1 = C_2 = C \tag{11.95}$$

then

$$C_3 = C_4 = KC, \tag{11.96}$$

where from Eq. (11.93)

$$K = \omega_0 T_c \tag{11.97}$$

For the case of equal time-constants the Q factor of the circuit in Fig. 11.37(a) is given by R_5/R_4. Thus the Q factor of the corresponding SC circuit in Fig. 11.37(b) is given by

$$Q = \frac{T_c/C_5}{T_c/C_4} \tag{11.98}$$

Thus C_5 should be selected from

$$C_5 = \frac{C_4}{Q} = \frac{KC}{Q} = \omega_0 T_c \frac{C}{Q} \tag{11.99}$$

Finally, the center-frequency gain of the bandpass function is given by

$$\text{Center-frequency gain} = \frac{C_6}{C_5} = Q\frac{C_6}{\omega_0 T_c C} \tag{11.100}$$

Exercise D11.31 Use $C_1 = C_2 = 20$ pF and design the circuit in Fig. 11.37(b) to realize a bandpass function with $f_0 = 10$ kHz, $Q = 20$, and unity center-frequency gain. Use a clock frequency $f_c = 200$ kHz. Find the values of C_3, C_4, C_5, and C_6.

Ans. 6.283 pF; 6.283 pF; 0.314 pF; 0.314 pF

A Final Remark

We have attempted to provide only an introduction to switched-capacitor filters. We have made many simplifying assumptions, the most important being the switched-capacitor–resistor equivalence (Eq. 11.90). This equivalence is correct only at $f_c = \infty$ and is approximately correct for $f_c \gg f$. Switched-capacitor filters are, in fact, sampled-data networks whose analysis and design can be carried out exactly using z-transform techniques. The interested reader is referred to the bibliography at the end of this chapter.

11.11 TUNED AMPLIFIERS

In this section we study a special kind of frequency-selective network, the LC-tuned amplifier. Figure 11.38 shows the general shape of the frequency response of a tuned amplifier. The techniques discussed apply to amplifiers with center frequencies in the range of a few hundred kHz to a few hundred MHz. Tuned amplifiers find application in the radio-frequency (RF) and intermediate-frequency (IF) sections of communications receivers and in a variety of other systems. It should be noted that the tuned-amplifier response of Fig. 11.38 is similar to that of the bandpass filter discussed in earlier sections.

As indicated in Fig. 11.38, the response is characterized by the center frequency ω_0, the 3-dB bandwidth B, and the *skirt selectivity,* which is usually measured as the ratio of the 30-dB bandwidth to the 3-dB bandwidth. In many applications, the 3-dB bandwidth is less than 5% of ω_0. This **narrow-band** property makes possible certain approximations that can simplify the design process, as will be explained later.

The tuned amplifiers studied in this section are small-signal voltage amplifiers in which the transistors operate in the class A mode. Tuned power amplifiers based on class C and other switching modes of operation are not studied in this book.

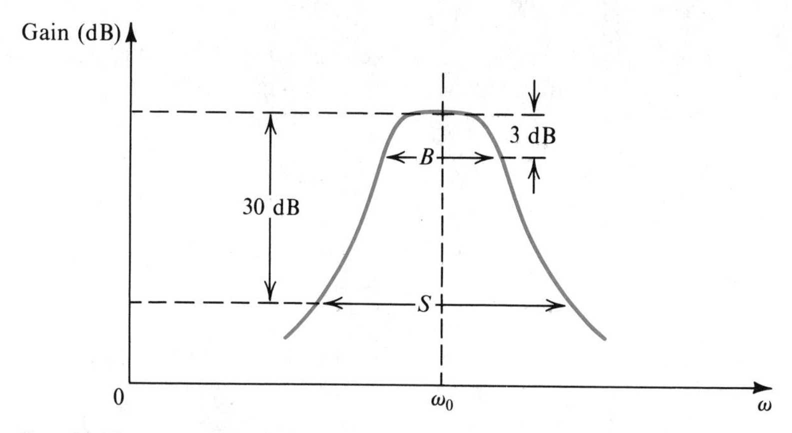

Fig. 11.38 Frequency response of a tuned amplifier.

The Basic Principle

The basic principle underlying the design of tuned amplifiers is the use of a parallel LCR circuit as the load, or at the input, of a BJT or a FET amplifier. This is illustrated in Fig. 11.39 with a MOSFET amplifier having a tuned-circuit load. For simplicity, the bias details are not included. Since this circuit uses a single tuned circuit, it is known as a **single-tuned amplifier.** The amplifier equivalent circuit is shown in Fig. 11.39(b). Here R denotes the parallel equivalent of R_L and the output resistance r_o of the FET, and C is the parallel equivalent of C_L and the FET output capacitance (usually very small). From the equivalent circuit we can write

$$V_o = \frac{-g_m V_i}{Y_L} = \frac{-g_m V_i}{sC + 1/R + 1/sL}$$

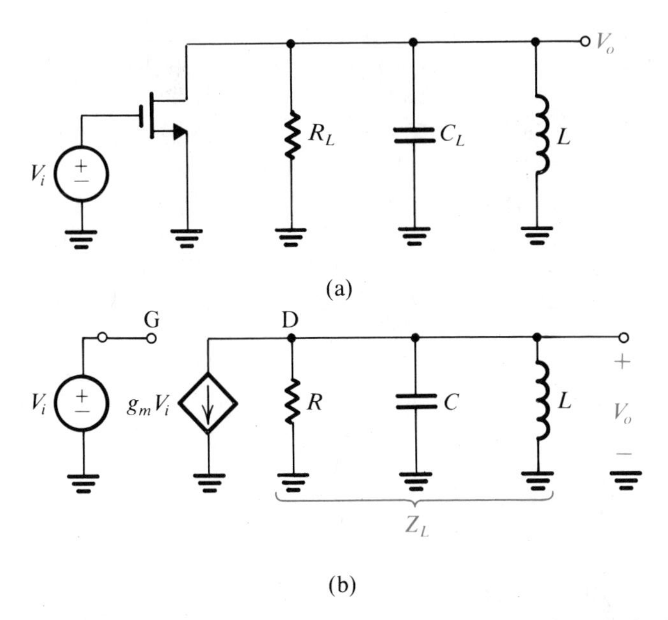

Fig. 11.39 The basic principle of tuned amplifiers is illustrated using a MOSFET with a tuned-circuit load. Bias details are not shown.

Thus the voltage gain can be expressed as

$$\frac{V_o}{V_i} = -\frac{g_m}{C} \frac{s}{s^2 + s(1/CR) + 1/LC} \tag{11.101}$$

which is a second-order bandpass function. Thus the tuned amplifier has a center frequency of

$$\omega_0 = 1/\sqrt{LC} \tag{11.102}$$

a 3-dB bandwidth of

$$B = \frac{1}{CR} \tag{11.103}$$

a Q factor of

$$Q \equiv \omega_0/B = \omega_0 CR \tag{11.104}$$

and a center-frequency gain of

$$\frac{V_o(j\omega_0)}{V_i(j\omega_0)} = -g_m R \tag{11.105}$$

Note that the expression for the center-frequency gain could have been written by inspection; at resonance the reactances of L and C cancel out and the impedance of the parallel LCR circuit reduces to R.

EXAMPLE 11.4 It is required to design a tuned amplifier of the type shown in Fig. 11.39, having $f_0 =$ 1 MHz, 3-dB bandwidth = 10 kHz, and center-frequency gain = −10 V/V. The FET available has at the bias point $g_m = 5$ mA/V and $r_o = 10$ kΩ. The output capacitance is negligibly small. Determine the values of R_L, C_L, and L.

Solution Center-frequency gain $= -10 = -5R$. Thus $R = 2$ kΩ. Since $R = R_L // r_o$, then $R_L =$ 2.5 kΩ.

$$B = 2\pi \times 10^4 = \frac{1}{CR}$$

Thus

$$C = \frac{1}{2\pi \times 10^4 \times 2 \times 10^3} = 7958 \text{ pF}$$

Since $\omega_0 = 2\pi \times 10^6 = 1/\sqrt{LC}$, we obtain

$$L = \frac{1}{4\pi^2 \times 10^{12} \times 7958 \times 10^{-12}} = 3.18\ \mu\text{H}.$$

Inductor Losses

The power loss in the inductor is usually represented by a series resistance r_s as shown in Fig. 11.40(a). However, rather than specifying the value of r_s, the usual practice is to specify the inductor Q factor at the frequency of interest,

$$Q_0 \equiv \frac{\omega_0 L}{r_s} \tag{11.106}$$

Typically, Q_0 is in the range of 50 to 200.

The analysis of a tuned amplifier is greatly simplified by representing the inductor loss by a parallel resistance R_p, as shown in Fig. 11.40(b). The relationship between R_p and Q_0 can be found by writing, for the admittance of the circuit in Fig. 11.40(a),

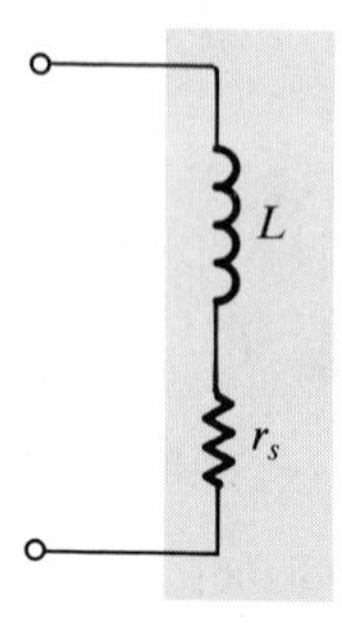

(a)

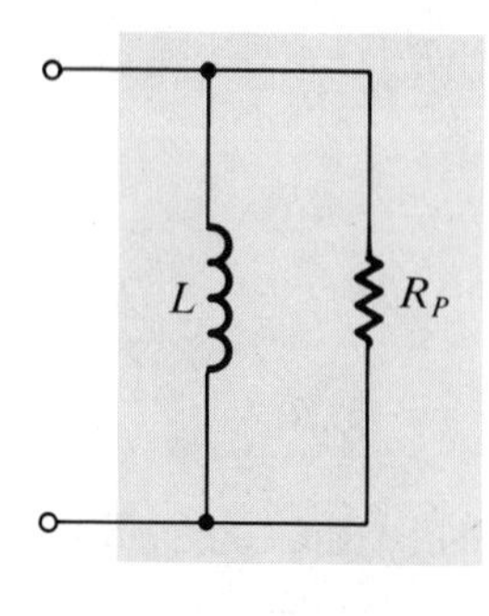

(b)

Fig. 11.40 Inductor equivalent circuits.

$$Y(j\omega_0) = \frac{1}{r_s + j\omega_0 L}$$

$$= \frac{1}{j\omega_0 L} \frac{1}{1 - j(1/Q_0)} = \frac{1}{j\omega_0 L} \frac{1 + j(1/Q_0)}{1 + (1/Q_0^2)}$$

For $Q_0 \gg 1$,

$$Y(j\omega_0) \simeq \frac{1}{j\omega_0 L}\left(1 + j\frac{1}{Q_0}\right) \tag{11.107}$$

Equating this to the admittance of the circuit in Fig. 11.40(b) gives

$$Q_0 = \frac{R_p}{\omega_0 L} \tag{11.108}$$

or, equivalently,

$$R_p = \omega_0 L Q_0 \tag{11.109}$$

Finally, it should be noted that the coil Q factor poses an upper limit on the value of Q achieved by the tuned circuit.

Exercise 11.32 If the inductor in Example 11.4 has $Q_0 = 150$, find R_p and then find the value to which R_L should be changed so as to keep the overall Q, and hence the bandwidth, unchanged.

Ans. 3 kΩ; 15 kΩ

Use of Transformers

In many cases it is found that the required value of inductance is not practical, in the sense that coils with the required inductance might not be available with the required high values of Q_0. A simple solution is to use a transformer to effect an impedance change. Alternatively, a tapped coil, known as an **autotransformer,** can be used, as shown in Fig. 11.41. Provided the two parts of the inductor are tightly coupled, which can be achieved by winding it on a ferrite core, the transformation relationships shown hold. The result is that the tuned circuit seen between terminals 1 and 1′ is equivalent to that in Fig. 11.39(b). For example, if a turns ratio $n = 3$ is used in the amplifier of Example 11.4, then a coil with inductance $L' = 9 \times 3.18 = 28.6\ \mu\text{H}$ and a capacitance $C' = 7{,}958/9 = 884$ pF will be required. Both of these values are more practical than the original ones.

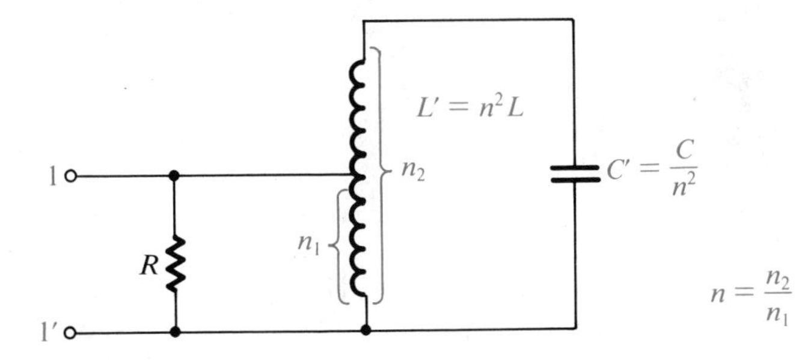

Fig. 11.41 A tapped inductor is used as an impedance transformer to allow using a higher inductance, L', and a smaller capacitance, C'.

In applications that involve coupling the output of a tuned amplifier to the input of another amplifier, the tapped coil can be used to raise the effective input resistance of the latter amplifier stage. In this way, one can avoid reduction of the overall Q. This point is illustrated in Fig. 11.42 and in the following exercises.

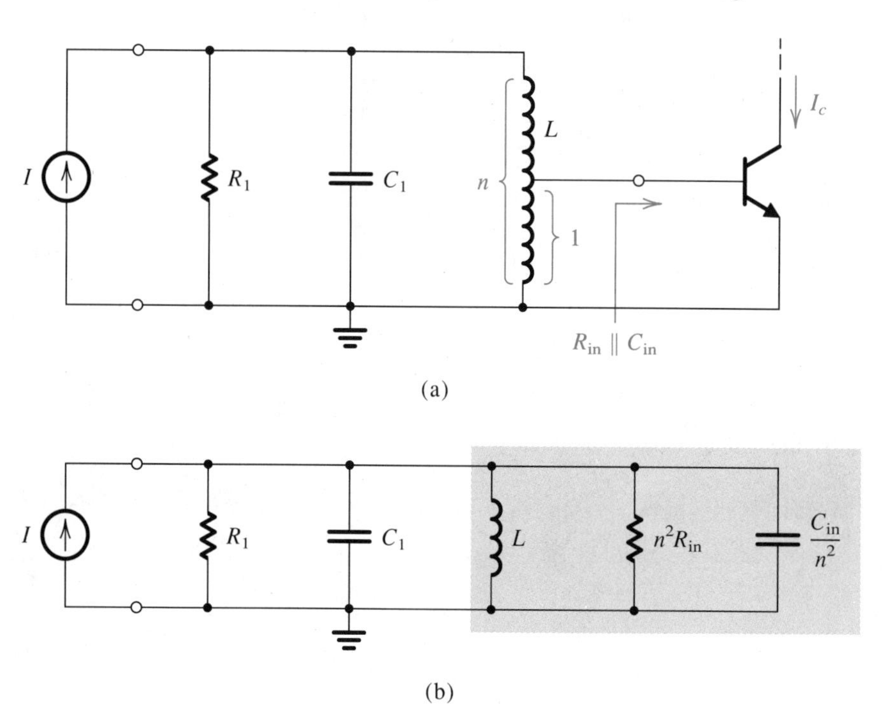

Fig. 11.42 **(a)** The output of a tuned amplifier is coupled to the input of another amplifier via a tapped coil, and **(b)** equivalent circuit. Note that the use of a tapped coil increases the effective input impedance of the second amplifier stage.

Exercises **11.33** Consider the circuit in Fig. 11.42(a), first without tapping the coil. Let $L = 5\ \mu$H and assume that R_1 is fixed at 1 kΩ. We wish to design a tuned amplifier with $f_0 = 455$ kHz and a 3-dB bandwidth of 10 kHz (this is the IF amplifier of an AM radio). If the BJT has $R_{in} = 1$ kΩ and $C_{in} = 200$ pF, find the actual bandwidth obtained and the required value of C_1.

Ans. 13 kHz; 24.27 nF

11.34 Since the bandwidth realized in Exercise 11.33 is greater than desired, find an alternative design utilizing a tapped coil as in Fig. 11.42(a). Find the value of n that allows the specifications to be just met. Also find the new required value of C_1 and the current gain I_c/I at resonance. Assume that at the bias point the BJT has $g_m = 40$ mA/V.

Ans. 1.36; 24.36 nF; 19.1 A/A

Amplifiers with Multiple Tuned Circuits

The selectivity achieved with the single-tuned circuit of Fig. 11.39 is not sufficient in many applications—for instance, in the IF amplifier of a radio or a TV receiver. Greater selectivity is obtained by using additional tuned stages. Figure 11.43 shows a BJT with

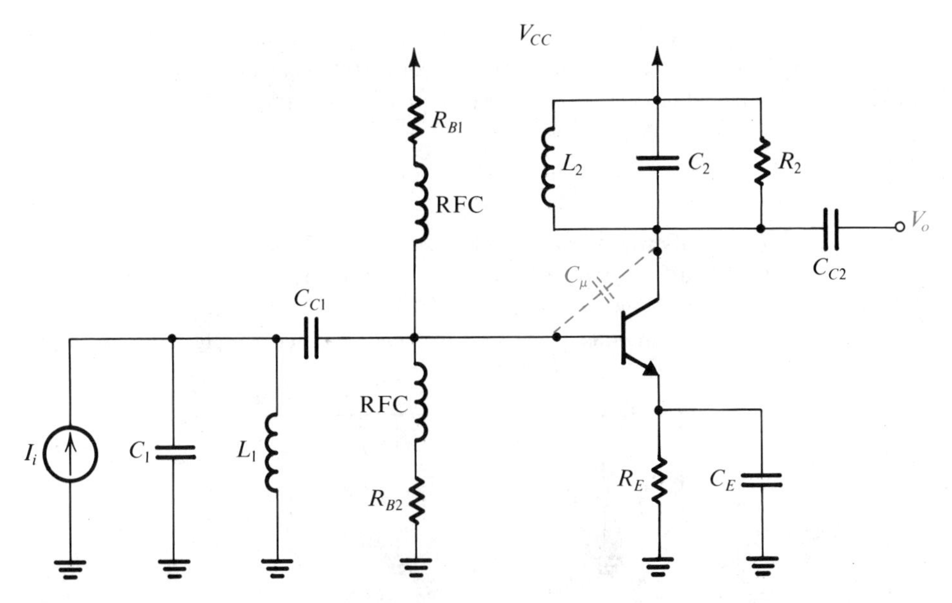

Fig. 11.43 A BJT amplifier with tuned circuits at the input and the output.

tuned circuits at both the input and the output.[9] In this circuit the bias details are shown, from which we note that biasing is quite similar to the classical arrangement employed in discrete-circuit design. However, in order to avoid the loading effect of the bias resistors R_{B1} and R_{B2} on the input tuned circuit, a **radio-frequency choke** (RFC) is inserted in series with each resistor. These chokes have high impedances at the frequencies of interest. The use of RFCs in biasing tuned RF amplifiers is common practice.

The analysis and design of the double-tuned amplifier of Fig. 11.43 is complicated by the Miller effect[10] due to capacitance C_μ. Since the load is not simply resistive, as was the case in the amplifiers studied in Chapter 7, the Miller impedance at the input will be complex. This reflected impedance will cause detuning of the input circuit as well as "skewing" of the response of the input circuit. Needless to say, the coupling introduced by C_μ makes tuning or aligning the amplifier quite difficult. Worse still, the capacitor C_μ can cause oscillations to occur (see Gray and Searle, 1969, and Problem 11.75).

Methods exist for **neutralizing** the effect of C_μ, using additional circuits so arranged as to feed back a current equal and opposite to that through C_μ. An alternative, and preferred, approach is to use circuit configurations that do not suffer from the Miller effect. These will be discussed below. Before leaving this section, however, we wish to point out that circuits of the type shown in Fig. 11.43 are usually designed utilizing the y-parameter model of the BJT (see Appendix B). This is done because here, in view of the fact that C_μ plays a significant role, the y-parameter model makes the analysis simpler (as

[9] Note that because the input circuit is a parallel resonant circuit, an input current source (rather than voltage source) signal is utilized.

[10] Here we use "Miller effect" to refer to the effect of the feedback capacitance C_μ in reflecting an input impedance that is a function of the amplifier load impedance.

compared to that using the hybrid-π model). Also, the y parameters can easily be measured at the particular frequency of interest, ω_0. For narrow-band amplifiers, the assumption is usually made that the y parameters remain approximately constant over the passband.

The Cascode and the CC-CB Cascade

From our study of amplifier frequency response in Chapter 7 we know that two amplifier configurations do not suffer from the Miller effect. These are the cascode configuration and the common-collector common-base cascade. Figure 11.44 shows tuned amplifiers based on these two configurations. The CC-CB cascade is usually preferred in IC imple-

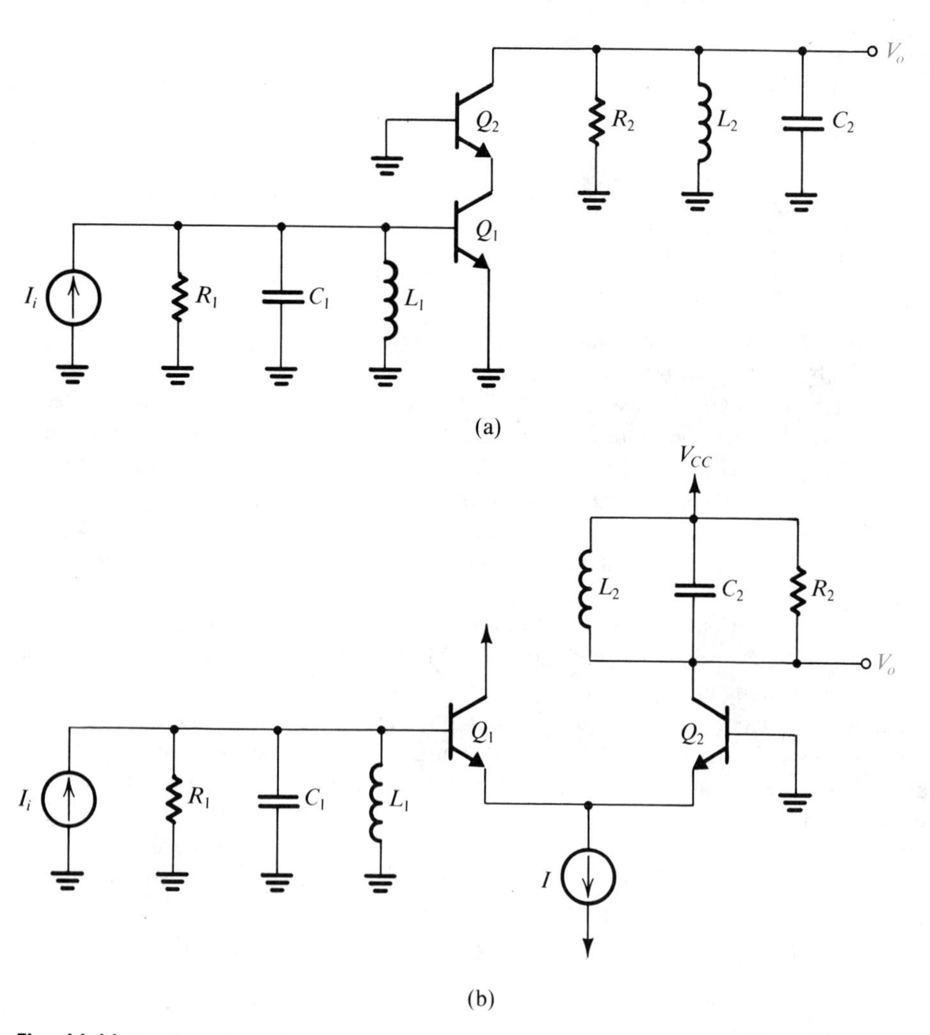

Fig. 11.44 Two tuned-amplifier configurations that do not suffer from the Miller effect: **(a)** cascode; **(b)** common-collector common-base cascade. (Note that bias details of the cascode circuit are not shown.)

mentations because its differential structure makes it suitable for IC biasing techniques. (Note that the biasing details of the cascode circuit are not shown in Fig. 11.43. Biasing can be done using arrangements similar to those discussed in earlier chapters.)

Synchronous Tuning

In the design of a tuned amplifier with multiple tuned circuits the question arises as to the frequency to which each circuit should be tuned. The object, of course, is for the overall response to exhibit high passband flatness and skirt selectivity. To investigate this question we shall assume that the overall response is the product of the individual responses; in other words, the stages do not interact. This can easily be achieved using circuits such as those in Fig. 11.44.

Consider first the case of N identical resonant circuits, known as the **synchronously tuned** case. Figure 11.45 shows the response of an individual stage and that of the cascade. Observe the bandwidth "shrinkage" of the overall response. The 3-dB bandwidth B of the overall amplifier is related to that of the individual tuned circuits, ω_0/Q, by (see Problem 11.77)

$$B = \frac{\omega_0}{Q}\sqrt{2^{1/N} - 1} \tag{11.110}$$

The factor $\sqrt{2^{1/N} - 1}$ is known as the bandwidth-shrinkage factor. Given B and N, we can use Eq. (11.110) to determine the bandwidth required of the individual stages.

Exercise 11.35 Consider the design of an IF amplifier for an FM radio receiver. Using two synchronously tuned stages with $f_0 = 10.7$ MHz, find the 3-dB bandwidth of each stage so that the overall bandwidth is 200 kHz. Using 3-μH inductors find C and R for each stage.

Ans. 310.8 kHz; 73.7 pF; 6.95 kΩ

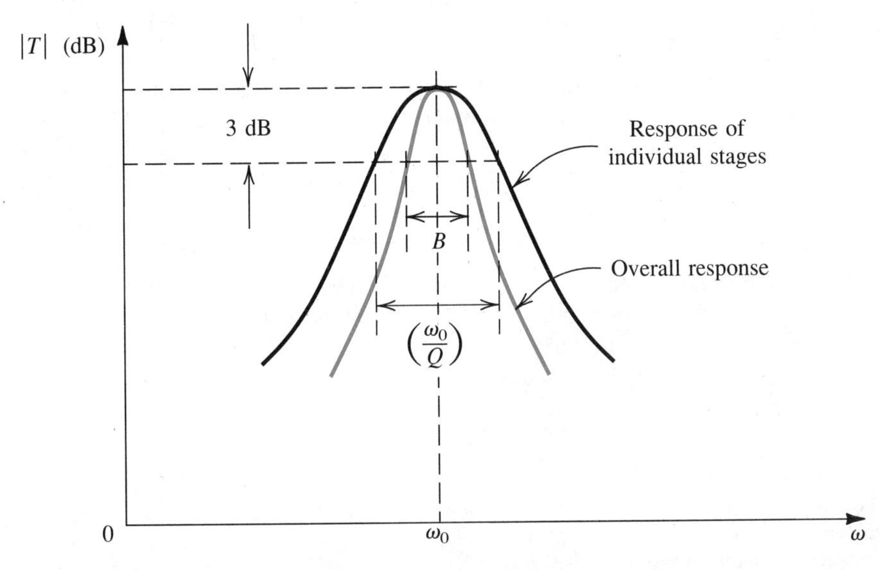

Fig. 11.45 Frequency response of a synchronously tuned amplifier.

Stagger-Tuning

A much better overall response is obtained by stagger-tuning the individual stages, as illustrated in Fig. 11.46. Stagger-tuned amplifiers are usually designed so that the overall response exhibits *maximal flatness* around the center frequency f_0. Such a response can be obtained by transforming the response of a maximally flat (Butterworth) low-pass filter up the frequency axis to ω_0. We show here how this can be done.

The transfer function of a second-order bandpass filter can be expressed in terms of its poles as

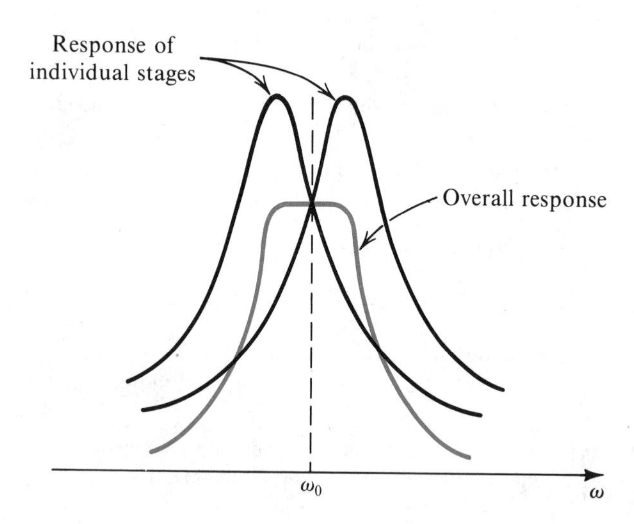

Fig. 11.46 Stagger-tuning the individual resonant circuits can result in an overall response with a passband flatter than that obtained with synchronous tuning (Fig. 11.45).

$$T(s) = \frac{a_1 s}{\left(s + \frac{\omega_0}{2Q} - j\omega_0\sqrt{1 - \frac{1}{4Q^2}}\right)\left(s + \frac{\omega_0}{2Q} + j\omega_0\sqrt{1 - \frac{1}{4Q^2}}\right)} \tag{11.111}$$

For a narrow-band filter $Q \gg 1$, and for values of s in the neighborhood of $+j\omega_0$ (see Fig. 11.47b) the second factor in the denominator is approximately $(2j\omega_0)$. Hence Eq. (11.111) can be approximated by

$$T(s) \simeq \frac{a_1/2}{s + \omega_0/2Q - j\omega_0} = \frac{a_1/2}{(s - j\omega_0) + \omega_0/2Q} \tag{11.112}$$

This is known as the **narrow-band approximation.**[11] Note that the magnitude response, for $s = j\omega$, has a peak value of $a_1 Q/\omega_0$ at $\omega = \omega_0$, as expected.

Now consider a first-order low-pass network with a single pole at $p = -\omega_0/2Q$ (we use p to denote the complex frequency variable for the low-pass filter). Its transfer function is

[11] The bandpass response is *geometrically symmetric* around the center frequency ω_0. That is, each two frequencies ω_1 and ω_2 at which the magnitude response is equal are related by $\omega_1\omega_2 = \omega_0^2$. For high Q, the symmetry becomes almost *arithmetic* for frequencies close to ω_0. That is, two frequencies with the same magnitude response are almost equally spaced from ω_0. The same is true for higher-order bandpass filters designed using the transformation presented in this section.

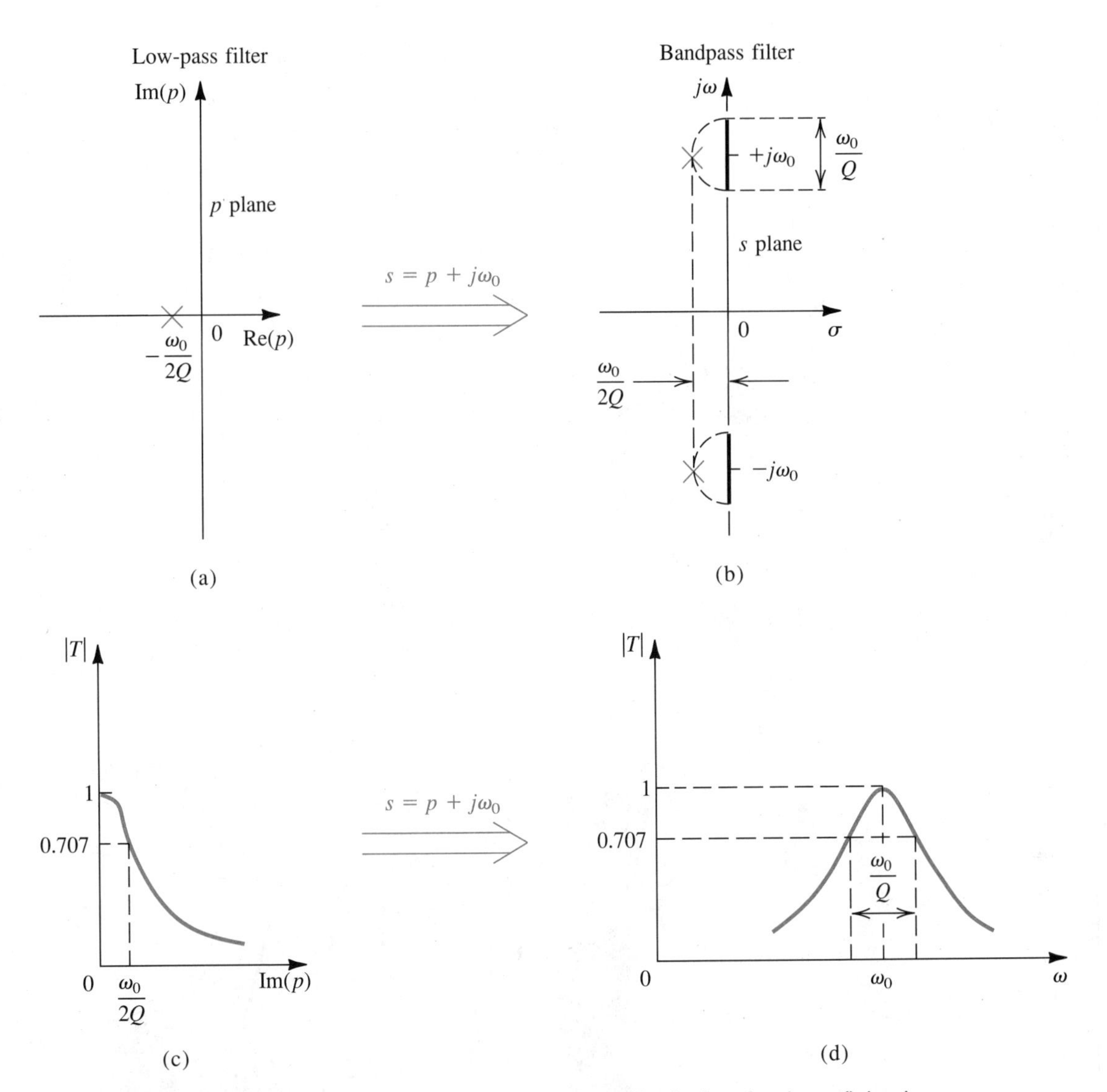

Fig. 11.47 Obtaining a second-order narrow-band bandpass filter by transforming a first-order low-pass filter. **(a)** Pole of the first-order filter in the p-plane. **(b)** Applying the transformation $s = p + j\omega_0$ and adding a complex conjugate pole results in the poles of the second-order bandpass filter. **(c)** Magnitude response of the first-order low-pass filter. **(d)** Magnitude response of the second-order bandpass filter.

$$T(p) = \frac{K}{p + \omega_0/2Q} \tag{11.113}$$

where K is a constant. Comparing Eqs. (11.112) and (11.113) we note that they are identical for $p = s - j\omega_0$ or, equivalently,

$$s = p + j\omega_0 \tag{11.114}$$

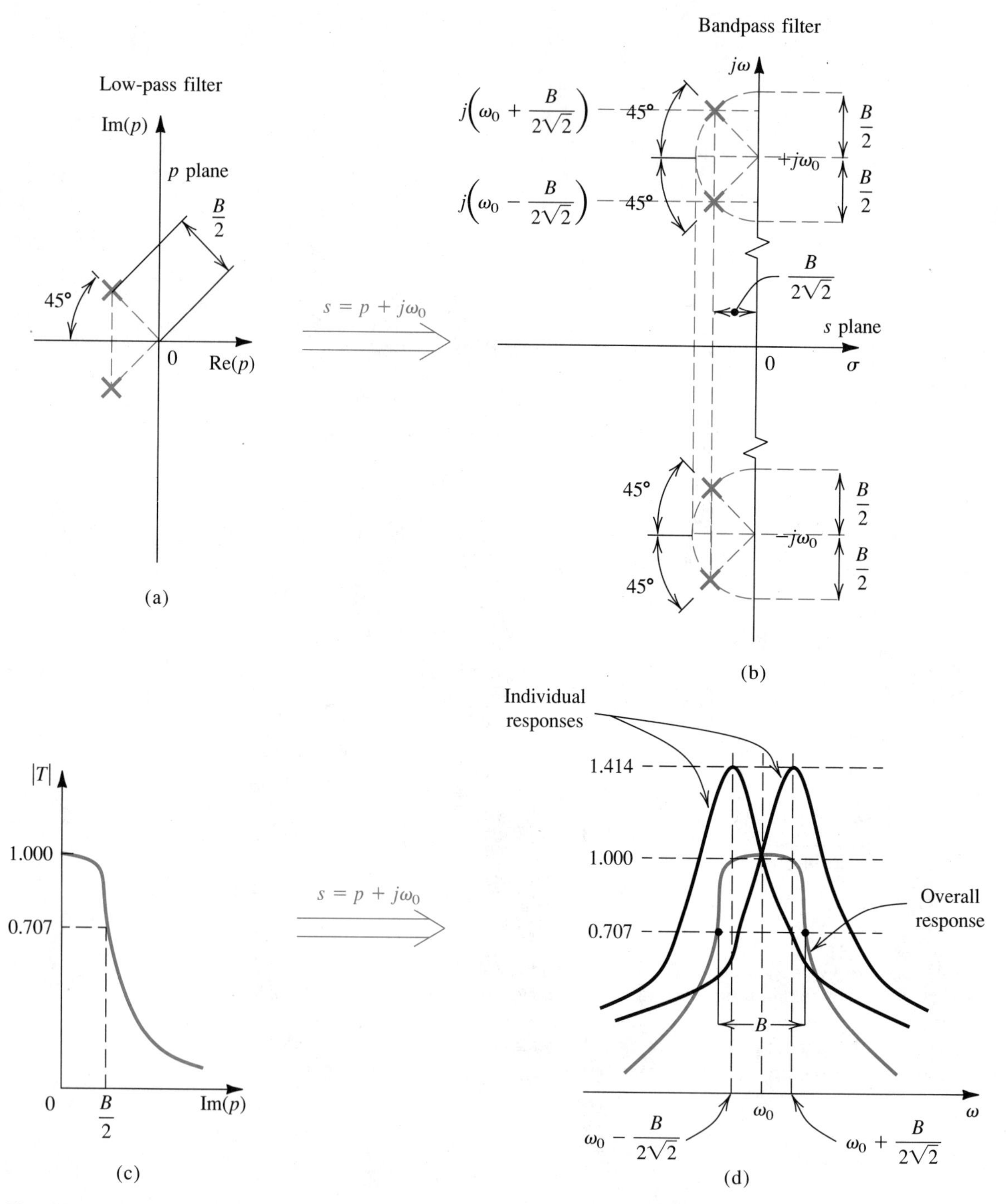

Fig. 11.48 Obtaining the poles and the frequency response of a fourth-order stagger-tuned narrow-band bandpass amplifier by transforming a second-order low-pass maximally flat response.

This result implies that the response of the second-order bandpass filter *in the neighborhood of its center frequency* $s = j\omega_0$ is identical to the response of a first-order low-pass filter with a pole at $(-\omega_0/2Q)$ *in the neighborhood of* $p = 0$. Thus the bandpass response can be obtained by shifting the pole of the low-pass prototype and adding the complex conjugate pole, as illustrated in Fig. 11.47. This is called a **low-pass to bandpass transformation** for *narrow-band* filters.

The transformation $p = s - j\omega_0$ can be applied to low-pass filters of order greater than one. For instance, we can transform a maximally flat second-order low-pass filter $(Q = 1/\sqrt{2})$ to obtain a maximally flat bandpass filter. If the 3-dB bandwidth of the bandpass filter is to be B rad/s, then the low-pass filter should have a 3-dB frequency and thus a pole frequency of $(B/2)$ rad/s, as illustrated in Fig. 11.48. The resulting fourth-order bandpass filter will be a stagger-tuned one, with its two tuned circuits (refer to Fig. 11.48) having

$$\omega_{01} = \omega_0 + \frac{B}{2\sqrt{2}} \qquad B_1 = \frac{B}{\sqrt{2}} \qquad Q_1 \simeq \frac{\sqrt{2}\omega_0}{B} \tag{11.115}$$

$$\omega_{02} = \omega_0 - \frac{B}{2\sqrt{2}} \qquad B_2 = \frac{B}{\sqrt{2}} \qquad Q_2 \simeq \frac{\sqrt{2}\omega_0}{B} \tag{11.116}$$

Note that in order for the overall response to have a normalized center-frequency gain of unity, the individual responses are shown in Fig. 11.48(d) to have equal center-frequency gains of $\sqrt{2}$. In practice, however, the individual responses need not have equal center-frequency gains, as illustrated in the following exercises.

Exercises 11.36 A stagger-tuned design for the IF amplifier specified in Exercise 11.35 is required. Find f_{01}, B_1, f_{02}, and B_2. Also give the value of C and R for each of the two stages. (Recall that 3-μH inductors are to be used.)

Ans. 10.77 MHz; 141.4 kHz; 10.63 MHz; 141.4 kHz; 72.8 pF; 15.5 kΩ; 74.7 pF; 15.1 kΩ

11.37 Using the fact that the voltage gain at resonance is proportional to the value of R, find the ratio of the gain at 10.7 MHz of the stagger-tuned amplifier designed in Exercise 11.36 and the synchronously tuned amplifier designed in Exercise 11.35. (*Hint:* For the stagger-tuned amplifier note that the gain at ω_0 is equal to the product of the gains of the individual stages at their 3-dB frequencies.)

Ans. 2.42

SUMMARY

- A filter is a linear two-port network with a transfer function $T(s) = V_o(s)/V_i(s)$. For physical frequencies, the filter transmission is expressed as $T(j\omega) = |T(j\omega)|e^{j\phi(\omega)}$. The magnitude of transmission can be expressed in decibels using either the gain function $G(\omega) \equiv 20 \log |T|$ or the attenuation function $A(\omega) \equiv -20 \log |T|$.

- The transmission characteristics of a filter are specified in terms of the edges of the passband(s) and the stopband(s); the maximum allowed variation in passband transmission, A_{max} (dB); and the minimum attenuation required in the stopband, A_{min} (dB). In some applications the phase characteristics are also specified.

- The filter transfer function can be expressed as the ratio of two polynomials in s; the degree of the denominator polynomial, N, is the filter order. The N roots of the denominator polynomial are the poles (natural modes).

- To obtain a highly selective response, the poles are complex and occur in conjugate pairs (except for one real pole when N is odd). The zeros are placed on the $j\omega$-axis in the stopband(s) including $\omega = 0$ and $\omega = \infty$.

▸ The Butterworth filter approximation provides a low-pass response that is maximally flat at $\omega = 0$. The transmission decreases monotonically as ω increases, reaching 0 (infinite attenuation) at $\omega = \infty$, where all N transmission zeros lie. Eq. (11.11) gives $|T|$, where ϵ is given by Eq. (11.14) and the order N is determined using Eq. (11.15). The poles are found using the graphical construction of Fig. 11.10, and the transfer function is given by Eq. (11.16).

▸ The Chebyshev filter approximation provides a low-pass response that is equiripple in the passband with the transmission decreasing monotonically in the stopband. All the transmission zeros are at $s = \infty$. Eq. (11.18) gives $|T|$ in the passband and Eq. (11.19) gives $|T|$ in the stopband, where ϵ is given by Eq. (11.21). The order N can be determined using Eq. (11.22). The poles are given by Eq. (11.23) and the transfer function by Eq. (11.24).

▸ Figures 11.13 and 11.14 provide a summary of first-order filter functions and their realizations.

▸ Figure 11.16 provides the characteristics of seven special second-order filtering functions.

▸ The second-order LCR resonator of Fig. 11.17(a) realizes a pair of complex conjugate poles with $\omega_0 = 1/\sqrt{LC}$ and $Q = \omega_0 CR$. This resonator can be used to realize the various special second-order filtering functions as shown in Fig. 11.18.

▸ By replacing the inductor of an LCR resonator with a simulated inductance obtained using the Antoniou circuit of Fig. 11.20(a), the op amp–RC resonator of Fig. 11.21(b) is obtained. This resonator can be used to realize the various second-order filter functions as shown in Fig. 11.22. The design equations for these circuits are given in Table 11.1.

▸ Biquads based on the two-integrator-loop topology are the most versatile and popular second-order filter realizations. There are two varieties: the KHN circuit of Fig. 11.24(a), which realizes the LP, BP, and HP functions simultaneously and can be combined with the output summing amplifier of Fig. 11.28(b) to realize the notch and all-pass functions; and the Tow–Thomas circuit of Fig. 11.25(b), which realizes the BP and LP functions simultaneously. Feedforward can be applied to the Tow–Thomas circuit to obtain the circuit of Fig. 11.26, which can be designed to realize any of the second-order functions [see Table 11.2].

▸ Single-amplifier biquads are obtained by placing a bridged-T network in the negative feedback path of an op amp. If the op amp is ideal, the poles realized are at the same locations as the zeros of the RC network. The complementary transformation can be applied to the feedback loop to obtain another feedback loop having identical poles. Different transmission zeros are realized by feeding the input signal to circuit nodes that are connected to ground. SABs are economic in their use of op amps but are sensitive to the op amp nonidealities and are thus limited to low-Q applications ($Q \leq 10$).

▸ The classical sensitivity function

$$S_x^y = \frac{\partial y/y}{\partial x/x}$$

is a very useful tool in investigating how tolerant a filter circuit is to the unavoidable inaccuracies in component values and to the nonidealities of the op amps.

▸ Switched-capacitor filters are based on the principle that a capacitor C, periodically switched between two circuit nodes at a high rate, f_c, is equivalent to a resistance $R = 1/Cf_c$ connecting the two circuit nodes. SC filters can be fabricated in monolithic form using CMOS IC technology.

▸ Tuned amplifiers utilize LC tuned circuits as loads, or at the input, of transistor amplifiers. They are used in the design of the tuner and the IF section of communication receivers. The cascode and the CC-CB cascade configurations are frequently used in the design of tuned amplifiers. Stagger-tuning the individual tuned circuits results in a flatter passband response (as compared to that obtained with all the tuned circuits synchronously tuned).

BIBLIOGRAPHY

P. E. Allen and E. Sanchez-Sinencio, *Switched-Capacitor Circuits,* New York: Van Nostrand Reinhold, 1984.

R. W. Brodersen, P. R. Gray, and D. A. Hodges, "MOS switched-capacitor filters," *Proceedings of the IEEE,* vol. 67, pp. 61–74, Jan. 1979.

K. K. Clarke and D. T. Hess, *Communication Circuits: Analysis and Design,* Chap. 6, Reading, Mass.: Addison Wesley, 1971.

M. S. Ghausi, *Electronic Devices and Circuits: Discrete and Integrated,* New York: Holt, Rinehart and Winston, 1985.

P. E. Gray and C. L. Searle, *Electronic Principles,* Chap. 17, New York: Wiley, 1969.

R. Gregorian and G. C. Temes, *Analog MOS Integrated Circuits for Signal Processing,* New York: Wiley-Interscience, 1986.

K. Martin, "Improved circuits for the realization of switched-capacitor filters," *IEEE Transactions on Circuits and Systems,* vol. CAS-27, no. 4, pp. 237–244, April 1980.

S. K. Mitra and C. F. Kurth, (eds.), *Miniaturized and Integrated Filters,* New York: Wiley-Interscience, 1989.

R. Schaumann, M. S. Ghausi, and K. R. Laker, *Design of Analog Filters,* Englewood Cliffs, N.J.: Prentice-Hall, 1990.

R. Schaumann, M. Soderstrand, and K. Laker, (eds.), *Modern Active Filter Design,* New York: IEEE Press, 1981.

A. S. Sedra, "Switched-capacitor filter synthesis," in *MOS VLSI Circuits for Telecommunications,* Y. Tsividis and P. Antognetti, (eds.), Englewood Cliffs, N.J.: Prentice-Hall, 1985.

A. S. Sedra and P. O. Brackett, *Filter Theory and Design: Active and Passive,* Portland, Ore.: Matrix, 1978.

A. S. Sedra, M. Ghorab, and K. Martin, "Optimum configurations for single-amplifier biquadratic filters," *IEEE Transactions on Circuits and Systems,* vol. CAS-27, no. 12, pp. 1155–1163, Dec. 1980.

W. M. Snelgrove, *FILTOR 2: A Computer-Aided Filter Design Package,* Dept. of Electrical Engineering, University of Toronto, 1981.

M. E. Van Valkenburg, *Analog Filter Design,* New York: Holt, Rinehart, and Winston, 1981.

A. I. Zverev, *Handbook of Filter Synthesis,* New York: Wiley, 1967.

PROBLEMS

Section 11.1: Filter Transmission, Types and Specification

11.1 The transfer function of a first-order low-pass filter (such as that realized by an RC circuit) can be expressed as $T(s) = \omega_0/(s + \omega_0)$, where ω_0 is the 3-dB frequency of the filter. Give in table form the values of $|T|$, ϕ, G, and A at $\omega = 0, 0.5\,\omega_0, \omega_0, 2\,\omega_0, 5\,\omega_0, 10\,\omega_0$, and $100\,\omega_0$.

*11.2 A filter has the transfer function $T(s) = 1/[(s + 1)(s^2 + s + 1)]$. Show that $|T| = \sqrt{1 + \omega^6}$ and find an expression for its phase response $\phi(\omega)$. Calculate the values of $|T|$ and ϕ for $\omega = 0.1$, 1, and 10 rad/s and then find the output corresponding to each of the following input signals:

(a) $2 \sin 0.1t$ (volts)

(b) $2 \sin t$ (volts)

(c) $2 \sin 10\,t$ (volts)

11.3 For the filter whose magnitude response is sketched (the colored curve) in Fig. 11.3 find $|T|$ at $\omega = 0$, $\omega = \omega_p$, and $\omega = \omega_s$. $A_{max} = 0.5$ dB, and $A_{min} = 40$ dB.

D11.4 A low-pass filter is required to pass all signals within its passband, extending from 0 to 4 kHz, with a transmission variation of at most 10% (i.e., the ratio of the maximum to minimum transmission in the passband should not exceed 1.1). The transmission in the stopband which extends from 5 kHz to ∞ should not exceed 0.1% of the maximum passband transmission. What are the values of A_{max}, A_{min}, and the selectivity factor for this filter?

11.5 A low-pass filter is specified to have $A_{max} = 1$ dB and $A_{min} = 10$ dB. It is found that its specifications can be just met with a single-time-constant RC circuit having a time constant of 1 s and a dc transmission of unity. What must ω_p and ω_s of this filter be? What is the selectivity factor?

11.6 Sketch transmission specifications for a high-pass filter having a passband defined by $f \geq 2$ kHz and a stopband defined by $f \leq 1$ kHz. $A_{max} = 0.5$ dB, and $A_{min} = 50$ dB.

11.7 Sketch transmission specifications for a bandstop filter that is required to pass signals over the bands $0 \leq f \leq 10$ kHz and 20 kHz $\leq f \leq \infty$ with A_{max} of 1 dB. The stopband extends from $f = 12$ kHz to $f = 16$ kHz, with a minimum required attenuation of 40 dB.

Section 11.2: The Filter Transfer Function

11.8 Consider a fifth-order filter whose poles are all at a radial distance from the origin of 10^3 rad/s. One pair of complex conjugate poles is at 18° angles from the $j\omega$-axis, and the other pair is at 54° angles. Give the transfer function in each of the following cases:

(a) The transmission zeros are all at $s = \infty$ and the dc gain is unity.

(b) The transmission zeros are all at $s = 0$ and the high-frequency gain is unity.

What type of filter results in each case?

11.9 A third-order low-pass filter has transmission zeros at $\omega = 2$ rad/s and $\omega = \infty$. Its natural modes are at $s = -1$ and $s = -0.5 \pm j0.8$. The dc gain is unity. Find $T(s)$.

11.10 Find the order N and the form of $T(s)$ of a bandpass filter having transmission zeros as follows: one at $\omega = 0$, one at $\omega = 10^3$ rad/s, one at 3×10^3 rad/s, one at 6×10^3 rad/s, and one at $\omega = \infty$. If this filter has a monotonically decreasing passband transmission with a peak at the center frequency of 2×10^3 rad/s, and equiripple response in the stopbands, sketch the shape of its $|T|$.

***11.11** Analyze the RLC network of Fig. P11.11 to determine its transfer function $V_o(s)/V_i(s)$ and hence its poles and zeros. (*Hint:* Begin the analysis at the output and work your way back to the input.)

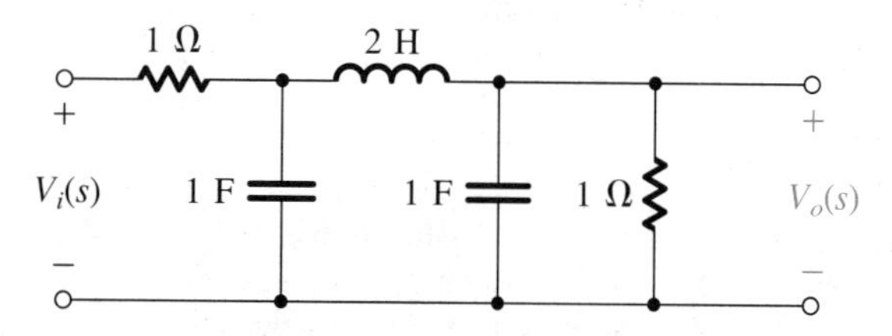

Fig. P11.11

Section 11.3: Butterworth and Chebyshev Filters

D11.12 Determine the order N of the Butterworth filter for which $A_{max} = 1$ dB, $A_{min} \geq 20$ dB, and the selectivity ratio $\omega_s/\omega_p = 1.3$. What is the actual value of minimum stopband attenuation realized? If A_{min} is to be exactly 20 dB, to what value can A_{max} be reduced?

11.13 Calculate the value of attenuation obtained at a frequency 1.6 times the 3-dB frequency of a seventh-order Butterworth filter.

11.14 Find the natural modes of a Butterworth filter with a 1-dB bandwidth of 10^3 rad/s and $N = 5$.

D11.15 Design a Butterworth filter that meets the following low-pass specifications: $f_p = 10$ kHz, $A_{max} = 2$ dB, $f_s = 15$ kHz, and $A_{min} = 15$ dB. Find N, the natural modes, and $T(s)$. What is the attenuation provided at 20 kHz?

***11.16** Sketch $|T|$ for a seventh-order low-pass Chebyshev filter with $\omega_p = 1$ rad/s and $A_{max} = 1$ dB. Use Eq. (11.18) to determine the values of ω at which $|T| = 1$ and the values of ω at which $|T| = 1/\sqrt{1+\epsilon^2}$. Indicate these values on your sketch. Use Eq. (11.19) to determine $|T|$ at $\omega = 2$ rad/s, and indicate this point on your sketch. For large values of ω, at what rate (in dB/octave) does the transmission decrease?

11.17 Contrast the attenuation provided by a fifth-order Chebyshev filter at $\omega_s = 2\,\omega_p$ to that provided by a Butterworth filter of equal order. For both, $A_{max} = 1$ dB. Sketch $|T|$ for both filters on the same axes.

D*11.18 It is required to design a low-pass filter to meet the following specifications: $f_p = 3.4$ kHz, $A_{max} = 1$ dB, $f_s = 4$ kHz, $A_{min} = 35$ dB.
(a) Find the required order of Chebyshev filter. What is the excess (above 35 dB) stopband attenuation obtained?
(b) Find the poles and the transfer function.

Section 11.4: First-Order and Second-Order Filter Functions

D11.19 Use the information displayed in Fig. 11.13 to design a first-order op amp–RC low-pass filter having a 3-dB frequency of 10 kHz, a dc gain magnitude of 10, and an input resistance of 10 kΩ.

D11.20 Use the information given in Fig. 11.13 to design a first-order op amp–RC high-pass filter with a 3-dB frequency of 100 Hz, a high-frequency input resistance of 100 kΩ, and a high-frequency gain magnitude of unity.

D*11.21 Use the information given in Fig. 11.13 to design a first-order op amp–RC spectrum-shaping network with a transmission zero frequency of 1 kHz, a pole frequency of 100 kHz, and a dc gain magnitude of unity. The low-frequency input resistance is to be 1 kΩ. What is the high-frequency gain that results? Sketch the magnitude of the transfer function versus frequency.

D*11.22 By cascading a first-order op amp–RC low-pass circuit with a first-order op amp–RC high-pass circuit one can design a wideband bandpass filter. Provide such a design for the case the midband gain is 12 dB and the 3-dB bandwidth extends from 100 Hz to 10 kHz. Select appropriate component values under the constraint that no resistors higher than 100 kΩ are to be used, and the input resistance is to be as high as possible.

D11.23 Derive $T(s)$ for the op amp–RC circuit in Fig. 11.14. We wish to use this circuit as a variable phase shifter by adjusting R. If the input signal frequency is 10^4 rad/s and if $C = 10$ nF, find the values of R required to obtain phase shifts of $-30°$, $-60°$, $-90°$, $-120°$, and $-150°$.

11.24 Show that by interchanging R and C in the op amp–

RC circuit of Fig. 11.14, the resulting phase shift covers the range 0 to 180° (with 0° at high frequencies and 180° at low frequencies).

11.25 Use the information in Fig. 11.16(a) to obtain the transfer function of a second-order low-pass filter with $\omega_0 = 10^3$ rad/s, $Q = 1$, and dc gain $= 1$. At what frequency does $|T|$ peak? What is the peak transmission?

D*11.26** Use the information in Fig. 11.16(a) to obtain the transfer function of a second-order low-pass filter that just meets the specifications defined in Fig. 11.3 with $\omega_p = 1$ rad/s and $A_{max} = 3$ dB. Note that there are two possible solutions. For each, find ω_0 and Q. Also, if $\omega_s = 2$ rad/s, find the value of A_{min} obtained in each case.

D11.27** Use two first-order op amp–RC all-pass circuits in cascade to design a circuit that provides a set of three-phase 60-Hz voltages, each separated by 120° and equal in magnitude, as shown in the phasor diagram of Fig. P11.27. These voltages simulate those used in three-phase power transmission systems. Use 1-μF capacitors.

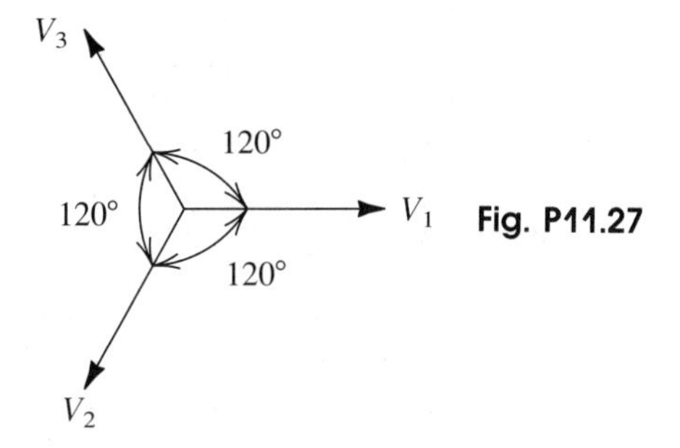

Fig. P11.27

11.28 Use the information given in Fig. 11.16(b) to find the transfer function of a second-order high-pass filter with natural modes at $-0.5 \pm j\sqrt{3}/2$ and a high-frequency gain of unity.

D11.29** (a) Show that $|T|$ of a second-order bandpass function is geometrically symmetric around the center frequency ω_0. That is, each pair of frequencies ω_1 and ω_2 for which $|T(j\omega_1)| = |T(j\omega_2|$ are related by $\omega_1\omega_2 = \omega_0^2$.
(b) Find the transfer function of the second-order bandpass filter that meets specifications of the form in Fig. 11.4 where $\omega_{p1} = 8100$ rad/s, $\omega_{p2} = 10{,}000$ rad/s, and $A_{max} = 1$ dB. If $\omega_{s1} =$ 3000 rad/s find A_{min} and ω_{s2}.

D*11.30 Use the result of Exercise 11.15 to find the transfer function of a notch filter that is required to eliminate a bothersome interference of 60-Hz frequency. Since the frequency of the interference is not stable, the filter should be designed to provide attenuation $\geq$ 20 dB over a 6-Hz band centered around 60 Hz. The dc transmission of the filter is to be unity.

11.31 Consider a second-order all-pass circuit in which errors in the component values result in the frequency of the zeros being slightly lower than that of the poles. Roughly sketch the expected $|T|$. Repeat for the case when the frequency of the zeros is slightly higher than the frequency of the poles.

11.32 Consider a second-order all-pass filter in which errors in the component values result in the Q factor of the zeros being greater than the Q factor of the poles. Roughly sketch the expected $|T|$. Repeat for the case the Q factor of the zeros is lower than the Q factor of the poles.

Section 11.5: The Second-Order LCR Resonator

D11.33 Design the LCR resonator of Fig. 11.17(a) to obtain natural modes with $\omega_0 = 10^4$ rad/s and $Q = 2$. Use $R = 10$ kΩ.

11.34 For the LCR resonator of Fig. 11.17(a) find the change in ω_0 that results from:
(a) increasing L by 1%
(b) increasing C by 1%
(c) increasing R by 1%

11.35 Derive an expression for $V_o(s)/V_i(s)$ of the high-pass circuit in Fig. 11.18(c).

D11.36 Use the circuit of Fig. 11.18(b) to design a low-pass filter with $\omega_0 = 10^5$ rad/s and $Q = 1\sqrt{2}$. Utilize a 0.1-μF capacitor.

D11.37 Modify the bandpass circuit of Fig. 11.18(d) to change its center-frequency gain from 1 to 0.5 without changing ω_0 or Q.

11.38 Consider the LCR resonator of Fig. 11.17(a) with node x disconnected from ground and connected to an input signal source V_x, node y disconnected from ground and connected to another input signal source V_y, and node z disconnected from ground and connected to a third input signal source V_z. Use superposition to find the voltage that develops across the resonator, V_o, in terms of V_x, V_y, and V_z.

11.39 Consider the notch circuit shown in Fig. 11.18(i). For what ratio of L_1 and L_2 does the notch occur at 0.9 ω_0? For this case, what is the magnitude of the transmission at frequencies $\ll \omega_0$? at frequencies $\gg \omega_0$?

Section 11.6: Second-Order Active Filters Based on Inductor Replacement

D11.40 Design the circuit of Fig. 11.20 (utilizing suitable component values) to realize an inductance of (a) 10 H, (b) 1 H, and (c) 0.1 H.

*11.41 Starting from first principles and assuming ideal op amps derive the transfer function of the circuit in Fig. 11.22(a).

*D11.42 It is required to design a fifth-order Butterworth filter having a 3-dB bandwidth of 10^4 rad/s and a unity dc gain. Use a cascade of two circuits of the type shown in Fig. 11.22(a) and a first-order op amp–RC circuit of the type shown in Fig. 11.13(a). Select appropriate component values.

D11.43 Design the circuit of Fig. 11.22(e) to realize an LPN function with $f_0 = 4$ kHz, $f_n = 5$ kHz, $Q = 10$ and a unity dc gain. Select $C_4 = 10$ nF.

D11.44 Design the all-pass circuit of Fig. 11.22(g) to provide a phase shift of 180° at $f = 1$ kHz and to have $Q = 1$. Use 1-nF capacitors.

11.45 Consider the Antoniou circuit of Fig. 11.20(a) with R_5 eliminated, a capacitor C_6 connected between node 1 and ground, and a voltage source V_2 connected to node 2. Show that the input impedance seen by V_2 is $R_2/s^2C_4C_6R_1R_3$. How does this impedance behave for physical frequencies ($s = j\omega$)? (This impedance is known as a **frequency-dependent negative resistance,** or FDNR.)

D11.46 Using the transfer function of the LPN filter, given in Table 11.1, derive the design equations given.

D11.47 Using the transfer function of the HPN filter, given in Table 11.1, derive the design equations given.

D**11.48 It is required to design a third-order low-pass filter whose $|T|$ is equiripple in both the passband and the stopband (in the manner shown in Fig. 11.3, (except that the response shown is for $N = 5$). The filter passband extends from $\omega = 0$ to $\omega = 1$ rad/s and the passband transmission varies between 1 and 0.9. The stopband edge is at $\omega = 1.2$ rad/s. The following transfer function was obtained using filter design tables:

$$T(s) = \frac{0.4508(s^2 + 1.6996)}{(s + 0.7294)(s^2 + s0.2786 + 1.0504)}$$

The actual filter realized is to have $\omega_p = 10^4$ rad/s.

(a) Obtain the transfer function of the actual filter by replacing s by $s/10^4$.

(b) Realize this filter as the cascade connection of a first-order LP op amp–RC circuit of the type shown in Fig. 11.13(a) and a second-order LPN circuit of the type shown in Fig. 11.22(e). Each section is to have a dc gain of unity. Select appropriate component values. (*Note:* A filter with an equiripple response in both the passband and the stopband is known as an **elliptic filter.**)

Section 11.7: Second-Order Active Filters Based on the Two-Integrator-Loop Topology

D11.49 Design the KHN circuit of Fig. 11.24(a) to realize a bandpass filter with a center frequency of 1 kHz and a 3-dB bandwidth of 50 Hz. Use 10-nF capacitors. Give the complete circuit and specify all component values. What value of center-frequency gain is obtained?

D11.50 (a) Using the KHN biquad with the output summing amplifier of Fig. 11.24(b) show that an all-pass function is realized by selecting $R_L = R_H = R_B/Q$. Also show that the flat gain obtained is $K\, R_F/R_H$.

(b) Design the all-pass circuit to obtain $\omega_0 = 10^4$ rad/s, $Q = 2$, and flat gain = 10. Select appropriate component values.

D11.51 Consider a notch filter with $\omega_n = \omega_0$ realized using the KHN biquad with an output summing amplifier. If the summing resistors used have 1% tolerances, what is the worst-case percentage deviation between ω_n and ω_0?

D11.52 Design the circuit of Fig. 11.26 to realize a low-pass notch filter with $\omega_0 = 10^4$ rad/s, $Q = 10$, dc gain = 1, and $\omega_n = 1.2 \times 10^4$ rad/s. Use $C = 10$ nF and $r = 20$ kΩ.

D11.53 In the all-pass realization using the circuit of Fig. 11.26, which component(s) does one need to trim to adjust only: (a) ω_z? (b) Q_z?

D**11.54 Repeat Problem 11.48 using the Tow–Thomas biquad of Fig. 11.26 to realize the second-order section in the cascade.

Section 11.8: Single-Amplifier Biquadratic Active Filters

D11.55 Design the circuit of Fig. 11.29 to realize a pair of poles with $\omega_0 = 10^4$ rad/s and $Q = 1/\sqrt{2}$. Use $C_1 = C_2 = 1$ nF.

11.56 Consider the bridged-T network of Fig. 11.28(a) with $R_1 = R_2 = R$ and $C_1 = C_2 = C$, and denote $CR = \tau$. Find the zeros and poles of the bridged-T network. If the network is placed in the negative-feedback path of an ideal infinite-gain op amp, as in Fig. 11.29, find the poles of the closed-loop amplifier.

**11.57 Consider the bridged-T network of Fig. 11.28(b) with $R_1 = R_2 = R$, $C_4 = C$, and $C_3 = C/16$. Let the network be placed in the negative-feedback path of an infinite-gain op amp and let C_4 be disconnected from ground and connected to the input signal source V_i. Analyze the resulting circuit to determine its transfer function $V_o(s)/V_i(s)$, where $V_o(s)$ is the voltage at the op amp output. Show that the filter realized is a bandpass and find its ω_0, Q, and center-frequency gain.

**11.58 Consider the bandpass circuit shown in Fig. 11.30. Let $C_1 = C_2 = C$, $R_3 = R$, $R_4 = R/4Q^2$, $CR = 2Q/\omega_0$, and $\alpha = 1$. Disconnect the positive input terminal of the op amp from ground and apply V_i through a voltage divider R_1, R_2 to the positive input terminal. Analyze the circuit to find its transfer function V_o/V_i. Find the voltage-divider ratio $R_2/(R_1 + R_2)$ so that the circuit realizes (a) an all-pass function and (b) a notch function. Assume the op amp to be ideal.

*11.59 Derive the transfer function of the circuit in Fig. 11.33(b) assuming the op amp to be ideal. Thus show that the circuit realizes a high-pass function. What is the high-frequency gain of the circuit? Design the circuit for a maximally flat response with a 3-dB frequency of 10^3 rad/s. Use $C_1 = C_2 = 10$ nF. (*Hint:* For a maximally flat response, $Q = 1/\sqrt{2}$ and $\omega_{3dB} = \omega_0$.)

D*11.60 Design a fifth-order Butterworth low-pass filter with a 3-dB bandwidth of 5 kHz and a dc gain of unity using the cascade connection of two Sallen-and-Key circuits (Fig. 11.34c) and a first-order section (Fig. 11.13a). Use a 10-kΩ value for all resistors.

11.61 The process of obtaining the complement of a transfer function by interchanging input and ground, as illustrated in Fig. 11.31, applies to any general network (not just RC networks as shown). Show that if the network n is a bandpass with a center-frequency gain of unity then the complement obtained is a notch. Verify this by using the RLC circuits of Figs. 11.18(d) and (e).

Section 11.9: Sensitivity

11.62 Evaluate the sensitivities of ω_0 and Q relative to R, L, and C of the bandpass circuit in Fig. 11.18(d).

*11.63 Verify the following sensitivity identities:
(a) If $y = uv$, then $S_x^y = S_x^u + S_x^v$.
(b) If $y = u/v$, then $S_x^y = S_x^u - S_x^v$.
(c) If $y = ku$, where k is a constant, then $S_x^y = S_x^u$.
(d) If $y = u^n$, where n is a constant, then $S_x^y = nS_x^u$.
(e) If $y = f_1(u)$ and $u = f_2(x)$, then $S_x^y = S_u^y \cdot S_x^u$.

*11.64 For the high-pass filter of Fig. 11.33(b), what are the sensitivities of ω_0 and Q to amplifier gain A?

*11.65 For the feedback loop of Fig. 11.34(a) use the expressions in Eqs. (11.77) and (11.78) to determine the sensitivities of ω_0 and Q relative to all passive components for the design in which $R_1 = R_2$.

11.66 For the op amp–RC resonator of Fig. 11.21(b) use the expressions for ω_0 and Q given in the top row of Table 11.1 to determine the sensitivities of ω_0 and Q to all resistors and capacitors.

Section 11.10: Switched-Capacitor Filters

11.67 For the switched-capacitor input circuit of Fig. 11.35(b), in which a clock frequency of 100 kHz is used, what input resistances correspond to C_1 capacitance values of 1 pF and 10 pF?

11.68 For a dc voltage of 1 V applied to the input of the circuit of Fig. 11.35(b), in which C_1 is 1 pF, what charge is transferred for each cycle of the two-phase clock? For a 100-kHz clock, what is the average current drawn from the input source? For a feedback capacitance of 10 pF, what change would you expect in the output for each cycle of the clock? For an amplifier that saturates at ±10 V and the feedback capacitor initially discharged, how many clock cycles would it take to saturate the amplifier? What is the average slope of the staircase output voltage produced?

D11.69 Repeat Exercise 11.31 for a clock frequency of 400 kHz.

D11.70 Repeat Exercise 11.31 for $Q = 40$.

D11.71 Design the circuit of Fig. 11.37(b) to realize, at the output of the second (noninverting) integrator, a maximally flat low-pass function with $\omega_{3dB} = 10^4$ rad/s and unity dc gain. Use a clock frequency $f_c = 100$ kHz and select $C_1 = C_2 = 10$ pF. Give the values of C_3, C_4, C_5, and C_6. (*Hint:* For a maximally flat response, $Q = 1/\sqrt{2}$ and $\omega_{3dB} = \omega_0$.)

Section 11.11: Tuned Amplifiers

*11.72 A voltage signal source with a resistance $R_s = 10$ kΩ is connected to the input of a common-emitter BJT amplifier. Between base and emitter is connected a tuned circuit with $L = 1\ \mu$H and $C = 200$ pF. The transistor is biased at 1 mA and has $\beta = 200$, $C_\pi = 10$ pF, and $C_\mu = 1$ pF. The transistor load is a resistance of 5 kΩ. Find ω_0, Q, the 3-dB bandwidth, and the center-frequency gain of this single-tuned amplifier.

11.73 A coil having an inductance of 10 μH is intended for applications around 1-MHz frequency. Its Q is specified to be 200. Find the equivalent parallel resistance R_p. What is the value of the capacitor required to produce resonance at 1 MHz? What additional parallel resistance is required to produce a 3-dB bandwidth of 10 kHz?

11.74 An inductance of 36 μH is resonated with a 1000-pF capacitor. If the inductor is tapped at one-third of its turns and a 1-kΩ resistor is connected across the third of the coil turns, find f_0 and Q of the resonator.

*11.75 Consider a common-emitter transistor amplifier loaded with an inductance L. Ignoring r_o and r_x, show that for $\omega C_\mu \ll 1/\omega L$, the input admittance is given by

$$Y_{in} \simeq \left(\frac{1}{r_\pi} - \omega^2 C_\mu L g_m\right) + j\omega(C_\pi + C_\mu)$$

Note: The real part of the input admittance can be negative. This can lead to oscillations.

*11.76 (a) Substituting $s = j\omega$ in the transfer function $T(s)$ of a second-order bandpass filter (see Fig. 11.16c), find $|T(j\omega)|$. For ω in the vicinity of ω_0 [that is, $\omega = \omega_0 + \delta\omega = \omega_0(1 + \delta\omega/\omega_0)$, where $\delta\omega/\omega_0 \ll 1$ so that $\omega^2 \simeq \omega_0^2(1 + 2\delta\omega/\omega_0)$] show that, for $Q \gg 1$,

$$|T(j\omega)| \simeq \frac{|T(j\omega_0)|}{\sqrt{1 + 4Q^2(\delta\omega/\omega_0)^2}}$$

(b) Use the result obtained in (a) to show that the 3-dB bandwidth B, of N synchronously tuned sections connected in cascade, is

$$B = (\omega_0/Q)\sqrt{2^{1/N} - 1}$$

**11.77 (a) Using the fact that for $Q \gg 1$ the second-order bandpass response in the neighborhood of ω_0 is the same as the response of a first-order low-pass with 3-dB frequency of $(\omega_0/2Q)$, show that the bandpass response at $\omega = \omega_0 + \delta\omega$, $\delta\omega \ll \omega_0$, is given by

$$|T(j\omega)| \simeq \frac{|T(j\omega_0)|}{\sqrt{1 + 4Q^2(\delta\omega/\omega_0)^2}}$$

(b) Use the relationship derived in (a) together with Eq. (11.110) to show that a bandpass amplifier with a 3-dB bandwidth B, designed using N synchronously tuned stages, has an overall transfer function given by

$$|T(j\omega)|_{\text{overall}} = \frac{|T(j\omega_0)|_{\text{overall}}}{[1 + 4(2^{1/N} - 1)(\delta\omega/B)^2]^{N/2}}$$

(c) Use the relationship derived in (b) to find the attenuation (in decibels) obtained at a bandwidth $2B$ for $N = 1$ to 5. Also find the ratio of the 30-dB bandwidth to the 3-dB bandwidth for $N = 1$ to 5.

*11.78 This problem investigates the selectivity of maximally flat stagger-tuned amplifiers derived in the manner illustrated in Fig. 11.48.

(a) The low-pass maximally flat (Butterworth) filter of 3-dB bandwidth $(B/2)$ and order N has the magnitude response

$$|T| = 1\Big/\sqrt{1 + \left(\frac{\Omega}{B/2}\right)^{2N}}$$

where $\Omega = \text{Im}(p)$, is the frequency in the low-pass domain. (This relationship can be obtained using the information provided in Section 11.3 on Butterworth filters.) Use this expression to obtain for the corresponding bandpass filter at $\omega = \omega_0 + \delta\omega$, $\delta\omega \ll \omega_0$,

$$|T| = 1\Big/\sqrt{1 + \left(\frac{\delta\omega}{B/2}\right)^{2N}}$$

(b) Use the transfer function of (a) to find the attenuation (in decibels) obtained at a bandwidth of $2B$ for $N = 1$ to 5. Also find the ratio of the 30-dB bandwidth to the 3-dB bandwidth for $N = 1$ to 5.

**11.79 Consider a sixth-order stagger-tuned bandpass amplifier with center-frequency ω_0 and 3-dB bandwidth B. The poles are to be obtained by shifting those of the third-order maximally flat low-pass filter, given in Fig. 11.10(c). For the three resonant circuits find ω_0, the 3-dB bandwidth, and Q.

CHAPTER 12

Signal Generators and Waveform-Shaping Circuits

INTRODUCTION

In the design of electronic systems the need frequently arises for signals having prescribed standard waveforms, for example, sinusoidal, square, triangular, pulse, and so on. Systems in which standard signals are required include: computer and control systems where clock pulses are needed for, among other things, timing; communication systems where signals of a variety of waveforms are utilized as information carriers; and test and measurement systems where signals, again of a variety of waveforms, are employed for testing and characterizing electronic devices and circuits. In this chapter we study signal-generator circuits.

There are two distinctly different approaches for the generation of sinusoids, perhaps the most commonly used of the standard waveforms. The first approach, studied in Sections 12.1–12.3, employs a *positive-feedback loop* consisting of an amplifier and an RC or an LC *frequency-selective network*. The amplitude of the generated sine waves is limited, or set, using a nonlinear mechanism, implemented either with a separate circuit or using the nonlinearities of the amplifying device itself. In spite of this, these circuits, which generate sine waves utilizing resonance phenomena, are known as *linear oscillators*. The name clearly distinguishes them from the circuits that generate sinusoids by way of the second approach. In these circuits, a sine wave is obtained by appropriately shaping a triangular waveform. We study waveform-shaping circuits in Section 12.9, following the study of triangular-waveform generators.

Circuits that generate square, triangular, pulse (etc.) waveforms, called nonlinear oscillators or function generators, employ circuit building blocks known as *multivibrators*. There are three types of multivibrators: The *bistable* (Section 12.4), the *astable* (Section 12.5), and the *monostable* (Section 12.7). The multivibrator circuits presented in this chapter employ op amps and are intended for precision analog applications. Multivibrator circuits using digital logic gates will be discussed in the next chapter.

A general and versatile scheme for the generation of square and triangular waveforms is obtained by connecting a bistable multivibrator and an op amp integrator in a feedback loop (Section 12.7). Similar results can be obtained using a commercially available, versatile IC chip, the 555 timer (Section 12.8). We conclude the chapter with a study of precision circuits that implement the rectifier functions introduced in Chapter 3. The circuits studied here, however, are intended for applications that demand precision, such as in instrumentation systems, including waveform generation.

12.1 BASIC PRINCIPLES OF SINUSOIDAL OSCILLATORS

In this section we study the basic principles on which the design of linear sine-wave oscillators is based. In spite of the name *linear oscillator,* some form of nonlinearity has to be employed to provide control of the amplitude of the output sine wave. In fact, all oscillators are essentially nonlinear circuits. This complicates the task of analysis and design of oscillators; no longer is one able to apply transform methods (s-plane) directly. Nevertheless, techniques have been developed by which the design of sinusoidal oscillators can be performed in two steps. The first step is a linear one, and frequency-domain methods of feedback circuit analysis can be readily employed. Subsequently, a nonlinear mechanism for amplitude control can be provided.

The Oscillator Feedback Loop

The basic structure of a sinusoidal oscillator consists of an amplifier and a frequency-selective network connected in a positive-feedback loop, such as that shown in block-diagram form in Fig. 12.1. Although in an actual oscillator circuit no input signal will be present, we include an input signal here to help explain the principle of operation. It is important to note that unlike the negative-feedback loop of Fig. 8.1, here the feedback signal x_f is summed with a *positive* sign. Thus the gain with feedback is given by

$$A_f(s) = \frac{A(s)}{1 - A(s)\beta(s)} \tag{12.1}$$

where we note the negative sign in the denominator.

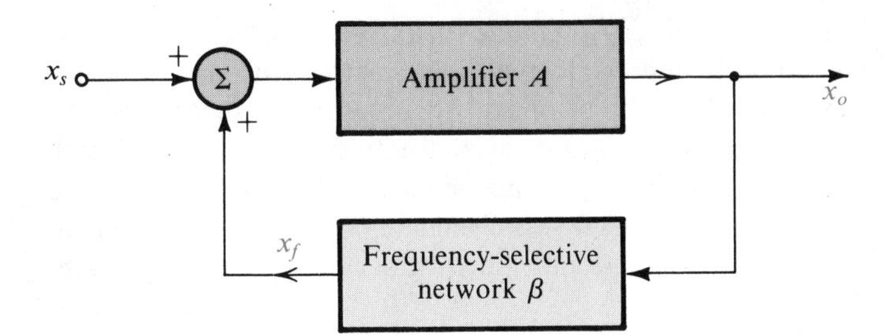

Fig. 12.1 The basic structure of a sinusoidal oscillator. A positive-feedback loop is formed by an amplifier and a frequency-selective network. In an actual oscillator circuit no input signal will be present; here an input signal x_s is employed to help explain the principle of operation.

According to the definition of loop gain in Chapter 8, the loop gain of the circuit in Fig. 12.1 is $-A(s)\beta(s)$. However, for our purposes here it is more convenient to drop the minus sign and define the loop gain $L(s)$ as

$$L(s) \equiv A(s)\beta(s) \tag{12.2}$$

The characteristic equation thus becomes

$$1 - L(s) = 0 \tag{12.3}$$

Note that this new definition of loop gain corresponds directly to the actual gain seen around the feedback loop of Fig. 12.1.

The Oscillation Criterion

If at a specific frequency f_0 the loop gain $A\beta$ is equal to unity, it follows from Eq. (12.1) that A_f will be infinite. That is, at this frequency the circuit will have a finite output for zero input signal. Such a circuit is by definition an oscillator. Thus the condition for the feedback loop of Fig. 12.1 to provide sinusoidal oscillations of frequency ω_0 is that

$$L(j\omega_0) \equiv A(j\omega_0)\beta(j\omega_0) = 1 \tag{12.4}$$

That is, *at ω_0 the phase of the loop gain should be zero and the magnitude of the loop gain should be unity*. This is known as the **Barkhausen criterion.** Note that for the circuit to oscillate at one frequency the oscillation criterion should be satisfied at one frequency only (that is, ω_0); otherwise the resulting waveform will not be a simple sinusoid.

An intuitive feeling for the Barkhausen criterion can be gained by considering once more the feedback loop of Fig. 12.1. For this loop to *produce* and *sustain* an output x_o with no input applied ($x_s = 0$), the feedback signal x_f,

$$x_f = \beta x_o$$

should be sufficiently large so that when multiplied by A it produces x_o,

$$Ax_f = x_o$$

that is,

$$A\beta x_o = x_o$$

which results in

$$A\beta = 1$$

It should be noted that the *frequency of oscillation* ω_0 is determined solely by the phase characteristics of the feedback loop; the loop oscillates at the frequency for which the phase is zero. It follows that the stability of the frequency of oscillation will be determined by the manner in which the phase $\phi(\omega)$ of the feedback loop varies with frequency. A "steep" function $\phi(\omega)$ will result in a more stable frequency. This can be seen if one imagines a change in phase $\Delta\phi$ due to a change in one of the circuit components. If $d\phi/d\omega$ is large, the resulting change in ω_0 will be small, as illustrated in Fig. 12.2.

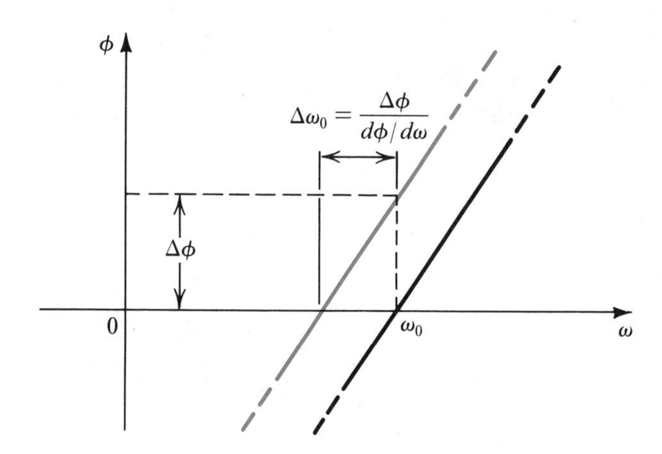

Fig. 12.2 Dependence of the oscillator frequency stability on the slope of the phase response. A steep phase response (that is, large $d\phi/d\omega$) results in a small $\Delta\omega_0$ for a given change in phase $\Delta\phi$ (resulting from a change in a circuit component).

An alternative approach to the study of oscillator circuits consists of examining the circuit poles, which are the roots of the *characteristic equation* (Eq. 12.3). For the circuit to produce *sustained oscillations* at a frequency ω_0 the characteristic equation has to have roots at $s = \pm j\omega_0$. Thus $1 - A(s)\beta(s)$ should have a factor of the form $s^2 + \omega_0^2$.

Exercise 12.1 Consider a sinusoidal oscillator formed of an amplifier with a gain of 2 and a second-order bandpass filter. Find the pole frequency and the center-frequency gain of the filter to produce sustained oscillations at 1 kHz.

Ans. 1 kHz; 0.5

Nonlinear Amplitude Control

The oscillation condition, the Barkhausen criterion, discussed above guarantees sustained oscillations in a mathematical sense. It is well known, however, that the parameters of any physical system cannot be maintained constant for any length of time. In other words, suppose we work hard to make $A\beta = 1$ at $\omega = \omega_0$; then the temperature changes and $A\beta$ becomes slightly less than unity. Obviously, oscillations will cease in this case. Conversely, if $A\beta$ exceeds unity, oscillations will grow in amplitude. We therefore need a mechanism for forcing $A\beta$ to remain equal to unity at the desired value of output amplitude. This task is accomplished by providing a nonlinear circuit for gain control.

Basically, the function of the gain-control mechanism is as follows: First, to ensure that oscillations will start, one designs the circuit such that $A\beta$ is slightly greater than unity. This corresponds to designing the circuit so that the poles are in the right half of the s-plane. Thus as the power supply is turned on, oscillations will grow in amplitude. When the amplitude reaches the desired level, the nonlinear network comes into action and causes the loop gain to be reduced to exactly unity. In other words, the poles will be "pulled back" to the $j\omega$-axis. This action will cause the circuit to sustain oscillations at this desired amplitude. If, for some reason, the loop gain is reduced below unity, the amplitude of the sine wave will diminish. This will be detected by the nonlinear network, which will cause the loop gain to increase to exactly unity.

As will be seen, there are two basic approaches to the implementation of the nonlinear amplitude-stabilization mechanism. The first approach makes use of a limiter circuit (see Chapter 3). Oscillations are allowed to grow until the amplitude reaches the level to which the limiter is set. Once the limiter comes into operation, the amplitude remains constant. Obviously, the limiter should be "soft" in order to minimize nonlinear distortion. Such distortion, however, is reduced by the filtering action of the frequency-selective network in the feedback loop. In fact, in one of the oscillator circuits studied in Section 12.2, the sine waves are hard-limited, and the resulting square waves are applied to a bandpass filter present in the feedback loop. The "purity" of the output sine waves will be a function of the selectivity of this filter. That is, the higher the Q of the filter, the less the harmonic content of the sine-wave output.

The other mechanism for amplitude control utilizes an element whose resistance can be controlled by the amplitude of the output sinusoid. By placing this element in the feedback circuit so that its resistance determines the loop gain, the circuit can be designed so that the loop gain reaches unity at the desired output amplitude. Diodes, or JFETs operated in the triode region, are commonly employed to implement the controlled-resistance element.

A Popular Limiter Circuit for Amplitude Control

We conclude this section by presenting a limiter circuit that is frequently employed for the amplitude control of op amp oscillators, as well as in a variety of other applications. The circuit is more precise and versatile than those presented in Chapter 3.

The limiter circuit is shown in Fig. 12.3(a), and its transfer characteristic is depicted in Fig. 12.3(b). To see how the transfer characteristic is obtained, consider first the case when the input signal v_I is small (close to zero) and the output voltage v_O is also small, so that v_A is positive and v_B is negative. It can be easily seen that both diodes D_1 and D_2 will be off. Thus all of the input current v_I/R_1 flows through the feedback resistance R_f, and the output voltage is given by

$$v_O = -(R_f/R_1)v_I \tag{12.5}$$

This is the linear portion of the limiter transfer characteristic in Fig. 12.3(b). We now can use superposition to find the voltages at nodes A and B in terms of $\pm V$ and v_O as

$$v_A = V\frac{R_3}{R_2 + R_3} + v_O\frac{R_2}{R_2 + R_3} \tag{12.6}$$

$$v_B = -V\frac{R_4}{R_4 + R_5} + v_O\frac{R_5}{R_4 + R_5} \tag{12.7}$$

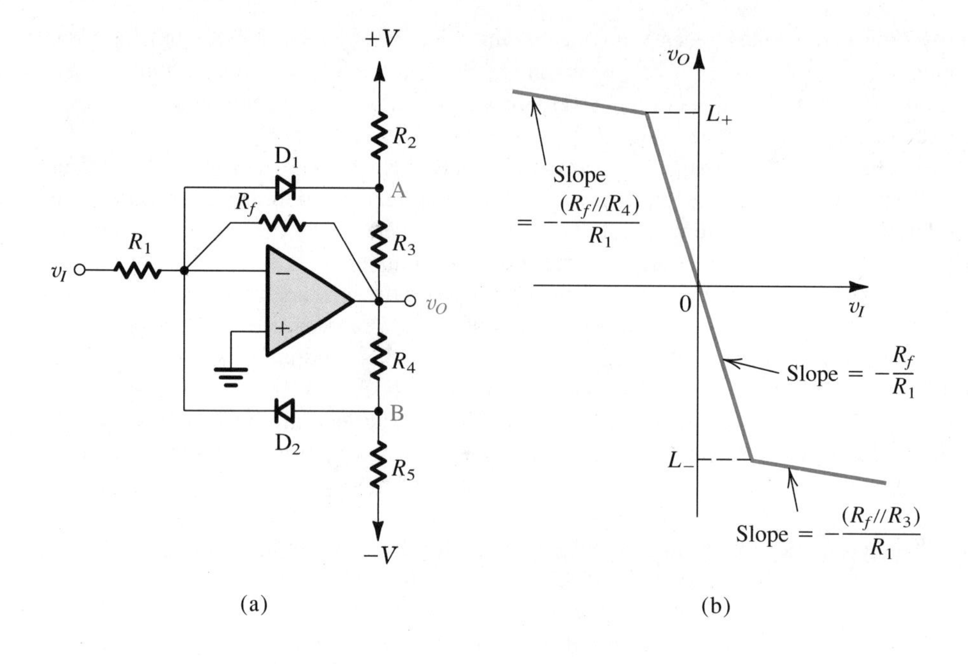

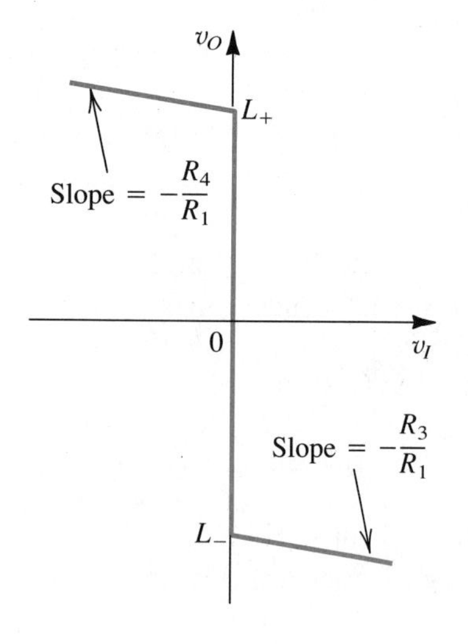

Fig. 12.3 (a) A popular limiter circuit. **(b)** Transfer characteristic of the limiter circuit; L_- and L_+ are given by Eqs. (12.8) and (12.9), respectively. **(c)** When R_f is removed the limiter turns into a comparator with the characteristics shown.

As v_I goes positive, v_O goes negative (Eq. 12.5), and we see from Eq. (12.7) that v_B will become more negative, thus keeping D_2 off. Equation (12.6) shows, however, that v_A becomes less positive. Then if we continue to increase v_I, a negative value of v_O will be reached at which v_A becomes -0.7 V or so and diode D_1 conducts. If we use the constant-voltage-drop model for D_1 and denote the voltage drop V_D, the value of v_O at which D_1

conducts can be found from Eq. (12.6). This is the negative limiting level, which we denote L_-,

$$L_- = -V\frac{R_3}{R_2} - V_D\left(1 + \frac{R_3}{R_2}\right) \tag{12.8}$$

The corresponding value of v_I can be found by dividing L_- by the limiter gain $-R_f/R_1$. If v_I is increased beyond this value, more current is injected into D_1 and v_A remains at approximately $-V_D$. Thus the current through R_2 remains constant, and the additional diode current flows through R_3. Thus R_3 appears in effect in parallel with R_f, and the incremental gain (ignoring the diode resistance) is $-(R_f \,//\, R_3)/R_1$. To make the slope of the transfer characteristic small in the limiting region, a low value should be selected for R_3.

The transfer characteristic for negative v_I can be found in a manner identical to that employed above. It can be easily seen that for negative v_I, diode D_2 plays an identical role to that played by diode D_1 for positive v_I. The positive limiting level L_+ can be found to be

$$L_+ = V\frac{R_4}{R_5} + V_D\left(1 + \frac{R_4}{R_5}\right) \tag{12.9}$$

and the slope of the transfer characteristic in the positive-limiting region is $-(R_f // R_4)/R_1$. We thus see that the circuit of Fig. 12.3(a) functions as a soft limiter, with the limiting levels L_+ and L_- independently adjustable by the selection of appropriate resistor values.

Finally we note that increasing R_f results in a higher gain in the linear region while keeping L_+ and L_- unchanged. In the limit, removing R_f altogether results in the transfer characteristic of Fig. 12.3(c), which is that of a comparator. That is, the circuit compares v_I with the comparator reference value of 0 V; $v_I > 0$ results in $v_O \simeq L_-$, and $v_I < 0$ yields $v_O \simeq L_+$.

Exercise 12.2 For the circuit of Fig. 12.3(a) with $V = 15$ V, $R_1 = 30$ kΩ, $R_f = 60$ kΩ, $R_2 = R_5 = 9$ kΩ, and $R_3 = R_4 = 3$ kΩ, find the limiting levels and the value of v_I at which the limiting levels are reached. Also determine the limiter gain and the slope of the transfer characteristics in the positive and negative limiting regions. Assume that $V_D = 0.7$ V.

Ans. ±5.93 V; ±2.97 V; −2; −0.095

12.2 OP AMP–RC OSCILLATOR CIRCUITS

In this section we shall study some practical oscillator circuits utilizing op amps and RC networks.

The Wien-Bridge Oscillator

One of the simplest oscillator circuits is based on the Wien bridge. Figure 12.4 shows a Wien-bridge oscillator without the nonlinear gain-control network. The circuit consists of an op amp connected in the noninverting configuration, with a closed-loop gain of $1 + R_2/R_1$. In the feedback path of this positive-gain amplifier an RC network is connected.

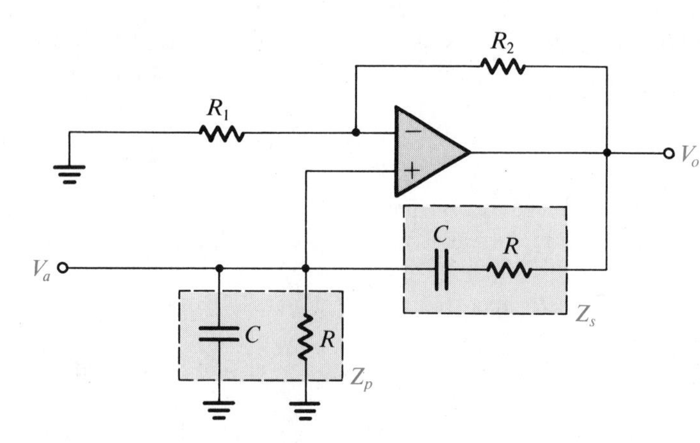

Fig. 12.4 Wien-bridge oscillator without amplitude stabilization.

The loop gain can be easily obtained by multiplying the transfer function $V_a(s)/V_o(s)$ of the feedback network by the amplifier gain,

$$L(s) = \left[1 + \frac{R_2}{R_1}\right] \frac{Z_p}{Z_p + Z_s}$$

Thus

$$L(s) = \frac{1 + R_2/R_1}{3 + sCR + 1/sCR} \tag{12.10}$$

Substituting $s = j\omega$ results in

$$L(j\omega) = \frac{1 + R_2/R_1}{3 + j(\omega CR - 1/\omega CR)} \tag{12.11}$$

The loop gain will be a real number (that is, the phase will be zero) at one frequency given by

$$\omega_0 CR = \frac{1}{\omega_0 CR}$$

That is,

$$\omega_0 = 1/CR \tag{12.12}$$

To obtain sustained oscillations at this frequency, one should set the magnitude of the loop gain to unity. This can be achieved by selecting

$$R_2/R_1 = 2 \tag{12.13}$$

To ensure that oscillations will start, one chooses R_2/R_1 slightly greater than 2. The reader can easily verify that if $R_2/R_1 = 2 + \delta$, where δ is a small number, the roots of the characteristic equation $1 - L(s) = 0$ will be in the right half of the s-plane.

The amplitude of oscillation can be determined and stabilized by using a nonlinear control network. Two different implementations of the amplitude-controlling function are shown in Figs. 12.5 and 12.6. The circuit in Fig. 12.5 employs a symmetrical feedback

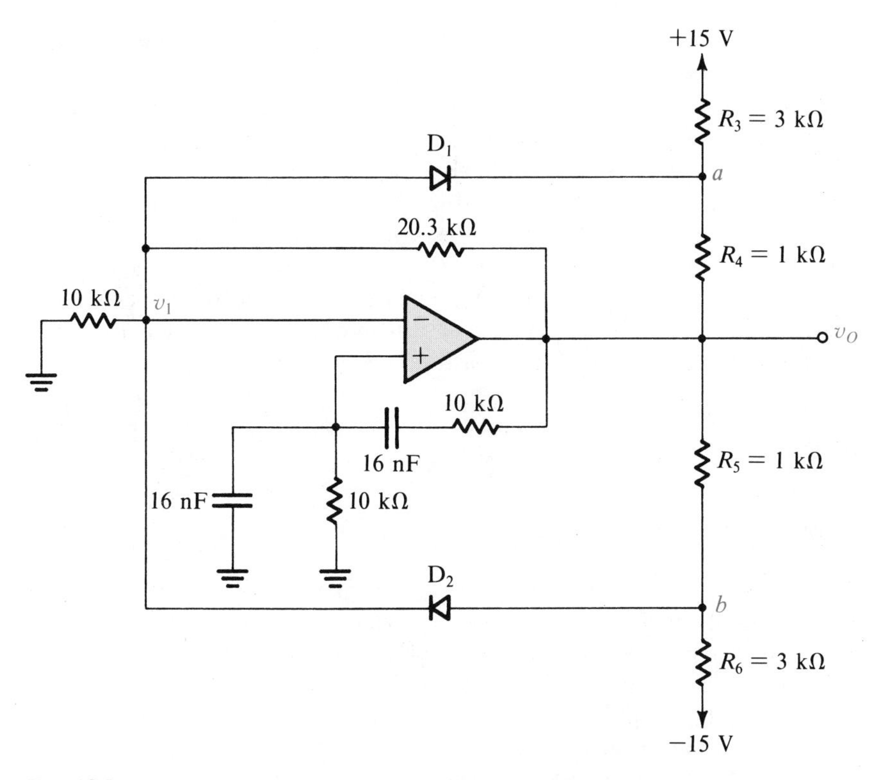

Fig. 12.5 A Wien-bridge oscillator with a limiter used for amplitude control.

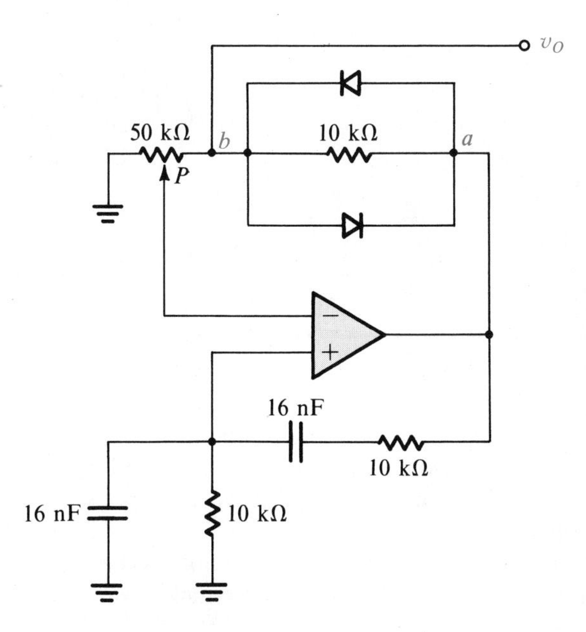

Fig. 12.6 A Wien-bridge oscillator with an alternative method for amplitude stabilization.

limiter of the type studied in Section 12.1. It is formed by diodes D_1 and D_2 together with resistors R_3, R_4, R_5, and R_6. The limiter operates in the following manner: At the positive peak of the output voltage v_O, the voltage at node b will exceed the voltage v_1 (which is about $\frac{1}{3}v_O$), and diode D_2 conducts. This will clamp the positive peak to a value determined by R_5, R_6, and the negative power supply. The value of the positive output peak can be calculated by setting $v_b = v_1 + V_{D2}$ and writing a node equation at node b while neglecting the current through D_2. Similarly, the negative peak of the output sine wave will be clamped to the value that causes diode D_1 to conduct. The value of the negative peak can be determined by setting $v_a = v_1 - V_{D1}$ and writing an equation at node a while neglecting the current through D_1. Finally, note that in order to obtain a symmetrical output waveform, R_3 is chosen equal to R_6, and R_4 equal to R_5.

Exercise 12.3 For the circuit in Fig. 12.5: (a) Disregarding the limiter circuit, find the location of the closed-loop poles, (b) Find the frequency of oscillation, (c) With the limiter in place, find the amplitude of the output sine wave (assume that the diode drop is 0.7 V).

Ans. $(10^5/16)(0.015 \pm j)$; 1 kHz; 21.36 V (peak-to-peak)

The circuit of Fig. 12.6 employs an inexpensive implementation of the parameter variation mechanism of amplitude control. Potentiometer P is adjusted until oscillations just start to grow. As the oscillations grow, the diodes start to conduct, causing the effective resistance between a and b to decrease. Equilibrium will be reached at the output amplitude that causes the loop gain to be exactly unity. The output amplitude can be varied by adjusting potentiometer P.

As indicated in Fig. 12.6, the output is taken at point b rather than at the op amp output terminal because the signal at b has lower distortion than that at a. To appreciate this point, note that the voltage at b is proportional to the voltage at the op amp input terminals and that the latter is a filtered (by the RC network) version of the voltage at node a. Node b, however, is a high-impedance node, and a buffer will be needed if a load is to be connected.

Exercise 12.4 For the circuit in Fig. 12.6 find the following: (a) The setting of potentiometer P at which oscillations just start. (b) The frequency of oscillation.

Ans. (a) 20 kΩ to ground; (b) 1 kHz

The Phase-Shift Oscillator

The basic structure of the phase-shift oscillator is shown in Fig. 12.7. It consists of a negative-gain amplifier ($-K$) with a three-section (third-order) RC ladder network in the feedback. The circuit will oscillate at the frequency for which the phase shift of the RC network is 180°. Only at this frequency will the total phase shift around the loop be 0 or 360°. Here we should note that the reason for using a three-section RC network is that three is the minimum number of sections (that is, lowest order) that is capable of producing a 180° phase shift at a finite frequency.

For oscillations to be sustained, the value of K should be equal to the inverse of the magnitude of the RC network transfer function at the frequency of oscillation. However,

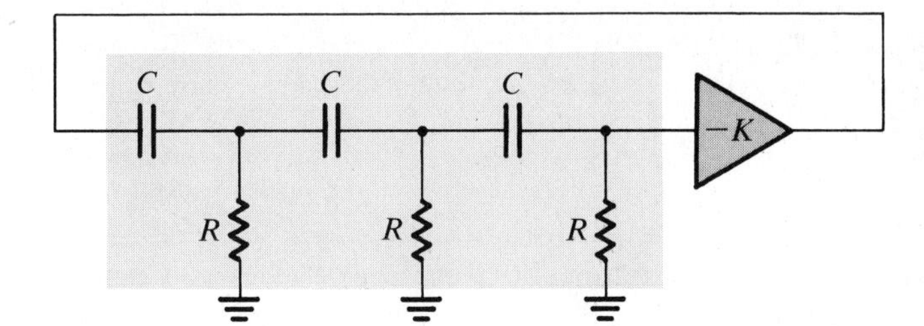

Fig. 12.7 Phase-shift oscillator.

in order to ensure that oscillations start, the value of K has to be chosen slightly higher than the value that satisfies the unity-loop-gain condition. Oscillations will then grow in magnitude until limited by some nonlinear control mechanism.

Figure 12.8 shows a practical phase-shift oscillator with a feedback limiter, consisting of diodes D_1 and D_2 and resistors R_1, R_2, R_3, and R_4 for amplitude stabilization. To start oscillations, R_f has to be made slightly greater than the minimum required value. Although the circuit stabilizes more rapidly, and provides sine waves with more stable amplitude, if R_f is made much larger than this minimum, the price paid is an increased output distortion.

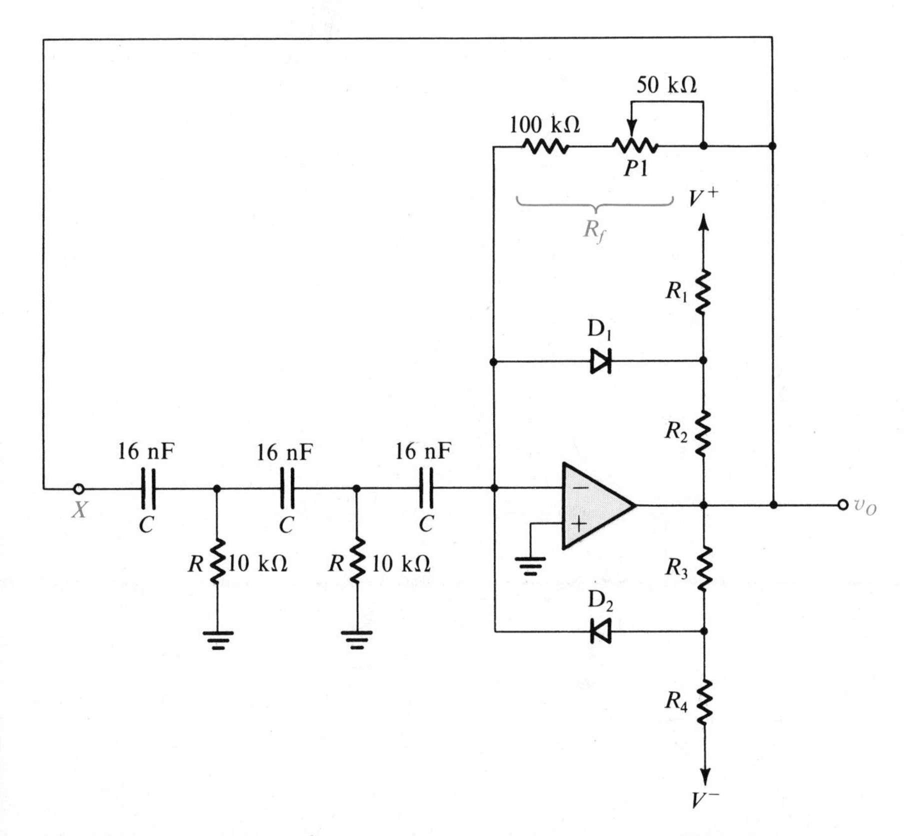

Fig. 12.8 Practical phase-shift oscillator with a limiter for amplitude stabilization.

Exercises 12.5 Consider the circuit of Fig. 12.8 *without* the limiter. Break the feedback loop at X and find the loop gain $A\beta \equiv V_o(j\omega)/V_x(j\omega)$. To do this it is easier to start at the output and work backward, finding the various currents and voltages, and eventually V_x in terms of V_o.

Ans. $\dfrac{\omega^2 C^2 R R_f}{4 + j(3\omega CR - 1/\omega CR)}$

12.6 Use the expression derived in Exercise 12.5 to find the frequency of oscillation f_0 and the minimum required value of R_f for oscillations to start in the circuit of Fig. 12.8.

Ans. 574.3 Hz; 120 kΩ

The Quadrature Oscillator

The **quadrature oscillator** is based on the two-integrator loop studied in Section 11.7. As an active filter, the loop is damped so as to locate the poles in the left half of the s-plane. Here no such damping will be used, since we wish to locate the poles on the $j\omega$-axis in order to provide sustained oscillations. In fact, to ensure that oscillations start, the poles are initially located in the right half-plane and then "pulled back" by the nonlinear gain control.

Figure 12.9 shows a practical quadrature oscillator. Amplifier 1 is connected as an inverting Miller integrator with a limiter in the feedback for amplitude control. Amplifier 2 is connected as a noninverting integrator (thus replacing the cascade connection of the

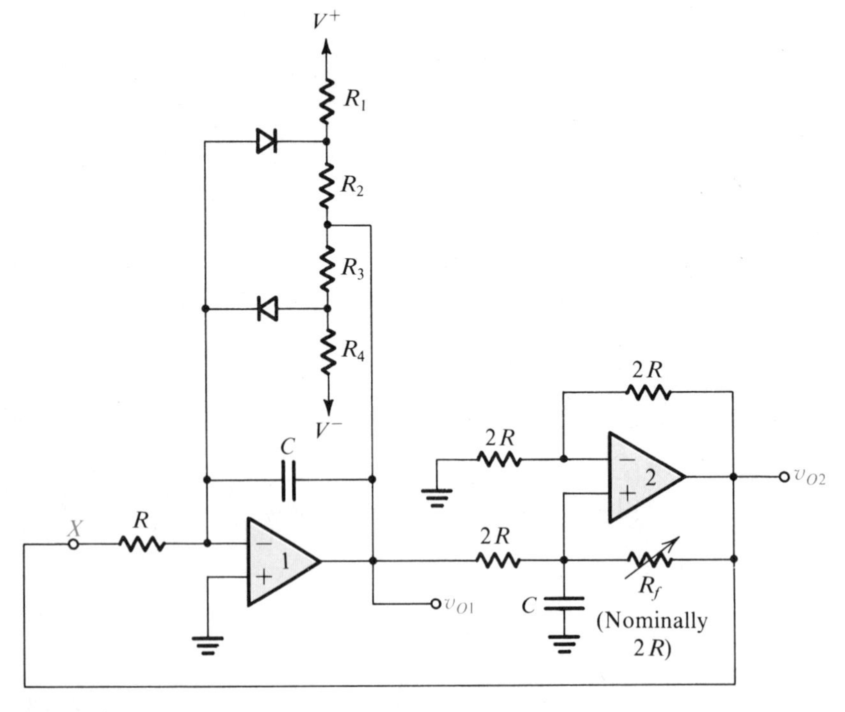

Fig. 12.9 A quadrature oscillator circuit.

Miller integrator and the inverter in the two-integrator loop of Fig. 11.25b). This noninverting integrator circuit is studied in Example 2.6.

The resistance R_f in the positive-feedback path of op amp 2 is made variable, with a nominal value of $2R$. Decreasing the value of R_f moves the poles to the right half-plane (Problem 12.19) and ensures that the oscillations start. Too much positive feedback, although it results in better amplitude stability, also results in higher output distortion (because the limiter has to operate ''harder''). In this regard, note that the output v_{O2} will be ''purer'' than v_{O1} because of the filtering action provided by the second integrator on the peak-limited output of the first integrator.

If we disregard the limiter and break the loop at X, the loop gain can be obtained as

$$L(s) \equiv \frac{V_{o2}}{V_x} = -\frac{1}{s^2C^2R^2} \tag{12.14}$$

Thus the loop will oscillate at frequency ω_0, given by

$$\omega_0 = \frac{1}{CR} \tag{12.15}$$

Finally, it should be pointed out that the name *quadrature oscillator* is used because the circuit provides two sinusoids with 90° phase difference. This should be obvious, since v_{O2} is the integral of v_{O1}. There are many applications for which quadrature sinusoids are required.

The Active-Filter Tuned Oscillator

The last oscillator circuit that we shall discuss is quite simple both in principle and in design. Nevertheless, the approach is general and versatile and can result in high-quality (that is, low-distortion) output sine waves. The basic principle is illustrated in Fig. 12.10. The circuit consists of a high-Q bandpass filter connected in a positive-feedback loop with a hard limiter. To understand how this circuit works, assume that oscillations have already

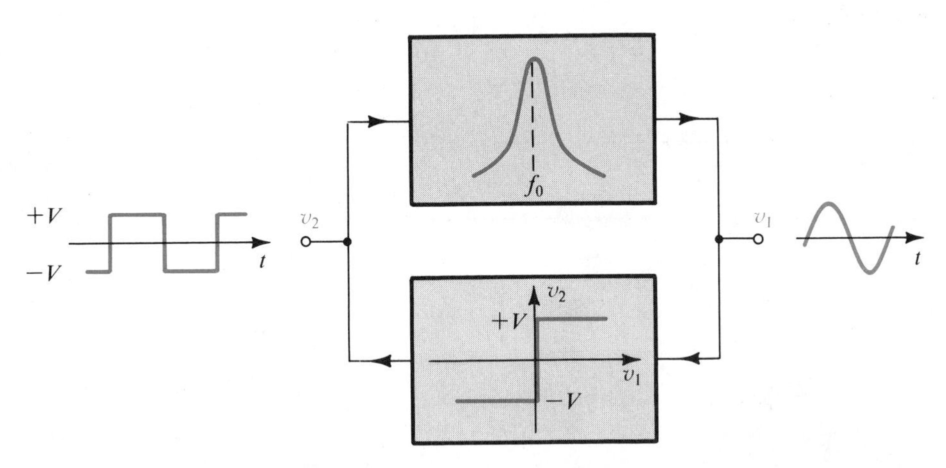

Fig. 12.10 Block diagram of the active-filter tuned oscillator.

started. The output of the bandpass filter will be a sine wave whose frequency is equal to the center frequency of the filter, f_0. The sine-wave signal v_1 is fed to the limiter, which produces at its output a square wave whose levels are determined by the limiting levels and whose frequency is f_0. The square wave in turn is fed to the bandpass filter, which filters out the harmonics and provides a sinusoidal output v_1 at the fundamental frequency f_0. Obviously, the purity of the output sine wave will be a direct function of the selectivity (or Q factor) of the bandpass filter.

The simplicity of this approach to oscillator design should be apparent. We have independent control of frequency and amplitude as well as of distortion of the output sinusoid. Any filter circuit with positive gain can be used to implement the bandpass filter. The frequency stability of the oscillator will be directly determined by the frequency stability of the bandpass-filter circuit. Also, a variety of limiter circuits (see Chapter 3) with different degrees of sophistication can be used to implement the limiter block.

Figure 12.11 shows one possible implementation of the active-filter tuned oscillator. This circuit uses a variation on the bandpass circuit based on the Antoniou inductance simulation circuit (see Fig. 11.22c). Here resistor R_2 and capacitor C_4 are interchanged. This makes the output of the lower op amp directly proportional to (in fact, twice as long as) the voltage across the resonator, and we can therefore dispense with the buffer amplifier K. The limiter used is a very simple one consisting of a resistance R_1 and two diodes.

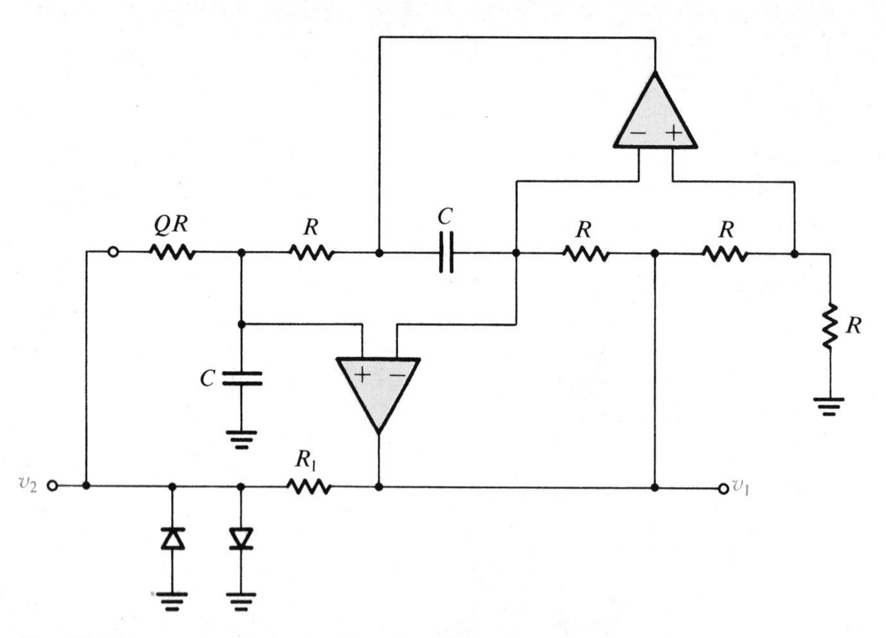

Fig. 12.11 Practical implementation of the active-filter tuned oscillator.

Exercise 12.7 Use $C = 16$ nF and find the value of R such that the circuit of Fig. 12.11 produces 1-kHz sine waves. If the diode drop is 0.7 V, find the peak-to-peak amplitude of the output sine wave. (*Hint:* A square wave with peak-to-peak amplitude of V volts has a fundamental component with $4V/\pi$ volts peak-to-peak amplitude.)

Ans. 10 kΩ; 3.6 V

A Final Remark

The op amp-RC oscillator circuits studied are useful for operation in the range 10 Hz to 100 kHz (or perhaps 1 MHz at most). Whereas the lower frequency limit is dictated by the size of passive components required, the upper limit is governed by the frequency-response and slew-rate limitations of op amps. For higher frequencies, circuits that employ transistors together with LC tuned circuits or crystals are frequently used.[1] These are discussed in Section 12.3.

12.3 LC AND CRYSTAL OSCILLATORS

Oscillators utilizing transistors (FETs or BJTs) and LC tuned circuits or crystals as feedback elements are used in the frequency range of 100 kHz to hundreds of MHz. They exhibit higher Q than the RC types. However, LC oscillators are difficult to tune over wide ranges, and crystal oscillators operate at a single frequency.

LC Tuned Oscillators

Figure 12.12 shows two commonly used configurations of LC tuned oscillators. They are known as (a) the **Colpitts oscillator** and (b) the **Hartley oscillator.** Both utilize a parallel LC circuit connected between collector and base (or between drain and gate if a FET is used) with a fraction of the tuned-circuit voltage fed to the emitter (the source in a FET). This feedback is achieved by way of a capacitive divider in the Colpitts oscillator and by way of an inductive divider in the Hartley circuit. Note that the bias details are not shown in order to focus attention on the oscillator's structure. In both circuits the resistor R models the losses of the inductors, the load resistance of the oscillator, and the output resistance of the transistor.

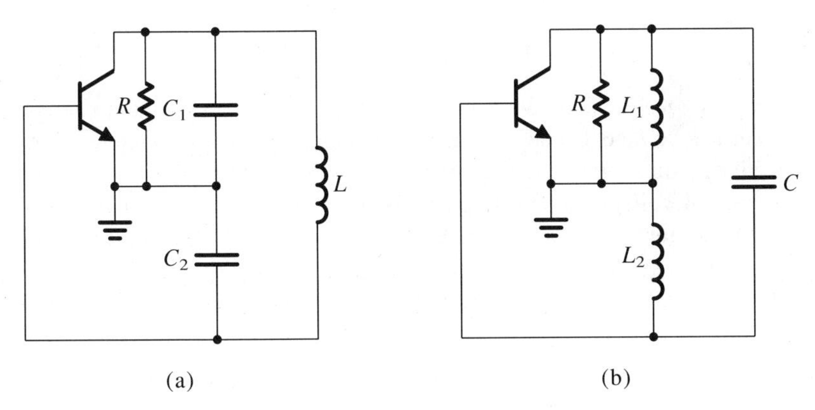

Fig. 12.12 Two commonly used configurations of LC-tuned oscillators: **(a)** Colpitts; **(b)** Hartley.

[1] Of course, transistors can be used in place of the op amps in the circuits just studied. At higher frequencies, however, better results are obtained with LC tuned circuits and crystals.

If the frequency of operation is sufficiently low that we can neglect the transistor capacitances, the frequency of oscillation will be determined by the resonance frequency of the parallel-tuned circuit (also known as a *tank circuit* because it behaves as a reservoir for energy storage). Thus for the Colpitts oscillator we have

$$\omega_0 = 1 \Big/ \sqrt{L\left(\frac{C_1 C_2}{C_1 + C_2}\right)} \tag{12.16}$$

and for the Hartley oscillator we have

$$\omega_0 = 1/\sqrt{(L_1 + L_2)C} \tag{12.17}$$

The ratio L_1/L_2 or C_1/C_2 determines the feedback factor and thus must be adjusted in conjunction with the transistor gain to ensure that oscillations will start. To determine the oscillation condition of the Colpitts oscillator we replace the transistor with its equivalent circuit, as shown in Fig. 12.13. To simplify the analysis we have neglected the transistor capacitance C_μ (C_{gd} for FET). Capacitance C_π (C_{gs} for FET), although not shown, can be considered to be a part of C_2. The input resistance r_π (infinite for FET) has also been neglected, assuming that at the frequency of oscillation $r_\pi \gg (1/\omega C_2)$. Finally, as mentioned earlier, the resistance R includes r_o of the transistor.

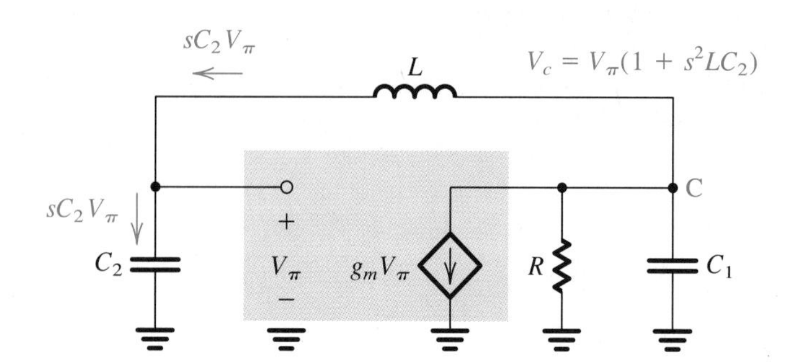

Fig. 12.13 Equivalent circuit of the Colpitts oscillator of Fig. 12.12(a). To simplify the analysis, C_μ and r_π are neglected. We can consider C_π to be part of C_2, and we can include r_o in R.

To find the loop gain we break the loop at the transistor base, apply an input voltage V_π, and find the returned voltage that appears across the input terminals of the transistor. We then equate the loop gain to unity. An alternative approach is to analyze the circuit and eliminate all current and voltage variables, and thus obtain one equation that governs circuit operation. Oscillations will start if this equation is satisfied. Thus the resulting equation will give us the conditions for oscillation.

A node equation at the transistor collector (node C) in the circuit of Fig. 12.13 yields

$$s\,C_2\,V_\pi + g_m\,V_\pi + \left(\frac{1}{R} + s\,C_1\right)(1 + s^2\,L\,C_2)\,V_\pi = 0$$

Since $V_\pi \neq 0$ (oscillations have started) it can be eliminated, and the equation can be rearranged in the form

$$s^3\,L\,C_1C_2 + s^2\,(L\,C_2/R) + s\,(C_1 + C_2) + \left(g_m + \frac{1}{R}\right) = 0 \tag{12.18}$$

Substituting $s = j\omega$ gives

$$\left(g_m + \frac{1}{R} - \frac{\omega^2 L C_2}{R}\right) + j\,[\omega(C_1 + C_2) - \omega^3 L C_1 C_2] = 0 \qquad (12.19)$$

For oscillations to start, both the real and imaginary parts must be zero. Equating the imaginary part to zero gives the frequency of oscillation as

$$\omega_0 = 1 \Big/ \sqrt{L\left(\frac{C_1 C_2}{C_1 + C_2}\right)} \qquad (12.20)$$

which is the resonance frequency of the tank circuit, as anticipated.[2] Equating the real part to zero together with using Eq. (12.20) gives

$$C_2/C_1 = g_m R \qquad (12.21)$$

which has a simple physical interpretation: For sustained oscillations, the magnitude of the gain from base to collector ($g_m R$) must be equal to the inverse of the voltage ratio provided by the capacitive divider, which from Fig. 12.12(a) can be seen to be $v_{eb}/v_{ce} = C_1/C_2$. Of course, for oscillations to start, the loop gain must be made greater than unity, a condition that can be stated in the equivalent form

$$g_m R > C_2/C_1 \qquad (12.22)$$

As oscillations grow in amplitude, the transistor's nonlinear characteristics reduce the loop gain to unity, thus sustaining the oscillations.

Analysis similar to that above can be carried out for the Hartley circuit (see Exercise 12.8). At high frequencies, more accurate transistor models must be used. Alternatively, the y parameters of the transistor can be measured at the intended frequency ω_0, and the analysis can then be carried out using the y-parameter model (see Appendix B). This is usually simpler and more accurate, especially at frequencies above about 30% of the transistor f_T.

As an example of a practical LC oscillator we show in Fig. 12.14 the circuit of a Colpitts oscillator, complete with bias details. Here the radio-frequency choke (RFC) provides a high reactance at ω_0 but a low dc resistance.

Finally, a few words are in order on the mechanism that determines the amplitude of oscillations in the LC tuned oscillators discussed above. Unlike the op amp oscillators that incorporate special amplitude-control circuitry, LC tuned oscillators utilize the nonlinear i_C–v_{BE} characteristics of the BJT (the i_D–v_{GS} characteristics of the FET) for amplitude control. Thus these LC tuned oscillators are known as *self-limiting oscillators*. Specifically, as the oscillations grow in amplitude, the effective gain of the transistor is reduced below its small-signal value. Eventually, an amplitude is reached at which the effective gain is reduced to the point that the Barkhausen criterion is satisfied exactly. The amplitude then remains constant at this value.

Reliance on the nonlinear characteristics of the BJT (or the FET) implies that the collector (drain) current waveform will be nonlinearly distorted. Nevertheless, the output

---

[2] If r_π is taken into account, the frequency of oscillation can be shown to shift slightly from the value given by Eq. (12.20).

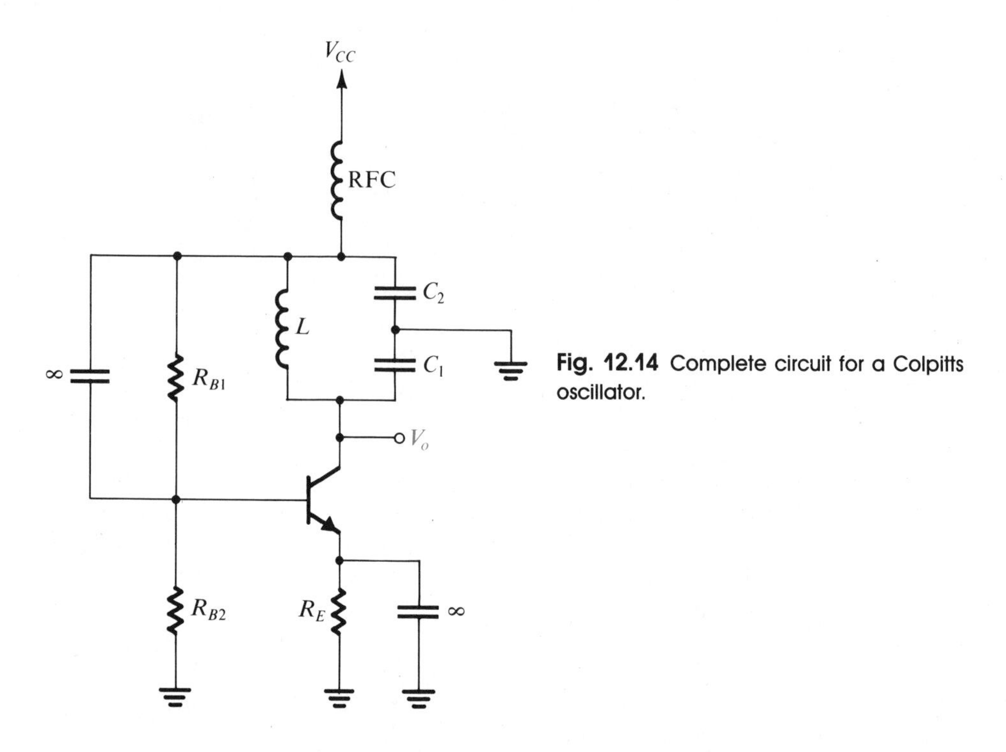

Fig. 12.14 Complete circuit for a Colpitts oscillator.

voltage signal will still be a sinusoid of high purity because of the filtering action of the LC tuned circuit. Detailed analysis of amplitude control, which makes use of nonlinear circuit techniques, is beyond the scope of this book. The interested reader is referred to Clarke and Hess, 1971.

Exercises 12.8 Show that for the Hartley oscillator of Fig. 12.12(b) the frequency of oscillation is given by Eq. (12.17) and that for oscillations to start $g_m R > (L_1/L_2)$.

D12.9 Using a BJT biased at $I_C = 1$ mA, design a Colpitts oscillator to operate at $\omega_0 = 10^6$ rad/s. Use $C_1 = 0.01\ \mu$F, and assume that the coil available has a Q of 100 (this can be represented by a resistance in parallel with C_1 given by $Q/\omega_0 C_1$). Also assume that there is a load resistance at the collector of 2 kΩ and that for the BJT, $r_o = 100$ kΩ. Find C_2 and L.

Ans. 0.35 μF; 100 μH (a somewhat smaller C_2 would be used to allow oscillations to grow in amplitude)

Crystal Oscillators

A piezoelectric crystal, such as quartz, exhibits electromechanical-resonance characteristics that are very stable (with time and temperature) and highly selective (having very high Q factors). The circuit symbol of a crystal is shown in Fig. 12.15(a), and the equivalent circuit model is given in Fig. 12.15(b). The resonance properties are characterized by a large inductance L (as high as hundreds of henrys), a very small series capacitance C_s (as small as 0.0005 pF), a series resistance r representing a Q factor $\omega_0 L/r$ that can be as high as a few hundred thousand, and a parallel capacitance C_p (a few picofarads). Capacitor C_p represents the electrostatic capacitance between the two parallel plates of the crystal. Note that $C_p \gg C_s$.

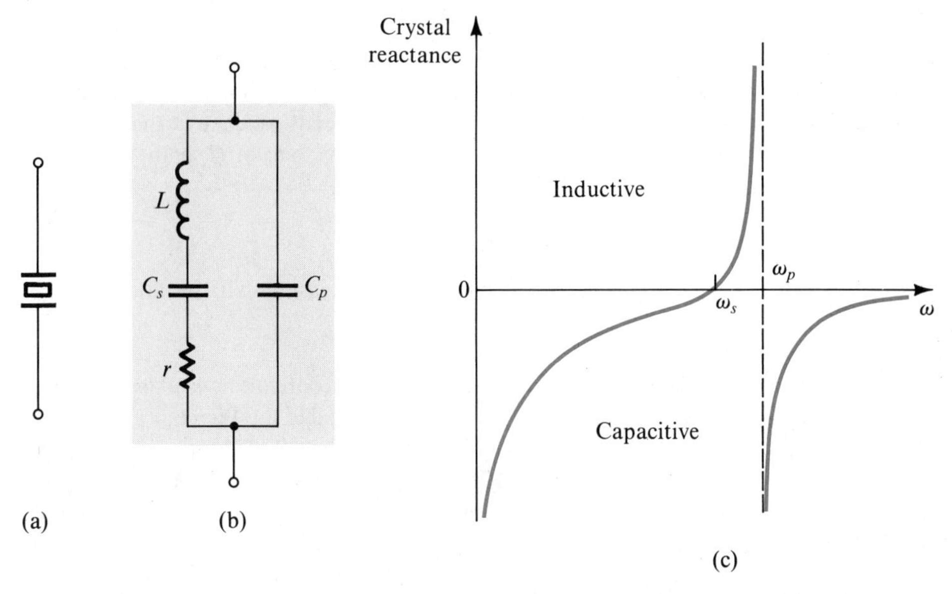

Fig. 12.15 A piezoelectric crystal: **(a)** circuit symbol; **(b)** equivalent circuit; **(c)** crystal reactance versus frequency [note that, neglecting the small resistance r, $Z_{\text{crystal}} = jX(\omega)$].

Since the Q factor is very high, we may neglect the resistance r and express the crystal impedance as

$$Z(s) = 1 \Big/ \left[sC_p + \frac{1}{sL + 1/sC_s} \right]$$

which can be manipulated to the form

$$Z(s) = \frac{1}{sC_p} \frac{s^2 + (1/LC_s)}{s^2 + [(C_p + C_s)/LC_sC_p]} \tag{12.23}$$

From Eq. (12.23) and from Fig. 12.15(b) we see that the crystal has two resonance frequencies: a series resonance at ω_s,

$$\omega_s = 1/\sqrt{LC_s} \tag{12.24}$$

and a parallel resonance at ω_p,

$$\omega_p = 1 \Big/ \sqrt{L\left(\frac{C_sC_p}{C_s + C_p}\right)} \tag{12.25}$$

Thus for $s = j\omega$ we can write

$$Z(j\omega) = -j\frac{1}{\omega C_p}\left(\frac{\omega^2 - \omega_s^2}{\omega^2 - \omega_p^2}\right) \tag{12.26}$$

From Eqs. (12.24) and (12.25) we note that $\omega_p > \omega_s$. However, since $C_p \gg C_s$, the two resonance frequencies are very close. Expressing $Z(j\omega) = jX(\omega)$, the crystal reactance

$X(\omega)$ will have the shape shown in Fig. 12.15(c). We observe that the crystal reactance is inductive over the very narrow frequency band between ω_s and ω_p. For a given crystal, this frequency band is well defined. Thus we may use the crystal to replace the inductor of the Colpitts oscillator (Fig. 12.12a). The resulting circuit will oscillate at the resonance frequency of the crystal inductance L with the series equivalent of C_s and

$$\left(C_p + \frac{C_1 C_2}{C_1 + C_2}\right) \tag{12.27}$$

Since C_s is much smaller than the three other capacitances, it will be dominant and

$$\omega_0 \simeq 1/\sqrt{LC_s} = \omega_s.$$

In addition to the basic Colpitts oscillator, a variety of configurations exist for crystal oscillators. Figure 12.16 shows a popular configuration (called the Pierce oscillator) utilizing a CMOS inverter (see Chapter 13) as amplifier. Resistor R_f determines a dc operating point in the high-gain region of the CMOS inverter. Resistor R_1 together with capacitor C_1 provides a low-pass filter that discourages the circuit from oscillating at a higher harmonic of the crystal frequency. Note that this circuit also is based on the Colpitts configuration.

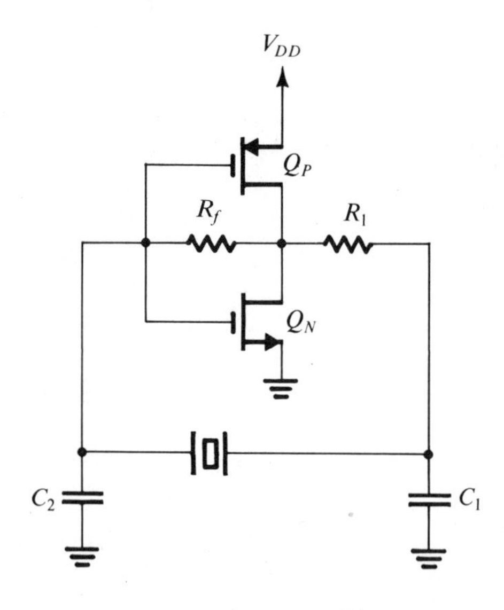

Fig. 12.16 A Colpitts (or Pierce) crystal oscillator utilizing a CMOS inverter as an amplifier.

The extremely stable resonance characteristics and the very high Q factors of quartz crystals result in oscillators with very accurate and stable frequencies. Crystals are available with resonance frequencies in the range of few kHz to hundreds of MHz. Temperature coefficients of ω_0 of 1 or 2 parts per million (ppm) per °C are achievable. Unfortunately, however, crystal oscillators, being mechanical resonators, are fixed-frequency circuits.

Exercise 12.10 A 2-MHz quartz crystal is specified to have $L = 0.52$ H, $C_s = 0.012$ pF, $C_p = 4$ pF, and $r = 120\ \Omega$. Find f_s, f_p, and Q.

Ans. 2.015 MHz; 2.018 MHz; 55,000

12.4 BISTABLE MULTIVIBRATORS

In this section we begin the study of the other type of waveform-generating circuits—nonlinear oscillators or function generators. These make use of a special class of circuits known as **multivibrators.** As mentioned earlier, there are three types of multivibrators: bistable, monostable and astable. This section is concerned with the first, the bistable multivibrator.

As its name indicates, the **bistable multivibrator** has *two stable states*. The circuit can remain in either stable state indefinitely and moves to the other stable state only when appropriately *triggered*.

The Feedback Loop

Bistability can be obtained by connecting an amplifier in a positive-feedback loop having a loop gain greater than unity. Such a feedback loop is shown in Fig. 12.17; it consists of an op amp and a resistive voltage divider in the positive-feedback path. To see how bistability is obtained, consider operation with the positive input terminal of the op amp near ground potential. This is a reasonable starting point since the circuit has no external excitation. Assume that the electrical noise that is inevitably present in every electronic circuit causes a small positive increment in the voltage v_+. This incremental signal will be amplified by the large open-loop gain A of the op amp, with the result that a much greater signal will appear in the op amp's output voltage v_O. The voltage divider R_1, R_2 will feed a fraction $\beta \equiv R_1/(R_1 + R_2)$ of the output signal back to the positive input terminal of the op amp. If $A\beta$ is greater than unity, as is usually the case, the fed-back signal will be greater than the original increment in v_+. This *regenerative* process continues until eventually the op amp saturates with its output voltage at the positive saturation level, L_+. When this happens the voltage at the positive input terminal, v_+, becomes $L_+R_1/(R_1 + R_2)$, which is positive and thus keeps the op amp in positive saturation. This is one of the two stable states of the circuit.

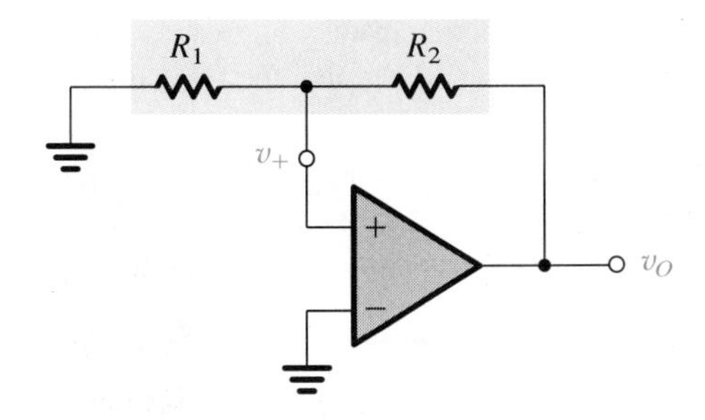

Fig. 12.17 A positive-feedback loop capable of bistable operation.

In the description above we assumed that when v_+ was near zero volts, a positive increment occurred in v_+. Had we assumed the equally probable situation of a negative increment, the op amp would have ended up saturated in the negative direction with $v_O = L_-$ and $v_+ = L_-R_1/(R_1 + R_2)$. This is the other stable state.

We thus conclude that the circuit of Fig. 12.17 has two stable states, one with the op amp in positive saturation and the other with the op amp in negative saturation. The circuit can exist in either of these two states indefinitely. We also note that the circuit cannot exist in the state for which $v_+ = 0$ for any length of time. This is a state of *unstable equilibrium;* any disturbance, such as that caused by noise, causes the bistable circuit to switch to one of its two stable states. This is in sharp contrast to the case when the feedback is negative, causing a virtual short circuit to appear between the op amp's input terminals and maintaining this virtual short circuit in the face of disturbances. A physical analogy for the operation of the bistable circuit is depicted in Fig. 12.18.

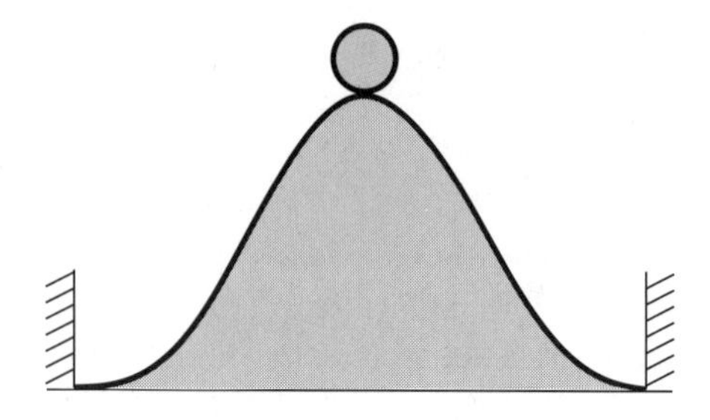

Fig. 12.18 A physical analogy for the operation of the bistable circuit. The ball cannot remain at the top of the hill for any length of time (a state of unstable equilibrium); the inevitably present disturbance will cause the ball to fall to one side or the other, where it can remain indefinitely (the two stable states).

Transfer Characteristics of the Bistable Circuit

The question naturally arises as to how we can make the bistable circuit of Fig. 12.17 change state. To help answer this crucial question, we derive the transfer characteristics of the bistable. Reference to Fig. 12.17 indicates that either of the two circuit nodes that are connected to ground can serve as an input terminal. We investigate both possibilities.

Figure 12.19(a) shows the bistable circuit with an input v_I applied to the inverting input terminal of the op amp. To derive the transfer characteristic v_O–v_I, assume that v_O is at one of its two possible levels, say L_+, and thus $v_+ = \beta L_+$. Now as v_I is increased from 0 V we can see from the circuit that nothing happens until v_I reaches a value equal to v_+ (that is, βL_+). As v_I begins to exceed this value, a net negative voltage develops between the input terminals of the op amp. This voltage is amplified by the open-loop gain of the op amp, and thus v_O goes negative. The voltage divider in turn causes v_+ to go negative, thus increasing the net negative input to the op amp and keeping the regenerative process going. This process culminates in the op amp saturating in the negative direction; that is, with $v_O = L_-$ and, correspondingly, $v_+ = \beta L_-$. It is easy to see that increasing v_I further has no effect on the acquired state of the bistable circuit. Figure 12.19(b) shows the transfer characteristic for increasing v_I. Observe that the characteristic is that of a comparator with a threshold voltage denoted V_{TH}, where $V_{TH} = \beta L_+$.

Next consider what happens as v_I is decreased. Since now $v_+ = \beta L_-$, we see that the circuit remains in the negative-saturation state until v_I goes negative to the point that it equals βL_-. As v_I goes below this value, a net positive voltage appears between the op amp's input terminals. This voltage is amplified by the op amp gain and thus gives rise to a positive voltage at the op amp's output. The regenerative action of the positive-feedback loop then sets in and causes the circuit eventually to go to its positive-saturation state, in which $v_O = L_+$ and $v_+ = \beta L_+$. The transfer characteristic for decreasing v_I is shown in Fig. 12.19(c). Here again we observe that the characteristic is that of a comparator, but with a threshold voltage $V_{TL} = \beta L_-$.

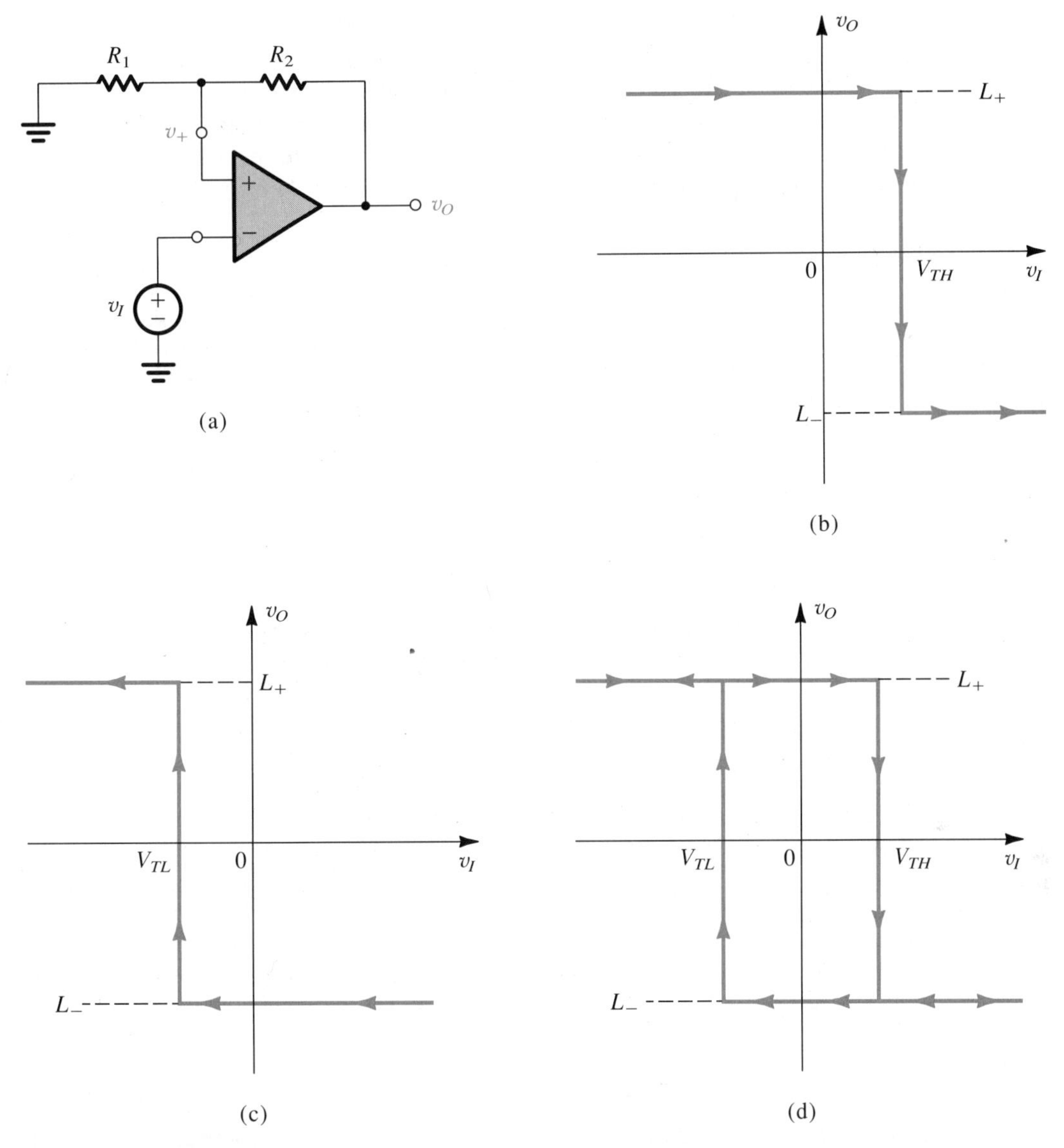

Fig. 12.19 (a) The bistable circuit of Fig. 12.17 with the negative input terminal of the op amp disconnected from ground and connected to an input signal v_I. **(b)** The transfer characteristic of the circuit in (a) for increasing v_I. **(c)** The transfer characteristic for decreasing v_I. **(d)** The complete transfer characteristics.

The complete transfer characteristics, v_O–v_I, of the circuit in Fig. 12.19(a) can be obtained by combining the characteristics in Figs. 12.19(b) and (c), as shown in Fig. 12.19(d). As indicated, the circuit changes state at different values of v_I, depending on whether v_I is increasing or decreasing. Thus the circuit is said to exhibit *hysteresis;* the width of the hysteresis is the difference between the high threshold V_{TH} and the low threshold V_{TL}. Also note that the bistable circuit is in effect a comparator with hysteresis. As will be shown shortly, adding hysteresis to a comparator's characteristics can be very beneficial in certain applications. Finally, observe that because the bistable circuit of Fig. 12.19 switches from the positive state ($v_O = L_+$) to the negative state ($v_O = L_-$) as v_I is increased past the positive threshold V_{TH}, the circuit is said to be *inverting*. A bistable with a *noninverting* transfer characteristics is presented shortly.

Triggering the Bistable Circuit

Returning now to the question of how to make the bistable circuit change state, we observe from the transfer characteristics of Fig. 12.19(d) that if the circuit is in the L_+ state it can be switched to the L_- state by applying an input v_I of value greater than $V_{TH} \equiv \beta L_+$. Such an input causes a net negative voltage to appear between the input terminals of the op amp, which initiates the regenerative cycle that culminates in the circuit switching to the L_- stable state. Here it is important to note that the input v_I merely initiates or *triggers* regeneration. Thus we can remove v_I with no effect on the regeneration process. In other words, v_I can be simply a pulse of short duration. The input signal v_I is thus referred to as a **trigger signal,** or simply a **trigger.**

The characteristics of Fig. 12.19(d) indicate also that the bistable circuit can be switched to the positive state ($v_O = L_+$) by applying a negative trigger signal v_I of magnitude greater than that of the negative threshold V_{TL}.

The Bistable Circuit as a Memory Element

We observe from Fig. 12.19(d) that for input voltages in the range $V_{TL} < v_I < V_{TH}$, the output can be either L_+ or L_-, *depending on the state that the circuit is already in*. Thus, for this input range, the output is determined by the *previous* value of the trigger signal (the trigger signal that caused the circuit to be in its current state). Thus the circuit exhibits *memory*. Indeed, the bistable multivibrator is the basic memory element of digital systems, as will be seen in Chapter 13. Finally, note that in analog circuit applications, such as the ones of concern to us in this chapter, the bistable circuit is also known as a **Schmitt trigger.**

A Bistable Circuit with Noninverting Transfer Characteristics

The basic bistable feedback loop of Fig. 12.17 can be used to derive a circuit with noninverting transfer characteristics by applying the input signal v_I (the trigger signal) to the terminal of R_1 that is connected to ground. The resulting circuit is shown in Fig. 12.20(a). To obtain the transfer characteristics we first employ superposition to the linear circuit formed by R_1 and R_2, thus expressing v_+ in terms of v_I and v_O as

$$v_+ = v_I \frac{R_2}{R_1 + R_2} + v_O \frac{R_1}{R_1 + R_2} \tag{12.28}$$

From this equation we see that if the circuit is in the positive stable state with $v_O = L_+$, positive values for v_I will have no effect. To trigger the circuit into the L_- state, v_I must be made negative and of such a value as to make v_+ decrease below zero. Thus the low

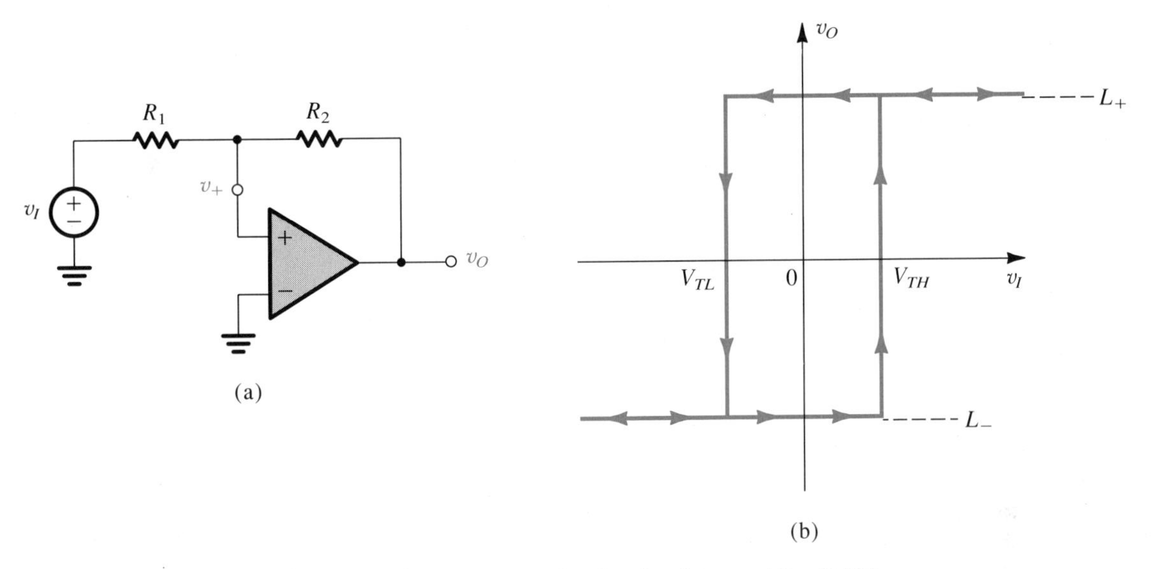

Fig. 12.20 (a) A bistable circuit derived from the positive-feedback loop of Fig. 12.17 by applying v_I through R_1. **(b)** The transfer characteristic of the circuit in (a) is noninverting. (Compare it to the inverting characteristic in Fig. 12.19d.)

threshold V_{TL} can be found by substituting in Eq. (12.28) $v_O = L_+$, $v_+ = 0$, and $v_I = V_{TL}$. The result is

$$V_{TL} = -L_+ \, (R_1/R_2) \tag{12.29}$$

Similarly, Eq. (12.28) indicates that when the circuit is in the negative-output state ($v_O = L_-$), negative values of v_I will make v_+ more negative with no effect on operation. To initiate the regeneration process that causes the circuit to switch to the positive state, v_+ must be made to go slightly positive. The value of v_I that causes this to happen is the high threshold voltage V_{TH}, which can be found by substituting in Eq. (12.28) $v_O = L_-$ and $v_+ = 0$. The result is

$$V_{TH} = -L_- \, (R_1/R_2) \tag{12.30}$$

The complete transfer characteristics of the circuit of Fig. 12.20(a) are displayed in Fig. 12.20(b). Observe that a positive triggering signal v_I (of value greater than V_{TH}) causes the circuit to switch to the positive state (v_O goes from L_- to L_+). Thus the transfer characteristic of this circuit is noninverting.

Application of the Bistable Circuit as a Comparator

The comparator is an analog-circuit building block that is used in a variety of applications ranging from detecting the level of an input signal relative to a preset threshold value, to the design of analog-to-digital (A/D) converters (see Section 10.11). Although one normally thinks of the comparator as having a single threshold value (see Fig. 12.21a), it is useful in many applications to add hysteresis to the comparator characteristics. If this is

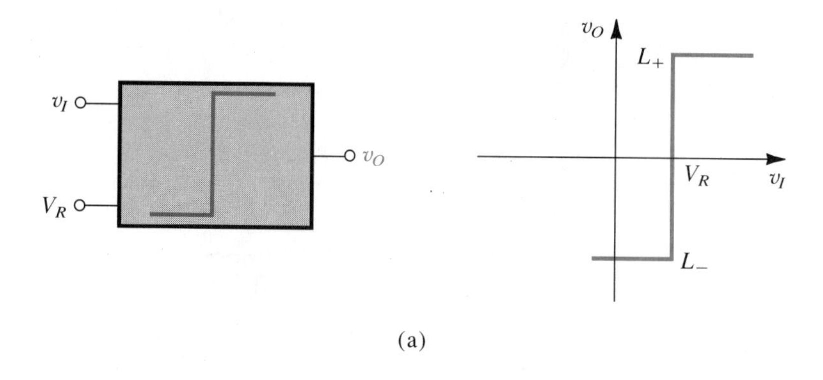

(a)

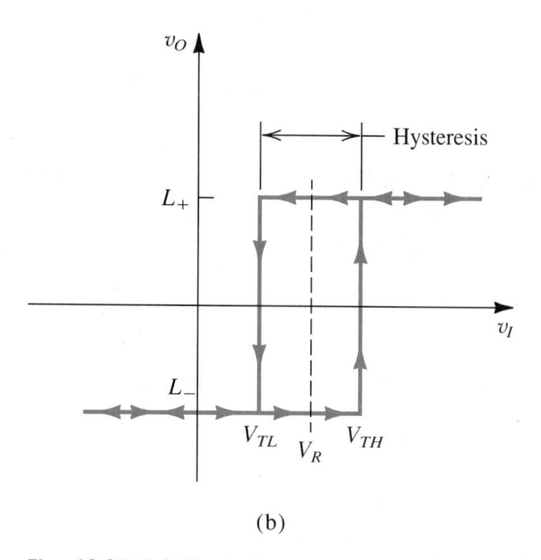

(b)

Fig. 12.21 (a) Block-diagram representation and transfer characteristic for a comparator having a reference, or threshold, voltage V_R. **(b)** Comparator characteristic with hysteresis.

done, the comparator exhibits two threshold values, V_{TL} and V_{TH}, symmetrically placed about the desired reference level, as indicated in Fig. 12.21(b). Usually V_{TH} and V_{TL} are separated by a small amount, say 100 mV.

To demonstrate the need for hysteresis we consider a common application of comparators. It is required to design a circuit that detects and counts the zero crossings of an arbitrary waveform. Such a function can be implemented using a comparator whose threshold is set to 0 V. The comparator provides a step change at its output every time a zero crossing occurs. Each step change can be used to generate a pulse, and the pulses are fed to a counter circuit.

Imagine now what happens if the signal being processed has—as it usually does have—interference superimposed on it, say of a frequency much higher than that of the signal. It follows that the signal might cross the zero axis a number of times around each of the zero-crossing points we are trying to detect, as shown in Fig. 12.22. The comparator would thus change state a number of times at each of the zero crossings, and our count would obviously be in error. If we have an idea of the expected peak-to-peak amplitude of the interference, the problem can be solved by introducing hysteresis of appropriate width in the comparator characteristics. Then, if the input signal is increasing in magnitude, the comparator with hysteresis will remain in the low state until the input level exceeds the high threshold V_{TH}. Subsequently the comparator will remain in the high state even if, owing to interference, the signal decreases below V_{TH}. The comparator will switch to the low state only if the input signal is decreased below the low threshold V_{TL}. The situation is illustrated in Fig. 12.22, from which we see that including hysteresis in the comparator characteristics provides an effective means for rejecting interference (thus providing another form of filtering).

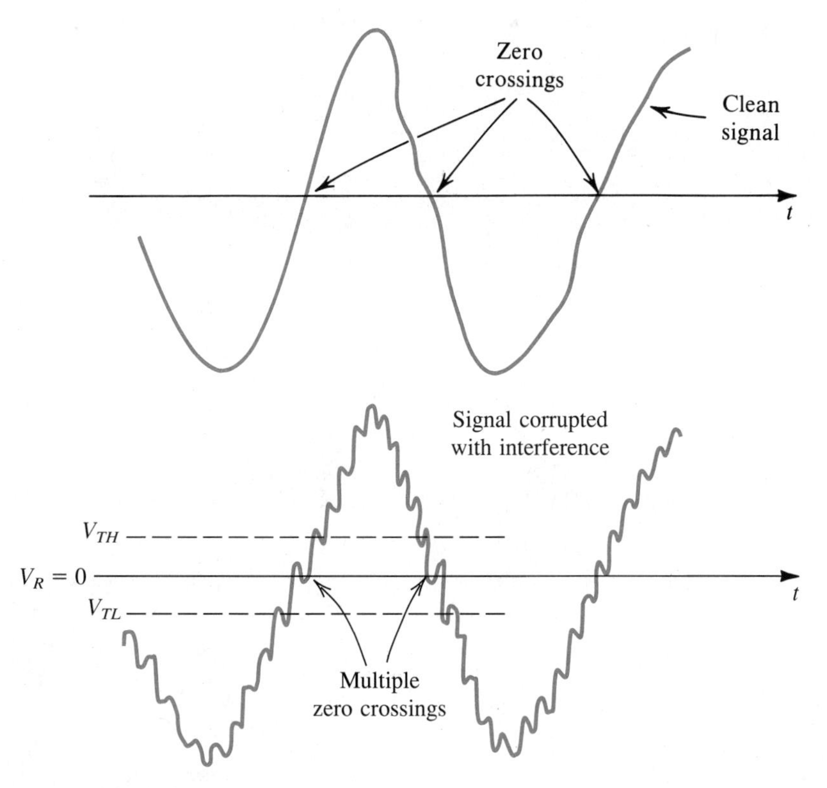

Fig. 12.22 Illustrating the use of hysteresis in the comparator characteristics as a means of rejecting interference.

Making the Output Levels More Precise

The output levels of the bistable circuit can be made more precise than the saturation voltages of the op amp are by cascading the op amp with a limiter circuit (see Section 3.7 for a discussion of limiter circuits). Two such arrangements are shown in Fig. 12.23.

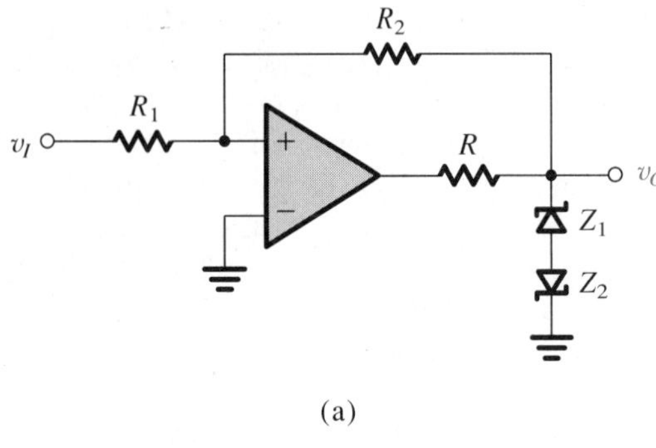

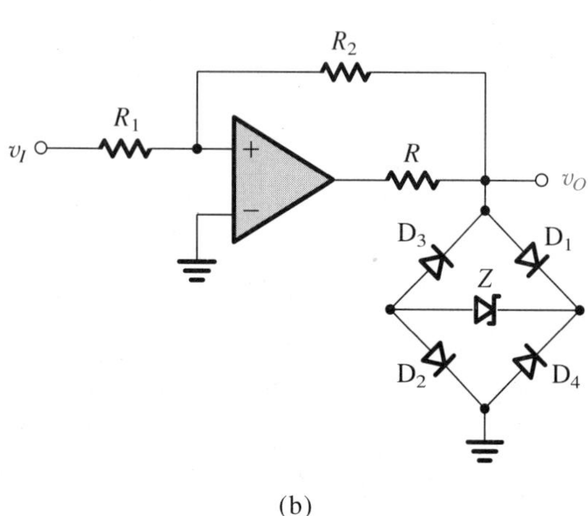

Fig. 12.23 Limiter circuits are used to obtain more precise output levels for the bistable circuit. In both circuits the value of R should be chosen to yield the current required for the proper operation of the zener diodes. **(a)** For this circuit $L_+ = V_{Z_1} + V_D$ and $L_- = -(V_{Z_2} + V_D)$, where V_D is the forward diode drop. **(b)** For this circuit $L_+ = V_Z + V_{D_1} + V_{D_2}$ and $L_- = -(V_Z + V_{D_3} + V_{D_4})$.

Exercises **D12.11** The op amp in the bistable circuit of Fig. 12.19(a) has output saturation voltages of ± 13 V. Design the circuit to obtain threshold voltages of ± 5 V. For $R_1 = 10$ kΩ find the value required for R_2.

Ans. 16 kΩ

D12.12 If the op amp in the circuit of Fig. 12.20(a) has ± 10-V output saturation levels, design the circuit to obtain ± 5-V thresholds. Give suitable component values.

Ans. Possible choice: $R_1 = 10$ kΩ and $R_2 = 20$ kΩ

12.13 Consider a bistable circuit with noninverting transfer characteristic, and let $L_+ = -L_- = 10$ V and $V_{TH} = -V_{TL} = 5$ V. If v_I is a triangular wave with a 0-V average, a 10-V peak amplitude, and a 1-ms period, sketch the waveform of v_O. Find the time interval between the zero crossings of v_I and v_O.

Ans. v_O is a square wave with 0-V average, 10-V amplitude, 1-ms period and is delayed by 125 μs relative to v_I.

12.14 Consider an op amp having saturation levels of ± 12 V used without feedback, with the inverting input terminal connected to $+3$ V and the noninverting input terminal connected to v_I. Characterize its operation as a comparator. What are L_+, L_-, and V_R, as defined in Fig. 12.21(a)?

Ans. $+12$ V; -12 V; $+3$ V

12.15 In the circuit of Fig. 12.20(a) let $L_+ = -L_- = 10$ V and $R_1 = 1$ kΩ. Find a value for R_2 that gives hysteresis of 100-mV width.

Ans. 200 kΩ

12.5 GENERATION OF SQUARE AND TRIANGULAR WAVEFORMS USING ASTABLE MULTIVIBRATORS

A square waveform can be generated by arranging for a bistable multivibrator to switch states periodically. This can be done by connecting the bistable multivibrator with an RC circuit in a feedback loop, as shown in Fig. 12.24(a). Observe that the bistable multivibrator has an inverting transfer characteristic and can thus be realized using the circuit of Fig. 12.19(a). This results in the circuit of Fig. 12.24(b). We shall show shortly that this circuit has no stable states and thus is appropriately named an **astable multivibrator.**

Operation of the Astable Multivibrator

To see how the astable multivibrator operates, refer to Fig. 12.24 and let the output of the bistable multivibrator be at one of its two possible levels, say L_+. Capacitor C will charge toward this level through resistor R. Thus the voltage across C, which is applied to the negative input terminal of the op amp and thus is denoted v_-, will rise exponentially toward L_+ with a time constant $\tau = CR$. Meanwhile the voltage at the positive input terminal of the op amp is $v_+ = \beta L_+$. This situation will continue until the capacitor voltage reaches the positive threshold V_{TH} at which point the bistable multivibrator will switch to the other stable state in which $v_O = L_-$ and $v_+ = \beta L_-$. The capacitor will then start discharging, and its voltage, v_-, will decrease exponentially toward L_-. This new state will prevail until v_- reaches the negative threshold V_{TL}, at which time the bistable multivibrator switches to the positive-output state, the capacitor begins to charge, and the cycle repeats itself.

From the above we see that the astable circuit oscillates and produces a square waveform at the output of the op amp. This waveform and the waveforms at the two input terminals of the op amp are displayed in Fig. 12.24(c). The period T of the square wave can be found as follows: During the charging interval T_1 the voltage v_- across the capacitor at any time t, with $t = 0$ at the beginning of T_1, is given by (see Appendix E)

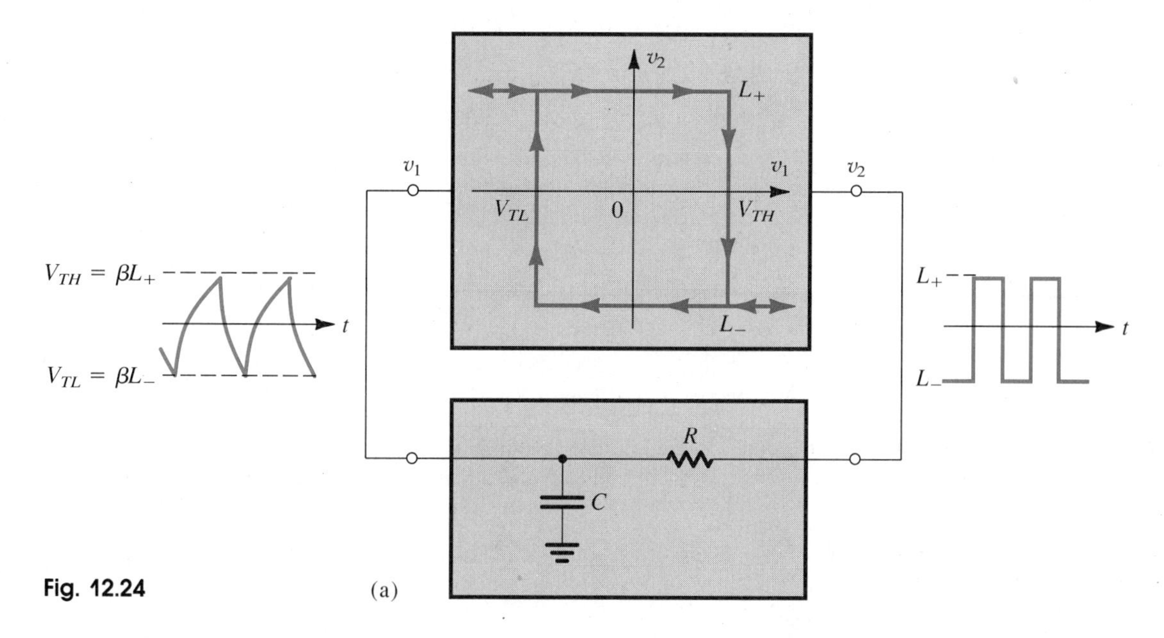

Fig. 12.24 (a)

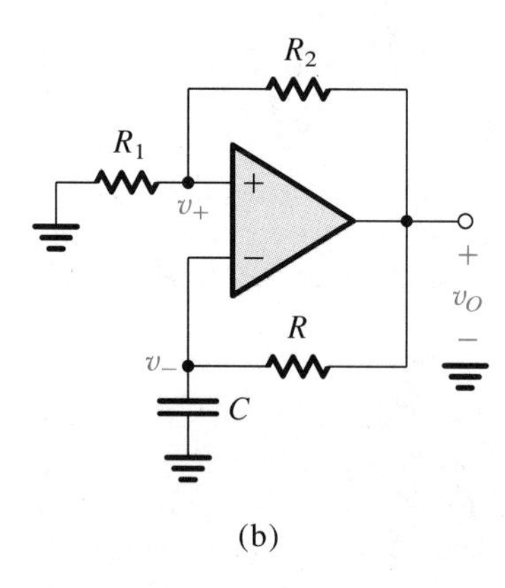

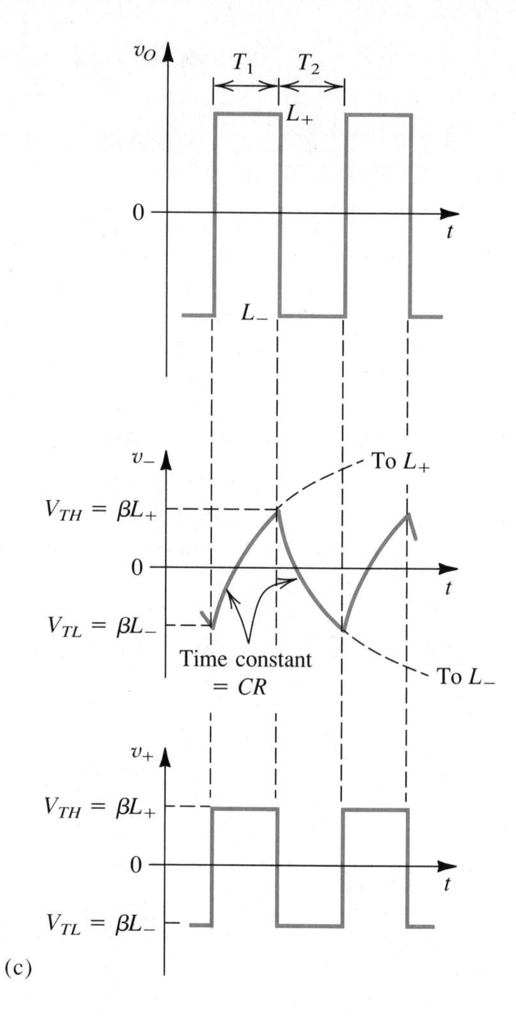

Fig. 12.24 (a) Connecting a bistable multivibrator with inverting transfer characteristics in a feedback loop with an RC circuit results in a square-wave generator. **(b)** The circuit obtained when the bistable multivibrator is implemented with the circuit of Fig. 12.19(a). **(c)** Waveforms at various nodes of the circuit in (b). This circuit is called an astable multivibrator.

$$v_- = L_+ - (L_+ - \beta L_-)e^{-t/\tau}$$

where $\tau = CR$. Substituting $v_- = \beta L_+$ at $t = T_1$ gives

$$T_1 = \tau \ln \frac{1 - \beta(L_-/L_+)}{1 - \beta} \tag{12.31}$$

Similarly, during the discharge interval T_2 the voltage v_- at any time t, with $t = 0$ at the beginning of T_2, is given by

$$v_- = L_- - (L_- - \beta L_+)e^{-t/\tau}$$

Substituting $v_- = \beta L_-$ at $t = T_2$ gives

$$T_2 = \tau \ln \frac{1 - \beta(L_+/L_-)}{1 - \beta} \tag{12.32}$$

Equations (12.31) and (12.32) can be combined to obtain the period $T = T_1 + T_2$. Normally, $L_+ = -L_-$, resulting in symmetrical square waves of period T given by

$$T = 2\tau \ln \frac{1 + \beta}{1 - \beta} \tag{12.33}$$

Note that this square-wave generator can be made to have variable frequency by switching different capacitors C (usually in decades) and by continuously adjusting R (to obtain continuous frequency control within each decade of frequency). Also, the waveform across C can be made almost triangular by using a small value for the parameter β. However, triangular waveforms of superior linearity can be easily generated using the scheme discussed next.

Before leaving this section, however, note that although the astable circuit has no stable states it has two *quasi-stable* states and remains in each for a time interval determined by the time constant of the RC network and the thresholds of the bistable multivibrator.

Exercises 12.16 For the circuit in Fig. 12.24(b), let the op amp saturation voltages be ± 10 V, $R_1 = 100$ kΩ, $R_2 = R = 1$ MΩ, and $C = 0.01$ μF. Find the frequency of oscillation.

Ans. 274.2 Hz

12.17 Consider a modification of the circuit of Fig. 12.24(b) in which R_1 is replaced by a pair of diodes connected in parallel in opposite directions. For $L_+ = -L_- = 12$ V, $R_2 = R = 10$ kΩ, $C = 0.1$ μF, and the diode voltage a constant denoted V_D, find an expression for frequency as a function of V_D. If $V_D = 0.70$ V at 25°C with a TC of -2 mV/°C, find the frequency at 0°C, 25°C, and 100°C. Note that the output of this circuit can be sent to a remotely connected frequency meter to provide a digital readout of temperature.

Ans. $f = 500/\ln[(12 + V_D)/(12 - V_D)]$ Hz; 3995 Hz, 4281 Hz, 4611 Hz, 5451 Hz

Generation of Triangular Waveforms

The exponential waveforms generated in the astable circuit of Fig. 12.24 can be changed to triangular by replacing the low-pass RC circuit with an integrator. (The integrator is, after all, a low-pass circuit with a corner frequency at dc.) The integrator causes linear charging and discharging of the capacitor, thus providing a triangular waveform. The resulting circuit is shown in Fig. 12.25(a). Observe that because the integrator is inverting it is necessary to invert the characteristics of the bistable circuit. Thus the bistable circuit required here is of the noninverting type and can be implemented using the circuit of Fig. 12.20(a).

We now proceed to show how the feedback loop of Fig. 12.25(a) oscillates and generates a triangular waveform v_1 at the output of the integrator and a square waveform v_2 at the output of the bistable circuit: Let the output of the bistable circuit be at L_+. A current equal to L_+/R will flow into the resistor R and through capacitor C, causing the output of the integrator to *linearly* decrease with a slope of $-L_+/CR$, as shown in Fig. 12.25(c). This will continue until the integrator output reaches the lower threshold V_{TL} of the bistable circuit, at which point the bistable circuit will switch states, its output becoming negative and equal to L_-. At this moment the current through R and C will reverse direction, and its value will become equal to $|L_-|/R$. It follows that the integrator output will start to increase linearly with a positive slope equal to $|L_-|/CR$. This will continue until the integrator output voltage reaches the positive threshold of the bistable circuit, V_{TH}. At this point the bistable circuit switches, its output becomes positive (L_+), the current into the integrator reverses direction, and the output of the integrator starts to decrease linearly, beginning a new cycle.

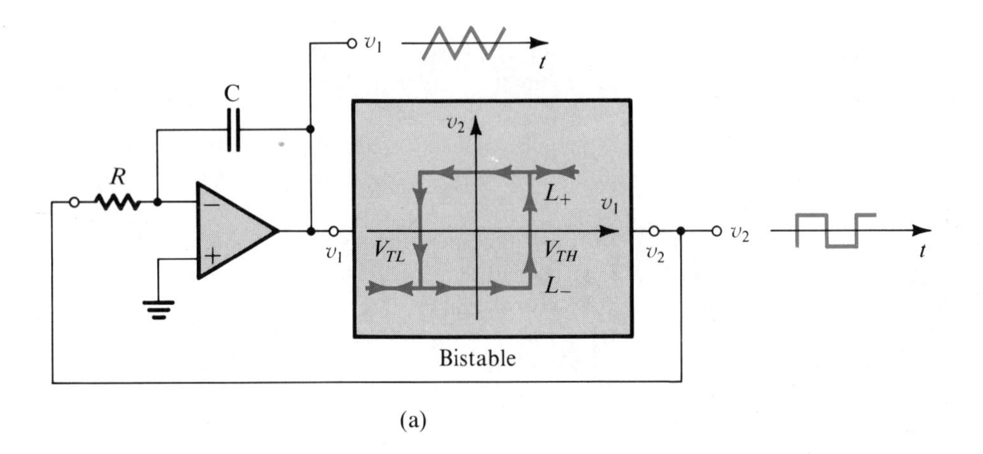

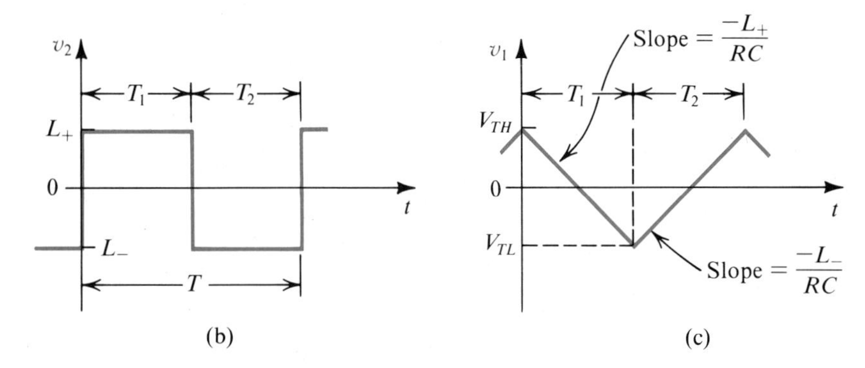

Fig. 12.25 General scheme for generating triangular and square waveforms.

From the above discussion it is relatively easy to derive an expression for the period T of the square and triangular waveforms. During the interval T_1 we have, from Fig. 12.25(c),

$$\frac{V_{TH} - V_{TL}}{T_1} = \frac{L_+}{CR}$$

from which we obtain

$$T_1 = CR\frac{V_{TH} - V_{TL}}{L_+} \tag{12.34}$$

Similarly, during T_2 we have

$$\frac{V_{TH} - V_{TL}}{T_2} = \frac{-L_-}{CR}$$

from which we obtain

$$T_2 = CR\frac{V_{TH} - V_{TL}}{-L_-} \tag{12.35}$$

Thus to obtain symmetrical square waves we design the bistable circuit to have $L_+ = -L_-$.

Exercise D12.18 Consider the circuit of Fig. 12.25(a) with the bistable circuit realized by the circuit in Fig. 12.20(a). If the op amps have saturation voltages of ± 10 V and if a capacitor $C = 0.01\ \mu$F and a resistor $R_1 = 10$ kΩ are used, find the values of R and R_2 (note that R_1 and R_2 are associated with the bistable circuit of Fig. 12.20a) such that the frequency of oscillation is 1 kHz and the triangular waveform has a 10-V peak-to-peak amplitude.

Ans. 50 kΩ; 20 kΩ

12.6 GENERATION OF A STANDARDIZED PULSE—THE MONOSTABLE MULTIVIBRATOR

In some applications the need arises for a pulse of known height and width generated in response to a trigger signal. Because the width of the pulse is predictable, its trailing edge can be used for timing purposes—that is, to initiate a particular task at a specified time. Such a standardized pulse can be generated by the third type of multivibrator, the **monostable multivibrator.**

The monostable multivibrator has one stable state in which it can remain indefinitely. It also has a quasi-stable state to which it can be triggered and in which it stays for a predetermined interval equal to the desired width of the output pulse. Once this interval expires, the monostable multivibrator returns to its stable state and remains there, awaiting another triggering signal. The action of the monostable multivibrator has given rise to its alternative name, the *one shot*.

Figure 12.26(a) shows an op amp monostable circuit. We observe that this circuit is an augmented form of the astable circuit of Fig. 12.24(b). Specifically, a clamping diode

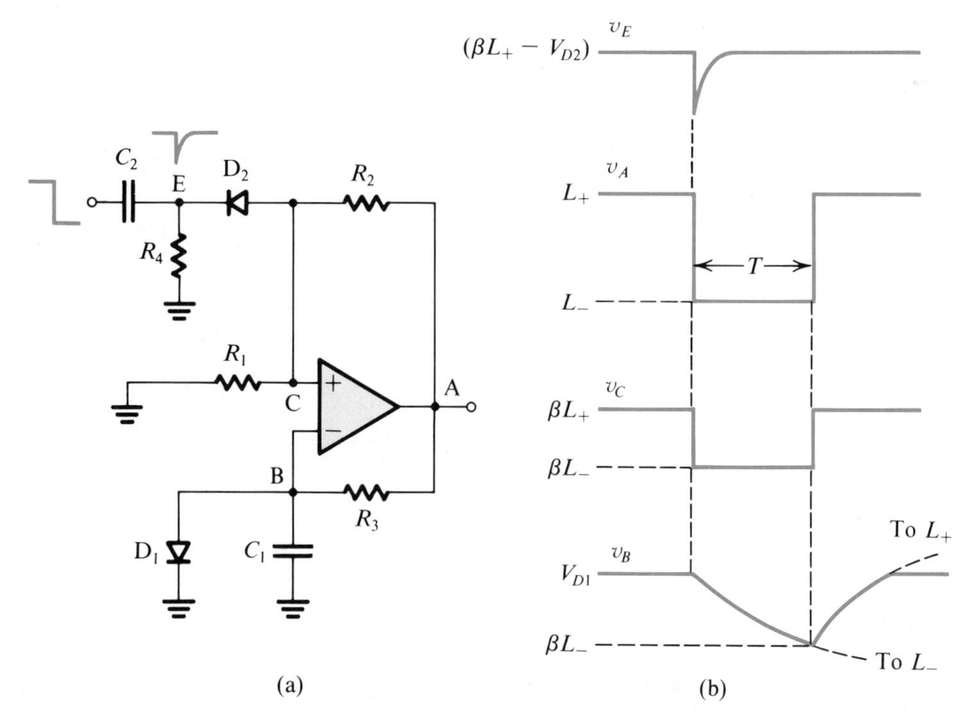

Fig. 12.26 **(a)** An op amp monostable circuit. **(b)** Signal waveforms in the circuit of (a).

D_1 is added across the capacitor C_1, and a trigger circuit composed of capacitor C_2, resistor R_4, and diode D_2 is connected to the noninverting input terminal of the op amp. The circuit operates as follows: In the stable state, which prevails in the absence of the triggering signal, the output of the op amp is at L_+ and diode D_1 is conducting through R_3 and thus clamping the voltage v_B to one diode drop above ground. We select R_4 much larger than R_1, so diode D_2 will be conducting a very small current and the voltage v_C will be very closely determined by the voltage divider R_1, R_2. Thus $v_C = \beta L_+$, where $\beta = R_1/(R_1 + R_2)$. The stable state is maintained because βL_+ is greater than V_{D1}.

Now consider the application of a negative-going step at the trigger input and refer to the signal waveforms shown in Fig. 12.26(b). The negative triggering edge will be coupled to the cathode of diode D_2 via capacitor C_2, and thus D_2 conducts heavily and pulls node C down. If the trigger signal is of sufficient height to cause v_C to go below v_B, the op amp will see a net negative input voltage and its output will switch to L_-. This in turn causes v_C to go negative to βL_-, which keeps the op amp in its newly acquired state. Note that D_2 now cuts off, thus isolating the circuit from any further changes at the trigger input terminal.

The negative voltage at A causes D_1 to cut off, and C_1 begins to discharge exponentially toward L_- with a time constant C_1R_3. The monostable multivibrator is now in its *quasi-stable state,* which will prevail until the declining v_B goes below the voltage at node C, which is βL_-. At this instant the op amp output switches back to L_+ and the voltage at node C goes back to βL_+. Capacitor C_1 then charges toward L_+ until diode D_1 turns on and the circuit returns to its stable state.

From Fig. 12.26(b) we observe that a negative pulse is generated at the output during the quasi-stable state. The duration T of the output pulse is determined from the exponential waveform of v_B,

$$v_B(t) = L_- - (L_- - V_{D1})e^{-t/C_1R_3}$$

by substituting $v_B(T) = \beta L_-$,

$$\beta L_- = L_- - (L_- - V_{D1})e^{-T/C_1R_3}$$

which yields

$$T = C_1R_3 \ln\left(\frac{V_{D1} - L_-}{\beta L_- - L_-}\right) \tag{12.36}$$

For $V_{D1} \ll |L_-|$, this equation can be approximated by

$$T \simeq C_1R_3 \ln\left(\frac{1}{1 - \beta}\right) \tag{12.37}$$

Finally, note that the monostable circuit should not be triggered again until capacitor C_1 has been recharged to V_{D1}; otherwise the resulting output pulse will be shorter than normal. This recharging time is known as the recovery period. Circuit techniques exist for shortening the recovery period.

Exercise 12.19 For the monostable circuit of Fig. 12.26(a) find the value of R_3 that will result in a 100-μs output pulse for $C_1 = 0.1\ \mu$F, $\beta = 0.1$, $V_D = 0.7$ V, and $L_+ = -L_- = 12$ V.

Ans. 6171 Ω

12.7 INTEGRATED-CIRCUIT TIMERS

Commercially available integrated-circuit packages exist that contain the bulk of the circuitry needed to implement monostable and astable multivibrators having precise characteristics. In this section we discuss the most popular of such ICs, the **555 timer.** Introduced in 1972 by the Signetics Corporation as a bipolar integrated circuit, the 555 is also available in CMOS technology and from a number of manufacturers.

The 555 Circuit

Figure 12.27 shows a block diagram representation of the 555 timer circuit (for the actual circuit, refer to Grebene 1984). The circuit consists of two comparators, an SR flip-flop, and a transistor Q_1 that operates as a switch. One power supply (V_{CC}) is required for operation, with the supply voltage typically 5 V. A resistive voltage divider, consisting of the three equal-valued resistors labeled R_1, is connected across V_{CC} and establishes the reference (threshold) voltages for the two comparators. These are $V_{TH} = \frac{2}{3}V_{CC}$ for comparator 1, and $V_{TL} = \frac{1}{3}V_{CC}$ for comparator 2.

We study SR flip-flops in Chapter 13. For our purposes here we note that an SR flip-flop (also called a latch) is a bistable circuit having complementary outputs, denoted Q and $\overline{Q}$. In the *set* state, the output at Q is ''high'' (approximately equal to V_{CC}) and that at $\overline{Q}$ is ''low'' (approximately equal to 0 V). In the other stable state, termed the *reset* state, the output at Q is low and that at $\overline{Q}$ is high. The flip-flop is set by applying a high

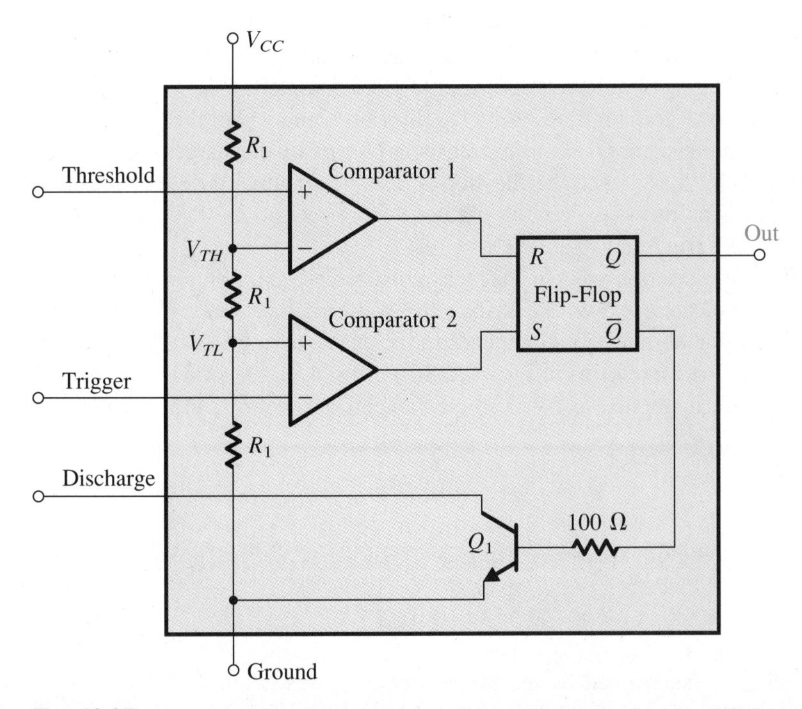

Fig. 12.27 Block diagram representation of the internal circuit of the 555 integrated-circuit timer.

level (V_{CC}) to its set input terminal, labeled S. To reset the flip-flop, a high level is applied to the reset input terminal, labeled R. Note that the reset and set input terminals of the flip-flop in the 555 circuit are connected to the outputs of comparator 1 and comparator 2, respectively.

The positive input terminal of comparator 1 is brought out to an external terminal of the 555 package, labeled Threshold. Similarly, the negative input terminal of comparator 2 is connected to an external terminal labeled Trigger, and the collector of transistor Q_1 is connected to a terminal labeled Discharge. Finally, the Q output of the flip-flop is connected to the output terminal of the timer package, labeled Out.

Implementing a Monostable Multivibrator Using the 555 IC

Figure 12.28(a) shows a monostable multivibrator implemented using the 555 IC together with an external resistor R and an external capacitor C. In the stable state the flip-flop will be in the reset state, and thus its $\overline{Q}$ output will be high, turning on transistor Q_1. Transistor Q_1 will be saturated, and thus v_C will be close to 0 V, resulting in a low level at the output of comparator 1. The voltage at the trigger input terminal, labeled v_{trigger}, is kept high (greater than V_{TL}), and thus the output of comparator 2 also will be low. Finally, note that since the flip-flop is in the reset state, Q will be low and thus v_O will be close to 0 V.

To trigger the monostable multivibrator, a negative input pulse is applied to the trigger input terminal. As v_{trigger} goes below V_{TL}, the output of comparator 2 goes to the high level, thus setting the flip-flop. Output Q of the flip-flop goes high, and thus v_O goes high, and output $\overline{Q}$ goes low, turning off transistor Q_1. Capacitor C now begins to charge up through resistor R, and its voltage v_C rises exponentially toward V_{CC}, as shown in Fig. 12.28(b). The monostable multivibrator is now in its quasi-stable state. This state prevails until v_C reaches, and begins to exceed, the threshold of comparator 1, V_{TH}, at which time the output of comparator 1 goes high, resetting the flip-flop. Output $\overline{Q}$ of the flip-flop now goes high and turns on transistor Q_1. In turn, transistor Q_1 rapidly discharges capacitor C, causing v_C to go to 0 V. Also, when the flip-flop is reset its Q output goes low, and thus v_O goes back to 0 V. The monostable multivibrator is now back in its stable state and is ready to receive a new triggering pulse.

From the above description we see that the monostable multivibrator produces an output pulse v_O as indicated in Fig. 12.28(b). The width of the pulse, T, is the time interval that the monostable multivibrator spends in the quasi-stable state; it can be determined by reference to the waveforms in Fig. 12.28(b) as follows. Denoting the instant at which the trigger pulse is applied as $t = 0$, the exponential waveform of v_C can be expressed as

$$v_C = V_{CC}(1 - e^{-t/CR}) \tag{12.38}$$

Substituting $v_C = V_{TH} = \frac{2}{3}V_{CC}$ at $t = T$ gives

$$T = CR \ln 3 \simeq 1.1CR \tag{12.39}$$

Thus the pulse width is determined by the external components C and R, which can be selected to have values as precise as desired.

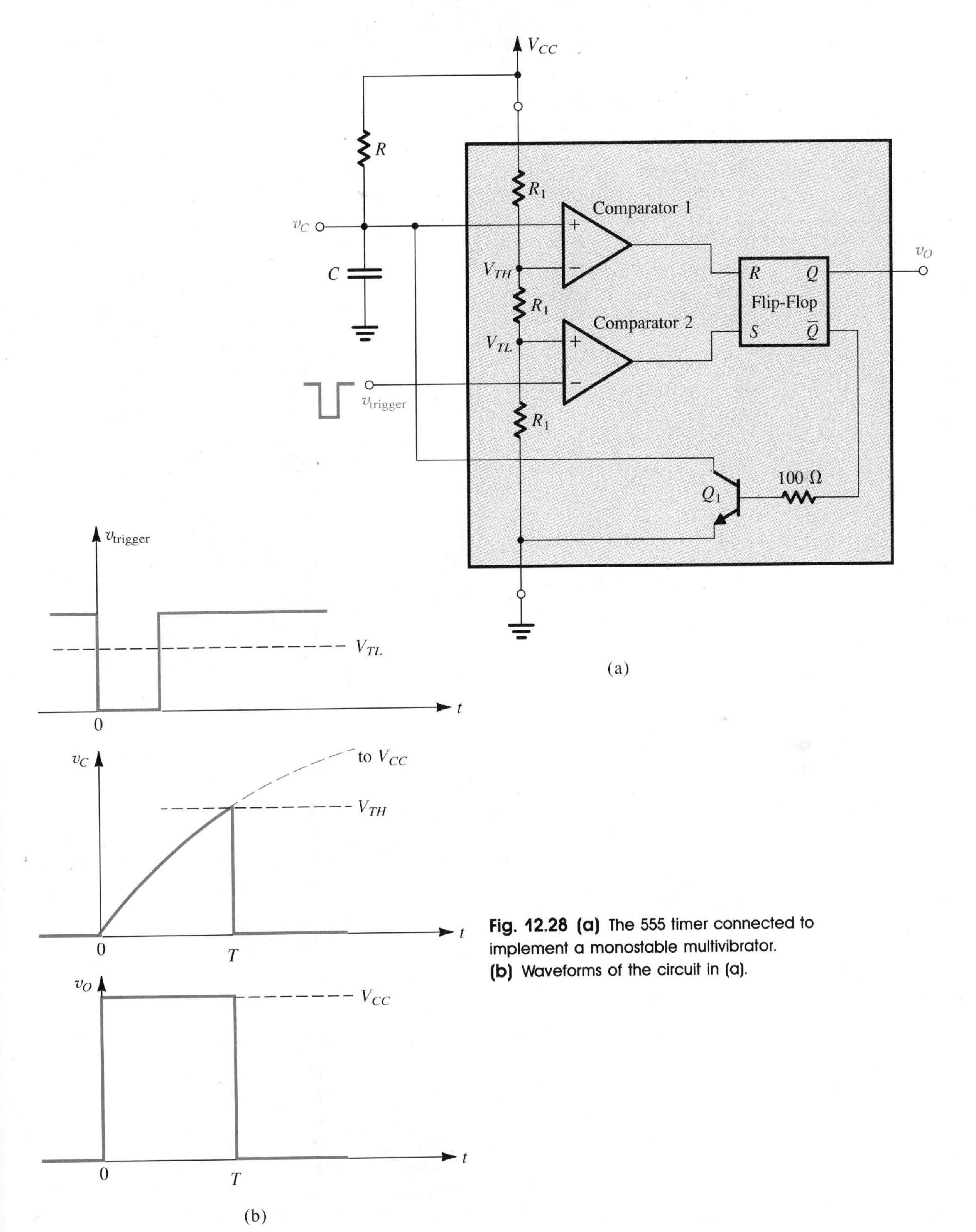

Fig. 12.28 (a) The 555 timer connected to implement a monostable multivibrator. **(b)** Waveforms of the circuit in (a).

An Astable Multivibrator Using the 555 IC

Figure 12.29(a) shows the circuit of an astable multivibrator employing a 555 IC, two external resistors, R_A and R_B, and an external capacitor C. To see how the circuit operates refer to the waveforms depicted in Fig. 12.29(b). Assume that initially C is discharged and the flip-flop is set. Thus v_O is high and Q_1 is off. Capacitor C will charge up through the series combination of R_A and R_B, and the voltage across it, v_C, will rise exponentially toward V_{CC}. As v_C crosses the level equal to V_{TL}, the output of comparator 2 goes low. This, however, has no effect on the circuit operation, and the flip-flop remains set. Indeed, this state continues until v_C reaches and begins to exceed the threshold of comparator 1, V_{TH}. At this instant of time, the output of comparator 1 goes high and resets the flip-flop. Thus v_O goes low, $\overline{Q}$ goes high, and transistor Q_1 is turned on. The saturated transistor Q_1 causes a voltage of approximately zero volts to appear at the common node of R_A and R_B. Thus C begins to discharge through R_B and the collector of Q_1. The voltage v_C decreases exponentially with a time constant CR_B toward 0 V. When v_C reaches the threshold of comparator 2, V_{TL}, the output of comparator 2, goes high and sets the flip-flop. The output v_O then goes high, and $\overline{Q}$ goes low, turning off Q_1. Capacitor C begins to charge through the series equivalent of R_A and R_B, and its voltage rises exponentially toward V_{CC} with a time constant $C\,(R_A + R_B)$. This rise continues until v_C reaches V_{TH}, at which time the output of comparator 1 goes high, resetting the flip-flop, and the cycle continues.

From the above description we see that the circuit of Fig. 12.29(a) oscillates and produces a square waveform at the output. The frequency of oscillation can be determined as follows. Reference to Fig. 12.29(b) indicates that the output will be high during the interval T_H, in which v_C rises from V_{TL} to V_{TH}. The exponential rise of v_C can be described by

$$v_C = V_{CC} - (V_{CC} - V_{TL})e^{-t/C(R_A+R_B)} \tag{12.40}$$

where $t = 0$ is the instant at which the interval T_H begins. Substituting $v_C = V_{TH} = \frac{2}{3}V_{CC}$ at $t = T_H$ and $V_{TL} = \frac{1}{3}V_{CC}$ results in

$$T_H = C(R_A + R_B)\ln 2 \simeq 0.69\ C(R_A + R_B) \tag{12.41}$$

We also note from Fig. 12.29(b) that v_O will be low during the interval T_L, in which v_C falls from V_{TH} to V_{TL}. The exponential fall of v_C can be described by

$$v_C = V_{TH}\ e^{-t/CR_B} \tag{12.42}$$

where we have taken $t = 0$ as the beginning of the interval T_L. Substituting $v_C = V_{TL} = \frac{1}{3}V_{CC}$ at $t = T_L$ and $V_{TH} = \frac{2}{3}V_{CC}$ results in

$$T_L = CR_B \ln 2 \simeq 0.69\ CR_B \tag{12.43}$$

Equations (12.41) and (12.43) can be combined to obtain the period T of the output square wave as

$$T = T_H + T_L = 0.69\ C(R_A + 2R_B) \tag{12.44}$$

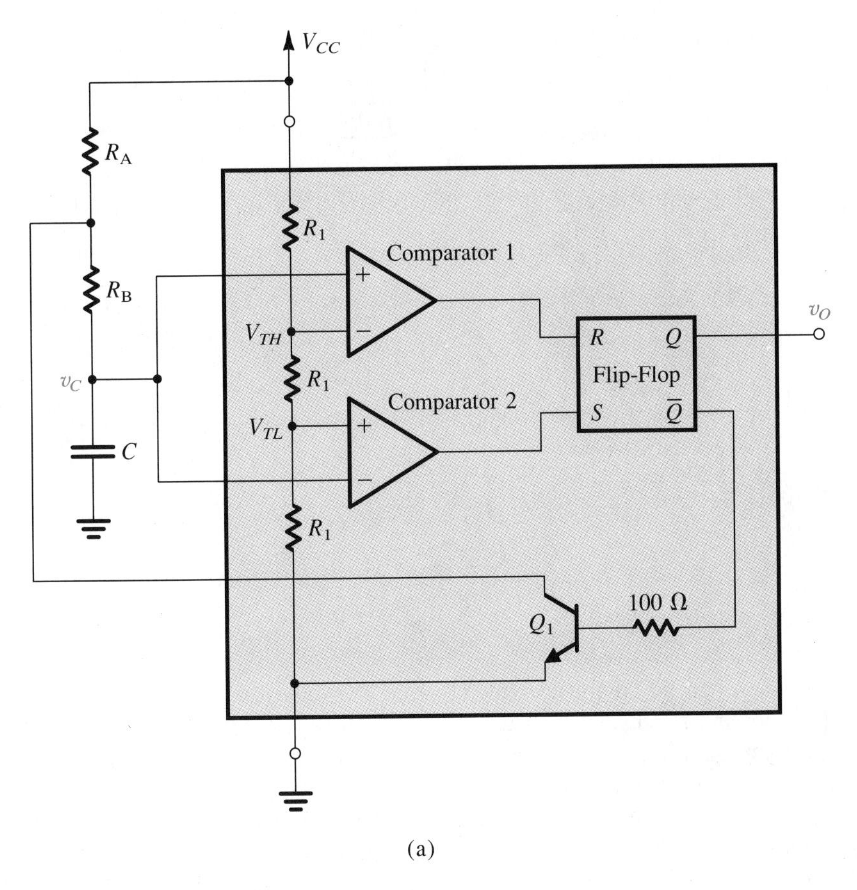

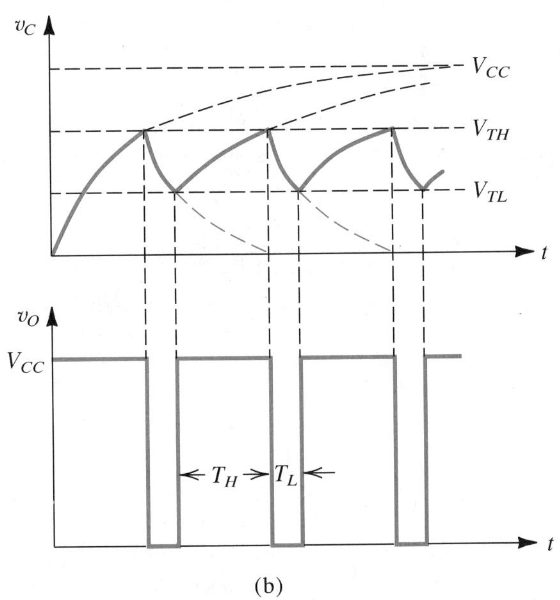

Fig. 12.29 (a) The 555 timer connected to implement an astable multivibrator. **(b)** Waveforms of the circuit in (a).

Also, the **duty cycle** of the output square wave can be found from Eqs. (12.41) and (12.43):

$$\text{Duty Cycle} \equiv \frac{T_H}{T_H + T_L} = \frac{R_A + R_B}{R_A + 2R_B} \tag{12.45}$$

Note that the duty cycle will always be greater than 0.5 (50%); it approaches 0.5 if R_A is selected much smaller than R_B.

Exercises 12.20 Using a 10-nF capacitor C, find the value of R that yields an output pulse of 100 μs in the monostable circuit of Fig. 12.28(a).

Ans. 9.1 kΩ

D12.21 For the circuit in Fig. 12.29(a), use a 1000-pF capacitor and find the values of R_A and R_B that result in an oscillation frequency of 100 kHz and a duty cycle of 75%.

Ans. 7.2 kΩ, 3.6 kΩ

12.8 NONLINEAR WAVEFORM-SHAPING CIRCUITS

Diodes or transistors can be combined with resistors to synthesize two-port networks having arbitrary nonlinear transfer characteristics. Such two-port networks can be employed in **waveform shaping**—that is, changing the waveform of an input signal in a prescribed manner to produce a waveform of a desired shape at the output. In this section we illustrate this application by a concrete example: the **sine-wave shaper.** This is a circuit whose purpose is to change the waveform of an input triangular-wave signal to a sine wave. Though simple, the sine-wave shaper is a practical building block used extensively in function generators. This method of generating sine waves should be contrasted to that using linear oscillators (Sections 12.1–12.3). Although linear oscillators produce sine waves of high purity, they are not convenient at very low frequencies. Also, linear oscillators are in general more difficult to tune over wide frequency ranges. In the following we discuss two distinctly different techniques for designing sine-wave shapers.

The Breakpoint Method

In the breakpoint method the desired nonlinear transfer characteristic (in our case the sine function shown in Fig. 12.30) is implemented as a piecewise linear curve. Diodes are utilized as switches that turn on at the various breakpoints of the transfer characteristic, thus switching into the circuit additional resistors that cause the transfer characteristic to change slope.

Consider the circuit shown in Fig. 12.31(a). It consists of a chain of resistors connected across the entire symmetric voltage supply $+V$, $-V$. The purpose of this voltage divider is to generate reference voltages that will serve to determine the breakpoints in the transfer characteristic. In our example these reference voltages are denoted $+V_2$, $+V_1$, $-V_1$, $-V_2$. Note that the entire circuit is symmetric, driven by a symmetric triangular wave and generating a symmetric sine wave. The circuit approximates each quarter-cycle

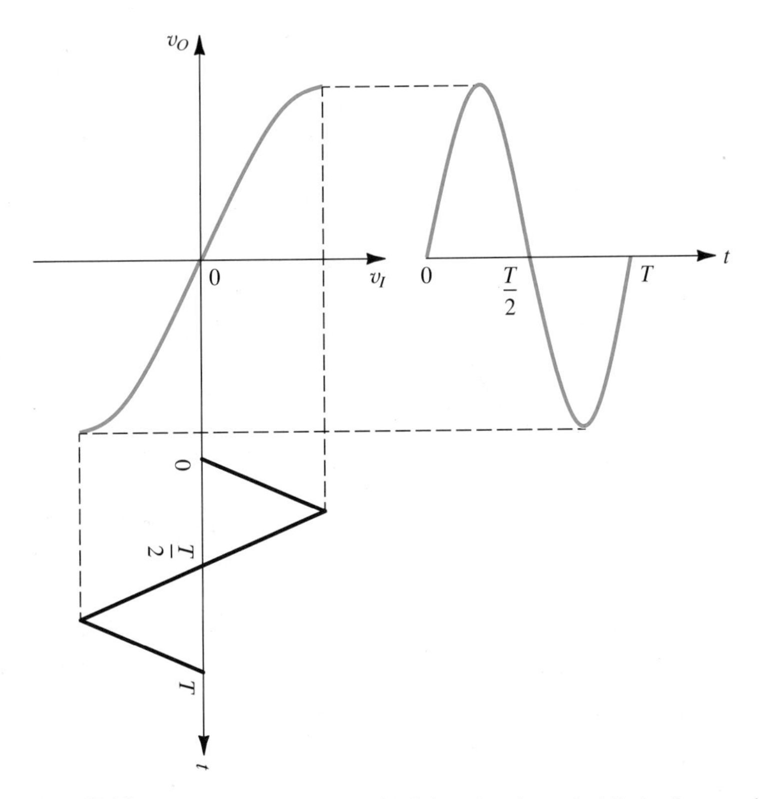

Fig. 12.30 Using a nonlinear (sinusoidal) transfer characteristic to shape a triangular waveform into a sinusoid.

of the sine wave by three straight-line segments; the breakpoints between these segments are determined by the reference voltages V_1 and V_2.

The circuit works as follows: Let the input be the triangular wave shown in Fig. 12.31(b), and consider first the quarter-cycle defined by the two points labeled 0 and 1. When the input signal is less in magnitude than V_1, none of the diodes conducts. Thus zero current flows through R_4, and the output voltage at B will be equal to the input voltage. But as the input rises to V_1 and above, D_2 (assumed ideal) begins to conduct. Assuming that the conducting D_2 behaves as a short circuit, we see that, for $v_I > V_1$,

$$v_O = V_1 + (v_I - V_1)\frac{R_5}{R_4 + R_5}$$

This implies that as the input continues to rise above V_1 the output follows but with a reduced slope. This gives rise to the second segment in the output waveform, as shown in Fig. 12.31(b). Note that in developing the above equation we have assumed that the resistances in the voltage divider are low-valued so as to cause the voltages V_1 and V_2 to be constant independent of the current coming from the input.

Next consider what happens as the voltage at point B reaches the second breakpoint determined by V_2. At this point D_1 conducts, thus limiting the output v_O *to* V_2 (plus, of course, the voltage drop across D_1 if it is not assumed ideal). This gives rise to the third

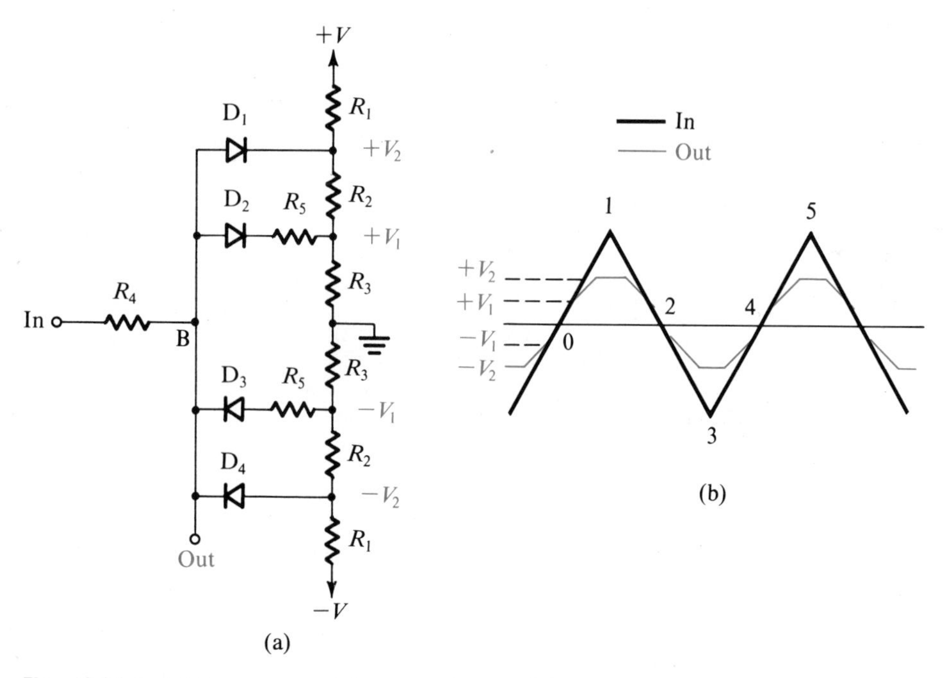

Fig. 12.31 (a) A three-segment sine-wave shaper. **(b)** The input triangular waveform and the output approximately sinusoidal waveform.

segment, which is flat, in the output waveform. The result is to ''bend'' the waveform and shape it into an approximation of the first quarter-cycle of a sine wave. Then, beyond the peak of the input triangular wave, as the input voltage decreases, the process unfolds, the output becoming progressively more like the input. Finally, when the input goes sufficiently negative, the process begins to repeat at $-V_1$ and $-V_2$ for the negative half-cycle.

Although the circuit is relatively simple, its performance is surprisingly good. A measure of goodness usually taken is to quantify the purity of the output sine wave by specifying the percentage **total harmonic distortion** (THD). This is the percentage ratio of the rms voltage of all harmonic components above the fundamental frequency (which is the frequency of the triangular wave) to the rms voltage of the fundamental. Interestingly, one reason for the good performance of the diode shaper is the beneficial effects produced by the nonideal i–v characteristics of the diodes—that is, the exponential knee of the junction diode as it goes into forward conduction. The consequence is a relatively smoothed transition from one line segment to the next.

Practical implementations of the breakpoint sine-wave shaper employ six to eight segments (as compared to the three used in the example above). Also, transistors are usually employed to provide more versatility in the design, with the goal being increased precision and lower THD. (See Grebene 1984, pages 592–595.)

The Nonlinear-Amplification Method

The other method we discuss for the conversion of a triangular wave into a sine wave is based on feeding the triangular wave to the input of an amplifier having a nonlinear transfer characteristic that approximates the sine function. One such amplifier circuit consists of a differential pair with a resistance connected between the two emitters, as shown in Fig. 12.32. With appropriate choice of the values of the bias current I and the resistance R, the differential amplifier can be made to have a transfer characteristic that

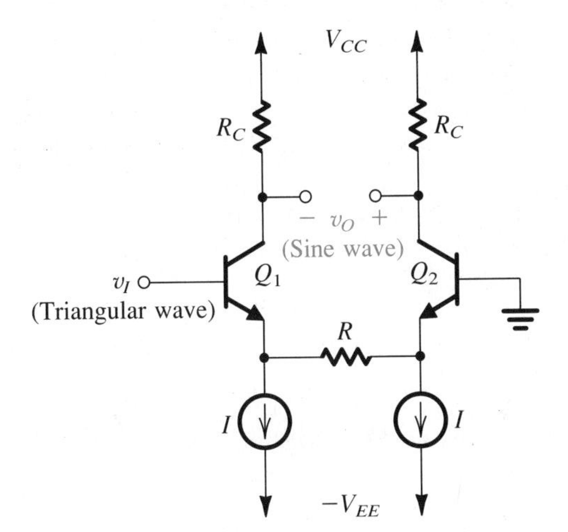

Fig.12.32 A differential pair with an emitter degeneration resistance used to implement a triangular-wave to sine-wave converter. Operation of the circuit can be graphically described by Fig. 12.30.

closely approximates that shown in Fig. 12.30. Observe that for small v_I the transfer characteristic of the circuit of Fig. 12.32 is almost linear, as a sine wave is near its zero crossings. At large values of v_I the nonlinear characteristics of the BJTs reduce the gain of the amplifier and cause the transfer characteristic to bend, approximating the sine wave as it approaches its peak. (More details on this circuit can be found in Grebene 1984, pages 595–597.)

Exercises D12.22 The circuit in Fig. E12.22 is required to provide a three-segment approximation to the nonlinear i–v characteristic, $i = 0.1v^2$, where v is the voltage in volts and i is the current in milliamperes. Find the values of R_1, R_2, and R_3 such that the approximation is perfect at $v = 2$ V, 4 V, and 8 V. Calculate the error in current value at $v = 3$ V, 5 V, 7 V, and 10 V. Assume ideal diodes.

Ans. 5 kΩ, 1.25 kΩ, 1.25 kΩ; −0.3 mA, +0.1 mA, −0.3 mA, 0

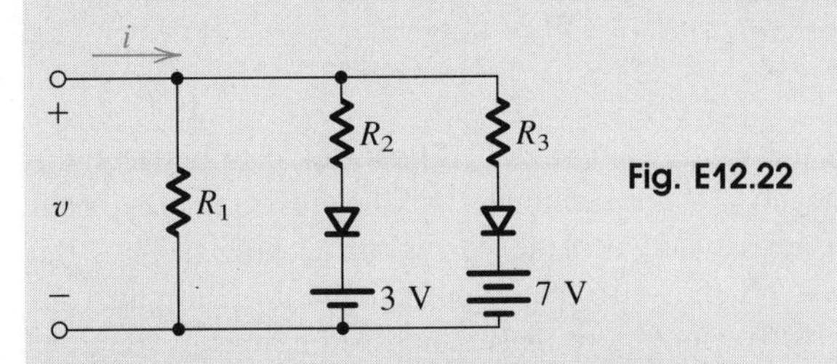

Fig. E12.22

12.23 A detailed analysis of the circuit in Fig. 12.32 shows that its optimum performance occurs when the values of I and R are selected so that $RI = 2.5\ V_T$, where V_T is the thermal voltage. For this design the peak amplitude of the input triangular wave should be 6.6 V_T, and the corresponding sine wave across R has a peak value of 2.42 V_T. For $I = 0.25$ mA and $R_C = 10$ kΩ, find the peak amplitude of the sine-wave output v_O. Assume $\alpha \simeq 1$.

Ans. 4.84 V

12.9 PRECISION RECTIFIER CIRCUITS

Rectifier circuits were studied in Chapter 3, where the emphasis was on their application in power-supply design. In such applications the voltages being rectified are usually much greater than the diode voltage drop, rendering the exact value of the diode drop unimportant to the proper operation of the rectifier. Other applications exist, however, where this is not the case. For instance, the signal to be rectified can be of a very small amplitude, say 0.1 V, making it impossible to employ the conventional rectifier circuits. Also, in instrumentation applications the need arises for rectifier circuits with very precise transfer characteristics.

In this section we study circuits that combine diodes and op amps to implement a variety of rectifier circuits with precise characteristics. Precision rectifiers, which can be considered a special class of wave-shaping circuits, find application in the design of instrumentation systems.

Precision Half-wave Rectifier—The "Superdiode"

Figure 12.33(a) shows a precision half-wave-rectifier circuit consisting of a diode placed in the negative-feedback path of an op amp, with R being the rectifier load resistance. The circuit works as follows: If v_I goes positive, the output voltage v_A of the op amp will go

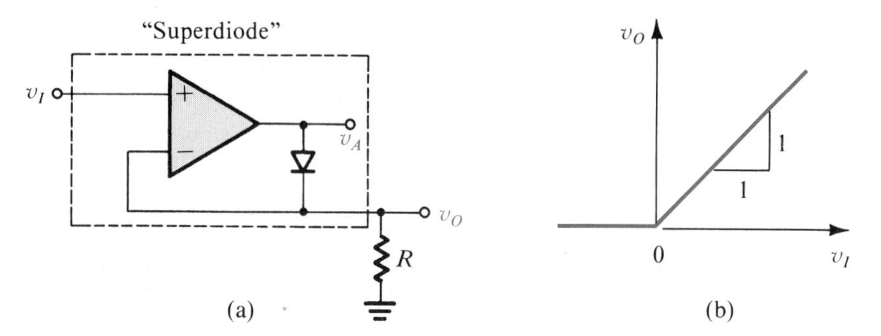

Fig. 12.33 The "superdiode" precision half-wave rectifier and its almost ideal transfer characteristic. Note that when $v_I > 0$ and the diode conducts, the op amp supplies the load current and the source is conveniently buffered, an added advantage.

positive and the diode will conduct, thus establishing a closed feedback path between the op amp's output terminal and the negative input terminal. This negative-feedback path will cause a virtual short circuit to appear between the two input terminals. Thus the voltage at the negative input terminal, which is also the output voltage v_O, will equal (to within a few millivolts) that at the positive input terminal, which is the input voltage v_I,

$$v_O = v_I \qquad v_I \geq 0$$

Note that the offset voltage ($\simeq 0.5$ V) exhibited in the simple half-wave rectifier circuit is no longer present. For the op amp circuit to start operation, v_I has to exceed only a negligibly small voltage equal to the diode drop divided by the op amp's open-loop gain.

In other words, the straight-line transfer characteristic v_O–v_I almost passes through the origin. This makes this circuit suitable for applications involving very small signals.

Consider now the case when v_I goes negative. The op amp's output voltage v_A will tend to follow and go negative. This will reverse-bias the diode, and no current will flow through resistance R, causing v_O to remain equal to 0 V. Thus for $v_I < 0$, $v_O = 0$. Since in this case the diode is off, the op amp will be operating in an open loop and its output will be at the negative saturation level.

The transfer characteristic of this circuit will be that shown in Fig. 12.33(b), which is almost identical to the ideal characteristic of a half-wave rectifier. The nonideal diode characteristics have been almost completely masked by placing the diode in the negative-feedback path of an op amp. This is another dramatic application of negative feedback. The combination of diode and op amp, shown in the dotted box in Fig. 12.33(a), is appropriately referred to as a "superdiode."

As usual, though, not all is well. The circuit of Fig. 12.33 has some disadvantages: When v_I goes negative and $v_O = 0$, the entire magnitude of v_I appears between the two input terminals of the op amp. If this magnitude is greater than few volts, the op amp may be damaged unless it is equipped with what is called "overvoltage protection" (a feature that most modern IC op amps have). Another disadvantage is that when v_I is negative, the op amp will be saturated. Although not harmful to the op amp, saturation should usually be avoided, since getting the op amp out of the saturation region and back into its linear region of operation requires some time. This time delay will obviously slow down circuit operation and limit the frequency of operation of the superdiode half-wave-rectifier circuit.

An Alternative Circuit

An alternative precision rectifier circuit that does not suffer from the disadvantages mentioned above is shown in Fig. 12.34. The circuit operates in the following manner: For positive v_I, diode D_2 conducts and closes the negative-feedback loop around the op amp. A virtual ground therefore will appear at the inverting input terminal, and the op amp's

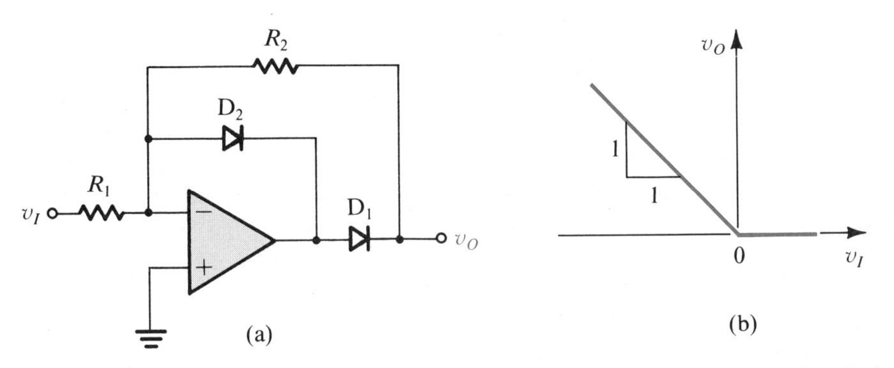

Fig. 12.34 (a) An improved version of the precision half-wave rectifier. Here diode D_2 is included to keep the feedback loop closed around the op amp during the off times of the rectifier diode D_1, thus preventing the op amp from saturating. **(b)** The transfer characteristic for $R_2 = R_1$.

output will be *clamped* at one diode drop below ground. This negative voltage will keep diode D_1 off, and no current will flow in the feedback resistance R_2. It follows that the rectifier output voltage will be zero.

As v_I goes negative, the voltage at the inverting input terminal will tend to go negative, causing the voltage at the op amp's output terminal to go positive. This will cause D_2 to be reverse-biased and hence cut off. Diode D_1, however, will conduct through R_2, thus establishing a negative-feedback path around the op amp and forcing a virtual ground to appear at the inverting input terminal. The current through the feedback resistance R_2 will be equal to the current through the input resistance R_1. Thus for $R_1 = R_2$ the output voltage v_O will be

$$v_O = -v_I \qquad v_I \leq 0$$

The transfer characteristic of the circuit is shown in Fig. 12.34(b). Note that unlike for the previous circuit, here the slope of the characteristic can be set to any desired value, including unity, by selecting appropriate values for R_1 and R_2.

As mentioned before, the major advantage of this circuit is that the feedback loop around the op amp remains closed at all times. Hence the op amp remains in its linear operating region, avoiding the possibility of saturation and the associated time delay required to "get out" of saturation. Diode D_2 "catches" the output voltage as it goes negative and clamps it to one diode drop below ground; hence D_2 is called a "catching diode."

An Application: Measuring AC Voltages

As one of the many possible applications of the precision rectifier circuits discussed in this section, consider the basic ac voltmeter circuit shown in Fig. 12.35. The circuit consists of a half-wave rectifier—formed by op amp A_1, diodes D_1 and D_2, and resistors R_1 and R_2—and a first-order low-pass filter—formed by op amp A_2, resistors R_3 and R_4, and capacitor C. For an input sinusoid having a peak amplitude V_p the output v_1 of the rectifier will consist of a half sine wave having a peak amplitude of V_pR_2/R_1. It can be shown using Fourier series analysis that the waveform of v_1 has an average value of $(V_p/\pi)(R_2/R_1)$ in

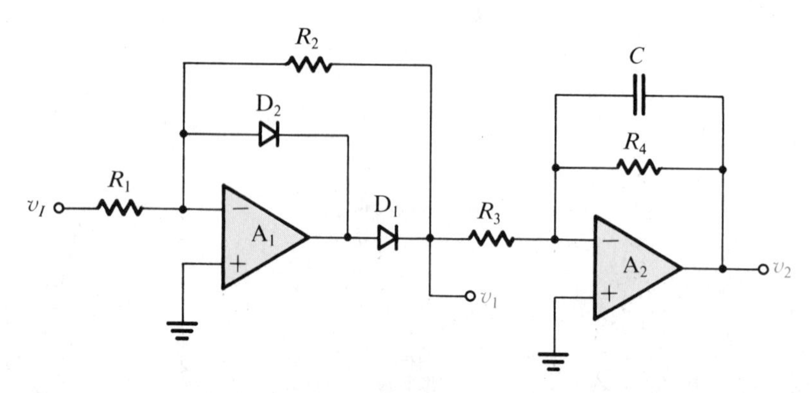

Fig. 12.35 A simple ac voltmeter consisting of a precision half-wave rectifier followed by a first-order low-pass filter.

addition to harmonics of the frequency ω of the input signal. To reduce the amplitudes of all of these harmonics to negligible levels, the corner frequency of the low-pass filter should be chosen much smaller than the lowest expected frequency $\omega_{\min}$ of the input sine wave. This leads to

$$\frac{1}{CR_4} \ll \omega_{\min}$$

Then the output voltage v_2 will be mostly dc, with a value

$$V_2 = -\frac{V_p}{\pi}\frac{R_2}{R_1}\frac{R_4}{R_3}$$

where R_4/R_3 is the dc gain of the low-pass filter. Note that this voltmeter essentially measures the average value of the negative parts of the input signal but can be calibrated to provide root-mean-square (rms) readings for input sinusoids.

Exercises 12.24 Consider the operational rectifier or superdiode circuit of Fig. 12.33(a), with $R = 1\ \text{k}\Omega$. For $v_I =$ 10 mV, 1 V, and -1 V, what are the voltages that result at the rectifier output and at the output of the op amp? Assume that the op amp is ideal and that its output saturates at ± 12 V. The diode has a 0.7-V drop at 1-mA current, and the voltage drop changes by 0.1 V per decade of current change.

Ans. 10 mV, 0.51 V; 1 V, 1.7 V; 0 V, -12 V

12.25 If the diode in the circuit of Fig. 12.33(a) is reversed, find the transfer characteristic v_O as a function of v_I.

Ans. $v_O = 0$ for $v_I \geq 0$; $v_O = v_I$ for $v_I \leq 0$

12.26 Consider the circuit in Fig. 12.34(a) with $R_1 = 1\ \text{k}\Omega$ and $R_2 = 10\ \text{k}\Omega$. Find v_O and the voltage at the amplifier output for $v_I = +1$ V, -10 mV, and -1 V. Assume the op amp to be ideal with saturation voltages of ± 12 V. The diodes have 0.7-V voltage drops at 1 mA, and the voltage drop changes by 0.1 V per decade of current change.

Ans. 0 V, -0.7 V; 0.1 V, 0.6 V; 10 V, 10.7 V

12.27 If the diodes in the circuit of Fig. 12.34(a) are reversed, find the transfer characteristic v_O as a function of v_I.

Ans. $v_O = -(R_2/R_1)v_I$ for $v_I \geq 0$; $v_O = 0$ for $v_I \leq 0$

12.28 Find the transfer characteristic for the circuit in Fig. E12.28.

Ans. $v_O = 0$ for $v_I \geq -5$ V; $v_O = -v_I - 5$ for $v_I \leq -5$ V

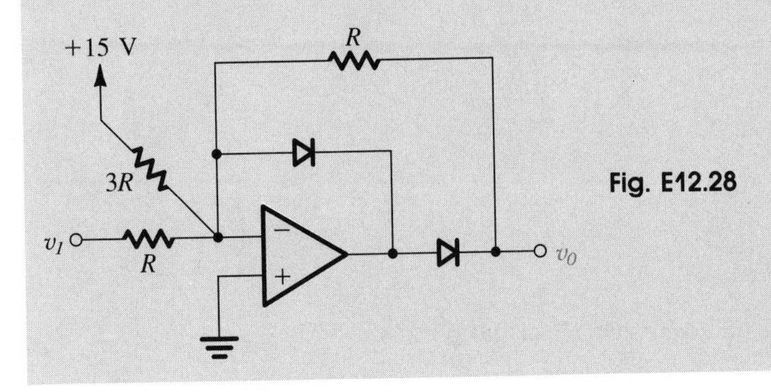

Fig. E12.28

Precision Full-Wave Rectifier

We now derive a circuit for a precision full-wave rectifier. From Chapter 3 we know that full-wave rectification is achieved by inverting the negative halves of the input-signal waveform and applying the resulting signal to another diode rectifier. The outputs of the two rectifiers are then joined to a common load. Such an arrangement is depicted in Fig. 12.36, which also shows the waveforms at various nodes. Now replacing diode D_A with a

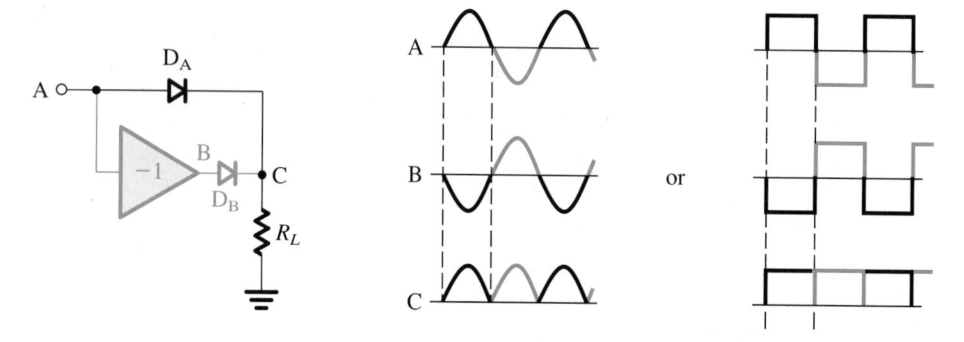

Fig. 12.36 Principle of full-wave rectification.

superdiode, and replacing diode D_B and the inverting amplifier with the inverting precision half-wave rectifier of Fig. 12.34 but without the catching diode, we obtain the precision full-wave rectifier circuit of Fig. 12.37(a).

To see how the circuit of Fig. 12.37 operates, consider first the case where the input at A is positive. The output of A_2 will go positive, turning D_2 on, which will conduct

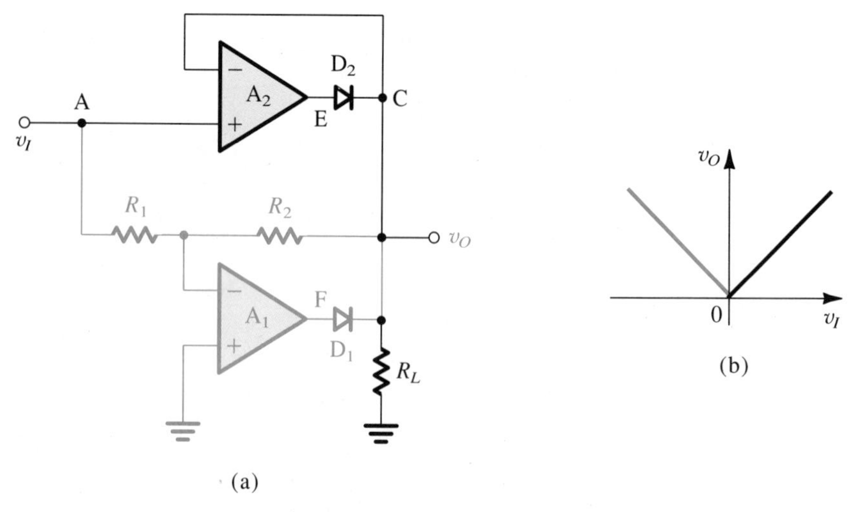

Fig. 12.37 (a) Precision full-wave rectifier based on the conceptual circuit of Fig. 12.36. **(b)** Transfer characteristic of the circuit in (a).

through R_L and thus close the feedback loop around A_2. A virtual short circuit will thus be established between the two input terminals of A_2, and the voltage at the negative input terminal, which is the output voltage of the circuit, will become equal to the input. Thus no current will flow through R_1 and R_2, and the voltage at the inverting input of A_1 will be equal to the input and hence positive. Therefore the output terminal (F) of A_1 will go negative until A_1 saturates. This causes D_1 to be turned off.

Next consider the case when A goes negative. The tendency for a negative voltage at the negative input of A_1 causes F to rise, making D_1 conduct to supply R_L and allowing the feedback loop around A_1 to be closed. Thus a virtual ground appears at the negative input of A_1, and the two equal resistances R_1 and R_2 force the voltage at C, which is the output voltage, to be equal to the negative of the input voltage at A and thus positive. The combination of positive voltage at C and negative voltage at A causes the output of A_2 to saturate in the negative direction, thus keeping D_2 off.

The overall result is perfect full-wave rectification, as represented by the transfer characteristic in Fig. 12.37(b). This precision is, of course, a result of placing the diodes in op amp feedback loops, thus masking their nonidealities. This circuit is one of many possible precision full-wave-rectifier or **absolute-value circuits.** Another related implementation of this function is examined in Exercise 12.30.

Exercises 12.29 In the full-wave rectifier circuit of Fig. 12.37(a) let $R_1 = R_2 = R_L = 10\ \text{k}\Omega$, and assume the op amps to be ideal except for output saturation at ± 12 V. When conducting a current of 1 mA each diode exhibits a voltage drop of 0.7 V, and this voltage changes by 0.1 V per decade of current change. Find v_O, v_E, and v_F corresponding to $v_I = +0.1$, $+1$, $+10$, -0.1, -1 and -10 volts.

Ans. +0.1 V, +0.6 V, −12 V; +1 V, +1.6 V, −12 V; +10 V, +10.7 V, −12 V; +0.1 V, −12 V, +0.63 V; +1 V, −12 V, +1.63 V; +10 V, −12 V, +10.73 V

D12.30 The block diagram shown in Fig. E12.30(a) gives another possible arrangement for implementing the absolute-value or full-wave-rectifier operation depicted symbolically in Fig. E12.30(b). As shown, the block diagram consists of two boxes: a half-wave rectifier, which can be implemented by the circuit in Fig. 12.34(a) after reversing both diodes, and a weighted inverting summer. Convince yourself that this block diagram does in fact realize the absolute-value operation. Then draw a complete circuit diagram, giving reasonable values for all resistors.

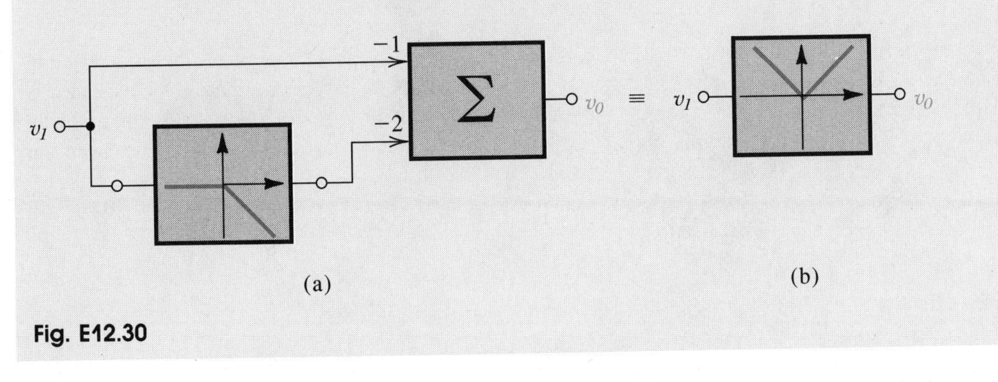

Fig. E12.30

A Precision Bridge Rectifier for Instrumentation Applications

The bridge rectifier circuit studied in Chapter 3 can be combined with an op amp to provide useful precision circuits. One such arrangement is shown in Fig. 12.38. This

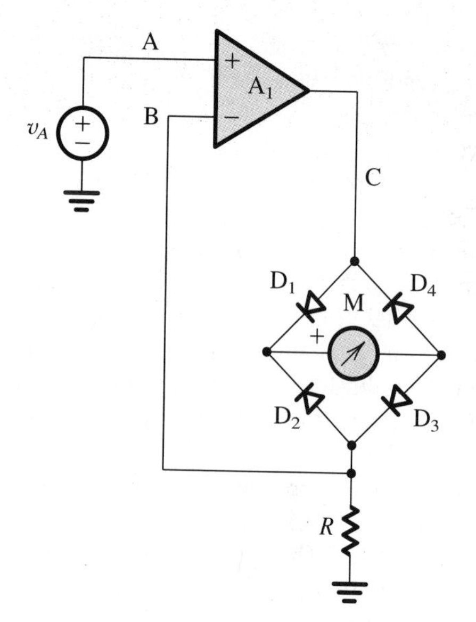

Fig. 12.38 Use of the diode bridge in the design of an ac voltmeter.

circuit causes a current equal to $|v_A|/R$ to flow through the moving-coil meter M. Thus the meter provides a reading that is proportional to the average of the absolute value of the input voltage v_A. All the nonidealities of the meter and of the diodes are masked by placing the bridge circuit in the negative feedback loop of the op amp. Observe that when v_A is positive, current flows from the op amp output through D_1, M, D_3, and R. When v_A is negative, current flows into the op amp output through R, D_2, M, and D_4. Thus the feedback loop remains closed for both polarities of v_A. The resulting virtual short circuit at the input terminals of the op amp causes a replica of v_A to appear across R. The circuit of Fig. 12.38 provides a relatively accurate high-input-impedance ac voltmeter using an inexpensive moving-coil meter.

Exercise D12.31 In the circuit of Fig. 12.38 find the value of R that would cause the meter to provide a full-scale reading when the input voltage is a sine wave of 5 V rms. Let meter M have a 1-mA, 50-Ω movement (that is, its resistance is 50 Ω, and it provides full-scale deflection when the average current through it is 1 mA). What are the approximate maximum and minimum voltages at the op amp's output? Assume that the diodes have constant 0.7-V drops when conducting.

Ans. 4.5 kΩ; +8.55 V; −8.55 V

Precision Peak Rectifiers

Including the diode of the peak rectifier studied in Chapter 3 inside the negative-feedback loop of an op amp, as shown in Fig. 12.39, results in a precision peak rectifier. The diode-op amp combination will be recognized as the superdiode of Fig. 12.33(a). Operation of the circuit in Fig. 12.39 is quite straightforward. For v_I greater than the output voltage, the op amp will drive the diode on, thus closing the negative-feedback path and

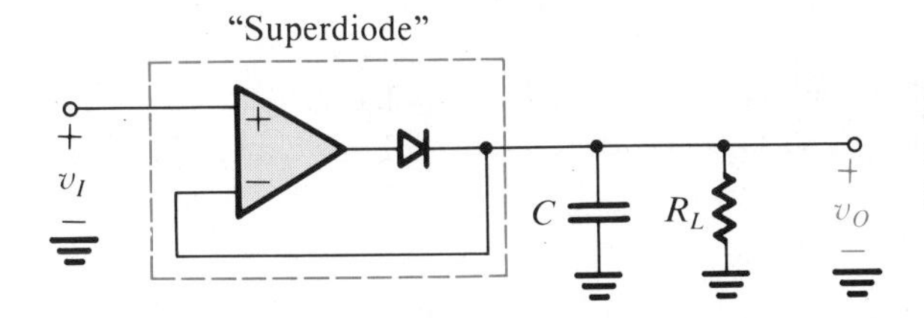

Fig. 12.39 A precision peak rectifier obtained by placing the diode in the feedback loop of an op amp.

causing the op amp to act as a follower. The output voltage will therefore follow that of the input, with the op amp supplying the capacitor-charging current. This process continues until the input reaches its peak value. Beyond the positive peak the op amp will see a negative voltage between its input terminals. Thus its output will go negative to the saturation level and the diode will turn off. Except for possible discharge through the load resistance, the capacitor will retain a voltage equal to the positive peak of the input. Inclusion of a load resistance is essential if the circuit is required to detect reductions in the magnitude of the positive peak.

A Buffered Precision Peak Detector

When the peak detector is required to hold the value of the peak for long times, the capacitor should be buffered, as shown in the circuit of Fig. 12.40. Here op amp A_2, which should have high input impedance and low input bias current, is connected as a voltage follower. The remainder of the circuit is quite similar to the half-wave-rectifier circuit of Figure 12.34. While diode D_1 is the essential diode for the peak rectification operation, diode D_2 acts as a catching diode to prevent negative saturation, and the associated delays, of op amp A_1. During the holding state, follower A_2 supplies D_2 with a small current through R. The output of op amp A_1 will then be clamped at one diode drop below the input voltage. Now if the input v_I increases above the value stored on C, which is equal to the output voltage v_O, op amp A_1 sees a net positive input that drives its output toward the positive saturation level and turns off diode D_2. Diode D_1 is then turned on and capacitor C is charged to the new positive peak of the input, after which time the circuit returns to the holding state. Finally, note that this circuit has a low-impedance output.

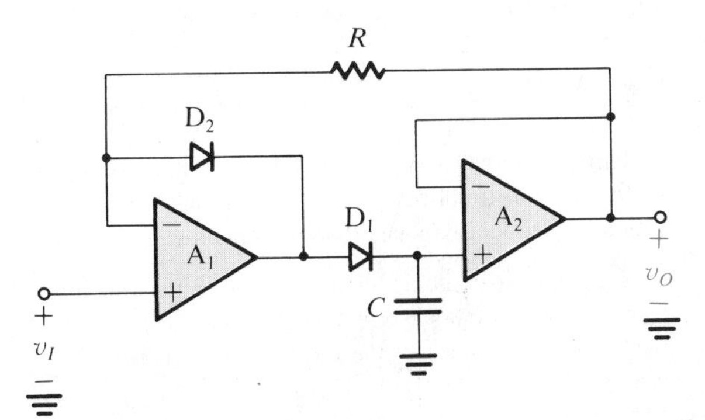

Fig. 12.40 A buffered precision peak rectifier.

A Precision Clamping Circuit

By replacement of the diode in the clamping circuit studied in Chapter 3 by a "super-diode," the precision clamp of Fig. 12.41 is obtained. Operation of this circuit should be self-explanatory.

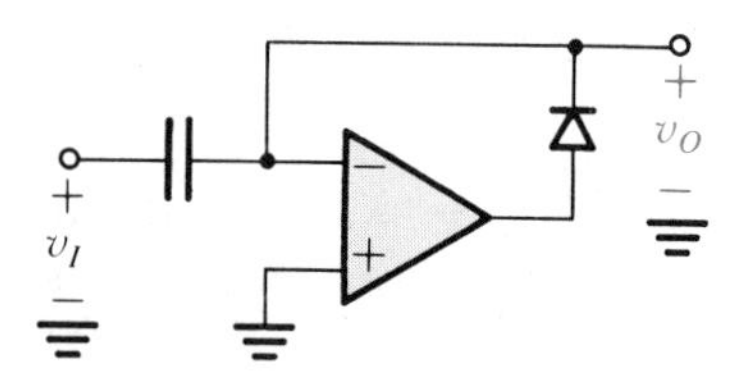

Fig. 12.41 A precision clamping circuit.

SUMMARY

▸ There are two distinctly different types of signal generators: linear oscillators, which utilize some form of resonance; and nonlinear oscillators or function generators, which employ a switching mechanism implemented with a multivibrator circuit.

▸ A linear oscillator can be realized by placing a frequency-selective network in the feedback path of an amplifier (an op amp or a transistor). The circuit will oscillate at the frequency at which the total phase shift around the loop is zero, provided that at this frequency the magnitude of loop gain is equal to, or greater than, unity.

▸ If in an oscillator the magnitude of loop gain is greater than unity, the amplitude will increase until a nonlinear amplitude-control mechanism is activated.

▸ The Wien-bridge oscillator, the phase-shift oscillator, the quadrature oscillator, and the active-filter-tuned oscillator are popular configurations for frequencies up to about 1 MHz. These circuits employ RC networks together with op amps or transistors. For higher frequencies, LC-tuned or crystal-tuned oscillators are utilized. A popular configuration is the Colpitts circuit.

▸ Crystal oscillators provide the highest possible frequency accuracy and stability.

▸ There are three types of multivibrators: bistable, monostable, and astable. Op amp circuit implementations of multivibrators are useful in analog circuit applications that require high precision. Implementations using digital logic gates will be studied in Chapter 13.

▸ The bistable multivibrator has two stable states and can remain in either state indefinitely. It changes state when triggered. A comparator with hysteresis is bistable.

▸ A monostable multivibrator has one stable state, in which it can remain indefinitely. When triggered, it goes into a quasi-stable state in which it remains for a predetermined interval, thus generating at the output a pulse of known width. It is also known as a one-shot multivibrator.

▸ An astable multivibrator has no stable state. It oscillates between two quasi-stable states, remaining in each for a predetermined interval. It thus generates a periodic waveform at the output.

▸ A feedback loop consisting of an integrator and a bistable multivibrator can be used to generate triangular and square waveforms.

▸ The 555 timer, a commercially available IC, can be used with external resistors and a capacitor to implement high-quality monostable and astable multivibrators.

▸ A sine wave can be generated by feeding a triangular waveform to a sine-wave shaper. A sine-wave shaper can be implemented either using diodes (or transistors) and resistors or using an amplifier having a nonlinear transfer characteristic that approximates the sine function.

▸ Diodes can be combined with op amps to implement precision rectifier circuits in which negative feedback serves to mask the nonidealities of the diode characteristics.

BIBLIOGRAPHY

G. B. Clayton, *Experimenting with Operational Amplifiers*, London: Macmillan, 1975.

G. B. Clayton, *Operational Amplifiers*, 2nd ed., London: Newnes-Butterworths, 1979.

S. Franco, *Design with Operational Amplifiers and Analog Integrated Circuits*, New York: McGraw-Hill, 1988.

M. S. Ghausi, *Electronic Devices and Circuits: Discrete and Integrated*, New York: Holt, Rinehart and Winston, 1985.

J. G. Graeme, G. E. Tobey, and L. P. Huelsman, *Operational Amplifiers: Design and Applications*. New York: McGraw-Hill, 1971.

P. E. Gray and C. L. Searle, *Electronic Principles*, Chap. 17, New York: Wiley, 1969.

A. B. Grebene, *Bipolar and MOS Analog Integrated Circuit Design*, New York: Wiley, 1984.

W. Jung, *IC Op Amp Cookbook*, Indianapolis, Ind.: Howard Sams, 1974.

Nonlinear Circuits Handbook, Norwood, Mass.: Analog Devices, Inc., 1976.

J. K. Roberge, *Operational Amplifiers: Theory and Practice*, New York: Wiley, 1975.

J. I. Smith, *Modern Operational Circuit Design*, New York: Wiley-Interscience, 1971.

L. Strauss, *Wave Generation and Shaping*, 2nd ed., New York: McGraw-Hill, 1970.

J. V. Wait, L. P. Huelsman, and G. A. Korn, *Introduction to Operational Amplifier Theory and Applications*, New York: McGraw-Hill, 1975.

PROBLEMS

Section 12.1: Basic Principles of Sinusoidal Oscillators

*12.1 Consider a sinusoidal oscillator consisting of an amplifier having a frequency-independent gain A (where A is positive) and a second-order bandpass filter with a pole frequency ω_0, a pole-Q denoted Q, and a center-frequency gain K.

(a) Find the frequency of oscillation, and find the condition that A and K must satisfy for sustained oscillations.

(b) Derive an expression for $d\phi/d\omega$ evaluated at $\omega = \omega_0$.

(c) Use the result of (b) to find an expression for the per unit change in frequency of oscillation resulting from a phase angle $\Delta\phi$ in the amplifier transfer function.

Hint: $\dfrac{d}{dx}(\tan^{-1} y) = \dfrac{1}{1+y^2}\dfrac{dy}{dx}$

12.2 For the oscillator described in Problem 12.1, show that independent of the value of A and K the poles of the circuit lie at a radial distance of ω_0. Find the value of AK that results in the poles (a) on the $j\omega$-axis, (b) in the right-half of the s-plane at a horizontal distance from the $j\omega$-axis of $\omega_0/(2Q)$.

D12.3 Sketch a circuit for a sinusoidal oscillator formed by an op amp connected in the noninverting configuration and a bandpass filter implemented by an RLC resonator (such as that in Fig. 11.18d). What should the amplifier gain be to obtain sustained oscillations? What is the frequency of oscillation? Find the percentage change in ω_0 resulting from a change of $+1\%$ in the value of (a) L, (b) C, (c) R.

12.4 An oscillator is formed by loading a transconductance amplifier having a positive gain with a parallel RLC circuit and connecting the output to the input directly (thus applying positive feedback with a factor $\beta = 1$). Let the transconductance amplifier have an input resistance of 10 kΩ and an output resistance of 10 kΩ. The LC resonator has $L = 10\ \mu$H, $C = 1000$ pF, and $Q = 100$. For what value of transconductance G_m will the circuit oscillate? At what frequency?

12.5 In a particular oscillator characterized by the structure of Fig. 12.1, the frequency-selective network exhibits a loss of 20 dB and a phase shift of 180° at ω_0. What is the minimum gain and the phase shift that the amplifier must have for oscillations to begin?

D12.6 Consider the circuit of Fig. 12.3(a) with R_f removed so as to realize the comparator function. Find suitable values for all resistors so that the comparator output levels are ± 6 V and so that the slope of the limiting characteristic is 0.1. Use power supply voltages of ± 10 V and assume the voltage drop of a conducting diode to be 0.7 V.

D12.7 Consider the circuit of Fig. 12.3(a) with R_f removed so as to realize the comparator function. Sketch the transfer characteristic. Show that by connecting a dc source V_B to the virtual ground of the op amp through a resistor R_B, the transfer characteristic is shifted along the v_I-axis to the point $v_I = -(R_1/R_B)V_B$. Utilizing available ±15-V dc supplies for $\pm V$ and for V_B, find suitable component values so that the limiting levels are ±5 V and the comparator threshold is at $v_I = +5$ V. Neglect the diode voltage drop (that is, assume $V_D = 0$). The input resistance of the comparator is to be 100 kΩ, and the slope in the limiting regions is to be ≤0.05 V/V. Use standard 5% resistors (see Appendix F).

12.8 Denoting the zener voltages of Z_1 and Z_2 by V_{Z1} and V_{Z2} and assuming that in the forward direction the voltage drop is approximately 0.7 V, sketch and clearly label the transfer characteristics v_O–v_I of the circuits in Fig. P12.8. Assume the op amps to be ideal.

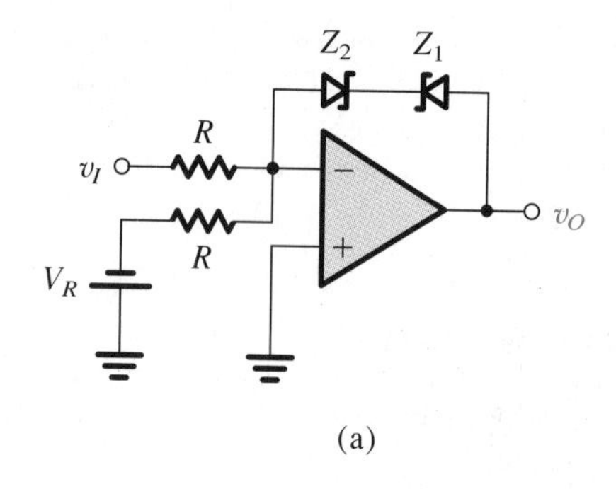

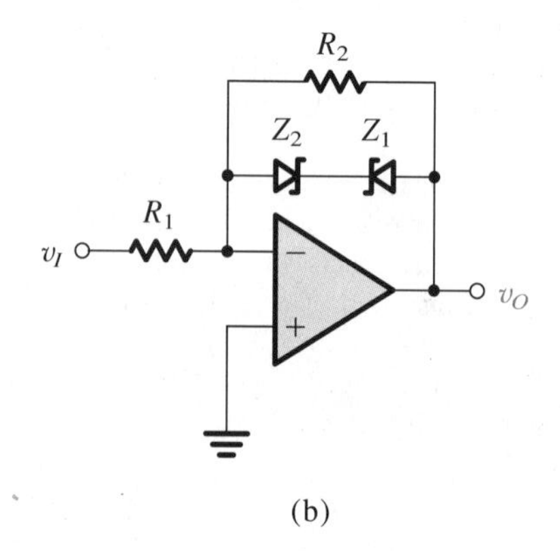

Fig. P12.8

Section 12.2: Op-Amp–RC Oscillator Circuits

12.9 For the Wien-bridge oscillator circuit in Fig. 12.4, show that the transfer function of the feedback network $[V_a(s)/V_o(s)]$ is that of a bandpass filter. Find ω_0 and Q of the poles, and find the center-frequency gain.

12.10 For the Wien-bridge oscillator of Fig. 12.4, let the closed-loop amplifier (formed by the op amp and the resistors R_1 and R_2) exhibit a phase shift of −0.1 rad in the neighborhood of $\omega = 1/CR$. Find the frequency at which oscillations can occur in this case, in terms of CR. (*Hint:* Use Eq. 12.11.)

12.11 For the Wien-bridge oscillator of Fig. 12.4, use the expression for loop gain in Eq. (12.10) to find the poles of the closed-loop system. Give the expression for the pole Q, and use it to show that to locate the poles in the right half of the s-plane, R_2/R_1 must be selected greater than 2.

D*12.12 Reconsider Exercise 12.3 with R_3 and R_6 increased to reduce the output voltage. What values are required for an output of 10 V peak to peak? What results if R_3 and R_6 are open circuited?

12.13 For the circuit in Fig. P12.13 find $L(s)$, $L(j\omega)$, the frequency for zero loop-phase, and R_2/R_1 for oscillation.

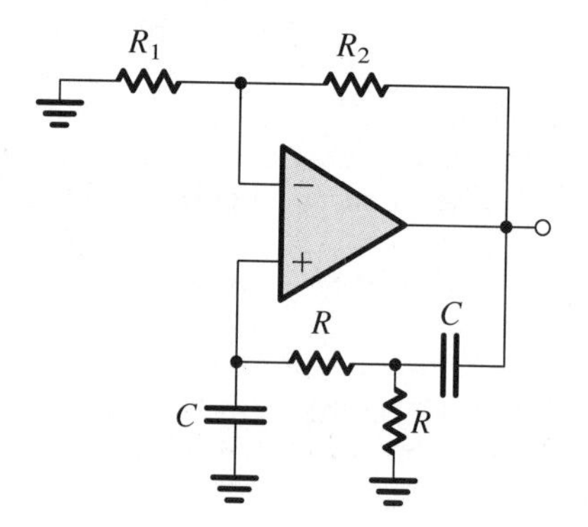

Fig. P12.13

12.14 Repeat Problem 12.13 for the circuit in Fig. P12.14.

***12.15** Consider the circuit of Fig. 12.6 with the 50-kΩ potentiometer replaced with two fixed resistors: 10 kΩ between the op amp's negative input and ground, and 18 kΩ. Modeling each diode as a 0.65-V battery in series with a 100-Ω resistance, find the peak-to-peak amplitude of the output sinusoid.

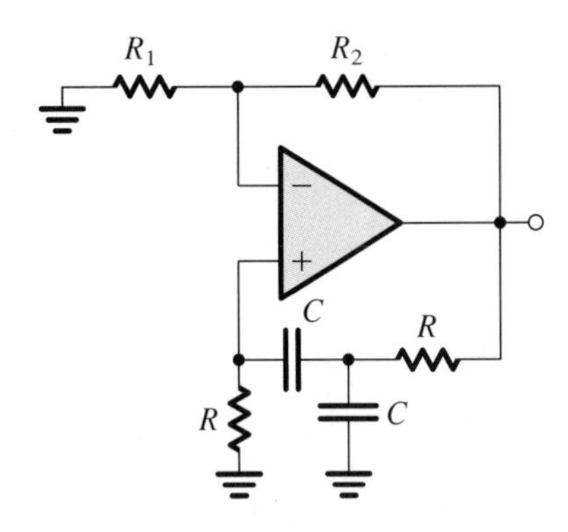

Fig. P12.14

D**12.16 Redesign the circuit of Fig. 12.6 for operation at 10 kHz using the same values of resistance. If at 10 kHz the op amp provides an excess phase shift (lag) of 5.7°, what will be the frequency of oscillation? (Assume that the phase shift introduced by the op amp remains constant for frequencies around 10 kHz.) To restore operation to 10 kHz, what change must be made in the shunt resistor of the Wien bridge? Also, to what must R_2/R_1 be changed?

*12.17 For the circuit of Fig. 12.8, connect an additional $R = 10\ \text{k}\Omega$ resistor in series with the rightmost capacitor C. For this modification (and ignoring the amplitude stabilization circuitry) find the loop gain $A\beta$ by breaking the circuit at node X. Find R_f for oscillation to begin, and find f_0.

D12.18 For the circuit in Fig. P12.18, break the loop at node X and find the loop gain (working backward for simplicity to find V_x in terms of V_o). For $R = 10\ \text{k}\Omega$, find C and R_f to obtain sinusoidal oscillations at 10 kHz.

*12.19 Consider the quadrature-oscillator circuit of Fig. 12.9 without the limiter. Let the resistance R_f be equal to $2R/(1 + \Delta)$, where $\Delta \ll 1$. Show that the poles of the characteristic equation are in the right-half s-plane and given by $s \simeq (1/CR)[(\Delta/4) \pm j]$.

*12.20 Assuming that the diode-clipped waveform in Exercise 12.7 is nearly an ideal square wave and that the resonator Q is 20, provide an estimate of the distortion in the output sine wave by calculating the magnitude (relative to the fundamental) of

(a) the second harmonic,
(b) the third harmonic,
(c) the fifth harmonic,
(d) the root mean square of harmonics to the tenth.

Note that a square wave of amplitude V and frequency ω is represented by the series

$$\frac{4V}{\pi}\left(\cos \omega t - \frac{1}{3}\cos 3\omega t + \frac{1}{5}\cos 5\omega t - \frac{1}{7}\cos 7\omega t + \cdots\right)$$

Section 12.3: LC and Crystal Oscillators

**12.21 Figure P12.21 shows four oscillator circuits of the Colpitts type, complete with bias detail. For each circuit derive an equation governing circuit operation, and find the frequency of oscillation and the gain condition that ensures that oscillations start.

**12.22 Consider the oscillator circuit in Fig. P12.22, and assume for simplicity that $\beta = \infty$.

(a) Find the frequency of oscillation and the minimum value of R_C (in terms of the bias current I) for oscillations to start.
(b) If R_C is selected equal to $(1/I)$ kΩ, where I is in milliamperes, convince yourself that oscillations will start. If oscillations grow to the point that V_o

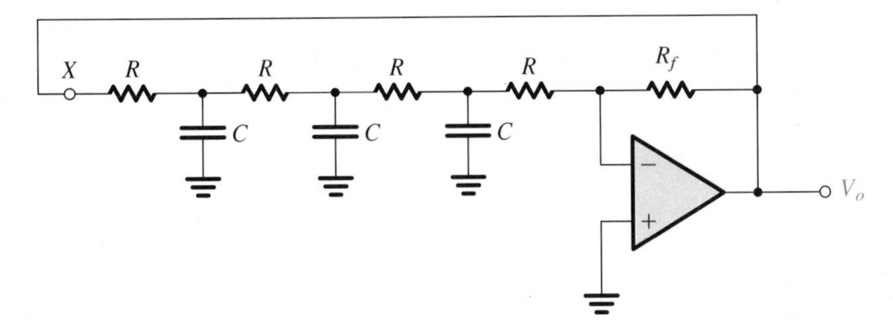

Fig. P12.18

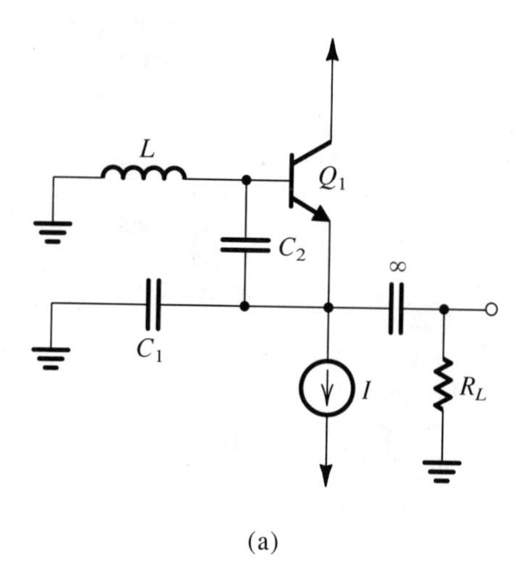

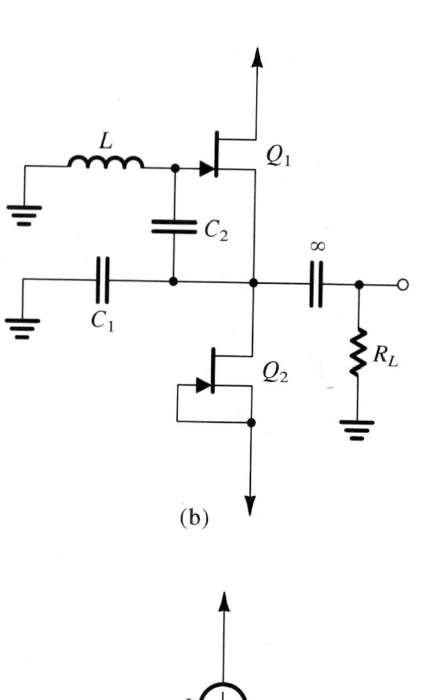

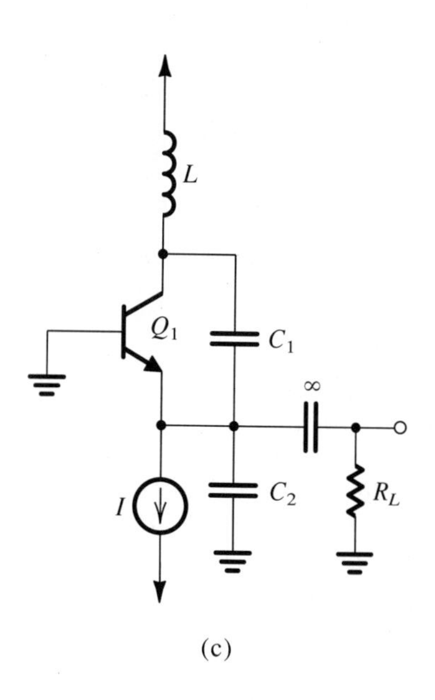

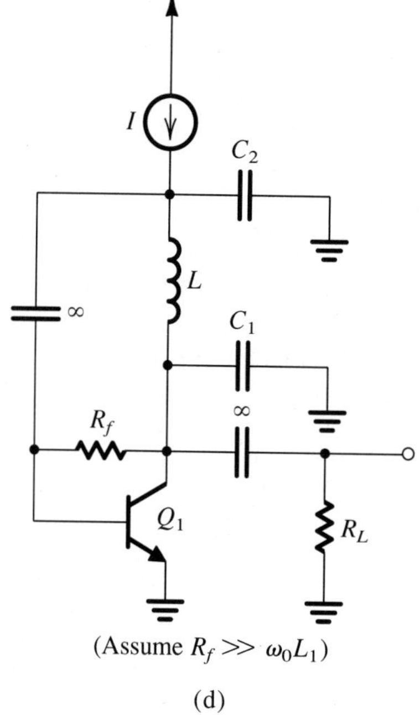

Fig. P12.21

is sufficiently large to turn the BJTs on and off, show that the voltage at the collector of Q_2 will be a square wave of 1 V peak to peak. Estimate the peak-to-peak amplitude of the output sine wave V_o.

12.23 Consider the Pierce crystal oscillator of Fig. 12.16 with the crystal as specified in Exercise 12.10. Let C_1 be variable in the range 1 to 10 pF, and let C_2 be fixed at 10 pF. Find the range over which the oscillation frequency can be tuned. (*Hint:* Use the result in the statement leading to the expression in Eq. 12.27.)

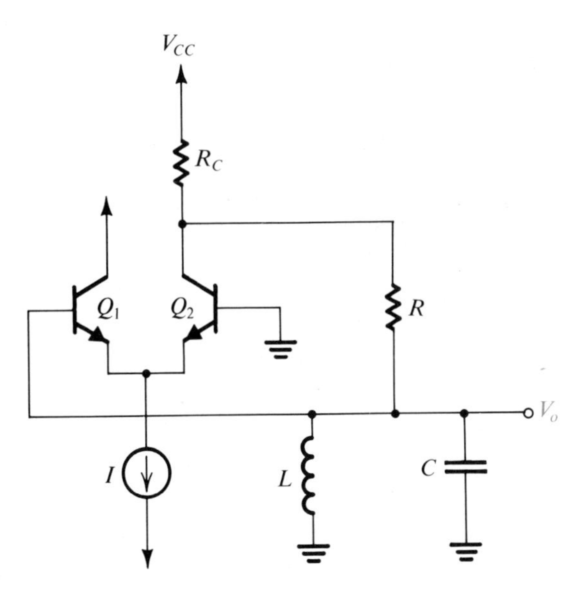

Fig. P12.22

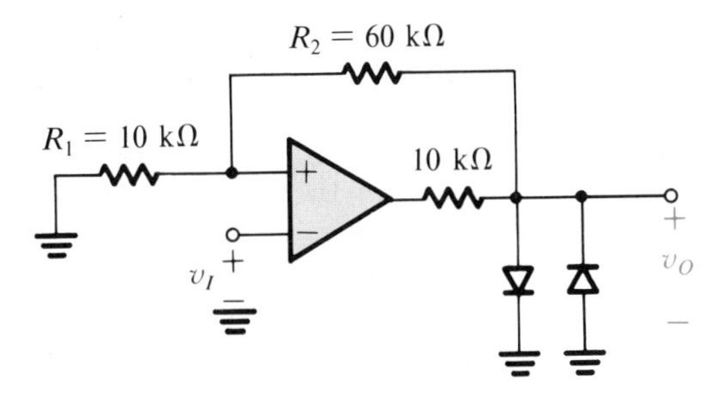

Fig. P12.26

Section 12.4: Bistable Multivibrators

12.24 Consider the bistable circuit of Fig. 12.19(a) with the op amp's positive input terminal connected to a positive voltage source V through a resistor R_3.

(a) Derive expressions for the threshold voltages V_{TL} and V_{TH} in terms of the op amp's saturation levels L_+ and L_-, R_1, R_2, R_3, and V.

(b) Let $L_+ = -L_- = 13$ V, $V = 15$ V and $R_1 = 10\ \text{k}\Omega$. Find the values of R_2 and R_3 that result in $V_{TL} = +4.9$ V and $V_{TH} = +5.1$ V.

12.25 Consider the bistable circuit of Fig. 12.20(a) with the op amp's negative input terminal disconnected from ground and connected to a reference voltage V_R.

(a) Derive expressions for the threshold voltages V_{TL} and V_{TH} in terms of the op amp's saturation levels L_+ and L_-, R_1, R_2, and V_R.

(b) Let $L_+ = -L_- = V$ and $R_1 = 10\ \text{k}\Omega$. Find R_2 and V_R that result in threshold voltages of 0 and $V/10$.

12.26 For the circuit in Fig. P12.26, sketch and label the transfer characteristic v_O–v_I. The diodes are assumed to have a constant 0.7-V drop when conducting, and the op amp saturates at ± 12 V. What is the maximum diode current?

12.27 Consider the circuit of Fig. P12.26 with R_1 eliminated and R_2 short circuited. Sketch and label the transfer characteristic v_O–v_I. Assume that the diodes have a constant 0.7-V drop when conducting and that the op amp saturates at ± 12 V.

*12.28 Consider a bistable circuit having a noninverting transfer characteristic with $L_+ = -L_- = 12$ V, $V_{TL} = -1$ V, and $V_{TH} = +1$ V.

(a) For a 0.5-V-amplitude sine-wave input having zero average, what is the output?

(b) Describe the output if a sinusoid of frequency f and amplitude of 1.1 V is applied at the input. By how much can the average of this sinusoidal input shift before the output becomes a constant value?

D12.29 Design the circuit of Fig. 12.23(a) to realize a transfer characteristic with ± 7.5-V output levels and ± 7.5-V threshold values. Design so that when $v_I = 0$ V a current of 0.1 mA flows in the feedback resistor and a current of 1 mA flows through the zener diodes. Assume that the output saturation levels of the op amp are ± 12 V. Specify the voltages of the zener diodes and give the values of all resistors.

Section 12.5: Generation of Square and Triangular Waveforms Using Astable Multivibrators

12.30 Find the frequency of oscillation of the circuit in Fig. 12.24(b) for the case $R_1 = 10\ \text{k}\Omega$, $R_2 = 16\ \text{k}\Omega$, $C = 10$ nF, and $R = 62\ \text{k}\Omega$.

D12.31 Augment the astable multivibrator circuit of Fig. 12.24(b) with an output limiter of the type shown in Fig. 12.23(b). Design the circuit to obtain an output square wave with 5-V amplitude and 1-kHz frequency using a 10-nF capacitor C. Use $\beta = 0.462$, and design for a current in the resistive divider approximately equal to the average current in the RC network over $\frac{1}{2}$ cycle. Assuming ± 13-V op amp saturation voltages, arrange that the zener operates at a current of 1 mA.

D12.32 Using the scheme of Fig. 12.25, design a circuit that provides square waves of 10 V peak to peak and triangular waves of 10 V peak to peak. The frequency is to be 1 kHz. Implement the bistable circuit with the circuit of Fig. 12.23(b). Use a 0.01-μF capacitor, and specify the values of all resistors and the required zener voltage. Design for a minimum zener current of 1 mA and for a maximum current in the resistive divider of 0.2 mA. Assume that the output saturation levels of the op amps are ± 13 V.

D*12.33 The circuit of Fig. P12.33 consists of an inverting bistable multivibrator with an output limiter and a noninverting integrator. (This integrator circuit was studied in Example 2.6.) Using equal values for all resistors except R_7 and a 0.5-nF capacitor, design the circuit to obtain a square wave at the output of the bistable multivibrator of 15-V peak-to-peak amplitude and 10-kHz frequency. Sketch and label the waveform at the integrator output. Assuming ± 13-V op amp saturation levels, design for a minimum zener current of 1 mA. Specify the zener voltage required, and give the values of all resistors.

Section 12.6: Generation of a Standardized Pulse—The Monostable Multivibrator

***12.34** Figure P12.34 shows a monostable multivibrator circuit. In the stable state, $v_O = L_+$, $v_A = 0$, and $v_B = -V_{ref}$. The circuit can be triggered by applying a positive input pulse of height greater than V_{ref}. For normal operation, $C_1R_1 \ll CR$. Show the resulting waveforms of v_O and v_A. Also, show that the pulse generated at the output will have a width T given by

$$T = CR \ln\left(\frac{L_+ - L_-}{V_{ref}}\right)$$

Note that this circuit has the interesting property that the pulse width can be controlled by changing V_{ref}.

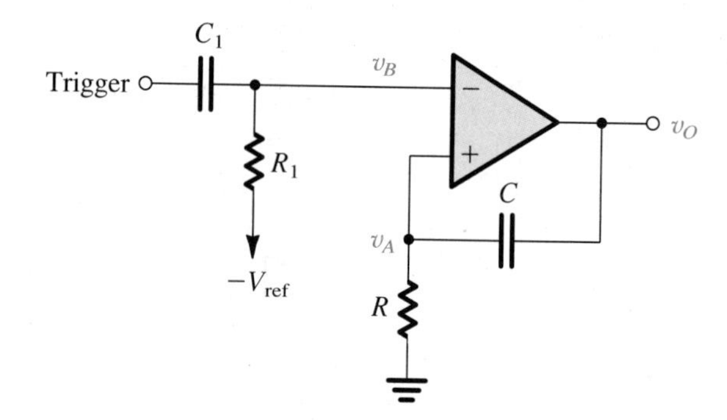

Fig. P12.34

12.35 For the monostable circuit considered in Exercise 12.19, calculate the recovery time.

D*12.36 Using the circuit of Fig. 12.26, with a nearly ideal op amp for which the saturation levels are ± 13 V, design a monostable multivibrator to provide a negative output pulse of 100-μs duration. Use capacitors of 0.1 nF and 1 nF. Wherever possible, choose resistors of 100 kΩ in your design. Diodes have a drop of

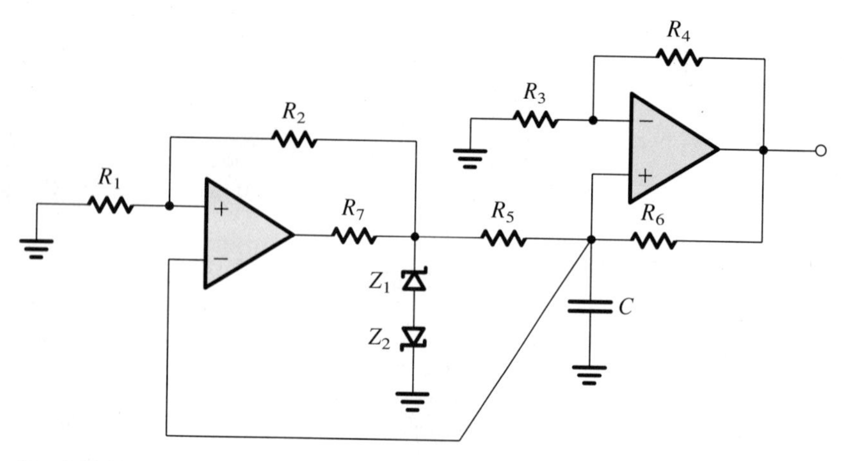

Fig. P12.33

0.7 V. What is the minimum input step size that will ensure triggering? How long does the circuit take to recover to a state in which retriggering is possible with a normal output?

Section 12.7: Integrated-Circuit Timers

12.37 Consider the 555 circuit of Fig. 12.27 when the Threshold and the Trigger input terminals are joined together and connected to an input voltage v_I. Verify that the transfer characteristic v_O–v_I is that of an inverting bistable circuit with thresholds $V_{TL} = \frac{1}{3}V_{CC}$ and $V_{TH} = \frac{2}{3}V_{CC}$ and output levels of 0 and V_{CC}.

12.38 (a) Using a 1-nF capacitor C in the circuit of Fig. 12.28, find the value of R that results in an output pulse of 10-μs duration.

(b) If the 555 timer used in (a) is powered with $V_{CC} = 15$ V, and assuming that V_{TH} can be varied externally (that is, it need not remain equal to $\frac{2}{3}$ V_{CC}), find its required value so that the pulse width is increased to 20 μs with other conditions the same as in (a).

D12.39 Using a 680-pF capacitor, design the astable circuit of Fig. 12.29(a) to obtain a square wave with a 50-kHz frequency and a 75% duty cycle. Specify the values of R_A and R_B.

***12.40** The node in the 555 timer at which the voltage is V_{TH} (that is, the inverting input terminal of comparator 1) is usually connected to an external terminal. This allows the user to change V_{TH} externally (that is, it no longer remains $\frac{2}{3}$ V_{CC}). Note, however, that whatever the value of V_{TH} becomes, V_{TL} always remains $\frac{1}{2}$ V_{TH}.

(a) For the astable circuit of Fig. 12.29, rederive the expressions for T_H and T_L, expressing them in terms of V_{TH} and V_{TL}.

(b) For the case $C = 1$ nF, $R_A = 7.2$ kΩ, $R_B = 3.6$ kΩ, and $V_{CC} = 5$ V, find the frequency of oscillation and the duty cycle of the resulting square wave when no external voltage is applied to the terminal V_{TH}.

(c) For the design in (b), let a sine-wave signal of a much lower frequency than that found in (b) and of 1-V peak amplitude be capacitively coupled to the circuit node V_{TH}. This signal will cause V_{TH} to change around its quiescent value of $\frac{2}{3}$ V_{CC}, and thus T_H will change correspondingly—a modulation process. Find T_H, and find the frequency of oscillation and the duty cycle at the two extreme values of V_{TH}.

Section 12.8: Nonlinear Waveform-Shaping Circuits

D*12.41 The two-diode circuit shown in Fig. P12.41 can provide a crude approximation to a sine-wave output when driven by a triangular waveform. To obtain a good approximation, we select the peak of the triangular waveform, V, so that the slope of the desired sine wave at the zero crossings is equal to that of the triangular wave. Also, the value of R is selected so that when v_I is at its peak the output voltage is equal to the desired peak of the sine wave. If the diodes exhibit a voltage drop of 0.7 V at 1-mA current, changing at the rate of 0.1-V per decade, find the values of V and R that will yield an approximation to a sine waveform of 0.7-V peak amplitude. Then find the angles θ (where $\theta = 90°$ when v_I is at its peak) at which the output of the circuit is 0.7, 0.65, 0.6, 0.55, 0.5, 0.4, 0.3, 0.2, 0.1, and 0 V. Use the angle values obtained to determine the values of the exact sine wave (that is, 0.7 sin θ), and thus find the percent error of this circuit as a sine shaper. Provide your results in tabular form.

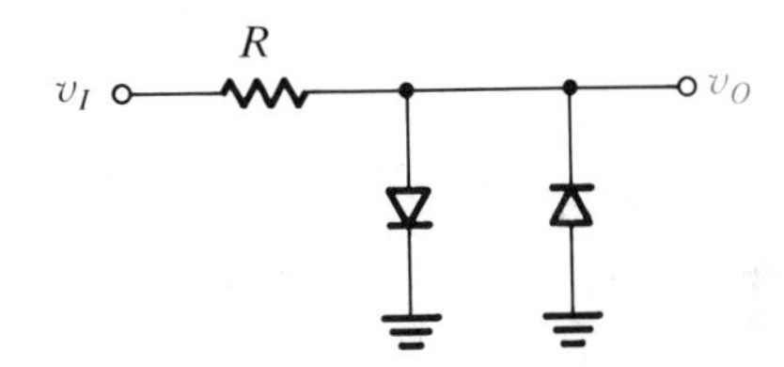

Fig. P12.41

D12.42 Design a two-segment sine-wave shaper using a 10-kΩ input resistor, two diodes, and two clamping voltages. The circuit, fed by a 10-V peak-to-peak triangular wave, should limit the amplitude of the output signal via a 0.7-V diode to a value corresponding to that of a sine wave whose zero-crossing slope matches that of the triangle. What are the clamping voltages you have chosen?

12.43 Show that the output voltage of the circuit in Fig. P12.43 is given by

$$v_O = -n\, V_T \ln\left(\frac{v_I}{I_S R}\right), \qquad v_I > 0$$

where I_S and n are the diode parameters and V_T is the thermal voltage. Since the output voltage is propor-

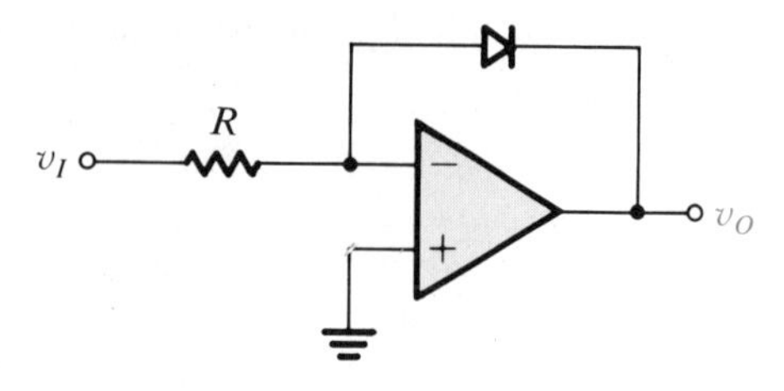

Fig. P12.43

tional to the logarithm of the input voltage, the circuit is known as a **logarithmic amplifier.** Such amplifiers find application in situations where it is desired to compress the signal range.

12.44 Verify that the circuit in Fig. P12.44 implements the transfer characteristic $v_O = v_1 v_2$ for $v_1, v_2 > 0$. Such a circuit is known as an analog multiplier. Check the circuit's performance for various combinations of input voltage of values, say, 0.5, 1, 2, and 3 volts. Assume all diodes to be identical, with 700-mV drop at 1-mA current and $n = 2$. Note that a *squarer* can easily be produced using a single input (for example, v_1) connected via a 0.5-kΩ resistor (rather than the 1-kΩ shown).

**12.45 Detailed analysis of the circuit in Fig. 12.32 shows that optimum performance (as a sine shaper) occurs when the values of I and R are selected so that $RI = 2.5\ V_T$, where V_T is the thermal voltage, and the peak amplitude of the input triangular wave is $6.6\ V_T$. If the output is taken across R (that is, between the two emitters), find v_I corresponding to $v_O = 0.25\ V_T$, $0.5\ V_T$, V_T, $1.5\ V_T$, $2\ V_T$, $2.4\ V_T$, and $2.42\ V_T$. Plot v_O–v_I and compare to the ideal curve given by

$$v_O = 2.42\ V_T \sin\left(\frac{v_I}{6.6\ V_T} \times 90°\right).$$

Section 12.9: Precision Rectifier Circuits

12.46 Two superdiode circuits connected to a common load resistor and having the same input signal have their diodes reversed, one with cathode to the load, the other with anode to the load. For a sine-wave input of 10 V peak to peak, what is the output waveform? Note that each half cycle of the load current is provided by a separate amplifier, and that while one

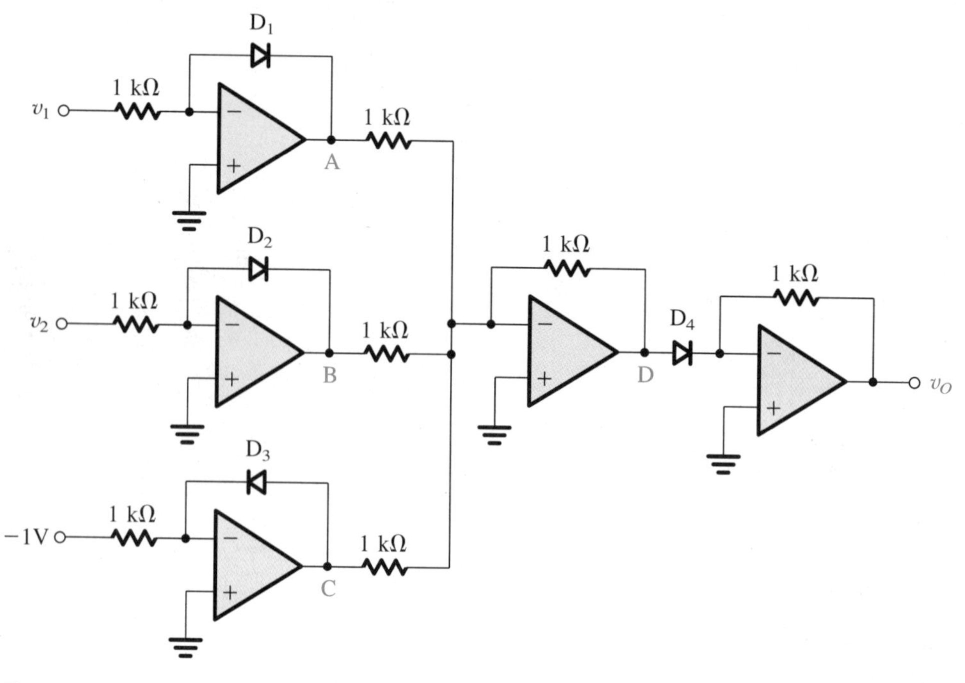

Fig. P12.44

amplifier supplies the load current, the other amplifier idles. This idea, called class B operation, is important in the implementation of power amplifiers.

D12.47 The superdiode circuit of Fig. 12.33(a) can be made to have gain by connecting a resistor R_2 in place of the short circuit between the cathode of the diode and the negative input terminal of the op amp, and a resistor R_1 between the negative input terminal and ground. Design the circuit for a gain of 2. For a 10-V peak-to-peak input sine wave, what is the average output voltage resulting?

D12.48 Provide a design of the inverting precision rectifier shown in Fig. 12.34(a) in which the gain is -2 for negative inputs and zero otherwise, and the input resistance is 100 kΩ. What values of R_1 and R_2 do you choose?

D*12.49 Provide a design for a voltmeter circuit similar to the one in Fig. 12.35, which is intended to function at frequencies of 10 Hz and above. It should be calibrated for sine-wave input signals to provide an output of +10 V for an input of 1 V rms. The input resistance should be as high as possible. To extend the bandwidth of operation, keep the gain in the ac part of the circuit reasonably small. As well, the design should be such as to reduce the size of the capacitor C required. The largest value of resistor available is 1 MΩ.

12.50 Plot the transfer characteristic of the circuit in Fig. P12.50.

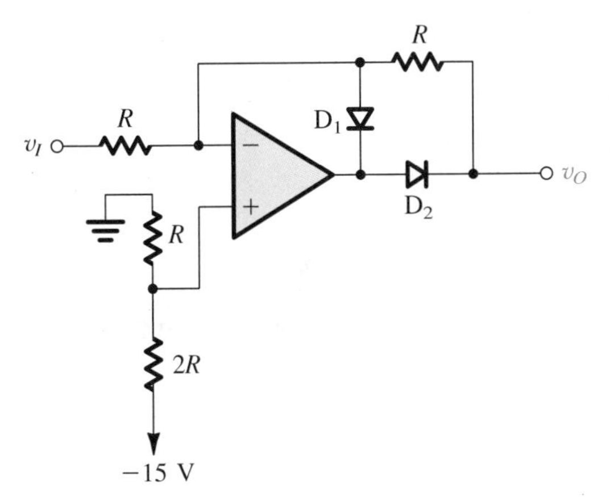

Fig. P12.50

12.51 Plot the transfer characteristics v_{O1}–v_I and v_{O2}–v_I of the circuit in Fig. P12.51.

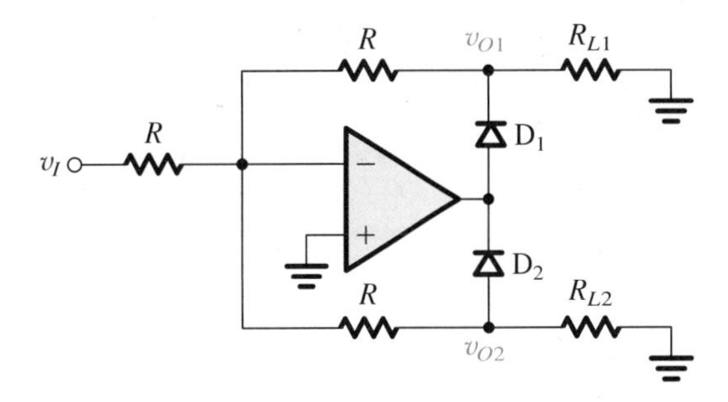

Fig. P12.51

12.52 Sketch the transfer characteristics of the circuit in Fig. P12.52.

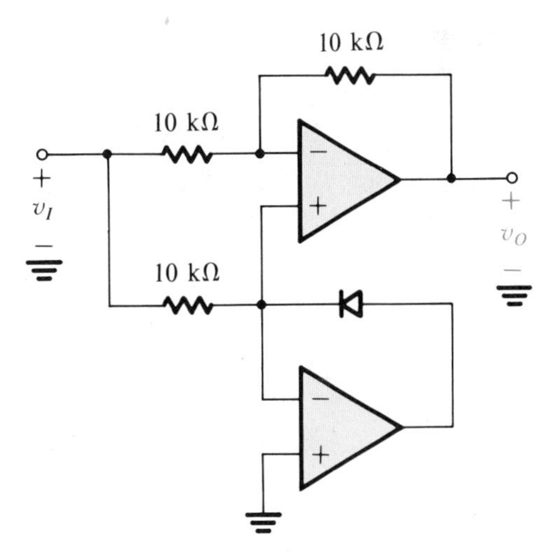

Fig. P12.52

D12.53 A circuit related to that in Fig. 12.38 is to be used to provide a current proportional to v_A($v_A \geq 0$) to a light-emitting diode (LED). The value of the current is to be independent of the diode's nonlinearities and variability. Indicate how this may be done easily.

***12.54** In the precision rectifier of Fig. 12.38, the resistor R is replaced by a capacitor C. What happens? For equivalent performance with a sine-wave input of 60-Hz frequency with $R = 1$ kΩ, what value of C should be used? What is the response of the modified circuit at 120 Hz? at 180 Hz? If the amplitude of v_A is kept fixed, what new function does this circuit perform? Now consider the effect of a waveform change on both circuits (the one with R and the one with C).

For a triangular-wave input of 60-Hz frequency that produces an average meter current of 1 mA in the circuit with R, what does the average meter current become when R is replaced with the C whose value was calculated above?

***12.55** A positive-peak rectifier utilizing a fast op amp and a junction diode in a superdiode configuration, and a 10-μF capacitor initially uncharged, is driven by a series of 10-V pulses of 10-μs duration. If the maximum output current that the op amp can supply is 10 mA, what is the voltage on the capacitor following one pulse? two pulses? ten pulses? How many pulses are required to reach 0.5 V? 1.0 V? 2.0 V?

D12.56 Consider the buffered precision peak rectifier shown in Fig. 12.40 when connected to a triangular input of 1-V peak-to-peak amplitude and 1000-Hz frequency. It utilizes an op amp whose bias current (directed into A_2) is 10 nA and diodes whose reverse leakage current is 1 nA. What is the smallest capacitor that can be used to guarantee an output ripple less than 1%?

PART 

DIGITAL CIRCUITS

INTRODUCTION

Part III is devoted to the study of digital circuits. In a digital circuit, signals take on a limited number of values. The most common digital systems employ two values and are said to be *binary* systems. In such a system, voltage signals are either "high" or "low" and the symbols 1 and 0 are used to denote the two possible levels.

Digital circuits operate on binary-valued input signals and produce binary-valued output signals. The operation of digital circuits can be described by a special kind of algebra called *Boolean algebra*. It is not our intention here to study Boolean algebra, a topic normally covered in digital systems texts. We shall in fact assume that the reader is familiar with the basic concepts of digital systems, and concentrate here on digital electronics.

Digital circuits play a very important role in today's electronic systems. They are employed in almost every facet of electronics, including communications, control, instrumentation, and, of course, computing. This widespread usage is due mainly to the availability of inexpensive integrated-circuit packages that contain powerful digital circuitry. The circuit complexity of a digital IC chip ranges from a small number of logic gates to a complete computer (a *microprocessor*) or 16 million bits of memory.

The conventional approach to designing digital systems consists of assembling the system using standard IC packages of various levels of complexity (and hence integration). As an alternative to using "off-the-shelf" components, the designer might opt for implementing part or all of the system using one or more customized very-large-scale-integrated (VLSI) circuits. However, custom IC design is usually justified only when the production volume is large (greater than about 100,000 parts). An intermediate approach, known as *semicustom design,* utilizes *gate array* chips. These are integrated circuits containing up to 10,000 unconnected logic gates. Their interconnection can be achieved by a final metallization step according to a pattern specified by the user so as to implement the user's particular functions. A more recently available type of gate arrays, known as *field programmable gate arrays* (FPGA), can, as their name indicates, be programmed directly by the user.

Whatever approach is taken in digital design, some familiarity with the various digital circuit technologies and design techniques is essential. The material to follow is intended to provide the reader with a basic understanding of digital electronics. It should also serve as an introduction to VLSI circuit design.

Prerequisite to the study of the topics covered in Part III is the material on basic electronic devices presented in Part I.

CHAPTER 13

MOS Digital Circuits

INTRODUCTION

This chapter is concerned with the study of MOS digital integrated circuits, by far the most popular technology for the implementation of digital systems. The small size, ease of fabrication, and low power dissipation of MOSFETs enable extremely high levels of integration of both logic and memory circuits.

Following an introduction to basic logic-circuit concepts (Section 13.1) that apply to both MOSFET and BJT circuits, the two families of MOS logic circuits are presented: NMOS in Sections 13.2–13.4, and CMOS in Sections 13.5 and 13.6. Although NMOS allows very high packing densities (that is, a very large number of gates per chip) and thus is particularly useful in VLSI design, CMOS is currently the most popular and versatile of all logic-circuit types and technologies. It is useful in almost every kind of application and is fabricated at all levels of integration, from SSI to VLSI. Its speed of operation, however, falls short of that achieved with bipolar circuits (Chapter 14) and with the emerging gallium-arsenide circuits presented in the concluding section of this chapter.

In addition to combinational logic circuits, this chapter presents such other essential digital-system building blocks as flip-flops and multivibrators, as well as providing an overview and introduction to memory circuits.

Finally, in preparation for the study of MOS digital circuits, the reader is urged to review the characteristics and basic circuits of MOSFETs (Chapter 5).

13.1 LOGIC CIRCUITS—SOME BASIC CONCEPTS

In this section we consider some basic concepts that underlie the design and specification of logic gate circuits. The material presented applies to both MOS and bipolar digital circuits.

Digital Signals

In binary digital circuits, two distinct voltage levels can represent the two values of binary variables. However, in order to allow for the inevitable component tolerances and miscellaneous other effects that change the signal voltage levels, two distinct voltage ranges are usually defined. As shown in Fig. 13.1, if the signal voltage lies in the range V_{L1} to V_{L2}, the signal is interpreted (by the digital circuit) as a logic 0. If the signal amplitude falls in the range V_{H1} to V_{H2}, it is interpreted as a logic 1. The two voltage bands are separated by a range in which the signal amplitude is not supposed to lie. This forbidden band represents an undefined or excluded region.

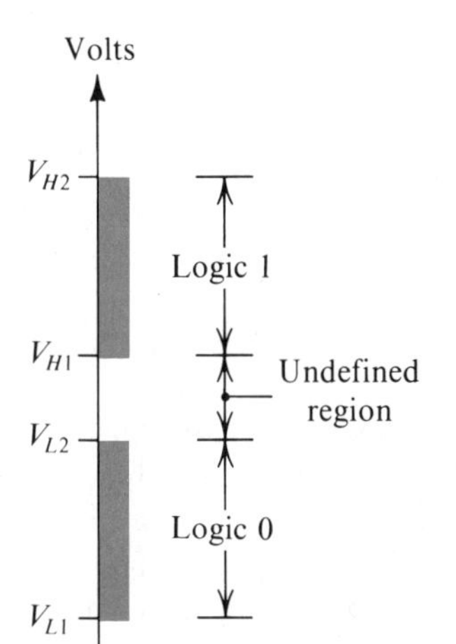

Fig. 13.1 Two distinct voltage ranges are used to represent the two values of binary variables.

Since the logic-1 voltages are higher than those that represent logic 0, the system illustrated in Fig. 13.1 is said to use **positive logic.** We can of course reverse our voltage band assignment, thus obtaining a **negative-logic** system. Throughout this book we shall assume the positive-logic convention. Furthermore, on many occasions we shall use the words "high" and "low" interchangeably with 1 and 0, respectively.

Logic Circuit Families

Integrated logic circuits are classified into a number of different families. Members of each family are made with the same technology, have similar circuit structure, and exhibit the same basic features. In this chapter we shall study two MOS families: NMOS and CMOS. The first utilizes *n*-channel MOSFETs exclusively; the latter uses both *n*- and *p*-channel transistors in complementary symmetric circuit configurations. In the next chapter we shall study two BJT logic families: transistor-transistor logic (TTL) and emitter-coupled logic (ECL). CMOS, TTL, and ECL are available as off-the-shelf components for conventional logic design, as well as being utilized in the design of special VLSI circuits. NMOS, on the other hand, is used only in the design of VLSI circuits such as microprocessors and memory chips.

It will be shown that each of the four logic circuit families offers a unique set of advantages and disadvantages. In designing a digital system one selects an appropriate logic family and attempts to implement as much of the system as possible using packages that belong to this family. In this way interconnection of the various packages is relatively straightforward. If, on the other hand, packages from more than one family have to be used, one has to design suitable *interface circuits*. The selection of a logic family is based on such considerations as logic flexibility, speed of operation, availability of complex functions, noise immunity, operating-temperature range, power dissipation, and cost. We will discuss some of these considerations in this chapter and the next.

Scale of Integration

Many types of logic functions are usually available within each logic family. Depending on the complexity of the circuit of the IC chip, the package can be classified as one of four types:

1. small-scale integrated (SSI) circuit
2. medium-scale integrated (MSI) circuit
3. large-scale integrated (LSI) circuit
4. very-large-scale integrated (VLSI) circuit

Although the boundaries between the different levels of integration are not very sharp, a rough guide, based on the number of "equivalent logic gates" on the chip, is as follows: SSI, 1 to 10 gates; MSI, 10 to 100 gates; LSI, 100 to 1000 gates; and VLSI, >1000 gates.

The basic circuit properties of a logic family can be established by studying the basic inverter circuit of that family. In the following we shall consider the general characteristics and structure of the logic inverter circuit.

The Basic Inverter

The logic inverter is basically a voltage-controlled switch such as that represented schematically in Fig. 13.2. As shown, the switch connected between terminals 2 and 3 is controlled by the input signal v_I applied between terminals 1 and 3 (terminal 3 is connected to the reference or ground point). The inverter output voltage v_O is taken across the switch—that is, between 2 and ground. When v_I is low (around 0 V) the switch is open and the output voltage v_O is high (equal to the supply voltage V^+). When v_I is high (above a specified threshold voltage, as explained below) the switch is closed and the output voltage is low (0 V). It should be obvious that this circuit realizes the logic inversion operation.

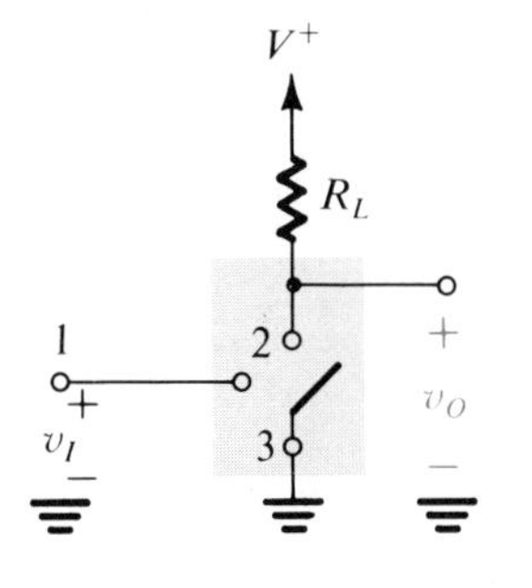

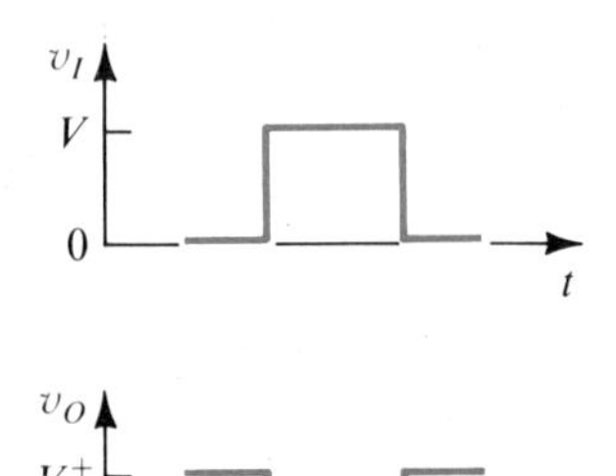

Fig. 13.2 A conceptual representation of the logic inverter as a voltage-controlled switch.

Practical logic inverter circuits differ from the conceptual circuit in Fig. 13.2 in a number of ways. First, the input terminal of the inverter usually draws some current from the driving source. Second, the switch is not ideal; specifically, when the switch is closed it does not behave as a short circuit but rather has a finite closure resistance (called **on resistance**) and sometimes an additional voltage drop (called **offset voltage**). Figure 13.3 shows an equivalent circuit of the switch in the closed position. As a result of these imperfections, the voltage v_O will not be zero in the on state. Third, the inverter switch may not switch instantaneously; rather, there may be a delay time between the application of the input change and the appearance of the output change. Even if the inverter switches instantaneously, the capacitance inevitably present between the output terminals will cause the output waveform to have finite rise and fall times. This point will be discussed in some detail shortly. Fourth, actual inverters may not exhibit a well-defined switching threshold as implied by the discussion of the conceptual circuit of Fig. 13.2. Transfer characteristics of logic inverters are discussed next.

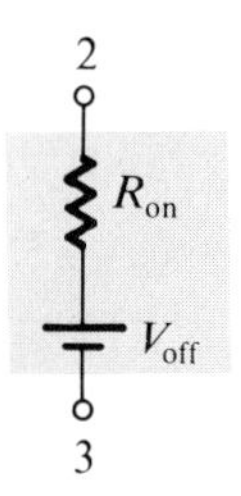

Fig. 13.3 Equivalent circuit for a typical switch in the on position.

Practical implementations of the logic inverter utilize a transistor (a MOSFET or a BJT) as the switching element and a resistor or another transistor for the load resistor R_L. Since MOS transistors require much smaller chip areas than medium- and high-valued resistors, MOS logic circuits almost always utilize MOSFETs as load elements.

The Inverter Transfer Characteristic

Figure 13.4 shows the ideal transfer characteristic for a logic inverter operated from a power supply V^+. As indicated, the inverter exhibits a threshold voltage $V_{th} = \frac{1}{2}V^+$. Input signals below this threshold voltage are interpreted as being low, and the inverter output is equal to the supply voltage V^+. Input signals above the threshold are interpreted as high,

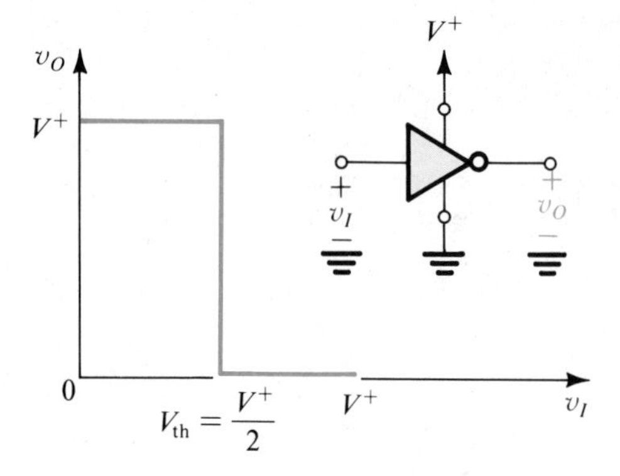

Fig. 13.4 Ideal transfer characteristic of the logic inverter.

and the inverter output is equal to 0 V. Note that this inverter is quite tolerant of errors in the value of the input signal. Also, with the threshold voltage at half the power supply, this tolerance is equally distributed between the two domains of input signal (low and high). Finally, the inverter characteristics are ideal in the sense that, independent of the exact value of input signal, the output is either V^+ or 0 V.

Actual logic inverters have transfer characteristics that only approximate the ideal one of Fig. 13.4. Figure 13.5 shows a typical inverter transfer characteristic. Note that the threshold voltage is no longer well defined and that there exists a *transition region* between the high and low states. Also, the high output (V_{OH}) and low output (V_{OL}) are no longer equal to V^+ and 0 V, respectively.

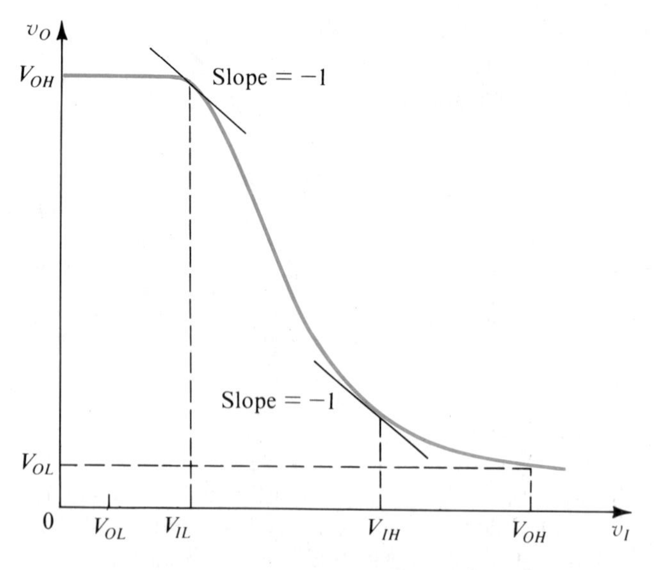

Fig. 13.5 Typical transfer characteristic of a logic inverter, illustrating the definition of the critical points.

Examining the inverter transfer characteristic in Fig. 13.5 in more detail, we note that there are three distinct regions:

1. The low-input region: $v_I < V_{IL}$
2. The transition (or uncertain) region: $V_{IL} \leq v_I \leq V_{IH}$
3. The high-input region: $v_I > V_{IH}$

Since the transition between one region and the next might not be sharp, it is customary to define V_{IL} and V_{IH} as the points at which the slope of the voltage transfer curve is -1, as indicated in Fig. 13.5. Of course, the slope in the transition region is much greater than unity. Input voltages less than V_{IL} are acknowledged by the gate as representing logic 0. Thus, *V_{IL} is the maximum allowable logic-0 value*. Similarly, input voltages greater than V_{IH} are acknowledged by the gate as representing logic 1. Thus, *V_{IH} is the minimum allowable logic-1 value*.

We should note in passing that the inverter transfer characteristic indicates that it is a grossly nonlinear device. In fact, digital circuits are extreme cases of the nonlinear circuits discussed in Chapters 3 and 12. If the inverter whose transfer characteristic is depicted in Fig. 13.5 were to be used as a linear amplifier, it would be biased at a point somewhere in the transition region, and the input signal swing would be kept small to restrict operation to a short, almost-linear segment in the transition region of the characteristic.

Noise Margins

A great advantage of binary digital circuits is their tolerance to variations in value of input signals. As long as the input signal is correctly interpreted as low or high, the accuracy of operation is not affected. This tolerance to variation in signal level can be considered as immunity to noise superimposed on the input signal. Here *noise* refers to extraneous signals that can be capacitively or inductively coupled into the digital circuit from other parts of the system or from outside the system. We shall now quantify the ability of a logic gate to reject noise.[1]

Consider once more the inverter transfer characteristic of Fig. 13.5, where the four parameters V_{IL}, V_{IH}, V_{OL}, and V_{OH} are indicated. On the horizontal axis we have also marked the points corresponding to the output levels V_{OL} and V_{OH}. In a logic system, one gate usually drives another. Thus a gate whose output is high, at V_{OH}, drives an identical gate whose specified minimum input logic-1 level is V_{IH}. The difference $V_{OH} - V_{IH}$ represents a margin of safety, for if noise were superimposed on the output signal of the driving gate (V_{OH}), the driven gate would not be bothered as long as the amplitude of the noise voltage was lower than $(V_{OH} - V_{IH})$. This difference is therefore called the **logic-1** or **"high" noise margin** and is denoted NM_H; that is,

$$NM_H \equiv V_{OH} - V_{IH} \tag{13.1}$$

The **logic-0** or **"low" noise margin,** NM_L, is similarly defined as

$$NM_L \equiv V_{IL} - V_{OL} \tag{13.2}$$

Due to the unavoidable variabilities in the values of circuit components and power-supply voltage, the manufacturer usually specifies *worst-case* values for the four parameters V_{OH}, V_{IH}, V_{OL}, and V_{IL}. These values are defined as follows:

V_{OH} the minimum voltage that will be available at a gate output when the output is supposed to be logic 1 (high)

[1] Unlike in analog circuits, where injected noise propagates throughout the system, in digital circuits, once a gate rejects noise at the input, the gate output will be correct and we need no longer concern ourselves with the input noise.

V_{IH} the minimum gate input voltage that will be unambiguously recognized by the gate as corresponding to logic 1

V_{OL} the maximum voltage that will be available at a gate output when the output is supposed to be logic 0 (low)

V_{IL} the maximum gate input voltage that will be unambiguously recognized by the gate as corresponding to logic 0

Instead of drawing the complete transfer characteristic, one usually is satisfied with the logic band diagram of Fig. 13.6. From this diagram it can be inferred that to maximize and equalize NM_L and NM_H, one ideally desires that $V_{IL} = V_{IH} =$ a value midway in the *logic swing* V_{OL} *to* V_{OH}. This implies that the transfer characteristic should switch abruptly; that is, it should exhibit high gain in the transition region. It also implies that switching should occur in the middle of the logic swing. To maximize this swing, V_{OL} should be as low as possible (ideally 0 V) and V_{OH} should be as high as possible (ideally equal to the power supply voltage V^+). Such an idealized transfer characteristic is shown in Fig. 13.4. As will be seen in Section 13.5, the CMOS logic family provides an excellent approximation to the ideal performance.

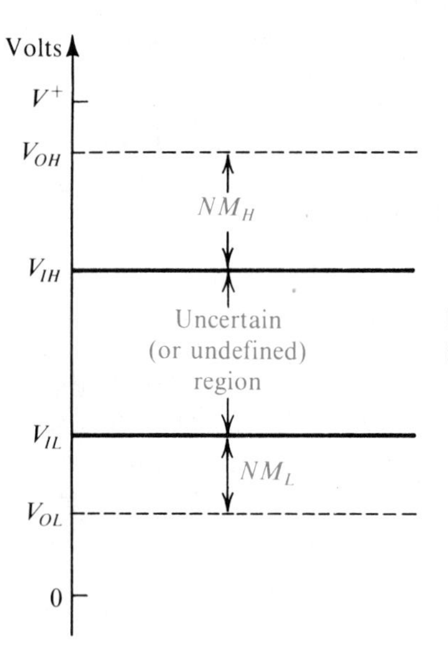

Fig. 13.6 Logic band diagram indicating the noise margins.

Power Dissipation

Of interest to a logic circuit designer is the amount of power consumed by the circuit. Knowledge of power dissipation enables the designer to determine the current that the power supply for a digital system must provide. The power dissipated in a logic circuit is composed of two components: static and dynamic. The **static power** is the power dissipated while the circuit is not changing states. Thus for the idealized inverter of Fig. 13.2 we see that when the output is high, the static power is zero; however, when the output is low, the static power is $(V^+/R_L)V^+$. If one assumes that on average a gate spends half the time in either state, then the average static power dissipation will be $(V^+)^2/2R_L$.

To visualize **dynamic-power dissipation,** consider the idealized inverter of Fig. 13.2 when driving a load capacitance C_L. This could be the input capacitance of another logic gate or the capacitance of the wire interconnection between the inverter and another part of the system. Let us assume that initially, at $t = 0-$, the input was high and the switch was closed. Thus the capacitor was initially discharged. Now let v_I change to low at $t = 0$. It follows that the switch will open, but since the capacitor voltage cannot change instantaneously, v_O will rise exponentially toward V^+. The charging current will be supplied through R_L, and thus power will be dissipated in R_L. If the instantaneous supply current is denoted by i, then the energy drawn from the supply will be $\int V^+ i\, dt = V^+ \int i\, dt = V^+ Q$, where Q is the charge supplied to the capacitor; that is, $Q = C_L V^+$. Thus the energy drawn from the supply is $C_L(V^+)^2$. Now, since the capacitor initially had zero energy and will finally have a stored energy of $\frac{1}{2}C_L(V^+)^2$, it follows that the energy dissipated in R_L is $\frac{1}{2}C_L(V^+)^2$. If we next let v_I go high, the switch will close and C_L will discharge through the switch resistance R_{on}, thus dissipating power through the switch. If we assume that V_{off} is zero, then the energy dissipated in R_{on} will be equal to the energy stored on the capacitor, $\frac{1}{2}C_L(V^+)^2$. It follows that in one cycle of operation the gate dissipates energy of $C_L(V^+)^2$. Thus if the inverter is switched on and off f times per second, the dynamic power dissipation will be

$$\text{Dynamic power dissipation} = f\, C_L(V^+)^2 \tag{13.3}$$

which is the sum of the power dissipation in R_L and R_{on}.

Fan-in and Fan-out

Fan-in of a gate is the number of its inputs. Thus a four-input NOR gate has a fan-in of 4. Fan-out is the maximum number of similar gates that a gate can drive while remaining within the guaranteed specifications. More will be said about fan-in and fan-out in the context of the various logic families.

Propagation Delay

Because of the dynamics involved in the action of a switching device such as a bipolar transistor, the switch in the inverter of Fig. 13.2 may not respond instantaneously to the control signal v_I. In addition, the inevitable load capacitance at the inverter output causes the waveform of v_O to depart from an ideal pulse. Figure 13.7 illustrates the typical response of an inverter to an input pulse with finite rise and fall times. The output pulse, of course, exhibits finite rise and fall times. In addition, there is a delay time between the input and output pulses. There are various ways for expressing this **propagation delay.** The usual way is to specify the times between the 50% points of the input and output waveforms at both the leading and trailing edges. These times are referred to as t_{PHL} (where HL indicates the high-to-low transition of the output) and t_{PLH} (where LH indicates the low-to-high transition of the output) in Fig. 13.7. The propagation time delay t_P can then be defined as the average of these two times,

$$t_P = \tfrac{1}{2}(t_{PHL} + t_{PLH}) \tag{13.4}$$

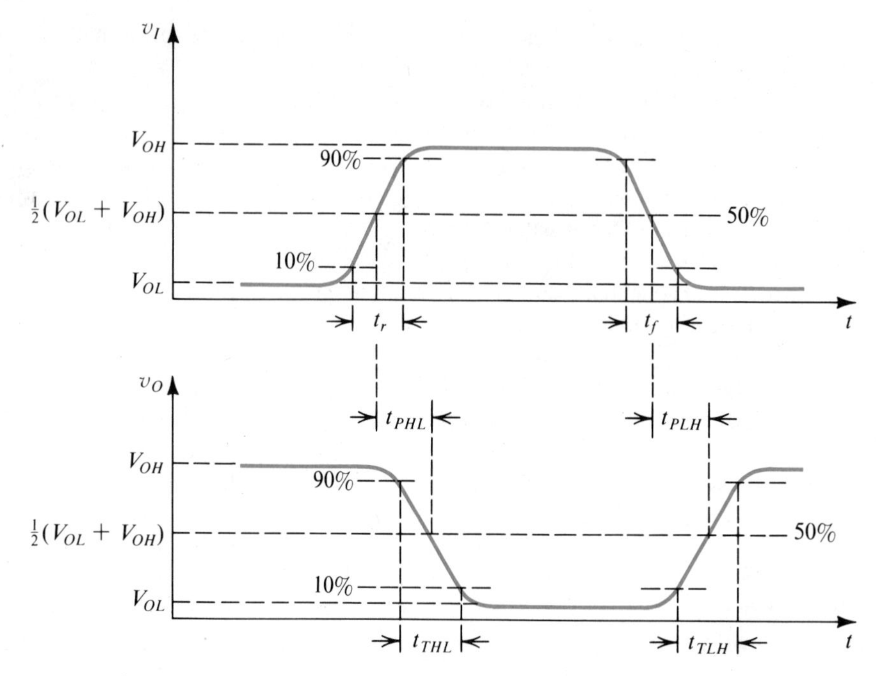

Fig. 13.7 Definitions of propagation delay and transition times of logic gates.

Note that the rise and fall times (or transition times) are specified using the 10% and 90% levels of the output excursion ($V_{OH} - V_{OL}$).

Delay-Power Product

One is usually interested in high-speed performance (low t_P) combined with low power dissipation. Unfortunately these two requirements are in conflict since, generally when designing a gate, if one attempts to reduce power dissipation by decreasing supply current, the gate delay will increase. It follows that a figure of merit for comparing logic families is the delay-power product *DP*, defined as

$$DP \equiv t_P P_D \tag{13.5}$$

where P_D is the power dissipation of the gate. Note that *DP* has the unit of joules. The lower the *DP* figure for a family, the more effective this logic family is.

Physical Packaging of Logic Circuits

Figure 13.8(a) shows a common physical package used to house IC logic circuits. This package is made of either plastic or ceramic and is called a *dual-in-line* (DIP) package. It has 14 leads, 7 brought out to each side. Other packages with 16, 24, and 40 pins exist. A functional diagram of a quad two-input NAND package is shown in Fig. 13.8(b).

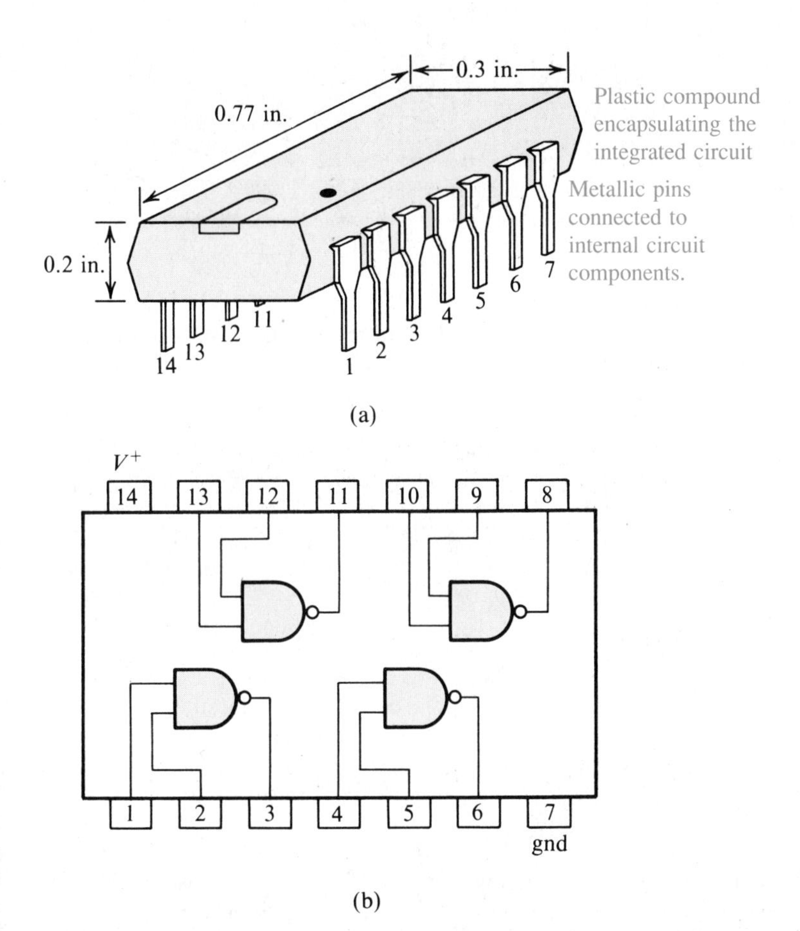

Fig. 13.8 A 14-pin integrated circuit. **(a)** Physical appearance. **(b)** Schematic of an integrated circuit providing 4 two-input NAND gates.

Exercises 13.1 The data sheet of the SN7400 quad 2-input NAND gate of the TTL family provides the following:

Logic-1 input voltage required at both input terminals to ensure a logic-0 level at the output: MIN (minimum) 2 V.
Logic-0 input voltage required at either input terminal to ensure a logic-1 level at the output: MAX (maximum) 0.8 V.
Logic-1 output voltage: MIN 2.4 V, TYP (typical) 3.3 V.
Logic-0 output voltage: TYP 0.22 V, MAX 0.4 V.
Logic-0-level supply current: TYP 12 mA, MAX 22 mA (for the entire package).
Logic-1-level supply current: TYP 4 mA, MAX 8 mA (for the entire package).
Propagation delay time to logic-0 level: TYP 7 ns, MAX 15 ns.
Propagation delay time to logic-1 level: TYP 11 ns, MAX 22 ns.

(a) Find the noise margin in both the 0 and 1 states. (b) Assuming that the gate is in the 1 state 50% of the time and in the 0 state 50% of the time, find the average static power dissipated in a typical gate. The power supply voltage is +5 V. (c) Assuming that the gate drives a capacitance $C_L = 45$ pF and is switched at a 1-MHz rate, find the dynamic power dissipation per gate using the typical values of the logic 1 and 0 levels at the output. (d) Find the typical value of the gate delay-power product (neglecting the dynamic power dissipation).

Ans. (a) 0.4 V, 0.4 V; (b) 10 mW; (c) 0.7 mW; (d) 90 picojoules

13.2 A logic inverter having a negligible static power dissipation is switched at the rate of 1 MHz. If the inverter is operated from a 10-V power supply and drives a 50-pF load capacitance, find the dynamic power dissipation and the average current drawn from the power supply. Assume that the output levels are close to 0 and 10 V.

Ans. 5 mW; 0.5 mA

13.3 Consider the inverter of Fig. 13.2 under the conditions that $V^+ = 5.5$ V, $R_L = 10$ kΩ, and the switch on-resistance $R_{on} = 1$ kΩ. Let $V_{off} = 0$. Find the values of V_{OH} and V_{OL}.

Ans. 5.5 V; 0.5 V

13.4 Let the inverter specified in Exercise 13.3 be fed with an ideal pulse having zero rise and fall times. Assuming that the switch operates instantaneously, find the propagation delays t_{PLH}, t_{PHL}, and t_P that result with a 50-pF load capacitance.

Ans. 347 ns; 32 ns; 189 ns

13.2 NMOS INVERTER WITH ENHANCEMENT LOAD

As mentioned above, the basic building block of a digital circuit family is the logic inverter. For NMOS, there are two basic inverter circuits: the enhancement-load inverter and the depletion-load inverter. In this section we study the first.

Static Characteristics

Figure 13.9(a) shows the enhancement-load inverter. The application of this circuit as an amplifier was studied in Section 5.9, where we derived the voltage transfer characteristic shown in Fig. 13.9(b). As a logic inverter, the circuit operates as follows: Logic-0 input signals, represented by voltages lower than the threshold voltage of Q_1, V_{t1}, cause Q_1 to be off, and thus the output voltage will be high at

$$V_{OH} = V_{DD} - V_{t2} \tag{13.6}$$

We note that the high output is lower than V_{DD} by the threshold voltage V_{t2}. This is a serious drawback of the enhancement-load inverter, resulting in reduced signal swing and noise margin.

With v_I high at the logic-1 level of $(V_{DD} - V_{t2})$, Q_1 will be in the triode region while Q_2 remains in the saturation region. The output will be V_{OL}.

In the transition region the voltage transfer characteristic is linear with a slope of $-\sqrt{K_1/K_2}$ (see Eqs. 5.77–5.79). The ratio of the conductance parameters K_1 and K_2, denoted K_R, is as follows:

$$K_R \equiv \frac{K_1}{K_2} = \frac{(W/L)_1}{(W/L)_2} \tag{13.7}$$

where we have substituted $K_1 = \frac{1}{2}\mu_n C_{OX}(W/L)_1$ and $K_2 = \frac{1}{2}\mu_n C_{OX}(W/L)_2$. The constant K_R is known as the *geometry ratio* or the *aspect ratio* of the inverter. In order to obtain a

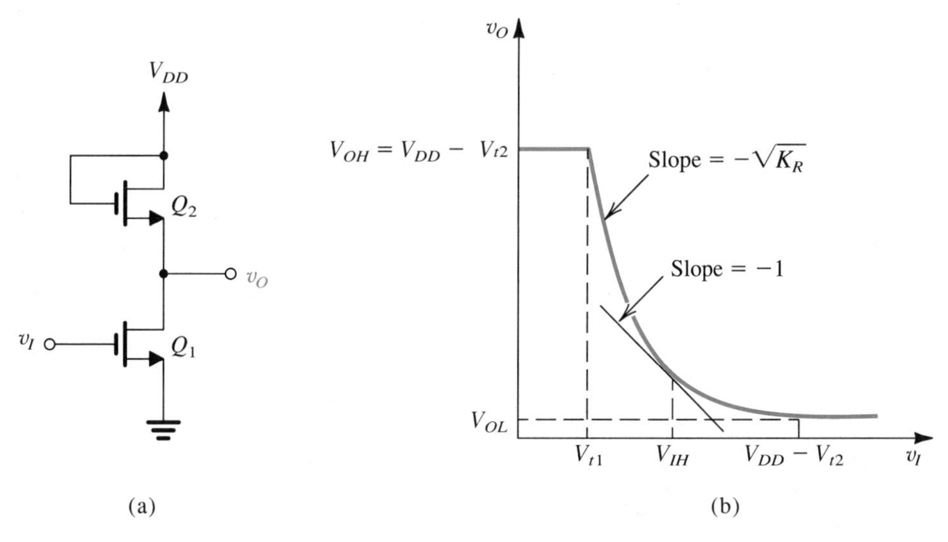

Fig. 13.9 (a) The enhancement-load inverter and **(b)** its voltage transfer characteristic.

reasonably sharp voltage transfer characteristic, and hence acceptable noise margins, K_R is usually greater than 8. Note, however, that the larger the value of K_R the greater is the silicon area occupied by the inverter, since there is a lower limit to the usable dimensions of a device in a given technology.

Exercise 13.5 An enhancement-load inverter having $(W/L)_1 = 3$ and $(W/L)_2 = \frac{1}{3}$ is fabricated with a technology for which the minimum practical transistor dimension (length or width) is 5 μm. Find the area occupied by this inverter and the value of its K_R. If K_R is to be quadrupled, find the ratio (W/L) for each device so that the silicon area required by the inverter is the minimum possible.

Ans. 150 μm^2; 9; $(W/L)_1 = 6$; $(W/L)_2 = \frac{1}{6}$

The Body Effect

Figure 13.10 shows the enhancement-load inverter with the substrate connections explicitly indicated. Note that while the source-to-body voltage (V_{SB}) of Q_1 is zero, that of Q_2 is equal to the output voltage v_O. The threshold voltage V_t of a MOSFET is related to V_{SB} via the relationship

$$V_t = V_{t0} + \gamma[\sqrt{V_{SB} + 2\phi_f} - \sqrt{2\phi_f}] \tag{13.8}$$

where V_{t0} is the threshold voltage at $V_{SB} = 0$, γ is a constant for the given fabrication process, and $2\phi_f$ is the equilibrium electrostatic potential of the p material forming the body. Typically, $V_{t0} = 1$ to 1.5 V, $\gamma = 0.3$ to 1 $V^{1/2}$, and $2\phi_f \simeq 0.6$ V. Using Eq. (13.8) we find that $V_{t1} = V_{t0}$ and

$$V_{t2} = V_{t0} + \gamma(\sqrt{v_O + 2\phi_f} - \sqrt{2\phi_f}) \tag{13.9}$$

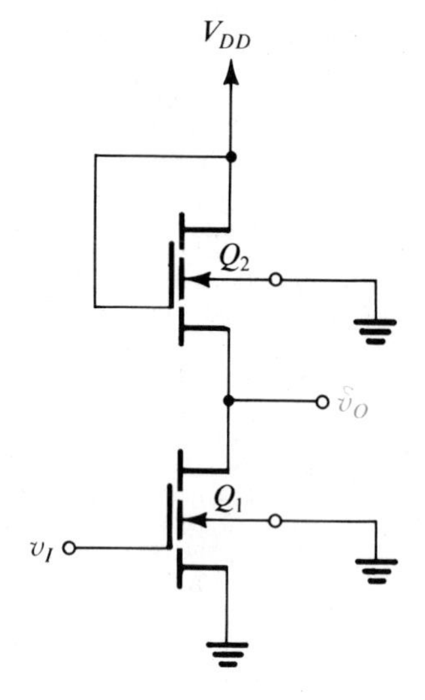

Fig. 13.10 The enhancement-load inverter with the body connections explicitly shown.

This relationship can be used together with the i–v relationship of the load transistor Q_2,

$$i_{D2} = K_2(V_{DD} - v_O - V_{t2})^2 \tag{13.10}$$

and the i_D–v_{DS} characteristics of the inverting transistor Q_1, to obtain a more accurate transfer characteristic for the enhancement-load inverter. This process, however, is quite tedious. Fortunately, such a detailed analysis is rarely necessary. It is sufficient to note that the body effect is significant only when v_O is high and results simply in a reduction in the value of V_{OH}.

EXAMPLE 13.1 Consider an enhancement-load inverter having $V_{t0} = 1$ V, $(W/L)_1 = 3$, $(W/L)_2 = \frac{1}{3}$, $\mu_n C_{OX} = 20\ \mu\text{A/V}^2$, $2\phi_f = 0.6$ V, $\gamma = 0.5\ \text{V}^{1/2}$, and $V_{DD} = 5$ V.

(a) Neglecting the body effect, find the critical points of the voltage transfer characteristic, and hence find the noise margins.

(b) Taking the body effect into account, find the modified values of V_{OH} and NM_H.

(c) Find the inverter current in both states, and hence find the average static power dissipation.

Solution

(a) Neglecting the body effect, $V_{t1} = V_{t2} = 1$ V. By reference to Fig. 13.9 we find that

$$V_{OH} = V_{DD} - V_t = 5 - 1 = 4 \text{ V}$$

The transfer characteristic in the transition region is a straight line with a slope of $-\sqrt{K_R} = -\sqrt{9} = -3$ V/V. Thus the slope changes abruptly from 0 to -3 at $v_I = V_t$. It follows that

$$V_{IL} = V_t = 1 \text{ V}$$

To find V_{OL} we assume that Q_1 is operating in the triode region and that Q_2 is operating in the saturation region, and we equate their drain currents to obtain

$$K_1[2(v_I - V_{t1})v_O - v_O^2] = K_2(V_{DD} - v_O - V_{t2})^2 \tag{13.11}$$

where we have neglected the finite MOSFET output resistance in saturation. Substituting $v_I = V_{OH} = V_{DD} - V_t$, $v_O = V_{OL}$, and $K_1 = 9K_2$ yields a quadratic equation in V_{OL} whose solution is

$$V_{OL} \simeq 0.3 \text{ V}$$

At this value of v_O, Q_1 will indeed be in the triode region, as assumed.

The value of V_{IH} is determined by assuming that Q_1 is in the triode region. Thus Eq. (13.11) applies. Differentiating both sides of Eq. (13.11) relative to v_I gives

$$K_1\left[2(v_I - V_{t1})\frac{dv_O}{dv_I} + 2v_O - 2v_O\frac{dv_O}{dv_I}\right] = -2K_2(V_{DD} - v_O - V_{t2})\frac{dv_O}{dv_I}$$

Substituting $v_I = V_{IH}$, $dv_O/dv_I = -1$, and $K_1 = K_R K_2$ results in

$$K_R[-(V_{IH} - V_{t1}) + 2v_O] = V_{DD} - v_O - V_{t2} \tag{13.12}$$

Substituting $v_I = V_{IH}$ in Eq. (13.11) and solving the resulting equation simultaneously with (13.12) results in

$$V_{IH} \simeq 2.2 \text{ V} \qquad \text{and} \qquad v_O \simeq 0.8 \text{ V}$$

From these values we can verify that Q_1 is indeed operating in the triode region.

The noise margins can now be found as

$$NM_H = V_{OH} - V_{IH} = 4 - 2.2 = 1.8 \text{ V} \qquad NM_L = V_{IL} - V_{OL} = 1 - 0.3 = 0.7 \text{ V}$$

(b) Taking the body effect into account, for V_{OH} we write

$$\begin{aligned} V_{OH} &= V_{DD} - V_{t2}\big|_{v_{SB} = v_{OH}} \\ &= V_{DD} - V_{t0} - \gamma(\sqrt{V_{OH} + 2\phi_f} - \sqrt{2\phi_f}) \end{aligned}$$

which for the given numerical values becomes

$$V_{OH} = 3.4 \text{ V}$$

Thus the body effect increases the threshold voltage of Q_2 to 1.6 V and thus decreases V_{OH} to 3.4 V and NM_H to 1.2 V.

(c) With v_O high, the current in the inverter is negligibly small, and hence the static power dissipation is negligible. For $v_O = V_{OL} = 0.3$ V, the inverter current can be determined from

$$\begin{aligned} I_{D2} &= K_2(V_{DD} - V_{OL} - V_t)^2 \\ &= \tfrac{1}{2} \times 20 \times \tfrac{1}{3} \times (5 - 0.3 - 1)^2 = 46 \ \mu\text{A} \end{aligned}$$

Thus in the low state the static power dissipation is

$$P_{DL} = 46 \times 5 = 230 \ \mu\text{W}$$

The average power dissipation of the inverter is

$$P_D = \tfrac{1}{2}(230 + 0) = 115\ \mu\text{W}$$

Dynamic Operation

We next consider the dynamic operation of the enhancement-load NMOS inverter. Figure 13.11(a) shows the inverter with a capacitive load C. It is assumed that C includes all the relevant MOSFET capacitances, the total input capacitance of all gates that are driven by the inverter under study, and the associated wiring (routing) capacitance. Lumping the various capacitive effects into a single load capacitance is, of course, an approximation, the purpose of which is to make the problem tractable for pencil-and-paper analysis. More accurate results can be obtained using more elaborate models together with a circuit analysis computer program such as SPICE (see Appendix C and, for more detail, the Spice Manual). Nevertheless, for gaining insight into circuit operation, there is no substitute for manual analysis.

Further simplification of the analysis is achieved by assuming that the input pulse is ideal, having zero rise and fall times, as shown in Fig. 13.11(b). This figure defines the propagation delay times that we wish to calculate. Figure 13.11(c) shows a sketch of i_{D2} versus v_O, which is the load curve, and a sketch of i_{D1} versus v_O for $v_I = V_{OH}$.

Before the application of the input pulse, $v_I = V_{OL}$ and $v_O = V_{OH}$. The inverter is operating at point B in Fig. 13.11(c) and the load capacitor C is charged to V_{OH}. When v_I goes high (to V_{OH}), transistor Q_1 turns on. However, since capacitor C cannot discharge instantaneously, the operating point jumps from B to D in Fig. 13.11(c). Transistor Q_1 will sink a relatively large current, thus discharging C. The operating point moves along the i_{D1} curve until it finally reaches point A, where $v_O = V_{OL}$. The propagation delay time t_{PHL} is the time for the operating point to move from D to N. The capacitor discharge current at any instant is the difference between i_{D1} and i_{D2} (see Fig. 13.11c). Although an exact solution is possible, we shall obtain an approximate estimate of t_{PHL} by finding an average value of the discharge current, denoted I_{HL}, as follows:

$$I_{HL} = \frac{i_{D1}(\text{D}) + i_{D1}(\text{N}) - i_{D2}(\text{M})}{2} \tag{13.13}$$

We can then compute t_{PHL} from

$$t_{PHL} = \frac{C\,[V_{OH} - \tfrac{1}{2}(V_{OH} + V_{OL})]}{I_{HL}} \tag{13.14}$$

As an example, for the inverter analyzed in Example 13.1 we have (neglecting the body effect) $V_{OH} = 4$ V, $V_{OL} = 0.3$ V, $\tfrac{1}{2}(V_{OH} + V_{OL}) = 2.15$ V, $i_{D1}(\text{D}) = 270\ \mu\text{A}$, $i_{D1}(\text{N}) = 250\ \mu\text{A}$ and $i_{D2}(\text{M}) = 11\ \mu\text{A}$. Thus $I_{HL} = 255\ \mu\text{A}$, and if we assume $C = 0.1$ pF (which is representative of on-chip capacitive loads) we obtain

$$t_{PHL} = \frac{0.1 \times 10^{-12} \times (4 - 2.15)}{255 \times 10^{-6}} = 0.7 \text{ ns}$$

Consider next the evaluation of t_{PLH}. As v_I goes low to V_{OL}, Q_1 turns off immediately and the circuit reduces to that in Fig. 13.11(d). Capacitor C is then charged up by the current

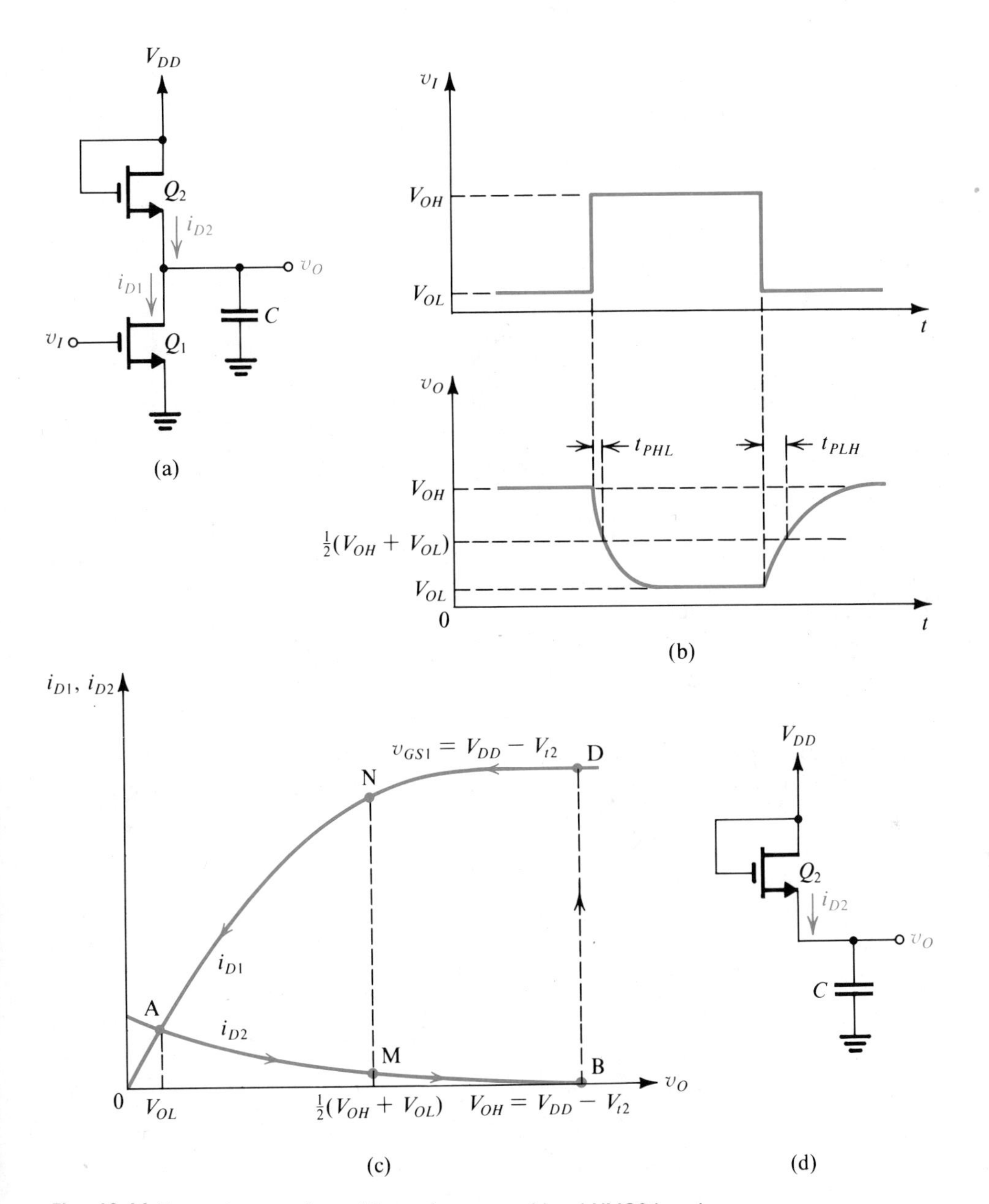

Fig. 13.11 Dynamic operation of the enhancement-load NMOS inverter.

supplied by Q_2, and the operating point moves along the load curve of Fig. 13.11(c) from A to, eventually, B. Because the available current is low, we expect t_{PLH} to be much longer than t_{PHL}. Although t_{PLH} can be found by solving the differential equation that describes the operation of the circuit in Fig. 13.11(d) (see Problem 13.18) we shall obtain an approximate value using an average charging current, computed as

$$I_{LH} = \tfrac{1}{2}[i_{D2}(\text{A}) + i_{D2}(\text{M})] \tag{13.15}$$

Thus

$$t_{PLH} = \frac{C\,[\frac{1}{2}(V_{OH} + V_{OL}) - V_{OL}]}{I_{LH}} \tag{13.16}$$

For example, for the inverter of Example 13.1 we have (neglecting the body effect)

$$i_{D2}(\text{A}) = 46\ \mu\text{A} \qquad \text{and} \qquad i_{D2}(\text{M}) = 11\ \mu\text{A}$$

Thus, $I_{LH} \simeq 29\ \mu$A and for $C = 0.1$ pF

$$t_{PLH} = \frac{0.1 \times 10^{-12} \times [\frac{1}{2}(4 + 0.3) - 0.3]}{29 \times 10^{-6}} = 6.4\ \text{ns}$$

The propagation delay of the inverter can be now computed as the average of t_{PHL} and t_{PLH}. For the inverter of Example 13.1 we obtain

$$t_P = \tfrac{1}{2}(0.7 + 6.4) \simeq 3.6\ \text{ns}$$

Here it should be noted that if the inverter is used to drive off-chip circuitry, then the load capacitance C will be two orders of magnitude larger than the value used in the above example. The propagation delay will increase proportionately and become hundreds of nanoseconds. This explains why enhancement-load NMOS is not used in SSI and MSI logic design.

Delay-Power Product

The delay-power product (DP) for enhancement-load NMOS can be found by multiplying the propagation delay by the static power dissipation. For the inverter in Example 13.1,

$$DP = 3.6 \times 10^{-9} \times 115 \times 10^{-6} = 0.4\ \text{pJ}$$

If we use capacitance values representative of those encountered in conventional discrete component design we obtain DP of the order of 10 to 100 pJ.

An approximate but useful expression for DP can be derived as follows: Assuming that $V_{OL} \simeq 0$ we can show that I_{LH} of Eq. (13.15) is approximately

$$I_{LH} \simeq \tfrac{5}{8}K_2(V_{DD} - V_t)^2 \tag{13.17}$$

Using this expression in Eq. (13.16) gives

$$t_{PLH} \simeq \frac{0.8\ C}{K_2(V_{DD} - V_t)} \tag{13.18}$$

Now, since $t_{PHL} \ll t_{PLH}$, we may evaluate t_P as

$$t_P \simeq \tfrac{1}{2}t_{PLH} = \frac{0.4C}{K_2(V_{DD} - V_t)} \tag{13.19}$$

For $V_{OL} \simeq 0$, the average static power dissipation is approximately

$$P_D \simeq \tfrac{1}{2}K_2(V_{DD} - V_t)^2 V_{DD} \tag{13.20}$$

Equations (13.19) and (13.20) can be combined to obtain

$$DP \simeq 0.2CV_{DD}(V_{DD} - V_t) \tag{13.21}$$

This expression shows that DP can be reduced by reducing either C or V_{DD}. Reducing V_{DD}, however, reduces the signal swing and the noise margins.

Exercises 13.6 For the enhancement-load inverter of Fig. 13.10, find K_R that results in $V_{OL} = 0.1$ V. Let $V_{t1} = V_{t2} = 1$ V and $V_{DD} = 5$ V, and neglect the body effect.

Ans. 26

13.7 Use the approximate expressions in Eqs. (13.19) and (13.21) to determine t_P and DP for the inverter of Example 13.1. Let $C = 0.1$ pF.

Ans. 3 ns; 0.4 pJ

13.8 Use the expression in Eq. (13.21) to determine DP for an enhancement-load NMOS inverter operated from $V_{DD} = 5$ V and having $V_t = 1$ V. Let $C = 10$ pF.

Ans. 40 pJ

13.3 NMOS INVERTER WITH DEPLETION LOAD

The use of a depletion MOSFET as the load element results in an inverter with higher gain having a sharper voltage transfer characteristic, and hence increased noise margins. Furthermore, the improved noise margins can be obtained while using a smaller geometry ratio, K_R, and hence smaller silicon area than required for the enhancement-load inverter. The price paid for the performance improvement is an extra fabrication step required to implant the channel of the depletion device.

Figure 13.12(a) shows the depletion-load inverter. We have studied the amplifier application of this circuit in Section 5.9. However, in deriving the voltage transfer characteristic in Section 5.9, we did not emphasize the body effect, which plays a significant role in the operation of the circuit as a logic inverter.[2] To be specific, we show in Fig. 13.12(b) plots of i_{D2} (normalized relative to K_2) versus v_O without (curve a) and with (curve b) the body effect taken into account. These are the curves representing the load of the inverting transistor Q_1. We note that without the body effect the depletion device behaves as a constant-current source over a wide range of v_O. This implies that the inverter transfer characteristic will be very sharp in the transition region. It also implies that a relatively large current is available to charge the load capacitance, resulting in a short t_{PLH}. Unfortunately, however, the body effect causes the depletion load to depart from constant-current operation. Thus, the inverter characteristics, though superior to those obtained with enhancement load, are not as good as was thought to be possible in the early stages of the development of this technology.

Analytically, the current in the depletion load is given by

$$i_{D2} = K_2 |V_{tD}|^2 \qquad \text{for } v_O \leq V_{DD} - |V_{tD}| \tag{13.22}$$

and

$$i_{D2} = K_2[2|V_{tD}|(V_{DD} - v_O) - (V_{DD} - v_O)^2] \qquad \text{for } v_O \geq V_{DD} - |V_{tD}| \tag{13.23}$$

[2] In Section 5.9 we did take the body effect into account, however, in determining the small-signal voltage gain of the depletion-load amplifier.

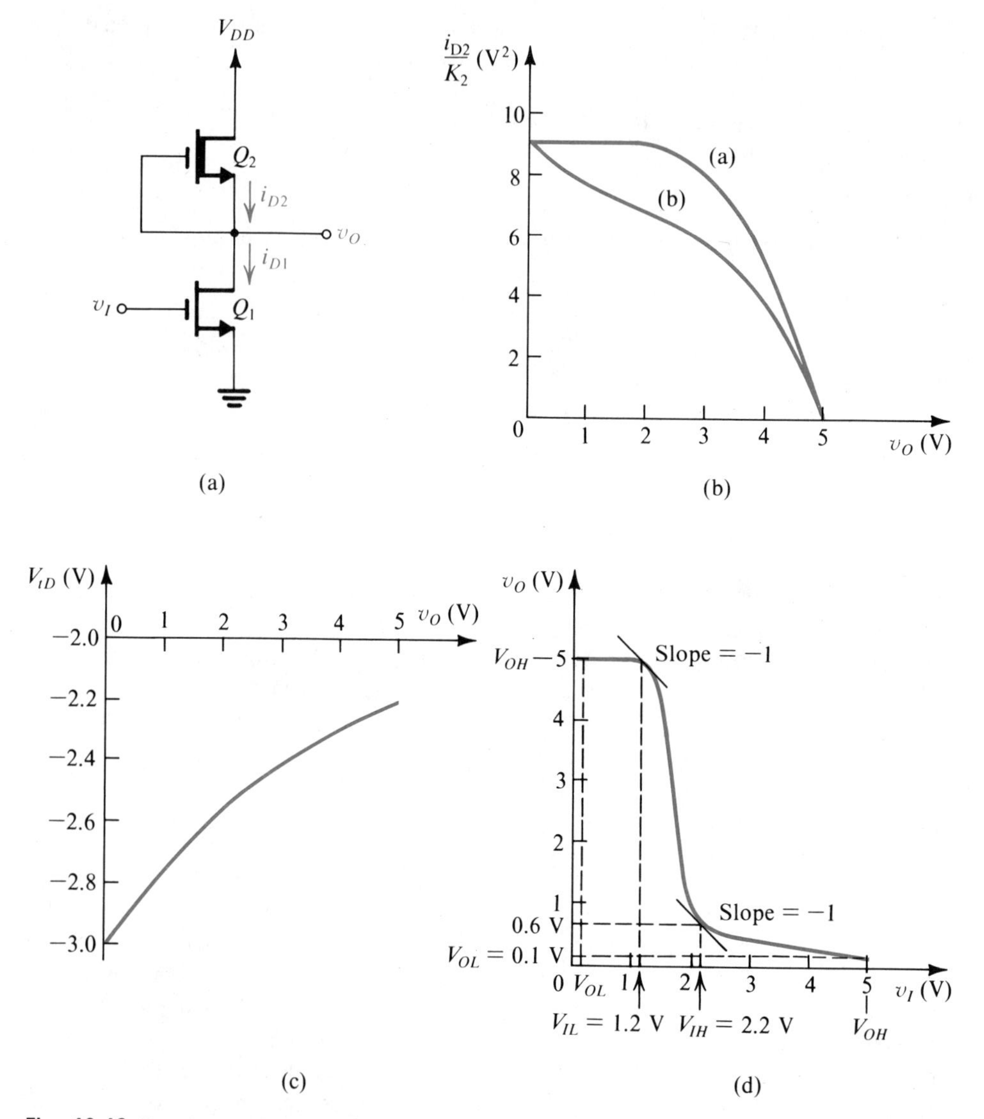

Fig. 13.12 The depletion-load inverter and its characteristics.

where V_{tD} is the threshold voltage of the depletion device. For $V_{SB}|_{Q2} = 0$, $V_{tD} = V_{tD0}$, which is typically -3 V. However, $V_{SB}|_{Q2} = v_O$ and V_{tD} is given by

$$V_{tD} = V_{tD0} + \gamma[\sqrt{v_O + 2\phi_f} - \sqrt{2\phi_f}] \tag{13.24}$$

Figure 13.12(c) shows a plot of V_{tD} versus v_O for $V_{tD0} = -3$ V, $\gamma = 0.5$ V$^{1/2}$, and $2\phi_f =$ 0.6 V. The plots in Fig. 13.12(b) are obtained using Eqs. (13.22)–(13.24) with the power supply $V_{DD} = 5$ V.

Static Characteristics

The characteristic curve of the depletion load (curve b in Fig. 13.12b) can be used in conjunction with the i_D–v_{DS} characteristics of the enhancement inverting transistor Q_1 to obtain the inverter transfer characteristic shown in Fig. 13.12(d). Observe that $V_{OH} = V_{DD}$, which is 1 to 1.5 V higher than the logic-1 level of the enhancement-load inverter. The following example illustrates the calculation of the critical points of the voltage transfer characteristic.

EXAMPLE 13.2

Consider a depletion-load inverter with $V_{tE} = 1$ V, $K_R = 9$, $V_{tD0} = -3$ V, $\mu_n C_{OX} = 20\ \mu A/V^2$, $2\phi_f = 0.6$ V, $\gamma = 0.5\ V^{1/2}$, and $V_{DD} = 5$ V. (Note that since this inverter has the same geometry ratio as the enhancement-load inverter of Example 13.1, the performance results can be easily compared.)

(a) Find the critical points of the voltage transfer characteristic, taking into account, where necessary, the body effect. Hence calculate the noise margins.

(b) Find $(W/L)_2$ so that the inverter current in the low-output state equals the corresponding value in the enhancement-load inverter of Example 13.1. Also find $(W/L)_1$.

(c) Find the average static power dissipation.

Solution

(a) $V_{OH} = V_{DD} = 5$ V. From Fig. 13.12(c) (or Eq. 13.24) we find that at $v_O = 5$ V, $V_{tD} = -2.2$ V. Thus Q_2 with $v_{GS} = 0$ can conduct a load current, if required. Of course, because Q_1 is off, the current in the inverter is zero.

To find V_{OL}, we let $v_I = V_{OH} = 5$ V and assume Q_1 to be in the triode region and Q_2 to be in the saturation region. Since we expect V_{OL} to be small, we may neglect the body effect and assume that $V_{tD} \simeq -3$ V. Thus

$$i_{D1} = K_1[2(5-1)V_{OL} - V_{OL}^2]$$

$$i_{D2} = K_2|V_{tD}|^2 = 9K_2$$

Equating i_{D1} and i_{D2}, substituting $K_1/K_2 = 9$, and solving the resulting quadratic equation, we get

$$V_{OL} \simeq 0.1 \text{ V}$$

To determine V_{IH} we assume that Q_1 is in the triode region while Q_2 is in saturation and equate their drain currents to obtain

$$K_1[2(v_I - 1)v_O - v_O^2] = 9K_2$$

where we have neglected the body effect. Substituting $K_2 = K_1/9$ gives

$$2(v_I - 1)v_O - v_O^2 = 1 \qquad (13.25)$$

Differentiating with respect to v_I yields

$$2(v_I - 1)\frac{dv_O}{dv_I} + 2v_O - 2v_O\frac{dv_O}{dv_I} = 0$$

Substituting $v_I = V_{IH}$ and $dv_O/dv_I = -1$ results in

$$v_O = \tfrac{1}{2}(V_{IH} - 1) \tag{13.26}$$

Substituting $v_I = V_{IH}$ in Eq. (13.25) and replacing v_O with the value given in Eq. (13.26) results in a quadratic equation in V_{IH} whose solution is $V_{IH} = 2.2$ V and from Eq. (13.26) the corresponding value of $v_O = 0.6$ V. Note that Q_1 is indeed in the triode region and Q_2 is in the saturation region, as assumed. Also note that the values obtained are approximate since we have assumed the body effect to be negligible.

To determine V_{IL} we assume that Q_1 is in saturation and Q_2 in the triode region and equate their currents. Since we expect v_O to be close to 5 V, we use $V_{tD} \simeq -2.2$ V (see Fig. 13.12c). The resulting equation is

$$(v_I - 1)^2 = (0.49)(5 - v_O) - \tfrac{1}{9}(5 - v_O)^2 \tag{13.27}$$

Differentiating relative to v_I, we obtain

$$2(v_I - 1) = -0.49\frac{dv_O}{dv_I} + \frac{2}{9}(5 - v_O)\frac{dv_O}{dv_I}$$

Substituting $v_I = V_{IL}$ and $dv_O/dv_I = -1$ gives

$$v_O = 9(V_{IL} - 0.69) \tag{13.28}$$

Substituting $v_I = V_{IL}$ in Eq. (13.27) and for v_O from Eq. (13.28) results in a quadratic equation whose solution yields V_{IL} and the corresponding v_O as

$$V_{IL} = 1.2 \text{ V} \quad v_O = 4.8 \text{ V}$$

The noise margins can now be calculated as follows:

$$NM_L = V_{IL} - V_{OL} = 1.2 - 0.1 = 1.1 \text{ V}$$

$$NM_H = V_{OH} - V_{IH} = 5 - 2.2 = 2.8 \text{ V}$$

Both of these values are larger than the corresponding values obtained for the enhancement-load inverter in Example 13.1 (0.7 V and 1.8 V, respectively).

(b) The enhancement-load inverter of Example 13.1 has at $v_O = V_{OL}$ a current of 46 μA. To establish an equal current in the depletion-load inverter we need to choose K_2 so that

$$K_2|V_{tD}|^2 = 46 \ \mu\text{A}$$

Thus,

$$K_2 = \tfrac{46}{9} = 5.1 \ \mu\text{A/V}^2$$

$$\tfrac{1}{2}(\mu_n C_{OX})\left(\frac{W}{L}\right)_2 = 5.1$$

which leads to

$$\left(\frac{W}{L}\right)_2 \simeq 0.5$$

With $K_R = 9$,

$$\left(\frac{W}{L}\right)_1 = 4.5$$

(c) Since the inverter is designed to have a current equal to that of the enhancement-load inverter of Example 13.1, the average static power dissipation will be the same; that is,

$$P_D = 115\ \mu\text{W}$$

The results of the above example indicate that a depletion-load inverter has markedly increased noise margins as compared to an enhancement-load inverter with the same geometry ratio, K_R. This implies that K_R of the depletion-load inverter can be reduced, thus reducing device sizes and silicon area while still maintaining good noise margins.

Exercise 13.9 If K_R of the inverter in Example 13.2 is reduced to 4 by reducing $(W/L)_1$ from 4.5 to 2, calculate the new values of V_{OH}, V_{OL}, V_{IH}, V_{IL}, NM_H, and NM_L.

Ans. 5 V; 0.3 V; 2.7 V; 1.5 V; 2.3 V; 1.2 V

Dynamic Operation

Figure 13.13 illustrates the dynamic operation of the depletion-load inverter in the presence of a load capacitance C. The propagation delay times t_{PLH} and t_{PHL} can be calculated by finding the average currents available to charge and discharge C, in the same manner used in the enhancement-load case. We shall not show the details of this analysis (see Exercise 13.10) but point out that the depletion load provides a relatively higher current

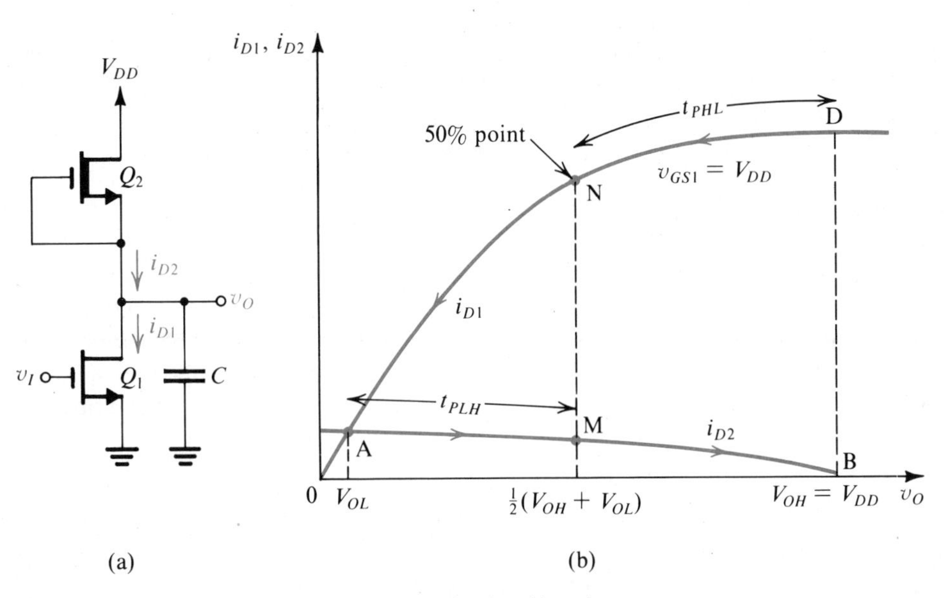

Fig. 13.13 Dynamic operation of the depletion-load inverter.

over a wider range of v_O than does the enhancement load. This results in a somewhat faster charging of the load capacitance, and correspondingly a slightly[3] shorter t_{PLH}.

Exercise 13.10 Consider the depletion-load inverter of Exercise 13.9 with $(W/L)_1 = 2$, $(W/L)_2 = 0.5$, and all the other parameters as specified in Example 13.2. Find the average values of the currents I_{HL} and I_{LH}. Also, for $C = 0.1$ pF calculate t_{PHL}, t_{PLH}, and t_P.

Ans. 300 μA; 38 μA; 0.8 ns; 6.2 ns; 3.5 ns

Delay-Power Product

An approximate expression can be derived for the delay-power product DP in a manner similar to that used for the enhancement-load inverter. The result (see Problem 13.29) is as follows:

$$DP \simeq \frac{1}{8\alpha} C V_{DD}^2 \tag{13.29}$$

where α is a fraction smaller than, but close to, 1, accounting for the variation of V_{tD} with v_O. This expression yields values somewhat smaller than those for enhancement-load circuits.

Exercise 13.11 For the depletion-load inverter of Exercises 13.9 and 13.10 let $\alpha \simeq 0.9$. Find the delay-power product first by multiplying P_D by t_P and then using the approximate expression in Eq. (13.29).

Ans. 0.4 pJ; 0.35 pJ

A Final Remark

Although it occupies a smaller chip area, the depletion-load NMOS inverter exhibits higher noise margins and a somewhat higher speed of operation than the enhancement-load inverter. For this reason, virtually all modern NMOS logic and memory circuits utilize the depletion-load technology.

13.4 NMOS LOGIC CIRCUITS

Figure 13.14(a) shows a two-input NOR gate in depletion-load NMOS technology. The logic gate operates as follows: If the voltage at any of the input terminals is high (at V_{DD}), then the corresponding transistor will be on and the output voltage will be low (at V_{OL}). The output voltage will be high only if the two inputs are simultaneously low. In this case both input transistors will be off and $v_Y = V_{DD}$. Thus the operation can be described by the Boolean expression

$$Y = \overline{A}\,\overline{B}$$

or, equivalently,

$$Y = \overline{A + B}$$

[3] The reduction in t_{PLH} is not as great because the voltage swing is higher in the depletion-load inverter.

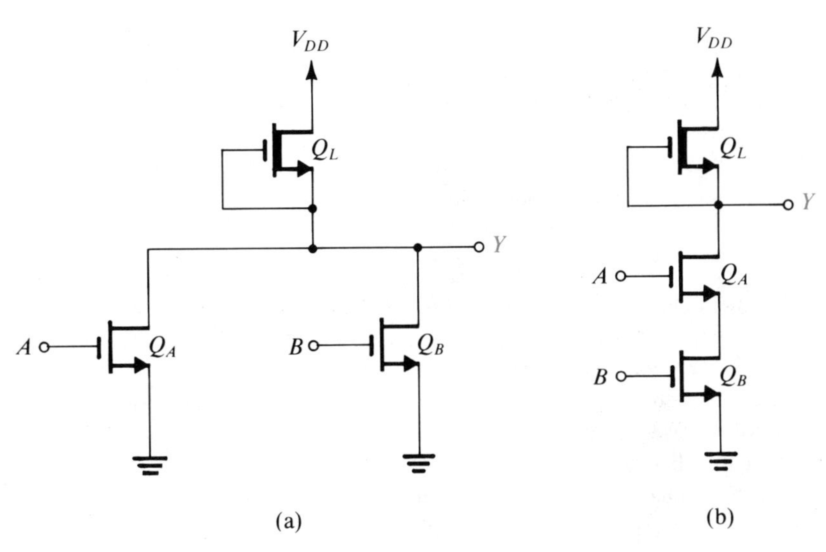

Fig. 13.14 (a) NOR gate and **(b)** NAND gate in depletion-load NMOS logic.

The input transistors are identical and have the same dimensions as the inverter transistor. Thus if either input is high, the output will be at V_{OL}. If both inputs are simultaneously high, Q_A and Q_B can be replaced by an equivalent transistor having the same length but twice the width of each of Q_A and Q_B, and analysis shows the output voltage to be lower than V_{OL}. The gate fan-in can be increased by adding additional input transistors.

Exercise 13.12 For the NOR gate in Fig. 13.14(a) let Q_A and Q_B be identical with $W = 12$ μm and $L = 6$ μm. The load transistor Q_L has $W = 6$ μm and $L = 12$ μm. Also let $V_{tE} = 1$ V, $V_{tD0} = -3$ V and $V_{DD} = 5$ V. Find the output voltage when **(a)** one input is high and **(b)** both inputs are high.

Ans. 0.29 V; 0.14 V

Whereas the NOR gate is formed by connecting input transistors in parallel, the NAND function is obtained by placing the input transistors in series. A two-input NAND gate is shown in Fig. 13.14(b). Here the output will be low only when both Q_A and Q_B are on, a situation obtained when both inputs are simultaneously high. Thus

$$\overline{Y} = AB$$

or

$$Y = \overline{AB} = \overline{A} + \overline{B}$$

When both Q_A and Q_B are conducting, the effective channel length between the output node and ground is double that of the inverter transistor. It follows that to keep the output voltage at the value of V_{OL} obtained in the inverter, each of the input transistors in the NAND gate should have double the width of the inverter transistor. In this way, the two series conducting transistors will have the same effective W/L ratio as the single inverter transistor. Of course, if we have N inputs, the width of each of the input transistors should be N times that of the inverter transistor. As a result, the silicon area required by a NAND gate is greater than that required by a NOR gate having the same number of inputs. This limits the application of the NAND implementation.

Exercise 13.13 If the optimum dimensions of a depletion-load NMOS inverter are $W/L = 12\ \mu m/6\ \mu m$ for the inverting transistor and $6\ \mu m/12\ \mu m$ for the load transistor, find the dimensions of the input transistors in a 3-input NAND gate.

Ans. $36\ \mu m/6\ \mu m$

As a final remark we note that NMOS logic circuits are simple to fabricate and require small chip area, and thus can be very densely packed on an IC chip. This permits very high levels of integration. Indeed, the most important application of NMOS is in the design of VLSI circuits such as microprocessors and random-access memory. In such applications, the low load-driving capability of NMOS is not a serious drawback. It is this low load-driving capability, however, that makes NMOS impractical for conventional digital system design. Thus NMOS logic is not available as SSI or MSI ''off-the-shelf'' components as is the case, for example, with CMOS or TTL.

13.5 THE CMOS INVERTER

Complementary MOS, or CMOS, is currently the most popular digital circuit technology. CMOS logic circuits are available as standard SSI and MSI packages for use in conventional digital system design. CMOS is also used in the design of general-purpose VLSI circuits such as memory and microprocessors. For custom and semicustom VLSI, CMOS is the technology of choice. Furthermore, CMOS is currently popular in analog-circuit applications as well (see Chapters 5, 6, and 10).

In this section we study the basic CMOS inverter, shown in Fig. 13.15(a). The inverter utilizes two matched enhancement-type MOSFETs: one, Q_N, with an n channel and the other, Q_P, with a p channel. As indicated, the body of each device is connected to

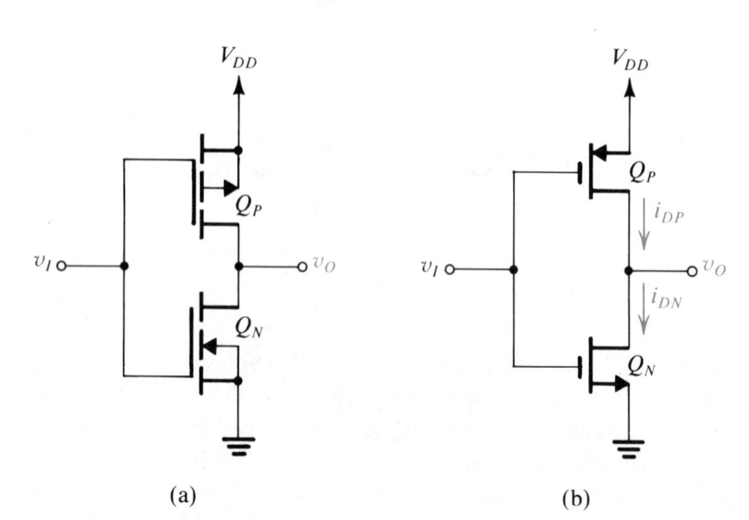

Fig. 13.15 (a) CMOS inverter. **(b)** Simplified circuit schematic for the inverter.

its source and thus no body effect arises. We shall therefore use the simplified circuit schematic diagram shown in Fig. 13.15(b).

Circuit Operation

We first consider the two extreme cases: when v_I is at logic-0 level, which is approximately 0 V, and when v_I is at logic-1 level, which is approximately V_{DD} volts. In both cases, we shall consider the n-channel device Q_N to be the driving transistor and the p-channel device Q_P to be the load. However, since the circuit is completely symmetric, this assumption is obviously arbitrary, and the reverse would lead to identical results.

Figure 13.16 illustrates the case when $v_I = V_{DD}$, showing the i_D–v_{DS} characteristic curve for Q_N with $v_{GSN} = V_{DD}$. (Note that $i_D = i$ and $v_{DSN} = v_O$). Superimposed on the Q_N characteristic curve is the load curve, which is the i_D–v_{SD} curve of Q_P for the case $v_{SGP} = 0$ V. Since $v_{SGP} < |V_t|$, the load curve will be a horizontal straight line at almost zero current level. The operating point will be at the intersection of the two curves, where we note that the output voltage is nearly zero (typically less than 10 mV) and the current through the two devices is also nearly zero. This means that the power dissipation in the circuit is very small (typically a fraction of a microwatt).

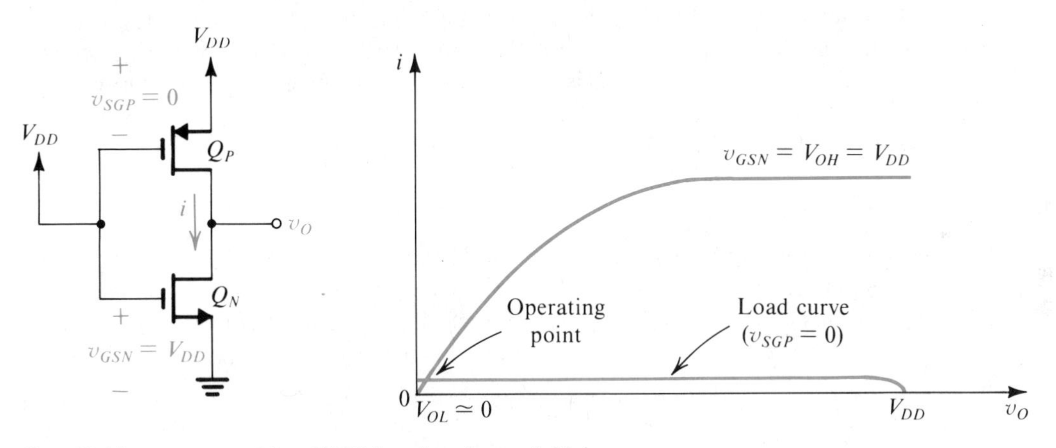

Fig. 13.16 Operation of the CMOS inverter when v_I is high.

The other extreme case, when $v_I = 0$ V, is illustrated in Fig. 13.17. In this case Q_N is operating at $v_{GSN} = 0$; hence its i_D–v_{DS} characteristic is almost a horizontal straight line at zero current level. The load curve is the i_D–v_{SD} characteristic of the p-channel device with $v_{SGP} = V_{DD}$. As shown, at the operating point the output voltage is almost equal to V_{DD} (typically less than 10 mV below V_{DD}), and the current in the two devices is still nearly zero. Thus the power dissipation in the circuit is very small in both extreme states.

From the above we conclude that the basic CMOS logic inverter behaves as an "ideal" logic element: The output voltage is almost equal to zero volts or V_{DD} volts, and the power dissipation is almost zero.

It should be noted, however, that in spite of the fact that the quiescent current is zero, the load-driving capability of the CMOS inverter is high. For instance, with the input

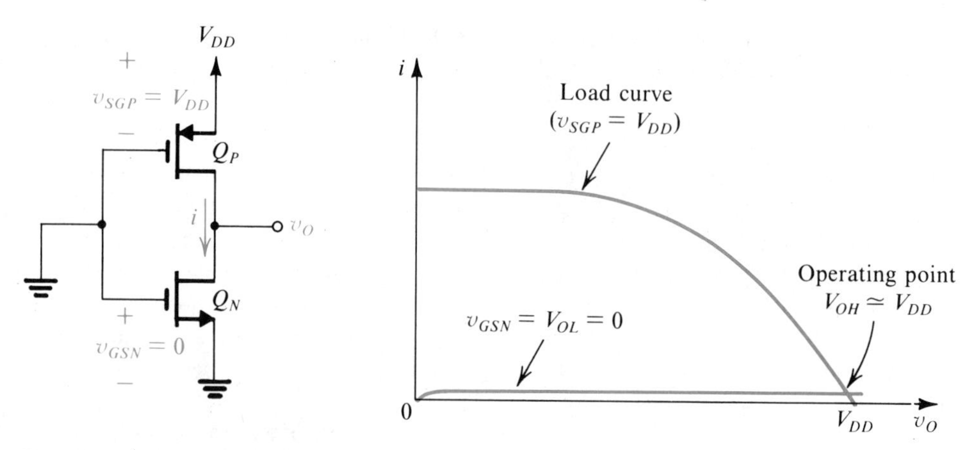

Fig. 13.17 Operation of the CMOS inverter when v_I is low.

high, as in the circuit of Fig. 13.16, transistor Q_N can sink a relatively large load current. This current can quickly discharge the load capacitance, as will be seen shortly. Because of its action in sinking load current and thus pulling the output voltage down toward ground, transistor Q_N is known as the "pull-down" device. Similarly, with the input low, as in the circuit of Fig. 13.17, transistor Q_P can source a relatively large load current. This current can quickly charge up a load capacitance, thus pulling the output voltage up toward V_{DD}. Hence, Q_P is known as the "pull-up" device.

The Voltage Transfer Characteristic

The complete voltage transfer characteristic of the CMOS inverter can be obtained by repeating the graphical procedure, used above in the two extreme cases, for all intermediate value of v_I. In the following, we shall calculate the critical points of the resulting voltage transfer curve. For this we need the i–v relationships of Q_N and Q_P. For Q_N,

$$i_{DN} = K_n[2(v_I - V_{tn})v_O - v_O^2] \qquad \text{for } v_O \leq v_I - V_{tn} \tag{13.30}$$

and

$$i_{DN} = K_n(v_I - V_{tn})^2 \qquad \text{for } v_O \geq v_I - V_{tn} \tag{13.31}$$

For Q_P,

$$i_{DP} = K_p[2(V_{DD} - v_I - |V_{tp}|)(V_{DD} - v_O) - (V_{DD} - v_O)^2] \qquad \text{for } v_O \geq v_I + |V_{tp}| \tag{13.32}$$

and

$$i_{DP} = K_p(V_{DD} - v_I - |V_{tp}|)^2 \qquad \text{for } v_O \leq v_I + |V_{tp}| \tag{13.33}$$

The CMOS inverter is usually designed to have $V_{tn} = |V_{tp}| = V_t$ and $K_n = K_p = K$. It should be noted that since μ_p is about half the value of μ_n, to make $K_p = K_n$ the width of the p-channel device is made about twice that of the n-channel device. With $K_n = K_p$, the inverter has equal current-driving capability in both directions (pull-up and pull-down).

With Q_N and Q_P matched, the CMOS inverter has the voltage transfer characteristic shown in Fig. 13.18. As indicated, the transfer characteristic has five distinct segments corresponding to the different modes of operation of Q_N and Q_P. The vertical segment BC is obtained when both Q_N and Q_P are operating in the saturation region. Because we are neglecting the finite output resistance in saturation, the inverter gain in this region is infinite. From symmetry, this vertical segment occurs at $v_I = V_{DD}/2$ and is bounded by $v_O(\mathrm{B}) = V_{DD}/2 + V_t$ and $v_O(\mathrm{C}) = V_{DD}/2 - V_t$.

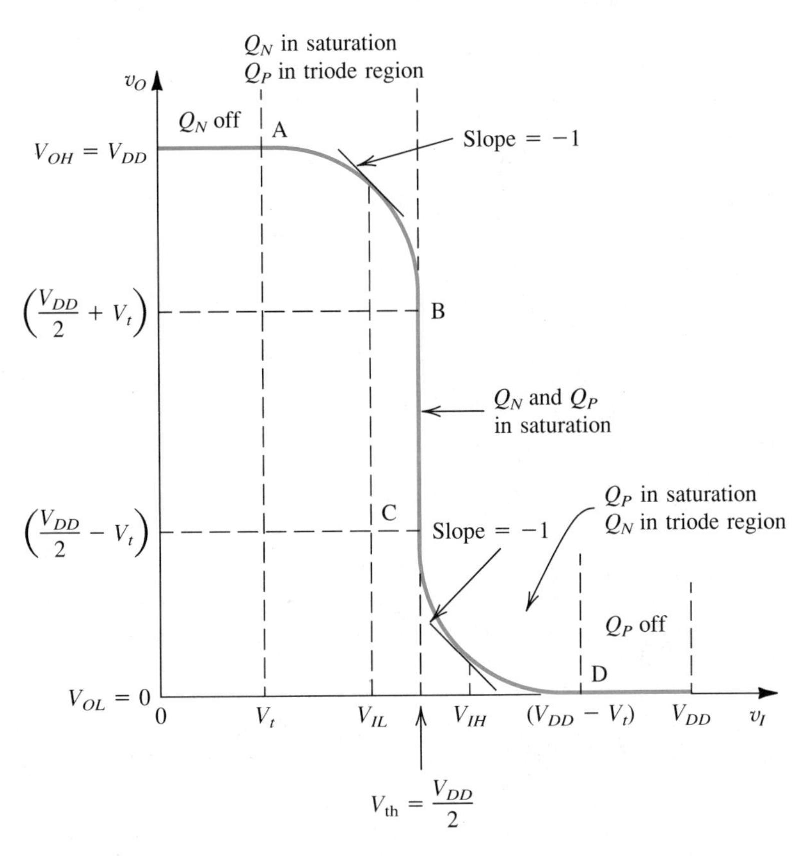

Fig. 13.18 The voltage transfer characteristic of the CMOS inverter.

To determine V_{IH} we note that Q_N is in the triode region, and thus its current is given by Eq. (13.30), while Q_P is in saturation and its current is given by Eq. (13.33). Equating i_{DN} and i_{DP}, and assuming matched devices, gives

$$2(v_I - V_t)v_O - v_O^2 = (V_{DD} - v_I - V_t)^2 \qquad (13.34)$$

Differentiating both sides relative to v_I results in

$$2(v_I - V_t)\frac{dv_O}{dv_I} + 2v_O - 2v_O\frac{dv_O}{dv_I} = -2(V_{DD} - v_I - V_t)$$

in which we substitute $v_I = V_{IH}$ and $dv_O/dv_I = -1$ to obtain

$$v_O = V_{IH} - \frac{V_{DD}}{2} \tag{13.35}$$

Substituting in Eq. (13.34) $v_I = V_{IH}$ and for v_O from Eq. (13.35) gives

$$V_{IH} = \tfrac{1}{8}(5V_{DD} - 2V_t) \tag{13.36}$$

V_{IL} can be determined in a manner similar to that used to find V_{IH}. Alternatively, we can use the symmetry relationship

$$V_{IH} - \frac{V_{DD}}{2} = \frac{V_{DD}}{2} - V_{IL}$$

together with V_{IH} from Eq. (13.36) to obtain

$$V_{IL} = \tfrac{1}{8}(3V_{DD} + 2V_t) \tag{13.37}$$

The noise margins can now be determined as follows:

$$\begin{aligned} NM_H &= V_{OH} - V_{IH} \\ &= V_{DD} - \tfrac{1}{8}(5V_{DD} - 2V_t) \\ &= \tfrac{1}{8}(3V_{DD} + 2V_t) \end{aligned} \tag{13.38}$$

$$\begin{aligned} NM_L &= V_{IL} - V_{OL} \\ &= \tfrac{1}{8}(3V_{DD} + 2V_t) - 0 \\ &= \tfrac{1}{8}(3V_{DD} + 2V_t) \end{aligned} \tag{13.39}$$

As expected, the symmetry of the voltage transfer characteristic results in equal noise margins. Of course, if Q_N and Q_P are not matched, the voltage transfer characteristic will no longer be symmetric and the noise margins will not be equal (see Problems 13.38 and 13.39).

Exercises 13.14 For a CMOS inverter with matched MOSFETs having $V_t = 1$ V find V_{IL}, V_{IH}, and the noise margins if $V_{DD} = 5$ V.

Ans. 2.1 V; 2.9 V; 2.1 V

13.15 Consider a CMOS inverter with $V_{tn} = |V_{tp}| = 2$ V, $(W/L)_N = 20$, $(W/L)_P = 40$, $\mu_n C_{OX} = 2\mu_p C_{OX} = 20\mu\text{A/V}^2$, and $V_{DD} = 10$ V. For $v_I = V_{DD}$ find the maximum current that the inverter can sink while v_O remains ≤ 0.5 V.

Ans. 1.55 mA

Current Flow and Power Dissipation

As the CMOS inverter is switched, current flows through the series connection of Q_N and Q_P. Figure 13.19 shows the inverter current as a function of v_I. We note that the current peaks at the switching point, $v_I = V_{DD}/2$. This current gives rise to dynamic power dissipation in the CMOS inverter. However, a more significant component of dynamic power

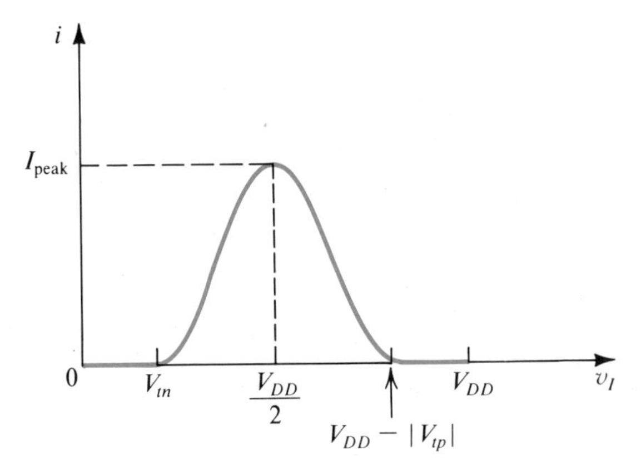

Fig. 13.19 The current in the CMOS inverter versus the input voltage.

dissipation results from the current that flows in Q_N and Q_P when the inverter is loaded by a capacitor C. This latter component of power dissipation, as given by Eq. (13.3), is

$$P_D = fCV_{DD}^2$$

where f is the frequency at which the inverter is switched.

Exercises 13.16 For the inverter specified in Exercise 13.15, find the peak current drawn from V_{DD} during switching.

Ans. 1.8 mA

13.17 Let the inverter specified in Exercise 13.15 be loaded by a 15-pF capacitance. Find the dynamic power dissipation that results when the inverter is switched at a frequency of 2 MHz. What is the average current drawn from the power supply?

Ans. 3 mW; 0.3 mA

Dynamic Operation

Unlike the NMOS inverters, the CMOS inverter features equally fast turn-on and turn-off times in the presence of capacitive loads. To illustrate, consider the capacitively loaded CMOS inverter of Fig. 13.20(a) driven by the ideal pulse (zero rise and fall times) shown in Fig. 13.20(b). Since the circuit is symmetric (assuming matched MOSFETs), the rise and fall times of the output waveform should be equal. It is sufficient, therefore, to consider either the turn-on or the turn-off process. In the following we consider the first.

Figure 13.20(c) shows the trajectory of the operating point obtained when the input pulse goes from $V_{OL} = 0$ to $V_{OH} = V_{DD}$ at time $t = 0$. Just prior to the leading edge of the input pulse (that is, at $t = 0-$) the output voltage equals V_{DD} and capacitor C is charged to this voltage. At $t = 0$, v_I rises to V_{DD}, causing Q_P to turn off immediately. From there on, the circuit is equivalent to that shown in Fig. 13.20(d) with the initial value of $v_O = V_{DD}$. Thus the operating point at $t = 0+$ is point E, at which it is seen that Q_N will be in the saturation region and conducting a large current. As C discharges, the current of Q_N remains constant until $v_O = V_{DD} - V_t$ (point F). Denoting this portion of the discharge

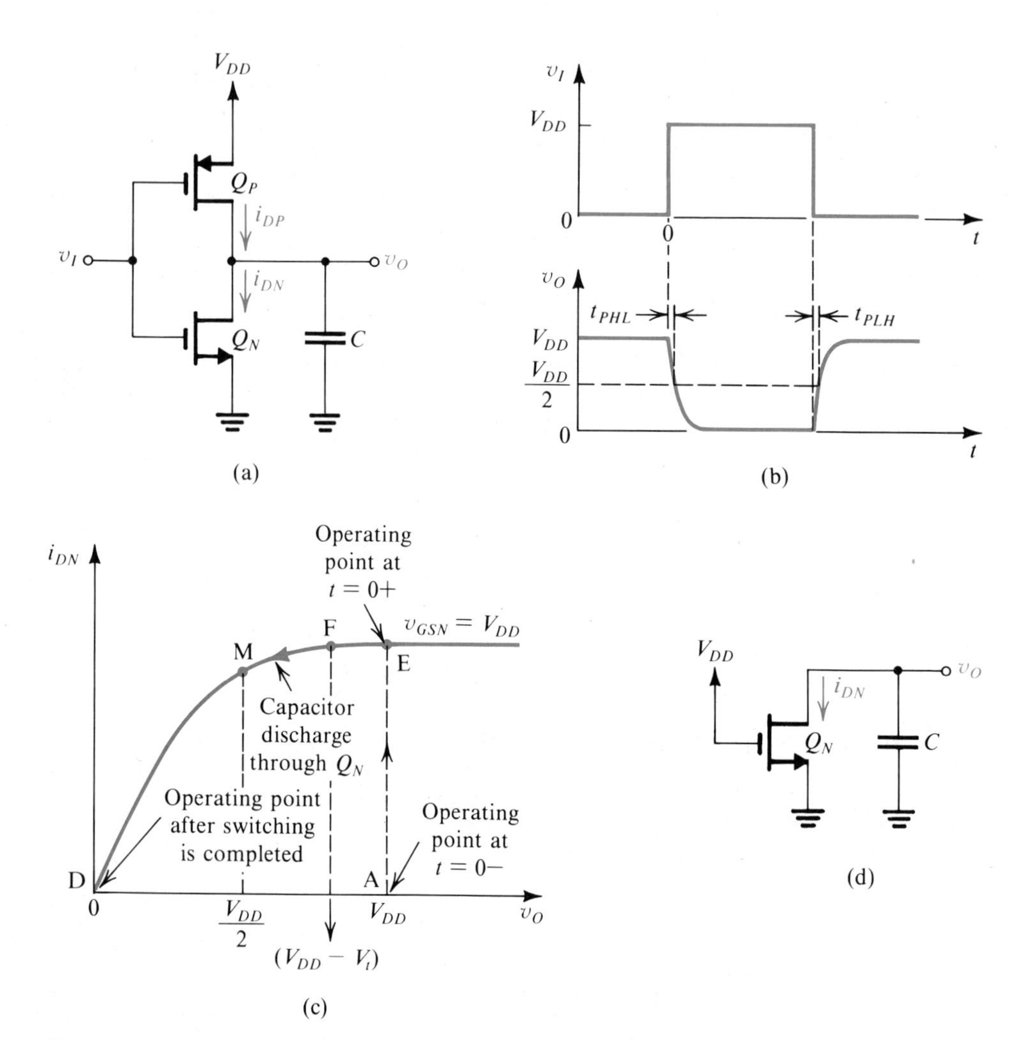

Fig. 13.20 Dynamic operation of a capacitively loaded CMOS inverter: **(a)** circuit; **(b)** input and output waveforms; **(c)** trajectory of the operating point as the input goes high and C discharges through Q_N; **(d)** equivalent circuit during the capacitor discharge.

interval t_{PHL1} we can write

$$t_{PHL1} = \frac{C\,[V_{DD} - (V_{DD} - V_t)]}{K_n(V_{DD} - V_t)^2}$$

$$= \frac{CV_t}{K_n(V_{DD} - V_t)^2} \tag{13.40}$$

Beyond point F, transistor Q_N operates in the triode region and thus its current is given by Eq. (13.30). This portion of the discharge interval can be described by

$$i_{DN}\,dt = -C\,dv_O$$

Substituting for i_{DN} from Eq. (13.30) and rearranging the differential equation, we obtain

$$-\frac{K_n}{C}dt = \frac{1}{2(V_{DD}-V_t)} \frac{dv_O}{\frac{1}{2(V_{DD}-V_t)}v_O^2 - v_O} \tag{13.41}$$

To find the component of the delay time t_{PHL} during which v_O decreases from $(V_{DD} - V_t)$ to the 50% point, $v_O = V_{DD}/2$, we integrate both sides of Eq. (13.41). Denoting this component of delay time t_{PHL2} we find that

$$-\frac{K_n}{C}t_{PHL2} = \frac{1}{2(V_{DD}-V_t)} \int_{v_O=V_{DD}-V_t}^{v_O=V_{DD}/2} \frac{dv_O}{\frac{1}{2(V_{DD}-V_t)}v_O^2 - v_O} \tag{13.42}$$

Using the fact that

$$\int \frac{dx}{ax^2 - x} = \ln\left(1 - \frac{1}{ax}\right)$$

enables us to evaluate the integral in Eq. (13.42) and thus obtain

$$t_{PHL2} = \frac{C}{2K_n(V_{DD}-V_t)} \ln\left(\frac{3V_{DD}-4V_t}{V_{DD}}\right) \tag{13.43}$$

The two components of t_{PHL} in Eqs. (13.40) and (13.43) can be added to obtain

$$t_{PHL} = \frac{C}{K_n(V_{DD}-V_t)}\left[\frac{V_t}{V_{DD}-V_t} + \frac{1}{2}\ln\left(\frac{3V_{DD}-4V_t}{V_{DD}}\right)\right] \tag{13.44}$$

For the usual case of $V_t \simeq 0.2V_{DD}$ this equation reduces to

$$t_{PHL} = \frac{0.8C}{K_n V_{DD}} \tag{13.45}$$

Similar analysis of the turn-off process yields an expression for t_{PLH} identical to that in Eq. (13.44) except for K_p replacing K_n.

Exercises 13.18 A CMOS inverter in a VLSI circuit operating from a 5-V supply has $(W/L)_n = 10\ \mu m/5\ \mu m$, $(W/L)_p = 20\ \mu m/5\ \mu m$, $V_{tn} = |V_{tp}| = 1$ V, $\mu_n C_{OX} = 2\ \mu_p C_{OX} = 20\ \mu A/V^2$. If the total effective load capacitance is 0.1 pF, find t_{PHL}, t_{PLH}, and t_P.

Ans. 0.8 ns; 0.8 ns; 0.8 ns

13.19 For the CMOS inverter of Exercise 13.15, which is intended for SSI and MSI circuit applications, find t_P if the load capacitance is 15 pF.

Ans. 6 ns

Delay-Power Product

The delay-power product of CMOS is obtained by multiplying the dynamic power dissipation by the propagation delay time. The resulting values range from much less than 1 pJ for VLSI circuits to about 10 pJ for commercially available SSI. The value of *DP* is directly proportional to the rate of switching and thus can be considerably reduced by operating at low rates. Also, like NMOS, *DP* for CMOS can be reduced by reducing the load capacitance and/or the power-supply voltage.

Exercise 13.20 Calculate the delay-power product of the CMOS inverter of Exercise 13.18 when it is operating at a switching rate of 50 MHz.

Ans. 0.1 pJ

13.6 CMOS GATE CIRCUITS

CMOS gate circuits are formed by extending the basic inverter circuit. To illustrate, we show a two-input NOR gate in Fig. 13.21(a). At the heart of the gate is the inverter Q_1, Q_2. The NOR gate circuit is obtained by adding a parallel *n*-channel device Q_3 and a series *p*-channel device Q_4. This process is repeated for each additional input. Since each additional input requires an additional pair of complementary MOSFETs, CMOS consumes more silicon area than NMOS, where, as we have seen in Section 13.4, each additional gate input requires the addition of only one transistor. The NAND gate of Fig. 13.21(b) is formed by a *dual* process: For each additional input we add a series *n*-channel device and a parallel *p*-channel device. It is interesting to note that if in one gate (say the NOR gate) we change each *n*-channel device to a *p*-channel device and vice versa and exchange the connections to V_{DD} and ground, we obtain the other gate.

Operation of the gate circuits in Fig. 13.21 is straightforward. The output of the circuit in Fig. 13.21(a) will be high (at V_{DD}) if both Q_2 and Q_4 are simultaneously on. This

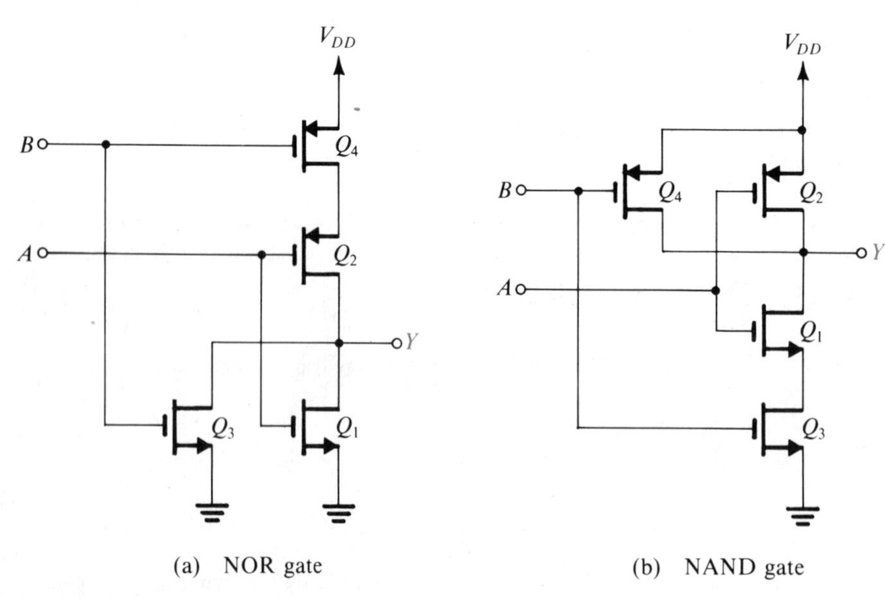

Fig. 13.21 CMOS logic gates.

will be achieved only when A and B are simultaneously low. Thus

$$Y = \overline{A}\,\overline{B}$$

or, equivalently,

$$Y = \overline{A + B}$$

which is the NOR logic function. On the other hand, the output of the circuit in Fig. 13.21(b) will be low (at 0 V) only when Q_1 and Q_3 are simultaneously on. This will occur only when A and B are simultaneously high. Thus

$$\overline{Y} = AB$$

or, equivalently,

$$Y = \overline{AB}$$

which is the NAND logic function.

Device Dimensions

It is desirable to design the CMOS gate circuits so as to provide equal output current-driving capability in both directions (pull-down and pull-up). For the two-input NOR gate of Fig. 13.21(a) we see that if the two inputs are tied together, then the pull-down current is the sum of the currents provided by Q_1 and Q_3. Since Q_1 and Q_3 are identical, the pull-down current is double that of the n-channel device. On the other hand, the pull-up current is provided by Q_2 and Q_4 in series and is thus equal to that of a single p-channel device. It follows that the output currents will be equal if each p-channel device is designed to have double the value of K of each of the n-channel devices; that is,

$$K_2 = K_4 = 2K_1 = 2K_3 \tag{13.46}$$

Now since $\mu_p \simeq \frac{1}{2}\mu_n$, we see that to satisfy the above condition the W/L ratio of each of the p-channel devices must be about four times the W/L ratio of each of the n-channel devices. In general, for an N-input NOR gate,

$$(W/L)_p \simeq 2N(W/L)_n \tag{13.47}$$

Applying the same reasoning to the NAND gate, we find that for the two-input NAND of Fig. 13.21(b) we should have $(W/L)_p \simeq (W/L)_n$. It follows that a two-input NAND requires smaller silicon area than a two-input NOR. In general, for an N-input NAND, one should use

$$\left(\frac{W}{L}\right)_n = \frac{N}{2}\left(\frac{W}{L}\right)_p \tag{13.48}$$

Exercise 13.21 Consider a CMOS technology in which the channel length of all devices is 5 μm and in which the minimum desired output current capability of the basic inverter is achieved with a $(W/L)_n = 2$. (a) For a two-input NOR gate find the width of each of the n-channel and p-channel devices and the total area occupied by all devices. (b) Repeat for a two-input NAND gate.

Ans. (a) NOR: $W_n = 10\ \mu$m, $W_p = 40\ \mu$m, area $= 500\ \mu\text{m}^2$; (b) NAND: $W_n = 20\ \mu$m, $W_p = 20\ \mu$m, area $= 400\ \mu\text{m}^2$

Gate Threshold

Even if the devices in a CMOS gate are designed according to the above considerations, the observed gate threshold will differ from the ideal value of $V_{DD}/2$ for a number of reasons. First, there are unavoidable mismatches. Second, the series devices (except for one) suffer from the body effect. Third, the gate threshold will depend on the signal values at the other inputs to the gate. The latter effect is illustrated in the following example.

EXAMPLE 13.3

Consider the two-input NOR gate of Fig. 13.21(a). Assume that $K_2 = K_4 = 2K_1 = 2K_3$ and that all devices have a V_t of 2 V. Calculate the value of the switching threshold of the gate V_{th} if (a) input terminal B is connected to ground and (b) input terminals A and B are joined together. Let the power supply voltage $V_{DD} = 10$ V and neglect the body effect in Q_2.

Solution

(a) Figure 13.22(a) shows a simplified circuit diagram of the NOR circuit with terminal B connected to ground. Since Q_3 will be off, we have eliminated it from the circuit altogether. The switching threshold V_{th} is the value of input voltage v_I at which both Q_1 and Q_2 are in the saturation region. (Figure 13.18 shows the threshold voltage obtained in the ideal situation.) For the no-load condition, the currents in Q_1 and Q_2 will be equal and given by

$$I = K_1(V_{th} - V_t)^2 = K_2(V_1 - V_{th} - V_t)^2 \tag{13.49}$$

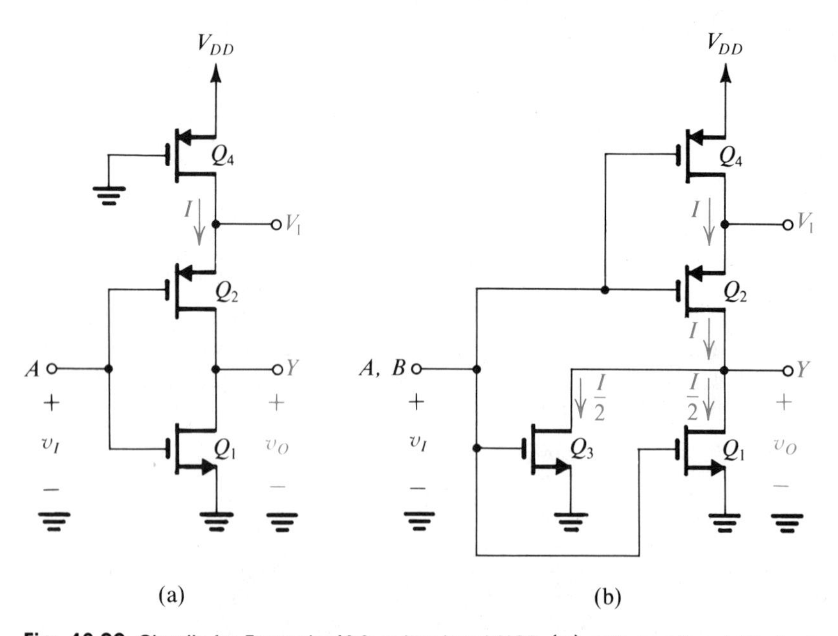

Fig. 13.22 Circuits for Example 13.3: a two-input NOR **(a)** with one input terminal grounded and **(b)** with the input terminals tied together.

where V_1 denotes the voltage at the source of Q_2. Since the value of V_{th} will be close to $V_{DD}/2$, V_1 will be higher than this value by at least V_t. Hence it is reasonable to expect that Q_4 will be operating in the triode region. We shall make this assumption and later check its validity. Thus we can write for Q_4

$$I = K_4[2(V_{DD} - V_t)(V_{DD} - V_1) - (V_{DD} - V_1)^2] \tag{13.50}$$

Eliminating I from Eqs. (13.49) and (13.50) and substituting $K_2 = K_4 = 2K_1$ gives two equations whose simultaneous solution yields $V_{th} = 5.31$ V and $V_1 = 9.65$ V. Thus Q_4 is indeed in the triode region, as assumed. Note that the value of V_{th} is slightly greater than the ideal value of $V_{DD}/2$, a result of the additional series device Q_4. At the switching threshold voltage V_{th} of 5.31 V the output voltage abruptly changes from $(5.31 + V_t)$ to $(5.31 - V_t)$—that is, from 7.31 V to 3.31 V.

(b) Next we consider the case where the two inputs are tied together as indicated in the circuit of Fig. 13.22(b). Again, we shall assume that with $v_I = V_{th}$ the value of V_1 will be such that Q_4 will be operating in the triode region. The other three transistors will be assumed to operate in saturation. Straightforward analysis gives $V_1 \simeq 9$ V and $V_{th} \simeq 4.5$ V. Thus we see that the gate threshold is 0.5 V below the ideal value of $V_{DD}/2 = 5$ V.

Transmission Gates

In addition to the basic CMOS gate circuits discussed above, the CMOS transmission gate, shown in Fig. 13.23, is an important building block in both digital and analog systems. We have studied the transmission gate in Section 5.10. An application of CMOS transmission gates will be presented in the next section.

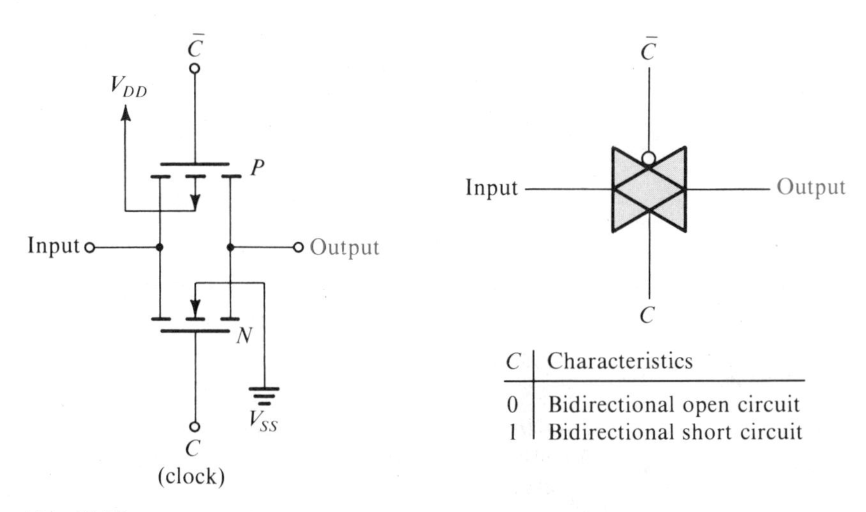

C	Characteristics
0	Bidirectional open circuit
1	Bidirectional short circuit

Fig. 13.23 Basic CMOS transmission gate.

Commercially Available CMOS Logic

As well as being used in VLSI circuit design, CMOS is available commercially in SSI, MSI, and LSI logic circuit packages for conventional digital system design. The oldest family of CMOS logic circuits, the 4000 series, utilizes metal gate technology (see Appendix A). These standard CMOS parts can operate over a supply voltage range of 3 to 18 V. (This should be contrasted to the typical allowed range for TTL of 4.75 to 5.25 V; see Chapter 14.)

Manufacturers of CMOS circuits usually specify minimum and maximum transfer characteristics in the manner of Fig. 13.24. All supplied CMOS gates are guaranteed to have transfer characteristics in the shaded area. The figure illustrates the definition of worst-case values for V_{IL} and V_{IH}—namely, as the input values at which v_O deviates from the corresponding ideal value (V_{DD} and 0) by a specified maximum, ΔV. (Note that these definitions differ from those we have been using and which are currently the standard.) Normally the deviation in output voltage ΔV is taken as $0.1V_{DD}$. As an example, Motorola specifies for standard CMOS logic:

$$V_{DD} = 5 \text{ V}: \quad V_{IL} = 1 \text{ V}, V_{IH} = 4 \text{ V}$$

$$V_{DD} = 10 \text{ V}: \quad V_{IL} = 2 \text{ V}, V_{IH} = 8 \text{ V}$$

$$V_{DD} = 15 \text{ V}: \quad V_{IL} = 2.5 \text{ V}, V_{IH} = 12.5 \text{ V}$$

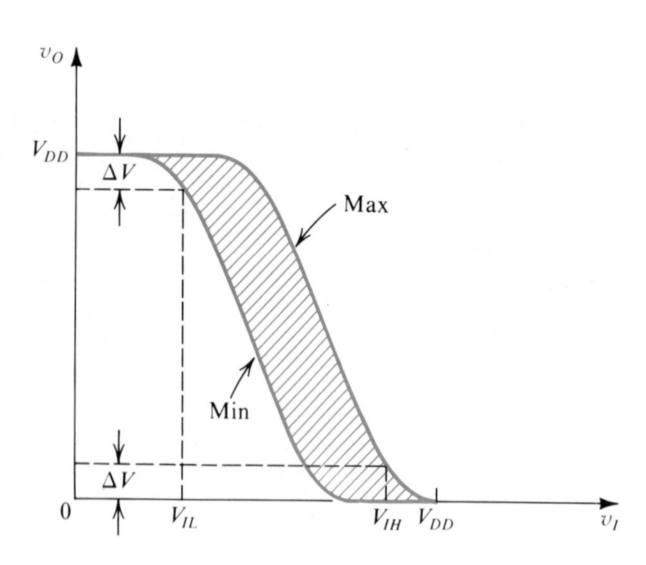

Fig. 13.24 The two limiting transfer characteristics specified for a CMOS gate.

These values can be used together with $\Delta V = 0.1V_{DD}$ to find the worst-case noise margins NM_H and NM_L:

$$V_{DD} = 5 \text{ V}: \quad NM_L = NM_H = 0.5 \text{ V}$$

$$V_{DD} = 10 \text{ V}: \quad NM_L = NM_H = 1 \text{ V}$$

$$V_{DD} = 15 \text{ V}: \quad NM_L = NM_H = 1 \text{ V}$$

The dynamic response of CMOS gates is specified in terms of the rise time, fall time, and propagation delay of the output pulse (see Fig. 13.7). As an example, for standard CMOS logic operated at $V_{DD} = 10$ V and driven with a pulse having 20-ns rise and fall times, Motorola specifies

$$t_{TLH} = (1.5 \text{ ns/pF})C_L + 15 \text{ ns}$$

$$t_{THL} = (0.75 \text{ ns/pF})C_L + 12.5 \text{ ns}$$

$$t_{PLH}, t_{PHL} = (0.66 \text{ ns/pF})C_L + 22 \text{ ns}$$

Note that the rise time is longer than the fall time because in these circuits the p device has a lower K than that of the n device. These formulas can be used to compute the switching times for a given load capacitance C_L. A standard SSI CMOS gate has an input capacitance of about 5 pF. Maximum fan-out of a CMOS gate is limited by the deterioration in its dynamic response due to the increased load capacitance. This should be contrasted to the case of TTL (Chapter 14), where maximum fan-out is limited by the degradation in noise margins.

An example of specifying dynamic power dissipation in CMOS logic is as follows: Motorola states that a standard CMOS gate operating with $V_{DD} = 10$ V and loaded with $C_L = 50$ pF draws an average dc supply current of 0.6 μA/kHz. Thus at a frequency of 1 MHz such a gate dissipates 6 mW. Finally, we note that although higher operating speeds are possible at higher supply voltages, these are obtained at the expense of increased dynamic power dissipation.

Improved-Performance CMOS

A variety of CMOS subfamilies exist with varying degrees of improved performance, including increased noise margins, higher output-driving capability, and higher speed of operation. Of special note is the buffered or B-series CMOS, which includes two cascaded CMOS inverters placed between the output node of the logic gate circuit and the external output terminal. The MOSFETs in the buffer inverters have W/L ratios that are much greater than those of the gate circuit proper. Thus the buffer provides an increased output drive capability. Also, the buffer provides additional voltage gain in the transition region. Since this results in a much sharper voltage transfer characteristic, the noise margins are increased.

Exercise 13.22 The buffered CMOS family is specified to have the following worst-case values. (a) At $V_{DD} = 5$ V, $V_{IL} = 1.5$ V and $V_{IH} = 3.5$ V. (b) At $V_{DD} = 10$ V, $V_{IL} = 3$ V and $V_{IH} = 7$ V. (c) At $V_{DD} = 15$ V, $V_{IL} = 4$ V and $V_{IH} = 11$ V.
By reference to Fig. 13.24 and using $\Delta V = 0.1\ V_{DD}$, calculate the corresponding worst-case noise margins.

Ans. (a) 1 V; (b) 2 V; (c) 2.5 V

(*Note:* Compare these values with the corresponding ones given in the text above for standard CMOS.)

Further improved versions of CMOS utilize devices with channel lengths in the submicron region ($< 1\ \mu$m). The use of *small-feature-size* devices results in increased speed of operation through the reduction of associated capacitances.

Some Practical Considerations

As mentioned in Chapter 5, because of the very high input resistance of MOS devices, static electricity can cause a charge to accumulate on the input capacitances during handling. Such charges can cause large voltages to appear at the MOSFET gate, which in turn can cause breakdown of the thin gate oxide. CMOS circuits include input diode networks, together with appropriate series resistors, for the purpose of limiting the voltages that may appear at the MOS device inputs (see Hodges and Jackson, 1988).

Since the input resistances of a CMOS gate are very high, a gate input left unconnected will float at an unspecified voltage. Usually, however, leakage currents of the input protection diodes are such that the input devices enter the active mode, allowing large currents to flow that cause overheating. Accordingly it is important that spare gate inputs be connected to an appropriate local power-supply pin or paralleled with another input (keeping in mind the effect of this on the gate switching threshold).

A Final Remark

CMOS logic circuits are highly versatile, finding application in diverse areas ranging from micropower circuits for watches and calculators to circuits that require high noise immunity, such as those used in automobiles and home appliances.

13.7 LATCHES AND FLIP-FLOPS

The logic circuits considered thus far are called **combinational** (or *combinatorial*). Their output depends only on the present value of the input. Thus these circuits do *not* have memory.

Memory is a very important part of digital systems. Its availability in digital computers allows for storing programs and data. Furthermore, it is important for temporary storage of the output produced by a combinational circuit for use at a later time in the operation of a digital system.

Logic circuits that incorporate memory are called **sequential circuits;** that is, their output depends not only on the present value of the input but also on the input's previous values. Such circuits require a timing generator (a *clock*) for their operation.[4]

In this section we shall study the basic memory element, the latch. We shall also examine two applications of latches, that of an NMOS **set/reset** (SR) flip-flop and a CMOS D (or data) flip-flop.

The Latch

The basic memory element, the latch, is shown in Fig. 13.25(a). It consists of two cross-coupled logic inverters, G_1 and G_2. The inverters form a positive-feedback loop. To investigate the operation of the latch we break the feedback loop at the input of one of

[4] Some combinational logic circuits also require a clock. Such circuits are called *dynamic logic* (see Hodges and Jackson, 1988). All the logic circuits we have studied fall into the static-logic category.

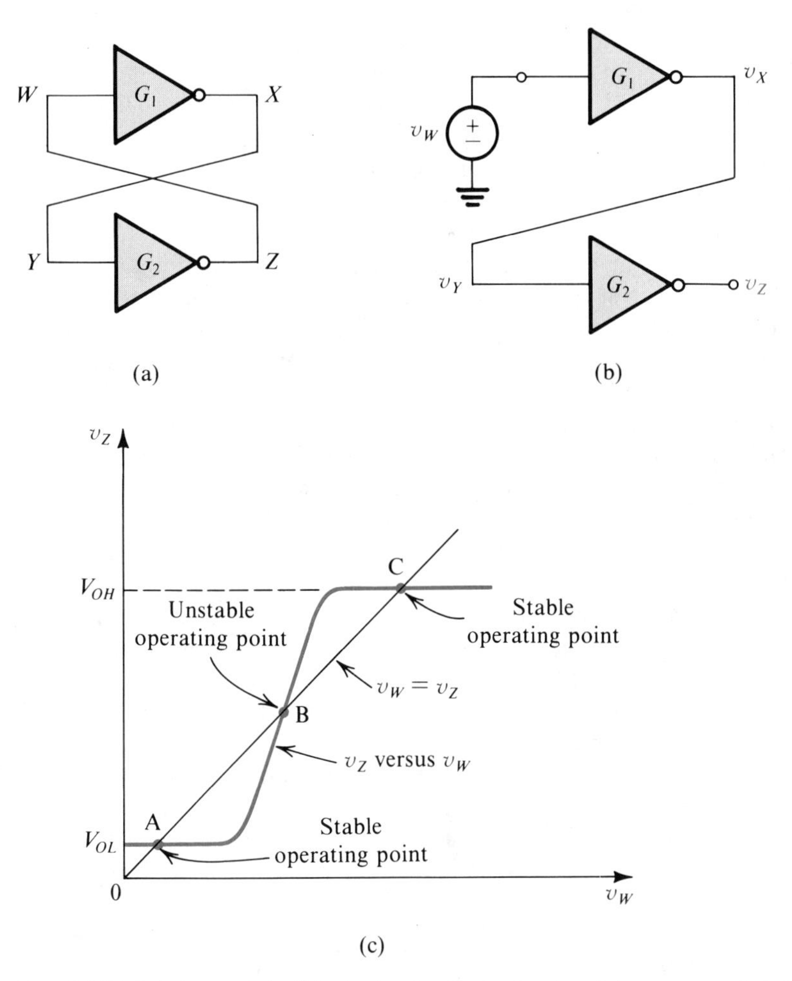

Fig. 13.25 (a) Basic latch. **(b)** The latch with the feedback loop opened. **(c)** Determining the operating point of the latch.

the inverters, say G_1, and apply an input signal, v_W in Fig. 13.25(b). Assuming that the input impedance of G_1 is large, breaking the feedback loop will not change the loop voltage transfer characteristic, which can be determined from the circuit of Fig. 13.25(b) by plotting v_Z versus v_W. This is the voltage transfer characteristic of two cascaded inverters and thus takes the shape shown in Fig. 13.25(c). Observe that the transfer characteristic consists of three segments, with the middle segment corresponding to the transition region of the inverters.

Also shown in Fig. 13.25(c) is a straight line with unity slope. This straight line represents the relationship $v_W = v_Z$ which is realized by reconnecting Z to W to close the feedback loop. As indicated, the straight line intersects the loop transfer curve at three points, A, B, and C. Thus any of these three points can serve as the operating point for the latch. We shall now show that while points A and C are stable operating points in the sense that the circuit can remain at either indefinitely, point B is an unstable operating point; the latch cannot operate at B for any period of time.

The reason point B is unstable can be seen by considering the latch circuit in Fig. 13.25(a) to be operating at point B, and taking account of the interference or noise that is inevitably present in any circuit. Let the voltage v_W increase by a small increment v_w. The voltage at X will increase (in magnitude) by a larger increment, equal to the product of v_w and the incremental gain of G_1 at point B. The resulting signal v_x is applied to G_2 and gives rise to an even larger signal at node Z. The voltage v_z is related to the original increment v_w by the loop gain at point B which is the slope of the v_Z versus v_W curve at point B. This gain is usually much greater than unity. Since v_z is coupled to the input of G_1, it will be further amplified by the loop gain. This regenerative process continues, shifting the operating point from B upward to point C. Since at C the loop gain is zero (or almost zero), no regeneration can take place.

In the description above, we assumed an initial positive voltage increment at W. Had we instead assumed a negative voltage increment we would have seen that the operating point moves downward from B to A. Again, since at point A the slope of the transfer curve is zero (or almost zero), no regeneration can take place. In fact for regeneration to occur the loop gain must be greater than unity, which is the case at point B.

The above discussion leads us to conclude that the latch has two stable operating points, A and C. At point C, v_W is high, v_X is low, v_Y is low, and v_Z is high. The reverse is true at point A. If we consider X and Z as the latch outputs, we see that in one of the stable states (say that corresponding to operating point A) v_X is high (at V_{OH}) and v_Z is low (at V_{OL}). In the other state (corresponding to operating point C) v_X is low (at V_{OL}) and v_Z is high (at V_{OH}). Thus the latch is a *bistable* circuit having two complementary outputs. In which of its two stable states the latch operates depends on the external excitation that forces it to the particular state. The latch then *memorizes* this external action by staying indefinitely in the acquired state. As a memory element the latch is capable of storing one bit of information. For instance, we can arbitrarily designate the state in which v_X is high and v_Z is low as corresponding to a stored logic 1. The other complementary state then designates a stored logic 0.

It now remains to devise a mechanism by which the latch can be *triggered* to change state. The latch together with the triggering circuitry forms a *flip-flop*. This will be discussed next. At this point, however, we wish to remind the reader that bistable circuits utilizing op amps were presented in Chapter 12.

The SR Flip-Flop

The simplest type of flip-flop is the set/reset (SR) flip-flop shown in Fig. 13.26(a). It is formed by cross-coupling two NOR gates, and thus it incorporates a latch. The second inputs of each NOR gate together serve as the trigger inputs of the flip-flop. These two inputs are labeled S (for set) and R (for reset). The outputs are labeled Q and $\overline{Q}$, emphasizing the fact that these are complementary. The flip-flop is considered set (that is, storing a logic 1) when Q is high and $\overline{Q}$ is low. When the flip-flop is in the other state (Q low, $\overline{Q}$ high), it is considered reset (storing a logic 0). In the rest or memory state (that is, when we do not wish to change the state of the flip-flop) both the S and R inputs should be low.

Consider the case when the flip-flop is storing a logic 0. Q will be low and thus both

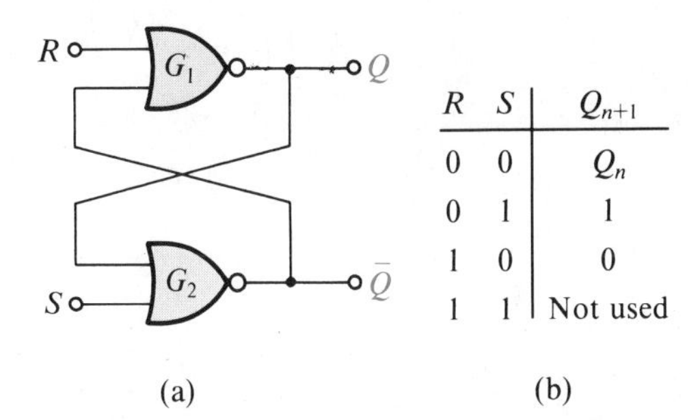

R	S	Q_{n+1}
0	0	Q_n
0	1	1
1	0	0
1	1	Not used

Fig. 13.26 The set/reset (SR) flip-flop and its truth table.

inputs to the NOR gate G_2 will be low. Its output will therefore be high. This high is applied to the input of G_1, causing its output Q to be low. To set the flip-flop we raise S to the logic-1 level while leaving R at 0. The 1 at the S terminal will force the output of G_2, $\overline{Q}$, to 0. Thus the two inputs to G_1 will be 0 and its output Q will go to 1. Now even if S returns to 0, the flip-flop remains in the newly acquired set state. Obviously, if we raise S to 1 again (with R remaining at 0) no change will occur. To reset the flip-flop we need to raise R to 1 while leaving $S = 0$. We can readily show that this forces the flip-flop into the reset state and that the flip-flop remains in this state even after R returns to 0. It should be observed that the trigger signal merely starts the regenerative action of the positive-feedback loop of the latch.

Finally, we inquire into what happens if both S and R are simultaneously raised to 1. The two NOR gates will cause both Q and $\overline{Q}$ to become 0 (note that in this case the complementary labeling of these two variables is incorrect). However, if R and S return to the rest state ($R = S = 0$) simultaneously, the state of the flip-flop will be undefined. In other words it will be impossible to predict the final state of the flip-flop. For this reason, this input combination is usually disallowed (that is, not used). Note, however, that this situation arises only in the idealized case, when both R and S return to 0 precisely simultaneously. In actual practice one of the two will return to 0 first, and the final state will be determined by the input that remains high longest.

The operation of the flip-flop is summarized by the *truth table* in Fig. 13.26(b), where Q_n denotes the value of Q at time t_n just before the application of the R and S signals, and Q_{n+1} denotes the value of Q at time t_{n+1} after the application of the input signals.

Rather than using two NOR gates, one can also implement an SR flip-flop by cross-coupling two NAND gates.

An SR flip-flop constructed using two two-input depletion-load NMOS NOR gates is shown in Fig. 13.27. This circuit is a direct implementation of the flip-flop shown in logic form in Fig. 13.26, and thus its operation should be self-evident.

A CMOS D Flip-Flop

A variety of flip-flop types exist. Most can be synthesized in terms of logic gates. The gates can then be replaced with their circuit implementations using the desired technology (NMOS, CMOS, TTL, etc.). For this reason we shall not present these flip-flop circuits here. An exception is a CMOS data, or D, flip-flop that utilizes transmission gates together with NOR gates in an interesting fashion.

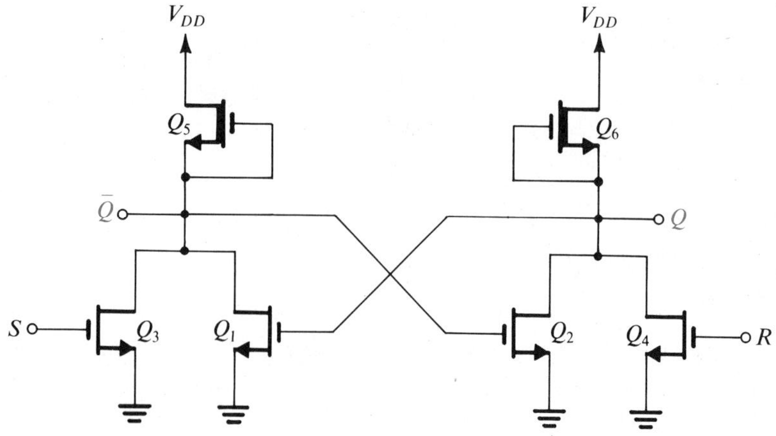

Fig. 13.27 An SR flip-flop formed by cross-coupling two depletion-load NMOS NOR gates.

The D flip-flop is shown in block-diagram form in Fig. 13.28. It has two inputs, the data input D and the clock input C. The complementary outputs are labeled Q and $\overline{Q}$. When the clock is low, the flip-flop is in the memory, or rest, state; signal changes on the D input line have no effect on the state of the flip-flop. As the clock goes high, the flip-flop acquires the logic level that existed on the D line just before the rising edge of the clock. Such a flip-flop is said to be **edge-triggered.** Some implementations of the D flip-flop include also direct set and reset inputs that override the clocked operation just described.

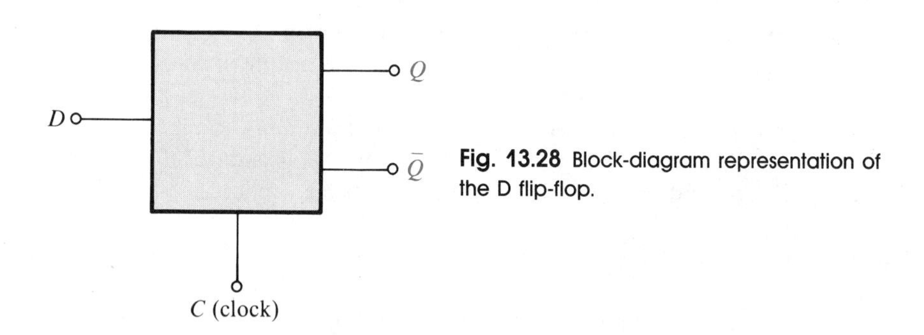

Fig. 13.28 Block-diagram representation of the D flip-flop.

Figure 13.29 shows a CMOS implementation of the D flip-flop including direct set and reset inputs. The flip-flop is composed of two SR flip-flops, called *master* and *slave*. Each consists of two NOR gates with a transmission gate inserted in the feedback loop. In addition, a transmission gate TG1 connects the D input to the master flip-flop, and another, TG3, connects the output of the master flip-flop to the input of the slave flip-flop. The outputs of the D flip-flop are buffered versions of the outputs of the slave section.

Triggering the flip-flop from the set and reset inputs is straightforward and needs no further discussion. Operation as an edge-triggered D flip-flop is best understood by con-

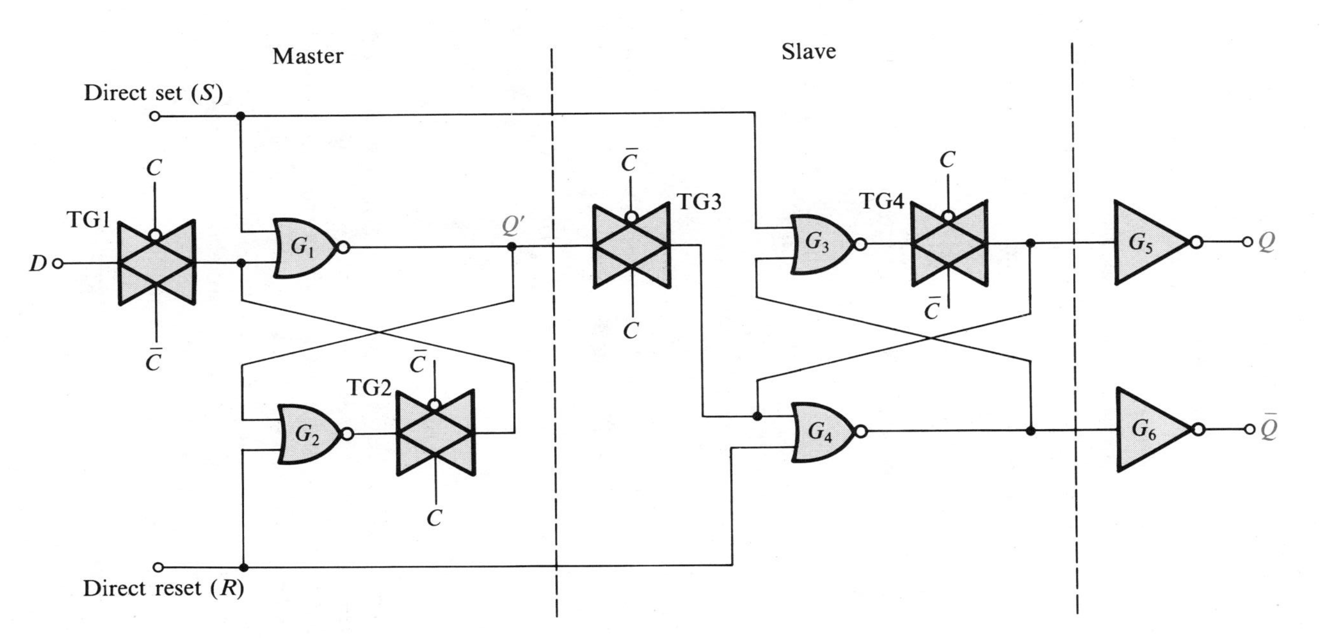

Fig. 13.29 Logic diagram of a CMOS D-type flip-flop.

sidering the transmission gates. When the clock is low, the transmission gates are in the following states:

$$\text{TG1} \rightarrow \text{ON}, \ \text{TG2} \rightarrow \text{OFF}, \ \text{TG3} \rightarrow \text{OFF}, \ \text{TG4} \rightarrow \text{ON}$$

Thus the slave is isolated from the master, and the feedback loop of the slave flip-flop is closed, making it retain its previous state. Meanwhile the feedback loop of the master flip-flop is open and the output of the master, Q', simply follows the complement of the input D.

When the clock input goes high, the transmission gates change to the following states:

$$\text{TG1} \rightarrow \text{OFF}, \ \text{TG2} \rightarrow \text{ON}, \ \text{TG3} \rightarrow \text{ON}, \ \text{TG4} \rightarrow \text{OFF}$$

The result is that the master is disconnected from input D and its feedback loop is closed. Thus its output Q' adopts the complement of the signal that existed on input line D just prior to the positive transition of the clock. Meanwhile the feedback loop of the slave is opened and its output Q adopts the complement of Q', which is the value of input D. The overall effect is that on the positive transition of the clock the output Q adopts the value of input D that existed just prior to the transition.

Exercise 13.23 For the SR flip-flop of Fig. 13.27 let $V_{DD} = 5$ V, $(W/L)_{1,2,3,4} = 2$, $(W/L)_{5,6} = 0.5$, $V_{tE} = 1$ V, $\mu_n C_{OX} = 20\ \mu\text{A/V}^2$, and $V_{tD} = -3$ V, and ignore the body effect. Assume that the total capacitance between each of the two output nodes and ground is 0.1 pF. Estimate the minimum required width of pulse at S or R to set or reset the flip-flop. (*Hint:* Use the results of Exercise 13.10.)

Ans. 7 ns

13.8 MULTIVIBRATOR CIRCUITS

As mentioned before, the flip-flop has two stable states and is called a bistable multivibrator. There are two other types of multivibrators: monostable and astable. The *monostable multivibrator* has one stable state in which it can remain indefinitely. It has another *quasi-stable* state to which it can be triggered. The monostable multivibrator can remain in the quasi-stable state for a predetermined interval T, after which it automatically reverts to the stable state. In this way the monostable multivibrator generates an output pulse of duration T. This pulse duration is in no way related to the details of the triggering pulse, as is indicated schematically in Fig. 13.30. The monostable multivibrator can therefore be used as a *pulse stretcher* or, more appropriately, a *pulse standardizer*. A monostable multivibrator is also referred to as a **one-shot.**

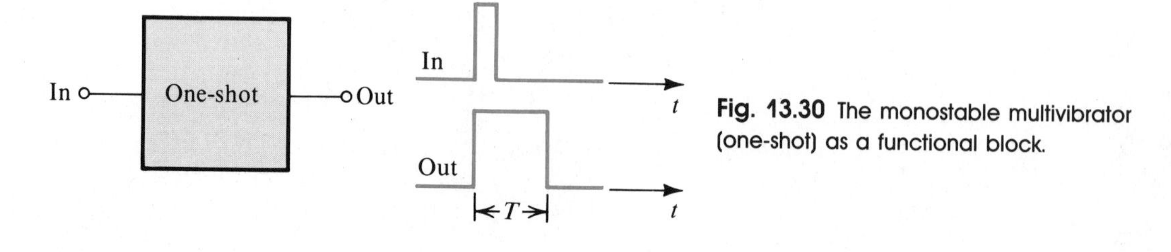

Fig. 13.30 The monostable multivibrator (one-shot) as a functional block.

The **astable multivibrator** has no stable states. Rather, it has two quasi-stable states, and it remains in each for predetermined intervals T_1 and T_2. Thus after T_1 seconds in one of the quasi-stable states the astable switches to the other quasi-stable state and remains there for T_2 seconds, after which it reverts back to the original state, and so on. The astable multivibrator thus oscillates with a period $T = T_1 + T_2$ or a frequency $f = 1/T$, and it can be used to generate periodic pulses such as those required for clocking.

In Chapter 12 we studied astable and monostable multivibrator circuits that use op amps. In the following we shall discuss monostable and astable circuits using logic gates.

A CMOS Monostable Circuit

Figure 13.31 shows a simple and popular circuit for a monostable multivibrator. It is composed of two two-input CMOS NOR gates, G_1 and G_2, a capacitor of capacitance C, and a resistor of resistance R. The input source v_I supplies the triggering pulses for the monostable multivibrator.

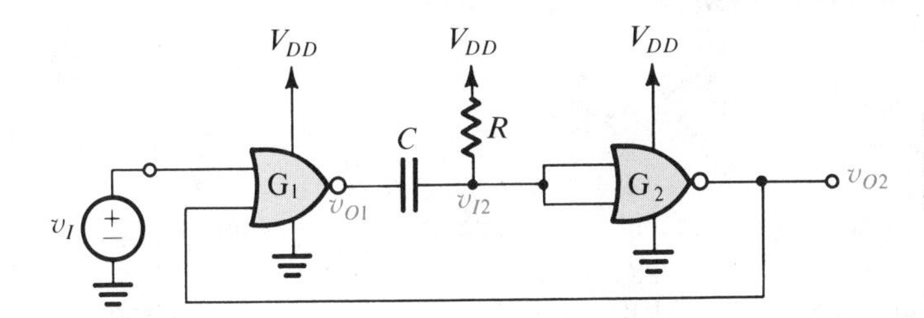

Fig. 13.31 Monostable circuit using CMOS NOR gates. Signal source v_I supplies the trigger pulses.

CMOS gates have a special arrangement of diodes connected at their input terminals, as indicated in Fig. 13.32(a). The purpose of these diodes is to prevent the input voltage signal from rising above the supply voltage V_{DD} (by more than one diode drop) and from falling below ground voltage (by more than one diode drop). These clamping diodes have an important effect on the operation of the monostable circuit. Specifically, we shall be interested in the effect of these diodes on the operation of the inverter-connected gate G_2. In this case, each pair of corresponding diodes appears in parallel, giving rise to the

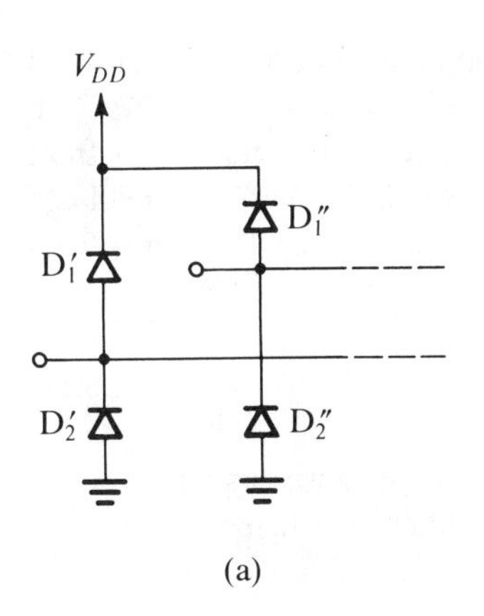

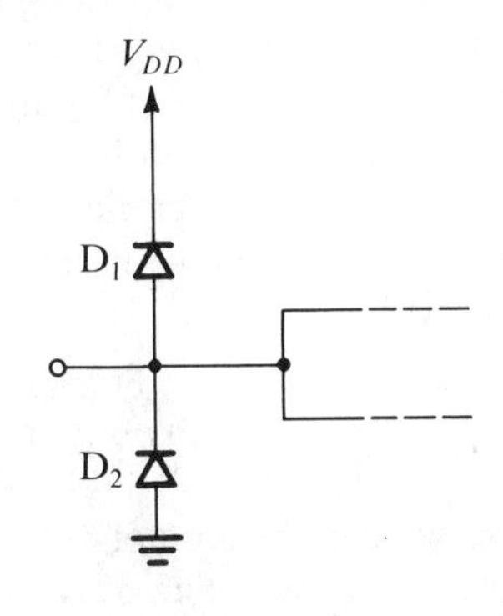

Fig. 13.32 (a) Diodes at the input of a two-input CMOS gate. **(b)** Equivalent diode circuit when the two inputs of the gate are shorted together.

equivalent circuit in Fig. 13.32(b). While the diodes provide a low-resistance path to the power supply for voltages exceeding the power supply limits, the input current for intermediate voltages is essentially zero.

To simplify matters we shall use the approximate equivalent output circuit of the gate, illustrated in Fig. 13.33. Figure 13.33(a) indicates that when the gate output is low, its output characteristics can be represented by a resistance R_{on} to ground, which is normally a few hundred ohms. In this state, current can flow from the external circuit into the output terminal of the gate; the gate is said to be *sinking* current. Similarly, the equivalent output circuit in Fig. 13.33(b) applies when the gate output is high. In this state, current can flow from V_{DD} through the output terminal of the gate into the external circuit; the gate is said to be *sourcing* current.

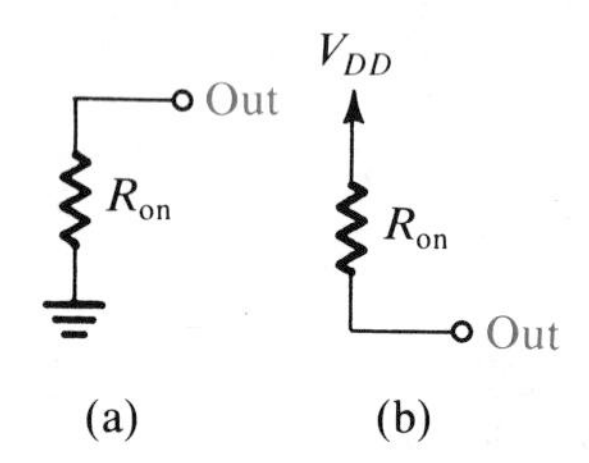

Fig. 13.33 Output equivalent circuit of CMOS gate when **(a)** the output is low and **(b)** the output is high.

To see how the monostable circuit of Fig. 13.31 operates, consider the timing diagram given in Fig. 13.34. Here a short triggering pulse of duration τ is shown in Fig. 13.34(a). In the following we shall neglect the propagation delays through G_1 and G_2. These delays, however, set a lower limit on the pulse width τ, $\tau > (t_{P1} + t_{P2})$.

Consider first the stable state of the monostable circuit—that is, the state of the circuit before the trigger pulse is applied. The output of G_1 is high at V_{DD}, the capacitor is discharged, and the input voltage to G_2 is high at V_{DD}. Thus the output of G_2 is low, at ground voltage. This low voltage is fed back to G_1; since v_I is low, the output of G_1 is high, as initially assumed.

Next consider what happens as the trigger pulse is applied. The output voltage of G_1 will go low. However, because G_1 will be sinking some current and because of its finite output resistance R_{on}, its output will not go all the way to 0 V. Rather, the output of G_1 drops by a value ΔV_1, which we shall shortly evaluate.

The drop ΔV_1 is coupled through C (which acts as a short circuit during the transient) to the input of G_2. Thus the input voltage of G_2 drops by an identical amount ΔV_1. Here we note that during the transient there will be an instantaneous current that flows from V_{DD} through R and C and into the output terminal of G_1 to ground. We thus have a voltage divider formed by R and R_{on} (note that the instantaneous voltage across C is zero) from which we can determine ΔV_1 as

$$\Delta V_1 = V_{DD}\frac{R}{R + R_{on}} \tag{13.51}$$

Returning to G_2, we see that the drop of voltage at its input causes its output to go high (to V_{DD}). This signal keeps the output of G_1 low even after the triggering pulse has disappeared. The circuit is now in the quasi-stable state.

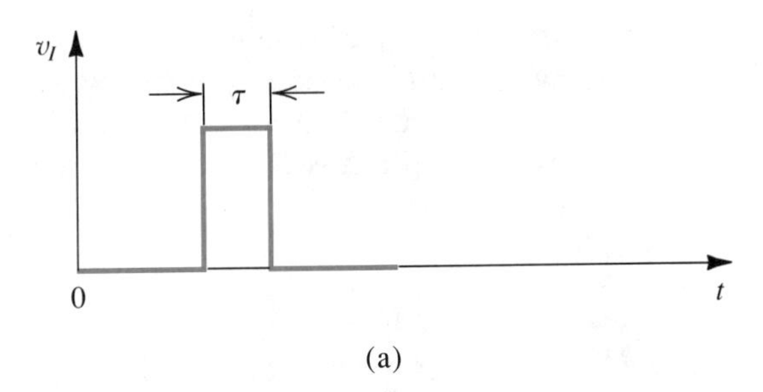

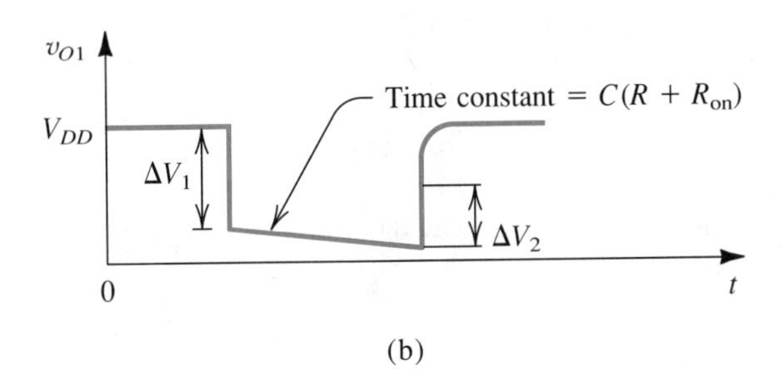

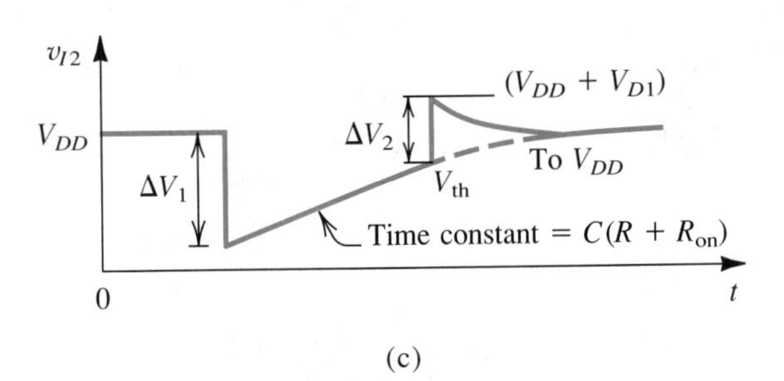

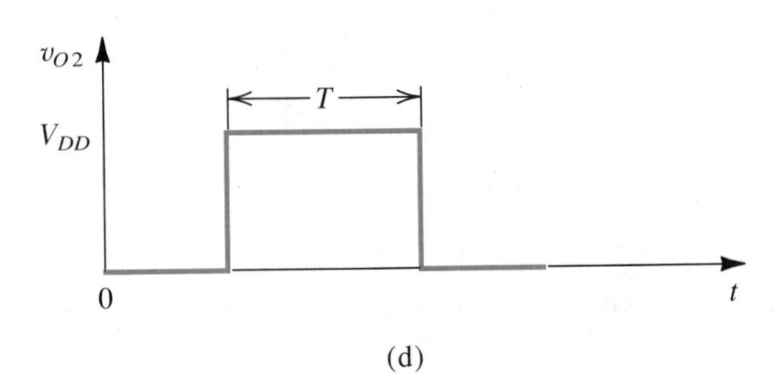

Fig. 13.34 Timing diagram for the monostable circuit in Fig. 13.31.

We next consider operation in the quasi-stable state. The current through R, C, and R_{on} causes C to charge, and the voltage v_{I2} rises exponentially toward V_{DD} with a time constant $C(R + R_{on})$, as indicated in Fig. 13.34(c). The voltage v_{I2} will continue to rise until it reaches the value of the threshold voltage V_{th} of inverter G_2. At this time G_2 will switch and its output v_{O2} will go to 0 V, which will in turn cause G_1 to switch. The output of G_1 will attempt to rise to V_{DD}, but, as will become obvious shortly, its instantaneous rise will be limited to an amount ΔV_2. This rise in v_{O1} is coupled faithfully through C to the input of G_2. Thus the input of G_2 will rise by an equal amount ΔV_2. Note here that because of diode D_1, between the input of G_1 and V_{DD}, the voltage v_{I2} can rise only to $V_{DD} + V_{D1}$, where V_{D1} (approximately 0.7 V) is the drop across D_1. Thus from Fig. 13.34(c) we see that

$$\Delta V_2 = V_{DD} + V_{D1} - V_{th} \tag{13.52}$$

Thus it is diode D_1 that limits the size of the increment ΔV_2.

Because now v_{I2} is higher than V_{DD} (by V_{D1}) current will flow from the output of G_1 through C and then through the parallel combination of R and D_1. This current discharges C until v_{I2} drops to V_{DD} and v_{O1} rises to V_{DD}. The charging circuit is depicted in Fig. 13.35, from which we note that the existence of the diode causes the discharging to be a nonlinear process. Although the details of the transient at the end of the pulse are not of immense interest, it is important to note that the monostable circuit should not be retriggered until the capacitor has been discharged, since otherwise the output obtained will not be the standard pulse which the one-shot is intended to provide. The capacitor discharge interval is known as the *recovery time*.

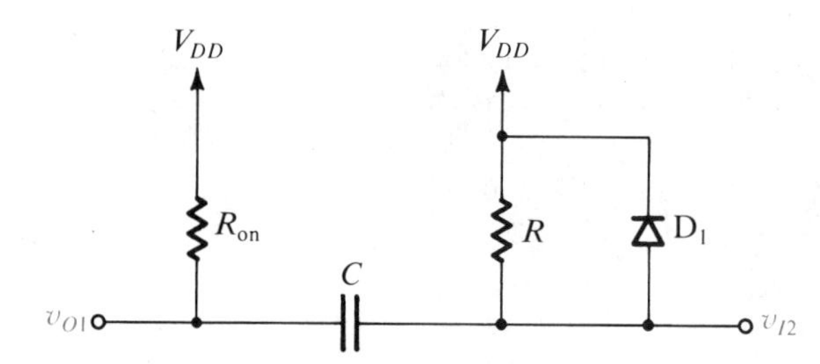

Fig. 13.35 Circuit that applies during the discharge of C (at the end of the monostable pulse interval *T*).

Exercise 13.24 Derive an expression for the pulse interval T of the monostable circuit in Fig. 13.34. (*Hint:* Use the information given in the timing diagram of Fig. 13.34(c) and the value of ΔV_1 given in Eq. (13.51).)

Ans.

$$T = C(R + R_{on}) \ln\left(\frac{R}{R + R_{on}} \frac{V_{DD}}{V_{DD} - V_{th}}\right)$$

An Astable Circuit

Figure 13.36(a) shows a popular astable circuit composed of two inverter-connected NOR gates, a resistor, and a capacitor. We shall consider its operation, assuming that the NOR gates are of the CMOS family. However, to simplify matters we shall make some further approximations: The finite output resistance of the CMOS gate will be neglected. Also, the clamping diodes will be assumed ideal (thus have zero voltage drop when conducting).

With these simplifying assumptions, the waveforms of Fig. 13.36(b) are obtained. The reader is urged to consider the operation of this circuit in a step-by-step manner and verify that the waveforms shown indeed apply.[5]

[5] Practical circuits often use a large resistance in series with the input to G_1. This limits the effect of diode conduction and allows v_{I1} to rise to a voltage greater than V_{DD} and, as well, to fall below zero.

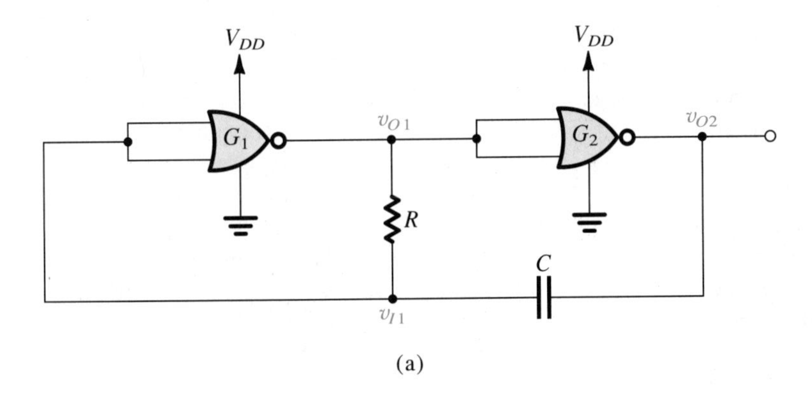

(a)

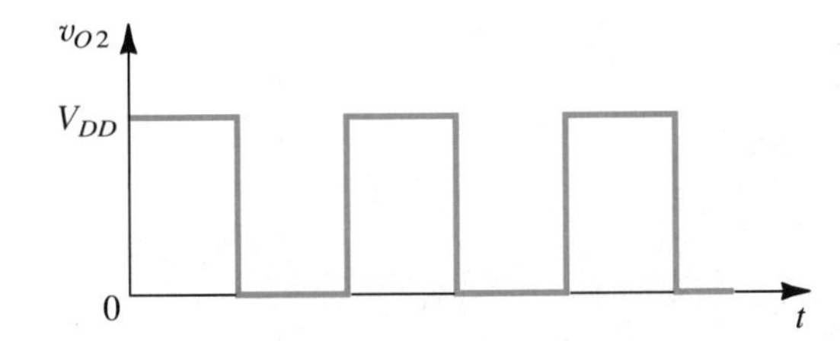

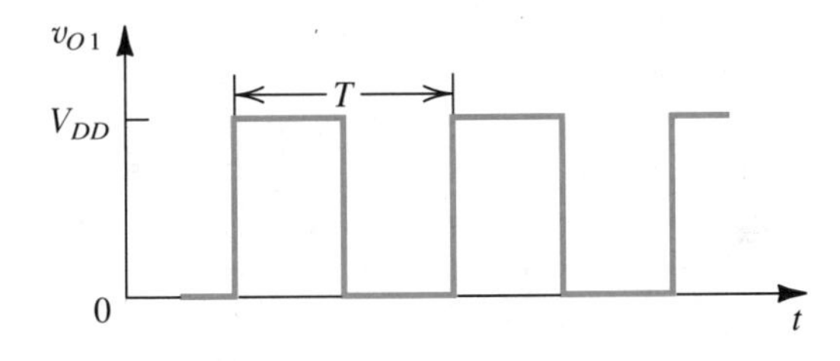

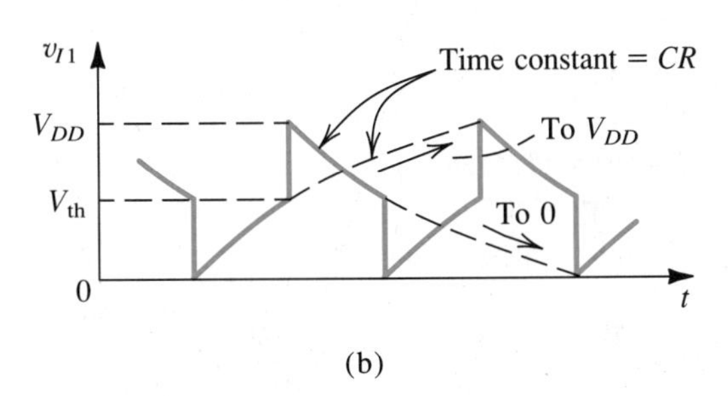

(b)

Fig. 13.36 (a) A simple astable multivibrator circuit using CMOS gates. **(b)** Waveforms for the astable circuit in (a). The diodes at the gate input are assumed ideal and thus limit the voltage v_{I1} to 0 and V_{DD}.

Exercise 13.25 Using the waveforms in Fig. 13.36(b), derive an expression for the period T of the astable multivibrator of Fig. 13.36(a).

Ans.

$$T = CR \ \ln\left(\frac{V_{DD}}{V_{DD} - V_{th}} \cdot \frac{V_{DD}}{V_{th}}\right)$$

13.9 RANDOM-ACCESS MEMORY (RAM)

A computer system, whether a large machine or a microcomputer, requires memory for storing data and program instructions. Furthermore, within a given computer system there usually are various types of memory utilizing a variety of technologies and having different *access times*. Broadly speaking, computer memory can be divided into two types: **main memory** and **mass-storage** memory. The main memory is usually the most rapidly accessible memory and the one from which most, often all, instructions in programs are executed. The main memory is usually of the random-access type. A **random-access memory** (RAM) is one in which the time required for storing (writing) information and for retrieving (reading) information is independent of the physical location (within the memory) in which the information is stored.

Random-access memories should be contrasted with *serial* or *sequential* memories, such as disks and tapes, from which data are available only in the same sequence in which the data were originally stored. Thus, in a serial memory the time to access particular information depends on the memory location in which the required information is stored, and the average access time is longer than the access time of random-access memory. In a computer system, serial memory is used for mass storage. Items not frequently accessed, such as large parts of the computer operating system, are usually stored in a *moving-surface memory* such as magnetic disk or tape.

Another important classification of memory is whether it is a *read/write* or a *read-only* memory. Read/write (R/W) memory permits data to be stored and retrieved at comparable speeds. Computer systems require random access, read/write memory for data and program storage.

Read-only memories (ROM) permit reading at the same high speeds as R/W memories (or perhaps higher), but restrict the writing operation. ROMs can be used to store a microprocessor operating-system program. They are also employed in operations that require table look-up, such as finding the values of mathematical functions. A popular application of ROMs is their use in video game cartridges. It should be noted that read-only memory is usually of the random-access type. Nevertheless, in the digital circuit jargon, the acronym RAM usually refers to read/write, random-access memory, while ROM is used for read-only memory.

The regular structure of memory circuits has made them an ideal application for VLSI circuit design. Indeed, at any moment, memory chips represent the state-of-the-art in packing density and hence integration level. At the present time chips containing 4M bits[6] are commercially available, while 16M-bit memory chips are being tested in research and development laboratories. In this section we shall study some of the basic circuits employed in VLSI RAM chips. Read-only memory circuits are studied in the next section.

Memory Chip Organization

The bits on a memory chip are either individually addressable, or addressable in groups of four or eight. As an example, a 64K-bit chip in which all bits are individually addressable

[6]The capacity of a memory chip to hold binary information as binary digits (or bits) is measured in K-bit units, where 1K bit = 1024 bits. Thus a 16K-bit chip contains 16,384 bits of memory, a 256K-bit chip contains 262,144 bits, and so on. Correspondingly, a 1M-bit chip contains 1,048,576 bits of memory.

is said to be organized as 64K words × 1 bit (or simply 64K × 1). Such a chip needs a 16-bit address (2^{16} = 65,536 = 64K). On the other hand, the 64K-bit chip can be organized as 16K words × 4 bits (16K × 4), in which case a 14-bit address is required. For simplicity we shall assume in our subsequent discussion that all the bits on a memory chip are individually addressable.

The bulk of a memory chip consists of the cells in which the bits are stored. Each memory cell is an electronic circuit capable of storing one bit. The storage cells on a chip are physically organized in a square matrix. As an example, Fig. 13.37 illustrates the organization of a 1K-bit chip. As indicated, the cell array has 32 rows and 32 columns. Each cell is connected to one of the 32 row lines, known rather loosely as **word lines,** and to one of the 32 column lines, known as **digit lines** or **bit lines.** A particular cell is selected by activating its word line and its digit line. This in turn is achieved by the row address decoder and the column address decoder. As indicated, 5 of the 10 address bits

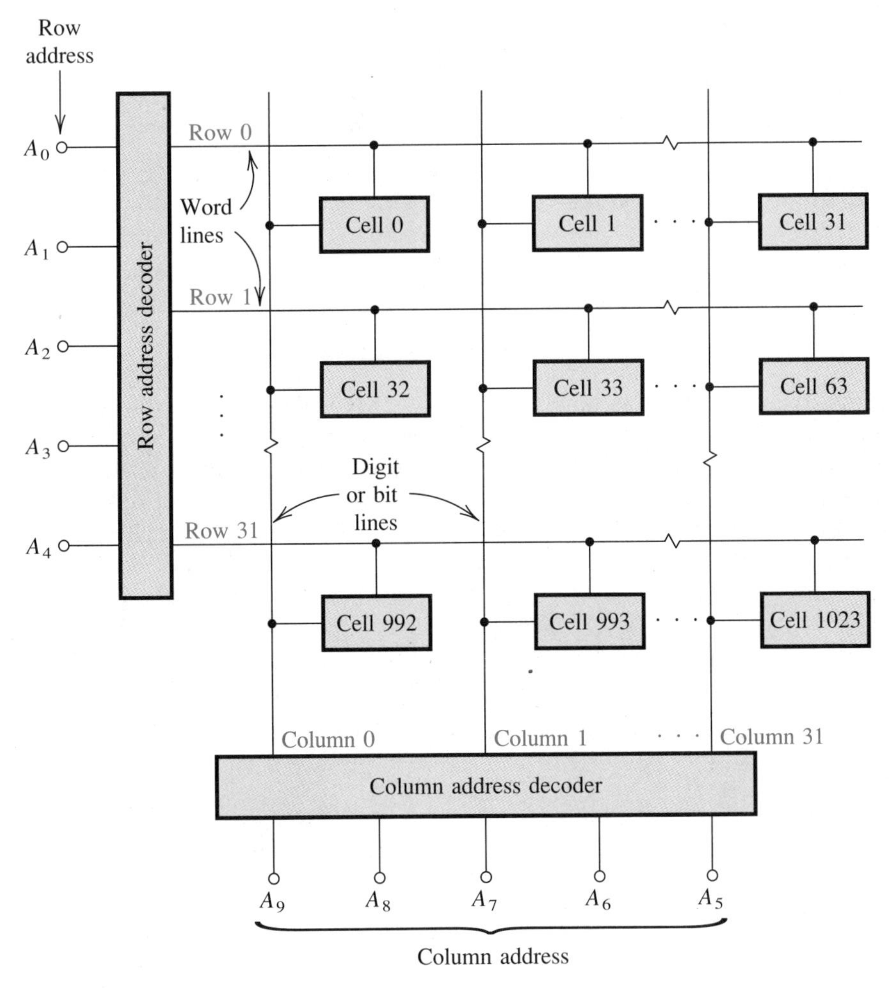

Fig. 13.37 Organization of a 1024-bit chip. Note that from a system point of view this chip is organized as 1024 words × 1 bit.

form the **row address,** and the other 5 address bits form the **column address.** The row address bits, labeled A_0 to A_4, are fed to the row address decoder, which selects one of the 32 word lines. Similarly, the column address bits, A_5 to A_9, are fed to the column address decoder, which selects one out of the 32 digit lines.

Address Buffers and Decoders

The address input terminals of a RAM chip are usually buffered using inverters. For instance, in an NMOS RAM the depletion-load circuits of Fig. 13.38 would be used. As indicated, the input buffers provide the true and the complement of the input address bits.

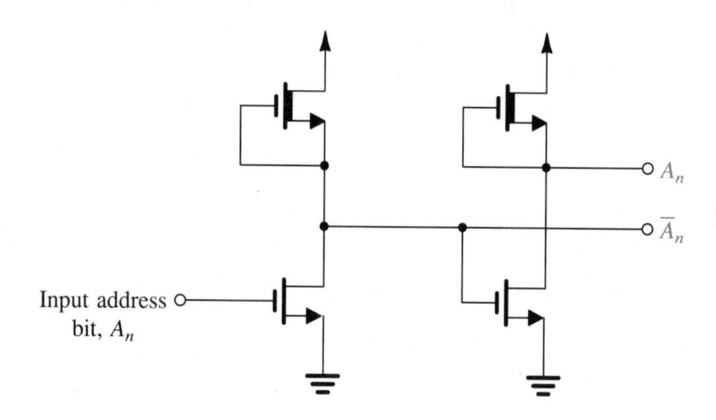

Fig. 13.38 A simple depletion-load address buffer.

Availability of the complements simplifies decoding. Each address decoder is usually a combinational circuit such as the NOR gate shown in Fig. 13.39. If, for instance, this NOR gate is used in the row address decoder, its output will be connected to one of the word lines and is inputs will be connected to the appropriate combination of address bits and their complements. This combination is selected so that the output of the NOR gate is high when we wish to select the particular word line to which it is connected.

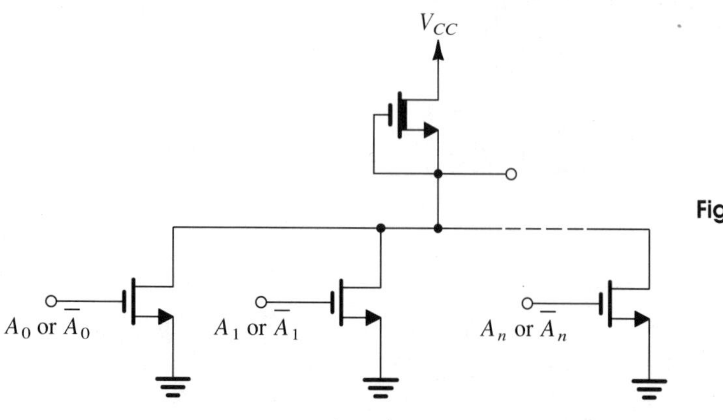

Fig. 13.39 A simple NOR decoder.

Exercise 13.26 For the 1K-bit RAM shown in Fig. 13.37, find the address bits that are to be connected to the inputs of the NOR gate whose output is connected to row 18.

Ans. A_0, $\overline{A}_1$, A_2, A_3, and $\overline{A}_4$.

The complete NOR address decoder is usually connected in array form, as illustrated in Fig. 13.40.

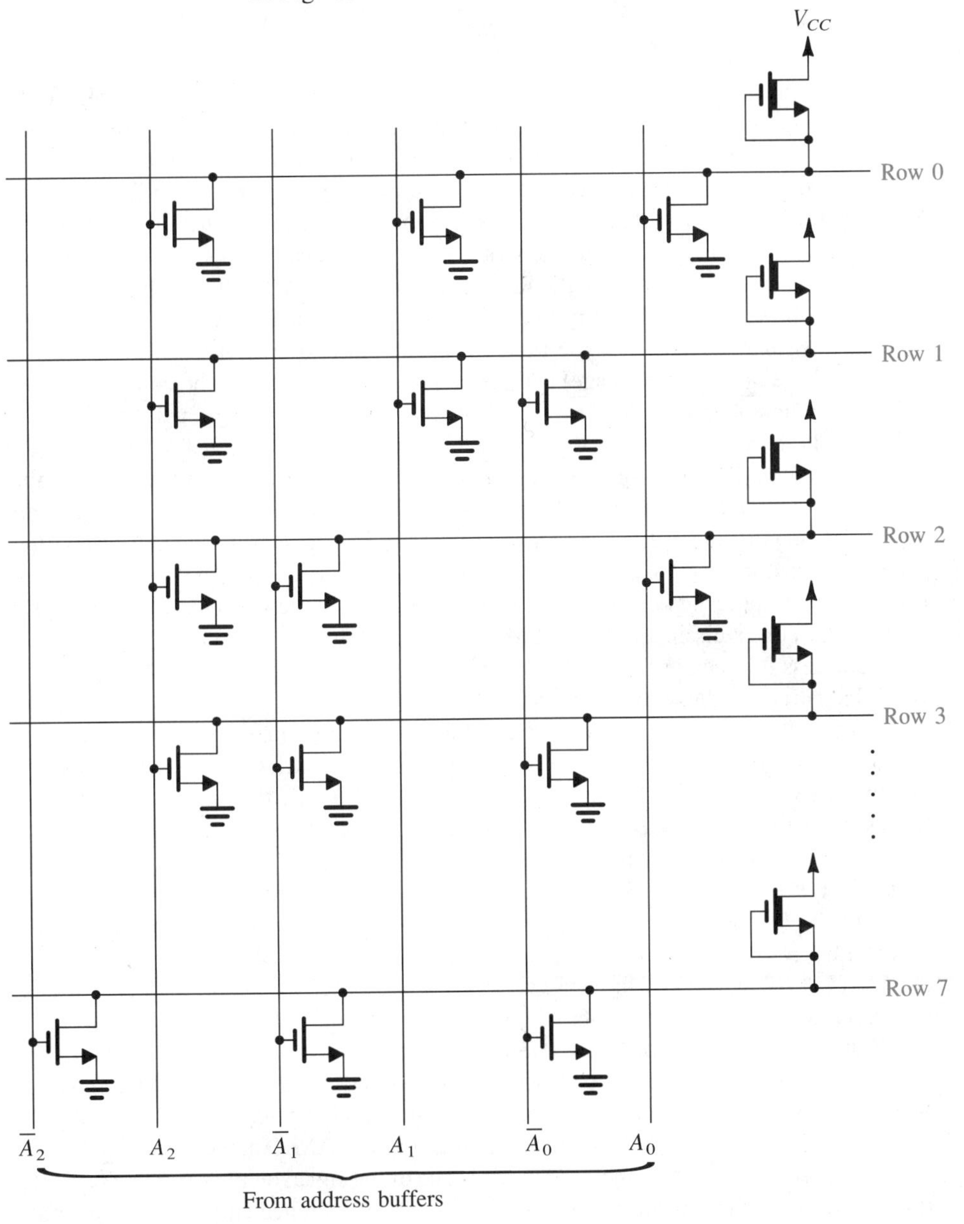

Fig. 13.40 A NOR address decoder in array form. One out of eight lines (row lines) is selected using a 3-bit address.

Memory Chip Timing

The **memory-access time** is the time between the initiation of a read operation and the appearance of the output data. The **memory-cycle time** is the minimum time allowed between two consecutive memory operations. To be on the conservative side, a memory operation is usually taken to include both read and write (in the same location). MOS memories have access and cycle times in the range of 10 to 300 ns.

Static Memory Cell

There are basically two types of MOS RAMs: static and dynamic. Static RAMs utilize flip-flops as the storage cells. As will be seen shortly, these flip-flops have to be as simple as possible in order to minimize the silicon area per cell. This is very important since the cell array constitutes by far the largest part of the memory chip. Dynamic RAMs on the other hand store the binary data on capacitors, resulting in further reduction in cell area at the expense of more elaborate read and write circuitry. We shall discuss static RAMs first.

Figure 13.41 shows typical static RAM cells in NMOS and CMOS technology. Each of the cells shown consists of a flip-flop formed by cross-coupling two inverters, and two *access transistors*, Q_5 and Q_6. The access transistors are turned on when the word line (row) is selected (raised in voltage) and they connect the flip-flop to the column (digit or D) line and the $\overline{\text{column}}$ ($\overline{\text{digit}}$ or $\overline{D}$) line. Note that here both the digit and $\overline{\text{digit}}$ lines are utilized. The access transistors act as transmission gates allowing bidirectional current flow between the flip-flop and the D and $\overline{D}$ lines. (To emphasize this point, their drains and sources are not distinguished.) The cell of Fig. 13.41(c) utilizes load resistors that are formed in the polysilicon layer via an additional processing step. Large-valued resistors can be obtained in this way, with the result that the power dissipation per cell is low. It is important to appreciate that one of the keys to realizing large-capacity memory chips is keeping the power dissipation per bit as small as possible.

To access the memory cells of Fig. 13.41 for reading or writing, the voltage of the word line is raised, thus turning on the access transistors (Q_5 and Q_6). In this way, one side of the cell flip-flop is connected to the D line and the other side is connected to the $\overline{D}$ line. Consider, as an example, the read operation of the cell in Fig. 13.41(a), and assume that the cell is storing a 0. In this case Q_1 is on and Q_2 is off. Before the read operation begins, the voltages of the D and $\overline{D}$ lines are equalized at about $V_{CC}/2$. When Q_5 and Q_6 are turned on, current flows from the D line through Q_5 and Q_1 to ground. This causes a drop in the D-line voltage. Simultaneously, current flows from V_{CC} through Q_4 and Q_6 and onto the $\overline{D}$ line, causing an increase in its voltage. The voltage signal that appears between the D and $\overline{D}$ lines is then fed to the column **sense amplifier** (there is one sense amplifier per column). Only the sense amplifier in the selected column will be active, and its output is connected to the data output line of the chip.

From the above description we infer that the currents conducted by the flip-flop transistors, together with the capacitances of the D and $\overline{D}$ lines (that is, between these lines and ground), determine the rise and fall times of the signals on the D and $\overline{D}$ lines. These times, in turn, contribute to the access time of the RAM. Another component of the access time is contributed by the nonzero rise time of the signal on the word line. This rise time is a result of the capacitance of the word line and the limited current-drive capability available from the output of the row decoder.

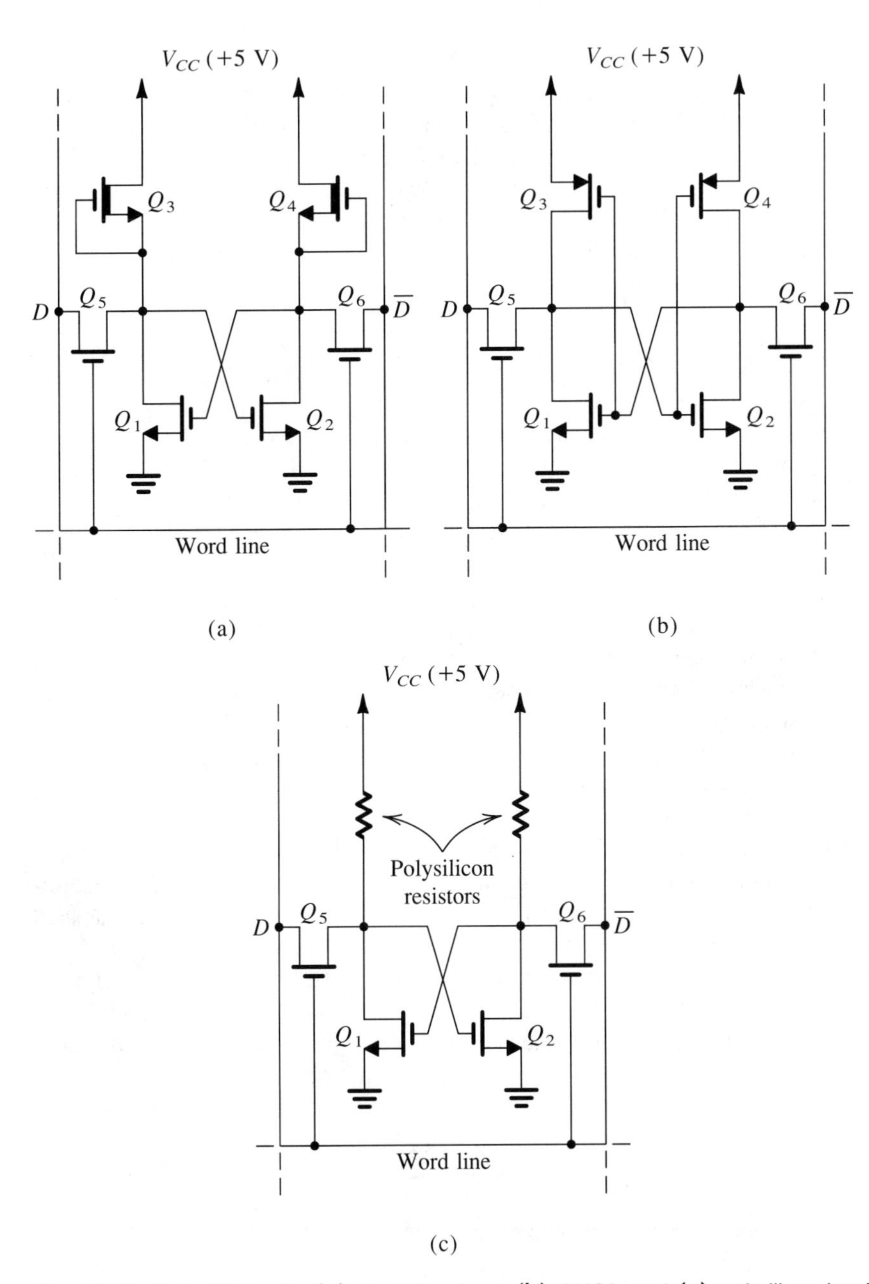

Fig. 13.41 Static RAM cells: **(a)** depletion load; **(b)** CMOS; and **(c)** polysilicon load.

To speed up RAM operation, both components of time delay mentioned above have to be reduced. This is the reason the D and $\overline{D}$ lines are usually *precharged* to $V_{CC}/2$ before the read operation. In this way the signal swing on each of the two lines is reduced. This in turn reduces the time taken by the column sense amplifier to reliably detect the voltage difference between D and $\overline{D}$.

To complete our discussion of cell access, consider the write operation. The data bit to be written and its complement are transferred to the D and $\overline{D}$ lines, respectively. Thus, if a 1 is to be written, the D line is raised to V_{CC} and the $\overline{D}$ line is lowered to ground. The conducting transistors Q_5 and Q_6 then (see Fig. 13.41a) cause the high voltage to appear at the gate of Q_2 and the low voltage to appear at the gate of Q_1. The flip-flop is then forced into the state in which the drain of Q_1 is high and that of Q_2 is low. This state, which denotes a stored 1, will be maintained indefinitely unless changed by another write operation.

Exercise 13.27 A (16K $\times$ 1) static RAM utilizing the cell shown in Fig. 13.41(c) requires 150 mW in standby (that is, when no reading or writing operations are carried out). If the standby current is nearly all required by the storage cells, what value (approximately) must the polysilicon resistors have if V_{CC} is +5 V? If V_{CC} of the cell is reduced to +2 V during standby, what must the resistance be? In each case, calculate the maximum current that can be drawn from the flip-flop at full voltage (5 V) without the loss of information that can occur if the drain of Q_1 or Q_2 is pulled below 1.5 V (V_t plus a safety margin). If the device K is designed to provide a low level $\leq$ 10 mV at the largest standing current and with full voltage applied, what is the greatest current the flip-flop can sink while retaining an output $\leq$ 0.5 V (V_t less a safety margin). $V_t = 1$ V.

Ans. 2.73 MΩ; 0.44 MΩ; 1.28 μA and 7.96 μA; 0.52 mA

We next consider the circuits for sensing and writing data. Figure 13.42 shows the Mth column of a static RAM. When this column is selected, the gate of Q_9 goes high and Q_9 turns on—establishing a low voltage at the sources of Q_7, Q_8, Q_{10}, and Q_{11}. During a read operation the voltage signals appearing on the D and $\overline{D}$ lines are applied to the gates of Q_7 and Q_8, which operate as common-source amplifiers, supplying amplified output signals at the chip Data out and $\overline{\text{Data out}}$ lines.

In a write operation, the input data bit and its complement are applied to the gates of transistors Q_{10} and Q_{11}, which operate as common-source amplifiers, supplying an amplified version of the input signal to D_M and $\overline{D}_M$. Finally, note that Q_{L1} and Q_{L2} act as load devices for the D and $\overline{D}$ lines.

Static RAMs can achieve very short access and cycle times. As an example, the Inmos IMS 1400, which is a 16K-bit static RAM fabricated in NMOS technology, has a 45-ns access time. The chip operates from a single 5-V supply and dissipates a maximum of 660 mW when operating. It has a "standby mode" in which it can hold its contents, but it is not available for read or write operation. The power dissipation in the standby mode is reduced to 110 mW.

As a final note, we observe that static RAMs can keep their contents indefinitely, as long as the power supply is connected. Although a clock is used for gating and synchronization, it is not essential for memory chip operation. Dynamic RAMs, on the other hand, require a clock for their operation, as explained next.

Dynamic RAMs

In a dynamic MOS RAM, binary data are stored in the form of charge on the cell capacitor. A logic 0 is represented by no charge and hence a voltage close to zero; a logic 1 is represented by a capacitor voltage of value close to the power supply. Due to the various leakage effects that are inevitably present, the capacitor charge will leak off. Thus, essen-

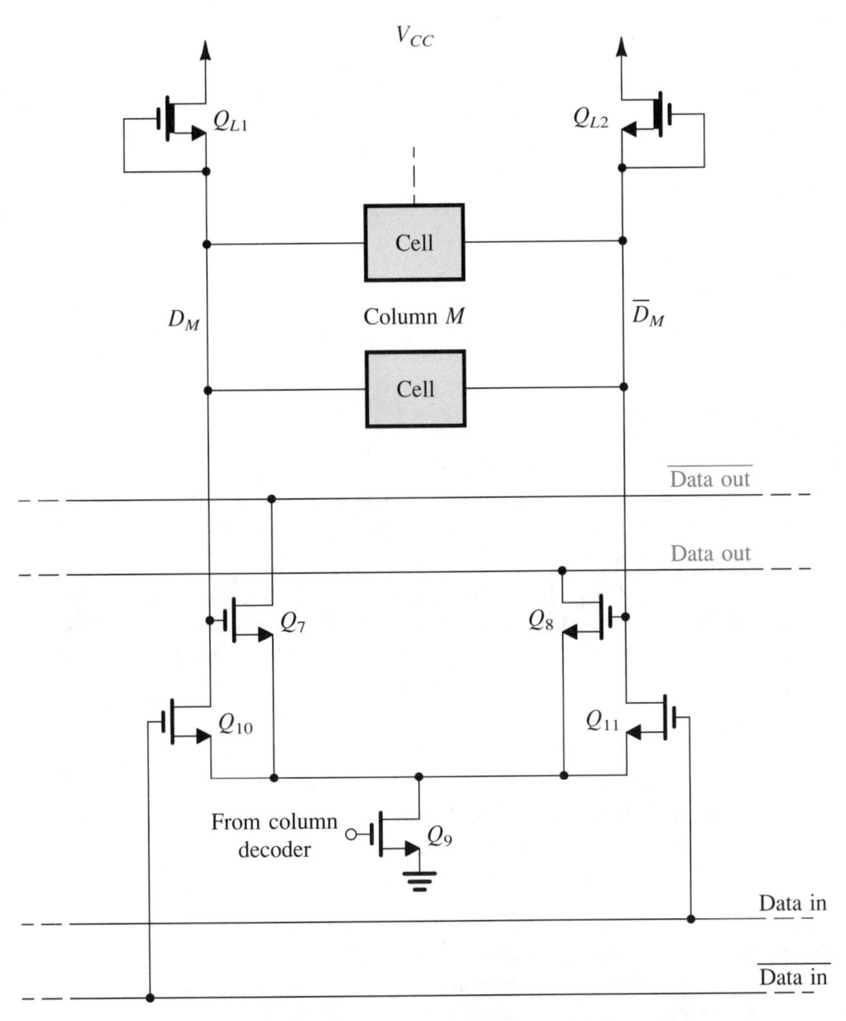

Fig. 13.42 Data-sensing and data-writing circuits for column M in a static RAM.

tial to the proper operation of dynamic RAMs is the **refresh** operation. During refresh, the cell content is read and the data bit is rewritten, thus restoring the capacitor voltage to its proper value. The refresh operation must be performed every 2 to 4 ms. The need for periodically refreshing the dynamic memory chip implies the necessity of having a clock.

The periodic refresh operation necessary in a dynamic RAM requires additional circuitry. Nevertheless, because the memory cell is very simple, as we will see very shortly, dynamic RAMs achieve much greater packing density than is possible with their static counterparts. Roughly speaking, at any given time in the past few years, the capacity of available dynamic RAM chips has been about four times that of static RAMs.

The most common storage cell employed in dynamic RAMs is shown in Fig. 13.43. The cell consists of a single enhancement-mode n-channel MOSFET, known as the *access*

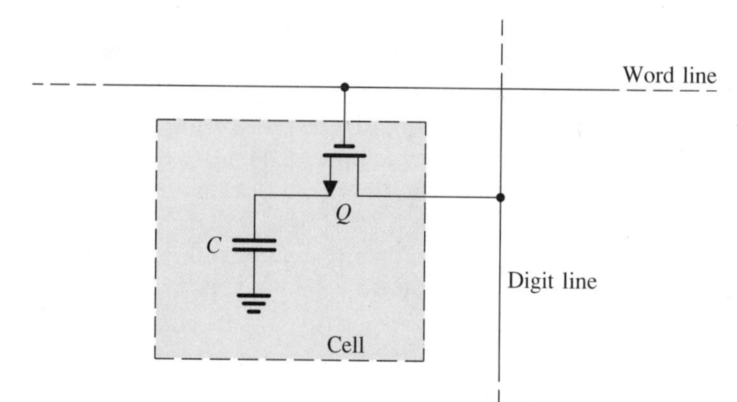

Fig. 13.43 The one-transistor dynamic RAM cell.

transistor, and a storage capacitor C. The cell is appropriately known as the **one-transistor cell.**[7] The gate of the transistor is connected to the word (or row) line, and its drain is connected to the digit (bit or column) line.

As in static RAMs, the row decoder selects a particular row by raising the voltage of its word line. This causes all the transistors in the selected row to become conductive, thereby connecting the storage capacitors of all the cells in the selected row to their respective digit lines. Thus the cell capacitor C is connected in parallel with the digit-line capacitance C_L as indicated in Fig. 13.44. Here it should be noted that typically C is about 0.05 pF and C_L is from 20 to 30 times larger. Now if the operation is a read and the cell is storing a logic 1, then the voltage on the cell capacitor C will cause a positive increment to appear across C_L. Since C_L is much greater than C, the voltage increment on C_L will be much smaller than the initial voltage on C. Obviously, if the cell is storing a logic 0, then no increment will appear on C_L.

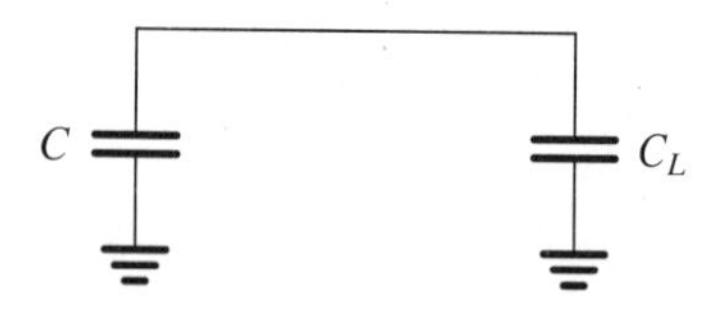

Fig. 13.44 When the voltage of the selected word line is raised, the transistor conducts, thus connecting the storage capacitor C to the digit-line capacitance C_L.

The change of voltage on the digit line is detected and amplified by the column sense amplifier. The amplified signal is then impressed on the storage capacitor, thus restoring its signal to the proper level. In this way, all the cells in the selected row are refreshed. Simultaneously, the signal at the output of the sense amplifier of the selected column is fed to the data output line of the chip.

The write operation proceeds similarly to the read operation except that the data bit to be written, which is impressed on the data input line, is applied by the column decoder to

[7]The name was originally used in order to distinguish this cell from earlier ones utilizing three transistors.

the selected digit line. This data bit is thus stored on the capacitor of the selected cell. Simultaneously, all the other cells in the selected row are simply refreshed.

Although the read and write operations result in automatic refreshing of the selected row, provision must be made for the periodic refreshing of the entire memory every 2 to 4 ms, as specified for the particular chip. The refresh operation is carried out in a burst mode, one row at a time. During refresh, the chip will not be available for read or write operations. This, however, is not a serious matter since the interval required to refresh the entire chip is typically less than about 2% of the time between refresh cycles. In other words, the memory chip remains available for normal operation for more than 98% of the time.

Exercises 13.28 In a particular one-transistor-per-cell dynamic MOS memory utilizing a 64×64 array, the storage cell capacitance is 0.05 pF, while the capacitance per cell on the digit line is 0.04 pF, and the input capacitance of the sense amplifier and associated circuitry is 0.5 pF. If, after degeneration due to leakage effects, the smallest signal allowed on the cell capacitance is 6 V, what is the corresponding signal available at the input of the sense amplifier when the access switch is closed?

Ans. 96.5 mV

13.29 A dynamic memory cell using a 0.05-pF capacitor and an MOS selection transistor employs 0- and 5-V levels for information storage. Sensing circuitry is adequate to permit the stored charge to decay to $1/e$ of its original value before refresh is required. The maximum allowed refresh interval for this design is 2 ms. What is the smallest equivalent resistance that can be allowed to shunt the storage capacitor? If the leakage phenomenon is best characterized as a current, what is the largest such current that can be tolerated?

Ans. 40×10^9 Ω; 79 pA

Because the signals available from dynamic RAM cells are very small, a critical part in the dynamic RAM is its sense amplifier.

13.10 READ-ONLY MEMORY (ROM)

As mentioned in the previous section, read-only memory (ROM) is memory that contains fixed data patterns. It is used in a variety of digital system applications. Currently, a very popular application of ROM is in microprocessor systems where it is used to store the instructions of the system operating program. ROM is particularly suited for such an application because it is nonvolatile; that is, it retains its contents when the power supply is switched off.

A ROM can be viewed as a combinational logic circuit for which the input is the collection of address bits of the ROM and the output is the set of data bits retrieved from the addressed location. This viewpoint leads to the application of ROMs in code conversion—that is, in changing the code of the signal from one code (say, binary) to another. Code conversion is employed, for instance, in secure communication systems, where the process is known as *scrambling*. It consists of feeding the code of the data to be transmitted to a ROM that provides corresponding bits in a supposedly secret code. The reverse process, which also uses a ROM, is applied at the receiving end.

In this section we will study various types of read-only memory. These include fixed ROM, which we refer to simply as ROM; programmable ROM (PROM); and erasable programmable ROM (EPROM).

A MOS ROM

Figure 13.45 shows a simplified 32-bit (or 8-word × 4-bit) MOS ROM. As indicated, the memory consists of an array of enhancement MOSFETs whose gates are connected to the word lines, whose sources are grounded, and whose drains are connected to the bit lines. Each bit line is connected to the power supply via a depletion-load device. A MOSFET exists in a particular cell if the cell is storing a 0; a cell storing a 1 has no MOSFET. This ROM can be thought of as 8 words of 4 bits each. The row decoder selects one of the 8 words by raising the voltage of the corresponding word line. The cell transistors connected to this word line will then conduct, thus pulling the voltage of the bit lines (to which transistors in the selected row are connected) down close to ground voltage (logic 0 level). The bit lines that are connected to cells (of the selected word) without transistors (that is, those cells that are storing 1s) will remain at the power-supply voltage (logic 1) because of the action of the pull-up depletion-load devices. In this way the bits of the addressed word can be read.

Mask-Programmable ROMs

The data stored in the ROMs discussed above is determined at the time of fabrication, according to the user's specifications. However, in order to avoid having to custom design each ROM from scratch (which would be an extremely costly process), ROMs are manufactured using a process known as **mask programming.** As explained in Appendix A, integrated circuits are fabricated on a wafer of silicon using a sequence of processing steps that include photomasking, etching, and diffusion. In this way, a pattern of junctions and interconnections is created on the surface of the wafer. One of the final steps in the fabrication process consists of coating the surface of the wafer with a layer of aluminum and then selectively (using a mask) etching away portions of the aluminum, leaving aluminum only where interconnections are desired. This last step can be used to program (that is, store a desired pattern in) a ROM. For instance, if the ROM is made of enhancement MOS transistors as in Fig. 13.45, then MOSFETs are included at all bit locations, but only the gates of those transistors where 0s are to be stored are connected to the word lines; the gates of transistors where 1s are to be stored are not connected. This pattern is determined by the mask, which is produced according to the user's specifications.

The economic advantages of the mask programming process should be obvious: All ROMs are fabricated similarly; customization occurs only during one of the final steps in fabrication.

Programmable ROMs (PROMs and EPROMs)

PROMs are ROMs that can be programmed by the user, but only once. A typical arrangement employed in BJT PROMs involves using polysilicon fuses to connect the emitter of each BJT to the corresponding digit line. Depending on the desired content of a ROM cell, the fuse can be either left intact or blown up using a large current. The programming process is obviously irreversible.

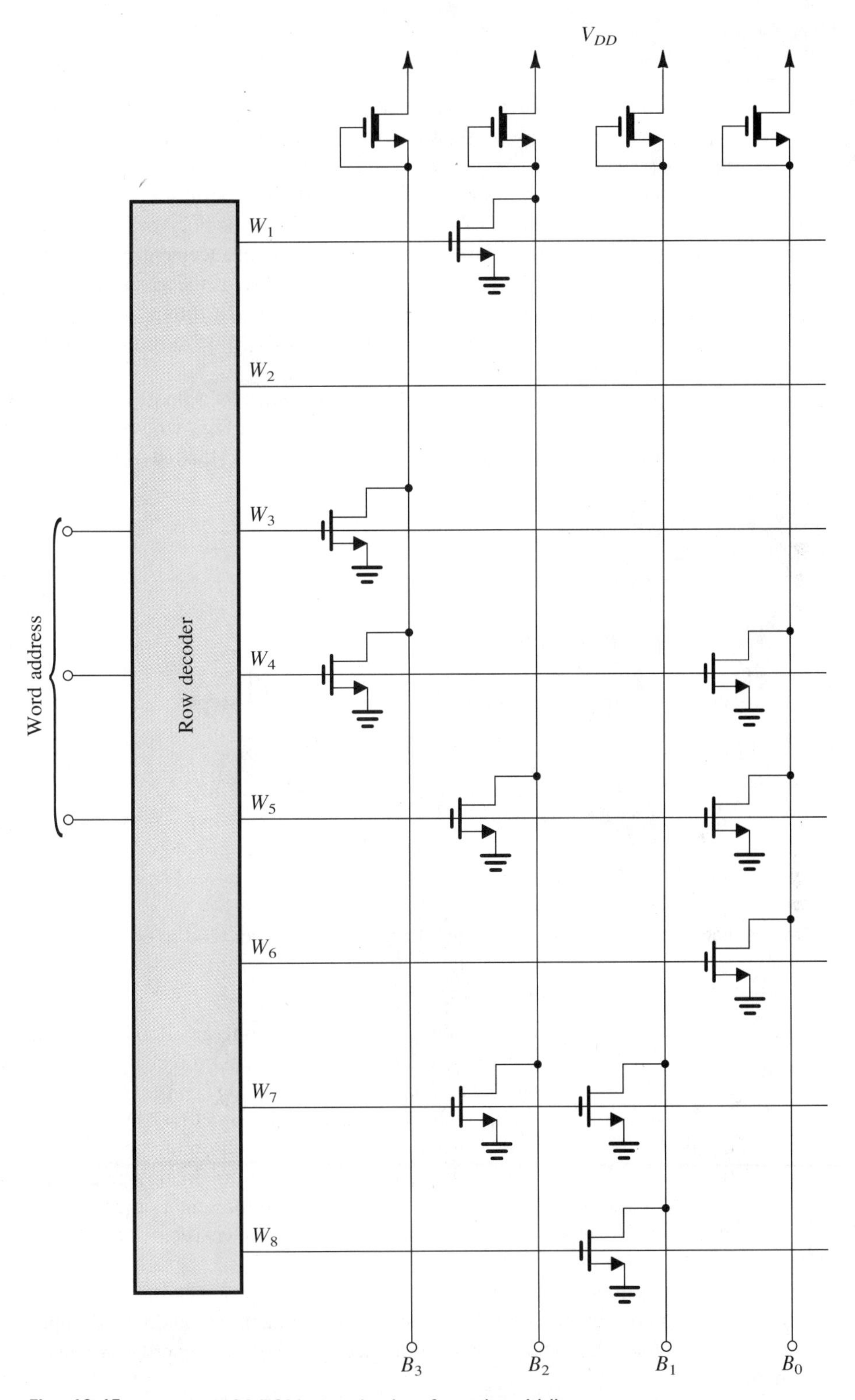

Fig. 13.45 A simple MOS ROM organized as 8 words × 4 bits.

An erasable programmable ROM, or EPROM, is a ROM that can be erased and reprogrammed as many times as the user wishes. It is therefore the most versatile type of read-only memory. It should be noted, however, that the process of erasure and reprogramming is time-consuming and is intended to be performed only infrequently.

State-of-the-art EPROMs use variants of the memory cell whose cross section is shown in Fig. 13.46(a). The cell is basically an enhancement-type *n*-channel MOSFET with two gates made of polysilicon material.[8] One of the gates is not electrically connected to any other part of the circuit; rather, it is left floating and is appropriately called a **floating gate.** The other gate, called a **select gate,** functions in the same manner as the gate of a regular enhancement MOSFET.

The MOS transistor of Fig. 13.46(a) is known as a **floating-gate transistor** and is given the circuit symbol shown in Fig. 13.46(b). In this symbol the broken line denotes the floating gate. The memory cell is known as the **stacked-gate cell.**

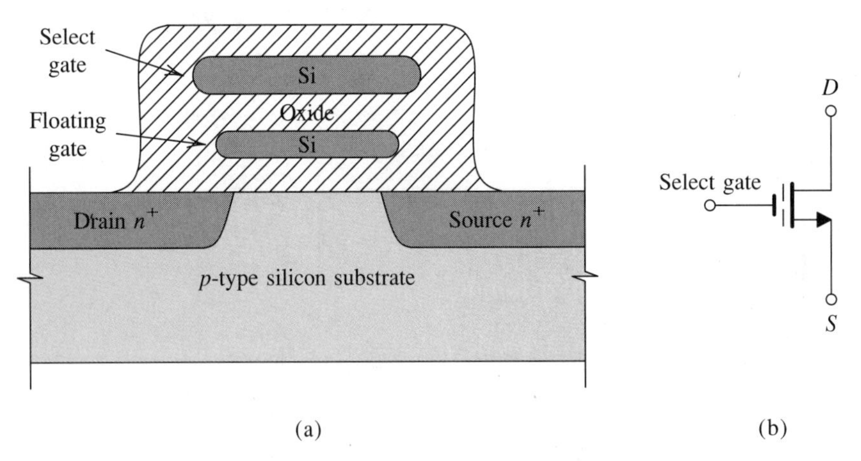

Fig. 13.46 (a) Cross section and **(b)** circuit symbol of the floating-gate transistor used as an EPROM cell.

Let us now examine the operation of the floating-gate transistor. Before the cell is programmed (we will shortly explain what this means), no charge exists on the floating gate and the device operates as a regular *n*-channel enhancement MOSFET. It thus exhibits the i_D–v_{GS} characteristic shown as curve (*a*) in Fig. 13.47. Note that in this case the threshold voltage (V_t) is rather low. This state of the transistor is known as the **not-programmed state.** It is one of two states in which the floating-gate transistor can exist. Let us arbitrarily take the not-programmed state to represent a stored 1. That is, a floating-gate transistor whose i_D–v_{GS} characteristic is that shown as curve (*a*) in Fig. 13.47 will be said to be storing a 1.

To program the floating-gate transistor, a large voltage (16–20 V) is applied between its drain and source. Simultaneously, a large voltage (about 25 V) is applied to its select gate. Figure 13.48 shows the floating-gate MOSFET during programming. In the absence

[8] See Appendix A for a description of silicon-gate technology.

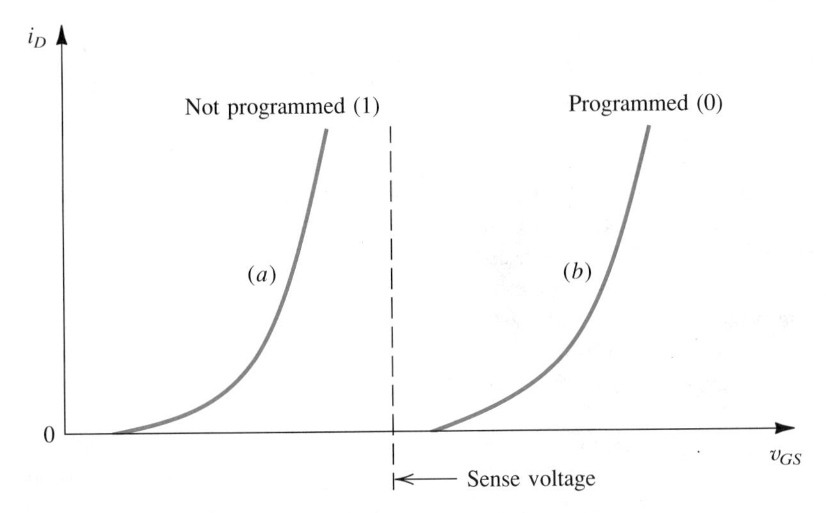

Fig. 13.47 Illustrating the shift in the i_D–v_{GS} characteristic of a floating-gate transistor as a result of programming.

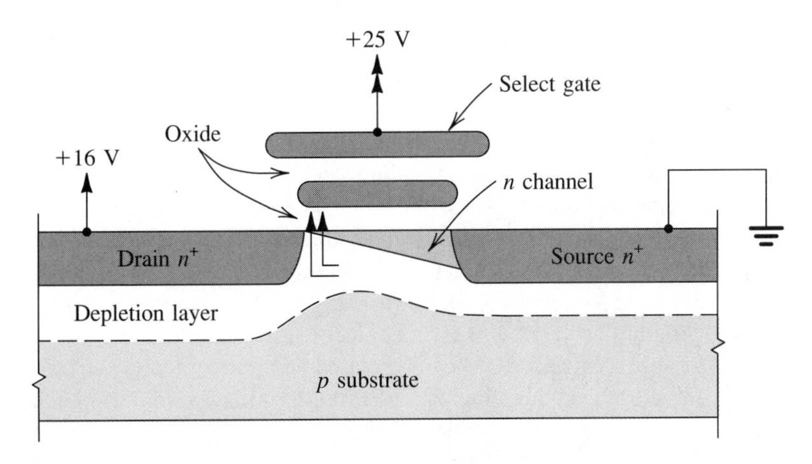

Fig. 13.48 The floating-gate transistor during programming.

of any charge on the floating gate the device behaves as a regular *n*-channel enhancement MOSFET. An *n*-type inversion layer (channel) is created at the wafer surface as a result of the large positive voltage applied to the select gate. Because of the large positive voltage at the drain, the channel has a tapered shape.

The drain-to-source voltage accelerates electrons through the channel. As these electrons reach the drain end of the channel they acquire sufficiently large kinetic energy and are referred to as *hot electrons*. The large positive voltage on the select gate (greater than the drain voltage) establishes an electric field in the insulating oxide. This electric field attracts the hot electrons and accelerates them toward the floating gate. In this way the floating gate is charged, and the charge that accumulates on it becomes trapped.

Fortunately, the process of charging the floating gate is self-limiting. The negative charge that accumulates on the floating gate reduces the strength of the electric field in the oxide to the point that it eventually becomes incapable of accelerating any more of the hot electrons.

Let us now inquire about the effect of the floating gate's negative charge on the operation of the transistor. The negative charge trapped on the floating gate will cause electrons to be repelled from the surface of the substrate. This implies that to form a channel, the positive voltage that has to be applied to the select gate will have to be greater than that required when the floating gate is not charged. In other words, the threshold voltage V_t of the programmed transistor will be higher than that of the not-programmed device. In fact, programming causes the i_D–v_{GS} characteristic to shift to that labeled (b) in Fig. 13.47. In this state, known as the *programmed state,* the cell is said to be storing a 0.

Once programmed, the floating-gate device retains its shifted i–v characteristic (curve *b*) even when the power supply is turned off. In fact, extrapolated experimental results indicate that the device can remain in the programmed state for as long as 100 years!

Reading the content of the stacked-gate cell is easy: A voltage V_{GS} somewhere between the low and high threshold values (see Fig. 13.47) is applied to the select gate. While a programmed device (one that is storing a 0) will not conduct, a not-programmed device (one that is storing a 1) will conduct heavily.

To return the floating-gate MOSFET to its not-programmed state, the charge stored on the floating gate has to be returned to the substrate. This *erasure* process can be accomplished by illuminating the cell with ultraviolet light of the correct wavelength (2537 Å) for a specified duration. The ultraviolet light imparts sufficient photon energy to the trapped electrons, allowing them to overcome the inherent energy barrier and thus to be transported through the oxide, back to the substrate. To allow this erasure process, the EPROM package contains a quartz window. Finally, it should be noted that the device is extremely durable and can be erased and programmed many times.

A more versatile programmable ROM is the electrically-erasable PROM (or EEPROM). As the name implies, an EEPROM can be erased and reprogrammed electrically without the need for ultraviolet illumination. EEPROMs utilize a variant of the floating-gate MOSFET.

13.11 GALLIUM-ARSENIDE DIGITAL CIRCUITS

We conclude this chapter with a brief discussion of digital logic circuits implemented using the emerging technology of gallium-arsenide. An introduction to this technology and its two basic devices, the MESFET and the Schottky-barrier diode (SBD), was given in Section 5.11. We urge the reader to review Section 5.11 before proceeding with the study of this section.

The major advantage that GaAs technology offers is a higher speed of operation than currently achievable using silicon devices. Gate delays of 10 to 100 ps have been reported for GaAs circuits. The disadvantages are a relatively high power dissipation per gate (1 to 10 mW); relatively small voltage swings and, correspondingly, narrow noise margins;

low packing density, mostly as a result of the high-power dissipation per gate; and low manufacturing yield. The present state of affairs is that a few specialized manufacturers produce SSI, MSI, and some LSI digital circuits performing relatively specialized functions, with a cost per gate considerably higher than that of silicon digital ICs. Nevertheless, the very high speeds of operation achievable in GaAs circuits make it a worthwhile technology whose applications will possibly grow.

Unlike the MOS logic-circuit families that we have studied in earlier sections of this chapter, and the bipolar logic families that we will study in Chapter 14, there are no standard GaAs logic-circuit families. The lack of standards extends not only to the topology of the basic gates but also to the power-supply voltages used. In the following we present examples of the most popular GaAs logic gate circuits.

Direct-Coupled FET Logic (DCFL)

Direct-coupled FET logic (DCFL) is the simplest form of GaAs digital logic circuits. The basic gate is shown in Fig. 13.49. The gate utilizes enhancement MESFETs, Q_1 and Q_2, for the input switching transistors, and a depletion MESFET for the load transistor Q_L. The gate closely resembles the depletion-load MOSFET circuit of Fig. 13.14(a). Like the MOSFET circuit, the GaAs circuit of Fig. 13.49 implements a two-input NOR function. Operation of the GaAs circuit, however, differs from that of the MOSFET circuit in a fundamental way.

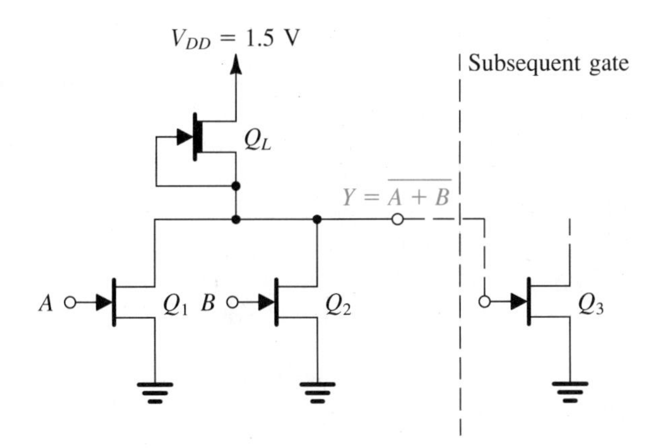

Fig. 13.49 A DCFL GaAs gate implementing a two-input NOR function. The gate is shown driving the input transistor Q_3 of another gate.

To see how the MESFET circuit of Fig. 13.49 operates, ignore input B and consider the basic inverter formed by Q_1 and Q_L. When the input voltage applied to node A, v_I, is lower than the threshold voltage of the enhancement MESFET Q_1, denoted V_{tE}, transistor Q_1 will be off. Recall that V_{tE} is positive and for GaAs MESFETs is typically 0.1 to 0.3 V. Now if the gate output Y is open circuited, the output voltage will be very close to V_{DD}. In practice, however, the gate will be driving another gate, as indicated in Fig. 13.49, where Q_3 is the input transistor of the subsequent gate. In such a case, current will flow from V_{DD} through Q_L and into the gate terminal of Q_3. Recalling that the gate to source of a GaAs MESFET is a Schottky-barrier diode that exhibits a voltage drop of about 0.7 V when conducting, we see that the gate conduction of Q_3 will clamp the output high voltage (V_{OH}) to about 0.7 V. This is in sharp contrast to the MOSFET case, where no gate conduction takes place.

Figure 13.50 shows the DCFL inverter under study with the input of the subsequent gate represented by a Schottky diode Q_3. With $v_I < V_{tE}$, $i_1 = 0$ and i_L flows through Q_3 resulting in $v_O = V_{OH} \simeq 0.7$ V. Since V_{DD} is usually low (1.2 to 1.5 V) and the threshold voltage of Q_L, V_{tD}, is typically -0.7 to -1 V, Q_L will be operating in the triode region. (To simplify matters, we shall ignore in this discussion the early-saturation effect exhibited by GaAs MESFETs.)

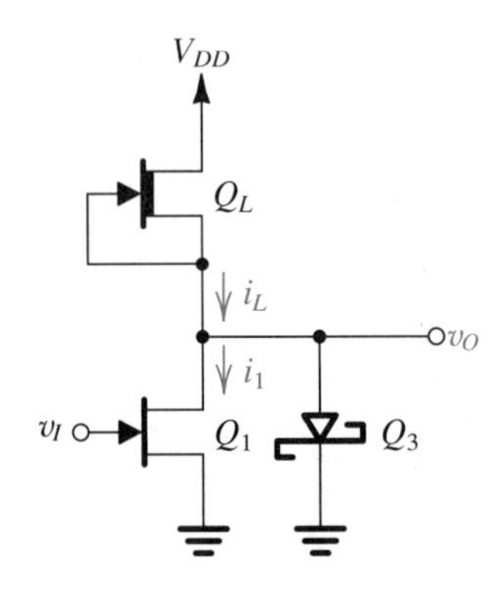

Fig. 13.50 The DCFL gate with the input of the subsequent gate represented by a Schottky diode Q_3.

As v_I is increased above V_{tE}, Q_1 turns on and conducts a current denoted i_1. Initially, Q_1 will be in the saturation region. Current i_1 subtracts from i_L, thus reducing the current in Q_3. The voltage across Q_3, v_O, decreases slightly. However, for the present discussion we shall assume that v_O will remain close to 0.7 V as long as Q_3 is conducting. This will continue until v_I reaches the value that results in $i_1 = i_L$. At this point, Q_3 ceases conduction and can be ignored altogether. Further increase in v_I results in i_1 increasing, v_O decreasing, and $i_L = i_1$. When $(V_{DD} - v_O)$ exceeds $|V_{tD}|$, Q_L saturates; and when v_O falls below v_I by V_{tE}, Q_1 enters the triode region. Eventually, when $v_I = V_{OH} = 0.7$ V, $v_O = V_{OL}$, which is typically 0.1 to 0.2 V.

From the description above we see that the output voltage swing of the DCFL gate is limited by gate conduction to a value less than 0.7 V (typically 0.5 V or so). Further details on the operation of the DCFL gate are illustrated by the following example.

EXAMPLE 13.4

Consider a DCFL gate fabricated in a GaAs technology for which $L = 1\ \mu$m, $V_{tD} = -1$ V, $V_{tE} = 0.2$ V, β (for 1-μm width) $= 10^{-4}$ A/V^2, and $\lambda = 0.1$ V^{-1}. Let the widths of the input MESFETs be 50 μm, and let the width of the load MESFET be 6 μm. $V_{DD} = 1.5$ V. Using a constant-voltage-drop model for the gate-source Schottky diode with $V_D = 0.7$ V, and neglecting the early-saturation effect of GaAs MESFETs (that is, using Eqs. 5.107 to describe MESFET operation), find V_{OH}, V_{OL}, V_{IH}, NM_H, NM_L, the static power dissipation, and the propagation delay for a total equivalent capacitance at the gate output of 30 fF.

Solution From the description above of the operation of the DCFL gate we found that $V_{OH} =$ 0.7 V. To obtain V_{OL}, we consider the inverter in the circuit of Fig. 13.50 and let $v_I =$ $V_{OH} = 0.7$ V. Since we expect $v_O = V_{OL}$ to be small, we assume Q_1 to be in the triode region and Q_L to be in saturation. (Q_3 is of course off.) Equating i_1 and i_L gives the equation

$$\beta_1\left[2(0.7-0.2)V_{OL} - V_{OL}^2\right](1 + 0.1\ V_{OL}) = \beta_L\left[0-(-1)\right]^2\left[1 + 0.1(1.5 - V_{OL})\right]$$

To simplify matters we neglect the terms 0.1 V_{OL} and substitute $\beta_L/\beta_1 = W_L/W_1 = 6/50$ to obtain a quadratic equation in V_{OL} whose solution gives $V_{OL} \simeq 0.17$ V.

Toward obtaining the value of V_{IL} we shall first find the value of v_I at which $i_1 = i_L$, the diode Q_3 turns off, and v_O begins to decrease. Since at this point $v_O = 0.7$ V we assume Q_1 is in saturation. Transistor Q_L has a v_{DS} of 0.8 V, which is less than $|V_{tD}|$ and is thus in the triode region. Equating i_1 and i_L gives

$$\beta_1(v_I - 0.2)^2(1 + 0.1 \times 0.7) = \beta_L[2(1)(1.5-0.7) - (1.5-0.7)^2]\ [1 + 0.1(1.5-0.7)]$$

Substituting $\beta_L/\beta_1 = W_L/W_1 = 6/50$ and solving the resulting equation yields $v_I =$ 0.54 V. Figure 13.51 shows a sketch of the transfer characteristic of the inverter. The slope dv_O/dv_I at point A can be found to be -14.2 V/V. We shall consider point A as the point at which the inverter begins to switch from the high-output state; thus $V_{IL} \simeq 0.54$ V.

To obtain V_{IH}, we find the co-ordinates of point B at which $dv_O/dv_I = -1$. This can be done using a procedure similar to that employed for the MOSFET inverters and assum-

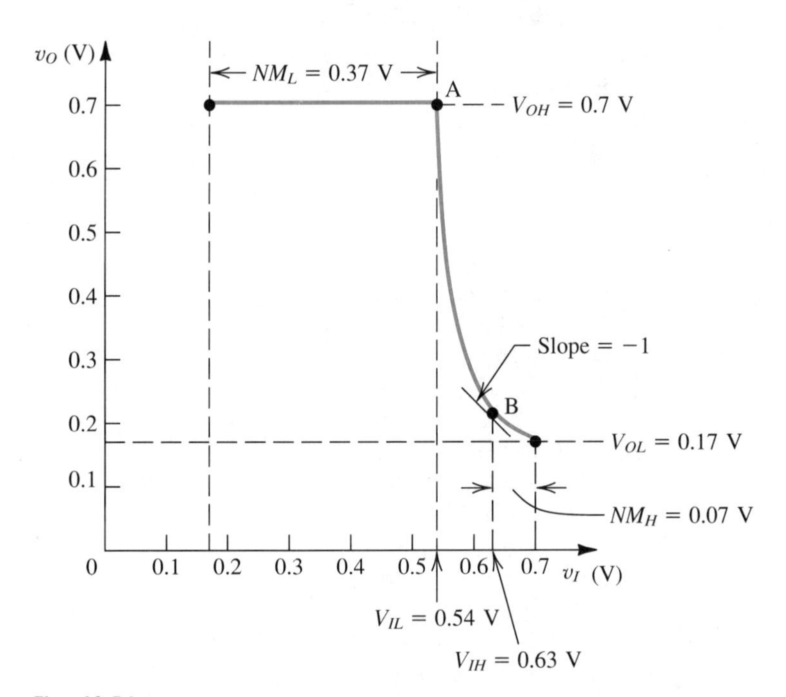

Fig. 13.51 Transfer characteristic of the DCFL inverter of Fig. 13.50.

ing Q_1 to be in the triode region and Q_L to be in saturation. Neglecting terms in $0.1\ v_O$, the result is $V_{IH} \simeq 0.63$ V. The noise margins can now be found as

$$NM_H \equiv V_{OH} - V_{IH} = 0.7 - 0.63 = 0.07 \text{ V}$$

$$NM_L \equiv V_{IL} - V_{OL} = 0.54 - 0.17 = 0.37 \text{ V}$$

The static power dissipation is determined by finding the supply current I_{DD} in the output-high and the output-low cases. When the output is high (at 0.7 V), Q_L is in the triode region and the supply current is

$$I_{DD} = \beta_L\Big[2(0+1)(1.5-0.7) - (1.5-0.7)^2\Big]\Big[1 + 0.1(1.5-0.7)\Big]$$

Substituting $\beta_L = 10^{-4} \times W_L = 0.6 \text{ mA/V}^2$ results in

$$I_{DD} = 0.61 \text{ mA}$$

When the output is low (at 0.17 V), Q_L is in saturation and the supply current is

$$I_{DD} = \beta_L(0+1)^2\Big[1 + 0.1(1.5 - 0.17)\Big] = 0.68 \text{ mA}$$

Thus the average supply current is

$$I_{DD} = \tfrac{1}{2}(0.61 + 0.68) = 0.645 \text{ mA}$$

and the static power dissipation is

$$P_D = 0.645 \times 1.5 \simeq 1 \text{ mW}$$

The propagation delay t_{PHL} is the time for the output voltage of the inverter to decrease from $V_{OH} = 0.7$ V to $\frac{1}{2}(V_{OH} + V_{OL}) = 0.435$ V. During this time v_I is at the high level of 0.7 V, and the capacitance C (assumed to be 30 fempto Farads [fF]) is discharged by $(i_1 - i_L)$; refer to Fig. 13.52(a). The average discharge current is found by calculating i_1 and i_L at the beginning and at the end of the discharge interval. The result is that i_1 changes from 1.34 mA to 1.28 mA and i_L changes from 0.61 mA to 0.66 mA.

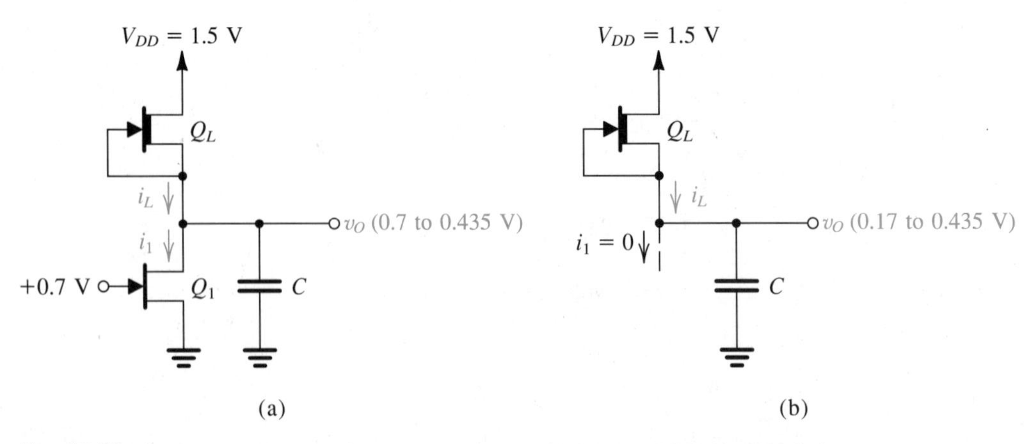

Fig. 13.52 Circuits for calculating the propagation delays of the DCFL inverter: **(a)** t_{PHL}; **(b)** t_{PLH}.

Thus the discharge current $(i_1 - i_L)$ changes from 0.73 mA to 0.62 mA for an average value of 0.675 mA. Thus

$$t_{PHL} = \frac{C\Delta V}{I} = \frac{30 \times 10^{-15}\,(0.7 - 0.435)}{0.675 \times 10^{-3}} = 11.8 \text{ ps}$$

To determine t_{PLH} we refer to the circuit in Fig. 13.52(b) and note that during t_{PLH}, v_O changes from $V_{OL} = 0.17\ V\ to\ \frac{1}{2}(V_{OH} + V_{OL}) = 0.435$ V. The charging current is the average value of i_L, which changes from 0.8 mA to 0.66 mA. Thus $i_{L|\text{average}} = 0.73$ mA and

$$t_{PLH} = \frac{30 \times 10^{-15} \times (0.435 - 0.17)}{0.73 \times 10^{-3}} = 10.9 \text{ ps}$$

The propagation delay of the DCFL gate can now be found as

$$t_P = \tfrac{1}{2}(t_{PHL} + t_{PLH}) = 11.4 \text{ ps}$$

As a final remark, we note that the analysis above was done using simplified device models; our objective is to show how the circuit works rather than to find accurate performance measures. These can be obtained using SPICE simulation with more elaborate models (see the SPICE manual).

Logic Gates Using Depletion MESFETs

The DCFL circuits studied above require both enhancement and depletion devices and thus are somewhat difficult to fabricate. Also, owing to the fact that the voltage swings and noise margins are rather small, very careful control of the value of V_{tE} is required in fabrication. As an alternative, we now present circuits that utilize depletion devices only.

Figure 13.53 shows the basic inverter circuit of a family of GaAs logic circuits known at FET logic (FL). The heart of the FL inverter is formed by the switching transistor Q_S and its load Q_L—both depletion-type MESFETs. Since the threshold voltage of a deple-

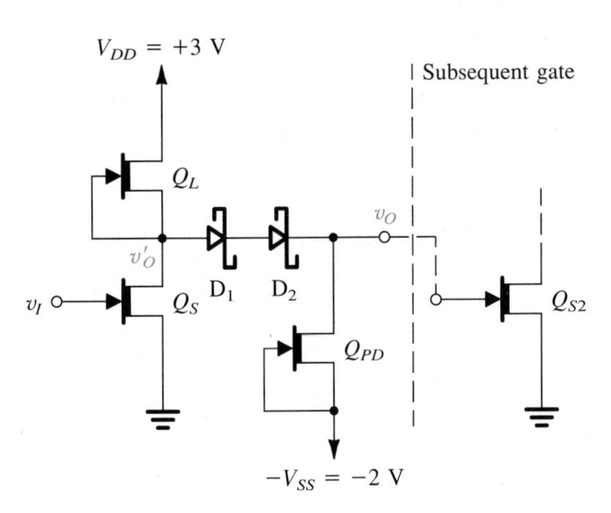

Fig. 13.53 An inverter circuit utilizing depletion-mode devices only. Schottky diodes are employed to shift the output logic levels to values compatible with the input levels required to turn the depletion MESFET Q_S on and off. This circuit is known as FET logic (FL).

tion MESFET, V_{tD}, is negative, a negative voltage $<V_{tD}$ is needed to turn Q_S off. On the other hand the output low voltage at the drain of Q_S will always be positive. It follows that the logic levels at the drain of Q_S are not compatible with the levels required at the gate input. The incompatibility problem is solved by simply shifting the level of the voltage v'_O down by two diode drops, that is, by approximately 1.4 V. This level shifting is accomplished by the two Schottky diodes D_1 and D_2. The depletion transistor Q_{PD} provides a constant-current bias for D_1 and D_2. To ensure that Q_{PD} operates in the saturation region at all times, its source is connected to a negative supply $-V_{SS}$, and the value of V_{SS} is selected to be equal to or greater than the lowest level of v_O (V_{OL}) plus the magnitude of the threshold voltage, $|V_{tD}|$. Transistor Q_{PD} also supplies the current required to discharge a load capacitance when the output voltage of the gate goes low, hence the name "pull-down" transistor and the subscript *PD*.

To see how the inverter of Fig. 13.53 operates, refer to its transfer characteristic, shown in Fig. 13.54. The circuit is usually designed using MESFETs having equal channel lengths (typically 1 μm) and having widths $W_S = W_L = 2W_{PD}$. The transfer characteristic shown is for the case $V_{tD} = -0.9$ V. For v_I lower than V_{tD}, Q_S will be off and Q_L will operate in saturation, supplying a constant current I_L to D_1 and D_2. Transistor Q_{PD} will also operate in saturation with a constant current $I_{PD} = \frac{1}{2}I_L$. The difference between the two currents will flow through the gate terminal of the input transistor of the next gate in the chain, Q_{S2}. Thus the input Schottky diode of Q_{S2} clamps the output voltage v_O to approximately 0.7 V, which is the output high level, V_{OH}. (Note that for this discussion we shall neglect the finite output resistance in saturation.)

As v_I is raised above V_{tD}, Q_S turns on. Since its drain is at +2.1 V, Q_S will operate in the saturation region and will take away some of the current supplied by Q_L. Thus the current flowing into the gate of Q_{S2} decreases by an equal amount. If we keep increasing v_I, a value is reached for which the current in Q_S equals $\frac{1}{2}I_L$, thus leaving no current to flow through the gate of Q_{S2}. This corresponds to the point labeled A on the transfer characteristic. A further slight increase in v_I will cause the voltage v'_O to fall to the point that Q_S enters the triode region. The segment AB of the transfer curve respresents the high-gain region of operation, having a slope equal to $-g_{ms}R$ where R denotes the total equivalent resistance at the drain node. Note that this segment is shown vertical in Fig. 13.54 because we are neglecting the output resistance in saturation.

The segment BC of the transfer curve corresponds to Q_S operating in the triode region. Here Q_L and Q_{PD} continue to operate in saturation and D_1 and D_2 remain conducting. Finally, for $v_I = V_{OH} = 0.7$ V, $v_O = V_{OL}$, which for the case $V_{tD} = -0.9$ V can be found to be -1.3 V.

Exercise 13.30 Verify that the co-ordinates of points A, B, and C of the transfer characteristic are as indicated in Fig. 13.54. Let $V_{tD} = -0.9$ V and $\lambda = 0$.

As indicated in Fig. 13.54, the FL inverter exhibits much higher noise margins than those for the DCFL circuit. The FL inverter, however, requires two power supplies.

The FL inverter can be used to construct a NOR gate by simply adding transistors with drain and source connected in parallel with those of Q_S.

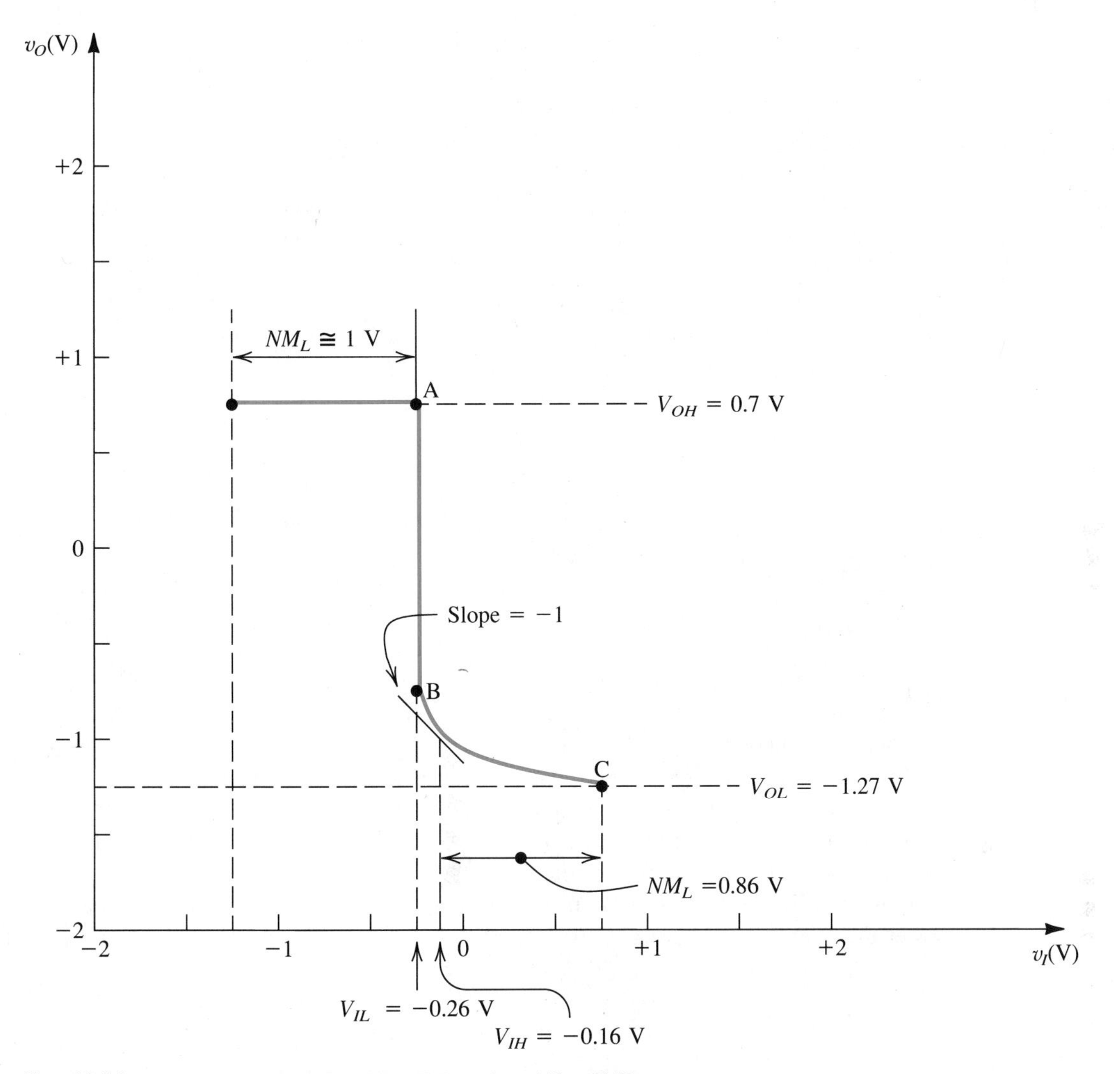

Fig. 13.54 Transfer characteristic of the FL inverter of Fig. 13.53.

Schottky Diode FET Logic (SDFL)

If the diode level-shifting network of the FL inverter is connected at the input side of the gate, rather than at the output side, we obtain the circuit shown in Fig. 13.55(a). This inverter operates in much the same manner as the FL inverter. The modified circuit, however, has a very interesting feature: The NOR function can be implemented by simply connecting additional diodes, as shown in Fig. 13.55(b). This logic form is known as Schottky diode FET logic (SDFL). SDFL permits higher packing density than other forms of MESFET logic because only an additional diode, rather than an additional transistor, is required for each additional input, and diodes require much smaller areas than transistors.

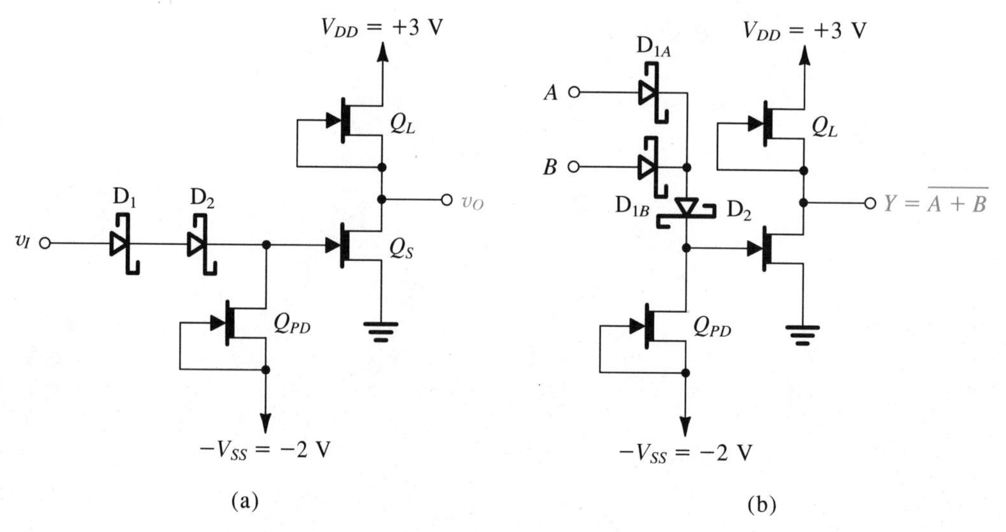

Fig. 13.55 (a) An SDFL inverter. **(b)** An SDFL NOR gate.

Buffered FET Logic (BFL)

Another variation on the basic FL inverter of Fig. 13.53 is possible. A source follower can be inserted between the drain of Q_S and the diode level-shifting network. The resulting gate, shown for the case of a two-input NOR, is depicted in Fig. 13.56. This form of GaAs logic circuit is known as buffered FET logic (BFL). The source-follower transistor Q_{SF} increases the output current-driving capability, thus decreasing the low-to-high propagation time. FL, BFL, and SDFL feature propagation delays of the order of 100 ps and power dissipation of the order of 10 mW/gate.

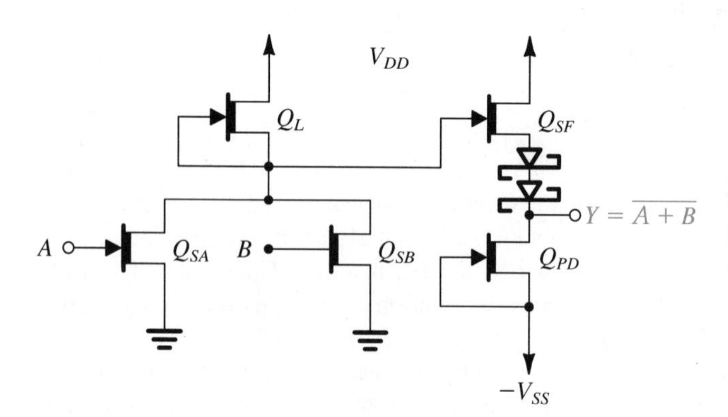

Fig. 13.56 A BFL two-input NOR gate. The gate is formed by inserting a source-follower transistor Q_{SF} between the inverting stage and the level-shifting stage.

SUMMARY

▶ The characteristics of a logic circuit family are determined by analyzing the circuit of its basic inverter.

▶ The static performance of a logic inverter is characterized by its voltage transfer characteristic, which in turn is characterized by the four parameters V_{OL}, V_{OH}, V_{IL}, and V_{IH} (see Fig. 13.5 for definitions). The noise margins are $NM_H = V_{OH} - V_{IH}$ and $NM_L = V_{IL} - V_{OL}$. An ideal inverter has $V_{OH} = V_{DD}$, $V_{OL} = 0$, $V_{IL} = V_{IH} = V_{DD}/2$, and thus $NM_H = NM_L = V_{DD}/2$.

▶ The dynamic performance of a logic inverter is characterized by the propagation delay t_P and the rise and fall times of the output waveform (see Fig. 13.7 for definitions).

▶ The two performance measures of a logic inverter—the average propagation delay t_P and the average power dissipation P_D—are combined to obtain the delay-power product $DP \equiv t_P P_D$, which serves as a figure of merit in comparing logic circuit families. Modern logic circuits have DP in the range 1 to 100 pJ.

▶ There are two MOS logic circuit families: NMOS and CMOS. NMOS has low output-driving capability, and thus its application is limited to VLSI circuit design. CMOS, on the other hand, is used in VLSI as well as in conventional design. At the present time CMOS is the most popular logic circuit technology. NMOS, however, occupies the least silicon area and thus allows higher levels of integration than CMOS.

▶ Of the two NMOS types, depletion-load circuits offer higher performance and smaller silicon area than enhancement-load circuits.

▶ Performance of depletion-load NMOS circuits is seriously degraded by the body effect on the load device.

▶ Although the static power dissipation of CMOS is negligibly small, dynamic power dissipation can be significant at high operating speeds. It is given by $fC_L V_{DD}^2$, where f is the switching frequency and C_L is the load capacitance.

▶ Flip-flops employ one or more latches. The basic latch is a bistable circuit (bistable multivibrator) implemented using two inverters connected in a positive-feedback loop. The latch can remain in either stable state indefinitely.

▶ A monostable multivibrator has one stable state, in which it can remain indefinitely, and one quasi-stable state, which it enters upon triggering and in which it remains for a predetermined interval T. Monostable circuits can be used to generate pulse signals of predetermined height and width.

▶ An astable multivibrator has no stable states. It has two quasi-stable states, between which it oscillates. The astable circuit is, in effect, a square-wave generator.

▶ A random-access memory (RAM) is one in which the time required for storing (writing) information and for retrieving (reading) information is independent of the physical location (within the memory) in which the information is stored.

▶ The bulk of a memory chip consists of the cells in which the bits are stored and which are organized in a square matrix. A cell is selected for reading or writing by activating its row, via the row-address decoder, and its column, via the column-address decoder. The sense amplifier detects the content of the selected cell and provides it to the data-output terminal of the chip.

▶ There are two kinds of MOS RAMs: static and dynamic. Static RAMs employ flip-flops as the storage cells. In a dynamic RAM, data are stored on a capacitor and thus must be refreshed periodically. Dynamic RAM chips provide the highest possible storage capacity.

▶ Read-only memory (ROM) contains fixed data patterns that are stored at the time of fabrication and cannot be changed by the user. On the other hand, the contents of an erasable programmable ROM (EPROM) can be changed by the user. The erasure and reprogramming is a time-consuming process and is performed only infrequently.

▶ Some EPROMS utilize floating-gate MOSFETs (Fig. 13.46) as the storage cells. The cell is programmed by applying a high voltage to the select gate. Erasure is achieved by illuminating the chip by ultraviolet light. The more versatile EEPROMs can be erased and reprogrammed electrically.

▶ GaAs technology provides logic circuits having very high speeds of operation, with gate delays in the 10 to 100 ps range. The power dissipation, however, is somewhat high (1 to 10 mW per gate), limiting the level of integration to LSI. Also, the low manufacturing yield results in a relatively high cost per gate.

▶ Direct-coupled FET logic (DCFL) utilizes enhancement-type MESFETs. The output voltage swing is limited to less than 0.7 V by gate conduction. Greater noise margins are obtained in circuits that employ depletion MESFETs. These, however, require level-shifting circuitry and two power supplies.

BIBLIOGRAPHY

M. I. Elmasry (Ed), *Digital MOS Integrated Circuits,* New York: IEEE Press, 1981.

D. A. Hodges and H. G. Jackson, *Analysis and Design of Digital Integrated Circuits,* Second Edition, New York: McGraw-Hill, 1988.

IEEE Journal of Solid-State Circuits. The October issue of each year has been devoted to digital circuits.

S. L. Long and S. L. Butner, *Gallium-Arsenide Digital Integrated Circuit Design,* New York: McGraw-Hill, 1990.

C. Mead and L. Conway, *Introduction to VLSI Systems,* Reading, Mass.: Addison-Wesley, 1980.

Motorola, *McMOS Handbook,* Phoenix, Ariz.: Motorola Inc., 1974.

Motorola, *Memory Data,* Pheonix, Ariz.: Motorola, Inc., 1989.

RCA, *COS/MOS Digital Integrated Circuits,* Publication No. SSD-203B, Somerville, N.J.: RCA Solid-State Division, 1974.

A. S. Sedra and K. C. Smith, *Microelectronic Circuits,* First Edition, Chapter 16, New York: Holt, Rinehart and Winston, 1982.

M. Shur, *GaAs Devices and Circuits,* New York: Plenum Press, 1987.

N. Weste and K. Eshraghian, *Principles of CMOS VLSI Design,* Reading, Mass.: Addison-Wesley, 1985.

PROBLEMS

Section 13.1: Logic Circuits—Some Basic Concepts

13.1 In the circuit of Fig. 13.2 let the input voltage v_I be relabeled as v_{I1}. Augment the circuit by connecting another switch between the output and ground. The new switch is to be controlled by a second input, v_{I2}. What happens to the output if either or both inputs are high? If both are low? This circuit is called a logic NOR.

13.2 A logic inverter modeled as in Fig. 13.2 employs a switch for which the offset voltage is 100 mV and the on resistance is 100 Ω. If the inverter load resistance is 1 kΩ and V^+ is 5 V, what are the two expected values of output voltage?

13.3 A particular logic inverter has a transfer characteristic for which the -1 slope points occur at $v_I = 1.4$ and 2.6 V. If V_{OL} and V_{OH} are 0.6 and 3.7 V, respectively, find the high and low noise margins.

13.4 Consider a logic inverter having a voltage transfer characteristic consisting of three straight-line segments, two horizontal ones at $v_O = V_{OH} = 4$ V and at $v_O = V_{OL} = 0.5$ V, and the third joining the points $v_I = V_{IL} = 1.5$ V and $v_I = V_{IH} = 2.5$ V. Find

(a) the noise margins,

(b) the value at which v_I and v_O are equal, and

(c) the voltage gain in the transition region.

13.5 For a particular logic family for which the supply voltage is V^+, $V_{OL} = 0.1\ V^+$, $V_{OH} = 0.8\ V^+$, $V_{IL} = 0.4\ V^+$, $V_{IH} = 0.6\ V^+$, what are the noise margins? What is the width of the transition region? For a minimum noise margin of 1 V what value of V^+ is required?

13.6 For a logic inverter with output levels of V_{OL} and V_{OH}, a power supply V^+, and a load capacitance C_L, show that when switched at the rate of f cycles per second the dynamic power dissipation is $fC_LV^+(V_{OH} - V_{OL})$. Evaluate the dynamic power dissipation for the case $V^+ = 5$ V, $V_{OL} = 0.5$ V, $V_{OH} = 4.5$ V, $C_L = 100$ pF, and $f = 1$ MHz.

***13.7** A logic inverter having the circuit of Fig. 13.2 with $V^+ = 5$ V and $R_L = 1$ kΩ, and the switch having an on-resistance of 100 Ω, is switched at a 10-MHz rate. The load capacitance is 10 pF, and the input remains high an average of 70% of the time. Calculate the static, dynamic, and total power dissipation in the gate. What is the power dissipated in the switch? in the load resistor?

13.8 Consider a logic gate for which t_{PLH}, t_{PHL}, t_{TLH}, and t_{THL} are 20 ns, 10 ns, 30 ns, and 15 ns, respectively. The rising and falling edges of the gate output can be approximated by linear ramps. Two such gates are connected in tandem and driven by an ideal input having zero rise and fall times. Calculate the time taken for the output voltage to complete 90% of its excursion for

(a) a rising input,

(b) a falling input.

What is the propagation time for the gate?

13.9 A particular logic gate has t_{PLH} and t_{PHL} of 50 and 70 ns, respectively, and dissipates 1 mW with output low and 0.5 mW with output high. Calculate the corresponding delay-power product (under the assumption of a 50% duty-cycle signal).

**13.10 An inverter that can be characterized by Fig. 13.2 has $V^+ = 5$ V, $R_L = 1$ kΩ, and $R_{on} = 100$ Ω. It is loaded by a similar inverter whose switching threshold is at 2 V, by means of a connection whose capacitance to ground is 50 pF. If both switches exhibit a pure delay of 10 ns from the moment their input signal threshold is crossed, how long does it take for an input step to open the switch of the second inverter? to close it?

**13.11 Consider the situation in which three identical inverting gates are connected in a closed ring. Using the ideas expressed in Fig. 13.7 and assuming that a rising edge occurs at the input of one of the inverters, sketch the signals that result at this and all other inputs. For simplicity, assume the rise and fall times to be zero, but assume that the propagation times are not zero. For a uniform propagation delay of 50 ns, what is the frequency of the resulting oscillator? What is it if $t_{PLH} = 60$ ns and $t_{PHL} = 40$ ns?

Section 13.2: NMOS Inverter with Enhancement Load

D13.12 For the circuit of Fig. 13.9 operating from a supply V_{DD} with devices for which $V_t = 0.2\, V_{DD}$, find an expression for K_R that results in $V_{OL} = x\, V_{DD}$, where $x < 1$. Ignore the body effect. Find the values of K_R required to obtain $x = 0.1$, 0.02, and 0.01.

13.13 Consider an enhancement-load inverter having $V_{t0} = 1$ V, $(W/L)_1 = 4$, $(W/L)_2 = \frac{1}{4}$, $\mu_n C_{OX} = 20\ \mu\text{A/V}^2$, $2\phi_f = 0.6$ V, $\gamma = 0.5\ \text{V}^{1/2}$, and $V_{DD} = 5$ V.

(a) Neglecting the body effect, find V_{OH}, V_{IL}, V_{OL}, V_{IH}, NM_H, and NM_L.

(b) Taking the body effect into account, find the modified values of V_{OH} and NM_H.

(c) Find the supply current in both states and hence find the static power dissipation.

13.14 For the inverter specified in Problem 13.13, neglecting the body effect, calculate the current available for discharging a load capacitor when $v_I = V_{OH}$ and $v_O = 4$ V, 3 V, and 1 V.

13.15 An n-channel enhancement MOS device, used in an inverter load operating from a 5-V supply, has V_{t0} ranging from 1 to 1.5 V and γ ranging from 0.3 to 1 $\text{V}^{1/2}$ and $2\phi_f = 0.6$ V. What is the lowest value of V_{OH} that will be found using transistors of this kind?

*13.16 Reconsider Example 13.1. Note that the calculations that result in $V_{IH} = 2.2$ V at $v_O = 0.8$ V and $V_{OL} = 0.3$ V are performed with the body effect neglected. Noting that with the body effect $V_{OH} = 3.4$ V, what does V_{OL} become? Now using the stated result for V_{IH}, iterate once by first finding the corresponding V_{t2} and thus V_{IH} and the corresponding v_O. Calculate the noise margins.

13.17 An enhancement-load inverter operated from a 5-V supply is found to have $V_{OH} = 3$ V. If $V_{t0} = 1$ V and $2\phi_f = 0.6$ V, what must γ be?

*13.18 Reconsider that analysis of t_{PLH} for the capacitively loaded enhancement-load inverter approximated by Eqs. (13.15) and (13.16), by writing and solving the differential equation that describes the operation of the circuit in Fig. 13.11(d). Ignore the body effect in your analysis. Contrast your result for the inverter of Example 13.1 and a 0.1-pF load with the value of 6.4 ns obtained by the approximate method.

D13.19 For the circuit of Example 13.1, for which t_{PHL} is approximately 0.7 ns and t_{PLH} is 6.4 ns with a 0.1-pF load, it is required to reduce t_{PLH} as much as possible by increasing the width of the load device. This, however, has the undesirable effect of raising V_{OL}. If the width is increased so that $V_{OL} = 1$ V, what does t_{PLH} become? Ignore the body effect.

13.20 Derive the result given in Eq. (13.17).

13.21 Use Eqs. (13.19), (13.20), and (13.21) to calculate the average propagation delay, the average power dissipation, and the delay-power product of an inverter for the following conditions.

$K_2(\mu\text{A/V}^2)$	V_t(V)	V_{DD}(V)	C(pF)
5	1	5	0.1
1	1	5	0.1
5	0.5	5	0.1
5	1	10	0.1
5	1	5	1.0
1	0.5	5	0.1

Section 13.3: NMOS Inverter with Depletion Load

13.22 For the NMOS inverter with depletion load (Fig. 13.12a) show that:

(a) When $v_I = V_{DD}$ and $v_O = V_{OL}$, the resistance between drain and source of Q_1 is given approximately by

$$r_{DS1} \simeq 1\Big/\left[\mu_n C_{OX}\,(W/L)_1\,(V_{DD} - V_{tE})\right]$$

where V_{OL} is assumed to be small.

(b) When v_O is close to V_{DD}, the drain-to-source resistance of Q_2 is given approximately by

$$r_{DS2} \simeq 1/[\mu_n C_{OX} (W/L)_2 |V_{tD}|]$$

Calculate the values of r_{DS1} and r_{DS2} for an inverter having $\mu_n C_{OX} = 50\ \mu A/V^2$, $V_{tE} = 0.8$ V, $V_{tD0} = -3$ V, $\gamma = 0.4\ V^{1/2}$, $2\phi_f = 0.6$ V, $(W/L)_1 = 10\ \mu m/2\ \mu m$, $(W/L)_2 = 2.5\ \mu m/2\ \mu m$, and $V_{DD} = 5$ V.

13.23 (a) Show that for the depletion-load inverter, V_{IH} is given by

$$V_{IH} = V_{tE} + 2\ |V_{tD}|/\sqrt{3K_R}$$

and the corresponding value of output voltage is

$$v_O = |V_{tD}|/\sqrt{3K_R}$$

(b) For the inverter whose parameters are specified in Problem 13.22, find V_{IH} and v_O by first neglecting the body effect. Then use the value of v_O to find V_{tD} and thus obtain a better estimate for V_{IH}.

D13.24 Use the expression for V_{IH} given in Problem 13.23 to show that for $NM_H = V_{DD}/2$ we must have

$$V_{tE} + 2\ |V_{tD}|/\sqrt{3K_R} = V_{DD}/2$$

For $V_{tE} = 0.8$ V, $V_{tD0} = -3$ V, and $V_{DD} = 5$ V, find the required value of K_R. Neglect the body effect.

D13.25 For the depletion-load inverter, ignoring the body effect, show that V_{OL} is given approximately by

$$V_{OL} \simeq V_{tD0}^2/2K_R(V_{DD} - V_{tE})$$

For $V_{DD} = 5$ V, $V_{tE} = 0.8$ V, $V_{tD0} = -3$ V, find K_R that results in $V_{OL} = 0.1$, 0.2, and 0.3 V.

13.26 (a) For the depletion-load inverter, show that

$$V_{IL} = V_{tE} + |V_{tD}|/K_R$$

and the corresponding output is

$$v_O \simeq V_{DD} - |V_{tD}|/2K_R$$

where it is assumed that v_O is very close to V_{DD}.

(b) Calculate the values of V_{IL} and v_O for the inverter whose parameters are specified in Problem 13.22.

(c) Use the value of V_{IL} obtained in (b) together with the values of V_{OL} found using the expression given in Problem 13.25 to determine the value of NM_L.

13.27 For the inverter whose parameters are specified in Problem 13.22, calculate the current when the output is low. Find the average static power dissipation.

13.28 Refer to Fig. 13.13 and let the inverter parameters be as specified in Problem 13.22. Determine the average current available to discharge the capacitor load, I_{HL}, and the average current available to charge the capacitor load, I_{LH}. For $C = 0.1$ pF calculate t_{PHL}, t_{PLH}, and t_P.

**13.29 The object of this problem is the derivation of the expression in Eq. (13.29) for the delay-power product, DP, of a depletion-load inverter. Refer to Fig. 13.13.

(a) Show that the average static power dissipation is given by $P_D = \frac{1}{2}K_2 V_{tD0}^2 V_{DD}$.

(b) Assuming that $V_{OL} \simeq 0$, $V_{tE} \ll V_{DD}$, and $i_{D2}(M) \ll i_{D1}(N)$, show that the average capacitor discharging current $I_{HL} \simeq \frac{7}{8}K_1 V_{DD}^2$ and thus

$$t_{PHL} \simeq \frac{4}{7}\frac{C}{K_1 V_{DD}}$$

(c) Assuming that Q_2 is still in saturation at point M (not necessarily a valid assumption but approximately true), show that the average capacitor charging current $I_{LH} \simeq \alpha K_2 V_{tD0}^2$, where α is a parameter that depends on the body effect. Find an approximate expression for α. Now show that $t_{PLH} \simeq CV_{DD}/2\alpha K_2 V_{tD0}^2$.

(d) Use the results of (b) and (c) to obtain an expression for t_P.

(e) Use the results of (a) and (d) to show that

$$DP = \frac{C\ V_{tD0}^2}{7K_R} + \frac{C\ V_{DD}^2}{8\alpha} \simeq \frac{1}{8\alpha} C\ V_{DD}^2$$

Section 13.4: NMOS Logic Circuits

D13.30 If the optimum dimensions of a depletion-load NMOS inverter are $W/L = 4\ \mu m/2\ \mu m$ for the inverting transistor and $2\ \mu m/4\ \mu m$ for the load transistor, find the dimensions of each input transistor in

(a) a 3-input NOR gate, and

(b) a 3-input NAND gate

in each case find the gate area.

D13.31 Find the logic function implemented by the circuit shown in Fig. P13.31. Give device sizes if the technology is that indicated in Problem 13.30.

D13.32 Show that the circuit in Fig. P13.32 implements the logic function $\overline{Y} = x_1 x_2 + \overline{x}_1 \overline{x}_2$. Find Y. (The latter form is known as the EXCLUSIVE-OR function.) Give device sizes if the technology is that indicated in Problem 13.30.

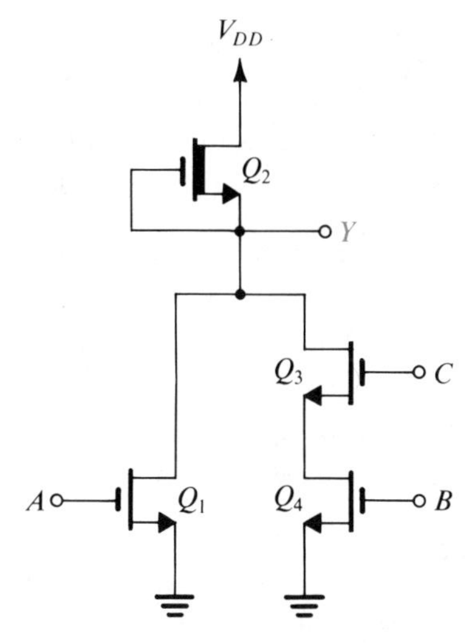

Fig. P13.31

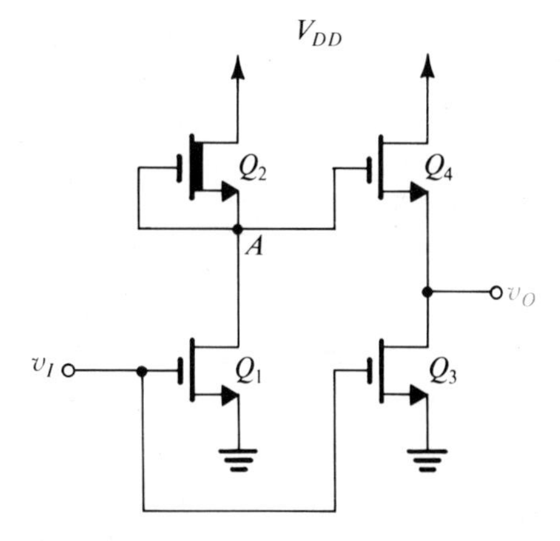

Fig. P13.33

**13.33 The circuit shown in Fig. P13.33 can be used in applications, such as output buffers and clock drivers, where capacitive loads require high current-drive capability. Let $V_{tE} = 1$ V, $V_{tD0} = -3$ V, $V_{DD} = 5$ V, $\mu_n C_{OX} = 20\ \mu\text{A/V}^2$, $W_1 = W_2 = 12\ \mu\text{m}$, $L_1 = L_2 = L_3 = L_4 = 6\ \mu\text{m}$, $W_3 = 120\ \mu\text{m}$, and $W_4 = 240\ \mu\text{m}$. To simplify matters we shall assume the body effect to be negligible.

(a) Find V_{AH}, V_{OH}, V_{AL}, and V_{OL} for v_I (high) = 5 V. (Note that V_{AH} denotes the high level at node A, etc.)

(b) Find the total static current with output high and with output low.

(c) As v_O goes low, find the peak current available to discharge a load capacitance $C = 10$ pF. Also find the discharge current for v_O at the 50% output swing point and thus find the average discharge current and t_{PHL}. Assume the capacitance at node A to be zero.

(d) Repeat (c) for v_O going high, thus finding t_{PLH}.

***13.34 The circuit shown in Fig. P13.34, called a bootstrap driver, is intended to provide a large output voltage swing and high output drive current with a reasonable standing current. Let $V_{DD} = 5$ V, $V_t = 1$ V, $K_1 = 40\ \mu\text{A/V}^2$, $K_2 = 10\ \mu\text{A/V}^2$, and $K_3 = 1\ \mu\text{A/V}^2$;

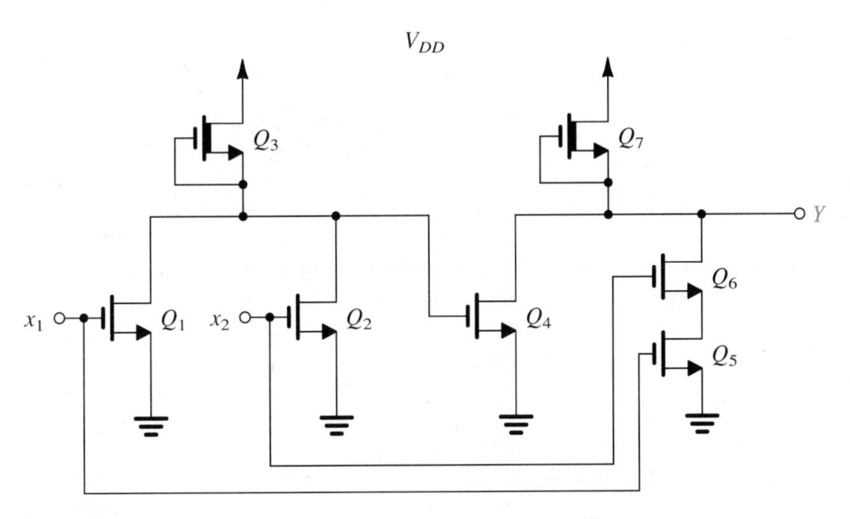

Fig. P13.32

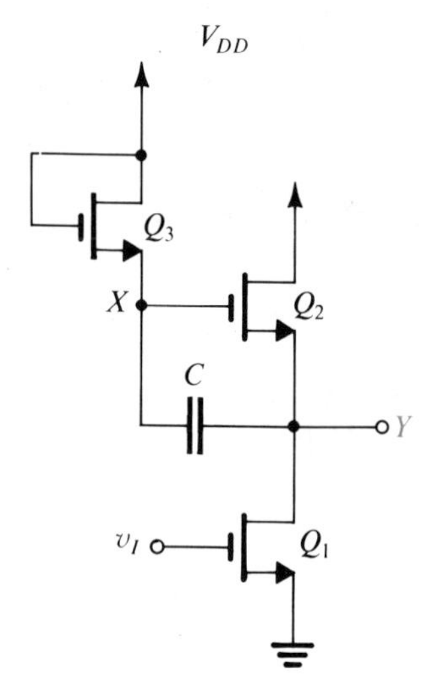

Fig. P13.34

and assume that C is large compared to any capacitances at nodes X and Y. Also, ignore the body effect.

(a) Find the voltages at nodes X and Y when v_I has been high (at 4 V) for some time.

(b) When v_I goes low to 0 V and Q_1 turns off, to what voltages do nodes X and Y rise?

(c) Find the current available to charge a load capacitance at Y as soon as v_I goes low and Q_1 turns off.

(d) As v_I goes high again, what current is immediately available to discharge the load capacitance at Y?

(e) What is that current shortly thereafter, when $v_Y = 4$ V?

Section 13.5: The CMOS Inverter

13.35 For the CMOS inverter, show that when v_I is high, the resistance between the output node and ground is approximately $1/[(\mu_n C_{OX})(W/L)_n (V_{DD} - V_{tn})]$. Also show that when the input is low, the resistance between the output node and the power supply is approximately $1/[(\mu_p C_{OX})(W/L)_p (V_{DD} - |V_{tp}|)]$. Evaluate both resistances for the case $\mu_n C_{OX} = 2\mu_p C_{OX} = 20\ \mu\text{A/V}^2$, $(W/L)_n = 10\ \mu\text{m}/5\ \mu\text{m}$, $(W/L)_p = 20\ \mu\text{m}/5\ \mu\text{m}$, $V_{tn} = -V_{tp} = 1$ V, and $V_{DD} = 5$ V.

13.36 What is the maximum current that the CMOS inverter specified in Problem 13.35 can sink with the output not exceeding 0.5 V? Also find the largest current that can be sourced with the output remaining within 0.5 V of V_{DD}.

*13.37 (a) In the transfer characteristic shown in Fig. 13.18, the segment BC is vertical because the Early effect is neglected. Taking the Early effect into account, use small-signal analysis to show that the slope of the transfer characteristic at $v_I = v_O = V_{DD}/2$ is

$$\frac{-2\,|V_A|}{(V_{DD}/2) - V_t}$$

where V_A is the Early voltage for Q_N and Q_P. Assume Q_N and Q_P to have equal values for K and $|V_t|$.

(b) A CMOS inverter with devices having $K_n = K_p$ is biased by connecting a resistor $R_G = 10$ MΩ between input and output. What is the dc voltage at input and output? What is the small-signal voltage gain and input resistance of the resulting amplifier? Assume the inverter to have the characteristics specified in Problem 13.35 with $|V_A| = 50$ V.

D13.38 Show that the threshold voltage of the CMOS inverter is given by

$$V_{th} = \frac{V_{DD} - |V_{tp}| + \sqrt{\dfrac{\mu_n (W/L)_n}{\mu_p (W/L)_p}}\,V_{tn}}{1 + \sqrt{\dfrac{\mu_n (W/L)_n}{\mu_p (W/L)_p}}}$$

Hence show that for $V_{tn} = -V_{tp}$, a threshold voltage of $V_{DD}/2$ is obtained by selecting

$$\frac{(W/L)_p}{(W/L)_n} = \frac{\mu_n}{\mu_p}$$

Find V_{th} for the case in which $V_{DD} = 5$ V, $V_{tn} = -V_{tp} = 0.8$ V, $\mu_n = 2.5\ \mu_p$, and $(W/L)_n = (W/L)_p$. If $L_n = L_p = 2\ \mu$m and $W_n = 4\ \mu$m, find the value of W_p that results in $V_{\text{th}} = V_{DD}/2$.

*13.39 A particular CMOS inverter uses n- and p-channel devices of identical sizes. If $\mu_n = 2\mu_p$, $|V_t| = 1$ V, and $V_{DD} = 5$ V, find V_{IL} and V_{IH} and hence the noise margins.

13.40 Repeat Exercise 13.14 for $V_{DD} = 10$ V and 15 V.

13.41 Repeat Exercise 13.14 for $V_t = 0.5$ V, 1.5 V, and 2 V.

D13.42 For a technology in which $V_{tn} = 0.2\ V_{DD}$, show that the maximum current that the CMOS inverter can sink while its low output level does not exceed $0.1\ V_{DD}$

is $0.075\,(\mu_n C_{OX})(W/L)_n\,V_{DD}^2$. For $V_{DD} = 5$ V, $\mu_n C_{OX} = 20\ \mu A/V^2$, $L_n = 5\ \mu m$, find the required transistor width to obtain a current of 1.5 mA.

13.43 For an inverter with $\mu_n C_{OX} = 2\,\mu_p C_{OX} = 50\ \mu A/V^2$, $V_{tn} = -V_{tp} = 0.8$ V, $(W/L)_n = 4\ \mu m/2\ \mu m$, $(W/L)_p = 8\ \mu m/2\ \mu m$, find the peak current drawn from a 5-V supply during switching.

13.44 If the inverter specified in Problem 13.43 is loaded with a 0.2-pF capacitance, find the dynamic power dissipation when the inverter is switched at a frequency of 20 MHz. What is the average current drawn from the power supply?

D13.45 Consider a CMOS inverter having $\mu_n C_{OX} = 50\ \mu A/V^2$, $V_{tn} = 0.8$ V, $L_n = 2\ \mu m$, and $V_{DD} = 5$ V. Find the minimum width required to obtain a propagation delay $t_{PHL} \leq 0.2$ ns when the inverter is loaded with a 0.2-pF capacitance. (*Hint:* Use the expression in Eq. 13.44.) What is the delay-power product exhibited by this inverter when it is switched at a frequency of 100 MHz?

13.46 Use Eq. (13.44) to explore the effect of variation of V_t on t_{PHL}. Derive expressions for t_{PHL} for $V_t = 0.1\ V_{DD}$, $0.2\ V_{DD}$, and $0.3\ V_{DD}$.

****13.47** Extend the analysis of the dynamic operation of the CMOS inverter to derive an expression for the fall time, t_{PHL}, which is the time for the output waveform to decrease from $0.9\ V_{DD}$ to $0.1\ V_{DD}$. Assume that the input is an ideal pulse. Evaluate t_{THL} for the inverter whose parameters are given in Exercise 13.18.

13.48 Reconsider Exercise 13.18 for an inverter in a new VLSI CMOS process in which all device dimensions are reduced by a factor of 2.5. Assume that C_{OX} is increased by the same factor, and find t_{PHL} for the cases in which

(a) $C_L = 0.1$ pF

(b) $C_L = 0.04$ pF

D13.49 A CMOS inverter for which $\mu_n C_{OX} = 50\ \mu A/V^2$, $V_{DD} = 5$ V, and $V_t = 0.2\ V_{DD}$ is loaded with a 0.1-pF capacitance. If the inverter is to be clocked at 100 MHz, find the value of (W/L) that will result in a delay-power product no larger than 0.1 pJ.

Section 13.6: CMOS Gate Circuits

13.50 Find the logic function implemented by the circuit shown in Fig. P13.50.

D13.51 Using the general idea introduced in Problem 13.50, and assuming the availability of complementary variables A, $\overline{A}$, B, $\overline{B}$, find a CMOS circuit to implement the function $Y = AB + \overline{A}\overline{B}$.

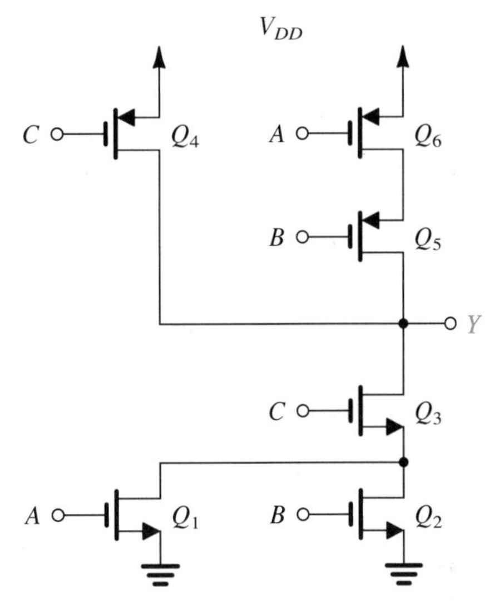

Fig. P13.50

D13.52 Considering a CMOS technology in which all devices have a channel length of 5 μm, and $(W/L)_n = 2$ in the basic inverter, find the appropriate sizes of the devices in the circuit of Problem 13.50 for output current capability the same in both directions and equal to that of the basic inverter. Assume $\mu_p \simeq \frac{1}{2}\mu_n$.

D13.53 Repeat Exercise 13.21 for a four-input NOR and a four-input NAND.

13.54 Repeat the analysis of the two-input NOR presented in Example 13.3 for the situation in which $V_{DD} = 5$ V and $|V_t| = 1$ V.

***13.55** What is the logic function implemented by the circuit in Fig. P13.55? Describe the output if the inputs are nondigital (continuous) signals in the range 0 to 5 V.

13.56 For standard CMOS gates whose propagation delay times are specified in the text for operation at 10 V, find the time of propagation through four levels of inverting (NAND) logic. Assume that each stage has a total fan-out of 4 (each with input capacitance of 5 pF) and that the wiring capacitance is 10 pF.

13.57 As specified in the text, a standard CMOS gate operating with $V_{DD} = 10$ V and loaded with $C_L = 50$ pF draws an average dc supply current of 0.6 μA/kHz. What is the component of this current that arises because of the repeated charge and discharge of C_L? Estimate the supply current for an unloaded gate ($C_L = 0$) in μA/kHz. Find the total gate dissipation for $V_{DD} = 10$ V, $f = 10$ MHz and $C_L = 100$ pF.

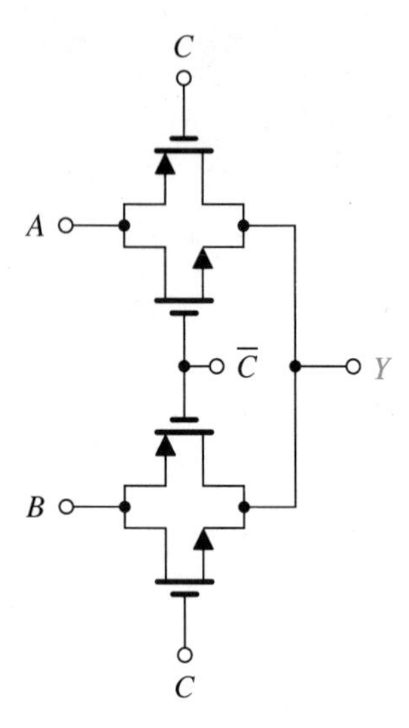

Fig. P13.55

13.58 A three-stage buffered CMOS inverter uses p-channel devices that are twice as wide as their n-channel counterparts. The output stage is the same size as the unbuffered gate, and each of the preceding stages has a W/L ratio that is one-tenth that of the stage that it drives. The input stage uses the smallest available n-channel device of unit area. What is the relative size of the buffered and unbuffered inverters? Note that all the channel lengths are the same.

Section 13.7: Latches and Flip-Flops

**13.59 Consider the latch of Fig. 13.25 as implemented in CMOS technology. Let $\mu_n C_{OX} = 2\ \mu_p C_{OX} = 20\ \mu A/V^2$, $W_p = 2W_n = 24\ \mu m$, $L_p = L_n = 6\ \mu m$, $|V_t| = 1$ V, and $V_{DD} = 5$ V.

(a) Plot the transfer characteristic of each inverter—that is, v_X versus v_W and v_Z versus v_Y. Determine the output of each inverter at input voltages of 1, 1.5, 2, 2.25, 2.5, 2.75, 3, 3.5, 4, and 5 volts.

(b) Use the characteristics in (a) to determine the loop voltage-transfer curve of the latch–that is, v_Z versus v_W. Find the coordinates of points A, B, and C as defined in Fig. 13.25(c).

(c) If the finite output resistance of the saturated MOSFET is taken into account, with $|V_A| = 100$ V, find the slope of the loop transfer characteristic at point B. What is the approximate width of the transition region?

13.60 Two CMOS inverters operating from a 5-V supply have V_{IH} and V_{IL} of 2.42 and 2.00 V and corresponding outputs of 0.4 V and 4.6 V, respectively, and are connected as a latch. Approximating the corresponding transfer characteristic of each gate by a straight line between the threshold points, sketch the latch open-loop transfer characteristic. What are the coordinates of point B? What is the loop gain at B?

D13.61 Sketch the logic gate symbolic representation of an SR flip-flop using NAND gates. Give the truth table that describes its operation.

D13.62 Sketch the circuit of the NAND SR flip-flop in Problem 13.61 using depletion-load NMOS. What is the rest state of the S and R inputs?

D13.63 Sketch the circuit of the NOR SR flip-flop of Fig. 13.26 using CMOS.

*13.64 Consider the SR flip-flop circuit shown in Fig. P13.64. Let $V_{DD} = 5$ V, $|V_t| = 1$ V, and $K_1 = K_2 = K_3 = K_4 = K$. Find the values of $K_5 = K_6$ so that the flip-flop switches state when a set or reset signal of $V_{DD}/2$ volts is applied.

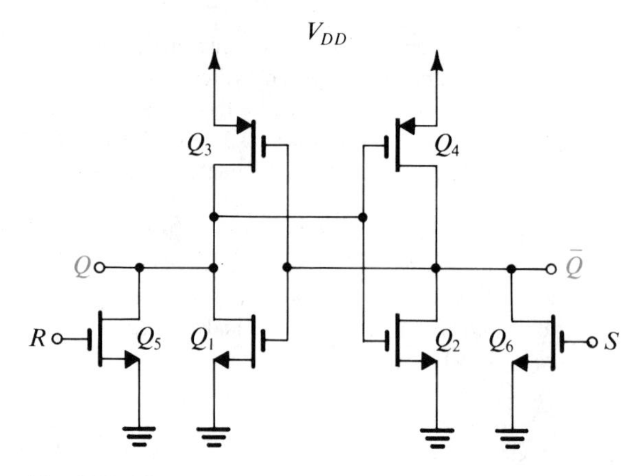

Fig. P13.64

Section 13.8: Multivibrator Circuits

D13.65 (a) For the monostable circuit of Fig. 13.31, find an approximate expression for the result of Exercise 13.24 under the conditions that $R_{on} \ll R$ and $V_{th} \simeq V_{DD}/2$.

(b) Use the approximate expression of (a) to find appropriate values for R and C so that $T = 1$ ms.

13.66 Consider the monostable circuit of Fig. 13.31 under the condition that $R_{on} \ll R$. What does the expression for T given in Exercise 13.24 become? If V_{th} is nominally 0.5 V_{DD} but can vary due to production variations in the range 0.4 V_{DD} to 0.6 V_{DD}, find the corresponding variation in T expressed as a percentage of the nominal value.

*13.67 The waveforms for the monostable circuit of Fig. 13.31 are given in Fig. 13.34. An expression for its period T is given in Exercise 13.24. Let $V_{DD} = 10$ V, $V_{th} = V_{DD}/2$, $R = 10$ kΩ, $C = 0.001$ μF, and $R_{on} = 200$ Ω. Find the values of T, ΔV_1, and ΔV_2. By how much does v_{O1} change during the quasi-stable state? What is the peak current that G_1 is required to sink? to source?

D13.68 Using the circuit of Fig. 13.31 (for which an expression for T is given in Exercise 13.24) design a monostable circuit with CMOS logic for which $R_{on} = 100$ Ω, $V_{DD} = 5$ V, and $V_{th} = 0.4$ V_{DD}. Use $C = 1$ μF to generate an output pulse of duration $T = 1$ s. What value of R should be used?

D13.69 (a) Use the expression given in Exercise 13.25 to find an expression for the frequency of oscillation f_O for the astable multivibrator of Fig. 13.36 under the condition that $V_{th} = V_{DD}/2$.
(b) Find suitable values for R and C to obtain $f_O = 100$ kHz.

13.70 Variations in manufacturing result in the CMOS gates used in implementing the astable circuit of Fig. 13.36 to have threshold voltages in the range 0.4 V_{DD} to 0.6 V_{DD} with 0.5 V_{DD} being the nominal value. Express the expected corresponding variation in the value of f_O (from nominal) as a percentage of the nominal value. (You may use the expression given in Exercise 13.25.)

*13.71 Consider a modification of the circuit of Fig. 13.36 in which a resistor equal to 10 R is inserted between the common node of C and R and the input node of G_1. This resistor allows the voltage labeled v_{I1} to rise above V_{DD} and below ground. Sketch the resulting modified waveforms of v_{I1} and show that the period T is now given by

$$T = CR \ln\left[\frac{2V_{DD} - V_{th}}{V_{DD} - V_{th}} \cdot \frac{V_{DD} + V_{th}}{V_{th}}\right]$$

Section 13.9: Random-Access Memory (RAM)

13.72 A 64 K-bit memory chip is organized in a square array and utilizes the simple NOR decoder of Fig. 13.39 for both row and column selection. **(a)** How many inputs would each decoder need? **(b)** What are the address bits connected to the inputs of the NOR gate whose output is connected to row 251?

13.73 Refer to the NOR address decoder of Fig. 13.40. Draw row 5 and clearly indicate the connections of its transistors.

D13.74 For the memory cell of Fig. 13.41(c) with polysilicon resistors of 1 MΩ and $V_{CC} = 5$ V, what is the W/L ratio required for Q_1 and Q_2 to ensure that $v_{DSon} \leq 0.1$ V? 0.2? What is the standby current and power in a collection of 16K such cells for a standby voltage of 2 V? Assume $V_t = 1$ V and $\mu_n C_{OX} = 20$ μA/V^2.

D*13.75 Consider the static RAM cell of Fig. 13.41(a). Let $V_{CC} = +5$ V, $K_1 = K_2$, $K_3 = K_4$, $K_5 = K_6$, $V_{tE} = 1$ V, and $V_{tD} = -3$ V, and ignore the body effect. **(a)** With Q_5 and Q_6 off and the cell storing a 1, the voltage at the drain of Q_2 is to be 0.1 V. Find K_4 in terms of K_2. **(b)** The D and $\overline{D}$ lines are precharged to 3 V, and the word line is raised to +5 V to read the stored 1. Transistor Q_5 turns on and current flows to the D line, raising its voltage by 0.25 V. The voltage at the drain of Q_1 correspondingly becomes approximately 3.25 V. Meanwhile Q_6 turns on and current flows from the $\overline{D}$ line, lowering its voltage to 2.75 V. The voltage at the drain of Q_2 rises. We wish to limit this voltage to 0.75 V (in order to avoid turning Q_1 on, which changes the state of the flip-flop). Find the required value of K_6 in terms of K_2. **(c)** The cell is storing a 0 and we wish to write a 1 into it. The word line is raised to +5 V and the $\overline{D}$ line is lowered to 0 V. Find the voltage that appears at the drain of Q_2 (before the flip-flop changes state and Q_2 turns on).

13.76 For a dynamic RAM cell utilizing a capacitance of 0.01 pF, refresh is required within 2 ms. If a signal loss on the capacitor of 1 V can be tolerated, what is the largest acceptable leakage current present at the cell?

D**13.77 Consider the one-transistor dynamic RAM cell in Fig. 13.43. For $V_{DD} = 5$ V, the digit-line voltages are either 0 or +5 V. To maximize the logic-1 level on C, arrangement is made to raise the word line from 0 to 7 V. $V_{tE} = 1$ V. For $C = 0.05$ pF, what value of K is required to ensure that C is charged to 4.5 V in 30 ns? [*Hint:* $\int dx/(ax^2 - x) = \ln(1 - 1/ax)$. You will also need the change of variables: $x = 5 - y$.]

Section 13.10: Read-Only Memory (ROM)

13.78 Give the eight words stored in the ROM of Fig. 13.45.

D13.79 Design the bit pattern to be stored in a (16 × 4) ROM that provides the 4-bit product of two 2-bit variables. Give a circuit implementation of the ROM array using a form similar to that in Fig. 13.45.

Section 13.11: Gallium-Arsenide Digital Circuits

13.80 For the DCFL gate in Example 13.4, verify that $V_{IH} = 0.63$ V, and find the corresponding value of v_O. For simplicity, use the fact that $v_O \ll 1$.

****13.81** For the DCFL gate in Example 13.4 with V_{tE} changed to 0.1 V, find V_{OH}, V_{OL}, V_{IL}, V_{IH}, NM_H, and NM_L. Compare the results with those found in Example 13.4, and comment on the effect of the value of V_{tE} on the noise margins.

D*13.82 Consider a DCFL gate fabricated in a GaAs technology for which $L = 1\ \mu$m, $V_{tD} = -1$ V, and $\lambda = 0.1\ \text{V}^{-1}$, and let $V_{DD} = 1.5$ V. Assume the on voltage of the Schottky diode to be 0.7 V.

(a) Derive expressions for V_{OL}, V_{OH}, V_{IL}, V_{IH}, NM_H, and NM_L in terms of V_{tE} and $m \equiv W_1/W_L$.

(b) For $V_{tE} = 0.2$ V find the value of m that yields $NM_H = 0.2$ V. What is the resulting value of NM_L?

(c) Repeat (b) for $V_{tE} = 0.1$ V.

13.83 For the FL gate in Fig. 13.53 let all channel lengths be equal at $1\ \mu$m, $W_S = W_L = 20\ \mu$m, $W_{PD} = 10\ \mu$m, β(per 1-μm width) $= 10^{-4}\ \text{A/V}^2$, $V_{tD} = -0.9$ V, and $\lambda = 0$. Calculate the static power dissipation of the gate.

***13.84** For the FL gate in Fig. 13.53 whose transfer characteristic is shown in Fig. 13.54, let $W_S = W_L = 20\ \mu$m, $W_{PD} = 10\ \mu$m, β(per 1-μm width) $= 10^{-4}\ \text{A/V}^2$, $V_{tD} = -0.9$ V, and $\lambda = 0.05\ \text{V}^{-1}$. Find an approximate value for the slope of the segment AB of the transfer characteristic, and use this value to estimate the width of the transition region. (*Hint:* The total resistance at the drain of Q_S is approximately the parallel equivalent of r_{oL}, r_{oS}, and r_{oPD}.)

CHAPTER 14

Bipolar Digital Circuits

INTRODUCTION

This is the second chapter of the two-chapter sequence devoted to digital circuits: In Chapter 13 we studied MOS digital circuits; here we shall study circuits implemented with bipolar junction transistors. A prerequisite for this material is a thorough familiarity with the BJT (Chapter 4). Also, it will be assumed that the reader is familiar with the general digital circuit concepts introduced in Section 13.1.

Our study of BJT digital circuits will begin with the Ebers–Moll model. This is a large-signal model whose application yields considerable insight into the operation of the BJT in the saturation region. We shall then discuss the dynamic operation of the BJT, relating its response times to the charge stored in its base.

Following a brief overview of early bipolar logic-circuit families, we study in detail two contemporary families: transistor-transistor logic (TTL) and emitter-coupled logic (ECL). For many years TTL has been the most popular circuit technology for implementing digital systems using SSI, MSI, and LSI packages. At the present time, TTL continues to be popular and is rivaled only by CMOS (Chapter 13). An important factor that contributed to the longevity of TTL is the continual performance improvement that this circuit technology has undergone over the years. Modern forms of TTL feature gate delays as low as 1.5 ns. As will be seen, in these improved circuits, the BJTs are prevented from saturation. This is done to avoid the time delay required to bring a transistor out of saturation. The other popular bipolar family, ECL, also avoids transistor saturation.

Except for the still emerging GaAs technology (Section 13.11), emitter-coupled logic is the fastest digital circuit technology available, featuring SSI and MSI gate delays of less than 1 ns, and even shorter delays in VLSI implementations. ECL finds application in digital communications circuits as well as in the high-speed circuits utilized in supercomputers.

A bipolar digital circuit technology that was popular a few years ago in VLSI applications is integrated injection logic (I^2L). It has, however, lost application ground to CMOS and will not be studied here.

The Chapter concludes with an introduction to a VLSI circuit technology that is becoming increasingly popular, BiCMOS. BiCMOS combines the advantages of both CMOS and bipolar circuits and provides the means for realizing very dense, low-power, high-speed integrated circuits.

Although the emphasis in this chapter is on logic circuits, other digital system building blocks such as flip-flops and multivibrators can be implemented in TTL and ECL following conventional approaches. Very-high-density memory chips, however, remain the exclusive domain of MOS technology.

14.1 THE BJT AS A DIGITAL CIRCUIT ELEMENT

We shall begin our study of BJT logic circuits with a summary of pertinent BJT characteristics. In addition, a popular large-signal model for the BJT will be introduced. Before proceeding with this material the reader is advised to review Chapter 4.

Saturating and Nonsaturating Logic

The most common usage of the BJT in digital circuits is to employ its two extreme modes of operation: cutoff and saturation. The resulting logic circuits are called **saturated** (or **saturating**) logic. The advantages of this mode of application include relatively large, well-defined logic swings and reasonably low power dissipation. The main disadvantage is the relatively slow response due to the long turnoff times of saturated transistors.

To obtain faster logic, one has to arrange the design such that the BJT does not saturate. We shall study two forms of nonsaturating BJT logic—namely, emitter-coupled logic (ECL), which is based on the differential pair studied in Section 6.1, and Schottky TTL, which is based on the use of special low-voltage-drop silicon diodes called Schottky diodes.

The Ebers–Moll (EM) Model

Although the simple large-signal transistor model developed in Chapter 4 is usually quite adequate for the approximate analysis of BJT digital circuits, more insight can be obtained from a more formal approach using a popular large-signal model of the BJT known as the **Ebers–Moll** (EM) model.

The EM model is a low-frequency (static) model based on the fact that the BJT is composed of two *pn* junctions, the emitter–base junction and the collector–base junction.

One can therefore express the terminal currents of the BJT as the superposition of the currents due to the two *pn* junctions, as shown in the following.

Figure 14.1 shows an *npn* transistor together with its EM model. The model consists of two diodes and two controlled sources. The diodes are D_E, the emitter–base junction

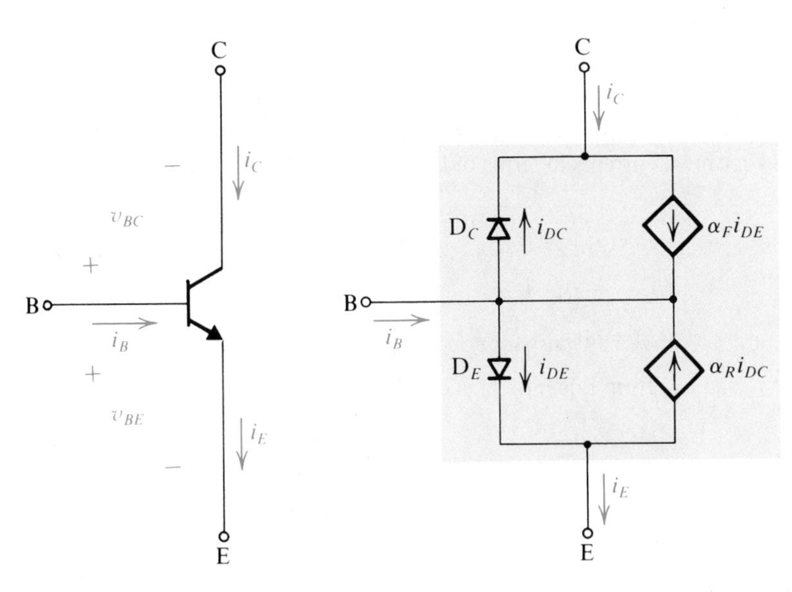

Fig. 14.1 An *npn* transistor and its Ebers–Moll (EM) model.

diode, and D_C, the collector–base junction diode. The diode currents i_{DE} and i_{DC} are given by the diode equation:

$$i_{DE} = I_{SE}(e^{v_{BE}/V_T} - 1) \quad (14.1)$$

$$i_{DC} = I_{SC}(e^{v_{BC}/V_T} - 1) \quad (14.2)$$

where I_{SE} and I_{SC} are the saturation or scale currents of the two diodes. Since the collector–base junction is usually of larger area than the emitter-base junction, I_{SC} is usually larger than I_{SE} (by a factor of 2 to 50).

As explained in Chapter 4, part of the emitter–base junction current i_{DE} reaches the collector and registers as collector current. It is this component that gives rise to the current source $\alpha_F i_{DE}$ in the model of Fig. 14.1. Here α_F denotes the **forward α** of the transistor (which is the parameter we simply called α previously). The value of α_F is usually very close to unity. Similarly, part of the collector–base junction current i_{DC} is transported across the base region and reaches the emitter. This component is represented in the EM model by the current source $\alpha_R i_{DC}$, where α_R denotes the **reverse α** of the transistor. Since the transistor structure is not physically symmetric but rather is optimized to have a large forward α, α_R is usually small (0.02 to 0.5).

A relationship exists (see Harris, Gray, and Searle, 1966) between the four parameters of the EM model and the transistor current scale I_S (see Chapter 4):

$$\alpha_F I_{SE} = \alpha_R I_{SC} = I_S \quad (14.3)$$

Since $\alpha_F \simeq 1$, we see that

$$I_{SE} \simeq I_S \tag{14.4}$$

Recall that for low-power (small-signal) transistors I_S is of the order of 10^{-14} to 10^{-15} A and is proportional to the area of the emitter–base junction.

The Transistor Terminal Currents

Having provided a qualitative physical justification for the EM model, we shall now use it to express the BJT terminal currents in terms of the junction voltages. From Fig. 14.1 we can write

$$i_E = i_{DE} - \alpha_R i_{DC} \tag{14.5}$$

$$i_C = -i_{DC} + \alpha_F i_{DE} \tag{14.6}$$

$$i_B = (1 - \alpha_F) i_{DE} + (1 - \alpha_R) i_{DC} \tag{14.7}$$

Substituting for i_{DE} and i_{DC} from Eqs. (14.1) and (14.2) and using the relationship in Eq. (14.3) gives

$$i_E = \frac{I_S}{\alpha_F}(e^{v_{BE}/V_T} - 1) - I_S(e^{v_{BC}/V_T} - 1) \tag{14.8}$$

$$i_C = I_S(e^{v_{BE}/V_T} - 1) - \frac{I_S}{\alpha_R}(e^{v_{BC}/V_T} - 1) \tag{14.9}$$

$$i_B = \frac{I_S}{\beta_F}(e^{v_{BE}/V_T} - 1) + \frac{I_S}{\beta_R}(e^{v_{BC}/V_T} - 1) \tag{14.10}$$

where β_F is the forward β and β_R is the reverse β,

$$\beta_F = \frac{\alpha_F}{1 - \alpha_F} \tag{14.11}$$

$$\beta_R = \frac{\alpha_R}{1 - \alpha_R} \tag{14.12}$$

While β_F is usually large, β_R is very small.

Exercise 14.1 A particular transistor is said to have $\alpha_F \simeq 1$ and $\alpha_R = 0.02$. Its emitter scale current is about 10^{-14} A. What is its collector scale current? What is the size of the collector junction relative to the emitter junction? What is the value of β_R?

Ans. 50×10^{-14} A; 50 times as large; 0.02

Application of the EM Model

We shall now consider the application of the EM model to characterize transistor operation in various modes.

The Normal Active Mode. Here the emitter–base junction is forward-biased and the collector–base junction is reverse-biased. The word **normal** is used to distinguish this mode from that in which the roles of the two junctions are interchanged (the *reverse* or *inverse* active mode). Since v_{BC} is negative and its magnitude is usually much greater than V_T, Eqs. (14.8) through (14.10) can be approximated as

$$i_E \simeq \frac{I_S}{\alpha_F}\, e^{v_{BE}/V_T} + I_S\left(1 - \frac{1}{\alpha_F}\right) \tag{14.13}$$

$$i_C \simeq I_S e^{v_{BE}/V_T} + I_S\left(\frac{1}{\alpha_R} - 1\right) \tag{14.14}$$

$$i_B \simeq \frac{I_S}{\beta_F}\, e^{v_{BE}/V_T} - I_S\left(\frac{1}{\beta_F} + \frac{1}{\beta_R}\right) \tag{14.15}$$

In each of these three equations one can normally neglect the second term on the right-hand side. This results in the familiar current–voltage relationships that characterize the active mode of operation.

Exercise 14.2 Use Eq. (14.8) to show that the i–v characteristic of the diode-connected transistor of Fig. E14.2 is given by

$$i = \frac{I_S}{\alpha_F}\,(e^{v/V_T} - 1) \simeq I_S e^{v/V_T}$$

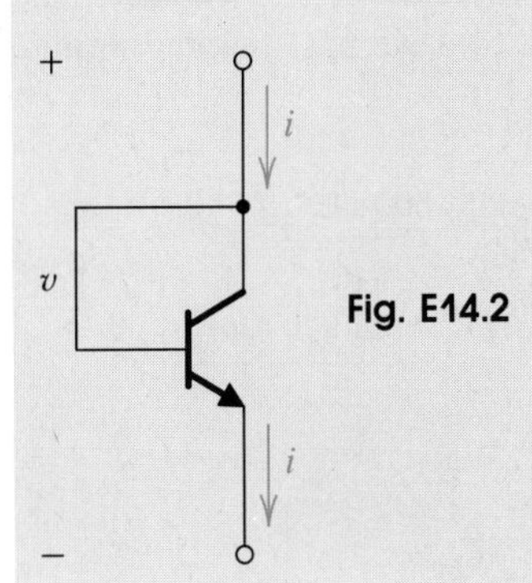

Fig. E14.2

The Saturation Mode. Consider first the normal (as opposed to reverse) saturation mode, as can be obtained in the circuit of Fig. 14.2. Assume that a current I_B is pushed into the base and that its value is sufficient to drive the transistor into saturation. Thus the collector current will be $\beta_{\text{forced}} I_B$, where $\beta_{\text{forced}} < \beta_F$. We wish to use the EM equations to derive an expression for $V_{CE\text{sat}}$.

In saturation both junctions are forward-biased. Thus V_{BE} and V_{BC} are both positive, and their values are much greater than V_T. Thus in Eqs. (14.9) and (14.10) we can assume that $e^{V_{BE}/V_T} \gg 1$ and $e^{V_{BC}/V_T} \gg 1$. Making these approximations and substituting $i_B = I_B$ and $i_C = \beta_{\text{forced}} I_B$ results in two equations that can be solved to obtain V_{BE} and V_{BC}. The

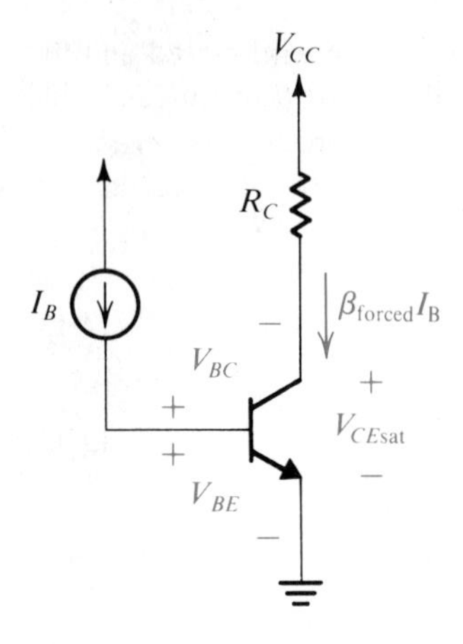

Fig. 14.2 Circuit in which the transistor can be operated in the normal saturation mode.

saturation voltage $V_{CE\text{sat}}$ can be then obtained as the difference between these two voltage drops:

$$V_{CE\text{sat}} = V_T \ln \frac{1 + (\beta_{\text{forced}} + 1)/\beta_R}{1 - \beta_{\text{forced}}/\beta_F} \tag{14.16}$$

It is instructive to use Eq. (14.16) to find $V_{CE\text{sat}}$ in a typical case. Table 14.1 provides numerical values for the case $\beta_F = 50$, $\beta_R = 0.1$, and various values of β_{forced}. Also, Fig. 14.3 shows a sketch of $V_{CE\text{sat}}$ versus β_{forced}. Since $i_C = \beta_{\text{forced}}\, I_B$ and I_B is constant, β_{forced} is proportional to i_C, and the curve in Fig. 14.3 is simply the v_{CE}–i_C characteristic for a constant base current I_B. From Table 14.1 and Fig. 14.3 we note that the infinite

Table 14.1 THE SATURATION VOLTAGE FOR THE EASE $\beta_F = 50$ AND $\beta_R = 0.1$

β_{forced}	50	48	45	40	30	20	10	1	0
$V_{CE\text{sat}}$ (mV)	∞	235	211	191	166	147	123	76	60

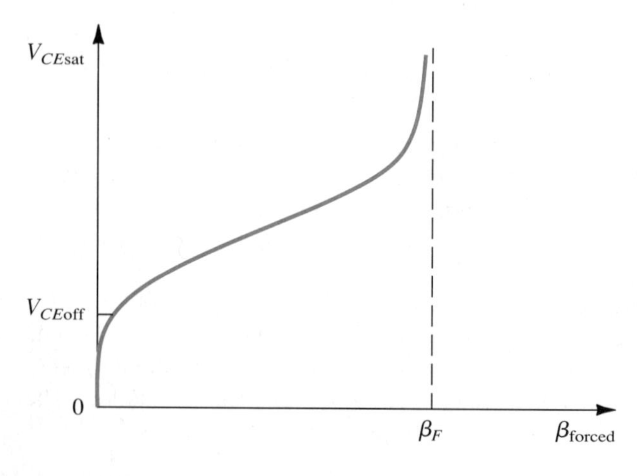

Fig. 14.3 Variation of $V_{CE\text{sat}}$ with forced β ($\beta_{\text{forced}} = I_c/I_B$). The vertical line obtained for $\beta_{\text{forced}} = \beta_F$ indicates that the transistor has left saturation and entered the active mode.

value of V_{CEsat} obtained at $\beta_{forced} = \beta_F$ is an indication that the transistor is at the boundary between saturation and active mode. Figure 14.3 illustrates this further by showing the independence of v_{CE} on β_{forced} (or i_C) in the active mode. As β_{forced} is reduced, the transistor is driven deeper into saturation, V_{BC} increases, and V_{CEsat} is reduced. Finally, for $\beta_{forced} = 0$, which corresponds to the collector being open-circuited, we obtain a small value of V_{CEsat}. This small value is almost equal to the *offset voltage* V_{CEoff} of the BJT switch, as defined in Fig. 4.61.

The numerical values of Table 14.1 suggest that for a saturated transistor $V_{CEsat} \simeq 0.1$–0.3 V. For approximate calculations we shall henceforth assume that for a transistor on the verge of saturation $V_{CEsat} = 0.3$ V; for a transistor "comfortably" saturated $V_{CEsat} = 0.2$ V; and for a transistor deep into saturation $V_{CEsat} = 0.1$ V.

Exercise 14.3 Use the entries for $\beta_{forced} = 1$ and 10 in Table 14.1 to calculate an approximate value for the collector–emitter saturation resistance, R_{CEsat}, of a transistor having $I_B = 1$ mA.

Ans. 5.2 Ω

The Inverse Mode. We shall next consider the operation of the BJT in the inverse or reverse mode. Figure 14.4 shows a simple circuit in which the transistor is used with its collector and emitter interchanged. Note that the currents indicated—namely, I_B, I_1, and I_2—have positive values. Thus since $i_C = -I_2$ and $i_E = -I_1$, both i_C and i_E will be negative.

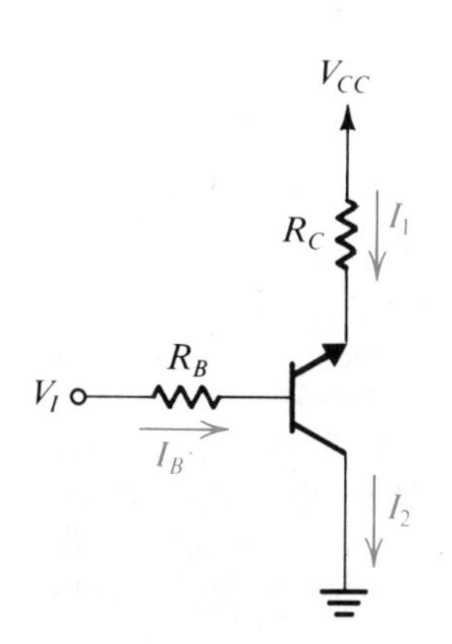

Fig. 14.4 Circuit in which the transistor is used in the reverse (or inverse) mode.

Since the roles of the emitter and collector are interchanged, the transistor in the circuit of Fig. 14.4 will operate in the active mode (called the **reverse active mode** in this case) when the emitter–base junction is reverse-biased. In such a case

$$I_1 = \beta_R I_B$$

Since β_R is usually very low, it makes little sense to operate the BJT in the reverse active mode.

The transistor in the circuit of Fig. 14.4 will saturate (that is, operate in the **reverse saturation mode**) when the emitter–base junction becomes forward-biased. In this case

$$\frac{I_1}{I_B} < \beta_R$$

We can use the EM equations to find an expression for $V_{EC\text{sat}}$ in this case. Such an expression can be directly obtained from Eq. (14.16) as follows: Replace β_{forced} by $-I_2/I_B$ and then replace I_2 by $I_1 + I_B$. The result is

$$V_{EC\text{sat}} = V_T \ln \frac{1 + \dfrac{1}{\beta_F} + \left(\dfrac{I_1}{I_B}\right)\left(\dfrac{1}{\beta_F}\right)}{1 - \left(\dfrac{I_1}{I_B}\right)\left(\dfrac{1}{\beta_R}\right)} \tag{14.17}$$

From this equation it can be seen that the minimum $V_{EC\text{sat}}$ is obtained when $I_1 = 0$. This minimum is very close to zero. Furthermore, we observe that the condition $I_1/I_B < \beta_R$ has to be satisfied in order that the denominator remain positive. This, of course, is the condition for the transistor to operate in the reverse saturation mode. Finally, note that since β_R is usually very low, I_1 has to be much smaller than I_B, with the result that $V_{EC\text{sat}}$ will be very small. This indeed is the reason for operating the BJT in the reverse saturation mode. Saturation voltages as low as a fraction of a millivolt have been reported. The disadvantage of the reverse saturation mode of operation is a relatively long turnoff time.

Exercise 14.4 For the circuit in Fig. 14.4, let $R_B = 1\ \text{k}\Omega$ and $V_I = V_{CC} = +5$ V. Assume that $V_{BC} = 0.6$ V, $\beta_R = 0.1$, and $\beta_F = 50$. Calculate approximate values for the emitter voltage in the following cases: $R_C = 1\ \text{k}\Omega$; $R_C = 10\ \text{k}\Omega$; and $R_C = 100\ \text{k}\Omega$.

Ans. +4.56 V; +0.6 V; +3.5 mV

Transistor Switching Times

Because of their internal capacitive effects, transistors do not switch in zero time. Figure 14.5 illustrates this point by displaying the waveform of the collector current i_C of a simple transistor inverter together with the waveforms for the input voltage v_I and the base current i_B. As indicated, when the input voltage v_I rises from the negative (or zero) level V_1 to the positive level V_2, the collector current does not respond immediately. Rather, a delay time t_d elapses before any appreciable collector current begins to flow. This delay time is required mainly for the EBJ depletion capacitance[1] to charge up to the forward-bias voltage V_{BE} (approximately 0.7 V). After this charging process is completed the collector current begins an exponential rise toward a final value of βI_{B2}, where I_{B2} is the current pushed into the base,[2] and is given as follows:

$$I_{B2} = \frac{V_2 - V_{BE}}{R_B} \tag{14.18}$$

The time constant of the exponential rise is determined by the junction capacitances. In fact, it is during the interval of the rising edge of i_C that the excess minority carrier charge is being stored in the base region (see Chapter 4).

Although the exponential rise of i_C is heading toward βI_{B2}, this value will never be

[1] Since during t_d the current is zero, the diffusion capacitance will be zero.
[2] We use β and β_F interchangeably.

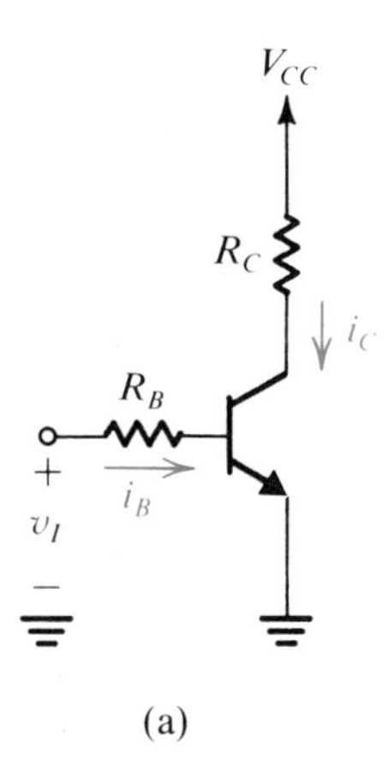

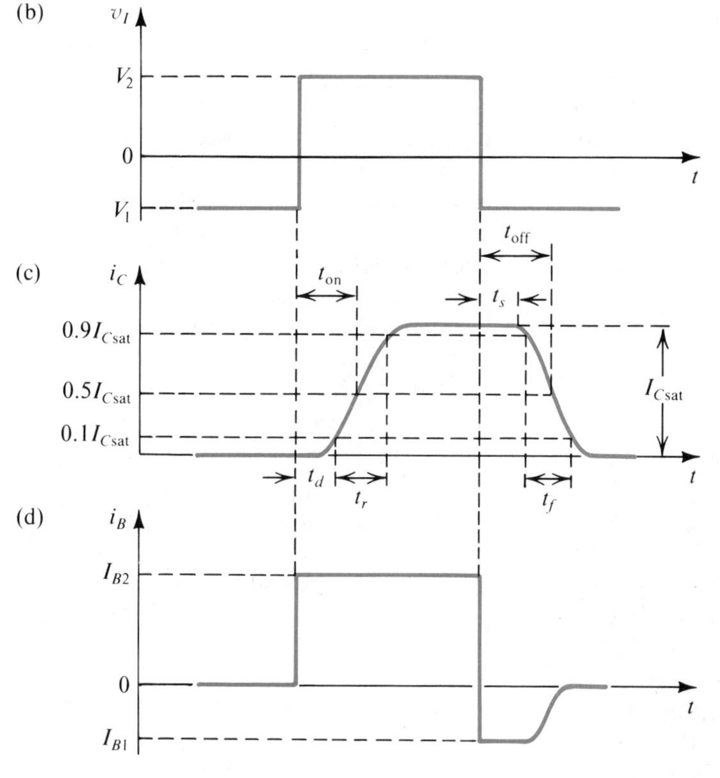

Fig. 14.5 Switching times of the BJT in the simple inverter circuit of (a) when the input v_I has the pulse waveform in (b). The effects of stored base charge following the return of v_I to V_1 are explained in conjunction with Eqs. (14.19) and (14.20).

reached, since the transistor will saturate and the collector current will be limited to I_{Csat}. A measure of the BJT switching speed is the **rise time** t_r indicated in Fig. 14.5(c). Another measure is the **turn-on time** t_{on}, also indicated in Fig. 14.5(c).

Figure 14.6(a) shows the profile of excess minority carrier charge stored in the base of a saturated transistor. Unlike the active mode case, the excess minority concentration is

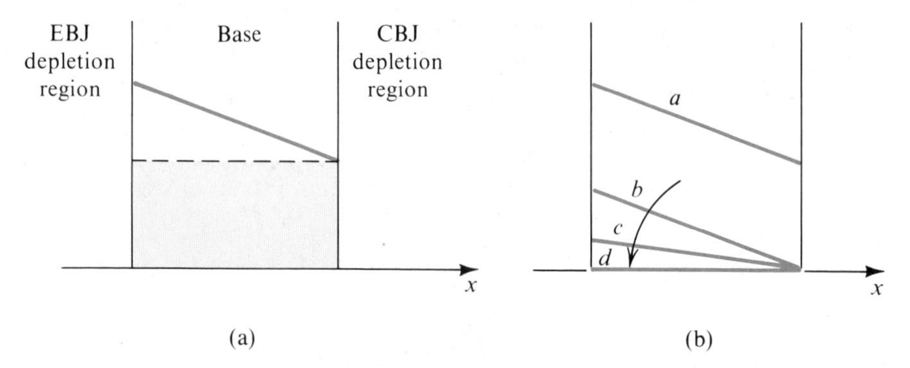

Fig. 14.6 (a) Profile of excess minority carriers in the base of a saturated transistor. The colored area represents the extra (or saturating) charge. **(b)** As the transistor is turned off, the extra stored charge has to be removed first. During this interval the profile changes from line *a* to line *b*. Then the profile decreases toward zero (line *d*), and the collector current falls exponentially to zero.

not zero at the edge of the CBJ. This is because the CBJ is now forward-biased. Of special interest here is the extra charge stored in the base and represented by the colored area in Fig. 14.6(a). Because this charge does not contribute to the slope of the concentration profile, it does not result in a corresponding collector-current component. Rather, this extra stored charge arises from the pushing of more current into the base than is required to saturate the transistor. The higher the overdrive factor used, the greater the amount of extra charge stored in the base. In fact the extra charge Q_s, called **saturating charge** or **excess charge,** is proportional to the excess base drive $I_{B2} - I_{Csat}/\beta$; that is,

$$Q_s = \tau_s(I_{B2} - I_{Csat}/\beta) \tag{14.19}$$

where τ_s is a transistor parameter known as the **storage time constant.**

Let us now consider the turn-off process. When the input voltage v_I returns to the low level V_1, the collector current does not respond but remains almost constant for a time t_s (Fig. 14.5c). This is the time required to remove the saturating charge from the base. During the time t_s, called the **storage time,** the profile of stored minority carriers will change from that of line *a* to that of line *b* in Fig. 14.6(b). As indicated in Fig. 14.5(d), the base current reverses direction because v_{BE} remains approximately 0.7 V while v_I is at the negative (or zero) level V_1. The reverse current I_{B1} helps to "discharge the base" and remove the extra stored charge; in the absence of the reverse base current I_{B1}, the saturating charge has to be removed entirely by recombination. It can be shown (see Millman and Taub, 1965) that the storage time t_s is given by

$$t_s = \tau_s \frac{I_{B2} - I_{Csat}/\beta}{I_{B1} + I_{Csat}/\beta} \tag{14.20}$$

Once the extra stored charge has been removed, the collector current begins to fall exponentially with a time constant determined by the junction capacitances. During the fall time, the slope of the excess-charge profile decreases toward zero, as indicated in Fig. 14.6(b). Finally, note that the reversed base current eventually decreases to zero as the EBJ capacitance charges up to the reverse-bias voltage V_1.

Typically t_d, t_r, and t_f are of the order of a few nanoseconds to a few tens of nanoseconds. The storage time t_s, however, is larger and usually constitutes the limiting factor on the switching speed of the transistor. As mentioned before, t_s increases with the overdrive factor (that is, with how deep the transistor is driven into saturation). It follows that if one desires high-speed digital circuits, then operation in the saturation region should be avoided. This is the idea behind the two nonsaturating forms of logic circuits we shall study—namely, Schottky TTL and ECL.

Exercise 14.5 We wish to use the transistor equivalent circuit of Fig. 7.22 to find an expression for the delay time t_d of a simple transistor inverter fed by a step voltage source with a resistance R_B. Since during the delay time the transistor is not conducting, the resistance r_π is infinite and C_π will consist entirely of the depletion capacitance C_{je}. As an approximation, C_{je} can be assumed to remain constant during t_d. Also, since the collector voltage does not change during t_d, the collector can be considered grounded. Assume that the two levels of v_1 are V_1 and V_2 and that the end of t_d can be taken as the time at which $v_\pi = 0.7$ V.

Ans. $t_d = (R_B + r_x)(C_{je} + C_\mu) \ln[(V_2 - V_1)/(V_2 - 0.7)]$

14.2 EARLY FORMS OF BJT DIGITAL CIRCUITS

In order to place the material of this chapter in proper perspective we shall examine briefly two early forms of bipolar logic circuit families.

The Basic BJT Inverter

Figure 14.7 shows the basic BJT logic inverter together with its voltage transfer characteristic. We have studied this circuit in detail in Chapter 4. Furthermore, in the previous section we studied BJT models that help in the analysis of the static and dynamic operation of the BJT inverter. Note that for a logic-0 input, $v_I \leq V_{IL}$, the transistor will be cut off and the output voltage will be equal to V_{CC}; that is, $V_{OH} = V_{CC}$. For a logic-1 input, $v_I \geq V_{IH}$, the BJT will be saturated and the output voltage will be equal to V_{CEsat}; that is, $V_{OL} = V_{CEsat} = 0.1$ to 0.2 V.

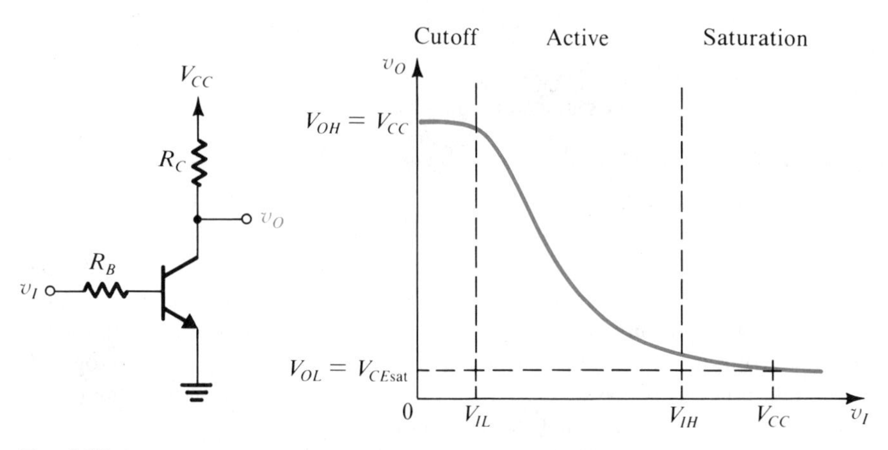

Fig. 14.7 The basic BJT inverter and its transfer characteristic.

Resistor-Transistor Logic (RTL)

By paralleling the outputs of two or more basic inverters we obtain the basic gate circuit of an early logic circuit family known as resistor-transistor logic (RTL). Figure 14.8 shows such a two-input NOR gate. The circuit operates as follows: If one of the inputs—say,

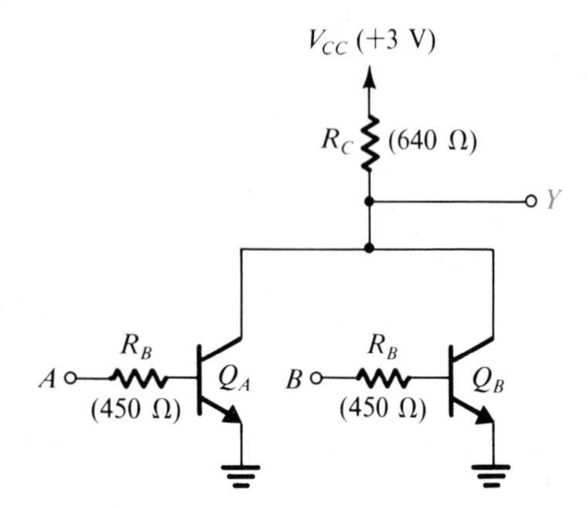

Fig. 14.8 A two-input NOR gate of the RTL family.

A—is high (logic 1), then the corresponding transistor (Q_A) will be on and saturated. This will result in $v_Y = V_{CE\text{sat}}$, which is low (logic 0). If the other input (*B*) is also high (logic 1), then the corresponding transistor (Q_B) will be on and saturated, thus also helping to keep the output low. It can be seen that for the output to be high ($v_Y = V_{CC}$), both Q_A and Q_B have to be off simultaneously. Clearly, this is obtained if both *A* and *B* are simultaneously low. That is, a logic 1 will appear at the output only in one case: *A* and *B* are low. We may therefore write the Boolean expression

$$Y = \bar{A}\bar{B}$$

which can be also written

$$Y = \overline{A + B}$$

which is a NOR function.

The fan-in of the RTL NOR gate can be increased by adding more input transistors. The resistor values and supply voltage indicated are those used in integrated-circuit RTL (circa the 1960s).

Although the high output level (V_{OH}) of a single gate is V_{CC}, this is not the case when the RTL gate is driving other similar gates. Since the input transistors of the driven gates will be turned on, the total base current will be supplied through resistor R_C of the driving gate. Thus the value of V_{OH} will be considerably lowered, to a value, depending on the fan-out, but closer to 1 V. Furthermore, this value will be reduced as the gate fan-out is increased. As a result, the noise margins of the RTL gate are rather narrow. This, together with the fact that RTL gates dissipate a rather large amount of power (the delay-power product is about 140 pJ), has resulted in the demise of RTL.

An RTL SR Flip-Flop

Before leaving RTL we wish to illustrate its application in the SR flip-flop circuit shown in Fig. 14.9. Obviously, this circuit can be formed by cross-coupling two two-input RTL NOR gates. However, the circuit was in fact quite popular in discrete-component circuit design before the advent of the integrated circuit in the early 1960s. Operation of the circuit is straightforward and follows closely the logic description of the SR flip-flop given in Section 13.7.

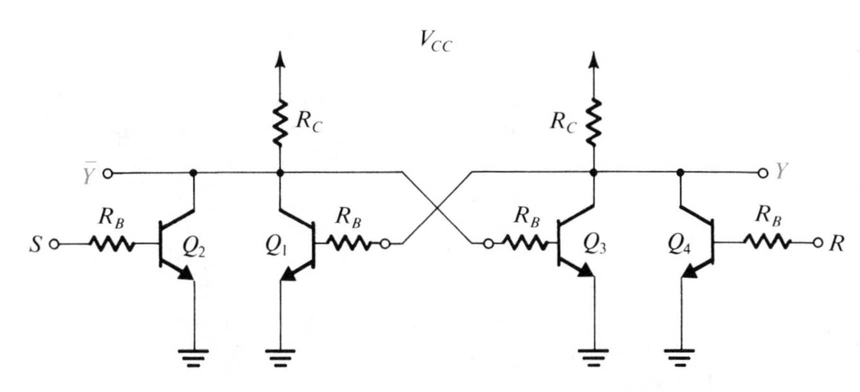

Fig. 14.9 An SR flip-flop formed by cross-coupling two NOR gates of the RTL family.

Diode-Transistor Logic (DTL)

Another early BJT logic circuit family is diode-transistor logic or DTL, exemplified by the two-input NAND gate shown in Fig. 14.10. DTL is of particular interest to us because, as we shall see in the next section, it is the circuit from which TTL has evolved.

The DTL circuit operates as follows: Let input B be left open. If a logic-0 signal ($\simeq$ 0 V) is applied to A, diode D_1 will conduct and the voltage at node X will be one diode

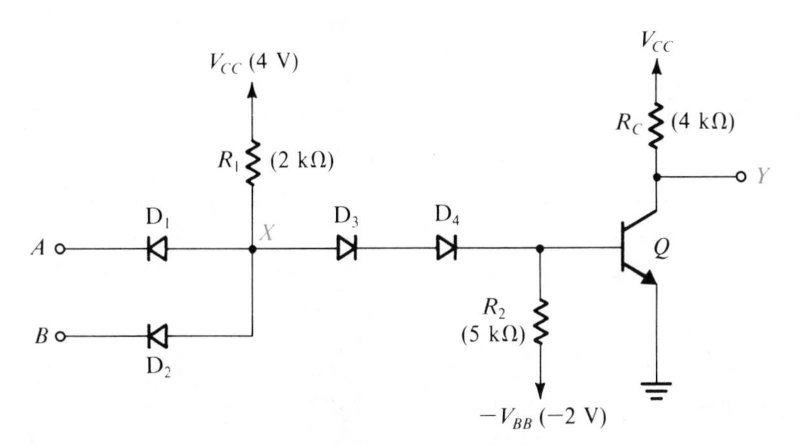

Fig. 14.10 A two-input NAND gate of the DTL family.

drop (0.7 V) above the logic-0 value. The two diodes D_3 and D_4 will be conducting, thus causing the base of transistor Q to be two diode drops below the voltage at node X. Thus the base will be at a small negative voltage, and hence Q will be off and $v_Y = V_{CC}$ (logic 1).

Consider now increasing the voltage v_A. It can be seen that diode D_1 will remain conducting and node X will keep rising in potential. Diodes D_3 and D_4 will remain conducting, and hence the base will also rise in potential. This situation will continue until the voltage at the base reaches about 0.5 V, at which point the transistor will start to conduct. This will occur when the voltage at A is

$$v_A \simeq 0.5 + V_{D4} + V_{D3} - V_{D1} \simeq 1.2 \text{ V}$$

Small increases in v_A above this threshold value will appear as increases in v_{BE} and hence in i_C. In this range the transistor will be in the active region. Eventually the voltage at the base will reach 0.7 V and the transistor will be fully conducting. At this point the voltage at X will be clamped to two diode drops above V_{BE}, and any further increases in v_A will reverse-bias D_1. It can be seen that the current in D_1 will begin to decrease when v_A reaches about 1.4 V. As D_1 stops conducting, all the current through R_1 will be diverted through D_3 and D_4 into the base of the transistor. The circuit is normally designed such that the current into the base will be sufficient to drive the transistor into saturation. Thus when A is at logic-1 level the transistor will be saturated and v_Y will be equal to V_{CEsat} ($\simeq$ 0.2 V), which is a logic-0 output.

Extrapolating from the above discussion, it can be seen that if either or both of the inputs is low the corresponding diode (D_1, D_2, or both) will be conducting, the transistor will be off, and the output Y will be high. The output will be low if the transistor is on, which will happen for only one particular input combination, when all the inputs are simultaneously high. We may therefore write the Boolean expression

$$\overline{Y} = AB$$

which can be rewritten

$$Y = \overline{AB}$$

which is a NAND function. This should come as no surprise, since the DTL circuit consists of a diode AND gate formed by D_1, D_2, and R_1 (see Section 3.1), followed by a transistor inverter. Finally, we note that because of their function in steering current into either R_2 or the transistor base, diodes D_3 and D_4 are known as "steering diodes."

DTL was popular in the 1960s and was implemented first using discrete components and subsequently in IC form. However, it was eventually replaced with transistor-transistor logic (TTL).

Exercises 14.6 Consider the RTL gate of Fig. 14.8 when driving N identical gates. Let both inputs to the driving gate be low. Convince yourself that the output voltage V_{OH} can be determined using the equivalent circuit shown in Fig. E14.6. Hence show that

$$V_{OH} = V_{CC} - R_C \frac{V_{CC} - V_{BE}}{R_C + R_B/N}$$

For $N = 5$, use the values given in Fig. 14.8, together with $V_{BE} = 0.7$ V to obtain the value of V_{OH}.

Ans. 1 V

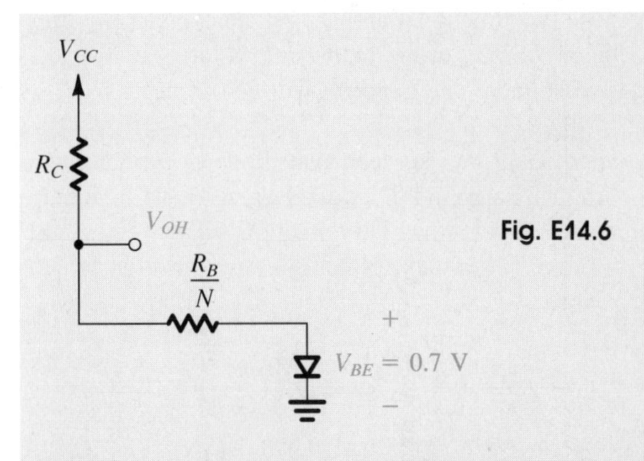

Fig. E14.6

14.7 For the DTL gate of Fig. 14.10 assume all conducting junctions to have a voltage drop of 0.7 V: (a) Find the current through D_1 when $v_A = 0.2$ V and $v_B = +4$ V. Also find the voltage at the base. (b) With $v_A = v_B = +4$ V find the transistor base current. If $V_{CEsat} = 0.2$ V find the value of β_{forced}.

Ans. (a) 1.25 mA, −0.5 V; (b) 0.41 mA, 2.3

14.3 TRANSISTOR-TRANSISTOR LOGIC (TTL OR T²L)

For more than two decades TTL has enjoyed immense popularity. Indeed, for the bulk of digital systems applications employing SSI and MSI packages, TTL is rivaled only by CMOS (Chapter 13).

We shall begin this section with a study of the evolution of TTL from DTL. In this way we shall explain the function of each of the stages of the complete TTL gate circuit. Characteristics of standard TTL gates will be studied in Section 14.4. Standard TTL, however, has now been virtually replaced with more advanced forms of TTL that feature improved performance. These will be discussed in Section 14.5.

Evolution of TTL From DTL

The basic DTL gate circuit in discrete form was discussed in the previous section (see Fig. 14.10). The integrated-circuit form of the DTL gate is shown in Fig. 14.11 with only one input indicated. As a prelude to introducing TTL, we have drawn the input diode as a diode-connected transistor (Q_1), which corresponds to how diodes are made in IC form.

This circuit differs from the discrete DTL circuit of Fig. 14.10 in two important aspects. First, one of the steering diodes is replaced by the base–emitter junction of a transistor (Q_2) that is either cut off (when the input is low) or in the active mode (when the input is high). This is done to increase the fan-out capability of the gate. A detailed explanation of this point, however, is not relevant to our study of TTL. Second, the resistance R_B is returned to ground rather than to a negative supply, as was done in the earlier discrete circuit. An obvious advantage of this is the elimination of the additional power supply. The disadvantage, however, is that the reverse base current available to remove the excess charge stored in the base of Q_3 is rather small. We shall elaborate on this point below.

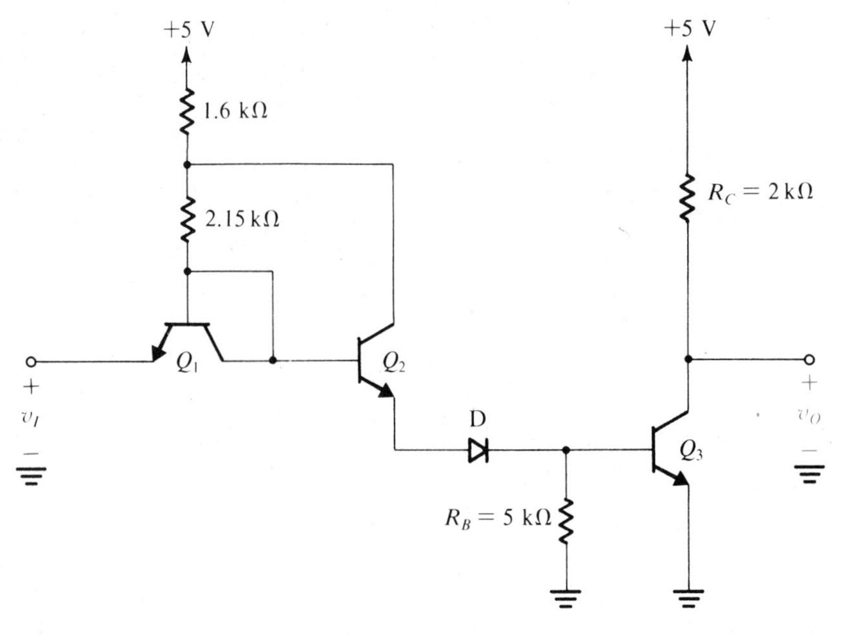

Fig. 14.11 IC form of the DTL gate with the input diode shown as a diode-connected transistor (Q_1). Only one input terminal is shown.

Exercise 14.8 Consider the DTL gate circuit shown in Fig. 14.11 and assume that $\beta(Q_2) = \beta(Q_3) = 50$. (a) When $v_I = 0.2$ V, find the input current. (b) When $v_I = +5$ V, find the base current of Q_3.

Ans. (a) 1.1 mA; (b) 1.6 mA

Reasons for the Slow Response of DTL

The DTL gate has relatively good noise margins and reasonably good fan-out capability. Its response, however, is rather slow. There are two reasons for this: first, when the input goes low and Q_2 and D turn off, the charge stored in the base of Q_3 has to leak through R_B to ground. The initial value of the reverse base current that accomplishes this "base discharging" process is approximately 0.7 V/R_B, which is about 0.14 mA. Because this current is quite small in comparison to the forward base current, the time required for the removal of base charge is rather long, which contributes to lengthening the gate delay.

The second reason for the relatively slow response of DTL derives from the nature of the output circuit of the gate, which is simply a common-emitter transistor. Figure 14.12 shows the output transistor of a DTL gate driving a capacitive load C_L. The capacitance C_L represents the input capacitance of another gate and/or the wiring and parasitic capacitances that are inevitably present in any circuit. When Q_3 is turned on, its collector voltage cannot instantaneously fall because of the existence of C_L. Thus Q_3 will not immediately saturate but rather will operate in the active region. The collector of Q_3 will therefore act as a constant-current source and will sink a relatively large current (βI_B). This large

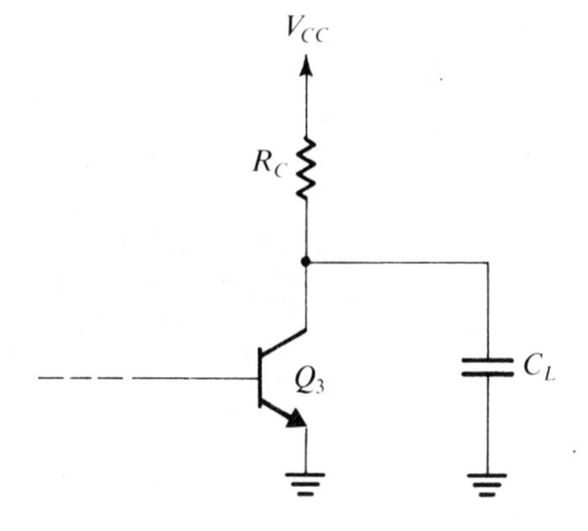

Fig. 14.12 The output circuit of a DTL gate driving a capacitive load C_L.

current will rapidly discharge C_L. We thus see that the common-emitter output stage features a short turn-on time. However, turnoff is another matter.

Consider next the operation of the common-emitter output stage when Q_3 is turned off. The output voltage will not rise immediately to the high level (V_{CC}). Rather, C_L will charge up to V_{CC} through R_C. This is a rather slow process, and it results in lengthening the DTL gate delay (and similarly the RTL gate delay).

Having identified the two reasons for the slow response of DTL, we shall see in the following how these problems are remedied in TTL.

Input Circuit of the TTL Gate

Figure 14.13 shows a conceptual TTL gate with only one input terminal indicated. The most important feature to note is that the input diode has been replaced by a transistor. One can think of this simply as if the short circuit between base and collector of Q_1 in Fig. 14.11 has been removed.

To see how the conceptual TTL circuit of Fig. 14.13 works, let the input v_I be high (say, $v_I = V_{CC}$). In this case current will flow from V_{CC} through R, thus forward-biasing the base–collector junction of Q_1. Meanwhile, the base–emitter junction of Q_1 will be reverse-biased. Therefore Q_1 will be operating in the **inverse active mode**—that is, in the

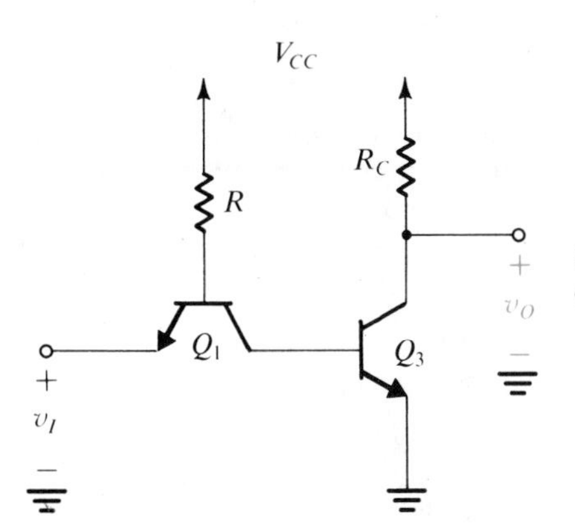

Fig. 14.13 Conceptual form of TTL gate. Only one input terminal is shown.

active mode but with the roles of emitter and collector interchanged. The voltages and currents will be as indicated in Fig. 14.14, where the current I can be calculated from

$$I = \frac{V_{CC} - 1.4}{R}$$

In actual TTL circuits Q_1 is designed to have a very low reverse β ($\beta_R \simeq 0.02$). Thus the gate input current will be very small, and the base current of Q_3 will be approximately equal to I. This current will be sufficient to drive Q_3 into saturation, and the output voltage will be low (0.1 to 0.2 V).

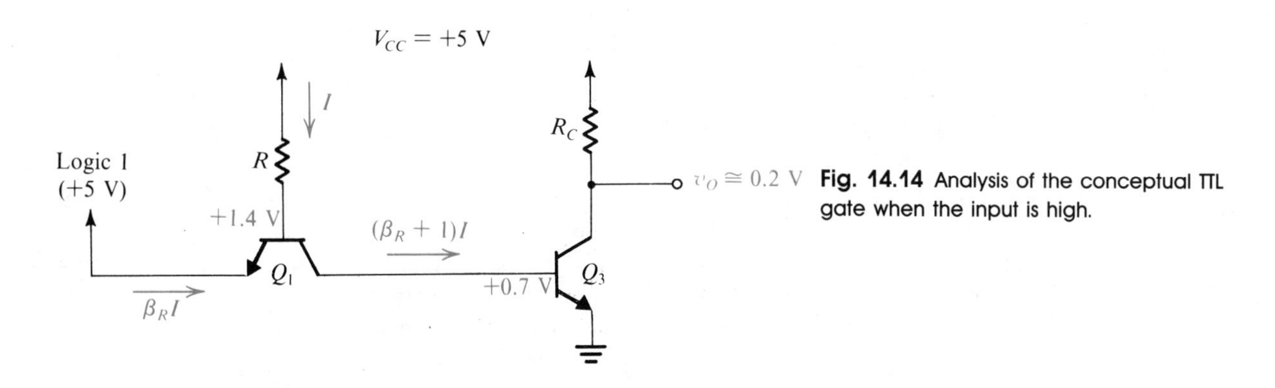

Fig. 14.14 Analysis of the conceptual TTL gate when the input is high.

Next let the gate input voltage be brought down to the logic-0 level (say, $v_I \simeq 0.2$ V). The current I will then be diverted to the emitter of Q_1. The base–emitter junction of Q_1 will become forward-biased, and the base voltage of Q_1 will therefore drop to 0.9 V. Since Q_3 *was* in saturation, its base voltage will remain at +0.7 V pending the removal of the excess charge stored in the base region. Figure 14.15 indicates the various voltage and current values immediately after the input is lowered. We see that Q_1 will be operating in the normal active mode[3] and its collector will carry a large current ($\beta_F I$).

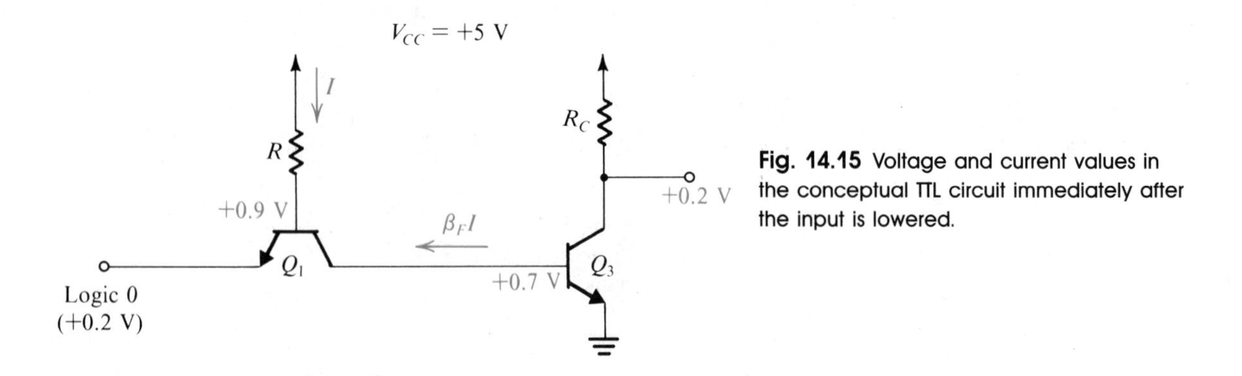

Fig. 14.15 Voltage and current values in the conceptual TTL circuit immediately after the input is lowered.

[3]Although the collector voltage of Q_1 is lower than its base voltage by 0.2 V, the collector–base junction will in effect be cut off and Q_1 will be operating in the active mode.

This large current rapidly discharges the base of Q_3 and drives it into cutoff. We thus see the action of Q_1 in speeding up the turn-off process.

As Q_3 turns off, the voltage at its base is reduced, and Q_1 enters the saturation mode. Eventually the collector current of Q_1 will become negligibly small, which implies that its V_{CEsat} will be approximately 0.1 V and the base of Q_3 will be at about 0.3 V, which keeps Q_3 in cutoff.

Output Circuit of the TTL Gate

The above discussion illustrates how one of the two problems that slow down the operation of DTL is solved in TTL. The second problem, the long rise time of the output waveform, is solved by modifying the output stage, as we shall now explain.

First, recall that the common-emitter output stage provides fast discharging of load capacitance but rather slow charging. The opposite is obtained in the emitter-follower output stage shown in Fig. 14.16. Here, as v_I goes high, the transistor turns on and provides a low output resistance (characteristic of emitter followers), which results in fast charging of C_L. On the other hand, when v_I goes low, the transistor turns off and C_L is then left to discharge slowly through R_E.

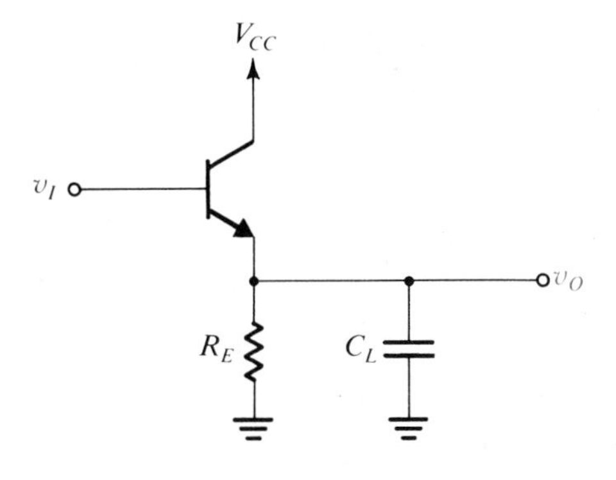

Fig. 14.16 An emitter-follower output stage with capacitive load.

It follows that an optimum output stage would be a combination of the common-emitter and the emitter-follower configurations. Such an output stage, shown in Fig. 14.17, has to be driven by two *complementary* signals v_{I1} and v_{I2}. When v_{I1} is high v_{I2} will be low, and in this case Q_3 will be on and saturated, and Q_4 will be off. The

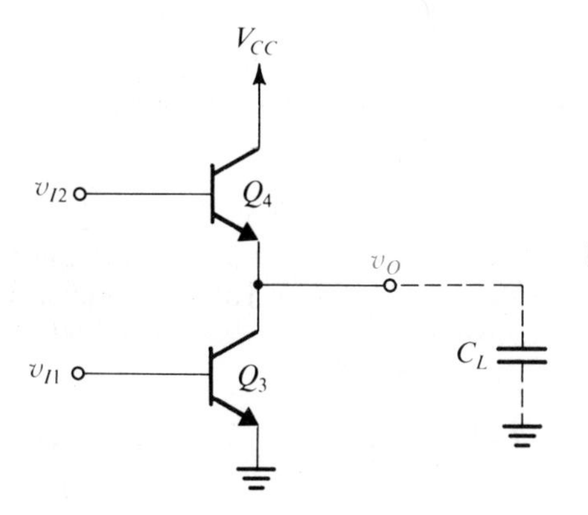

Fig. 14.17 The totem-pole output stage.

common-emitter transistor Q_3 will then provide the fast discharging of load capacitance and in steady state provide a low resistance (R_{CEsat}) to ground. Thus when the output is low, the gate can *sink* substantial amounts of current through the saturated transistor Q_3.

When v_{I1} is low and v_{I2} is high, Q_3 will be off and Q_4 will be conducting. The emitter follower Q_4 will then provide fast charging of load capacitance. It also provides the gate with a low output resistance in the high state and hence with the ability to *source* a substantial amount of load current.

Because of the appearance of the circuit in Fig. 14.17, with Q_4 stacked on top of Q_3, the circuit has been given the name **totem-pole output stage.** Also, because of the action of Q_4 in *pulling up* the output voltage to the high level, Q_4 is referred to as the **pull-up transistor.** Since the pulling up is achieved here by an active element (Q_4), the circuit is said to have an **active pull-up.** This is in contrast to the **passive pull-up** of RTL and DTL gates. Finally, note that a special **driver circuit** is needed to generate the two complementary signals v_{I1} and v_{I2}.

EXAMPLE 14.1

We wish to analyze the circuit shown together with its driving waveforms in Fig. 14.18 to determine the waveform of the output signal v_O. Assume that Q_3 and Q_4 have $\beta = 50$.

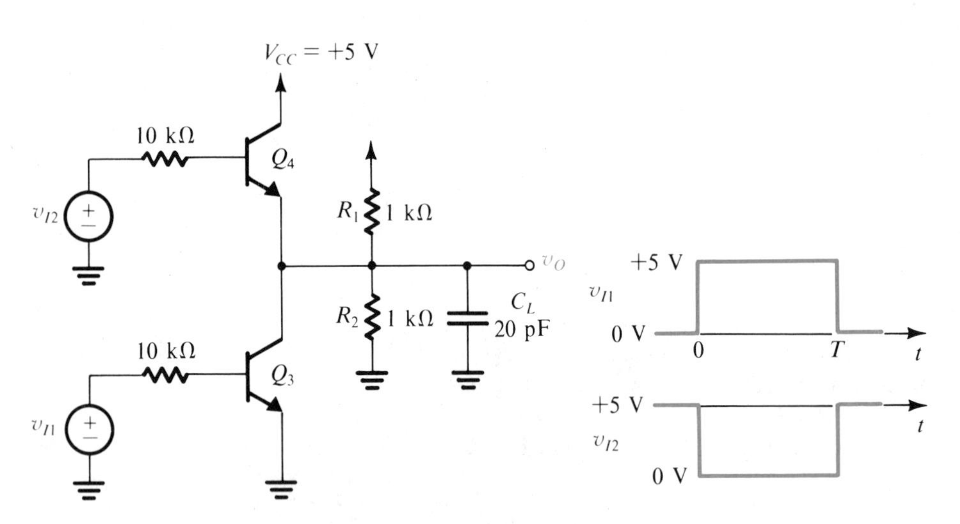

Fig. 14.18 Circuit and input waveforms for Example 14.1.

Solution Consider first the situation before v_{I1} goes high—that is, at time $t < 0$. In this case Q_3 is off and Q_4 is on, and the circuit can be simplified to that shown in Fig. 14.19. In this simplified circuit we have replaced the voltage divider (R_1, R_2) by its Thévenin equivalent. In the steady state, C_L will be charged to the output voltage v_O, whose value can be obtained as follows:

$$5 = 10 \times I_B + V_{BE} + I_E \times 0.5 + 2.5$$

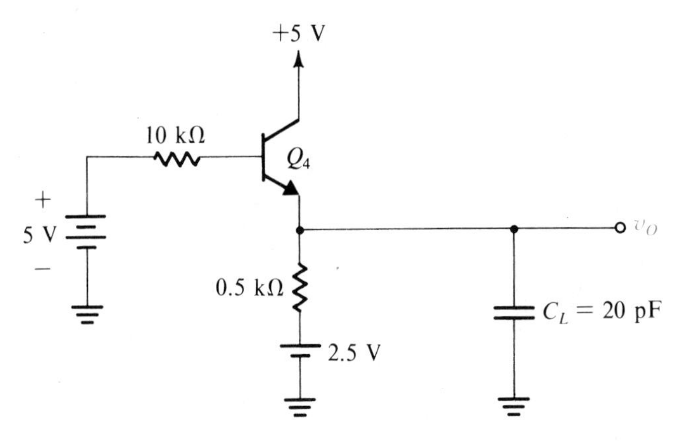

Fig. 14.19 The circuit of Fig. 14.18 when Q_3 is off.

Substituting $V_{BE} \simeq 0.7$ V and $I_B = I_E/(\beta + 1) = I_E/51$ gives $I_E = 2.59$ mA. Thus the output voltage v_O is given by

$$v_O = 2.5 + I_E \times 0.5 = 3.79 \text{ V}$$

We next consider the circuit as v_{I1} goes high and v_{I2} goes low. Transistor Q_3 turns on and transistor Q_4 turns off, and the circuit simplifies to that shown in Fig. 14.20. Again we have used the Thévenin equivalent of the divider (R_1, R_2). We shall also assume that the switching times of the transistors are negligibly small. Thus at $t = 0+$ the base current of Q_3 becomes

$$I_B = \frac{5 - 0.7}{10} = 0.43 \text{ mA}$$

Since at $t = 0$ the collector voltage of Q_3 is 3.79 V, and since this value cannot change instantaneously because of C_L, we see that at $t = 0+$ transistor Q_3 will be in the active mode. The collector current of Q_3 will be βI_B, which is 21.5 mA, and the circuit will have the equivalent shown in Fig. 14.21(a). A simpler version of this equivalent circuit, obtained using Thévenin's theorem, is shown in Fig. 14.21(b).

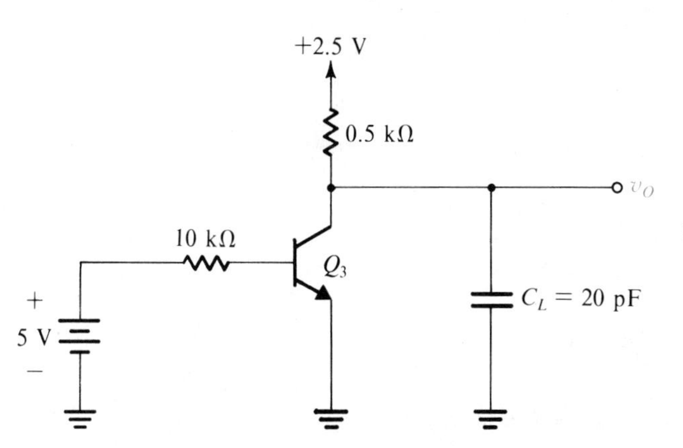

Fig. 14.20 The circuit of Fig. 14.18 when Q_4 is off.

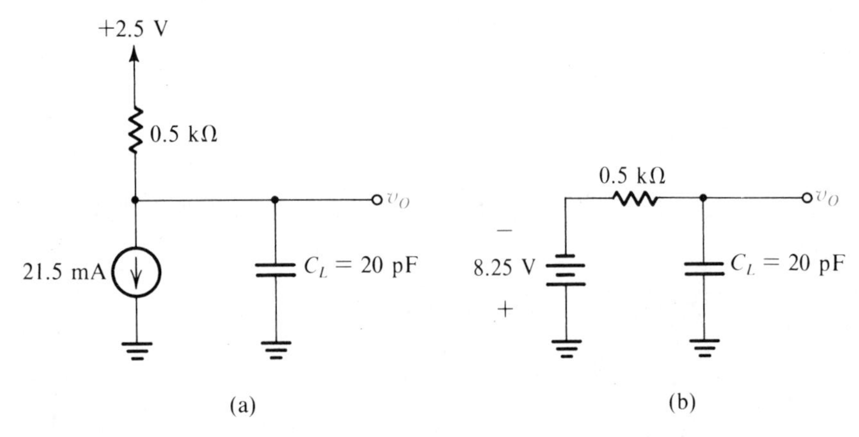

Fig. 14.21 (a) Equivalent circuit for the circuit in Fig. 14.20 when Q_3 is in the active mode. **(b)** Simpler version of the circuit in (a) obtained using Thévenin's theorem.

The equivalent circuit of Fig. 14.21 applies as long as Q_3 remains in the active mode. This condition persists while C_L is being discharged and until v_O reaches about +0.3 V, at which time Q_3 enters saturation. This is illustrated by the waveform in Fig. 14.22. The time for the output voltage to fall from +3.79 V to +0.3 V, which can be considered the **fall time** t_f, can be obtained from

$$-8.25 - (-8.25 - 3.79)e^{-t_f/\tau} = 0.3$$

which results in

$$t_f \simeq 0.34\tau$$

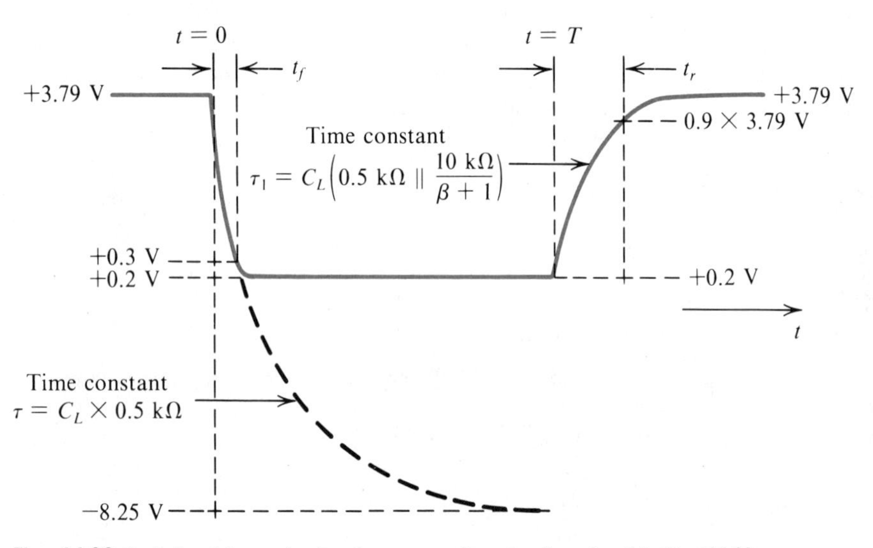

Fig. 14.22 Details of the output voltage waveform for the circuit in Fig. 14.18.

where

$$\tau = C_L \times 0.5 \text{ k}\Omega = 10 \text{ ns}$$

Thus $t_f = 3.4$ ns.

After Q_3 enters saturation, the capacitor discharges further to the final steady-state value of V_{CEsat} ($\simeq 0.2$ V). The transistor model that applies during this interval is more complex; since the interval in question is quite short, we shall not pursue the matter further.

Consider next the situation as v_{I1} goes low and v_{I2} goes high at $t = T$. Transistor Q_3 turns off as Q_4 turns on. We shall assume that this occurs immediately, and thus at $t = T+$ the circuit simplifies to that in Fig. 14.19. We have already analyzed this circuit in the steady state and thus know that eventually v_O will reach $+3.79$ V. Thus v_O rises exponentially from $+0.2$ V toward $+3.79$ V with a time constant of $C_L\{0.5 \text{ k}\Omega // [10 \text{ k}\Omega/(\beta + 1)]\}$, where we have neglected the emitter resistance r_e. Denoting this time constant τ_1 we obtain $\tau_1 = 2.8$ ns. Defining the rise time t_r as the time for v_O to reach 90% of the final value, we obtain $3.79 - (3.79 - 0.2)e^{-t_r/\tau_1} = 0.9 \times 3.79$, which results in $t_r = 6.4$ ns. Figure 14.22 illustrates the details of the output voltage waveform.

The Complete Circuit of the TTL Gate

Figure 14.23 shows the complete TTL gate circuit. It consists of three stages: the input transistor Q_1, whose operation has already been explained, the driver stage Q_2, whose function is to generate the two complementary voltage signals required to drive the totem-pole circuit, which is the third (output) stage of the gate. The totem-pole circuit in the TTL

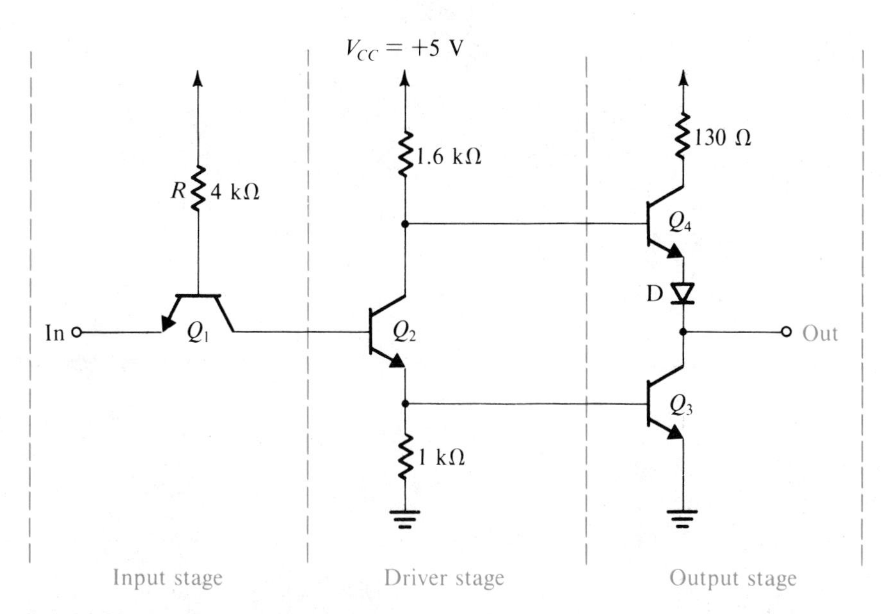

Fig. 14.23 The complete TTL gate circuit with only one input terminal indicated.

gate has two additional components: the 130-Ω resistance in the collector circuit of Q_4 and the diode D in the emitter circuit of Q_4. The function of these two additional components will be explained shortly. Notice that the TTL gate is shown with only one input terminal indicated. Inclusion of additional input terminals will be considered in Section 14.4.

Because the driver stage Q_2 provides two complementary (that is, out-of-phase) signals, it is known as a **phase splitter.**

We shall now provide a detailed analysis of the TTL gate circuit in its two extreme states: one with the input high and one with the input low.

Analysis when the Input Is High

When the input is high (say, +5 V), the various voltages and currents of the TTL circuit will have the values indicated in Fig. 14.24. The analysis illustrated in Fig. 14.24 is quite straightforward, and the order of the steps followed is indicated by the circled numbers. As expected, the input transistor is operating in the inverse active mode, and the input current, called the **input high current** I_{IH}, is small; that is,

$$I_{IH} = \beta_R I \simeq 15\ \mu\text{A}$$

where we assume that $\beta_R \simeq 0.02$.

The collector current of Q_1 flows into the base of Q_2, and its value is sufficient to saturate the phase-splitter transistor Q_2. The latter supplies the base of Q_3 with sufficient current to drive it into saturation and lower its output voltage to V_{CEsat} (0.1 to 0.2 V). The voltage at the collector of Q_2 is $V_{BE3} + V_{CEsat}$ (Q_2), which is approximately +0.9 V. If

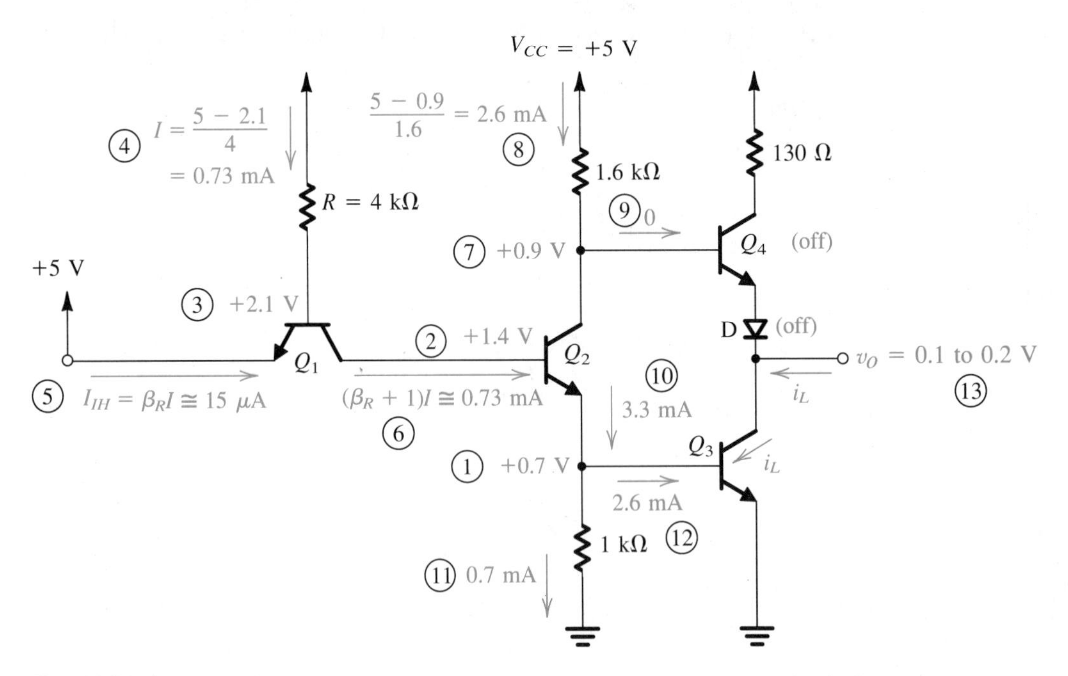

Fig. 14.24 Analysis of the TTL gate with the input high. The circled numbers indicate the order of the analysis steps.

diode D were not included, this voltage would be sufficient to turn Q_4 on, which is contrary to the proper operation of the totem-pole circuit. Including diode D ensures that both Q_4 and D remain off. The saturated transistor Q_3 then establishes the low output voltage of the gate (V_{CEsat}) and provides a low impedance to ground.

In the low-output state the gate can sink a load current i_L, provided that the value of i_L does not exceed $\beta \times 2.6$ mA, which is the maximum collector current that Q_3 can sustain while remaining in saturation. Obviously the greater the value of i_L, the greater the output voltage will be. To maintain the logic-0 level below a certain specified limit, a corresponding limit has to be placed on the load current i_L. As will be seen shortly, it is this limit that determines the maximum fan-out of the TTL gate.

Figure 14.25 shows a sketch of the output voltage v_O versus the load current i_L of the TTL gate when the output is low. This is simply the v_{CE}–i_C characteristic curve of Q_3 measured with a base current of 2.6 mA. Note that at $i_L = 0$, v_O is the offset voltage, which is about 100 mV.

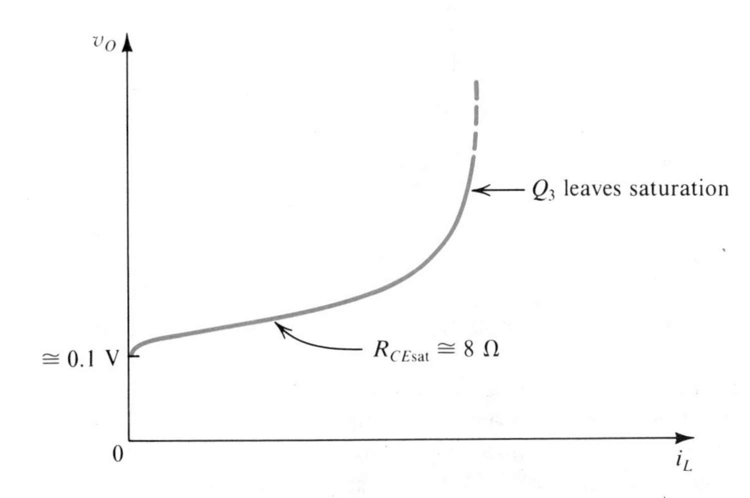

Fig. 14.25 The v_O–i_L characteristic of the TTL gate when the output is low.

Exercise 14.9 Assume that the saturation portion of the v_O–i_L characteristic shown in Fig. 14.25 can be approximated by a straight line (of slope = 8 Ω) that intersects the v_O axis at 0.1 V. Find the maximum load current that the gate is allowed to sink if the logic-0 level is specified to be ≤ 0.3 V.

Ans. 25 mA

Analysis when the Input Is Low

Consider next the operation of the TTL gate when the input is at the logic-0 level ($\simeq 0.2$ V). The analysis is illustrated in Fig. 14.26, from which we see that the base–emitter junction of Q_1 will be forward-biased and the base voltage will be approximately +0.9 V. Thus the current I can be found to be approximately 1 mA. Since 0.9 V is insufficient to forward-bias the series combination of the collector-base junction of Q_1 and the base–emitter junction of Q_2 (at least 1.2 V would be required), the latter will be off.

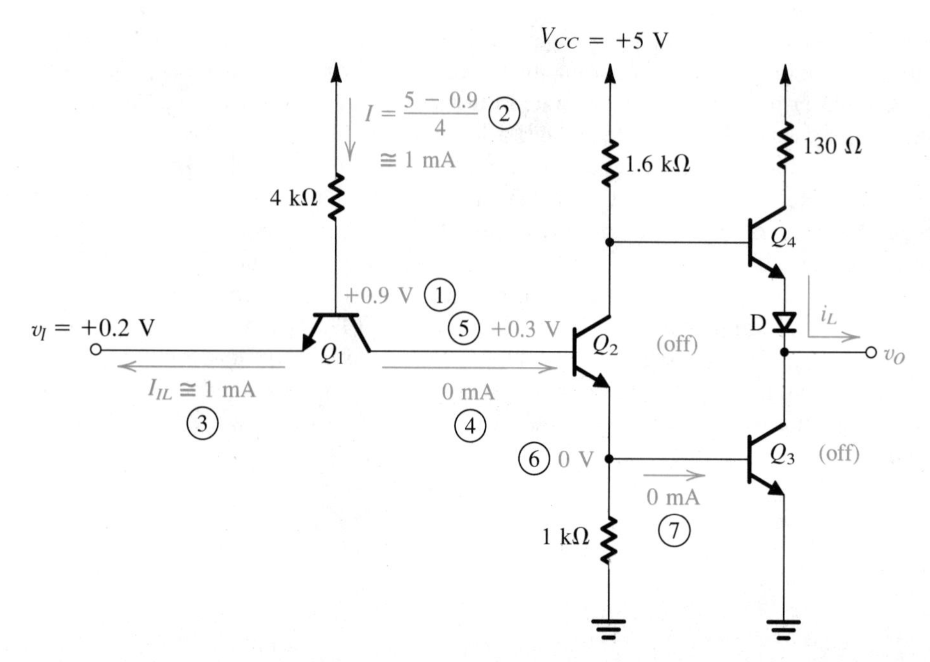

Fig. 14.26 Analysis of the TTL gate when the input is low. The circled numbers indicate the order of the analysis steps.

Therefore the collector current of Q_1 will be almost zero and Q_1 will be saturated, with $V_{CEsat} \simeq 0.1$ V. Thus the base of Q_2 will be at approximately +0.3 V, which is indeed insufficient to turn Q_2 on.

The gate input current in the low state, called **input-low current** I_{IL}, is approximately equal to the current I ($\simeq 1$ mA) and flows out of the emitter of Q_1. If the TTL gate is driven by another TTL gate, the output transistor Q_3 of the driving gate should sink this current I_{IL}. Since the output current that a TTL gate can sink is limited to a certain maximum value, the maximum fan-out of the gate is directly determined by the value of I_{IL}.

Exercises 14.10 Consider the TTL gate analyzed in Exercise 14.9. Find its maximum allowable fan-out using the value of I_{IL} calculated above.

Ans. 25

14.11 Use Eq. (14.16) to find V_{CEsat} of transistor Q_1 when the input of the gate is low (0.2 V). Assume that $\beta_F = 50$ and $\beta_R = 0.02$.

Ans. 98 m V

Let us continue with our analysis of the TTL gate. When the input is low, we see that both Q_2 and Q_3 will be off. Transistor Q_4 will be on and will supply (source) the load current i_L. Depending on the value of i_L, Q_4 will be either in the active mode or in the saturation mode.

With the gate output terminal open, the current i_L will be very small (mostly leakage) and the two junctions (base–emitter junction of Q_4 and diode D) will be barely conducting. Assuming that each junction has a 0.65-V drop and neglecting the voltage drop across the 1.6-kΩ resistance, we find that the output voltage will be

$$v_O \simeq 5 - 0.65 - 0.65 = 3.7 \text{ V}$$

As i_L is increased, Q_4 and D conduct more heavily, but for a range of i_L, Q_4 remains in the active mode, and v_O is given by

$$v_O = V_{CC} - \frac{i_L}{\beta + 1} \times 1.6 \text{ k}\Omega - V_{BE4} - V_D \tag{14.21}$$

If we keep increasing i_L, a value will be reached at which Q_4 saturates. Then the output voltage becomes determined by the 130-Ω resistance according to the approximate relationship

$$v_O \simeq V_{CC} - i_L \times 130 - V_{CE\text{sat}}(Q_4) - V_D \tag{14.22}$$

Function of the 130-Ω Resistance

At this point the reason for including the 130-Ω resistance should be evident: It is simply to limit the current that flows through Q_4, especially in the event that the output terminal is accidentally short-circuited to ground. This resistance also limits the supply current in another circumstance, namely, when Q_4 turns on while Q_3 is still in saturation. To see how this occurs, consider the case where the gate input was high and then is suddenly brought down to the low level. Transistor Q_2 will turn off relatively fast because of the availability of a large reverse current supplied to its base terminal by the collector of Q_1. On the other hand, the base of Q_3 will have to discharge through the 1-kΩ resistance, and thus Q_3 will take some time to turn off. Meanwhile Q_4 will turn on, and a large current pulse will flow through the series combination of Q_4 and Q_3. Part of this current will serve the useful purpose of charging up any load capacitance to the logic-1 level. The magnitude of the current pulse will be limited by the 130-Ω resistance to about 30 mA.

The occurrence of these current pulses of short duration (called current spikes) raises another important issue. The current spikes have to be supplied by the V_{CC} source and, because of its finite source resistance, will result in voltage spikes (or "glitches") superimposed on V_{CC}. These voltage spikes could be coupled to other gates and flip-flops in the digital system and thus might produce false switching in other parts of the system. This effect, which might loosely be called **crosstalk,** is a problem in TTL systems. To reduce the size of the voltage spikes, capacitors (called bypass capacitors) should be connected to ground at frequent locations on the supply rail. These capacitors lower the impedance of the supply-voltage source and hence reduce the magnitude of the voltage spikes. Alternatively, one can think of the bypass capacitors as supplying the impulsive current spikes.

Exercises 14.12 Assuming that Q_4 has $\beta = 50$ and that at the verge of saturation $V_{CE\text{sat}} = 0.3$ V, find the value of i_L at which Q_4 saturates.

Ans. 4.16 mA

14.13 Assuming that at a current of 1 mA the voltage drops across the emitter–base junction of Q_4 and the diode D are each 0.7 V, find v_O when i_L = 1 mA and 10 mA. (Note the result of the previous exercise.)

Ans. 3.6 V; 2.7 V

14.14 Find the maximum current that can be sourced by a TTL gate while the output high level (V_{OH}) remains greater than the minimum guaranteed value of 2.4 V.

Ans. 12.3 mA

14.4 CHARACTERISTICS OF STANDARD TTL

Because of its popularity and importance, TTL will be studied further in this and the next sections. In this section we shall consider some of the important characteristics of standard TTL gates. Special improved forms of TTL will be dealt with in Section 14.5.

Transfer Characteristic

Figure 14.27 shows the TTL gate together with a sketch of its voltage transfer characteristic drawn in a piecewise-linear fashion. The actual characteristic is, of course, a smooth curve. We shall now explain the transfer characteristic and calculate the various breakpoints and slopes. It will be assumed that the output terminal of the gate is open.

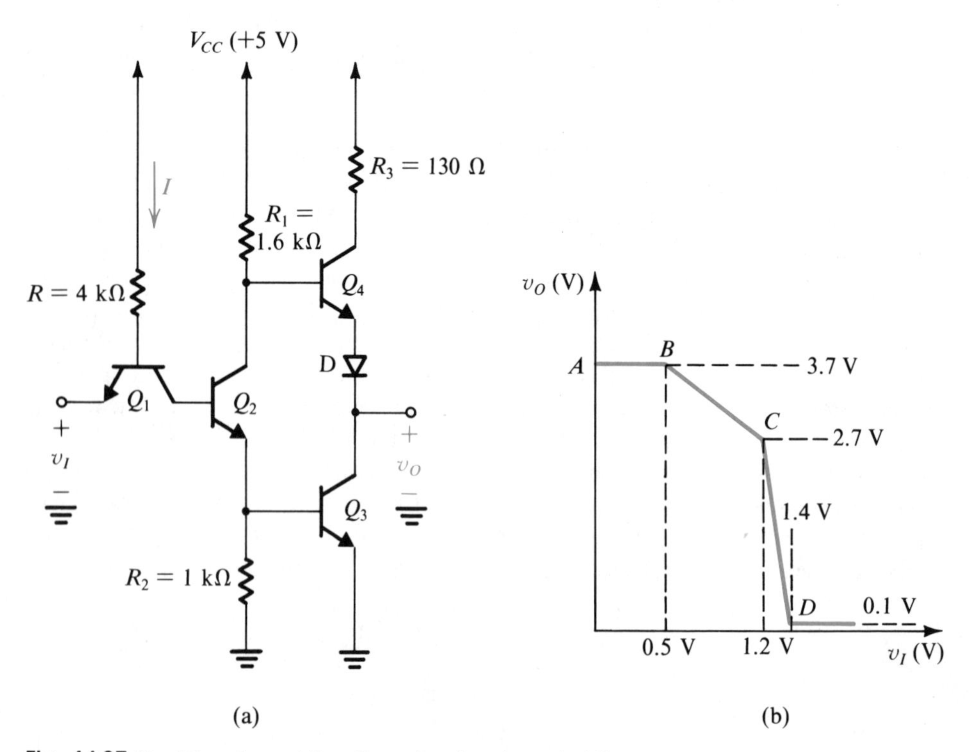

Fig. 14.27 The TTL gate and its voltage transfer characteristic.

Segment AB is obtained when transistor Q_1 is saturated, Q_2 and Q_3 are off, and Q_4 and D are on. The output voltage is approximately two diode drops below V_{CC}. At point B the phase splitter (Q_2) begins to turn on because the voltage at its base reaches 0.6 V (0.5 V + V_{CEsat} of Q_1).

Over segment BC, transistor Q_1 remains saturated, but more and more of its base current I gets diverted to its base–collector junction and into the base of Q_2, which operates as a linear amplifier. Transitor Q_4 and diode D remain on, with Q_4 acting as an emitter follower. Meanwhile the voltage at the base of Q_3, although increasing, remains insufficient to turn Q_3 on (less than 0.6 V).

Let us now find the slope of segment BC of the transfer characteristic. Let the input v_I increase by an increment Δv_I. This increment appears at the collector of Q_1, since the saturated Q_1 behaves (approximately) as a three-terminal short circuit as far as signals are concerned. Thus at the base of Q_2 we have a signal Δv_I. Neglecting the loading of emitter follower Q_4 on the collector of Q_2, we can find the gain of the phase splitter from

$$\frac{v_{c2}}{v_{b2}} = \frac{-\alpha_2 R_1}{r_{e2} + R_2} \tag{14.23}$$

The value of r_{e2} will obviously depend on the current in Q_2. This current will range from zero (as Q_2 begins to turn on) to the value that results in a voltage of about 0.6 V at the emitter of Q_2 (the base of Q_3). This value is about 0.6 mA and corresponds to point C on the transfer characteristic. Assuming an average current in Q_2 of 0.3 mA, we obtain $r_{e2} \simeq 83\ \Omega$. For $\alpha = 0.98$, Eq. (14.23) results in a gain value of 1.45. Since the gain of the output follower Q_4 is close to unity, the overall gain of the gate, which is the slope of the BC segment, is about -1.45.

As already implied, breakpoint C is determined by Q_3 starting to conduct. The corresponding input voltage can be found from

$$\begin{aligned} v_I(C) &= V_{BE3} + V_{BE2} - V_{CEsat}(Q_1) \\ &= 0.6 + 0.7 - 0.1 = 1.2 \text{ V} \end{aligned}$$

At this point the emitter current of Q_2 is approximately 0.6 mA. The collector current of Q_2 is also approximately 0.6 mA; neglecting the base current of Q_4, the voltage at the collector of Q_2 is

$$v_{C2}(C) = 5 - 0.6 \times 1.6 \cong 4 \text{ V}$$

Thus Q_2 is still in the active mode. The corresponding output voltage is

$$v_O(C) = 4 - 0.65 - 0.65 = 2.7 \text{ V}$$

As v_I is increased past the value of $v_I(C) = 1.2$ V, Q_3 begins to conduct and operates in the active mode. Meanwhile, Q_1 remains saturated, and Q_2 and Q_4 remain in the active mode. The circuit behaves as an amplifier until Q_2 and Q_3 saturate and Q_4 cuts off. This occurs at point D on the transfer characteristic, which corresponds to an input voltage $v_I(D)$ obtained from

$$\begin{aligned} v_I(D) &= V_{BE3} + V_{BE2} + V_{BC1} - V_{BE1} \\ &= 0.7 + 0.7 + 0.7 - 0.7 = 1.4 \text{ V} \end{aligned}$$

Note that we have in effect assumed that at point D transistor Q_1 is still saturated, but with $V_{CEsat} \simeq 0$. To see how this comes about, note that from point B on, more and more of the base current of Q_1 is diverted to its base–collector junction. Thus while the drop across the base–collector junction increases, that across the base–emitter junction decreases. At point D these drops become almost equal. For $v_I > v_I(D)$ the base–emitter junction of Q_1 cuts off; thus Q_1 leaves saturation and enters the inverse active mode.

Calculation of gain over the segment CD is a relatively complicated task. This is due to the fact that there are two paths from input to output: one through Q_3 and one through Q_4. A simple but gross approximation for the gain of this segment can be obtained from the coordinates of points C and D in Fig. 14.27(b), as follows:

$$\text{Gain} = -\frac{v_O(C) - v_O(D)}{v_I(D) - v_I(C)}$$

$$= -\frac{2.7 - 0.1}{1.4 - 1.2} = -13 \text{ V/V}$$

From the transfer curve of Fig. 14.27(b) we can determine the critical points and the noise margins as follows: $V_{OH} = 3.7$ V; V_{IL} is somewhere in the range of 0.5 V to 1.2 V, and thus a conservative estimate would be 0.5 V; $V_{OL} = 0.1$ V; $V_{IH} = 1.4$ V; $NM_H = V_{OH} - V_{IH} = 2.3$ V; and $NM_L = V_{IL} - V_{OL} = 0.4$ V. It should be noted that these values are computed assuming that the gate is not loaded and without taking into account power-supply or temperature variations.

Exercise 14.15 Taking into account the fact that the voltage across a forward-biased *pn* junction changes by about -2 mV/°C, find the coordinates of points *A*, *B*, *C*, and *D* of the gate transfer characteristic at -55°C and at $+125$°C. Assume that the characteristic in Fig. 14.27(b) applies at 25°C, and neglect the small temperature coefficient of V_{CEsat}.

Ans. At -55°C: (0, 3.38), (0.66, 3.38), (1.52, 2.16), (1.72, 0.1); at $+125$°C: (0, 4.1), (0.3, 4.1), (0.8, 3.46), (1.0, 0.1)

Manufacturers' Specifications

Manufacturers of TTL usually provide curves for the gate transfer characteristic, the input i–v characteristic, and the output i–v characteristic, measured at the limits of the specified operating temperature range. In addition, guaranteed values are usually given for the parameters V_{OL}, V_{OH}, V_{IL}, and V_{IH}. For standard TTL (known as the 74 series) these values are $V_{OL} = 0.4$ V, $V_{OH} = 2.4$ V, $V_{IL} = 0.8$ V, and $V_{IH} = 2$ V. These limit values are guaranteed for a specified tolerance in power-supply voltage and for a maximum fan-out of 10. From our discussion in Section 14.3 we know that the maximum fan-out is determined by the maximum current that Q_3 can sink while remaining in saturation and while maintaining a saturation voltage lower than a guaranteed maximum ($V_{OL} = 0.4$ V). Calculations performed in Section 14.3 indicate the possibility of a maximum fan-out of 20 to 30. Thus the figure specified by the manufacturer is appropriately conservative.

The parameters V_{OL}, V_{OH}, V_{IL}, and V_{IH} can be used to compute the noise margins as follows:

$$NM_H = V_{OH} - V_{IH} = 0.4 \text{ V}$$

$$NM_L = V_{IL} - V_{OL} = 0.4 \text{ V}$$

Exercises 14.16 In Section 14.3 we found that when the gate input is high, the base current of Q_3 is approximately 2.6 mA. Assume that this value applies at 25°C and that at this temperature $V_{BE} \simeq 0.7$ V. Taking into account the −2-mV/°C temperature coefficient of V_{BE} and neglecting all other changes, find the base current of Q_3 at −55°C and at +125°C.

Ans. 2.2 mA; 3 mA

14.17 Figure E14.17 shows sketches of the i_L–v_O characteristics of a TTL gate when the output is low. Use these characteristics together with the results of Exercise 14.16 to calculate the value of β of transistor Q_3 at −55°C, +25°C, and +125°C.

Ans. 16; 25; 28

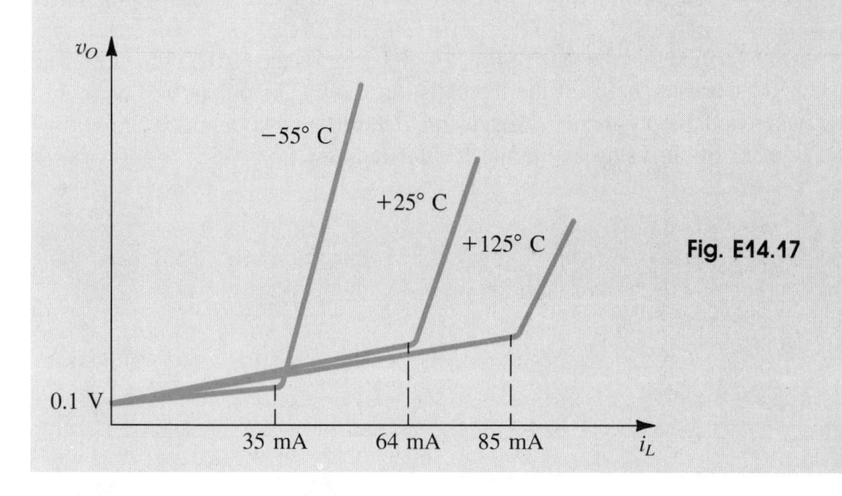

Fig. E14.17

Propagation Delay

The propagation delay of TTL gates is defined conventionally as the time between the 1.5-V points of corresponding edges of the input and output waveforms. For standard TTL (also known as *medium-speed* TTL) t_P is typically about 10 ns.

As far as power dissipation is concerned it can be shown (see Exercise 14.18 below) that when the gate output is high the gate dissipates 5 mW, and when the output is low the dissipation is 16.7 mW. Thus the average dissipation is 11 mW, resulting in a delay-power product of about 100 pJ.

Exercise 14.18 Calculate the value of the supply current (I_{CC}), and hence the power dissipated in the TTL gate, when the output terminal is open and the input is (a) low at 0.2 V (see Fig. 14.26) and (b) high at +5 V (see Fig. 14.24).

Ans. (a) 1 mA, 5 mW; (b) 3.33 mA, 16.7 mW

Dynamic Power Dissipation

In Section 14.3 the occurrence of supply current spikes was explained. These spikes give rise to additional power drain from the V_{CC} supply. This **dynamic power** is also dissipated in the gate circuit. It can be evaluated by multiplying the average current due to the spikes by V_{CC}, as illustrated by the solution of Exercise 14.19.

Exercise 14.19 Consider a TTL gate that is switched on and off at the rate of 1 MHz. Assume that each time the gate is turned off (that is, the output goes high) a supply-current pulse of 30-mA amplitude and 2-ns width occurs. Also assume that no current spike occurs when the gate is turned on. Calculate the average supply current due to the spikes, and the dynamic power dissipation.

Ans. 60 μA; 0.3 mW

The TTL NAND Gate

Figure 14.28 shows the basic TTL gate. Its most important feature is the **multiemitter transistor** Q_1 used at the input. Figure 14.29 shows the structure of the multiemitter transistor.

It can be easily verified that the gate of Fig. 14.28 performs the NAND function. The output will be high if one (or both) of the inputs is (are) low. The output will be low in only one case: when both inputs are high. Extension to more than two inputs is straightforward and is achieved by diffusing additional emitter regions.

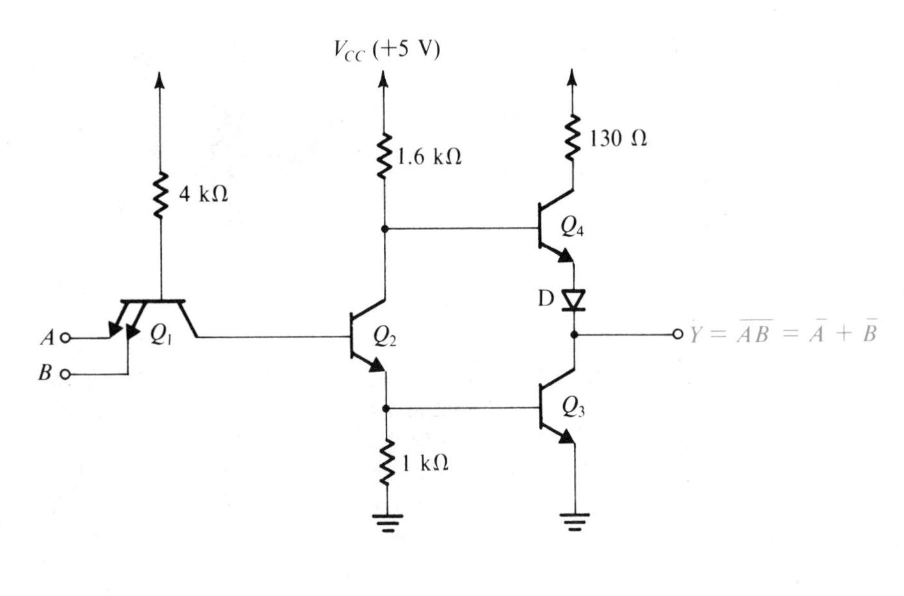

Fig. 14.28 The TTL NAND gate.

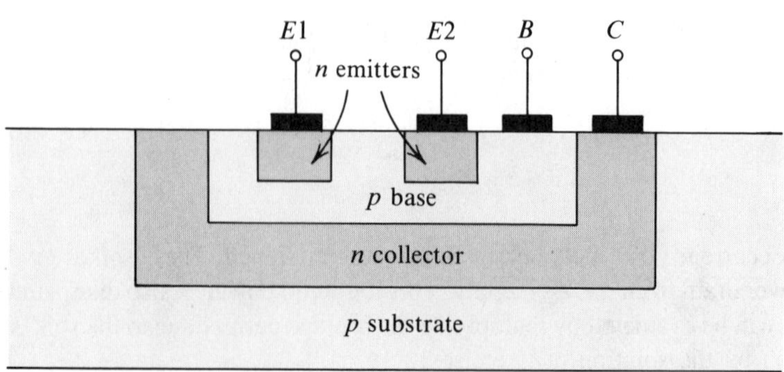

Fig. 14.29 Structure of the multiemitter transistor Q_1.

Although theoretically an unused input terminal may be left open-circuited, this is generally not a good practice. An open-circuit input terminal acts as an "antenna" that "picks up" interfering signals and thus could cause erroneous gate switching. An unused input terminal should therefore be connected to the positive power supply *through a resistance* (of, say, 1 kΩ). In this way the corresponding base-emitter junction of Q_1 will be reverse-biased and thus will have no effect on the operation of the gate. The series resistance is included in order to limit the current in case of breakdown of the base–emitter junction due to transients on the power supply.

Other TTL Logic Circuits

On a TTL MSI chip there are many cases in which logic functions are implemented using "stripped-down" versions of the basic TTL gate. As an example we show in Fig. 14.30 the TTL implementation of the AND-OR-INVERT function. As shown, the phase-splitter transistors of two gates are connected in parallel, and a single output stage is used. The reader is urged to verify that the logic function realized is as indicated.

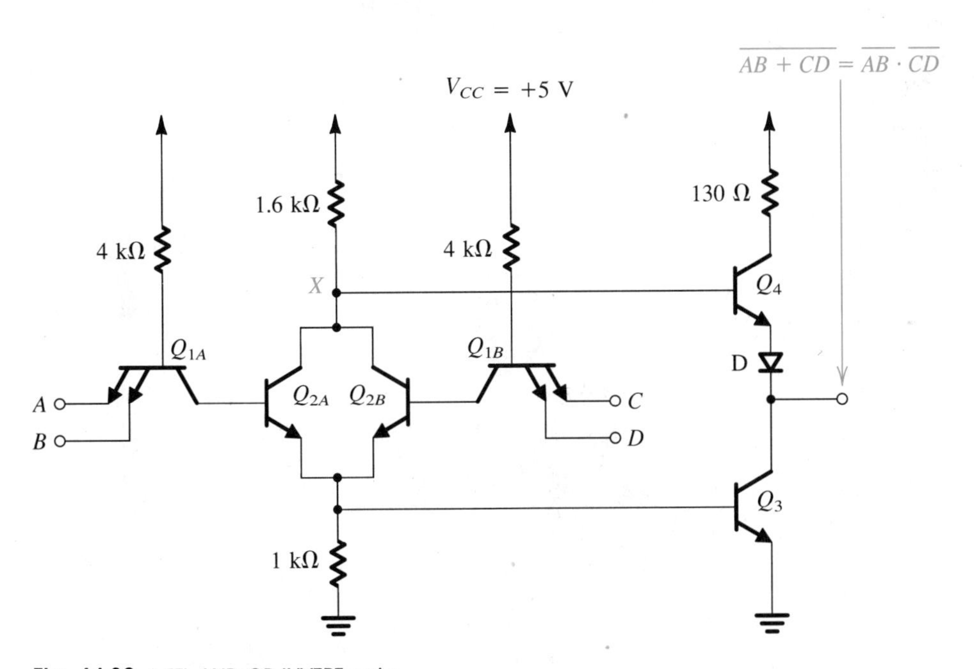

Fig. 14.30 A TTL AND-OR-INVERT gate.

At this point it should be noted that the totem-pole output stage of TTL does *not* allow connecting the output terminals of two gates to realize the AND function of their outputs (known as the wired-AND connection). To see the reason for this, consider two gates whose outputs are connected together, and let one gate have a high output and the other have a low output. Current will flow from Q_4 of the first gate through Q_3 of the second gate. The current value will fortunately be limited by the 130-Ω resistance. Obviously, however, no useful logic function is realized by this connection.

The lack of wired-AND capability is a drawback of TTL. Nevertheless, the problem is solved in a number of ways, including doing the paralleling at the phase-splitter stage, as illustrated in Fig. 14.30. Another solution consists of deleting the emitter-follower transistor altogether. The result is an output stage consisting solely of the common-emitter transistor Q_3 without even a collector resistance. Obviously, one can connect the outputs of such gates together to a common collector resistance and achieve a wired-AND capability. TTL gates of this type are known as **open-collector TTL.** The obvious disadvantage is the slow rise time of the output waveform.

Another useful variant of TTL is the **tristate** output arrangement explored in Exercise 14.20.

Exercise 14.20 The circuit shown in Fig. E14.20 is called tristate TTL. Verify that when the terminal labeled Third state is high, the gate functions normally and that when this terminal is low, both transistors Q_3 and Q_4 cut off and the output of the gate is an open circuit. The latter state is the third state, or the high-output-impedance state.

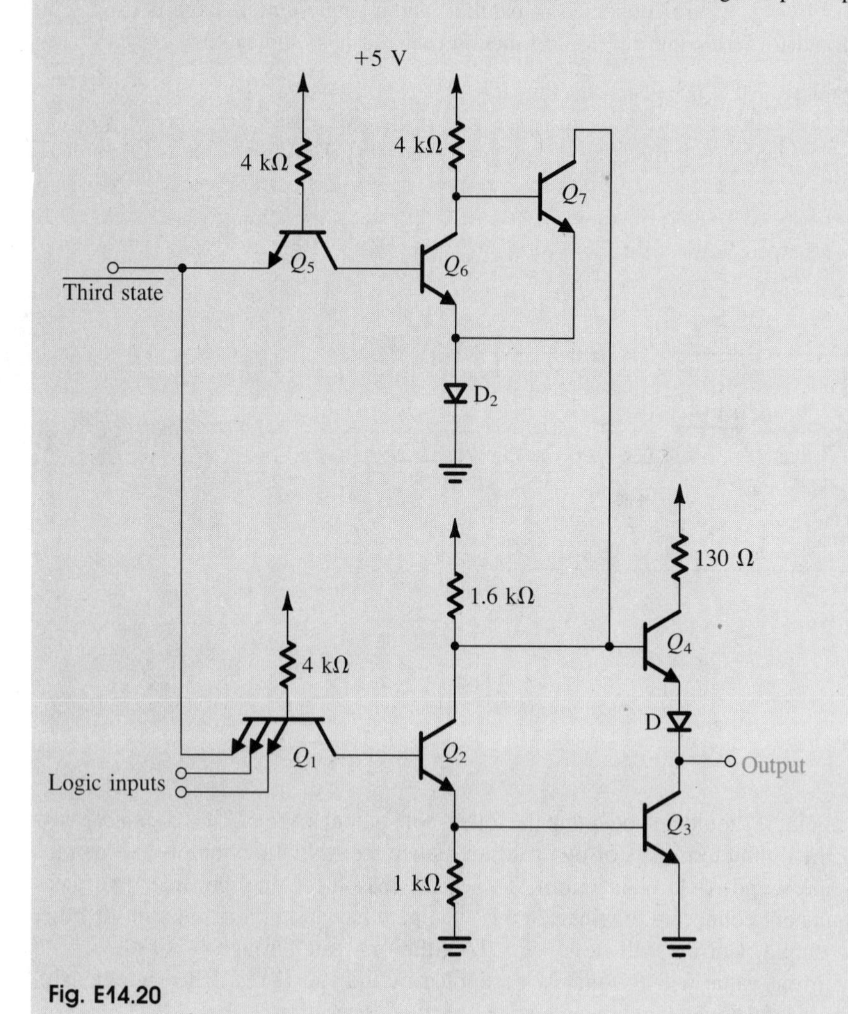

Fig. E14.20

Tristate TTL enables the connection of a number of TTL gates to a common output line (or *bus*). At any particular time the signal on the bus will be determined by the one TTL gate that is *enabled* (by raising its third-state input terminal). All other gates will be in the third state and thus will have no control of the bus.

14.5 TTL FAMILIES WITH IMPROVED PERFORMANCE

The standard TTL circuits studied in the two previous sections were introduced in the mid-1960s. Since then, several improved versions have been developed. In this section we shall discuss some of these improved TTL subfamilies. As will be seen the improvements are in two directions: increasing speed and reducing power dissipation.

The speed of the standard TTL gate of Fig. 14.28 is limited by two mechanisms: first, transistors Q_1, Q_2, and Q_3 saturate, and hence we have to contend with their finite storage time. Although Q_2 is discharged reasonably quickly because of the active mode of operation of Q_1, as already explained, this is not true for Q_3, whose base charge has to leak out through the 1-kΩ resistance in its base circuit. Second, the resistances in the circuit, together with the various transistor and wiring capacitances, form relatively long time constants, which contribute to lengthening the gate delay.

It follows that there are two approaches to speeding up the operation of TTL. The first is to prevent transistor saturation and the second is to reduce the values of all resistances. Both approaches are utilized in the Schottky TTL circuit family.

Schottky TTL

In Schottky TTL, transistors are prevented from saturation by connecting a low-voltage-drop diode between base and collector, as shown in Fig. 14.31. These diodes, formed as a metal-to-semiconductor junction, are called Schottky diodes and have a forward voltage drop of about 0.5 V. Schottky diodes[4] are easily fabricated and do not increase chip area.

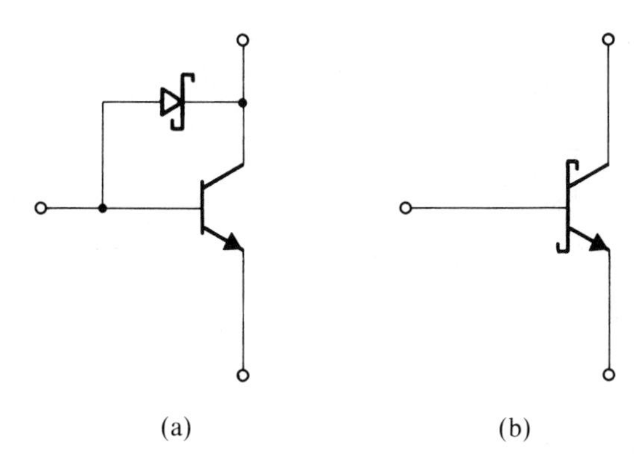

Fig. 14.31 (a) A transistor with a Schottky diode clamp. **(b)** Circuit symbol for the connection in (a), known as a Schottky transistor.

[4]Note that silicon Schottky diodes exhibit voltage drops of about 0.5 V, whereas GaAs Schottky diodes (Section 5.11) exhibit voltage drops of about 0.7 V.

In fact, the Schottky TTL fabrication process has been designed to yield transistors with smaller areas and thus higher β and f_T than those produced by the standard TTL process. Figure 14.31 also shows the symbol used to represent the combination of a transistor and a Schottky diode, referred to as a Schottky transistor.

The Schottky transistor does not saturate since some of its base current drive is shunted away from the base by the Schottky diode. The latter then conducts and clamps the base-collector junction voltage to about 0.5 V. This voltage is smaller than the value required to forward-bias the collector-base junction of these small-sized transistors. In fact, a Schottky transistor begins to conduct when its v_{BE} is about 0.7 V, and the transistor is fully conducting when v_{BE} is about 0.8 V. With the Schottky diode clamping v_{BC} to about 0.5 V, the collector-to-emitter voltage is about 0,.3 V and the transistor is still operating in the active mode. By avoiding saturation, the Schottky-clamped transistor exhibits a very short turnoff time.

Figure 14.32 shows a Schottky-clamped, or simply Schottky, TTL NAND gate. Comparing this circuit to that of standard TTL shown in Fig. 14.28 reveals a number of variations. First and foremost, Schottky clamps have been added to all transistors except

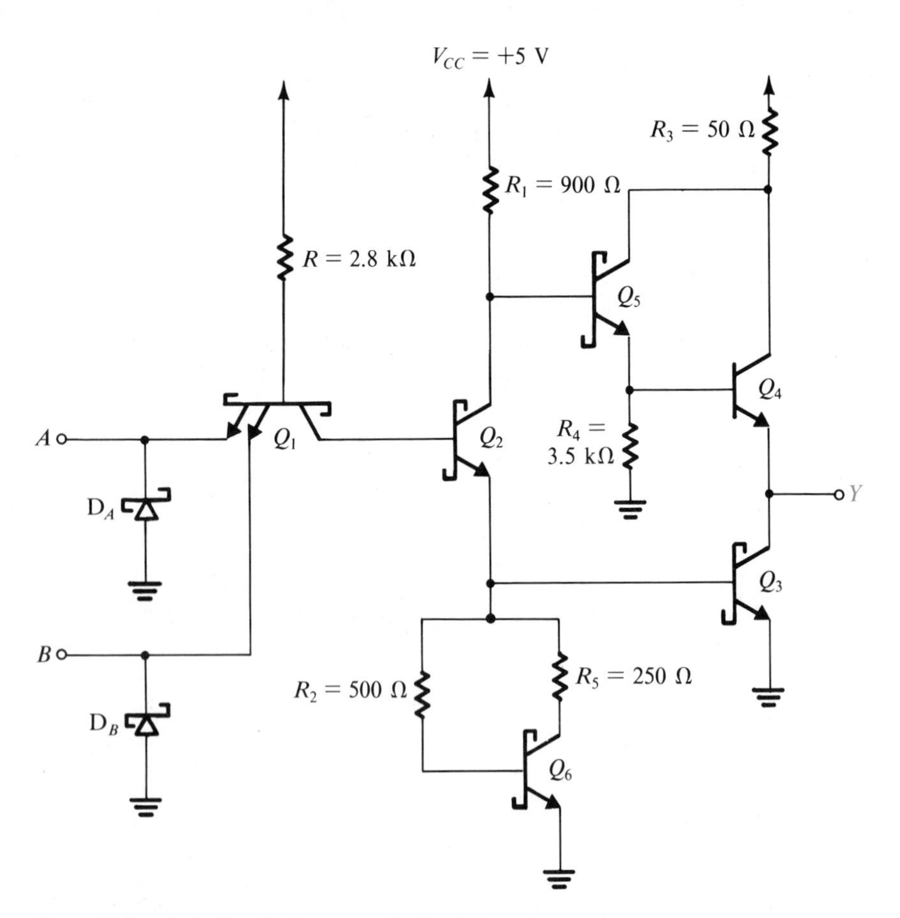

Fig. 14.32 A Schottky TTL (known as STTL) NAND gate.

Q_4. As will be seen shortly, transistor Q_4 can never saturate and thus does not need a Schottky clamp. Second, all resistances have been reduced to almost half the values used in the standard circuit. These two changes result in a much shorter gate delay. The reduction in resistance values, however, increases the gate power dissipation (by a factor of about 2).

The Schottky TTL gate features three other circuit techniques that further improve performance. These are as follows:

1. The diode D needed to prevent Q_4 from conducting when the gate output is low is replaced by transistor Q_5, which together with Q_4 forms a Darlington pair. This Darlington stage provides increased current gain and hence increased current-sourcing capability. This, together with the lower output resistance of the gate (in the output-high state), yields a reduction in the time required to charge the load capacitance to the high level. Note that transistor Q_4 never saturates because

$$\begin{aligned} V_{CE4} &= V_{CE5} + V_{BE4} \\ &\simeq 0.3 + 0.8 = 1.1 \text{ V} \end{aligned}$$

2. Input clamping diodes, D_A and D_B, are included.[5] These diodes conduct only when the input voltages go below ground level. This could happen due to ''ringing'' on the wires connecting the input of the gate to the output of another gate. Ringing occurs because such connecting wires behave as *transmission lines* that are not properly terminated. Without the clamping diodes, ringing can cause the input voltage to transiently go sufficiently negative so as to cause the substrate-to-collector junction of Q_1 (see Fig. 14.29) to become forward-biased. This in turn would result in improper gate operation. Also, ringing can cause the input voltage to go sufficiently positive to result in false gate switching. The input diodes clamp the negative excursions of the input ringing signal (to -0.5 V). Their conduction also provides a power loss in the transmission line, which results in *damping* of the ringing waveform and thus a reduction in its positive-going part.

3. The resistance between the base of Q_3 and ground has been replaced by a nonlinear resistance realized by transistor Q_6, and two resistors, R_2 and R_5. This nonlinear resistance is known as an **active pull-down,** in analogy to the active pull-up provided by the emitter-follower part of the totem-pole output stage. This feature is an ingenious one, and we shall discuss it in more detail.

Active Pull-Down. Figure 14.33 shows a sketch of the i–v characteristic of the nonlinear network composed of Q_6 and its two associated resistors, R_2 and R_5. Also shown for comparison is the linear i–v characteristic of a resistor that would be connected between the base of Q_3 and ground if the active pull-down were not used.

The first characteristic to note of the active pull-down is that it conducts negligible current (and thus behaves as a high resistance) until the voltage across it reaches a V_{BE} drop. Thus the gain of the phase splitter (as a linear amplifier) will remain negligibly small until

[5] Some standard TTL circuits also include input clamping diodes. Note, however, that in Schottky TTL, the input clamping diodes are of the Schottky type.

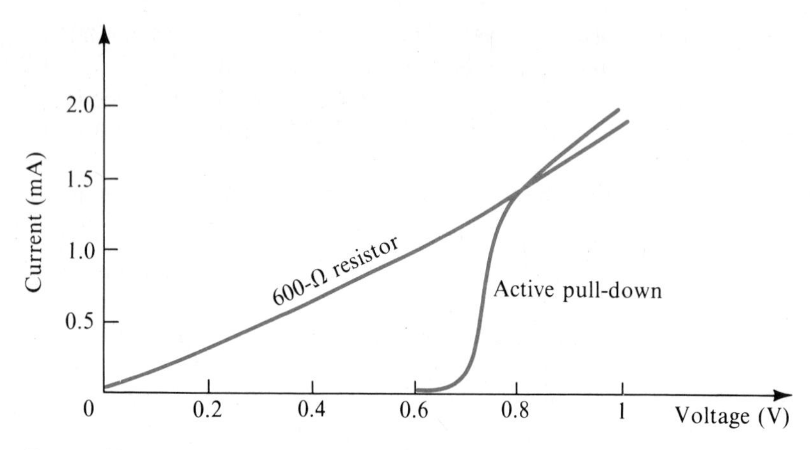

Fig. 14.33 Comparison of the i–v characteristic of the active pull-down with that of a 600-Ω resistor.

a V_{BE} drop develops between its emitter and ground, which is the onset of conduction of Q_3. In other words, Q_2, Q_3, and Q_6 will turn on almost simultaneously. Thus segment BC of the transfer characteristic (see Fig. 14.27b) will be absent and the gate transfer characteristic will become much sharper, as shown in Fig. 14.34. Since the active pull-down circuit causes the transfer characteristic to become "squarer," it is also known as a "squaring circuit." The result is an increase in noise margins.

The active pull-down circuit also speeds up the turn-on and turnoff of Q_3. To see how this comes about observe that the active pull-down draws a negligible current over a good

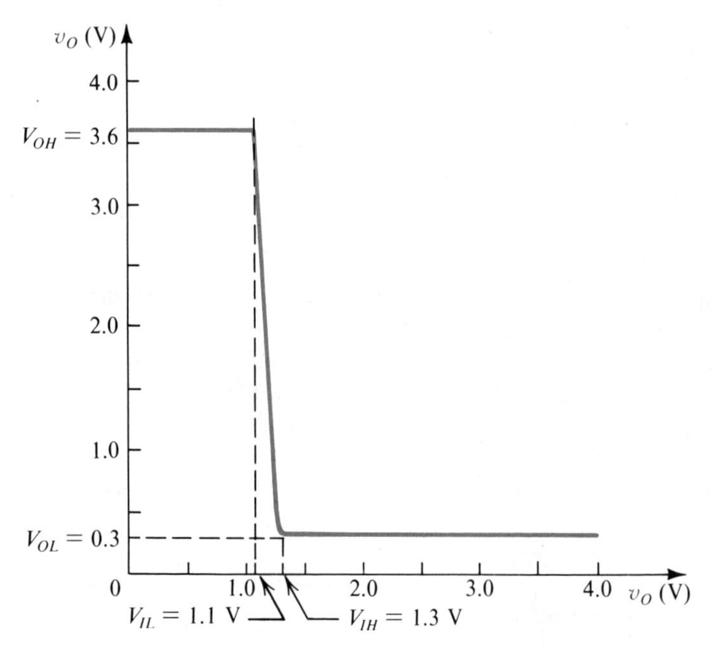

Fig. 14.34 Voltage transfer characteristic of the Schottky TTL gate.

part of its characteristic. Thus the current supplied by the phase splitter will initially be diverted into the base of Q_3, thereby speeding up the turn-on of Q_3. On the other hand, during turnoff the active pull-down transistor Q_6 will be operating in the active mode. Its large collector current will flow through the base of Q_3 in the reverse direction, thus quickly discharging the base-emitter capacitance of Q_3 and turning it off fast.

Finally, the function of the 250-Ω collector resistance of Q_6 should be noted; without it the base-emitter voltage of Q_3 would be clamped to V_{CE6}, which is about 0.3 V.

Exercises 14.21 Show that the values of V_{OH}, V_{IL}, V_{IH}, and V_{OL} of the Schottky TTL gate are as given in Fig. 14.34. Assume that the gate output current is small and that a Schottky transistor conducts at $v_{BE} = 0.7$ V and is fully conducting at $v_{BE} = 0.8$ V.

14.22 Calculate the input current of a Schottky TTL gate with the input voltage low (at 0.3 V).

Ans. 1.4 mA

14.23 For the Schottky TTL gate calculate the current drawn from the power supply with the input low at 0.3 V (remember to include the current in the 3.5-kΩ resistor) and with the input high at 3.6 V. Hence calculate the gate power dissipation in both states and the average power dissipation.

Ans. 2.6 mA; 5.4 mA; 13 mW; 27 mW; 20 mW

Performance Characteristics. Schottky TTL (known as Series 74S) is specified to have the following worst-case parameters.

$$V_{OH} = 2.7 \text{ V} \qquad V_{OL} = 0.5 \text{ V}$$

$$V_{IH} = 2.0 \text{ V} \qquad V_{IL} = 0.8 \text{ V}$$

$$t_P = 3 \text{ ns} \qquad P_D = 20 \text{ mW}$$

Note that the delay-power product is 60 pJ, as compared with about 100 pJ for standard TTL.

Exercise 14.24 Use the values given above to calculate the noise margins NM_H and NM_L for a Schottky TTL gate.

Ans. 0.7 V; 0.3 V

Low-Power Schottky TTL

Although Schottky TTL achieves a very low propagation delay, the gate power dissipation is rather high. This limits the number of gates that can be included per package. The need therefore arose for a modified version that achieves a lower gate dissipation, possibly at the expense of an increase in gate delay. This is obtained in the low-power Schottky (LS) TTL subfamily, represented by the two-input NAND gate shown in Fig. 14.35.

The low-power Schottky circuit of Fig. 14.35 differs from that of the regular Schottky TTL of Fig. 14.32 in several ways: most important, note that the resistances used are about ten times larger, with the result that the power dissipation is only about a tenth of that of the Schottky circuit (2 mW versus 20 mW). However, as expected, the use of larger resistances is accompanied by a reduction in the speed of operation. To help compensate for this speed reduction, some circuit design innovations are employed: First, the

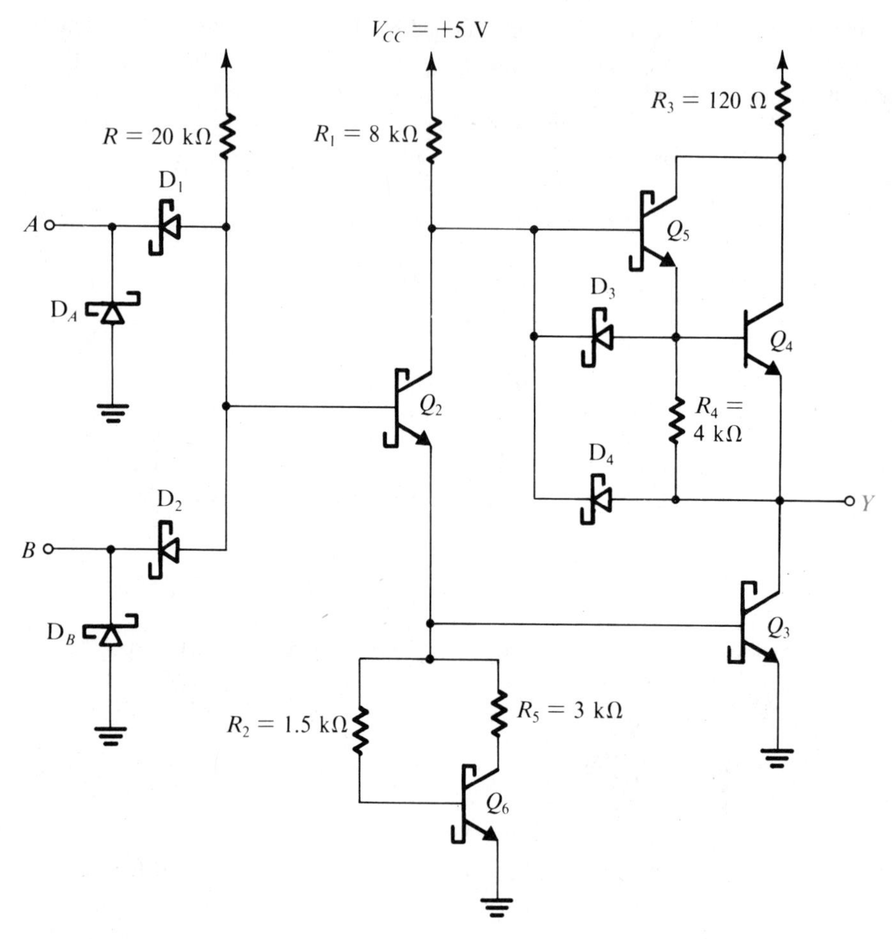

Fig. 14.35 A low-power Schottky TTL (known as LSTTL) gate.

input multiemitter transistor has been eliminated in favor of Schottky diodes, which occupy a smaller silicon area and hence have smaller parasitic capacitances. In this regard it should be recalled that the main advantage of the input multiemitter transistor is that it rapidly removes the charge stored in the base of Q_2. In Schottky-clamped circuits, however, transistor Q_2 does not saturate, eliminating this aspect of the need for Q_1.

Second, two Schottky diodes, D_3 and D_4, have been added to the output stage to help speed up the turnoff of Q_4 and the turn-on of Q_3 and hence the transition of the output from high to low. Specifically, as the gate input is raised and Q_2 begins to turn on, some of its collector current will flow through diode D_3. This current constitutes a reverse base current for Q_4 and thus aids in the rapid turnoff of Q_4. Simultaneously, the emitter current of Q_2 will be supplied to the base of Q_3, thus causing it to turn on faster. Some of the collector current of Q_2 will also flow through D_4. This current will help discharge the gate load capacitance and thus shorten the transition time from high to low. Both D_3 and D_4 will be cut off under static conditions.

Finally, observe that the other terminal of the emitter resistor of Q_5 is now connected to the gate output. It follows that when the gate output is high, the output current will be first supplied by Q_5 through R_4. Transistor Q_4 will turn on only when a 0.7-V drop develops across R_4, and thereafter it sources additional load current. However, when the gate is supplying a small load current, Q_4 will be off and the output voltage will be approximately.

$$V_{OH} = V_{CC} - V_{BE5}$$

which is higher than the value obtained in the regular Schottky TTL gate.

Exercises 14.25 The voltage transfer characteristic of the low-power Schottky TTL gate of Fig. 14.35 has the same shape as that of the regular Schottky TTL gate, shown in Fig. 14.34. Calculate the values of V_{OH}, V_{IL}, V_{IH}, and V_{OL}. Assume that a Schottky transistor conducts at $v_{BE} = 0.7$ V and is fully conducting at $v_{BE} = 0.8$ V, and that a Schottky diode has a 0.5-V drop. Also assume that the gate output current is very small.

Ans. 4.3 V; 0.9 V; 1.1 V; 0.3 V

14.26 For the low-power Schottky TTL gate, using the specifications given in Exercise 14.25, calculate the supply current in both states. Hence calculate the average power dissipation.

Ans. 0.21 mA; 0.66 mA; 2 mW

Although the guaranteed voltage levels and noise margins of low-power Schottky TTL (known as Series 74LS) are similar to those of the regular Schottky TTL, the gate delay and power dissipation are

$$t_P = 10 \text{ ns} \qquad P_D = 2 \text{ mW}$$

Thus although the power dissipation has been reduced by a factor of 10, the delay has been increased by only a factor of 3. The result is a delay-power product of only 20 pJ.

Further-Improved TTL Families

There are other TTL families with further improved characteristics. Of particular interest is the advanced Schottky (Series 74AS) and the advanced low-power Schottky (Series 74ALS). We shall not discuss the circuit details of these families here. Table 14.2 provides a comparison of the TTL subfamilies based on gate delay and power dissipation.

Table 14.2 PERFORMANCE COMPARISON OF TTL FAMILIES

	Standard TTL (Series 74)	Schottky TTL (Series 74S)	Low-Power Schottky TTL (Series 74LS)	Advanced Schottky TTL (Series 74AS)	Advanced Low-Power Schottky TTL (Series 74ALS)
t_P, ns	10	3	10	1.5	4
P_D, mW	10	20	2	20	1
DP, pJ	100	60	20	30	4

Observe that advanced low-power Schottky offers a very small delay-power product. In conclusion, we note that at the present time TTL is still a popular logic circuit family in systems assembled using SSI and MSI chips. Although standard TTL is no longer used in new designs, the advanced circuit types are frequently employed. The speed of the advanced Schottky family is rivaled only by that achieved in emitter-coupled logic (ECL).

14.6 EMITTER-COUPLED LOGIC (ECL)

Emitter-coupled logic (ECL) is the fastest logic circuit family. High speed is achieved by operating all transistors out of saturation, thus avoiding storage time delays, and by keeping the logic signal swings relatively small (about 0.8 V), thus reducing the time required to charge and discharge the various load and parasitic capacitances.

Unlike Schottky TTL, where saturation is prevented by diverting the excess base current drive into the Schottky clamp, saturation in ECL is avoided by using the BJT differential pair as a current switch. An introduction to the BJT differential pair was given in Sections 6.1 and 6.2, which we urge the reader to review before proceeding with the study of ECL.

ECL Families

Currently there are two popular forms of ECL—namely, ECL 10K and ECL 100K. The ECL 100K Series features gate delays of the order of 0.75 ns and dissipates about 40 mW/gate, for a delay-power product of 30 pJ. Although its power dissipation is relatively high, the 100K Series provides the shortest available gate delay.

The ECL 10K Series is slightly slower; it features a gate propagation delay of 2 ns and a power dissipation of 25 mW for a delay-power product of 50 pJ. Although the value of *DP* is higher than that obtained in the 100K Series, the 10K Series is easier to use. This is due to the fact that the rise and fall times of the pulse signals are deliberately made longer, thus reducing signal coupling, or crosstalk, between adjacent signal lines. ECL 10K has an "edge speed" of about 3.5 ns, as compared with the approximately 1 ns of ECL 100K. In the following we shall study the popular ECL 10K in some detail.

In addition to its usage in SSI and MSI circuit packages, ECL is also employed in LSI and VLSI applications. More recently, a variant of ECL known as **current-mode logic** (CML) has become popular in VLSI applications (see Treadway, 1989, and Wilson, 1990).

The Basic Gate Circuit

The basic gate circuit of the ECL 10K family is shown in Fig. 14.36. The circuit consists of three parts. The network composed of Q_1, D_1, D_2, R_1, R_2, and R_3 generates a reference voltage V_R whose value at room temperature is -1.32 V. As will be shown below, the value of this reference voltage is made to change with temperature in a predetermined manner so as to keep the noise margins almost constant. Also, the reference voltage V_R is made relatively insensitive to variations in the power supply voltage V_{EE}.

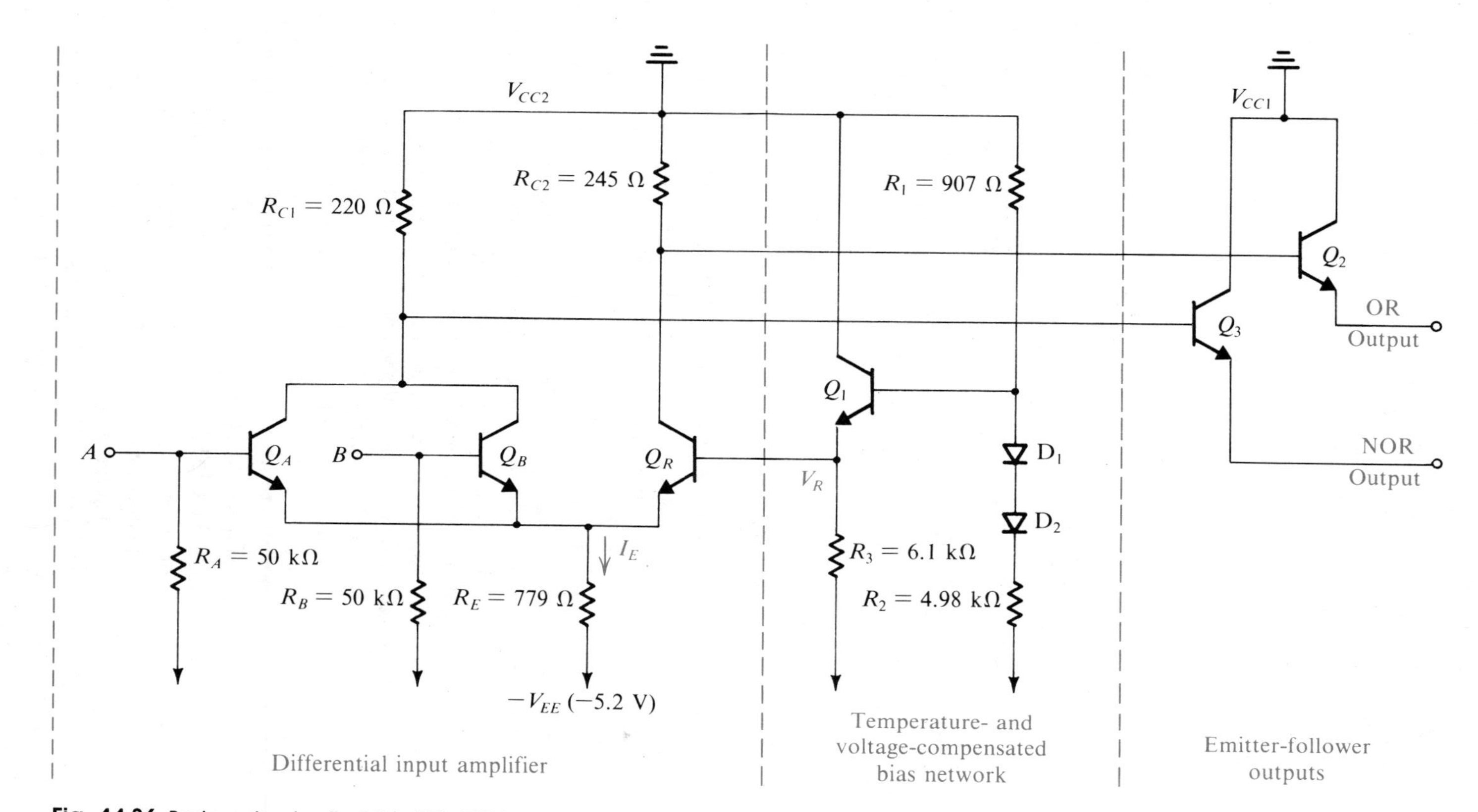

Fig. 14.36 Basic gate circuit of the ECL 10K family.

Exercise 14.27 Figure E14.27 shows the circuit that generates the reference voltage V_R. Assuming that the voltage drop across each of D_1, D_2, and the base–emitter junction of Q_1 is 0.75 V, calculate the value of V_R. Neglect the base current of Q_1.

Ans. −1.32 V

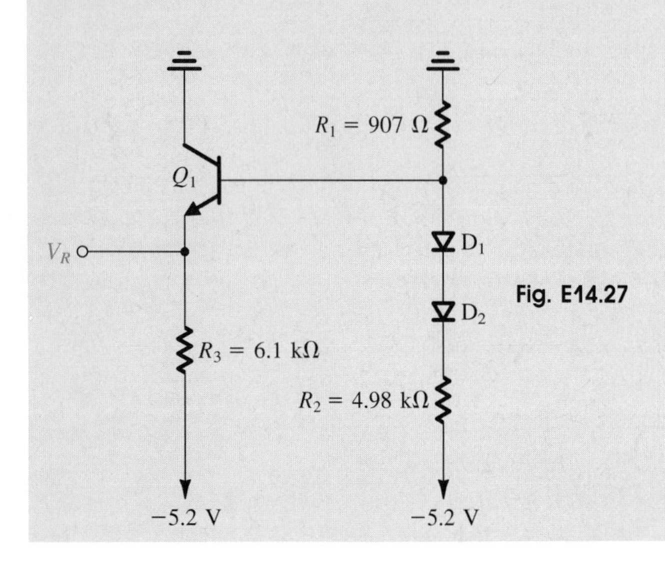

Fig. E14.27

The second part, and the heart of the gate, is the differential amplifier formed by Q_R and either Q_A or Q_B. This differential amplifier is biased not by a constant-current source, as was done in the circuits of Chapter 6, but with a resistance R_E connected to the negative supply $-V_{EE}$. One side of the differential amplifier consists of the reference transistor Q_R, whose base is connected to the reference voltage V_R. The other side consists of a number of transistors (two in the case shown) connected in parallel, with separated bases connected to the gate inputs. If the voltages applied to A and B are at the logic-0 level, which, as we will soon find out, is about 0.4 V below V_R, both Q_A and Q_B will be off and the current I_E in R_E will flow through the reference transistor Q_R. The resulting voltage drop across R_{C2} will cause the collector voltage of Q_R to be low.

On the other hand, when the voltage applied to A or B is at the logic-1 level, which, as we will show shortly, is about 0.4 V above V_R, transistor Q_A or Q_B, or both, will be on and Q_R will be off. Thus the current I_E will flow through Q_A or Q_B, or both, and an almost equal current flows through R_{C1}. The resulting voltage drop across R_{C1} will cause the collector voltage to drop. Meanwhile, since Q_R is off, its collector voltage rises. We thus see that the voltage at the collector of Q_R will be high if A or B, or both, is high, and thus at the collector of Q_R the OR logic function, $A + B$, is realized. On the other hand, the common collector of Q_A and Q_B will be high only when A and B are simultaneously low. Thus, at the common collector of Q_A and Q_B the logic function $\bar{A}\bar{B} = \overline{A + B}$ is realized. We therefore conclude that the two-input gate of Fig. 14.36 realizes the OR function and its complement, the NOR function. The availability of complementary outputs is an important advantage of ECL; it simplifies logic design and avoids the use of additional inverters with associated time delay.

It should be noted that the resistance connecting each of the gate input terminals to the negative supply enables the user to leave an unused input terminal open. An open input terminal will then be *pulled down* to the negative supply voltage, and its associated transistor will be off.

Exercise 14.28 With input terminals A and B left open, find the current I_E through R_E. Also find the voltages at the collector of Q_R and at the common collector of the input transistors Q_A and Q_B. Use $V_R = -1.32$ V, V_{BE} of $Q_R \simeq 0.75$ V, and assume that β of Q_R is very high.

Ans. 4 mA; −1 V; 0 V

The third part of the ECL gate circuit is composed of the two emitter followers, Q_2 and Q_3. The emitter followers do not have on-chip loads, since in most applications of high-speed logic circuits the gate output drives a transmission line terminated at the other end, as indicated in Fig. 14.37.

The emitter followers have two purposes. First, they shift the level of the output signals by one V_{BE} drop. Thus, using the results of Exercise 14.28 above, we see that the output levels become approximately −1.75 V and −0.75 V. These shifted levels are centered approximately around the reference voltage ($V_R = -1.32$ V), which means that one gate can drive another. This compatibility of logic levels at input and output is always an essential requirement in the design of gate circuits.

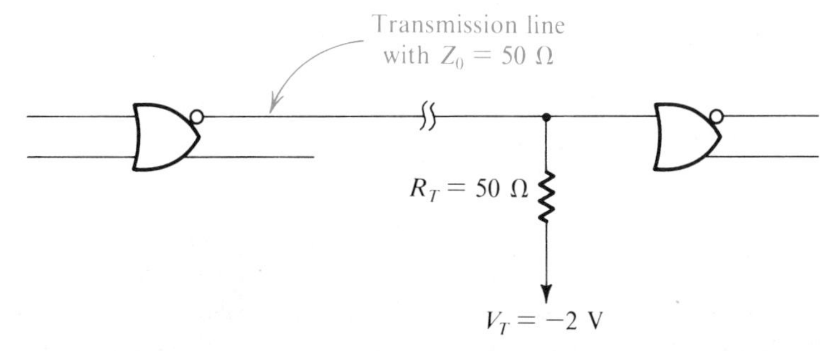

Fig. 14.37 The proper way to connect high-speed logic gates such as ECL. Properly terminating the transmission line connecting the two gates eliminates "ringing" that would otherwise corrupt the logic signals.

The second function of the output emitter followers is that they provide the gate with low output resistances and with the large output currents required for charging load capacitances. Since these large transient currents can cause spikes on the power-supply line, the collectors of the emitter followers are connected to a power-supply terminal V_{CC1} separate from that of the differential amplifier and the reference-voltage circuit, V_{CC2}. Here we note that the supply current of the differential amplifier and the reference circuit remains almost constant. The use of separate power-supply terminals prevents the coupling of power-supply spikes from the output circuit to the gate circuit, and thus lessens the likelihood of false gate switching. Both V_{CC1} and V_{CC2} are of course connected to the same system ground external to the chip.

Voltage Transfer Characteristics

Having provided a qualitative description of the operation of the ECL gate we shall now derive its voltage transfer characteristics. This will be done under the conditions that the outputs are terminated in the manner indicated in Fig. 14.37. Assuming that the B input is low and thus Q_B is off, the circuit simplifies to that shown in Fig. 14.38. We wish to analyze this circuit to determine v_{OR} versus v_I and v_{NOR} versus v_I (where $v_I \equiv v_A$).

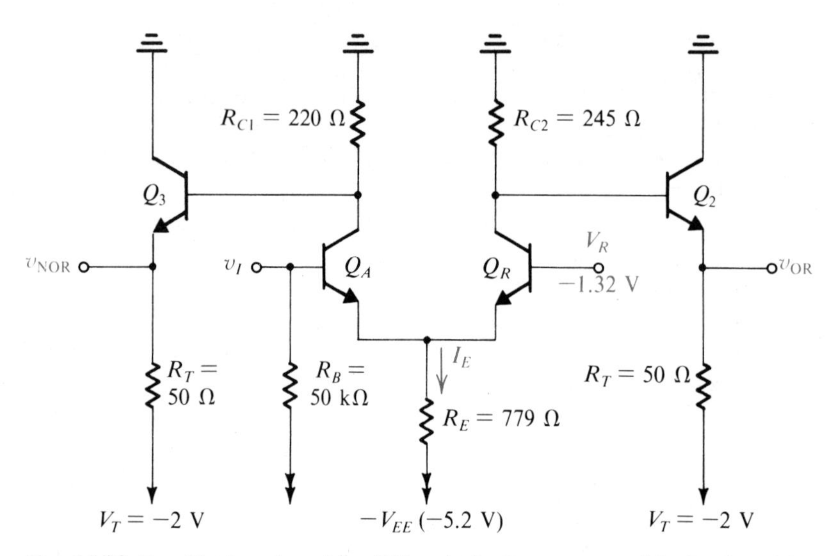

Fig. 14.38 Simplified version of the ECL gate for the purpose of finding transfer characteristics.

In the analysis to follow we shall make use of the exponential i_C–v_{BE} characteristic of the BJT. Since the BJTs used in ECL circuits have small areas (in order to have small capacitances and hence high f_T), their scale currents I_S are small. We will therefore assume that at an emitter current of 1 mA an ECL transistor has a V_{BE} drop of 0.75 V.

The OR Transfer Curve

Figure 14.39 shows a sketch of the OR transfer characteristic, v_{OR} versus v_I, with the parameters V_{OL}, V_{OH}, V_{IL}, and V_{IH} indicated. However, in order to simplify the calculation of V_{IL} and V_{IH} we shall use an alternative to the unity-gain definition. Specifically, we shall assume that at point x, transistor Q_A is conducting 1% of I_E while Q_R is conducting 99% of I_E. The reverse will be assumed for point y. Thus at point x we have

$$\frac{I_E|_{Q_R}}{I_E|_{Q_A}} = 99$$

Using the exponential i_E–v_{BE} relationship we obtain

$$V_{BE}|_{Q_R} - V_{BE}|_{Q_A} = V_T \ln 99 = 115 \text{ mV}$$

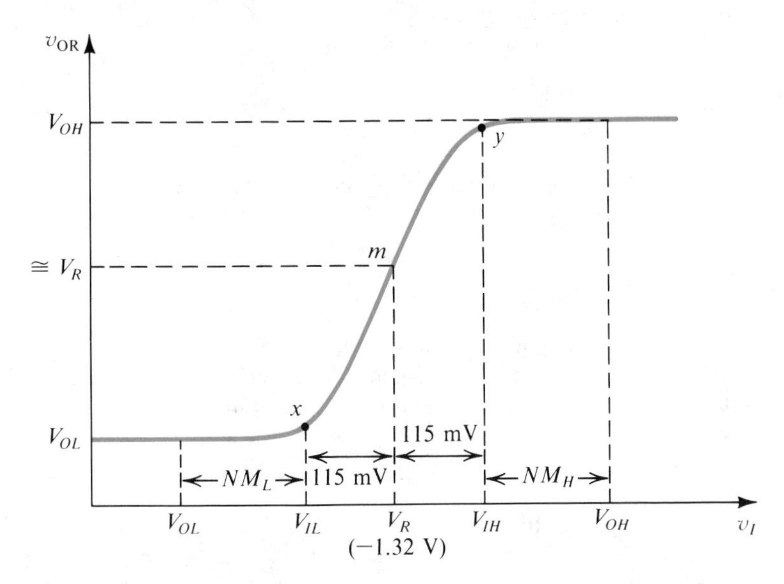

Fig. 14.39 The OR transfer characteristic v_{OR} versus v_I, for the circuit in Fig. 14.38.

which gives

$$V_{IL} = -1.32 - 0.115 = -1.435 \text{ V}$$

Assuming Q_A and Q_R to be matched, we can write

$$V_{IH} - V_R = V_R - V_{IL}$$

which can be used to find V_{IH} as

$$V_{IH} = -1.205 \text{ V}$$

To obtain V_{OL} we note that Q_A is off and Q_R carries the entire current I_E, given by

$$\begin{aligned} I_E &= \frac{V_R - V_{BE}\big|_{Q_R} + V_{EE}}{R_E} \\ &= \frac{-1.32 - 0.75 + 5.2}{0.779} \\ &\simeq 4 \text{ mA} \end{aligned}$$

(If we wish, we can iterate to determine a better estimate of $V_{BE}\big|_{Q_R}$ and hence of I_E.) Assuming that Q_R has a high β so that its $\alpha \simeq 1$, then its collector current will be approximately 4 mA. If we neglect the base current of Q_2, we obtain for the collector voltage of Q_R

$$V_C\big|_{Q_R} \simeq -4 \times 0.245 = -0.98 \text{ V}$$

Thus a first approximation for the value of the output voltage V_{OL} is

$$V_{OL} = V_C\big|_{Q_R} - V_{BE}\big|_{Q_2}$$
$$\simeq -0.98 - 0.75 = -1.73 \text{ V}$$

We can use this value to find the emitter current of Q_2 and then iterate to determine a better estimate of its base-emitter voltage. The result is $V_{BE2} \simeq 0.79$ V and, correspondingly,

$$V_{OL} \simeq -1.77 \text{ V}$$

At this value of output voltage, Q_2 supplies a load current of about 4.6 mA.

To find the value of V_{OH} we assume that Q_R is completely cut off (because $v_I > V_{IH}$). Thus the circuit for determining V_{OH} simplifies to that in Fig. 14.40. Analysis of this circuit assuming $\beta_2 = 100$ results in $V_{BE2} \simeq 0.83$ V, $I_{E2} = 22.4$ mA, and

$$V_{OH} \simeq -0.88 \text{ V}$$

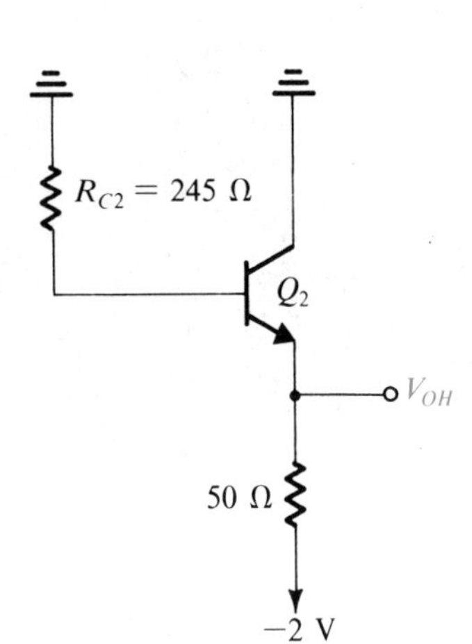

Fig. 14.40 Circuit for determining V_{OH}.

Exercise 14.29 For the circuit in Fig. 14.38, determine the values of I_E obtained when $v_I = V_{IL}$, V_R, and V_{IH}. Also, find the value of v_{OR} corresponding to $v_I = V_R$. Assume that $v_{BE} = 0.75$ V at a current of 1 mA.

Ans. 3.97 mA; 4.00 mA; 4.12 mA; -1.31 V

Noise Margins

The results of Exercise 14.29 indicate that the bias current I_E remains approximately constant. Also, the output voltage corresponding to $v_I = V_R$ is approximately equal to V_R. Notice further that this is also approximately the midpoint of the logic swing; specifically,

$$\frac{V_{OL} + V_{OH}}{2} = -1.325 \simeq V_R$$

Thus the output logic levels are centered around the midpoint of the input transition band. This is an ideal situation from the point of view of noise margins, and it is one of the reasons for selecting the rather arbitrary-looking numbers $V_R = -1.32$ V and $V_{EE} =$ 5.2 V.

The noise margins can now be evaluated as follows:

$$NM_H = V_{OH} - V_{IH} \qquad\qquad NM_L = V_{IL} - V_{OL}$$

$$= -0.88 - (-1.205) = 0.325 \text{ V} \qquad\qquad = -1.435 - (-1.77) = 0.335 \text{ V}$$

Note that these values are approximately equal.

The NOR Transfer Curve

The NOR transfer characteristic, which is v_{NOR} versus v_I for the circuit in Fig. 14.38, is sketched in Fig. 14.41. The values of V_{IL} and V_{IH} are identical to those found above for the OR characteristic. To emphasize this we have labeled the threshold points x and y, the same letters used in Fig. 14.39.

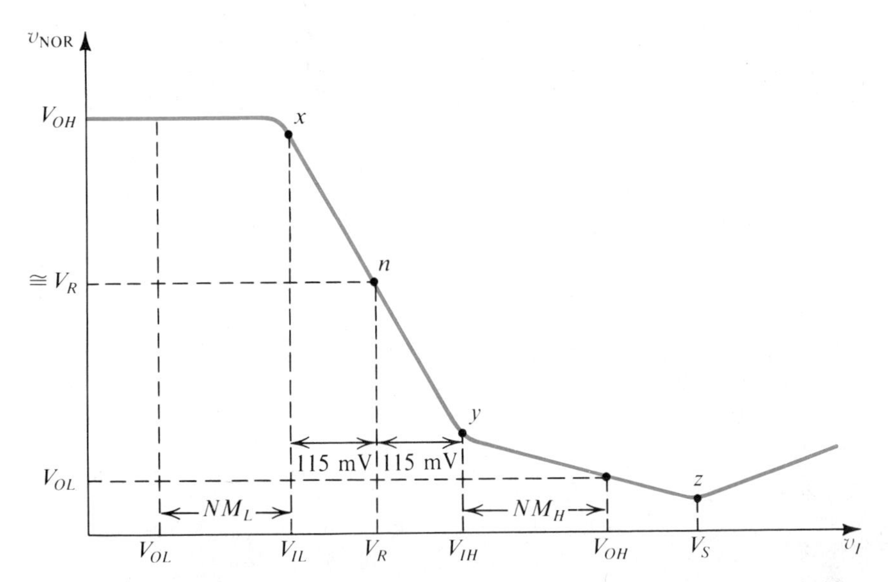

Fig. 14.41 The NOR transfer characteristic, v_{NOR} versus v_I, for the circuit in Fig. 14.38.

For $v_I < V_{IL}$, Q_A is off and the output voltage v_{NOR} can be found by analyzing the circuit composed of R_{C1}, Q_3, and its 50-Ω termination. Except that R_{C1} is slightly smaller than R_{C2}, this circuit is identical to that in Fig. 14.40. Thus the output voltage will be only slightly greater than the value V_{OH} found previously. In the sketch of Fig. 14.41 we have assumed that the output voltage is approximately equal to V_{OH}.

For $v_I > V_{IH}$, Q_A is on and is conducting the entire bias current. The circuit then simplifies to that in Fig. 14.42. This circuit can be easily analyzed to obtain v_{NOR} versus v_I for the range $v_I \geq V_{IH}$. A number of observations are in order. First, note that $v_I = V_{IH}$

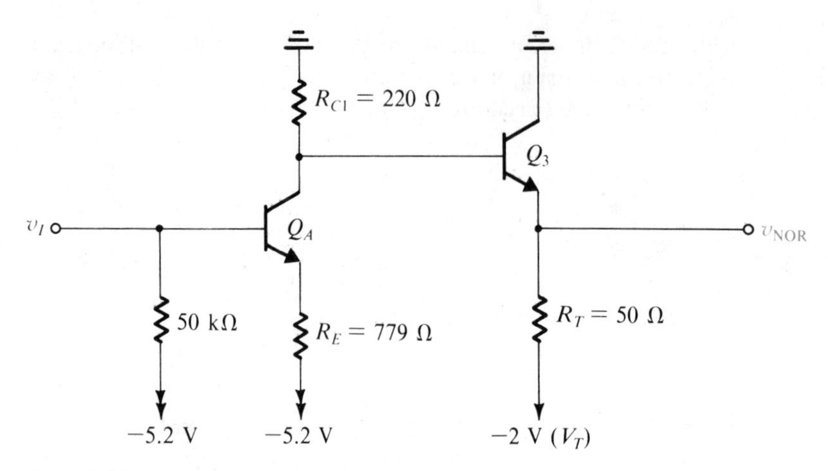

Fig. 14.42 Circuit for finding v_{NOR} versus v_I for the range $v_I > V_{IH}$.

results in an output voltage slightly higher than V_{OL}. This is because R_{C1} is smaller than R_{C2}. In fact, R_{C1} is chosen lower in value than R_{C2} so that with v_I equal to the normal logic-1 value (that is, V_{OH}, which is approximately -0.88 V) the output will be equal to the V_{OL} value previously found for the OR output.

Second, note that as v_I exceeds V_{IH}, transistor Q_A operates in the active mode and the circuit of Fig. 14.42 can be analyzed to find the gain of this amplifier, which is the slope of the segment *yz* of the transfer characteristic. At point *z* transistor Q_A saturates. Further increments in v_I (beyond the point $v_I = V_S$) cause the collector voltage and hence v_{NOR} to increase. The slope of the segment of the transfer characteristic beyond point *z*, however, is not unity but is about 0.5 because as Q_A is driven deeper into saturation a portion of the increment in v_I appears as an increment in the base–collector forward-bias voltage. The reader is urged to solve Exercise 14.30, which is concerned with the details of the NOR transfer characteristic.

Exercise 14.30 Consider the circuit in Fig. 14.42. (a) For $v_I = V_{IH} = -1.205$ V, find v_{NOR}. (b) For $v_I = V_{OH} = -0.88$ V, find v_{NOR}. (c) Find the slope of the transfer characteristic at the point $v_I = V_{OH} = -0.88$ V. (d) Find the value of v_I at which Q_A saturates (that is, V_S). Assume that $V_{BE} = 0.75$ V at a current of 1 mA, $V_{CEsat} \simeq 0.3$ V, and $\beta = 100$.

Ans. (a) -1.70 V; (b) -1.79 V; (c) -0.24 V/V; (d) -0.58 V

Manufacturers' Specifications

ECL manufacturers supply gate transfer characteristics of the form shown in Figs. 14.39 and 14.41. The manufacturer usually provides such curves measured at a number of temperatures. In addition, at each relevant temperature, worst-case values for the parameters V_{IL}, V_{IH}, V_{OL}, and V_{OH} are given. These worst-case values are specified with the inevitable component tolerances taken into account. As an example, Motorola specifies that for MECL 10,000 at 25°C the following worst-case values apply:[6]

$$V_{IL\max} = -1.475 \text{ V} \qquad V_{IH\min} = -1.105 \text{ V}$$

$$V_{OL\max} = -1.630 \text{ V} \qquad V_{OH\min} = -0.980 \text{ V}$$

These values can be used to determine worst-case noise margins,

$$NM_L = 0.155 \text{ V} \qquad NM_H = 0.125 \text{ V}$$

which are about half the *typical* values previously calculated.

For additional information on MECL specifications the interested reader is referred to the Motorola (1988, 1989) publications listed in the bibliography at the end of this chapter.

Fan-Out

When the input signal to an ECL gate is low, the input current is equal to the current that flows in the 50-kΩ pull-down resistor. Thus

$$I_{IL} = \frac{-1.77 + 5.2}{50} \simeq 69 \ \mu\text{A}$$

When the input is high, the input current is greater because of the base current of the input transistor. Thus, assuming a transistor β of 100, we obtain

$$I_{IH} = \frac{-0.88 + 5.2}{50} + \frac{4}{101} \simeq 126 \ \mu\text{A}$$

Both of these current values are quite small, which, coupled with the fact that the output resistance of the ECL gate is very small, ensures that little degradation of logic signal levels results from the input currents of fan-out gates. It follows that the fan-out of ECL gates is not limited by logic-level considerations but rather by the degradation of the circuit speed (rise and fall times). This latter effect is due to the capacitance that each fan-out gate presents to the driving gate (approximately 3 pF). Thus while the *dc fan-out* can be as high as 90 and thus does not represent a design problem, the *ac fan-out* is limited by considerations of circuit speed to 10 or so.

Speed

The speed of operation of a logic family is measured by the delay of its basic gate and by the rise and fall times of the output waveforms. Typical values of these parameters for ECL have already been given. Here we should note that because the output circuit is an emitter follower the rise time of the output signal is shorter than its fall time, since on the rising edge of the output pulse the emitter follower functions and provides the output current required to charge up the load and parasitic capacitances. On the other hand, as the signal at the base of the emitter follower falls, the emitter follower cuts off and the load

[6] MECL is the trade name used by Motorola for its ECL.

capacitance discharges through the combination of load and pull-down resistances. This point was explained in detail in Section 14.3.

Signal Transmission

In order to take full advantage of the very high speed of operation possible with ECL, special attention should be paid to the method of interconnecting the various logic gates in a system. To appreciate this point we shall briefly discuss the problem of signal transmission.

ECL deals with signals whose rise times may be 1 ns or even less, the time it takes for light to travel only 30 cm or so. For such signals a wire and its environment becomes a relatively complex circuit element along which signals propagate with finite speed (perhaps half the speed of light—that is, 15 cm/ns). Unless special care is taken, energy that reaches the end of such a wire is not absorbed but rather returns as a *reflection* to the transmitting end, where (without special care) it may be re-reflected. The result of this process of reflection is what can be observed as **ringing,** a damped oscillatory excursion of the signal about its final value.

Unfortunately ECL is particularly sensitive to ringing because the signal levels are so small. Thus it is important that transmission of signals be well controlled and surplus energy absorbed to prevent reflections. The accepted technique is to limit the nature of connecting wires in some way. One way is to insist that they be very "short," where short is taken with respect to the signal rise time. The reason for this is that if the wire connection is so short that reflections return while the input is still rising, the result becomes only a somewhat slowed and "bumpy" rising edge.

If, however, the reflection returns *after* the rising edge, it produces not simply a modification of the initiating edge but an *independent second event*. This is clearly bad! The restriction is thus made that the time taken for a signal to go from one end of a line and back should be less than the rise time of the driving signal by some factor—say, 5. Thus for a signal with a 1-ns rise time and for propagation at the speed of light (30 cm/ns), a double path of only 0.2 ns equivalent length, or 6 cm, would be allowed, representing in the limit a wire only 3 cm from end to end.

Such is the restriction on ECL 100K. However, ECL 10K has an intentionally slower rise time of about 3.5 ns. Using the same rules, wires can accordingly be as long as about 10 cm for ECL 10K.

If greater lengths are needed, then transmission lines must be used. These are simply wires in a controlled environment in which the distance to a ground reference plane or second wire is highly controlled. Thus they might simply be twisted pairs of wires, one of which is grounded, or parallel ribbon wires, every second of which is grounded, or so-called microstrip lines on a printed-circuit (PC) board. The last are simply copper strips of controlled geometry on one side of a printed-circuit board, the other side of which consists of a grounded plane.

Such transmission lines have a *characteristic impedance* R_0 which ranges from a few tens of ohms to hundreds of ohms. Signals propagate on such lines somewhat slower than the speed of light, perhaps half as fast. When a transmission line is terminated at its receiving end in a resistance equal to its characteristic impedance R_0, all the energy sent on the line is absorbed at the receiving end and no reflections occur. Thus signal integrity is maintained. Such transmission lines are said to be *properly terminated*. A properly terminated line appears at its sending end as a resistor of value R_0. The followers of ECL 10K with their open emitters and low output resistances (specified to be 7 Ω maximum) are ideally suited for driving transmission lines. ECL is also good as a line receiver. The simple gate with its high (50 kΩ) pull-down input resistor represents a very high resistance to the line. Thus a few such gates can be connected to a terminated line with little difficulty.

Much more on the subject of logic-signal transmission in ECL can be found in Taub and Schilling (1977) and Motorola (1988).

Power Dissipation

Because of the differential-amplifier nature of ECL, the gate current remains approximately constant and is simply steered from one side of the gate to the other depending on the input logic signals. Thus, unlike TTL, the supply current and hence the gate power dissipation of unterminated ECL remain relatively constant independent of the logic state of the gate. It follows that no voltage spikes are introduced on the supply line. Such spikes are a dangerous source of noise in a digital system, as explained in connection with TTL. It follows that in ECL the need for supply-line bypassing is not as great as in TTL. This is another advantage of ECL.

At this point we should mention that although an ECL gate would operate with $V_{EE} = 0$ and $V_{CC} = +5.2$ V, the selection of $V_{EE} = -5.2$ V and $V_{CC} = 0$ V is recommended because in the circuit all signal levels are referenced to V_{CC}, and ground is certainly an excellent reference.

Exercise 14.31 For the ECL gate in Fig. 14.36 calculate an approximate value for the power dissipated in the circuit under the condition that all inputs are low and that the emitters of the output followers are left open. Assume that the reference circuit supplies four identical gates, and hence only a quarter of the power dissipated in the reference circuit should be attributed to a gate.

Ans. 22.4 mW

Thermal Effects

In our analysis of the ECL gate of Fig. 14.36, we found that at room temperature the reference voltage V_R is -1.32 V. We have also shown that the midpoint of the output logic swing is approximately equal to this voltage, which is an ideal situation in that it results in equal high and low noise margins. In Example 14.2 below we shall derive expressions for the temperature coefficients of the reference voltage and of the output low and high voltages. In this way, it will be shown that the midpoint of the output logic swing varies with temperature at the same rate as the reference voltage. As a result, although the magnitudes of the 1 and 0 noise margins change with temperature, their values remain equal. This is an added advantage of ECL and is a demonstration of the degree of design optimization of this gate circuit.

EXAMPLE 14.2 We wish to determine the temperature coefficient of the reference voltage V_R and of the midpoint between V_{OL} and V_{OH}.

Solution To determine the temperature coefficient of V_R, consider the circuit in Fig. E14.27 and assume that the temperature changes by +1°C. Denoting the temperature coefficient of the diode and transistor voltage drops by δ, where $\delta \simeq -2$ mV/°C, we obtain the equivalent circuit shown in Fig. 14.43. In the latter circuit the changes in device voltage drops are considered as signals, and hence the power supply is shown as a signal ground.

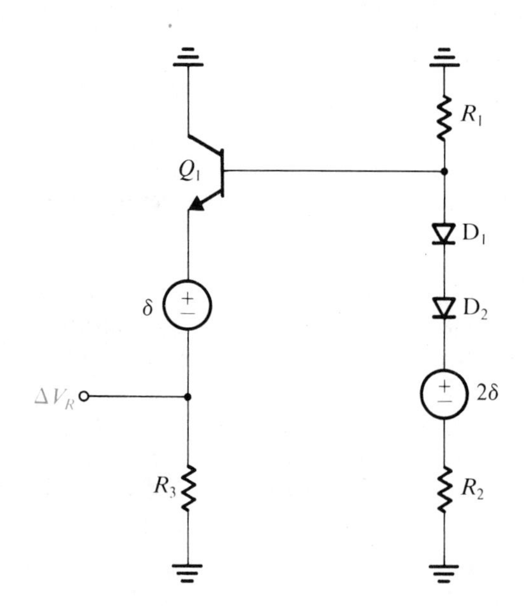

Fig. 14.43 Equivalent circuit for determining the temperature coefficient of the reference voltage V_R.

In the circuit of Fig. 14.43 we have two signal generators, and we wish to analyze the circuit to determine ΔV_R, the change in V_R. We shall do so using the principle of superposition. Consider first the branch R_1, D_1, D_2, 2δ, and R_2, and neglect the signal base current of Q_1. The voltage signal at the base of Q_1 can be easily obtained from

$$v_{b1} = \frac{2\delta \times R_1}{R_1 + r_{d1} + r_{d2} + R_2}$$

where r_{d1} and r_{d2} denote the incremental resistances of diodes D_1 and D_2, respectively. The dc bias current through D_1 and D_2 is approximately 0.64 mA, and thus $r_{d1} = r_{d2} =$ 39.5 Ω. Hence $v_{b1} \simeq 0.3\delta$. Since the gain of the emitter-follower Q_1 is approximately unity, it follows that the component of ΔV_R due to the generator 2δ is approximately equal to v_{b1}, that is, $\Delta V_{R1} = 0.3\delta$.

Consider next the component of ΔV_R due to the generator δ. Reflection of the total resistance of the base circuit, $[R_1 \| (r_{d1} + r_{d2} + R_2)]$, into the emitter circuit by dividing it by $\beta + 1$ ($\beta \simeq 100$) results in the following component of ΔV_R:

$$\Delta V_{R2} = -\frac{\delta \times R_3}{[R_B/(\beta + 1)] + r_{e1} + R_3}$$

where R_B denotes the total resistance in the base circuit and r_{e1} denotes the emitter resistance of Q_1 ($\simeq 40\ \Omega$). This calculation yields $\Delta V_{R2} \simeq -\delta$. Adding this value to that due to the generator 2δ gives $\Delta V_R \simeq -0.7\delta$. Thus for $\delta = -2$ mV/°C the temperature coefficient of V_R is $+1.4$ mV/°C.

We next consider determination of the temperature coefficient of V_{OL}. The circuit on which to perform this analysis is shown in Fig. 14.44. Here we have three generators whose contributions can be considered separately and the resulting components of ΔV_{OL} summed. The result is

$$\Delta V_{OL} \simeq \Delta V_R \frac{-R_{C2}}{r_{eR} + R_E} \frac{R_T}{R_T + r_{e2}} - \delta \frac{-R_{C2}}{r_{eR} + R_E} \frac{R_T}{R_T + r_{e2}} - \delta \frac{R_T}{R_T + r_{e2} + R_{C2}/(\beta + 1)}$$

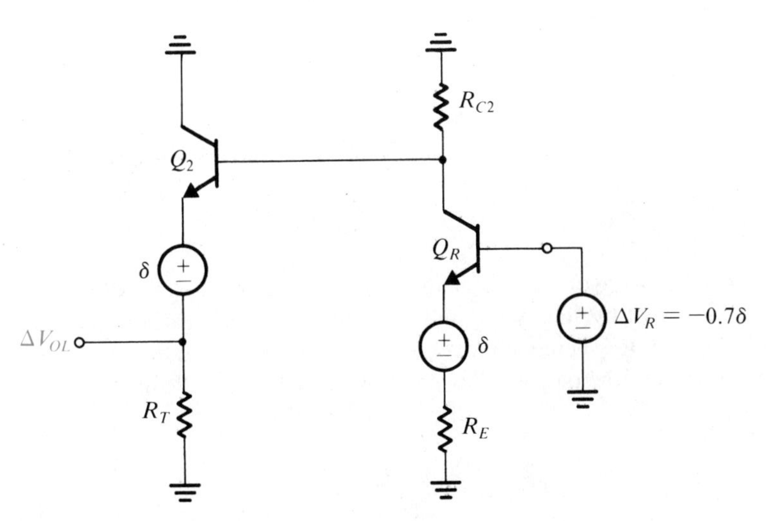

Fig. 14.44 Equivalent circuit for determining the temperature coefficient of V_{OL}.

Substituting the values given and those obtained throughout the analysis of this section, we find $\Delta V_{OL} \simeq -0.43\delta$.

The circuit for determining the temperature coefficient of V_{OH} is shown in Fig. 14.45, from which we obtain

$$\Delta V_{OH} = -\delta \frac{R_T}{R_T + r_{e2} + R_{C2}/(\beta + 1)} \simeq -0.93\delta$$

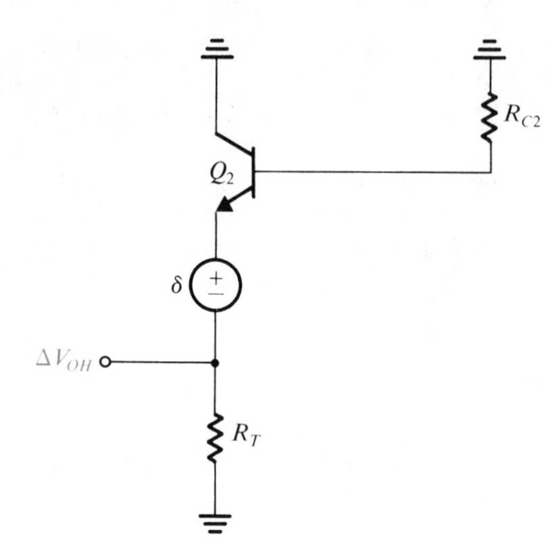

Fig. 14.45 Equivalent circuit for determining the temperature coefficient of V_{OH}.

We now can obtain the variation of the midpoint of the logic swing as

$$\frac{\Delta V_{OL} + \Delta V_{OH}}{2} = -0.68\delta$$

which is approximately equal to that of the reference voltage $V_R(-0.7\delta)$.

The Wired-OR Capability

The emitter-follower output stage of the ECL family allows an additional level of logic to be performed at very low cost by simply wiring the outputs of several gates in parallel. This is illustrated in Fig. 14.46 where the outputs of two gates are wired together. Note that the base-emitter diodes of the output followers provide a positive OR function: This

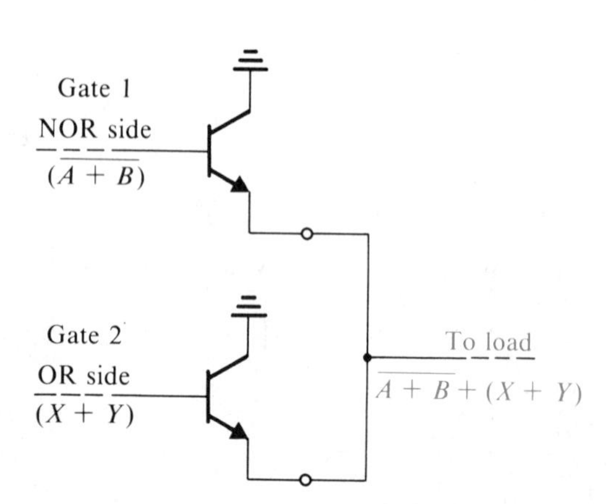

Fig. 14.46 The wired-OR capability of ECL.

wired-OR connection can be used to provide gates with high fan-in as well as to increase the flexibility of ECL in logic design.

A Final Remark

ECL is an important logic family that has been successfully applied in the design of high-speed digital communications systems as well as computer systems. It is also currently applied in LSI and VLSI circuit design.

14.7 BiCMOS DIGITAL CIRCUITS

We conclude our study of digital circuits with a brief introduction to a VLSI circuit technology that is becoming increasingly popular, BiCMOS. As its name implies, BiCMOS technology combines *Bi*polar and *CMOS* circuits on the same IC chip. The resulting circuits retain the low-power, high input-impedance, and wide noise margins of CMOS (Chapter 13) and the high current-driving capability and high speed of operation of bipolar transistors. The result is a circuit technology that is capable of implementing very dense, high-speed, and low-power digital integrated circuits. Furthermore, since the technology of BiCMOS is well suited for the implementation of high-performance analog circuits (Sections 6.8 and 10.8), it makes possible the realization of both analog and digital functions on the same IC chip, making the ''system on a chip'' an attainable goal.

The basic BiCMOS logic inverter is shown in Fig. 14.47. It consists of a CMOS inverter formed by transistors Q_1 and Q_2 and a bipolar output stage formed by transistors Q_3 and Q_4 together with resistors R_1 and R_2. To see how the BiCMOS inverter operates, consider first the case when the input v_I is low at 0 V. Transistor Q_2 will be off, and no current will flow into the base of Q_4. Resistor R_2 causes the voltage at the base of Q_4 to be zero, and thus Q_4 will be off. The PMOS transistor Q_1 will be on, establishing a low-impedence conducting channel between the base of Q_3 and V_{DD}. Now, if the output terminal of the gate is open circuited, no current will flow through Q_3 or R_1, and thus no current will flow through Q_1. Thus the voltage drop across Q_1 will be nearly zero, and

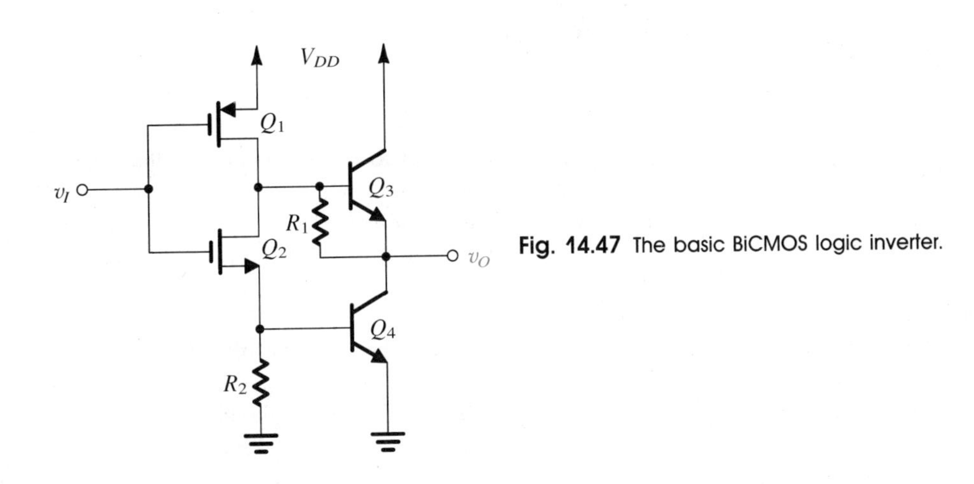

Fig. 14.47 The basic BiCMOS logic inverter.

resistor R_1 will cause the voltage at the gate output to be equal to V_{DD}. Resistor R_1 thus functions as a pull-up resistor and results in $V_{OH} = V_{DD}$.

If, however, the output terminal of the gate is required to source current to a load, as would be required, for instance, to charge a load capacitance during a low-to-high transition at the output, transistor Q_3 turns on and acts as an emitter follower, providing the gate with a low output resistance and a high current-driving capability. In this case, transistor Q_1 supplies Q_3 with its required base current, which is amplified by the current gain β of Q_3. The resulting large current available from Q_3 serves to charge the load capacitance very quickly, making the gate delay t_{PLH} very short.

Consider next the situation as v_I goes high to V_{DD}. Transistor Q_1 turns off and Q_2 turns on. The drain current of Q_2 serves to remove the base charge of Q_3, which turns off quickly. Transistor Q_2 supplies Q_4 with base current; thus Q_4 turns on. Now, since the load capacitance at the output terminal will maintain the collector voltage of Q_4 high, Q_4 will operate in the active region, thus providing a large output current that quickly discharges the load capacitance, resulting in a short delay t_{PHL}. With the load capacitance discharged, and assuming no dc load, the currents in Q_4 and Q_2 will be nearly zero; and we see that the path through R_1, the conducting channel of Q_2, and R_2, pulls the voltage of the gate output terminal down to ground level, resulting in $V_{OL} = 0$. Now, if v_I goes low again, Q_2 turns off, the base charge of Q_4 leaks out to ground through R_2, and the gate returns to the high-output state described earlier.

We thus see that the BiCMOS inverter has the large output voltage swing characteristic of CMOS and the high output current-driving capability and the correspondingly short propagation delays characteristic of BJTs. Also, the gate has a nearly zero power dissipation in its two states and an almost infinite input resistance. The BiCMOS circuit exhibits a higher speed of operation than obtainable in CMOS and a lower power dissipation than achievable in bipolar circuits. BiCMOS thus combines the best features of its two constituent technologies. The only drawback of BiCMOS is its somewhat complex fabrication process, making the cost per gate higher than that of CMOS circuits. The increased cost is, however, outweighed by the possibility of realizing circuit functions of sophistication and performance levels not otherwise attainable.

With a fan-out of zero, the delay of a BiCMOS gate is typically 1 ns, which is somewhat larger than that of CMOS. The speed advantage of BiCMOS, however, becomes evident when the gate is required to drive others. For instance, at a fan-out of 10, the time delay of a BiCMOS gate is still close to 1 ns, whereas that of a CMOS gate is about twice as long.

Before leaving the basic inverter circuit two comments are in order. First, the above description of gate operation is oversimplified in order to focus attention on the salient features of BiCMOS. In particular we have assumed that the bipolar devices operate in the active mode during the gate turn on and turnoff. In actual practice, the intrinsic resistance of the collector region of each BJT (about 100 Ω) causes it to saturate, thus limiting its collector current to a value less than βI_B. Second, in an actual BiCMOS gate, R_1 and R_2 are implemented using MOS transistors (see Deierling, 1989).

Since the logic function in the BiCMOS inverter circuit is performed by the CMOS inverter formed by Q_1 and Q_2, logic gate circuits can be formed in the manner used for CMOS circuits in Section 13.6. As an example, Fig. 14.48 shows the circuit for a two-input NAND gate.

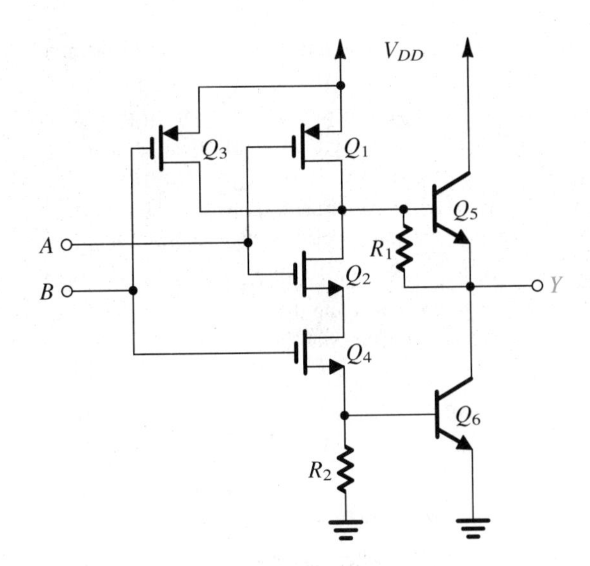

Fig. 14.48 BiCMOS NAND gate. Output Y is low only when A and B are simultaneously high: $\bar{Y} = AB$, which can be written $Y = \overline{AB}$.

BiCMOS technology is currently being applied in a variety of products including microprocessors, static RAMs, and gate arrays (see Alvarez, 1989).

Exercise D14.32 The threshold voltage of the BiCMOS inverter of Fig. 14.47 is the value of v_I at which both Q_1 and Q_2 are conducting equal currents and operating in the saturation region. At this value of v_I, Q_4 will be on, causing the voltage at the source of Q_2 to be approximately 0.7 V. It is required to design the circuit so that the threshold voltage is equal to $V_{DD}/2$. For $V_{DD} = 5$ V, $|V_t| = 0.6$ V, and assuming equal channel lengths for Q_1 and Q_2 and that $\mu_n \simeq 2.5\ \mu_p$, find the required ratio of widths, W_2/W_1.

Ans. 1

SUMMARY

▶ The Ebers–Moll model is based on the fact that the BJT is composed of two *pn* junctions: the emitter–base junction, having a scale current I_{SE}, and the collector-base junction, having a scale current I_{SC}. The current I_{SC} is usually 2 to 50 times greater than I_{SE}.

▶ Although the forward α, α_F, is close to unity, the reverse α, α_R, is 0.02 to 0.5.

▶ For small collector currents, a saturated transistor has $V_{CEsat} \simeq 0.1$ V.

▶ Saturation voltages as low as a fraction of a millivolt can be obtained when the transistor is operated in the reverse saturation mode.

▶ Before a saturated transistor begins to turn off, the extra charge stored in its base must be removed. The storage time can be shortened by avoiding deep saturation and by arranging for a reverse base current to flow. For high-speed operation, saturation should be avoided altogether.

▶ Transistor-transistor logic (TTL) evolved from diode-transistor logic (DTL).

▶ The TTL gate consists of three sections: the input stage, which implements the AND function and utilizes either a multiemitter transistor (in standard TTL) or Schottky diodes (in modern forms of TTL); the phase splitter, which generates a pair of complementary signals to drive the output stage; and the output stage, which utilizes the totem-pole configuration and provides logic inversion. The basic gate implements the NAND function.

▶ In standard TTL, when the gate input goes low, the input multiemitter transistor operates in the normal active mode and supplies a large collector current to discharge the base of the phase splitter rapidly.

▸ The totem-pole output stage consists of a common-emitter transistor, which can sink large load currents and thus rapidly discharge the load capacitance, and an emitter follower, which can source large load currents and thus rapidly charge the load capacitance.

▸ To increase the speed of TTL, transistors are prevented from saturation. This is achieved by connecting a Schottky diode between the base and collector. Schottky diodes are formed as metal-to-semiconductor junctions and exhibit low forward voltage drops. The Schottky diode shunts some of the base current drive of the BJT and thus keeps it out of saturation.

▸ The state of the art in TTL circuits is represented by advanced Schottky TTL, having $t_P = 1.5$ ns, $P_D = 20$ mW, and $DP = 30$ pJ; and advanced low-power Schottky TTL, having $t_P = 4$ ns, $P_D = 1$ mW, and $DP = 4$ pJ.

▸ Emitter-coupled logic (ECL) is the fastest logic circuit family. It achieves its high speed of operation by avoiding transistor saturation and by utilizing small logic-signal swings.

▸ In ECL the input signals are used to steer a constant bias current between a reference transistor and an input transistor. The basic gate configuration is that of a differential amplifier.

▸ There are two popular ECL types: ECL 10K, having $t_P = 2$ ns, $P_D = 25$ mW, and $DP = 50$ pJ; and ECL 100K, having $t_P = 0.75$ ns, $P_D = 40$ mW, and $DP = 30$ pJ. ECL 10K is easier to use because the rise and fall times of its signals are deliberately made long (about 3.5 ns).

▸ Because of the very high operating speeds of ECL, care should be taken in connecting the output of one gate to the input of another. Transmission line techniques are usually employed.

▸ The design of the ECL gate is optimized so that the noise margins are equal and remain equal as temperature changes.

▸ The ECL gate provides two complementary outputs, realizing the OR and NOR functions.

▸ The outputs of ECL gates can be wired together to realize the OR function of the individual output variables.

▸ BiCMOS combines the low-power and wide noise margins of CMOS with the high current-driving capability and thus the short gate delays of BJTs to obtain a technology that is capable of implementing very dense, low-power, high-speed VLSI circuits that can also include analog functions.

BIBLIOGRAPHY

A. R. Alvarez, (Ed.), *BiCMOS Technology and Applications,* Boston: Kluwer Academic Publishers, 1989.

K. Deierling, "Digital design," Chapter 5 in *BiCMOS Technology and Applications,* A. R. Alvarez, (Ed.), Boston: Kluwer Academic Publishers, 1989.

Fairchild, *TTL Data Book,* Mountain View, Calif.: Fairchild Camera and Instrument Corp., Dec. 1978.

L. S. Garrett, "Integrated-circuit digital logic families," a three-part article published in *IEEE Spectrum,* Oct., Nov., and Dec. 1970.

I. Getreu, *Modeling the Bipolar Transistor,* Beaverton, Ore.: Tektronix Inc., 1976.

J. N. Harris, P. E. Gray, and C. L. Searle, *Digital Transistor Circuits,* Vol. 6 of the SEEC Series, New York: Wiley, 1966.

D. A. Hodges and H. G. Jackson, *Analysis and Design of Digital Integrated Circuits,* 2nd ed., New York: McGraw-Hill, 1988.

IEEE Journal of Solid-State Circuits. The October issue of each year has been devoted to digital circuits.

J. Millman and H. Taub, *Pulse, Digital, and Switching Waveforms,* chap. 20, New York: McGraw-Hill, 1965.

Motorola, MECL System Design Handbook, Phoenix, Ariz.: Motorola Semiconductor Products Inc., 1988.

Motorola, *MECL Device Data,* Phoenix, Ariz.: Motorola Semiconductor Products Inc., 1989.

L. Strauss, *Wave Generation and Shaping,* 2nd ed., New York: McGraw-Hill, 1970.

H. Taub and D. Schilling, *Digital Integrated Electronics,* New York: McGraw-Hill, 1977.

Texas Instruments Staff, *Designing with TTL Integrated Circuits,* New York: McGraw-Hill, 1971.

R. L. Treadway, "DC analysis of current-mode logic," *IEEE Circuits and Devices,* vol. 5, no. 2, pp. 21–35, Mar. 1989.

G. R. Wilson, "Advances in bipolar VLSI," Proceedings of the IEEE, Vol. 78, No. 11, pp. 1707–1719, November 1990.

PROBLEMS

Section 14.1: The BJT as a Digital Circuit Element

14.1 Repeat Exercise 14.1 for a transistor for which $\alpha_R = 0.5$.

14.2 A transistor characterized by the Ebers–Moll model shown in Fig. 14.1 is operated with both emitter and collector grounded and a base current of 1 mA. If the collector junction is 10 times larger than the emitter junction and $\alpha_F \simeq 1$, find i_C and i_E.

14.3 Derive expressions for the i–v characteristic of the diode-connected transistors shown in Fig. P14.3, in terms of I_S, α_F, and α_R. If the two transistors are identical, and when the currents i are made equal to a value I, it is found that the voltage v is 0.7 V for the diode in (a) and 0.6 V for the diode in (b), find the relative sizes of the emitter–base and collector–base junctions.

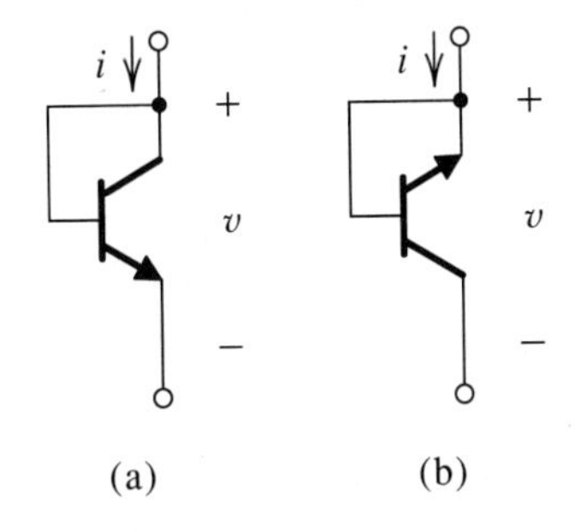

Fig. P14.3

14.4 For the diode-connected transistor shown in Fig. P14.4, find expressions for i_E and i in terms of v, I_S, β_R, and β_F. Assume $\beta_F \gg \beta_R$.

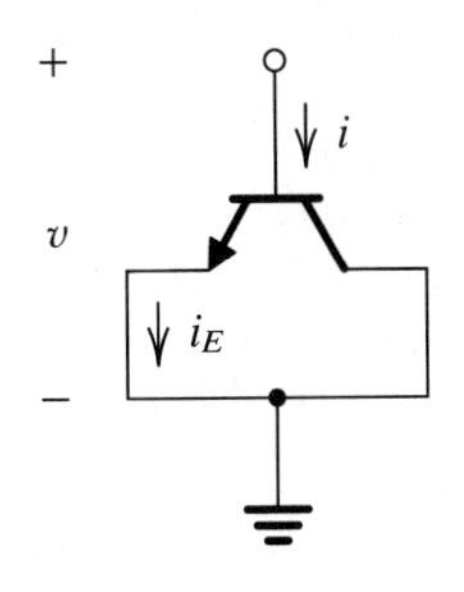

Fig. P14.4

14.5 A BJT for which $\beta_F = 100$ and $\alpha_R = 0.2$ operates with a constant base current but with the collector open. What value of V_{CEsat} would you measure?

*14.6 A BJT for which I_B is 0.5 mA has $V_{CEsat} = 140$ mV at $I_C = 10$ mA, and $V_{CEsat} = 170$ mV at $I_C = 20$ mA. Estimate the value of its saturation resistance. What are β_F and β_R? What is its offset voltage?

**14.7 Consider a BJT operated in saturation with a constant base current I_B. The v_{CE}–i_C characteristic is described by Eq. (14.16) with β_{forced} replaced by i_C/I_B (see also Fig. 14.2). Find an expression for the incremental resistance $\partial v_{CE}/\partial i_C$ in saturation and simplify the resulting expression by assuming that $\beta_F \gg 1$. Then show that the minimum value of the incremental resistance R_{CEsat} is obtained by operating the BJT at a collector current of $\beta_F I_B/2$ (that is, at $\beta_{forced} = \beta_F/2$) and that the value of the minimum resistance is approximately $4V_T/\beta_F I_B$. By extrapolating the straight-line tangent corresponding to this value of minimum incremental resistance to $i_C = 0$, show that an estimate of the offset voltage for this case is V_T $[\ln(\beta_F/\beta_R) - 2]$. Calculate the minimum value of R_{CEsat}, and the offset voltage, for $I_B = 1$ mA, $\beta_F = 50$, and $\beta_R = 0.1$.

*14.8 Using the data given in Table 14.1, plot V_{CEsat} versus i_C for a transistor operated at a constant base current $I_B = 1$ mA. Estimate the minimum value of its incremental saturation resistance $R_{CEsat} \equiv \partial v_{CEsat}/\partial i_C$. Convince yourself that this minimum occurs at $i_C \simeq \beta_F I_B/2$. Extrapolate the straight-line tangent of slope equal to the minimum R_{CEsat} to obtain an estimate for the offset voltage of the BJT switch. Compare your results to those obtained in Problem 14.7.

*14.9 For a grounded-emitter *npn* transistor for which $\beta_F = 100$ and $\beta_R = 1$ in a circuit in which $I_B = 1$ mA and $\beta_{forced} = 10$:

(a) Calculate and label all the currents in the branches of the EM model shown in Fig. 14.1(b).

(b) If $I_S = 10^{-14}$ A, find the voltages across the two junctions and V_{CEsat}.

(c) Verify the value of V_{CEsat} found in (b) using Eq. (14.16).

(d) If the collector lead is cut while the base connection remains, find the new values of V_{BE}, V_{BC}, and V_{CEsat}.

*14.10 A BJT with fixed base current has V_{CEsat} of 60 mV with the emitter grounded and the collector open-circuited. When the collector is grounded and the emitter is open-circuited, V_{CEsat} becomes -1 mV. Estimate values for β_R and β_F for this transistor.

14.11 A particular BJT for which the storage time-constant is 10 ns and β is 50 is operated in a circuit for which $I_{Csat} = 10$ mA and the base turn-on current, I_{B2}, is 2 mA. Calculate the storage delay under conditions that the base turnoff current, I_{B1}, is
(a) 0 mA, (b) 1 mA, (c) 2 mA.

14.12 Use the result of Exercise 14.5 to find $(C_{je} + C_{\mu})$ for the BJT in a circuit for which $V_1 = 0$ V, $V_2 = +3$ V, $R_B = 1$ kΩ, and $t_d = 3$ ns. Assume that $r_x = 50$ Ω.

14.13 A BJT, when operated in the circuit of Fig. 14.5 with $R_B = 1$ kΩ, $R_C = 1$ kΩ, $V_{CC} = V_2 = 3$ V, and $V_1 = 0$ V with $V_{CEsat} \simeq 0$ V, has a storage delay of 10 ns. What storage delay would you expect if $V_{CC} = V_2$ is raised to 5 V? Assume $V_{BE} = 0.7$ V and $\beta = 50$.

14.14 A transistor whose β is 100 and whose storage time-constant is 20 ns is operated with a collector current of 10 mA and a forced base current of 1 mA. What is the excess saturating base charge under these conditions? What does it become if the base drive is reduced to 0.11 mA? In both cases find the storage time, assuming that the turnoff base current is 0.1 mA.

Section 14.2: Early Forms of BJT Digital Circuits

***14.15** In this problem we wish to find the critical points of the voltage transfer characteristic, and hence the noise margins, of the RTL NOR gate of Fig. 14.8. Consider the situation where $v_B = 0$ V, and thus Q_B is cut off. The voltage transfer characteristic is v_Y versus v_A. For this purpose let the transistor have $v_{BE} = 0.7$ V at $i_C = 1$ mA, $\beta_F = 50$, and $\beta_R = 0.1$, and note that (1) V_{OH} was calculated in Exercise 14.6, under the conditions of a fan-out $N = 5$, and was found to be 1 V; and (2) V_{IL} is the value of v_A at which Q_A begins to conduct. Let it be approximately 0.6 V.
(a) V_{IH} is the value of v_A at which Q_A saturates with, say, $\beta_{forced} = \beta_F/2$. Use Eq. (14.16) to calculate V_{CEsat}. Then find I_C, V_{BE}, I_B, and V_{IH}.
(b) $V_{OL} = V_{CEsat}$ that is obtained for $v_A = V_{OH} = 1$ V. Find the value of V_{OL}.
(c) Sketch the voltage transfer characteristic.
(d) Calculate NM_H and NM_L.

14.16 Consider the RTL gate shown in Fig. 14.8.
(a) Find the current drawn from the dc supply when $v_Y = V_{OL} = 0.1$ V. Then find the power dissipation in the gate in this state (neglect the power dissipated due to the base drive of the BJTs).
(b) With the transistor cut off and the gate driving other gates so that $v_Y = V_{OH} = 1$ V, find the current drawn from the dc supply and also the gate power dissipation in this state.
(c) Use the results of (a) and (b) to compute the average power dissipation of the RTL gate.
(d) If the propagation delay of the RTL gate is 10 ns, find its delay-power product.

14.17 What is the logic function implemented by the circuit shown in Fig. P14.17?

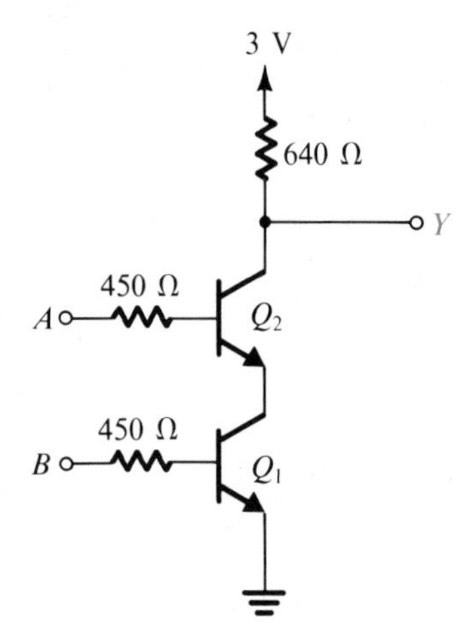

Fig. P14.17

14.18 Consider the circuit in Fig. 14.9. If $V_{CC} = 5$ V, $R_C = R_B = 1$ kΩ, $V_{CEsat} = 0.1$ V, and $V_{BE} = 0.7$ V, what are the voltage levels at Y and $\overline{Y}$ when the set and reset inputs are inactive but following an interval in which S was high while R was low?

14.19 For the DTL gate of Fig. 14.10 whose operation is described in the text, find V_{OH}, V_{OL}, V_{IL}, V_{IH}, and the noise margins. Assume that the voltage drop of a conducting junction is a constant 0.7 V. V_{IL} can be taken as the input voltage at which Q begins to conduct (V_{BE} reaches 0.5 V), and V_{IH} can be taken as the input value at which the voltage across the input diode reaches 0.5 V. To find V_{OL}, calculate β_{forced} and use Eq. (14.16) to calculate V_{CEsat}, assuming that $\beta_F = 50$ and $\beta_R = 0.1$.

14.20 For the DTL gate of Fig. 14.10 calculate the total current in each supply and also the gate power dissipation for the two cases: v_Y high and v_Y low. Then find the average power dissipation in the DTL gate.

D14.21 Following the general direction suggested by the two-input NAND DTL gate in Fig. 14.10, sketch a circuit (using one transistor, and diodes and resistors) to implement the function $Y = \overline{AB + CD}$.

Section 14.3: Transistor-Transistor Logic (TTL or T²L)

D14.22 Consider the DTL gate of Fig. 14.11 when the output is low and the gate is driving N identical gates. Let Q_3 have $\beta_F = 50$ and $\beta_R = 0.1$. Utilize the results of Exercise 14.8 to determine
(a) the output voltage for $N = 0$, and
(b) the maximum allowed fan-out N under the constraint that the output voltage does not exceed twice the value found in (a).

***14.23** For the DTL gate of Fig. 14.11 let $\beta = 100$ and $V_{BE} = V_D = 0.7$ V. Calculate the base current supplied to Q_3 when v_I goes high. When v_I goes low, what is the value of reverse current that flows through the base of Q_3 to remove the saturating charge? If $\tau_s = 10$ ns, compute the storage delay t_s using Eq. (14.20).

D*14.24 For the circuit in Fig. 14.11, what is the minimum β of Q_2 and Q_3 (assumed matched) that ensures that Q_3 saturates with a base overdrive factor of 5? Assume $V_{BE} = V_D = 0.7$ V and $V_{CEsat} = 0.2$ V. (Recall that the base overdrive factor is the ratio of the base current to the minimum base current required to saturate the transistor.)

14.25 Consider the output stage of a DTL gate, consisting of a transistor with $\beta_F = 50$ and a forced β of 10, a load resistance of 2 kΩ connected to a +5-V supply. If the load capacitance is 10 pF, calculate the 10% to 90% rise and fall times of the output voltage waveform. Assume $V_{CEsat} = 0.2$ V.

14.26 Consider the circuit of Fig. 14.13 with a 5-V supply, $R = 4$ kΩ, $R_C = 2$ kΩ, $V_{BE} = 0.7$ V, V_{CEsat} $(Q_3) \simeq 0.2$ V, $\beta_F = 20$, and $\beta_R = 0.1$. What input current flows with input high (≥ 1.4 V)? with input low (at 0.2 V)? What is the value of V_{OH} with no load? For what fan-out of similar circuits does V_{OH} decrease by 2 V?

14.27 Consider the circuit of Fig. 14.13 with $V_{CC} = 3$ V, $R = 3$ kΩ, and $R_C = 1$ kΩ, as the input rises slowly from 0 V. If V_{CEsat} of Q_1 is 0.1 V, and if Q_3 turns on when its V_{BE} reaches 0.6 V, at what value of input voltage does Q_3 begin to conduct? This is a good estimate of V_{IL}.

***14.28** A variant of the T²L gate shown in Fig. 14.23 is being considered in which all resistances are tripled. For input high, estimate all node voltages and branch currents with $\beta_F = 30$, $\beta_R = 0.01$, $V_{BE} = 0.7$ V, and a load of 1 kΩ connected to the 5-V supply.

***14.29** Repeat the analysis of the circuit suggested in Problem 14.28 with input low (at +0.2 V) and a resistor of 1 kΩ connected from the output to ground.

***14.30** Two TTL gates of the type described in Problem 14.28, one with input high and one with input low, have their outputs accidentally joined. What output voltage results? What current flows in the short circuit?

***14.31** A transistor for which $\beta_F = 50$ and $\beta_R = 5$ is used for Q_3 in Fig. 14.24. For a base current of 2.5 mA, what is V_{CEsat} for $i_L = 0$, 1, 10, and 100 mA? Estimate R_{CEsat} at 0.5, 5, and 50 mA.

14.32 Consider the output circuit of the gate in Fig. 14.26. What is the output voltage when a current of 2 mA is extracted? What is the (small-signal) output resistance at this current level? Use $\beta = 50$.

14.33 Consider the output circuit of the gate in Fig. 14.26. For $\beta = \infty$ and $V_{CEsat} = 0.2$ V, at what output current does Q_4 saturate? For $\beta = 20$, at what current does saturation occur?

14.34 If the output of the circuit in Fig. 14.26 is short-circuited to ground, what current flows? Assume high β, $V_{CEsat} = 0.2$ V, and $V_{BE} = V_D = 0.7$ V. What is the minimum value of β for which your analysis holds?

Section 14.4: Characteristics of Standard TTL

D14.35 For the TTL gate shown with its transfer characteristic in Fig. 14.27, consider the effect of changing the value of R_2 to 0.5 kΩ and to 2 kΩ. Sketch and label the original and two modified characteristics. Which do you like best? Why? The trouble is that as R_2 is raised, the storage delay of the gate is increased.

*****14.36** Consider the circuit of Fig. 14.27 with a 200-Ω resistor connected between the output and the input to bias the inverter circuit in its linear region.
(a) Perform a dc analysis on the circuit to determine all currents and voltages. Assume that Q_1 saturates with V_{CEsat} of approximately 0.1 V and that all other transistors are in the active mode with $\beta = 50$ and $V_{BE} = 0.7$ V. (Note that Q_4 will remain in the active mode even though its collector voltage will be slightly lower than its base voltage.)

(b) Using the simple T model for each active transistor, find an approximate value for the small-signal voltage gain of the inverter.

14.37 Using the data provided in the answers to Exercise 14.15, find the noise margins of the T^2L gate at $-55°C$ and $+125°C$.

14.38 Analysis of the TTL gate circuit of Fig. 14.27 when v_O is low shows that the base current of Q_3 at $-55°C$, $+25°C$, and $+125°C$ is 2.2 mA, 2.6 mA, and 3 mA, respectively. If at these three temperatures, β_F of Q_3 is 13, 20, and 25, respectively, find the maximum current that the gate output can sink at each of the three temperatures, while Q_3 remains in saturation.

D14.39 A designer, considering the possibility of raising the input threshold of the TTL gate shown in Fig. 14.27, adds two diodes, one in series with the emitter of Q_2 and one in series with D. Why is the second diode necessary? Sketch the transfer characteristic and find V_{OH}, V_{OL}, V_{IH}, V_{IL}, and the noise margins.

14.40 For an eight-input TTL NAND gate resembling that in Fig. 14.28, β_R of Q_1 is 0.04. For all inputs high, what is the additional current supplied to Q_2 as a result of Q_1 operating in the inverted mode?

D14.41 Following the direction indicated by Fig. 14.30, sketch the circuit of a three-input NOR gate.

14.42 Analyze the circuit of Fig. 14.30 to determine all currents and voltages for the following three cases:
(a) $v_A = v_B = v_C = v_D = +5$ V.
(b) $v_A = v_C = v_D = +5$ V and $v_B = +0.2$ V.
(c) $v_A = v_C = +5$ V and $v_B = v_D = +0.2$ V.
Use $V_{BE} = 0.7$ V and $V_{CEsat} = 0.2$ V.

D14.43 Sketch a circuit that implements the logic function $Y = \overline{AB + CD}$ using two two-input NAND gates of the open-collector type and a resistor.

14.44 Consider the tristate gate shown in Fig. E14.20 as the voltage on the tristate input terminal is raised from the "low" value. What is the voltage just required to turn off Q_7 and thus release the 1.6-kΩ resistor to drive Q_4 or Q_3 via Q_2?

Section 14.5: TTL Families with Improved Performance

14.45 Consider a Schottky transistor whose emitter is grounded, whose base is connected to a +5-V input signal via a 10-kΩ resistor, and whose collector is connected to a +5-V supply via a 1-kΩ resistor. Assume that $V_{BE} = 0.8$ V, the voltage drop of the Schottky diode is 0.5 V, and $\beta = 50$. Find the base and collector currents of the intrinsic BJT and the current through the Schottky diode.

14.46 For the Schottky TTL NAND gate of Fig. 14.32, find the current that flows in an output short-circuit to ground when
(a) the inputs are both low, and
(b) the inputs are both high.
Assume $V_{BE} = 0.8$ V and $V_D = 0.5$ V.

*14.47 The BJT in the circuit shown in Fig. P14.47 begins to conduct at $v_{BE} = 0.7$ V and is fully conducting at $v_{BE} = 0.8$ V. The same applies to D_7. The Schottky diodes have a voltage drop of 0.5 V.
(a) What is the logic function performed?
(b) Find V_{OL} and V_{OH}.
(c) Find V_{IL} and V_{IH}.
(d) Find the noise margins.
(e) Find the current drawn from the supply when A is high, B is high, and C is low.

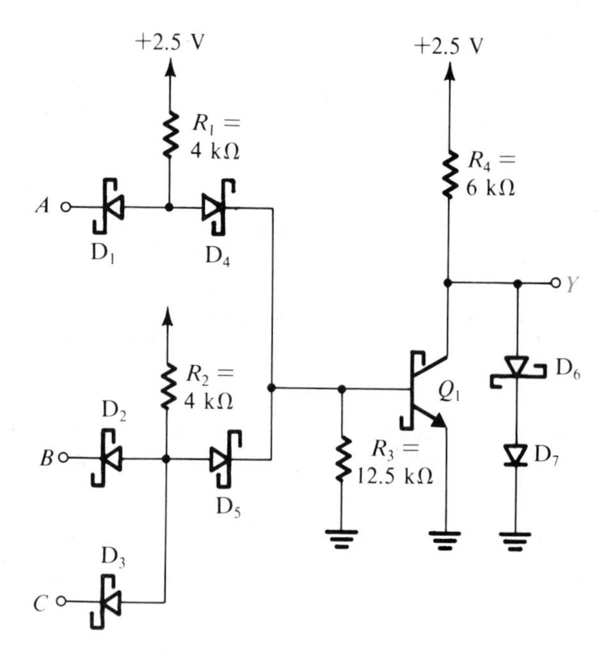

Fig. P14.47

Section 14.6: Emitter-Coupled Logic (ECL)

D14.48 For the ECL circuit in Fig. P14.48, the transistors exhibit V_{BE} of 0.75 V at an emitter current I and have very high β.
(a) Find V_{OH} and V_{OL}.

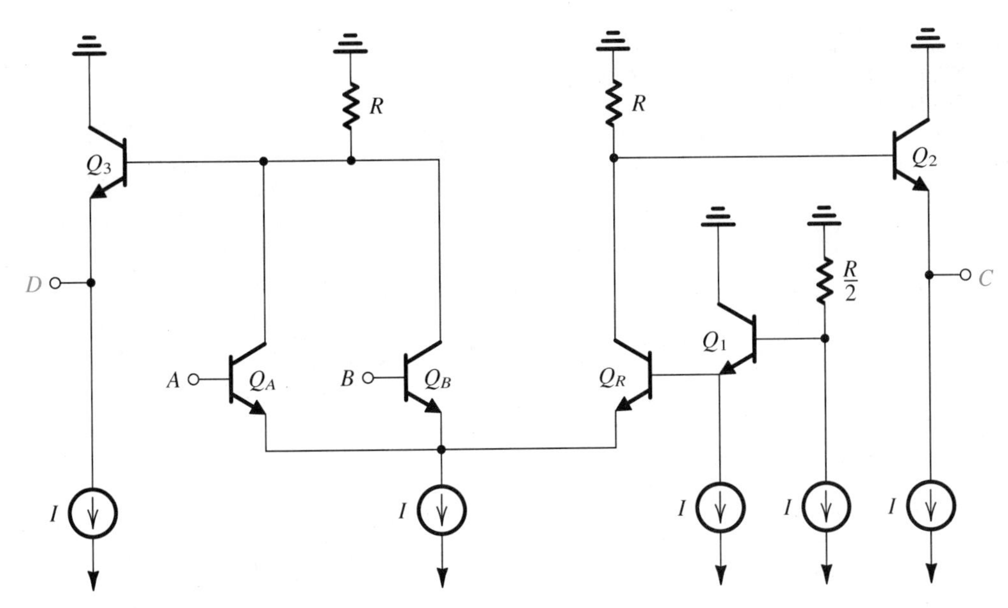

Fig. P14.48

(b) For the input at B sufficiently negative for Q_B to be cut off, what voltage at A causes a current of $I/2$ to flow in Q_R?
(c) Repeat (b) for a current in Q_R of $0.99I$.
(d) Repeat (c) for a current in Q_R of $0.01I$.
(e) Use the results of (c) and (d) to specify V_{IL} and V_{IH}.
(f) Find NM_H and NM_L.
(g) Find the value of IR that makes the noise margins equal to the width of the transition region, $V_{IH} - V_{IL}$.
(h) Using the IR value obtained in (g), give numerical values for V_{OH}, V_{OL}, V_{IH}, V_{IL}, and V_R for this ECL gate.

*14.49 Three logic inverters are connected in a ring. Specifications for this family of gates indicates a typical propagation delay of 3 ns for high-to-low output transitions and 7 ns for low-to-high transitions. Assume that for some reason the input to one of the gates undergoes a low-to-high transition. By sketching the waveforms at the outputs of the three gates and keeping track of their relative positions, show that the circuit functions as an oscillator. What is the frequency of oscillation of this **ring oscillator?** In each cycle, how long is the output high? low?

*14.50 Following the idea of a ring oscillator introduced in Problem 14.49, consider an implementation using a ring of five ECL 100K inverters. Assume that the inverters have linearly rising and falling edges (and thus the waveforms are trapezoidal in shape). Let the 0 to 100% rise and fall times be equal to 1 ns. Also, let the propagation delay (for both transitions) be equal to 1 ns. Provide a labeled sketch of the five output signals, taking care that relevant phase information is provided. What is the frequency of oscillation?

*14.51 Using the logic and circuit flexibility of ECL indicated by Figs. 14.36 and 14.46, sketch an ECL logic circuit that realizes the exclusive OR function, $Y = \bar{A}B + A\bar{B}$. (*Hint:* $\bar{A}B + A\bar{B} = \overline{\overline{A + \bar{B}} + \overline{\bar{A} + B}}$.)

*14.52 For the circuit in Fig. 14.38, whose transfer characteristic is shown in Fig. 14.39, calculate the incremental voltage gain from input to the OR output at points x, m, and y of the transfer characteristic. Assume $\beta = 100$. Use the results of Exercise 14.29, and let the output at x be -1.77 V and that at y be -0.88 V. Hint: Recall that x and y are defined by a 1%, 99% current split.

14.53 For the circuit in Fig. 14.38, whose transfer characteristic is shown in Fig. 14.39, find V_{IL} and V_{IH} if x and y are defined as the points at which
(a) 90% of the current I_E is switched.
(b) 99.9% of the current I_E is switched.

14.54 For the symmetrically loaded circuit of Fig. 14.38

and for typical output signal levels ($V_{OH} = -0.88$ V and $V_{OL} = -1.77$ V), calculate the power lost in both load resistors R_T and both output followers. What then is the total power dissipation of a single ECL gate including its symmetrical output terminations?

14.55 Considering the circuit of Fig. 14.40, what is the value of β of Q_2, for which the high noise margin (NM_H) is reduced by 50%?

***14.56** Consider an ECL gate whose inverting output is terminated in a 50-Ω resistance and connected to a load capacitance C. As the input of the gate rises, the output emitter follower cuts off and the load capacitance C discharges through the 50-Ω load (until the emitter follower conducts again). Find the value of C that will result in a discharge time of 1 ns. Assume that the two output levels are -0.88 V and -1.77 V.

14.57 For signals whose rise and fall times are 3.5 ns, what length of unterminated gate-to-gate wire interconnect can be used if a ratio of rise time to return time of 5 to 1 is required? Assume the environment of the wire to be such that the signal propagates at two-thirds the speed of light (which is 30 cm/ns).

***14.58** For the circuit in Fig. P14.58 let the levels of the inputs A, B, C, and D be 0 and +5 V. For all inputs low at 0 V, what is the voltage at E? If A and C are raised to +5 V, what is the voltage at E? Assume $|V_{BE}| = 0.7$ V and $\beta = 50$. Express E as a logic function of A, B, C, and D.

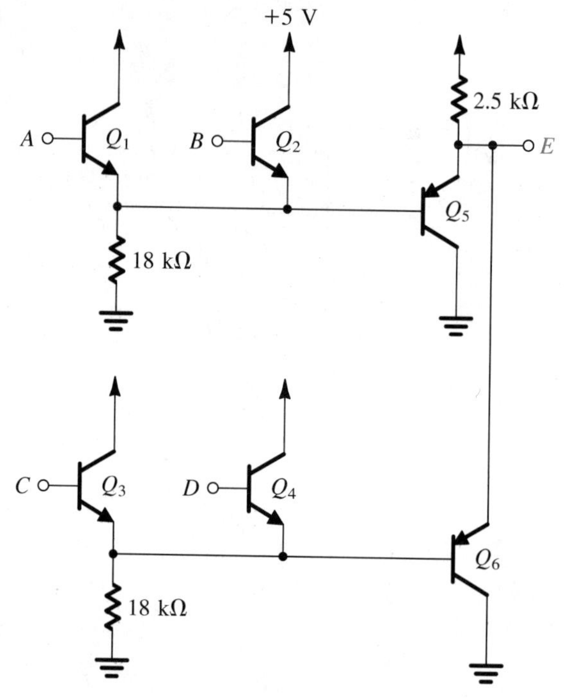

Fig. P14.58

APPENDIX 

Integrated-Circuit Fabrication Technology

INTRODUCTION

The purpose of this appendix is to familiarize the reader with integrated-circuit terminology. An explanation is given of standard integrated-circuit processes. The characteristics of the devices available in integrated-circuit form are also presented. This is done to aid the reader in understanding those aspects of integrated-circuit design that are distinct from discrete-circuit design. In order to take proper advantage of the economics of integrated circuits, designers have had to overcome some serious device limitations (such as poor resistor tolerances) while exploiting device advantages (such as good component matching). An understanding of device characteristics is therefore essential in designing good integrated circuits, and also helps when applying commercial integrated circuits to system design.

This appendix will consider only silicon technology. Although germanium and gallium arsenide are also used to make semiconducting devices, silicon is still the most popular material and will remain so for some time. The physical properties of silicon make it suitable for fabricating active devices with good electrical characteristics. In addition, silicon can easily be oxidized to form an excellent insulating layer (glass). This insulator is used to make capacitor structures and allows the construction of field-controlled devices. It also serves as a good mask against foreign impurities, which could diffuse into the high-purity silicon material. This masking property allows the formation of integrated circuits; active and passive circuit elements can be built together on the same piece of material (substrate) at the same time, and they can be interconnected to form a complete circuit function.

INTEGRATED-CIRCUIT PROCESSES

The basic processes involved in the fabrication of integrated circuits will be described in the following sections.

Wafer Preparation

The starting material for modern integrated circuits is very-high-purity silicon. The material is grown as a single crystal. It takes the shape of a solid cylinder 10 to 12.5 cm in diameter and 1 m in length, which is steel gray in color. This crystal is then sawed (like a loaf of bread) to produce wafers 10 to 12.5 cm in diameter and 200 μm thick (a micrometer or micron is a millionth of a meter). The surface of the wafer is then polished to a mirror finish.

The basic electrical and mechanical properties of the wafer depend on the direction in which the crystal was grown (crystal orientation) and the number and type of impurities present. Both variables are strictly controlled during crystal growth. Impurities can be added on purpose to the pure silicon in a process known as **doping.** Doping allows controlled alteration of the electrical properties of the silicon, in particular the resistivity. It is also possible to control the type of carrier used to produce electrical conduction, being either holes (in *p*-type silicon) or electrons (in *n*-type silicon). If a large number of impurity atoms is added, then the silicon is said to be heavily doped. When designating relative doping concentrations on diagrams of devices, it is common to use + and − symbols. Thus a heavily doped (low-resistivity) *n*-type silicon wafer would be referred to as n^+ material. This ability to control the doping of silicon permits the formation of diodes, transistors, and resistors in integrated circuits.

Oxidation

Oxidation refers to the chemical process of silicon reacting with oxygen to form silicon dioxide. To speed up the reaction, it is necessary to heat the wafers to the 1000 to 1200°C range. The heating is performed in special high-temperature furnaces. To avoid the introduction of even small quantities of contaminants (which could significantly alter the electrical properties of the silicon), it is necessary to maintain an ultraclean environment for the processing. This is true for all processing steps involved in the fabrication of an integrated circuit. Specially filtered air is circulated in the processing area, and all personnel must wear special lint-free clothing.

The oxygen used in the reaction can be introduced either as a high-purity gas (referred to as a "dry oxide") or as water vapor (forming a "wet oxide"). In general, a wet oxide has a faster growth rate, but a dry oxide has better electrical characteristics. The oxide layer grown has excellent electrical insulation properties. It has a dielectric constant of about 3.5, and it can be used to form excellent capacitors. It also serves as a good mask against many impurities. It can therefore be used to protect the silicon surface from contaminants. It can also be used as a masking layer, allowing the introduction of dopants into the silicon only in regions that are not covered with oxide. This masking property is what permits the convenient fabrication of integrated circuits.

The silicon dioxide layer is a thin, transparent film, and the silicon surface is highly reflective. If white light is incident on the oxidized wafer, constructive and destructive interference effects occur in the oxide, causing certain colors to be absorbed strongly. The wavelengths absorbed depend on the thickness of the oxide layer. This absorption produces different colors in the different regions of a processed wafer. The colors can be quite vivid, ranging from blue to greens and reds, and are immediately obvious when a finished chip is viewed under a microscope. It should be remembered, though, that the color is due to an optical effect. The oxide layer is transparent, and the silicon underneath it is steel-gray in color.

Diffusion

Diffusion is the process by which atoms move through the crystal lattice. In fabrication, it relates to the introduction of impurity atoms (dopants) into silicon to change its doping. The rate at which dopants diffuse in silicon is a strong function of temperature. This allows us to introduce the impurities at a high temperature (1000 to 1200°C) to obtain the desired doping. The slice is then cooled to room temperature, and the impurities are essentially ''frozen'' in position. The diffusion process is performed in furnaces similar to those used for oxidation. The depth to which the impurities diffuse depends on both the temperature and time allowed.

The two most common impurities used as dopants are boron and phosphorus. Boron is a *p*-type dopant, and phosphorus is an *n*-type dopant. Both dopants are effectively masked by thin silicon dioxide layers. By diffusing boron into an *n*-type substrate, a *pn* junction is formed (diode). A subsequent phosphorus diffusion will produce an *npn* structure (transistor). In addition, if the doping concentration is heavy, the diffused layer can be used as a conductor.

Ion Implantation

In addition to diffusion, impurities can be introduced into silicon by the use of an ion implanter. An ion implanter produces ions of the desired impurity, accelerates them by an electric field, and allows them to strike the silicon surface. The ions become embedded in the silicon. The depth of penetration is related to the energy of the ion beam, which can be controlled by the accelerating-field voltage. The quantity of ions implanted can be controlled by varying the beam current (flow of ions). Since both voltage and current can be accurately measured and controlled, ion implantation results in much more accurate and reproducible impurity profiles than can be obtained by diffusion. In addition, ion implantation can be performed at room temperature. Ion implantation normally is used when accurate control of the dopant is essential for device operation.

Chemical Vapor Deposition

Chemical vapor deposition (CVD) is a process by which gases or vapors are chemically reacted, leading to the formation of a solid on a substrate. In our case, CVD can be used to deposit silicon dioxide on a silicon substrate. For instance, if silane gas and oxygen are mixed above a silicon substrate, silicon dioxide deposits as a solid on the silicon. The oxide layer formed is not as good as a thermally grown oxide, but it is good enough to act as an electrical insulator. The advantage of a CVD layer is that the oxide deposits at a faster rate and a lower temperature (below 500°C).

If silane gas alone is used, then a silicon layer deposits on the wafer. If the reaction temperature is high enough (above 1000°C), then the layer is deposited as a crystalline layer (assuming the substate is crystalline silicon). This is because the atoms have enough energy to align themselves in the proper crystal directions. Such a layer is said to be an epitaxial layer, and the deposition process is referred to as **epitaxy** instead of CVD. At lower temperatures, or if the substrate is not single-crystal silicon, the atoms are not all aligned along the same crystal direction. Such a layer is called **polycrystalline silicon,** since it consists of many small crystals of silicon aligned in various directions. Such layers are normally doped very heavily to form a high conductivity region that can be used for interconnecting devices.

Metallization

The purpose of metallization is to interconnect the various components of the integrated circuit (transistors, resistors, etc.) to form the desired circuit. Metallization involves the deposition of a metal (aluminum) over the entire surface of the silicon. The required interconnection pattern is then selectively etched. The aluminum is deposited by heating it in vacuum until it vaporizes. The vapors then contact the silicon surface and condense to form a solid aluminum layer.

Packaging

A finished silicon wafer may contain from 100 to 1000 finished circuits or chips. Each chip contains between 10 and 10^7 transistors and is rectangular in shape, typically between 1 and 10 mm on each edge. The circuits are first tested electrically (while still in wafer form) using an automatic probing station. Bad circuits are marked for later identification. The circuits are then separated from each other (by dicing) and the good circuits (dies) are mounted in packages (headers). Fine gold wires are traditionally used to interconnect the pins of the package to the metallization pattern on the die. Finally, the package is sealed under vacuum or in an inert atmosphere.

Photolithography

The surface geometry of the various integrated-circuit components is defined photographically: The silicon surface is coated with a photosensitive layer and then exposed to light through a master pattern on a photographic plate. The layer is then developed to reproduce the pattern on the wafer. Very fine surface geometries can be reproduced accurately by this technique. The resulting layer is not attacked by the chemical etchants used for silicon dioxide or aluminum and so forms an effective mask. This allows ''windows'' to be etched in the oxide layer in preparation for subsequent diffusion processes. This process is used to define transistor regions and to isolate one transistor from another.

INTEGRATED-CIRCUIT STRUCTURES

The structure of the basic integrated-circuit components available to the circuit designer will be presented in this section. Only a standard bipolar junction-isolated process will be considered initially. The processing steps involved are outlined in Fig. A.1 along with the corresponding transistor cross section. The process requires six masking steps, four diffusions, and one epitaxial layer growth. The possible components produced include *npn* and *pnp* bipolar transistors, *p*-channel junction field-effect transistors, resistors, and capacitors. Each of the components will be described in the following sections.

Resistors

Resistors can be made from the collector layer (*n*-type), base region (*p*-type), emitter region (n^+-type), or the ''pinched-base'' region (base region between the n^+ diffused region and the *n*-type epitaxial region). The value of resistance obtained is related to the surface geometry and the region resistivity (doping). Figure A.2 gives the surface and cross-sectional views of the possible resistor structures. Table A.1 gives typical characteristics of the resulting resistors. Note that it is difficult to obtain high resistor values. Also note that although the tolerance of the resistor value is very poor (20% to 50%), matching of two similar resistor values is quite good (5%). Thus circuit designers should design circuits that exploit resistor matching and should avoid all circuits that require a specific resistor value. Also note that diffused resistors have a significant temperature coefficient.

Capacitors

Two types of capacitor structure are available in bipolar integrated circuits. Figure A.3 shows cross sections of those structures. The junction capacitor makes use of one of the junctions available by diffusion. Either the collector–base or the emitter–base junction can be used, but the low breakdown voltage of the emitter–base junction (6 to 7 V) makes it less popular. The capacitance is determined by geometry and doping levels, but normally it is restricted to values less than 20 pF. The tolerance is 20%. Junction capacitors must always be reverse-biased and have a large voltage dependence.

A capacitor can also be formed using the oxide as a dielectric. The aluminum metallization and the emitter region form a parallel-plate structure. This structure makes an excellent capacitor with very low voltage and temperature coefficients, low leakage, and high breakdown voltage. Although the tolerance is 10%, capacitors on the same chip can be matched to better than 1%. Again the values are restricted to a few tens of picofarads.

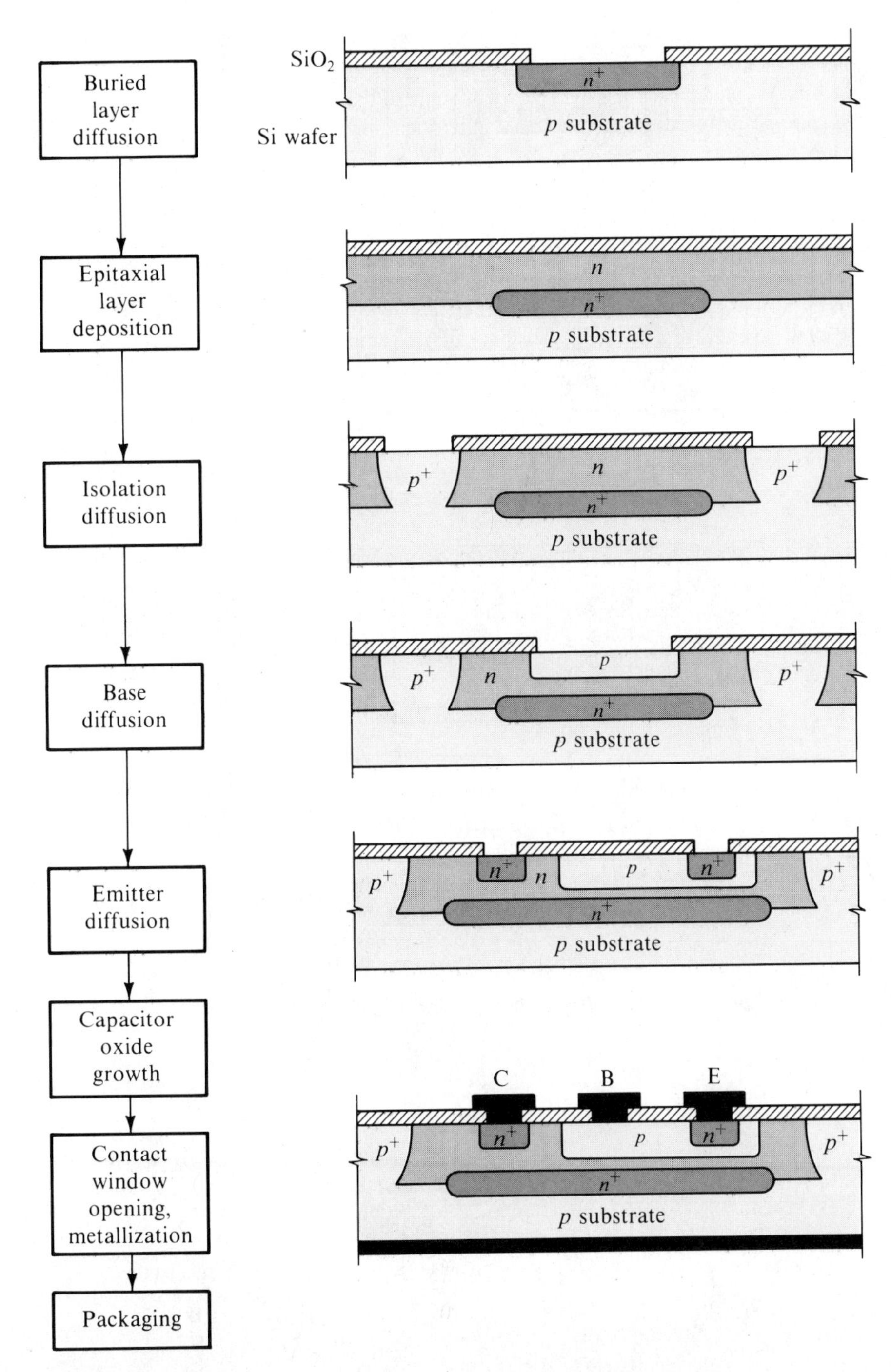

Fig. A.1 Standard bipolar integrated-circuit process forming an *npn* BJT.

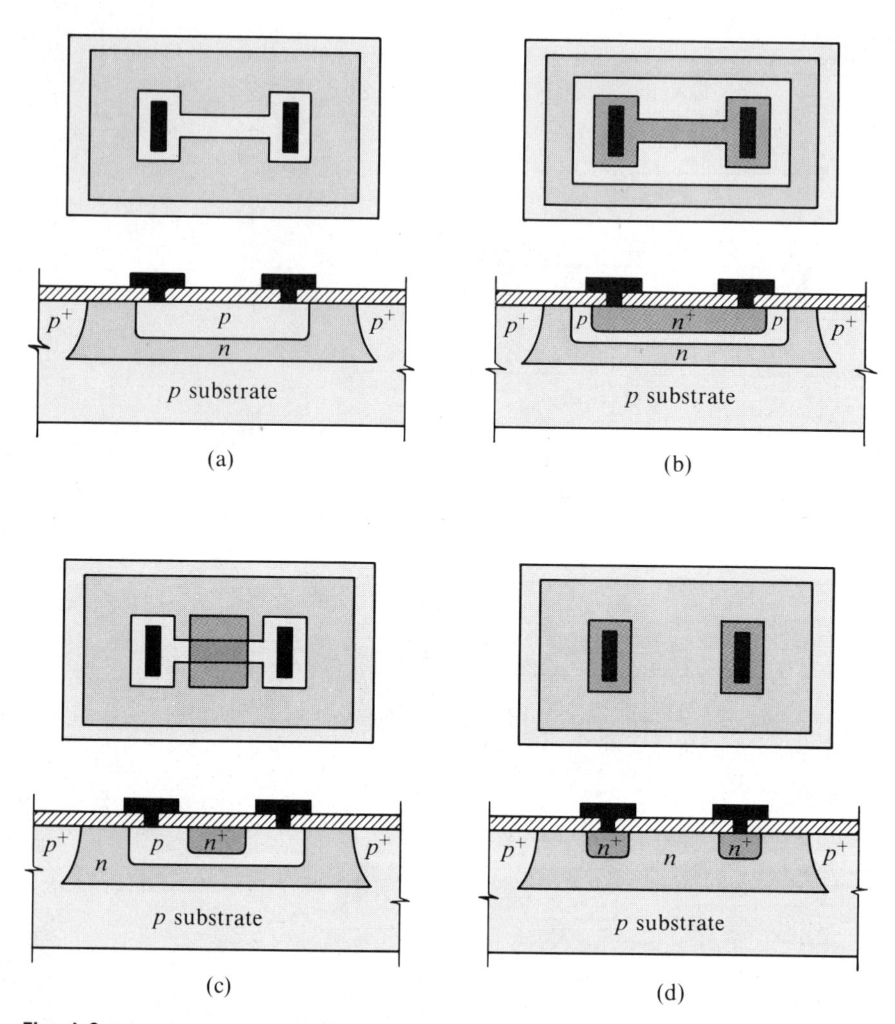

Fig. A.2 Integrated-circuit resistor structures. **(a)** Base resistor. **(b)** Emitter resistor. **(c)** Pinched-base resistor. **(d)** Collector resistor.

Table A.1 DIFFUSED RESISTOR CHARACTERISTICS

	Resistor Type			
	Base	Emitter	Pinched Base	Collector
Range (ohms)	50–50k	5–100	10k–500k	1k–10k
Tolerance	20%	20%	50%	50%
Matching	5%	5%	5%	5%
Temperature coefficient	0.1%/°C	0.2%/°C	0.5%/°C	0.8%/°C
Breakdown voltage (V)	40	6	6	70

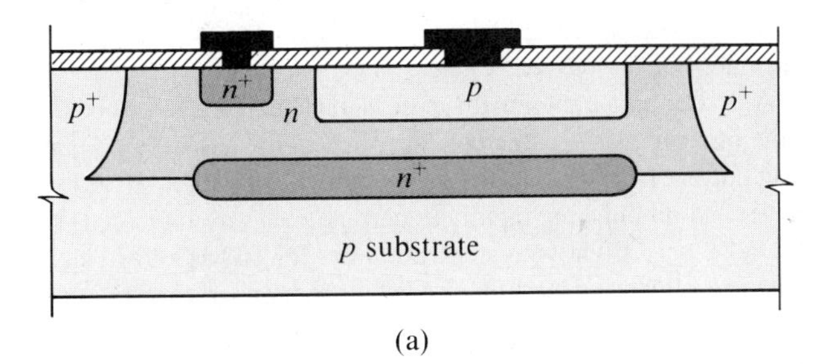

(a)

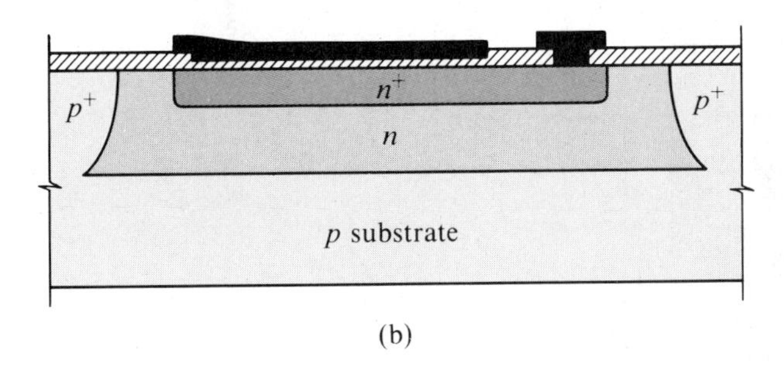

(b)

Fig. A.3 Integrated-circuit capacitor structures. **(a)** Collector–base junction capacitor. **(b)** Oxide capacitor.

Bipolar Junction Transistors

Three basic types of bipolar transistors are easily available: *npn, lateral pnp,* and *substrate pnp*. The *npn* structure is repeated in cross-section form in Fig. A.4. It has a beta typically of 100 to 500, a collector breakdown voltage of 40 V, and a cutoff frequency of 500 MHz. The normal operating-

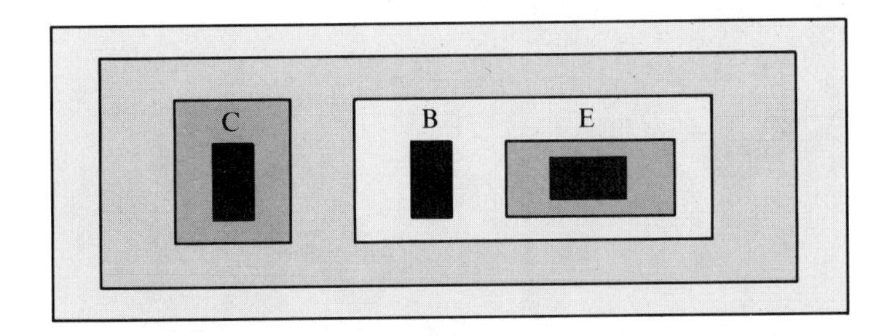

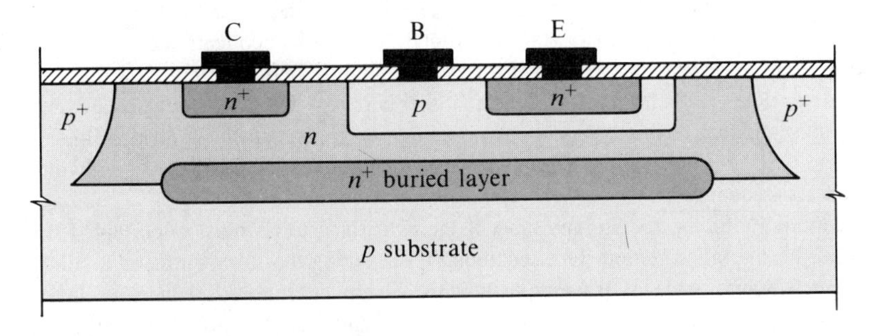

Fig. A.4 An *npn* bipolar transistor.

current range is from a few microamperes to a few tens of milliamperes. Higher operating currents are obtained by increasing the surface area of the device.

The lateral *pnp* is shown in Fig. A.5. The base region in this case is the lateral area between the two *p* diffusions, hence the name (the other transistor structures are vertical devices). This is the most commonly used *pnp* structure. It suffers from very low beta, being typically about 20 at low collector currents (a few microamperes) and diminishing rapidly as current levels are increased. At milliampere current levels, its gain may be only slightly greater than unity. It also has a low cutoff frequency, around 1 MHz. Its breakdown voltage is over 40 V. Despite its poor characteristics, the lateral *pnp* has been the best *pnp* transistor available in integrated circuits until recently.

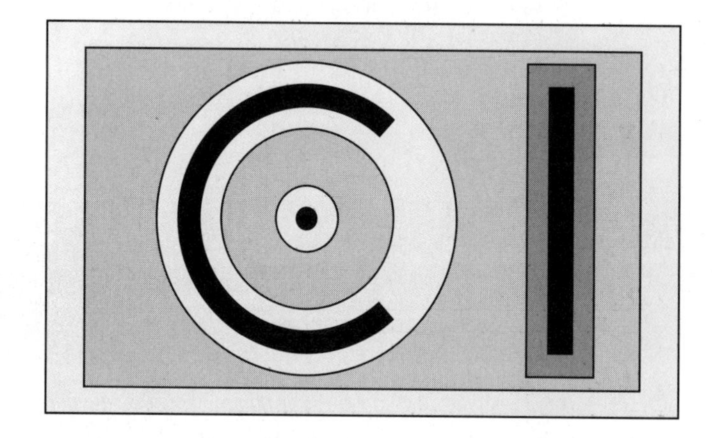

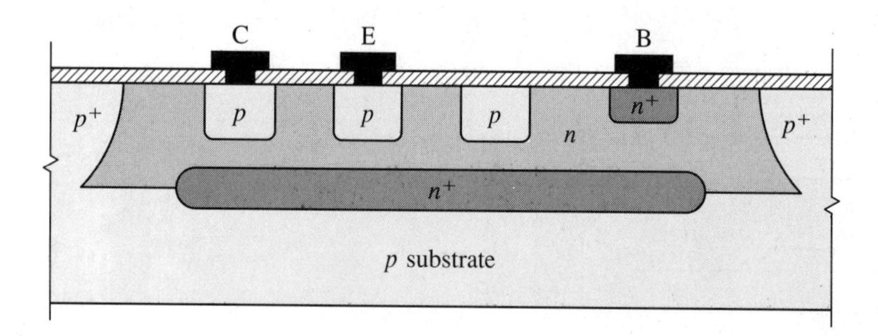

Fig. A.5 Lateral *pnp* bipolar transistor.

The substrate *pnp* structure is shown in Fig. A.6. In this case, the collector is electrically connected to the substrate, which is normally a signal ground. This limits application of the device to common-collector configurations. It does, however, have a slightly better current gain than the lateral *pnp* and exhibits somewhat better performance at higher currents. The characteristics are only slightly improved, so integrated-circuit designers avoid using *pnp* transistors, if at all possible. If needed, they are best used in unity-gain configurations to avoid serious degradation of bandwidth due to their low cutoff frequency.

For digital integrated circuits, the storage time of the *npn* transistor is most important if the device is operated in the saturation region. One technique for reducing the storage time is to allow gold to diffuse into the silicon. Gold acts as an impurity in the silicon and serves to reduce the carrier lifetime, hence the storage time. It also reduces the current gain and breakdown voltage, but these reductions are not important in digital circuits. It is also not possible to make corresponding *pnp*

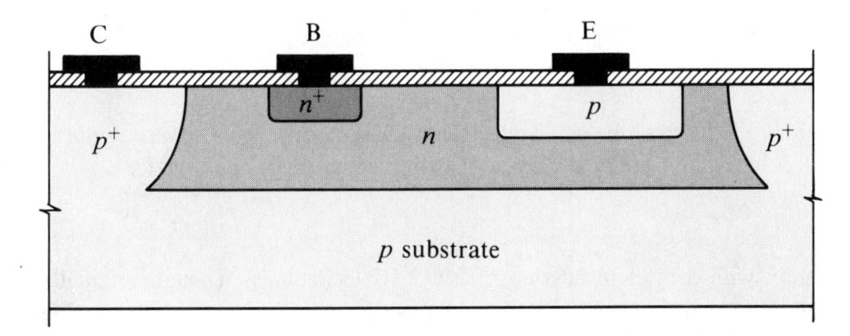

Fig. A.6 Substrate *pnp* bipolar transistor.

transistor structures with gains greater than unity due to the low lifetime, but again the *pnp* is not needed in usual digital circuits. An alternative to gold doping is to avoid saturation region operation by Schottky clamping the collector–base junction. This will not affect any transistor parameters, and yet the storage time can be eliminated. Fortunately, a suitable Schottky diode can be formed by using aluminum as the metal and the collector *n* region as the semiconductor. This leads to the very simple structure shown in Fig. A.7.

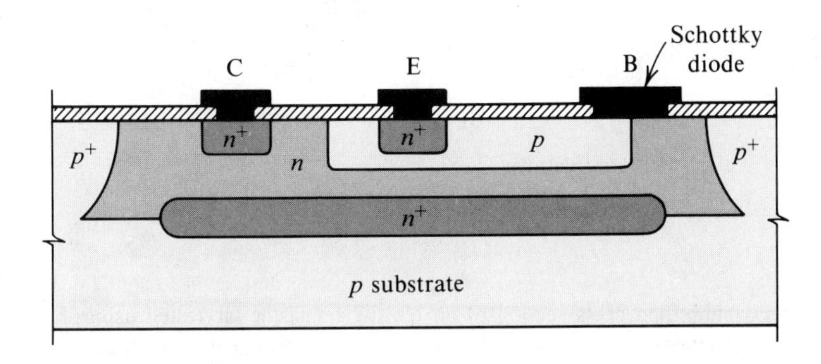

Fig. A.7 Schottky-clamped *npn* bipolar transistor.

Junction Field-Effect Transistors

A *p*-channel JFET can be formed using the bipolar process. The corresponding cross section is shown in Fig. A.8. This structure suffers from a low gate-drain breakdown voltage of 6 V and poor matching of pinch-off voltages. The use of ion implantation as an extra processing step, however, permits much improvement in pinch-off matching and control. The structure can be used to functionally replace *pnp* bipolar devices in circuits. It is capable of providing increased gain and cutoff

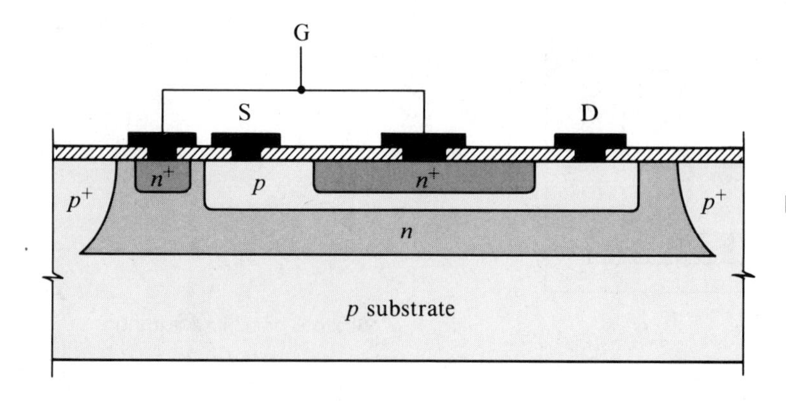

Fig. A.8 *p*-channel JFET.

frequency, with values similar to those of the *npn* device. Thus the *p*-channel JFET is very useful in analog integrated-circuit design and is to be preferred over the use of lateral *pnp* devices.

MOS Technology

We conclude this appendix with a brief discussion of MOS IC technology. Though originally employed for the fabrication of digital circuits, MOS technology is currently used also for analog circuits, and for VLSI chips containing both digital and analog circuits. Initially, MOS IC technology was based on the PMOS transistor, whose cross section is shown in Fig. A.9. The simplicity of the structure should be noted. Indeed, the PMOS process is simple and economical. The performance of the resulting circuits, however, falls considerably short of that currently available from modern NMOS and CMOS processes. The PMOS process, exemplified by the transistor of Fig. A.9, utilizes aluminum gates. Most modern NMOS and CMOS processes employ polycrystalline silicon (poly) to form the gate electrodes.

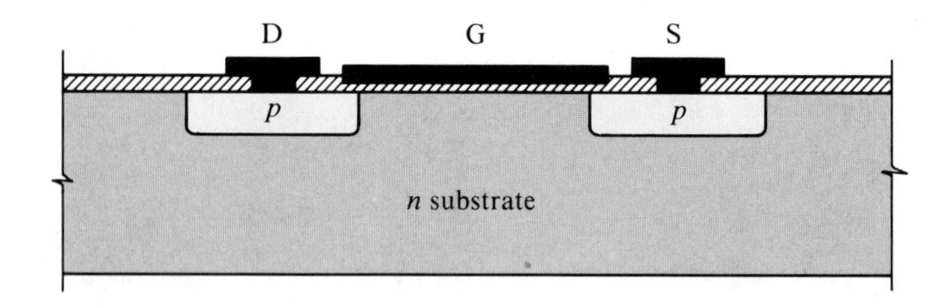

Fig. A.9 PMOS transistor (aluminum gate).

Figure A.10 shows a simplified cross section of an NMOS transistor fabricated using the silicon-gate process. Here also the structure is simple, with the result that a large number of devices can be fabricated on the same chip. Indeed, NMOS technology provides the highest possible integration density.

The NMOS transistor shown in Fig. A.10 is of the enhancement type. Depletion devices can be formed by using ion implantation to implant a channel at the substrate surface. Modern NMOS technology employs depletion transistors as load devices.

Compared to NMOS, CMOS technology requires additional processing steps but allows much greater flexibility in circuit design. Also, the resulting circuits, both analog and digital, exhibit improved performance over those fabricated in NMOS technology.

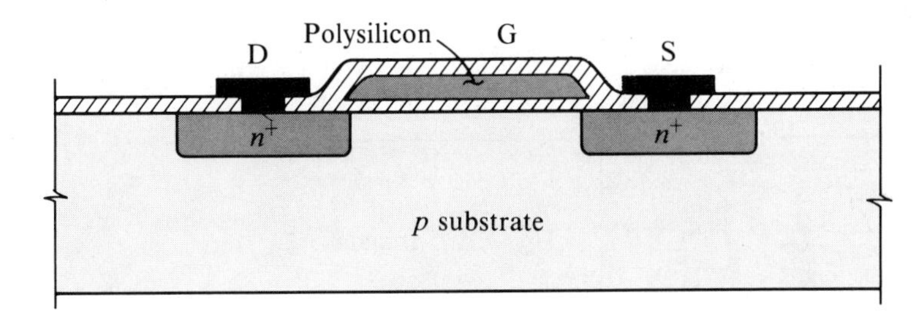

Fig. A.10 NMOS transistor (silicon gate).

Although CMOS technology can utilize either the metal gate or the silicon gate technique, most modern CMOS processes are of the silicon gate variety. Figure A.11 shows a somewhat simplified cross section of silicon gate CMOS transistors. As indicated, the p-channel transistor is formed in an n well. For this reason, the process illustrated is known as an n-well process (p-well processes are possible and are equally popular). The gate electrodes are formed of high-conductivity n^+ polycrystalline silicon that is deposited using a CVD process. The poly is then covered with a thick oxide layer that is deposited also using CVD. For the n-well CMOS structure shown, the p substrate is tied to ground while the n well is tied to V_{DD}.

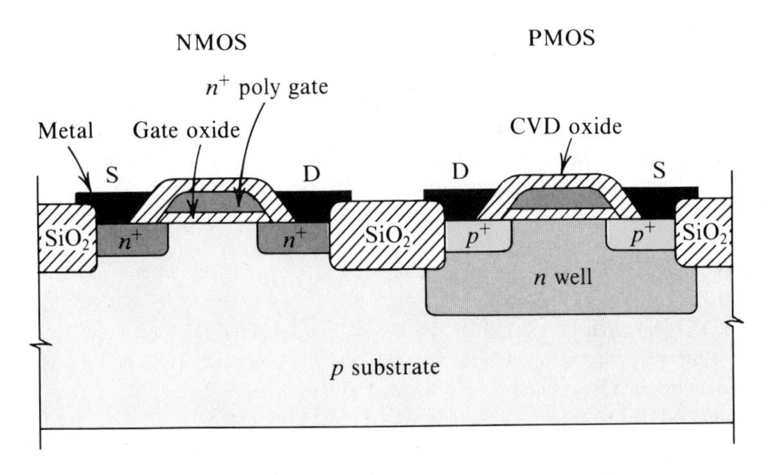

Fig. A.11 CMOS transistors in silicon-gate, n-well technology.

The CMOS process illustrated in Fig. A.11 utilizes a single layer of polysilicon and is thus known as a single-poly process. More advanced processes employ two layers of polysilicon. These double-poly processes provide increased flexibility. For instance, high-quality capacitors for analog circuit applications can be implemented using the two poly layers as electrodes and the silicon dioxide as the dielectric. Also, because of their high conductivity, the polysilicon layers can be used for forming interconnection patterns, thus enabling greater levels of integration.

BIBLIOGRAPHY

A. B. Grebene, *Bipolar and MOS Analog Integrated Circuit Design,* New York: Wiley, 1984.

D. J. Hamilton and W. G. Howard, *Basic Integrated Circuit Engineering,* New York: McGraw-Hill, 1975.

APPENDIX B

Two-Port Network Parameters

INTRODUCTION

At various points throughout the text, we make use of some of the different possible ways to characterize linear two-port networks. A summary of this topic is presented in this appendix.

CHARACTERIZATION OF LINEAR TWO-PORT NETWORKS

A two-port network (Fig. B.1) has four port variables: V_1, I_1, V_2, and I_2. If the two-port network is linear, we can use two of the variables as excitation variables and the other two as response variables. For instance, the network can be excited by a voltage V_1 at port 1 and a voltage V_2 at port 2, and the two currents, I_1 and I_2, can be measured to represent the network response. In this case V_1 and V_2 are independent variables and I_1 and I_2 are dependent variables, and the network operation can be described by the two equations

$$I_1 = y_{11}V_1 + y_{12}V_2 \quad \text{(B.1)}$$

$$I_2 = y_{21}V_1 + y_{22}V_2 \quad \text{(B.2)}$$

Here the four parameters y_{11}, y_{12}, y_{21}, and y_{22} are admittances, and their values completely characterize the linear two-port network.

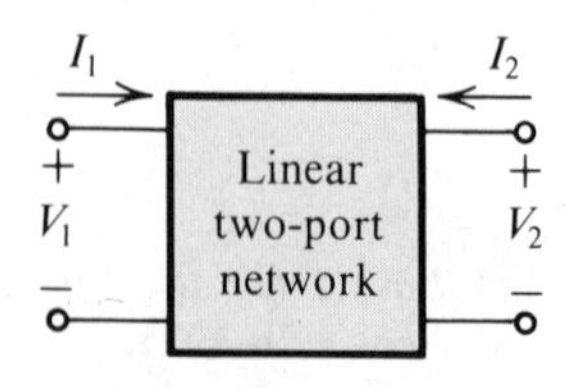

Fig. B.1 The reference directions of the four port variables in a linear two-port network.

Depending on which two of the four port variables are used to represent the network excitation, a different set of equations (and a correspondingly different set of parameters) is obtained for characterizing the network. In the following we shall present the four parameter sets commonly used in electronics.

y Parameters

The short-circuit admittance (or *y*-parameter) characterization is based on exciting the network by V_1 and V_2, as shown in Fig. B.2(a). The describing equations are Eqs. (B.1) and (B.2). The four admittance parameters can be defined according to their roles in Eqs. (B.1) and (B.2).

Specifically, from Eq. (B.1) we see that y_{11} is defined as

$$y_{11} = \left.\frac{I_1}{V_1}\right|_{V_2=0} \tag{B.3}$$

Thus y_{11} is the input admittance at port 1 with port 2 short-circuited. This definition is illustrated in Fig. B.2(b), which also provides a conceptual method for measuring the input short-circuit admittance y_{11}.

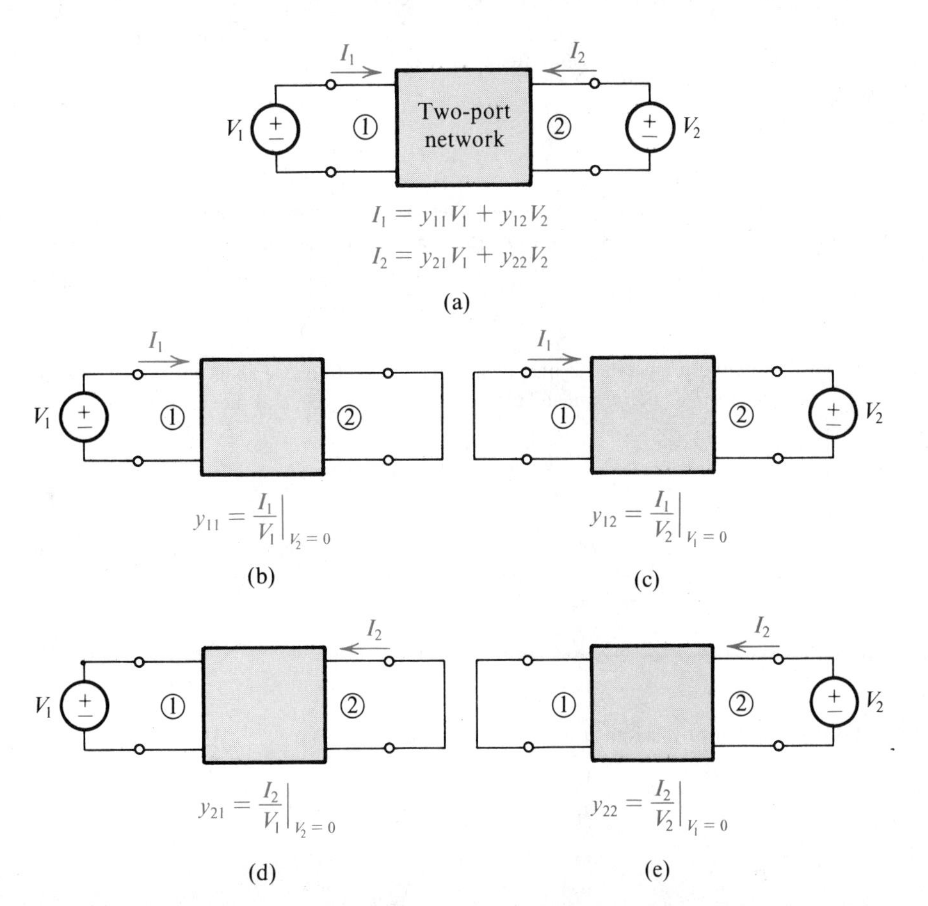

Fig. B.2 Definition and conceptual measurement circuits for *y* parameters.

The definition of y_{12} can be obtained from Eq. (B.1) as

$$y_{12} = \left.\frac{I_1}{V_2}\right|_{V_1=0} \tag{B.4}$$

Thus y_{12} represents transmission from port 2 to port 1. Since in amplifiers, port 1 represents the input port and port 2 the output port, y_{12} represents internal *feedback* in the network. Figure B.2(c) illustrates the definition and the method for measuring y_{12}.

The definition of y_{21} can be obtained from Eq. (B.2) as

$$y_{21} = \left.\frac{I_2}{V_1}\right|_{V_2=0} \tag{B.5}$$

Thus y_{21} represents transmission from port 1 to port 2. If port 1 is the input port and port 2 the output port of an amplifier, then y_{21} provides a measure of the forward gain or transmission. Figure B.2(d) illustrates the definition and the method for measuring y_{21}.

The parameter y_{22} can be defined, based on Eq. (B.2), as

$$y_{22} = \left.\frac{I_2}{V_2}\right|_{V_1=0} \tag{B.6}$$

Thus y_{22} is the admittance looking into port 2 while port 1 is short-circuited. For amplifiers, y_{22} is the output short-circuit admittance. Figure B.2(e) illustrates the definition and the method for measuring y_{22}.

z Parameters

The open-circuit impedance (or *z*-parameter) characterization of two-port networks is based on exciting the network by I_1 and I_2, as shown in Fig. B.3(a). The describing equations are

$$V_1 = z_{11}I_1 + z_{12}I_2 \tag{B.7}$$

$$V_2 = z_{21}I_1 + z_{22}I_2 \tag{B.8}$$

Owing to the duality between the *z*- and *y*-parameter characterizations we shall not give a detailed discussion of *z* parameters. The definition and method of measuring each of the four *z* parameters is given in Fig. B.3.

h Parameters

The hybrid (or *h*-parameter) characterization of two-port networks is based on exciting the network by I_1 and V_2, as shown in Fig. B.4(a) (note the reason behind the name *hybrid*). The describing equations are

$$V_1 = h_{11}I_1 + h_{12}V_2 \tag{B.9}$$

$$I_2 = h_{21}I_1 + h_{22}V_2 \tag{B.10}$$

from which the definition of the four *h* parameters can be obtained as

$$h_{11} = \left.\frac{V_1}{I_1}\right|_{V_2=0} \qquad h_{21} = \left.\frac{I_2}{I_1}\right|_{V_2=0}$$

$$h_{12} = \left.\frac{V_1}{V_2}\right|_{I_1=0} \qquad h_{22} = \left.\frac{I_2}{V_2}\right|_{I_1=0}$$

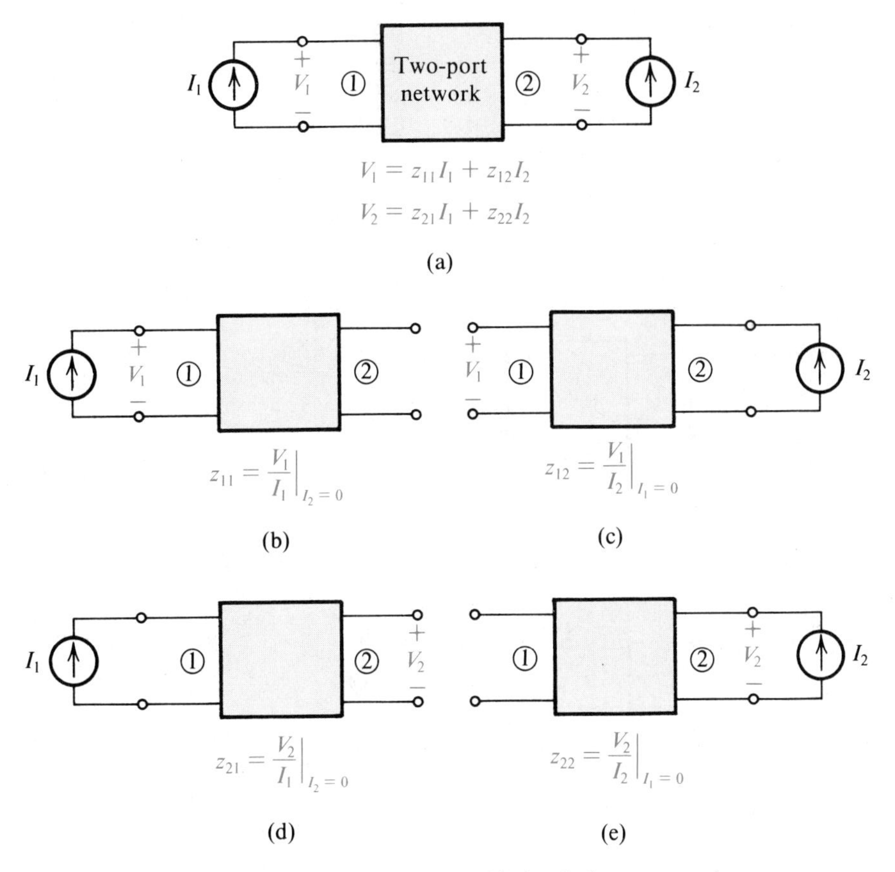

Fig. B.3 Definition and conceptual measurement circuits for *z* parameters.

Thus h_{11} is the input impedance at port 1 with port 2 short-circuited. The parameter h_{12} represents the reverse or feedback voltage ratio of the network, measured with the input port open-circuited. The forward-transmission parameter h_{21} represents the current gain of the network with the output port short-circuited; for this reason h_{21} is called the *short-circuit current gain*. Finally, h_{22} is the output admittance with the input port open-circuited.

The definitions and conceptual measuring setups of the h parameters are given in Fig. B.4.

g Parameters

The inverse-hybrid (or g-parameter) characterization of two-port networks is based on excitation of the network by V_1 and I_2, as shown in Fig. B.5(a). The describing equations are

$$I_1 = g_{11}V_1 + g_{12}I_2 \tag{B.11}$$

$$V_2 = g_{21}V_1 + g_{22}I_2 \tag{B.12}$$

The definitions and conceptual measuring setups are given in Fig. B.5.

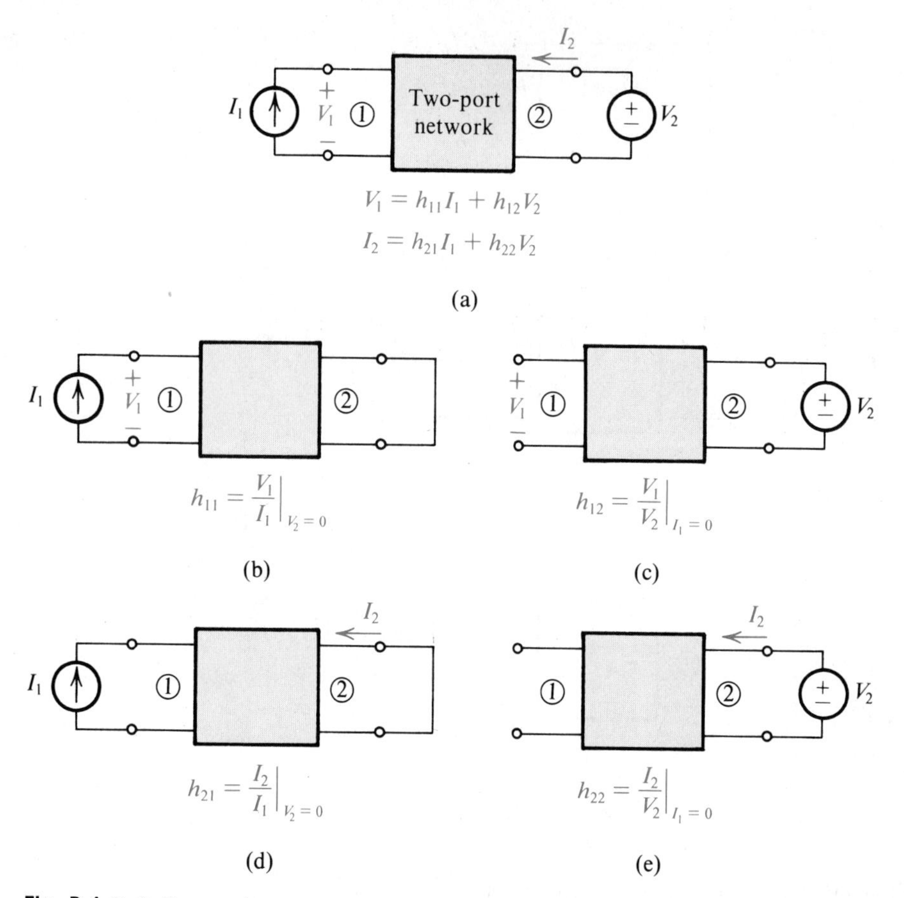

Fig. B.4 Definition and conceptual measurement circuits for h parameters.

Equivalent Circuit Representation

A two-port network can be represented by an equivalent circuit based on the set of parameters used for its characterization. Figure B.6 shows four possible equivalent circuits corresponding to the four parameter types discussed above. Each of these equivalent circuits is a direct pictorial representation of the corresponding two equations describing the network in terms of the particular parameter set.

Finally, it should be mentioned that other parameter sets exist for characterizing two-port networks, but these are not discussed or used in this book.

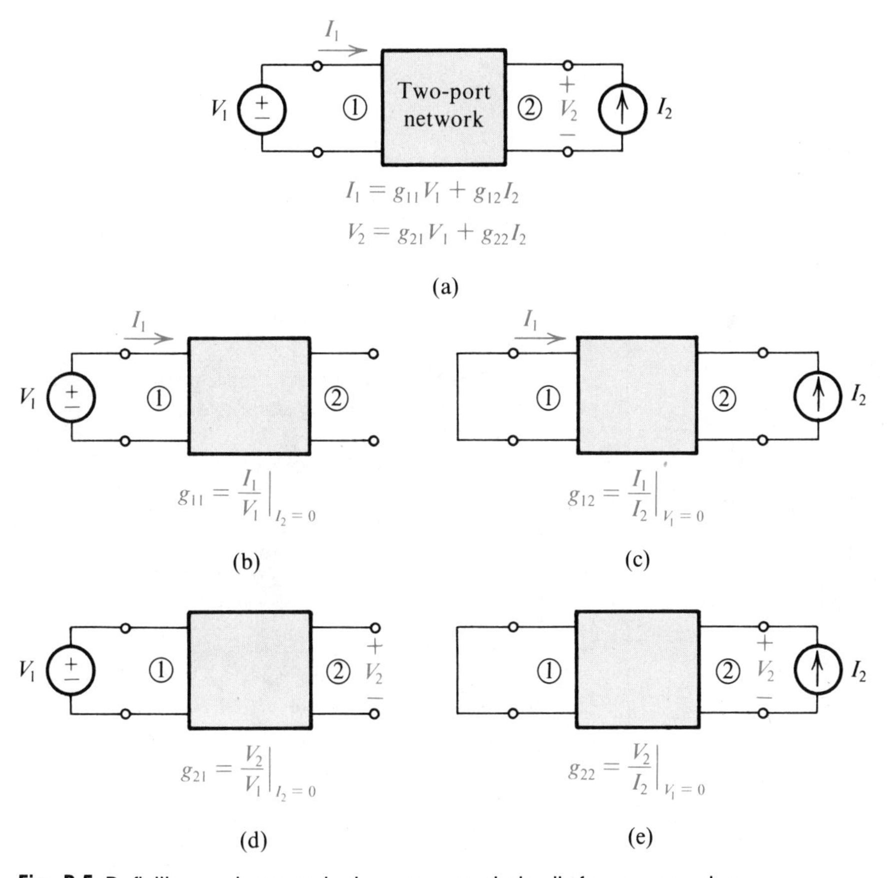

Fig. B.5 Definition and conceptual measurement circuits for g parameters.

Exercise B.1 Figure EB.1 shows the small-signal equivalent circuit model of a transistor. Calculate the values of the h parameters.

Ans. $h_{11} \simeq 2.6\ \text{k}\Omega$; $h_{12} \simeq 2.5 \times 10^{-4}$; $h_{21} \simeq 100$; $h_{22} \simeq 2 \times 10^{-5}\ \mho$

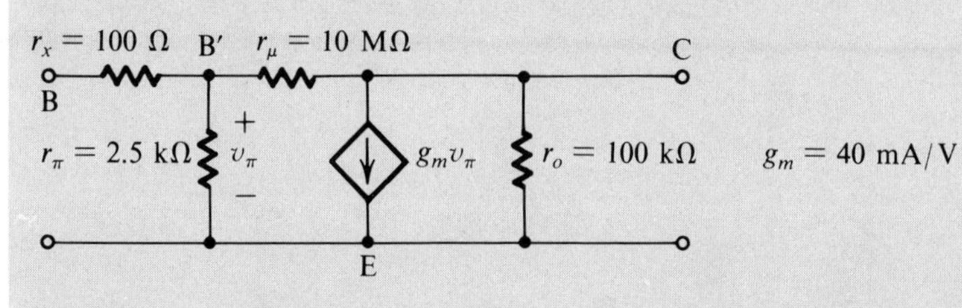

Fig. EB.1

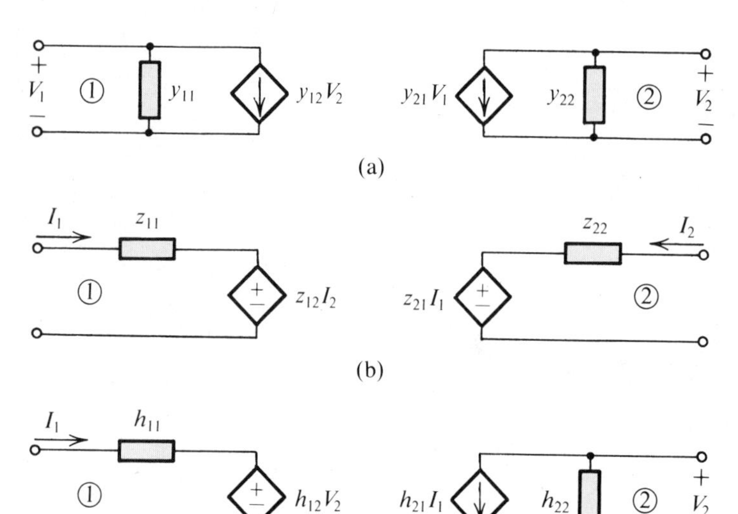

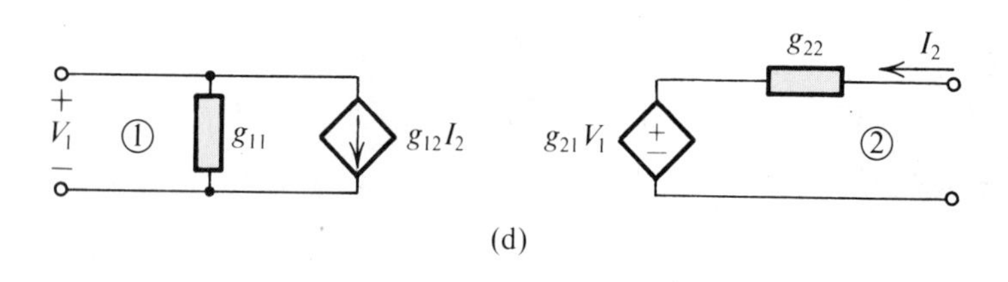

Fig. B.6 Equivalent circuits for **(a)** *y*, **(b)** *z*, **(c)** *h*, and **(d)** *g* parameters.

PROBLEMS

B.1 **(a)** An amplifier characterized by the *h*-parameter equivalent circuit of Fig. B.6(c) is fed with a source having a voltage V_s and a resistance R_s, and is loaded in a resistance R_L. Show that its voltage gain is given by

$$\frac{V_2}{V_s} = \frac{-h_{21}}{(h_{11} + R_s)(h_{22} + 1/R_L) - h_{12}h_{21}}$$

(b) Use the expression derived in (a) to find the voltage gain of the transistor in Exercise B.1 for $R_s = 1\ \text{k}\Omega$ and $R_L = 10\ \text{k}\Omega$.

B.2 The terminal properties of a two-port network are measured with the following results: With the output short-circuited and an input current of 0.01 mA, the output current is 1.0 mA and the input voltage is 26 mV. With the input open-circuited and a voltage of 10 V applied to the output, the current in the output is 0.2 mA and the voltage measured at the input is 2.5 mV. Find values for the *h* parameters of this network.

B.3 Figure PB.3 shows the high-frequency equivalent circuit of a BJT. (For simplicity, r_x has been omitted.) Find the *y* parameters.

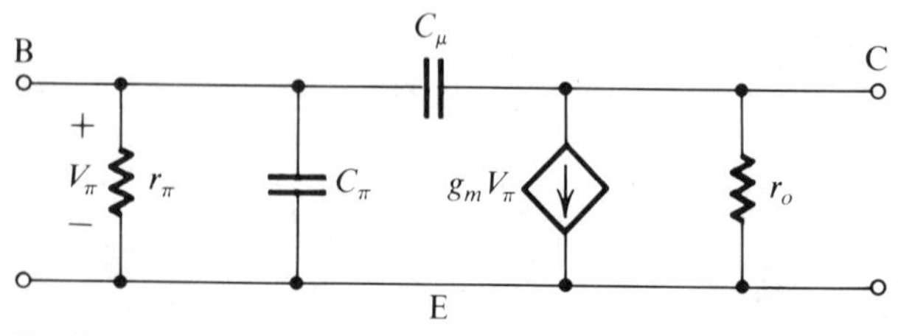

Fig. PB.3

APPENDIX 

Computer Aids for Electronic Circuit Design

INTRODUCTION

Earlier in the book we have indicated situations where computer aids may be appropriate in the process of circuit design. In this appendix we wish to expand upon these comments; describe more generally the role we see for such aids; and finally, by means of an example, show the benefits that can be derived from their use. Readers requiring a much more detailed exposition to computer-aided analysis and design of electronic circuits should refer to the SPICE supplement to this text.

THE CURRENT SITUATION

Computer aids for those working with electronic circuits, whether linear or digital, are becoming increasingly popular. Although design and analysis aids were once available only on relatively large machines on a time-sharing basis, the present trend is toward the provision of such tools at lower and lower cost on more accessible machines. Many such tools are already available (often in somewhat limited form) for use with personal computers of relatively modest cost. Development of computer aids is continuing at a rapid pace.

Computer aids for the process of development of electronic products take many forms: some relatively simple and low-cost but useful, such as printed-circuit-board (PCB) layout programs; and some quite sophisticated, such as parameterized circuit simulators, a class of programs that have not yet reached their potential even on large machines, and certainly not in a low-cost form. Since the number and variety of such aids is quite large, particularly if design is contemplated to the level of custom VLSI, we will focus here on one of the most important classes of programs—circuit simulation.

The simulator we have chosen for illustrative purposes is SPICE 2, an evolved (intermediate) circuit-simulator program developed at the Electronics Research Laboratory of the University of California, Berkeley, as part of a long-term research effort in electronic circuit design. While there are other, more sophisticated, more complete, and more parameterized proprietary products available (some of which have origins in earlier SPICE developments), none is as accessible as SPICE. As a result of the generous distribution policies of the Berkeley Group, SPICE 2 is installed in a large number of academic and industrial organizations throughout the world. Thus circuit simulation by SPICE 2 at modest (but not negligible) cost is a reality for everyone interested in electronics.

WHEN TO USE SPICE

Before we proceed with the details of SPICE data specification and application, it is important to review its place among the tools and techniques related to the study of electronics.

In spite of the fact that SPICE and similar tools are popularly associated with design, as implied by the popular acronym CAD (Computer-Aided Design), this connection is really somewhat indirect. At present SPICE and other such tools are simply circuit simulators, and thus they are actually tools for performing analysis. While we have already seen that analysis is an essential part of the design process, it is certainly not all there is, nor even the most important part. But most critically, as this text has demonstrated again and again, the process of design is not unique: There are many choices, each representing a particular balance in a cost–benefit trade-off.

Thus we have seen that at the very early stages of a design a very simple, perhaps approximate, analysis is appropriate. The challenge then is to discover rapidly whether a conjectured solution to a circuit's problem is likely to work at all. Thus, very simple and rapid analysis is the key to "good design," particularly at the stage of selection of circuit topology. For without some degree of this ability the designer is not likely to find the circuit whose intrinsic strengths match the specifications of the problem to be solved.

On the other hand, at later stages of the design process, a more thorough analysis is justified. Often it is the only means by which one of several competing designs may be selected; frequently it is the final step before a commitment to production is made. Thorough analysis is particularly important in the process of design for VLSI, since the potential loss of time and money resulting from a design fault can be very great indeed.

So a circuit simulator such as SPICE is best used very near the end of the design process, both as an independent verification of operation and as a means by which relatively subtle parameter optimization can be performed. In general it is *not* a technique for selection of a basic circuit topology.

Certainly at present there is no design aid capable of the conceptualization, winnowing, and selection of topologies which characterize the techniques used by the best circuit designers. Nor in a general sense is there likely to be. What *is* becoming available, however, are parameterized design aids in which, for a given topology, ranges of component values are tested and even optimized. In this regard, in order to avoid the combinatorial explosion (and corresponding cost of computer time) implied by simultaneous variation of many parameters, random sampling techniques are employed to "cover" the corresponding multidimensional parameter space. (In view of the implied wager placed on the adequacy of coverage, these techniques are referred to as Monte Carlo!) While certainly of great importance in the past for special purposes, and likely to be more generally applied in the future, such techniques are beyond our present need and will not be discussed further here.

SOME SPECIFICS ON SPICE

"SPICE is a general-purpose circuit simulation program for nonlinear dc, nonlinear transient, and linear ac analysis. Circuits may contain resistors, capacitors, inductors, mutual inductors, independent voltage and current sources, four types of dependent sources, transmission lines, and the four most common semiconductor devices: diodes, BJTs, JFETs, and MOSFETs.'' (See Vladimirescu, Newton, and Pederson, 1980.)

SPICE provides a hierarchy of built-in models for semiconductor devices. If model-parameter data are available, the more sophisticated models can be invoked; otherwise a simpler model is used by default. Thus for BJTs, although the integral-charge model of Gummel and Poon (which we have not studied in this book) is potentially available, the Ebers–Moll model is used in default mode. Correspondingly there is a hierarchy of three MOS models that can include second-order effects such as channel-length modulation (short-channel effects) and subthreshold conduction.

TYPES OF ANALYSIS

DC Analysis

The dc-analysis part of SPICE determines the dc operating point of the circuit with inductors shorted and capacitors open. It is performed automatically prior to transient analysis and small-signal analysis to determine the operating point of each device, and thus to establish appropriate device-model parameters.

The dc-analysis part of SPICE can also be used to generate dc transfer curves as an independent source is stepped over a range, and as corresponding dc output variables are calculated. Also available are dc small-signal sensitivities of specified output variables with respect to circuit parameters.

AC Small-Signal Analysis

The ac small-signal part of SPICE computes ac output variables over a specifiable range of frequencies on the basis of linearized small-signal models established from dc bias-point data. Typical transfer functions (gain, transimpedance, etc.) are available. Techniques are also available for evaluating noise and distortion characteristics.

Transient Analysis

The transient analysis part of SPICE computes transient output variables as a function of time over a specifiable interval. Such an analysis is useful for determining, for example, the step and pulse responses of logic-gate circuits.

Convergence

It is interesting and important to note that SPICE uses an iterative process to obtain both dc and transient solutions. Iteration is terminated when branch currents and node voltages converge to within a tolerance of 0.1% or 1 pA or 1 μV, whichever is larger. Although SPICE algorithms are found to be quite reliable, convergence is not guaranteed. In the event of failure to converge, the program terminates with no guarantee of relevance of data provided. This may be the case, for example, when dc analysis of regenerative circuits, such as flip-flops and Schmitt triggers, is attempted.

Input Format

The input format for SPICE is of the (relatively) free-format type in which labeled entries are separated by blanks, a comma, an equal sign, or a parenthesis. Extra spaces are ignored. Number fields may be either integer or floating point, using either decimal or scientific notation or mnemonic scale factors (for example, G for 10^9, MEG for 10^6, M for 10^{-3}, U for 10^{-6}, N for 10^{-9}, and so on). Letters following a number are ignored if not scale factors, as are other letters after scale factors. Thus 10, 10V, 10 VOLTS, and 10 HZ all represent the same values and M, MA, MSEC, and MMHOS represent the same scale factor.

CIRCUIT DESCRIPTION

The circuit to be analyzed is described to SPICE by a sequence of lines entered at a computer terminal, workstation, or personal computer. There are element lines which define the topology and element values, and control lines which set model parameters and measurement nodes. The first line must be a title and the last .END. In between, the order is arbitrary, except that continued lines must follow immediately.

Each element in the circuit is specified by an element line containing the element name, connected nodes, and electrical parameter value(s). The first letter of an element name denotes element type (for example, R for resistor), and the name can be from one to eight characters. Nodes are specified by nonnegative integers but need not be numbered sequentially. The datum node (ground) must be numbered 0. Every node must have two connections, except for MOSFET substrate nodes (and unterminated transmission lines).

Comments can be interspersed by prefacing them with an asterisk (*). Comments are *very important* for later understanding (by both the originating designer and other users) and should be used liberally.

Basic (but incomplete) specification formats are given in the table below:

Component	Name	Nodes and value		
Resistor	Rxxxxxxx	N+	N−	VALUE
Capacitor	Cxxxxxxx	N+	N−	VALUE
Inductor	Lxxxxxxx	N+	N−	VALUE
VCCS	Gxxxxxxx	N+	N−	NC + NC − VALUE
VCVS	Exxxxxxx	N+	N−	NC + NC − VALUE
CCCS	Fxxxxxxx	N+	N−	VNAM VALUE
CCVS	Hxxxxxxx	N+	N−	VNAM VALUE
Voltage source	Vxxxxxxx	N+	N−	QUAL
Current source	Ixxxxxxx	N+	N−	QUAL

where:

1. The component name begins with a particular letter as indicated, and is from one to eight alphanumeric characters long.
2. N+ and N− indicate the nodes connected, the first being positive (if that matters).
3. VALUE is in units of ohms, farads, henries, mhos, V/V, A/A, ohms, respectively, for the first seven components above.
4. NC+ and NC− are nodes across which the controlling voltage appears.
5. VNAM is the voltage source through which the controlling current flows.
6. QUAL is a set of qualifiers of the source, whether DC or transient (including pulse, sinusoid, exponential or piecewise-linear) with amplitudes and other qualifiers, or AC with magnitude and phase.

7. Note that a voltage source of zero volts is used in SPICE as a means to indicate the location of a current measurement.

While the previous elements are simple enough to allow for relatively rudimentary descriptions, often on a single line, that is not the case for semiconductor devices for which named separate definitions using .MODEL are required. Subsequent semiconductor device specifications refer to the named model definition (for example, MName). Thus for the four semiconductor devices, we have

Device	Name	Nodes and models		
Diode	Dxxxxxxx	N+	N−	MNAME AREA
BJT	Qxxxxxxx	NC	NB	NE NS MNAME AREA
JFET	Jxxxxxxx	ND	NG	NS MNAME AREA
MOSFET	Mxxxxxxx	ND	NG	NS NB MNAME L W

where

1. MNAME refers to a user- or system-defined device model.
2. AREA is an (optional) area-scaling factor.
3. NC, NB, NE, NS are nodes to which the collector, base, emitter, and (optionally) substrate are connected.
4. ND, NG, NS are nodes to which the drain, gate, and source are connected.
5. ND, NG, NS, NB are nodes to which the drain, gate, source, and substrate are connected.
6. L, W are the channel length and width in meters.

A description of model creation and the MODEL construct is beyond our present need and will not be included here. *Note again* that the preceding description of specification-line formats, while essentially correct, is not complete and is intended only to provide an introductory appreciation of the example that follows.

EXAMPLE C.1

The requirement is to provide a SPICE analysis of the two-stage CMOS op amp presented originally in Fig. 10.23, for two sets of device models: a simple one and a considerably more completely specified one characteristic of the 5-micron process historically available at Bell Northern Research (BNR). The simple model was presented originally in Example 10.2 where (in SPICE notation) VTO corresponds to V_t, KP to (μC_{OX}) and LAMBDA to $1/|V_A|$ with values of 1 V, 20 μA/V^2, 0.04 V^{-1}, and -1 V, 10 μA/V^2, 0.04 V^{-1}, for *n*- and *p*-channel devices, respectively. For the BNR model, the devices are similar, though not identical. In particular $|V_A|$ is higher, as we shall see when comparing gain and output-resistance results.

Solution Figure C.1 repeats Fig. 10.23 but with nodes labeled for SPICE input purposes. It also includes the W/L specification for each of the devices employed.

Figure C.2 is an exact version of the input presented via an interactive terminal to a version of SPICE, namely SPICE 2G.5 (10 Aug 81), installed in a DEC VAX 11/780 at the University of Toronto.

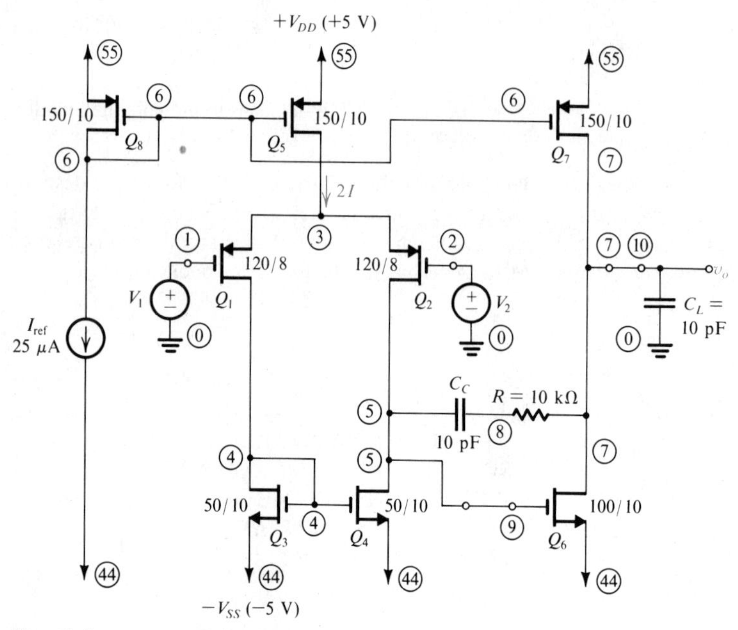

Fig. C.1 Two-stage CMOS op amp for Example C.1.

```
p-channel input cmos opamp     * opamp.spi *
.options gmin=1e-8 nomod
.width out=80
* transistor models
.model   mn   nmos (level=2  vto=1 kp=20u lambda=0.04)
* ciruit description
.model   mp   pmos (level=2  vto=-1 kp=10u lambda=0.04)
m1 4 1 3 55 mp l=8u w=120u
m2 5 2 3 55 mp l=8u w=120u
m3 4 4 44 44 mn l=10u w=50u
m4 5 4 44 44 mn l=10u w=50u
m5 3 6 55 55 mp l=10u w=150u
m6 7 9 44 44 mn l=10u w=100u
m7 7 6 55 55 mp l=10u w=150u
m8 6 6 55 55 mp l=10u w=150u
* frequency compensation
cc 5 8 10pF
rc 8 7 10k
*
* load
cl 10 0 10pf
*
*
vdd 55 0 dc 5
vss 44 0 dc -5
v2 2 0  dc 0 ac
v1 1 0 dc 0
vtmp 9 5 dc 0
ibias 6 44 dc 25u
*
* ammeters
vampo 7 10 dc 0
*
* initialize node voltages
.nodeset v(3)=1.2 v(4)=-3.2 v(5)=-3.2 v(6)=3.5 v(7)=0 v(8)=0
*
* small signal gain
.tf v(7) v2
*
.end
```

Fig. C.2 SPICE input for the circuit of Fig. C.1 utilizing the simple MOS device model.

The following list is a commentary on the input on nearly a line-by-line basis. Note that input is in lowercase letters; it will be reproduced in uppercase format in the output.

1. Line 1 is the title that will appear directly on the output.
2. Line 2 is an options-control list where gmin sets the value of minimum conductance to 1×10^{-8} mhos and nomod suppresses the output of model parameters.
3. Line 3 is a (printer) width control that includes 80-column output.
4. Line 4 is a comment.
5. Line 5 is a model specification for mn, an NMOS device using the MOS2 model with $V_t = 1$ V, $\mu C_{OX} = 20\ \mu A/V^2$, and $V_A = 1/\lambda = 25$ V.
6. Line 6 is a premature comment with a typo.
7. Line 7 is a model specification for mp, a PMOS device using the MOS2 model with $V_t = 1$ V, $\mu C_{OX} = 10\ \mu A/V^2$, and $V_A = 1/\lambda = 25$ V.
8. Lines 8 through 15 provide a device specification and wiring list for eight MOS devices where in terms of line 8, m1 is Q_1, connected with drain at node 4, gate at node 1, source at node 3, and substrate at node 55, for which model mp applies, with length of 8 microns and width of 120 microns.
9. Line 16 is a comment.
10. Line 17 describes a capacitor cc connected from node 5 to node 8 of value 10 pF. Note the typo F (only the p is used).
11. Line 18 describes a resistor rc connected from node 8 to node 7 of value 10 kΩ.
12. Line 19 is a (blank) comment.
13. Line 20 is a comment.
14. Line 21 describes a capacitor cl connected from node 10 to node 0 (ground) of value 10 pF.
15. Lines 22 and 23 are (blank) comments.
16. Line 24 describes an independent voltage source vdd connected between node 55 (and thus defining it) and node 0 (ground) of type dc and value +5 V.
17. Line 25 establishes −5 V on node 44.
18. Line 26 describes an independent voltage source v2 connected from node 2 [the (+)input] to node 0 (ground) with a dc component of 0 and an ac component of unit magnitude and 0 phase (by default).
19. Line 27 describes an independent voltage source v1 connected from node 1 [the (−)input] to node 0 (ground) of type dc and value 0.
20. Line 28 describes an independent voltage source vtmp connected from node 9 to node 5 of type dc and value 0 (to be used for current measurement).
21. Line 29 describes an independent current source ibias connected from node 6 to node 44 of type dc and value 25 μA.
22. Lines 29 and 30 are comments.

23. Line 31 describes an independent voltage source vampo connected from node 7 to node 10 of type dc and value 0 (to be used for current measurement).
24. Lines 32 and 33 are comments.
25. Line 34 is an initializing control to establish voltages at nodes 3, 4, 5, 6, 7, 8 prior to iteration.
26. Lines 35 and 36 are comments.
27. Line 37 is an output-control which requests a small-signal analysis with output at node 7 and input at node 2.
28. Line 38 is a comment.
29. Line 39 is an end-control signifying the end of the process in line 1.

Figure C.3 provides the output results. Note that the input data are first reproduced complete with typos, then a small-signal bias solution is presented. This includes dc node voltages and supply currents with total power dissipation. Then operating-point information for each of the transistors is presented. Note that in the latter part of the table all capacitances except C_{gs} are zero, a consequence of the simple model specified. Finally note that the gain and input and output resistances are provided at the end.

For the complex MOS model, Fig. C.4 shows the input model data (which replaces lines 5 and 7 of Fig. C.2). Figure C.5 provides the output corresponding to Fig. C.3 for the simple model. Note here the relative completeness of the operating-point information, including more nonzero capacitances. Finally note that the computer CPU time taken has increased to 4.32 units.

Finally, by way of a summary of all we now know about this op amp, Table C.1 presents a tabulation of results. While a lot can be said about these results, only a few points will be made here. Possibly the most reassuring is that the results are quite similar. Probably the most interesting is that the hand-calculated results lie somewhere between the extremes of SPICE using the simple model and the more complex one. That the results are strongly dependent on the detail of the diverse models is noteworthy as well.

Table C.1 COMPARISON OF THE RESULTS OF THE ANALYSIS OF THE CIRCUIT OF FIG. C.1 BY HAND AND USING SPICE WITH TWO DIFFERENT MODELS

	Units	Hand Calculation	SPICE 2	
			Simple	Complex
First stage gain	V/V	−62.5	−55.1	−50.4
First stage output resistance	kΩ	—	779	714
Second stage gain	V/V	−50.0	−38.9	−68.2
Second stage output resistance	kΩ	—	321	596
Open-loop gain	V/V	3125	2145	3435
Output resistance	kΩ	—	321	596
Input resistance	Ω	—	1×10^{20}	1×10^{20}
Unity-gain frequency	MHz	1.0	0.972	0.955
Phase margin	degrees	57.8	68.3	63.4
Positive slew rate	V/μs	—	1.37	1.25
Negative slew rate	V/μs	−2.5	−3.07	−2.90

```
1*******07/04/86 ********  SPICE 2G.5 (10AUG81)  ********00:06:51*****

0P-CHANNEL INPUT CMOS OPAMP    * OPAMP.SPI *

0****      INPUT LISTING                    TEMPERATURE =    27.000 DEG C

0**********************************************************************


 .OPTIONS GMIN=1E-8 NOMOD
 .WIDTH OUT=80
 * TRANSISTOR MODELS
 .MODEL  MN  NMOS (LEVEL=2  VTO=1 KP=20U LAMBDA=0.04)
 * CIRUIT DESCRIPTION
 .MODEL  MP  PMOS (LEVEL=2  VTO=-1 KP=10U LAMBDA=0.04)
 M1 4 1 3 55 MP L=8U W=120U
 M2 5 2 3 55 MP L=8U W=120U
 M3 4 4 44 44 MN L=10U W=50U
 M4 5 4 44 44 MN L=10U W=50U
 M5 3 6 55 55 MP L=10U W=150U
 M6 7 9 44 44 MN L=10U W=100U
 M7 7 6 55 55 MP L=10U W=150U
 M8 6 6 55 55 MP L=10U W=150U
 * FREQUENCY COMPENSATION
 CC 5 8 10PF
 RC 8 7 10K
 *
 * LOAD
 CL 10 0 10PF
 *
 *
 VDD 55 0 DC 5
 VSS 44 0 DC -5
 V2 2 0  DC 0 AC
 V1 1 0 DC 0
 VTMP 9 5 DC 0
 IBIAS 6 44 DC 25U
 *
 * AMMETERS
 VAMPO 7 10 DC 0
 *
 * INITIALIZE NODE VOLTAGES
 .NODESET V(3)=1.2 V(4)=-3.2 V(5)=-3.2 V(6)=3.5 V(7)=0 V(8)=0
 *
 * SMALL SIGNAL GAIN
 .TF V(7) V2
 *
 .END
1*******07/04/86 ********  SPICE 2G.5 (10AUG81)  ********00:06:51*****

0P-CHANNEL INPUT CMOS OPAMP    * OPAMP.SPI *

0****      SMALL SIGNAL BIAS SOLUTION       TEMPERATURE =    27.000 DEG C

0**********************************************************************

 NODE   VOLTAGE      NODE   VOLTAGE      NODE   VOLTAGE      NODE   VOLTAGE


 (  1)     .0000     (  2)     .0000     (  3)    1.3841     (  4)   -3.4904

 (  5)   -3.4904     (  6)    3.4411     (  7)   -1.0221     (  8)   -1.0221

 (  9)   -3.4904     ( 10)   -1.0221     ( 44)   -5.0000     ( 55)    5.0000
```

Fig. C.3 SPICE output obtained in response to the input of Fig. C.2.

VOLTAGE SOURCE CURRENTS

NAME	CURRENT
VDD	-8.359e-05
VSS	8.359e-05
V2	0.000e+00
V1	0.000e+00
VTMP	0.000e+00
VAMPO	0.000e+00

TOTAL POWER DISSIPATION   6.25e-04 WATTS

1*******07/04/86 ******** SPICE 2G.5 (10AUG81)   ********00:06:51*****

0P-CHANNEL INPUT CMOS OPAMP   * OPAMP.SPI *

0****   OPERATING POINT INFORMATION   TEMPERATURE = 27.000 DEG C

0***

0

0**** MOSFETS

0	M1	M2	M3	M4	M5	M6	M7
0MODEL	MP	MP	MN	MN	MP	MN	MP
ID	-1.37e-05	-1.37e-05	1.38e-05	1.38e-05	-2.74e-05	3.09e-05	-3.09e-05
VGS	-1.384	-1.384	1.510	1.510	-1.559	1.510	-1.559
VDS	-4.875	-4.875	1.510	1.510	-3.616	3.978	-6.022
VBS	3.616	3.616	.000	.000	.000	.000	.000
VTH	-1.000	-1.000	1.000	1.000	-1.000	1.000	-1.000
VDSAT	-.384	-.384	.510	.510	-.559	.510	-.559
GM	7.16e-05	7.16e-05	5.42e-05	5.42e-05	9.80e-05	1.21e-04	1.10e-04
GDS	6.83e-07	6.83e-07	5.88e-07	5.88e-07	1.28e-06	1.47e-06	1.63e-06
GMB	0.00e+00	0.00e+00	0.00e+00	0.00e+00	0.00e+00	0.00e+00	0.00e+00
CBD	0.00e+00	0.00e+00	0.00e+00	0.00e+00	0.00e+00	0.00e+00	0.00e+00
CBS	0.00e+00	0.00e+00	0.00e+00	0.00e+00	0.00e+00	0.00e+00	0.00e+00
CGSOVL	0.00e+00	0.00e+00	0.00e+00	0.00e+00	0.00e+00	0.00e+00	0.00e+00
CGDOVL	0.00e+00	0.00e+00	0.00e+00	0.00e+00	0.00e+00	0.00e+00	0.00e+00
CGBOVL	0.00e+00	0.00e+00	0.00e+00	0.00e+00	0.00e+00	0.00e+00	0.00e+00
CGS	2.21e-13	2.21e-13	1.15e-13	1.15e-13	3.45e-13	2.30e-13	3.45e-13
CGD	0.00e+00	0.00e+00	0.00e+00	0.00e+00	0.00e+00	0.00e+00	0.00e+00
CGB	0.00e+00	0.00e+00	0.00e+00	0.00e+00	0.00e+00	0.00e+00	0.00e+00

0	M8
0MODEL	MP
ID	-2.50e-05
VGS	-1.559
VDS	-1.559
VBS	.000
VTH	-1.000
VDSAT	-.559
GM	8.94e-05
GDS	1.07e-06
GMB	0.00e+00
CBD	0.00e+00
CBS	0.00e+00
CGSOVL	0.00e+00
CGDOVL	0.00e+00
CGBOVL	0.00e+00
CGS	3.45e-13
CGD	0.00e+00
CGB	0.00e+00

0****   SMALL-SIGNAL CHARACTERISTICS

0   V(7)/V2   = 2.145e+03

0   INPUT RESISTANCE AT V2   = 1.000e+20

0   OUTPUT RESISTANCE AT V(7)   = 3.210e+05

0

JOB CONCLUDED

0   TOTAL JOB TIME   2.57

Fig. C.3 *(continued)*

```
* transistor models
*
.model  mn  nmos  (level=2  vto=1  nsub=1e16  tox=8.5e-8  uo=750
+                  cgso=4e-10  cgdo=4e-10  cgbo=2e-10
+                  ucrit=5e4  uexp=.14  utra=0  vmax=5e4  rsh=15
+                  cj=4e-4  mj=2  pb=.7  cjsw=8e-10  mjsw=2
+                  js=1e-6  xj=1u  ld=.7u)
.model  mp  pmos  (level=2  vto=-1  nsub=2e15  tox=8.5e-8  uo=250
+                  cgso=4e-10  cgdo=4e-10  cgbo=2e-10
+                  ucrit=1e4  uexp=.03  utra=0  vmax=3e4  rsh=75
+                  cj=1.8e-4  mj=2  pb=.7  cjsw=6e-10  mjsw=2
+                  js=1e-6  xj=.9u  ld=.6u)
*
```

Fig. C.4 Input model data (to replace lines 5 and 7 of Fig. C.2) for the more elaborate MOS model.

```
 * SMALL SIGNAL GAIN
 .TF V(7) V2
 *
 .END
1*******07/04/86 ********  SPICE 2G.5 (10AUG81)  ********00:06:04*****

0P-CHANNEL INPUT CMOS OPAMP   * OPAMP.SPI *

0****     SMALL SIGNAL BIAS SOLUTION        TEMPERATURE =   27.000 DEG C

0***********************************************************************

  NODE   VOLTAGE     NODE   VOLTAGE     NODE   VOLTAGE     NODE   VOLTAGE

 (  1)     .0000    (  2)     .0000    (  3)    1.8834    (  4)   -3.5384

 (  5)   -3.5384    (  6)    3.4244    (  7)    -.9032    (  8)    -.9032

 (  9)   -3.5384    ( 10)    -.9032    ( 44)   -5.0000    ( 55)    5.0000



     VOLTAGE SOURCE CURRENTS

     NAME        CURRENT

     VDD       -8.173e-05

     VSS        8.173e-05

     V2         0.000e+00

     V1         0.000e+00

     VTMP       1.059e-22

     VAMPO      0.000e+00

     TOTAL POWER DISSIPATION   6.07e-04  WATTS
1*******07/04/86 ********  SPICE 2G.5 (10AUG81)  ********00:06:04*****

0P-CHANNEL INPUT CMOS OPAMP   * OPAMP.SPI *

0****     OPERATING POINT INFORMATION       TEMPERATURE =   27.000 DEG C

0***********************************************************************
```

Fig. C.5 SPICE output for the circuit in Example C.1 utilizing the more elaborate MOS device model specified in Fig. C.4.

```
0**** MOSFETS

0              M1          M2          M3          M4          M5          M6          M7
0MODEL         MP          MP          MN          MN          MP          MN          MP
 ID     -1.35e-05  -1.35e-05   1.35e-05   1.35e-05  -2.69e-05   2.96e-05  -2.96e-05
 VGS       -1.883      -1.883      1.462      1.462     -1.576      1.462     -1.576
 VDS       -5.422      -5.422      1.462      1.462     -3.117      4.097     -5.903
 VBS        3.117       3.117       .000       .000       .000       .000       .000
 VTH       -1.515      -1.515       .952       .952      -.959       .941      -.952
 VDSAT      -.318       -.318       .284       .284      -.456       .292      -.462
 GM      7.18e-05    7.18e-05   5.37e-05   5.37e-05   8.64e-05   1.15e-04   9.43e-05
 GDS     7.87e-07    7.87e-07   6.08e-07   6.08e-07   1.09e-06   7.76e-07   8.92e-07
 GMB     8.55e-06    8.55e-06   3.90e-05   3.90e-05   2.63e-05   8.25e-05   2.83e-05
 CBD     0.00e+00    0.00e+00   0.00e+00   0.00e+00   0.00e+00   0.00e+00   0.00e+00
 CBS     0.00e+00    0.00e+00   0.00e+00   0.00e+00   0.00e+00   0.00e+00   0.00e+00
 CGSOVL  4.80e-14    4.80e-14   2.00e-14   2.00e-14   6.00e-14   4.00e-14   6.00e-14
 CGDOVL  4.80e-14    4.80e-14   2.00e-14   2.00e-14   6.00e-14   4.00e-14   6.00e-14
 CGBOVL  1.36e-15    1.36e-15   1.72e-15   1.72e-15   1.76e-15   1.72e-15   1.76e-15
 CGS     2.21e-13    2.21e-13   1.16e-13   1.16e-13   3.58e-13   2.33e-13   3.58e-13
 CGD     0.00e+00    0.00e+00   0.00e+00   0.00e+00   0.00e+00   0.00e+00   0.00e+00
 CGB     0.00e+00    0.00e+00   0.00e+00   0.00e+00   0.00e+00   0.00e+00   0.00e+00

0              M8
0MODEL         MP
 ID     -2.50e-05
 VGS       -1.576
 VDS       -1.576
 VBS         .000
 VTH        -.964
 VDSAT      -.451
 GM      8.08e-05
 GDS     1.42e-06
 GMB     2.48e-05
 CBD     0.00e+00
 CBS     0.00e+00
 CGSOVL  6.00e-14
 CGDOVL  6.00e-14
 CGBOVL  1.76e-15
 CGS     3.58e-13
 CGD     0.00e+00
 CGB     0.00e+00

0****       SMALL-SIGNAL CHARACTERISTICS

0      V(7)/V2                                   =   3.435e+03
0      INPUT RESISTANCE AT V2                    =   1.000e+20
0      OUTPUT RESISTANCE AT V(7)                 =   5.960e+05
0
          JOB CONCLUDED
0           TOTAL JOB TIME              4.32
```

Fig. C.5 *(continued)*

Now, having analyzed the op amp by several means, we are in a particularly good position to find anomalous behavior, and possibly to correct for it in a redesign. We see, for example, that the slew rate is somewhat less than desired and asymmetric as well. Motivated by this result we consider the slew rate computation and realize that SR+ is essentially $I_{D7}/C_C + C_L$ where C_C and C_L are about the same size, and in particular that the result is a slew rate of about

$$\frac{I}{C} = \frac{30 \times 10^{-6}}{(10 + 10) \times 10^{-12}}$$

or 1.5 V/μs, about as the SPICE results show.

To improve this, our only option is to raise I_{D7}, by perhaps a factor of 2, expecting that SR$-$, controlled primarily by I_{D1}, will remain the same. We note, however regrettably, that the gain of stage 2 will reduce somewhat. Our logical next step is to evaluate this by hand computation and then select a modified design, which we would check by SPICE analysis.

CONCLUDING REMARKS

SPICE is an extremely valuable circuit-simulation program. But it does not provide an alternative to hand analysis; it is usually used at a later stage in the design process to help in design optimization.

We hope that the reader has access to SPICE. If so, considerable benefit can be gained by using it to analyze some of the circuits described in the examples, exercises, and problems in this text. To aid in this process, an ancillary for this book is available. It contains a detailed description of when and how to use SPICE. It also contains examples and problems on the use of SPICE which are keyed to the sections of this book.

BIBLIOGRAPHY

A. Vladimirescu, A. R. Newton, and D. O. Pederson, "SPICE Version 2G.1 User's Guide," Berkeley: University of California, Department of Electrical Engineering and Computer Science, 1980.

APPENDIX 

Some Useful Network Theorems

INTRODUCTION

In this appendix we review three network theorems that are useful in simplifying the analysis of electronic circuits: Thévenin's theorem, Norton's theorem, and the source-absorption theorem.

THÉVENIN'S THEOREM

Thévenin's theorem is used to represent a part of a network by a voltage source V_t and a series impedance Z_t, as shown in Fig. D.1. Figure D.1(a) shows a network divided into two parts, A and B. In Fig. D.1(b) part A of the network has been replaced by its Thévenin equivalent: a voltage source V_t and a series impedance Z_t. Figure D.1(c) illustrates how V_t is to be determined: Simply open-circuit the two terminals of network A and measure (or calculate) the voltage that appears between these two terminals. To determine Z_t we reduce all external (that is, independent) sources in network A to zero by short-circuiting voltage sources and open-circuiting current sources. The impedance Z_t will be equal to the input impedance of network A after this reduction has been performed, as illustrated in Fig. D.1(d).

NORTON'S THEOREM

Norton's theorem is the *dual* of Thévenin's theorem. It is used to represent a part of a network by a current source I_n and a parallel impedance Z_n, as shown in Fig. D.2. Figure D.2(a) shows a network divided into two parts, A and B. In Fig. D.2(b) part A has been replaced by its Norton's equivalent: a current source I_n and a parallel impedance Z_n. The Norton's current source I_n can be measured (or calculated) as shown in Fig. D.2(c). The terminals of the network being reduced (network A) are shorted, and the current I_n will be equal simply to the short-circuit current. To

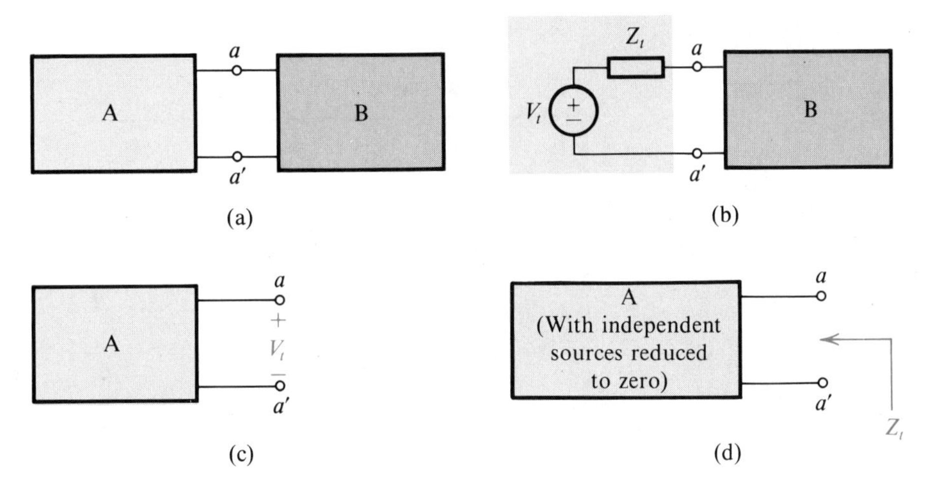

Fig. D.1 Thévenin's theorem.

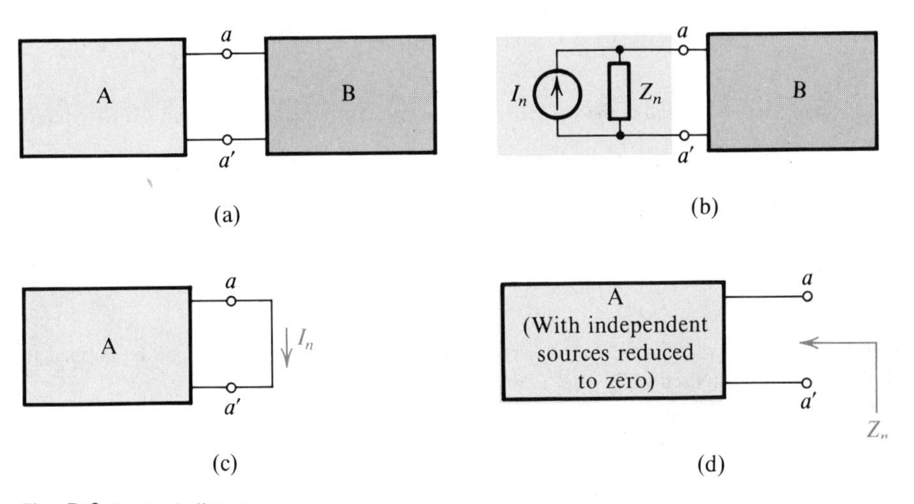

Fig. D.2 Norton's theorem.

determine the impedance Z_n we first reduce the external excitation in network A to zero, that is, short-circuit independent voltage sources and open-circuit independent current sources. The impedance Z_n will be equal to the input impedance of network A after this source-elimination process has taken place. Thus the Norton impedance Z_n is equal to the Thévenin impedance Z_t. Finally, note that $I_n = V_t/Z$, where $Z = Z_n = Z_t$.

EXAMPLE D.1

Figure D.3(a) shows a bipolar junction transistor circuit. The transistor is a three-terminal device with the terminals labeled E (emitter), B (base), and C (collector). As shown, the base is connected to the dc power supply V^+ via the voltage divider composed of R_1 and R_2. The collector is connected to the dc supply V^+ through R_3 and to ground through R_4. To simplify the analysis we wish to reduce the circuit through application of Thévenin's theorem.

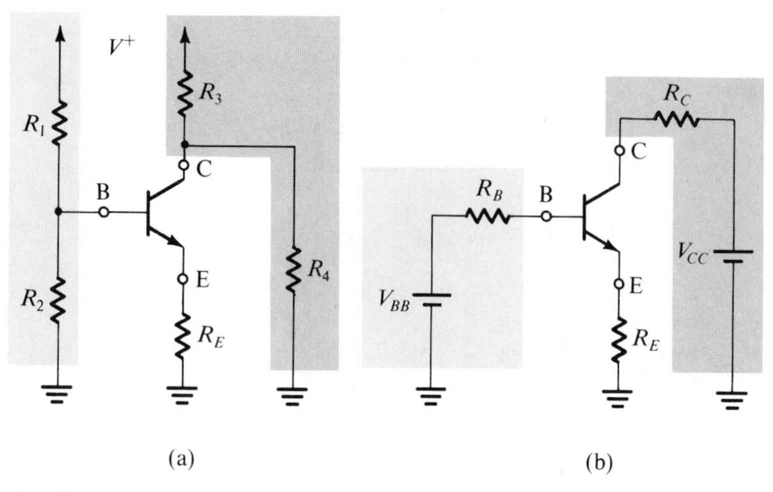

Fig. D.3 Thévenin's theorem applied to simplify the circuit of **(a)** to that in **(b)**. (See Example D.1.)

Solution Thévenin's theorem can be used at the base side to reduce the network composed of V^+, R_1, and R_2 to a dc voltage source V_{BB},

$$V_{BB} = V^+ \frac{R_2}{R_1 + R_2}$$

and a resistance R_B,

$$R_B = R_1 // R_2$$

where // denotes ''in parallel with.'' At the collector side Thévenin's theorem can be applied to reduce the network composed of V^+, R_3, and R_4 to a dc voltage source V_{CC},

$$V_{CC} = V^+ \frac{R_4}{R_3 + R_4}$$

and a resistance R_C,

$$R_C = R_3 // R_4$$

The reduced circuit is shown in Fig. D.3(b).

SOURCE-ABSORPTION THEOREM

Consider the situation shown in Fig. D.4. In the course of analyzing a network we find a controlled current-source I_x appearing between two nodes whose voltage difference is the controlling voltage V_x. That is, $I_x = g_m V_x$ where g_m is a conductance. We can replace this controlled source by an impedance $Z_x = V_x / I_x = 1/g_m$, as shown in Fig. D.4, because the current drawn by this impedance will be equal to the current of the controlled source that we have replaced.

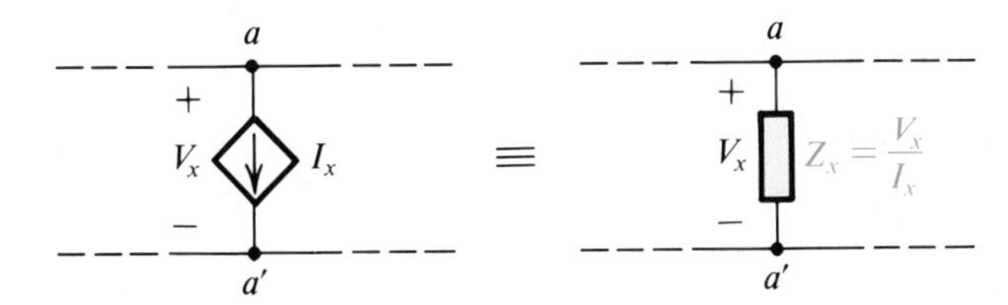

Fig. D.4 The source-absorption theorem.

EXAMPLE D.2

Figure D.5(a) shows the small-signal equivalent circuit model of a transistor. We want to find the resistance R_{in} "looking into" the emitter terminal E—that is, between the emitter and ground—with the base B and collector C grounded.

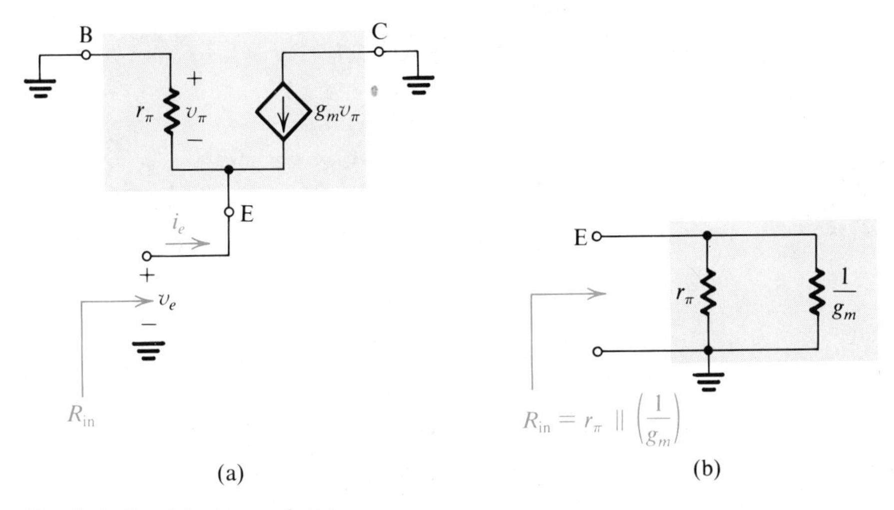

Fig. D.5 Circuit for Example D.2.

Solution From Fig. D.5(a) we see that the voltage v_π will be equal to $-v_e$. Thus looking between E and ground we see a resistance r_π in parallel with a current source drawing a current $g_m v_e$ away from terminal E. The latter source can be replaced by a resistance $(1/g_m)$, resulting in the input resistance R_{in} given by

$$R_{in} = r_\pi // (1/g_m)$$

as illustrated in Fig. D.5(b)

Exercises D.1 A source is measured and found to have a 10-V open-circuit voltage and to provide 1 mA into a short circuit. Calculate its Thévenin and Norton equivalent source parameters.

Ans. $V_t = 10$ V; $Z_t = Z_n = 10$ kΩ; $I_n = 1$ mA

D.2 In the circuit shown in Fig. ED.2 the diode has a voltage drop $V_D \simeq 0.7$ V. Use Thévenin's theorem to simplify the circuit and hence calculate the diode current I_D.

Ans. 1 mA

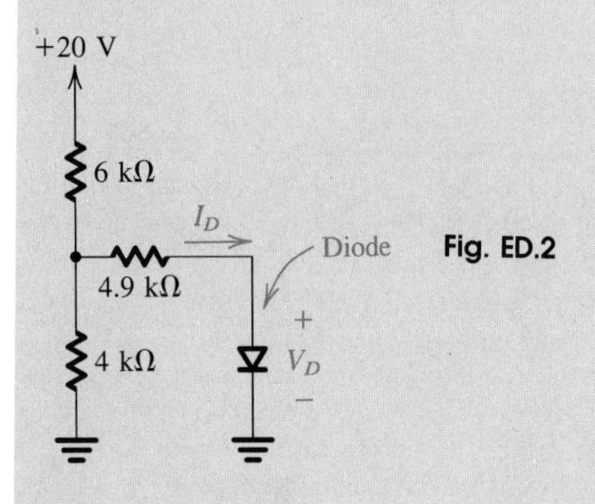

Fig. ED.2

D.3 The two-terminal device M in the circuit of Fig. ED.3 has a current $I_M \simeq 1$ mA independent of the voltage V_M across it. Use Norton's theorem to simplify the circuit and hence calculate the voltage V_M.

Ans. 5 V

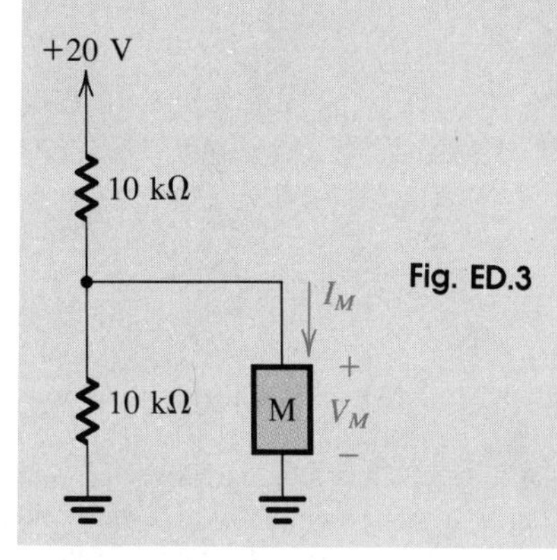

Fig. ED.3

PROBLEMS

D.1 Consider the Thévenin equivalent circuit characterized by V_t and Z_t. Find the open-circuit voltage V_{oc} and the short-circuit current (that is, the current that flows when the terminals are shorted together) I_{sc}. Express Z_t in terms of V_{oc} and I_{sc}.

D.2 Repeat Problem D.1 for a Norton equivalent characterized by I_n and Z_n.

D.3 A voltage divider consists of a 9-kΩ resistor connected to +10 V and a resistor of 1 kΩ connected to ground. What is the Thévenin equivalent of this voltage divider? What output voltage results if it is loaded with 1 kΩ? Calculate this two ways: directly and using your Thévenin equivalent.

D.4 Find the output voltage and output resistance of the

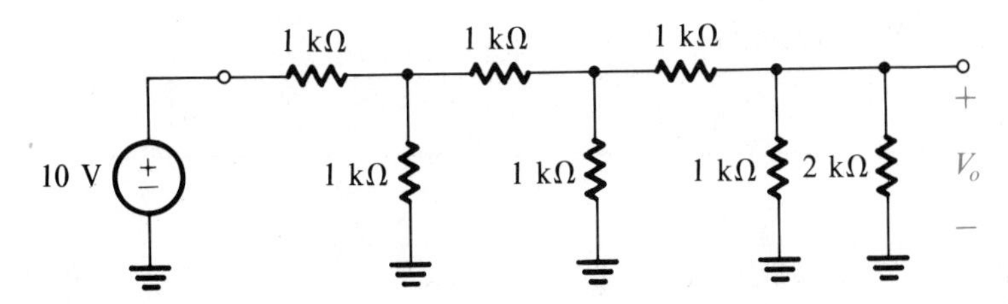

Fig. PD.4

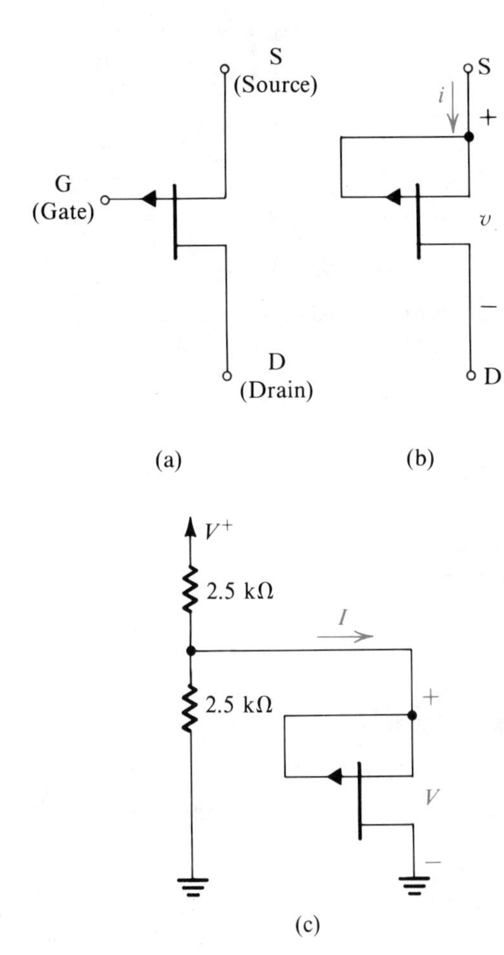

Fig. PD.6

circuit shown in Fig. PD.4 by considering a succession of Thévenin equivalent circuits.

D.5 Repeat Example D.2 with a resistance R_B connected between B and ground in Fig. D.5 (that is, rather than grounding the base B as indicated in Fig. D.5).

***D.6** Figure PD.6(a) shows the circuit symbol of the *p*-channel junction field-effect transistor (JFET), which is studied in Chapter 5. As indicated, the JFET has three terminals. When the gate terminal G is connected to the source terminal S, the two-terminal device shown in Fig. PD.6(b) is obtained. Its i–v characteristic is given by

$$i = I_{DSS}\left[2\frac{v}{V_P} - \left(\frac{v}{V_P}\right)^2\right] \qquad \text{for } v \leq V_P$$

$$i = I_{DSS} \qquad \text{for } v \geq V_P$$

where I_{DSS} and V_P are constants for the particular JFET. Now consider the circuit shown in Fig. PD.6(c) and let $V_P = 2$ V and $I_{DSS} = 2$ mA. For $V^+ = 10$ V show that the JFET is operating in the constant-current mode and find the voltage across it. What is the minimum value of V^+ for which this mode of operation is maintained? For $V^+ = 2$ V find the values of I and V.

APPENDIX 

Single-Time-Constant Circuits

INTRODUCTION

Single-time-constant (STC) circuits are those circuits that are composed of, or can be reduced to, one reactive component (inductance or capacitance) and one resistance. An STC circuit formed of an inductance L and a resistance R has a time constant $\tau = L/R$. The time constant τ of an STC circuit composed of a capacitance C and a resistance R is given by $\tau = CR$.

Although STC circuits are quite simple, they play an important role in the design and analysis of linear and digital circuits. For instance, the analysis of an amplifier circuit can usually be reduced to the analysis of one or more STC circuits. For this reason we will review in this appendix the process of evaluating the response of STC circuits to sinusoidal and other input signals such as step and pulse waveforms. The latter signal waveforms are encountered in some amplifier applications but are more important in switching circuits, including digital circuits.

EVALUATING THE TIME CONSTANT

The first step in the analysis of an STC circuit is to evaluate its time constant τ.

EXAMPLE E.1 Reduce the circuit in Fig. E.1(a) to an STC circuit, and find its time constant.

Solution The reduction process is illustrated in Fig. E.1 and consists of repeated applications of Thévenin's theorem. The final circuit is shown in Fig. E.1(c), from which we obtain the time constant as

$$\tau = C\{R_4//[R_3 + (R_1//R_2)]\}$$

Rapid Evaluation of τ

In many instances it will be important to be able to evaluate rapidly the time constant τ of a given STC circuit. A simple method for accomplishing this goal consists first of reducing the excitation to zero; that is, if the excitation is by a voltage source, short it, and if by a current source, open it. Then if the circuit has one reactive component and a number of resistances, ''grab hold'' of the two

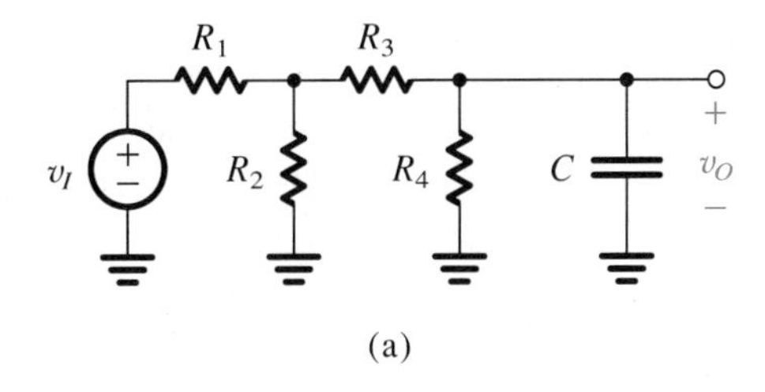

(a)

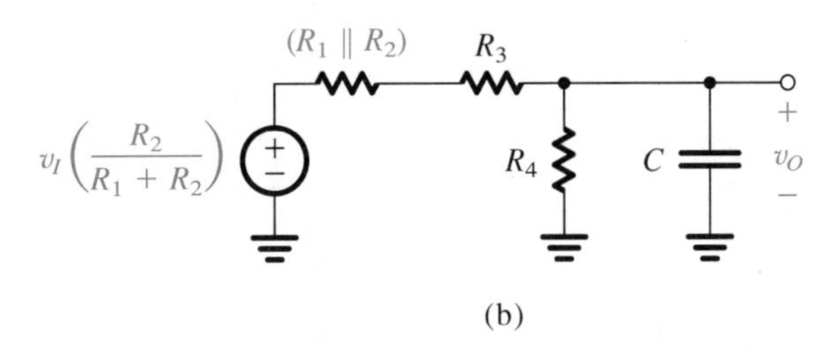

(b)

$R_4 \parallel [R_3 + (R_1 \parallel R_2)]$

$v_I \left(\frac{R_2}{R_1 + R_2}\right)\left(\frac{R_4}{R_4 + R_3 + (R_1 \parallel R_2)}\right)$

C v_O

(c)

Fig. E.1 The reduction of the circuit in **(a)** to the STC circuit in **(c)** through the repeated application of Thévenin's theorem.

terminals of the reactive component (capacitance or inductance) and find the equivalent resistance R_{eq} seen by the component. The time constant is then either L/R_{eq} or CR_{eq}. As an example, in the circuit of Fig. E.1(a) we find that the capacitor C ''sees'' a resistance R_4 in parallel with the series combination of R_3 and (R_2 in parallel with R_1). Thus

$$R_{eq} = R_4//[R_3 + (R_2//R_1)]$$

and the time constant is CR_{eq}.

In some cases it may be found that the circuit has one resistance and a number of capacitances or inductances. In such a case the procedure should be inverted; that is, ''grab hold'' of the resistance terminals and find the equivalent capacitance C_{eq}, or equivalent inductance L_{eq}, seen by this resistance. The time constant is then found as $C_{eq}R$ or L_{eq}/R. This is illustrated in Example E.2.

EXAMPLE E.2

Find the time constant of the circuit in Fig. E.2.

Solution

After reducing the excitation to zero by short-circuiting the voltage source, we see that the resistance R ''sees'' an equivalent capacitance $C_1 + C_2$. Thus the time constant τ is given by

$$\tau = (C_1 + C_2)R$$

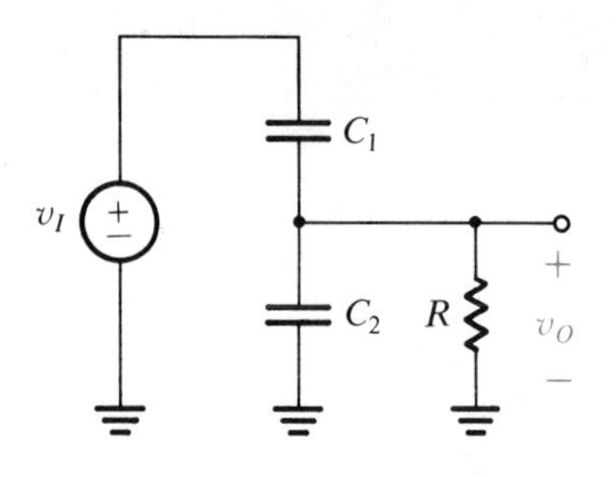

Fig. E.2 Circuit for Example E.2.

Finally, there are cases where an STC circuit has more than one resistance and more than one capacitance (or more than one inductance). In such a case some initial work must be performed to simplify the circuit, as illustrated by Example E.3.

EXAMPLE E.3

Here we show that the response of the circuit in Fig. E.3(a) can be obtained using the method of analysis of STC circuits.

Solution

The analysis steps are illustrated in Fig. E.3. In Fig. E.3(b) we show the circuit excited by two separate but equal voltage sources. The reader should convince himself or herself of the equivalence of the circuits in Fig. E.3(a) and Fig. E.3(b). The ''trick'' employed to obtain the arrangement in Fig. E.3(b) is a very useful one.

Application of Thévenin's theorem to the circuit to the left of the line XX' and then to the circuit to the right of that line results in the circuit of Fig. E.3(c). Since this is a linear circuit, the response may be obtained using the principle of superposition. Specifically, the output voltage v_O will be the sum of the two components v_{O1} and v_{O2}. The first component, v_{O1}, is the output due to the left-hand-side voltage source with the other voltage source reduced to zero. The circuit for calculating v_{O1} is shown in Fig. E.3(d). It is an STC circuit with a time constant given by

$$\tau = (C_1 + C_2)(R_1 // R_2)$$

Similarly, the second component v_{O2} is the output obtained with the left-hand-side voltage source reduced to zero. It can be calculated from the circuit of Fig. E.3(e), which is an STC circuit with a time constant equal to that given above.

Finally, it should be observed that the fact that the circuit is an STC one can also be ascertained by setting the independent source v_I in Fig. E.3(a) to zero. Also, the time constant is then immediately obvious.

CLASSIFICATION OF STC CIRCUITS

STC circuits can be classified into two categories, *low-pass* (LP) and *high-pass* (HP) types, with each of the two categories displaying distinctly different signal responses. The task of finding whether an STC circuit is of LP or HP type may be accomplished in a number of ways, the simplest of which uses the frequency-domain response. Specifically, low-pass circuits pass dc (that is, signals with zero frequency) and attenuate high frequencies, with the transmission being zero at $\omega = \infty$. Thus we can test for the circuit type either at $\omega = 0$ or at $\omega = \infty$. At $\omega = 0$ capacitors should be replaced by open circuits ($1/j\omega C = \infty$) and inductors should be replaced by short circuits ($j\omega L =$

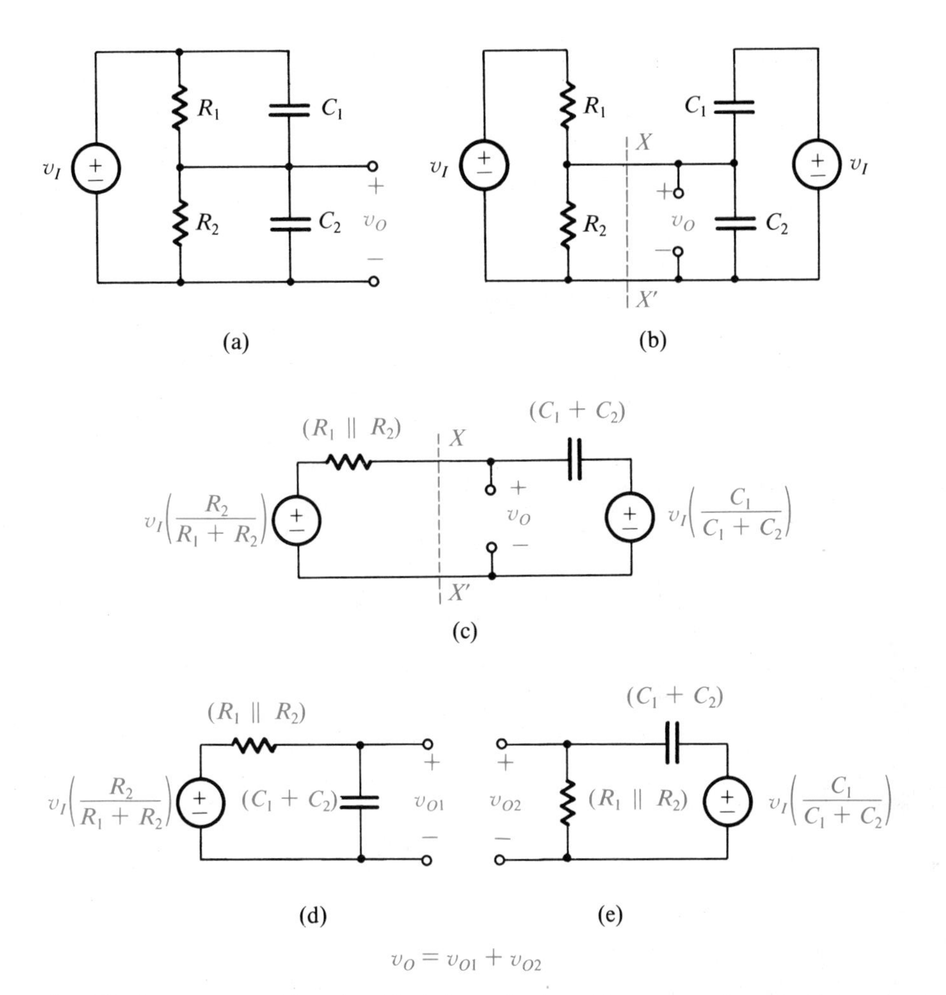

Fig. E.3 The response of the circuit in **(a)** can be found by superposition, that is by summing the responses of the circuits in **(d)** and **(e)**.

0). Then if the output is zero, the circuit is of the high-pass type, while if the output is finite, the circuit is of the low-pass type. Alternatively, we may test at $\omega = \infty$ by replacing capacitors by short circuits ($1/j\omega C = 0$) and inductors by open circuits ($j\omega L = \infty$). Then if the output is finite, the circuit is of the HP type, whereas if the output is zero, the circuit is of the LP type. Table E.1 provides a summary of these results (s.c., short circuit; o.c., open circuit).

Table E.1 RULES FOR FINDING THE TYPE OF STC CIRCUIT

Test at	Replace	Circuit is LP if	Circuit is HP if
$\omega = 0$	C by o.c. L by s.c.	Output is finite	Output is zero
$\omega = \infty$	C by s.c. L by o.c.	Output is zero	Output is finite

Figure E.4 shows examples of low-pass STC circuits, and Fig. E.5 shows examples of high-pass STC circuits. For each circuit we have indicated the input and output variables of interest. Note that a given circuit can be of either category, depending on the input and output variables. The reader is urged to verify, using the rules of Table E.1, that the circuits of Figs. E.4 and E.5 are correctly classified.

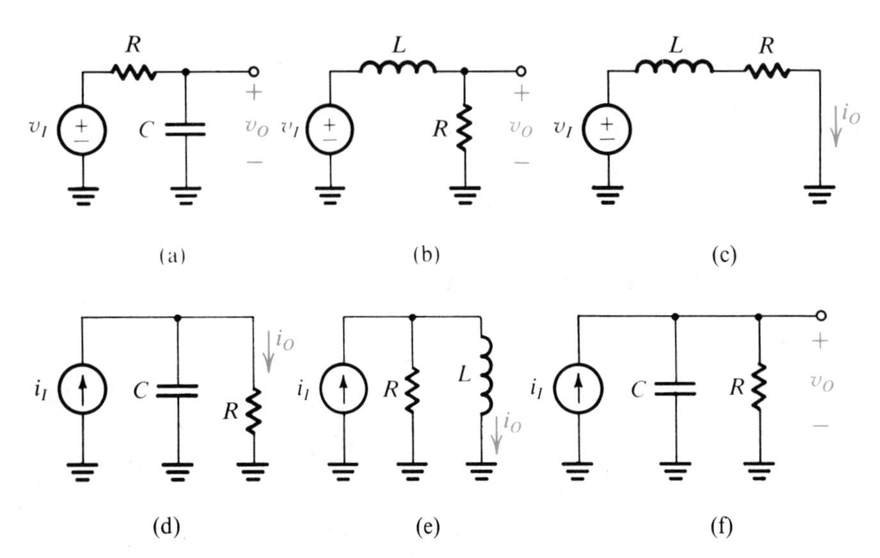

Fig. E.4 STC circuits of the low-pass type.

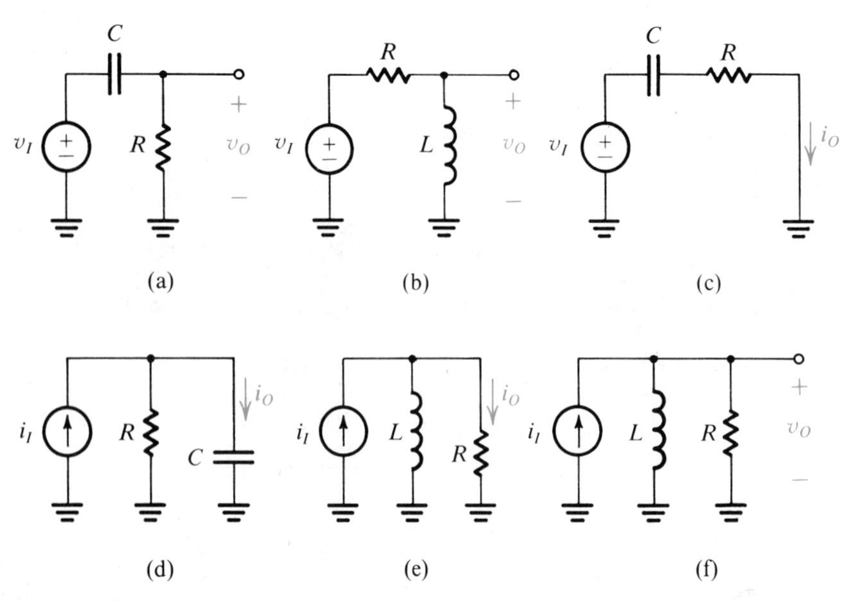

Fig. E.5 STC circuits of the high-pass type.

Exercises E.1 Find the time constants for the circuits shown in Fig. EE.1.

Ans. (a) $\frac{(L_1//L_2)}{R}$; (b) $\frac{(L_1//L_2)}{(R_1//R_2)}$

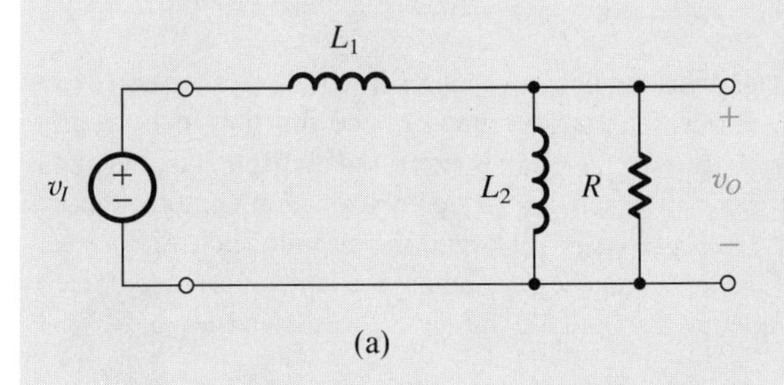

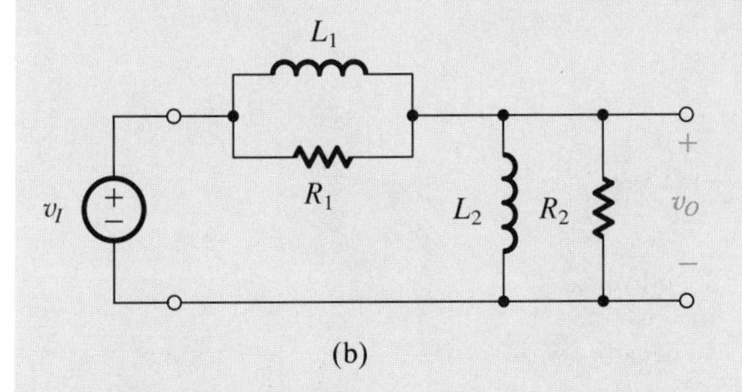

Fig. EE.1

E.2 Classify the following circuits as STC high-pass or low-pass: Fig. E.4(a) with output i_O in C to ground; Fig. E.4(b) with output i_O in R to ground; Fig. E.4(d) with output i_O in C to ground; Fig. E.4(e) with output i_O in R to ground; Fig. E.5(b) with output i_O in L to ground; and Fig. E.5(d) with output v_O across C.

Ans. HP; LP; HP; HP; LP; LP

FREQUENCY RESPONSE OF STC CIRCUITS

Low-Pass Circuits

The transfer function $T(s)$ of an STC low-pass circuit always can be written in the form

$$T(s) = \frac{K}{1 + (s/\omega_0)} \tag{E.1}$$

which, for physical frequencies, where $s = j\omega$, becomes

$$T(j\omega) = \frac{K}{1 + j(\omega/\omega_0)} \tag{E.2}$$

where K is the magnitude of the transfer function at $\omega = 0$ (dc) and ω_0 is defined by

$$\omega_0 = 1/\tau$$

with τ being the time constant. Thus the magnitude response is given by

$$|T(j\omega)| = \frac{K}{\sqrt{1 + (\omega/\omega_0)^2}} \tag{E.3}$$

and the phase response is given by

$$\phi(\omega) = -\tan^{-1}(\omega/\omega_0) \tag{E.4}$$

Figure E.6 shows sketches of the magnitude and phase responses for an STC low-pass circuit. The magnitude response shown in Fig. E.6(a) is simply a graph of the function in Eq. (E.3). The magnitude is normalized with respect to the dc gain K and is expressed in dB, that is, the plot is for $20 \log|T(j\omega)/K|$, with a logarithmic scale used for the frequency axis. Furthermore, the frequency variable has been normalized with respect to ω_0. As shown, the magnitude curve is closely defined by two straight-line asymptotes. The low-frequency asymptote is a horizontal straight line at 0 dB. To find the slope of the high-frequency asymptote consider Eq. (E.3) and let $\omega/\omega_0 \gg 1$, resulting in

$$|T(j\omega)| \simeq K \frac{\omega_0}{\omega}$$

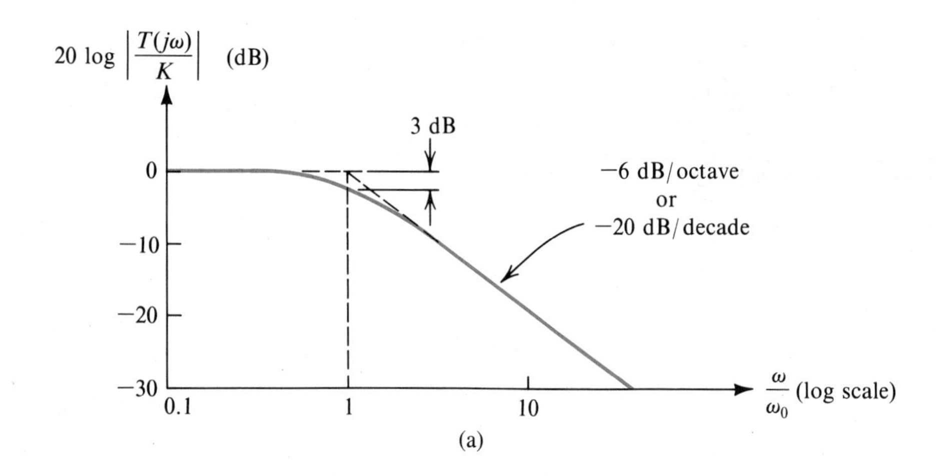

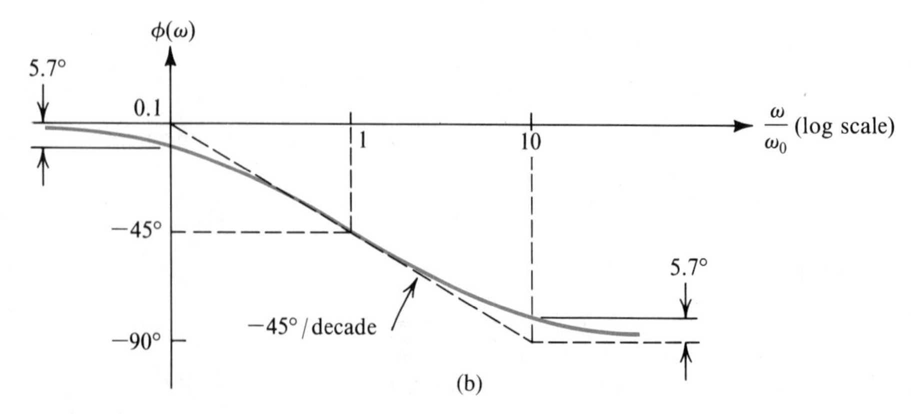

Fig. E.6 (a) Magnitude and **(b)** phase response of STC circuits of the low-pass type.

It follows that if ω doubles in value, the magnitude is halved. On a logarithmic frequency axis, doublings of ω represent equally spaced points, with each interval called an *octave*. Halving the magnitude function corresponds to a 6-dB reduction in transmission ($20 \log 0.5 = -6$ dB). Thus the slope of the high-frequency asymptote is -6 dB/octave. This can be equivalently expressed as -20 dB/decade, where a decade refers to an increase in frequency by a factor of 10.

The two straight-line asymptotes of the magnitude-response curve meet at the "corner frequency" or "break frequency" ω_0. The difference between the actual magnitude-response curve and the asymptotic response is largest at the corner frequency, where its value is 3 dB. To verify that this value is correct, simply substitute $\omega = \omega_0$ in Eq. (E.3) to obtain

$$|T(j\omega_0)| = K/\sqrt{2}$$

Thus at $\omega = \omega_0$ the gain drops by a factor of $\sqrt{2}$ relative to the dc gain, which corresponds to a 3-dB reduction in gain. The corner frequency ω_0 is appropriately referred to as the 3-dB frequency.

Similar to the magnitude response, the phase-response curve, shown in Fig. E.6(b), is closely defined by straight-line asymptotes. Note that at the corner frequency the phase is $-45°$, and that for $\omega \gg \omega_0$ the phase approaches $-90°$. Also note that the $-45°$/decade straight line approximates the phase function, with a maximum error of 5.7°, over the frequency range $0.1\omega_0$ to $10\omega_0$.

EXAMPLE E.4

Consider the circuit shown in Fig. E.7(a), where an ideal voltage amplifier of gain $\mu = -100$ has a small (10-pF) capacitance connected in its feedback path. The amplifier is fed by a voltage source having a source resistance of 100 kΩ. Show that the frequency response V_o/V_s of this amplifier is equivalent to that of an STC circuit, and sketch the magnitude response.

Solution

Direct analysis of the circuit in Fig. E.7(a) results in the transfer function

$$\frac{V_o}{V_s} = \frac{\mu}{1 + sRC_f(-\mu + 1)}$$

which can be seen to be that of a low-pass STC circuit with a dc gain $\mu = -100$ (or, equivalently, 40 dB) and a time constant $\tau = RC_f(-\mu + 1) = 100 \times 10^3 \times 10 \times 10^{-12} \times 101 \simeq 10^{-4}$ s, which corresponds to a frequency $\omega_0 = 1/\tau = 10^4$ rad/s. The magnitude response is sketched in Fig. E.7(b).

High-Pass Circuits

The transfer function $T(s)$ of an STC high-pass circuit always can be expressed in the form

$$T(s) = \frac{Ks}{s + \omega_0} \tag{E.5}$$

which for physical frequencies $s = j\omega$ becomes

$$T(j\omega) = \frac{K}{1 - j\omega_0/\omega} \tag{E.6}$$

where K denotes the gain as s or ω approaches infinity and ω_0 is the inverse of the time constant τ,

$$\omega_0 = 1/\tau$$

The magnitude response

$$|T(j\omega)| = \frac{K}{\sqrt{1 + (\omega_0/\omega)^2}} \tag{E.7}$$

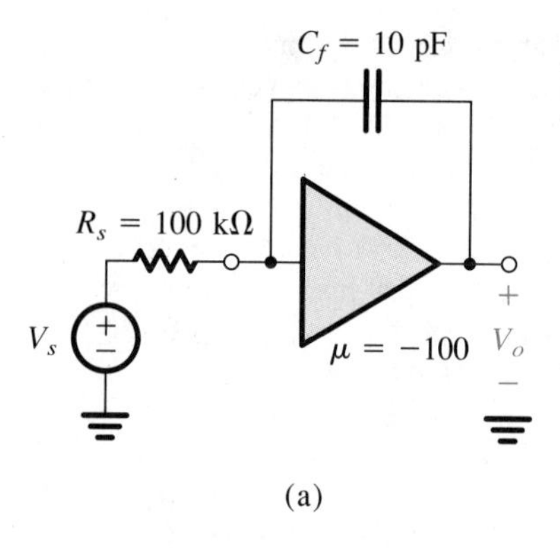

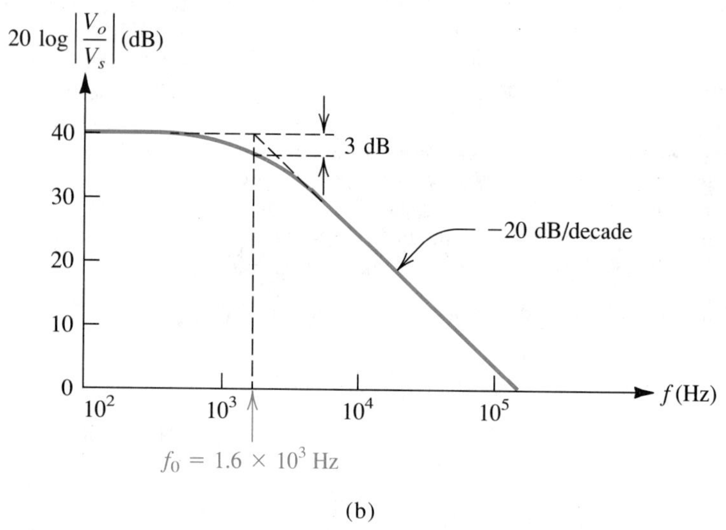

Fig. E.7 (a) An amplifier circuit and **(b)** a sketch of the magnitude of its transfer function.

and the phase response

$$\phi(\omega) = \tan^{-1}(\omega_0/\omega) \tag{E.8}$$

are sketched in Fig. E.8. As in the low-pass case, the magnitude and phase curves are well defined by straight-line asymptotes. Because of the similarity (or, more appropriately, duality) with the low-pass case, no further explanation will be given.

Exercises E.3 Find the dc transmission, the corner frequency f_0, and the transmission at $f = 2$ MHz for the low-pass STC circuit shown in Fig. EE.3.

Ans. −6 dB; 318 kHz; −22 dB

E.4 Find the transfer function $T(s)$ of the circuit in Fig. E.2. What type of STC network is it?

Ans. $T(s) = \dfrac{C_1}{C_1 + C_2} \dfrac{s}{s + [1/(C_1 + C_2)R]}$; HP

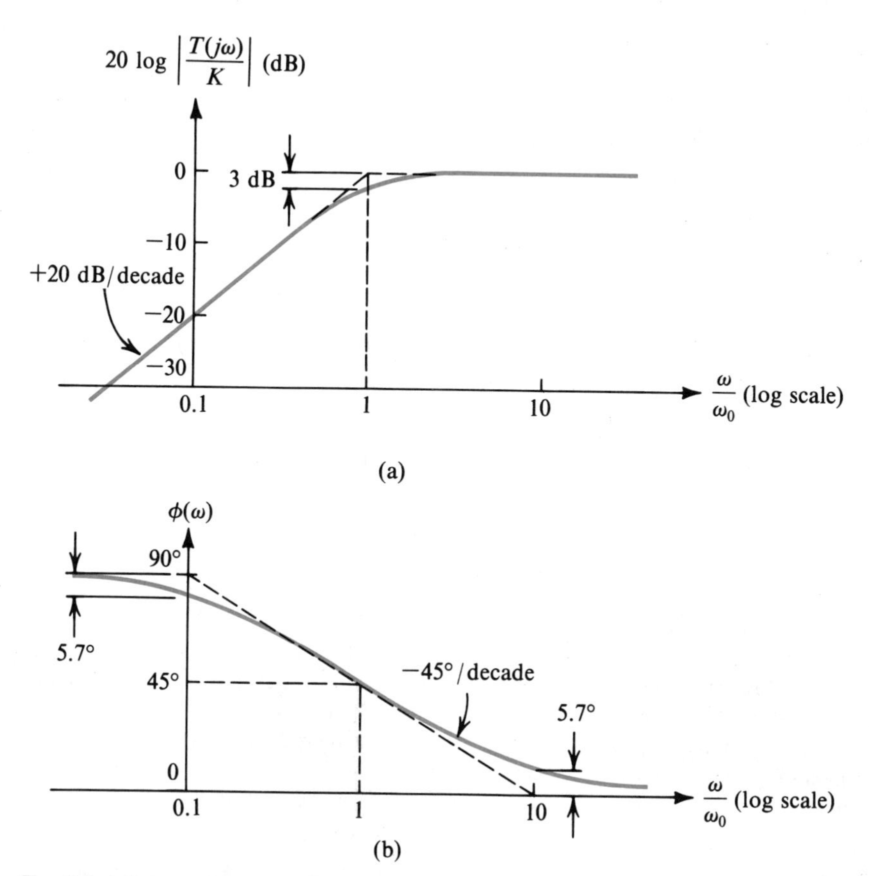

Fig. E.8 **(a)** Magnitude and **(b)** phase response of STC circuits of the high pass type.

E.5 For the circuit of Exercise E.4, if $R = 10\ \text{k}\Omega$, find the capacitor values that result in the circuit having a high-frequency transmission of 0.5 V/V and a corner frequency $\omega_0 = 10$ rad/s.

Ans. $C_1 = C_2 = 5\ \mu\text{F}$

E.6 Find the high-frequency gain, the 3-dB frequency f_0, and the gain at $f = 1$ Hz of the capacitively coupled amplifier shown in Fig. EE.6. Assume the voltage amplifier to be ideal.

Ans. 40 dB; 15.9 Hz; 16 dB

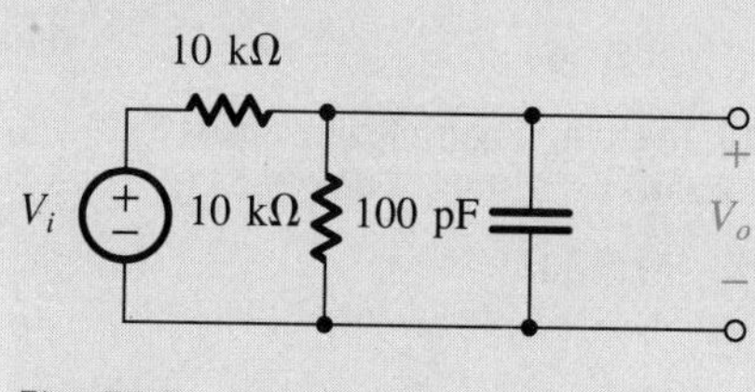

Fig. EE.3

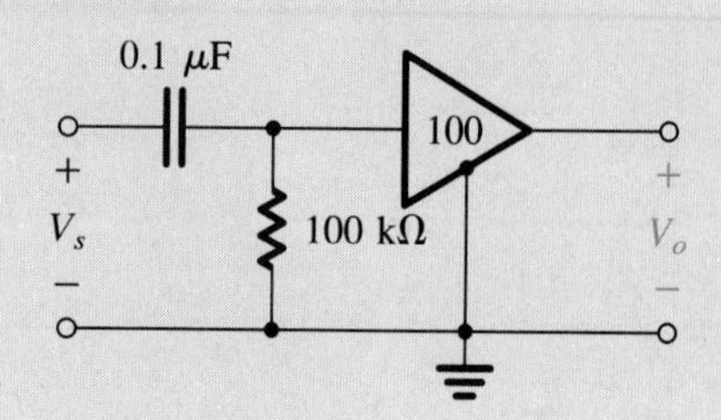

Fig. EE.6

STEP RESPONSE OF STC CIRCUITS

In this section we consider the response of STC circuits to the step-function signal shown in Fig. E.9. Knowledge of the step response enables rapid evaluation of the response to other switching signal waveforms, such as pulses and square waves.

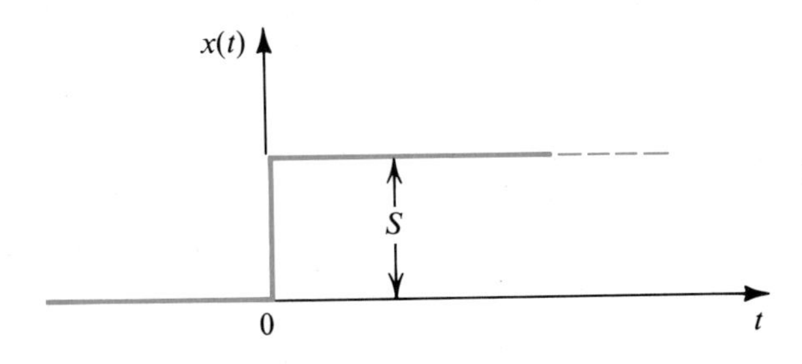

Fig. E.9 A step-function signal of height S.

Low-Pass Circuits

In response to an input step signal of height S, a low-pass STC circuit (with a dc gain $K = 1$) produces the waveform shown in Fig. E.10. Note that while the input rises from 0 to S at $t = 0$, the output does not respond to this transient and simply begins to rise exponentially toward the *final* dc value of the input, S. In the long term—that is, for $t \gg \tau$—the output approaches the dc value S, a manifestation of the fact that low-pass circuits faithfully pass dc.

The equation of the output waveform can be obtained from the expression

$$y(t) = Y_\infty - (Y_\infty - Y_{0+})e^{-t/\tau} \tag{E.9}$$

where Y_∞ denotes the *final* value or the value toward which the output is heading and Y_{0+} denotes the value of the output immediately after $t = 0$. This equation states that the output at any time t is equal to the difference between the final value Y_∞ and a gap whose initial value is $Y_\infty - Y_{0+}$ and which is "shrinking" exponentially. In our case $Y_\infty = S$ and $Y_{0+} = 0$; thus

$$y(t) = S(1 - e^{-t/\tau}) \tag{E.10}$$

The reader's attention is drawn to the slope of the tangent to $y(t)$ at $t = 0$, which is indicated in Fig. E.10.

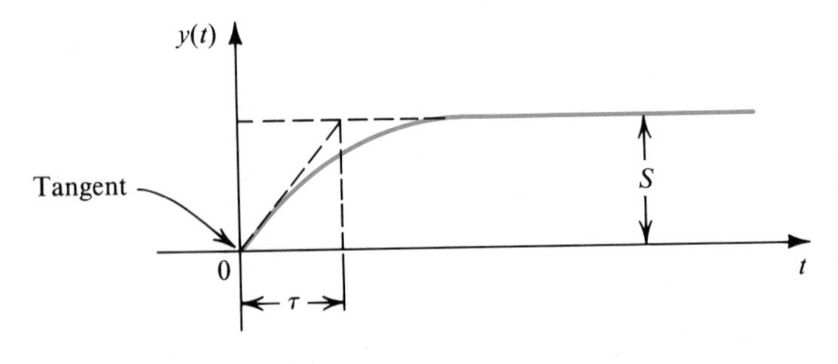

Fig. E.10 The output $y(t)$ of a low-pass STC circuit excited by a step of height S.

High-Pass Circuits

The response of an STC high-pass circuit (with a high-frequency gain $K = 1$) to an input step of height S is shown in Fig. E.11. The high-pass circuit faithfully transmits the transient of the input signal (the step change) but blocks the dc. Thus the output at $t = 0$ follows the input,

$$Y_{0+} = S$$

and then it decays toward zero,

$$Y_{\infty} = 0$$

Substituting for Y_{0+} and Y_{∞} in Eq. (E.9) results in the output $y(t)$,

$$y(t) = Se^{-t/\tau} \tag{E.11}$$

The reader's attention is drawn to the slope of the tangent to $y(t)$ at $t = 0$, indicated in Fig. E.11.

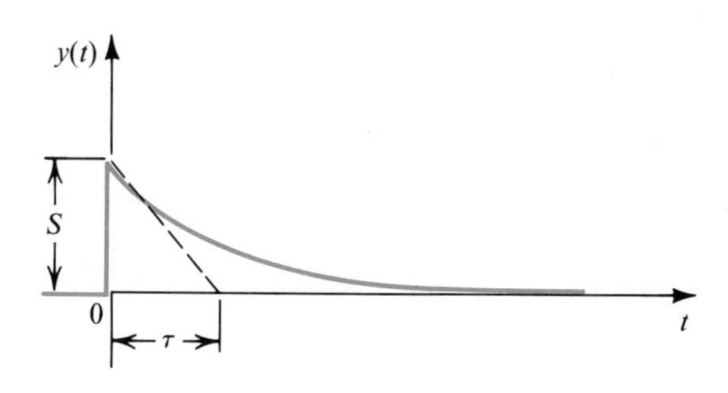

Fig. E.11 The output $y(t)$ of a high-pass STC circuit excited by a step of height S.

EXAMPLE E.5 This example is a continuation of the problem considered in Example E.3. For an input v_I that is a 10-V step, find the condition under which the output v_O is a perfect step.

Solution Following the analysis in Example E.3, which is illustrated in Fig. E.3, we have

$$v_{O1} = k_r[10(1 - e^{-t/\tau})]$$

where

$$k_r \equiv \frac{R_2}{R_1 + R_2}$$

and

$$v_{O2} = k_c(10e^{-t/\tau})$$

where

$$k_c \equiv \frac{C_1}{C_1 + C_2}$$

and

$$\tau = (C_1 + C_2)(R_1 // R_2)$$

Thus

$$v_O = v_{O1} + v_{O2}$$
$$= 10k_r + 10e^{-t/\tau}(k_c - k_r)$$

It follows that the output can be made a perfect step of height $10k_r$ volts if we arrange that

$$k_c = k_r$$

that is, if the resistive voltage-divider ratio is made equal to the capacitive voltage-divider ratio.

This example illustrates an important technique, namely, that of the "compensated attenuator." An application of this technique is found in the design of the oscilloscope probe. The oscilloscope probe problem is investigated in Problem E.3.

Exercises E.7 For the circuit of Fig. E.4(f) find v_O if i_I is a 3-mA step, $R = 1$ kΩ, and $C = 100$ pF.

Ans. $3(1 - e^{-10^7 t})$

E.8 In the circuit of Fig. E.5(f) find $v_O(t)$ if i_I is a 2-mA step, $R = 2$ kΩ, and $L = 10$ μH.

Ans. $4e^{-2\times 10^8 t}$

E.9 The amplifier circuit of Fig. EE.6 is fed with a signal source that delivers a 20-mV step. If the source resistance is 100 kΩ, find the time constant τ and $v_O(t)$.

Ans. $\tau = 2 \times 10^{-2}$ s; $v_O(t) = 1 \times e^{-50t}$

E.10 For the circuit in Fig. E.2 with $C_1 = C_2 = 0.5$ μF, $R = 1$ MΩ, find $v_O(t)$ if $v_I(t)$ is a 10-V step.

Ans. $5e^{-t}$

E.11 Show that the area under the exponential of Fig. E.11 is equal to that of the rectangle of height S and width τ.

PULSE RESPONSE OF STC CIRCUITS

Figure E.12 shows a pulse signal whose height is P and whose width is T. We wish to find the response of STC circuits to input signals of this form. Note at the outset that a pulse can be considered as the sum of two steps: a positive one of height P occurring at $t = 0$ and a negative one of height P occurring at $t = T$. Thus the response of a linear circuit to the pulse signal can be obtained by summing the responses to the two step signals.

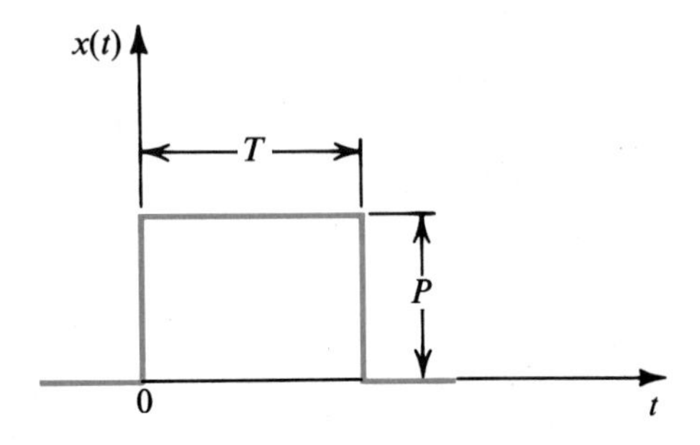

Fig. E.12 A pulse signal with height P and width T.

Low-Pass Circuits

Figure E.13(a) shows the response of a low-pass STC circuit (having unity dc gain) to an input pulse of the form shown in Fig. E.12. In this case we have assumed that the time constant τ is in the same range as the pulse width T. As shown, the LP circuit does not respond to the step change at the leading edge of the pulse; rather, the output starts to rise exponentially toward a final value of P. This exponential rise, however, will be stopped at time $t = T$, that is, at the trailing edge of the pulse when the input undergoes a negative step change. Again the output will respond by starting an exponential decay toward the final value of the input, which is zero. Finally, note that the area under the output waveform will be equal to the area under the input pulse waveform, since the LP circuit faithfully passes dc.

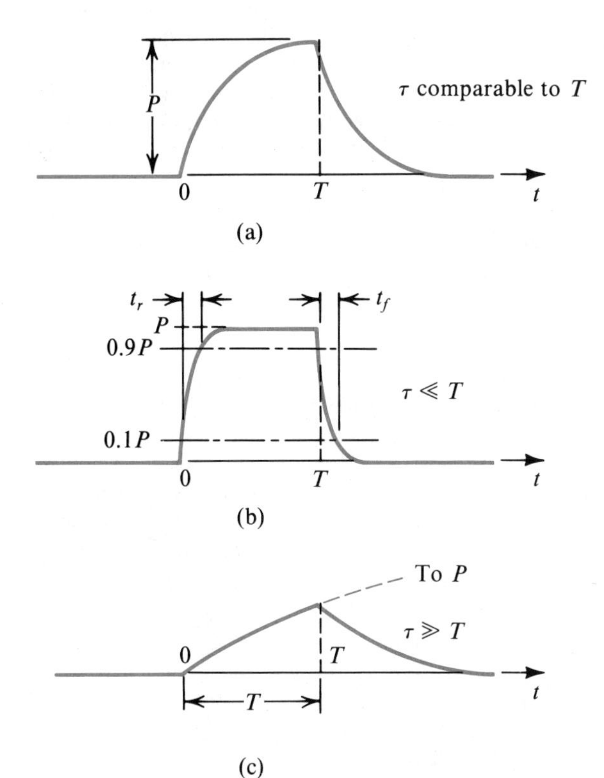

Fig. E.13 Pulse response of STC low-pass circuits.

In connecting a pulse signal from one part of an electronic system to another, a low-pass effect usually occurs. The low-pass circuit in this case is formed by the output resistance (Thévenin's equivalent resistance) of the system part from which the signal originates and the input capacitance of the system part to which the signal is fed. This unavoidable low-pass filter will cause distortion—of the type shown in Fig. E.13(a)—of the pulse signal. In a well-designed system such distortion is kept to a low value by arranging that the time constant τ be much smaller than the pulse width T. In this case the result will be a slight rounding of the pulse edges, as shown in Fig. E.13(b). Note, however, that the edges are still exponential.

The distortion of a pulse signal by a parasitic (that is, unwanted) low-pass circuit is measured by its *rise time* and *fall time*. The rise time is conventionally defined as the time taken by the amplitude to increase from 10% to 90% of the final value. Similarly, the fall time is the time during which the pulse amplitude falls from 90% to 10% of the maximum value. These definitions are illustrated in Fig. E.13(b). By use of the exponential equations of the rising and falling edges of the output waveform it can be easily shown that

$$t_r = t_f \simeq 2.2\tau \tag{E.12}$$

which can be also expressed in terms of $f_0 = \omega_0/2\pi = 1/2\pi\tau$ as

$$t_r = t_f \simeq \frac{0.35}{f_0} \tag{E.13}$$

Finally, we note that the effect of the parasitic low-pass circuits that are always present in a system is to "slow down" the operation of the system: In order to keep the signal distortion within acceptable limits one has to use a relatively long pulse width (for a given low-pass time constant).

The other extreme case—namely, when τ is much larger than T—is illustrated in Fig. E.13(c). As shown, the output waveform rises exponentially toward the level P. However, since $\tau \gg T$, the value reached at $t = T$ will be much smaller than P. At $t = T$ the output waveform starts its exponential decay toward zero. Note that in this case the output waveform bears little resemblance to the input pulse. Also note that because $\tau \gg T$ the portion of the exponential curve from $t = 0$ to $t = T$ is almost linear. Since the slope of this linear curve is proportional to the height of the input pulse, we see that the output waveform approximates the time integral of the input pulse. That is, a low-pass network with a large time constant approximates the operation of an *integrator*.

High-Pass Circuits

Figure E.14(a) shows the output of an STC HP circuit (with unity high-frequency gain) excited by the input pulse of Fig. E.12, assuming that τ and T are comparable in value. As shown, the step transition at the leading edge of the input pulse is faithfully reproduced at the output of the HP circuit. However, since the HP circuit blocks dc, the output waveform immediately starts an exponential decay toward zero. This decay process is stopped at $t = T$ when the negative step transition of the input occurs and the HP circuit faithfully reproduces it. Thus at $t = T$ the output waveform exhibits an *undershoot*. Then it starts an exponential decay toward zero. Finally, note that the area of the output waveform above the zero axis will be equal to that below the axis for a total average area of zero, consistent with the fact that HP circuits block dc.

In many applications an STC high-pass circuit is used to couple a pulse from one part of a system to another part. In such an application it is necessary to keep the distortion in the pulse shape as small as possible. This can be accomplished by selecting the time constant τ to be much longer than the pulse width T. If this is indeed the case, the loss in amplitude during the pulse period T will be very small, as shown in Fig. E.14(b). Nevertheless, the output waveform still swings negatively, and the area under the negative portion will be equal to that under the positive portion.

Consider the waveform in Fig. E.14(b). Since τ is much larger than T, it follows that the portion of the exponential curve from $t = 0$ to $t = T$ will be almost linear and that its slope will be equal to the slope of the exponential curve at $t = 0$, which is P/τ. We can use this value of the slope to determine the loss in amplitude ΔP as

$$\Delta P \simeq \frac{P}{\tau} T \tag{E.14}$$

The distortion effect of the high-pass circuit on the input pulse is usually specified in terms of the per-unit or percentage loss in pulse height. This quantity is taken as an indication of the "sag" in the output pulse,

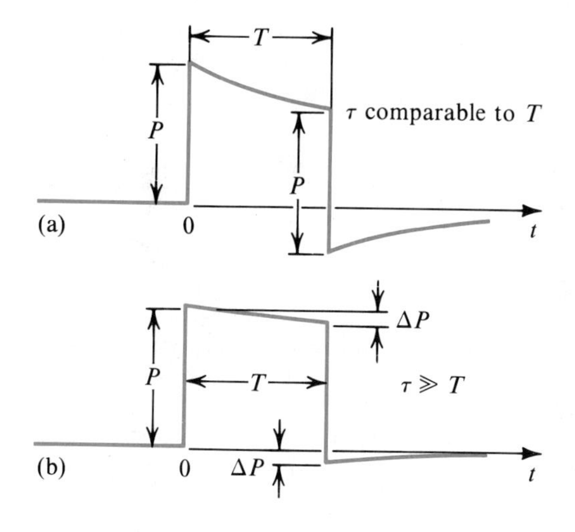

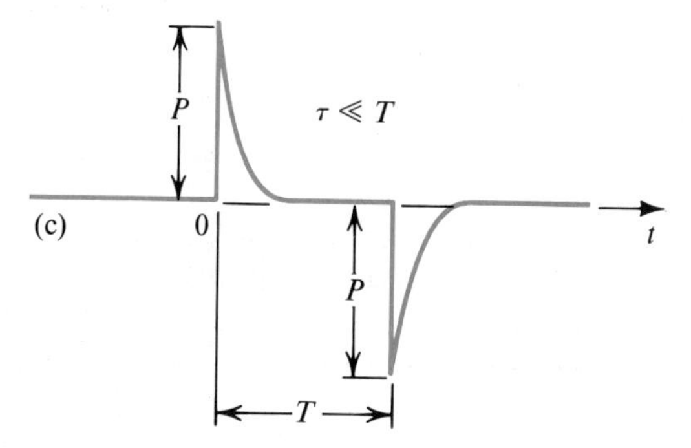

Fig. E.14 Pulse response of STC high-pass circuits.

$$\text{Percentage sag} \equiv \frac{\Delta P}{P} \times 100 \tag{E.15}$$

Thus

$$\text{Percentage sag} = \frac{T}{\tau} \times 100 \tag{E.16}$$

Finally, note that the magnitude of the undershoot at $t = T$ is equal to ΔP.

The other extreme case—namely, $\tau \ll T$—is illustrated in Fig. E.14(c). In this case the exponential decay is quite rapid, resulting in the output becoming almost zero shortly beyond the leading edge of the pulse. At the trailing edge of the pulse the output swings negatively by an amount almost equal to the pulse height P. Then the waveform decays rapidly to zero. As seen from Fig. E.14(c), the output waveform bears no resemblance to the input pulse. It consists of two spikes: a positive one at the leading edge and a negative one at the trailing edge. Note that the output waveform is approximately equal to the time derivative of the input pulse. That is, for $\tau \ll T$ an STC high-pass circuit approximates a *differentiator*. However, the resulting differentiator is not an ideal one; an ideal differentiator would produce two impulses. Nevertheless, high-pass STC circuits with short time constants are employed in some applications to produce sharp pulses or spikes at the transitions of an input waveform.

Exercises E.12 Find the rise and fall times of a 1-μs pulse after it passes through a low-pass RC circuit with a corner frequency of 10 MHz.

Ans. 35 ns

E.13 Consider the pulse response of a low-pass STC circuit, as shown in Fig. E.13(c). If $\tau = 100T$ find the output voltage at $t = T$. Also find the difference in the slope of the rising portion of the output waveform at $t = 0$ and $t = T$ (expressed as a percentage of the slope at $t = 0$).

Ans. $0.01P$; 1%

E.14 The output of an amplifier stage is connected to the input of another stage via a capacitance C. If the first stage has an output resistance of 10 kΩ and the second stage has an input resistance of 40 kΩ, find the minimum value of C such that a 10-μs pulse exhibits less than 1% sag.

Ans. 0.02 μF

E.15 A high-pass STC circuit with a time constant of 100 μs is excited by a pulse of 1-V height and 100-μs width. Calculate the value of the undershoot in the output waveform.

Ans. 0.632 V

BIBLIOGRAPHY

R. Littauer, *Pulse Electronics*, New York: McGraw-Hill, 1965.

J. Millman and H. Taub, *Pulse, Digital, and Switching Waveforms*, New York: McGraw-Hill, 1965.

PROBLEMS

E.1 Consider the circuit of Fig. E.3(a) and the equivalent shown in (d) and (e). There, the output, $v_O = v_{O1} + v_{O2}$, is the sum of outputs of a low-pass and a high-pass circuit, each with the time constant $\tau = (C_1 + C_2)(R_1//R_2)$. What is the condition that makes the contribution of the low-pass circuit at zero frequency equal to the contribution of the high-pass circuit at infinite frequency? Show that this condition can be expressed as $C_1R_1 = C_2R_2$. If this condition applies, sketch $|V_o/V_i|$ versus frequency for the case $R_1 = R_2$.

E.2 Use the voltage-divider rule to find the transfer function $V_o(s)/V_i(s)$ of the circuit in Fig. E.3(a). Show that the transfer function can be made independent of frequency if the condition $C_1R_1 = C_2R_2$ applies. Under such condition the circuit is called a *compensated attenuator*. Find the transmission of the compensated attenuator in terms of R_1 and R_2.

D**E.3 The circuit of Fig. E.3(a) is used as a compensated attenuator (see Problems E.1 and E.2) for the oscilloscope probe. The object is to reduce the signal voltage applied to the input amplifier of the oscilloscope, with the signal attenuation independent of frequency. The probe itself includes R_1 and C_1, while R_2 and C_2 model the oscilloscope input circuit. For an oscilloscope having an input resistance of 1 MΩ and an input capacitance of 30 pF design a compensated "10 to 1 probe"—that is, a probe that attenuates the input signal by a factor of 10. Find the input impedance of the probe when connected to the oscilloscope, which is the impedance seen by v_I in Fig. E.3(a). Show that this impedance is 10 times higher than that of the oscilloscope itself. This is the great advantage of the 10:1 probe.

E.4 In the circuits of Fig. E.4 and E.5 let $L = 10$ mH, $C = 0.01$ μF, and $R = 1$ kΩ. At what frequency does a phase angle of 45° occur?

*E.5 Consider a voltage amplifier with an open-circuit voltage gain $A_{vo} = -100$ V/V, $R_o = 0$, $R_i =$ 10 kΩ, and an input capacitance C_i (in parallel with R_i) of 10 pF. The amplifier has a feedback capacitance (a capacitance connected between output and input) $C_f = 1$ pF. The amplifier is fed with a voltage source V_s having a resistance $R_s = 10\ \text{k}\Omega$. Find the amplifier transfer function $V_o(s)/V_s(s)$ and sketch its magnitude response versus frequency (dB versus frequency on a log axis).

E.6 For the circuit in Fig. PE.6 assume the voltage amplifier to be ideal. Derive the transfer function $V_o(s)/V_i(s)$. What type of STC response is this? For $C = 0.01\ \mu\text{F}$ and $R = 100\ \text{k}\Omega$ find the corner frequency.

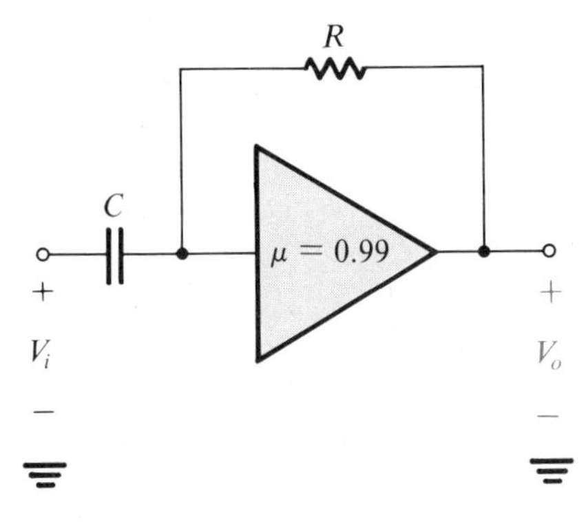

Fig. PE.6

E.7 For the circuits of Figs. E.4(b) and E.5(b) find $v_O(t)$ if v_I is a 10-V step, $R = 1\ \text{k}\Omega$, and $L = 1\ \text{mH}$.

E.8 Consider the exponential response of an STC low-pass circuit to a 10-V step input. In terms of the time constant τ find the time taken for the output to reach 5 V, 9 V, 9.9 V, and 9.99 V.

E.9 The high-frequency response of an oscilloscope is specified to be like that of an STC LP circuit with a 100-MHz corner frequency. If this oscilloscope is used to display an ideal step waveform, what rise time (10% to 90%) would you expect to observe?

E.10 An oscilloscope whose step response is like that of a low-pass STC circuit has a rise time of t_s seconds. If an input signal having a rise time of t_w seconds is displayed, the waveform seen will have a rise time t_d seconds, which can be found using the empirical formula $t_d = \sqrt{t_s^2 + t_w^2}$. If $t_s = 35$ ns, what is the 3-dB frequency of the oscilloscope? What is the observed rise time for a waveform rising in 100 ns, 35 ns, and 10 ns? What is the actual rise time of a waveform whose displayed rise time is 49.5 ns?

E.11 A pulse of 10-ms width and 10-V amplitude is transmitted through a system characterized as having an STC high-pass response with a corner frequency of 10 Hz. What undershoot would you expect?

E.12 An RC differentiator having a time constant τ is used to implement a short-pulse detector. When a long pulse with $T \gg \tau$ is fed to the circuit, the positive and negative peak outputs are of equal magnitude. At what pulse width does the negative output peak differ from the positive one by 10%?

E.13 A high-pass STC circuit with a time constant of 1 ms is excited by a pulse of 10-V height and 1-ms width. Calculate the value of the undershoot in the output waveform. If an undershoot of 1 V or less is required, what is the time constant necessary?

DE.14 A capacitor C is used to couple the output of an amplifier stage to the input of the next stage. If the first stage has an output resistance of 2 kΩ and the second stage has an input resistance of 3 kΩ, find the value of C so that a 1-ms pulse exhibits less than 1% sag. What is the associated 3-dB frequency?

DE.15 An RC differentiator is used to convert a step voltage change V to a single pulse for a digital logic application. The logic circuit that the differentiator drives distinguishes signals above $V/2$ as "high" and below $V/2$ as "low." What must the time constant of the circuit be to convert a step input into a pulse that will be interpreted as "high" for 10 μs?

DE.16 Consider the circuit in Fig. E.7(a) with $\mu = -100$, $C_f = 100$ pF, and the amplifier ideal. Find the value of R so that the gain $|V_o/V_s|$ has a 3-dB frequency of 1 kHz.

APPENDIX 

Standard Resistance Values

Discrete resistors are available only in standard values. The table below provides the multipliers for the standard values of 5%-tolerance and 1%-tolerance resistors. Thus in the kΩ range of 5% resistors one finds resistances of 1.0, 1.1, 1.2, 1.3, 1.5, . . . , kΩ. In the same range, one finds 1% resistors of values of 1.00, 1.02, 1.05, 1.07, 1.10, . . . , kΩ.

5% resistor values	1% resistor values			
10	100	178	316	562
11	102	182	324	576
12	105	187	332	590
13	107	191	340	604
15	110	196	348	619
16	113	200	357	634
18	115	205	365	649
20	118	210	374	665
22	121	215	383	681
24	124	221	392	698
27	127	226	402	715
30	130	232	412	732
33	133	237	422	750
36	137	243	432	768
39	140	249	442	787
43	143	255	453	806
47	147	261	464	825
51	150	267	475	845
56	154	274	487	866
62	158	280	499	887
68	162	287	511	909
75	165	294	523	931
82	169	301	536	953
91	174	309	549	976

APPENDIX 

Answers to Selected Problems

Chapter 1

1.1 1 kΩ **1.2** **(a)** $v = 10 \sin(6.28 \times 10^4 t)$; **(b)** $v = 170 \sin(377t)$; **(c)** $v = 0.1 \sin(10^3 t)$; **(d)** $v = 0.1 \sin(6.28 \times 10^3 t)$ **1.5** 1.041 V; 0.922 V; −1.041 V; −0.922 V
1.7 200 V/V, 46 dB; 100 A/A, 40 dB; 2×10^4 W/W, 43 dB **1.9** 26 mW
1.10 91.9 mV **1.12** 1.2 V; −10 V/V; 0.5 V peak **1.14** 3; 2.07 V
1.16 19.05 V/V **1.17** $R_{in} = R_i/(1 + g_m R_i)$ **1.19** $v_o/v_s = g_m R_L/(1 + R_s/r_\pi)$; 100 V/V
1.20 0.4 mH **1.22** 880 V/V, 58.9 dB; 2200 A/A, 66.8 dB; 1.94 W/μW, 62.9dB
1.24 $v_o = g_m R(v_1 - v_2)$; 0 V; 10 V
1.26 $V_i(s)/V_s(s) = [R_i/(R_i + R_s)]/[1 + sC_i R_i R_s/(R_i + R_s)]$; low-pass STC with $K = R_i/(R_i + R_s)$ and $\omega_0 = 1/C_i(R_i//R_s)$; 1.75 MHz
1.28 $0.434/RC$ rad/s **1.30** 40 kΩ; 11.4 kΩ; 18.9 mA/V
1.32 $V_o(s)/V_i(s) = R_2(R_1 C_1 s + 1)/[s(R_1 R_2)(C_1 + C_2) + R_1 + R_2]$; $T = R_2/(R_1 + R_2)$
1.33 20.2 Hz; 4.96 kHz; 0.174 dB; 20.2 Hz; 4.96 kHz

Chapter 2

2.1 5; 14 **2.2** 1001 V/V **2.5** −10 V/V **(a)** −5 V/V; **(b)** −20 V/V
2.8 $R_2 = 10$ MΩ, $R_1 = 200$ kΩ; 200 kΩ **2.10** ±10 mV **2.12** $R_{in} = R_1 + R_2/(1 + A)$
2.14 $A = (1 + R_2/R_1)(k - 1)/(1 - x/100)$; 2×10^4 V/V **2.16** −3 V/V; −8 V/V
2.17 **(a)** −9.89 V/V, 10.1 kΩ; **(b)** −9.89 V/V, 10.1 kΩ **(c)** −9.79 V/V, 10.1 kΩ; **(d)** −9.89 V/V, 10.1 kΩ; **(e)** 0 V/V, 10 kΩ; **(f)** −1000 V/V, 100 Ω.
2.19 **(a)** 15.9 Hz; **(b)** −90° (90° lagging); **(c)** rises by 10 times; **(d)** −90°
2.22 $V_o(s)/V_i(s) = -(R_2/R_1)/(1 + sR_2C)$; $R_1 = 1$ kΩ; $R_2 = 100$ kΩ; $C = 0.398$ nF; 398 kHz
2.24 1.59 kHz; 10 V (peak-to-peak)
2.26 $V_o(s)/V_i(s) = -R_2 Cs/(1 + R_1 C_s)$; $R_1 = 1$ kΩ; $R_2 = 100$ kΩ; $C = 1.59$ μF; 1.0 Hz
2.28 $v_O = v_1 - v_2/2$; −1.5 V **2.30** $R_{v_1} = 20$ kΩ; $R_{v_2} = 120$ kΩ; $R_f = 60$ kΩ
2.34 $v_O = 10(v_2 - v_1)$; $v_O = 2 \sin(2\pi \times 1000t)$ **2.35** $v_O/v_I = 1/x$; +1 to +∞; add 1 kΩ in series with grounded end of pot

2.37 $v_o/v_i = 1/(1 + 1/A)$; 0.999, −0.1%; 0.990, −1.0%; 0.909, −9.1%
2.39 2.5 kΩ; 1.275 V. **2.40** $v_O = v_2 - v_1$; R; $2R$; $2R$; R
2.44 R_1 = 100 kΩ pot plus 1 kΩ fixed; R_2 = 100 kΩ; R_3 = 200 kΩ; R_4 = 100 kΩ
2.45 0 V, 0 V; 0.1 V, 0.2 V; 1 V, 2 V; 2 V, 4 V; 6.5 kΩ
2.47 (a) 1.5 V (peak-to-peak), 1.5 V (peak-to-peak) of opposite phase, 3.0 V (peak-to-peak); (b) 3 V/V; (c) 56 V (peak-to-peak), 19.8 V (rms)
2.49 4.2×10^4 V/V; 181 Hz; 7.6×10^6 Hz
2.51 47.6 kHz; 19.9 V/V; 1.99 V/V **2.53** (a) $(\sqrt{2} - 1)^{1/2} f_1$; (b) 10 kHz; (c) 64.4 kHz, about six times greater
2.54 (a) $f_t/(1 + K)$, $Kf_t/(1 + K)$; (b) f_t/K, f_t; noninverting preferred at low gains
2.56 For each, $f_{3dB} = f_t/3$ **2.58** 380 V/V; $\frac{1}{20}$ V/V; 19 V/V; 20
2.59 3.33×10^6 V/V, 1.06 MHz; 1.33×10^6 V/V, 3.18 MHz; 1.0×10^6 V/V, 106 MHz; 1.12×10^6 V/V, 15.9 MHz
2.60 9.19 mV **2.62** 0.5 μs; triangular **2.63** 80 V/μs **2.66** (a) 31.8 kHz; (b) 0.795 V; (c) 0 to 200 kHz; (d) 1 V peak
2.67 0.01 V/V; −250 V/V; 25×10^3, 88 dB
2.70 20.2 V (peak-to-peak)
2.72 159 pF in series with 10^4 Ω all paralleled by 10^8 Ω; 10^5 Ω
2.75 1.75 mH in series with 1.1 Ω all paralleled by 1 kΩ; 1.1 Ω; 1.56 Ω; 110 Ω; 10^3 Ω
2.76 4.95 mV **2.78** 8 mV; 12 mV **2.80** 10 V; 5 V; 15 mV
2.82 (a) 100 mV; (b) 0.2 V; (c) 10 kΩ, 10 mV; (d) 110 mV

Chapter 3

3.1 −5 V, 1 mA; +5 V, 0 mA; +5 V, 1 mA; −5 V, 0 mA **3.3** 25 kΩ
3.5 0 V, 1 mA; −3.33 V, 0 mA; **3.7** 1.7 kΩ; 170 V **3.9** 1/2; 60 mA
3.11 red on; neither on; green on **3.13** 346 mV; 1.2×10^6 I_S **3.15** 3.8 mA; 22.9 mV
3.17 1.022; 1.78×10^{-15} A **3.20** 0.336 mA; 0.664 V **3.22** R = 1.624 kΩ
3.25 0.0 V; +3.57 V
3.26 −4.3 V, 0.93 mA; +5 V, 0.0 mA; 4.3 V, 0.93 mA; −5 V, 0.0 mA;
3.28 0.0 V, 0.86 mA; −3.57 V, 0.0 mA
3.30 R = 1690 Ω; 170 V; essentially the same, since supply is very large
3.31 116.4°; 111 mA peak; 36 V (peak inverse) **3.34** 5 Ω
3.36 (b) 10 mA, 930 Ω, 10 mA diode;
(c) $\Delta V_O/I_L = -(mnV_T/I_D)((V^+ - 0.7m)/(V^+ - 0.7m + mnV_T))$
3.38 Best regulation for 15 mA supply: 24.3 mV change in output
3.40 >2.19 kΩ; 2.125 kΩ; R = 390 Ω **3.42** 65 mV
3.44 (1) 110 Ω, 43.5 mV/V; (2) 9.2 kΩ, 75.4 mV/V
3.46 (c) 2.83 V; (d) 9.3 mA; (e) 10 V **3.48** 1.994 V **3.50** 49.6 V; 4.65 V; 0.22 V
3.52 (a) 9.475 to 1; (b) 1.065 to 1 **3.54** 49.3 V; 48.6 V
3.56 (a) 23.6 V; (b) 417 μV; (c) 32.7 V; (d) 0.68 A; (e) 1.30 A.
3.58 8.865 V; 22 μs; 4.12 A; 8.24 A
3.65 0.51; 0.70; 1.70; 10.8; 0.0; −0.51; −0.70; −1.70; −10.8 V; (semi-)hard limiter; 1 V/V; 0.8 V; −0.8 V
3.67 14.14 V

Chapter 4

4.1 1.34×10^{-16} A; 1.12×10^{-14} A; the second is 84 times larger **4.2** 125; 0.992
4.5 11×; 1000× **4.8** 0.5 μA **4.10** $0.8\dot{3}$; 5 **4.12** −0.553 V; −6.0 V; 0.41 mA
4.14 0.980 **4.16** 100; 99; 19.5 **4.18** R_C = 7.5 kΩ; R_E = 14.2 kΩ **4.20** 4.02 μA
4.22 200 kΩ; 2 MΩ **4.24** 49 V; 49 kΩ **4.26** 4.85 V
4.28 0.30 V; 15 μA; 800 μA; 785 μA; $52.\dot{3}$; 0.981

4.30 $R_C = 10$ kΩ; $R_E = 13$ kΩ; $R_B = 165$ kΩ; 0.863 to 1.008 mA; 6.37 to 4.92 V
4.32 1.075 kΩ; saturation **4.35** 80 mA/V, 1.875 kΩ, 12.4 Ω; 8 mA/V, 18.75 kΩ, 124 Ω
4.36 3.17×10^{-16} A **4.38** 3.33 kΩ; 0.50 kΩ
4.40 Gain $= -(V_{CC} - V_{BE} - \hat{V}_{be})/(V_T + \hat{V}_{be})$; +2.25 V; 1.55 $V_{(peak)}$; −310 V/V
4.42 3 V; 40 mA/V; $v_o/v_i = g_m R_C = 80$ V/V **4.44** 68 μA; 1.18 V **4.46** r_e
4.48 1 mA; 0.996 V/V; −0.630 V/V **4.50** −4000 V/V
4.52 $i_C = 5(1 + v_{CE}/100)$; $v_{CE} = 10 - i_C$; 4.76 V; 5.24 mA; for +30 μA, 2.91 mA and −2.91 V; for −30 μA, −3.08 mA and +3.08 V
4.54 (a) 5.15; (b) $V_{BB} = 0.35\ V_{CC} + 0.7$; (c) 20.5 kΩ, 14.8 kΩ, 1.67 kΩ; (d) 1.85 kΩ
4.56 4.3 kΩ; 4.0 kΩ; nominally, i_C increases by 6%. Actual increase from 0.980 to 1.04 mA and v_O reduced by 0.16 V, or 16%
4.58 1.94 mA; 1.14 kΩ; −77.6 mA/V; 2.11 kΩ; −8.13 V/V; −45.3 A/A **4.60** 10.1 V/V
4.62 (a) 14.3 kΩ; (b) 10 kΩ; (c) −64.5 V/V
4.64 12.75 kΩ; −11.0 V/V; 17.84 mV; −196 mV **4.67** (a) 2.43; (b) 0.631 kΩ; (c) 0.166 **4.69** 50 Ω; 9.9 V/V
4.71 (a) 1.44 mA, 1.44 V, 2.14 V; 5.54 mA, 5.54 V, 6.24 V; (b) 9.80 kΩ, 50.7 kΩ; (c) 0.478 V/V, 0.828 V/V **4.73** 119 Ω; 0.998 V/V; 0.892 V/V; −2.50 V; 3.45 V
4.75 0.784 V/V **4.77** (a) 2.0 mA, 29.95 μA, 3.785 V, 3.085 V; (b) 0.9876 V/V, 203.5 kΩ; (c) 0.996 V/V, 448 kΩ; (d) 0.818 V/V (e) 0.796 V/V
4.78 2.7 kΩ; 2.95
4.80 2.16 V, 2.86 V, 2.86 V; 2.39 V, 2.69 V, 3.09 V; 2.97 V, 3.27 V, 3.67 V
4.82 1.30 V, 2; 4.29 V, 3; 3.07 V, 3.07 V, 1.25, 3.04 **4.85** 0.138 mA
4.87 0.178 V, 198 μs **4.89** 43.3 V, 13.3 V, 12.6 V

Chapter 5

5.1 3 mA; 4 mA **5.3** 1/16 mA/V^2; 1 V **5.5** Available ratios: 9:1, 3:1, 1:1, 1:3, 1:9
5.7 0.5 mA/V^2; 0.536 V **5.9** 50 kΩ, 5 kΩ; $0.2V_{Ds}$% change in each case
5.11 1 V to 2.16 V **5.13** (a) $\partial i_D/\partial T = (\partial K/\partial T)(v_{GS} - V_t)^2 - 2K(v_{GS} - V_t)(\partial V_t/\partial T)$; (b) 0.001 per unit/°C **5.15** 3 V; 4 mA **5.17** 0.586 V **5.19** 20 mA; −2 V
5.21 ∞; V_A/I_{DSS} **5.23** −2 V; 1 mA/V^2; 1 V; 1 mA; 1.1 mA
5.25 0.096 V; −0.106 V; square-law characteristic **5.27** ≤ −3 V; 2.25 mA
5.29 (a) 12.93 kΩ; (b) 75 μm **5.31** 200 μm; 50 μm; 15 kΩ **5.33** 5 V; 0.125 V
5.35 3 kΩ; 10 kΩ **5.37** 0.125 mA; +6 V
5.39 0.5 mA; 8 V; 5 V; 3.3 V, 6.67 V; 5 V; 0.5 mA; 5 V; 0.293 V
5.41 4 mA; 0.0 V; 500 Ω **5.43** 2 mA; 0.205 mA; −0.195 mA; 2 mA/V; same
5.46 1 kΩ **5.47** 500 μm; 2 V **5.49** 0 V; 2 kΩ
5.51 4 V; 1 mA; 99 kΩ **5.53** 7.93 kΩ, for which use 8.2 kΩ; 0.043 mA to 0.146 mA
5.55 2 V; 1.34 V; 1.34 mA **5.57** $V_{GG} = 8$ V; 2 kΩ; 4 mA; 5.63 mA
5.59 0.1 mA; 3 V; 1.85 V peak **5.61** (a) −3 V; (b) −3.41 V; (c) −4 V
5.63 −6 V/V **5.65** 200 Ω; 3.57 V/V; 100 Ω; 4.76 V/V
5.67 (a) 35 kΩ, 50 kΩ, 10 MΩ; (b) 0.4 mA/V, 400 kΩ; (c) −7.66 V/V; (d) 0.934 V/V, 2.32 kΩ; (e) 0.456 V
5.69 (a) 6 V, 2.97 mA, −2.01 V, 12 V; (b) 2.99 mA/V, 101 kΩ; (d) 0.42 MΩ, 0.808 V/V (e) −3.98 V/V (f) −3.22 V/V
5.71 $R_G = 1$ MΩ; $R_S = 1.28$ kΩ; 0.86 V/V **5.73** −10 V/V **5.75** −10 V/V
5.77 (a) 6 V, 5.87 V, 3 V; (b) 6 V, 4.45 V, 0.205 V; (c) 6 V, 0.008 V, 0.002 V
5.80 20 μm; 100 μm
5.81 225 μA; 225 μA; $A_{gd} = -3000$ V/V; with r_o, $A_{gd} = -115.4$ V/V, $R_{in} = 85.9$ kΩ; −9.13 V/V; ±2 V
5.83 6.67 kΩ; 1.538 kΩ; 660 kΩ **5.85** 48.2 kHz **5.87** 50 μm; 2 kΩ
5.89 15 mA; −21 V/V **5.91** 4 V; −8 V/V; 10 MΩ; 2.2 MΩ

Chapter 6

6.1 -2.7 V; 3.02 V; 3.02 V **6.3** $+0.5$ V, -0.5 V, -5.0 V
6.5 (a) $V_{CC} - R_C I/2$; (b) $R_C I/2$; (c) 4.0 V; (d) 0.404 mA, 9.9 kΩ **6.7** 0 V; -5 V
6.9 2 mA; 4 mA; 17.3 mV **6.11** 2 mA/V; 201 kΩ
6.13 (a) 0.4 mA, 10 mV; (b) 1.40 mA, 0.60 mA; (c) -2.0 V, $+2.0$ V; (d) 40 V/V
6.15 39.6 V/V; 50.5 kΩ **6.18** (a) 20 V/V; (b) 0.231 V/V; (c) $86.6 \equiv 38.7$ dB;
(d) $v_o = 0.2 \sin(2\pi \times 1000t) - 0.023 \sin(2\pi \times 60t)$
6.19 $R_E = 25\ \Omega$; $R_C = 10\ \text{k}\Omega$; $R_o \geq 50\ \text{k}\Omega$; $R_{i_{cm}} = 5\ \text{M}\Omega$; $\pm 12\ V$ would do, ± 15 V would be better.
6.21 2% mismatch, for example ±1% resistors **6.23** 74.4 MΩ **6.25** 2.5 mV
6.27 0.125 mV **6.29** 1.32 μA; 0.33 μA; 0.99 μA
6.31 $I/3$; $2I/3$; $R_C I/3$; 16.7 mV; 17.3 mV; 0.495 μA; 0.5 μA; 0.33 μA
6.33 $I_O = \alpha(R_2 V_{CC} + (R_1 - R_2)V_{BE})/(R_E(R_1 + R_2))$; $R_1 = R_2 = R_E(1 - 2V_{BE}/V_{CC})$;
$R_E = 7.5\ \text{k}\Omega$; $R_1 = R_2 = 6.8\ \text{k}\Omega$; $+8.2$ V
6.35 4.3 kΩ; 0 V ideally, 0.4 or 0.5 in practice
6.36 (a) $i_{C2} = 2$ mA, $i_{C3} = 3$ mA; (b) $i_{C1} = 0.5$ mA, $i_{C3} = 3/2$ mA;
(c) $i_{C1} = 0.33$ mA, $i_{C2} = 0.67$ mA
6.39 $I_O/I_{\text{REF}} = n/(1 + (n + 1)/\beta)$; $\beta = 20(n + 1)$ **6.41** $R = 8.63\ \text{k}\Omega$; 8.63 kΩ
6.43 $V_{B1} = 9.3$ V; $V_{B2} = -9.3$ V; $I_{R1} = 1.86\ \text{mA} = I_{C3} = I_{C4}$; $V_{C3} = 3.72$ V;
$V_{C4} = V_{C5} = V_{B5} = 0.7$ V; $I_{C6} = 1.86$ mA; $V_{C6} = 3.14$ V;
$I_{C2} = I_{C7} = I_{C8} = I_{C9} = I_{C10} = I_{C11} = 1.86$ mA; $I_{R4} = 3.72$ mA; $V_{C8} = V_{C7} = -3.72$ V;
$V_{C9} = V_{C10} = V_{B10} = V_{B11} = 4.3$ V; $V_{C11} = 1.86$ V
6.47 Replicate Q_2; $I_O/I_{\text{REF}} = 1/(1 + (n + 1)/(\beta(\beta + 1)))$; 9 **6.50** 2 μA; 0.2%
6.51 (a) $I_O/I_{\text{REF}} = 0.5(1/(1 + 2/(\beta^2 + 2\beta)))$ (b) Replace Q_3, Q_4 by seven identical transistors: one, separately, provides 1 mA; two joined, 2 mA; four joined, 4 mA; 999.2 μA, 1998.4 μA, 3996.8 μA
6.53 6.61 MΩ **6.55** v_O stays same; v_O lowers by 0.43 V
6.57 Use two-transistor mirror with $R = 146.5\ \text{k}\Omega$; $R_i = 100\ \text{k}\Omega$; $R_o = 500\ \text{k}\Omega$;
$A_{vo} = 2000$ V/V; $I_B = 0.5\ \mu$A; -13.6 V to $+14.3$ V (for small output signals); 67 MΩ
6.59 -1 V/V **6.61** CMRR $= \beta_P RI/(2V_T)$; $10^5 \equiv 100$ dB
6.63 12.85 V/V; 10^{-3} V/V; $12.9 \times 10^3 \equiv 82.2$ dB; 15.6 mV
6.65 $R = 586\ \Omega$; $R_D = 16\ \text{k}\Omega$; 16 V/V; -14 V **6.67** (a) 1.2 V, 1.2 V, -1.2 V, 0 V;
(b) 1.245 V, 1.141 V, -1.141 V, $+0.104$ V;
(c) 1.283 V, 1.0 V, -1.0 V, $+0.283$ V; 0.283 V
6.69 $W/L = 25$; 20 μA **6.71** 150 mV; V_t; 60%; 24% **6.73** 136.5 μA
6.76 80 μA; 100 μA **6.78** 50 μA **6.80** 58.5 kΩ **6.81** 4000; very large
6.85 (b) $I = 10\ \mu$A; 32 μm; 1.25 V; 0.95 V **6.87** 100 μm; 1 kΩ
6.89 1 mA; 0.553 V; 4.55 V; 447 kΩ **6.91** 10 mA; 2 mA; -100 V/V **6.93** 12.47 V/V
6.95 R_5; reduce to 7.37 kΩ; 4104 V/V; reduce R_4 to 1.12 kΩ
6.97 Raises to 3 kΩ; 5581 V/V; 3378 V/V; removal of internal loading is important

Chapter 7

7.1 $V_o(s)/V_i(s) = RC_1 s/(1 + sR(C_1 + C_2))$; STC with $C_{\text{eq}} = C_1 // C_2$; high-pass; zero at 0 Hz; pole at 15.9 Hz
7.3 0 dB, $-90°$; $+0.04$ dB, $-95.0°$ **7.4** 1 kHz; 510 Hz; 105 Hz
7.7 (a) $V_o(s)/V_i(s) = G_m R_L R_i/[(R_i + R_s)(1 + 1/((R_i + R_s)C_C s))(1 + C_L R_L s)]$;
(b) $A_M = G_m R_L R_i/(R_i + R_s)$, $F_L(s) = (1 + 1/((R_1 + R_2)C_C s))^{-1}$, $F_H(s) = 1/(1 + C_L R_L s)$;
(c) 1.1 mA/V; (d) 0.144 μF (e) 15.9 pF **7.9** (a) 2.18; (b) 7.05
7.11 (a) 10 Hz; (b) 10.49 Hz **7.13** 5.67×10^6 rad/s
7.15 (a) $A_M = -R_L/(1/g_m + R_s)$;

(b) $R_{gs} = (R + R_S)/(1 + g_m R_S)$, $R_{gd} = R_L(1 + g_m R/(1 + g_m R_s)) + R$;
(c) −20 V/V, 0.45 Mrad/s, 9 Mrad/s; −14.3 V/V, 0.623 Mrad/s, 8.9 Mrad/s; −10 V/V, 0.866 Mrad/s, 8.7 Mrad/s
7.17 −10 V/V; 18.6 μF; 1.43 Hz; $V_o(s)/V_i(s) = -10(s + 2\pi \times 1.43)/(s + 2\pi \times 10)$; −10/7
7.19 0.7 μF; 13.5 Hz; 22.7 Hz
7.21 g_m = 110 μA/V; g_{mb} = 17 μA/V; r_o = 0.5 MΩ; C_{sb} = 0.05 pF; C_{db} = 0.03 pF; C_{gb} = 0.05 pF; C_{gs} = 0.08 pF; C_{gd} = 0.01 pF 7.22 195 MHz 7.25 2.33 MΩ
7.27 1 kΩ; −500 V/V; 5 kΩ
7.29 $V_o(s)/V_i(s) = -g_m R_L(1 - s(g_m/C_{gd})^{-1})/(1 + s(R_L C_{gd} + C_L))$; $\omega_P = -1/(R_L(C_{gd} + C_L))$; $\omega_Z = g_m/C_{gd}$; 80 Mrad/s; 8 Grad/s
7.31 21.2 kHz; 25.5 MHz 7.33 35 MHz; 3.18 Hz
7.35 (a) 5 kΩ, 5.26 kΩ; 200 A/A, 212.6 A/A; (b) 5.21 kΩ, 24.4 Ω, 40.8 MA/V, 1.025 mA;
(c) 2×10^{-4} V/V, avoid effect of C_μ; (d) 26.0 MΩ, 6.12 kHz; (e) 16×10^{-6} A/V, 128 kΩ, 131 V
7.37 3.64×10^9 rad/s; 24.3 Mrad/s 7.39 0.54 pF; 20 mA/V; 7.5 kΩ; 33.3 MHz
7.41 10 ω_β 7.43 5.40 kΩ; −17.4 V/V; 1.67 MHz
7.45 C_E = 15.6 μF; C_{C_2} = 5.96 μF; C_{C_1} = 1.08 μF 7.47 193
7.49 −4000 V/V; 265 kHz; 1.06 GHz
7.51 $I_o(s)/I_i(s) = (1 - sC_\mu/g_m)/(1 + s(2C_\pi + C_\mu)/g_m)$; 214 MHz; 3.18 GHz
7.53 −15 V/V; 20.3 MHz, 707 MHz, 124 MHz; 20 MHz
7.55 (c) −44.7 V/V; 1.14 MHz, 87.3 MHz; 2.03 GHz
7.58 C_2 = 2.8 μF; C_1 = 0.32 μF; 0.922 V/V; 31.5 MHz 7.59 0.963 V/V; 816 MHz
7.61 (a) 2.51 MΩ, −4008 V/V; (b) 107.8 kHz, C_L, then C_{μ_2} directly and multiplied;
(c) dominant effect is reduction of r_o and output time constant by a factor of 10, and thus f_H increases by a factor of 10
7.63 92.5 V/V; 0.327 MHz; 30.3 MHz
7.65 74.6 V/V; 0.397 MHz; 29.6 MHz 7.68 15.9 kHz; 500 kHz
7.69 46.07 V/V; 6.98 MHz
7.71 (a) −66.7 V/V, 117 kHz; (b) −66 V/V, 3.8 MHz; (c) 49.5 V/V, 4.6 MHz;
(d) −192 V/V, 1.60 MHz; (e) −66 V/V, 3.8 MHz; (f) 49.5 V/V, 4.6 MHz

Chapter 8

8.1 9.99×10^{-3}; 91; −9% 8.3 (b) 1110; (c) 20 dB; (d) 10 V, 9 mV, 1 mV; (e) 2.44%
8.5 $\beta = 9/A$; 1.0 8.7 10; (a) 989, 990; (b) 937.5, 960 8.9 100 kHz; 10 Hz
8.11 0.997 V/V; 300 V/V; 49.5 dB
8.13 β = 0.08; five segments with gains of 0, 11.11 V/V, 12.34 V/V, 11.11 V/V, and 0 V/V and break points (v_s, v_o) at (−1.17 V, −14 V), (−0.81 V, −10 V), (+8.1 V, +10 V), (+1.17 V, +14 V) 8.15 1 A/mA; 9 mA/A 8.17 $\beta = (1/\alpha)(R_{E1}R_{E2})/(R_{E1} + R_{E2} + R_f)$
8.19 1 V/μA; 9 μA/V 8.21 lower; 99; 10 kΩ
8.23 (a) $h_{11} = R_1R_2/(R_1 + R_2)$ V/A, $h_{12} = R_2/(R_1 + R_2)$ V/V, $h_{22} = 1/(R_1 + R_2)$ A/V, $h_{21} = -R_2/(R_1 + R_2)$ A/A; (b) 10 Ω, 0.01 V/V, 0.99×10^{-3} ℧, −0.01 A/A
8.25 76.3 V/V; 1.0 V/V; 0.987 V/V; 1.54 MΩ; 2.02 Ω
8.27 0.0 V; 0.7 V; 31.3 V/V; 0.1 V/V; 7.6 V/V; ∞; 163 Ω
8.29 (a) R_{D1} = 1.65 kΩ, R_{E2} = 88 Ω, R_L = 100 Ω;
(b) 3.05 V/V, 1.0 V/V, 0.75 V/V, ∞, 24.7 Ω
8.31 R_E = 1 kΩ; 0.988 mA/V; 258 MΩ; 9.67 MΩ 8.33 0.1 mA/V; 0.098 mA/V
8.35 7.99 V/mA; 171 Ω; 106 Ω 8.37 (a) shunt-series; (b) series-series;
(c) shunt-shunt 8.39 −192 V/V; 29.9 Ω; 29.5 Ω 8.41 9.09 A/A; 110 Ω; 90.9 kΩ

8.43 $V_{E2} = 1.41$ V; $V_{B2} = 2.11$ V; $V_O = 5$ V; $I_{E1} = 100\ \mu$A; $I_{E2} = 10.1$ mA; 2.54 V/V; 1.45 kΩ **8.44** 3.72 V/V; 141 Ω **8.48** 0.0594 V/V **8.50** 23.4 V/V; 175 Ω
8.51 10^4 rad/s; 0.002 V/V; 500 V/V **8.53** 8×10^{-3} V/V
8.55 99 V/V; 1.01×10^6 Hz; 10^7 Hz; 101 **8.57** 9×10^{-3}; 17.94 kHz
8.59 2 V/V; 17.3 MHz **8.61** 7.86 kHz; 51.8°; 0.01414 **8.63** 4.9×10^{-5}; 1.69×10^4
8.65 2.48×10^4 V/V; 6.54×10^3 V/V **8.67** 10^3 Hz; 2000 **8.69** 10 Hz; 15.9 nF
8.70 59 pF; 239×10^6 Hz

Chapter 9

9.1 Upper limit (same in all cases): 4.7 V, 5.4 V; lower limits: −4.3 V, −3.6 V; −2.15 V, −1.45 V; −4.7 V, −4.0 V
9.4 152 Ω; 0.978 V/V; 0.996 V/V; 0.998 V/V; 2% **9.6** IV_{CC}; IV_{CC}; IV_{CC}; IV_{CC}
9.7 10%; 13.3%; 16%; 20% **9.9** 7 V **9.11** 4 V; 12.8%; 11.1 kΩ
9.13 5.0 V peak; 3.18 V peak; 3.38 Ω; 4.83 Ω; 3.65 W; 2.59 W
9.15 $\hat{V}_o^2/R_L$; $V_{SS}\hat{V}_o/R_L$; $\hat{V}_o/V_{SS}$; 100%; V_{SS}; V_{SS}^2/R_L; $V_{SS}/2$; 50% **9.17** 2.5 V **9.19** 12.5
9.21 20.7 mA; 788 mW; 7.9°C; 38.7 mA **9.23** 1.34 kΩ; 1.04 kΩ **9.25** 50 W; 2.5 A
9.27 140°C; 0.570 V **9.29** 100 W; 0.4°C/W **9.31** 0.85 Ω
9.33 0 mA, 0 mA; 20 μA, 22.5 μA; −20 μA; −22.5 μA
9.35 1.96 mA; 38.4 μA; out of base 1 and into base 2; 3.4 μA; 277 kΩ; 0.94 V/V
9.37 0.064 v_i; 64.1 mA/V; −64.1 V/V; 14 kΩ
9.39 $R_1 = 300$ kΩ; $R_2 = 632$ kΩ; 9.48 V; −10.65 V
9.41 13 Ω; 433 mV; 0.33 μA **9.43** $R_1 = 60$ kΩ; $R_2 = 5$ kΩ; 9.7 nA
9.45 $I_{E1} = I_{E2} = 17\ \mu$A; $I_{E3} = I_{E4} = 358\ \mu$A; $I_{E5} = I_{E6} = 341\ \mu$A; 10.3 V
9.47 14 V; 1.9 W; 11 V **9.49** $R_3 = R_4 = 40$ Ω; $R_1 = R_2 = 2.2$ kΩ
9.51 40 kΩ; 50 kΩ **9.53** $L = \mu_n(v_{GS} - V_t)/U_{sat}$; 3 μm; 3 A; 1 A/V

Chapter 10

10.2 36.3 μA
10.3 625 mV; for A, 7.3 mA/V, 134 Ω, 6.85 kΩ, 274 kΩ; for B, 21.9 mA/V, 44.7 Ω, 2.28 kΩ, 91.3 kΩ **10.5** $(I_3/I_1)^{1/2} = [(1/K_1)^{1/2} + (1/K_2)^{1/2}]/[(1/K_3)^{1/2} + (1/K_4)^{1/2}]$; 100 μA
10.7 **(a)** 0.73 mA; **(b)** 0.586 mA, 0.786 mA; **(c)** 3.4 V **10.9** 616 mV; 535 mV; 4.05 kΩ
10.11 4.75 μA; 625 mV; 551 mV; 1.94 kΩ **10.13** 56.5 kΩ; 9.353 μA
10.15 226 to 250; ±5% **10.17** +14.7 V to −12.9 V **10.19** 6.37 kΩ; 270 μA; 270 μA
10.21 1.68 mA; 50.4 mW **10.23** Raise R₁, R_2 to 4.63 kΩ **10.26** 0.96 mV
10.27 $R_{o9} = [1 + 50R/(65.8 + R)]2.63$ MΩ; 18.2 kΩ; 15.6 MΩ **10.30** 19.5 MΩ; 498 MΩ
10.31 3.10 MΩ; 9.38 mA/V **10.33** 4.2 V to −3.6 V **10.35** 21 mA
10.37 2.5×10^5 V/V; 62 Ω; 1.91×10^5 V/V; ±4 V **10.39** 11.4 MHz **10.41** 637 kΩ
10.43 159 kHz; 15.9 MHz
10.45 **(a)** to $\frac{1}{2}$ previous value; increased by a factor of 2 to 125 μA/V, 31.25 μA/V; **(b)** increase to −125 V/V and 6250 V/V; **(c)** part due to V_t is not changed; other parts reduced to $\frac{1}{2}$ previous; **(d)** C_C doubled; slew rate halved
10.47 $V_{OS} = g_{m3}\ \Delta V_t/G_{m1}$; 1.6 mV **10.49** **(a)** 19.9 pF **(b)** 31.8 pF
10.51 9.95 pF; 16.5 kΩ; 2.5 V/μs **10.52** 10 pF; 1.40 V; 15.6 μm/μm
10.56 61.3 MΩ; 3750 V/V **10.57** 7.85 μA; 25.1 **10.59** 27.6×10^6 Hz; 15.4 pF
10.60 six bits; 0.159 V; seven bits; seven bits; 0.118 V; 0.059 V **10.63** $I/16$; $I/8$; $I/4$; $I/2$
10.65 Use op amp with $R/2$ input and $50R$ feedback to drive V_{ref}; 15 sine-wave amplitudes, from 0.625 V peak to 9.375 V peak; an output of 10 V (peak-to-peak) corresponds to a digital input of (1000).
10.67 8.19 ms; 4.095 ms; 9.90 V; no, stays the same!

Chapter 11

11.1 1 V/V, 0°, 0 dB, 0 dB
0.894 V/V, −26.6°, −0.97 dB, 0.97 dB
0.707 V/V, −45.0°, −3.01 dB, 3.01 dB
0.447 V/V, −63.4°, −6.99 dB, 6.99 dB
0.196 V/V, −78.7°, −14.1 dB, 14.1 dB
0.100 V/V, −84.3°, −20.0 dB, 20.0 dB
0.010 V/V, −89.4°, −40.0 dB, 40.0 dB
11.3 1.000; 0.944; 0.010 **11.5** 0.509 rad/s; 3 rad/s; 5.90
11.8 $T(s) = 10^{15}/[(s + 10^3)(s^2 + 618s + 10^6)(s^2 + 1618s + 10^6)]$, low-pass;
$T(s) = s^5/[(s + 10^3)(s^2 + 618s + 10^6)(s^2 + 1618s + 10^6)]$, high-pass
11.9 $T(s) = 0.2225(s^2 + 4)/[(s + 1)(s^2 + s + 0.89)]$
11.11 $T(s) = 0.5/[(s + 1)(s^2 + s + 1)]$; poles at $s = -1, -\frac{1}{2} \pm j\sqrt{3}/2$, 3 zeros at $s = \infty$
11.13 28.6 dB
11.15 N = 5; f_0 = 10.55 kHz, at −108°, −144°, −180°, −216°, −252°;
$p_1 = -20.484 \times 10^3 + j63.043 \times 10^3$ (rad/s), $p_2 = -53.628 \times 10^3 + j38.963 \times 10^3$ (rad/s),
$p_3 = -\omega_0 = -66.288 \times 10^3$ rad/s, $p_4 = -53.628 \times 10^3 - j38.963 \times 10^3$ (rad/s),
$p_5 = -20.484 \times 10^3 - j63.043 \times 10^3$ (rad/s);
$T(s) = \omega_0^5/[(s + \omega_0)(s^2 + 1.618\omega_0 s + \omega_0^2)(s^2 + 0.618\omega_0 s + \omega_0^2)]$; 27.8 dB
11.19 R_1 = 10 kΩ; R_2 = 100 kΩ; C = 159 pF
11.21 R_1 = 1 kΩ; R_2 = 1 kΩ; C_1 = 0.159 μF; C_2 = 1.59 nF;
High-frequency gain = −100 V/V
11.23 $T(s) = (1 - RCs)/(1 + RCs)$; 2.68 kΩ, 5.77 kΩ, 10 kΩ, 17.3 kΩ, 37.3 kΩ
11.25 $T(s) = 10^6/(s^2 + 10^3 s + 10^6)$; 707 rad/s; 1.16 V/V
11.27 R = 4.59 kΩ; R_1 = 10 kΩ **11.28** $T(s) = s^2/(s^2 + s + 1)$
11.30 $T(s) = (s^2 + 1.42 \times 10^5)/(s^2 + 375s + 1.42 \times 10^5)$
11.33 L = 0.5 H; C = 20 nF **11.35** $V_o(s)/V_i(s) = s^2/(s^2 + s/RC + 1/LC)$
11.37 Split R into two parts, leaving $2R$ in its place and adding $2R$ from the output to ground.
11.39 $L_1/L_2 = 0.235$; $|T| = L_2/(L_1 + L_2)$; $|T| = 1$.
11.40 For all resistors = 10 kΩ, C_4 is (a) 0.1 μF, (b) 0.01 μF, (c) 1000 pF; For R_5 = 100 kΩ and $R_1 = R_2 = R_3$ = 10 kΩ, C_4 is (a) 0.01 μF, (b) 1000 pF, (c) 100 pF
11.43 $R_1 = R_2 = R_3 = R_5$ = 3979 Ω; R_6 = 39.79 kΩ; C_{61} = 6.4 nF; C_{62} = 3.6 nF
11.44 $C_4 = C_6$ = 1 nF; $R_1 = R_2 = R_3 = R_5 = R_6 = r_1 = r_2$ = 159 kΩ
11.48 (a) $T(s) = 0.451 \times 10^4(s^2 + 1.70 \times 10^8)/[(s + 0.729 \times 10^4)(s^2 + 0.279 \times 10^4 s + 1.05 \times 10^8)]$; (b) For LP section: C = 10 nF, $R_1 = R_2$ = 13.7 kΩ; For LPN section: C = 10 nF, $R_1 = R_2 = R_3 = R_5$ = 9.76 kΩ, R_6 = 35.9 kΩ, C_{61} = 6.18 nF, C_{62} = 3.82 nF
11.49 C = 10 nF; R = 15.9 kΩ; $R_1 = R_f$ = 10 kΩ; R_2 = 10 kΩ; R_3 = 390 kΩ; 39 V/V
11.51 ±1%
11.53 (a) For only ω_z, change C_1 and r or R_3, or change R_2 and r or R_3; R_2 and R_3 preferred; (b) For only Q_z, change only r, or only R_3
11.55 R_3 = 141.4 kΩ; R_4 = 70.7 kΩ
11.57 $T(s) = -(16\, s/RC)/[s^2 + 2\, s/RC + 16/(RC)^2]$; Bandpass; $\omega_0 = 4/RC$; Q = 2; Center-frequency gain = 8 V/V
11.59 $T(s) = s^2/[s^2 + (C_1 + C_2)s/R_3C_1C_2 + 1/R_4R_3C_1C_2]$; High-pass; High-frequency gain = 1 V/V; R_3 = 141.4 kΩ; R_4 = 70.7 kΩ
11.60 For first-order section: C_1 = 3.18 nF; For one S and K section, the grounded and floating capacitors are, respectively, C_2 = 984 pF and C_3 = 10.3 nF; For the other S and K section, corresponding capacitors are C_4 = 2.57 nF and C_5 = 3.93 nF, respectively.
11.62 Sensitivities of ω_0 to R, L, C are 0, $-\frac{1}{2}$, $-\frac{1}{2}$ respectively, and of Q are 1, $-\frac{1}{2}$, $\frac{1}{2}$, respectively.

Chapter 12

12.1 (a) $\omega = \omega_0$, $AK = 1$; (b) $d\phi/d\omega$ at $\omega = \omega_0$ is $-2Q/\omega_0$; (c) $\Delta\omega_0/\omega_0 = -\Delta\phi/2Q$.
12.3 To non-inverting input, connect LC to ground and R to output; $A = 1 + R_2/R_1 \geq 1.0$; Use $R_1 = 10$ kΩ, $R_2 = 100$ Ω (say); $\omega_0 = 1/\sqrt{LC}$ (a) $-\frac{1}{2}\%$; (b) $-\frac{1}{2}\%$; (c) 0%.
12.5 Minimum gain is 20 dB; phase shift is 180°.
12.6 Use $R_2 = R_5 = 10$ kΩ; $R_3 = R_4 = 5$ kΩ; $R_1 = 50$ kΩ
12.9 $V_a(s)/V_o(s) = (s/RC)/[s^2 + 3s/RC + 1/R^2C^2]$; with magnitude zero at $s = 0$, $s = \infty$; $\omega_0 = 1/RC$; $Q = \frac{1}{3}$; Gain at $\omega_0 = \frac{1}{3}$.
12.10 $\omega = 1.16/CR$. **12.12** $R_3 = R_6 = 6.5$ kΩ; $v_O = 2.08$ $\text{V}_{\text{(peak-to-peak)}}$
12.13 $L(s) = (1 + R_2/R_1)(s/RC)/[s^2 + s3/RC + 1/R^2C^2]$; $L(j\omega) = (1 + R_2/R_1)/[3 - j(1/\omega RC - \omega RC)]$; $\omega = 1/RC$; for oscillation, $R_2/R_1 = 2$.
12.15 18.5 V. **12.17** $A\beta(s) = -(R_f/R)/[1 + 6/RCs + 5/R^2C^2s^2 + 1/R^3C^3s^3]$; $R_f = 29R$; $f_0 = 0.065/RC$
12.20 (a) 0.00; (b) 0.01875; (c) 0.0104; (d) 0.0233; THD $\approx$ 2%
12.21 For circuits (a), (b), (d), characteristic equation is: $C_1C_2Ls^3 + (C_2L/R_L)s^2 + (C_1 + C_2)s + 1/R_L + g_m = 0$; $\omega_0 = [(C_1 + C_2)/C_1C_2L]^{1/2}$; $g_mR_L = C_2/C_1$; For circuit (c): $LC_1C_2s^3 + (C_1L/R_L)s^2 + (C_1 + C_2)s + 1/R_L + g_m = 0$; $\omega_0 = [(C_1 + C_2)/C_1C_2L]^{1/2}$; $g_mR_L = C_1/C_2$.
12.23 From 2.01612 MHz to 2.01724 MHz.
12.25 (a) $V_{TL} = V_R(1 + R_1/R_2) - L_+R_1/R_2$, $V_{TH} = V_R(1 + R_1/R_2) - L_-R_1/R_2$; (b) $R_2 = 200$ kΩ, $V_R = 0.0476V$.
12.28 (a) Either +12 V or −12 V; (b) Symmetrical square wave of frequency f and amplitude ±12 V, and lags the input by 65.4°. Maximum shift of average is 0.1 V.
12.29 $V_Z = 6.8$ V; $R_1 = R_2 = 37.5$ kΩ; $R = 4.1$ kΩ
12.31 $V_Z = 3.6$ V, $R_2 = 6.67$ kΩ; $R = 50$ kΩ; $R_1 = 24$ kΩ, $R_2 = 27$ kΩ.
12.33 $V_Z = 6.8$ V; $R_1 = R_2 = R_3 = R_4 = R_5 = R_6 = 100$ kΩ; $R_7 = 5.0$ kΩ; Output is a symmetric triangle with half period of 50 μs and ±7.5 V peaks.
12.35 96 μs **12.36** $R_1 = R_2 = 100$ kΩ; $R_3 = 134.1$ kΩ; $R_4 = 470$ kΩ 6.5 V; 61.8 μs.
12.38 (a) 9.1 kΩ; (b) 13.3 V **12.39** $R_A = 21.3$ kΩ; $R_B = 10.7$ kΩ
12.41 V = 1.0996 V; $R = 400$ Ω; Table rows, for v_O, θ, 0.7 sin θ, error % are:
0.70 V, 90°, 0.700 V, 0%;
0.65 V, 63.6°, 0.627 V, 3.7%;
0.60 V, 52.4°, 0.554 V, 8.2%;
0.55 V, 46.1°, 0.504 V, 9.1%;
0.50 V, 41.3°, 0.462 V, 8.3%;
0.40 V, 32.8°, 0.379 V, 5.6%;
0.30 V, 24.6°, 0.291 V, 3.1%;
0.20 V, 16.4°, 0.197 V, 1.5%;
0.10 V, 8.2°, 0.100 V, 0%;
0.00 V, 0°, 0.0 V, 0%.
12.42 ±2.5 V
12.45 Table rows; circuit v_O/V_T, circuit v_I/V_T, ideal v_O/V_T, and error as a % of ideal are:
0.250, 0.451, 0.259, −3.6%
0.500, 0.905, 0.517, −3.4%
1.000, 1.847, 1.030, −2.9%
1.500, 2.886, 1.535, −2.3%
2.000, 4.197, 2.035, −1.7%
2.400, 6.292, 2.413, −0.6%
2.420, 6.539, 2.420, 0.0%
12.47 $R_1 = R_2 = 10$ kΩ (say); 3.18 V

12.49 $R_1 = 1\ \text{M}\Omega$; $R_2 = 1\ \text{M}\Omega$; $R_3 = 45\ \text{k}\Omega$; $R_4 = 1\ \text{M}\Omega$; $C = 0.16\ \mu\text{F}$ for a corner frequency of 1 Hz.
12.53 Use op amp circuit with v_A connected to positive input, LED between output and negative input and resistor R between negative input and ground; $I_{\text{LED}} = v_A/R$.
12.54 $i_M = C\ |dv/dt|$; $C = 2.65\ \mu\text{F}$; $i_{M120} = 2\ i_{M60}$; $i_{M180} = 3\ i_{M60}$; Acts as a linear frequency meter for fixed input amplitude; with C, has a dependence on waveform rate of change; 1.272 mA.
12.55 10 mV, 20 mV, 100 mV; 50 pulses, 100 pulses, 200 pulses

Chapter 13

13.1 For either or both inputs high, $v_O = 0$; for both inputs low, $v_O = V^+$
13.3 $NM_L = 0.8$ V; $NM_H = 1.1V$
13.5 $NM_L = 0.3V^+$; $NM_H = 0.2V^+$; Transition width $= 0.2V^+$; $V^+ = 5$ V
13.7 15.9 mW, 2.3 mW, 18.2 mW; 2.6 mW; 15.6 mW **13.9** 45 pJ
13.10 24.9 ns; 40.8 ns
13.13 **(a)** 4.0 V, 1.0 V, 0.158 V, 1.93 V, 2.07 V, 0.84 V; **(b)** 3.40 V, 1.47 V; **(c)** For v_O high, $I_S = 0$; For v_O low, $I_S = 36.9\ \mu\text{A}$; For 50% duty cycle, $P_D = 92.3\ \mu\text{W}$
13.15 2.51 V **13.17** 0.891 $\text{V}^{1/2}$ **13.19** 1.6 ns
13.21 2 ns, 0.2 mW, 0.4 pJ; 10 ns, 0.04 mW, 0.4 pJ; 1.78 ns, 0.253 mW, 0.45 pJ; 0.89 ns, 2.03 mW, 1.80 pJ; 20 ns, 0.20 mW, 4 pJ; 8.89 ns, 0.051 mW, 0.45 pJ
13.24 4.15 **13.25** 10.7; 5.4; 3.6 **13.27** 281 μA; 0.702 mW
13.28 1.66 mA; 247 μA; 0.15 ns; 1.01 ns; 0.58 ns
13.31 $Y = \overline{A + BC}$; $(W/L)_1 = 4/2$; $(W/L)_2 = 2/4$; $(W/L)_3 = (W/L)_4 = 8/2$
13.33 **(a)** 5 V, 4 V, 1.35 V, 0.03 V; **(b)** 0 mA, 221 μA; **(c)** 3.2 mA, 2.4 mA, 2.8 mA, 7.1 ns
13.36 75 μA; 75 μA **13.37** **(b)** $V_{DD}/2$, -66.7 V/V, 146 kΩ
13.39 $V_{IL} = 1.80$ V; $V_{IH} = 2.54$ V; $NM_L = 1.80$ V; $NM_H = 2.46$ V
13.41 2.0 V, 3.0 V, 2.0 V; 2.25 V, 2.75 V, 2.25 V; 2.375 V, 2.625 V, 2.375 V
13.43 144.5 μA **13.45** 11.81 μm; 0.10 pJ **13.48** **(a)** 0.32 ns; **(b)** 0.128 ns
13.49 $(W/L)_n \geq 1.6$; $(W/L)_p \geq 3.2$
13.51 8 transistors total, 4 PMOS, 4 NMOS; From V_{DD}, 2 series-connected PMOS pairs, joined in parallel to Y and controlled by A, B and $\overline{A}$, $\overline{B}$, respectively; From Y, 2 parallel-connected NMOS pairs, joined in series to ground and controlled by A, B and $\overline{A}$, $\overline{B}$ respectively
13.53 NOR: $W_n = 10\ \mu\text{m}$, $W_p = 80\ \mu\text{m}$, area $= 1800\ \mu\text{m}^2$; NAND: $W_n = 40\ \mu\text{m}$, $W_p = 20\ \mu\text{m}$, area $= 1200\ \mu\text{m}^2$
13.55 $Y = A \cdot \overline{C} + B \cdot C$; Y is continuous, being the same as A when C is low, and the same as B when C is high
13.57 0.5 μA/kHz; 0.1 μA/kHz; 110 mW
13.58 Buffered inverter is 1.11 times the area of the unbuffered (with 1% of its input capacitance!)
13.61 Cross-coupled NAND with truth table columns: ($\overline{S}$, $\overline{R}$, Q, $\overline{Q}$) and corresponding rows (1, 1, 0 or 1, 1 or 0), (1, 0, 0, 1), (0, 1, 1, 0), and (0, 0, 1, 1), the latter being the doubly-gated, normally-unused state
13.64 $K_5 = K_6 = 6.11\ K$ **13.65** **(a)** $T = 0.693\ CR$; **(b)** $C = 0.1\ \mu\text{F}$, $R = 14.4\ \text{k}\Omega$; $C = 0.01\ \mu\text{F}$, $R = 144\ \text{k}\Omega$
13.67 6.87 μs; 9.8 V; 5.7 V; $\approx$0.1 V; 0.98 mA; The source current can be as large as 21 mA (for $R_{\text{on}} = 200\ \Omega$), but is clearly limited by K_P of G_1 to a much smaller value
13.69 **(a)** $f_O = 0.72/CR$; **(b)** $C = 1000$ pF, $R = 7.2\ \text{k}\Omega$; $C = 100$ pF, $R = 72\ \text{k}\Omega$
13.70 $\pm 2.8\%$

13.74 For $v_{DS_{on}} = 0.1$ V, $(W/L)_{1,2} = 0.62$, $I_S = 30.1$ mA, $P_D = 60.3$ mW; For $v_{DS_{on}} = 0.2$ V, $(W/L)_{1,2} = 0.31$, $I_S = 27.5$ mA, $P_D = 55.1$ mW
13.75 (a) $K_4 = 0.0878\, K_2$; (b) $K_6 = 0.225\, K_2$; (c) 0.47 V **13.76** 5 pA
13.78 Word n ($B_3\ B_2\ B_1\ B_0$): 1(1 0 1 1), 2(1 1 1 1), 3(0 1 1 1), 4(0 1 1 0), 5(1 0 1 0), 6(1 1 1 0), 7(1 0 0 1), 8(1 1 0 1)
13.81 For $V_{tE} = 0.2$ V, 0.1 V; $V_{OH} = 0.7$ V, 0.7 V; $V_{OL} = 0.17$ V, 0.129 V; $V_{IL} = 0.54$ V, 0.441 V; $V_{IH} = 0.63$ V, 0.529 V; $NM_H = 0.07$ V, 0.17 V; $NM_L = 0.37$ V, 0.312 V; NM become more balanced as V_{tE} lowers.
13.83 For fanout ≥ 1 (and $v_O = 0.7$ V), $P_D = 5.91$ mW or 7.05 mW for v_I low and high respectively, or $P_D = 6.48$ mW on average.

Chapter 14

14.1 2×10^{-14} A; twice as large; 1.00
14.3 $i_a \approx (I_S/\alpha_F)e^{v/V_T}$; $i_b \approx (I_S/\alpha_R)e^{v/V_T}$; the collector–base junction is 54.6 times larger than the emitter–base junction **14.5** 40.2 mV **14.6** 3 Ω; 69.7; 0.11; 110 mV
14.9 (a) $i_{DE} = 11.88$ mA, $i_{DC} = 1.76$ mA, $\alpha_F i_{DE} = 11.76$ mA, $\alpha_R i_{DC} = 0.88$ mA; (b) $V_{BE} = 694.8$ mV, $V_{BC} = 630$ mV; $V_{CE_{sat}} = 64.8$ mV; (c) 64.8 mV; (d) 650 mV, 632.7 mV, 17.3 mV
14.11 (a) 90 ns; (b) 15 ns; (c) 8.2 ns **14.13** 17.8 ns
14.15 (a) $V_{CE_{sat}} = 156$ mV, $I_C = 4.44$ mA, $V_{BE} = 737$ mV, $I_B = 0.178$ mA, $V_{IH} = 817$ mV; (b) $V_{OL} = 116$ mV; (d) $NM_H = 0.183$ V, $NM_L = 0.484$ V
14.17 NAND
14.19 $V_{OH} = 4$ V; $V_{OL} = 90$ mV; $V_{IL} = 1.2$ V; $V_{IH} = 1.6$ V; $NM_H = 2.4$ V; $NM_L = 1.1$ V
14.22 (a) 82.8 mV; (b) 41 **14.23** 1.77 mA; 0.14 mA; 105.8 ns
14.25 $t_r = 43.9$ ns; $t_f = 3.57$ ns **14.27** 0.5 V **14.30** 0.193 V; 10.7 mA
14.31 4.56 mV; 6.37 mV; 19.4 mV; 95.7 mV; 1.81 Ω; 1.45 Ω; 0.85 Ω
14.33 3.85 mA; 10.5 mA **14.34** 31.5 mA; $\beta > 14$
14.37 At −55°C, $NM_H = 1.66$ V, $NM_L = 0.56$ V; At 125°C, $NM_H = 3.1$ V, $NM_L = 0.2$ V
14.39 4-segment characteristic with breakpoints at $(v_I, v_O) = (1.1$ V, 3.0 V), (1.9 V, 2.0 V), (2.1 V, 0.1 V); $V_{OH} = 3.0$ V; $V_{OL} = 0.1$ V; $V_{IH} = 2.1$ V; $V_{IL} = 1.1$ V; $NM_H = 0.9$ V; $NM_L = 1.0$ V
14.40 0.232 mA **14.44** 1.3 V **14.45** $I_B = 0.10$ mA; $I_C = 5.02$ mA; $I_D = 0.32$ mA
14.47 (a) $Y = \overline{A + B \cdot C}$; (b) 0.3 V, 1.3 V; (c) 0.7 V, 0.8 V; (d) $NM_L = 0.4$ V, $NM_H = 0.5$ V (e) 1.092 mA
14.49 33.3 MHz; high for 13 ns; low for 17 ns
14.52 0.329 V/V; 8.94 V/V; 0.651 V/V
14.53 (a) −1.375 V, −1.265 V; (b) −1.493 V, −1.147 V
14.55 21.2 **14.57** 7 cm

Appendix B

B.2 $h_{11} = 2.6$ kΩ; $h_{12} = 2.5 \times 10^{-4}$; $h_{21} = 100$; $h_{22} = 2 \times 10^{-5}$ ℧
B.3 $y_{11} = 1/r_\pi + s(C_\pi + C_\mu)$; $y_{12} = -sC_\mu$; $y_{21} = -sC_\mu + g_m$; $y_{22} = 1/r_o + sC_\mu$

Appendix D

D.1 $Z_t = V_{oc}/I_{sc}$ **D.3** 1 V, 0.90 kΩ; 0.526 V
D.5 $R_{in} = (r_\pi + R_B)/(1 + g_m r_\pi)$

Appendix E

E.2 $V_o(s)/V_i(s) = R_2/(R_1 + R_2)$ **E.4** 10^5 rad/s **E.6** HP; 10 rad/s
E.7 $v_O(t) = 10(1 - e^{-t/10^{-6}})$; $v_o(t) = 10\, e^{-10^6 t}$ **E.9** 3.5 ns **E.11** −4.67 V
E.13 −6.32 V; 9.5 ms **E.15** 14.4 μs

Index

Boldface page numbers signify either the presence of a definition or of an important concept. The page numbers followed by "n" signify information included in a footnote on that page.

G

H

I